International Association of Fire Chiefs

National Fire Protection Association

Fundamentals of Fire Fighter Skills

Evidence-Based Practices

ENHANCED THIRD EDITION

JONES & BARTLETT LEARNING

Jones & Bartlett Learning
World Headquarters
5 Wall Street
Burlington, MA 01803
978-443-5000
info@jblearning.com
www.jblearning.com

National Fire Protection Association
1 Batterymarch Park
Quincy, MA 02169-7471
www.NFPA.org

International Association of Fire Chiefs
4025 Fair Ridge Drive
Fairfax, VA 22033
www.IAFC.org

Jones & Bartlett Learning books and products are available through most bookstores and online booksellers. To contact Jones & Bartlett Learning directly, call 800-832-0034, fax 978-443-8000, or visit our website, www.jblearning.com.

Substantial discounts on bulk quantities of Jones & Bartlett Learning publications are available to corporations, professional associations, and other qualified organizations. For details and specific discount information, contact the special sales department at Jones & Bartlett Learning via the above contact information or send an email to specialsales@jblearning.com.

13653-1

Production Credits

General Manager, Safety and Trades: Doug Kaplan
General Manager, and Executive Publisher: Kimberly Brophy
VP, Manufacturing and Inventory Control: Therese Connell
VP of Sales, Public Safety Group: Matthew Maniscalco
Director of Sales, Public Safety Group: Patricia Einstein
Executive Editor: Bill Larkin
Senior Acquisitions Editor: Janet Maker
Senior Development Editor: Janet Morris
Senior Editor: Jennifer Deforge-Kling
Production Manager: Jenny L. Corriveau
Associate Production Editor: Kristen Rogers
Senior Marketing Manager: Brian Rooney
Composition: diacriTech
Cover Design: Kristin E. Parker
Rights & Photo Research Associate: Lian Bruno
Cover Image: Photo by David Traiforos, luvfireimages.com
Printing and Binding: RR Donnelley
Cover Printing: RR Donnelley

To order this product, use ISBN: 978-1-284-12284-8

The Library of Congress has cataloged the first printing as:

Library of Congress Cataloging-in-Publication Data unavailable at time of printing.

6048

Printed in the United States of America
20 19 18 17 16 10 9 8 7 6 5 4 3 2 1

Brief Contents

Contents

CHAPTER 3 Personal Protective Equipment and Self-Contained Breathing Apparatus 42

CHAPTER 4 Fire Service Communications 90

Skill Drills

CHAPTER 16

CHAPTER 17

CHAPTER 18

CHAPTER 19

CHAPTER 21

CHAPTER 22

CHAPTER 25 Not Included

CHAPTER 26

CHAPTER 27

CHAPTER 32

CHAPTER 33

CHAPTER 34

CHAPTER 36

Resource Preview

Fundamentals of Fire Fighter Skills: Evidence-Based Practices, Enhanced Third Edition

With the release of the *Enhanced Third Edition*, Jones & Bartlett Learning, the National Fire Protection Association, and the International Association of Fire Chiefs have joined forces to raise the bar for the fire service once again. *Fundamentals of Fire Fighter Skills: Evidence-Based Practices, Enhanced Third Edition* presents data based on the extensive research conducted by UL and NIST, offering completely revised content in the areas of fire behavior, ventilation, suppression, fire attack and foam, and search and rescue. This one-volume text meets and exceeds the Fire Fighter I and Fire Fighter II professional qualifications levels as outlined in the 2013 edition of NFPA 1001, *Standard for Fire Fighter Professional Qualifications*. It also covers all of the Job Performance Requirements (JPRs) at the awareness and operations levels of the 2013 edition of NFPA 472, *Standard for Competence of Responders to Hazardous Materials/Weapons of Mass Destruction Incidents*.

New and Exciting Changes

With each new edition of *Fundamentals of Fire Fighter Skills*, we strive to enhance the effectiveness of this critically important Fire Fighter I and Fire Fighter II curriculum package. We understand how critical it is to prepare future fire fighters for a safe and successful career. It is with this in mind that we offer *Fundamentals of Fire Fighter Skills: Evidence-Based Practices, Enhanced Third Edition*, which includes complete revisions of the following chapters:

- Fire Behavior
- Search and Rescue
- Ventilation
- Fire Attack and Foam
- Suppression

In addition, the *Enhanced Third Edition* continues to offer the following:

- Fire Fighter Safety and Personal Protective Equipment and Self-Contained Breathing Apparatus as standalone chapters to increase student comprehension and improve lesson delivery.
- Water Supply with a single emphasis on providing water (static or municipal water supply) to the fire apparatus.
- Fire Fighter II content identified throughout each chapter with a green margin bar so there is a clear distinction between Fire Fighter I and Fire Fighter II level material. Chapters that are completely devoted to Fire Fighter II material are identified as well.
- Fire Fighter II in Action, to promote critical thinking in potential leaders. It presents challenging situations and asks the Fire Fighter II candidate, "What would you do?"

Chapter Resources

With the *Enhanced Third Edition*, safety is fundamental. It features a laser-like focus on fire fighter injury prevention, including dedicated chapters on safety, personal protective equipment, and survival. Reducing fire fighter injuries and deaths requires the dedicated efforts of every fire fighter, of every fire department, and of the entire service working together. Because experience can be the best instructor, actual National Fire Fighter Near-Miss Reporting System cases are discussed to illustrate critical lessons about safety.

Fundamentals of Fire Fighter Skills: Evidence-Based Practices, Enhanced Third Edition features reinforce and expand on fundamental information and reinforce rapid information retrieval. These features include:

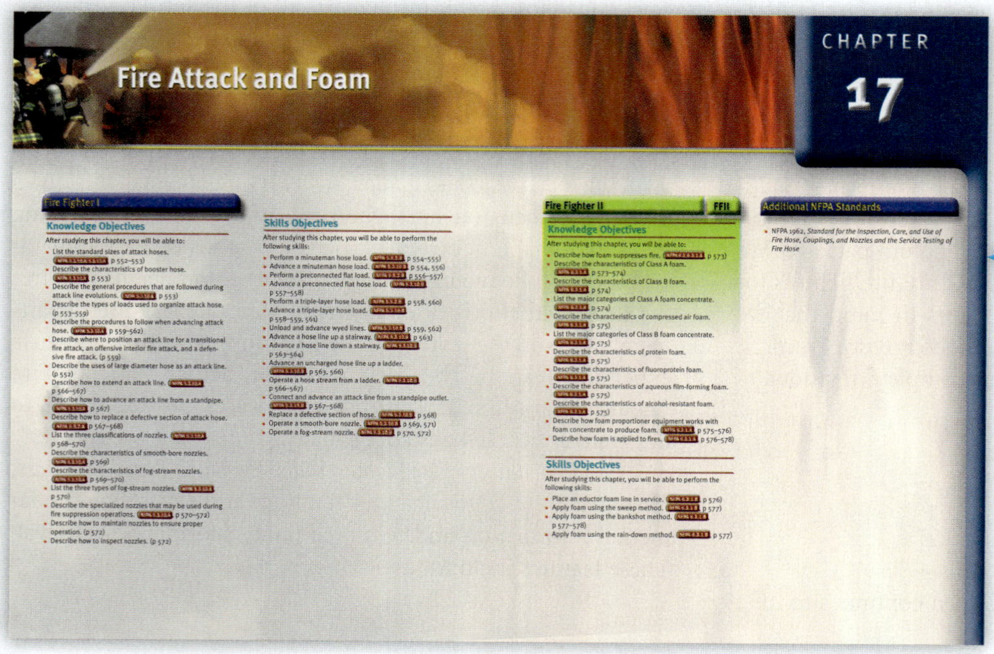
CHAPTER 17

Fire Attack and Foam

Fire Fighter I

Knowledge Objectives

After studying this chapter, you will be able to:

- List the standard sizes of attack hoses. (p 552–553)
- Describe the characteristics of booster hose. (p 553)
- Describe the general procedures that are followed during attack line evolutions. (p 553)
- Describe the types of loads used to organize attack hose. (p 553–559)
- Describe the procedures to follow when advancing attack hose. (p 559–562)
- Describe where to position an attack line for a transitional fire attack, an offensive interior fire attack, and a defensive fire attack. (p 559)
- Describe the uses of large diameter hose as an attack line. (p 552)
- Describe how to extend an attack line. (p 566–567)
- Describe how to advance an attack line from a standpipe. (p 567)
- Describe how to replace a defective section of attack hose. (p 567–568)
- List the three classifications of nozzles. (p 568–570)
- Describe the characteristics of smooth-bore nozzles. (p 569)
- Describe the characteristics of fog-stream nozzles. (p 569–570)
- List the three types of fog-stream nozzles. (p 570)
- Describe the specialized nozzles that may be used during fire suppression operations. (p 570–572)
- Describe how to maintain nozzles to ensure proper operation. (p 572)
- Describe how to inspect nozzles. (p 572)

Skills Objectives

After studying this chapter, you will be able to perform the following skills:

- Perform a minuteman hose load. (p 554–555)
- Advance a minuteman hose load. (p 554, 556)
- Perform a preconnected flat load. (p 556–557)
- Advance a preconnected flat hose load. (p 557–558)
- Perform a triple-layer hose load. (p 558, 560)
- Advance a triple-layer hose load. (p 558–559, 561)
- Unload and advance wyed lines. (p 559, 562)
- Advance a hose line up a stairway. (p 563)
- Advance a hose line down a stairway. (p 563–564)
- Advance an uncharged hose line up a ladder. (p 563, 566)
- Operate a hose stream from a ladder. (p 566–567)
- Connect and advance an attack line from a standpipe outlet. (p 567–568)
- Replace a defective section of hose. (p 568)
- Operate a smooth-bore nozzle. (p 569, 571)
- Operate a fog-stream nozzle. (p 570, 572)

Fire Fighter II FFII

Knowledge Objectives

After studying this chapter, you will be able to:

- Describe how foam suppresses fire. (p 573)
- Describe the characteristics of Class A foam. (p 573–574)
- Describe the characteristics of Class B foam. (p 574)
- List the major categories of Class A foam concentrate. (p 574)
- Describe the characteristics of compressed air foam. (p 575)
- List the major categories of Class B foam concentrate. (p 575)
- Describe the characteristics of protein foam. (p 575)
- Describe the characteristics of fluoroprotein foam. (p 575)
- Describe the characteristics of aqueous film-forming foam. (p 575)
- Describe the characteristics of alcohol-resistant foam. (p 575)
- Describe how foam proportioner equipment works with foam concentrate to produce foam. (p 575–576)
- Describe how foam is applied to fires. (p 576–578)

Skills Objectives

After studying this chapter, you will be able to perform the following skills:

- Place an eductor foam line in service. (p 576)
- Apply foam using the sweep method. (p 577)
- Apply foam using the bankshot method. (p 577–578)
- Apply foam using the rain-down method. (p 577)

Additional NFPA Standards

- NFPA 1962, *Standard for the Inspection, Care, and Use of Fire Hose, Couplings, and Nozzles and the Service Testing of Fire Hose*

Chapter Objectives

NFPA 1001 and 472 is correlated to the Knowledge Objectives and Skill Objectives listed at the beginning of each chapter.

- Page references are included for quick reference to content.
- Knowledge Objectives outline the most important topics covered in the chapter.
- Skills Objectives map the skill drills provided in the chapter.

You Are the Fire Fighter

Each chapter opens with a case study intended to stimulate classroom discussion, capture students' attention, and provide an overview on the key topics in the chapter.

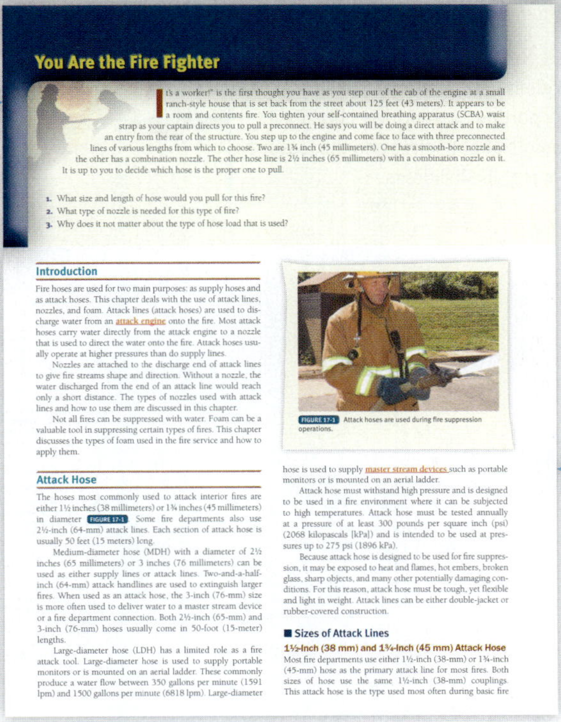

You Are the Fire Fighter

"It's a worker!" is the first thought you have as you step out of the cab of the engine at a small ranch-style house that is set back from the street about 125 feet (43 meters). It appears to be a room and contents fire. You tighten your self-contained breathing apparatus (SCBA) waist strap as your captain directs you to pull a preconnect. He says you will be doing a direct attack and to make an entry from the rear of the structure. You step up to the engine and come face to face with three preconnected lines of various lengths from which to choose. Two are 1¾ inch (45 millimeters). One has a smooth-bore nozzle and the other has a combination nozzle. The other hose line is 2½ inches (65 millimeters) with a combination nozzle on it. It is up to you to decide which hose is the proper one to pull.

1. What size and length of hose would you pull for this fire?
2. What type of nozzle is needed for this type of fire?
3. Why does it not matter about the type of hose load that is used?

Introduction

Fire hoses are used for two main purposes: as supply hoses and as attack hoses. This chapter deals with the use of attack lines, nozzles, and foam. Attack lines (attack hoses) are used to discharge water from an attack engine onto the fire. Most attack hoses carry water directly from the attack engine to a nozzle that is used to direct the water onto the fire. Attack hoses usually operate at higher pressures than do supply lines.

Nozzles are attached to the discharge end of attack lines to give fire streams shape and direction. Without a nozzle, the water discharged from the end of an attack line would reach only a short distance. The types of nozzles used with attack lines and how to use them are discussed in this chapter.

Not all fires can be suppressed with water. Foam can be a valuable tool in suppressing certain types of fires. This chapter discusses the types of foam used in the fire service and how to apply them.

FIGURE 17-1 Attack hoses are used during fire suppression operations.

Attack Hose

The hoses most commonly used to attack interior fires are either 1½ inches (38 millimeters) or 1¾ inches (45 millimeters) in diameter FIGURE 17-1. Some fire departments also use 2½-inch (64-mm) attack lines. Each section of attack hose is usually 50 feet (15 meters) long.

Medium-diameter hose (MDH) with a diameter of 2½ inches (65 millimeters) or 3 inches (76 millimeters) can be used as either supply lines or attack lines. Two-and-a-half-inch (64-mm) attack handlines are used to extinguish larger fires. When used as an attack hose, the 3-inch (76-mm) size is more often used to deliver water to a master stream device or a fire department connection. Both 2½-inch (65-mm) and 3-inch (76-mm) hoses usually come in 50-foot (15-meter) lengths.

Large-diameter hose (LDH) has a limited role as a fire attack tool. Large-diameter hose is used to supply portable monitors or is mounted on an aerial ladder. These commonly produce a water flow between 350 gallons per minute (1591 lpm) and 1500 gallons per minute (6818 lpm). Large-diameter hose is used to supply master stream devices such as portable monitors or is mounted on an aerial ladder.

Attack hose must withstand high pressure and is designed to be used in a fire environment where it can be subjected to high temperatures. Attack hose must be tested annually at a pressure of at least 300 pounds per square inch (psi) (2068 kilopascals [kPa]) and is intended to be used at pressures up to 275 psi (1896 kPa).

Because attack hose is designed to be used for fire suppression, it may be exposed to heat and flames, hot embers, broken glass, sharp objects, and many other potentially damaging conditions. For this reason, attack hose must be tough, yet flexible and light in weight. Attack lines can be either double-jacket or rubber-covered construction.

Sizes of Attack Lines

1½-Inch (38 mm) and 1¾-Inch (45 mm) Attack Hose

Most fire departments use either 1½-inch (38-mm) or 1¾-inch (45-mm) hose as the primary attack line for most fires. Both sizes of hose use the same 1½-inch (38-mm) couplings. This attack hose is the type used most often during basic fire

Hot Terms

Hot Terms are easily identifiable within the chapter and define key terms that the student must understand. A comprehensive glossary of Hot Terms also appears in the Wrap-Up.

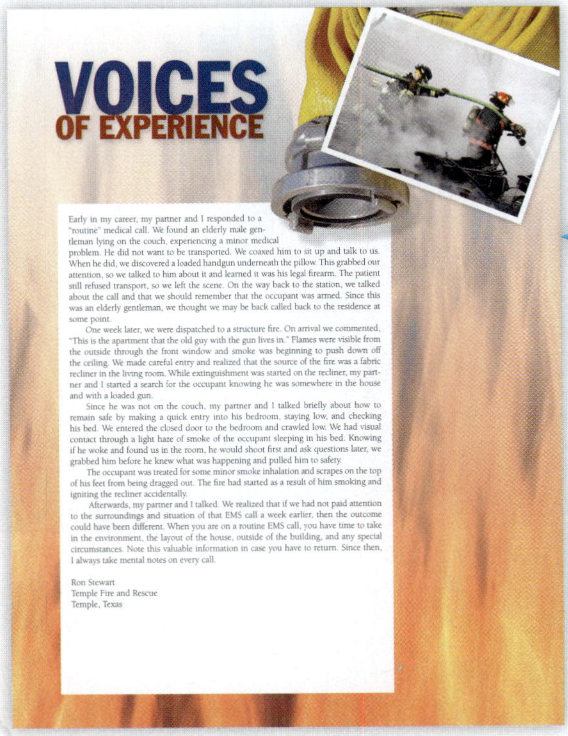

VOICES OF EXPERIENCE

Early in my career, my partner and I responded to a "routine" medical call. We found an elderly male gentleman lying on the couch, experiencing a minor medical problem. He did not want to be transported. We coaxed him to sit up and talk to us. When he did, we discovered a loaded handgun underneath the pillow. This grabbed our attention, so we talked to him about it and learned it was his legal firearm. The patient still refused transport, so we left the scene. On the way back to the station, we talked about the call and that we should remember that the occupant was armed. Since this was an elderly gentleman, we thought we may be back called back to the residence at some point.

One week later, we were dispatched to a structure fire. On arrival we commented, "This is the apartment that the old guy with the gun lives in." Flames were visible from the outside through the front window and smoke was beginning to push down off the ceiling. We made careful entry and realized that the source of the fire was a fabric recliner in the living room. While extinguishment was started on the recliner, my partner and I started a search for the occupant knowing he was somewhere in the house and with a loaded gun.

Since he was not on the couch, my partner and I talked briefly about how to remain safe by making a quick entry into his bedroom, staying low, and checking his bed. We entered the closed door to the bedroom and crawled low. We had visual contact through a light haze of smoke of the occupant sleeping in his bed. Knowing if he woke and found us in the room, he would shoot first and ask questions later, we grabbed him before he knew what was happening and pulled him to safety.

The occupant was treated for some minor smoke inhalation and scrapes on the top of his feet from being dragged out. The fire had started as a result of him smoking and igniting the recliner accidentally.

Afterwards, my partner and I talked. We realized that if we had not paid attention to the surroundings and situation of that EMS call a week earlier, then the outcome could have been different. When you are on a routine EMS call, you have time to take in the environment, the layout of the house, outside of the building, and any special circumstances. Note this valuable information in case you have to return. Since then, I always take mental notes on every call.

Ron Stewart
Temple Fire and Rescue
Temple, Texas

Voices of Experience

Veteran fire fighters share their accounts of memorable incidents while offering advice and encouragement. These essays highlight what it is truly like to be a fire fighter.

Near Miss Reports

Utilizing incident data, National Fire Fighter Near Miss Reporting System cases highlight important points about safety and the lessons learned from real-life incidents.

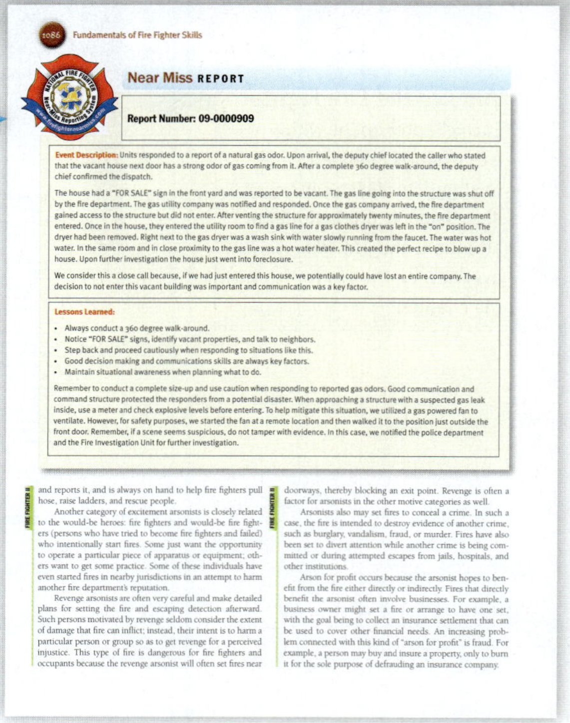

1086 Fundamentals of Fire Fighter Skills

Near Miss REPORT

Report Number: 09-0000909

Event Description: Units responded to a report of a natural gas odor. Upon arrival, the deputy chief located the caller who stated that the vacant house next door has a strong odor of gas coming from it. After a complete 360 degree walk-around, the deputy chief confirmed the dispatch.

The house had a "FOR SALE" sign in the front yard and was reported to be vacant. The gas line going into the structure was shut off by the fire department. The gas utility company was notified and responded. Once the gas company arrived, the fire department gained access to the structure but did not enter. After venting the structure for approximately twenty minutes, the fire department entered. Once in the house, they entered the utility room to find a gas line for a gas clothes dryer was left in the "on" position. The dryer had been removed. Right next to the gas dryer was a wash sink with water slowly running from the faucet. The water was hot water. In the same room and in close proximity to the gas line was a hot water heater. This created the perfect recipe to blow up a house. Upon further investigation the house just went into foreclosure.

We consider this a close call because, if we had just entered this house, we potentially could have lost an entire company. The decision to not enter this vacant building was important and communication was a key factor.

Lessons Learned:

- Always conduct a 360 degree walk-around.
- Notice "FOR SALE" signs, identify vacant properties, and talk to neighbors.
- Step back and proceed cautiously when responding to situations like this.
- Good decision making and communications skills are always key factors.
- Maintain situational awareness when planning what to do.

Remember to conduct a complete size-up and use caution when responding to reported gas odors. Good communication and command structure protected the responders from a potential disaster. When approaching a structure with a suspected gas leak inside, use a meter and check explosive levels before entering. To help mitigate this situation, we utilized a gas powered fan to ventilate. However, for safety purposes, we started the fan at a remote location and then walked it to the position just outside the front door. Remember, if a scene seems suspicious, do not tamper with evidence. In this case, we notified the police department and the Fire Investigation Unit for further investigation.

FIRE FIGHTER II

and reports it, and is always on hand to help fire fighters pull hose, raise ladders, and rescue people.

Another category of excitement arsonists is closely related to the would-be heroes: fire fighters and would-be fire fighters (persons who have tried to become fire fighters and failed) who intentionally start fires. Some just want the opportunity to operate a particular piece of apparatus or equipment; others want to get some practice. Some of these individuals have even started fires in nearby jurisdictions in an attempt to harm another fire department's reputation.

Revenge arsonists are often very careful and make detailed plans for setting the fire and escaping detection afterward. Such persons motivated by revenge seldom consider the extent of damage that fire can inflict; instead, their intent is to harm a particular person or group so as to get revenge for a perceived injustice. This type of fire is dangerous for fire fighters and occupants because the revenge arsonist will often set fires near doorways, thereby blocking an exit point. Revenge is often a factor for arsonists in the other motive categories as well.

Arsonists also may set fires to conceal a crime. In such a case, the fire is intended to destroy evidence of another crime, such as burglary, vandalism, fraud, or murder. Fires have also been set to divert attention while another crime is being committed or during attempted escapes from jails, hospitals, and other institutions.

Arson for profit occurs because the arsonist hopes to benefit from the fire either directly or indirectly. Fires that directly benefit the arsonist often involve businesses. For example, a business owner might set a fire or arrange to have one set, with the goal being to collect an insurance settlement that can be used to cover other financial needs. An increasing problem connected with this kind of "arson for profit" is fraud. For example, a person may buy and insure a property, only to burn it for the sole purpose of defrauding an insurance company.

Fire Fighter II Content

All Fire Fighter II content is highlighted clearly within the chapter.

574 Fundamentals of Fire Fighter Skills

FIRE FIGHTER II

prevent its ignition. Class A foam concentrate can be added to water streams and applied with several types of nozzles.

Class B foam is used to fight Class B fires—that is, fires involving flammable and combustible liquids. The various types of Class B foam are formulated to be effective on different types of flammable liquids. Note that some liquids are incompatible with different foam formulations and will destroy the foam before the foam can control the fire.

In a flammable-liquid fire, the liquid itself does not burn. Only the flammable vapors that are evaporating from the surface of the liquid and mixing with air can burn. Depending on the temperature and the physical properties of the liquid, the amount of vapors being released from the surface of a liquid varies. For example, gasoline produces flammable vapors down to a temperature of −45°F (−42.7°C). Some vapors are lighter than air and rise up into the atmosphere. Other vapors are heavier than air and flow across the surface and along the ground, collecting in low spots.

Foam extinguishes flammable-liquid fires by separating the fuel from the fire. When a blanket of foam completely covers the surface of the liquid, the release of flammable vapors stops. Preventing the production of additional vapors eliminates the fuel source for the fire, which extinguishes the fire.

Once a foam blanket has been applied, it must not be disturbed. The fuel under the foam blanket is still capable of producing flammable vapors. Thus, if the foam blanket is disturbed by wind, by someone walking through the liquid, or by hose streams breaking up the foam blanket, flammable vapors will be released and could be reignited easily.

When using foam to extinguish a flammable-liquid fire, it is critically important to apply enough foam to cover the liquid surface fully. If not enough foam is available to completely cover the surface at one time, the fire will have the ability to continue burning and the suppression efforts will be ineffective. The rate of foam application must also be high enough to cover the surface and maintain a blanket on top of the liquid. If the application rate is too low, thermal updrafts caused by the fire will keep the foam from covering the surface, and the heat of the fire will destroy the foam that has already been applied.

Class B foam can also be applied to a flammable-liquid spill in an effort to prevent a fire. A foam blanket floating on the surface of the liquid will inhibit the production of vapors that might otherwise be ignited. In this situation, the rate of foam application is not as critical because there is no fire working to destroy the foam while it is being applied.

Fire fighters must ensure that they use the proper foam for the situation that is encountered. For example, they must use an alcohol-resistant foam if the incident involves a polar solvent. Polar solvents are chemicals (such as alcohols) that readily mix with water. An ordinary foam would be broken down quickly if it came in contact with this type of product. Make sure to use the proper foam given the product involved in the incident.

Class A foams are not designed to resist hydrocarbons or to self-seal on a liquid surface. If Class A foam is used on a flammable liquid, fire fighters would find it difficult to extinguish a fire and the foam blanket would not mend itself if it became disrupted. In this case, the risk of reignition could create a dangerous situation.

Foam Concentrates

Foam concentrate is the product that is mixed with water in different ratios to produce foam solution. The foam solution is the product that is actually applied to extinguish a fire or to cover a spill.

Class A Foams

Class A foams are usually formulated to be mixed with water in ratios from 0.1 percent (1 gallon of concentrate to 999 gallons of water, or 1 liter of concentrate to 999 liters of water) to 1 percent (1 gallon of concentrate to 99 gallons of water, or 1 liter of concentrate to 99 liters of water). The end product can be tweaked to have different properties by varying the percentage of foam concentrate in the mixture and the application method. It is possible to produce "wet" foam that has good penetrating properties, for example, or "drier" foam that is more effective for applying a protective layer of foam onto a building.

Class B Foams

Most Class B foam concentrates are designed to be used in strengths of either 3 percent or 6 percent. A 3 percent foam mixes 3 gallons of foam concentrate with 97 gallons of water (11 liters of concentrate to 367 liters of water) to produce 100 gallons (100 liters) of foam solution. A 6 percent foam mixes 6 gallons of foam concentrate with 94 gallons of water (22 liters of concentrate to 356 liters of water) to produce 100 gallons (378 liters) of foam solution. Some foams are designed to be used at 3 percent for ordinary hydrocarbons and at 6 percent for polar solvents. Fire fighters must determine which type of fuel is involved in the incident so that they can select the correct proportioning rate.

The compatibility of foam agents with other extinguishing agents needs to be considered as well. For example, some combinations of dry chemical extinguishing agents and foam agents can cause an adverse reaction. The compatibility data needed are available from the manufacturers of the agents.

Fire fighters must not mix different types of foam concentrate or even different brands of the same type of foam concentrate unless they are known to be compatible. Some concentrates have been known to react with other types of foams, causing a congealing of the concentrate in the storage containers or in the proportioning system. This type of reaction can plug a foam system and render it useless.

Fire Fighter Safety Tips

Although most foam concentrates are somewhat corrosive, they pose few health risks to fire fighters if normal precautions are taken. In the past, some fire fighters have experienced minor skin irritations from coming into contact with foam concentrate. As always, before using any equipment, fire fighters must read and follow the manufacturer's directions when applying foam.

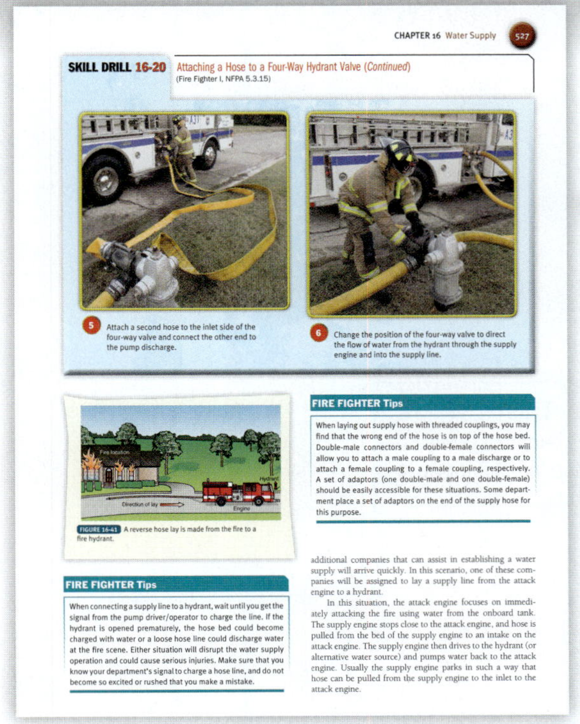

CHAPTER 16 Water Supply 527

SKILL DRILL 16-20 Attaching a Hose to a Four-Way Hydrant Valve *(Continued)*
(Fire Fighter I, NFPA 5.3.15)

5 Attach a second hose to the inlet side of the four-way valve and connect the other end to the pump discharge.

6 Change the position of the four-way valve to direct the flow of water from the hydrant through the supply engine and into the supply line.

FIGURE 16-41 A reverse hose lay is made from the fire to a fire hydrant.

FIRE FIGHTER Tips

When laying out supply hose with threaded couplings, you may find that the wrong end of the hose is on top of the hose bed. Double-male connectors and double-female connectors will allow you to attach a male coupling to a male discharge or to attach a female coupling to a female coupling, respectively. A set of adaptors (one double-male and one double-female) should be easily accessible for these situations. Some departments place a set of adaptors on the end of the supply hose for this purpose.

FIRE FIGHTER Tips

When connecting a supply line to a hydrant, wait until you get the signal from the pump driver/operator to charge the line. If the hydrant is opened prematurely, the hose bed could become charged with water or a loose hose line could discharge water at the fire scene. Either situation will disrupt the water supply operation and could cause serious injuries. Make sure that you know your department's signal to charge a hose line, and do not become so excited or rushed that you make a mistake.

additional companies that can assist in establishing a water supply will arrive quickly. In this scenario, one of these companies will be assigned to lay a supply line from the attack engine to a hydrant.

In this situation, the attack engine focuses on immediately attacking the fire using water from the onboard tank. The supply engine stops close to the attack engine, and hose is pulled from the bed of the supply engine to an intake on the attack engine. The supply engine then drives to the hydrant (or alternative water source) and pumps water back to the attack engine. Usually the supply engine parks in such a way that hose can be pulled from the supply engine to the inlet to the attack engine.

Skill Drills

Written step-by-step explanations and visual summaries of key skills. This clear, concise format enhances student comprehension of complex procedures. Each Skill Drill identifies if the skill is Fire Fighter I or Fire Fighter II with the corresponding NFPA job performance requirement.

FIRE FIGHTER Tips

This series of tips provides advice from masters of the trade.

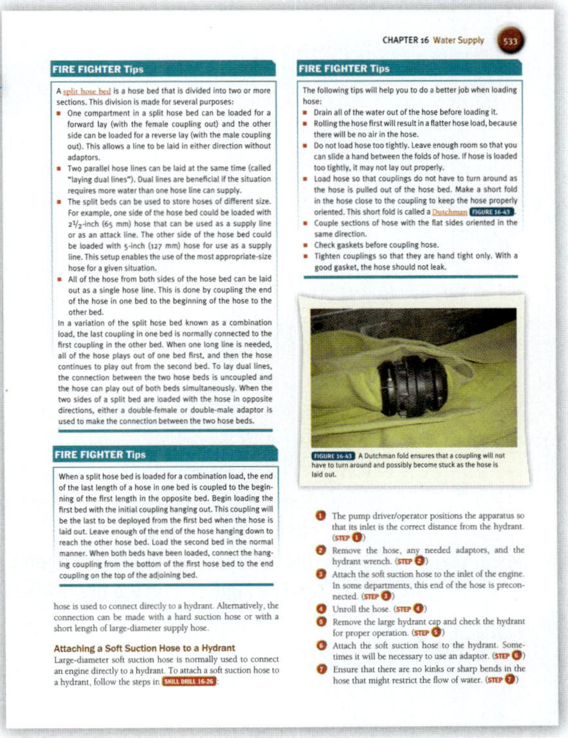

CHAPTER 16 Water Supply 533

FIRE FIGHTER Tips

A split hose bed is a hose bed that is divided into two or more sections. This division is made for several purposes:

- One compartment in a split hose bed can be loaded for a forward lay (with the female coupling out) and the other side can be loaded for a reverse lay (with the male coupling out). This allows a line to be laid in either direction without adaptors.
- Two parallel hose lines can be laid at the same time (called "laying dual lines"). Dual lines are beneficial if the situation requires more water than one hose line can supply.
- The split beds can be used to store hoses of different size. For example, one side of the hose bed could be loaded with 2½-inch (65 mm) hose that can be used as a supply line or as an attack line. The other side of the hose bed could be loaded with 5-inch (127 mm) hose for use as a supply line. This setup enables the use of the most appropriate-size hose for a given situation.
- All of the hose from both sides of the hose bed can be laid out as a single hose line. This is done by coupling the end of the hose in one bed to the beginning of the hose to the other bed.

In a variation of the split hose bed known as a combination load, the last coupling in one bed is normally connected to the first coupling in the other bed. When one long line is needed, all of the hose plays out of one bed first, and then the hose continues to play out from the second bed. To lay dual lines, the connection between the two hose beds is uncoupled and the hose can play out of both beds simultaneously. When the two sides of a split bed are loaded with the hose in opposite directions, either a double-female or double-male adaptor is used to make the connection between the two hose beds.

FIRE FIGHTER Tips

When a split hose bed is loaded for a combination load, the end of the last length of a hose in one bed is coupled to the beginning of the first length in the opposite bed. Begin loading the first bed with the initial coupling hanging out. This coupling will be the last to be deployed from the first bed when the hose is laid out. Leave enough of the end of the hose hanging down to reach the other hose bed. Load the second bed in the normal manner. When both beds have been loaded, connect the hanging coupling from the bottom of the first hose bed to the end coupling on the top of the adjoining bed.

hose is used to connect directly to a hydrant. Alternatively, the connection can be made with a hard suction hose or with a short length of large-diameter supply hose.

Attaching a Soft Suction Hose to a Hydrant

Large-diameter soft suction hose is normally used to connect an engine directly to a hydrant. To attach a soft suction hose to a hydrant, follow the steps in SKILL DRILL 16-26.

FIRE FIGHTER Tips

The following tips will help you to do a better job when loading hose:

- Drain all of the water out of the hose before loading it.
- Rolling the hose first will result in a flatter hose load, because there will be no air in the hose.
- Do not load hose too tightly. Leave enough room so that you can slide a hand between the folds of hose. If hose is loaded too tightly, it may not lay out properly.
- Load hose so that couplings do not have to turn around as the hose is pulled out of the hose bed. Make a short fold in the hose close to the coupling to keep the hose properly oriented. This short fold is called a Dutchman FIGURE 16-43.
- Couple sections of hose with the flat sides oriented in the same direction.
- Check gaskets before coupling hose.
- Tighten couplings so that they are hand tight only. With a good gasket, the hose should not leak.

FIGURE 16-43 A Dutchman fold ensures that a coupling will not have to turn around and possibly become stuck as the hose is laid out.

1. The pump driver/operator positions the apparatus so that its inlet is the correct distance from the hydrant. (STEP 1)
2. Remove the hose, any needed adaptors, and the hydrant wrench. (STEP 2)
3. Attach the soft suction hose to the inlet of the engine. In some departments, this end of the hose is preconnected. (STEP 3)
4. Unroll the hose. (STEP 4)
5. Remove the large hydrant cap and check the hydrant for proper operation. (STEP 5)
6. Attach the soft suction hose to the hydrant. Sometimes it will be necessary to use an adaptor. (STEP 6)
7. Ensure that there are no kinks or sharp bends in the hose that might restrict the flow of water. (STEP 7)

CHAPTER 2 Fire Fighter Safety 31

greatly reduce the possibility that vehicle occupants will be ejected from the vehicle.

In some departments where fire fighters are not on duty at the fire station, it may be necessary for members to respond to the station or to the emergency scene in their private vehicles. Driving a private vehicle during an emergency response requires a special degree of safety. Such emergency driving requires added considerations and represents a much different type of driving from the routine driving.

Motor vehicle crashes are the second most common cause of fire fighter deaths. Although some of these crashes involve fire department apparatus, all too many involve fire fighters driving their private vehicles to an emergency. The ensuing collisions sometimes result in injury or death to the fire fighter and to civilians. Even when no one is injured, crashes cause fire fighters to be delayed in getting to the emergency.

Many collisions occur when the motor vehicle operator loses control of the vehicle. These crashes may be caused by driving too fast for the prevailing conditions, braking inappropriately, changing directions too abruptly, or tracking around a curve too fast. In addition, many collisions involving emergency vehicles occur on open roads, where excessive speed is often cited as a primary cause. Intersections are common sites of collisions involving emergency vehicles; most of these collisions occur when the emergency vehicle operator fails to stop at an intersection and ensure that other traffic has stopped before proceeding through the intersection.

A motor vehicle collision actually consists of a series of separate collision events. The first collision occurs when the vehicle collides with a second vehicle or with a stationary object; the second collision occurs when the occupants of the vehicle collide with the interior of the vehicle.

FIGURE 2-4 Protective clothing should be properly positioned so that you can quickly don it.

FIRE FIGHTER II Tip FFII

Developing situational awareness on the emergency scene is essential to your safety and survival. This concept can be summed up by saying that you must have a "big picture" view of the emergency scene and avoid "tunnel vision." This ability comes only through experience and with well-developed competency in basic fire fighter skills. The less you have to concentrate on the components of your task, the more you are able to observe your surroundings. This quality emerges through hours and hours of practice and skill repetition. Do not stop practicing until your basic skills become muscle memories. If you can do your job with your eyes closed, you will be able to observe much more when they are open!

Fire Fighter Safety Tips

Always fasten your seat belt each and every time you get into a motor vehicle. It is vital that you follow this rule each and every time you climb into a fire engine.

Laws and Regulations Governing Emergency Vehicle Operation

Four general principles govern emergency vehicle operation:

- Emergency vehicle operators are subject to all traffic regulations unless a specific exemption is made. A specific exemption is a statement that appears in a statute, such as "The driver of an authorized vehicle may exceed the maximum speed limits so long as he or she does not endanger life or property."
- Exceptions are legal only when the vehicle is operating in emergency mode.
- Even with an exemption, the emergency vehicle operator can be found criminally or civilly liable if involved in a crash.
- An exemption does not relieve the operator of an authorized emergency vehicle from the duty to drive with reasonable care for all persons using the highway.

Safe Driving Practices

Vehicle occupants who are not wearing seat belts are much more likely to suffer serious injuries or death than are occupants who have their seat belts properly fastened. Air bags are most effective when seat belts are properly fastened; they are not effective without properly applied seat belts. Seat belts also

Fire Fighter Safety Tips

Fire Fighter Safety Tips reinforce safety-related concerns for both fire fighters and civilians.

720 Fundamentals of Fire Fighter Skills

When static water sources are used, the preincident plan must identify drafting sites (locations where an engine can draft water directly from the static source) or the locations of the nearest dry hydrants (an arrangement of pipes that is permanently connected to a static water supply). It is also important to measure the distance from the water source to the fire building to determine whether a large-diameter hose can be used and whether additional engines will be needed. Finally, the preincident plan should outline the operation that would be required to deliver water to the fire.

If a tanker (or tender) shuttle (a system of tankers [or tenders] transporting water from a water supply to the fire scene) will be used to deliver water, the preincident plan must include several additional details. For instance, sites for filling the tankers (or tenders) and for discharging their loads must be identified FIGURE 23-15. The preincident plan should also identify how many tankers (or tenders) will be needed based on the total distance they must travel, the quantity of water each vehicle can transport, and the total time it takes to empty and refill the vehicle.

FIGURE 23-15 The points where tankers (or tenders) can be filled and where they can discharge their loads must be identified in the preincident plan.

FIRE FIGHTER II Tips FFII

A private water system may not be able to simultaneously supply both sprinklers and fire hoses effectively. In one incident, a commercial property in Bluffton, Indiana, was destroyed when the responding fire department hooked its pumpers to private fire hydrants. The subsequent reduction in water flow rendered the building's sprinklers ineffective.

A large industrial or commercial complex may have its own water supply system that provides water for automatic sprinkler systems, standpipes, and private hydrants. These systems usually include storage tanks or reservoirs as well as fixed fire pumps to deliver the water under pressure. The details and arrangement of the private water supply system must be determined during the preincident survey.

The survey should indicate how much water is stored on the property and where the tanks are located. If a fixed fire pump is available, its location, capacity (in gallons or liters per minute), and power source (electricity or a diesel engine) should be noted. It is also important to confirm that these private water systems are being maintained in good operating condition. For example, the private water supply system on an abandoned property may be useless if it is no longer properly maintained.

In many cases, the same private water main provides water for both the sprinkler system and the private hydrants. If the fire department uses the private hydrants, the water supply to the sprinklers may be compromised. The preincident plan for these sites should note whether public, off-site hydrants should be used instead of the private hydrants.

Utilities

During an emergency incident, it may be necessary to turn off utilities such as electricity or natural gas as a safety measure. The preincident survey should note the locations of shut-offs for electricity, natural gas, propane gas, fuel oil, and other energy sources. These sites, as well as contact information for the appropriate utility company, should be included in the preincident plan. In some cases, special knowledge or equipment may be required to disconnect utilities. This information must be noted in the preincident plan, along with the procedure for contacting the appropriate individual or organization.

Electrical wires pose particular problems during firefighting operations. Electricity may be supplied by overhead wires or through underground cables. Overhead wires can be deadly if a ground ladder or aerial ladder comes in contact with them. The preincident plan should show the locations of high-voltage electrical lines and equipment that could prove dangerous to fire fighters. If electricity is supplied by underground cables, the shut-off may be located inside the basement of a building or in an underground vault. These locations should also be noted on the preincident plan.

If propane gas or fuel oil is stored on the property, the preincident plan should show the location and note the capacity of each tank. The presence of an emergency generator should

FIRE FIGHTER II Tips FFII

Identifying the available water supply for a particular building is important. Consider the following issues:

- How much water will be needed to fight a fire in the structure?
- Where are the nearest hydrants located?
- How much water is available and at what pressure?
- Where are hydrants that are on different water mains located?
- How many feet of hose would be needed to deliver the water to the fire?

If there are no hydrants, consideration should be given to these issues:

- What is the nearest water supply?
- Is the water readily available year-round? Is it seasonal?
- How many tanker (or tender) trucks will be needed to deliver the water?
- How long will it take to establish a tanker (or tender) shuttle?
- What is the total travel time from the scene to the fill site and then from the fill site back to the scene?

FIRE FIGHTER II Tips

This series of tips provides targeted advice for the Fire Fighter II on the fireground.

Wrap-Up

Chief Concepts

Highlight critical information from the chapter in a bulleted format to help students prepare for exams.

Hot Terms

A collection of key terms and definitions from the chapter.

Fire Fighter in Action

This feature promotes critical thinking through the use of case studies and provides instructors with discussion points for the classroom.

Fire Fighter II in Action

This feature promotes critical thinking in potential leaders. It presents challenging situations and asks the Fire Fighter II candidate, "What would you do?"

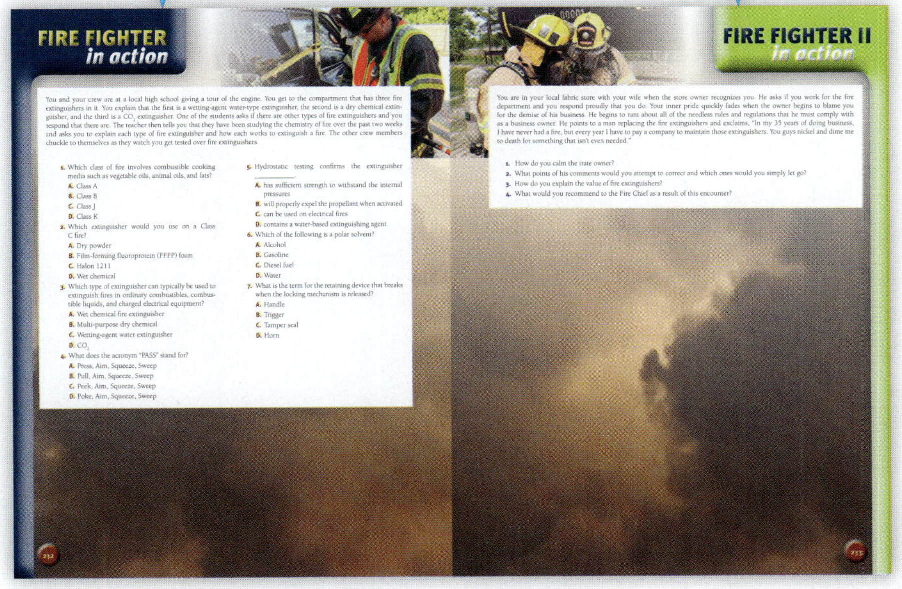

Instructor Resources

Instructor's ToolKit

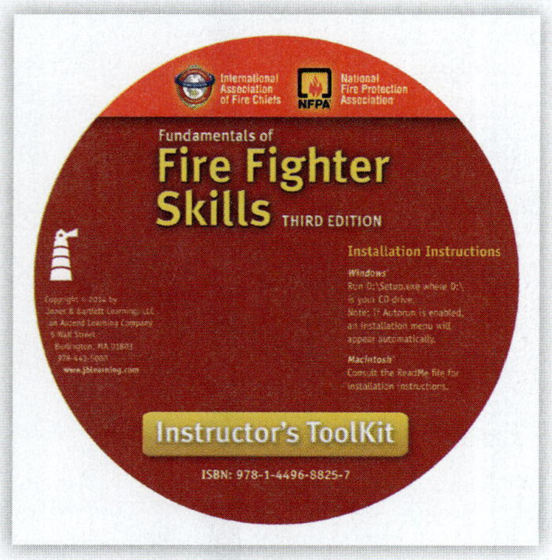

Preparing for class is easy with the resources on this CD. Instructors can choose resources for the Fire Fighter I, Fire Fighter II, or Fire Fighter I & II combined. The CD includes the following resources:

- Adaptable PowerPoint Presentations
- Detailed Lesson Plans
- Image and Table Bank
- Skills Sheets
- Student Workbook Answers

Instructor's Test Bank

The *Instructor's Test Bank* contains over 2,000 multiple-choice questions, and allows instructors to create tailor-made classroom tests and quizzes quickly and easily by selecting, editing, organizing, and printing a test along with an answer key, including page references to the text.

The questions found in the *Instructor's Test Bank* are separated by level so you can create exams for Fire Fighter I, Fire Fighter II, or Fire Fighter I and II combined. Each question includes the corresponding NFPA 1001 and/or NFPA 472 objective being tested.

Student Resources

Student Workbook

This resource is designed to encourage critical thinking and aid comprehension of the course materials. The student workbook uses the following activities to enhance mastery of the material.

- Case studies and corresponding questions
- Figure-labeling exercises
- Matching, fill-in-the-blank, short-answer, and multiple-choice questions
- Skill Drill activities
- Answers to the workbook activities are located on the *Instructor's ToolKit*.

Acknowledgments

Jones & Bartlett Learning, the National Fire Protection Association, and the International Association of Fire Chiefs would like to thank the current editors, contributors, and reviewers of *Fundamentals of Fire Fighter Skills: Evidence-Based Practices, Enhanced Third Edition*.

Editorial Board

Thomas McGowan
National Fire Protection Association
Quincy, Massachusetts

Glen D. Rudner
Virginia Department of Emergency Management, retired
North Chesterfield, Virginia

Kenneth Willette
National Fire Protection Association
Quincy, Massachusetts

Dave Schottke
Fairfax Volunteer Fire Department
Fairfax, Virginia

Shawn Kelley
International Association of Fire Chiefs
Fairfax, Virginia

Contributors and Reviewers

W. Chris Allan
Albemarle Fire and Rescue
Albemarle, North Carolina

Bryan Altman
Worth County Fire Rescue
Sylvester, Georgia

Benjamin Andersen
Loveland Fire and Rescue
Loveland, Colorado

Dr. Michael Armstrong
Augusta County Fire-Rescue
County of Augusta, Virginia

Rick Armstrong
Regional Education Service Agencies V/Public Service Training
Parkersburg, West Virginia

Shane Baird
Durango Fire Rescue
Durango, Colorado

Greg Barton
Beverly Hills Fire Department
Beverly Hills, California

Bernard W. Becker III
Cleveland State University
Cleveland, Ohio

Patrick Beckley
Ohio Fire Academy
Reynoldsburg, Ohio

Deward Beeler
Tri Township Fire District
St. Charles, Michigan

Dave Belcher
Violet Township Fire Department
Pickerington, Ohio

Bill Benjamin
Portland Community College
Portland, Oregon

Brian C. Bonner
Homewood Fire Department
Homewood, Alabama

Bryan C. Bowling
Butler Tech Public Safety Education Complex
Hamilton, Ohio

Scott Breeze
Independence Fire District
Independence, Kentucky

Britt Brinson
Georgia Firefighter Standard and Training Council
Forsyth, Georgia

Debra L. Burch
Connecticut Fire Academy
Windsor Locks, Connecticut

Jason Cairney
Justice Institute of British Columbia Fire & Safety Division
Westminster, British Columbia, Canada

Mike Camelo Sr.
Southwest Florida Public Safety Academy: Fire Academy Division
Fort Myers, Florida

Jim Campbell
Pike Township Fire Department
Indianapolis, Indiana

Lawrence Carter
Apex Fire Department
Apex, North Carolina

Jeffery C. Cash
City of Cherryville Fire Department
Cherryville, North Carolina

Ben Celius
Evergreen Fire/Rescue
Evergreen, Colorado

Wayne Cheek
City of Durham Fire Department
Hillsborough, North Carolina

Glenn C. Clapp
Fire Training Commander
High Point, North Carolina Fire Department

Cari A. Coll
Vashon Island Fire & Rescue
Vashon, Washington

Dean Colthorp
Justice Institute of British Columbia
Fire & Safety Division
Westminster, British Columbia, Canada

Christopher Connor
Bucksport Fire Rescue
Bucksport, Maine

Tom Crowder
Clackamas Fire District #1
Clackamas, Oregon

Stephan M. Curry
Columbia Fire Department
Columbia, South Carolina

Kevin B. Czarnecki
Connecticut Fire Academy
Windsor Locks, Connecticut

Jason D'Eliso
Scottsdale Fire Department
Scottsdale, Arizona

Brian D. Daake
Beatrice Fire/Rescue
Beatrice, Nebraska

Ed Davies
Mount Lebanon Fire Department
Pittsburgh, Pennsylvania

John T. Dean
Phoenix Fire Department
Phoenix, Arizona

Sean DeCrane
Cleveland Division of Fire
Cleveland, Ohio

Tucker Dempsey
Delaware State Fire School
Dover, Delaware

Todd Dettman
Wake Technical Community College
Raleigh, North Carolina

Bill Dicken
Central Pierce Fire & Rescue
Tacoma, Washington

Todd Eddy
Alabama Fire College
Tuscaloosa, Alabama

Christopher J. Edwards
Honolulu Fire Department Training and
Research Bureau
Honolulu, Hawaii

William B. Eisenhardt
Berea Fire Department
Berea, Ohio

Julie Ann Evert
Salt River Fire Department
Scottsdale, Arizona

Michael J. Fagel
University of Chicago-Master of Threat Risk
Program
Chicago, Illinois

David Feichtner
Farmington Hills Fire Department
Farmington Hills, Michigan

R.C. Fellows
Pleasant Valley Joint Fire District
Galloway, Ohio

Tim Fitts
Guilford Technical Community College
Jamestown, North Carolina

Steve Flaherty
Grand Rapids Fire Department
Grand Rapids, Minnesota

David Wm. Fogerson
East Fork Fire and Paramedic Districts
Gardnerville, Nevada

Phil Fritts
Jonesborough Fire Department
Jonesborough, Tennessee

Mark L. Froelich
Vincennes University
Vincennes, Indiana

Neil R. Fulton
Norwich Fire Department
Norwich, Vermont

Charles Garrity
Berkshire Community College
Pittsfield, Massachusetts

Rob Gaylor
City of Westfield Fire Department
Westfield, Indiana

Anthony Gianantonio
Palm Bay Fire Rescue
Palm Bay, Florida

Steve R. Goheen
Clark State Community College
Springfield, Ohio

Jason Goodale
Lovedale Fire Department
Johnstown, Colorado

Robert W. Green
Culpeper County Office of Emergency
Services
Culpeper, Virginia

Andrew S. Gurwood
Bucks County Public Safety Training Center
Northampton, Pennsylvania

Robert A. Halpin
Highpoint Fire Department
High Point, North Carolina

Jack Hamilton
Odessa Fire/Rescue
Odessa, Texas

Jim Harmes
Grand Blanc Fire
Grand Blanc, Michigan

Chris Harner
City of Pueblo Fire Department
Pueblo, Colorado

Howard Harper
Cincinnati State College
Cincinnati, Ohio

Thomas L. "Tommy" Harper
Jefferson College of Health Sciences
Roanoke, Virginia

Jeffry J. Harran
Buckskin Fire Department
Parker, Arizona

Jeff Harris
Durango Fire and Rescue Authority
Durango, Colorado

Robert Havens
Lamar Institute of Technology/Port Arthur Fire Department
Beaumont, Texas

Lee Hedgepeth
Bossier Parish Fire District 1
Haughton, Louisiana

Justin R. Heim
Eagle Fire Department
Eagle, Wisconsin

Andy Hernandez
Coolidge Fire Department
Coolidge, Arizona

Jacob Hrywniak
Bergen County Law & Public Safety Institute
Mahwah, New Jersey

Wesley Hutchins
Forsyth Technical Community College
Winston-Salem, North Carolina

Darin Keith
Rock Island Arsenal Fire & Emergency Services
Rock Island, Illinois

J. Nathan Kempfer
City of Lawrence Fire Department
Lawrence, Indiana

Stephen Knight
Regional Education Service Agencies V
Parkersburg, West Virginia

Robert Knuth
North Dakota Firefighters Association
Bismarck, North Dakota

Richard Kosmoski
Middlesex County Fire Academy
Sayreville, New Jersey

Kevin Kupietz
Halifax Community College
Weldon, North Carolina

Roger Land
Lake Superior State University
Sault Ste. Marie, Michigan

John Lankford
Auburn Fire Division
Auburn, Alabama

John J. Lewis
City of Passaic Fire Department
Passaic, New Jersey

Robert A. Lindstedt
Southern Maine Community College
South Portland, Maine

Joshua Livermore
Bullhead City Fire Department
Bullhead City, Arizona

R. Paul Long
York County Department of Fire and Life Safety
York County, Virginia

Daniel Madrzykowski
National Institute of Standards and Technology (NIST), U.S. Department of Commerce
Gaithersburg, Maryland

F. Patrick Marlatt
Maryland Fire and Rescue Institute
College Park, Maryland

Mark Martin
Perry Township Fire Department
Masillon, Ohio

Al Martinez
Praire State College
Chicago Heights, Illinois

Paul D. Matheis
Southern California Training Officers Association
Irvine, California

Randy McCreadie
Cincinnati State Technical and Community College
Cincinnati, Ohio

Bill McGrath
CMEs Training
Fort Lauderdale, Florida

Matt Mead
Shepherd Tri-Township Fire Department
Shepherd, Michigan

Tony Mecham
CAL FIRE and the San Diego County Fire Department
San Diego, California

Todd Mesick
Maynard Fire Department
Utica, New York

Allen Michel
Nebraska State Fire Marshal – Training Division
Grand Island, Nebraska

Jon Mik
Muskegon County Hazmat Team
Muskegon, Michigan

William R Montrie
Springfield Township Fire Department
Holland, Ohio

Matthew C. Morgan
Great Oaks Public Safety Services
Cincinnati, Ohio

Nicholas Morgan
St. Louis Fire Department
St. Louis, Missouri

Michael L. New
Lower Columbia College
Longview, Washington

David Nix
Wilburton Fire Department
Wilburton, Oklahoma

Tom O'Brien
Newark Ohio Fire Department
Newark, Ohio

Patrick M. O'Connor
Providence Fire Department
Providence, Rhode Island

Shannon D. Orndorff
North Carolina Office of State Fire Marshal
Raleigh, North Carolina

Walter Osellame
Fire Services Training, College of the Rockies
Cranbrook, British Columbia, Canada

Nick Palmer
Louisiana State University Fire and Emergency Training Institute at Pine Country
Minden, Louisiana

Mike Pannell
DeKalb County Fire Rescue
Douglasville, Georgia

Scott Pascu
University of Akron
Akron, Ohio

Gary E. Patrick
Silverhill Volunteer Fire Department
Silverhill, Alabama

Dana Peloso
Oxford Fire Department
Oxford, Massachusetts

Michael D. Penders
Massachusetts Firefighting Academy
Stow, Massachusetts

Bill Pfeifer
Nebraska State Fire Marshals, Training Division
Randolph, Nebraska

Rick Picard
Peoria Fire Department
Peoria, Arizona

Lee Poteat
Dekalb County Fire Rescue
Tucker, Georgia

Jason Powell
Johnson City Fire Department
Johnson City, Tennessee

Joseph Ramsey
Winston-Salem Fire Department
Winston-Salem, North Carolina

Carl Raymond
Southern Crescent Technical College
Griffin, Georgia

David Reeves
Guilford Technical Community College
Jamestown, North Carolina

Joe Ridgeway
Lone Star College – CyFair
Cypress, Texas

Michael Ridley
Wilton Fire Department
Wilton, California

Todd Rielage
Chesterfield-Union Township Fire Department
Chesterfield, Indiana

Jeff Roberts
Durham Fire Department Training Division / Special Operations
Durham, North Carolina

Gerald F. Robinson
Ohio Fire Academy
Reynoldsburg, Ohio

Cary R. Roccaforte
TEEX/ESTI at Brayton Fire Training Field
College Station, Texas

Mark Romer
Division Chief, Retired
Lincoln, California

Derek Rosenquist
Estes Valley Fire Protection District
Estes Park, Colorado

Jose V. Salazar
Loudon County Fire-Rescue
Leesburg, Virginia

Mark Schuman
North Washington Fire Protection District
Denver, Colorado

John Shafer
Green Maltese LLC
Greencastle Fire Department
Greencastle, Indiana

Peter Silva, Jr.
Wisconsin Technical College System
Madison, Wisconsin

Ryan Skabroud
Lakeshore Technical College
Clevland, Wisconsin

Kenneth Smarr
DeKalb County Fire Rescue
Covington, Georgia

Jeffrey Smith
Pueblo Magnet High School
Tucson, Arizona

Joshua Smith
Mitchell Community College,
Statesville, North Carolina

Rich Solomon
Sable Altura Fire Rescue
Arapahoe County, Colorado

Brian Staska
Riverland Community College
Austin, Minnesota

Ron Stewart
Temple Fire and Rescue
Temple, Texas

Lori Stoney
Homewood Fire & Rescue Service
Homewood, Alabama

Gary Styers
Mooresville Fire-Rescue
Mooresville, North Carolina

David Swain
Durham Fire Department
Durham, North Carolina

Jimmy Talton
Reece City Volunteer Fire Department
Attalla, Alabama

Brent Thomas
Hamilton Emergency Services - Fire
Hamilton, Ontario, Canada

Matthew B. Thorpe
North Carolina Office of State Fire Marshal
King, North Carolina

Ian Toms
Raleigh Fire Department
Raleigh, North Carolina

James Utsey
Arkansas Fire Academy
Camden, Arkansas

Phil Vossmeyer
Cincinnati State Technical and Community College
Cincinnati, Ohio

Ronald J. Walls
San Bernardino County Fire Department
San Bernardino, California

James Weston
Sunset Utah Sunset Fire and Rescue
Clinton, Utah

W. Douglas Whittaker
Onondaga Community College Public Safety Training Center
Syracuse, New York

Charles A. Williams
King George County Department of Fire, Rescue, & Emergency Services
King George County, Virginia

Brent Willis
Martinez Columbia Fire Rescue
Martinez, Georgia

David Winter
College Place Fire Department
College Place, Washington

James Woolf
Twinsburg Fire Department
Twinsburg, Ohio

James Yazloff
San Pasqual Reservation Fire Academy
Valley Center, California

Gray Young
Louisiana State University Fire and Emergency Training Institute
Minden, Louisiana

Alan Zygmunt
Connecticut Fire Academy
Windsor Locks, Connecticut

Photographic Contributors

We would like to extend a huge thank you to Glen E. Ellman who was the photographer for this project. Glen Ellman is a commercial photographer and fire fighter based in Fort Worth, Texas. His expertise and professionalism are unmatched!

We also need to thank Frank Becerra from the Fort Worth Fire Department, Jennifer Schottke and Dave Schottke from the Fairfax City Fire Department, and Forest Reeder from Des Plaines Fire Department for their technical assistance with the photo shoots. We could not have captured so many complex and technically challenging skills without your expertise.

Finally, thank you to following people and organizations that opened up their facilities for these photo shoots:

Burleson Fire Department, Texas
Gary Wisdom, Fire Chief
David Franks
Mike Jones
Eddie Limon
James Pribble

Chicago Fire Department, Illinois
José A. Santiago, Fire Commissioner
Peter Van Dorpe, Chief of the Training Division
Michael DeJesus
Romeo King
Billy Latham
Jamar Sullivan
Joseph Vidinich

Fort Worth Fire Department, Texas
J.W. Brunson, Captain
David Johns, Captain
Martin Alcantar, Lieutenant
Frank Becerra, Fire Fighter
Mitchell Berry, Fire Fighter
Kevin Kouns, Fire Fighter
Dennis Scott Jr., Fire Fighter
Patrick Vega, Fire Fighter

Des Plaines Fire Department, Illinois
Alan Wax, Fire Chief
Forest Reeder, Division Chief
Nick Ciaburri
Joseph Ciraulo
Jeff Gove
Dan Hauser
Chris Jannusch
Mike Kiszkowski
Ted Lee
James Mulvihill
Ryan Petty
Jeff Schuck
Lukasz Szerlag
Bob Ward
Anna Watson
Tony Zulfer

San Antonio College, Regional Fire Academy, San Antonio, Texas
Martin D. Davila, Chairperson
Daniel Reese, Course Coordinator
Joe Martinez, Chief
Justin Flores, Recruit
Jorge Castilla, Recruit
Pete Rodriquez, Recruit
Jose Mejia, Recruit
Eduardo Flores, Recruit

Fairfax City Fire Department, Virginia
Dave Rohr, Fire Chief
Andrew Vita, Assistant Chief
Mark Ciarrocca, Battalion Chief
Andrea Clark, Captain
Jason Gorres, Master Technician
Michael Carpenter, Firemedic
Derek Sivsher, Firemedic

Virginia Department of Emergency Management (VDEM), Virginia
Tom Jordan, Technological Hazards Division

York County Department Fire and Life Safety, Virginia
Christopher Sadler, Battalion Chief
Richard Burgess, Lieutenant
Wade Dunlap, Hazardous Materials Technician
Robert Kudley, Hazardous Materials Technician
Reggie Rivers, Hazardous Materials Technician
Mark Shields, Hazardous Materials Technician
Brian Williams, Hazardous Materials Technician

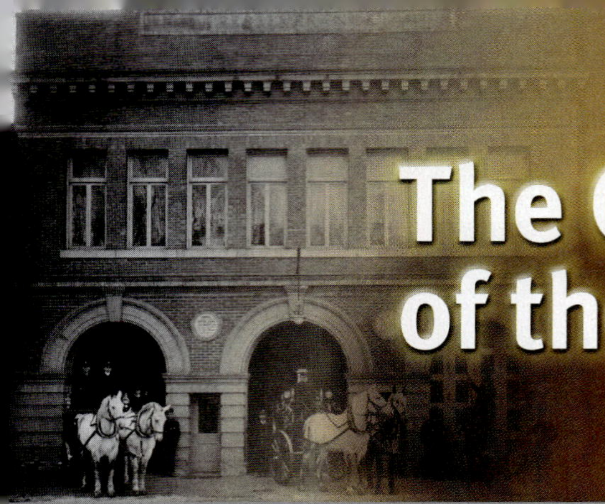

The Orientation and History of the Fire Service

Fire Fighter I

Knowledge Objectives

After studying this chapter, you will be able to:

- List five guidelines for successful fire fighter training. (p 4)
- Describe the general requirements for becoming a fire fighter. (p 5)
- Outline the roles and responsibilities of a Fire Fighter I. (NFPA5.1, p 5–6)
- Describe the common positions of fire fighters within the fire department. (NFPA5.1.1, p 6–7)
- Describe the specialized response roles within the fire department. (p 7)
- Explain the concept of governance and describe how the fire department's regulations, policies, and standard operating procedures affect it. (NFPA5.1.1, p 8)
- Locate information in departmental documents and standard operating procedures. (NFPA5.1.2, p 8)
- List the different types of fire department companies and describe their functions. (NFPA5.1.1, p 8–10)
- Describe how to organize a fire department in terms of staffing, function, and geography. (NFPA5.1.1, p 10–11)
- Explain the basic structure of the chain of command within the fire department. (NFPA5.1.1, p 11)
- Define the four basic management principles used to maintain organization within the fire department. (p 12)
- Explain the evolution of the methods and tools of firefighting from colonial days to the present. (p 13–17)
- Explain how building codes prevent the loss of life and property. (p 13–14)
- Describe the evolution of funding for fire department services. (p 17–18)

Skills Objectives

There are no skill objectives for Fire Fighter I candidates.

CHAPTER 1

Fire Fighter II FFII

Knowledge Objectives

- Outline the responsibilities of a Fire Fighter II. (NFPA 6.1, p 6)
- Describe the roles of a Fire Fighter II within the fire department. (NFPA 6.1.1, p 6)

Skills Objectives

There are no skill objectives for Fire Fighter II candidates. NFPA 1001 contains no Fire Fighter II Job Performance Requirements for this chapter.

Additional NFPA Standards

- NFPA 1500, *Standard on Fire Department Occupational Safety and Health Program*
- NFPA 1582, *Standard on Comprehensive Operational Medical Program for Fire Departments*

You Are the Fire Fighter

It is your first day in the station after having completed four months of recruit academy. Your new crew welcomes you. As you are being shown around the station, the doorbell rings—and you are told your first official duty is to answer it. At the door, you find a young woman and her young son. She sheepishly says that her son has been begging her to meet the fire fighters because he wants to be one when he grows up.

As you begin the tour, you show the pair the living quarters and then move to the engine room and the boot room. Like most youngsters, the visiting boy is full of questions and asks, "Where are the rest of the fire fighters for all those coats? Do you drive the fire truck? Why does that truck have a big ladder on top?" The mother tells him to give you a chance to explain before asking any more questions. You smile as you ponder your answers.

1. How do you explain the difference between the number of turnout gear ensembles and the number of personnel at the station?
2. Explain how the duties of the captain, the engineer, and the fire fighters differ.
3. Explain the types of apparatus used in the fire service and their functions.

Introduction

Training to become a fire fighter is challenging. The art and science of extinguishing fires is much more complex than most people imagine. You will be challenged—both physically and mentally—during this course. You must keep your body in excellent condition so you can complete your assignments; you must also remain mentally alert to cope with the various conditions you will encounter. Fire fighter training will expand your understanding of fire suppression—that is, the various activities involved in controlling and extinguishing fires.

By the time you complete this course, you will be well equipped to continue a centuries-old tradition of preserving lives and property threatened by fires **FIGURE 1-1**. You will have an understanding of the many duties and responsibilities of the fire service, and you should have a feeling of personal satisfaction about your accomplishment.

This chapter first offers five guidelines for you to remember as you progress through this course and take your place as a fire fighter in your community. It next presents the responsibilities of a Fire Fighter I and a Fire Fighter II. The chapter then explains the organization of the fire service and describes the chain of command that connects all members of a department. Finally, it reviews the history and development of the modern-day fire service.

FIGURE 1-1 When you join the fire service, you join a profession with a long and noble history of protecting and serving the community. This memorial to fallen fire fighters is located at the National Fire Academy in Emmittsburg, Maryland.

Fire Fighter Guidelines

During this course of study, you will need to practice and work hard. Do your best. The five guidelines listed here will help both to keep you on target to become a proud and accomplished fire fighter and throughout your career as a fire fighter.

1. *Be safe*. Safety should always be uppermost in your mind. Keep yourself safe. Keep your teammates safe. Keep the public you serve safe.
2. *Follow orders*. Your supervisors have more training and experience than you do. If you can be counted on to follow orders, you will become a dependable member of the team.
3. *Work as a team*. Fighting fires requires the coordinated efforts of all department members. Teamwork is essential to success.
4. *Think!* Lives will depend on the choices you make. Put your brain in gear. Think about what you are studying.
5. *Follow the Golden Rule*. Treat each team member, victim, or citizen as an important person or as you would treat a member of your own family. Everyone is an important person or family member to someone, and everyone deserves your best efforts.

FIRE FIGHTER Tips

The mission of the fire department is to save lives and protect property through prevention, education, suppression, and rescue activities.

Fire Fighter Qualifications

Not everyone can become a fire fighter. Those who do succeed in achieving this status understand the vital mission of the fire department: to save lives and protect property through prevention, education and suppression. A fire fighter must be healthy and in good physical condition, assertive enough to enter a dangerous situation, yet mature enough to work as a member of a team **FIGURE 1-2**. The job requires a person who has the desire to learn, the will to practice, and the ability to apply the skills of the trade. A fire fighter is constantly learning as the body of knowledge about fires expands and the technology used in fighting fires evolves.

FIGURE 1-2 Firefighting requires a special person.

The training and performance qualifications for fire fighters are specified in NFPA 1001, *Standard for Fire Fighter Professional Qualifications*. Age, education requirements, medical requirements, and other criteria are established locally.

Age Requirements

Most career fire departments require that candidates be at least 18 years of age, although some have specified a minimum age of 21. Candidates should possess a valid driver's license, have a clean driving record, have no criminal record, and be drug free.

Volunteer fire departments usually have different age requirements, which often depend on insurance considerations. Some volunteer fire departments allow "junior" members to join at age 16 but restrict their activities until they reach age 18.

Education Requirements

Most career fire departments require that applicants have at least a high school diploma or equivalent. Some require candidates to have completed more advanced college-level courses in a field related to fire or Emergency Medical Services (EMS). All departments require fire fighters who wish to be considered for promotion or extra responsibility to take additional courses.

Medical Requirements

Because firefighting is both stressful and physically demanding, fire fighters must undergo a medical evaluation before their training begins. This medical evaluation will identify most medical conditions or physical limitations that could increase the risk of injury or illness either to the candidate or to other fire fighters. Medical requirements for fire fighters are specified in NFPA 1582, *Standard on Comprehensive Operational Medical Program for Fire Departments*.

Physical Fitness Requirements

Physical fitness requirements are established to ensure that fire fighters have the strength and stamina needed to perform the tasks associated with firefighting and emergency operations. Several performance-testing scenarios have been validated by qualified organizations. NFPA 1001 allows individual fire departments to choose the fitness-testing method that will be used for fire fighter candidates.

Emergency Medical Care Requirements

Delivering emergency medical care is an important function of most fire departments. NFPA 1001 allows individual fire departments to specify the level of emergency medical care training required for entry-level personnel. At a minimum, you will need to understand infection control procedures and be able to perform cardiopulmonary resuscitation (CPR), control bleeding, and manage shock. These skills are discussed in more detail in the Fire and Emergency Medical Care chapter. Many departments require fire fighters to become certified as an Emergency Medical Responder, an Emergency Medical Technician (EMT)–Basic, or a Paramedic.

Roles and Responsibilities of Fire Fighter I and Fire Fighter II

The first step in understanding the organization of the fire service is to learn your roles and responsibilities as a Fire Fighter I or Fire Fighter II. As you progress through this text, you will learn what to do and how to do it so that you can take your place confidently among the ranks.

Roles and Responsibilities for Fire Fighter I

- Don (put on) and doff (take off) personal protective equipment (PPE) properly.
- Hoist hand tools using appropriate ropes and knots.
- Understand and correctly apply appropriate communication protocols.
- Use self-contained breathing apparatus (SCBA).
- Respond on apparatus to an emergency scene.
- Establish and operate safely in emergency work areas.
- Force entry into a structure.
- Exit a hazardous area safely as a team.
- Set up ground ladders safely and correctly.

- Attack a passenger vehicle fire, an exterior Class A fire, and an interior structure fire.
- Conduct search and rescue in a structure.
- Perform ventilation of an involved structure.
- Overhaul a fire scene.
- Conserve property with salvage tools and equipment.
- Connect a fire department engine to a water supply.
- Extinguish incipient Class A, Class B, Class C, and Class D fires.
- Illuminate an emergency scene.
- Turn off utilities.
- Combat a ground cover fire.
- Perform fire safety surveys.
- Clean and maintain equipment.

Per manufacturer recommendation

FIRE FIGHTER II

Roles and Responsibilities for Fire Fighter II

- Prepare reports.
- Communicate the need for assistance.
- Coordinate an interior attack line team.
- Extinguish an ignitable liquid fire.
- Control a flammable gas cylinder fire.
- Protect evidence of fire cause and origin.
- Assess and disentangle victims from motor vehicle accidents.
- Assist special rescue team operations.
- Perform a fire safety survey.
- Present fire safety information.
- Maintain fire equipment.
- Perform annual service tests on fire hoses.

Every member of the fire service will interact with the public. In addition to interacting with citizens at the scene of incidents, fire fighters will encounter people who visit the fire station requesting a tour or asking questions about specific fire safety issues FIGURE 1-3. Fire fighters should be prepared to assist these visitors and use this opportunity to provide them with additional fire safety information. Use every contact with the public to deliver positive public relations and an educational message.

FIGURE 1-3 Any contact with the public should be used as an educational opportunity.

Roles Within the Fire Department

General Roles

A fire fighter may be assigned any task from placing hose lines to extinguishing fires. Generally, the fire fighter is not responsible for any command functions and does not supervise other personnel, except on a temporary basis when promoted to an acting officer. Through additional training and promotion, a fire fighter may fulfill many roles during a career, particularly in smaller departments. Some of the more common positions include the following:

- Fire apparatus driver/operator: Often called an engineer or technician, the driver is responsible for getting the fire apparatus to the scene safely, as well as setting up and running the pump or operating the aerial ladder once it arrives on the scene. This function is a full-time role in some departments; in other departments, it is rotated among the fire fighters.
- Company officer: The company officer is usually a lieutenant or captain who is in charge of an apparatus. This person leads the company both on scene and at the station. The company officer is responsible for initial firefighting strategy, personnel safety, and the overall activities of the fire fighters on their apparatus. Once command is established, the company officer focuses on tactics.
- Safety officer: The safety officer watches the overall operation for unsafe practices. He or she has the authority to halt any firefighting activity, allowing the activity to resume only when it can be done safely and correctly. The senior ranking officer may act as safety officer until the appointed safety officer arrives or until another officer is delegated those duties.
- Training officer: The training officer is responsible for updating the training of current fire fighters and for training new fire fighters. He or she must be aware of the most current techniques of firefighting and EMS. The training officer must coordinate various aspects of the fire department's training program and maintain documentation of all training activities.
- Incident commander (IC): The incident commander is responsible for the management of all incident operations. This position focuses on the overall strategy of the incident and is often assumed by the battalion/district chief.
- Fire marshal/fire inspector/fire investigator: The fire marshal inspects businesses and enforces public safety laws and fire codes. He or she may also respond to fire scenes to help investigate the cause of a fire. The investigator may have full police powers to investigate and arrest suspected arsonists and people causing false alarms.

- Fire and life safety education specialist: The fire and life safety education specialist educates the public about fire safety and injury prevention and presents juvenile fire safety programs.
- 911 dispatcher/telecommunicator: From the communications center, the dispatcher takes calls from the public, sends appropriate units to the scene, assists callers with emergency medical information (treatment that can be performed until the medical unit arrives), and assists the IC with obtaining needed resources.
- Fire apparatus maintenance personnel: Apparatus mechanics repair and service fire and EMS vehicles, keeping them ready to respond to emergencies. These individuals are usually trained by equipment manufacturers to repair vehicle engines, lights, and all parts of the fire pump and aerial ladders.
- Fire police: Fire police are usually fire fighters who control traffic and secure the scene from public access. Many fire police are also sworn peace officers.
- Information management: "Info techs" are fire fighters or civilians who take care of a department's computer and networking systems. Maintaining computerized fire reports, e-mail, and a functional computer network requires the services of properly trained computer personnel.
- Public information officer: The public information officer serves as a liaison between the IC and the news media.
- Fire protection engineer: The fire protection engineer reviews plans and works with building owners to ensure that their fire suppression and detection systems will meet the applicable codes and function as needed. Some fire protection engineers actually design these systems, and most hold an engineering degree.

Specialized Response Roles

Many assignments require specialized training. Indeed, most large departments have teams of fire fighters who respond to specific types of calls. Members of these teams are usually required to be fire fighters before they begin additional training. Specialist positions include the following:

- Aircraft/crash rescue fire fighter (ARFF): Aircraft rescue fire fighters are based on military and civilian airports and receive specialized training in aircraft fires, extrication of victims on aircraft, and extinguishing agents. They wear special PPE and respond in specialized fire apparatus that protects them from high-temperature fires caused by substances such as jet fuel.
- Hazardous materials technician: "Hazmat" technicians have training and certification in chemical identification, leak control, decontamination, and clean-up procedures.
- Technical rescue technician: "Tech rescue" technicians are trained in special rescue techniques for incidents involving structural collapse, trench rescue, swiftwater rescue, confined-space rescue, high-angle rescue, and other unusual situations. The units they work in are sometimes called urban search and rescue teams.
- SCUBA dive rescue technician: Many fire departments, especially those located near waterways, lakes, or an ocean, use SCUBA technicians who are trained in rescue, recovery, and search procedures in both water and under-ice situations. (SCUBA stands for "self-contained underwater breathing apparatus.")
- Emergency Medical Services (EMS) personnel: EMS personnel administer prehospital care to people who are sick or injured. Prehospital calls account for the majority of responses in many departments, so fire fighters are often cross-trained with EMS personnel. EMS training levels are normally divided into three categories: Emergency Medical Technician–Basic, Advanced Emergency Medical Technician, and Emergency Medical Technician–Paramedic.
  1. Emergency Medical Technicians (EMTs): Most EMS providers are EMTs. They have training in basic emergency care skills, including oxygen therapy, bleeding control, CPR, automated external defibrillation, basic airway devices, and assisting patients with certain medications.
  2. Advanced Emergency Medical Technician (AEMT): AEMTs can perform more procedures than EMT-Basics, but they are not yet Paramedics. They have training in specific aspects of advanced life support, such as intravenous (IV) therapy, interpretation of cardiac rhythms, defibrillation, and airway intubation.
  3. Paramedic: A Paramedic has completed the highest level of training in EMS. These personnel have extensive training in advanced life support, including IV therapy, administering drugs, cardiac monitoring, inserting advanced airways (endotracheal tubes), manual defibrillation, and other advanced assessment and treatment skills.

Working with Other Organizations

Fire departments are a vital part of their communities. To fulfill its mission, each fire department must interact with other organizations in the community. Consider the response to a motor vehicle crash FIGURE 1-4. In an area with a centralized 911 call center, the fire department, EMS, law enforcement officials, and tow-truck operators will all be notified of this event. If the crash affects utility service, the utility company will be notified as well. If EMS is not part of the fire department, the EMS provider will be contacted. In addition, fire fighters frequently interact with hospital personnel who assume care for ill or injured victims who are transported to emergency rooms.

When multiple agencies work together at an incident, a unified command must be established as part of the incident command system (ICS). A unified command system eliminates multiple command posts, establishes a single set of incident goals and objectives, and ensures mutual communication and cooperation. Although there can be only one IC at each scene, each agency must have input in handling an emergency. The

FIGURE 1-4 Many agencies work together at the scene of a motor vehicle crash.

ICS is discussed more thoroughly in the Incident Command System chapter.

Other organizations may be involved in different types of incidents. For example, utility companies, the Salvation Army, and the Red Cross may be part of the team responding to structure fires. Wildland fires may involve representatives from various government agencies and jurisdictions. Large-scale incidents may call upon several agencies, including the following:

- Public works
- School administrators
- Medical examiners or coroners
- Funeral directors
- Government officials (mayor, city manager)
- Federal Bureau of Investigation
- Military
- Federal Emergency Management Agency
- Search and rescue teams
- Fire investigators
- Various state agencies

Communications centers keep a list of contact names and phone numbers so these parties may be notified without delay when their presence at an incident is required.

Fire Department Governance

Governance is the process by which an organization exercises authority and performs the functions assigned to it. The governance of a fire department depends on regulations, policies, and standard operating procedures (SOPs). Each of these concepts is discussed in this section.

Regulations are developed by various government or government-authorized organizations to implement a law that has been passed by a government body. For example, federal occupational health and safety laws are adopted as regulations by some states. These regulations may apply to activities within the fire department.

Fire dept. can't make people leave their house

Policies are developed to provide definitive guidelines for present and future actions. Fire department policies, for example, outline what is expected in stated conditions. These policies often require personnel to make judgments and to determine the best course of action within the stated policy. Policies governing parts of a fire department's operations may be developed by other government agencies, such as personnel policies that cover all employees of a city or county.

Standard operating procedures (SOPs) provide specific information on the actions that should be taken to accomplish a certain task **FIGURE 1-5**. SOPs are developed within the fire department, are approved by the chief of the department, and ensure that all members of the department perform a given task in the same manner. They provide a uniform way of dealing with emergency situations, enabling fire fighters from different stations or companies to work together smoothly, even if they have never encountered one another before. These procedures are vital because they enable everyone in the department to function properly and know what is expected for each task. Fire fighters must learn and frequently review departmental SOPs. An example would be your policy and procedure manuals.

In some fire departments, suggested operating guidelines (SOGs) are utilized and are not as strict as SOPs, because conditions may dictate that the fire fighter or officer use his or her personal judgment in completing the procedure. This flexibility allows the responder to deviate from a set procedure, yet still be held accountable for that action.

A practical way to organize a department's SOP manual is with removable pages collected in a three-ring binder, which ensures that regular updates can be added easily. Each department member should have an SOP manual and must update it as needed. This manual should be organized in sections—such as administration, safety, scene operations, apparatus and equipment, station duties, uniforms, and miscellaneous—and use a simple numbering system based on section and policy numbers. Many fire departments maintain their SOPs on a computer network, which simplifies the process of providing all employees with up-to-date SOPs.

The Organization of the Fire Service

Company Types

A fire department includes many different types of companies, each of which has its own job at the scene of an emergency. Companies may be composed of various combinations of people and equipment; in smaller departments, a company may fill many roles. The most common types of companies and their roles are described here:

- Engine company: An engine company is responsible for securing a water source, deploying handlines, conducting search and rescue operations, and putting water on the fire **FIGURE 1-6**. Each fire engine has a pump, carries hoses, and maintains a booster tank of water. Engines also carry a limited quantity of ladders and hand tools.

Anytown Fire Department

Standard Operating Procedure

Date: 1/1/08
Section: 1 – Administration SOP 01-01 Page 1 of 1
Maintaining Station Logbooks

Purpose

This guideline is provided to ensure that information documented in station logbooks is maintained in a consistent manner, station to station, shift to shift, throughout the department. The station commander shall have discretion in formatting the information.

Scope

This guideline shall be followed by all authorized department personnel entering information into the station logbooks. Any deviations from this guideline will be the responsibility of the individual making the deviation.

Policy

The station log is an important component of the fire department's record-keeping system. All entries must be legible, written in black ink, and the writer identified by name and computer identification number. All members should treat the logbook as a legal document and record all of the station's activities as well as other information deemed important by the station commander.

Procedure

The following guidelines should be used when making logbook notations:

1. The day, date, and shift noted on the top line of the page.
2. A list of each person assigned to that shift to include their computer identification numbers, and the apparatus to which they are assigned for the shift. Personnel on leave should be identified with the type of leave (ie, annual, sick, leave without pay, funeral, military, training) noted. A notation should be made by the name of any member temporarily assigned to another station during that shift.
3. Daily safety talks should be recorded prior to the section dedicated to emergency responses. For more information on safety talks, refer to SOP 31.3.
4. The section of the log dedicated to recording chronological activity should have two (2) columns on the left side of the book: the far left should note time and the second column the incident number. A column to the far right should be used to note the identification number of the person making the entry.
5. Incident numbers for medic unit responses should be noted in blue ink.
6. Incident numbers for fire suppression responses should be noted in red ink.
7. Units responding, the address to which the response is being made, and the type of situation found should be noted behind the message number (ie, E-5 responded to 304 Albemarle Drive/Gas Leak).
8. The term fill-in should be used in the logbook to denote one station standing by in another station to cover a company that has responded to an incident.
9. Entries for medic units should include the hospital to which the patient was transported.
10. Station maintenance/repairs and apparatus maintenance should be documented in the section of the logbook dedicated to that purpose. This documentation may also be maintained in a separate logbook dedicated to those types of entries.

Although the information requirements provided here must be maintained in the station's logbook, the format used for ensuring that documentation is left to the station commander's discretion.

Approved: ______________________________ Date: ______________________
Fire Chief

FIGURE 1-5 A sample standard operating procedure.

FIGURE 1-6 An engine company secures a water source and extinguishes the fire.

- Truck company: A truck company (also called a ladder company) specializes in forcible entry, ventilation, roof operations, search and rescue, and deployment of ground ladders FIGURE 1-7. Trucks carry several ground ladders, ranging from 8 feet to 50 feet in length, as well as an extensive quantity of tools. Trucks are also equipped with an aerial device, such as an aerial ladder, tower ladder, or ladder/platform. These aerial devices can be raised and positioned above a roof to provide fire fighters with a stable, safe work zone.
- Rescue company: A rescue company usually is responsible for rescuing victims from fires, confined spaces, trenches, and high-angle situations. Rescue companies carry an extensive array of regular and specialized tools.
- Wildland/brush company: A wildland/brush company is dispatched to wildland and brush fires that larger engines cannot reach. Because they often work in rough terrain, wildland/brush companies use four-wheel drive vehicles. They carry a tank of water and a pump that enables them to pump water while the truck is moving. These companies also carry special firefighting equipment such as portable pumps, rakes, and shovels.
- Hazardous materials company: A hazardous materials company responds to and controls scenes involving spilled or leaking hazardous chemicals. These companies have special equipment, PPE, and training to handle most emergencies involving chemicals.
- Emergency Medical Services company: An EMS company may include medical units such as ambulances or first-response vehicles. These companies respond to and assist in transporting medical and trauma victims to medical facilities for further treatment. They often carry medications, defibrillators, and other equipment that can stabilize a critical patient during transport. Engine or truck companies may be staffed with EMS providers who act as first responders until a transport ambulance arrives FIGURE 1-8.

Other Views of Fire Service Organization

There are several other ways to look at the organization of a fire department, including in terms of staffing, function, and geography.

Staffing

A fire department must have sufficient trained personnel available to respond to a fire at any hour of the day, every day of the year. Staffing issues affect all fire departments—career departments, combination departments, and volunteer departments. In volunteer departments, it is particularly important to ensure that enough responders are available at all times, particularly during the day. In the past, when many people worked in or near the communities where they lived, volunteer response time was not an issue. Today, however, many people have longer commutes and work longer hours, so the number of people available to respond during the day may be limited. Some volunteer departments have been forced to hire full-time fire fighters during daytime hours to ensure sufficient personnel are available to respond to an incident.

FIGURE 1-7 A truck company provides aerial support at a fire.

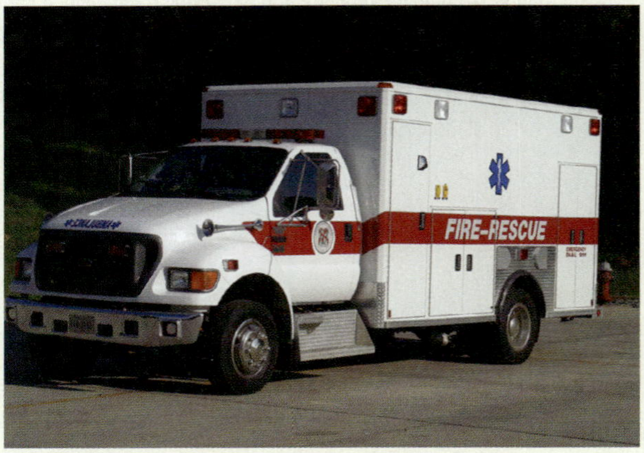

FIGURE 1-8 An EMS company delivers medical services.

Function

Fire departments can be organized along functional lines. For example, the training division is responsible for leading and coordinating department-wide training activities; truck companies have certain functional responsibilities at a fire; hazardous materials squads have different functional responsibilities that support the overall mission of the fire department.

Geography

Each fire department is responsible for a specific geographic area, and each station in the department is assigned the primary responsibility for a geographic area within the community. Fire stations are distributed throughout a community in a manner intended to ensure a rapid response time to every location in that community. This design enables the fire department to distribute and use specialized equipment efficiently throughout the community.

Chain of Command

Should manage 3–7 people

The organizational structure of a fire department consists of a chain of command. Although the precise ranks may vary in different departments, the basic concept remains the same across the fire service. The chain of command creates a structure for managing the department and the fire-ground operations.

Fire fighters usually report to a lieutenant, who is responsible for a single fire company (such as an engine company) on a single shift. Lieutenants can provide a number of practical skills and tips to new recruits.

The next level in the chain of command is the captain. Captains are responsible not only for managing a fire company on their shift, but also for coordinating the company's activities with other shifts. A captain may be in charge of the activities of a fire station.

Captains report directly to chiefs. Several levels of chiefs are designated. Battalion chiefs (also known as district chiefs) are responsible for coordinating the activities of several fire companies in a defined geographic area such as a station or district. A battalion chief is usually the officer in charge of a single-alarm working fire.

Above battalion chiefs are assistant or division chiefs. Assistant or division chiefs are usually in charge of a functional area, such as training, within the department. These officers report directly to the chief of the department.

The chief of the department has overall responsibility for the administration and operations of the fire department. The chief can delegate responsibilities to other members of the department but is still responsible for ensuring that these activities are carried out properly.

The fire service's chain of command is used to implement department policies. This organizational structure enables a fire department to determine the most efficient and effective way to fulfill its mission and to communicate this information to all members of the department FIGURE 1-9. Adhering to the chain of command ensures that a given task is carried out in a uniform manner. A variety of documents, including the previously discussed regulations, policies, procedures, and SOPs, are used to achieve this goal.

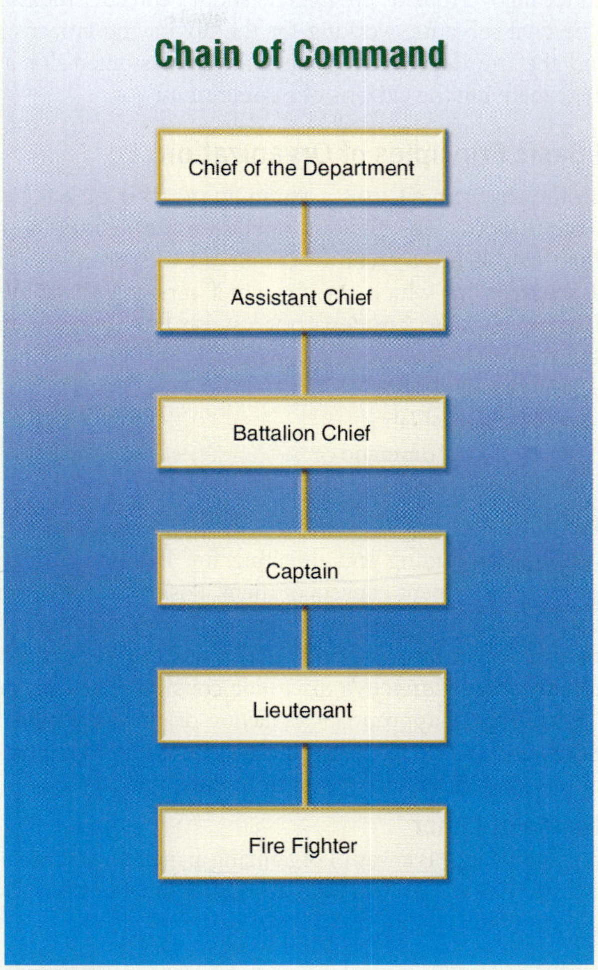

FIGURE 1-9 The chain of command ensures that the department's mission is carried out efficiently and effectively.

Source of Authority

Governments—whether municipal, state, provincial, or national—are charged with protecting the welfare of the public against common threats. Fire is one such peril, because an uncontrolled fire threatens everyone in the community. Citizens accept certain restrictions on their behavior and pay taxes to protect themselves and the common good. As a consequence, people charged with protecting the public may be given certain privileges to enable them to perform effectively. For example, fire departments can legally enter a locked home without permission to extinguish a fire and protect the public.

The fire service draws its authority from the governing entity responsible for protecting the public from fire—whether it is a town, a city, a county, a township, or a special fire district. Federal and state governments also grant authority to fire departments. In some states, private corporations have contracts to provide fire protection to municipalities or government agencies. The head of the fire department (the fire chief) is accountable to the leaders of the governing body, such as the city council, the county commission, the mayor, or the city manager. Because of the relationship between a fire department and local government, fire fighters should consider themselves to be civil servants, working for the tax-paying citizens who fund the fire department. The ultimate customers for a fire department are the citizens of a community.

Basic Principles of Organization

The fire department uses a paramilitary style of leadership. In other words, fire fighters operate under a rank system, which establishes a chain of command (discussed earlier in this chapter) in which the fire chief serves as head of the department. Most fire departments use four basic management principles:

- Discipline
- Division of labor
- Unity of command
- Span of control

Discipline

Discipline is guiding and directing fire fighters to do what their fire department expects of them. Positive discipline consists of providing guidelines for the right way of doing things. Examples of positive discipline are policies, SOPs, training, and education. Corrective discipline consists of actions taken to discourage inappropriate behavior or poor performance. Examples of corrective discipline are counseling sessions, formal reprimands, or suspension from duty.

Division of Labor

Division of labor is a way of organizing an incident by breaking down the overall strategy into a series of smaller tasks. Some fire departments are divided into units based on function. For example, engine companies establish water supplies and pump water; truck companies perform forcible entry, rescue, and ventilation functions. Each of these functions can be divided into multiple assignments, which can then be assigned to individual fire fighters. With division of labor, the specific assignment of a task to an individual makes that person responsible for completing the task and prevents duplication of job assignments.

Unity of Command

Unity of command is the concept that each fire fighter answers to only one supervisor, each supervisor answers to only one boss, and so on FIGURE 1-10. In this way, the chain of command ensures that everyone is answerable to the fire chief and establishes a direct route of responsibility from fire chief to fire fighter.

At a fire ground, all functions are assigned according to incident priorities. A fire fighter with more than one supervising officer during an emergency may be overwhelmed with various assignments, and the incident priorities may not be accomplished in a timely and efficient manner. The concept of unity of command is designed to avoid such conflicts and lead to more effective firefighting.

Span of Control

Span of control is the number of people whom one person can supervise effectively. Most experts believe that span of control should extend to no more than five people in a complex or rapidly changing environment. This number can change, however, depending on the assignment or task to be completed.

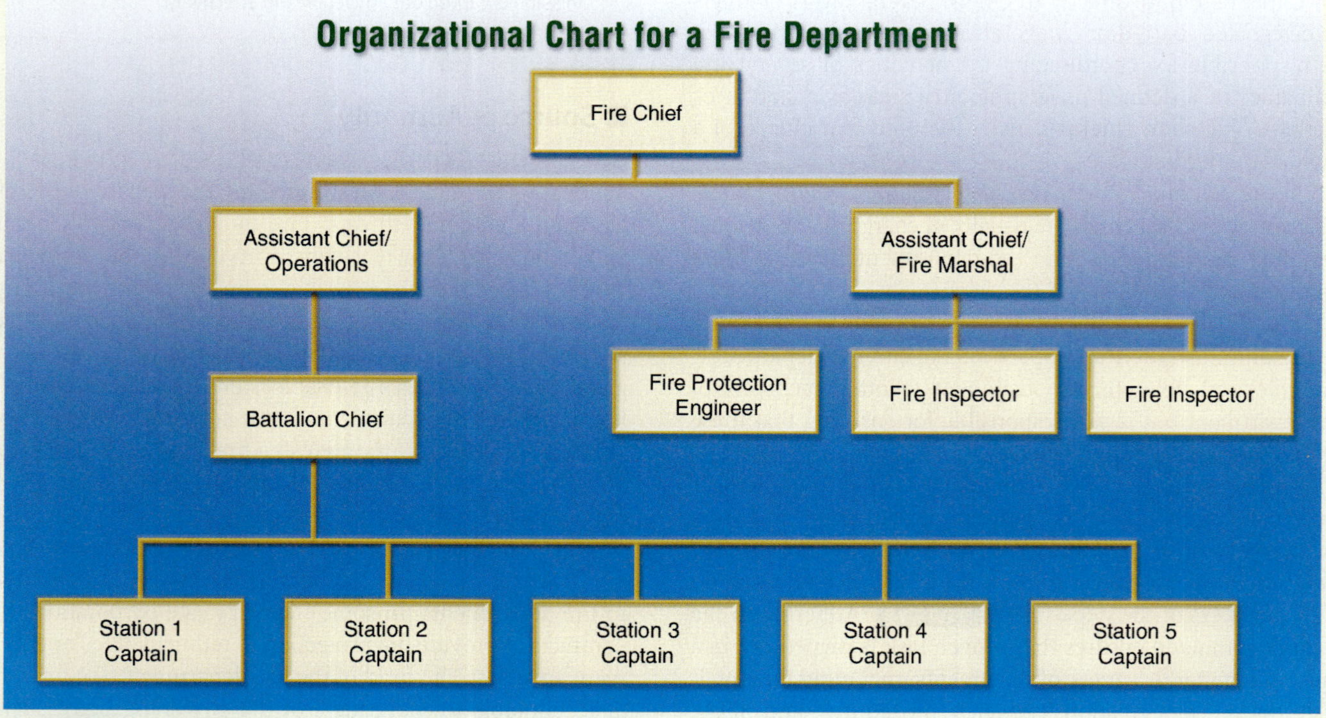

FIGURE 1-10 The organization of a typical fire department.

The History of the Fire Service

Since prehistoric times, controlled fire has been a source of comfort and warmth, but uncontrolled fire has brought death and destruction. Historical accounts from the ancient Roman Empire describe community efforts to suppress uncontrolled fire. In 24 B.C., the Roman emperor Augustus Caesar created what was probably the first fire department. Called the Familia Publica, it was composed of approximately 600 slaves who were stationed around the city and charged with watching for and fighting fires. Of course, because the Familia Publica consisted of slaves, these conscripts had little interest in preserving the homes of their masters and little desire to take risks, so fires continued to be a problem.

In about 60 A.D., under the emperor Nero, the Corps of Vigiles was established as the Roman Empire's fire protectors. This group of 7000 free men was responsible for firefighting, fire prevention, and building inspections. The Corps of Vigiles adopted the formal rank structure of the Roman military, which continues to be used by today's fire departments.

The American Fire Service

The first documented structure fire in North America occurred in Jamestown, Virginia, in 1607. This fire started in the community blockhouse and almost burned down the entire settlement. At that time, most structures were built entirely of combustible materials such as straw and wood. Local ordinances soon required the use of less flammable building materials and mandated that fires be banked (covered over) throughout the night. In 1630, the city of Boston established the first fire regulations in North America when it banned wood chimneys and thatched roofs. In the Dutch colony of New York in 1647, Governor Peter Stuyvesant not only banned wood chimneys and thatched roofs but also required that chimneys be swept out regularly. Fire wardens imposed fines on homeowners who did not obey these regulations; the money collected was used to pay for firefighting equipment.

The first paid fire department in the United States was established in 1679 in Boston, which also had the first fire stations and fire engines. The first volunteer fire company began in Philadelphia in 1735, under the leadership of Benjamin Franklin. There, citizens were required to keep buckets filled with water outside their doors, readily available to fight fires. Franklin recognized the many dangers of fire and continually sought ways to prevent it. For example, he developed the lightning rod to help draw lightning strikes (a common cause of fires) away from homes. Another early volunteer fire fighter, George Washington, imported one of the first hand-powered fire engines from England. He donated this engine to the Alexandria, Virginia, Fire Department in 1765.

In 1871, two major fires significantly affected the development of both the fire service and fire codes. At the time, the city of Chicago was a "boom town" with 60,000 buildings—40,000 of which were constructed of wood, with roofs of tar and felt or wooden shingles. Given the lax construction regulations and no rain for three weeks, the city was primed for a major fire. On October 8, 1871, a fire started in a barn on the west side of the city. The fire department was already exhausted from fighting a four-block fire earlier in the day. Errors in judging the location of the fire and in signaling the alarm resulted in a delayed response time. The Great Chicago Fire burned through the city for three days. When it was over, more than 2000 acres and 17,000 homes had been destroyed, the city had suffered more than $200 million in damage, and 300 people were dead and 90,000 homeless.

At the same time, another major fire was raging just 262 miles north of Chicago in Peshtigo, Wisconsin. This fire was not as highly publicized as the Great Chicago Fire, but proved even more deadly. Throughout the summer, the north woods of Wisconsin had experienced drought-like conditions, and logging operations had left pine branches carpeting the forest floor. A flash forest fire created a "tornado of fire" more than 1000 feet high and 5 miles wide. More than 2400 square miles of forest land burned, several small communities were destroyed, and more than 2200 people lost their lives. The Peshtigo firestorm even jumped the 60-mile-wide Green Bay to destroy several hundred more square miles of land and settlements on Wisconsin's northeast peninsula.

As a result of the Great Chicago Fire, but also in response to the Peshtigo Fire, communities began to enact strict building and fire codes **FIGURE 1-11**. The development of water pumping systems, advances in firefighting equipment, and improvements in communications and alarm systems helped ensure that such tragedies did not recur.

Building Codes

Throughout history, the threat of fire has served as an impetus for communities to establish building codes. Although the first building codes, which were developed in ancient Egypt, focused on preventing building collapse, building codes were quickly recognized as an effective means of preventing, limiting, and containing fires.

Colonial communities had few building codes. The first settlers had a difficult time erecting even primitive shelters, which often were constructed of wood with straw-thatched roofs. The fireplaces used for cooking and heating may have had chimneys constructed of smaller logs. The all-wood construction and use of open fires meant that fires were a constant threat to early settlers in America. As communities developed, they enacted codes restricting the hours during which open fires were permitted and the materials that could be used for roofs and chimneys. In 1678, Boston required that "tyle" or slate be used for all roofs. After British forces burned Washington, D.C., in 1814 (during the War of 1812), codes prohibited the

FIRE FIGHTER Tips

Benjamin Franklin organized the first fire insurance company in the United States and coined the phrase, "An ounce of prevention is worth a pound of cure"—an apt motto for fire safety. Early insurance companies marked the homes of their policy holders with a plaque or fire mark that showed the name or logo of the insurance company. The insurance company paid fire companies to respond to those buildings displaying their fire mark. Sadly, others were left to burn.

FIGURE 1-11 The Great Chicago Fire of 1871 caused the deaths of 300 people and led to changes in firefighting operations in the United States.

building of wooden houses in the area. Some building codes required the construction of a fire-resistive wall, or firewall, of brick or mortar between two buildings.

Today's building codes not only govern construction materials, but also frequently require built-in fire prevention and safety measures. Required fire detection equipment notifies both building occupants and the fire department. Built-in fire suppression and sprinkler systems help contain fires to a small area and prevent small fires from growing into major fires. Fire escapes, stairways, and doors that unlock when the alarms sound and that open outward enable occupants to escape a burning building safely. Without modern building code requirements, high-rise buildings and large shopping centers could not be built safely.

FIRE FIGHTER Tips

Modern building codes help to ensure that fire departments receive prompt notification of fires, and limit or extinguish fires that might otherwise overwhelm local firefighting resources.

At first, each community established its own building codes. As larger government jurisdictions were established, however, a more uniform code was adopted across the United States, reflecting a minimum standard. Local communities were permitted to make this code stricter as needed.

Today, U.S. codes and standards are written by national organizations such as the National Fire Protection Association (NFPA). Volunteer committees of citizens and representatives of businesses, insurance companies, and government agencies research and develop proposals, which are in turn debated and reviewed by various groups. The final document, known as a consensus document, is then presented to the public. Many states adopt selected NFPA codes and standards as law.

Training and Education

Fire fighter training and education have also come a long way over the years. The first fire fighters simply needed sufficient muscular strength and endurance to pass buckets or operate a hand pumper. As equipment became more complex, however, the importance of formalized training and good judgment increased.

Today's fire fighters operate high-tech, costly equipment, including million-dollar apparatus, radios, thermal imaging cameras (high-tech devices that use infrared technology to find objects giving off a heat signature), and self-contained breathing apparatus (SCBA; air packs used by fire fighters to enter a hazardous atmosphere) FIGURE 1-12. These tools, as well as better fire detection devices, have greatly increased the safety and effectiveness of modern day fire fighters. The most important "machines" on the fire scene, however, remain the intelligent, knowledgeable, well-trained, physically capable fire fighters who have the ability and determination to attack the fire. A thermal imaging camera may be able to find someone trapped in a burning building, but it takes a smart, able-bodied fire fighter to remove the victim safely.

The increasing complexity of both the world and the science of firefighting requires that fire fighters continually sharpen their skills and increase their knowledge of potential hazards. That need for ongoing education is why training courses such as this one are just as important as good physical fitness to today's fire fighters.

FIGURE 1-12 A thermal imaging camera is one of the many tools available to modern day fire fighters.

VOICES OF EXPERIENCE

In order to appreciate today's fire service, the aspiring fire fighter should know the evolutionary progression of the fire service. While the fire service in the United States dates back to Benjamin Franklin, the global fire service can trace its origins back to the bucket brigades of the ancient Egyptians. The fire service has developed from a small group of neighbors throwing water on a fire to a cultural force and a family that works tirelessly for the betterment of society.

Change often occurs due to incidents from the Great Chicago Fire of 1871, to the Hotel Vendome Fire in Boston in 1972, to the Station Nightclub fire in Rhode Island in 2003, to the World Trade Center Attack on September 11, 2001. The fire service has grown and progressed due to these past events and sacrifices. In order to embrace the future, the fire service must fully understand the importance of these incidents and the bravery shown at these sites. Never forget the past, because it tells us how we became who we are today. To honor those sacrifices, always strive for a better tomorrow working within this great service.

Firefighting is not just an occupation—it is a way of life rich with great pride and tradition. The fire service has developed and changed over the years and will continue to do so long after you have retired. It is your job to help write the pages of history of the fire service with the pride and integrity that so many others have done before you.

Dana Peloso
Oxford Fire Department
Oxford, Massachusetts

FIRE FIGHTER Tips

You will find many resources on the history of firefighting at the public library, in local bookstores, and on the Internet. These resources make for interesting reading and learning about how things have—or have not—changed over the years.

Fire Equipment

Today's equipment and apparatus evolved over a number of years as new inventions were adapted to the needs of the fire service. Colonial-era fire fighters, for example, had only buckets, ladders, and fire hooks (tools used to pull down burning structures) at their disposal. Homeowners were required to keep buckets filled with sand or water and to bring them to the scene of the fire. Some towns also required that ladders be available so that fire fighters could access the roof to extinguish small fires. If all else failed, the fire hook was used to pull down a burning building and prevent the fire from spreading to nearby structures. The "hook-and-ladder truck" evolved from this early equipment.

Buckets gave way to hand-powered pumpers in 1720, when Richard Newsham developed the first such pumper in London, England. As many as 16 strong men were used to power the pump, making it possible to propel a steady stream of water from a safe distance. In 1829, more powerful steam-powered pumpers replaced the hand-powered pumpers. Unfortunately, many volunteer fire fighters felt threatened by these steam engines and fought against their use. Steam engines were heavy machines that were pulled to the fire by a trained team of horses. They required constant attention, which limited their use to larger cities that could bear the costs of maintaining the horses and the steamers.

The advent of the internal combustion engine in the early 1900s greatly changed the fire service and enabled even small towns to have machine-powered pumpers. Today, both staffed and unstaffed firehouses keep fire engines ready to respond at any hour of the day or night. Although they require regular maintenance, current equipment does not require the constant attention that horses or steam engines did. Modern fire apparatus carry water, a pumping mechanism, hoses, equipment, and personnel. In this sense, a single apparatus has replaced several single-function vehicles from the past.

The progress in fire protection equipment extends beyond trucks. Without an adequate water supply, modern-day apparatus would be helpless. The advent of municipal water systems provides large quantities of water to extinguish major fires.

Romans developed the first municipal water systems, just as they had developed the first fire companies. It was not until the 1800s, however, that water distribution systems were applied to fire suppression efforts. George Smith, a fire fighter in New York City, developed the first fire hydrants in 1817. He realized that using a valve to control access to the water in the pipes would enable fire fighters to tap into the system whenever a fire occurred. These valves, or fireplugs, were used with both above-ground and below-ground piping systems.

Because small fires are more easily controlled, the sooner a fire department is notified, the more likely it will be able to extinguish the fire and minimize losses. The introduction of public call boxes in Washington, D.C., in 1860 represented a major advance in this direction. The call boxes, which were placed around the city, enabled citizens to send a coded telegraph signal to the fire department, which received the message as a series of bells. The fire department could determine the location of the fire alarm box being used by the number of bells in the signal. When fire fighters arrived at the alarm box, the caller could then direct them to the exact location of the fire. Similar units are still used in many areas but are being replaced by more immediate and effective communications systems.

FIRE FIGHTER Tips

Buckets made of leather could not hold water for any length of time, so in some parts of the United States, fire buckets were kept filled with sand. After the sand was thrown on the fire, the bucket would be refilled with water and used in a bucket brigade.

Communications

Good communication is vital for effective firefighting. When a fire is discovered, fire fighters must be summoned and citizens alerted to the danger. During the firefight, officers must be able to communicate with fire fighters or summon additional resources. Not surprisingly, then, improvements in communication systems are tied to improvements in the fire service.

During the colonial period, fire wardens or night watchmen patrolled neighborhoods and sounded the alarm if a fire was discovered. Some towns, including Charleston, South Carolina, built a series of fire towers where wardens would watch for fires. In many towns, ringing the community fire bell or church bells alerted citizens to a fire.

In the late 1800s, telegraph fire alarm systems were installed in large cities. These systems enabled more rapid reporting of fires and made it possible for officers to request additional resources or to let dispatchers know that the fire was extinguished. In small towns, telegraph fire alarm systems were gradually replaced by community sirens mounted on poles or tall rooftops to signal a fire. Today most fire departments rely on pagers or two-way radios to summon part-time or volunteer fire fighters to emergencies.

Many of the early communications systems have also been replaced by hard-wired and cellular telephones that enable citizens to report an emergency from almost anywhere **FIGURE 1-13**. Use of telephones has greatly reduced the time and difficulty of reporting a fire. The introduction of computer-aided dispatch facilities has likewise improved response times, because the closest available fire units can quickly be sent to the site of the emergency.

FIGURE 1-13 Modern technology allows citizens to report an emergency from almost any location.

FIGURE 1-14 The chief's trumpet was once used to amplify this commander's voice. Today it serves as a symbol of authority in the fire service.

Communications during the firefight also have improved over the years, from simply shouting loud enough to be heard over the chaos to using two-way radios. Before electronic amplification became available, the chief officer shouted commands through his trumpet. The chief's trumpet, or bugle, eventually became a symbol of authority FIGURE 1-14. Although chief officers no longer use trumpets for communicating, the use of multiple trumpets to symbolize the rank of chief signifies this person's need to communicate as well as to lead. Today's two-way radios enable fire units and individual fire fighters to remain in contact with one another at all times.

FIRE FIGHTER Tips

The historic symbol of the chief officer's trumpet is still used on a chief's badge. This series of crossed trumpets is one of the cherished traditions of the fire service.

Paying for Fire Service

As previously noted, volunteer fire departments were common in colonial America and are still used today in many areas. The first fire wardens were employed by communities and paid from community funds. The introduction of hand-powered pumpers hastened the development of a permanent fire service. The question of who would pay for the equipment and the fire fighters, however, was not settled for many years.

Fire insurance companies were established in England soon after the Great Fire of London in 1666 to help victims cope with the financial loss from fires. Such companies collected fees (premiums) from homeowners and businesses, in return pledging to repay the owner for any losses resulting from fire. The first fire insurance company in America was established in 1736 in Charleston, South Carolina, but it folded shortly after a major fire occurred in the city in 1740.

Because the insurance companies could save money if a fire was put out before much damage was done, they agreed to pay fire companies for trying to extinguish fires. Houses that had insurance were designated with a fire mark FIGURE 1-15. Most fire companies were loosely governed and organized, and more than one company might show up to fight a fire. If two fire

FIGURE 1-15 A fire mark indicated the homeowner had insurance that would pay the fire company for extinguishing the blaze.

companies arrived at a fire, however, a dispute might arise over which company would collect the money. Consequently, municipalities began assuming the role of providing fire protection.

Today almost all the fire protection in the United States is funded directly or indirectly through tax dollars. Some jurisdictions fund the fire department as a public service; others contract with an independent fire department. Volunteer fire departments frequently conduct fund-raising activities. Ultimately, however, all funds for operating fire departments come from the citizens of the community.

Fire Service in the United States Today

The fire service in the United States today is the product of evolution occurring over the past 400 years. As a beginning fire fighter, it is helpful for you to learn from the past as well as to study the modern-day U.S. fire service.

According to the NFPA, there are approximately 1.1 million fire fighters in the United States. Of this number, 27 percent are full-time, career fire fighters and 73 percent are volunteers. Three out of four career fire fighters work in communities with populations of 25,000 or more. More than half of all volunteer fire fighters work in fire departments that protect small, rural communities with populations of 2500 or less. There are approximately 30,000 fire departments throughout the United States. These departments may be composed of all career fire fighters, all volunteer fire fighters, or a mix of career and volunteer fire fighters, depending on the community's needs and available resources **FIGURE 1-16**.

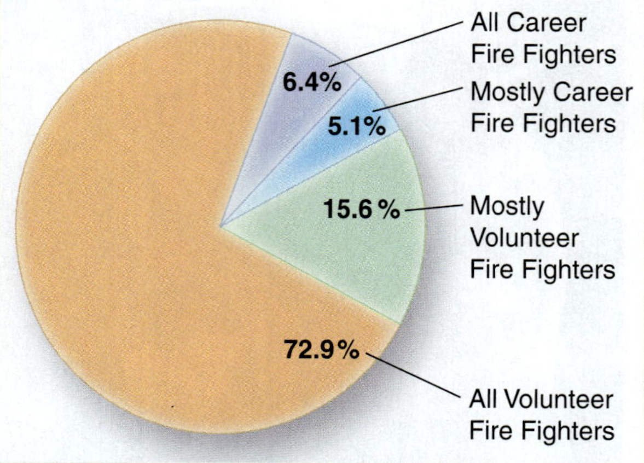

FIGURE 1-16 The majority of fire departments consist of fire fighters who are volunteers.

- All career—All of the members of the department are paid, full-time fire fighters.
- Combination departments:
  - Combination—These departments include both paid, full-time fire fighters and either on-call fire fighters or volunteers.
  - Mostly volunteer—More than half of the members in these departments are volunteers, although they do include paid, full-time fire fighters as well.
- All volunteer—All of the members of the fire department are volunteer fire fighters.

Fire Fighter Safety Tips

Fire Statistics for the United States, 2010

- U.S. fire departments responded to 1,331,500 fires.
- There were 3120 civilian fire fatalities.
- One civilian death from fire occurred every 169 minutes.
- Property loss was $11,593,000,000.
- Home fires caused 2640 civilian fire fatalities (85 percent of all civilian fire-related deaths).
- Fires accounted for 5 percent of the 28,205,000 calls for fire department assistance.
- Eight percent of the calls were false reports.
- Sixty-six percent of the calls were for aid such as EMS.

Half of the volunteers serve rural areas

Wrap-Up

Chief Concepts

- Throughout your training and career, keep in mind the five fire fighter guidelines:
  - *Be safe*. Safety for yourself, your teammates, and the public should be your number one priority.
  - *Follow orders*. Remember that your supervisors have more training and experience than you do.
  - *Work as a team*. Teamwork is essential to safety and successful fire suppression.
  - *Think!* Always keep your mind sharp and focused.
  - *Follow the Golden Rule*. Treat each team member, victim, or citizen as you would like to be treated. Every life is important.
- The training and performance qualifications for fire fighters are specified in NFPA 1001, *Standard for Fire Fighter Professional Qualifications*. Age, education requirements, medical requirements, and other criteria are established locally.
- A Fire Fighter I works in a team under direct supervision to suppress fires.
- A Fire Fighter II works in a team under general supervision. A Fire Fighter II may assume command, transfer command, and coordinate command within the incident command system.
- Throughout your career, you may assume several roles in the fire department. Each role requires additional training:
  - The fire apparatus driver/operator is responsible for getting the fire apparatus to the scene safely, as well as setting up and running the pump or operating the aerial ladder.
  - The company officer leads the company both on the scene and at the station.
  - The safety officer watches the overall operation for unsafe practices.
  - The training officer is responsible for updating the training of current fire fighters and for training new fire fighters.
  - The incident commander is responsible for the management of all incident operations.
  - The fire marshal/fire inspector/fire investigator inspects businesses and enforces public safety laws and fire codes. He or she may also respond to fire scenes to help investigate the cause of a fire.
  - The fire and life safety education specialist educates the public about fire safety and injury prevention.
  - The 911 dispatcher/telecommunicator takes calls from the public and dispatches appropriate units to an emergency.
  - Fire apparatus maintenance personnel repair and service fire and EMS vehicles, keeping them ready to respond to emergencies.
  - Fire police control traffic and secure the scene from the public.
  - Information management professionals are fire fighters or civilians who take care of a fire department's computer network system.
  - The public information officer serves as a liaison between the incident commander and the news media.
  - The fire protection engineer reviews plans and works with building owners to ensure that their fire suppression and detection systems will meet the applicable codes and function as needed.
- Many emergencies require specialized skills. Most large fire departments have teams of specialized fire fighters who can respond to specific emergencies:
  - An aircraft/crash rescue fire fighter (ARFF) works at military and civilian airports and has specialized training in aircraft fires, extricating victims from aircraft, and extinguishing agents.
  - A hazardous materials technician is trained to identify chemicals, control leaks, decontaminate a scene, and clean up a scene.
  - A technical rescue technician is trained in special rescue techniques for incidents involving structural collapse, trench rescue, swiftwater rescue, confined-space rescue, high-angle rescue, and other situations.
  - A SCBA dive rescue technician is trained in water rescue, recovery, and search procedures.
  - EMS personnel are trained to administer prehospital medical care to victims.
- When multiple agencies, such as police, fire, and EMS, work together at an incident, a unified command must be established as part of the incident command system. A unified command establishes a single set of incident goals under a single leader and ensures mutual communication and cooperation.
- Governance is the process by which an organization exercises authority and performs the functions assigned to it. The governance of a fire department depends on regulations, policies, and SOPs.
  - Regulations are developed by various government or government-authorized organizations to implement a law that has been passed by a government body.
  - Policies are developed to provide definitive guidelines for present and future actions.
  - SOPs provide specific information on the actions that should be taken to accomplish a certain task.

- A fire department includes many different types of companies to perform specific tasks at the scene of an emergency:
  - The engine company is responsible for securing a water source, deploying handlines, conducting search and rescue operations, and putting water on the fire.
  - The truck company is responsible for forcible entry, ventilation, roof operations, search and rescue, and ground ladder deployment.
  - The rescue company is responsible for rescuing victims from fires, confined spaces, trenches, and high-angle situations.
  - A wildland/brush company is dispatched to wildland and brush fires.
  - A hazardous materials company responds to and controls scenes involving spilled or leaking hazardous chemicals.
  - An EMS company includes ambulances and responds to medical emergencies.
- The chain of command may differ from fire department to fire department, but the basic concept remains the same across the fire service. The chain of command, from lowest rank to highest, is:
  - Fire fighter
  - Lieutenant
  - Capitan
  - Battalion chief
  - Assistant or division chief
  - Chief of the department
- Four basic management principles apply to the fire service:
  - Discipline comprises the set of guidelines that a fire department establishes for fire fighters. Regulations, policies, and procedures are all forms of discipline.
  - Division of labor is a way of organizing an incident by breaking down an overall strategy into a series of smaller tasks.
  - Unity of command is the concept that each fire fighter answers to only one supervisor.
  - Span of control is the number of people whom one person can supervise effectively.
- Highly destructive fires such as the Great Chicago Fire and the Peshtigo, Wisconsin, fire of 1871 spurred communities to enact strict building and fire codes in an effort to prevent large loss of life and property. Today's building codes not only govern construction materials, but also frequently require built-in fire prevention and safety measures such as fire detectors.
- Fire equipment has evolved from leather buckets and wooden ladders in colonial times to today's thermal imaging cameras.
- Communications have evolved from simple shouting to using two-way radios.
- The fire service in the United States today is the product of an evolution over the past 400 years. As a beginning fire fighter, it is helpful for you to learn from the past and to study the fire service in the United States today.

Hot Terms

911 dispatcher/telecommunicator From the communications center, the dispatcher takes the calls from the public, sends appropriate units to the scene, assists callers with treatment instructions until the EMS unit arrives, and assists the incident commander with needed resources.

Advanced Emergency Medical Technician (AEMT) A member of EMS who can perform limited procedures that usually fall between those provided by an EMT-Basic and those provided by an EMT-Paramedic, including IV therapy, interpretation of cardiac rhythms, defibrillation, and airway intubation.

Aircraft/crash rescue fire fighter (ARFF) An individual with specialized training in aircraft fires, extrication of victims from aircraft, and extinguishing agents. These fire fighters wear special types of turnout gear and respond in fire apparatus that protect them from high-temperature fires.

Assistant or division chief A midlevel chief who often has a functional area of responsibility, such as training, and who answers directly to the fire chief.

Banked Covering a fire to ensure low burning.

Battalion chief Usually the first level of fire chief; also called a district chief. These chiefs are often in charge of running calls and supervising multiple stations or districts within a city. A battalion chief is usually the officer in charge of a single-alarm working fire.

Captain The second rank of promotion in the fire service, between the lieutenant and the battalion chief. Captains are responsible for managing a fire company and for coordinating the activities of that company among the other shifts.

Chain of command A rank structure, spanning the fire fighter through the fire chief, for managing a fire department and fire-ground operations.

Chief of the department The top position in the fire department. The fire chief has ultimate responsibility for the fire department and usually answers directly to the mayor or other designated public official.

Chief's trumpet An obsolete amplification device that enabled a chief officer to give orders to fire fighters during an emergency. It was a precursor to a bullhorn and portable radios.

Company officer The officer, or any other position of comparable responsibility in the department, in charge of a fire department company or station. (NFPA 1143)

Consensus document A code document developed through agreement between people representing different organizations and interests. NFPA codes and standards are consensus documents.

Discipline The guidelines that a department sets for fire fighters to work within.

Division of labor Breaking down an incident or task into a series of smaller, more manageable tasks and assigning personnel to complete those tasks.

Doff To take off an item of clothing or equipment.

Don To put on an item of clothing or equipment.

Emergency Medical Services (EMS) company A company that may be made up of medical units and first-response vehicles. Members of this company respond to and assist in the transport of medical and trauma victims to medical facilities. They often have medications, defibrillators, and paramedics who can stabilize a critical patient.

Emergency Medical Services (EMS) personnel Personnel who are responsible for administering prehospital care to people who are sick and injured. Prehospital calls make up the majority of responses in most fire departments, and EMS personnel are cross-trained as fire fighters.

Emergency Medical Technicians (EMT) EMS personnel who account for most of the EMS providers in the United States. EMTs have training in basic emergency care skills, including oxygen therapy, bleeding control, CPR, automated external defibrillation, use of basic airway devices, and assisting patients with certain medications.

Engine company A group of fire fighters who work as a unit and are equipped with one or more pumping engines that have rated capacities of 2840 L/min (750 gpm) or more. (NFPA 1410)

Fire and life safety education specialist A member of the fire department who deals with the public on education, fire safety, and juvenile fire safety programs.

Fire apparatus driver/operator A fire department member who is authorized by the authority having jurisdiction to drive, operate, or both drive and operate fire department vehicles. (NFPA 1451)

Fire apparatus maintenance personnel The people who repair and service the fire and EMS vehicles, ensuring that they are always ready to respond to emergencies.

Fire Fighter I A person, at the first level of progression as defined in Chapter 5, who has demonstrated the knowledge and skills to function as an integral member of a firefighting team under direct supervision in hazardous conditions. (NFPA 1001)

Fire Fighter II A person, at the second level of progression as defined in Chapter 6, who has demonstrated the skills and depth of knowledge to function under general supervision. (NFPA 1001)

Fire hook A tool used to pull down burning structures.

Fire mark Historically, an identifying symbol on a building informing fire fighters that the building was insured by a company that would pay them for extinguishing the fire.

Fire marshal/fire inspector/fire investigator A member of the fire department who inspects businesses and enforces laws that deal with public safety and fire codes. A fire investigator may also respond to fire scenes to help incident commanders investigate the cause of a fire. Investigators may have full police powers of arrest and deal directly with investigations and arrests.

Fireplug A valve installed to control water accessed from wooden pipes.

Fire police Members of the fire department who protect fire fighters by controlling traffic and securing the scene from public access. Many fire police are sworn peace officers as well as fire fighters.

Fire protection engineer A member of the fire department who is responsible for reviewing plans and working with building owners to ensure that the design of and systems for fire detection and suppression will meet applicable codes and function as needed.

Fire wardens Individuals who were charged with enforcing fire regulations in colonial America.

Governance The process by which an organization exercises authority and performs the functions assigned to it.

Hazardous materials company A fire company that responds to and controls scenes where hazardous materials have spilled or leaked. Responders wear special suits and are trained to deal with most chemicals.

Hazardous materials technician A person who responds to hazardous materials/weapons of mass destruction incidents using a risk-based response process by which they analyze the problem at hand, select applicable decontamination procedures, and control a release while using specialized protective clothing and control equipment. (NFPA 472)

Incident commander (IC) The person who is responsible for all decisions relating to the management of the incident and is in charge of the incident site. (NFPA 1500)

Incident command system (ICS) The combination of facilities, equipment, personnel, procedures, and communications operating within a common organizational structure that has responsibility for the management of assigned resources to effectively accomplish stated objectives pertaining to an incident or training exercise. (NFPA 1670)

Information management Fire fighters or civilians who take care of the computer and networking systems that a fire department needs to operate.

Lieutenant A company officer who is usually responsible for a single fire company on a single shift; the first in line among company officers.

Paramedic EMS personnel with the highest level of training in EMS, including cardiac monitoring, administering drugs, inserting advanced airways, manual defibrillation, and other advanced assessment and treatment skills.

Policies Formal statements that provide guidelines for present and future actions. Policies often require personnel to make judgments.

Public information officer An individual who has demonstrated the ability to conduct media interviews and prepare news releases and media advisories. (NFPA 1035)

Regulations Rules, usually issued by a government or other legally authorized agency, that dictate how something must be done. Regulations are often developed to implement a law.

Rescue company A group of fire fighters who work as a unit and are equipped with one or more rescue vehicles. (NFPA 1410)

Safety officer An individual appointed by the authority having jurisdiction as qualified to maintain a safe working environment at all live fire training evolutions. (NFPA 1403)

SCUBA dive rescue technician A responder who is trained to handle water rescues and emergencies, including recovery and search procedures, in both water and under-ice situations. (SCUBA stands for "self-contained underwater breathing apparatus.")

Self-contained breathing apparatus (SCBA) An atmosphere-supplying respirator that supplies a respirable air atmosphere to the user from a breathing air source that is independent of the ambient environment and designed to be carried by the user. (NFPA 1981)

Span of control The maximum number of personnel or activities that can be effectively controlled by one individual (usually three to seven). (NFPA 1006)

Standard operating procedure (SOP) A written organizational directive that establishes or prescribes specific operational or administrative methods to be followed routinely for the performance of designated operations or actions. (NFPA 1521)

Suggested operating guideline (SOG) Another term for standard operating procedure.

Technical rescue technician A fire fighter who is trained in special rescue techniques for incidents involving structural collapse, trench rescue, swiftwater rescue, confined-space rescue, and other unusual rescue situations.

Thermal imaging camera Electronic devices that detect differences in temperature based on infrared energy and then generate images based on those data. These devices are commonly used in obscured environments to locate victims.

Training officer The person designated by the fire chief as having authority for overall management and control of the organization's training program. (NFPA 1401)

Truck company A group of fire fighters who work as a unit and are equipped with one or more pieces of aerial fire apparatus. (NFPA 1410)

Unity of command The concept according to which each person within an organization reports to one, and only one, designated person. (NFPA 1026)

Wildland/brush company A fire company that is dispatched to woods and brush fires where larger engines cannot gain access. Wildland/brush companies have four-wheel drive vehicles and special firefighting equipment.

FIRE FIGHTER *in action*

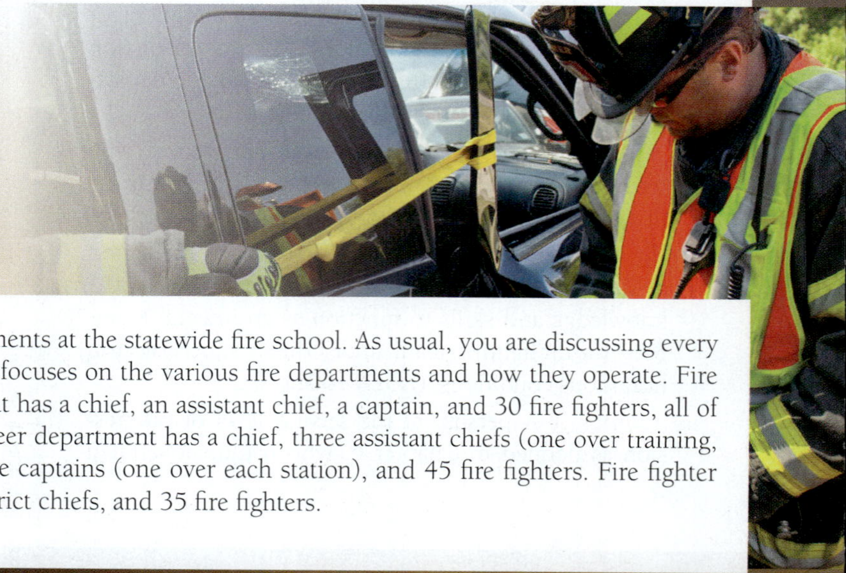

You are with a group of fire fighters from several departments at the statewide fire school. As usual, you are discussing every topic under the sun; right now, however, the discussion focuses on the various fire departments and how they operate. Fire fighter Tom Smith explains that his volunteer department has a chief, an assistant chief, a captain, and 30 fire fighters, all of whom report to the chief. You respond that your volunteer department has a chief, three assistant chiefs (one over training, one over operations, and one over fire prevention), three captains (one over each station), and 45 fire fighters. Fire fighter Brad Smith says that his department has a chief, five district chiefs, and 35 fire fighters.

FIRE FIGHTER *in action* (continued)

As the fire fighters continue to discuss the differences in the structures of their departments, you find that there are many differences in department organization, even though they are all small, volunteer departments that respond to comparable types and quantities of incidents.

1. Which of the following is not typically a type of incident to which fire departments respond?
   - **A.** Motor vehicles collisions
   - **B.** Hazardous materials spills
   - **C.** Vicious dogs on the loose
   - **D.** Emergency medical requests
2. What are the three most common staffing methods of fire departments?
   - **A.** Public safety officer
   - **B.** Volunteer
   - **C.** Career
   - **D.** Combination
3. Which role would most likely fall under the Fire Prevention Division?
   - **A.** Fire police
   - **B.** Fire marshal
   - **C.** Incident commander
   - **D.** Company officer
4. If there are an equal number of fire fighters assigned to each station at your department, what is the span of control for each captain?
   - **A.** 1 to 1
   - **B.** 3 to 1
   - **C.** 5 to 1
   - **D.** 15 to 1
5. At fire fighter Brad Smith's department, fire fighters are report to a district chief as well as the fire chief. Which organizational principle does this violate?
   - **A.** Span of control
   - **B.** Unity of command
   - **C.** Division of labor
   - **D.** Chain of command
6. Which position would fire fighter Brad Smith's department most likely add if it wanted a supervisory position between the rank of fire fighter and district chief?
   - **A.** Division chief
   - **B.** Battalion chief
   - **C.** Fire fighter
   - **D.** Assistant chief

FIRE FIGHTER II *in action*

You are thrilled to learn that you will be assigned to Ladder Company 34, which is in one of the busiest districts in the city. It covers multiple low-income high-rise apartments in the downtown area and is known to routinely have a working fire each day. It is served by an engine company, a truck company, a rescue company, and a battalion chief.

Generally, it takes years of experience and proving one's skills before getting this coveted assignment. However, your excitement turns to concern as you wonder how well you will perform. Are you up to it? Can you step up to a leadership role within the organization, while still a fire fighter? These and other questions race through your mind as you pack your gear to head for the new assignment.

1. What is the benefit of division of labor through the use of engine companies, truck companies, rescue companies, and chiefs?
2. Why are regulations, policies, and standard operating procedures important for you to understand and follow?
3. What responsibility do you have for fire prevention in your new district?
4. How can you develop a positive image with your new supervisors and crews?

Fire Fighter Safety

Fire Fighter I

Knowledge Objectives

After studying this chapter, you will be able to:

- List the major causes of death and injury in fire fighters. (p 26–27)
- Explain how to submit a Near Miss Report. (p 28)
- List the three groups whom fire fighters must always consider when ensuring safety at the incident scene. (p 27)
- List and describe the four components of a fire fighter safety program. (p 27–28)
- Describe the 16 fire fighter life safety initiatives. (p 29)
- Describe the connection between physical fitness and fire fighter safety. (NFPA5.1.1, p 28–30)
- Describe the components of a well-rounded physical fitness program. (NFPA5.1.1, p 28–30)
- Explain the practices fire fighters should take to promote optimal physical and mental health. (NFPA5.1.1, p 28–31)
- Describe the purpose of an employee assistance program. (NFPA5.1.1, p 30)
- Explain how fire fighter candidates, instructors, and veteran fire fighters work together to ensure safety during training. (p 30)
- Describe the steps to ensure safety when responding to an emergency. (p 30–31)
- Describe the steps to ensure safety when driving to an emergency incident. (p 31–33)
- List the four general principles that govern emergency vehicle operation. (p 31–33)
- List the guidelines for safe emergency vehicle response. (p 33)
- Explain how the teamwork concept is applied during every stage of an emergency incident to ensure the safety of all fire fighters. (p 33–35)
- Describe how the personnel accountability system is implemented during an emergency incident. (p 35–36)
- List the common hazards at an emergency incident. (p 36)
- Describe the measures fire fighters follow to ensure electrical safety at an emergency incident. (p 36)
- Describe how to lift and move objects safely. (p 36)
- Explain how rehabilitation is used to protect the safety of fire fighters during an emergency incident. (p 36–38)
- Explain the role of a critical incident stress debriefing in preserving the mental well-being of fire fighters. (p 38)
- Describe how to ensure safety at the fire station. (p 38)
- Describe how to ensure safety outside of the workplace. (p 38)

Skills Objectives

There are no skill objectives for Fire Fighter I candidates.

CHAPTER 2

Fire Fighter II — FFII

Knowledge Objectives

There are no knowledge objectives for Fire Fighter II candidates. NFPA 1001 contains no Fire Fighter II Job Performance Requirements for this chapter.

Skills Objectives

There are no skill objectives for Fire Fighter II candidates. NFPA 1001 contains no Fire Fighter II Job Performance Requirements for this chapter.

Additional NFPA Standards

- NFPA 1250, *Recommended Practice in Emergency Service Organization Risk Management*
- NFPA 1451, *Standard for a Fire Service Vehicle Operations Training Program*
- NFPA 1500, *Standard on Fire Department Occupational Safety and Health Program*
- NFPA 1582, *Standard on Comprehensive Operational Medical Program for Fire Departments*

You Are the Fire Fighter

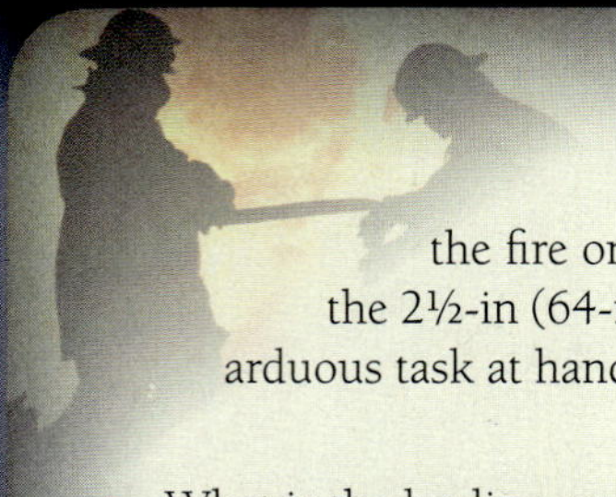

You and your crew are at a large fire in an apartment that started on the third floor, extended into the attic, and is now venting through the roof. Your captain gave the incident commander (IC) the accountability tags, and you were assigned to Division D to confine the fire on the third floor. You quickly ran through an air cylinder as you pulled ceilings after having stretched the 2½-in (64-millimeter) hose to the third floor. Your crew quickly changes cylinders and gets ready to continue the arduous task at hand. Then you get the word from the IC: "Engine 213, report to rehab."

1. What is the leading cause of fire fighter injury and death?
2. What are the safety measures taken during this incident?
3. What are the potential hazards during this incident?

Introduction

This chapter covers the topics of injury prevention, means of reducing fire fighter injuries and deaths, and safety and health measures needed during all activities performed by fire fighters, from training to the fireground to fire station duties. Firefighting, by its very nature, is dangerous. Each individual fire fighter must learn safe methods of confronting the risks presented during training exercises, on the fire ground, and at other emergency scenes.

Every fire department must do everything it can to reduce the hazards and dangers of the job and help prevent fire fighter injuries and deaths. Each fire department must have a strong commitment to fire fighter safety and health, with fire fighters taking the lead for their health and safety. Safety must be fully integrated into every activity, procedure, and job description.

Appropriate safety measures must be applied routinely and consistently. During incidents, safety officers are responsible for evaluating the hazards of various situations and recommending appropriate safety measures to the incident commander (IC). Each accident, injury, or near miss must be thoroughly investigated to learn why it happened and how it can be avoided in the future.

Advances in technology and equipment require fire departments to review and revise their safety policies and procedures regularly. Information reviews and research by designated health and safety officers can identify new hazards as well as appropriate risk-management measures. In addition, reports of accidents, fatalities, and near misses from other fire departments can help identify common problems and lead to the development of effective preventive actions.

Causes of Fire Fighter Deaths and Injuries

The National Fire Protection Association (NFPA) reports that each year, on average, 80 to 100 fire fighters are killed in the line of duty in the United States. These deaths occur at emergency incident scenes, in fire stations, during training, and while responding to or returning from emergency situations. Approximately the same number of fire fighter deaths occur on the fire ground or emergency scene as during training or while performing other nonemergency duties. Approximately one fourth of all deaths occur while fire fighters are responding to or returning from alarms **FIGURE 2-1**. Heart attack and stroke are the highest cause of fire fighter death due to injury.

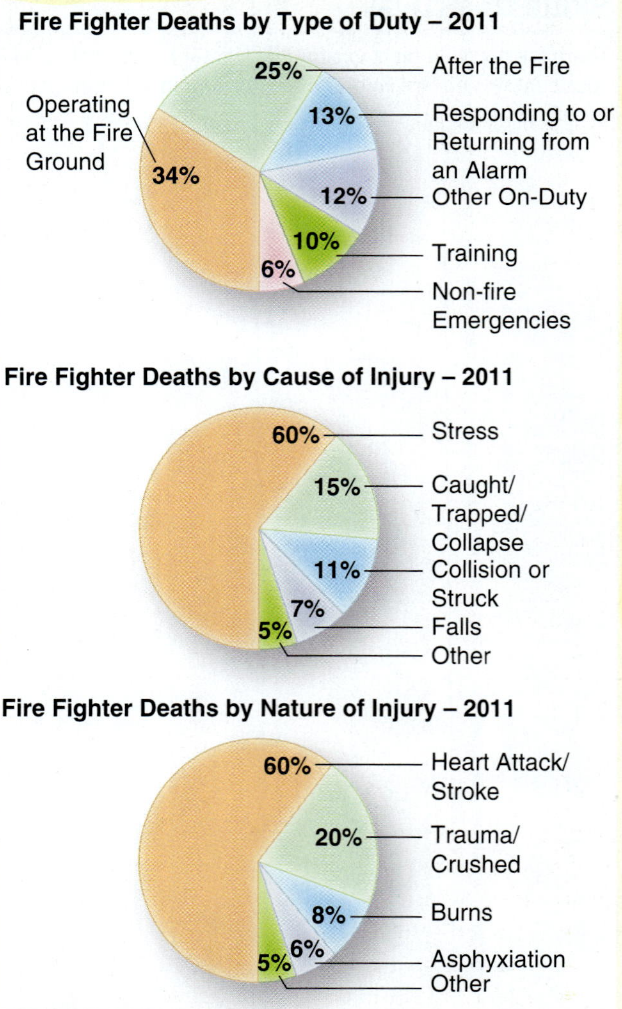

FIGURE 2-1 Fire fighter deaths in the United States.

Vehicle collisions are a major cause of fire fighter fatalities. For every 1000 emergency responses, it is estimated that one vehicle collision involving an emergency vehicle occurs FIGURE 2-2. One study found that 39 percent of the fire fighters who died in those incidents were not using seat belts. Fire fighters should never overlook basic safety procedures, such as always fastening seat belts.

The NFPA estimates that 71,875 fire fighters were injured in the line of duty in 2010. Fewer than half of these injuries occurred while fighting fires. The most common injuries were strains, sprains, and soft-tissue injuries. Burns accounted for only 10 percent of the total injuries. Smoke and inhalation injuries accounted for 5 percent of all fire-ground injuries TABLE 2-1.

FIGURE 2-2 Motor vehicle collisions are a leading cause of death for fire fighters.

TABLE 2-1 Fire Fighter Injuries

Types of Injury	Percentage of Total Injuries
Strains and sprains	53%
Soft-tissue injuries and fractures	17%
Burns	10%
Smoke and inhalation injuries	5%
Other	15%

Injury Prevention

Injury prevention is a responsibility shared by each member of the firefighting team. Fire fighters must always consider three groups when ensuring safety at the scene:

- Their personal safety
- The safety of other team members
- The safety of everyone present at an emergency scene

To reduce the risks of accidents, injuries, occupational illnesses, and fatalities, a successful safety program must include four major components:

- Standards and procedures
- Personnel
- Training
- Equipment

Standards and Procedures

Because safety is such a high priority, several organizations set the standards intended to ensure a safe working environment for the fire service. NFPA 1500, *Standard on Fire Department Occupational Safety and Health Program*, provides a template for implementing a comprehensive health and safety program. Additional NFPA standards focus on specific subjects directly related to health and safety—for example, NFPA 1582, *Standard on Comprehensive Operational Medical Program for Fire Departments*.

The federal Occupational Safety and Health Administration (OSHA), along with a variety of state and provincial health and safety agencies, develops and enforces government regulations on workplace safety. NFPA standards often are incorporated by reference in government regulations.

Every fire department should have a set of standard operating procedures (SOPs) or standard operating guidelines (SOGs), which outline how to perform various functions and operations. SOPs and SOGs cover a range of topics—from uniforms and grooming to emergency scene operations. These procedures should incorporate safe practices and policies. Each fire fighter is responsible for understanding and following these procedures.

The fire department chain of command also enforces safety goals and procedures. In particular, the command structure keeps everyone working toward common goals in a safe manner. An incident command system (ICS) is a nationally recognized plan to establish command and control of emergency incidents. The ICS is flexible enough to meet the needs of any emergency situation, so it should be implemented at every emergency scene—from a routine auto accident to a major disaster involving responders from numerous agencies. More information on the incident command system is presented in the Incident Command System chapter.

Many fire departments have a health and safety committee that is responsible for establishing policies and monitoring fire fighter safety. Members of this committee should include representatives from every area, component, and level within the department, from fire fighters to chief officers. The health and safety officer and the fire department physician also should be members of the committee.

Personnel

A safety program is only as effective as the individuals who implement it. A majority of safety issues occur during emergency operations, this also accounts for the largest portion of most fire department operations and budgets.

Teamwork is an essential element of safe emergency operations. On the fire ground and during any hazardous activity, fire fighters must work together to get the job done. The lives of citizens and fire fighters alike depend on compliance with basic safety concepts and principles of operation.

An overall plan also is essential to coordinate the activities of every team, crew, or unit involved in the operation. The ICS coordinates and tracks the location and function of every individual or work group involved in an operation.

Freelancing is acting independently of a superior's orders or the fire department's SOPs. Freelancing has no place on the fire ground; it poses a danger to the fire fighter who acts independently and every other fire fighter on the emergency scene. A fire fighter who freelances can easily get into trouble by being in the wrong place at the wrong time or by doing the wrong thing. For example, a fire fighter who enters a burning structure without informing a superior may be trapped by rapidly changing conditions. By the time the fire fighter is missed, it may be too late to perform a rescue. Searching for a missing fire fighter exposes others to unnecessary risk.

Incident Safety Officer

Incident safety officers are members of the fire department whose primary responsibility is safety. At the emergency scene, the designated incident safety officer reports directly to the IC and has the authority to correct or stop any action that is judged to be unsafe. Incident safety officers observe operations and conditions, evaluate risks, and work with the IC to identify hazards and ensure the safety of all personnel. They also determine when fire fighters can work without self-contained breathing apparatus (SCBA) after a fire is extinguished.

Incident safety officers can enhance safety in the workplace, at emergency incidents, and at training exercises. Even so, it is important to remember that each and every member of the fire department shares the responsibility for promoting safety, both as an individual and as a member of the team.

Training

Adequate training is essential for fire fighter safety. The initial fire fighter training covers the potential hazards of each skill and outlines the steps necessary to avoid injury. Fire fighters must avoid sloppy practices or shortcuts that might potentially contribute to injuries and learn how to identify hazards and unsafe conditions.

The knowledge and skills developed during training classes are essential to maintain safety at actual emergency scenes. The initial training course is just the beginning—fire fighters must continually seek out additional courses to keep their skills current.

Equipment

A fire fighter's equipment ranges from power and hand tools to personal protective equipment (PPE) and electronic instruments. Fire fighters must know how to use equipment in the correct manner and then operate it safely at all times. Equipment also must be properly maintained. Poorly maintained equipment can create additional hazards to the user or fail to operate when needed.

Manufacturers usually supply operating instructions and safety procedures for their equipment. These instructions cover proper use of the equipment, its limitations, and warnings about potential hazards. Fire fighters must read and heed these warnings and instructions. In addition, new equipment must meet applicable standards to ensure that it can perform under the difficult and dangerous conditions often encountered on the fire ground.

Reducing Fire Fighter Injuries and Deaths

Reducing fire fighter injuries and deaths requires the dedicated efforts of every fire fighter, of every fire department, and of the entire fire community working together. In 1992, Congress created the National Fallen Firefighters Foundation to lead a nationwide effort to remember the U.S. fallen fire fighters. In the years since then, this foundation has expanded its programs to sponsor the annual National Fallen Firefighters Memorial Weekend, offer support programs for family members and other survivors, award scholarships to fire service members' survivors, and work to prevent line-of-duty injuries and deaths.

Most fire fighter injuries and deaths are the result of preventable situations. Recognizing this fact, the National Fallen Firefighters Foundation has developed programs with the goal of reducing line-of-duty deaths. For example, the National Fire Fighter Near-Miss Reporting System provides a method for reporting situations that could have resulted in injuries or deaths. This system, which is accessible over the Internet, provides a means for all fire fighters to learn from situations that occur both rarely and frequently. Representative near-miss reports are included in most chapters of this text so that you can benefit from the experiences of others without getting in harm's way.

In an effort to do more to prevent line-of-duty deaths and injuries, the National Fallen Firefighters Foundation has developed a fire fighter safety initiative called Everyone Goes Home. The goal of this program is to raise awareness of life safety issues, improve safety practices, and allow everyone to return home at the end of their shift. In particular, the Everyone Goes Home program has developed "The 16 Firefighter Life Safety Initiatives," which describe steps that need to be taken to change the current culture of the fire service to help make it a safer place for all TABLE 2-2.

Safety and Health

Safety and well-being are directly related to personal health and physical fitness. Although fire departments regularly monitor and evaluate the health of fire fighters, each department member is responsible for his or her own personal health, conditioning, and nutrition. To be an effective fire fighter, you should eat a healthy diet, maintain a healthy weight, and exercise regularly.

Good health requires that you get an adequate amount of uninterrupted sleep to maintain alertness, prevent stress, and avoid illnesses and injuries. Because it is not always possible to get adequate sleep during long duty shifts, it is important that you get adequate amounts of uninterrupted sleep during your off-duty time. Establish a consistent sleep schedule and sleep routine, such as turning off all electronic devices a half-hour before bedtime to allow your mind to wind down and prepare for sleep.

All fire fighters—whether career or volunteer—should spend at least an hour each day in physical fitness training.

TABLE 2-2 16 Firefighter Life Safety Initiatives

1. Define and advocate the need for a cultural change within the fire service relating to safety and incorporating leadership, management, supervision, accountability, and personal responsibility.
2. Enhance personal and organizational accountability for health and safety throughout the fire service.
3. Focus greater attention on the integration of risk management with incident management at all levels, including strategic, tactical, and planning responsibilities.
4. Empower all fire fighters to stop unsafe practices.
5. Develop and implement national standards for training, qualifications, and certification (including regular recertification) that are equally applicable to all fire fighters based on the duties they are expected to perform.
6. Develop and implement national medical and physical fitness standards that are equally applicable to all fire fighters based on the duties they are expected to perform.
7. Create a national research agenda and a data collection system that are related to the initiatives.
8. Take advantage of available technology whenever it can produce higher levels of health and safety.
9. Thoroughly investigate all fire fighter fatalities, injuries, and near-misses.
10. Implement grant programs to support the implementation of safe practices and/or mandate safe practices as an eligibility requirement.
11. Develop and champion national standards for emergency response policies and procedures.
12. Develop and champion national protocols for response to violent incidents.
13. Ensure that fire fighters and their families have access to counseling and psychological support.
14. Provide more resources for public education, and champion this kind of education as a critical fire and life safety program.
15. Advocate for the enforcement of codes and installation of home sprinklers.
16. Make safety a primary consideration in the design of apparatus and equipment.

Fire fighters should be examined by either a personal or departmental physician before beginning any new workout routine. An exercise routine that includes weight training, cardiovascular workouts, and stretching with a concentration on job-related exercises is ideal. For example, many fire fighters use a stair-climbing machine to focus on the muscle groups used for climbing. This type of exercise builds cardiovascular endurance for the fire ground, but other muscle groups should not be neglected. Physical fitness must be a career-long activity; it is not something to be left at the fire academy when you graduate. Firefighting is a stressful activity that demands you maintain a good fitness level throughout your career.

Fire Fighter Safety Tips

Maintaining proper hydration is essential to performing at your peak physical level.

FIRE FIGHTER II Tip FFII

Maintain your physical fitness, because your life and the lives of your crew depend on it!

Hydration is an important part of every workout. A good guideline is to consume 8 to 10 ounces (0.2 to 0.3 liter) of water for every 5 to 10 minutes of physical exertion. Do not wait until you feel thirsty to start rehydrating. In fact, fire fighters should drink up to a gallon of water each day to keep properly hydrated. Proper hydration enables muscles to work longer and reduces the risk of injuries at the emergency scene. More information about hydration is presented in the Fire Fighter Rehabilitation chapter.

Diet is another important aspect of physical fitness. A healthy menu includes fruits, vegetables, low-fat foods, whole grains, and lean protein. Pay attention to portion sizes—unfortunately, most people eat larger portions than their bodies need. Substitute healthy choices (such as fruit) for high-calorie desserts.

Heart disease is the leading cause of death both in the United States as a whole and among fire fighters in particular. A healthy lifestyle that includes a balanced diet, weight training, and cardiovascular exercises can help reduce many risk factors for heart disease and enable fire fighters to meet the physical demands of the job FIGURE 2-3.

FIGURE 2-3 Regular exercise will help you to stay healthy and perform your job effectively.

Many fire departments have adopted policies that prohibit the use of tobacco products by fire fighters, both on duty and off duty. Smoking is a major risk factor in cardiovascular disease, reduces the efficiency of the body's respiratory system, and increases the risk of lung and other types of cancer. Fire fighters should avoid tobacco products entirely for both health and insurance reasons.

Alcohol is another substance that fire fighters should avoid. Alcohol is a mood-altering substance that can be abused. Excessive alcohol use can damage the body and affect performance. Fire fighters who have consumed alcohol within the previous 8 hours must not be permitted to engage in training or emergency operations.

Drug use has absolutely no place in the fire service. Many fire departments have drug-testing programs to ensure that fire fighters do not use or abuse drugs. The illegal use of drugs endangers your life, the lives of your team members, and the public you serve.

Of course, everyone is subject to an occasional illness or injury. You should not try to work when ill or injured. Operating safely as a member of a team requires both fitness and concentration. Do not compromise the safety of the team or your personal health by trying to work while you are ill or injured.

FIRE FIGHTER II Tip **FFII**

Find time for yourself and your family as a mental buffer from the stressors of the job.

Employee Assistance Programs

Employee assistance programs (EAPs) provide confidential help with a wide range of problems that might affect performance. Many fire departments have established EAPs so that fire fighters can get counseling, support, or other assistance in dealing with physical, financial, emotional, or substance abuse problems. A fire officer may refer a fire fighter to an EAP if a problem begins or may begin to affect the individual's job performance. Fire fighters who take advantage of an EAP can do so with complete confidentiality and without fear of retribution.

Safety During Training

During training, fire fighters learn and practice the actual skills that they will use later under emergency conditions. Typically, the patterns that develop during training will continue during actual emergency incidents. Thus developing the proper working habits during training courses helps ensure safety later.

Many of the skills covered during training can be dangerous if they are not performed correctly. According to the NFPA, each year approximately 10 percent of fire fighter deaths occur during training exercises. Use of proper protective gear and good teamwork are as important during training as they are on the fire ground.

Instructors and veteran fire fighters are more than willing to share their experiences and advice. They can explain and demonstrate every skill and point out the safety hazards involved because they have performed these skills hundreds of times and know what to do. But here, too, safety is a shared responsibility. Do not attempt anything you feel is beyond your ability or knowledge. If you see something that you believe is an unsafe practice, bring it to the attention of your instructors or a designated safety officer.

As mentioned earlier, freelancing is a dangerous practice, both on the fire ground and during training exercises. Wait for specific instructions or orders before beginning any task. Do not assume that something is safe and act independently. Follow instructions and learn to work according to the proper procedures.

Teamwork is also important during training exercises. Assignments are given to firefighting teams during most live fire exercises. Teams must stay together. If any member of the team becomes fatigued, is in pain or discomfort, or needs to leave the training area for any reason, notify the instructor or safety officer. Emergency Medical Services (EMS) personnel should be available to perform an examination and to transport ill or injured personnel to further treatment if necessary. A fire fighter who is injured during training should not return until medically cleared for duty.

Safety During the Emergency Response

When a company is dispatched to an emergency, fire fighters need to get to the apparatus and don the appropriate PPE quickly before mounting the vehicle and proceeding to the incident. Walk quickly to the apparatus; do not run. Take care not to slip and become injured before you reach the apparatus. Be careful getting into and out of fire apparatus.

Personal protective gear should be properly positioned so that you can don it quickly before getting into the apparatus **FIGURE 2-4**. In addition, make sure that seat belts are properly fastened before the fire apparatus begins to move. All personnel responding on fire apparatus must be seated with seat belts fastened, and seat belts should remain fastened until the fire apparatus comes to a complete stop. Fire fighters can don SCBA while seated in some vehicles. Learn how to do this without compromising safety.

Drivers of fire apparatus have a great responsibility. They must get the fire apparatus and the crew members to the emergency scene without having or causing a traffic accident en route. They must know the streets in their first-due area and any target hazards. They must be able to operate the vehicle skillfully and keep it under control at all times. In addition, they must anticipate all responses from other drivers who might not see or hear an approaching emergency vehicle or know what to do when confronted by one. Prompt response is a goal, but safe response is a much higher priority. Before you are qualified to drive any piece of fire department apparatus, you must undergo special training that covers the principles of emergency driving, applicable motor vehicle regulations, and characteristics of the equipment you will be driving. You also need to practice driving with a qualified instructor. You need to demonstrate competency in both required knowledge and skills. These driving skills are not required for the Fire Fighter I and Fire Fighter II courses and are beyond the scope of this textbook.

FIGURE 2-4 Protective clothing should be properly positioned so that you can quickly don it.

FIRE FIGHTER II Tip FFII

Developing situational awareness on the emergency scene is essential to your safety and survival. This concept can be summed up by saying that you must have a "big picture" view of the emergency scene and avoid "tunnel vision." This ability comes only through experience and with well-developed competency in basic fire fighter skills. The less you have to concentrate on the components of your task, the more you are able to observe your surroundings. This quality emerges through hours and hours of practice and skill repetition. Do not stop practicing until your basic skills become muscle memories. If you can do your job with your eyes closed, you will be able to observe much more when they are open!

Safe Driving Practices

Vehicle occupants who are not wearing seat belts are much more likely to suffer serious injuries or death than are occupants who have their seat belts properly fastened. Air bags are most effective when seat belts are properly fastened; they are not effective without properly applied seat belts. Seat belts also greatly reduce the possibility that vehicle occupants will be ejected from the vehicle.

In some departments where fire fighters are not on duty at the fire station, it may be necessary for members to respond to the station or to the emergency scene in their private vehicles. Driving a private vehicle during an emergency response requires a special degree of safety. Such emergency driving requires added considerations and represents a much different type of driving from the routine driving.

Motor vehicle crashes are the second most common cause of fire fighter deaths. Although some of these crashes involve fire department apparatus, all too many involve fire fighters driving their private vehicles to an emergency. The ensuing collisions sometimes result in injury or death to the fire fighter and to civilians. Even when no one is injured, crashes cause fire fighters to be delayed in getting to the emergency.

Many collisions occur when the motor vehicle operator loses control of the vehicle. These crashes may be caused by driving too fast for the prevailing conditions, braking inappropriately, changing directions too abruptly, or tracking around a curve too fast. In addition, many collisions involving emergency vehicles occur on open roads, where excessive speed is often cited as a primary cause. Intersections are common sites of collisions involving emergency vehicles; most of these collisions occur when the emergency vehicle operator fails to stop at an intersection and ensure that other traffic has stopped before proceeding through the intersection.

A motor vehicle collision actually consists of a series of separate collision events. The first collision occurs when the vehicle collides with a second vehicle or with a stationary object; the second collision occurs when the occupants of the vehicle collide with the interior of the vehicle.

Fire Fighter Safety Tips

Always fasten your seat belt each and every time you get into a motor vehicle. It is vital that you follow this rule each and every time you climb into a fire engine.

Laws and Regulations Governing Emergency Vehicle Operation

Four general principles govern emergency vehicle operation:

- Emergency vehicle operators are subject to all traffic regulations unless a specific exemption is made. A specific exemption is a statement that appears in a statute, such as "The driver of an authorized vehicle may exceed the maximum speed limits so long as he or she does not endanger life or property."
- Exceptions are legal only when the vehicle is operating in emergency mode.
- Even with an exemption, the emergency vehicle operator can be found criminally or civilly liable if involved in a crash.
- An exemption does not relieve the operator of an authorized emergency vehicle from the duty to drive with reasonable care for all persons using the highway.

Laws governing emergency vehicle operation vary from one state to another. You must follow the laws and regulations of your state regarding emergency vehicle operations. Some states permit private vehicles to operate as emergency vehicles in responding to a fire station or to the scene of an emergency; other states outline a limited use of emergency equipment and dictate certain restrictions. You must understand and follow the specific laws and regulations of your state as well as your department's SOPs.

Standard Operating Procedures for Personal Vehicles

The use of personal vehicles to respond to fire and EMS calls constitutes a fire department function. For you to use your personal vehicle in a situation requiring an emergency response, this use must be permitted by state laws and regulations, and this operation must also be permitted by your local fire department. If your local fire department does not permit the use of personal vehicles for emergency responses, then you are not permitted to engage in this behavior regardless of what is permitted by your state's laws. Your fire department should have SOPs that spell out whether private vehicles can be used for emergency response. These procedures typically address which kind of training you must complete before you are permitted to use your personal vehicle for emergency response. This training will usually include a course in defensive driving and functioning as an emergency vehicle operator.

Safe Driving Begins with You

It has often been stated that the biggest cause of vehicle crashes is "a loose nut behind the wheel." In other words, a major factor in crashes is the attitude and ability of the vehicle operator. A competent emergency vehicle operator needs to have a confident, but not cocky, attitude. The emergency vehicle operator should have good judgment, mental fitness, maturity, physical fitness, and driving habits. It is important to maintain a clean driving record, because a person who has been cited for multiple moving violations in his or her personal vehicle will usually be a poor risk as the driver of an emergency vehicle. To a good driver, it is important to know the state and local laws relating to motor vehicle operations. In addition, an emergency vehicle operator needs to understand reaction time, braking distance, and stopping distance of the vehicle.

Emergency driving requires good reactions and alertness. It is important for the emergency vehicle operator to be in good physical condition and to have good eyesight. Driving while impaired is simply asking for trouble. Impairment can result from many different sources—for example, using some prescribed medicines, using some over-the-counter medicines, and being overly fatigued. Anyone who has been drinking alcoholic beverages should never think of driving. Driving while eating, texting, or talking on a communications device are all forms of distracted driving. Distracted driving during routine driving or when responding to an emergency is to be avoided—it is the cause of many collisions and deaths.

In many cases, one of the best predictors of future performance is past behavior. Your driving record is an important consideration in your career in a fire department. A person who has past moving violations for speeding, reckless operation, aggressive driving, chargeable crashes, or driving under the influence of alcohol or drugs may be excluded from driving any emergency vehicles. Your driving record, both on-duty and off-duty, is important to your career as a fire fighter.

Vehicle Collision Prevention

Safe driving practices will prevent most vehicle collisions. If you are using your personal vehicle to respond to an emergency, it is important to take the characteristics of your vehicle into account while driving. For example, four-wheel-drive pickup trucks and large sport utility vehicles have a higher center of gravity than sleek sports cars, and sudden changes in direction are more likely to result in rollover crashes when a vehicle has a high center of gravity. For this reason, it is critical to learn the characteristics and limitations of your vehicle.

Anticipate the road and road conditions. If you regularly travel the same roads to report to your fire station, learn the characteristics of that strip of roadway and be able to anticipate when you need to slow down or stop. Always drive at a safe speed for that road: Traveling on a limited-access highway is much different from traveling through a school zone, for example. Observe the traffic conditions, and slow down when more traffic congestion is present. Expect that motorists around you may do anything at any time. Upon the approach of an emergency vehicle, for example, some motorists will slow down, some will stop, some will speed up, some will pull to the right, and others will turn left in front of you. Expect the unexpected.

Make allowances for weather conditions. On a clear day, it might seem that you can see forever—but you really cannot. In conditions of rain, snow, fog, dust, or darkness, the distance over which you can see is greatly reduced; reduce your speed to compensate for the limited visibility. At night, your vision is limited by the distance your headlights reach, so reduce your speed accordingly. Recognize that you cannot see objects outside the projection of the headlight beams. Understand that the goal of an emergency response is not to drive as fast as you can, but rather to arrive on the scene as quickly as you can while maintaining safety.

Adjust your speed of response to accommodate any storm conditions. Rainstorms reduce visibility and produce slippery road surfaces. When water collects on a roadway, your vehicle can hydroplane on the thin film of water that separates your tires and the road surface—and you have no control over your vehicle when hydroplaning occurs. During stormy weather, slow down and be prepared for slippery road surfaces. Be aware of high winds from tornados, hurricanes, or other windstorms that can topple trees onto the roadway. During heavy rainstorms, watch for water flowing over the road. You cannot see how deep the water is, and you cannot tell whether the flood water has washed out part of the roadway. Slow down when driving following an earthquake or tornado because you will not know where the roadway, bridges, and overpasses have been blocked, damaged, or destroyed.

When operating an emergency vehicle, you are not exempt from the laws of physics. The laws of physics tell us that when the speed of a vehicle doubles, the force exerted by that vehicle increases by a factor of four. Thus, when the speed of a vehicle increases from 20 miles per hour to 40 miles per hour, the force exerted by that vehicle is increased by a factor of four. Higher speeds require more braking power and a longer distance to bring the vehicle to a stop.

To drive safely, you need to understand where the greatest risks are located. As mentioned earlier, studies have shown that many crashes involving emergency vehicles occur at intersections FIGURE 2-5. Most fire departments require all emergency vehicles to come to a full stop when the emergency vehicle operator encounters a stop sign or a red traffic light. If you enter an intersection without stopping, you may not be able to see approaching traffic, and many motorists do not drive defensively. They assume that if there is no stop sign or red light, they can proceed through the intersection without looking for oncoming vehicles.

In addition, many emergency vehicle crashes occur on open stretches of straight roads during daylight hours. The primary cause of these crashes is excessive speed. Most of us have limited experience with high-speed driving, so it is important to keep within the limits of your driving skills and within the limits of the vehicle you are operating. Many departments limit the speed during emergency responses to 5 miles per hour over the posted speed limit. Multiple studies have shown that little time is saved by driving at high rates of speed.

During an emergency response, be alert for the presence of other emergency vehicles. These vehicles may carry other fire fighters responding from their homes, or fire, EMS, or law enforcement personnel responding to the same incident. It is difficult to hear the sounds of other apparatus when you are on an emergency response.

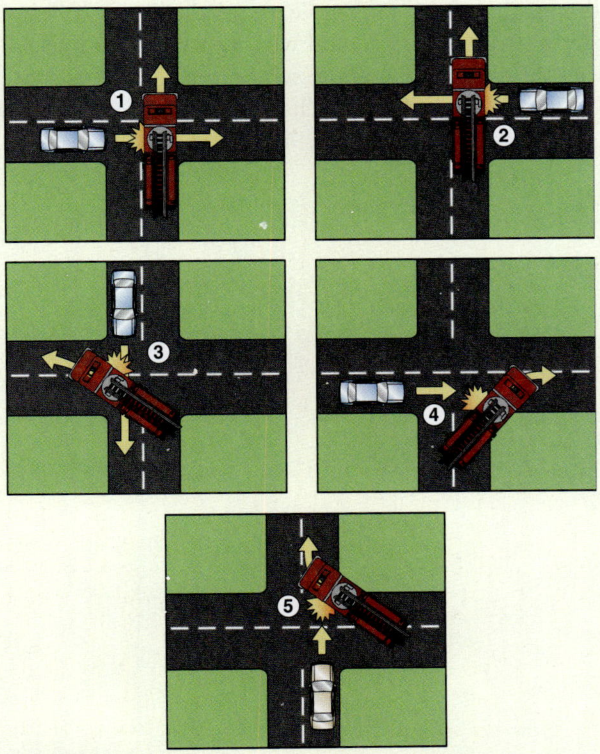

FIGURE 2-5 Every two-lane intersection has five potential crash points, making intersections very dangerous.

Drive with a cushion of safety around you—that is, in such a manner that even if a motorist does the unexpected, you still have enough time and distance to avoid a collision. Remember that the sound of a siren carries only a limited distance. Sometimes in an urban environment motorists may not be able to determine the location of an approaching emergency vehicle because of surrounding buildings, which distort the sound of the siren. Similarly, emergency lights have limitations and cannot be depended on to clear a path of travel for you.

■ The Importance of Vehicle Maintenance

Do not underestimate the importance of proper vehicle maintenance for both fire department vehicles and your private vehicle. It is important to perform regular maintenance of the engine, transmission, and other equipment to ensure dependable starting and running. Keep the brakes and tires in proper condition to ensure dependable stopping. In addition, note that tire tread depth for emergency vehicles should exceed the minimum required for nonemergency vehicles. The suspension system and steering system need to be properly maintained to ensure safe handling of the vehicle. Likewise, the windshield wipers and windshield washers must be kept in good condition to ensure clear vision. Regularly check all headlights, taillights, and turn signals for proper operation. Private vehicles that are used for emergency response should be maintained at a higher level than vehicles that are not subject to the stresses of emergency operation.

TABLE 2-3 summarizes the guidelines when driving to respond to an emergency call.

Safety at Emergency Incidents

At the emergency scene, fire fighters should never charge blindly into action. The officer in command will "size up" the situation, carefully evaluating the conditions to determine whether the burning building is safe to enter or what needs to be done to make the scene safe for action to take place. Wait for instructions and follow directions for the specific tasks to be performed. Do not freelance or act independently of command.

TABLE 2-3	Guidelines for Safe Emergency Vehicle Response
1.	Drive defensively.
2.	Follow agency policies in regard to posted speed limits.
3.	Always maintain a safe distance. Use the "4-second rule": Stay at least 4 seconds behind another vehicle in the same lane.
4.	Maintain an open space or cushion in the lane next to you as an escape route in case the vehicle in front of you stops suddenly.
5.	Always assume that other drivers will not hear your siren or see your emergency lights.
6.	Select the shortest and least congested route to the scene at the time of dispatch.
7.	Visually clear all directions of an intersection before proceeding.
8.	Go with the flow of traffic.
9.	Watch carefully for bystanders and pedestrians. They may not move out of your way or could move the wrong way.

VOICES OF EXPERIENCE

I was the Instructor-in-Charge of a live fire exercise. During one of our evolutions, the fire was growing as expected and the students were instructed on the essential fire behaviors being exhibited. Everything was textbook perfect. When the fire began to rollover, the student on the nozzle was instructed to open up and chase the fire back into the room of origin by opening the nozzle up high at the ceiling. When he did, the nozzle was on a wide fog pattern instead of the proper pattern. This caused the fire to have a tremendous amount of air pushed into it and intensify toward us.

Instead of adjusting his pattern, the student on the nozzle lowered his nozzle and attempted to duck under the fire. This allowed the fire to travel closer toward us. I reached up and helped him to adjust his nozzle stream. We redirected the appropriate stream into the fire to extinguish it. When we exited the structure, I advised the instructor with the backup team to overhaul the room while I inspected the students on the hose team and myself to make certain that we were not injured. We escaped unscathed. Why?

Because we took the time to make certain that everyone had the appropriate PPE. By following the strict guidelines of NFPA 1403, *Standard for Live Fire Training Evolutions*, no one was injured. In addition, the students learned valuable lessons about nozzle manipulation, proper hose stream selection, and why we wear the proper PPE during training and emergencies. It was an important lesson for the instructors as well because by following NFPA 1403, we ensured that we had the proper safety precautions in place for the students and instructional staff.

W. Chris Allan
Albemarle Fire and Rescue
Albemarle, North Carolina

Teamwork

On the fire ground or emergency scene, a firefighting team should always consist of at least two fire fighters who work together and are in constant communication with each other. Some departments call this scheme the buddy system FIGURE 2-6. In some cases, fire fighters work directly with the company officer, and all crew members function as a team. In other situations, two individual fire fighters may form a team that is assigned to carry out a specific task. In either case, the company officer must always know where teams are and what they are doing.

Partners or assigned team members should enter together, work together, and leave together. If one member of a team must leave the fire building for any reason, the entire team must leave together, regardless of whether it is a two-person team or an entire crew working as a team.

Before entering a burning building to perform interior search and rescue or fire suppression operations, fire fighters must be properly equipped with approved and appropriate PPE. Partners should check each other's PPE to ensure it has been donned and is working correctly before they enter a hazardous area.

When working in a hazardous area, team members should maintain visual, vocal, or physical contact with one another at all times. At least one member of each team should have a portable radio to maintain contact with the IC or a designated individual in the chain of command who remains outside the hazardous area. This radio can be used to relay pertinent information and to summon help if the team becomes disoriented, trapped, or injured. In such a situation, the IC can contact the team with new instructions or an evacuation order.

Any fire fighters operating in a hazardous area require backup personnel. The backup team must be able to communicate with the entry team, either by sight or by radio, and should be ready to provide assistance. During the initial stages of an incident, at least two fire fighters must remain outside the hazardous area, properly equipped to respond immediately if the entry team needs to be rescued.

Once crews are assigned to enter the hazardous area, a designated rapid intervention crew (also known as a rapid intervention team) should be established and positioned outside the hazardous area. This team's sole responsibility is to be prepared to provide emergency assistance to crews working inside the hazardous area.

Accountability

Every fire department must have a personnel accountability system to track personnel and assignments on the emergency scene. This system should record the individuals assigned to each company, crew, or entry team; the assignments for each team; and the team's current activities. Several kinds of accountability systems are acceptable, ranging from paper assignments or display boards to laptop computers and electronic tracking devices.

Some departments use a "passport" system. In this scheme, each company officer carries a small magnetic board called a passport. Each crew member on duty with that company has a magnetic name tag on the board FIGURE 2-7. At the incident scene, the company's passport is given to a designated individual, who uses it to track the assignment and location of every company at the incident.

Other fire departments employ an "accountability tag" system. With this approach, each fire fighter carries a name tag and turns it in or places it in a designated location on the apparatus when the individual is on the scene. The tags are then collected and used to track crews who are working together as a unit. This system works well when crews are organized based on the available personnel.

Both passport and accountability-tag systems provide an up-to-date accounting of everyone who is working at the incident and how they are organized. At set intervals, an accountability check is performed to account for everyone. Usually, a company officer reports on the status of each crew. The company officer should always know exactly where each crew is and what it is doing. If a crew splits into two or more teams, the company officer should be in contact with at least one member of each team.

An accountability check is also performed when the operational strategy changes or when a situation occurs that could endanger fire fighters. If an accountability check is needed, the list of personnel and assignments should be readily available at the command post.

FIGURE 2-6 A firefighting team should consist of at least two members who work together.

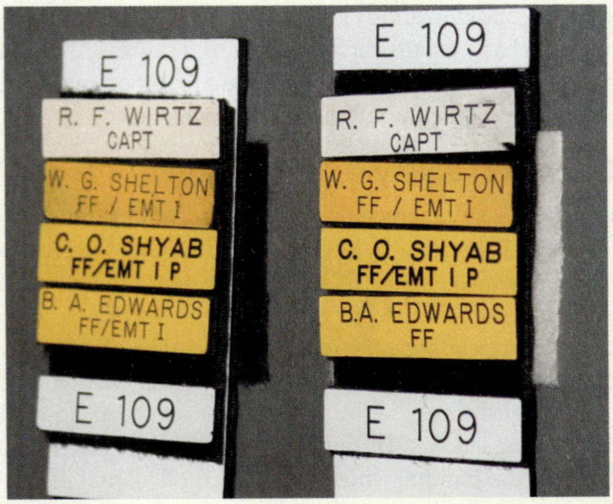

FIGURE 2-7 A passport lists each fire fighter assigned to a crew.

Fire fighters must learn which kind of accountability system their department uses, how to work within it, and how this system works within the ICS. They are responsible for complying with this system and staying in contact with a company officer or assigned supervisor at all times. Most importantly, fire fighters must always remember that teams must stay together.

The accountability system is valuable for identifying fire fighters who are lost or trapped in a fire or other emergency situation. The Fire Fighter Survival chapter covers many of these self-survival skills and the steps for calling for help ("mayday").

Incident Scene Hazards

Fire fighters must be aware of their surroundings when performing their assigned tasks at an emergency scene. At an incident, make a safe exit from the fire apparatus and look at the building or situation for safety hazards such as traffic, downed utility wires, and adverse environmental conditions. An incident on a street or highway must first be secured with proper traffic- and scene-control devices. Flares, traffic cones, or barrier tape are all measures that can help keep the public at a safe distance from the scene. Emergency vehicles can be placed to block traffic and protect the incident scene. Always operate within established boundaries and protected work areas.

Changing fire conditions will also affect safety. During the overhaul phase and while picking up equipment, watch out for falling debris, smoldering areas of fire, and sharp objects. If a safety officer is not on the scene, another qualified person should be assigned to monitor the atmosphere to ensure it is safe to remove SCBA or enter the area without SCBA. Because the chance for injury increases when you are tired, do not let down your guard even though the main part of the firefighting operation is over.

Using Tools and Equipment Safely

Fire fighters must learn how to use tools and equipment properly and safely before attempting to use them at an actual emergency incident. Follow the proper procedures and safety precautions both in training and at an incident scene. In addition, don protective gear such as PPE, safety glasses, and hearing protection when they are required.

Proper maintenance of tools includes sharpening, lubricating, and cleaning each tool. Equipment should always be in excellent condition and ready for use. All fire fighters should be able to do basic repairs such as changing a saw blade or a handlight battery. Practice these tasks at the fire station until you can perform them quickly and safely on the emergency scene.

Electrical Safety

Electricity is an emergency scene hazard that must always be respected. Electricity is in every building and involved with many vehicle collisions. In some cases, energized power lines may be present on the fire ground. These wires may be energized, yet nearly invisible when lying on the ground. This is especially true in conditions of limited visibility such as darkness, fog, and smoke. Always check for overhead power lines when raising ladders. During any major fire, the electric power supply to the building should be turned off as part of the fireground task called "controlling the utilities."

Fire departments are often called to electrical emergencies, such as those involving downed power lines, fires, arcing wires, and transformer fires. Always have the power disconnected or turn off the power if you are trained and permitted by department policy.

Fire apparatus should be parked outside the area and away from power lines when responding to a call for an electrical emergency. A downed power line should be considered energized until the power company confirms that it is dead. Secure the area around the power line and keep the public at a safe distance. Never drive fire apparatus over a downed line or attempt to move it using tools. If a sparking power line causes a brush fire, if it is safe to do so attempt to contain the fire from a safe distance but do not use water near the line.

Lifting and Moving

Lifting and moving objects is part of a fire fighter's daily duties. Do not try to move something that is too heavy alone—ask for help. Never bend at the waist to lift an object; instead, bend at the knees and use your legs to lift the weight. If you must move objects over a long distance, use equipment such as handcarts, hand trucks, and wheelbarrows.

In addition to moving inanimate objects, fire fighters must often move sick or injured victims. Don't move an injured person, unless their life is in danger, until an EMT or other appropriate medical personnel are on scene to stabilize the victim. Discuss and evaluate the options before moving a victim, and then proceed very carefully. If necessary, request help. Never be afraid to call for additional resources, such as another engine or truck company, to assist in lifting and moving a heavy victim.

Working in Adverse Weather Conditions

In adverse weather conditions, fire fighters must dress appropriately. A turnout coat, pants, boots, and helmet can keep you warm and dry in rain, snow, or ice. Firefighting gloves and knit caps will also help retain body heat and keep you warm. If conditions are icy, make smaller movements, watch your step, and keep your balance. During the hot summer months, it is difficult to stay cool in PPE. Many departments have SOPs covering PPE use in summer months. Check your department's SOPs for guidance on acceptable PPE use in adverse weather.

Rehabilitation

Rehabilitation is a systematic process that provides periods of rest and recovery for emergency workers during an incident. It is usually conducted in a designated area away from the hazards of the emergency scene. The rehabilitation area, or "rehab," is usually staffed by fire and EMS personnel.

A fire fighter who is sent to rehabilitation should be accompanied by the other members of the crew. The company officer should inform the IC of their change in location. While in rehabilitation, fire fighters should take advantage of the opportunity to rest, rehydrate, have their vital signs checked and receive treatment for minor injuries FIGURE 2-8. Rehabilitation gives fire fighters the chance to cool off in hot weather and to warm up in cold weather.

Rehabilitation time can be used to replace SCBA cylinders, obtain new batteries for portable radios, and make repairs or adjustments to tools or equipment. Firefighting teams can discuss recently completed assignments and plan their next work cycle. When a crew is released from rehabilitation, its members

FIGURE 2-8 In the rehabilitation area, fire fighters can rest and rehydrate.

should be rested, refreshed, and ready for another work cycle. If the crew is too exhausted or unable to return to work or have not been medically cleared to return to work, they should be replaced and released from the incident.

Never be afraid or embarrassed to admit you need a break when on the emergency scene. Heat exhaustion is a common condition at firefighting incidents and is characterized by profuse sweating, dizziness, confusion, headache, nausea, and cramping. If a fire fighter shows signs or symptoms of heat exhaustion, the company officer should be notified immediately. The company officer will then request approval for rehabilitation so the problem can be treated.

Heat exhaustion is usually not life-threatening and, if identified early, can be remedied by rehydration, cooling, and rest. Left untreated, however, heat exhaustion can progress quickly to heat stroke, which is a life-threatening emergency. A lack of sweating, low blood pressure, shallow breathing, and seizures are

Near Miss REPORT

Report Number: 05-0000535

Synopsis: Inadequate protective clothing contributes to fire fighter injury.

Event Description: At a structure fire in a metal frame and metal siding workshop, I was on a 2.5-inch (6 centimeter) line with a fire fighter from another department. At that time, we were conducting an exterior attack on the fire from the front of the building, in front of a metal roll-up door. The door had fallen down and was lying in the way of us advancing the hose line into the shop to extinguish hot spots. The door was also positioned in such a way that it was difficult for me to get the stream where I wanted it inside the shop. I shut down the nozzle and moved the metal door, which was all the way on the ground now and not connected to the building. We had no problem directing our fire stream after that, and we had a clear access point into the shop.

About 10 minutes later, I was picking up an object to move it out of the way (something smaller than the door) with my right hand, which had just a regular leather work glove on it the whole time, when I noticed blood streaming down my wrist and in my glove. I reported to the ambulance and got it flushed, cleaned, and wrapped; I then went back to work.

If the cut had been any deeper, it would have nicked one of my veins and I could have been in bad shape. I never felt the metal door cut into me. I was wearing full PPE the whole time, with the exception of my leather work gloves. I'll have a pretty good scar to remind me. If I had been wearing my fire gloves, which have extended cuffs on them, and had asked for a hand moving the door, I probably would not have been injured. However, at the time, there were not many fire fighters on scene to help me, and coordination between the four or five departments that were there seemed somewhat nonexistent to me. Four other fire fighters were taken out for rehab, two had IVs started on scene, and two or three were transported to the hospital. The contents of the shop were pretty much a total loss.

Lessons Learned: Wear your PPE so that no skin is exposed. Ask for help when moving heavy objects—don't try to fight the fire by yourself.

Fewer trucks and more personnel on the scene can sometimes lead to less chaos. Likewise, better coordination of tasks and teams leads to more effective operations.

Better training and equipment for volunteer fire fighters would also be helpful. Some fire fighters who were on scene just stood around not knowing what was going on and couldn't do much because they didn't have the knowledge or the equipment to assist safely. Some groups of fire fighters with no PPE were walking around the shop with small portable foam units applying it to the fire through holes in the wall, even as others were washing the foam away with the hose lines from the front of the building and negating what the others were doing.

some of the signs of heat stroke. If you or a member of your team experiences these symptoms during training or on the fire ground, place a "mayday" or "fire fighter in trouble" call. You must report to your superior and seek immediate medical attention.

A fire fighter who experiences chest pain or discomfort should stop and seek medical attention immediately. As noted earlier, heart attack is the leading cause of death among fire fighters. More information on fire fighter rehabilitation is presented in the Fire Fighter Rehabilitation chapter.

Violence at the Scene

Fire fighters are sometimes dispatched to calls involving guns, domestic disputes, injuries from an assault, or other violent scenes. In such incidents, the staging area and the fire apparatus should be located at some distance from the scene until the police arrive, investigate, and declare the area safe. Only then should fire fighters proceed to enter the emergency scene. A fire fighter's personal safety should always be paramount. If there is any threat to personal safety, slowly back away from the emergency scene to a safe distance and request the police to secure the scene. Do not become a victim.

If you are confronted with a potentially violent situation, do not respond violently. Remain calm, speak quietly, and attempt to gain the person's trust and call for police assistance. You might consider taking additional classes to increase your understanding and develop appropriate skills for these situations.

Mental Well-being

Some emergency calls are particularly difficult and emotionally traumatic. After experiencing such an incident, fire fighters who were involved may be encouraged to meet with a counselor or attend a critical incident stress debriefing (CISD). Usually, a stress debriefing is held as soon as possible after a traumatic call. It provides a forum for firefighting and EMS personnel to discuss the anxieties, stress, and emotions triggered by a difficult call. Follow-up sessions can be arranged for individuals who continue to experience stressful or emotional responses after a challenging incident.

Some fire departments have qualified counselors available 24 hours a day. The initial counseling may consist of a group session for all fire fighters and rescuers; alternatively, it can also be done on a one-on-one basis or in smaller groups. Everyone handles stress differently, and new fire fighters need time to develop the personal resources necessary to deal with difficult situations. If any call proves emotionally disturbing to you, ask your superior officer for a referral to a qualified counselor. If you are having trouble sleeping or find yourself having difficulty dealing with your thoughts or feelings, tell a fire officer or contact an Employee Assistance Program.

Safety at the Fire Station

The fire station is just as much a workplace as the fire ground. Indeed, fire fighters will spend much of their time during a 12- or 24-hour shift at the fire station. While there, be careful when working with power tools, ladders, electrical appliances, pressurized cylinders, and hot surfaces. Injuries that occur at the firehouse can be just as devastating as those that occur at an emergency incident scene.

Safety Outside Your Workplace

Continue to follow safe practices when you are off duty. An accident or injury, regardless of where it happens, can end your career as a fire fighter. For example, if you are using a ladder while off duty, follow the same safety practices that you would use on duty. Keep your seat belt fastened in your personal vehicle, just as you are required to do when you are on duty FIGURE 2-9.

Fire Fighter Safety Tips

Remember these guidelines to stay safe—both on and off the job:

- You are personally responsible for safety. Keep yourself safe. Keep your teammates safe. Keep citizens—your customers—safe.
- Work as a team. The safety of the entire firefighting unit depends on the efforts of each member. Become a dependable member of the team.
- Follow orders. Freelancing can endanger other fire fighters as well as yourself.
- Think! Before you act, think about what you are doing. Many people are depending on you.

FIGURE 2-9 Always use your seat belt.

Wrap-Up

Chief Concepts

- Every fire fighter and fire department must have a strong commitment to safety and health. Safety must be fully integrated into every activity, procedure, and job description.
- Each year, on average, 80 to 100 fire fighters are killed at emergency incident scenes, in fire stations, during training, and while responding or returning from an emergency.
- The majority of all fire fighter deaths are caused by heart attacks or strokes.
- Motor vehicle collisions are a major cause of fire fighter fatalities. Always wear your seat belt each and every time you are in a motor vehicle.
- Fire fighters must always consider three groups when ensuring safety at the scene:
  - Their personal safety
  - The safety of other team members
  - The safety of everyone at the emergency scene
- A successful safety program must have four major components:
  - Standards and procedures—Regulations, ranging from NFPA standards to OSHA regulations to local laws, govern the fire department's SOPs and SOGs on ensuring a safe work environment.
  - Personnel—From teams working together to safety officers, personnel ensure that a fire department's safety program is implemented correctly.
  - Training—Training does not end with this course. To ensure that you attain and maintain a high level of skills and knowledge, you must seek out additional courses and training opportunities throughout your career.
  - Equipment—To ensure safety, equipment must be properly maintained and fire fighters must be trained how to use it correctly.
- The national Fallen Firefighters Foundation's "16 Firefighter Life Safety Initiatives" describes the steps that need to be taken to change the current culture of the fire service to help make it a safer place for all fire fighters.
- Safety and well-being are directly related to personal health and fitness. You should eat a healthy diet, maintain a healthy weight, exercise regularly, and sleep for 7 to 8 hours whenever possible.
- A well-rounded fitness routine includes weight training, cardiovascular workouts, and stretching.
- Employee assistance programs are available to provide fire fighters with confidential counseling, support, or assistance in dealing with a physical, financial, emotional, or substance abuse problem.
- Each year, 10 percent of fire fighter fatalities occur during training exercises. Do not attempt anything you feel is beyond your ability or knowledge. If you see something you feel is unsafe, say something.
- Four general principles govern emergency vehicle operation:
  - Emergency vehicle operators are subject to all traffic regulations unless a specific exemption is made.
  - Exceptions are legal only when operating in emergency mode.
  - Even with an exemption, the emergency vehicle operator can be found criminally or civilly liable if involved in a crash.
  - An exemption does not relieve the operator of an authorized emergency vehicle from the duty to drive with reasonable care for all persons using the highway.
- Emergency driving requires good reactions and alertness. Driving while impaired or distracted is unacceptable. Your driving record, both on-duty and off-duty, will impact your career as a fire fighter.
- Safe driving practices will prevent most vehicle collisions. Anticipate the road and road conditions. Make allowances for weather conditions. Learn where the greatest risks are located on your routes. Be alert for the presence of other emergency vehicles.
- The personnel accountability system tracks personnel and assignments at the emergency scene. Several types of systems are acceptable, ranging from paper assignments to display boards to laptop computers and electronic tracking devices.
- The electric power supply should be turned off to safely secure the scene for fire fighters.
- Do not try to move a heavy object alone—ask for help.
- A fire fighter who experiences chest pain or discomfort should stop and seek medical attention immediately.
- A critical incident stress debriefing is a forum in which firefighting and EMS personnel can discuss anxieties, stress, and emotions triggered by a difficult incident.
- The fire station is just as much a workplace as the fire ground. Be careful when working with power tools, ladders, electrical appliances, pressurized cylinders, wet floors, and hot surfaces.
- An accident or injury, regardless of when or where it happens, can end your firefighting career.

Hot Terms

Critical incident stress debriefing (CISD) A postincident meeting designed to assist rescue personnel in dealing with psychological trauma as the result of an emergency. (NFPA 1006)

Employee assistance programs (EAPs) Fire service programs that provide confidential help to fire fighters with personal issues.

Freelancing The dangerous practice of acting independently of command instructions.

Occupational Safety and Health Administration (OSHA) The federal agency that regulates worker safety and, in some cases, responder safety. OSHA is part of the U.S. Department of Labor.

Personnel accountability system A system that readily identifies both the locations and the functions of all members operating at an incident scene. (NFPA 1500)

Rapid intervention crew (RIC) A minimum of two fully equipped personnel who are on site, in a ready state, for immediate rescue of injured or trapped fire fighters. Also known as a rapid intervention team.

Rehabilitation An intervention designed to mitigate against the physical, physiological, and emotional stress of fire fighting so as to sustain a member's energy, improve performance, and decrease the likelihood of on-scene injury or death. (NFPA 1521)

FIRE FIGHTER *in action*

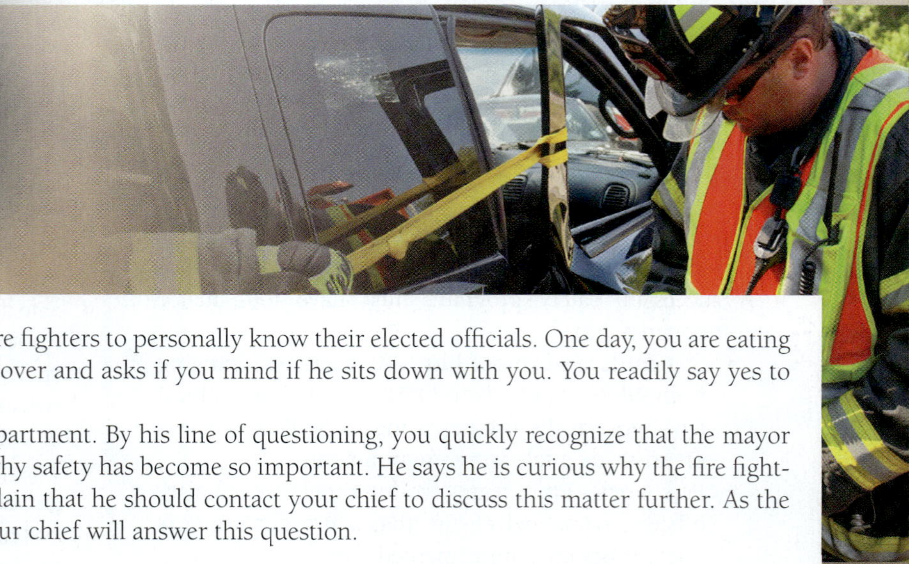

You live in a small town, so it is common for the fire fighters to personally know their elected officials. One day, you are eating at the only diner in town when the mayor comes over and asks if you mind if he sits down with you. You readily say yes to the opportunity to eat with the mayor.

The conversation quickly turns to the fire department. By his line of questioning, you quickly recognize that the mayor does not understand the risks of firefighting and why safety has become so important. He says he is curious why the fire fighters need new safety gear and equipment. You explain that he should contact your chief to discuss this matter further. As the conversation shifts gears, you think about how your chief will answer this question.

1. Approximately how many fire fighters are killed each year in the United States in the line of duty?
   - **A.** 40–60
   - **B.** 80–100
   - **C.** 130–150
   - **D.** 200–210
2. What is the cause of most fire fighter deaths?
   - **A.** Burns
   - **B.** Building collapse
   - **C.** Heart attack
   - **D.** Smoke inhalation
3. Out of all fire fighter injuries in the line of duty, what percentage occurred while fighting fires?
   - **A.** 25 percent
   - **B.** 45 percent
   - **C.** 65 percent
   - **D.** 85 percent
4. What is the most important action you can take to ensure your safety while responding to an emergency incident?
   - **A.** Remain seated in the fire apparatus.
   - **B.** Fasten your seat belt.
   - **C.** Fasten your self-contained breathing apparatus (SCBA) straps.
   - **D.** Wear your helmet.
5. How much water should be consumed to maintain proper hydration for each 5–10 minutes of physical exertion?
   - **A.** 4–6 ounces (0.1 to 0.17 liter)
   - **B.** 8–10 ounces (0.2 to 0.3 liter)
   - **C.** 12–16 ounces (0.35 to 0.4 liter)
   - **D.** 20–24 ounces (0.6 to 0.7 liter)

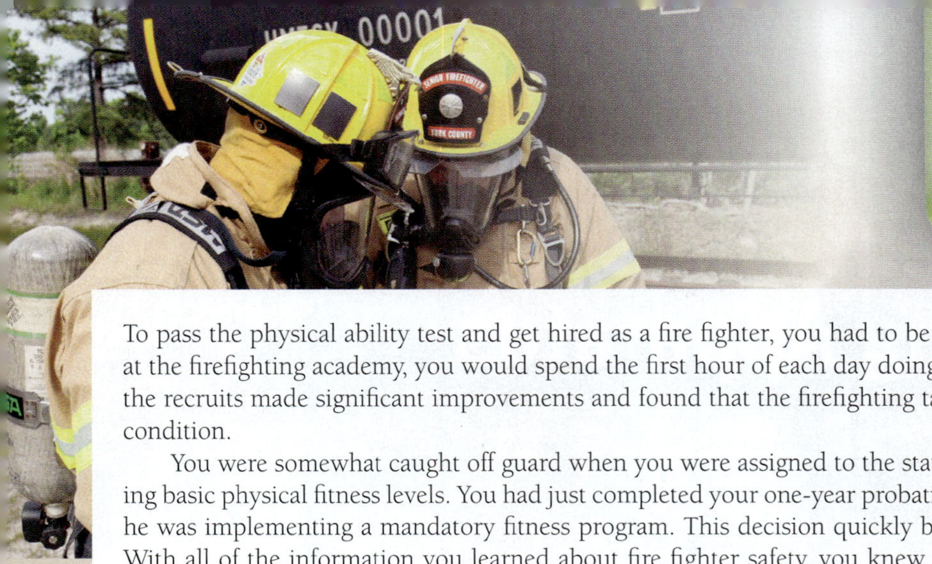

FIRE FIGHTER II *in action*

To pass the physical ability test and get hired as a fire fighter, you had to be in fairly good shape. Even so, during your time at the firefighting academy, you would spend the first hour of each day doing some form of a physical fitness activity. Each of the recruits made significant improvements and found that the firefighting tasks became even easier with the better physical condition.

You were somewhat caught off guard when you were assigned to the station and found many of the crew members lacking basic physical fitness levels. You had just completed your one-year probationary period when the fire chief announced that he was implementing a mandatory fitness program. This decision quickly became the hot topic around the firehouse table. With all of the information you learned about fire fighter safety, you knew the benefits of being physically fit, so you were shocked to find that the crew was opposed to it.

1. What would you say to the other crew members as they discuss this topic around the table?
2. What would you do to create a change in the attitude of the crew?
3. How would you handle it if you were asked by another fire fighter to stop working out while on-duty because it was making the rest of them look bad?
4. When should you "go with the flow" and when should you "stand up for your principles"?

Personal Protective Equipment and Self-Contained Breathing Apparatus

Fire Fighter I

Knowledge Objectives

After studying this chapter, you will be able to:

- List the components of personal protective equipment (PPE) or the structural firefighting ensemble. (NFPA 5.1.2, p 44–49)
- Describe the type of protection provided by the structural firefighting ensemble. (p 44–45)
- Explain how each design element of a fire helmet works to protect the head, face, and eyes. (p 45–46)
- Explain why protective hoods are a part of the structural firefighting ensemble. (p 46–47)
- Explain how each design element of a turnout coat works to protect the upper body. (p 46–47)
- Describe how each design element of boots works to protect the feet. (p 48)
- Describe how each design element of gloves works to protect the hands and wrist. (p 48)
- Explain how a personal alert safety system (PASS) helps to ensure fire fighter safety. (NFPA 5.3, p 49)
- List the limitations of PPE. (p 50)
- Explain the role of the fighter's work uniform as part of the PPE ensemble. (NFPA 5.1.2, p 50–51)
- Describe how to inspect the condition of PPE. (NFPA 5.3, p 51, 54)
- Describe how to properly maintain PPE. (NFPA 5.5.1, p 51, 54)
- Describe the specialized protective clothing required for vehicle extrication and wildland fires. (p 54)
- List the respiratory hazards posed by smoke and fire. (NFPA 5.3.1A, p 54–56)
- List the conditions that require respiratory protection or self-contained breathing apparatus (SCBA). (NFPA 5.3.1.A, p 56)
- Describe the differences between open-circuit breathing apparatus and closed-circuit breathing apparatus. (p 56)
- Describe when a supplied-air respirator is used. (p 56)
- Describe the limitations of SCBA. (NFPA 5.3.1A, p 58–59)
- Describe the physical and psychological limitations of an SCBA user. (NFPA 5.3.1.A, p 58–59)
- List and describe the major components of SCBA. (NFPA 5.3.1.A, p 59–64)
- Describe the devices on an SCBA that can assist the user in air management. (NFPA 5.3.1, p 62)
- Describe the pathway that air travels through an SCBA. (p 62)
- Explain the skip-breathing technique. (NFPA 5.3.1, 5.3.1B, p 62)
- Explain how to inspect SCBA to ensure that it is operation ready. (NFPA 5.5.1, p 73)
- List the complete sequence of donning PPE. (NFPA 5.1.2, 5.3.1A, 5.3.1, p 73)
- Describe the importance of SCBA inspections and SCBA operational testing. (NFPA 5.5.1, p 73–78)
- Explain the procedures for refilling SCBA cylinders. (p 78–81)

Skills Objectives

After studying this chapter, you will be able to perform the following skills:

- Don approved personal protective clothing. (NFPA 5.1.2, p 51–52)
- Doff approved personal protective clothing. (NFPA 5.1.2, p 51, 53)
- Don an SCBA from an apparatus seat mount. (NFPA 5.3.1B, p 64–65)
- Don an SCBA from a compartment mount. (NFPA 5.3.1B, p 64)
- Don an SCBA from a storage case using the over-the-head method. (NFPA 5.3.1B, p 66–67)
- Don an SCBA from a storage case using the coat method. (NFPA 5.3.1B, p 66, 68)
- Don an SCBA from a seat-mounted position with a safety latch. (NFPA 5.3.1B, p 66, 69)
- Don a face piece. (NFPA 5.3.1B, p 67–72)
- Doff an SCBA. (p 73–74)
- Perform a visible inspection of an SCBA. (NFPA 5.5.1, p 73–74)
- Perform an operational inspection of an SCBA. (NFPA 5.5.1, p 75–78)
- Replace an SCBA cylinder. (NFPA 5.3.1B, p 75–76, 79)
- Replace an SCBA cylinder on another fire fighter. (NFPA 5.3.1B, p 78, 80)
- Refill an SCBA cylinder from a cascade system. (p 78–79, 81)
- Clean an SCBA. (NFPA 5.5.1, p 81–82)

CHAPTER 3

Fire Fighter II — FFII

Knowledge Objectives

There are no knowledge objectives for Fire Fighter II candidates. NFPA 1001 contains no Fire Fighter II Job Performance Requirements for this chapter.

Skills Objectives

There are no skill objectives for Fire Fighter II candidates. NFPA 1001 contains no Fire Fighter II Job Performance Requirements for this chapter.

Additional NFPA Standards

- NFPA 1404, *Standard for Fire Service Respiratory Protection Training*
- NFPA 1500, *Standard on Fire Department Occupational Safety and Health Program*
- NFPA 1582, *Standard on Comprehensive Operational Medical Program for Fire Departments*
- NFPA 1851, *Standard on Selection, Care, and Maintenance of Protective Ensembles for Structural Fire Fighting and Proximity Fire Fighting*
- NFPA 1852, *Standard on Selection, Care, and Maintenance of Open-Circuit Self-Contained Breathing Apparatus (SCBA)*
- NFPA 1971, *Standard on Protective Ensembles for Structural Fire Fighting and Proximity Fire Fighting*
- NFPA 1975, *Standard on Station/Work Uniforms for Emergency Services*
- NFPA 1977, *Standard on Protective Clothing and Equipment for Wildland Fire Fighting*
- NFPA 1981, *Standard on Open-Circuit Self-Contained Breathing Apparatus (SCBA) for Emergency Services*
- NFPA 1982, *Standard on Personal Alert Safety Systems (PASS)*

You Are the Fire Fighter

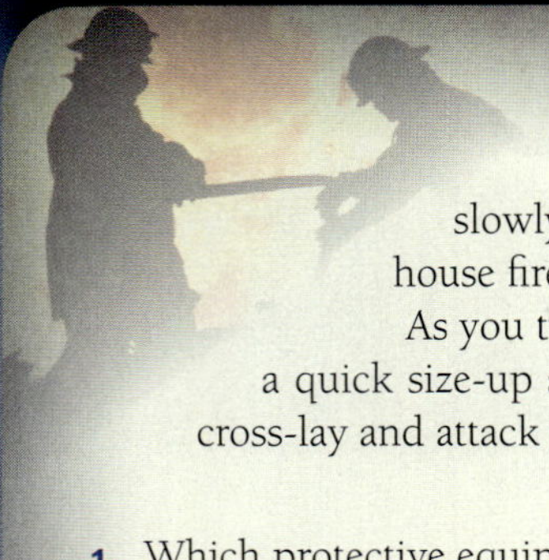

One cool fall evening, you are dispatched for a smoke odor in the area. It seems like a fairly routine call because residents are beginning to use their fireplaces for the first time of the year and the weather conditions tend to cause the smoke to hang low. As the engine drives slowly through the neighborhood of bungalow-style houses, you notice the distinct smell of a working house fire rather than that of firewood.

As you turn the corner, there it is—smoke drifting from the basement windows of a house. Your captain does a quick size-up and requests a full assignment from dispatch. She then tells you and the other fire fighter to pull a cross-lay and attack the fire.

1. Which protective equipment would you need for this situation?
2. How is the most efficient way to put on an SCBA?
3. Once the fire is over, what is the proper way to care for your protective equipment?

Introduction

Two safety components used by the fire fighter require special consideration: personal protective equipment (PPE) and self-contained breathing apparatus (SCBA). For you to remain safe and to do your job effectively, you need to understand the purposes for which this equipment is made, what it can do, and what it cannot do. PPE is designed to protect the body against a *limited* amount of heat. If you exceed the protection offered by this equipment, then severe injuries and death can occur. SCBA enables fire fighters to safely enter some smoky and toxic environments. It provides respiratory protection for a limited period of time. When fire fighters exceed the limits of SCBA, however, the consequences can be deadly.

Personal Protective Equipment

Personal protective equipment is an essential component of a fire fighter's safety system. It enables a person to survive under conditions that might otherwise result in death or serious injury. Different PPE ensembles are designed for specific hazardous conditions, such as structural firefighting, wildland firefighting, airport rescue and firefighting, hazardous materials operations, and emergency medical operations.

The various PPE ensembles provide specific protections, so an understanding of their designs, applications, and limitations is critical. For example, a structural firefighting ensemble will protect the wearer from the heat, smoke, and toxic gases present in building fires, but it cannot provide long-term protection from extreme weather conditions and it limits range of motion **FIGURE 3-1**. The more you know about the protection your PPE can provide, the better you will be able to judge conditions that exceed its limitations.

A fire fighter's PPE must provide full-body coverage and protection from a variety of hazards. To be effective, the entire ensemble must be worn whenever potential exposure to those hazards exists. In addition, PPE must be cleaned, maintained, and inspected regularly to ensure that it will continue to provide the intended degree of protection when it is needed. Worn or damaged articles must be repaired or replaced.

FIGURE 3-1 A protective ensemble for structural firefighting provides protection from multiple hazards.

Structural Firefighting Ensemble

Structural firefighting PPE enables fire fighters to enter burning buildings and work in areas with elevated temperatures and concentrations of toxic gases. Without PPE, fire fighters would

Fire Fighter Safety Tips

A structural firefighting ensemble is designed only for structural firefighting—it is not designed for other functions such as hazardous materials, water rescue, or wildland firefighting. Instead, each of these specialized functions requires a specific set of PPE. If your department performs specialized operations, you need the appropriate PPE for that activity.

TABLE 3-1	Protection Furnished by Personal Protective Equipment

- Provides thermal protection
- Repels water
- Provides impact protection
- Protects against cuts and abrasions
- Furnishes padding against injury
- Provides respiratory protection

be unable to conduct search and rescue operations or perform fire suppression activities. A structural firefighting ensemble is designed to cover every inch of the body. It provides protection from the fire, keeps water away from the body, and helps reduce trauma from cuts or falls. This type of PPE is designed to be worn with self-contained breathing apparatus (SCBA), which provides respiratory protection.

The structural firefighting ensemble consists of a protective coat, trousers or coveralls, a helmet, a hood, boots, and gloves. The helmet must have a face shield, goggles, or both. The clothing is worn with SCBA and a personal alert safety system (PASS) device. All of these elements must be worn together to provide the necessary level of protection FIGURE 3-2.

Protection Provided

A structural firefighting ensemble is designed to provide full-body coverage and offers several different types of protection TABLE 3-1. The coat and pants have outer shells that can withstand elevated temperatures, repel water, and provide protection from abrasions and sharp objects. The knees may be reinforced with pads for greater protection when crawling. Reflective trim adds visibility in dark or smoky environments. Insulating layers of fire-resistant materials protect the skin from high temperatures. A moisture barrier between different layers of PPE keeps liquids and vapors, such as hot water or steam, from reaching the skin.

The helmet protects the head from falling debris; it includes ear coverings to protect the ears as well. The face shield helps protect the eyes. A fire-retardant hood covers any exposed skin between the coat collar and the helmet. Gloves protect the hands from heat, cuts, and abrasions. Boots protect the feet and ankles from the fire, keep them dry, prevent puncture injuries, and protect the toes from crushing injuries.

An SCBA provides respiratory protection by giving the fire fighter an independent, limited air supply. In this way, the fire fighter's respiratory system remains safe from toxic products and hot gases present in the atmosphere.

Helmet

Fire helmets are manufactured in several designs and shapes using a variety of materials. Nevertheless, each design must meet the requirements specified in NFPA 1971, *Standard on Protective Ensembles for Structural Fire Fighting and Proximity Fire Fighting*. The hard outer shell of the helmet is lined with energy-absorbing material and has a suspension system to provide impact protection against falling objects FIGURE 3-3. The helmet shell also repels water, protects against steam, and creates a thermal barrier against heat and cold. Its shape helps to deflect water away from the head and neck.

Face and eye protection can be provided by a face shield, goggles, or both. These components must be attached to the helmet and are used when SCBA is not needed or when the

FIGURE 3-2 The complete structural firefighting ensemble consists of a helmet, coat, trousers, protective hood, gloves, boots, SCBA, and PASS device.

FIGURE 3-3 A helmet is constructed with multiple layers and components.

SCBA face piece is not in place. A chin strap is also required and must be worn to maintain the helmet in the proper position. In addition, this strap helps to keep the helmet on the fire fighter's head during an impact.

Fire helmets have an inner liner for added thermal protection. This liner also provides protection for the wearer's ears and neck. When entering a burning building, the fire fighter should pull down the ear tabs to ensure maximum protection.

Helmets are manufactured with adjustable inner suspension systems that hold the head away from the shell and cushion it against impacts. This suspension system must be adjusted to fit the individual, who should try on the helmet with the SCBA face piece and hood in place. Because helmet shells come in a variety of sizes, fire fighters must be sure to wear the correct-size shell.

In many departments, helmet shells are color-coded according to the fire fighter's rank and function. They often carry company numbers, rank insignia, or other markings. Some fire departments use a shield mounted on the front of the helmet to identify the fire fighter's rank and company. Bright, reflective materials are applied to make the fire fighter more visible in all types of lighting conditions.

NFPA 1971 requires that helmets approved for structural firefighting have a label permanently attached to the inside of the shell. This label lists the manufacturer, model, date of manufacture, weight, size, and recommended cleaning procedures.

Protective Hood

Although the helmet's ear tabs cover the ears and neck, this area is still at risk for burns when the head is turned or the neck is flexed. Protective hoods provide additional thermal protection for these areas. Such a hood, which is constructed of a flame-resistant material such as Nomex®, PBI®, or carbon fiber, covers the whole head and neck, except for that part of the face protected by the SCBA face piece FIGURE 3-5. The lower part of the hood, which is called the bib, drapes down inside the turnout coat.

Protective hoods are worn over the face piece but under the helmet. After securing the face piece straps, the fire fighter should carefully fit the hood around the face piece so that no areas of bare skin are left exposed. The hood must fit snugly around the clear area of the face piece so that vision is not compromised and hot gases cannot leak between the face piece and the hood.

Turnout Coat

The coats used for structural firefighting are generally called bunker coats or turnout coats FIGURE 3-6. Only coats that meet NFPA 1971 requirements should be used for structural firefighting.

Turnout coats consist of three layers. The outer layer or shell is constructed of a sturdy, flame-resistant, water-repellant material such as Nomex, Kevlar®, or PBI®. Fluorescent reflective material applied to the light-colored outer shell makes the fire fighter more visible in smoky conditions and at night. The light-colored fabric also makes it easier to identify contaminants such as hydrocarbons, blood, and body fluids on the coat.

The second layer of the coat is the moisture barrier, which usually consists of a flexible membrane attached to a thermal barrier material (the third layer). The moisture barrier helps prevent the transfer of water, steam, and other fluids to the skin. It is critical because water applied to a fire generates large amounts of superheated steam, which can engulf fire fighters and burn unprotected skin.

The third layer of the turnout coat comprises a thermal barrier, which is made of a multilayered or quilted material that insulates the body from external temperatures. It enables fire fighters to operate in elevated temperatures generated by a fire and keeps the body warm during cold weather.

The front of the turnout coat has an overlapping flap to provide a secure seal. The inner closure is secured first, and

Standards for Personal Protective Equipment

The PPE for fire fighters must be manufactured according to exacting standards. Each item must have a permanent label verifying that the particular item meets the requirements of the standard FIGURE 3-4. Usage limitations as well as cleaning and maintenance instructions should also be provided.

The requirements for firefighting PPE are outlined in these standards:

- NFPA 1971, *Standard on Protective Ensembles for Structural Fire Fighting and Proximity Fire Fighting*
- NFPA 1977, *Standard on Protective Clothing and Equipment for Wildland Fire Fighting*

The requirements for SCBA and PASS devices are outlined in two standards:

- NFPA 1981, *Standard on Open-Circuit Self-Contained Breathing Apparatus (SCBA) for Emergency Services*
- NFPA 1982, *Standard on Personal Alert Safety Systems (PASS)*

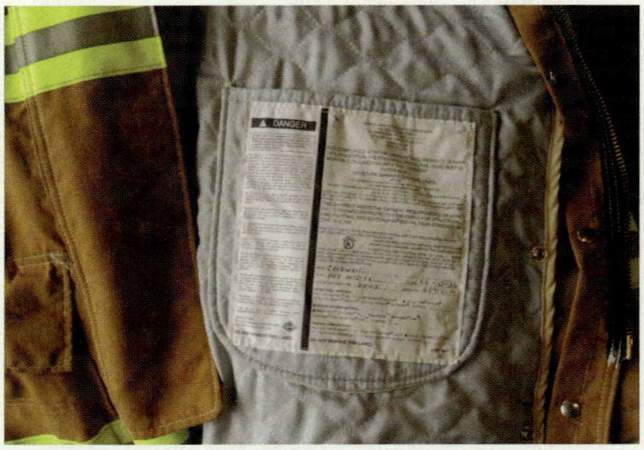

FIGURE 3-4 The label provides important information about each item of PPE.

FIGURE 3-5 A protective hood.

FIGURE 3-6 A bunker or turnout coat.

then the outer flap is secured, creating a double seal. Several different combinations of D-rings, snaps, zippers, and Velcro can be used to secure the inner and outer closures.

The collar of the turnout coat works with the hood to protect the neck. The collar has snaps or a Velcro closure system in front to keep it in a raised position. The coat's sleeves feature wristlets that prevent liquids or hot embers from getting between the sleeves and the skin. They also prevent the sleeves from riding up on the wrists, so no skin is exposed between the gloves and the sleeves, which could result in wrist burns.

Turnout coats come in two lengths—long and short. Both will protect the body if the matching style of pants or coveralls is worn simultaneously. The coat must be long enough that you can raise your arms over your head without exposing your midsection. The sleeve length should not hinder arm movement, and the coat should be large enough that it does not interfere with breathing or other movements.

Pockets in the coat can be used for carrying small tools or extra gloves. Additional pockets or loops can be installed to hold radios, microphones, flashlights, or other accessories.

Fire Fighter Safety Tips

The inner thermal liner of most turnout coats can be removed while the outer shell is being cleaned, but the turnout coat should never be used without the thermal liner. Severe injury could occur if the coat is used without the liner.

Bunker Pants

The protective trousers included as part of the structural firefighting ensemble are also called bunker pants or turnout pants FIGURE 3-7. They can be constructed in a waist-length design or

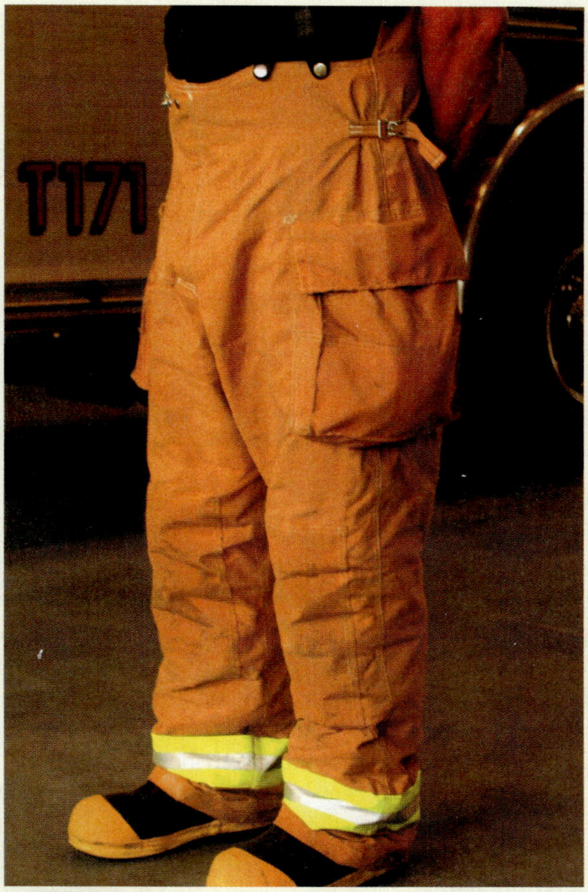

FIGURE 3-7 Bunker pants.

a bib-overall configuration. Like turnout coats, bunker pants must satisfy the conditions specified in NFPA 1971.

Also like turnout coats, bunker pants are constructed with three layers. The outer shell resists abrasions and repels water. The second layer is a moisture barrier that protects the skin from liquids and steam burns. The third, inner layer is a quilted, thermal barrier that protects the body from elevated temperatures. Bunker pants are reinforced around the ankles and knees with leather or extra padding.

Bunker pants are manufactured with a double-fastener system at the waist, similar to the front flap of a turnout coat. Florescent or reflective stripes around the ankles provide added visibility. Suspenders hold the pants up. Pants should be large enough that you can don them quickly. They should be big enough to allow you to crawl and bend your knees easily, but they should not be bigger than necessary.

Boots

Structural firefighting boots can be constructed of rubber or leather; they are available in a variety of lengths. Rubber firefighting boots come in a step-in style without laces FIGURE 3-8. Leather firefighting boots are available in a knee-length, pull-on style or in a shorter version with laces FIGURE 3-9. Many fire fighters install a zipper on the laced boots to aid in quick donning and doffing of this component of PPE.

Both boot styles must meet the requirements specified in NFPA 1971. The outer layer of these boots repels water and must be both flame and cut resistant. Boots must also have a heavy sole with a slip-resistant design, a puncture-resistant sole, and a reinforced toe to prevent injury from falling objects. An inner liner constructed of a material such as Nomex or Kevlar adds thermal protection.

Boots must be the correct size for the firefighter's feet. The foot should be secure within the boot to prevent ankle injuries and enable secure footing on ladders or uneven surfaces. Improperly sized boots will cause blisters and other problems.

FIGURE 3-8 Rubber firefighting boots.

FIGURE 3-9 Leather firefighting boots.

Fire Fighter Safety Tips

Fire boots must be worn with approved bunker pants. Boots worn without bunker pants leave the legs unprotected and exposed to injury.

Gloves

Gloves are an important part of the firefighting ensemble, because most fire suppression tasks require the use of the hands FIGURE 3-10. Gloves must provide adequate protection, yet permit the manual dexterity needed to accomplish tasks. NFPA 1971 specifies that gloves must be resistant to heat, liquid absorption, vapors, cuts, and penetration. These items must have a wristlet to prevent skin at the wrist from being exposed during normal firefighting activities.

Firefighting gloves are usually constructed of heat-resistant leather. The wristlets are usually made of knitted Nomex or Kevlar. The liner adds thermal protection and serves as a moisture barrier.

Many fire fighters carry a second set of gloves in their bunker gear or on the fire apparatus so they can change gloves if one pair becomes wet or is damaged. Do not wring or twist wet gloves, as this motion can tear or damage the inner liners.

Although gloves furnish needed protection, they inevitably reduce manual dexterity. For this reason, fire fighters need to practice manual skills while wearing gloves to become accustomed to them and to learn to adjust their movements accordingly.

Fire Fighter Safety Tips

Plain leather work gloves or plastic-coated gloves should never be used for structural firefighting. Gloves used for firefighting must be labeled as such and meet the specifications given by NFPA 1971.

Respiratory Protection

The SCBA is an essential component of the PPE used for structural firefighting. Without adequate respiratory protection, fire fighters would be unable to mount an attack. The design, parts, operation, use, and maintenance of this complex piece of equipment are covered later in this chapter.

FIRE FIGHTER Tips

Fire fighters have personal preferences about the tools and equipment they carry in their pockets. Observe which items seasoned members of your department carry in their pockets. Their choices will help you decide whether you need to carry similar equipment.

FIGURE 3-10 Firefighting gloves.

Personal Alert Safety System

A personal alert safety system (PASS) is an electronic device that sounds a loud audible signal when a fire fighter becomes trapped or injured. This device will sound automatically if a fire fighter is motionless for a set time period. It can also be manually activated to notify other fire fighters that the user needs assistance.

A PASS device can be either separate from or integrated into the SCBA unit FIGURE 3-11. Most PASS devices turn on automatically when the SCBA is activated. The separate PASS devices are often worn on the SCBA harness and must be turned on manually. All fire fighters must check that their PASS devices are on and working properly before they enter a burning building or hazardous area.

The PASS device combines an electronic motion sensor with an alarm system. If the user remains motionless for 30 seconds, it will produce a low warning tone before sounding a full alarm. The user can reset the device by moving during this warning period. Newer PASS devices may also include a radio transmitter that sends a signal to the command post when the alarm sounds.

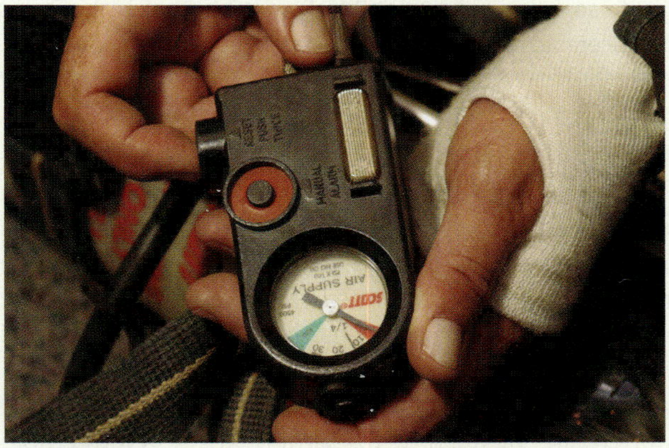

FIGURE 3-11 An integrated PASS device can save a fire fighter's life.

Additional Personal Protective Equipment

Some eye protection is provided by the face shield mounted on fire helmets. When additional eye protection is needed, such as when they are using power saws or hydraulic rescue tools, fire fighters can use approved goggles or safety glasses. Goggles can be carried easily in the pockets of the turnout coat.

Fire fighters are often exposed to loud noises such as sirens and engines. Because hearing loss is cumulative, it is important to limit exposure to loud sounds. An intercom system on the apparatus can provide hearing protection FIGURE 3-12. A small microphone is incorporated into large earmuffs located at each riding and operating position. Fire fighters don the earmuffs, which reduce engine and siren noise; at the same time, the microphone enables crew members to talk to each other in a normal tone of voice and to hear the apparatus radio. Flexible ear plugs are useful in other situations involving loud sounds. Fire fighters should use the hearing protection supplied by their departments to prevent hearing loss when using power equipment or other equipment that produces noise.

A fire fighter should always carry a hand light, given that most interior firefighting takes place in near-dark, zero-visibility conditions. A good working hand light can illuminate your surroundings, mark your location, and help you find your way under difficult conditions.

Two-way radios link the members of a firefighting team. At least one member of each team working inside a burning building or in any hazardous area should always have a radio. Some fire departments provide a radio for every on-duty fire fighter. Follow your department's standard operating procedure (SOP)

FIGURE 3-12 Some fire apparatus are equipped with an intercom system.

on radio use. A radio should be considered part of PPE and carried with you whenever it is appropriate.

Any time you are operating in a roadway or are close to traffic, you are required to wear a bright colored reflective safety vest that meets the latest iteration of ANSI Standard 207. This standard requires the vest to contain 450 square inches (11 meters) of high-visibility fabric and 201 square inches (5 meters) of retro-reflective material.

The final piece of safety equipment is the drag rescue device. The drag rescue device is a fabric handle integrated within the turnout coat that is designed to aid in the rescue of an incapacitated fire fighter. A rescuer can grab onto the drag rescue device and drag the incapacitated fire fighter to safety.

FIRE FIGHTER Tips

Each item of PPE clothing is designed to have overlapping material from one item to the next. For example, boots overlap the pants, coats overlap pants, and coats overlap gloves. This overlapping ensures that the body is protected at all times.

Limitations of the Structural Firefighting Ensemble

The structural firefighting ensemble protects a fire fighter from the hostile environment of a fire. To provide complete protection, each component must be properly donned and worn continually. Unfortunately, even today's advanced PPE has drawbacks and limitations. Understanding those limitations will help you avoid situations that could result in serious injury or death.

First, this ensemble is not easy to don. It includes several components, all of which must be put on in the proper order and correctly secured. You must be able to don your equipment quickly and correctly, either at the fire station before you respond to an emergency or after you arrive at the scene. Properly donning and doffing PPE takes practice FIGURE 3-13.

PPE is also heavy, accounting for nearly 50 pounds (23 kilogram) of extra weight. This increased weight means that everything you do—even walking—requires more energy and strength. Tasks such as advancing an attack line up a stairway or using an axe to ventilate a roof can be difficult, even for a fire fighter in excellent physical condition.

FIGURE 3-13 It takes practice to don the full set of PPE.

Because PPE retains body heat and perspiration, wearing this equipment makes it difficult for the body to cool itself. Perspiration is retained inside the protective clothing rather than being released through evaporation to cool you. As a consequence, fire fighters in full protective gear can rapidly develop elevated body temperatures, even when the ambient temperature is cool. The problem of overheating is more acute when surrounding temperatures are high, which is one reason why fire fighters must undergo regular rehabilitation and fluid replacement.

Another drawback of PPE is that it limits mobility. Full turnout gear not only limits the range of motion, but also makes movements awkward and difficult. This constraint increases energy expenditure and adds stress.

Wearing PPE also decreases normal sensory abilities. The sense of touch is reduced by wearing heavy gloves, for example. Turnout coats and pants protect skin but reduce its ability to determine the temperature of hot air. Sight is restricted when you wear SCBA. The plastic face piece reduces peripheral vision, and the helmet, hood, and coat make turning the head difficult. Both the earflaps and the protective hood over the ears limit hearing. Speaking becomes muffled and distorted by the SCBA face piece, even if it is equipped with a special voice amplification system.

For these reasons, fire fighters must become accustomed to wearing and using PPE. Practicing skills while wearing PPE will help you become comfortable with its operation and limitations.

Fire Fighter Safety Tips

Do not don protective clothing inside the apparatus while en route to an emergency incident. Instead, stay in your assigned seat, properly secured by a seat belt or safety harness, while the vehicle is in motion. Either don protective clothing in the firehouse before mounting the apparatus, or put it on at the scene after you arrive.

FIRE FIGHTER Tips

Practice donning your protective clothing ensemble until you can do so quickly and smoothly. Remember that each piece of the ensemble must be in place for maximum protection.

Work Uniforms

A fire fighter's work uniform is also part of the personal protective package. Certain synthetic fabrics can melt at relatively low temperatures and cause severe burns, even when worn under a complete personal protective ensemble. NFPA 1975, *Standard on Station/Work Uniforms for Emergency Services*, defines criteria for selecting appropriate fabrics for work uniforms.

Clothing containing nylon or polyester, even if these materials are blended with natural fibers, should not be worn in a firefighting environment. Clothing made of natural fibers, such as cotton or wool, is generally safer to wear.

Nylon and polyester will melt, but cotton and wool may char or even burn without melting. Special synthetic fibers such as Nomex and PBI®, which are used in turnout clothing, have excellent resistance to high temperatures.

Donning Personal Protective Clothing

Donning (i.e., putting on) protective clothing must be done in a specific order to obtain maximum protection; it must also be done quickly. Your goal should be to don PPE properly.

Following a set pattern of donning PPE helps reduce the time it takes to dress. This exercise does not include donning SCBA, however. First become proficient in donning personal protective clothing, and then add the SCBA. Follow the steps in SKILL DRILL 3-1 to don personal protective clothing:

1. Prepare your equipment for donning. Place equipment in a logical order for donning. (STEP 1)
2. Place the legs of your bunker pants over your boots, and fold the pants down around them.
3. Place your protective hood over the top of your boots to remind you to put on your protective hood. (STEP 2)
4. Lay out your coat, helmet, and gloves.
5. Place your protective hood over your head and down around your neck.
6. Step into your boots and pull up your bunker pants. Place the suspenders over your shoulders, and secure the front of the pants using the closure system. (STEP 3)
7. Put on your turnout coat, and secure the inner and outer closures. (STEP 4)
8. Place your helmet on your head with the ear tabs extended, and adjust the chin strap securely. Turn up your coat collar and secure it in front. (STEP 5)
9. Put on your gloves. (STEP 6)
10. Check all clothing to be sure it is properly secured and in the correct position.
11. Have your partner check your clothing. (STEP 7)

Doffing Personal Protective Clothing

To doff (i.e., remove) your personal protective clothing, reverse the procedure used in getting dressed. Follow the steps in SKILL DRILL 3-2 to doff personal protective clothing:

1. Remove your gloves. (STEP 1)
2. Open the collar of your turnout coat. (STEP 2)
3. Release the helmet chin strap and remove your helmet. (STEP 3)
4. Remove your turnout coat. (STEP 4)
5. Remove your bunker pants and boots. (STEP 5)
6. Remove your protective hood. (STEP 6)

PPE should be cleaned anytime it becomes dirty—by being exposed to smoke, blood or body fluids, or chemicals, for example—to ensure your safety and health. Smoke contains carcinogens, substances capable of causing cancer. Body fluids can transmit diseases and chemicals can cause burns.

Once PPE has been properly cleaned, it should be placed in a convenient location until the next response. Such equipment may be kept close to the fire apparatus, on the fire apparatus, or in an equipment locker. Wherever it is stored, personal protective clothing must be properly maintained, organized, and ready for the next emergency response.

Care of Personal Protective Clothing

Approved personal protective clothing is built to exacting standards but requires proper care if it is to continue to afford maximum protection. Avoid unnecessary wear on turnout clothing. A complete set of approved turnout clothing (excluding SCBA) costs more than $2000. Keep this expensive equipment in good shape for its intended use—structural firefighting.

Check the condition of your PPE on a regular basis. Clean it when necessary; repair worn or damaged PPE at once. PPE that is worn or damaged beyond repair must be replaced immediately because it will not be able to protect you.

Avoid unnecessary cuts or abrasions on the outer material. This material already meets NFPA standards—do not look for new opportunities to test its effectiveness. If the fabric is damaged, it must be properly repaired to retain its protective qualities. Follow the manufacturer's instructions for repairing or replacing PPE.

You must also keep personal protective clothing clean so that it will maintain its protective properties. Dirt will build up in the clothing fibers from routine use and exposure to fire environments. Smoke particles will also become embedded in the outer shell material. The interior layers will frequently be soaked with perspiration. Regular cleaning should remove most of these contaminants.

Other contaminants are formed from the by-products of burnt plastics and synthetic products. These residues, which are flammable, can become trapped between the fibers or build up on the outside of PPE, damaging the materials and reducing their protective qualities. A fire fighter who is wearing contaminated PPE is, in essence, bringing additional fuel into the fire on the clothing.

PPE that has been badly soiled by exposure to smoke, other products of combustion, melted tar, petroleum products, or other contaminants needs to be cleaned as soon as possible. Items that have been exposed to chemicals or hazardous materials may have to cleaned by a professional cleaning company or be impounded for decontamination or disposal.

Cleaning instructions are listed on the tag attached to the clothing. Always follow the manufacturer's cleaning instructions. Failure to do so may reduce the effectiveness of the garment and create an unsafe situation for the wearer.

Some fire departments have special washing machines that are approved for cleaning turnout clothing. Other departments contract with an outside firm to clean and repair protective clothing. In either case, the manufacturer's instructions for cleaning and maintaining the garment must be followed. If you wash your turnout gear in house, make sure that you follow the manufacturer's instructions for drying these garments properly.

SKILL DRILL 3-1 Donning Personal Protective Clothing

(Fire Fighter I, NFPA 5.1.2)

1. Place your equipment in a logical order for donning.

2. Place your protective hood over your head and down around your neck.

3. Put on your boots and pull up your bunker pants. Place the suspenders over your shoulders and secure the front of the pants.

4. Put on your turnout coat and close the front of the coat.

5. Place your helmet on your head and adjust the chin strap securely. Turn up your coat collar and secure it in front.

6. Put on your gloves.

7. Have your partner check your clothing.

SKILL DRILL 3-2

Doffing Personal Protective Clothing
(Fire Fighter I, NFPA 5.1.2)

1 Remove your gloves.

2 Open the collar of your turnout coat.

3 Release the helmet chin strap and remove your helmet.

4 Remove your turnout coat.

5 Remove your bunker pants and boots.

6 Remove your protective hood.

FIRE FIGHTER Tips

Follow your fire department's policies in regard to cleaning contaminated PPE (i.e., PPE that comes in contact with body fluids such as blood). Your fire department may require that the PPE be bagged in a biohazard bag and professionally cleaned. Never clean your PPE or work uniform in home washers to prevent cross-contamination with your home laundry.

Fire Fighter Safety Tips

In hot weather, remove as much of your PPE as possible when you arrive at rehab. This step will help you to cool down quickly. Some fire departments even place fans and misting equipment in the rehabilitation area to provide additional cooling.

Fire Fighter Safety Tips

Make sure your PPE is dry before using it on the fire ground. If wet protective clothing is exposed to the high temperatures of a structural fire, the water trapped in the liner materials will turn into steam and be trapped inside the moisture barrier. The result may be painful steam burns.

Other PPE may also require regular cleaning and maintenance. For example, the outer shell of your helmet should be cleaned with a mild soap as recommended by the manufacturer. The inner parts of helmets should be removed and cleaned according to the manufacturer's instructions. The chin strap and suspension system must be properly adjusted and all parts of the helmet kept in good repair.

Protective hoods and gloves get dirty quickly and should be cleaned according to the manufacturer's instructions. Most hoods can be washed with appropriate soaps or detergents. Repair or discard gloves or hoods that have holes in them; do not use them. Even a small cut or opening in these items can result in a burn injury.

Boots should be maintained according to the manufacturer's instructions. Rubber boots need to be kept in a place that does not result in damage to the boot. Leather boots must be properly maintained to keep them supple and in good repair. Any type of boot needs to be repaired or replaced if the outer shell is damaged. Do not store these items or any other PPE in direct sunlight.

Correctly storing your PPE is another important step in the proper care and maintenance of your PPE. PPE should only be stored in the proper PPE storage areas within the fire station. PPE should not be stored in living areas or in cars.

Specialized Protective Clothing

Vehicle Extrication

Due to the risk of fire at the scene of a vehicle extrication incident, members of the emergency team will wear full turnout gear. The officer in charge may also designate one or more members to don SCBA and stand by with a charged hose line. A structural firefighting ensemble also protects against many of the hazards present at a vehicle extrication incident, such as broken glass and sharp metal objects.

Some protective clothing, such as special gloves and coveralls or jumpsuits, has been specifically designed for vehicle extrication. These items are generally lighter in weight and more flexible than structural firefighting PPE, although they may be constructed of the same basic materials.

Fire fighters performing vehicle extrication must always be aware of the possibility of contact with blood or other body fluids. Latex gloves should be worn when providing patient treatment. Eye protection should be worn, given the possibilities of breaking glass, contact with body fluids, metal debris, and accidents with tools. Proper cleaning of extrication gear is especially important because many of the fluids encountered during vehicle extrication are oil based—for example, gasoline, diesel fuel, transmission fluid, and brake fluid. Because of the flammability of these fluids, they pose a hazard for fire fighters who enter a burning building while wearing oil-contaminated turnout gear.

Wildland Fires

Gear that is designed specifically for fighting wildland or brush fires must meet the requirements outlined in NFPA 1977, *Standard on Protective Clothing and Equipment for Wildland Fire Fighting*. In this type of firefighting ensemble, the jacket and pants are made of fire-resistant materials, such as Nomex or specially treated cotton, which are designed for comfort and maneuverability while working in the wilderness. Wildland fire fighters wear a helmet of a thermoresistant plastic, eye protection, and pigskin or leather gloves. The boots are designed to provide comfort and sure footing while hiking to a remote fire scene and to provide protection against crush and puncture injuries. Structural firefighting gear is not designed for extended wildland firefighting. It may produce high body temperatures resulting in heat exhaustion or heat stroke.

SCBA self contained breathing apparatus

Respiratory Protection

Respiratory protection equipment is an essential component of the structural firefighting personal protection ensemble. An SCBA is both expensive and complicated; using one confidently requires practice. You must be proficient in using SCBA before you engage in interior fire suppression activities.

The interior atmosphere of a burning building is considered to be IDLH (immediately dangerous to life and health). Attempting to work in such an atmosphere without proper respiratory protection can cause serious injury or death. Never enter or operate in a fire atmosphere without first ensuring that you have appropriate respiratory protection.

Respiratory Hazards of Fires

Fire fighters need respiratory protection because a fire involves a complex series of chemical reactions that can rapidly affect the atmosphere in unpredictable ways. The most readily evident by-product of a fire is smoke. The visible smoke produced by a fire contains many different substances, some of which are dangerous if inhaled. In addition, smoke contains invisible, highly toxic products of combustion. The process of combustion consumes oxygen and can lower the oxygen concentration

FIRE FIGHTER Tips

Recent research conducted by the Underwriters Laboratories demonstrated that residual smoke on turnout gear retains many heavy metals, including arsenic, cobalt, chromium, lead, mercury, phosphorus, and phthalates. Exposure to these chemicals has been linked to increased risk of cancer. Dirty personal protective gear is a danger to your health.

in the atmosphere below the level necessary to support life. In addition, the atmosphere of a fire may become so hot that one unprotected breath can result in fatal respiratory burns.

These respiratory hazards require fire fighters to use respiratory protection in all fire environments, regardless of whether the environment is known to be contaminated, suspected of being contaminated, or could possibly become contaminated without warning. The use of SCBA allows fire fighters to enter and work in a fire atmosphere with a safe, independent air supply.

Smoke

Most fires do not have a supply of oxygen sufficient to allow them to consume all of the available fuel. As a consequence, most fires result in incomplete combustion and produce a variety of by-products, which are then released into the atmosphere. Many of these by-products are extremely toxic. Collectively, the airborne products of combustion are called smoke. Smoke includes three major components: particles, vapors, and gases.

Fire Fighter Safety Tips

Smoky environments and oxygen-deficient atmospheres are deadly! It is essential to use SCBA at all times when operating in a smoky environment.

Smoke Particles

Smoke particles consist of unburned, partially burned, and completely burned substances. These particles are lifted in the thermal column produced by the fire and are usually readily visible. The completely burned particles are primarily ash; the unburned and partially burned smoke particles can include a variety of substances. The concentration of unburned or partially burned particles depends on the amount of oxygen that was available to fuel the fire.

Many smoke particles are so small that they can pass through the natural protective mechanisms of the respiratory system and enter the lungs. Some are toxic to the body and can result in severe injuries or death if they are inhaled. These particles also can prove extremely irritating to the eyes and digestive system.

Smoke Vapors

Smoke also contains small droplets of liquids. These smoke vapors are similar to fog, which consists of small water droplets suspended in the air. When oil-based compounds burn, they produce small hydrocarbon droplets that become part of the smoke. If inhaled or ingested, these compounds can affect the respiratory and circulatory systems. Some of the toxic droplets in smoke can cause poisoning if they are absorbed through the skin.

Fire Fighter Safety Tips

Dirty turnout clothing is dangerous. It may be saturated with flammable substances, cancer-causing chemicals, or infectious material, for example. Turnout clothing must be properly cleaned and maintained to provide maximum protection.

When water is applied to a fire, it creates steam and water droplets that in turn become part of the smoke. These water droplets can absorb some of the toxic substances contained in the smoke.

Toxic Gases

A fire also produces several types of gases. The amount of oxygen available to the fire and the type of fuel being burned determine precisely which gases are produced. A fire fueled by wood, for example, produces a different mixture of gases than one fueled by petroleum-based products. (Many common products—such as plastics—are made from petroleum-based compounds.) The concentrations of these gases can change rapidly as the oxygen supply is consumed or as fresh oxygen is introduced to the combustion process.

Many of the gases produced by residential or commercial fires are highly toxic. Carbon monoxide (CO), hydrogen cyanide, and phosgene are three of many gases often present in smoke.

CO is deadly in small quantities. When inhaled, this gas quickly replaces the oxygen in the bloodstream because it binds with hemoglobin molecules (which normally carry oxygen to the body's cells) in the blood 200 times more readily than oxygen does. Even a small concentration of CO can quickly disable and kill a fire fighter. CO is odorless, colorless, and tasteless, which also makes it dangerous. Because this gas cannot be detected without special instruments, fire fighters must always assume that it is present in the atmosphere around a fire.

Hydrogen cyanide is formed when plastic products—such as the polyvinyl chloride pipe used in residential construction—burn. This poisonous gas is quickly absorbed by the blood and interferes with cellular respiration. Just a small amount of hydrogen cyanide can easily render a person unconscious. Recent studies have indicated that cyanide poisoning in fire fighters may be more common than previously recognized.

Phosgene gas is formed from incomplete combustion of many common household products, including vinyl materials. This gas can affect the body in several ways. At low levels, it causes itchy eyes, a sore throat, and a burning cough. At higher levels, phosgene gas can cause pulmonary edema (fluid retention in the lungs) and death. Phosgene gas was used as a weapon in World War I to disable and kill soldiers.

Fires produce other gases that may affect the human body in harmful ways. Among these toxic gases are hydrogen chloride and several compounds containing different combinations of nitrogen and oxygen. Carbon dioxide, which is produced when sufficient oxygen is available to allow for complete combustion, is not toxic, but it can displace oxygen from the atmosphere and cause asphyxiation. Inhalation of these substances should be avoided.

Oxygen Deficiency

Oxygen is required to sustain human life. Normal outside or room air contains approximately 21 percent oxygen. A decrease in the amount of oxygen in the air, however, may drastically affect an individual's ability to function TABLE 3-2. An atmosphere with an oxygen concentration of 19.5 percent or less is considered oxygen deficient. If the oxygen level drops below 17 percent, people can experience disorientation, an inability to control their muscles, and irrational thinking, which can make escaping a fire much more difficult.

TABLE 3-2 Physiological Effects of Reduced Oxygen Concentration

Oxygen Concentration	Effect
21%	Normal breathing air
17%	Judgment and coordination impaired; lack of muscle control
12%	Headache, dizziness, nausea, fatigue
9%	Unconsciousness
6%	Respiratory arrest, cardiac arrest, death

During compartment fires (i.e., fires burning within enclosed areas), oxygen deficiency occurs in two ways. First, the fire consumes large quantities of the available oxygen, thereby decreasing the concentration of oxygen in the atmosphere. Second, the fire produces large quantities of other gases, which decrease the oxygen concentration by displacing the oxygen that would otherwise be present inside the compartment.

Increased Temperature

Heat is also a respiratory hazard. The temperature of the gases generated during a fire varies, depending on the fire conditions and the distance traveled by the hot gases. Inhaling superheated gases produced by a fire can immediately cause severe burns of the respiratory tract. If the gases are hot enough, a single inhalation can cause fatal respiratory burns. More information about fire behavior and products of combustion appears in the Fire Behavior chapter.

Other Toxic Environments

Not all hazardous atmospheric conditions are caused by fires. Indeed, fire fighters may encounter toxic gases or oxygen-deficient atmospheres in numerous types of emergency situations. Respiratory protection is just as important in these situations as in a fire suppression operation.

For example, toxic gases may be released at hazardous materials incidents from leaking storage containers or industrial equipment, from chemical reactions, or from the normal decay of organic materials. CO can be produced by internal combustion engines or improperly operating heating appliances. Exposure to CO or another toxic gas should be suspected whenever fire fighters find one or more unconscious patients with no evident cause.

Toxic gases can quickly fill confined spaces or below-grade structures. Any confined space or below-grade area must be treated as a hazardous atmosphere until it has been tested to ensure that the concentration of oxygen is adequate and that no hazardous or dangerous gases are present.

Conditions That Require Respiratory Protection

More fire deaths are caused by smoke inhalation rather than burns. Clearly, then, adequate respiratory protection is essential to your safety as a fire fighter. In fact, the products of combustion from house fires and commercial fires may be so toxic that just a few breaths can result in death—which is why you must always take respiratory protection precautions before entering the fire. When you first arrive at the scene of a fire, you do not have any way to measure the immediate danger to your life and your health posed by that fire. You must use full PPE and approved breathing apparatus if you will enter and operate within this atmosphere.

Whenever you are in an area where smoke is present, you must use SCBA—including when you respond to outside fires such as vehicle and dumpster fires. Use your SCBA at the fire scene and continue to wear it during overhaul until the air has been tested and proven to be safe by the safety officer. Do not remove your SCBA just because a fire has been knocked down.

SCBA must also be used in any situation where there is a possibility of toxic gases being present or oxygen deficiency, such as in a confined space. Always assume that the atmosphere is hazardous until it has been tested and proven to be safe.

Types of Breathing Apparatus

The basic respiratory protection hardware used by the fire service is the SCBA. The term "self-contained" in the acronym for this equipment refers to the requirement that the apparatus be the sole source of the fire fighter's air supply. In other words, SCBA is an independent air supply that will last for a predictable duration.

The two main types of SCBA are open-circuit and closed-circuit devices. Open-circuit breathing apparatus is typically used for structural firefighting. A tank of compressed air provides the breathing air supply for the user, and exhaled air is released into the atmosphere through a one-way valve FIGURE 3-14. Approved open-circuit SCBA comes in several different models, designs, and options.

Closed-circuit breathing apparatus recycles the user's exhaled air. The air passes through a mechanism that removes carbon dioxide and adds oxygen within a closed system, where the oxygen is generated from a chemical canister FIGURE 3-15.

Many closed-circuit SCBA units include a small oxygen tank as well as chemically generated oxygen. Closed-circuit breathing apparatus is often used for extended operations, such as mine rescue work, and operations in long tunnels where breathing apparatus must be worn for a long time.

A supplied-air respirator (SAR) uses an external source for the breathing air FIGURE 3-16. In this type of device, a hose line is connected to a breathing-air compressor or to compressed air cylinders located outside the hazardous area. The user breathes air through the line and exhales through a one-way valve, just as with an open-circuit SCBA.

Although SARs are commonly used in industrial settings, they are not used by fire fighters for structural firefighting. Hazardous materials teams and confined-space rescue teams sometimes use SARs for specialized operations. Some fire service SCBA units can be adapted for use as SARs.

FIGURE 3-14 Open-circuit SCBA.

FIGURE 3-15 Closed-circuit SCBA.

FIGURE 3-16 A supplied-air respirator may be needed for special rescue operations.

Fire Fighter Safety Tips

Always use SCBA before entering a confined space or a below-ground structure. Continue to use respiratory protection until the air is tested and deemed safe by the safety officer. Do not attempt confined-space operations without special training.

FIRE FIGHTER Tips

Do not confuse SCBA and SCUBA. SCBA is used by fire fighters during fire suppression activities. Self-contained underwater breathing apparatus (SCUBA) is used by divers while swimming underwater.

SCBA Standards and Regulations

In the United States, the National Institute for Occupational Safety and Health (NIOSH) sets the design, testing, and certification requirements for SCBA. NIOSH is a federal agency that researches, develops, and implements occupational safety and health programs. It also investigates fire fighter fatalities and serious injuries, and makes recommendations on how to prevent accidents from recurring. NFPA 1981 requires that all SCBA equipment used in the emergency service must meet the design, testing, and certification requirements established by NIOSH.

Occupational Safety and Health Administration (OSHA) and state agencies are responsible for establishing and enforcing regulations for respiratory protection programs. In some states and in Canadian provinces, individual occupational safety and health agencies establish and enforce these regulations.

The NFPA has developed three standards directly related to SCBA:

- NFPA 1500, *Standard on Fire Department Occupational Safety and Health Program*, includes the basic requirements for SCBA use and program management.
- NFPA 1404, *Standard for Fire Service Respiratory Protection Training*, sets the requirements for an SCBA training program within a fire department.
- NFPA 1981, *Standard on Open-Circuit Self-Contained Breathing Apparatus (SCBA) for Emergency Services*, includes requirements for the design, performance, testing, and certification of open-circuit SCBA for the fire service.

These standards specify approved SCBA equipment, initial training for each fire fighter, and annual retraining. Each fire department must follow applicable standards and regulations to ensure safe working conditions for all personnel.

Uses and Limitations of Self-Contained Breathing Apparatus

SCBA is designed to provide pure breathing air to fire fighters who are working in the hostile environment of a fire. It must meet rigid manufacturing specifications so that it can function in the increased temperatures and smoke-filled environments that fire fighters encounter. When properly maintained, such equipment will provide sufficient quantities of air to enable fire fighters to perform rigorous tasks.

Using SCBA requires that fire fighters develop unique skills, including different breathing techniques. This equipment limits normal sensory awareness; the senses of smell, hearing, and sight are all affected by the apparatus. Proficiency in the use of SCBA and other PPE requires ongoing training and practice.

Like any type of equipment, SCBA also has its limitations. Some of these limitations apply to the equipment; others apply to the user's physical and psychological abilities.

FIRE FIGHTER Tips

Some states are governed by OSHA regulations; fire departments in these states are required to comply with the OSHA respiratory protection regulations, such as OSHA 1910.134. OSHA 1910.134 covers the fit testing, medical screening, and training of all fire fighters in the use of SCBA.

Differences cannot be more than 10% off

Limitations of the Equipment

Because an SCBA carries its own air supply in a pressurized cylinder, its use is limited by the amount of air in the cylinder. SCBA for structural firefighting must carry enough air for a minimum of 30 minutes; cylinders rated for 45 minutes and 60 minutes are also available. These duration ratings, however, are based on ideal laboratory conditions. The realistic useful life of an SCBA cylinder for firefighting operations is usually much less than the rated duration, and actual use time will depend on the size of the user, his or her physical fitness and conditioning, the amount of physical exertion, and the user's degree of calm. An SCBA cylinder will generally have a realistic useful life of no more than 50 percent of the rated time. For example, an SCBA cylinder rated for 30 minutes can be expected to last for a maximum of 15 minutes during strenuous firefighting.

Fire fighters must manage their working time while using SCBA so that they have enough time to exit from the hazardous area before exhausting the air supply. To properly manage the air supply, a fire fighter must consider the following factors:

- The time and the effort it will take to reach the task destination. Climbing stairs will take more energy and air than walking across a flat floor, for example.
- The amount of air that will be available upon reaching the task destination.
- The amount of time necessary to complete the task and the air that will be used during that period. Some tasks will take more energy and air.
- The amount of time it will take to reach a safe area. At the end of this time, the firefighter must have a reserve of air for unexpected emergencies.

SCBA provides a very limited window of time for firefighting and a safe exit from the hazardous conditions of the fire. It is essential that you have a margin of safety built into your air supply for the unexpected. In some cases, you may need to begin exiting from the fire scene before half of your air supply is exhausted.

The weight of an SCBA varies, based on the manufacturer and the type and size of the cylinder. Generally, an SCBA weighs at least 25 pounds (11 kilogram). The size of the unit also makes it more difficult for the user to fit into small places **FIGURE 3-17**. The added weight and bulk decrease the user's flexibility and mobility, and shift the user's center of gravity.

FIGURE 3-17 SCBA expands a user's profile, making it more difficult to pass through tight spaces.

The design of the SCBA face piece limits the fire fighter's vision—particularly his or her peripheral vision. The face piece may fog up under some conditions, further limiting visibility. SCBA also may affect the user's ability to communicate, depending on the type of face piece and any additional hardware provided, such as voice amplification and radio microphones. The equipment is noisy during inhalation and exhalation, which may limit the user's hearing as well.

Physical Limitations of the User

Conditioning is important for SCBA users. An out-of-shape fire fighter will consume the air supply from this equipment more quickly and will have to exit the fire building long before a well-conditioned fire fighter does so. Overweight or poorly conditioned fire fighters are also at greater risk for heart attack due to physical stress.

Altogether, the protective clothing and SCBA that must be worn when fighting fires weigh more than 50 pounds. Moving with this extra weight requires additional energy, which in turn increases air consumption and body temperature. Taken collectively, this activity places additional stress on a fire fighter's body. A person with ideal body weight and in good physical condition will be able to perform more work per cylinder of air than a person who is overweight or out of shape.

The weight and bulk of the complete PPE ensemble limits a fire fighter's ability to walk, climb ladders, lift objects, and crawl through restricted spaces. Fire fighters must become accustomed to these limitations and learn to alter their movements accordingly. Practice and conditioning are key to becoming proficient in wearing and using PPE while fighting fires.

Psychological Limitations of the User

In addition to the physical limitations, the user must make mental adjustments when wearing an SCBA. Breathing through an SCBA is different from normal breathing and can be very stressful. Covering your face with a face piece, hearing the air rushing in, hearing valves open and close, and exhaling against positive pressure are all foreign sensations. The surrounding environment, which is often dark and filled with smoke, is foreign as well.

Fire fighters must adjust so that they can operate effectively under these stressful conditions. Practice in donning PPE, breathing through SCBA, and performing firefighting tasks in darkness helps to build confidence, not only in the equipment, but also in the fire fighter's personal skills.

Training generally introduces one skill at a time. Practice each skill as it is introduced and try to become proficient in that skill. As your skills improve, you will be able to tackle tasks characterized by increasing levels of difficulty.

FIRE FIGHTER Tips

The goal of training is to bring you to a level of comfort and proficiency in using your equipment. Start with a friendly environment until you are used to your equipment. Then add stressors such as darkness, smoke, and heat, one at a time.

Components of Self-Contained Breathing Apparatus

SCBA consists of four main parts: the backpack and harness, the air cylinder assembly, the regulator assembly, and the face piece assembly. Although the basic features and operations of all models are similar, you need to become familiar with the specific SCBA used by your department.

Backpack and Harness

The backpack provides the frame for mounting the other working parts of the SCBA **FIGURE 3-18**. It is usually constructed of a lightweight metal or composite material.

FIGURE 3-18 SCBA backpacks come in a variety of models.

The SCBA harness consists of the straps and fasteners used to attach the SCBA to the fire fighter. Most harnesses have two adjustable shoulder straps and a waist belt. Depending on the specific model of SCBA, the waist belt and shoulder straps will carry different proportions of the weight. The procedures for tightening and adjusting the straps also vary based on the model. The harness must be secure enough to keep the SCBA firmly fastened to the user, but not so tight that it interferes with breathing or movements. The waist strap must be tight enough to keep the SCBA from moving from side to side or getting caught on obstructions. Some SCBA models are equipped with a reinforced harness that can be used to help drag a fallen fire fighter out of danger.

Air Cylinder Assembly

A compressed air cylinder holds the breathing air for an SCBA. This removable cylinder is attached to the backpack harness and can be changed quickly in the field. An experienced fire fighter should be able to remove and replace the cylinder in complete darkness.

Fire fighters should be familiar with the type of cylinders used in their departments. Cylinders are marked with the materials used in their construction, the working pressure, and the rated duration.

The air pressure in filled SCBA cylinders ranges from 2200 to 4500 pounds per square inch (psi) (15,168 to 31,026 kilopascals). The greater the air pressure, the more air that can be stored in the cylinder.

Low-pressure cylinders, which are pressurized at 2200 psi, (15,168 kilopascals) can be constructed of steel or aluminum and are usually rated for 30 minutes of use. Composite cylinders are generally constructed of an aluminum shell wrapped with carbon, Kevlar, or glass fibers. They are significantly lighter in weight, can be pressurized up to 4500 psi (31,026 kilopascals), and are rated for 30, 45, or 60 minutes of use.

As previously noted, the rated duration times are established under laboratory conditions. A working fire fighter can quickly use up the air because of exertion, so the ratings should be viewed with caution. Generally, the working time available for a particular cylinder is half the rated duration.

The neck of an air cylinder is equipped with a hand-operated shut-off valve. The pressure gauge is located near the shut-off valve; it shows the amount of pressure currently in the cylinder. Be careful not to damage the threads or let any dirt get into the outlet of the cylinder.

FIRE FIGHTER Tips

The pressure gauge on the cylinder is important for checking the amount of air in the cylinder. This gauge cannot be viewed by the user while wearing the SCBA.

Regulator Assembly

SCBA regulators may be mounted on the waist belt or shoulder strap of the harness or attached directly to the face piece FIGURE 3-19. The regulator controls the flow of air to the user. Inhaling decreases the air pressure in the face piece. This change in pressure opens the regulator, which in turn releases air from the cylinder into the face piece. When inhalation stops, the regulator shuts off the air supply. Exhaling opens a second valve (the exhalation valve), thereby expelling the exhaled air into the atmosphere. SCBA regulators are capable of delivering large volumes of air to support the strenuous activities required in firefighting. Some units are equipped with a dual-path pressure reducer, a feature that automatically provides a backup method for air to be supplied to the regulator if the primary passage malfunctions.

SCBA regulators maintain a slight positive air pressure to the face piece in relation to the ambient air pressure outside the face piece. This feature helps to prevent the hazardous atmosphere outside the face piece from leaking into the face piece during inhalation. If any leakage occurs in the area where the face piece and the face make the seal, the positive-pressure breathing air inside the face piece will keep the hazardous atmosphere from entering the device. Regardless of the positive pressure, a proper face piece-to-face seal must always be maintained. Breathing with this slight positive pressure may require some practice. New fire fighters often report that it takes more energy to breathe when first using positive-pressure SCBA, but note that this sensation gradually decreases.

To activate the SCBA, it is necessary to open the air cylinder valve, don the SCBA, attach the regulator to the face piece, and breathe. Some SCBA models require that the user turn a yellow-colored valve to activate the unit. If the air supply is partially or completely cut off during use, you can fully open the purge valve to secure an emergency supply of air. This action will create a constant flow of air that will rapidly deplete the remaining air supply in the air cylinder. If it is necessary to open the purge valve, you should immediately exit from the IDLH area. During normal use, the purge valve can be momentarily opened to remove condensed air from the face piece.

Like the cylinder, the regulator has a gauge that indicates the pressure of the breathing air remaining in the cylinder. This gauge enables the user to monitor the amount of air remaining in the cylinder. The regulator pressure gauge can be mounted directly on the regulator, or it may be located on a separate hose so it can be attached to a shoulder strap for easier viewing. If the pressure gauges are working correctly, the readings on the regulator and cylinder pressure gauges should be within 100 psi (689 kilopascals) of each other. The cylinder pressure gauge must be visible to the user during SCBA use.

SCBA is also required to have a second heads-up air pressure display, which must be visible in the face piece. Some SCBAs contain a total of four light-emitting diode (LED) displays that indicate whether the air cylinder is full, three-fourths full, one-half full, or one-fourth full. Such heads-up displays enable the user to constantly monitor the amount of air in the air cylinder. Other LEDs on the display may provide additional information—for example, low batteries or other problems FIGURE 3-20.

NFPA standards require that SCBA include an end-of-service-time-indicator (EOSTI), or low-air alarm. This warning device tells the user that the end of the breathing air supply is approaching. NFPA 1500 requires that an exit strategy be practiced when the SCBA cylinder reaches a level of 600 liters or more. This alarm may take the form of a bell or whistle, a vibration, or a flashing LED. SCBAs are required to have two types of low-air alarms that operate independently of each other and activate different senses. For example, one alarm might ring a bell, whereas the second alarm might vibrate or flash an LED.

Fire fighters should never ignore the EOSTI warning. Most fire departments require fire fighters to exit from the IDLH area before the EOSTI alarm sounds because the low-air alarm does not sound until three-fourths of the air supply has

FIGURE 3-19 SCBA regulator.

FIGURE 3-20 A heads-up display.

been exhausted. A reserve of one-fourth of the air supply will not be adequate for a safe escape.

Many SCBA models include integrated PASS devices. A PASS is designed to help colleagues locate a downed fire fighter by sending out a loud audible signal. This device can be manually activated by a fire fighter in distress. It is also automatically activated if a fire fighter remains motionless for 30 seconds. Turning on the air supply automatically activates this PASS device, which ensures that a fire fighter does not forget to turn the PASS device on when entering a hazardous area. If your SCBA is equipped with a PASS device, learn how to activate it and how to turn it off FIGURE 3-21.

Communications between fire fighters wearing SCBAs is difficult. To facilitate communication, SCBAs are required to be equipped with a voice communication system. This functionality may be as simple as mechanical voice diaphragm, or it may be as sophisticated as an electronic system. Some SCBAs are also equipped with voice communications systems that utilize two-way radios.

SCBAs are required to be equipped with a rapid intervention crew/company universal air connection (RIC UAC). This device is connected to the SCBA and used to refill a cylinder if a trapped fire fighter is running out of air. A universal air connection is attached to a hose from a full air cylinder and brought to the downed fire fighter by a rapid intervention team, who refill the downed fire fighter's cylinder. This connection should not be used for the normal refilling of a cylinder.

Fire department SCBAs are required to be certified to provide protection against certain chemical, biological, radiological, and nuclear agents—that is, agents that could be released as a result of a terrorist attack. SCBAs provide protection against these agents by preventing chemical fumes, disease-causing biological organisms, and radioactive particles from being inhaled and entering the fire fighter's respiratory system. However, an SCBA does not protect from contamination by other means of transmission.

Many accessories are available for SCBAs, including data logging, unit IDs, tracking devices, special rescue harness connections, corrective eye lenses for face pieces, and electronic communications devices. Data logging and unit IDs aid in keeping track of the use of each SCBA and checking the proper functioning of each unit. Exit locator devices attempt to give the fire fighter some idea of how to exit a hazardous environment. Some of these tracking devices work with by sending out sounds of varying pitch. Special rescue harnesses are reinforced straps that enable the SCBA to be used as a rescue harness. Corrective lenses for face pieces enable SCBA users to use their face pieces as corrective eyewear. Electronic communications systems integrate radio communications between SCBA users. If your SCBA is equipped with any of these accessories, you must become competent in the operation of these extra devices. To ensure proper functioning of your SCBA, do not add any devices or accessories that are not approved by the manufacturer of the equipment.

Although many models of SCBA regulators exist, fire fighters must learn how to operate only the particular model that is used in their department. They should be able to operate the regulator in the dark and with gloves on.

Face Piece Assembly

The face piece delivers breathing air to the fire fighter and protects the face from high temperatures and smoke FIGURE 3-22. It consists of a face mask with a clear lens, an exhalation valve, and—on models with a harness-mounted regulator—a flexible low-pressure supply hose. In some models, the regulator is attached directly to the face piece.

The face piece covers the entire face. The part that comes in contact with the skin is made of special rubber or silicon, because these materials provide for a tight seal. The clear lens allows for better vision. Exhaled air is expelled from the face piece through the one-way exhalation valve, which has a spring mechanism to maintain positive pressure inside the face piece. Because it is difficult to communicate through a face piece, a voice amplification device or mechanical diaphragm is used to facilitate communications. A mechanical diaphragm uses a vibrating air-tight membrane to transmit the fire fighter's voice without the use of electricity.

Face pieces are equipped with nose cups to help prevent fogging of the clear lens. In addition, nose cups prevent the build-up of carbon dioxide (CO_2) by directing exhaled air toward the exhalation valve. Fogging is a greater problem

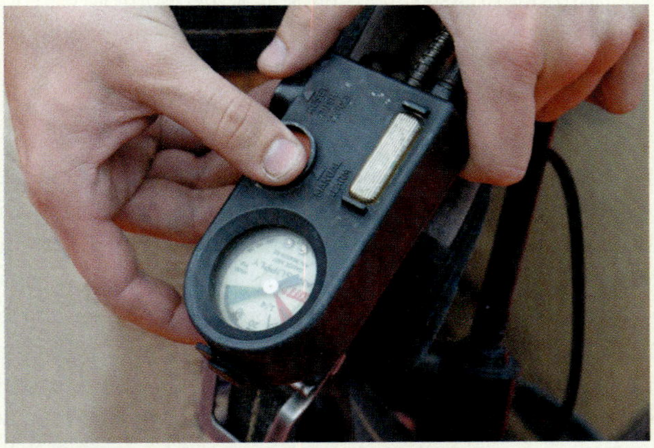

FIGURE 3-21 A PASS device can be integrated into an SCBA.

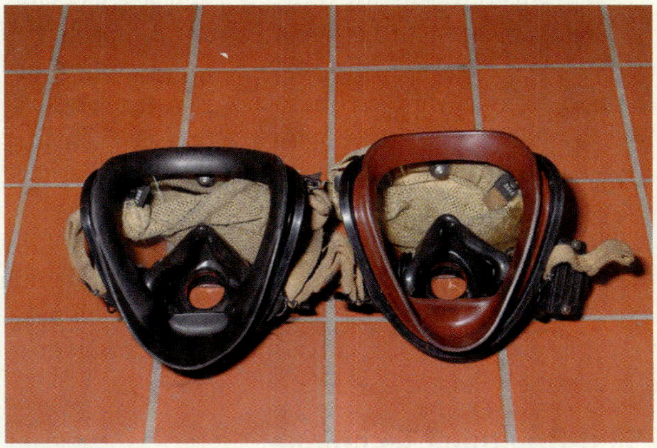

FIGURE 3-22 SCBA face pieces come in several sizes.

in colder climates. It occurs because the compressed air you breathe is dry, but the air you exhale is moist. The purge valve can be opened slightly for a second or two to clear condensation from the eyepieces. The flow of dry air from the regulator helps to prevent fogging of the lens.

The face piece is held in place with a weblike series of straps, or a net and straps. Face pieces should be stored with the straps in the longest position to make them easier to don. Pull the end of the straps toward the back of the head (not out to the sides) to tighten them and ensure a snug fit.

A leak in the face piece seal may result from an improperly sized face piece, an improper donning procedure, or facial hair around the edge of the face piece. In particular, the following factors can affect the seal on a SCBA face piece:

- Facial hair, sideburns, or beard
- A low hairline that interferes with the sealing surface
- Ponytails or buns that interfere with the smooth and close fit on the head harness
- A skull cap that projects under the face piece or temple pieces
- The absence of teeth
- Improper size face piece

An improperly fitted face piece may lead to exposure to a hazardous environment. Such leaks are dangerous for two reasons. First, a large leak could overcome the positive pressure in the face piece and allow contaminated air to enter the face piece. Second, a leak of any size will deplete the breathing air and reduce the amount of time available for firefighting.

Face pieces are manufactured in several sizes. NFPA 1500 requires that all fire fighters must have their face pieces fit-tested annually to ensure that they are wearing the proper size. Some departments issue individual face pieces to each fire fighter; others provide a selection of sizes on each apparatus. NFPA 1500 also requires that the sealing surface of the face piece be in direct contact with the user's skin; that is, hair or a beard cannot be in the seal area.

Pathway of Air Through an SCBA

In an SCBA, the breathing air is stored under pressure in the cylinder. This air passes through the cylinder shut-off valve into the high-pressure air line (hose), which then takes it to the regulator. The regulator reduces the high pressure to low pressure.

The regulator opens when the user inhales, reducing the pressure on the downstream side. In an SCBA unit with a face piece-mounted regulator, the air next goes directly into the face piece. In units with a harness-mounted regulator, the air travels from the regulator through a low-pressure hose into the face piece. From the face piece, the air is inhaled through the user's air passages and into the lungs.

When the user exhales, used air is returned to the face piece. The exhaled air is exhausted from the face piece through the exhalation valve. This cycle repeats with every inhalation. As the pressure in the face piece drops, the exhalation valve closes and the regulator opens.

Skip-Breathing Technique

The skip-breathing technique helps conserve air while using an SCBA in a firefighting situation. In this technique, the fire fighter takes a short breath, holds it, takes a second short breath (without exhaling in between breaths), and then relaxes with a long exhale. Each breath should take 5 seconds.

Fire Fighter Safety Tips

Make sure that your face piece is properly fitted and is the correct size for your face.

FIRE FIGHTER Tips

Controlled-Breathing Technique

Controlled breathing helps extend the SCBA air supply. It consists of a conscious effort to inhale naturally through the nose and to force exhalation from the mouth. Practicing controlled breathing during training will help you to maximize the efficient use of air while you are working.

A simple drill can demonstrate the benefits of skip breathing. In this exercise, one fire fighter dons PPE and an SCBA with a full air cylinder, and walks in a circle around a set of traffic cones, around the track at the local school, or, if safety permits, around the parking lot at the fire station. A second fire fighter times how long it takes for the fire fighter to completely deplete the air in the SCBA. After the first fire fighter is completely rested, the air cylinder is replaced, and the same drill is repeated using the skip-breathing technique. A comparison of the times after completion of both cycles should confirm that skip breathing conserves the air in the cylinder.

Keep in mind that as you are able to decrease the rate and depth of your breathing, you will increase the amount of time you can breathe from a single cylinder of air. Keep calm, perform your assignments efficiently, and do not start breathing from your air supply until you need to.

Mounting Breathing Apparatus

SCBA should be located so that fire fighters can don it quickly when they arrive at the scene of a fire. Seat-mounted brackets enable fire fighters to don SCBA en route to an emergency scene, without unfastening their seat belts or otherwise endangering themselves. This approach enables fire fighters to begin work as soon as they arrive.

Several types of apparatus seat-mounting brackets are available. Some hold the SCBA with the friction of a clip. Others are equipped with a mechanical hold-down device that must be released to remove the SCBA. Regardless of which mounting system is employed, it must hold the SCBA securely in the bracket; a collision or sudden stop should not dislodge the SCBA from the brackets, because a loose SCBA can be a dangerous projectile. The fire fighter who dons SCBA from a seat-mounted bracket should not tighten the shoulder straps while seated, so as not to dislodge the SCBA in a sudden-stop situation. Fire fighters must remain securely restrained by a seat belt or combination seat belt/shoulder harness anytime the apparatus is moving.

VOICES OF EXPERIENCE

I was the officer on Ladder 1 in downtown Hamilton's central station. We were dispatched around 01:00 hours to an18-story high-rise. We had been to the same building twice for alarm bells already during this shift, both malicious alarm activations.

On arrival, the building superintendent waved us in and was very excited. I exited the apparatus and approached him in my full PPE and SCBA. The building superintendent stated that there was light smoke on the 10th floor and that he did not know which unit was on fire. As I radioed dispatch to upgrade the call to a full first alarm–structure fire, my three crew members met me with full PPE, SCBA, forcible entry tools, medical bag, high-rise kit, and extra SCBA cylinders—everything we bring in for a high-rise call.

We were able to get to the 10th floor, find the fire unit, remove the occupant, and provide medical care in the safety of the stairwell below the fire floor. We also connected to the hose cabinet on the fire floor and knocked down the fire in the room of origin. Thanks to our efforts, the fire was contained to the room of origin with no impact on the rest of the building. The occupant was treated for minor smoke inhalation and released from the hospital.

It is easy to become complacent when you respond to the same location over and over again. Had my crew and I not responded as if this were a "legitimate" call, we would have had to don proper PPE and SCBA and collect our high-rise equipment upon arrival at the scene. I am confident that if we had to take the time to perform these tasks, it would have allowed the fire to reach the fully developed phase and spread beyond the room of origin to the entire unit. Heat and smoke would most likely have caused permanent (or worse) injury to the occupant. The increase in fire intensity and smoke generation would have reduced visibility on the fire floor to the point that it would have taken longer to extend hose lines from the floor below, find the fire, and extinguish it. Treat each alarm as if it is "the big one."

Brent Thomas
Hamilton Emergency Services—Fire
Hamilton, Ontario, Canada

Compartment-mounted SCBA units also can be donned quickly. These units are used by fire fighters who arrive in vehicles without seat-mounted SCBA or whose seats were not equipped with them. Driver/operators and fire fighters who arrive in their private vehicles often use these types of SCBA. The mounting brackets should be positioned high enough to allow for easy donning of the equipment. Some mounting brackets allow the fire fighter to lower the SCBA without removing it from the mounting bracket. Older apparatus may have the brackets mounted on the exterior of the vehicle. Any exterior-mounted SCBA should be protected from weather and dirt by a secure cover.

SCBA also may be kept in a storage case. This method is most appropriate for transporting extra SCBA units. It should not be used to transport SCBA that will be used during the initial phase of operations at a fire scene.

The SCBA should be stored on the apparatus in ready-for-use condition, with the main cylinder valve closed.

Donning Self-Contained Breathing Apparatus

Donning SCBA is an important skill. Fire fighters should be able to don and activate SCBA in 1 minute; both the fire fighter's personal safety and the effectiveness of the firefighting operation depend on mastery of this skill. Fire fighters must already be wearing full PPE before they don SCBA. Before beginning the actual donning process, fire fighters must carefully check the SCBA to ensure it is ready for operation:

- Check whether the air cylinder is at least 90 percent of its rated pressure.
- If the SCBA has an air saving/donning switch, confirm that it is activated.
- Open the cylinder valve two or three turns, listen for the low-air alarm to sound, and then open the valve fully.
- Check the pressure gauges on both the regulator and the cylinder. The reading for these gauges should be within 10 percent of each other.
- Check all harness straps to ensure that they are fully extended.
- Check all valves to ensure that they are in the correct position. (An open purge valve will waste air.)

Donning SCBA from an Apparatus Seat Mount

Donning SCBA while en route to an emergency can save valuable time. However, this maneuver requires that you don all of your protective clothing before mounting the apparatus. Place your arms through the shoulder straps as you sit down, and then fasten your seat belt. Alternatively, you can fasten your seat belt first, and then slide one arm at a time through the shoulder straps of the SCBA harness. You can partially tighten the shoulder straps while you are seated.

When you arrive at the emergency scene, release your seat belt, activate the bracket release, and exit the apparatus. Be sure to take a face piece with you. Face pieces should be either kept in a storage bag close to each seat-mounted SCBA or attached to the harness. After exiting from the apparatus, attach the waist strap of the SCBA unit, and then tighten and adjust the shoulder and waist straps.

Follow the steps in SKILL DRILL 3-3 to don SCBA from a seat-mounted bracket. Before beginning this skill drill, inspect the SCBA to ensure it is ready for service.

1. Don your hood, bunker pants, boots, and turnout coat. Safely mount the apparatus and sit in the seat, placing your arms through the SCBA shoulder straps. (STEP 1)
2. Fasten your seat belt. Partially tighten the shoulder straps; do not fully tighten them at this time. When the apparatus comes to a complete stop at the emergency scene, release your seat belt and release the SCBA from the mounting bracket.
3. Carefully exit the apparatus. (STEP 2)
4. Attach the waist belt of the harness and cinch it down. (STEP 3)
5. Adjust shoulder straps until they are tight. (STEP 4)
6. Open the main cylinder valve. Activate the air saver valve if needed. (STEP 5)
7. Remove or loosen your helmet and pull the hood back. Don the face piece and check for leaks. Pull the protective hood up over your head, put the helmet on, and secure the chin strap. (STEP 6)
8. If necessary, connect the regulator to the face piece. (STEP 7)
9. Activate the air flow and PASS alarm. (STEP 8)

Donning SCBA from a Compartment Mount

To don a compartment-mounted SCBA, slide one arm through the shoulder harness strap, and then slide the other arm through the other shoulder strap. Release the SCBA from the mounting bracket. Adjust the shoulder straps so that you can carry the SCBA fairly high on your back. Attach the ends of the waist strap and tighten the waist strap.

Follow the steps in SKILL DRILL 3-4 (Fire Fighter I, NFPA 5.3.1) to don SCBA from a side-mounted compartment or bracket. Before starting, confirm that the SCBA has been inspected and is ready for service. Examine the equipment to see how it is mounted in the compartment. Check the release mechanism and determine how it operates. If the SCBA is mounted on an exterior bracket, remove the protective cover before beginning the donning sequence.

1. Don PPE and hood. Stand in front of the SCBA bracket and fully open the main cylinder valve.
2. Turn your back toward the SCBA, slide your arms through the shoulder straps, and partially tighten the straps.
3. Release the SCBA from the bracket and step away from the apparatus.
4. Attach the waist belt and tighten it.
5. Adjust the shoulder straps.
6. Don the face piece and check for an adequate seal.
7. Pull the protective hood into position, don your helmet, and secure the chin strap.
8. If necessary, connect the regulator to the face piece.
9. Activate the air flow and PASS alarm.

SKILL DRILL 3-3

Donning SCBA from an Apparatus Seat Mount
(Fire Fighter I, NFPA 5.3.1)

1 Don your bunker pants, boots, and turnout coat. Place your hood around your neck prior to mounting the fire apparatus. Safely mount the apparatus and sit in the seat, placing your arms through the SCBA shoulder straps.

2 Fasten your seat belt. Partially tighten the shoulder straps. When the apparatus stops, release the seat belt and release the SCBA from its brackets. Exit the apparatus.

3 Attach the waist belt of the harness and cinch it down.

4 Adjust shoulder straps until they are tight.

5 Open the main cylinder valve. Active the air saver valve if needed.

6 Remove or loosen your helmet and pull the hood back. Don the face piece and check for leaks. Replace the protective hood and helmet, and secure the chin strap.

7 If necessary, connect the regulator to the face piece.

8 Activate the air flow and PASS alarm.

Donning SCBA from the Ground, the Floor, or a Storage Case

Fire fighters must sometimes don an SCBA that is stored in a case or on the ground. Two methods can be used in these situations: the over-the-head method and the coat method.

Over-the-Head Method

To don an SCBA using the over-the-head method, place the SCBA on the ground or on the floor with the cylinder valve facing away from you. Lay the shoulder straps out to each side of the backpack. Grasp the backplate with both hands, and lift the SCBA over your head. Let the backpack slide down your back; the straps will slide down your arms. Balance the unit on your back. First attach and tighten the waist strap, and then tighten the shoulder straps.

Follow the steps in SKILL DRILL 3-5 to don SCBA using the over-the-head method. Before starting, confirm that the SCBA has been inspected and is ready for service. Don your bunker pants, boots, and turnout coat. Place your protective hood over your neck. Place your helmet and gloves close to the SCBA case.

1. Open the protective case and lay out the SCBA so that the cylinder valve is away from you and the shoulder straps are to the sides. (STEP 1)
2. Fully open the main cylinder valve. Activate the air saver valve if needed. (STEP 2)
3. Bend down and grasp the SCBA backplate with both hands. Using your legs, lift the SCBA over your head. Once the SCBA clears your head, rotate it 180 degrees so that the waist straps are pointed toward the ground. (STEP 3)
4. Slowly slide the pack down your back. Make sure that your arms slide into the shoulder straps. Once the SCBA is in place, tighten the shoulder straps and secure the waist strap. (STEP 4)
5. Make sure your hood is pulled down around your neck. Don the face piece and check for an adequate seal. Pull your protective hood into position, don your helmet, and secure the chin strap. (STEP 5)
6. If necessary, connect the regulator to the face piece. Activate the air flow and PASS alarm. (STEP 6)

Coat Method

To don an SCBA using the coat method, place the SCBA on the ground or on the floor with the cylinder valve facing toward you. Spread out and extend the shoulder straps. Use your left hand to grasp the left shoulder strap close to the backplate; use your right hand to grasp the right shoulder strap farther away from the backplate. Swing the SCBA over your left shoulder. Release your right arm and slide it through the right shoulder harness strap. Tighten both shoulder straps. Attach and tighten the waist belt.

Follow the steps in SKILL DRILL 3-6 to don SCBA using the coat method. Before starting, confirm that the SCBA has been inspected and is ready for service. Don your bunker pants, boots, and turnout coat. Place your protective hood over your neck. Place your helmet and gloves close to the SCBA case.

1. Open the protective case and lay out the SCBA so that the cylinder valve is facing you and the straps are laid out to the sides. Fully open the main cylinder valve. Open the air saver valve if needed. Place your dominant hand on the opposite shoulder strap. For safety reasons, be sure to grasp the strap as close to the backplate as possible. (STEP 1)
2. Lift the SCBA and swing it over your dominant shoulder, being mindful of people or objects around you. (STEP 2)
3. Slide your other hand between the SCBA cylinder and the corresponding shoulder strap. (STEP 3)
4. Tighten the shoulder straps. (STEP 4)
5. Attach the waist belt and adjust its tightness. (STEP 5)
6. Make sure your hood is pulled down around your neck. Don the face piece and check for an adequate seal. (STEP 6)
7. Pull the protective hood into position, don your helmet, and secure the chin strap. If necessary, connect the regulator to the face piece. Activate the air flow and PASS alarm. (STEP 7)

These instructions will have to be modified for different SCBA models. In particular, the sequence for adjusting the shoulder straps and the waist belt varies with different models. Modifications must also be made for SCBA models with waist-mounted regulators. Refer to the specific manufacturer's instructions supplied with each unit. Follow the SOPs for your department.

To don SCBA from a seat-mounted position with a safety latch, follow the steps in SKILL DRILL 3-7:

1. With the SCBA straps fully extended, place your dominant arm between the shoulder strap and the SCBA bottle. Repeat the same process with the other arm. (STEP 1)
2. Tighten the shoulder straps. (STEP 2)
3. Attach the waist belt and adjust its tightness. (STEP 3)
4. If the apparatus has a SCBA locking device, detach the device from the SCBA. *Note*: Fire fighters should never take off their seat belts to perform this action. Stay seated and buckled up while the apparatus is in motion. (STEP 4)

Donning the Face Piece

Your face piece keeps contaminated air outside and pure breathable air inside. To perform properly, it must be the correct size and must be adjusted to fit your face. Make sure you have been tested to determine the proper size for you. The requirements for face piece fit testing are described in NFPA 1500. Fit testing needs to be performed before fire fighters are permitted to use SCBA. To ensure a proper fit, such testing should be repeated every year.

No facial hair can be present in the seal area. Eyeglasses that pass through the seal area cannot be worn with a face piece, because they can cause leakage between the face piece and your skin. Your face piece must match your SCBA—you cannot interchange a face piece from a different SCBA model.

Face pieces for various brands and models of SCBAs may differ slightly. Some have the regulator mounted on the face piece; others have it mounted on the harness straps. Fire

SKILL DRILL 3-5 Donning SCBA Using the Over-the-Head Method

(Fire Fighter I, NFPA 5.3.1)

1. Open the protective case and lay out the SCBA so that the cylinder valve is away from you and the shoulder straps are to the sides.

2. Fully open the main cylinder valve. Activate the air saver valve if needed.

3. Bend down and grasp the SCBA backplate with both hands. Using your legs, lift the SCBA over your head. Once the SCBA clears your head, rotate it 180 degrees so that the waist straps are pointed toward the ground.

4. Slowly slide the pack down your back. Make sure that your arms slide into the shoulder straps. Once the SCBA is in place, tighten the shoulder straps and secure the waist strap.

5. Make sure your hood is pulled down around your neck. Don the face piece and check for an adequate seal. Pull your protective hood into position, don your helmet, and secure the chin strap.

6. If necessary, connect the regulator to the face piece. Activate the air flow and PASS alarm.

fighters must learn about the specific face pieces used by their departments.

Follow the steps in SKILL DRILL 3-8 to don a face piece. Before beginning this procedure, make sure you have donned your PPE and protective hood. Remove your helmet.

1. Fully extend the straps on the face piece. (STEP 1)
2. Rest your chin in the chin pocket at the bottom of the mask. (STEP 2)
4. Fit the face piece to your face, bringing the straps or webbing over your head. (STEP 3)
4. Tighten the lowest two straps. To tighten them, pull the straps straight back, rather than out and away from your head. Check the harness net to make sure it is lying flat against the back of the head. (STEP 4)

SKILL DRILL 3-6

Donning SCBA Using the Coat Method
(Fire Fighter I, NFPA 5.3.1)

1 Open the protective case and lay out the SCBA so that the cylinder valve is facing you and the straps are laid out to the sides. Fully open the main cylinder valve. Open the air saver valve if needed. Place your dominant hand on the opposite shoulder strap. For safety reasons, be sure to grasp the strap as close to the backplate as possible.

2 Lift the SCBA and swing it over your dominant shoulder, being mindful of people or objects around you.

3 Slide your other hand between the SCBA cylinder and the corresponding shoulder strap.

4 Tighten the shoulder straps.

5 Attach the waist belt and adjust its tightness.

6 Make sure your hood is pulled down around your neck. Don the face piece and check for an adequate seal.

7 Pull the protective hood into position, don your helmet, and secure the chin strap. If necessary, connect the regulator to the face piece. Activate the air flow and PASS alarm.

SKILL DRILL 3-7 Donning SCBA from a Seat-Mounted Position with a Safety Latch

(Fire Fighter I, NFPA 5.3.1)

1 With the SCBA straps fully extended, place your dominant arm between the shoulder strap and the SCBA bottle. Repeat the same process with the other arm.

2 Tighten the shoulder straps.

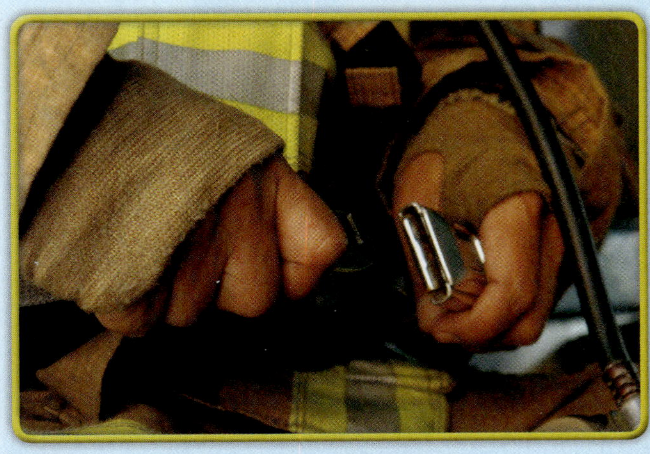

3 Attach the waist belt and adjust its tightness.

4 If the apparatus has a SCBA locking device, detach the device from the SCBA. *Note*: Fire fighters should never take off their seat belts to perform this action. Stay seated and buckled up while the apparatus is in motion.

5. Tighten the pair of straps at your temple, if these straps are present.
6. If your model has additional straps, tighten the top strap(s) last. (STEP 5)
7. Check for a proper seal. This process depends on the model and type of face piece you use. (STEP 6)
8. Confirm that your nose fits in the chin cup.
9. Pull the protective hood up so that it covers all bare skin. Make sure it does not get under your face piece or obscure your vision. (STEP 7)
10. Replace your helmet and secure the chin strap. (STEP 8)
11. Attach the regulator to your face piece or attach the low-pressure air supply hose to the regulator. (STEP 9)

Fire Fighter Safety Tips

The following method is one way to check the seal of a face mask:

1. Don the SCBA and begin breathing cylinder air.
2. Completely close the air cylinder valve.
3. Breathe on the respirator until all air stops flowing from the breathing regulator.
4. Inhale slowly and hold your breath. Your face piece should be slightly drawn to your face.
5. Listen and feel for air leakage around the face piece seal. Check that the negative pressure in the face piece does not change.
6. If no change occurs, no leaks are present.

Safety Precautions for Self-Contained Breathing Apparatus

As you practice using your SCBA, remember that this equipment is your protection against serious injury or death in hazardous conditions. Practice safe procedures from the beginning.

Before you enter a hazardous environment, make sure that your PASS device is activated and that you are properly logged into your personnel accountability system. Always work in teams of two in hostile environments. In addition, always have at least two fire fighters outside at the ready whenever two fire fighters are working in a hostile environment.

SCBA Use During Emergency Operations

As a fire fighter, your job is to practice using SCBA until you are confident that you can carry out a variety of tasks in hazardous conditions while depending on your equipment to supply you with safe air to breathe. Because hostile environments are often unpredictable, fire fighters must be prepared to react if an emergency situation occurs while they are using SCBA. In emergencies, follow simple guidelines. Most importantly, keep calm, stop, and think. Panic increases air consumption. Try to control your breathing by maintaining a steady rate of respirations. A calm person has a greater chance of surviving an emergency.

If you experience a problem with your SCBA, try to exit the IDLH area to a safe environment. If your cylinder contains air but no air comes out of your regulator, many SCBAs have a purge value that can be opened slightly to release a constant supply of air. This measure will rapidly empty your cylinder, however. In this circumstance, you must immediately exit from the hazardous environment. If you are in danger, follow the steps for self-survival and calling a "mayday." This topic is discussed in the Fire Fighter Survival chapter.

FIRE FIGHTER Tips

Restricted Spaces

The size and shape of SCBA may make it difficult for you to fit through tight openings when wearing this equipment. Several techniques may help you navigate these spaces:

- Change your body position: Rotate your body by 45 degrees and try again.
- Loosen one shoulder strap and change the location of the SCBA on your back.
- If you have no other choice, you may have to remove your SCBA. In this case, do not let go of the backpack and harness for any reason. Keep the unit in front of you as you navigate through the tight space. Reattach the harness as soon as you are through the restricted space. *Note*: This is an absolutely "last resort" procedure!

You will be using your SCBA in a variety of conditions that most people would consider to be emergencies. Because you are a fire fighter, these activities are predictable for you. Therefore, you need to master a wide variety of firefighting skills while wearing SCBA. Practice these tasks first in conditions of good visibility, and then progress to doing the same tasks in conditions of limited visibility. You need to practice most of the skills you will perform as a fire fighter—for example, advancing hose lines, climbing ladders, crawling through windows, performing rescues, providing medical assistance, and crawling through confined spaces—while wearing SCBA.

Exercises for Mastering SCBA

The following are three exercises that may be practiced in order to assist in the mastery of using SCBA properly.

The Conducting a Primary Search skill drill in the Search and Rescue chapter describes how to perform search and rescue while wearing SCBA. Another skill you need to practice is passing through a restricted space while wearing an SCBA. This activity is illustrated in the Opening a Wall to Escape skill drill in the Fire Fighter Survival chapter.

A third activity to familiarize you with your SCBA simulates the activities you would perform when ventilating a roof. Once you have mastered donning and doffing your PPE ensemble, you may perform this activity to help you master the use of your SCBA. In doing so, you will work with at least one other team member.

Bring an axe and other appropriate ventilation tools with you to the roof. Don your face piece and activate the air supply. With the axe, assess the construction of the building for safety and sound the rafters for safety.

Next, have a second team member support you. Chop on the roof or simulated roof using short, powerful, controlled strokes. Then switch places with your team member. Be sure to monitor your air supply during this activity. Exit the hazard area in time to be out of hazards before your low-air alarm activates.

FIRE FIGHTER Tips

Many fire departments need to take precautions to ensure proper operation of SCBAs in cold temperatures. SCBAs should be stored in a location that is kept at a temperature greater than 32°F (0°C). After use, this equipment should be dried thoroughly to avoid retained moisture that might freeze moving parts. When SCBA units are doffed in below-freezing temperatures, the face piece and the regulator should be placed inside the turnout coat to keep it warm and prevent any moisture from freezing.

FIRE FIGHTER Tips

Refilling a cylinder using the RIC UAC in below-freezing conditions presents two concerns. The first concern relates to the protective cap on the RIC UAC coupling. Water on the cover of the coupling may freeze, preventing a connection between the filling hose and the universal air connection. The second concern is that if the universal air connection is used to fill a SCBA cylinder in below-freezing temperatures, and the SCBA is then taken into a warmer temperature, the difference in temperature could cause the pressure in the cylinder to expand to unsafe levels. If there is excess pressure, open the air cylinder valve and the purge valve to release the excess pressure.

SKILL DRILL 3-8 Donning a Face Piece
(Fire Fighter I, NFPA 5.3.1)

1 Fully extend the straps on the face piece.

2 Rest your chin in the chin pocket at the bottom of the mask.

3 Fit the face piece to your face, bringing the straps or webbing over your head.

4 Tighten the lowest two straps. To tighten them, pull the straps straight back, rather than out and away from your head. Check the harness net to make sure it is lying flat against the back of the head.

(Continued)

SKILL DRILL 3-8 Donning a Face Piece (*Continued*)

(Fire Fighter I, NFPA 5.3.1)

5 Tighten the pair of straps at your temple, if these straps are present. If your model has additional straps, tighten the top strap(s) last.

6 Check for a proper seal. This process depends on the model and type of face piece you use. Confirm that your nose fits in the chin cup.

7 Pull the protective hood up so that it covers all bare skin. Make sure it does not get under your face piece or obscure your vision.

8 Replace your helmet and secure the chin strap.

9 Attach the regulator to your face piece or attach the low-pressure air supply hose to the regulator.

Doffing Self-Contained Breathing Apparatus

The procedure for doffing SCBA depends on which model you use and whether it has a face piece–mounted regulator or a harness-mounted regulator. Follow the procedures recommended by the manufacturer and your department's SOPs.

In general, when doffing your SCBA, you should reverse the steps you followed to don your SCBA. Follow the steps in SKILL DRILL 3-9 to doff your SCBA:

1. Remove your gloves. Remove the regulator from your face piece or disconnect the low-pressure air supply hose from the regulator. (STEP 1)
2. Shut off the air-supply valve or fully depress the air saver/donning switch to stop the flow of air.
3. Remove your helmet and pull your protective hood down around your neck. (STEP 2)
4. Loosen the straps on your face piece. (STEP 3)
5. Remove your face piece. (STEP 4)
6. Release your waist belt. (STEP 5)
7. Loosen the shoulder straps and remove the SCBA. (STEP 6)
8. Shut off the air-cylinder valve. (STEP 7)
9. Bleed the air pressure from the regulator by opening the purge valve. (STEP 8)
10. If you have an integrated PASS device, turn it off.
11. Place the SCBA in a safe location where it will not get dirty or damaged. (STEP 9)

Putting It All Together: Donning the Entire PPE Ensemble

The complete PPE ensemble consists of both personal protective clothing and respiratory protection (SCBA). Although donning personal protective clothing and donning and operating SCBA can be learned and practiced separately, you must be able to integrate these skills to have a complete PPE ensemble. Each part of the complete ensemble must be in the proper place to provide whole-body protection.

The steps for donning a complete PPE are summarized here:

1. Place the protective hood over your head and bring it down around your neck.
2. Put on your bunker pants and boots. Adjust the suspenders and secure the front flap of the pants.
3. Put on your turnout coat and secure the front.
4. Open the air-cylinder valve on your SCBA and check the air pressure. Press the air saver valve to prevent air flow if needed.
5. Put on your SCBA.
6. Tighten both shoulder straps of the SCBA harness.
7. Attach the waist belt of the harness and tighten it. Tighten the chest straps, if present.
8. Fit the face piece to your face.
9. Tighten the face piece straps, beginning with the lowest straps.
10. Check the face piece for a proper seal. (Follow the manufacturer's instructions.)
11. Pull the protective hood up so that it covers all bare skin but does not obscure your vision.
12. Place your helmet on your head with the ear tabs extended, and secure the chin strap.
13. Turn up your coat collar and secure it in front.
14. Put on your gloves.
15. Check your clothing to be sure it is properly secured.
16. Make sure your PASS device is turned on.
17. Attach your regulator or turn it on to start the flow of breathing air.
18. Work safely!

SCBA Inspection and Maintenance

SCBA must be properly cleaned, inspected, and prepared for the next use each time it is used, whether in an actual emergency incident or as part of a training exercise. The air cylinder must be changed or refilled, the face piece and regulator must be sanitized according to the manufacturer's instructions, and the unit must be cleaned, inspected, and checked for proper operation. It is the user's responsibility to ensure that the SCBA is in good working order and a ready-to-use condition before it is returned to the fire apparatus.

Each SCBA must be checked on a regular basis to ensure that it meets these requirements. Different procedures can be followed for inspection and for operational testing. Inspection and operational testing should be conducted after a unit has been used, and on a regular schedule. In career departments, inspection and testing are done at the beginning of each shift. In volunteer departments, this step is commonly performed on a weekly schedule.

If an SCBA inspection reveals any problems that cannot be remedied by routine maintenance, the SCBA must be removed from service for repair. Only properly trained and certified personnel are authorized to repair SCBA.

FIRE FIGHTER Tips

Follow the manufacturer's recommendations and avoid the use of aerosol cleaners or any alcohol-containing cleaner, as these materials degrade the rubber material in the face piece. Also, face pieces should be air dried or wiped with a soft, nonabrasive cloth to avoid scratching the lens.

Inspection of SCBA

The purpose of a SCBA inspection is to identify any parts of the SCBA that are visibly damaged and need to be repaired or replaced to ensure continued safe operation. This visible inspection can be done in conjunction with the operational testing sequence described here. Follow the steps in SKILL DRILL 3-10 for the SCBA inspection:

1. Visually inspect the air cylinder and valve assembly for dents and gouges. Look for black or discolored areas that indicate exposure to flame. (STEP 1)

SKILL DRILL 3-9

Doffing SCBA
(Fire Fighter I, NFPA 5.3.1)

1 Remove your gloves. Remove the regulator from your face piece or disconnect the low-pressure air supply hose from the regulator.

2 Shut off the air-supply valve or fully depress the air saver/donning switch to stop the flow of air. Remove your helmet and pull your protective hood down around your neck.

3 Loosen the straps on your face piece.

4 Remove your face piece.

5 Release your waist belt.

6 Loosen the shoulder straps and remove the SCBA.

7 Shut off the air-cylinder valve.

8 Bleed the air pressure from the regulator by opening the purge valve.

9 If you have an integrated PASS device, turn it off. Place the SCBA in a safe location where it will not get dirty or damaged.

2. Check the cylinder for the current hydrostatic test date and date of manufacture. Note that fiber-wrapped cylinders must be replaced every 15 years.
3. Check the cylinder pressure gauge to be sure it is full. (STEP 2)
4. Inspect hoses and rubber parts for damage or deterioration. (STEP 3)
5. Inspect the harness, webbing, buckles, fasteners, and cylinder retention system for damage. (STEP 4)

6. Verify that the SCBA has been cleaned according to the manufacturer's and department's recommendations.
7. Inspect the regulator for intact gaskets and other visible damage. (STEP 5)
8. Inspect the face piece for damage and warn components. Look for damage to the lenses and check for the presence of a nose cup. (STEP 6)
9. Inspect the head harness to confirm that all parts are present and working properly. (STEP 7)
10. Check the quick disconnects and the universal air connection to make sure they are not damaged and that they are operating properly. (STEP 8)

Operational Testing

The operational testing sequence is designed to check the function of the many parts of the SCBA to ensure safe use of the device. It concentrates on the working parts of the SCBA. Operational testing should be done after each use as well as at the beginning of each shift in career departments or on a set schedule in volunteer departments. Follow the steps in SKILL DRILL 3-11 for operational testing of SCBAs.

1. Check the regulator purge valve to be sure it is closed. (STEP 1)
2. Depress the air save switch, if present. (STEP 2)
3. Slowly open the air cylinder valve. Check for proper operation of the heads-up display and of the low-battery indicator. Confirm that the low-air warning and PASS devices are working. (STEP 3)
4. Check the remote pressure gauge for proper operation. Its reading should be within 10 percent of the air-cylinder pressure gauge reading. (STEP 4)
5. Don the face piece. Adjust it to obtain a good seal. Inhale sharply to start the flow of air. Breathe normally to check for proper operation. (STEP 5)
6. Remove the regulator or face piece; air should flow freely. (STEP 6)
7. Depress the air saver/donning switch to stop the flow of air. (STEP 7)
8. Open the purge valve to check for air flow. (STEP 8)
9. Close the purge valve to stop the flow of air. (STEP 9)
10. Rotate the air cylinder valve to close it. (STEP 10)
11. Open the purge valve slightly to vent residual air pressure from the system. Watch the heads-up display to verify its proper operation as the air pressure is exhausted. (STEP 11)
12. Once the air flow stops, close the purge valve. Complete any reporting that is required. (STEP 12)

Annual Inspection

A complete annual inspection and maintenance procedure must be performed on each SCBA unit. The annual inspection must be performed by a certified manufacturer's representative or a person who has been trained and certified to perform this work. SCBA requires regular inspection and maintenance to ensure that it will perform as intended.

Servicing SCBA Cylinders

A pressurized SCBA cylinder contains a tremendous amount of potential energy. Not only does the air within the cylinder exert considerable pressure on its walls, but the cylinder itself is used under extreme conditions on the fire ground. If the cylinder ruptures and suddenly releases this energy, it can cause serious injury or death. For this reason, cylinders must be regularly inspected and tested to ensure they are safe. These devices must be visually inspected during regularly scheduled inspections. A more detailed inspection is required if a cylinder has been exposed to excessive heat, has come into contact with flame, has been exposed to chemicals, or has been dropped.

The U.S. Department of Transportation requires hydrostatic testing for SCBA cylinders on a periodic basis and limits the number of years that a cylinder can be used. For example, composite-fiber overwrapped cylinders must be replaced after 15 years. Hydrostatic testing seeks to identify any defects or damage that might render the cylinder unsafe. Any cylinder that fails a hydrostatic test should be immediately taken out of service and cannot be used.

Cylinders constructed of different materials have different testing requirements. Aluminum, steel, and carbon-fiber cylinders must be hydrostatically tested every 5 years. Cylinders constructed of composite materials such as Kevlar or fiberglass fibers must be tested every 3 years. Fire fighters must know which types of cylinders are used by their departments and must check each cylinder for a current hydrostatic test date before filling it.

Replacing SCBA Cylinders

A used air cylinder can be quickly replaced with a full cylinder in the field to enable you to continue firefighting activities. A fire fighter who is working alone must doff his or her SCBA to replace the air cylinder; two fire fighters who are working together can change each other's cylinders without removing their SCBA. The steps listed in this section outline how a single person makes a cylinder change. This procedure may vary slightly depending on the model of SCBA being used. Follow the procedure recommended by the SCBA manufacturer and by your department's SOPs.

Practice changing air cylinders until you become proficient at this task. A fire fighter should be able to change cylinders in the dark and while wearing gloves if necessary. Follow the steps in SKILL DRILL 3-12 to replace an SCBA cylinder:

1. Place the SCBA on the floor or a bench. (STEP 1)
2. Turn off the cylinder valve. (STEP 2)
3. Bleed off the pressure by opening the purge valve. (STEP 3)
4. Disconnect the high-pressure supply hose. Keep the ends clean. (STEP 4)
5. Release the cylinder from the backpack and remove cylinder. (STEP 5)
6. Slide a full cylinder into the backpack. Align the outlet to connect the supply hose. Lock the cylinder in place. (STEP 6)
7. Check that the "O" ring is present and in good shape. (STEP 7)

SKILL DRILL 3-10 Visible SCBA Inspection

(Fire Fighter I, NFPA 5.3.1)

1 Visually inspect the air cylinder and valve assembly for dents and gouges. Look for black or discolored areas that indicate exposure to flame.

2 Check the cylinder for the current hydrostatic test date and date of manufacture. Check the cylinder pressure gauge to be sure it is full.

3 Inspect hoses and rubber parts for damage or deterioration.

4 Inspect the harness, webbing, buckles, fasteners, and cylinder retention system for damage.

5 Verify that the SCBA has been cleaned according to the manufacturer's and department's recommendations. Inspect the regulator for intact gaskets and other visible damage.

6 Inspect the face piece for damage and warn components. Look for damage to the lenses and check for the presence of a nose cup.

7 Inspect the head harness to confirm that all parts are present and working properly.

8 Check the quick disconnects and the universal air connection to make sure they are not damaged and that they are operating properly.

8 Connect the high-pressure hose to the cylinder. Hand-tighten only. (STEP 8)

9 Open the cylinder valve. Check the regulator gauge or remote gauge. Its reading should be within 10 percent of the reading on the cylinder gauge. (STEP 9)

Replacing an SCBA Cylinder on Another Fire Fighter

If you need to quickly reenter the fire scene, a second person can replace your air cylinder while you continue to wear your SCBA harness. Be sure that you are physically able to fight a second round with the fire. It is better to allow yourself a few minutes in rehab than to get back into the fire immediately and require rescue from other crew members. Do not overtax

SKILL DRILL 3-11 SCBA Operational Inspection
(Fire Fighter I, NFPA 5.3.1)

1. Check the regulator purge valve to be sure it is closed.

2. Depress the air save switch, if present.

3. Slowly open the air cylinder valve. Check for proper operation of the heads-up display and of the low-battery indicator. Confirm that the low-air warning and PASS devices are working.

4. Check the remote pressure gauge for proper operation. Its reading should be within 10 percent of the air-cylinder pressure gauge reading.

5. Don the face piece. Adjust it to obtain a good seal. Inhale sharply to start the flow of air. Breathe normally to check for proper operation.

6. Remove the regulator or face piece; air should flow freely.

7. Depress the air saver/donning switch to stop the flow of air.

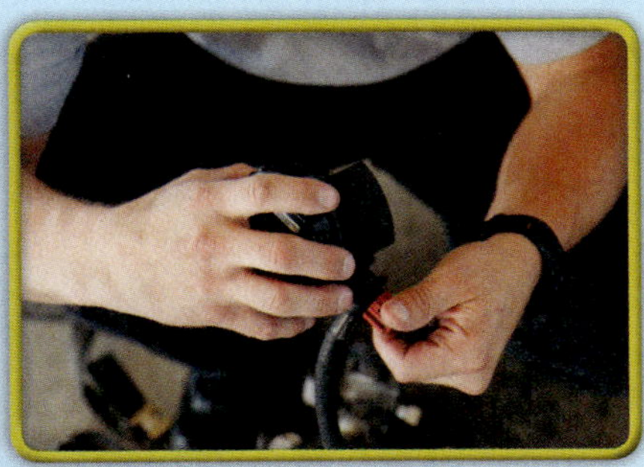

8. Open the purge valve to check for air flow.

9. Close the purge valve to stop the flow of air.

(Continued)

SKILL DRILL 3-11 SCBA Operational Inspection (*Continued*)

(Fire Fighter I, NFPA 5.3.1)

10 Rotate the air cylinder valve to close it.

11 Open the purge valve slightly to vent residual air pressure from the system. Watch the heads-up display to verify its proper operation as the air pressure is exhausted.

12 Once the air flow stops, close the purge valve. Complete any reporting that is required.

yourself by replacing the cylinder and going back to work without adequate rest when you need it.

Follow the steps in SKILL DRILL 3-13 to replace a SCBA cylinder on another fire fighter:

1. Remove the regulator from the face piece or remove the face piece so that you can breathe ambient air. (STEP 1)
2. Turn off the cylinder valve on the used cylinder. (STEP 2)
3. Open the purge valve to bleed off pressure from the high-pressure supply line. (STEP 3)
4. Disconnect the high-pressure supply hose. (STEP 4)
5. Release the SCBA cylinder from the backpack assembly and set it aside. (STEP 5)
6. Slide the full SCBA cylinder into the backpack. (STEP 6)
7. Check for the presence and satisfactory condition of the "O" ring. (STEP 7)
8. Lock the SCBA cylinder into place. (STEP 8)
9. Connect the high-pressure hose to the SCBA cylinder. (STEP 9)
10. Open the cylinder valve and state the pressure reading of the SCBA cylinder to the fire fighter.
11. Notify the fire fighter that the SCBA cylinder change is complete. (STEP 10)

Refilling SCBA Cylinders

Compressors and cascade systems are used to refill SCBA cylinders. A compressor or a cascade system can be permanently located at a maintenance facility or at a firehouse, or it can be mounted on a truck or a trailer for mobile use. Mobile filling units are often brought to the scene of a large fire.

Compressor systems filter atmospheric air, compress it to a high pressure, and transfer it to the SCBA cylinders FIGURE 3-23. Cascade systems have several large storage cylinders of compressed breathing air connected by a high-pressure manifold system. The empty SCBA cylinder is connected to the cascade system, and compressed air is transferred from the storage tanks to the cylinder. The storage cylinder valves must be opened and closed, one at a time, to fill the SCBA cylinder to the recommended pressure. This is done by opening the lowest-pressure SCBA cylinder first and then filling it from cylinders containing more pressure. The specific steps for filling SCBA cylinders vary with different systems. It is important to follow the manufacturer's directions when refilling SCBA cylinders.

Follow the steps in SKILL DRILL 3-14 to safely fill SCBA cylinders from a cascade system:

1. Complete the fill record form, including the date, hydrostatic test date, and cylinder serial number. (STEP 1)
2. Ensure that the cylinder is safe to fill by checking its date of manufacture and the hydrostatic test date. (STEP 2)
3. Check the cylinder for visible damage. (STEP 3)

FIGURE 3-23 Air compressors are installed at many fire stations to refill SCBA cylinders.

SKILL DRILL 3-12 Replacing an SCBA Cylinder

(Fire Fighter I, NFPA 5.3.1)

1 Place the SCBA on the floor or a bench.

2 Turn off the cylinder valve.

3 Open the purge valve to bleed off the pressure.

4 Disconnect the high-pressure supply hose.

5 Release the cylinder from the backpack and remove cylinder.

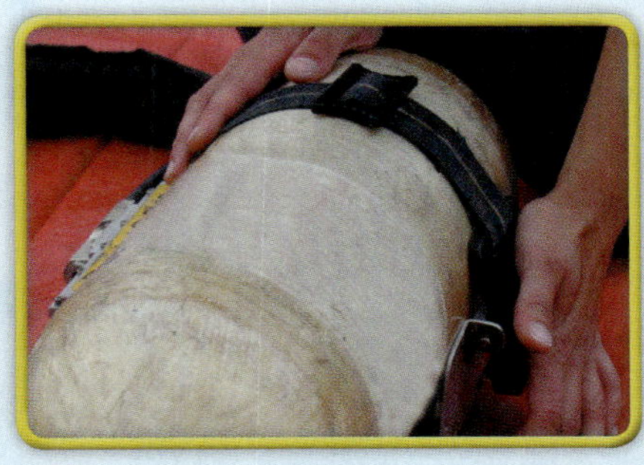

6 Slide a full cylinder into the backpack. Align the outlet to the supply hose. Lock the cylinder in place.

7 Check that the "O" ring is present and in good shape.

8 Connect the high-pressure hose to the air cylinder.

9 Open the cylinder valve. Check the gauge reading.

4. Follow the cascade system filling procedures.
5. Secure the cascade system after use according to system procedures. (STEP 4)

Proper training is required to fill SCBA cylinders. Whether your department has an air compressor or a cascade system, only those fire fighters who have been trained on the safe use of this equipment should refill air cylinders.

Cleaning and Sanitizing SCBA

Most SCBA manufacturers provide specific instructions for the care and cleaning of their models. The first step in cleaning the SCBA is to rinse the entire unit using a hose with clean water. The harness assembly and cylinder can be cleaned with a mild soap-and-water solution. If additional cleaning is needed, the unit can be scrubbed with a stiff brush. After scrubbing, the SCBA harness and cylinder should be rinsed with clean water.

After a fire, face pieces and regulators can be cleaned with a mild soap and warm water or with a disinfectant cleaning solution. The face piece should be fully submerged in the cleaning solution. If additional cleaning is needed, a soft brush can be used to scrub the face piece. During the cleaning process, avoid scratching the lens or damaging the exhalation

SKILL DRILL 3-13

Replacing an SCBA Cylinder on Another Fire Fighter
(Fire Fighter I, NFPA 5.3.1)

1. Remove the regulator from the face piece or remove the face piece so that you can breathe ambient air.

2. Turn off the cylinder valve on the used cylinder.

3. Open the purge valve to bleed off pressure from the high-pressure supply line.

4. Disconnect the high-pressure supply hose.

5. Release the SCBA cylinder from the backpack assembly and set it aside.

6. Slide the full SCBA cylinder into the backpack.

7. Check for the presence and satisfactory condition of the "O" ring.

8. Lock the SCBA cylinder into place.

9. Connect the high-pressure hose to the SCBA cylinder.

10. Open cylinder valve and notify the fire fighter that the SCBA cylinder change is complete.

valve. The regulator can be cleaned with the same solution, but should not be submerged. The face piece and regulator should then be rinsed with clean water.

Fire Fighter Safety Tips

The rapid intervention crew/company universal air connection (RIC UAC) is designed to be used only for emergency refilling of an air cylinder. It should never be used for routine filling of an air cylinder.

Allow the SCBA time to dry completely before returning it to service. Also check for any damage before returning this equipment to service. Follow the steps in **SKILL DRILL 3-15** to clean and sanitize an SCBA:

1. Inspect the SCBA for any damage that might have occurred before cleaning. (**STEP 1**)
2. Remove the face piece from the regulator. On some models, the regulator also can be removed from the harness. (**STEP 2**)
3. Detach the SCBA cylinder from the harness. (**STEP 3**)
4. Rinse all parts of the SCBA with clean water. Water from a garden hose can be used for this step. (**STEP 4**)
5. Using a stiff brush, along with a soap-and-water solution, scrub the SCBA cylinder and harness. Rinse and set these pieces aside to dry. (**STEP 5**)
6. In a 5-gallon (19 liter) bucket, make a mixture of mild soap and water; alternatively, use the manufacturer's recommended cleaning and disinfecting solution and water. (**STEP 6**)
7. Submerge the SCBA face piece into the soapy water or cleaning solution. For heavier cleaning, allow the face piece to soak. (**STEP 7**)

SKILL DRILL 3-14 Refilling SCBA Cylinders from a Cascade System

(Fire Fighter I, NFPA 5.3.1)

1 Complete the fill record form, including the date, hydrostatic test date, and cylinder serial number.

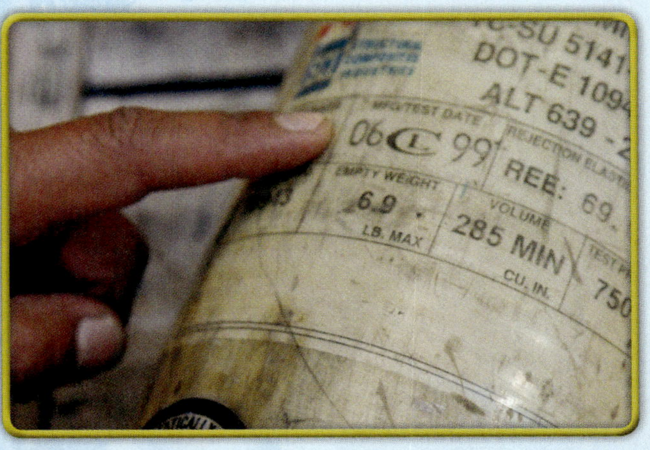

2 Ensure the cylinder is safe to fill by checking its date of manufacture and the hydrostatic test date.

3 Check the cylinder for visible damage.

4 Follow the cascade system filling procedures. Secure the cascade system after use according to system procedures.

8. Clean the regulator with the soapy water or cleaning solution, following the manufacturer's instructions.
9. Use a soft brush, if necessary, to scrub contaminants from the face piece and regulator. (**STEP 8**)
10. Completely rinse the face piece and the regulator with clean water. Set them aside and allow them to dry.
11. Reassemble and inspect the entire SCBA before placing it back in service. (**STEP 9**)

SKILL DRILL 3-15

Cleaning SCBA

(Fire Fighter I, NFPA 5.3.1)

1. Inspect the SCBA for any damage that might have occurred before cleaning.

2. Remove the face piece from the regulator. On some models, the regulator also can be removed from the harness.

3. Detach the SCBA cylinder from the harness.

4. Rinse all parts of the SCBA with clean water. Water from a garden hose can be used for this step.

5. Using a stiff brush, along with a soap-and-water solution, scrub the SCBA cylinder and harness. Rinse and set these pieces aside to dry.

6. In a 5-gallon (19 liter) bucket, make a mixture of mild soap and water; alternatively, use the manufacturer's recommended cleaning and disinfecting solution and water.

7. Submerge the SCBA face piece into the soapy water or cleaning solution. For heavier cleaning, allow the face piece to soak.

8. Clean the regulator with the soapy water or cleaning solution, following the manufacturer's instructions.

9. Completely rinse the face piece and the regulator with clean water. Set them aside and allow them to dry. Reassemble and inspect the entire SCBA before placing it back in service.

Near Miss REPORT

Report Number: 09-0000331

Synopsis: SCBA fails during maze exercise.

Event Description: During a Fire Fighter 1, Breathing Apparatus 1 class, a student suffered an SCBA failure. The student was in line for the maze exercise. While in line, the fire fighter heard an air leak and loss of air in the mask. The student continued to attempt to breathe in the mask. These attempts led to hyperventilation and a shift in the fire fighter's level of consciousness. The fire fighter's fellow students supported her and removed her mask. Instructors provided initial care (patient assessment and oxygen) and called EMS. The fire fighter was observed and released by the local Emergency Department Division Chiefs of Training. The Health and Safety Division were notified and an investigation initiated within the hour.

The SCBA failed at the intermediate pressure line between the first stage regulator and the mask mounted second stage regulator. A product change in the SCBA placed a swivel nut connection in the middle of the intermediate pressure line. In previous models this had been a one piece line. The swivel nut became loose and came undone, causing the air leak at 90 psi (620 kilopascal) and loss of air pressure to the mask.

The fire fighter's SCBA and mask were impounded by Health and Safety. All 40 sets of SCBA at the training academy were inspected. One additional SCBA was found to have the swivel nut assembly loose. The Air Maintenance Shop was notified. A Safety Alert was sent to all Stations. The manufacturer was contacted.

The swivel nut assembly is on the shoulder strap of the SCBA harness. It is often hidden from view by a cloth channel the line runs through. In attempts to recreate the failure it was found that under pressure, no air leak is heard until the swivel nut is on the last thread. When separation occurs the "O" ring is expelled or damaged. Even if the fire fighter could reassemble the line, leakage and loss of integrity would occur. The manufacturer has since developed torque standards for this component.

Lessons Learned:

All product "improvements" need to be critically looked at to determine risk, need for procedural changes, etc. We have added a step in the daily SCBA checklist which addressed the need to visually inspect the intermediate pressure line swivel nut assembly and attempt to disassemble using finger pressure. Should a swivel nut assembly be loose, the SCBA is to be removed from service and the Air Maintenance Shop notified. SCBA assigned to the Training Division, because of their high utilization, will receive twice yearly inspections and testing by the Air Maintenance Shop.

Fire Fighter Safety Tips

Refilling SCBA cylinders requires special precautions because of the high pressures involved in this procedure. The SCBA cylinder must be placed in a shielded container while it is being refilled FIGURE 3-24. Such a container is designed to prevent injury if the cylinder ruptures. In addition, the hydrostatic test date must be checked before the cylinder is refilled to ensure that its certification has not expired. Special procedures must be followed to ensure that the air used to fill the SCBA cylinder is not contaminated.

FIGURE 3-24 SCBA cylinders are refilled in a protective enclosure.

Wrap-Up

Chief Concepts

- Personal protective equipment is an essential component of a fire fighter's safety system. It enables a person to survive under conditions that might otherwise result in death or serious injury.
- Structural firefighting PPE enables fire fighters to enter burning buildings and work in areas with elevated temperatures and concentrations of toxic gases. It provides protection from the fire, keeps water away from the body, and helps reduce trauma from cuts or falls. This type of PPE is designed to be worn with SCBA.
- The components of structural PPE include the following items:
  - Bunker coat and pants—Have tough outer shells to withstand temperature extremes and repel water, and insulating layers to protect the skin.
  - Helmet—Protects the entire head and face from trauma; includes ear coverings and a face shield.
  - Protective hood—Constructed of a flame-resistant material to provide additional protection to the ears and neck.
  - Gloves—Constructed of heat-resistant leather to protect the hands from heat, water, vapors, cuts, and puncture wounds.
  - Boots—May be constructed of rubber or leather; designed to protect the feet from heat and trauma.
  - Work uniform—Must be made of natural fibers or special synthetic fibers to prevent burns.
  - SCBA—Provides essential respiratory protection that enables fire fighters to enter the fire atmosphere with a safe, independent air supply.
  - PASS—A device that will automatically sound if a fire fighter is motionless for a set time period.
  - Additional equipment—Goggles, earmuffs, ear plugs, hand light, and two-way radios.
- Structural PPE has several limitations:
  - Weight—PPE adds 50 pounds (23 kilogram) of extra weight to the fire fighter.
  - Overheating—PPE retains body heat and perspiration, which raises the risk of overheating.
  - Mobility—Full turnout gear limits the range of motion and makes movements awkward and difficult.
- Fire fighters should be able to don PPE in 1 minute or less. Following a set pattern of donning PPE helps reduce the time it takes to dress.
- Check the condition of PPE on a regular basis. Clean it when necessary, and get it repaired at once if it is worn or damaged. Damaged PPE will not protect you.
- Keep PPE clean to maintain its protective properties. The flammable by-products of burnt plastics and synthetic products can become trapped between the fibers, damaging the PPE.
- Cleaning instructions are listed on the tag attached to the clothing. Always follow the manufacturer's cleaning instructions.
- The specialized protective equipment used during vehicle extrication includes special gloves and coveralls or jumpsuits. Latex gloves may be used if the fire fighter provides emergency medical care.
- The PPE for wildland fires is designed for comfort and maneuverability. It includes a jacket and pants made of fire-resistant materials, a helmet, eye protection, and pigskin or leather gloves. The boots are designed to provide comfort and sure footing.
- Respiratory hazards associated with fires include smoke, smoke particles, smoke vapors, toxic gases, oxygen deficiency, and increased temperatures.
- The two main types of SCBA are open-circuit and closed-circuit devices. Open-circuit breathing apparatus is typically used for structural firefighting. A tank of compressed air provides the breathing air supply for the user, and exhaled air is released into the atmosphere through a one-way valve. Closed-circuit breathing apparatus recycles the user's exhaled air and is often used for extended operations, such as mine rescue work.
- One of the biggest limitations of SCBA is the limited amount of air in the cylinder. Fire fighters must manage their working time while using SCBA so that they have enough time to exit from the hazardous area before exhausting the air supply.
- Physical conditioning is important for all SCBA users. A fire fighter with an ideal body weight and in good physical condition will be able to perform more work per cylinder of air than a person who is overweight or out of shape.
- Breathing through an SCBA is different from normal breathing and can be very stressful. Covering your face with a face piece, hearing the air rushing in, hearing valves open and close, and exhaling against a positive pressure are all foreign sensations. Practicing breathing through SCBA in darkness helps to build confidence in the equipment and your skills.
- SCBA consists of four main parts: the backpack and harness, the air cylinder assembly, the regulator assembly, and the face piece assembly. A compressed air cylinder is removable from the backpack harness and can be changed quickly in the field. SCBA regulators may be mounted on the waist belt or shoulder strap of the

harness or attached directly to the face piece; they control the flow of air to the user. The face piece delivers breathing air to the fire fighter and protects the face from high temperatures and smoke.

- The pathway of air through an SCBA begins in the cylinder. The air passes through the cylinder shut-off valve into the high-pressure air line that takes it to the regulator. The regulator opens when the user inhales, reducing the pressure on the downstream side. In an SCBA unit with a face piece–mounted regulator, the air next goes directly into the face piece. In units with a harness-mounted regulator, the air travels from the regulator through a low-pressure hose into the face piece. From the face piece, the air is inhaled through the user's air passages and into the lungs. When the user exhales, used air is returned to the face piece. The exhaled air is exhausted from the face piece through the exhalation valve.
- The skip-breathing technique helps conserve air while using an SCBA in a firefighting situation. In this technique, the fire fighter takes a short breath, holds it, takes a second short breath without exhaling in between breaths, and then relaxes with a long exhale. Each breath should take 5 seconds.
- Each SCBA must be checked on a regular basis to ensure that it is ready for use. Inspection and operational testing should be conducted after a unit has been used, and on a regular schedule. In career departments, inspection and testing is done at the beginning of each shift. In volunteer departments, it is commonly done on a weekly schedule.
- Compressors and cascade systems are used to refill SCBA cylinders. A compressor or a cascade system can be permanently located at a maintenance facility or at a firehouse, or it can be mounted on a truck or a trailer for mobile use. Mobile filling units are often brought to the scene of a large fire.
- The steps for donning a complete PPE are as follows:
  1. Place the protective hood over your head and bring it down around your neck.
  2. Put on your bunker pants and boots. Adjust the suspenders and secure the front flap of the pants.
  3. Put on your turnout coat and secure the front.
  4. Open the air-cylinder valve on your SCBA and check the air pressure. Press the air saver valve to prevent air flow if needed.
  5. Put on your SCBA.
  6. Tighten both shoulder straps of the SCBA harness.
  7. Attach the waist belt of the harness and tighten it. Tighten the chest straps, if present.
  8. Fit the face piece to your face.
  9. Tighten the face piece straps, beginning with the lowest straps.
  10. Check the face piece for a proper seal. (Follow the manufacturer's instructions.)
  11. Pull the protective hood up so that it covers all bare skin but does not obscure your vision.
  12. Place your helmet on your head with the ear tabs extended, and secure the chin strap.
  13. Turn up your coat collar and secure it in front.
  14. Put on your gloves.
  15. Check your clothing to be sure it is properly secured.
  16. Make sure your PASS device is turned on.
  17. Attach your regulator or turn it on to start the flow of breathing air.
  18. Work safely!

Hot Terms

Air cylinder The component of the SCBA that stores the compressed air supply.

Air line The hose through which air flows, either within an SCBA or from an outside source to a supplied air respirator.

Backpack The harness of the SCBA, which supports the components worn by a fire fighter.

Bunker coat The protective coat worn by a fire fighter for interior structural firefighting; also called a turnout coat.

Bunker pants The protective trousers worn by a fire fighter for interior structural firefighting; also called turnout pants.

Carbon monoxide (CO) A toxic gas produced through incomplete combustion.

Carcinogen A cancer-causing substance that is identified in one of several published lists, including, but not limited to, *NIOSH Pocket Guide to Chemical Hazards*, *Hazardous Chemicals Desk Reference*, and the *ACGIH 2007 TLVs* and *BEIs*. (NFPA 1851)

Cascade system A method of piping air tanks together to allow air to be supplied to the SCBA fill station using a progressive selection of tanks, each with a higher pressure level. (NFPA 1901)

Closed-circuit breathing apparatus SCBA designed to recycle the user's exhaled air. This system removes carbon dioxide and generates fresh oxygen.

Compressor A device used for increasing the pressure and density of a gas. (NFPA 853)

Doff To take off an item of clothing or equipment.

Don To put on an item of clothing or equipment.

Dual-path pressure reducer A feature that automatically provides a backup method for air to be supplied to the regulator of an SCBA if the primary passage malfunctions.

End-of-service-time-indicator (EOSTI) A warning device on a SCBA that alerts the user that the end of the breathing air is approaching.

Face piece A component of SCBA that fits over the face.

Fire helmet Protective head covering worn by fire fighters to protect the head from falling objects, blunt trauma, and heat.

Hand light A small, portable light carried by fire fighters to improve visibility at emergency scenes; it is often powered by rechargeable batteries.

Heads-up display A visual display of information and system conditions status that is visible to the wearer of the SCBA.

Hydrogen cyanide A toxic gas produced by the combustion of materials containing cyanide.

Hydrostatic testing Pressure testing of the extinguisher to verify its strength against unwanted rupture. (NFPA 10)

Immediately dangerous to life and health (IDLH) Any condition that would pose an immediate or delayed threat to life, cause irreversible adverse health effects, or interfere with an individual's ability to escape unaided from a hazardous environment. (NFPA 1670, 2004)

Incomplete combustion A burning process in which the fuel is not completely consumed, usually due to a limited supply of oxygen.

Kevlar® A strong synthetic material used in the construction of protective clothing and equipment.

Light-emitting diode (LED) An electronic semiconductor that emits a single-color light when activated.

National Institute for Occupational Safety and Health (NIOSH) The U.S. federal agency responsible for research and development on occupational safety and health issues.

Nomex® A fire-resistant synthetic material used in the construction of personal protective equipment for firefighting.

Nose cups An insert inside the face piece of an SCBA that fits over the user's mouth and nose.

Occupational Safety and Health Administration (OSHA) The U.S. federal agency that regulates worker safety and, in some cases, responder safety. It is part of the U.S. Department of Labor.

Open-circuit breathing apparatus SCBA in which the exhaled air is released into the atmosphere and is not reused.

Oxygen deficiency Any atmosphere where the oxygen level is less than 19.5 percent. Low oxygen levels can have serious effects on people, including adverse reactions such as poor judgment and lack of muscle control.

PBI® A fire-retardant synthetic material used in the construction of personal protective equipment.

Personal alert safety system (PASS) A device that continually monitors for lack of movement of the wearer and, if no movement is detected, automatically activates an alarm signal indicating the wearer is in need of assistance. The device can also be manually activated to trigger the alarm signal. (NFPA 1982)

Personal protective equipment (PPE) The basic protective equipment for wildland fire suppression includes a helmet, protective footwear, gloves, and flame-resistant clothing as defined in NFPA 1977, *Standard on Protective Clothing and Equipment for Wildland Fire Fighting*. (NFPA 1051)

Phosgene A chemical agent that causes severe pulmonary damage; it is a by-product of incomplete combustion.

Pounds per square inch (psi) The standard unit for measuring pressure.

Pressure gauge A device that measures and displays pressure readings. In an SCBA, the pressure gauges indicate the quantity of breathing air that is available at any time.

Protective hood A part of a fire fighter's personal protective equipment that is designed to be worn over the head and under the helmet; it provides thermal protection for the neck and ears.

Rapid intervention crew/company universal air connection (RIC UAC) A system that allows emergency replenishment of breathing air to the SCBA of disabled or entrapped fire or emergency services personnel. (NFPA 1981)

Respirator A device that provides respiratory protection for the wearer. (NFPA 1994)

SCBA harness The part of SCBA that allows fire fighters to wear it as a "backpack."

SCBA regulator The part of the SCBA that reduces the high pressure in the cylinder to a usable lower pressure and controls the flow of air to the user.

Self-contained breathing apparatus (SCBA) A respirator with an independent air supply that is used by fire fighters to enter toxic or otherwise dangerous atmospheres.

Self-contained underwater breathing apparatus (SCUBA) A respirator with an independent air supply that is used by underwater divers.

Smoke particles Airborne solid materials consisting of ash and unburned or partially burned fuel released by a fire.

Supplied-air respirator (SAR) An atmosphere-supplying respirator for which the source of breathing air is not designed to be carried by the user. (NFPA 1404)

Turnout coat Protective coat that is part of a protective clothing ensemble for structural firefighting; also called a bunker coat.

Turnout pants Protective trousers that are part of a protective clothing ensemble for structural firefighting; also called bunker pants.

Two-way radio A portable communication device used by fire fighters. Every firefighting team should carry at least one radio to communicate distress, progress, changes in fire conditions, and other pertinent information.

FIRE FIGHTER *in action*

Your crew was first-in to the fully involved house, and you were the lead on the nozzle as you worked to put down the flames. After completing rehab, you were tasked with overhaul, but you did not mind; you were getting to fight fire. You think to yourself, "This is the greatest job in the world."

You are still pumped from the excitement as you pull back into the station. As you step off the engine, you notice the soot covering your gear and your SCBA. You think to yourself that it will be a while before you get to bed.

1. Which method should routinely be used to clean the SCBA?
- **A.** Soap and water
- **B.** Clean water only
- **C.** Nonsolvent degreaser
- **D.** Solvent-based degreaser

2. When refilling SCBA cylinders from a cascade system, how should it be done?
- **A.** One cascade cylinder at a time, starting with the one with the lowest pressure
- **B.** One cascade cylinder at a time, starting with the one with the highest pressure
- **C.** One cascade cylinder at a time, but the order does not matter
- **D.** Open all cascade cylinders at the same time

3. In career departments, how frequently should an SCBA unit be operationally tested?
- **A.** Annually
- **B.** Quarterly
- **C.** Monthly
- **D.** Daily

4. Pressure readings on the cylinder and the regulator should be ____ each other.
- **A.** the same as
- **B.** within 10 psi of (69 kilopascal)
- **C.** within 50 psi of (345 kilopascal)
- **D.** within 100 psi of (690 kilopascal)

5. Full PPE will weigh nearly ____ pounds.
- **A.** 25 (11 kilogram)
- **B.** 35 (16 kilogram)
- **C.** 50 (23 kilogram)
- **D.** 70 (32 kilogram)

6. A(n) ____ is an SCBA designed to recycle the user's exhaled air.
- **A.** supplied-air respirator
- **B.** open-circuit breathing apparatus
- **C.** closed-circuit breathing apparatus
- **D.** cascade system

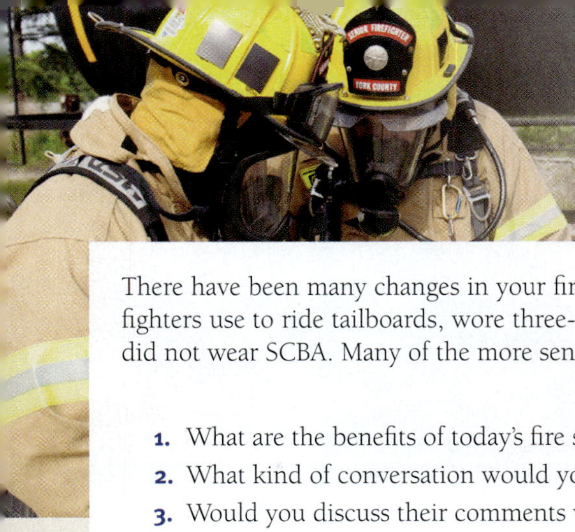

FIRE FIGHTER II *in action*

There have been many changes in your fire department over the past 20 years. You have heard all the stories about how fire fighters use to ride tailboards, wore three-quarter boots rather than bunker pants, used ¾″ booster lines for house fires, and did not wear SCBA. Many of the more senior guys talk fondly about those days and how much better it was.

1. What are the benefits of today's fire service over the fire service of yesteryear when it comes to safety?
2. What kind of conversation would you have with the senior fire fighters?
3. Would you discuss their comments with other recruits?
4. How will you promote today's safety standards?

Fire Service Communications

Fire Fighter I

Knowledge Objectives

After studying this chapter, you will be able to:

- Describe the role of the communications center in the fire service. (NFPA 5.2, p 92–93)
- Describe the role and responsibilities of a telecommunicator. (NFPA 5.2.1, p 93)
- List the requirements of a communications center. (p 93–94)
- Describe the equipment used in a communications center. (NFPA 5.2.1, p 94)
- Describe how computer-aided dispatch assists in dispatching the correct resources to an emergency incident. (NFPA 5.2.1, p 94)
- Describe the basic services provided by the communications center. (NFPA 5.2.1, p 95–96)
- List the five major steps in processing an emergency incident. (NFPA 5.2.1A, p 96)
- Describe how telecommunicators conduct a telephone interrogation. (NFPA 5.2.2, p 96–97)
- Describe how municipal fire alarm systems, private and automatic fire alarm systems, and citizens can activate the emergency response system. (NFPA 5.2.1, p 98–99)
- Describe how location validation systems operate. (NFPA 5.2.1, p 98–99)
- Describe the three types of fire service radios. (NFPA 5.2.3, p 103–104)
- Describe how two-way radio systems operate. (NFPA 5.2.3, p 103–104)
- Explain how a repeater system works to enhance fire service communications. (NFPA 5.2.3, p 105–106)
- Explain how a trunking system works to enhance fire service communications. (NFPA 5.2.3, p 106)
- Describe the basic principles of effective radio communication. (NFPA 5.2.3, p 106–108)
- Describe when to use plain language and how ten-codes are implemented in fire service communications. (NFPA 5.2.3A, p 107–108)
- Outline the information provided in arrival and progress reports. (p 108–109)
- Describe fire department procedures for answering nonemergency business and personal telephone calls. (NFPA 5.2.2, 5.2.2A, p 111)

Skills Objectives

After studying this chapter, you will be able to perform the following skills:

- Initiate a response to a simulated emergency. (NFPA 5.2.1B, p 102)
- Observe the operation of a communications center. (p 103)
- Display how to use a portable radio. (NFPA 5.2.3B, p 107)
- Operate and answer the fire station telephone. (NFPA 5.2.1B, 5.2.2B, p 111)

CHAPTER 4

Fire Fighter II | FFII

Knowledge Objectives

After studying this chapter, you will be able to:

- Define emergency traffic. (NFPA 6.2.2, 6.2.2A, p 109)
- Explain how to initiate a mayday call. (NFPA 6.2.2, 6.2.2A, p 109)
- Describe common evacuation signals. (NFPA 6.2.2, 6.2.2A, p 109)
- Explain the importance of an incident report to the entire fire service. (NFPA 6.2, 6.2.1, p 109–110)
- Describe how to collect the necessary information for a thorough incident report. (NFPA 6.2.1, 6.2.1A, p 109–110)
- Describe the resources that list the codes utilized in incident reports. (NFPA 6.2.1A, p 109–110)
- Explain the consequences of an incomplete or inaccurate incident report. (NFPA 6.2.1A, p 109–110)

Skills Objectives

After studying this chapter, you will be able to perform the following skills:

- Display how to use a portable radio. (NFPA 6.2.2B, p 107)
- Describe how to use the National Fire Incident Reporting System Data Entry Tool. (NFPA 6.2.1B, p 110)

Additional NFPA Standards

- NFPA 901, *Standard Classifications for Incident Reporting and Fire Protection Data*
- NFPA 902, *Fire Reporting Field Incident Guide*
- NFPA 1061, *Standard for Professional Qualifications for Public Safety Telecommunicator*
- NFPA 1221, *Standard for the Installation, Maintenance, and Use of Emergency Services Communications Systems*

You Are the Fire Fighter

At 3:04 a.m., you are dispatched to a report of a fire at 3256 West Madison. Your captain asks if any additional details are available, and the dispatcher says that the caller said there was smoke coming from the eaves of the building and then hung up. You are getting excited about the prospect of a working fire. Then the dispatcher comes back and says that this may be a false report, because the caller used a cell phone from the east side of town. Your aggravation rises at the prospect of being woken up for another false call—at least until the dispatcher comes back with a frantic sound in her voice and says that a home security company just reported a fire alarm at 3256 *East* Madison. Thoughts plow through your mind, wondering what is going on.

1. What is the process for receiving 911 calls in an emergency communications center?
2. How would the dispatcher know where the cell phone call originated?
3. How are alarm systems monitored and then reported to the communications center?

Introduction

Rapidly developing technology and advanced communications systems are having a tremendous effect on fire department communications. As a fire fighter, you must be familiar with the communications systems, equipment, and procedures used in your department. This chapter provides a basic guide to help you understand how fire department communications systems work and how commonly used systems are configured.

Every fire department depends on a functional communications system. When a citizen requests assistance or an alarm sounds, the communications center dispatches the appropriate units to the incident, and it continues to maintain communication with those units throughout the entire duration of the incident. The communications system is the link between fire fighters on the scene, the rest of the organization, and others. The communications center monitors everything that happens at the incident scene and processes all requests for assistance or special resources.

At the scene, fire fighters need to communicate with one another so that the incident commander (IC) can manage the operation efficiently based on progress reports or requests for assistance from fire fighters. The incident command system (ICS) depends on the presence of a functional on-site communications system. During incidents, fire fighters must be able to communicate not only with one another, but also with other emergency response agencies.

The communications center does more than simply dispatch units and communicate with them during emergency operations. It also tracks the location and status of every other fire department unit. The communications center must always know which units can be dispatched to an incident, and it must be able to contact those units promptly. The communications center is also responsible for redeploying units to maintain adequate coverage for all areas.

In addition to these special communications requirements, a fire department must have a communications infrastructure that allows it to function as an effective organization. Basic administration and day-to-day management require an efficient communications network, including telephone and data links with every fire station and work site.

The Communications Center

The communications center is the hub of the fire department emergency response system. It serves as the central processing point for all information relating to an emergency incident and all of the information relating to the location, status, and activities of fire department units. It connects and controls all of the department's communications systems. The communications center functions much like the human brain—that is, information comes in via the nerves, is processed, and is then sent back out to be acted upon by the various parts of the body.

The communications center is a physical location, whose size and complexity will vary depending on the needs of the department. The communications center for one department may be a small room in the fire station; another department may have a specially designed, highly sophisticated facility with advanced technological equipment. The fire department in a small community may need a simple system, whereas the public safety agencies in a metropolitan area may require a large facility. Some distinct differences between rural and urban dispatch centers are apparent—specifically, the number of dispatchers needed at any given time, the type of training that is required, and, to some extent, the type of equipment used. No matter what their size, however, all communications centers perform the same basic functions FIGURE 4-1.

FIGURE 4-1 No matter what their size, all communications centers perform the same basic functions.

Many fire departments operate stand-alone communications centers, which serve only a single agency. Others operate or are served by a communications center that works with several fire departments, and sometimes with all of the fire departments in a county or region. In many areas, the fire department communications center is co-located with other public safety agencies, including law enforcement and separate Emergency Medical Services (EMS) providers. A joint facility may house separate personnel and independent systems for each service delivery agency; alternatively, the entire operation may be integrated, with all employees being cross-trained to receive calls and dispatch responders to any type of emergency incident.

Telecommunicators

The employees who staff a communications center are known as dispatchers or telecommunicators. Telecommunicators have been professionally trained to work in a public safety communications environment. Their completion of advanced training and professional certification programs ensures that these skilled, competent individuals can fulfill their critical roles in the public safety system.

Just as special qualities can make an individual a good fire fighter, so a telecommunicator must possess certain qualities to be successful. The job of a telecommunicator can be complicated, demanding, and extremely stressful. Even in the face of these challenges, the successful telecommunicator is able to understand and follow complicated procedures, perform multiple tasks effectively, memorize information, and make decisions quickly.

One of the telecommunicator's most important skills is the ability to communicate effectively with citizens to obtain critical information, even when the caller is highly stressed or in extreme personal danger. Even if an emotional caller criticizes or insults the telecommunicator, the telecommunicator must respond professionally and focus on obtaining the essential information. Voice control and the ability to maintain composure under pressure are important qualities; the telecommunicator must always be clear, calm, and in control.

A telecommunicator must be skilled in operating all of the systems and equipment in the communications center. He or she must understand and follow the fire department's operational procedures, particularly those relating to dispatch policies and protocols, radio communications, and incident management. The telecommunicator must keep track of the status and location of each unit at all times, and monitor the overall deployment and availability of resources throughout the system. NFPA 1061, *Standard for Professional Qualifications for Public Safety Telecommunicator*, contains a complete list of qualifications for telecommunicator candidates.

Communications Facility Requirements

The fire department communications center must be designed and operated to ensure that its critical mission can be performed with a very high degree of reliability. The performance requirements in NFPA 1221, *Standard for the Installation, Maintenance, and Use of Emergency Services Communications Systems*, govern the design and construction of a fire department or public safety communications center. These requirements apply whether the communications center serves a small community with only one or two fire stations or a metropolitan area with dozens of stations.

The communications center should be well protected against natural threats such as floods, and it should be able to withstand predictable damaging forces such as floods, earthquakes, snowstorms, tornados, hurricanes, and other severe storms. It should be located so that it can continue to function in times of civil unrest. In addition, this center should be able to operate at maximum capacity, without interruption, even when other community services are severely affected. The building should be equipped with emergency generators and other systems so that it can continue to operate for several days in even the most challenging conditions.

NFPA 1221 also requires backup systems for all of the critical equipment in a communications center, so that the failure of a single component or system will not disable the entire operation. For example, there must be more than one way of transmitting a dispatch message from the communications center to each fire station. A backup radio transmitter should be available, and the telephone system must be able to receive calls even if part of the system is damaged. The design of the facility should minimize its vulnerability to a fire originating inside the building as well as to nearby fires. The center must also be secured to prevent unauthorized entry.

A backup communications center at a different location should be established. If some unanticipated situation makes it impossible to operate from the primary location, this backup location can be activated to ensure ongoing operation of the communications function.

Plans need to be in place for the reporting of emergencies in the event that the emergency communications center fails. Some communities have plans in place to locate fire department and law enforcement vehicles at strategic locations so citizens can report emergencies. Fire station personnel may plan to staff a watch desk to report emergencies from citizens who walk into the fire station.

Communications Center Equipment

A communications center is usually equipped with several types of communications systems and equipment. Although the specific requirements depend on the size of the operation and the configuration of the local communications systems, most centers will have the following equipment:

- Dedicated 911 telephones
- Public telephones
- Direct-line phones to other agencies
- Equipment to receive alarms from public or private fire alarm systems
- Computers and/or hard-copy files and maps to locate addresses and select units to dispatch
- Equipment for alerting and dispatching units to emergency calls
- Two-way radio system(s)
- Recording devices to record phone calls and radio traffic
- Backup electrical generators
- Records and record management system

Computer-Aided Dispatch

Computers are used in almost all communications centers. Most large communications centers use computer-aided dispatch (CAD) systems; many smaller centers have smaller-scale versions of CAD systems.

As the name suggests, a CAD system is designed to assist a telecommunicator by performing specific functions more quickly and efficiently than they can be done manually. Many variations in CAD systems exist, ranging from very simple versions to highly sophisticated systems that perform many different functions. All of the functions performed by a CAD system can be performed manually by trained and experienced telecommunicators; the benefit of such a system derives from the fact that the computer enables the telecommunicator to perform these functions more quickly and accurately FIGURE 4-2.

A well-designed CAD system can shorten the time it takes for a telecommunicator to receive and dispatch calls. In this way, a CAD system helps meet the most important objective in processing an emergency call: sending the appropriate units to the correct location as quickly as possible. The increased efficiency provided by a CAD system is particularly important in large communications centers that process a high volume of calls.

Because a computer can manage large amounts of information efficiently and accurately, a CAD system can be used to track the status of a large number of units. It will know which units are available to respond to a call, which units are assigned to incidents, and which units are temporarily assigned to cover different areas. The most advanced type of CAD system utilizes global positioning system (GPS) devices in the department's vehicles, allowing it to track the exact location of each unit. A CAD system can also look up addresses and determine the closest fire stations in order of response for any location. It can then select those units that can respond quickly to an alarm, even if some of the units that would normally respond to the call are currently unavailable.

Some CAD systems transmit dispatch information directly to mobile data terminals, or to computers that are located in the fire station or on the apparatus FIGURE 4-3. In addition, CAD systems can provide immediate access to information such as preincident plans, hazardous materials lists, lockbox locations, and information such as whether persons with limited mobility reside at the address of the emergency call. By linking the CAD system to other data files, fire fighters gain access to even more useful information such as travel route instructions, maps, and reference materials.

Voice Recorders and Activity Logs

Almost everything that happens in a communications center is recorded, either by a voice recording system or by an activity logging system. Most communications centers can automatically record everything that is said over the telephone or radio, 24 hours a day. In addition, most have an instant playback unit that allows the telecommunicator to replay conversations for the previous 10 to 15 minutes at the touch of a button. This

FIGURE 4-2 A CAD system enables a telecommunicator to work more quickly and efficiently.

FIGURE 4-3 Some CAD systems transmit dispatch information directly to terminals in fire stations and mobile data terminals in the apparatus.

feature is particularly valuable if the caller talks very quickly, has an unusual accent, hangs up, or is disconnected. The telecommunicator can replay the message several times, if necessary, to determine exactly what was said.

The logging system keeps a detailed record of every incident and activity that occurs. These records include every call that is entered, every unit that is dispatched, and every significant event that occurs in relation to an emergency incident. Times are recorded when a call is received, when the units are dispatched, when they report that they are en route, when they arrive at the scene, when the incident is under control, and when the last unit leaves the scene. One major advantage of a CAD system is that it automatically captures and stores every event as it occurs. Before such systems were developed, someone had to write down and time-stamp every transaction, including every radio transmission; all of these hard-copy logs then had to be retained for future reference.

Voice recorders and activity logs are maintained for several reasons. First, they serve as legal records of the official delivery of a government service by a public agency (i.e., the fire department). These records may be required for legal proceedings, sometimes years after the incident occurred. They may be needed to defend the fire department's actions when questions are raised about an unfortunate outcome. The records provided by voice recorders and activity logs accurately document the events and can often demonstrate that the organization and its employees performed ethically, responsibly, and professionally. They also make it difficult to hide an error, if a mistake was made.

Second, records are valuable in reviewing and analyzing information about department operations. Good record keeping enables a fire department to examine what happened on a particular call as well as to measure workloads, system performance, activity trends, and other factors as part of its planning and budget preparation. For example, analysts and planners can use data from CAD systems to study deployment strategies and to make the most efficient use of fire department resources.

Call Response and Dispatch

Critical functions performed by most CAD systems include verifying an address and determining which units should respond to an alarm. This is usually a simple task in a small fire department with only two or three fire stations, but it becomes much more complicated in a large urban system with dozens of fire stations and individual units. When the telecommunicator enters an address and an incident description code into a CAD system, the system can make a recommendation on the appropriate units to dispatch in less than a second.

Data links that provide the location of the caller and automatically enter the address can also save time. If duplicate addresses might exist or if the address entered is not a valid location, the CAD system will prompt the telecommunicator to ask the caller for more information. Hard-copy files of the most essential information should always be maintained for backup, so that calls can be dispatched even if the system goes down (i.e., is out of service).

The telecommunicator's first responsibility is to obtain the information that is required to dispatch the appropriate units to the correct location **FIGURE 4-4**. At that point, the incident must be processed according to standard protocols. The telecommunicator must decide which agencies or units should respond and transmit the necessary information to them. The generally accepted performance objective, from the time a call reaches the communications center until the units are dispatched, is 1 minute or less.

FIGURE 4-4 Telecommunicators must obtain information and relay it accurately to the appropriate responders.

Most requests for fire department response are made by telephone. Although a few communities use a regular seven-digit telephone number or "0" to report an emergency, most areas in the United States and Canada have implemented the 911 system. According to the National Emergency Number Association, 93 percent of the population in the United States and Canada has access to some type of 911 system to report an emergency.

All calls to 911 are automatically directed to a designated public safety answering point (PSAP) for that community or jurisdiction. If all public safety communications are located in the same facility, the call can be answered, processed, and dispatched immediately. If fire and EMS communications operate in separate facilities, those calls must be transferred to a fire/EMS-trained telecommunicator. The call-taking process is described in more detail later in this chapter.

Communications Center Operations

Several different functions are performed in a fire department or public safety communications center. All of these activities must be performed accurately and efficiently, even in the most challenging circumstances. For example, the activity level in a communications center can ratchet up from calm to extreme in less than a minute during an emergency, but the chaos outside must not affect the operations inside. If the communications center fails to perform its mission, the fire department will not be able to deliver much-needed emergency services.

A communications center performs the following basic functions:

- Receiving calls for emergency incidents and dispatching fire department units
- Supporting the operations of those fire department units delivering emergency services
- Coordinating fire department operations with other agencies
- Keeping track of the status of each fire department unit at all times
- Monitoring the level of coverage and managing the deployment of available units
- Notifying designated individuals and agencies of particular events and situations
- Maintaining records of all emergency-related activities
- Maintaining information required for dispatch purposes

Receiving and Dispatching Emergency Calls

Most fire department responses begin when a call comes in to the communications center, which then dispatches one or more units. Even if a citizen reports an emergency directly to a fire station or a crew discovers a situation, the communications center is immediately notified to initiate the emergency response process.

There are five major steps in processing an emergency incident:

1. Call receipt
2. Location validation
3. Classification and prioritization
4. Unit selection
5. Dispatch

Call receipt refers to the process of receiving an initial call and obtaining the necessary information to initiate a response. It includes telephone calls from the general public as well as other notification methods, such as automatic fire alarm systems, public fire alarm boxes, requests from other agencies, calls reported directly to fire stations, and calls initiated via radio from fire department units or other public safety agencies.

Location validation ensures that the information received is adequate to dispatch units to the correct location. Potential duplicate addresses—such as two streets with the same or very similar names—must be eliminated from consideration. For example, it is important to differentiate 123 Main Street from 123 Main Road. The information must point to a valid location on a map or in a street index system, and that location must be within the geographic jurisdiction of the potential dispatch units. Without a valid address, a communications center will be unable to send units to the proper location.

Classification and prioritization is the process of assigning a response category, based on the nature of the reported problem. Most fire departments respond to many kinds of situations, ranging from outside vegetation fires and fires in high-rise buildings to heart attacks and multiple-casualty incidents. The nature of the call dictates which units or combination of units should be dispatched. Dispatch policies are usually spelled out in the department's standard operating practices (SOPs) and dispatch protocols, which state the numbers and types of units to dispatch for each type of situation.

Unit selection is the process of determining exactly which unit or units to dispatch, based on the location and classification of the incident. The usual policy is to dispatch the closest available unit that can provide the necessary assistance. When all units are available and in their fire stations, this determination can usually be made quickly based on preprogrammed information. Unit selection becomes more complicated when some units are out of position or unavailable. This step often requires quick decision-making skills, even when the CAD system is programmed to follow set policies for various situations. Some dispatch centers are equipped with automatic vehicle locator systems that track apparatus by using GPS devices. These systems enable dispatchers to dispatch the closest units that are in service, thereby increasing the efficiency of the system.

Dispatch is the important step of actually alerting the selected units to respond and transmitting the information to them. Fire departments use a variety of dispatch systems, ranging from telephone lines to radio systems. The communications center must have at least two separate ways of notifying each fire station.

FIRE FIGHTER Tips

Dispatchers must be careful about which information is given to the public. Information such as availability of individual fire departments, apparatus placement, and available equipment should be monitored. An SOP should clarify which information is allowed to be broadcast and which information is not.

Call Receipt

Telephones

The public generally uses telephones to report emergency incidents. In most communities, calls to 911 connect the caller with a public service answering point. The PSAP can take the information immediately or transfer the call to the appropriate agency, based on the nature of the emergency. In some systems, calling 911 connects the caller directly with a telecommunicator, who obtains the required information.

FIRE FIGHTER Tips

Some communities have implemented a 311 system to handle nonemergency calls. This system is often used to link calls with health and human services or other community resources that can assist with a problem when it might not necessarily be life threatening in nature. The underlying premise of the 311 system is to reduce the number of nonemergency calls to 911.

The communications center has a seven-digit nonemergency telephone number. Some people may use that number to report an emergency because they are not sure that their particular situation is serious enough to be considered a "true emergency."

Any number that is published as a fire department telephone number should be answered at all times, including nights and weekends. The emergency number should always be pronounced as "nine-one-one," not "nine-eleven."

The telecommunicator who takes the call must conduct a telephone interrogation, asking the caller questions to obtain the required information. Initially, the telecommunicator will need to know the location of the emergency and the nature of the situation. Many 911 systems can automatically provide the location of the telephone where a call originates. This information may be unavailable or inaccurate, however, if the call was made on a wireless phone or from someplace other than the location of the emergency incident. The exact location must be obtained so that units can respond directly to the incident scene.

The telecommunicator must also interpret the nature of the problem from the caller's description. The caller may be distressed, excited, and unable to organize his or her thoughts. There may be language barriers; the caller might speak a different language or might not know the right words to explain the situation. Telecommunicators must follow SOPs and use active listening to interpret the information. Many communications centers use a structured set of questions to obtain and classify information on the nature of the situation. The telecommunicator must remember that the caller thinks the situation is an emergency, and he or she must treat every call as such until it is determined that no emergency exists.

Telecommunicators cannot allow gaps of silence to occur while questioning the caller. If the caller suddenly becomes silent, something may have happened to him or her; the caller might be in personal danger or extremely upset. Conversely, if the telecommunicator is silent, the caller might think that the telecommunicator is no longer on the line or no longer listening.

Disconnects are another problem, because callers to 911 might hang up accidentally or be disconnected after initially reaching the communications center. A telecommunicator who is unable to return the call and reach the original caller will usually dispatch a police officer to the location to determine if more help is needed.

A telecommunicator should never argue with a caller. Although callers may raise their voices, scream, or shout, the telecommunicator must continue to speak calmly and remain professional at all times. He or she should remember that the caller is simply reacting to the emergency situation.

With just two critical pieces of information—the location and nature of the problem—the telecommunicator can initiate a response. Local SOPs, however, may require the telecommunicator to obtain additional information. Getting the caller's name and contact phone number is useful in case it is necessary to call back for additional information. If the caller is in danger or distress, the telecommunicator will try to keep the line open and remain in contact with the caller until help arrives. In many communities, telecommunicators are trained to provide self-help instructions for callers, advising them on what to do until the fire department arrives. For example, if a building fire is being reported, the telecommunicator will advise the occupants to evacuate and wait outside; if the caller is reporting a medical incident, the telecommunicator may provide first-aid instructions, based on the patient's symptoms.

TDD/TTY/Text Phones

Speech- and/or hearing-impaired citizens can communicate by telephone using special devices that display text rather than transmitting audio **FIGURE 4-5**. Three such systems are TDD (telecommunications device for the deaf), TTY (teletype), and text phones. The Americans with Disabilities Act requires that communications centers be able to receive calls via text messages as well as by voice communication. Telecommunicators must know how to use this equipment to communicate with speech- and hearing-impaired persons. Many communications centers have a special telephone line with a seven-digit number set aside for such calls, but the 911 lines must also accommodate text messaging. Any incoming calls without voice contact at the other end should be checked for TDD/TTY/text phone systems.

FIGURE 4-5 TDD/TTY/text phone systems enable speech- and/or hearing-impaired persons to communicate over phone lines.

FIRE FIGHTER Tips

Next generation dispatch systems will permit dispatch centers to receive texts, video clips, and other communications from the general public.

Direct-Line Telephones

A direct-line (or ring-down) telephone connects two predetermined points. Picking up the phone at one end causes an immediate ring at the other end. Direct-line connections often link police and fire communications centers or two fire communications centers that serve adjacent areas. Direct lines also may connect hospitals, private alarm companies, utility companies, airports, and similar facilities with the fire department communications center. These lines are often used in both directions, so that the communications center can both receive calls for assistance and send notifications and requests for response.

The communications center may be connected by direct lines to each fire station in its jurisdiction. A direct line also can be linked to the station's public address speakers to announce dispatch messages.

Municipal Fire Alarm Systems

Some communities have installed fire alarm boxes or emergency telephones on street corners or in public places. A fire alarm box transmits a coded signal to the communications center. Although this signal indicates an alarm and identifies the location of the box, it does not indicate what kind of emergency is occurring. The major drawbacks to this type of reporting system are the high rate of false alarms and the inability to determine the nature of the call. Public fire alarm boxes may also be connected to automatic fire alarm systems in buildings. When the building's alarm system is activated, the public fire alarm box transmits an alarm directly to the communications center.

Public fire alarm boxes are usually connected by networks of dedicated cables that transmitted the alarm signals to a fire department facility and to individual fire stations. In many cases, these hard-wired boxes have been replaced by radio-operated fire alarm boxes. In recent years, many communities have eliminated their public fire alarm systems because of increasing numbers of false alarms, high maintenance costs, and development of alternative notification systems, including private and public hard-wired and cellular telephones. Other communities have converted their public fire alarm systems into call box systems.

A call box connects a person directly to a telecommunicator. The caller can request a full range of emergency assistance—from police, fire, or emergency medical assistance to a tow truck. Call boxes can be directly connected to a wireless or hard-wired telephone network or can operate on a radio system. Emergency call boxes are often located along major highways and in bridges, tunnels, subways, large complexes such as universities, and in other locations where nearby phones are not available and where mobile phones do not work. In locations without electric or telephone service, call boxes may be solar powered FIGURE 4-6.

FIGURE 4-6 Wireless call boxes are often found in places without other kinds of telephone service.

Private and Automatic Fire Alarm Systems

Private and automatic fire alarm systems use several types of arrangements to transmit alarms to the local fire department communications center. Many commercial, industrial, and residential buildings have automatic fire alarm systems, which use heat detectors, smoke detectors, or other devices to initiate an alarm, such as water flow alarms on automatic sprinkler systems. The Fire Detection, Protection, and Suppression Systems chapter discusses these systems in detail.

The connection used to transmit an alarm from a private system to a public fire communications center depends on many factors. These systems may be monitored by a privately operated central station alarm service or other alarm monitoring service, which relays the alarm to the fire communications center. A direct line may connect a central alarm service and the fire department communications center. Some communications centers provide monitoring services. Finally, private alarm systems may be connected to a public fire alarm box. No matter how the alarm reaches the communications center, the result is the creation of an incident report and the dispatch of fire department apparatus.

Walk-ins

Although most emergencies are reported by telephone, some people actually come to a fire station seeking assistance FIGURE 4-7. When a walk-in occurs, the station should contact and advise the communications center of the situation immediately. The communications center will create an incident record and dispatch any needed additional assistance. Even if the units at the station can handle the situation, they should notify the communications center that they are occupied with an incident.

Citizens should be able to come to a fire station and report an emergency at all times, even when the station is unoccupied. Many departments install a direct-line telephone to the communications center just outside each fire station. These phones should be marked with a simple sign stating, "If the station is vacant, pick up the telephone in the red box to report an emergency."

Location Validation

Enhanced 911 Systems

Most enhanced 911 systems have features that can help the telecommunicator obtain identifying information. For example, automatic number identification (ANI) shows the telephone

FIGURE 4-7 A citizen might walk into the fire station and report an emergency.

number where the call originated. Automatic location identification (ALI) queries a database to show the location of the telephone, the subscriber's name, and other details FIGURE 4-8. These features, when combined with an accurate database, will generate a computer display listing all of the information when the call is received. Another feature in an enhanced 911 system ensures that each call will be directed to the appropriate PSAP for that location.

Using enhanced 911, the telecommunicator can often pinpoint the location of an incident, even if the caller is unable to provide that information. Even so, the telecommunicator should always confirm that this information is correct and refers to the actual location of the emergency. If the caller has recently moved, the database may contain the old address.

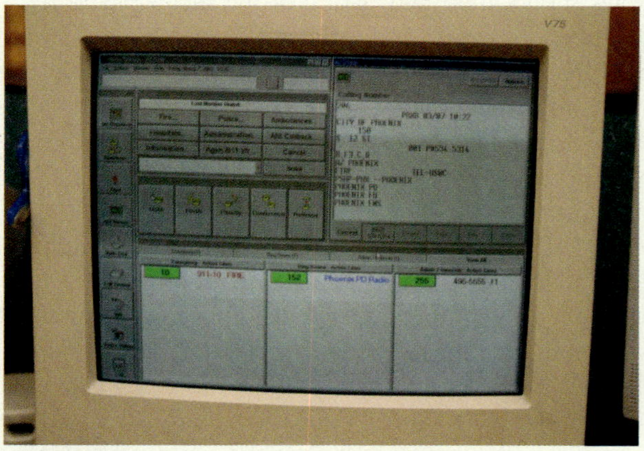

FIGURE 4-8 An automatic locator identification (ALI) system provides information about the origin of a 911 call.

Also, sometimes a billing address for the telephone service is provided rather than the physical location of the telephone.

Enhanced 911 features can save both time and lives, but only if they are used in conjunction with an accurate, up-to-date database. Telephone service providers, state regulatory agencies, and local emergency service agencies must cooperate to keep the database current and accurate.

With the large use of wireless (cell) phones, calls from wireless telephones are now as common as those from hard-wired telephones FIGURE 4-9.

Although wireless phones have certainly enhanced public access to emergency services, they have also created new challenges. The ANI and ALI enhancements to 911 systems were developed for hard-wired telephone systems. The Federal Government has mandated that wireless telephone companies work to update their systems to be able to pinpoint the geographic coordinates of a 911 call from a cellular phone and transmit that information to dispatch centers upon request.

FIGURE 4-9 Calls from wireless telephones are now as common as those from hard-wired telephones.

Determining the location of the incident may also prove problematic, because people who call 911 on their cellular telephones may not know their exact location. In addition, the dispatch center may receive numerous calls for the same incident from callers calling from different locations, thus making it even more difficult to determine the correct location.

Initiating a Response to a Simulated Emergency

Fire fighters should understand the steps followed by the personnel who staff the communications center as they receive emergency calls, solicit needed information from a caller, and dispatch appropriate units to the emergency. Follow the steps in SKILL DRILL 4-1 to initiate a response to a simulated emergency:

1. Identify your agency.
2. Ask whether there is an emergency.

3. Organize your questions to get the following information:
   a. Incident location
   b. Type of incident
   c. When incident occurred
4. Obtain the following information:
   a. Caller's name
   b. Location of the caller, if different from the incident
   c. Callback number (**STEP 1**)
5. Record the information needed.
6. Initiate an alarm following the protocols of your communications center. This may be done by dialing 911 if you are at the fire station, using a mobile phone, or using your mobile or portable radio to transmit the information to the dispatch center. The protocols in your department may vary from the steps listed here. Follow the protocols of the agency having jurisdiction for your department's communications. (**STEP 2**)

Call Classification and Prioritization

The next step in processing an emergency incident is classifying and prioritizing each call as it is received. The nature of the problem—what is happening at the scene—determines the urgency of the call and its priority. Although most fire department calls are dispatched immediately, it is sometimes necessary to delay dispatch to a lower-priority call if a more urgent situation arises or if several calls come in at the same time. Some calls qualify for emergency response (red lights and siren); others are considered nonemergency. Consult your local SOPs to learn the difference in your jurisdiction.

The call classification determines the number and types of units that are dispatched. The standard response to different incidents is established by the fire department's SOPs and can range from a single unit to an initial assignment of 10 or more units and dozens of fire fighters.

Unit Selection

After verifying the location and classifying the situation, a telecommunicator must select the specific units to dispatch to an incident. Generally, the standard assignment to each type of incident in each geographic area is stored on a system of run cards. Run cards are prepared in advance and either entered into the CAD system or stored in hard-copy form. They list units in the proper order of response, based on response distance or estimated response time, and often specify the units that would be dispatched through several levels of multiple alarms.

The run card assignments can be used as long as all of the units that would normally respond on a call are available. If any units are unavailable, the telecommunicator must adjust the assignments by selecting substitute units in their order of response. In a large, busy system with many simultaneous incidents, the status of individual units is constantly changing—a factor that complicates the assignment process unless the system is equipped with an automatic vehicle locator system that is integrated into the computerized dispatch protocols.

Most CAD systems are programmed to select the units for an incident automatically, based on the location, call classification, and actual status of all units. Such a system will recommend a dispatch assignment, which the telecommunicator can then accept or adjust, based on circumstances or special information. The same process is used to dispatch any additional units to an incident, whether it is a multiple alarm or a request for particular units or capabilities.

FIRE FIGHTER Tips

On occasion, the communications center may receive calls about issues that cannot be handled by the fire department or other agencies participating in the communications system. In these cases, the telecommunicator should make every effort to accommodate the caller. If possible, transfer the caller to the proper agency. In addition, provide the caller with the proper contact information in case the transfer fails. The caller will appreciate the assistance and retain a positive image of the department's service.

Dispatch

The telecommunicator's next step is to transmit the dispatch information to the assigned units quickly and accurately. This information can be transmitted in many different ways, but every department must have at least two separate methods for sending a dispatch message from the communications center to each fire station. The primary connection can be a hard-wired circuit, a telephone line, a computer-based data link, a microwave system, or a radio system.

Most fire departments dispatch verbal messages to the appropriate fire stations. The dispatch message is broadcast over speakers in the fire station, so that everyone immediately knows the location and nature of the incident and the units that should respond. Radio transmissions are used to contact a unit that is out of the station. The fire station or each vehicle must confirm that the message was received and the units are responding. If the communications center does not receive the confirmation within a set time period, it must dispatch substitute units.

A CAD system can be programmed to alert the appropriate fire stations automatically. Such a system can send the dispatch information to computer terminals or printers, sound distinctive tones, turn on lights and public address speakers, turn off the stove, and perform additional functions. The dispatch message also can go directly to each individual vehicle that has a computer terminal as well as to the station. The CAD notification often is accompanied by a verbal announcement over the radio and the fire station speakers.

Fire departments with volunteer responders must be able to reach them with a dispatch message. Some departments issue pagers to their individual volunteers. Some pagers can receive a vocal dispatch message; others display an alphanumeric message. Some CAD systems permit text messages to be sent to wireless phones with incident information. Volunteer fire departments in some communities may also rely on outdoor sirens, horns, or whistles to notify their members

VOICES OF EXPERIENCE

One winter evening, my department was called to assist in covering a neighboring area with a heavy rescue truck since most of their apparatus was working another incident. Not more than 10 minutes later, we were dispatched to a working residence fire. As we pulled up to the scene, flames were showing from the C-side of the structure. To my surprise, we were the first due company, but in a truck without water. I quickly did a scene size-up and observed fire venting through a back door and noted the structure had a basement. My partner grabbed a handline from an arriving engine and we made entry.

As we made our way into the structure, we encountered heavy fire in the rear. We were able to knock down a majority of the fire in seconds. We continued to search for fire and decided that we needed to find the basement stairs. As soon as the decision was made, the fire fighter behind me fell over to my left. He was alright and stated that he had found the basement stairs.

I proceeded cautiously to where my fellow fire fighter had just been. As I moved forward, I felt the floor starting to sag a foot or so. It was as if the plywood was on a spring. I knew I had felt this feeling before. I had a flashback to three years prior, in my fire recruit class, when during a simulated search and rescue operation, I entered a prop designed to demonstrate the signs of floor collapse. Having learned my lesson in class, I quickly backed up and told my crew to exit quickly. As our hose team was making their egress, I looked to my left and observed a large amount of flames coming from the heating registers on the floor. We were over a basement fire.

There is no doubt in my mind that the training I had in recruit class saved my life and the lives of my company. Fire simulations and drills are developed for a reason: to save fire fighters' lives. I often wonder what would have happened if I didn't pay attention during that evolution. This is why you have to make the most of EVERY training opportunity, big or small. Someday when you least expect it, your life might depend on it.

Todd Rielage
Chesterfield-Union Township Fire Department
Chesterfield, Indiana

SKILL DRILL 4-1 Initiating a Response to a Simulated Emergency

(Fire Fighter I, NFPA 5.2.1)

1 Identify your agency. Ask whether there is an emergency. Organize your questions to get the following information: incident location, type of incident, and when the incident occurred. Obtain the following information: caller's name, location of the caller, if different from the incident, and callback number.

2 Record the information needed. Initiate an alarm following the protocols of your communications center.

of an emergency. These audible devices usually can be activated by remote control from the communications center. Volunteers then call the communications center by radio or telephone to receive specific instructions.

Operational Support and Coordination

After the communications center dispatches the units, it begins to provide incident support and coordination. Someone in the communications center must remain in contact with the responding units throughout the entirety of the incident. The telecommunicator must confirm that the dispatched units actually received the alarm, record their en route times, provide any additional or updated information, and record their on-scene arrival times.

Operational support and coordination encompass all communications between the units and the communications center during an entire incident. Two-way radios, laptop computers, mobile communications devices, wireless phones, and mobile computer terminals may all be used to exchange information.

Generally, the IC will communicate with a telecommunicator operating a radio in the communications center. Progress and incident status reports, requests for additional units or release of extra units, notifications, and requests for information or outside resources are examples of incident communications. The communications center closely monitors the radio and provides any needed support for the incident.

Each part of the public safety network—fire, EMS, and police—must be aware of what other agencies are doing at the incident. The communications center serves as the hub of the network that supports units operating at emergency incidents, and it coordinates the fire department's activities and requirements with other agencies and resources. For example, the telecommunicator might need to notify nonemergency resources such as gas, electric, or telephone companies of the incident and request that they take certain actions. The communications center should have accurate, current telephone numbers and contact information for every relevant agency.

Status Tracking and Deployment Management

The communications center must know the location and status of every fire department unit at all times. Units should never get "lost" in the system, whether they are available for dispatch, assigned to an incident, or in the repair shop. As previously stated, the communications center must always know which units are available and which units are not available for dispatch. The changing conditions at an incident may require frequent reassignment of units, including ambulances. In addition, units may be needed from outside the normal response area or from other districts.

Tracking the status of the various units is difficult if the telecommunicator must rely on radio reports and colored magnets or tags on a map or status board. CAD systems make this job much easier, because status changes can be entered through digital status units or computer terminals.

Communications centers must continually monitor the availability of units in each geographic area and redeploy units when an area has insufficient coverage. Many fire departments list both unit relocations and multiple response units on the run cards. This information is only valid, however, if major incidents occur one at a time and when all units are available. If a department has a large volume of routine incidents, it may need to redeploy units to balance coverage, even when no major incidents are in progress.

Usually, a supervisor in the communications center is responsible for determining when and where to redeploy units as well as for requesting coverage from surrounding jurisdictions. For example, units from many different jurisdictions may be redeployed to respond to large-scale incidents under regional or statewide plans. These plans must include a system for tracking every unit and a designated communications center for maintaining contact with all units.

Touring the Communications Center

It is helpful to new fire fighters to tour the emergency communications center. Observing the actual operation of this center will give you a much better understanding of the role of the telecommunicator. SKILL DRILL 4-2 (Fire Fighter I, NFPA 5.2.2) lists the steps for touring your local communications center:

1. Arrange for a tour.
2. Conduct yourself in a professional manner.
3. Observe the use of equipment FIGURE 4-10.
4. Observe the receipt of a reported emergency.
5. Differentiate the needs of fire, police, and EMS personnel.
6. Understand the telecommunicator's job.

FIRE FIGHTER Tips

Vehicle locator systems and mobile tracking devices are valuable tools to accurately and effectively keep command staff and communication dispatchers up-to-date on the status of all units. These systems are especially useful in large departments and at large emergency incidents.

Radio Systems

Fire department communications systems depend on two-way radio systems. Radios link the communications center and individual units; they also link units at an incident scene. A radio system is an integral component of the ICS because it links all of the units on an incident—both up and down the chain of command and across the organization chart. Usually, every fire department vehicle has a mobile radio, and at least one—if not every—member of a team carries a portable radio during an emergency incident. A radio may be the fire fighter's only link to the incident organization and the only means to call for help in a dangerous situation. Radios also are used to transmit dispatch information to fire stations, to page volunteer fire fighters, and to link mobile computer terminals.

Fire departments use many different types of radios and radio systems. Technological advances are rapidly adding new features and system configurations, making it impossible to describe all of the possible features, principles of operation, and systems here. Instead, this section describes common systems and operating features. As a fire fighter, you must know how to operate your assigned radio and learn your own department's radio procedures.

FIGURE 4-10 Observing the actual operation of a call center will give you a much better understanding of the role of telecommunicators.

Radio Equipment

Three types of fire service radios are distinguished: the portable radio, the mobile radio, and the base station.

A portable radio is a hand-held two-way radio that is small enough for a fire fighter to carry at all times FIGURE 4-11. The radio body contains an integrated speaker and microphone, an on/off switch or knob, a volume control, channel select switch, and a "push-to-talk" (PTT) button.

A portable radio must have an antenna to receive and transmit signals. A popular optional attachment is an extension microphone/speaker unit that can be clipped to a collar or shoulder strap, while the radio remains in a pocket or pouch.

A portable radio is usually powered by a rechargeable battery, which should be checked at the beginning of each shift or prior to each use. Because the battery has a limited capacity, it must be recharged or replaced after extended operations. Battery-operated portable radios also have limited transmitting power. The signal can be heard only within a certain range and is easily blocked or overpowered by a stronger signal.

Mobile radios are more powerful two-way radios that are permanently mounted in vehicles and powered by the vehicle's

FIRE FIGHTER Tips

It can be frustrating to the fire fighter on the street when he or she has to deal with an inadequate dispatch center. Always remember that someone on the other end of the radio transmission is attempting to do his or her job with whatever is available to the telecommunicator. Stay professional, and remember that many citizens with scanners are listening to everything you say.

FIGURE 4-11 A portable radio should be carried by each individual fire fighter.

FIGURE 4-12 A mobile radio is permanently mounted in a vehicle.

FIRE FIGHTER Tips

To enable fire fighters to communicate more effectively while wearing an SCBA mask, several manufacturers have developed speaker/microphone systems that interface a portable radio with SCBA face pieces. These devices enable fire fighters to hear messages more clearly while wearing SCBA and to transmit information over their portable radios more easily, clearly, and simply. If your department uses this type of device, learn how to use it effectively.

electrical system FIGURE 4-12. Both mobile and portable radios share similar features, but mobile radios usually have a fixed speaker and an attached, hand-held microphone on a coiled cord. The PTT button is on the microphone, and the antenna is usually mounted on the exterior of the vehicle. Fire apparatus often include headsets with a combined intercom/radio system that enables crew members to talk to each other and hear the radio.

Base station radios are permanently mounted in a building, such as a fire station, communications center, or remote transmitter site FIGURE 4-13. These kinds of radios are more powerful than either portable or mobile radios. The antenna of the base station is often mounted on a radio tower so the transmissions have a wide coverage area FIGURE 4-14. Public safety radio systems often use multiple base stations at different locations to cover large geographic areas. The communications center operates these stations by remote control.

Mobile data terminals are computer devices that transmit data by radio. Some departments use them to track unit status, transmit dispatch messages, and exchange different types of information. Some terminals simply include buttons to send status messages such as "responding," "on the scene," and "back in service"; others comprise advanced laptop computers FIGURE 4-15.

■ Radio Operation

In the United States, the design, installation, and operation of two-way radio systems are regulated by the Federal Communications Commission (FCC). The FCC has established strict limitations governing the assignment of

FIGURE 4-13 Base station radios are installed at fixed locations.

FIGURE 4-15 Mobile computers exchange data over a radio channel.

FIGURE 4-14 A base station radio antenna is often mounted on a radio tower to provide maximum coverage.

frequencies to ensure that all users have adequate access. Every system must be licensed and operated within these guidelines.

Radios work by broadcasting electronic signals on certain frequencies, which are designated in units of megahertz (MHz). A frequency is an assigned space on the radio spectrum; only those radios tuned to that specific frequency can hear the message. The FCC licenses an agency to operate on one or more specific frequencies. A particular agency may be licensed to operate on several frequencies. These frequencies are often programmed into the radio and can be adjusted only by a qualified technician.

A radio channel uses either one frequency (simplex channel) or two frequencies (duplex channel). With a simplex channel, each radio transmits signals and receives signals on the same frequency, so a message goes directly from one radio to every other radio that is set to receive that frequency. With a duplex channel, each radio transmits signals on one frequency and receives on the other frequency. Duplex channels are used with repeater systems, which are described later in this chapter.

U.S. fire service frequencies are located in several different ranges, including 33 to 46 MHz (VHF low band); 150 to 174 MHz (VHF high band); 450 to 460 MHz (UHF band); 700 MHz; and 800 MHz. Additional groups of frequencies are allocated in specific geographic areas with a high demand for public safety agency radio channels. Each band has certain advantages and disadvantages relating to geographic coverage, topography (hills and valleys), and penetration into structures. Some bands have a large number of different users, which can cause interference problems, particularly in densely populated metropolitan areas.

Generally, one radio can be programmed to operate on several frequencies in a particular band. This limitation may become a problem if neighboring fire departments or police, fire, and EMS systems within the same jurisdiction operate on different bands. When different agencies working at the same incident communicate via different bands, they must make complicated arrangements with cross-band repeaters and multiple radios in command vehicles so that they can communicate with one another.

Repeater Systems

Radio messages can be broadcast over only a limited distance, for two reasons: (1) The signal weakens as it travels farther from the source; and (2) buildings, tunnels, and topography create interference. These problems are most significant for hand-held portable radios, which have limited transmitting power. Mobile radios that operate in systems covering large geographic areas face similar problems. To compensate for these shortcomings, fire departments may use radio repeater systems FIGURE 4-16.

In a repeater system, each radio channel uses two separate frequencies—one to transmit and the other to receive. When a low-power radio transmits over the first frequency, the signal is received by a repeater unit that automatically rebroadcasts it on the second frequency over a more powerful radio. All radios set on the designated channel receive the boosted signal on the second frequency. This approach enables the transmission to reach a wider coverage area.

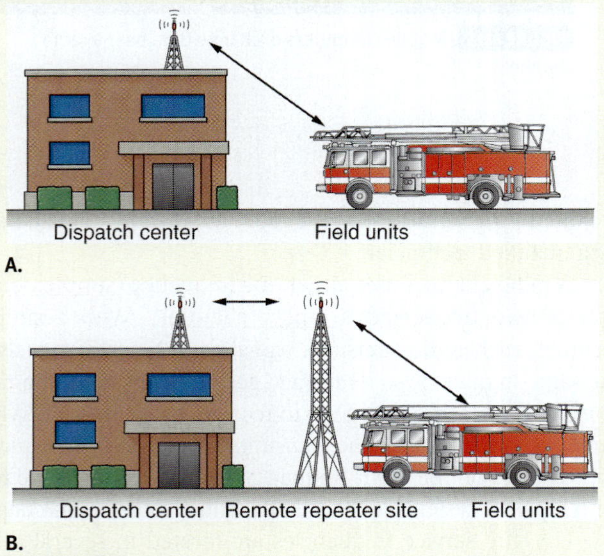

FIGURE 4-16 Direct and repeater channels send and receive transmissions in different ways. **A.** Direct channel. **B.** Repeater channel.

Many public safety radio systems have multiple receivers (voter receivers), which are geographically distributed over the service area to capture weak signals. The individual receivers forward their signals to a device that selects the strongest signal and rebroadcasts it over the system's base station radio(s) and transmission tower(s). With this configuration, messages that originate from a mobile or hand-held portable radio have as strong a signal as a base station radio at the communications center. As long as the original signal is strong enough to reach one of the voter receivers, the system is effective. If the signal does not reach a repeater, however, no one will hear it.

Some fire departments switch from a duplex channel to a simplex channel for on-scene communications. This configuration is sometimes called a talk-around channel, because it bypasses the repeater system. In this case, the radios transmit and receive on the same frequency, and the signal is not repeated. A talk-around channel often works well for short-distance communications, such as from the command post to crews inside a house fire or from one unit to another inside a building. Conversations on a talk-around channel cannot be monitored by the communications center; instead, the IC must use a more powerful radio on a repeater channel to maintain contact with the communications center.

Mobile repeater systems also can boost signals at the incident scene by creating a localized, on-site repeater system. The mobile repeater can be permanently mounted in a vehicle or set up at the command post. It captures the weak signal from a portable radio for rebroadcast. Some fire departments use cross-band repeaters, which boost the power of a weak signal and transmit it over a different radio band. In this way, a fire fighter using a UHF portable radio inside a building can communicate with the IC who is outside on a VHF radio. Similar systems may be used in large buildings or underground structures so crews working inside can communicate with units on the outside or with the communications center.

Trunking Systems

Radio systems based on newer technology and digital communications are being introduced to take advantage of the additional frequencies being allocated by the FCC for public safety use. Unfortunately, these new systems are generally incompatible with older radios and require expensive infrastructure changes.

The new radio technologies use a trunking system, a group of shared frequencies controlled by a computer. The computer allocates the frequencies for each transmission. The radio operator sets the radio on a talk group and communicates with the computer on a control frequency. When a user presses the transmit button, the computer assigns a frequency for that message and directs all of the radios in the talk group to receive the message on that frequency. In this way, a trunking system makes more efficient use of available frequencies and can provide coverage over extensive networks. Using digital technology also is more efficient than using analog technology, and enables more users to communicate at the same time in a limited portion of the radio spectrum. Trunking systems also allow different radios to be tied together for a given incident. With a trunking system, it is possible to link different agencies on the same system. For instance, public works, the highway department, fire department, and the police department could all be linked for a specific incident if they needed to share radio communications.

Using a Radio

As a fire fighter, you must learn how to operate any radio assigned to you and how to work with the particular radio systems used by your fire department. You must know when to use "Channel 3" or "Tac 5" or "Charlie 2," which buttons to press, and which knobs to turn. Study the training materials and SOPs provided by your department to learn this information.

When a radio is assigned to you, make sure you know how to operate and maintain it. Check the battery to be sure it is fully charged. Learn the protocol for changing and recharging portable radio batteries. On the fire ground, your radio is your link to the outside world. It must work properly—because your life depends on it.

The Future of Public Safety Communications

Currently, public safety agencies in the United States operate on 10 widely separated radio frequency bands. This configuration limits the ability of one agency to talk to another agency, especially when they are using widely separated frequencies. With the expansion of fire, law enforcement, and EMS agencies, there has been an increased need for additional frequencies, located close to each other, on which to operate public safety radios. Unfortunately, the demand for radio frequencies has outstripped the number of frequencies available. This problem is especially difficult to remedy because only a limited number of radio frequencies exist.

The problem of overcrowding is exacerbated by the increase in the number of functions for which radios are being used. Early radios operated on an analog system that could transmit only voice communications. By contrast, newer radios use a digital system that can be used to transmit voice, data, or video communications. Data such as dispatch information can be sent to mobile data terminals mounted on fire apparatus, for example. Information such as building plans, maps, inspection reports, and hazardous materials data can be sent over digital radio systems as well. Digital technology also produces better voice quality and allows for a greater number of users to operate on one frequency.

Trunked radio systems provide an efficient means for multiple agencies in the same geographic area to share a single radio system on the same frequencies. These systems rely on a computer to select and use an open frequency each time someone pushes a microphone button to talk; individual users are placed in different talk groups based on the types of functions they perform. Trunked radio systems greatly reduce idle time on each radio frequency. They also allow different agencies to talk to one another during mass-casualty events or natural disasters.

Some public safety agencies now operate radio systems on the 800 MHz frequency band. Because of the limited number of frequencies available, users on nearby frequencies often cause interference on public safety radios. Currently, some 800 MHz users are being moved to other frequencies to avoid this problem. In addition, recent Congressional action has made frequencies in the 700 MHz band available to public safety agencies. Because some of these frequencies are currently being used by broadcast TV stations, these TV stations must modify their broadcast equipment before the 700 MHz band becomes available for use by public safety agencies.

Public safety leaders hope that these efforts will result in future communications systems that enable effective transmission of data, voice, and video signals. Ideally, these systems will operate without interference from adjoining radio frequencies and will support communications within agencies and between agencies both during day-to-day operations and during events involving mass casualties and disasters.

To use a radio, follow the steps in **SKILL DRILL 4-3** (Fire Fighter I, NFPA 5.2.3, Fire Fighter II, NFPA 6.2.3):

1. Listen to determine that the channel is clear of any other traffic. Depress the "push-to-talk" (PTT) button and wait at least 2 seconds before speaking. This delay enables the system to capture the channel without cutting off the first part of the message. Some systems sound a distinctive tone when the channel is ready.
2. Speak across the microphone at a 45-degree angle and hold the microphone 1 to 2 inches (2.5 to 5 centimeters) from your mouth. Never speak on the radio if you have anything in your mouth **FIGURE 4-17**.
3. Know what you are going to say before you start talking. Speak clearly and keep the message brief and to the point.
4. Release the PTT button only after you have finished speaking.

If you hold a portable radio perpendicular to the ground with the antenna pointing toward the sky, you will get better transmission and reception. Range and transmission quality also will improve if you remove the radio from the radio pocket or belt clip before you use it.

You should be familiar with all departmental SOPs governing the use of radios. All radio transmissions should be pertinent to the situation, with unnecessary radio traffic being avoided. Remember that radio communications are automatically recorded. They also may be heard by everyone with a radio scanner, including the news media and the general public. Given this fact, you should avoid saying anything of a sensitive nature or anything you might later regret. Radio tapes provide a complete record of everything that is said and can be admissible as evidence in a legal case.

Most fire departments use plain English for radio communications, but some use codes for standard messages. A few departments have developed intricate systems of radio codes to fit any situation. Ten-codes, a system of coded messages that begin with the number 10, were once widely used. They can be problematic, however, when units from different jurisdictions have to communicate with one another. To be effective, codes must be understood by all parties and mean the same thing to all users. NIMS and NFPA standards recommend using plain English rather than codes. If your

FIGURE 4-17 Speak across the microphone at a 45-degree angle and hold the microphone 1 to 2 inches from your mouth.

department uses radio codes, you must know all of the codes and understand how they are used. Even plain English may include some generally accepted terminology within a system, such as "code blue" for a cardiac arrest. The objective is to be clearly and easily understood, without having to explain every message in detail. If your department has a few codes or special words for special situations, they should be specified in the SOPs.

FIRE FIGHTER Tips

Ten-codes are not approved by NIMS.

Arrival and Progress Reports

Fire fighters and telecommunicators must know how to transmit clear, accurate size-up and progress reports. The first-arriving unit at an incident—whether a fire fighter, a company officer, or a chief—should always give a brief initial radio report and establish command as described in the Incident Command System chapter. This initial report should convey a preliminary assessment of the situation and give other units a sense of what is happening so that they can anticipate what their assignments may be when they arrive at the scene. The communications center should repeat the initial size-up information.

FIRE FIGHTER Tips

Many departments have adopted a regional numbering system for apparatus to avoid confusion at major incidents. This numbering often involves assigning a three-digit number for each piece of apparatus. The first number identifies the fire department to which that apparatus belongs. The second and third numbers identify the number of that unit in the department. For example, all apparatus belonging to Everytown Fire Department might start with the number "4". Everytown Engine 3, then, is identified as Engine 403; Everytown Engine 33 is identified by the number 433. When units are dispatched by a regional dispatch center, this numbering system avoids the need to identify the name of each unit. When fire apparatus from several different departments are dispatched to the same fire, it also avoids the confusion that occurs when two "Engine 3" apparatus are on the same scene. With a three-digit numbering system, Everytown Engine 3 is designated as Engine 433, whereas Anothertown Engine 3 is designated as Engine 633.

For example, on arrival at the scene, a company might make the following report:

> Engine 410 to Communications. Engine 410 is on location at the intersection of High Street and Chambersburg Road. We have a two-story, wood-frame dwelling with fire showing from two windows on the second floor, side A. Engine 410 is assuming High Street Command. This will be an offensive operation.

The communications center would respond by announcing:

> Engine 410, I have you on the scene assuming High Street Command. Fire showing from the second floor, side A, of a two-story dwelling. Offensive operation.

This initial report should be followed by a more detailed report after the IC has had time to gather more information and make initial assignments. An example of such a report follows:

> **IC:** High Street Command to Communications.

> **Communications Center:** Go ahead, High Street Command.

> **IC:** At 176 High Street, we are using all companies on the first alarm for a fire on the second floor, side A, of a two-story, wood-frame dwelling, approximately 30 feet by 40 feet. Exposure Alpha is the street, Exposure Baker is a similar dwelling, Exposure Charlie is an alley, and Exposure Delta is a vacant lot. Companies are engaged in offensive operations and the fire appears to be confined. We have an all clear at this time. Battalion Chief 406 is in Command.

Communications Center: Copied High Street Command. You have an all-clear for a fire in a dwelling at 176 High Street. The fire appears to be confined to the second floor. Battalion 406 is IC.

FIRE FIGHTER Tips

Verifying the information received over the radio is vital. Always restate an important message or instruction to confirm that it has been received and understood.

IC: Command to Ladder 207. We need you to open the roof for ventilation, directly over the fire area.

Ladder 207: Ladder 207 copies. We are going to the roof to ventilate directly over the fire.

For the duration of an incident, the IC should provide regular updates to the communications center. Information about certain events—such as command transfer, "all clear," and "fire under control"—should be provided when they occur. Reports also should be made when the situation changes significantly and at least every 10 to 20 minutes during a working incident.

In some fire departments, the communications center prompts the IC to report at set intervals, known as time marks. Time marking allows the IC to assess the progress of the incident and decide whether the strategy or tactics should be changed. All progress reports are noted in the activity log to document the incident.

If additional resources, such as apparatus, equipment, or personnel, are needed, the IC transmits these requests to the communications center. The communications center will advise the IC of additional resources that are dispatched.

FIRE FIGHTER II

Emergency Messages

Sometimes the telecommunicator must interrupt normal radio transmissions for emergency traffic. Emergency traffic is an urgent message that takes priority over all other communications. When a unit needs to transmit emergency traffic, the telecommunicator generates a distinctive alert tone to notify everyone on the frequency to stand by, so the channel is available for the emergency communication. Once the emergency message is complete, the telecommunicator notifies all units to resume normal radio traffic.

The most important emergency traffic is a fire fighter's call for help. Most departments use mayday to indicate that a fire fighter is lost, is missing, or requires immediate assistance. If a mayday call is heard on the radio, all other radio traffic should stop immediately. The fire fighter making the mayday call should describe the situation, location, and help needed. Fire fighters should study and practice the procedure for responding to a mayday call. An example of a mayday call follows:

Fire fighter: MAYDAY ... MAYDAY ... MAYDAY.

[All radio traffic stops.]

FIRE FIGHTER II

IC: Unit calling MAYDAY, go ahead.

Fire fighter: This is Engine 403. We are on the second floor and running out of air. Fire has cut off our escape route. We request a ladder to the window on the Charlie side of the building so we can evacuate.

IC: Command copied. Engine 403, your escape route cut off by fire. I am sending the rapid intervention crew to Charlie side with a ladder.

The procedure for responding to a mayday call—which is used when fire fighters are in imminent danger—should be studied and practiced frequently. The steps for calling a mayday are explained more fully in the Fire Fighter Survival chapter.

Many fire department portable radios have an emergency button that can be pressed to transmit an emergency signal to the dispatcher. This transmission identifies which radio is sending the emergency signal. If your radios are equipped with this feature, learn the procedure for using it.

Another emergency traffic message is "evacuate the building," which warns all units inside a structure to abandon the building immediately. Most fire departments have specified a standard evacuation signal to warn all personnel to pull back to a safe location. The evacuation signal that is commonly used is a sequence of three blasts on an apparatus air horn, repeated several times, or sirens sounded on "high-low" for 15 seconds. An evacuation warning should be announced at least three times to ensure that everyone hears it; the warning should also be announced on the radio by the IC. Because no universal evacuation signal has been established, fire fighters must learn their department's SOP for emergency evacuation.

After the evacuation, the radio airwaves should remain clear so the IC can do a roll call of all units. This step ensures that all personnel have safely exited the building. After the roll call, the IC will allow the resumption of normal radio traffic.

Records and Reporting

The final communications step in an emergency situation is the incident report. Every time a fire department unit responds to an incident, someone must complete the proper reports. The incident report includes where and when the incident occurred, who was involved, and what happened. Incident reports for fires should include details about the origin of the fire, the extent of damage, and any injuries or fatalities. Such reports can be completed and submitted on paper, although many fire departments now enter them on computers and store the information in a computerized database. Several different incident-reporting software packages are available, and many large fire departments use custom-designed software.

In the United States, the National Fire Incident Reporting System (NFIRS) is used to compile and analyze incident reports at the local, state, and national levels. This information is used to help reduce the loss of life and property by fire. It also gives a clear picture of the fire problem in the United States. NFIRS 5.0 was developed by the United States Fire Administration in partnership with the National Fire Information Council. It has been offered to all states and territories to help them develop a more usable database of fire statistics. The NFIRS 5.0 system includes code improvements and additional modules such as EMS, Department Apparatus, Wildland, and Personnel that address the increasingly diverse demands faced by fire departments. In this way, it expands data collection beyond fires to encompass the full range of fire department activities.

Obtaining the Necessary Information

After each incident, the officer in charge will need to complete an incident report. The property owner and/or occupant is a primary source of information for the report. Any bystanders or eyewitnesses should also be questioned about what they observed. The officer preparing the report should share information with any authorities—such as a fire marshal, law enforcement personnel, or insurance investigator—who also are looking into the cause of the fire. The model number and serial number of any equipment damaged by or involved in the fire should be recorded as well.

Required Coding Procedures

Codes are often used in incident reports to indicate incident type, actions taken, and property use. The NFIRS Reference Guide identifies these information codes, as does the Quick Reference Guide, a condensed version that does not include graphics or examples. The Coding Questions and Answers Guide also provides coding information for NFIRS 5.0 incident reports. Fire administrators in all states have been invited to participate in NFIRS 5.0, although participation is not mandatory. If your department has its own incident reports and coding procedures, you should follow department SOPs.

Consequences of Incomplete and Inaccurate Reports

From a legal standpoint, records and reports are vital parts of the emergency. Information must be complete, clear, and concise because these records can become admissible evidence in a court case. Improper or inadequate documentation can have long-term negative consequences. Fire reports are considered to be public records under the Freedom of Information Act, so they may be viewed by an attorney, an insurance company, the news media, or the public. If a fatality or loss occurs, incomplete or inaccurate reports may be used to prove that the fire department was negligent. The department, the fire chief, and others may be held accountable.

Using the NFIRS Data Entry Tool

To use the NFIRS Data Entry Tool, follow the steps in **SKILL DRILL 4-4** (Fire Fighter II, NFPA 6.2.1):

1. Connect to the Internet. To start the NFIRS Data Entry Tool, go to the Windows Start menu and move the cursor to **Programs**, then **NFIRSv521**, then **Data Entry Tool**.
2. A connection will be established, and the NFIRS Data Entry Tool will open. Beside the red fire fighter's hat is your FDID (fire department identification) and department name, which is displayed within your state or county level. Under the Fire Dept menu, select **Open Fire Dept** and enter your department, personnel, and apparatus information. Click **OK** when you are done.
3. To enter a new incident, select **New Incident** under the Incident menu, or click the **New** button under Incident on the right-hand menu. Type in your FDID number, the incident date (in the format "mm/dd/yyyy"), and the incident number, if this information does not appear by default. (All yellow fields must be completed.) Click **OK** when you are done.
4. Once you have entered the key information, you must complete the required basic module. Click **Basic Module** to select it, and then click the **Open** button under the module on the right-hand menu.
5. Sections B–E of the basic module will open. After you complete them, click the **Sections F–J** tab at the top of your screen, or click the **Next Tab** button on the lower right.
6. Continue through the forms, completing each one. Use the tabs at the top or the tabs to the lower right to advance through the forms. When you are finished, save the information by selecting **Save Incident** under the Incident menu or click the **Save** button under Incident on the right-hand menu.
7. If you have not completed all required fields or have entered an invalid code, the "Critical Validation Errors Exist" prompt will alert you. The Validation Errors and Warning screen gives details of the form, field, error message, and error level (critical or warning). You must correct critical errors to save the incident as valid (V), but you can also save the incident as invalid (I) and edit it to correct the errors at a later date.
8. To search for incidents, select **Open Incident** under the Incident menu, or click the **Open** button under Incident on the right-hand menu, and then click **Search**.
9. For a general search, complete the fields under Address Search Criteria. For a more detailed search, fill in the incident number, exposure number, status (valid or invalid), incident type, incident date range, and/or property use under Key Search Criteria.
10. To exit the NFIRS Data Entry Tool, select **Exit NFIRS** from the Incident menu. Click **Yes** when the pop-up asks, "Exit NFIRS Data Entry Tool?"

Taking Calls: Emergency, Nonemergency, and Personal Calls

One of the first things you should learn when assigned to a fire station is how to use the telephone and intercom systems. You must be able to use them to answer a call or to announce a response. Keep your personal calls to a minimum, so incoming phone lines are open to receive emergency calls.

A fire fighter who answers the telephone in a fire station, fire department facility, or communications center is a representative of the fire department. Use your department's standard greeting when you answer the phone: "Good afternoon, Pleasant Town Fire Department. Captain Smith speaking. How may I help you?" Be prompt, polite, professional, and concise.

Detailed SOPs for obtaining information and processing calls should be provided to any fire fighter assigned to answer incoming emergency telephone lines. An emergency call can come in on any fire department telephone line. As mentioned earlier, someone may call your fire station directly instead of dialing 911 or another published emergency number. If this happens, you are responsible for ensuring that the caller receives the appropriate emergency assistance. Your department's SOPs should outline exactly which steps you should take in this situation, such as whether you should take the information yourself or connect the caller directly to the communications center, if possible.

SKILL DRILL 4-5 (Fire Fighter I, NFPA 5.2.1) lists the steps in obtaining essential information from a caller for an emergency response:

1. When you answer a call on any incoming line, follow your department's SOPs. Answer promptly and professionally. Identify yourself and the department or location. For example, you might say, "Good afternoon. Fire Station 403, Fire Fighter Jones speaking."
2. Determine immediately whether the caller has an emergency FIGURE 4-18.
3. If the call involves an emergency, follow your department's SOPs. You may have to transfer the call to the communications center or take the information yourself and relay it to the communications center.
4. Take the information from the caller, be sure to get as much information as possible:
   a. Type of incident/situation
   b. Location of the incident
   c. Cross streets or identifying landmarks
   d. Indication of scene safety
   e. Caller's name
   f. Caller's location, if different from the incident location
   g. Caller's callback number
5. If your station or your unit will be responding to the call, always advise the communications center immediately.
6. Be prepared to take accurate information or messages on all calls. Include the date, time, name of the caller, caller's phone number, message, and call taker's name. Have questions already formed and organized to maintain control of the conversation. Never leave someone on hold for any longer than necessary. Always terminate the call in a courteous manner. Let the caller hang up first.

FIGURE 4-18 Determine whether the caller has an emergency.

FIRE FIGHTER Tips

Today most people carry mobile cell phones, most of which have the ability to take pictures. Taking a picture of a person in an emergency situation while you are a member of a fire department can violate the privacy rights of that person and subject you and your department to possible legal action. Become familiar with the policy of your department regarding the use of private cell phones for taking pictures of people in emergency situations. Do not ever consider posting any such pictures on the Internet.

Wrap-Up

Chief Concepts

- Every fire department depends on a functional communications system.
- The communications center is the central processing point for all information relating to an emergency incident and all information relating to the location, status, and activities of fire department units. Depending on the size of the fire department, the communications center may be a small room or a large building. The communications center may serve one fire department or several public service agencies.
- Telecommunicators receive information from citizens and process that information to correctly dispatch resources to an emergency incident.
- The following equipment is found in communications centers:
  - Dedicated 911 telephones
  - Public telephones
  - Direct-line phones to other agencies
  - Equipment to receive alarms from public or private fire alarm systems
  - Computers and/or hard-copy files and maps to locate addresses and select units to dispatch
  - Equipment for alerting and dispatching units to emergency calls
  - Two-way radio system(s)
  - Recording devices to record phone calls and radio traffic
  - Backup electrical generators
- Computer-aided dispatch (CAD) enables telecommunicators to work more effectively. It tracks the status of units and assists the telecommunicator in quickly dispatching units to an emergency incident. Some CAD systems transmit dispatch information directly to mobile data terminals.
- Most communications centers automatically record everything that is said over the telephone or radio. This feature can prove valuable if the caller talks very quickly, has a strong accent, hangs up, or is disconnected. The recordings also serve as legal records for the fire department. They can assist in reviewing and analyzing information about department operations as well.
- A communications center performs the following basic functions:
  - Receiving calls for emergency incidents and dispatching fire department units
  - Supporting the operations of fire department units delivering emergency services
  - Coordinating fire department operations with other agencies
  - Keeping track of the status of each fire department unit at all times
  - Monitoring the level of coverage and managing the deployment of available units
  - Notifying designated individuals and agencies of particular events and situations
  - Maintaining records of all emergency-related activities
  - Maintaining information required for dispatch purposes
- There are five major steps in processing an emergency incident:
  - Call receipt—The process of receiving an initial call and gathering information.
  - Location validation—Ensuring that the address is valid.
  - Classification and prioritization—Assigning a response category based on the nature of the problem.
  - Unit selection—The process of determining exactly which unit or units to dispatch based on the location and classification of the incident.
  - Dispatch—The alerting of the selected units to respond and transmit information to them.
- Calls may be received via telephone, municipal fire alarm systems, private and automatic fire alarm systems, and walk-ins.
- Enhanced 911 systems may be able to display information about where a call originated and even the name and address of the caller.
- Fire department communications systems depend on two-way radio systems. A radio system is an integral component of the ICS because it links all of the units on an incident—both up and down the chain of command and across the organization chart.
- Three types of fire service radios may be used:
  - Portable radio—A hand-held two-way radio that the fire fighter carries at all times. The battery of such a radio should be checked at the beginning of each shift.
  - Mobile radio—Two-way radios permanently mounted in vehicles and powered by the vehicle's electrical system.
  - Base station—Radios permanently mounted in a building, such as a fire station, communications center, or remote transmitter site. Public safety radio systems often use multiple base stations at different locations to cover large areas.
- Radios work by broadcasting electronic signals on certain frequencies. These frequencies are often programmed into the radio and can be adjusted only by a qualified technician.
- A radio channel uses either one frequency (simplex channel) or two frequencies (duplex channel). With a simplex channel, each radio transmits signals and receives signals on the same frequency, so a message goes directly from one radio to every other radio set to that frequency. With a duplex channel, each radio transmits signals on one frequency and receives messages on another frequency.
- In a repeater system, each radio channel uses two separate frequencies—one to transmit and the other to receive. When a low-power radio transmits over the first frequency, the signal is received by a repeater unit that automatically rebroadcasts it on the second frequency over a more powerful radio. All radios set on the designated channel receive the boosted signal on the second

frequency. This approach enables the transmission to reach a wider coverage area.

- With a trunking system, a group of shared frequencies are controlled by a computer. The computer allocates the frequency for each transmission. The radio operator sets the radio on a talk group and communicates with the computer on a control frequency. When a user presses the transmit button, the computer assigns a frequency for that message and directs all of the radios in the talk group to receive the message on that frequency.
- The first-arriving unit at an incident should always give a brief initial radio report. This report should convey a preliminary assessment of the situation and give other units a sense of what is happening so that they can anticipate what their assignments may be when they arrive at the scene.
- Emergency traffic is an urgent message that takes priority over all other communications.
- When a unit needs to transmit emergency traffic, the telecommunicator generates a distinctive alert tone to notify everyone on the frequency to stand by, so the channel is available for the emergency communication. Once the emergency message is complete, the telecommunicator notifies all units to resume normal radio traffic.
- The most important emergency traffic is a fire fighter's call for help. Most departments use "mayday" to indicate that a fire fighter is lost, is missing, or requires immediate assistance. If a mayday call is heard on the radio, all other radio traffic should stop immediately. The fire fighter making the mayday call should describe the situation, location, and help needed.
- An incident report describes where and when the incident occurred, who was involved, and what happened. Incident reports for fires should include details about the origin of the fire, the extent of damage, and any injuries or fatalities. Such reports can be completed and submitted on paper, although many fire departments enter them on computers and store the information in a computerized database.
- From a legal standpoint, records and reports are vital parts of the emergency. Information must be complete, clear, and concise because these records can become admissible evidence in a court case.
- A fire fighter who answers the telephone in a fire station, fire department facility, or communications center is a representative of the fire department. Use your department's standard greeting when you answer the phone. Be prompt, polite, professional, and concise.

Hot Terms

Activity logging system A device that keeps a detailed record of every incident and activity that occurs.

Automatic location identification (ALI) An enhanced 911 service feature that displays where the call originated or where the phone service is billed.

Automatic number identification (ANI) An enhanced 911 service feature that shows the calling party's phone number on the display screen at the telecommunicator's terminal.

Base station A stationary radio transceiver with an integral AC power supply. (NFPA 1221)

Call box A system of telephones connected by phone lines, radio equipment, or cellular technology to a communications center or fire department.

Communications center A building or portion of a building that is specifically configured for the primary purpose of providing emergency communications services or public safety answering point services to one or more public safety agencies under the authority or authorities having jurisdiction. (NFPA 1221)

Computer-aided dispatch (CAD) A combination of hardware and software that provides data entry, makes resource recommendations, notifies and tracks those resources before, during, and after fire service alarms, preserving records of those alarms and status changes for later analysis. (NFPA 1221)

Direct line A telephone that connects two predetermined points.

Dispatch To send out emergency response resources promptly to an address or incident location for a specific purpose. (NFPA 450)

Duplex channel A radio system that uses two frequencies per channel; one frequency transmits and the other receives messages. Such a system uses a repeater site to transmit messages over a greater distance than is possible with a simplex system.

Emergency traffic An urgent message, such as a call for help or evacuation, transmitted over a radio that takes precedence over all normal radio traffic.

Evacuation signal A distinctive signal intended to be recognized by the occupants as requiring evacuation of the building. (NFPA 72)

Federal Communications Commission (FCC) The federal regulatory authority that oversees radio communications in the United States.

Mayday A code indicating that a fire fighter is lost, missing, or trapped, and requires immediate assistance.

Mobile data terminal Technology that allows fire fighters to receive information while in the fire apparatus or at the station.

Mobile radio A two-way radio that is permanently mounted in a fire apparatus.

National Fire Incident Reporting System (NFIRS) A system used by fire departments to report and maintain computerized records of fires and other fire department incidents in a uniform manner.

Portable radio A battery-operated, hand-held transceiver. (NFPA 1221)

Public safety answering point (PSAP) A facility in which 911 calls are answered. (NFPA 1221)

Radio repeater system A radio system that automatically retransmits a radio signal on a different frequency.

Run cards Cards used to determine a predetermined response to an emergency.

Simplex channel A radio system that uses one frequency to transmit and receive all messages.

Talk-around channel A simplex channel used for on-site communications.

TDD/TTY/text phone system User devices that allow speech- and/or hearing-impaired citizens to communicate over a telephone system. TDD stands for telecommunications device for the deaf; TTY stands for teletype; text phones visually display text. The displayed text is the equivalent of a verbal conversation between two hearing persons.

Telecommunicator A trained individual who is responsible for answering requests for emergency and nonemergency assistance from citizens. This individual assesses the need for a response and alerts responders to the incident.

Telephone interrogation The phase in a 911 call during which the telecommunicator asks questions to obtain vital information such as the location of the emergency.

Ten-codes A system of predetermined coded messages, such as "What is your 10-20?", used by responders over the radio.

Time marks Status updates provided to the communications center every 10 to 20 minutes. Such an update should include the type of operation, the progress of the incident, the anticipated actions, and the need for additional resources.

Trunking system A radio system that uses a shared bank of frequencies to make the most efficient use of radio resources.

Voice recording system Recording devices or computer equipment connected to telephone lines and radio equipment in a communications center to record telephone calls and radio traffic.

FIRE FIGHTER *in action*

Today is the day you study dispatching in your recruit class. You see that 4 hours is allocated for this topic. You wonder how it can take 4 hours to show how a dispatcher answers a phone, asks some questions, and then pushes a button to dispatch the fire units.

The director of the emergency communications center arrives and tells the class that you will be going over to the communications center. You are asked to sit next to a man with a headset on and five computer screens in front of him. He hands you a headset so you can listen in. Within minutes, you are amazed at how hard this job really is.

1. ______________________ is an enhanced 911 service feature that displays where the call originated or where the phone service is billed.

A. Automatic location identification (ALI)

B. Automatic number identification (ANI)

C. Call box

D. TDD/TTY/text phone system

2. A ______________________ is a radio system that uses a shared bank of frequencies to make the most efficient use of radio resources.

A. trunked system

B. simplex system

C. radio repeater system

D. mobile data system

3. A ______________________ is a radio system that uses one frequency to transmit and receive all messages.

A. trunked system

B. simplex system

C. radio repeater system

D. mobile data system

FIRE FIGHTER *in action* (continued)

4. A(n) ________________________ is a code indicating that a fire fighter is lost, missing, or trapped, and requires immediate assistance.
   - **A.** emergency evacuation signal
   - **B.** ten-code
   - **C.** box call
   - **D.** mayday
5. Two-way radios that are permanently mounted in vehicles and powered by the vehicle's electrical system are called ________________________.
   - **A.** portable radios
   - **B.** mobile radios
   - **C.** base radios
   - **D.** simplex radios
6. Which step comes first in the processing of an emergency incident in a communications center?
   - **A.** Classification and prioritization
   - **B.** Unit selection
   - **C.** Location validation
   - **D.** Dispatch

FIRE FIGHTER II *in action*

Your engine is dispatched to the report of a fire alarm in a building. It is the third call for the same building during this shift, and the second one since you went to bed. It's difficult to get too excited when you know the building is having trouble with its system. When you reach the site, you get off the engine with your gear, the captain checks the alarm panel, and you catch the elevator to the sixth floor.

As you step off the elevator, you smell smoke and see a haze hanging down approximately 2 feet (0.6 meter) from the ceiling. Clearly, this call is not an alarm malfunction. Your captain tells you to go to the stairwell and connect to the standpipe, but to not charge it, while she checks for the source. She says she thinks the source is food on a stove, so the captain says don't have dispatch wake anyone else up. You get connected to the standpipe but your captain isn't back. You open the door and find the hallway filled with smoke. You try to call your captain on the radio but do not receive a response. You get a pit in your stomach as you realize this isn't good.

1. Would this be a situation where you would initiate emergency traffic?
2. How would you activate a mayday?
3. Which actions would this message precipitate at the communications center?
4. Which information should you relay to the communications center?

Incident Command System

Fire Fighter I

Knowledge Objectives

There are no knowledge objectives for Fire Fighter I candidates. NFPA 1001 contains no Fire Fighter I Job Performance Requirements for this chapter.

Skills Objectives

There are no skill objectives for Fire Fighter I candidates. NFPA 1001 contains no Fire Fighter I Job Performance Requirements for this chapter.

Fire Fighter II FFII

Knowledge Objectives

After studying this chapter, you will be able to:

- Describe the characteristics of the incident command system. (NFPA 6.1.1, p 119–122)
- Explain the organization of the incident command system. (NFPA 6.1.1, p 122–126)
- Function within an assigned role within the incident command system. (NFPA 6.1.1, p 122–128)
- Organize and coordinate an incident command system until command is transferred. (NFPA 6.1.1, p 128–131)
- Transfer command within an incident command system. (NFPA 6.1.1, p 131–134)

Skills Objectives

After studying this chapter, you will be able to perform the following skills:

- Operate within the incident command system. (NFPA 6.1.2, p 130)
- Assume command. (NFPA 6.1.2, p 131)
- Transfer command. (NFPA 6.1.2, p 134)

Note: To be NIMS compliant, in this chapter we will refer to the incident management system as the incident command system.

FFII

CHAPTER 5

Additional NFPA Standards

- NFPA 1026, *Standard for Incident Management Personnel Professional Qualifications*
- NFPA 1500, *Standard on Fire Department Occupational Safety and Health Program*
- NFPA 1521, *Standard for Fire Department Safety Officer*
- NFPA 1561, *Standard on Emergency Services Incident Management System*

You Are the Fire Fighter

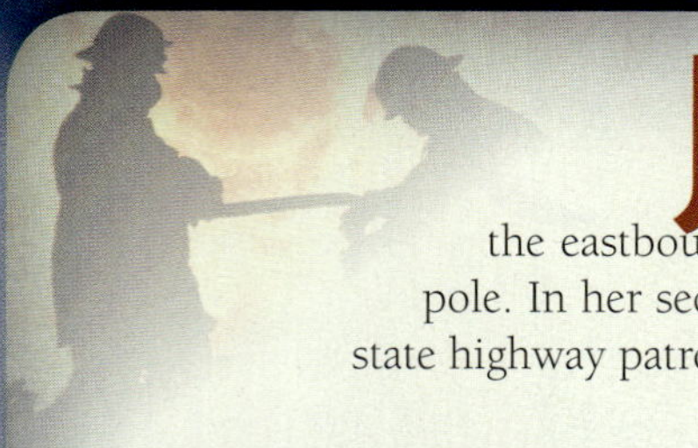

Just before lunch, you hear the unmistakable sound of the alert tones in the station. Your engine company and EMS unit are dispatched to U.S. Highway 20 close to County Road 270 for a motor vehicle collision. The dispatcher informs you that one vehicle is on its side on the eastbound lanes of Route 20 and a second vehicle has gone off the westbound lanes and hit a utility pole. In her second transmission, the dispatcher informs you that two EMS units from Essex County and the state highway patrol are also responding. Your pulse quickens when you find yourself the first to arrive on scene.

1. Why is there a need for an incident command system?
2. Who is in charge of a situation like this one, where several different agencies are operating at the same scene?
3. What will your role be?

Introduction

All fire department emergency operations and training exercises should be conducted within the framework of an incident command system (ICS). Using an ICS ensures that operations are coordinated and conducted safely and effectively. This type of system provides a standard approach, structure, and operational procedure to organize and manage any operation—from an emergency to a training session. The same principles apply whether the situation involves a single engine company or hundreds of emergency responders from dozens of different agencies. These principles also can be applied to many nonemergency events—such as large-scale public events—that require planning, coordination, and communication.

Planning, supervision, and communication are key components of an ICS. Model procedures for incident management have been developed and widely adopted to provide a standard approach that can be used by many different agencies. These standards and guidelines are defined at a national level in the National Incident Management System (NIMS). The command structure discussed in this chapter is a critical component of NIMS. Within a single fire department or a group of fire departments and other public safety agencies (such as police or Emergency Medical Services [EMS]) that routinely respond together, the use of a standard system for managing an incident is essential. Enhanced coordination results when every organization that could potentially be involved in a major incident is familiar with and has practiced with the same ICS.

Some fire departments use procedures that vary from the ICS model or other model systems in certain aspects. As a fire fighter, it is essential for you to become intimately familiar with the system that is used within your jurisdiction. Where local variations exist, the overall system usually remains similar in structure and application to the ICS model.

NIMS integrates all facets of response into a comprehensive national framework. This system was developed as an outgrowth of a presidential directive issued on February 28, 2003. It provides a consistent nationwide template to enable federal, state, and local governments as well as private-sector and nongovernmental organizations to work together effectively and efficiently to prepare for, prevent, respond to, and recover from domestic incidents, regardless of those incidents' cause, size, or complexity, including acts of terrorism and natural disasters. Along with the basic concepts of flexibility and standardization, the NIMS principles are now taught in every incident management course.

History of the Incident Command System

Prior to the 1970s, each individual fire department had its own method for commanding and managing incidents. Often, the organization established to direct operations at an incident scene depended on the style of the chief on duty. Perhaps not surprisingly, this individualized approach did not work well when units from different districts or mutual aid companies responded to a major incident. What was adequate for routine incidents became ineffective and confusing with large-scale incidents, rapidly changing situations, and units that did not normally work together.

Such a fragmented approach to managing emergency incidents is no longer considered acceptable. Over the past 30 years, formal ICS organizations (or incident management systems) have been developed and refined. Today's ICS structures comprise an organized system of roles, responsibilities, and standard operating procedures (SOPs) that are widely used to manage and direct emergency operations. As a result, the same basic approaches, organizational structures, and terminology are used by thousands of fire departments and emergency response agencies across the United States.

The move to develop a standard system began approximately 40 years ago, after several large-scale wildland fires in southern California proved disastrous for both the fire service and residents of the region. A number of fire-related agencies at the local, state, and federal levels decided that better organization was necessary to effectively combat these costly fires. Collectively, these agencies established an organization known as FIRESCOPE (**FI**re **RES**ources of **C**alifornia **O**rganized for **P**otential **E**mergencies) to develop solutions for a variety of problems associated with large, complex emergency incidents during wildland fires. These problems related to the following issues:

- Command and control procedures
- Resource management
- Terminology
- Communications

FIRESCOPE developed the first standard ICS. Originally, ICS was intended only for large, multijurisdictional or multiagency incidents involving more than 25 resources or operating units. It proved so successful, however, that it was applied to structural firefighting and eventually became an accepted system for managing all emergency incidents FIGURE 5-1.

At the same time, the fire-ground command (FGC) system was developed and adopted by many fire departments. The concepts underlying the FGC system were similar to those that formed the foundation of the ICS, albeit with some differences in terminology and organizational structure. The FGC system was initially designed for day-to-day fire department incidents involving fewer than 25 fire suppression companies, but it could be expanded to meet the needs of larger-scale incidents.

During the 1980s, the ICS developed by FIRESCOPE was adopted by all federal and most state wildland firefighting agencies. The National Fire Academy (NFA) also used the FIRESCOPE ICS as the model fire service ICS for all of its courses. Additional federal agencies, including the Federal Emergency Management Agency (FEMA) and the Federal Bureau of Investigation, adopted the same model for use during major disasters or terrorist events. All federal agencies could now learn and use the same basic system.

Several federal regulations and consensus standards—including NFPA 1500, *Standard on Fire Department Occupational Safety and Health Program*—were adopted during the 1980s. These standards mandated the use of an ICS at emergency incidents. NFPA 1561, *Standard on a Fire Department Incident Management System*, which was issued in 1990, identified the key components of an effective system and described the importance of using such a system at all emergency incidents. Fire departments could use either ICS or FGC to meet the requirements of NFPA 1561.

In the following years, users of different systems from across the United States formed the National Fire Service Incident Management System Consortium to develop "model procedure guides" for implementing effective incident management systems at various types of incidents. The resulting system, which blended the best aspects of both ICS and FGC, is now known formally as ICS—the incident command system. An ICS can be used at any type or size of emergency incident and by any type or size of department or agency. Reflecting this change, NFPA 1561 is now called *Standard on Emergency Services Incident Management System*. NFPA 1026, *Standard for Incident Management Personnel Professional Qualifications*, was developed to define the various job performance requirements for each of the positions classified within the NIMS model.

FIGURE 5-1 ICS was first developed to coordinate efforts during large-scale wildland fires.

Characteristics of the Incident Command System

An ICS provides a standard, professional, organized approach to managing emergency incidents. The use of an ICS enables a fire department and any other type of emergency response agency to operate both more safely and more effectively. A standardized approach facilitates and coordinates the use of resources from multiple agencies, ensuring that they work toward common objectives. It also eliminates the need to develop a special approach for each situation.

Effective management of incidents requires an organizational structure to provide both a hierarchy of authority and responsibility and formal channels for communications. Under the ICS, the specific responsibilities and authority of everyone in the organization are clearly stated, and all relationships are well defined. Important characteristics of an ICS include the following:

- Recognized jurisdictional authority and responsibility
- Applicable to all risk and hazard situations
- Applicable to both day-to-day operations and major incidents
- Unity of command
- Span of control
- Modular organization
- Common terminology
- Integrated communications
- Consolidated incident action plans
- Designated incident facilities
- Resource management

Jurisdictional Authority

Jurisdictional authority is usually not a problem at an incident with a single focus or threat such as a structure fire, where the highest-ranking or designated officer from the local fire department assumes incident command. Matters can become more complicated, however, when several jurisdictions are involved or when multiple agencies within a single jurisdiction have authority for various aspects of the incident. For example, suppose a military aircraft crashes in a national park and starts a wildland fire that spreads across the park's boundaries into an adjoining state. There, the fire causes a hazardous material to leak into a navigable waterway and approaches a county jail located inside the limits of an incorporated city. Such a situation involves both military and civilian agencies as well as multiple levels of government. Chaos would ensue if each affected jurisdiction claimed to be in charge of this incident and attempted to issue orders to the other jurisdictions.

When responsibilities of responders overlap, the ICS may employ a unified command (UC). This structure brings representatives of different agencies together to work on one plan and ensures that all actions are fully coordinated. "Command," although the chosen term of the ICS system, is perhaps misleading. It is important to remember that incidents are *managed*; personnel are *commanded*. Incident command (IC), whether conducted by an individual or through UC, is a management and leadership position. This role is responsible for setting strategic objectives, maintaining a comprehensive understanding of the impact of an incident, and identifying the strategies required to manage that impact effectively. The Command function is structured in one of two ways: single or unified.

Single command is the most traditional perception of the Command function and is the genesis of the term *incident commander*. When an incident occurs within a single jurisdiction, and when there is no jurisdictional or functional agency overlap, a single incident commander should be identified and designated as having overall incident management responsibility by the appropriate jurisdictional authority. This does not mean that other agencies do not respond or do not have a role in supporting the management of the incident.

Single command is best used when a single discipline in a single jurisdiction is responsible for the strategic objectives associated with managing the incident. Single command is also appropriate in the later stages of an incident that was initially managed through UC. Over time, as many incidents become stabilized, the strategic objectives become increasingly focused on a single jurisdiction or discipline. In this situation, it is appropriate to transition from unified to single command.

It is also acceptable, if all agencies and jurisdictions agree, to designate a single incident commander in multiagency and multijurisdictional incidents. In this situation, however, Command personnel should be carefully chosen. The incident commander is responsible for developing the strategic incident objectives on which the incident action plans (IAPs) will be based. IAPs are oral or written plans containing general objectives reflecting the overall strategy for managing an incident. The incident commander is responsible for the IAP and all requests pertaining to the ordering and releasing of incident resources.

The introduction of NIMS sparked tremendous discussion and debate related to the broader emphasis given to the concept of unified command. UC is a critical evolution of the ICS system that recognizes an important reality of incident management: A multiagency or multijurisdictional response to an incident is a routine occurrence. UC provides a framework that allows agencies with different legal, geographic, and functional responsibilities to coordinate, plan, and interact effectively. UC, through a consensus-based approach, addresses the challenges that were encountered nationwide when multiple organizations responding to the same incident established separate, but concurrent, incident management structures. This approach, which is still routine in many jurisdictions, leads to inefficiency and duplication of effort that results in ineffective incident management, on-scene conflict, and substantial safety issues. UC is crafted specially to address these issues.

The concept of UC represents a clear departure from the traditional view of incident command. Unfortunately, it is frequently misunderstood and difficult for some organizations to implement. To be effective in implementing UC, agencies and individuals need to worry less about the structural change and more about the changes required in organizational culture and interagency relationships. As incidents become increasingly complex, the likelihood of any one individual having the knowledge and expertise required to effectively develop, implement, and evaluate strategic objectives for a major incident decreases rapidly.

The lesson of UC is to be concerned less with who is in charge and more with what is required to safely and effectively manage the incident. For UC to be effective, the management of the incident depends on a collaborative process to establish incident objectives and designate priorities that accommodate those objectives. This approach must yield a single IAP that clearly defines the various agency and jurisdictional responsibilities for incident management. UC also yields a single, integrated set of incident objectives. Moreover, the use of UC reinforces those other aspects of ICS that make it effective for managing incidents. For example, because Command is provided using the UC model, the incident is managed using a single organizational structure, thereby limiting duplication and freeing more resources for overall incident management duties. In addition, UC ensures that the incident will be managed from a single incident command post (ICP). Finally, UC allows for a single planning section, which helps prevent multiple, redundant requests for similar items and the inefficient use of limited supplies and resources.

FIRE FIGHTER II Tips **FFII**

Incoming units and personnel need to be patient and give the incident commander enough time to perform a proper and thorough size-up. This step is essential in creating an effective IAP. If all personnel are not patient and do not wait for the IAP, then freelancing could occur. Freelancing can lead to confusion and danger at the scene.

■ All-Risk, All-Hazard System

The ICS has evolved into an all-risk, all-hazard system that can be applied to manage resources at fires, floods, tornadoes, plane crashes, earthquakes, hazardous materials incidents, or any other type of emergency situation FIGURE 5-2. This kind

FIGURE 5-2 ICS can be used to manage different types of emergency incidents involving several agencies.

of system has also been used to manage many nonemergency events, such as large-scale public events, that have similar requirements for command, control, and communication. The flexibility of the ICS enables the management structure to expand as needed, using whatever components are necessary. The operations of multiple agencies and organizations can be integrated smoothly in the management of the incident.

Everyday Applicability

An ICS can and should be used for everyday operations as well as major incidents. Command should be established at every incident, whether it is a trash bin fire, an automobile collision, or a building fire FIGURE 5-3. Regular use of the system ensures familiarity with standard procedures and terminology. It also increases the users' confidence in the system. Frequent use of ICS for routine situations makes it easier to apply to larger incidents.

Unity of Command

Unity of command is a management concept in which each person has only one direct supervisor. All orders and assignments come directly from that supervisor, and all reports are made to the same supervisor. This approach eliminates the confusion that can result when a person receives orders from multiple bosses. Unity of command reduces delays in solving problems as well as the potential for life and property losses. By ensuring that each person has only one supervisor, unity of command can increase overall accountability, prevent freelancing, improve the flow of communication, assist with the coordination of operational issues, and enhance the safety of the entire situation.

An ICS is not necessarily a rank-oriented system. The best-qualified person should be assigned at the appropriate level for each situation, even if that means a lower-ranking individual is temporarily assigned to a higher-level position. This concept is critical for the effective application of the system and must be embraced by all participants. Additionally, a critical and developing component of NIMS will be a national credentialing standard for ICS positions such as Command and Section Chiefs in Operations, Plans, Logistics, and Finance/Administration.

FIGURE 5-3 Because of effective fire prevention measures, fire fighters respond to more car accidents than fires.

Span of Control

Span of control refers to the number of subordinates who report to one supervisor at any level within the organization. Span of control relates to all levels of ICS—from the strategic level, to the operational/tactical level, to the task level (individual companies or crews).

In most situations, one person can effectively supervise only three to seven people or resources. Because of the dynamic nature of emergency incidents, an individual who has command or supervisory responsibilities in an ICS normally should not directly supervise more than five people. The actual span of control should depend on the complexity of the incident and the nature of the work being performed. For example, at a complex incident involving hazardous materials, the span of control might be only three; during less intense operations, the span of control could be as high as seven.

Modular Organization

The ICS is designed to be flexible and modular. The ICS organizational structure—Command, Operations, Planning, Logistics, and Finance/Administration—is predefined, ready to be staffed and made operational as needed. Indeed, an ICS has often been characterized as an organizational toolbox, where only the tools needed for the specific incident are used. In an ICS, these tools consist of position titles, job descriptions, and an organizational structure that defines the relationships between positions. Some positions and functions are used frequently, whereas others are needed only for complex or unusual situations. Any position can be activated simply by assigning someone to it.

For example, a small structure fire can usually be managed by Command, in the form of a person who directly supervises four or five company officers. Each company officer supervises his or her respective company fire fighters. At a larger fire, when more companies respond, the ICS would expand to include divisions of up to five companies. At the same time, officers would be assigned to perform specific functions, such as safety and planning. At even more complex incidents, the additional levels and positions within the ICS structure would be filled in the same manner. The organization can expand as much as needed to manage the incident.

Common Terminology

The ICS promotes the use of common terminology both within an organization and among all agencies involved in emergency incidents. Common terminology means that each word has a single definition, and no two words used in managing an emergency incident have the same definition. Everyone uses the same terms to communicate the same thoughts, so everyone understands what is meant. Each job comes with one set of responsibilities, and everyone knows who is responsible for each duty.

Common terminology is particularly important for radio communications. For example, in many fire departments, the term "tanker" refers to a mobile water supply apparatus. To wildland firefighting agencies, however, "tanker" means a fixed-wing aircraft used to drop an extinguishing agent on a fire. Without common terminology, a request for a "tanker" could result in the Operations Section Chief getting an airplane instead of the truck he or she actually requested from Command.

FIRE FIGHTER II Tips **FFII**

NIMS Guide 0001, National Resource Typing Criteria, is the NIMS document that lists the proper terminology for the common resources used at large-scale emergencies.

Integrated Communications

Integrated communications ensures that everyone at an emergency can communicate with both supervisors and subordinates. The ICS must support communication up and down the chain of command at every level. A message must be able to move efficiently through the system from Command down to the lowest level, and from the lowest level up to the Command **FIGURE 5-4**. Within a fire department, the primary means of communicating at the incident scene is by radio.

FIGURE 5-4 Integrated communications is essential for successful operations during an emergency.

Consolidated Incident Action Plans

An ICS ensures that everyone involved in the incident is following one overall plan. Different components of the organization may perform different functions, but all of their efforts contribute to meeting the same goals and objectives. Everything that occurs is coordinated with everything else. At smaller incidents, Command develops an action plan and communicates the incident priorities, objectives, strategies, and tactics to all of the operating units. Representatives from all participating agencies meet regularly to develop and update this plan. In both large and small incidents, all personnel involved in the incident understand what their specific roles are and how they fit into the overall plan.

Designated Incident Facilities

Designated incident facilities are assigned locations where specific functions are always performed. For example, Command will always be based at the ICP. The staging area, rehabilitation area, casualty collection point, treatment area, base of operations, and helispot are all designated areas where particular functions take place. The facilities required for the specific incident are established according to the specific IAP or a predefined ICS plan.

Resource Management

Resource management entails the use of a standard system of assigning and keeping track of the resources involved in the incident. In structural firefighting, the basic units are companies. Engine companies, ladder companies, and other units are dispatched to the incident and assigned specific functions. In some cases, groups from several different companies are assigned to task forces and strike teams. The Resource Management System of the ICS keeps track of the various company assignments.

At large-scale incidents, units are often dispatched to a staging area, rather than going directly to the incident location. A staging area is a location close to the incident scene where a number of units can be held in reserve, ready to be assigned if needed.

Personnel are the most vital resource the fire service has. Under the ICS, the incident commander uses a personnel accountability system to track each and every member at the scene. The Fire Fighter Survival chapter discusses the various personnel accountability systems in detail.

The ICS Organization

The ICS structure identifies a full range of duties, responsibilities, and functions that are performed at emergency incidents; it also defines the relationships among those components. Some components are used at almost every incident, whereas others apply to only the largest and most complex situations. The five major components of an ICS organization are Command, Operations, Planning, Logistics, and Finance/Administration.

An ICS organization chart may be quite simple or very complex. Each block on an ICS organization chart refers to a function area or a job description. Positions are staffed as they are needed. The only position that must be filled at every incident is Incident Command. Incident Command decides which additional components are needed for the given situation and activates those positions by assigning someone to perform those tasks.

Fire fighters must understand the overall structure of ICS as well as the basic roles and responsibilities of each position within the ICS organization. As an emergency develops, a fire fighter could start in Logistics, move to Operations, and eventually assume a Command position. Knowing how ICS works enables fire fighters to see how different roles and responsibilities work together and relate to one another. This grasp of the "big picture" makes it easier for fire fighters to focus on a particular assignment, instead of being overwhelmed by the incident.

Command

On an ICS organization chart, the first component is Command FIGURE 5-5. Command is the only position in the ICS that must always be filled for every incident, because there must always be someone in charge. Command is established when the first unit arrives on the scene and is maintained until the last unit leaves the scene.

In the ICS structure, Command (either single or unified) is ultimately responsible for managing an incident and has the necessary authority to direct all activities at the incident scene. Command is directly responsible for the following tasks:

- Determining strategy
- Selecting incident tactics
- Setting the action plan
- Developing the ICS organization
- Managing resources
- Coordinating resource activities
- Providing for scene safety
- Releasing information about the incident
- Coordinating with outside agencies

Unified Command

When multiple agencies with overlapping jurisdictions or legal responsibilities respond to the same incident, UC provides several advantages. In this approach, representatives from each agency cooperate to share command authority. They work together and are directly involved in the decision-making process. UC helps ensure cooperation, avoids confusion, and guarantees agreement on goals and objectives FIGURE 5-6.

Incident Command Post

The incident command post (ICP) is the headquarters location for the incident. Command functions are centered in the ICP; thus Command and all direct support staff should always be located at this site. The location of the ICP should be broadcast to all units as soon as it is established.

Relative to the incident scene, the ICP should be in a nearby, protected location. Often, the ICP for a major incident is located in a special vehicle or building. This choice of location enables the command staff to function without needless distractions or interruptions. For large incidents that are geographically spread out, the command post may be some distance from part of the emergency incident.

Command Staff

Individuals on the Command Staff perform functions that report directly to Command and cannot be delegated to work in other major sections of the organization FIGURE 5-7. The safety officer, liaison officer, and public information officer are always part of the Command Staff. In addition, aides, assistants, and

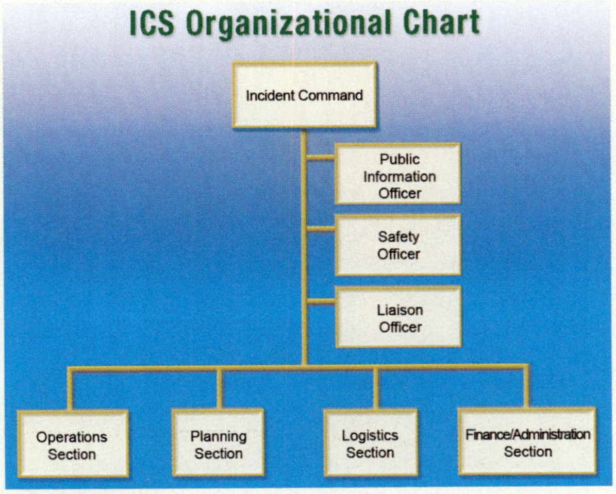

FIGURE 5-5 The ICS organization chart.

FIGURE 5-6 A unified command involves many agencies directly in the decision-making process for a large incident.

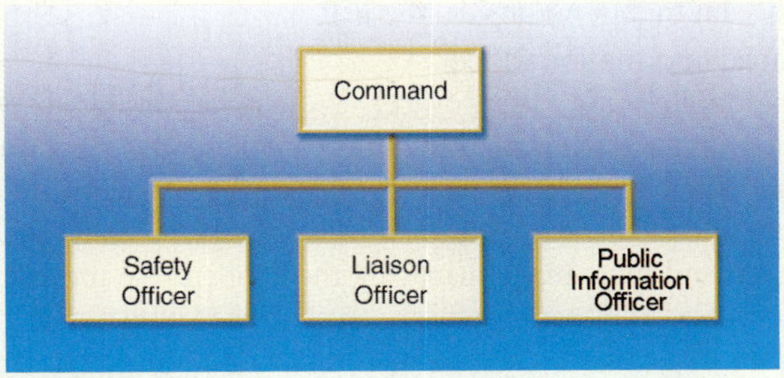

FIGURE 5-7 The Command Staff report directly to Command.

advisors may be assigned to work directly for members of the Command Staff. An aide is a fire fighter (sometimes an officer) who serves as a direct assistant to a member of the Command or general staff.

Safety Officer

The safety officer is responsible for ensuring that safety issues are managed effectively at the incident scene. He or she serves as the eyes and ears of Command in terms of safety—identifying and evaluating hazardous conditions, watching out for unsafe practices, and ensuring that safety procedures are followed. The safety officer is appointed early during an incident. As the incident becomes more complex and the number of resources present at the scene increases, additional qualified personnel can be assigned as assistant safety officers.

Although the safety officer is an advisor to Command, he or she has the authority to stop or suspend operations when unsafe situations occur. This authority is clearly stated in national standards, including NFPA 1500, *Standard on Fire Department Occupational Safety and Health Program*; NFPA 1521, *Standard for Fire Department Safety Officer*; NFPA 1561, *Standard on Emergency Services Incident Management System* and NFPA 1026. Several state and federal regulations require the assignment of a safety officer at incidents involving hazardous materials and at certain technical rescue incidents.

At a fire scene, the safety officer should be an individual who is knowledgeable in fire behavior, building construction and collapse potential, firefighting strategy and tactics, hazardous materials, rescue practices, and departmental safety rules and regulations. He or she should also have considerable experience in incident response and specialized training in occupational safety and health. Many larger fire departments have full-time safety officers who perform administrative functions relating to health and safety when they are not responding to emergency incidents. The NFA's Incident Safety Officer and Advanced Safety Operations and Management courses are excellent resources for people who are interested in serving in this capacity.

Liaison Officer

The liaison officer is Command's representative to a point of contact for representatives from outside agencies. This member of the Command Staff is responsible for exchanging information with representatives from those agencies. During an active incident, Command may not have time to meet directly with everyone who comes to the ICP. The liaison officer functions as the representative of Command under these circumstances, obtaining and providing information, or directing people to the proper location or authority. The liaison area should be adjacent to, but not inside, the ICP.

Public Information Officer

The public information officer (PIO) is responsible for gathering and releasing incident information to the news media and other appropriate agencies FIGURE 5-8. At a major incident, the public will want to know what is being done. Because Command must make managing the incident the top priority, the public information officer serves as the contact person for media requests, which frees up Command to concentrate on the incident. A media headquarters should be established near—but not in—the ICP. The information presented to the media by the public information officer needs to be approved by the IC.

FIGURE 5-8 The public information officer (PIO) is responsible for gathering and releasing incident information to the media and other appropriate agencies.

General Staff Functions

Command has the overall responsibility for the entire incident command organization, although some elements of Command's responsibilities can be handled by the Command Staff. When the incident is too large or too complex for just one person to manage effectively, Command may appoint someone to oversee parts of the operation. Everything that occurs at an emergency incident can be divided among the major functional components within ICS FIGURE 5-9:

- Operations
- Planning
- Logistics
- Finance/Administration

The staff of these four sections are known as the ICS general staff. Command decides which (if any) of these four positions need to be activated, when to activate them, and who should be placed in each position. (Recall that the blocks on the ICS organizational chart refer to functional areas or job descriptions, not to positions that must always be staffed.)

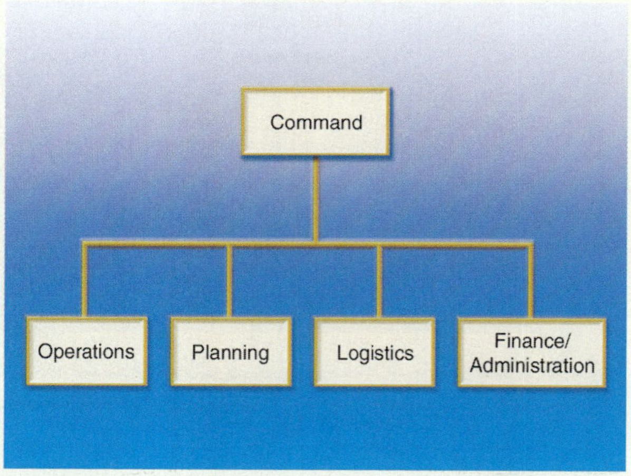

FIGURE 5-9 Major functional components of ICS.

The four Section Chiefs on the ICS general staff, when they are assigned, may run their operations from the main ICP, although this structure is not required. At a large incident, the four functional organizations may operate from different locations, but will always be in direct contact with Command.

Operations

The Operations Section is responsible for taking direct action to control the incident **FIGURE 5-10**. The Operations Section fights the fire, rescues any trapped individuals, treats any injured victims, and does whatever else is necessary to alleviate the emergency situation.

For most structure fires, Command directly supervises the functions of the Operations Section. At complex incidents, a separate Operations Section Chief takes on this responsibility so that Command can focus on overall strategy while the Operations Section Chief focuses on the tactics that are required to get the job done.

Operations are conducted in accordance with an IAP that outlines what the strategic objectives are and how emergency operations will be conducted. At most incidents, the IAP is relatively simple and can be expressed in a few words or phrases. For a large-scale incident, however, it can be a lengthy document that is regularly updated and used for daily briefings of the Command Staff.

Planning

The Planning Section is responsible for the collection, evaluation, dissemination, and use of information relevant to the incident **FIGURE 5-11**. The Planning Section works with pre-incident plans, building construction drawings, maps, aerial photographs, diagrams, reference materials, and status boards. It is also responsible for developing and updating the IAP. The Planning Section develops what needs to be done by whom and identifies which resources are needed.

Command activates the Planning Section when information needs to be obtained, managed, and analyzed. The Planning Section Chief reports directly to Command. Individuals assigned to Planning examine the current situation, review available information, predict the probable course of events, and prepare recommendations for strategies and tactics. The Planning Section also keeps track of resources at large-scale incidents and provides Command with regular situation and resource status reports.

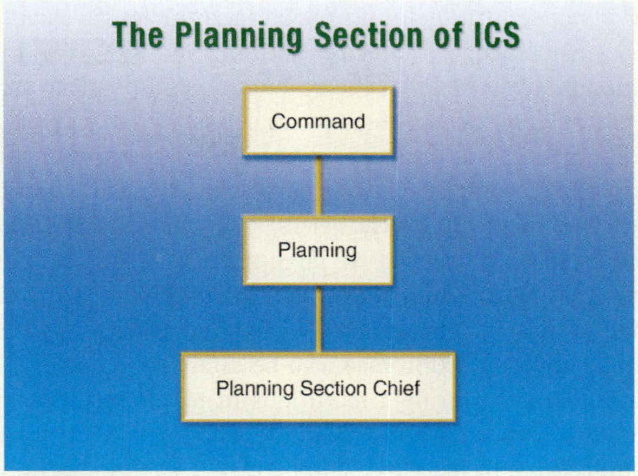

FIGURE 5-11 The organizational structure of the Planning Section.

Logistics

The Logistics Section is responsible for providing supplies, services, facilities, and materials during the incident **FIGURE 5-12**. The Logistics Section Chief reports directly to Command and serves as the supply officer for the incident. Among the responsibilities of this section are keeping apparatus fueled, providing food and refreshments for fire fighters, obtaining the foam concentrate needed to fight a flammable liquids fire, and arranging for specialized equipment to remove a large pile of debris.

In many fire departments, these logistical functions are routinely performed by support services personnel. These groups work in the background to ensure that the members of the Operations Section have whatever they need to get the job done. Resource-intensive or long-duration situations may require assignment of a Logistics Section Chief, however, because service and support requirements may be so complex or so extensive that they need their own management component.

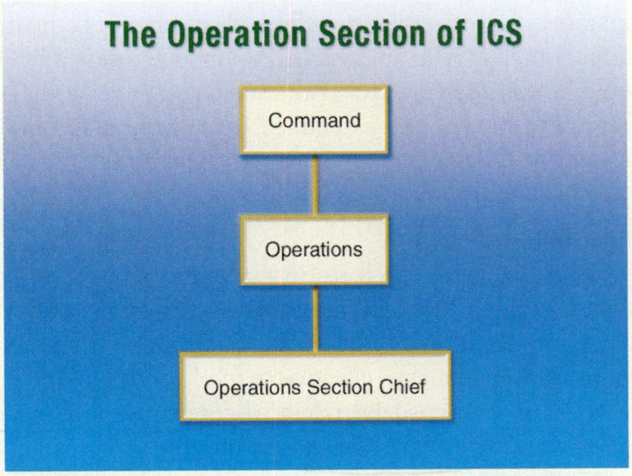

FIGURE 5-10 The organizational structure of the Operations Section.

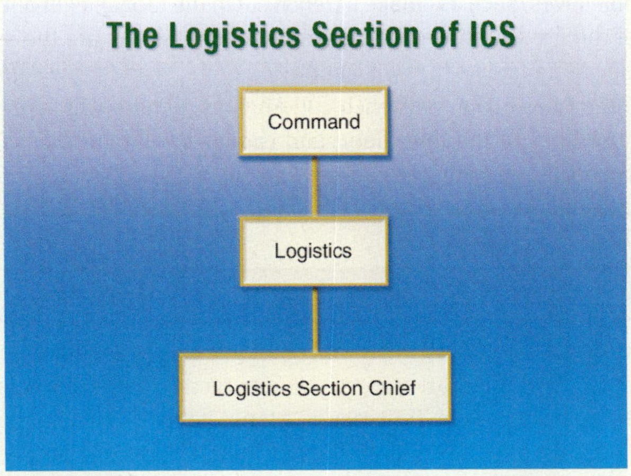

FIGURE 5-12 The organizational structure of the Logistics Section.

Finance/Administration

The Finance/Administration Section is the fourth major ICS component managed directly by Command FIGURE 5-13. This section is responsible for the accounting and financial aspects of an incident, as well as any legal issues that may arise in its aftermath. This function is not staffed at most incidents, because cost and accounting issues are typically addressed after the incident. Nevertheless, a Finance/Administration Section may be needed at large-scale and long-term incidents that require immediate fiscal management, particularly when outside resources must be procured quickly. A Finance/Administration Section may also be established during a natural disaster or during a hazardous materials incident where reimbursement may come from the shipper, carrier, chemical manufacturer, or insurance company.

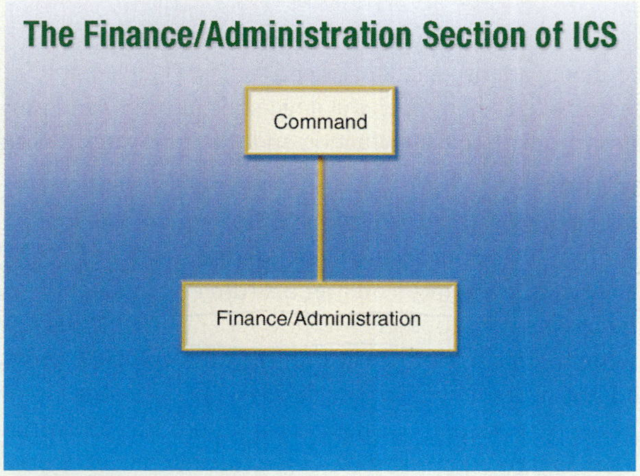

FIGURE 5-13 The organizational structure of the Finance/Administration Section.

FIRE FIGHTER II Tips **FFII**

Strategy is the "big picture" plan of what has to be done: "Stop the fire from extending into Exposure D" is a strategic objective. *Tactics* are the steps that are taken to achieve the strategic objectives, such as "Make a trench cut in the roof and get ahead of the fire with hose lines on the top floor." *Tasks* are the specific assignments that will get the job done: "Ladder 1 will go to the roof to make the trench cut; Engines 3 and 5 will take the hose lines to the third floor; and Ladder 4 will pull ceilings to provide access into the cockloft" are all tasks.

Standard ICS Concepts and Terminology

Fire departments and other emergency response agencies implement ICS in an effort to organize emergency scene activities in a standard manner. Emergency scenes tend to be chaotic, so organizing operations at an incident often poses a serious challenge, particularly if the agencies involved use different terms to describe certain concepts and resources. As mentioned earlier, one of the strengths of ICS is its use of standard terminology. In ICS, specific terms apply to various parts of an incident organization. Understanding these basic concepts and terminology is the first step in understanding the system.

Some fire departments may use slightly different terminology. For this reason, fire fighters must know both the terminology used by their department and the standard ICS terminology. This section defines important ICS terms and examines their use in organizing and managing an incident.

Single Resources and Crews

A single resource is an individual vehicle and its assigned personnel FIGURE 5-14. For example, an engine and its crew would be a single resource; a ladder company would be a second single resource. A *company officer* is the individual in charge of a company. A company operates as a work unit, with all crew members working under the supervision of the company officer. Companies are assigned to perform tasks such as search and rescue, attacking the fire with a hose line, forcible entry, and ventilation.

FIGURE 5-14 A single resource consists of an individual company and the personnel that arrive on that unit.

A crew is a group of fire fighters who are working without apparatus. For example, members of an engine company or ladder company who are assigned to operate inside a building would be considered a crew. Additional personnel at the scene of an incident who are assembled to perform a specific task may also be called a crew. A crew must have an assigned leader or company officer.

Divisions and Groups

Organizational units such as divisions and groups are established to aggregate single resources and/or crews under one supervisor. The primary reason for establishing divisions and groups is to maintain an effective span of control.

- A division usually refers to companies and/or crews working in the same geographic area.
- A group usually refers to companies and/or crews working on the same task or function, although not necessarily in the same location.

The flexibility of the ICS enables organizational units to be created as needed, depending on the size and scope of the incident. In the early stages of an incident, individual companies are often assigned to work in different areas or perform different tasks. As the incident grows and more companies are assigned to it, Command can establish divisions and groups. An individual is assigned to supervise each division or group as it is created.

These organizational units are particularly useful when several resources are working near one another, such as on the same floor inside a building, or are doing similar tasks, such as ventilation. The assigned supervisor can directly observe and coordinate the actions of several crews.

A division includes those resources responsible for operations within a defined geographic area. This area could be a floor inside a building, the rear of the fire building, or a section of a wildland/brush fire. Divisions are most often employed during routine fire department emergency operations. By assigning all of the units in one area to one supervisor, Command is better able to coordinate their activities and ensure that all resources are working to further a similar strategy.

The division structure provides effective coordination of the tactics being employed by different companies working in the same area. For example, the division supervisor would coordinate the actions of a crew that is advancing a hose line into the area, a crew that is conducting search and rescue operations in the same area, and a crew that is performing horizontal ventilation FIGURE 5-15.

An alternative way of organizing resources is by function (rather than by location). A group includes those resources assigned to a specific function, such as ventilation, search and rescue, or water supply. Groups are responsible for performing an assignment, wherever it may be required, and often work across division lines. For example, the officer assigned to supervise the ventilation group would use the radio designation "Ventilation Group."

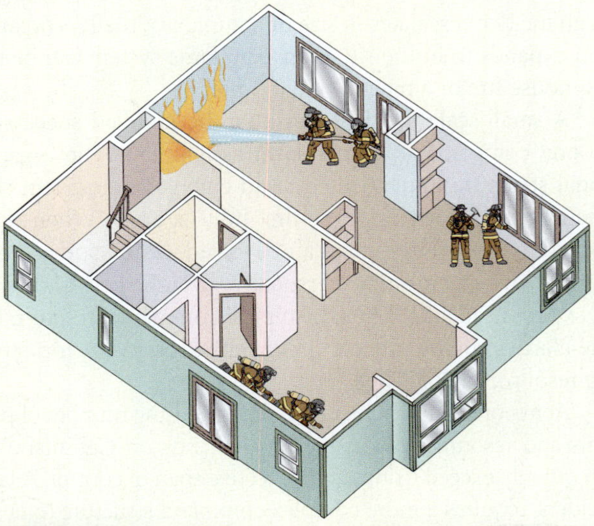

FIGURE 5-15 A division refers to companies or crews working in the same geographic area.

Division and group supervisors have the same rank within ICS. Divisions do not report to groups, and groups do not report to divisions. Instead, the supervisors are required to work together—that is, to coordinate their actions and activities with one another. For example, a group supervisor must coordinate with the division supervisor when the group enters the division's geographic area, particularly if the group's assignment will affect the division's personnel, operations, or safety. The division supervisor, in turn, must be aware of everything that is happening within that area. Effective communication between divisions and groups is critical during emergency operations.

Branches

A branch is a higher level of combined resources than divisions and groups. At a major incident, several different activities may occur in separate geographic areas or involve distinct functions. The span of control might still be a problem, even after the establishment of divisions and groups. In these situations, Command can establish branches to place a higher-level supervisor (a branch director) in charge of a number of divisions or groups.

Location Designators

The ICS uses a standard system to identify the different parts of a building or a fire scene. Every fire fighter must be familiar with this terminology.

The exterior sides of a building are generally known as Sides A, B, C, and D. The front of the building is Side A, with Sides B, C, and D following in a clockwise direction around the building. The companies working in front of the building are assigned to Division A, and the radio designation for their supervisor is "Division A." Similar terminology is used for companies working at the sides and rear of the building.

The areas adjacent to a burning building are called exposures. Exposures take the same letter as the adjacent side of the building. A fire fighter facing Side A can see the adjacent building on the left (Exposure B), and the building to the right (Exposure D). If the burning building is part of a row of buildings, the buildings to the left are called Exposures B, B1, B2, and so on. The buildings to the right are Exposures D, D1, D2, and so on FIGURE 5-16.

Within a building, divisions commonly take the number of the floor on which they are working. For example, fire fighters working on the fifth floor would be in Division 5, and the radio designation for the chief assigned to that area would be "Division 5." Crews doing different tasks on the fifth floor would all be part of this division.

Task Forces and Strike Teams

Task forces and strike teams are groups of single resources that have been assigned to work together for a specific purpose or for a certain period of time. Grouping resources reduces the span of control by placing several units under a single supervisor.

A task force includes two to five single resources, such as different types of units assembled to accomplish a specific task. For example, a task force may be composed of two engines and

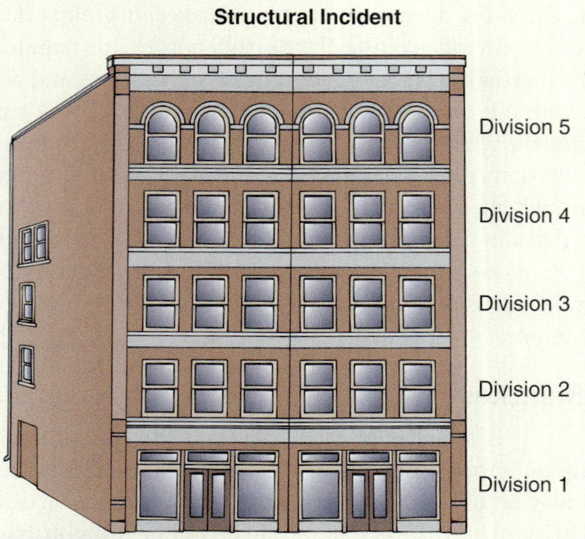

FIGURE 5-16 Location designators in ICS.

one truck company, two engines and two brush units, or one rescue company and four ambulances. A task force operates under the supervision of a task force leader. All communications for the separate units in the task force are directed to the task force leader.

Task forces are often part of a fire department's standard dispatch philosophy. In the Los Angeles City Fire Department, for example, a task force consists of two engines and one ladder truck, staffed by a total of 10 fire fighters. Some departments create task forces consisting of one engine and one brush unit for responses during wildland fire season. The brush unit responds with the engine company wherever it goes.

A strike team is five units of the same type with an assigned leader. A strike team could be five engines (engine strike team), five trucks (truck strike team), or five ambulances (EMS strike team) FIGURE 5-17. A strike team operates under the supervision of a strike team leader.

FIGURE 5-17 A strike team is five units of the same type under one leader.

Strike teams are commonly used to combat wildland fires—incidents to which dozens or hundreds of companies may respond. During wildland fire seasons, many departments establish strike teams of five engine companies that will be dispatched and work together on major wildland fires FIGURE 5-18. The assigned companies rendezvous at a designated location and then respond to the scene together. Each engine has an officer and fire fighters, but only one officer is designated as the strike team leader. All communications for the strike team are directed to the strike team leader.

EMS strike teams, which consist of five ambulances and a supervisor, are often organized to respond to multiple-casualty incidents or disasters. Rather than requesting 15 ambulances and establishing an organizational structure to supervise 15 single resources, Command can request three EMS strike teams and can coordinate with three strike team leaders.

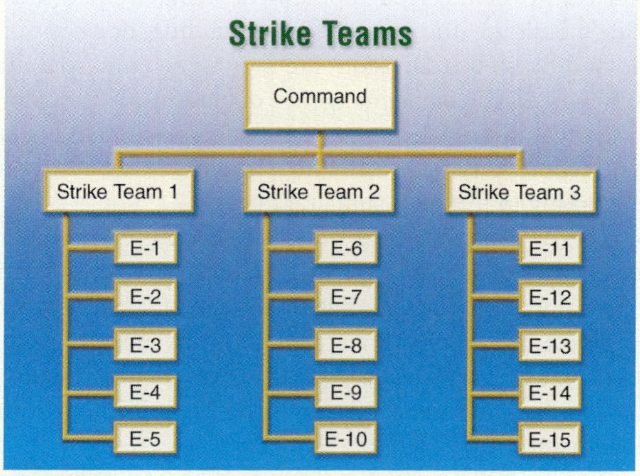

FIGURE 5-18 The organization of a strike team.

Implementing ICS

The ICS helps to organize every incident scene in a standard, consistent manner. Fire departments develop SOPs, and then train and practice using ICS to ensure consistency at emergency incidents. This approach increases safety and efficiency. As an incident escalates in size or complexity, the ICS organization expands to fit the situation. The same system can be used at a house fire or a major wildland fire.

A small-scale incident can often be handled successfully by one company or a first-alarm assignment. The organizational structure implemented at an emergency incident starts small, with the arrival of the first unit or units. When only a limited number of resources are involved, Command can often manage the organization personally or with the assistance of an experienced aide. Thus the typical command structure for a one-alarm structure fire consists of only Command and reporting resources FIGURE 5-19.

At a more complex incident, the increasing number of problems and resources places greater demands on Command and can quickly exceed Command's effective span of control. A larger incident requires a more complex command structure to ensure that no details are overlooked or personnel safety compromised.

The modular design of ICS allows the organization to expand, based on the needs of the incident, by activating predetermined components. Command can delegate specific

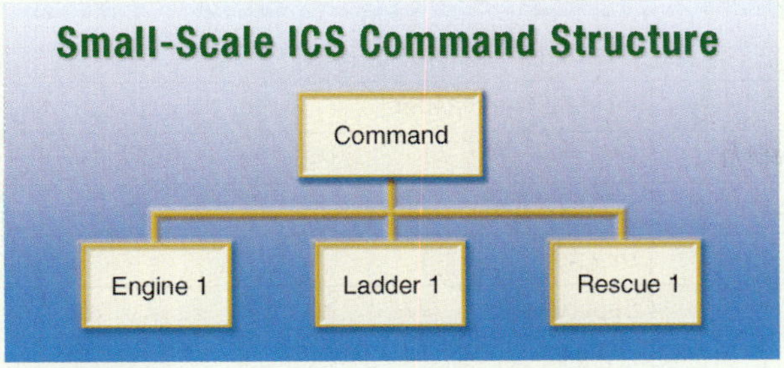

FIGURE 5-19 At a small-scale incident, the typical ICS command structure may consist of only Command and reporting resources.

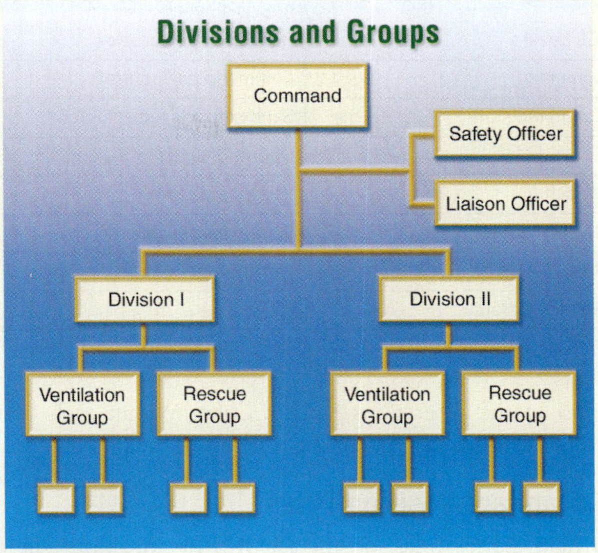

FIGURE 5-20 Divisions and groups are organized to manage the span of control and to supervise and coordinate units working together.

responsibilities and authority to other personnel, thereby creating an effective incident organization. An individual who receives an assignment knows the basic responsibilities of the job, because they are defined in advance.

The most frequently used ICS components in structural firefighting are divisions and groups FIGURE 5-20. These components place several single resources under one supervisor, effectively reducing Command's span of control. Command can also assign individuals to special jobs, such as safety officer and liaison officer, to establish a more effective organization for the incident.

In the largest and most complex incidents, other ICS components can be activated. Command can then delegate responsibility for managing major components of the operation, such as operations or logistics, to general staff positions. Each of these positions has a wide range of responsibilities related to a major component of the organization.

If necessary, branches can be established, creating an additional level within the organization FIGURE 5-21. For example, a structural collapse during a major fire may result in several trapped victims and many victims who need medical treatment and transportation. In such a situation, the Operations Section Chief would have multiple responsibilities that could exceed his or her span of control. Activating a Fire Suppression Branch, a Rescue Branch, and a Casualty Branch would address this problem. One officer would be responsible for each branch and

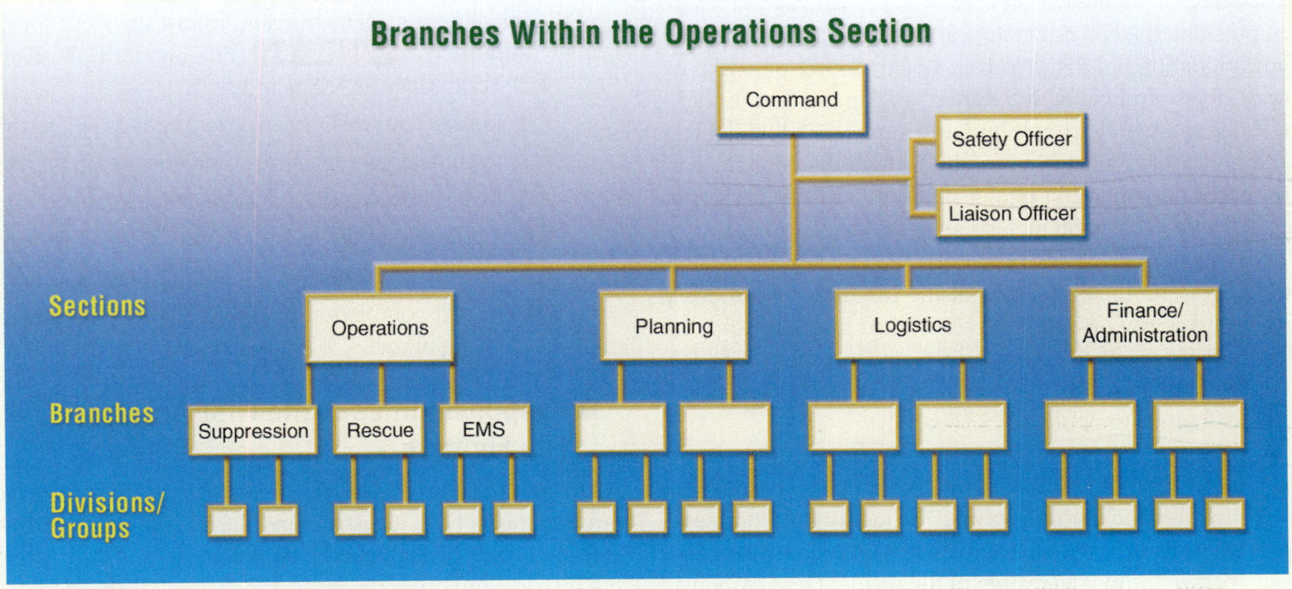

FIGURE 5-21 Creating branches within the Operations Section is one way to manage the span of control during a large incident.

TABLE 5-1 Levels of an ICS Organization

ICS Level	ICS Function/Location	Position Designator
Command	Command and control	Incident Commander
Command staff	Safety, liaison, information	Officer
General staff	Operations, Planning, Logistics, Finance/Administration	Section Chief
Branch	Varies (e.g., EMS)	Director
Division/group	Varies (e.g., Division A)	Supervisor
Unit/crew/strike	Varies (e.g., Rehab)	Leader
Team/task force		(Company officer)

report to the Operations Section Chief. Within each branch, several divisions or groups would report to the branch director.

Standard Position Titles

To help clarify roles within the ICS organization, standard position titles are used for personnel within the organization. Each level of the organization has a different designator for the individual in charge. The position title typically includes the functional or geographic area of responsibility, followed by a specific designator TABLE 5-1.

Division and group supervisors are usually chief officers, but company officers may be assigned to these positions as well. Typically, the company officer from the first unit assigned to a division or group will serve as the supervisor until a chief officer is available. A chief officer with a strong background in operations is usually assigned to the position of Operations Section Chief. The other Section Chief positions (Planning, Logistics, and Finance/Administration) are staffed as needed by Command. These individuals should understand their roles and be able to handle the responsibilities associated with these positions.

Working Within the Incident Command System

To an outsider, the ICS might appear to be a large, complicated organizational model that involves a complex set of SOPs. By contrast, to the individual fire fighter working within this system, ICS is really quite simple and uncomplicated. Fire fighters should understand what ICS is and how it works. More importantly, each individual fire fighter must understand his or her place and role in the system.

This section discusses how every emergency incident is conducted. Certain components of the ICS are used on every incident and at every training exercise. In addition, fire fighters generally have specific responsibilities and procedures to follow in most situations.

Three basic components always apply:

- Command is established at every incident and is maintained from the time that the first unit arrives until the time that the last unit leaves. The identity of Command may change, but there is always one single function (person) who is in charge of the incident and responsible for everything that happens. SOPs may dictate who will be Command at any time.
- Each fire fighter always reports to one supervisor. A fire fighter's supervisor will usually be a company officer. The company officer directly supervises a small group of fire fighters, such as an engine company or ladder company, who work together. At an incident scene, the company officer provides instructions and must always know where each fire fighter is and what he or she is doing. If the company officer assigns two fire fighters to work together away from the rest of the company, both fire fighters remain under the supervision of the company officer. The company officer could be an acting officer (a fire fighter temporarily designated as a "fill-in" officer), or a fire fighter could be assigned to work temporarily under the supervision of a different officer.
- The company officer reports to Command. If only one company is present at the scene, the company officer is Command, at least until someone else arrives and assumes that role. At a small incident, the company officer may report directly to Command. At a large incident, several layers of supervision may separate a company officer and Command.

To make the ICS work, every fire fighter must know how to function within this system. To fulfill your role as a member of the incident management team, follow the steps for operating within the ICS SKILL DRILL 5-1 (Fire Fighter II, NFPA 6.1.2):

1. Verify that the ICS is in use.
2. Repeat your assignment over the radio to verify it.
3. Complete personal size-up to ensure safety.
4. Account for yourself and for other team members.
5. Update your supervising officer regularly.
6. Provide personnel accountability reports as necessary.
7. Report completion of each assignment to your supervising officer.

Responsibilities of the First-Arriving Fire Fighters

The first fire fighters to arrive at an emergency incident are the foundation of the ICS structure. Rarely is a high-ranking chief sent to a fire to evaluate the situation, design the organization structure, and order the resources that will be needed. Instead, the ICS builds its organization from the bottom up, around the units that take initial action.

The officer in charge of the first-arriving unit is responsible for taking initial action and becomes Command, with all of the authority that accompanies that position. This officer is responsible for managing the operation until he or she is relieved by a senior officer. If there is no officer on the first-arriving unit, the fire fighter with the greatest seniority is in charge until an officer arrives and assumes command. The position of Command must have an unbroken line of succession from the moment the first unit arrives on the scene until the incident is considered finished.

Establishing Command

The ICS establishes that the Command function is initiated when the first unit arrives and is transferred as required for the duration of the incident. The individual who is in charge of the first-arriving unit automatically establishes command of the incident until a superior takes over that role.

The individual who initially establishes command must formally announce that fact over the radio. This announcement eliminates any possible confusion over who is in charge. The initial report should include the following information SKILL DRILL 5-2 (Fire Fighter II, NFPA 6.1.2):

1. A size-up report.
2. Command designation.
3. The unit or individual who is assuming command.
4. An initial situation report.
5. The initial action being taken.

For example, the first-arriving unit at a single-family dwelling fire might deliver the following radio report:

> Engine 4 is on the scene at 309 Central Avenue, with smoke showing from Side A of a single-family dwelling. Engine 4 is laying a supply line and will make an offensive attack through the front door. Engine 4 is assuming Central Avenue Command.

Usually, the first-arriving unit at an incident scene is an engine company or some other unit that takes direct action. Regardless of the type of emergency or the type of unit, the first officer on the scene is in charge. This officer must assess the situation, determine incident priorities, and provide direction for his or her own crew as well as any other units that are arriving. Any units coming in behind the first unit know that they will be taking their orders from Command until a higher-level officer assumes command.

The officer who initially assumes command must decide whether to take action and directly supervise the initial attack crew or whether to concentrate on managing the incident as Command. This decision depends on many factors, including the nature of the situation, the resources that are on the scene or expected to arrive quickly, and the ability of the crew to work safely without direct supervision.

If the incident is large and complicated, the best option for Command is to establish an ICP and focus on sizing up the situation, directing incoming units, and requesting additional resources. Command's own crew can be assigned to work with an acting officer or to join forces with another company officer and crew. If the situation is less critical, Command might be able to function as both a company officer and Command simultaneously, at least temporarily.

If the first-arriving unit is a chief officer, the chief officer automatically assumes command and begins executing Command responsibilities. If a company officer had previously assumed command, the company officer would transfer Command authority to the first-arriving chief officer. The officer relinquishing command should provide an assessment of the situation to the incoming personnel, as described in the "Transfer of Command" section later in this chapter.

Most fire departments have written procedures that specify who will assume command in certain situations. If multiple units respond from the same station and arrive together, there should be an established protocol for which officer assumes command. If a company officer arrives only seconds ahead of a chief officer, the chief officer should assume command from the outset.

Confirming the Command

The initial announcement of Command definitely confirms that Command has been established at an incident. If no one announces that he or she is in command, all responders should realize that Command has not been established. The announcement of Command also reinforces Command's personal commitment to the position through a conscious personal act and a standard organizational act.

Identifying the Incident

Individual fire department procedures may vary in terms of the specific protocol they use for naming an incident. The first officer to assume command should establish an identity that clearly identifies the location of the incident, such as "Engine 10 will be Seventh Avenue Command." This announcement reduces confusion on the radio and establishes a continuous identity for Command, regardless of who holds that position during the incident.

There can be only one "Seventh Avenue Command" at a time, so there is no confusion about who is in charge of the incident. When anyone needs to talk to Command, a call to "Seventh Avenue Command" should be answered by the individual who is in command.

Transfer of Command

Transfer of command occurs whenever one person relinquishes command of an incident and another individual becomes Command. For example, the first-arriving company officer would transfer command to the first-arriving chief officer, who would later transfer command to a higher-ranking officer during a major incident. Some departments require transfer of command when a higher-ranking officer arrives at an incident; others give the higher-ranking officer the option of assuming command or leaving the existing personnel in the Command assignment.

When a higher-ranking officer arrives at the scene of an emergency incident and takes charge, that officer assumes the moral and legal responsibility for managing the overall operation. Established procedures must be followed whenever command is transferred. One of the most important requirements

Near Miss REPORT

Report Number: 08-0000111

Synopsis: Poor communications create opposing handlines.

Event Description: My department responded to a report of a garage fire in a large subdivision in our fire district. The structure was a three-stall garage attached to a roughly 4000-square-foot (1219-meter) house. We responded in an engine with a 1250-gpm (4700 lpm), 1000-gallon (3800-liter) tank. There were five people on board. The ranks of the five were as follows: Assistant Chief, who was the driver (career); Chief, who was riding in the passenger seat (career); 1 fire fighter (volunteer); a captain (career); and me, a firefighter (career).

On arrival, we found the garage well involved, as well as two vehicles inside the garage. We stretched a 1¾-inch (45 mm) line to the "D" side to go into the garage. An attempt to knock down the fire was made with little success. The next entry was made by me and the captain through the front door on the "A" side. We encountered heavy smoke and heat. We eventually made our way to the door that went into the garage, which was left open by the occupants upon discovering the fire. At this time positive pressure ventilation (PPV) was in place and conditions improved. The captain and I went into the garage and began attacking the fire. At the same time someone sprayed from an outside line and in return pushed heat and smoke onto us inside the garage. The captain immediately got on the radio and told Command to shut down that line and conditions deteriorated rapidly. As a result we had to pull back into the house and regroup.

Lessons Learned: It is vital to have a good and competent incident commander with a strong command presence. In this case, the incident commander did not have any accountability system set in place, and several fire fighters were freelancing. Incident command must keep in constant communication with all fire fighters who are inside a burning structure, and in return they must keep incident command updated on their status.

of command transfer is the accurate and complete exchange of incident information FIGURE 5-22. The officer who is relinquishing command needs to give incoming Command a current situation status report that includes the following information:

- Tactical priorities
- Action plans
- Hazardous or potentially hazardous conditions
- Accomplishments
- Assessment of effectiveness of operations
- Current status of resources:
  - Assigned or working
  - Available for assignment
  - En route
  - Additional resource requirements

If transfer of command occurs very early during an incident, the transfer of information may be brief. For example, the first-arriving company officer might have been in command for only a few minutes and have little information to report when command transfers to the first-arriving chief officer. The chief may have heard all of the exchanges over the radio and know what the current situation is. Conversely, if the company officer has been Command for several minutes, there may be a significant amount of information to report, such as the current assignments of all first-alarm companies. Whether the information is minimal or substantial, the transfer must be accurate and complete.

FIGURE 5-22 The incoming Command needs to be briefed before assuming this role.

VOICES OF EXPERIENCE

I was acting as a lieutenant when we received a call about a possible structure fire: smoke was visible and residents were evacuating. While en-route my mind was racing, thinking of the decisions the lieutenant needs to make. Our engine was getting closer when I saw smoke blowing from the roof line of the house. I told the crew to prepare for an attack upon arrival.

We arrived on scene to find smoke coming from the roof line and the chimney. The front door was open and people were gathered on the front driveway. I gave my radio report and initiated a fast attack. I quickly questioned the occupants about any possible victims and possible locations of the fire. They stated that the fire was in the rear corner of the house and that all occupants were out.

I proceeded inside with my crew. The smoke was dark and low. We made our way toward the rear corner of the structure. My nozzle man found a door, tried to advance into it, and became stuck momentarily. I backed up and moved into an opening and that was when I felt the heat. It was quite considerable, but not too uncomfortable. I directed my nozzle man and another crew member to follow me into this opening. It was dark and hot. I heard on the radio that positive pressure ventilation was ready at the door and that a fire fighter was positioned at the rear corner for window removal. I got down on my hands and knees to see any fire in front of us. I moved my interior crew into position and radioed to take the window and start the positive pressure ventilation fan. We directed the stream onto the fire. The attack in coordination with the ventilation cleared the room of heat and smoke quickly.

It was later found that a mattress was completely burnt and most contents in the room had been consumed by the fire. But we made certain that the fire never advanced into the rest of the structure. Teamwork and communication is essential—especially when a new or acting position is filled.

Anthony Gianantonio
Palm Bay Fire Rescue
Palm Bay, Florida

Each fire department establishes specific procedures for transferring command. In some cases, the transfer of command may be done via radio. The most effective method, however, is face-to-face communication. Standard command worksheets and a status board are valuable tools for tracking and transferring information at an ICP.

Command is always maintained for the entire duration of an incident. In the later stages of an incident, after the situation is under control, it might be transferred to a lower-level commander. A downward transfer of command requires the same type of briefing and exchange of information as an upward command transfer. The officer in charge of the last company remaining on the scene would be Command. When that company leaves the scene, the Command function is terminated.

Command Transfer Rationale

Command may be transferred at different points during an emergency incident for several important reasons. A first-arriving company officer can usually direct the initial operations of two or three additional companies, but as situations become more complex, the problems of maintaining control increase rapidly. A company officer's primary responsibility is to supervise one crew and ensure that the crew members operate safely. When three or more companies are operating at an incident, it is better to have a chief officer assume command.

As more companies are assigned to an incident, the Command structure must also expand. That is, the organization must grow to maintain an effective span of control. Additional chief officers may be assigned to the incident, and a higher-ranking officer may assume the Command role. A command transfer may be required if the situation is beyond the training and experience of the current Command. A more experienced officer may have to assume command to ensure proper management of the incident.

All fire fighters should know the steps involved in transferring command during an emergency incident SKILL DRILL 5-3 (Fire Fighter II, NFPA 6.1.2):

1. Assume command.
2. Follow established procedures for transferring command.
3. Transfer command in a face-to-face meeting if possible. If not possible, transfer command over the radio.
4. Communicate to the incoming Command personnel the tactical priorities, action plans, hazardous conditions, effectiveness of operations, status of resources, and need for additional resources.
5. Formally announce the transfer of command over the radio.

FIRE FIGHTER II Tips **FFII**

The goal of the National Incident Management System (NIMS) is to train all emergency responders to be able to operate within this national-scale plan. To make it easier for responders all over the United States to access this material and to complete this training, a variety of course offerings are available through the U.S. Department of Homeland Security, FEMA's Web site, and Jones & Bartlett Learning. The introductory course, ICS 100: Introduction to the Incident Command System, introduces the ICS and provides the foundation for higher levels of training in incident management. IS-700: Introduction to NIMS is also required to meet the requirements of the presidential directive. Check with your training officer to determine which level of incident management training you need as a beginning fire fighter.

Wrap-Up

Chief Concepts

- An incident command system (ICS) provides a standard, professional, and organized approach to managing emergency incidents. This standard approach provides common objectives and coordination of resources from multiple agencies.
- Several characteristics are critical to an ICS:
  - Organized approach: An ICS imposes "order on chaos" on the fire ground and enables a safer and more efficient operation than would be possible if personnel and units worked independently of one another.
  - Terminology: The ICS uses a standard terminology for effective communications.
  - All-risk: The ICS can be used at any type of emergency incident.
  - Jurisdictional authority: The ICS enables different jurisdictions, agencies, and organizations to work cooperatively on a single incident.
  - Span of control: The ICS maintains the desired span of control through flexible levels of organization.

- Everyday applicability: The ICS can, and should, be used at every single incident, every single time.
- Modular: The ICS is based on standard modules that are activated as needed to manage an incident.
- Integrated communications: Everyone responding to the incident can communicate up and down the chain of command as needed.
- Incident action plan: Every incident has a plan that outlines the strategic objectives. Large incidents will have formal plans.
- Designated incident facilities: Standardized facilities—such as an incident command post, staging area, and rehabilitation area—are established as needed.
- Resource management: A standard system is used for assigning and keeping track of incident resources.

- Five major functions are part of ICS:
  - Command: Responsible for the entire incident. This is the only function that is always staffed.
  - Operations: Responsible for most fire-ground functions, including suppression, search and rescue, and ventilation.
  - Planning: Responsible for developing the incident action plan.
  - Logistics: Responsible for obtaining the resources needed to support the incident.
  - Finance/Administration: Responsible for tracking expenditures and managing the administrative functions at the incident.
- The command staff assists Command at the incident:
  - The safety officer is responsible for the overall safety of the incident. He or she has the authority to stop any action or operation if it creates a safety hazard on the scene.
  - The liaison officer is responsible for coordinating operations between the fire department and other agencies that may be involved in the incident.
  - The public information officer is responsible for coordinating media activities and providing the necessary information to the various media organizations.
- An example of a single resource would be an engine company or a ladder company.
- Single resources can be combined into task forces or strike teams.
- Other organizational units that can be established under ICS include divisions, groups, branches, task forces, and strike teams.
- The ICS can be expanded infinitely to accommodate any size of incident. Branches can be established to aggregate similar functions, such as suppression, EMS, or hazardous materials.
- Three basic components always apply when working within the ICS:
  - At every incident, Command must always be established.
  - Each fire fighter always reports to one supervisor.
  - The company officer reports to Command.
- As the incident grows or continues, it may be necessary to transfer command to another officer. Transfer of command must be done in a seamless manner to ensure continuity of command.

Hot Terms

Branch The organizational level having functional, geographical, or jurisdictional responsibility for major aspects of incident operations. (NFPA 1026)

Branch director A person in a supervisory level position in either the operations or logistics function to provide a span of control. (NFPA 1561)

Command The first component of the ICS. It is the only position in the ICS that must always be staffed.

Command Staff The public information officer, safety officer, and liaison officer, all of whom report directly to the incident commander and are responsible for functions in the incident management system that are not a part of the function of the line organization. (NFPA 1561)

Crew A team of two or more fire fighters. (NFPA 1500)

Designated incident facilities Assigned locations where specific functions are always performed.

Division A supervisory level established to divide an incident into geographic areas of operations. (NFPA 1561)

Division supervisor A person in a supervisory-level position who is responsible for a specific geographic area of operations at an incident. (NFPA 1561)

Finance/Administration Section Section responsible for all costs and financial actions of the incident or planned event, including the time unit, procurement unit, compensation/claims unit, and the cost unit. (NFPA 1026)

Fire-ground command (FGC) An incident management system developed in the 1970s for day-to-day fire department incidents (generally handled with fewer than 25 units or companies).

FIRESCOPE (Fire Resources of California Organized for Potential Emergencies) An organization of agencies established in the early 1970s to develop a standardized system for managing fire resources at large-scale incidents such as wildland fires.

Group A supervisory level established to divide an incident into functional areas of operation. (NFPA 1561)

Group supervisor A person in a supervisory-level position who is responsible for a functional area of operation. (NFPA 1561)

ICS general staff The chiefs of each of the four major sections of ICS: Operations, Planning, Logistics, and Finance/Administration.

Incident action plan (IAP) The objectives reflecting the overall incident strategy, tactics, risk management, and member safety that are developed by the incident commander. Incident action plans are updated throughout the incident. (NFPA 1500)

Incident command post (ICP) The field location at which the primary tactical-level, on-scene incident command functions are performed. (NFPA 1026)

Incident command system (ICS) The combination of facilities, equipment, personnel, procedures, and communications operating within a common organizational structure that has responsibility for the management of assigned resources to effectively accomplish stated objectives pertaining to an incident or training exercise. (NFPA 1670)

Integrated communications The ability of all appropriate personnel at the emergency scene to communicate with their supervisor and their subordinates.

Liaison officer A member of the Command Staff who serves as the point of contact for assisting or coordinating agencies. (NFPA 1026)

Logistics Section Section responsible for providing facilities, services, and materials for the incident or planned event, including the communications unit, medical unit, and food unit within the service branch and the supply unit, facilities unit, and ground support unit within the support branch. (NFPA 1026)

Logistics Section Chief The general staff position responsible for directing the logistics function. It is generally assigned on complex, resource-intensive, or long-duration incidents.

National Incident Management System (NIMS) A system mandated by Homeland Security Presidential Directive 5 (HSPD-5) that provides a systematic, proactive approach guiding government agencies at all levels, the private sector, and nongovernmental organizations to work seamlessly to prepare for, prevent, respond to, recover from, and mitigate the effects of incidents, regardless of cause, size, location, or complexity, so as to reduce the loss of life or property and harm to the environment. (NFPA 1026)

Operations Section Section responsible for all tactical operations at the incident or planned event, including up to 5 branches, 25 divisions/groups, and 125 single resources, task forces, or strike teams. (NFPA 1026)

Operations Section Chief The general staff position responsible for managing all operations activities. It is usually assigned when complex incidents involve more than 20 single resources or when Command cannot be involved in the details of tactical operations.

Planning Section Section responsible for the collection, evaluation, dissemination, and use of information related to the incident situation, resource status, and incident forecast. (NFPA 1026)

Planning Section Chief The general staff position responsible for planning functions. It is assigned when Command needs assistance in managing information.

Public information officer (PIO) A member of the Command Staff who is responsible for interacting with the public and media or with other agencies with incident-related information requirements. (NFPA 1026)

Resource management Under NIMS, includes mutual-aid agreements; the use of special federal, state, local, and tribal teams; and resource mobilization protocols. (NFPA 1026)

Safety officer A member of the Command Staff who is responsible for monitoring and assessing safety hazards and unsafe situations, and for developing measures for ensuring personnel safety. (NFPA 1561)

Single command A command structure in which a single individual is responsible for all of the strategic objectives of the incident. It is typically used when an incident is within a single jurisdiction and is managed by a single discipline.

Single resource An individual, a piece of equipment and its personnel, or a crew or team of individuals with an identified supervisor that can be used on an incident or planned event. (NFPA 1026)

Staging area A prearranged, strategically placed area, where support response personnel, vehicles, and other equipment can be held in an organized state of readiness for use during an emergency. (NFPA 424)

Strike team Specified combinations of the same kind and type of resources, with common communications and a leader. (NFPA 1026)

Strike team leader The person in charge of a strike team. This individual is responsible to the next higher level in the incident organization and serves as the point of contact for the strike team within the organization.

Task force Any combination of single resources assembled for a particular tactical need, with common communications and a leader. (NFPA 1051)

Task force leader The person in charge of a task force. This individual is responsible to the next higher level in the incident organization and serves as the point of contact for the task force within the organization.

Transfer of command The formal procedure for transferring the duties of an incident commander at an incident scene. (NFPA 1026)

Unified command An application of the incident command system that allows all agencies with jurisdictional responsibility for an incident or planned event, either geographical or functional, to manage an incident or planned event by establishing a common set of incident objectives and strategies. (NFPA 1561)

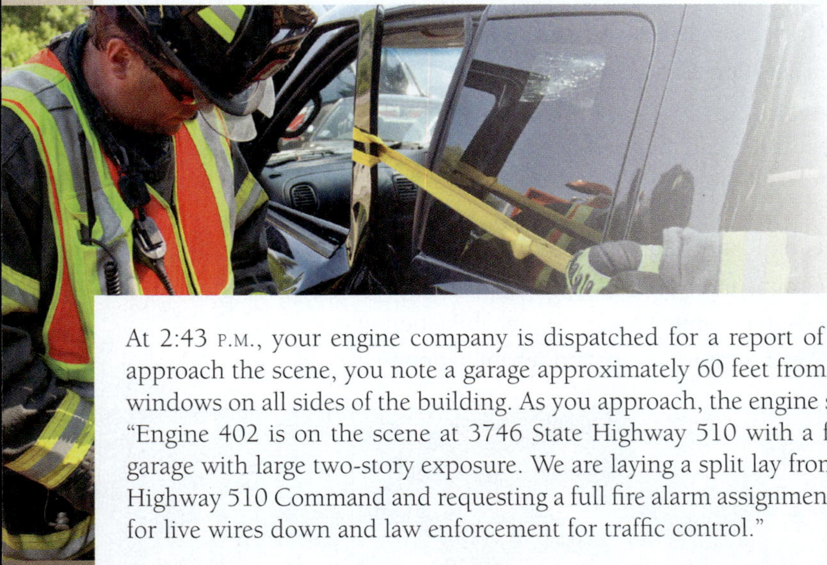

FIRE FIGHTER *in action*

At 2:43 P.M., your engine company is dispatched for a report of smoke in the 3700 block of State Highway 510. As you approach the scene, you note a garage approximately 60 feet from a large two-story farmhouse that has fire coming from the windows on all sides of the building. As you approach, the engine stops on the road and you hear your captain's initial report. "Engine 402 is on the scene at 3746 State Highway 510 with a fully involved 40 by 60 feet (12 by 18 meters), one-story garage with large two-story exposure. We are laying a split lay from the road to protect exposures. Engine 402 is establishing Highway 510 Command and requesting a full fire alarm assignment, and a strike team of water tankers. Request Lewis Electric for live wires down and law enforcement for traffic control."

1. What is the best command structure for handling this incident?
   - **A.** An incident command system using all major functional components
   - **B.** A unified command
   - **C.** A single command
   - **D.** No incident command is needed for a single-building fire
2. When is it necessary to establish Command at a fire or EMS scene?
   - **A.** Only when a chief officer is dispatched
   - **B.** At all incidents
   - **C.** Whenever more than one agency is involved
   - **D.** When the incident requires more resources than the initial response
3. Which phase of the incident command system was the captain's report?
   - **A.** Transfer of command
   - **B.** Termination of command
   - **C.** Establishing command
   - **D.** Confirming command

FIRE FIGHTER *in action* *(continued)*

4. Which of the following ICS components must be implemented for this emergency?
   1. Operations
   2. Planning
   3. Logistics
   4. Finance/Administration
   5. Command
      - **A.** 1, 2, 3, 4, and 5
      - **B.** 1, 2, and 5
      - **C.** 2, 3, and 5
      - **D.** 5
5. What is the benefit of requesting a strike team of water tankers rather than five individual water tankers?
   - **A.** A strike team responds under the direction of a team leader.
   - **B.** All water tankers from the strike team will come from a single jurisdiction.
   - **C.** Individual water tankers are not able to communicate with each other.
   - **D.** A strike team will provide its own rehabilitation unit.
6. The full alarm includes two additional engines, two truck companies, a rescue company, two chief officers, and a rehabilitation unit, in addition to the strike team of water tankers. Which of the following would be an appropriate organizational structure?
   - **A.** IC, Safety, Division 3, Division B, Water Supply Group, and rehabilitation unit
   - **B.** IC, Safety, Division 3, Division 1, Water Supply Group, and rehabilitation unit
   - **C.** IC, Safety, Exposure Division, Fire Attack Group, Water Supply Group, and rehabilitation unit
   - **D.** IC, Safety, Exposure Group, Fire Attack Group, Water Supply Group, and rehabilitation unit

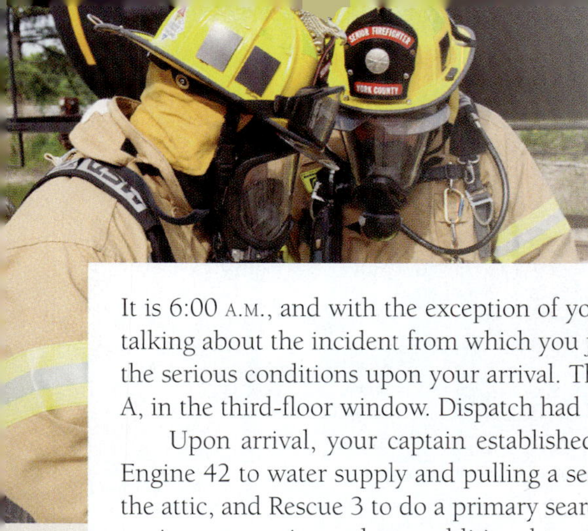

FIRE FIGHTER II *in action*

It is 6:00 A.M., and with the exception of your captain, you and the rest of the crew of Engine 413 are sitting on the tailboard talking about the incident from which you just returned. Everyone felt good about having just made a really good stop, given the serious conditions upon your arrival. The incident involved a three-story apartment complex with fire showing from Side A, in the third-floor window. Dispatch had reported people trapped in the adjoining units.

Upon arrival, your captain established Command and assigned you to confine and attack the fire. He then assigned Engine 42 to water supply and pulling a second attack line, Truck 16 to set up ventilation and then check for extension into the attic, and Rescue 3 to do a primary search of the third-floor units. A second alarm was requested, bringing two additional engine companies and two additional truck companies. The two engines were assigned to exposure protection, one truck company was assigned to the rapid intervention crew (RIC), and the other truck was assigned to salvage.

The discussion eventually turned to complaints by the crew that they could have used the help of the captain rather than having him stand in the front yard barking orders. Another chimed in about the waste of crew members being assigned to RIC rather than doing "some real work."

1. You know the importance of having a strong command structure, particularly in serious fires such as this incident. How do you explain this point to the other crew members?
2. Leadership often requires someone to stand up for what is right, even when it is not popular. How do you get the buy-in for the use of RIC when others say, "We are paid to risk our lives"?
3. The ICS delegates authority and responsibility down the organizational structure. How do you motivate subordinates to ensure the overall goal is achieved?
4. Why is a maximum span of control of five to one generally recommended when on the fire ground?

Fire Behavior

Fire Fighter I

Knowledge Objectives

After studying this chapter, you will be able to:

- Describe the chemistry of fire. (p 142–143)
- List the three states of matter. (NFPA5.3.10.A, p 143)
- List the five forms of energy. (p 143–144)
- Explain the concept of the fire triangle. (NFPA5.3.11.A, p 144)
- Explain the concept of the fire tetrahedron. (NFPA5.3.11.A, p 144)
- Describe the chemistry of combustion. (NFPA5.3.11.A, p 144–145)
- Describe the by-products of combustion. (NFPA5.3.11.A, p 145)
- Explain how fires are spread by conduction, convection, and radiation. (NFPA5.3.12.A, p 146–147)
- Define flow path and describe how it influences the growth of a building fire. (p 147)
- Describe the four methods of extinguishing fires. (p 148)
- Define Class A, B, C, D, and K fires. (p 148–149)
- Describe the importance of the following characteristics in solid-fuel fires: composition of fuel, amount of fuel, and configuration of fuel. (NFPA5.3.11.A, p 149–150)
- Describe the four stages of fire development: incipient stage, growth stage, fully developed stage, and decay stage. (NFPA5.3.11.A, p 151–154)
- Define the following terms: thermal layering, rollover, backdraft, flashover, rapid fire growth, and ventilation-limited fires. (p 152–153)
- Describe the conditions that cause thermal layering. (NFPA5.3.12.A, p 152)
- Describe the conditions that lead to rollover. (NFPA5.3.12.A, p 152)
- Describe the conditions that lead to a backdraft. (NFPA 5.3.11.A, p 153)
- Describe the conditions that lead to flashover. (NFPA5.3.12.A, p 152)
- Describe the conditions that lead to rapid fire growth. (p 153)
- Describe the conditions that lead to a ventilation-limited fire. (p 153–154)
- Describe how fire behaves in modern structures. (p 155)
- Describe how the wind effect impacts fire behavior. (p 156)
- Describe the characteristics of liquid-fuel fires. (p 156)
- Define the characteristics of gas-fuel fires. (p 157)
- Explain the concept of vapor density. (p 157)
- Explain the concept of flammability limits. (p 157)
- Describe the causes and effects of a boiling liquid/expanding vapor explosion (BLEVE). (p 157)
- Describe the process of reading smoke. (p 157–160)

Skills Objectives

There are no skill objectives for Fire Fighter I candidates. NFPA 1001 contains no Fire Fighter I Job Performance Requirements for this chapter.

CHAPTER 6

Fire Fighter II — FFII

Knowledge Objectives

There are no knowledge objectives for Fire Fighter II candidates. NFPA 1001 contains no Fire Fighter II Job Performance Requirements for this chapter.

Skills Objectives

There are no skill objectives for Fire Fighter II candidates. NFPA 1001 contains no Fire Fighter II Job Performance Requirements for this chapter.

You Are the Fire Fighter

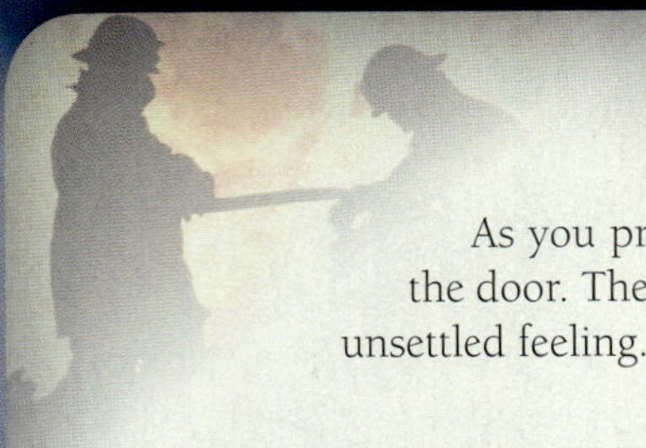

It is 3:46 am and you arrive at a house with a fire reported by a neighbor. You pull an 1¾″ (4 centimeter) hoseline and advance it to the front door. Even though there is little visible smoke, you smell the familiar scent of fire. You notice the windows are stained black with soot. As you prepare to force entry, you notice light brown smoke pushing through between the threshold and the door. There is a crack across the living room window. The hair on the back of your neck rises as you get an unsettled feeling.

1. Why would this situation make you uneasy?
2. What type of conditions may be present in this structure?
3. What type of event may be likely to occur?

Introduction

This chapter describes the behavior of fire.

Since the days when someone discovered how to use fire for cooking food and keeping warm, fire has fueled our lives. Although we have progressed from having open fires for cooking and for warmth, we continue to depend on fires to fuel the combustion of our society. Most of our electric power is created by burning fossil fuels. Our modern gasoline- and diesel-powered vehicles depend on small explosions (internal combustion) to propel them. Most of our homes are heated by open flames in well-designed furnaces.

Unfortunately, destruction of lives and property by uncontrolled fires has also been occurring since ancient times. Despite our advanced technology for detecting and combating fires, we continue to wage an ongoing battle with fire. The United States has a higher mortality rate from fires than many other developed countries that keep fire statistics. According to the National Fire Protection Association (NFPA), in the United States in 2014, there were 3,275 civilian fire deaths and 15,775 civilian fire injuries. Public fire departments responded to 1,298,000 fires. These fires resulted in a property loss of 11.6 billion dollars. Of these 1,298,000 fires, 494,000, or 38 percent, were structure fires. Of the total number of structure fires, 386,000, or 78.2 percent, were residential structure fires.

This chapter explores the relationship between various types of fuels and the combustion process. It describes the four stages of fire growth. It emphasizes recent research studies that have demonstrated how fire growth is influenced by changes in ventilation and how it can be controlled by limiting ventilation or the application of water. It describes how the composition of the fuel, the amount of fuel, and the configuration of the fuel influence the combustion process with solid fuels.

FIRE FIGHTER Tips

When studying fire behavior, some basic units of measure must be used. In the United States, most people use the British system of units. Every other country uses the International System of Units, also known as the SI or metric system. Whichever system is used, it is important to use it consistently.

In addition, the following terms are defined: thermal layering, rollover, backdraft, flashover, rapid fire growth, ventilation-limited fires, fuel-limited fires, and smoke explosions. This chapter describes how to assess the volume, velocity, density, and color of smoke. By relating the key attributes of smoke to the size of the building, weather conditions, and the rate of change of the smoke, it becomes possible to determine the fire's location and its stage of development. This chapter also describes the characteristics of liquid fuel fires.

This book uses the British system and provides metric conversion factors or equivalent values as necessary. In the British system, distance is measured in feet and inches, liquid volume is measured in gallons, temperature is measured in degrees Fahrenheit, and pressure is measured in pounds per square inch. In the metric system, distance is measured in meters, liquid volume is measured in liters, temperature is measured in degrees Celsius, and pressure is measured in pascals or kilopascals.

The Chemistry of Fire

To be a safe and effective fire fighter, you need to understand the conditions needed for a fire to ignite and grow. Understanding these conditions will increase your effectiveness as a fire fighter. A well-trained and experienced fire fighter can put out more fire with less water because of his or her understanding of the chemistry of fire.

■ What Is Fire?

Fire is a rapid chemical process that produces heat and usually light. Everyone can identify fire when they see it, but few can explain the process that is taking place to produce it. Fire is characterized by the production of a flame, which can be seen in many different colors, depending on what is burning and how much heat is being produced. Fire requires fuel that is in the form of combustible vapors in order for the chemical process of combustion to occur. This energy can be stored in a solid or liquid form. For a fuel to burn, it must first be changed into a flammable vapor. Materials that remain in a solid or liquid state will not burn while in that state. Solid fuels such as wood, liquid fuels such as gasoline, and gaseous fuels such as propane will all burn after they have changed into a vapor.

States of Matter

Matter is made up of atoms and molecules. Matter exists in three states: solid, liquid, and gas FIGURE 6-1. An understanding of these states of matter is helpful in understanding fire behavior.

FIGURE 6-1 Matter can be classified as solids, liquids, or gases.

We all know what makes an object solid: It has a definite shape. Most uncontrolled fires are fed by solid fuels. In structure fires, the building and most of the contents are solids. A solid has a definite size and shape. Some solids, such as wax or plastic, may change to a liquid or a gaseous state when heated. Cold makes most solids more brittle, whereas heat makes them more flexible. Because a solid is rigid, only a limited number of molecules are present on its surface; the majority of molecules are cushioned or insulated by the outer surface of the solid.

A liquid will assume the shape of the container in which it is placed. Most liquids expand when heated and will turn into gases when sufficiently heated. Liquids, for all practical purposes, cannot be compressed. This characteristic allows fire fighters to pump water for long distances through pipelines or hoses. A liquid has no independent shape, but it does have a definite volume. The liquid with which fire fighters are most concerned is water.

A gas is a type of fluid that has neither independent shape nor independent volume, but rather tends to expand indefinitely. The gas we most commonly encounter is air, the mixture of invisible odorless, tasteless gases that surrounds the earth. The mixture of gases in air maintains a constant composition—21 percent oxygen, 78 percent nitrogen, and 1 percent argon and trace amounts of other gases such as carbon dioxide, neon, methane, helium, krypton, hydrogen, and xenon. Oxygen is required for us to live, but it is also required for fires to burn. We will explore the reaction of fuels with oxygen as we look at the chemistry of burning.

Fuels

Fuels are materials that store energy. Think of the vast amount of heat that is released during a large fire. The energy released in the form of heat and light has been stored in the fuel before it is burned. The release of the energy in a gallon of gasoline, for example, can move a car miles down the road FIGURE 6-2. Many common household materials, such as laptop computers, furniture, and carpets, consist of fuels that will burn under the right conditions.

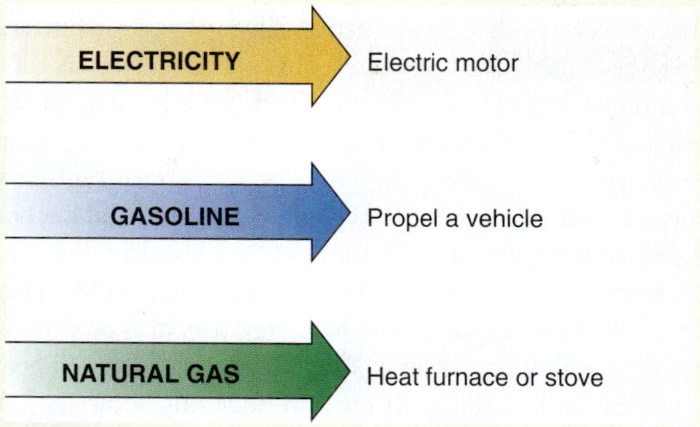

FIGURE 6-2 Energy being converted to work.

Types of Energy

Energy exists in the following forms: chemical, mechanical, electrical, light, and nuclear. Regardless of the form in which the energy is stored, it can be changed from one form to another. For example, electrical energy can be converted to heat or to light. Similarly, mechanical energy can be converted to electrical energy through a generator.

Chemical Energy

Chemical energy is the energy created by a chemical reaction. Some chemical reactions produce heat (exothermic); others absorb heat (endothermic). The combustion process is an exothermic reaction, because it releases heat energy. Most chemical reactions occur because bonds are established between two substances or bonds are broken as two substances are chemically separated. Heat is also produced whenever oxygen combines with a combustible material. If the reaction occurs slowly in a well-ventilated area, the heat is released harmlessly into the air. If the reaction occurs very rapidly or within an enclosed space, the mixture can be heated to its ignition temperature and can begin to burn. Fire is an example of energy being released as a result of a chemical reaction. An example of this occurs when a bundle of rags soaked with linseed oil releases enough heat through oxidization, causing the rags to ignite spontaneously.

Mechanical Energy

Mechanical energy is converted to heat when two materials rub against each other and create friction. For example, a fan belt rubbing against a seized pulley produces heat. Heat is also produced when mechanical energy is used to compress air in a compressor. Water falling over a dam is another example of mechanical energy.

Electrical Energy

Electrical energy is converted to heat energy in several different ways. For example, electricity produces heat when it flows through a wire or any other conductive material. The greater the flow of electricity and the greater the resistance of the material, the greater the amount of heat produced. Examples of electrical energy that can produce enough heat to start a fire include electric heating elements, overloaded wires, electrical arcs, and lightning. Electrical energy is carried through the electrical wires inside homes and can be stored in batteries that convert chemical energy to electrical energy.

Light Energy

Light energy is caused by electromagnetic waves packaged in discrete bundles called photons. This energy travels as thermal radiation, a form of heat. When light energy is hot enough, it can sometimes be seen in the form of visible light. If it is of a frequency that we cannot see, the energy may be felt as heat but not seen as visible light. One example of light energy is the radiant energy we receive from the sun. We think of candles, fires, light bulbs, and lasers as common forms of light energy. We should recognize that, while these do produce light energy, they also produce heat. They emit both heat and light. They transfer most of their heat via convection and radiation. If they are touching something, they would be using conduction to transfer heat as well.

Nuclear Energy

Nuclear energy is created by splitting the nucleus of an atom into two smaller nuclei (nuclear fission) or by combining two small nuclei into one large nucleus (fusion). Nuclear reactions release large amounts of energy in the form of heat. These reactions can be controlled, as in a nuclear power plant, or uncontrolled, as in an atomic bomb explosion. In a nuclear power plant, the nuclear reaction releases carefully controlled amounts of heat, which then convert water to steam so as to power a steam turbine generator. Both uncontrolled explosions and controlled reactions release radioactive material, which can cause injury or death. Nuclear energy is stored in radioactive materials and converted to electricity by nuclear power-generating stations.

Conservation of Energy

The law of conservation of energy states that energy cannot be created or destroyed by ordinary means. Energy can, however, be converted from one form to another. Think of an automobile. Chemical energy in the gasoline is converted to mechanical energy when the car moves down the road. If you apply the brakes to stop the car, the mechanical energy is converted to heat energy by the friction of the brakes. Similarly, in a house fire, the stored chemical energy in the wood of the house is converted into heat and light energy during the fire.

Conditions Needed for Fire

To understand the behavior of fire, you need to consider the three basic elements needed for a combustion to occur: fuel, oxygen, and heat. First, a combustible fuel must be present. Second, oxygen must be available in sufficient quantities. Third, a source of ignition (heat) must be present. If we graphically place these three components together, the result is the fire triangle FIGURE 6-3.

A fourth factor, a chemical chain reaction, must result from the first three elements to produce and maintain a self-sustaining fire. That is, the fuel, oxygen, and heat must interact in such a way as to create a self-sustaining chemical chain reaction to keep the combustion process going.

One way of visualizing this process is to show the chemical chain reaction joining the elements of the fuel, oxygen, and heat. This depiction reflects the central role that the chemical reaction plays in maintaining the process of flaming combustion. This relationship is sometimes characterized as the fire tetrahedron FIGURE 6-4. A tetrahedron is a four-sided, three-dimensional figure. Each side of the fire tetrahedron represents one of the four elements needed for a fire to occur.

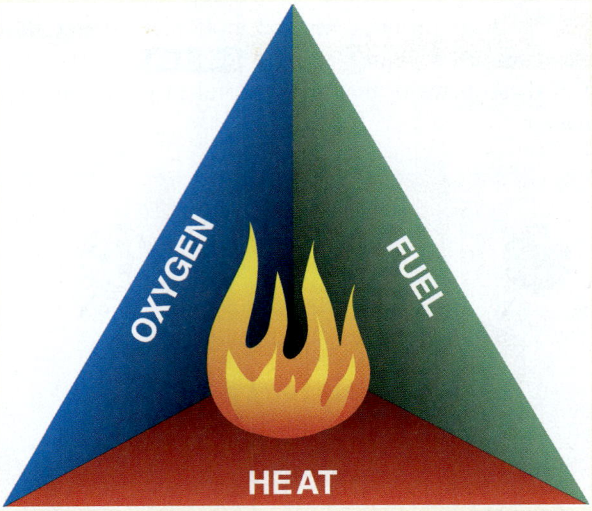

FIGURE 6-3 The fire triangle consists of fuel, oxygen, and heat.

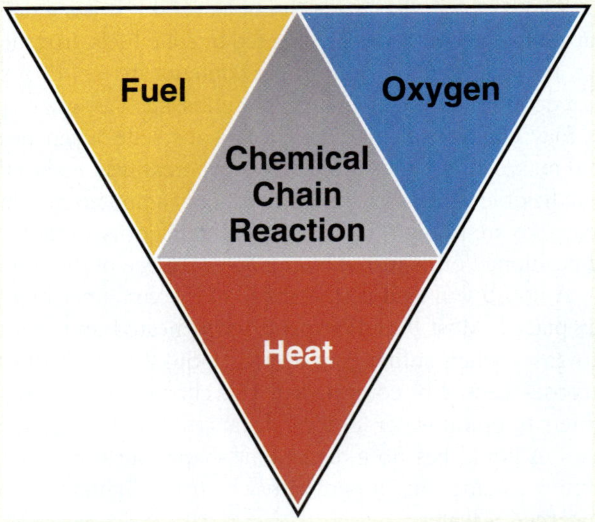

FIGURE 6-4 A chemical chain reaction unites fuel, oxygen, and heat in the fire triangle.

The key point to remember is that fuel, oxygen, and heat must be present for a fire to start and to continue burning. If you remove any of these elements, the fire will go out.

Chemistry of Combustion

The smallest unit of matter is an atom. Some atoms exist in nature as pairs. For example, two atoms of oxygen are paired up in air. The formula for the oxygen pair is written O_2.

When atoms of one element combine chemically with atoms of another element, they produce a compound that is made up of molecules. For example, when an atom of oxygen (O) combines chemically with two atoms of hydrogen (H_2), the resulting compound is water (H_2O).

Almost all fuels consist of hydrogen (H) and carbon (C) atoms; hence they are called hydrocarbons. Fuels may contain a wide variety of other elements, but for our purpose we can describe the self-sustaining chemical reaction of combustion simply by considering the fuel to contain carbon and hydrogen.

When this fuel combines with oxygen, it produces two by-products, water and carbon dioxide:

Hydrogen–Carbon Fuel + Oxygen = Water and Carbon Dioxide

Graphically, this reaction is depicted as follows:

$$\text{H—C (hydrocarbon fuels)} + O_2 = H_2O + CO_2$$

Of course, the reactions that occur in most actual fires are not so simple as this basic reaction suggests. First, hydrocarbon fuels are very complex molecules that contain a wide variety of atoms besides hydrogen and carbon. As a consequence, the combustion process produces numerous toxic by-products. Second, the fires encountered by fire fighters burn in the presence of a limited amount of oxygen. The resulting incomplete combustion produces significant quantities of deadly gases and compounds. Also, the production of large quantities of heat and light is an integral part of combustion. Thus the reaction for combustion should be rewritten as follows:

$$\text{H—C (hydrocarbon fuels)} + O_2 = H_2O + CO_2 + \text{Light} + \text{Heat}$$

To understand the reactions involved in fire, it is helpful to differentiate between oxidation, combustion, and pyrolysis. Oxidation is the process in which oxygen combines chemically with another substance to create a new compound. For example, steel that is exposed to oxygen results in rust. The process of oxidation can be very slow; indeed, it can take years for oxidation to become evident. Slow oxidation does not produce easily measurable heat. Combustion, by contrast, is a rapid chemical process in which the combination of a substance with oxygen produces heat and light. For fire fighters' purposes, the terms "combustion" and "fire" can be used interchangeably. Pyrolysis is the decomposition of a solid material caused by heat, which breaks down the material to produce a flammable vapor.

Products of Combustion

The classic reaction for burning that results in water and carbon dioxide does not occur in most fire situations that fire fighters encounter. Building fires burn with a limited supply of oxygen, and result in incomplete combustion and produce a variety of toxic by-products. These by-products are collectively called smoke. Smoke includes three major components: particles (which are solids), vapors (which are finely suspended liquids—that is, aerosols), and gases FIGURE 6-5.

Smoke particles include unburned, partially burned, and completely burned substances. The unburned particles are lifted in the thermal column produced by the fire. Some partially burned particles become part of the smoke because inadequate oxygen is available to allow for their complete combustion. Completely burned particles are primarily ash. Some particles in smoke are small enough that they can get past the protective mechanisms of the respiratory system and enter the lungs, and most of them are very toxic to the body.

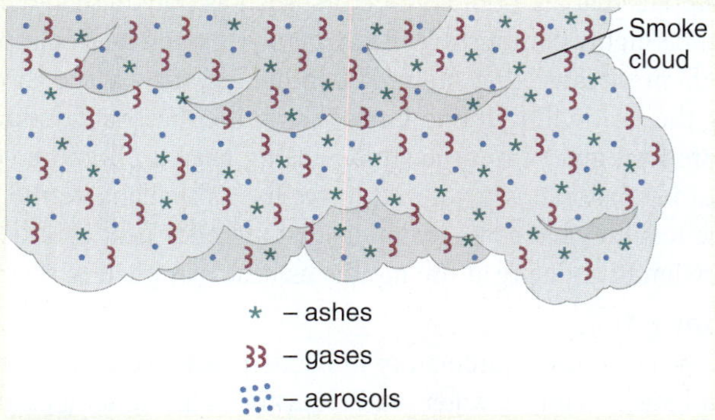

FIGURE 6-5 Smoke is composed of solids (ashes), vapors (aerosols, or small droplets), and gases.

Smoke may also contain small droplets of liquids called aerosols. The fog that is formed on a cool night consists of small water droplets suspended in the air. When water is applied to a fire, small droplets may also be suspended in the smoke or haze that forms. Similarly, when oil-based compounds burn, they produce small oil-based droplets that become part of the smoke. Oil-based or lipid compounds can cause great harm to a person when they are inhaled. In addition, some toxic droplets cause poisoning if absorbed through the skin.

Smoke contains a wide variety of gases. The composition of gases in smoke will vary greatly, depending on the amount of oxygen available to the fire at that instant. The composition of the substance being burned also influences the composition of the smoke. In other words, a fire fueled by wood will produce a different composition of gases than a fire fueled by petroleum-based fuels, including plastics.

Because smoke is the product of incomplete combustion and it contains unburned hydrocarbons, we need to remember that it is a form of fuel. A column of hot black smoke coming in contact with an adequate supply of oxygen and an ignition source can ignite suddenly and violently.

Almost all of the gases produced by a fire are toxic to the body, including carbon monoxide, hydrogen cyanide, and phosgene. Carbon monoxide is deadly in small quantities and was used to kill people in gas chambers. Hydrogen cyanide was also used to kill convicted criminals in gas chambers. Phosgene gas was used in World War I as a poisonous gas to disable soldiers. Even carbon dioxide, which is an inert gas, can displace oxygen and cause hypoxia.

A discussion of the by-products of combustion would be incomplete without considering heat. Because smoke is the result of fire, it is hot. The temperature of smoke will vary depending on the conditions of the fire and the distance the smoke travels from the fire. Injuries from smoke may occur because of the inhalation of the particles, droplets, and gases that make up smoke. The inhalation of hot gases in smoke may also cause injuries in the form of severe burns of the skin and the respiratory tract.

Fire Spread and Heat Transfer

Fire fighters work hard to prevent fires and to extinguish them. Fire fighters' enemy is fire—and one way to outsmart this enemy is by learning to think like the enemy. By learning how fires grow and spread, fire fighters can better

Fire Fighter Safety Tips

Smoke consists mainly of unburned forms of hydrocarbon fuels. As a consequence, when smoke is combined with adequate oxygen and heat, it will burn. Smoke contains large amounts of potential energy that can be released violently under the right conditions.

Fire Fighter Safety Tips

Smoky environments are deadly! It is essential to use self-contained breathing apparatus (SCBA) whenever you are operating in a smoky environment. This rule applies whether you are dealing with a structure fire, a dumpster fire, a car fire, or any other type of fire that generates smoke. Toxic gases are still present during the overhaul stage of a fire, even when there is little smoke visible. A recent UL smoke study determined that 97% of the fire's particulates are too small to be visible to the human eye.

determine how to use their suppression forces most effectively. The primary principle needed to understand fire spread is the principle of heat transfer. By studying heat transfer, fire fighters can understand how a fire grows, how to predict the size and growth rate of the fire, and determine the most effective means by which to attack the fire. Heat transfer occurs when there is a difference in temperature between two objects. Heat transfer occurs from a hotter object to a cooler object until they reach equal temperatures. When two objects have the same temperature heat transfer does not occur.

The rate of heat transfer is dependent on two factors: the difference in temperature between the two substances, and the ability of the materials to conduct heat. The bigger the temperature difference in the two substances, the greater the rate of heat transfer. There are three primary mechanisms by which heat is transferred: conduction, convection, and radiation. The term used to describe heat transfer is heat flux. Heat flux is the measure of the rate of heat transfer from one surface to another. For example, if a sofa is actively burning and heat is moving to the ceiling, the heat flux would indicate how much heat was being transferred to the ceiling. Temperature is the degree of sensible heat of a body as measured by a thermometer or similar instrument. Higher temperatures reflect greater energy of a material. Temperature is measured on the Celsius scale or the Fahrenheit scale.

Conduction

Conduction is the process of transferring heat to and through one solid to another. This transfer occurs when two solids are in contact with each other and one has a higher temperature than the other. This heat transfer occurs because of the kinetic energy of the molecules within the solid materials FIGURE 6-6. Conduction transfers energy directly from one molecule to another, much as a billiard ball transfers energy from one billiard ball to the next. Objects vary in their ability to conduct energy. Metals generally have a greater ability to conduct heat than wood does. Objects that have more tightly packed molecules are more efficient in conducting heat than objects that are less densely constructed. For example, heat applied to a copper pipe will be readily conducted along the pipe. TABLE 6-1 shows a comparison of the thermal conductivity of some common materials.

Heat transfer by conduction is dependent on three separate factors. First is the thermal conductivity of the material being heated. Better conductors will transfer more heat. The second factor is the area being heated. If all other factors remain the same, heating a small area will result in less heat transfer than heating a larger area. The third factor in determining the amount of heat transfer is the difference in temperature between the heated object and the second object that is being heated. The greater the temperature difference or energy difference between the two materials, the greater the heat transfer will be.

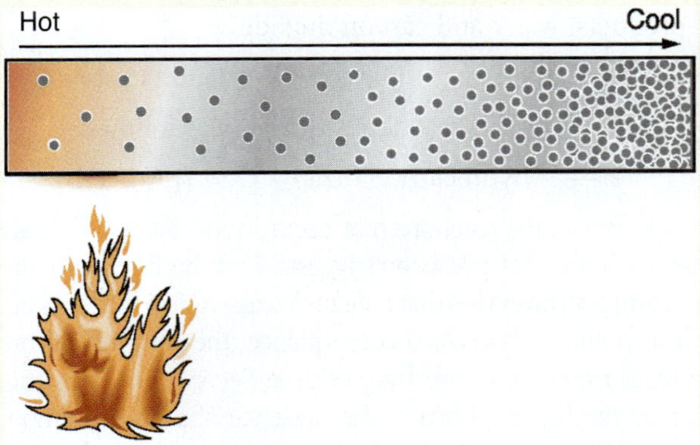

FIGURE 6-6 Conduction through a solid object.

TABLE 6-1 Thermal Conductivity of Common Building Materials and Air

Material	Thermal Conductivity (W/mK)
Copper	385
Carbon steel	36.0
Concrete	0.8
Brick	0.5
Air	0.024
Mineral insulation	0.04
Wood, oven dried	0.04

(Measured at 20°C)

Insulating materials are designed to limit the transfer of heat. Insulating materials have little ability to conduct heat due to the air trapped within the insulating material. The air or other gas trapped in small packets prevents the degree of heat transfer that occurs in more densely constructed materials, and gases are not efficient conductors of heat from one solid to another because the gas molecules are farther apart than they are in denser materials. The result is that most insulating materials limit the rate of heat transfer by conduction. However, some materials that are poor conductors, such as polyurethane foam, can be highly combustible. They pyrolyze when heat is applied to them. Because they are poor conductors, the surface absorbing the heat will pyrolyze faster than if the material conducted more heat away from the surface being heated. As more combustible material is pyrolyzed, more fuel is available to the fire. The result is that some materials that are poor heat conductors may ignite and spread fire rapidly as heat is applied to them.

Convection

Convection is the circulatory movement that occurs in a gas or fluid such as air or water, with areas of differing temperatures being present in the same medium owing to the variation of the substance's density of the liquid or the pressure exerted by the gases and the action of gravity FIGURE 6-7. To understand

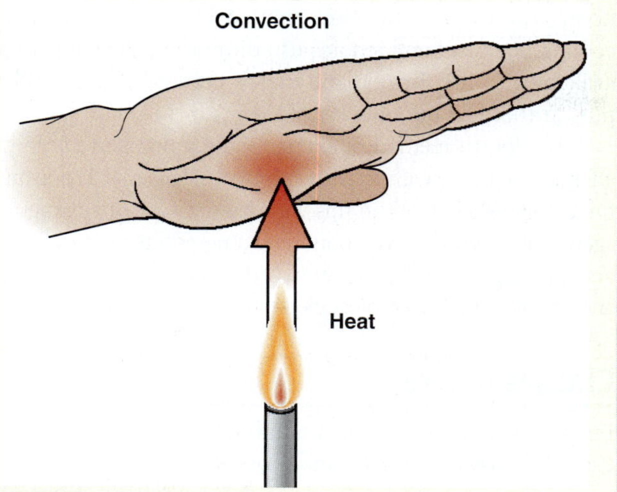

FIGURE 6-7 Convection.

the movement of gases in a fire, consider a container of water being heated on a stove. If you put a drop of food coloring in one part of the water and start to heat the water, you can readily see the circulation of the water within the container. As the water heats up, it expands and becomes lighter than the surrounding water. This causes a column of water to rise to the top; the cooler water then falls to the bottom. Thus a cycle of warmer water rises to the top, pushing the cooler water to the bottom. The convection currents have the same effect of a small pump pushing the water around in a set pathway.

A simple definition of convection is the transfer of heat by the flow of gases or fluid from hotter areas to cooler areas. Convection in a fire involves primarily smoke and hot gases generated by the fire. This transfer occurs from a hotter gas to a cooler surface. The heat of the fire warms the gases and particles in the smoke. The hotter and less dense column of gases rises and displaces cooler, denser gases downward. A large fire burning in the open can generate a plume of heated gases and smoke that rises high in the air. This convection stream can carry smoke and large bands of burning fuel for several blocks before the gases cool and fall back to the earth. If winds are present during a large fire, they will influence the direction in which the convection currents travel.

When a fire occurs in a building, the convection currents generated by the fire rise in the room and travel along the ceiling. These currents carry hot gases, which may ultimately heat combustible room surfaces and contents enough to ignite them FIGURE 6-8. If the fire room has openings, convection may then carry the fire outside the room of origin and to other parts of the building.

As a fire grows, it creates increasing volumes of hot gases and smoke. These hot gases first create a plume, which rises to the upper levels of the fire compartment. If the fire continues to grow and the pressure in the fire compartment is sufficient, the hot gases will push laterally outside the room where the fire started. Because the pressure in the lower parts of the fire compartment is lower, cooler air will be drawn into the lower levels of the fire compartment. The volume of space in which the cooler air is being drawn into the fire and the volume of space where the hotter gases and smoke are escaping from the fire are called the flow path. A flow path

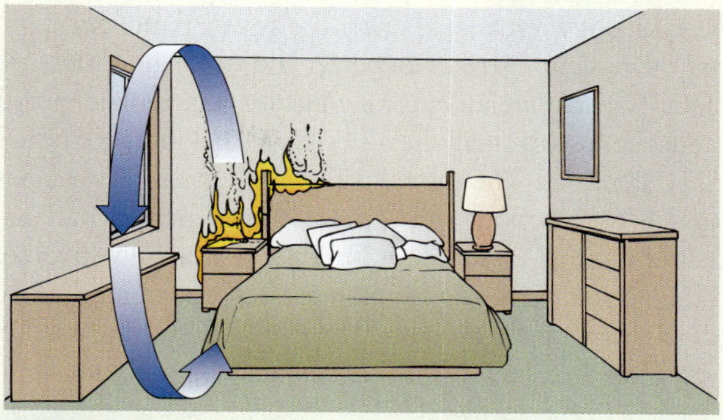

FIGURE 6-8 Convection currents in a room-and-contents fire.

is the boundary for the flow of gases within the fire building. It is the volume of space in which the movement of cooler air enters through an inlet vent and the space where hot gases and smoke exit through an exhaust outlet. The flow path transfers hot gases and smoke from the higher pressure within the fire area toward the lower pressure areas accessible through doorways, window openings, and roof openings. The fire growth will progress along the exhaust portion of the flow path. Flow paths are dependent on the location of higher and lower pressure within the building and external factors such as ventilation and wind direction. In addition, convection currents are influenced by the layout of the building. For example, air will flow quickly through a structure with an open floor plan versus a structure made up of small, closed rooms. The *Ventilation* chapter contains more information on flow paths.

Radiation

Radiation is the transfer of heat through the emission of energy in the form of invisible waves FIGURE 6-9. The sun radiates energy to the earth over the vast miles of outer space; this electromagnetic radiation readily travels through the vacuum of space. When this energy is absorbed, it becomes converted to heat—that is, the heat you feel as the sun touches your body on a warm day. Of course, the direction in which the radiation travels can be changed or redirected, as when a mirror reflects the sun's rays and bounces the energy in another direction.

FIGURE 6-9 Radiation.

Thermal radiation from a fire travels in all directions. The effect of thermal radiation, however, is not seen or felt until the radiation strikes an object and heats the surface of the object. Thermal radiation is a significant factor in the growth of a campfire from a small flicker of flame to a fire hot enough to ignite large logs. The growth of a small fire in a wastebasket to a full-blown room-and-contents fire is due in part to the effect of thermal radiation. A building that is fully involved in fire radiates a tremendous amount of energy in all directions. Indeed, the radiant heat from a large building fire can travel several hundred feet to ignite an unattached building.

FIRE FIGHTER Tips

Radiant heat can travel through the air gaps in a water spray.

Methods of Extinguishment

There are many variations on the methods used to extinguish fires exist; however, they boil down to four main methods: cooling the burning material, excluding oxygen from the fire, removing fuel from the fire, and interrupting the chemical reaction with a flame inhibitor FIGURE 6-10. There are many variations on the way that these methods can be implemented, and sometimes a combination of these methods is used to achieve suppression of fires.

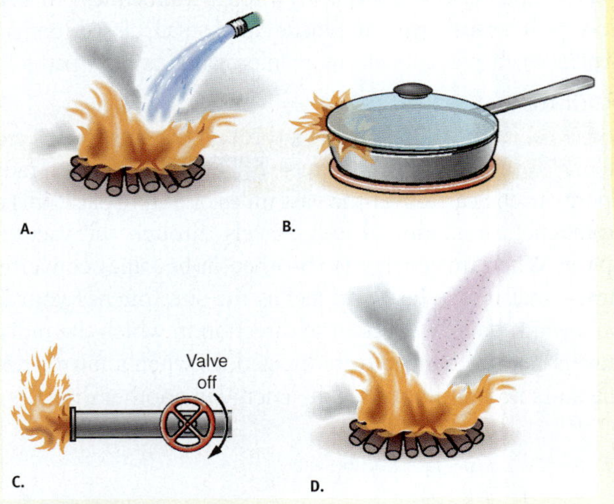

FIGURE 6-10 The four basic methods of fire extinguishment. **A.** Cool the burning material. **B.** Exclude oxygen from the fire. **C.** Remove fuel from the fire. **D.** Interrupt the chemical reaction with a flame inhibitor.

The method most commonly used to extinguish fires is to cool the burning material. The actions fire fighters take to set up a water supply, lay hose lines to the fire, and apply water to the fire are all steps in implementing this method of fire extinguishment.

A second method of extinguishing a fire is to exclude oxygen from the fire. One simple way to do so is to place the lid on an unvented charcoal grill or to close the door to a fire room. Reducing the amount of air reaching a fire will retard the growth. Applying foam to a petroleum fire covers flammable vapors.

Removing fuel from a fire will also extinguish the fire. For example, shutting off the supply of natural gas to a fire being fueled by this gas will extinguish the natural gas fire but not necessarily the resulting fires caused by the gas. Reducing or cooling hot gases and smoke also diminishes the supply of fuel available to a fire. In wildland fires, a firebreak cut around a fire puts further fuel out of reach of the fire.

The fourth method of extinguishing a fire is to interrupt the chemical reaction with a flame inhibitor. Halon- or Halotron-type fire extinguishers work in this way. These clean, fire extinguishing agents can be applied with portable extinguishers or through a fixed system designed to flood an enclosed space. They leave no residue and are often used to protect electronics and computer systems.

Classes of Fire

Fires are generally categorized into one of five classes: Class A, Class B, Class C, Class D, and Class K.

Class A Fires

Class A fires involve ordinary solid combustible materials such as wood, paper, plastics, and cloth FIGURE 6-11. Natural vegetation such as the grass that burns in ground cover fires is also considered to be part of this group of materials. The method most commonly used to extinguish Class A fires is to cool the fuel with water to a temperature that is below the ignition temperature or a combination of limiting ventilation and applying water.

FIGURE 6-11 A Class A fire involves wood, paper, or other ordinary combustibles.

Class B Fires

Class B fires involve flammable or combustible liquids such as gasoline, kerosene, diesel fuel, and motor oil FIGURE 6-12. Fires involving gases such as propane or natural gas are also classified

FIGURE 6-12 A Class B fire involves flammable liquids such as gasoline.

as Class B fires. These fires can be extinguished by shutting off the supply of fuel or by using foam to exclude oxygen from the fuel.

Class C Fires

Class C fires involve energized electrical equipment FIGURE 6-13. They are listed in a separate class because of the electrical hazard they present. Incorrectly attacking a Class C fire with an extinguishing agent that conducts electricity can result in injury or death to the fire fighter. Once the power is cut to a Class C fire, the fire is treated as a Class A or Class B fire depending on the type of material that is burning.

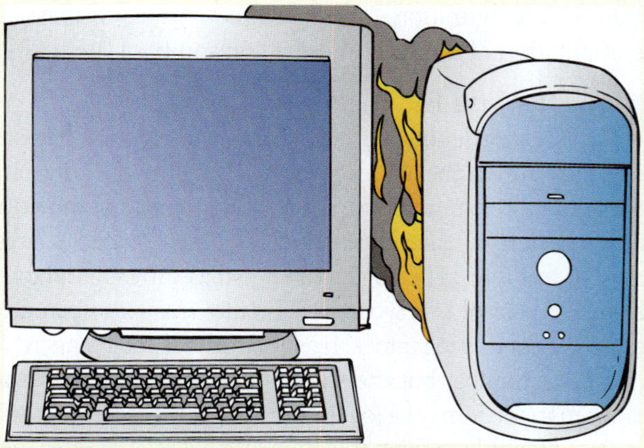

FIGURE 6-13 A Class C fire involves energized electrical equipment.

Class D Fires

Class D fires involve combustible metals such as sodium, magnesium, and titanium FIGURE 6-14. These fires are assigned to a special class because the application of water to fires involving these metals will result in violent explosions. Instead, these fires must be attacked with special agents to prevent explosions and to smother the fire and remove the supply of oxygen to the fire.

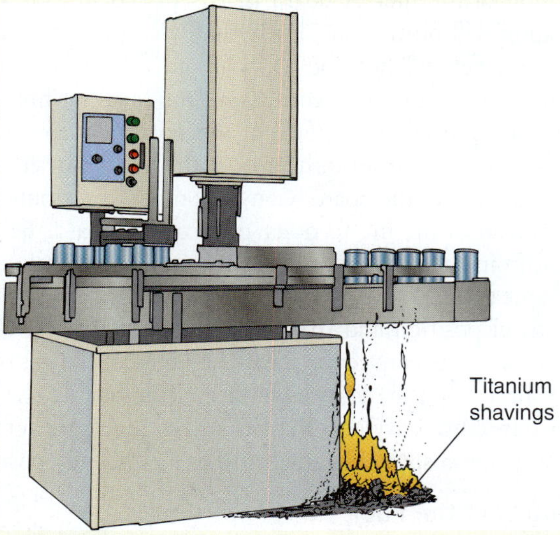

FIGURE 6-14 A Class D fire involves metals such as sodium or titanium.

Class K Fires

Class K fires involve combustible cooking oils and fats in kitchens FIGURE 6-15. Heating vegetable or animal fats or oils in appliances such as deep-fat fryers can result in serious fires that are difficult to fight with ordinary fire extinguishers. Special Class K extinguishers are available to handle this type of fire. These extinguishers contain a wet agent that combines with the cooking oils to produce a soap-like substance. These agents also absorb heat from the fire. Yellow fire extinguisher

FIGURE 6-15 A Class K fire involves a deep-fat fryer in a kitchen.

Some fires may fit into more than one class. For example, a fire involving a wood building could also involve petroleum products. Likewise, a fire involving energized electrical circuits—a Class C fire—might also involve Class A or Class B materials. If a fire involves live electrical sources, it should be treated as a Class C fire until the source of electricity has been disconnected.

Fire Fighter Safety Tips

Why do metal fires explode when water is applied to them? When water is applied to certain heated metals, the water (H_2O) reacts with the heated metal and produces hydrogen (H_2), which burns with an explosive force.

Characteristics of Solid-Fuel Fires

Most building fires that fire fighters encounter involve solid fuels. Solid fuels have a definite size and shape. A variety of solid fuels are found in most buildings. These include wood and wood-based products, fabrics, paper, carpeting, and many petroleum-based fuels, such as plastics and petroleum-based foams. Wood is a commonly encountered building material. Building contents may contain a variety of materials. Older furniture was commonly constructed of solid-wood frames, and natural materials such as cotton were used for padding and upholstery. Wall coverings in older houses were often paper. Modern furniture is more frequently constructed from petroleum-based foams and fabrics and uses press board and plastic frames. Today, wall coverings are more likely to be latex-based paint or other petroleum-based materials. Solid fuels have certain characteristics that influence how a fire progresses. The building materials and the building contents will influence how a fire burns.

Solid fuels burn when they are heated sufficiently to change them into flammable vapors. When fuels are heated, they begin to change chemically and decompose, thereby releasing flammable vapors. This process is called pyrolysis. Some fuels change from a solid form directly into a vapor. Other solid fuels first change into a liquid before becoming a vapor. Plastics and petroleum-based materials pyrolyze faster than wood-based products. This means that it requires less heating to decompose and ignite many of these fuels than it does to ignite wood.

Heating wood products also causes them to pryloyze. This decomposition results in the release of flammable vapors. However, it requires more heat to pyrolyze wood than it does to pyrolyze plastic-based substances. Wood contains varying amounts of moisture. When wood is heated, the first change that occurs is that the water is vaporized and escapes from the wood. This causes the wood to begin to char. This charring is a natural mechanism of the wood in an attempt to protect itself. As the wood chars, it begins to lose mass, which can compromise the load carrying capabilities. As the wood is heated to about 425°F (218°C), pyrolysis begins. The energy required to vaporize the water delays the beginning of pyrolysis as the wood is heated.

The characteristics of solid fuels that influence their combustion can be broken into three major categories: the composition of the fuel, the amount of the fuel, and the configuration of the fuel.

Composition of Fuel

The chemical composition of the fuel will have a significant impact on how the fuel burns. Wood is composed primarily of a natural fiber called cellulose. Cellulose is combustible. Most of the modern room contents including synthetic foams, upholstery coverings, various types of wall coverings, and plastic furniture are manufactured from petroleum products. Petroleum-based products generally contain more potential heat energy than products made from natural products. This means that fires fueled by modern petroleum-based products have a much higher heat release rate than fires fueled by natural products such as wood and cotton.

Another factor that influences how the fire burns and the amount of heat that is released is the amount of moisture contained in the fuel. The water contained in a fuel serves as a cooling agent. It requires heat to vaporize the water from the fuel before the fuel can be pyrolyzed to create a flammable vapor. For example, green fire wood from a recently cut tree contains a lot of moisture and will be hard to burn in a fire place, whereas fire wood that has been allowed to dry will ignite more easily and burn faster with a higher heat release rate.

Amount of Fuel

The second characteristic of fuel that determines how it burns is the amount of fuel available to the fire. In a fuel-limited fire if all other factors are the same, when more fuel is available, there is a higher heat release rate than when less fuel is available. Under identical conditions, a fire burning with four wooden pallets will have a much lower heat release rate than a fire fueled with 40 wood pallets.

Configuration of Fuel

The third characteristic of fuel that determines how it burns is the configuration of the fuel. There are three factors that contribute to configuration. These are the surface-to-mass ratio, the orientation of the fuel, and the continuity of the fuel.

Surface-to-Mass Ratio

Solid fuels have a definite size and shape. The size and shape of the fuel greatly impact the ability of the fuel to ignite, the time it takes the fuel to be consumed, and the heat release rate of the burning fuel. One example of a common fuel is wood. Consider a large log that is 10 feet long. This log has a large mass or weight; however, the surface area of this wood is relatively small compared to its mass. If we cut this log into 4" by 4" beams, the total mass remains the same as the log, but the surface area is much greater than it was when in the form of a log. If the original log were cut into thin shingles, it would maintain the same original total mass as the log, but would have a far greater total surface area than the original log or the 4" by 4" wood beams. The energy required for ignition of the low surface-to-mass ratio log is much greater than the energy required to ignite the higher surface-to-mass ratio shingles. The higher the surface-to-mass ratio is, the less energy that is required to ignite the fuel. This is due to lesser energy being required to heat the material to a temperature sufficient to burn. For example, holding a match to a large log will not usually result in ignition of the log, but holding a match to a thin piece of shingle from the same source will usually result in ignition of the shingle.

Orientation of the Fuel

The second factor that determines how a fuel burns is the orientation of the fuel. A board that is placed in a horizontal position will burn more slowly than an equal-sized board placed in a vertical position. A board in a horizontal position that is ignited on one corner will produce hot fire gases that rise away from the surface of the board. As convection currents carry heat upward, heat from the fire is carried away from the surface of the board. Convection will contribute little to the spread of the fire to other parts of the board. Because the heat from the convection currents rises, it has little effect on the spread of the fire. However, if a similar board is placed in a vertical position and the board is ignited on one of the lower corners, the convection created by the hot fire gases will transfer heat along the vertical surface of the board. The heat absorbed by convective heat transfer will be much greater than the heat that is absorbed by the board in a horizontal position.

Continuity of the Fuel

The third factor that determines how a fuel burns is the continuity of the fuel. Continuity refers to the closeness of one piece of fuel to another. The closer the fuel is the easier and more quickly it will ignite. Continuity can occur in a horizontal direction along floors, ceilings, or horizontal surfaces of building contents. It can also occur in a vertical direction along walls or the vertical surfaces of building contents. Fuel that is in contact with other fuel will ignite faster and reach a peak heat release rate more quickly. This rate of growth will be partly due to spread by radiation, convection, and in some cases, close proximity conduction.

FIRE FIGHTER Tips

Fire fighters must remember that smoke is one of the most common forms of fuel that we encounter. Hot smoke is a powerful and potentially deadly fuel and is present at almost every fire we encounter.

Solid-Fuel Fire Development

A fire progresses through four classic stages of growth: the incipient stage, the growth stage, the fully developed stage, and the decay stage FIGURE 6-16.

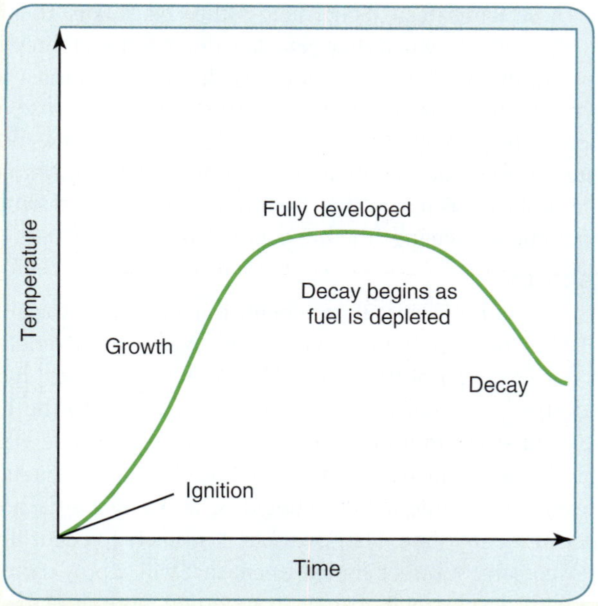

FIGURE 6-16 Illustration of a traditional fire growth curve with a fuel-limited fire.

1. Incipient Stage

The first stage of fire development is called the incipient stage. Incipient means beginning to come into being. The incipient stage occurs when there is an adequate supply of fuel, oxygen, and heat or ignition FIGURE 6-17. Ignition is defined as the action of setting something on fire. The ignition of a fire can occur in many different ways. One simple way to ignite a fire is by lighting some paper with a match. At the incipient stage a fire is small and confined to the initial fuel that was ignited. Because the fire is small, the initial growth is largely dependent on the type of fuel, and how much fuel can be pyrolyzed into a vapor. Radiant heat from the fire begins to pyrolyze, increasing amounts of fuel into flammable gases. The fire at this stage consumes relatively small amounts of oxygen, so at this stage the fire is fuel dependent.

FIGURE 6-17 The incipient stage of a fire.

As the fire begins to produce more heat, a plume of hot gases and smoke begins to rise from the fire. As this plume rises, it is cooled by the surrounding cooler air. As increasing amounts of heat are generated, the plume carries more heat and hot gases upward. If it reaches the ceiling, it will spread out in circle that increases in size. These hot gases begin to heat the surfaces with which they make contact. In the incipient stage, because the amount of heat is limited, there is little increase in the temperature of the room. The amount of smoke and combustion gases in the compartment is usually not high enough to pose a serious threat to occupants if they exit quickly. Many fires in the incipient stage can be safely extinguished with a portable fire extinguisher.

It is important to understand the stages of fire growth do not have a clear dividing line between them. A fire can progress rapidly from the incipient stage to a growth stage in a very short period of time. Because of the rapid fire growth, building occupants and fire fighters who are not equipped with a complete ensemble of personal protective equipment (PPE) including SCBA should exit from even a small fire.

2. Growth Stage

The second stage of fire development is the growth stage. A fire in the growth stage produces more interaction and is more dependent on the environment in the compartment around it than a fire in the incipient stage FIGURE 6-18. Especially important are the composition of the compartment surfaces and contents, and the placement and configuration of the compartment contents. The amount of ventilation is also an important factor.

FIGURE 6-18 The growth stage of a fire.

As a fire begins to grow it produces larger amounts of heat, producing a higher heat release rate. In other words, the fire is generating a higher heat release rate. The hot gases and smoke start to rise because they become lighter as they are heated. This hot column of gases and smoke is called a plume. A plume is defined as an elongated and moving column of gases and smoke. The location of the fuel in a compartment influences the fire development. If the fuel is in the center of the compartment, it can draw air from four sides. The hot gases and smoke in the plume mix with cooler air. This cooling reduces the rate of growth, the length of the flame, and the length of the plume, thereby reducing vertical extension of the fire. These factors help to slow the rate of fire growth. If the burning fuel is located close to a corner, the plume can entrain (draw along and transport) air from only two sides. This has the effect of drawing in less cool air at the base of the fire. As a result, the combustion zone increases the height. In this situation, the growth of the fire will be faster than if the burning fuel were in the middle of the room. If the fuel is located close to a wall, it can entrain air on three sides. This situation will result in a growth rate slower than a fire located in a corner, but faster than a fire with fuel in the middle of the room. A higher fire plume will increase the hot gases at the ceiling level of the compartment. More hot gases at the ceiling will radiate more heat back to the room surfaces, the room contents, and to the burning fuel, thereby increasing the rate of fire growth.

The following paragraphs describe specific types of fire conditions. These include thermal layering, rollover, flashover, backdraft, rapid fire growth, and the behavior of ventilation-limited fires. These conditions are not limited to the growth stage of fire development, but are introduced here because many of them are likely to occur in the growth stage.

Thermal Layering

As a fire continues to grow, more hot gases are generated. Hot gases are lighter and buoyant and therefore tend to seek a higher level in the compartment. Since cooler gases are heavier, they tend to be at lower levels in the compartment. The phenomenon of gases forming into layers according to temperature is called thermal layering. It is also referred to as heat stratification. Anytime the ventilation of the fire is changed, it will change the thermal layering.

As the fire grows, the amount of hot gases increase. As the hot fire gases rise, they increase the rate of heat transfer back to the contents and walls of the fire compartment. The layer of hot gases in the upper level of the compartment creates an increase in pressure. This layer of hot gases tends to push out through the upper level of any available openings such as doors or windows. Because the cooler gases at the bottom of the room exert less pressure than the hotter gases, cooler air is drawn into the fire compartment usually at lower levels.

The common boundary or interface between the hot gases and the cooler gases is called the neutral plane. The neutral plane is the level in a compartment opening at which the pressure of the hot gases and smoke leaving the compartment and the pressure of the cooler gases entering the compartment are equal. In order for a neutral plane to exist, there must be a flow of cooler gases entering the compartment and a flow of hot gases exiting the compartment. Firefighters should avoid working in a layer of hot gases. Stay low whenever possible to remain under the layer of hot gases in the upper levels of the fire compartment. You will learn more about steps that can be taken to ventilate and control the fire to reduce the amount of hot gases and to increase the level of the neutral plane.

As the fire continues to grow and the amount of heat generated increases, you may notice small flames "dancing" in the hot gas layer. These isolated flames are an indication that the gases in the hot layer are within their flammable range. This indicates that this layer is hot enough to cause ignition. Often, these flames will be around the edges of the hot plume because this area has sufficient oxygen for combustion. These isolated flames may indicate that the fire is ventilation limited. There is not sufficient oxygen to ignite additional quantities of hot fuel. This may also indicate that conditions are approaching a point where a fire may soon flashover.

Rollover

Rollover (also known as flameover) is the spontaneous ignition of hot gases in the upper levels of a room or compartment. During the growth stage of the fire, the hottest gases rise to the top of the room. If some of these gases reach ignition temperature, they will ignite from the radiation of heat coming from the flaming fire or from direct contact with open flames. In other words, the upper layer of flammable vapor catches fire. These flames can flicker across the ceiling and then go out. Alternatively, when they get a few degrees hotter, they can extend throughout the room at ceiling level. Rollover is a sign that the temperature is rising—and if it continues to rise, the temperature throughout the room will soon reach the point where the room and contents will spontaneously and rapidly ignite. Rollover is a sign that flashover is imminent if actions are not taken to change the fire conditions.

Flashover

Flashover is not a specific moment, but rather the transition from a fire that is growing by igniting one type of fuel to another, to a fire where all of the exposed surfaces have ignited. It is a rapid change or transition from the growth stage to the fully developed stage. In a fire, it may be hard to determine which stage a fire is in. Fires do not always follow the four classic stages of development. Fire fighters should be aware that a flashover can occur at any time. Flashovers are less likely to occur if the fire is in a large room or compartment that will supply more air during fire development. Larger areas entrain more air in the fire plume, which tends to cool the plume, and reduce the radiation and convection of heat to the room surface and contents.

The critical temperature for a flashover to occur is approximately 1000°F. Once this temperature is reached, all fuels in the room are involved in the fire, including the floor coverings. So the temperature at the floor level may be as high as the temperature of the ceiling before flashover. Fire fighters, even with full PPE, cannot survive for more than a few seconds in a flashover.

Flashover may not occur if the fire remains ventilation controlled. A ventilation-controlled fire will produce a limited amount of heat energy. If the supply of oxygen is not increased, the fire may enter the decay stage. During the decay stage, heat will continue to pyrolyze fuels in the compartment. This pyrolysis produces additional vaporized fuel in the form of smoke. A fire in this decay stage can flashover quickly if oxygen is added. Keeping doors and windows to the fire compartment closed will help to reduce the amount of oxygen available to the fire and may help prevent a flashover. It is important to note that even with increased ventilation, many compartment fires will remain ventilation controlled rather than advancing to a free-burning state because there is not sufficient oxygen to produce a free-burning state. FIGURE 6-19 illustrates the ventilation-limited stage of a fire.

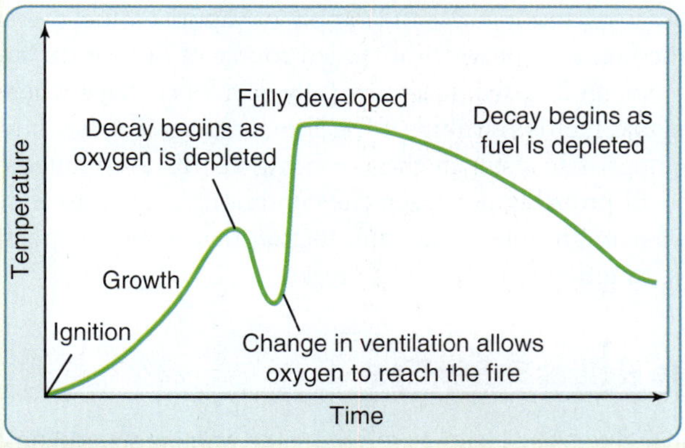

FIGURE 6-19 Illustration of a ventilation-limited fire, which often occurs in houses with lightweight construction and petroleum-based contents.

FIRE FIGHTER Tips

Flashover can be prevented by cooling the fuel, removing the fuel, or reducing the amount of oxygen present by controlling ventilation.

FIRE FIGHTER Tips

A searching fire fighter may have difficulty recognizing potential flashover conditions at the upper levels of the room. Fire fighters should also be alert for indications of potential flashover at lower levels of the room such as off-gassing of furniture or carpeting. Visible indications of steam or smoke indicate these materials are pre-heating, signaling imminent flashover.

Backdraft

A backdraft is caused by the introduction of oxygen, a change of the ventilation profile, into an enclosure where the superheated gases and contents are already hot enough for ignition but do not have sufficient oxygen to combust. The development of a backdraft requires a unique set of conditions. When a fire generates quantities of combustible gases, these gases can become heated above their ignition temperatures. Backdrafts require a "closed box"—that is, a room or building that is ventilation limited. When the fire chamber has a limited amount of oxygen, combustion is reduced, yet superheated fuel in the form of smoke still fills the chamber. If a supply of new oxygen is then introduced into the room by intentional ventilation, such as opening the front door or a window breaking because of excess heat, explosive combustion can occur owing to the presence of the superheated flammable gases. This kind of explosive combustion may exert enough force to cause severe injury or death to fire fighters. Signs and symptoms of an impending backdraft include the following:

- Any confined fire with a large heat build-up
- Little visible flame from the exterior of the building
- A "living fire," where the building appears to be breathing due to smoke puffing out of and then being sucked back into the building
- Smoke that seems to be pressurized
- Smoke-stained windows (an indication of a significant fire)
- No smoke showing
- Turbulent smoke
- Thick yellowish smoke (containing sulfur compounds)

It is important for fire fighters to stay alert for the conditions that signal a possible backdraft. What triggers a backdraft when the conditions are right is the introduction of oxygen into a room that was lacking a sufficient oxygen supply to support a fire. A backdraft can occur when a window breaks because of excess heat, when a fire fighter opens a closed door, or when a fire fighter tries to ventilate a fire by opening a door or breaking a window. It is important to reduce the possibility of a backdraft developing. The chapters *Ventilation* and *Fire Attack and Foam* explain specific ways to limit the supply of oxygen to the fire and to cool a ventilation-limited fire.

Rapid Fire Growth

Introducing air into a ventilation-limited fire can result in explosive fire growth. Research has demonstrated that fire fighters making entry through the front door can introduce enough air into the fire area to produce rapid fire growth and flashover. This change can occur in less than two minutes and advance so fast that it is not possible to escape this deadly environment. Fire fighters need to carefully consider what constitutes ventilation. Opening any door, window, skylight, or roof introduces oxygen into a burning building. The fire service has not traditionally taught that making entry into a building is part of ventilation, yet studies have shown that opening the front door has a profound impact on the growth of the fire. Repeated experiments produced a violent flashover or backdraft after the front door is opened. Rapid fire growth can be prevented by cooling the fuel, removing the fuel or controlling the amount of oxygen present. During rapid growth, a fire can progress to its maximum heat release rate unless actions are taken to limit the supply of oxygen or apply water to the fire.

Behavior of Ventilation-Limited Fires

Recent experiments conducted by Underwriters Laboratories (UL), the National Institute for Standards and Technology (NIST), and the Fire Department of New York (FDNY) have demonstrated that many building fires become ventilation limited because of a limited supply of oxygen. Newer houses are tightly sealed with added insulation and caulking, double-paned windows, and storm windows. A fire in a building during the growth stage can consume enough oxygen to reduce the concentration of oxygen below the level needed to maintain a growing fire. When this occurs the fire is said to become ventilation limited. A ventilation-limited fire is restricted because there is not sufficient oxygen available for the fire to burn as rapidly as it would with an unlimited supply of oxygen. Assessing a ventilation-limited fire presents a challenge to fire fighters. A fire in this condition may initially appear to be a small fire in the incipient stage, but it contains a large amount of energy in the form of hot gases and smoke, which are vaporized fuel. Often flames are not visible. If the supply of oxygen is increased, a ventilation-limited fire can suddenly progress to a flashover in less than two minutes. Research indicates that fires in modern residential occupancies are likely to enter a ventilation-limited decay stage prior to the arrival of the first due engine company.

From the outside, a ventilation-limited fire may be deceiving and may look like a small fire. All that is needed is more oxygen to produce explosive fire growth. Be aware that a fire that appears to be "small and relatively harmless" may in fact be a ventilation-limited fire containing enormous amounts of energy that needs only a healthy dose of oxygen to rapidly grow.

3. Fully Developed Stage

The third stage of fire development is the fully developed stage. During this stage, the fire is consuming the maximum amount of fuel possible and it achieves the maximum heat release rate possible for the fuel supply and oxygen present FIGURE 6-20. During the fully developed stage, the fire may be ventilation-limited or fuel-limited. When a fire has an unlimited supply of oxygen, the fire is fuel limited. For example, consider a building under construction that has been framed with wood and combustible sheathing but lacks windows and doors. If this building catches fire, the fully developed fire may be fuel limited because it has an abundant supply of oxygen. Most structure fires that fire fighters encounter have limited openings from the fire compartment to the outside. These fires will be limited by the amount of oxygen available, since there are limited openings available to admit oxygen and to exhaust the hot fire gases and smoke.

FIGURE 6-20 The fully developed stage of a fire.

During the fourth stage of fire development, as the fire releases large quantities of heat, the energy pyrolyzes large amount of fuel, generating large quantities of smoke and fire gases. The smoke and fire gases generate increased pressure in the fire compartment and force hot fire gases from the fire compartment. The heat release rate will be dependent on the size of the compartment openings, which provide a supply of oxygen and openings for hot fire gases.

When the hot fire gases are exhausted from the fire building, if they are above the ignition temperature of the gases, they may ignite upon mixing with a fresh supply of oxygen. This is what produces flames from compartment openings during this active burning state. If these hot fire gases travel to another room or compartment, they can transfer enough heat to spread the fire to rooms or compartments adjacent to the fire compartment. The supply of oxygen may be greater in adjacent areas. As hot vaporized fuel travels to adjacent areas, it can ignite when it reaches a new supply of oxygen. The large quantities of heat transferred to adjacent areas will also pyrolyze fuel in these areas and increase the supply of vaporized fuel available to the fire. This provides an added source of fuel for the fire.

Not all fires will reach the fully developed stage where all available fuel is burning. If the fire compartment has limited openings that result in the fire being ventilation-limited, the fire will probably not reach the conditions needed to achieve the maximum rate of heat production that would be possible with an unlimited supply of oxygen.

Fire Fighter Safety Tips

Fire fighters must learn to recognize the warning signs of flashover and use the reach of their hose stream to cool as close to the fire as possible, thereby reducing the chance of a flashover. A fire fighter—even one who is wearing full turnout gear—has little chance of avoiding severe injury or death if he or she is caught in a flashover.

4. Decay Stage

The fourth and final stage in fire development is the decay stage. This condition can occur because of a decreasing fuel supply or because of a limited oxygen supply FIGURE 6-21.

One example of a fuel-controlled fire occurs in cases where an outdoor fire consumes the fuel supply. It can also occur in a room or compartment fire when the fuel supplied by the compartment contents is being consumed and the fire has not extended beyond the compartment because of fire-resistant construction. During the decay stage, active flaming combustion decreases or stops. This change is fuel controlled and not ventilation controlled. The heat will continue to pyrolyze the fuel and create flammable gases and vapors. These flammable fuel products will continue to burn, but the rate of

FIGURE 6-21 The decay stage of a fire.

combustion will continue to decrease and eventually stop when the fuel supply is exhausted.

A second type of situation occurs when a fire with an available fuel supply becomes oxygen limited. The type of fire can also enter a decay stage, but for a very different reason. In oxygen-limited decay, the rate of combustion slows and visible flames decrease or disappear. Because there is still significant heat in and around the fire and the fire compartment, fuels will continue to pyrolze and create additional flammable vapors and gases. The rate of pyrolysis will slow as the rate of combustion slows, but large quantities of flammable fuel may still be present. This is decay caused by a ventilation-limited fire. If additional oxygen is introduced into the fire compartment, rapid or violent fire growth can develop quickly.

Smoke Explosion

A smoke explosion occurs when a mixture of flammable gases and oxygen are present, usually in a void or other area separate from the fire compartment. Smoke may travel some distance from the fire, and be below its ignition temperature. When it comes in contact with a source of ignition, the flammable mixture ignites often in a violent manner. The conditions needed to produce a smoke explosion often include: the presence of void spaces, combustible building materials, a ventilation-controlled fire that produces unburned fuel, and a relatively cool temperature. Smoke explosions can occur during the decay stage of a fire, but they are not limited to the stage. They sometimes result in a violent explosion during final extinguishment or overhaul. In a smoke explosion, there is no change to the ventilation profile such as an open door or window; rather it occurs from the travel of smoke within the structure to an ignition source.

Fire Behavior in Modern Structures

Modern construction techniques and synthetic contents have altered fire behavior in today's fires. Houses built in the last 20 years or 30 years are constructed to be energy and cost efficient. They contain insulation to prevent the heat loss in the winter and loss of cool air in the summer. In addition, they are sealed to prevent air exchange through the walls and around the doors and windows. As a consequence, when a fire occurs in the house, only a limited quantity of air can enter the structure when the doors and windows are closed. In many parts of the country, doors and windows are kept closed most of the year because buildings are being heated or cooled by heating and cooling systems.

Many newer houses are constructed with lightweight manufactured building components. These manufactured building components contain many highly combustible products such as glues and coatings, most of which are made from petroleum products. They contain much less material (less mass) than houses built with dimensional lumber. Petroleum-based products pyrolyze faster and contain more potential heat energy. This means that they have a higher heat release rate than fires in older houses constructed primarily of wood. These products are not only flammable but also burn hotter and faster than an older house that is constructed primarily of wood. The lightweight frames of newer houses fail much faster than heavier wood framing of older homes. The lightweight construction also provides greater surface area of structural members and provide greater void spaces between floors.

The second major change in houses relates to the widespread use of plastics and petroleum-based products for furniture and accessories. These petroleum-based products contain much more energy than wood and natural fiber furnishings. They catch fire easily, reach high temperatures quickly, have a high heat release rate, and release huge quantities of thick, dark smoke when burning. This rapid combustion requires huge amounts of oxygen to maintain the growth stage (i.e., to remain burning) **FIGURE 6-22**.

These changes in building construction materials and techniques and changes in the building contents have, in turn, changed the dynamics of fire behavior inside a structure. Recent tests by Underwriters Laboratories documented a pattern of fire behavior that had been previously observed by veteran fire fighters.

Fires in modern structures progress to the fully developed stage quickly. In most cases, these fires then enter a ventilation-limited stage because there is a limited supply of oxygen supplied from the openings to the outside. During this ventilation-limited stage, the heat from the fire continues to pyrolyze available fuels, thereby increasing the supply of vaporized fuel. All that is needed is an increased supply of oxygen to change the fire into a rapid growth stage, often in the form of a flashover.

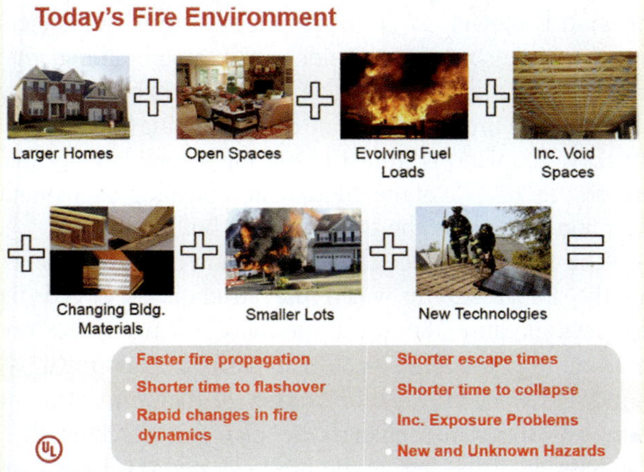

FIGURE 6-22 Modern fire environment.

FIRE FIGHTER Tips

The intensity and the speed with which a fire burns depend on the concentration of oxygen available to the fire. During the ignition stage, there is usually sufficient oxygen available for the fire to grow. In the growth, fully developed, and decay stages, a fire in an enclosed compartment may be regulated by the amount of oxygen available to it. If the atmosphere contains 16 to 21 percent oxygen, the fire will typically burn with an open flame. If the concentration of oxygen drops to less than 16 percent, open-flame burning will decrease. If the concentration of oxygen drops to less than 5 percent of the atmosphere, burning will stop. If oxygen is introduced into a dying fire, however, the conditions can change quickly and result in a dangerous situation for fire fighters. Properly coordinated ventilation and fire attack procedures are designed to prevent these conditions from occurring.

Often, the fire department arrives on the scene while the fire is in a ventilation-limited decay stage. One of the first actions taken by the fire department is to open the front door to gain access to the house. Unfortunately, opening the door has the unintentional effect of introducing a fresh supply of oxygen to the fire. When a fresh supply of oxygen comes in contact with superheated flammable fuel, the result is rapid growth of the fire often in the form of a flashover.

To prevent this scenario from occurring, fire fighters must limit the ventilation of the fire by keeping doors and windows closed until crews are ready to apply water as close to the fire as possible. This resetting the fire temporarily cools the fire until entry can be made to perform final extinguishment. This requires coordinated timing in implementing building entry, ventilation, and application of water to the fire. More information on these topics is presented in the *Forcible Entry; Ventilation; Fire Attack and Foam*; and *Fire Suppression* chapters.

Wind Effect

One factor that greatly influences the behavior of a fire is the wind. When wind is present, the behavior of the fire changes drastically. For example, when the air is relatively still, the best approach to a fire may be through the "A" side of a building. Conversely, if the wind is blowing at 20 to 25 miles (32 to 40 kilometers per hour) per hour from the "C' side of the same building, entering on the "A" side may be a deadly miscalculation because the wind is pushing the fire to the "A" side of the building FIGURE 6-23. The impact of the wind is similar to placing several large ventilation fans on the "C" side of the building and pushing huge quantities of air and fire toward the "A" side of the building.

The potential effect of the wind may not be evident during the initial size-up when the structure is closed tightly. However, because the wind can have a huge effect on fire behavior, the initial size-up of a structure fire must always include an evaluation of the impact of the wind. This evaluation will sometimes change the plan for attacking the fire. For example, if the seat of the fire is located on the "C" side of the building and a strong wind is blowing from that direction, the incident commander might choose not to begin ventilation operations on the "C" side windows if the rescue crew needs to enter through the "A" side of the building. Opening up a window on the "C" side could cause the fire to spread rapidly toward the "A" side. NIST research has shown that wind speeds as low as 10 mph can create wind-driven fire conditions inside the structure.

Characteristics of Liquid-Fuel Fires

Fires involving liquid or gaseous fuels have some different characteristics from fires that involve solid fuels. To understand the behavior of these fires, you need to understand the characteristics of liquid and gaseous fuels.

Recall that solid fuels do not burn in the solid state but instead must be converted to a vapor before they will burn. Liquids share the same characteristic: They must be converted to a vapor and be mixed with oxygen in the proper concentration before they will burn. Three conditions must be present for a vapor and air mixture to ignite:

- The fuel and air must be present at a concentration within a flammable range.
- There must be an ignition source with enough energy to start ignition.
- The ignition source and the fuel mixture must make contact for long enough to transfer the energy to the air–fuel mixture.

As liquids are heated, the molecules in the material become more active and the speed of vaporization of the molecules increases. Most liquids will eventually reach their boiling points during a fire. As the boiling point is reached, the amount of flammable vapor generated increases significantly FIGURE 6-24. Because most liquid fuels are a mixture of compounds (e.g., gasoline contains approximately 100 different compounds), however, the fuel does not have a single boiling point. Instead, the flammability of the mixture is determined by the compound with the lowest ignition temperature—the ignition temperature is the temperature at which that fuel will spontaneously ignite.

The amount of liquid that will be vaporized is also related to the volatility of the liquid. The higher the temperature, the more liquid that will evaporate. Liquids that have a lower molecular weight will tend to vaporize more readily than liquids with a higher molecular weight. As more of the liquid vaporizes, the mixture may reach a point where enough vapor is present in the air to create a flammable vapor–air mixture.

Two additional terms are used to describe the flammability of liquids: flash point and flame point. The flash point is the lowest temperature at which a liquid produces a flammable vapor. It is measured by determining the lowest temperature at which a liquid

FIGURE 6-23 An interior structure fire can be affected by an exterior wind.

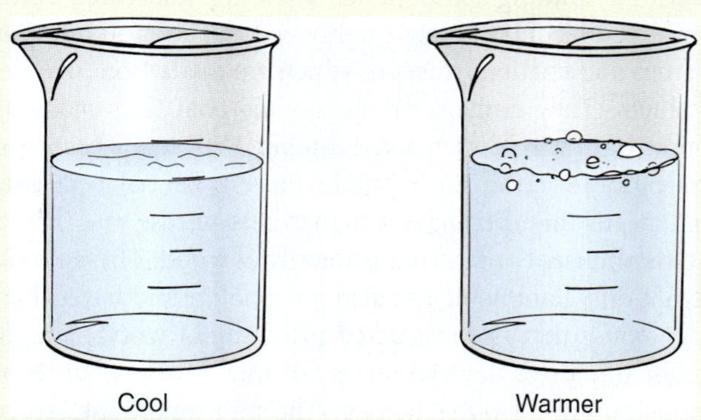

FIGURE 6-24 As liquids are heated, the molecules become more active and the speed of vaporization increases.

will produce enough vapor to support a small flame for a short period of time (the flame may go out quickly) TABLE 6-2. The flame point (also known as the fire point) is the lowest temperature at which a liquid produces enough vapor to sustain a continuous fire.

TABLE 6-2 Flash Point and Flame Point

Fuel	Flash Point (°F/°C)
Gasoline	−45/−43
Ethanol	54/12
Diesel fuel	104–131/40–55
SAE N 10 motor oil	340/171

Source: Principles of Fire Protection Chemistry and Physics. NFPA, 1998, p. 115.

Characteristics of Gas-Fuel Fires

By learning about the characteristics of flammable gas fuels, you can help to prevent injuries or deaths in emergency situations and work to mitigate the conditions causing the problem. Two terms are used to describe the characteristics of flammable vapors: vapor density and flammability limits.

Vapor Density

Vapor density refers to the weight of a gas fuel and measures the weight of the gas compared to air TABLE 6-3. The weight of air is assigned the value of 1. A gas with a vapor density less than 1 will rise to the top of a confined space or rise in the atmosphere. For example, hydrogen gas, which has a vapor density of 0.07, is a very light gas. Conversely, a gas with a vapor density greater than 1 is heavier than air and will settle close to the ground. For example, propane gas has a vapor density of 1.51 and will settle to the ground. By comparison, carbon monoxide has a vapor density of 0.97—almost the same as that of air—so it mixes readily with all layers of the air. In situations where a flammable gas is present, fire fighters need to recognize the vapor density of the escaping fuel so that they can take actions to prevent the ignition of the fuel and allow the gaseous fuel to safely escape into the atmosphere.

TABLE 6-3 Vapor Density of Common Gases

Gaseous Substance	Vapor Density
Gasoline	> 3.0
Ethanol	1.6
Methane	0.55
Propane	1.6

Flammability Limits

Mixtures of flammable gases and air will burn only when they are mixed in certain proportions. If too much fuel is present in the mixture, there will not be enough oxygen to support the combustion process; if too little fuel is present in the mixture, there will not be enough fuel to support the combustion process. The range of gas–air mixtures that will burn varies from one fuel to another. Carbon monoxide will burn when mixed with air in concentrations between 12.5 percent and 74 percent. By contrast, natural gas will burn only when it is mixed with air in concentrations between 4.5 percent and 15 percent.

The terms flammability limits and explosive limits are used interchangeably because under most conditions, if the flammable gas–air mixture will not explode, it will not ignite. The lower flammable limit (LFL) refers to the minimum amount of gaseous fuel that must be present in a gas–air mixture for the mixture to be flammable. In the case of carbon monoxide, the lower limit is 12.5 percent. The upper flammable limit (UFL) of carbon monoxide is 74 percent. Test instruments are available to measure the percentage of fuels in gas–air mixtures and to determine when an emergency scene is safe.

Boiling Liquid/Expanding Vapor Explosions

One potentially deathly set of circumstances involving liquid and gaseous fuels is a boiling liquid/expanding vapor explosion (BLEVE). A BLEVE occurs when a liquid fuel is stored in a vessel under pressure. If the vessel is filled with propane, for example, the bottom part of the vessel would contain liquid propane and the upper part of the vessel would contain vaporized propane FIGURE 6-25.

If this sealed container is subjected to heat from a fire, the pressure that builds up from the expansion of the liquid will prevent the liquid from evaporating. Normally, evaporation cools the liquid and allows it to maintain its temperature. If heating continues, the temperature inside the vessel reaches a level that exceeds the boiling point of the liquid. At some point, the vessel will fail, releasing all of the heated fuel in a massive explosion. The released fuel instantly becomes vaporized and ignited as a huge fireball.

The key to preventing a BLEVE is to cool the top of the tank, which contains the vapor. This action will prevent the fuel from building up enough pressure to cause a catastrophic rupture of the container. Prevention of BLEVEs is covered in more detail in the Fire Suppression chapter.

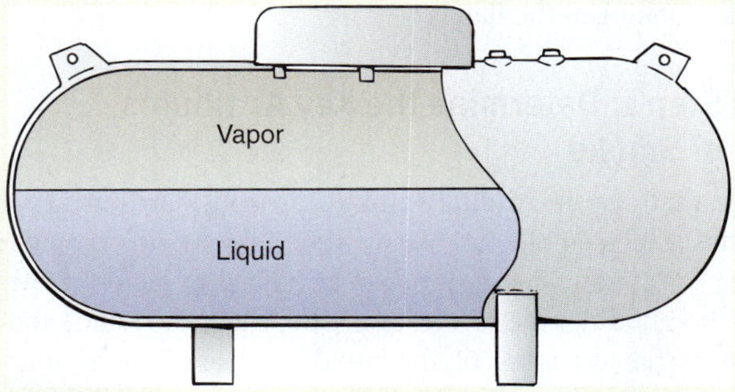

FIGURE 6-25 A propane tank contains both liquid and vapor.

Smoke Reading

Learning the principles underlying fire behavior helps fire fighters to understand the rules that govern the way fires burn. One practical application of these principles is learning how to "read" the smoke at a fire. Being able to read smoke potentially enables you to learn where the fire is, how big it is, and where it is moving. At a fire, most untrained people look at flames. Flames indicate where the fire is now, but they do not tell you how big the fire is and where it is moving. Fires are dynamic events—what you see this minute will probably change quickly. Before developing a plan of attack, fire fighters want to anticipate where the fire will be in a few minutes. At a structure fire, there are often no visible flames, because the fire is occurring inside the building. The ability to read smoke gives fire fighters information that they need to mount a more effective attack on a fire and may help to save either their lives or the lives of the building's occupants.

It is helpful to think about smoke as being a fuel. Fuels behave in predictable ways, when they are combined with the right mixture of oxygen and heat. For example, flashovers occur when fuels in a room or confined space are heated until they ignite simultaneously. Most of the fuel in a flashover is in the form of smoke. A backdraft is a sudden, explosive ignition of fire gases that occurs when oxygen is introduced into a superheated space with limited oxygen. As in a flashover, these fire gases are really smoke. Both flashovers and backdrafts are fed by superheated fuel, in the form of smoke. Thus, as you study the art of smoke reading, think of this exercise as studying the fuel that is all around you at a fire.

As mentioned earlier, smoke is composed of three major components: particles, vapors (aerosols), and gases. The solids consist of carbon, soot, dust, and fibers; the vapors contain suspended hydrocarbons and water; and the gases consist of carbon monoxide and a wide variety of other gases. Taken together, these three components of smoke largely represent fuels that will burn when heated sufficiently in the presence of oxygen. Therefore, hot smoke is extremely flammable and will ultimately dictate the fire's behavior. Smoke today is not the smoke of yesterday. Most of today's building materials and building contents are plastics, which are made from petroleum products. As a consequence, these materials give off more toxic gases and burn at higher temperatures than the materials used for construction and contents in earlier times.

The best place to observe patterns of smoke is from the outside of the fire building. Compare the smoke coming out of openings in different parts of the building. It will tell you more than looking at the flames.

Step 1: Determine the Key Attributes of Smoke

The first step in reading smoke is to consider four key attributes of the smoke: smoke volume, smoke velocity, smoke density, and smoke color. You also need to factor in the size of the building, the layout of the building, the wind, and any other characteristics of the building that might change the appearance of these key attributes.

The volume of smoke coming from the fire gives some idea of how much fuel is being heated to the point that it gives off gases (off-gassing). As you assess the smoke volume, also consider the size of the burning building (the box). It takes relatively little smoke to fill up a small building, but a lot of smoke is required to fill up a large building. Thus you must consider the size of the box when considering the amount of smoke coming from it. Assessing the volume of smoke alone will not provide a complete picture of the fire, but it sets the stage for better understanding the fire.

The velocity (speed) at which smoke is leaving the building suggests how much pressure is accumulating in the building. Smoke is pushed both by heat and by volume. When smoke is pushed by heat, it will rise and then slow down gradually. When smoke is pushed by volume, it will slow down immediately. Assess the smoke velocity to determine whether it is being pushed by heat or by volume.

Next, consider whether the smoke has a laminar flow or a turbulent flow. Laminar smoke flow is a smooth or streamlined flow. It indicates that the box (the building and its contents) are absorbing heat and the pressure in the box is not too high FIGURE 6-26. By contrast, turbulent smoke flow is agitated, boiling, or angry. This type of flow is caused by rapid molecular expansion of the gases within the smoke and the restrictions created by the building (the box) FIGURE 6-27. This expansion occurs when the box cannot absorb any more heat. In such circumstances, the radiant heat that the box has absorbed is radiated back into the smoke. Turbulent smoke contains an immense amount of energy. When this energy reaches a point where the smoke is heated to its ignition temperature, flashover will occur. Whenever you see turbulent smoke, be aware that flashover is likely to occur very soon.

Next, compare the velocity or speed of the smoke leaving different openings of the building. By comparing the speed and type of flow from similar-sized openings, fire fighters can get a good idea of where the fire is located. A similar-sized opening with a higher velocity of smoke is closer to the location of the fire. Look for the fastest smoke coming from the most restrictive opening. Studying the velocity of smoke provides valuable information as fire fighters prepare to attack the fire.

Evaluate the thickness of the smoke. Smoke density suggests how much fuel is contained in the smoke. The denser the smoke is, the more fuel it contains. Dense smoke can produce a powerful flashover when conditions are right. In fact, thick

FIGURE 6-26 Laminar smoke flow.

FIGURE 6-27 Turbulent smoke flow.

smoke can flash over even without turbulent flow because the supply of fuel reaches from the heart of the fire to the location where the smoke is issuing from the building.

When you think of smoke as fuel, you quickly realize that fire fighters who are surrounded by dense smoke are working in a pool of flammable fuel. While no one would send a fire fighter into a pool of petroleum, we are doing just that when we send a fire fighter into dense smoke.

Moreover, dense smoke contains many poisonous substances. When fire fighters are in an environment of dense smoke that has banked down to the floor, they are in an environment that will not support life for a building occupant. Just a few breaths of this toxic brew will result in death. Also, if a flashover occurs, this environment will immediately change to a deadly environment for fire fighters—even those in full turnout gear and SCBA. There is no sense in a fire fighter risking his or her life to try to save someone who is already dead.

The color of smoke may give you some indication of which stage the fire is in and which substances are burning. When a single substance is burning, the color of the smoke may provide valuable clues about what that substance is. Of course, most of the fires involve multiple substances burning, so using smoke color alone to determine what is burning may not always yield a clear answer.

When first-arriving fire fighters evaluate smoke color, they can determine the stage of heating and obtain information about the location of the fire in the building. Some solid materials such as wood will emit a white-colored smoke when they are first heated. This white color is primarily a result of moisture being released from the material. The same effect is apparent when fire fighters apply water to a fire: The smoke goes from black to white. As material dries out, however, the color of the smoke emitted changes. Smoke from burning wood, for example, changes to tan or brown. Plastics and painted or stained surfaces emit a gray smoke, which is a combination of black from the hydrocarbons and white from the escaping moisture. When synthetic materials are heated, the white vapor created is uncombusted fuel.

The color of smoke can also help fire fighters determine the location of the fire. As smoke travels, its heat evaporates moisture from some of the materials it passes through. This added moisture tends to change black smoke to lighter smoke the farther it travels from the fire. In addition, the carbon suspended in the black smoke settles out and is filtered by the materials it passes through. Therefore, carbon-rich black smoke often becomes lighter in color as it travels farther from the initial fire site, because of the loss of carbon from the smoke.

How do you differentiate between white smoke caused by early heating and white smoke from a hot fire that has traveled a distance? White smoke that is lazy or slow indicates early heating. White smoke that has its own pressure indicates smoke from a hot fire that has traveled for some distance.

Black Fire

Black fire is a high-volume, high-velocity, turbulent, ultra-dense, black smoke. You can think of it as a form of fire, because in many ways it is. Black fire is so hot (temperatures up to 1000°F [540°C]) that it creates a deadly environment. Black fire can produce charring as well as heat damage to steel and concrete. Its presence indicates impending auto ignition and flashover. Thus black fire is often just seconds away from becoming a deadly environment even for fire fighters wearing full PPE. Because black fire is not tenable for fire fighters in full protective equipment, it certainly is not tenable for unprotected occupants of a building. When black fire occurs, there are usually no lives to be saved in that location.

FIRE FIGHTER Tips

Smoke rising from a fire can help you determine the energy being given off by the fire. A fire that is generating large amounts of energy will emit quantities of smoke that are actively mixing or fluctuating. This angry mixing is called turbulent flow. As the smoke travels away from the fire and radiates some of its energy, the flow smooths out. This smooth flow is called laminar flow and indicates smoke that has less energy either because the fire is small or because the smoke has travelled some distance and has lost the high energy of turbulent smoke.

Step 2: Determine What Is Influencing the Key Attributes

Consider the size of the structure that is on fire. How big is the box in which the fire is contained? A small amount of smoke rapidly fills up a small building, so it does not take a very big fire to fill a small building with smoke. Conversely, in a large building (such as a "big box" store or a home improvement center), it takes a large fire to produce enough smoke to fill the whole building and pressurize the smoke coming out of the building. A fire fighter who pulls up to a large building that has signs of smoke from the outside should realize that only a significant fire would produce enough smoke to vent to the outside. Always consider the size of the box when examining smoke for clues about the fire.

Also consider how wind, thermal balance, fire streams, ventilation openings, and an operating fire sprinkler system might affect the normal characteristics of smoke. Wind can change the direction in which the smoke travels. The normal thermal balance in a room or building can be affected by the heating and air-conditioning systems, especially in a high-rise building. Ventilation openings are made with the intention of changing the normal flow of smoke, so the characteristics of the smoke would naturally be expected to change as the building is ventilated. An operating sprinkler system will cause smoke to be cooled; this cool smoke will then hang close to the bottom of the room, making it difficult to see.

Step 3: Determine the Rate of Change

To complete the smoke reading, determine the rate of change for the event. Remember that flames indicate what is happening now, whereas smoke gives a more complete picture of the characteristics of the fire and where it is going. Are the volume of smoke, the velocity of smoke, the density of smoke, and the color of smoke changing? How are they changing? How rapidly are these changes occurring? What do these changes suggest about the progression of the fire?

Step 4: Predict the Event

To assess the size and location of a fire, it is necessary to first consider the four key attributes of smoke—volume, velocity, density, and color. Then, fire fighters need to consider what might be influencing these key attributes. Next, fire fighters should try to determine the rate of change: Is it getting better or is it getting worse? Finally, it is time to put these pieces of the puzzle together. The previously mentioned information should help fire fighters to better determine the location of the fire, the size of the fire, and the potential for a hostile fire event such as a backdraft or a flashover. Communicate the key parts of these observations to the company officer. With practice, this assessment process will help you systematically evaluate the information the smoke signals are giving you about the fire.

One way to become more proficient in smoke reading is to review videos of fires. Assess the smoke at the beginning of the tape and try to identify the location of the fire and the stage of burning. You will be surprised at how much information you can obtain from a short video clip. This information on smoke reading might not have been taught to fire fighters who completed their initial training in the past, but today's smoke is deadly. If you want to stay alive, you need to understand it.

Smoke Reading Through a Door

When you see indications of a hot fire, such as darkened windows, but little visible smoke is coming from around closed doors and from around windows, you may be dealing with a fire that is in a decay stage because it does not have sufficient oxygen. This is a sign of great danger: As soon as this fire receives a fresh supply of oxygen, it will likely produce a violent backdraft. Fires can be dangerous even when little smoke is showing.

When you open a door, watch what the smoke does. If smoke exits through the top half of the door and clean air enters through the bottom half of the opening, then the fire is probably on the same level. This phenomenon is sometimes called smoke that has found balance.

If smoke rises and the opening clears out, fresh air is being pulled into the building. This indicates the fire is probably above the level of the opening.

If smoke thins when the door is opened, but smoke still fills the door, the fire is probably below the level of the opening.

Wrap-Up

Chief Concepts

- Fire is a rapid chemical process that produces heat and light.
- Matter is made up of atoms and molecules.
- Matter exists in three states: solid, liquid, and gas.
    - A solid has definite capacity for resisting forces and under ordinary conditions retains a definite size and shape.
    - A liquid assumes the shape of the container in which it is placed.
    - A gas is a type of liquid that has neither independent shape nor independent volume, but rather tends to expand indefinitely.
- Fuels are materials that store energy. Energy exists in many forms, including chemical, mechanical, electrical, light, or nuclear.
    - Chemical energy is the energy created by a chemical reaction.
    - Mechanical energy is converted to heat when two materials rub against each other and create friction.
    - Electrical energy is converted to heat energy when it flows through a conductive material. Light energy is caused by electromagnetic waves packaged in photons and travels as thermal radiation.
    - Nuclear energy is stored in radioactive materials and converted to electricity by nuclear power-generating stations.
- The three basic conditions needed for a fire to occur are fuel, oxygen, and heat. A fourth factor, a chemical chain reaction, is required to maintain a self-sustaining fire.
- The by-product of fire is smoke. Smoke includes three major components: particles, vapors, and gases. Smoke consists mainly of unburned forms of hydrocarbon fuels.
- Fire may be spread by conduction, convection, and radiation.
- Direct contact is a flame touching a fuel.

- Conduction is the transfer of heat through matter, like heat traveling up a metal spoon.
- Convection is the circulatory movement that occurs in a gas or fluid. Convection currents in a fire involve hot gases generated by the fire that rise because they are lighter, creating a higher pressure. Cooler gases are denser and move to the lower areas of the compartment. Convection pushes hot gases from the fire compartment to other areas of the building and is instrumental in spreading the fire beyond the room of origin.
- Radiation is the transfer of heat through the emission of energy in the form of invisible waves.
- The four principal methods of fire extinguishment are cooling the fuel, excluding oxygen, removing the fuel, and interrupting the chemical reaction.
- Fires are categorized as Class A, Class B, Class C, Class D, and Class K. These classes reflect the type of fuel that is burning and the type of hazard that the fire represents.
- Class A fires involve ordinary solid combustible materials such as wood, paper, and cloth.
- Class B fires involve flammable or combustible liquids such as gasoline, kerosene, diesel fuel, and motor oil.
- Class C fires involve energized electrical equipment.
- Class D fires involve combustible metals such as sodium, magnesium, and titanium.
- Class K fires involve combustible cooking oils and fats in kitchens.
- Most fires encountered by fire fighters involve solid fuels. Solid fuels do not actually burn in a solid state. Instead, they must be heated or pyrolyzed to decompose into a vapor before they will burn.
- Free-burning
  - Three primary factors that influence the combustion of solid fuel fires are: the composition of the fuel, the amount of fuel, and the configuration of the fuel. The factors that influence the fuel configuration are the surface-to-mass ratio, the orientation of the fuel, and the continuity of the fuel.
- There are four stages of fire development: the incipient stage, the growth stage, the fully developed stage, and the decay stage.
- The incipient stage occurs as the fuel starts to burn, but the fire is confined to the area of origin and does not significantly affect the atmosphere in the fire compartment.
- The growth stage occurs as the fire begins to involve fuels beyond the ignition point. During this stage, the fire begins to create a plume of hot gases that begin to act on the fire compartment. The temperature within the fire compartment begins to rise and to pyrolyze fuels close to the fire. When pressure from the burning gases reaches a higher pressure, the fire may begin to spread beyond the room of origin.
- The fully developed stage occurs when all available fuel has ignited and heat is being produced at the maximum rate. Some fires will not reach a fully developed stage because they are ventilation limited.
- During the decay stage, the rate of burning slows down because less fuel is available.
- The growth of room-and-contents fires depends on the characteristics of the room and the contents of the room. Synthetic products are widely used in today's homes. The by-products of heated plastics are not only flammable, but also toxic.
- Special conditions within the fire compartment include: thermal layering of gases, rollover, backdraft, flashover, rapid fire growth, and ventilation-limited fires.
- Thermal layering is the property of gases in an enclosed space in which they form layers according to their temperature. The hottest gases travel by convection currents to the top level of the room.
- Rollover is the ignition of the hot, unburned gases that have accumulated at the top of the fire compartment.
- Backdraft is caused by a change of the ventilation profile, permitting the introduction of oxygen into an enclosure where superheated gases and contents are hot enough for ignition, but the fire does not have sufficient oxygen to cause their combustion.
- Flashover is the near-simultaneous ignition of most of the exposed combustible materials in an enclosed area.
- Modern structures tend to be more tightly sealed, be constructed of lighter-weight materials, and contain more plastics. These characteristics can lead to a greater risk of backdrafts when a fire occurs in such a structure.
- Liquid-fuel fires require the proper mixture of fuel and air, an ignition source, and contact between the fuel mixture and the ignition.
- The characteristics of flammable vapors can be described in terms of vapor density and flammability limits. Vapor density reflects the weight of a gas compared to air. Flammability limits vary widely for different fuels.
- A boiling liquid/expanding vapor explosion (BLEVE) is a catastrophic explosion in a vessel containing both a boiling liquid and a vapor.
- Assessment of smoke volume, velocity, density, and color enables fire fighters to predict the location of a fire and its stage of development.
- Smoke reading requires fire fighters to evaluate the effect of the building, the weather, and ventilation on the smoke.

Hot Terms

Aerosols Small solid particles or liquid droplets that are suspended in air.

Atom The smallest particle of an element, which can exist alone or in combination.

Backdraft A phenomenon that occurs when a fire takes place in a confined area, such as a sealed aircraft fuselage, and burns undetected until most of the oxygen within is consumed. The heat continues to produce flammable gases, mostly in the form of carbon monoxide. These gases are heated above their ignition temperature and when a supply of oxygen is introduced, as when normal entry points are opened, the gases could ignite with explosive force. (NFPA 402)

Black fire A hot, high-volume, high-velocity, turbulent, ultra-dense black smoke that indicates an impending flashover or auto ignition.

Boiling liquid/expanding vapor explosion (BLEVE) An explosion that occurs when a tank containing a volatile liquid at the bottom of the tank and a flammable gas at the top of the tank is heated to the point where the tank ruptures.

Box A burning structure.

Chemical energy Energy that is created or released by the combination or decomposition of chemical com-pounds.

Class A fire A fire in ordinary combustible materials, such as wood, cloth, paper, rubber, and many plastics. (NFPA 10)

Class B fire A fire in flammable liquids, combustible liquids, petroleum greases, tars, oils, oil-based paints, solvents, lacquers, alcohols, and flammable gases. (NFPA 10)

Class C fire A fire that involves energized electrical equipment. (NFPA 10)

Class D fire A fire in combustible metals, such as magnesium, titanium, zirconium, sodium, lithium, and potassium. (NFPA 10)

Class K fire A fire in a cooking appliance that involves combustible cooking media (vegetable or animal oils and fats). (NFPA 10)

Combustion A chemical process of oxidation that occurs at a rate fast enough to produce heat and usually light in the form of either a glow or a flame. (NFPA 101)

Compartment A space completely enclosed by walls and a ceiling. The compartment enclosure is permitted to have openings in walls to an adjoining space if the openings have a minimum lintel depth of 8 inches (203 mm) from the ceiling and the openings do not exceed 8 feet (2.44 m) in width. A single opening of 36 inches (914 mm) or less in width without a lintel is permitted when there are no other openings to adjoining spaces. (NFPA 13)

Conduction Heat transfer to another body or within a body by direct contact. (NFPA 921)

Convection Heat transfer by circulation within a medium such as a gas or a liquid. (NFPA 921)

Decay stage The stage of fire development in which the fire has consumed either the available fuel or oxygen and is starting to die down.

Electrical energy Heat that is produced by electricity.

Endothermic Reactions that absorb heat or require heat to be added.

Exothermic Reactions that result in the release of energy in the form of heat.

Fire A rapid, persistent chemical reaction that releases both heat and light.

Flow path The volume in a structure between an inlet and an outlet that allows the movement of heat and smoke from the higher pressure within the fire area toward the lower-pressure areas accessible via doorways, halls, stairs, and window openings.

Fire tetrahedron A geometric shape used to depict the four components required for a fire to occur: fuel, oxygen, heat, and chemical chain reactions.

Fire triangle A geometric shape used to depict the three components of which a fire is composed: fuel, oxygen, and heat.

Flame point (fire point) The lowest temperature at which a substance releases enough vapors to ignite and sustain combustion.

Flammability limits (explosive limits) The upper and lower concentration limits (at a specified temperature and pressure) of a flammable gas or vapor in air that can be ignited, expressed as a percentage of the fuel by volume.

Flashover A transition stage in the development of a compartment fire in which surfaces exposed to thermal radiation reach ignition temperature more or less simultaneously and fire spreads rapidly throughout the space, resulting in full room involvement or total involvement of the compartment or enclosed space. (NFPA 921)

Flash point The lowest temperature at which a liquid produces a flammable vapor.

Fuel A material that will maintain combustion under specified environmental conditions. (NFPA 53)

Fully developed stage The stage where the fire is consuming the maximum amount of fuel possible. It achieves the maximum heat release rate possible for the fuel supply and oxygen present. It may be ventilation limited or fuel limited.

Gas A material that has a vapor pressure greater than 300 kPa absolute (43.5 psia) at 50°C (122°F) or is completely gaseous at 20°C (68°F) at a standard pressure of 101.3 kPa absolute (14.7 psia). (NFPA 30A)

Growth stage The stage of fire development in which the fire is spreading beyond the point of origin and beginning to involve other fuels in the immediate area.

Heat flux The measure of the rate of heat transfer from one surface to another. It is measured in kilowatts per meter squared (kW/m^2) or Btu/ft^2.

Hypoxia A state of inadequate oxygenation of the blood and tissue sufficient to cause impairment of function. (NFPA 99B)

Ignition The action of setting something on fire.

Ignition temperature Minimum temperature a substance should attain to ignite under specific test conditions. (NFPA 921)

Incipient stage The stage of fire development where the fire has not progressed beyond a size that can be extinguished with a portable fire extinguisher. The fire is not big enough to influence the fire compartment. At this stage, the fire does not produce a sufficient quantity of smoke to require protective clothing or self-contained breathing apparatus for respiratory protection.

Laminar smoke flow Smooth or streamlined movement of smoke, which indicates that the pressure in the building is not excessively high.

Liquid Any material that (1) has a fluidity greater than that of 300 penetration asphalt when tested in accordance with ASTM D5, *Standard Test Method for Penetration of Bituminous Materials*, or (2) is a viscous substance for which a specific melting point cannot be determined but that is determined to be a liquid in accordance with ASTM D4359, *Standard Test Method for Determining Whether a Material Is a Liquid or a Solid*. (NFPA 30)

Lower flammable limit (LFL) That concentration of a combustible material in air below which ignition will not occur; also known as the lower explosive limit (LEL). Mixtures below this limit are said to be "too lean." (NFPA 329)

Matter A substance made up of atoms and molecules.

Mechanical energy A form of potential energy that can generate heat through friction.

Neutral plane The interface at a vent, such as a doorway or a window opening, between the hot gas flowing out of a fire compartment and the cool air flowing into the compartment where the pressure difference between interior and exterior is zero.

Oxidation Reaction with oxygen either in the form of the element or in the form of one of its compounds. (NFPA 53)

Plume The column of hot gases, flames, and smoke rising above a fire; also called convection column, thermal updraft, or thermal column. (NFPA 921)

Pyrolysis The destructive distillation of organic compounds in an oxygen-free environment that converts the organic matter into gases, liquids, and char. (NFPA 820)

Radiation The emission and propagation of energy through matter or space by means of electromagnetic disturbances that display both wave-like and particle-like behavior. (NFPA 801)

Rollover The condition where unburned fuel (pyrolysate) from the originating fire has accumulated in the ceiling layer to a sufficient concentration (i.e., at or above the lower flammable limit) that it ignites and burns. This can occur without ignition of, or prior to, the ignition of other fuels separate from the origin. (NFPA 921)

Smoke The airborne solid and liquid particulates and gases evolved when a material undergoes pyrolysis or combustion, together with the quantity of air that is entrained or otherwise mixed into the mass. (NFPA 318)

Smoke color The attribute of smoke that reflects the stage of burning of a fire and the material that is burning in the fire.

Smoke density The thickness of smoke. Because it has a high mass per unit volume, smoke is difficult to see through.

Smoke explosion A violent release of confined energy that occurs when a mixture of flammable gases and oxygen are present, usually in a void or other area separate from the fire compartment, and come in contact with a source of ignition. In a smoke explosion, there is no change to the ventilation profile, such as an open door or window; rather it occurs from the travel of smoke within the structure to an ignition source.

Smoke velocity The speed of smoke leaving a burning building.

Smoke volume The quantity of smoke, which indicates how much fuel is being heated.

Solid One of the three stages of matter; a material that has three dimensions and is firm in substance.

Temperature The degree of sensible heat of a body as measured by a thermometer or similar instrument.

Thermal column A cylindrical area above a fire in which heated air and gases rise and travel upward.

Thermal layering The stratification (heat layers) that occurs in a room as a result of a fire.

Thermal radiation The means by which heat is transferred to other objects.

Turbulent smoke flow Agitated, boiling, angry-movement smoke, which indicates great heat in the burning building. It is a precursor to flashover.

Upper flammable limit (UFL) The highest concentration of a combustible substance in a gaseous oxidizer that will propagate a flame. (NFPA 68)

Vapor density The weight of an airborne concentration (vapor or gas) as compared to an equal volume of dry air.

Ventilation-limited A fire in an enclosed building that is restricted because there is insufficient oxygen available for the fire to burn as rapidly as it would with an unlimited supply of oxygen. Increasing the supply of oxygen to a ventilation-limited fire may result in a rapid flashover.

Volatility The ability of a substance to produce combustible vapors.

FIRE FIGHTER *in action*

You are spending your evening searching the Web for firefighting clips. Hours pass by as you watch video clip after video clip. Then you run across a video clip that makes you drop your mouse. The video clip shows two fire fighters entering a two-story house with fire coming out of a side window. Smoke is pouring out of the upper half of the front door, and you can see that the upper floor is full of smoke.

As the clip continues, the smoke on the upper floor gets darker and darker and begins boiling out of the window. To your surprise, a fire fighter emerges in the turbulent smoke of the window. He is clearly in trouble.

1. What stage is this fire in?
- **A.** Incipient stage
- **B.** Growth stage
- **C.** Fully developed stage
- **D.** Decay stage

2. How is the heat from this fire traveling from one room to another on the second floor?
- **A.** Convection
- **B.** Direct contact
- **C.** Conduction
- **D.** Radiation

3. Which is not a part of the fire triangle?
- **A.** Fuel
- **B.** Chemical chain reaction
- **C.** Oxygen
- **D.** Ignition source

4. ____________ is the speed of the smoke leaving a burning building.
- **A.** Smoke velocity
- **B.** Smoke density
- **C.** Smoke volume
- **D.** Smoke color

5. ____________ is a hot, high-volume, high-velocity, turbulent, ultra-dense black smoke that indicates an impending flashover or autoignition.
- **A.** Black fire
- **B.** Laminar flow
- **C.** BLEVE
- **D.** Pyrolysis

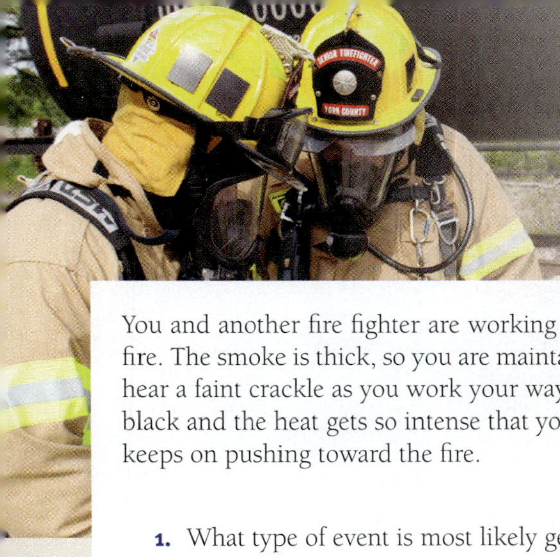

FIRE FIGHTER II *in action*

You and another fire fighter are working your way through dirty brown smoke in a house fire trying to find the seat of the fire. The smoke is thick, so you are maintaining physical contact with each other. You can feel the heat on your back, and you hear a faint crackle as you work your way to the back of the structure. Before you can reach the sound, the smoke turns jet-black and the heat gets so intense that you are forced onto your stomach. You know this is a bad situation, but your partner keeps on pushing toward the fire.

1. What type of event is most likely getting ready to occur?
2. What type of protective actions do you need to take?
3. What do you say, if anything, to your partner?
4. If you feel that you need to retreat, but your partner does not, what would you do?

Building Construction

Fire Fighter I

Knowledge Objectives

After studying this chapter, you will be able to:

- Explain how occupancy classifications affect fire suppression operations. (p 168)
- Explain how the contents of a structure affect fire suppression operations. (p 168–169)
- Describe the characteristics of masonry building materials. (p 169)
- Describe the characteristics of concrete building materials. (p 169–170)
- Describe the characteristics of steel building materials. (p 170–171)
- Describe the characteristics of glass building materials. (p 171)
- Describe the characteristics of gypsum building materials. (p 171–172)
- Describe the characteristics of wood building materials. (p 172–173)
- Describe the characteristics of engineered wood building materials. (p 172)
- Describe the characteristics of plastic building materials. (p 173)
- List the five types of building construction. (p 173)
- Describe the characteristics of Type I construction. (NFPA 5.3.12.A, p 174–175)
- Describe the effects of fire on Type I construction. (NFPA 5.3.12.A, p 174–175)
- Describe the characteristics of Type II construction. (NFPA 5.3.12.A, p 175–176)
- Describe the effects of fire on Type II construction. (NFPA 5.3.12.A, p 175–176)
- Describe the characteristics of Type III construction. (NFPA 5.3.12.A, p 176)
- Describe the effects of fire on Type III construction. (NFPA 5.3.12.A, p 176)
- Describe the characteristics of Type IV construction. (NFPA 5.3.12.A, p 176–177)
- Describe the effects of fire on Type IV construction. (NFPA 5.3.12.A, p 176–177)
- Describe the characteristics of Type V construction. (NFPA 5.3.12.A, p 177–180)
- Describe the effects of fire on Type V construction. (NFPA 5.3.12.A, p 177–180)
- Describe the characteristics of balloon-frame construction. (NFPA 5.3.12.A, p 178)
- Describe the effects of fire on balloon-frame construction. (NFPA 5.3.12.A, p 178)
- Describe the characteristics of platform-frame construction. (NFPA 5.3.12.A, p 180)
- Describe the effects of fire on platform-frame construction. (NFPA 5.3.12.A, p 180)
- Describe the purpose of a foundation in a structure. (p 180)
- List the warning signs of foundation collapse. (p 180)
- Explain how floor construction affects fire suppression operations. (p 180–181)
- Describe the characteristics of fire-resistive floors. (p 180–181)
- Describe the characteristics of wood-supported floors. (p 181)
- Describe the characteristics of ceiling assemblies. (p 181)
- List the three primary types of roofs. (p 181–184)
- Describe the characteristics of trusses. (p 184–185)
- List the types of trusses. (p 184–185)
- Describe the effects of fires on trusses. (p 185–186)
- Describe the characteristics of walls. (p 186–187)
- List the common types of walls in structures. (p 186–187)
- Describe the characteristics of door assemblies. (p 188)
- Describe the characteristics of window assemblies. (p 188)
- Describe the characteristics of fire doors. (p 188–189)
- Describe the characteristics of fire windows. (p 188–189)
- Explain the effect that interior finishes have on fire suppression operations. (p 189)
- Describe the hazards that buildings under construction or demolition pose to fire fighters. (p 190)
- Describe how building construction factors into preincident planning and incident size-up. (p 190–191)

Skills Objectives

There are no skill objectives for Fire Fighter I candidates. NFPA 1001 contains no Fire Fighter I Job Performance Requirements for this chapter.

CHAPTER

7

Fire Fighter II — FFII

Knowledge Objectives

There are no knowledge objectives for Fire Fighter II candidates. NFPA 1001 contains no Fire Fighter II Job Performance Requirements for this chapter.

Skills Objectives

There are no skill objectives for Fire Fighter II candidates. NFPA 1001 contains no Fire Fighter II Job Performance Requirements for this chapter.

Additional NFPA Standards

- NFPA 80, *Standard for Fire Doors and Other Opening Protectives*
- NFPA 220, *Standard on Types of Building Construction*

You Are the Fire Fighter

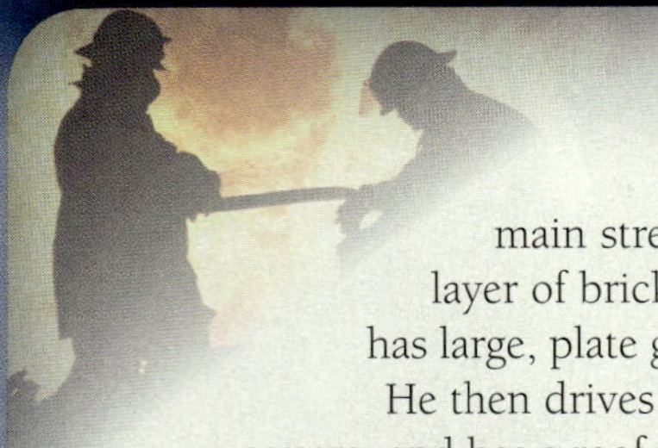

Your crew is driving around your district looking at the various buildings. The engineer drives through a historic residential district and points to a two-story house with a wrap-around porch, tall narrow windows, and a stone foundation. He then cuts over to the old main street and points to a commercial building. It is three stories tall and made of brick. Every seventh layer of brick is laid crosswise. It has a flat roof and is about 30' (9 meters) wide and 90' (27 meters) deep. It has large, plate glass windows on the first floor, and there are normal size windows on the second and third floor. He then drives to a residential part of town built in the 1940s. Each house is one story, is about 40' (12 meters) square, and has a roof with asphalt shingles. Each home has wooden clapboard siding and a small porch on the front. He then drives to a newer commercial area where there is a stand-alone convenience store. It was built in 1998 and has a flat roof and metal siding. He turns to you and says, "We just looked at four different buildings. Tell me about each one."

1. Why do fire fighters need to understand building construction?
2. What effect does building construction have on fires?
3. Which building would most likely collapse first if exposed to fire?

Big Chapter for test questions

Introduction

Knowing the basic types of building construction is vital for fire fighters because building construction affects how fires grow and spread. Fire fighters must be able to recognize and understand different types of building construction. They must understand how each type of building construction reacts when exposed to the effects of heat and fire. This knowledge will help them anticipate the fire's behavior and respond accordingly. An understanding of building construction will also help fire fighters determine when it is safe to enter a burning building and when it is necessary to evacuate a building. Your safety, the safety of your team members, and the safety of the building's occupants—all depend on your knowledge of building construction.

Construction, however, is merely one component of a complex relationship. Fire fighters also must consider the occupancy of the building and the building contents. Fire risk factors vary depending on how a building is used. To establish a course of action at a fire, it is necessary to consider the interactions between three factors: the building construction, the occupancy of the building, and the contents of the building. The following sections relate to occupancy and contents, and the rest of this chapter focuses on those aspects of building construction that are important to a fire fighter.

Occupancy

The term occupancy refers to how a building is used. Based on a building's occupancy classification, a fire fighter can predict who is likely to be inside the building, how many people are likely to be there, and what they are likely to be doing. Occupancy also suggests the types of hazards and situations that might be encountered in the building.

Occupancy classifications are used in conjunction with building and safety codes to establish regulatory requirements, including the types of construction that can be used to construct buildings of a particular size, use, or location. Regulations classify building occupancies into major categories—such as residential, health care, business, and industrial—based on common characteristics.

Occupancy classifications can be used to predict the number of occupants who are likely to be at risk in a fire. For example, hospitals and nursing homes are occupied 24 hours a day by persons who will probably need assistance to evacuate. An office building probably has a large population during the day, but most of the workers should be able to evacuate the site without assistance. A daycare center might be filled with young children during the day but is likely to be empty at night. A nightclub is probably empty during the day and crowded at night.

Fire hazards in different types of occupancies also vary greatly. A computer repair shop has different characteristics from an auto transmission shop, for example. A factory could be used to manufacture cast-iron pipe fittings or airplanes. The materials used could be flammable, toxic, or fire resistant. Fire fighters must consider each of these factors when responding to a particular building.

Building codes require that certain types of building construction be used for specific types of occupancies. In most communities, however, buildings that were built for one reason are often being used for some other purpose. Buildings constructed before modern building codes were adopted may present significant hazards for fire fighters.

Contents

Contents also must be considered when responding to a building. Building contents vary widely, but are usually closely related to the occupancy of a building. Some buildings have noncombustible contents that would not feed a fire, whereas others could be so dangerous that fire fighters could not safely attack any fire at the site. For example, an automobile repair shop generally contains many toxic and flammable substances.

Whenever possible, fire fighters should prepare a prefire or preincident plan that accounts for these possibilities when buildings present special hazards.

Even similar occupancies can pose different levels of risk, however. For example, three warehouses, although identical from the exterior, could have very different contents that would either increase or reduce the risks to fire fighters. One might contain ceramic floor tiles, the second might be filled with wooden furniture, and the third might serve as a storehouse for swimming pool chemicals. The tiles will not burn, the furniture will burn easily, and the chemicals create toxic products of combustion and contaminated runoff.

When evaluating the risks at any alarm, consider the occupancy, the contents, and the building construction. They represent an interwoven package of risks to fire fighters.

Types of Construction Materials

An understanding of building construction begins with the materials used. Building components are usually made of different materials. The properties of these materials and the details of their construction determine the basic fire characteristics of the building itself.

Function, appearance, price, and compliance with building and fire codes are all considerations when selecting building materials and construction methods. Architects often place a priority on functionality and aesthetics when selecting materials, whereas builders are often more concerned about price and ease of construction. For their part, building owners are usually interested in durability and maintenance expenses, as well as the materials' initial cost and aesthetics.

Fire fighters, however, use a completely different set of factors to evaluate construction methods and materials. Their chief concern is the behavior of the building under fire conditions. In particular, a building material or construction method that is attractive to architects, builders, and building owners might have the potential to create serious problems or deadly hazards for fire fighters.

The most commonly used building materials are wood, engineered wood products, masonry, concrete, steel, aluminum, glass, gypsum board, and plastics. Within these basic categories, hundreds of variations are possible. The key factors that affect the behavior of each of these materials under fire conditions are outlined here:

- Combustibility: Whether or not a material will burn determines its combustibility. Materials such as wood will burn when they are ignited, releasing heat, light, and smoke, until they are completely consumed by the fire. Concrete, brick, and steel are noncombustible materials that cannot be ignited and are not consumed by a fire.
- Thermal conductivity: This characteristic describes how readily a material will conduct heat. Heat flows very readily through metals such as steel and aluminum. By contrast, brick, concrete, and gypsum board are poor conductors of heat.
- Decrease in strength at elevated temperatures: Many materials lose strength at elevated temperatures. For example, steel loses strength and will bend or buckle when exposed to the high temperatures associated with fires, and aluminum melts in a fire. Some engineered building products are constructed of wood products held together by glue or plastics; the glue in these products begins to fail long before they are hot enough to burn. By contrast, bricks and concrete can generally withstand high temperatures for extended periods of time.
- Thermal expansion when heated: Some materials—steel, in particular—expand significantly when they are heated. A steel beam exposed to a fire will stretch (elongate); if it is restrained so that it cannot elongate, it will sag, warp, or twist. As a general rule, a steel beam will elongate at a rate of 1 inch (2.5 centimeters) per 10 feet (3 meters) of length at a temperature of 1000°F (540°C). This expansion can push out the walls of buildings and cause collapse.

Masonry

Masonry includes stone, concrete blocks, and brick. The individual components are usually bonded together into a solid mass with mortar, which is produced by mixing sand, lime, water, and Portland cement. A concrete–masonry unit (CMU) is a preassembled wall section of brick or concrete blocks that contains steel reinforcing rods. A complete CMU can be delivered to a construction site, ready to be erected without further preparation.

Masonry materials are inherently fire resistive FIGURE 7-1; that is, they do not burn or deteriorate at the temperatures normally encountered in building fires. Masonry is also a poor conductor of heat, so it will limit the passage of heat from one side through to the other side. For these reasons, masonry is often used to construct fire walls to protect vulnerable materials. A fire wall helps prevent the spread of a fire from one side of the wall to the other.

All masonry walls are not necessarily fire walls, however. A single layer of masonry may be placed over a wood-framed building to make it appear more substantial. If a masonry wall contains unprotected openings, a fire can often spread through them. If the mortar has deteriorated or the wall has been exposed to fire for a prolonged time, a masonry wall can collapse during a fire.

A masonry structure also can collapse under fire conditions if the roof or floor assembly collapses. In such a case, the masonry falls because of the mechanical action of the collapse. Likewise, aged, weakened mortar can contribute to a collapse. Regardless of the cause, a collapsing masonry wall can be a deadly hazard to fire fighters.

Concrete

Like masonry, concrete is a naturally fire-resistive material. Because it does not burn or conduct heat well, it is often used to insulate other building materials from fire. Concrete does not have a high degree of thermal expansion—that is, it does not expand greatly when exposed to heat and fire—nor will it lose strength when exposed to high temperatures.

A.

B.

FIGURE 7-1 Masonry materials are inherently fire resistive. **A.** Solid masonry wall. **B.** Brick veneer wall.

Concrete is made from a mixture of Portland cement and aggregates such as sand and gravel. Different formulations of concrete can be produced for specific building purposes. Concrete is inexpensive and easy to shape and form. It is often used for foundations, columns, floors, walls, roofs, and exterior pavement.

Under compression, concrete is strong and can support a great deal of weight; in contrast, under tension, it is weak. When this material is used in building construction, steel reinforcing rods are often embedded in the concrete to strengthen it under tension. In turn, the concrete insulates the steel reinforcing rods from heat.

Although it is inherently fire resistive, concrete can be damaged by exposure to a fire. For example, a fire may convert trapped moisture in the concrete to steam. As the steam expands, it creates internal pressure that can cause sections of the concrete surface to break off in a process called spalling. Severe spalling in reinforced concrete can expose the steel reinforcing rods to the heat of the fire. If the fire is hot enough, the steel might weaken, resulting in a structural collapse, though this is a rare event.

Fire Fighter Safety Tips

Heat transfer can occur through exposed steel reinforcing rods. This mode of heat transfer is known as conduction.

Steel

Steel is the strongest building material in widespread use. It is considered very strong in terms of both tension and compression. Steel can also be produced in a wide variety of shapes and sizes, ranging from heavy beams and columns to thin sheets. This material is often used in the structural framework of a building to support floor and roof assemblies. It is resistant to aging and does not rot, although most types of steel will rust unless they are protected from exposure to air and moisture.

Steel is an alloy of iron and carbon. Other metals may be added to the mix to produce steel with special properties, such as stainless steel or galvanized steel. When considered by itself, steel is not fire resistive. It will melt at extremely high temperatures, although these extraordinary temperatures are rarely encountered at structure fires **FIGURE 7-2**. Steel conducts heat well, so it tends to expand and lose strength as the temperature increases. For this reason, other materials—such as masonry, concrete, or layers of gypsum board—are often used to protect steel from the heat of a fire. Sprayed-on coatings of mineral or cement-like materials are also used to insulate steel members. The amount of heat absorbed by steel depends on the mass of the object and the amount of protection surrounding it. Smaller, lighter pieces of steel absorb heat more easily than larger and heavier pieces.

FIGURE 7-2 Steel will lose strength at temperatures of more than 1000°F (540°C).

Steel both expands and loses strength as it is heated. As a consequence, an unprotected steel roof beam directly exposed to a fire may elongate sufficiently to cause a supporting wall to collapse. Heated steel beams will often sag and twist, whereas steel columns tend to buckle as they lose strength. The bending and distortion are caused by the uneven heating that occurs in actual fire situations. Many residential occupancies are now constructed using metal studs instead of wood.

Failure of a steel structure depends on three factors: the mass of the steel components, the loads placed upon them, and the methods used to connect the steel pieces. Unfortunately, there are no accurate indicators that would enable fire fighters to predict when a steel beam is ready to fail. Any sign of bending, sagging, or stretching of steel structural members should be considered a warning of an immediate risk of failure. If this sign is noticed, all fire fighters should immediately clear the area because of the potential for imminent collapse.

Other Metals

A variety of other metals, including aluminum, copper, and zinc, are also used in building construction. Aluminum is often used for siding, window frames, door frames, and roof panels. Copper is used primarily for electrical wiring and piping; it is sometimes used for decorative roofs, gutters, and downspouts as well. Zinc is used primarily as a coating to protect metal parts from rust and corrosion.

Aluminum is occasionally used as a structural material in building construction. It is more expensive and not as strong as steel, so its use is generally limited to light-duty applications such as awnings and sunshades. Aluminum expands more extensively than steel when heated and loses strength quickly when exposed to a fire. This material has a lower melting point than steel, so it will often melt and drip in a fire.

Glass

Almost all buildings contain glass—in windows, doors, skylights, and sometimes walls. Ordinary glass breaks very easily, but glass can be manufactured to resist breakage and to withstand impacts or high temperatures.

Although ordinary glass will usually break when exposed to fire, specially formulated glass fire rated can be used as a fire barrier in certain situations. The thermal conductivity of glass is rarely a significant factor in the spread of fire.

Many different types of glass are available for use in building construction:

- Ordinary window glass will usually break with a loud pop when heat exposure to one side causes it to expand and creates internal stresses that fracture the glass. The broken glass forms large shards, which usually have sharp edges.
- Tempered glass is much stronger than ordinary glass and more difficult to break. Some tempered glass can be broken with a spring-loaded center punch; it will shatter into small pieces that do not have the sharp edges of ordinary glass. Tempered or laminated glass is usually required in installations such as storm doors to prevent injuries from shards of broken glass.
- Laminated glass is manufactured so that a thin sheet of plastic is placed between two sheets of glass. The resulting product is much stronger than ordinary glass, is difficult to break with ordinary hand tools, and will usually deform instead of breaking. When exposed to a fire, laminated glass windows are likely to crack, but remain in place. Laminated glass is sometimes used in buildings to help soundproof areas.
- Glass blocks are thick pieces of glass similar to bricks or tiles. They are designed to be built into a wall with mortar so that light can be transmitted through the wall. Glass blocks have limited strength and are not intended to be used as part of a load-bearing wall, but they can usually withstand a fire. Some glass blocks are approved for use with fire-rated masonry walls.
- Wired glass is made by molding tempered glass with a reinforcing wire mesh FIGURE 7-3. When wired glass is subjected to heat, the wire holds the glass together and prevents it from breaking. This material is often used in fire doors and windows designed to prevent fire spread.

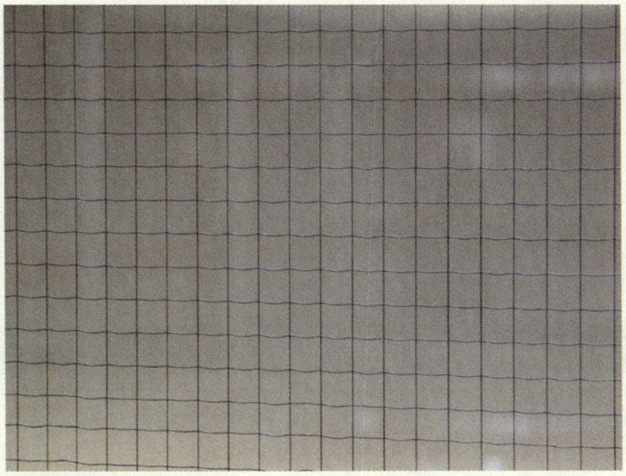

FIGURE 7-3 Wired glass.

Gypsum Board

Gypsum is a naturally occurring mineral composed of calcium sulfate and water molecules that is used to make plaster of Paris. Gypsum is a good insulator and noncombustible; it will not burn even in atmospheres of pure oxygen.

Gypsum board (also called drywall, sheetrock, or plasterboard) is commonly used to cover the interior walls and ceilings of residential living areas and commercial spaces. It is also required in some applications, such as residential garages, to create a rated fire-resistant barrier. Gypsum board is manufactured in large sheets consisting of a layer of compacted gypsum sandwiched between two layers of specially produced paper. In a building, these sheets are nailed or screwed in place on a framework of wood or metal studs, and the edges are secured with a special tape. The nail or screw heads and tape are then covered with a thin layer of plaster.

Gypsum board has limited combustibility, because the paper covering will burn slowly when exposed to a fire. It does not conduct or release heat to an extent that would contribute to fire spread, so this material is often used to create a firestop or to protect building components from fire. Notably, gypsum blocks are sometimes used as protective insulation around steel members or to create fire-resistive enclosures or interior fire walls.

When gypsum board is heated, some of the water present in the calcium sulfate will evaporate, causing the board to deteriorate. If it is exposed to a fire for a long time, this material will fail. Water will also weaken and permanently damage gypsum board.

Even if the gypsum board retains its integrity after a fire, the sections that were directly exposed to the fire should be replaced after the incident. The sections that were heated will probably be unable to serve as effective fire barriers in the future.

Although gypsum board is a good finishing material and an effective fire barrier, it is not a strong structural material. It must be properly mounted on and supported by wood or steel studs. Gypsum board mounted on wood framing will protect the wood from fire for a limited time.

■ Wood

Wood is probably the most commonly used building material in today's environment. It is inexpensive to produce, is easy to use, and can be shaped into many different forms, ranging from heavy structural supports to thin strips of exterior siding. Both soft woods (such as pine) and hard woods (such as oak) are used for building construction.

FIRE FIGHTER Tips

Older buildings were built with lumber that was true to its stated dimensions. A two-by-four stud actually measured 2 inches by 4 inches (5 by 10 centimeters). In buildings built today, a two-by-four stud measures $1^1/_2$ inches by $3^3/_8$ inches (4 by 8 centimeters) or less. In a fire, the true two-by-four stud will resist fire longer than the $1^1/_2$ inches by $3^3/_8$ inches (4 by 8 centimeters) stud.

Solid lumber is squared and cut into uniform lengths. Examples of solid lumber range from the heavy timbers used in mills and barns, to smaller two-by-four studs, to wood siding, interior wood trim, wood floors, and wood shingles on roofs. Solid wood can be used in a wide variety of applications in a building. This type of lumber may be found in multiple sizes in either older or legacy construction or newer or contemporary construction.

Engineered Wood Products

Engineered wood products are also called manufactured board, manmade wood, or composite wood. They include plywood, fiberboard, oriented strand board, and particle board—materials that are used in roofs, walls, floors, countertops, doors, I-joists, trusses, and cabinets. These products are manufactured from small pieces of wood that are held together with glue or adhesive. The adhesives include urea-formaldehyde resins, phenol-formaldehyde resins, melamine-formaldehyde resins, and polyurethane resins.

Engineered wood products are preferred over natural wood for a variety of reasons. In particular, they are usually less expensive than large solid planks of natural wood. Such products can also be fabricated in longer lengths and for specific applications that are not possible with natural wood. They can maximize the natural characteristics of wood, thereby producing greater strength and stiffness. These products are easy to work with using ordinary tools, and they can readily be shaped into curved surfaces. With all of these advantages, it is not surprising that the use of engineered wood products has greatly increased in recent years.

Unfortunately, some drawbacks also arise with engineered wood products. One disadvantage is that some of these products may be prone to warping in conditions of high humidity. A second disadvantage is that these products may contain toxic products, such as formaldehyde, which are introduced during manufacturing. The biggest disadvantage—at least from fire fighters' perspective—is that these products burn quickly and fail quickly and without warning under fire conditions and produce large void spaces.

Fire Fighter Safety Tips

The glue that is used to create laminated wood is flammable.

Implications for Fire Fighters

For fire fighters, the most important characteristic of wood and manufactured wood products is their high combustibility. Wood acts as fuel in a structural fire and can provide a path for the fire to spread. It ignites at fairly low temperatures and is gradually consumed by the fire, weakening and eventually leading to the collapse of the structure. Great quantities of heat and hot gases are created by the wood burning process, until eventually all that remains after the fuel is consumed is a small quantity of residual ash.

Wood components weaken as they are consumed by fire. A burning wood structure becomes weaker every minute, making it imperative that fire fighters always operate on the side of safety in such environments.

The rate at which wood ignites, burns, and decomposes depends on several factors:

- Ignition: A small ignition source contains less energy and will take a longer time to ignite a fire. Using an accelerant will greatly speed this process.
- Moisture: Damp or moist wood takes longer to ignite and burn. New lumber usually contains more moisture than lumber that has been in a building for many years. A higher relative humidity in the atmosphere also makes wood more difficult to ignite.
- Density: Heavy, dense wood is harder to ignite than lighter or less dense wood.
- Preheating: The more the wood is preheated, the faster it ignites.
- Size and form: The rate of combustion is directly related to the surface area of the wood. During the combustion process, high temperatures release flammable gases from the surface of the wood. A large solid beam is difficult to ignite and burns relatively slowly; by contrast, a lightweight truss ignites more easily and is consumed rapidly.

Exposure to the high temperatures generated by a fire can also decrease the strength of wood through the process of pyrolysis. This chemical change occurs when wood or other materials are heated to a temperature high enough to release some of the volatile compounds in the wood, albeit without igniting these gases. The wood then begins to decompose without combustion.

Fire-Retardant–Treated Wood

Wood cannot be treated to make it completely noncombustible. Nevertheless, impregnating wood with mineral salts makes it more difficult to ignite and slows the rate of burning. Fire-retardant treatment can significantly reduce the fire hazards of wood construction. On the downside, the treatment process can also reduce the strength of the wood.

In some cases, fire-retardant–treated wood can pose a danger for fire fighters. Some fire-retardant chemicals used to treat plywood roofing panels, for example, will cause the wood to rapidly deteriorate and weaken. The plywood may then fail, necessitating the replacement of these panels. Fire fighters should know whether fire-retardant–treated plywood is used in the community. If it is, they should avoid standing on roofs made with these panels during a fire, because the extra weight could cause the plywood to collapse.

Fire Fighter Safety Tips

Avoid standing on roofs that have been constructed with fire-retardant–treated wood or with trusses. Both types of roofs can collapse in a short period of time.

Plastics

Plastics are synthetic materials that are used in a wide variety of products today FIGURE 7-4. Plastics may be transparent or opaque, stiff or flexible, tough or brittle. They are often combined with other materials to produce building products.

FIGURE 7-4 Plastic building materials can be used for siding, window frames, and interior finishes.

Although plastics are rarely used for structural support, they may be found throughout a building. Building exteriors may include vinyl siding, plastic window frames, and plastic panel skylights. Foam plastic materials may be used as exterior or interior insulation. Plastic pipe and fittings, plastic tub and shower enclosures, and plastic lighting fixtures are commonly used. Even carpeting and floor coverings often contain plastics.

The combustibility of plastics varies greatly. Some plastics ignite easily and burn quickly, whereas others will burn only when an external source of heat is present. Some plastics will withstand high temperatures and fire exposure without igniting.

Many plastics produce quantities of heavy, dense, dark smoke and release high concentrations of toxic gases as they burn. This smoke resembles smoke from a petroleum fire—not surprising given that most plastics are made from petroleum products.

Thermoplastic materials melt and drip when exposed to high temperatures, even those as low as 500°F (260°C). Although heat can be used to shape these materials into a variety of desired forms, the dripping, burning plastic can rapidly spread a fire. By contrast, thermoset materials are fused by heat and will not melt as the plastic burns, although their strength will decrease dramatically.

Types of Construction

Fire fighters need to understand and recognize five types of building construction—designated as Types I through V. These classifications are directly related to fire protection and fire behavior in the building.

Buildings are classified based on the combustibility of the structure and the fire resistance of its components. Buildings using Type I or Type II construction are assembled with primarily noncombustible materials and limited amounts of wood and other materials that will burn. This does not mean that Type I and Type II buildings cannot be damaged by a fire. In fact, if the contents of the building burn, the fire can seriously damage or destroy the structure.

In buildings using Type III, Type IV, or Type V construction, both the structural components and the building contents will burn. Wood and wood products are used to varying degrees in these buildings. If the wood ignites, the fire will weaken and consume both the structure and its contents. Structural elements of these buildings can be damaged or destroyed in a very short time.

Fire resistance refers to the length of time that a building or building component can withstand a fire before igniting. Fire resistance ratings are stated in hours, based on the time that a test assembly withstands or contains a standard test fire. For example, walls are rated based on whether they stop the progress of the test fire for periods ranging from 20 minutes to 4 hours. A floor assembly is rated based on whether it supports a load for 1 hour or 2 hours. Because ratings are based on a standard test fire but an actual fire could be more or less severe than the test fire, ratings are considered merely guidelines; that is, an assembly with a 2-hour rating will not necessarily withstand a fire for 2 hours.

Building codes specify the type of construction to be used, based on the height, area, occupancy classification, and location of the building. Other factors, such as the presence of automatic sprinklers, will also affect the required construction classification. NFPA 220, *Standard on Types of Building Construction*, provides detailed requirements for each type of building construction.

Near Miss REPORT

Report Number: 07-0000726

Synopsis: Truss roof collapse causes near miss at church fire.

Event Description: We were involved in fire suppression activities on a working attic fire at a church. Company crew members were attempting to perform a trench cut to ventilate and confine the fire to an area adjacent to the fire wall. Upon laddering the roof, and while crew members were still operating on and from the aerial apparatus, a sudden collapse of the truss roof occurred, endangering the crew because of fire and radiant heat. The collapse involved approximately 40 feet of the roof line, and a fireball extended upward approximately 30 feet (9 meters) into the air. The ventilation crew had to make an immediate descent down the aerial to escape harm. The collapse occurred about 10 minutes after arrival of first-in companies. All firefighters involved were in full personal protective equipment (PPE) and self-contained breathing apparatus (SCBA). Awareness of wood truss construction must always be considered during fire suppression activities.

Lessons Learned: Fire ground accountability, awareness of wood truss construction, the fire's impact because of direct impingement, and the duration of the fire must always be considered during fire suppression activities. In this case, had the crew actually been actively ventilating or on the roof, serious injuries and most likely firefighter fatalities would have occurred.

Awareness training for wood truss construction should be included in any ventilation and structural fire suppression training. Command and control, fire-ground accountability, and fire-ground safety officer training should also be included in such education.

We must always keep in mind that the fire ground is a dynamic, ever-changing scene, and we must adapt to those conditions and stay aware of them. Remember—no structure is worth our lives. Everyone goes home.

Know 5 Types of Construction

Type I Construction: Fire Resistive

Type I construction (also called fire-resistive construction) is the most fire-resistive category of building construction. It is used for buildings designed for large numbers of people, buildings with a high life-safety hazard, tall buildings, large-area buildings, and buildings containing special hazards. Type I construction is commonly found in schools, hospitals, and high-rise buildings FIGURE 7-5.

FIGURE 7-5 Type I construction is commonly found in schools, hospitals, and high-rise buildings.

Building with Type I construction can withstand and contain a fire for a specified period of time. The fire resistance and combustibility of all construction materials are carefully evaluated in such cases, and each building component must be engineered to contribute to fire resistance of the entire building to warrant this classification.

All of the structural members and components used in Type I construction must be made of noncombustible materials, such as steel, concrete, or gypsum board. In addition, the structure must be constructed or protected so that it has at least 2 hours of fire resistance. Building codes specify the fire-resistance requirements for different components. For example, columns and load-bearing walls in multistory buildings could have a fire-resistance requirement of 3 or 4 hours, whereas a floor might be required to have a fire-resistance rating of only 2 hours. Some codes allow Type I buildings to use a limited amount of combustible materials as part of the interior finish.

If a Type I building exceeds specific height and area limitations, codes generally require the use of fire-resistive walls and/or floors to subdivide it into compartments. A compartment might consist of a single floor in a high-rise building or a part of a floor in a large-area building. In any event, a fire in one compartment should not spread to any other parts of the building. To ensure that fire is contained, stairways, elevator shafts, and utility shafts should be enclosed in construction that prevents fire from spreading from floor to floor or from compartment to compartment.

Type I buildings typically use reinforced concrete and protected steel-frame construction. As mentioned earlier, concrete is noncombustible and provides thermal protection around the steel reinforcing rods. Reinforced concrete can fail, however, if it is subjected to a fire for a long period of time or if the building contents create an extreme fire load. Given its tendency to lose strength in the face of high temperatures, structural steel framing must be protected from the heat of a fire. In Type I construction, the structural steel members are generally encased in concrete, shielded by a fire-resistive ceiling, covered with multiple layers of gypsum board, or protected by a sprayed-on insulating material **FIGURE 7-6**. An unprotected steel beam exposed to a fire can fail and cause the entire building to collapse.

FIGURE 7-6 Sprayed-on fireproofing materials are often used to protect structural steel.

Type I building materials should not provide enough fuel by themselves to create a serious fire. Thus it is the contents of the building that determine the severity of a Type I building fire. In theory, a fire could consume all of the combustible contents of a Type I building, yet leave the structure basically intact.

Although Type I construction might provide the highest degree of safety, it does not eliminate all of the risks to occupants and fire fighters. Serious fires can still occur in fire-resistive buildings, because the burning contents may produce copious quantities of heat and smoke. Even though the structure is designed to give fire fighters time to conduct an interior attack, it can be very difficult to extinguish a fire in a Type I building. A fire that is fully contained within fire-resistive construction can be very hot and difficult to ventilate. For this reason, fire-resistive construction is typically combined with an automatic fire detection system and automatic sprinklers in modern high-rise buildings.

Sometimes, the burning contents of a building provide sufficient fuel and generate enough heat to overwhelm the fire-resistant properties of the construction. In these cases, the fire can escape from its compartment and damage or destroy the structure. Similarly, inadequate or poorly maintained construction can undermine the fire-resistive properties of the building materials. Under extreme fire conditions, a Type I building can collapse. Even so, these buildings have a much lower potential for structural failure than do other buildings.

Type II Construction: Noncombustible

Type II construction is also referred to as noncombustible construction **FIGURE 7-7**. All of the structural components in a Type II building must be made of noncombustible materials. The fire-resistive requirements, however, are less stringent for Type II construction than for Type I construction. In some cases, no fire-resistance requirements apply to the building. In other cases, known as protected noncombustible construction, a fire-resistance rating of 1 or 2 hours may be required for certain elements.

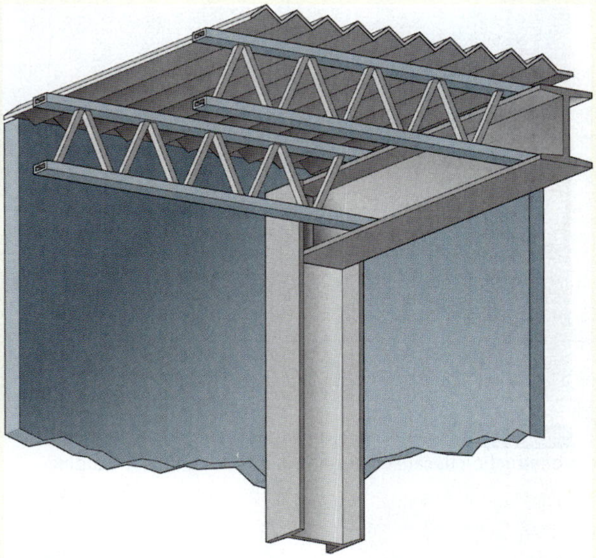

FIGURE 7-7 Type II construction is also referred to as noncombustible construction.

Noncombustible construction is most common in single-story warehouse or factory buildings, where vertical fire spread is not an issue. Unprotected construction of this type is generally limited to a maximum of two stories. Some multistory buildings are constructed using protected noncombustible construction.

Steel is the most common structural material in Type II buildings. Insulating materials can be applied to the steel when fire resistance is required. A typical example of Type II construction is a large-area, single-story building with a steel frame, metal or concrete block walls, and a metal roof deck. Fire walls are sometimes used to subdivide these large-area buildings and prevent catastrophic losses. Undivided floor areas must be protected with automatic sprinklers to limit the fire risk.

Fire severity in a Type II building is determined by the contents of the building because the structural components contribute little or no fuel and interior finish materials are limited. If the building contents provide a high fuel load, a fire can collapse and destroy the structure. In this kind of construction, automatic sprinklers should be used to protect combustible and valuable contents. Fires in buildings equipped with monitored fire detection systems are more likely to be reported earlier than fires in buildings without monitored fire detection systems.

Type III Construction: Ordinary

Type III construction is also referred to as ordinary construction because it is used in a wide variety of buildings, ranging from commercial strip malls to small apartment buildings FIGURE 7-8. Ordinary construction is usually limited to buildings of no more than four stories, but it can sometimes be found in buildings as tall as six or seven stories.

FIGURE 7-8 Type III construction is referred to as ordinary construction because it is used in a wide variety of buildings.

Type III buildings have masonry exterior walls, which support the floors and the roof structure. The interior structural and nonstructural members—including the walls, floor, and roof—are all constructed of wood. In most ordinary construction, gypsum board or plaster is used as an interior finish material, covering the wood framework and providing minimal fire protection.

Fire-resistance requirements for interior construction of Type III buildings are limited. The gypsum or plaster coverings over the interior wood components provide some fire resistance. Key interior structural components may be required to have fire resistance ratings of 1 or 2 hours, or there may be no requirements at all. Some Type III buildings use interior masonry load-bearing walls to meet the requirements for fire-resistant structural support.

A building of Type III construction has two separate fire loads: the contents and the combustible building materials used. Even a vacant building will contain a sufficient quantity of wood and other combustible components to produce a large fire. Given these materials' combustibility, a fire involving both the contents and the structural components can quickly destroy the building.

Fortunately, most fires originate with the building contents and are extinguished before the flames ignite the Type III structure itself. Once a fire extends into the structure itself and begins to consume the fuel within the walls and above the ceilings, it becomes much more difficult to control and prone to roof or floor collapse.

Type III construction presents several problems for fire fighters. For example, an electrical fire can begin inside the void spaces within the walls, floors, and roof assemblies and extend to the contents. The void spaces also allow a fire to extend vertically and horizontally, spreading from room to room and from floor to floor. In such a case, fire fighters will have to open the void spaces to fight the fire. An uncontrolled fire within the void spaces is likely to destroy the building. Fire fighters need to be especially aware that fires in void spaces under floors or under the roof can quickly destroy the strength of these components. Do not be surprised by these "hidden" fires; instead, anticipate them.

The fire resistance of interior structural components often depends on the age of the building and on local building codes. Older Type III buildings were built from solid dimensional lumber, which can contain or withstand a fire for a limited time. Newer or renovated buildings might contain lightweight assemblies that can be damaged much more quickly and are prone to early failure. Floors and roofs in these buildings collapse suddenly and without warning. Because fire fighters cannot always determine the type of structure during a fire, you should assume that lightweight components may be present in any Type III building.

Exterior walls also can collapse if a fire causes significant damage to the interior structure. Because the exterior walls, the floors, and the roof are all connected in a stable building, the collapse of the interior structure could make the free-standing masonry walls unstable and prone to collapse.

FIRE FIGHTER Tips

Fire-resistive structures with sprinklers are a great asset in assisting fire fighters in controlling the burning contents. When they activate, however, the sprinklers produce steam when the water hits the fire. As the sprinklers continue to operate, the steam is cooled and condenses; the condensed water vapor may then fill an enclosed area from floor to ceiling. Without proper ventilation, fire fighters may find themselves with virtually no visibility. Recall that 1 gallon (4 liters) of water weighs 8.33 pounds (3.77 kilograms)—an important reason why fire fighters must calculate the total water load on the building.

Type IV Construction: Heavy Timber

Type IV construction is also known as heavy timber construction. A heavy timber building has exterior walls that consist of masonry construction, and interior walls, columns, beams, floor assemblies, and roof structure that are made of wood. The exterior walls are usually brick and are extra thick to support the weight of the building and its contents.

The wood used in Type IV construction is much heavier than that used in Type III construction. It is more difficult to ignite and will withstand a fire for a much longer time before the building collapses. A typical heavy timber building might have 8-inch-square wood posts (20 centimeters) supporting 14-inch-deep wood beams (35 centimeters) and 8-inch floor joists (20 centimeters). The floors could be constructed of solid wood planks, 2 or 3 inches (5 or 8 centimeters) thick, with a top layer of wood serving as the finished floor surface.

Heavy timber construction, if it meets building codes, has no concealed spaces or voids. This structure helps reduce the horizontal and vertical fire spread that often occurs in ordinary construction buildings. Unfortunately, many heavy timber buildings do contain vertical openings for elevators, stairs, or machinery, which can provide a path for a fire to travel from one floor to another.

In the past, heavy timber construction was used to construct buildings as tall as six to eight stories with open spaces suitable for manufacturing and storage occupancies. Today, modern buildings of this style are usually built with either fire-resistive (Type I) or noncombustible (Type II) construction. New buildings of Type IV construction are rare, except for special structures that feature the construction components as architectural elements.

The heavy, solid wood columns, support beams, floor assemblies, and roof assemblies used in heavy timber construction will withstand a fire much longer than the smaller wood members used in ordinary and lightweight combustible construction. Nevertheless, once involved in a fire, the structure of a heavy timber building can burn for many hours. A fire that ignites the combustible portions of heavy timber construction is likely to burn until it runs out of fuel and the building is reduced to a pile of rubble. As the fire consumes the heavy timber support members, the masonry walls will become unstable and collapse.

Mill construction was common during the 1800s in the United States, especially in northeastern states. In particular, large mill buildings were used as factories and warehouses. In many communities, dozens of these large buildings were clustered together in industrial areas, creating the potential for huge fires involving multiple multistory buildings. The radiant heat from a major fire in one of these buildings could spread the fire to nearby buildings, resulting in the loss of several surrounding structures.

Mill construction was state-of-the-art for its time. Automatic sprinkler systems were developed to protect buildings of mill construction; as long as the sprinklers are properly maintained, these buildings have a good safety record. Without sprinklers, however, a heavy timber mill building becomes a major fire hazard.

Only a few of these original mill buildings are still used for manufacturing. Many more have been converted into small shops, galleries, office buildings, and residential occupancies. These conversions tend to divide the open spaces into smaller compartments and create void spaces within the structure. The revamped buildings may contain either noncombustible contents or highly flammable contents, depending on their use. Occupancies may range from unoccupied storage facilities to occupied loft apartments and office space. Appropriate fire protection and life-safety features, such as modern sprinkler systems, must be built into the conversions to ensure the safety of occupants.

Type V Construction: Wood Frame

In Type V construction, all of the major components are constructed of wood or other combustible materials **FIGURE 7-9**. Type V construction is often called wood-frame construction and is the most common type of construction used today. Wood-frame construction is used not only in one- and two-family dwellings and small commercial buildings, but also in larger structures such as apartment and condominium complexes and office buildings up to four stories in height.

FIGURE 7-9 Type V (wood-frame) construction is the most common type of construction used today.

Many wood-frame buildings do not have any fire-resistive components. Some codes require a 1-hour fire resistance rating for limited parts of wood-frame buildings, particularly those of more than two stories. Plaster or gypsum board barriers are often used to achieve this rating, but fire fighters should not assume that these barriers will be present or effective in all Type V buildings.

Because all of their structural components are combustible, wood-frame buildings can rapidly become totally involved in a fire. In addition, Type V construction usually creates voids and channels that allow a fire within the structure to spread quickly. A fire that originates in a Type V building can easily extend to other nearby buildings.

Wood-frame buildings often collapse and suffer major destruction from fires. Smoke detectors are essential to warn building residents early if a fire occurs. Although compartmentalization can help limit the spread of a fire, fire detection devices and automatic sprinklers are the most effective way of protecting lives and property in Type V buildings.

Wood-frame buildings are constructed in a variety of ways. Older wood-frame construction was assembled from solid lumber, which relied on its mass to provide strength.

To reduce costs and create the largest building with the least amount of material, lightweight construction makes extensive use of engineered wood products such as wooden I-beams and wooden trusses. With this construction technique, structural assemblies are engineered to be just strong enough to carry the required load. As a result, there is little built-in safety margin, and these buildings can collapse early, suddenly, and completely during a fire.

The presence of flammable building contents may greatly increase the severity of a fire and accelerate the speed with which the fire destroys a wood-frame building. Indeed, the structural elements of such a building can be damaged or destroyed in a very short time. Whenever a fire has been burning under a floor or roof, you should assume that the structure is unsafe and may collapse. Fires that start to consume structural parts of a Type V building present a high risk of building collapse. From a fire fighter's point of view, lightweight wood-frame construction is a campfire waiting to happen.

Wood-frame buildings might be covered with wood siding, vinyl siding, aluminum siding, brick veneer, or stucco. The covering on a building does not reflect the type of construction of that building. Just because you see a brick covering does not mean that the building is constructed of solid brick walls.

Two systems are used to assemble wood-frame buildings: balloon-frame construction and platform-frame construction. Because these systems developed in different eras, fire fighters can anticipate the type of construction based on the age of a building.

During a structural fire, the brick veneer is likely to collapse or peel away as the wall behind it burns or collapses. A single thickness of bricks is rarely strong enough to stand independently. Fire fighters should be aware of buildings constructed in this manner and maintain a safe distance from potentially falling bricks.

Fire Fighter Safety Tips

Lightweight wood-frame construction is a campfire waiting to happen. A fire in this type of construction will spread rapidly throughout the building and to adjacent exposures.

Fire Fighter Safety Tips

Many buildings have a brick veneer on the exterior walls of wood-frame construction. A thin layer of brick or stone might be applied to enhance the appearance of the building or to reduce the risk of fire spread. Unfortunately, this practice can give the impression that a building is Type III (ordinary) construction when it is really Type V (wood-frame) construction.

Balloon-Frame Construction

Balloon-frame construction was popular between the late 1800s and the mid-1900s. In a balloon-frame building, the exterior walls are assembled with wood studs that run continuously from the basement to the roof FIGURE 7-10. In a two-story building, the floor joists that support the first and second floors are nailed to these continuous studs. As a result, an open channel between each pair of studs extends from the foundation to the attic. Each of these channels provides a path that enables a fire to spread from the basement to the attic without being visible on the first- or second-floor levels. Fire fighters must anticipate that a fire originating on a lower level will quickly extend through these voids, and they should open the void spaces to check for hidden fires and to prevent rapid vertical extension.

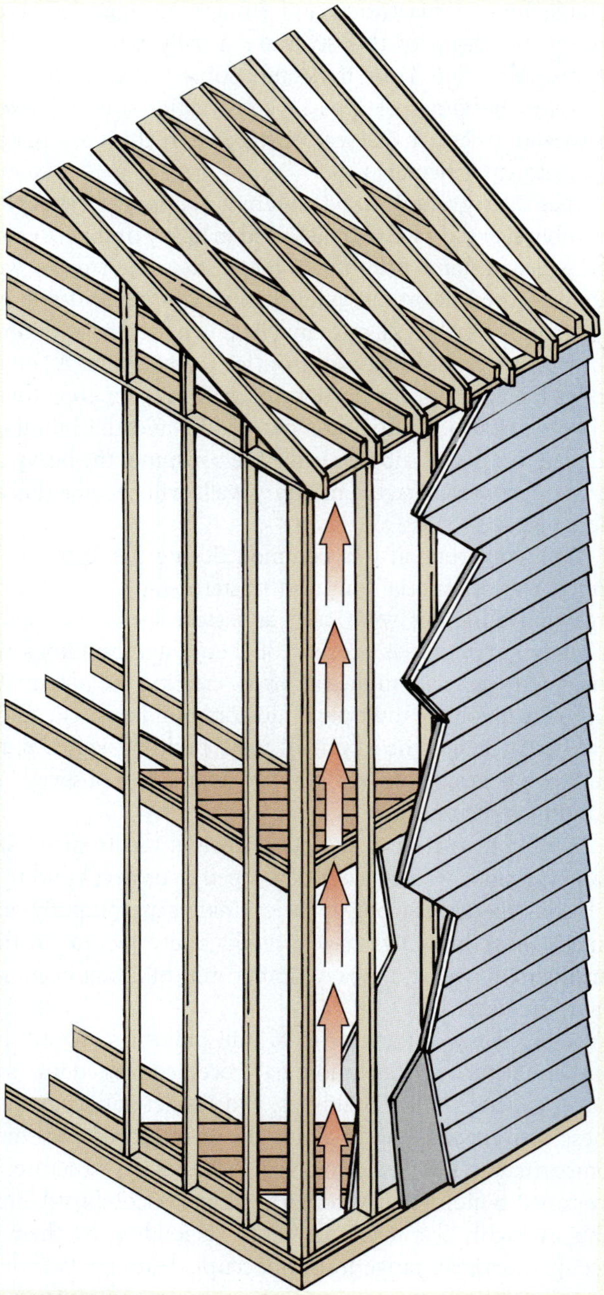

FIGURE 7-10 In a balloon-frame building, the exterior walls create channels that enable a fire to spread from the basement to the attic.

VOICES OF EXPERIENCE

It was early Monday morning, the day after the Super Bowl. It was 34 degrees below zero with the wind chill factor. The fire was two houses east of my house and was the second house from the corner. As I rounded the front of my house, I could see a well-established fire in the basement in the corner of sectors A and D. I knew that the upstairs occupants had escaped the fire and they were supposed to be the only ones in the house.

As soon as I saw the fire in the basement, I knew we were going to lose the building, an old balloon-frame two-story. I knew a fire in the basement was going straight up the exterior walls into the second floor void spaces and into the attic. An aerial ladder was set up in front in an attempt to cover the burn building and protect the exposure (a house less than ten feet from the burn building).

As the crews began to fight the fire, we were informed by an upstairs occupant that sometimes the landlord let a young man stay in the basement. She thought that he might be there. Three separate crews attempted entry into the basement without success. The fire did exactly what we thought it was going to do; it went straight up to the roof.

It was so cold that three engines and one ladder truck sustained substantial damage. One of them froze to the ground and had to be thawed out the next day. Most of the water we poured into the building settled into the basement and froze. We didn't find the victim until the next day. Not only was everything frozen, but the house had suffered severe damage and was unsafe to enter.

A young man lost his life that night, vehicles were damaged, and a building was destroyed. But all of the fire fighters working that night went home to their families without injury because we knew what the fire was going to do and how the building was going to react.

Al Martinez
Prairie State College
Chicago Heights, Illinois

Platform-Frame Construction

Platform-frame construction is used for almost all modern wood-frame construction. In a building with platform construction, the exterior wall studs are not continuous. Instead, the first floor is constructed as a platform, and the studs for the exterior walls are erected on top of it. The first set of studs extends only to the underside of the second-floor platform, which blocks any vertical void spaces. The studs for the second-floor exterior walls are erected on top of the second-floor platform. At each level, then, the floor platform blocks the path of any fire rising within the void spaces. Platform framing prevents fire from spreading from one floor to another through continuous stud spaces. Although a fire can eventually burn through the wood platform, the platform will slow the fire spread.

Building Components

The construction classification system gives fire fighters a general idea of the materials used in building and an indication of how the materials will react to fire. In addition, each building also has several different components that should be recognized. A fire fighter who understands how these components function will have a better understanding of the risks involved with a burning building. Some building construction features are safer for fire fighters, as this section explains.

The following major components of a building are discussed in this section:

- Foundations
- Floors and ceilings
- Roofs
- Trusses
- Walls
- Doors and windows
- Interior finishes and floor coverings

Foundations

The primary purpose of a building foundation is to transfer the weight of the building and its contents to the ground FIGURE 7-11. The weight of the building itself is called the dead load; the weight of the building's contents is called the live load. The foundation ensures that the base of the building is planted firmly in a fixed location, which helps keep all other components connected.

FIGURE 7-11 The foundation supports the entire weight of the building.

Modern foundations are usually constructed of concrete or masonry, although wood pilings may be used in some areas. Foundations can be either shallow or deep, depending on the type of building and the soil composition. Some buildings are built on concrete footings or piers; others are supported by steel piles or wooden posts driven into the ground.

As long as the foundation remains stable and intact, it is usually not a major concern for fire fighters. If the foundation shifts or is in poor repair, however, it can become a critical concern. A burning building with a weak foundation or inadequate lateral bracing could be in imminent danger of collapse.

Most foundation problems are caused by circumstances other than a fire, such as improper construction, shifting soil conditions, or earthquakes. Although fires can damage timber post and other wood foundations, most foundations remain intact even after a severe fire has damaged the rest of the building.

When examining a building, take a close look at the foundation. Look for cracks that indicate movement of the foundation. If the building has been modified or remodeled, look for areas where the support could be compromised.

Floors and Ceilings

Floor construction is very important to fire fighters for three major reasons. First, fire fighters who are working inside a building must rely on the integrity of the floor to support their weight. A floor that fails could drop the fire fighters into a fire on a lower level. Second, in a multistory structure, fire fighters may be working below a floor (or roof), which would fall on them if it collapsed. Third, the floor system influences whether a fire spreads vertically from floor to floor within a building or whether it remains contained on a single level. Some floor systems are designed to resist fires, whereas others have no fire-resistant capabilities.

Fire-Resistive Floors

In multistory buildings, floors and ceilings are generally considered a combined structural system—that is, the structure that supports a floor also supports the ceiling of the story below. In a fire-resistive building, this system is designed to prevent a fire from spreading vertically and to prevent a collapse when a fire occurs in the space below the floor–ceiling assembly. Fire-resistive floor–ceiling systems are rated in terms of hours of fire resistance based on a standard test fire.

Concrete floors are common in fire-resistive construction. The concrete can be cast in place or assembled from panels or beams of precast concrete, which are made at a factory and then transported to the construction site for placement. The thickness of the concrete floor depends on the load that the floor needs to support.

Concrete floors can be either self-supporting or supported by a system of steel beams or trusses. In the latter case, the steel can be protected by sprayed-on insulating materials or covered with concrete or gypsum board. If the ceiling is part of the fire-resistive rating, it provides a thermal barrier to protect the steel members from a fire in the area below the ceiling.

The ceiling below a fire-resistant floor can be constructed from plaster or gypsum board, or it can be a system of tiles suspended from the floor structure. In many cases, a void space between the ceiling and the floor above it contains building systems and equipment such as electrical or telephone wiring, heating and air-conditioning ducts, and plumbing and fire sprinkler system pipes FIGURE 7-12. If the space above the ceiling is not subdivided by fire-resistant partitions or protected by automatic sprinklers, a fire can quickly extend horizontally across a large area.

FIGURE 7-12 The space between the ceiling and the underside of the floor above it often holds electrical and communications wiring and other building systems.

Wood-Supported Floors

Wood floor structures are common in non-fire-resistive construction. Such wooden floor systems range from heavy timber construction, which is often found in old mill buildings, to modern, lightweight construction.

Although heavy timber construction provides a huge fuel load for a fire, it can also contain and withstand a fire for a considerable length of time without collapsing. Heavy timber construction uses posts and beams that are at least 8 inches (20 centimeters) on the smallest side and often as large as 14 inches (35 centimeters) in depth. The floor decking is often assembled from solid wood boards, 2 or 3 inches (5 to 8 centimeters) thick, which are covered by an additional 1 inch of finished wood flooring. The depth of the wood in this system will often contain a fire for an hour or more before the floor fails or burns through.

Conventional wood flooring, which was widely used for many decades, is much lighter than heavy timber but uses solid lumber as beams, floor joists, decking, and finished flooring. It burns readily when exposed to a fire but generally takes approximately 20 minutes to burn through or reach structural failure. This time estimate is only a general, unscientific guideline; the actual burning rate depends on many other factors.

Modern lightweight construction uses structural elements that are much less substantial than conventional lumber. For example, lightweight wooden trusses or engineered wooden I-beams are often used as supporting structures. Thin sheets of plywood are used as decking, and the top layer often consists of a thin layer of concrete or wood covered by carpet. This floor construction provides little resistance to fire. The lightweight structural elements can fail or the fire can burn through the decking quickly. In Department of Health and Human Services National Institute for Occupational Safety and Health (NIOSH) Publication No. 2009-114, NIOSH reported that a recent test demonstrated that it took 19 minutes for a traditional unprotected residential floor assembly to fail. Under the same test conditions, an unprotected lightweight engineered I-joist failed in only 6 minutes.

Unfortunately, it is impossible to tell how a floor is constructed by looking at it from above. The important information about a floor can be observed only from below, if it is visible at all. The building's age and local construction methods can provide significant clues to a floor's stability, but you should be aware that many older buildings have been renovated using modern lightweight systems and materials. Fire fighters should use preincident surveys and other planning activities to gather essential structural information about buildings. In some reconstructions of office or computer rooms, the floor has been replaced with removable floor panels, so as to hide the different types of cables under the floor.

Ceiling Assemblies

From a structural standpoint, the ceiling is considered to be part of the floor assembly. Ceilings are mainly included in buildings for appearance reasons. Their primary function is to hold light fixtures and to diffuse light. Ceilings also conceal heating, ventilation, and air-conditioning (HVAC) distribution systems. In addition, they conceal wiring and fire sprinkler systems.

Ceilings can be part of the fire-resistive package. For example, they can be covered with plaster or gypsum board, or they can consist of dropped ceilings covered with mineral tiles. Some ceilings are fire rated as part of the floor assembly.

Fire fighters should be aware that hollow spaces between the floor and the ceiling can contribute to the horizontal spread of fire. Penetrations in the floor can also contribute to the vertical spread of fire.

Roofs

Safe interior firefighting operations depend on a stable roof. If the roof collapses, the fire fighters inside the building may be injured or killed. Interior fires generate hot gases that accumulate under the roof. Thus fire fighters who are performing ventilation activities on the roof to release the heated gases also must depend on the structural integrity of the roof.

The primary purpose of a roof is to protect the inside of a building from the weather. Often, roofs are not designed to be as strong as floors, especially in warmer climates that do not experience heavy snowfall. If the space under the roof is used for storage, or if extra HVAC equipment is mounted on the roof, the load can exceed the designer's expectations. Adding the weight of several fire fighters to an overloaded roof can be disastrous. Also, this space may contain different types of insulation to help protect the structure from climate changes.

In older homes, for example, this insulation often contains ground-up paper that can lead to smoldering and fire spread.

Several methods and materials are used for roof construction. Roofs are constructed in three primary designs: pitched, curved, and flat. The major components of a roof assembly are the supporting structure, the roof deck, and the roof covering.

Pitched Roofs

A pitched roof has sloping or inclined surfaces. Pitched roofs are used on many houses and some commercial buildings. The pitch or angle of the roof can vary depending on local architectural styles and climate. Variations of pitched roofs include gable, hip, mansard, gambrel, and lean-to roofs FIGURE 7-13.

The slope of a pitched roof can present either a minor inconvenience or a complete lack of secure footing, depending on its angle. Fire fighters working on a pitched roof are always in danger of falling, but the risk is particularly high when the roof is wet or icy, when the roof covering is not secure, or when smoke or darkness obscures vision. Roof ladders and aerial apparatus are used to provide a secure working platform for fire fighters working on a pitched roof.

Pitched roofs are usually supported by either rafters or trusses. Rafters are solid wood joists mounted in an inclined position. Pitched roofs supported by rafters usually have solid wood boards as the roof decking. Fire fighters can sometimes detect the impending failure of a roof with solid decking supported by rafters by a spongy feeling underfoot.

Modern lightweight construction uses manufactured wood trusses to construct most pitched roofs. Many lightweight trusses are manufactured using gusset plates that penetrate the wood no more than $^{3}/_{8}$ inch (1 centimeter). The decking usually consists of thin plywood or a sheeting material such as wood particleboard. When these trusses and decking are exposed to heat and fire, they often fail after a short period of time with no prior warning. Fire fighters cannot work safely on these roofs when the supporting structures become involved in a fire.

Steel trusses also are used to support pitched roofs. The fire resistance of steel trusses is directly related to how well the steel is protected from the heat of the fire.

Several roof-covering materials are used on pitched roofs, usually in the form of shingles or tiles FIGURE 7-14. Shingles are usually made from felt or mineral fibers impregnated with asphalt, although metal and fiberglass shingles are also used. Wood shingles, often made from cedar, are popular in some

A. B. C. D.

FIGURE 7-13 Examples of pitched roofs: **A.** A gable roof. **B.** A hip roof. **C.** A mansard roof. **D.** A gambrel roof.

areas. In older construction, wood shingles were often mounted on individual wood slats instead of continuous decking; these types of shingles burn rapidly in dry weather conditions.

Shingles are generally durable, economical, and easily repaired. A shingle roof that has aged and deteriorated should be removed completely and replaced. Some older buildings may have newer layers of shingles on top of older layers, which can make it difficult to cut an opening for ventilation.

Roofing tiles are usually made from clay or concrete products. Clay tiles, which have been used for roofing since ancient times, can be either flat or rounded. Rounded clay tiles are sometimes called mission tiles. Clay tiles are both durable and fire resistant.

Slate tiles are produced from thin sheets of rock. Slate is an expensive, long-lasting roofing material, but it is very brittle and becomes slippery when wet. Slate tiles also pose a threat to fire fighters below when a pitched roof is being ventilated, because falling tiles can become airborne missiles.

Metal panels of galvanized steel, copper, and aluminum may be used on pitched roofs as well. Expensive metals such as copper are often used for their decorative appearance, whereas lower-cost galvanized steel is used on barns and industrial buildings. Corrugated, galvanized metal panels are strong enough to be mounted on a roof without roof decking. Metal roof coverings will not burn, but they can conduct heat to the roof decking.

FIGURE 7-14 Several roof-covering materials are used on pitched roofs. **A.** A tile roof. **B.** A wood shingle roof. **C.** A slate roof.

Curved Roofs

Curved roofs are often used for supermarkets, warehouses, industrial buildings, arenas, auditoriums, bowling alleys, churches, airplane hangars, and other buildings that require large, open interiors. These kinds of roofs are usually supported by steel or wood bowstring trusses or arches.

The decking on curved roof buildings can range from solid wooden boards or plywood to corrugated steel sheets. The type of material in use must be identified before ventilation openings can be made. Often, the roof covering consists of layers that include felt, mineral fibers, and asphalt, although some curved roofs are covered with foam plastics or plastic panels.

Flat Roofs

Flat roofs are found on houses, apartment buildings, shopping centers, warehouses, factories, schools, and hospitals **FIGURE 7-15**. Most flat roofs have a slight slope so that water can drain off the structure. If the roof does not have the proper slope or if the drains are not maintained, water may pool on the roof, overloading the structure and causing a collapse.

The support systems for most flat roofs are constructed of either wood or steel. A wood support structure uses solid wooden beams and joists; laminated wood beams may be included to provide extra strength or to span long distances. Lightweight construction uses wood trusses or wooden I-beams as the supporting members. Such lightweight assemblies are much less fire resistant and collapse more quickly than do solid-wood systems.

FIGURE 7-15 Most flat roofs have a slight slope so that water can drain off.

Steel may also be used to support flat roofs. Open-web steel trusses (sometimes called bar joists) can span long distances and remain stable during a fire if they are not subjected to excessive heat. Automatic sprinklers can protect the steel from the heat; without such protection, however, the steel over a fire will soon weaken and sag. Heated steel support members can even elongate enough to push out the exterior walls of the building. Cracks in the outside walls at the roof level are a warning indicator of this condition.

Flat roof decks can be constructed of wood planking, plywood, corrugated steel, gypsum, or concrete. The material used depends on the building's age and size, the climate, the cost of materials, and the type of roof covering.

The roof covering is applied on top of the deck. Most flat roofs are covered with multiple-layer, built-up roofing systems that help prevent leakage. A typical built-up roof covering contains five layers: a vapor barrier, thermal insulation, a waterproof membrane, a drainage layer, and a wear course. The outside of the flat roof reveals only the top layer, which is often gravel; it serves as the wear course and protects the underlying layers from wear and tear. Cutting through all of the layers to open a ventilation hole can be quite a challenge.

Most flat roof coverings contain highly combustible materials, including asphalt, roofing felt, tarpaper, rubber or plastic membranes, and plastic insulation layers. These materials can be difficult to ignite. Once lit, however, they burn readily, releasing great quantities of heat and black smoke in the process.

Flat roofs can present unique problems during vertical ventilation operations. A roof with a history of leaks could have several patches where additional layers of roofing material make for an extra-thick covering. Sometimes a whole new roof is constructed on top of the old one. After opening a hole in the new roof, fire fighters might discover the old roof and have to cut into that as well. Fire fighters might also discover fire burning in the void space between the two roofs.

A similar problem may be encountered in remodeled buildings or buildings with additions. An old flat roof may be found under a new pitched section, resulting in two separate void spaces. The old roof may also provide a large supply of fuel for the fire. Such a situation is difficult to predict unless the incident commander has the building plan in advance.

Fire Fighter Safety Tips

When working on a flat roof, avoid walking on skylights and other openings that will not support your weight. Stay away from the edges of the roof.

Fire Fighter Safety Tips

Flat roofs with parallel chord trusses and corrugated metal decks have been known to roll down into ventilation holes during roof operations. Some fire departments no longer perform vertical ventilation in buildings with this type of roof.

Trusses

A truss is a structural component that is composed of smaller pieces joined to form a triangular configuration or a system of triangles. Trusses are used extensively in support systems for both floors and roofs. They are common in modern construction for several reasons. The triangular geometry creates a strong, rigid structure that can support a load much greater than its own mass. For example, both a solid beam and a simple truss with the same overall dimensions can support the same load. The truss has several advantages over the solid beam, however. It requires much less material than the beam, is much lighter, and can span a long distance without supports. Trusses are often prefabricated and transported to the building location for installation there.

Trusses are frequently used in residential construction, apartment buildings, small office buildings, commercial buildings, warehouses, fast-food restaurants, airplane hangars, churches, and even firehouses. They may be clearly visible, or they may be concealed within the construction. Trusses are widely used in new construction and often replace heavier solid beams and joists in renovated or modified older buildings.

The strength of a truss depends on both its members and the connections between them. A properly assembled and installed truss is strong. If any member or connector begins to fail, however, the strength of the entire truss will be compromised.

Trusses can be used for many different purposes. In building construction, those made of wood, steel, or combinations of wood and steel are used primarily to support roofs and floors. Three of the most commonly used types of trusses are the parallel chord truss, the pitched chord truss, and the bowstring truss.

A parallel chord truss has two parallel horizontal members connected by a system of diagonal and sometimes vertical members **FIGURE 7-16**. The top and bottom members are called the chords, and the connecting pieces are called the web members. Parallel chord trusses are typically used to support flat roofs or floors. In lightweight construction, an engineered wood truss is often assembled with wood chords and either wood or light steel web members. A steel bar joist is another example of a parallel chord truss.

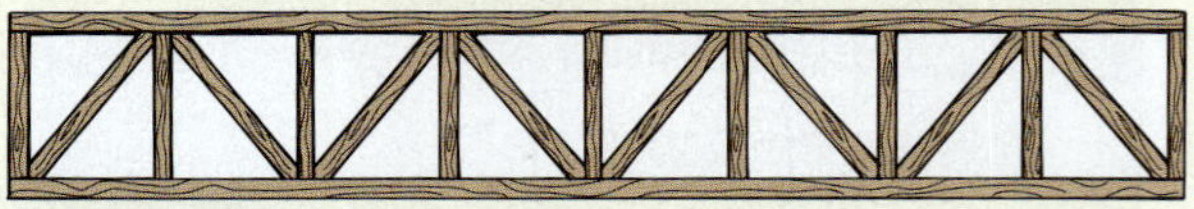

FIGURE 7-16 A parallel chord truss.

A pitched chord truss is typically used to support a sloping roof **FIGURE 7-17**. Most modern residential construction uses a series of prefabricated wood pitched chord trusses to support the roof. The roof deck is supported by the top chords, and the ceilings of the occupied rooms are attached to the bottom chords. In this way, the trusses define the shape of the attic.

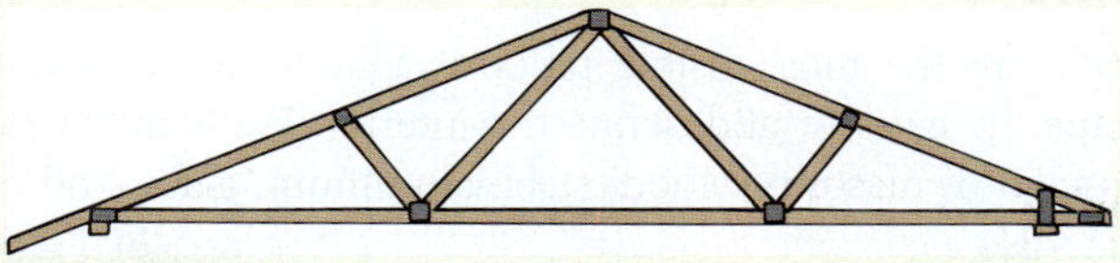

FIGURE 7-17 A pitched chord truss.

A bowstring truss has the same shape as an archery bow, where the top chord represents the curved bow and the bottom chord represents the straight bowstring FIGURE 7-18. The top chord resists compressive forces and the bottom chord resists tensile forces. Bowstring trusses are usually quite large and widely spaced in the structure. They were popular in warehouses, supermarkets, and similar buildings with large, open floor areas. The roof of a building with bowstring trusses has a distinctive curved shape.

Trusses are everywhere, so it is important that you learn how to recognize them. A building with a bowstring truss roof will have a distinctive curved roof FIGURE 7-19. In some buildings, the trusses are exposed and readily visible from the inside of the building. In other buildings, the floor or roof trusses are not visible because they are enclosed in attic or floor spaces FIGURE 7-20.

Roof off limits if a single truss fails

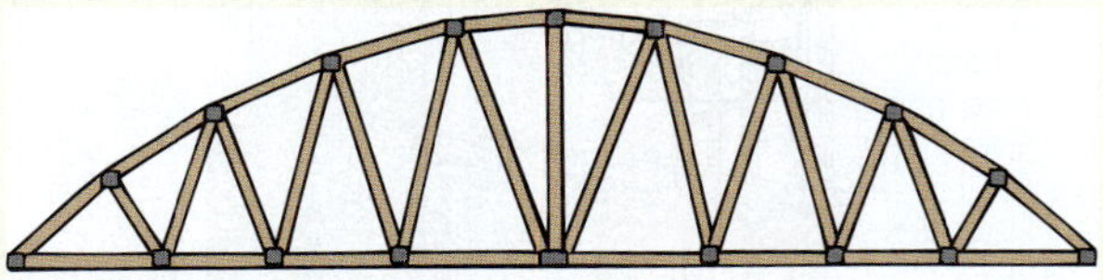

FIGURE 7-18 A bowstring truss.

FIGURE 7-19 A building with a bowstring truss.

Fire Fighter Safety Tips

Checking the floor or roof to see if it is spongy does not ensure safety with modern lightweight construction.

FIGURE 7-20 This building may have hidden trusses.

Trusses Under Fire

Under normal circumstances, a properly engineered and installed truss will do a satisfactory job of supporting the roof or floor assembly for which it was designed. However, like many other building components, trusses display different characteristics when exposed to heat and fire. Because these components are engineered to work with the least amount of wood or steel needed to support a given weight, they provide little margin of safety for fire fighters operating during a fire.

Trusses consist of a small cross-section of wood with a high surface-to-mass ratio. As a consequence, a fire burning for a very short period of time can destroy enough wood to weaken a truss. All it takes is a failure at one point in a truss to produce a failure in the entire truss. Wooden trusses constructed with metal gusset plates are especially prone to failure. These gusset plates tie the chords and members of the truss together. They have sharp pointed indentions that are pounded into the wood to hold the chords and web members together. Because the gusset plates are embedded in the wood at a depth of only $^3/_8$ inch (1 centimeter), heating the metal gusset plate will decompose the wood fibers, quickly leading to failure of the truss. Likewise, wooden trusses held together with steel pins can fail in a fire. When heated, the steel pins conduct heat into the wood, which in turn causes decomposition of the wood. When the wood weakens, failure initially occurs at that point, followed by failure of the entire truss. Fire can also weaken the glue in glued trusses, causing the truss to weaken.

Because trusses are constructed with a small margin of safety, failure of one truss usually results in overload and subsequent failure of adjoining trusses. For this reason, failure of a single truss usually results in catastrophic failure of the floor or roof it supports.

In the past, fire fighters were taught that sounding the floor or checking the roof to see if it was "spongy" would help determine if that part of the structure was safe. This technique has proved unreliable when dealing with structural members built from trusses. In fact, fire fighters have reported cases where roofs and floors have collapsed with no warning signs.

Steel 100' truss can elongate as much as 9" at 1000 °F

Steel trusses are also prone to failure during fires. When thin tubular steel is heated enough to cause a tube in the truss to weaken, the entire truss can fail. If steel trusses are protected from heat with thermal insulation, they may withstand the heat from a fire for a longer period of time. Many steel trusses do not have any protection from heat, however, so they often fail during a major fire. Steel has a tendency to sag before it fails completely, which may give a slight warning of its impending failure.

During a fire, steel trusses and I-beams can fail because of elongation and twisting. Francis Brannigan noted that a 100-foot-long (30 meters) beam or truss can elongate by as much as 9 inches (23 centimeters) when heated to 1000°F (540°C). Elongated steel beams can rotate and drop wooden trusses—with deadly consequences. Elongation can also push down masonry walls.

In some cases, trusses may fail because of overload. As buildings are modified, or when the use of buildings changes, more weight may be placed on floors or roofs than they were intended to carry. For example, air conditioners may be placed on roofs and attic spaces may be used for storage purposes. This extra weight might exceed the weight for which the trusses were designed. In the event of a fire, the added weight can contribute to the rapid failure of a truss-supported floor or roof.

Truss spaces can also contribute to the spread of fires. Spaces under floors occupied by trusses can serve as conduits for the spread of fires. These fires are often difficult to detect and may spread rapidly, much as fires spread quickly through any open spaces. Because truss-supported floors can be covered with many types of material (including cement), it may be difficult to determine that a fire burning under the floor has weakened the floor. Be alert for the possibility of a fire burning under you. Fires that occur in the space above a truss-supported ceiling are also challenging to detect. Whenever a truss-supported ceiling has fire above it, be prepared for the possibility of a rapid collapse of the trusses supporting the roof.

Trusses exist in many types of buildings—an important fact for fire fighters to recognize. Identifying the presence of trusses will enable you to better predict the behavior of a particular building under fire conditions. Understanding that buildings with trusses give fire fighters few signs of impending failure enables you to evaluate the risk and benefit of different approaches to fire suppression in that building.

FIRE FIGHTER Tips

A new component in lightweight building construction is the truss joist I-beam (or TGI beam). This framing material is made up of three wooden components. Channels may be drilled through TGI beams, thus providing another avenue for fire to spread.

Engineered lumber and laminated beams are being used more often in modern construction. These products are made up of smaller pieces of wood bound together with an adhesive process. Exposure to fire may break the bonds of the adhesive process and lead to lumber or beam failure.

Walls

Walls are the most visible parts of a building because they shape the exterior and define the interior. Walls may be constructed of masonry, wood, steel, aluminum, glass, and other materials.

Fire Fighter Safety Tips

Trusses can and will fail without warning early in a fire. Fire fighters should stay off of roofs where fire is known or suspected to be in the truss loft.

Walls are either load bearing or nonbearing. Load-bearing walls provide structural support FIGURE 7-21. A load-bearing wall supports a portion of both the building's weight (dead load) and its contents (live load), transmitting that load down to the building's foundation. Damaging or removing a load-bearing wall can result in a partial or total collapse of the building. Load-bearing walls can be either exterior or interior walls.

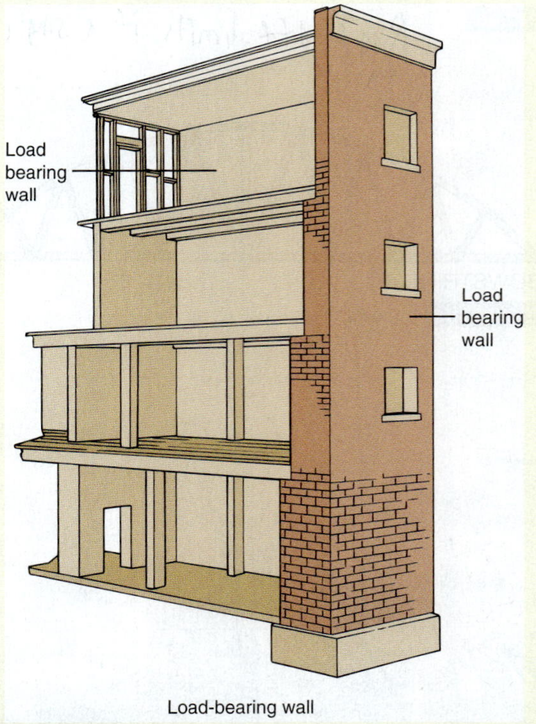

FIGURE 7-21 A load-bearing wall provides structural support.

Nonbearing walls, by contrast, support only their own weight FIGURE 7-22. For this reason, most nonbearing walls can be breached or removed without compromising the structural integrity of the building. Many nonbearing walls are interior partitions that divide the building into rooms and spaces. The exterior walls of a building can also be nonbearing, particularly when a system of columns supports the building.

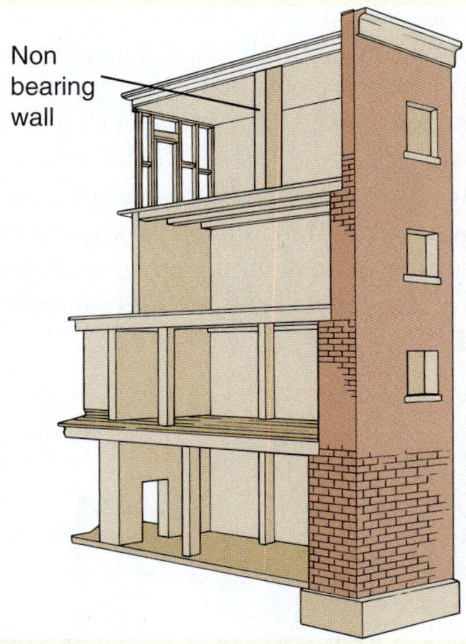

FIGURE 7-22 A nonbearing wall supports only its own weight.

In addition to load-bearing and nonbearing walls, several specialized walls may be used in construction:

- Party walls are constructed on the line between two properties and are shared by a building on each side of the line. They are almost always load-bearing walls. A party wall is often—but not always—constructed as a fire wall between the two properties.
- Fire walls are designed to limit the spread of fire from one side of the wall to the other side. A fire wall might divide a large building into sections or separate two attached buildings. Fire walls usually extend from the foundation up to and through the roof of a building. They are constructed of fire-resistant materials and may be fire rated.
- Fire partitions are interior walls that extend from a floor to the underside of the floor above. They often enclose fire-rated interior corridors or divide a floor area into separate fire compartments.
- Fire enclosures are fire-rated assemblies that enclose interior vertical openings, such as stairwells, elevator shafts, and chases for building utilities. A fire enclosure prevents fire and smoke from spreading from floor to floor via the vertical opening. In multistory buildings, fire enclosures also protect the occupants when they are using the exit stairways.
- Curtain walls are nonbearing exterior walls attached to the outside of the building. These walls often serve as the exterior skin on a steel-frame high-rise building.

Solid, load-bearing masonry walls, at least 6 to 8 inches thick, can be used for buildings up to six stories high. In contrast, nonbearing masonry walls can be almost any height. Masonry walls provide a durable, fire-resistant outer covering for a building, so they are often used as fire walls. A well-designed masonry fire wall is often completely independent of the structures on either side. Even if the building on one side burns completely and collapses, the fire wall should prevent the fire from spreading to the building on the other side.

Older buildings often have masonry load-bearing walls that are several feet thick at the bottom and then decrease in thickness as the building's height increases. Modern masonry walls are typically reinforced with steel rods or concrete to provide a more efficient and more durable structural system.

When properly constructed and maintained, masonry walls are strong and can withstand a vigorous assault by fire. If the interior structure begins to collapse and exerts unanticipated forces on the exterior walls, however, solid masonry walls may fail during a fire.

Even though a building has an outer layer of brick or stone, fire fighters should never assume that the exterior walls are masonry. Many buildings that look like masonry are actually constructed using wood-frame techniques and materials. A single layer of brick or stone, called a veneer layer, is applied to the exterior walls to give the appearance of a durable and architecturally desirable outer covering. If the wood structure is damaged during a fire, this veneer is likely to collapse. For this reason, it is imperative that a collapse zone be set up around a burning building.

Fire Fighter Safety Tips

To determine if a brick wall is a solid masonry wall or a brick veneer, look at the pattern of the bricks. If every seventh course of bricks is turned sideways so that the end of the brick is visible (kings row pattern), it is usually a solid masonry wall. If all of the courses of brick are the same, it is probably a brick veneer.

Wood framing is used to construct the walls in most houses and many small commercial buildings. The wood framing used for exterior walls can be covered with a variety of materials, including wooden siding, vinyl siding, aluminum siding, stucco, and masonry veneer. Moisture barriers, wind barriers, and other types of insulation are usually applied to the outside of the vertical studs before the outer covering is applied.

Vertical wooden studs support the walls and partitions inside the building. If fire resistance is critical, steel studs are used to frame walls. The wood framing is usually covered with gypsum board and any of a variety of interior finish materials. The space between the two wall surface coverings may be empty, it may contain thermal and sound insulating materials, or it may contain electrical wiring, telephone wires, and plumbing. These spaces often provide pathways through which fire can spread.

Fire Fighter Safety Tips

The United States Fire Administration (USFA) has partnered with the American Forest and Paper Association to develop educational materials to enhance fire fighter awareness of the fire performance of different types of lightweight construction components. The fire fighter educational material developed under this partnership is available, free of charge from the USFA web site.

Doors and Windows

Doors and windows are important components of any building. Although they generally have different functions—doors provide entry and exit, whereas windows provide light and ventilation—in an emergency, doors and windows are almost interchangeable. A window can serve as an entry or an exit, for example, whereas a door can provide light and ventilation.

Hundreds of door and window designs exist, with many different applications. Of particular concern to fire fighters are fire doors and fire windows. Even wood doors provide some barrier to fire spread. Closing a door during a fire, then, may give you a few additional minutes to search a room.

Door Assemblies

Most doors are constructed of either wood or metal. Hollow-core wooden doors are often used inside buildings. A typical hollow-core door has an internal framework whose outer surfaces are covered by thin sheets of wood. Because hollow-core doors can be easily opened with simple hand tools, they should not be used where security is a concern. A fire can usually burn through a hollow-core door in a short time.

Solid-core wooden doors are used where a more substantial door is required. These types of doors are manufactured from solid panels or blocks of wood and are more difficult to force open. A solid-core door provides some fire resistance and can often keep a fire contained within a room for 20 minutes or more, giving fire fighters a chance to arrive on the scene and mount an effective attack.

Metal doors are more durable and fire resistant than wooden doors. Some metal doors have a solid wood core that is covered on both sides by a thin sheet of metal; others are constructed entirely of metal and reinforced for added security. The interior of a metal door can be either hollow or filled with wood, sound-deadening material, or thermal insulation. Although most approved fire doors are metal, not all metal doors should be considered fire doors.

FIRE FIGHTER Tips

Manufactured (mobile) homes use lightweight building components throughout the structure to reduce weight. As a result, most parts of these structures are combustible. Such homes typically have few doors and small windows, making entry for fire suppression or rescue difficult. Once a fire starts, especially in an older manufactured (mobile) home, it can destroy the entire structure within a few minutes.

Window Assemblies

Fire fighters frequently use windows as entry points to attack a fire and as emergency exits. Windows also provide ventilation during fires, allowing smoke and heat to escape the building and cooler, fresh air to enter it.

Windows come in many shapes, sizes, and designs for different buildings and occupancies FIGURE 7-23. Fire fighters must become familiar with the specific types of windows found in local occupancies, learn to recognize them, and understand how they operate. It is especially important to know if a particular type of window is very difficult to open or cannot be opened. Today's windows are being built with more energy-conserving features. As part of this trend, double- and triple-pane windows are common in some parts of the country. Window coatings and more durable types of glass may make forcible entry more difficult during firefighting operations. The Forcible Entry chapter contains more information on window construction.

FIGURE 7-23 Windows can open in many different ways. This is just one example.

Fire Doors and Fire Windows

Fire doors and fire windows are constructed to prevent the passage of flames and heat through an opening during a fire. They must be tested and meet the standards specified in NFPA 80, *Standard for Fire Doors and Other Opening Protectives*. Fire doors and fire windows come in many different shapes and sizes, however, and they provide different levels of fire resistance. For example, fire doors can swing on hinges, slide down or across an opening, or roll down to cover an opening.

The fire rating on a door or window covers the actual door or window, the frame, the hinges or closing mechanism, the latching hardware, and any other equipment that is required to operate the door or window. All of these items must be tested and approved as a combined system. All fire doors must have a mechanism that keeps the door closed or automatically closes the door when a fire occurs. Doors that are normally open can be closed by the release of a fusible link, by a smoke or heat detector, or by activation of the fire alarm system.

Fire doors and fire windows are rated for a particular duration of fire resistance to a standard test fire. This is similar to the fire-resistance rating system used for building construction assemblies. A 1-hour rating, however, does not guarantee that the door will resist any fire for 60 minutes; rather, it simply establishes that the door will resist the standard test fire for 60 minutes. In any given fire situation, a door rated at 1 hour will probably last twice as long as a door rated 30 minutes.

Fire windows are used when a window is needed in a required fire-resistant wall. These windows are often made of wired glass, which is designed to withstand exposures to high temperatures without breaking. Fire-resistant glass without wires is available for some applications; in such cases, special steel window frames are required to keep the glass firmly in place.

Wired glass is also used to provide vision panels in or next to fire doors FIGURE 7-24. Vision panels allow a person to view conditions on the opposite side of a door without opening the door, or to make sure no one is standing in front of the door before opening it. In these configurations, the components making up the entire assembly—including the door, the window, and the frame—must be tested and approved together.

When a window is required only for light passage, glass blocks can sometimes be used instead of wired glass. Glass blocks will resist high temperatures and remain in place during a fire (assuming they are properly installed). The size of the opening is limited and depends on the fire-resistance rating of the wall.

FIGURE 7-24 Wired glass is used to provide vision panels in or next to fire doors.

Interior Finishes and Floor Coverings

The term interior finish is typically used to refer to the exposed interior surfaces of a building. These materials affect how a particular building or occupancy reacts when a fire occurs. Considerations related to interior finishes include whether a material will ignite easily or resist ignition, how quickly a flame will spread across the material surface, how much energy the material will release when it burns, how much smoke it will produce, and which components the smoke will contain (e.g., toxic particles or gases).

A room with a bare concrete floor, concrete block walls, and a concrete ceiling has no interior finishes that will increase the fire load. But when the same room has an acrylic carpet with rubber padding on the floor, wooden baseboards, varnished wood paneling on the walls, and foam plastic acoustic insulation panels on the ceiling, the situation will be vastly different: These interior finishes ignite quickly, spread flames quickly across other surfaces, and release significant quantities of heat, flames, and toxic smoke.

Different interior finish materials contribute in various ways to a building fire. Each individual material has certain characteristics, and fire fighters must evaluate the particular combination of materials in a room or space. Typical wall coverings include painted plaster or gypsum board, wallpaper or vinyl wall coverings, wood paneling, and many other surface finishes. Floor coverings might include different types of carpet, vinyl floor tiles, or finished wood flooring. All of these products will burn to some extent, and each one involves a different set of fire risk factors. Fire fighters must understand the hazards posed by different interior finish materials if they are to operate safely at the scene of a fire. Keep in mind that most of the furnishings and interior finishes in a house consist of some type of plastic that is derived from petroleum products.

Manufactured Housing

Manufactured housing includes both mobile homes and modular homes. These homes are constructed with lightweight building components throughout the structure to reduce weight. Most parts of manufactured houses are combustible. Many manufactured houses lack large windows that can be used for rescue purposes and fire suppression, and they often contain few doors. Once a fire has started in a mobile home, it can destroy the entire structure within just a few minutes. As a consequence, the death rate in mobile home fires is three times as high as the death rate in fires involving other types of single-family homes. People who live in mobile homes without a working smoke alarm are at an especially high risk of dying from a fire.

Buildings Under Construction or Demolition

Buildings that are under construction or renovation or in the process of demolition present a variety of problems and extra hazards for fire fighters. In many cases, the fire protection features found in finished buildings are missing from these structures. For example, automatic sprinklers might not have been installed yet or might have been disconnected. Smoke detectors might not have been installed or might be inoperable. There may be no coverings on the walls, leaving the entire wood framework fully exposed. Missing doors and windows can provide an unlimited supply of fresh air to feed the fire and allow the flames to spread rapidly. Fire-resistive enclosures may be missing, leaving critical structural components unprotected. Sprinkler and standpipe systems may be inoperative. In addition, construction sites and demolition sites often have large quantities of building materials stored close to buildings. These materials can add a huge additional fire load and contribute to the rapid spread of fire often from one building to another. All of these factors can enhance fire spread, leading to quick structural failure.

Many fires at construction and demolition sites are inadvertently caused by workers using torches to weld, cut, or take apart pieces of the structure. Tanks of flammable gases and piles of highly combustible construction materials might be left in locations where they could add even more fuel to a fire. Buildings under construction or demolition are often left unoccupied for many hours, resulting in delayed discovery and reporting of fires. In some cases, it might prove difficult for fire apparatus to approach the structure or for fire fighters to access working hydrants. All of these problems must be anticipated when considering the fire risks associated with a construction or demolition site.

Fire Fighter Safety Tips

Buildings under construction or demolition are at high risk for a major fire. Such was the case at the Deutsche Bank building in New York City, where two fire fighters lost their lives in 2007. This building was damaged in the terrorist attacks of September 11, 2001, and was set for demolition. The Fire Department of New York responded to a working fire in the building and attempted to use the standpipe system to fight the fire. A part of the standpipe system had been removed, however, and was inoperable.

Fire Fighter Safety Tips

The majority of fire fighter deaths and injuries occur at residential fires from structural collapse. The most frequent cause of death and injury is due to the collapse of floors, followed by the collapse of ceilings and walls. Recent tests by Underwriters Laboratories revealed that common items found in modern construction, such as finger-joined parallel chord trusses, can fail in as little time as 4 minutes.

Preincident Planning and Incident Size-Up

When a building is on fire, how do you determine which type of construction was used to build the building? Even experienced fire fighters need to assess a building carefully to make this determination.

The best way to gather this information is to conduct preincident planning. Preincident planning allows fire department personnel to identify the type of building construction as well as to survey the specific characteristics of that building. This type of investigation helps you prepare to quickly, safely, and effectively attack a fire in that building. Preincident planning allows you the luxury of being at the building before a fire breaks out.

During preincident planning, you can document the characteristics of a building before a fire starts there **FIGURE 7-25**. In your examination of the building, consider the following questions: Which type of building construction is present? Which type of occupancy is this building? Which type of contents does it have? Which type of support do the floors have? Is it safe to climb onto the roof? Does the building contain roof trusses, floor trusses, or manufactured wood I-joists that might collapse suddenly? Which types of fire detection and fire suppression systems are installed in the building? The importance of preplanning cannot be overemphasized. An excellent way to identify building construction is to inspect the building multiple times while it is being built. Your preplanning will be greatly advanced by this exercise. More detailed information about preplanning is presented in the Preincident Planning chapter.

Of course, it is not possible to do preincident planning for every property in your department. In cases where you have not conducted a preincident inspection, it will be necessary to rely on an incident size-up to determine the type of construction present in the fire building **FIGURE 7-26**. By learning the general characteristics of the building types in your first-due area, you will get some idea of the type of construction in a given neighborhood. In a neighborhood where the houses were built in the 1920s, for example, you are likely to find balloon construction. In a neighborhood populated by recently completed houses, you are likely to encounter lightweight construction with wooden I-beams and wooden trusses.

FIGURE 7-25 Preincident inspections help you determine the type of building construction before a fire occurs there.

FIGURE 7-26 A thorough incident size-up can help identify hazards at a fire.

Use the information you have learned about building construction to make firefighting safer both for you and for building residents. Your knowledge of building construction, preincident inspections, and incident size-up will help you gather the information you need to remain a safe and effective fire fighter.

Over the course of your career as a fire fighter, it is also important to continue to keep up with the changes in building construction. Spend some time at construction sites. Review the latest fire studies from places like Underwriters Laboratories, which are typically available online. The information you gain may be valuable at your next fire scene.

FIRE FIGHTER Tips

It is important to know which types of construction are found in your response area. When a building is under construction, contact the builder or site foreperson and arrange for a tour. Ask questions about the structure and its construction. Inquire about any special features and its fire protection system.

Wrap-Up

Chief Concepts

- It is vital for fire fighters to understand the basic types of building construction and to recognize how building construction affects fire growth and spread.
- Based on a structure's occupancy classification, a fire fighter can predict who is likely to be inside the building. For example, a bank is unlikely to have occupants at 3 am.
- Building contents are usually related to the occupancy of the structure. For example, a warehouse is likely to store goods or chemicals, whereas an apartment building is likely to be filled with human occupants and furniture.
- The most commonly used building materials are wood, engineered wood products, masonry, concrete, steel, aluminum, glass, gypsum board, and plastics. The key factors that affect the behavior of each of these materials under fire conditions are combustibility, thermal conductivity, decrease in strength at elevated temperatures, and thermal expansion when heated.
- Masonry includes stone, concrete blocks, and brick. These types of components are usually bonded together with mortar. Masonry is fire resistive and a poor conductor of heat. Even so, a masonry structure can collapse under fire conditions if the roof or floor assembly collapses.
- Concrete is a fire-resistive material that does not conduct heat well. It is often used to insulate other building materials from fire. Concrete does not lose strength when exposed to high temperatures. Steel reinforcing rods are often embedded within this material to strengthen it. Under high heat from fire, a fire may convert trapped moisture in the concrete to steam and create spalling; severe spalling can, in turn, expose steel reinforcing rods to heat.
- Steel is often used in the structural framework of buildings. It is strong and resistant to aging, but can rust. Steel is not fire resistive and conducts heat. Steel both expands and loses strength when heated. The risk of failure of a steel structure depends on the mass of the steel components, the loads placed on them, and the methods used to connect the components. Any sign of bending, sagging, or stretching of steel structural members should be viewed as a warning of imminent collapse.
- Glass is noncombustible but not fire resistive. Ordinary glass will usually break when exposed to fire. Tempered glass is stronger and more difficult to break. Laminated glass is likely to crack and remain in place when exposed to fire. Glass blocks have limited strength and are not intended to be used as part of a load-bearing wall. Wired glass incorporates wires that hold the glass together and prevent it from breaking when exposed to heat.
- Gypsum board is commonly used to cover the interior walls and ceilings of residences. It is manufactured in large sheets consisting of a layer of compacted gypsum sandwiched between two layers of specially produced paper. Gypsum board has limited combustibility and does not conduct or release heat to contribute to fire spread. It may fail if exposed to fire for extended periods.
- Engineered wood products are also called manufactured board, manmade wood, or composite wood. These products are manufactured from small pieces of wood that are held together with glue or adhesives (e.g., urea-formaldehyde resins, phenol-formaldehyde resins, melamine-formaldehyde resins, and polyurethane resins). Engineered wood products may warp under high humidity and release toxic fumes during a fire.
- The most important characteristic of wood and engineered wood products is their high combustibility. The rate at which wood ignites, burns, and decomposes depends on several factors: ignition, moisture, density, preheating, and the size and form.
- Wood cannot be treated to make it completely noncombustible, but it can be made to be more difficult to ignite and burn through application of a fire-retardant treatment.
- Plastics are rarely used for structural support but may be found throughout a building. The combustibility of plastics varies greatly. Many plastics produce quantities of heavy, dense, dark smoke and release high concentrations of toxic gases.
- There are five types of building construction:
  - Type I construction: fire resistive—Can withstand and contain a fire for a specified period of time. All of the structural members and components used in this construction must be made of noncombustible materials and constructed so that it provides at least 2 hours of fire resistance. Under extreme fire conditions, a Type I building can collapse.
  - Type II construction: noncombustible—Includes structural components made of noncombustible materials. The fire severity is determined by the contents in the building.
  - Type III construction: ordinary—Used in a wide variety of buildings, ranging from strip malls to small apartment buildings. These buildings have masonry exterior walls and interior structural members constructed of wood. The resulting structure may have void spaces that can allow fire spread.
  - Type IV construction: heavy timber—Structure with masonry construction and interior walls, columns, beams, floor assemblies, and a roof structure made of heavy wood. It has no concealed spaces or voids. The columns, support beams, floor assemblies, and roof assemblies will withstand a fire much longer than the smaller wood members used in ordinary construction.

- Type V construction: wood frame—All of the major components are constructed of wood or other combustible materials. Type V is the most common type of construction used today. Many of these structures do not have any fire-resistive components. Fire detection devices and automatic sprinklers are the most effective way of protecting lives in these buildings.

- In a balloon-frame building, an open channel extends from the foundation to the attic that can enable hidden fire to spread throughout the structure.
- Platform-frame construction is used in most modern construction. In this building technique, each level is constructed as a separate platform. At each level, the floor platform blocks the path of any fire rising within the void spaces.
- Every building contains the following major components:
  - Foundation—Usually constructed of concrete or masonry and designed to transfer the weight of the building and its contents to the ground.
  - Floors and ceilings—Might be designed to resist fires and might have no fire-resistant capabilities. In a fire-resistive building, the floor is designed to prevent a fire from spreading vertically and to prevent a collapse when a fire occurs in the space below the floor–ceiling assembly. Wood floor structures are common in non-fire-resistive construction. The ceiling is considered to be part of the floor assembly.
  - Roof—Designed to protect the inside of the building from the weather. Roofs are constructed in three primary designs: pitched, curved, and flat. The major components of a roof assembly are the supporting structure, the roof deck, and the roof covering.
  - Trusses—A structural component composed of smaller pieces joined to form a triangular configuration or a system of triangles. Trusses are used as support systems for both floors and roofs. The strength of a truss depends on its members as well as the connections between them. The types of trusses include parallel chord truss, pitched chord truss, and bowstring truss. Depending on their construction, trusses may fail when exposed to heat and fire.
  - Walls—Either load bearing or nonbearing. Load-bearing walls provide structural support. Nonbearing walls support only their own weight.
  - Doors—Typically constructed of either wood or metal. Hollow-core wooden doors are often used inside buildings and can be easily opened. Metal doors are more durable and fire resistant.
  - Windows—Used frequently by fire fighters as entry points to attack a fire and as emergency exits.
  - Interior finished and floor coverings—The exposed interior surfaces of a building. Different interior finish materials contribute in various ways to a structure fire. For example, plastics emit toxic gases when they burn.
- Buildings under construction or demolition often lack fire protection features and present additional hazards to fire fighters.
- The preincident plan should contain information on the structure's type of building construction.

Hot Terms

Balloon-frame construction An older type of wood frame construction in which the wall studs extend vertically from the basement of a structure to the roof without any fire stops.

Bowstring truss A truss that is curved on the top and straight on the bottom.

Combustibility The property describing whether a material will burn and how quickly it will burn.

Contemporary construction Buildings constructed since about 1970 that incorporate lightweight construction techniques and engineered wood components. These buildings exhibit less resistance to fire than older buildings.

Curtain wall Nonbearing walls that separate the inside and outside of the building but are not part of the support structure for the building.

Curved roof A roof with a curved shape.

Dead load The weight of the aerial device structure and all materials, components, mechanisms, or equipment permanently fastened thereto. (NFPA 1901)

Fire enclosure A fire-rated assembly used to enclose a vertical opening such as a stairwell, elevator shaft, and chase for building utilities.

Fire partition An interior wall extending from the floor to the underside of the floor above.

Fire wall A fire division assembly with a fire resistance rating of 3 test hours or longer, built to permit complete burnout and collapse of the structure on one side without extension of fire through the fire wall or collapse of the fire wall. (NFPA 901)

Fire window A window assembly rated in accordance with NFPA 257 and installed in accordance with NFPA 80. (NFPA 5000)

Flat roof A horizontal roof; often found on commercial or industrial occupancies.

Glass blocks Thick pieces of glass that are similar to bricks or tiles.

Gypsum A naturally occurring material consisting of calcium sulfate and water molecules.

Gypsum board The generic name for a family of sheet products consisting of a noncombustible core primarily of gypsum with paper surfacing. (NFPA 5000)

Interior finish The exposed surfaces of walls, ceilings, and floors within buildings. (NFPA 101)

Laminated glass Safety glass. The lamination process places a thin layer of plastic between two layers of glass, so that the glass does not shatter and fall apart when broken.

Laminated wood Pieces of wood that are glued together.

Legacy construction An older type of construction that used sawn lumber and was built before about 1970.

Live load The load produced by the use and occupancy of the building or other structure, which does not include construction or environmental loads such as wind load, snow load, rain load, earthquake load, flood load, or dead load. Live loads on a roof are those produced (1) during maintenance by workers, equipment, and materials; and (2) during the life of the structure by movable objects such as planters and by people. [ASCE/SEI 7:4.1] (NFPA 5000)

Load-bearing wall A wall that is designed to provide structural support for a building.

Manufactured (mobile) home A factory-assembled structure or a structure transportable in one or more sections that is built on a permanent chassis and designed to be used as a dwelling without a permanent foundation when connected to the required utilities, including the plumbing, heating, air-conditioning, and electric systems contained therein.

Masonry Built-up unit of construction or combination of materials such as clay, shale, concrete, glass, gypsum, tile, or stone set in mortar. (NFPA 5000)

Nonbearing wall Any wall that is not a bearing wall. [ASCE/SEI 7:11.2] (NFPA 5000)

Occupancy The purpose for which a building or other structure, or part thereof, is used or intended to be used. (NFPA 5000)

Parallel chord truss A truss in which the top and bottom chords are parallel.

Party wall A wall constructed on the line between two properties.

Pitched chord truss A type of truss typically used to support a sloping roof.

Pitched roof A roof with sloping or inclined surfaces.

Platform-frame construction Construction technique for building the frame of the structure one floor at a time. Each floor has a top and bottom plate that acts as a firestop.

Pyrolysis The destructive distillation of organic compounds in an oxygen-free environment that converts the organic matter into gases, liquids, and char. (NFPA 820)

Rafters Joists that are mounted in an inclined position to support a roof.

Spalling Chipping or pitting of concrete or masonry surfaces. (NFPA 921)

Tempered glass A type of safety glass that is heat treated so that, under stress or fire, it will break into small pieces that are not as dangerous.

Thermal conductivity A property that describes how quickly a material will conduct heat.

Thermoplastic material Plastic material capable of being repeatedly softened by heating and hardened by cooling and, that in the softened state, can be repeatedly shaped by molding or forming. (NFPA 5000)

Thermoset material Plastic material that, after having been cured by heat or other means, is substantially infusible and cannot be softened and formed. (NFPA 5000)

Truss A collection of lightweight structural components joined in a triangular configuration that can be used to support either floors or roofs.

Type I construction (fire resistive) Buildings with structural members made of noncombustible materials that have a specified fire resistance.

Type II construction (noncombustible) Buildings with structural members made of noncombustible materials without fire resistance.

Type III construction (ordinary) Buildings with the exterior walls made of noncombustible or limited-combustible materials, but with interior floors and walls made of combustible materials.

Type IV construction (heavy timber) Buildings constructed with noncombustible or limited-combustible exterior walls, and interior walls and floors made of large-dimension combustible materials.

Type V construction (wood frame) Buildings with exterior walls, interior walls, floors, and roof structures made of wood.

Wired glass A glazing material with embedded wire mesh. (NFPA 80)

Wooden beam Load-bearing member assembled from individual wood components.

Wood truss An assembly of small pieces of wood or wood and metal.

FIRE FIGHTER *in action*

A tornado has ripped through your community in the early hours of the morning damaging numerous buildings. Your crew has been tasked with conducting a damage assessment of a two block area and reporting your findings to the incident commander.

The first house is slightly damaged with no risk of collapse. The next house has significant damage. It is a wood-frame house with several collapsed load-bearing walls. It is unsafe to enter without shoring. The third wood-frame home is completely destroyed. The fourth building is a noncombustible commercial structure with block walls and steel trusses. It is partially collapsed. As you continue your search, you realize the magnitude of this event.

1. What type of construction is the commercial structure?
   - **A.** Type II
   - **B.** Type III
   - **C.** Type IV
   - **D.** Type V
2. What type of construction were the houses?
   - **A.** Type II
   - **B.** Type III
   - **C.** Type IV
   - **D.** Type V
3. A ___ is a wall that is designed to provide structural support for a building.
   - **A.** non-load bearing
   - **B.** load bearing
   - **C.** curtain wall
   - **D.** fire wall
4. A ____ is usually constructed of concrete or masonry and designed to transfer the weight of the building and its contents to the ground.
   - **A.** foundation
   - **B.** roof
   - **C.** beam
   - **D.** column
5. Type III construction is also known as ____.
   - **A.** heavy timber
   - **B.** noncombustible
   - **C.** ordinary
   - **D.** fire resistive
6. ___ are mounted in an inclined position to support a roof.
   - **A.** Trusses
   - **B.** Rafters
   - **C.** Joists
   - **D.** Beams

FIRE FIGHTER II *in action*

You arrive at an old two-story, wood-frame farmhouse with wood siding and shake shingles. It has very tall, narrow windows. The windows on the second floor are directly above the windows on the first floor. The house is about 40' by 40' and has a basement. There is turbulent brown smoke coming from the basement windows, and there is light brown, laminar flow smoke coming out the gable end of the roof. You do not see any smoke through the front window. Your captain directs you to pull a 1 ¾" hoseline from the engine and go to the front door to attack the fire in the basement.

1. What type of construction do you suspect this building is?
2. Where is the fire going to travel?
3. What hazards may be associated with this type of construction?
4. What are the sings of a potential building collapse?

Portable Fire Extinguishers

Fire Fighter I

Knowledge Objectives

After studying this chapter, you will be able to:

- State the primary purposes of fire extinguishers. (NFPA 5.3.16.A, p 199)
- Define Class A fires. (NFPA 5.3.16.A, p 200)
- Define Class B fires. (NFPA 5.3.16.A, p 200–201)
- Define Class C fires. (NFPA 5.3.16.A, p 201)
- Define Class D fires. (NFPA 5.3.16.A, p 201)
- Define Class K fires. (NFPA 5.3.16.A, p 201–202)
- Explain the classification and rating system for fire extinguishers. (NFPA 5.3.16.A, p 202–206)
- Describe the types of agents used in fire extinguishers. (NFPA 5.3.16.A, p 206–209)
- Describe the types of operating systems in fire extinguishers. (NFPA 5.3.16.A, p 209–216)
- Describe the basic steps of fire extinguisher operation. (NFPA 5.3.16, p 216–218)
- Explain the basic steps of inspecting, maintaining, recharging, and hydrostatic testing of fire extinguishers. (NFPA 5.3.16, p 225–227)
- Select the proper class of fire extinguisher. (NFPA 5.3.16.B, p 216)

Skills Objectives

After studying this chapter, you will be able to perform the following skills:

- Transport the extinguisher to the location of the fire. (NFPA 5.3.16.B, p 218–219)
- Attack a Class A fire with a stored-pressure water-type fire extinguisher. (NFPA 5.3.16.B, p 218, 220)
- Attack a Class A fire with a multipurpose dry-chemical fire extinguisher. (NFPA 5.3.16.B, p 219)
- Attack a Class B flammable liquid fire with a dry-chemical fire extinguisher. (NFPA 5.3.16.B, p 220–221)
- Attack a Class B flammable liquid fire with a stored-pressure foam fire extinguisher. (NFPA 5.3.16.B, p 221–222)
- Operate a carbon dioxide extinguisher. (NFPA 5.3.16.B, p 222–223)
- Use a halogenated agent-type extinguisher. (NFPA 5.3.16.B, p 223–224)
- Use dry-powder fire-extinguishing agents. (NFPA 5.3.16.B, p 224–225)
- Use a wet-chemical (Class K) fire extinguisher. (NFPA 5.3.16.B, p 225–226)

CHAPTER 8

Fire Fighter II — FFII

Knowledge Objectives

There are no knowledge objectives for Fire Fighter II candidates. NFPA 1001 contains no Fire Fighter II Job Performance Requirements for this chapter.

Skills Objectives

There are no skill objectives for Fire Fighter II candidates. NFPA 1001 contains no Fire Fighter II Job Performance Requirements for this chapter.

Additional NFPA Standards

- NFPA 10, *Standard for Portable Fire Extinguishers*
- NFPA 11, *Standard for Low-, Medium-, and High-Expansion Foam*

You Are the Fire Fighter

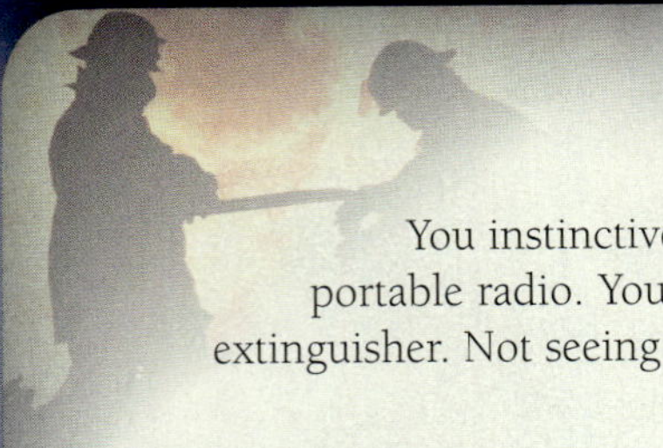

You are the designated cook for the day and the kitchen is bare. You go into the grocery store while the rest of the crew waits in the apparatus. As you make your way through the produce section, an employee runs up to you and says that there is a fire in the back room. You instinctively run to the back room while instructing another employee to run out and notify your crew by portable radio. You find a fire in a trash can that is starting to spread. You quickly scan the room looking for a fire extinguisher. Not seeing one, you ask the employee where the closest one is located.

1. What type of fire extinguisher do you need for this fire?
2. How large of a fire will it put out?
3. How do fire inspectors ensure the fire extinguisher will work when needed?

Introduction

Portable fire extinguishers are required in many types of occupancies as well as in commercial vehicles, boats, aircraft, and various other locations. Fire prevention efforts encourage citizens to keep fire extinguishers in their homes, particularly in their kitchens. Fire extinguishers are used successfully to put out hundreds of fires every day, preventing millions of dollars in property damage as well as saving lives. Most fire extinguishers are easy to operate and can be used effectively by an individual with only basic training.

Fire extinguishers range in size from models that can be operated with one hand to large, wheeled models that contain several hundred pounds of extinguishing agent (material used to stop the combustion process) **FIGURE 8-1**. Extinguishing agents include water, water with different additives, dry chemicals, wet chemicals, dry powders, and gaseous agents. Each agent is suitable for specific types of fires.

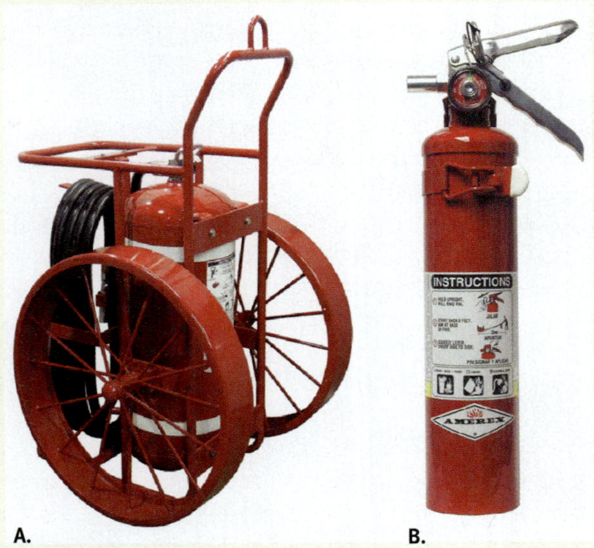

A.

B.

FIGURE 8-1 Portable fire extinguishers can be large or small.
A. A wheeled extinguisher. **B.** A one-hand fire extinguisher.

Fire extinguishers are designed for different purposes and involve different operational methods. As a fire fighter, you must know which is the most appropriate kind of extinguisher to use for different types of fires, which kinds must not be used for certain fires, and how to use any fire extinguisher you may encounter, especially those used by your department and carried on your apparatus. The selection of the proper fire extinguisher builds on the information presented in the Fire Behavior chapter.

Fire fighters use fire extinguishers to control small fires that do not require the use of a hose line. In addition, a portable backpack-type fire extinguisher can be used to control and overhaul a wildland fire located beyond the reach of hoses. Special types of fire extinguishers are appropriate for situations where the application of water would be dangerous, ineffective, or undesirable. For example, applying water to a fire that involves expensive electronic equipment is both dangerous and costly. Using the correct fire extinguisher could control the fire without causing additional damage to the equipment.

This chapter also covers the operation and maintenance of the most common types of portable fire extinguishers. These principles will enable you to use fire extinguishers correctly and effectively, thereby reducing the risk of personal injury and property damage.

FIRE FIGHTER Tips

Fire fighters are often called out to do fire extinguisher training for the public. It is essential that you understand the characteristics and operations of each type of fire extinguisher.

Fire Fighter Safety Tips

Do not place yourself in a dangerous situation by trying to fight a large fire with a small fire extinguisher. You cannot fight a fire or protect yourself with an empty extinguisher, an under-sized extinguisher, or the inappropriate type of extinguisher.

Fire Fighter Safety Tips

Proper technique in using a portable fire extinguisher is important for both safety and effectiveness. Practice using a portable fire extinguisher under the careful supervision of a trained instructor.

Purposes of Fire Extinguishers

Portable fire extinguishers have two primary uses: to extinguish incipient-stage fires (those that have not spread beyond the area of origin), and to control fires where traditional methods of fire suppression are not recommended.

Fire extinguishers are placed in many locations so that they will be available for immediate use on small, incipient-stage fires, such as a fire in a wastebasket. A trained individual with a suitable fire extinguisher could usually control this type of fire FIGURE 8-2. As flames spread beyond the wastebasket to other contents of the room, however, such a fire becomes increasingly difficult to control with only a portable fire extinguisher.

FIGURE 8-2 A trained individual with a suitable fire extinguisher can usually control an incipient-stage fire.

Fire extinguishers are also used to control fires in situations where traditional extinguishing methods are not recommended. For example, using water on fires that involve energized electrical equipment increases the risk of electrocution to fire fighters. Applying water to a fire in a computer or electrical control room could cause extensive damage to the electrical equipment. In these cases, it would be better to use a fire extinguisher filled with the appropriate extinguishing agent. Special extinguishing agents are also required for fires that involve flammable liquids, cooking oils, and combustible metals. Extinguishers are designed to provide an effective discharge pattern for fires involving specific types of fuels. Using an improper type of extinguisher can spread burning material. An appropriate type of fire extinguisher needs to be available in areas containing flammable materials.

Use of Portable Fire Extinguishers in Incipient-Stage Fires

Most fire department vehicles carry at least one fire extinguisher, and many vehicles carry two or more extinguishers of different types. Fire fighters often use these portable extinguishers to control incipient-stage fires quickly. At times, a fire fighter may even use an extinguisher from the fire-site premises to control an incipient fire.

One advantage of fire extinguishers is their portability. It may take less time to control a fire with a portable extinguisher than it would to advance and charge a hose line. Fire department vehicles that are not equipped with water or fire hoses usually carry at least one multipurpose fire extinguisher. If you arrive at an incipient-stage fire in one of these vehicles, you might be able to control the fire with the portable extinguisher. When you take this course of action, make sure you are not placing yourself in a dangerous situation by using a fire extinguisher.

The primary disadvantage of fire extinguishers is that they are "one-shot" devices. In other words, once the contents of a fire extinguisher have been discharged, the device is no longer effective in fighting fires until it is recharged. If the extinguisher does not control the fire, some other device or method must be employed. This is a serious limitation when compared to a fire hose with a continuous water supply.

Special Extinguishing Agents

As mentioned earlier, some types of fires require special extinguishing agents. As a fire fighter, you must know which fires require special extinguishing agents, which type of extinguisher should be used, and how to operate the different types of special-purpose extinguishers.

Special wet-chemical extinguishing agents are used for commercial kitchen fires that involve cooking oils in deep-fat fryers. A dry chemical is used for residential kitchen fires involving vegetable and peanut oils. Specific dry chemicals are used for combustible metal fires, and certain "clean" agents are used for fires in electronic equipment. Using water or an unsuitable fire-extinguishing chemical to fight these types of fires can cause unnecessary damage and may pose a danger to the fire extinguisher operator.

Portable extinguishers are sometimes used in combination with other techniques. For example, with a pressurized-gas fire, water may be used to cool hot surfaces and prevent reignition, while a dry chemical agent is used to extinguish the flames FIGURE 8-3. Certain types of portable extinguishers can be helpful in overhauling a fire. The extinguishing agents contained in these devices break down the surface tension of the water, allowing water to penetrate the materials and reach deep-seated fires.

FIGURE 8-3 Water can cool hot surfaces.

FIGURE 8-4 Most ordinary combustible materials are included in the definition of Class A fires.

Classes of Fires

It is essential to match the appropriate type of extinguisher to the type of fire. Fires and fire extinguishers are grouped into classes according to their characteristics. Some extinguishing agents work more efficiently than others on certain types of fires. In some cases, selecting the proper extinguishing agent will mean the difference between extinguishing a fire and being unable to control it.

It is dangerous to apply the wrong extinguishing agent to a fire. Using a water extinguisher on an electrical fire, for example, can cause an electrical shock as well as a short-circuit in the equipment. Likewise, a water extinguisher should never be used to fight a grease fire. Burning grease is generally hotter than 212°F (100°C), so when water comes in contact with it, the water converts to steam, which expands very rapidly. If the water penetrates the surface of the grease, the steam is produced within the grease. As the steam expands, the hot grease erupts like a volcano and splatters over everything and everyone nearby, potentially resulting in burns or injuries to people and spreading the fire.

Before selecting a fire extinguisher, ask yourself, "Which class of fire am I fighting?" There are five classes of fires, each of which affects the choice of extinguishing equipment.

Class A Fires

Class A fires involve ordinary combustibles such as wood, paper, cloth, rubber, household rubbish, and some plastics FIGURE 8-4 . Natural vegetation, such as grass and trees, is also Class A material. Water is the most commonly used extinguishing agent for Class A fires, although several other agents can be used effectively.

FIRE FIGHTER Tips

Most residential fires involve Class A materials. Even though the source of the ignition could be Class B or Class C in nature, the resulting fire frequently involves normal combustibles. A fire that begins with an unattended pot of grease on the stove, for example, might ignite the wooden cabinets and other kitchen contents. A short-circuit in an electrical outlet might ignite ordinary combustibles in the immediate vicinity. In both of these cases, the fire would involve primarily Class A materials.

Class B Fires

Class B fires involve flammable or combustible liquids, such as gasoline, oil, grease, tar, lacquer, oil-based paints, and some plastics FIGURE 8-5 . Fires involving flammable gases, such as propane or natural gas, are also categorized as Class B fires.

FIGURE 8-5 Class B fires involve flammable liquids and gases.

Examples of Class B fires include a fire in a pot of molten roofing tar, a fire involving splashed fuel on a hot lawnmower engine, and burning natural gas that is escaping from a gas meter struck by a vehicle. Several different types of extinguishing agents are approved for use in Class B fires.

Class C Fires

Class C fires involve energized electrical equipment, which includes any device that uses, produces, or delivers electrical energy FIGURE 8-6. A Class C fire could involve building wiring and outlets, fuse boxes, circuit breakers, transformers, generators, or electric motors. Power tools, lighting fixtures, household appliances, and electronic devices such as televisions, radios, and computers could be involved in Class C fires as well. The equipment must be plugged in or connected to an electrical source, but not necessarily operating, for the fire to be classified as Class C.

Electricity does not burn, but electrical energy can generate tremendous heat that could ignite nearby Class A or B materials. As long as the equipment is energized, the incident must be treated as a Class C fire. Agents that will not conduct electricity, such as dry chemicals or carbon dioxide (CO_2), must be used on Class C fires.

Class D Fires

Class D fires involve combustible metals such as magnesium, titanium, zirconium, sodium, lithium, and potassium. Special techniques and extinguishing agents are required to fight combustible metals fires FIGURE 8-7. Normally used extinguishing agents can react violently—even explosively—if they come in contact with burning metals. Violent reactions also can occur when water strikes burning combustible metals.

Class D fires are most often encountered in industrial occupancies, such as machine shops, repair shops, and metal recycling plants, as well as in fires involving aircraft and automobiles. Magnesium and titanium—both combustible metals—are used to produce automotive and aircraft parts because they combine high strength with light weight. Sparks from cutting, welding, or grinding operations could ignite a Class D fire, or the metal items could become involved in a fire that originated elsewhere.

Because of the chemical reactions that could occur during a Class D fire, it is important to select the proper extinguishing agent and application technique for this kind of incident. Choosing the correct fire extinguisher for a Class D fire requires expert knowledge and experience.

Class K Fires

Class K fires involve combustible cooking oils and fats FIGURE 8-8. This is a relatively new classification; cooking oil fires were previously classified as Class B combustible liquid fires. The introduction of high-efficiency, modern deep-fat fryers and the trend toward using vegetable oils instead of animal fats to fry foods in recent years have resulted in higher cooking temperatures. These higher temperatures

FIGURE 8-7 Combustible metals in Class D fires require special extinguishing agents.

FIGURE 8-6 Class C fires involve energized electrical equipment or appliances.

FIGURE 8-8 Class K fires involve cooking oils and fats.

required the development of a new class of wet-chemical extinguishing agents to combat fires involving cooking oils. Many restaurants continue to use extinguishing agents that were approved for Class B fires, however. Class B extinguishers are not as effective for cooking oil fires as are Class K extinguishers.

Classification of Fire Extinguishers

Portable fire extinguishers are classified and rated based on their characteristics and capabilities. This information is important for selecting the proper extinguisher to fight a particular fire TABLE 8-1. It is also used to determine which types of fire extinguishers should be placed in a given location so that incipient-stage fires can be controlled quickly.

In the United States, Underwriters Laboratories, Inc. (UL) is the organization that developed the standards, classification, and rating system for portable fire extinguishers. This system rates fire extinguishers for both safety and effectiveness. Each fire extinguisher has a specific rating that identifies the classes of fires for which it is both safe and effective.

TABLE 8-1	Types of Fires
Class A	Ordinary combustibles
Class B	Flammable or combustible liquids
Class C	Energized electrical equipment
Class D	Combustible metals
Class K	Kitchen fires involving oils and fats

The classification system for fire extinguishers uses both letters and numbers. The letters indicate the classes of fire for which the extinguisher can be used, and the numbers indicate its effectiveness. Fire extinguishers that are safe and effective for more than one class will be rated with multiple letters. For example, an extinguisher that is safe and effective for Class A fires will be rated with an "A"; one that is safe and effective for Class B fires will be rated with a "B"; and one that is safe and effective for both Class A and Class B fires will be rated with both an "A" and a "B."

Class A and Class B fire extinguishers also include a number, indicating the relative effectiveness of the fire extinguisher in the hands of a nonexpert user. On Class A extinguishers, this number reflects the amount of water it contains. An extinguisher that is rated 1-A contains the equivalent of 1.25 gallons (5.56 liters) of water. A typical Class A extinguisher contains 2.5 gallons (9 liters) of water and has a 2-A rating. The higher the number, the greater the extinguishing capability of the extinguisher. An extinguisher that is rated 4-A should be able to extinguish approximately twice as much fire as one that is rated 2-A.

The effectiveness of Class B extinguishers is based on the approximate area (measured in square feet) of burning fuel that these devices are capable of extinguishing. A 10-B rating indicates that a nonexpert user should be able to extinguish a fire in a pan of flammable liquid that is 10 square feet (3 meters) in surface area. An extinguisher rated 40-B should be able to control a flammable liquid pan fire with a surface area of 40 square feet.

Numbers are used to rate an extinguisher's effectiveness only for Class A and Class B fires. If the fire extinguisher can also be used for Class C fires, it contains an agent proven to be nonconductive to electricity and safe for use on energized electrical equipment. For instance, a fire extinguisher that carries a 2-A:10-B:C rating can be used on Class A, Class B, and Class C fires. It has the extinguishing capabilities of a 2-A extinguisher when applied to Class A fires, has the capabilities of a 10-B extinguisher when applied to Class B fires, and can be used safely on energized electrical equipment.

Standard test fires are used to rate the effectiveness of fire extinguishers. Such testing may involve different agents, amounts, application rates, and application methods. Fire extinguishers are rated for their ability to control a specific type of fire as well as for the extinguishing agent's ability to prevent rekindling. Some agents can successfully suppress a fire but are unable to prevent the material from reigniting. A rating is given only if the extinguisher completely extinguishes the standard test fire and prevents rekindling.

Labeling of Fire Extinguishers

Fire extinguishers that have been tested and approved by an independent laboratory are labeled to clearly designate the classes of fire the unit is capable of extinguishing safely. This traditional lettering system has been used for many years and is still found on many fire extinguishers. More recently, a universal pictograph system, which does not require the user to be familiar with the alphabetic codes for the different classes of fires, has been developed.

Traditional Lettering System

The traditional lettering system uses the following labels FIGURE 8-9:

- Extinguishers suitable for use on Class A fires are identified by the letter "A" on a solid green triangle. The triangle has a graphic relationship to the letter "A."
- Extinguishers suitable for use on Class B fires are identified by the letter "B" on a solid red square. Again, the shape of the letter mirrors the graphic shape of the box.
- Extinguishers suitable for use on Class C fires are identified by the letter "C" on a solid blue circle, which also incorporates a graphic relationship between the letter "C" and the circle.
- Extinguishers suitable for use on Class D fires are identified by the letter "D" on a solid yellow, five-pointed star.
- Extinguishers suitable for use on Class K (combustible cooking oil) fires are identified by a pictograph showing a fire in a frying pan. Because the Class K designation is new, there is no traditional-system alphabet graphic for it.

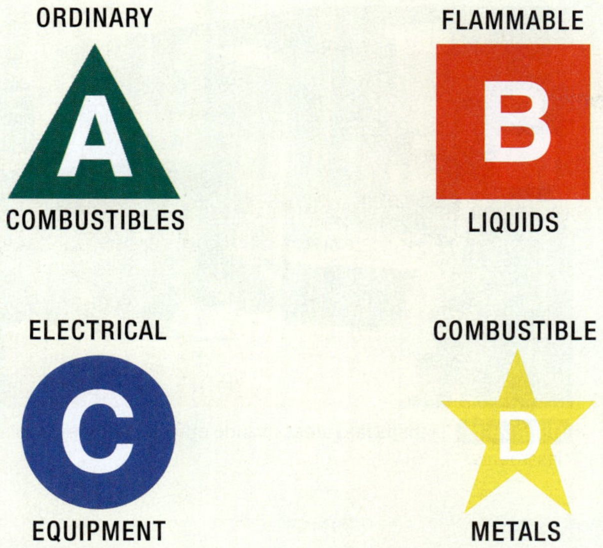

FIGURE 8-9 Traditional letter labels on fire extinguishers often incorporate a shape as well as a letter.

Pictograph Labeling System

The pictograph system, such as described for Class K fire extinguishers, uses symbols rather than letters on the labels. This system also clearly indicates whether an extinguisher is inappropriate for use on a particular class of fire. The pictographs are all square icons, each of which is designed to represent a certain class of fire FIGURE 8-10. The icon for Class A fires is a burning trashcan beside a wood fire. The Class B fire extinguisher icon is a flame and a gasoline can; the Class C icon is a flame and an electrical plug and socket. As noted previously, extinguishers rated for fighting Class K fires are labeled with an icon showing a fire in a frying pan.

Under this pictograph labeling system, the presence of an icon indicates that the extinguisher has been rated for that class of fire. A missing icon indicates that the extinguisher has not been rated for that class of fire. A red slash across an icon indicates that the extinguisher must not be used on that type of fire, because doing so would create additional risk.

FIGURE 8-10 The icons for Class A, B, C, and K fires.

For example, an extinguisher rated for Class A fires only would show all three icons, but the icons for Class B and Class C would have red diagonal lines through them. This three-icon array signifies that the extinguisher contains a water-based extinguishing agent, making it unsafe to use on flammable liquid or electrical fires.

Certain extinguishers labeled as appropriate for Class B and Class C fires do not include the Class A icon, but may be used to put out small Class A fires. The fact that they have not been rated for Class A fires indicates that they are less effective in extinguishing a common combustible fire than a comparable Class A extinguisher would be.

FIRE FIGHTER Tips

The safest and surest way to extinguish a Class C fire is to turn off the power and treat it like a Class A or B fire. If you are unable to turn off the power, you should be prepared for reignition, because the electricity could reignite the fire after it is extinguished.

Fire Extinguisher Placement

Fire codes and regulations require the installation of fire extinguishers in many areas so that they will be available to fight incipient-stage fires. NFPA 10, *Standard for Portable Fire Extinguishers*, lists the requirements for placing and mounting portable fire extinguishers as well as the appropriate mounting heights.

The regulations for each type of occupancy specify the maximum floor area that can be protected by each extinguisher, the maximum travel distance from the closest extinguisher to a potential fire, and the types of fire extinguishers that should be provided. Two key factors must be considered when determining which type of extinguisher should be placed in each area: the class of fire that is likely to occur and the potential magnitude of an incipient fire.

Extinguishers should be mounted so they are readily visible and easily accessible FIGURE 8-11. Heavy extinguishers should not be mounted high on a wall. If the extinguisher is mounted too high, a smaller person might be unable to lift it off its hook or could be injured in the attempt.

According to NFPA 10, the recommended mounting heights for the placement of fire extinguishers are as follows:

- Fire extinguishers weighing up to 40 pounds (18 kilograms) should be mounted so that the top of the extinguisher is not more than 5 feet (2 meters) above the floor.
- Fire extinguishers weighing more than 40 pounds (18 kilograms) should be mounted so that the top of the extinguisher is not more than 3 feet (1 meters) above the floor.
- The bottom of an extinguisher should be at least 4 inches (10 centimeters) above the floor.

FIGURE 8-11 Extinguishers should be mounted in locations with unobstructed access and visibility.

Classifying Area Hazards

Areas are divided into three risk classifications—light, ordinary, and extra hazard—based on the amount and type of combustibles that are present, including building materials, contents, decorations, and furniture. The quantity of combustible materials present is sometimes called a building's fire load and is measured as the average weight of combustible materials per square foot (or per square meter) of floor area. The larger the fire load, the larger the potential fire.

The occupancy use category does not necessarily determine the building's hazard classification. The recommended hazard classifications for different types of occupancies are simply guidelines based on typical situations. The hazard classification for each area should be based on the actual amount and type of combustibles that are present.

Light (Low) Hazard

Light (low) hazard locations are areas where the majority of materials are noncombustible or arranged so that a fire is not likely to spread. Light hazard environments usually contain limited amounts of Class A combustibles, such as wood, paper products, cloth, and similar materials. A light hazard environment might also contain some Class B combustibles (flammable liquids and gases), such as copy machine chemicals or modest quantities of paints and solvents, but all Class B materials must be kept in closed containers and stored safely. Examples of common light hazard environments include most offices, classrooms, churches, assembly halls, and hotel guest rooms FIGURE 8-12.

FIGURE 8-12 Light hazard areas include offices, churches, and classrooms.

Ordinary (Moderate) Hazard

Ordinary (moderate) hazard locations contain more Class A and Class B materials than do light hazard locations. Typical examples of ordinary hazard locations include retail stores with on-site storage areas, light manufacturing facilities, auto showrooms, parking garages, research facilities, and workshops or service areas that support light hazard locations, such as hotel laundry rooms or restaurant kitchens FIGURE 8-13.

FIGURE 8-13 Auto showrooms, hotel laundry rooms, and parking garages are classified as ordinary hazard areas.

Ordinary hazard areas also include warehouses that contain Class I and Class II commodities. Class I commodities include noncombustible products stored on wooden pallets or in corrugated cartons that are shrink-wrapped or wrapped in paper. Class II commodities include noncombustible products stored in wooden crates or multilayered corrugated cartons.

Extra (High) Hazard

Extra (high) hazard locations contain more Class A combustibles and/or Class B flammables than do ordinary hazard locations. Typical examples of extra hazard areas include woodworking shops; service and repair facilities for cars, aircraft, or boats; and many kitchens and other cooking areas that have deep fryers, flammable liquids, or gases under pressure FIGURE 8-14. In addition, areas used for manufacturing processes such as painting, dipping, or coating, as well as facilities used for storing or handling flammable liquids, are classified as extra hazard environments. Warehouses containing products that do not meet the definitions of Class I and Class II commodities are also considered extra hazard locations.

Determining the Most Appropriate Placement of Fire Extinguishers

Several factors must be considered when determining the number and types of fire extinguishers that should be placed in each area of an occupancy. Among these factors are the types of fuels found in the area and the quantities of those materials.

Some areas may need extinguishers with more than one rating or more than one type of fire extinguisher. Environments that include Class A combustibles require an extinguisher rated for Class A fires; those with Class B combustibles require an extinguisher rated for Class B fires. Areas that contain both Class A and Class B combustibles, however, require either an extinguisher that is rated for both types of fires or a separate extinguisher for each class of fire.

FIGURE 8-14 Kitchens, woodworking shops, and auto repair shops are considered possible extra hazard locations.

Most buildings require extinguishers that are suitable for fighting Class A fires because ordinary combustible materials—such as furniture, partitions, interior finish materials, paper, and packaging products—are so common. Even where other classes of products are used or stored, there is still a need to defend the facility from a fire involving common combustibles.

A single multipurpose extinguisher is generally less expensive than two individual fire extinguishers and eliminates the problem of selecting the proper extinguisher for a particular fire. However, it is sometimes more appropriate to install Class A extinguishers in general-use areas and to place extinguishers that are especially effective in fighting Class B or Class C fires near those specific hazards.

In some facilities, a variety of conditions are present. In these occupancies, each area must be individually evaluated and the extinguisher installation tailored to its particular circumstances. A restaurant is a good example of this situation. The dining areas contain common combustibles, such as furniture, tablecloths, and paper products, that would require an extinguisher rated for Class A fires. In the restaurant's kitchen, where the risk of fire involves cooking oils, a Class K extinguisher would provide the best defense.

Similarly, within a hospital, extinguishers for Class A fires would be appropriate in hallways, offices, lobbies, and patient rooms. Class B extinguishers should be mounted in laboratories and areas where flammable anesthetics are stored or handled. Electrical rooms should have extinguishers that are approved for use on Class C fires, whereas hospital kitchens would need Class K extinguishers.

Methods of Fire Extinguishment

Understanding the nature of fire is key to understanding how extinguishing agents work and how they differ from one another. All fires require three basic ingredients: fuel, heat, and oxygen. In scientific terminology, burning is called rapid oxidation. This chemical process occurs when a fuel is combined with oxygen, resulting in the formation of ash or other waste products and the release of energy as heat and light.

The combustion process begins when the fuel is heated to its ignition point—the temperature at which it begins to burn. The energy that initiates this process can come from many different sources, including a spark or flame, friction, electrical energy, or a chemical reaction. Once a substance begins to burn, it will generally continue burning as long as adequate supplies of oxygen and fuel to sustain the chemical reaction are present, unless something interrupts the process.

Most extinguishers stop the burning by cooling the fuel below its ignition point, by cutting off the supply of oxygen, or by performing both actions. Some extinguishing agents interrupt the complex system of molecular chain reactions that occur between the heated fuel and the oxygen. Modern portable fire extinguishers contain agents that use one or more of these methods.

Cooling the Fuel

If the temperature of the fuel falls below its ignition temperature, the combustion process will stop. Water extinguishes a fire using this method.

Cutting Off the Supply of Oxygen

Creating a barrier that interrupts the flow of oxygen to the flames will also extinguish a fire. Putting a lid on a pan of burning food is an example of this technique **FIGURE 8-15**; applying a blanket of foam to the surface of a burning liquid is another example. Surrounding the fuel with a layer of CO_2 can also cut off the supply of oxygen necessary to sustain the burning process.

FIGURE 8-15 Covering a pan of burning food with a lid will extinguish a fire by cutting off the supply of oxygen.

Interrupting the Chain of Reactions

Some extinguishing agents work by interrupting the molecular chain reactions required to sustain combustion. In some cases, just a very small quantity of the agent can accomplish this objective.

Types of Extinguishing Agents

An extinguishing agent is the substance contained in a portable fire extinguisher that puts out a fire. A variety of chemicals, including water, are used in portable fire extinguishers. The best extinguishing agent for a particular hazard depends on several factors, including the types of materials involved and the anticipated size of the fire. Portable fire extinguishers use seven basic types of extinguishing agents:

- Water
- Dry chemicals
- CO_2
- Foam
- Halogenated agents
- Dry powder
- Wet chemicals

Water

Water is an efficient, plentiful, and inexpensive extinguishing agent. When it is applied to a fire, it is quickly converted from a liquid into steam, absorbing great quantities of heat in the process. As the heat is removed from the combustion process, the fuel cools below its ignition temperature and the fire stops burning.

Water is an excellent extinguishing agent for Class A fires. Many Class A fuels will absorb liquid water, which further lowers the temperature of the fuel. This also prevents rekindling.

Water is a much less effective extinguishing agent for other classes of fires. Applying water to hot cooking oil, for example, can cause explosive splattering, which can spread the fire and endanger the operator of the fire extinguisher. Many burning flammable liquids will simply float on top of water. Because water conducts electricity, it is dangerous to apply a stream of water to any fire that involves energized electrical equipment. If water is applied to a burning combustible metal, a violent reaction can occur. Because of these limitations, plain water is used only in Class A fire extinguishers.

One notable disadvantage of water is that it freezes at 32°F (0°C). In areas that are subject to below-freezing temperatures, loaded-stream fire extinguishers can be used to counteract this limitation. These extinguishers combine an alkali metal salt with water. The salt lowers the freezing point of water, so the extinguisher can be used in much colder areas.

Wetting agents can also be added to the water in a fire extinguisher. These agents reduce the surface tension of the water, allowing it to penetrate more effectively into many fuels, such as baled cotton and fibrous materials.

Another method of applying water is through the use of water mist. Water-mist fire extinguishers contain distilled water, which is discharged from the device as a fine spray. These extinguishers are used where regular extinguishers might cause excessive damage. Typical uses include museums, rare book collections, and hospital operating rooms.

Dry Chemicals

Dry-chemical fire extinguishers deliver a stream of very finely ground particles onto a fire. Different chemical compounds are used to produce extinguishers of varying capabilities and characteristics. The dry-chemical extinguishing agents work in two ways. First, the dry chemicals interrupt the chemical chain reactions that occur as part of the combustion process. Second, the tremendous surface area of the finely ground particles allows them to absorb large quantities of heat.

Dry chemical extinguishing agents offer several advantages over water extinguishers:

- They are effective on Class B (flammable liquids and gases) fires.
- They can be used on Class C (energized electrical equipment) fires, because the chemicals are nonconductive.
- They are not subject to freezing.

The first dry-chemical extinguishers were introduced during the 1950s and were rated only for Class B and C fires. The industry term for these B:C-rated units is "ordinary dry chemical" extinguishers.

During the 1960s, multipurpose dry-chemical fire extinguishers were introduced. These extinguishers are rated for Class A, B, and C fires. The chemicals in these extinguishers form a crust over Class A combustible fuels, thereby preventing rekindling FIGURE 8-16.

FIGURE 8-16 Multipurpose dry-chemical extinguishers can be used for Class A, B, and C fires.

Multipurpose dry-chemical extinguishing agents take the form of fine particles that are treated with other chemicals to help maintain an even flow when the extinguisher is being used. These additives are intended to prevent the particles from absorbing moisture, which could cause packing or caking and interfere with the extinguisher's discharge.

One disadvantage of dry-chemical extinguishers is that the chemicals—particularly the multipurpose dry chemicals—are corrosive and can damage electronic equipment, such as computers, telephones, and copy machines. The fine particles are carried in air and settle like a fine dust inside the equipment. Over a period of months, this residue can corrode metal parts, causing considerable damage. If electronic equipment is exposed to multipurpose dry-chemical extinguishing agents, it should be cleaned professionally within 48 hours after exposure.

Five compounds are used as the primary dry-chemical extinguishing agents:

- Sodium bicarbonate (rated for Class B and C fires only)
- Potassium bicarbonate (rated for Class B and C fires only)
- Urea-based potassium bicarbonate (rated for Class B and C fires only)
- Potassium chloride (rated for Class B and C fires only)
- Ammonium phosphate (rated for Class A, B, and C fires)

Sodium bicarbonate is often used in small household extinguishers. Potassium bicarbonate, potassium chloride, and urea-based potassium bicarbonate all have greater fire-extinguishing capabilities (per unit volume) for Class B fires than does sodium bicarbonate. Potassium chloride is more corrosive than the other dry-chemical extinguishing agents.

Ammonium phosphate is the only dry-chemical extinguishing agent that is rated as suitable for use on Class A fires. Although ordinary dry-chemical extinguishers can be used against Class A fires, a water dousing is also needed to extinguish any smoldering embers and prevent rekindling.

The selection of which dry-chemical extinguisher to use depends on the compatibility of different agents with one another and with any products that they might contact. Some dry-chemical extinguishing agents cannot be used in combination with particular types of foam.

Carbon Dioxide

Carbon dioxide is a gas that is 1.5 times heavier than air. When CO_2 is discharged on a fire, it forms a dense cloud that displaces the air surrounding the fuel. This effect interrupts the combustion process by reducing the amount of oxygen that can reach the fuel. The placement of a blanket of CO_2 over the surface of a liquid fuel can also disrupt the fuel's ability to vaporize.

In portable carbon dioxide (CO_2) fire extinguishers, CO_2 is stored under pressure as a colorless and odorless liquid. It is discharged through a hose and expelled on the fire through a horn. The CO_2 is very cold when it is released; it forms a visible cloud of "dry ice" when moisture in the air freezes as it comes into contact with the CO_2.

CO_2 fire extinguishers are rated for Class B and C fires only. This extinguishing agent does not conduct electricity and has two significant advantages over dry chemical agents: It is not corrosive and it does not leave any residue.

CO_2 also has several limitations and disadvantages:

- Weight: CO_2 extinguishers are heavier than similarly rated extinguishers that use other extinguishing agents FIGURE 8-17.
- Range: CO_2 extinguishers have a short discharge range, which requires the operator to be close to the fire, increasing the risk of personal injury.
- Weather: CO_2 does not perform well at temperatures below 0°F (–18°C) or in windy or drafty conditions, because it dissipates before it reaches the fire.
- Confined spaces: When used in confined areas, CO_2 dilutes the oxygen in the air. If it is diluted enough, people in the space can begin to suffocate.
- Suitability: CO_2 extinguishers are not suitable for use on fires involving pressurized fuel or on cooking grease fires.

FIGURE 8-17 Carbon dioxide extinguishers are heavy due to the weight of the container and the large quantity of agent needed to extinguish a fire. They also have a large discharge nozzle, making them easily identifiable.

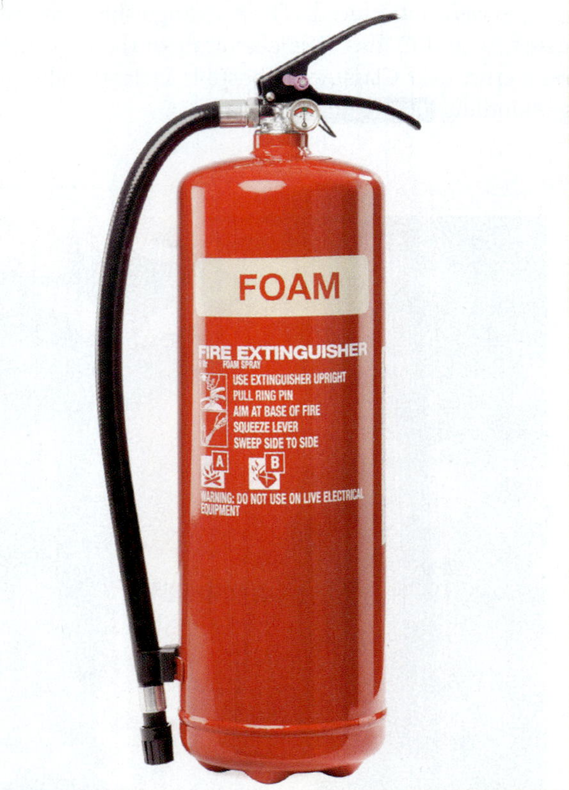

FIGURE 8-18 An AFFF extinguisher produces an effective foam for use on Class B fires.

Foam

Foam fire extinguishers discharge a water-based solution to which a measured amount of foam concentrate has been added. The nozzles on foam extinguishers are designed to introduce air into the discharge stream, thereby producing a foam blanket. Foam extinguishing agents are formulated for use on either Class A or Class B fires.

Class A foam extinguishers for ordinary combustible fires extinguish fires in the same way that water extinguishes fires. This type of extinguisher can be produced by adding Class A foam concentrate to the water in a standard 2.5-gallon (11 liter), stored-pressure extinguisher. The foam concentrate reduces the surface tension of the water, allowing for better penetration of the foam into the burning materials.

Class B foam extinguishers discharge a foam solution that floats across the surface of a burning liquid, creating a blanket that separates the fuel from oxygen. This blanket prevents the fuel from vaporizing. It forms a barrier between the fuel and the oxygen, extinguishing the flames and preventing reignition. These foaming agents are not suitable for Class B fires that involve pressurized fuels or cooking oils, however.

The most commonly encountered Class B additives are aqueous film-forming foam (AFFF) and film-forming fluoroprotein (FFFP) foam **FIGURE 8-18**. Both concentrates produce very effective foams. Which one should be used depends on the product's compatibility with a particular flammable liquid and other extinguishing agents that could be used on the same fire.

Some Class B foam extinguishing agents are approved for use on polar solvents—that is, water-soluble flammable liquids such as alcohols, acetone, esters, and ketones. Only extinguishers that are specifically labeled for use with polar solvents should be used if these products are present.

Although they are not specifically intended to extinguish Class A fires, most Class B foams can also be used on ordinary combustibles. The reverse is not true, however: Class A foams are *not* effective on Class B fires. Foam extinguishers are not suitable for use on Class C fires and cannot be stored or used at freezing temperatures.

Wet Chemicals

Wet-chemical fire extinguishers are the only type of extinguisher to qualify under the new Class K rating requirements. They use wet-chemical extinguishing agents, which are chemicals applied as water solutions. Before Class K extinguishing agents were developed, most fire-extinguishing systems for kitchens used dry chemicals. The minimum requirement for a commercial kitchen was a 40-B-rated sodium bicarbonate or potassium bicarbonate extinguisher. These systems required extensive clean-up after their use, which often resulted in serious business interruptions.

All new fixed extinguishing systems in restaurants and commercial kitchens now use wet-chemical extinguishing agents. These agents are specifically formulated for use in commercial kitchens and food-product manufacturing facilities, especially where food is cooked in a deep fryer. The fixed systems discharge the agent directly over the cooking surfaces. There is no numeric rating of their efficiency in portable fire extinguishers.

The Class K wet chemical agents include aqueous solutions of potassium acetate, potassium carbonate, and potassium citrate, either singly or in various combinations. These wet agents convert the fatty acids in cooking oils or fats to a soap or foam, a process known as saponification.

When wet chemical agents are applied to burning vegetable oils, they create a thick blanket of foam that quickly smothers the fire and prevents it from reigniting while the hot oil cools. Such agents are discharged as a fine spray, which reduces the risk of splattering. They are very effective at extinguishing cooking oil fires, and clean-up afterward is much easier, allowing a business to reopen sooner.

Halogenated Agents

Halogenated extinguishing agents are produced from a family of liquefied gases, known as halogens, that includes fluorine, bromine, iodine, and chlorine. Hundreds of different formulations can be produced from these elements; these myriad versions have many different properties and potential uses. Although several of these formulations are very effective for extinguishing fires, only a few of them are commonly used as extinguishing agents.

Halogenated extinguishing agents are commonly called clean agents because they leave no residue and are ideally suited for areas that contain computers or sensitive electronic equipment. Per pound, they are approximately twice as effective at extinguishing fires as is CO_2.

Two categories of halogenated extinguishing agents are distinguished: Halons and halocarbons. A 1987 international agreement, known as the Montreal Protocol, limited Halon production because these agents damage the earth's ozone layer. Halons have since been replaced by a new family of extinguishing agents, halocarbons.

The halogenated agents are stored as liquids and are discharged under relatively high pressure. They release a mist of vapor and liquid droplets that disrupts the molecular chain reactions within the combustion process, thereby extinguishing the fire. These agents dissipate rapidly in windy conditions, as does CO_2, so their effectiveness is limited in outdoor locations. Because halogenated agents also displace oxygen, they should be used with care in confined areas.

Halon 1211 (bromochlorodifluoromethane) should be used judiciously and only in situations where its clean properties are essential, owing to its detrimental environmental impact. Small Halon 1211 extinguishers are rated for Class B and C fires, but are unsuited for use on fires involving pressurized fuels or cooking grease. Larger Halon extinguishers are also rated for Class A fires.

Currently, four types of halocarbon agents are used in portable extinguishers: hydrochlorofluorocarbon, hydrofluorocarbon, perfluorocarbon, and fluoroiodocarbon.

Dry Powder

Dry-powder extinguishing agents are chemical compounds used to extinguish fires involving combustible metals (Class D fires). These agents are stored in fine granular or powdered form and are applied to smother the fire. They form a solid crust over the burning metal, which both blocks out oxygen (the fuel for the fire) and absorbs heat.

The most commonly used dry-powder extinguishing agent is formulated from finely ground sodium chloride (table salt) plus additives to help it flow freely over a fire. A thermoplastic material mixed with the agent binds the sodium chloride particles into a solid mass when they come into contact with a burning metal.

Another dry powder agent is produced from a mixture of finely granulated graphite powder and phosphorus-containing compounds. This agent cannot be expelled from fire extinguishers; instead, it is produced in bulk form and applied by hand, using a scoop or a shovel. When applied to a metal fire, the phosphorus compounds release gases that blanket the fire and cut off its supply of oxygen; the graphite absorbs heat from the fire, allowing the metal to cool below its ignition point. Other specialized dry-powder extinguishing agents are available for fighting specific types of metal fires. For details, see NFPA's *Fire Protection Handbook*.

Class D agents must be applied very carefully so that the molten metal does not splatter. No water should come in contact with the burning metal, because even a trace quantity of moisture can cause a violent reaction.

Fire Extinguisher Design

All portable fire extinguishers use pressure to expel their contents onto a fire, and many rely on pressurized gas to eject the extinguishing agent. This gas can be stored either with the extinguishing agent in the body of the extinguisher or externally in a separate cartridge or cylinder. With external storage, the extinguishing agent is put under pressure only as it is used.

Some extinguishing agents, such as CO_2, are called self-expelling agents. Most of these agents are normally gases, but are stored as liquids under pressure in the fire extinguisher. When the confining pressure is released, the agent rapidly expands, causing it to self-discharge. Hand-operated pumps are used to expel the agent when water or water with additives is the extinguishing agent.

Portable Fire Extinguisher Components

Most hand-held portable fire extinguishers have six basic parts FIGURE 8-19:

- A cylinder or container that holds the extinguishing agent
- A carrying handle

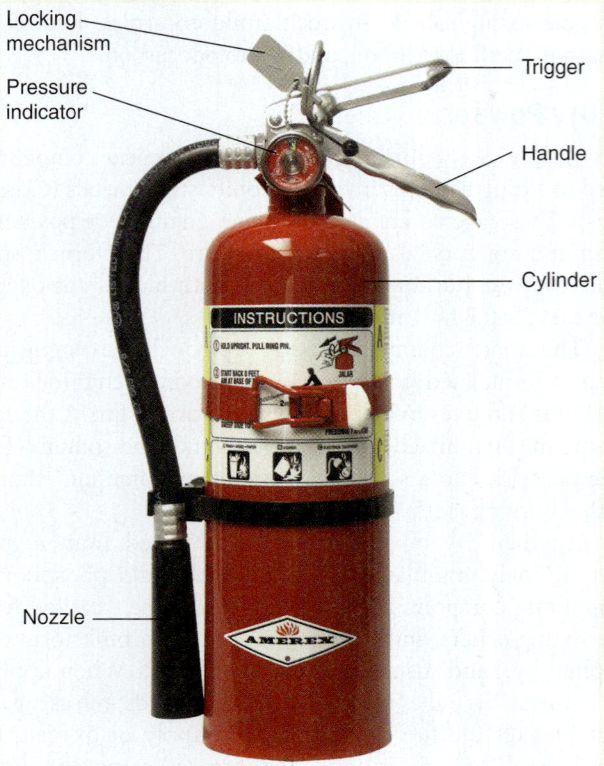

FIGURE 8-19 Portable fire extinguishers have six basic parts.

- A nozzle or horn
- A trigger and discharge valve assembly
- A locking mechanism to prevent accidental discharge
- A pressure indicator

Cylinder or Container

The body of the extinguisher, known as the cylinder or container, holds the extinguishing agent. Nitrogen, compressed air, or CO_2 can be used to pressurize the cylinder to expel the agent. Stored-pressure fire extinguishers hold both the extinguishing agent, in wet or dry form, and the expeller gas under pressure in the cylinder. Cartridge/cylinder fire extinguishers rely on an external cartridge of pressurized gas, which is released only when the extinguisher is to be used.

Handle

The handle is used to carry a portable fire extinguisher and, in many cases, to hold it during use. The actual design of the handle varies from model to model, but all extinguishers that weigh more than 3 pounds have handles. In many cases, the handle is located just below the trigger mechanism.

Nozzle or Horn

The extinguishing agent is expelled through a nozzle or horn. In some extinguishers, the nozzle is attached directly to the valve assembly at the top of the extinguisher. In other models, the nozzle is found at the end of a short hose.

Foam extinguishers have a special aspirating nozzle that introduces air into the extinguishing agent, thereby creating the foam. Carbon dioxide extinguishers have a tubular or conical horn, which is often mounted at the end of a short hose.

Pump tank fire extinguishers, which are nonpressurized manually operated water extinguishers, usually have a nozzle at the end of a short hose. The manually operated pump may be mounted directly on the cylinder or it may be part of the nozzle assembly.

Trigger

The trigger is the mechanism that is squeezed or depressed to discharge the extinguishing agent. In some models, the trigger is a button positioned just above the handle. On most portable fire extinguishers, however, the trigger is a lever located above the handle. To release the extinguishing agent, the operator lifts the extinguisher by the handle and simultaneously squeezes down on the discharge lever.

Cartridge/cylinder extinguisher models usually have a two-step operating sequence. First, a handle or lever is pushed to pressurize the stored agent; then, a trigger-type mechanism incorporated in the nozzle assembly is used to control the discharge.

Locking Mechanism

The locking mechanism is a simple quick-release device that prevents accidental discharge of the extinguishing agent. The simplest form of locking mechanism is a stiff pin, which is inserted through a hole in the trigger to prevent it from being depressed. The pin usually has a ring at the end so that it can be removed quickly.

A special plastic tie, called a tamper seal, is used to secure the pin. This seal is designed to break easily when the pin is pulled. Removing the pin and tamper seal is best accomplished with a twisting motion. The tamper seal makes it easy to see whether the extinguisher has been used and not recharged. It also discourages people from playing or tinkering with the extinguisher.

Pressure Indicator

The pressure indicator or gauge shows whether a stored-pressure extinguisher has sufficient pressure to operate properly. Over time, the pressure in an extinguisher may dissipate. Checking the gauge first will tell you whether the extinguisher is ready for use.

Pressure indicators vary in terms of both design and sophistication. Most extinguishers use a needle gauge. Pressure may be shown in pounds per square inch (psi) or on a three-step scale (too low, proper range, too high). Pressure gauges are usually color-coded, with a green area indicating the proper pressure zone. CO_2 extinguishers do not have pressure indicators or gauges, but rather are weighted to determine the remaining agent. Likewise, extinguishers that are pressurized by a cartridge lack a pressure gauge.

Some disposable fire extinguishers intended for home use have an even simpler pressure indicator—that is, a plastic pin built into the cylinder. Pressing on the pin tests the pressure within the extinguisher. If the pin pops back up, the extinguisher has enough pressure to operate; if it remains depressed, the pressure has dropped below an acceptable level.

Wheeled Fire Extinguishers

Wheeled fire extinguishers are large units that are mounted on wheeled carriages and typically contain between 150 and 350 pounds (68 and 158 kilograms) of extinguishing agent. The wheeled design lets one person transport the extinguisher to the fire. If a wheeled extinguisher is intended for indoor use, doorways and aisles must be wide enough to allow for its passage to every area where it could be needed.

Wheeled fire extinguishers usually have long delivery hoses, so the unit can stay in one spot as the operator moves around to attack the fire from more than one side. Usually, a separate cylinder containing nitrogen or some other compressed gas provides the pressure necessary to operate the extinguisher.

Wheeled fire extinguishers are most often installed in special hazard areas, such as at military bases, airports, and industrial settings. Models with rubber tires or wide-rimmed wheels are available for outdoor installations.

Fire Extinguisher Characteristics

Portable fire extinguishers vary according to their extinguishing agent, capacity, effective range, and the time it takes to completely discharge the extinguishing agent. They also have different mechanical designs. This section describes the basic characteristics of seven types of extinguishers, which are organized by type of extinguishing agent:

- Water extinguishers
- Dry-chemical extinguishers
- CO_2 extinguishers
- Class B foam extinguishers
- Halogenated-agent extinguishers
- Dry-powder extinguishing agents
- Wet-chemical extinguishers

Water Extinguishers

Water extinguishers are used to cool the burning fuel below its ignition temperature. They are intended for use primarily on Class A fires. Class B foam extinguishers, which are a specific type of water extinguisher, are intended for fires involving flammable liquids. Water extinguishers include stored-pressure, loaded-stream, and wetting-agent models.

Stored-Pressure Water Extinguishers

The most popular kind of stored-pressure water fire extinguisher is the 2.5-gallon (9 liter) model with a 2-A rating FIGURE 8-20. Many fire department vehicles carry this type of fire extinguisher for use on incipient-stage Class A fires. Such an extinguisher expels water in a solid stream with a range of 35 to 40 feet (9 to 12 meters) through a nozzle at the end of a short hose. The discharge time is approximately 55 seconds if the extinguisher is used continuously. A full extinguisher weighs about 30 pounds (14 kilograms).

Because the contents of these extinguishers can freeze, these devices should not be installed in areas where the temperature is expected to drop below 32°F (0°C). Antifreeze models

FIGURE 8-20 Most fire departments use stored-pressure water extinguishers.

of stored-pressure water extinguishers, called loaded-stream extinguishers, are available. The loaded-stream agent will not freeze at temperatures as low as −40°F (−40°C).

The recommended procedure for operating a stored-pressure water extinguisher is to set it on the ground, grasp the handle with one hand, and pull out the ring pin or release the locking latch with the other hand. At this point, the extinguisher can be lifted and used to douse the fire. Use one hand to aim the stream at the fire, and squeeze the trigger with the other hand. The stream of water can be turned into a spray by putting a thumb at the end of the nozzle; this technique is often used after the flames have been extinguished as part of the effort to thoroughly soak the fuel.

Stored-pressure water extinguishers can be recharged at any location that provides water and a source of compressed air. Follow the manufacturer's instructions to ensure proper and safe recharging.

Loaded-Stream Water Extinguishers

Loaded-stream water extinguishers discharge a solution of water containing an alkali metal salt that prevents freezing at temperatures as low as −40°F (−40°C). The most common model is the 2.5-gallon (9 liter) unit, which is identical to a typical stored-pressure water extinguisher. Hand-held models are available with capacities of 1 to 2.5 gallons (4 to 9 liters) of water and are rated from 1-A to 3-A.

Larger units, including a 17-gallon (64 liter) unit rated 10-A and a 33-gallon (124 liter) unit rated 20-A, are also available. Pressure for these extinguishers is supplied by a separate cylinder of CO_2.

Wetting-Agent and Class A Foam Water Extinguishers

Wetting-agent water fire extinguishers expel water that contains a solution intended to reduce its surface tension (the physical property that causes water to bead or form a puddle on a flat surface). Reducing the surface tension allows water to spread over the fire and penetrate more efficiently into Class A fuels.

Class A foam extinguishers contain a solution of water and Class A foam concentrate. This agent has foaming properties as well as the ability to reduce surface tension.

Both wetting-agent and Class A foam extinguishers are available in the same configurations as water extinguishers, including hand-held stored-pressure models and wheeled units. These extinguishers should not be exposed to temperatures below 40°F (4°C).

Pump Tank Water Extinguishers

Pump tank water fire extinguishers come in sizes ranging from 1-A-rated, 1.5-gallon (6 liter) units to 4-A-rated, 5 gallon (19 liter) units. The water in these devices is not stored under pressure. Instead, the pressure needed to expel the water is provided by a hand-operated, double-acting, vertical piston pump, which moves water out through a short hose on both the up and the down strokes. This type of extinguisher sits upright on the ground during use. A small bracket at the bottom allows the operator to steady the extinguisher with one foot while pumping.

Pump tank extinguishers can be used with antifreeze agents. The manufacturer should be consulted for details, because some antifreeze agents (such as common salt) can corrode the extinguisher or damage the pump. Extinguishers with steel shells corrode more easily than those with copper or nonmetallic shells.

Backpack Water Extinguishers

Backpack water extinguishers are used primarily outdoors for fighting brush and grass fires FIGURE 8-21. Most of these units have a tank capacity of 5 gallons (19 liters) and weigh approximately 50 pounds (23 kilograms) when full. Backpack extinguishers are listed by UL but do not carry numeric ratings.

The water tank can be made of fiberglass, stainless steel, galvanized steel, nylon, canvas, or brass. Backpack extinguishers are designed to be refilled easily in the field, such as from a lake or a stream, through a wide-mouth opening at the top. A filter keeps dirt, stones, and other contaminants from entering the tank. Antifreeze agents, wetting agents, or other special water-based extinguishing agents can also be used with backpack water extinguishers.

Most backpack extinguishers are operated via hand pumps. The most common design has a trombone-type, double-acting piston pump located at the nozzle, which is attached to the tank by a short rubber hose. To discharge the extinguisher, the operator holds the pump in both hands and moves the piston back and forth.

FIGURE 8-21 A backpack extinguisher can be easily refilled from a natural water source such as a lake or river.

Some models have a compression pump built into the side of the tank. On these devices, it takes about 10 strokes of the pump handle to build up the initial pressure, which is maintained through continuous slow strokes. The operator uses the other hand to control the discharge. A lever-operated shut-off nozzle is provided at the end of a short hose.

Dry-Chemical Extinguishers

Dry-chemical extinguishers contain a variety of chemical extinguishing agents in granular form. Hand-held dry-chemical extinguishers are available with capacities ranging from 1 to 30 pounds (0.45 to 14 kilograms) of agent. Wheeled fire extinguishers are available with capacities up to 350 pounds (159 kilograms) of agent. Ordinary dry-chemical models can be used to extinguish Class B and C fires, whereas multipurpose dry-chemical units are rated for use on Class A, B, and C fires. Some large dry-chemical extinguishers are mounted on fire apparatus to deal with special risks.

All dry-chemical extinguishing agents can be used on Class C fires that involve energized electrical equipment. Unfortunately, the residue left by the dry chemical can be very damaging to computers, electronic devices, and electrical equipment.

Stored-pressure units expel the dry chemical agent in the same manner as stored-pressure water extinguishers. The dry chemical in a cartridge/cylinder extinguisher is not stored

VOICES OF EXPERIENCE

I once was assigned to conduct a search at a reported structure fire. My crew and I made entry with a portable fire extinguisher to find a mattress in a bedroom on fire. We extinguished the fire with the portable fire extinguisher while the engine company was stretching a hose line. Since the fire was in its incipient stage, it was easily contained with the portable fire extinguisher and the hose line was not needed. We were able to limit the damage to the room of origin. We overhauled the bedroom, packed up, and went home.

That one act of using the portable fire extinguisher saved the property owner the added expense of having to repair any water damage from the hose lines. The insurance agent gave a donation to the fire department out of his own pocket to thank us for using the portable fire extinguisher instead of automatically dousing the whole place with water from our hose line.

Rob Gaylor
City of Westfield Fire Department
Westfield, Indiana

under pressure. Instead, these extinguishers have a sealed, pressurized cartridge connected to the storage cylinder. They are activated by pushing down on a lever that punctures the cartridge and pressurizes the cylinder.

Most small, hand-held dry-chemical extinguishers are designed to discharge their contents completely in as little as 8 to 20 seconds. Larger units may discharge for as long as 30 seconds.

Depending on the extinguisher's size, the horizontal range of the discharge stream can be from 5 to 30 feet (2 to 9 meters). Some models have special nozzles that allow for a longer range. The long-range nozzles are useful when the fire involves burning gas or a flammable liquid under pressure or when the operator is working in a strong wind.

The trigger allows this type of extinguisher to be discharged intermittently, starting and stopping the agent flow. Releasing the trigger stops the flow of agent; however, this does not mean that the extinguisher can be put aside and used again later. These extinguishers do not retain their internal pressure for extended periods, because the granular dry chemical causes the valve to leak.

Dry-chemical fire extinguishers can be stored and used in areas where temperatures fall below the freezing point.

Ordinary Dry-Chemical Extinguishers

Ordinary dry-chemical extinguishers are available in hand-held models with ratings up to 160-B:C. Larger, wheeled units carry ratings up to 640-B:C.

Fire Fighter Safety Tips

Discharging a dry-chemical fire extinguisher in a confined space can create a cloud of very fine dust that can impair vision and cause difficulty breathing. Self-contained breathing apparatus (SCBA) should be used to protect fire fighters from both toxic gases from the fire and the dry chemical dust discharged from the fire extinguisher.

FIRE FIGHTER Tips

A dry-chemical extinguisher will usually continue to lose pressure after a partial discharge. Pressure loss can occur even when only a very small amount of agent has been discharged and the pressure gauge still indicates that the extinguisher is properly charged. This loss occurs because the agent leaves residue in the valve assembly and allows the stored pressure to leak out slowly.

Multipurpose Dry-Chemical Extinguishers

Ammonium phosphate is commonly called a multipurpose dry chemical agent because it can be used on Class A, B, and C fires. When it is used on Class A fires, the chemical coats the surface of the fuel to prevent continuing combustion.

Multipurpose dry-chemical extinguishers are available in hand-held models with ratings ranging from 1-A to 20-A and from 10-B:C to 120-B:C. Larger, wheeled models have ratings ranging from 20-A to 40-A and from 60-B:C to 320-B:C.

Multipurpose dry-chemical extinguishers should never be used on cooking oil (Class K) fires, such as those involving deep-fat fryers located in commercial kitchens. The ammonium phosphate–based extinguishing agent is acidic and will not react with cooking oils to produce the smothering foam needed to extinguish this type of fire. Even worse, the acid will counteract the foam-forming properties of any alkaline extinguishing agent that is applied to the same fire.

Carbon Dioxide Extinguishers

CO_2 extinguishers are rated to fight Class B and C fires. Carbon dioxide extinguishes a fire by enveloping the fuel in a cloud of inert gas, which reduces the oxygen content of the surrounding atmosphere and smothers the flames. Because the CO_2 discharge is very cold, it also helps to cool the burning materials and surrounding areas.

CO_2 gas is 1.5 times heavier than air, colorless, odorless, nonconductive, and inert. It is also noncorrosive and does not leave any residue. These factors are important in areas where costly electronic components or computer equipment must be protected. CO_2 is also used around food preparation areas and in laboratories.

CO_2 is both an expelling agent and an extinguishing agent. It is stored under a pressure of 823 psi (1178 Kpa), which keeps the CO_2 in liquid form at room temperature. When the pressure is released, the liquid CO_2 rapidly converts to a gas, and the expansion of the gas forces the agent out of the container.

The CO_2 is discharged through a siphon tube that reaches to the bottom part of the storage cylinder. It is forced through a hose to a horn or cone-shaped applicator that is used to direct the flow of the agent. When discharged, the agent is very cold and contains a mixture of CO_2 gas and solid CO_2, which is quickly converted to a gas.

Compared to other types of extinguishers, CO_2 extinguishers have a relatively short discharge range, 3 to 8 feet (1 to 2 meters). This limitation presents a safety issue because the operator must be close to the fire. Depending on their size, CO_2 extinguishers can discharge completely in 8 to 30 seconds. CO_2 extinguishers are not toxic when used around food, and their use does not require clean-up because the CO_2 gas dissipates into the air.

CO_2 fire extinguishers are not recommended for outdoor use or for use in locations with strong air currents. In such settings, the gas will be rapidly dissipated and will not be effective in smothering the flames.

CO_2 extinguishers have a trigger mechanism that can be operated intermittently to preserve the remaining agent. The pressurized CO_2 will remain in the extinguisher, but the extinguisher must be recharged after use. The extinguisher can be weighed to determine how much agent is left in the storage cylinder.

The smaller CO_2 extinguishers contain from 2 to 5 pounds (1 to 2 kilograms) of agent. These units are designed to be

Fire Fighter Safety Tips

Do not aim the CO_2 extinguisher discharge at anyone or allow it to come in contact with exposed skin, because frostbite could result. CO_2 discharged into a confined space will reduce the oxygen level in that space. For this reason, anyone entering the confined area must wear SCBA.

operated with one hand. The horn is attached directly to the discharge valve on the top of the extinguisher by a hinged metal tube. In larger models, the horn is attached at the end of a short hose. These models require two-handed operation.

Class B Foam Extinguishers

Class B foam extinguishers are very similar in appearance and operation to water extinguishers. Instead of plain water, however, they discharge a solution of water and either AFFF or FFFP foam concentrate. The agent is discharged through an aspirating nozzle, which mixes air into the stream. The result is a foam blanket that will float over the surface of a flammable liquid.

Class B foam agents are very effective in fighting Class A fires but are not suitable for Class C fires or for fires involving flammable liquids or gases under pressure. They are not intended for use on cooking oil fires, and only certain foam extinguishers can be used on fires involving polar solvents. Detailed information on the use of AFFF and FFFP is available in NFPA 11, *Standard for Low-, Medium-, and High-Expansion Foam*.

AFFF and FFFP hand-held stored-pressure extinguishers are available in two sizes: 1.6 gallon (6 liter), rated 2-A:10-B, and 2.5 gallon (9 liter), rated 3-A:20-B. A wheeled model with a 33-gallon (125 liter) capacity is rated 20-A:160-B.

Foam extinguishing agents are not effective at freezing temperatures. Consult the extinguisher manufacturer for information on using foam agents effectively at low temperatures.

Wet-Chemical Extinguishers

Wet-chemical extinguishers are used to protect Class K (kitchen) installations, which are equipped with deep-fat fryers, cooking oils, and grills. Many commercial cooking installations use fixed, automatic fire-extinguishing systems as their first line of defense. Portable Class K wet-chemical extinguishers are currently available in two sizes, 1.5 gallons (6 liters) and 2.5 gallons (9 liters). There are no numerical ratings for these extinguishers.

FIRE FIGHTER Tips

Dry chemical agents and dry powder agents have very different meanings in relation to fire extinguishers. Dry-powder fire extinguishers are designed for use in suppressing Class D fires. Dry-chemical fire extinguishers are rated for Class B and C fires or for Class A, B, and C fires. Do not confuse these two terms.

Halogenated-Agent Extinguishers

Halogenated-agent fire extinguishers include both Halon agents and halocarbon agents. Because Halon agents can destroy the Earth's protective ozone layer, their use is strictly controlled. The halocarbons, however, are not subject to the same environmental restrictions. Both types of agents are available in hand-held extinguishers rated for Class B and C fires. Larger-capacity models are also rated for use on Class A fires.

In this type of extinguisher, the agent is discharged as a streaming liquid, which can be directed at the base of a fire. The discharges from halogenated-agent extinguishers have a horizontal stream range of 9 to 15 feet (3 to 4 meters).

The halogenated agents are nonconductive and leave no residue that can damage electrical equipment.

Halon 1211 (bromochlorodifluoromethane) is available in hand-held stored-pressure extinguishers with capacities that range from 1 pound (0.45 kilogram), rated 1-B:C, to 22 pounds (10 kilograms), rated 4-A:80-B:C. Wheeled Halon 1211 models are available with capacities up to 150 pounds (68 kilograms) with a rating of 30-A:160-B:C. The wheeled fire extinguishers use a nitrogen booster charge from an auxiliary cylinder to expel the agent.

Dry-Powder Extinguishing Agents

Dry-powder extinguishing agents are intended for fighting Class D fires involving combustible metals. The extinguishing agents and techniques required to put out Class D fires vary greatly, depending on the specific fuel, the quantity involved, and the physical form of the fuel (e.g., grindings, shavings, or solid objects). The agent and the application method must be suited to the particular situation.

Each dry-powder extinguishing agent is listed for use on specific combustible-metal fires. This information and instructions related to the recommended application method are printed on the container. Consult the manufacturer's recommendations for information about Class D agents and extinguishers, as well as NFPA's *Fire Protection Handbook*.

Dry-Powder Fire Extinguishers

Dry-powder fire extinguishers using sodium chloride–based agents are available with 30-pound (14 kilogram) capacity in either stored-pressure or cylinder/cartridge models. Wheeled models are available with 150- and 350-pound (68 to 160 kilogram) capacities.

Dry-powder extinguishers have adjustable nozzles that allow the operator to vary the flow of the extinguishing agent. When the nozzle is fully opened, the hand-held models have a range of 6 to 8 feet (2 to 2.5 meter). Extension wand applicators are available to direct the discharge from a more distant position.

Bulk Dry Powder Agents

Bulk dry powder agents are available in 40- and 50-pound (18- and 23-kilogram) pails and 350-pound (159 kilogram) drums. The same sodium chloride–based agent that is used in portable fire extinguishers can be stored in bulk form and applied by hand.

Another dry-powder extinguishing agent for Class D fires, graded granular graphite mixed with phosphorus-containing compounds, cannot be expelled from a portable fire extinguisher. Instead, this agent must be applied manually from a pail or other container using a shovel or scoop.

Fire Fighter Safety Tips

Before deciding to use a fire extinguisher, size up the fire to ensure that the extinguisher has an adequate capacity and holds the proper extinguishing agent.

Fire Fighter Safety Tips

Always test a fire extinguisher before using it. Activate the trigger long enough to ensure that the agent discharges properly.

Use of Fire Extinguishers

Fire extinguishers should be simple to operate. Indeed, an individual with only basic training should be able to use most fire extinguishers safely and effectively. Every portable extinguisher should be labeled with printed operating instructions.

There are six basic steps in extinguishing a fire with a portable fire extinguisher:

1. Locate the fire extinguisher.
2. Select the proper classification of extinguisher.
3. Transport the extinguisher to the location of the fire.
4. Activate the extinguisher to release the extinguishing agent.
5. Apply the extinguishing agent to the fire for maximum effect.
6. Ensure your personal safety by having an exit route.

Although these steps are not complicated, practice and training are essential for effective fire suppression with a fire extinguisher. In particular, tests have shown that the effective use of Class B portable fire extinguishers depends heavily on user training and expertise. A trained expert can extinguish a fire up to twice as large as a nonexpert can, using the same extinguisher.

As a fire fighter, you should be able to operate any fire extinguisher that you might be required to use, whether it is carried on your fire apparatus, hanging on the wall of your firehouse, or placed in some other location in your community.

Locating a Fire Extinguisher

As a fire fighter, you should know which types of fire extinguishers are carried on your department's apparatus and where each type of extinguisher is located. You should also know where fire extinguishers are located in and around the fire station and other workplaces. You should have at least one fire extinguisher in your home and another in your personal vehicle, and you should know exactly where they are located. Knowing the exact locations of extinguishers can save valuable time in an emergency.

Selecting the Proper Fire Extinguisher

Selecting the proper extinguisher requires an understanding of the classification and rating system for fire extinguishers. Knowing which types of agents are available, how they work, which ratings the fire extinguishers carried on your fire apparatus have, and which extinguisher is appropriate for a particular fire situation is also important.

Fire fighters should be able to assess a fire quickly, determine whether the fire can be controlled by an extinguisher, and identify the appropriate extinguisher. Using an extinguisher with an insufficient rating may not completely extinguish the fire, which can place the operator in danger of being burned or otherwise injured. If the fire is too large for the extinguisher, consider other options, such as obtaining additional extinguishers or making sure that a charged hose line is ready to provide backup.

Understanding the fire extinguisher rating system and the different types of agents will enable a fire fighter who must use an unfamiliar extinguisher to determine whether it is suitable for a particular fire situation. A quick look at the label should be all that is needed in such a case.

Fire fighters should also be able to determine the most appropriate type of fire extinguisher to place in a given area, based on the types of fires that could occur and the hazards that are present TABLE 8-2. In some cases, one type of extinguisher might be preferred over another. An extinguishing agent such as CO_2 or a halogenated agent is better for a fire involving electronic equipment, for example, because it leaves no residue. A dry-chemical extinguisher is generally more appropriate than a CO_2 extinguisher to fight an outdoor fire, because wind will quickly dissipate CO_2. A dry-chemical extinguisher would also be the best choice for a fire involving a flammable liquid leaking under pressure from a pipe, whereas foam would be a better choice for a fire involving a liquid spill on the ground.

Transporting a Fire Extinguisher

The best method of transporting a hand-held portable fire extinguisher depends on the extinguisher's size, weight, and design. Hand-held portable models can weigh as little as 1 pound (0.45 kilogram) or as much as 50 pounds (23 kilograms). The ability to handle the heavier extinguishers depends on an individual operator's personal strength.

Extinguishers with a fixed nozzle should be carried in the favored or stronger hand. This approach enables the operator to depress the trigger and direct the discharge easily. Extinguishers that have a hose between the trigger and the nozzle should be carried in the weaker or less-favored hand so that the favored hand can grip and aim the nozzle.

Heavier extinguishers may have to be carried as close as possible to the fire and placed upright on the ground. The operator can then depress the trigger with one hand, while holding the nozzle and directing the stream with the other hand.

TABLE 8-2 Fire Extinguisher Types

Extinguishing Agent	Method of Operation	Capacity	Horizontal Range of Stream	Approximate Time of Discharge	Protection Required below 40°F (4°C)	UL or ULC Classifications[a]
Water	Stored-pressure Stored-pressure or pump Pump Pump	6L 2½ gal 4 gal 5 gal	30 to 40 ft 30 to 40 ft 30 to 40 ft 30 to 40 ft	40 sec 1 min 2 min 2 to 3 min	Yes Yes Yes Yes	1-A 2-A 3-A 4-A
Water (wetting agent)	Stored-pressure Stored-pressure Stored-pressure Stored-pressure	1½ gal 25 gal (wheeled) 45 gal (wheeled) 60 gal (wheeled)	20 ft 35 ft 35 ft 35ft	30 sec 1 ½ min 2 min 2 ½ min	Yes Yes Yes Yes	2-A 10-A 30-A 40-A
Loaded stream	Stored-pressure Stored-pressure	2½ gal 33 gal (wheeled)	30 to 40 ft 50 ft	1 min 3 min	No No	2-A 20-A
Water mist	Stored-pressure Stored-pressure	6L 2½ gal	NA NA	60 to 70 sec 70 to 80 sec	Yes Yes	2-A:C 2-A:C
AFFF, FFFP	Stored-pressure Stored-pressure Nitrogen cylinder	2½ gal 6L 33 gal	20 to 25 ft 20 to 25 ft 30 ft	50 sec 50 sec 1 min	Yes Yes Yes	3-A:20 to 40-B 2A:10B 20-A:160-B
Carbon dioxide[b]	Self-expelling Self-expelling Self-expelling Self-expelling	2½ to 5 lb 10 to 15 lb 20 lb 50 to 100 lb (wheeled)	3 to 8 ft 3 to 8 ft 3 to 8 ft 3 to 10 ft	8 to 30 sec 8 to 30 sec 10 to 30 sec 10 to 30 sec	No No No No	1 to 5-B:C 2 to 10-B:C 10-B:C 10 to 20-B:C
Regular dry chemical (sodium bicarbonate)	Stored-pressure Cartridge or stored-pressure Cartridge or stored-pressure Stored-pressure Nitrogen cylinder or stored-pressure	1 to 2½ lb 2¾ to 5 lb 6 to 30 lb 50 lb (wheeled) 75 to 350 lb (wheeled)	5 to 8 ft 5 to 20 ft 5 to 20 ft 20 ft 15 to 45 ft	8 to 12 sec 8 to 25 sec 10 to 25 sec 35 sec 20 to 105 sec	No No No No No	2 to 10-B:C 5 to 20-B:C 10 to 160-B:C 160:B-C 40 to 320-B:C
Purple K dry chemical (potassium bicarbonate)	Cartridge or stored-pressure Cartridge or stored-pressure Cartridge or stored-pressure Cartridge or stored-pressure Nitrogen cylinder or stored-pressure	2 to 5 lb 5½ to 10 lb 16 to 30 lb 48 to 50 lb (wheeled) 125 to 315 lb (wheeled)	5 to 12 ft 5 to 20 ft 10 to 20 ft 20 ft 15 to 45 ft	8 to 10 sec 8 to 20 sec 8 to 25 sec 30 to 35 sec 30 to 80 sec	No No No No No	5 to 30-B:C 10 to 80-B:C 40 to 120-B:C 120 to 160-B:C 80 to 640-B:C
Super K dry chemical (potassium chloride)	Cartridge or stored-pressure Cartridge or stored-pressure Cartridge or stored-pressure Cartridge or stored-pressure Cartridge or stored-pressure	2 to 5 lb 5 to 9 lb 9½ to 20 lb 19½ to 30 lb 125 to 200 lb (wheeled)	5 to 8 ft 8 to 12 ft 10 to 15ft 5 to 20 ft 15 to 45 ft	8 to 10 sec 10 to 15 sec 15 to 20 sec 10 to 25 sec 30 to 40 sec	No No No No No	5 to 10-B:C 20 to 40-B:C 40 to 60-B:C 60 to 80-B:C 160-B:C
Multipurpose/ ABC dry chemical (ammonium phosphate)	Stored-pressure Stored-pressure or cartridge Stored-pressure or cartridge Stored-pressure or cartridge Stored-pressure or cartridge Nitrogen cylinder or stored-pressure	1 to 5 lb 2½ to 9 lb 9 to 17 lb 17 to 30 lb 45 to 50 lb (wheeled) 110 to 315 lb (wheeled)	5 to 12 ft 5 to 12 ft 5 to 20 ft 5 to 20 ft 20 ft 15 to 45 ft	8 to 10 sec 8 to 15 sec 10 to 25 sec 10 to 25 sec 25 to 35 sec 30 to 60 sec	No No No No No	1 to 3-A[c] and 2 to 10-B:C 1 to 4-A and 10 to 40-B:C 2 to 20-A and 10 to 80-B:C 3 to 20-A and 30 to 120-B:C 20 to 30-A and 80 to 160-B:C 20 to 40-A and 60 to 320-B:C
Dry chemical (foam-compatible)	Cartridge or stored-pressure Cartridge or stored-pressure Cartridge or stored-pressure Nitrogen cylinder or stored-pressure	4¾ to 9 lb 9 to 27 lb 18 to 30 lb 150 to 350 lb (wheeled)	5 to 20 ft 5 to 20 ft 5 to 20 ft 15 to 45 ft	8 to 10 sec 10 to 25 sec 10 to 25 sec 20 to 150 sec	No No No No	10 to 20-B:C 20 to 30-B:C 40 to 60-B:C 80 to 240-B:C
Dry chemical (potassium bicarbonate urea based)	Stored-pressure Stored-pressure Nitrogen cylinder or stored-pressure	5 to 11 lb 9 to 23 lb 175 lb (wheeled)	11 to 22 ft 15 to 30 ft 70 ft	18 sec 17 to 33 sec 62 sec	No No No	40 to 80-B:C 60 to 160-B:C 480-B:C
Wet chemical	Stored-pressure Stored-pressure Stored-pressure	3L 6L 2½ gal	8 to 12 ft 8 to 12 ft 8 to 12 ft	30 sec 35 to 45 sec 75 to 85 sec	No No No	K K K

(Continued)

TABLE 8-2 Fire Extinguisher Types (*Continued*)

Extinguishing Agent	Method of Operation	Capacity	Horizontal Range of Stream	Approximate Time of Discharge	Protection Required below 40°F (4°C)	UL or ULC Classifications[a]
Halon 1211 (bromochloro-difluromethane)	Stored-pressure Stored-pressure Stored-pressure Stored-pressure Stored-pressure Stored-pressure	0.9 to 2 lb 2 to 3 lb 5½ to 9 lb 13 to 22 lb 50 lb 150 lb (wheeled)	6 to 10 ft 6 to 10 ft 9 to 15 ft 14 to 16 ft 35 ft 20 to 35 ft	8 to 10 sec 8 to 10 sec 8 to 15 sec 10 to 18 sec 30 sec 30 to 44 sec	No No No No No No	1 to 2-B:C 5-B:C 1-A:10-B:C 2 to 4-A and 20 to 80-B:C 10-A:120-B:C 30-A:160 to 240-B:C
Halon 1211/1301 (bromochlorodi-fluromethane and bromotrifluoro-methane) mixtures	Stored-pressure or self-expelling Stored-pressure	0.9 to 5 lb 9 to 20 lb	3 to 12 ft 10 to 18 ft	8 to 10 sec 10 to 22 sec	No No	1 to 10-B:C 1-A:10-B:C to 4-A:80-B:C
Halocarbon type	Stored-pressure	1.4 to 150 lb	6 to 35 ft	9 to 23 sec	No	1B:C to 10A:80-B:C

Note: Halon should be used only where its unique properties are deemed necessary.
[a]UL and ULC ratings checked as of July 24, 1987. Readers concerned with subsequent ratings should review the pertinent lists and supplements issued by these laboratories: Underwriters Laboratories, Inc., 333 Pfingsten Road, Northbrook, IL 60062, or Underwriters Laboratories of Canada, 7 Crouse Road, Scarborough, Ontario, Canada M1R 3A9.
[b]Carbon dioxide extinguishers with metal horns do not carry a C classification.
[c]Some small extinguishers containing ammonium phosphate–based dry chemical do not carry an A classification.
NA = Not available.
Source: National Fire Protection Association. *Fire Protection Handbook*, Vol. 2, 20th ed. Quincy, MA, 2008.

To transport a fire extinguisher, follow the steps in **SKILL DRILL 8-1**:

1. Locate the closest appropriate fire extinguisher. (**STEP 1**)
2. Assess that the extinguisher is safe and effective for the type of fire being attacked. Release the mounting bracket straps. (**STEP 2**)
3. Lift the extinguisher using good body mechanics. Lift small extinguishers with one hand, and large extinguishers with two hands. (**STEP 3**)
4. Walk briskly—do not run—toward the fire. If the extinguisher has a hose and nozzle, carry the extinguisher with one hand and grasp the nozzle with the other hand. (**STEP 4**)

Never turn your back to the fire

Basic Steps of Fire Extinguisher Operation

Activating a fire extinguisher to apply the extinguishing agent is a single operation that involves four steps. The PASS acronym is a helpful way to remember these steps:

1. Pull the safety pin.
2. Aim the nozzle at the base of the flames.
3. Squeeze the trigger to discharge the agent.
4. Sweep the nozzle across the base of the flames.

Most fire extinguishers have very simple operation systems. Practice discharging different types of extinguishers in training situations to build confidence in your ability to use them properly and effectively.

Ensure Your Personal Safety

When using a fire extinguisher, always approach the fire with an exit behind you. If the fire suddenly expands or the extinguisher fails to control it, you must have a planned escape route. Never let the fire get between you and a safe exit. After suppressing a fire, do not turn your back on it. Always watch it, and be prepared for it to rekindle until the fire has been fully overhauled.

When ordinary civilians use fire extinguishers on incipient fires, they are probably wearing their normal clothing. As a fire fighter, however, you should wear your personal protective clothing and use appropriate personal protective equipment (PPE). Take advantage of the protection they provide.

If you must enter an enclosed area where an extinguisher has been discharged, wear full PPE and use SCBA. The atmosphere within the enclosed area will probably contain a mixture of combustion products and extinguishing agents, and the oxygen content within the space may be dangerously low.

To attack a Class A fire with a stored-pressure water fire extinguisher, follow the steps in **SKILL DRILL 8-2**:

1. Size up the fire to determine whether a stored-pressure water extinguisher is safe and effective for this fire. Ensure that extinguisher is large enough to be safe and effective. (**STEP 1**)
2. Ensure your safety. Make sure you have an exit route from the fire. Do not turn your back on a fire. (**STEP 2**)
3. Remove the hose and nozzle. Quickly check the pressure gauge to verify that the extinguisher is adequately charged. (**STEP 3**)
4. Pull the pin to release the extinguisher control valve. You must be within 35 to 40 feet (11 to 12 meters) of the fire to be effective. (**STEP 4**)
5. Aim the nozzle and sweep the water stream at the base of the flames. (**STEP 5**)
6. Overhaul the fire; take steps to prevent rekindling, break apart tightly packed fuel, and summon additional help if needed. (**STEP 6**)

Note: This fire extinguisher is safe and effective to use on Class A fires. It is not effective on Class B fires, and it is not safe to use on Class C fires.

SKILL DRILL 8-1 Transporting a Fire Extinguisher

(Fire Fighter I, NFPA 5.3.16)

1 Locate the closest fire extinguisher.

2 Assess that the extinguisher is safe and effective for the type of fire being attacked. Release the mounting bracket straps.

3 Lift the extinguisher using good body mechanics. Lift small extinguishers with one hand, and large extinguishers with two hands.

4 Walk briskly—do not run—toward the fire. If the extinguisher has a hose and nozzle, carry the extinguisher with one hand and grasp the nozzle with the other hand.

To attack a Class A fire with a multipurpose dry-chemical fire extinguisher, follow the steps in SKILL DRILL 8-3 (Fire Fighter I, 5.3.16):

1. Size up the fire to determine whether a multipurpose dry-chemical fire extinguisher is safe and effective for this fire. Ensure that extinguisher is large enough to be safe and effective.
2. Ensure your safety. Make sure you have an exit route from the fire. Do not turn your back on a fire.
3. Remove the hose and nozzle. Quickly check the pressure gauge to verify that the extinguisher is adequately charged.
4. Pull the pin to release the extinguisher control valve. You must be within 10 to 30 feet (3 to 9 meters) of the fire to be effective.
5. Aim the nozzle and sweep the dry-chemical discharge at the base of the flames. Coat the burning fuel with dry chemical.
6. Overhaul the fire; take steps to prevent rekindling, break apart tightly packed fuel, and summon additional help if needed.

Note: This fire extinguisher is safe and effective on Class A, B, and C fires.

SKILL DRILL 8-2 Attacking a Class A Fire with a Stored-Pressure Water-Type Fire Extinguisher

(Fire Fighter I, NFPA 5.3.16)

1. Size up the fire to determine whether a stored-pressure water extinguisher is safe and effective for this fire. Ensure that extinguisher is large enough to be safe and effective.
2. Ensure your safety. Make sure you have an exit route from the fire. Do not turn your back on a fire.
3. Remove the hose and nozzle. Quickly check the pressure gauge to verify that the extinguisher is adequately charged.

4. Pull the pin to release the extinguisher control valve. You must be within 35 to 40 feet (11 to 12 meters) of the fire to be effective.
5. Aim the nozzle and sweep the water stream at the base of the flames.
6. Overhaul the fire; take steps to prevent rekindling, break apart tightly packed fuel, and summon additional help if needed.

To attack a Class B flammable liquid spill or pool fire with a dry-chemical extinguisher, follow the steps in SKILL DRILL 8-4:

1. Size up the fire to determine whether a dry-chemical fire extinguisher is safe and effective for this fire. Ensure that extinguisher is large enough to be safe and effective.
2. Ensure your safety. Make sure you have an exit route from the fire. Do not turn your back on a fire.
3. Remove the hose and nozzle. Quickly check the pressure gauge to verify that the extinguisher is adequately charged. (STEP 1)
4. Pull the pin to release the extinguisher control valve. You must be within 10 to 30 feet (3 to 9 meters) of the fire to be effective. (STEP 2)
5. Aim the nozzle and sweep the dry-chemical discharge across the surface of the burning liquid. Start at the

SKILL DRILL 8-4 Attacking a Class B Flammable Liquid Fire with a Dry-Chemical Fire Extinguisher

(Fire Fighter I, NFPA 5.3.16)

1 Check the pressure gauge.

2 Pull the pin to release the extinguisher value.

3 Aim the nozzle and sweep the dry-chemical discharge across the surface of the burning liquid. Start at the near edge of the fire and work toward the back.

4 Overhaul the fire; take steps to prevent rekindling, keep a blanket of dry chemical over the fuel, and summon additional help if needed.

near edge of the fire and work toward the back. (STEP 3)

6. Overhaul the fire; take steps to prevent rekindling, keep a blanket of dry chemical over the fuel, and summon additional help if needed. (STEP 4)

Note: This fire extinguisher is safe to use on Class B and C fires. It is effective on Class B fires, but is not rated for Class A fires.

To attack a Class B flammable liquid spill or pool fire with a stored-pressure fire extinguisher containing an AFFF or FFFP extinguishing agent, follow the steps in SKILL DRILL 8-5:

1. Size up the fire to determine whether an AFFF or FFFP fire extinguisher is safe and effective for this fire. Ensure that extinguisher is large enough to be safe and effective.
2. Ensure your safety. Make sure you have an exit route from the fire. Do not turn your back on a fire.
3. Remove the hose and nozzle. Quickly check the pressure gauge to verify that the extinguisher is adequately charged.
4. Pull the pin to release the extinguisher control valve. You must be within 20 to 25 feet (6 to 8 meters) of the fire to be effective. (STEP 1)
5. Aim the nozzle and discharge the stream of foam so the foam drops gently onto the surface of the burning liquid at the front or the back of the

SKILL DRILL 8-5 Attacking a Class B Flammable Liquid Fire with a Stored-Pressure Foam Fire Extinguisher (AFFF or FFFP)

(Fire Fighter I, NFPA 5.3.16)

1. Size up the fire and ensure your safety. Remove the hose and nozzle. Quickly check the pressure gauge to verify that the extinguisher is adequately charged. Pull the pin to release the extinguisher control valve. You must be within 20 to 25 feet (6 to 8 meters) of the fire to be effective.

2. Aim the nozzle and discharge the stream of foam so the foam drops gently onto the surface of the burning liquid at the front or the back of the container. Let the foam blanket flow across the surface of the burning liquid. Avoid splashing foam on the burning liquid, because it can cause the burning fuel to splatter.

3. Overhaul the fire; keep a thick blanket of foam intact over the hot liquid, reapply foam over any hot spots, and summon additional help if needed.

container. Let the foam blanket flow across the surface of the burning liquid. Avoid splashing foam on the burning liquid, because it can cause the burning fuel to splatter. (STEP 2)

6. Overhaul the fire; keep a thick blanket of foam intact over the hot liquid, reapply foam over any hot spots, and summon additional help if needed. (STEP 3)

Note: This fire extinguisher is safe to use on Class A, B, and C fires. It is effective on Class B fires by forming a foam blanket. It is effective on Class A fires by cooling and penetrating the fuel.

To operate a CO_2 extinguisher, follow the steps in SKILL DRILL 8-6:

1. Size up the fire to determine whether CO_2 is a safe and effective agent for this fire. Ensure that extinguisher is large enough to be safe and effective.
2. Ensure your safety. Make sure you have an exit route from the fire. Do not turn your back on a fire.
3. Remove the horn or nozzle. Quickly squeeze to verify that the extinguisher is charged; CO_2 extinguishers do not have pressure gauges. (STEP 1)

SKILL DRILL 8-6 Operating a Carbon Dioxide Extinguisher

(Fire Fighter I, NFPA 5.3.16)

1 Size up the fire to determine whether CO_2 is a safe and effective agent for this fire. Ensure that extinguisher is large enough to be safe and effective. Ensure your safety. Make sure you have an exit route from the fire. Do not turn your back on a fire. Remove the horn or nozzle. Quickly squeeze to verify that the extinguisher is charged; CO_2 extinguishers do not have pressure gauges.

2 Pull the pin to release the extinguisher control valve. You must be within 5 to 8 feet (2 to 2.5 meters) of the fire to be effective.

3 Aim the horn or nozzle and sweep at the base of the flames.

4 Overhaul the fire; take steps to prevent rekindling and summon additional help if needed.

4. Pull the pin to release the extinguisher control valve. You must be within 5 to 8 feet (2 to 2.5 meters) of the fire to be effective. (STEP 2)
5. Aim the horn or nozzle and sweep at the base of the flames. (STEP 3)
6. Overhaul the fire; take steps to prevent rekindling and summon additional help if needed. (STEP 4)

Note: The fire extinguisher is most effective when used in areas without wind. It is designed for Class B and C fires only.

To apply a halogenated extinguishing agent to a fire in an electrical equipment room, follow the steps in SKILL DRILL 8-7 (Fire Fighter I, 5.3.16):

1. Size up the fire to determine whether a halogenated stored-pressure fire extinguisher is safe and effective for this fire. Ensure that extinguisher is large enough to be safe and effective.
2. Ensure your safety. Turn off electricity if possible. Make sure you have an exit route from the fire. Do not turn your back on a fire.

3. Remove the hose and nozzle. Quickly check the pressure gauge to verify that the extinguisher is adequately charged.
4. Pull the pin to release the extinguisher control valve. You must be within 3 to 18 feet (1 to 5 meters) of the fire to be effective.
5. Aim the nozzle at the base of the flames to sweep the flames off the surface starting at the near edge of the flames.
6. Overhaul the fire; take steps to prevent rekindling, continue to apply the extinguishing agent to cool the fuel, and summon additional help if needed.

Note: This fire extinguisher is safe to use on Class A, B, and C fires. It is effective on Class B fires, but requires large quantities to be effective on Class A fires.

To apply a dry-powder extinguishing agent to an incipient-stage fire involving combustible metal filings or shavings (Class D fire), follow the steps in SKILL DRILL 8-8:

1. Size up the fire to determine whether a dry-powder fire extinguisher is safe and effective for the metal involved in this fire. Ensure that extinguisher is large enough to be safe and effective. Avoid the use of water and other extinguishing agents that may react explosively.
2. Ensure your safety. Burning metals produce very high temperatures and can burn explosively. Make sure you have an exit route from the fire. Do not turn your back on a fire.
3. Remove the hose and nozzle. Quickly check the pressure gauge to verify that the extinguisher is adequately charged.
4. Pull the pin to release the extinguisher control valve. You must be within 6 to 8 feet (2 to 2.5 meters) of the fire to be effective.
5. Aim the nozzle and fully open the valve to provide a maximum range, and then reduce the valve to produce a soft, heavy flow down to completely cover the burning metal. The method of application may vary depending on the type of metal burning and the extinguishing agent. (STEP 1)
6. Overhaul the fire; take steps to prevent rekindling, continue to place a thick layer of the extinguishing

SKILL DRILL 8-8 Using a Dry-Powder Fire Extinguisher (Class D)

(Fire Fighter I, NFPA 5.3.16)

1 Aim the nozzle and fully open the valve to provide a maximum range, and then reduce the valve to produce a soft, heavy flow down to completely cover the burning metal. The method of application may vary depending on the type of metal burning and the extinguishing agent.

2 Overhaul the fire; take steps to prevent rekindling, continue to place a thick layer of the extinguishing agent over the hot metal to form an airtight blanket, allow the hot metal to cool, and summon additional help if needed.

agent over the hot metal to form an airtight blanket, allow the hot metal to cool, and summon additional help if needed. (STEP 2)

Note: This fire extinguisher is safe to use only on Class D fires. It is effective only on the metals for which it is intended.

To apply a wet-chemical (Class K) extinguishing agent on a fire in a deep-fat fryer, follow the steps in SKILL DRILL 8-9:

1. Size up the fire to determine whether a wet-chemical (Class K) fire extinguisher is safe and effective for this fire. Ensure that extinguisher is large enough to be safe and effective.
2. Ensure your safety. Make sure you have an exit route from the fire. Do not turn your back on a fire.
3. Remove the hose and nozzle. Quickly check the pressure gauge to verify that the extinguisher is adequately charged.
4. Pull the pin to release the extinguisher control valve. You must be within 8 to 12 feet (2 to 4 meters) of the fire to be effective.
5. Aim the nozzle and discharge the stream of wet chemical drops gently into the surface of the burning liquid at the front or the back of the deep-fat fryer. (STEP 1)
6. Let the deep foam blanket flow across the surface of the burning liquid. Avoid splashing foam on the burning liquid, as it can cause the burning fuel to splatter. (STEP 2)
7. Continue to apply the agent until the foam blanket has extinguished all of the flames. (STEP 3)
8. Do not disturb the foam blanket even after all of the flames have been suppressed. If reignition occurs, repeat these steps.
9. Overhaul the fire; keep a thick blanket of foam intact over the hot liquid, reapply wet chemical over any hot spots or thin spots, and summon additional help if needed. (STEP 4)

Note: This fire extinguisher is designed to be used on Class K fires—that is, fires in deep-fat fryers and on grills. It is not rated for other classes of fire.

The Care of Fire Extinguishers

Fire extinguishers must be regularly inspected and properly maintained to ensure that they will be available for use in an emergency. Records must be kept to confirm that the required inspections and maintenance have been performed on schedule. The individuals assigned to perform these functions must be properly trained and must always follow the manufacturer's recommendations for inspecting, maintaining, recharging, and testing the equipment. Persons performing maintenance or recharging of portable fire extinguishers should be qualified according to NFPA 10, *Standard for Portable Fire Extinguishers*.

Inspection

According to NFPA 10, an inspection is a "quick check" to verify that a fire extinguisher is available and ready for immediate use. Fire extinguishers on fire apparatus should be inspected as part of the regular equipment checks mandated by your department. Inspection should take place each month. The fire fighter charged with inspecting the extinguishers should perform the following tasks:

- Ensure that all tamper seals are intact.
- Determine fullness by weighing or "hefting" the extinguisher.
- Examine all parts for signs of physical damage, corrosion, or leakage.
- Check the pressure gauge to confirm that it is in the operable range.
- Ensure that the extinguisher is properly identified by type and rating.
- Check the hose and nozzle for damage or obstruction by foreign objects.
- Check the hydrostatic test date of the extinguisher.

If an inspection reveals any problems, the extinguisher should be removed from service until the required maintenance procedures are performed. Spare extinguishers should be used until the problem is corrected.

The pressure gauge on a stored-pressure extinguisher indicates whether the pressure is sufficient to expel the entire agent. Some stored-pressure extinguishers use compressed air, whereas others use compressed nitrogen. The weight of the extinguisher and the presence of an intact tamper seal should indicate that the unit is full of extinguishing agent.

Cartridge-type extinguishers contain a predetermined quantity of a pressurizing gas that expels the agent. This gas is released only when the cartridge is punctured. A properly charged extinguisher will be full of extinguishing agent and will have a cartridge that has not been punctured.

Self-expelling agents, such as CO_2 and some halogenated agents, do not require a separate gas cartridge. The only way to determine whether a CO_2 extinguisher is properly charged is to weigh it. The proper weight should be indicated on the outside of the extinguisher.

Maintenance

The maintenance requirements and intervals for various types of fire extinguishers are outlined in NFPA 10. Maintenance includes an internal inspection as well as any repairs that may be required. These procedures must be performed periodically, depending on the type of extinguisher. An inspection may also reveal the need to perform maintenance procedures.

Maintenance procedures must always be performed by a qualified person. Some procedures can be performed only at a properly licensed facility. The specific qualifications and training requirements are determined by the manufacturer and the jurisdictional authority. Untrained personnel should never be allowed to perform fire extinguisher maintenance.

Common indications that an extinguisher needs maintenance include the following findings:

- The pressure gauge reading is outside the normal range.
- The inspection tag is out-of-date.

SKILL DRILL 8-9 Using a Wet-Chemical (Class K) Fire Extinguisher

(Fire Fighter I, NFPA 5.3.16)

1. Size up the fire. Make sure you have an exit route from the fire. Remove the hose and nozzle. Check the pressure gauge. Pull the pin to release the extinguisher control valve. Aim the nozzle and discharge the stream of wet chemical drops gently into the surface of the burning liquid at the front or the back of the deep-fat fryer.

2. Let the deep foam blanket flow across the surface of the burning liquid. Avoid splashing foam on the burning liquid.

3. Continue to apply the agent until the foam blanket has extinguished all of the flames.

4. Do not disturb the foam blanket even after all of the flames have been suppressed. If reignition occurs, repeat these steps.

- The tamper seal is broken, especially in extinguishers with no pressure gauge.
- The extinguisher does not appear to be full of extinguishing agent.
- The hose or nozzle assembly is obstructed.
- There are signs of physical damage, corrosion, or rust.
- There are signs of leakage around the discharge valve or nozzle assembly.

Recharging

A fire extinguisher must be recharged after each and every use, even if it was not completely discharged. The only exceptions are nonrechargeable extinguishers, which should be replaced after any use. All performance standards assume that an extinguisher will be fully charged when it has to be used. Immediately after any use, the extinguisher should be taken out of service until it has been properly recharged. This guideline also applies to any extinguisher that leaks or has a pressure gauge reading above or below the proper operating range. Most rechargeable extinguishers must be recharged by qualified personnel.

When an extinguisher is recharged, the extinguishing agent is refilled and the system that expels the agent is fully charged. Both the quantity of the agent and the pressurization must be verified. After recharging is complete, a tamper seal is installed; this seal provides assurance that the extinguisher has not been fully or even partially discharged since the last time it was recharged.

Typical 2.5-gallon (9 liter), stored-pressure water extinguishers can be recharged by fire fighters using water and a source of compressed air. Before the extinguisher is refilled, all remaining stored pressure must be discharged so that the valve assembly can be safely removed. Water is then added up to the water-level indicator, and the valve assembly is replaced. Finally, compressed air is introduced to raise the pressure to the level indicated on the gauge.

Hydrostatic Testing

Most fire extinguishers are pressurized vessels, meaning that they are designed to hold a steady internal pressure. The ability of an extinguisher to withstand this internal pressure is measured by periodic hydrostatic testing. The hydrostatic testing requirements for fire extinguishers are established by the U.S. Department of Transportation and can be found in NFPA 10 TABLE 8-3. Such tests are conducted in a special test facility and involve filling the extinguisher with water and applying above-normal pressure.

Each extinguisher has an assigned maximum interval between hydrostatic tests, usually 5 or 12 years, depending on the construction material and vessel type TABLE 8-4. The date of the most recent hydrostatic test must be indicated on the outside of the extinguisher. An extinguisher may not be refilled if the date of the most recent hydrostatic test is not within the prescribed limit. Any extinguisher that is out-of-date should be removed from service and sent to the appropriate maintenance facility for hydrostatic testing.

TABLE 8-3 Hydrostatic Test Pressure Requirements

Carbon dioxide extinguishers Carbon dioxide and nitrogen cylinders (used with wheeled extinguishers)	5/3 service pressure stamped on cylinder
Carbon dioxide extinguishers with cylinder specification ICC3	3000 psi (20,685 kPa)
All stored-pressure and bromochlorofluoromethane (Halon 1211)	Factory test pressure not to exceed three times the service pressure
Carbon dioxide hose assemblies	1250 psi (8619 kPa)
Dry-chemical and dry-powder hose assemblies	300 psi (2068 kPa) or at service pressure, whichever is higher

TABLE 8-4 Hydrostatic Test Interval for Extinguishers

Extinguisher Type	Test Interval (Years)
Stored-pressure water-loaded steam and/or antifreeze agent	5
Wetting agent	5
AFFF (aqueous film-forming foam)	5
FFFP (film-forming fluoroprotein foam)	5
Dry chemical with stainless-steel shells	5
Carbon dioxide	5
Wet chemical	5
Dry chemical, stored pressure, with mild steel shells, brazed brass shells, or aluminum shells	12
Dry chemical, cartridge or cylinder operated, with mild steel shells	12
Halogenated agents	12
Dry powder, stored pressure, cartridge or cylinder operated, with mild steel shells	12

Wrap-Up

Chief Concepts

- Fire extinguishers are used successfully to put out hundreds of fires every day, preventing millions of dollars in property damage and saving uncounted lives.
- Fire fighters use fire extinguishers to control small fires that do not require the use of a hose line.
- Portable fire extinguishers have two primary uses: to extinguish incipient-stage fires and to control fires where traditional methods of fire suppression are not recommended.
- The primary disadvantage of fire extinguishers is that they are "one-shot" devices.
- The five classes of fires affect the choice of extinguishing equipment:
  - Class A fires involve ordinary combustibles such as wood, paper, cloth, rubber, household rubbish, and some plastics.
  - Class B fires involve flammable or combustible liquids, such as gasoline, oil, grease, tar, lacquer, oil-based paints, and some plastics.
  - Class C fires involve energized electrical equipment, which includes any device that uses, produces, or delivers electrical energy.
  - Class D fires involve combustible metals such as magnesium, titanium, zirconium, sodium, lithium, and potassium. Special techniques and extinguishing agents are required to fight combustible-metal fires.
  - Class K fires involve combustible cooking oils and fats.
- Portable fire extinguishers are classified and rated based on their characteristics and capabilities.
- The classification system for fire extinguishers uses both letters and numbers. The letters indicate the classes of fire for which the extinguisher can be used, and the numbers indicate its effectiveness. Fire extinguishers that are safe and effective for more than one class will be rated with multiple letters. Numbers are used to rate an extinguisher's effectiveness for only Class A and Class B fires.
- The traditional lettering system uses the following labels:
  - Extinguishers suitable for use on Class A fires are identified by the letter "A" on a solid green triangle. The triangle has a graphic relationship to the letter "A."
  - Extinguishers suitable for use on Class B fires are identified by the letter "B" on a solid red square. The shape of the letter mirrors the graphic shape of the box.
  - Extinguishers suitable for use on Class C fires are identified by the letter "C" on a solid blue circle, which also incorporates a graphic relationship between the letter "C" and the circle.
  - Extinguishers suitable for use on Class D fires are identified by the letter "D" on a solid yellow, five-pointed star.
  - Extinguishers suitable for use on Class K (combustible cooking oil) fires are identified by a pictograph showing a fire in a frying pan.
- The pictograph system uses symbols rather than letters on the labels. Under this system, the presence of an icon indicates that the extinguisher has been rated for that class of fire.
- NFPA 10 lists the requirements for placing and mounting portable fire extinguishers as well as the appropriate mounting heights:
  - Fire extinguishers weighing up to 40 pounds (18 kilograms) should be mounted so that the top of the extinguisher is not more than 5 feet (2 meters) above the floor.
  - Fire extinguishers weighing more than 40 pounds (18 kilograms) should be mounted so that the top of the extinguisher is not more than 3 feet (1 meter) above the floor.
  - The bottom of an extinguisher should be at least 4 inches (10 centimeters) above the floor.
- Areas are divided into three risk classifications based on the fire risks associated with the materials in those areas:
  - In light (low) hazard locations, the majority of materials are noncombustible or arranged to limit fire spread, such as in offices.
  - Ordinary (moderate) hazard locations contain more Class A and Class B materials than do light hazard locations, such as auto showrooms and parking garages.
  - Extra (high) hazard locations contain more Class A combustibles and/or Class B flammables than do ordinary hazard locations, such as woodworking shops.
- Most extinguishers stop fires by cooling the fuel below its ignition point, by cutting off the supply of oxygen, or by combining the two techniques.
- Portable fire extinguishers use seven basic types of extinguishing agents:
  - Water—Used to extinguish Class A fires.
  - Dry chemicals—Used mainly to extinguish Class B and C fires. This type of extinguisher can be used to suppress Class A fires, but water is also needed to fully extinguish any embers.
  - Carbon dioxide—Used to extinguish Class B and C fires only.
  - Foam—Used to extinguish Class A or B fires. Be sure to double-check the type of foam in the extinguisher. Most Class B foams can be used on Class A fires, but Class A foams are not effective on Class B fires.
  - Halogenated agents—Include Halons and halocarbons. Halons are limited-use chemicals due to

their propensity to damage the environment. These agents are used to extinguish Class B and C fires.
  - Dry powders—Chemical compounds used to extinguish Class D fires.
  - Wet chemicals—Used to extinguish Class K fires.
- Most hand-held portable fire extinguishers have six basic parts:
  - A cylinder or container that holds the extinguishing agent
  - A carrying handle
  - A nozzle or horn
  - A trigger and discharge valve assembly
  - A locking mechanism to prevent accidental discharge
  - A pressure indicator or gauge that shows whether a stored-pressure extinguisher has sufficient pressure to operate properly
- Portable fire extinguishers vary according to their extinguishing agent, capacity, and effective range, along with the time it takes to completely discharge the extinguishing agent. They also have different mechanical designs:
  - Stored-pressure water-type fire extinguishers—Expel water in a solid stream with a range of 35 to 40 feet (11 to 12 meters) through a nozzle at the end of a short hose.
  - Loaded-stream water extinguishers—Discharge a solution of water containing a chemical to prevent freezing.
  - Wetting-agent water-type fire extinguishers—Expel water that contains a solution to reduce its surface tension, which enables water to penetrate more efficiently into Class A fuels.
  - Pump tank water-type fire extinguishers—Do not store water under pressure, but instead expel it through use of a hand-operated, double-acting, vertical piston pump.
  - Backpack water extinguishers—Five-gallon water tanks used primarily outdoors for fighting brush and grass fires.
  - Ordinary dry-chemical extinguishers—Available in hand-held models and larger, wheeled units to extinguish Class B and C fires.
  - Multipurpose dry-chemical extinguishers—Available in hand-held models and larger, wheeled units to extinguish Class A, B, and C fires.
  - Carbon dioxide extinguishers—Include a siphon tube and horn or cone-shaped applicator that is used to direct the flow of the agent. These extinguishers have a relatively short discharge range of 3 to 8 feet (1 to 2 meters).
  - Class B foam extinguishers—Very similar in appearance and operation to water extinguishers, except these extinguishers have an aspirating nozzle.
  - Wet-chemical extinguishers—Available in two sizes, 1.5 gallons (6 liters) and 2.5 gallons (9 liters).
  - Halogenated-agent extinguishers—Available in hand-held extinguishers for Class B and C fires and larger-capacity models for use on Class A fires.
  - Dry-powder fire extinguishers—Rely on sodium chloride–based agents. Models with 30-pound (14-kilogram) capacity are available in either stored-pressure or cylinder/cartridge versions. Wheeled models are available with 150- and 350-pound (68- to 159-kilogram) capacities.
- There are six basic steps in extinguishing a fire with a portable fire extinguisher:
  - Locate the fire extinguisher.
  - Select the proper classification of extinguisher.
  - Transport the extinguisher to the location of the fire.
  - Activate the extinguisher to release the extinguishing agent.
  - Apply the extinguishing agent to the fire for maximum effect.
  - Ensure your personal safety by having an exit route.
- Activating a fire extinguisher to apply the extinguishing agent is a single operation that involves four steps. The PASS acronym is a helpful way to remember these steps:
  - Pull the safety pin.
  - Aim the nozzle at the base of the flames.
  - Squeeze the trigger to discharge the agent.
  - Sweep the nozzle across the base of the flames.
- When using a fire extinguisher, always approach the fire with an exit behind you. If you are outside, make sure the wind is at your back.
- Fire extinguishers must be regularly inspected and properly maintained to ensure that they will be available for use in an emergency.
- The fire fighter charged with inspecting the extinguishers should perform the following tasks:
  - Ensure that all tamper seals are intact.
  - Determine fullness by weighing or "hefting" the extinguisher.
  - Examine all parts for signs of physical damage, corrosion, or leakage.
  - Check the pressure gauge to confirm that it is in the operable range.
  - Ensure that the extinguisher is properly identified by type and rating.
  - Shake dry-chemical extinguishers to mix or redistribute the agent.
  - Check the hose and nozzle for damage or obstruction by foreign objects.
  - Check the hydrostatic test date of extinguisher.
- Common indications that an extinguisher needs maintenance include the following findings:
  - The pressure gauge reading is outside the normal range.
  - The inspection tag is out-of-date.
  - The tamper seal is broken, especially in extinguishers with no pressure gauge.

- The extinguisher does not appear to be full of extinguishing agent.
- The hose or nozzle assembly is obstructed.
- There are signs of physical damage, corrosion, or rust.
- There are signs of leakage around the discharge valve or nozzle assembly.

- A fire extinguisher must be recharged after each and every use, even if it was not completely discharged. The only exceptions are nonrechargeable extinguishers, which should be replaced after any use.
- Most fire extinguishers are pressurized vessels, meaning that they are designed to hold a steady internal pressure. The ability of an extinguisher to withstand this internal pressure is measured by periodic hydrostatic testing. These tests, which are conducted in a special test facility, involve filling the extinguisher with water and applying above-normal pressure.

Hot Terms

Ammonium phosphate An extinguishing agent used in dry-chemical fire extinguishers that can be used on Class A, B, and C fires.

Aqueous film-forming foam (AFFF) A concentrated aqueous solution of one or more hydrocarbon and/or fluorochemical surfactants that forms a foam capable of producing a vapor-suppressing, aqueous film on the surface of hydrocarbon fuels. (NFPA 403)

Carbon dioxide (CO_2) fire extinguisher A fire extinguisher that uses carbon dioxide gas as the extinguishing agent. It is rated for use on Class B and C fires.

Cartridge/cylinder fire extinguisher A fire extinguisher that has the expellant gas in a separate container from the extinguishing agent storage container. The storage container is pressurized by a mechanical action that releases the expellant gas.

Class A fire A fire in ordinary combustible materials, such as wood, cloth, paper, rubber, and many plastics. (NFPA 10)

Class B fire A fire in flammable liquids, combustible liquids, petroleum greases, tars, oils, oil-based paints, solvents, lacquers, alcohols, and flammable gases. (NFPA 10, 2007)

Class C fire A fire that involves energized electrical equipment. (NFPA 10)

Class D fire A fire in combustible metals, such as magnesium, titanium, zirconium, sodium, lithium, and potassium. (NFPA 10)

Class K fire A fire in a cooking appliance that involves combustible cooking media (vegetable or animal oils and fats). (NFPA 10)

Clean agent Electrically nonconducting, volatile, or gaseous fire extinguishant that does not leave a residue upon evaporation. (NFPA 2001)

Cylinder The body of the fire extinguisher where the extinguishing agent is stored.

Dry-chemical fire extinguisher An extinguisher that uses a mixture of finely divided solid particles to extinguish fires. The agent is usually sodium bicarbonate, potassium bicarbonate, or ammonium phosphate based, with additives being included to provide resistance to packing and moisture absorption and to promote proper flow characteristics. These extinguishers are rated for use on Class B and C fires, although some are also rated for Class A fires.

Dry-powder extinguishing agent An extinguishing agent used in putting out Class D fires. Common examples include sodium chloride and graphite-based powders.

Dry-powder fire extinguisher A fire extinguisher that uses an extinguishing agent in powder or granular form. It is designed to extinguish Class D combustible-metal fires by crusting, smothering, or heat-transferring means.

Extinguishing agent A material used to stop the combustion process. Extinguishing agents may include liquids, gases, dry chemical compounds, and dry powder compounds.

Extra (high) hazard locations Occupancies where the total amounts of Class A combustibles and Class B flammables are greater than expected in occupancies classed as ordinary (moderate) hazards.

Film-forming fluoroprotein (FFFP) foam A protein-based foam concentrate incorporating fluorinated surfactants that forms a foam capable of producing a vapor-suppressing, aqueous film on the surface of hydrocarbon fuels. (NFPA 403)

Fire load The weight of combustibles in a fire area (measured in ft^2 or m^2) or on a floor in buildings and structures, including either contents or building parts, or both. (NFPA 914)

Halogenated-agent fire extinguisher An extinguisher that uses a halogenated extinguishing agent.

Halogenated extinguishing agent A liquefied gas extinguishing agent that puts out fires by chemically interrupting the combustion reaction between the fuel and oxygen.

Halon 1211 A halogenated agent whose chemical name is bromochlorodifluoromethane, $CBrClF_2$; it is a multipurpose, Class ABC-rated agent effective against flammable liquid fires. (NFPA 408)

Handle The grip used for holding and carrying a portable fire extinguisher.

Horn The tapered discharge nozzle of a carbon dioxide–type fire extinguisher.

Hydrostatic testing Pressure testing of a fire extinguisher to verify its strength against unwanted rupture. (NFPA 10)

Ignition point The minimum temperature at which a substance will burn.

Incipient stage The initial or beginning stage of a fire, in which it can be controlled or extinguished by portable extinguishers or small amounts of dry extinguishing agents, without the need for protective clothing or breathing apparatus. (NFPA 484)

Light (low) hazard locations Occupancies where the total amount of combustible materials is less than expected in an ordinary hazard location.

Loaded-stream fire extinguisher A water-based fire extinguisher that uses an alkali metal salt as a freezing-point depressant.

Locking mechanism A device that locks an extinguisher's trigger to prevent its accidental discharge.

Multipurpose dry-chemical fire extinguisher A fire extinguisher rated to fight Class A, B, and C fires.

Nozzle A constricting appliance attached to the end of a fire hose or monitor to increase the water velocity and form a stream. (NFPA 1965)

Ordinary (moderate) hazard locations Occupancies that contain more Class A and Class B materials than are found in light hazard locations.

PASS Acronym for the steps involved in operating a portable fire extinguisher: Pull pin, Aim nozzle, Squeeze trigger, Sweep across burning fuel.

Polar solvent A water-soluble flammable liquid such as alcohol, acetone, ester, and ketone.

Pressure indicator A gauge on a pressurized portable fire extinguisher that indicates the internal pressure of the expellant.

Pump tank fire extinguisher A nonpressurized, manually operated water extinguisher that is rated for use on Class A fires. Discharge pressure is provided by a hand-operated, double-acting piston pump.

Rapid oxidation A chemical process that occurs when a fuel is combined with oxygen, resulting in the formation of ash or other waste products and the release of energy as heat and light.

Saponification The process of converting the fatty acids in cooking oils or fats to soap or foam; the action caused by a Class K fire extinguisher.

Self-expelling fire extinguisher A fire extinguisher in which the agents have sufficient vapor pressure at normal operating temperatures to expel themselves. (NFPA 10)

Stored-pressure fire extinguisher A fire extinguisher in which both the extinguishing material and the expellant gas are kept in a single container, and that includes a pressure indicator or gauge. (NFPA 10)

Stored-pressure water-type fire extinguisher A fire extinguisher in which water or a water-based extinguishing agent is stored under pressure.

Tamper seal A retaining device that breaks when the locking mechanism is released.

Trigger The button or lever used to discharge the agent from a portable fire extinguisher.

Underwriters Laboratories, Inc. (UL) The U.S. organization that tests and certifies that fire extinguishers (among many other products) meet established standards. The Canadian equivalent is Underwriters Laboratories of Canada.

Wet-chemical extinguishing agent An extinguishing agent for Class K fires. It commonly consists of solutions of water and potassium acetate, potassium carbonate, potassium citrate, or any combination thereof.

Wet-chemical fire extinguisher A fire extinguisher for use on Class K fires that contains a wet-chemical extinguishing agent.

Wetting-agent water-type fire extinguisher An extinguisher that expels water combined with a chemical or chemicals to reduce its surface tension; it is used for Class A fires.

Wheeled fire extinguisher A portable fire extinguisher equipped with a carriage and wheels, which is intended to be transported to the fire by one person. (NFPA 10)

FIRE FIGHTER *in action*

You and your crew are at a local high school giving a tour of the engine. You get to the compartment that has three fire extinguishers in it. You explain that the first is a wetting-agent water-type extinguisher, the second is a dry chemical extinguisher, and the third is a CO_2 extinguisher. One of the students asks if there are other types of fire extinguishers and you respond that there are. The teacher then tells you that they have been studying the chemistry of fire over the past two weeks and asks you to explain each type of fire extinguisher and how each works to extinguish a fire. The other crew members chuckle to themselves as they watch you get tested over fire extinguishers.

1. Which class of fire involves combustible cooking media such as vegetable oils, animal oils, and fats?

A. Class A
B. Class B
C. Class J
D. Class K

2. Which extinguisher would you use on a Class C fire?

A. Dry powder
B. Film-forming fluoroprotein (FFFP) foam
C. Halon 1211
D. Wet chemical

3. Which type of extinguisher can typically be used to extinguish fires in ordinary combustibles, combustible liquids, and charged electrical equipment?

A. Wet chemical fire extinguisher
B. Multi-purpose dry chemical
C. Wetting-agent water extinguisher
D. CO_2

4. What does the acronym "PASS" stand for?

A. Press, Aim, Squeeze, Sweep
B. Pull, Aim, Squeeze, Sweep
C. Peek, Aim, Squeeze, Sweep
D. Poke, Aim, Squeeze, Sweep

5. Hydrostatic testing confirms the extinguisher ____________.

A. has sufficient strength to withstand the internal pressures
B. will properly expel the propellant when activated
C. can be used on electrical fires
D. contains a water-based extinguishing agent

6. Which of the following is a polar solvent?

A. Alcohol
B. Gasoline
C. Diesel fuel
D. Water

7. What is the term for the retaining device that breaks when the locking mechanism is released?

A. Handle
B. Trigger
C. Tamper seal
D. Horn

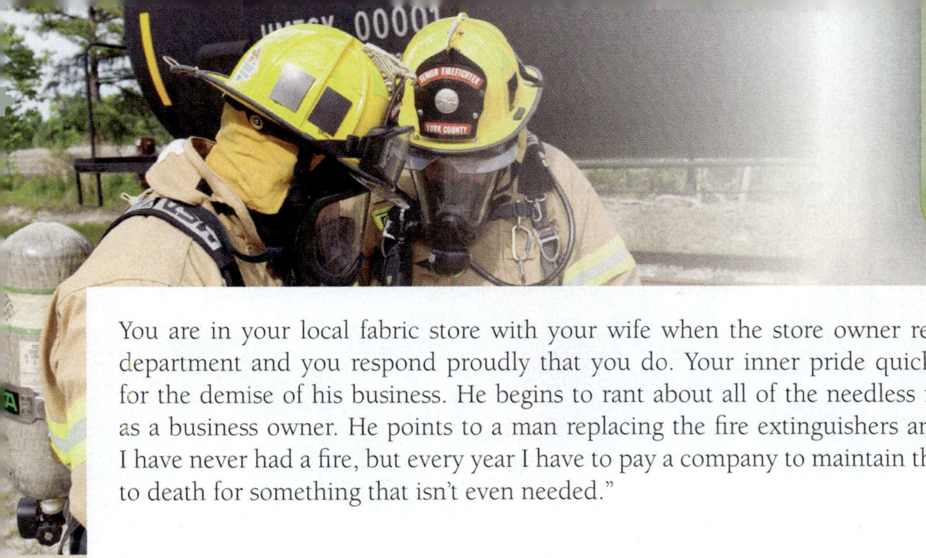

FIRE FIGHTER II *in action*

You are in your local fabric store with your wife when the store owner recognizes you. He asks if you work for the fire department and you respond proudly that you do. Your inner pride quickly fades when the owner begins to blame you for the demise of his business. He begins to rant about all of the needless rules and regulations that he must comply with as a business owner. He points to a man replacing the fire extinguishers and exclaims, "In my 35 years of doing business, I have never had a fire, but every year I have to pay a company to maintain those extinguishers. You guys nickel and dime me to death for something that isn't even needed."

1. How do you calm the irate owner?
2. What points of his comments would you attempt to correct and which ones would you simply let go?
3. How do you explain the value of fire extinguishers?
4. What would you recommend to the Fire Chief as a result of this encounter?

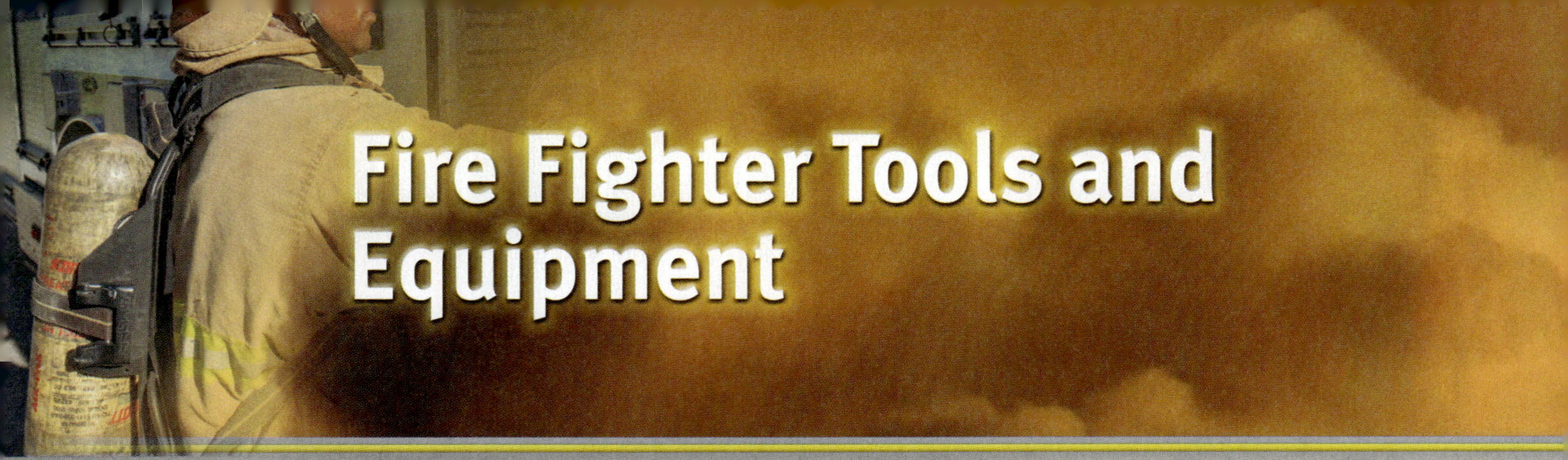

Fire Fighter Tools and Equipment

Fire Fighter I

Knowledge Objectives

After studying this chapter, you will be able to:

- Describe the general purposes of tools and equipment. (NFPA 5.5.1.B, p 236–237)
- Describe the safety considerations for the use of tools and equipment. (NFPA 5.5.1.B, p 236)
- Describe why it is important to use tools and equipment effectively. (NFPA 5.5.1.B, p 236–237)
- Describe why it is important for you to know where tools are stored. (NFPA 5.5.1.B, p 237)
- List and describe tools and equipment that are used for rotating. (p 237–239)
- List and describe tools and equipment that are used for pushing or pulling. (p 239–240)
- List and describe tools and equipment that are used for prying or spreading. (p 240–242)
- List and describe tools and equipment that are used for striking. (p 242–244)
- List and describe tools and equipment that are used for cutting. (p 244–248)
- Describe the tools used in response and scene size-up activities. (p 248)
- Describe the tools used in a forcible entry. (NFPA 5.3.4, p 248–249)
- Describe the tools used during an interior attack. (NFPA 5.3.8, p 249–251)
- Describe ventilation tools. (NFPA 5.3.11, p 250–251)
- Describe the hand tools needed during an overhaul assignment. (NFPA 5.3.13, p 251)
- Explain how tools and equipment are staged for rapid access. (NFPA5.5.1B, p 252)
- Describe the importance of properly maintaining tools and equipment. (NFPA 5.5.1, 5.5.1.A, p 252–253)
- Describe how to clean and inspect hand tools. (NFPA 5.5.1, 5.5.1.A, 5.5.1.B, p 253)

Skills Objectives

There are no skill objectives for Fire Fighter I candidates. NFPA 1001 contains no Fire Fighter I Job Performance Requirements for this chapter.

CHAPTER 9

Fire Fighter II — FFII

Knowledge Objectives

After studying this chapter, you will be able to:

- Describe the tools used in search and rescue operations. (NFPA 6.4.2.A, 6.4.2.B, p 250)
- Explain how tools and equipment are staged for rapid access. (NFPA 6.5.4.B, p 252)
- Describe how to maintain power equipment and power tools. (NFPA 6.5.4, 6.5.4.A, p 252–253)

Skills Objectives

There are no skill objectives for Fire Fighter II candidates. NFPA 1001 contains no Fire Fighter II Job Performance Requirements for this chapter.

You Are the Fire Fighter

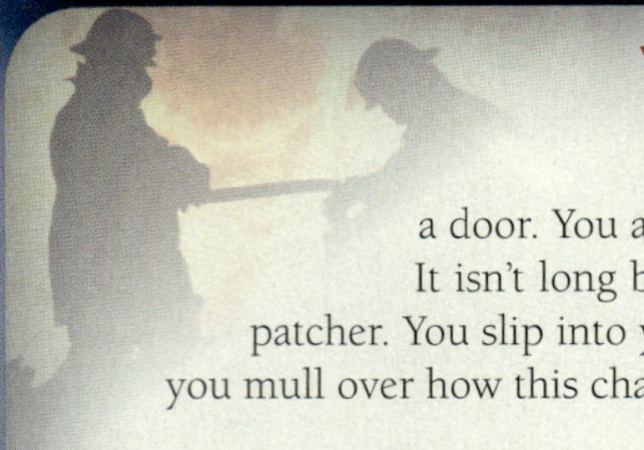

You are excited about getting to travel to another station to work on an engine company. Ever since you were in recruit class, engine work has appealed to you. You have seen the engine crew dismount the engine with their tools in hand, heading to the structure to force open a door. You are pumped knowing that today that fire fighter could be you.

It isn't long before the alarm sounds, "Person trapped, motor vehicle collision, 4th and Elm," directs the dispatcher. You slip into your turnout gear, take your seat, and put on your seatbelt. As the engine pulls from the station, you mull over how this changes your plan for what you would be doing today.

1. What is considered "engine work"?
2. What kinds of tools will you need for this job?
3. What other kind of knowledge will you need?

Introduction

Fire fighters use tools and equipment to perform a wide range of activities. A fire fighter must know how to use tools effectively, efficiently, and safely, even when it is dark or visibility is limited. This chapter provides an overview of the general functions of the most commonly used tools and equipment and discusses how they are used during fire suppression and rescue operations. As you will see, the same tools may be used in different ways during each phase of fire suppression or rescue operations. This chapter also explains how to maintain tools and equipment so that they will always be ready for emergency use.

General Considerations

Tools and equipment are used in almost all fire suppression and rescue operations. As you progress through your training, you will learn how to operate the different types of tools and equipment used by your department.

Hand tools are used to extend or multiply the actions of your body and to increase your effectiveness in performing specific functions. Most of these tools operate using simple machine principles: A pike pole extends your reach, allows you to penetrate through a ceiling, and enables you to apply force to pull down ceiling material; an axe multiplies the cutting force you can exert on a given area. By contrast, power tools and equipment use an external source of power, such as an electric motor or an internal combustion engine. In certain cases, they are faster and more efficient than hand tools.

Safety

Safety is a prime consideration when using any kind of tool or equipment. You need to operate the equipment so that you, your fellow fire fighters, victims, and bystanders are not accidentally injured. Personal safety also requires the use of the proper personal protective equipment (PPE):

- Approved helmet
- Firefighting hood
- Eye protection
- Face shield
- Firefighting gloves
- Turnout coat
- Bunker pants
- Boots
- Self-contained breathing apparatus (SCBA)
- Personal alert safety system (PASS)

Conditions of Use/Operating Conditions

The best way to learn how to use tools and equipment properly is under optimal conditions of visibility and safety. In the beginning, you must be able to see what you are doing and practice without endangering yourself and others. As you become more proficient, you should practice using tools and equipment under more difficult working conditions. Eventually, you should be able to use tools and equipment safely and effectively even when darkness or smoke decreases visibility. You must be able to work safely in hazardous areas, or from a ladder or pitched roof, where you are surrounded by noise and other activities, while wearing all of your protective clothing and using your SCBA. For this reason, some departments require fire fighters to practice certain skills and evolutions in total darkness or with their face masks covered to simulate the darkness of actual fires.

Effective Use

Effective, efficient use of tools and equipment means using the least amount of energy to accomplish the task. Being effective means that you achieve the desired goal; being efficient means that you produce the desired effect without wasting time or energy. When you are assigned a task on the fire ground, your objective is to complete that task safely and quickly. If you waste energy by working inefficiently, you will not be able to perform additional tasks. If you know which tools and equipment are needed for each phase of firefighting, you will be able to achieve the desired objective quickly and have the energy needed to complete the remaining tasks.

New fire fighters are often surprised by the strength and energy required to perform many tasks. An aggressive, continuous program of physical fitness will enable you to maintain your body in the optimal state of readiness. It does little good to practice using a tool for hours if you are not in good physical condition.

As your training continues, you will learn which tools and equipment are used during different phases of fire-ground operations. For example, the tools needed for forcible entry differ from the tools needed for overhaul. Knowing which tools are needed for the work that must be done will help you prepare for the different tasks that unfold on the scene of a fire. The specific steps for properly using tools and equipment are presented in this chapter and throughout this text; your fire department instructors will demonstrate them as well.

Most fire departments have standard operating procedures (SOPs) or standard operating guidelines (SOGs) that specify which tools and equipment are needed in various situations. As a fire fighter, you must know where every tool and piece of equipment is carried on your apparatus. Knowing how to use a piece of equipment does you no good if you cannot find it quickly. Your company officer and your department's SOPs will assist you in knowing which tools to bring to different situations.

Some fire fighters carry a selection of small tools and equipment in the pockets of their turnout coats or bunker pants. Check whether your department requires you to carry certain tools and equipment at all times. Ask senior fire fighters for recommendations about which tools and equipment you should carry.

Functions

An engine or truck company carries a number of tools and different types of equipment. Often, the easiest way to learn and remember these tools is to group them by the function that each performs TABLE 9-1 and TABLE 9-2. Most of the tools carried by fire departments fit into the following functional categories:

- Rotating (assembly or disassembly)
- Prying or spreading
- Pushing or pulling
- Striking
- Cutting
- Multiple use

Rotating Tools

Rotating tools apply a rotational force to make something turn. The most commonly used rotating tools are screwdrivers, wrenches, and pliers, which are used to assemble (fit together)

TABLE 9-1	General Tool Functions
• Rotate, assemble, or disassemble FIGURE 9-1 • Pry or spread FIGURE 9-2 • Push or pull FIGURE 9-3 • Strike FIGURE 9-4 • Cut FIGURE 9-5 • A combination of functions FIGURE 9-6	

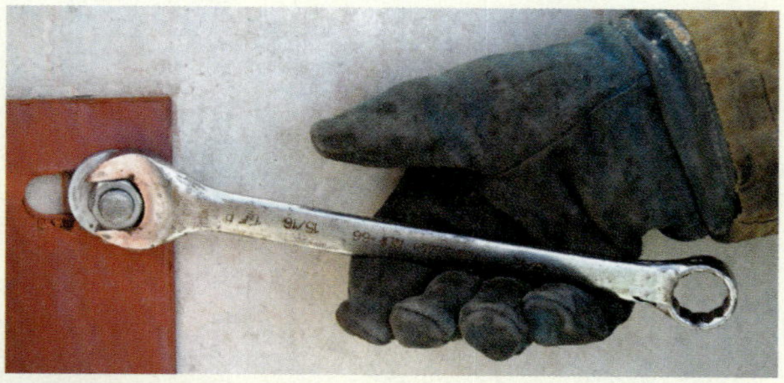

FIGURE 9-1 Rotate, assemble, or disassemble.

FIGURE 9-2 Pry or spread.

FIGURE 9-3 Push or pull.

FIGURE 9-4 Strike.

FIGURE 9-5 Cut.

or disassemble (take apart) parts that are connected with threaded fasteners TABLE 9-3.

Assembling and disassembling are basic mechanical skills that are routinely employed by fire fighters to solve problems. Most fire apparatus carries a tool kit with a selection of screwdrivers with different heads, open-end wrenches, box wrenches, socket wrenches, pliers, adjustable wrenches, and pipe wrenches.

FIGURE 9-6 A combination of functions.

TABLE 9-2 Tool Functions

Function	Tool
Rotate	Box-end wrench Gripping pliers Hydrant wrench Open-end wrench Pipe wrench Screwdriver Socket wrench Spanner wrench
Push or pull	Ceiling hook Clemens hook Drywall hook K tool Multipurpose hook Pike pole Plaster hook Roofman's hook San Francisco hook
Pry or spread	Claw bar Crowbar Flat bar Halligan tool Hux bar Hydraulic spreader Kelly tool Pry bar Rabbet tool
Strike	Battering ram Chisel Flat-head axe Hammer Mallet Maul Pick-head axe Sledgehammer Spring-loaded center punch
Cut	Axe Bolt cutter Chainsaw Cutting torch Hacksaw Handsaw Hydraulic shears Reciprocating saw Rotary saw Seat belt cutter

TABLE 9-3	Tools for Assembly and Disassembly
• Box-end wrenches • Pipe wrenches • Gripping pliers • Screwdrivers • Hydrant wrenches • Socket wrenches • Open-end wrenches • Spanner wrenches	

Various sizes and types of screw heads are available, including slotted head, Phillips head, Roberts head, and others. A screwdriver with interchangeable heads is sometimes more useful than a selection of different screwdrivers. Pliers and wrenches also come in a variety of shapes and sizes for different applications.

A spanner wrench is used to tighten or loosen fire hose couplings (one set or pair of connection devices that allow a fire hose to be interconnected with additional lengths of hose). If the couplings are not too tight, they can be loosened by hand, without the use of a spanner wrench. A hydrant wrench is used to open or close a hydrant by rotating the valve stem, and to remove the caps from the hydrant outlets.

Examples of rotating tools include the following items FIGURE 9-7:

- Box-end wrench: A hand tool with a closed end that is used to tighten or loosen bolts. Some styles of box-end wrenches have ratchets for easier use.
- Gripping pliers: A hand tool with a pincer-like working end that can also be used to bend wire or hold smaller objects.
- Hydrant wrench: A hand tool used to operate the valves on a hydrant. Some versions are plain wrenches, whereas others have a ratchet (a mechanism that will rotate only one way) feature FIGURE 9-8.
- Open-end wrench: A hand tool with an open end that is used to tighten or loosen bolts.
- Combination wrench: A hand tool with an open-end wrench on one end and a box-end wrench on the other.

FIGURE 9-8 A hydrant wrench in action.

FIGURE 9-7 Hydrant wrench, spanner wrench, pipe wrench, open-end wrench, box-end wrench, gripping pliers, screwdriver, and socket wrench.

- Pipe wrench: A wrench with one fixed grip and one movable grip that can be adjusted to fit securely around pipes and other tubular objects.
- Screwdriver: A tool that is used to turn screws.
- Socket wrench: A wrench that fits over a nut or bolt and uses the action of an attached handle to tighten or loosen the nut or bolt.
- Spanner wrench: A special wrench used to tighten or loosen hose couplings. Spanners come in several sizes.

Fire Fighter Safety Tips

When you are learning to operate tools and equipment, use them in areas without hazards and with adequate light. Gradually increase the amount of PPE worn and decrease the light to the levels typically found on the fire ground. Eventually, you should be so familiar with your tools and equipment that you can use them under adverse conditions while wearing full PPE.

Pushing/Pulling Tools

A second type of tools is those used for pushing or pulling. Such tools can extend the reach of the fire fighter as well as increase the power the fire fighter can exert upon an object TABLE 9-4. Pushing and pulling tools have many different uses in fire department operations.

An example of a tool that extends the fire fighter's reach is a pike pole FIGURE 9-9. This wood or fiberglass pole has a metal head attached to one end. It is used primarily to pull down a ceiling in an effort to get to the seat of a fire burning above. The metal head has a sharpened point that can be punched through the ceiling and a hook that can grab and pull the ceiling down. The proper use of a pike pole is discussed in depth in the Salvage and Overhaul chapter.

Pike poles come in several different sizes and with a variety of heads FIGURE 9-10. The most common length of 4 to 6 feet (1 to 2 meter) enables a fire fighter to stand on a floor and pull down a 10-feet-high (3 meter) ceiling. Closet hooks, which are intended for use in tight spaces, are commonly 2 to 4 feet (0.6 to 1 meter) long. Some pike poles are equipped with handles as long as 12 or 14 feet (3 to 4 meter) for use in rooms with very high ceilings; others may have a "D"-type handle for better pulling power. It is important to bring the right-sized pike pole to a fire. It the pike pole is too short, you will not be able to reach the ceiling. If it is too long, you will not be able to use it in a room with a low ceiling.

TABLE 9-4	Tools for Pushing and Pulling
• Ceiling hook • Multipurpose hook • Roofman's hook • Clemens hook • Pike pole • San Francisco hook • Drywall hook • Plaster hook	

FIGURE 9-9 A pike pole in action.

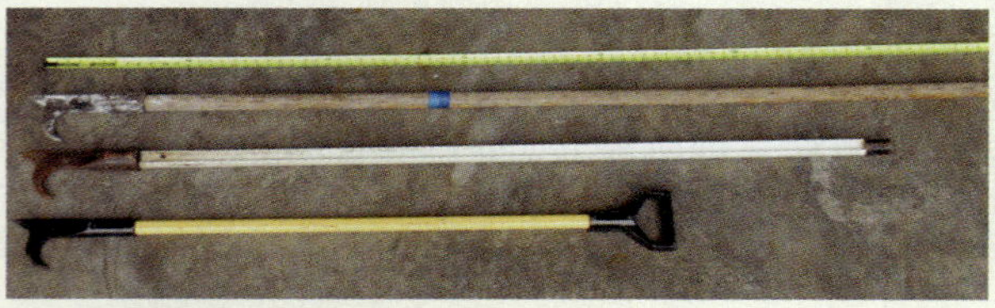

FIGURE 9-10 Pike poles come in several different sizes.

The different head designs of pike poles are intended for different types of ceilings and come in a variety of configurations. Many fire departments use one type of pike pole for plaster ceilings and another type for drywall ceilings.

Commonly encountered types of pulling poles and hooks include the following:

- Ceiling hook: A tool consisting of a long wood or fiberglass pole and a metal point with a spur at right angles that can be used to probe ceilings and pull down plaster lath material.
- Clemens hook: A multipurpose tool that can be used for forcible entry and ventilation applications because of its unique head design.
- Drywall hook: A specialized version of a pike pole designed to remove drywall.
- Multipurpose hook: A long pole with a wooden or fiberglass handle and a metal hook.
- Plaster hook: A long pole with a pointed head and two retractable cutting blades on the side.
- Roofman's hook: A long pole with a solid metal hook.
- San Francisco hook: A multipurpose tool used for forcible entry and ventilation applications. It includes a built-in gas shut-off and directional slot.

A K tool is another type of pushing or pulling tool FIGURE 9-11. It is used to pull the lock cylinder out of a door, exposing the locking mechanism so it can be unlocked easily.

Prying/Spreading Tools

A third type of tools is those used for prying or spreading. These tools may be as simple as a pry bar or as mechanically complex as a hydraulic spreader TABLE 9-5. They come in several versions that are designed for different applications.

FIGURE 9-11 Components of a K tool.

TABLE 9-5 Tools for Prying and Spreading

- Claw bar
- Halligan tool
- Kelly tool
- Crowbar
- Hux bar
- Pry bar
- Flat bar
- Hydraulic spreader
- Rabbet tool

A simple pry bar consists of a hardened steel rod with a tapered end that can be inserted into a small area. The bar acts as a lever to multiply the force that a person can exert to bend or pry objects apart. A properly positioned pry bar can apply an enormous amount of force.

Fire departments use a wide range of prying tools in different sizes and with different features. One of the most popular is a Halligan tool, which was designed by a New York City fire fighter. This tool incorporates a sharp pick, a flat prying surface, and a forked claw. It can be used for forcible entry applications, as can several other prying tools. The specific uses and applications of these tools are covered in the chapters on the various phases of fire suppression.

The following tools are used for prying FIGURE 9-12:

- Claw bar: A tool with a pointed claw-hook on one end and a forked- or flat-chisel pry on the other end that can be used for forcible entry.
- Crowbar: A straight bar made of steel or iron with a forked-like chisel on the working end.
- Flat bar: A specialized prying tool made of flat steel with prying ends suitable for performing forced entry.
- Halligan tool: A prying tool designed for use in the fire service that also incorporates a pick, an adz, and a fork; sometimes known as a Hooligan tool. The difference is that a Halligan tool is a forged piece of steel.
- Hux bar: A multipurpose tool that can be used for forcible entry and ventilation applications because of its unique design. A Hux bar also can be used as a hydrant wrench.
- Kelly tool: A steel bar with two main features—a large pick and a large chisel or fork.
- Pry bar: A specialized prying tool made of a hardened steel rod with a tapered end that can be inserted into a small area FIGURE 9-13.

Hydraulic spreaders are an example of machine-powered rescue tools that can be used for prying and spreading FIGURE 9-14. The use of hydraulic power enables you to apply

FIGURE 9-13 A pry bar in action.

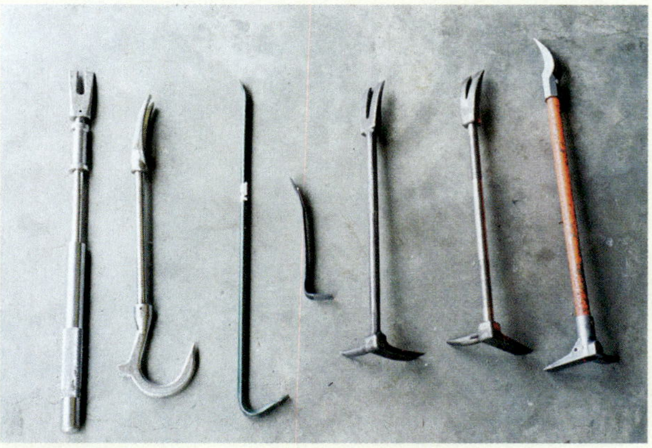

FIGURE 9-12 Tools used for prying and spreading.

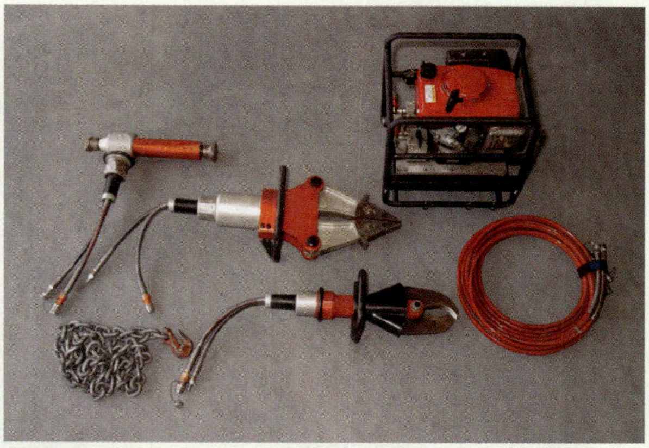

FIGURE 9-14 The components of a hydraulic rescue tool.

several tons of force on a very small area. You must have special training to operate these machines safely, however. Fire and rescue departments most commonly use this equipment for extrication of victims from motor vehicles and machinery FIGURE 9-15.

Hand-powered hydraulic tools are also used for prying and spreading. One of these, the rabbet tool, is designed to quickly open doors FIGURE 9-16.

Striking Tools

Striking tools are used to apply an impact force to an object TABLE 9-6. They are often employed to gain entrance to a building or a vehicle or to make an opening in a wall or roof. This equipment can also be used to force the end of a prying tool into a small opening. Specific use of these tools is covered in the Forcible Entry chapter.

Striking tools include the following items FIGURE 9-17:

- Hammer: A hand tool constructed of solid material with a long handle and a head affixed to the top of the handle, with one side of the head used for striking and the other side used for prying.
- Mallet: A short-handled hammer with a round head.
- Sledgehammer: A long, heavy hammer that requires the use of both hands.
- Maul: A specialized striking tool (weighing 6 pounds (3 kilograms) or more) with an axe on one end and a sledgehammer on the other end.
- Flat-head axe: A tool with a head that has an axe blade on one end and a flat head that can be used for striking objects on the opposite end FIGURE 9-18A.

Fire Fighter Safety Tips

Some fire fighters carry knives, combination tools, a small length of rope, or wooden wedges in their turnout gear. The tools and equipment that you carry will depend on your department's requirements, the type of area served by your department, and your personal preference.

TABLE 9-6	Striking Tools
• Battering ram • Hammer • Pick-head axe • Chisel • Mallet • Sledgehammer • Flat-head axe • Maul • Spring-loaded center punch	

FIGURE 9-15 A hydraulic spreader is often used for extrication of victims from vehicles.

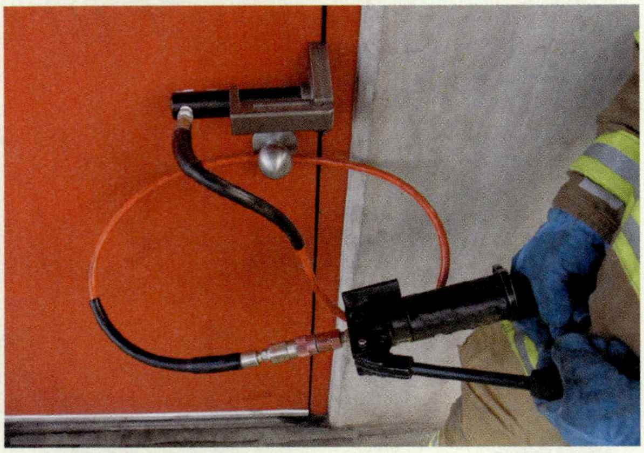

FIGURE 9-16 A rabbet tool can open doors quickly.

FIGURE 9-17 Striking tools (from top): hammer, mallet, maul, sledgehammer.

A.

B.

FIGURE 9-18 Two types of axes: **A.** Flat-head axe. **B.** Pick-head axe.

- Pick-head axe: A tool with a head that has an axe blade on one end and a pointed "pick" on the opposite end FIGURE 9-18B.
- Battering ram: A heavy metal bar used to break down doors FIGURE 9-19.
- Chisel: A metal tool with one sharpened end that can be used to break apart material when used in conjunction with a hammer, mallet, or sledgehammer FIGURE 9-20.
- Spring-loaded center punch: A spring-loaded punch that is used to break tempered automobile glass.

Among the most frequently used tools in the fire service are axes, including both pick-head axes and flat-head axes. Both types of axes have a wide cutting blade that can be used to chop into a wall, roof, or door. A pick-head axe has a point or pick that can be used for puncturing, pulling, and prying FIGURE 9-21. A flat-head axe can be used as a striking tool for forcible entry, usually in combination with a prying tool, such as a Halligan tool. Together, the flat-head axe and the Halligan tool are sometimes referred to as "the irons"; this combination is very effective in most forcible-entry situations FIGURE 9-22. Forcible entry is covered in depth in the Forcible Entry chapter.

FIGURE 9-19 Using a battering ram.

FIGURE 9-20 Using a chisel with a hammer.

FIGURE 9-21 A pick-head axe can be used to pry up boards.

FIGURE 9-22 A flat-head axe can be used with a Halligan tool to force open a door.

FIGURE 9-23 A spring-loaded center punch can be used to break a car window safely.

FIRE FIGHTER Tips

Many fire departments carry a striking tool and a prying bar strapped together, a combination sometimes referred to as "the irons." A flat-head axe and a Halligan tool are carried together on many vehicles and are taken into a fire building by one of the crew members.

Fire Fighter Safety Tips

Some tools require a considerable amount of movement and room to operate. Always confirm that no one is in danger of being injured before you use these tools. Look around and make sure you can operate an axe or sledgehammer safely and effectively.

The spring-loaded center punch is a striking tool used primarily on cars FIGURE 9-23. It can exert a large amount of force on a pinpoint-size portion of tempered automobile glass. This action disrupts the integrity of tempered glass and causes the window to shatter into small, uniform-sized pieces. A spring-loaded center punch is often used in vehicular crashes to gain access to a victim who needs care. Vehicle extrications are covered in depth in the Vehicle Rescue and Extrication chapter.

Cutting Tools

Cutting tools have sharp edges that sever objects. They come in several forms and are used to cut a wide variety of substances FIGURE 9-24. Cutting tools used by fire fighters range from knives and wire cutters that can be carried in the pockets of turnout coats to seat belt cutters, bolt cutters (a scissors-like tool used to cut through items such as chains or padlocks), saws, cutting torches (a torch that produces a high-temperature flame capable of melting metal), and hydraulic shears TABLE 9-7. Each of these tools is designed to work on certain types of materials. Fire fighters can be injured and cutting tools can be ruined if the tools are used incorrectly.

Bolt cutters are most often used to cut through chains or padlocks to open doors or gates. By concentrating the cutting force on a small area, it is possible to break through many chains in just a few seconds.

Fire departments often carry several different types of saws. This equipment can be classified into two main categories, based on the power source. Handsaws are manually powered, whereas mechanical saws are usually powered by electric motors or gasoline-powered engines. Handsaws include hacksaws, carpenter's handsaws, keyhole saws, and coping saws.

Hacksaws are designed to cut metal FIGURE 9-25. Different blades can be used, depending on the type of metal being cut. Hacksaws are useful when metal needs to be cut under closely controlled conditions.

Carpenter's handsaws are designed for cutting wood. Saws with large teeth are effective in cutting large timber or tree branches; they are often useful at motor-vehicle crashes where tree limbs may hamper the rescue effort. Saws with finer teeth are designed for cutting finished lumber. A coping saw is used to cut curves in wood; it consists of a handsaw with a narrow blade set between the ends of a U-shaped frame. A keyhole saw, a specialty saw, is narrow and slender and can be used to cut keyholes in wood.

VOICES OF EXPERIENCE

One of the best sets of tools that a fire fighter can grab is the irons: the flathead axe and halligan tool strapped together. These tools are married together using a heavy duty Velcro or rubber strap. They are connected by placing the fork end of the Halligan onto the head of the axe between the tip and the handle. This enables the two tools to be carried with one hand by the fire fighter. Working together, the irons can make quick work of forcing open conventional doors.

Using the irons inside a structure is beneficial as well. For example, when entry is made into the structure for fire suppression or search and rescue activities, remember to bring the irons with you. When conducting search and rescue operations, the flat-head axe is a good tool to assist in searching for victims.

There is great emphasis put on forcible entry, but forcible exit must also be considered. Always ask the question, "If things go bad, what can I carry in that will allow me to breach a wall and get out?" The irons could be the tool that saves your life by allowing you to breach a wall and exit the structure.

Joseph Ramsey
Winston-Salem Fire Department
Winston-Salem, North Carolina

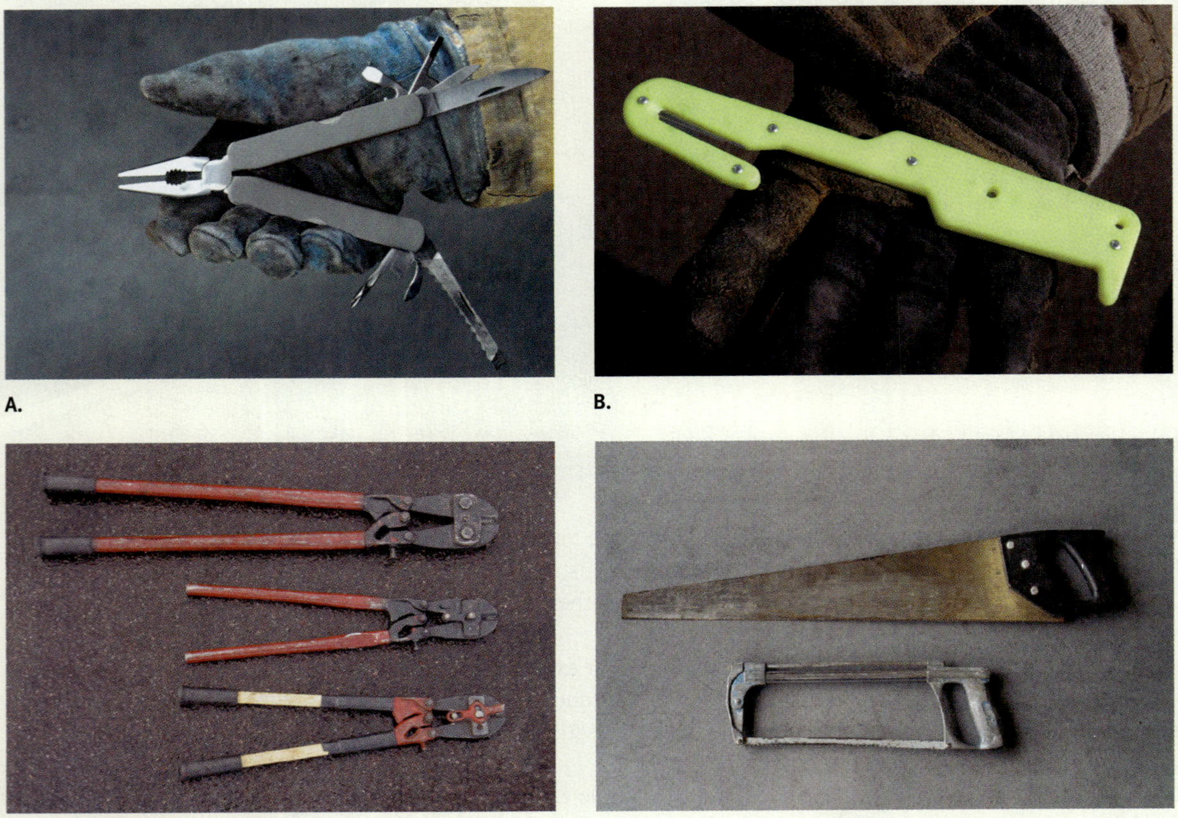

A. B. C. D.

FIGURE 9-24 Cutting tools. **A.** Combination tool. **B.** Seat belt cutter. **C.** Bolt cutters. **D.** Handsaws.

TABLE 9-7 Cutting Tools

- Axes
- Bolt cutters
- Chainsaws
- Cutting torches
- Hacksaws
- Handsaws
- Hydraulic shears
- Reciprocating saws
- Rotary saws
- Seat belt cutters

FIGURE 9-25 A fire fighter using a hacksaw to cut through metal.

Three primary types of mechanical saws are chainsaws, rotary saws, and reciprocating saws. Although handsaws have a valuable role in firefighting operations, power saws have the advantage of accomplishing more work in a shorter period of time. They also enable fire fighters to conserve energy, resulting in less fatigue. Because mechanical saws are powerful, they must be used only by trained operators. Nevertheless, they offer some distinct disadvantages relative to handsaws. Specifically, power saws are heavy to carry and sometimes can be difficult to start. They may also require an electrical connection, although cordless models are becoming more widely available.

Most people are familiar with the gasoline-powered or electric chainsaws commonly used to cut wood, particularly trees. Fire fighters often use saws with special chains to cut ventilation openings in roofs constructed of wood, metal, tar, gravel, or insulating materials **FIGURE 9-26**.

Rotary saws may be powered by either electric motors or gasoline engines **FIGURE 9-27**. In some rotary saws, the cutting part of the saw is a round metal blade with teeth, with different blades being used depending on the type of material being cut. Other rotary saws use a flat, abrasive disk for cutting. The disks are made of composite materials and are designed to wear down as they are used. It is important to match the appropriate saw blade or saw disk to the material being cut.

Reciprocating saws are powered by either an electric motor or a battery motor that rapidly pulls a saw blade back and forth **FIGURE 9-28**. As with rotary saws, different blades are

FIGURE 9-26 A fire fighter can use a chainsaw to cut through a roof.

FIGURE 9-27 A fire fighter using a rotary saw.

used to cut different materials. Reciprocating saws are most commonly used to cut metal during extrication of a victim from a motor vehicle.

FIRE FIGHTER Tips

Ventilation saws have a depth gauge on the bar and are specifically designed for roof ventilation.

The cutting tools that require the most training are hydraulic shears and cutting torches. Hydraulic shears are often used along with hydraulic spreaders and rams in extrication of victims from automobiles; the same hydraulic power source can be used with all three types of tools. Hydraulic shears can cut quickly through metal posts and bars.

Cutting torches produce an extremely high-temperature flame and are capable of heating steel until it melts, thereby cutting through an object FIGURE 9-29. Because these torches

FIGURE 9-29 A cutting torch can be used to cut through a metal door.

A.

B.

FIGURE 9-28 **A.** The components of a reciprocating saw. **B.** A reciprocating saw being used during the extrication of a victim from a vehicle.

produce such high temperatures (5700°F [3148°C]), operators must be specially trained before using them. Cutting torches are sometimes used for rescue situations and for cutting through heavy steel objects. One drawback of this equipment is that it cannot be used in situations where flammable fuels are present.

Multiple-Function Tools

Certain tools are designed to perform multiple functions, thereby reducing the total number of tools needed to achieve a goal. For example, a flat-head axe can be used as either a cutting tool or a striking tool. Some combination tools can be used to cut, to pry, to strike, and to turn off utilities.

Special-Use Tools

Some fire situations require special-use tools that perform other functions. For example, fire departments located in areas where brush and ground fires occur frequently may need to carry rakes, brooms, shovels, and combination tools that can be used for raking, chopping, cutting, and leaf blowing. These tools are described in the Wildland and Ground Fires chapter.

Rescue squads may also use specialized equipment such as jacks and air bags for lifting heavy objects, come alongs (small, hand-operated winches) for lifting or moving heavy objects or bending objects, and tripods FIGURE 9-30. You can learn more about the proper use of this special equipment by taking special rescue courses or during in-service training.

FIGURE 9-30 Heavy-duty air bags can be used to lift vehicles in rescue situations.

Phases of Use

The process of extinguishing a fire usually involves a sequence of steps or stages. Each phase of a fire-ground operation may require the use of certain types of tools and equipment.

The basic steps of fire suppression are summarized here:

- Response/size-up: This phase begins when the emergency call is received and continues as the units travel to the incident scene. The last part of this phase involves the initial observation and evaluation of factors used to determine which strategy and tactics will be employed.
- Forcible entry: This phase begins when entry to buildings, vehicles, aircraft, or other confined areas is locked or blocked, requiring fire fighters to use special techniques to gain access.
- Interior attack: During this phase, a team of fire fighters is assigned to enter the fire structure, locate the fire, and extinguish the fire.
- Search and rescue: This phase involves searching for any victims trapped by the fire and extricating them from the building.
- Rapid intervention: During this phase, a rapid intervention company/crew (RIC) provides immediate assistance to injured or trapped fire fighters.
- Ventilation: This step involves changing air within a compartment by natural or mechanical means.
- Overhaul: The final phase is to ensure that all hidden fires are extinguished after the main fire has been suppressed.

Response/Size-Up

The response and size-up phase enables you to anticipate emergency situations. At this time, you should consider the information from the dispatcher along with preincident plan information about the location. This information can provide you with an idea of the nature and possible gravity of the situation, as well as the types of problems that might arise during firefighting operations. For example, an automobile fire on the highway will present different problems and require different tools than a call for smoke coming from a single-family house. A different thinking process occurs when you are dispatched at midnight to a house fire that may have trapped a family inside than when you respond to a report of a kitchen stove fire at suppertime. Even though information about the incident is limited at this point, the response and size-up phase is the time to start thinking about the types of tools and equipment that you might need.

Most fire departments have SOPs or SOGs that specify the tools and equipment required for different types of fires. Each crew member is expected to bring specific tools and equipment from the apparatus. These requirements take into account the roles that different units have within the fire department. A member of an engine company, for example, will not bring the same tools that a member of a truck company will bring. Likewise, the tools carried for an interior fire attack differ from those carried for an outside or defensive attack.

Upon arrival at the scene, the company officers will size up the situation and develop the action plans for each company, following SOPs and SOGs. As a crew member, you are responsible for following the directions of your company officer.

Forcible Entry

Gaining entrance to a locked building or structure can present a challenge to even the most seasoned fire fighter. Buildings are often equipped with security devices designed to keep unwanted people out, but these same devices can make it very difficult for fire fighters to gain access to the building. Forcible entry is the process of entering a building by overcoming these barriers. The skills involved in this phase are discussed in depth in the Forcible Entry chapter.

Several types of tools can be used for forcible entry, including an axe, a prying tool, or a K tool. A flat-head axe and a Halligan tool, collectively called the irons, are used in combination to pry open a door, although they may permanently damage both the door and the frame. Prying tools used for gaining access include pry bars, crowbars, Halligan tools, Hux bars, and the hydraulic-powered rabbet tool.

A K tool can be used to pull out a cylinder lock mounted in a wood or heavy metal door, so that the lock can be released FIGURE 9-31. This comparatively nondestructive process leaves the door and most of the locking mechanism undamaged. The building owner can then have the lock cylinder replaced at a relatively low cost.

A variety of striking tools can be used for forcible entry when brute force is needed to break into a building. These items include flat-head axes, hammers, sledgehammers, and battering rams.

Sometimes the easiest—or only—way to gain access is to use cutting tools. An axe can be used to cut out a door panel. A power saw can be used to cut through a wood wall. Bolt cutters can be used to remove a padlock. Cutting torches or rotary power saws can be used to cut through metal security bars or the hinges on metal security doors. Although cutting is a destructive process, it is justified when required to save lives or property.

FIGURE 9-31 Use a K tool to pull out and release a cylinder lock.

Many techniques and types of tools can be used to gain entry into secured structures TABLE 9-8. The exact tool needed will depend on the method of entry and the type of obstacle. Because having greater experience usually suggests the best way to gain entry in each situation, rely on the orders and advice of your officer and fellow fire fighters. More techniques used for forcible entry are described in the Forcible Entry chapter.

Interior Firefighting Tools and Equipment

The process of fighting a fire inside a building involves several tasks that are usually performed simultaneously or in rapid succession by teams of fire fighters. While one crew is advancing a hose line to attack the fire, another crew may be searching for occupants and a third crew may be performing ventilation tasks. This is referred to as combined operations (combined ops). An additional company may be standing by as a RIC, in case a fire fighter needs to be rescued.

Some basic tools and equipment should be carried by every crew working inside a burning building. Crews may also carry specialized tools and equipment needed for their particular assignment. The basic tools enable them to solve problems they may encounter while performing interior operations. For example, crew members may encounter obstacles such as locked doors, or they may need to open an emergency escape route. They may need to establish horizontal ventilation by forcing, opening, or breaking a window. They may have to gain access to the space above the ceiling by using a pike pole or making a hole in a wall or floor with an axe. A powerful light is also important, because smoke can quickly reduce interior visibility to just a few inches.

The basic set of tools for interior firefighting includes the following items:

- A prying tool, such as a Halligan tool
- A striking tool, such as a flat-head axe or a sledgehammer

TABLE 9-8 Forcible Entry Tools

Type of Tool	Use in Forcible Entry
Prying tools	Prying tools can include a Halligan tool, a flat bar, and a crowbar, among other options. They can be used to break windows or force open doors.
Axe, flat-head axe, pick-head axe	An axe can be used to cut through a door or break a window. A flat-head axe can be used in conjunction with a Halligan tool to force open a door.
Sledgehammer	A sledgehammer can be used to breach walls or break a window. It can also be used in conjunction with other tools, such as a Halligan tool, to force open a door or break off a padlock.
Hammer, mallet	These tools may be used in conjunction with other tools such as chisels or punches to force entry through windows or doors.
Chisel, punch	These tools can be used to make small openings through doors or windows.
K tool	The K tool provides a "through the lock" method of opening a door, thereby minimizing damage to the door.
Rotary saw, chainsaw, reciprocating saw	Power saws can cut openings through obstacles including doors, walls, fences, gates, security bars, and other barriers.
Bolt cutter	A bolt cutter can be used to cut off padlocks or cut through obstacles such as fences.
Battering ram	A battering ram can be used to breach walls.
Hydraulic door opener	A hydraulic door opener can be used to force open a door.
Hydraulic rescue tool	A hydraulic rescue tool can be used in a variety of situations to break, breach, or force openings in doors, windows, walls, fences, or gates.

- A cutting tool, such as an axe
- A pushing/pulling tool, such as a pike pole
- A hand light or portable light

The specific tools that must be carried by each crew are usually defined in the fire department's training manuals and SOPs. These requirements are based on local conditions and preferences and may differ depending on the type of company and the assignment.

The interior attack team is responsible for advancing a hose line, finding the fire, and applying water to extinguish the flames. The members of this team need the basic tools that will enable them to reach the seat of the fire.

FIRE FIGHTER II

Search and Rescue Tools and Equipment

Search and rescue needs to be carried out quickly, shortly after arrival on the fire ground. A search team should carry the same basic hand tools as the interior attack team:

- Pushing tool (short pike pole)
- Prying tool (Halligan tool)
- Striking tool (sledgehammer or flat-head axe)
- Cutting tool (axe)
- Portable hand light

In addition to being equipped for forcible entry and emergency exit, a search and rescue team may use tools to probe under beds for unconscious victims. A short pike pole or closet hook is relatively light and may reduce the time needed to search an area by extending the fire fighter's reach. An axe handle can also be used for this purpose.

Other types of tools used for search and rescue include thermal imaging devices, portable lights, and life lines FIGURE 9-32, FIGURE 9-33, and FIGURE 9-34. The techniques of search and rescue are discussed in the Search and Rescue chapter.

FIGURE 9-32 A thermal imaging device.

FIGURE 9-33 A fire fighter using a thermal imaging device.

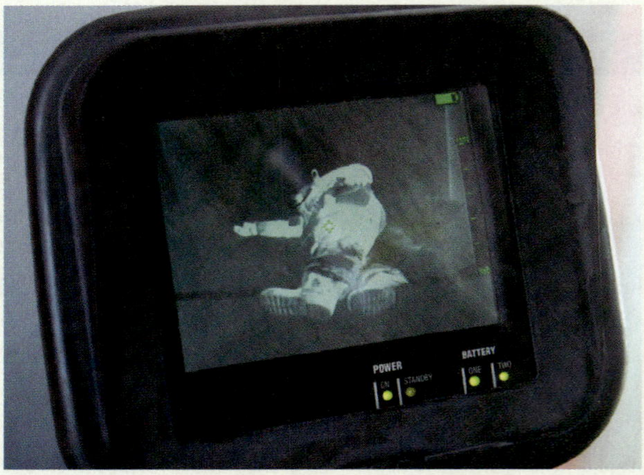

FIGURE 9-34 An image produced by a thermal imaging device.

Rapid Intervention Tools and Equipment

The RIC is designated to stand by to provide immediate assistance to any fire fighters who become lost, trapped, or injured during an incident or training exercise. Members of this team should carry the standard set of tools for interior firefighting as well as extra tools and equipment that are particularly important for search and rescue tasks. The extra tools and equipment should help the RIC find and gain access to a fire fighter who is in trouble, extricate a fire fighter who is trapped under debris, provide breathing air for a fire fighter who has experienced an SCBA failure or run out of air, and remove an injured or unconscious fire fighter from the building. All of this equipment should be gathered and staged with the RIC, so that it will be immediately available if needed.

An RIC should carry the following special equipment:

- Thermal imaging device
- Additional portable lighting
- Life lines
- Prying tools
- Striking tools
- Cutting tools, including a power saw
- SCBA and spare air cylinders with Rapid Intervention Team universal air connection
- Litter or patient packaging device

Ventilation Tools and Equipment

The objective of ventilation is to provide openings so that fresh air can enter a burning structure and hot gases and products of combustion can escape from the building. See the Ventilation chapter for more detailed information on ventilation operations. Many of the same tools used for forcible entry are also applied to provide ventilation. For example, power saws and axes are commonly used to cut through roofs and vent combustion by-products.

Fans are used to remove smoke from a building or to introduce fresh air into a structure. With positive-pressure ventilation, fresh air is blown *into* a building through selected openings to force contaminated air out through other openings. This kind of ventilation allows fire fighters to work quickly, safely, and effectively to locate and suppress the fire.

Negative-pressure ventilation uses fans placed at selected openings to draw contaminated air *out of* a building. This technique is used when there are no suitable openings for positive-pressure ventilation or when introducing pressurized air could accelerate the fire. Negative-pressure ventilation may also be the best option when positive-pressure ventilation might potentially spread the products of combustion.

Ventilation fans can be powered either by electric or gasoline motors or by water pressure FIGURE 9-35. A gasoline-powered fan may not be suitable in some situations, because it can introduce carbon monoxide into a structure if an exhaust hose is not available. If the building atmosphere contains potentially dangerous levels of flammable gases, a water-powered fan may be the best choice.

Horizontal ventilation usually involves opening outer doors and windows to allow fresh air to enter and to remove contaminated air. When ventilation fans are used, they are placed at these openings. Positive-pressure fans blow in fresh air, whereas negative-pressure fans draw out smoke and contaminants. To provide for adequate ventilation, unlocked or easily released windows and doors should be opened normally. Locked or jammed windows and doors may have to be broken or forced open using basic interior firefighting tools.

It may also be necessary to create interior openings within the building so that contaminated air can reach the exterior openings. Opening interior doors is the easiest, quickest, and often most effective way to ventilate interior spaces. A series of electrically powered fans may be placed throughout a large structure to help move smoke in the desired direction.

In vertical ventilation, openings in the roof or the highest part of a building allow smoke and hot gases to escape.

FIGURE 9-35 Different types of fans used in ventilation.

Whenever possible, existing openings such as doors, windows, roof hatches, and skylights should be used for vertical ventilation. It may be necessary to force them open or to break them using forcible entry tools.

In some circumstances, it may be necessary to cut through a roof to make an effective vertical ventilation opening. Cutting tools such as axes and power saws are used for this purpose. Pike poles will also be needed to pull down ceilings after the roof covering is opened.

The following special equipment is needed for ventilation:

- Positive-pressure fans
- Negative-pressure (exhaust) fans
- Pulling and pushing tools (long pike poles)
- Cutting tools (power saws and axes)

Overhaul Tools and Equipment

The purpose of overhaul is to examine the fire scene carefully and ensure that all hidden fires are extinguished. During this phase, burned debris must be removed and potential hot spots in enclosed spaces behind walls, above ceilings, and under floors exposed. Both tasks can be accomplished using simple hand tools.

As part of overhaul, you need a tool that can shut off activated sprinkler heads so as to avoid excess water damage. In buildings equipped with sprinkler systems, fire fighters should shut off any sprinkler heads that have been activated by the fire to reduce water damage and to restore the fire protection system in the building. Whenever possible, a sprinkler system should be restored to service. A single sprinkler head can be shut off with a wooden wedge—a technique that leaves the sprinkler system with an inoperable sprinkler head when it is turned on. A sprinkler head can also be shut off with a special tool that contains a fusible link. This shut-off tool allows an activated sprinkler head to be placed back in service until a replacement head can be installed.

Pike poles are commonly used for pulling down ceilings and opening holes in walls. Axes, and sometimes power saws, are used to open walls and floors. Prying and striking tools are also used to open closed spaces. Shovels, brooms, rakes, and buckets are used to clear away debris. Water vacuums are used to remove water from buildings.

The widespread introduction of infrared thermal imaging devices has made it possible to "see" hot spots behind walls without physically cutting into them FIGURE 9-36. This technology has reduced the risk of missing dangerous hot spots as well as curtailed the amount of time and effort it takes to overhaul a fire scene. The process of overhauling a fire scene and the tools used to accomplish this task are covered more fully in the Salvage and Overhaul chapter.

The following tools are used during overhaul operations:

- Pushing tools (pike poles of varying lengths)
- Prying tools (Halligan tool)
- Striking tools (sledgehammer, flat-head axe, hammer, mallet)
- Cutting tools (axes, power saws)
- Debris-removal tools (shovels, brooms, rakes, buckets, carryalls)
- Water-removal equipment (water vacuums)
- Ventilation equipment (electric-, gas-, or water-powered fans)

FIGURE 9-36 Thermal imaging devices can see hot spots.

- Portable lighting
- Sprinkler shut-off tools
- Thermal imaging devices

Fire Fighter Safety Tips

There is a valuable saying in the fire service: "Hand tools—never leave your apparatus without them."

Fire Fighter Safety Tips

Keep the manufacturers' manuals for all of the department's tools and equipment in a safe and easily accessible location, and refer to them whenever you have any questions about that device.

FIRE FIGHTER II

Tool Staging

Many fire departments have SOPs for staging necessary equipment nearby during a fire. This procedure often involves placing a tarp or salvage cover on the ground at a designated location and laying out commonly used tools and equipment where they can be accessed readily. A similar procedure may be used for rescue operations, where tools that are likely to be needed can be laid out, ready for use. This kind of tool staging saves valuable time because fire fighters do not have to return to their own apparatus or search several different vehicles to find a particular tool.

A department's SOPs usually specify the types of tools and equipment to be staged. The tool staging area can be located outside the building or, in the case of a high-rise or large building, at a convenient location inside the structure. Additional personnel may be directed to bring particular items to the tool staging location at working fires.

Maintenance

All tools and equipment must be properly maintained so that they will be ready for use when they are needed. That means you must keep equipment clean and free from rust, keep cutting blades sharpened, and keep fuel tanks filled. Every tool and piece of equipment must be ready for use before you respond to an emergency incident. Follow the manufacturer's instructions for cleaning and maintaining each piece of equipment. In addition, use equipment only for its intended purposes. For example, a pike pole is made for pushing and pulling; it is not a lever and will break if used inappropriately. Use the right tool for the job.

Cleaning and Inspecting Hand Tools

All hand tools should be completely cleaned and inspected and their conditions documented after each use TABLE 9-9. Remove all dirt and debris from the tools. If appropriate, use water streams to remove the debris and mild soap to clean the equipment thoroughly. Learn how to safely use the cleaning solutions that your department and the manufacturer specify for cleaning tools and equipment.

To prevent rust, metal tools must be dried completely, either by hand or by air, before being returned to the apparatus. Coating unpainted metal surfaces with a light film of oil will help prevent rusting.

Cutting tools should be sharpened after each use. Before any tool is placed back into service, it should be inspected for damage.

Avoid painting tools, because paint may hide defects or visible damage. Keep the number of markings on a tool to a minimum.

Wooden and fiberglass handles should be inspected for cracks, splinters, gouges, and other defects. Repair or replace any damaged handle.

FIRE FIGHTER II

Cleaning and Inspecting Power Equipment and Tools

Power equipment and tools are used for lighting, ventilation, salvage, and overhaul. Use power equipment and tools only after you have received training in their use. Read and heed the instructions supplied with the equipment.

TABLE 9-9 Cleaning and Inspecting Hand Tools

Metal parts	All metal parts should be clean and dry. Remove rust with steel wool. Do not oil the striking surface of metal tools, as this treatment may cause them to slip.
Wood handles	Inspect for damage such as cracks and splinters. Sand the handle if necessary. Do not paint or varnish the handle; instead, apply a coat of boiled linseed oil. Check that the tool head is tightly fixed to the handle.
Fiberglass handles	Clean with soap and water. Inspect for damage. Check that the tool head is tightly fixed to the handle.
Cutting edges	Inspect for nicks or other damage. File and sharpen as needed. Power grinding may weaken some tools, so hand sharpening may be required.

FIRE FIGHTER II

Test power equipment and tools frequently, and have these items serviced regularly by a qualified shop. Keep records of all inspections and maintenance performed on power equipment and tools. In addition, preventive maintenance will help ensure that equipment and tools operate properly when needed. If a power tool or equipment is defective, report this fact to your officer.

Fill each tool with the proper fuel, remembering that some tools operate on gasoline whereas others use various types of gasoline and oil mixtures. Many small engines operate on gasoline, for example, but others are designed to burn diesel fuel. Fuels have a limited storage life, especially fuels containing ethanol. If the fuel is not used in a certain period of time, it may be necessary to drain and refill equipment with fresh fuel. Check the manufacturer's recommendations before fueling a particular piece of equipment.

After returning from a fire, clean, inspect, and record maintenance data for all of your overhaul tools to ensure that they are in a "ready state."

Steps for Cleaning and Inspecting Power Tools

All power equipment should be left in a "ready state" for immediate use at the next incident:

- All debris should be removed, and the tool should be clean and dry.

FIRE FIGHTER II

- All fuel tanks should be filled completely with fresh fuel.
- Any dull or damaged blades and chains should be replaced.
- Belts should be inspected to ensure they are tight and undamaged.
- All guards should be securely in place.
- All hydraulic hoses should be cleaned and inspected.
- All power cords should be inspected for damage.
- All hose fittings should be cleaned, inspected, and tested to ensure a tight fit.
- Tools should be started to ensure that they operate properly.
- Tanks on water vacuums should be emptied, washed, cleaned, and dried.
- Hoses and nozzles on water vacuums should be cleaned and dried.

It is essential to read the manufacturer-provided manuals and follow all instructions on the care and inspection of power tools and equipment. Keep all manufacturers' manuals in a safe and easily accessible location, and refer to these documents when cleaning and inspecting the tools and equipment.

Learn the proper procedure for reporting a problem with a power tool and taking it out of service. Remember—your safety depends on the quality of your tools and equipment.

Near Miss REPORT

Report Number: 06-0000499

Synopsis: Hydraulic ram fails during field check.

Event Description: There was a dramatic failure of a long [brand deleted] hydraulic ram at a station today during routine equipment exercising. The weld down the spine of the ram split open, which caused hydraulic fluid to spray far and wide from a longitudinal crack. The excursion of the ram's outer housing caused the heads to shear off the two Allen head machine bolts holding the bracket for the front handgrip of the ram. The fire fighter involved was wearing full PPE and eye protection. He was uninjured, and his person was not contaminated.

Risk management paperwork was completed and submitted by the lieutenant for the damaged gear and ram. The fire fighter turned in that gear and received replacement turnouts on the same day from our warehouse. Digital photographs have been taken of the mechanical equipment by a deputy fire marshal. The deputy chief will receive copies as well as the members of the safety committee made up of management and union members. The logistics utility worker was instructed to take the ram and the new power plant back to the vendor for evaluation.

My observation was that this ram looked very old—almost all of its paint was worn off of it. I never even contemplated the fact that these housings are not one milled piece. They appear to be two halves welded together; the weld is then ground down flush and the finish applied to give a uniform appearance. None I'd ever seen before looked like this one. We have dozens of pieces of such equipment of different ages and states of wear.

Lessons Learned: All personnel should be reminded to always wear turnout PPE whenever operating hydraulic power equipment under all circumstances. Eye protection is paramount. Routinely, all of our hydraulic power tools should be carefully looked over before operation. Obviously worn equipment should be removed from service and replacements sought through appropriate channels. Command staff as well as rank and file personnel should be sensitive to this issue.

Wrap-Up

Chief Concepts

- Tools and equipment are used in almost all fire suppression and rescue operations.
- Hand tools are used to extend or multiply the actions of your body and to increase your effectiveness in performing specific functions.
- Power tools and equipment use an external source of power and are faster and more efficient than hand tools.
- Always wear PPE when using tools or equipment.
- Most tools fit into the following functional categories:
  - Rotating tools—Apply a rotational force to make something turn and used to assemble and disassemble items. Rotating tools include screwdrivers, wrenches, and pliers.
  - Prying or spreading tools—Used to pry and spread. May be as simple as a pry bar or as mechanically complex as a hydraulic spreader. Prying and spreading tools include claw bar, crowbar, and Halligan tool.
  - Pushing or pulling tools—Extend your reach and increase the power you can exert upon an object. Pushing and pulling tools include pike poles, closet hooks, K tools, and ceiling hooks.
  - Striking tools—Used to apply an impact force to an object. Striking tools include hammers, mallets, axes, and spring-loaded center punches.
  - Cutting tools—Tools with sharp edges that sever objects. Cutting tools include bolt cutters, chainsaws, cutting torches, and handsaws.
  - Multiple-use tools—Designed to perform multiple functions, thereby reducing the total number of tools needed to achieve a goal. Some combination tools can be used to cut, to pry, to strike, and turn off utilities.
- Forcible entry tools are used to gain entrance to a locked building:
  - Prying tools—Used to break windows or force open doors.
  - Axes—Used to cut through a door or break a window.
  - Sledgehammer—Used to breach walls or break a window.
  - Hammer or mallet—May be used with other tools such as chisels or punches to force entry through windows or doors.
  - Chisel or punch—Used to make small openings through doors or windows.
  - K tool—Provides a "through the lock" method of opening a door, minimizing damage to the door.
  - Saws—Can cut openings through obstacles including doors, walls, fences, gates, security bars, and other barriers.
  - Bolt cutter—Used to cut off padlocks or cut through obstacles such as fences.
  - Battering ram—Used to breach walls.
  - Hydraulic door opener—Used to force open a door.
  - Hydraulic rescue tool—Used in a variety of situations to break, breach, or force openings in doors, windows, walls, fences, or gates.
- The interior attack team is responsible for advancing a hose line, finding the fire, and applying water to extinguish the flames. The members of this team need the basic tools that will allow them to reach the seat of the fire:
  - A prying tool, such as a Halligan tool
  - A striking tool, such as a flat-head axe or a sledgehammer
  - A cutting tool, such as an axe
  - A pushing/pulling tool, such as a pike pole
  - A strong hand light or portable light
- A search and rescue team should carry the same basic hand tools as the interior attack team, as well as a short pike pole, thermal imaging devices, portable lights, and life lines.
- An RIC should carry the following special equipment:
  - Thermal imaging device
  - Additional portable lighting
  - Life lines
  - Prying tools
  - Striking tools
  - Cutting tools, including a power saw
  - SCBA or spare air cylinders with Rapid Intervention Team universal air connection
- The objective of ventilation is to provide openings so that fresh air can enter a burning structure and hot gases and products of combustion can escape from the building. The following special equipment is needed for ventilation:
  - Positive-pressure fans
  - Negative-pressure (exhaust) fans
  - Pulling and pushing tools (long pike poles)
  - Cutting tools (power saws and axes)
- The purpose of overhaul is to examine the fire scene carefully and ensure that all hidden fires are extinguished.
- The following tools are used during overhaul operations:
  - Pushing tools (pike poles of varying lengths)
  - Prying tools (Halligan tool)
  - Striking tools (sledgehammer, flat-head axe, hammer, mallet)
  - Cutting tools (axes, power saws)
  - Debris-removal tools (shovels, brooms, rakes, buckets, carryalls)
  - Water-removal equipment (water vacuums)
  - Ventilation equipment (electric-, gas-, or water-powered fans)
  - Portable lighting
  - Sprinkler shut-off tools
  - Thermal imaging devices

- Tool staging often involves placing a tarp or salvage cover on the ground at a designated location and laying out commonly used tools and equipment so that they can be readily accessed.
- All tools and equipment must be properly maintained so that they will be ready for use when they are needed. That means you must keep equipment clean and free from rust, keep cutting blades sharpened, and keep fuel tanks filled.
- Test power equipment and tools frequently, and have them serviced regularly.
- Read and follow the manufacturer's manuals, and follow all instructions on the care and inspection of power tools and equipment.

Hot Terms

Battering ram A tool made of hardened steel with handles on the sides used to force doors and to breach walls. Larger versions may be used by as many as four people; smaller versions are made for one or two people.

Bolt cutter A cutting tool used to cut through thick metal objects such as bolts, locks, and wire fences.

Box-end wrench A hand tool used to tighten or loosen bolts. The end is enclosed, as opposed to an open-end wrench. Each wrench is a specific size, and most have ratchets for easier use.

Carpenter's handsaw A saw designed for cutting wood.

Ceiling hook A tool with a long wooden or fiberglass pole that has a metal point with a spur at right angles at one end. It can be used to probe ceilings and pull down plaster lath material.

Chainsaw A power saw that uses the rotating movement of a chain equipped with sharpened cutting edges. It is typically used to cut through wood.

Chisel A metal tool with one sharpened end that is used to break apart material in conjunction with a hammer, mallet, or sledgehammer.

Claw bar A tool with a pointed claw-hook on one end and a forked- or flat-chisel pry on the other end. It is often used for forcible entry.

Clemens hook A multipurpose tool that can be used for several forcible entry and ventilation applications because of its unique head design.

Closet hook A type of pike pole intended for use in tight spaces, commonly 2 to 4 feet (0.6 to 1 meter) in length.

Come along A hand-operated tool used for dragging or lifting heavy objects that uses pulleys and cables or chains to multiply a pulling or lifting force.

Coping saw A saw designed to cut curves in wood.

Coupling One set or pair of connection devices attached to a fire hose that allow the hose to be interconnected to additional lengths of hose or adapters and other fire-fighting appliances. (NFPA 1963)

Crowbar A straight bar made of steel or iron with a forked-like chisel on the working end that is suitable for performing forcible entry.

Cutting torch A torch that produces a high-temperature flame capable of heating metal to its melting point, thereby cutting through an object. Because of the high temperatures (5700°F [3148°C]) that these torches produce, the operator must be specially trained before using this tool.

Drywall hook A specialized version of a pike pole that can remove drywall more effectively because of its hook design.

Flat bar A specialized type of prying tool made of flat steel with prying ends suitable for performing forcible entry.

Flat-head axe A tool that has a head with an axe on one side and a flat head on the opposite side.

Forcible entry Techniques used by fire personnel to gain entry into buildings, vehicles, aircraft, or other areas of confinement when normal means of entry are locked or blocked. (NFPA 402)

Gripping pliers A hand tool with a pincer-like working end that can be used to bend wire or hold smaller objects.

Hacksaw A cutting tool designed for use on metal. Different blades can be used for cutting different types of metal.

Halligan tool A prying tool that incorporates a pick and a fork, specifically designed for use in the fire service.

Hammer A striking tool.

Hand light A small, portable light carried by fire fighters to improve visibility at emergency scenes. It is often powered by rechargeable batteries.

Handsaw A manually powered saw designed to cut different types of materials. Examples include hacksaws, carpenter's handsaws, keyhole saws, and coping saws.

Hux bar A multipurpose tool that can be used for several forcible entry and ventilation applications because of its unique design. It may also be used as a hydrant wrench.

Hydrant wrench A hand tool that is used to operate the valves on a hydrant; it may also be used as a spanner wrench. Some models are plain wrenches, whereas others have a ratchet feature.

Hydraulic shears A lightweight, hand-operated tool that can produce up to 10,000 pounds (4,500 kilogram) of cutting force.

Hydraulic spreader A lightweight, hand-operated tool that can produce up to 10,000 pounds (4,500 kilogram) of prying and spreading force.

Interior attack The assignment of a team of fire fighters to enter a structure and attempt fire suppression.

Irons A combination of tools, usually consisting of a Halligan tool and a flat-head axe, that are commonly used for forcible entry.

Kelly tool A steel bar with two main features: a large pick and a large chisel or fork.

Keyhole saw A saw designed to cut circles in wood for keyholes.

K tool A tool that is used to remove lock cylinders from structural doors so the locking mechanism can be unlocked.

Life line A rope secured to a fire fighter that enables the fire fighter to retrace his or her steps out of a structure.

Mallet A short-handled hammer.

Maul A specialized striking tool, weighing 6 pounds (3 kilogram) or more, with an axe on one end and a sledgehammer on the other end.

Mechanical saw A saw that is usually powered by an electric motor or a gasoline engine. The three primary types of mechanical saws are chainsaws, rotary saws, and reciprocating saws.

Multipurpose hook A long pole with a wooden or fiberglass handle and a metal hook on one end used for pulling.

Open-end wrench A hand tool that is used to tighten or loosen bolts. The end is open, as opposed to a box-end wrench. Each wrench is a specific size.

Overhaul The process of final extinguishment after the main body of a fire has been knocked down. All traces of fire must be extinguished at this time. (NFPA 402, 2002)

Pick-head axe A tool that has a head with an axe on one side and a pointed end ("pick") on the opposite side.

Pike pole A pole with a sharp point ("pike") on one end coupled with a hook. It is used to make openings in ceilings and walls. Pike poles are manufactured in different lengths for use in rooms of different heights.

Pipe wrench A wrench having one fixed grip and one movable grip that can be adjusted to fit securely around pipes and other tubular objects.

Plaster hook A long pole with a pointed head and two retractable cutting blades on the side.

Pry bar A specialized prying tool made of a hardened steel rod with a tapered end that can be inserted into a small area.

Rabbet tool A hydraulic spreading tool designed to pry open doors that swing inward.

Rapid intervention company/crew (RIC) A minimum of two fully equipped personnel on site, in a ready state, who are assigned to immediate rescue of injured or trapped fire fighters. Also called a rapid intervention team (RIT).

Reciprocating saw A saw that is powered by an electric motor or a battery motor, and whose blade moves back and forth.

Response Immediate and ongoing activities, tasks, programs, and systems to manage the effects of an incident that threatens life, property, operations, or the environment. (NFPA 1600)

Roofman's hook A long pole with a solid metal hook used for pulling.

Rotary saw A saw that is powered by an electric motor or a gasoline engine, and that uses a large rotating blade to cut through material. The blades can be changed depending on the material being cut.

San Francisco hook A multipurpose tool that can be used for several forcible entry and ventilation applications because of its unique design, which includes a built-in gas shut-off and directional slot.

Screwdriver A tool used for turning screws.

Search and rescue The process of searching a building for a victim and extricating the victim from the building.

Seat belt cutter A specialized cutting device that cuts through seat belts.

Size-up The observation and evaluation of existing factors that are used to develop objectives, strategy, and tactics for fire suppression. (NFPA 1051)

Sledgehammer A long, heavy hammer that requires the use of both hands.

Socket wrench A wrench that fits over a nut or bolt and uses the ratchet action of an attached handle to tighten or loosen the nut or bolt.

Spanner wrench A type of tool used to couple or uncouple hoses by turning the rocker lugs on the connections.

Spring-loaded center punch A spring-loaded punch used to break automobile glass.

Thermal imaging device An electronic device that detects differences in temperature based on infrared energy and then generates images based on those data. It is commonly used in smoke-filled environments to locate victims as well as to search for hidden fire during size-up and overhaul.

Ventilation The changing of air within a compartment by natural or mechanical means. Ventilation can be achieved by introduction of fresh air to dilute contaminated air or by local exhaust of contaminated air. (NFPA 302, 2004)

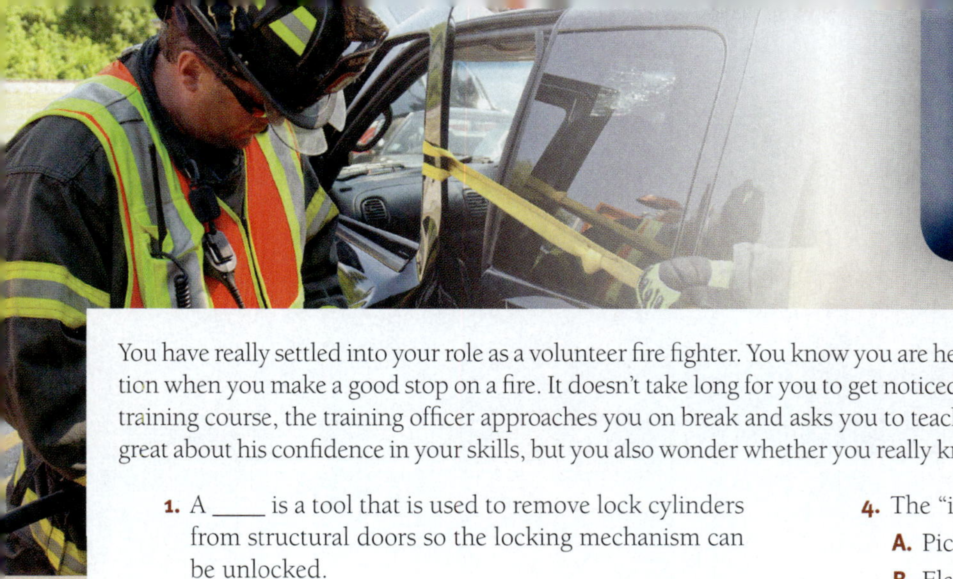

FIRE FIGHTER *in action*

You have really settled into your role as a volunteer fire fighter. You know you are helping your community and you feel the satisfaction when you make a good stop on a fire. It doesn't take long for you to get noticed by others in the department. In fact, at a recent training course, the training officer approaches you on break and asks you to teach a class on basic tools and equipment. You feel great about his confidence in your skills, but you also wonder whether you really know enough about the tools to teach it to others.

1. A _____ is a tool that is used to remove lock cylinders from structural doors so the locking mechanism can be unlocked.
   A. K-tool
   B. kelly tool
   C. rabbet tool
   D. keyhole saw

2. A ___ is a prying tool that incorporates a pick and a fork, designed for use in the fire service.
   A. ceiling hook
   B. pry bar
   C. Halligan tool
   D. crowbar

3. A ___ is a tool made of hardened steel with handles on the sides used to force doors and to breach walls.
   A. battering ram
   B. Halligan tool
   C. maul
   D. mallet

4. The "irons" usually consists of what two tools?
   A. Pick-head axe and halligan
   B. Flat-head axe and halligan
   C. Flat-head axe and Clemens hook
   D. Pick-head axe and Clemens hook

5. Gripping pliers are included in which functional category?
   A. Cutting
   B. Pushing or pulling
   C. Prying or spreading
   D. Rotating

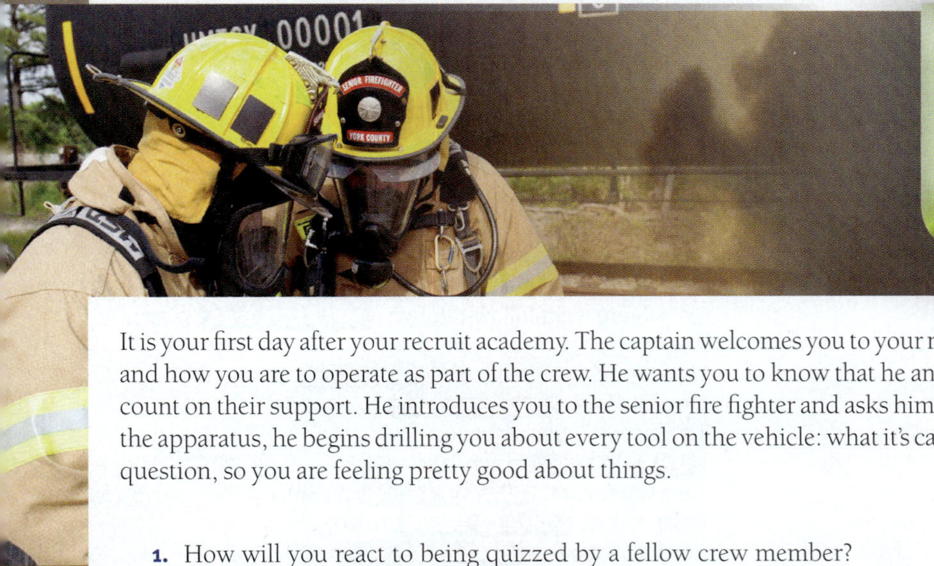

FIRE FIGHTER II *in action*

It is your first day after your recruit academy. The captain welcomes you to your new home. He goes over his expectations with you and how you are to operate as part of the crew. He wants you to know that he and the crew want you to be successful and you can count on their support. He introduces you to the senior fire fighter and asks him to show you around the station. When you get to the apparatus, he begins drilling you about every tool on the vehicle: what it's called and what it's for. You are able to answer every question, so you are feeling pretty good about things.

1. How will you react to being quizzed by a fellow crew member?
2. Will you have any questions for your fellow crew member?
3. Does this introduction to the fire station match your expectations of the first day on the job?
4. If you were unable to answer any of the questions, how would you react?

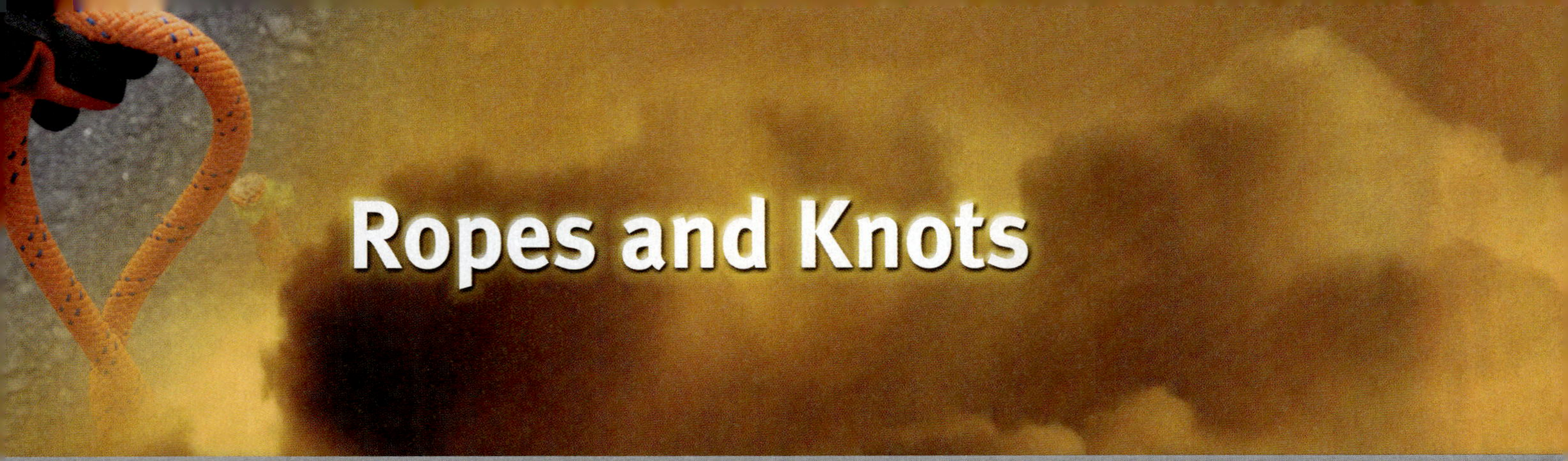

Ropes and Knots

Fire Fighter I

Knowledge Objectives

After studying this chapter, you will be able to:

- Describe the three primary types of fire service rope. (NFPA 5.3.20.A, p 260)
- List the two types of life safety rope and their minimum breaking strength. (NFPA 5.3.20.A, p 260–261)
- Describe the characteristics of escape rope. (NFPA 5.3.20.A, p 261)
- Describe the characteristics of utility ropes. (NFPA 5.3.20.A, p 261–262)
- List the advantages of synthetic fiber ropes. (NFPA 5.3.20.A, p 262–263)
- List the disadvantages of synthetic fiber ropes. (NFPA 5.3.20.A, p 263)
- List the types of synthetic fibers that are used in fire service rope. (NFPA 5.3.20.A, p 263)
- Describe how twisted ropes are constructed. (NFPA 5.3.20.A, p 263–264)
- Describe how braided ropes are constructed. (NFPA 5.3.20.A, p 264)
- Describe how kernmantle ropes are constructed. (NFPA 5.3.20.A, p 264–265)
- Explain the differences between dynamic kernmantle rope and static kernmantle rope. (NFPA 5.3.20.A, p 265)
- List the four components of the rope maintenance formula. (NFPA 5.3.20.A, p 268)
- Describe how to preserve rope strength and integrity. (NFPA 5.3.20.A, p 268–270)
- Describe how to clean rope. (NFPA5.5.1, p 268–269)
- Describe how to inspect rope. (NFPA5.5.1, p 269–270)
- Describe how to keep an accurate rope record. (NFPA5.5.1, p 270)
- Describe how to store rope properly. (NFPA5.5.1, p 270–271)
- List the terminology used to describe the parts of a rope when tying knots. (NFPA 5.3.20.A, p 271)
- List the common types of knots that are used in the fire service. (NFPA 5.3.20.A, p 272)
- Describe the characteristics of a safety knot. (NFPA 5.3.20, p 272–273)
- Describe the characteristics of a hitch. (NFPA 5.3.20, p 272–278)
- Describe the characteristics of a half hitch. (NFPA 5.3.20, p 272)
- Describe the characteristics of a clove hitch. (NFPA 5.3.20, p 272–273)
- Describe the characteristics of a figure eight knot. (NFPA 5.3.20, p 279)
- Describe the characteristics of a bowline knot. (NFPA 5.3.20, p 279)
- Describe the characteristics of a bend. (NFPA 5.3.20, p 279, 286)

Skills Objectives

After studying this chapter, you will be able to perform the following skills:

- Care for life safety ropes. (NFPA 5.5.1.A, 5.5.1.B, p 268)
- Clean fire department ropes. (NFPA 5.5.1.A, p 268–269)
- Inspect fire department ropes. (NFPA5.5.1, p 270)
- Place a life safety rope in a rope bag. (NFPA5.5.1, p 270–271)
- Tie a safety knot. (NFPA 5.1.2, 5.3.20.B, p 272–273)
- Tie a half hitch. (NFPA 5.1.2, 5.3.20.B, p 272, 274)
- Tie a clove hitch in the open. (NFPA 5.1.2, 5.3.20.B, p 273, 276–277)
- Tie a clove hitch around an object. (NFPA 5.1.2, 5.3.20.B, p 273–274, 277–278)
- Tie a figure eight knot. (NFPA 5.1.2, 5.3.20.B, p 279–280)
- Tie a figure eight on a bight. (NFPA 5.1.2, 5.3.20.B, p 279, 281)
- Tie a figure eight follow-through. (NFPA 5.1.2, 5.3.20.B, p 279, 282)
- Tie a figure eight bend. (NFPA 5.1.2, 5.3.20.B, p 279, 283)
- Tie a bowline. (NFPA 5.1.2, 5.3.20.B, p 279, 284)
- Tie a sheet or Becket bend. (NFPA 5.1.2, 5.3.20.B, p 279, 285–286)
- Tie a water knot. (NFPA 5.1.2, 5.3.20.B, p 286)
- Hoist an axe. (NFPA 5.1.2, 5.3.20.B, p 287–288)
- Hoist a pike pole. (NFPA 5.1.2, 5.3.20.B, p 288–289)
- Hoist a ladder. (NFPA 5.1.2, 5.3.20.B, p 289–290)
- Hoist a charged hose line. (NFPA 5.1.2, 5.3.20.B, p 289, 291)
- Hoist an uncharged hose line. (NFPA 5.1.2, 5.3.20.B, p 289, 292)
- Hoist an exhaust fan or power tool. (NFPA 5.1.2, 5.3.20.B, p 290–291, 293)

CHAPTER 10

Fire Fighter II FFII

Knowledge Objectives

After studying this chapter, you will be able to:

- Describe the hardware components used during a rope rescue. (NFPA 6.4.2, p 265–268)
- Describe the characteristics of a carabiner. (NFPA 6.4.2, p 265–266)
- Describe the characteristics of a harness. (NFPA 6.4.2, p 266)
- List the types of incidents that might require a rope rescue. (p 266–268)

Skills Objectives

There are no separate Fire Fighter II skill objectives for this chapter.

Additional NFPA Standard

- NFPA 1983, *Standard on Life Safety Rope and Equipment for Emergency Services*

You Are the Fire Fighter

You and your partner are on the fourth floor of a multistory apartment building and find yourself trapped by a rapidly progressing fire. You get to a bedroom at the front of the building and know your only escape route is out the window. You don't have time to wait for the aerial ladder to get to you, so you quickly consider your training for self-rescue. You pull out your rope and begin the most important task you have ever done in your life.

1. What type of rope would you want to use for this type of activity?
2. What rope construction method would be used for this rope?
3. What type of knot would you use?

Introduction

In the fire service, ropes are widely used to hoist or lower tools, appliances, or people; to pull a person to safety; or to serve as a life line in an emergency. A rope might be your only means of accessing a trapped person or your only way of escaping from a fire.

Learning about ropes and knots is an important part of your training as a fire fighter. This chapter gives you a basic understanding of the importance of ropes and knots in the fire service. You can then build on this foundation as you develop skills in handling ropes and tying knots. You must be able to tie simple knots accurately without hesitation or delay.

This chapter discusses different types of rope construction and the materials used in making ropes. It covers the care, cleaning, inspection, and storage of ropes. It also shows how to tie eight essential knots and explains how to secure tools and equipment so that they can be raised or lowered using ropes. Finally, it discusses additional uses for ropes in various rescue situations.

FIGURE 10-1 A life safety rope is a critical tool for fire fighters.

Types of Rope

Three primary types of rope are used in the fire service, each of which is dedicated to a distinct function. Life safety rope is used solely for supporting people. It must be used whenever a rope is needed to support a person, whether during training or during firefighting, rescue, or other emergency operations FIGURE 10-1. Escape rope is a single-purpose, emergency self-escape, self-rescue rope. Utility rope is used in most other cases, when it is not necessary to support the weight of a person, such as when hoisting or lowering tools or equipment.

Life Safety Rope

The life safety rope is a critical tool used *only* for life-saving purposes; it must never be used for utility purposes FIGURE 10-2. Life safety rope must be used in every situation where the rope must support the weight of one or more persons. In these situations, rope failure could result in serious injury or death.

Because a fire fighter's equipment must be extremely reliable, the criteria for design, construction, and performance of life safety rope and related equipment are specified in NFPA 1983, *Standard on Life Safety Rope and Equipment for Emergency Services*.

NFPA 1983 lists very specific standards for the construction of life safety rope. This standard also requires the rope manufacturer to include detailed instructions for the proper use, maintenance, and inspection of the life safety rope, including the conditions for removing the rope from service. In addition, the manufacturer must supply a list of criteria that must be reviewed before a life safety rope that has been used in the field can be used again. If the rope does not meet all of the criteria, it must be retired from service.

Types of Life Safety Ropes

NFPA 1983 defines the performance requirements for two types of life safety rope: technical use life safety rope and general use life safety rope. Technical use life safety rope by definition has a diameter that is 3/8" (9.5 mm) or greater, but is less than 1/2" (12.5mm). In addition to its smaller diameter, technical use life safety ropes are also weaker than general use life safety ropes. With a minimum breaking strength of 4496 pounds force (lbf) (20 kN), technical use life safety ropes are used by highly trained rescue teams that deploy to very technical environments such as mountainous and/or wilderness terrain TABLE 10-1.

TABLE 10-1 Life Safety Rope Strength

Rope Classification	Life Safety Rope Strength
Escape rope (NFPA 1983)	3304 lbf (13.5 kN)
Technical use life safety rope (NFPA 1983)	4496 lbf (20 kN)
General use life safety rope (NFPA 1983)	8992 lbf (40 kN)

FIGURE 10-2 A life safety rope.

The most common life safety rope carried by the fire service in the United States is the general use life safety rope. With a minimum breaking strength of 8992 lbf (40 kN), the general use life safety rope allows the technical rescuer a much greater margin of safety. With respect to its diameter, a general use life safety rope can be no larger than 5/8" (16 mm) and no smaller than 7/16" (11 mm).

FIGURE 10-3 An escape rope is designed to be used by only one person.

Escape Ropes

An escape rope is intended to be used by a fire fighter only for self-rescue from an extreme situation. This rope is designed to carry the weight of only one person and to be used only one time FIGURE 10-3. Its purpose is to provide the fire fighter with a method of escaping from a life-threatening situation. The escape rope should be replaced by a new rope if the escape rope is exposed to an immediate danger to life and health environment.

When you are fighting a fire, you should always have a safe way to get out of a situation and reach a safe location. You might be able to go back through the door that you entered, or you might have another exit route, such as through a different door, through a window, or down a ladder. If conditions suddenly change for the worse, having an escape route can save your life.

Sometimes, however, you can find yourself in a situation where conditions deteriorate so quickly that you cannot use your planned exit route. For example, the stairway you used might collapse behind you, or the room you are in might suddenly flash over (a phase in the development of a contained fire in which exposed surfaces reach ignition temperature more or less simultaneously and fire spreads rapidly throughout the space), blocking your planned route out. In such a situation, you might need to take extreme measures to get out of the building. The escape rope was developed specifically for this type of emergency self-rescue situation. It can support the weight of one person and fits easily into a small packet or pouch FIGURE 10-4.

FIGURE 10-4 A fire fighter can easily carry an escape rope.

FIRE FIGHTER Tips

Escape ropes are not classified as life safety ropes.

Utility Rope

Utility rope is used when it is not necessary to support the weight of a person. Fire department utility rope is used for hoisting or lowering tools or equipment, for making

FIGURE 10-5 Utility ropes are used for hoisting and lowering tools.

ladder halyards (rope used on extension ladders to raise a fly section), for marking off areas, and for stabilizing objects FIGURE 10-5. Utility ropes also require regular inspection.

Utility ropes must not be used in situations where life safety rope is required. Conversely, life safety rope must not be used for utility applications. A fire fighter must be able to recognize instantly the type of rope from its appearance and markings.

Fire Fighter Safety Tips

Many fire departments use color coding or other visible markings to identify different types of rope. This allows a fire fighter to determine very quickly if a rope is a life safety rope or a utility rope. The length of each rope should also be clearly marked by a tag or a label on the rope bag.

Rope Materials

Ropes can be made from many different types of materials. The earliest ropes were made from naturally occurring vines or fibers that were woven together. Today, ropes are made of synthetic materials such as nylon or polypropylene. Because ropes have many different uses, certain materials can work better than others in particular situations.

Natural Fibers

In the past, fire departments used ropes made from natural fibers, such as manila, because there were no alternatives. In these ropes, the natural fibers are twisted together to form strands. A strand can contain hundreds of individual fibers of different lengths. Today, ropes made from natural fibers are still used as utility ropes but are no longer acceptable as life safety ropes TABLE 10-2. Natural fiber ropes can be weakened by mildew and deteriorate with age, even when properly stored. A wet manila rope can absorb 50 percent of its weight in water, making it very susceptible to deterioration. A wet natural fiber rope is also very difficult to dry.

TABLE 10-2	Drawbacks to Using Natural Fiber Ropes
• Lose their load-carrying ability over time • Subject to mildew • Absorb 50 percent of their weight in water • Degrade quickly	

Synthetic Fibers

Since the introduction of nylon in 1938, synthetic fibers have been used to make ropes. In addition to nylon, several newer synthetic materials—such as polyester, polypropylene, and polyethylene—have been used in rope construction FIGURE 10-6. Synthetic fibers have several advantages over natural fibers TABLE 10-3. For example, synthetic fibers are generally stronger than natural fibers, so it is possible to use a smaller diameter rope without sacrificing strength. Synthetic materials can also produce very long fibers that run the full length of a rope to provide greater strength and added safety.

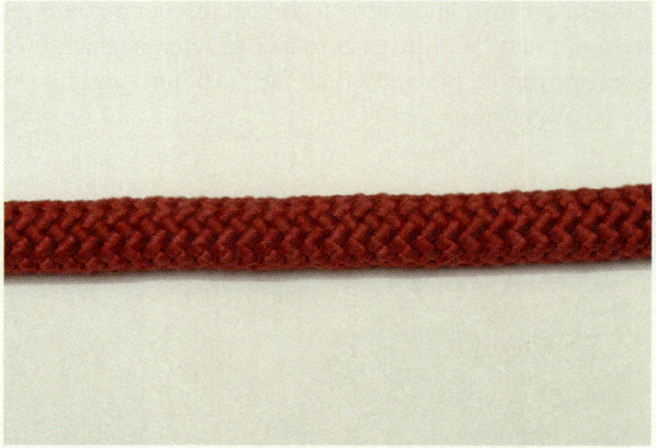

FIGURE 10-6 Synthetic fibers are generally stronger than natural fibers are.

Synthetic ropes are more resistant to rotting and mildew than natural fiber ropes are and do not degrade as rapidly. Depending on the material, they might also provide more resistance to melting and burning than natural fiber ropes do. In addition, synthetic ropes absorb much less water when wet and can be washed and dried. Some types of synthetic rope can float on water, which is a major advantage in water rescue situations.

TABLE 10-3	Advantages to Using Synthetic Fiber Ropes

- Strength-to-diameter ratio
- Difficult to meet block creel construction requirement with natural fibers
- Longevity over natural fibers

However, ropes made from synthetic fibers do have some drawbacks. Prolonged exposure to ultraviolet light as well as exposure to strong acids or alkalis can damage some synthetic ropes and decrease its life expectancy.

Life safety ropes are always made of synthetic fibers. Before any rope can be used for life safety purposes, it must meet the manufacturing requirements outlined in the current NFPA 1983. These standards specify that life safety rope must be woven of block creel construction (without knots or splices in the yarns, ply yarns, strands, braids, or rope). Rope of any other material or construction may not be used as a life safety rope.

The most common synthetic fiber used in life safety ropes is nylon. It has a high melting temperature with good abrasion resistance and is strong and lightweight. Nylon ropes are also resistant to most acids and alkalis. Polyester is the second most common synthetic fiber used for life safety ropes. Some life safety ropes are made of a combination of nylon and polyester or other synthetic fibers.

FIRE FIGHTER Tips

Ropes used for water rescue are often kept in special throw bags. The rescuer holds onto one end of the rope and throws the bag for the victim to catch. The rescuer then uses the rope to pull the victim to shore.

FIGURE 10-7 Polypropylene rope is often used in water rescues.

Polypropylene is the lightest of the synthetic fibers. Because it does not absorb water and floats, polypropylene rope is often used for water rescue situations FIGURE 10-7. However, polypropylene is not as suitable as nylon for fire department life-safety uses because it is not as strong, it is hard to knot, and it has a low melting point.

Rope Construction

Several types of rope construction are possible. The best choice of rope construction depends on the specific application.

Twisted and Braided Rope

Twisted ropes, which are also called laid ropes, are made of individual fibers twisted into strands. The strands are then twisted together to make the rope FIGURE 10-8A. This method of rope construction has been used for hundreds of years. Both natural and synthetic fibers can be used to make twisted rope.

This method of construction does have a disadvantage in that it exposes all of the fibers to the outside of the rope, where they are subject to abrasion. Abrasion can damage the rope

Near Miss REPORT

Report Number: 08-0000031
Report Date: 01/17/2008

Event Description: During the stabilization for performing extrication, a fire fighter used a bungee cord to tie open the passenger door of a SUV that was lying on its side. I (EMT/Fire Fighter) was performing patient care and had been leaning into the vehicle to work with the patient. The door broke loose from the bungee cord and slammed my dominant hand between the door and frame. As the door bounced back up, I pulled my hand out and the fire fighter who had improperly tied off the door had his thumb shut in the door. After regaining my composure, I returned to patient care until the ambulance arrived. We were both put out of service once the ambulance arrived with a crew to continue patient care. We both had to have x-rays. No broken bones, but I am still under doctor's care.

At our next fire meeting, I took the fire fighter out to our medical unit and opened the compartment which holds our c-collars (which we had used on the scene) and showed him that there were two sections of rope lying right there in plain sight. Had I been leaning in the vehicle door when the door broke free, my neck would have been smashed in the door instead, and I would most likely be dead.

Lessons Learned: The lesson learned for the fire fighter is that bungee cords do NOT replace ropes. Additional extrication training is being scheduled along with additional training in ICS and the need for appointing a safety officer on every scene.

A.

B.

FIGURE 10-8 **A.** Twisted rope. **B.** Braided rope.

fibers and reduce rope strength. Twisted ropes tend to stretch and are prone to unraveling when a load is applied.

Braided ropes are constructed by weaving or intertwining strands—typically synthetic fibers—together in the same way that hair is braided FIGURE 10-8B. This method of construction also exposes all of the strands to the outside of the rope where they are subject to abrasion. Braided rope stretches under a load, but it is not prone to twisting. A double-braided rope has an inner braided core covered by a protective braided sleeve so that only the fibers in the outer sleeve are exposed to the outside; the inner core remains protected from abrasion in this construction.

Kernmantle Rope

Kernmantle rope consists of two distinct parts: the kern and the mantle. The kern is the center or core of the rope. The mantle, or sheath, is a braided covering that protects the core from dirt and abrasion. Although both parts of a kernmantle rope are made with synthetic fibers, different fibers can be used for the kern and the mantle.

Each fiber in the kern extends for the entire length of the rope without knots or splices. This block creel construction is required under NFPA 1983 for all life safety ropes. The continuous filaments produce a core that is stronger than one constructed of shorter fibers that are twisted or braided together.

Kernmantle construction produces a very strong and flexible rope that is relatively thin and lightweight FIGURE 10-9. This construction is well suited for rescue work and is very popular for life safety rope.

FIGURE 10-9 The parts of a kernmantle rope.

Dynamic and Static Rope

A rope can be either dynamic or static, depending on how it reacts to an applied load. A dynamic rope is designed to be elastic and stretch when it is loaded. A static rope has a limited range of elasticity. The differences between dynamic and static ropes result from both the fibers used and the construction method.

Dynamic rope is typically used in safety lines for mountain climbing because it stretches and cushions the shock if a climber falls for a long distance. A static rope is more suitable for most fire rescue situations, where falls from great heights are not anticipated. Teams that specialize in rope rescue often carry both static and dynamic ropes for use in different situations.

Dynamic and Static Kernmantle Ropes

Kernmantle ropes can be either dynamic or static. A dynamic kernmantle rope is constructed with overlapping or woven fibers in the core. When the rope is loaded, the core fibers are pulled tighter, which gives the rope its elasticity.

In the core of a static kernmantle rope, all of the fibers are laid parallel to each other. Such a rope has very little elasticity and limited elongation under an applied load. Most fire department life safety ropes use static kernmantle construction. This type of rope is well suited for lowering a person and can be used with a pulley system for lifting individuals. It can also be used to create a bridge between two structures. In this type of rope, the manufacturer includes an imbedded trailer. This trailer includes the name of the manufacturer, model number, make number, serial number, and date of manufacture.

FIRE FIGHTER II

Technical Rescue Hardware

During technical rescue incidents, ropes are often used to access and extricate individuals. In addition to the rope itself, several hardware components might be used. Fire fighters, for example, often use a carabiner, FIGURE 10-10. This device

FIGURE 10-10 Carabiners.

Near Miss REPORT

Report Number: 09-0000302
Report Date: 03/20/2009

Event Description: We were wrapping up training at the local drill tower by packing and stowing ropes. A fire fighter dropped a full 200' rope bag from the second story walkway of the drill tower. It landed by two other fire fighters who were packing another rope bag. It could have broken a fire fighter's neck.

Lessons Learned: Teamwork; safety; concern for fellow fire fighters; equipment care; and common sense need to be more important than personal convenience.

FIRE FIGHTER II

is used to connect one rope to another rope or to other hardware such as an anchor plate, swivel, or pulley. Several types of carabiners are available, and you should know how to operate the type used by your department.Only a few are recognized for use in the fire service and rescue. Refer to NFPA 1983.

Harnesses

A harness is a piece of rescue or safety equipment made of webbing and worn by a person. It is used to secure the person to a rope or to a solid object. Two types of harnesses—Class II and Class III—can be used by rescuers, depending on the circumstances encountered:

- Class II harness (seat harness) fastens around the rescuer's waist and legs and has a design load of 600 lbs (272 kg). It is used to support a fire fighter, particularly in rescue situations FIGURE 10-11.
- Class III harness (full body harness) fastens around the rescuer's waist and thighs as well as secures the rescuer's waist and shoulders. It is the most secure type of harness and is often used to support a fire fighter who is being raised or lowered on a life safety rope FIGURE 10-12.

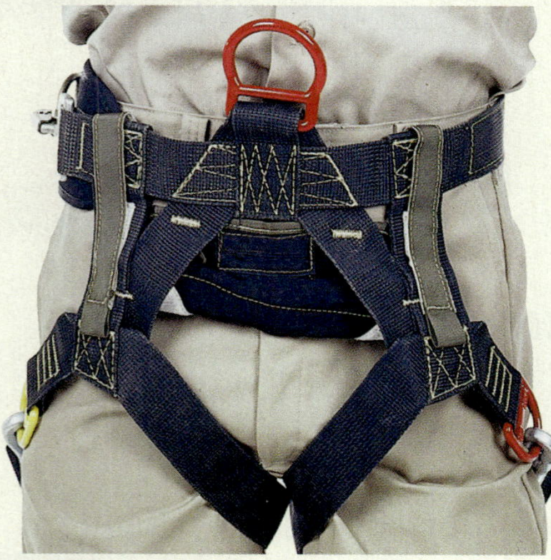

FIGURE 10-11 Class II harness (seat harness).

Harnesses need to be cleaned and inspected regularly, just as life safety ropes do. Follow the manufacturer's instructions for cleaning and inspecting harnesses.

Rope Rescue

Rope rescue involves raising and lowering rescuers to access injured or trapped individuals, as well as raising or lowering victims who are rescued so that they can be given appropriate medical treatment. This chapter outlines the basics of rope rescue and gives you a foundation for learning the more complex parts of rope rescue. An approved rope rescue course is required to attain true proficiency in rope rescue skills.

Rope rescue courses cover the technical skills needed to raise or lower people using mechanical advantage systems and to remove someone from a rock ledge or a confined space

FIRE FIGHTER II

FIGURE 10-12 Class III harness (full body harness).

FIGURE 10-13 Rope rescues require intense technical training.

FIGURE 10-13. They also cover the equipment and skills needed to accomplish these rescues safely.

Rope Rescue Incidents

Most rope rescue incidents involve people who are trapped in normally inaccessible locations such as a mountainside or the outside of a building FIGURE 10-14. Rescuers often have to

FIGURE 10-14 Ropes are invaluable when a person is trapped in an inaccessible area, such as outside a high building.

lower themselves using a system of anchors, webbing, ropes, carabiners, and other devices to reach the trapped person. Once rescuers reach the person, they then have to stabilize him or her and determine how to get the person to safety. Sometimes the victim has to be lowered or raised to a safe location. Extreme cases could involve more complicated operations, such as transporting the victim in a basket lowered by a helicopter.

The type and number of ropes used in a rope rescue depend on the situation. There is almost always a primary rope that bears the weight of the rescuer (or rescuers) as he or she attempts to reach the victim. The rescuers often have a second line attached to them, known as a belay line, which serves as a backup if the main line fails. Additional lines might be needed to raise or lower the trapped individual, depending on the circumstances.

Trench Rescue

Rescues in collapsed trenches often are complicated and involve a number of different skills, such as shoring, air quality monitoring, confined-space operations, and rope rescue. Ropes are often used to stabilize and remove the trapped person. After the rescuers shore the walls of the trench and remove the dirt covering the victim, they place the person in a Stokes basket or on a backboard and lift him or her to the surface. If the trench is deep, ropes might be used to raise the victim to the surface.

Confined-Space Rescue

A confined-space rescue can take place in locations as varied as tanks, silos, underground electrical vaults, storm drains, and similar structures. It is often very difficult to extricate an unconscious or injured person from these locations because of the poor ventilation and limited entry or exit area. For this reason, ropes are often used to remove an injured or unconscious victim FIGURE 10-15.

FIGURE 10-15 Ropes are often used to remove an injured or unconscious victim from a confined space.

Water Rescue

Ropes can be used in a variety of ways during water rescue operations. The simplest situation involves a rescuer on the shore throwing a rope to a person in the water and pulling that individual to shore. A more complicated situation could involve a rope stretched across a stream or river FIGURE 10-16. A boat

FIGURE 10-16 Ropes ensure rescuer safety during water rescues.

might be tethered to the rope, and rescuers on shore might maneuver the boat using a series of ropes and pulleys.

Rope Maintenance

All ropes—but especially life safety ropes—need proper care to perform in an optimal manner. Maintenance is necessary for all kinds of equipment and all types of rope, and it is absolutely essential for life safety ropes. Your life and the lives of others depend on the proper maintenance of your life safety ropes.

There are four parts to the maintenance formula:

- Care
- Clean
- Inspect
- Store

Care for the Rope

You must follow certain principles to preserve the strength and integrity of rope TABLE 10-4:

- Protect the rope from sharp and abrasive surfaces. Use edge protectors when the rope must pass over a sharp or unpadded surface.
- Protect the rope from rubbing against another rope or webbing. Friction generates heat, which can damage or destroy the rope.
- Protect the rope from heat, chemicals, and flames.
- Protect the rope from prolonged exposure to sunlight. Ultraviolet radiation can damage rope.

TABLE 10-4 Principles to Preserve Strength and Integrity of Rope

- Protect the rope from sharp, abrasive surfaces.
- Protect the rope from heat, chemicals, and flames.
- Protect the rope from rubbing against another rope.
- Protect the rope from prolonged exposure to sunlight.
- Follow the manufacturer's recommendations.

FIRE FIGHTER Tips

A shock load can occur when a rope is suddenly placed under unusual tension—for example, when someone attached to a life safety rope falls until the length of the rope or another rescuer stops the drop. A utility rope can be shock-loaded in a similar manner if a piece of equipment that is being raised or lowered drops suddenly.

Any rope that has been shock-loaded should be inspected and might have to be removed from service. Although there might not be any visible damage, shock loading can cause damage that is not immediately apparent. Repeated shock loads can severely weaken a rope so that it can no longer be used safely. Keeping accurate rope records helps identify potentially damaged rope.

- Do not step on a rope! Your footstep could force shards of glass, splinters, or abrasive particles into the core of the rope, damaging the rope fibers.
- Follow the manufacturer's recommendations for rope care.

To care for life safety ropes properly, follow the steps in SKILL DRILL 10-1 (Fire Fighter I, NFPA 5.5.1):

1. Protect the rope from sharp and abrasive edges; use edge protectors.
2. Protect the rope from rubbing against other ropes.
3. Protect the rope from heat, chemicals, flames, and sunlight.
4. Avoid stepping on a rope.

Clean the Rope

Many ropes made from synthetic fibers can be washed with a mild soap and water. In addition, a special rope washer can be attached to a garden hose FIGURE 10-17. Some manufacturers recommend placing the rope in a mesh bag and washing it in a front-loading washing machine.

When washing a rope, use a mild detergent. Do not use bleach because it can damage rope fibers. Follow the manufacturer's recommendations for specific care of your rope. Do not pack or store wet or damp rope. Air drying is usually recommended. The drying rope should be suspended and should not lie on the floor. The use of mechanical drying devices is not usually recommended. Rope should never be dried or stored in direct sunlight.

FIGURE 10-17 Some fire departments use a rope washer to clean their ropes.

Follow the steps in SKILL DRILL 10-2 to clean fire department ropes:

1. Wash the rope with mild soap and water. (STEP 1)
2. Use a rope washer or machine, if recommended by the rope's manufacturer. (STEP 2)
3. Air dry the rope out of direct sunlight.
4. Inspect the rope and replace it in the rope bag so that it is ready for use. (STEP 3)

SKILL DRILL 10-2 Cleaning Fire Department Ropes

(Fire Fighter I, NFPA 5.5.1)

1 Wash the rope with mild soap and water.

2 Use a rope washer or machine if recommended by the rope's manufacturer.

3 Air dry the rope out of direct sunlight. Inspect the rope and replace it in the rope bag so that it is ready for use.

Inspect the Rope

Life safety ropes must be inspected after each use, whether the rope was used for an emergency incident or in a training exercise TABLE 10-5. Unused rope should also be inspected on a regular schedule. Some departments inspect all rope, including life safety and utility ropes, every three months. Obtain the inspection criteria from the rope manufacturer.

Inspect the rope visually, looking for cuts, frays, or other damage as you run it through your fingers. Make sure that your grasping hand is gloved to protect yourself from sharp objects imbedded in the rope. Because you cannot see the inner core of a kernmantle rope, feel for any depressions (flat spots or lumps on the inside). Examine the sheath for any discolorations, abrasions, or flat spots FIGURE 10-18. If you have any doubt about whether the rope has been damaged, consult your company officer TABLE 10-6.

TABLE 10-5 Questions to Consider When Inspecting Life Safety Ropes

- Has the rope been exposed to heat or flame?
- Has the rope been exposed to abrasion?
- Has the rope been exposed to chemicals?
- Has the rope been exposed to shock loads?
- Are there any depressions, discolorations, or lumps in the rope?

FIGURE 10-18 Rope inspection is a critical step.

TABLE 10-6	Signs of Possible Rope Deterioration

- Discoloration
- Shiny markings from heat or friction
- Damaged sheath
- Core fibers poking through the sheath

A life safety rope that is no longer usable must be pulled from service and either destroyed or marked as a utility rope. A downgraded rope must be clearly marked so that it cannot be confused with a life safety rope.

Follow the steps in SKILL DRILL 10-3 (Fire Fighter I, NFPA 5.5.1) to inspect fire department ropes:

1. Inspect the rope after each use and at regular intervals.
2. Examine the core by looking for depressions, flat spots, or lumps.
3. Examine the sheath by looking for discolorations, abrasions, flat spots, and imbedded objects.
4. Remove the rope from service if it is damaged.

Rope Record

Each piece of rope must be marked for identification. In addition, a rope record must be kept for each piece of life safety rope. This record should include a history of when the rope was purchased, when it was used, how it was used, and what kinds of loads were applied to it. Each inspection should also be recorded. Fire departments should maintain rope records for both utility ropes and life safety ropes.

Store the Rope

Proper care ensures a long life for your rope and reduces the chance of equipment failure and accident. Store ropes away from temperature extremes, out of sunlight, and in areas where there is some air circulation. Avoid placing ropes where fumes from gasoline, oils, or hydraulic fluids might damage them. Apparatus compartments used to store ropes should be separated from compartments used to store any oil-based products or machinery powered by gasoline or diesel fuel. Do not place any heavy objects on top of the rope FIGURE 10-19.

Rope bags are used to protect and store ropes. Each bag should hold only one rope. Rope may also be coiled for storage FIGURE 10-20. Very long pieces of rope are sometimes stored on spools FIGURE 10-21.

Follow the steps in SKILL DRILL 10-4 to place a life safety rope into a rope bag:

1. Inspect rope to be certain it is fit for service.
2. Carefully load the life safety rope into the rope bag. (STEP 1)
3. Do not try to coil the rope in the bag because this action will cause the rope to kink and become tangled when it is pulled out.
4. Return the rope and the rope bag to its proper location for storage or service. (STEP 2)

FIGURE 10-19 Ensure that ropes are stored safely.

FIGURE 10-20 Ropes may be coiled for storage.

FIGURE 10-21 Long sections of rope are sometimes stored on spools.

SKILL DRILL 10-4 Placing a Life Safety Rope in a Rope Bag

(Fire Fighter I, NFPA 5.5.1)

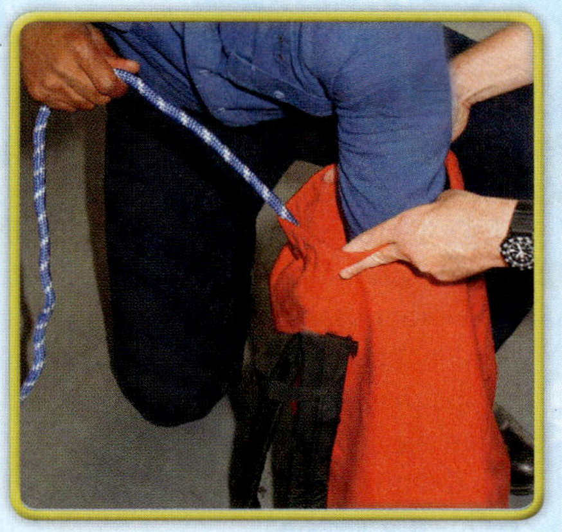

1. Inspect rope to be certain it is fit for service. Carefully load the life safety rope into the rope bag.

2. Do not try to coil the rope in the bag because this action will cause the rope to kink and become tangled when it is pulled out. Return the rope and the rope bag to its proper location for storage or service.

Knots

Knots are prescribed ways of fastening lengths of rope or webbing to objects or to each other. As a fire fighter, you must know how to tie certain knots and when to use each type of knot. Knots can be used for one or more particular purposes. Hitches, such as the clove hitch, are used to attach a rope around an object. Knots, such as the figure eight and the bowline, are used to form loops. Bends, such as the sheet bend or Becket bend, are used to join two ropes together. Safety knots, such as the overhand knot, are used to secure the ends of ropes to prevent them from coming untied.

Any knot reduces the load-carrying capacity of the rope in which it is placed by a certain percentage **TABLE 10-7**. You can avoid an unnecessary reduction in rope strength if you know which type of knot to use and how to tie the knot correctly.

TABLE 10-7 Effect of Knots on Rope Strength

Group	Knot	Reduction in Strength
Loop knots	Figure eight on a bight	20%
	Figure eight follow-through	19%
	Bowline	33%

Terminology

Specific terminology is used to refer to the parts of a rope in describing how to tie knots **FIGURE 10-22**:

- The working end is the part of the rope used for forming the knot.
- The running end is the part of the rope used for lifting or hoisting.
- The standing part is the rope between the working end and the running end.
- A bight is formed by reversing the direction of the rope to form a U bend with two parallel ends **FIGURE 10-23**.
- A loop is formed by making a circle in the rope **FIGURE 10-24**.
- A round turn is formed by making a loop and then bringing the two ends of the rope parallel to each other **FIGURE 10-25**.

FIGURE 10-22 The sections of a rope used in tying knots.

FIGURE 10-23 A bight.

FIGURE 10-24 A loop.

FIGURE 10-25 A round turn.

A fire fighter should know how to tie and properly use these simple knots:

- Safety knot (overhand knot)
- Half hitch
- Clove hitch
- Figure eight
- Figure eight on a bight
- Figure eight follow-through
- Figure eight bend
- Bowline
- Bend (sheet or Becket bend)

Safety Knot

A safety knot (also referred to as an overhand knot or a keeper knot) is used to secure the leftover working end of the rope. It provides a degree of safety to ensure that the primary knot will not become undone. A safety knot should always be used to finish the other basic knots.

A safety knot is simply an overhand knot in the loose end of the rope that is made around the standing part of the rope. It secures the loose end and prevents it from slipping back through the primary knot.

Follow the steps in SKILL DRILL 10-5 to tie a safety knot:

1. Take the loose end of the rope, beyond the knot, and form a loop around the standing part of the rope. (STEP 1)
2. Pass the loose end of the rope through the loop. (STEP 2)
3. Tighten the safety knot by pulling on both ends at the same time.
4. Test whether you have tied a safety knot correctly by sliding it on the standing part of the rope. A knot that is tied correctly will slide. (STEP 3)

Hitches

Hitches are knots that wrap around an object, such as a pike pole or a fencepost. They are used to secure the working end of a rope to a solid object or to tie a rope to an object before hoisting it.

Half Hitch

The half hitch is not a secure knot by itself, which is why it is used only in conjunction with other knots. For example, when hoisting an axe or pike pole, you use the half hitch to keep the hoisting rope aligned with the handle. On long objects, you might need to use several half hitches.

Follow the steps in SKILL DRILL 10-6 to tie a half hitch:

1. Grab the rope with your palm facing away from you. (STEP 1)
2. Rotate your hand so that your palm is facing you. This will make a loop in the rope. (STEP 2)
3. Pass the loop over the end of the object. (STEP 3)
4. Finish the half-hitch knot by positioning it and pulling tight. (STEP 4)

Clove Hitch

A clove hitch is used to attach a rope firmly to a round object, such as a tree or a fencepost. It can also be used to tie a hoisting rope around an axe or pike pole. A clove hitch can be tied anywhere in a rope and will hold equally well if tension is applied to either end of the rope or to both ends simultaneously.

There are two different methods of tying this knot. A clove hitch tied in the open is used when the knot can be formed and then slipped over the end of an object, such as an axe or pike

SKILL DRILL 10-5 Tying a Safety Knot

(Fire Fighter I, NFPA 5.5.2)

1 Take the loose end of the rope, beyond the knot, and form a loop around the standing part of the rope.

2 Pass the loose end of the rope through the loop.

3 Tighten the safety knot by pulling on both ends at the same time. Test whether you have tied a safety knot correctly by sliding it on the standing part of the rope. A knot that is tied correctly will slide. Note: When complete, the safety knot should sit directly next to any accompanying knots.

pole. It is tied by making two consecutive loops in the rope. Follow the steps in SKILL DRILL 10-7 to tie a clove hitch in the open:

1. Starting from left to right on the rope, grab the rope with crossed hands with the left positioned higher than the right. (STEP 1)
2. Holding onto the rope, uncross your hands. This will create a loop in each hand. (STEP 2)
3. Slide the right-hand loop behind the left-hand loop. (STEP 3)
4. Slide both loops over the object. (STEP 4)
5. Pull in opposite directions to tighten the clove hitch.
6. Tie a safety knot in the working end of the rope. (STEP 5)

If the object is too large or too long to slip the clove hitch over one end, the same knot can be tied around the object. Follow the steps in SKILL DRILL 10-8 to tie a clove hitch around an object:

1. Place the working end of the rope over the object. (STEP 1)
2. Make a complete loop around the object, working end down. (STEP 2)

FIRE FIGHTER Tips

When tying a safety knot in a clove hitch, the end of the rope should pass around the object the clove hitch is being tied to and the overhand safety knot is tied onto the rope.

SKILL DRILL 10-6 Tying a Half Hitch

(Fire Fighter I, NFPA 5.5.2)

1 Grab the rope with your palm facing away from you.

2 Rotate your hand so that your palm is facing you. This will make a loop in the rope.

3 Pass the loop over the end of the object.

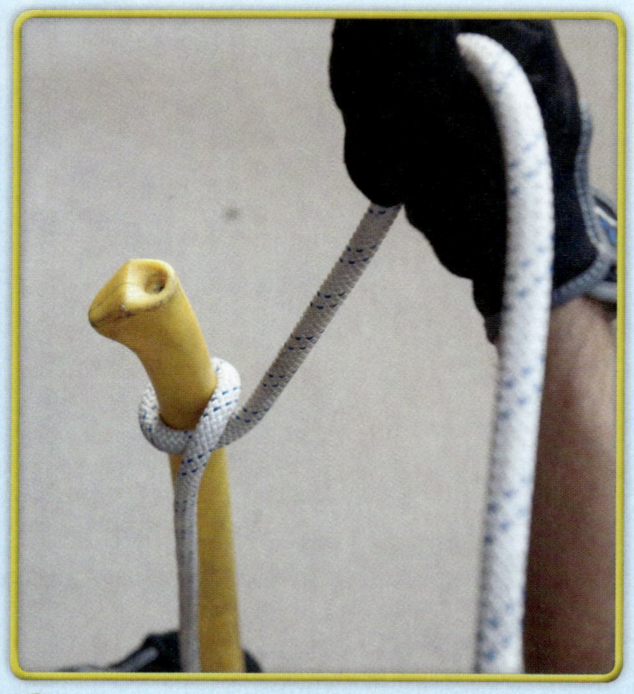

4 Finish the half-hitch knot by positioning it and pulling tight.

3. Make a second loop around the object a short distance above the first loop.
4. Pass the working end of the rope under the second loop, above the point where the second loop crosses over the first loop. (STEP 3)
5. Tighten the knot and secure it by pulling on both ends. (STEP 4)
6. Tie a safety knot in the working end of the rope. (STEP 5)

VOICES OF EXPERIENCE

Sunday mornings in the summer are fun. As a lieutenant and training officer for my company, if I'm not teaching at the training academy, you will find me leading an entertaining drill at the station with the fire fighters who randomly show up that day. Our township has a mixture of unique features, which demand a solid core of all-around skills. Coming up with skills to review on a Sunday morning is not difficult.

When the tones dropped, we were in the middle of reviewing tool tying. The dispatcher announced, "Rescue box 73-88, the rod and gun club, a water rescue." As I boarded the apparatus, I never could have anticipated what we would find on scene.

The rod and gun club is situated on hilly terrain. Hidden in the back, behind the two-story club house, is a normally small stream. On that day it was a torrent. Heavy rain had caused a deluge of water to pour down from the northern counties, turning what was a 30 foot-wide, shallow trickling brook, into a 12 mile-per-hour, 4 foot deep section of rapids and trapping three 12-year-old children on a narrow island in the middle.

The action plan developed by our chiefs called for us to erect a modified Tyrolean traverse (water highline). The plan was to establish a secure line onto which the rescuers, along with their victims, could be attached and assisted to the safe bank of the span. I would become the master rigger for the evolution. I sent four fire fighters down stream with a 100 foot static kernmantle lifeline and orders to find a way to get across and establish a catch line should any of our team or the children become caught in the water and propelled downstream. Once the down stream team had completed their task, I attached one end of a 100 foot static kernmantle lifeline to a 3 foot diameter tree down stream from the rescue sight at a safe place on the bank using a tensionless hitch. I then ordered two fire fighters upstream to attempt a crossing, realizing that even if they made it, they would need to compensate for the water flow to intersect the island. I had them take rescue harnesses, webbing, Prusik cords, and carabineers, anticipating their success. This way the children could be properly packaged and the rescuers safely connected to the highline during the retrieval phase.

Once the rescuers reached the island, they secured the other end of the rope to a similar sized tree using a clove hitch on a bight and began the task of packaging the children. While this was going on, I ordered the fashioning of a simple Z-rig and the placement of a change of direction pulley on a tree directly in front of the island rescue point. We then tensioned the line, holding it fast with progress capture Prusik cords. With the highline in place, the children harnessed and fastened securely to the rescuers and the rescuers properly fastened to the highline using long Prusik cords, the rescuers were able to safely carry the children to the bank.

The next day I went back to the site of the rescue. I walked across the babbling brook, staying well above the water using the stones, thinking, this is exactly how those children crossed. As I gathered the ropes from the other side, I took a moment to reflect on how my practiced skills had helped three youngsters escape a flash flood on a sunny Sunday morning without a cloud in the sky. I felt satisfied.

Andrew S. Gurwood
Bucks County Public Safety Training Center
Northampton, Pennsylvania

SKILL DRILL 10-7 Tying a Clove Hitch in the Open

(Fire Fighter I, NFPA 5.5.2)

1 Starting from left to right on the rope, grab the rope with crossed hands with the left positioned higher than the right.

2 Holding onto the rope, uncross your hands. This will create a loop in each hand.

3 Slide the right-hand loop behind the left-hand loop.

(Continued)

SKILL DRILL 10-7 Tying a Clove Hitch in the Open (*Continued*)

(Fire Fighter I, NFPA 5.5.2)

4 Slide both loops over the object.

5 Pull in opposite directions to tighten the clove hitch. Tie a safety knot in the working end of the rope.

SKILL DRILL 10-8 Tying a Clove Hitch Around an Object

(Fire Fighter I, NFPA 5.5.2)

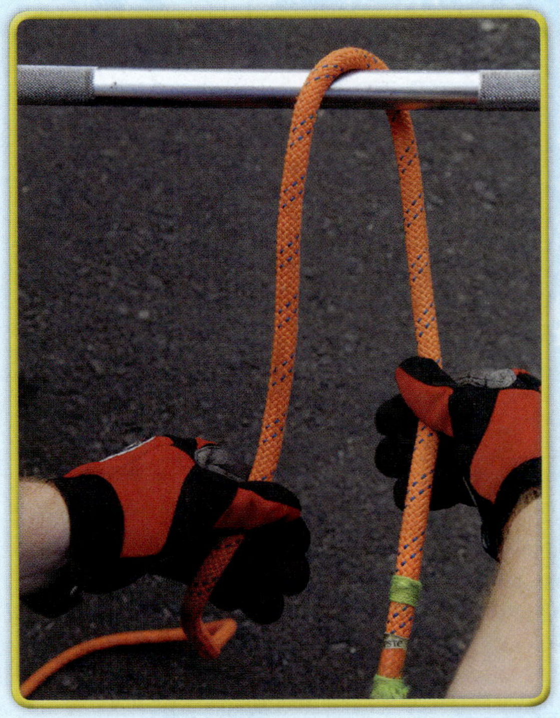

1 Place the working end of the rope over the object.

2 Make a complete loop around the object, working end down.

(*Continued*)

SKILL DRILL 10-8 Tying a Clove Hitch Around an Object (*Continued*)

(Fire Fighter I, NFPA 5.5.2)

3 Make a second loop around the object a short distance above the first loop. Pass the working end of the rope under the second loop, above the point where the second loop crosses over the first loop.

4 Tighten the knot and secure it by pulling on both ends.

5 Tie a safety knot in the working end of the rope.

Near Miss REPORT

Report Number: 10-1038

Synopsis: Knot overlooked during rope training.

Event Description: Our rescue unit was staffed one day with untrained personnel. The apparatus was scheduled for rope training and the untrained member was going to rappel. All safety equipment was checked and appropriate rigging and knots were tied and checked for safe operation. As the untrained member stepped into position to rappel, someone noticed that the knot wasn't properly attached to the member's harness. The member was pulled away from the edge and the equipment was checked again.

Lessons Learned: Untrained personnel should not be allowed to participate in training if they are not educated in all areas of the training.

Loop Knots

Loop knots are used to form a loop in the end of a rope. These loops can be used for hoisting tools, for securing a person during a rescue, for securing a rope to a fixed object, or for identifying the end of a rope stored in a rope bag. When tied properly, these knots will not slip, yet are easy to untie.

Figure Eight Knot

A figure eight is a basic knot used to produce a family of other knots, including the figure eight on a bight, the figure eight follow-through, and the figure eight bend. Follow the steps in SKILL DRILL 10-9 to tie a figure eight knot:

1. Form a bight in the rope. (STEP 1)
2. Loop the working end of the rope completely around the standing end of the rope. (STEP 2)
3. Thread the working end through the opening of the bight. (STEP 3)
4. Tighten the knot by pulling on both ends simultaneously. When you pull the knot tight, it will have the shape of a figure eight. (STEP 4)

Figure Eight on a Bight

The figure eight on a bight knot creates a secure loop at the working end of a rope. This loop can be used to attach the end of the rope to a fixed object or a piece of equipment or to tie a life safety rope around a person. Follow the steps in SKILL DRILL 10-10 to tie a figure eight on a bight:

1. Double over a section of the rope to form a bight. The closed end of the bight becomes the working end of the rope. (STEP 1)
2. Hold the two sides of the bight as if they were one rope.
3. Form a loop in the doubled section of the rope. (STEP 2)
4. Pass the working end of the bight through the loop. (STEP 3)
5. Pull the knot tight. Pulling the knot tight locks the neck of the bight and forms a secure loop. (STEP 4)
6. Use a safety knot to tie the loose end of the rope to the standing part. (STEP 5)

Figure Eight Follow-Through

A figure eight follow-through knot creates a secure loop at the end of the rope when the working end must be wrapped around an object or passed through an opening before the loop can be formed. It is very useful for attaching a rope to a fixed ring or a solid object with an "eye." Follow the steps in SKILL DRILL 10-11 to tie a figure eight with a follow-through:

1. Tie a simple figure eight in the standing part of the rope, far enough back to make a loop. Leave this knot loose. (STEP 1)
2. Thread the working end through the opening or around the object, and bring it back through the original figure eight knot in the opposite direction. (STEP 2)
3. Once the working end has been threaded through the knot, pull the knot tight. (STEP 3)
4. Secure the loose end with a safety knot. (STEP 4)

Figure Eight Bend

The figure eight bend or tracer 8 is used to join two ropes together. Follow the steps in SKILL DRILL 10-12 to tie a figure eight bend:

1. Tie a figure eight near the end of one rope. (STEP 1)
2. Thread the end of the second rope completely through the knot from the opposite end.
3. Pull the knot tight. (STEP 2)
4. Tie a safety knot on the loose end of each rope to the standing part of the other. (STEP 3)

Bowline

A bowline knot also can be used to form a loop. This type of knot is frequently used to secure the end of a rope to an object or anchor point. It can also be used to hoist equipment. Follow the steps in SKILL DRILL 10-13 to tie a bowline:

1. Make the desired sized loop and bring the working end back to the standing part. (STEP 1)
2. Form another small loop in the standing part of the rope with the section close to the working end on top.
3. Thread the working end up through this loop from the bottom. (STEP 2)
4. Pass the working end over the loop, around and under the standing part, and back down through the same opening. (STEP 3)
5. Tighten the knot by holding the working end and pulling the standard part of the rope backward. (STEP 4)
6. Tie a safety knot in the working end of the rope. (STEP 5)

Bends

Bends are used to join two ropes together. The sheet bend or Becket bend can be used to join two ropes of unequal size. A sheet bend knot also can be used to join rope to a chain or to pull up a rope of a different diameter. It should not be relied upon to support a human load. Follow the steps in SKILL DRILL 10-14 to tie a sheet or Becket bend:

1. Form a bight in the working end of one rope. If the ropes are of unequal size, the bight should be made in the larger rope. (STEP 1)
2. Pass the end of the second rope up through the opening of the bight, between the two parallel sections of the first rope. (STEP 2)
3. Loop the second rope completely around both sides of the bight.
4. Thread the second rope under itself and on top of the two sides of the bight. (STEP 3)

SKILL DRILL 10-9 Tying a Figure Eight Knot

(Fire Fighter I, NFPA 5.5.2)

1 Form a bight in the rope.

2 Loop the working end of the rope completely around the standing part of the rope.

3 Thread the working end back through the bight.

4 Tighten the knot by pulling on both ends simultaneously.

SKILL DRILL 10-10 Tying a Figure Eight on a Bight
(Fire Fighter I, NFPA 5.5.2)

1 Form a bight and identify the closed end of the bight as the working end of the rope.

2 Holding both sides of the bight together, form a loop.

3 Feed the working end of the bight back through the loop.

4 Pull the knot tight.

5 Secure the loose end of the rope with a safety knot.

SKILL DRILL 10-11 Tying a Figure Eight Follow-Through

(Fire Fighter I, NFPA 5.5.2)

1 Tie a simple figure eight in the standing part of the rope, far enough back to make a loop. Leave this knot loose.

2 Thread the working end through the opening or around the object, and bring it back through the original figure eight knot in the opposite direction.

3 Once the working end has been threaded through the knot, pull the knot tight.

4 Secure the loose end with a safety knot.

SKILL DRILL 10-12 Tying a Figure Eight Bend
(Fire Fighter 1, NFPA 5.1.2)

1. Tie a figure eight near the end of one rope.

2. Thread the end of the second rope completely through the knot from the opposite end. Pull the knot tight.

3. Tie a safety knot on the loose end of each rope to the standing part of the other.

SKILL DRILL 10-13 Tying a Bowline

(Fire Fighter I, NFPA 5.5.2)

1 Make the desired sized loop and bring the working end back to the standing part.

2 Form another small loop in the standing part of the rope with the section close to the working end on top. Thread the working end up through this loop from the bottom.

3 Pass the working end over the loop, around and under the standing part, and back down through the same opening.

4 Tighten the knot by holding the working end and pulling the standard part of the rope backward.

5 Tie a safety knot in the working end of the rope.

SKILL DRILL 10-14 Tying a Sheet or Becket Bend

(Fire Fighter I, NFPA 5.5.2)

1 Using your left hand, form a bight at the working end of the first (larger) rope.

2 Thread the working end of the second (smaller) rope up through the bight.

3 Loop the second (smaller) rope completely around both sides of the bight. Pass the working end of the second (smaller) rope between the original bight and under the second rope.

4 Tighten the knot.

5 Tie a safety knot in the working end of each rope.

SKILL DRILL 10-15 Tying a Water Knot
(Fire Fighter I, NFPA 5.5.2)

1 In one end of the webbing, approximately 6 inches (152 mm) from the end, tie an overhand knot.

2 With the other end of the webbing, start retracing from the working end through the knot until approximately 6 inches (152 mm) is left on the other end.

3 Tie an overhand knot on each tail as a safety.

5. To tighten the knot, hold the first rope firmly while pulling back on the second rope. (STEP 4)
6. Use a safety knot to secure the loose ends of both ropes. (STEP 5)

Water Knot

The water knot or ring bend is used to join webbing of the same or different sizes together. When a single piece of webbing is used and the opposite ends are tied to each other, a loop or sling is created. Follow the steps in SKILL DRILL 10-15 to tie a water knot:

1. In one end of the webbing, approximately 6 inches (152 mm) from the end, tie an overhand knot. (STEP 1)
2. With the other end of the webbing, start retracing from the working end through the knot until approximately 6 inches (152 mm) is left on the other end. (STEP 2)
3. Tie an overhand knot on each tail as a safety. (STEP 3)

There are many ways to tie each of these knots. Find one method that works for you and use it all the time. In addition, your department might require that you learn how to tie other knots. It is important to become proficient in tying knots. With practice, you should be able to tie these knots in the dark, with heavy gloves on, and behind your back.

A knot should be properly "dressed" by tightening and removing twists, kinks, and slack from the rope. The finished knot is then firmly fixed in position. The configuration of a properly dressed knot should be evident so that it can be easily inspected.

Knot-tying skills can be quickly lost without adequate practice. Practice tying knots while you are on the telephone or watching television FIGURE 10-26. You never know when you will need to use these skills in an emergency situation. For

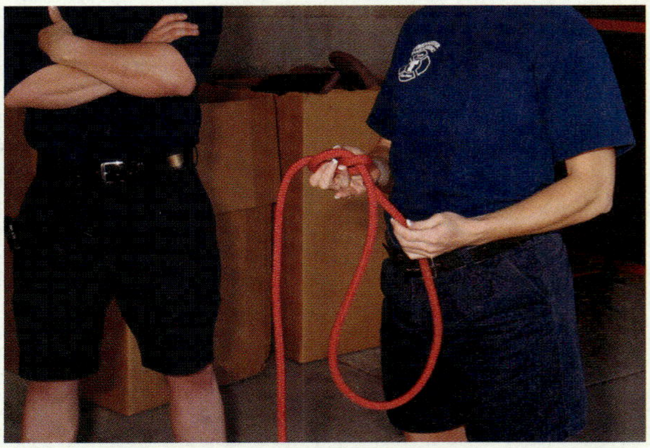

FIGURE 10-26 To maintain your knot-tying skills, practice tying different knots frequently.

added practice, try tying these knots with your gloves on or in darkness. These are conditions you will encounter often as a fire fighter.

Hoisting

Tying knots is not an idle exercise but rather a practical skill that you will frequently use on the job. In emergency situations, you might have to raise or lower a tool to other fire fighters. It is important for you to learn how to raise and lower an axe, a pike pole, a ladder, a charged hose line, an uncharged hose line, and an exhaust fan.

It is essential to ensure that the rope is tied securely to the object being hoisted so that the tool does not fall. Additionally, your co-workers must be able to remove the rope and place the tool into service quickly. When you are hoisting or lowering a tool, make sure no one is standing under the object. Keep the scene clear of people to avoid any chance of an accident.

■ Hoisting an Axe

An axe should be hoisted in a vertical position with the head of the axe down. After receiving the end of the utility rope from the fire fighter above, follow the steps in SKILL DRILL 10-16 to hoist an axe:

1. Tie the end of the hoisting rope around the handle near the head using either a figure eight on a bight or a clove hitch. Slip the knot down the handle from the end to the head. (STEP 1)

SKILL DRILL 10-16 Hoisting an Axe
(Fire Fighter I, NFPA 5.5.2)

1 Tie the end of the hoisting rope around the handle near the head using either a figure eight on a bight or a clove hitch. Slip the knot down the handle from the end to the head.

2 Loop the standing part of the rope under the head.

3 Place the standing part of the rope parallel to the axe handle.

4 Use one or two half hitches along the axe handle to keep the handle parallel to the rope. Communicate with the fire fighter above that the axe is ready to raise.

2. Loop the standing part of the rope under the head. (STEP 2)
3. Place the standing part of the rope parallel to the axe handle. (STEP 3)
4. Use one or two half hitches along the axe handle to keep the handle parallel to the rope.
5. Communicate with the fire fighter above that the axe is ready to raise. (STEP 4)

To release the axe, hold the middle of the handle, release the half hitches, and slip the knot up and off.

Hoisting a Pike Pole

A pike pole should be hoisted in a vertical position with the head at the top. After receiving the end of the utility rope from the fire fighter above, follow the steps in SKILL DRILL 10-17 to hoist a pike pole:

1. Place a clove hitch over the bottom of the handle and secure it close to the bottom of the handle.
2. Leave enough length of rope below the clove hitch for a tag line while raising the pike pole. (STEP 1)

SKILL DRILL 10-17 Hoisting a Pike Pole
(Fire Fighter I, NFPA 5.5.2)

1 Place a clove hitch over the bottom of the handle and secure it close to the bottom of the handle. Leave enough length of rope below the clove hitch for a tag line while raising the pike pole.

2 Place a half hitch around the handle above the clove hitch to keep the rope parallel to the handle.

3 Slip a second half hitch over the handle and secure it near the head of the pike pole.

4 Communicate with the fire fighter above that the pike pole is ready to raise.

3. Place a half hitch around the handle above the clove hitch to keep the rope parallel to the handle. (STEP 2)
4. Slip a second half hitch over the handle and secure it near the head of the pike pole. (STEP 3)
5. Communicate with the fire fighter above that the pike pole is ready to raise. (STEP 4)

Hoisting a Ladder

A ladder should be hoisted in a vertical position. A tag line should be attached to the bottom to keep the ladder under control as it is hoisted. If it is a roof ladder, the hooks should be in the retracted position. After receiving the end of the utility rope from the fire fighter above, follow the steps in SKILL DRILL 10-18 to hoist a ladder:

1. Tie a figure eight on a bight to create a loop, approximately 3 or 4 feet (1–1.3 meters) in diameter, that is large enough to fit around both ladder beams. (STEP 1)
2. Pass the rope between two beams of the ladder, three or four rungs from the top.
3. Pull the end of the loop under the rungs and toward the tip at the top of the ladder. (STEP 2)
4. Place the loop around the top tip of the ladder and pull on the running end of the rope to remove the slack from the rope. (STEP 3)
5. Attach a tag line to the bottom rung of the ladder to stabilize it as it is being hoisted.
6. Communicate with the fire fighter above that the ladder is ready to be raised.
7. Stabilize the bottom of the ladder as it is being hoisted by holding on to the tag line. (STEP 4)

Hoisting a Charged Hose Line

It is almost always preferable to hoist a dry hose line because water adds considerable weight to a charged line. Water weighs 8.33 pounds (3.79 kilograms) per gallon, which can make hoisting a charged line much more difficult. However, there can be occasions when it is necessary to hoist a charged hose line. After receiving the end of the utility rope from the fire fighter above, follow the steps in SKILL DRILL 10-19 to hoist a charged hose line:

1. Make sure that the nozzle is completely closed and secure. A charged hose line should have the nozzle secured in a closed position as it is hoisted because an unsecured handle (known as the bale) could get caught on something as the hose is being hoisted. The nozzle could open suddenly, resulting in an out-of-control hose line suspended in midair, with the potential for causing serious injuries and damage.
2. Use a clove hitch, 1 or 2 feet (0.3–0.6 meter) behind the nozzle, to tie the end of the hoisting rope around a charged hose line.
3. Use a safety knot to secure the loose end of the rope below the clove hitch. (STEP 1)
4. Make a bight in the rope even with the nozzle shut-off handle, which must be in the forward (completely off) position.
5. Insert the bight through the handle opening and slip it over the end of the nozzle. When the bight is pulled tight, it will create a half hitch and secure the handle in the off position while the charged hose line is hoisted.
6. Communicate with the fire fighter above that the hose line is ready to hoist. (STEP 2)

The knot can be released after the line is hoisted by removing the tension from the rope and slipping the bight back over the end of the nozzle.

Hoisting an Uncharged Hose Line

Before hoisting a dry hose line, you should fold the hose back on itself and place the nozzle on top of the hose. This action ensures that water will not reach the nozzle if the hose is accidentally charged while being hoisted. It also eliminates an unnecessary stress on the couplings by ensuring that the rope pulls on the hose and not directly on the nozzle. After receiving the end of the utility rope from the fire fighter above, follow the steps in SKILL DRILL 10-20 to hoist an uncharged hose line:

1. Fold about 3 feet (1 meter) of hose back on itself and place the nozzle on top of the hose. (STEP 1)
2. Make one or two half hitches in the rope and slip it over the nozzle. Move the half hitch along the hose and secure it about 6 inches (15 centimeters) from the fold. (STEP 2)
3. Tie a clove hitch near the end of the rope, wrapping the rope securely around both the nozzle and the hose. The clove hitch should hold both the nozzle and the hose. (STEP 3)
4. Communicate with the fire fighter above that the hose line is ready to hoist.
5. Hoist the hose with the fold at the top and the nozzle pointing down.
6. Before releasing the rope, the fire fighters at the top must pull up enough hose so that the weight of the hanging hose does not drag down the hose. (STEP 4)

FIRE FIGHTER II Tips **FFII**

Ropes and knots are challenging in the best of circumstances. To be proficient and to work in an effective and efficient manner when tying knots and moving equipment, you should consistently work on your rope and knot skills. Keep a short piece of rope close at hand so that you can practice your skills during down times.

Hoisting an Exhaust Fan or Power Tool

Several types of tools and equipment—including an exhaust fan, a chainsaw, a circular saw, or any other object that has a strong closed handle—can be hoisted using the same

SKILL DRILL 10-18 Hoisting a Ladder
(Fire Fighter I, NFPA 5.5.2)

1. Tie a figure eight on a bight to create a loop, approximately 3 or 4 feet (1–1.3 meters) in diameter, that is large enough to fit around both ladder beams.

2. Pass the rope between two beams of the ladder, three or four rungs from the top. Pull the end of the loop under the rungs and toward the tip at the top of the ladder.

3. Place the loop around the top tip of the ladder.

4. Pull on the running end of the rope to remove the slack from the rope. Attach a tag line to the bottom rung of the ladder to stabilize it as it is being hoisted. Communicate with the fire fighter above that the ladder is ready to be raised. Stabilize the bottom of the ladder as it is being hoisted by holding on to the tag line.

technique. In this situation, the hoisting rope is secured to the object by passing the rope through the opening in the handle. A figure eight with a follow-through knot is then used to close the loop.

Some types of equipment require that you use additional half hitches to balance the object in a particular position while it is being hoisted. For example, power saws are hoisted in a level position to prevent the fuel from leaking out.

You should practice hoisting the actual tools and equipment used in your department. You should be able to perform this task automatically and in adverse conditions.

Remember that you always use utility rope for hoisting tools. You do not want to get oil or grease on designated life safety ropes. If a life safety rope becomes oily or greasy, it should be taken out of service and destroyed so that it will not be mistakenly used again as a life safety rope. The damaged rope can be cut into short lengths and used for utility rope.

After receiving the end of the utility rope from the fire fighter above, follow the steps in SKILL DRILL 10-21 to hoist an exhaust fan:

1. Tie a figure eight knot in the rope about 3 feet (1 meter) from the working end of the rope. (STEP 1)

SKILL DRILL 10-19 Hoisting a Charged Hose Line
(Fire Fighter I, NFPA 5.5.2)

1 Make sure that the nozzle is completely closed and secure. Use a clove hitch, 1 or 2 feet (0.3–0.6 meters) behind the nozzle, to tie the end of the hoisting rope around a charged hose line. Use a safety knot to secure the loose end of the rope below the clove hitch.

2 Make a bight in the rope even with the nozzle shut-off handle. Insert the bight through the handle opening and slip it over the end of the nozzle. When the bight is pulled tight, it will create a half hitch and secure the handle in the off position while the charged hose line is hoisted. Communicate with the fire fighter above that the hose line is ready to hoist.

2 Loop the working end of the rope around the fan handle and back to the figure eight knot. (STEP 2)

3 Secure the rope by tying a figure eight with a follow-through by threading the working end back through the first figure eight in the opposite direction.

4 Attach a tag line to the fan for better control.

5 Communicate with the fire fighter above that the exhaust fan is ready to hoist. (STEP 3)

- Twisted
- Braided
- Kermantle

SKILL DRILL 10-20 Hoisting an Uncharged Hose Line

(Fire Fighter I, NFPA 5.5.2)

1 Fold about 3 feet (1 meter) of hose back on itself and place the nozzle on top of the hose.

2 Make one or two half hitches in the rope and slip it over the nozzle. Move the half hitch along the hose and secure it about 6 inches (15 centimeters) from the fold.

3 Tie a clove hitch near the end of the rope, wrapping the rope securely around both the nozzle and the hose. The clove hitch should hold both the nozzle and the hose.

4 Communicate with the fire fighter above that the hose line is ready to hoist. Hoist the hose with the fold at the top and the nozzle pointing down. Before releasing the rope, the fire fighters at the top must pull up enough hose so that the weight of the hanging hose does not drag down the hose.

SKILL DRILL 10-21 Hoisting an Exhaust Fan or Power Tool

(Fire Fighter I, NFPA 5.5.2)

1 Tie a figure eight knot in the rope about 3 feet (1 meter) from the working end of the rope.

2 Loop the working end of the rope around the fan handle and back to the figure eight knot.

3 Secure the rope by tying a figure eight with a follow-through by threading the working end back through the first figure eight in the opposite direction. Attach a tag line to the fan for better control. Communicate with the fire fighter above that the exhaust fan is ready to hoist.

Wrap-Up

Chief Concepts

- The three primary types of rope in the fire service are:
  - *Life safety rope.* Used to support people during a rescue. This type of rope is used only for life-saving purposes.
  - *Escape rope.* Used as a single-purpose emergency self-escape rope.
  - *Utility rope.* Used to perform all other tasks, such as hoisting equipment.
- Life safety ropes are rated as either technical use life safety rope or general use life safety rope.
- An escape rope is designed to be used once by one fire fighter.
- Ropes can be made of natural or synthetic fibers. Natural fibers can be weakened by mildew and age, so they cannot be used to make life safety ropes. Synthetic fibers are generally stronger than natural fibers and are used in life safety ropes.
- Synthetic fibers can be damaged by ultraviolet light.
- Nylon is the most common synthetic fiber used in life safety ropes, followed by polyester and polypropylene. Polypropylene does not absorb water, so it is often used for water rescues.
- Common rope construction includes the following types:
  - *Twisted ropes.* Individual fibers twisted into strands that are twisted together to form the rope.
  - *Braided ropes.* Constructed by weaving or intertwining strands together in the same way that hair is braided.
  - *Kernmantle ropes.* Consists of the kern and the mantle. The kern is the core of the rope and the mantle is the braided covering that protects the core. These ropes can be either dynamic or static.
- During technical rescue incidents, several hardware components are used in addition to rope to rescue victims:
  - *Carabiner.* Connects one rope to another rope or a harness
  - *Harness.* Webbing that secures a person to a rope or solid object
- Most rope rescue incidents involve people who are trapped in normally inaccessible locations such as trenches, confined spaces, and open water.
- All ropes need proper care to perform in an optimal manner. The four parts to the rope maintenance formula are the following:
  - Care
  - Clean
  - Inspect
  - Store
- The principles of caring for a rope include the following:
  - Protect the rope from sharp and abrasive surfaces.
  - Protect the rope from rubbing against another rope or webbing.
  - Protect the rope from heat, chemicals, and flames.
  - Protect the rope from prolonged exposure to sunlight.
- When inspecting life safety rope, consider these questions:
  - Has the rope been exposed to heat or flame?
  - Has the rope been exposed to abrasion?
  - Has the rope been exposed to chemicals?
  - Has the rope been exposed to shock loads?
  - Are there any depressions, discolorations, or lumps in the rope?
- A rope record for a life safety rope includes a history of when the rope was purchased, when it was used, how it was used, each inspection, and what kinds of loads were applied to it.
- Ropes should be stored away from temperature extremes, out of sunlight, and in areas with good air circulation. Rope bags can be used to protect and store ropes.
- Knots can be used for one or more particular purposes. Hitches, such as the clove hitch, are used to attach a rope around an object. Knots, such as the figure eight and the bowline, are used to form loops. Bends, such as the sheet bend or Becket bend, are used to join two ropes together. Safety knots, such as the overhand knot, are used to secure the ends of ropes to prevent them from coming untied.
- Specific terminology is used to refer to the parts of a rope in describing how to tie knots:
  - The working end is the part of the rope used for forming the knot.
  - The running end is the part of the rope used for lifting or hoisting.
  - The standing part is the rope between the working end and the running end.
  - A bight is formed by reversing the direction of the rope to form a U bend with two parallel ends.
  - A loop is formed by making a circle in the rope.
  - A round turn is formed by making a loop, and then bringing the two ends of the rope parallel to each other.
- The knots that a fire fighter should know how to tie are as follows:
  - *Safety knot.* Used to finish other basic knots
  - *Half hitch.* Knots that wrap around an object
  - *Clove hitch.* Used to attach a rope to a round object

- *Figure eight.* Basic knot used to produce a family of other knots
- *Figure eight on a bight.* Used to create a secure loop at the working end of a rope
- *Figure eight follow-through.* Used to create a secure loop at the end of the rope when the working end must be wrapped around an object or passed through an opening before the loop can be formed.
- *Bowline.* Used to create a loop
- *Water Knot.* Used to join the ends of webbing together.
- The bends that a fire fighter should know how to create in order to join two ropes together are the sheet bend and the Becket bend.

Hot Terms

Bend A knot used to join two ropes together.

Bight A U shape created by bending a rope with the two sides parallel.

Block creel construction Rope constructed without knots or splices in the yarns, ply yarns, strands or braids, or rope. (NFPA 1983)

Braided rope Rope constructed by intertwining strands in the same way that hair is braided.

Carabiner An auxiliary equipment system item; load-bearing connector with a self-closing gate used to join other components of life safety rope. (NFPA 1983)

Depressions Indentations felt on a kernmantle rope that indicate damage to the interior (kern) of the rope.

Dynamic rope A rope generally made from synthetic materials that is designed to be elastic and stretch when loaded. Dynamic rope is often used by mountain climbers.

Escape rope A system component; a single-purpose, one-time use, emergency self-escape (self-rescue) rope; not classified as a life safety rope. (NFPA 1983)

Harness A piece of equipment worn by a rescuer that can be attached to a life safety rope.

Hitch A knot that attaches to or wraps around an object so that when the object is removed, the knot will fall apart. (NFPA 1670, *Standard on Operations and Training for Technical Search and Rescue Incidents*)

General use life safety rope A life safety rope that is no larger than 5/8" (16 mm) and no smaller than 7/16" (11 mm), with a minimum breaking strength of 8992 lbf (40 kN).

Kernmantle rope Rope made of two parts—the kern (interior component) and the mantle (the outside sheath).

Knot A fastening made by tying together lengths of rope or webbing in a prescribed way. (NFPA 1670)

Ladder halyards Rope used on extension ladders to raise a fly section.

Life safety rope Rope dedicated solely for the purpose of supporting people during rescue, firefighting, other emergency operations, or during training evolutions. (NFPA 1983)

Loop A piece of rope formed into a circle.

Rope bag A bag used to protect and store rope so that the rope can be easily and rapidly deployed without kinking.

Rope record A record for each piece of rope that includes a history of when the rope was placed in service, when it was inspected, when and how it was used, and which types of loads were placed on it.

Round turn A piece of rope looped to form a complete circle with the two ends parallel.

Running end The part of a rope used for lifting or hoisting.

Safety knot A knot used to secure the leftover working end of the rope.

Shock load An instantaneous load that places a rope under extreme tension, such as when a falling load is suddenly stopped as the rope becomes taut.

Standing part The part of a rope between the working end and the running end.

Static rope A rope generally made out of synthetic material that stretches very little under load.

Technical use life safety rope A life safety rope with a diameter that is 3/8" (9.5 mm) or greater, but is less than 1/2" (12.5mm), with a minimum breaking strength of 4496 lbf (20 kN). Used by highly trained rescue teams that deploy to very technical environments such as mountainous and/or wilderness terrain.

Twisted rope Rope constructed of fibers twisted into strands, which are then twisted together.

Utility rope Rope used for securing objects, for hoisting equipment, or for securing a scene to prevent bystanders from being injured. Utility rope must never be used in life safety operations.

Water Knot A knot used to join the ends of webbing together.

Working end The part of the rope used for forming a knot.

FIRE FIGHTER *in action*

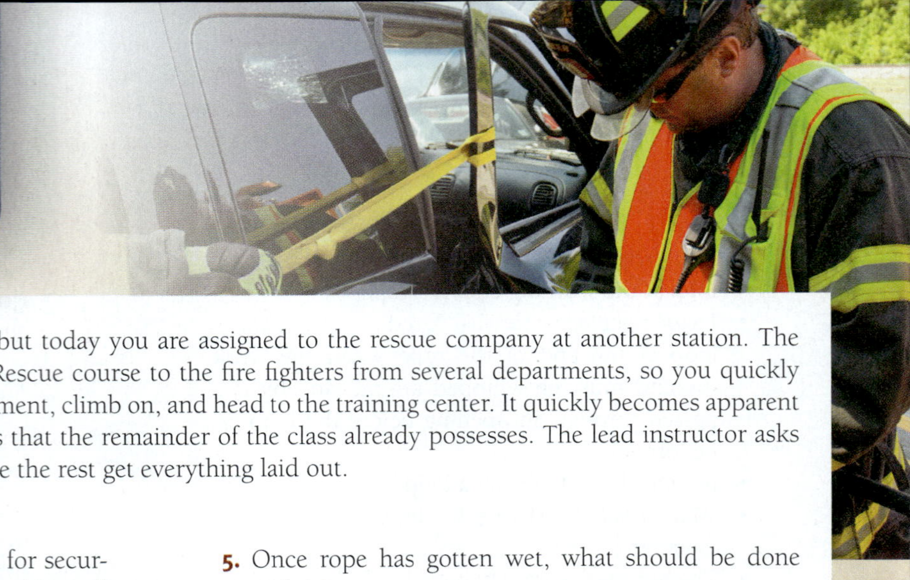

You are normally assigned to a truck company, but today you are assigned to the rescue company at another station. The captain is scheduled to teach the Intro to Tech Rescue course to the fire fighters from several departments, so you quickly familiarize yourself with the location of the equipment, climb on, and head to the training center. It quickly becomes apparent that one particular student lacks basic rope skills that the remainder of the class already possesses. The lead instructor asks you to review the basics with the fire fighter while the rest get everything laid out.

1. What type of rope is designed to be used for securing objects, hoisting equipment, and blocking off scenes?
   - **A.** Technical use life safety rope
   - **B.** General use life safety rope
   - **C.** Utility rope
   - **D.** Escape rope
2. The part of a rope used for lifting or hoisting is called the ________.
   - **A.** Standing end
   - **B.** Running end
   - **C.** Bight
   - **D.** Round turn
3. The knot used to secure the leftover working end of the rope.
   - **A.** Safety
   - **B.** Figure eight
   - **C.** Bowline
   - **D.** Half hitch
4. A ________ rope is a rope generally made out of synthetic material that stretches very little under load.
   - **A.** Dynamic
   - **B.** Twisted
   - **C.** Braided
   - **D.** Static
5. Once rope has gotten wet, what should be done with it?
   - **A.** Lay it in the sunlight so that it does not mold.
   - **B.** Air dry it out of direct sunlight.
   - **C.** Place it in a clothes dryer on high heat.
   - **D.** The rope can no longer be used.
6. Which knot is used to create a secure loop at the end of the rope when the working end must be wrapped around an object or passed through an opening before the loop can be formed?
   - **A.** Figure eight follow-through
   - **B.** Figure eight on a bight
   - **C.** Half hitch
   - **D.** Clove hitch
7. A ________ is when a piece of rope is looped to form a complete circle with the two ends parallel.
   - **A.** Bight
   - **B.** Hitch
   - **C.** Round turn
   - **D.** Loop

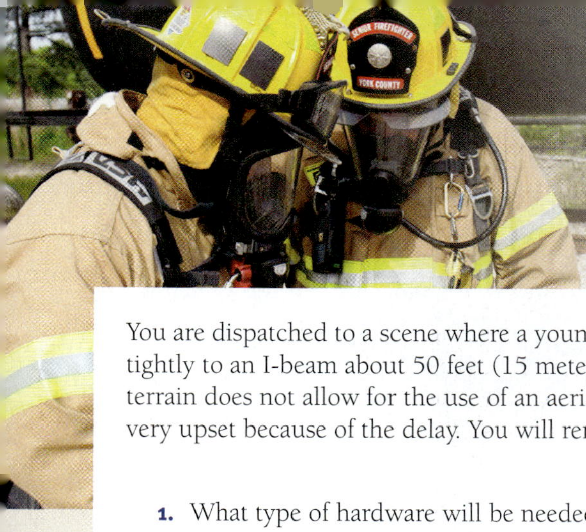

FIRE FIGHTER II *in action*

You are dispatched to a scene where a young boy has climbed up on a cell tower. On your arrival, you find that he is clinging tightly to an I-beam about 50 feet (15 meters) up the tower and the parents of the boy are on-scene and they are frantic. The terrain does not allow for the use of an aerial ladder, so the decision is made to use a high-angle rescue. The crowd is getting very upset because of the delay. You will remain on the ground while the rope rescue technician makes the ascent.

1. What type of hardware will be needed for this rescue?
2. What type of harness will be needed for this rescue?
3. How will the parents be reassured?
4. How will emotions of the crowd be manged**FIGURE 10-9** **B.** This close-up shows the core and mantle construction.

Response and Size-Up

Fire Fighter I

Knowledge Objectives

After studying this chapter, you will be able to:

- Explain how fire stations receive and process dispatch information. (p 301)
- Identify the information in a typical dispatch message. (p 301)
- Describe how to safely ride a fire apparatus to an emergency scene. (NFPA 5.3.2, p 301–302)
- List the NFPA standards that require fire fighters to wear safety belts while riding in a fire apparatus (NFPA 5.3.2, p 302).
- List the prohibited practices when riding in a fire apparatus to an emergency scene. (NFPA5.3.2.A, p 303)
- Describe how to manage traffic safely at an emergency scene. (NFPA 5.3.3, 5.3.3.A, p 303–304)
- Describe the actions fire fighters must take when arriving at an emergency scene. (NFPA 5.3.3, p 304)
- Explain how to shut off a structure's electrical service. (NFPA 5.3.3, 5.3.18, 5.3.18.A, 5.3.18.B, p 305–306)
- Explain how to shut off a structure's gas service. (NFPA 5.3.3, 5.3.18, 5.3.18.A, 5.3.18.B, p 306)
- Explain how to shut off a structure's water service. (NFPA 5.3.3, 5.3.18, 5.3.18.A, 5.3.18.B, p 306)
- Describe the process of performing an initial size-up. (p 306–311)
- List the two basic categories of information used in the size-up process. (p 308–310)
- Explain how the size-up process determines the resources required at the emergency incident. (p 310–311)
- List the five objectives of an incident action plan. (p 311–312)
- Describe the two methods of attack used to extinguish a fire. (p 313)

Skills Objectives

After studying this chapter, you will be able to perform the following skills:

- Mount an apparatus safely. (NFPA 5.3.2.A, 5.3.2.B, p 302)
- Dismount from an apparatus safely. (NFPA 5.3.2.B, 5.3.3.B, p 303)

CHAPTER 11

Fire Fighter II

FFII

Knowledge Objectives

There are no knowledge objectives for Fire Fighter II candidates. NFPA 1001 contains no Fire Fighter II Job Performance Requirements for this chapter.

Skills Objectives

There are no skill objectives for Fire Fighter II candidates. NFPA 1001 contains no Fire Fighter II Job Performance Requirements for this chapter.

Additional NFPA Standards

- NFPA 1500, *Standard on Fire Department Occupational Safety and Health Program*

You Are the Fire Fighter

You are dispatched to a report of a house fire. As you pull out of the station, the dispatcher reports multiple calls and that occupants are trapped in the building. Upon your arrival, there is black smoke boiling out of a back window, with brown smoke coming from the eaves and upstairs windows. As the fire officer gives a size-up of the building and conditions, a neighbor runs up and says that he heard someone screaming from the upstairs. Without missing a beat, the fire officer develops a mental incident action plan and makes assignments to the crews. You are amazed by how she is able to calmly coordinate such a chaotic scene.

1. What type of attack will be made on this fire?
2. What is the highest priority for this fire?
3. What objectives must be assigned for this incident?

Introduction

Response involves a series of actions that begin when a crew is dispatched to an alarm and end with the crew's arrival at the emergency incident. Response actions include receiving the alarm, donning protective clothing and equipment, mounting the apparatus, and transporting equipment and personnel to the emergency incident quickly and safely.

Because fire fighters must be ready to react immediately to an alarm, preparations for response begin long before the alarm is sounded. These preparations include checking personal equipment, ensuring that the fire apparatus is ready, and making sure that all equipment carried on the apparatus is ready for use. Fire fighters also should be familiar with their response district, know the buildings under their protection, and understand their department's standard operating procedures (SOPs).

Response actions for the apparatus driver also include considering road and traffic conditions, determining the best route to the incident, identifying nearby hydrant locations or water sources, and selecting the best position for the apparatus at the incident scene.

Size-up is a systematic process of information gathering and situation evaluation that begins when an alarm is received. It continues during the response and includes the initial observations made upon arrival at the incident scene. Size-up information is essential for determining the appropriate strategy and tactics for each situation.

The incident commander (IC) and company officers are ultimately responsible for obtaining the necessary information to manage the emergency incident. Each fire fighter should also be involved in the process of gathering and processing information. The observations made by individual fire fighters will help them anticipate which actions might be necessary at the scene and will provide company officers with critical information.

Response

Trained fire fighters must be ready to respond to an emergency at any time during their tour of duty. This process begins by ensuring that your personal protective equipment (PPE) is complete, ready for use, and in good condition. At the beginning of each tour of duty, place your PPE in its designated location, which depends on your assigned riding position on the apparatus **FIGURE 11-1**.

You should also conduct a daily inspection of the self-contained breathing apparatus (SCBA) for your riding position at the beginning of each tour of duty **FIGURE 11-2**. The air

FIGURE 11-1 Check your PPE and place it in the designated location for your riding position.

cylinder should be full, the face piece clean, and the personal alert safety system (PASS) operable. Also check the availability and operation of your hand light and any hand tools you might require, based on your assigned position on the apparatus FIGURE 11-3. Recheck your PPE and tools thoroughly when you return from each emergency response and clean them whenever they are used. This routine will help ensure that your gear and equipment are fully functional when the next alarm comes.

Alarm Receipt

The emergency response process begins when an alarm is received at the fire station. Fire fighters should be familiar with the dispatch method or methods used by their departments. In many fire departments, a local or regional communications center dispatches individual units. In smaller departments, the dispatcher may be located at the fire station or other location. Departments should have both a primary method and a backup method of transmitting alarms to stations.

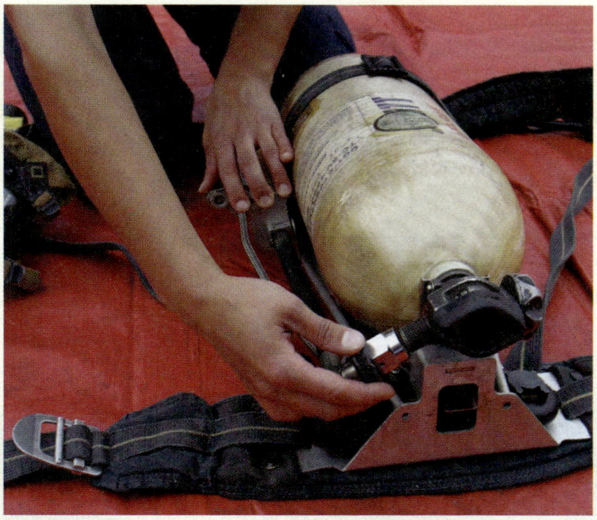

FIGURE 11-2 Conduct the daily check of your SCBA at the beginning of each tour of duty.

FIGURE 11-3 Check the tools assigned to your riding position.

Although radio, telephone, and public address systems are often used to transmit information to fire stations, the use of computer terminals and printers to transmit dispatch messages is increasing. Some fire departments still use a system of bells to transmit alarms. Although volunteer or rural departments may use outdoor sirens or horns to summon fire fighters to an emergency, most volunteer fire fighters receive dispatch messages over pagers or cell phones.

Dispatch messages contain varying amounts of useful information, based on what the dispatcher learns from the person initially reporting the incident. At a minimum, the dispatch information will include the location of the incident, the type of emergency (e.g., vehicle fire, structure fire, medical call), and the units that are due to respond. Computer-aided dispatch systems often provide additional information about the building or premises on the dispatch printout.

As more information becomes available, the telecommunicator will include it in dispatch messages to later-responding units or advise the responding units by radio while they are en route. Each additional piece of information can help responding fire fighters plan their strategies and prepare themselves for the incident.

When an alarm is received, the response should be prompt and efficient. Responding fire fighters should walk briskly to the apparatus. There is no need to run; the objective is to respond quickly, without injuring anyone or causing any damage. Follow established procedures to ensure that stoves, faucets, and other appliances at the station are shut off. Wait until the apparatus bay doors are fully open before leaving the station.

FIRE FIGHTER Tips

Volunteer fire fighters who are not assigned to specific tours of duty or riding positions should check their PPE, SCBA, and associated tools and equipment on a regular basis to ensure that these items are ready for use whenever an emergency response becomes necessary. After each use, all PPE should be carefully cleaned and checked before it is put away.

Riding the Apparatus

Don your PPE before mounting the apparatus; do not attempt to dress while the apparatus is on the road. Your seat belt should be fastened whenever the apparatus is moving. Wait until you dismount at the incident scene to don any protective clothing that was not donned prior to mounting the apparatus. Don SCBA only after the apparatus stops at the scene, unless the SCBA is seat mounted.

All equipment should be properly mounted, stowed, or secured on the fire apparatus. Unsecured equipment in the crew compartment can prove dangerous if the apparatus must stop or turn quickly, because a flying tool, map book, or PPE can seriously injure a fire fighter.

Be careful when mounting and dismounting apparatus, because the steps on fire apparatus are often high and can be slippery. Follow the steps in **SKILL DRILL 11-1** to mount an apparatus properly:

1. When mounting (climbing aboard) fire apparatus, always have at least one hand firmly grasping a handhold and at least one foot firmly placed on a foot surface. Maintain the one hand and one foot placement until you are seated. (**STEP 1**)
2. Fasten your seat belt, and leave it fastened until the apparatus is stopped at its destination. Don any other required safety equipment for the response, such as hearing protection and intercom. (**STEP 2**)

All fire fighters must be seated in their assigned riding positions with seat belts and/or harnesses fastened before the apparatus begins to move. NFPA 1500, *Standard on Fire Department Occupational Safety and Health Program* and NFPA 1001, *Standard for Fire Fighter Professional Qualifications*, requires all fire fighters to be in their seats, with seat belts secured, whenever the vehicle is in motion. Do not unbuckle your seat belt to don any clothing or equipment while the apparatus is en route to an incident.

The noise produced by sirens and air horns can have long-term, damaging effects on a fire fighter's hearing. For this reason, your department should provide hearing protection for personnel riding on fire apparatus. Use it. Some of these devices include radio and intercom capabilities so that fire fighters can talk to one another and hear information from the dispatcher or IC.

During transport, limit conversation to the exchange of pertinent information. Listen for instructions from the IC, for instructions from your company officer, and for additional information about the incident over the radio. The vehicle operator's attention should be focused on driving the apparatus safely to the scene of the incident.

The ride to the incident is a good time to consider any relevant factors that could affect the situation. These factors could include the time of day or night, the temperature, the presence of precipitation or wind, the type of occupancy, the type of construction, and the location and type of incident. Using this time to think ahead will help you mentally prepare for the various possibilities that you might encounter at the scene.

Emergency Response

The fire apparatus driver/operator must always exercise caution when driving to an incident. Fire apparatus can be very large, heavy, and difficult to maneuver. Operating an emergency vehicle without the proper regard for safety can endanger the lives of both the fire fighters on the vehicle and any civilian drivers and pedestrians encountered along the way. Although the impulse to respond as speedily as possible is understandable, safety should never be compromised for a faster response time.

Fire fighters who drive emergency vehicles must have special driver training and know the laws and regulations that

SKILL DRILL 11-1 Mounting Apparatus
(Fire Fighter I, NFPA 5.3.2)

1 When mounting (climbing aboard) fire apparatus, always have at least one hand firmly grasping a handhold and at least one foot firmly placed on a foot surface. Maintain the one hand and one foot placement until you are seated.

2 Fasten your seat belt and then don any other required safety equipment for the response, such as hearing protection.

apply to emergency responses. Many jurisdictions require a special driver's license to operate fire apparatus. The rules that apply to emergency vehicles are very specific, and the driver/operator is legally responsible for the safe operation of the vehicle and the safety of the vehicle occupants at all times.

An emergency vehicle must always be operated with due regard for the safety of everyone on the road. Although most states and provinces permit drivers of emergency vehicles to take exception to specific traffic regulations when responding to emergency incidents, driver/operators must always consider the potential actions of other drivers before making such a decision. For example, traffic laws require other drivers to yield the right of way to an emergency vehicle. There is no assurance, however, that other drivers will do so when an emergency vehicle approaches. The driver/operator must also anticipate which routes other units responding to the same incident will take.

Fire fighters who respond to emergency incidents in their personal vehicles must follow the specific laws and regulations of their state or province and follow departmental SOPs related to this issue. Some jurisdictions allow fire fighters to equip their privately owned vehicles with warning devices and to operate them as emergency vehicles; others grant no special status to privately owned vehicles driven during an emergency response. In some areas, volunteer fire fighters responding to an emergency incident can use colored lights to request the right of way from other drivers. Additional information about emergency driving is presented in the Fire Fighter Safety chapter.

Prohibited Practices

For safety's sake, SOPs often prohibit specific actions during response. As noted previously, fire fighters must remain seated with their seat belts securely fastened while the emergency vehicle is in motion. Never unfasten your seat belt to retrieve or don equipment. Do not dismount the apparatus until the vehicle comes to a complete stop.

In addition, you should never stand up while riding on apparatus. Do not hold on to the side of a moving vehicle or stand on the rear step. When a vehicle is in motion, everyone aboard must be seated and belted in an approved riding position.

Dismounting a Stopped Apparatus

When the apparatus arrives at the incident scene, the driver/operator will park it in a location that is both safe and functional. Wait until the vehicle comes to a complete stop before dismounting. Always check for traffic before opening the doors or stepping out of the apparatus. During the dismount, watch for other hazards—for example, ice and snow, downed power lines, uneven terrain, or hazardous materials—that could be present.

Be careful when dismounting apparatus. The increased weight of PPE and adverse conditions can contribute to slips, strains, and sprains. Use handrails when mounting or dismounting the apparatus. Follow the steps in **SKILL DRILL 11-2** to dismount an apparatus safely:

1. Become familiar with your riding position and the safest way to dismount. (**STEP 1**)
2. Keep one hand grasping a handhold and one foot placed firmly on a flat surface when leaving the apparatus, especially on wet or potentially icy roadway surfaces. (**STEP 2**)

Traffic Safety on the Scene

An emergency incident scene presents several risks to fire fighters in addition to the hazards of fighting fires and performing other duties. One of these dangers is traffic, particularly when the incident scene is on a street or highway. Traffic safety should be a major concern for the first-arriving units

SKILL DRILL 11-2 Dismounting a Stopped Apparatus
(Fire Fighter I, NFPA 5.3.2, 5.3.3)

1 Become familiar with your riding position and the safest way to dismount.

2 Maintain the one hand and one foot placement when leaving the apparatus, especially on wet or potentially icy roadway surfaces.

because approaching drivers might not see emergency workers or realize how much room fire fighters need to work safely.

The first unit or units to arrive at the incident scene have a dual responsibility. Not only must the fire fighters focus on the emergency situation facing them, but they must also consider approaching traffic, including other emergency vehicles, and other, less obvious hazards. Always check for traffic before opening doors and dismounting the apparatus, and watch out for traffic when working in the street. Follow departmental SOPs to close streets quickly and to block access to areas where operations are being conducted.

One of the most dangerous work areas for fire fighters is on the scene of a highway incident, where traffic can be approaching at high speeds. Place traffic cones, flares, emergency scene signage, the traffic safety officer, and other warning devices far enough away from the incident to slow approaching traffic and direct it away from the work area. Placement of emergency vehicles on the scene is also critical. With proper placement, such vehicles can act as a barrier between oncoming traffic and the scene. All fire fighters working at highway incidents should wear high-visibility safety vests in addition to their normal PPE. These vests should meet the ANSI 207 Standard for Public Safety Vests. Many fire departments have specific SOPs covering required safety procedures for these incidents.

FIRE FIGHTER Tips

- Do not attempt to mount or dismount a moving vehicle.
- Do not remove your seat belt until the apparatus comes to a complete stop.
- Do not stand directly behind an apparatus that is backing up. Instead, stand off to one side, where the driver can see you in the rear-view mirror. All fire apparatus should have working, audible alarms when in reverse gear.

Arrival at the Incident Scene

From the moment that fire fighters arrive at an emergency incident scene, their department's SOPs and the structured incident command system (ICS) must guide all of their actions. Fire fighters should always work in assigned teams (companies or crews) and be guided by a strategic plan for the incident. Teamwork and disciplined action are essential to provide for the safety of all fire fighters and the effective, efficient conduct of operations.

The command structure plans, coordinates, and directs operations. Fire fighters responding to an incident on an apparatus constitute the crew assigned to that vehicle and take direction from their company officer, who ensures that their actions are coordinated with the overall plan. A fire fighter who arrives at the scene independently of an apparatus must report to the IC and receive an assignment to a company or crew under the supervision of a company officer.

Fire fighters who take action on their own, without regard to SOPs, the command structure, or the strategic plan, are freelancing. This type of activity—whether it is done by individual fire fighters, groups of fire fighters, or full companies—is unacceptable. Freelancing cannot be tolerated at any emergency incident. The safety of every person on the scene can be compromised by fire fighters who do not work within the system.

Personnel must not respond to an emergency incident scene unless they have been dispatched or have an assigned duty to respond. Unassigned units and individual personnel arriving on the scene can overload the IC's ability to manage the incident effectively. Individuals who simply show up and find something to do are likely to compromise their own safety and create more problems for the command staff. All personnel must operate within the established system, reporting to a designated supervisor under the direction of the IC.

Personnel Accountability System

A personnel accountability system should be used to track every fire fighter at every incident scene. This system maintains an updated list of the fire fighters assigned to each vehicle or crew and tracks each crew's assignment at the fire scene. If any fire fighters are reported missing, lost, or injured, the accountability system can identify who is missing and what his or her last assignment was.

Different fire departments use different types of accountability systems. Consult your agency's SOPs for more specific information about the system being used by your department and always follow the required procedures.

Many fire departments use personnel accountability tags (PATs) to track individual fire fighters. An accountability tag generally includes the fire fighter's name, ID number, and photograph and may include additional information, such as an individual's medical history, medications, allergies, and other factors. When this system is used, fire fighters who respond to an incident on an apparatus deposit their PATs in a designated location on the vehicle FIGURE 11-4. The PATs are then collected from each vehicle and taken to the command post. Fire fighters who respond directly to the scene must report to the command post to receive an assignment and deposit their PATs.

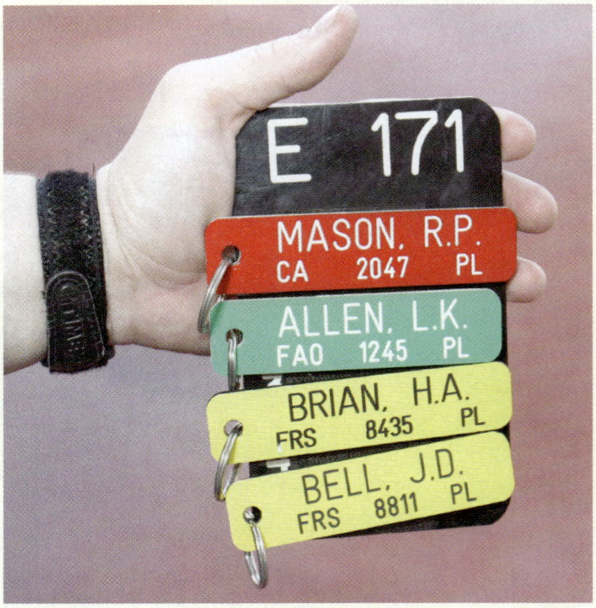

FIGURE 11-4 There should be a designated location on the fire apparatus for the fire fighters to deposit their PATs.

Fire Fighter Safety Tips

Teamwork is a hallmark of effective fire-ground operations. Training and teamwork produce effective, coordinated operations. Unassigned individual efforts and freelancing can disrupt the operations and endanger the lives of both fire fighters and civilians.

Controlling Utilities

Controlling utilities is one of the first tasks that must be accomplished at many working structure fires. Most departments have written SOPs that define when the utilities are to be shut off. Although this responsibility is often assigned to a particular company or crew, all fire fighters should know how to shut off a building's electrical, gas, and water service.

Electrical and gas utilities to a burning building may be disconnected for any of several reasons. For example, if faulty electrical equipment or a gas appliance caused the fire, shutting off the supply will help alleviate the problem. Disconnecting electrical service can also prevent problems such as short circuits and electrical arcing that could result from fire or water damage. In addition, disconnecting electrical service reduces the risk of electrical injury to fire fighters. Shutting down gas service eliminates the potential for explosion due to damaged, leaking, or ruptured gas piping.

Controlling utilities is particularly important if fire fighters need to open walls or ceilings to look for fire in the void spaces, to cut ventilation holes in roofs, or to penetrate through floors. Because these spaces often contain both electrical lines and gas pipes, the danger of electrocution or explosion exists unless these utilities are disconnected.

For example, a fire fighter using an axe to open a void space could be electrocuted if the tool contacts energized electrical wires. Fire fighters operating a hose could be injured if the water conducts electricity back to the nozzle. Fire fighters might also be electrocuted if they come in contact with wires that have been pulled loose or exposed by a structural collapse.

Gas pipes in walls, ceilings, roofs, or floors can also be inadvertently damaged by fire fighters. A power saw slicing through a gas pipe can cause a rapid release of gas and create the potential for an explosion. A structural collapse can rupture gas lines or create leaks in the stressed piping.

Controlling utilities also alleviates the risk of an additional fire or explosion. Shutting off electrical service, for example, eliminates potential ignition sources that might cause an explosion if leaking gas has accumulated inside the building; shutting off the gas supply will prevent any further leakage. The electrical supply must be disconnected at a location outside the area where gas might be present. Interrupting the power at a remote location alleviates the risk of an electrical arc that could cause a gas explosion.

If a serious water leak has occurred inside the building, shutting off the water supply may prevent electrical problems and help minimize additional water damage to the structure and contents. Shutting off the water supply to a fire-damaged building is often necessary to control leaking pipes, which could cause additional damage to the building.

Electrical Service

Electric delivery service arrangements depend on the utility company providing that service and the age of the system. The most common installation is a service drop from above-ground utility wires to the electric meter, which is typically mounted on the outside of the building **FIGURE 11-5**. Other types of electric service installations include underground connections, inside meters, and multiple services or separate electric meters for different tenants or areas inside a building.

Fire departments must work with their local electric utility companies to identify the different types of electric services and arrangements used in the area and to determine the proper procedures for dealing with each type. Often, a main disconnect switch can be operated to interrupt the power. Most utility companies will train fire fighters to identify and operate shutoff devices on typical installations. Other utility companies request that the fire department call a company representative to shut off the power. Shutting off large systems that involve high-voltage equipment should be done by a company technician or a trained individual from the premises.

A utility company representative should be called to interrupt power from a remote location, such as a utility pole. This step may be necessary if the outside wires have been damaged by fire, if fire fighters are working with ladders or aerial apparatus, or if an explosion is possible. In urban areas, the electric utility company can usually dispatch a qualified technician or crew to respond quickly to a fire or emergency incident. In many cases, qualified technicians are dispatched automatically to all working fires. Some utility companies can interrupt the service to an area from a remote location in response to a fire department request.

In general, to turn off an electrical service to a structure, you must first identify the circuit breaker box that supplies the

FIGURE 11-5 The electric service often has an exterior meter connected to above-ground utility wires.

building or part of the building in which you need to disconnect the power. You can then turn off the main circuit breaker that supplies the whole breaker box. It is important to leave a fire fighter at the circuit breaker box, or to use a lockout system that tags the box, to ensure that the power does not get turned on accidentally.

Gas Service

Both natural gas and liquefied petroleum gas (LPG) are used for heating and cooking. Generally, natural gas is delivered through a network of underground pipes. By contrast, LP gas is usually delivered by a tank truck and stored in a container on the premises. In some areas, LPG is distributed from one large storage tank through a local network of underground pipes to several customers.

A single valve usually controls the natural gas supply to a building. This valve is generally located outside the building at the entry point of the gas piping; natural gas service has a distinctive piping arrangement. In older buildings, the shut-off valve for a natural gas system may be located in the basement.

The shut-off valve for a natural gas system is usually a quarter-turn valve with a locking device so it can be secured in the off position **FIGURE 11-6**. When the handle is in line with the pipe, the valve is open; when the handle is at a right angle to the pipe, the valve is closed. A special key, an adjustable wrench, or a spanner wrench can be used to turn the handle.

The valve for an LPG system is usually located at the storage tank. The more common type of LPG valve, however, has a distinctive handle that indicates the proper rotation direction to open or close the valve. To shut off the flow of gas, rotate the handle to the fully closed position.

A gas valve that has been shut off must not be reopened until the system piping has been inspected by a qualified person. Air must be purged from the system and any pilot lights must be reignited to prevent gas leaks.

Water Service

Water service to a building can usually be shut off by closing one valve at the entry point. Many communities permit the water service to be turned off at the connection between the utility pipes and the building's system. This underground valve, which is often accessed through a curb box, is located outside the building and can be operated with a special wrench or key. In most cases, another valve is found inside the building, usually in the basement (if there is one), where the water line enters. In warmer climates, water supply valves are sometimes located above ground.

FIGURE 11-6 A typical natural gas service includes a pressure-reducing valve and a gas meter along with the shut-off valve.

Size-Up

Size-up is the process of initially evaluating an emergency situation to determine which actions need to be taken and which actions can be undertaken safely. It is always the first step in making plans to bring the situation under control. The initial size-up of an incident is conducted when the first unit arrives on the scene and identifies the appropriate actions for that unit or units. Size-up is also used to determine whether additional resources are needed. Size-up is one of the most important steps that need to be taken at an emergency scene. It is often the key factor in determining whether an emergency operation is a success or a failure.

Fire Fighter Safety Tips

Size-up is often a key factor in determining whether an emergency operation is a success or a failure.

The initial size-up is often conducted by the first-arriving company officer, who serves as the IC until a higher-ranking officer arrives on the scene and assumes command. The IC uses the size-up results to develop an initial plan and to set the stage for the actions that follow.

As more complete and detailed information becomes available, the IC will improve both the initial size-up and the initial plan. At a major incident, the size-up process might continue through several stages. The ongoing size-up must consider the effectiveness of the initial plan, the impact that fire fighters are having on the problem, and any changing circumstances at the incident.

Although officers usually perform size-up, all fire fighters must understand how to formulate an operational plan, how to gather and process information, and how to use this information to change plans during the operation. If no officer is present on the first-arriving unit, a fire fighter becomes responsible for assuming command and conducting the preliminary size-up until an officer arrives. Individual fire fighters are often asked to obtain information and to report their observations for ongoing size-up. They should routinely make observations during incidents to maintain their personal awareness of the situation and to develop their personal competence. One factor in an effective size-up relies on the appropriate use of the senses of each fire fighter.

Managing Information

The size-up process requires a systematic approach to managing information. Emergency incidents are often complicated, chaotic, and rapidly changing. The IC must look at a

VOICES OF EXPERIENCE

I have come to appreciate that when it comes to being aware of your surroundings at all times, even experienced fire fighters can lose focus of the dangers that lie right in front of them. As the IC of a working structure fire at a two-story single family dwelling, one of my more senior fire fighters was walking toward me to receive further orders.

I was standing in the middle of the yard near an electrical service line that had burned away from the house. I was trying to protect our personnel from coming in contact with it until we could set some warning cones in place to identify the hazard. As he continued to approach me, he focused on the flames and smoke coming from the structure and not on the electrical line. He did not hear me yelling at him to watch out for the electrical line on the ground.

He was about four feet from me when he stopped short and realized that he was about to step directly on the line. His eyes were as big as silver dollars and he nearly dropped the tools in his hands. For me, it was like watching a scene in slow motion. Although the electrical line was not arcing and was probably dead, the power company had not verified that this was the case.

Even seasoned fire fighters can get distracted. Even in broad daylight they can miss cues and clues that can get them injured or killed if they fail to maintain constant situational awareness.

Steve Flaherty
Grand Rapids Fire Department
Grand Rapids, Minnesota

complicated situation, identify the key factors that apply to the specific incident, and develop an action plan based on known facts, observations, realistic expectations, and certain assumptions. As the operation unfolds, the IC must continually revise, update, supplement, and reprocess the size-up information to ensure that the plan is still valid or to identify when the plan needs to be changed.

Size-up relies on two basic categories of information: facts and probabilities. Facts are data elements that are accurate and based on prior knowledge, a reliable source of information, or an immediate, on-site observation. Probabilities are factors that can be reasonably assumed, predicted, or expected to occur but are not necessarily accurate. It is often necessary to rely on probabilities because fire fighters rarely have all the facts that are necessary to create an action plan based on facts alone.

Often, the initial size-up is based on a limited amount of factual information and a larger amount of probable information. It is refined as more factual information is obtained and as probable information is either confirmed or revised. Effective size-up requires a combination of training, experience, and good judgment.

FIRE FIGHTER Tips

Building maintenance personnel can often provide valuable information about a structure and its mechanical systems.

Facts

Facts are bits of accurate information obtained from various sources. For example, the communications center will provide some facts about the incident during dispatch. A preincident plan will contain many facts about the structure. Maps, manuals, and other references could provide additional information. Mobile computer data terminals, laptop computers, and tablets have the ability to call up previously entered facts at a touch of a few buttons. Individuals with specific training, such as a building engineer or a utility representative, can add even more specific information. In addition, a seasoned officer will have built up a knowledge bank of valuable information based on experience, training, and direct observations.

The initial dispatch information will contain facts such as the location and nature of the situation. This information could be very general ("a building reported on fire in the vicinity of Central Avenue and Main Street") or very specific ("a smoke alarm activation in room 3102 of the First National Bank Building at 173 South Main Street").

The time of day, temperature, and weather conditions are other factors that can easily be determined and incorporated into the initial size-up. Based on these basic facts, an officer might have certain expectations about the incident. For example, whether a building is likely to be occupied or unoccupied, whether the occupants are likely to be awake or sleeping, and whether traffic will delay the arrival of additional units may all be inferred from the time of day. Weather conditions such as snow and ice might delay the arrival of fire apparatus, create operational problems with equipment, and require additional fire fighters to perform basic functions such as stretching hose lines and raising ladders. Strong winds can cause rapid extension (spread) of a fire to exposed buildings or push a fire in unexpected ways. High heat and humidity will affect fire fighters' performance and may cause heat casualties.

An experienced fire officer will also have a basic knowledge of the community and the department's available resources. The officer may remember the types of occupancies, construction characteristics of typical buildings, and specific information about particular buildings from previous incidents or preincident planning visits.

A preincident plan can be especially helpful during size-up because it contains significant information about a structure. A preincident plan generally provides details about a building's construction, layout, contents, special hazards, and fire protection systems. The Preincident Planning chapter contains specific information about the contents and development of such plans. As you study the Preincident Planning chapter, consider how preincident plans feed into the size-up process.

Basic facts about a building's size, layout, construction, and occupancy can often be observed upon arrival, if they are not known in advance. In particular, the officer must consider the size, height, and construction of the building during the size-up. The action plan for a single-story, wood-frame dwelling on an acre (4,047 square meters) of land will be quite different than that for a steel-frame, high-rise tower on a crowded urban street.

The age of the building is often an important consideration in size-up because building and fire codes change over time. Older wooden buildings may have utilized balloon-frame construction, which can provide a path for fire to spread rapidly into uninvolved areas. Floor and roof systems in newer buildings typically use trusses for support; these trusses are susceptible to early failure and collapse. Lightweight and composite engineered building systems tend to fail quickly and without warning signs under fire conditions. In addition, the weight of air-conditioning or heating units on the roof may contribute to early structural collapse. Knowledge of building construction is vital when performing an initial size-up. Review the Building Construction chapter for construction details that will enable you to perform an accurate size-up.

A plan for rescuing occupants and attacking a fire must also incorporate information about the building layout, such as the number, locations, and construction of stairways. The plan for a building with an open stairway that connects several floors might identify keeping the fire out of the stairway as a priority. This focus enables fire fighters to use the stairway for rescue and prevents the fire from spreading to other floors. Ladder placement, use of aerial or ground ladders, and emergency exit routes all depend on the building layout.

Any special factors that will assist or hinder operations must be identified during size-up as well. For example, bars on the windows will limit access and complicate the rescue of trapped victims. Fire detection and suppression systems play a vital part in initial size-up. A fire in a building that contains a properly designed and installed fire detection system with external notification will be reported sooner to the fire department than an unoccupied building with no alarm system. Likewise, an operational sprinkler system may confine or extinguish the fire. The presence of a standpipe system makes getting a hose line in operation on an upper floor of a burning building much quicker and easier. The presence of firewalls may also confine a fire to the location of origin. The information presented in the Fire Detection, Protection, and Suppression Systems chapter will help you become more skillful in performing a size-up.

The occupancy of a building also represents critical information. An apartment building, a restaurant, an automotive repair garage, an office building, a warehouse filled with concrete blocks (often referred to as a concrete–masonry unit), and a chemical manufacturing plant all present fire fighters with different sets of problems that must be addressed.

It is important for the IC to visualize the building from the side where the IC is located—but it is equally important to try to visualize the other three sides of the building. The IC should initially perform a size-up by walking completely around all four sides of the building, a process often referred to as a "walk-around" or "doing a 360." By completing a loop around a building, the IC can observe fire and smoke conditions on all sides of the building. In addition, the IC can look for signs of occupancy, locations of windows and doors, building construction, fire suppression devices, exposures, and locations of utility entrances such as electrical and gas. It will not always be possible for the IC to complete this lap around the fire building. In those cases where the commander cannot complete a walk-around, the companies assigned to the sides and back can give a limited size-up report to the IC, who will usually be located on the front ("A" side) of the building.

The size of the fire and its location within the building often dictate hose line size, placement, and ventilation sites. The fire's location is also critical in determining which occupants are in immediate need of rescue. For example, someone directly above a burning apartment should be rescued immediately, whereas a person at the opposite end of the building, far from the fire, is probably not in immediate danger.

Direct visual observations will give the best information about the size and location of a fire, particularly when combined with information about the building. Although visible flames can indicate where the fire is located and how intensely it is burning, they might not tell the whole story. Flames issuing from only one window suggest that the fire is confined to that room, but it could also be spreading through void spaces to other parts of the building.

Often, the location of a fire within a building cannot be determined from the exterior, particularly when visibility is obscured by smoke FIGURE 11-7. An experienced fire fighter will observe where smoke is visible, how much is apparent, what color it is, and how it moves. Smoke reading, as described in the Fire Behavior chapter, is a vital part on the scene size-up because this process helps the IC predict where the fire might go. Sometimes even the smell of smoke from afar can be a clue. A burning pot of food, an overheated electrical motor, and burning wood all have distinctive odors, for example.

Inside a building, fire fighters can use their observations and sensations to help them work safely and effectively. If smoke limits visibility, a crackling sound or the sensation of heat coming from one direction may indicate the seat of the fire

FIGURE 11-7 It can be difficult to determine the location and extent of a fire inside the building.

(i.e., the main area of the fire). Blistering paint and smoke seeping through cracks can lead fire fighters to a hidden fire burning within a wall. Thermal imaging cameras are an effective device for locating fires, as they can sometimes pinpoint the location of a fire that otherwise would be hard to find. In this way, thermal imagers help fire fighters determine the location of a fire quickly and without putting fire fighters at much risk.

The IC needs to gather as much factual information as possible about a fire. Because the IC often is located at a command post outside the building, company officers will be requested to report their observations from different locations. A company that is operating a hose line inside the building can report on interior conditions, whereas the ventilation crew on the roof will have a very different perspective. The IC may request that an officer or a fire fighter prepare a reconnaissance report, which entails the inspection and exploration of a specific area to gather information for the IC. The IC assembles, interprets, and bases decisions on this information.

Regular progress reports from companies working in different areas will offer updated information about the situation. Progress reports enable an IC to judge whether an operational plan is effective or needs to be changed.

Probabilities

Probabilities refer to events and outcomes that can be predicted or anticipated, based on facts, observations, common sense, and previous experiences. Fire fighters frequently use probabilities to anticipate or predict what is likely to happen in various situations. The attack plan is also based on probabilities, predicting where the fire is likely to spread and anticipating potential problems.

Near Miss REPORT

Report Number: 10-0000966

Synopsis: Good entry practice alerts crew to danger.

Event Description: Our department responded to a large four-story (three up and one down) home involved in fire and smoke. The home was in a very rural part of our district, which significantly delayed our arrival due to both weather and travel time.

I was a member of the first unit to arrive. Upon arrival, I found a female occupant trapped on the second-floor balcony. I threw up a ground ladder from the first-arriving engine company, and the shift captain accessed the victim and assisted her down the ladder.

With the victim removed, I completed another 360-degree inspection of the home. I could see a glow in the basement, but no fire. There was now turbulent smoke coming from the eaves of the home. I instructed the shift captain to enter from the front door on side "A" and assess fire activity and the percentage of involvement. I told him what I had observed from the exterior and to be aware of fire activity in the basement.

The captain forced the door and began to enter with three other members. He was crawling on the floor even though there were only mild to moderate smoke conditions on the first floor. Visibility was not an issue. The shift captain was carrying a pick-head axe and sounded the floor as he crawled. Each of the other team members carried a tool. In total, the crew had two pike poles, a drywall hook, and the axe carried by the captain into the home.

Upon sounding the floor a few times, the captain reported that one of the floor boards had given way under the stress of the axe. He immediately ordered his team to retreat and evacuate. He then notified me that when the floor gave way after sounding, he could see fire throughout the basement. Soon after evacuating, the entire first floor became involved in fire.

Lessons Learned: In my opinion, this is a situation that went right. We communicated as a team and used our training and experience to ensure everyone went home. Due to the clear visibility in the first floor, it could have been a time when complacency led us to walk in and put out the fire. However, we crawled in, and, as a result, we crawled out. We could have very easily been carried out in a bag. Our shift captain recognized the integrity of the floor had been compromised and retreated to safety with his team. We later learned that the home had a dumbwaiter system that spread the fire rapidly from the basement to the attic area. A good size-up, communication, and training allowed us to clock out at the end of shift.

An IC must be able to identify the probabilities that apply to a given situation quickly. For example, a fire in an apartment building in the middle of the night will probably involve occupants who need to be rescued. In such circumstances, the IC would assign additional crews to search for potential victims, even if no factual information indicates that any occupants need to be rescued. Similarly, a fire burning on the top floor of a structure in a row of attached buildings has a high probability of spreading to adjoining buildings through the cockloft. In this case, the IC's plan would include opening the roof above the fire and sending crews into the exposed occupancies to check for fire extension in the cockloft.

The concepts of convection, conduction, and radiation enable an IC to predict how a fire will extend in a particular situation. By observing a particular combination of smoke and fire conditions in a particular type of building, the IC can identify a range of possibilities for fire extension within the structure or to other exposed buildings. Using these probabilities, the IC can predict what is likely to happen and develop a plan to control the situation effectively. Effective size-up requires a good knowledge of fire behavior to evaluate the probabilities for the spread of a fire.

The IC must also evaluate the potential for collapse of a burning structure. The building's construction, the location and intensity of the fire, and the length of time the structure has been burning are all factors that must be considered in making this determination. Houses built in the last 30 years using lightweight construction techniques, for example, tend to collapse in a much shorter period of time than houses built in earlier eras. When the possibility of collapse exists, risk management dictates that the IC not permit fire fighters to enter the building or immediately order all fire fighters to exit the building.

Resources

Resources include all of the means that are available to fight a fire or conduct emergency operations at any other type of emergency incident. Resource requirements depend on the size and type of incident. Resource availability depends on the capacity of a fire department to deliver fire fighters, fire apparatus, equipment, water, and other items that can be used at the scene of an incident.

A fire department's basic resources consist of its personnel and apparatus. Firefighting resources are usually defined as the number of engine companies, ladder companies, special units, and command officers required to control a particular fire. An IC should be able to request the required number of companies and know that each unit will arrive with the appropriate equipment and the necessary fire fighters to perform a standard set of functions at the emergency scene. The IC usually will have a good idea of how many and which types of companies are available to respond, how they are staffed and equipped, and how long they should take to arrive. This information might have to be updated at the incident, particularly in areas served by volunteer fire fighters, because the number of fire fighters available to respond may vary at different times.

Water supply is another critical resource. It is rarely a problem if the fire occurs in a district that has hydrants and a strong, reliable municipal water system with adequate-sized water mains. In an area without hydrants, however, water supply may be severely limited. Even when a static water supply is available, it takes time to establish a water supply from such a source. If the water supply is limited to tankers or tenders, the amount of water that can be delivered by a tanker shuttle is limited. In this situation, the IC would need to call for an additional supply of tankers.

Sometimes, resources include more than fire fighters, fire apparatus, and a water source. A fire in a flammable liquids storage facility will require large quantities of foam and the equipment to apply it effectively. A hazardous materials incident might require special monitoring equipment, chemical protective clothing, and bulk supplies to neutralize or absorb a spilled product. A building collapse might require heavy equipment to move debris and a structural engineer to determine where fire fighters can work safely. The fire department must have these supplies available or be able to obtain them quickly when they are needed. In addition, resources for a large-scale incident must include food and fluids for the rehydration of fire fighters, fuel for the apparatus, and other supplies. The Red Cross, Salvation Army, and law enforcement agencies often provide support resources.

The size-up process enables the IC to determine which resources will be needed to control the specific situation and to ensure their availability at an appropriate time. An action plan to control an incident can be effective only if the necessary resources can be assembled on a timely basis. If a delay occurs, the IC must anticipate how much the fire will grow and where it will spread. If the desired resources cannot be obtained, the IC must develop a realistic plan using the available resources to gain eventual control of the situation.

Ideally, a fire department will be able to dispatch enough fire fighters and apparatus to control any situation within its jurisdiction. More practically, most departments have established mutual aid agreements with surrounding jurisdictions that require them to assist one another if a situation requires more resources than the local community can provide. In some areas, hundreds of fire fighters and apparatus can be assembled to respond to a large-scale incident. In more remote areas, the resources available to fight a fire can be very limited.

If resources are insufficient or delayed, a fire can become too large to be controlled by available personnel. For example, if only 20 fire fighters and four apparatus are available, they will have to bring the fire under control before it gets too big, limit actions to protecting exposures, or wait until the fire burns itself down to a manageable size before tackling it. The IC must always determine which actions can be taken safely with the resources that are currently available.

Resources, in turn, must be organized to support efficient emergency operations. Fire fighters must be properly equipped and organized in companies. Equipment and procedures must be standardized. The ICS must be designed to manage all of the resources that could be used at a large-scale incident, and the communications system must enable the IC to coordinate operations effectively.

Incident Action Plan

Based on information gathered during size-up, the IC develops an incident action plan that outlines the steps needed to control the situation. The initial IC develops a basic plan for beginning operations. If the situation expands or becomes more complicated, this plan may be revised and expanded as additional information is obtained, more resources become available, and the ICS structure grows.

The incident action plan should be based on the five basic fire-ground objectives, which are listed here in order of priority:

1. Save lives. This includes keeping fire fighters safe and rescuing victims.
2. Protect exposures.
3. Confine the fire.
4. Extinguish the fire.
5. Salvage property and overhaul the fire.

This system of priorities clearly establishes that the highest priority in any emergency situation is saving lives. The remaining priorities focus on saving property. Making sure that the fire does not spread to any exposure (an area adjacent to the fire that may become involved if not protected) is a higher priority than confining the fire within the burning building. After the fire is confined, the next priority is to extinguish it. The final priority is to protect property from additional damage and make sure the fire is completely out.

These priorities are not separate and exclusive, of course. That is, one objective does not have to be accomplished before fire fighters tackle the next one. Often, more than one objective can be addressed simultaneously, and certain activities help achieve more than one objective. For example, if a direct attack on the fire will bring it under control very quickly, the objectives of protecting exposures, confining the fire, and extinguishing the fire might all be addressed simultaneously. Extinguishing the fire might also be the best way to protect the lives of building occupants. Frequently, salvage crews may be working on lower floors even as fire suppression crews continue attacking a fire on an upper floor.

RECEO VS

Fire departments and their incident commanders employ different strategies and tactics to meet incident objectives. Lloyd Layman was first credited with developing a tactical model that identified seven basic tactics that some fire departments follow on the fireground. This model is commonly referred to by the acronym RECEO VS:

- Rescue—The removal of victims from a life-threatening situation.
- Exposures—The protection of surrounding structures from a fire.
- Confinement—The containment of a fire to one area or the prevention of further areas from becoming involved in the fire.
- Extinguishment—The suppression of the fire.
- Overhaul—The process of ensuring that the fire is suppressed completely.
- Ventilation—The process of removing smoke, heat, and the products of combustion from a structure.
- Salvage—The removal or protection of property that could be damaged during firefighting or overhaul operations.

Fire Officer: Principles and Practice, second edition, covers the strategy and tactics employed by fire officers when creating an incident action plan in greater detail. Note that the steps in this acronym are not listed in the order that they are performed on the fire ground. It is only intended to aid in recalling what needs to be done—not when it needs to be done.

These priorities guide the IC in making difficult decisions, particularly if not enough resources are available to address every priority. If a decision must be made between saving lives and saving property, saving lives comes first. After rescue is completed and if the fire is still spreading, the IC should then place exposure protection ahead of salvage and overhaul.

A similar set of priorities can be established for any emergency situation: Saving lives is always more important than protecting property. The IC must always place a higher priority on bringing the problem under control than on cleaning up after the problem.

Rescue

Protecting lives is the first consideration at a fire or any other emergency incident. The need for rescue depends on many circumstances. Notably, the number of people in danger is likely to vary based on the type of occupancy and the time of day. A commercial building that is crowded during the workday might have few, if any, occupants at night. At night, rescue is more likely needed in a residential occupancy, such as a house, an apartment building, or a hotel.

The degree of risk to the lives of the building's occupants must also be evaluated. A fire that involves one apartment on the 10th floor of a high-rise apartment building could threaten the lives of both the residents on that floor and the residents who live directly above the fire. In contrast, residents below the fire are probably not in any danger, and residents several floors above the fire might be safer staying in their apartments until the fire is extinguished instead of walking down smoke-filled stairways.

If the size-up suggests that people may be trapped in a burning building, the IC must make a determination of whether it is safe to send fire fighters into the burning building. The IC should use survivability profiling to make this decision. Survivability profiling entails examining the situation and making an informed decision based on the presenting circumstances of whether civilians could survive the existing fire and smoke conditions. Through this process, the IC decides whether to commit fire fighters to life-saving and interior suppression operations to save live victims. Fire fighters should engage in interior operations only if there is a reasonable probability that trapped civilians might be alive.

The upper limit in which people can survive has been shown to be 212°F (100°C). Death occurs at temperatures of 350°F (175°C) in approximately 3 minutes. Lethal first-degree respiratory burns occur in 230 seconds (less than 4 minutes). An analysis of burn victims has demonstrated that respiratory burns from heat kill victims first, followed closely by toxic smoke.

In today's fires, temperatures can rise higher than 500°F (160°C) in 3 to 4 minutes. Flashovers occur at 1100°F (590°C). We must always remember that if an environment is not tenable for a fire fighter in full PPE, it is unlikely that trapped victims will survive either. It is not acceptable to risk the lives of fire fighters to try to save the lives of already dead victims.

Fire Fighter Safety Tips

It is not acceptable to risk the lives of fire fighters to try to save the lives of already dead victims.

Often, the best way to protect lives is to extinguish the fire quickly. For this reason, efforts to control the fire are usually initiated at the same time as rescue operations. For example, hose lines may be used to protect exit paths and keep the fire away from victims during search and rescue operations. The Search and Rescue chapter discusses specific tactics and techniques for these operations.

Exposure Protection

The secondary objective of protecting property encompasses multiple priorities. The first priority in protecting property is to keep the fire from spreading beyond the area of origin or involvement when the fire department arrives. The IC must start by making sure the fire is not expanding. If the fire is burning in only one room, the objective should be to ensure that it does not spread beyond that room. If more than one room is involved, the objective might be to contain the fire to one apartment or one floor level. Sometimes the objective is to confine the fire to one building, particularly if multiple buildings are attached or closely spaced to the site of the fire.

In some cases, the IC must look ahead of the fire and identify a place to stop its spread. If flames are extending quickly through the cocklofts in a row of attached buildings, for example, the IC might place companies with hose lines ahead of the fire to stop its progress.

The IC must sometimes weigh potential losses when deciding where to attack a fire. If a fire in a vacant building threatens to spread to an adjacent occupied building, the IC will usually assign companies to protect the exposure before attacking the main body of fire. If, however, a fire in an occupied building might spread to a vacant building, the IC's decision might be to attack the fire first and worry about controlling the spread later.

Confinement

After ensuring that the fire is not actively extending into any exposed areas, the IC will focus on confining it to a specific area. To do so, the IC will define a perimeter and plan operations so that the fire does not expand beyond that area. Fire fighters on the perimeter must be alert to any indications that the fire is spreading to those limits.

Thermal imaging devices, which can detect sources of heat, are valuable tools for finding fires in void spaces and hence for containing the spread of fire. The principles and use of thermal imaging devices are covered more fully in the Search and Rescue chapter.

Extinguishment

Depending on the size of the fire and the risk involved, the IC will mount either an offensive attack or a defensive strategy to extinguish a fire. An offensive attack is used with most small fires. With this approach, fire fighters advance into the fire building with hose lines or other extinguishing agents and overpower the fire. If the fire is not too large and the attacking fire fighters can apply enough extinguishing agent, the fire can usually be terminated quickly and efficiently. Extinguishing the fire in this way often satisfies several priorities at the same time, including exposure protection and confinement.

At regular intervals during an offensive attack, the IC must evaluate the progress being made. If the progress is inadequate, the IC must either adjust the tactics or alter the overall strategy. For example, the fire might produce more heat than the water from hose lines can absorb or fire fighters might be unable to penetrate far enough to make a direct attack on the fire.

Sometimes the problem can be resolved by adding resources and intensifying the offensive attack. Larger hose lines or more fire fighters with additional hose lines, for instance, might be able to extinguish the fire successfully. Coordinating ventilation with an offensive attack might enable fire fighters to get close enough to extinguish the fire. Special extinguishing agents might also be used to extinguish the fire.

When the fire is too large or too dangerous to extinguish with an offensive attack, the IC will implement a defensive attack. In these situations, all fire fighters are ordered out of the building, and heavy streams are operated from outside the fire building. A defensive strategy is required when the IC determines that the risk to fire fighters' lives is excessive, when rescue is impossible, and when structural collapse is possible. The IC who adopts a defensive strategy has determined that there is no property left to save or that the potential for saving property does not justify the risk to fire fighters. Sometimes a defensive strategy is effective in extinguishing the fire; at other times, it simply keeps the fire from spreading to exposed properties FIGURE 11-8.

FIGURE 11-8 A defensive strategy involves an exterior fire attack with heavy streams.

In some situations, all fire fighters may be withdrawn from the area and the fire allowed to burn itself out. These situations generally involve potentially explosive or hazardous materials that represent an extreme danger to fire fighters.

Each strategy poses some risk to fire fighters, so it is important that the IC consider both the risks and the benefits of each strategy. Sending fire fighters into an unoccupied building may pose a large risk to fire fighters, with the only potential benefit being to save property. Sending fire fighters into an occupied building to save lives has great benefit, but only if there is a chance to save the occupants and the operation does not pose a huge risk to the fire fighters. Each IC must evaluate the risk versus benefit of each action to be taken at an emergency scene.

Ventilation

Ventilation operations include all activities to remove smoke, heat, and the products of combustion from the interior of the structure. Ventilation can be a stand-alone strategy to prevent the spread of heat and smoke in the structure, or it may be used in combination with other tactics to support the safe removal of occupants. Ventilation operations are covered in depth in the Ventilation chapter.

Salvage and Overhaul

Salvage operations are conducted to save property by preventing avoidable property losses. Salvage is the removal or protection of property that could be damaged during firefighting or overhaul operations FIGURE 11-9. Salvage operations are often aimed at reducing smoke and water damage to the structure and contents once the fire is under control.

FIGURE 11-9 Salvage operations are conducted to save property by preventing avoidable property losses.

The overhaul process is conducted after a fire is under control, with the goal being to completely extinguish any remaining pockets of fire FIGURE 11-10. The IC is responsible for ensuring that the fire is completely out before terminating operations. Floors, walls, ceilings, and attic spaces should be checked for signs of heat, smoke, or fire. Window casings, wooden door jambs, baseboards, electrical outlets, and heating or air-conditioning vents can often hide small, smoldering fires. To mitigate this risk, debris from the burned contents of the structure should be removed and thoroughly doused to reduce the potential for rekindle, or a reignition of the fire. For large fires, a fire department unit may be assigned to stay at the scene, or to return to the scene every few hours to check for indications of residual fire. These operations are covered in depth in the Salvage and Overhaul chapter.

FIGURE 11-10 Overhaul is conducted after a fire is under control to completely extinguish any remaining pockets of fire.

Wrap-Up

Chief Concepts

- Response actions include receiving the alarm, donning protective clothing and equipment, mounting the apparatus, and transporting equipment and personnel to the emergency incident quickly and safely.
- Size-up is the systematic process of gathering information and evaluating the incident. It is essential for determining the appropriate strategy and tactics to handle the incident.
- You must be prepared to respond to an emergency at all times. Inspect your PPE and SCBA daily to ensure that it is ready for operations.
- The response process begins when the alarm is received at the fire station. Dispatch messages will include the location of the incident, the type of emergency, and the units that are due to respond.

- Don your PPE before mounting the apparatus. While riding in the apparatus, your seat belt should be secure, and you should be concentrating on mentally preparing for the emergency. Consider any relevant factors that could affect the emergency, such as the weather and the time of day.
- Upon arrival at the scene, traffic safety should be a major concern. Always check for traffic before exiting the apparatus. Follow departmental SOPs to close streets quickly and block access for civilian vehicles to the incident.
- Fire fighters should always work in assigned teams and be guided by a strategic plan for the incident. Teamwork and disciplined action are essential to provide for the safety of all fire fighters and the effective, efficient conduct of operations.
- Upon arriving at the scene, check into the personnel accountability system.
- To protect the safety of fire fighters, controlling utilities is one of the first tasks to be accomplished. The electrical service and gas supply should be shut off and locked.
- Size-up is the first step in making plans to bring the emergency incident under control.
    - The initial size-up is often conducted by the first-arriving company officer, who serves as the IC until a higher-ranking officer arrives at the scene and assumes command.
    - The IC uses the initial size-up information to develop an initial plan. This plan is revised as additional information is gathered.
- Size-up relies on two basic categories of information: facts and probabilities.
    - Facts are data elements that are accurate and based on prior knowledge, a reliable source of information, or an immediate, on-site observation.
    - Probabilities are factors that can be reasonably assumed, predicted, or expected to occur, but are not necessarily accurate.
- The preincident plan contains facts that can be essential in creating a plan. It provides details about a building's construction, layout, contents, special hazards, and fire protection systems. This information can be used in determining how to rescue occupants and attack a fire.
- The IC often operates out of a command post outside the structure and relies on company officers to provide reconnaissance reports. Progress reports from companies enable an IC to judge whether an operational plan is effective or needs to be changed.
- Probabilities refer to events and outcomes that can be predicted or anticipated, based on facts, observations, common sense, and previous experiences. Fire fighters frequently use probabilities to anticipate or predict what is likely to happen in various situations. The attack plan is also based on probabilities, predicting where the fire is likely to spread and anticipating potential problems.
- Resources for fire fighters include all of the means that are available to fight a fire or conduct emergency operations at any other type of emergency incident. Resource requirements reflect the size and type of incident. Resource availability depends on the capacity of a fire department to deliver fire fighters, fire apparatus, equipment, water, and other items that can be used at the scene of an incident.
- Based on information gathered during size-up, the IC develops an incident action plan that outlines the steps needed to control the situation. The incident action plan should be based on the five basic fireground objectives:
    - Save lives—Protecting lives is the first consideration at a fire or any other emergency incident. Often the best way to protect lives is to quickly extinguish the fire.
    - Protect exposures—The first priority in protecting property is to keep the fire from spreading.
    - Confine the fire—Once the exposures are protected, the IC will focus on confining the fire to a specific area.
    - Extinguish the fire—Depending on the size of the fire and the risk involved, the IC will mount either an offensive attack or a defensive strategy to extinguish a fire.
    - Salvage property and overhaul the fire—Salvage is the removal or protection of property that could be damaged during firefighting operations. Overhaul operations completely extinguish any remaining pockets of fire.

Hot Terms

Balloon-frame construction An older type of wood frame construction in which the wall studs extend vertically from the basement of a structure to the roof without any fire stops.

Defensive attack Exterior fire suppression operations directed at protecting exposures.

Exposure The heat effect from an external fire that might cause ignition of, or damage to, an exposed building or its contents. (NFPA 80A)

Extension Fire that moves into areas not originally involved in the incident, including walls, ceilings, and attic spaces; also, the movement of fire into uninvolved areas of a structure.

Freelancing The dangerous practice of acting independently of command instructions.

Offensive attack An advance into the fire building by fire fighters with hose lines or other extinguishing agents that are intended to overpower the fire.

Overhaul The process of final extinguishment after the main body of a fire has been knocked down. All traces of fire must be extinguished at this time. (NFPA 402)

Wrap-Up, continued

Personnel accountability system A system that readily identifies both the locations and the functions of all members operating at an incident scene. (NFPA 1500)

Personnel accountability tag (PAT) An identification card used to track the location of a fire fighter on an emergency incident.

Preincident plan A written document resulting from the gathering of general and detailed information to be used by public emergency response agencies and private industry for determining the response to reasonable anticipated emergency incidents at a specific facility.

Reconnaissance report The inspection and exploration of a specific area for the purpose of gathering information for the incident commander.

Rekindle A return to flaming combustion after apparent but incomplete extinguishment. (NFPA 921)

Response The deployment of an emergency service resource to an incident. (NFPA 901)

Salvage A firefighting procedure for protecting property from further loss following an aircraft accident or fire. (NFPA 402)

Seat of the fire The main area of the fire.

Size-up The observation and evaluation of existing factors that are used to develop objectives, strategy, and tactics for fire suppression. (NFPA 1051)

Survivability profiling The process of using available data and observations to determine if there is a reasonable chance that victims trapped inside might still be alive. This process is used to decide whether to send fire fighters into a fire or other hazardous environment.

Thermal imaging devices Electronic devices that detect differences in temperature based on infrared energy and then generate images based on those data. These devices are commonly used in obscured environments to locate victims.

FIRE FIGHTER *in action*

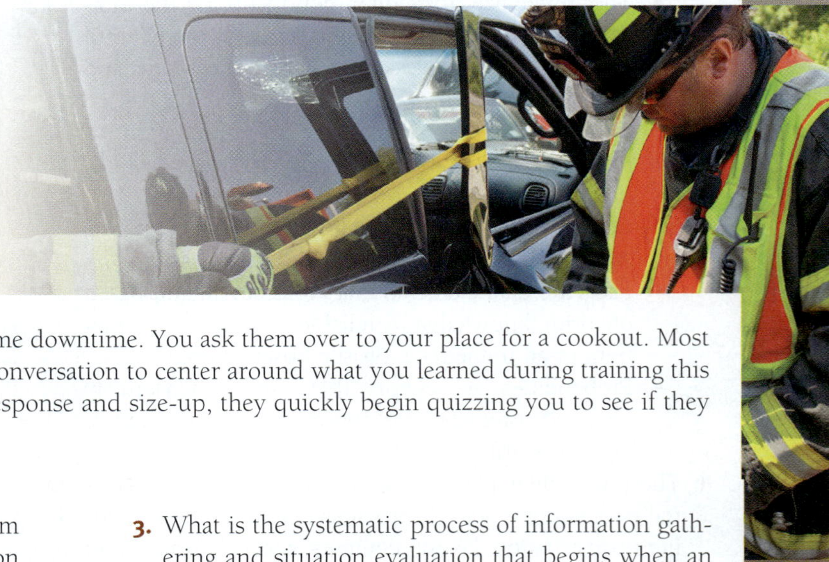

It is the weekend. You and your buddies are ready for some downtime. You ask them over to your place for a cookout. Most of them are fire fighters, so it does not take long for the conversation to center around what you learned during training this week. When you tell them that you are learning about response and size-up, they quickly begin quizzing you to see if they can stump you.

1. A(n) _______ is a written document resulting from the gathering of general and detailed information to be used by public emergency response agencies and private industry for determining the response to reasonable anticipated emergency incidents at a specific facility.
   - **A.** preincident plan
   - **B.** reconnaissance report
   - **C.** incident action plan
   - **D.** size-up
2. _______ is conducted after a fire is under control, with the goal being to completely extinguish any remaining pockets of fire.
   - **A.** Overhaul
   - **B.** Salvage
   - **C.** Exposure protection
   - **D.** Extinguishment
3. What is the systematic process of information gathering and situation evaluation that begins when an alarm is received?
   - **A.** Preincident plan
   - **B.** Reconnaissance report
   - **C.** Incident action plan
   - **D.** Size-up
4. When should you don your PPE?
   - **A.** Prior to mounting the apparatus
   - **B.** While en route to the scene
   - **C.** While en route to the scene, but only if you can remain in your seat belt
   - **D.** Upon arrival at the scene

5. How should fire fighters cut the electricity to a building?
   - **A.** Turn off the main circuit breaker serving the building.
   - **B.** Pull the electrical meter to the building.
   - **C.** Cut the loops at the electrical pole feeding the building.
   - **D.** Only electrical system personnel should ever cut any electricity to a building or area.

6. When does the emergency response process begin?
   - **A.** When the dispatcher answers the phone.
   - **B.** When the fire chief receives information from the dispatcher.
   - **C.** When an alarm is received at the fire station.
   - **D.** When the apparatus leaves the fire station.

FIRE FIGHTER II *in action*

Your crew is the first to arrive at a one-car rollover on a major highway. The car is in a ditch, so the driver/operator pulls the fire engine onto the shoulder at a slight angle. You see the driver slumped over the steering wheel, blood dripping down his face. Despite the presence of the engine with lights flashing, vehicles still zoom past the scene at full speed.

1. Who is responsible for setting out safety cones?
2. What is the first item of PPE every member of your crew should don?
3. What additional resources can assist in maintaining your safety?
4. What is your first priority at this incident?

Forcible Entry

Fire Fighter I

Knowledge Objectives

After studying this chapter, you will be able to:

- Describe the situations and circumstances that require forcible entry into a structure. (NFPA 5.3.4, p 320–321)
- List the general safety rules to follow when utilizing forcible entry tools. (p 321)
- List the general carrying tips when utilizing forcible entry tools. (NFPA 5.3.4.B, p 321)
- List the general maintenance tips when utilizing forcible entry tools. (p 321–322)
- List the types of tools used in forcible entry. (p 322)
- List the striking tools used in forcible entry. (p 322)
- Describe the tasks that striking tools are used for in forcible entry. (p 322)
- List the prying and spreading hand tools used in forcible entry. (p 322–323)
- Describe the tasks that prying and spreading hand tools are used for in forcible entry. (p 322–323)
- List the cutting tools used in forcible entry. (p 323–324)
- Describe the tasks that cutting tools are used for in forcible entry. (p 323–324)
- List the lock and specialty tools used in forcible entry. (p 324–325)
- Describe the tasks that lock and specialty tools are used for in forcible entry. (p 324–325)
- Describe the basic components of a door. (NFPA 5.3.4.A, p 325)
- Explain the differences between a solid-core and a hollow-core door. (NFPA 5.3.4.A, p 326)
- Describe the basic classifications of doors. (NFPA 5.3.4.A, p 327)
- Explain how the door classification affects forcible entry operations. (NFPA 5.3.4.A, p 327–331)
- Describe the basic configurations of window construction. (NFPA 5.3.4.A, p 331–333)
- Describe the common styles of window frames. (NFPA 5.3.4.A, p 333–338)
- Explain how the style of window frame affects forcible entry operations. (NFPA 5.3.4.A, p 333–338)
- Describe the major components of a door lock. (NFPA 5.3.4.A, p 338–339)
- Describe the major components of a padlock. (NFPA 5.3.4.A, p 339–340)
- Describe the four major types of locks. (NFPA 5.3.4.A, p 339–342)
- Explain how the type of lock affects forcible entry operations. (NFPA 5.3.4.A, p 339–342)
- Describe the tools used to force entry through locks. (p 340–343)
- Describe how to force entry through security gates and windows. (p 342–344)
- Explain the differences between load-bearing and nonbearing walls. (NFPA 5.3.4.A, p 344)
- Describe the materials used in exterior and interior walls. (NFPA 5.3.4.A, p 344)
- List the basic steps and considerations in forcible entry operations. (p 347–349)
- Describe how forcible entry operations affect salvage operations. (NFPA 5.3.14.A, p 349)

Skills Objectives

After studying this chapter, you will be able to:

- Force entry into an inward-opening door. (NFPA 5.3.4.B, p 327–328)
- Force entry into an outward-opening door. (NFPA 5.3.4.B, p 329–330)
- Open an overhead garage door using the triangle method. (NFPA 5.3.4.B, p 331–332)
- Force entry through a wooden double-hung window. (NFPA 5.3.4.B, p 334)
- Force entry through a casement window. (NFPA 5.3.4.B, p 337)
- Force entry through a projected window. (NFPA 5.3.4.B, p 338)
- Force entry using a K tool. (NFPA 5.3.4.B, p 341–342)
- Force entry using an A tool. (NFPA 5.3.4.B, p 342–343)
- Force entry by unscrewing a lock. (NFPA 5.3.4.B, p 342–343)
- Breach a wall frame. (NFPA 5.3.4.B, p 344–345)
- Breach a masonry wall. (NFPA 5.3.4.B, p 345–346)
- Breach a metal wall. (NFPA 5.3.4.B, p 345–347)
- Breach a floor. (NFPA 5.3.4.B, p 346–348)

CHAPTER

12

Fire Fighter II FFII

Knowledge Objectives

After studying this chapter, you will be able to:

- Describe the connection between tools and forcible entry tasks. (NFPA 6.3.2.A, 6.3.2.B, p 322-325)

Skills Objectives

There are no skill objectives for Fire Fighter II candidates. NFPA 1001 contains no Fire Fighter II Job Performance Requirements for this chapter.

You Are the Fire Fighter

There is a working fire on the second floor of a local hotel. Your crew is assigned to search for occupants who may still be in the building. The hotel is a three-story, wood-frame building with a hallway down the center. It was built in 1981. As you dismount the fire apparatus, your captain tells you to grab the tools for forcible entry.

1. What kind of access issues would you expect for this type of structure?
2. What kind of tools would you want to take with you?
3. How would the forcible entry challenges for this structure differ from other types of structures?

Introduction

One of a fire fighter's most dynamic and challenging tasks, forcible entry, is defined as gaining access to a structure when the normal means of entry are locked, secured, obstructed, blocked, or unable to be used for some other reason. This chapter examines forcible entry into structures.

Forcible entry requires strength, knowledge, proper techniques, and skill. Because it often causes damage to the property, fire fighters must consider the results of using different forcible entry methods and select the one appropriate for the situation. If rapid entry is needed to save a life or prevent a more serious loss of property, it is appropriate to use maximum force. When the situation is less urgent, fire fighters can take more time and use less force.

Another factor to consider when making forcible entry is the need to secure the premises after operations are completed. Fire fighters must never leave the premises in a condition that would allow unauthorized entry. If a door or a lock is destroyed during an urgent forcible entry, the building owner should be called to secure the building. When the situation is not urgent, consider using entry methods that result in less damage and can be repaired more easily.

A fire fighter who is skilled in forcible entry should be able to get the job done quickly, with as little damage as possible. As part of this task, the fire fighter must consider the type of building construction, possible entry points, the types of securing devices that are present, and the best tools and techniques for the specific situation. Selecting the right tool and using the proper technique save valuable time—and could save lives.

Doors, windows, locks, and security devices can be combined in countless variations to prevent easy entry. Given this diversity, fire fighters must be familiar with new technology, including new styles of windows, doors, locks, and security devices that are common in the local response area. In addition, they must understand how these devices operate. Learn to recognize different types of doors, windows, and locks. The best time to examine these components is during inspection and preincident planning visits. Touring buildings under construction or renovation is also an excellent way to learn about building construction and to examine different devices. Talking with construction workers and locksmiths can provide valuable information about how to best approach a particular lock, window, or door.

This chapter reviews door, window, and lock construction. In addition, it discusses the tools and techniques used for forcible entry through doors, windows, walls, and floors. The skill drills cover the most common forcible entry techniques.

Forcible Entry Situations

Forcible entry is usually required at emergency incidents where time is a critical factor. A rapid forcible entry, using the right tools and methods, might result in a successful rescue or allow an engine company to make an interior attack and control a fire before it extends farther into the structure or to adjoining structures. Fire fighters must study and practice forcible entry techniques so that their proficiency with these skills increases with every experience.

Company officers usually select both the point of entry and the method to be used. They ensure that the efforts of different companies are properly coordinated for safe, effective operations. For example, forcible entry actions must be coordinated with hose teams because the entry must be made before engine companies can get inside the structure to attack a fire. Opening a door before the hose lines are in place and ready to advance allows fresh air to enter the structure, possibly resulting in rapid fire spread or a backdraft. Making an opening at the wrong location can undermine a well-planned attack. Anytime you open a door or window, that action may provide ventilation for a fire. Consider forcible entry to be part of ventilation, and make sure it is coordinated with the rest of the fire attack.

Before beginning forcible entry, remember this simple rule: "Try before you pry." In other words, always check doors or windows to confirm that forcible entry is truly needed. Be absolutely sure that you are at the address of the emergency incident. There is nothing harder to explain than why you broke down the door of a house located three doors down from the structure that is on fire. Check for the presence of a lockbox if your community has a lockbox program. The presence of a lockbox will give you immediate access to a key to open that door. An unlocked door requires no force; a window that can be opened does not need to be broken. Checking first takes only a few seconds, but it could save several minutes of effort and unnecessary property damage. It is equally important to look for alternative entry points. Do not spend time working on a locked door when, for example, a nearby window provides easy access to the same room.

Unusually difficult forcible entry situations may require the expertise of specialty companies such as technical rescue teams. These companies have experience working with specialized tools and equipment to gain access to almost any area.

FIRE FIGHTER Tips

Many communities have installed lockbox systems. With these systems, a master key carried by fire department members opens a small lockbox at the incident structure. This lockbox contains the keys needed to open the structure's door. If your community has a lockbox system, always look for a lockbox before beginning any type of forcible entry. The fastest way to gain access to a building is with a key.

Forcible Entry Tools

Choosing the right forcible entry tool can be a very important decision. Fire departments use many different types of forcible entry tools, ranging from basic cutting, prying, and striking tools to sophisticated mechanical and hydraulic equipment. Fire fighters must be familiar with the following information:

- Which tools the department uses
- What the uses and limitations of each tool are
- How to select the proper tool for the job
- How to safely operate each tool
- How to carry the tools safely
- How to inspect and maintain each tool, thereby ensuring its readiness for service and safe operation

General Tool Safety

Any tool that is used incorrectly or is maintained improperly can be dangerous to both the operator and other persons in the vicinity. Anyone who uses a tool should understand the proper operating and maintenance procedures before beginning the task. Always follow the manufacturer's recommendations for operation and maintenance.

General safety tips for using tools include the following points:

1. Always wear the appropriate protective equipment. When conducting forcible entry during fire suppression operations, this includes a full set of structural firefighting personal protective equipment (PPE).
   - Goggles or approved eye protection are required when working with cutting or striking tools to prevent eye injuries from pieces of metal, glass, or other materials.
   - Gloves provide protection from sharp cutting blades, sharp edges, and broken glass. Good gloves can mean the difference between a minor injury and a severed finger.
   - A helmet provides essential protection from falling debris.
   - A turnout coat and pants help to protect skin from sharp metal edges or broken glass.
   - Boots protect the feet from sharp objects and from falling objects.
2. Learn to recognize the materials used in building and lock construction, and become familiar with the appropriate tools and techniques used for each kind of material. For example, many locks are made of case-hardened steel, which combines carbon and nitrogen to produce a very hard outer coating that cannot be penetrated by most ordinary cutting tools. Using the wrong tool to cut case-hardened steel could break the tool and potentially injure the user and other fire fighters.
3. Keep all tools clean, properly serviced according to the manufacturer's guidelines, and ready for use. Immediately report any tool that is damaged or broken and take it out of service.
4. Do not leave tools lying on the ground or floor; instead, return them to a tool staging area or the apparatus. Tools that are left lying around are tripping hazards that may increase the number of injuries on the fire ground.

General Carrying Tips

Tools can cause injuries even when they are not in use. Carrying a tool improperly, for example, can result in muscle strains, abrasions, or lacerations. The following general carrying tips apply to all tools:

- Do not try to carry a tool or piece of equipment that is too heavy or designed to be used by more than one person **FIGURE 12-1**. Instead, request assistance from another fire fighter to carry this type of tool.
- Always use your legs—not your back—when lifting heavy tools.
- Keep all sharp edges and points away from your body at all times. Cover or shield them with a gloved hand to protect those around you.
- Carry long tools with the head down toward the ground. Be aware of overhead obstructions and wires, especially when using pike poles and ladders.

General Maintenance Tips

All tools should be kept in a "ready state"—that is, the tool should be in proper working order, in its proper storage place, and ready for immediate use. Forcible entry tools are generally stored in a designated compartment on the apparatus. Hand tools should be clean, and cutting blades should be sharp. Power tools should be completely fueled and treated with a

FIGURE 12-1 Request assistance to help carry a heavy load.

fuel-stabilizing product, if necessary, to ensure easy starting. Also remember that metal cutting blades can be damaged by gasoline vapors. Every crew member should be able to locate the right tool immediately, confident that it is ready for use FIGURE 12-2.

FIGURE 12-2 All tools should be kept in a neat and ready state.

All tools require regular maintenance and cleaning to ensure that they will be ready for use in an emergency. Thorough, conscientious daily or weekly checks of these items should be performed, particularly with infrequently used tools. Always follow the manufacturer's instructions and guidelines, and store maintenance manuals in an easily accessible location. Keep proper records to track maintenance, repairs, and any warranty work that is performed on tools.

Types of Forcible Entry Tools

Forcible entry tools include hand tools and power tools. They can be classified into the following categories:

- Striking tools
- Prying/spreading tools
- Cutting tools
- Lock tools

Many of the basic firefighting tools are described and pictured in the Fire Fighter Tools and Equipment chapter.

Striking Tools

Striking tools generate an impact force directly on an object or another tool. They are generally hand tools that are powered by human energy. The head of a striking tool is usually made of hardened steel.

FIRE FIGHTER Tips

The forcible entry tools covered in this chapter are not the only ones available. Different fire departments use different tools, and regional preferences for a particular piece of equipment may influence the items that are made available. You must become familiar with the tools used in your department.

Fire Fighter Safety Tips

Before swinging a striking tool, make sure there is a clear area of at least the length of the tool handle. Station another fire fighter outside the swing area as a safety person to keep others from entering the zone. Make sure that you have checked for potential signs of backdraft before using any striking tool.

Flat-Head Axe

The flat-head axe was one of the first tools developed by humans and is still widely used for a variety of purposes. One side of the axe head is a cutting blade; the other side is a flat striking surface. Fire fighters often use the flat side to strike a Halligan tool and drive a wedge into an opening. Most fire apparatus carry both flat-head axes and pick-head axes.

Battering Ram

The battering ram is used to force doors and breach walls. Originally, it was a large log used to smash through enemy fortifications. Today's battering rams are usually made of hardened steel and have handles; two to four people are needed to use a battering ram.

Sledgehammer

Sledgehammers come in a variety of weights and sizes. The head of the hammer can weigh from 2 to 20 pounds (1 to 9 kilograms). The handle may be short like a carpenter's hammer or long like an axe handle. A sledgehammer can be used by itself to break down a door or in conjunction with other striking tools such as the Halligan bar.

Prying/Spreading Hand Tools

Hand tools designed for prying and spreading are often used by fire fighters to force entry into buildings. This section describes the most commonly used forcible entry tools.

Halligan Tool (Bar)

The Halligan tool or bar is widely used by the fire service FIGURE 12-3. Pairing a Halligan tool with a flat-head axe creates a tool set often referred to collectively as the irons. These two tools are commonly used to perform forcible entry. The Halligan tool incorporates three different tools—the adz, the pick, and the claw. The adz end is used to pry open doors and windows; the pick end is used to make holes or break glass; and the claw is used to pull nails and to pry apart wooden slats. A fire fighter can use a flat-head axe to strike the Halligan bar into place, thereby creating a better bite (a small opening that allows better tool access in forcible entry) between a door or window and its frame.

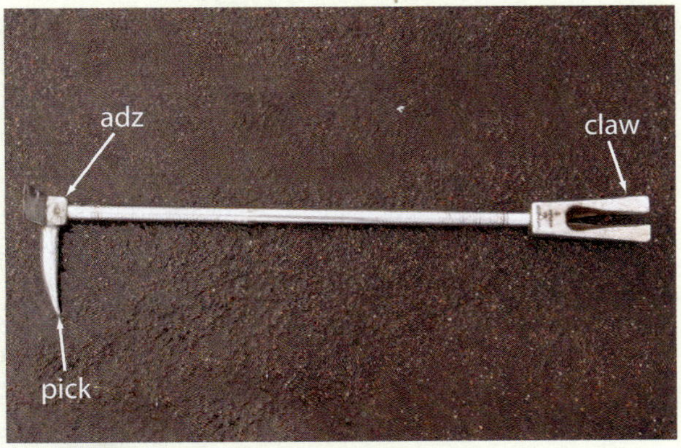

FIGURE 12-3 A Halligan tool has multiple purposes.

Pry Bar/Hux Bar/Crowbar

Pry bars, Hux bars, and crowbars are made from hardened steel cast into a variety of shapes and sizes. These tools are most commonly used to force doors and windows, to remove nails, or to separate building materials. The different shapes allow fire fighters to exert different amounts of leverage in diverse situations.

Pry Axe

The pry axe is a multipurpose tool that can be used both to cut and to force open doors and windows FIGURE 12-4. It includes an adz, a pick, and a claw. The tool consists of two parts: the body, which has the adz and pick and looks similar to a miniature pick axe, and a handle with a claw at the end, which slides into the body. The handle of the pry axe can be extended to provide extra leverage when prying, or it may be removed and inserted into the head of the adz to provide rotational leverage. Extreme caution should be used when handling this tool. Over time, the mechanism that locks the handle into position may become worn, allowing the handle to slip out and potentially injure fire fighters.

Hydraulic Tools

Hydraulic-powered tools such as spreaders, cutters, and rams are often used for forcible entry. These tools require hydraulic pressure, which can be provided by a high-pressure, motor-operated pump or a hand pump.

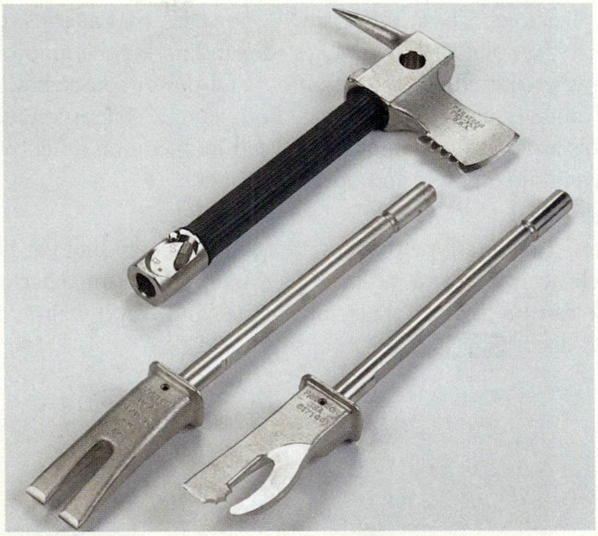

FIGURE 12-4 The pry axe is a multipurpose tool that can be used to cut and pry.

Hydraulic cutters and spreaders are typically used in vehicle extrication, but can also be used in some forcible entry situations FIGURE 12-5. Hydraulic rams come in a variety of lengths and sizes and can be used to apply a powerful force in one direction. The operation of these tools is covered in the Vehicle Rescue and Extrication chapter.

A rabbet tool is a small hydraulic spreader operated by a hand-powered pump. This tool is designed with teeth that will fit into a door jamb or rabbet (a type of door frame that has the door stop cut into the frame). As the spreader opens, it applies a powerful force that can open many doors.

Cutting Tools

Cutting tools are primarily used for cutting doors, roofs, walls, and floors. Although they do not work as quickly as power tools, hand-operated cutting tools are proven and reliable in many situations. Because they do not require a power source, these items often can be deployed more quickly than power tools.

FIGURE 12-5 A hydraulic spreader.

Several power cutting tools are available for different applications. These tools can be powered by batteries, electricity, gasoline, or hydraulics, depending on the amount of power required. Each type of tool and power source has both advantages and disadvantages. For example, battery-powered tools are portable and can be placed in operation quickly, but have limited power and operating times.

Axe

Many different types of axes are available, including flat-head, pick-head, pry, and multipurpose axes. The cutting edge of an axe can be used to break into plaster and wood walls, roofs, and doors FIGURE 12-6.

FIGURE 12-6 The cutting edge of an axe can be used to break into plaster and wood walls, roofs, and doors.

Although a flat-head axe was described earlier in this chapter as a striking tool, it is also classified as a cutting tool. The pick-head axe is similar to a flat-head axe, but has a pick instead of a striking surface opposite the blade. This pick can be used to make an entry point or a small hole if needed.

Specialty axes such as the pry axe and the multipurpose axe can also be used for purposes other than cutting. The multipurpose axe, for example, can be used for cutting, striking, or prying. It includes a pick, a nail puller, a hydrant wrench, and a gas main shut-off wrench.

Bolt Cutters

Bolt cutters are used to cut metal components such as bolts, padlocks, chains, and chain-link fences. These tools are available in several different sizes, based on the blade opening and the handle length. The longer the handle, the greater the cutting force that can be applied. Note that bolt cutters may not be able to cut into some heavy-duty padlocks that are constructed from case-hardened metal.

Circular Saw

Most fire departments use gasoline-powered circular saws both for making forcible entry and for cutting ventilation holes. These tools are light, powerful, and easy to use, with blades that can be changed quickly. The different blades enable the saw to cut several materials.

Fire departments generally carry several different circular saw blades so that the proper blade can be used in different situations.

- Carbide-tipped blades are specially designed to cut through hard surfaces or wood. They stay sharp for long periods, so more cuts can be made before the blade needs to be changed.
- Metal-cutting blades consist of a composite material made with aluminum oxide. They are used to cut metal doors, locks, or gates.
- Masonry-cutting blades are abrasive and made of a composite material that includes silicon carbide or steel. They can cut concrete, masonry, and similar materials. Because masonry-cutting blades resemble metal-cutting blades, the operator must check the label on the blade before using it. Blades with missing labels should be discarded. Do not store masonry-cutting blades near gasoline, because the gasoline vapors will cause the composite materials in the blade to decompose. In such a case, when the blade is later used, it could disintegrate.

FIRE FIGHTER Tips

Almost all cutting tools have built-in safety features. Always take advantage of the safety features on every tool. To avoid injury, never remove safety guards from tools—not even for "just one cut."

FIRE FIGHTER Tips

Caution should be used when using gas-powered equipment in flammable atmospheres because they are likely to produce sparks. Caution should also be used when using such equipment in enclosed spaces because these tools produce carbon monoxide in their exhaust fumes.

Lock Tools/Specialty Tools

Lock and specialty tools are used to disassemble the locking mechanism on a door. These devices will cause minimal damage to the door and the door frame. If they are used properly, the door and frame should be undamaged, although the lock generally must be replaced. An experienced user can usually gain entry in less than a minute with these tools.

K Tool, A Tool, and J Tool

The K tool is designed to cut into a lock cylinder FIGURE 12-7A. To shear the cylinder from a lock, you place the cutting edge

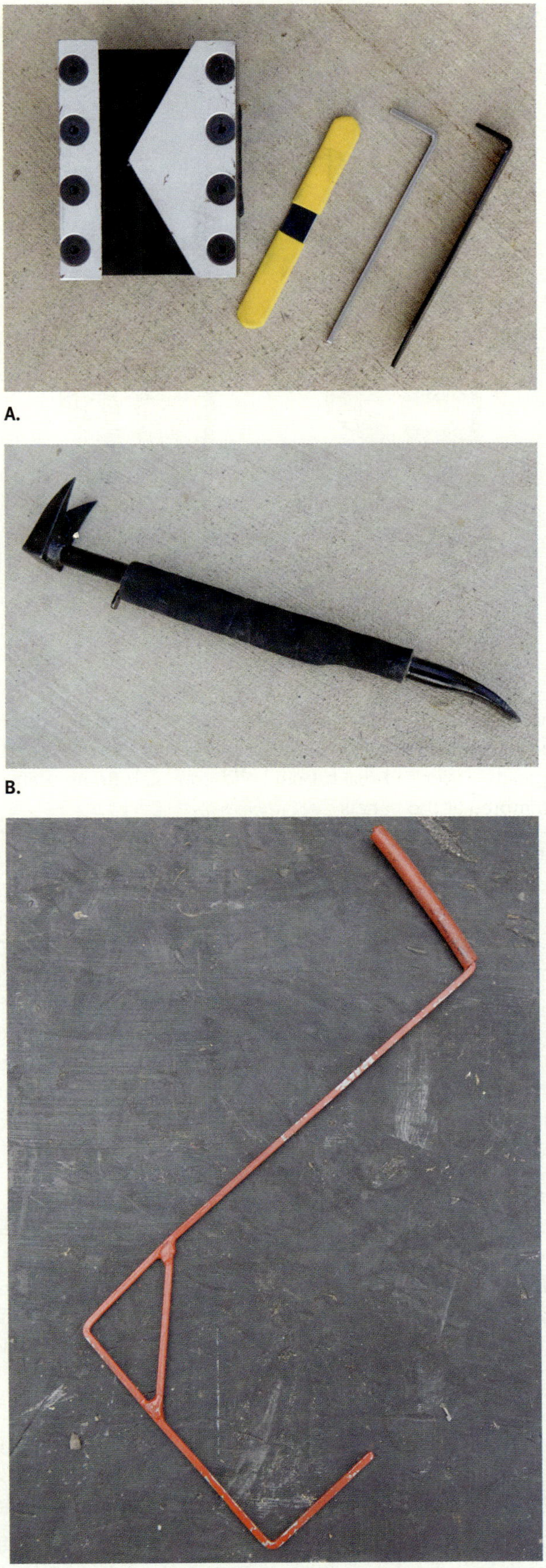

A. B. C.

FIGURE 12-7 **A.** K tool. **B.** A tool. **C.** J tool.

of the K tool on the top of the lock, insert a pry bar into the K tool, and strike the pry bar with a flat-head axe. If the lock has a protective ring, the K tool might not be able to cut through it. Once the lock cylinder is removed, another tool can then be used to open the locking mechanism.

The A tool is similar to the K tool except that the pry bar is built into the cutting part of the tool FIGURE 12-7B. This tool gets its name from its A-shaped cutting edges. To use an A tool, you put the cutting head on the lock cylinder and use a striking tool to force it down into the cylinder until the lock can be forced out of the door. Once the lock cylinder is removed, you would then use another tool to open the locking mechanism.

A J tool will fit between double doors that have panic bars FIGURE 12-7C. Slide the J tool between the doors and pull to engage the panic bars.

Shove Knife

A shove knife is an old tool, yet is frequently used by modern-day fire fighters. Slipping the knife between the door and the frame at the latch forces the latch back and opens the door. Most new doors have latches that will not respond to this tool.

Duck-Billed Lock Breakers

To open a padlock using duck-billed lock breakers, drive the point into the shackles (the U-shaped part) of the lock. The increasing size of the point forces the shackles apart until they break. While this causes no damage to the lock body, the lock will be inoperable after use of the duck-billed lock breaker.

Locking Pliers and Chain

The locking pliers and chain are used to clamp a padlock securely in place so that the shackles can be cut safely with a circular saw or cutting torch. To use this tool, one fire fighter clamps the pliers to the lock body while a second fire fighter maintains a steady tension on the chain as the lock is being cut.

Bam-Bam Tool

The bam-bam tool contains a case-hardened screw, which is inserted and secured in the keyway of a lock. After the screw is in place, you drive the handle of the tool into the keyway and pull the tumbler out of the lock.

Doors

In most cases, the best point to attempt forcible entry to a structure is the door or a window. Doors and windows are constructed as entry points, so they are made generally of weaker materials than are walls or roofs. An understanding of the basic construction of doors will help you select the proper tool and increase your likelihood of successfully gaining entry.

Basic Door Construction

Doors can be categorized both by their construction material and by the way they open. Both interior and exterior doors have the same basic components FIGURE 12-8:

- Door (the entryway itself)
- Jamb (the frame)
- Hardware (the handles, hinges, and other components)
- Locking mechanism

FIGURE 12-8 The parts of a door.

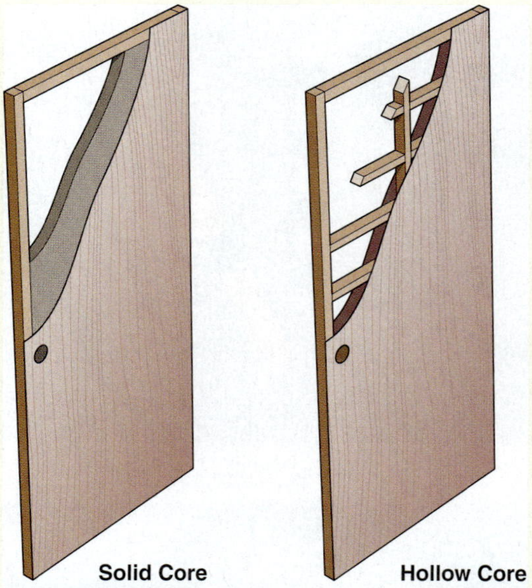

FIGURE 12-9 A slab door may have either solid-core or hollow-core construction.

Construction Materials

Doors are generally constructed of wood, metal, or glass. The design of the door and the construction material used will determine the difficulty in forcing entry through this point.

Wood

Wood doors are commonly used in residences and may be found in some commercial buildings. Three types of wood swinging doors are distinguished: slab, ledge, and panel. Slab doors come in solid-core or hollow-core designs and are attached to wood-frame construction with normal hardware FIGURE 12-9.

Solid-core doors are constructed of solid wood core blocks covered by a face panel. Some have a fire rating. Such doors are typically used for entrance doors. Solid-core doors are heavy and may be difficult to force, but their construction enables them to contain fire better than hollow-core doors do.

Hollow-core doors have a lightweight, honeycomb interior, which is covered by a face panel. They are often used as interior doors, such as for bedrooms. Hollow-core doors are easy to force and will burn through quickly.

Ledge doors are simply wood doors with horizontal bracing. These doors, which are often constructed of tongue-and-groove boards, may be found on warehouses, sheds, and barns.

Panel doors are solid wood doors that are made from solid planks to form a rigid frame with solid wood panels set into the frame. Panel doors are used as both exterior and interior doors and may be made from a variety of types of wood. These doors resist fire longer than hollow-core slab doors do and are typically easier to breach than solid-core slab doors if entry is attempted at the panels.

Metal

Metal doors may be either decorative (for residential use) or utilitarian (for warehouses and factories). Like wood doors, they may have either a hollow-core or a solid-core construction. Hollow-core metal doors have a metal framework interior so they are as lightweight as possible. By contrast, solid-core metal doors have a foam or wood interior that is intended to reduce the door's weight without affecting its strength. Metal doors may be set in either a wood frame or a metal frame. Residential metal doors may appear to be panel doors and are often used as entry doors.

Glass

Glass doors generally have a steel frame with tempered glass; alternatively, they may be simply tempered glass and not require a frame, but have metal supports to attach hardware. Glass doors are easy to force, but can be dangerous owing to the large number of small broken pieces that are produced when glass is broken.

Fire Fighter Safety Tips

Opening a door can allow air to enter the structure, potentially resulting in fire spread or a backdraft. Be careful and listen to instructions when performing forcible entry. Everything needs to be coordinated!

A.

B.

FIGURE 12-10 **A.** An outward-opening door will have hinges showing. **B.** A door with the hinges not showing will open inward.

Types of Doors

Doors are also classified based on how they open. The five most common ways that doors open are inward, outward, sliding, revolving, and overhead. Overhead, sliding, and revolving doors are the most readily identified. Inward-opening and outward-opening doors can be differentiated based on whether the hinges are visible. If you can see the hardware, the door will swing toward you (outward) FIGURE 12-10A. If the hinges are not visible, the door will swing away from you (inward) FIGURE 12-10B.

Door frames may be constructed of either wood or metal. Wood-framed doors come in two styles: stopped and rabbet. Stopped door frames have a piece of wood attached to the frame that stops the door from swinging past the latch. Rabbeted door frames are constructed with the stop cut built into the frame so it cannot be removed.

Metal-framed doors are more difficult to force than wood-framed doors. Metal frames have little flexibility, so when a metal frame is used with a metal door, forcing entry will be very difficult. Metal frames look like rabbeted door frames.

Inward-Opening Doors

Design

Inward-opening doors of wood, steel, or glass can be found in most structures. They have an exterior frame with a stop or rabbet that keeps the door from opening past the latch. The locking mechanisms may range from standard doorknob locks to deadbolt locks or sliding latches.

Forcing Entry

A simple solution to gaining entry through inward-opening doors may be to break a small window on the door or adjacent to it, reach inside, and operate the locking mechanism. Remember—"try before you pry."

If no window is available and the door is locked, stronger measures are required. To force an inward-opening door, first ascertain which type of frame it has. If the door has a stopped frame, use a prying tool near the locking mechanism to pry the stop away from the frame. After removing the stop, reinsert the prying tool near the latch and pry the door away from the frame. Once the latch clears the strike plate, push the door inward. Use a striking tool to force the prying tool farther into the jamb.

To force entry into an inward-opening door, follow the steps in SKILL DRILL 12-1:

1. Look for any safety hazards as you evaluate the door. Inspect the door for the location and number of locks and their mechanisms. (STEP 1)
2. Place the adz of the Halligan tool into the door frame, between the door jamb and the door stop, near the lock, with the beveled end against the door. (STEP 2)
3. Once the Halligan tool is in position, have your partner, on your command, drive the tool farther into the gap between the rabbeted jamb or stop and the door. Make sure that the tool is not driven into the door jamb itself. (STEP 3)
4. Once the tool is past the stop but between the door and the jamb, push the Halligan tool toward the door to force it open.

SKILL DRILL 12-1 Forcing Entry into an Inward-Opening Door

(Fire Fighter I, NFPA 5.3.4)

1. Size up the door, looking for any safety hazards. Inspect the door for the location and number of locks and mechanisms.

2. Place the adz of the Halligan tool into the door frame between the door jamb and the door stop, near the lock, with the beveled end of the tool against the door.

3. Once the Halligan tool is in position, have your partner, on your command, drive the tool farther into the gap between the rabbeted jamb or stop and the door. Make sure that the tool is not driven into the door jamb itself.

4. Once the tool is past the stop but between the door and the jamb, push the Halligan tool toward the door to force it open. If more leverage is needed, your partner can slide the axe head between the bevel of the Halligan tool and the door. It may be necessary to push in on the door. Secure the door to prevent it from closing behind you.

5. If more leverage is needed, your partner can slide the axe head between the bevel of the Halligan tool and the door. It may be necessary to push in the door.
6. Once the door is open, secure it to prevent it from closing behind you. (STEP 4)

Outward-Opening Doors

Design

Outward-opening doors are used in commercial occupancies and for most exits FIGURE 12-11. They are designed so that people can leave a building quickly during an emergency. Outward-opening doors may be constructed of wood, metal, or glass. They usually have exposed hinges, which may present an entry opportunity. More frequently, however, these hinges will be sealed so that the pins cannot be removed. Several types of locks, including handle-style locks and deadbolts, may be used with these doors.

Forcing Entry

Before forcing entry to an outward-opening door, check the hinges to see if they can be disassembled or the pins removed. If that would take too long or cannot be done, place the adz

FIGURE 12-11 An outward-opening door enables rapid exit from a building.

FIRE FIGHTER Tips

Self-closing doors may work via a means as simple as a mechanism above the door or as complex as a magnet or electrically charged release that functions by closing the doors when the fire alarm sounds. Either way, when these doors are forced, the mechanism that closes the doors should be disabled, along with using some type of door chock or wedge to keep the doors from closing behind you once entry is made.

end of a prying tool into the door frame near the locking mechanism. Use a striking tool to drive it farther into the door jamb and get a good bite on the door. Then leverage the tool to force the door outward away from the jamb.

To force entry into an outward-opening door, follow the steps in SKILL DRILL 12-2:

1. Size up the door, looking for any safety hazards. Determine the number and location of locks and the locking mechanism.
2. Place the adz end of the Halligan tool between the door and the frame, near the locking mechanism, or between the mechanism and a secondary lock. (STEP 1)
3. Once the Halligan tool is in position, give the command to your partner to strike the Halligan tool and drive the adz end farther into the gap. (STEP 2)

SKILL DRILL 12-2 Forcing Entry into an Outward-Opening Door

(Fire Fighter I, NFPA 5.3.4)

1 Size up the door, looking for any safety hazards. Place the adz end of the Halligan tool between the door and the frame either near the locking mechanism, or between the mechanism and a secondary lock.

2 Once the Halligan tool is in position, have your partner strike the Halligan tool on your command and drive the adz end farther into the gap.

3 Pry in a downward direction with the fork end of the tool and then force the door outward. Secure the door to prevent it from closing behind you.

4. Pry down with the fork end of the tool and then force the door outward.
5. Always secure the door to prevent it from closing behind you. (STEP 3)

Sliding Doors

Design

Most sliding doors are constructed of tempered glass in a wooden or metal frame. Such doors are commonly found in residences and hotel rooms that open onto balconies or patios FIGURE 12-12. Sliding doors generally have two sections and a double track; one side is fixed in place, while the other side slides. A weak latch on the frame of the door secures the movable side. Many people prop a wood or metal rod in the track to provide additional security. If this has been done, look for another means of entry if possible to avoid destroying the door and leaving the premise open.

FIGURE 12-12 A reinforced sliding glass door may be difficult to open without breaking the glass.

Forcing Entry

Forcing open sliding glass doors may be either very easy or very difficult. If the doors are not reinforced with a rod in the track, they should be easy to force. The locking mechanisms are not strong, and any type of prying tool can be used. Place the tool into the door frame near the locking mechanism and force the door away from the mechanism. If a rod in the track prevents the door from moving and there is no other way to enter, the only choice is to break the glass to gain forcible entry. This may create a hazard if fire fighters must drag hose lines or victims through the sharp glass debris. Clean all broken glass from the opening to avoid such problems.

Revolving Doors

Design

Revolving doors are most commonly found in upscale buildings and buildings in large cities FIGURE 12-13. They are usually made of four glass panels with metal frames. The panels are designed to collapse outward with a certain amount of pressure to allow for rapid escape during an emergency. Revolving doors are generally secured by a standard cylinder lock or slide latch lock.

FIGURE 12-13 Building codes may require outward-opening doors next to a revolving door.

Because of Life Safety Code®, fire and building code requirements, standard outward-opening doors are often found adjacent to the revolving doors. It may be easier to attempt to force entry through these doors instead of going through the revolving doors. Repairing any damage to the outward-opening doors will be less expensive for the building owner, too.

Forcing Entry

Forcible entry through revolving doors should be avoided whenever possible. Even if the doors can be forced open, the opening created will not be large enough to allow many people to exit quickly and easily. Forcible entry through a revolving door can be achieved by attacking the locking mechanism directly or by breaking the glass.

Overhead Doors

Design

Overhead doors come in many different designs, ranging from standard residential garage doors to high-security commercial roll-up doors FIGURE 12-14. Most residential garage doors have three or four panels, which may or may not include windows. Some residential overhead garage doors come in a single section and tilt rather than roll up. These doors, which may be made of wood or metal, usually have a hollow core that is filled with insulation or foam. By contrast, commercial security overhead doors are made of metal panels or hardened steel rods. They may use solid-core or hollow-core construction, depending on the amount of security needed. Overhead doors can be secured with cylinder-style locks, padlocks, or automatic garage door openers.

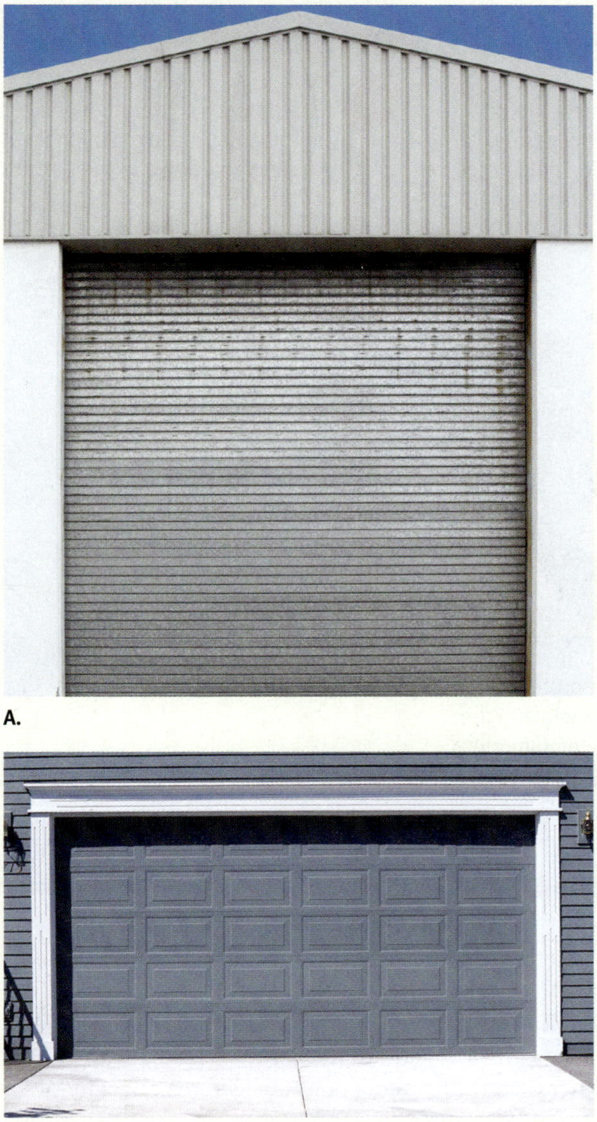

A.

B.

FIGURE 12-14 **A.** Commercial overhead doors are made of metal panels or steel rods. **B.** Residential overhead garage doors may tilt or roll up.

Forcing Entry

Before forcing entry through an overhead door, perform a careful size-up of the door. Most residential garage doors are not very sturdy; breaking a window or panel and manually operating the door lock or pulling the emergency release on the automatic opener from the inside may be all that is needed. If the fire is located behind the overhead door, however, it might have weakened the door springs, making it impossible to raise the door. In such a case, either the door must be cut or another means of entry found.

Secure any raised door with a pike pole or other support to ensure that it does not close on either fire fighters or their attack lines. Put the prop under the door near the track. If the door has an emergency release cord, it might be able to serve as a second safety measure.

The quickest way to force entry through a security roll-up door is to cut the door with a torch or saw. Make the cut in the shape of a triangle, with the point at the top. Pad this opening to prevent injury to those entering and exiting through the opening. To open an overhead garage door using the triangle method, follow the steps in SKILL DRILL 12-3:

1. Before cutting, check for any safety hazards during the size-up of the garage door.
2. Select the appropriate tool to make the cut. The best choice is a power saw with a metal-cutting blade.
3. Wearing full turnout gear and eye protection, start the saw and ensure it is in proper working order. (STEP 1)
4. Be aware of the environment behind the door. If necessary, cut a small inspection hole, large enough to insert a hose nozzle. (STEP 2)
5. Starting at a center high point in the door, make a diagonal cut to the right, down to the bottom of the door. (STEP 3)
6. From the same starting point, make a second diagonal cut to the left, down to the bottom of the door.
7. Fold the door down to form a large triangle. (STEP 4)
8. If necessary, pad or protect the cut edges of the triangle and the bottom panel to prevent injuries as fire fighters enter or leave the premises. (STEP 5)

Fire Fighter Safety Tips

Always secure doors in the open position before entering a building. Prop open overhead doors with a pike pole. Use the emergency release cord if one is available. If this step is not taken, the overhead doors could come down on fire fighters and on their attack lines, trapping them and cutting off their water supply. This problem could complicate the attack and lead to injuries or deaths.

Windows

Windows provide both air flow and light to the inside of buildings; they also can provide emergency entry or exits. Windows are often easier to force than doors. Understanding how to force entry into a window requires an understanding of both window-frame construction and glass construction. Window frames are made of the same materials used in doors—wood, metal, vinyl, or combinations of these materials—and will often match the door construction of a building.

SKILL DRILL 12-3 Opening an Overhead Garage Door Using the Triangle Method

(Fire Fighter I, NFPA 5.3.4)

1. Before cutting, check for any safety hazards. Select the appropriate tool to make the cut. Wearing full turnout gear and eye protection, start the saw and ensure it is in proper working order.

2. Be aware of the environment behind the door. If necessary, cut a small inspection hole, large enough to insert a hose nozzle.

3. Starting at a center high point in the door, make a diagonal cut to the right, down to the bottom of the door.

4. From the same starting point, make a second diagonal cut to the left, down to the bottom of the door. Fold the door down to form a large triangle.

5. If necessary, pad or protect the cut edges of the triangle and the bottom panel to prevent injuries as fire fighters enter or leave the premises.

This section reviews window construction, glass construction, and forcible entry techniques. The goal of forcible entry using windows is to gain entry with minimal or no damage to the window. As with doors, you should try to open the window normally before using any force. Although breaking the glass is the easiest way to force entry, it is also dangerous. Glass may be the least expensive part of the window, but replacement costs are increasing as energy-efficient windows are being used more widely.

Safety

Always take proper protective measures, including wearing full PPE, when forcing entry through windows. Ensure that the area around the window, both inside and outside the building and below the window, is clear of other personnel. Broken pieces of wood, shattered glass, and sharp metal can cause serious injury.

Remember that forcing entry through windows during a fire situation might cause the fire to spread. Do not attempt forcible entry through a window unless a proper fire attack is in place. Always stand to the windward side, with your hands higher than the breaking point, when breaking windows. This positioning ensures that the broken glass will fall away from your hands and body. Placing the tip of the tool in the corner of the window will give you more control in breaking the window. After the window is broken, clear glass from the entire frame, so that no glass shards will stick out and cause injury to fire fighters who enter or exit through the window.

Glass Construction

The glazed (transparent) part of the window is most commonly made of glass. Window glass comes in several configurations: regular glass, double/triple-pane glass, plate glass (for large windows), laminated glass, and tempered glass. Plastics such as Plexiglas and Lexan may also be used in windows. The window may contain one or more panes of glass. Insulated glass, for example, usually consists of two or more pieces of glass in the window and will be discussed in more detail later.

FIRE FIGHTER Tips

Lexan glass is used as a security measure because it cannot be broken with conventional tools and methods. It looks like glass, but if you take an axe to it, the axe will bounce back. For windows made of Lexan glass, cut the window frame in order to gain entry.

Regular or Annealed Glass

Single-pane, regular or annealed glass is often used in construction because it is relatively inexpensive; larger pieces are called plate glass. This type of glass is easily broken with a pike pole. When broken, plate glass creates long, sharp pieces called shards, which can penetrate helmets, boots, and other protective gear, causing severe lacerations and other injuries.

Double/Triple-Pane Glass (Insulated Windows)

Double/triple-pane glass is used in many homes because it improves home insulation by using two panes of glass with an air pocket between them. Some double/triple-pane windows may include an inert gas such as argon between the panes for additional insulation value. These windows are sealed units, which makes them more expensive to replace. However, replacing the glass alone is less expensive than replacing the entire window assembly. Forcing entry through insulated windows is basically the same as forcing entry through single-pane windows, except that the two panes may need to be broken separately. These kinds of windows also will produce dangerous glass shards.

Plate Glass

Commercial plate glass is a stronger, thicker glass used in large window openings. Although it is being replaced by tempered glass in modern construction for safety reasons, commercial plate glass can still be found in older large buildings, storefronts, and residential sliding doors. It can easily be broken with a sharp object such as a Halligan tool or a pike pole. When broken, commercial plate glass will create many large, sharp pieces.

Laminated Glass

Laminated glass, also known as safety glass, is used to prevent windows from shattering and causing injury. Laminated glass is molded with a sheet of plastic between two sheets of glass. This type of glass is most commonly used in vehicle windshields, but it may also be found in other applications such as doors or building windows.

Tempered Glass

Tempered glass is specially heat treated, making it four times stronger than regular glass. This type of glass is commonly found in side and rear windows in vehicles, in commercial doors, in newer sliding glass doors, and in other locations where a person might accidentally walk into the glass and break it. Tempered glass breaks into small pellets without sharp edges to help prevent injury during accidents. The best way to break tempered glass is by using a sharp, pointed object in the corner of the frame. During vehicle extrication, a center punch is often used to break a vehicle window.

Wired glass is tempered glass that has been reinforced with wire. This kind of glass may be clear or frosted, and it is often used in fire-rated doors that require a window or sight line from one side of the door to the other. Wired glass is difficult to break and force.

Frame Designs

Window frames come in many styles. This chapter covers the most common ones, but fire fighters must be familiar with both the types in their response area and the forced entry techniques for each type.

Double-Hung Windows

Design

Double-hung windows contain two movable sashes, usually made of wood or vinyl, that move up and down **FIGURE 12-15**. These kinds of windows are frequently found in residences and have wood, plastic, or metal runners. Newer double-hung window sashes may be removed or swung in for cleaning. They may have either one locking mechanism that is found in the center of the window or two locks on each side of the lower sash that prevent the sashes from moving up or down.

FIGURE 12-15 Double-hung windows allow the inner and outer sashes to move freely up and down.

Forcing Entry

Forcing entry through double-hung windows involves opening or breaking the locking mechanism. Place a prying tool under the lower sash and force it up to break the lock or remove it from the track. This technique must be done carefully; otherwise, it may cause the glass to shatter. Because forcing the locks causes extensive damage to the window, breaking the glass and opening the locks may be a less expensive method of gaining entry.

To force entry through a wooden double-hung window, follow the steps in SKILL DRILL 12-4:

1. Size up the window for any safety hazards and locate the locking mechanism. (STEP 1)
2. Place the pry end of the Halligan tool under the bottom sash in line with the locking mechanism. (STEP 2)
3. Pry the bottom sash upward to displace the locking mechanism.
4. Secure the window in the open position so that it cannot close. (STEP 3)

Fire Fighter Safety Tips

When forcing entry through older, double-hung windows, try sliding a hacksaw blade or similar blade up between the sashes. The teeth on the blade will grip the locking mechanism and can often move it into the unlocked position. This technique will not work on newer double-hung windows because of their newer latch design.

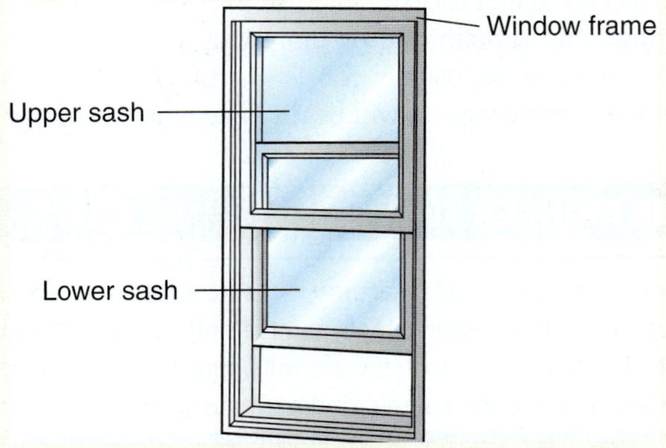

FIGURE 12-16 A single-hung window has only one movable sash.

Single-Hung Windows

Design

Single-hung windows are similar to double-hung windows, except that the upper sash is fixed and only the lower sash moves FIGURE 12-16. The locking mechanism is the same as on double-hung windows as well. It may be difficult to distinguish between single-hung and double-hung windows from the exterior of a building.

Forcing Entry

Forcing entry through a single-hung window involves the same technique as forcing entry through a double-hung window. Place a prying tool under the lower sash and force it up, which will usually break the lock or remove the sash from the track. Be careful because the glass will probably shatter. Breaking the glass and opening the window is generally easier with a single-hung window.

Jalousie Windows

Design

A jalousie window is made of adjustable sections of tempered glass encased in a metal frame that overlap each other when closed FIGURE 12-17. This type of window is often found in mobile homes and is operated by a small hand-wheel or crank located in the corner of the window.

Forcing Entry

Forcing entry through jalousie windows can be difficult because they are made of tempered glass and have several panels. Forcing open the panels or breaking a lower one to operate the crank is possible but does not leave a large-enough

SKILL DRILL 12-4 Forcing Entry Through a Wooden Double-Hung Window

(Fire Fighter I, NFPA 5.3.4)

1. Size up the window for any safety hazards and locate the locking mechanism.

2. Place the pry end of the Halligan tool under the bottom sash in line with the locking mechanism.

3. Pry the bottom sash upward to displace the locking mechanism. Secure the window so that it does not close.

VOICES OF EXPERIENCE

When I was a fire fighter assigned to a four person engine company, we received a call for smoke inside a structure. While traveling to the incident scene, dispatch reported that it appeared that the structure was on fire, but access could not be made due to a locked gate. When we arrived, I exited the apparatus and promptly retrieved some bolt cutters. I was making my way to the gate to cut the lock when I heard the fire officer order the driver/operator to ram the gate.

The driver/operator asked if the fire officer really wanted to do that and the fire officer repeated the command. I stepped back from the engine for my safety. The driver/operator complied and at a low speed tried to force the gate open with the fire engine. The first attempt failed. The driver/operator backed the fire engine up and sped up in an attempt to break through the locked gate. This time the gate gave way at the hinges and flew open. Upon entering the structure, a small fire in the kitchen area of the business was found and quickly extinguished.

After the incident it was discovered that the action of ramming the gate had caused damage to the bumper, front right corner of the cab, and the chrome bell of the fire engine. In addition, there was considerable damage to the fence and gate of the business. We determined through a Post Incident Analysis that if the fire officer had allowed the lock to be cut by the bolt cutters, no time would have been lost and hundreds of dollars in damage could have been avoided. It all came down to using the proper tool for the job.

Mark Schuman
North Washington Fire Protection District
Adams County, Colorado

FIGURE 12-17 Jalousie windows are opened and closed with a small hand crank.

opening for a person to enter the building. Removing or breaking the panels one at a time is time-consuming and does not always leave a clean entry point with the framing system still in place. The best strategy is to avoid these windows.

Awning Windows

Design

Awning windows are similar in operation to jalousie windows, except that they usually have one large or two medium-sized glass panels instead of many small panes FIGURE 12-18. Awning windows are operated by a hand crank located in the corner or in the center of the window. These kinds of windows can be found in residential, commercial, and industrial structures. Residential awning windows may be framed in wood, vinyl, or metal, whereas commercial and industrial windows will usually have metal frames.

FIGURE 12-18 An awning window has larger panels than a jalousie window does.

Commercial and industrial awning windows often use a lock and a notched bar to hold the window open, rather than a crank. These so-called projected windows are discussed later in this chapter.

Forcing Entry

Forcing entry through an awning window requires the same technique as is used for jalousie windows—that is, break or force the lower panel and operate the crank, or break out all the panels. Depending on the size of the panels and the window frame, it may be easier to access an awning window. The larger opening of an awning window makes entry easier as well.

Horizontal-Sliding Windows

Design

Horizontal-sliding windows are similar to sliding doors FIGURE 12-19. The latch is similar as well and attaches to the window frame. People often place a rod or pole in the track to prevent break-ins. Newer sliding windows have latches between the windows, similar to those found on double-hung windows.

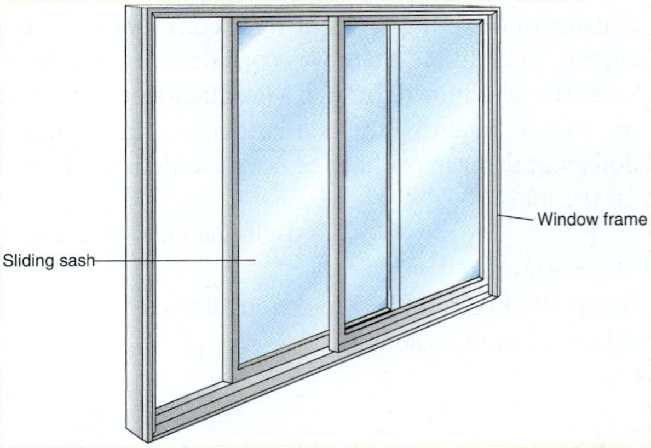

FIGURE 12-19 Horizontal-sliding windows work like sliding doors.

Forcing Entry

Forcing entry through sliding windows is just like forcing entry through sliding doors: Place a pry bar near the latch to break the latch or the plate. If a rod has been inserted into the track, look for another entry point or break the glass, although the latter step should be the last resort.

Casement Windows

Design

Casement windows have a steel or wood frame and open away from the building with a crank mechanism FIGURE 12-20. Although they are similar to jalousie or awning windows, casement windows have a side hinge, rather than a top hinge. Several types of locking mechanisms can be used with these windows. Like jalousie windows, these windows should be avoided when forcible entry is necessary because they are difficult to force open.

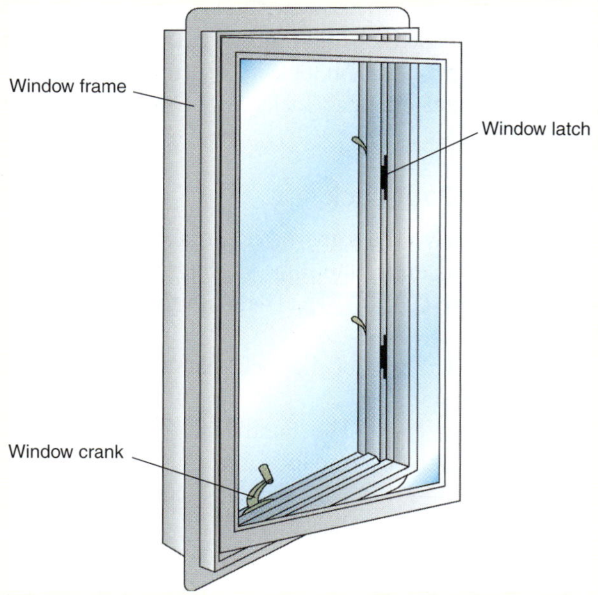

FIGURE 12-20 Casement windows open to the side with a crank mechanism.

Forcing Entry

The best way to force entry through a casement window is to break out the glass from one or more of the panes, locate the locking mechanisms, and crank them manually. To force entry through a casement window, follow the steps in SKILL DRILL 12-5:

1. Size up the window to check for any safety hazards and locate the locking mechanism.
2. Select an appropriate tool to break out a windowpane.
3. Stand to the windward side of the window and break out the pane closest to the locking mechanism, keeping the tool lower than your hands. (STEP 1)
4. Remove all of the broken glass in the pane to prevent injuries. (STEP 2)
5. Reach in and manually operate the window crank to open the window. (STEP 3)

Projected Windows

Design

Projected windows (also called factory windows) are usually found in older warehouse or commercial buildings FIGURE 12-21. They can project inward or outward on an upper hinge. Screens are rarely used with these windows, but forcing entry may not be easy, depending on the integrity of the frame, the type of locking mechanism used, and the window's distance off the ground. These windows may have fixed, metal-framed wire glass panes above them.

Forcing Entry

Avoid forcing entry through a projected window when possible. These windows are often difficult to force open, and they may be difficult for a person to enter. To force entry, access one

SKILL DRILL 12-5 Forcing Entry Through a Casement Window

(Fire Fighter I, NFPA 5.3.4)

1 Size up the window to check for any safety hazards and locate the locking mechanism. Select an appropriate tool to break out a windowpane. Stand to the windward side of the window and break out the pane closest to the locking mechanism, keeping the tool lower than your hands.

2 Remove all of the broken glass in the pane to prevent injuries.

3 Reach in and manually operate the window crank to open the window.

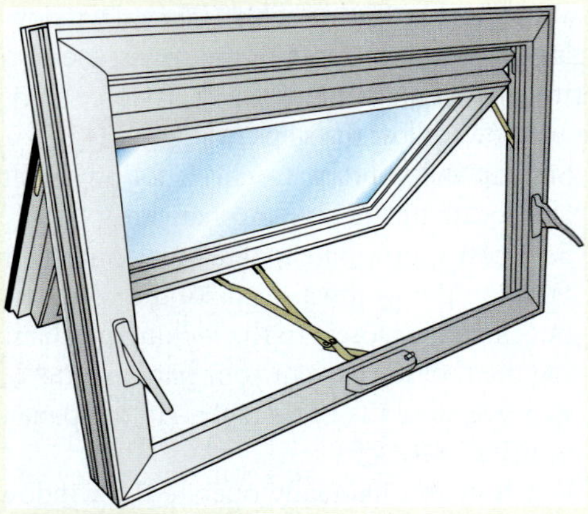

FIGURE 12-21 Projected windows may open inward or outward.

pane, unlock the mechanism, and open the window by hand. If the opening created is not large enough, break out the entire window assembly.

To force entry through a projected or factory window, follow the steps in SKILL DRILL 12-6:

1. Size up the window to check for any safety hazards and locate the locking mechanism.
2. Select an appropriate tool, such as a pike pole, to break a pane of glass.
3. Stand to the windward side of the window and break the pane closest to the locking mechanism, keeping the tool head lower than your hands. (STEP 1)
4. Remove all of the broken glass in the frame to prevent injuries. (STEP 2)
5. Reach in and manually open the locking mechanism and the window.
6. If necessary, use a cutting tool, such as a torch, to remove the window frame and enlarge the hole. (STEP 3)

Locks

Locks have been used for hundreds of years. They range from the very simple push-button locks found in most homes to the complex computer-operated locks found in banks and other high-security areas.

Parts of a Door Lock

All door locks have the same basic parts. Gaining entry will be easier with an understanding of what these parts are and how they operate FIGURE 12-22. There are three major parts of a door lock:

- Latch—The part of the lock that "catches" and holds the door frame
- Operator lever—The handle, doorknob, or keyway that turns the latch to lock it or unlock it
- Deadbolt—A second, separate latch that locks and reinforces the regular latch

Parts of a Padlock

All padlocks, like all door locks, have similar parts FIGURE 12-23:

- Shackles—The U-shaped top of the lock that slides through a hasp and locks in the padlock itself

SKILL DRILL 12-6 Forcing Entry Through a Projected Window

(Fire Fighter I, NFPA 5.3.4)

1 Size up the window to check for any safety hazards and locate the locking mechanism. Select an appropriate tool, such as a pike pole, to break a pane of glass. Stand to the windward side of the window and break the pane closest to the locking mechanism, keeping the tool head lower than your hands.

2 Remove all of the broken glass in the frame to prevent injuries.

3 Reach in and manually open the locking mechanism and the window. If necessary, use a cutting tool, such as a torch, to remove the window frame and enlarge the hole.

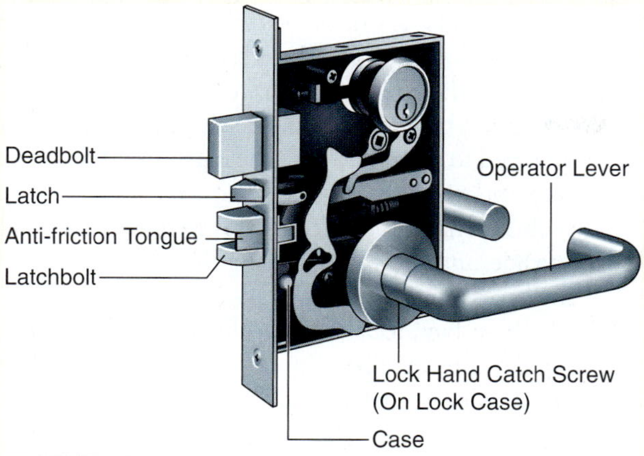

FIGURE 12-22 All door locks have the same basic parts.

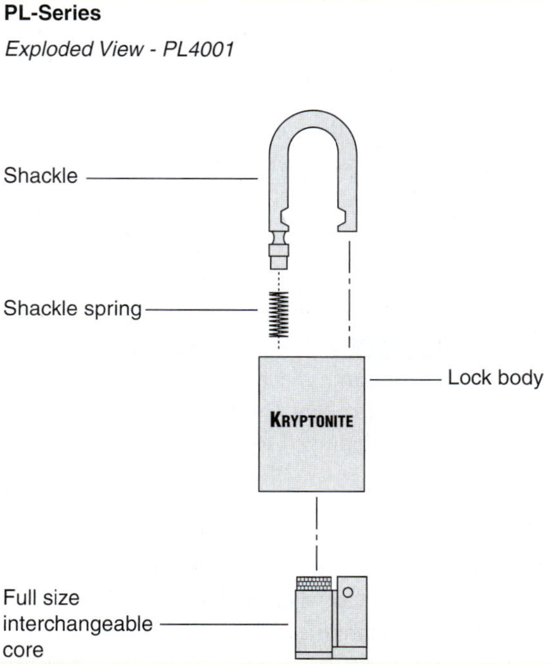

FIGURE 12-23 All padlocks have four basic parts.

- Unlocking mechanism—The keyway, combination wheels, or combination dial used to open the padlock
- Lock body—The main part of the padlock that houses the locking mechanisms and the retention part of the lock

Safety

Tool safety is an important consideration when executing forcible entry through a door or padlock. The tools used for cutting the cylinders must be kept sharp, and fire fighters must be careful to avoid being cut when using these tools. Care also must be taken when using striking tools when forcing a lock to avoid pinching or crushing fingers or hands. Fire fighters must wear proper protective equipment such as gloves and use tools carefully to minimize the risk of injury.

Types of Locks

The four major lock categories are cylindrical locks, padlocks, mortise locks, and rim locks.

Cylindrical Locks

Design

Cylindrical locks are the most common fixed lock in use today FIGURE 12-24. They are relatively inexpensive and are standard in most doors. The locks and handles are set into predrilled holes in the doors. One side of the door usually has a key-in-the-knob lock; the other side will have a keyway, a button, or some type of locking/unlocking mechanism.

Forcing Entry

Forcing entry into a cylindrical lock is usually very easy because most mechanisms are not very strong. Place a pry bar near the locking mechanism and lever it to force the lock.

Padlocks

Design

Padlocks are the most common locks on the market today and come in many designs and strengths. Both regular-duty and heavy-duty padlocks are available. Padlocks come with a variety of unlocking mechanisms, including keyways, combination wheels, or combination dials. Operating the unlocking

FIGURE 12-24 A cylindrical lock.

mechanism opens one side of the lock to release the shackle and allow entry. Shackles for regular padlocks generally have a diameter of ¼ inch (6 millimeters) or less and are not made of case-hardened metal. Shackles for heavy-duty padlocks are ¼ inch (6 millimeters) or larger in diameter and made of case-hardened steel or some other case-hardened metal. The design of some padlocks hides the latch, making forced entry difficult.

One of these locks is the American Series 2000, also referred to as the hockey puck lock FIGURE 12-25. These types of locks have hidden shackles and cannot be forced through conventional methods. A large pipe wrench with a cheater bar can be used to twist the locks from their mounting tabs. However, if the lock has a mounting bracket, this technique will be ineffective. In such a case, you will need a rotary saw or a cutting torch to force entry. When using this method, you should focus your effort approximately two-thirds of the way up from the keyway to cut the pin in half.

Forcing Entry

Several techniques can be used to force entry through padlocks without causing extensive property damage. Before breaking the padlock, consider cutting the shackle or hasp first. This approach saves the lock and makes securing the building easier. Breaking the shackle is the best method for forcing entry through padlocked doors. If the padlock is made of case-hardened steel, however, many conventional methods of breaking the lock will prove ineffective.

The tools most commonly used to force entry through a padlock are bolt cutters, duck-billed lock breakers, a bam-bam tool, locking pliers and chain, and a torch:

- Bolt cutters can quickly and easily break regular-duty padlocks, but cannot be used on heavy-duty, case-hardened steel padlocks. To use bolt cutters on a padlock, open the jaws as wide as possible. Close the jaws around one side of the lock shackles to cut through the shackle. Once one shackle of a regular-duty lock is cut, the other side will spin freely and allow access.
- The duck-billed lock breaker has a large metal wedge attached to a handle. Place the narrow end of the wedge into the center of the shackle and force it through with another striking tool. The wedge will spread the shackle until it breaks, freeing the padlock and allowing access to the building.
- The bam-bam tool can pull the lock cylinder out of a regular-duty padlock, but it will not work on higher-end padlocks that include security rings to hold the cylinder in place. This tool contains a case-hardened screw that is placed in the keyway. Once the screw is set, the sliding hammer pulls the tumblers out of the padlock so that the trip-lever mechanism inside the lock can be opened manually.
- The locking pliers and chain are attached to a padlock to secure it in a fixed position so that the lock can be cut. Use a rotary saw with a metal-cutting blade or a torch to cut the padlock after it has been secured. Securing the padlock is necessary because it is too dangerous to hold the lock with a gloved hand or to cut into it while it is loose.

FIGURE 12-25 The Pro Series Hidden Shackle Padlock, also referred to as the hockey puck lock.

Mortise Locks

Design

Mortise locks, like cylindrical locks, are designed to fit in predrilled openings inside a door; they are commonly found in hotel rooms FIGURE 12-26. Mortise locks have both a latch and a bolt built into the same mechanism, each of which operates independently of the other. The latch will lock the door, but the bolt can also be deployed for added security.

Forcing Entry

The design and construction of mortise locks make them difficult to force. A door with a mortise lock may be forced with conventional means but will probably require the use of a through-the-lock technique.

Rim Locks and Deadbolts

Design

Rim locks and deadbolts are different types of locks that can be surface mounted on the interior of the door frame FIGURE 12-27. They are commonly found in residences as secondary locks to support the through-the-handle locks. Such locks can be identified from the outside by the keyway that has been bored into the door. These locks have a bolt that extends at least 1 inch (2.5 centimeters) into the door frame, making the door more difficult to force.

Forcing Entry

Rim locks and deadbolts are the most challenging locks to break. They can be very difficult to force with conventional methods, such that through-the-lock methods may be the only option. Skill Drills 12-7 through 12-9 present various through-the-lock techniques.

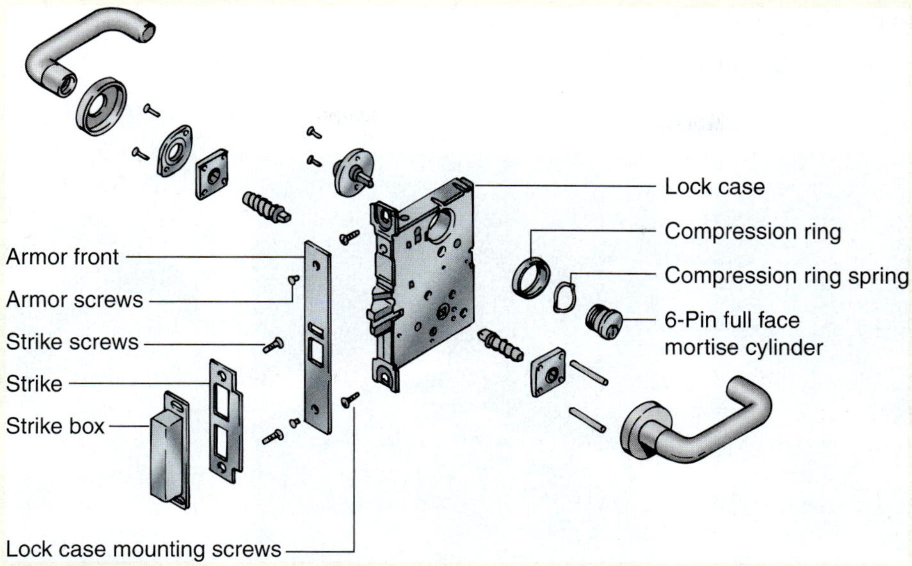

FIGURE 12-26 Mortise locks have a latch and a bolt, which operate independently of each other.

FIGURE 12-27 Rim locks and deadbolts are mounted on the inside of a residence door.

To force entry using a K tool, follow the steps in SKILL DRILL 12-7:

1. Size up the lock area, checking for any safety hazards. Determine which type of lock is used, whether it is regular or heavy-duty construction, and whether the cylinder has a case-hardened collar, which may hamper proper cutting.
2. Place the K tool over the face of the cylinder, noting the location of the keyway.
3. Using a Halligan or similar pry tool, tap the K tool onto the cylinder. (STEP 1)
4. Place the adz end of the Halligan tool into the slot on the K tool, and strike the end of the Halligan tool with a flat-head axe to drive the K tool farther into the lock cylinder. (STEP 2)
5. Pry up on the Halligan tool to pull out the lock and expose the locking mechanism.
6. Using the small tools that come with the K tool, turn the mechanism to open the lock.
7. The lock should release, allowing you to open the door. (STEP 3)

FIRE FIGHTER Tips

If the K tool does not fit over a residential lock, try an A tool.

SKILL DRILL 12-7 Forcing Entry Using a K Tool

(Fire Fighter I, NFPA 5.3.4)

1 Place the K tool over the face of the cylinder, noting the location of the keyway. Using a Halligan or similar pry tool, tap the K tool onto the cylinder.

2 Place the adz end of the Halligan tool into the slot on the K tool, and strike the end of the Halligan tool with a flat-head axe to drive the K tool farther into the lock cylinder.

3 Pry up on the Halligan tool to pull out the lock and expose the locking mechanism. Using the small tools that come with the K tool, turn the mechanism to open the lock. The lock should release, allowing you to open the door.

To force entry using an A tool, follow the steps in SKILL DRILL 12-8:

1. Size up the door and lock area, checking for any safety hazards.
2. Place the cutting edges of the A tool at the top of the cylinder, between the cylinder and the door frame. (STEP 1)
3. Using a flat-head axe or similar tool, drive the A tool into the cylinder.
4. Pry up on the A tool to remove the cylinder from the door.
5. Insert a key tool into the hole to manipulate the locking mechanism and open the door. (STEP 2)

To force entry by unscrewing the lock, follow the steps in SKILL DRILL 12-9:

1. Size up the door and lock area, checking for any safety hazards.
2. Using a set of vise grips, lock the pliers onto the outer housing of the lock with a good grip. (STEP 1)
3. Unscrew the housing until it can be fully removed.
4. Manipulate the locking mechanism with lock tools to disengage the arm, and then open the door. (STEP 2)

■ Forcing Entry Through Security Gates and Windows

Many homes are equipped with metal security gates and security bars over the windows, which are intended to provide protection from break-ins for the building residents. Security gates are usually equipped with a lock that must be unlocked using a key from both the inside and the outside. As a consequence, if a key is not available to the building occupants, they can become trapped in their own house. Likewise, security bars should be equipped with a locking mechanism that enables the residents to quickly open the security bars from the inside and use the window for an exit in case of a fire. Unfortunately, many security bars are permanently mounted into the window frame and cannot be removed quickly.

FIRE FIGHTER Tips

Electromagnetic locks use powerful magnets to keep doors shut. Doors with electromagnetic locks may be forced open with prying tools or may require hydraulic tools. Follow the SOPs of your fire department.

Thus these security devices present a dilemma: The same devices that were designed to keep criminals out can, in the event of a fire, make it difficult for the building occupants and fire fighters to escape from the fire. Such devices also complicate fire fighters' efforts to gain access to the building to perform search and rescue operations and fire suppression activities. Given these facts, it is important for fire fighters to know how to rapidly force entry when these devices are in place.

SKILL DRILL 12-8 Forcing Entry Using an A Tool

(Fire Fighter I, NFPA 5.3.4)

1 Place the cutting edges of the A tool at the top of the cylinder, between the cylinder and the door frame.

2 Using a flat-head axe or similar tool, drive the A tool into the cylinder. Pry up on the A tool to remove the cylinder from the door. Insert a key tool into the hole to manipulate the locking mechanism and open the door.

SKILL DRILL 12-9 Forcing Entry by Unscrewing the Lock

(Fire Fighter I, NFPA 5.3.4)

1 Using a set of vise grips, lock the pliers onto the outer housing of the lock with a good grip.

2 Unscrew the housing until it can be removed. Manipulate the locking mechanism with lock tools to disengage the arm, and then open the door.

FIRE FIGHTER Tips

Some gates that secure driveways may be siren activated. In such a case, no forcible entry technique other than sounding the siren on the fire apparatus is required to open the gates.

One method of breaching a security gate is to remove the lock cylinder using a K tool, an A tool, or a bam-bam tool. If the lock cylinder can be accessed and removed, it is a simple matter to turn the locking mechanism and open the gate. If the lock cylinder has been protected to minimize the chance of break-in, however, it may be necessary to cut through part of the gate to open it. This can be done by using a circular saw that is equipped with an appropriate blade or, in some cases, by using a hydraulic cutter. An alternative is to use a hydraulic spreader to force the anchor from the masonry or wood to which it is attached. Be sure to wear PPE, including adequate eye protection, when operating these tools. Plan your cuts to minimize the amount of metal that needs to be cut to open the gate.

Breaching security bars on a window can sometimes be accomplished by removing the lock cylinder, if one is present on the outside of the security bars. If no lock cylinder is present, it will be necessary to cut through the bars using either a circular saw or hydraulic cutters. If the bars are not well anchored, you might be able to pry them from the window. A hydraulic spreader can sometimes be used to force the anchor from the masonry or wood to which it is attached. This method will not always work, however, when the bars are securely anchored in masonry or concrete.

Look for the types of security gates and bars that are present in your community. Talking with the installers of these products will give you more knowledge about how to gain access through them.

Breaching Walls and Floors

On occasion, forced entry through a window, lock, or door might not be possible or might take too long. In these cases, consider the option of breaching a wall or floor. Breaching a wall can also be an option for removal of injured people or for emergency escape. An understanding of basic construction concepts is required to execute this technique safely and quickly.

Load-Bearing/Nonbearing Walls

Before breaching a wall, first consider whether the wall is load bearing. A load-bearing wall supports the building's ceiling and/or rafters, so removing or damaging such a wall could cause the building or wall to collapse. By contrast, a nonbearing wall can be removed safely and without danger. Nonbearing walls are also called partition walls or simply partitions. Refer to the Building Construction chapter for a review of wall construction.

Exterior Walls

Exterior walls can be constructed of one or more materials. Many residences have both wood and brick, aluminum siding, or masonry block construction FIGURE 12-28. Commercial buildings usually have concrete, masonry, or metal exterior walls. Commercial buildings also may have steel I-beam construction; older commercial structures may have heavy timber construction. Because there are exceptions to every rule, it is important to know the construction methods and materials used for buildings in your specific response area.

Deciding whether to breach an exterior wall is a difficult choice. Masonry, metal, and brick are formidable materials, and breaking through them can be very difficult. The best tools to use in breaching a concrete or masonry exterior wall are a battering ram, a sledgehammer, and a rotary saw with a concrete blade.

FIGURE 12-28 Residences may have exterior walls of multiple materials, such as wood and masonry block.

Interior Walls

Interior walls in residences are usually constructed of wood or metal studs covered by plaster, gypsum, or sheetrock. Some newer residential construction contains a laminate sheetrock, which is extremely difficult to penetrate. Commercial buildings may have concrete block interior walls. Breaching an interior wall can be dangerous to fire fighters for several reasons. For example, many interior walls contain electrical wiring, plumbing, cable wires, and phone wires—all of which present hazards to fire fighters. Interior walls may also be load bearing. Load-bearing walls can be breached without undue hazard as long as studs are not removed. Extreme care should be taken, however, if any studs are removed.

After determining whether the wall is load bearing, sound it to locate a stud away from any electrical outlets or switches. Tap on the wall; the area between studs will make a hollow sound compared to the solid sound directly over the stud.

After locating an appropriate site, make a small hole (which can also serve as an inspection hole to check for fire) to check for any obstructions. If the area is clear, expand the opening to reveal the studs. Walls should be breached as close as possible to the studs because this choice makes a large opening and cutting is easier. If possible, enlarge the opening by removing at least one stud to enable quick escapes.

To breach a wall frame, follow the steps in SKILL DRILL 12-10:

1. Size up the wall, checking for safety hazards such as electrical outlets, wall switches, or any signs of plumbing. Inspect the overall scene to ensure that the wall is not load bearing.
2. Using a striking or cutting tool, sound the wall to locate any studs, and then make a hole between the studs. (STEP 1)
3. Cut the sheetrock as close to the studs as possible. (STEP 2)
4. Enlarge the hole by extending it from stud to stud and as high as necessary. (STEP 3)

SKILL DRILL 12-10 Breaching a Wall Frame

(Fire Fighter I, NFPA 5.3.4)

1 Size up the wall, checking for safety hazards such as electrical outlets, wall switches, or any signs of plumbing. Inspect the overall scene to ensure that the wall is not load bearing. Using a striking or cutting tool, sound the wall to locate any studs, and then make a hole between the studs.

2 Cut the sheetrock as close to the studs as possible.

3 Enlarge the hole by extending it from stud to stud and as high as necessary.

To breach a masonry wall, you create an upside-down V in the wall. Follow the steps in SKILL DRILL 12-11:

1. Size up the wall, checking for any safety hazards such as electrical outlets, wall switches, and plumbing. Inspect the overall area to ensure that the wall is not load bearing.
2. Select a row of masonry blocks at 2 to 3 feet (61 to 91 centimeters) above the floor.
3. Using a sledgehammer, knock two holes each in five masonry blocks along the selected row. Each hole should pierce into the hollow core of the masonry blocks.
4. Repeat the process on four masonry blocks above the first row.
5. Repeat the process on three masonry blocks above the second row, on two masonry blocks above the third row, and on one masonry block above the fourth row. (STEP 1)
6. An upside-down V has now been created. (STEP 2)
7. Begin knocking out the remaining portion of the masonry blocks by hitting the masonry blocks parallel to the wall rather than perpendicular to it. (STEP 3)
8. Clear any reinforcing wire using bolt cutters, and enlarge the hole as needed. (STEP 4)

To breach a metal wall, follow the steps in SKILL DRILL 12-12:

1. Size up the wall, checking for any safety hazards such as electrical outlets, wall switches, or plumbing. Inspect the overall area to ensure that the wall is not load-bearing.
2. Select an appropriate tool, such as a rotary saw with a metal-cutting blade or a torch.
3. Open the wall using a diamond-shaped or inverted V-cut. (STEP 1)
4. Make the cut so the opening piece falls or is bent into the building so no sharp bottom edges are exposed. (STEP 2)

Floors

The two most popular floor materials found in residences and commercial buildings are wood and poured concrete. Both are very resilient and may be difficult to breach. Conduct a thorough size-up of the site before considering forcible entry through this route; breaching a floor should be a last-resort measure. A rotary saw with an appropriate blade is the best

SKILL DRILL 12-11 Breaching a Masonry Wall

(Fire Fighter I, NFPA 5.3.4)

1. Select a row of masonry blocks at 2 to 3 feet (61 to 91 centimeters) above the floor. Using a sledgehammer, knock two holes each in five masonry blocks along the selected row. Each hole should pierce into the hollow core of the masonry blocks. Repeat the process on four masonry blocks above the first row. Repeat the process on three masonry blocks above the second row, on two masonry blocks above the third row, and on one masonry block above the fourth row.

2. An upside-down V has now been created.

3. Begin knocking out the remaining portion of the masonry blocks by hitting the masonry blocks parallel to the wall rather than perpendicular to it.

4. Clear any reinforcing wire using bolt cutters, and enlarge the hole as needed.

tool to breach a floor; a chainsaw may be a better choice when used on a wood floor.

To breach a floor, follow the steps in SKILL DRILL 12-13:

1. Size up the floor area, checking for hazards in the area to be cut.
2. Sound the floor with an axe or similar tool to locate the floor joists. (STEP 1)
3. Use an appropriate cutting tool to cut one side of the hole.
4. Cut the opposite side. (STEP 2)
5. Remove any flooring, such as carpet, tile, or floorboards, that has been loosened.

SKILL DRILL 12-12 Breaching a Metal Wall

(Fire Fighter I, NFPA 5.3.4)

1 Select an appropriate tool, such as a rotary saw with a metal-cutting blade or a torch. Open the wall using a diamond-shaped or inverted V-cut.

2 Make the cut so the opening piece falls or is bent into the building so no sharp bottom edges are exposed.

6. Cut a similar opening into the subfloor, until the proper size of hole is achieved. (STEP 3)
7. Secure the area around the hole to prevent others working in the area from falling through the hole. (STEP 4)

Vehicle Entry

Some fire departments receive training from a locksmith in the use of slim jims and other devices to unlock vehicles. If you are not trained and equipped to gain entry into cars, in an emergency you can quickly gain entry by breaking a side window as far away from the trapped child or pet as possible. This is easily done using a spring-loaded center punch or other sharp object as described in the Vehicle Rescue and Extrication chapter.

FIRE FIGHTER Tips

Fire departments are regularly called because people have locked their car keys in their cars. Many fire departments do not respond to these routine calls. Instead, they may refer individuals to locksmiths or emergency lockout companies that can promptly respond to such requests. These routine calls are very different from calls where a child or a pet has been locked in a vehicle, however—this type of situation represents an emergency situation that requires a fire department response.

Systematic Forcible Entry

The first step in forcible entry is to think. Several issues need to be evaluated before deciding on a course of action. Double-check the address for the incident—it is embarrassing to try to explain to homeowners why you have destroyed their backdoor when the burning food on the stove was reported three houses down the street. Finding the right address can be a challenge, especially if rear entrances off an alley are not clearly marked.

Second, look for the presence of a lockbox. Lockboxes are attached to buildings, usually close to a main entrance. These boxes can be unlocked with a master key that is carried by the fire department. The master key is coded to open all lockboxes in that jurisdiction. Once entry is made into the lockbox, the fire department officer can remove keys that will open the specific building. With a lockbox system, rapid entry can be obtained by the fire department into any building equipped with a lockbox and the building can be secured after the incident by the fire department. To maintain security over the lockbox master keys, some systems are equipped with a device that keeps the master key locked in the fire apparatus unless released by the emergency dispatcher. This approach creates a record of each time the

SKILL DRILL 12-13 Breaching a Floor

(Fire Fighter I, NFPA 5.3.4)

1 Sound the floor with an axe or similar tool to locate the floor joists.

2 Use an appropriate cutting tool to cut one side of the hole. Cut the opposite side.

3 Remove any flooring, such as carpet, tile, or floorboards, that has been loosened. Cut a similar opening into the subfloor, until the proper size of hole is achieved.

4 Secure the area around the hole to prevent others working in the area from falling through the hole.

master key is used, allows the dispatcher to record the time and place the key was used, and identifies the officer requesting the use of the master key. You need to determine if your department has a lockbox program, and if it does, how it works.

Third, evaluate the threat level in the incident. When flames and smoke are visible, the need for forcible entry is much higher than when a fire alarm activation occurs in the middle of a summer thunderstorm with no sign of fire. Life safety is a major consideration. A fire alarm activation in a closed business in the middle of the night is much less potentially life threatening than a call where people are reportedly in distress or trapped.

Fourth, consider how to make entry with the least amount of damage. This objective must be balanced with the severity of the emergency. Generally speaking, you want to make entry with the least damage in the shortest amount of time, but this assessment must be balanced with the type of emergency present.

Do not forget that you need to work in a coordinated fashion with other members of the team. When a forcible entry team opens a door or window, the breach produces as much ventilation to the fire as if the door or window were opened in ventilation operations.

Forcible Entry and Salvage

Before Entry

The decision of which type of forcible entry to use will have a direct impact on salvage operations. In a nonemergency situation, through-the-lock methods will minimize damage, thereby saving the property from excessive repairs. In an emergency situation, more urgent methods of forcible entry may cause massive amounts of destruction, requiring expensive repairs to a door, window, or wall.

After Entry

When the operation is over and normal salvage and overhaul have begun, the structure must be secured before fire fighters leave the scene. The use of conventional forcible entry techniques and specialized tools requires that doors and windows be secured to prevent break-ins or theft after the call. If through-the-lock methods are used, a hasp and padlock may be all that are needed to secure a site. A resourceful crew might also screw a forced door shut from the inside and leave from a door that has not been forced. Broken windows or doors can be boarded up with sheets of plywood. Police will often increase patrols around a damaged building when needed to deter potential thieves. The building owner should be called to secure the building.

FIRE FIGHTER Tips

Many building owners are grateful to fire departments for explaining to them the types of assistance that board-up companies can provide. These companies are available in most communities to secure buildings after fire, water, or storm damage. The average homeowner has few opportunities to use such a company, and many people are not aware of the role these companies play in reducing the damage and in helping to prevent further damage to the building. Some fire departments offer building owners a list of licensed or approved companies that provide emergency board-up services in their community.

The fire department is charged not only with saving lives, but also with protecting property. Do everything possible to reduce the amount of damage caused by both the fire and your own actions. Homeowners will be especially appreciative of support from the fire department after the fire is extinguished.

Wrap-Up

Chief Concepts

- Forcible entry is required at emergency incidents where time is a critical factor. Company officers usually select both the point of entry and the method to be used. They also ensure that the efforts of different companies are coordinated for safe and effective operations.
- Before beginning forcible entry operations, remember to "try before you pry."
- Fire departments use many different types of forcible entry tools, ranging from basic cutting, prying, and striking tools to sophisticated mechanical and hydraulic equipment.
- Four types of forcible entry tools are used:
  - Striking tools—Generate an impact force directly on an object or another tool. They include the flat-head axe, battering ram, and sledgehammer.
  - Prying/spreading tools—Designed for prying and spreading. They include the Halligan tool, pry bar/hux bar/crowbar, pry axe, spreaders, cutters, and rams.
  - Cutting tools—Primarily used for cutting doors, roofs, walls, and floors. They include the axe, bolt cutters, and circular saw.
  - Lock tools and specialty tools—Used to disassemble the locking mechanism on a door. They include the K tool, A tool, J tool, shove knife, duck-billed lock breakers, locking pliers and chain, and bam-bam tool.
- Every door contains four major components:
  - Door—The entryway itself
  - Jamb—The frame
  - Hardware—The handles, hinges, and other components
  - Locking mechanism
- Doors are generally constructed of wood, metal, or glass. Wood and metal doors can be solid-core or hollow-core types.
- Doors are classified by how they open:
  - Inward—Can be made of wood, steel, or glass and can be found in most structures. The locking mechanisms range from standard doorknob locks to deadbolt locks or sliding latches.
  - Outward—Used in commercial occupancies and for most exits. The hinges are often exposed. Several types of locks, including handle-style locks and deadbolts, may be used with these doors.
  - Sliding—Constructed of tempered glass in a wooden or metal frame. A weak latch on the frame of the door secures the movable side.
  - Revolving—Made of four glass panels with metal frames. Generally secured by a standard cylinder lock or slide latch lock.
  - Overhead—Range from standard residential garage doors to high-security commercial roll-up doors. Made be secured with cylinder-style locks, padlocks, or automatic garage door openers.
- Window construction includes glazed glass, regular or annealed glass, double-pane glass, plate glass, laminated glass, and tempered glass.
- Window frame designs include the following types:
  - Double-hung windows—Made of wood or vinyl and two movable sashes.
  - Single-hung windows—Similar to double-hung windows except that the upper sash is fixed and only the lower sash moves.
  - Jalousie windows—Made of adjustable sections of tempered glass encased in a metal frame that overlap each other when closed.
  - Awning windows—Operate like jalousie windows except that awning windows are one large or two medium-sized glass panels instead of many small panes.
  - Horizontal-sliding windows—Similar to sliding doors.
  - Casement windows—Have a steel or wood frame and open away from the building with a crank mechanism.
  - Projected windows—Usually found in older warehouses and project either outward or inward on a hinge.
- Locks range in sophistication from the very simple push-button locks to complex computer-operated locks.
- There are three major parts of a door lock:
  - Latch—The part of the lock that "catches" and holds the door frame
  - Operator lever—The handle, doorknob, or keyway that turns the latch to lock it or unlock it
  - Deadbolt—A second, separate latch that locks and reinforces the regular latch
- All padlocks, like all door locks, have similar parts:
  - Shackles—The U-shaped top of the lock that slides through a hasp and locks in the padlock itself
  - Unlocking mechanism—The keyway, combination wheels, or combination dial used to open the padlock
  - Lock body—The main part of the padlock that houses the locking mechanisms and the retention part of the lock
- Locks can be classified into four major categories:
  - Cylindrical locks—Most common fixed lock in use. One side of the door usually has a key-in-the-knob lock; the other side will have a keyway, a button, or some type of locking mechanism.
  - Padlocks—Most common locks on the market. Available with a variety of unlocking mechanisms, including keyways, combination wheels, or combination dials.

- Mortise locks—Designed to fit in predrilled openings and have both a latch and a bolt into the same mechanism, each of which operates independently of the other.
- Rim locks—Include deadbolts that can be surface mounted on the interior of the door frame. These locks have a bolt that extends at least 1 inch into the door frame.

- On occasion, breaching a wall or floor may be necessary. Before breaching a wall, first consider whether the wall is load bearing. Removing or damaging this type of wall could cause the building or wall to collapse.
- The steps in systematic forcible entry are as follows:
  - Think.
  - Look for a lockbox.
  - Evaluate the situation.
  - Make entry with the least amount of damage possible.
  - Recognize that forcible entry operations will require you to take measures to secure the property during salvage operations.

Hot Terms

Adz The prying part of the Halligan tool.

Annealed The process of forming standard glass.

A tool A cutting tool with a pry bar built into the cutting part of the tool.

Awning windows Windows that have one large or two medium-size panels, which are operated by a hand crank from the corner of the window.

Bam-bam tool A tool with a case-hardened screw, which is secured in the keyway of a lock and used to remove the keyway from the lock.

Battering ram A tool made of hardened steel with handles on the sides, which is used to force doors and to breach walls. Larger versions may be used by as many as four people; smaller versions are made for one or two people.

Bite A small opening made to enable better tool access in forcible entry.

Bolt cutter A cutting tool used to cut through thick metal objects, such as bolts, locks, and wire fences.

Case-hardened steel Steel created in a process that uses carbon and nitrogen to harden the outer core of a steel component, while the inner core remains soft. Case-hardened steel can be cut only with specialized tools.

Casement windows Windows in a steel or wood frame that open away from the building via a crank mechanism.

Claw The forked end of a tool.

Cutting tools Tools that are designed to cut into metal or wood.

Cylindrical locks The most common fixed locks in use today. The locks and handles are placed into predrilled holes in the doors. One side of the door will usually have a key-in-the-knob lock; the other will have a keyway, a button, or some other type of locking/unlocking mechanism.

Deadbolt Surface- or interior-mounted lock on or in a door with a bolt that provides additional security.

Door An entryway; the primary choice for forcing entry into a vehicle or structure.

Double-hung windows Windows that have two movable sashes that can go up and down.

Double-pane glass A window design that traps air or inert gas between two pieces of glass to help insulate a house.

Duck-billed lock breaker A tool with a point that can be inserted into the shackles of a padlock. As the point is driven farther into the lock, it gets larger and forces the shackles apart until they break.

Exterior wall A wall—often made of wood, brick, metal, or masonry—that makes up the outer perimeter of a building. Exterior walls are often load bearing.

Forcible entry Techniques used by fire personnel to gain entry into buildings, vehicles, aircraft, or other areas of confinement when normal means of entry are locked or blocked. (NFPA 402)

Glazed Glass or transparent or translucent plastic sheet used in windows, doors, skylights, or curtain walls. [ASCE/SEI 7:6.2] (NFPA 5000)

Hardware The parts of a door or window that enable it to be locked or opened.

Hockey puck lock A type of lock with hidden shackles that cannot be forced open through conventional methods.

Hollow-core door A door made of panels that are honeycombed inside, creating an inexpensive and lightweight design.

Horizontal-sliding windows Windows that slide open horizontally.

Interior wall A wall inside a building that divides a large space into smaller areas; also known as a partition.

Irons A combination tool, normally consisting of the Halligan tool plus a flat-head axe.

Jalousie windows Windows made of small slats of tempered glass, which overlap each other when the window is closed. Often found in trailers and mobile homes, jalousie windows are held together by a metal frame and operated by a small hand wheel or crank found in the corner of the window.

Jamb The part of a doorway that secures the door to the studs in a building.

J tool A tool that is designed to fit between double doors equipped with panic bars.

K tool A tool that is used to remove lock cylinders from structural doors so the locking mechanism can be unlocked.

Wrap-Up, continued

Laminated glass Safety glass. The lamination process places a thin layer of plastic between two layers of glass, so that the glass does not shatter and fall apart when broken.

Latch A spring-loaded latch bolt or a gravity-operated steel bar that, after release by physical action, returns to its operating position and automatically engages the strike plate when it is returned to the closed position. (NFPA 80)

Lock body The part of a padlock that holds the main locking mechanisms and secures the shackles.

Locking mechanism A standard doorknob lock, deadbolt lock, or sliding latch.

Mortise locks Door locks with both a latch and a bolt built into the same mechanism; the two locking mechanisms operate independently of each other. Mortise locks are often found in hotel rooms.

Operator lever The handle of a door that turns the latch to open it.

Padlocks The most common types of locks on the market today, built to provide regular-duty or heavy-duty service. Several types of locking mechanisms are available, including keyways, combination wheels, and combination dials.

Partition A nonstructural interior wall that spans horizontally or vertically from support to support. The supports may be the basic building frame, subsidiary structural members, or other portions of the partition system. [ASCE/SEI 7:11.2] (NFPA 5000)

Pick The pointed end of a pick axe, which can be used to make a hole or bite in a door, floor, or wall.

Plate glass A type of glass that has additional strength so it can be formed in larger sheets but will still shatter upon impact.

Projected windows Windows that project inward or outward on an upper hinge; also called factory windows. They are usually found in older warehouses or commercial buildings.

Pry axe A specially designed hand axe that serves multiple purposes. Similar to a Halligan bar, it can be used to pry, cut, and force doors, windows, and many other types of objects. Also called a multipurpose axe.

Rabbet A type of door frame in which the stop for the door is cut into the frame.

Rim locks Surface- or interior-mounted locks located on or in a door with a bolt that provide additional security.

Shackles The U-shaped part of a padlock that runs through a hasp and then is secured back into the lock body.

Solid-core door A door design that consists of wood filler pieces inside the door. This construction creates a stronger door that may be fire rated.

Striking tools Tools designed to strike other tools or objects such as walls, doors, or floors.

Tempered glass A type of safety glass that is heat treated so that it will break into small pieces that are not as dangerous.

Unlocking mechanism A keyway, combination wheel, or combination dial.

FIRE FIGHTER *in action*

You are on the first-in engine for an alarm in a large industrial building. The company manufactures rubber belts and hoses. It is early Sunday morning, so you are pretty sure this is going to be an accidental alarm trip. Dispatch advises that the key holder is en route with a 25-minute estimated time of arrival. There is no evidence of fire showing from the outside of the structure. Your captain has you set a ground ladder to a first floor window to take a look across the factory floor. From the ladder, you can see a thick haze across the factory floor.

1. The window you peer through is called a factory window and a __________ window.

A. casement
B. projected
C. double-hung
D. single-hung

2. The front door to the office area is a glass door. What type of glass would most likely be used in this door?

A. Tempered
B. Plate
C. Double-pane
D. Laminate

3. Which type of door is most commonly used as a front exit?
   - **A.** Inward swinging
   - **B.** Outward swinging
   - **C.** Sliding
   - **D.** Revolving
4. On a __________ door, you can check the hinges to see if they can be disassembled or the pins removed before more destructive methods are used.
   - **A.** Inward swinging
   - **B.** Outward swinging
   - **C.** Sliding
   - **D.** Revolving
5. Which tool is designed to fit between double doors equipped with panic bars so it can be forced open?
   - **A.** Halligan tool
   - **B.** Bam-bam tool
   - **C.** K-tool
   - **D.** J-tool
6. A __________ is a second, separate latch that locks and reinforces the regular latch.
   - **A.** deadbolt
   - **B.** jamb
   - **C.** latch
   - **D.** rabbet

FIRE FIGHTER II *in action*

You have just been transferred into a new station. You are assisting another fire fighter as he is cleaning out the compartments on the fire apparatus. He pulls a variety of equipment from the forcible entry compartment. As he pulls out the K tool, he says, "This is covered with grime from sitting in this tray for years. I can't remember when I last used it. It's faster to use the sledge to break the door down."

1. How would you address the issue of the K tool being covered with grime?
2. How do you get others to do what is best, rather than what is fastest?
3. How do you determine the best method to force entry into a structure?
4. What should you say to this fire fighter if he asks, "Don't you agree?"

Ladders

Fire Fighter I

Knowledge Objectives

After studying this chapter, you will be able to:

- List and describe the parts of a ladder. (NFPA 5.3.6.A, p 357–359)
- Categorize the different types of ladders. (NFPA 5.3.6.A, p 359–362)
- Inspect ladders. (NFPA 5.5.1, p 362–363)
- Maintain ladders. (NFPA 5.5.1, p 363–364)
- Clean ladders. (NFPA 5.5.1, p 364–365)
- Describe when, where, and who performs service testing on ladders. (NFPA 5.5.1, p 365–366)
- Specify the hazards associated with ladders. (NFPA 5.3.6, p 366–367)
- Itemize the measures fire fighters should take to ensure safety when working with and on ladders. (NFPA 5.3.6, p 366–368)
- Cite the factors and guidelines used to select the appropriate ladder from the fire apparatus. (NFPA 5.3.6.A, p 369–370)
- Describe how to remove a ladder from the apparatus. (p 370)
- Describe how to lift ladders. (p 370–371)

Skills Objectives

After studying this chapter, you will be able to perform the following skills:

- Inspect, clean, and maintain a ladder. (NFPA 5.5.1, p 364–365)
- Carry a portable ladder using the one-fire fighter carry. (NFPA 5.3.6.B, p 371–372)
- Carry a portable ladder using the two-fire fighter shoulder carry. (NFPA 5.3.6.B, p 372–373)
- Carry a portable ladder using the three-fire fighter shoulder carry. (NFPA 5.3.6.B, p 372–373)
- Carry a portable ladder using the two-fire fighter suitcase carry. (NFPA 5.3.6.B, p 374)
- Carry a portable ladder using the three-fire fighter suitcase carry. (NFPA 5.3.6.B, p 374–375)
- Carry a portable ladder using the three-fire fighter flat carry. (NFPA 5.3.6.B, p 374, 376)
- Carry a portable ladder using the four-fire fighter flat carry. (NFPA 5.3.6.B, p 375, 378)
- Carry a portable ladder using the three-fire fire fighter flat-shoulder carry. (NFPA 5.3.6.B, p 376, 379)
- Carry a portable ladder using the four-fire fighter flat-shoulder carry. (NFPA 5.3.6.B, p 378–380)
- Raise a portable ladder using the one-fire fighter rung raise for ladders shorter than 14 feet. (NFPA 5.3.6.B, p 381–382)
- Raise a portable ladder using the one-fire fighter rung raise for ladders taller than 14 feet. (NFPA 5.3.6.B, p 381, 383)
- Tie the halyard. (NFPA 5.3.6.B, p 381–382, 384)
- Raise a portable ladder using the four-fire fighter beam raise. (NFPA 5.3.6.B, p 382, 385)
- Raise a portable ladder using the two-fire fighter rung raise. (NFPA 5.3.6.B, p 384–387)
- Raise a portable ladder using the three-fire fighter rung raise. (NFPA 5.3.6.B, p 387–388)
- Raise a portable ladder using the four-fire fighter rung raise. (NFPA 5.3.6.B, p 387, 390)
- Climb a ladder. (NFPA 5.3.12.B, p 391–392)
- Use a leg lock to work from a ladder. (NFPA 5.3.6.B, p 393–394)
- Deploy a roof ladder. (NFPA5.3.12.B, p 393–395)
- Inspect a chimney. (NFPA 5.3.6.B, p 396)

CHAPTER 13

Fire Fighter II FFII

Knowledge Objectives

There are no knowledge objectives for Fire Fighter II candidates. NFPA 1001 contains no Fire Fighter II Job Performance Requirements for this chapter.

Skills Objectives

There are no skill objectives for Fire Fighter II candidates. NFPA 1001 contains no Fire Fighter II Job Performance Requirements for this chapter.

Additional NFPA Standards

- NFPA 1901, *Standard for Automotive Fire Apparatus*
- NFPA 1931, *Standard for Manufacturer's Design of Fire Department Ground Ladders*
- NFPA 1932, *Standard on Use, Maintenance, and Service Testing of In-Service Fire Department Ground Ladders*
- NFPA 1983, *Standard on Life Safety Rope and Equipment for Emergency Services*

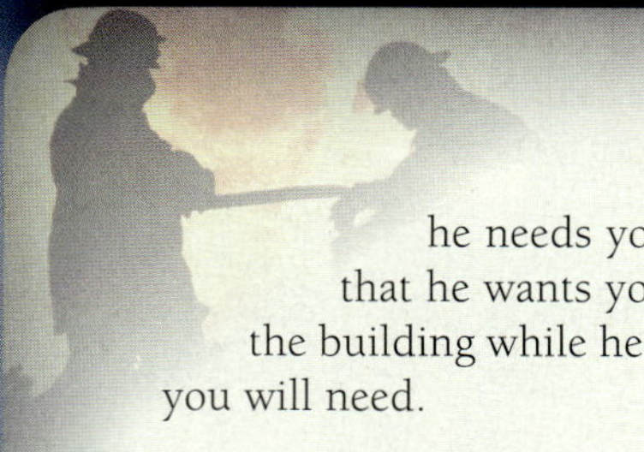

You Are the Fire Fighter

Along with your crew, you find yourself at the scene of a fast-moving fire in an apartment building. When your crew reports to the command post, the incident commander (IC) tells your captain that a truck company was sent inside to conduct a primary search so he needs your crew to set ground ladders to the building. The IC points to the aerial ladder and tells you that he wants you and the other fire fighters to ladder the first, second, and third floors of the B and D sides of the building while he does a 360-degree walk-around. You head to the truck, quickly making estimates of the ladders you will need.

1. Which types of ladders will work best for this situation?
2. How would you estimate the ladder lengths you will need?
3. What concerns do you have when you are laddering the building?

Introduction

Despite the many technological advances that have occurred in the fire service, ladders remain a fundamental piece of equipment that have not changed much. Ladders are one of the fire fighter's basic tools, carried on nearly every piece of structural firefighting apparatus.

Portable ladders are among the most functional, versatile, durable, inexpensive, easy-to-use, and rapidly deployable tools used by fire fighters. No advanced technology can substitute for a ladder as a means of rapid, safe vertical access for fire rescue, ventilation, suppression, and escape. Every fire fighter must be proficient in the basic skills of working with ladders and physically capable of performing ladder evolutions.

Functions of a Ladder

Ladders provide a vertical path, either up or down, from one level of a structure to another. They can provide access to an area or egress (a method to exit or escape) from an area FIGURE 13-1. At times, a ladder can be used as a work platform so that a fire fighter can perform various functions in locations that could not be reached otherwise.

FIGURE 13-1 Ladders can provide access to or exit from a structure.

Ladders are most often used to provide access to and exit from areas above grade (the level at which the ground intersects the foundation of a structure). When used outside a building, they can enable fire fighters to reach the roof for ventilation operations, enter a window for an interior search, or rescue a victim. Exterior ladders also provide an emergency exit for crews working inside a structure. When used within a building, ladders can allow fire fighters to access the attic or cockloft or may provide a safe path between floors, enabling fire fighters to avoid a damaged stairway. A response team can use a ladder to access and exit from the top of a rail car during a hazardous materials incident.

Fire fighters also use ladders to access and exit areas below grade. Lowering a ladder into a trench or manhole allows them to mark the location of a victim in case of secondary collapse while the rescue operation is in progress, to reach the level of an incident, or to escape from a below-grade site safely. In addition, a ladder can be used to reach an injured person in a ravine or to access a highway accident from an overpass.

Ladders are useful even at grade level. They can be deployed to create a bridge across a small opening, for example, or to enable fire fighters to climb over a fence or obstruction to reach an emergency scene.

Finally, ladders can be used as work platforms. For example, a fire fighter who has climbed a ladder to a window can stand on the ladder to open the window for ventilation or to direct a hose stream into the building or when working on a roof. Ladders are also used during overhaul operations, where they may serve as a platform for a fire fighter to remove exterior trim in a wood-frame building.

Secondary Functions of Ladders

Ladders serve several other secondary functions in the fire service. Roof ladders provide stable footing and distribute the weight of fire fighters during pitched roof operations. A ladder gin is an A-shaped structure formed with two ladder sections. Ladder gins are used to lower a rescuer into a trench or manhole or to raise a victim safely from a below-grade site.

Ladders provide elevated platforms for equipment as well as for fire fighters. They are especially useful when elevation would extend the reach or enhance the operations of the equipment. For example, a hose line attached to a ladder section can be deployed to protect exposures or confine a fire. A smoke ejector may be mounted on a ladder to provide ventilation. Portable lighting also may be secured to ladders to illuminate working areas.

Ladders can be used as ramps for moving equipment, or for hoisting or lowering victims. They can help shore up a damaged wall, support a hose line over an opening, or guide a rope up, down, or over an obstacle. By using a tarp or salvage cover, fire fighters can turn a ladder into a channel or chute for water and debris.

Ladder Construction

In its most basic design, a ladder consists of two beams connected by a series of parallel rungs. Fire service ladders, however, are specialized tools with several different parts. To use and maintain ladders properly, fire fighters must be familiar with the various types of ladders, as well as with the different parts and terms used to describe them.

Basic Ladder Components

The basic components of a straight ladder are found in most other types of ladders as well FIGURE 13-2.

Beams

A beam is one of the two main structural components that run the entire length of most ladders or ladder sections. The beams support the rungs and carry the load of a person from the rungs down to the ground.

Three basic types of ladder beam construction are used:

- Trussed beam: A trussed beam ladder has a top rail and a bottom rail, which are joined by a series of smaller pieces called truss blocks. The rungs are attached to the truss blocks. Trussed beams are usually constructed of aluminum or wood FIGURE 13-3A. This construction is often used for longer extension ladders.
- I-beam: An I-beam ladder has thick sections at the top and the bottom, which are connected by a thinner section. The rungs are attached to the thinner section of the beam. This type of beam is usually made from fiberglass FIGURE 13-3B.
- Solid beam: A solid beam ladder has a simple rectangular cross-section. Many wooden ladders have solid beams. Rectangular aluminum beams, which are usually hollow or C-shaped, are also classified as solid beams.

Rail

The rail is the top or bottom section of a trussed beam. Each trussed beam has two rails. The term "rail" can also be used to refer to the top and bottom surfaces of an I-beam.

Truss Block

A truss block is a piece or assembly that connects the two rails of a trussed beam. The rungs are attached to the truss blocks. Truss blocks can be made from metal or wood.

Rung

A rung is a crosspiece that spans the two beams of a ladder. The rungs serve as steps and transfer the weight of the user to the beams. Most portable ladders used by fire departments have aluminum rungs, but wooden ladders are still constructed with wood rungs.

Tie Rod

A tie rod is a metal bar that runs from one beam of the ladder to the other and keeps the beams from separating. Tie rods are typically found in wooden ladders.

Tip

The tip is the very top of the ladder.

Butt

The butt is the end of the ladder that is placed against the ground when the ladder is raised. It is sometimes called the heel or base.

Butt Spurs

Butt spurs are metal spikes that are attached to the butt of a ladder. They prevent the butt from slipping out of position.

Butt Plate

A butt plate is an alternative to a simple butt spur. This swiveling plate is attached to the butt of the ladder and incorporates both a spur and a cleat or pad.

Roof Hooks

Roof hooks are spring-loaded, retractable, curved metal pieces that are attached to the tip of a roof ladder. These

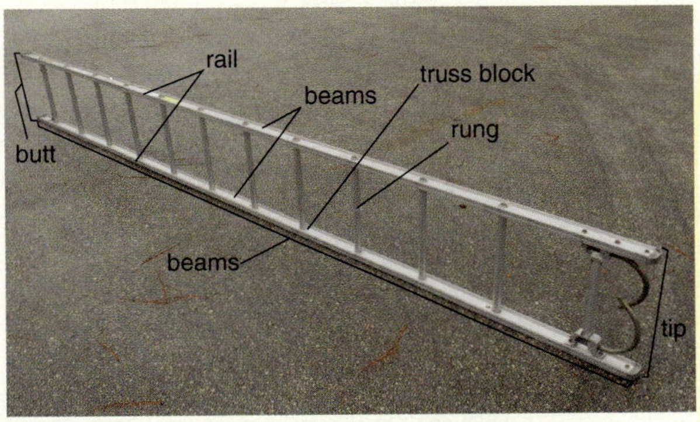

FIGURE 13-2 The basic components of a straight ladder.

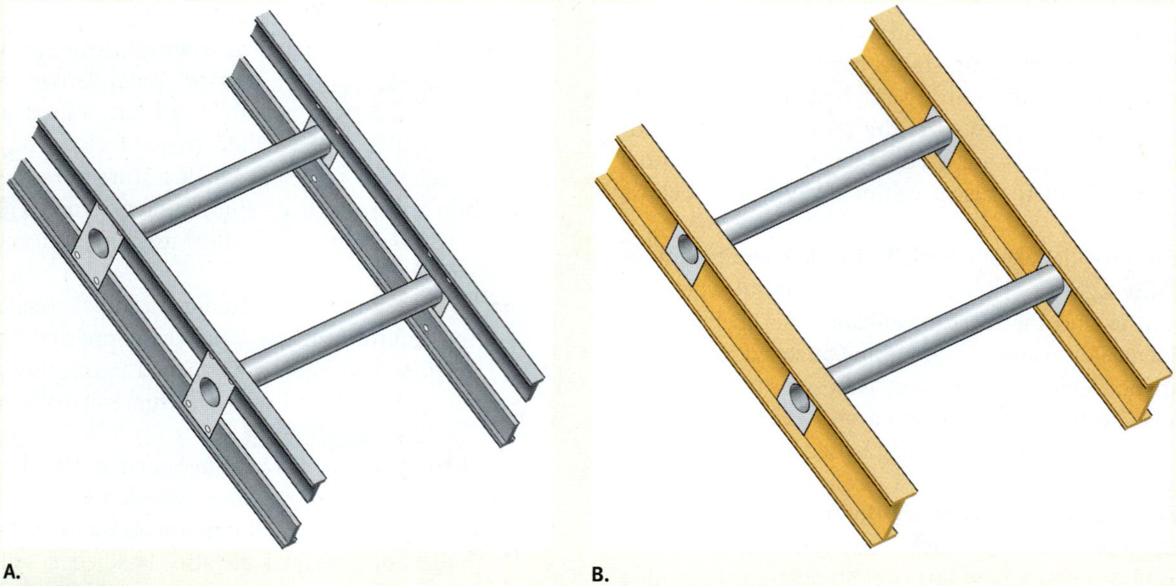

FIGURE 13-3 Two types of beam construction found in ladders: **A.** Trussed beam. **B.** I-beam.

hooks are used to secure the tip of the ladder to the peak of a pitched roof.

Heat Sensor Label

A heat sensor label identifies when the ladder has been exposed to a specific amount of heat conditions that could damage its structural integrity; such a label changes colors when it is exposed to a particular temperature.

Protection Plates

Protection plates are reinforcing pieces that are placed on a ladder at chaffing and contact points to prevent damage from friction or contact with other surfaces.

Extension Ladder Components

An extension ladder is an assembly of two or more ladder sections that fit together and can be extended or retracted to adjust the ladder's length. Extension ladders have additional parts FIGURE 13-4.

Bed (Base) Section

The bed (base) section is the widest section of an extension ladder. It serves as the base; all other sections are raised from the bed section. The bottom of the bed section rests on the supporting surface.

Fly Section

A fly section is the part of an extension ladder that is raised or extended from the bed section. Extension ladders often have more than one fly section, each of which extends from the previous section.

Guides

Guides are strips of metal or wood that guide a fly section of an extension ladder as it is being extended. Channels or slots in the bed or fly section may also serve as guides.

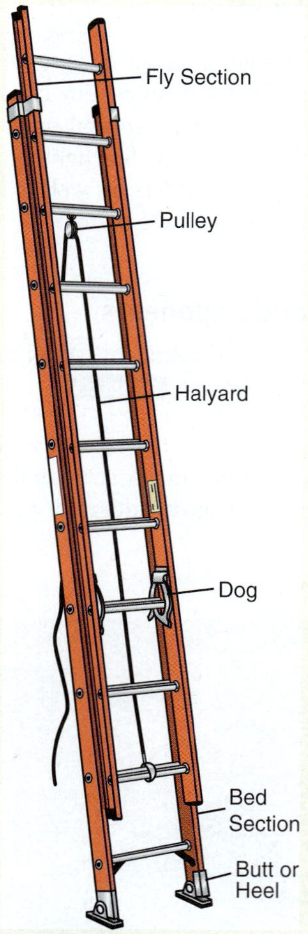

FIGURE 13-4 Extension ladders have additional parts that are not found in straight ladders.

Halyard

The halyard is the rope or cable used to extend or hoist the fly sections of an extension ladder. The halyard runs through the pulley.

Pawls

Pawls are the mechanical locking devices that are used to secure the extended fly sections of an extension ladder. They are sometimes called dogs, ladder locks, or rung locks.

Pulley

The pulley is a small grooved wheel that is used to change the direction of the halyard pull. A downward pull on the halyard creates an upward force on the fly sections, extending the ladder.

Stops

Stops are pieces of wood or metal that prevent the fly sections of a ladder from overextending and collapsing the ladder. They are also referred to as stop blocks.

Staypole

Staypoles (also called tormentors) are long metal poles that are attached to the top of the bed section and that help stabilize the ladder as it is being raised and lowered. One pole is attached to each beam of a long (40 feet (12 m) or longer) extension ladder with a swivel joint. Each pole has a spur on the other end. Ladders with staypoles are typically referred to as Bangor ladders. In some parts of the United States, Bangor ladders may be referred to as tormentor ladders.

Types of Ladders

Ladders used in the fire service can be classified into two broad categories: aerial and portable. Aerial ladders are permanently mounted and operated from a piece of motorized fire apparatus. Portable ladders are carried on fire apparatus but are designed to be removed and used in other locations.

Aerial Apparatus

Aerial apparatus vary greatly in terms of their design and function. This discussion is limited to a brief overview of the basic styles of aerial apparatus. Detailed requirements for aerial apparatus are documented in NFPA 1901, *Standard for Automotive Fire Apparatus*. This standard sets minimum performance requirements for apparatus referred to as either "aerial ladders" or "elevating platforms."

Aerial ladders are permanently mounted, power-operated ladders with a working length of between 50 feet (15 m) and 137 feet (41 m). Aerial ladders contain two or more sections FIGURE 13-5. Most aerial ladders have a permanently mounted waterway and monitor nozzle. Those built on a straight chassis are often referred to as straight-stick aerials; those built on a tractor-trailer chassis are called tillered aerials or tiller trucks. A tillered ladder truck requires a second operator at the back of the trailer to steer the back of the trailer. Tillered ladder trucks can be maneuvered into tight places that straight chassis trucks cannot reach.

FIGURE 13-5 An example of an aerial ladder.

An elevating platform apparatus includes a passenger-carrying platform (bucket) attached to the tip of a ladder or boom FIGURE 13-6. The ladder or boom must have at least two sections, which may be telescoping or articulating (jointed). An elevating platform apparatus that is 110 feet (33 m) or less in length must also have a prepiped waterway and a permanently mounted monitor nozzle. An elevated platform supported by a boom may have an aerial ladder for continuous access to the platform.

Portable Ladders

Most fire apparatus carry portable ladders (also called ground ladders). Portable ladders are designed to be removed from the apparatus and used in different locations. They may take

FIGURE 13-6 An elevated platform serves as a secure working platform for fire fighters.

the form of either general-purpose ladders, such as straight or extension ladders, or specialized ladders such as attic ladders or roof ladders.

Fire department engines are required to carry at least one roof ladder, one extension ladder, and one folding ladder. Most fire department engines carry 24- or 28-feet (7 or 8 m) ladders, which can reach the roof of a typical two-story building. Ladder trucks are required to carry a total of 115 feet (35 m) of ladders. They must carry at least one roof ladder, two straight ladders, and two extension ladders. Ladder trucks usually carry 35- or 40-feet (11 or 12 m) ladders, which can reach the roofs of most three-story buildings.

Generally, fire service portable ladders are limited to a maximum length of 50 feet (15 m). If a building's height exceeds the capabilities of portable ladders, aerial apparatus must be used.

Straight Ladder

A straight ladder is a single-section, fixed-length ladder. Straight ladders may also be called wall ladders or single ladders. These lightweight ladders can be raised quickly, and can reach the windows and roofs of one- and two-story structures. Straight ladders are commonly 12 to 14 feet (4 to 4.2 m) long, but can be as much as 20 feet (6 m) long.

> **FIRE FIGHTER Tips**
>
> Fire service portable ladders should be constructed and certified as compliant with the most recent edition of NFPA 1931, *Standard for Manufacturer's Design of Fire Department Ground Ladders*.

Roof Ladder

A roof ladder (sometimes called a hook ladder) is a straight ladder that is equipped with retractable hooks at one end. The hooks secure the tip of the ladder to the peak of a pitched roof, when the ladder lies flat on the roof. A roof ladder provides stable footing and distributes the weight of fire fighters and their equipment, thereby helping reduce the risk of structural failure in the roof assembly. Roof ladders can be used on flat roofs as a work platform to distribute weight. Roof ladders are usually 12 to 18 feet (4 to 5 m) long FIGURE 13-7.

FIGURE 13-7 Roof ladders are commonly 12 to 18 feet (4 to 5 m) long.

> **Fire Fighter Safety Tips**
>
> Roof ladders are not free-hanging ladders. The roof hooks will not support the full weight of the ladder and anyone on it when the ladder is in a vertical position.

Extension Ladder

An extension ladder is an adjustable-length ladder with multiple sections. The bed or base section supports one or more fly sections. Pulling on the halyard extends the fly sections along a system of brackets or grooves FIGURE 13-8. The fly sections, in turn, lock in place at set intervals so that the rungs are aligned to facilitate climbing.

An extension ladder is usually heavier than a straight ladder of the same length and requires more than one person to set it up. Because the length is adjustable, an extension ladder can replace several straight ladders, and it can be stored in places where a longer straight ladder would not fit.

Bangor Ladder

Bangor ladders (tormentor) are extension ladders with staypoles that offer added stability during raising, lowering, and climbing operations. Staypoles are required on ladders of 40 feet (12 m) or greater length, but can often be found on 35-foot (11 m) ladders. The poles help keep these heavy

FIGURE 13-8 An extension ladder can be used at any length, from fully retracted to fully extended.

FIGURE 13-9 Staypoles are used to stabilize a Bangor ladder while it is being raised or lowered.

ladders under control while fire fighters are maneuvering them into place FIGURE 13-9. When the ladder is positioned correctly, the staypoles are planted in the ground on either side for additional stability.

Combination Ladder

A combination ladder can be converted from a straight ladder to a stepladder configuration (A-frame), or from an extension ladder to a stepladder configuration. Such ladders are convenient for indoor use and for maneuvering in tight spaces. They are generally 6 to 10 feet (2 to 3 m) in length in the A-frame configuration and 10 to 15 feet (3 to 5 m) in length in the extension configuration FIGURE 13-10.

Folding Ladder

A folding ladder (also called an attic ladder) is a narrow, collapsible ladder that is designed to allow access to attic scuttle holes and confined areas FIGURE 13-11. The two beams fold in to enhance the ladder's portability. Folding ladders are commonly available in 8- to 14-feet (2 to 2.4 m) lengths.

Fresno Ladder

A Fresno ladder is a narrow, two-section extension ladder that is designed to provide attic access. The Fresno ladder is

FIGURE 13-10 Combination ladders can be used in several different configurations. This is a combination stepladder and straight ladder.

A. B.

FIGURE 13-11 Folding ladders are designed to be used in narrow spaces or restrictive passages. **A.** In the portable position. **B.** Fully extended.

generally short—just 10 to 14 feet (3 to 4 m)—so it has no halyard; it is extended manually **FIGURE 13-12**. A Fresno ladder can be used in tight space applications, such as bridging over a damaged section of an interior stairway.

Inspection, Maintenance, and Service Testing of Portable Ladders

Portable ladders used by the fire service must be able to withstand extreme conditions. NFPA 1931, *Standard for Manufacturer's Design of Fire Department Ground Ladders*, establishes requirements for the construction of new ladders. These requirements are based on probable conditions during emergency operations.

Ladders can be dropped, overloaded, or exposed to temperature extremes during use. NFPA 1932, *Standard on Use, Maintenance, and Service Testing of In-Service Fire Department Ground Ladders*, provides general guidance for ground ladder users. Portable ladders must be regularly inspected, maintained, and service-tested, following the NFPA standard. In addition, ladders should always be inspected and maintained in accordance with the manufacturer's recommendations.

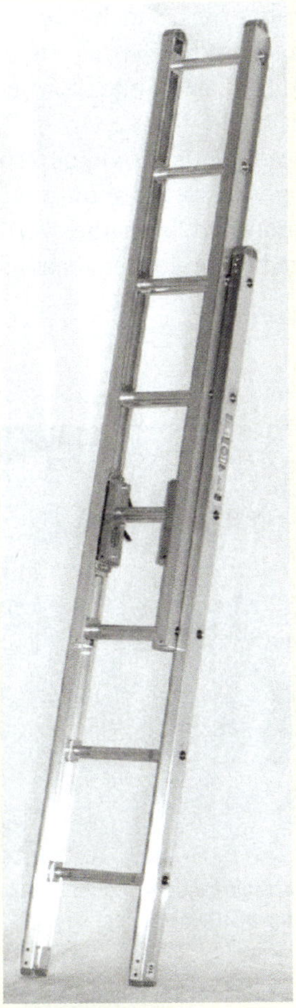

FIGURE 13-12 A Fresno ladder is extended manually.

Inspection

Ground ladders must be visually inspected at least monthly. A ladder should also be inspected after each use. A general visual inspection should, at a minimum, address the following items:

- Beams: Check the beams for cracks, splintering, breaks, gouges, checks, wavy conditions, or deformations.
- Rungs: Check all rungs for snugness, tightness, punctures, wavy conditions, worn serrations, splintering, breaks, gouges, checks, or deformations.
- Halyard: Check the halyard for fraying or kinking; ensure that it moves smoothly through the pulleys.
- Wire-rope halyard extensions: Check the wire-rope halyard extensions on three- and four-section ladders for snugness. This check should be performed when the ladder is in the bedded position to ensure that the upper sections will align properly during operation.
- Ladder slides: Check the ladder slide areas for chafing. Also check for adequate wax, if the manufacturer requires wax.
- Pawls: Check the pawl (dog) assemblies for proper operation.
- Butt spurs: Check the butt spurs for excessive wear or other defects.
- Heat sensors: Check the labels to see whether the sensors indicate that the ladder has been exposed to excessive heat.
- Bolts and rivets: Check all bolts and rivets for tightness; bolts on wood ladders should be snug and tight without crushing the wood.
- Welds: Check all welds on metal ladders for cracks or apparent defects.
- Roof hooks: Check the roof hooks for sharpness and proper operation.
- Metal surfaces: Check metal surfaces for signs of surface corrosion.
- Fiberglass and wood surfaces: Check fiberglass ladders for loss of gloss on the beams. Check for damage to the varnish finish on wooden ground ladders. Check for signs of wood rot—that is, dark, soft spots.

If the inspection reveals any deficiencies, the ladder must be removed from service until repairs are made. Minor repairs that require simple maintenance can often be performed by properly trained fire fighters at the fire station. By contrast, repairs involving the structural or mechanical components of a ladder must be performed only by qualified personnel at a properly equipped repair facility.

Ladders that have been exposed to excessive heat must be removed from service and tested before being used again. A ladder that fails the structural stability test should be permanently removed from service.

Fire Fighter Safety Tips

Fire fighters should perform routine ladder maintenance. Repairs to portable ladders should be performed only by trained repair personnel.

Maintenance

All fire fighters should be able to perform routine ladder maintenance. Maintenance is simply the regular process of keeping the ladder in proper operating condition. Fundamental maintenance tasks include the following:

- Clean and lubricate the dogs, following the manufacturer's instructions FIGURE 13-13.
- Clean and lubricate the slides on extension ladders in accordance with the manufacturer's recommendations.
- Replace worn halyards and wire rope on extension ladders when they fray or kink FIGURE 13-14.
- Clean and lubricate hooks. Remove rust and other contaminants, and lubricate the folding roof hook assemblies on roof ladders to keep them operational FIGURE 13-15.

FIGURE 13-13 Extension ladder dogs (pawls) must operate smoothly.

FIGURE 13-14 Replace the halyard if it is worn or damaged.

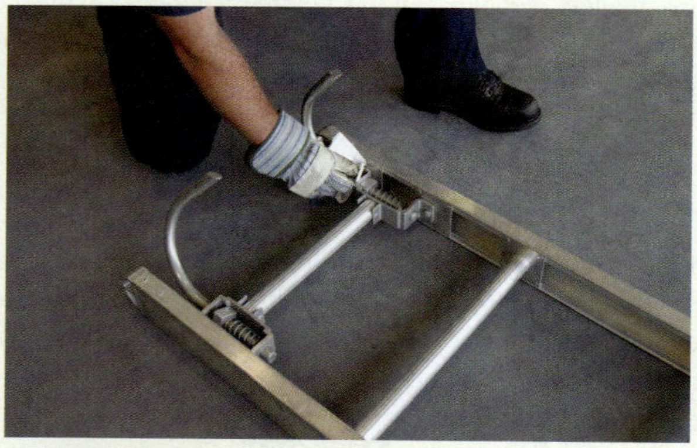

FIGURE 13-15 Roof hooks must operate smoothly.

- Check the heat sensor labels. Replace the sensors when they reach their expiration date. Remove a ladder that has been exposed to high temperatures from service for testing.
- Maintain the finish on fiberglass and wooden ladders in accordance with the manufacturer's recommendations.
- Ensure that portable ladders are *not* painted except for the top and bottom 18 inches (0.47 m) of each section, because paint can hide structural defects in the ladder. The tip and butt are painted for purposes of identification and visibility.
- Maintenance records should be kept.

Ladders that are in storage should be placed on racks or in brackets and protected from the weather. Fiberglass ladders can be damaged by prolonged exposure to direct sunlight FIGURE 13-16.

FIGURE 13-16 Portable ladders should be stored on racks or in brackets, out of the weather or direct sunlight.

Cleaning

Ladders must be cleaned regularly to remove any road grime and dirt that have built up on the apparatus during storage. Ladders should be cleaned before each inspection to ensure that any hidden faults can be observed. They should also be cleaned after each use to remove dirt and debris.

Use a soft-bristle brush and water to clean ladders. A mild, diluted detergent may be used, if allowed by the manufacturer's recommendations. Remove any tar, oil, or grease deposits with a safety solvent as recommended by the manufacturer.

Rinse and dry the cleaned ladder before placing it back on the apparatus.

Fire Fighter Safety Tips

Do not get any solvent on the halyard of an extension ladder. Contact with solvents can damage halyard ropes.

The steps for inspecting, cleaning, maintaining, and storing ladders are listed in SKILL DRILL 13-1:

1. Clean all components following manufacturer and national standards.
2. Visually inspect the ladder for wear and damage. (STEP 1)

3. Lubricate the ladder dogs, guides, and pulleys using the recommended material. (STEP 2)
4. Perform a functional check of all components. (STEP 3)
5. Complete the maintenance record for the ladder. Tag and remove the ladder from service if deficiencies are found.
6. Return the ladder to the apparatus or storage area. (STEP 4)

Service Testing

Service testing is performed periodically to evaluate the continued usefulness of a ladder during its life. Service tests should be performed on both new ladders and ladders that have been in use for some time. Such tests measure the structural integrity of a portable ladder, ensuring that ladders are safe to use for their intended purposes. Service testing of portable ladders must follow NFPA 1932, *Standard on Use,*

SKILL DRILL 13-1 Inspect, Clean, and Maintain a Ladder

(Fire Fighter I, NFPA 5.5.1)

1. Clean all components following manufacturer and national standards. Visually inspect the ladder for wear and damage.

2. Lubricate the ladder dogs, guides, and pulleys using the recommended material.

3. Perform a functional check of all components.

4. Complete the maintenance record for the ladder. Tag and remove the ladder from service if deficiencies are found. Return the ladder to the apparatus or storage area.

Maintenance, and Service Testing of In-Service Fire Department Ground Ladders.

Service tests should be conducted before a new ladder is used and annually while the ladder remains in service. A ladder that has been exposed to extreme heat, has been overloaded, has been impact or shock loaded, is visibly damaged, or is suspected of being unsafe for any other reason must be removed from service until it has passed a service test. A repaired ladder must also undergo a service test before it can be returned to service, unless a halyard replacement was the only repair.

One part of the ladder service test is the horizontal-bending test, which evaluates the structural strength of a ladder. To perform this test, the ladder is placed in a horizontal position across a set of supports. A weight is then put on the ladder, and the amount of deflection or bending caused by the weight is measured to evaluate the ladder's strength **FIGURE 13-17**. Additional tests are performed on the extension hardware of extension ladders. The hooks on roof ladders are tested as well.

FIGURE 13-17 A horizontal-bending test evaluates the structural strength of a ladder.

Service testing of ground ladders and aerial ladders requires special training and equipment and must be conducted only by qualified personnel. Many fire departments use outside contractors to perform these tests. The results of all service tests must be recorded and kept for future reference.

Ladder Safety

Several potential hazards are associated with ladder use, but unfortunately these risks are easily overlooked during emergency operations. As a consequence, many fire fighters have been seriously injured or killed in ladder accidents, both on the fire ground and during training sessions.

Ladders must be used with caution; follow standard procedures and regularly reinforce your skills through training. Many safety precautions should be followed from the time a ladder is removed from the apparatus until it is returned to the apparatus. Basic safety issues include the following concerns:

- General safety
- Lifting and moving ladders
- Placement of ground ladders
- Working on a ladder
- Rescue operations
- Ladder damage

These issues are explained in general here and will be revisited in the discussions of using specific types of ladders.

General Safety Requirements

In any firefighting operation, full turnout clothing and protective equipment are essential for working with ground ladders. Turnout gear provides protection from mechanical injuries as well as from fire and heat injuries. Helmets, coats, pants, gloves, and protective footwear provide protection from falling debris, impact injuries, and pinch injuries. Fire fighters must be able to work with and on ladders while wearing self-contained breathing apparatus (SCBA). You should practice working with ladders while wearing all of your personal protective equipment (PPE), including gloves and eye protection.

Lifting and Moving Ladders

Teamwork is essential when working with ladders. Some ladders are heavy and can be very awkward to maneuver, particularly when they are extended. Crew members must use proper lifting techniques and coordinate all movements.

Never attempt to lift or move a ladder that weighs more than you are capable of lifting safely. Because of their shape, ladders may be awkward to carry, especially over uneven terrain, through gates, or over snow and ice. It is better to ask for help in moving a ladder than to risk injury or delay the placement of the ladder.

Placement of Ground Ladders

Fire fighters working on the fire ground should survey the area where a portable ladder will be used before placing the ladder or beginning to raise it. If possible, inspect the area before retrieving the ladder from the apparatus. If you note any hazards, consider changing the position of the ladder. Work in the safest available location.

Sometimes a ladder must be placed in a potentially hazardous location. If you are aware of the hazard, you can take corrective actions and implement special precautions to avoid accidents.

The most important check assesses the location of overhead utility lines. If a ladder comes in contact with or close to an electrical wire, the fire fighters who are handling it can be electrocuted. If the line falls, it can electrocute other fire fighters as well. Avoid placing a ladder against a surface that has been energized by a damaged or fallen power line.

Do not assume that it is safe to use wood or fiberglass ladders around power lines. Metal ladders will certainly conduct electricity, but a wet or dirty wood or fiberglass ladder can also conduct electricity.

If possible, do not raise ladders anywhere near overhead wires. At a minimum, keep ladders at least 10 feet (3 m) away from any power lines while using or raising them. If a power line is nearby, make sure that enough fire fighters are present to keep the ladder under control as it is raised. One fire fighter should watch to make sure that the ladder does not come too close to the line. If the ladder falls into a power line, the results can be deadly.

Other types of overhead obstructions may also be hazards during a ladder raise. If the ladder hits something as it is being raised, the weight and momentum of the ladder will shift, which can cause fire fighters to lose control and drop the ladder.

Ladders must be placed on stable and relatively level surfaces FIGURE 13-18. The shifting weight of a climber can easily tip a ladder placed on a slope or unstable surface. Always evaluate the stability of the surface before placing the ladder, and check it again during the course of the firefighting operations. A ladder placed on snow or mud, for example, may become unstable as the snow melts or as the base of the ladder sinks into the mud. In addition, water from the firefighting operations can make soft ground even more unstable.

It is also important to ensure that the tip of the ladder is placed against a stable surface. Placing the tip of a ladder against an unstable surface can result in the ladder falling forward into a fire or being pushed backward. The failure of an unstable wall, for example, can occur during firefighting operations or during overhaul. Avoid injury or death—do not place the tip of a ladder against any unstable surface.

Ladders should not be exposed to direct flames or extreme temperatures, because excessive heat can cause permanent damage to or catastrophic failure of such equipment. Check the heat sensor labels immediately after a ladder has been exposed to high temperatures.

Working on a Ladder

Before climbing a ladder, make sure it is set at the proper climbing angle (approximately 75 degrees) for maximum load capacity and safety. To accomplish this, position the base of the ladder out from the building at a distance equal to one-fourth of the ladder's height FIGURE 13-19. For example, if a ladder is being raised to a window 20 feet (6 m) above the ground, the base of the ladder should be placed 5 feet (1.5 m) from the building. Achieving a proper angle is important to your safety and the safety of people being rescued.

When using an extension ladder, ensure that the pawls (dogs) are locked and the halyard tied before anyone climbs the ladder.

To prevent slipping, another fire fighter should secure the base of the ladder by heeling (footing) it FIGURE 13-20. This technique uses the fire fighter's weight to keep the base of the ladder from slipping. The base of the ladder can also be mechanically secured to a solid object. An alternative to securing the base is to use a rope or strap to secure the tip of the ladder to the building.

All ladders have weight limits. Most fire department portable ladders are designed to support a weight of 750 pounds (340 kg). In a rescue situation, this weight limit is equivalent to one victim and two fire fighters wearing full protective clothing

Fire Fighter Safety Tips

Always maintain at least 10 feet (3 m) of clearance between a ladder and utility lines to prevent electrocution.

FIGURE 13-18 Ladders must be placed on stable and relatively level surfaces.

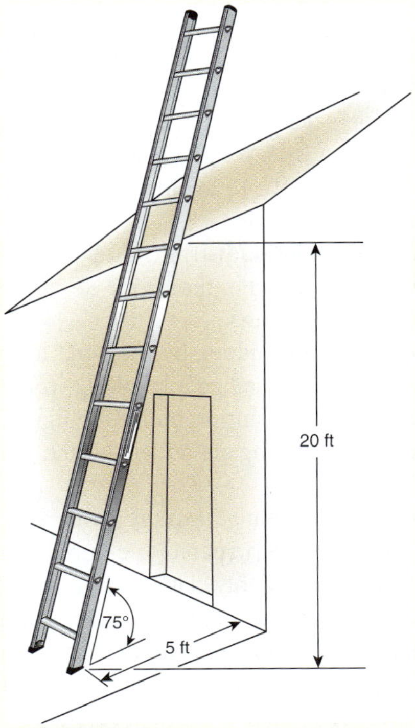

FIGURE 13-19 Place the base of a ladder out from the building at a distance equal to one-fourth of the ladder's height to achieve a climbing angle of 75 degrees.

FIGURE 13-20 To prevent a ladder from slipping, fire fighters should secure the base by heeling the ladder.

Fire Fighter Safety Tips

Many ladders used for construction work and home repair are rated for a working load of only 225 to 300 pounds (102 to 136 kg)—significantly less than the 750 pounds (340 kg) rating of approved fire department ladders. Remember the difference between these two types of ladders. If you encounter a situation where a construction ladder is already present, you still need to set up and use a fire department–approved ladder to avoid the possibility of a catastrophic ladder failure.

and equipment. It would also accommodate two fire fighters climbing or working on the ladder along with their equipment. In contrast, ladders designed for everyday use may be rated for as little as 225 pounds (102 kg). Given the potential difference in ratings, it is important to avoid using unapproved ladders for fire department operations.

The weight should be distributed along the length of the ladder. Only one fire fighter should be on each section of an extension ladder at any time.

While climbing the ladder, be prepared for falling debris, misguided hose streams, or people falling from the building. Knowing these risks, it is important for fire fighters to wear full turnout gear when working on ladders—such gear will help to protect you.

If fire conditions change rapidly while you are working on a ladder, you must be prepared to climb down quickly. For example, if the fire suddenly flashes over or if flames break through a window located near the ladder, you must be prepared to move quickly out of danger. Turnout gear will not protect you from direct exposure to flames for more than a few seconds.

A fire fighter who is working from a ladder is in a less stable position than one who is working on the ground. While on a ladder, you must constantly adjust your balance, especially when swinging a hand tool or reaching for a trapped occupant. There is always the danger of falling as well as a risk to people below if something falls or drops. To minimize your risk of falling, use a safety belt or a leg lock to secure yourself to the ladder.

■ Rescue

Ladders can be an essential tool for rescue. For example, portable ladders are often used to reach and remove trapped occupants from the upper stories of a building. When positioning a ladder for rescue, place the tip of the ladder at or slightly below the window sill. This position provides a direct pathway for the fire fighter entering the window and a direct route for exiting the structure. The ladder should not occlude any part of the window opening. Fire fighters need to address several important safety concerns before going up a ladder to rescue someone.

A person who is in extreme danger may not wait to be rescued. Jumpers not only risk their own lives, but may also endanger the fire fighters trying to rescue them. Fire fighters have been seriously injured by persons who jumped before a rescue could be completed.

A trapped person might try to jump onto the tip of an approaching ladder or to reach out for anything or anyone nearby. As a fire fighter, you might be pulled or pushed off the ladder by the person you are trying to rescue. If several people are trapped, they might all try to climb down the ladder at the same time.

It is important to make verbal contact with any person you are trying to rescue. You must remain in charge of the situation and not let the individual panic. Tell the person to remain calm and wait to be rescued. If enough fire fighters are present, one could maintain contact with the person while others raise the ladder.

After you reach the person, you still have to get him or her down to the ground safely. Try to make the person realize that any erratic movements could cause the ladder to tip over. It is often helpful to have one fire fighter assist the person, while a second fire fighter acts as a guide and backup to the rescuer. Although modern fire ladders are designed to support this load, a ladder that is overloaded during a rescue operation must be removed from service for inspection and service testing. See the Search and Rescue chapter for specific ladder rescue techniques.

■ Ladder Damage

Ladders are easily damaged while in use. For example, a ladder might be overloaded or used at a low angle during a rescue. An unexpected shift in fire conditions could bring the ladder in direct contact with flames. Whenever a ladder is used outside of its recommended limits, it should be taken out of service for inspection and testing, even if there is no visible damage.

Using Portable Ladders

Portable ladders are often urgently needed during emergency incidents. Because an accident or error in handling or using a ladder can result in death or serious injury, all fire fighters must know how to work with ladders.

Using a ladder requires that fire fighters complete a series of consecutive tasks. The first step is to select the best ladder for the job from those available. Fire fighters must then remove the ladder from the apparatus and carry it to the location where it will be used. The next step is to raise and secure the ladder. At the end of the operation, the ladder must be lowered and returned to the apparatus. Each of these tasks is important to the safe and successful completion of the overall objective.

Selecting the Ladder

The first step in using a portable ladder is to select the appropriate ladder from those available. Fire fighters must be familiar with all of the ladders carried on their apparatus. Engine and ladder company apparatus usually carry several portable ladders of various lengths. Many other types of apparatus, such as tankers (water tenders) and rescue units, often carry additional portable ladders. NFPA 1901, *Standard for Automotive Fire Apparatus*, identifies minimum requirements for portable ladders for each type of apparatus **TABLE 13-1**.

Selecting an appropriate ladder requires that fire fighters estimate the heights of windows and rooflines. General guidelines have been developed for estimating the heights for residential and commercial buildings **TABLE 13-2**.

Ladder placement will also affect the size and length of the ladder needed. When the ladder is used to access a roof, for example, the tip of the ladder should extend several feet above the roofline to provide a convenient handhold and secure footing for fire fighters as they mount and dismount. This extra length also makes the tip of the ladder visible to fire fighters who are working on the roof **FIGURE 13-21**. A common rule of thumb is to ensure that at least five ladder rungs show above the roofline.

Ladders used to provide access to windows must be longer than those used in rescue operations. During access operations, the ladder is placed next to the window, with the ladder tip even with the top of the window opening **FIGURE 13-22**. During rescue operations, however, the tip of the ladder should be immediately below the windowsill, which prevents the ladder from obstructing the window opening while a trapped occupant is removed **FIGURE 13-23**.

The final factor in determining the correct length of ladder is the angle formed by the ladder and the placement surface (ground). A portable ladder should be placed at an angle of approximately 75 degrees for maximum strength and stability. To satisfy this criterion, the ladder must be slightly longer than the vertical distance between the ground and the target

TABLE 13-1 Minimum Ladder Complement for Apparatus as Specified in NFPA 1901, *Standard for Automotive Fire Apparatus*

Pumper	1 folding ladder 1 roof ladder, straight and equipped with hooks 1 extension ladder
Quick attack	1 extension ladder must be 12 feet (4 m) or longer
Aerial/ladder	Portable ladders that have a total length of 115 feet (35 m) or more and contain a minimum of: • 1 folding ladder • 2 roof ladders, straight and equipped with hooks • 2 extension ladders
Quint	Portable ladders that have a total length of 85 feet (26 m) or more and contain a minimum of: • 1 folding ladder • 1 roof ladder, straight and equipped with hooks • 1 extension ladder

TABLE 13-2 Approximate Distances for Residential and Commercial Construction

Residential floor-to-floor height	8–10 feet (2–3 m)
Residential floor-to-windowsill height	3 feet (1 m)
Commercial floor-to-floor height	10–12 feet (3–4 m)
Commercial floor-to-windowsill height	4 feet (1 m)

FIGURE 13-21 The tip of the ladder should extend above the roofline during roof operations.

FIGURE 13-22 To provide access through a window, the tip of the ladder should be on the upwind side of the window and even with the top of the window opening.

FIGURE 13-23 For entry and rescue operations, the tip of the ladder should be just below the windowsill.

point. Generally, a ladder requires an additional 1 foot (.5 m) in length for every 15 feet (5 m) of vertical height.

Thus, to reach a window that is 30 feet (9 m) above grade level, a ladder would have to be at least 32 feet (10 m) long. Because ladders used in roof operations need to extend at least above the roofline, accessing a roof that is 30 feet (9 m) above grade level requires a ladder at least 35 feet (11 m) long.

Removing the Ladder from the Apparatus

Ladders are mounted on apparatus in various ways. Such equipment should be stored in locations where it will not be exposed to excessive heat, engine exhaust, or mechanical damage. Fire fighters must be familiar with how ladders are mounted on their apparatus and practice removing them safely and quickly.

Ladders are often nested one inside another and mounted on brackets on the side of the engine. Fire fighters should note the nesting order and location of the ladders relative to the brackets when removing or replacing them. Ladders that are not needed should not be placed on the ground in front of an exhaust pipe because they could be damaged by the hot exhaust.

Ladders can also be stored on overhead hydraulic lifts FIGURE 13-24. The hydraulic mechanism keeps them out of the way until they are needed, when they can be lowered to a convenient height. It is important to review the operation of the hydraulic lift used on your fire department's apparatus so that you can operate it efficiently and safely during an emergency situation.

On some vehicles, portable ladders are stored in compartments under the hose bed or aerial device. The ladders may lie flat or vertically on one beam FIGURE 13-25. To remove or replace them, fire fighters slide the ladders out the rear of the vehicle. For this reason it is important to leave adequate room behind such apparatus so that ladders can be removed at a fire scene.

Lifting Ladders

Many ladders are heavy and awkward to handle. To prevent lifting injuries while handling ladders, fire fighters must work together to lift and carry long or heavy ladders.

FIGURE 13-24 This hydraulic mechanism lowers the ladders to a convenient height when they are needed.

FIGURE 13-25 Ladders in rear compartments can be stored either flat or vertically. This configuration requires adequate space behind the truck to remove ladders at an emergency scene.

When fire fighters are working as a team to lift or carry a ladder, one fire fighter must act as the leader, providing direction and coordinating the actions of all team members. The lead fire fighter needs to call out the intended movements clearly, using standard terminology and clear two-step commands. For example, when lowering a ladder, the leader should say, "Prepare to lower," followed by the command, "Lower." Team members must communicate in a clear and concise fashion, using the same terminology for commands; there should be no confusion about the specific meaning of each command. Because terminology may differ among jurisdictions, fire fighters who are newly or temporarily assigned to an apparatus should verify which commands are used and what they mean.

A prearranged method should be in place for determining the lead fire fighter for ladder lifts. In some fire departments, the fire fighter on the right side at the butt end of the ladder is always the leader. In other departments, the fire fighter at the tip of the ladder on the right side may be the leader. The departmental policy should be consistent for all ladder operations.

Near Miss REPORT

Report Number: 07-854

Synopsis: Fire fighters trapped on a roof without a second means of egress.

Event Description: We were called to the scene of a large fire in a two-story apartment building in mid-December. It was cold out and while we normally run four fire fighters, this particular night we had only three. The roof was slick from the water and ice. We were assigned to the ventilation sector by the IC. We threw our ground ladder up as well as placed a roof ladder on the roof. We made a ventilation hole that after a few seconds allowed the fire to come through. The roof conditions had deteriorated so badly by this point that we were unable to make it back to our ladders. An aerial operator from a neighboring department saw that we were having trouble and moved the stick over just below where we were. Our only option at this point was to slide down the roof and hope we hit the ladder. Luckily, all three of us made it to the aerial without injury.

Lessons Learned: From that fire on, we constantly train on putting up two means of egress. It also made us firm believers in having a four-fire fighter truck crew. We believe if we would have had four firefighters, the extra fire fighter could have been throwing other ladders, giving us second and third points of egress.

Additionally, you must use good lifting techniques when handling ladders. When bending to pick up a ladder, bend at the knees and keep the back straight FIGURE 13-26. Lift and lower the load with the legs, rather than the back. Take care to avoid twisting motions during lifting and lowering, as these motions often lead to back strains.

FIGURE 13-26 Bend your knees and keep your back straight when lifting or lowering a ladder.

Carrying Ladders

Once a ladder has been removed from the apparatus or lifted from the ground, it must be carried to the placement site. Ladders can be carried at shoulder height or at arm's length. They can be carried either on edge or flat. The most common carries are explained in the following sections.

One-Fire Fighter Carry

Most straight ladders and roof ladders less than 18 feet (5 m) long can be safely carried by a single fire fighter. The steps involved in the one-fire fighter carry are outlined in SKILL DRILL 13-2:

1. Start with the ladder mounted in a bracket or standing on one beam. Locate the center of the ladder. (STEP 1)
2. Place an arm between two rungs of the ladder just to one side of the middle rung.
3. Lift the top beam of the ladder and rest it on your shoulder. (STEP 2)
4. Walk carefully with the butt end first. Carry roof ladders with the hook end in the front. (STEP 3)

A straight ladder is carried with the butt end first and pointed slightly downward; a roof ladder is carried tip first. Most ladders are carried with the butt end forward, because the placement of the butt determines the positioning of the ladder. Roof ladders are carried with the hooks toward the front. A roof ladder usually must be carried up another ladder to reach the roof. The roof ladder is then usually pushed up the pitched roof slope from the butt end until the hooks engage the peak.

Two-Fire Fighter Shoulder Carry

The two-fire fighter shoulder carry is generally used with extension ladders up to 35 feet (11 m) long. It can also be used to carry straight ladders or roof ladders that are too long for one person to handle. To perform the two-fire fighter shoulder carry, follow the steps in SKILL DRILL 13-3:

SKILL DRILL 13-2 One-Fire Fighter Carry
(Fire Fighter I, NFPA 5.3.6)

1 Start with the ladder mounted in a bracket or standing on one beam. Locate the center of the ladder. Place an arm between two rungs of the ladder just to one side of the middle rung.

2 Lift the top beam of the ladder and rest it on your shoulder.

3 Walk carefully with the butt end first.

1. Start with the ladder mounted in a bracket or standing on one beam. Both fire fighters are positioned on the same side of the ladder.
2. Facing the butt end of the ladder, one fire fighter is positioned near the butt end of the ladder and the second fire fighter is positioned near the tip of the ladder. (STEP 1)
3. Both fire fighters place one arm between two rungs and, on the leader's command, lift the ladder onto their shoulders.
4. The ladder is carried butt end first. (STEP 2)
5. The fire fighter closest to the butt covers the sharp butt spur with a gloved hand to prevent injury to other fire fighters in the event of a collision. (STEP 3)

Three-Fire Fighter Shoulder Carry

Three fire fighters may be needed to carry a heavy ladder. This carry is similar to the two-fire fighter shoulder carry, with the additional fire fighter positioned at the middle of the ladder. All three fire fighters stand on the same side of the ladder. To perform the three-fire fighter shoulder carry, follow the steps in SKILL DRILL 13-4:

1. Start with the ladder mounted in a bracket or standing on one beam. Three fire fighters stand on the same side of the ladder.
2. Facing the butt end of the ladder, one fire fighter is positioned at the butt end of the ladder, one fire fighter is at the middle of the ladder, and one fire fighter is at the tip of the ladder. (STEP 1)

SKILL DRILL 13-3 Two-Fire Fighter Shoulder Carry

(Fire Fighter I, NFPA 5.3.6)

1. Start with the ladder mounted in a bracket or standing on one beam. Both fire fighters are positioned on the same side of the ladder. Facing the butt end of the ladder, one fire fighter is positioned near the butt end of the ladder and a second fire fighter is positioned near the tip of the ladder.
2. Both fire fighters place one arm between two rungs and, on the leader's command, lift the ladder onto their shoulders. The ladder is carried butt end first.
3. The butt spurs are covered with a gloved hand while the ladder is transported.

SKILL DRILL 13-4 Three-Fire Fighter Shoulder Carry

(Fire Fighter I, NFPA 5.3.6)

1. All three fire fighters are positioned on the same side of the ladder. Facing the butt end of the ladder, one fire fighter is positioned at the butt end of the ladder, one fire fighter is at the middle of the ladder, and one fire fighter is at the tip of the ladder.
2. Each fire fighter places an arm between two rungs and, on the leader's command, all fire fighters hoist the ladder onto their shoulders.
3. The ladder is carried butt end first. The fire fighter closest to the butt covers the sharp butt spur with a gloved hand.

3. Each fire fighter places an arm between two rungs and, on the leader's command, all fire fighters hoist the ladder onto their shoulders. (STEP 2)
4. The ladder is carried butt end first. The fire fighter closest to the butt covers the sharp butt spur with a gloved hand to prevent injury to other fire fighters in the event of a collision. (STEP 3)

Two-Fire Fighter Suitcase Carry

The two-fire fighter suitcase carry is commonly used with straight and extension ladders. With this technique, the ladder is carried at arm's length. To perform the two-fire fighter suitcase carry, follow the steps in SKILL DRILL 13-5:

1. Start with the ladder resting on the ground on one beam. Both fire fighters are positioned on the same side of the ladder.
2. Facing the butt end of the ladder, one fire fighter is positioned at the butt end of the ladder and the other fire fighter is positioned at the tip of the ladder. (STEP 1)
3. Each fire fighter reaches down and grasps the upper beam of the ladder. (STEP 2)
4. On the leader's command, both fire fighters pick the ladder up from the ground and carry it with the butt end forward. The fire fighter closest to the butt covers the sharp butt spur with a gloved hand. (STEP 3)

Three-Fire Fighter Suitcase Carry

A three-fire fighter suitcase carry can be used for heavier ladders. This technique is similar to the two-fire fighter suitcase carry, with the addition of a third fire fighter at the center of the ladder. All three fire fighters remain on the same side of the ladder. To perform the three-fire fighter suitcase carry, follow the steps in SKILL DRILL 13-6:

1. Start with the ladder resting on the ground on one beam. All three fire fighters are positioned on the same side of the ladder.
2. Facing the butt end of the ladder, one fire fighter is positioned at the butt end of the ladder, one fire fighter is at the middle of the ladder, and one fire fighter is at the tip of the ladder. (STEP 1)
3. All three fire fighters reach down and grasp the upper rung of the ladder. (STEP 2)
4. On the leader's command, the fire fighters pick the ladder up from the ground and carry it at arm's length. (STEP 3)
5. The fire fighter closest to the butt covers the sharp butt spur with a gloved hand. (STEP 4)

Three-Fire Fighter Flat Carry

The three-fire fighter flat carry is typically used with extension ladders up to 35 feet (11 m) long. To perform the three-fire fighter flat carry, follow the steps in SKILL DRILL 13-7:

1. The carry begins with the bed section of the ladder flat on the ground. Two fire fighters stand on one side of the ladder, and the third fire fighter stands on the opposite side of the ladder.
2. Facing the butt end of the ladder, one fire fighter is positioned at the butt end of the ladder with the second fire fighter at the tip end of the ladder. The third fire fighter is positioned at the center of the ladder on the opposite side. (STEP 1)
3. All three fire fighters kneel down and grasp the closer beam at arm's length. (STEP 2)
4. On the leader's command, all three fire fighters pick up the ladder and carry it butt end forward. The fire fighter closest to the butt covers the sharp butt spur with a gloved hand. (STEP 3)

SKILL DRILL 13-5 Two-Fire Fighter Suitcase Carry

(Fire Fighter I, NFPA 5.3.6)

1. Start with the ladder resting on the ground on one beam. Both fire fighters are positioned on the same side of the ladder. Facing the butt end of the ladder, one fire fighter is positioned at the butt end of the ladder and the other fire fighter is positioned at the tip of the ladder.
2. Each fire fighter reaches down and grasps the upper beam of the ladder.
3. On the leader's command, both fire fighters pick the ladder up from the ground and carry it with the butt end forward. The fire fighter closest to the butt covers the sharp butt spur with a gloved hand.

SKILL DRILL 13-6 Three-Fire Fighter Suitcase Carry
(Fire Fighter I, NFPA 5.3.6)

1 Start with the ladder resting on the ground on one beam. All three fire fighters are positioned on the same side of the ladder. Facing the butt end of the ladder, one fire fighter is positioned at the butt end of the ladder, one fire fighter is at the middle of the ladder, and one fire fighter is at the tip of the ladder.

2 All three fire fighters reach down and grasp the upper beam of the ladder.

3 On the leader's command, the fire fighters pick the ladder up from the ground and carry it at arm's length.

4 The fire fighter closest to the butt covers the sharp butt spur with a gloved hand.

Four-Fire Fighter Flat Carry

The four-fire fighter flat carry is similar to the three-fire fighter flat carry described in the previous section, except that two fire fighters are positioned on each side of the ladder. On each side, one fire fighter is at the tip and the other is at the butt end of the ladder. To perform the four-fire fighter flat carry, follow the steps in SKILL DRILL 13-8:

1. The carry begins with the bed section of the ladder flat on the ground. Two fire fighters stand on each side of the ladder.
2. All four fire fighters face the butt end of the ladder. One fire fighter is positioned at each corner of the ladder, with two fire fighters at the butt end of the ladder and two fire fighters at the tip end of the ladder. (STEP 1)
3. On the leader's command, all four fire fighters kneel down and grasp the closer beam at arm's length. (STEP 2)
4. On the leader's command, all four fire fighters pick up the ladder and carry it butt end forward. (STEP 3)
5. The fire fighter closest to the butt covers the sharp butt spur with a gloved hand. (STEP 4)

SKILL DRILL 13-7 Three-Fire Fighter Flat Carry

(Fire Fighter I, NFPA 5.3.6)

1 The bed section of the ladder is flat on the ground. Two fire fighters stand on one side of the ladder, and the third fire fighter stands on the opposite side of the ladder. Facing the butt end of the ladder, one fire fighter is positioned at the butt end of the ladder with the second fire fighter at the tip end of the ladder. The third fire fighter is positioned at the center of the ladder on the opposite side.

2 All three fire fighters kneel down and grasp the closer beam at arm's length.

3 On the leader's command, all three fire fighters pick up the ladder and carry it butt end forward. The fire fighter closest to the butt covers the sharp butt spur with a gloved hand.

Three-Fire Fighter Flat Shoulder Carry

The three-fire fighter flat shoulder carry is similar to the three-fire fighter flat carry, except that the ladder is carried on the shoulders instead of at arm's length. The fire fighters face the tip of the ladder as they lift it, but then pivot into the ladder as they raise it to shoulder height.

This carry is useful when fire fighters must carry the ladder over short obstacles. Raising the ladder to shoulder height increases the potential for back strain, however, so fire fighters must follow proper lifting techniques.

To perform the three-fire fighter flat shoulder carry, follow the steps in SKILL DRILL 13-9:

1. The carry begins with the bed section of the ladder flat on the ground. Two fire fighters are positioned on one side of the ladder, and the third fire fighter is on the opposite side of the ladder. Facing the tip end of the ladder, one fire fighter is positioned at the butt end of the ladder and the second fire fighter is at the tip end of the ladder. The third fire fighter is at the center of the ladder on the opposite side. (STEP 1)
2. All three fire fighters kneel and grasp the closer beam. (STEP 2)
3. On the leader's command, the fire fighters stand, raising the ladder. (STEP 3)
4. As the ladder approaches chest height, the leader instructs the fire fighters to pivot into the ladder. (STEP 4)
5. The fire fighters place the beam of the ladder on their shoulders. (STEP 5)
6. The ladder is carried in this position with the butt moving forward. The fire fighter closest to the butt covers the sharp butt spur with a gloved hand. (STEP 6)

Fire Fighter Safety Tips

When fire fighters of different heights are working together and carrying ladders, you can make these operations safer and smoother by considering the placement of fire fighters. When performing three-person carries, place shorter fire fighters at the butt end of the ladder and taller fire fighters at the tip end of the ladder. This strategy prevents a shorter fire fighter from being placed between two taller fire fighters and having to carry the ladder at an awkward and unsafe height. When performing four-person carries, placing shorter fire fighters at the butt end of the ladder prevents the ladder from being tipped to one side while it is being carried.

VOICES OF EXPERIENCE

While some may undervalue the importance of ladder training and drills, when the time comes and ground ladders need to be placed rapidly, fire fighters will fall back on the training and evolutions they practiced until they became second nature. For example, my fire department once received a call for a structure fire in a building in an older section of the downtown area. Upon arrival, several occupants were leaning out of the second story windows. The fire had trapped them in their apartments.

Using an aerial ladder was not an option due to the power lines in front of the building. Ground ladders were the safest option. We positioned the ground ladders and accomplished four rescues in a matter of minutes from when we arrived at the scene. Relying on our training enabled us to "ladder the building" in a safe and efficient manner and quickly perform the rescues. This was a clear example of training reinforcing the necessary skills needed in the entire team for a successful conclusion to an incident.

Chris Harner
City of Pueblo, Colorado Fire Department
Pueblo, Colorado

SKILL DRILL 13-8 Four-Fire Fighter Flat Carry
(Fire Fighter I, NFPA 5.3.6)

1. Two fire fighters stand on each side of the ladder. All four fire fighters face the butt end of the ladder. One fire fighter is positioned at each corner of the ladder, with two fire fighters at the butt end of the ladder and two fire fighters at the tip end of the ladder.

2. On the leader's command, all four fire fighters kneel down and grasp the closer rung at arm's length.

3. On the leader's command, all four fire fighters pick up the ladder and carry it butt end forward.

4. The fire fighter closest to the butt covers the sharp butt spur with a gloved hand.

Four-Fire Fighter Flat Shoulder Carry

The four-fire fighter flat shoulder carry is similar to the three-fire fighter flat shoulder carry described in the previous section, except that two fire fighters are positioned on each side of the ladder. On each side, one fire fighter is at the tip and the other at the butt end of the ladder. To perform the four-fire fighter flat shoulder carry, follow the steps in SKILL DRILL 13-10:

1. The carry begins with the bed section of the ladder flat on the ground. Two fire fighters are positioned on each side of the ladder.
2. All four fire fighters face the tip end of the ladder. One fire fighter is positioned at each corner of the ladder, with two fire fighters at the butt end and two fire fighters at the tip end of the ladder. (STEP 1)
3. On the leader's command, all four fire fighters kneel and grasp the closer beam. (STEP 2)
4. On the leader's command, the fire fighters stand, raising the ladder. (STEP 3)
5. As the ladder approaches chest height, the fire fighters all pivot toward the ladder. (STEP 4)

SKILL DRILL 13-9 Three-Fire Fighter Flat Shoulder Carry

(Fire Fighter I, NFPA 5.3.6)

1. Two fire fighters are positioned on one side of the ladder, and the third fire fighter is on the opposite side of the ladder. Facing the tip end of the ladder, one fire fighter is positioned at the butt end of the ladder and the second fire fighter is at the tip end of the ladder. The third fire fighter is at the center of the ladder on the opposite side.

2. All three fire fighters kneel and grasp the closer beam.

3. On the leader's command, the fire fighters stand, raising the ladder.

4. As the ladder approaches chest height, the leader instructs the fire fighters to pivot into the ladder.

5. The fire fighters place the beam of the ladder on their shoulders.

6. The ladder is carried in this position with the butt moving forward. The fire fighter closest to the butt covers the sharp butt spur with a gloved hand.

6. The fire fighters place the beam of the ladder on their shoulders. All four fire fighters face the butt of the ladder.
7. The ladder is carried in this position, with the butt moving forward. The fire fighter closest to the butt covers the sharp butt spur with a gloved hand. (STEP 5)

Placing a Ladder

The first step in raising a ladder is selecting the proper location for the ladder. Generally, an officer or senior fire fighter will select the general area for ladder placement, and the fire fighter at the butt of the ladder will determine the exact site at which to position it. Both the officer ordering the ladder and the fire fighter placing the ladder need to consider several factors in their decisions.

A raised ladder should be positioned at an angle of approximately 75 degrees to provide the best combination of strength, stability, and vertical reach. This angle also creates a comfortable climbing angle. When a fire fighter is standing on a rung, the rung at shoulder height will be approximately one arm's length away. The ratio of the ladder height (vertical reach) to the distance from the structure should be 4:1. For example, if the vertical reach of a ladder is 20 feet (6 m), the butt of the ladder should be placed 5 feet (1.5 m) out from the wall. Most new ladders have an inclination guide on their beams to help with ladder placement.

When calculating vertical reach, remember to add extra footage for rooftop operations. If the butt of the ladder is placed too close to the building, the climbing angle will be too steep. In such a case, the tip of the ladder could pull away from the

SKILL DRILL 13-10 Four-Fire Fighter Flat Shoulder Carry

(Fire Fighter I, NFPA 5.3.6)

1. Two fire fighters are positioned on each side of the ladder. All four fire fighters face the tip end of the ladder. One fire fighter is positioned at each corner of the ladder, with two fire fighters at the butt end and two fire fighters at the tip end of the ladder.

2. On the leader's command, all four fire fighters kneel and grasp the closer beam.

3. On the leader's command, the fire fighters stand, raising the ladder.

4. As the ladder approaches chest height, the fire fighters all pivot toward the ladder.

5. The fire fighters place the beam of the ladder on their shoulders. All four fire fighters face the butt of the ladder. The ladder is carried in this position, with the butt moving forward. The fire fighter closest to the butt covers the sharp butt spur with a gloved hand.

building as the fire fighter climbs, making the ladder unstable. Conversely, if the butt is too far from the structure, the climbing angle will be too shallow, reducing the load capacity of the ladder and increasing the risk that the butt could slip out from under the ladder.

The ladder should be placed on a stable and relatively level surface, as ladders placed on uneven surfaces are prone to tipping FIGURE 13-27. As a fire fighter climbs the ladder, the moving weight will cause the load to shift back and forth between the two beams. Unless both butt ends are firmly placed on solid ground, this uneven load can start a rocking motion that will tip the ladder over.

Ladders should not be placed on top of unstable structures such as manholes or trap doors. The combined weight of the ladder, fire fighters, and equipment could cause the cover or door to fail, injuring the fire fighters.

As mentioned earlier, ladders should be placed in areas with no overhead obstructions. If a ladder comes into contact with overhead utility lines—and particularly electric power lines—both the fire fighters working on the ladder and those stabilizing it could be injured or killed. This is true for all ladders, regardless of their composition (fiberglass, wood, or metal).

Be aware that a ladder can become energized even if it does not actually touch an electric line. For example, a ladder that enters the electromagnetic field surrounding a power line can become energized. For this reason, ladders should be placed at least 10 feet (3 m) away from energized power lines.

FIGURE 13-27 A ladder placed on an uneven surface is unsafe because it can easily tip over.

Finally, portable ladders should not be placed in high-traffic areas unless no other alternative is available. For example, because the main entrance to a structure is heavily used, a ladder should not be placed where it would obstruct the door.

Fire Fighter Safety Tips

Ladders can become energized without making contact with overhead electric lines. An electric force field surrounds electric lines, with the size of this field increasing in tandem with both the voltage in the line and the relative humidity in the air. Always keep ladders at least 10 feet (3 m) away from electric lines.

Raising a Ladder

Once the position has been selected, the ladder must be raised for climbing. Two commonly used techniques for raising portable ladders are the beam raise and the rung raise. A beam raise is typically used when the ladder must be raised parallel to the target surface. A rung raise is often used when the ladder can be raised from a position perpendicular to the target surface.

The number of fire fighters required to raise a ladder depends on the length and weight of the ladder, as well as on the available clearance from obstructions. A single fire fighter can safely raise many straight ladders and lighter extension ladders. Two or more fire fighters are required to raise longer and heavier ladders.

One-Fire Fighter Rung Raises

There are two variations of the one-fire fighter rung raise. One is used by a single fire fighter to raise a small, straight ladder, typically 14 feet (4 m) or less in length. Always check for overhead hazards before raising a ladder. To perform this raise, follow the steps in SKILL DRILL 13-11:

1. Place the ladder flat on the ground. Position the heel of the ladder where it will be when the ladder is raised.
2. Stand at the tip of the ladder and check for overhead hazards.
3. Raise the ladder to hip level.
4. Walk hand-over-hand down the rungs until the ladder is vertical. (STEP 1)
5. Place one foot against the beam of the ladder and lean it into place.
6. Check the ladder for a 75-degree climbing angle. Reposition it if necessary. Check the tip of the ladder and the heel of the ladder to ensure safety before climbing. (STEP 2)

The other variation of the one-fire fighter rung raise is generally used with straight ladders longer than 14 feet (4 m) and with extension ladders that can be safely handled by one fire fighter. Each fire fighter will have a different safety limit, depending on his or her strength and the weight of the ladder. To perform this variation of the one-fire fighter rung raise, follow the steps in SKILL DRILL 13-12:

1. Place the butt of the ladder on the ground directly against the structure so that both spurs contact the ground and the structure. Lay the ladder on the ground. If the ladder is an extension ladder, place the base section on the ground. (STEP 1)
2. Stand at the tip of the ladder and check for overhead hazards.
3. Take hold of a rung near the tip, bring that end of the ladder to chest height, and then step beneath the ladder. (STEP 2)
4. Raise the ladder using a hand-over-hand motion as you walk toward the structure until the ladder is vertical and against the structure. (STEP 3)
5. If an extension ladder is being used, hold the ladder vertical against the structure and extend the fly section by pulling the halyard smoothly, with a hand-over-hand motion, until the desired height is reached and the dogs are locked. (STEP 4)
6. Pull the butt of the ladder out from the structure to create a 75-degree climbing angle. To move the butt away from the structure, grip a lower rung and lift slightly while pulling outward. At the same time, apply pressure to an upper rung to keep the tip of the ladder against the structure. (STEP 5)
7. If the ladder is an extension ladder, it will be necessary to rotate the ladder so the fly section is out.
8. The halyard should be tied as described in the Tying the Halyard skill drill. (STEP 6)
9. Check the tip and the heel of the ladder to ensure safety before climbing. (STEP 7)

Tying the Halyard

The halyard of an extension ladder should always be tied after the ladder has been extended and lowered into place. A tied halyard stays out of the way and provides a safety backup to the dogs for securing the fly section. To tie the halyard, follow the steps in SKILL DRILL 13-13:

1. Wrap the excess halyard rope around two rungs of the ladder and pull the rope tight across the upper of the two rungs. (STEP 1)

SKILL DRILL 13-11 One-Fire Fighter Rung Raise for Ladders Less Than 14 Feet Long

(Fire Fighter I, NFPA 5.3.6)

1 Place the ladder flat on the ground. Position the heel of the ladder where it will be when the ladder is raised. Stand at the tip of the ladder and check for overhead hazards. Raise the ladder to hip level. Walk hand-over-hand down the rungs until the ladder is vertical.

2 Place one foot against the beam of the ladder and lean it into place. Check the ladder for a 75-degree climbing angle. Reposition it if necessary. Check the tip of the ladder and the heel of the ladder to ensure safety before climbing.

2. Tie a clove hitch around the upper rung and the vertical section of the halyard. Refer to the Ropes and Knots chapter to review how to tie a clove hitch. (STEP 2)
3. Pull the clove hitch tight. (STEP 3)
4. Place an overhand safety knot as close to the clove hitch as possible to prevent slipping. (STEP 4)

Two-Fire Fighter Beam Raise

The two-fire fighter beam raise is used with midsized extension ladders up to 35 feet (11 m) long. To perform the two-fire fighter beam raise, follow the steps in SKILL DRILL 13-14:

1. The two-fire fighter beam raise begins with a shoulder or suitcase carry. One fire fighter stands near the butt of the ladder, and the other fire fighter stands near the tip. (STEP 1)
2. The fire fighter at the butt of the ladder places that end of the ladder on the ground, while the fire fighter at the tip of the ladder holds that end. (STEP 2)
3. The fire fighter at the butt of the ladder places a foot on the butt of the beam that is in contact with the ground and grasps the upper beam. (STEP 3)
4. The fire fighter at the tip of the ladder checks for overhead hazards and then begins to walk toward the butt, while raising the lower beam in a hand-over-hand fashion until the ladder is vertical. (STEP 4)
5. The two fire fighters pivot the ladder into position as necessary. (STEP 5)
6. The fire fighters face each other, one on each side of the ladder, and heel the ladder by each placing the toe of one boot against the opposing beams of the ladder. (STEP 6)
7. One fire fighter extends the fly section by pulling the halyard smoothly with a hand-over-hand motion until the fly section is at the height desired and the pawls (dogs) are locked.
8. The other fire fighter stabilizes the ladder by holding the outside of the two beams, so that if the fly comes down suddenly it will not strike the fire fighter's hands. (STEP 7)
9. The fire fighter facing the structure places one foot against one beam of the ladder, and then both fire fighters lean the ladder into place.
10. The halyard is tied.
11. The fire fighters check the ladder for a 75-degree climbing angle and for stability at the top and the bottom of the ladder. (STEP 8)

SKILL DRILL 13-12

One-Fire Fighter Rung Raise for Ladders More Than 14 Feet Long

(Fire Fighter I, NFPA 5.3.6)

1 Place the butt of the ladder on the ground directly against the structure so that both spurs contact the ground and the structure. Lay the ladder on the ground. If the ladder is an extension ladder, place the base section on the ground. Stand at the tip of the ladder and check for overhead hazards.

2 Take hold of a rung near the tip, bring that end of the ladder to chest height, and then step beneath the ladder.

3 Raise the ladder using a hand-over-hand motion as you walk toward the structure until the ladder is vertical and against the structure.

4 If an extension ladder is being used, hold the ladder vertical against the structure and extend the fly section by pulling the halyard smoothly, with a hand-over-hand motion, until the desired height is reached and the dogs are locked.

5 Pull the butt of the ladder out from the structure to create a 75-degree climbing angle. To move the butt away from the structure, grip a lower rung and lift slightly while pulling outward. At the same time, apply pressure to an upper rung to keep the tip of the ladder against the structure.

6 If the ladder is an extension ladder, it will be necessary to rotate the ladder so the fly section is out. The halyard should be tied as described in the Tying the Halyard skill drill.

7 Check the tip and the heel of the ladder to ensure safety before climbing.

SKILL DRILL 13-13 Tying the Halyard

(Fire Fighter I, NFPA 5.3.6)

1 Wrap the excess halyard rope around two rungs of the ladder and pull the rope tight across the upper of the two rungs.

2 Tie a clove hitch around the upper rung and the vertical section of the halyard.

3 Pull the clove hitch tight.

4 Place an overhand safety knot as close to the clove hitch as possible to prevent slipping.

Two-Fire Fighter Rung Raise

The two-fire fighter rung raise is also commonly used with midsized extension ladders up to 35 feet (11 m) long. To perform the two-fire fighter rung raise, follow the steps in SKILL DRILL 13-15:

1. The two-fire fighter rung raise begins from a shoulder carry or suitcase carry, with one fire fighter near the butt of the ladder and the other near the tip.
2. The fire fighter at the butt of the ladder places the butt of the lower beam on the ground, while the fire fighter at the tip holds the other end.
3. The fire fighter at the tip rotates the ladder so that both beams are in contact with the ground. (STEP 1)
4. The fire fighter at the butt of the ladder stands on the bottom rung, grasps a higher rung with both hands, crouches down, and leans backward to heel the ladder.
5. The fire fighter at the tip of the ladder checks for overhead hazards, then swings under the ladder and walks toward the butt, advancing down the ladder and lifting the rungs in a hand-over-hand fashion until the ladder is vertical. (STEP 2)
6. The two fire fighters stand on opposite sides of the ladder and pivot it into position as necessary. (STEP 3)
7. The fire fighters face each other, one on each side of the ladder, and heel the ladder by each placing the toe

SKILL DRILL 13-14 Two-Fire Fighter Beam Raise
(Fire Fighter I, NFPA 5.3.6)

1 The two-fire fighter beam raise begins with a shoulder or suitcase carry. One fire fighter stands near the butt of the ladder, and the other fire fighter stands near the tip.

2 The fire fighter at the butt of the ladder places that end of the ladder on the ground, while the fire fighter at the tip of the ladder holds that end.

3 The fire fighter at the butt of the ladder places a foot on the butt of the beam that is in contact with the ground and grasps the upper beam.

4 The fire fighter at the tip of the ladder checks for overhead hazards and then begins to walk toward the butt, while raising the lower beam in a hand-over-hand fashion until the ladder is vertical.

5 The two fire fighters pivot the ladder into position as necessary.

6 The fire fighters face each other, one on each side of the ladder, and heel the ladder by each placing the toe of one boot against the opposing beams of the ladder.

7 One fire fighter extends the fly section by pulling the halyard smoothly with a hand-over-hand motion until the fly section is at the height desired and the pawls (dogs) are locked. The other fire fighter stabilizes the ladder by holding the outside of the two beams, so that if the fly comes down suddenly it will not strike the fire fighter's hands.

8 The fire fighter facing the structure places one foot against one beam of the ladder, and then both fire fighters lean the ladder into place. The halyard is tied. The fire fighters check the ladder for a 75-degree climbing angle and check for stability at the top and the bottom of the ladder.

SKILL DRILL 13-15 Two-Fire Fighter Rung Raise

(Fire Fighter I, NFPA 5.3.6)

1 The two-fire fighter rung raise begins from a shoulder carry or suitcase carry, with one fire fighter near the butt of the ladder and the other near the tip. The fire fighter at the butt of the ladder places the butt of the lower beam on the ground, while the fire fighter at the tip holds the other end. The fire fighter at the tip rotates the ladder so that both beams are in contact with the ground.

2 The fire fighter at the butt of the ladder stands on the bottom rung, grasps a higher rung with both hands, crouches down, and leans backward to heel the ladder. The fire fighter at the tip of the ladder checks for overhead hazards, then swings under the ladder and walks toward the butt, advancing down the ladder and lifting the rungs in a hand-over-hand fashion until the ladder is vertical.

3 The two fire fighters stand on opposite sides of the ladder and pivot it into position as necessary.

4 The fire fighters face each other, one on each side of the ladder, and heel the ladder by each placing the toe of one boot against the opposite beams of the ladder.

5 If using an extension ladder, one fire fighter extends the fly section by pulling the halyard smoothly with a hand-over-hand motion until the tip is at the desired height and the ladder locks (dogs) are locked. The other fire fighter stabilizes the ladder by holding the outside of the base section's beams, so that if the fly comes down suddenly it will not strike the fire fighter's hands.

6 The fire fighter facing the structure places one foot against one beam of the ladder, and then both fire fighters lean the ladder into place. The halyard is tied. The fire fighters check the ladder for a 75-degree climbing angle and for stability at the bottom and at top of the ladder.

of one boot against the opposite beams of the ladder. (**STEP 4**)

8. If using an extension ladder, one fire fighter extends the fly section by pulling the halyard smoothly with a hand-over-hand motion until the tip is at the desired height and the ladder locks (dogs) are locked. The other fire fighter stabilizes the ladder by holding the outside of the base section's beams, so that if the fly comes down suddenly it will not strike the fire fighter's hands. (**STEP 5**)

9. The fire fighter facing the structure places one foot against one beam of the ladder, and then both fire fighters lean the ladder into place.
10. The halyard is tied.
11. The fire fighters check the ladder for a 75-degree climbing angle and for stability at the bottom and top of the ladder. (STEP 6)

Three- and Four-Fire Fighter Rung Raises

The three- and four-fire fighter rung raises are used for very heavy ladders. The basic steps in these raises are similar to the two-fire fighter rung raise. In a three-fire fighter rung raise, the third fire fighter assists with the hand-over-hand raising of the ladder. When four fire fighters are raising the ladder, two anchor the butt of the ladder while the other two raise the ladder. Each of the two fire fighters at the butt places his or her inside foot on the bottom rung of the ladder and bends over, grasping a rung in front of him or her.

To perform the three-fire fighter rung raise, follow the steps in SKILL DRILL 13-16:

1. The raise begins from a shoulder carry or suitcase carry, with one fire fighter near the butt of the ladder, one fire fighter in the middle, and one fire fighter near the tip.
2. The fire fighter at the butt of the ladder places the butt of the lower beam on the ground, while the fire fighter at the tip holds that end. The fire fighter in the middle moves to the tip. The fire fighters at the tip rotate the ladder so that both butts are in contact with the ground.
3. The fire fighter at the butt of the ladder stands on the bottom rung, grasps a higher rung with both hands, crouches down, and leans backward to heel the ladder. (STEP 1)
4. The fire fighters at the tip of the ladder check for overhead hazards, then begin to walk toward the butt, advancing down the ladder and lifting the rungs in a hand-over-hand fashion until the ladder is vertical. (STEP 2)
5. All three fire fighters pivot the ladder into position as necessary. (STEP 3)
6. Two fire fighters face each other, one on each side of the ladder, grasp the outsides of the beams, and heel the ladder by each placing the toe of one boot against the opposite beams of the ladder. (STEP 4)
7. The third fire fighter extends the fly section by pulling the halyard smoothly with a hand-over-hand motion until the tip is at the desired height and the dogs are locked. (STEP 5)
8. One fire fighter heels the ladder while the other two lean the ladder into place.
9. The halyard is tied.
10. The fire fighters check the ladder for a 75-degree climbing angle and for security at the butt end and at the tip of the ladder. (STEP 6)

To perform the four-fire fighter rung raise, follow the steps in SKILL DRILL 13-17:

1. The raise begins with a four-fire fighter flat carry. Two fire fighters are at the butt of the ladder, and two fire fighters are at the tip. (STEP 1)
2. The two fire fighters at the butt of the ladder place the butt end of the ladder on the ground, while the two fire fighters at the tip hold that end of the ladder. (STEP 2)
3. The two fire fighters at the butt of the ladder stand side-by-side, facing the ladder. Each fire fighter places the inside foot on the bottom rung and the other foot on the ground outside the beam. Both crouch down, grab a rung and the beam, and lean backward. (STEP 3)
4. The two fire fighters at the tip of the ladder check for overhead hazards and then begin to walk toward the butt of the ladder, advancing down the rungs in a hand-over-hand fashion until the ladder is vertical. (STEP 4)
5. The fire fighters pivot the ladder into position, as necessary. (STEP 5)
6. Two fire fighters heel the ladder by placing a boot against each beam. Each fire fighter places the toe of one boot against one of the beams. The third fire fighter stabilizes the ladder by holding it on the outside of the rails. (STEP 6)
7. The fourth fire fighter extends the fly section by pulling the halyard smoothly with a hand-over-hand motion until the tip reaches the desired height and the dogs are locked. (STEP 7)
8. The two fire fighters facing the structure each place one foot against one beam of the ladder while the other two fire fighters lower the ladder into place.
9. The halyard is tied.
10. The fire fighters check the ladder for a 75-degree climbing angle and for security at the butt end and at the tip of the ladder. (STEP 8)

Fire Fighter Safety Tips

When raising the fly section of an extension ladder, do not wrap the halyard around your hand. If the ladder falls or if the fly section retracts unexpectedly, your hand could be caught in the halyard and you could be seriously injured. Also, do not place your foot under the fly sections. If the rope slips, your foot will be crushed.

Fly Section Orientation

When raising extension ladders, fire fighters must know whether the fly section should be placed toward the building (fly in) or away from the building (fly out). The ladder manufacturer and department standard operating procedures (SOPs) will specify whether the fly should be facing in or out. In general, manufacturers of fiberglass and metal ladders recommend that the fly sections be placed away (fly out) from the structure. In contrast, wooden fire-service ladders are often designed to be used with the fly in.

SKILL DRILL 13-16 Three-Fire Fighter Rung Raise

(Fire Fighter I, NFPA 5.3.6)

1 The fire fighter at the butt of the ladder places the butt of the lower beam on the ground, while the fire fighter at the tip holds that end. The fire fighter in the middle moves to the tip. The fire fighters at the tip rotate the ladder so that both butts are in contact with the ground. The fire fighter at the butt of the ladder stands on the bottom rung, grasps a higher rung with both hands, crouches down, and leans backward to heel the ladder.

2 The fire fighters at the tip of the ladder check for overhead hazards, then begin to walk toward the butt, advancing down the ladder and lifting the rungs in a hand-over-hand fashion until the ladder is vertical.

3 All three fire fighters pivot the ladder into position as necessary.

4 Two fire fighters face each other, one on each side of the ladder, grasp the outsides of the beams, and heel the ladder by each placing the toe of one boot against the opposite beams of the ladder.

5 The third fire fighter extends the fly section by pulling the halyard smoothly with a hand-over-hand motion until the tip is at the desired height and the dogs are locked.

6 One fire fighter heels the ladder while the other two lean the ladder into place. The halyard is tied. The fire fighters check the ladder for a 75-degree climbing angle and for security at the butt end and at the tip of the ladder.

Securing the Ladder

Several approaches may be used to prevent a ladder from moving once it is in place. One option is to have a fire fighter stand between the ladder and the structure, grasp the beams, and lean back to pull the ladder into the structure—a process called heeling the ladder FIGURE 13-28. A fire fighter on the outside of the ladder, facing the structure, can also heel the ladder by placing a boot against the beam of the ladder FIGURE 13-29.

Each of these methods has its own advantages and disadvantages:

- Heeling from under the ladder: The ladder protects the heeling fire fighter from debris that may fall from the structure, but this approach can cause an eye injury

FIGURE 13-28 The fire fighter's weight can pull the ladder against the structure to secure the ladder.

FIGURE 13-29 A fire fighter can push against the ladder and place his or her foot on the butt end of the ladder to heel it.

if the fire fighter looks up during his or her partner's climb (owing to falling debris).

- Heeling from outside the ladder: This technique allows the heeling fire fighter to survey the building and warn the climber of possible hazards and avoid falling debris.

A rope, a rope–hose tool, or webbing also can be used to secure a ladder in place. The lower part of the ladder can be tied to any solid object to keep the base from kicking out; in fact, the base should always be secured if the ladder is used at a low angle. The tip of the ladder can be tied to a secure object near the top to keep it from pulling away from the building. The best method is to secure both the tip and the base.

Climbing the Ladder

Climbing a ladder on the fire or emergency scene should be done in a deliberate and controlled manner. Always make sure the ladder is secure (tied or heeled). Before climbing an extension ladder, verify that the dogs are locked and the halyard is secured.

The proper climbing angle should be checked as well. See the section Placing a Ladder in this chapter for more information. A raised ladder should be positioned at an angle of approximately 75 degrees to provide the best combination of strength, stability, and vertical reach. This angle also creates a comfortable climbing angle. When a fire fighter is standing on a rung, the rung at shoulder height will be approximately one arm's length away. The ratio of the ladder height (vertical reach) to the distance from the structure should be 4:1. For example, if the vertical reach of a ladder is 20 feet (6 meter), the butt of the ladder should be placed 5 feet (1.5 meter) out from the wall. Some ladder beams include level guide stickers (placed there by the ladder's manufacturer), so that the bottom line of the indicator will be parallel to the ground when the ladder is positioned properly. Standing at the base of the ladder and extending the arms straight out is another way to check the climbing angle **FIGURE 13-30**. The hands should comfortably reach the beams or rungs if the angle is appropriate. Finally, the angle can be measured by ensuring that the butt of the ladder is one-fourth of the working height out from the base of the structure.

Keep any bouncing and shifting to a minimum while climbing the ladder. Your eyes should be focused forward, with only occasional glances upward. This approach will prevent debris such as falling glass from injuring your face or eyes. Lower the protective face shield for additional protection.

SKILL DRILL 13-17 Four-Fire Fighter Rung Raise

(Fire Fighter I, NFPA 5.3.6)

1 The raise begins with a four-fire fighter flat carry. Two fire fighters are at the butt of the ladder, and two fire fighters are at the tip.

2 The two fire fighters at the butt of the ladder place the butt end of the ladder on the ground, while the two fire fighters at the tip hold that end of the ladder.

3 The two fire fighters at the butt of the ladder stand side-by-side, facing the ladder. Each fire fighter places the inside foot on the bottom rung and the other foot on the ground outside the beam. Both crouch down, grab a rung and the beam, and lean backward.

4 The two fire fighters at the tip of the ladder check for overhead hazards and then begin to walk toward the butt of the ladder, advancing down the rungs in a hand-over-hand fashion until the ladder is vertical.

5 The fire fighters pivot the ladder into position, as necessary.

6 Two fire fighters heel the ladder by placing a boot against each beam. Each fire fighter places the toe of one boot against one of the beams. The third fire fighter stabilizes the ladder by holding it on the outside of the rails.

7 The fourth fire fighter extends the fly section by pulling the halyard smoothly with a hand-over-hand motion until the tip reaches the desired height and the dogs are locked.

8 The two fire fighters facing the structure each place one foot against one beam of the ladder while the other two fire fighters lower the ladder into place. The halyard is tied. The fire fighters check the ladder for a 75-degree climbing angle and for security at the butt end and at the tip of the ladder.

Use a hand-over-hand motion on the rungs of the ladder, or slide both hands along the underside of the beams while climbing. Sliding the hands along the beams is more secure because it maintains three contact points with the ladder at all times (both hands and one foot).

If tools must be moved up or down, it is better to hoist them by ropes than to carry them up a ladder. Carrying tools on a ladder reduces the fire fighter's grip and increases the potential for injury if a tool slips or falls. Sometimes you will need to carry a tool up a ladder, however, so you should practice doing it safely. To climb a ladder while holding a tool, hold the tool against one beam with one hand and maintain contact with the opposite beam with the other hand. Tools such as pike poles can be hooked onto the ladder and moved up every few rungs. One method of climbing a ladder while carrying a tool is described in SKILL DRILL 13-18:

1. Check the ladder to ensure that it is at a safe angle to climb and is secure.
2. Grasp the tool comfortably and securely in one hand, and hold that hand and the tool against the beam of the ladder. (STEP 1)
3. Wrap the other hand around the opposite beam and begin climbing. (STEP 2)
4. Climb smoothly and safely by maintaining contact between your free hand and the beam and by sliding the tool along the opposite beam.
5. Be sure not to overload the ladder. No more than two fire fighters should be on a ladder at the same time. As mentioned earlier, a properly placed ladder should be able to support two rescuers (with their protective clothing and equipment) and one victim. (STEP 3)

Dismounting the Ladder

When a ladder is used to reach a roof or a window entry, the fire fighter will have to dismount the ladder. Fire fighters can minimize their risk of slipping and falling by making sure that the surface is stable—that is, testing its stability with a tool—before dismounting. For example, fire fighters who dismount from a ladder onto a roof typically sound the roof with an axe before stepping onto the surface FIGURE 13-31.

FIGURE 13-30 The climbing angle can be checked by standing on the bottom rung and holding the arms straight out.

FIGURE 13-31 Check the stability of the roof before dismounting from the ladder.

SKILL DRILL 13-18 Climbing a Ladder While Carrying a Tool

(Fire Fighter I, NFPA 5.3.12 B)

1 Check the ladder to ensure that it is at a safe angle to climb and is secure. Grasp the tool comfortably and securely in one hand, and hold that hand and the tool against the beam of the ladder.

2 Wrap the other hand around the opposite beam and begin climbing.

3 Climb smoothly and safely by maintaining contact between your free hand and the beam and by sliding the tool along the opposite beam.

Try to maintain contact with the ladder at three points when dismounting. For example, a fire fighter who steps onto a roof should keep two hands and one foot on the ladder while checking the footing. This consideration is particularly important if the surface slopes or is covered with rain, snow, or ice. Do not shift your weight onto the roof until you have tested the footing.

It is also important to check the footing when entering a window from a ladder positioned to the side of the window. Before stepping into the window, confirm that the interior surface is structurally sound and offers secure footing. The three points of contact rule also applies in this situation.

Under heavy fire and smoke conditions, fire fighters sometimes dismount from a ladder by climbing over the tip and sliding over the windowsill into the building. Sound the floor inside the window before entering to confirm that it is solid and stable. To remount a ladder under these conditions, back out the window feet first and rest your abdomen on the sill until you can feel the ladder under your feet. Under better conditions, sit on the windowsill with legs out and roll onto the ladder.

Working from a Ladder

Fire fighters must often work from a ladder. To avoid falling while doing so, the fire fighter must be secured to the ladder. Fire fighters use two different methods to secure themselves to a ladder.

The first method involves the use of a ladder belt—a piece of equipment specifically designed to secure a fire fighter to a ladder or elevated surface FIGURE 13-32. Fire fighters should use only ladder belts designed and certified under the requirements specified by NFPA 1983, *Standard on Life Safety Rope and Equipment for Emergency Services*. Utility belts designed to carry tools should never be used as ladder belts.

The second method used to secure a fire fighter to a ladder is to apply a leg lock. The leg lock is simple and secure, and requires no special equipment.

To apply a leg lock, follow the steps in SKILL DRILL 13-19:

1. Climb to the desired work height and step up to one more rung. (STEP 1)
2. Note the side of the ladder where the work will be performed. Extend your leg between the rungs on the side opposite the side you will be working. (STEP 2)
3. Once your leg is between the rungs, bend your knee and bring your foot back under the rung and through to the climbing side of the ladder. (STEP 3)
4. Secure your foot against the next lower rung or the beam of the ladder. Use your thigh for support and step down one rung with the opposite foot. (STEP 4)
5. The use of a leg lock enables you to have two hands free for a variety of tasks, including placing a roof ladder or controlling a hose stream. (STEP 5)

FIGURE 13-32 A ladder belt has a large hook that is designed to secure a fire fighter to a ladder.

Placing a Roof Ladder

Several methods can be used to properly position a roof ladder on a sloping roof. One commonly used method is described in SKILL DRILL 13-20:

1. Carry the roof ladder to the base of the climbing ladder that is already in place to provide access to the roofline. (STEP 1)
2. Place the butt end of the roof ladder on the ground, and rotate the hooks of the roof ladder to the open position. (STEP 2)
3. Use a one-fire fighter beam or rung raise to lean the roof ladder against one beam of the other ladder with the hooks oriented outward. (STEP 3)
4. Climb the lower ladder until you reach the midpoint of the roof ladder, and then slip one shoulder between two rungs of the roof ladder and shoulder the ladder. (STEP 4)
5. Climb to the roofline of the structure, carrying the roof ladder on one shoulder.
6. Secure yourself to the ladder by using a ladder belt or applying a ladder leg lock. (STEP 5)
7. Place the roof ladder on the roof surface with hooks down. Push the ladder up toward the peak of the roof with a hand-over-hand motion. (STEP 6)
8. Once the hooks have passed the peak, pull back on the roof ladder to set the hooks and check that they are secure. (STEP 7)
9. To remove a roof ladder from the roof, reverse the process just described. After releasing the hooks from the peak, it may be necessary to turn the ladder on one of its beams or turn it so the hooks are pointing up to slide the ladder down the roof without catching the hooks on the roofing material. Carrying a roof ladder in this manner requires strength and practice.

SKILL DRILL 13-19 Use a Leg Lock to Work from a Ladder

(Fire Fighter I, NFPA 5.3.6)

1 Climb to the desired work height and step up to one more rung.

2 Note the side of the ladder where the work will be performed. Extend your leg between the rungs on the side opposite the side you will be working.

3 Once your leg is between the rungs, bend your knee and bring your foot back under the rung and through to the climbing side of the ladder.

4 Secure your foot against the next lower rung or the beam of the ladder. Use your thigh for support and step down one rung with the opposite foot.

5 The use of a leg lock enables you to have two hands free for a variety of tasks.

SKILL DRILL 13-20 Deploy a Roof Ladder

(Fire Fighter I, NFPA 5.3.12 B)

1 Carry the roof ladder to the base of the climbing ladder that is already in place to provide access to the roofline.

2 Place the butt end of the roof ladder on the ground, and rotate the hooks of the roof ladder to the open position.

3 Use a one-fire fighter beam or rung raise to lean the roof ladder against one beam of the other ladder with the hooks oriented outward.

4 Climb the lower ladder until you reach the midpoint of the roof ladder, and then slip one shoulder between two rungs of the roof ladder and shoulder the ladder.

5 Climb to the roofline of the structure, carrying the roof ladder on one shoulder. Secure yourself to the ladder by using a ladder belt or applying a ladder leg lock.

6 Place the roof ladder on the roof surface with hooks down. Push the ladder up toward the peak of the roof with a hand-over-hand motion.

7 Once the hooks have passed the peak, pull back on the roof ladder to set the hooks and check that they are secure.

Fire Fighter Safety Tips

When deploying a roof ladder, make sure that the hooks are pointed away from you as you raise and lower the ladder on the roof. Should the roof ladder slip or fall from the roof, the hooks could strike you if they are positioned toward you.

Inspect a Chimney

Each year, fire departments receive many calls for chimney fires and other chimney-related problems. Many times it is necessary to inspect the chimney to determine whether a problem is truly present. To access the chimney, it is necessary to safely ladder the building and proceed to the roof. The steps for inspecting a chimney are described in **SKILL DRILL 13-21**:

1. Make sure you are equipped with a complete set of PPE, including SCBA, and are operating under the direction of your officer.
2. Remove the proper-size extension ladder and the roof ladder from the apparatus.
3. Carry the ladders to the building and raise the extension ladder so that five rungs project above the roofline.
4. Confirm that the roof is safe to climb on. If it is not, use an aerial ladder.
5. When possible, place the roof ladder on the upwind side of the chimney.
6. Working with a partner, climb to the chimney while carrying the appropriate equipment.
7. Inspect the chimney, and report your findings to the incident commander.
8. Remove the tools and ladder when the task is complete.

As you work on learning the skills needed to perform vital ladder operations, do not lose sight of the dangers inherent in working with ladders. As you practice the skills described in this chapter, follow these safety rules:

- Wear appropriate PPE. This equipment should include SCBA when you are operating on the fire ground.
- Choose the proper ladder for the job.
- Choose the proper number of people for the job. Move smoothly, safely, and efficiently, keeping the ladder under your control at all times.
- Lift with your legs, keeping your back straight.
- Use proper hand and foot placement during carries, raises, climbing, and lowering of ladders.
- Look overhead for wires and obstructions before raising or lowering a ladder. Communicate the results of this check by stating "Clear overhead" to your partner.
- Allow the heel person to take charge of the processes of raising and lowering the ladder.
- After extending an extension ladder, check the pawls (ladder locks) to confirm that they are set and make sure the halyard is tied off.
- Check the ladder for the proper 75-degree angle.
- Ensure that the wall or roof will support the ladder before climbing it.
- Remain aware of hazards to yourself and to others.
- Maintain hand contact with the ladder. Watch your hands when climbing up, and watch your legs when climbing down.
- Use an appropriate leg lock or ladder belt when working on a ladder.

Wrap-Up

Chief Concepts

- The primary function of ladders is to provide safe access to and egress from otherwise inaccessible areas.
- Ladders can be used for several secondary functions, including channeling debris, serving as a lift point, and holding other firefighting equipment.
- The two main structural components of the ladder are the beam and the rung. The beams support the rungs and carry the load of the fire fighter from the rungs down to the ground.
- There are three basic types of ladder beam construction: trussed beam, I-beam, and solid beam. Trussed beams are usually constructed of aluminum or wood. I-beam ladders are usually made from fiberglass.
- Many wooden ladders have solid beams. Rectangular aluminum beams, which are usually hollow or C-shaped, are also classified as solid beam ladders.
- The rung is a crosspiece that spans the two beams of the ladder. Most portable ladders in fire departments have aluminum rungs, but wooden rungs are also in use.
- Additional parts of a typical portable ladder include the following components:
  - Rail—The top or bottom section of a trussed beam or the top and bottom surfaces of an I-beam.
  - Truss block—A piece or assembly that connects the two rails of a trussed beam.
  - Tie rod—A metal bar that connects one beam to the other.
  - Tip—The very top of the ladder.
  - Butt—The end of the ladder.
  - Butt spurs—Metal spikes designed to help stabilize the ladder in place.
  - Butt plate—An alternative to butt spurs; a swiveling plate that incorporates both a spur and a cleat.
  - Protection plates—Reinforcing plates placed at chafing or contact points.
- An extension ladder is an assembly of two or more ladder sections that fit together and can be extended or retracted to adjust the length of the ladder. Additional parts of an extension ladder include the following components:
  - Bed section—The base of the extension ladder.
  - Fly section—The section of the ladder that is raised from the bed section.
  - Guides—Strips of metal or wood or channels that guide the fly section during extension.
  - Halyard—The rope or cable used to extend the fly sections.
  - Pawls—Mechanical locking devices that secure the fly sections.
  - Pulley—A small wheel used to change the direction of the halyard pull.
  - Stops—Pieces of wood or metal that prevent the fly sections from becoming overextended.
  - Staypoles—Long metal poles used to stabilize the extension ladder.
- There are two general categories of ladders: aerial and portable. Aerial ladders are permanently mounted and operated from a fire apparatus. Portable ladders are carried on fire apparatus and removed from the apparatus to be used in other locations.
- Portable ladders include general-purpose ladders such as straight or extension ladders and specialized ladders such as attic or roof ladders.
- Fire apparatus are required to carry at least one roof ladder, one extension ladder, and one folding ladder.
- A straight ladder is a single-section, fixed-length ladder.
- A roof ladder is a straight ladder with retractable hooks at one end.
- An extension ladder is an adjustable-length ladder with multiple sections.
- Bangor ladders are extension ladders with staypoles for added stability.
- A combination ladder can be converted from a straight ladder to a stepladder configuration, or from an extension ladder to a stepladder configuration.
- A folding ladder is a narrow, collapsible ladder designed to allow access to attic scuttle areas and confined areas.
- A Fresno ladder is a narrow, two-section extension ladder designed to provide attic access.
- Portable ladders must be regularly inspected, maintained, and service tested following NFPA 1932.
- A ladder should be inspected visually after every use. If the inspection reveals any deficiencies, the ladder must be removed from service until repairs are made.
- All fire fighters should be able to perform routine ladder maintenance, including cleaning the ladders to remove any road grime and dirt. Ladders should be cleaned after every use.
- Service testing of portable ladders must follow NFPA 1932.
- Safety precautions should be followed from the time a ladder is removed from the apparatus until it is returned to the apparatus. Full turnout gear and PPE are essential

for working with portable and aerial ladders. Additional general safety rules include the following:
- Lift with your legs, keeping your back straight.
- Look overhead for wires and obstructions before raising or lowering a ladder.
- After extending an extension ladder, check the pawls (ladder locks) to confirm that they are set and make sure the halyard is tied off.
- Check the ladder for the proper 75-degree angle. Ensure that the wall or roof will support the ladder before climbing it.

- When positioning a ladder for rescue, place the tip of the ladder at or slightly below the window sill. This provides a direct pathway to enter and exit the window.
- Communication is key to coordinating efforts when working with ladders. One leader should be designated, and his or her commands should be followed during carries and raises.

Hot Terms

Aerial ladder A self-supporting, turntable-mounted, power-operated ladder of two or more sections permanently attached to a self-propelled automotive fire apparatus and designed to provide a continuous egress route from an elevated position to the ground. (NFPA 1901)

Bangor ladder A ladder equipped with tormentor poles or staypoles that stabilize the ladder during raising and lowering operations.

Base (bed) section The lowest or widest section of an extension ladder. (NFPA 1931)

Beam The main structural side of a ground ladder. (NFPA 1931)

Butt The end of the beam that is placed on the ground, or other lower support surface, when ground ladders are in the raised position. (NFPA 1931)

Butt plate An alternative to a simple butt spur; a swiveling plate with both a spur and a cleat or pad that is attached to the butt of the ladder.

Butt spurs That component of ground ladder support that remains in contact with the lower support surface to reduce slippage. (NFPA 1931)

Combination ladder A ground ladder that is capable of being used both as a stepladder and as a single or extension ladder. (NFPA 1931)

Egress To go or come out; to exit from an area or a building.

Extension ladder A non-self-supporting ground ladder that consists of two or more sections traveling in guides, brackets, or the equivalent arranged so as to allow length adjustment. (NFPA 1931)

Fly section Any section of an aerial telescoping device beyond the base section. (NFPA 1901)

Folding ladder A single-section ladder with rungs that can be folded or moved to allow the beams to be brought into a position touching or nearly touching each other. (NFPA 1931)

Fresno ladder A narrow, two-section extension ladder that has no halyard. Because of its limited length, it can be extended manually.

Grade A measurement of the angle used in road design and expressed as a percentage of elevation change over distance. (NFPA 1901)

Guides Strips of metal or wood that serve to guide a fly section during extension. Channels or slots in the bed or fly section may also serve as guides.

Halyard Rope used on extension ladders for the purpose of raising a fly section(s). (NFPA 1931)

Heat sensor label A label that changes color at a preset temperature to indicate a specific heat exposure. (NFPA 1931)

I-beam A ladder beam constructed of one continuous piece of I-shaped metal or fiberglass to which the rungs are attached.

Ladder belt A compliant equipment item that is intended for use as a positioning device for a person on a ladder. (NFPA 1983)

Ladder gin An A-shaped structure formed with two ladder sections. It can be used as a makeshift lift when raising a trapped person. One form of the device is called an A-frame hoist.

Pawls Devices attached to a fly section(s) to engage ladder rungs near the beams of the section below for the purpose of anchoring the fly section(s). (NFPA 1931)

Portable ladder A ladder carried on fire apparatus but designed to be removed from the apparatus and deployed by fire fighters where needed.

Protection plates Reinforcing material placed on a ladder at chafing and contact points to prevent damage from friction and contact with other surfaces.

Pulley A device with a free-turning, grooved metal wheel (sheave) used to reduce rope friction. Side plates are available for a carabiner to be attached. (NFPA 1670)

Rail The top or bottom piece of a trussed beam assembly used in the construction of a trussed ladder. Also, the top and bottom surfaces of an I-beam ladder. Each beam has two rails.

Roof hooks The spring-loaded, retractable, curved metal pieces that allow the tip of a roof ladder to be secured to the peak of a pitched roof. The hooks fold outward from each beam at the top of a roof ladder.

Roof ladder A single ladder equipped with hooks at the top end of the ladder. (NFPA 1931)

Rung The ladder crosspieces, on which a person steps while ascending or descending. (NFPA 1931)

Solid beam A ladder beam constructed of a solid rectangular piece of material (typically wood), to which the ladder rungs are attached.

Staypoles Poles attached to each beam of the base section of extension ladders, which assist in raising the ladder and help provide stability of the raised ladder. (NFPA 1931)

Stop A piece of material that prevents the fly sections of a ladder from becoming overextended, leading to collapse of the ladder.

Tie rod A metal rod that runs from one beam of the ladder to the other to keep the beams from separating. Tie rods are typically found in wood ladders.

Tip The very top of the ladder.

Truss block A piece of wood or metal that ties the two rails of a trussed beam ladder together and serves as the attachment point for the rungs.

Trussed beam A ladder beam constructed of top and bottom rails joined by truss blocks that tie the rails together and support the rungs.

FIRE FIGHTER *in action*

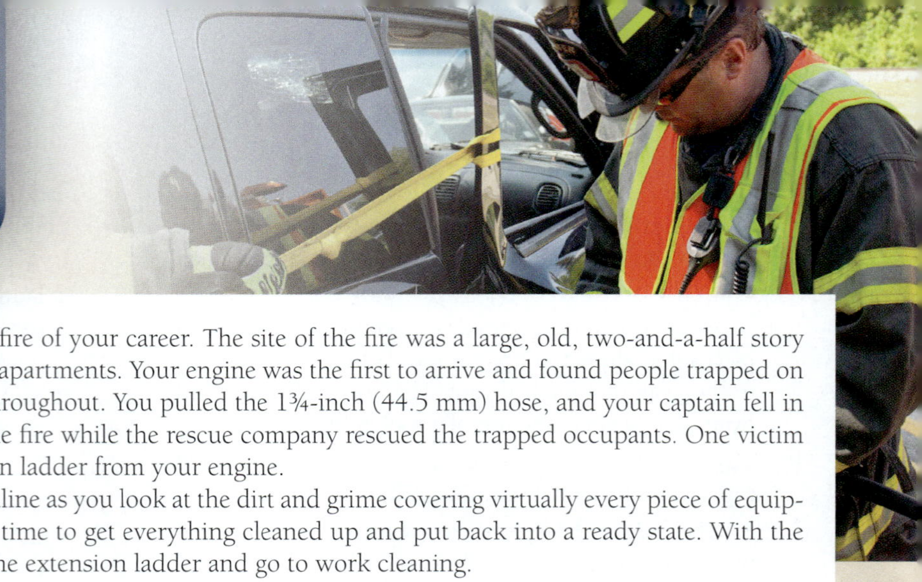

You just got back to the station from the largest fire of your career. The site of the fire was a large, old, two-and-a-half story wood frame house that had been converted into apartments. Your engine was the first to arrive and found people trapped on the second floor with heavy smoke conditions throughout. You pulled the 1¾-inch (44.5 mm) hose, and your captain fell in behind you. You made the stairs and confined the fire while the rescue company rescued the trapped occupants. One victim was rescued down the 35-foot (11 mm) extension ladder from your engine.

You can still feel the aftereffects of the adrenaline as you look at the dirt and grime covering virtually every piece of equipment on the engine. You know it will take some time to get everything cleaned up and put back into a ready state. With the help of your crewmate, you start by pulling off the extension ladder and go to work cleaning.

1. When possible, which is the preferred method of cleaning the ladder?
   - **A.** Stiff-bristle brush, mild detergent, and water
   - **B.** Soft-bristle brush and water
   - **C.** Stiff-bristle brush and water
   - **D.** Soft-bristle brush, mild solvent, and water
2. How should wire-rope halyard extensions be checked?
   - **A.** With the fly section fully extended
   - **B.** With the fly section partially extended
   - **C.** With the fly section fully bedded
   - **D.** The position of the fly section does not matter
3. In addition to after each use, when should a ladder normally be visually inspected?
   - **A.** Daily
   - **B.** Weekly
   - **C.** Monthly
   - **D.** Yearly
4. What are the spring-loaded, retractable, curved metal pieces that allow the tip of a roof ladder to be secured to the peak of a pitched roof?
   - **A.** Halyard
   - **B.** Tie rod
   - **C.** Pawls
   - **D.** Roof hooks
5. What are the two main structural components that run the entire length of most ladders or ladder sections?
   - **A.** Rails
   - **B.** Truss blocks
   - **C.** Beams
   - **D.** Guides

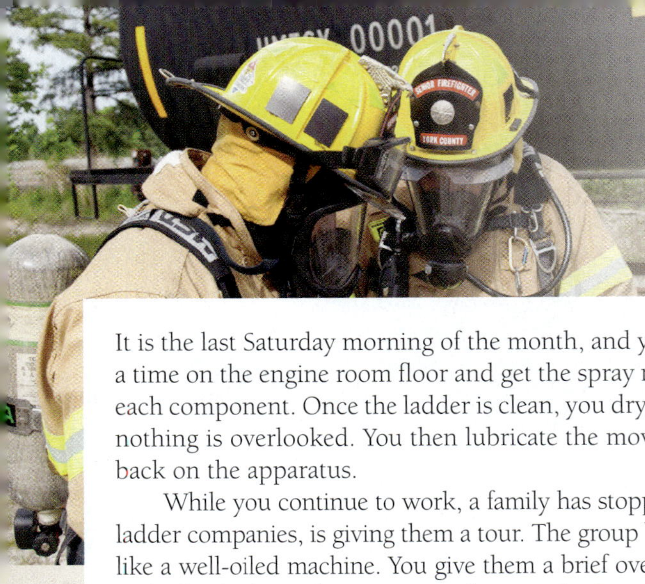

FIRE FIGHTER II *in action*

It is the last Saturday morning of the month, and you are performing your monthly visual inspection. You place one ladder at a time on the engine room floor and get the spray nozzle. You collect your soft-bristle brush and go to work carefully cleaning each component. Once the ladder is clean, you dry off the moisture and then carefully inspect each part of the ladder, ensuring nothing is overlooked. You then lubricate the moving parts as recommended by the manufacturer, before placing the ladder back on the apparatus.

While you continue to work, a family has stopped by the station. Fire Fighter Smith, who is filling in from one of the busier ladder companies, is giving them a tour. The group briefly stops to observe you working diligently to ensure your ladders operate like a well-oiled machine. You give them a brief overview of what you are doing, and the visitors tell you that they are surprised at the level of care that goes into cleaning. As they move on, you overhear Smith comment to the family that you work at a slow station and that real fire fighters do not have time to spend their days cleaning ladders.

1. Would you address the comment with the guests?
2. How would you address the issue with Smith?
3. Is this an issue you would address with your captain?
4. If you were transferred to a busy station, would you continue to clean, inspect, and maintain the ladders the same as you do at the slow station?

Search and Rescue

Fire Fighter I

Knowledge Objectives

After studying this chapter, you will be able to:

- Describe the mission of search operations. (NFPA 5.3.9, p 404)
- Describe the mission of rescue operations. (NFPA 5.3.9, p 404)
- Explain how search and rescue operations are coordinated with other fire suppression operations. (NFPA 5.3.9, p 404–407)
- Identify the factors to evaluate during a search and rescue size-up. (NFPA 5.3.9, p 405–407)
- Describe how to perform a risk–benefit analysis. (NFPA 5.3.9, p 405)
- Describe the factors that determine the level of risk faced by occupants. (NFPA 5.3.9.A, p 405–406)
- Explain how search operations are coordinated. (NFPA 5.3.9, p 407)
- List the priorities of search operations. (NFPA 5.3.9, p 407)
- Describe the objectives of a primary search. (NFPA 5.3.9.A, p 407–408)
- Describe the search patterns commonly used in search operations. (NFPA 5.3.9.A, p 409–410)
- Explain how thermal imagers are used during search operations. (NFPA 5.3.9.A, p 409)
- Describe how and when search ropes are utilized during search operations. (NFPA 5.3.9.A, p 413)
- Describe the role of a fire officer during search operations. (NFPA 5.3.9.A, p 414)
- Explain how a vent–entry–search is commonly performed. (NFPA 5.3.9.A, p 413)
- Describe the objectives of a secondary search. (NFPA 5.3.9.A, p 414)
- Describe three types of searches that are used in residential fires. (p 410–413)
- Describe one type of search that is used in large commercial fires. (p 413–414)
- Describe the advantages and disadvantages of conducting a standard search. (p 410)
- Describe the advantages and disadvantages of conducting an oriented search. (p 412)
- Describe the advantages and disadvantages of the oriented-vent-enter-isolate-search sequence (O-VEIS). (p 413)
- Describe the advantages and disadvantages of the team search. (p 414)
- Explain how fire fighters maintain safety through risk management. (NFPA 5.3.9, p 415–416)
- List the tools and equipment used in search and rescue operations. (NFPA 5.3.9.A, p 416)
- Describe the methods fire fighters use to determine whether an area is tenable. (NFPA 5.3.9.A, p 416)
- List the major types of rescue. (NFPA 5.3.9.A, p 417)
- Describe the concept of sheltering-in-place. (NFPA 5.3.9.A, p 417)
- Describe how to assist a victim to an exit. (NFPA 5.3.9.A, p 418)
- List the common types of simple victim carries performed during rescue operations. (NFPA 5.3.9.A, p 418)
- List the five emergency drags performed during rescue operations. (NFPA 5.3.9.A, p 420–427)
- Describe the conditions that may require a ground ladder rescue. (NFPA 5.3.9.A, p 428)

Skills Objectives

After studying this chapter, you will be able to:

- Conduct a standard search. (p 410–411)
- Conduct an oriented search. (p 412–413)
- Conduct an oriented-vent-enter-isolate search (O-VEIS). (p 413–414)
- Conduct a team search. (NFPA 5.3.9B, p 414)
- Conduct a secondary search. (NFPA 5.3.9B, p 415)
- Perform a one-person walking assist. (NFPA 5.3.9B, p 418–419)
- Perform a two-person walking assist. (NFPA 5.3.9B, p 418, 420)
- Perform a two-person extremity carry. (NFPA 5.3.9B, p 418, 421)
- Perform a two-person seat carry. (NFPA 5.3.9B, p 418, 422)
- Perform a two-person chair carry. (NFPA 5.3.9B, p 419, 423)
- Perform a cradle-in-arms carry. (NFPA 5.3.9B, p 419, 423)
- Perform a clothes drag. (NFPA 5.3.9B, p 421, 424)
- Perform a blanket drag. (NFPA 5.3.9B, p 421-422, 425)
- Perform a standing drag. (NFPA 5.3.9B, p 422, 426)
- Perform a webbing sling drag. (NFPA 5.3.9B, p 424, 427)
- Perform a fire fighter drag. (NFPA 5.3.9B, p 424, 427)
- Perform a one-person emergency drag from a vehicle. (NFPA 5.3.9B, p 424, 428)
- Perform a long backboard rescue. (NFPA 5.3.9B, p 424-425, 429)
- Rescue a conscious victim from a window. (NFPA 5.3.9B, p 429–432)
- Rescue an unconscious victim from a window. (NFPA 5.3.9B, p 431, 433)
- Rescue an unconscious child or small adult from a window. (NFPA 5.3.9B, p 432, 434)
- Rescue a large adult. (NFPA 5.3.9B, p 432, 434–435)

CHAPTER

14

Fire Fighter II — FFII

Knowledge Objectives

There are no knowledge objectives for Fire Fighter II candidates. NFPA 1001 contains no Fire Fighter II Job Performance Requirements for this chapter.

Skills Objectives

There are no skill objectives for Fire Fighter II candidates. NFPA 1001 contains no Fire Fighter II Job Performance Requirements for this chapter.

Additional NFPA Standards

- NFPA 1500, *Standard on Fire Department Occupational Safety and Health Program*

You Are the Fire Fighter

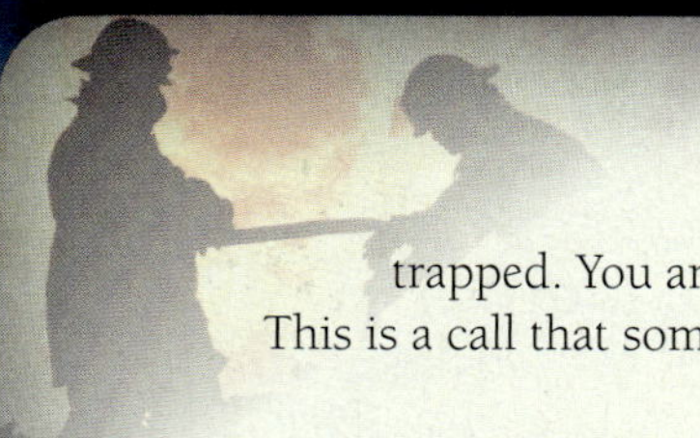

You arrive on the scene of a fast-moving apartment fire. The structure is a three-story, center-hallway, ordinary-constructed apartment building with fire on the second floor. The second- and third-floor hallways are smoke charged. Dispatch is reporting occupants trapped. You are assigned to conduct a primary search on the third floor while another crew begins the fire attack. This is a call that some fire fighters may not experience in their careers.

1. How will you go about making the search?
2. What will you do if you find a victim in the first unit you search?
3. Would you take a hose line with you on the search?

Introduction

The mission of the fire department is to save lives and protect property. Saving lives is always the highest priority at a fire scene, which explains why search and rescue are such important operations. The first fire fighters to arrive must always consider the possibility that lives could be in danger and act accordingly.

Saving lives remains the highest priority until it is determined that every occupant who was in danger has been found and moved to a safe location, or until it is no longer possible to rescue anyone successfully. The first situation occurs when a thorough search is performed and confirms that there are no occupants remaining to be rescued. The second situation occurs when fire conditions make it unlikely or impossible that any occupants could still be alive to be rescued.

Search and rescue are almost always performed in tandem, yet they are actually separate actions. The purpose of search operations is to locate living victims. The purpose of rescue operations is to physically remove an occupant or victim from a dangerous environment. Rescue occurs when a fire fighter leads an occupant to an exit or carries an unconscious victim out of a burning building or rescues the victim using a ladder. Both are examples of rescues because both the occupant and the victim were physically removed from imminent danger through the actions of a fire fighter.

Any fire department company or unit can be assigned to search and rescue operations. Many fire departments routinely assign this responsibility to ladder (or truck) companies, rescue companies, or to designated engine companies. All fire fighters must be trained and prepared to perform search and rescue functions.

Search and rescue operations must be conducted quickly and efficiently. A systematic approach will ensure that every occupant who can possibly be saved is successfully located and removed from danger. Fire fighters should practice the specific search and rescue procedures used by their departments.

Often, the first-arriving fire fighters will not know whether any occupants are inside a burning building or how many occupants might be inside. Fire fighters should never assume that a building is unoccupied. The only way to confirm that every occupant has safely evacuated a building is to conduct a thorough search. The decision to enter a burning building to search for living victims is one that may be made by the incident commander (IC), the officer in charge, or by a fire fighter. However, fire fighters need to recognize that a room that has flashed over is not survivable even for a fire fighter in full personnel protective equipment for more than a few seconds. A room that contains a ventilation-limited fire may, at first glance, appear to be a small incipient fire, but in reality it may be a larger fire in a ventilation-limited state. Fire fighters should not risk their lives when the conditions are not survivable for occupants or fully protected fire fighters.

Search and Rescue Operations

■ Coordinating Search and Rescue Operations with Fire Suppression

Although search and rescue is a top priority, it is rarely the only action taken by first-arriving fire fighters. Instead, the IC and fire fighters must plan and coordinate all fire suppression operations—from ventilation to fire attack—to support the search and rescue priority.

At times, the most effective initial step that can be taken to achieve an effective search and rescue effort is to slow down the growth of the fire or to extinguish it. Because smoke is fuel, cooling the stream of hot gases, particles, and aerosols in the room of origin can result in a dramatic slowing of the fire growth and a dramatic reduction in the temperature of a large part of the fire building. The application of a solid stream into the room of origin can result in dramatic cooling of not only the room of origin, but also of other parts of the building. This technique is called a transitional attack. It is an offensive fire attack initiated by an exterior, indirect handline operation into the fire compartment to initiate cooling while transitioning into an interior attack. The use of a transitional fire attack can be an important part of an effective search and rescue operation. You will learn how to conduct a transitional fire attack in the chapter, *Fire Attack and Foam*.

Any rescue effort must take into account the flow path of the fire. Any operation conducted in the flow path increases the risks posed to the fire fighter. Getting caught in the exhaust flow path that is carrying hot gases away from the fire can be deadly in a matter of seconds. Any change in any opening in the fire building, whether it is the intentional opening of a door, window, or roof, or an unintentional change produced by glass failing or a change in the wind direction can be deadly to both rescuers and persons trapped in the fire.

At times, fire fighters must take action to confine or control a fire before search and rescue operations can begin. For example, it might be necessary to position hose lines to keep the fire away from potential victims or to protect the entry and exit paths, so that the victims can be found and safely removed. The closing of doors may also limit fire growth and protect trapped occupants. In some cases, the best way to save lives is to control the fire with water application and eliminate the danger quickly.

Other fire scene activities also must be coordinated with search and rescue. For example, forcible entry is sometimes needed to provide entrances and exits for search and rescue teams. Sometimes, well-placed ventilation may be indicated. At other times, it may be best to limit ventilation by closing doors or by not opening windows. Portable lighting can also provide valuable assistance to interior search crews.

Often, the search for potential victims provides valuable information about the location and extent of the fire within a building. In essence, the searchers act as a reconnaissance team to determine which areas are involved and where a fire might spread. They report this information to the IC, who develops the overall fire suppression plan.

Search and Rescue Size-Up

The size-up process at every fire should include a specific evaluation of the critical factors for search and rescue—namely, the number of occupants in the building, their location, the degree of risk to their lives, and their ability to evacuate by themselves. Usually, this information is not immediately available, so fire fighters' actions must be based on a combination of observations and expectations. The type of occupancy; the size, construction, and condition of the building; the apparent smoke and fire conditions; and the time of day and the day of the week are all important observations that could indicate whether and how many people may need to be rescued.

A search and rescue plan can then be developed based on this information. Such a plan identifies the areas to be searched, the priorities for searching different areas, the number of search teams required, and any additional actions needed to support the search and rescue activities. One search team might be sufficient for a small building, whereas multiple teams might be needed to search a large building.

FIRE FIGHTER Tips

Although search teams both search for and rescue victims, the preferred term for these teams is simply "search team." During a technical rescue incident, teams that search and rescue are called "search and rescue teams."

Risk–Benefit Analysis

Plans for search and rescue must take into consideration both the risks and the benefits of the operation. In some situations, search and rescue operations must be limited or cannot be performed at all. For example, if a building demonstrates potential backdraft conditions, the risk to fire fighters might be too high and the possibility of saving any occupants inside the structure might be too low to justify sending a search and rescue team into the building FIGURE 14-1. A similar decision might be made if the fire occurs in an abandoned building or a lightweight

FIGURE 14-1 The IC must weigh the risk to the fire fighters against the possibility of saving any occupant inside before authorizing an interior search.

construction building in danger of structural collapse. In other situations, it might not be possible to conduct an interior search until the fire has been extinguished.

Occupancy Factors

When a building is known or believed to be occupied, fire fighters should first rescue the occupants who are in the most immediate danger, followed by those who are in less danger. Search teams should be assigned on the basis of these priorities. Several factors determine the level of risk faced by the occupants of a burning building—namely, the location of the fire within the building, the direction of the fire spread, the volume and intensity of the fire, and smoke conditions in different areas of the structure.

In establishing search priorities, fire fighters should also consider where occupants are most likely to be located. Occupants who are close to the fire, above the fire, or in the path of fire spread are usually at greater risk than occupants who are located farther away from the fire. An occupant in the apartment where the fire started is probably in immediate danger, whereas the occupants of a lower or upper floor may be relatively safe. This decision will be somewhat dependent on the type of construction used in the fire building.

Building occupants who are at the windows or on balconies calling for help obviously realize that they are in danger and want to be rescued. There may be other occupants in the building who cannot be seen from the exterior, however—perhaps because they are asleep, unconscious, incapacitated, or trapped. A search must be conducted to locate these individuals before they can be rescued.

Occupants who are asleep are at greater risk than those who are awake. Young children and elderly people, who are more likely to need assistance to evacuate, are at greater risk than adolescents and adults. Individuals who are confined to a bed or wheelchair are at greater risk than occupants who can move freely. Consequently, more fire fighters will be needed for search and rescue operations in a nursing home than in an office building, where the occupants can usually make their way out unassisted.

A fire in a single-family dwelling generally places the members of one family at risk, whereas a fire in a large apartment building may endanger several families. Generally, the life risk in residential occupancies is higher at night and on weekends, when more people are at home. Risk is greatest late at night, when the occupants are probably asleep. A fire that occurs in a residential occupancy on a weekday afternoon would present a significantly different rescue problem from a fire in the same location at 2:00 a.m.

An office building that is fully occupied on a weekday afternoon is typically unoccupied or has a low occupancy at midnight. A warehouse fire probably presents a direct risk to few occupants. Conversely, a nightclub fire on a Friday or Saturday night could endanger hundreds of lives. The risk posed to victims may also increase depending on the building construction: Occupants of an unprotected, wood-frame building are in greater danger than those in a building with fire-resistive construction and automatic sprinklers.

Fire Fighter Safety Tips

Search and rescue size-up considerations include the following factors:

- Occupancy
- Size of the building
- Construction of the building
- Time of day and day of week
- Number of occupants
- Degree of risk to the occupants presented by the fire
- Ability of occupants to exit on their own

Known and obvious rescue

Observations

Fire fighters should never automatically assume that a building is unoccupied. An observant fire fighter notices clues that indicate whether a building is occupied and how many occupants are likely to be present. The initial 360-degree size-up of an incident can provide valuable information on the probability of finding occupants, the number of occupants, and the most likely location of any occupants.

For example, the first-responding unit at a residential fire might see two cars in the driveway and numerous toys on the front lawn. These signs indicate that the house is probably occupied by a family with children. In such a situation, the IC would likely assign several search teams and emphasize the importance of quickly searching every room where victims might be located. Conversely, an absence of cars, locked doors, and an overflowing mailbox would suggest that the house is vacant. If there is a chance that savable victims might be present, the IC might order a search, in case someone is inside, but the assignment might go to a single search team FIGURE 14-2.

Similar observations can be made for other structures. Although an empty parking lot does not guarantee that a building is empty, the presence of a car probably indicates that at least one maintenance worker or security guard is inside—but it might also mean that a car is simply parked in the lot. Boarded-up windows and doors on a building surrounded by a chain-link fence are indications that the building is probably unoccupied.

Occupant Information

Fire fighters must make sure that everyone is accounted for. An important question to ask is, "Are you sure that no one else was home?" The dispatcher may relay information on the status of the occupants. Neighbors are also another source of information about the status of the occupants.

It is often difficult to obtain accurate information from occupants who have just escaped from a burning building, especially if they believe that family members, close friends, or even a beloved pet might still be inside. Occupants may be too emotional to speak or even to think clearly. If someone says that he or she thinks occupants are still inside, fire fighters should obtain as much information as possible. Ask specific questions, such as "Who is still inside?" or "Where is your son's room located?" and "Where was he when you last saw him?" It is always easier to find someone if you know for whom you are looking and where to look. Many fire fighters can tell stories about searching for a missing baby, only to discover that "Baby" is the family dog.

Building Size and Arrangement

The size and arrangement of the building are other important factors to consider in planning and conducting a search. A small, one-story house is simpler to search than a larger building with multiple units or a large commercial building. A large building

A. B.

FIGURE 14-2 Exterior observations can often provide a good indication of whether a building is occupied. **A.** Parked cars indicate the residence is occupied. **B.** Boarded-up windows indicate the residence is unoccupied.

FIRE FIGHTER Tips

Do not consider all of the information gathered from neighbors and bystanders to be accurate. In the stress of the moment, well-intentioned people may make incorrect statements. Do not rush into an empty burning house because of a bad tip from a bystander.

with many rooms must be searched in a systematic fashion. Access to an interior layout or floor plan is often helpful when planning and assigning teams to search a building. To ensure that each area is searched promptly and that no areas are omitted, specific areas must be assigned to different search teams in an organized manner. Assignments are often based on the stairway locations, corridor arrangements, and the apartment or room numbering system. Teams must report when they have completed searching each area. Even though the complexity of the search increases as a building gets larger, searchers need to be aware of the fire flow path in the building and avoid actions that could place them in a dangerous position. Practice good door control at the point of entry into the building. Close the door to each room after you have completed searching that room.

Because it is difficult to determine this information in the midst of an emergency, fire fighters often conduct preincident surveys of buildings such as apartment complexes and offices. They assemble information about the building during the preincident survey and use it to prepare preincident plans, so they will be prepared when an emergency occurs.

Preincident plans can include a variety of valuable information:

- Corridor layouts
- Exit locations
- Stairway locations
- Apartment layouts
- Number of bedrooms in apartments
- Locations of handicapped residents' apartments
- Special function rooms or areas

Fire fighters should note how the floors of a building are numbered. Buildings constructed on a slope may appear to have a different number of levels when viewed from the "A" or front side than when viewed from the "C" or back side. For example, a fire may appear to be on the third floor to a fire fighter standing in front of the building, while a fire fighter at the rear of the building might report that it is on the fifth floor. Without knowing how floors are numbered, fire fighters can be sent to work above a fire instead of below it. This situation resulted in multiple fire fighter deaths in Pittsburgh several years ago.

Search Coordination

The overall plan for the incident must focus on the life-safety priority as long as search and rescue operations are still under way. As soon as all searches are complete, the priority can shift to controlling and extinguishing the fire.

The IC makes search assignments and designates a fire officer to be in charge of the search effort (search officer). The search officer keeps the IC informed of the status and results of the search effort. As the search team completes its search of each area, the search officer must notify the IC of the results. An "All clear" report indicates that an area has been searched and all victims have been removed.

If fire fighters who have been assigned to perform some other task discover a victim or come upon a critical rescue situation, they must notify the IC immediately. Because life safety is always the highest priority, the IC may send a rescue team to meet the search team so that the search team does not lose the continuity of its search.

Another critical aspect of search coordination is keeping track of everyone who was rescued or who escaped without assistance. This information should be tracked at the command post so that reports of missing occupants can be matched to reports of rescued victims. Fire fighters should also conduct an exterior search for any missing occupants. An occupant who escaped from the fire may be lying unconscious on the ground nearby or in the care of neighbors, for example. An occupant who jumped from a window could be injured and unable to move.

Search Priorities

A search begins in the areas where victims are at the greatest risk. One or two search teams can usually go through all of the rooms in a single-family dwelling in 15 minutes or less. Multiple search teams and a systematic division of the building are needed to cover larger structures such as apartment buildings. Area search assignments should be based on a system of priorities:

- The first priority is to search those areas where live victims may be located immediately around the fire, and then the rest of the fire floor.
- The second priority is to search the area directly above the fire and the rest of that floor.
- The next priority is higher-level floors, working from the top floor down, because smoke and heat are likely to accumulate in these areas.
- Generally, areas below the fire floor are a lower priority.

These priorities are guidelines and need to be modified based on occupancy, construction type, and the circumstances at that incident.

At a high-rise building fire, the IC might assign two or more search teams to each floor. The search teams must work together closely and coordinate their searches to ensure that all areas are covered.

Primary Search

Two types of searches are performed in buildings: primary and secondary. A primary search is a quick attempt to locate any potential victims who are in danger. This search should be as thorough as time permits and should cover any places where victims are likely to be found. Fire conditions might make it impossible to conduct a primary search in some areas, or they may limit the time available for an exhaustive search. A primary search should be completed in 15 minutes or less.

A secondary search is conducted after the fire has been suppressed. At this point, any remaining occupants have most likely been exposed to conditions that cannot support life, such as flashover. The main purpose of the secondary search is to find deceased victims. If possible, it should be conducted by a different search team, so that each area of the building is examined with a fresh set of eyes. The secondary search is covered in more detail later in this chapter.

During the primary search, fire fighters rapidly search the accessible areas of a burning structure where conditions may still permit human life to exist. The objective of this search is to find any potential victims as quickly as possible and to remove them from danger. When fire fighters complete the primary search, they have gone as far as they can and have removed anyone who was rescued. The phrase "Primary all clear" is used to report that the primary search has been completed.

By necessity, the primary search is conducted quickly and gives priority to those areas where victims are most likely to be located. Time is a critical concern, because fire fighters must reach potential victims before they are burned, overcome by smoke and toxic gases, or trapped by a structural collapse. Search teams may have only a few minutes to conduct a primary search. In that limited amount of time, fire fighters must try to find anyone who could be in danger and remove those individuals to a safe area. Active fire conditions may limit the areas that can be searched quickly as well as the time that can be spent in each area.

Fire fighters should try to check all of the areas where victims might be, such as beds, cribs, and sofas. Adults who tried to escape on their own are often found near doors or windows. Some people—particularly children—may try to hide in a closet, in a bathtub or shower, or under a piece of furniture. Be aware that young children may be frightened by the appearance of a fire fighter wearing full personal protective gear.

The primary search is frequently conducted in conditions that expose both fire fighters and victims to the risks presented by heavy smoke, heat, structural collapse, and entrapment by the fire. Search teams must often work in conditions of zero visibility and may have to crawl along the floor to stay below layers of hot gases. Because the beam from a powerful hand light might be visible only a few inches inside a smoke-filled building, search teams must practice searching for victims in total darkness. They must know how to keep track of their location and how to get back to their entry point. Practicing these skills in a controlled environment will enable fire fighters to perform confidently under similar conditions at a real emergency incident.

Fire fighters must rely on their senses when they search a building:

- Sight: Can you see anything?
- Sound: Can you hear someone calling for help, moaning, or groaning?
- Touch: Do you feel a victim's body?

If smoke and fire restrict visibility, fire fighters must use sound and touch to find victims. Every few seconds a member of the search team should yell out, "Is anyone in here? Can anyone hear me?" Then listen. Hold your breath to quiet your breathing regulator and stop moving. People who have suffered burns or are semiconscious may not be able to speak intelligibly, so you will need to listen for guttural sounds, cries, faint voices, groans, and moans. Focus on the direction of any sounds you hear.

As you search, feel around for the hands, legs, arms, and torsos of potential victims. Use side crawls to extend your reach sideways; sweep ahead and to the sides as you crawl. Practice until you can tell the difference between a piece of furniture and a human body. The type of furniture you encounter may also provide clues for finding victims. Spindle-type legs in a bedroom may indicate a crib that needs to be searched. An unusually low bed may be the bottom berth of a children's bunk bed.

Zero-visibility conditions can be disorienting. In such circumstances, fire fighters must follow walls and note turns and doorways to avoid becoming lost. The use of a thermal imager can be beneficial in improving the speed and efficiency of a search. After locating a victim or completing a search of an area, the search team must be able to retrace its path and return to its entry point. Search teams must also identify a secondary escape route in case fire conditions change and block their planned exit route. Thus, during the search, fire fighters should note the locations of stairways, doors, and windows. Always be aware of the nearest exit and an alternate exit. It is also important to know the location of interior doors in the case that fire fighters need to isolate themselves from rapid fire growth.

If the situation deteriorates rapidly, a window can serve as an emergency exit from a room. Fire fighters should keep track of the exterior walls and window locations, even reaching up to feel for the windows if conditions force them to crawl.

Fire fighters must remain in voice contact with someone outside the room and must be able to state their location at all times, particularly if they need assistance. The search team members must be able to give their location, including the building section and floor, to the IC if they need assistance. If the search team needs a ladder for the rescue, the IC will need to know where the ladder should be placed (which side of the building) and how long a ladder will be needed (which floor). Fire fighters should use standard incident command system terminology to describe their location over the radio.

Fire Fighter Safety Tips

While searching for victims, you also need to focus on your safety. Continuously monitor your air supply and the conditions of the fire and the structure. You should also be aware of the status of other crews, such as the fire attack crew and the ventilation crew.

General Search Techniques

Fire fighters should employ standard techniques to search assigned areas quickly, efficiently, and safely. They should operate in teams of two or more and should always stay close together **FIGURE 14-3**. Partners must remain in direct visual, voice, or physical contact with each other.

At least one member of each search team should also have a radio to maintain contact with the command post or someone outside the building. The search team can use this radio to call for help if they become disoriented, trapped by fire, or need assistance. If the search team finds a victim, they can notify the IC so that help will be available to help remove the victim from the building and begin medical treatment. The

FIGURE 14-3 Search teams always include at least two members.

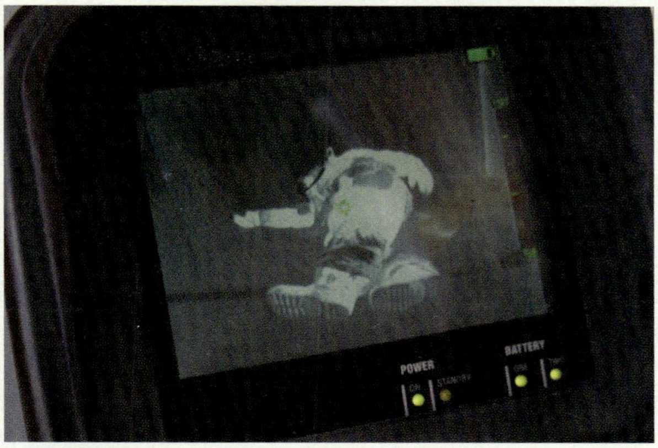

FIGURE 14-4 A thermal imager can capture an image of a person through thick smoke.

search team must notify the IC when the search of each area has been completed so the IC can make informed decisions about which steps to take next.

Rescue teams perform two types of searches, a primary search and a secondary search. A primary search is a quick attempt to locate any potential victims who are in danger. The primary search should be efficient and as thorough as possible. It should cover any places that can be safely searched where a victim might be located. Fire conditions may make it impossible or unsafe to search in some parts of a building. Changing fire conditions will limit the time available for a primary search.

Thermal Imagers

A thermal imager (TI) is a valuable tool for conducting a search in a smoke-filled building. This piece of equipment is similar to a video camera, except that it displays images of heat emitted from surfaces instead of visible light images. The images appear on a display screen and show the relative amounts of heat being radiated by different objects. It is important that you learn how to properly set any controls on your thermal imager and practice operating it in order to be proficient in its use during an emergency situation.

A major benefit of using a thermal imager is that it can "see" the heat signature of a person in conditions of total darkness or through smoke that totally obscures normal vision FIGURE 14-4. Fire fighters will be able to identify the shape of a human body with a thermal image scan because the body will be either warmer or cooler than its surroundings.

Temperature differences also mean that the thermal imager can show furniture, walls, doorways, and windows. This information may enable a fire fighter to navigate through the interior of a smoke-filled building. The TI can sometimes be used to locate a fire in a smoke-filled building or behind walls or ceilings. A thermal image scan of the exterior of a building can sometimes locate the fire source and the direction of fire spread. Scanning a door before opening it may indicate whether the room is safe to enter.

Several types of thermal imagers are available. Some are hand-held devices, similar to a hand-held computer with a built-in monitor. Others are helmet-mounted devices that produce an image in front of the face piece of the user's self-contained breathing apparatus (SCBA). Fire fighters need training and practice to become proficient in using these devices, interpreting the images, and maneuvering confidently through a building while wearing a thermal imager.

One of the limitations of TIs is that they reveal their images only to the person who is viewing them. This means that the viewer has a visual image that is not imprinted on the minds of the other search team members. This places the other members of the team at a disadvantage. The person operating the TI communicates this information to the search team members by telling them what is visible on the camera screen. This can create a difficult situation if the TI fails and leaves the team members without this special set of eyes. It can also create a dangerous situation if the TI and the route of egress is not clear to the team members because they have been relying on the TI for their directions. It is important to use good search methods and to encourage all team members to remain aware of their surroundings.

Remember that the TI does not "see" fire. It measures the heat flux; that is, the heat that is given off by an object. It shows differences in the amount of heat given off by different objects or from different parts of the room. By itself, it does not show you the integrity of a structure. Be alert for areas that show small heat signatures (areas of higher temperature), which may indicate that a floor is not safe to walk on or an area of roof that is unsafe to be under. If the TI gives any indication that might indicate an unsafe condition, make sure that a thorough exterior size-up has been conducted.

FIRE FIGHTER Tips

Not all rooms have just four walls; some may have five or more. To determine how many walls a room has, check for walls on each side of the door as you enter the room. If there is a wall parallel to the door, on the right of the door, and to the left of the door, then you will need to search five walls. If there is a parallel wall only to the right or to the left of the door, then the room has four walls. Noting the number of walls in a room about to be searched is a good way to ensure that the entire room has been searched and to maintain orientation.

Four Standard Methods of Searching

Four standard methods of searching are described here. They are the standard search, the oriented search, the oriented-vent-enter-isolate-search (O-VEIS), and the team search. It is important to understand how each is conducted, the advantages and disadvantages of each, and where each is most appropriately used.

Although the search methods differ, the tasks of the search team remain the same. There are several factors to be addressed by the search team beyond locating victims. First and foremost, the team members must keep themselves safe. They need to carefully monitor their air supplies to ensure they know when to exit in order to be out of the IDLH zone before their low-air alarm sounds. They also need to monitor changes in the fire conditions, to make sure their position is not in danger of an immediate flashover, and to monitor changes in the level and characteristics of smoke, the heat level, and any changes in the location of flames. Possible indications of structural changes, such as unusual sounds or sagging, must be noted and reported to the IC. Changes in the fire suppression activities are very important, too. The search team needs to know if the fire attack team is making progress, or if they are having difficulty in establishing a water supply or advancing the attack line. The search team members must have a plan for conducting a search that will maximize their efforts. Perhaps most importantly, they need to keep track of the progress they are making in completing the search.

FIGURE 14-5 To conduct a left-handed search pattern, locate the wall with your left hand as you enter the room. Turn right at each corner. This will keep the wall on your left as you search around the room. Keep your body at a right angle to the wall you are searching.

FIGURE 14-6 To conduct a right-handed search pattern, locate the wall with your right hand as you enter the room. Turn left at each corner. This will keep the wall on your right as you search around the room. Keep your body at a right angle to the wall you are searching.

■ Standard Search

The standard search is conducted with a search team of two or more members. All members of the search team cover the same area together. Two fire fighters enter a room together. The first fire fighter proceeds to the left, around the outside wall while keeping his or her left shoulder or hand close to that wall. This is called a left-hand search **FIGURE 14-5**. At the same time the second fire fighter proceeds to the right and keeps his or her right shoulder or hand close to the wall on the right. This is called a right-hand search **FIGURE 14-6**. The two searchers meet halfway around the room. Each searcher has covered about one half of the perimeter of the room. At this point the searchers can continue around the rest of the way around the room and do a quick re-search of the area initially searched by the other searcher or each one can each retrace their steps to return to their original starting point. Once they have returned to the door by which they entered, they proceed to the next room. If the room is large, it may be necessary for each rescuer to perform side crawls to check the center parts of the room. This is done by keeping your body at right angles to the wall being searched at all times, and moving away from the wall with two or three side crawls. The searcher then crawls back to the wall and moves forward.

Some departments advocate that each searcher should carry a tool to extend their reach. Others believe that the tool reduces the searcher's ability to feel the texture and composition of an object. The advantages and disadvantages of using a tool when searching are discussed later in this chapter.

The standard search is the most commonly taught search method used by fire fighters. The primary use of the standard search is in residential fires, where two fire fighters should be able to search an average-sized house in 15 minutes or less.

An advantage to the standard search is that it keeps two searchers together and in voice contact of each other. Disadvantages of the standard search include the following: each searcher must be responsible for monitoring his or her own air supply, and monitoring his or her safety as related to changes in the fire situation, structural stability, and other fireground operations. This has the tendency to divide the searchers' attention. It is likely that these searchers may not notice deteriorating conditions. Also, they may fail to monitor the status of their own air supplies. Another disadvantage is that if a victim is located, the searchers will exit with that victim and the continuity of the search is more likely to be lost than if using the oriented-team search method. This means that it is hard to make and follow a systematic plan for searching that does not miss areas that need to be searched.

Fire Fighter Safety Tips

There are two additions to the classic vent–enter–search method that can make this procedure safer:

1. Isolate the fire. Be sure to close the door between the room and the hallway as soon as you enter the room. According to recent National Institute of Technology and Standards studies, this step will help to keep the room more tenable for a few minutes while you perform a quick search of the room.
2. Use a team of two fire fighters when performing O-VEIS. One fire fighter enters the room to isolate it and perform the search. The second fire fighter remains outside the window through which entry was made; this fire fighter monitors the fire conditions and can warn the inside fire fighter of changing fire conditions. The second fire fighter can also assist in rescuing any victims who are found.

Conducting a Standard Search

Follow the steps in SKILL DRILL 14-1 to perform a standard search of a residential structure and rescue victims:

1. Don your personal protective equipment (PPE), including SCBA, and enter the personnel accountability system.
2. Bring hand tools, a hand light, a radio, a thermal imaging device, and search ropes if indicated.
3. Notify command that the search is starting, and indicate the area to be searched and the direction of the search.
4. Use hand tools or ground ladders if needed to gain access to the site.
5. Conduct a quick and systematic search by staying on an outside wall and searching from room to room.
6. Maintain contact with an outside wall. (STEP 1)
7. Maintain team integrity using visual, voice, or direct contact.
8. Use the most efficient movement based on the hazard encountered: duck walk, crawl, stand only when you can see your feet and it is not hot, and use tools to extend your reach.
9. Clear each room visually or by touch, then close the door.
10. Search the area, including stairs up to the landing on the next floor. (STEP 2)
11. Periodically listen for victims and sounds of fire.
12. Communicate the locations of doors, windows, and inside corners to the other team members.
13. Observe fire, smoke, and heat conditions; update command on this information.
14. Locate and remove victims; notify the IC.
15. When the search is complete, conduct a personnel accountability report.
16. Report the results of the search to your officer. (STEP 3)

SKILL DRILL 14-1 Conducting a Standard Search

(Fire Fighter I, NFPA 5.3.9)

1 Notify command that the search is starting, and indicate the area to be searched and the direction of the search. Conduct a quick and systematic search by staying on an outside wall and searching from room to room. Maintain contact with an outside wall.

2 Maintain team integrity using visual, voice, or direct contact. Use the most efficient movement based on the hazard encountered: duck walk, crawl, stand only when you can see your feet and it is not hot, and use tools to extend your reach. Clear each room visually or by touch, then close the door. Search the area, including stairs up to the landing on the next floor.

3 Periodically listen for victims and sounds of fire. Communicate the locations of doors, windows, and inside corners to other team members. Observe fire, smoke, and heat conditions; update command on this information. Locate and remove victims; notify the IC. When the search is complete, conduct a personnel accountability report. Report the results of the search to your officer.

Oriented Search

An oriented search is conducted using a search team consisting of an officer and one to three searchers. The officer remains outside the individual rooms that are being searched and a single searcher systematically searches one room at a time. At first glance this seems like it would decrease the efficiency of the search; however, consider that the responsibilities of the search team include all of the following activities: maintaining their safety, monitoring their air supplies, monitoring the progression of the fire and the fire suppression efforts, developing a systematic search plan, assessing the progress of the search effort, monitoring the activities on the rest of the fire scene, and coordinating activities with the incident commander. The oriented search method assigns all the above responsibilities to the team leader or officer. It assigns the search of one room at a time to each of the other team members. It provides a framework for accomplishing a systematic and efficient search. The oriented method of searching is well suited to searching a residential setting.

The safety of the crew is maintained by the officer monitoring changes in the search area. By staying outside the rooms being searched, the officer is in a better position to note these changes than the crew members who are searching inside individual rooms. The officer can watch for changes in the fire conditions and in the structural integrity of the building and periodically monitor the status of the air supply of all team members. The officer also keeps track of the progress of the search, which rooms have been searched, and which are yet to be searched. The officer must maintain voice contact with each member of the search team. This method enables the fire fighters who are searching to concentrate all their attention on the task of searching their assigned area. The officer controls the door and monitors the means of exit for crew members. If possible, there should be a primary exit and a second one that can be used if the first one becomes untenable.

Follow the steps in SKILL DRILL 14-2 to conduct an oriented search:

1. Assemble an oriented search team consisting of one officer and one to three searchers.
2. Don your personal protective equipment (PPE), including SCBA and hand tools.
3. Enter personal accountability system.
4. Notify the incident commander that search is starting.
5. Proceed to the area to be searched.
6. The officer remains outside the room to be searched to monitor safety conditions, air supplies, and the status of the fire. The officer maintains a systematic search pattern and coordinates activities with the incident commander.
7. The officer directs each searcher to proceed to one room to be searched.
8. Searchers use a left-handed or a right-handed search pattern and perform two to three side crawls to extend the search into the room.
9. Upon completion of the search of each room, the officer directs searchers to the next rooms to be searched, and closes the door.
10. If a victim is found, the officer notifies the incident commander and requests a second team to help remove the victim. The oriented search team moves the victim toward the exit and turns over the care and removal of the victim to the second team wherever they meet.
11. The oriented search team then returns to the last location searched and continues the systematic search of the building.
12. When search is complete, the officer conducts a personal accountability report.
13. The officer reports results of search and the personal accountability report to the incident commander.

As a team member enters a room to be searched, the searcher begins a right-handed or left-handed search of the room. It is important to remain perpendicular to the wall as you search. When necessary, the searcher can side crawl two or three crawls from the wall, then side crawl back to the wall and move forward to repeat this process again. This method enables a single searcher to conduct a quick search of most bedrooms in a short period of time.

The searched room or apartment should be marked so other personnel will know that it has been searched. Chalk, crayons, felt tip markers, or masking tape can be used to mark the door for this purpose. Some fire departments use a two-part marking system to indicate when a search is in progress and when it has been completed: In this system, a slash ("/") indicates that a search is in progress, and an "X" indicates that the search has been completed FIGURE 14-7. Other fire departments place an object in the doorway or attach a tag or latch strap to the doorknob to indicate that the room has been searched.

Continuity of the search can be easily lost when a victim is found and removed from the hazardous environment. When this happens, it is hard to resume the search from the same location. By the time a second team has been assembled and arrives at the location where the victim was found, it is hard to pick up the search without duplication of efforts

FIGURE 14-7 In large buildings, doors should be marked after the rooms are searched.

and wasting valuable time. In switching from one team to another, it is easy to miss areas that need to be searched. The most efficient way to maintain the continuity of the search is for the search officer to notify command that a victim has been found, move toward the exit with the victim, and be met by a team that can remove the victim the rest of the way out of the building and begin emergency medical care. This approach enables the search team to return to the location they were searching and continue the search for additional victims without losing much time and without losing the continuity of the search. This operation requires planning, lots of practice and a well-coordinated fire suppression operation.

The advantages of an oriented search are that it provides a greater degree of safety for all crew members, and it has been shown by the studies done in Toledo, Ohio to take less time than using the standard search method. An oriented search requires a well-trained search officer who concentrates on the planning and implementation of the search plan, and searchers who have practiced this technique until they are proficient.

Oriented Vent Enter Isolate Search (O-VEIS)

The original vent-enter-search method was developed for situations where there was a porch located in front of bedroom windows and a person in one of these bedrooms needed to be rescued. In the beginning, vent, enter, and search was used in extreme situations in which there was a shortage of personnel and one fire fighter could quickly place a ladder to the porch roof, quickly open or break the bedroom window, enter the bedroom, perform a quick search of the room, and exit back onto the porch roof. Many times it was possible to access a second bedroom from the same porch, thereby searching two bedrooms in a short period of time. This evolution is very dangerous for several reasons. Opening a bedroom window is a type of ventilation, which adds oxygen to the fire, contributing to a flashover or directing the hot fire gases toward the open window. Vent, enter, search violates the two-in/two-out rule, and because it is often done before any hose line is in place, the fire may expand rapidly. VES is a very dangerous procedure and should not be used. A safer modification of VES is the oriented-vent-enter-isolate-search (O-VEIS). This method requires a team of two rescuers and stresses the need to quickly close the door between the room to be searched and the rest of the house.

To conduct an oriented-vent-enter-isolate-search, a team of two fire fighters places a ladder in front of the window leading to the room or to the porch in front of the room to be searched. They then climb to the porch roof and assess the situation. After assessing the room to determine if it appears to be tenable for a victim and for the fire fighter, they open or remove the window. The officer or person in charge remains outside to continue to assess the situation in the room, maintain radio contact, and monitor activities in the other parts of the fire ground FIGURE 14-8.

The second fire fighter quickly enters the room. As soon as possible, this fire fighter locates the door to the hallway and immediately closes it. This action, according to recent NIST studies, isolates the room to be searched from the fire flow path. Even a thin hollow core door will provide some protection from the flow path for a short period of time. This isolation reduces the heat and smoke in that room for a few minutes and decreases the chance that the room will quickly become untenable for the rescuer and any victims present. It prevents unplanned ventilation, which may rapidly accelerate the fire growth; it also improves visibility in the bedroom and reduces the chance for flashover.

FIGURE 14-8 In an oriented-vent-enter-isolate-search (O-VEIS), the leader remains outside to monitor the fire and safety conditions and to assist the inside searcher.

If a victim is found, the officer can assist in removing the victim to the porch. If a victim is found and there is another bedroom to be searched from the same porch, it may be more efficient for the officer to request a second rescue crew to remove the victim from the porch roof so the original search team can begin to search the second bedroom. An oriented-vent-enter-isolate-search is somewhat safer and more efficient than the original vent-enter-search scenario. This is not to say that it is a safe procedure. It should be considered only when the room is tenable for a potential victim. It should not be attempted if the room is fully charged with smoke and in danger of flashing over when additional oxygen is introduced. O-VEIS should be used only in dire emergencies in which a sizeable risk has a large potential benefit. If no porch is present, the oriented-vent-enter-isolate-search method can be performed by placing a ladder to the window in the room to be searched. In this case, the second rescuer remains on the ladder to monitor the situation.

Team Search

A team search, also called a rope search, is a method of searching that uses a search rope or hose to provide searchers with an anchor point to keep them in contact with each other and with the egress route from a burning building. The rope plays out as the team progresses and keeps them in contact with a route out of the building. It is anchored about 10 feet outside the exit door to give the searchers an identifiable route out of the building. If the direction of travel inside the building needs to be changed, the rope is anchored to a stationary object at a height of 1 1/2 to 2 feet above the floor to make it easy to find and to follow.

This method of searching is more complicated than the three methods above. Its use requires a lot of training and coordination. Team searching is well suited for locating and removing lost occupants or injured fire fighters. It is a valuable tool for a RIT working to locate a missed, downed, or trapped fire fighter especially in a large space. Its use enables other fire fighters to locate the search team quickly. It can be a valuable method to locate a fire in large warehouse and other large open structures when the visibility is severely limited. It helps fire fighters keep in contact with the exit in a building that contains cold smoke. Cold smoke is generated when a fire sprinkler system is activated during a fire. The sprinkler cools the smoke and causes it to bank down to the floor. Cool smoke and condensed water droplets severely limit visibility.

A rope search is not often used in a residential fire unless the house has large open spaces or unless the house is very large. It is a time-consuming method. When a rope is used as a search tool, knots are sometimes placed every 20 feet to enable searchers to know how far they have progressed into the building. The number of knots indicates the number of feet from the end of the rope. For example, four knots would equal 80 feet of travel into the building. Some fire departments prefer to use a fire hose because it is larger and easier to locate when visibility is limited. Also fire hose is always available. In cases of emergency, a fire fighter can determine which direction leads to the exit by identifying the male and female couplings.

The team consists of an anchor person who ties off the line to a stationary object that is located outside the building. The anchor person feeds the line out as the team moves into the structure, and lets the team members know how far they have progressed into the building. This person keeps slight tension on the line at a comfortable height above the floor (about 18 to 20 inches), and secures the rope to a stationary object if the team changes direction within the building.

The officer of the search team is responsible for developing a search plan, keeping the team safe, and communicating with the incident command. The officer stays in contact with the rope and maintains voice contact with the searchers. This helps to provide them with an orientation. The officer ensures that each member is monitoring his or her air supply. This officer is usually responsible for operating the thermal imaging camera and directing the team based on the images seen on the screen. The officer must also assess any changing conditions around the search team, and communicate these changes to the incident commander.

The members of the team are responsible for following the rope into the building and searching off the rope. This is done by placing one searcher on each side of the rope. When searching large areas, each of the searchers performs two or three side crawls at a right angle to the rope. The searcher on the left side of the rope side crawls to the left of the rope and the searcher on the right side of the rope side crawls to the right of the rope. Some departments use a short lateral rope that clips to the main rope. This enables them to move a limited distance from the main search rope.

When the search team is ready to exit, the officer should assume the anchor position on the search line and maintain tension on the line as the team works its way along the line to the building exit. The officer assumes this position in order to remain accountable for the other members of the team.

The advantages of using a team search include the following factors. It keeps searchers together, preventing individuals from getting lost or separated. It keeps searchers oriented to the exit and provides a safe and direct route out of the building. It provides a safe way for searching large areas, especially where there are limited landmarks to provide directional guidance. It allows a rescue team to find the searchers if they need to enter to remove a victim who has been located by the search team. And it provides a quick exit any time it is needed by the search team.

There are several disadvantages to the team search. One is that it relies solely on the search rope. This may result in more time spent holding onto the rope than engaging in active searching. It is also hard to maintain the continuity of the search if a victim is found and removed. If the search team locates a victim and leaves to remove the victim, it will be hard for a new crew to return to the same place to continue a systematic search.

All search methods require a thorough training program that includes sufficient practice in order for you to perform efficient and effective searches. And you need regular practice to maintain proficiency.

Secondary Search

A secondary search is conducted as soon as the fire is under control or fully extinguished and there are sufficient resources available **FIGURE 14-9**. The secondary search is a slower systematic search to locate deceased victims. It is conducted slowly and methodically to ensure that no areas are overlooked. Areas that could not be covered during the primary search are covered during the secondary search. There are fewer risks to fire fighters

FIGURE 14-9 A secondary search is conducted after the fire is under control.

during the secondary search than during the primary search. Once the fire is under control, conditions during the secondary search should be much different, ventilation should have cleared out most of the smoke, and visibility should be greatly improved. Lights can be brought and set up to make it easier to see the area that is being searched. It is important to remain alert for conditions indicating an unstable structure. Watch for signs of a fire rekindle. Continue to use SCBA until the air quality has been thoroughly tested. Air that looks clear can contain high concentrations of poisonous gases and small suspended particles of toxic byproducts of the fire. One recent UL study showed that 97% of these particulates were invisible to the human eye.

The secondary search should be conducted in an orderly, systematic manner to ensure that all areas have been searched. The search team should check all places where someone—especially a child—could hide, including closets, bathtubs, shower enclosures, behind doors, under windows, under beds and furniture, and inside toy chests.

The oriented search method can be helpful in ensuring that all areas are covered. Many departments try to have a different crew conduct the secondary search. This ensures that all areas are covered with a fresh set of eyes. Fire fighters can search better without having to crawl under layers of hot gases. When the secondary search has been completed, the team leader should notify the IC by reporting, "Secondary search is clear."

Follow the steps in SKILL DRILL 14-3 (Fire Fighter, I NFPA 5.3.9) to perform a secondary search:

1. Don PPE, including SCBA, and enter the personnel accountability system.
2. Bring hand tools, a hand light, a radio, a thermal imaging device, and search ropes if indicated.
3. Notify command that the search is starting, and indicate the area to be searched and the direction of the search.
4. Conduct a systematic, thorough search for any possible victims.
5. Check for hidden fire, salvage needs, and indicators of the fire's cause, its origin, or arson.
6. Protect any evidence.
7. Maintain team integrity using visual, voice, or direct contact.
8. Notify command of the results of the search and personnel accountability report.

Fire Fighter Safety Tips

Even though the fire is under control, safety remains a top consideration during a secondary search. SCBA should be used until the air is tested and pronounced safe to breathe because levels of carbon monoxide and other poisonous gases often remain high for a long time after a fire is extinguished.

Search Safety

Search and rescue operations present a high risk to fire fighters. During searches, fire fighters are exposed to the same hazards that may endanger the lives of potential victims. Even though fire fighters have the advantages of protective clothing, protective equipment, training, teamwork, and SOPs, they can still be seriously or fatally injured during these operations. Safety is an essential consideration in all search and rescue operations.

Risk Management

Search and rescue situations require a very special type of risk management. Although every emergency operation involves a degree of unavoidable, inherent risk to fire fighters, during search and rescue operations fire fighters will probably encounter situations that involve a significantly higher degree of personal risk. This level of risk is acceptable only if there is a reasonable probability of saving a life.

The IC is responsible for managing the level of risk during emergency operations. He or she must determine which actions will be taken in each situation. Actions that present a high level of risk to the safety of fire fighters are justified only if lives can be saved. In contrast, only a limited risk level is acceptable to save property. When there is no possibility of saving either lives or property, no risk is acceptable.

The acceptance of unusual levels of risk can be justified only when victims are known or believed to be in immediate danger, and there is a reasonable probability that they can be rescued. It is not acceptable to risk the lives of fire fighters when rescuing any victims is not a possibility. For example, if a building is fully involved in flames, a room has flashed over, or the smoke has banked all the way down to the floor, there is little or no chance that any occupants can still be alive, so there is no reason to send fire fighters inside to search for them.

It is important to:

- conduct a thorough 360-degree size-up
- use a thermal imager
- evaluate the possibility of a ventilation-limited fire
- practice good door control to limit the supply of oxygen to the fire
- consider the use of a transitional fire attack to cool the fire before starting the search process

The risk involved in conducting search and rescue operations must always be weighed against the probability of finding someone who is savable—that is, who is still alive to be rescued.

A primary search is conducted during the active stages of a fire, when time is a critical factor. During a primary search, it may be necessary to accept a high level of risk to save a life. A secondary search, which is conducted after the fire is extinguished, is designed to locate any deceased victims and should not expose fire fighters to a high level of risk.

The IC must decide whether to conduct a primary search based on a risk–benefit evaluation of the situation. In making this decision, the IC must consider the stage of the fire, the condition of the building, and the presence of any other hazards. These risks must be weighed against the probability of finding any occupants who could be rescued. Such an on-scene evaluation is often guided by established policies. For example, many fire departments have policies that prohibit fire fighters from entering abandoned structures known to be in poor structural condition. This kind of policy reflects the fact that such a building has a high risk of collapse and the low probability that anyone would be inside.

The IC may decide not to conduct a primary search because the risk to fire fighters is too great or because the possibility of making a successful rescue is too remote. Such a decision might be based on advanced fire conditions, potential backdraft conditions, imminent structural collapse, or other circumstances **FIGURE 14-10**. The IC may be able to identify these conditions from the exterior, or he or she may learn of them from a team assigned to conduct a search. A search team that encounters conditions that make entry impossible should report their findings back to the IC. In these situations, a secondary search is conducted when it is safe to enter.

Search and Rescue Equipment

To perform search and rescue properly, fire fighters must have the appropriate equipment:

- PPE
- Portable radio
- Hand light or flashlight
- Forcible entry (exit) tools
- Hose lines
- Thermal imaging devices
- Ladders
- Long rope(s)
- A piece of tubular webbing or short rope (16 to 24 feet) (5m to 7m)

Full PPE, which is always required for structural firefighting, is essential for interior search and rescue operations. The proper attire includes a helmet, protective hood, bunker coat, turnout pants, boots, and gloves. Each fire fighter must use SCBA and carry a flashlight or hand light. Before entering the building, fire fighters must activate their personal alert safety system devices.

At least one member of each search team should be equipped with a portable radio. Ideally, each individual should have a radio. If a fire fighter gets into trouble, the radio is the best means to obtain assistance.

Each fire fighter assigned to search and rescue should carry at least one forcible entry hand tool, such as an axe, Halligan tool, or short ceiling hook. These tools can be used both to open an area for a search and, if necessary, to open an emergency exit path. A hand tool can also be used to extend the fire fighter's reach during a sweep for unconscious victims.

FIGURE 14-10 A primary search cannot be safely conducted in a building that is fully involved in a fire.

A search team that is working close to the fire should carry a hose line or be accompanied by another team with a hose line. A hose line can protect the fire fighters and enable them to search a structure more efficiently. This measure is essential when a search team is working close to or directly above the fire. The hose line can be used to knock down the fire, to protect a means of egress (stairway or corridor), and to protect the victims as they are escaping.

Methods to Determine Whether an Area Is Tenable

Fire fighters must make rapid, accurate, and ongoing assessments about the safety of the building for fire fighters while working at a structure fire. When they arrive, fire fighters must quickly determine the type of structure involved, the stability of the structure and the possibility of collapse, and the life-safety risk involved. Try to determine how long the structure has been on fire. Are there reports of fire showing before your arrival? Consider the possibility that if no fire is showing, this could indicate a ventilation-limited fire condition. Dark black and turbulent smoke, blackened windows, the appearance of a "breathing" building, soot around windows or doors indicating smoke travel, puffing or little to no smoke showing, and intense heat are signs of possible flashover or backdraft. Sagging walls, chipped or cracked mortar or cement, warped or failing structural steel, and any partial collapse are significant indicators of an impending collapse.

Even after the decision has been made to enter a burning structure, fire fighters must continually reevaluate the safety of the operation. Those working on the search team rely on fire officers, team members, and the IC to notify them of changing fire conditions. Fire officers must be alert for changing conditions inside the building and for any information about changing conditions that is from the other parts of the fire scene. The IC may know or see things that fire fighters working inside might not know, for example, and may call for an evacuation as the fire situation changes. For their part, fire fighters in the interior of the building should check and recheck the surfaces they are working on. Floors that have burned through, that are sagging or that are warm, furniture or carpeting off-gasing, or flame "rollover" should be avoided. A rapid rise in the amount of heat or flame "rollover" may indicate the potential for flashover. Evaluate the floor before entering the building. Since sounding the floor is not a valid measure of stability, a 360-degree size-up is essential to determine where the fire is located. Sounding floors to determine if they are stable is a dangerous practice to be discouraged. Floors that are supported by lightweight wooden trusses and other manufactured components fail quickly when exposed to fire. Entering a burning building when there is a fire burning beneath you is an act of great risk. Underwriters Laboratories studies have indicated all residential floor systems are at risk of sudden catastrophic collapse without warning. Extreme caution should be used when operating on any floor above a fire.

Only in imminent life-threatening situations where immediate action can prevent the loss of life or serious injury should fire fighters take actions at a higher level of risk. The initial IC must evaluate the situation and determine whether this level of risk is justified. The IC also must be prepared to explain his or her decision.

Fire Fighter Safety Tips

During search and rescue operations, fire fighters should follow these guidelines:

- Work from a single plan.
- Maintain radio contact with the IC, both through the chain of command and via portable radios.
- Monitor fire conditions during the search.
- Coordinate ventilation operations with search and rescue activities.
- Adhere to the personnel accountability system.
- Stay with a partner.

FIRE FIGHTER Tips

The IC must always balance the risks involved in an emergency operation with the potential benefits:

- Actions that present a high level of risk to the safety of fire fighters are justified only if there is a potential to save lives.
- Only a limited level of risk is acceptable to save valuable property.
- It is not acceptable to risk the safety of fire fighters when there is no chance of saving lives or property.

Rescue Techniques

Rescue is the removal of a located person who is unable to escape from a dangerous situation. Fire fighters rescue people not only from fires but also from a wide variety of accidents and mishaps. Although this section refers primarily to rescuing occupants from burning buildings, many of the techniques discussed here can be used for other types of situations. As a fire fighter, you must learn and practice the various types of assists, carries, drags, and other techniques used to rescue people from fires. After mastering these techniques, you will be able to rescue victims from life-threatening situations.

FIRE FIGHTER Tips

Specific safety requirements for search and rescue operations are defined in NFPA 1500, *Standard on Fire Department Occupational Safety and Health Program,* and in regulations enforced by the Occupational Safety and Health Administration. The NFPA requirements state that a team of at least two fire fighters must enter together, and at least two other fire fighters must remain outside the danger area, ready to rescue the fire fighters who are inside the building. This policy is sometimes called the two-in/two-out rule.

Rescue is the second component of search and rescue. When you locate a victim during a search, you must direct, assist, or carry that person to a safe area. Rescue can be as simple as verbally directing an occupant toward an exit, or it can be as demanding as extricating a trapped, unconscious victim and physically carrying that person out of the building. The term "rescue" is generally applied to situations where the rescuer physically assists or removes the victim from the dangerous area.

Most people who realize that they are in a dangerous situation will attempt to escape on their own. Children and elderly persons, physically or developmentally handicapped persons, impaired people, and ill or injured persons, however, may be unable to escape and will need to be rescued. People who are sleeping, are under the influence of alcohol or drugs, or have diminished mental capacity may not become aware of the danger in time to escape from the structure on their own. Toxic gases may incapacitate even healthy individuals before they can reach an exit. Victims also may become trapped when a rapidly spreading fire, an explosion, or a structural collapse cuts off potential escape routes.

Because fires threaten the lives of both victims and rescuers, the first priority is to remove the victim from the fire building or dangerous area as quickly as possible. It is sometimes better to move the victim to a safe area first and then provide any necessary medical treatment. The assists, lifts, and carries described in this chapter should not be used if you suspect that the victim has a spinal injury, however, unless there is no other way to remove the person from the life-threatening situation.

Always use the safest and most practical means of egress when removing a victim from a dangerous area. A building's normal exit system, such as interior corridors and stairways, should be used if it is open and safe. If the regular exits cannot be used, an outside fire escape, a ladder, or some other method of egress must be found. Ladder rescues, which are covered in this chapter, can be both difficult and dangerous, whether the victim is conscious and physically fit or is unconscious and injured FIGURE 14-11.

Shelter-in-Place

In some situations, the best option is to shelter the occupants in place instead of trying to remove them from a fire building. This option should be considered when the occupants are conscious and are found in a part of the building that is adequately protected from the fire by fire-resistive construction or fire suppression systems. If smoke and fire conditions block the exits, victims might be safer staying in the sheltered location than attempting to evacuate through a hazardous environment.

Such a situation occurs in a high-rise apartment building when a fire is confined to one apartment or one floor. In this scenario, the stairways and corridors may be filled with smoke, but the occupants who are remote from the fire would be very safe on their balconies or in their apartments with the windows open to provide fresh air. They would be exposed to greater levels of risk if they attempted to exit than if they remained in their apartments until the fire is extinguished. When occupants remain in place, it is important to remember how important it is for them to isolate the room where they are from the fire. This can be done by practicing good door control or by determining whether they are safer with the windows open or with the windows closed. Open windows can serve as an entry point for smoke from the outside or actually draw the fire to the occupant. The decision to follow a shelter-in-place strategy in this case must be made by the rescue officer or IC. Potential locations to shelter-in-place occupants should be identified on the preincident plan.

Fire Fighter Safety Tips

To exit from a fire, follow the hose line in the direction of the male coupling.

FIGURE 14-11 Bringing an unconscious victim or a person who needs assistance down a ladder can be difficult and dangerous.

Exit Assist

The simplest rescue is the exit assist, in which the victim is responsive and able to walk without assistance or with very little assistance. The fire fighter may simply need to guide the person to safety or to provide a minimal level of physical support. Even if the victim can walk without assistance, the fire fighter should take the person's arm or use the one-person walking assist (discussed later in this section) to make sure that the victim does not fall or become separated from the rescuer.

Two types of assists can be used to help responsive victims exit a fire situation:

- One-person walking assist
- Two-person walking assist

One-Person Walking Assist

The one-person walking assist can be used if the person is capable of walking. To perform a one-person walking assist, follow the steps in SKILL DRILL 14-4:

1. Help the victim stand next to you, facing the same direction. (STEP 1)
2. Have the victim place his or her arm behind your back and around your neck.
3. Hold the victim's wrist as it drapes over your shoulder.
4. Put your free arm around the victim's waist and help the victim to walk. (STEP 2)

Two-Person Walking Assist

The two-person walking assist is useful if the victim cannot stand and bear weight without assistance. With this assist technique, the two rescuers completely support the victim's weight. It may be difficult to walk through doorways or narrow passages using this type of assist. To perform a two-person walking assist, follow the steps in SKILL DRILL 14-5:

1. Two fire fighters stand facing the victim, one on each side of the victim. (STEP 1)
2. Both fire fighters assist the victim to a standing position. (STEP 2)
3. Once the victim is fully upright, place the victim's right arm around the neck of the fire fighter on the right side. Place the victim's left arm around the neck of the fire fighter on the left side. The victim's arms should drape over the fire fighters' shoulders. The fire fighters hold the victim's wrist in one hand. (STEP 3)
4. Both fire fighters put their free arms around the victim's waist, grasping each other's wrists for support and locking arms together behind the victim. (STEP 4)
5. Both fire fighters slowly assist the victim to walk. Fire fighters must coordinate their movements and move slowly. (STEP 5)

Simple Victim Carries

Four simple carry techniques can be used to move a victim who is conscious and responsive, but incapable of standing or walking:

- Two-person extremity carry
- Two-person seat carry
- Two-person chair carry
- Cradle-in-arms carry

Two-Person Extremity Carry

The two-person extremity carry requires no equipment and can be performed in tight or narrow spaces, such as the corridors of mobile homes, small hallways, and narrow spaces between buildings. The focus of this carry is on the victim's extremities. To perform a two-person extremity carry, follow the steps in SKILL DRILL 14-6:

1. Two fire fighters help the victim to sit up. (STEP 1)
2. One fire fighter kneels behind the victim, reaches under the victim's arms, and grasps the victim's wrists. (STEP 2)
3. The second fire fighter backs in between the victim's legs, reaches around, and grasps the victim behind the knees. (STEP 3)
4. At the command of the first fire fighter, both fire fighters stand up and carry the victim away, walking straight ahead. The fire fighters must coordinate their movements. (STEP 4)

Two-Person Seat Carry

The two-person seat carry is used with victims who are disabled or paralyzed. This type of carry requires the assistance of two fire fighters, and moving through doors and down stairs may be difficult. To perform a two-person seat carry, follow the steps in SKILL DRILL 14-7:

1. Two fire fighters kneel near the victim's hips, one on each side of the victim. (STEP 1)

SKILL DRILL 14-4 Performing a One-Person Walking Assist

(Fire Fighter I, NFPA 5.3.9)

1 Help the victim stand.

2 Have the victim place his or her arm around your neck, and hold on to the victim's wrist, which should be draped over your shoulder. Put your free arm around the victim's waist and help the victim to walk.

2. Both fire fighters raise the victim to a sitting position and link arms behind the victim's back. (STEP 2)
3. The fire fighters place their free arms under the victim's knees and link them together. (STEP 3)
4. If possible, the victim puts his or her arms around the fire fighters' necks for additional support. (STEP 4)

Two-Person Chair Carry

The two-person chair carry is particularly suitable when a victim must be carried through doorways, along narrow corridors, or up or down stairs. In this technique, two rescuers use a chair to transport the victim. A folding chair should not be used for this purpose, because folding chairs can collapse when used in this way to carry a victim. The chair must be strong enough to support the weight of the victim while he or she is being carried. The victim should feel much more secure with this carry than with the two-person seat carry, and he or she should be encouraged to hold on to the chair. To perform a two-person chair carry, follow the steps in SKILL DRILL 14-8:

1. Tie the victim's hands together or have the victim grasp his or her hands together.
2. One fire fighter stands behind the seated victim, reaches down, and grasps the back of the chair. (STEP 1)
3. The fire fighter tilts the chair slightly backward on its rear legs so that the second fire fighter can step back between the legs of the chair and grasp the tips of the chair's front legs. The victim's legs should be between the legs of the chair. (STEP 2)
4. When both fire fighters are correctly positioned, the fire fighter behind the chair gives the command to lift and walk away.
5. Because the chair carry may force the victim's head forward, watch the victim for airway problems. (STEP 3)

Cradle-in-Arms Carry

The cradle-in-arms carry can be used by one fire fighter to carry a child or a small adult. With this technique, the fire fighter should be careful of the victim's head when moving through doorways or down stairs. To perform the cradle-in-arms carry, follow the steps in SKILL DRILL 14-9:

1. Kneel beside the child, and place one arm around the child's back and the other arm under the thighs. (STEP 1)
2. Lift slightly and roll the child into the hollow formed by your arms and chest. (STEP 2)
3. Be sure to use your leg muscles to stand. (STEP 3)

SKILL DRILL 14-5 Performing a Two-Person Walking Assist

(Fire Fighter I, NFPA 5.3.9)

1. Two fire fighters stand facing the victim, one on each side of the victim.

2. The fire fighters assist the victim to a standing position.

3. Once the victim is fully upright, drape the victim's arms around the necks and over the shoulders of the fire fighters, each of whom holds one of the victim's wrists.

4. Both fire fighters put their free arm around the victim's waist, grasping each other's wrists for support and locking their arms together behind the victim.

5. Assist walking at the victim's speed.

■ Emergency Drags

The most efficient method to remove an unconscious or unresponsive victim from a dangerous location is a drag. Five types of emergency drags can be used to remove unresponsive victims from a fire situation:

- Clothes drag
- Blanket drag
- Webbing sling drag
- Fire fighter drag
- Emergency drag from a vehicle

SKILL DRILL 14-6 Performing a Two-Person Extremity Carry

(Fire Fighter I, NFPA 5.3.9)

1 Two fire fighters help the victim to sit up.

2 The first fire fighter kneels behind the victim, reaches under the victim's arms, and grasps the victim's wrists.

3 The second fire fighter backs in between the victim's legs, reaches around, and grasps the victim behind the knees.

4 The first fire fighter gives the command to stand and carry the victim away, walking straight ahead. Both fire fighters must coordinate their movements.

When using an emergency drag, the rescuer should make every effort to pull the victim in line with the long axis of the body to provide as much spinal protection as possible. The victim should be moved head first to protect the head.

Clothes Drag

The clothes drag is used to move a victim who is on the floor or the ground and is too heavy for one rescuer to lift and carry alone. In this technique, the rescuer drags the person by pulling on the clothing in the neck and shoulder area. The rescuer should grasp the clothes just behind the collar, use the arms to support the victim's head, and drag the victim away from danger. To perform the clothes drag, follow the steps in SKILL DRILL 14-10:

1. Crouch behind the victim's head, grab the shirt or jacket around the collar and shoulder area, and support the head with your arms. (STEP 1)
2. Lift with your legs until you are fully upright. Walk backward, dragging the victim to safety. (STEP 2)

Blanket Drag

The blanket drag can be used to move a victim who is not dressed or who is dressed in clothing that is too flimsy for the clothes drag (for example, a nightgown). This procedure requires the use of a large sheet, blanket, curtain, or rug. Place

SKILL DRILL 14-7 Performing a Two-Person Seat Carry
(Fire Fighter I, NFPA 5.3.9)

1 Kneel beside the victim near the victim's hips.

2 Raise the victim to a sitting position and link arms behind the victim's back.

3 Place your free arms under the victim's knees and link arms.

4 If possible, the victim puts his or her arms around your necks for additional support.

the item on the floor and roll the victim onto it, and then pull the victim to safety by dragging the sheet or blanket.

To perform a blanket drag, follow the steps in SKILL DRILL 14-11:

1. Lay the victim supine (face up) on the ground. Stretch out the material for dragging next to the victim. (STEP 1)
2. Roll the victim onto the right or left side. Neatly bunch one-third of the material against the victim so the victim will lie approximately in the middle of the material. (STEP 2)
3. Lay the victim back down (supine) on the material. Pull the bunched material out from underneath the victim and wrap it around the victim. (STEP 3)
4. Grab the material at the head and drag the victim backward to safety. (STEP 4)

The standing drag is performed by following the steps in SKILL DRILL 14-12:

1. Kneel at the head of the supine victim. (STEP 1)
2. Raise the victim's head and torso by 90 degrees, so that the victim is leaning against you. (STEP 2)
3. Reach under the victim's arms, wrap your arms around the victim's chest, and lock your arms. (STEP 3)
4. Stand straight up using your legs.
5. Drag the victim out. (STEP 4)

SKILL DRILL 14-8

Performing a Two-Person Chair Carry
(Fire Fighter I, NFPA 5.3.9)

1 One fire fighter stands behind the seated victim, reaches down, and grasps the back of the chair.

2 The fire fighter tilts the chair slightly backward on its rear legs so that the second fire fighter can step back between the legs of the chair and grasp the tips of the chair's front legs. The victim's legs should be between the legs of the chair.

3 When both fire fighters are correctly positioned, the fire fighter behind the chair gives the command to lift and walk away. Because the chair carry may force the victim's head forward, watch the victim for airway problems.

SKILL DRILL 14-9

Performing a Cradle-in-Arms Carry
(Fire Fighter I, NFPA 5.3.9)

1 Kneel beside the child, and place one arm around the child's back and the other arm under the thighs.

2 Lift slightly and roll the child into the hollow formed by your arms and chest.

3 Be sure to use your leg muscles to stand.

SKILL DRILL 14-10 Performing a Clothes Drag
(Fire Fighter I, NFPA 5.3.9)

1 Crouch behind the victim's head, and grab the shirt or jacket around the collar and shoulder area.

2 Lift with your legs until you are fully upright. Walk backward, dragging the victim to safety.

Webbing Sling Drag

The webbing sling drag provides a secure grip around the upper part of a victim's body, allowing for a faster removal from the dangerous area. In this drag, a sling is placed around the victim's chest and under the armpits, then used to drag the victim. The webbing sling helps support the victim's head and neck. A webbing sling can be rolled and kept in a turnout coat pocket. A carabiner can be attached to such a sling to secure the straps under the victim's arms and provide additional protection for the victim's head and neck.

To perform the webbing sling drag, follow the steps in SKILL DRILL 14-13:

1. Using a prepared webbing sling, place the victim in the center of the loop so the webbing is behind the victim's back in the area just below the armpits. (STEP 1)
2. Take the large loop over the victim and place it above the victim's head. Reach through, grab the webbing behind the victim's back, and pull through all the excess webbing. This creates a loop at the top of the victim's head and two loops around the victim's arms. (STEP 2)
3. Adjust your hand placement to protect the victim's head while dragging. (STEP 3)

Fire Fighter Drag

The fire fighter drag can be used if the victim is heavier than the rescuer because it does not require lifting or carrying the victim. To perform the fire fighter drag, follow the steps in SKILL DRILL 14-14:

1. Tie the victim's wrists together with anything that is handy—a cravat (a folded triangular bandage), gauze, a belt, or a necktie. (STEP 1)
2. Get down on your hands and knees and straddle the victim. (STEP 2)
3. Pass the victim's tied hands around your neck, straighten your arms, and drag the victim across the floor by crawling on your hands and knees. (STEP 3)

Emergency Drag from a Vehicle

An emergency drag from a vehicle is performed when the victim must be quickly removed from a vehicle to save his or her life. The drags described in this section might be used, for example, if the vehicle is on fire or if the victim requires cardiopulmonary resuscitation.

One Rescuer

There is no effective way for one person to remove a victim from a vehicle without some movement of the neck and spine. Preventing excess movement of the victim's neck, however, is important. To perform an emergency drag from a vehicle with only one rescuer, follow the steps in SKILL DRILL 14-15:

1. Grasp the victim under the arms and cradle his or her head between your arms. (STEP 1)
2. Gently pull the victim out of the vehicle. (STEP 2)
3. Lower the victim down into a horizontal position in a safe place. (STEP 3)

Long Backboard Rescue

If four or more fire fighters are present, one fire fighter can support the victim's head and neck, while the second and third fire fighters can move the victim by lifting under the victim's arms. The victim can then be moved in line with the long axis of the body, with the head and neck stabilized in a neutral position. Whenever possible, a long backboard should be

SKILL DRILL 14-11 Performing a Blanket Drag
(Fire Fighter I, NFPA 5.3.9)

1 Stretch out the material you are using next to the victim.

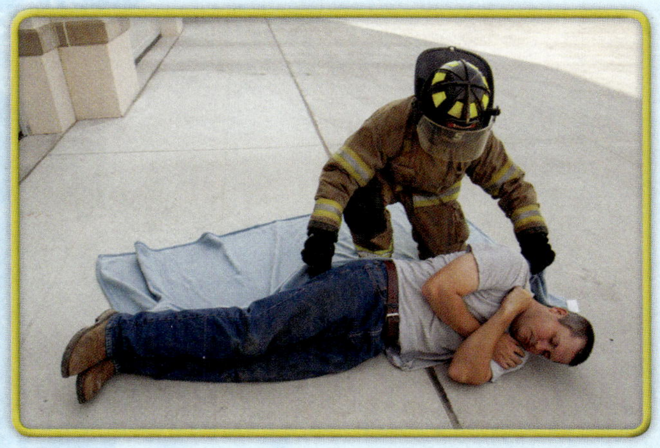

2 Roll the victim onto one side. Neatly bunch one-third of the material against the victim's body.

3 Lay the victim back down (supine). Pull the bunched material out from underneath the victim and wrap it around the victim.

4 Grab the material at the head and drag the victim backward to safety.

used to remove a victim from the vehicle. Follow the steps in SKILL DRILL 14-16 to perform this type of rescue:

1. The first fire fighter supports the victim's head and cervical spine from behind. Support may be applied from the side, if necessary, by reaching through the driver's-side doorway. (STEP 1)
2. The second fire fighter serves as team leader and, as such, gives commands until the patient is supine on the backboard. Because the second fire fighter lifts and turns the victim's torso, he or she must be physically capable of moving the victim. The second fire fighter works from the driver's-side doorway. If the first fire fighter is also working from that doorway, the second fire fighter should stand closer to the door hinges toward the front of the vehicle. The second fire fighter applies a cervical collar. (STEP 2)
3. The second fire fighter provides continuous support of the victim's torso until the victim is supine on the backboard. Once the second fire fighter takes control of the torso, usually in the form of a body hug, he or she should not let go of the victim for any reason. Some type of cross-chest shoulder hug usually works well, but you will have to decide which method works best for you with any given victim. You cannot simply reach into the car and grab the victim, because this will merely twist the victim's torso. Instead, you must rotate the victim as a unit.
4. The third fire fighter works from the front passenger's seat and is responsible for rotating the victim's legs

SKILL DRILL 14-12 Performing a Standing Drag
(Fire Fighter I, NFPA 5.3.9)

1 Kneel at the head of the supine victim.

2 Raise the victim's head and torso by 90 degrees so that the victim is leaning against you.

3 Reach under the victim's arms, wrap your arms around the victim's chest, and lock your arms.

4 Stand straight up using your legs. Drag the victim out.

and feet as the torso is turned, ensuring that they remain free of the pedals and any other obstruction. With care, the third fire fighter should first move the victim's nearer leg laterally without rotating the victim's pelvis and lower spine. The pelvis and lower spine rotate only as the third fire fighter moves the second leg during the next step. Moving the nearer leg early makes it much easier to move the second leg in concert with the rest of the body. After the third fire fighter moves the legs together, both legs should be moved as a unit. (STEP 3)

5 The victim is rotated 90 degrees so that his or her back is facing out the driver's door and his or her feet are on the front passenger's seat. This coordinated movement is done in three or four "eighth turns." The second fire fighter directs each quick turn by saying, "Ready, turn" or "Ready, move." Hand position changes should be made between moves.

6 In most cases, the first fire fighter will be working from the back seat. At some point—either because the door post is in the way or because he or she cannot reach farther from the back seat—the first fire fighter will be unable to follow the torso rotation. At that time, the third fire fighter should assume temporary support of the victim's head and neck until the first fire fighter can regain control of

SKILL DRILL 14-13 Performing a Webbing Sling Drag

(Fire Fighter I, NFPA 5.3.9)

1 Place the victim in the center of the loop so the webbing is behind the victim's back.

2 Take the large loop over the victim and place it above the victim's head. Reach through, grab the webbing behind the victim's back, and pull through all the excess webbing. This creates a loop at the top of the victim's head and two loops around the victim's arms.

3 Adjust your hand placement to protect the victim's head while dragging.

SKILL DRILL 14-14 Performing a Fire Fighter Drag

(Fire Fighter I, NFPA 5.3.9)

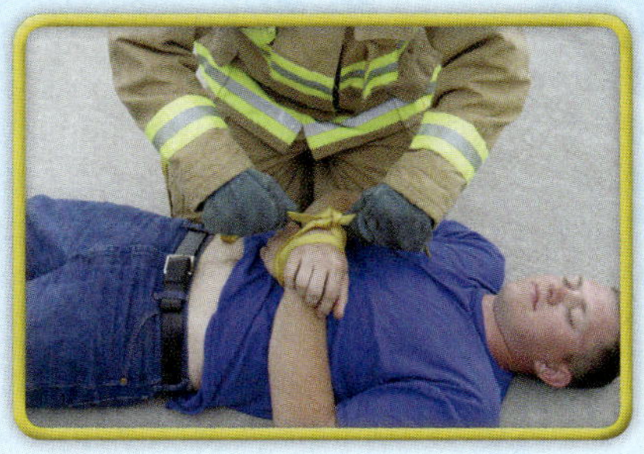

1 Tie the victim's wrists together with anything that is handy.

2 Get down on your hands and knees and straddle the victim.

3 Pass the victim's tied hands around your neck, straighten your arms, and drag the victim across the floor by crawling on your hands and knees.

the head from outside the vehicle. If a fourth fire fighter is present, he or she stands next to the second fire fighter. The fourth fire fighter takes control of the victim's head and neck from outside the vehicle without involving the third fire fighter. As soon as the change has been made, the rotation can continue. (**STEP 4**)

7 Once the victim has been fully rotated, the backboard is placed against the victim's buttocks on the seat. Do not try to wedge the backboard under the victim. If only three fire fighters are present, place the backboard within arm's reach of the driver's door before the move so that the board can be pulled into place when needed. In such cases, the far end of the board can be left on

SKILL DRILL 14-15 Performing a One-Person Emergency Drag from a Vehicle
(Fire Fighter I, NFPA 5.3.9)

1 Grasp the victim under the arms and cradle his or her head between your arms.

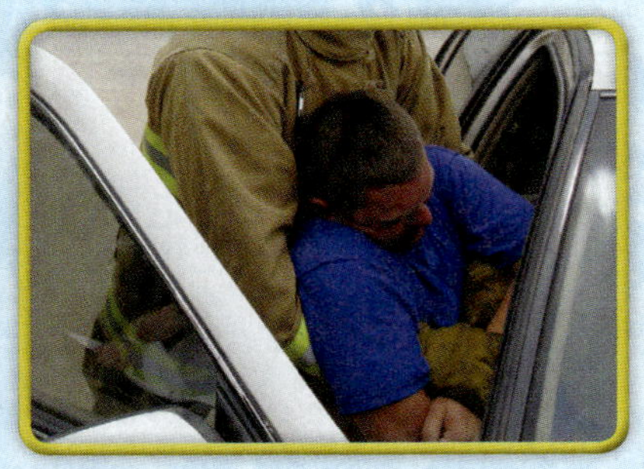

2 Gently pull the victim out of the vehicle.

3 Lower the victim down into a horizontal position in a safe place.

the ground. When a fourth fire fighter is available, the first fire fighter exits the rear seat of the car, places the backboard against the victim's buttocks, and maintains pressure in toward the vehicle from the far end of the board. (Note: When the door opening allows, some fire fighters prefer to insert the backboard onto the car seat before rotating the victim.)

8. As soon as the victim has been rotated and the backboard is in place, the second and third fire fighters lower the victim onto the board while supporting the head and torso so that neutral alignment is maintained. The first fire fighter holds the backboard until the victim is secured. (STEP 5)
9. The third fire fighter moves across the front seat to be in position at the victim's hips. If the third fire fighter stays at the victim's knees or feet, he or she will be ineffective in helping to move the body's weight, because the knees and feet follow the hips.
10. The fourth fire fighter maintains support of the head and now takes over giving the commands. The second fire fighter maintains direction of the extrication. This fire fighter stands with his or her back to the door, facing the rear of the vehicle.
11. The backboard should be immediately in front of the third fire fighter. The second fire fighter grasps the victim's shoulders or armpits. Then, on command, the second and third fire fighters slide the victim 8 to 12 inches along the backboard, repeating this slide until the victim's hips are firmly on the backboard. (STEP 6)
12. The third fire fighter gets out of the vehicle and moves to the opposite side of the backboard, across from the second fire fighter. The third fire fighter now takes control at the victim's shoulders, and the second fire fighter moves back to take control of the hips. On command, these two fire fighters move the victim along the board in 8- to 12-inch (20 cm to 30 cm) slides until the victim is placed fully on the board. (STEP 7)
13. The first (or fourth) fire fighter continues to maintain support of the victim's head. The second and third fire fighters grasp their side of the board, and then carry it and the victim away from the vehicle and toward the prepared cot nearby. (STEP 8)

These steps must be considered a general procedure to be adapted as needed based on the type of vehicle, type of victim, and rescue crew available. Two-door cars differ from four-door models; larger cars differ from smaller compact models; pickup trucks differ from full-size sedans and four-wheel-drive vehicles. Likewise, you will handle a large, heavy adult differently from a small adult or child. Every situation will be different—a different car, a different patient, and a different crew. Your resourcefulness and ability to adapt are necessary elements to successfully perform this rescue technique.

Assisting a Person Down a Ground Ladder

Using a ground ladder to rescue a trapped occupant is one of the most critical, stressful, and demanding tasks performed by fire fighters. Assisting someone down a ladder carries a considerable risk of injury to both fire fighters and victims. Whenever it is possible to use a stairway, fire escape, or aerial tower for rescue, these options should be considered before using ground ladders. Fire fighters must use proper techniques to safely accomplish a ladder rescue. In addition, they must have the physical strength and stamina needed to accomplish the rescue without causing injury to anyone involved. Although circumstances could require an individual fire fighter to work alone, at least two fire fighters should work as a team to rescue a victim whenever possible.

Time is a critical factor in many rescue situations—someone waiting to be rescued is often in immediate danger and may be preparing to jump. As a consequence, fire fighters may have only a limited time to work. They must quickly and efficiently raise a ladder and assist the victim to safety.

Ladder rescue begins with proper placement of the ladder. As described in the Ladders chapter, a ladder used to rescue a person from a window should have its tip placed just below the windowsill. This positioning makes it easier for the victim to mount the ladder. If possible, one or more fire fighters in the interior of the building should help the victim onto the ladder, and one fire fighter should stay on the ladder to assist the individual down.

Any ladder used for rescue should be heeled or tied in. The weight of an occupant and one or two fire fighters, all moving on the ladder at the same time, can easily destabilize a ladder that is not adequately secured.

Rescuing a Conscious Person from a Window

When a rescue involves a conscious person, fire fighters should establish verbal contact as quickly as possible to reassure the victim that help is on the way. Many people have jumped to

SKILL DRILL 14-16 Performing a Long Backboard Rescue

(Fire Fighter I, NFPA 5.3.9)

1 The first fire fighter provides in-line manual support of the victim's head and cervical spine.

2 The second fire fighter gives commands and applies a cervical collar.

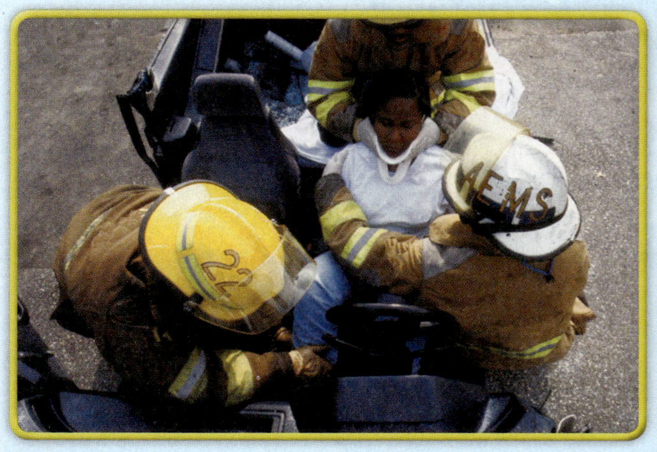

3 The third fire fighter frees the victim's legs from the pedals and moves the legs together without moving the victim's pelvis or spine.

4 The second and third fire fighters rotate the victim as a unit in several short, coordinated moves. The first fire fighter (relieved by the fourth fire fighter as needed) supports the victim's head and neck during rotation (and later steps).

(*Continued*)

SKILL DRILL 14-16 Performing a Long Backboard Rescue (*Continued*)

(Fire Fighter I, NFPA 5.3.9)

5 The first (or fourth) fire fighter places the backboard on the seat against the victim's buttocks. The second and third fire fighters lower the victim onto the long backboard.

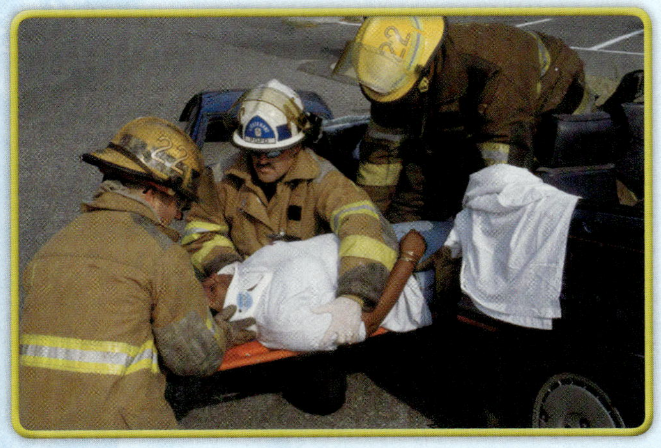

6 The third fire fighter moves to an effective position for sliding the victim. The second and third fire fighters slide the victim along the backboard in coordinated, 8- to 12-inch (20 to 30 cm) moves until the victim's hips rest on the backboard.

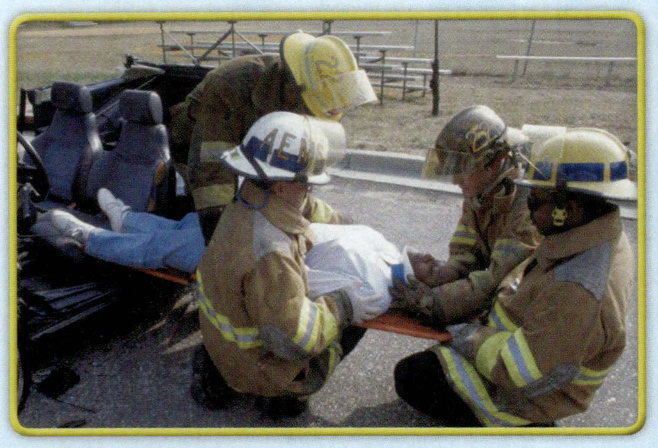

7 The third fire fighter exits the vehicle and moves to the backboard opposite the second fire fighter. Working together, they continue to slide the victim until the victim is fully on the backboard.

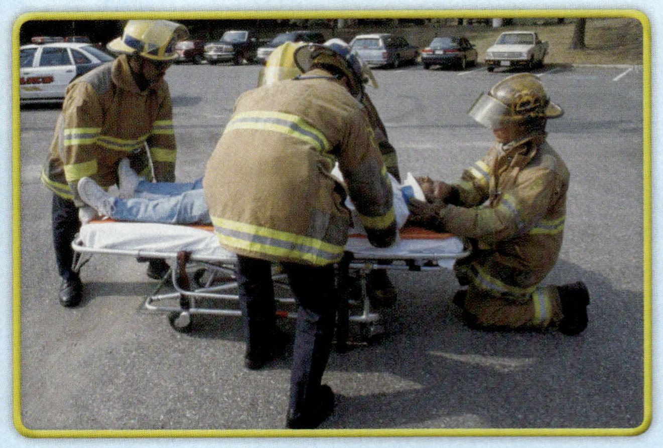

8 The first (or fourth) fire fighter continues to stabilize the victim's head and neck while the second, third, and fourth fire fighters carry the victim away from the vehicle.

their deaths just seconds before a ladder could be raised to a window.

All ladder rescues should be performed with two fire fighters whenever possible. **SKILL DRILL 14-17** presents a technique that could, if necessary, be performed by a single fire fighter:

1. The rescue team places the ladder into the rescue position, with the tip of the ladder just below the windowsill, and secures the ladder in place. (**STEP 1**)
2. The first fire fighter climbs the ladder, makes contact with the victim, and climbs inside the window to assist the victim. Make contact as soon as possible to calm the victim, and encourage the victim to stay at the window until the rescue can be performed. (**STEP 2**)
3. The second fire fighter climbs up to the window, standing one rung below the windowsill. This leaves at least one rung available for the victim. The fire fighter on the ladder should apply a leg lock or use a safety belt. When ready, the fire fighter advises the victim to slowly come out onto the ladder, feet first and facing the ladder. (**STEP 3**)
4. The fire fighter forms a semi-circle around the victim, with both hands on the beams of the ladder. (**STEP 4**)

5. The fire fighter and victim proceed slowly down the ladder, one rung at a time, with the fire fighter always staying one rung below the victim.
6. If the victim slips or loses his or her footing, the fire fighter's legs should keep the victim from falling.
7. The fire fighter can take control of the victim at any time by leaning in toward the ladder and squeezing the victim against the ladder. It is important to talk to the person being rescued. This helps to reassure and calm him or her. Verbalize each step and encourage the person to keep his or her gaze forward. (STEP 5)

Rescuing an Unconscious Person from a Window

If the trapped person is unconscious, one or more fire fighters will have to climb inside the building and pass the person out of the window to a fire fighter on the ladder. Caution should be used when lowering an unconscious victim down a ladder, because it is very easy for the victim's arms or legs to get caught in the ladder. To rescue an unconscious person via a ladder, follow the steps in SKILL DRILL 14-18:

1. The rescue team places and secures the ladder in the rescue position, with the tip of the ladder just below the windowsill. (STEP 1)
2. One fire fighter climbs up the ladder and into the window to assist from the inside. The second fire fighter climbs up to the window opening and waits for the victim. (STEP 2)
3. The fire fighter on the ladder keeps a firm grip on the ladder with both hands on the rungs. One leg should be straight and the other should be bent so that the thigh is horizontal to the ground, with the knee at a 90-degree angle. The foot of the straight leg should be one rung below the foot of the bent leg.
4. When both fire fighters are ready, the interior fire fighter passes the victim out through the window and onto the ladder. The victim's back should be toward the ladder, so that the victim is face-to-face with the fire fighter on the ladder. (STEP 3)
5. The victim is lowered so that his or her groin rests on the horizontal leg of the fire fighter. The fire fighter places his or her arms under the arms of the victim and holds on to the rungs. It is important to keep the balls of both feet on the rungs of the ladder, because it is much more difficult to move your feet in this position if the heels are close to the rungs.

SKILL DRILL 14-17 Rescuing a Conscious Victim from a Window

(Fire Fighter I, NFPA 5.3.9)

1 The rescue team places the ladder into the rescue position, with the tip of the ladder just below the windowsill, and secures the ladder in place.

2 The first fire fighter climbs the ladder, makes contact with the victim, and climbs inside the window to assist the victim. Make contact as soon as possible to calm the victim, and encourage the victim to stay at the window until the rescue can be performed.

3 The second fire fighter climbs up to the window, standing one rung below the windowsill. This leaves at least one rung available for the victim. When ready, the fire fighter advises the victim to slowly come out onto the ladder, feet first and facing the ladder.

(Continued)

SKILL DRILL 14-17 Rescuing a Conscious Victim from a Window (*Continued*)

(Fire Fighter I, NFPA 5.3.9)

4 The fire fighter forms a semi-circle around the victim, with both hands on the beams of the ladder.

5 The fire fighter and victim proceed slowly down the ladder, one rung at a time, with the fire fighter always staying one rung below the victim. If the victim slips or loses his or her footing, the fire fighter's legs should keep the victim from falling. The fire fighter can take control of the victim at any time by leaning in toward the ladder and squeezing the victim against the ladder.

6. The fire fighter climbs down the ladder slowly, one rung at a time. The victim is always supported at the groin by one of the fire fighter's legs. The fire fighter's arms remain under the victim's arms to support the upper torso. Optionally, the fire fighter in the interior of the building may tie the victim's hands together and place them over the neck of the fire fighter on the ladder. (STEP 4)

Rescuing an Unconscious Child or a Small Adult from a Window

Small adults and children can be cradled across a fire fighter's arms during a rescue. To use this rescue technique, the child must be light enough that the fire fighter can descend safely, using only arm strength to support the victim. To carry an unconscious child down a ladder, follow the steps in SKILL DRILL 14-19:

1. The rescue team sets up and secures the ladder in the rescue position, with the tip of the ladder just below the windowsill. (STEP 1)
2. The first fire fighter climbs the ladder and enters the window to rescue from the interior. The second fire fighter climbs the ladder to the window opening and waits for the victim. Both of the second fire fighter's arms should be level with his or her hands on the beams. (STEP 2)
3. When ready, the interior fire fighter passes the victim to the fire fighter on the ladder so the victim is cradled across the fire fighter's arms. (STEP 3)
4. The fire fighter climbs down the ladder slowly, with the victim being held in his or her arms. The fire fighter's hands should slide down the beams. (STEP 4)

Rescuing a Large Adult from a Window

Three fire fighters using two ladders may be needed to rescue very tall or heavy adults. To rescue a large adult using a ladder, follow the steps in SKILL DRILL 14-20:

SKILL DRILL 14-18 Rescuing an Unconscious Victim from a Window

(Fire Fighter I, NFPA 5.3.9)

1 Place the tip of the ladder just below the windowsill.

2 One fire fighter enters to rescue the victim. The second fire fighter climbs to the window.

3 The fire fighter waiting on the ladder places both hands on the rungs, with one leg straight and the other horizontal to the ground with the knee at an angle of 90 degrees. The interior fire fighter passes the victim through the window and onto the ladder, keeping the victim's back toward the ladder.

4 The victim is lowered so that he or she straddles the fire fighter's leg. The fire fighter's arms should be positioned under the victim's arms, holding on to the rungs. Step down one rung at a time, transferring the victim's weight from one leg to the other. The victim's arms can also be secured around the fire fighter's neck.

1. The rescue team places and secures two ladders, side-by-side, in the rescue position. The tips of the two ladders should be just below the windowsill. (STEP 1)
2. Multiple fire fighters may be required to enter the window to assist from the inside. (STEP 2)
3. Two fire fighters, one on each ladder, climb up to the window opening and wait for the victim. (STEP 3)
4. When ready, the victim is lowered down across the arms of the fire fighters, with one fire fighter supporting the victim's legs and the other supporting the victim's arms. Once in place, the fire fighters can slowly descend the ladder, using both hands to hold on to the ladder rungs. (STEP 4)

Removal of Victims by Ladders

Ladders should be used to remove victims only when it is not possible to use interior stairways, fire escapes, or aerial towers. A ladder rescue is often frightening to conscious victims. Rescuing an unconscious victim by ladder is dangerous and difficult, but it may be the best way to save a life.

SKILL DRILL 14-19 Rescuing an Unconscious Child or a Small Adult from a Window
(Fire Fighter I, NFPA 5.3.9)

1 Place the ladder in rescue position, with the tip below the windowsill.

2 One fire fighter enters the window to assist the victim. The second fire fighter stands on the ladder to receive the victim, with both arms level and hands on beams.

3 The victim is placed in the fire fighter's arms.

4 The fire fighter descends, keeping arms level and sliding the hands down the beams.

SKILL DRILL 14-20 Rescuing a Large Adult from a Window

(Fire Fighter I, NFPA 5.3.9)

1 The rescue team places and secures two ladders, side-by-side, in the rescue position. The tips of the two ladders should be just below the windowsill.

2 Multiple fire fighters may be required to enter the window to assist from the inside.

3 Two fire fighters, one on each ladder, climb up to the window opening and wait for the victim.

4 When ready, the victim is lowered down across the arms of the fire fighters, with one fire fighter supporting the victim's legs and the other supporting the victim's arms. Once in place, the fire fighters can slowly descend the ladder, using both hands to hold on to the ladder rungs.

Aerial Ladders and Platforms

An aerial ladder or platform can be used for rescue operations. The same basic rescue techniques are used with both aerial and ground ladders, but aerial ladders offer several advantages over ground ladders. Aerial ladders are much stronger and have a longer reach. In addition, they are wider and more stable than ground ladders, with side rails that provide greater security for both rescuers and victims FIGURE 14-12.

Aerial platforms are even more suitable than aerial ladders for rescue operations. These devices reduce the risk of slipping and falling because the victim is lowered to the ground mechanically. An aerial platform is usually preferred for rescue work if one is available.

Ground Ladders

Before a ground ladder can be used in a rescue, it must be properly positioned and secured. Positioning and securing ground ladders are covered in the Ladders chapter. Additional personnel will be needed to secure the ladder and to assist in bringing the victim down the ladder. It is not often that fire fighters need to perform a rescue using a ground ladder. This type of rescue is physically taxing and requires carefully executed skills. It is important to practice the skills to be prepared for this event, as those skills may be needed at any time.

FIGURE 14-12 An aerial ladder is stronger and more stable than a ground ladder.

Chief Concepts

- Search and rescue are almost always performed in tandem, yet they are separate operations. The purpose of search operations is to locate living victims. The purpose of rescue operations is to physically remove an occupant or victim from a dangerous environment.
- The IC and fire fighters must plan and coordinate all fire suppression operations, from ventilation to fire attack, to support the search and rescue priority.
- Sometimes fire fighters must confine or control a fire before search and rescue operations can begin.
- The size-up process at every fire should include a specific evaluation of the critical factors for search and rescue—the number of occupants in the building, their location, the degree of risk to their lives, and their ability to evacuate by themselves. The type of occupancy; the size, construction, and condition of the building; the apparent smoke and fire conditions; and the time of day and the day of the week are all important observations that could indicate whether and how many people may need to be rescued. A search and rescue plan can then be developed based on this information.
- Plans for search and rescue must consider the risks and benefits of the operations. In some cases, such as when conditions are ripe for a backdraft, search and rescue operations may not be performed.
- Search and rescue size-up considerations include the following factors:
  - Occupancy
  - Size of the building
  - Construction of the building
  - Time of day and day of week
  - Number of occupants
  - Degree of risk to the occupants presented by the fire
  - Ability of occupants to exit on their own
- Never assume that a building is occupied or unoccupied. Look for clues, such as cars parked in the driveway or a light on in the window, that indicate the presence of an occupant. However, the decision to perform search and rescue operations should not be made solely on these observations.
- Dispatch, neighbors, bystanders, and occupants who escaped on their own are sources of information on the status of a structure's occupants.
- A search begins in the areas where victims are at the greatest risk. Area search assignments should be based on a system of priorities:
  - First priority—Search the areas where live victims may be located immediately around the fire, and then the rest of the fire floor.
  - Second priority—Search the area directly above the fire and the rest of that floor.
  - Third priority—Search the higher-level floors, working from the top floor down, because smoke and heat are likely to accumulate in these areas.
  - Lowest priority—Search areas below the fire floor.
- A primary search is a quick attempt to locate victims who are in danger. Fire fighters must rely on their senses such as touch, hearing, and sight as they search for victims in smoky conditions.
- When searching a room, fire fighters should use the walls to orient themselves and as guides around the room. The room should be searched in a systematic pattern.
- Thermal imaging devices may be used to search for victims in darkness or smoke-obscured conditions. The thermal imaging device detects the temperature differential of the victim's body from the room.
- Search ropes may be used in large, open areas so fire fighters may maintain their orientation while searching for victims.
- The vent-entry-isolate-search method uses a window as the entry and exit point to a room.
- A secondary search is conducted after the fire is under control or fully suppressed. It is a slower search that is designed to locate any deceased victims and search any areas not covered in the primary search.
- Although every emergency operation involves a degree of unavoidable, inherent risk to fire fighters, during search and rescue operations fire fighters will probably encounter situations that involve a significantly higher degree of personal risk. Assuming this level of risk is acceptable only if there is a reasonable probability of saving a life. In some cases, the IC may decide that the risk to fire fighters is too great and a primary search may not be conducted.
- Sagging walls, chipped or cracked mortar or cement, warped or failing structural steel, and any partial collapse are significant indicators of an impending collapse. Dark black turbulent smoke, blackened windows, the appearance of a "breathing" building, and intense heat are signs of possible flashover or backdraft.
- Rescue is the removal of a located person who is unable to escape from a dangerous situation; it is the second component of search and rescue. When you locate a victim during a search, you must direct, assist, or carry that person to a safe area. Rescue can be as simple as verbally directing an occupant toward an exit, or it can be as demanding as extricating a trapped, unconscious victim and physically carrying that person out of the building.
- Children and elderly persons, physically or developmentally handicapped persons, impaired people, and ill or injured persons may be unable to escape from a structure and, therefore, may need to be rescued. Victims also may become trapped when a rapidly spreading fire, an explosion, or a structural collapse cuts off potential escape routes. Because fires threaten the lives of both victims and rescuers, the first priority is to remove the victim from the fire building or dangerous area as quickly as possible.

- In some situations, the best option is to shelter the occupants in place instead of trying to remove them from a fire building. This option should be considered when the occupants are conscious and are found in a part of the building that is adequately protected from the fire by fire-resistive construction and/or fire suppression systems. Such a situation occurs in a high-rise apartment building when a fire is confined to one apartment or one floor.
- The simplest rescue is the exit assist, in which the victim is responsive and able to walk without assistance or with very little assistance. Two types of assists can be used to help responsive victims exit a fire situation:
  - One-person walking assist
  - Two-person walking assist
- Four simple carry techniques can be used to move a victim who is conscious and responsive, but incapable of standing or walking:
  - Two-person extremity carry
  - Two-person seat carry
  - Two-person chair carry
  - Cradle-in-arms carry
- The most efficient method to remove an unconscious or unresponsive victim from a dangerous location is a drag. Five emergency drags can be used to remove unresponsive victims from a fire situation:
  - Clothes drag
  - Blanket drag
  - Webbing sling drag
  - Fire fighter drag
  - Emergency drag from a vehicle
- When using an emergency drag, make every effort to pull the victim in line with the long axis of the body to provide as much spinal protection as possible. The victim should be moved head first to protect the head.
- Assisting someone down a ground ladder carries a considerable risk of injury to both fire fighters and victims. Whenever it is possible to use a protected stairway, fire escape, or aerial tower for rescue, these options should be considered before using ground ladders.

Hot Terms

Primary search A rapid search to locate living victims.

Rekindle A return to flaming combustion after apparent but incomplete extinguishment. (NFPA 921)

Rescue Those activities directed at locating endangered persons at an emergency incident, removing those persons from danger, treating the injured, and providing for transport to an appropriate healthcare facility. (NFPA 402)

Search The process of locating victims who are in danger.

Search rope A guide rope used by fire fighters that allows them to maintain contact with a fixed point.

Secondary search A methodical search designed to locate deceased victims after the fire has been suppressed, searching areas not covered during the primary search.

Thermal imager (TI) An electronic device that detects differences in temperature based on infrared energy and then generates images based on those data. This device is used in smoky environments to locate victims.

Transitional attack An offensive fire attack initiated by an exterior, indirect handline operation into the fire compartment to initiate cooling while transitioning into interior direct fire attack in coordination with ventilation operations.

Two-in/two-out rule A safety procedure that requires a minimum of two personnel to enter a hazardous area and a minimum of two backup personnel to remain outside the hazardous area during the initial stages of an incident.

FIRE FIGHTER *in action*

You are working the night shift on a truck company at a dual-company station. Shortly after 1:15 a.m., the gong sounds, indicating a fire in a house located just down the street from the station. You slide down the pole and quickly mount the apparatus when the dispatcher indicates that neighbors are reporting they hear screaming coming from the house. When you arrive, you find a two-story, wood-frame house with heavy smoke showing. The engine begins pulling handlines, while two members of your crew are assigned ventilation duties and you and the other fire fighter are assigned primary search responsibility.

1. The main purpose of __________ is to find deceased victims.
   A. search
   B. rescue
   C. primary search
   D. secondary search
2. When rescuing a victim from a window, the ladder tip should be placed __________.
   A. just below the windowsill
   B. just above the windowsill
   C. beside the window
   D. above the window header
3. Where should a search begin?
   A. Search the areas where live victims may be located immediately around the fire.
   B. Search the area directly above the fire and the rest of that floor.
   C. Search the higher-level floors, working from the top floor down.
   D. Search the fire floor, working toward your way toward the fire.
4. The minimum number of rescuers required to properly remove a victim from a vehicle and onto a longboard is __________.
   A. 2
   B. 3
   C. 4
   D. 5
5. The most efficient method to remove an unconscious or unresponsive victim from a dangerous location is a __________.
   A. two-person walking assist
   B. clothes drag
   C. two-person extremity carry
   D. two-person chair carry

FIRE FIGHTER II *in action*

You are at the training center waiting for the class to begin. A number of the fire fighters are standing around outside discussing search and rescue. One of the more outspoken individuals makes the comment, "We should search every structure regardless whether it is vacant." He goes on to say, "As far as I am concerned, we should search no matter what because we might save a life."

1. What would you do if every other person there agreed with his comments?
2. What does the fire service generally recommend for firefighting in vacant structures?
3. Is this fire fighter's suggestion in these circumstances correct?
4. How would you address this situation?

Ventilation

Fire Fighter I

Knowledge Objectives

After studying this chapter, you will be able to:

- Describe the importance of including ventilation in a size-up. (p 443)
- Describe signs of a ventilation-limited fire. (p 443)
- Describe the impact of door control on ventilation. (p 444)
- Describe the impact of ventilation location. (p 444)
- Describe the impact of ventilation hole size. (p 445)
- Describe the impact of exterior suppression on fire behavior. (p 445)
- List ventilation priorities. (p 445)
- Describe the importance of the timing and coordination of ventilation and suppression. (p 445–446)
- Describe the effect of wind to the fire flow path. (p 446)
- Explain why keeping the lid on a fire helps to control the fire growth. (p 446)
- Describe how modern construction contributes to flashover. (p 446)
- Describe steps that can be taken to minimize backdrafts and flashovers. (p 447)
- Explain how fire research aids in effective ventilation practices. (p 448)
- Describe the role of mechanical ventilation in fire suppression. (p 449)
- Explain how fire-resistive construction affects ventilation operations. (NFPA 5.3.11A, p 449)
- Describe how the characteristics of ordinary construction affect ventilation operations. (NFPA 5.3.11A, p 449–450)
- Describe how the characteristics of wood-frame construction affect ventilation operations. (NFPA 5.3.11A, p 450)
- List the two basic types of ventilation. (NFPA 5.3.11A, p 451)
- Explain how horizontal ventilation removes contaminated atmosphere from a structure. (NFPA 5.3.11A, p 452)
- List the two methods of horizontal ventilation. (NFPA 5.3.11, p 452–460)
- Explain how natural ventilation removes contaminated atmosphere from a structure. (NFPA 5.3.11A, p 452)
- Describe the techniques used to provide natural ventilation to a structure. (NFPA 5.3.11, p 452–457)
- Explain how mechanical ventilation removes contaminated atmosphere from a structure. (NFPA 5.3.11A, p 457–460)
- Describe the techniques used to provide mechanical ventilation to a structure. (NFPA 5.3.11, p 457–460)
- Describe how negative-pressure ventilation removes contaminated atmosphere from a structure. (NFPA 5.3.11, p 457–458)
- Describe the techniques used to provide negative-pressure ventilation to a structure. (NFPA 5.3.11, p 457–458)
- Describe how positive-pressure ventilation removes contaminated atmosphere from a structure. (NFPA 5.3.11A, p 458–460)
- Describe the advantages of positive-pressure ventilation. (NFPA 5.3.11A, p 458–460)
- Describe the disadvantages of positive-pressure ventilation. (NFPA 5.3.11A, p 458–460)
- Describe the techniques used to provide positive-pressure ventilation to a structure. (NFPA 5.3.11, p 459–460)
- Describe how hydraulic ventilation removes contaminated atmosphere from a structure. (NFPA 5.3.11A, p 461)
- Describe how vertical ventilation removes contaminated atmosphere from a structure. (NFPA 5.3.12, p 461–464)
- Describe how to ensure fire fighter safety during vertical ventilation operations. (NFPA 5.3.12, p 461–463)
- Identify the warning signs of roof collapse. (NFPA 5.3.12A, p 464)
- Describe the components and characteristics of roof support structures. (NFPA 5.3.12A, p 464)
- Describe the components and characteristics of roof coverings. (NFPA 5.3.12A, p 464)
- Explain how roof construction affects fire resistance. (NFPA 5.3.12A, p 465)
- List the differences in solid-beam construction and truss construction in roofs. (NFPA 5.3.12A, p 465–466)
- List the basic types of roof design. (NFPA 5.3.12A, p 466–468)
- Describe the characteristics of flat roofs. (NFPA 5.3.12A, p 466)
- Describe the characteristics of pitched roofs. (NFPA 5.3.12A, p 466)
- Describe the characteristics of arched roofs. (NFPA 5.3.12A, p 466)
- Describe the techniques of vertical ventilation. (NFPA 5.3.12, p 468-469)
- List the tools utilized in vertical ventilation. (NFPA 5.3.12, p 469)
- List the types of roof cuts utilized in vertical ventilation operations. (NFPA 5.3.12A, p 469–476)
- Describe the characteristics of a rectangular or square cut. (NFPA 5.3.12A, p 471–472)
- Describe the characteristics of a louver cut. (NFPA 5.3.12A, p 472–473)

CHAPTER

15

- Describe the characteristics of a triangular cut. (NFPA 5.3.12A, p 473–474)
- Describe the characteristics of a peak cut. (NFPA 5.3.12A, p 474)
- Describe the characteristics of a trench cut. (NFPA 5.3.12A, p 474, 476)
- Describe the special considerations in ventilating concrete roofs. (NFPA 5.3.12A, p 477)
- Describe the special considerations in ventilating metal roofs. (NFPA 5.3.12A, p 477)
- Describe the special considerations in ventilating basements. (NFPA 5.3.12A, p 476)
- Describe the special considerations in ventilating high-rise buildings. (NFPA 5.3.12A, p 477)
- Describe the special considerations in ventilating windowless buildings. (NFPA 5.3.12A, p 478)
- Describe the special considerations in ventilating large buildings. (NFPA 5.3.12A, p 478)
- Explain how to ensure that ventilation equipment is in a state of readiness. (p 479–480)

Skills Objectives

After studying this chapter, you will be able to:

- Break glass with a hand tool. (NFPA 5.3.11 B, p 453–454)
- Break a window with a ladder. (NFPA 5.3.11 B, p 453, 455)
- Break a window with a Halligan tool. (NFPA 5.3.11 B, p 456)
- Deliver negative-pressure ventilation. (NFPA 5.3.11 B, p 458)
- Deliver positive-pressure ventilation. (NFPA 5.3.11 B, p 460)
- Perform hydraulic ventilation. (NFPA 5.3.11 B, p 461)
- Sound a roof. (NFPA 5.3.12 B, p 463–464)
- Operate a power saw. (NFPA 5.3.12 B, p 469–470)
- Perform a rectangular cut. (NFPA 5.3.12 B, p 471)
- Make a louver cut. (NFPA 5.3.12 B, p 472)
- Make a triangular cut. (NFPA 5.3.12 B, p 473)
- Make a peak cut. (NFPA 5.3.12 B, p 474–475)
- Make a trench cut. (NFPA 5.3.12 B, p 474)
- Perform a readiness check on a power saw. (NFPA 5.3.12 B, p 479–480)
- Maintain a power saw. (NFPA 5.3.12 B, p 479, 481)

Fire Fighter II FFII

Knowledge Objectives

After studying this chapter, you will be able to:

- List the tactical priorities in structural firefighting operations and how the tactical priorities affect ventilation operations. (NFPA 6.3.2, 6.3.2.B, p 450)
- Identify the warning signs of roof collapse. (NFPA 6.3.2.A, p 464)

Skills Objectives

There are no skill objectives for Fire Fighter II candidates. NFPA 1001, *Standard for Fire Fighter Professional Qualifications*, contains no Fire Fighter II Job Performance Requirements for this chapter.

You Are the Fire Fighter

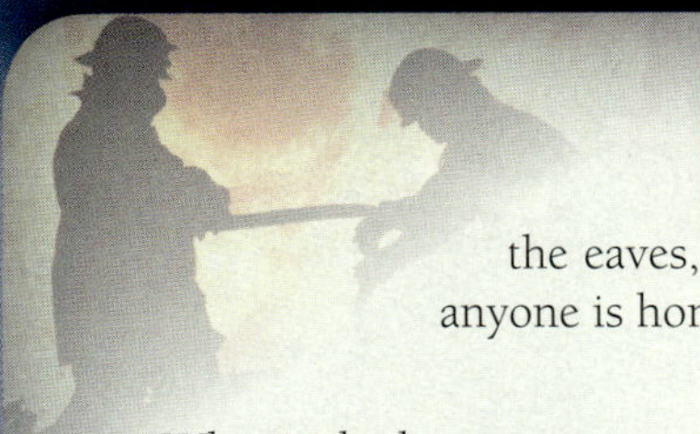

It is a routine day at the grocery store where you work full-time. You are busy stocking shelves when your pager goes off, saying there is a fire in a house only a few blocks away. On your arrival, you find a one-story, wood-frame house with dirty brown smoke pushing out through the eaves, siding, and other cracks. The windows are stained black. The neighbors say they do not think anyone is home because the occupants work during the day.

1. What is the best way to ventilate this structure?
2. Which concerns do you have about ventilating this structure?
3. When would you want to ventilate this structure?

Should be a coordinated attack with

Introduction

This chapter explains the purpose and importance of ventilation. It describes when to perform ventilation, where to ventilate, how to perform ventilation operations, and stresses the importance of coordinating ventilation with other fire-suppression activities.

Ventilation is defined as the fire-fighting operation of removing smoke and heat from a structure by opening windows and doors or making holes in the roof. One aspect that the definition does not cover is the coordination of ventilation. Clarifying this importance would provide clarity to the operating fire fighter. For our fire fighters operating in the field, maybe the definition should read:

> *Planned, systematic, and coordinated removal of heat, smoke, and fire gases (fire effluent) and their replacement with fresh air.*

Fire Fighter Safety Tips

Effective ventilation removes smoke, gases, and heat, but it also brings in cooler, cleaner, and oxygen-rich air. If ventilation is not coordinated with fire attack operations, the hazards from the fire can increase.

Understanding how and when to perform ventilation is vital to conducting safe and effective fire suppression. Properly performed and coordinated, ventilation can help to remove hot gases from the fire compartment, make it easier to locate the seat of the fire, improve visibility, and contribute to faster and safer knockdown, more effective fire suppression, and improved efficiency of searching. Poorly performed and poorly timed ventilation can quickly add oxygen to a fire and turn a ventilation-limited fire in a decay stage into an actively growing fire. Recently, tactical considerations for ventilation have changed because of the findings from the live fire experiments conducted by Underwriters Laboratories. The information and techniques presented in this chapter incorporates the results of those experiments.

Fire service ventilation can be looked at as a two-step process. The first step, which can be done during size-up, is a determination of the need for ventilation, an assessment of the location and amount of ventilation needed, and the coordination of ventilation operations with other parts of the fire-suppression operations. Planning and assessment of ventilation needs are discussed in the first sections of this chapter.

The second step of ventilating a fire is the mechanical operations of actually opening or closing doors and windows, opening skylights, and cutting openings in the roof. The information on conducting ventilation operations is discussed in the second part of this chapter.

Before fire fighters open a door, take out a window, or put a hole in the roof, they should ask themselves the following questions:

- Why am I ventilating? Is it to provide a tenable atmosphere for a potential occupant? Am I creating a more tenable atmosphere for possible occupants and fire fighters?
- Where do I want to accomplish the ventilation? Do I want to perform horizontal ventilation, vertical ventilation, or positive pressure ventilation?
- When do I want to perform the ventilation? I have to coordinate my efforts with the actions of the suppression team and the search and rescue teams, or I could be making conditions worse.

FIRE FIGHTER Tips

According to tests performed by the National Institute of Standards and Technology (NIST) and Underwriters Laboratories (UL), the amount of time needed for an underventilated fire to transition through flashover could be as fast as tens of seconds or as long as a few minutes depending on the size of the structure and the fire conditions at the time of ventilation. In some cases, a backdraft can occur in as little as 10 seconds after ventilation.

Size-Up and Ventilation

Size-up should include consideration of the tactics needed to ventilate the fire structure. Ventilation is not a one size fits all–type of operation where you perform exactly the same actions at every fire. It is necessary to evaluate multiple factors at each fire in order to determine which actions need to be taken and in which order these actions need to be initiated.

One of the most important aspects of size-up is the determination of where the fire is located, how big the fire is, and what stage the fire is in. Remember that a fire where there is little or no smoke showing may indicate a small fire in the incipient stage, a fire that can easily be extinguished with a small amount of water and little fire damage. However, a fire that initially has little or no smoke showing may also be a ventilation-limited fire. This type of fire contains large quantities of hot unburned fire gases in an oxygen-deficient environment. These hot gases may ignite and spread quickly if supplied with additional oxygen. If the fire is in a ventilation-limited decay stage, ventilation that is not coordinated with the application of water will result in sudden and rapid fire growth.

A thorough size-up of each fire is important to develop a plan not only for rescue and suppression, but also for ventilation. This size-up should include consideration of the following factors: the fire department arrival time, the size of the building, the type of building construction, the geometry or shape of the building, the potential for rescue, the potential for building collapse, the amount of fuel in the building contents, the impact of modern synthetic contents versus old natural-based contents, and the type of fire attack that can be used. You also need to assess possible places to ventilate, and evaluate how these ventilation options will impact the fire flow path and the spread of hot fire gases and smoke.

Recognizing Ventilation-Limited Fires

When fire fighters arrive at a fire that is in an underventilated stage, there may be no visible flames and very little smoke and they may incorrectly assume they are confronting a small fire, or a fire that has exhausted most of its fuel supply. Underventilated fires often appear to be in a state of decay. In fact, an underventilated fire is like a sleeping giant waiting for a supply of oxygen to transform it into an out-of-control inferno. Oxygen will be supplied if heat breaks a window or when a fire fighter forces open the front door of the residence. These are two of the many actions that can convert a fire from an underventilated state to a sudden flashover, or a backdraft. An underventilated fire contains large amounts of thermal energy. It also has plenty of fuel in the form of suspended flammable smoke particles and heated flammable gases. All it needs is a supply of oxygen to rapidly convert it to an active-burning state. In some cases, a backdraft can occur in as little as 10 seconds after ventilation. As building construction materials and techniques have changed and the fuel load of building contents is increased, fire behavior in residential fires has changed. This requires us to modify some of the ventilation procedures that have been used in the past. These changes have occurred because of the observations of many fire fighters at today's fires and controlled tests conducted by NIST and UL. These studies are available at www.nist.gov/fire, or at www.ULfirefightersafety.com.

Ventilation and Fire Behavior

Changes in ventilation may change fire behavior. Before ventilating, it is important to understand how a given action will affect the fire behavior. Because many of our building fires today involve modern synthetic contents, fires today grow much faster than fires with older contents. Today, many of the building fires we encounter will be in a ventilation-limited decay stage upon our arrival. This means that any action we take that results in the introduction of added oxygen to the fire is likely to change the fire from a ventilation-limited decay stage to a fuel-limited growth stage. An action as simple as opening the front door of the building can quickly change the fire from a decay stage to a growth stage.

The chapter on fire behavior describes the flow path as the volume in a structure between an inlet and an outlet that allows the movement of heat and smoke from the higher pressure within the fire area towards the lower pressure areas accessible via doorways, halls, stairs, and window openings. The flow path can function as a single uni-directional flow, between the fire and an opening to the outside acting as an exhaust. It also may act as bi-directional flows, acting as an inlet for fresh air and an outlet, or exhaust. In a bi-directional flow, a fire fighter may identify the neutral plane that defines the inlet and outlet. The neutral plane is the interface at a vent, such as a doorway or a window opening, between the hot gas flowing out of a fire compartment and the cool air flowing into the compartment where the pressure difference between interior and exterior is zero.

The flow path will be determined by the building design and which doors and windows are open to the outside. Any operations that are conducted in the exhaust portion of the flow path place fire fighters at significant risk for injury or death because of the large amount of heat and smoke that travel along the flow path.

Fire Fighter Safety Tips

Fire fighters operating in the exhaust portion of the flow path are at a significantly increased risk for injury or death due to the flow of hot fire gases and smoke moving over them.

There are several factors that you need to consider before ventilating a fire. These will influence the effectiveness of the ventilation and the speed with which the ventilation changes the fire behavior. These include: the impact of door control, the impact of the ventilation location, the impact of the ventilation hole size, and the impact of exterior suppression on fire behavior.

Fire Fighter Safety Tips

Some important points to remember regarding ventilation and fire flow paths are:

- Forcing the front door is ventilation and must be thought of as ventilation. While forcing entry is necessary to fight the fire, it must also trigger the thought that air is being fed to the fire and the clock is ticking before either the fire gets extinguished or it grows until an untenable condition exists, jeopardizing the safety of everyone in the structure.
- Once the front door is opened, attention should be given to the flow through the front door. A rapid rush of air or a tunneling effect could indicate a ventilation-limited fire.
- Every new ventilation opening provides a new flow path to the fire and vice versa. This could create very dangerous conditions when there is a ventilation-limited fire. You never want to be between where the fire is and where it wants to go without water or a door to close.
- Fire showing does not mean the fire is vented—it means it is venting and additional ventilation points will grow the fire if water is not applied.
- "Taking the lid off" does not guarantee positive results. Vertical ventilation is the most efficient type of natural ventilation. It allows for the most hot gases to exit the structure; however, it also allows the most air to be entrained into the structure. Coordination of vertical ventilation must occur with fire attack—just like with horizontal ventilation.
- Interior fire attack is still the most common and important fire attack method, but emphasis needs to be put on controlling ventilation and cooling fire gases. Apply water to the smoke to cool it as you move through it.

The Impact of Door Control

One of the first actions taken by most fire departments at a structure fire is to gain access to the building by opening the front door. This action changes the ventilation profile of the fire. It results in additional cool air being supplied to the fire. This can rapidly change a ventilation-limited fire into a fuel-controlled fire. If you limit the air inlet, you will limit the fire's ability to grow. The simple act of closing the front door after forcing entry will limit the air supply to the fire and slow the fire growth until a crew is ready to make access to the building and perform a coordinated fire attack. Sometimes the most effective act in ventilation is to limit ventilation until a coordinated fire attack can be made. When making an interior fire attack or rescue effort, try to maintain control of the front door, keeping it closed as much as possible to limit the flow of air. When making a fire attack, you can keep the front door open just enough to allow the hose line to be advanced to the fire. It may be necessary to station a firefighter at the front door to maintain this partial door closure.

Fire growth may also be limited by closing interior doors. Closing the door after searching a room will potentially limit the supply of air available to the fire. It will also help to reduce the risk of the fire transitioning into that room or other areas connected to that room. When closing the door to the room of origin or the involved room, it can help to contain a room-and-contents fire to the room until water can be applied.

Fire Fighter Safety Tips

If you limit air inlets to the fire, you can limit the ability of the fire to grow. Sometimes the best ventilation is less ventilation.

The Impact of the Ventilation Location

The location of the ventilation opening has an impact on the growth of the fire. The location of ventilation openings will influence how fast a fire reacts to changes in ventilation, and will determine how effective the opening is at removing hot gases and smoke from the fire location.

Ventilating directly over the fire will produce the fastest impact on the behavior of the fire, and will exhaust the greatest amount of combustion products. However, if this ventilation is done without coordinated water application, the result will mostly likely be added fire growth because of the added supply of oxygen to the fire. In other words, ventilating over the fire is helpful only if the fire attack is coordinated with ventilation. If the ventilation openings are farther from the fire, it will take longer for any changes in ventilation to impact the fire. These more distant openings will be less effective in removing smoke and hot gases from the fire due to the limitation of the door opening restricting airflow. The application of water absorbs some of the heat produced by the fire. This water application can help to remove more energy than the fire is creating. Once the fire growth has been slowed, a ventilation opening can be more effective in removing hot gases and smoke. This will improve the conditions in the fire compartment, reducing the air temperature and improving visibility.

Other factors that will affect how fast the fire will respond to oxygen include: the stage the fire is in, whether the fire is vent limited or fuel limited, the number and size of the ventilation inlets and outlets, the shape of the ventilation openings, the temperature of the fire room, the configuration of the walls, and the amount and type of contents in the fire flow path. The information in the following list is taken from the "Fire Service Summary Report: Study of the Effectiveness of Fire Service Vertical Ventilation and Suppression Tactics in Single Family Homes," published by the Firefighter Safety Research Institute of Underwriters Laboratories:

- The fire does not react to additional oxygen instantaneously. A ventilation action may appear to be positive at first, as the air is entrained into the ventilation-limited fire; two minutes later, conditions could become deadly without water application.
- The higher the interior temperatures, the faster the fire reacts. If the fire is showing on arrival, the interior temperatures are higher than if the house is closed. This means that additional ventilation openings are going to create more burning in a shorter period of time.
- The closer the air is to the fire, the faster the fire reacts. Venting the fire room will more quickly increase burning; it will also let the hot gases out faster after water is applied.

- The higher the ventilation, the faster the fire reacts. Faster and more efficient ventilation means faster air entrainment, which means more burning and higher temperatures. It also means better ventilation after water is applied.
- The more air, the faster the fire reacts. Also, the more exhaust, the more air that can be entrained into the fire. A bigger ventilation hole in the roof means that more air will be entrained into the fire. If the fire is fuel limited, this is good, but if the fire is ventilation limited, this could be bad.

Fire Fighter Safety Tips

Limiting the flow path by limiting inlet and exhaust openings until water is ready to be applied is important to limit the supply of oxygen to the fire and to reduce the amount of heat release and temperature rise within the structure.

The Impact of the Ventilation Hole Size

Recent fire studies conducted by UL demonstrated that larger-sized vertical ventilation openings did not localize the growth of the fire. In addition, vertical ventilation alone did not reduce the temperatures in the fire building. Their studies compared the impact of cutting a 4' x 4' roof ventilation hole with the effect of cutting a 4' x 8' roof ventilation hole. Neither the small hole nor the larger hole had the effect of lowering the temperature in the fire area when only ventilation was performed.

When performing ventilation in coordination with an exterior fire attack, the larger hole had a more positive effect for any potential victims and for fire fighters operating inside the fire building. It appears that vertical ventilation by itself does not usually have a positive effect on a ventilation-limited fire. However, vertical ventilation in coordination with an exterior or interior application of water as close to the fire as possible has the effect of improving visibility, reducing the temperature in the fire compartment, and temporarily limiting the growth of the fire. Therefore, when conducting a coordinated attack in conjunction with any water application, it appears that a larger ventilation hole is more effective in lowering the temperature of the fire compartment and other parts of the building. After water application, it also improves visibility and reduces the temperature, making conditions better for fire fighters and victims inside, although this improvement was for a very short time. Although it is important to understand that with today's fires, the fire service simply cannot vent enough to vent the enormous amounts of smoke and particulates.

The Impact of Exterior Suppression on Fire Behavior

The results of the ventilation tests conducted by UL seem to indicate that ventilation by itself has a negative impact on improving fire conditions. With fires in a ventilation-limited state, which is most of the fires we encounter today, ventilation alone has the result of introducing more oxygen into the fire therefore resulting in fire growth. It does not appear that you can create large enough ventilation openings to reduce the temperature of the fire compartment or the surrounding areas. For the greatest reduction in production of hot gases and smoke, use a coordinated fire attack by performing ventilation and applying water to the fire from a safe location as close to the fire as possible. This removes large amounts of hot fuel, results in a reduction in internal temperatures, and improves the internal visibility. Ventilation, then, is a vitally important part of fire suppression, but to be effective, it must be carried out in close coordination with the application of water as close to the fire as possible. Providing ventilation without this coordination can contribute to the growth of the fire. Applying water to the fire as quickly as possible from either the exterior or interior can help to improve conditions in the entire fire structure. Even a small amount of water improves the conditions within the residence. This also improves fire fighter safety and the safety of any potential victims. Ventilation without water application may lead to very dangerous conditions.

Ventilation Priorities

The second part of this chapter describes the steps needed to perform various types of ventilation. The type of ventilation used will depend on several factors. One is the ease of executing (performing) that type of ventilation. Another consideration is the effectiveness of each type of ventilation. For the purposes of this discussion, consider ventilating to include opening or closing doors, windows, skylights, and roofs.

Usually the door is the easiest of these to use for ventilation. It is possible to open some doors without tools. The door may have been left open by the building residents. If the door is locked, it is usually not too difficult to open with simple tools. And they often need to be opened to provide access to or egress from the building. Windows are the next easiest to open. These can sometimes be opened manually or they can be opened quickly by removing the glass and frame. First-floor windows may sometimes be removed without the use of ladders. Second-floor windows are usually readily accessible using ground ladders. Ventilating skylights and roofs usually requires gaining access using ground ladders or aerial devices. Sloped roofs may require the use of roof ladders for performing ventilation.

The effectiveness of each type of ventilation depends on the size of the opening and its location. Opening a door will usually create a larger ventilation opening than opening a window. Some windows are easily opened while others are constructed with reinforced materials that are difficult to remove or break. Opening a skylight is often easier and faster than cutting a hole in the roof. Roofing materials vary greatly as noted in the chapter *Building Construction*. Even after a roof has been cut open, there may be additional layers of ceiling materials that need to be removed to create an exhaust opening.

Timing and Coordination of Ventilation and Suppression

Remember that if you limit the amount of air entering the fire compartment, you limit the fire's ability to grow. Keeping the front door and other ventilation openings closed until fire fighters are prepared to apply water to the fire will reduce the speed of the fire growth. This is a temporary action until water can be applied to the fire. When considering ventilation options and timing, remember that "taking the lid off" does not guarantee positive results. To make sure the fire does not get larger, you must first change the fire from a ventilation-limited stage to a fuel-limited stage by applying water. Once the water has begun to cool the fire, more energy is being absorbed by the water than is being generated by the fire. At this point, the ventilation will begin to work as intended. Remember that a vital part of a coordinated fire attack is ventilation. Because it takes time to get a charged hose line in place, you should sometimes delay actually opening a vent until a charged hose line is in place and the crew is ready to apply water to the fire from a close and safe location. Whenever you think about ventilation, think about the *Three Ws* (when, where, and why) of ventilation, which includes timing. Effective ventilation requires an incident command function to coordinate timing and a fire attack.

The Effect of Wind on Fire Flow Path

Wind is a powerful force that can rapidly change the direction and speed of a fire and its flow path. Any time a window or door is opened on the side of a building that faces the wind, known as the windward side, a huge amount of oxygen under high pressure is introduced into the fire. This action may be accidental, such as a glass window or patio slider door breaking because of heat from the fire, or it can be an intentional act of ventilation. A sudden increase in oxygen combining with hot flammable smoke (fuel) from the fire will result in rapid fire growth, and can also produce a sudden change in the direction of the flow path due to the pressure imposed by the wind. Think of the wind as a giant positive pressure fan forcing huge quantities of oxygen into a ventilation-limited fire **FIGURE 15-1**. Remember to keep the wind at your back during a fire attack and avoid ventilating on the windward or leeward side of a fire unless it is part of a well-organized suppression effort. The National Institute of Standards and Technology (NIST) has

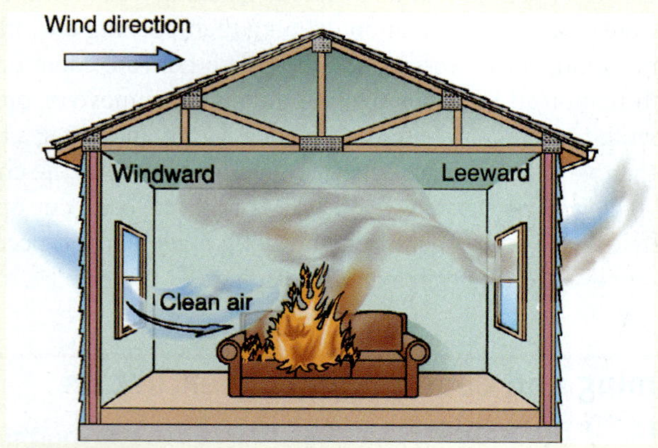

FIGURE 15-1 On a windy day, proper ventilation helps to remove smoke and heat from a structure.

conducted experiments on wind-driven fires. More information about these studies is available at the NIST website. These tests were conducted using high-rise structures but wind is also a very real threat in one- and two-family structure fires.

Keeping the Lid on Ventilation-Limited Fires

One means of controlling a ventilation-limited fire is to limit the amount of oxygen that is available to the fire. Understanding how flow paths function helps aid in this control. Limiting the oxygen available to the fire until other suppression activities can be implemented may help to keep the fire contained. Opening the front door of a house supplies the fire with oxygen and may quickly produce a violent flashover. Opening the door must be considered an act of ventilation. While ventilation needs to be considered as a critical part of our fire-suppression activities, it should always be conducted in coordination with effective and sustained suppression efforts. Leaving the front door closed as long as possible aids significantly in controlling a fire by limiting the amount of oxygen supplied to it, as does stationing a fire fighter at the front door to keep it partly closed as a hose line is advanced. This fire fighter can also assist in advancing the hose line as they control the door. Key points regarding the fire flow path and ventilation drawn from the fire experiments conducted by NIST and the FDNY are shown in "Key Points" box in the section on *Ventilation and Fire Behavior*.

Fire Fighter Safety Tips

Many structure fires today are ventilation-limited or ventilation-controlled fires and are discovered in the ventilation-controlled stage.

Modern Construction and Flashover

Consider today's modern construction, which is built with fast-growth lumber and/or lightweight and manufactured building components. The structure and contents contain resins and glues, and houses are filled with petroleum-based furnishings. This recipe gives us more fuel, fuels that reach flashover faster, fuels that burn hotter, and building components that fail sooner. Add to this the fact that most recently built residences are sealed from the outside with tight widows, vapor barriers, and insulation, and now you have buildings that may catch fire and quickly and reach high temperatures. The wide use of heating and air conditioning systems means that in many parts of the country our houses are tightly closed for most of the year. The result is that today, many fires reach a point where they have a limited supply of oxygen if the doors and windows remain closed and temperatures can reach the point of flashover in a few minutes. At this point, many fires use up enough oxygen that the amount of oxygen in the house begins to drop. When this happens, the fire, which is still very hot and has an abundance of fuel, becomes underventilated, or a ventilation-limited fire.

Minimizing Backdrafts and Flashovers

Ventilation is a major factor in two significant fire-ground phenomena: backdraft and flashover. Both can be deadly situations, and fire fighters should exercise great caution when conditions indicate that either is possible. A transitional fire attack reduces the chance of a backdraft or flashover. Apply water to the fire from a safe location as close to the fire as possible. This will reduce heat and fuel in the form of hot gases and smoke from the fire compartment. It will reduce the temperature of the space, ideally converting the ventilation-limited fire to a fuel-limited fire. Ventilation will then be more effective in removing hot gases and smoke from the fire space and allow entry into the interior space for final fire suppression FIGURE 15-2.

Backdrafts and Ventilation

A backdraft can occur when a building is charged with hot gases and most of the available oxygen has been consumed. Few flames might be apparent, but the hot gases can contain rich amounts of unburned or partially burned fuel. If clean air is introduced to the mixture, the fuel can ignite and explode. To reduce the chance of a backdraft, fire fighters try to release as much heat and unburned products of combustion as possible.

Placing a ventilation opening as high as possible within the building or area can help to eliminate potential backdraft conditions. A roof opening can draw the hot gases up and relieve the interior pressure. As the mixture rises into the open atmosphere and is exposed to oxygen, it might ignite. The ventilation crew should ensure that a charged hose line is in place to attack the fire before they ventilate the building.

The attack crews should charge their hose lines outside the building. The ventilation crew should not ventilate until the attack crew is ready to advance the attack line. It is very important to remember that opening the front door of a structure to begin a fire attack is an act of ventilation. The door should remain closed until the attack team is ready to advance the hose line into the building.

Once the attack team sees flames inside the structure, they should open their hose streams to extinguish the fire and cool the interior atmosphere as quickly as possible. In cases where flames are beginning to move across the ceiling, a stream of water on the ceiling helps to cool the upper area of the room without upsetting the thermal balance.

Flashovers and Ventilation

Flashover is a process that occurs when the air in a room is superheated and exposed combustibles in the space are near their ignition point TABLE 15-1. In a typical room-and-contents fire, the ceiling temperature can reach 1200°F (648°C). Because the potential harm from a flashover is so great, you need to be able to recognize any conditions that might indicate the possibility of a flashover and take steps to prevent it whenever possible. Do not enter an environment if signs of an impending flashover are present. Use a transitional attack to cool the fire compartment, vent, and then enter.

TABLE 15-1 Chemical Properties of Fire Gases

Chemical Name	Ignition Temperature	Flammable Range
Propane	896°F/480°C	2.1 to 9.5 percent
Methane	1076°F/580°C	5 to 15 percent
Carbon Monoxide	1128°F/609°C	12.5 to 74 percent
Hydrogen Cyanide	1000°F/538°C	5.6 to 40 percent

A.

B.

FIGURE 15-2 **A.** Building structure and fire conditions should be evaluated for signs of backdraft before any type of ventilation opening is made. **B.** Improperly venting a fire in an oxygen-deficient atmosphere can cause a backdraft.

Fire Fighter Safety Tips

There are no reliable signs or symptoms to warn you of an impending flashover in a ventilation-controlled fire. If a fire fighter recognizes the potential signs of pending flashover, either effect egress or apply water to the upper areas of the compartment immediately.

Fire Fighter Safety Tips

During a flashover, the flow of hot gases has been measured at speeds between 12 and 20 miles per hour with no wind present. This is faster than a fire fighter can run to try to escape the path of the flashover.

Combining Research and Practice

The recent UL and NIST fire research studies have demonstrated that our traditional assumptions about the effects of ventilation need to be modified. Ventilation has often been defined as the systematic removal of heat, smoke, and fire gases from a building, and replacement of those elements with cool, fresh air. Although this definition is and always has been technically true, it can lead to a misapplication of the principle. What research has repeatedly demonstrated is that when dealing with ventilation-limited fires, any cooling effect produced by ventilation alone is minimal and short-lived. Because of the extremely fuel-rich environment found on today's fire ground, ventilation that is not preceded by and concurrent with or immediately followed by effective suppression will introduce enough oxygen to rapidly bring the fire area to flashover. TABLE 15-2 shows the conclusions for residential fire behavior that were drawn from the fire experiments conducted by NIST, UL, and the FDNY.

TABLE 15-2 Residential Fire Behavior

- Increasing the air flow to a ventilation-limited structure fire by opening doors, windows, or roof openings will increase the fire hazard from the fire for both building occupants and fire fighters.
- Increasing the air flow to a ventilation-limited structure may lead to a rapid transition to flashover.
- There is a need for coordinated fire attack to coordinate the activities of size-up, entry, ventilation, search and rescue, and application of water.

Courtesy of NIST.

It is important to understand the results of recent fire research on fire ventilation, and then be able to put these findings into practice at fire scenes. We have noted that ventilation must be closely coordinated with all other activities being carried out at a fire scene. Well-planned ventilation, when coordinated with the proper application of water, has several benefits. It greatly cools the fire building and removes large quantities of highly flammable fuels in the form of superheated gases and smoke. This greatly reduces the chance for flashovers or backdrafts. Ventilation helps to improve the visibility within the fire building, thereby making it easier to search for fire victims. Proper ventilation helps firefighters maintain their orientation within the building. Reduced temperatures make it safer for fire fighters and increases the potential for the survival of building occupants. When properly performed, ventilation can assist in locating the seat of the fire; on the other hand, improperly performed ventilation can turn a manageable fire into an out-of-control inferno FIGURE 15-3.

As noted in the preceding section, applying water through a window or door from a safe location as close to the fire as possible just before or at the same time ventilation occurs is called a transitional attack. This offensive exterior fire attack must occur prior to entry, search, and suppression. Surprisingly, the UL/NIST research demonstrated the application of water from the outside, when properly applied, did NOT push the fire from one place to another. The water introduced into the fire from the outside produced cooling, not just in the fire compartment, but also in other areas of the fire building. Significant cooling was achieved even when the water was applied to a body of fire on the interior of the structure at a distance from the seat of the fire. While this application of water did not totally extinguish the fire, it cooled the fire gases throughout most parts of the fire building, making it safer for fire fighters to enter and complete the suppression of the fire. It also greatly increased the chance for survival of any fire victims. It is important to stress that the transitional fire

A. B.

FIGURE 15-3 Proper ventilation enables fire fighters to control a fire rapidly. **A.** Vented structure. **B.** Unvented structure.

attack does not reduce the importance of proper ventilation, so it requires close coordination between the fire attack team and the ventilation crew at a fire. More information about mounting an effective and safe transitional fire attack is presented in the chapter *Fire Suppression*.

FIRE FIGHTER Tips

Concerns about saving energy have led to increased use of insulated thermal glass, both in new construction and as a replacement for conventional glass in older windows. Thermopane windows have multiple layers of glass sealed together, with a small airspace between the layers. These windows are much more difficult to break than conventional glass windows are. In fact, a fire fighter might have to use multiple strikes and additional force to break a thermopane window.

The Role of Mechanical Ventilation

In addition to making or controlling openings to influence ventilation, fire fighters can use mechanical ventilation to direct the flow of combustion gases. Fans can be used to draw or pull smoke through openings (negative-pressure ventilation) or to introduce fresh air to displace smoke and other products of combustion (positive-pressure ventilation). Hose streams can be used to create a hydraulic ventilation to remove hot gases and smoke from a room. Some buildings have ventilation systems that are designed to remove smoke or prevent smoke from entering certain areas.

How Building Construction Affects Ventilation

The way a building is constructed affects ventilation operations. Each construction type presents its own set of problems.

Fire-Resistive Construction

Fire-resistive construction refers to a building in which all of the structural components are made of noncombustible or limited-combustible materials. This type of construction generally has spaces divided into compartments, which limit potential fire spread. A fire is most likely to involve the combustible contents of an interior space within such a building.

Although fire-resistive construction is intended to confine a fire, it does contain potential avenues for fire spread. Openings for mechanical systems such as heating and cooling ducts, plumbing and electrical chases, elevator shafts, and stairwells all provide paths for fire spread. A fire can also spread from one floor to another through exterior windows, a phenomenon called vertical fire extension (or auto-exposure) FIGURE 15-4.

Smoke can travel through a fire-resistive building using the same routes. Stairways, elevator shafts, and heating, ventilation, and air-conditioning (HVAC) systems are common pathways. Because the windows in these buildings are often sealed and are not broken easily, there can be limited opportunities for creating ventilation openings. In some cases, one

FIGURE 15-4 Vertical fire extension occurs when the fire jumps from floor to floor through exterior windows.

stairway can be used as an exhaust shaft, while other stairways are cleared for rescue and access. Ventilation fans are often required to direct the flow of smoke. Some windowless buildings contain a number of glass panels with an indicator (sometimes a Maltese cross) in a lower corner, indicating that the panel is made of tempered safety glass and can be easily broken with a sharp, pointed object.

In many of these buildings, engineered designs are installed to assist fire fighters with ventilation or controlling the movement of hot gases and smoke throughout the building. Smoke and heat vents are installed in the ventilation systems, stairwells and elevator shafts are designed with positive pressure to limit smoke migration, and HVAC systems are designed to be used by arriving fire fighters to pressurize floors or areas to limit fire spread. In some cases, the most valuable person on scene may be the building engineer who can assist in the operations of many of these systems from the fire command center.

The roof on a fire-resistive building is usually supported by a steel or concrete roof deck. It can be difficult or impossible to make vertical ventilation holes in such a roof. These roofs might, however, have openings for skylights or for HVAC ducts. Similar roofing materials can be found in buildings with noncombustible construction.

Ordinary Construction

Ordinary construction buildings have exterior walls made of noncombustible or limited-combustible materials that support the roof and floor assemblies. The interior walls and floors are usually wood construction; the roof usually has a wood deck and a wood structural support system. In this type of construction, the roof can be cut with power saws or axes to open vertical ventilation holes. Ordinary construction buildings usually have windows and doors that can be used for horizontal ventilation.

In this type of construction, the walls and floors might include numerous openings for plumbing and electrical chases. If a fire breaks through the interior finish materials (plaster or drywall), it can spread undetected within these void spaces to other portions of the structure. Interior stairwells generally allow a fire to extend vertically within the building. Often, ordinary construction buildings have been remodeled and have multiple roof ceilings, making ventilation more difficult.

Although ventilation can be performed either horizontally or vertically in such a case, it can be hard to ventilate in the area close to where the fire is burning.

The vertical openings in ordinary construction frequently provide a path for fire to extend into an attic or cockloft (open space between the ceiling of the top floor and the underside of the roof). Heat and smoke can accumulate within these spaces and then spread laterally between sections of a large building or into attached buildings. Coordinated suppression and vertical ventilation are essential to extinguish attic fires FIGURE 15-5.

Wood-Frame Construction

Wood-frame construction has many of the same features as ordinary construction. The primary difference is that the exterior walls in a wood-frame building are not required to be constructed of masonry or noncombustible materials.

Wood-frame buildings often contain many void spaces where fire can spread, including attics or cocklofts. Modern, fast-growth lumber; lightweight wood-truss roofs; and manufactured I-beam floors, which can fail quickly under fire conditions, are common in newer buildings.

Older buildings of wood-frame construction were often assembled with balloon-frame construction. This type of construction includes direct vertical channels within the exterior walls, so a fire can spread very quickly from a lower level to the attic or cockloft in such a building. These buildings were constructed with readymade "ventilation ducts" that can carry fire gases and smoke to the attic. Heated smoke and gases can accumulate under the roof, requiring rapid vertical roof ventilation.

FIGURE 15-5 Coordinated suppression and vertical ventilation are essential to extinguish fires.

Modern wood-frame construction uses platform-frame techniques. In these buildings, the structural frame is built one floor at a time. Between each floor a plate at the floor and the ceiling acts as a fire stop. Its presence limits the fire from spreading up and helps contain the fire on a single floor.

Modern wood-frame construction typically uses lightweight components to save money and provide more longer and open areas, which presents problems when a fire occurs. Lightweight trusses are subject to failure after only a few minutes of exposure to a fire. Wooden I-beams contain much less wood than the solid beams used in older construction and burn more rapidly and might fail quickly. The fire service cannot become complacent with modern dimensional lumber as new lumber is engineered to "grow quickly." In some cases, a tree may be only 18 years of age when harvested. UL studies have shown that modern dimensional lumber may fail as quickly as engineered products.

FIRE FIGHTER II

Tactical Priorities

Ventilation is directly related to the two major tactical priorities in structural firefighting operations: life safety and fire containment and property conservation.

Venting for Life Safety and Containment

Life safety and containment are the primary goals of the fire service, whether it involves rescuing civilians or protecting fire department personnel. Good ventilation practices help to clear smoke, heat, and toxic gases from the structure, which gives occupants a better chance to survive. It provides firefighting crews with increased visibility and makes the structure more tenable, enabling more rapid searches for victims.

Ventilation in coordination with an effective transitional fire attack cools the interior of the building, limits fire spread, and allows fire fighters to advance hose lines more safely and rapidly to attack the fire. A coordinated fire attack and ventilation effort must consider the timing and the direction of the attack, as well as the type and location of possible ventilation openings. The swift control of the fire reduces the risk to fire fighters as well as to building occupants.

Venting for Property Conservation

Another priority of the fire service is property conservation, which involves reducing the losses caused by means other than direct involvement in the fire. Since much of the damage from a fire can be smoke damage, ventilation can play a significant role in limiting property damage. If a structure is ventilated rapidly and correctly, the damages associated with smoke, heat, water, and overhaul operations can all be reduced.

Location and Extent of Smoke and Fire Conditions

Fire fighters must be able to recognize when ventilation is needed and where it should be provided, based on the circumstances of each fire situation. Many factors must be considered, including the size of the fire, the stage of combustion, the location of the fire within a building, and the available ventilation options. An experienced fire fighter learns to recognize these significant factors immediately and to interpret each situation quickly.

Ventilation should occur as close to the fire as possible. The most desirable options are to provide a ventilation opening directly over the seat of the fire or to open a door or window that will release heat and smoke from the fire directly to the exterior. If it is not possible to create a ventilation opening in the immediate area of the fire, fire fighters must be able to predict how ventilation from another location will affect the fire. Fire travels toward a ventilation opening, unless influenced by outside forces such as strong winds along the fire flow path.

The color, location, movement, and amount of smoke can provide valuable clues about the fire's size, intensity, and fuel **FIGURE 15-6**.

Smoke movement is a good indicator of the fire's temperature. A very hot fire produces smoke that moves quickly, rolling and forcing its way out through an opening. The hotter the fire, the more the gases expand, the greater the pressure in the structure, the faster the smoke's movement. Cooler smoke moves more slowly and gently. On a cool, damp day with very little wind, this type of smoke might hang low to the ground (a phenomenon known as smoke inversion).

FIGURE 15-6 The color, location, and amount of smoke can provide valuable clues for fire fighters.

FIRE FIGHTER Tips

In many recent fires where fire fighters have been injured, the initial crews described the smoke as being light and light colored. This seemingly benign appearance should not fool fire fighters: Such smoke can be just as dangerous as thick black smoke. Always remember that these unburned products of combustion are producing toxic gases and vapors, no matter what the smoke's color.

A different phenomenon occurs in buildings with automatic sprinkler systems. The water discharged from the sprinklers cools the smoke and produces a cold smoke that may hardly move within the building. This type of smoke can fill a large warehouse space from floor to ceiling with an opaque mixture that behaves much like fog on a damp day. Because this cold smoke has a lower temperature than the surrounding atmosphere, it sits at the lower levels of the room, making it hard to see and hard to locate the seat of a fire. Mechanical ventilation is often needed to clear this cold smoke from the building. A similar situation occurs when smoke becomes trapped within a building long enough to cool it to the ambient temperature.

Types of Ventilation

To remove the products of combustion and other airborne contaminants from a structure, fire fighters use two basic types of ventilation: horizontal and vertical.

Horizontal ventilation takes advantage of the doors, windows, and other openings at the same level as the fire. In some cases, fire fighters might make additional openings in a wall to provide horizontal ventilation.

Vertical ventilation involves making openings in roofs or floors so that heat, smoke, and toxic gases can escape from the structure in a vertical direction. Pathways for vertical ventilation can include ceilings, stairwells, exhaust vents, and roof openings such as skylights, scuttles, or monitors. Additional openings can be created by cutting holes in the roof or the floor and making sure that the opening extends through every layer of the roof or floor. Vertical ventilation typically incorporates horizontal openings, such as doors and windows, to provide an intake for the make-up air to replace the smoke that is being exhausted through the vertical vent.

Ventilation can be either natural or mechanical. Natural ventilation depends on convection currents and other natural forces, such as the wind, to move heat and smoke out of a building and to allow clean air to enter. Mechanical ventilation uses fans or other powered equipment to introduce clean air or to exhaust heat and smoke.

Ventilation can also be classified as intentional or accidental. Intentional ventilation occurs when the ventilation is planned and done on purpose. Accidental ventilation occurs when a window or door fails because of the fire or a window or door is left open by mistake. Any ventilation may quickly change the direction or intensity of a fire, but unexpected ventilation may jeopardize fire fighters operating inside a structure due to rapid changes in the fire's behavior.

When referring to ventilation techniques, fire fighters use the term *contaminated atmosphere* to describe the products of combustion that must be removed from a building and the term *clean air* to refer to the outside air that replaces them. The contaminated atmosphere can include any combination of heat, smoke, and gases produced by combustion, including toxic gases. A contaminated atmosphere also can be defined as a dangerous or undesirable atmosphere that was not caused by a fire. For example, natural gas, carbon monoxide, ammonia, and many other products can contaminate the atmosphere as a result of a leak, spill, or similar event. The same basic techniques of ventilation can be applied to many of these situations as well.

Horizontal Ventilation

Horizontal ventilation uses horizontal openings in a structure, such as windows and doors, and can be employed in many situations, particularly in small fires **FIGURE 15-7**. Horizontal ventilation is commonly used in residential fires, room-and-contents fires, and fires that can be controlled quickly by the attack team.

Horizontal ventilation can be a rapid, generally easy way to clear a contaminated atmosphere. Often, outlets for horizontal ventilation can be made simply by opening a door or window.

FIGURE 15-7 Windows are frequently used in horizontal ventilation.

In other cases, fire fighters might need to break a window or to use forcible entry techniques. Search and attack teams operating inside a structure as well as fire fighters operating outside the building use horizontal ventilation to reduce heat and smoke conditions.

Horizontal ventilation is most effective when the opening goes directly into the room or space where the fire is located. Opening an exterior door or window, for example, allows heat and smoke to flow directly outside and eliminates the problems that can occur when these products travel through the interior spaces of a building.

Horizontal ventilation is more difficult if there are no direct openings to the outside or if the openings are inaccessible. In these situations, it might be necessary to direct the air flow through other interior spaces or to use vertical ventilation. For example, fire fighters need to use horizontal ventilation if an inaccessible or damaged roof makes vertical ventilation impossible.

Horizontal ventilation also can be used in less urgent situations if the building can be ventilated without additional structural damage. Low-urgency situations might include residual smoke caused by a stovetop fire or a small natural gas leak.

Horizontal ventilation tactics include both natural and mechanical methods.

Fire Fighter Safety Tips

Closing a door provides a temporary barrier to the spread of fire. This can protect parts of a building from fire and smoke.

Natural Ventilation

Natural ventilation depends on convection currents, wind, and other natural air movements to allow a contaminated atmosphere to flow out of a structure. The heat of a fire creates convection currents that move smoke and gases up toward the roof or ceiling and out and away from the fire source. Opening or breaking a window or door allows these products of combustion to escape through natural ventilation.

Natural ventilation can be used only when the natural convection air currents or wind are adequate to move the contaminated atmosphere out of the building and replace it with fresh air. Mechanical devices (discussed later in this chapter) can be used if natural forces do not provide adequate ventilation.

Natural ventilation is often used when quick ventilation is needed, such as when attacking a room-and-contents fire or a first-floor residential fire with people trapped on the second floor. For search and attack teams to enter these buildings and act quickly, smoke and heat must be ventilated immediately.

Wind speed and direction play an important role in natural ventilation. If possible, windows on the downwind side of a building should be opened first so that the contaminated atmosphere flows out. Openings on the upwind side can then be used to bring in clean air. Conversely, opening a window on the upwind side first could push the fire into uninvolved areas of the structure, particularly on a windy day.

FIRE FIGHTER Tips

Opening windows on the upwind side of a building before opening downwind-side windows can force the fire into uninvolved areas of the structure.

Breaking Glass

Natural ventilation uses existing or created openings in a building. The methods used to create horizontal openings employ a variety of tools and techniques. For example, some windows can simply be opened by hand. If the window cannot be opened and the need for ventilation is urgent, however, fire fighters should not hesitate to break the glass. Breaking a window is a fast and simple way to create a ventilation opening.

When breaking glass, the fire fighter should always use a hand tool (Halligan tool, axe, or pike pole) and keep his or her hands above or to the side of the falling glass. This tactic prevents pieces of glass from sliding down the tool and potentially injuring the fire fighter. The tool should then be used to clear the entire opening of all remaining pieces of glass, thereby creating the largest opening possible and providing a way for fire fighters to enter or exit through the window in the event of an emergency.

Clearing the broken glass also reduces the risk of injury to anyone who might be in the area and eliminates the danger of sharp pieces falling out later. Before breaking glass—and particularly if the window is above the ground floor—fire fighters must look out to ensure that no one will be struck by the falling glass.

To break glass on the ground floors with a hand tool, follow the steps in SKILL DRILL 15-1:

1. Wear full personal protective equipment (PPE), including eye protection and SCBA.
2. Select a hand tool and position yourself to the side of the window. (STEP 1)
3. With your back facing the wall, swing backward forcefully with the tip of the tool, striking the top one-third of the glass. (STEP 2)
4. Clear the remaining glass from the opening with the hand tool. (STEP 3)

Breaking a Window from a Ladder

A similar technique can be used to break a window on an upper floor, with the fire fighter working from a ladder. The ladder should be positioned to the upwind (or windward) side of the window so that smoke is carried away from the fire fighter. The ladder should be positioned to the upwind (or windward) side of the window with the tip even with the top of the window. The fire fighter should climb to a position level with the window and lock into the ladder for safety. Using a hand tool, the fire fighter should strike and break the window, and then clear the opening completely. The fire fighter should not be positioned below the window, where glass could slide down the handle of the tool FIGURE 15-8.

FIGURE 15-8 When breaking an upper-story window, the fire fighter should place the ladder to the side of the window.

Breaking a Window with a Ladder

Fire fighters on the ground can use the tip of a ladder to break and clear a window when immediate ventilation is needed on an upper floor. Dropping the tip of the ladder into the upper half of the glass breaks it. Moving the tip of the ladder back and forth inside the opening usually clears the remaining glass.

This technique requires proper ladder selection. Usually, second-floor windows can be reached by 16- (5 m) or 20-foot (6 m) roof ladders, as well as by bedded 24- (7 m) or 28-foot (8.5 m) extension ladders. Extension ladders are needed to break windows on the third or higher floors.

To perform this maneuver, the ladder can be raised directly into the top half of the window or it can be raised next to the window to determine the proper height for the tip. The ladder is then rolled into the window, drawn back at the tip, and forcibly dropped into the top third of the window. The objective is to push the broken glass into the window opening, although there is always a risk that some of the glass will fall outward. For this reason, fire fighters working below the window must wear full PPE to shield themselves from falling glass. A word of caution: The higher the window, the greater the danger of falling glass. It is also important to remember that this form of ventilation must still be used in coordination with suppression operations and the incident's tactical objectives.

When a ladder is used to break a window, shards of glass might be left hanging from the edges of the opening. If the opening will be used for access, it needs to be cleared with a tool before anyone enters through it.

To break a window with a ladder, follow the steps in SKILL DRILL 15-2:

1. Wear full PPE, including eye protection.
2. Select the proper size of ladder for the job.

SKILL DRILL 15-1 Breaking Glass with a Hand Tool
(Fire Fighter I, NFPA 5.3.11)

1 Position yourself to the side of the window.

2 With your back facing the wall, swing backward forcefully with the tip of the tool striking the top one-third of the glass.

3 Clear the remaining glass from the opening with the hand tool.

SKILL DRILL 15-2 Breaking Windows with a Ladder

(Fire Fighter I, NFPA 5.3.11)

1 Check for overhead lines. Use standard procedures for performing a ladder raise.

2 Raise the ladder next to the window. Extend the tip so that it is even with the top third of the window. If a roof ladder is used, extend the hooks toward the window.

3 Position the ladder in front of the window.

4 Forcibly drop the ladder into the window. Exercise caution—falling glass can cause serious injury.

5 Raise the ladder from the window and move it to the next window to be ventilated.

3. Check for overhead lines. Use standard procedures for performing a ladder raise using one, two, or three fire fighters, depending on the size of the ladder. (**STEP 1**)
4. Raise the ladder next to the window. Extend the tip so that it is even with the top third of the window. If a roof ladder is used, extend the hooks toward the window. (**STEP 2**)
5. Coordinate operations with the incident's tactical objectives.
6. Position the ladder in front of the window. (**STEP 3**)
7. Forcibly drop the ladder into the window.
8. Exercise caution—falling glass can cause serious injury. Glass might slide down the beams of the ladder. (**STEP 4**)
9. Raise the ladder from the window and move it to the next window to be ventilated. Either carry the ladder vertically or pivot the ladder on its feet. (**STEP 6**)

Breaking a Window from Above

In some instances, fire fighters might be located above the window that needs to be broken. For example, fire fighters could be on the roof, performing roof ventilation; they could be on a floor above the fire; or they might not have a ladder available that can reach the window.

One way to break a window from above is to attach a Halligan tool to a rope or chain, toss it over the roof edge, and allow it to swing into the window. An alternative method is to use a long-handled tool, such as a pike pole, and to reach down from the edge of the roof, from a balcony, or from a window on the floor above the fire. Smoke and heat will be released when the window is broken, so be sure to stand to the side to keep from being enveloped by these products of combustion.

To break a window from above with a Halligan tool, follow the steps in **SKILL DRILL 15-3**:

1. Look over the roof edge to locate the window to be broken. Always be sure to check if crews are operating below you when you do this maneuver because glass and debris might fall down from the broken window.
2. Attach the Halligan tool securely to the rope or chain using a clip or hook.
3. Lower the Halligan tool over the edge of the roof until the tool is located in the middle of the window. Mark this spot on the rope or chain; it is the pivot point for the toss. (**STEP 1**)
4. Raise the tool back to the roof. (**STEP 2**)
5. Toss the Halligan tool over the roofline and out approximately the distance down to the window. This action allows the tool to swing into the window with enough force to break the glass.
6. Be sure to be in a safe position because this technique puts you above and in the direct path of the products of combustion that are being vented. Always be aware of the location of the roof edge as well as your safety zone on the roof.
7. If the tool becomes lodged in the window, give it a few shakes and pulls. It should free itself, breaking whatever it is caught on. Always check whether crews are operating below you when you perform this maneuver because glass and debris might fall down. (**STEP 3**)

SKILL DRILL 15-3 Breaking Windows on Upper Floors: Halligan Toss
(Fire Fighter I, NFPA 5.3.11)

1 Lower the Halligan tool over the edge of the roof to measure the distance.

2 Raise the Halligan tool back to the roof.

3 Toss the Halligan tool over the roofline so that it swings back to break the glass.

Opening Doors

Like windows, door openings can be used for natural ventilation operations. Doorways have certain advantages when used for ventilation. Specifically, they are large openings—often twice the size of an average window. In fact, some doorways can be large enough to drive a truck through. A locked door can be opened using forcible entry techniques.

The problem with using doorways as ventilation openings is that their use for this purpose compromises entry and exit for interior fire attack teams as well as search teams. If heat and smoke are pouring through a doorway, fire fighters cannot enter or exit through that portal. Because protecting exit paths is a priority, doorways are more suitable as entries for clean air. The contaminated atmosphere should, therefore, be discharged through a different opening.

FIRE FIGHTER Tips

Watch for glass sliding down the beams of the ladder. Wear full protective clothing and SCBA (on air) and gloves when performing this type of ventilation because this procedure can cause injury to an unwary fire fighter. Coordinate operations with the incident's tactical objectives.

Fire fighters often approach a fire through the interior of a building to attack from the unburned side. Opening an exterior door that leads directly to the fire can create an opening for oxygen-rich air to enter the building or a vent that pushes the heat and smoke out of the building. Treat the opening of a door as a major act of ventilation. It should be opened only when the hose line is charged and the attack team is ready to advance.

Before opening an exterior door, examine the area around the door. A small amount of smoke can be the sign of a small fire or a sign of an underventilated fire. Proceed with extreme caution. Once the exterior door is opened, note how the air moves through the exterior door. A rapid movement of air into the structure can indicate an underventilated fire. If the fire is actively burning, you might see clean air entering through the bottom half of the exterior door and smoke escaping through the top half (bi-directional flow).

Doorways are also good places to put mechanical ventilation devices when natural ventilation needs to be augmented. Fans can be used to push clean air in (positive-pressure ventilation) or to pull contaminated air out (negative-pressure ventilation).

Fire Fighter Safety Tips

Opening an exterior door is an act of ventilation. In an underventilated fire, this act can quickly produce a backdraft.

Mechanical Ventilation

Mechanical ventilation uses large high-powered fans to augment natural ventilation; that is, it involves the use of fans or other powered equipment. There are three different methods of mechanical ventilation. Negative-pressure ventilation uses fans called smoke ejectors to exhaust smoke and heat from a structure. Positive-pressure ventilation uses fans to introduce clean air into a structure and push the contaminated atmosphere out. Hydraulic ventilation moves air by using fog or broken-pattern fire streams to create a pressure differential behind and in front of the nozzle.

In some buildings, HVAC systems can be used to provide mechanical ventilation. To perform this function, such a system needs to be configured so that the zones can be controlled correctly and effectively. The building engineer or maintenance staff might have the specific knowledge required to assist in this effort. Some HVAC systems can be used to exhaust smoke or supply fresh air if a fire occurs, whereas others should be shut down immediately. Information about HVAC systems should be obtained during preincident planning surveys.

Negative-Pressure Ventilation

The basic principles of air flow are used in negative-pressure ventilation. Fire fighters locate the source of the fire and, using a fan or blower, exhaust the products of combustion out through a window or door. The fan creates a negative pressure that draws the heat, smoke, and fire gases out of the building. In turn, fresh, clean air must enter the structure, to replace the exhausted contaminated air.

Smoke ejectors are usually 16 to 24 inches (40 to 60 cm) in diameter, although larger models are available. They can be powered by electricity, gasoline, or water pressure FIGURE 15-9.

Negative-pressure ventilation can be used to clear smoke out of a structure after a fire, particularly if natural ventilation would be too slow or if no natural cross-ventilation exists. In large buildings, negative-pressure ventilation can be used to pull heat and smoke out while fire fighters are working, although it is not generally used before or during a fire attack.

FIGURE 15-9 A smoke ejector pulls products of combustion out of a structure. Gasoline-powered smoke ejectors might have problems running because of reduced oxygen or carbon particles in the smoke.

Although this type of ventilation is usually associated with horizontal ventilation, an ejector can be used in a roof-vent operation to pull heated gases out of an attic.

Negative-pressure ventilation has several limitations related to its positioning, power source, maintenance, and air flow control. Often, fire fighters must enter the heated and smoke-filled environment to set up the ejector, and they might need to use braces and hangers to position it properly. Because most ejectors run on electricity, a power source and a cord are required for them to function. The ejectors also get very dirty while they are running, so they must be taken apart and cleaned after each use.

Air flow control with a negative-pressure system is also difficult. The fan must be completely sealed so that the exhausted air is not immediately drawn back into the building. This phenomenon, which is called recirculation, reduces the effectiveness of the ejector. Recirculation can be eliminated by completely blocking the opening around the fan.

Smoke ejectors usually have explosion-proof motors, which makes them excellent choices for venting flammable or combustible gases and other hazardous environments.

To perform negative-pressure ventilation, follow the steps in SKILL DRILL 15-4:

1. Determine the area to be ventilated and the outside wind direction.
2. Place the smoke ejector so that it is not on the upwind side of the building.
3. Hang the fan as high in the selected opening as possible.
4. Remove any obstructions from the area used to ventilate smoke, including curtains, screens, and other debris.
5. Use a salvage cover to prevent recirculation if the structure's windows are not intact. (STEP 1)
6. Provide an opening on the windward side of the structure to provide cross-ventilation. (STEP 2)

FIRE FIGHTER Tips

Negative-pressure ventilation is not frequently used in initial fire attacks. However, you should understand how it works because it can be a useful tool to exhaust contaminated air whether that is caused by smoke or other toxic products.

Positive-Pressure Ventilation

Positive-pressure ventilation uses large, powerful fans to force fresh air into a structure. The fans create a positive pressure inside the structure, which displaces the contaminated atmosphere and pushes heat and other products of combustion out. Positive-pressure ventilation can be used to reduce interior temperatures and smoke conditions in coordination with an initial fire attack or used to clear a contaminated atmosphere after a fire has been extinguished.

Positive-pressure fans are usually set up at exterior doorways, often at the same opening used by the attack team FIGURE 15-10. In an average-sized residence, these fans may be most effective in front of the exterior door. This creates a cone of air that improves the flow of fresh air entering the fire building. This pushes clean air in the same direction as the advancing hose line so that fire fighters move toward the fire enveloped in a stream of clean, cool air.

SKILL DRILL 15-4 Delivering Negative-Pressure Ventilation

(Fire Fighter I, NFPA 5.3.11)

1 Hang the fan in the upper part of the opening. Use salvage covers to prevent churning.

2 Provide openings on the windward side for cross-ventilation.

FIGURE 15-10 Positive-pressure fans are usually set up at exterior doorways, often at the same opening used by the attack team.

While the fan pushes fresh air into the building, it is equally important to have an opening on the other side of the building to exhaust the superheated air and smoke. This opening should be as close to the fire as possible. Improper placement of the fan or the exhaust opening can help to spread the fire. When performed properly, positive-pressure ventilation can push heat and smoke out of the building through an exhaust vent on the opposite side of the building.

For this kind of system to work effectively, the integrity of the building must be intact and ventilation openings kept to a minimum. Positive-pressure ventilation does not work properly if there are too many openings. There must be an exhaust opening near the seat of the fire to allow the heat and smoke to escape **FIGURE 15-11**. The entry and exit openings should be approximately the same size to create the desired positive pressure within the structure. If there is no exhaust opening or if it is too small, the heat and smoke can flow into other areas of the building or back toward the attack team. If the exhaust opening is too large, it does not create the build-up of pressure inside the structure that would otherwise create the velocity needed to remove the smoke effectively.

FIGURE 15-11 There must be an opening near the seat of the fire to allow heat and products of combustion to exhaust.

To increase the efficiency of positive-pressure ventilation and to prevent smoke and fire from reaching other parts of the building, fire fighters should close doors to unaffected areas of the structure. This is an effective action to maintain a tenable environment for people trapped in other parts of the building.

Fire Fighter Safety Tips

According to studies by the National Institute of Technology and Standards, when positive-pressure ventilation is performed as part of the initial fire attack, the fan increases the heat of the fire as a result of the addition of oxygen-rich air for a few minutes. To ensure fire fighter safety, the attack crew should wait until the heat release is stabilized before advancing toward the fire source.

Fire Fighter Safety Tips

The importance of close coordination among members of the ventilation crew and the attack team cannot be overstressed. Good communication and proper coordination of operations can prevent backdraft.

When planning positive-pressure ventilation, the force and direction of the wind need to be taken into consideration. Positive-pressure fans should never be placed in a position where they are blowing against the wind. Check for the direction and speed of the wind before placing fans and monitor the wind for any changes in direction or intensity.

Because smaller areas can be ventilated rapidly, venting one room at a time or one section of the building at a time can quickly clear trapped smoke out of a building after a fire. In a high-rise building, positive-pressure fans can blow fresh air up through a stair shaft. By opening the doors, fire fighters can clear one floor at a time. Very large structures can be ventilated by placing multiple fans side by side.

FIRE FIGHTER Tips

In high-rise fires, a large number of people need to be safely evacuated, often by a single stairway. The attack stairway and evacuation stairways must be established and announced early in the event. The evacuation stairway must be free of smoke and toxic gases. Positive-pressure fans can be used to keep smoke and products of combustion from entering the stairwell. In emergencies, they can also be used to introduce clean, oxygen-rich air into the potentially toxic atmosphere of the stairwell. When using positive-pressure fans for this purpose, keep in mind that the engines can increase carbon monoxide levels, so ensure that the fans are properly ventilated.

Positive-pressure ventilation has several advantages over negative-pressure ventilation. For instance, a single fire fighter can set up a fan very quickly. Because the fan is positioned outside the structure, the fire fighter does not have to enter a hazardous environment, and the fan does not interfere with interior operations. Positive pressure is both quick and efficient because the entire space within the structure is under pressure.

When used with a well-placed exhaust opening, positive-pressure ventilation can help keep a fire confined to a smaller area and increase safety for interior operating crews, as well as any building occupants. In addition, positive-pressure fans do not require as much cleaning and maintenance as negative-pressure fans because the products of combustion never move through the fan.

Positive-pressure ventilation does have some disadvantages, however. The most important concern is the potential for improper use of positive pressure to spread a fire. If the fire is burning in a structural space such as in a pipe chase or above a ceiling, positive pressure can push it into unaffected areas of the structure. If the fire is located in structural void spaces, positive-pressure ventilation should not be used until access to these spaces is available and attack crews are in place. Ineffective positive-pressure ventilation occurs if positive pressure is used and the exhaust gases do not have openings to release the positive pressure from the building.

Positive-pressure fans operate at high velocity and can be very noisy. Most are powered by internal combustion engines, so they can increase carbon monoxide levels if they are run for significant periods of time after the fire is extinguished. Natural ventilation should be used after the structure is cleared to prevent carbon monoxide build-up. Because positive-pressure fan motors can get hot, they are unsafe to use if combustible vapors are present.

SKILL DRILL 15-5 outlines the steps in performing positive-pressure ventilation:

1. Determine the location of the fire within the building and the direction of attack.
2. Place the fan in front of the opening to be used for the fire attack. The exact position of the fan will depend on the size of the opening, the size of the fan, and the direction of any wind. (**STEP 1**)
3. Provide an exhaust opening at or near the fire. This opening can be made either before the fan is started or when the fan is started. (**STEP 2**)
4. Check for interior openings that could allow the products of combustion to be pushed into unwanted areas.
5. Start the fan and check the cone of air produced; it should completely cover the opening. This can be checked by running a hand around the door frame to feel the direction of air currents.
6. Allow smoke to clear—usually 30 seconds to 1 minute depending on the size of the area to be ventilated and the smoke conditions before entering the building. (**STEP 3**)
7. Remember: if conditions worsen because of the ventilation, turn the fan off.

FIRE FIGHTER Tips

Doors used for ventilation must be kept open to ensure that ventilation is sustained. Depending on the time available and the type of door and lock, fire fighters might remove the door, disable the closure device, or place a wedge in the hinge side or under the door. Overhead doors must be blocked in the open position, thereby ensuring that they cannot accidentally roll down and close. Check the manufacturer's recommendations.

On some fans, if the fan was tipped down while getting it out of the apparatus, the oil sensor will cause the fan not to start. Fire fighters need to know the procedure to use for the fan to get it started.

SKILL DRILL 15-5 Delivering Positive-Pressure Ventilation

(Fire Fighter I, NFPA 5.3.12)

1 Place the fan in front of the opening to be used for the fire attack.

2 Provide an exhaust opening at or near the fire.

3 Start the fan and allow the smoke to clear.

Hydraulic Ventilation

Hydraulic ventilation uses the water stream from the hose line to exhaust smoke and heated gases from a structure. To use this ventilation technique, the fire fighter working the hose line directs a narrow fog stream or a broken stream from a smoothbore nozzle out of the building through an opening, such as a window or doorway. The contaminated atmosphere is drawn into a low-pressure area behind the nozzle. An induced draft created by the high-pressure stream of water then pulls the smoke and gas out through the opening FIGURE 15-12. A well-placed fog stream can move a tremendous volume of air through an opening.

Hydraulic ventilation is most useful in clearing smoke and heat out of a room after the fire is under control. Hydraulic ventilation can move several thousand cubic feet of air per minute and is effective in clearing both heat and smoke from a fire room. Because it does not require any specialized equipment, it can be performed by the attack team with the same hose line that was used to control the fire.

Some disadvantages are also associated with the use of hydraulic ventilation. Fire fighters must enter the heated, toxic environment and remain in the path of the products of combustion as they are being exhausted. If the water supply is limited, using water for ventilation must be balanced with using water for fire attack. This technique can cause excessive water damage if it is used improperly or for long periods of time. Also, in cold climates, ice build-up can occur on the ground, creating a safety hazard.

Ideally, the ventilation opening should be created before the hose line advances into the fire area. The hose stream also can be used to break a window before it is directed on the fire. The steam created when the water hits the fire pushes the heat and smoke out through the vent. Hydraulic ventilation can then be used to clear the remaining heat and smoke from the room.

To perform hydraulic ventilation, follow the steps in SKILL DRILL 15-6:

1. Enter the room and remain close to the ventilation opening.
2. Place the nozzle through the opening and open the nozzle to a narrow fog or broken-spray pattern.
3. Keep directing the stream outside and back into the room until the fog pattern almost fills the opening. The nozzle should be 2 to 4 feet (0.6 to 1.2 meters) inside the opening.
4. Stay low, out of the heat and smoke, or to one side to keep from partially obstructing the opening.

FIGURE 15-12 Hydraulic ventilation can be used to draw smoke out of a building after the fire is controlled.

Vertical Ventilation

Vertical ventilation refers to the release of smoke, heat, and other products of combustion into the atmosphere in a vertical direction. Vertical ventilation occurs naturally, owing to the fire's flow path (convection), if an opening is available above a fire. Convection causes the products of combustion to rise and flow through the opening. In some situations, vertical ventilation can be assisted by mechanical means such as fans or hose streams.

Although vertical ventilation refers to any opening that allows the products of combustion to travel up and out, this term is most often applied to describe operations on the roof of a structure. The roof openings can be existing features, such as skylights or bulkheads, or holes created by fire fighters who cut through the roof covering. The choice of roof openings depends primarily on the building's roof construction.

A vertical ventilation opening should be made as close as possible to the seat of the fire. Smoke issuing from the roof area, melted asphalt shingles, or steam coming from the roof surface are all signs that fire fighters can use to identify the hottest point. In order for a vertical exhaust vent to be effective, there must also be a horizontal intake vent to admit air. In addition, if horizontal vents are made on a floor above the fire, they can produce the effect of vertically venting the fire.

Safety Considerations in Vertical Ventilation

Vertical ventilation should be performed only when it is necessary and can be done safely. Rooftop operations involve significant inherent risks. If horizontal ventilation can sufficiently control the situation, vertical ventilation might be unnecessary. Before performing vertical ventilation, fire fighters must evaluate all pertinent safety issues and avoid unnecessary risks. In particular, they should assess the roof for roof scuttles, heat vents, plumbing vents, louver ventilation, translucent roof panels, cell phone towers, photovoltaic panels, and fan shafts.

The most obvious risk in a rooftop operation is the possibility that the roof will collapse. Many fire fighters have died during vertical ventilation attempts because the fire had compromised the structural integrity of the roof support system. If the structural system supporting the roof fails, both the roof and the fire fighters standing on it will fall into the fire. Fire fighters also can fall through an area of the roof deck that has been weakened by the fire.

Determine the type of roof construction before beginning roof ventilation operations. If the roof is constructed with lightweight trusses, ventilate the roof from an aerial ladder or from an aerial basket. Roofs built using lightweight manufactured trusses can fail quickly and without warning.

Near Miss REPORT

Report Number: 10-0000745

Event Description: Brackets [] denote reviewer deidentification.

Tower [1] responded to a residential structure fire. Tower [1] was the first arriving unit to the scene with light smoke showing from the front door. The homeowner reported the fire in the basement of the home. Crews opted for an interior attack with a handline from Tower [1]. Crews located the basement access from the kitchen and proceeded to make their way to the basement of the home. Upon entering the basement, fire was found in the laundry room of the home. Interior attack teams extinguished the visible flames and then called for ventilation to help with low visibility. After requesting ventilation, heat intensified at a rapid rate downstairs, forcing attack teams to pull out of the basement to the main level of the home. After reaching the top of the stairs, interior crews noted extreme heat now on the main level. Heavy, thick smoke conditions blinded attack teams, causing them to take shelter in the kitchen of the home. Fire was observed to be above firefighters. Original egress was no longer an option because of dark smoke and fire. The interior teams' attack line was burned and rendered not usable. Crews recognized the rapid change and bailed out the kitchen window because of extreme heat and smoke conditions. Two of the three crew members suffered minor injuries. Approximately 20 seconds after crews bailed out, the kitchen floor where we were standing collapsed into the basement. The flashover was believed to be caused by extreme heat building in a couch on the main level of the home. When the ventilation fan was placed to the front door, it is possible that the rush of fresh air helped to fuel the smoldering couch and ignited a fire.

Lessons Learned: Our department teaches us to recognize rapid change. In this instance, our bailout training helped with confidence not to hesitate and to bail out. Another important point is that the flashover occurred approximately five minutes after fire was deemed "out." Crews were venting the structure when the flashover occurred.

Fire Fighter Safety Tips

Ventilate roofs constructed with lightweight construction from an aerial ladder or aerial tower FIGURE 15-13. Do not attempt to stand on these roofs if there are signs of heat or fire underneath.

FIGURE 15-13 Fire fighters operating from the safety of aerial apparatus.

Falling from a roof—either off of the building or through a ventilation opening, open shaft, or skylight into the building—are risks fire fighters encounter when operating on roofs. Smoke reduces visibility and can cause disorientation, increasing the risk of falling off the edge of a roof. Factors such as darkness, snow, ice, or rain can also create hazardous situations. Not surprisingly, a sloping roof presents even greater risks than a flat roof. Safe operations on steep roofs require fire fighters to operate from an aerial ladder or basket or to use a roof ladder.

Effective vertical ventilation assists with prompt fire control and reduces the risks to fire fighters operating inside the building. In some situations, prompt, proper, and coordinated vertical ventilation can save the lives of building occupants by relieving interior conditions, opening exit paths, and allowing fire fighters to enter and perform search and rescue operations.

Vertical ventilation should always be performed as quickly and efficiently as possible. If they provide an adequate opening, skylights, ventilators, bulkheads, and other existing openings should be opened first. If fire fighters must cut through the roof covering to create a hole, the location should be identified and the operation should be performed promptly and efficiently. It must be done in coordination with the other fire fighters operating on the fire scene.

Fire fighters who are working on the roof should always have two safe exit routes. A second ground ladder or aerial device should be positioned to provide a quick alternative exit, in case conditions on the roof deteriorate rapidly. The two escape routes should be separate from each other, preferably in opposite directions from the operation site. The team working on the roof should always know where these exit routes are located, and they should plan ventilation operations to work toward these locations.

The ventilation opening should never be located between the fire fighters and their escape route. A ventilation opening releases smoke and hot gases from the fire. If this exhaust flow path of hot gases and smoke separates the fire fighters from their exit from the roof, they will be in serious danger FIGURE 15-14.

Once a ventilation opening has been made, fire fighters should withdraw to a safe location. There is no good reason to stand around the vent hole and look at it; this behavior simply exposes the fire fighters to an unnecessary risk.

Before fire fighters climb or walk onto any roof, they should test the roof to ensure that it will support their weight. Older roofs constructed of heavy beams can become soft or spongy before they fail. Modern roofs constructed with lightweight trusses can fail suddenly and without warning.

Modern lightweight construction frequently uses thin plywood for roof decking, with a variety of waterproof and insulating coverings then being applied to the plywood. A relatively minor fire under this type of roof can cause the plywood to delaminate, with no visible indication of damage or weakness from above. The weight of a fire fighter on a damaged area can be enough to open a hole and cause the fire fighter to fall through the roof.

Sounding is not a safe method to test the condition of a roof. It can give fire fighters a false sense of security on some types of roofs. For example, slate and tile roofs can appear solid even if the supporting structure is ready to fail, owing to the rigid surface of the tiles. Sounding lightweight construction truss roofs is not a reliable test of roof safety. Always look for other visible indicators of possible collapse and take the roof construction into account.

FIGURE 15-14 Be prepared for smoke and flames that are released when the roof is opened.

Determine the type of roof construction before beginning roof ventilation operations. If the roof is constructed with lightweight trusses, operate from an aerial ladder or from an aerial basket. To sound a roof for safety, follow the steps in SKILL DRILL 15-7:

1. Before stepping onto a roof, use a hand tool (such as an axe, pike pole, or sledge) to strike the roof decking with considerable force. Listen and feel for rebound. Be aware that sounding lightweight construction truss roofs is not a reliable test of roof safety. (STEP 1)
2. If a tool penetrates, the roof is not safe.
3. Locate support members by listening to the sound and watching the bounce of the tool. Support members create a solid sound and make the tool bounce from the roof. By contrast, the spaces between supports will sound hollow, and the tool will not bounce.
4. Continue to sound the area around you periodically to monitor conditions. (STEP 2)
5. Sound the roof as you leave the area. You are not completely safe until your feet are back on solid ground. (STEP 3)

Fire Fighter Safety Tips

Sounding a roof to see if it is stable is not a safe test with a truss or lightweight roof. Most roofs constructed in the last 20 years are of lightweight construction.

A fire fighter's path to a proposed vent-hole site should follow the areas of greatest support and strength. These locations include the roof edges, which are supported by bearing walls, and the hips and valleys of the roof, where structural materials are doubled for strength. These areas should be stronger than interior sections of the roof.

Fire fighters should always be aware of their surroundings when working on a roof. Knowing where the roof's edge is located, for example, can help avoid accidentally walking or falling off the roof. Maintaining this knowledge is especially important—although difficult—in darkness, adverse weather, or heavy smoke conditions. It is equally important to watch for openings in the roof, such as skylights and ventilation shafts.

The order of the cuts in making a ventilation opening should be planned carefully. Fire fighters should be upwind, have a clear exit path, and be standing on a firm section of the roof or using a roof ladder. If fire fighters are upwind from the open hole, the wind will push the heat and smoke away from them; thus, the escaping heat and smoke should not block their exit path. To prevent an accidental fall into the building, fire fighters should stand on a portion of the roof that is firmly supported, particularly when making the final cuts.

SKILL DRILL 15-7 Sounding a Roof
(Fire Fighter I, NFPA 5.3.12)

1 Use a hand tool to check the roof before stepping onto it.

2 Use the tools to sound ahead and on both sides as you walk. Locate the support members by sound and rebound. Check conditions around your work area periodically.

3 Sound the roof along your exit path.

Basic Indicators of Roof Collapse

Roof collapse poses the greatest risk to fire fighters performing vertical ventilation. Some roofs—particularly truss roofs—give little or no warning that they are about to collapse.

Fire fighters who are assigned to vertical ventilation tasks should always be aware of the condition of the roof. They should immediately retreat from the roof if they notice any of the following signs:

- Any visible indication of sagging roof supports
- Any indication that the roof assembly is separating from the walls, such as the appearance of fire or smoke near the roof edges
- Any structural failure of any portion of the building, even if it is some distance from the ventilation operation
- Any sudden increase in the intensity of the fire from the roof opening
- High heat indicators on a thermal imaging camera

Fire Fighter Safety Tips

Signs of imminent roof collapse include fire showing around roof vents, melting snow, and evaporating or steaming water. These signs indicate that the roof is being compromised by fire below it and is not safe.

Roof Construction

Roofs can be constructed of many different types of materials in several configurations. Nevertheless, all roofs have two major components—a support structure and a roof covering.

Roof Support Structures

The roof support system provides the structural strength to hold the roof in place. It must also be able to bear the weight of any rain or snow accumulation, any rooftop machinery or equipment, and any other loads placed on the roof, such as fire fighters or other people walking on the roof. The support system can be constructed of solid beams of wood, steel, or concrete or a system of trusses. Trusses are produced in several configurations and are made of wood, steel, or a combination of wood and steel.

Roof Coverings

The roof covering constitutes the weather-resistant surface of the roof and can have several layers. Roof covering materials include shingles and composite materials, tar and gravel, rubber, foam plastics, and metal panels. The roof decking is a rigid layer made of wooden boards, plywood sheets, or metal panels. In addition to the decking, the roof covering typically includes a waterproof membrane and insulation to retain heat in winter and limit solar heating in the summer.

Vertical ventilation operations often involve cutting a hole through the roof covering. The selection of tools, the technique used, the time required to make an opening, and the personnel requirements all depend on the roofing materials and the layering configurations.

Effects of Roof Construction on Fire Resistance

Fire can very quickly burn through some roofs, whereas others retain their integrity for long periods. The difference arises because each type of material used in roof construction is affected differently by fire. These reactions either increase or decrease the time available to perform roof operations. For example, solid beams are generally more fire resistant than truss systems are because the former materials are larger and heavier (more mass). Wooden beams are affected by fire more quickly than concrete beams are. Steel, unlike wood, does not burn but elongates and loses strength when heated.

Most roofs eventually fail as a result of fire exposure— some very quickly and others more gradually. In some situations, the fire weakens or burns through the supporting structure so that it collapses while the roof covering remains intact. Structural failure usually results in sudden and total collapse of the roof. A structural failure involving other building components could also cause the roof to collapse.

In other cases, the fire burns through the roof covering, even though the structure is still sound. This type of roof failure usually begins with a "burn through" close to the seat of the fire or at a high point above the fire. As combustible products in the roof covering become involved in the fire, the opening spreads, eventually causing the roof to collapse.

The inherent strength and fire resistance of a roof are often determined by local climate conditions. In areas that receive winter snow loads, roof structures can usually support a substantial weight and might include multiple layers of insulating material. A serious fire could burn under this type of roof with very little visible evidence from above. Heavily constructed roofs, supported by solid structural elements, can burn for hours without collapsing.

In warmer climates, however, roof construction is often very light. Roofs in these geographic areas might simply function as umbrellas. The supporting structure can be just strong enough to support a thin deck with a waterproof, weather-resistant membrane. This type of roof can burn through very quickly, and the supporting structure can collapse after a short fire exposure.

Solid-Beam vs. Truss Construction

The two major structural support systems for roofs use either solid-beam or truss construction. It can be impossible to determine which system is used simply by looking at the roof from the exterior of the building, particularly when the building is on fire. Given this fact, information on the roof support system should be obtained during preincident planning surveys.

The basic difference between solid-beam and truss construction is in the way individual load-bearing components are made. Solid-beam construction uses solid components, such as girders, beams, and rafters. Trusses are assembled from smaller, individual components. In most cases, it makes no difference whether solid-beam or truss construction is used. For fire fighters, however, there is an important difference: A truss system can collapse quickly and suddenly and without warning when it is exposed to a fire even for a short period of time.

Trusses are constructed by assembling relatively small and lightweight components in a series of triangles. The resulting system can efficiently span long distances while supporting a load. In most cases, a solid beam that spanned an equal distance and carried an equivalent load would be much larger and heavier than a truss FIGURE 15-15. The disadvantage of truss construction is that some trusses fail completely when only one of the smaller components is weakened or when only one of the connections between components fails.

FIGURE 15-15 A lightweight truss can carry as much weight as a much heavier solid beam but might fail quickly and without warning during a fire.

A truss is not necessarily bad or inherently weak construction. Indeed, some trusses are very strong and are assembled with substantial components that can resist fire just as well as solid beams. Many trusses, however, are made of lightweight materials and are designed to carry as much load as possible with as little mass as possible. These trusses have been shown to fail more quickly when exposed to fire.

Truss construction can be used with almost any type of roof or floor, including flat roofs, pitched roofs, arched roofs, or overhanging roofs. Fire fighters should assume that any modern construction uses truss construction for the roof support system until proven otherwise.

The individual components of a lightweight truss used for roof support are often wood 20 × 40 (2-inch by 4-inch) (50.8 mm by 101.6 mm) sections or a combination of wood and lightweight steel bars or tubes. Because these components are small, they can be weakened quickly by a fire. Even more critical, however, are the points where the individual components join together. These connection points are often the weakest portion of the truss and the most common point of failure.

For example, the components of a truss constructed with 20 × 40 wood pieces can be connected by heavy-duty staples or by gusset plates (connecting plates made of wood or lightweight metal) FIGURE 15-16. Due to pyrolysis and the loss of wood mass, these connections can fail quickly in a fire. When one connection fails, the truss loses its ability to support a load.

If only one truss fails, other trusses in the system might be able to absorb the additional load, but the overall strength of the system is still compromised. A fire that causes one truss to fail probably weakens additional trusses in the same area. These trusses also can fail, sometimes as soon as the additional load is transferred. In most cases, a series of trusses fails in rapid succession, resulting in a total collapse of the roof. Sometimes just the weight of a fire fighter walking on the roof is enough to trigger a collapse.

FIGURE 15-16 Gusset plates can fail in a fire, destroying the truss.

Some trusses are made of metal, usually individual steel bars or angle sections that are welded together. Lightweight steel trusses, known as bar joists, often support flat roofs on commercial or industrial buildings. When exposed to the heat of a fire, these metal trusses expand and lose strength. A roof supported by bar joists will probably sag before it collapses, a warning that fire fighters should heed. The expansion can stretch the trusses, which causes the supporting walls to collapse suddenly. Horizontal cracks in the upper part of a wall may indicate that the steel roof supports are pushing outward.

Roof Designs

Fire fighters must be able to identify the various types of roofs and the materials used in their construction. Creating ventilation openings requires the use of many different tools and techniques, depending on the type of construction and roof decking.

Flat Roofs

Flat roofs can be constructed with many different support systems, roof decking systems, and materials. Although they are classified as flat, most roofs have at least a slight slope so that water can flow to roof drains or scuppers.

Flat roof construction is generally very similar to floor construction. The roof structure can be supported by solid components, such as beams and rafters (the elements that support the roof), or by trusses. The horizontal beams or trusses often run from one exterior load-bearing wall (a wall that supports the weight of a floor or roof) to another load-bearing wall. In some buildings, the roof is supported by a system of vertical columns and/or interior load-bearing walls.

The roof deck is usually constructed of multiple layers, beginning with wooden boards, plywood sheets, or metal decking. Then come one or more layers of roofing paper, insulation, tar and gravel, rubber, gypsum, lightweight concrete, or foam plastic. Some roof decks have only a single layer, consisting of metal sheets or precast concrete sections.

Flat roofs often have vents, skylights, scuttles (small openings or hatches with movable lids), or other features that penetrate the roof deck. Removing the covers from these openings provides vertical ventilation without the need to make cuts through the roof deck.

Flat roofs also can have parapet walls, freestanding walls that extend above the normal roofline. A parapet can be an extension of a firewall, a division wall between two buildings, or a decorative addition to the exterior wall.

Pitched Roofs

Pitched roofs have a visible slope that provides for rain, ice, and snow runoff. The pitch (angle) of the roof usually depends on both local weather conditions and the aesthetics of construction. A pitched roof can be supported either by trusses or by a system of rafters and beams. The rafters usually run from one load-bearing wall up to a center ridge pole and back to another load-bearing wall.

Most pitched roofs have a layer of solid sheeting, which can be metal, plywood, or wooden boards. This layer is often covered by a weather-resistant membrane and an outer covering such as shingles, slate, or tiles. Some pitched roofs have a system of laths (thin, parallel strips of wood) instead of solid sheeting to support the outer covering.

As with a flat roof, the type of roof construction material will dictate how to ventilate a pitched roof. For example, a slate or tile pitched roof can be opened by breaking the tiles and pushing them through the supporting laths. A tin roof can be cut and "peeled" back like the lid on a can. Opening a wooden roof usually requires cutting, chopping, or sawing.

Pitched roofs have either steep or gradual slopes. Roof ladders should be used to provide a stable support while working on this kind of roof. A ground or aerial ladder can be used to access the lower part of the roof; in such a case, the roof ladder is placed on the sloping surface and hooked over the center ridge pole.

Arched Roofs

Arched roofs are generally found in commercial structures because they create large open spans without requiring the use of columns FIGURE 15-17. Arched roofs are common in warehouses, supermarkets, bowling alleys, and similar buildings. These roofs are supported by trusses or by other types of construction, including large wood, steel, or concrete arches.

Arched roofs are often supported by bowstring trusses, which give the roof its distinctive curved shape. Bowstring trusses are usually constructed of wood and spaced 6 to 20 feet

FIGURE 15-17 An arched roof is commonly used for warehouses, supermarkets, and bowling alleys.

VOICES OF EXPERIENCE

On a typical weekday morning, our department was dispatched to a reported fire in an assisted living apartment building under construction. Upon our arrival, the construction crew advised that they had extinguished the fire with a dry chemical extinguisher. I entered the apartment as part of a four-person engine crew. Upon approaching the storage room, we could hear that a natural gas line feeding the furnace had been compromised and we advised the incident commander that the gas needed to be shut off. There was no active fire at this point and only light visible smoke was showing from the ceiling area of the room.

After the gas was shut down, I entered the room with an attic ladder and began assessing the damage. As I was up at the ceiling, I noticed a steady plume of light smoke coming from the damaged area indicating a smoldering fire somewhere. This accompanied by the dry chemical from the extinguishers got me thinking about my future health and how bad inhaling this stuff would be. Having been in the fire service for eighteen years, I know I've inhaled some nasty chemicals. This was enough for me to get down from the ladder and put my SCBA mask on and go on air.

I re-entered the room and went to the top of the ladder again to search for the smoldering fire. The incident commander stated that I was only in there for about thirty seconds when the entire room erupted into an orange fireball. Apparently the trapped gas that escaped from the burst line found the smoldering fire before I did. I just remember seeing an orange fireball hitting me in the face and then being on my back at the opposite end of the patio.

Taking that extra minute to mask up and put my gloves on saved my life. Our department has discussed the situation many times and has come to the same conclusion: when complacency takes the place of precaution, the results can be deadly.

James Woolf
Twinsburg Fire Department
Twinsburg, Ohio

(1.8 to 6 meters) apart. They support a roof deck of wooden boards or plywood sheeting and a covering of waterproof membrane. Although these roofs are quite distinctive when seen from above or outside, their structure might not be evident from inside the building because a flat ceiling is attached to the bottom chords of the trusses. This creates a huge attic space that is often used for storage. A hidden fire within this space can severely and quickly weaken the bowstring trusses.

The collapse of a bowstring truss roof is usually very sudden. A large area can collapse owing to the long spans and wide spaces between trusses. Fire fighters should be wary of any fire involving the truss space in these buildings. Many fire fighters have been killed when a bowstring truss roof collapsed.

Buildings with this type of roof construction should be identified and documented during preincident planning surveys. Parapet walls, smoke, and other obstructions make it difficult to recognize the distinctive arched shape of a bowstring truss roof during an incident.

Vertical Ventilation Techniques

Roof Ventilation

The objective of any roof ventilation operation is simple: to provide the largest opening in the appropriate location, using the least amount of time and the safest technique. Among the different roof openings that can provide vertical ventilation are built-in roof openings, inspection openings to locate the optimal place to vent, primary expandable openings located directly over the fire, and defensive secondary openings intended to prevent fire spread.

Before starting any vertical ventilation operation, fire fighters must make an initial assessment. They should note construction features and indications of possible fire damage, establish safety zones and exit paths, and identify built-in roof openings that can be used immediately. The sooner a building is ventilated, the safer it becomes for hose teams working inside the structure.

Ventilation operations must not be conducted in unsafe locations. A less efficient opening in a safe location is better than an optimal opening in a location that jeopardizes the lives of the ventilation team. In some cases, it might not be safe to conduct any rooftop operations, unless they can be performed from the safety of an aerial ladder or an aerial basket. If vertical ventilation cannot be performed, fire fighters have to rely on horizontal ventilation techniques.

Vertical ventilation is most effective when the opening is placed at the highest point, directly over the fire. Information on the fire's location as well as visible clues from the roof can be used to pinpoint the best location to vent. The incident commander (IC) or attack crews might be able to provide some direction, and the ventilation team should look for signs such as smoke issuing through the roof, tar bubbling up, or other indications of heat. If the roof is wet, rising steam or melted snow could indicate the hottest area.

Examination holes allow the team members to evaluate conditions under the roof and to verify the proper location for a ventilation opening. Sometimes a single slit cut (a kerf cut) into the roof with a power saw gives fire fighters some indication of the location of the fire. If a single cut is not effective, a triangular examination hole can be created very quickly by making three small cuts in the roof. Examination holes are used to determine how large an area is involved, whether a fire is spreading, and in which direction it is moving FIGURE 15-18.

Once the spot directly over the fire is located, the ventilation team should determine the most appropriate type of opening to make. Built-in rooftop openings provide readily available ventilation openings FIGURE 15-19. Skylights, rooftop stairway exit doors, louvers, and ventilators can quickly be transformed into ventilation openings simply by removing a cover or an obstruction. This approach also results in less property damage to the building. Many residential roofs have a vent installed along the ridge line of the roof to keep the attic space ventilated. Often, these can be removed quickly because they are attached with small nails.

FIGURE 15-18 A triangular cut can be used as an examination hole.

FIGURE 15-19 Built-in rooftop openings often can provide vertical ventilation very quickly.

If the roof must be cut to provide a ventilation opening, cutting one large hole is better than making several small ones. Some departments recommend starting with a rectangular opening about 4 feet long by 4 feet wide (1.2 meters long by 1.2 meters wide). If necessary, this hole can be expanded by continuing the cuts to make a larger opening FIGURE 15-20. If the crews inside the building report that smoke conditions are lifting and temperature levels are dropping, the vent is effective and probably large enough. If interior crews do not see a difference, the opening might be obstructed or need to be expanded.

Once the roof opening is made, a hole of the same size should be made in the ceiling material below to allow heat and smoke to escape from the interior of the building. If the ceiling is not opened, only the heat and smoke from the space between the ceiling and the roof will escape through the roof opening. The blunt end of a pike pole or hook can be used to push down as much of the ceiling material as possible. A roof opening is much less effective if there is no ceiling hole or if the ceiling hole is too small.

Tools Used in Vertical Ventilation

Several tools can be used in roof ventilation. Fire fighters commonly use power saws to cut vent openings but also use many hand tools as well. For example, axes, Halligan tools, pry bars, tin cutters, pike poles, and other hooks can all be used to remove coverings from existing openings, cut through the roof decking, remove sections of the roof, and punch holes in the interior ceiling. Usually the specific tools needed can be determined by looking at the building and the roof construction features.

Ventilation team members should always carry a standard set of tools as well as a utility rope for hoisting additional equipment if needed. All personnel involved in roof operations should use SCBA and wear full PPE. In this scenario, ground ladders or aerial apparatus can provide access to the roof.

Power saws effectively cut through most roof coverings. A rotary saw with a wood- or metal-cutting blade or a chain saw can be used for this purpose, for example. Special carbide-tipped saw blades can cut through typical roof construction materials. To operate a power saw, follow the steps in SKILL DRILL 15-8:

1. Make sure that the saw is checked during the daily apparatus inspection.
2. Confirm that the cutting device or blade is appropriate for the material anticipated. If it is not, put the correct blade on before going to the roof.
3. Briefly inspect the blade or chain for obvious damage.
4. Ensure that your proper protective gear is in place, including eye protection, hearing protection, SCBA, and full PPE. (STEP 1)
5. Before going to the roof, start the saw to ensure that it runs properly. Set the choke to the halfway position, or as recommended by the manufacturer. (STEP 2)
6. Stay clear of any moving parts of the saw, use a foot or knee to anchor the saw to the ground, and pull the starter cord as recommended to start the saw. (STEP 3)
7. Run the saw briefly at full throttle to verify its proper operation. (STEP 4)
8. Shut down the saw, wait for the blade to stop completely, and then carry the saw to the roof. (STEP 5)
9. If possible, work off of a roof ladder or aerial platform for added safety. (STEP 6)
10. Start the saw in an area slightly away from where you intend to cut. (STEP 7)
11. Always run the saw at maximum throttle when cutting. The saw should be running at full speed before the blade touches the roof decking. Keep the throttle fully open while cutting and removing the blade from the cut to reduce the tendency for the blade to bind. (STEP 8)

Types of Roof Cuts

Roof construction is a major consideration in determining which type of cut to use. Some roofs are thin and easy to cut with an axe or a power saw; others have multiple layers that are difficult to cut. Being familiar with the roof types in the local area and the best methods for opening them increases your efficiency and effectiveness in ventilation operations.

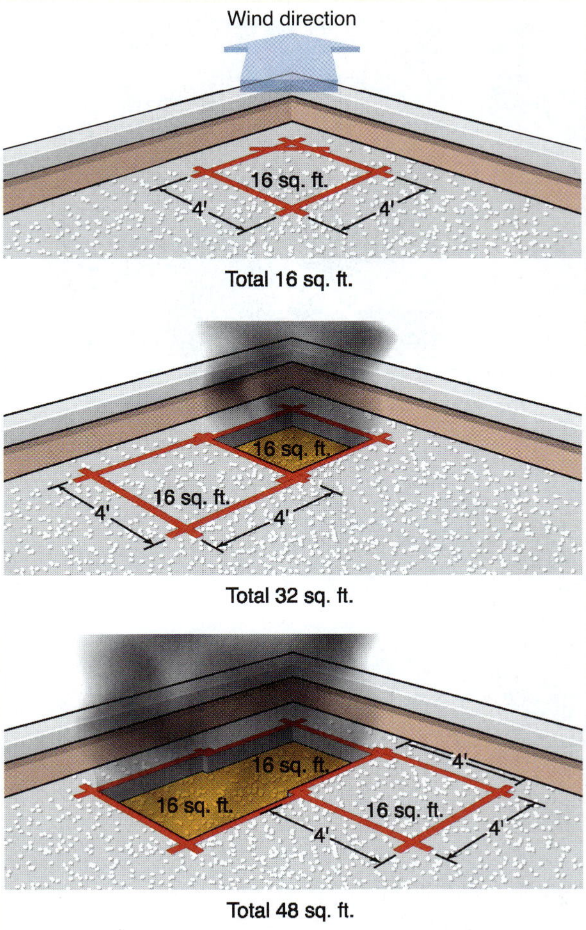

FIGURE 15-20 A ventilation hole can be quickly expanded by extending the cuts.

SKILL DRILL 15-8 Operating a Power Saw

(Fire Fighter I, NFPA 5.3.12)

1 Confirm that the cutting device or blade is appropriate for the material anticipated. Briefly inspect the blade or chain for obvious damage. Ensure that your proper protective gear is in place.

2 Before going to the roof, start the saw to ensure that it runs properly. Set the choke to the halfway position, or as recommended by the manufacturer.

3 Stay clear of any moving parts of the saw, use a foot or knee to anchor the saw to the ground, and pull the starter cord as recommended to start the saw.

4 Run the saw briefly at full throttle to verify its proper operation.

5 Shut down the saw, wait for the blade to stop completely, and then carry the saw to the roof.

6 If possible, work off of a roof ladder or aerial platform for added safety.

7 Start the saw in an area slightly away from where you intend to cut.

8 Always run the saw at maximum throttle when cutting. The saw should be running at full speed before the blade touches the roof decking. Keep the throttle fully open while cutting and removing the blade from the cut to reduce the tendency for the blade to bind.

Fire Fighter Safety Tips

Always carry and handle tools and equipment in a safe manner. When carrying tools up a ladder, hold the beam of the ladder instead of the rungs and run the hand tool up the beam. This technique works for most hand tools, including axes, pike poles and hooks, and tin roof cutters. Hoist tools using a rope with an additional tagline to keep them away from the building. Carry power saws and equipment in a sling across your back. Always turn off power equipment before you carry it.

Rectangular or Square Cut

A 4-foot by 4-foot (1.2-meter by 1.2-meter) rectangular or square cut is the most common vertical ventilation opening. It requires the fire fighter to make four cuts completely through the roof decking, using an axe or power saw. When using a power saw, the fire fighter must carefully avoid cutting through the structural supports. After the four cuts are completed, a section of the roof deck can be removed.

The fire fighter should stand upwind of the opening, with an unobstructed exit path. The first and last cuts should be made parallel to and just inside the roof supports. The fire fighter making the cuts must always stand on a solid portion of the roof. A triangular cut in one corner of the planned opening can be used as a starting point for prying the deck up with a hand tool.

Depending on the roof construction, it might be possible to lift out the entire section at once. If several layers of roofing material are present, fire fighters might have to peel them off in layers. The decking itself could consist of plywood sheets or individual boards that have to be removed one at a time.

To perform the rectangular cut, follow the steps in SKILL DRILL 15-9:

1. Check the roof for soundness. Use a tool to locate the roof supports.
2. Make the first cut parallel to a roof support. The cut should be made so that the support is outside the area that will be opened. (STEP 1)
3. Make a small triangular cut at the first corner of the opening. (STEP 2)
4. Make two cuts perpendicular to the roof supports. Do not cut through roof supports. Rock the saw over them to avoid damaging the integrity of the roof structure. (STEP 3)

SKILL DRILL 15-9 Making a Rectangular or Square Cut
(Fire Fighter I, NFPA 5.3.12)

1 Locate the roof supports by sounding. Make the first cut parallel to the roof support.

2 Make a triangular cut at the first corner.

3 Make two cuts perpendicular to the roof supports, and then make the final cut parallel to another roof support.

4 Pull out or push in the triangle cut.

5 Punch out the ceiling below. Be wary of a sudden updraft of hot gases or flames.

5. Make the final cut parallel to and slightly inside a roof support. Be sure to stand on the solid portion of the roof when making this cut.
6. Use a hand tool to pull out the corner triangle or push it through to create a small starter hole. (STEP 4)
7. If possible, use the hand tool to pull the entire roof section free and flip it over onto the solid roof. It might be necessary to pull the decking out, one board at a time.
8. After opening the roof deck, use a pike pole to punch out the ceiling below. This hole should be the same size as the opening in the roof decking. Be wary of a sudden updraft of hot gases or flames. (STEP 5)

Louver Cut

Another common cut is the louver cut, which is particularly suitable for flat or sloping roofs with plywood decking. To make a louver cut, fire fighters use a power saw or axe to make two parallel cuts, approximately 4 feet (1.2 meters) apart, perpendicular to the roof supports. The fire fighters then make cuts parallel to the roof supports, approximately halfway between each pair of supports. Using each roof support as a fulcrum, the cut sections are tilted to create a series of louvered openings.

Louver cuts can quickly create a large opening. Continuing the same cutting pattern in any direction creates additional louver sections. To make a louver cut, follow the steps in SKILL DRILL 15-10:

1. Sound the roof with a tool to locate the roof supports. (STEP 1)
2. Make two parallel cuts perpendicular to the roof supports. Do not cut through the roof supports. Rock the saw over them to avoid damaging the integrity of the roof structure. (STEP 2)

SKILL DRILL 15-10 Making a Louver Cut

(Fire Fighter I, NFPA 5.3.12)

1 Locate the roof supports by sounding.

2 Make two parallel cuts perpendicular to the roof supports.

3 Cut parallel to the supports and between pairs of supports in a rectangular pattern.

4 Tilt the panel to a vertical position.

3. Make cuts parallel to the supports and between pairs of supports in a rectangular pattern. (STEP 3)
4. Strike the nearest side of each section of the roofing material with an axe or maul, pushing it down on one side; use the support at the center of each panel as a fulcrum. Tilt the panel over the middle support.
5. Moving horizontally along the roof, make additional louver openings.
6. Open the interior ceiling area below the opening by using the butt end of a pike pole. This hole should be the same size as the opening made in the roof decking. (STEP 4)

Triangular Cut

The triangular cut works well on metal roof decking because it prevents the decking from rolling away as it is cut. Using saws or axes, the fire fighter removes a triangle-shaped section of decking. Smaller triangular cuts can be made between supports (so that the decking falls into the opening) or over supports (to create a louver effect). Because triangular cuts are generally smaller than other types of roof ventilation openings, several might be needed to create an adequately sized vent.

To make a triangular cut, follow the steps in SKILL DRILL 15-11:

1. Locate the roof supports. (STEP 1)
2. Make the first cut from just inside a support member in a diagonal direction toward the next support member. (STEP 2)
3. Begin the second cut at the same location as the first, and make it in the opposite diagonal direction, forming a V shape. (STEP 3)
4. Make the final cut along the support member so as to connect the first two cuts. Cutting from this

SKILL DRILL 15-11 Making a Triangular Cut

(Fire Fighter I, NFPA 5.3.12)

1 Locate the roof supports.

2 Make the first cut from just inside a support member in a diagonal direction toward the next support member.

3 Begin the second cut at the same location as the first, and make it in the opposite diagonal direction, forming a V shape.

4 Make the final cut along the support member so as to connect the first two cuts. Cutting from this location allows fire fighters the full support of the member directly below them while performing ventilation.

location allows fire fighters the full support of the member directly below them while performing ventilation. (STEP 4)

Peak Cut

Peak cuts are limited to peaked roofs sheeted with plywood. In these structures, the 4-foot by 8-foot (1.2-meter by 2.4-meter) sheets of plywood are applied starting at the bottom edge of the roof and working toward the top. A partial sheet is usually placed at the top, along the roof peak. A small gap is left between the two sides of the roof to allow the plywood sheeting to expand and contract with temperature changes.

To make a peak cut, the fire fighter first uses a hand tool to clear the outer covering along the peak (roof cap), which reveals the roof supports through the gap in the sheeting. Using a power saw or axe, the fire fighter makes a series of vertical cuts between the supports from the top to the bottom of the plywood sheet. The individual panels are then struck with an axe and louvered or pried up with a hook.

Repeating this technique on both sides of the roof essentially removes the entire peak. Unless multiple layers of roof decking are present, there is no need to make horizontal cuts. The opening can be expanded either vertically or horizontally with a minimum of cutting.

To make a peak cut, follow the steps in SKILL DRILL 15-12:

1. Sound the roof with a tool to locate the roof supports. (STEP 1)
2. Clear the roofing materials away from the roof cap. (STEP 2)
3. Make the first cut vertically, at the farthest point away. Start at the roof peak in the area between the support members and cut down to the bottom of the first plywood panel. (STEP 3)
4. Make parallel downward cuts between supports, moving horizontally along the roofline to make additional ventilation openings. (STEP 4)
5. Strike the nearest side of the roofing material with an axe or maul, pushing it in, using the support located at the center as a fulcrum. This causes one end of the roofing material to go downward into the opening and the other to rise up.
6. If necessary, repeat this process on both sides of the peak, horizontally across the peak, or vertically toward the roof edge. (STEP 5)
7. Open the interior ceiling area below the opening by using the butt end of a pike pole. This hole should be the same size as the vent opening made in the roof decking. (STEP 6)

Trench Cut

Trench cut (or strip cut) ventilation is used to stop fire spread in long narrow buildings, such as strip malls and small storage complexes. A trench cut creates a large opening ahead of the fire, removing a section of fuel and letting heated smoke and gases flow out of the building. Essentially, it is a firebreak in the roof. Water is sometimes applied through the trench cut.

Trench cuts are a defensive ventilation tactic intended to stop the progress of a large fire, particularly one that is advancing through an attic or cockloft. The IC who chooses this tactic is "writing off" part of the building and identifying a point where crews will be able to stop the fire.

A trench cut is made from one exterior wall across to the other. It begins with two parallel cuts, spaced 2 to 4 feet (0.6 to 1.2 meters) apart. Approximately every 4 feet (1.2 meters), fire fighters make short perpendicular cuts between the two parallel cuts. They can then lift the roof covering out in sections, completely opening a section of the roof. On a pitched roof, the trench should run from the peak down. On a sloping roof, it should start at the higher end and work toward the lower end.

A trench cut is a secondary cut, used to limit the fire spread, rather than a primary cut, located over the seat of the fire. A primary vent should still be made before crews start working on the trench cut.

A trench cut must be made far in advance of the fire. The ventilation crew must be able to complete the cut and the interior crew get into position before the fire passes the trench. Inspection holes should be made to ensure that the fire has not already passed the chosen site before the trench cut is completed. The crew members' lives will be in danger and the tactic is useless if the fire advances beyond the trench before it is opened.

Although trench cuts are effective, they require both time and personnel to complete. As with all types of vertical ventilation, careful coordination between the ventilation crew and the interior attack crew is essential. While the trench cut is being made, attack teams should be deployed inside the building to defend the area in front of the cut.

To make a trench cut, follow the steps in SKILL DRILL 15-13 (Fire Fighter I, NFPA 5.3.12) FIGURE 15-21:

1. After a primary cut has been made over the seat of the fire, cut a number of small inspection holes to identify a point sufficiently far ahead of the fire travel.
2. Make two parallel cuts, 2 to 4 feet (0.6 to 1.2 meters) apart, across the entire roof, starting at the ridge pole

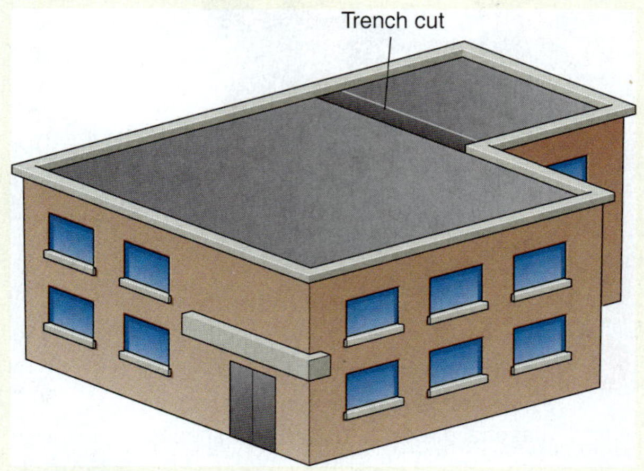

FIGURE 15-21 Making a trench cut.

SKILL DRILL 15-12 Making a Peak Cut

(Fire Fighter I, NFPA 5.3.12)

1 Locate the roof supports.

2 Clear the roofing materials away from the roof cap.

3 Make the first cut vertically, at the farthest point away. Start at the roof peak in the area between the support members and cut down to the bottom of the first plywood panel.

4 Make parallel downward cuts between supports, moving horizontally along the roofline to make additional ventilation openings.

5 Strike the nearest side of the roofing material with an axe or maul, pushing it in, using the support located at the center as a fulcrum. This causes one end of the roofing material to go downward into the opening and the other to rise up. If necessary, repeat this process on both sides of the peak, horizontally across the peak, or vertically toward the roof edge.

6 Open the interior ceiling area below the opening by using the butt end of a pike pole. This hole should be the same size as the vent opening made in the roof decking.

(for peaked roofs) or a load-bearing wall (for flat roofs).

3. Cut between the two long cuts to make a row of rectangular sections.
4. Remove the rectangular panels to open the trench.

Special Considerations

Many obstacles can be encountered during ventilation operations. Fire fighters must be creative and remember the basic objectives of ventilation—to create high openings as rapidly as possible so that hose lines can be advanced into the building.

Poor access or obstructions such as trees, fences, electric lines, or tight exposures can prevent fire fighters from getting close enough to place ladders. Many residential and commercial roofs have multiple layers; some even have a new roof built on top of an older one. Window openings in abandoned buildings might be boarded or sealed.

Steel bars, shutters, and other security features can also hinder ventilation efforts. A building that requires forcible entry is also likely to present ventilation challenges. To enhance their security, some commercial buildings have roofs made of steel plating. In these buildings, fire fighters should not attempt vertical ventilation through the roof but rather should use alternative ventilation techniques.

Ventilating Basements

Research involving basement fires has provided valuable information about safe and effective ways to ventilate and extinguish basement fires. Many vertical voids within a building originate or terminate in the basement, providing ample opportunities for fire gases to spread throughout the building. In many parts of the country, especially in newer subdivisions with lightweight construction, basement ceilings are left unfinished when the home is sold. Under these conditions, fires originating in the basement will quickly involve the floor system, resulting in a failure of the floor over the fire and early collapse, with the potential for causing injury or death to fire fighters entering the building with conditions of limited visibility.

Fire Fighter Safety Tips

Remember that basement fires often do not look like basement fires. Because of the many voids between the basement and other parts of the building, a fire that started in the basement may at first appear to be a kitchen fire or a fire on the first or second floor of the building. As a safety measure, operate as though the floor, your working platform, has been weakened by a fire underneath until you can prove that there is no fire beneath you.

The traditional tactic of pushing down the interior stairway to the seat of the fire should no longer be considered a safe option. One of the findings of recent research on basement fires was that water applied via the interior stairs had a limited effect on cooling the basement or extinguishing the fire. However, water applied through an external window or external door quickly darkened down the fire and reduced temperatures throughout the building FIGURE 15-22 .Water applied through an external window or door did not "push" fire or hot gases up the interior stairs. This cooling effect lasted for several minutes before the fire grew back to the size it was before the application of the water TABLE 15-3.

When attacking basement fires, the use of interior stairs for ventilation should not be considered a safe option because of the potential for failure of the floor system and the danger of fire fighters being in the exhaust flow path of hot gases from the fire. The use of the interior stairs for flowing water onto the fire is unsafe and is ineffective in reducing the temperature of the fire compartment and the rest of the building.

In some basement fires, the temperature reaches untenable levels even for fire fighters in full PPE. Recent UL tests have shown that in some basement fires, the temperature at the floor level is not significantly lower than the temperatures in other parts of the basement. This seems to indicate that it may not always be a safe idea to send fire fighters into basement fires before the compartment has been cooled.

Whenever possible, basement fires should be ventilated through exterior windows or doors. This ventilation will result in oxygen being added to a ventilation-limited fire and may cause rapid growth of the fire unless a hose stream is quickly directed into the volume of fire. The application of a straight stream as close to the fire as possible for about 20 to 30 seconds is effective in cooling the fire compartment and the rest of the building. Once this offensive exterior attack is completed, it will be safer

FIGURE 15-22 Figure showing water being applied to a basement fire through a basement window.

TABLE 15-3 Basement Fires

- Fire flows from basement fires developed in locations other than the stairs, as the floor assembly often failed close to the location where the fire started.
- Flowing water at the top of the interior stairs had limited impact on basement fires.
- Offensive exterior attack through a basement window was effective in cooling the fire compartment.
- Offensive fire attack through an exterior door was effective in cooling the fire compartment.

Courtesy of NIST.

and easier to enter the basement to complete extinguishment of the remaining fire. More information about extinguishing basement fires is presented in the chapter, *Fire Suppression*.

Ventilating Concrete Roofs

Some commercial or industrial structures have concrete roofs. Concrete roofs can be constructed with "poured-in-place" concrete, with precast concrete sections of roof decking placed on a steel or concrete supporting structure, or with T-beams. Such roofs are generally flat and difficult to breach. The roof decking is usually very stable, but fire conditions underneath could weaken the supporting structural components or load-bearing walls, leading to failure and collapse.

There are few options for ventilating concrete roofs. Even special concrete-cutting saws are generally ineffective against these structures. In such cases, fire fighters should use alternative ventilation openings such as vents, skylights, and other roof penetrations or horizontal ventilation.

Ventilating Metal Roofs

Metal roofs and metal roof decks also present many challenges for the ventilation crew. Because metal conducts heat more quickly than other roofing materials do, discoloration and warping of this material can indicate the seat of the fire. Tin-cutter hand tools can be used to slice through thin metal coverings, whereas special saw blades might be needed to cut through metal roof decking. In many cases, the metal is on the bottom and supports a built-up or composite roof covering.

Metal roof decking is often supported by lightweight steel bar joists, which can sag or collapse when exposed to a fire. Because the metal decking is lightweight, the supporting structure can be relatively weak, with widely spaced bar joists. The resulting assembly can fail quickly with only limited fire exposure.

As the fire heats the metal deck, the tar roof covering can melt and leak through the joints into the building, where it releases flammable vapors. When this sequence of events occurs, it can quickly spread the fire over a wide area under the roof decking. Fire fighters should look for indications of dripping or melting tar and begin rapid ventilation to dissipate the flammable vapors before they can ignite. Hose streams should be used to cool the roof decking from below to stop the tar from melting and producing vapors.

When a metal roof deck is cut, the metal can roll down and create a dangerous slide directly into the opening. The triangular cut prevents the decking from rolling away as easily, so it is the preferred option, even though several cuts can be needed to create an adequately sized vent.

Ventilating High-Rise Buildings

Ventilating a high-rise building can be challenging. A high-rise building resembles a stack of individual floor compartments connected by stairways, elevator shafts, and other vertical passages. Most high-rise buildings have sealed windows that are difficult to break. In addition, high-rise buildings have unique patterns of smoke movement, such that smoke might be trapped on individual floors or move up or down within the vertical shafts.

Many newer high-rise buildings have incorporated smoke management capabilities into their HVAC systems. Use of this type of system enables different areas to be pressurized with fresh air and contaminated air to be exhausted directly to the outside. If the HVAC system does not have this capacity, it can complicate problems by circulating smoke to different areas of the building.

A phenomenon called the stack effect can occur in high-rise structures. The stack effect is a response to the differences in temperature inside and outside a building. A cold outer atmosphere and a heated interior cause smoke to rise quickly through stairways, elevator shafts, and other vertical openings, filling the upper levels of the building **FIGURE 15-23**.

The opposite situation can occur on a hot summer day, when the interior temperature is much cooler than the outside atmosphere. In this scenario, the heavy cooler air pushes the smoke down the vertical openings, toward a lower-level exit **FIGURE 15-24**.

The situation can change if the fire produces enough heat to alter the temperature profile within the building. The air currents

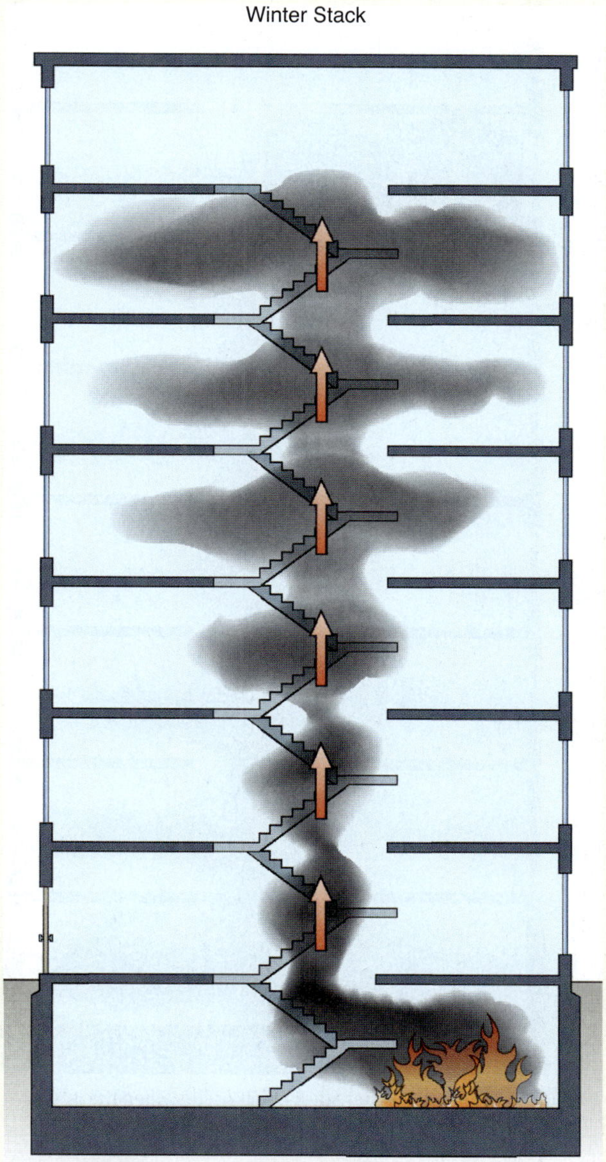

FIGURE 15-23 A cold outer atmosphere and a heated interior create a winter stack effect.

within a tall building might, for example, be stronger than the convection currents that normally govern the flow of smoke and other products of combustion. A strong wind through an open or broken window can change the direction of smoke. Contaminated air might suddenly move toward the opening or be pushed back as a strong draft of fresh air enters the building.

After smoke mixes with fresh air or is hit by water from sprinklers or hose streams, it cools and can "sit" in one location. This cold smoke can fill several floors and usually needs to be cleared by mechanical means.

A key objective in ventilating a high-rise building is to manage the air movement in stairways and elevator shafts. Positive-pressure ventilation can help keep stairways clear of smoke and ventilate individual floors in high-rise structures.

FIGURE 15-24 A summer stack effect occurs when the interior temperature is much cooler than the outside air.

At least one stairwell should be designated as the evacuation stairwell and announced to the operating companies. This stairwell should be kept well ventilated and designated as an occupant and rescue route. These stairs must be used by rescuers and escaping victims only. Some newer buildings have smoke-proof stair towers or pressurized stair shafts that are designed to keep smoke out of the stairway. Otherwise, positive-pressure fans can be used in a stairway to keep smoke out. Place the positive-pressure fans at a ground-floor doorway, positioning them so that fresh air blows into the stairway. Make certain that the engines of the positive-pressure fans are well ventilated because they can increase carbon monoxide levels.

A pressurized stairway also can be used to clear smoke from a floor. Opening a door from the stairway allows fresh air to enter the floor. The contaminated atmosphere can then be vented out through a window or another stairway.

Ventilating Windowless Buildings

Many structures do not have windows FIGURE 15-25. Some buildings are designed without windows; others have bricked-up or covered windows. These buildings pose two significant risks to fire fighters: Heat and products of combustion are trapped, and fire fighters have no secondary exit route.

Windowless buildings are similar to basements in terms of the ventilation approach to be used. Any ventilation needs to be as high as possible and probably requires mechanical assistance. Using existing rooftop openings, cutting openings in the roof, reopening boarded-up windows or doors, and making new openings in exterior walls are all possible ways to ventilate windowless structures.

Ventilating Large Buildings

Providing adequate ventilation is more difficult in large buildings than in smaller ones. In a large building, a ventilation hole placed in the wrong location can draw the fire toward the opening, spreading the fire to an area that had not previously been involved. This underscores the importance of coordinating ventilation operations with the overall fire attack strategy.

FIGURE 15-25 Structures without windows pose significant problems in ventilation.

Smoke cools as it travels into unaffected portions of a large building. A fire sprinkler system also cools the smoke, causing it to stratify. As the cold smoke fills the area, it becomes more difficult to clear.

If possible, fire fighters should use interior walls and doors to create several smaller compartments in a large building, thereby limiting the spread of heat and smoke. The smaller areas can be cleared one at a time with positive-pressure fans. Several fans can be used in a series or in parallel lines to clear smoke from a large area.

Equipment Maintenance

It is important that all equipment used for ventilation is kept in good repair and ready to operate at peak efficiency. It is important to read and follow the manufacturer's instructions for properly maintaining power equipment at regular and properly documented intervals. Fuel must be rotated regularly if not used. Fuel tanks need to be filled to the recommended levels. All fire fighters who use ventilation tools need to practice using them regularly to ensure safe and effective operation when needed.

To perform a readiness check on a power saw, see SKILL DRILL 15-14:

1. Don the proper PPE.
2. Follow the manufacturer's and your department's recommendations.
3. Make certain that the fuel tank is full. Make certain that the bar and chain oil reservoirs are full. (STEP 1)
4. Check the throttle trigger for smooth operation. (STEP 2)
5. Ensure that the saw, blade, air filter, and chain brake are clean and working.
6. Inspect the blade for even wear, and lubricate the sprocket tip if needed. (STEP 3)
7. Check the chain for wear, missing teeth, or other damage. Check the chain end for proper tension.
8. Check the chain catcher. (STEP 4)
9. Check for loose nuts and screws, and tighten them if needed.
10. Check the starter and starter cord for wear. (STEP 5)
11. Inspect the spark plugs. (STEP 6)
12. Start the saw.
13. Make certain that both the bar and the chain are being lubricated while the saw is running.
14. Check the chain brake. (STEP 7)
15. Make certain that the stop switch functions.
16. Record the results of the inspection. (STEP 8)

To maintain a power saw, see SKILL DRILL 15-15:

1. Don the proper PPE.
2. Follow the manufacturer's and your department's recommendations.
3. Remove the clutch cover, bar, and chain, and inspect them for damage and wear. Replace these components if necessary. (STEP 1)
4. Clean and inspect the clutch cover, drive, bar, chain, and power head using a cloth, cleaning solvent, and compressed air.
5. Inspect the air filter, and clean or replace it as needed. (STEP 2)
6. Coat the components with a substance recommended by the manufacturer to prevent rusting.
7. Place the bar and chain back on, flipping the bar over each time to help wear the bar evenly.
8. Replace the clutch cover. (STEP 3)
9. Adjust the chain tension (make sure the bar and chain are cool before adjusting them). (STEP 4)
10. Tighten the clutch cover nuts and other nuts and screws.
11. Fill the power saw with fuel. Fill the bar and chain oil reservoirs.
12. Perform a periodic check.
13. Document the maintenance on the departmental log. (STEP 5)

SKILL DRILL 15-14 Performing a Readiness Check on a Power Saw

(Fire Fighter I, NFPA 5.3.12)

1 Make certain that the fuel tank is full. Make certain that the bar and chain oil reservoirs are full.

2 Check the throttle trigger for smooth operation.

3 Ensure that the saw, blade, air filter, and chain brake are clean and working. Inspect the blade for even wear, and lubricate the sprocket tip if needed.

4 Check the chain for wear, missing teeth, or other damage. Check the chain end for proper tension. Check the chain catcher.

5 Check for loose nuts and screws, and tighten them if needed. Check the starter and starter cord for wear.

6 Inspect the spark plugs.

7 Start the saw. Make certain that both the bar and the chain are being lubricated while the saw is running. Check the chain brake.

8 Make certain that the stop switch functions. Record the results of the inspection.

SKILL DRILL 15-15 Maintaining a Power Saw

(Fire Fighter I, NFPA 5.3.12)

1 Remove, clean, and inspect the clutch cover, bar, and chain for damage and wear. Replace if necessary.

2 Inspect the air filter and clean/replace as needed.

3 Lubricate components as recommended by the manufacturer. Place the bar and chain back on, flipping the bar over each time to help wear bar evenly. Replace the clutch cover.

4 Adjust the chain tension (make sure the bar and chain cool before adjusting).

5 Fill the power saw with fuel. Fill the bar and chain oil reservoirs.

Wrap-Up

Chief Concepts

- Ventilation is the process of the systematic and coordinated removal of smoke, heat, and toxic gases from a burning building that also admits a supply of oxygen-rich air into the fire. The ventilation process has two different components. The first is the removal of smoke, toxic gases, and hot air. The second is the addition of cooler, cleaner, and oxygen-rich air.
- Fire size-up needs to include the type of ventilation that is appropriate for the fire conditions and a plan to implement ventilation operations in coordination with other fire-suppression activities.
- Many building fires in buildings today are ventilation limited. It is important to be able to recognize this condition, since uncoordinated ventilation can cause rapid fire growth.
- Factors that influence the effectiveness of ventilation operations include: door control, the ventilation location relative to the fire, the ventilation hole size, and the coordination of ventilation with fire suppression.
- Ventilation practices should take into consideration the advantages and disadvantages of using doors, windows, and roofs.
- Effective and safe ventilation requires close coordination between ventilation activities and fire-suppression activities.
- It is important to recognize how wind conditions can impact ventilation and fire flow paths.
- In some circumstances, the best initial ventilation practice is to keep the building openings closed until hose lines are in place to apply water.
- Modern construction practices contribute to ventilation-limited fires and rapid heat build-up.
- Backdrafts and flashovers can be minimized through the use of a transitional fire attack that is coordinated with appropriate ventilation.
- All of the structural components in fire-resistive construction are made of noncombustible or limited-combustible materials. It generally has spaces divided into compartments, which limit potential fire spread. However, the products of combustion can spread through stairways and HVAC systems. The roof is usually supported by a steel or concrete roof deck that is difficult to impossible to cut vertical ventilation holes in.
- Ordinary construction buildings have exterior walls made of noncombustible or limited-combustible materials that support the roof and floor assemblies. The walls and floors are usually wood construction. The roof can be cut with power saws or axes to open vertical ventilation holes. These buildings usually have windows and doors that can be used for horizontal ventilation. Vertical openings frequently provide a path for fire to extend into an attic or cockloft.
- Wood-frame buildings often contain many void spaces. Modern fast-growth lumber and lightweight construction components, found in newer buildings, can fail quickly under fire conditions.
- Horizontal ventilation takes advantage of the doors, windows, and other openings at the same level as the fire. In some cases, fire fighters make additional openings in a wall to provide horizontal ventilation. Horizontal ventilation is commonly used in residential fires, room-and-contents fires, and fires that can be controlled quickly by the attack team.
- Vertical ventilation refers to any opening that allows the products of combustion to travel up and out. It involves making openings in roofs or floors so that heat, smoke, and toxic gases can escape from the structure in a vertical direction. Pathways for vertical ventilation can include ceilings, stairwells, exhaust vents, and roof openings such as skylights, scuttles, or monitors. Additional openings can be created by cutting holes in the roof or the floor and making sure that the opening extends through every layer of the roof or floor. The choice of roof openings depends primarily on the building's roof construction.
  - Mechanical ventilation includes negative-pressure ventilation, positive-pressure ventilation, and hydraulic ventilation.
  - Negative-pressure ventilation uses smoke ejectors to exhaust smoke and heat from a structure. It can be used to move smoke out of a structure after a fire.
  - Positive-pressure ventilation uses fans to introduce clean air into a structure and push the contaminated atmosphere out. It can be used to reduce interior temperatures and smoke conditions in coordination with a fire attack or clear a contaminated atmosphere after a fire has been extinguished.
  - Hydraulic ventilation moves air using fog or broken-pattern fire streams to create a pressure differential behind and in front of the nozzle. It is most useful in clearing smoke and heat out of a room after the fire is under control.
- Before performing vertical ventilation, fire fighters must evaluate all pertinent safety issues and avoid unnecessary risks. The biggest risk is roof collapse. Assess the roof for roof scuttles, heat vents, plumbing vents, louver ventilation, and fan shafts to prevent tripping or falling from the roof.
- When working on a roof, have two safe exit routes. A second ground ladder or aerial device should be positioned to provide a quick escape route. The ventilation opening should never be located between the exit route and the ventilation crew.

- Fire fighters who are assigned to vertical ventilation tasks should always be aware of the condition of the roof. They should immediately retreat from the roof if they notice any of the following signs:
  - Any visible indication of sagging roof supports
  - Any indication that the roof assembly is separating from the walls, such as the appearance of fire or smoke near the roof edges
  - Any structural failure of any portion of the building, even if it is some distance from the ventilation operation
  - Any sudden increase in the intensity of the fire from the roof opening
  - High heat indicators on a thermal imaging camera
- The two major components of roof construction are roof support structures and roof coverings. The roof support system provides the structural strength to hold the roof in place. The structural system is either solid-beam or truss construction. The roof covering is the weather-resistant surface and may consist of many layers.
- Roof designs include flat roofs, pitched roofs, and arched roofs.
  - Flat roof construction is similar to floor construction. It can be supported by solid components or by trusses. Flat roofs often have vents, skylights, scuttles, or other features that penetrate the roof deck. Removing the covers from these openings provides vertical ventilation without the need to make cuts through the roof deck.
  - Pitched roofs have a visible slope. They can be supported by trusses or a system of rafters and beams. Many of these roofs have a layer of solid sheeting covered by a weather-resistant membrane and outer covering. The roof construction material dictates how to ventilate the roof.
  - Arched roofs create large open spans without the use of columns. They are often supported with bowstring trusses. The collapse of a bowstring truss is usually very sudden.
- The types of vertical ventilation openings include the following:
  - Built-in roof openings
  - Inspection openings
  - Primary (expandable) openings
  - Secondary (defensive) openings
- Some commercial or industrial structures have concrete roofs. There are few options for ventilating these structures. Use alternative ventilation openings such as vents or skylights.
- Metal roofs conduct heat and are often supported by lightweight steel metal joists.
- Both horizontal and vertical ventilation can be required to vent a basement.
- HVAC systems may be used to ventilate high-rise buildings.
- To ensure successful ventilation operations, all equipment and tools must be in a ready state and properly maintained.

Hot Terms

Arched roof A rounded roof usually associated with a bowstring truss.

Chase Open space within walls for wires and pipes.

Cockloft The concealed space between the top-floor ceiling and the roof of a building.

Ejectors Also called smoke ejectors, electric fans used in negative-pressure ventilation.

Fire-resistive construction Buildings that have structural components of noncombustible materials with a specified fire resistance. Materials can include concrete, steel beams, and masonry block walls. Fire-resistive construction is also known as Type I building construction, as defined in NFPA 220, *Standard on Types of Building Construction*.

Flow path The volume in a structure between an inlet and an outlet that allows the movement of heat and smoke from the higher pressure within the fire area towards the lower pressure areas accessible via doorways, halls, stairs, and window openings.

Fuel-limited A fire that has sufficient oxygen for fire growth, but has a limited amount of fuel available for burning.

Gusset plate A connecting plate used in trusses, typically made of wood or lightweight metal.

Horizontal ventilation The process of making openings on the same level as the fire so that smoke, heat, and gases can escape horizontally from a building through openings such as doors and windows.

Hydraulic ventilation Ventilation that relies on the movement of air caused by a fog stream that is placed 2 to 4 feet (0.6 to 1.2 meters) in front of the open window.

Kerf cut A cut that is only the width and depth of the saw blade. It is used to inspect cockloft spaces from the roof.

Lath Thin strips of wood used to make the supporting structure for roof tiles.

Louver cut A cut that is made using power saws and axes to cut along and between roof supports so that the sections created can be tilted into the opening.

Mechanical ventilation A process of removing heat, smoke, and gases from a fire area by using exhaust fans, blowers, air-conditioning systems, or smoke ejectors. (NFPA 402, *Guide for Aircraft Rescue and Fire-Fighting Operations*)

Natural ventilation The flow of air or gases created by the difference in the pressures or gas densities between the outside and inside of a vent, room, or space. (NFPA

Wrap-Up, continued

853, *Standard for the Installation of Stationary Fuel Cell Power Systems*)

Negative-pressure ventilation Ventilation that relies on electric fans to pull or draw the air from a structure or area.

Neutral plane The interface at a vent, such as a doorway or a window opening, between the hot gas flowing out of a fire compartment and the cool air flowing into the compartment where the pressure difference between interior and exterior is zero.

Ordinary construction Buildings whose exterior walls are made of noncombustible or limited-combustible materials, but whose interior floors and walls are made of combustible materials. Ordinary construction is also known as Type III building construction, as defined in NFPA 220, *Standard on Types of Building Construction*.

Parapet walls Walls on a flat roof that extend above the roofline.

Peak cut A ventilation opening that runs along the top of a peaked roof.

Positive-pressure ventilation Ventilation that relies on fans to push or force clean air into a structure.

Primary cut The main ventilation opening made in a roof to allow smoke, heat, and gases to escape.

Roof covering The membrane, which can also be the roof assembly, that resists fire and provides weather protection to the building against water infiltration, wind, and impact. (NFPA 5000, *Building Construction and Safety Code*)

Roof decking The rigid component of a roof covering.

Secondary cut An additional ventilation opening made for the purpose of creating a larger opening or limiting fire spread.

Smoke inversion The condition in which smoke hangs low to the ground because of the presence of cold air.

Sounding The process of striking a roof with a tool to determine where the roof supports are located.

Stack effect The vertical air flow within buildings caused by the temperature-created density differences between the building interior and exterior or between two interior spaces. (NFPA 92B, *Standard for Smoke Management Systems in Malls, Atria, and Large Spaces*)

Strip cut Another term for a trench cut.

Transitional attack An offensive fire attack initiated by an exterior, indirect handline operation into the fire compartment to initiate cooling while transitioning into interior direct fire attack in coordination with ventilation operations.

Trench cut A roof cut that is made from one load-bearing wall to another load-bearing wall and that is intended to prevent horizontal fire spread in a building.

Triangular cut A triangle-shaped ventilation cut in the roof decking that is made using a saw or an axe.

Underventilated fire A fire that has an adequate supply of fuel available but is not receiving an adequate supply of oxygen.

Ventilation The fire-fighting operation of the systematic and coordinated removal of smoke and heat from a structure by opening windows and doors or making holes in the roof. (NFPA 923)

Vertical ventilation The process of making openings so that smoke, heat, and gases can escape vertically from a structure.

Wood-frame construction Buildings whose exterior walls, interior walls, floors, and roofs are made of combustible wood materials. Wood-frame construction is also known as Type V building construction, as defined in NFPA 220, *Standard on Types of Building Construction*.

FIRE FIGHTER *in action*

You arrive at work to find that you are being assigned to the truck company for the day. Your captain hardly gets the words out of his mouth when you are dispatched to a fire in a building. The structure is a newer commercial building constructed of metal-clad, structural steel. The caller indicated that a construction crew was digging a trench when it hit a gas main. The gas ignited, impinged on the building, and caught it on fire. The IC indicated that this would be an offensive attack with an objective to limit fire spread from the current suite that is involved. Your crew is assigned to ventilate the structure to support fire control.

1. You decide to vertically ventilate the structure. What is your greatest concern?
   - **A.** Falling from the roof
   - **B.** The roof collapsing
   - **C.** Cutting the hole in the wrong location
   - **D.** Crews putting a hoseline through the ventilation hole
2. What is a trench cut?
   - **A.** A cut that is only the width and depth of the saw blade; it is used to inspect cockloft spaces from the roof
   - **B.** A ventilation opening that runs along the top of a peaked roof
   - **C.** An additional ventilation opening made for the purpose of creating a larger opening or limiting fire spread
   - **D.** A roof cut that is made from one load-bearing wall to another load-bearing wall and that is intended to prevent horizontal fire spread in a building
3. What is hydraulic ventilation?
   - **A.** The process of making openings on the same level as the fire so that smoke, heat, and gases can escape horizontally from a building through openings such as doors and windows
   - **B.** A process of removing heat, smoke, and gases from a fire area by using exhaust fans, blowers, air-conditioning systems, or smoke ejectors
   - **C.** Ventilation that relies on fans to push or force clean air into a structure
   - **D.** Ventilation that relies on the movement of air caused by a fog stream that is placed 2 to 4 feet (0.6 to 1.2 m) in front of the open window
4. What is smoke inversion?
   - **A.** The condition in which smoke hangs low to the ground due to the presence of cold air
   - **B.** The vertical airflow within buildings caused by temperature-created density differences between the building interior and exterior or between two interior spaces
   - **C.** The process in which rising smoke, heat, and gases encounter a horizontal barrier such as a ceiling and begin to move out and back down
   - **D.** Recirculation of exhausted air that is drawn back into a negative-pressure fan in a circular motion
5. A cut that is made using power saws and axes to cut along and between roof supports so that the sections created can be tilted into the opening is a ______.
   - **A.** louvered cut
   - **B.** kerf cut
   - **C.** triangle cut
   - **D.** secondary cut
6. When setting up positive-pressure ventilation, how far should the fan be placed in front of the opening?
   - **A.** Inside the building, 3 to 4 feet (0.9 to 1.2 m) from the doorway
   - **B.** In the doorway
   - **C.** 3 to 4 feet (0.9 to 1.2 m) in front of the doorway
   - **D.** 8 to 10 feet (2.4 to 3.1 m) in front of the doorway

FIRE FIGHTER II *in action*

You are dispatched to 256 East Main Street for the report of a fire in a two-story house. The residential development is about 20 years old. Your dispatch center reports that it has received multiple calls for this incident. Two callers reported seeing smoke coming from around the front door and windows. One caller reported seeing flames in the living room.

You arrive on the scene in six minutes. From the A side, the only signs of fire you see are dirty windows and some smoke stains around the front door and windows. After conducting a 360-degree size-up, your officer reports slight smoke on the A and B sides of the house. She orders the second due engine to establish a water supply to your engine.

1. Which stage of fire development do you think this scenario represents?
2. Why does your officer instruct you to force the front door and then to keep it closed until the hose crew is ready for action?
3. Why is close coordination between fire suppression personnel and the ventilation crew important?
4. When attempting a rescue using VEIS, why is it important to isolate the door of the room being searched?

Water Supply

Fire Fighter I

Knowledge Objectives

After studying this chapter, you will be able to:

- Describe the equipment and procedures that are used to access static sources of water. (NFPA 5.3.15.A, p 489)
- Describe the characteristics of a mobile water supply apparatus. (NFPA 5.3.15.A, p 490)
- Describe the advantages of a portable tank system. (NFPA 5.3.15.A, p 490–491)
- Describe how municipal water systems supply water to communities. (p 492–495)
- List the types of fire hydrants. (p 495–496)
- Describe the characteristics of wet-barrel hydrants. (NFPA 5.3.15.A, p 495)
- Describe the characteristics of dry-barrel hydrants. (NFPA 5.3.15.A, p 495–496)
- Describe the common guidelines that govern the location of fire hydrants. (p 496–497)
- Describe how to inspect a fire hydrant. (p 497–500)
- Describe how to test a fire hydrant. (p 500–504)
- Explain the principles of fire hydraulics. (p 503–504)
- Describe how water flow is measured. (p 503–504)
- Describe how water pressure is measured. (p 504)
- Explain how friction loss affects water pressure. (p 504)
- Explain how elevation pressure affects water pressure. (p 504)
- Describe how to prevent water hammer. (NFPA 5.3.1.B, p 504)
- List the two types of fire hose. (NFPA 5.3.15.A, p 504)
- Describe the characteristics of small-diameter hose. (NFPA 5.3.15.A, p 505)
- Describe the characteristics of medium-diameter hose. (NFPA 5.3.15.A, p 505)
- Describe the characteristics of large-diameter hose. (NFPA 5.3.15.A, p 505)
- Explain how fire hose is constructed. (NFPA 5.3.15.A, p 505)
- Describe the characteristics of double-jacket hose. (NFPA 5.3.15, p 505)
- Describe the characteristics of rubber-jacket hose. (NFPA 5.3.15, p 505)
- Describe the characteristics of couplings. (NFPA 5.3.15, p 505–506)
- List the common types of couplings. (NFPA 5.3.15, p 506)
- Describe the two types of supply hose. (NFPA 5.3.15, p 510–512)
- List the common types of hose damage. (NFPA 5.5.2A, p 512–513)
- Describe how to clean and maintain hose. (NFPA 5.5.2, p 513)
- Describe how to perform a hose inspection. (NFPA 5.5.2, p 513–515)
- List the common hose appliances used in conjunction with fire hoses. (NFPA 5.3.15, p 515–517)
- Describe the characteristics of wyes. (NFPA 5.3.15, p 515)
- Describe the characteristics of water thieves. (NFPA 5.3.15, p 515)
- Describe the characteristics of Siamese connections. (NFPA 5.3.15, p 515–516)
- Describe the characteristics of reducers. (NFPA 5.3.15, p 516)
- Describe the characteristics of hose jackets. (NFPA 5.3.15, p 516–517)
- Describe the characteristics of hose rollers. (NFPA 5.3.15, p 517)
- Describe the characteristics of hose clamps. (NFPA 5.3.15, p 517)
- Describe the types of valves used to control water in pipes or hose lines. (NFPA 5.3.15, p 517)
- List the common types of hose rolls used to organize supply hose. (NFPA 5.3.15, p 517–519)
- Describe the procedures used to connect supply lines to a fire hydrant. (NFPA 5.3.15, p 522–528)
- Describe the common types of supply line evolutions. (NFPA 5.3.15, p 522–528)
- Describe the common techniques used to load supply hose. (NFPA 5.3.15, p 528–532)
- Describe the common techniques used to carry and advance supply hose. (NFPA 5.3.15, p 535–539)
- Describe the two types of standpipe systems. (p 539)
- Describe how to unload a fire hose. (NFPA 5.3.15, p 544)

Skills Objectives

After studying this chapter, you will be able to perform the following skills:

- Set up a portable tank. (NFPA 5.3.15.B, p 491)
- Operate a fire hydrant. (NFPA 5.3.15.A, p 497–498)
- Shut down a fire hydrant. (NFPA 5.3.15.B, p 497, 499)
- Test a fire hydrant. (p 502–503)
- Replace the swivel gasket on a fire hose. (NFPA 5.3.15.B, p 507)

CHAPTER

16

- Perform the one-fire fighter foot-tilt method of coupling a fire hose. (NFPA 5.3.15, p 508–509)
- Perform the two-fire fighter method of coupling a fire hose. (NFPA 5.3.15, p 508–509)
- Perform the one-fire fighter knee-press method of uncoupling a fire hose. (NFPA 5.3.15, p 508, 510)
- Perform the two-fire fighter stiff-arm method of uncoupling a fire hose. (NFPA 5.3.10.B, p 508, 510)
- Uncouple a fire hose with a spanner wrench. (NFPA 5.3.10.B, p 508, 511)
- Connect two lines with a damaged coupling. (p 510)
- Clean and maintain fire hose. (NFPA 5.5.2B, p 513–514)
- Mark a defective hose. (NFPA 5.5.2A, p 513)
- Perform a straight hose roll. (p 518–519)
- Perform a single-doughnut roll. (p 518, 520)
- Perform a twin-doughnut roll. (p 518, 521)
- Perform a self-locking twin-doughnut roll. (p 518–519, 522)
- Perform a forward hose lay. (NFPA 5.3.15B, p 523–525)
- Attach a fire hose to a four-way hydrant valve. (NFPA 5.3.15B, p 523–527)
- Perform a reverse hose lay. (NFPA 5.3.15B, p 528–529)
- Perform a split hose lay. (NFPA 5.3.15B, p 528)
- Perform a flat hose load. (NFPA 5.3.15B, p 529–531)
- Perform a horseshoe hose load. (NFPA 5.3.15B, p 531–532)
- Perform an accordion hose load. (NFPA 5.3.15B, p 531–532, 534)
- Attach a soft suction hose to a fire hydrant. (NFPA 5.3.15B, p 533, 535–536)
- Attach a hard suction hose to a fire hydrant. (NFPA 5.3.15B, p 534–537)
- Perform a working hose drag. (NFPA 5.3.10.B, p 537–538)
- Perform a shoulder carry. (NFPA 5.3.10.B, p 539–540)
- Advance an accordion load. (NFPA 5.3.10.B, p 539, 541)
- Connect a hose line to supply a fire department connection. (p 540, 542)
- Replace a damaged hose line. (NFPA 5.5.2A, p 543)
- Drain a fire hose. (p 543)

Fire Fighter II FFII

Knowledge Objectives

After studying this chapter, you will be able to:

- Describe how to perform a service test on a fire hose. (NFPA 6.5.5.A, p 513–514)
- List the information that should be noted on a hose record. (NFPA 6.5.5.A, p 515)

Skills Objectives

After studying this chapter, you will be able to perform the following skills:

- Perform an annual service test. (NFPA 6.5.5.B, p 513–514)

Additional NFPA Standards

- NFPA 24, *Standard for the Installation of Private Fire Service Mains and Their Appurtenances*
- NFPA 1142, *Standard on Water Supplies for Suburban and Rural Fire Fighting*
- NFPA 1962, *Standard for the Inspection, Care, and Use of Fire Hose, Couplings, and Nozzles and the Service Testing of Fire Hose*

You Are the Fire Fighter

You arrive at a fire in a barn that is fully involved. The attack is clearly going to be a defensive operation, but with a house and other buildings nearby, aggressive actions will be required to prevent their loss. The house and other buildings are located one-fourth mile away from the barn, down a narrow, tree-lined lane. While the other crews are setting up exposure protection, you are assigned to be water supply officer. The incident commander says he needs a 500 gallons per minute (gpm) (2273 liters per minute [lpm]) flow to be established.

1. Which water sources could you consider?
2. Which components need to be included to establish a water supply?
3. How will you deal with the narrow lane?

Water Supply

One primary objective at a fire is to get water on the fire—the action that cools the fire and extinguishes it. To accomplish this objective, many factors must come together. This chapter covers several topics related to water supply. It describes the hydraulics or behavior of flowing water, rural and municipal water supplies, the operation of fire hydrants, the construction and care of fire hose, the roles of hose appliances, and supply line evolutions. Thus this chapter covers the water flow during the first half of its journey, from the source to the fire pumper. The Fire Attack and Foam chapter covers the second half of the journey, from the attack pumper to the fire.

The importance of a dependable and adequate water supply for fire-suppression operations is self-evident. The hose line is not only the primary weapon for fighting fire, but also the fire fighter's primary defense against being burned or driven out of a burning building. The basic plan for fighting most fires depends on having an adequate supply of water to safely confine, control, and extinguish the fire.

If the water supply is interrupted while crews are working inside a building, fire fighters can be trapped, injured, or killed. Fire fighters entering a burning building need to be confident that their water supply is both reliable and adequate to operate hose lines for their protection and to extinguish the fire.

Ensuring a dependable water supply is a critical fire-ground operation that must be accomplished as soon as possible. A water supply should be established at the same time as other initial fire-ground operations are conducted. At many fire scenes, size-up, forcible entry, raising ladders, search and rescue, ventilation, and establishing a water supply all occur concurrently.

Fire fighters can obtain water from one of two sources. Municipal water systems furnish water under pressure through fire hydrants FIGURE 16-1. In contrast, rural areas may depend on static water sources such as lakes and streams. These bodies of water serve as drafting sites for fire department apparatus to obtain and deliver water to the fire scene.

Often, the water that is carried in a tank on one of the first-arriving vehicles is used in the initial attack. Although many fires are successfully controlled using tank water, this tactic does not ensure the adequacy and reliability of the water supply. The establishment of an adequate, continuous water supply then becomes the primary objective to support the fire attack FIGURE 16-2. The operational plan must ensure that an adequate and reliable water supply is available before the tank becomes empty.

FIGURE 16-1 The water that comes from a hydrant is provided by a municipal or private water system.

FIGURE 16-2 Mobile water supply apparatus can deliver limited quantities of water to the scene of a fire.

FIGURE 16-3 Any accessible body of water can be used as a static source.

Rural Water Supplies

Many fire departments protect areas that are not serviced by municipal water systems. In these areas, residents usually depend on individual wells or cisterns to supply water for their domestic uses. Because there are no fire hydrants in these areas, fire fighters must depend on water from other sources for firefighting activities. Fire fighters in rural areas must know how to get water from the sources that are available.

Static Sources of Water

Each County needs 20ft of "hard tube"

Several potential static water sources can be used for fighting fires in rural areas. Both natural and human-made bodies of water such as rivers, streams, lakes, ponds, oceans, canals, reservoirs, swimming pools, and cisterns can be used to supply water for fire suppression FIGURE 16-3. Some areas have many different static sources, whereas others have few or none at all.

Water from a static source can be used to fight a fire directly, if it is close enough to the fire scene. Otherwise, it must be transported to the fire using long hose lines, engine relays, or mobile water supply tankers/tenders.

To be useful, static water sources must be accessible to a fire engine or portable pump. If a road or hard surface is located within 20 feet (6 m) of the water source, a fire engine can drive close enough to draft water directly into the pump through a hard suction hose. Rural fire departments should identify these areas and practice establishing drafting operations at all of these locations. Some fire departments construct special access points so engines can approach the water source.

Dry hydrants, also called drafting hydrants, provide quick and reliable access to static water sources. A dry hydrant is a pipe with a strainer on one end and a connection for a hard suction hose on the other end. The strainer end should be located below the water's surface and away from any silt or potential obstructions. The other end of the pipe should be accessible to fire apparatus, with the connection at a convenient height for an engine hook-up FIGURE 16-4. When a hard suction hose is connected to a drafting hydrant, the engine can draft water from the static source.

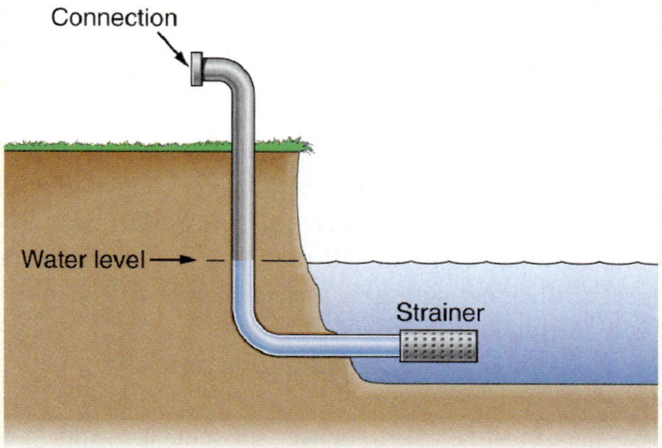

FIGURE 16-4 A dry hydrant or drafting hydrant can be placed at an accessible location near a static water source.

Dry or drafting hydrants are often installed in lakes and rivers and close to clusters of buildings where there is a recognized need for fire protection. They may also be installed in farm cisterns or connected to swimming pools on private property to make water available for the local fire department. In some areas, dry or drafting hydrants are used to enable fire fighters to reach water under the frozen surface of a lake or river. NFPA 1142, *Standard on Water Supplies for Suburban and Rural Fire Fighting*, has more information about dry hydrants.

The portable pump is an alternative means of accessing water in areas that are inaccessible to fire apparatus FIGURE 16-5. The portable pump can be hand-carried or transported by an off-road vehicle to the water source. Portable pumps can deliver as much as 500 gallons of water per minute (2273 lpm).

FIGURE 16-5 A portable pump can be used if the water source is inaccessible to a fire department engine.

Mobile Water Supply Apparatus

Mobile water supply apparatus (also called tankers or water tenders) also can deliver water to fight fires in rural areas. These trucks are designed specifically to carry large volumes of water. Fire department engines usually carry at least 500 gallons (2273 L) of water in the booster tank, whereas fire department tankers generally carry 1000 to 3500 gallons of water (4546 to 13,638 L) FIGURE 16-6. Some tankers can transport as much as 5000 gallons (22,730 L) of water.

If a mobile water supply apparatus is the only source of water for fighting a structural fire, the attack must be carefully planned. Enough water must be available on the scene to supply the required hose lines. If the water supply is exhausted before the fire is extinguished, the attack team will be in serious danger and the building will probably be lost. Conversely, if water use is too conservative, fire-suppression efforts will probably be unsuccessful. To be both safe and successful, the fire department must be able to deliver adequate amounts of water and to maintain a reliable supply of water to the fire scene.

Some fire departments begin the attack using water from the tank of the first-arriving unit. Tankers then pump additional water directly into the attack pumper to keep it full. In rural areas, several tankers may be dispatched for a structural fire. These tankers can deliver water to the attack engine in sequence, refilling their own supplies at a remote location and returning to the fire scene as needed.

FIGURE 16-6 A fire department tanker or tender.

Portable Tanks

Portable tanks carried on fire apparatus can be quickly set up at a fire scene. These tanks typically hold between 600 and 5000 gallons (2727 and 22,730 liters) of water and should be placed so that they can be accessed from multiple directions. With this supply method, one engine drafts water from the tank, using a hard suction hose just as it would if the water came from any other static source FIGURE 16-7. A tanker is used to fill the portable tank. The pump driver/operator primes the pump and begins drafting water out of the portable tank, while the tanker leaves to get another load of water.

Speed is a primary advantage when using a portable tank system, in that tankers do not have to hook up to the attack pumper and slowly transfer the water. Instead, a dump valve enables the tankers to offload as much as 3000 gallons (13,638 L) of water in 1 minute into a portable tank FIGURE 16-8.

FIGURE 16-7 An engine may be set up to draft water from a portable tank.

FIGURE 16-8 A dump valve allows a tanker to discharge water into the portable tank quickly.

The faster the tanker can offload its water, the more quickly it can return to the fill site for another load.

Another advantage of the portable tank system is its ability to expand rapidly. Additional portable tanks can be set up and linked together to increase water storage capacity, additional pumpers can be used to draft water from the portable tanks, and additional tankers can be used to deliver more water at a faster rate. A series of several tankers may also be used as shuttles, dumping water either simultaneously or in sequence. When using multiple tanks, a jet siphon assists in keeping the water level at the maximum capacity in the tank closest to the pumper.

To set up a portable tank, follow the steps in **SKILL DRILL 16-1**:

1. Two fire fighters lift the portable tank off the apparatus. This tank may be mounted on a side rack or on a hydraulic rack that lowers it to the ground. (**STEP 1**)
2. Place the portable tank on level ground beside the engine. The pump driver/operator will indicate the best location. (**STEP 2**)
3. Expand the tank (metal-frame type) or lay it flat (self-expanding type). (**STEP 3**)
4. Place a strainer on the end of the suction hose, put the suction hose into the tank, and connect it to the engine. (**STEP 4**)
5. A second fire fighter helps the tanker driver discharge water into the portable tank. If the tank is self-expanding, the fire fighters may need to hold the collar until the water level is high enough for the tank to support itself. (**STEP 5**)

SKILL DRILL 16-1 Setting Up a Portable Tank
(Fire Fighter I, NFPA 5.3.15)

1 Two fire fighters lift the portable tank off the apparatus. This tank may be mounted on a side rack or on a hydraulic rack that lowers it to the ground.

2 Place the portable tank on level ground beside the engine. The pump driver/operator will indicate the best location.

3 Expand the tank (metal-frame type) or lay it flat (self-expanding type).

4 One fire fighter helps the pump operator place the strainer on the end of the suction hose, put the suction hose into the tank, and connect it to the engine.

5 The second fire fighter helps the tanker driver discharge water into the portable tank. If the tank is self-expanding, the fire fighters may need to hold the collar until the water level is high enough for the tank to support itself.

Tanker Shuttles

When a large volume of water is needed for an extended period at a fire, a tanker shuttle can be used to deliver water from a fill site to the fire scene. The number of tankers needed will depend on the distance between the fill site and the fire scene, the time it takes the tankers to dump and refill, and the flow rate required at the fire scene.

All of the components of a tanker shuttle must be set up so water moves efficiently from the fill site to the fire scene. At the fill site, the tankers must be refilled without delay. The routes in both directions must be planned so that tankers make efficient round trips, without having to back up or make U-turns. At the fire scene, the tankers should be able to drive up, dump their water into the portable tanks, and immediately return to the fill site. The portable tanks, in turn, must be large enough to receive the full load of each tanker as it arrives. An effective tanker shuttle can deliver several hundred gallons of water per minute without interruption.

A separate component of the incident command structure—titled "water supply"—is usually established to coordinate a tanker-shuttle operation. Departments must practice these operations to ensure proper coordination and effective water delivery.

Fire Fighter Safety Tips

Never get between two apparatus or tanks!

Municipal Water Systems

Municipal water systems make clean water available to people in populated areas and provide water for fire protection. As the name suggests, most municipal water systems are owned and operated by a local government agency, such as a city, county, or special water district. Some municipal water systems are privately owned; however, the basic design and operation of both private and government systems are very similar. Municipal water is supplied to homes, commercial establishments, and industries.

Hydrants make the same water supply available to the fire department. In addition, most automatic sprinkler systems and many standpipe systems are connected directly to a municipal water source. A municipal water system has three major components: the water source, the treatment plant, and the distribution system.

Water Sources

Municipal water systems can draw water from wells, rivers, streams, lakes, or human-made storage facilities called reservoirs. The source will depend on the geographic and hydrologic features of the area. Many municipal water systems draw water from several sources to ensure a sufficient supply. Underground pipelines or open canals supply some cities with water from sources that are many miles away.

The water source for a municipal water system needs to be large enough to meet the total demands of the service area. Most of these systems include large storage facilities, thereby ensuring that they will be able to meet the community's water supply demands if the primary water source becomes unavailable for some reason. The backup supply for some systems can provide water for several months or years. By contrast, in some other systems, the supply may last only a few days.

Water Treatment Facilities

Municipal water systems also include a water treatment facility, where impurities are removed from the water **FIGURE 16-9**. The nature of the treatment system depends on the quality of the untreated source water. Water that is clean and clear from the source requires little treatment. Other systems must use extensive filtration to remove impurities and foreign substances. Some treatment facilities use chemicals to remove impurities and improve the water's taste. All of the water in the system must be suitable for drinking.

Chemicals and ultraviolet (UV) radiation are used to kill bacteria and harmful organisms and to keep the water pure as it moves through the distribution system to individual homes or businesses. After the water has been treated, it enters the distribution system. As of 2014 just over 5000 hydrants

Water Distribution System

The distribution system delivers water from the treatment facility to the end users and fire hydrants through a complex network of underground pipes, known as water mains. In most cases, the distribution system also includes pumps, storage tanks, control valves, reservoirs, and other necessary components to ensure that the required volume of water can be delivered, where and when it is needed, at the required pressure.

Water pressure requirements differ, depending on how the water will be used. Generally, water pressure ranges from 20 pounds per square inch (psi) to 80 psi (138 to 551 kPa) at the delivery point. The recommended minimum pressure for water coming from a fire hydrant is 20 psi (138 kPa), but it is possible to operate with lower hydrant pressures under some circumstances.

Most water distribution systems rely on an arrangement of pumps to provide the required pressure, either directly or indirectly. In systems that use pumps to supply direct pressure, if the

FIGURE 16-9 Impurities are removed at the water treatment facility.

VOICES OF EXPERIENCE

I had only been out of the training academy three weeks and I knew everything there was to know about being a hard-charging, smoke-eating fire fighter. We were wrapping up an EMS call when I heard the radio crackling for a structure fire in our first-due area. The fire was in an industrial park attached to an airport. The business made pumps and used a large heated oil vat to waterproof their shipping crates. The oil had overflowed, igniting on the heater, and was spreading across the property.

As we arrived on scene, there was a smoke column so big that it was as black as night. We had a three-person crew and we were going to be first engine—*this is* what it was all about. When we pulled up to the gate my heart was racing, as were the thoughts in my head: remember the hydrant wrench, do not forget a tool, which hose will we pull? Then my captain turned to me, "Take the plug." I was off. This is what I had trained for. This was my chance to prove myself. I wrapped the hydrant in record time, I shrieked, "Go!" at the top of my lungs, and the engine disappeared into a wall of black smoke.

I was a machine. Fueled by pure adrenaline and training, I knew that I had made those connections in record time. The wrench landed on the hydrant with a solid thud and was immediately spinning at light speed. That's when it happened. Nothing. That wrench spun, and spun—but the hose lay there like a limp noodle. My first real fire—this can't be happening! I turned to look toward my crew only to see the unbroken wilted yellow hose disappearing around the corner into that dense wall of smoke. What do I do now? Nobody mentioned *this* in the academy.

Back then, only the captain had a radio. I looked left, then looked right—no flashing lights, no red trucks. I was all alone, so I ran in full PPE for the entire 1000 feet of laid hose. When I got to the engine, I was so out of breath that I could barely talk. But through a combination of breathless nouns, flailing gestures, and some kind of sign language, I was able to tell the engineer that there was no water. He stopped, looked me square in the eye, and stared at me for a moment. Then he shrugged, said, "Okay," and went back to work.

Ultimately, the next engine laid another supply line and we got the water we needed from them. The fire went out, nobody got hurt, and everyone went home. I also learned that day that I didn't know quite as much about fighting fires as I thought I did.

In the decades since that day, I have been to dozens of fires where an adequate water supply was an issue. Broken hydrants, burst hose lines, nozzles filled with gravel from the water supply—the bottom line was always the same. We could not do our job until we fixed *that* problem. Water is the most important suppression tool a fire fighter has. We take it for granted until it is not there.

Rick Picard
Peoria Fire Department
Peoria, Arizona

pumps stop operating, the pressure will be lost and the system will be unable to deliver adequate water to the end users or to hydrants. Most municipal systems maintain multiple pumps and backup power supplies to reduce the risk of a service interruption due to a pump failure. These extra pumps can sometimes be used to boost the flow for a major fire or during a high-demand period.

In a pure gravity-feed system, the water source, treatment plant, and storage facilities are located on high ground while the end users live in lower-lying areas, such as a community in a valley **FIGURE 16-10**. This type of system may not require any pumps, because gravity, through the elevation differentials, provides the necessary pressure to deliver the water. In some systems, the elevation pressure is so high that pressure-control/reduction devices are needed to keep parts of the system from being subjected to excessive pressures.

Most municipal water supply systems use both pumps and gravity to deliver water. Pumps can be used to deliver water from the treatment plant to elevated water storage towers or to reservoirs located on hills or high ground. These elevated storage facilities then maintain the desired water pressure in the distribution system, ensuring that water can be delivered under pressure even if the pumps are not operating **FIGURE 16-11**. When the elevated storage facilities need refilling, large supply pumps are used. Additional pumps may be installed to increase the pressure in particular areas, such as a booster pump that provides extra pressure for a hilltop neighborhood.

A combination pump-and-gravity-feed system must maintain enough water in the elevated storage tanks and reservoirs to meet anticipated demands. If more water is being used than the pumps can supply, or if the pumps are out of service, some systems will be able to operate for several days by relying solely on their elevated storage reserves. Other systems may be able to function for only a few hours in such scenarios.

The underground water mains that deliver water to the end users come in several different sizes. Large mains, known as primary feeders, carry large quantities of water to a section of the town or city. Smaller mains, called secondary feeders, distribute water to a smaller area. The smallest pipes, called distributors, carry water to the users and hydrants along individual streets.

The size of the water mains required depends on the amount of water needed both for normal consumption and for fire protection in that location. Most jurisdictions specify the

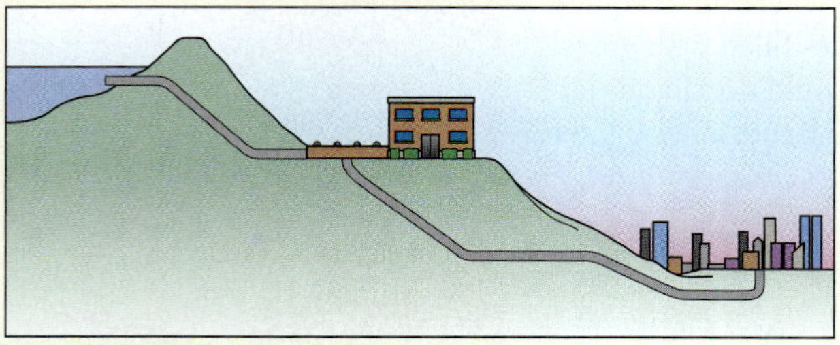

FIGURE 16-10 A gravity-feed system can deliver water to a low-lying community without the need for pumps.

FIGURE 16-11 Water that is stored in an elevated tank can be delivered to the end users under pressure.

minimum-size main that can be installed in a new municipal water system to ensure an adequate flow. Some municipal water systems, however, may include undersized water mains in older areas of the community. The volume of water delivered through a water main may decrease if the pipe becomes corroded or partly filled with sediment. Fire fighters must know the arrangement and capacity of the water systems in their response areas.

Water mains in a well-designed system will follow a grid pattern. A grid arrangement provides water flow to a fire hydrant from two or more directions and establishes multiple paths from the source to each area. This kind of organization helps to ensure an adequate flow of water for firefighting. The grid design also helps to minimize downtime for the other portions of the system if a water main breaks or needs maintenance work. With a grid pattern, the water flow can be diverted around the affected section.

Older water distribution systems may have dead-end water mains, which supply water from only one direction. Such dead-end water mains may still be found in the outer reaches of a municipal system. Hydrants on this type of main will have a limited water supply. If two or more hydrants on the same dead-end main are used to fight a fire, the upstream hydrant will have more water and water pressure than the downstream hydrants.

Control valves installed at intervals throughout a water distribution system allow different sections of the water supply to be turned off or isolated. These valves are used when a water main breaks or when work must be performed on a section of the system.

Shut-off valves are located at the connection points where the underground mains meet the distributor pipes. These valves control the flow of water to individual customers or to individual fire hydrants FIGURE 16-12. If the water system in a building or to a fire hydrant becomes damaged, the shut-off valves can be closed to prevent further water flow.

The fire department should notify the water department when fire operations will require prolonged use of large quantities of water. The water department may, in turn, be able to increase the normal volume and/or pressure by starting additional pumps. In some systems, the water department can open valves to increase the flow to a certain area in response to fire department operations at major fires.

FIGURE 16-12 A shut-off valve controls the water supply to an individual user or fire hydrant.

Types of Fire Hydrants

Fire hydrants provide water for firefighting purposes. Public hydrants are part of the municipal water distribution system and draw water directly from the public water mains. Hydrants are also installed on private water systems supplied by the municipal water system or from a separate source. The water source as well as the adequacy and reliability of the supply to private hydrants must be noted and identified to ensure that sufficient water supplies will be available when fighting fires.

Most fire hydrants consist of an upright steel casing (barrel) attached to the underground water distribution system. The two major types of municipal hydrants are the dry-barrel hydrant and the wet-barrel hydrant. Hydrants are equipped with one or more valves to control the flow of water through the hydrant. In addition, one or more outlets are provided to connect fire department hoses to the hydrant. These outlets are sized to fit the 2½-inch (65 mm) or larger fire hoses used by the local fire department. Threads utilized on fire hydrants are usually of the national standard type, although some jurisdictions have their own thread type.

Wet-Barrel Hydrants

Wet-barrel hydrants are used in locations where temperatures do not drop below freezing. These hydrants always have water in the barrel and do not have to be drained after each use.

Wet-barrel hydrants usually have separate valves that control the flow to each individual outlet FIGURE 16-13. A fire fighter can hook up one hose line and begin flowing water, and later attach a second hose line and open the valve for that outlet, without shutting down the hydrant.

Dry-Barrel Hydrants

Dry-barrel hydrants are used in climates where temperatures fall below freezing. The valve that controls the flow of water into the barrel of the hydrant is located at the base, several feet below ground (below the frost line), to keep the hydrant from freezing FIGURE 16-14. The length of the barrel depends on the climate and the depth of the valve. Water enters the barrel of the hydrant only when it is needed. Turning the nut on the top of the hydrant rotates the operating stem, which opens the valve so that water flows up into the barrel of the hydrant.

Whenever this type of hydrant is not in use, the barrel must remain dry. If the barrel contains standing water, it will freeze in cold weather and render the hydrant inoperable. After each use, the water drains out through an opening at the bottom of the barrel. This drain is fully open when the hydrant valve is fully shut. When the hydrant valve is opened, the drain closes, thereby preventing water from being forced out of the drain when the hydrant is under pressure. If the drain becomes clogged, however, the hydrant may not drain. In this case, it may be necessary to pump water out of the hydrant before it freezes.

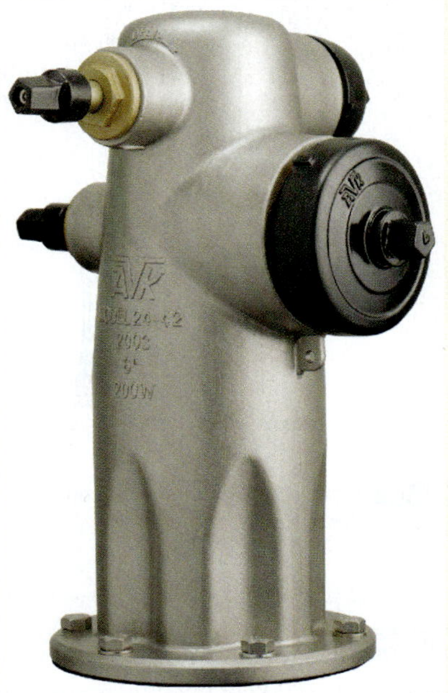
FIGURE 16-13 A wet-barrel hydrant has a separate valve for each outlet.

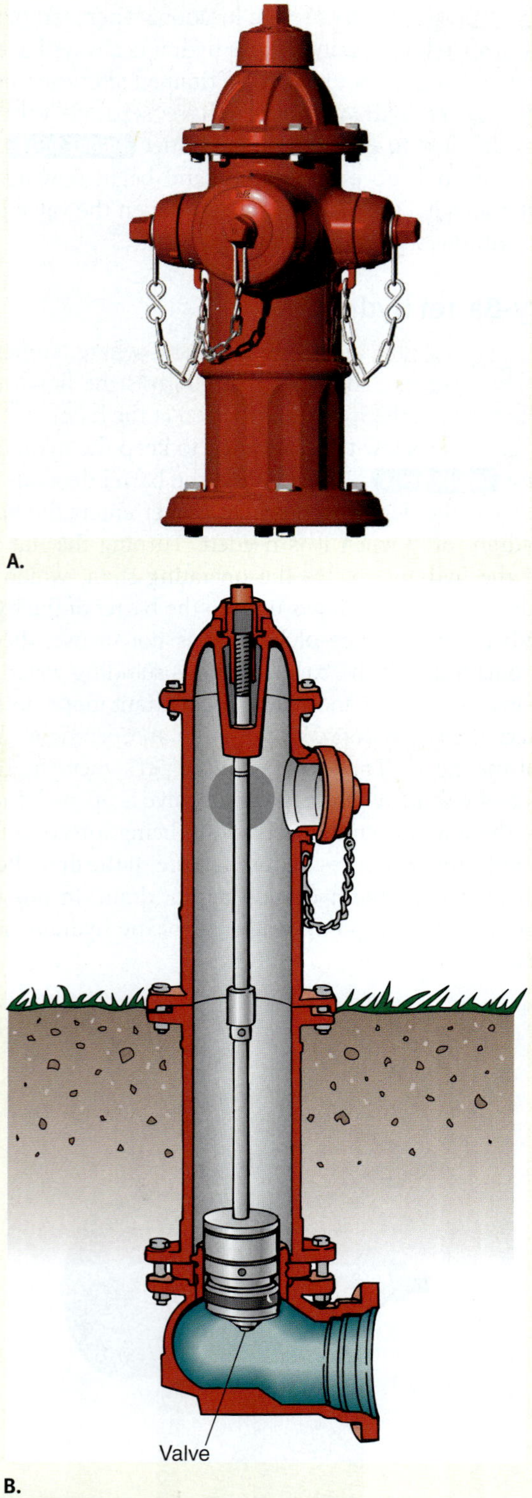

FIGURE 16-14 A dry-barrel hydrant (A) is controlled by an underground valve (B).

A partially opened valve means that the drain is also partially open, so pressurized water can flow out. This leakage can erode (undermine) the soil around the base of the hydrant and may damage the hydrant. For this reason, a hydrant should always be either fully opened or fully closed. A fully opened hydrant also makes the maximum flow available to fight a fire.

FIRE FIGHTER Tips

The terms "dry-barrel hydrant" and "dry hydrant" are very easy to confuse, but they are very different. Dry-barrel hydrants are connected to a pressurized municipal water system. Dry hydrants (also known as drafting hydrants) are a permanent piping system that is connected to a static water source such as a stream, a pond, or a lake. To get water from a dry or drafting hydrant, you must draft (pump) the water using a fire pumper or a portable pump.

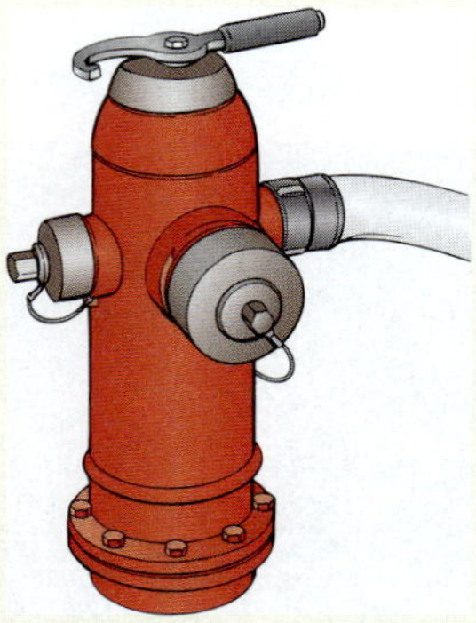

FIGURE 16-15 Most dry-barrel hydrants have only one large valve at the bottom of the barrel that controls the flow of water.

Most dry-barrel hydrants contain only one large valve that controls the flow of water FIGURE 16-15. Each outlet must be connected to a hose or an outlet valve, or have a hydrant cap firmly in place before the valve is turned on. Many fire departments use special hydrant valves that allow additional connections to be made after water is flowing through the first hose. If additional outlets are needed later, separate outlet valves can be connected before the hydrant is opened.

In some areas, vandals may dispose of trash or place foreign objects in the empty barrels of dry-barrel hydrants. These materials can obstruct the water flow or damage a fire department pumper if they are drawn into the pump. It is a good practice to check the operation of the hydrant and flush out debris before connecting a hose to a hydrant. To do so, the fire fighter making the connection opens the large outlet cap and then opens the hydrant valve just enough to ensure that water flows into the hydrant and flushes out any foreign matter. This step takes only a few seconds. The fire fighter then closes the hydrant valve, connects the hose, and reopens the valve all the way. Fire departments should ensure that inspections and tests to keep hydrants operating smoothly are performed regularly.

Fire Hydrant Locations

Fire hydrants are located according to local standards and nationally recommended practices. They may be placed a certain distance apart, perhaps every 500 feet (152 m) in residential areas and every 300 feet (91 m) in high-value commercial and industrial areas. In many communities, hydrants are located at every street intersection, with mid-block hydrants being installed if the distance between intersections exceeds a specified limit.

In some cases, the requirements for locating hydrants are based on the occupancy, construction, and size of a building. A builder may be required to install additional hydrants when a new building is constructed so that no part of the building will be more than a specified distance from the closest hydrant.

Knowing the plan for installing fire hydrants makes them easier to find in emergency situations. Fire fighters who perform fire inspections or develop preincident plans should identify the locations and flow rates of nearby fire hydrants for each building or group of buildings.

Fire Hydrant Operation

Operate a Hydrant

Fire fighters must be proficient in operating fire hydrants. SKILL DRILL 16-2 outlines the steps in getting water from a dry-barrel hydrant efficiently and safely. These same steps, with the modifications noted, apply to wet-barrel hydrants as well.

1. Remove the cap from the outlet you will be using. (STEP 1)
2. Quickly look inside the hydrant opening for any objects that may have been thrown into the hydrant. (STEP 2) (Omit this step for a wet-barrel hydrant.)
3. Check that the remaining hydrant caps are snugly attached. (STEP 3) (Omit this step for a wet-barrel hydrant.)
4. Place the hydrant wrench on the stem nut. Check the top of the hydrant for the arrow indicating which direction to turn the nut to open the hydrant valve. (STEP 4)
5. Open the hydrant just enough to determine that there is a good flow of water and to flush out any objects that may have been put into the hydrant. (STEP 5) (Omit this step for a wet-barrel hydrant.)
6. Shut off the flow of water. (STEP 6) (Omit this step for a wet-barrel hydrant.)
7. Attach the hose or valve to the hydrant outlet. (STEP 7)
8. When instructed to do so by your officer or the pump driver/operator, start the flow of water. Turn the hydrant wrench to fully open the valve. This may take 12 to 14 turns, depending on the type of hydrant. (STEP 8)
9. Open the hydrant slowly to avoid a pressure surge. Once the flow of water has begun, you can open the hydrant valve more quickly. Make sure that you open the hydrant valve completely. If the valve is not fully opened, the drain hole will remain open. (STEP 9)

Note: This skill drill applies to a dry-barrel hydrant. For a wet-barrel hydrant, omit steps 2, 3, 5, and 6. Open the valve for the particular outlet that will be used.

Individual fire departments may have their own variations on this procedure. For example, some departments specify that the wrench be left on the hydrant. Other departments require that the wrench be removed and returned to the fire apparatus so that an unauthorized person cannot interfere with the operation. Always follow the standard operating procedures (SOPs) for your department.

Shutting Down a Hydrant

Shutting a hydrant down properly is just as important as opening a hydrant properly. If the hydrant is damaged during shutdown, it cannot be used until it has been repaired. To shut down a hydrant efficiently and safely, follow the steps in SKILL DRILL 16-3:

1. Turn the hydrant wrench slowly until the valve is closed. (STEP 1)
2. Allow the hose to drain by opening a drain valve or disconnecting a hose connection downstream. Slowly disconnect the hose from the hydrant outlet, allowing any remaining pressure to escape. (STEP 2)
3. On dry-barrel hydrants, leave one outlet open until the water drains from the hydrant. (STEP 3)
4. Replace the hydrant cap. (STEP 4)

Note: Do not leave or replace the caps on a dry-barrel hydrant until you are sure that the water has completely drained from the barrel. If you feel suction on your hand when you place it over the opening, the hydrant is still draining. In very cold weather, you may have to use a hydrant pump to remove all of the water and prevent freezing.

Maintaining Fire Hydrants

Inspecting Fire Hydrants

Because hydrants are essential to fire suppression efforts, fire fighters must understand how to inspect and maintain them. Hydrants should be checked on a regular schedule—no less frequently than once a year—to ensure that they are in proper operating condition. During inspections, fire fighters may encounter some common problems and should know how to correct them.

The first factors to check when inspecting hydrants are visibility and accessibility. Hydrants should always be visible from every direction, so they can be easily spotted. A hydrant should not be hidden by tall grass, brush, fences, debris, dumpsters, or any other obstructions FIGURE 16-16. In winter,

FIGURE 16-16 Hydrants should not be hidden or obstructed.

SKILL DRILL 16-2 Operating a Fire Hydrant

(Fire Fighter I, NFPA 5.3.15)

1 Remove the cap from the outlet you will be using.

2 Quickly look inside the hydrant opening for foreign objects (dry-barrel hydrant only).

3 Check that the remaining caps are snugly attached (dry-barrel hydrant only).

4 Attach the hydrant wrench to the stem nut. Check for an arrow indicating the direction to turn to open.

5 Open the hydrant enough to verify flow and flush the hydrant (dry-barrel hydrant only).

6 Shut off the flow of water (dry-barrel hydrant only).

7 Attach the hose or valve to the hydrant outlet.

8 When instructed, turn the hydrant wrench to fully open the valve.

9 Open the hydrant slowly to avoid a pressure surge.

hydrants must be clear of snow. No vehicles should be allowed to park in front of a hydrant.

In many communities, hydrants are painted in bright reflective colors for increased visibility. The bonnet (the top of the hydrant) and or caps may also be color coded to indicate the available flow rate of a hydrant TABLE 16-1. Colored reflectors are sometimes mounted next to hydrants or placed in the pavement in front of them to make them more visible at night. In areas that experience lots of snow, hydrants are sometimes marked with a small flag mounted on top of a pole that is attached to the hydrant.

Hydrants should be installed at an appropriate height above the ground. Their outlets should not be so high or so low that fire fighters have difficulty connecting hose lines

SKILL DRILL 16-3 Shutting Down a Hydrant

(Fire Fighter I, NFPA 5.3.15)

1 Turn the wrench to slowly close the hydrant valve.

2 Drain the hose line. Slowly disconnect the hose from the hydrant outlet.

3 Leave one hydrant outlet open until the hydrant is fully drained (dry-barrel hydrant only).

4 Replace the hydrant cap.

TABLE 16-1 Fire Hydrant Colors

Class	Flow Available at 20 psi (138 kPa)	Color
Class C	Less than 500 gpm (2273 lpm)	Red
Class B	500–999 gpm (2273–4541 lpm)	Orange
Class A	1000–1499 gpm (4546–6814 lpm)	Green
Class AA	1500 gpm and higher (6819 lpm)	Light blue

NFPA 24 recommends that fire hydrants be color coded to indicate the water flow available from each hydrant at 20 psi. It is recommended that the top bonnet and the hydrant caps be painted according to the above system. The colors give you an idea of how much water can be obtained from a hydrant during a fire.

to them. NFPA 24, *Standard for the Installation of Private Fire Service Mains and Their Appurtenances*, requires a minimum of 18 inches (45 cm) from the center of a hose outlet to the finished grade. Hydrants should be positioned so that the connections—especially the large steamer connection—are facing the street.

During a hydrant inspection, check the exterior of the hydrant for signs of damage. Open the steamer port of dry-barrel hydrants to ensure that the barrel is dry and free of debris. Make sure that all caps are present and that the outlet hose threads are in good working order FIGURE 16-17.

FIGURE 16-17 All hydrants should be checked at least annually.

FIRE FIGHTER Tips

Before you leave the fire scene, make sure that dry-barrel hydrants are completely drained, even if the weather is warm. During winter, any water left in the hydrant can freeze. Fire fighters may lose valuable time connecting a hose to a frozen hydrant, only to discover that it will not operate. If this happens, fire fighters will be without water until they can locate a working hydrant and can reposition and reconnect the hose lines.

The second part of the inspection ensures that the hydrant works properly. Open the hydrant valve just enough to confirm that water flows out and flushes any debris out of the barrel. After flushing, shut down the hydrant. Leave the cap off dry-barrel hydrants to ensure that they drain properly. A properly draining hydrant will create suction against a hand placed over the outlet opening. When the hydrant is fully drained, replace the cap.

If the threads on the discharge ports need cleaning, use a steel brush and a small triangular file to remove any burrs in the threads. Also check the gaskets in the caps to make sure they are not cracked, broken, or missing. Replace worn gaskets with new ones, which should be carried on each apparatus. Follow the manufacturer's recommendations for maintaining any parts that require lubrication.

Testing Fire Hydrants

The amount of water available to fight a fire at a given location is a crucial factor in planning an attack. Will the hydrants deliver enough water at the needed pressure to enable fire fighters to control a fire? If not, what can be done to improve the water supply? How can fire fighters obtain additional water if a fire does occur?

Fire-suppression companies are often assigned to test the flow from hydrants in their districts. The procedures for testing hydrants are relatively simple, but a basic understanding of the concepts of hydraulics and careful attention to detail are required. This section explains some of the basic theory and terminology of hydraulics and describes how the tests are conducted and the results are recorded.

Flow and Pressure

To understand the procedures for testing fire hydrants, fire fighters must remember the terminology that is used. The flow or quantity of water moving through a pipe, hose, or nozzle is measured in terms of its volume, usually specified in units of gallons (or liters) per minute. Water pressure refers to an energy level and is measured in units of psi (or kilopascals [kPa]). Volume and pressure are two different, but mathematically related, measurements.

Water that is not moving has potential (static) energy. When the water is moving, it has a combination of potential energy and kinetic (in motion) energy. Both the quantity of water flowing and the pressure under a specific set of conditions must be measured as part of testing any water system, including hydrants.

Static pressure is the pressure in a system when the water is not moving. Static pressure is potential energy, because it would cause the water to move if there were some place the water could go. This kind of pressure causes the water to flow out of an opened fire hydrant. If no static pressure is present, nothing happens when a fire fighter opens a hydrant.

Static pressure is generally created by elevation pressure or pump pressure, or both. An elevated storage tank creates elevation pressure in the water mains. Gravity also creates elevation pressure (sometimes referred to as head pressure) in a water system as the water flows from a hilltop reservoir to the water mains in the valley below. Pumps create pressure by bringing the energy from an external source into the system.

To measure static pressure in a water distribution system, place a pressure gauge on a hydrant port and open the hydrant valve. No water can be flowing out of the hydrant when static pressure is measured.

When measured in this way, the static pressure reading assumes that there is no flow in the system. Of course, because municipal water systems deliver water to hundreds or thousands of users, there is almost always at least some water flowing within the system. Thus, in most cases, a static pressure reading actually measures the normal operating pressure of the system.

Normal operating pressure refers to the amount of pressure in a water distribution system during a period of normal consumption. In a residential neighborhood, for example, people are constantly using water to care for lawns, wash clothes, bathe, and perform other normal household activities. In an industrial or commercial area, normal consumption occurs during a normal business day as water is used for various purposes. The water distribution system uses some of the static pressure to deliver this water to residents and businesses. A pressure gauge connected to a hydrant during a period of normal consumption will indicate the normal operating pressure of this system.

Fire fighters need to know how much pressure will be in the system when a fire occurs. Because the regular users of the system will be drawing off a normal amount of water even during firefighting operations, the normal operating pressure is sufficient for measuring available water. (*Note*: The normal operating pressure may change according to the time of day in some areas owing to high demand during certain hours.)

Residual pressure is the amount of pressure that remains in the system when water is flowing. When fire fighters open a hydrant and start to draw large quantities of water out of the system, some of the potential energy of still water is converted to the kinetic energy of moving water. However, not all of the potential energy turns into kinetic energy—some of it is used to overcome friction in the pipes. The pressure remaining while the water is flowing constitutes the residual pressure.

Residual pressure is important because it provides the best indication of how much more water is available in the system. The more water that is flowing, the less residual pressure in the system. In theory, when the maximum amount of water is flowing, the residual pressure is zero, and there is no more potential energy to push more water through the system. In reality, 20 psi (138 kPa) is considered the minimum usable residual pressure necessary to reduce the risk of damage to underground water mains or pumps.

At the scene of a fire, the pump driver/operator uses the difference between static pressure and residual pressure to determine how many more attack lines or appliances can be operated from the available water supply. The pump driver/operator can refer to a set of tables to calculate the maximum amount of available water; these tables are based on the static and residual pressure readings taken during hydrant testing. Flow pressure measures the quantity of water flowing through an opening during a hydrant test. When a stream of water flows out through an opening (known as an orifice), all of the pressure is converted to kinetic energy. To calculate the volume of water flowing, fire fighters measure the pressure at the center of the water stream as it passes through the opening, and then factor in the size and flow characteristics of the orifice. A Pitot gauge is used to measure flow pressure in psi (or kPa) and to calculate the flow in gallons (or liters) per minute.

Knowing the static pressure, the flow in gallons (liters) per minute, and the residual pressure enables fire fighters to calculate the amount of water that can be obtained from a hydrant or a group of hydrants on the same water main. The test procedure is described in the next section.

Hydrant Testing Procedure

The procedure for testing hydrant flows requires two adjacent hydrants, a Pitot gauge, and an outlet cap with a pressure gauge. Fire fighters measure static pressure and residual pressure at one hydrant and then open the other hydrant to let water flow out. The two hydrants should be connected to the same water main and preferably at approximately the same elevation FIGURE 16-18.

The cap gauge is placed on one of the outlets of the first hydrant. The hydrant valve is then opened to allow water to fill the hydrant barrel. The initial pressure reading on this gauge is recorded as the static pressure.

At the second hydrant, fire fighters remove one of the discharge caps and open the hydrant. They put the Pitot gauge into the middle of the stream and take a reading, which is recorded as the Pitot pressure. At the same time, fire fighters at the first hydrant record the residual pressure reading.

Using the size of the discharge opening (usually 2½ inches) (65 mm) and the Pitot pressure, fire fighters can calculate the flow in gpm or look it up in a table. Such a table usually incorporates factors to adjust for the shape of the discharge opening. Fire fighters can use special graph paper or computer software to plot the static pressure and the residual pressure at the test flow rate. The line defined by these two points shows the number of gallons (liters) per minute that is available at any residual pressure. The flow available for fire suppression is usually defined as the number of gallons (liters) per minute available at 20 psi (138 kPa) residual pressure.

Several special devices are available to simplify the process of taking accurate Pitot readings. Some outlet attachments have smooth tips and brackets that hold the Pitot gauge in the exact

FIGURE 16-18 Testing hydrant flow requires two hydrants on the same water main.

required position **FIGURE 16-19**. The flow can also be measured with an electronic flow meter instead of a Pitot gauge.

To test the operability and flow of a fire hydrant, follow the steps in **SKILL DRILL 16-4**:

1. Place a cap gauge on one of the outlets of the first hydrant. (**STEP 1**)
2. Open the first hydrant valve to fill the hydrant barrel.
3. Record the initial pressure reading on the gauge as the static pressure. (**STEP 2**)
4. Move to the second hydrant, remove one of the discharge caps, and open the second hydrant. (**STEP 3**)
5. Place the Pitot gauge one-half the diameter of the orifice away from the opening.
6. Record this pressure as the Pitot pressure.
7. Record the pressure on the first hydrant as the residual pressure.
8. Use the recorded pressures to calculate or look up the flow rates at 20 psi (138 kPa) residual pressure.
9. Document your findings. (**STEP 4**)

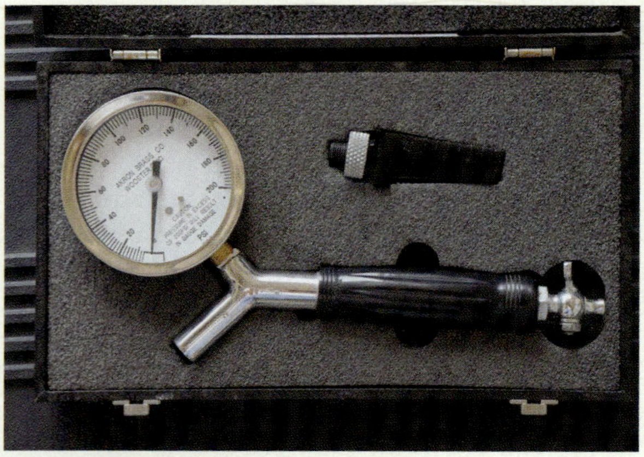

FIGURE 16-19 The Pitot gauge.

FIRE FIGHTER Tips

You have learned about the initial steps needed to establish a water supply. Now you can appreciate how important these tasks are for the entire firefighting team and understand how identifying the locations of fire hydrants and static water sources during preincident planning can save time and lives.

Near Miss REPORT

Report Number: 11-0000246

Synopsis: Hydrant cap blows off during checks.

Event Description: I was sent to a local apartment complex to investigate reported damage to a fire hydrant. Engine [number removed] was on the air for area familiarization and was in the apartment complex. The apartment complex consists of 32 12-unit, three-story, lightweight construction apartment buildings with limited access. The buildings are all equipped with an NFPA-13R compliant sprinkler system. The complex is served by a private fire hydrant system that is maintained by the complex owners.

While in the complex, fire fighters stopped at several hydrants to see which pressures and flows were available. After checking one hydrant, the crew moved to an adjacent hydrant and was charging the hydrant when the 2½-inch (65 mm) discharge cap came off. The cap was stopped by the attached chain. The fire fighter opening the hydrant was not injured, as he was standing to the rear of the hydrant, out of the way of the cap.

An examination of the hydrant revealed damage to the threads that prevented the cap from being replaced properly. The fire fighters stated they noticed the cap appeared to have been backed off the threads prior to charging the hydrant. Both fire fighters stated neither one of them attempted to tighten the cap prior to charging the hydrant. Both fire fighters stated they noticed the damage to the threads after they attempted to replace the cap when the cap would not screw back onto the connection.

Lessons Learned: This example demonstrates the need to slow down and check equipment prior to using it—especially privately owned and maintained equipment. Some of the private hydrant systems are not being properly maintained due to budgetary issues faced by their owners. Municipal budget cuts are also affecting the enforcement of regulations related to not maintaining fire protection equipment by property owners.

SKILL DRILL 16-4 Testing a Fire Hydrant

1 Place a cap gauge on one of the outlets of the first hydrant.

2 Open the hydrant valve to fill the hydrant barrel. Record the initial pressure reading on the gauge as the static pressure.

3 Move to the second hydrant, remove one of the discharge caps, and open the hydrant.

4 Place the Pitot gauge one-half the diameter of the orifice away from the opening and record this pressure as the Pitot pressure. Record the pressure on the first hydrant as the residual pressure. Use the recorded pressures to calculate or look up the flow rates at 20 psi (138 kPa) residual pressure. Document your findings.

Fire Hydraulics

To understand how a water supply system works, it is necessary to first understand the principles of water movement. Water is a liquid, which means that it cannot be compressed. Fire hydraulics deals with the properties of energy, pressure, and water flow as related to fire suppression. When operating supply lines or attack hose lines at a fire, it is important to understand some basic principles of hydraulics—namely, friction loss in different sizes of hose lines, elevation-related changes in pressure, and the development of water hammers. Fire fighters who advance to the position of pump driver/operator will learn more about fire service hydraulics.

Flow

Flow refers to the volume of water that is being moved through a pipe or hose. In fire hydraulics, flow is measured in gpm. Canadian fire departments use the metric system and measure flow in liters per minute (lpm). The conversion factor is 1 gal = 3.785 L.

Pressure

The amount of energy in a body or stream of water is measured as pressure. In fire hydraulics, pressure is measured in psi. Pressure is required to push water through a hose, to expel water through a nozzle, or to lift water up to a higher level. A pump adds energy to a water stream, causing an increase in pressure. In the metric system, pressures are measured in kPa. The conversion factor is 1 psi = 6.894 kPa.

Friction Loss

Friction loss is the decrease in pressure that occurs as water moves through a pipe or hose. This loss of pressure represents the energy required to push the water through the hose. Friction loss is influenced by the diameter of the hose, the volume of water traveling through the hose, and the distance the water travels. In a given size hose, a higher flow rate produces more friction loss. In a 2½-inch (65 mm) hose, for example, a 300-gpm (1364 lpm) flow causes much more friction loss than a 200-gpm (909 lpm) flow. At a given flow rate, the smaller the diameter of the hose, the greater the friction loss. At a flow of 500 gpm (2,273 lpm), the friction loss in a 2½-inch (65 mm) hose is much greater than the friction loss in a 4-inch (100 mm) hose. With any combination of flow and diameter, the friction loss is directly proportional to the distance. At a flow of 250 gpm (1136 lpm) in a 2½-inch hose (65 mm), for example, the friction loss in 200 feet (61 m) of hose is double the friction loss in 100 feet (30 m) of hose.

Elevation Pressure

Elevation affects water pressure. An elevated water tank supplies pressure to a municipal water system because of the difference in height between the water in the water tank and the underground delivery pipes. If a fire hose is laid down a hill, the water at the bottom will have additional pressure owing to the change in elevation. Conversely, if a fire hose is advanced upstairs to the third floor of a building, it will lose pressure as a result of the energy that is required to lift the water. The pump driver/operator must take elevation changes into account when setting the discharge pressure.

Water Hammer

Water hammer is a surge in pressure caused by suddenly stopping the flow of a stream of water. A fast-moving stream of water has a large amount of kinetic energy. If the water suddenly stops moving when a valve is closed, all of the kinetic energy is converted to an instantaneous increase in pressure. Because water cannot be compressed, the additional pressure is transmitted along the hose or pipe as a shock wave. Water hammer can rupture a hose, cause a coupling to separate, or damage the plumbing on a piece of fire apparatus. Severe water hammer can even damage an underground piping system. In addition, fire fighters have been injured by equipment that was damaged by a water hammer.

A similar situation can occur if a valve is opened too quickly and a surge of pressurized water suddenly fills a hose. The surge in pressure can damage the hose or cause the fire fighter at the nozzle to lose control of the stream.

Fire Fighter Safety Tips

Open and close hydrants and nozzles slowly to prevent water hammer.

To prevent water hammer, always open and close fire hydrant valves slowly. Pump driver/operators also need to open and close the valves on fire engines slowly. When you are operating the nozzle on an attack line, open the nozzle slowly. Most importantly, when you close the shut-off valve on an attack line, do it slowly.

Fire Hoses

Functions of Fire Hoses

Fire hoses are used for two main purposes: as supply hoses and as attack hoses. Supply hoses (or supply lines) are used to deliver water from a static source or from a fire hydrant to an attack engine. The water can come directly from a hydrant, or it can come from another engine that is being used to provide a water supply for the attack engine. Supply line sizes are 2½ inches (65 mm), 3 inches (76 mm), 4 inches (100 mm), 5 inches (127 mm), and 6 inches (152 mm). Supply hoses are designed to carry larger volumes of water at lower pressures than attack lines.

Attack hoses (or attack lines) are used to discharge water from an attack engine onto the fire. Most attack hoses carry water directly from the attack engine to a nozzle that is used to direct the water onto the fire. In some cases, an attack line is attached to a deck gun, an aerial device, or some other type of master stream appliance. Attack hoses usually operate at higher pressures than do supply lines. These lines can also be attached to the outlet of a standpipe system inside a building. The steps for using attack lines are covered in the Fire Attack and Foam chapter.

Sizes of Hose

Fire hoses range in size from 1 inch to 6 inches (25 mm to 152 mm) in diameter FIGURE 16-20. The nominal hose size refers to the inside diameter of the hose when it is filled with water.

FIGURE 16-20 Fire hose comes in a wide range of sizes for different uses and situations.

Smaller-diameter hoses are used as attack lines; larger-diameter hoses are almost always used as supply lines. Medium-diameter hoses can be used as either attack lines or supply lines.

Small-diameter hose (SDH) ranges in size from 1 inch (25 mm) to 2 inches (50 mm) in diameter. Many fire department vehicles are equipped with a reel of 1-inch (25 mm) hard rubber hose called a booster hose (or booster line), which is used for fighting small outdoor fires. A lightweight, collapsible, 1-inch (25 mm) hose, known as forestry hose, is often used to fight brush fires.

The hoses that are most commonly used to attack interior fires are either 1½ inches (38 mm) or 1¾ inches (45 mm) in diameter. They are usually connected directly to a handline nozzle. Some fire departments also use 2-inch (50 mm) attack lines. Each section of attack hose is usually 50 feet (15 m) long.

Medium-diameter hose (MDH) has a diameter of 2½ inches (65 mm) or 3 inches (76 mm). Hoses in this size range can be used as either supply lines or attack lines. Large handline nozzles are often attached to 2½-inch (65 mm) hose to attack larger fires. When used as an attack hose, the 3-inch (76 mm) size is more often used to deliver water to a master stream device or a fire department connection. These hose sizes also come in 50-feet (15 m) lengths.

Large-diameter hose (LDH) has a diameter of 3½ inches (88 mm) or more. Standard LDH sizes include 4-inch (100 mm) and 5-inch (127 mm) diameters, which are used as supply lines by many fire departments. The largest LDH size is 6 inches (152 mm) in diameter. Standard lengths of either 50 feet (15 m) or 100 feet (30 m) are available for LDH.

Fire hose is designed to be used as either attack hose or supply hose. Attack hose must withstand higher pressures and is designed to be used in a fire environment where it can be subjected to high temperatures, sharp surfaces, abrasion, and other potentially damaging conditions. Large-diameter supply hose is constructed to operate at lower pressures than attack hose and in less severe operating conditions; however, it must still be durable and resistant to external damage. Attack hose can be used as supply hose, but large-diameter supply hose must never be used as attack hose.

Attack hose must be tested annually at a pressure of at least 300 psi (2068 kPa) and is intended to be used at pressures up to 275 psi (1896 kPa). Supply hose must be tested annually at a pressure of at least 200 psi (1379 kPa) and is intended to be used at pressures up to 185 psi (1275 kPa). Most LDH is constructed as supply line; however, some fire departments use special LDH that can withstand higher pressures.

Hose Construction

3½ Supply
Smaller is attack

Most fire hose is constructed with an inner waterproof liner surrounded by either one or two outer layers. The outer layers provide the strength needed to withstand the high pressures exerted by the water inside the hose. This strength is provided by a woven mesh made from high-strength synthetic fibers (such as nylon) that are resistant to high temperatures, mildew, and many chemicals. These fibers can also withstand some mechanical abrasion.

Double-jacket hose is constructed with two layers of woven fibers. The outer layer serves as a protective covering, while the inner layer provides most of the strength. The tightly woven outer jacket can resist abrasion, cutting, hot embers, and other external damage. The woven fibers are treated to resist water and provide added protection from many common hazards that are likely to be encountered at the scene of a fire.

Instead of a double jacket, some fire hoses are constructed with a durable rubber-like compound as the outer covering. This material is bonded to a single layer of strong woven fibers that provides the strength needed to keep the hose from rupturing under pressure. This type of construction is called rubber-covered hose or rubber-jacket hose FIGURE 16-21.

Both types of hose are designed to be stored flat and to fold easily when there is no water inside the hose. This allows a much greater length of hose to be stored in the hose compartments on a fire apparatus.

The hose liner (also called the hose inner jacket) is the inner part of the hose FIGURE 16-22. This liner prevents water from leaking out of the hose and provides a smooth inside surface for water to move against. Without this smooth surface, excessive friction would arise between the moving water and inside of the hose, reducing the amount of pressure that could reach the nozzle. The inner liner is usually made of a synthetic rubber compound or a thin flexible membrane material that can be flexed and folded without developing leaks. In a double-jacket hose, it is bonded to the inner woven jacket. In a rubber-covered hose, the inner and outer layers are usually bonded together and the woven fibers are contained within these layers.

Hose Couplings

Couplings are used to connect individual lengths of fire hose together. They are also used to connect a hose line to a hydrant, to an intake or discharge valve on an engine, or to a variety of

FIGURE 16-21 Rubber-covered hose.

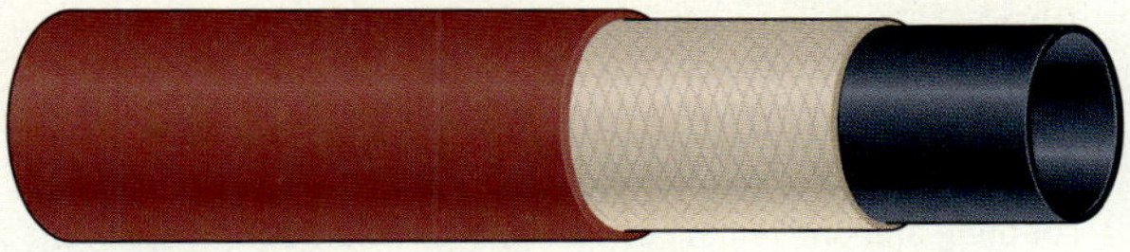

FIGURE 16-22 The liner inside a fire hose can be made from synthetic rubber or a variety of other membrane materials.

nozzles, fittings, and appliances. A coupling is permanently attached to each end of a section of fire hose. The two most common types of fire hose couplings are threaded hose couplings or Storz-type (nonthreaded) hose couplings.

Fire Fighter Safety Tips

Gloves should always be worn when handling hose. Metal shavings, glass shards, or other sharp objects may potentially become embedded in the fibers of the hose jackets. These objects could easily sever a muscle or tendon in a bare hand—and end a career.

Threaded Couplings

Threaded couplings are used on most hoses up to 3 inches (76 mm) in diameter and on both soft suction hose and hard suction hose. A set of threaded couplings consists of a male coupling, which has the threads on the outside, and a female coupling, which has matching threads on the inside FIGURE 16-23. A length of fire hose has a male coupling on one end and a female coupling on the other end. The female coupling has a swivel, so the male and female ends can be attached together without twisting the hose. Threads are male

Male hose couplings are manufactured in a single piece, whereas female hose couplings are manufactured in two pieces. Thus the combination of the male and female hose couplings together is sometimes called a three-piece coupling set. Occasionally, when fire fighters need to reduce the size of a hose coupling to make it fit a smaller coupling or when they need to attach two hoses that possess different threads, a reducer or adaptor will be added to both the male and female couplings. This arrangement is called a five-piece coupling set.

Most fire departments use standardized hose threads, which allows fire hose from different departments to be connected together. The use of standardized hose threads is not universal among all jurisdictions, however, so some fire departments carry special adaptors on their apparatus for use during mutual aid firefighting operations.

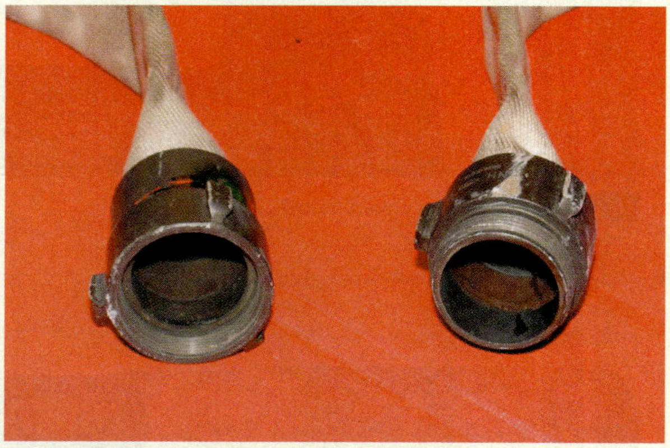

FIGURE 16-23 A set of threaded couplings includes one male coupling and one female coupling. The female coupling (on the left) has threads on the inside of the coupling. The male coupling (on the right) has exposed threads on the outside of the coupling.

When connecting fire hoses with threaded couplings, the threads must be properly aligned so that the male and female couplings will engage fully. When the couplings are properly aligned, the two ends should attach together with minimal resistance. The swivel on the female coupling should be turned until the connection is snug, but only hand tight, so that the couplings can be easily disconnected.

If any leakage occurs after the hose is filled with water, further tightening may be needed. You can use a spanner wrench to gently tighten the couplings until the leakage is stopped. Spanner wrenches are also used to connect and disconnect hose couplings FIGURE 16-24. In particular, you may need to use a spanner wrench to uncouple the hose after it has been pressurized with water. Normally two spanner wrenches are used together to rotate the two couplings in opposing directions.

Fire hose couplings are constructed with lugs to aid in the coupling and uncoupling of the hose. Three types of lugs are used for this purpose: pin lugs, rocker lugs, and recessed lugs FIGURE 16-25. Pin lugs look like small cylinders that extend

FIGURE 16-24 Spanner wrenches are used to tighten or loosen hose couplings.

FIGURE 16-25 Pin lugs, recessed lugs, and rocker lugs.

outward from the hose coupling. They are rarely found on fire hose today because the pins tend to snag as the hose is being pulled over rough surfaces. Instead, pin lungs have largely been replaced by rocker lugs. Rocker lugs (rocker pins) are rectangular-shaped extensions on hose couplings. The edges of rocker lugs are beveled to prevent them from catching on objects as the hose is dragged. Recessed lugs are circular indentations in hose couplings that are shaped to hold a specially designed spanner wrench. They are usually found on the couplings for ¾-inch (19 mm) or 1-inch (25 mm) booster hose.

Higbee indicators (sometimes called Higbee notches or Higbee cuts) show the position where the ends of the threads on a pair of couplings are properly aligned with each other. Using the Higbee indicators will help you to couple hose more quickly. When the indicators on the male and female couplings are aligned, the two couplings should connect quickly and easily FIGURE 16-26.

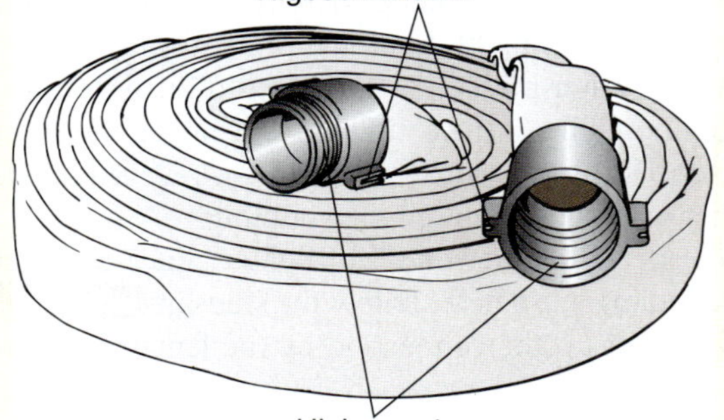

FIGURE 16-26 Higbee indicators show the position where the threads on a pair of couplings are properly aligned with each other.

An important part of a threaded coupling is the rubber gasket. The gasket is an O-shaped piece of rubber that sits inside the swivel section of the female coupling. When the male coupling is tightened down against it, a seal is formed that stops water from leaking. If the gasket is damaged or missing, the coupling will leak. These gaskets can deteriorate with time. Also, using a wrench to tighten couplings on an empty hose or to overtighten couplings on a filled hose can damage the gaskets and cause them to leak. The gaskets must be changed periodically as part of the maintenance on the hose.

Although a leaking coupling is not a critical problem during most firefighting operations, it can result in unnecessary water damage. During cold weather, a leaking coupling can cause ice to form and create a significant safety hazard. The best way to prevent leaks is to make sure the gaskets are in good condition and to replace any gaskets that are missing or damaged. To replace the swivel gasket, follow the steps in SKILL DRILL 16-5:

1. Fold the new gasket, bringing the thumb and the forefinger together and creating two loops. (STEP 1)
2. Place either of the two loops into the coupling and against the gasket seat. (STEP 2)
3. Using the thumb, push the remaining unseated portions into the coupling until the entire gasket is properly positioned against the coupling seat. (STEP 3)

Storz-Type Couplings

Storz-type hose couplings are designed so that the couplings on both ends of a length of hose are the same. In other words, there is no male or female end to the hose. When this system is used, each coupling can be attached to any other coupling of the same diameter FIGURE 16-27. Storz-type couplings are made for all hose sizes; however, in North America they are most often used on LDH.

SKILL DRILL 16-5 Replacing the Swivel Gasket

(Fire Fighter I, NFPA 5.5.4)

1 Fold the new gasket, bringing the thumb and the forefinger together and creating two loops.

2 Place either of the two loops into the coupling and against the gasket seat.

3 Using the thumb, push the remaining unseated portions into the coupling until the entire gasket is properly positioned against the coupling seat.

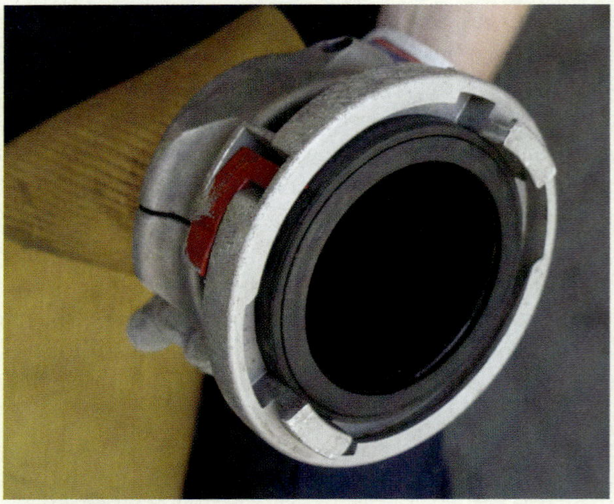

FIGURE 16-27 Storz-type couplings are designed so that the couplings on both ends of a length of hose are the same.

Storz-type couplings are connected by mating the two couplings face-to-face and then turning them clockwise one-third of a turn. To disconnect a set of couplings, the two parts are rotated counterclockwise one-third of a turn. A spanner wrench can be used to tighten a leaking coupling or to release a connection that cannot be loosened by hand. Some Storz-type couplings have a small pin or lever that must be released before uncoupling the hose—a design that prevents accidental uncoupling of the hose.

Adaptors are used to connect Storz-type couplings to threaded couplings or to connect couplings of different sizes together. Many fire departments use LDH with Storz-type couplings as a supply line between a hydrant and an engine.

Storz-type couplings can also be found on fire department connections. They can be used to supply a standpipe or sprinkler system. In addition, some hydrant manufacturers place Storz-type connections on hydrants—a practice that allows fire departments to rapidly make hydrant connections.

Coupling and Uncoupling Hose

Several techniques are used for attaching and releasing hose couplings. Depending on the circumstances, one technique may be more effective than another. A fire fighter should learn how to perform each skill. To perform the one-fire fighter foot-tilt method of coupling fire hose, follow the steps in SKILL DRILL 16-6:

1. Place one foot on the hose behind the male coupling.
2. Push down with your foot to tilt the male coupling upward. (STEP 1)
3. Place one hand behind the female coupling and grasp the hose. (STEP 2)
4. Place the other hand on the coupling swivel.
5. Bring the two couplings together and align the Higbee indicators. Rotate the swivel in a clockwise direction to connect the hoses. (STEP 3)

To perform the two-fire fighter method for coupling a fire hose, follow the steps in SKILL DRILL 16-7:

1. Pick up the male end of the coupling. Grasp it directly behind the coupling and hold it tightly against the body. (STEP 1)
2. The second fire fighter holds the female coupling firmly with both hands. (STEP 2)
3. The second fire fighter brings the female coupling to the male coupling. (STEP 3)
4. The second fire fighter aligns the female coupling with the male coupling. Use the Higbee indicators for easy alignment. (STEP 4)
5. The second fire fighter turns the female coupling counterclockwise until it clicks, which indicates that the threads are aligned. (STEP 5)
6. Turn the female coupling clockwise to couple the hoses. (STEP 6)

Charged hose lines should never be disconnected while the water inside the hose is under pressure. The loosened couplings can flail around wildly and cause serious injury to fire personnel or bystanders in the vicinity. Instead, always shut off the water supply and bleed off the pressure before uncoupling hoses. The water pressure will make it difficult to uncouple a charged hose line. If the coupling resists an attempt to uncouple it, check to make sure the pressure is relieved before using spanner wrenches to loosen the coupling.

To perform the one-fire fighter knee-press method of uncoupling a fire hose, follow the steps in SKILL DRILL 16-8:

1. Pick up the connection by the female coupling end. (STEP 1)
2. Turn the connection upright, resting the male coupling on a firm surface. (STEP 2)
3. Place a knee on the female coupling and press down on it with your body weight (which compresses the gasket).
4. Turn the female swivel counterclockwise and loosen the coupling. (STEP 3)

To perform the two-fire fighter stiff-arm method of uncoupling a hose, follow the steps in SKILL DRILL 16-9:

1. Two fire fighters face each other and firmly grasp their respective coupling. (STEP 1)
2. With elbows locked straight, they push toward each other. (STEP 2)
3. While pushing toward each other, the fire fighters turn the coupling counterclockwise, loosening the coupling. (STEP 3)

To uncouple a hose with spanner wrenches, follow the steps in SKILL DRILL 16-10:

1. With the connection on the ground, straddle the connection above the female coupling. (STEP 1)
2. Place one spanner wrench on the female coupling, with the handle of the wrench to the left. (STEP 2)
3. Place the second spanner wrench on the male coupling, with the handle of the wrench to the right. (STEP 3)
4. Push both spanner wrench handles down toward the ground, loosening the connection. (STEP 4)

SKILL DRILL 16-6

Performing the One-Fire Fighter Foot-Tilt Method of Coupling a Fire Hose
(Fire Fighter I, NFPA 5.3.10)

1. Place one foot on the hose behind the male coupling. Push down with your foot to tilt the male coupling upward.

2. Place one hand behind the female coupling and grasp the hose.

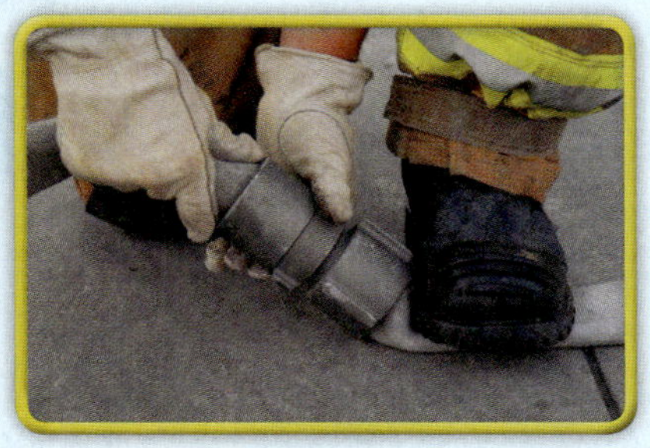

3. Place the other hand on the coupling swivel. Bring the two couplings together and align the Higbee indicators. Rotate the swivel in a clockwise direction to connect the hoses.

SKILL DRILL 16-7

Performing the Two-Fire Fighter Method of Coupling a Fire Hose
(Fire Fighter I, NFPA 5.3.10)

1. Pick up the male end of the coupling. Grasp it directly behind the coupling and hold it tightly against the body.

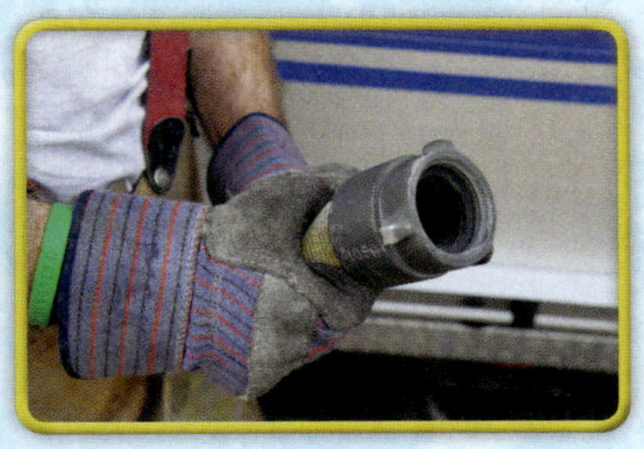

2. The second fire fighter holds the female coupling firmly with both hands.

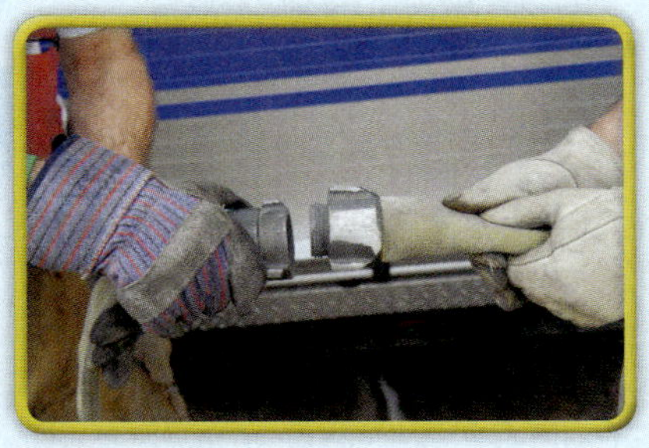

3. The second fire fighter brings the female coupling to the male coupling.

4. The second fire fighter aligns the female coupling with the male coupling. Use the Higbee indicators for easy alignment.

5. The second fire fighter turns the female coupling counterclockwise until it clicks, which indicates that the threads are aligned.

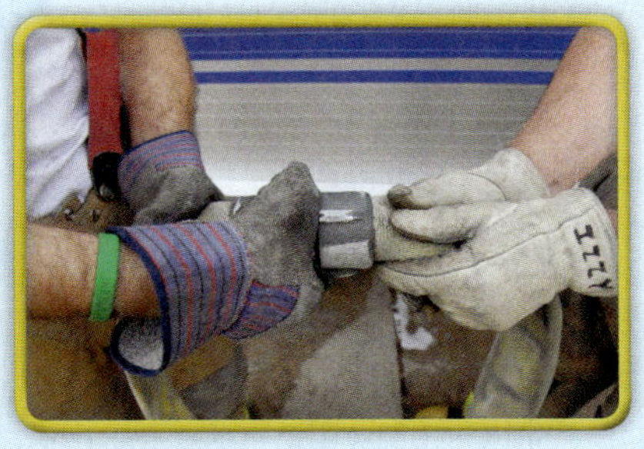

6. Turn the female coupling clockwise to couple the hoses.

SKILL DRILL 16-8 Performing the One-Fire Fighter Knee-Press Method of Uncoupling a Fire Hose

(Fire Fighter I, NFPA 5.3.10)

1. Pick up the connection by the female coupling end.

2. Turn the connection upright, resting the male coupling on a firm surface.

3. Place a knee on the female coupling and press down on it with your body weight. Turn the female swivel counterclockwise and loosen the coupling.

SKILL DRILL 16-9 Performing the Two-Fire Fighter Stiff-Arm Method of Uncoupling a Fire Hose

(Fire Fighter I, NFPA 5.3.10)

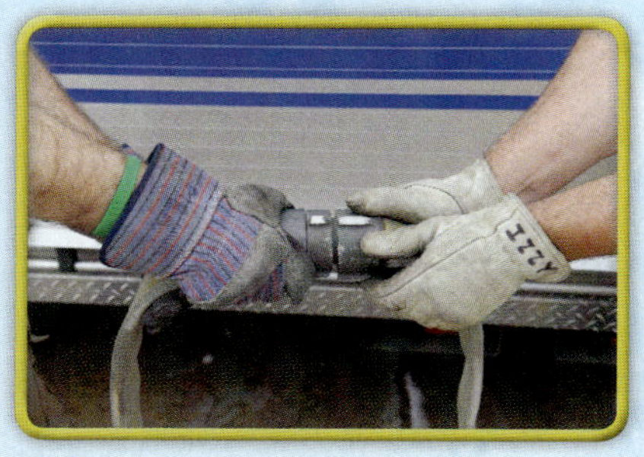

1. Two fire fighters face each other and firmly grasp their respective coupling.

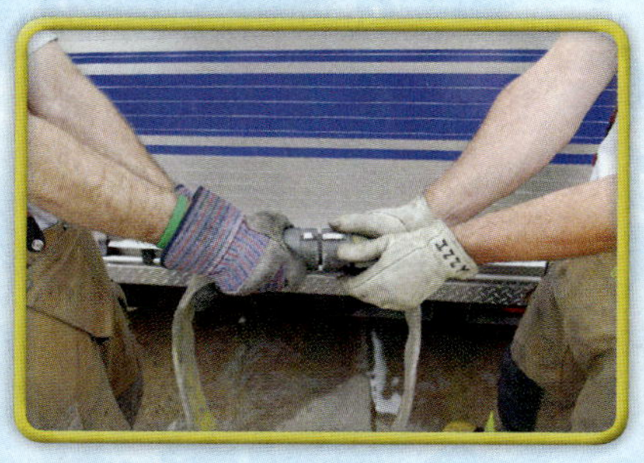

2. With elbows locked straight, they push toward each other.

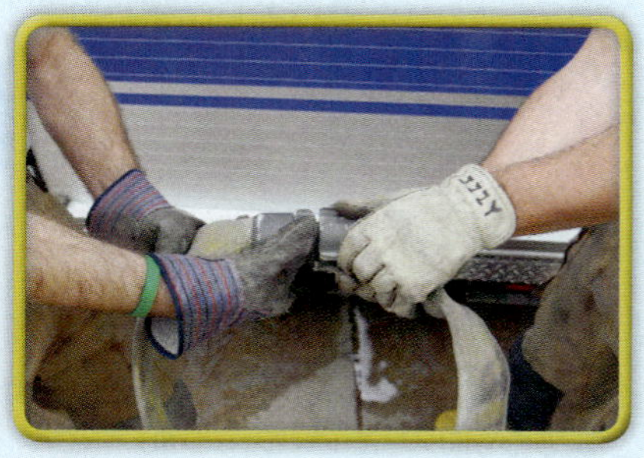

3. While pushing toward each other, the fire fighters turn the coupling counterclockwise, loosening the coupling.

To connect two lines with a damaged coupling, follow the steps in SKILL DRILL 16-11:

1. Using a hose jacket, open the hose jacket and place the damaged coupling in one end.
2. Place the second coupling in the other end of the jacket.
3. Close the hose jacket, ensuring that the latch is secure. Slowly bring the hose line up to pressure, allowing the gaskets to seal around the hose ends.

Fire Fighter Safety Tips

Never attempt to uncouple charged hose lines.

Attack Hose

Attack hose is designed to be used for fire suppression. This type of hose carries the water from the attack pumper to the fire. The common sizes of attack hose are 1½ inch (38 mm) or 1¾ inch (45 mm) lines, 1-inch (25 mm) booster lines, and 1-inch (25 mm) or 1½-inch (38 mm) forestry lines. Attack hose is discussed in detail in the Fire Attack and Foam chapter.

Supply Hose

Supply hose is used to deliver water to an attack engine from a pressurized source, such as a fire hydrant or from a supply engine. Supply lines range from 2½ inches (65 mm) to

SKILL DRILL 16-10 Uncoupling a Hose with Spanner Wrenches

(Fire Fighter I, NFPA 5.3.10)

1 With the connection on the ground, straddle the connection above the female coupling.

2 Place one spanner wrench on the female coupling, with the handle of the wrench to the left.

3 Place the second spanner wrench on the male coupling, with the handle of the wrench to the right.

4 Push both spanner wrench handles down toward the ground, loosening the connection.

6 inches (152 mm) in diameter. The choice of diameter is based on the preferences and operating requirements of each fire department. It also depends on the amount of water needed to supply the attack engine, the distance from the source to the attack engine, and the pressure that is available at the source.

Fire department engines are normally loaded with at least one bed of hose that can be laid out as a supply line. When threaded couplings are used, this hose can be laid out from the hydrant to the fire (known as a straight lay) or from the fire to the hydrant (known as a reverse lay). Sometimes engines are loaded with two beds of hose so they can easily drop a supply line in either direction. If Storz-type couplings are used or the necessary adaptors are provided, hose from the same bed can be laid in either direction.

A 2½-inch (65 mm) hose may be used as either a supply line or an attack line. Such hose has a limited flow capacity as supply hose, but it can be effective at low to moderate flow rates and over short distances. Sometimes two parallel lines of 2½-inch (65 mm) hose are used to provide a more effective water supply.

Large-diameter supply lines are much more efficient than 2½-inch (65 mm) hose for moving larger volumes of water over longer distances. Given this fact, many fire departments use 4-inch (100 mm) or 5-inch (127 mm) hose as their standard supply line. A single 5-inch (127 mm) supply line can deliver flows exceeding 1500 gpm (6819 lpm) under some conditions. LDH is heavy and difficult to move after it has been charged with water, however. This hose comes in 50-feet (15 m) and 100-feet (30 m) lengths. A typical fire engine may carry anywhere from 750 feet (228 m) to 1250 feet (381 m) of supply hose.

Soft Suction Hose

A soft suction hose is a short section of LDH that is used to connect a fire department engine directly to the large steamer outlet on a hydrant FIGURE 16-28. This type of hose is used to allow as much water as possible to flow from the hydrant to the pump through a single line. A soft suction hose has a female connection on each end, with one end matching the local hydrant threads and the other end matching the threads on a large-diameter inlet to the engine. The couplings have

FIGURE 16-28 Soft suction hose.

large handles to allow for quick tightening by hand. The hose can range from 4 inches (100 mm) to 6 inches (152 mm) in diameter and is usually between 10 feet (3 m) and 25 feet (7.6 m) in length.

Hard Suction Hose

A hard suction hose is a special type of supply hose that is used to draft water from a static source such as a river, lake, or portable drafting basin FIGURE 16-29. The water is drawn through this hose into the pump on a fire department engine or into a portable pump. This type of line is called a hard suction hose because it is designed to remain rigid and will not collapse when a vacuum is created in the hose to draft the water into the pump. It can also be used to carry water from a fire hydrant to the pumper.

Hard suction hose normally comes in 10-feet (3 m) or 20-feet (6 m) sections. The diameter is based on the capacity of the pump but can be as large as 6 inches (152 mm). Hard suction hose can be made from either rubber or plastic; however, the newer plastic versions are much lighter and more flexible.

Long handles are provided on the female couplings of hard suction hose to assist in tightening the hose. To draft water, it is essential to have an airtight connection at each coupling. Sometimes it may be necessary to gently tap the handles on the female couplings with a rubber mallet to tighten the hose or to disconnect it. Tapping these handles with anything metal, however, could cause damage to the handles or the coupling. To create an airtight seal between couplings, it is important to have a flexible gasket to facilitate the seal. These gaskets need to be replaced regularly.

FIGURE 16-29 Hard suction hose.

Hose Care, Maintenance, and Inspection

Fire hose should be regularly inspected and tested following the procedures in NFPA 1962, *Standard for the Inspection, Care and Use of Fire Hose, Couplings, and Nozzles and the Service Testing of Fire Hose*. Hoses that are not properly maintained can deteriorate over time and eventually burst. In addition, the gaskets in female couplings need to be checked regularly and replaced when they are worn or damaged.

Causes and Prevention of Hose Damage

Fire hose is a lifeline for fire fighters. Every time fire fighters respond to a fire, they rely on fire hose to deliver the water needed to attack the fire and protect themselves from it. Fire hose is a highly engineered product designed to perform well under adverse conditions. Fire fighters must be careful to prevent damage to the hose that could result in premature or unexpected failure. The factors that most commonly damage fire hose include mechanical causes, UV radiation, chemicals, heat, cold, and mildew.

Mechanical Damage

Mechanical damage can occur from many sources. Hose that is dragged over rough objects or along a roadway can be damaged by abrasion, for example. Broken glass and sharp objects can cut through the hose. Be especially careful if you need to place a hose line through a broken window; remove any protruding sharp edges of glass first. Particles of grit caught in the fibers can damage the jacket or puncture holes in the liner. Reloading dirty hose can cause damage to the fibers in the hose jacket.

Fire hose is likely to be damaged if it is run over by a vehicle. For this reason, hose ramps should be used if traffic must drive over a hose that is in the roadway. Hose couplings can also be damaged by dropping them on the ground. In particular, the exposed threads on male couplings are easily damaged if they are dropped. Avoid dragging hose couplings, as this practice can cause damage to the threads and to the swivels.

Heat and Cold

Hoses can be damaged by heat and cold as well as by prolonged exposure to sunlight. Heat is an obvious concern when fighting a fire. A hose that is directly exposed to a fire can burn through and burst quickly. Burning embers and hot coals can also damage the hose, causing small leaks or weakening the hose so that it is likely to burst under pressure. Always visually inspect any hose that has come in direct contact with a fire.

Avoid storing a hose in places where it will come in contact with hot surfaces, such as a heating unit or the exhaust pipe on a vehicle. If the apparatus is parked outside, use a hose cover to protect the hose from sunlight.

In cold weather, freezing is a threat to hoses. Freezing can rupture the inner liner and break fibers in the hose jacket. When fire fighters are working in below-freezing temperatures, water should be kept flowing through the hose to prevent freezing. If a line must be shut down temporarily, the nozzle should be left partly open to keep the water moving and the stream should be directed to a location where it will not cause additional water damage. When a line is no longer needed, the hose should be drained and rolled before it freezes.

Hose that is frozen or encased in ice can often be thawed out with a steam generator. Another option is to carefully chop the hose out using an axe, being careful not to cut the hose itself. The hose can then be transported back to the fire station to thaw. Do not attempt to bend a section of frozen hose. In situations where the hose is frozen solid, it may be necessary to transport the hose back to the fire station on a flatbed truck.

Chemicals

Many chemicals can damage fire hoses. Such chemicals may be encountered at incidents in facilities where chemicals are manufactured, stored, or used as well as in locations where their presence is not anticipated. Most vehicles contain a wide variety of chemicals that can damage fire hose, including battery acid, gasoline, diesel fuel, antifreeze, motor oil, and transmission fluid. Hose may come in contact with these chemicals at vehicle fires or at the scene of a collision where chemicals are spilled on the roadway. In particular, supply hoses often come in contact with residues from these chemicals when lines are laid in the roadway. Fire fighters should remove chemicals from the hose as soon as possible and wash the hose with an approved detergent, thoroughly rinse it, and let it dry completely.

Mildew

Mildew is a type of fungus that can grow on fabrics and materials in warm, moist conditions. A fire hose that has been packed away while it is still wet and dirty is a natural breeding ground for mildew. This fungus feeds on nutrients found in many natural fibers, which can cause the fibers to rot and deteriorate. In the days when cotton fibers were used in fire hose jackets, mildew was a major problem. Hose had to be washed and completely dried after every use before it could be placed back on the apparatus.

Modern fire hose is made from synthetic fibers that are resistant to mildew, and most types can be repacked without drying. Nevertheless, mildew may still grow on exposed fibers if they are soiled with contaminants that will provide mildew with the necessary nutrients. The fibers in rubber-covered hose are protected from mildew.

Fire Fighter Safety Tips

Whenever a fire hose has suffered possible damage, it should be thoroughly inspected and tested according to NFPA 1962, *Standard for the Inspection, Care and Use of Fire Hose, Couplings, and Nozzles and the Service Testing of Fire Hose*, before it is returned to service.

Cleaning and Maintaining Hoses

To clean hose that is dirty or contaminated, follow the steps in SKILL DRILL 16-12:

1. Lay the hose out flat.
2. Rinse the hose with water. (STEP 1)
3. Gently scrub the hose with mild detergent, paying attention to soiled areas. (STEP 2)
4. Turn over the hose and repeat steps 1 and 2.
5. Give a final rinse to the hose with water.
6. Lay out or hang the hose and allow it to dry before properly storing it. (STEP 3)

FIRE FIGHTER Tips

To prevent UV radiation damage, do not dry hose in sunlight.

Fire Fighter Safety Tips

Hose washing machines are unable to sufficiently clean couplings on jacketed hose.

Hose Inspections

FIRE FIGHTER II

Each length of hose should be tested at least annually, according to the procedures listed in NFPA 1962, *Standard for the Inspection, Care and Use of Fire Hose, Couplings, and Nozzles and the Service Testing of Fire Hose*. The hose testing procedures are complicated and require special equipment. This equipment must be operated according to the manufacturer's instructions.

Visual hose inspections should be performed at least quarterly. A visual inspection should also be performed after each use, either while the hose is being cleaned and dried or when it is reloaded onto the apparatus. If any defects are found, that length of hose should be immediately removed from service and tagged with a description of the problem. Hose that has not been used in 30 days should be unpacked, inspected, cleaned, and reloaded. The appropriate notifications must be made to have the hose repaired.

To clearly mark a defective hose, follow the steps in SKILL DRILL 16-13 FIGURE 16-30:

1. Inspect the hose for defects.
2. Upon finding a defect, mark the area on the hose and remove the hose from service.
3. Tag the hose as defective and provide a description of the defect, take it out of service, and notify your superiors.

FIRE FIGHTER II

To perform an annual service test on fire hose, follow the steps in SKILL DRILL 16-14:

1. Don turnout gear.
2. Connect up to 300 feet (91 m) of hose to a hose testing gate valve on the discharge valve of a fire department pumper or hose tester.

SKILL DRILL 16-12 Cleaning and Maintaining Hoses

(Fire Fighter I, NFPA 5.5.2)

1. Lay the hose out flat. Rinse the hose with water.
2. Gently scrub the hose with mild detergent, paying attention to soiled areas.
3. Turn over the hose and repeat steps 1 and 2. Give a final rinse to the hose with water. Lay out or hang the hose and allow it to dry before properly storing it.

FIGURE 16-30 Upon finding a defect, mark the area on the hose and remove the hose from service.

FIRE FIGHTER II

3. Attach a nozzle to the end of the hose.
4. Slowly fill each hose with water at 50 psi (344 kPa), and remove kinks and twists in the hose.
5. Open the nozzle to purge air from the hose, discharging the water away from the test area. Close the nozzle.
6. Measure and record the length of each section of hose.
7. Mark the position of each hose coupling on the hose.
8. Check each coupling for leaks. When leaks are found, remove the hose from service if the leak is behind the coupling. If the leak is in front of the coupling, tighten the leaking couplings. Replace gaskets if necessary after shutting down the hose line.
9. Close the hose testing gate valve.
10. Ensure that all fire fighters are clear of the test area.
11. Increase the pressure on the hose to the pressure required by NFPA 1962, and maintain that pressure for 5 minutes.
12. Monitor the hose and couplings for leaks as the pressure increases during the test.
13. Shut the gates and open the nozzle to bleed off the pressure.
14. Uncouple the hose and drain it.
15. Inspect the marks placed on the hose jacket near the couplings to determine whether slippage occurred.
16. Tag hose that failed.
17. Mark hose that passed.
18. Record the results in the departmental logs.

FIRE FIGHTER II

FIRE FIGHTER II

Hose Records

A hose record is a written history of each individual length of fire hose. Each length of hose should be identified with a unique number stenciled or painted on it. A hose record will contain information such as the following:

- Hose size, type, and manufacturer
- Date when the hose was manufactured
- Date when the hose was purchased
- Dates when the hose was tested
- Any repairs that have been made to the hose

Some fire departments keep hose records on cards or paper files. Others keep the records in a database in a computer system.

Hose Appliances

A hose appliance is any device used in conjunction with a fire hose for the purpose of delivering water. You should be familiar with wyes, water thieves, Siamese connections, double-male and double-female adaptors, reducers, hose clamps, hose jackets, and hose rollers. It is important to learn how to use those hose appliances and tools required by your fire department. In particular, you should understand the purpose of each device and be able to use each appliance correctly. Some hose appliances are used primarily with supply lines, whereas others are most often used with attack lines; many hose appliances have applications with both supply lines and attack lines. Both types are discussed in this section.

Wyes

A wye is a device that splits one hose stream into two hose streams. The word "wye" refers to a Y-shaped part or object. When threaded couplings are used, a wye has one female connection and two male connections.

The wye that is most commonly used in the fire service splits one 2½-inch (65 mm) hose line into two 1½-inch (38 mm) hose lines. This appliance is used primarily on attack lines. A gated wye is equipped with two quarter-turn ball valves so that the flow of water to each of the split lines can be controlled independently FIGURE 16-31. A gated wye enables fire fighters to initially attach and operate one hose line and then to add a second hose later if necessary. The use of a gated wye avoids the need to shut down the hose line supplying the wye so as to attach the second line. Some gated wyes are used on LDH. These wyes have an inlet for LDH (4 inch (100 mm) or 5 inch (127 mm)) and several 2½-inch (65 mm) outlets.

Water Thief

A water thief is similar to a gated wye, but includes an additional 2½-inch (65 mm) outlet FIGURE 16-32. It is used primarily on attack lines and larger ones are used with LDH. The water that comes from a single 2½-inch (65 mm) inlet can be directed to two 1½-inch (38 mm) outlets and one 2½-inch (65 mm) outlet. Under most conditions, it will not be possible to supply all three outlets at the same time because the capacity of the supply hose is limited.

A water thief can be placed near the entrance to a building to provide the water for interior attack lines. One or two 1½-inch (38 mm) attack lines can be used in such a case. If necessary, they can then be shut down and a 2½-inch (65 mm) line substituted for them. Sometimes the 2½-inch (65 mm) line is used to knock down a fire and the two 1½-inch (38 mm) lines are used for overhaul.

FIGURE 16-31 A gated wye is used to split one 2½-inch (65 mm) hose line into two 1½-inch (38 mm) lines.

FIGURE 16-32 A water thief.

Siamese Connection

A Siamese connection is a hose appliance that combines two hose lines into one. The most commonly used type of Siamese connection combines two 2½-inch (65 mm) hose lines into a single line FIGURE 16-33. This scheme increases the flow of water on the outlet side of the Siamese connection. A Siamese connection that is used with threaded couplings has two female connections on the inlets and one male connection on the outlet. It can be used with supply lines and with some attack lines.

FIGURE 16-33 A typical Siamese connection has two female inlets and a single outlet.

A Siamese connection is sometimes used on an engine inlet to allow water to be received from two different supply lines. These kinds of connections are also used to supply master stream devices and ladder pipes. Siamese connections are commonly installed on the fire department connections that are used to supply water to standpipe and sprinkler systems in buildings. Fire fighters should go over their department's policies on the correct procedures for supplying fire department Siamese connections.

Adaptors

Adaptors are used for connecting hose couplings that have the same diameter but dissimilar threads. Dissimilar threads could be encountered when different fire departments are working together or in industrial settings where the hose threads of the building's equipment do not match the threads of the municipal fire department. Some private fire hydrants may have different threads from the municipal system and require an adaptor. Adaptors are also used to connect threaded couplings to Storz-type couplings. They are useful for both supply lines and attack lines.

Adaptors can also be used when it is necessary to connect two female couplings or two male couplings. A double-female adaptor is used to join two male hose couplings. A double-male adaptor is used to join two female hose couplings. Double-male and double-female adaptors are often employed when performing a reverse hose lay FIGURE 16-34.

Reducers

A reducer is used to attach a smaller-diameter hose to a larger-diameter hose FIGURE 16-35. Usually the larger end has a female connection and the smaller end has a male connection. One type of reducer is used to attach a 2½-inch (65 mm) hose line to a 1½-inch (38 mm) attack line. Many 2½-inch (65 mm) nozzles are constructed with a built-in reducer, so that a 1½-inch (38 mm) line can be attached for overhaul. Reducers are also used to attach a 2½-inch (65 mm) supply line to a larger suction inlet on a fire pumper.

Hose Jacket

A hose jacket is a device that is placed over a leaking section of hose to stop a leak FIGURE 16-36. The best way to handle a leak in a section of hose is to replace the defective section of hose.

FIGURE 16-34 Double-female and double-make adaptors are used to join two couplings of the same sex.

FIGURE 16-35 A reducer is used to connect a smaller-diameter hose line to the end of a larger-diameter line.

FIGURE 16-36 A hose jacket is used to repair a leaking hose line.

A hose jacket can provide a temporary fix until the section of hose can be replaced. This device should be used only in cases where it is not possible to quickly replace the leaking section of hose. It can be used for both supply lines and attack lines.

The hose jacket consists of a split metal cylinder that fits tightly over the outside of a hose line. This cylinder is hinged on one side to allow it to be placed over the leak. A fastener is then used to clamp the cylinder tightly around the hose. Gaskets on each end of the hose jacket prevent water from leaking out the ends of the hose jacket.

Hose Roller

A hose roller is used to protect a hose line that is being hoisted over the edge of a roof or over a windowsill **FIGURE 16-37**. This appliance keeps the hose from chafing or kinking at the sharp edge. A hose roller is sometimes called a hose hoist because it makes it easier to raise (i.e., hoist) a hose over the edge of the building. Hose rollers are also used to protect ropes when hoisting an object over the edge of a building and during rope rescue operations. They are typically used with attack lines.

Hose Clamp

A hose clamp is used to temporarily stop the flow of water in a hose line. These devices are often applied to supply lines, allowing a hydrant to be opened before the line is hooked up to the intake of the attack engine. The fire fighter at the hydrant does not have to wait for the pump operator to connect the line to the pump intake before opening the hydrant. As soon as the line is connected, the clamp is released. A hose clamp can also be used to stop the flow in a line if a hose ruptures or if an attack line needs to be connected to a different appliance **FIGURE 16-38**.

Valves

Valves are used to control the flow of water in a pipe or hose line. Several different types of valves are used on fire hydrants, fire apparatus, standpipe and sprinkler systems, and attack hose lines. The important thing to remember when opening and closing any valve or nozzle is to do it s-l-o-w-l-y so as to prevent water hammer.

Commonly encountered valves include the following types:

FIGURE 16-37 A hose roller is used to protect a hose when it is hoisted over a sharp edge of a roof or a windowsill.

FIGURE 16-38 A hose clamp is use to temporarily interrupt the flow of water in a hose line.

- Ball valves: Ball valves are used on nozzles, gated wyes, and engine discharge gates. They consist of a ball with a hole in the middle. When the hole is lined up with the inlet and the outlet, water flows through it. As the ball is rotated, the flow of water is gradually reduced until it is shut off completely **FIGURE 16-39A**.
- Gate valves: Gate valves are found on hydrants and on sprinkler systems. Rotating a spindle causes a gate to move slowly across the opening. The spindle is rotated by turning it with a wrench or a wheel-type handle **FIGURE 16-39B**.
- Butterfly valves: Butterfly valves are often found on the large pump intake connections where a hard suction hose or soft suction hose is connected. They are opened or closed by rotating a handle one-quarter turn **FIGURE 16-39C**.
- Four-way hydrant valve: This appliance, which is attached to a fire hydrant, enables water to flow directly from the hydrant. It can boost pressure in a supply line without interrupting the flow of water. The use of a four-way hydrant valve is covered in detail later in this chapter.
- Remote-controlled hydrant valve: Remote-controlled hydrant valves are attached to a fire hydrant. They allow the operator to turn the hydrant on without flowing water into the hose line. These valves are operated by the pump operator using a radio control. They free up the hydrant person for other tasks and give the pump operator control over the hydrant from a distance.

Hose Rolls

An efficient way to transport a single section of fire hose is in the form of a roll. Rolled hose is both compact and easy to manage. A fire hose can be rolled many different ways, depending on

FIGURE 16-39 **A.** Ball valve

FIGURE 16-39 **B.** Gate valve

FIGURE 16-39 **C.** Butterfly valve

how it will be used. Follow the SOPs of your department when rolling hose.

Straight or Storage Hose Roll

The straight roll is a simple and frequently used hose roll. It is used for general handling and transporting of hose, as well as for rack storage of hose. To perform a straight roll, follow the steps in SKILL DRILL 16-15:

1. Lay the length of hose to be rolled flat and straight. (STEP 1)
2. Begin by folding the male coupling over on top of the hose. (STEP 2)
3. Roll the hose to the female coupling. (STEP 3)
4. Set the hose roll on its side, and tap any protruding hose flat with a foot. (STEP 4)

Single-Doughnut Hose Roll

The single-doughnut roll is used when the hose will be put into use directly from the rolled state. With this arrangement, the hose has both couplings on the outside of the roll. The hose can be connected and extended by one fire fighter. As the hose is extended, it unrolls. To perform a single-doughnut roll, follow the steps in SKILL DRILL 16-16:

1. Place the hose flat and in a straight line. (STEP 1)
2. Locate the midpoint of the hose. (STEP 2)
3. From the midpoint, move 5 feet (1.5 m) toward the male coupling end.
4. Start rolling the hose toward the female coupling. (STEP 3)
5. At the end of the roll, wrap the excess hose of the female end over the male coupling to protect the threads. (STEP 4)

Twin-Doughnut Hose Roll

The twin-doughnut roll is used primarily to make a small compact roll that can be carried easily. To perform a twin-doughnut hose roll, follow the steps in SKILL DRILL 16-17:

1. Lay the hose flat and in a straight line. (STEP 1)
2. Bring the male coupling alongside the female coupling. (STEP 2)
3. Fold the far end over and roll toward the couplings, creating a double roll. (STEP 3)
4. The roll can be carried by hand, by a rope, or by a strap. (STEP 4)

Self-Locking Twin-Doughnut Hose Roll

The self-locking twin-doughnut hose roll is similar to the twin-doughnut roll, except that it forms its own carry loop. To perform a self-locking twin-doughnut roll, follow the steps in SKILL DRILL 16-18:

1. Lay the hose flat and bring the couplings alongside each other. (STEP 1)
2. Move one side of the hose over the other, creating a loop. This creates the carrying shoulder loop. (STEP 2)
3. Bring the loop back toward the couplings to the point where the hose crosses. (STEP 3)
4. From the point where the hose crosses, begin to roll the hose toward the couplings, with the loop as its center. This creates a loop on each side of the roll. (STEP 4)

SKILL DRILL 16-15 Performing a Straight Hose Roll

(Fire Fighter I, NFPA 5.5.2)

1 Lay the length of hose to be rolled flat and straight.

2 Begin by folding the male coupling over on top of the hose.

3 Roll the hose to the female coupling.

4 Set the hose roll on its side, and tap any protruding hose flat with a foot.

5. On completion of the rolling, position the couplings on the top of the rolls. (STEP 5)
6. Position the loops so that one is larger than the other. Pass the larger loop over the couplings and through the smaller loop, which secures the rolls together and forms the shoulder loop. (STEP 6)

Fire Hose Evolutions

Fire hose evolutions are standard methods of working with fire hose to accomplish different objectives in a variety of situations. Most fire departments set up their equipment and conduct regular training so that fire fighters will be prepared to perform a set of standard hose evolutions. Hose evolutions involve specific actions that are assigned to specific members of a crew, depending on their riding positions on the apparatus. Every fire fighter should know how to perform all of the standard evolutions quickly and proficiently. When an officer calls for a particular evolution to be performed, each crew member should know exactly what to do.

Hose evolutions are divided into supply line operations and attack line operations. Supply line operations involve laying hose lines and making connections between a water supply source and an attack engine. Attack line operations

SKILL DRILL 16-16 Performing a Single-Doughnut Roll

(Fire Fighter I, NFPA 5.5.2)

1 Place the hose flat and in a straight line.

2 Locate the midpoint of the hose.

3 From the midpoint, move 5 feet (1.5 m) toward the male coupling end. Start rolling the hose toward the female coupling.

4 At the end of the roll, wrap the excess hose of the female end over the male coupling to protect the threads.

SKILL DRILL 16-17 Performing a Twin-Doughnut Roll

(Fire Fighter I, NFPA 5.5.2)

1 Lay the hose flat and in a straight line.

2 Bring the male coupling alongside the female coupling.

3 Fold the far end over and roll toward the couplings, creating a double roll.

4 The roll can be carried by hand, by a rope, or by a strap.

SKILL DRILL 16-18 Performing a Self-Locking Twin-Doughnut Roll

(Fire Fighter I, NFPA 5.5.2)

1 Lay the hose flat and bring the couplings alongside each other.

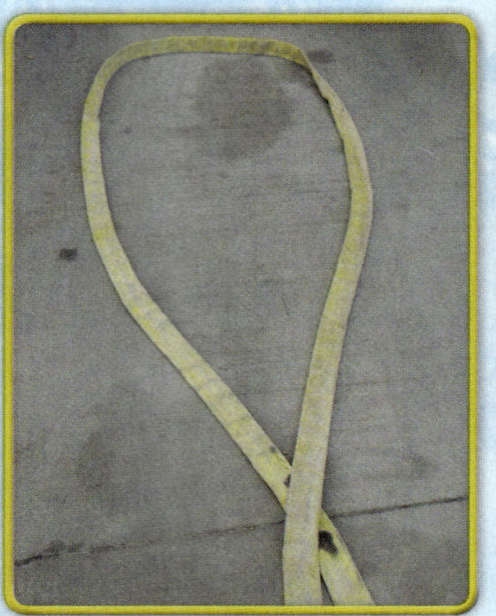

2 Move one side of the hose over the other, creating a loop. This creates the carrying shoulder loop.

3 Bring the loop back toward the couplings to the point where the hose crosses.

4 From the point where the hose crosses, begin to roll the hose toward the couplings, with the loop as its center. This creates a loop on each side of the roll.

5 On completion of the rolling, position the couplings on the top of the rolls.

6 Position the loops so that one is larger than the other. Pass the larger loop over the couplings and through the smaller loop.

Fire Fighter Safety Tips

When performing a forward lay, the fire fighter who is connecting the supply hydrant must not stand between the hose and the hydrant. When the apparatus starts to move off, the hose could become tangled and suddenly be pulled taut. Anyone standing between the hose and the hydrant could be seriously injured.

involve advancing hose lines from an attack engine to apply water onto the fire. Attack line evolutions are covered in the Fire Attack and Foam chapter.

Supply Line Evolutions

The objective of laying a supply line is to deliver water from a hydrant or a static water source to an attack engine. In most cases, this operation involves laying a continuous hose line out of the bed of the fire apparatus as the vehicle drives forward. It can be done using either a forward lay or a reverse lay. A forward lay starts at the hydrant and proceeds toward the fire; the hose is laid in the same direction as the water flows—from the hydrant to the fire. A reverse lay involves laying the hose from the fire to the hydrant; the hose is laid in the opposite direction to the water flow. Each fire department will determine its own preferred methods and procedures for supply line operations based on available apparatus, water supply, and regional considerations.

Forward Hose Lay

The forward hose lay is most often used by the first-arriving engine company at the scene of a fire FIGURE 16-40. This method allows the engine company to establish a water supply without assistance from another company. A forward lay also places the attack engine close to the fire, allowing access to additional hoses, tools, and equipment that are carried on the apparatus.

A forward hose lay can be performed using MDH (2½ inches (65 mm) or 3 inches (76 mm)) or LDH (3½ inches (88 mm) and larger). The larger the diameter of the hose, the more water that can be delivered to the attack engine through a single supply line. When MDH is used and the beds are arranged to lay dual lines, a company can lay two parallel lines from the hydrant to the fire.

FIGURE 16-40 A forward hose lay is made from the hydrant to the fire.

If the fire hydrant is located close to the fire, it may supply a sufficient quantity of water from the hydrant alone. For example, a 5-inch (127 mm) hose can supply 700 gpm (3,182 lpm) over a distance of 500 feet (152 m), losing only some 20 pounds (137 kPa) of pressure due to friction loss.

To perform a forward hose lay, follow the steps in SKILL DRILL 16-19:

1. Stop the fire apparatus 10 feet (3 m) from the hydrant. (STEP 1)
2. Grasp enough hose to reach to the hydrant and to loop around the hydrant. Step off the apparatus, carrying the hydrant wrench and all necessary tools.
3. Loop the end of the hose around the hydrant or secure the hose as specified in the local SOP. Do not stand between the hose and the hydrant. Never stand on the hose. (STEP 2)
4. Signal the pump driver/operator to proceed to the fire once the hose is secured. (STEP 3)
5. Once the apparatus has moved off and a length of supply line has been removed from the apparatus and is lying on the ground, remove the appropriate-size hydrant cap nearest to the fire. Follow the local SOP for checking the operating condition of the hydrant. (STEP 4)
6. Attach the supply hose to the outlet. An adaptor may be needed if a large-diameter hose with Storz-type couplings is used. (STEP 5)
7. Attach the hydrant wrench to the hydrant. (STEP 6)
8. The pump driver/operator uncouples the hose and attaches the end of the supply line to the pump inlet or clamps the hose close to the pump, depending on the local SOP.
9. When the pump driver/operator signals to charge the hose by prearranged hand signal, radio, or air horn, open the hydrant slowly and completely. (STEP 7)
10. Follow the hose back to the engine and remove any kinks from the supply line. (STEP 8)

Four-Way Hydrant Valve

In cases where long supply lines are needed or when MDH is used, it is often necessary to place a supply engine at the hydrant. The supply engine pumps water through the supply line to increase the pressure and volume of water to the attack engine. Some departments use a four-way hydrant valve to connect the supply line to the hydrant, ensuring that the supply line can be charged with water immediately, yet still allowing for a supply engine to connect to the line later.

When the four-way valve is placed on the hydrant, the water flows initially from the hydrant through the valve to the supply line, which delivers the water to the attack engine. The second engine can hook up to the four-way valve and redirect the flow by changing the position of the valve. The water then flows from the hydrant to the supply engine. The supply engine increases the pressure and discharges the water into the supply line, boosting the flow of water to the attack engine. This operation can be accomplished without uncoupling any lines or interrupting the flow.

To use a four-way valve, follow the steps in SKILL DRILL 16-20:

1. Stop the attack engine 10 feet (3 m) past the hydrant to be used.
2. Grasp the four-way valve, the attached hose, and enough hose to reach to and loop around the hydrant. Carry the four-way valve from the apparatus, along with the hydrant wrench and any other needed tools.
3. Loop the end of the hose around the hydrant or secure the hose with a rope as specified in the local SOP. Do not stand between the hydrant and the hose.
4. Signal the attack engine driver/operator to proceed to the fire. (STEP 1)
5. Once enough hose has been removed from the apparatus and is lying in the street, remove the steamer port from the fire hydrant. Follow the local SOP for checking the operating condition of the hydrant.

6. Attach the four-way valve to the hydrant outlet (an adaptor may be needed).
7. Attach the hydrant wrench to the hydrant.
8. The attack engine driver/operator uncouples the hose and attaches the end of the supply line to the pump inlet.
9. The attack engine driver/operator signals by prearranged hand signal, radio, or air horn to charge the supply line.
10. Open the hydrant slowly and completely. (**STEP 2**)

SKILL DRILL 16-19 Performing a Forward Hose Lay
(Fire Fighter I, NFPA 5.3.15)

1. Stop the fire apparatus 10 feet (3 m) from the hydrant.

2. Grasp enough hose to reach to the hydrant and to loop around the hydrant. Step off the apparatus, carrying the hydrant wrench and all necessary tools. Loop the end of the hose around the hydrant or secure the hose as specified in the local SOP. Do not stand between the hose and the hydrant. Never stand on the hose.

3. Signal the pump driver/operator to proceed to the fire once the hose is secured.

4. Once the apparatus has moved off and a length of supply line has been removed from the apparatus and is lying on the ground, remove the appropriate-size hydrant cap nearest to the fire. Follow the local SOP for checking the operating condition of the hydrant.

(Continued)

SKILL DRILL 16-19 Performing a Forward Hose Lay (*Continued*)

(Fire Fighter I, NFPA 5.3.15)

5 Attach the supply hose to the outlet. An adaptor may be needed if a large-diameter hose with Storz-type couplings is used.

6 Attach the hydrant wrench to the hydrant.

7 The pump driver/operator uncouples the hose and attaches the end of the supply line to the pump inlet or clamps the hose close to the pump, depending on the local SOP. When the pump driver/operator signals to charge the hose by prearranged hand signal, radio, or air horn, open the hydrant slowly and completely.

8 Follow the hose back to the engine and remove any kinks from the supply line.

11. Initially, the attack engine is supplied with water from the hydrant. When the supply engine arrives at the fire scene, the supply engine driver/operator stops at the hydrant with the four-way valve. (STEP 3)
12. The supply engine driver/operator attaches a hose from the four-way valve outlet to the intake side of the engine. (STEP 4)
13. Attach a second hose to the inlet side of the four-way valve and connect the other end to the pump discharge. (STEP 5)
14. Change the position of the four-way valve to direct the flow of water from the hydrant through the supply engine and into the supply line. (STEP 6)

Reverse Hose Lay

The reverse hose lay is the opposite of the forward lay FIGURE 16-41. In the reverse lay, the hose is laid out from the fire to the hydrant, in the direction opposite to the flow of the water. This evolution can be used when the attack engine arrives at the fire scene without a supply line. It may be a standard tactic in areas where sufficient hydrants are available and

SKILL DRILL 16-20 Attaching a Hose to a Four-Way Hydrant Valve

(Fire Fighter I, NFPA 5.3.15)

1 Stop the attack engine 10 feet (3 m) past the hydrant to be used. Grasp the four-way valve, the attached hose, and enough hose to reach to and loop around the hydrant. Carry the four-way valve from the apparatus, along with the hydrant wrench and any other needed tools. Loop the end of the hose around the hydrant or secure the hose with a rope as specified in the local SOP. Signal the attack engine driver/operator to proceed to the fire.

2 Once enough hose has been removed from the apparatus and is lying in the street, remove the steamer port from the fire hydrant. Follow the local SOP for checking the operating condition of the hydrant. Attach the four-way valve to the hydrant outlet. Attach the hydrant wrench to the hydrant. The attack engine driver/operator uncouples the hose and attaches the end of the supply line to the pump inlet. The attack engine driver/operator signals charge the supply line. Open the hydrant slowly and completely.

3 Initially, the attack engine is supplied with water from the hydrant. When the supply engine arrives at the fire scene, the supply engine driver/operator stops at the hydrant with the four-way valve.

4 The supply engine driver/operator attaches a hose from the four-way valve outlet to the intake side of the engine.

(Continued)

SKILL DRILL 16-20 Attaching a Hose to a Four-Way Hydrant Valve (*Continued*)

(Fire Fighter I, NFPA 5.3.15)

5 Attach a second hose to the inlet side of the four-way valve and connect the other end to the pump discharge.

6 Change the position of the four-way valve to direct the flow of water from the hydrant through the supply engine and into the supply line.

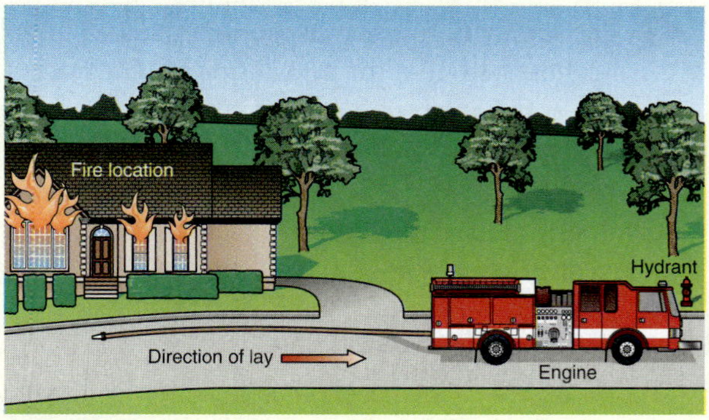

FIGURE 16-41 A reverse hose lay is made from the fire to a fire hydrant.

FIRE FIGHTER Tips

When connecting a supply line to a hydrant, wait until you get the signal from the pump driver/operator to charge the line. If the hydrant is opened prematurely, the hose bed could become charged with water or a loose hose line could discharge water at the fire scene. Either situation will disrupt the water supply operation and could cause serious injuries. Make sure that you know your department's signal to charge a hose line, and do not become so excited or rushed that you make a mistake.

FIRE FIGHTER Tips

When laying out supply hose with threaded couplings, you may find that the wrong end of the hose is on top of the hose bed. Double-male connectors and double-female connectors will allow you to attach a male coupling to a male discharge or to attach a female coupling to a female coupling, respectively. A set of adaptors (one double-male and one double-female) should be easily accessible for these situations. Some department place a set of adaptors on the end of the supply hose for this purpose.

additional companies that can assist in establishing a water supply will arrive quickly. In this scenario, one of these companies will be assigned to lay a supply line from the attack engine to a hydrant.

In this situation, the attack engine focuses on immediately attacking the fire using water from the onboard tank. The supply engine stops close to the attack engine, and hose is pulled from the bed of the supply engine to an intake on the attack engine. The supply engine then drives to the hydrant (or alternative water source) and pumps water back to the attack engine. Usually the supply engine parks in such a way that hose can be pulled from the supply engine to the inlet to the attack engine.

To perform a reverse lay, follow the steps in SKILL DRILL 16-21:

1. Pull sufficient hose to reach from the supply engine to the inlet of the attack engine.
2. Anchor the hose to a stationary object if possible. Wrapping a portion of the hose around the front wheels of the attack engine is one possible strategy; wrapping it around a tree is another. Do not stand between the hose and the stationary object. (STEP 1)
3. The supply engine driver/operator drives away, laying out hose from the attack engine to the fire hydrant or static water source. (STEP 2)
4. Connect the supply line to the inlet of the attack engine. (STEP 3)
5. The supply engine driver/operator uncouples the supply hose from the hose remaining in the hose bed and attaches the supply hose to the discharge side of the supply engine. (STEP 4)
6. The supply engine driver/operator connects the supply engine to the hydrant or water source. The four-way hydrant valve can be used here if needed to boost the supply pressure.
7. Upon the signal from the attack engine, the supply engine driver/operator charges the supply line. (STEP 5)

Split Hose Lay

A split hose lay is performed by two engine companies in situations where hose must be laid in two different directions to establish a water supply FIGURE 16-42. This evolution could be used when the attack engine must approach a fire either along a dead-end street with no hydrant or down a long driveway. To perform a split hose lay, the attack engine drops the end of its supply hose at the corner of the street and performs a forward lay toward the fire. The supply engine stops at the same intersection, pulls off enough hose to connect to the end of the supply line that is already there, and then performs a reverse hose lay to the hydrant or static water source. With the two lines connected together, the supply engine can pump water to the attack engine.

A split lay often requires coordination by two-way radio, because the attack engine must advise the supply engine of the plan and indicate where the end of the supply line is being dropped. In many cases, the attack engine is out of sight when the supply engine arrives at the split point.

A split hose lay does not necessarily require split hose beds. It can be performed with or without split beds if the necessary adaptors are used.

To perform a split hose lay, follow the steps in SKILL DRILL 16-22 (Fire Fighter I, NFPA 5.3.15):

1. The attack engine driver/operator stops at the intersection or driveway entrance.
2. A fire fighter from the attack engine removes the end of the supply line from the hose bed and anchors it.
3. The attack engine driver/operator lays out a hose line while proceeding slowly toward the fire.
4. The attack engine fire fighter either proceeds by foot to the fire or waits at the intersection for the supply engine, according to local SOPs.

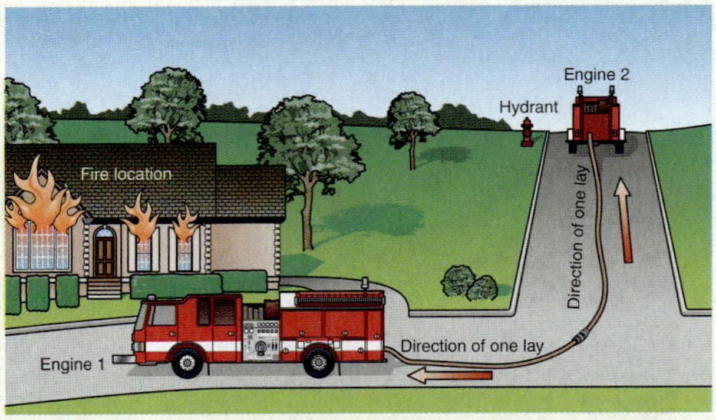

FIGURE 16-42 Two engines performing a split hose lay. Engine 1 performs a forward hose lay from the corner to the fire. Engine 2 performs a reverse hose lay from the hose at the corner of a hydrant.

5. When the supply engine arrives at the intersection, it stops and connects the end of its supply hose to the hose end laid in the street by the attack engine. If threaded couplings are used, a double-male adaptor may be required.
6. A fire fighter anchors the end of the supply line from the supply engine company and lays out a hose line to the hydrant or static water source.
7. The attack engine driver/operator can start pumping from the booster tank and then switch over to the supply line when the supply engine is ready to pump.
8. The supply engine driver/operator positions the apparatus at the water supply according to the local SOP.
9. The supply engine driver/operator pulls off hose from the hose bed until the next coupling. The hose is broken at this connection and is connected to the pump discharge or the four-way hydrant valve following the local SOP.
10. Upon the signal from the attack engine, the supply engine driver/operator charges the supply line.

Loading Supply Hose

This section describes the basic procedures for loading supply line hose into the hose beds on a fire apparatus. Hose can be loaded in several different ways, depending on how it will be laid out at the fire scene. The hose must be easily removable from the hose bed, without kinks or twists, and without the possibility of becoming caught or tangled. The ideal hose load would be easy to load, avoid wear and tear on the hose, have few sharp bends, and allow the hose to play out of the hose bed smoothly and easily.

You must learn the specific hose loads used by your fire department. When loading hose, always remember that the time and attention that go into loading the hose properly will prove valuable when it becomes necessary to use the hose at a fire. You should wear appropriate safety gear when loading hose—at a minimum, your helmet, gloves, and boots. Remember that many modern hose beds are several feet above the ground, and that a fall from a hose bed could result in significant injuries.

Three basic hose loads are commonly used for supply line hose: the flat load, the horseshoe load, and the accordion load. Any one of these methods can be used to load hose for either a straight lay or a reverse lay.

Flat Hose Load

The flat hose load is the easiest loading technique to implement and can be used for any size of hose, including LDH. Because the hose is placed flat in the hose bed, it should lay out flat without twists or kinks. With this type of load, wear and tear on the edges of the hose from the movement and vibration of the vehicle during travel are minimized. The flat hose load can be used with a single hose bed or a split bed. Many variations of the flat hose load exist, so follow your department's SOPs.

To load hose in a flat load, follow the steps in SKILL DRILL 16-23:

1. If you are loading hose with threaded couplings, determine whether the hose will be used for a forward lay or a reverse lay.
2. To set up the hose for a forward lay, place the male hose coupling in the hose bed first. To set up the hose for a reverse lay, place the female hose coupling in the hose bed first. (STEP 1)

SKILL DRILL 16-21 Performing a Reverse Hose Lay

(Fire Fighter I, NFPA 5.3.15)

1. Pull sufficient hose to reach from the supply engine to the inlet of the attack engine. Anchor the hose to a stationary object if possible. Do not stand between the hose and the stationary object.

2. The supply engine driver/operator drives away, laying out hose from the attack engine to the fire hydrant or other static water source.

3. Connect the supply line to the inlet of the attack engine.

4. The supply engine driver/operator uncouples the supply hose from the hose remaining in the hose bed and attaches the supply hose to the discharge side of the supply engine.

5. The supply engine driver/operator connects the engine to the hydrant or water source. The four-way hydrant valve can be used here if needed to boost the supply pressure. Upon the signal from the attack engine, the supply engine driver/operator charges the supply line.

SKILL DRILL 16-23 Performing a Flat Hose Load
(Fire Fighter I, NFPA 5.5.2)

1 To set up the hose for a forward lay, place the male hose coupling in the hose bed first. To set up the hose for a reverse lay, place the female hose coupling in the hose bed first.

2 Start the hose load with the coupling at the front end of the hose compartment.

3 Fold the hose back on itself at the rear of the hose bed.

4 Run the hose back to the front end on top of the previous length of hose. Fold the hose back on itself so that the top of the hose is on the previous length.

5 While laying the hose back to the rear of the hose bed, angle the hose to the side of the previous fold.

6 Continue to lay the hose in neat folds until the whole hose bed is covered with a layer of hose. Continue to load the layers of hose until the required amount of hose is loaded.

3. Start the hose load with the coupling at the front end of the hose compartment. (STEP 2)
4. Fold the hose back on itself at the rear of the hose bed. (STEP 3)
5. Run the hose back to the front end on top of the previous length of hose.
6. Fold the hose back on itself so that the top of the hose is on the previous length. (STEP 4)
7. While laying the hose back to the rear of the hose bed, angle the hose to the side of the previous fold. (STEP 5)
8. Continue to lay the hose in neat folds until the whole hose bed is covered with a layer of hose.
9. To make this hose load neat, make every other layer of hose slightly shorter or alternate the folds. This keeps the ends from getting too high at the folds.
10. Continue to load the layers of hose until the required amount of hose is loaded. (STEP 6)

• When a hole: Shut down or clamp off • Replace damaged section of hose and replace with two sections of hose

Horseshoe Hose Load

The horseshoe hose load is accomplished by placing the hose on its edge and positioning it around the perimeter of the hose bed in a U-shape. At the completion of the first U-shape, the hose is folded inward to form another U-shape in the opposite direction. This continues until a complete layer is filled; then another layer is started above the first. When the hose load is completed, the hose in each layer is in the shape of a horseshoe. A major advantage of the horseshoe load is that it contains fewer sharp bends than the other hose loads.

A horseshoe load cannot be used for LDH because the hose tends to fall over when it stands on edge. This kind of load also leads to more wear on the hose because the weight of the hose is supported by the edges. Also, when laying out a horseshoe load, the hose tends to lay out in a wave-like manner from one side of the street to the other.

Many variations of the horseshoe hose load exist, so follow your department's SOPs. To perform a horseshoe hose load, follow the steps in SKILL DRILL 16-24:

1. If you are loading threaded hose, determine whether you want to load the hose for a forward lay or a reverse lay. For a forward lay, start with the male coupling in the rear corner of the hose bed. For a reverse lay, start with the female coupling in the rear corner of the hose bed. (STEP 1)
2. Lay the first length of hose on its edge against the right or left wall of the hose bed. (STEP 2)
3. At the front of the hose bed, lay the hose across the width of the bed and continue down the opposite side toward the rear. (STEP 3)
4. When the hose reaches the rear of the hose bed, fold the hose back on itself and continue laying it back toward the front of the hose bed. Keep the hose tight to the previous row of hose around the hose bed until it is back to the rear on the starting side. Fold the hose back on itself again, and continue packing the hose tight to the previous row. (STEP 4)
5. Continue to pack the hose on the first layer. Each fold of hose will decrease the amount of space available inside the horseshoe. Once the center of the horseshoe is filled in, begin a second layer by bringing the hose from the rear of the hose bed and laying it around the perimeter of the hose bed. Complete additional layers using the same pattern as used for the first layer. Finish the hose load with any adaptors or appliances used by your department. (STEP 5)

FIRE FIGHTER Tips

When referring to the hose bed, the end closest to the cab is called the front of the hose bed. The end closest to the tailboard is called the rear of the hose bed.

Accordion Hose Load

The accordion hose load is performed with the hose placed on its edge. In this loading technique, the hose is laid side-to-side in the hose bed. One advantage of the accordion load is that it is easy to implement in the hose bed. One layer is loaded from left to right, and then the next layer is loaded above it from left to right.

The accordion hose load also has some disadvantages. Because the hose is stacked on its side, the hose experiences more wear than with a flat load. The accordion hose load is not recommended for LDH because this kind of hose tends to collapse when placed on its side.

Many variations of the accordion hose load exist, so follow your department's SOPs. To perform the accordion hose load, follow the steps in SKILL DRILL 16-25:

1. Determine whether the hose will be used for a forward lay or a reverse lay.
2. To set up the hose for a forward lay, place the male hose coupling in the hose bed first. To set up the hose for a reverse lay, place the female hose coupling in the hose bed first.
3. Start the hose lay with the coupling at the front end of the hose compartment.
4. Lay the first length of hose in the hose bed on its edge against the side of the hose bed. (STEP 1)
5. Double the hose back on itself at the rear of the hose bed. Leave the female end extended so that the two hose beds can be cross-connected. (STEP 2)
6. Lay the hose next to the first length and bring it to the front of the hose bed.
7. Fold the hose at the front of the hose bed so that the bend is even to the edge of the hose bed. Continue to lay folds of hose across the hose bed. (STEP 3)

SKILL DRILL 16-24 Performing a Horseshoe Hose Load

(Fire Fighter I, NFPA 5.5.2)

1 For a forward lay, start with the male coupling in the rear corner of the hose bed. For a reverse lay, start with the female coupling in the rear corner of the hose bed.

2 Lay the first length of hose on its edge against the right or left wall of the hose bed.

3 At the front of the hose bed, lay the hose across the width of the bed and continue down the opposite side toward the rear.

4 When the hose reaches the rear of the hose bed, fold the hose back on itself and continue laying it back toward the front of the hose bed. Keep the hose tight to the previous row of hose around the hose bed until it is back to the rear on the starting side. Fold the hose back on itself again, and continue packing the hose tight to the previous row.

5 Continue to pack the hose on the first layer. Each fold of hose will decrease the amount of space available inside the horseshoe. Once the center of the horseshoe is filled in, begin a second layer by bringing the hose from the rear of the hose bed and laying it around the perimeter of the hose bed. Complete additional layers using the same pattern as used for the first layer. Finish the hose load with any adaptors or appliances used by your department.

8. Alternate the length of the hose folds at each end to allow more room for the folded ends.
9. When the bottom layer is completed, angle the hose upward to begin the second tier.
10. Continue the second layer by repeating the steps used to complete the first layer. (STEP 4)

Connecting a Fire Department Engine to a Water Supply

When an engine is located at a hydrant, a supply hose must be used to deliver the water from the hydrant to the engine. This special type of supply line is intended to deliver as much water as possible over a short distance. In most cases, a soft suction

FIRE FIGHTER Tips

A split hose bed is a hose bed that is divided into two or more sections. This division is made for several purposes:

- One compartment in a split hose bed can be loaded for a forward lay (with the female coupling out) and the other side can be loaded for a reverse lay (with the male coupling out). This allows a line to be laid in either direction without adaptors.
- Two parallel hose lines can be laid at the same time (called "laying dual lines"). Dual lines are beneficial if the situation requires more water than one hose line can supply.
- The split beds can be used to store hoses of different size. For example, one side of the hose bed could be loaded with $2^1/_2$-inch (65 mm) hose that can be used as a supply line or as an attack line. The other side of the hose bed could be loaded with 5-inch (127 mm) hose for use as a supply line. This setup enables the use of the most appropriate-size hose for a given situation.
- All of the hose from both sides of the hose bed can be laid out as a single hose line. This is done by coupling the end of the hose in one bed to the beginning of the hose to the other bed.

In a variation of the split hose bed known as a combination load, the last coupling in one bed is normally connected to the first coupling in the other bed. When one long line is needed, all of the hose plays out of one bed first, and then the hose continues to play out from the second bed. To lay dual lines, the connection between the two hose beds is uncoupled and the hose can play out of both beds simultaneously. When the two sides of a split bed are loaded with the hose in opposite directions, either a double-female or double-male adaptor is used to make the connection between the two hose beds.

FIRE FIGHTER Tips

When a split hose bed is loaded for a combination load, the end of the last length of a hose in one bed is coupled to the beginning of the first length in the opposite bed. Begin loading the first bed with the initial coupling hanging out. This coupling will be the last to be deployed from the first bed when the hose is laid out. Leave enough of the end of the hose hanging down to reach the other hose bed. Load the second bed in the normal manner. When both beds have been loaded, connect the hanging coupling from the bottom of the first hose bed to the end coupling on the top of the adjoining bed.

hose is used to connect directly to a hydrant. Alternatively, the connection can be made with a hard suction hose or with a short length of large-diameter supply hose.

Attaching a Soft Suction Hose to a Hydrant

Large-diameter soft suction hose is normally used to connect an engine directly to a hydrant. To attach a soft suction hose to a hydrant, follow the steps in SKILL DRILL 16-26:

FIRE FIGHTER Tips

The following tips will help you to do a better job when loading hose:

- Drain all of the water out of the hose before loading it.
- Rolling the hose first will result in a flatter hose load, because there will be no air in the hose.
- Do not load hose too tightly. Leave enough room so that you can slide a hand between the folds of hose. If hose is loaded too tightly, it may not lay out properly.
- Load hose so that couplings do not have to turn around as the hose is pulled out of the hose bed. Make a short fold in the hose close to the coupling to keep the hose properly oriented. This short fold is called a Dutchman FIGURE 16-43.
- Couple sections of hose with the flat sides oriented in the same direction.
- Check gaskets before coupling hose.
- Tighten couplings so that they are hand tight only. With a good gasket, the hose should not leak.

FIGURE 16-43 A Dutchman fold ensures that a coupling will not have to turn around and possibly become stuck as the hose is laid out.

1. The pump driver/operator positions the apparatus so that its inlet is the correct distance from the hydrant. (STEP 1)
2. Remove the hose, any needed adaptors, and the hydrant wrench. (STEP 2)
3. Attach the soft suction hose to the inlet of the engine. In some departments, this end of the hose is preconnected. (STEP 3)
4. Unroll the hose. (STEP 4)
5. Remove the large hydrant cap and check the hydrant for proper operation. (STEP 5)
6. Attach the soft suction hose to the hydrant. Sometimes it will be necessary to use an adaptor. (STEP 6)
7. Ensure that there are no kinks or sharp bends in the hose that might restrict the flow of water. (STEP 7)

SKILL DRILL 16-25 Performing an Accordion Hose Load

(Fire Fighter I, NFPA 5.5.2)

1 Lay the first length of hose in the hose bed on its edge against the side of the hose bed.

2 Double the hose back on itself at the rear of the hose bed. Leave the female and extended so that the two hose beds can be cross-connected.

3 Lay the hose next to the first length and bring it to the front of the hose bed. Fold the hose at the front of the hose bed so that the bend is even to the edge of the hose bed. Continue to lay folds of hose across the hose bed.

4 Alternate the length of the hose folds at each end to allow more room for the folded ends. When the bottom layer is completed, angle the hose upward to begin the second tier. Continue the second layer by repeating the steps used to complete the first layer.

8. Open the hydrant slowly when so indicated by the pump driver/operator. Check all connections for leaks. Tighten the couplings if necessary. (STEP 8)
9. Where required place chafing blocks under the hose where it contacts the ground to prevent mechanical abrasion. (STEP 9)

Attaching a Hard Suction Hose to a Hydrant

Although it is not commonly done, a hard suction hose can be used to connect an engine to a hydrant. Implementing this connection can be a difficult task because the hard suction is heavy and has limited flexibility, so additional personnel may be needed to lift and attach the hose. The apparatus must be carefully positioned to make this connection properly.

To attach a hard suction hose to a hydrant, follow the steps in SKILL DRILL 16-27:

1. The pump driver/operator positions the apparatus so that the intake on the apparatus is located the correct distance from the hydrant. (STEP 1)
2. Remove the pump inlet cap on the apparatus.
3. Remove the hydrant steamer outlet cap, which is the largest one. Check the hydrant for proper operation.
4. Place an adaptor on the hydrant if needed.
5. With a partner's help, remove a section of hard suction hose from the engine. (STEP 2)
6. Connect the hard suction hose to the large intake on the engine. (STEP 3)

7. Connect the opposite end to the hydrant, using a double-female adaptor with the hydrant thread on one side and the suction hose thread on the other side.
8. The pump driver/operator may need to move the apparatus slightly to position the apparatus to make the final connection.
9. Slowly open the hydrant when instructed to do so by the pump driver/operator. (STEP 4)

Supply Hose Carries and Advances

Several different techniques are used to carry and advance supply hose. The best technique for a particular situation will depend on the size of the hose, the distance over which it must be moved, and the number of fire fighters available to perform the task. The same techniques can be used for both supply lines and attack lines.

Fire Fighter Safety Tips

When loading hose on an apparatus, always use caution in climbing up and down on the apparatus. If you are loading hose at a fire scene, watch out for wet, slippery surfaces, ice, or other hazards. Also, wet hose can be heavy and you may need to reach, stretch, or lift the hose to get it into the hose bed. Use caution!

SKILL DRILL 16-26 Attaching a Soft Suction Hose to a Fire Hydrant

Fire Fighter I (NFPA 5.3.15)

1. The pump driver/operator positions the apparatus so that its inlet is the correct distance from the hydrant.

2. Remove the hose, any needed adaptors, and the hydrant wrench.

3. Attach the soft suction hose to the inlet of the engine if it is not already attached to the engine intake.

4. Unroll the hose.

(Continued)

SKILL DRILL 16-26 Attaching a Soft Suction Hose to a Fire Hydrant (*Continued*)

Fire Fighter I (NFPA 5.3.15)

5 Remove the large hydrant cap. Check the hydrant for proper operation.

6 Attach the soft suction hose to the hydrant.

7 Ensure that there are no kinks or sharp bens in the hose that might restrict the flow of water.

8 Open the hydrant slowly when so indicated by the driver/operator. Check all connections for leaks. Tighten the couplings if necessary.

9 Where required, placing chafing blocks under the hose where it contacts the ground to prevent mechanical abrasion.

SKILL DRILL 16-27 Attaching a Hard Suction Hose to a Fire Hydrant

Fire Fighter I (NFPA 5.3.15)

1 The pump driver/operator positions the apparatus so that the intake on the apparatus is located the correct distance from the hydrant.

2 Remove the pump inlet cap. Remove the hydrant steamer outlet cap. Place an adaptor on the hydrant if needed. With a partner's help, remove a section of hard suction hose from the engine.

3 Connect the hard suction hose to the large intake on the engine.

4 Connect the opposite end to the hydrant. Slowly open the hydrant when instructed to do so by the pump driver/operator.

Whenever possible, a hose line should be laid out and positioned as close as possible to the location where it will be operated before it is charged with water. A charged line is much heavier and more difficult to maneuver than a dry hose line. A suitable amount of extra hose should be available to allow for maneuvering after the line is charged.

Working Hose Drag

The working hose drag technique is used to deploy hose from a hose bed and advance the line over a relatively short distance to the desired location. Depending on the size and length of the hose, several fire fighters may be required to perform this task.

To perform a working hose drag, follow the steps in SKILL DRILL 16-28:

1. Place the end of the hose over the shoulder. (STEP 1)
2. Hold onto the coupling with the hand. (STEP 2)
3. Walk in the direction you want to advance the hose. (STEP 3)
4. As the next hose coupling is ready to come off the hose bed, have a second fire fighter grasp the coupling and place the hose over the shoulder. (STEP 4)
5. Continue this process until enough hose has been pulled out of the hose bed. (STEP 5)

SKILL DRILL 16-28 Performing a Working Hose Drag

Fire Fighter I (NFPA 5.3.10)

1. Place the end of the hose over the shoulder.

2. Hold onto the coupling with the hand.

3. Walk in the direction you want to advance the hose.

4. As the next hose coupling is ready to come off the hose bed, a second fire fighter grasps the coupling and places the hose over the shoulder.

5. Continue this process until enough hose has been pulled out of the hose bed.

Shoulder Carry

The shoulder carry is used to transport full lengths of hose over a longer distance than it is practical to drag the hose. It is also useful when a hose line must be advanced around obstructions. For example, this technique can be used to stretch an attack line from the front of a building, around to the rear entrance, and up to the second floor. The shoulder carry can also be employed to stretch a supply line to an attack engine in a location where the hose cannot be laid out by another engine.

This technique requires practice and good teamwork to be successful. By working together to complete tasks efficiently, fire fighters can achieve their goal of extinguishing the fire in the shortest period of time. To perform a shoulder carry, follow the steps in SKILL DRILL 16-29:

1. Stand at the tailboard of the engine. Grasp the end of the hose and place it over your shoulder so that the coupling is at chest height.
2. Have a second fire fighter place additional hose on your shoulder so that the ends of the folds reach to about knee level. Continue to place folds on your shoulder, but only as much as you can safely carry. (STEP 1)
3. Hold the hose to prevent it from falling off your shoulder. Continue to hold the hose and move forward about 15 feet (4.5 m).
4. A second fire fighter stands at the tailboard to receive a load of hose and then move forward. (STEP 2)
5. When enough fire fighters have received hose loads, the hose can be uncoupled from the hose bed. (STEP 3)
6. The pump driver/operator connects the coupling to the engine.
7. All of the fire fighters start walking toward the fire, coordinating their movements. The last fire fighter should not start offloading hose from his or her shoulder until all of the hose has been laid out. The next fire fighter in line then starts laying out hose from his or her shoulder.
8. Each fire fighter lays out his or her supply of hose until the entire length is laid out. (STEP 4)

To advance an accordion load using a shoulder carry, follow the steps in SKILL DRILL 16-30:

1. Find the end of the accordion load, whether it is a nozzle or a coupling.
2. Using two hands, grasp the end of the load and the number of folds it will take to make an adequate shoulder load. (STEP 1)
3. Pull the accordion load about one-third of the way off the apparatus. (STEP 2)
4. Twist the folds so that they become flat, with the end of the accordion load (nozzle or coupling) on the bottom of the now flat shoulder load. (STEP 3)
5. Transfer the hose to the opposite shoulder while turning so that you face in the direction you will walk. (STEP 4)
6. Place the shoulder load over your shoulder and grasp it tightly with both hands. (STEP 5)
7. Walk away from the apparatus, pulling the shoulder load out of the hose bed. (STEP 6)
8. If additional fire fighters are needed, they may follow steps 1–6 to assist in removing the required amount of hose.

Connecting Supply Hose Lines to Standpipe and Sprinkler Systems

Another water supply evolution is furnishing water to standpipe and sprinkler systems. Fire department connections on buildings are provided so that the fire department can pump water into standpipe and sprinkler systems. This setup is considered to be a supply line because it supplies water to the standpipe and sprinkler systems. What is different is that this supply line is connected to the discharge side of the attack engine.

The function of the hose line in this case is to provide either a primary or secondary water supply for the sprinkler or standpipe system. The same basic techniques are used to connect the hose lines to either type of system.

Standpipe systems are used to provide a water supply for attack lines that will be operated inside the building. Outlets are provided inside the building where fire fighters can connect attack lines. The fire fighters inside the building must then depend on fire fighters outside the building to supply the water to the fire department connection FIGURE 16-44.

Two types of standpipe systems exist:

- A dry standpipe system depends on the fire department to provide all of the water.
- A wet standpipe system has a built-in water supply, but the fire department connection is provided to deliver a higher flow or to boost the pressure.

The pressure requirements for standpipe systems depend on the height where the water will be used inside the building.

The fire department connection for a sprinkler system is also used to supplement the normal water supply. The required pressures and flows for different types of sprinkler

FIGURE 16-44 A standpipe connection.

SKILL DRILL 16-29 Performing a Shoulder Carry

Fire Fighter I (NFPA 5.3.10)

1 Stand at the tailboard of the engine. Grasp the end of the hose and place it over your shoulder so that the coupling is at chest height. Have a second fire fighter place additional hose on your shoulder so that the ends of the folds reach about knee level. Continue to place folds on your shoulder, but only as much as you can safely carry.

2 Hold the hose to prevent it from falling off your shoulder. Continue to hold the hose and move forward about 15 feet (4.5 m). A second fire fighter then stands at the tailboard to receive a load of hose and then move forward.

3 When enough fire fighters have received hose loads, the hose can be uncoupled from the hose bed.

4 All of the fire fighters start walking toward the fire. The last fire fighter should not start offloading hose from his or her shoulder until all of the hose is laid out. The next fire fighter then starts laying out hose from his or her shoulder. Each fire fighter lays out his or her supply of hose until the entire length is laid out.

systems can vary significantly. As a guideline, sprinkler systems should be fed at 150 psi (1034 kPa) unless more specific information is available.

To connect a hose line to supply a fire department connection, follow the steps in **SKILL DRILL 16-31**:

1. Locate the fire department connection to the standpipe or sprinkler system.
2. Extend a hose line from the engine discharge to the fire department connection using the size of hose required by the fire department's SOPs. Some fire departments use a single hose line, whereas others call for two or more lines to be connected. (**STEP 1**)
3. Remove the caps on the standpipe inlet. Some caps are threaded into the connections and must be unscrewed. Other caps are designed to break away when struck with a tool such as a hydrant wrench or spanner wrench. (**STEP 2**)

SKILL DRILL 16-30 Advancing an Accordion Load

Fire Fighter I (NFPA 5.3.10)

1. Using two hands, grasp the end of the load and the number of folds it will take to make an adequate shoulder load.

2. Pull the accordion load about one-third of the way off the apparatus.

3. Twist the folds so that they become flat, with the end of the accordion load (nozzle or coupling) on the bottom of the now flat shoulder load.

4. Transfer the hose to the opposite shoulder while turning so that you face in the direction you will walk.

5. Place the shoulder load over your shoulder and grasp it tightly with both hands.

6. Walk away from the apparatus, pulling the shoulder load out of the hose bed.

SKILL DRILL 16-31 Connecting a Hose Line to Supply a Fire Department Connection

Fire Fighter I (NFPA 5.3.15)

1. Locate the fire department connection to the standpipe or sprinkler system. Extend a hose line from the engine discharge to the fire department connection using the size of hose required by the fire department's SOPs. Some fire departments use a single hose line, whereas others call for two or more lines to be connected.

2. Remove the caps on the standpipe inlet. Some caps are threaded into the connections and must be unscrewed. Other caps are designed to break away when struck with a tool such as a hydrant wrench or spanner wrench.

3. Visually inspect the interior of the connection to ensure that it does not contain any debris that might obstruct the water flow. Never stick your hand or fingers inside the connections; fire fighters have been injured from broken glass or needles inside these connections. Attach the hose line to the connection. Notify the pump driver/operator when the connection has been completed.

4. Visually inspect the interior of the connection to ensure that it does not contain any debris that might obstruct the water flow. Do not stick your hand or fingers inside the connections; fire fighters have been injured from broken glass or needles inside these connections.
5. Attach the hose line to the connection.
6. Notify the pump driver/operator when the connection has been completed. (STEP 3)

Replacing a Defective Section of Hose

With proper maintenance and testing, the risk of fire hose failure should be low—but even so, it is always a possibility. Fire fighters need to know what to do if a section of hose bursts or develops a major leak while it is being used. Every fire fighter should know how to quickly replace a length of defective hose and restore the flow.

A burst hose line should be shut down as soon as possible. If the line cannot be shut down at the pump, at the fire hydrant, or at a control valve, a hose clamp can be used to stop the flow. Place the hose clamp on an undamaged section of hose upstream from the problem. After the water flow has been shut off, quickly remove the damaged section of hose and replace it with two sections of hose. Using two sections of hose will ensure that the replacement hose is long enough to replace the damaged section.

To replace a hose section, follow the steps in SKILL DRILL 16-32:

1. Shut down or clamp off the damaged line.
2. Remove the damaged section of the hose.
3. Replace the damaged section with two new sections to ensure that the hose's length will be adequate.
4. Restore the water flow.

Fire Fighter Safety Tips

Some departments use LDH with Storz-type couplings to connect to fire department connections. Hose that is rated for use as attack hose should be used to connect to a standpipe system.

Large-diameter supply line hose is rated for a safe working pressure of 185 psi (1275 kPa). Standpipe systems generally require at least 150 psi (1034 kPa) to be provided at the fire department connection, but a water hammer can cause a spike of much higher pressure. The excess pressure could burst the hose and place fire fighters who are working inside the building in danger.

Many fire department SOPs require two hose lines to be connected to a fire department connection. If one line breaks or if the flow is interrupted, water will still be delivered to the system through the other line.

Draining and Picking Up Hose

To put a hose back into service, it must be drained of water. To do so, lay the hose straight on a flat surface, and then lift one end of the hose to shoulder level. Gravity will allow the water to flow to the lower portion of the hose and eventually out of the hose. As you proceed down the length of hose, fold the hose back and forth over your shoulder. When you reach the end of the section, you will have the whole section of hose on your shoulder.

To drain a hose, follow the steps in SKILL DRILL 16-33:

1. Lay the section of hose straight on a flat surface. (STEP 1)
2. Starting at one end of the section, lift the hose to shoulder level. (STEP 2)

SKILL DRILL 16-33 Draining a Hose
Fire Fighter I (NFPA 5.5.2)

1 Lay the section of hose straight on a flat surface.

2 Starting at one end of the section, lift the hose to shoulder level.

3 Move down the length of hose, folding it back and forth over the shoulder.

4 Continue down the length until the entire section is on the shoulder.

3. Move down the length of hose, folding it back and forth over the shoulder. (STEP 3)
4. Continue down the length until the entire section is on the shoulder. (STEP 4)

Unloading Hose

There are times other than a fire when fire fighters will need to unload the hose from an engine. For example, hose should be unloaded and reloaded on a regular basis to place the bends in different portions of the hose, because leaving bends in the same locations for long periods of time is likely to cause weakened areas. In addition, hose might have to be unloaded to change out apparatus. Sometimes all of the equipment from one engine must be transferred to another vehicle. It could also be necessary to offload the hose for annual testing to be conducted.

The following procedure should be used to unload a hose from a hose bed:

1. Use a large area such as a parking lot for this procedure. Be sure it is clean.
2. Disconnect any gate valves or nozzles from the hose before you begin.
3. Grasp the end of the hose, and pull it off the engine in a straight line.
4. When a coupling comes off the engine, disconnect the hose and pull off the next section of hose.
5. When all of the hose has been removed from the hose bed, use a broom to brush off any dirt or debris on both sides of the hose jacket.
6. Sweep out any debris or dirt from the hose bed.
7. Store hose rolls off the floor on a rack, in a cool dry area.

Wrap-Up

Chief Concepts

- In rural areas, fire departments often depend on water from static sources to maintain their water supply. Water may be used from the static source directly or transported via a mobile water supply tanker/tender. These trucks are designed to carry large volumes of water, ranging from 500 gallons (2,273 liters) to 5000 gallons (22,730 liters).
  - Dry or drafting hydrants have a strainer on one end and connect to a fire apparatus to supply water from a static source.
  - Portable tanks are carried on fire apparatus and can hold between 600 (2,727 liters) and 5000 gallons (22,730 liters) of water. These tanks can be set up quickly and linked together to increase the water storage capacity.
  - A tanker shuttle can deliver water from a static source fill site to the fire scene.
- Municipal water systems make clean water available to people in populated areas and provide water for fire protection. Hydrants make this water supply available to the fire department.
- Municipal water systems can draw water from wells, rivers, or lakes; they carry the water via pipelines or canals to a water treatment facility, and then to the water distribution system, a complex network of underground pipes.
  - Most municipal water supply systems use both pumps and gravity to deliver water.
  - Underground water mains come in several sizes. Large mains, or primary feeders, carry large quantities of water to a section of a city. Smaller mains, or secondary feeders, distribute water to a smaller area. The smallest pipes, or distributors, carry water to the users and hydrants along individual streets.
  - Shut-off valves are located at the connection points where the underground mains meet the distributor pipes. These valves can be used to prevent water flow if the water system in the building or the fire hydrant is damaged.
- The two types of fire hydrants are wet-barrel hydrants and dry-barrel hydrants.
  - Wet-barrel hydrants are used in areas where temperatures do not drop below freezing. These hydrants do not have to be drained after each use.
  - Dry-barrel hydrants are used in areas where temperatures drop below freezing. When this type of hydrant is not in use, the barrel must be dry.
- Fire hydrants are located according to local standards and nationally recommended practices. In many communities, hydrants are located at every street intersection.
- Fire fighters must be proficient in operating fire hydrants, including the tasks of turning on the hydrant, shutting off the hydrant, and inspecting the hydrant.

- To understand fire hydrant testing procedures, you must understand some basic concepts:
  - The flow or quantity of water moving through a pipe, hose, or nozzle is described in terms of its volume, usually specified in units of gallons (or liters) per minute.
  - Water pressure refers to an energy level and is measured in units of psi (or kilopascals). Volume and pressure are two different, but mathematically related, measurements.
  - Water that is not moving has potential (static) energy. When the water is moving, it has a combination of potential energy and kinetic (in motion) energy. Both the quantity of water flowing and the pressure under a specific set of conditions must be measured when testing any water system, including hydrants.
  - Static pressure is created by elevation pressure and/or pump pressure.
  - Normal operating pressure refers to the amount of pressure in a water distribution system during a period of normal consumption.
  - Residual pressure is the amount of pressure that remains in the system when water is flowing.
  - Flow pressure measures the quantity of water flowing through an opening. A Pitot gauge is used to measure flow pressure in psi (kPa) and to calculate the flow in gallons per minute (liters per minute).
  - Knowing the static pressure, the flow in gallons per minute, and the residential pressure enables fire fighters to calculate the amount of water that can be obtained from a hydrant or a group of hydrants on the same water main.
- Fire hoses are used as supply hoses and as attack hoses. Supply hoses are used to deliver water from a static source or from a fire hydrant to an attack engine. Fire hoses range in size from 1 inch to 6 inches (25 mm to 152 mm).
  - Small-diameter hose (SDH) ranges in size from 1 inch to 2 inches (25 mm to 50 mm) in diameter. It is commonly used for attack lines.
  - Medium-diameter (MDH) hose has a diameter of 2½ inches or 3 inches (65 mm or 76 mm). It is used for both attack and supply lines.
  - Large-diameter hose (LDH) has a diameter of 3½ inches (88 mm) or more. It is used for supply lines.
- Fire hose is constructed with an inner waterproof liner surrounded by either one or two outer layers.
- Couplings are used to connect individual lengths of fire hose together. They are also used to connect a hose line to a hydrant, to an intake or discharge valve on an engine, or to a variety of nozzles, fittings, and appliances. A coupling is permanently attached to each end of a section of fire hose.
- The two types of couplings are threaded couplings and Storz-type couplings.
  - Threaded couplings are used on most hoses up to 3 inches (76 mm) in diameter and on soft suction hose and hard suction hose. Threaded couplings consist of a male coupling and a female coupling.
  - Storz-type couplings are the same on both ends.
- Supply hose can either be soft suction or hard suction.
  - Soft suction hose is a short section of LDH that is used to connect a fire department engine directly to the large steamer outlet on a hydrant. It has a female connection on each end, with one end matching the local hydrant threads and the other end matching the threads on a large-diameter inlet to the engine.
  - A hard suction hose is a special type of supply hose that is used to draft water from a static source such as a river, lake, or portable drafting basin. It is designed to remain rigid and will not collapse when a vacuum is created in the hose to draft the water into the pump. It can also be used to carry water from a fire hydrant to the pumper.
- Fire hose should be inspected and tested following the procedures in NFPA 1962.
- The most common causes of hose damage are mechanical damage, heat and cold damage, chemical damage, and mildew damage.
- Visual hose inspections should be performed at least quarterly. A visual inspection of hose should be performed after each use. If any defects are found, the length of hose should be marked and removed from service.
- Hose appliances include wyes, gated wyes, water thieves, Siamese connections, adaptors, reducers, hose jackets, hose rollers, and hose clamps.
- Rolled hose is compact and easy to manage. A fire hose can be rolled many different ways, depending on how it will be used. Follow the SOPs of your department when rolling hose.
- The objective of laying a supply line is to deliver water from a hydrant or a static water source to an attack engine. In most cases, this operation involves laying a continuous hose line out of the bed of the fire apparatus as the vehicle drives forward. It can be done using either a forward lay or a reverse lay.
  - A forward lay starts at the hydrant and proceeds toward the fire; the hose is laid in the same direction as the water flows—from the hydrant to the fire.
  - A reverse lay involves laying the hose from the fire to the hydrant; the hose is laid in the opposite direction to the water flow.
  - A split hose lay is performed by two engine companies in situations where hose must be laid in two different directions to establish a water supply. The attack engine drops the end of its supply hose at the corner of the street and performs a forward lay toward the fire. The supply engine stops at the same

intersection, pulls off enough hose to connect to the end of the supply line that is already there, and then performs a reverse hose lay to the hydrant or static water source. With the two lines connected together, the supply engine can pump water to the attack engine.

- Hose can be loaded in several different ways, depending on how it will be laid out at the fire scene.
  - The hose must be easily removable from the hose bed, without kinks or twists, and without the possibility of becoming caught or tangled.
  - The ideal hose load would be easy to load, avoid wear and tear on the hose, have few sharp bends, and allow the hose to play out of the hose bed smoothly and easily.
- Several different techniques are used to carry and advance supply hose.
  - The best technique for a particular situation will depend on the size of the hose, the distance over which it must be moved, and the number of fire fighters available to perform the task.
  - The same techniques can be used for both supply lines and attack lines.
- Whenever possible, a hose line should be laid out and positioned as close as possible to the location where it will be operated before it is charged with water.
  - A charged line is much heavier and more difficult to maneuver than a dry hose line.
  - A suitable amount of extra hose should be available to allow for maneuvering after the line is charged.

Hot Terms

Accordion hose load A method of loading hose on a vehicle that results in a hose appearance that resembles accordion sections. It is achieved by standing the hose on its edge, then placing the next fold on its edge, and so on.

Adaptor A device that joins hose couplings of the same type, such as male to male or female to female.

Attack engine An engine from which attack lines have been pulled.

Attack hose (attack line) Hose designed to be used by trained fire fighters and fire brigade members to combat fires beyond the incipient stage. (NFPA 1961)

Ball valves Valves used on nozzles, gated wyes, and engine discharge gates. They consist of a ball with a hole in the middle of the ball.

Booster hose (booster line) A noncollapsible hose used under positive pressure having an elastomeric or thermoplastic tube, a braided or spiraled reinforcement, and an outer protective cover. (NFPA 1962)

Butterfly valves Valves that are found on the large pump intake valve where the suction hose connects to the inlet of the fire pump.

Distributors Relatively small-diameter underground pipes that deliver water to local users within a neighborhood.

Double-female adaptor A hose adaptor that is used to join two male hose couplings.

Double-jacket hose A hose constructed with two layers of woven fibers.

Double-male adaptor A hose adaptor that is used to join two female hose couplings.

Dry-barrel hydrant A type of hydrant used in areas subject to freezing weather. The valve that allows water to flow into the hydrant is located underground, and the barrel of the hydrant is normally dry.

Dry hydrant An arrangement of pipe permanently connected to a water source other than a piped, pressurized water supply system that provides a ready means of water supply for firefighting purposes and that utilizes the drafting (suction) capability of a fire department pump. (NFPA 1142)

Dump valve A large opening from the water tank of a mobile water supply apparatus for unloading purposes. (NFPA 1901)

Dutchman A short fold placed in a hose when loading it into the bed; the fold prevents the coupling from turning in the hose bed.

Elevated water storage tower An above-ground water storage tank that is designed to maintain pressure on a water distribution system.

Elevation pressure The amount of pressure created by gravity. Also known as head pressure.

Fire hydraulics The physical science of how water flows through a pipe or hose.

Flat hose load A method of putting a hose on a vehicle in which the hose is laid flat and stacked on top of the previous section.

Flow pressure The amount of pressure created by moving water.

Forward lay A method of laying a supply line where the line starts at the water source and ends at the attack engine.

Four-way hydrant valve A specialized type of valve that can be placed on a hydrant and that allows another engine to increase the supply pressure without interrupting flow.

Friction loss The reduction in pressure resulting from the water being in contact with the side of the hose. This contact requires force to overcome the drag that the wall of the hose creates.

Gate valves Valves found on hydrants and sprinkler systems.

Gated wye A valved device that splits a single hose into two separate hoses, allowing each hose to be turned on and off independently.

Gravity-feed system A water distribution system that depends on gravity to provide the required pressure. The system storage is usually located at a higher elevation than the end users.

Hard suction hose A hose used for drafting water from static supplies (lakes, rivers, wells, and so forth). It can also be used for supplying pumps on fire apparatus from hydrants if designed for that purpose. The hose contains a semirigid or rigid reinforcement. (NFPA 1963)

Higbee indicators Indicators on the male and female threaded couplings that indicate where the threads start. These indicators should be aligned before fire fighters start to thread the couplings together.

Horseshoe hose load A method of loading hose in which the hose is laid into the bed along the three walls of the bed, so that it resembles a horseshoe.

Hose appliance A piece of hardware (excluding nozzles) generally intended for connection to fire hose to control or convey water. (NFPA 1962)

Hose clamp A device used to compress a fire hose so as to stop water flow.

Hose jacket A device used to stop a leak in a fire hose or to join hoses that have damaged couplings.

Hose liner (hose inner jacket) The inside portion of a hose that is in contact with the flowing water.

Hose roller A device that is placed on the edge of a roof and is used to protect hose as it is hoisted up and over the roof edge.

Large-diameter hose (LDH) A hose 3.5 inches (89 mm) or larger that is designed to move large volumes of water to supply master stream appliances, portable hydrants, manifolds, standpipe and sprinkler systems, and fire department pumpers from hydrants and in relay. (NFPA 1410)

Master stream device A large-capacity nozzle that can be supplied by two or more hose lines of fixed piping. It can flow 300 gallons per minute (1363 lpm). These devices include deck guns and portable ground monitors.

Medium-diameter hose (MDH) Hose with a diameter of 2½ or 3 inches (65 mm or 76 mm).

Mildew A fungus that can grow on hose if the hose is stored wet. Mildew can damage the jacket of a hose.

Mobile water supply apparatus A vehicle designed primarily for transporting (pickup, transporting, and delivering) water to fire emergency scenes to be applied by other vehicles or pumping equipment. (NFPA 1901)

Municipal water system A water distribution system that is designed to deliver potable water to end users for domestic, commercial, industrial, and fire protection purposes.

Normal operating pressure The observed static pressure in a water distribution system during a period of normal demand.

Pitot gauge A type of gauge that is used to measure the velocity pressure of water that is being discharged from an opening. It is used to determine the flow of water from a hydrant or nozzle.

Portable tanks Folding or collapsible tanks that are used at the fire scene to hold water for drafting.

Primary feeders The largest-diameter pipes in a water distribution system, which carry the greatest amounts of water.

Private water system A privately owned water system that operates separately from the municipal water system.

Reducer A fitting used to connect a small hose line or pipe to a larger hose line or pipe. (NFPA 1142)

Remote-controlled hydrant valve A valve that is attached to a fire hydrant to allow the operator to turn the hydrant on without flowing water into the hose line.

Reservoir A water storage facility.

Residual pressure The pressure that exists in the distribution system, measured at the residual hydrant at the time the flow readings are taken at the flow hydrants. (NFPA 24)

Reverse lay A method of laying a supply line where the supply line starts at the attack engine and ends at the water source.

Rocker lugs (rocker pins) Fittings on threaded couplings that aid in coupling the hoses.

Rubber-covered hose (rubber-jacket hose) Hose whose outside covering is made of rubber, which is said to be more resistant to damage.

Secondary feeders Smaller-diameter pipes that connect the primary feeders to the distributors.

Shut-off valve Any valve that can be used to shut down water flow to a water user or system.

Siamese connection A device that allows two hoses to be connected together and flow into a single hose.

Small-diameter hose (SDH) Hose with a diameter ranging from 1 to 2 inches (25 mm to 50 mm).

Soft suction hose A large-diameter hose that is designed to be connected to the large port on a hydrant (steamer connection) and into the engine.

Spanner wrench A type of tool used to couple or uncouple hoses by turning the rocker lugs on the connections.

Split hose bed A hose bed arranged to enable the engine to lay out either a single supply line or two supply lines simultaneously.

Split hose lay A scenario in which the attack engine forward lays a supply line from an intersection to the fire, and the supply engine reverse lays a supply line from the hose left by the attack engine to the water source.

Static pressure The pressure that exists at a given point under normal distribution system conditions measured at the residual hydrant with no hydrants flowing. (NFPA 24)

Static water source A water source such as a pond, river, stream, or other body of water that is not under pressure.

Steamer port The large-diameter port on a hydrant.

Storz-type (nonthreaded) hose coupling A hose coupling that has the property of being both the male and the female coupling. It is connected by engaging the lugs and turning the coupling a one-third turn.

Supply hose (supply line) Hose designed for the purpose of moving water between a pressurized water source and a pump that is supplying attack lines. (NFPA 1961)

Tanker shuttle A method of transporting water from a source to a fire scene using a number of mobile water supply apparatus.

Threaded hose coupling A type of coupling that requires a male fitting and a female fitting to be screwed together.

Water hammer The surge of pressure that occurs when a high-velocity flow of water is abruptly shut off. The pressure exerted by the flowing water against the closed system can be seven or more times that of the static pressure. (NFPA 1962)

Water main A generic term for any underground water pipe.

Water supply A source of water for firefighting activities. (NFPA 1144)

Water thief A device that has a 2½-inch (65 mm) inlet and a 2½-inch (65 mm) outlet in addition to two 1½-inch (38 mm) outlets. It is used to supply many hoses from one source.

Wet-barrel hydrant A hydrant used in areas that are not susceptible to freezing. The barrel of the hydrant is normally filled with water.

Wye A device used to split a single hose into two separate lines.

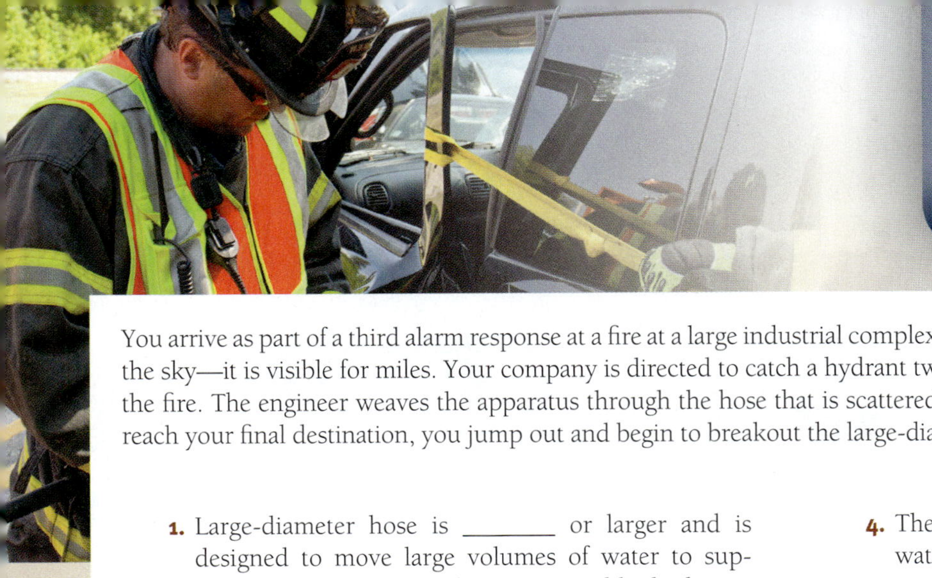

FIRE FIGHTER *in action*

You arrive as part of a third alarm response at a fire at a large industrial complex. A huge column of black smoke is billowing into the sky—it is visible for miles. Your company is directed to catch a hydrant two blocks away and set up master streams to cool the fire. The engineer weaves the apparatus through the hose that is scattered everywhere like spaghetti on a plate. Once you reach your final destination, you jump out and begin to breakout the large-diameter hose before setting up the master streams.

1. Large-diameter hose is _______ or larger and is designed to move large volumes of water to supply master stream appliances, portable hydrants, manifolds, standpipe and sprinkler systems, and fire department pumpers from hydrants and in relay.

A. 2.5 inches (65 mm)
B. 3 inches (76 mm)
C. 3.5 inches (88 mm)
D. 4 inches (100 mm)

2. The _______ hose load is achieved by standing the hose on its edge, then placing the next fold on its edge, and so on.

A. flat
B. forward
C. reverse
D. accordion

3. _______ is the pressure that exists at a given point under normal distribution system conditions measured at the residual hydrant with no hydrants flowing.

A. Static pressure
B. Flow pressure
C. Residual pressure
D. Elevation pressure

4. The _______ are the largest-diameter pipes in a water distribution system, carrying the greatest amounts of water.

A. distributors
B. primary feeders
C. secondary feeders
D. large-diameter hose

5. A _______ is a device that has a 2½-inch (65 mm) inlet and a 2½-inch (65 mm) outlet in addition to two 1½-inch (38 mm) outlets. It is used to supply many hoses from one source.

A. water thief
B. gated wye
C. four-way hydrant valve
D. dump valve

6. A _______ is a hose adaptor that is used to join two female hose couplings.

A. hose jacket
B. double-male
C. double-female
D. Siamese connection

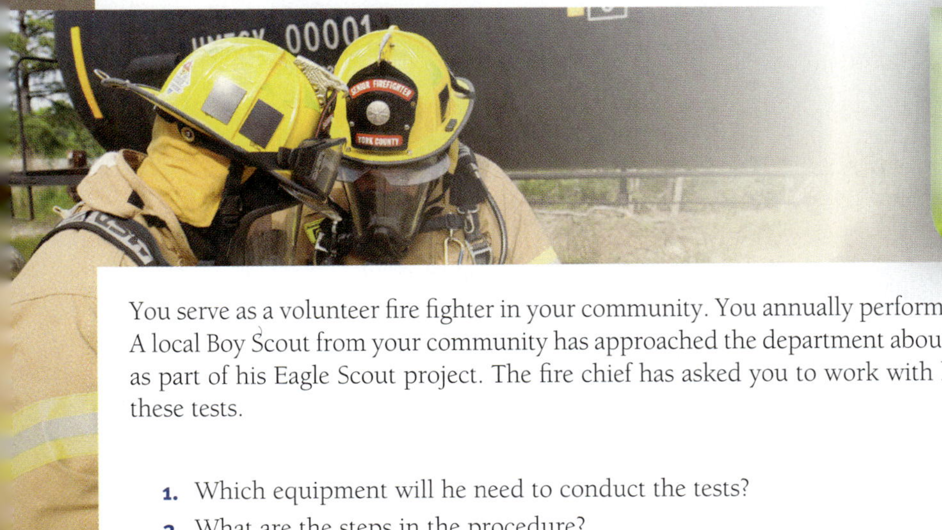

FIRE FIGHTER II *in action*

You serve as a volunteer fire fighter in your community. You annually perform hydrant test and inspection in your community. A local Boy Scout from your community has approached the department about assisting with testing all of the hydrants in town as part of his Eagle Scout project. The fire chief has asked you to work with him on how to properly conduct and document these tests.

1. Which equipment will he need to conduct the tests?
2. What are the steps in the procedure?
3. What does he need to develop as a record of his tests?
4. Is this permitted by your insurance company? If so, to what level can he assist?

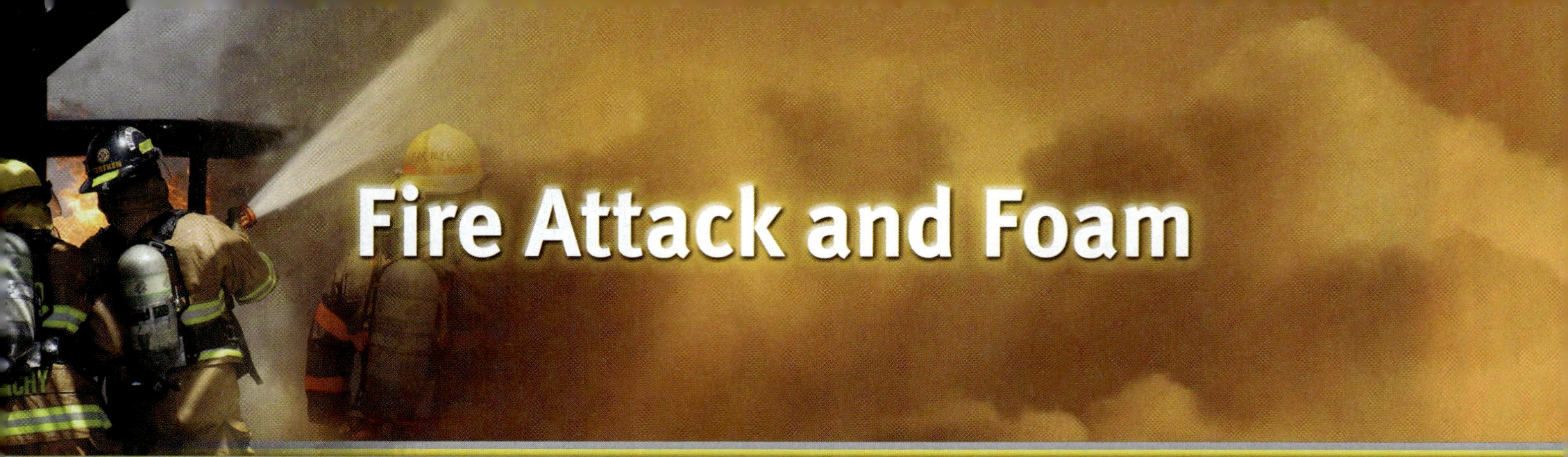

Fire Attack and Foam

Fire Fighter I

Knowledge Objectives

After studying this chapter, you will be able to:

- List the standard sizes of attack hoses. (NFPA 5.3.10.A, 5.3.13.A, p 552–553)
- Describe the characteristics of booster hose. (NFPA 5.3.10.A, p 553)
- Describe the general procedures that are followed during attack line evolutions. (NFPA 5.3.10.A, p 553)
- Describe the types of loads used to organize attack hose. (p 553–559)
- Describe the procedures to follow when advancing attack hose. (NFPA 5.3.10.A, p 559–562)
- Describe where to position an attack line for a transitional fire attack, an offensive interior fire attack, and a defensive fire attack. (p 559)
- Describe the uses of large diameter hose as an attack line. (p 552)
- Describe how to extend an attack line. (NFPA 5.3.10.A, p 566–567)
- Describe how to advance an attack line from a standpipe. (NFPA 5.3.10.A, p 567)
- Describe how to replace a defective section of attack hose. (NFPA 5.5.2.A,p 567–568)
- List the three classifications of nozzles. (NFPA 5.3.10.A, p 568–570)
- Describe the characteristics of smooth-bore nozzles. (NFPA 5.3.10.A, p 569)
- Describe the characteristics of fog-stream nozzles. (NFPA 5.3.10.A, p 569–570)
- List the three types of fog-stream nozzles. (NFPA 5.3.10.A, p 570)
- Describe the specialized nozzles that may be used during fire suppression operations. (NFPA 5.3.10.A, p 570–572)
- Describe how to maintain nozzles to ensure proper operation. (p 572)
- Describe how to inspect nozzles. (p 572)

Skills Objectives

After studying this chapter, you will be able to perform the following skills:

- Perform a minuteman hose load. (NFPA 5.5.2.B, p 554–555)
- Advance a minuteman hose load. (NFPA 5.3.10.B, p 554, 556)
- Perform a preconnected flat load. (NFPA 5.5.2.B, p 556–557)
- Advance a preconnected flat hose load. (NFPA 5.3.10.B, p 557–558)
- Perform a triple-layer hose load. (NFPA 5.5.2.B, p 558, 560)
- Advance a triple-layer hose load. (NFPA 5.3.10.B, p 558–559, 561)
- Unload and advance wyed lines. (NFPA 5.3.10.B, p 559, 562)
- Advance a hose line up a stairway. (NFPA 5.3.10.B, p 563)
- Advance a hose line down a stairway. (NFPA 5.3.10.B, p 563–564)
- Advance an uncharged hose line up a ladder. (NFPA 5.3.10.B, p 563, 566)
- Operate a hose stream from a ladder. (NFPA 5.3.10.B, p 566–567)
- Connect and advance an attack line from a standpipe outlet. (NFPA 5.3.15.B, p 567–568)
- Replace a defective section of hose. (NFPA 5.3.10.B, p 568)
- Operate a smooth-bore nozzle. (NFPA 5.3.10.B, p 569, 571)
- Operate a fog-stream nozzle. (NFPA 5.3.10.B, p 570, 572)

CHAPTER 17

Fire Fighter II — FFII

Knowledge Objectives

After studying this chapter, you will be able to:

- Describe how foam suppresses fire. (NFPA 6.3, 6.3.1.A, p 573)
- Describe the characteristics of Class A foam. (NFPA 6.3.1.A, p 573–574)
- Describe the characteristics of Class B foam. (NFPA 6.3.1.A, p 574)
- List the major categories of Class A foam concentrate. (NFPA 6.3.1.A, p 574)
- Describe the characteristics of compressed air foam. (NFPA 6.3.1.A, p 575)
- List the major categories of Class B foam concentrate. (NFPA 6.3.1.A, p 575)
- Describe the characteristics of protein foam. (NFPA 6.3.1.A, p 575)
- Describe the characteristics of fluoroprotein foam. (NFPA 6.3.1.A, p 575)
- Describe the characteristics of aqueous film-forming foam. (NFPA 6.3.1.A, p 575)
- Describe the characteristics of alcohol-resistant foam. (NFPA 6.3.1.A, p 575)
- Describe how foam proportioner equipment works with foam concentrate to produce foam. (NFPA 6.3.1.A, p 575–576)
- Describe how foam is applied to fires. (NFPA 6.3.1.A, p 576–578)

Skills Objectives

After studying this chapter, you will be able to perform the following skills:

- Place an eductor foam line in service. (NFPA 6.3.1.B, p 576)
- Apply foam using the sweep method. (NFPA 6.3.1.B, p 577)
- Apply foam using the bankshot method. (NFPA 6.3.1.B, p 577–578)
- Apply foam using the rain-down method. (NFPA 6.3.1.B, p 577)

Additional NFPA Standards

- NFPA 1962, *Standard for the Inspection, Care, and Use of Fire Hose, Couplings, and Nozzles and the Service Testing of Fire Hose*

You Are the Fire Fighter

"It's a worker!" is the first thought you have as you step out of the cab of the engine at a small ranch-style house that is set back from the street about 125 feet (43 meters). It appears to be a room and contents fire. You tighten your self-contained breathing apparatus (SCBA) waist strap as your captain directs you to pull a preconnect. He says you will be doing a direct attack and to make an entry from the rear of the structure. You step up to the engine and come face to face with three preconnected lines of various lengths from which to choose. Two are 1¾ inch (45 millimeters). One has a smooth-bore nozzle and the other has a combination nozzle. The other hose line is 2½ inches (65 millimeters) with a combination nozzle on it. It is up to you to decide which hose is the proper one to pull.

1. What size and length of hose would you pull for this fire?
2. What type of nozzle is needed for this type of fire?
3. Why does it not matter about the type of hose load that is used?

Introduction

Fire hoses are used for two main purposes: as supply hoses and as attack hoses. This chapter deals with the use of attack lines, nozzles, and foam. Attack lines (attack hoses) are used to discharge water from an attack engine onto the fire. Most attack hoses carry water directly from the attack engine to a nozzle that is used to direct the water onto the fire. Attack hoses usually operate at higher pressures than do supply lines.

Nozzles are attached to the discharge end of attack lines to give fire streams shape and direction. Without a nozzle, the water discharged from the end of an attack line would reach only a short distance. The types of nozzles used with attack lines and how to use them are discussed in this chapter.

Not all fires can be suppressed with water. Foam can be a valuable tool in suppressing certain types of fires. This chapter discusses the types of foam used in the fire service and how to apply them.

FIGURE 17-1 Attack hoses are used during fire suppression operations.

Attack Hose

The hoses most commonly used to attack interior fires are either 1½ inches (38 millimeters) or 1¾ inches (45 millimeters) in diameter FIGURE 17-1. Some fire departments also use 2½-inch (64-mm) attack lines. Each section of attack hose is usually 50 feet (15 meters) long.

Medium-diameter hose (MDH) with a diameter of 2½ inches (65 millimeters) or 3 inches (76 millimeters) can be used as either supply lines or attack lines. Two-and-a-half-inch (64-mm) attack handlines are used to extinguish larger fires. When used as an attack hose, the 3-inch (76-mm) size is more often used to deliver water to a master stream device or a fire department connection. Both 2½-inch (65-mm) and 3-inch (76-mm) hoses usually come in 50-foot (15-meter) lengths.

Large-diameter hose (LDH) has a limited role as a fire attack tool. Large-diameter hose is used to supply portable monitors or is mounted on an aerial ladder. These commonly produce a water flow between 350 gallons per minute (1591 lpm) and 1500 gallons per minute (6818 lpm). Large-diameter hose is used to supply master stream devices such as portable monitors or is mounted on an aerial ladder.

Attack hose must withstand high pressure and is designed to be used in a fire environment where it can be subjected to high temperatures. Attack hose must be tested annually at a pressure of at least 300 pounds per square inch (psi) (2068 kilopascals [kPa]) and is intended to be used at pressures up to 275 psi (1896 kPa).

Because attack hose is designed to be used for fire suppression, it may be exposed to heat and flames, hot embers, broken glass, sharp objects, and many other potentially damaging conditions. For this reason, attack hose must be tough, yet flexible and light in weight. Attack lines can be either double-jacket or rubber-covered construction.

Sizes of Attack Lines

1½-Inch (38 mm) and 1¾-Inch (45 mm) Attack Hose

Most fire departments use either 1½-inch (38-mm) or 1¾-inch (45-mm) hose as the primary attack line for most fires. Both sizes of hose use the same 1½-inch (38-mm) couplings. This attack hose is the type used most often during basic fire training. Handlines of this size can usually be operated by one fire fighter, although a second person on the line makes it much easier to advance and control the hose. This hose is often stored on fire apparatus as a preconnected attack line in lengths ranging from 150 feet to 350 feet (45 to 106 meters), ready for immediate use.

The primary difference between 1½-inch (38-mm) and 1¾-inch (45-mm) hose is the amount of water that can flow though the hose. Depending on the pressure in the hose and the type of nozzle used, a 1½-inch (38-mm) hose can generally flow between 60 and 125 gpm (273 and 568 lpm). An equivalent 1¾-inch (45-mm) hose can flow between 120 and 180 gpm (946 and 818 lpm). This difference is important because the amount of fire that can be extinguished is directly related to the amount of water that is applied to it. A 1¾-inch (45-mm) hose can deliver much more water and is only slightly heavier and more difficult to advance than a 1½-inch (38-mm) hose line.

2½-Inch (65 mm) Attack Hose

A 2½-inch (65-mm) hose is used as an attack line for fires that are too large to be controlled by a 1½-inch (38-mm) or 1¾-inch (45-mm) hose line. A 2½-inch (65-mm) handline hose is generally considered to deliver a flow of approximately 250 gpm (946 lpm). It takes at least two fire fighters to safely control a 2½-inch (65-mm) handline hose owing to the weight of the hose and the water and the nozzle reaction force. A 50-foot (15-meter) length of dry 2½-inch (65-mm) hose weighs about 30 pounds (14 kilograms). When the hose is charged and filled with water, however, it can weigh as much as 140 pounds (64 kilograms) per length. A 2½-inch (65-mm) hose is most often used for interior attacks in large buildings and for defensive exterior attacks.

Higher flows—up to approximately 350 gpm (1591 lpm)—can be achieved with higher pressures and larger nozzles. It is difficult to operate a handline hose at these high flow rates, however. Such flows are more likely to be used to supply a master stream device.

Booster Hose

A booster hose (booster line) is usually carried on a hose reel that holds 150 feet or 200 feet (45 to 61 meters) of rubber hose. Booster hose contains a steel wire that gives it a rigid shape. This rigid shape allows the hose to flow water without pulling all of the hose off the reel. Booster hose is light in weight and can be advanced quickly by one person.

The disadvantage of booster hose is its limited flow. The normal flow from a 1-inch (38-mm) booster hose is in the range of 40 to 50 gpm (182 to 227 lpm), which is not an adequate flow for extinguishing structure fires. As a consequence, the use of booster hose is typically limited to small outdoor fires and trash dumpster fires. This type of hose should not be used for structural firefighting.

Forestry Lines

Small-diameter hose lines, typically 1 inch (25-mm) or 1½ inches (38-mm) in diameter, are often used for fighting wildland and ground fires. Large volumes of water are not often needed for these fires, and the small diameter provides much better maneuverability through brush and trees. These hose lines are sometimes extended for hundreds of feet.

Attack Line Evolutions

Attack lines are used to deliver water from an attack engine to a nozzle, which discharges the water onto the fire. Attack line evolutions are standard methods of working with attack lines to accomplish different objectives in a variety of situations. Most fire departments set up their equipment and conduct regular training so that fire fighters will be prepared to perform a set of standard hose evolutions. Hose evolutions involve specific actions that are assigned to specific members of a crew, depending on their riding positions on the apparatus. Every fire fighter should know how to perform all of the standard evolutions quickly and proficiently. When a fire officer calls for a particular evolution to be performed, each crew member should know exactly what to do.

Attack lines are usually stretched from an attack engine to the fire. The attack engine is usually positioned close to the fire, and attack lines are stretched manually by fire fighters. In some situations, an attack engine will drop an attack line at the fire and drive from the fire to a hydrant or water source. This procedure is similar to the reverse lay evolution for supply lines, except that the hose will be used as an attack line.

Most engines are equipped with preconnected attack lines, which provide a predetermined length of attack hose that is already equipped with a nozzle and connected to a pump discharge outlet. An additional supply of attack hose that is not preconnected is usually carried in another hose bed or compartment. To create an attack line with this hose, the desired length of hose is removed from the bed; a coupling is then disconnected and attached to a pump discharge outlet. This hose can also be used to extend a preconnected attack line or to attach to a wye or a water thief.

Attack hose is loaded in such a manner that it can be quickly and easily deployed. There are many ways to load attack lines into a hose bed, and this section presents only a few of the most common hose loads. Your department might use a variation of one of these techniques. It is important for you to master the hose loads used by your department.

FIRE FIGHTER II Tips **FFII**

When advancing a hose line into a building, the officer should be the second or third person in the "stack." This allows him or her to monitor the progress of the fire attack as well as provide direction to the fire fighter operating the nozzle. Being positioned at the front also allows the officer to help other fire fighters control hose movement, fix kinks, control the door, and thereby control airflow reaching the fire. The officer also remains mobile to continue to check fire and building conditions.

Preconnected Attack Lines

Preconnected hose lines are intended for immediate use as attack lines. A preconnected hose line has a predetermined length of hose with the nozzle already attached and is connected to a discharge outlet on the fire engine. The most commonly used attack lines are 1¾-inch (45-mm) hoses, generally ranging from 150 feet to 250 feet (45.7 to 76 meters) in length. Many engines are also equipped with preconnected 2½-inch (65-mm) hose lines to enable them to make a quick attack on larger fires.

Attack lines should be loaded in the hose bed in a manner that ensures they can be quickly stretched from the attack engine to the fire. It should be possible for one or two fire fighters to remove the hose quickly from the hose bed and advance the hose to the fire. In many cases, an attack line is stretched in two stages. First, the hose is laid out from the attack engine to the building entrance or to the location where the fire attack will begin. When performing a transitional attack, the hose should be positioned in a location to cool the fire environment from the safest location, where the hose stream will be directed as a straight or solid stream, an indirect attack, to the ceiling of the room, as close to the body of the fire as possible. This straight or solid stream should be directed at the ceiling, simulating a sprinkler system. The stream should remain steady through the top of the window frame to avoid sealing the window outlet. This permits the higher pressure to continue to use the window as an exhaust. When performing an interior fire attack, the hose is advanced into the building to reach the fire. Extra hose should be flaked out at the entrance to the fire building.

Whichever hose load technique is used, the hose should not become tangled as it is being removed from the hose bed and advanced. Laying out the hose should not require multiple trips between the engine and the fire attack point. It should be easy to lay the hose around obstacles and corners. If additional fire fighters are available, they can remain behind to help move hose line around any obstacles. It should also be possible to repack the hose quickly and with minimal personnel. There is no perfect hose load that works well for every situation. The three most common hose loads for preconnected attack lines are the minuteman load, the flat load, and the triple-layer load. Variations in circumstances can make one type of hose load preferable for your community.

Because they face different types of fire situations, most departments load attack lines of different lengths. An engine could be equipped with a 150-foot (45.7-meter) preconnected line and a 250-foot (76.2-meter) preconnected line, for example. Fire fighters should pull an attack line that is long enough to reach the fire, but not so long that an excess of hose might slow down the operation and become tangled.

Preconnected hose lines can be placed in several different locations on a fire engine. For example, a section of a divided hose bed at the rear of the apparatus can be loaded with a preconnected attack line. Transverse hose beds are installed above the pump on many engines and loaded so that the hose can be pulled off from either side of the apparatus. Preconnected lines can be loaded into special trays that are mounted on the side of fire apparatus.

Many engines include a special compartment in the front bumper that can store a short preconnected hose line. This line is often used for vehicle fires and dumpster fires, where the apparatus can drive up close to the incident and a longer hose line is not needed.

Booster hose is another type of preconnected attack line. Booster reels holding ¾-inch (19-mm) or 1-inch (25-mm) hose can be mounted in a variety of locations on fire apparatus. Booster lines should not be used for interior fires, but may be useful for an initial exterior attack.

Minuteman Load

To prepare a 200-foot (61-meter) minuteman hose load, connect the female coupling to the preconnect discharge. Lay the first two sections in the hose bed using a flat load. After the first loop of hose, make short pulling handles at the end of the hose bed. Leave the male end of the second section of hose at the side of the hose compartment. Attach the remaining two sections of hose together, and place the nozzle on the male end of the last section of hose. Place these sections of hose in the hose bed, starting with the nozzle end. Leave long pulling loops in the first loops attached to the nozzle. When you reach the female end of the last section of hose, attach it to the male end of the second section of hose placed into the hose bed.

> **FIRE FIGHTER Tips**
>
> When laying out an attack line, with threaded couplings, you might find that the wrong end of the hose is on top of the hose bed. Double-male connectors and double-female connectors enable you to attach a male coupling to a male discharge or to attach a female coupling to a female coupling, respectively. A set of adaptors (one double-male and one double-female) should be easily accessible for these situations. Some departments place a set of adaptors on the end of the hose for this purpose.

To perform a minuteman load, follow the steps in **SKILL DRILL 17-1**:

1. Connect the female end of the first length of hose to the preconnect discharge outlet. (**STEP 1**)
2. Flat load the hose even to the edges of the hose bed. At the second fold, make a loop or ear. This loop will create a handle to be used when advancing the hose load. (**STEP 2**)
3. Flat load the rest of the 100 feet (30 meters) of hose. Place the male hose coupling to the side of the hose bed. (**STEP 3**)
4. Couple the remaining two hose sections and attach the nozzle. (**STEP 4**)
5. Place the nozzle in the hose bed facing outward. Feed one layer of hose under the nozzle. Wrap the first layer of hose around the tip of the nozzle. (**STEP 5**)
6. Continue flat loading the remaining two sections of hose. (**STEP 6**)
7. Connect the female coupling from the last section of hose to the male coupling from the first section of hose. (**STEP 7**)

To advance a minuteman hose load, follow the steps in **SKILL DRILL 17-2**:

1. Grasp the nozzle and the folds on top of it.
2. Pull this top part of the load approximately one-third out of the hose bed. (**STEP 1**)
3. Turn away from the hose bed, and place this top part of the hose load on your shoulder. Walk away from the apparatus until the rest of the top section of hose drops from the hose bed. (**STEP 2**)
4. Turn back around toward the hose bed and grasp the loop or ear from the first section of hose. (**STEP 3**)
5. Walk away from the apparatus until all hose is clear from the hose bed.
6. Continue walking away, allowing the rest of the hose to deploy from the top of the load on your shoulder. (**STEP 4**)

SKILL DRILL 17-1 Performing a Minuteman Hose Load

(Fire Fighter I, NFPA 5.5.2)

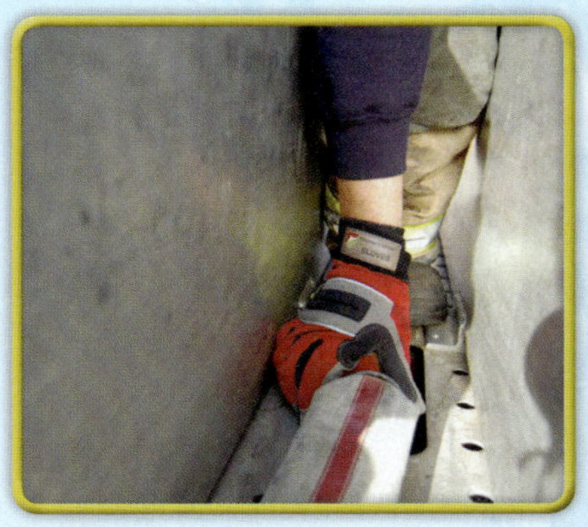

1 Connect the female end of the first length of hose to the preconnect discharge outlet.

2 Flat load the hose even to the edges of the hose bed. At the second fold, make a loop or ear. This loop will create a handle to be used when advancing the hose load.

3 Flat load the rest of the 100 feet (30 meters) of hose. Place the male hose coupling to the side of the hose bed.

4 Couple the remaining two hose sections and attach the nozzle.

5 Place the nozzle in the hose bed facing outward. Feed one layer of hose under the nozzle. Wrap the first layer of hose around the tip of the nozzle.

6 Continue flat loading the remaining two sections of hose.

7 Connect the female coupling from the last section of hose to the male coupling from the first section of hose.

SKILL DRILL 17-2 Advancing a Minuteman Hose Load

(Fire Fighter I, NFPA 5.3.10)

1. Grasp the nozzle and the folds on top of it. Pull this top part of the load approximately one-third out of the hose bed.

2. Turn away from the hose bed, and place this top part of the hose load on your shoulder. Walk away from the apparatus until the rest of the top section of hose drops from the hose bed.

3. Turn back around toward the hose bed and grasp the loop or ear from the first section of hose.

4. Walk away from the apparatus until all hose is clear from the hose bed. Continue walking away, allowing the rest of the hose to deploy from the top of the load on your shoulder.

Fire Fighter Safety Tips

When loading hose on an apparatus, always use caution in climbing up and down on the apparatus. If you are loading hose at a fire scene, watch out for wet, slippery surfaces, ice, or other hazards. Also, wet hose can be heavy and you might need to reach, stretch, or lift the hose to get it into the hose bed. Use caution!

Preconnected Flat Load

To prepare the preconnected flat load, attach the female end of the hose to the preconnect discharge. Begin placing the hose flat in the hose bed. When about one-third of the hose is in the bed, make an 8-inch (203-mm) loop at the end of the hose bed. This loop will be used as a pulling handle. When two-thirds of the hose is loaded, make a second pulling loop about twice the size of the first loop. Finish loading the hose, attach the nozzle, and place it on top of the hose bed. The preconnected flat load is now ready for use.

To perform a preconnected flat hose load, follow the steps in SKILL DRILL 17-3:

1. Attach the female end of the hose to the preconnect discharge. (STEP 1)
2. Begin flat loading the hose in the hose bed. (STEP 2)

SKILL DRILL 17-3 Performing a Preconnected Flat Load

(Fire Fighter I, NFPA 5.5.2)

1 Attach the female end of the hose to the preconnect discharge.

2 Begin flat loading the hose in the hose bed.

3 After the first layer is loaded, make the first fold even with the edge of the hose bed. On the second layer, make a loop or ear.

4 Flat load the remainder of the hose in the hose bed. Attach the nozzle to the male coupling at the end of the hose.

3. After the first layer is loaded, make the first fold even with the edge of the hose bed. On the second layer, make a loop or ear. (STEP 3)
4. Flat load the remainder of the hose in the hose bed. Attach the nozzle to the male coupling at the end of the hose. (STEP 4)

To advance a preconnected flat hose load, follow the steps in SKILL DRILL 17-4:

1. Grasp the top section of hose and place it over your shoulder. (STEP 1)
2. Walk away from the apparatus until the top load drops onto your shoulder. (STEP 2)
3. Turn back to the hose bed and grasp the lower loop or ear. Pull the remainder of the hose from the hose bed. (STEP 3)
4. Turn away from the apparatus and continue walking, letting the hose lay out off of your shoulder. (STEP 4)

FIRE FIGHTER Tips

When referring to the hose bed, the end closest to the cab is called the front of the hose bed. The end closest to the tailboard is called the rear of the hose bed.

Triple-Layer Load

To prepare the triple-layer load, attach the female end of the hose to the preconnect discharge. Connect the sections of hose together. Extend the hose directly from the hose bed. Pick up the hose two-thirds of the distance from the discharge to the

SKILL DRILL 17-4 Advancing a Preconnected Flat Hose Lead

(Fire Fighter I, NFPA 5.3.10)

1 Grasp the top section of hose and place it over your shoulder.

2 Walk away from the apparatus until the top load drops onto your shoulder.

3 Turn back to the hose bed and grasp the lower loop or ear. Pull the remainder of the hose from the hose bed.

4 Turn away from the apparatus and continue walking, letting the hose lay out off of your shoulder.

hose nozzle. Carry the hose back to the apparatus, forming a three-layer loop. Pick up the entire length of folded hose (this will take several people). Lay the tripled-folded hose in the hose bed in an S shape with the nozzle on top.

To perform a triple-layer hose load, follow the steps in SKILL DRILL 17-5:

1. Attach the female end of the hose to the preconnect discharge. (STEP 1)
2. Connect the sections of hose together. (STEP 2)
3. Extend the hose directly from the hose bed. Pick up the hose two-thirds of the distance from the discharge to the hose nozzle. (STEP 3)
4. Carry the hose back to the apparatus, forming a three-layer loop. (STEP 4)
5. Pick up the entire length of folded hose. (This will take several people.) (STEP 5)
6. Lay the tripled-folded hose in the hose bed in an S shape with the nozzle on top. (STEP 6)

FIRE FIGHTER Tips

The triple-layer load requires several fire fighters and must be practiced often.

To advance a triple-layer hose load, follow the steps in SKILL DRILL 17-6:

1. Grasp the nozzle and the top fold. (STEP 1)
2. Turn away from the hose bed and place the hose on the shoulder. (STEP 2)
3. Walk away from the vehicle until the entire load is out of the bed. (STEP 3)

4. When the load is out of the bed, drop the fold. (STEP 4)
5. Extend the nozzle the remaining distance. (STEP 5)

Wyed Lines

To reach a fire that is some distance from the engine, it can be necessary first to advance a larger-diameter line, such as a 2½-inch (65-mm) hose line, and then split it into two 1¾-inch (45-mm) attack lines. This is accomplished by attaching a gated wye or a water thief to the end of the 2½-inch (65-mm) line, and then attaching the two attack lines to the gated outlets.

To unload and advance wyed lines, follow the steps in SKILL DRILL 17-7:

1. Grasp one of the attack lines and pull it from the bed. (STEP 1)
2. Pull the second attack line from the bed and place it far enough from the first line so that you can walk between the hose lines. This will keep the lines from becoming entangled. (STEP 2)
3. Grasp the gated wye and pull it from the bed.
4. The apparatus will deploy the remaining hose from the wye back to a water source. Place the gated wye so that one attack line is on the other side. (STEP 3)
5. The individual attack lines can now be extended to the desired positions. (STEP 4)

FIRE FIGHTER Tips

To prepare to advance the hose, form an upright loop about 4 feet (1.3 meters) in diameter outside the door to the fire building. If more hose is needed, one person can roll this loop toward the nozzle. This will yield several additional feet of hose at the nozzle.

Advancing Attack Lines

Attack lines are used for three different types of evolutions. These are transitional fire attack, offensive interior fire attack, and defensive fire attack. A transitional fire attack is an offensive exterior attack that occurs in coordination with entry, search, and tactical ventilation. When performing a transitional attack, the hose should be positioned in a safe location as close to the fire as possible to cool the fire environment. The hose stream should be directed as a straight or solid stream, an indirect attack, to the ceiling of the room, as close to the body of the fire as possible. This straight stream acts somewhat like a sprinkler system. Hold the straight stream steady or sweep side to side through the top of the window frame to avoid sealing the window outlet; this permits the higher pressure to continue to use the window as an exhaust opening for smoke and hot gases. The hose should be flaked out so it will not become tangled when the hose line is charged. The transitional attack line helps to temporarily darken the fire down, cool the fire gases, and reduce the production of hot gases. This helps to provide a safer environment for the fire fighters entering the building and makes the environment more survivable for an occupant trapped in the fire building.

An offensive interior fire attack requires the hose line to be placed in position in front of the door where the firefighters will make entry into the fire building. A defensive fire attack is used when it is unsafe to enter the fire building and the goal of fire fighters is to contain the fire to the building of origin. The placement of the hose line will be different for each of these evolutions. It is important to know which type of fire attack is to be used before starting to advance a hose line. More information about these evolutions is presented in Chapter 22, *Fire Suppression*.

To perform an interior fire attack, an attack line is usually advanced in two stages. The first stage involves laying out the hose to the building entrance. The second stage is to advance the line into the building to the location of the fire, where it will be operated.

When the attack line has been laid out to the entry point, the extra hose that will be advanced into the building should be flaked out in a serpentine pattern, ensuring that it will not become tangled when it is charged. Specifically, the hose should be flaked out with lengths of hose running parallel to the front of the fire building so that it can be easily advanced into the building FIGURE 17-2. It should also be set back from the doorway so that it does not obstruct the entry and exit path.

Make sure that you flake out the hose before it is charged with water. Once the line is charged, the hose becomes much more difficult to maneuver and advance. It is also important to verify that the hose length will reach the location where the line is needed inside the building.

While fire fighters are flaking out the hose and preparing to enter the building, the company officer will be completing the size-up. Other members of the team will be carrying out other tasks to support the operation. For example, they might be forcing entry, getting into position for ventilation, establishing a water supply, and performing search and rescue. All of these tasks must be performed in sequence to maintain a safe environment and to extinguish the fire efficiently.

Once the hose is flaked out, signal the driver/operator to charge the line. Open the nozzle slowly to bleed out any trapped air and to make sure the hose is operating properly. If you are using an adjustable nozzle, make sure the nozzle is set to deliver the appropriate stream. For an interior attack, this setting will usually be a straight stream pattern. Once this is done, slowly close the nozzle.

FIGURE 17-2 Hose should be flaked out in a serpentine pattern outside the building entrance.

SKILL DRILL 17-5 Performing a Triple-Layer Hose Load

(Fire Fighter I, NFPA 5.5.2)

1. Attach the female end of the hose to the preconnect discharge.

2. Connect the sections of hose together.

3. Extend the hose directly from the hose bed. Pick up the hose two-thirds of the distance from the discharge to the hose nozzle.

4. Carry the hose back to the apparatus, forming a three-layer loop.

5. Pick up the entire length of folded hose.

6. Lay the triple-folded hose in the hose bed in an S shape with the nozzle on top.

SKILL DRILL 17-6 Advancing a Triple-Layer Hose Load

(Fire Fighter I, NFPA 5.3.10)

1 Grasp the nozzle and the top fold.

2 Turn away from the hose bed and place the hose on the shoulder.

3 Walk away from the vehicle until the entire load is out of the bed.

4 When the load is out of the bed, drop the fold.

5 Extend the nozzle the remaining distance.

SKILL DRILL 17-7 Unloading and Advancing Wyed Lines

(Fire Fighter I, NFPA 5.3.10)

1 Grasp one of the attack lines and pull it from the bed.

2 Pull the second attack line from the bed and place it far enough from the first line so that you can walk between the hose lines. This will keep the lines from becoming entangled.

3 Grasp the gated wye and pull it from the bed. The apparatus will deploy the remaining hose from the wye back to a water source. Place the gated wye so that one attack line is on the other side.

4 The individual attack lines can now be extended.

Quickly recheck all parts of your personal protective equipment (PPE). Make sure your coat is fastened and your collar is turned up and fastened in front. Check your gloves and make sure you have a hand light. Check your partner's equipment and have your partner check your equipment. If you have time, catch your breath while you are breathing ambient air. Be ready to start breathing air from your SCBA and to advance the charged hose line as soon as your officer directs you to do so.

When you are given the command to advance the hose, keep safety as your number one priority. Make sure the other members of the nozzle team are ready. Do not stand in front of the door as it is opened: You do not know what might happen.

As you move inside the building, stay low to avoid the greatest amount of heat and smoke. If you cannot see because of the dense smoke, use your hands to feel the pathway in front of you. Feel in front of you so that you do not fall into a hole or other opening. Look for the glow of fire, and check for the sensation of heat coming through your face piece. Communicate with the other members of the nozzle team as you advance.

As you advance the hose line, you need to have enough hose to enable you to move forward. Ideally, a hose line crew consists of at least three members at the nozzle and a fourth member at the door. As resistance is encountered in advancing the line, the fire fighter at the nozzle can help to pull more hose, while the fire fighter at the door is responsible for feeding more hose into the building and controlling the door, thereby controlling the airflow into the building. This door control will limit the ability of the fire to grow rapidly as fire fighters get into position to apply the hose stream.

Charged hose lines are not easy to advance through a house or other building. For example, the hose line can become caught on a door frame. It is only with good teamwork that efficient hose line advancement can occur.

Advancing an Attack Line up a Stairway

When advancing a hose line up stairs, arrange to have an adequate amount of extra hose close to the bottom of the stairs. Utilizing a shoulder carry with the minuteman hose load allows for easy advancement. Make sure all members of the team are ready to move on command. It is difficult to move a charged hose line up a set of stairs while flowing water through the nozzle; shutting down the hose line while you are moving up the stairs will often allow you to get to the top of the stairs more quickly and safely. Follow the directions of your officer.

To advance an uncharged hose line up a stairway, follow the steps in SKILL DRILL 17-8:

1. Use a shoulder carry to advance up the stairs. (STEP 1)
2. When ascending the stairway, lay the hose against the outside of the stairs to avoid sharp bends and kinks and to reduce tripping hazards. (STEP 2)
3. Arrange excess hose so that it is available to fire fighters entering the fire floor. (STEP 3)

Advancing an Attack Line down a Stairway

Advancing a charged hose line down a stairway is also difficult and dangerous. The smoke and flames from the fire tend to travel up the stairway of a residential structure. This means fire fighters will be operating in the exhaust portion of the flow path of a basement fire. These operations will increase the risks these operating members will encounter. Due to the risks of operating in a potential flow path, fire officers should consider cooling the fire environment from the safest location possible, the exterior. Applying a straight or solid stream into a basement window prior to entry will cool the fire environment, limit fire damage to the floor system, and reduce the thermal assault to the operating members. All of these improve the operating environment for the fire company. If the application of an exterior line is not possible and an interior attack must be made, get down the stairway and position yourself below the heat and smoke as quickly as you can. Keep as low as possible to avoid the worst of the heat and smoke. When you advance a hose line down a stairway, you have a major advantage on your side: Gravity is working with you to bring the hose line down the stairs. You should never advance toward a fire unless your hose line is charged and ready to flow water. Wearing PPE and SCBA changes your center of gravity. If you try to crawl down a stairway headfirst, you are likely to find yourself tumbling head over heels. Instead, move down the stairway feet first, using your feet to feel for the next step. Move carefully but as quickly as possible to get below the worst of the heat and smoke.

To advance a hose line down a stairway, follow the steps in SKILL DRILL 17-9:

1. Advance forward with the charged hose line. (STEP 1)
2. Descend stairs feet first. (STEP 2)
3. Position fire fighters at areas where hose lines could snag. (STEP 3)

Advancing an Attack Line up a Ladder

If a hose line has to be advanced up a ladder, it should be positioned correctly before the line is charged. To do so, place the hose on the left side of the ladder. Place the hose across your chest, with the nozzle draped over your right shoulder. Climb up the ladder with the uncharged hose line in this position. With the hose

SKILL DRILL 17-8 Advancing a Hose Line up a Stairway
(Fire Fighter I, NFPA 5.3.10)

1 Use a shoulder carry to advance up the stairs.

2 When ascending the stairway, lay the hose against the outside of the stairs to reduce tripping hazards. Avoid sharp bends.

3 Arrange excess hose so that it is available to fire fighters entering the fire floor.

Near Miss REPORT

Report Number: 05-0000281

Synopsis: Early in the afternoon, a commercial structure fire call was sent out. Ten varying units were dispatched.

Event Description: Early in the afternoon, a commercial structure fire call was sent out. Ten varying units were dispatched. While en route, dispatch advised numerous calls. I was on a rescue unit that was second on the scene. The first engine reported nothing showing and entered the structure. With no following radio traffic, the first engine exited the building and reentered with a preconnect. They eventually reported fire on the top floor of the two-story apartment. Rescue was assigned search. When we entered the structure, other units were receiving other assignments or arriving at the scene. The first engine was Fire Attack One. They had advanced no further than the top of the stairs. We, Search, entered and went upstairs to find high heat and blackout conditions. We eventually were trying to do fire attack duties and search. Thermal imaging showed no victims in the room of ignition, but high heat temps were obvious. Once a window was located, I knocked it out. A rush of heat was felt and flames became quickly visible and growing. When we made our way back to the door of the room, the flames were rolling over us. My partner started a combination pattern with the fog nozzle. The fire quickly subsided. At the time, we didn't feel as if we were in much danger. Once we exited the building other fire fighters stated a large fireball rushed out of the window we had broken out to vent. It was then we realized what the conditions were. It was a closer call than I realized.

Lessons Learned: Never underestimate fire conditions, whether they are improving or not. We probably should have worked together on venting and having a nozzle ready simultaneously. Communication before and after we made entry would have been helpful. We also took on too many tasks for a two-man crew.

SKILL DRILL 17-9 Advancing a Hose Line down a Stairway

(Fire Fighter I, NFPA 5.3.10)

1 Advance the charged hose line.

2 Descend stairs feet first. If there is smoke, position yourself underneath it.

3 Position fire fighters at areas where hose lines could snag.

arranged in this way, if the line is mistakenly charged while you are climbing, it will not push you away from the ladder.

Additional fire fighters should pick up the hose about every 25 feet (7 meters) and help to advance it up the ladder. The nozzle is then passed over the top rung of the ladder and into the fire building. Additional hose should be fed up the ladder until sufficient hose is inside the building to reach the fire. The hose should be secured to the ladder with a hose strap or other rope or webbing to keep it from becoming dislodged.

To advance a hose line up a ladder, follow the steps in SKILL DRILL 17-10:

1. If a hose line needs to be advanced up a ladder, it should be advanced before it is charged.
2. Advance the hose line to the ladder. (STEP 1)

VOICES OF EXPERIENCE

As a young fire fighter, I responded to a large industrial fire involving a number of mutual aid engines and trucks due to the intensity, size, and type of the incident. I drove the department squad car behind the two engines from headquarters station. The need for good communication, quick planning, and precise action was paramount.

When my department's engine stopped in front of the fire, the captain issued commands, and the crew pulled our fire department's dual hose lays. It was clear to me that the industrial fire was a prime situation for a dual 2 ½-inch attach streams. I pulled the squad car to the side of the road and ran directly to the wye and loosened it to replace it with a 1 ½-inch nozzle. I couldn't find the nozzle so I sped to another department's engine, but nothing looked familiar. I then realized that the captain had not given the order for the dual 2 ½-inch attack streams. There I was, with an open lay hose lying in the street. I had to quickly replace the wye before water arrived.

Fire fighters should remember that the incident commander makes appropriate decisions and directs the crew based on his or her assessment of the scene. No matter how insightful and skilled fire fighters may be, only the incident commander can be expected to make the call for action. Then, and only then, should fire fighters react. Any action other than that which is called for can add chaos to an already chaotic situation. An open hose, under pressure, in the wake of the need for attack lines, only slows down a process that is designed for speed and efficiency. In this case, both would have been sacrificed and a scramble to recoup would have been precious time wasted.

Roger Land
Lake Superior State University
Sault Ste. Marie, Michigan

SKILL DRILL 17-10 Advancing an Uncharged Hose Line up a Ladder
(Fire Fighter I, NFPA 5.3.10)

1. Advance the hose line to the ladder. Pick up the nozzle; place the hose across the chest, with the nozzle draped over the shoulder. Climb up the ladder with the uncharged hose line.

2. Once the first fire fighter reaches the first fly section of the ladder, a second fire fighter shoulders the hose to assist advancing the hose line up the ladder. To avoid overloading of the ladder, enforce a limit of one fire fighter per fly section. The nozzle is placed over the top rung of the ladder and advanced into the fire area.

3. Additional hose can be fed up the ladder until sufficient hose is in position. The hose can be secured to the ladder with a hose strap to support its weight and keep it from becoming dislodged.

3. Pick up the nozzle; place the hose across the chest, with the nozzle draped over the shoulder.
4. Climb up the ladder with the uncharged hose line.
5. Once the first fire fighter reaches the first fly section of the ladder, a second fire fighter shoulders the hose to assist advancing the hose line up the ladder. To avoid overloading of the ladder, enforce a limit of one fire fighter per fly section.
6. The nozzle is placed over the top rung of the ladder and advanced into the fire area. (STEP 2)
7. Additional hose can be fed up the ladder until sufficient hose is in position.
8. The hose can be secured to the ladder with a hose strap to support its weight and keep it from becoming dislodged. (STEP 3)

Operating an Attack Line from a Ladder

A hose stream can be operated from a ladder and directed into a building through a window or other opening. To operate a fire hose from a ladder, follow the steps in SKILL DRILL 17-11:

1. Climb the ladder with a hose line to the height at which the line will be operated.
2. Apply a leg lock or use a ladder belt. (STEP 1)
3. Place the hose between two rungs and secure the hose to the ladder with a rope hose tool, rope, or piece of webbing. (STEP 2)
4. Carefully operate the hose stream from the ladder. Be careful when opening and closing nozzles and redirecting the stream because of the nozzle backpressure. This force could destabilize the ladder. (STEP 3)

Extending an Attack Line

When choosing a preconnected hose line or assembling an attack line, it is better to have too much hose than not enough. With a hose that is longer than necessary, you can flake out the excess hose. With a hose that is too short, you cannot advance it to the seat of the fire without shutting down the line and taking the time to extend it.

In some circumstances, it can be necessary to extend a hose line by adding additional lengths of hose—for example, if the fire is farther from the apparatus than initially estimated. More hose line could also be needed to reach the burning area or to complete extinguishment.

There are two basic ways to extend a hose line. First, fire fighters can disconnect the hose from the discharge gate on the attack engine and add the extra hose at that location. This method requires advancing the full length of the attack line to take advantage of the extra hose, which could take time and considerable effort.

Alternatively, fire fighters can add the hose to the discharge end of the hose. This addition can be achieved if the nozzle is a breakaway type nozzle that can be separated between the shut-off and the tip. With this type of nozzle, the valve is shut down, the nozzle tip is removed, and the extra hose is attached to the shut-off valve. The nozzle tip can then be installed on the male end of the added hose. This evolution allows fire fighters to lengthen the attack hose without shutting it down at the pump. A standpipe kit can be used to supply the added section of hose.

Advancing an Attack Line from a Standpipe Outlet

Standpipe outlets inside a building are provided for fire fighters to connect attack hose lines. Their availability eliminates the need to advance hose lines all the way from the attack engine on the street outside to an upper floor or a fire deep inside a large area building. In tall buildings, it would be impossible to advance hose lines up the stairways in a reasonable time and to supply sufficient pressure to fight a fire on an upper floor.

Standpipe outlets are often located in stairways, and standard operating procedures (SOPs) generally require attack lines to be connected to an outlet one floor below the fire. The working space in the stairway and around the outlet valve is usually limited. Before opening the door, it is important to flake out the hose line properly so that it will be ready to advance into the fire floor. Before charging the hose line, the hose line should be flaked out on the stairs going up from the fire floor. When the hose line is charged and advanced into the fire floor, gravity will help to move the line forward. This technique is much easier than trying to pull the charged hose line up the stairs.

To connect and advance an attack line from a standpipe outlet, follow the steps in SKILL DRILL 17-12:

1. Carry a standpipe hose bundle to the standpipe connection below the fire. Remove the cap from the standpipe. (STEP 1)
2. Attach the proper adaptor or an appliance such as a gated wye. (STEP 2)
3. Flake the hose up the stairs to the floor above the fire. (STEP 3)
4. Extend the hose to the fire floor and prepare for the fire attack. (STEP 4)

Replacing a Defective Section of Hose

With proper maintenance and testing, the risk of fire hose failure should be low—but even so, it is always a possibility. Fire fighters need to know what to do if a section of hose bursts or develops a major leak while it is being used. Every fire fighter

SKILL DRILL 17-11 Operating a Hose Stream from a Ladder
(Fire Fighter I, NFPA 5.3.10)

1. Climb the ladder with a hose line to the height at which the line will be operated. Apply a leg lock or use a ladder belt.
2. Place the hose and secure the hose to the ladder with a rope hose tool, rope, or piece of webbing.
3. Carefully operate the hose stream from the ladder. Be careful when opening and closing nozzles and redirecting the stream because of the nozzle backpressure. This force could destabilize the ladder.

should know how to quickly replace a length of defective hose and restore the flow.

A burst hose line should be shut down as soon as possible. If the line cannot be shut down at the pump or at a control valve, a hose clamp can be used to stop the flow in an undamaged section of hose upstream from the problem. After the water flow has been shut off, quickly remove the damaged section of hose and replace it with two sections of hose. Using two sections of hose line ensures that the replacement hose is long enough to replace the damaged section.

To replace a hose section, follow the steps in SKILL DRILL 17-13 (Fire Fighter I, NFPA 5.3.10):

1. Shut down or clamp off the damaged hose line.
2. Remove the damaged section of the hose line.
3. Replace the damaged section with two new sections to ensure that the hose line's length will be adequate. Restore the water flow.

Nozzles

Nozzles are attached to the discharge end of attack lines to give fire streams shape and direction. Without a nozzle, the water discharged from the end of a hose would reach for only a short distance. Nozzles are used on all sizes of handlines as well as on master stream devices.

SKILL DRILL 17-12 Connecting and Advancing an Attack Line from a Standpipe Outlet
(Fire Fighter I, NFPA 5.3.15)

1 Carry a standpipe hose bundle to the standpipe connection below the fire. Remove the cap from the standpipe.

2 Attach the proper adaptor or an appliance such as a gated wye.

3 Flake the hose up the stairs to the floor above the fire.

4 Extend the hose to the fire floor and prepare for the fire attack.

Nozzles can be classified into three groups:

- Low-volume nozzles flow 40 gpm (182 lpm) or less. They are primarily used for booster hoses; their use is limited to small outside fires.
- Handline nozzles are used on hose lines ranging from 1½ inches (38 mm) to 2½ inches (65 mm) in diameter. Handline streams usually flow between 60 and 350 gpm (273 and 1591 lpm).
- Master stream nozzles are used on deck guns, portable monitors, and ladder pipes that flow more than 350 gpm (1591 lpm).

Low-volume and handline nozzles incorporate a shut-off valve that is used to control the flow of water. The control valve for a master stream is usually separate from the nozzle itself. All nozzles have some type of device or mechanism to direct the water stream into a certain shape. Some nozzles also incorporate a mechanism that can automatically adjust the flow based on the water's volume and pressure.

Nozzle Shut-Offs

The nozzle shut-off or bale enables the fire fighter at the nozzle to start or stop the flow of water. The handle that controls this valve is called a bale. Some nozzles incorporate a rotary control valve that is operated by rotating the nozzle in one direction to open it and in the opposite direction to shut off the flow of water.

Two different types of nozzles are manufactured for the fire service: smooth-bore nozzles and fog-stream nozzles. Smooth-bore nozzles produce a solid column of water, whereas fog-stream nozzles separate the water into droplets. The size of the water droplets and the discharge pattern can be varied by adjusting the nozzle setting. Nozzles must have an adequate volume of water and an adequate pressure to produce a good fire stream. The volume and pressure requirements vary according to the type and size of nozzle.

FIGURE 17-3 Smooth-bore nozzle.

Smooth-Bore Nozzles

The simplest smooth-bore nozzle consists of a shut-off valve and a smooth-bore tip that gradually decreases the diameter of the stream to a size smaller than the hose diameter **FIGURE 17-3**. Smooth-bore nozzles are manufactured to fit both handlines and master stream devices. Those that are used for master streams and ladder pipes often consist of a set of stacked tips, where each successive tip in the stack has a smaller-diameter opening. Tips can be quickly added or removed to provide the desired stream size. This approach allows different sizes of streams to be produced under different conditions.

Using a smooth-bore nozzle offers several advantages. For example, a smooth-bore nozzle has a longer reach than a combination fog nozzle operating at a straight stream setting. Smooth-bore nozzles also operate at lower pressures than adjustable-stream nozzles do. Most smooth-bore nozzles are designed to operate at 50 psi (344 kPa), whereas adjustable-stream nozzles generally require pressures of 75 to 100 psi (517 to 689 kPa). Lower nozzle pressure makes it easier for a fire fighter to handle the nozzle.

A straight or solid stream extinguishes a fire with less air movement and less disturbance of the thermal layering than does a fog stream, which in turn makes the heat conditions less intense for fire fighters during an interior attack. It is also easier for the operator to see the pathway of a solid stream than a fog stream.

There are also some disadvantages associated with smooth-bore nozzles. These nozzles are not as effective for hydraulic ventilation. A fire fighter cannot change the setting of a smooth-bore nozzle to produce a fog pattern; in contrast, a fog nozzle can be set to produce a straight stream.

To operate a smooth-bore nozzle, follow the steps in **SKILL DRILL 17-14**.

1. Select the desired tip size and attach it to the nozzle shut-off valve. (**STEP 1**)
2. Attain a stable stance. (**STEP 2**)
3. Slowly open the valve, allowing water to flow. (**STEP 3**)
4. Open the valve completely to achieve maximum effectiveness. (Failure to open the valve fully results in reduced flow, depriving the line of the necessary gallons per minute or liters per minute.) (**STEP 4**)
5. Direct the stream to the desired location. (**STEP 5**)

Fog-Stream Nozzles

Fog-stream nozzles produce fine droplets of water **FIGURE 17-4**. These are sometimes called spray nozzles. The advantage of creating these droplets of water is that they absorb heat much more quickly and efficiently than does a solid column of water. A fog nozzle can be used to reduce the room temperature to prevent a flashover. The smaller droplets will absorb more heat per gallon

FIGURE 17-4 Fog-stream nozzle.

than water from a straight stream nozzle. Discharging 1 gallon of water in 100 cubic feet (3 liters in 2 cubic meters) of involved interior space can extinguish a fire in 30 seconds. Fog nozzles can produce a variety of stream patterns, ranging from a straight stream to a narrow fog cone of less than 45 degrees to a wide-angle fog pattern that is close to 90 degrees.

The straight streams produced by fog nozzles have openings in the center. Therefore, a fog nozzle cannot produce a solid stream. The straight stream from a fog-stream nozzle breaks up faster and does not have the reach of a solid stream. A straight stream also is affected more dramatically by wind than a solid stream is.

The use of fog-stream nozzles offers several advantages. First, these nozzles can be used to produce a variety of stream patterns by rotating the tip of the nozzle. In addition, fog streams are effective at absorbing heat and can be used to create a water curtain to protect fire fighters from extreme heat.

Fog nozzles move large volumes of air along with the water. This can be an advantage or a disadvantage, depending on the situation. A fog stream can be used to exhaust smoke and gases through hydraulic ventilation. Unfortunately, this air movement can also result in sudden heat inversion in a room, which then pushes hot steam and gases down onto the fire fighters. If used incorrectly, a fog pattern can push the fire into unaffected areas of a building.

To produce an effective stream, nozzles must be operated at the pressure recommended by the manufacturer. For many years, the standard operating pressure for fog-stream nozzles was 100 psi (689 kPa). In recent years, some manufacturers have produced low-pressure nozzles that are designed to operate at 50 or 75 psi (344 or 517 kPa). The advantage of low-pressure nozzles is that they produce less reaction force, which makes them easier to control and advance. Lower nozzle pressure also decreases the risk that the nozzle will get out of control.

To operate a fog-stream nozzle, follow the steps in SKILL DRILL 17-15:

1. Select the desired nozzle.
2. Attain a stable stance (if standing). (STEP 1)
3. Slowly open the valve, allowing water to flow. (STEP 2)
4. Open the valve completely. (Failure to open the valve fully restricts water flow, reducing the necessary gallons per minute/liters per second rate). (STEP 3)
5. Select the desired water pattern by rotating the bezel of the nozzle.
6. Apply water where needed. (STEP 4)

Three types of fog-stream nozzles are available, with the difference among the types related to the water delivery capability:

- A fixed-gallonage fog nozzle delivers a preset flow at the rated discharge pressure. The nozzle could be designed to flow 30, 60, or 100 gpm (136, 273, or 455 lpm).
- An adjustable-gallonage fog nozzle allows the operator to select a desired flow from several settings by rotating a selector bezel to adjust the size of the opening. For example, a nozzle could have the options of flowing 60, 95, or 125 gpm (273, 432, 568 lpm). Once the setting is chosen, the nozzle delivers the rated flow only as long as the rated pressure is provided at the nozzle.
- An automatic-adjusting fog nozzle can deliver a wide range of flows. As the pressure at the nozzle increases or decreases, its internal spring-loaded piston moves in or out to adjust the size of the opening. The amount of water flowing through the nozzle is adjusted to maintain the rated pressure and produce a good stream. A typical automatic nozzle could have an operating range of 90 to 225 gpm (409 to 1023 lpm) while maintaining 100 psi (689 kPa) discharge pressure.

Other Types of Nozzles

Several other types of nozzles are used for special purposes. If your fire department has other types of specialty nozzles, you need to become proficient in their use and operation. Piercing nozzles, for example, are used to make a hole in automobile sheet metal, aircraft, or building walls or floors so as to extinguish fires behind these surfaces FIGURE 17-5.

Cellar nozzles and Bresnan distributor nozzles are used to fight fires in cellars and other inaccessible places such as attics and cocklofts FIGURE 17-6. These nozzles discharge water in a wide circular pattern as the nozzle is lowered

SKILL DRILL 17-14 Operating a Smooth-Bore Nozzle

(Fire Fighter I, NFPA 5.3.10)

1 Select the desired tip size and attach it to the nozzle shut-off valve.

2 Attain a stable stance (if standing).

3 Slowly open the valve, allowing water to flow.

4 Open the valve completely to achieve maximum effectiveness.

5 Direct the stream to the desired location.

SKILL DRILL 17-15 Operating a Fog-Stream Nozzle
(Fire Fighter I, NFPA 5.3.10)

1 Attain a stable stance (if standing).

2 Slowly open the valve, allow water to flow.

3 Open the valve completely.

4 Select the desired water pattern by rotating the bezel of the nozzle. Apply water where needed.

vertically through a hole into the cellar. They work like a large sprinkler head.

Water curtain nozzles are used to deliver a flat screen of water that then forms a protective sheet (curtain) of water on the surface of an exposed building FIGURE 17-7. The water curtains must be directed onto the exposed building because radiant heat can pass through the air gaps in the water curtain.

Nozzle Maintenance and Inspection

Nozzles should be inspected on a regular basis, along with all of the equipment on every fire department vehicle. In particular, nozzles should be checked after each use before they are placed back on the apparatus. They should be kept clean and clear of debris. Debris inside the nozzle will affect the performance of the nozzle, possibly reducing flow. Dirt and grit can also interfere with the valve operation and prevent the nozzle from opening and closing fully. A light grease on the valve ball keeps it operating smoothly.

On fog nozzles, inspect the fingers on the face of the nozzle. Make sure all fingers are present and that the finger ring can spin freely. Any missing fingers or failure of the ring to spin drastically affects the fog pattern. Any problems noted should be referred to a competent technician for repair.

FIRE FIGHTER Tips

Some nozzles are made so that the tip of the nozzle can be separated from the shut-off valve. Such a breakaway nozzle allows fire fighters to shut off the flow, unscrew the nozzle tip, and then add more lengths of hose to extend the hose line without shutting off the valve at the engine.

FIGURE 17-5 Piercing nozzle.

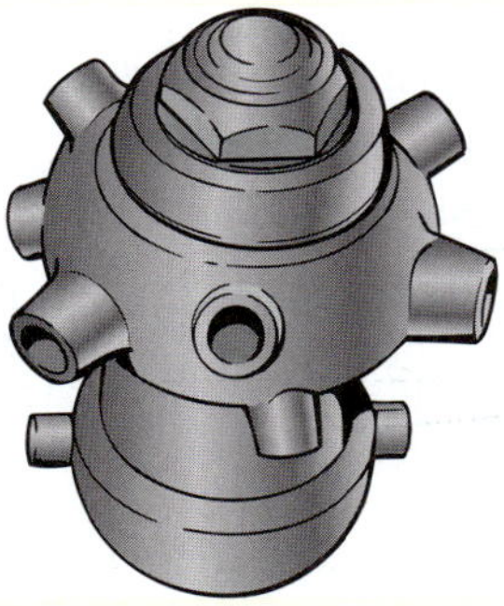
FIGURE 17-6 Bresnan distributor nozzle.

FIRE FIGHTER II Tips FFII

The common names that are used to identify mobile water supply apparatus vary in different parts of the country. In the western parts of the United States where aerial water tankers are used commonly in fighting wildland fires, the term tanker is used to identify aerial mobile water supply apparatus and the term tender is used to describe truck-mounted mobile water supply apparatus. In parts of the country where aerial apparatus is not used commonly, the term tanker is used to refer to truck-mounted mobile water supply apparatus. Be sure you understand the terminology used in your department to avoid confusion on the fire scene.

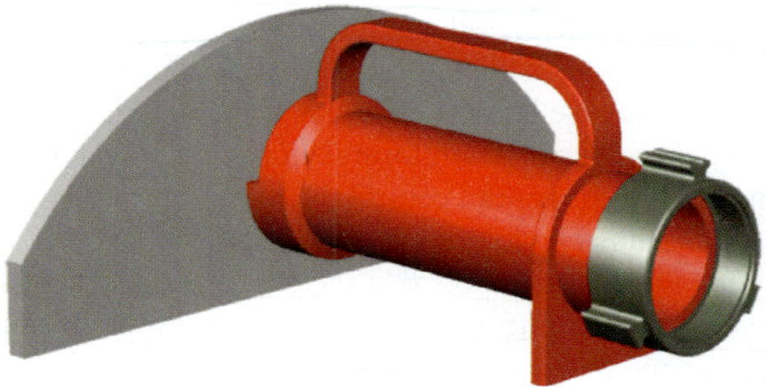
FIGURE 17-7 Water curtain nozzle.

FIRE FIGHTER II

Foam

Firefighting foam can be used to fight multiple types of fires and to prevent the ignition of materials that could become involved in a fire. The use of foam is increasing as many new types of foam have become available and efficient systems for applying them have been developed. Foams also have been developed for use in neutralizing hazardous materials and decontamination. Many fire departments use several different types of foam for a variety of situations.

Firefighting foam is produced by mixing foam concentrate with water to produce a solution that can serve as an effective extinguishing agent. Several different types of foam are used for fires involving different types of fuels. Each type of foam requires the appropriate type of concentrate, the proper equipment to mix the concentrate with water in the required proportions, and the proper application equipment and techniques to mix the foam solution with air to generate the appropriate finished foam for the specific application. Fire fighters must become familiar with the specific types of foam used by their fire department and the proper techniques for using them. It is particularly important to learn where and when to use each type of foam that is available.

Foam Classifications

The basic classifications of firefighting foams are either Class A or Class B. There are many types of foam within each classification.

Class A foam is used to fight fires involving ordinary combustible materials, such as wood, paper, and textiles. It is also effective on organic materials such as hay and straw. Class A foam is particularly useful for protecting buildings in rural areas during forest and brush fires when the supply of water is limited.

Class A foam increases the effectiveness of water as an extinguishing agent by reducing the surface tension of water. This allows the water to penetrate dense materials instead of running off the surface and allows more heat to be absorbed. The foam also keeps water in contact with unburned fuel to

FIRE FIGHTER II

prevent its ignition. Class A foam concentrate can be added to water streams and applied with several types of nozzles.

Class B foam is used to fight Class B fires—that is, fires involving flammable and combustible liquids. The various types of Class B foam are formulated to be effective on different types of flammable liquids. Note that some liquids are incompatible with different foam formulations and will destroy the foam before the foam can control the fire.

In a flammable-liquid fire, the liquid itself does not burn. Only the flammable vapors that are evaporating from the surface of the liquid and mixing with air can burn. Depending on the temperature and the physical properties of the liquid, the amount of vapors being released from the surface of a liquid varies. For example, gasoline produces flammable vapors down to a temperature of –45°F (–42.7°C). Some vapors are lighter than air and rise up into the atmosphere. Other vapors are heavier than air and flow across the surface and along the ground, collecting in low spots.

Foam extinguishes flammable-liquid fires by separating the fuel from the fire. When a blanket of foam completely covers the surface of the liquid, the release of flammable vapors stops. Preventing the production of additional vapors eliminates the fuel source for the fire, which extinguishes the fire.

Once a foam blanket has been applied, it must not be disturbed. The fuel under the foam blanket is still capable of producing flammable vapors. Thus, if the foam blanket is disturbed by wind, by someone walking through the liquid, or by hose streams breaking up the foam blanket, flammable vapors will be released and could be reignited easily.

When using foam to extinguish a flammable-liquid fire, it is critically important to apply enough foam to cover the liquid surface fully. If not enough foam is available to completely cover the surface at one time, the fire will have the ability to continue burning and the suppression efforts will be ineffective. The rate of foam application must also be high enough to cover the surface and maintain a blanket on top of the liquid. If the application rate is too low, thermal updrafts caused by the fire will keep the foam from covering the surface, and the heat of the fire will destroy the foam that has already been applied.

Class B foam can also be applied to a flammable-liquid spill in an effort to prevent a fire. A foam blanket floating on the surface of the liquid will inhibit the production of vapors that might otherwise be ignited. In this situation, the rate of foam application is not as critical because there is no fire working to destroy the foam while it is being applied.

Fire fighters must ensure that they use the proper foam for the situation that is encountered. For example, they must use an alcohol-resistant foam if the incident involves a polar solvent. Polar solvents are chemicals (such as alcohols) that readily mix with water. An ordinary foam would be broken down quickly if it came in contact with this type of product. Make sure to use the proper foam given the product involved in the incident.

Class A foams are not designed to resist hydrocarbons or to self-seal on a liquid surface. If Class A foam is used on a flammable liquid, fire fighters would find it difficult to extinguish a fire and the foam blanket would not mend itself if it became disrupted. In this case, the risk of reignition could create a dangerous situation.

Foam Concentrates

Foam concentrate is the product that is mixed with water in different ratios to produce foam solution. The foam solution is the product that is actually applied to extinguish a fire or to cover a spill.

Class A Foams

Class A foams are usually formulated to be mixed with water in ratios from 0.1 percent (1 gallon of concentrate to 999 gallons of water, or 1 liter of concentrate to 999 liters of water) to 1 percent (1 gallon of concentrate to 99 gallons of water, or 1 liter of concentrate to 99 liters of water). The end product can be tweaked to have different properties by varying the percentage of foam concentrate in the mixture and the application method. It is possible to produce "wet" foam that has good penetrating properties, for example, or "drier" foam that is more effective for applying a protective layer of foam onto a building.

Class B Foams

Most Class B foam concentrates are designed to be used in strengths of either 3 percent or 6 percent. A 3 percent foam mixes 3 gallons of foam concentrate with 97 gallons of water (11 liters of concentrate to 367 liters of water) to produce 100 gallons (100 liters) of foam solution. A 6 percent foam mixes 6 gallons of foam concentrate with 94 gallons of water (22 liters of concentrate to 356 liters of water) to produce 100 gallons (378 liters) of foam solution. Some foams are designed to be used at 3 percent for ordinary hydrocarbons and at 6 percent for polar solvents. Fire fighters must determine which type of fuel is involved in the incident so that they can select the correct proportioning rate.

The compatibility of foam agents with other extinguishing agents needs to be considered as well. For example, some combinations of dry chemical extinguishing agents and foam agents can cause an adverse reaction. The compatibility data needed are available from the manufacturers of the agents.

Fire fighters must not mix different types of foam concentrate or even different brands of the same type of foam concentrate unless they are known to be compatible. Some concentrates have been known to react with other types of foams, causing a congealing of the concentrate in the storage containers or in the proportioning system. This type of reaction can plug a foam system and render it useless.

Fire Fighter Safety Tips

Although most foam concentrates are somewhat corrosive, they pose few health risks to fire fighters if normal precautions are taken. In the past, some fire fighters have experienced minor skin irritations from coming into contact with foam concentrate. As always, before using any equipment, fire fighters must read and follow the manufacturer's directions when applying foam.

FIRE FIGHTER II

Some constituents in Class B foam concentrates that have been widely used in the past are being phased out now because of environmental concerns. Newer concentrates have been developed that are equally effective but lack the undesirable properties. The major categories of Class B foam concentrate are as follows:

- Protein foam
- Fluoroprotein foam
- Film-forming fluoroprotein (FFFP)
- Aqueous film-forming foam (AFFF)
- Alcohol-resistant foam

Standard personal protective clothing provides appropriate protection when working with foam agents. Fire fighters are advised to rinse the skin after coming into contact with foam and to flush all equipment with clear water after foam use. Protective clothing should be rinsed with plain water.

Protein Foam

Protein foams are made from animal by-products. They are effective on Class B hydrocarbon fires and are applied in 3 percent or 6 percent delivery rates.

Fluoroprotein Foam

Fluoroprotein foams are made from the same base materials as protein foam but include additional fluorochemical surfactant additives. The additives allow this foam to produce a fast-spreading film.

Film-forming fluoroprotein (FFFP) agents are composed of protein plus film-forming fluorinated surface-active agents, which make them capable of forming water solution films on the surface of most flammable hydrocarbons and of conferring a fuel-shedding property to the foam generated. FFFP foams have a fast-spreading and leveling characteristic. Like other foams, they act as surface barriers that exclude air and prevent vaporization.

Aqueous Film-Forming Foam

Aqueous film-forming foam (AFFF) is a synthetic-based foam that is particularly suitable for spill-related fires involving gasoline and light hydrocarbon fuels. It can form a seal across a surface quickly and has excellent vapor suppression capabilities.

Alcohol-Resistant Foam

Alcohol-resistant foam has properties similar to AFFF; however, it is formulated so that alcohols and other polar solvents do not dissolve the foam. Regular foams cannot be used on these products.

Compressed Air Foam Systems

A compressed air foam systems (CAFS) is a relatively new method of making class A foam.

Compressed air foam (CAF) is produced by injecting compressed air into a stream of water that has been mixed with 0.1% to 1.0% foam. This results in a small, highly compacted foam. CAF adheres to most surfaces and was thought to absorb more heat than water. The CAF coats the surface of a fuel and may reduce rekindling and smoldering of unburned fuel.

FIRE FIGHTER II

CAF is produced by a combination of a centrifugal fire pump, a foam metering device to provide a 0.1% to 1.0% concentration of foam that is injected on the discharge side of the pump, and an air compressor to inject compressed air bubbles into the water-foam combination. This mixture is said to be scrubbed by its movement through the attack line. As it exits the nozzle, the compressed air expands, producing a finished foam product. Hose lines containing CAF are lighter than lines completely filled with water, because the CAF consists of a mixture of water and a significant amount of air. The finished foam product is usually discharged at a rate of between 40 to 125 gallons of water per minute as measured before expansion.

Foams of different consistencies can be produced by adjusting the ratios of air to water. Drier foams are produced by increasing the ratio of air to water. Wetter foams are produced by decreasing the ratio of air to water. Drier foams have longer drain times. The bubbles do not burst as quickly, so it is a longer lasting product. Wetter foams will drain more quickly in the presence of heat. Because CAF will adhere to most surfaces, it is used to coat the sides and roofs of buildings during a wildland fire as a precautionary measure. When prepared as a drier foam, it will remain in place for a fairly long period of time. When it is used for interior structural fire fighting, a wetter foam is often used. It is also used to protect exposures during structure fires.

In spite of earlier claims that the use of CAF resulted in faster knockdown and more rapid cooling, current research conducted by NIST indicated that a given amount of water applied to a fire produces similar cooling and structural knockdown to using a similar mass flow of compressed air foam. These tests also indicated that there did not appear to be significant differences in the post-extinguishment smoke between using water and using foam.

If your department uses CAF, you need to complete special training, including instruction for pump operators and theoretical and practical training for other fire fighters. This training should include practice with live burn situations.

Foam Equipment

Foam equipment includes the proportioning equipment used to mix foam concentrate and water to produce foam solution, as well as the nozzles and other devices that are used to apply the foam. Many different types of proportioning and application systems are available. Most engine companies carry the necessary equipment to place at least one foam attack line into operation. Structural firefighting apparatus can also be designed with built-in foam-proportioning systems and on board tanks of foam concentrate to provide greater capabilities. Many fire departments specify that new apparatus include both Class A and Class B integrated foam systems or have special foam apparatus available for situations when large quantities of foam are needed.

FIRE FIGHTER II

Foam Proportioning Equipment

A foam proportioner is the device that mixes the foam concentrate into the fire stream in the proper percentage. The two types of proportioners—eductors and injectors—are available in a wide range of sizes and capacities. Foam solution can also be produced by batch mixing or premixing.

Foam Eductors

A foam eductor draws foam concentrate from a container or storage tank into a moving stream of water. An eductor can be built into the plumbing of an engine, or a portable eductor can be inserted in an attack hose line. A foam eductor is usually designed to work at a predetermined pressure and flow rate. A metering valve can be adjusted to set the percentage of foam concentrate that is educted into the stream.

The most common type of portable in-line eductor used by fire departments is sized to work with a 1½-inch (38-mm) or 1¾-inch (44-mm) attack line. This type of eductor requires 200 psi (1379 kPa) of water pressure to draw foam concentrate from a portable container into the stream. An attack line with 150 feet (36 meters) of 1½-inch (38-mm) hose can be connected on the discharge side of the eductor to deliver the foam to a nozzle.

Foam Injectors

Foam injectors add the foam concentrate to the water stream under pressure. Most injector-based proportioning systems can work across a range of flow rates and pressures. A metering system measures the flow rate and pressure of the water and adjusts the injector to add the proper amount of foam concentrate. This type of system is often installed on special foam apparatus.

To place an eductor foam line in service, follow the steps in SKILL DRILL 17-16 (Fire Fighter II, NFPA 6.3.1):

1. Make sure all necessary equipment is available, including an in-line foam eductor and an air-aspirating nozzle. Ensure that enough foam concentrate is available to suppress the fire.
2. Don all PPE.
3. Procure an attack line, remove the nozzle, and replace it with the air-aspirating nozzle.
4. Place the in-line eductor in the hose line no more than 150 feet (36 meters) from the nozzle.
5. Place the foam concentrate container next to the eductor, check the percentage at which the foam concentrate should be used (found on container label), and set the metering device on the eductor accordingly.
6. Place the pickup tube from the eductor into the foam concentrate, keeping both items at similar elevations to ensure sufficient induction of foam concentrate.
7. Charge the hose line with water, ensuring there is a minimum of 200 psi (1379 kPa) at the eductor.
8. Flow water through the hose line until foam starts to come out of the nozzle. The hose line is now ready to be advanced onto burning flammable liquids.
9. Apply foam using one of the three application methods (sweep technique, bankshot technique, or rain-down technique—discussed later in this chapter) depending on the situation.

FIRE FIGHTER II

Batch Mixing

Foam concentrate can be poured directly into an apparatus booster tank to produce foam solution, a technique called batch mixing. If the booster tank has a capacity of 500 gallons (1893 liters), 15 gallons (56 liters) of 3 percent foam concentrate should be added. If 6 percent foam concentrate is used, 30 gallons (113 liters) of foam concentrate should be added to the booster tank. It might be necessary to drain sufficient water from the tank first to make room for the foam concentrate. After the concentrate has been added, the solution should be mixed by circulating the water through the pump before it is discharged.

Premixing

Premixed foam is commonly used in 2½-gallon (9.5-liter) portable fire extinguishers. Foam fire extinguishers are filled with premixed foam solution and pressurized with compressed air or nitrogen. Some vehicles are equipped with large tanks holding 50 or 100 gallons (189 or 379 liters) of premixed foam, which operate in the same manner.

Foam Application

Foam can be applied to a fire or spill through portable extinguishers, handlines, master stream devices, or a variety of fixed systems for special applications. Most fire departments apply foam through either handlines or master stream devices on fire apparatus. Several types of nozzles are used, with the goal of producing different types of foam. The major difference between these nozzles is the manner in which air is introduced into the stream of foam solution to produce the desired consistency of foam and air bubbles.

Foam can be applied with a wide range of expansion rates, depending on the amount of air that is mixed into the stream and the size of the bubbles that are produced. Low-expansion foam has little entrained air and a small bubble structure. This type of foam is often produced with standard adjustable fog nozzles, where the air is entrained by the flowing stream and mixed into the foam solution. This type of nozzle is often used to apply AFFF or Class A foam.

Medium-expansion foam is produced with special aerating nozzles that are designed to introduce more air into the stream and produce a consistent bubble structure (aeration). This type of nozzle is generally used with protein and fluoroprotein foams to produce a thicker blanket of foam. It is also recommended for use with alcohol-resistant foams to produce a thicker foam blanket.

High-expansion foam contains a much higher proportion of air and large bubbles. Such a system uses a high-expansion foam generator to introduce large quantities of air into the discharge stream. High-expansion foam is sometimes used in automatic systems that are designed to completely fill a large space with foam. These systems are most likely to be found in aircraft hangars or other large storage areas.

Foam Application Techniques

When applying foam from a handline, the correct application techniques must be used to produce the desired quality of

FIRE FIGHTER II

foam and successfully blanket the surface of a burning liquid or spill. These techniques can be direct or indirect and include the sweep, bankshot, and rain-down methods of application.

Sweep Method

The sweep (or roll-on) method should be used only on a pool of flammable product that is on open ground. With this technique, the fire fighter sweeps the stream along the ground just in front of the target to produce a quantity of foam and then uses the energy of the stream to push the foam blanket across the surface. The stream is moved back and forth in a slow, steady horizontal motion to push the foam forward gently until the area is covered. It is important to push the foam gingerly so that the blanket is not broken. The fire fighter might need to move to different positions to be sure that the entire surface of the product is covered by the foam blanket.

To perform the sweep method of applying foam, follow the steps in SKILL DRILL 17-17:

1. Open the nozzle and test to ensure that foam is being produced.
2. Move within a safe range of the product and open the nozzle.
3. Direct the stream of foam onto the ground just in front of the pool of product.
4. Allow the foam to roll across the top of the pool of product until it is completely covered.
5. Be aware that the fire fighters might have to change positions along the spill so as to adequately cover the entire pool.

Bankshot Method

The bankshot (or bank-down) method is used at fires where the fire fighter can use an object to deflect the foam stream and let it flow down onto the burning surface. This method could be used to apply foam to an open-top storage tank or a rolled-over transport vehicle, for example. The fire fighter should sweep the foam back and forth against the object while the foam flows down and spreads back across the surface. As with all foam application methods, it is important to let the foam blanket flow gently on the surface of the flammable liquid to form a blanket.

To perform the bankshot method of applying foam, follow the steps in SKILL DRILL 17-18:

1. Open the nozzle and test to ensure that foam is being produced. (STEP 1)
2. Move within a safe range of the product and open the nozzle.
3. Direct the stream of foam onto a solid structure such as a wall or metal tank so that the foam is directed off the object and onto the pool of product.
4. Allow the foam to flow across the top of the pool of product until it is completely covered.
5. Be aware that the fire fighters might have to bank the foam off several areas of the solid object so as to extinguish the burning product. (STEP 2)

FIRE FIGHTER II

Rain-Down Method

The rain-down application method consists of lofting the foam stream into the air above the fire and letting it fall down gently onto the surface. The stream should be broken so that the amount of foam falling in the same area does not cause splashing or break up the blanket that has already been applied. Carefully observe how the foam blanket is building up and direct your stream so that the entire surface is covered.

FIRE FIGHTER II Tips **FFII**

Utilizing either Class A or Class B foam on a brush fire after extinguishment can allow for additional penetration and reduce the likelihood of rekindle.

To perform the rain-down method of applying foam, follow the steps in SKILL DRILL 17-19 (Fire Fighter II, NFPA 6.3.1):

1. Open the nozzle and test to ensure that foam is being produced.
2. Move within a safe range of the product and open the nozzle.
3. Direct the stream of foam into the air so that the foam breaks apart in the air and falls onto the pool of product.
4. Allow the foam to flow across the top of the pool of product until it is completely covered.
5. Be aware that the fire fighters might have to move to several locations and shoot the foam into the air so as to extinguish the burning product.

Backup Resources

When attempting to use foam to extinguish a flammable-liquid fire, it is critically important that enough foam concentrate be available to complete the job. If the flow of foam must be interrupted while additional foam supplies are obtained, the fire will destroy the foam that has already been applied. It is better to wait until an adequate supply of foam concentrate is on hand than to waste the limited supply that is immediately available.

Foam manufacturers provide specific formulas that enable fire fighters to calculate how much foam is required to extinguish fires of a certain size. Most fire departments have contingency plans to deliver quantities of foam to the scene of a major incident. This foam could be located on designated vehicles or in storage at fire stations, for example. Backup sources often include airport crash vehicles and petroleum facilities that have their own foam apparatus and often keep large quantities of foam in storage. Manufacturers of foam products also have emergency programs to deliver large quantities of foam to the scene of exceptionally large-scale incidents.

SKILL DRILL 17-18 Apply Foam with the Bankshot Method

(Fire Fighter II, NFPA 6.3.1)

1. Open the nozzle and test to ensure that foam is being produced.

2. Move within a safe range of the product and open the nozzle. Direct the stream of foam onto a solid structure such as a wall or metal tank so that the foam is directed off the object and onto the pool of product. Allow the foam to flow across the top of the pool of product until it is completely covered. Be aware that the fire fighters might have to bank the foam off several areas of the solid object so as to extinguish the burning product.

FIRE FIGHTER II

■ Foam Apparatus

Some fire departments operate apparatus that is specifically designed to produce and apply foam. The most common examples are found at airports, where these apparatus are used for aircraft rescue and firefighting. These large vehicles carry the foam concentrate and water on board and are designed to quickly apply large quantities of foam to a flammable-liquid fire. Remote-control monitors can be used to apply foam while the vehicles are in motion. If there is an airport close to your department, you should practice with them and know the best way to access them, before an emergency occurs.

Wrap-Up

Chief Concepts

- Most attack hoses carry water directly from the attack engine to a nozzle that is used to direct the water onto the fire.
- Attack hoses usually operate at higher pressures than supply lines do.
- The hoses most commonly used to attack interior fires are either 1½ inches (38 millimeters) or 1¾ inches (45 millimeters) in diameter. Each section of attack hose is usually 50 feet (15 meters) long.
- Attack hose must be tested annually at a pressure of at least 300 psi (2068 kPa) and is intended to be used at pressures up to 275 psi (1896 kPa).
- Depending on the pressure in the hose and the type of nozzle used, a 1½-inch (38-mm) hose can generally flow between 60 and 125 gpm (273 and 568 lpm). An equivalent 1¾-inch (44-mm) hose can flow between 120 and 180 gpm (545 and 818 lpm).
- A 2½-inch (65-mm) hose is used as an attack line for large fires and generally delivers a flow of approximately 250 gpm (1136 lpm). It takes at least two fire fighters to control a 2½-inch (65-mm) handline safely because when the hose is charged with water it can weigh as much as 140 pounds (64 kilograms).
- Booster hose contains a steel wire that gives it a rigid shape so that the hose can flow water without pulling all of the hose off the reel. The normal flow from a 1-inch (25-mm) booster hose is in the range of 40 to 50 gpm (182 to 227 lpm). This type of hose should not be used for structural firefighting.
- Attack lines are usually stretched from an attack engine to the fire. The attack engine is usually positioned close to the fire, and attack lines are stretched manually by fire fighters. In some situations, an engine will drop an attack line at the fire and drive from the fire to a hydrant or water source. This procedure is similar to the reverse lay evolution, except that the hose will be used as an attack line.
- Attack hose is loaded so that it can be quickly and easily deployed. The three most common hose loads used for preconnected attack lines are the minuteman load, the flat load, and the triple-layer load.
- Preconnected hose lines can be placed in several different locations on a fire engine:
  - A section of a divided hose bed at the rear of the apparatus can be loaded with a preconnected attack line.
  - Transverse hose beds are installed above the pump on many engines and loaded so that the hose can be pulled off from either side of the apparatus.
  - Preconnected lines can be loaded into special trays that are mounted on the side of fire apparatus.
  - Many engines include a special compartment in the front bumper that can store a short preconnected hose line.
- To reach a fire that is some distance from the engine, it might be necessary to first advance a larger-diameter line, and then split it into two 1¾-inch (45-mm) attack lines. This is accomplished by attaching a gated wye or a water thief to the end of the 2½-inch (65-mm) line, and then attaching the two attack lines to the gated outlets.
- A basic attack evolution looks like this:
  - Attack line is stretched in two stages. First, the hose is laid out from the attack engine to the building entrance or to a location close to the fire. Then, the hose is advanced into the building to reach the fire. Extra hose should be deposited at the entrance to the fire building. Make sure you flake out the hose before it is charged with water.
  - Once the hose is flaked out, signal the pump driver/operator to charge the line.
  - Once the hose is flaked out, check your PPE and your partner's PPE.
  - When you are given the command to advance the hose, keep safety as your number one priority.
  - As you move inside the building, stay low to avoid the greatest amount of heat and smoke. Communicate with the other members of the nozzle team as you advance.
  - As you advance the hose line, you need to have enough hose to enable you to move forward.
  - Charged hose lines are not easy to advance. It is only with good teamwork that efficient hose line advancement can occur.
- Nozzles are attached to the discharge end of attack lines to give fire streams shape and direction. Nozzles are used on all sizes of handlines as well as on master stream devices.
- Nozzles can be classified into three groups:
  - Low-volume nozzles flow 40 gpm (182 lpm) or less. They are primarily used for booster hoses; their use is limited to small outside fires.
  - Handline nozzles are used on hose lines ranging from 1½ inches (38 millimeters) to 2½ inches (65 millimeters) in diameter. Handline streams usually flow between 60 and 350 gpm (273 and 1591 lpm).
  - Master stream nozzles are used on deck guns, portable monitors, and ladder pipes that flow more than 350 gpm (1591 lpm).
- Nozzle shut-off enables the fire fighter at the nozzle to start or stop the flow of water.

- Two different types of nozzles are manufactured for the fire service:
  - *Smooth-bore nozzles.* Produce a solid column of water.
  - *Fog-stream nozzles.* Separate the water into droplets. The size of the water droplets and the discharge pattern can be varied by adjusting the nozzle setting.
- Specialized nozzles include these:
  - *Piercing nozzles.* Used to make a hole in sheet metal or building walls to extinguish fires.
  - *Cellar nozzles and Bresnan distributor nozzles.* Used to fight fires in cellars and other inaccessible places such as attics and cocklofts.
  - *Water curtain nozzles.* Used to deliver a flat screen of water that forms a protective sheet of water on the surface of an exposed building.
- Firefighting foam can be used to fight multiple types of fires and to prevent the ignition of additional fuels. Several different types of foam are used for fires involving different types of fuels.
- Foams are either Class A or Class B.
  - Class A foam is used to fight fires involving ordinary combustible materials, such as wood, paper, and textiles.
  - Class B foam is used to fight Class B fires, or fires involving flammable and combustible liquids.
- Foam extinguishes flammable-liquid fires by separating the fuel from the fire. When a blanket of foam completely covers the surface of the liquid, the release of flammable vapors stops. Preventing the production of additional vapors eliminates the fuel source for the fire, which extinguishes the fire.
- Foam concentrate is the product that is mixed with water in different ratios to produce foam solution. The foam solution is the product that is actually applied to extinguish a fire or to cover a spill.
- The major categories of Class B foam concentrate are protein foam, fluoroprotein foam, film-forming fluoroprotein, aqueous film-forming foam, and alcohol-resistant foams.
- Compressed air foam systems are a new method of making Class A foam. Compressed air foam is produced by injecting compressed air into a stream of water that has been mixed with 0.1 percent to 1.0 percent foam.
- A foam proportioner is the device that mixes the foam concentrate into the fire stream in the proper percentage. The two types of proportioners—eductors and injectors—are available in a wide range of sizes and capacities.
- Foam solution can also be produced by batch mixing or premixing.
  - Batch mixing is a technique where foam concentrate is poured directly into an apparatus booster tank to produce foam solution.
  - Premixed foam is commonly used in 2½-gallon (9.5-liter) portable fire extinguishers.

Hot Terms

Adjustable-gallonage fog nozzle A nozzle that allows the operator to select a desired flow from several settings.

Aerating The process of inducing air into the foam solution, which expands and finishes the foam.

Alcohol-resistant foam Used for fighting fires involving water-soluble materials or fuels that are destructive to other types of foams. Some alcohol-resistant foams are capable of forming a vapor-suppressing aqueous film on the surface of hydrocarbon fuels. (NFPA 412)

Aqueous film-forming foam (AFFF) A concentrated aqueous solution of one or more hydrocarbon and/or fluorochemical surfactants that forms a foam capable of producing a vapor-suppressing, aqueous film on the surface of hydrocarbon fuels. (NFPA 403)

Attack engine An engine from which attack lines have been pulled.

Automatic-adjusting fog nozzle A nozzle that can deliver a wide range of water stream flows. It operates by means of an internal spring-loaded piston.

Bankshot (or bank-down) method A method that applies the stream onto a nearby object, such as a wall, instead of directly aiming at the fire.

Batch mixing Pouring foam concentrate directly into the fire apparatus water tank, thereby mixing a large amount of foam at one time.

Booster hose (booster line) A noncollapsible hose used under positive pressure having an elastomeric or thermoplastic tube, a braided or spiraled reinforcement, and an outer protective cover. (NFPA 1962)

Breakaway type nozzle A nozzle with a tip that can be separated from the shut-off valve.

Bresnan distributor nozzle A nozzle that can be placed in confined spaces. The nozzle spins, spreading water over a large area.

Cellar nozzle A nozzle used to fight fires in cellars and other inaccessible places. The device works by spreading water in a wide pattern as the nozzle is lowered through a hole into the cellar.

Class A foam Foam intended for use on Class A fires. (NFPA 1150)

Class B foam Foam intended for use on Class B fires. (NFPA 1901)

Compressed air foam (CAF) Class A foam produced by injecting compressed air into a stream of water that has been mixed with 0.1 percent to 1.0 percent foam.

Compressed air foam system (CAFS) A foam system that combines air under pressure with foam solution to create foam. (NFPA 1901)

Film-forming fluoroprotein (FFFP) A protein-based foam concentrate incorporating fluorinated surfactants that forms a foam capable of producing a vapor-suppressing aqueous film on the surface of hydrocarbon fuels. This foam can show an acceptable level of compatibility with dry chemicals and might be suitable for use with those agents. (NFPA 412)

Fixed-gallonage fog nozzle A nozzle that delivers a set number of gallons per minute (liters per minute) as per the nozzle's design, no matter what pressure is applied to the nozzle.

Fluoroprotein foam A protein-based foam concentrate to which fluorochemical surfactants have been added. This has the effect of giving the foam a measurable degree of compatibility with dry chemical extinguishing agents and an increase in tolerance to contamination by fuel. (NFPA 402)

Foam concentrate A concentrated liquid foaming agent as received from the manufacturer. (NFPA 11)

Foam eductor A device placed in the hose line that draws foam concentrate from a container and introduces it into the fire stream.

Foam injector A device installed on a fire pump that meters out foam by pumping or injecting it into the fire stream.

Foam proportioner A device or method to add foam concentrate to water to make foam solution. (NFPA 1901)

Foam solution A homogeneous mixture of water and foam concentrate in the proper proportions. (NFPA 1901)

Fog-stream nozzle A nozzle that is placed at the end of a fire hose and separates water into fine droplets to aid in heat absorption.

Handline nozzle A nozzle with a rated discharge of less than 1591 L/min (350 gpm). (NFPA 1964)

Low-volume nozzle A nozzle that flows 40 gallons per minute (182 lpm) or less.

Master stream device A large-capacity nozzle that can be supplied by two or more 2½-inch (64-mm) hose lines or large-diameter hose or fixed piping. It can flow between 350 and 1500 gallons per minute (1591 and 6818 lpm). These devices include deck guns and portable ground monitors.

Master stream nozzle A nozzle with a rated discharge of 1591 lpm (350 gpm) or greater. (NFPA 1964)

Nozzle A constricting appliance attached to the end of a fire hose or monitor to increase the water velocity and form a stream. (NFPA 1965)

Nozzle shut-off A device that enables the fire fighter at the nozzle to start or stop the flow of water.

Piercing nozzle A nozzle that can be driven through sheet metal or other material to deliver a water stream to that area.

Premixed foam Solution produced by introducing a measured amount of foam concentrate into a given amount of water in a storage tank. (NFPA 11)

Protein foam A protein-based foam concentrate that is stabilized with metal salts to make a fire-resistant foam blanket. (NFPA 402)

Smooth-bore nozzle A nozzle that produces a straight stream that is a solid column of water.

Smooth-bore tip A nozzle device that is a smooth tube, which is used to deliver a solid column of water.

Sweep (or roll-on) method A method of applying foam that involves sweeping the stream just in front of the target.

Triple-layer load A hose loading method in which the hose is folded back onto itself to reduce the overall length to one-third before loading the hose in the bed. This load method reduces deployment distances.

Water curtain nozzle A nozzle used to deliver a flat screen of water that forms a protective sheet of water to protect exposures from fire.

FIRE FIGHTER *in action*

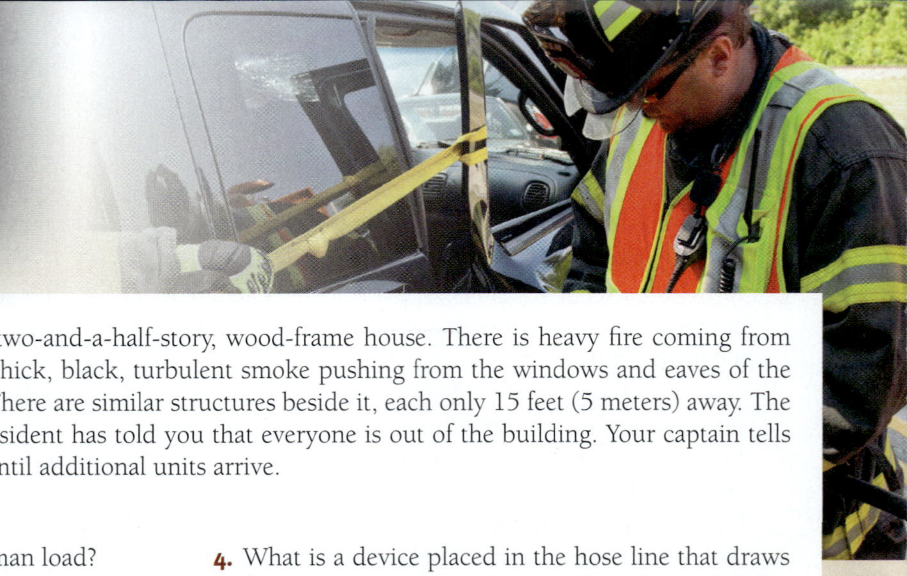

You are on the scene of a fire involving a large, two-and-a-half-story, wood-frame house. There is heavy fire coming from the second floor at the rear of the structure and thick, black, turbulent smoke pushing from the windows and eaves of the remainder of the second floor and the attic area. There are similar structures beside it, each only 15 feet (5 meters) away. The siding on one of them is already smoking. The resident has told you that everyone is out of the building. Your captain tells you to pull a preconnect to protect an exposure until additional units arrive.

1. What steps will you take to pull a minuteman load?
  - **A.** While facing the engine, pull the hose one-third of the way out, place the hose on your shoulder, turn, and then walk toward the proper location.
  - **B.** While facing away from the engine, pull the hose one-third of the way out, turn, place the hose on your shoulder, and then walk toward the proper location.
  - **C.** Place your arm though the larger loop, and then grasp the smaller loop with that hand. Pull the hose and walk toward the proper location.
  - **D.** Place your arm through the larger loop, and then grasp the smaller loop with the opposite hand. Pull the hose and walk toward the proper location.

2. Hose with a diameter of 2½ or 3 inches (65 or 76 millimeters) is considered ________.
  - **A.** small diameter
  - **B.** medium diameter
  - **C.** intermediate diameter
  - **D.** large diameter

3. A nozzle that can deliver a wide range of water stream flows and that operates by means of an internal spring-loaded piston is called a/an ________.
  - **A.** adjustable-gallonage fog nozzle
  - **B.** Bresnan distributing nozzle
  - **C.** fog stream nozzle
  - **D.** automatic-adjusting fog nozzle

4. What is a device placed in the hose line that draws foam concentrate from a container and introduces it into the fire stream called?
  - **A.** Foam injector
  - **B.** Aerator
  - **C.** Foam eductor
  - **D.** Batch mixer

5. What is the method that applies the stream onto a nearby object, such as a wall, instead of directly aiming at the fire called?
  - **A.** Roll-on
  - **B.** Sweep
  - **C.** Rain-down
  - **D.** Bank-down

6. ________ foam increases the effectiveness of water as an extinguishing agent by reducing the surface tension of water.
  - **A.** Class A
  - **B.** Class B
  - **C.** Class C
  - **D.** Class D

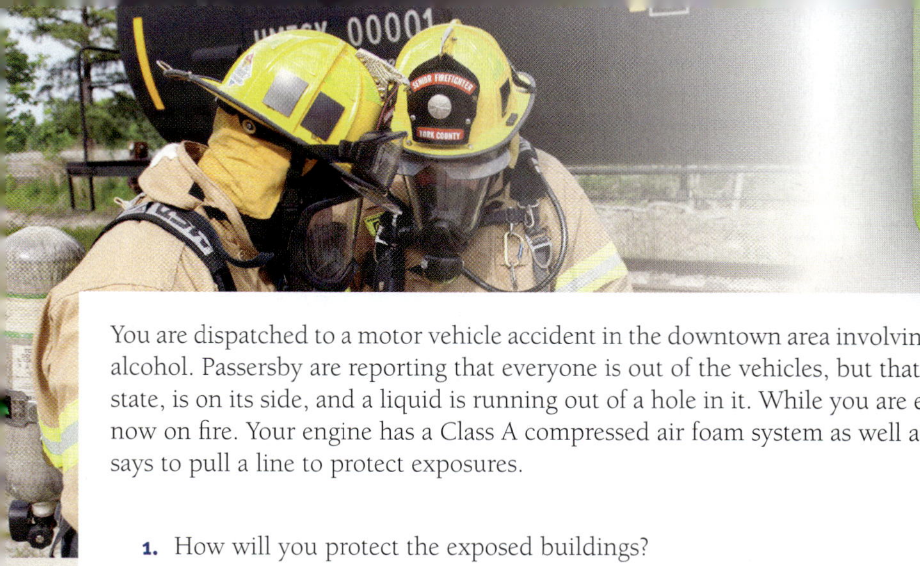

FIRE FIGHTER II *in action*

You are dispatched to a motor vehicle accident in the downtown area involving a small passenger car and a tanker transporting alcohol. Passersby are reporting that everyone is out of the vehicles, but that the tanker rolled over the guardrail of the interstate, is on its side, and a liquid is running out of a hole in it. While you are en route, the dispatcher reports that the tanker is now on fire. Your engine has a Class A compressed air foam system as well as a Class B foam system with AFFF. Your captain says to pull a line to protect exposures.

1. How will you protect the exposed buildings?
2. Do you have the proper foam to extinguish the fire?
3. Will CAFs be of any value in this scenario?
4. What concerns do you have with this incident?

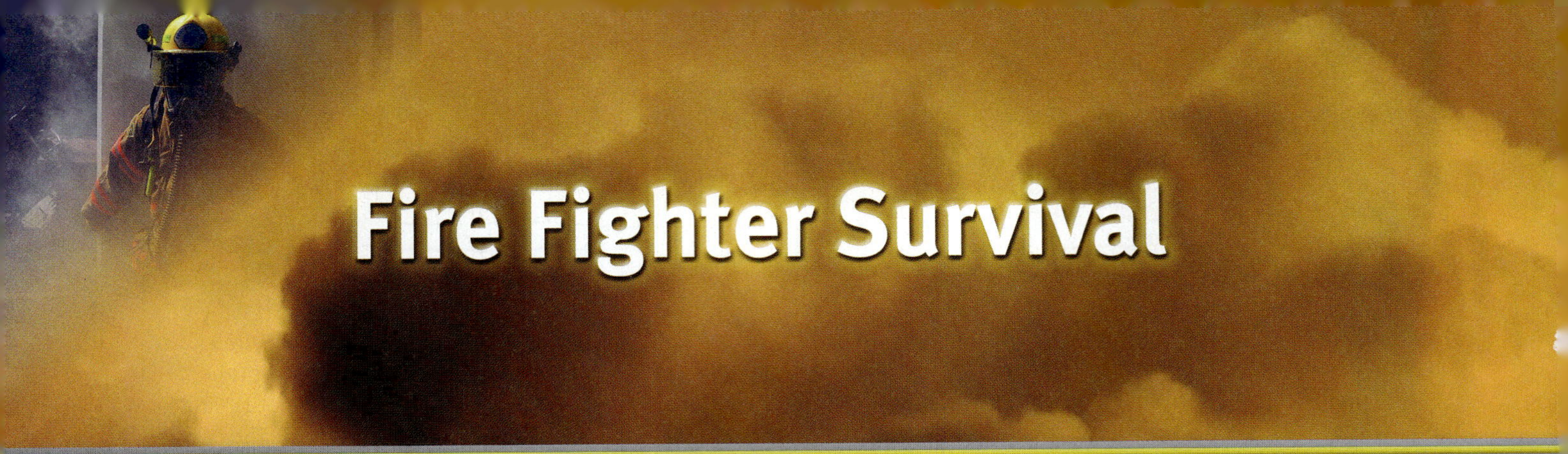

Fire Fighter Survival

Fire Fighter I

Knowledge Objectives

After studying this chapter, you will be able to:

- Describe how to apply a risk–benefit analysis to an emergency incident. (p 586–587)
- List the common hazard indicators that should alert fire fighters to a potentially life-threatening situation. (NFPA 5.3.5.A, p 587)
- List the 11 Rules of Engagement for Fire Fighter Survival. (p 588)
- Explain how to maintain team integrity during emergency operations. (NFPA 5.3.5, p 588–589)
- Define personnel accountability system. (NFPA 5.3.5.A, p 589–590)
- Describe the types of personnel accountability systems and how they function. (NFPA 5.3.5.A, p 590)
- Explain how a personnel accountability report is taken. (p 590)
- Describe how to initiate emergency communications procedures. (NFPA 5.3.5.A, p 590–591)
- Describe the methods used for maintaining orientation. (p 593)
- Describe common self-rescue techniques. (NFPA 5.3.5.A, p 593)
- Describe how to find a safe location while awaiting rescue. (NFPA 5.3.5, 5.3.5.A, p 596–600)
- Describe air management procedures. (NFPA 5.3.1.B, 5.3.5.A, 5.3.5.B, p 600)
- Describe how rapid intervention crews provide an air supply to a trapped fire fighter. (NFPA 5.3.9B, p 601)
- Explain the importance of the rehabilitation process. (p 602–603)
- List the common causes of critical incident stress. (p 603–605)
- Explain the goal of critical incident stress management. (p 603–605)

Skills Objectives

After studying this chapter, you will be able to:

- Initiate a mayday call. (NFPA 5.2.4, 5.2.4.A, 5.2.4.B, p 591–592)
- Perform a self-rescue using a hose line. (NFPA 5.3.5.B, p 594)
- Perform a self-rescue by locating a door or window for an emergency exit. (p 594–595)
- Perform a self-rescue by opening a wall to escape. (NFPA 5.3.1.B, 5.3.9B, p 596–597)
- Perform a self-rescue by removing an entanglement. (p 596, 598)
- Rescue a downed fire fighter using the fire fighter's SCBA straps. (NFPA 5.3.9B, p 601–602)
- Rescue a downed fire fighter using a drag rescue harness. (NFPA 5.3.9B, p 601, 603)
- Rescue a downed fire fighter as a two-person team. (NFPA 5.3.5B, p 601–602, 604)

CHAPTER 18

Fire Fighter II FFII

Knowledge Objectives

There are no knowledge objectives for Fire Fighter II candidates. NFPA 1001 contains no Fire Fighter II Job Performance Requirements for this chapter.

Skills Objectives

There are no skill objectives for Fire Fighter II candidates. NFPA 1001 contains no Fire Fighter II Job Performance Requirements for this chapter.

Additional NFPA Standards

- NFPA 1500, *Standard on Fire Department Occupational Safety and Health Program*

You Are the Fire Fighter

It is the middle of August, and you are dispatched to a fire in a barn in a neighboring jurisdiction. As you leave your station, you can see a large column of black smoke in the distance. Upon your arrival, you find that the barn is fully involved, the plastic on a recreational vehicle (RV) that is parked nearby is smoking heavily, and embers have ignited a ground fire across the road in grass dried out by the recent drought. The incident commander (IC) has assigned crews to protect the RV and other exposures, as well as crews to address the grass fire. A water shuttle has been established, so the IC directs your captain to take over personnel accountability and orders you to establish a rehabilitation (rehab) unit. You quickly search for the optimal location for the rehab unit and begin to pull the essential items together.

1. What should be considered when establishing a rehab unit?
2. Why is it important to establish rehab early in an incident such as this one?
3. Why should personnel be assigned to personnel accountability?

Introduction

Fire fighter survival is the most important outcome in any successful fire department operation. Fire fighters often perform their duties in environments that are inherently dangerous. Survival is a positive outcome that can be achieved only by working and training in a manner that recognizes risks and hazards and consistently applies safe operating policies and procedures. "Everybody goes home" should be the goal of every member of the team. This chapter introduces the actions, attitudes, and systems that are important in achieving that goal.

Fire fighters often encounter situations where their survival depends on making the right decisions and taking the appropriate actions. Many of the factors that cause fire fighter deaths and injuries appear again and again in postincident studies as risks that are often present at fires and other emergency incidents. Fire fighters must learn to recognize dangerous situations and to take actions to prevent more dangerous situations from occurring.

Risk–Benefit Analysis

A risk–benefit analysis approach to emergency operations can limit the risk of fire fighter deaths and injuries. This kind of analysis weighs the positive results that can be achieved against the probability and severity of potential negative consequences. A standard approach to risk–benefit analysis should be incorporated into the standard operating procedures (SOPs) for all fire departments. NFPA 1500, *Standard on Fire Department Occupational Safety and Health Program*, is an excellent reference guide to this approach as it applies to fire department operations.

In the fire service, risk–benefit analysis should be practiced and implemented by everyone on the fire ground. The incident commander (IC) is responsible for the analysis of the overall scene, while company officers and fire fighters need to process the risks and benefits of the strategies implemented by command and be cognitive of changes on the fire ground.

The IC must always assess the risks and benefits before committing crews to the interior of a burning structure. If the potential benefits are marginal and the risks are too high, fire fighters should not be engaged to conduct an interior attack. The IC must also reassess the risks and benefits periodically during an operation. There should be no hesitation to terminate an interior attack and move to an exterior defensive operation at any time if the risks associated with an offensive operation change so that they outweigh the benefits of this strategy.

The risk–benefit method attempts to predict the potential risks that fire fighters might face when entering a burning structure and weighs them against the results that might be achieved by undertaking such a mission. If a building is known to be unoccupied and has no value, there is no justification for risking the lives of fire fighters to save it. Similarly, if a fire has already reached the stage where no occupants could survive and no property of value can be saved, there is no justification for risking the lives of fire fighters in an interior attack **FIGURE 18-1**.

FIGURE 18-1 If a fire has already reached the stage where no occupants could survive and no property of value can be saved, there is no justification for risking the lives of fire fighters in an interior attack.

In a situation where no lives are at stake but there is a reasonable expectation that property could be saved, this policy allows for fire fighters to be committed to an interior attack. It recognizes that a standard set of hazards is anticipated in an interior fire attack situation, but that the combination of personal protective clothing and equipment, training, and SOPs is designed to allow fire fighters to work safely in this type of environment. All standard approaches to operational safety must be followed without compromise. No property is ever worth the life of a fire fighter.

It is permissible to risk the life of a fire fighter only in situations where there is a reasonable and realistic possibility of saving a life. The determination that a risk is acceptable in a particular situation does not justify taking unsafe actions, however; it merely justifies taking actions that involve a higher level of risk.

Company officers and safety officers engage in risk–benefit analysis on an ongoing basis. If they perceive that the risk–benefit balance has shifted, it is their responsibility to report their observations and recommendations to the IC. The IC reevaluates the risk–benefit balance whenever conditions change and decides whether the strategy must be changed.

In addition, each individual fire fighter involved in the operation should conduct a risk–benefit analysis from his or her own perspective. A report from an observant fire fighter to his or her company officer can be crucial to the safety of an operation.

Hazard Indicators

As you progress in your career as a fire fighter, it will become evident that fire suppression and other types of emergency operations involve many different types of hazards. Fire fighters must be capable of working safely in an environment that includes a wide range of inherent hazards. While the danger of firefighting should never be taken for granted or perceived as routine, a fire fighter must learn to routinely follow safe SOPs. Some hazards are encountered at almost every incident, whereas others are rare occurrences. A fire fighter should recognize the various types of hazardous conditions and react to them appropriately.

An example of a commonly encountered hazard would be the presence of smoke inside a structure. Fires produce smoke, and breathing smoke is well recognized as dangerous. The proper response to this hazard is to always wear appropriate personal protective equipment (PPE) for the hazard at hand. At the same time, you should recognize that turbulent smoke and black fire indicate that the atmosphere may soon prove deadly even for fire fighters who are in complete PPE including self-contained breathing apparatus (SCBA). (The Fire Behavior chapter has more information on smoke reading.) This is a situation when hazards are recognized and standard solutions are applied.

Hazardous conditions may or may not be evident by simple observation. For example, smoke and flames are usually visible, whereas hazards related to the construction details of a building might not be so easily detectable. You should be aware of the indicators of hazards that present themselves in a less obvious manner. Making this kind of mental connection requires knowledge of what can be hazardous, given a particular set of circumstances. Hazard recognition becomes easier through study and experience:

- Building construction: Knowledge of building construction can help in anticipating fire behavior in different types of buildings and recognizing the potential for structural collapse. Is the building designed to withstand a fire for some period of time, or is it likely to weaken quickly? Has the building been modified from its original use? Does the structure incorporate void spaces that would allow a fire to spread quickly? Is it constructed using lightweight construction techniques? Does the building incorporate truss roofs or floors or lightweight construction that could fail in a short period of time?
- Weather conditions: Inclement weather can turn a routine incident into a hazardous one. Rain and snow make operations on a roof even more dangerous than usual. Raising a ladder close to power lines during windy weather involves an increased level of risk. Wind-driven fires can rapidly change direction, or wind can rapidly increase the intensity of a fire, often with deadly results.
- Occupancy: A name such as "Central Plating Company" on a building should cause fire fighters to anticipate that hazardous materials could be present. A warning placard outside would be a more specific hazardous materials indicator FIGURE 18-2. Many common commercial buildings such as hardware stores and home improvement centers contain huge amounts of highly flammable and hazardous materials.

These are just a few examples of observable factors that might indicate a hazard. At each and every incident, fire fighters should look for what could be critical indications of a hazard.

FIGURE 18-2 The NFPA 704 diamond indicates that hazardous materials are present.

Safe Operating Procedures

SOPs define the manner in which a fire department conducts operations at an emergency incident. Many of these procedures were developed to protect fire fighters' health and safety.

SOPs must be learned and practiced before they can be implemented. Training is the only way for individuals to become proficient at performing any set of skills. To be effective, the fire fighter must apply the same skills routinely and consistently at every training session and at every emergency incident. When under great pressure, most people will revert to habit; thus, when self-survival is at stake, well-practiced habits can be a lifesaver.

Rules of Engagement for Fire Fighter Survival

To help define safe practices of engaging structural firefighting, the International Association of Fire Chiefs (IAFC) and National Volunteer Fire Council (NVFC) developed the Rules of Engagement for Firefighter Survival. These 11 rules are summarized here:

1. **Size up your tactical area of operation**. Fire fighters and company officers need to pause for a moment and look over their area of responsibility. During this size-up, you need to evaluate your individual risk exposure and determine a safe approach to complete your assigned objectives.
2. **Determine the occupant survival profile**. Fire fighters and company officers need to consider the fire conditions to determine where an occupant could be alive in the conditions that present themselves. They must continue an ongoing assessment of these conditions.
3. ***Do not* risk your life for lives or property that cannot be saved**. Do not engage in high-risk search and rescue and firefighting activities when the fire conditions prevent occupant survival or if the property destruction is inevitable.
4. **Extend *limited* risk to protect *savable* property**. Limit your risk to a reasonable, cautious, and conservative level when trying to save a building.
5. **Extend *vigilant* and *measured* risk to protect and rescue *savable* lives**. Engage in search and rescue and firefighting operations in a calculated, controlled, and safe manner, while remaining alert to changing conditions during high-risk search and rescue operations, only in conditions where lives can be saved.
6. **Go in together, stay together, and come out together.** Ensure that you and your fellow fire fighters always enter burning buildings as a team of two or more and that no fire fighter is allowed to be alone at any time while entering, operating in, or exiting a hazardous building.
7. **Maintain continuous awareness of your air supply, situation, location, and fire conditions.** Maintain constant situational awareness of your SCBA air supply, your location in the building, and all that is happening in your area of operation and elsewhere on the fire ground that may affect your risk and safety. Monitor your air supply and ensure that you are outside the hazard zone before your low-air alarm activates.
8. **Constantly monitor fire-ground communications for critical radio reports.** Maintain constant awareness of all fire-ground radio communications on their assigned channels for progress reports, critical messages, and other information that may affect your risk and safety.
9. **Report unsafe practices or conditions that can harm you. Stop, evaluate, and decide**. Report unsafe practices and exposure to unsafe conditions to your supervisor. There should be no penalty for this reporting.
10. **Abandon your position and retreat before deteriorating conditions can harm you.** Be aware of unsafe fire conditions and exit immediately to a safe area when you are exposed to deteriorating conditions, unacceptable risk, or a life-threatening condition.
11. **Declare a mayday as soon as you *think* you are in danger.** Declare a mayday as soon as you are faced with a life-threatening situation; declare a mayday as soon as you think you are in trouble. Failure to declare a mayday as soon as possible puts your life and the lives of fellow fire fighters in even more danger.

Team Integrity

Teamwork is an essential requirement in firefighting FIGURE 18-3. A team is defined as two or more individuals attempting to achieve a common goal. Given the intense physical labor necessary to pull hose or raise ladders, team efforts are imperative in firefighting. The most important reason for maintaining team integrity in an emergency operation is for safety: Team members keep track of one another and provide immediate assistance to other team members when needed. The standard operational team in firefighting operations is a company, usually consisting of three to five fire fighters working under the direct supervision of a company officer. Career fire fighters may arrive as a company on the same apparatus, whereas on-call or volunteer fire fighters should assemble in companies upon arrival.

Individual companies are assigned as a group to perform tasks at the scene of an emergency incident. The IC keeps track of the companies involved in the response and communicates with company officers. Every company officer must have a

FIGURE 18-3 A team attempts to achieve a common goal.

Near Miss REPORT

Report Number: 07-0000726

Synopsis: Truss roof collapse causes near miss at church fire.

Event Description: We were involved in fire suppression activities on a working attic fire at a church. Company crew members were attempting to perform a trench cut to ventilate and confine the fire to an area adjacent to the fire wall. Upon laddering the roof, and while crew members were still operating on and from the aerial apparatus, a sudden collapse of the truss roof occurred, endangering the crew due to fire and radiant heat. The collapse involved approximately 40 feet (12 meter) of the roof line, and a fireball extended upward approximately 30 feet (9 meter) in the air. The ventilation crew had to make an immediate descent down the aerial apparatus to escape harm. The collapse occurred about 10 minutes after arrival of the first-in companies. All fire fighters involved were in full PPE and SCBA. Awareness of wood truss construction must always be considered during fire suppression activities.

Lessons Learned: Fire-ground accountability, awareness of wood truss construction, the fire's impact due to direct impingement, and the duration of the fire must always be considered during fire suppression activities. Had the crew actually been actively ventilating or on the roof, there would have been serious injuries and most likely fire fighter fatalities.

Awareness training for wood truss construction should be included in any ventilation and structural fire suppression training. Command and control, fire-ground accountability and fire-ground safety officer training should also be included as part of the training activities.

We must always keep in mind that the fire ground is a dynamic, ever-changing scene, and we must adapt to those conditions and stay aware. Remember—no structure is worth our lives. Everyone goes home.

portable radio to maintain contact with the IC. The company officer is responsible for knowing the location of every company member at all times; he or she should also know exactly what each individual fire fighter is doing at all times.

Team integrity means that a company arrives at a fire together, works together, and leaves together. The members of a company should always be oriented to one another's location, activities, and condition. In a hazardous atmosphere, such as inside a structure, a fire fighter should always be able to contact his or her company members by voice, sight, or touch. When air cylinders need to be replaced, all of the company members should leave the building together, change cylinders together, and return to work together. If any members require rehabilitation, the full company should go to rehabilitation together and return to action together.

In some cases, a company will be divided into teams of two to perform specific tasks. For example, a ladder company might be split into a roof team and an interior search team. The initial attack on a room-and-contents fire in a single-family dwelling might be made by two fire fighters, while two other fire fighters remain outside, following the two-in/two-out rule. For safety and survival reasons, fire fighters should always use a buddy system, working in teams of at least two. If one team member should become incapacitated, his or her partner can provide help immediately and call for assistance. The two team members must remain in direct contact with each other. At least one member of every team must have a portable radio to maintain contact with the IC; in many fire departments, every member carries a portable radio.

Personnel Accountability System

A personnel accountability system provides a systematic way to keep track of the locations and functions of all personnel operating at the scene of an incident **FIGURE 18-4**. The IC has the responsibility to account for each fire fighter involved in an emergency incident. A personnel accountability system SOP

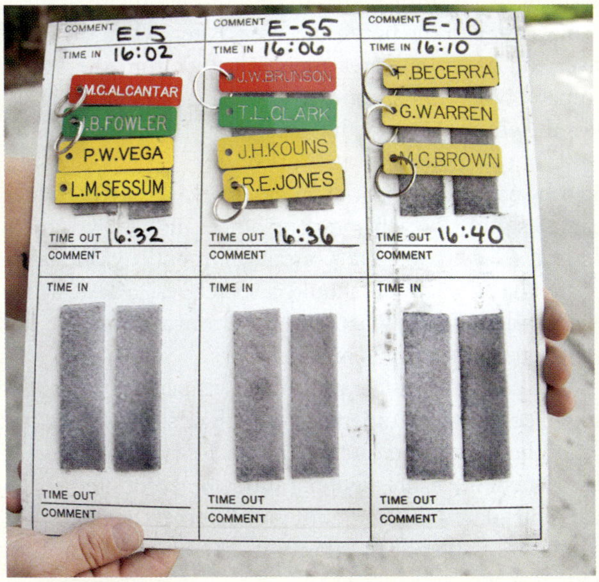

FIGURE 18-4 Personnel accountability system.

should be designed to establish standard procedures for performing this function and managing the necessary information.

A personnel accountability system must keep track of every company or team of fire fighters working at the scene of the incident, from the time they arrive until the time they are released. In addition, it must be able to identify each individual fire fighter who is assigned to a particular company or team. The system assumes that all of the company members will be together—that is, working as a unit under the supervision of the company officer—unless the IC has been advised of a change. At any point in the operation, the IC should be able to identify the location and function of each company and, within seconds, determine which individual fire fighters are assigned to that company.

The personnel accountability systems used by fire departments take many different forms. Whatever type of system is used, the only way for it to be truly effective is if every fire fighter takes responsibility for its use at all incidents. This rule includes even the smallest of incidents. Some fire departments start with a written roster of the team members assigned to each company; others use a computer database tied to a daily staffing program. Lists, roll calls, or electronic tracking databases should be used to account for all personnel at the beginning of every tour of duty or when changes are made to assignments throughout the tour.

Each piece of apparatus should carry a list of assigned company members. One popular personnel accountability system uses a Velcro or magnetic nametag for each company member. The tags for all members assigned to the apparatus are affixed to a special board carried in the cab. This board, which is sometimes called a passport, should be kept up-to-date throughout the tour of duty and should denote any change in the assigned team. In the case of a volunteer company, the tags can be affixed to the board at the time of an alarm.

At the scene of an incident, these boards are physically left with a designated person at the command post or at the entry point to a hazardous area. Depositing the board and the tags indicates that the company and those individual fire fighters have entered the hazardous area. Later, when the company leaves the hazardous area or is released from the scene, the company officer must retrieve the board—an action that indicates the entire company has safely left the hazardous area.

A personnel accountability report (PAR) is a roll call taken by each supervisor at an emergency incident. When the IC requests a PAR, each company officer physically verifies that all assigned members are present and confirms this information to the IC. The IC should also request a PAR at tactical benchmarks, such as when a change from an offensive strategy to a defensive strategy is made.

The company officer must be in visual or physical contact with all company members to verify their status. The company officer then communicates with the IC by radio to report a PAR. Whenever a fire fighter cannot be accounted for, he or she is considered missing until proven otherwise. A report of a missing fire fighter always becomes the highest priority at the incident scene.

If unusual or unplanned events occur at an incident, a PAR should always be performed. For example, if there is a report of an explosion, a structural collapse, or a fire fighter missing or in need of assistance, a PAR should immediately take place so that command can determine if any personnel are missing, identify any missing individuals, and establish their last known locations and assignments. These locations would then serve as the starting points for any search and rescue crews.

Emergency Communications Procedures

Communication is a critical part of any emergency operation, but it is especially important in relation to fire fighter safety and survival **FIGURE 18-5**. Breakdowns in the communication process are a major contributor to fire fighter injuries and deaths. Conversely, good communications can save the lives of fire fighters in a dangerous situation.

The Fire Service Communications chapter discussed the proper radio communications procedures. The following is a model of good radio communication:

> "Ladder 203 to Central IC."
> "This is Central IC. Go ahead, Ladder 203."
> "Ladder 203 has completed roof ventilation and is available for another assignment."
> "IC copies. Roof ventilation completed and Ladder 203 available for reassignment."

Such standard communications procedures ensure that the message is clearly stated and then repeated back as confirmation that it has been received and understood. Note that some

FIGURE 18-5 Communication is a critical part of any emergency operation.

fire departments might open radio communications differently. For example, "Central IC, this is Ladder 203." Consult your department's SOPs.

When a fire fighter is missing or in trouble, or when an imminent safety hazard is detected, it is even more important to communicate clearly and to ensure that the message is understood. For this reason, much of fire fighters' communications have been standardized. There are standard methods to transmit emergency information, standard information items to convey, and standard responses to emergency messages. Reserved phrases, sounds, and signals for emergency messages should be a part of every fire department's SOPs. Likewise, a separate set of standard phrases should be used for conveying information about emergency situations and hazardous conditions. These phrases should be known and practiced by everyone in the department. The same procedures and terminology should be used by all of the fire departments that can be expected to work together at emergency incidents. In many areas, these procedures are coordinated regionally.

The word "mayday" is used to indicate that a fire fighter is in trouble and requires immediate assistance. It could be used if a fire fighter is lost, is trapped, has an SCBA failure, or is running out of air. A fire fighter can transmit a mayday message to request assistance, or another fire fighter could transmit a mayday message to report that a fire fighter is missing or in trouble. A mayday call takes precedence over all other radio communications.

The term "emergency traffic" is used to indicate an imminent fire-ground hazard, such as a potential explosion or structural collapse. This kind of message would also be used to order fire fighters to immediately withdraw from interior offensive attack positions and switch to a defensive strategy. An emergency traffic message takes precedence over all other radio communications with the exception of a mayday call. Any use of the term "mayday" should bring an immediate response. The IC will interrupt any other communications and direct the person reporting the mayday to proceed. That person will then state the emergency, including all pertinent information, as in the following example:

> "This is Fire Fighter Jones with Engine 212. I'm running low on air and I'm separated from my crew. I'm on the third floor and I believe that I'm in the C side of the building."

The IC will repeat this information back and then initiate procedures to rescue Fire Fighter Jones. The next actions could include committing the rapid intervention crew (RIC) to the rescue effort, calling for additional resources, and redirecting other teams to support a search and rescue operation.

In many fire departments, the communications center sounds a special tone over the radio to alert all members that an emergency situation is in progress. It then repeats the information to be certain that everyone at the incident scene heard the message correctly. Some departments sound a horn in such circumstances.

Similar procedures are followed for emergency traffic situations. All imminent hazards and emergency instructions should be announced in a manner that captures the attention of everyone at the incident scene and ensures that the message is clearly understood.

Initiating a Mayday

A mayday message, as described previously, is used to indicate that a fire fighter is in trouble and needs immediate assistance. Analysis of fire fighter fatalities and serious injuries has shown that fire fighters often wait too long before calling for assistance. Instead of initiating a mayday call when they first get into trouble, all too often fire fighters wait until the situation is absolutely critical before requesting help. This delay could reflect a fear of embarrassment if the situation turns out to be less severe than anticipated, combined with a hope that the fire fighter will be able to resolve the problem without assistance.

The failure to act promptly can be fatal in many situations. Systems have been designed to do everything possible to protect fire fighters and to rescue fire fighters from dangerous situations—but those systems are effective only when fire fighters act appropriately when they find themselves in trouble. Do not hesitate to call for help when you think you need it.

To initiate a mayday call, follow your fire department's SOPs. In most cases, this will involve transmitting the message "Mayday, mayday, mayday" over the radio to initiate the process. When this transmission is acknowledged by the IC, give a LUNAR (location, unit number, name, assignment, and resources needed) report. Manually activate your personal alert safety system (PASS) device so that other fire fighters can hear the signal and come to your location. Also activate the emergency alert button on your portable radio, if it has this feature. All of these actions will increase the probability that the RIC or other fire fighters will be able to locate you. It is also important to remember that you should continue to work to free yourself from the dangerous situation.

To initiate a mayday call, follow the steps in SKILL DRILL 18-1:

1. Use your radio to call, "Mayday, mayday, mayday."
2. Give a LUNAR report: (STEP 1)
   - a. Location
   - b. Unit number
   - c. Name
   - d. Assignment
   - e. Resources needed
3. Activate your PASS device.
4. Attempt self-rescue.
5. If you are able to move, identify a safe location where you can await rescue. (STEP 2)
6. Lie on your side in a fetal position with your PASS device pointing out so that it can be heard. (STEP 3)
7. Point your flashlight toward the ceiling.
8. Slow your breathing as much as possible to conserve your air supply. (STEP 4)

Rapid Intervention Companies/Crews

RICs are established for the sole purpose of rescuing fire fighters who are operating at emergency incidents. In some fire departments, the term "rapid intervention team" or "fire fighter assistance and safety team" is used to describe this assignment. A RIC is a company or crew that is assigned to stand by at the incident scene, fully dressed in PPE and SCBA and equipped

SKILL DRILL 18-1 Initiating a Mayday Call
(Fire Fighter I, NFPA 5.2.4)

1. Use your radio to call, "Mayday, mayday, mayday." Give a LUNAR report: your location, unit number, name, assignment, and resources needed.

2. Activate your PASS device. Attempt self-rescue. If you are able to move, identify a safe location where you can await rescue.

3. Lie on your side in a fetal position with your PASS device pointing out so it can be heard.

4. Point your flashlight toward the ceiling. Slow your breathing as much as possible to conserve your air supply.

for action, and that is ready to deploy immediately when assigned to do so by the IC.

A RIC is an extension of the two-in/two-out rule. During the early stages of an interior fire attack, a minimum of two fire fighters are required to establish an entry team and a minimum of two additional fire fighters are required to remain outside the hazardous area. The two fire fighters who remain outside the hazardous area can perform other functions, but they must have donned full PPE and SCBA, so that they can immediately enter and assist the entry team if the interior fire fighters need to be rescued. The two fire fighters who remain outside the hazardous area fulfill the first stage of a RIC procedure. The second stage involves a dedicated RIC, specifically assigned to stand by for fire fighter rescue assignments.

NFPA 1500, *Standard on Fire Department Occupational Safety and Health Program*, includes a complete description of RIC procedures and provides examples of their use at typical scenes. A number of options are available for establishing and organizing RIC crews or teams, including dispatching one or more additional companies to incidents with this specific assignment. Some small fire departments rely on an automatic response from an adjoining community to provide a dedicated RIC at working incidents.

SOPs should dictate when a RIC is required, how it is assigned, and where it should be positioned at an incident scene. A RIC should be in place at any incident where fire fighters are operating in conditions that are immediately dangerous to life and health (IDLH), which includes any structure fire where fire fighters are operating inside a building and using SCBA. Many fire departments specify a special set of rescue equipment, including extra SCBA air cylinders, that is brought to the standby location. All RIC members must wear full PPE and SCBA, and be ready for action at an instant.

The IC should quickly deploy the RIC to any situation where a fire fighter needs immediate assistance. Situations that call for the use of the RIC would include a lost or missing fire fighter, an injured fire fighter who has to be removed from a hazardous location, and a fire fighter trapped by a structural collapse. The U.S. federal regulation that establishes requirements for using respiratory protection (OSHA 29 CFR 1910.134) mandates the use of backup teams at incidents where operations are conducted in an unsafe atmosphere (potentially IDLH).

Fire Fighter Survival Procedures

Some of the most important procedures you need to learn are directly related to your personal safety and to the safety of the fire fighters who will be working with you. These procedures are designed to keep you from entering into dangerous situations and to guide you in situations where you could be in immediate danger. Your personal safety will depend on learning, practicing, and consistently following these procedures. However, do not lose sight of the fact that preventing life-threatening situations is preferable to utilizing survival procedures.

Maintaining Orientation

As you practice working with SCBA, you will discover that it is very easy to become disoriented in a dark, smoke-filled building. Sometimes the visibility inside a burning building can be measured in inches. In these conditions, it is extremely important to stay oriented. If you become lost, you could run out of air before you find your way out. The atmosphere outside your face piece could be fatal in one or two breaths. Even if you call for assistance, rescuers might not be able to find you. For all these reasons, safety and survival inside a fire building are directly related to remaining oriented within the building.

Several methods can help fire fighters stay oriented inside a smoke-filled building. Before entering a building, look at it from the outside to get an idea of the size, shape, arrangement, and number of stories. After entering the structure, paint a picture in your mind of your surroundings. Given that most of your senses will be disrupted by dense smoke, touch may become your only means of moving through a structure. By feeling out your surroundings, you will be able to tell which type of room you have entered. Feel for changes in floor coverings, and always look for alternative means of egress.

Preplanning helps with orientation. By reviewing the preincident plan before entering the structure you will have a general idea of the structure's floor plan. See the Preincident Planning chapter for more information.

One of the most basic methods to remain oriented is to always stay in contact with a hose line—for example, by always keeping a hand or foot on the hose line. To find your way out, feel for the couplings. Place one hand on the male coupling and the other hand on the female coupling. To exit the building, travel in the direction of the hand holding the male coupling.

Team integrity is an important factor in maintaining orientation. Everyone works as a team to stay oriented. When the team members cannot see one another, they must stay in direct physical contact or within the limits of verbal contact. Standard procedures are employed to ensure that fire fighters work efficiently as a team, such as having one member remain at the doorway of the room while the other members perform a systematic search inside. The searchers then work their way around the room in one direction until they get back to the fire fighter at the door.

A guideline can be used for orientation when inside a structure. A guideline is a rope that is attached to an object on the exterior or a known fixed location; it is stretched out as a team enters the structure. The fire fighters can follow the guideline back out, or another team can follow the guideline in to find them. Some fire departments use a series of knots to mark distances along the rope to indicate how far it is stretched. Be aware that using the guideline technique requires intense practice.

Training sessions in conditions with no visibility will build confidence. This type of training can be conducted inside any building, while using an SCBA face piece with the lens covered. More realistic conditions can be simulated by using smoke machines inside training facilities. Fire fighters should practice navigating around furniture, following charged hose lines, climbing and descending stairs, and fitting through narrow spaces. Distracting noises, such as operating fans, can add a realistic feel. Team integrity must always be maintained in these practice sessions.

Self-Rescue

If you become separated from your team, or if you become lost, disoriented, or trapped inside a structure, you can use several techniques for self-rescue. The first step is always to call for assistance. Do not wait until it is too late for anyone to find you before making it known that you are in trouble. As described in the section "Initiating a Mayday," time is critical when your safety is at risk. You need to initiate the process as soon as you *think* you are in trouble, not when you are absolutely *sure* you are in grave danger.

If you are simply separated from your team, the best option is to take a normal route of egress to the exterior. Follow a hose line back to an open doorway, descend a ladder, or locate a window or door—all of these options can provide a direct exit and require no special tools or techniques. You must immediately communicate with your company officer or the IC to inform them that you have exited safely.

To rescue yourself by locating and following a hose line to an exit, follow the steps in **SKILL DRILL 18-2**:

1. Initiate a mayday.
2. Stay calm and control your breathing. (**STEP 1**)
3. Systematically search the room to locate a hose line.
4. Follow the hose line to a hose coupling.
5. Identify the male and female ends of the coupling.
6. Move from the female coupling to the male coupling. (**STEP 2**)
7. Follow the hose out.
8. Exit the hazard area.
9. Notify command of your location. (**STEP 3**)

To locate a door or window that can be used for an exit, follow the steps in **SKILL DRILL 18-3**:

1. Initiate a mayday.
2. Stay calm and control your breathing. (**STEP 1**)
3. Systematically locate a wall. (**STEP 2**)
4. Use a sweeping motion on the wall to locate an alternative exit.
5. Identify the opening as a window, interior door, or external door. (**STEP 3**)
6. If the first opening identified is not adequate for an exit, continue to search.
7. Maintain your orientation and stay low.
8. Exit the room safely if possible.
9. If unable to exit, assume the downed fire fighter position in a safe location or find refuge.
10. Keep command informed of your situation. (**STEP 4**)

In addition to these actions, fire fighters can use many more complicated techniques to escape from dangerous predicaments, such as breaching a wall. These methods must be learned and practiced regularly so that you will be ready to

SKILL DRILL 18-2 Performing Self-Rescue

(Fire Fighter I, NFPA 5.3.5)

1 Initiate a mayday. Stay calm and control your breathing.

2 Systematically search the room to locate a hose line. Follow the hose line to a hose coupling. Identify the male and female ends of the coupling. Move from the female coupling to the male coupling.

3 Follow the hose out. Exit the hazard area. Notify command of your location.

SKILL DRILL 18-3 Locating a Door or Window for Emergency Exit

(Fire Fighter I, NFPA 5.3.5)

1. Initiate a mayday. Stay calm and control your breathing.

2. Systematically locate a wall.

3. Use a sweeping motion on the wall to locate an alternative exit. Identify the opening as a window, interior door, or external door.

4. If the first opening identified is not adequate for an exit, continue to search. Maintain your orientation and stay low. Exit the room safely if possible. If unable to exit, assume the downed fire fighter position in a safe location or find refuge. Keep command informed of your situation.

perform them without hesitation if you are ever in a situation that requires them.

To escape through a wall that is constructed with wood or metal studs, you need to expose the studs and then reduce the size of your profile to get your body and your SCBA through the narrow opening. Sometimes you can get through a narrow opening by loosening your waist strap and one strap of your SCBA and then repositioning your SCBA as you twist through the limited opening. At other times you may have to remove your SCBA backpack while keeping your face piece on. You will then have to push your SCBA in front of you. Hold on to your backpack carefully to prevent it from becoming separated from your face piece. The technique you need to use will depend on your size. A small person may be able to squeeze through a limited-size opening without removing SCBA; a larger person trying to escape through the same small opening may have to remove the SCBA backpack to fit through it.

One method of opening a wall and using a reduced profile to escape through a wall is shown in **SKILL DRILL 18-4**:

1. Identify deteriorating conditions that require exiting through a wall.
2. Initiate a mayday.
3. Use a hand tool or your feet to open a hole in the wall between two studs.
4. If using a hand tool, drive the tool completely through the wall to check for obstacles on the other side. (**STEP 1**)
5. Enlarge the hole.
6. Enter the hole head first to check the floor and fire conditions on the other side of the wall. (**STEP 2**)
7. Loosen the SCBA waist strap and remove one shoulder strap.
8. Sling the SCBA to one side to reduce your profile. (**STEP 3**)
9. Escape through the opening in the wall. Adjust the SCBA straps to their normal position. (**STEP 4**)
10. Assist others through the opening in the wall.
11. Report your status to command. (**STEP 5**)

Disentanglement is an important skill that fire fighters need to learn and practice. When crawling in a smoke-filled building, a fire fighter can easily become entangled in wires, cables, and debris. Above most dropped ceilings is a huge and varied mass of electrical wires and conduit, computer and TV cables, plumbing and sewage pipes, telephone cables, and heating and air-conditioning tubing and ducts. Because many dropped ceilings are held up only with wires, it does not take much fire or water damage to bring this huge tangle of utilities crashing down. If these wires fall on your back while you are wearing an SCBA, you could be easily trapped.

Many fire fighters carry small tools in the pockets of their PPE that can be used to cut through wires or small cables. This procedure can be very difficult if poor-visibility conditions prevent the fire fighter from seeing and identifying the entangling material.

To escape from an entanglement, follow the steps in **SKILL DRILL 18-5**:

1. Initiate a mayday.
2. Stay calm and control your breathing. (**STEP 1**)
3. Change your position—back up and turn on your side to try to free yourself. (**STEP 2**)
4. Use the swimmer stroke to try to free yourself. (**STEP 3**)
5. Loosen the SCBA straps, remove one arm, and slide the air pack to the front of your body to try to free the SCBA.(**STEP 4**)
6. Cut the wires or cables causing the entanglement. Be aware of any possible electrocution risk.
7. If you are unable to disentangle yourself, notify command of your situation.
8. If you are able to exit, notify command that you are out of danger. (**STEP 5**)

Some additional self-rescue methods involve using tools and equipment in ways that they were not designed for. These so-called last-resort methods should be taught only by experienced and qualified instructors and must be practiced with strict safety measures in place. Some of these methods are quite controversial. Some fire departments consider them unacceptable owing to the high risk of injury when they are used or even practiced; other fire departments provide training in them because these measures could save the life of a fire fighter in an extreme situation. This policy decision must be made by each fire department.

Safe Locations

A safe location is a temporary location that provides refuge while awaiting rescue or finding a method of self-rescue from an extremely hazardous situation. The term "safe" in this case is relative: The location may not be particularly safe, but it is less dangerous than the alternatives. Wildland fire fighters carry lightweight reflective emergency shelters as a form of safe location. When overrun by a fire, taking refuge in the shelter might be the fire fighter's only chance for survival.

Safe locations are important when situations become critical and few alternatives are available. When a critical situation occurs, fire fighters should know where to look for a safe location and recognize one when it presents itself.

If fire fighters become trapped inside a burning building, a room with a door and a window could be a safe location. Closing the door could keep the fire out of the room for at least a few minutes, and breaking the window could provide fresh air and an escape route. In this case, the safe location ideally provides time for a rescue team to reach the fire fighters, a ladder to be raised to the window, or another group of fire fighters to control the fire.

A roof or floor collapse often leaves a void adjacent to an exterior wall. If trapped fire fighters can reach this void, it could be the best place to go. Breaching the wall might be the only way out of the void or the only way for a rescue team to reach the trapped personnel.

SKILL DRILL 18-4 Opening a Wall to Escape

(Fire Fighter I, NFPA 5.3.5)

1 Identify deteriorating conditions that require exiting through a wall. Initiate a mayday. Use a hand tool or your feet to open a hole in the wall between two studs. If using a hand tool, drive the tool completely through the wall to check for obstacles on the other side.

2 Enlarge the hole. Enter the hole head first to check the floor and fire conditions on the other side of the wall.

3 Loosen the SCBA waist strap and remove one shoulder strap. Sling the SCBA to one side to reduce your profile.

4 Escape through the opening in the wall. Adjust the SCBA straps to their normal position.

5 Assist others through the opening in the wall. Report your status to command.

SKILL DRILL 18-5 Escaping from an Entanglement

(Fire Fighter I, NFPA 5.3.5)

1. Initiate a mayday. Stay calm and control your breathing.

2. Change your position—back up and turn on your side to try to free yourself.

3. Use the swimmer stroke to try to free yourself.

4. Loosen the SCBA straps, remove one arm, and slide the air pack to the front of your body to try to free the SCBA.

5. Cut the wires or cables causing the entanglement. Be aware of any possible electrocution risk. If you are unable to disentangle yourself, notify command of your situation. If you are able to exit, notify command that you are out of danger.

VOICES
OF EXPERIENCE

It was the first call of the night—a possible structure fire. We arrived to see dark black smoke rolling out from the basement windows of a single-story house. My stomach dropped and it seemed that everything went into slow motion. A voice yelled, "My husband is in the basement! Can you hear me? I said my husband is in the basement!" This was an older home, with an outside basement entrance. I kept thinking, "We need to find this guy. He's going to be okay if we can find him."

Just inside the door to the basement, my partner and I found the husband. He had made it up the stairs and had almost made it to safety before he passed out in front of the basement door. My partner and I dragged him out of the stairway and handed him off to the paramedics.

Then my heart skipped a beat—my partner was gone. I followed the hoseline back into the basement. He was not there. I began to feel panic. Where could he be? I had to find him. So I pulled the attack line even farther into the basement. The further I went in, the worse the conditions became. I was trying to find my partner as I crawled down the stairway, getting closer and closer to the fire.

A second team from a mutual aid department came down the hoseline to help. I tried to communicate to them that I had lost my partner. We went in deeper. It was hot and we could not find the fire. We could hear the fire, but where was it? And where was my partner? It sounded like a freight train above us. The roar of the fire became more and more intense.

All of a sudden something said it was time to get out. Conditions were changing rapidly, we couldn't find the fire, and it was getting louder and hotter. Even though we could not find my partner, we had to leave and we had to leave immediately. We scrambled up the stairs, rushing to get out. We crawled out of the doorway to the outside. Within seconds the entire basement flashed, pushing flames out of the stairway that we had just come out of. Panic stricken, I looked around and found my partner standing in front of me, completely unconcerned about being lost. Later I learned that he had stayed with the victim we had extricated.

We were lucky at that fire. Basement fires can be extremely dangerous. Not knowing where my partner was and going alone to look for him was a critical error. I was lucky that the other team went in with me. I was so focused on trying to find my partner that I would not have noticed the warning signs and would have died looking for a guy who wasn't even there.

David Winter
College Place Fire Department
College Place, Washington

Maintaining team integrity is important when escaping to a safe location. In some cases, fire fighters have survived in safe locations long enough to work together to create an opening. The ability to use a radio to call for help and to describe the location where fire fighters are trapped can be critical.

Like all of the emergency procedures described in this chapter, these activities require good instruction and practice. Always follow your department's operating guidelines when it comes to self-rescue techniques. Follow the SOPs used by your department for these evolutions.

Air Management

Air management is important to all fire fighters when using SCBA. Air management reflects a basic fact: Air equals time. How long a fire fighter can survive in a dangerous atmosphere depends on how much air is in the SCBA and how quickly the fire fighter uses it. You have to manage your air supply to manage your time in a hazardous area.

The time in the hazardous atmosphere must take into account the time it takes to enter and get to the location where the fire fighter plans to operate and the time it takes to exit safely. The work time is limited by the amount of air in the SCBA cylinder, after allowing for entry and exit time. If it will take 6 minutes to reach a safe atmosphere and the fire fighter has only 3 minutes of air left, the results could be disastrous.

Recall that the time rating for an SCBA is based on a standard rate of consumption for a typical adult under low-exertion conditions in a test laboratory. Fire fighters typically consume air much more rapidly than occurs in the standard test. In fact, fire fighters often consume a 45-minute-rated air supply in just 20 to 25 minutes.

The actual rate of air consumption varies significantly among individual fire fighters and depends on the activities that they are performing. Chopping holes with an axe or pulling ceilings with a pike pole, for example, will cause a fire fighter to consume air more rapidly than will conducting a survey with a thermal image camera. Even given the same workload, no two individuals will use exactly the same amount of air from their SCBA cylinders. Differences in body size, fitness levels, and even heredity factors cause air consumption rates to be faster or slower than the average.

Air management must be a team effort as well as an individual responsibility. Fire fighters will always be working as part of a team when using SCBA, and team members inevitably use air at different rates. The member who uses the air supply most rapidly determines the working time for the team.

A good way to determine your personal air usage rate is to participate in an SCBA consumption exercise. This drill puts fire fighters through a series of exercises and obstacles to learn how quickly each fire fighter uses air when performing different types of activities. Repeated practice can help you learn to use air more efficiently by building your confidence and learning to control your breathing rate. Keeping physically fit results in a more efficient use of oxygen and allows you to work for longer periods with a limited air supply.

Practice sessions should take place in a nonhazardous atmosphere. A typical skill-building exercise might include the following components:

- While wearing full PPE and breathing air from SCBA, perform typical activities such as carrying hose packs up stairs, hitting an object with an axe or sledgehammer, or crawling along a hallway.
- Record the following information: how much air is in the cylinder when you start; how long you work; whether the physical labor is light or heavy; when the low-air alarm sounds; and when the cylinder is completely empty.

This type of exercise will allow you to determine your personal rate of consumption and work on improving your times. With practice, you will be able to predict how long your air supply should last under a given set of conditions.

Knowledge of your team members' physical conditions and their workloads can help you look out for their safety as well. Because many tasks are divided up in any given operation, one member may be working to his or her maximum level while another team member is acting only in a supportive role. This difference in exertion levels could cause the first team member to use up his or her air supply much faster without realizing it.

When using SCBA at an emergency incident, be aware of the limitations of the device. Do not enter a hazardous area unless your air cylinder is full. Keep track of your air supply by using the heads-up display to determine how much air you have left. Always follow the rule of air management: Know how much air you have in your SCBA before you go in, and manage that amount so that you leave the hazardous environment *before* your SCBA low-air alarm activates. Do not wait until the low-pressure alarm sounds to start thinking about leaving the hazardous area; you should be out of the hazardous area before this alarm sounds.

Fire Fighter Safety Tips

Follow the rule of air management: Know how much air you have in your SCBA before you go in, and manage that amount so that you leave the hazardous environment *before* your SCBA low-air alarm activates.

Even when fire fighters employ the best possible air management practices, emergency situations can occur. An SCBA can malfunction or fire fighters may be trapped inside a building by a structural collapse or an unexpected situation. In these circumstances, remaining calm and knowing how to use all of the emergency features of the SCBA can be critically important. Remaining calm, controlling your breathing rate, and taking shallow breaths can slow air consumption. All of these techniques require practice to build confidence.

Fire Fighter Safety Tips

Before entering a hazard zone:

- Ensure that all crew members have portable radios and appropriate hand tools.
- Keep crews together.
- Work under the incident command system.
- Follow guidelines for managing your air supply.

Rescuing a Downed Fire Fighter

One of the most critical and demanding situations in the fire service arises when fire fighters have to rescue another fire fighter who is in trouble. Air management has critical implications in a fire fighter rescue situation. Specifically, air management must be considered for both the rescuers and the fire fighter who is in trouble.

When you reach a downed fire fighter, the first step is to assess the fire fighter's condition. Is the fire fighter conscious and breathing? Does he or she have a pulse? Is the fire fighter trapped or injured? Make a rapid assessment and notify the IC of your situation and location. Have the RIC deployed to your location and quickly determine the additional resources needed.

The most critical decision at this point is whether the fire fighter can be moved out of the hazardous area quickly and easily or whether considerable time and effort will be required to bring the fire fighter to safety. If it will take more than a minute or two to remove the fire fighter from the building, air supply will be an important consideration. You need to provide enough air to keep the fire fighter breathing for the duration of the rescue operation.

Fire fighters must know and understand all aspects of their SCBA in this type of situation. If the downed fire fighter has a working SCBA, determine how much air remains in the cylinder. A fire fighter who is breathing and has an adequate air supply is not in immediate, life-threatening danger. If the SCBA contains very little air or no air, however, air management is a critical priority. In this case, rescuers need to move the fire fighter out of the hazardous area immediately or provide an additional air supply. A spare SCBA air cylinder should be brought to the location and connected to the downed fire fighter's SCBA through the rapid intervention crew/company universal air connection (RIC UAC). This procedure provides for rapid emergency refilling of breathing air to an SCBA for the downed or trapped fire fighter. Many RIC teams bring a complete replacement SCBA as part of their rescue effort in case it is needed.

If the fire fighter's SCBA is not working, check whether a simple fix can make it operational. Did a hose or regulator come off? Was the cylinder valve accidentally closed? If regulator failure is a possibility, open the bypass valve.

When it is necessary to quickly remove a downed fire fighter, you can use the fire fighter's SCBA straps to create a rescue harness. To rescue a downed fire fighter by making a rescue harness from the fire fighter's SCBA straps, follow the steps in **SKILL DRILL 18-6**:

1. Locate the downed fire fighter.
2. Activate the mayday procedure, if that step has not already been taken.
3. Quickly assess the condition of the downed fire fighter and the situation.
4. Shut off the PASS device as needed to aid in communication. (**STEP 1**)
5. Position yourself at the legs of the downed fire fighter.
6. Lift one leg of the downed fire fighter up onto your shoulder. (**STEP 2**)
7. Locate and loosen the waist straps and shoulder straps (if needed) of the downed fire fighter's SCBA harness. (**STEP 3**)
8. Unbuckle the downed fire fighter's waist strap, and buckle the waist strap under the lifted leg.
9. Tighten the straps.
10. Remove the downed fire fighter from the hazard area to a safe area. (**STEP 4**)

To rescue a downed fire fighter who is wearing PPE and SCBA from the immediate hazard using a drag rescue device, follow the steps in **SKILL DRILL 18-7**:

1. Locate the downed fire fighter.
2. Activate the mayday procedure, if that step has not already been taken.
3. Shut off the PASS device to aid in communication.
4. Assess the situation and the condition of the downed fire fighter. (**STEP 1**)
5. Use the rapid intervention crew/company universal air connection (RIC–UAC) to fill the downed fire fighter's air supply cylinder, if needed. (**STEP 2**)
6. Access the fire fighter's drag rescue device. (**STEP 3**)
7. Remove the downed fire fighter from the hazard area to a safe area. (**STEP 4**)

As a member of a two-person team, you can also remove a downed fire fighter with PPE and SCBA from a hazardous area by having one fire fighter pull on the downed fire fighter's SCBA straps and the second fire fighter support the legs of the downed fire fighter. To do so, follow the steps in **SKILL DRILL 18-8**:

1. Locate the downed fire fighter.
2. Activate the mayday procedure, if that step has not already been taken.
3. Shut off the PASS device to aid in communication.
4. Assess the situation and the condition of the downed fire fighter.
5. Use the RIC–UAC to fill the downed fire fighter's air supply cylinder, if needed. (**STEP 1**)
6. The second fire fighter converts the downed fire fighter's SCBA harness into a rescue harness. (**STEP 2**)

SKILL DRILL 18-6 Rescuing a Downed Fire Fighter Using SCBA Straps as a Rescue Harness

(Fire Fighter I, NFPA 5.3.5)

1 Locate the downed fire fighter. Activate the mayday procedure, if that step has not already been taken. Quickly assess the condition of the downed fire fighter and the situation. Shut off the PASS device as needed to aid in communication.

2 Position yourself at the legs of the downed fire fighter. Lift one leg of the downed fire fighter up onto your shoulder.

3 Locate and loosen the waist straps and shoulder straps (if needed) of the downed fire fighter's SCBA harness.

4 Unbuckle the downed fire fighter's waist strap, and buckle the waist strap under the lifted leg. Tighten the straps. Remove the downed fire fighter from the hazard area to a safe area.

7. The first fire fighter grabs the shoulder straps or uses the webbing to create a handle to pull the downed fire fighter. (STEP 3)
8. The second fire fighter stays at the downed fire fighter's legs and supports the legs of the downed fire fighter. (STEP 4)
9. Remove the downed fire fighter from the hazard area to a safe area. (STEP 5)

Rehabilitation

The Fire Fighter Rehabilitation chapter covers fire fighter rehabilitation in detail. The purpose of rehabilitation is to reduce the effects of fatigue during an emergency operation FIGURE 18-6. Firefighting involves very demanding physical labor. When combined with extremes of weather and the mental stresses associated with emergency incidents, it can

SKILL DRILL 18-7

Rescuing a Downed Fire Fighter Using a Drag Rescue Device

(Fire Fighter I, NFPA 5.3.5)

1 Locate the downed fire fighter. Activate the mayday procedure, if that step had not already been taken. Shut off the PASS device to aid in communication. Assess the situation and the condition of the downed fire fighter.

2 Use the rapid intervention crew/company universal air connection (RIC–UAC) to fill the downed fire fighter's air supply cylinder, if needed.

3 Access the fire fighter's drag rescue device.

4 Remove the downed fire fighter from the hazard area to a safe area.

challenge the strength and endurance of even fire fighters who are physically fit and well prepared. Rehabilitation helps fire fighters retain the ability to perform at the current incident and restores their capacity to work at later incidents.

At the simplest of incidents, a rehabilitation area can be set up on the tailboard of an apparatus with a water cooler. At larger incidents, a complete rehabilitation operation should be established, including personnel to monitor the vital signs of fire fighters and provide first aid. In whatever size or form it takes, rehabilitation is integral to fire fighter safety and survival. The personnel accountability system must continue to track fire fighters who are assigned to report to rehabilitation and note when they are released from rehabilitation for another assignment.

Counseling and Critical Incident Stress

Fire fighters are often exposed to stressful situations. Sometimes they are involved in situations that are described as critical incidents; these incidents challenge the capacity of

SKILL DRILL 18-8 Rescuing a Downed Fire Fighter as a Two-Person Team

(Fire Fighter I, NFPA 5.3.5)

1. Locate the downed fire fighter. Activate the mayday procedure, if that step has not already been taken. Shut off the PASS device to aid in communication. Assess the situation and the condition of the downed fire fighter. Use the RIC–UAC to fill the downed fire fighter's air supply cylinder, if needed.

2. The second fire fighter converts the downed fire fighter's SCBA harness into a rescue harness.

3. The first fire fighter grabs the shoulder straps or uses the webbing to create a handle to pull the downed fire fighter.

4. The second fire fighter stays at the downed fire fighter's legs and supports the legs of the downed fire fighter.

5. Remove the downed fire fighter from the hazard area to a safe area.

FIGURE 18-6 Rehabilitation reduces the effects of fatigue produced by an operation.

FIGURE 18-7 Group stress debriefings are sometimes used to alleviate the stress reactions generated by high-stress emergency situations.

most individuals to deal with stress. Examples of critical incidents include the following:

- Line-of-duty deaths (police, fire/rescue, Emergency Medical Services)
- Suicide of a colleague
- Serious injury to a colleague
- Situations that involve a high level of personal risk to fire fighters
- Events in which the victim is known to the fire fighters
- Multiple-casualty/disaster/terrorism incidents
- Events involving death or life-threatening injury or illness to a victim, especially a child
- Events that are prolonged or end with a negative or unexpected outcome

This list is not complete, nor is it necessarily a fact that any of these situations will seriously trouble every individual. Normal coping mechanisms help many fire fighters to handle many situations. Some individuals have a high capacity to deal with stressful situations through exercise, talking to friends and family, or turning to their religious beliefs. These are healthy, nondestructive ways to deal with the pressures of being exposed to a critical incident.

Sometimes individuals react to critical incidents in ways that are not positive—such as alcohol or drug abuse, depression, the inability to function normally, or a negative attitude toward life and work. These symptoms can occur in anyone, even individuals who normally have healthy coping skills. Reactions will vary from one individual to the next, both in type and severity. Many times fire fighters do not realize they are affected in a deeply negative manner. A somewhat routine incident, however, may trigger negative reactions from a critical incident that occurred in the past. Critical incident stress can also be cumulative, building up over time. This condition, which is called burnout, cannot be traced to any one incident.

The aim of counseling and critical incident stress management (CISM) is to prevent these reactions from having a negative impact on the fire fighter's work and life, over both the short term and the long term FIGURE 18-7. Fire fighters should understand the counseling resources and CISM systems available to them and know how they can access them. Maintaining good emotional health is a simple but important part of fire fighter survival.

The recognized stages of emotional reaction experienced by fire fighters and other rescue personnel after a stressful incident can include the following:

1. Anxiety
2. Denial/disbelief
3. Frustration/anger
4. Inability to function logically
5. Remorse
6. Grief
7. Reconciliation/acceptance

These stages can occur within minutes or hours, or they can take several days or months to unfold. Not all of the steps will occur for every event, and they do not necessarily occur in the order given here.

Counseling aims to help fire fighters recognize and deal with these reactions in the most positive manner possible. Although the ways in which this outcome is accomplished may vary, the purpose remains the same.

Your fire department should have some type of counseling program in place. All department members should know how it is activated and be supportive of its intent and function. Fire fighters should realize that their emotional and mental health must be protected, just as much as their physical health and safety.

Wrap-Up

Chief Concepts

- Risk–benefit analysis weighs the positive results that can be achieved against the probability and severity of potential negative consequences. It should be practiced and implemented by everyone on the fire ground. While the IC is responsible for the analysis of the overall scene, company officers and fire fighters need to process the risks and benefits of the strategies implemented by command.
- No property is ever worth the life of a fire fighter.
- Smoke is an obvious hazard of fire. You should also be familiar with the less obvious hazard indicators, including inclement weather conditions and the likelihood that certain occupancies might hold hazardous materials.
- Safe operating procedures define the manner in which a fire department conducts operations at an emergency incident.
- The 11 rules of engagement identified by the IAFC and NVFC are as follows:
  - Size up your tactical area of operation.
  - Determine the occupant survival profile.
  - Do not risk your life for lives or property that cannot be saved.
  - Extend limited risk to protect savable property.
  - Extend vigilant and measured risk to protect and rescue savable lives.
  - Go in together, stay together, and come out together.
  - Maintain continuous awareness of your air supply, situation, location, and fire conditions.
  - Constantly monitor fire-ground communications for critical radio reports.
  - Report unsafe practices or conditions that can harm you. Stop, evaluate, and decide.
  - Abandon your position and retreat before deteriorating conditions can harm you.
  - Declare a mayday as soon as you think you are in danger.
- Team integrity means that a company arrives at a fire together, works together, and leaves together. Personnel accountability systems help to ensure team integrity by tracking all of the locations and functions of the personnel on the scene.
- A personnel accountability report is a roll call taken by each supervisor at an emergency incident. These headcounts are made at regular, predetermined intervals throughout the incident, generally every 10 minutes.
- Mayday is used if a fire fighter is lost, is trapped, has an SCBA failure, or is running out of air. A fire fighter can transmit a mayday message to request assistance, or another fire fighter can transmit a mayday message to report that a fire fighter is missing or in trouble.
- A mayday call takes precedence over all other radio communications.
- Emergency traffic is used to indicate an imminent fire-ground hazard. This kind of message would also be used to order fire fighters to immediately withdraw from interior offensive attack positions and switch to a defensive strategy. An emergency traffic message takes precedence over all other radio communications except a mayday call.
- The sole purpose of RICs is to rescue fire fighters operating at emergency incidents. An RIC is an extension of the two-in/two-out rule. This team should be in place at any incident where fire fighters are operating in conditions that are immediately dangerous to life and health and should be deployed to any situation where a fire fighter needs immediate assistance.
- One of the most basic methods to remain oriented in a dark, smoke-filled environment is to always stay in contact with a hose line or guideline.
- The first step in attempting a self-rescue is to initiate a mayday.
- A safe location is a temporary location that provides refuge while awaiting rescue or finding a method of self-rescue. In this case, "safe" means less hazardous.
- Always follow the rule of air management: Know how much air you have left in your SCBA before you enter a building, and manage that amount so that you leave the hazardous environment before your low-air alarm sounds.
- SCBA straps or a drag rescue harness may be used to help drag a downed fire fighter to a safe location.
- Critical incident stress is a known hazard that emergency personnel can learn to handle. This hazard can affect fire fighters' emotional and mental health. Numerous resources are available to help fire fighters maintain good emotional and mental health.

Hot Terms

Air management The use of a limited air supply in such a way as to ensure that it will last long enough to enter a hazardous area, accomplish needed tasks, and return safely.

Critical incident stress management (CISM) A program designed to reduce both acute and chronic effects of stress related to job functions. (NFPA 450)

Drag rescue device (DRD) A component integrated within the protective coat element to aid in the rescue of an incapacitated fire fighter. (NFPA 1851)

Guideline A rope used for orientation when fire fighters are inside a structure where there is low or no visibility. The line is attached to a fixed object outside the hazardous area.

Personnel accountability report (PAR) Periodic reports verifying the status of responders assigned to an incident or planned event. (NFPA 1026)

Risk–benefit analysis An assessment of the risk to rescuers versus the benefits that can be derived from their intended actions. (NFPA 1006)

Safe location A location remote or separated from the effects of a fire so that such effects no longer pose a threat. (NFPA 101)

Self-rescue Escaping or exiting a hazardous area under one's own power. (NFPA 1006)

FIRE FIGHTER *in action*

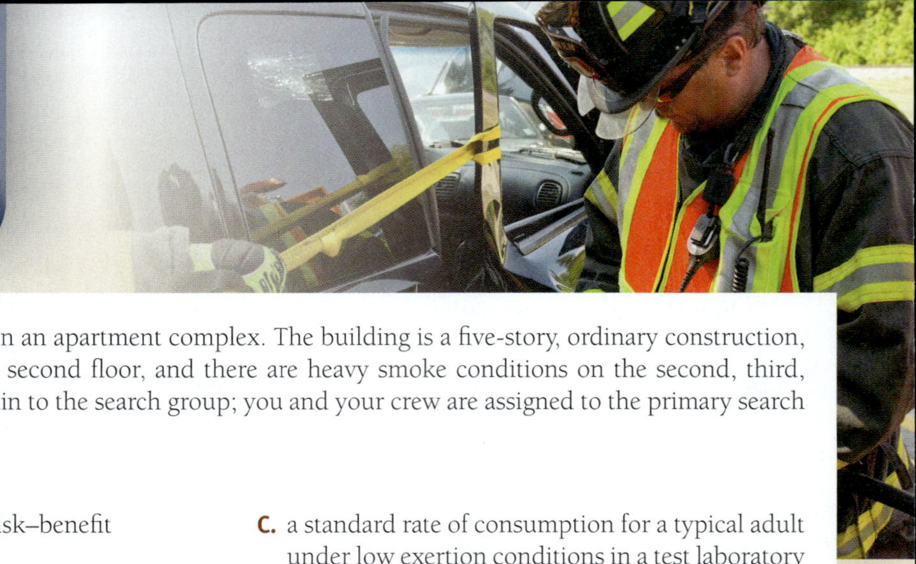

It is 2:00 a.m. and you are on the scene of a fire in an apartment complex. The building is a five-story, ordinary construction, with a center hallway design. The fire is on the second floor, and there are heavy smoke conditions on the second, third, fourth, and fifth floors. The IC assigns your captain to the search group; you and your crew are assigned to the primary search of the third floor.

1. Who is responsible for conducting a risk–benefit analysis during this fire?
- **A.** Incident commander
- **B.** Company officer
- **C.** Fire fighter
- **D.** All of the above

2. Which of the following is *not* part of the 11 Rules of Engagement?
- **A.** Constantly monitor fire-ground communications for critical radio reports.
- **B.** Report unsafe practices or conditions that can harm you.
- **C.** Do not abandon your position and retreat before deteriorating conditions can harm you unless directed to do so by the IC.
- **D.** Declare a mayday as soon as you *think* you are in danger.

3. If you become lost during your search, what should you do?
- **A.** Continue searching until you find something that is familiar
- **B.** Switch the hand that is in contact with the wall and try to find your way out
- **C.** Transmit a mayday communication over the radio
- **D.** Transmit an "emergency traffic" communication over the radio

4. The time rating for an SCBA is based on ______.
- **A.** a standard rate of consumption for a typical fire fighter under high exertion conditions in a test laboratory
- **B.** a standard rate of consumption for a typical adult under high exertion conditions in a test laboratory
- **C.** a standard rate of consumption for a typical adult under low exertion conditions in a test laboratory
- **D.** a standard rate of consumption for a typical fire fighter under low exertion conditions in a test laboratory

5. When is a personnel accountability report not necessary?
- **A.** When changing from an offensive to a defensive strategy
- **B.** When an emergency evacuation has been declared
- **C.** Every 10 minutes
- **D.** Upon transfer of command

6. What is the purpose of critical incident stress management?
- **A.** To prevent the public's reaction to critical incidents from having a negative impact on their work and life
- **B.** To prevent a fire fighter's reaction to critical incidents from having a negative impact on the fire fighter's work and life
- **C.** To prevent fire fighters from having an emotional response to critical situations
- **D.** To prevent the public from having an emotional response to critical situations

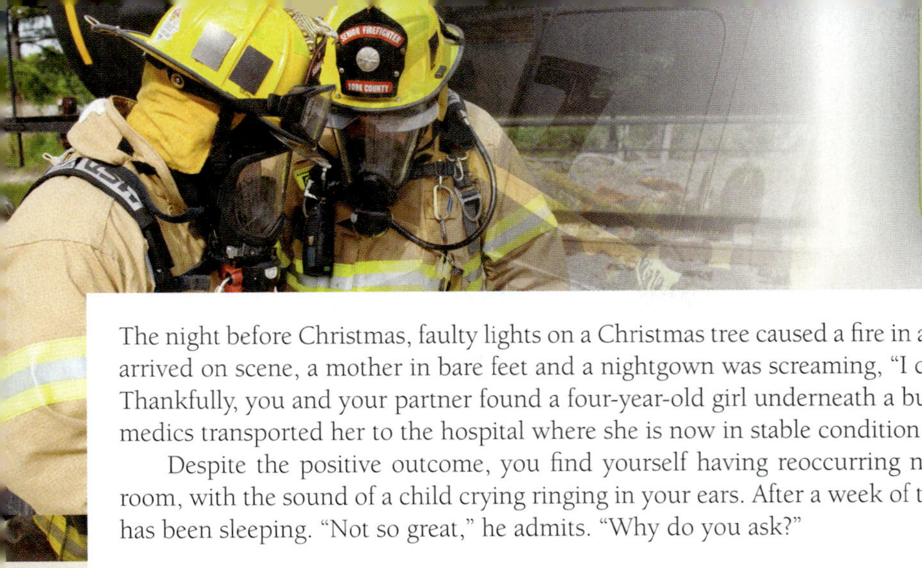

FIRE FIGHTER II *in action*

The night before Christmas, faulty lights on a Christmas tree caused a fire in a multi-unit apartment building. When your crew arrived on scene, a mother in bare feet and a nightgown was screaming, "I can't find my daughter, I can't find my daughter!" Thankfully, you and your partner found a four-year-old girl underneath a bunk bed in the second floor apartment. The paramedics transported her to the hospital where she is now in stable condition.

Despite the positive outcome, you find yourself having reoccurring nightmares about being trapped in a pitch black room, with the sound of a child crying ringing in your ears. After a week of the same nightmare, you ask your partner how he has been sleeping. "Not so great," he admits. "Why do you ask?"

1. What would your response be?
2. What would your next step be to resolve this issue?
3. Would you inform your captain about this issue?
4. If your partner is reluctant to seek help, how would you convince him of his best interests?

Salvage and Overhaul

Fire Fighter I

Knowledge Objectives

After studying this chapter, you will be able to:

- Describe the equipment used to illuminate an emergency scene. (NFPA 5.3.17, p 612)
- Describe the safety precautions to take when working with lighting equipment. (NFPA 5.3.17, 5.3.17.A, p 612)
- Describe the types of lights used to illuminate exterior and interior scenes. (NFPA 5.3.17, p 613)
- Describe how to operate lighting equipment to light exterior and interior scenes. (NFPA 5.3.17, 5.3.17.A, p 614–615)
- Explain the purpose of salvage operations. (NFPA 5.3.14.A, p 616–617)
- List the tasks involved in a salvage operation. (p 616–617)
- Describe the safety precautions that need to be considered when performing salvage. (p 617)
- List the tools used to perform salvage operations. (p 617)
- Describe the salvage techniques commonly used to prevent water damage. (p 617–618)
- Describe the general procedures for preventing excess water damage from fire sprinklers. (NFPA 5.3.14.A, p 618–623)
- List the equipment used to shut down fire sprinklers. (NFPA 5.3.14.A, p 618)
- Describe the identifying characteristics of a main control valve of a fire sprinkler system. (NFPA 5.3.14.A, p 619)
- Describe the general procedures and equipment used to remove excess water from a structure. (NFPA 5.3.14.B, p 622–626)
- Describe the general procedures and equipment used to limit smoke and heat damage. (p 626)
- Describe how to maintain salvage covers. (NFPA 5.5.1, p 628)
- Explain when fire investigators should become involved in salvage operations. (NFPA 5.3.14.B, p 631)
- Describe the purpose of overhaul operations. (NFPA 5.3.13, 5.3.14, p 632)
- List the concerns that must be addressed to ensure the safety of fire fighters who are performing overhaul. (NFPA 5.3.13.A, p 633)
- List the indicators of possible structural collapse. (NFPA 5.3.13.A, p 633)
- Explain how to preserve structural integrity during overhaul. (NFPA 5.3.13, p 633)
- Describe how to coordinate overhaul operations with fire investigators. (NFPA 5.3.13.A, 5.3.13.B, p 633)
- Explain how fire fighters determine overhaul locations. (NFPA 5.3.13, p 634)
- List the tools that are used for overhaul operations. (NFPA 5.3.13, 5.3.13.A, p 635–636)
- Describe the general techniques used in overhaul operations. (NFPA 5.3.13, 5.3.14, p 636)

Skills Objectives

After studying this chapter, you will be able to perform the following skills:

- Illuminate an emergency scene. (NFPA 5.3.17.B, p 614–615)
- Use a sprinkler stop to shut down a sprinkler head. (NFPA 5.3.14.B, p 618–619)
- Use a sprinkler wedge to shut down a sprinkler head. (NFPA 5.3.14.B, p 618, 620)
- Close and reopen a main control valve. (NFPA 5.3.14.B, p 619–621)
- Close and open a post indicator valve. (NFPA 5.3.14.B, p 620–622)
- Construct a water chute. (NFPA 5.3.14.B, p 623)
- Construct a water catch-all. (NFPA 5.3.14.B, p 624)
- Fold a salvage cover for one-fire fighter deployment. (NFPA 5.3.14.B, p 626–627)
- Fold a salvage cover for two-fire fighter deployment. (NFPA 5.3.14.B, p 626–628)
- Fold and roll a salvage cover. (NFPA 5.3.14.B, p 627–629)
- Perform a one-fire fighter salvage cover roll. (NFPA 5.3.14.B, p 627, 630)
- Perform a shoulder toss. (NFPA 5.3.14.B, p 628, 631)
- Perform a balloon toss. (NFPA 5.3.14.B, p 628, 632)
- Open a ceiling to check for fire using a pike pole. (NFPA 5.3.8.B, 5.3.13.B, p 636–637)
- Open an interior wall to check for fire. (NFPA 5.3.8.B, 5.3.13.B, p 636, 638)

CHAPTER 19

Fire Fighter II | FFII

Knowledge Objectives

After studying this chapter, you will be able to:

- Describe the types of generators used to power lighting equipment. (p 614)
- Describe how generators operate. (p 614)
- Describe how to clean and maintain lighting equipment. (NFPA 6.5.4.A, p 615)
- Describe how to maintain generators. (NFPA 6.5.4, 6.5.4.A, p 615)

Skills Objectives

After studying this chapter, you will be able to perform the following skill:

- Conduct a weekly/monthly generator test. (NFPA 6.5.4.B, p 615–616)

You Are the Fire Fighter

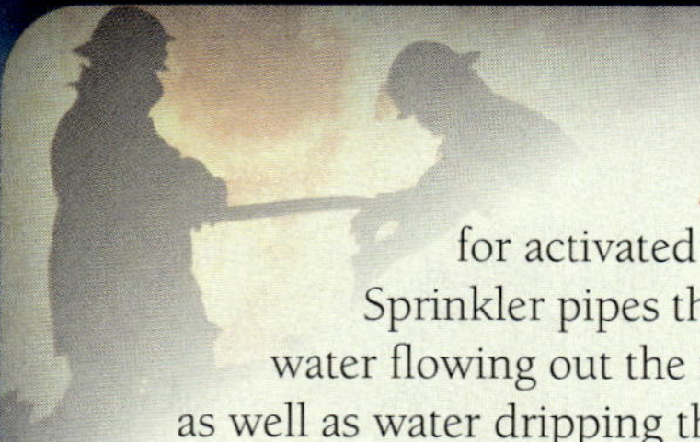

After a bitterly cold and snowy winter with several sub-zero days with strong winds, the sun is finally shining. Today the temperature is climbing to near record highs. The sun and unexpected warmth have brightened the collective mood of the fire station when the calls for activated fire alarms come rolling in.

Sprinkler pipes throughout the community are breaking due to the thaw. You arrive at a four-story warehouse with water flowing out the front door. You find several broken pipes on various floors that are flowing onto the merchandise as well as water dripping through the office ceiling.

1. How would you go about protecting the contents of the office?
2. Where would you begin in order to protect the merchandise?
3. What equipment will you need to protect the office and merchandise from water damage?

Introduction

The basic goals of firefighting are first to save lives, second to control the fire, and third to protect property. Salvage and overhaul operations help meet that third goal by limiting and reducing property losses resulting from a fire. Salvage efforts protect property and belongings from damage, particularly from the effects of smoke and water. Overhaul ensures that a fire is completely extinguished by finding and exposing any smoldering or hidden pockets of fire in an area that has been burned.

Salvage and overhaul are usually conducted in close coordination with each other and have a lower priority than conducting search and rescue operations or controlling the fire. If a department has enough personnel on scene to address more than one priority, salvage can be performed concurrently with fire suppression. Overhaul, however, can begin only when the fire is under control.

Throughout the salvage and overhaul process, fire fighters must attempt to preserve evidence related to the cause of a fire, particularly when working near the suspected area of fire origin and when the fire cause is not obvious. In general, disturb as little as possible in or near the area of origin until a fire investigator determines the cause of the fire and gives permission to clear the area. Knowing what caused a fire can help prevent future fires and enable victims to recoup losses under their insurance coverage. This subject is covered in greater depth in the Fire Cause Determination chapter.

Fire fighters performing salvage and overhaul must be able to see where they are going, what they are doing, and which potential hazards, if any, are present. The smoke might have cleared, but electrical power might still be unavailable. Portable lights should be set up to fully illuminate areas during salvage and overhaul.

Lighting

Lighting is an important concern at many fires and other emergency incidents. Because many emergency incidents occur at night, lighting is required to illuminate the scene and enable safe, efficient operations. Inside fire buildings, heavy smoke can completely block out natural light. In addition, electrical service is often interrupted or disconnected for safety purposes. As a consequence, setting up interior lighting is essential for fire fighters to conduct search and rescue, ventilation, fire suppression, salvage, and overhaul operations.

Most fire departments use different types of lighting equipment depending on the particular situation. Spotlights, for example, project a narrow concentrated beam of light. Floodlights project a more diffuse light over a wide area. Lights can be portable or permanently mounted on fire apparatus.

Although the primary responsibility for fire-scene lighting often belongs to truck (ladder) companies or support units, any company can be assigned this duty. Lighting equipment can be mounted on any apparatus, and some fire departments equip special vehicles with high-powered lights to illuminate incident scenes.

Safety Principles and Practices

The lighting and power equipment used at a fire scene generally operates on 110-volt AC (alternating current), which is the same as standard household current. Some systems require higher voltage. All electrical cords, junction boxes, lights, and power tools must be maintained properly and handled carefully to avoid electrical shocks. In addition, all electrical equipment must be properly grounded. Electrical cords must be well insulated, without cuts or defects, and properly sized to handle the required amperage.

Generators should have ground-fault interrupters (GFI) to prevent a fire fighter from receiving a potentially fatal electric shock. A GFI senses when there is a problem with an electrical ground and interrupts the current in such a case, shutting down both the power source and the equipment it is feeding.

Some portable generators are equipped with a grounding rod that must be inserted into the ground. Always use a grounding device if one is provided. Avoid areas of standing or flowing water when placing power cords and junction boxes at a fire scene, and put electrical equipment on higher ground whenever possible.

Lighting Equipment

Portable lights can be taken into buildings to illuminate the interior. They can also be set up outside the structure to illuminate the fire or emergency incident scene FIGURE 19-1. Portable lights usually range from 300 watts to 1500 watts and can use several types of bulbs, including quartz bulbs, halogen bulbs, and light-emitting diodes.

Portable light fixtures are connected to the generator with electrical cords. These electrical cords should be stored neatly coiled or on permanently mounted reels attached to the fire apparatus. The cord is then pulled from the reel to the place where the power is needed. Portable reels can be taken from the apparatus into the fire scene.

FIGURE 19-1 Portable emergency lights come in various sizes and levels of brightness.

Junction boxes are used as mobile power outlets. They are placed in convenient locations so that cords for individual lights and electrical equipment can be attached to them. Junction boxes used by fire departments are protected by waterproof covers and are often equipped with small lights so that they can be easily located in the dark.

The connectors and plugs used for fire department lighting have special connectors that attach with a slight clockwise twist. This setup keeps the power cords from becoming unplugged during fire department operations.

Lights also can be permanently mounted on fire apparatus to illuminate incident scenes FIGURE 19-2. In such a case, the vehicle operator can immediately illuminate the emergency scene simply by pressing the generator starter button. Some vehicle-mounted lights can be manually raised to illuminate a larger area. Mechanically operated light towers, which can be raised and rotated by remote control, can create near-daylight conditions at an incident scene.

FIGURE 19-2 An apparatus-mounted light tower can provide near-daylight conditions at an incident scene.

Battery-Powered Lights

Battery-powered lights generally are used by individual fire fighters to find their way in dark areas or to illuminate their immediate work area. These kinds of lights are most often used during the first few critical minutes of an incident because they are lightweight, are easily transported, do not require power cords, and can be used immediately FIGURE 19-3. The lights are powered by either disposable or rechargeable batteries; thus they have a limited operating time before the batteries need to be recharged or replaced.

Powerful lights are needed to pierce the smoke inside a fire building. Large hand lights that project a powerful beam of light are preferred for search and rescue activities and interior

FIGURE 19-3 Hand lights manufactured for use by fire fighters have light outputs ranging from 3500 to 75,000 candlepower.

fire suppression operations when fire fighters must penetrate smoke-filled areas quickly. These lights can be equipped with a shoulder strap for easy transport. Every crew member entering a fire building should be equipped with a high-powered hand light.

A personal flashlight is another type of battery-operated light used by fire fighters. Always carry a flashlight as part of your personal protective equipment (PPE); it could be a lifesaver if your primary light source fails. Flashlights are not as powerful as the larger hand lights, however, and they will not operate for as long on one set of batteries. A fire fighter's flashlight should be rugged and project a strong light beam.

FIRE FIGHTER II

Electrical Generators

More powerful lighting equipment requires a separate source of electricity, generally a 110-volt AC delivered through power cords. The electricity can be supplied by a generator, an inverter, or a building's electrical system, if the power has not been interrupted.

Power inverters convert 12-volt DC (direct current) from a vehicle's electrical system to 110-volt AC power. An inverter can provide a limited amount of AC current and is typically used to power one or two lights mounted directly on the apparatus. Some vehicles have additional outlets for operating portable lights or small electrically powered tools. Most power inverters do not produce enough power to operate high-intensity lighting equipment, large power tools, or ventilation fans. Be aware that connecting devices that draw too much current to an inverter can seriously damage it.

Electrical generators are powered by gasoline or diesel engines; they can be either portable or permanently mounted on fire apparatus. Portable generators are small enough to be removed from the apparatus and carried to the fire scene. These devices come in various sizes and can produce as much as 6000 watts (6 kilowatts) of power. A portable generator has a small gasoline- or diesel-powered motor with its own fuel tank.

Apparatus-mounted generators have much larger power capacities, sometimes exceeding 20 kilowatts. The smaller units, which are similar to portable generators, can be permanently mounted in a compartment on the apparatus. By contrast, larger generators are usually mounted directly on the vehicle, often have diesel engines, and draw fuel directly from the vehicle's fuel tank. Permanently mounted generators can also be powered by the apparatus engine through a power take-off and a hydraulic pump.

Another potential power source for lighting is a building's normal power supply. It is usually not an option if a serious fire has occurred in the building because the current will be disconnected for safety reasons. In contrast, if the fire is relatively minor, the power might not have to be interrupted and can be used for lighting. Alternatively, power can sometimes be obtained from a nearby building. When drawing power from a building, be aware that you might not know the capacity of the circuit from which you are obtaining electricity. Do not overload a household electrical circuit.

FIRE FIGHTER II

Fire Fighter Safety Tips

Never run a portable generator in a confined space or an area without adequate ventilation unless you and everyone else in the area are wearing self-contained breathing apparatus (SCBA). The exhaust from the generator contains carbon monoxide, which can kill you.

Lighting Methods

Effective lighting improves the efficiency and safety of all crews at an emergency incident, but it requires practice to set up quickly and efficiently. Departmental standard operating procedures (SOPs) or standard operating guidelines (SOGs), or the incident commander (IC), will dictate the responsibility for setting up lighting.

Exterior Lighting

Exterior lighting should be provided at all incident scenes during hours of darkness so that fire fighters can see what they are doing, recognize hazards, and find victims who need rescuing. Scene lighting also makes fire fighters more visible to drivers who are approaching the scene or maneuvering emergency vehicles. Powerful exterior lights can often be seen through the windows and doors of a dark, smoke-filled building and can guide disoriented fire fighters or victims to safety.

Apparatus operators should turn on their apparatus-mounted floodlights and position them to illuminate as much of the area as possible. Position vehicles with light towers to use their lights for maximum effectiveness. The entire area around a fire-struck building should be illuminated. If some areas cannot be covered with apparatus-mounted lights, use portable lights.

Interior Lighting

Portable lights can provide interior lighting at a fire scene. During interior operations, quickly extend a power cord and set up a portable light at the entry point to the structure. Fire fighters looking for the building exit can then use this light at the door as a beacon. Extend additional lights into the building to illuminate interior areas as needed and as time permits.

Interior operations often progress through the search and rescue phase and the fire suppression stage quickly, with these steps sometimes being completed before effective interior lighting can be established. If interior operations continue for a lengthy period, portable lighting should be used to provide as much light as possible in the areas where fire fighters are working.

When operations reach the salvage and overhaul phase, take extra time to provide adequate interior lighting in all areas so that crews can work safely. Proper lighting enables fire fighters to see what they are doing and to observe any dangerous conditions that need to be addressed. Although exterior lighting is generally needed only at night, interior lighting may be needed even when it is a bright day outside.

Follow the steps in SKILL DRILL 19-1 (Fire Fighter I, NFPA 5.3.17) to illuminate an emergency scene (FIGURE 19-4):

1. Wear PPE. Depending on condition, you may or may not wear SCBA.
2. Inspect all equipment while setting it up.

3. Start the power plant, or engage the inverter.
4. Connect cords, plug adaptors, GFIs, and lighting equipment.
5. Ensure proper grounding and GFI use.
6. Ensure that the scene is adequately and safely illuminated.

Fire Fighter Safety Tips

A good rule to remember: Light early, light often, and light safely.

FIGURE 19-4 Inspect all equipment while setting it up.

FIRE FIGHTER II

Cleaning and Maintenance

It is important to properly maintain lighting equipment and generators to keep them in good operating condition. To clean lighting equipment, follow the manufacturer's instructions for cleaning. In general, the directions specify how to clean the floodlights and search lights with a mild soap or detergent. Always use a soft cloth to clean and polish the lens. Avoid the use of strong solvents, as they may damage plastic or rubber components.

When cleaning and maintaining a generator, also follow the manufacturer's instructions. Avoid overfilling the fuel tank on portable generators. When a fuel tank is too full, it can overflow onto the hot engine and start a fire. An overfull tank can also leak fuel as the fuel expands, especially if the tank was filled in a cool environment and then brought into a warmer place. Spilled fuel can damage paint under certain circumstances.

Clean the external parts of the generator with a mild soap or detergent as recommended by the manufacturer. Remove any dirt, vegetation, or charred materials on the device. Avoid the use of strong solvents, as they can damage rubber and plastic parts of the generator and can remove paint.

If a solvent is recommended by the manufacturer for any cleaning operation, follow the directions for its use on the product label. Also read the material safety data sheet before starting to work with any solvent. Check whether the product is flammable. If it is, be sure to follow the safety recommendations for its use. Always use flammable solvents in a well-ventilated area that is not close to any flame or heat source.

FIRE FIGHTER II

Fire Fighter Safety Tips

Solvents can irritate or burn the skin, eyes, and lungs. Some of these agents are known to cause cancer and should be avoided. Solvents can be absorbed through the lungs as well as through the skin. For solvents that can be absorbed through skin, check whether chemically resistant gloves are required during their use. Also check whether eye protection or respiratory protection is required.

Test and run generators on a weekly or monthly basis to confirm that they will start, run smoothly, and produce power. Run gasoline-powered generators for 15 to 30 minutes to reduce any deposit build-up that could foul the spark plugs and make the generator hard to start.

At the same time, inspect and test other electrical equipment on the apparatus. Take junction boxes out and check them for cracked covers or broken outlets. Plug them in to make sure the GFI works properly. Check power cords for tears in the protective covering, mechanical damage, fraying, heat damage, or burns. Inspect plugs for loose or bent prongs. Inspect and test the operation of all power tools and equipment on the apparatus. Check lighting fixtures to confirm that they are in good repair and that the bulbs are working.

After shutting down the generator, refill the fuel tank. Follow the manufacturer's recommendations for generator engine maintenance, including regular oil changes. To conduct a weekly/monthly generator test, follow the steps in SKILL DRILL 19-2:

1. Remove the generator from the apparatus compartment or open all doors as needed for ventilation. Install the grounding rod if needed. (STEP 1)
2. Check the oil and fuel levels, and start the generator.
3. Connect the power cord or junction box to the generator, connect a load such as a fan or lights, and make sure the generator attains the proper speed. Check the voltage and amperage gauges to confirm efficient operation. (STEP 2)
4. Run the generator under load for 15 to 30 minutes.
5. Turn off the load and listen as the generator slows down to idle speed. Allow the generator to idle for approximately 2 minutes before turning it off.
6. Disconnect all power cords and junction boxes, and remove the grounding rod if present.
7. Clean all power cords, plugs, adaptors, GFIs, and tools, and replace them in proper storage areas. Allow the generator to cool for 5 minutes.
8. Refill the generator with fuel and oil as needed, and return the generator to its compartment. Fill out the appropriate paperwork. (STEP 3)

SKILL DRILL 19-2 Conducting a Weekly/Monthly Generator Test FFII

(Fire Fighter II, NFPA 6.5.4)

1. Remove the generator from the apparatus compartment or open all doors as needed for ventilation. Install the grounding rod, if needed.

2. Check the oil and fuel levels, and start the generator. Connect the power cord or junction box to the generator; connect a load such as a fan or lights, and make sure the generator attains the proper speed. Check the voltage and amperage gauges to confirm efficient operation.

3. Run the generator under load for 15 to 30 minutes. Turn off the load and listen as the generator slows down to idle speed. Allow the generator to idle for approximately 2 minutes before turning it off. Disconnect all power cords and junction boxes; clean all power cords, plugs, adaptors, GFIs, and tools, and replace them in proper storage areas. Allow the generator to cool for 5 minutes. Refill the generator with fuel and oil as needed, and return the generator to its compartment. Fill out the appropriate paperwork.

Salvage Overview

Salvage operations are conducted to save property from a fire and to reduce the damage that results from a fire. Often, the damage caused by smoke and water can be more extensive and costly to repair or replace than the property that is burned. Salvage operations are also conducted in other situations that endanger property, such as floods or water leaking from burst pipes.

Salvage efforts usually are aimed at preventing or limiting secondary losses that result from smoke and water damage, fire suppression efforts, and other causes. They must be performed promptly to preserve the building and protect the contents from avoidable damage. Salvage operations include ejecting smoke, removing heat, controlling water runoff, removing water from the building, securing a building after a fire, covering broken windows and doors, and providing temporary patches for ventilation openings in the roof to protect the structure and contents.

Fire fighters must help protect property from avoidable loss or damage as part of their professional responsibilities. At the scene of a fire or any other type of emergency, the victims' property is entrusted to the care of fire fighters, who are expected to protect that property to the best of their ability.

Salvage efforts at residential fires often focus on protecting personal property. These items often have personal implications for the victims' lives. Fire fighters may be able to identify and protect critical personal belongings, such as wallets, purses, car keys, or medications. The value of some items, such as photographs and family heirlooms, cannot be measured in dollars, but these items are certainly treasured by victims and are often irreplaceable.

FIRE FIGHTER Tips

One way many fire departments help homeowners after a fire is by having a list of board-up and fire restoration companies that they can give to a homeowner. Board-up companies are prepared to respond quickly and secure a property to protect it from rain, snow, or damage from vandals. Fire restoration companies are skilled in removing smoke and water from a building and its contents before more damage occurs. Because many building owners are not aware of these companies, your recommendations are often greatly appreciated by the owner or occupant of a damaged building at this highly stressful time.

At commercial or industrial fires, the salvage priorities are quite different from those at residential fires. The preincident plan should identify critical items and enable fire fighters to focus their salvage efforts appropriately. For example, a law firm might identify its files as the most valuable property, whereas a library might place more importance on books and computer files than on furniture. A preincident plan should note whether the contents of a warehouse consist of valuable electronic components or industrial steel pipe. Factory machinery could also be a higher priority than the finished products stored on premises.

Fire Fighter Safety Tips

At the emergency scene, the IC should know where every crew is working and what those crew members are doing at all times. The IC relies on your company officer to know your location and activities and to relay this information to the command post.

Safety Considerations During Salvage Operations

Safety is a primary concern during salvage operations. The PPE required will depend on the site of the salvage operations, the stage of the fire, and any hazards present. Salvage operations often begin while the fire is still burning and continue for several hours after it has been extinguished. During firefighting operations and immediately afterward, fire fighters should wear a full set of protective clothing and equipment, including SCBA. After the fire is out, the area must be ventilated, and the atmosphere tested and determined to be safe by the safety officer before fire fighters can work without SCBA.

If salvage is conducted in a different part of the building than where the fire occurred, several floors below the fire, or in a neighboring building, fire fighters may have other protective clothing requirements. Hazards in these areas might be quite different from those in the fire area. Fire fighters conducting salvage operations in remote areas can wear partial PPE (boots, turnout pants, helmet, and gloves) with the approval of the safety officer.

Structural collapse is always a possibility during salvage operations because of fire damage to the building's structural components. For example, lightweight trusses can collapse quickly when damaged by a fire. Heavy objects, such as air-conditioning units and appliances, can crash through roofs or ceilings weakened by fire damage.

Even the water used to douse the fire can contribute to structural collapse. Recall that each gallon of water weighs 8.3 pounds (4 kilograms). Extinguishing a major fire can, therefore, load the structure with tons of extra water weight. If the structure is already weakened by the fire, the extra water weight can cause a catastrophic structural collapse, with the potential to injure or kill fire fighters. For example, a ladder pipe discharging 800 gallons (2,903 kilograms) of water per minute into a building is adding 6400 pounds (3.2 tons) (2.9 tonnes) of weight to the building every minute.

Sometimes this danger may be masked by absorbent materials such as bolts of cloth or bales of rags stored in the building. Uneven absorption of water also presents a danger. For example, stacked rolls of newsprint in a printing factory can collapse if the bottom layer becomes waterlogged and, therefore, unable to support the stack.

Water can also create potential hazards for fire fighters conducting salvage operations on floors below the fire. Water used to extinguish a fire inevitably leaks down through the building and can cause major damage. If it becomes trapped in the void space above a ceiling, it may cause sections of the ceiling to collapse. Fire fighters should watch for warning signs such as water leaking through ceiling joints and sagging areas of ceilings. Water that accumulates in a basement can hide a stairway and present additional hazards if utility connections are also located in the basement.

Damage to the building's utility systems presents significant risks to fire fighters who are conducting salvage operations. Gas and electrical services should always be shut off to eliminate the potential hazards of electrocution or explosion stemming from gas leaks.

Salvage Tools

Fire fighters conducting salvage operations use a variety of special tools and equipment TABLE 19-1. Salvage covers and other protective items shield and cover building contents. Other salvage tools remove smoke, water, and other products that can damage property FIGURE 19-5.

TABLE 19-1	Salvage Tools
	• Salvage covers: treated canvas or plastic • Box cutter for cutting plastic • Floor runners • Wet/dry vacuums • Squeegees • Submersible pumps and hose • Sprinkler shut-off kit • Ventilation fans, power blowers • Small tool kit • Pike poles to construct water chutes

Using Salvage Techniques to Prevent Water Damage

The best way to prevent water damage at a fire scene is to limit the amount of water used to fight the fire. In other words, use only enough water to knock down the fire quickly; do not use

FIGURE 19-5 Salvage tools.

FIGURE 19-6 Most fires are controlled by the activation of only one or two sprinkler heads.

more water than is needed. Do not continue to douse a fire that is already out, because this action merely creates unnecessary water damage. Turn off hose nozzles when they are not in use. If a nozzle must be left partly open to prevent freezing during cold weather, direct the flow through a window. Take the time to tighten any leaky hose couplings after the fire has been knocked down, because this measure prevents additional water damage during overhaul. Look for leaking water pipes in the building; if you find any, shut off the building's water supply.

Deactivating Sprinklers

Buildings equipped with sprinkler systems require special steps to limit water damage. Sprinklers can control a fire efficiently, but will keep flowing until the activated sprinklers are shut off or the entire system is shut down. This ongoing flow can cause significant water damage.

Sprinklers should be shut down as soon as the IC declares that the fire is under control. Do not shut down sprinklers prematurely; the fire may rekindle, causing major damage. Between the time when sprinklers are shut down and when overhaul is completed, a fire fighter with a portable radio should stand by and be ready to reactivate the sprinklers if necessary.

Sprinkler systems are usually designed so that only those sprinkler heads directly over or close to the fire will be activated. Indeed, most fires can be contained or fully extinguished by the release of water from only one or two sprinkler heads FIGURE 19-6. Although movies may show all sprinkler heads activating at once, this does not occur in real life except in specially designed deluge sprinkler systems. More information on sprinkler systems is presented in the Fire Detection, Protection, and Suppression Systems chapter.

If only one or two sprinklers have been activated, inserting a sprinkler wedge or a sprinkler stop can quickly stop the flow. This action keeps the rest of the system operational during overhaul.

A sprinkler wedge is a small triangular piece of wood. Inserting two wedges on opposite sides, between the orifice and the deflector of the sprinkler, and pushing them together effectively plugs the opening, stopping the flow from standard upright and pendant sprinkler heads.

A sprinkler stop is a more sophisticated mechanical device with a rubber stopper that can be inserted into a sprinkler head. Several types of sprinkler stops are available, including some that work only with specific sprinkler heads. Some types of sprinkler stops such as a Shutgun™ are manufactured with a fusible link so they can shut off a sprinkler head and return that sprinkler head to temporary service.

Not all sprinklers can be shut off with a wedge or a sprinkler stop. Recessed sprinklers, which are often installed in buildings with finished ceilings, are usually difficult to shut off. If the individual heads cannot be shut off, stop the water flow by closing the sprinkler control valve.

To stop the flow of water from a sprinkler using a sprinkler stop, follow the steps in SKILL DRILL 19-3:

1. Have a sprinkler stop in hand. (STEP 1)
2. Place the flat-coated part of the sprinkler stop over the sprinkler head orifice and between the frame of the sprinkler head. (STEP 2)
3. Push the lever to expand the sprinkler stop until it snaps into position. (STEP 3)

To stop the flow of water from a sprinkler using a pair of sprinkler wedges, follow the steps in SKILL DRILL 19-4:

1. Hold one wedge in each hand. (STEP 1)
2. Insert the two wedges, one from each side, between the discharge orifice and the sprinkler head deflector. (STEP 2)
3. Bump the wedges securely into place to stop the water flow. (STEP 3)

If individual sprinklers cannot be plugged or if several sprinklers have been activated, you can stop the flow by closing the sprinkler control valve. After the valve is closed, the water trapped in the piping system will drain out through the activated sprinklers for several minutes. If a drain valve is found near the control valve, open it to drain the system quickly. This step also directs the water flow to a location where it will not cause additional damage.

SKILL DRILL 19-3 Using a Sprinkler Stop
(Fire Fighter I, NFPA 5.3.14)

1 Have a sprinkler stop in hand.

2 Place the flat-coated part of the sprinkler stop over the sprinkler orifice and between the frame of the sprinkler.

3 Push the lever to expand the sprinkler stop until it snaps into position.

The main control valve for a sprinkler system is usually an outside stem and yoke (OS&Y) valve or a post indicator valve (PIV). An OS&Y valve is typically found in a mechanical room in the basement or on the ground-floor level of a building. A PIV is located outside the building or on an exterior wall. Some sprinkler systems also have zone valves that control the flow of water to sprinklers in different areas of the building. High-rise buildings, for example, usually have a zone valve for each floor. Closing a zone valve stops the flow to sprinklers in the zone; in the rest of the building, however, the sprinklers remain operational. The locations of the main control valves and zone valves should be identified during preincident planning visits to a building.

Sprinkler control valves should always be locked in the open position, usually with a chain and padlock, and should be equipped with a tamper alarm that is tied into the fire alarm system. Fire fighters who are sent to shut off a sprinkler control valve should take a pair of bolt cutters to remove the lock if the key is not available.

To close and reopen the main control valve, follow the steps in SKILL DRILL 19-5:

1. Locate the OS&Y valve as indicated on the preincident plan. It may be either inside the building or close by outside the building. The stem of an OS&Y valve protrudes from the valve handle in the open position.
2. Identify the valve that controls sprinklers in the fire area. Smaller systems will have only one OS&Y valve; large buildings may have either multiple sprinkler systems or one main control valve and additional zone valves controlling the flow to different areas.

SKILL DRILL 19-4 Using Sprinkler Wedges
(Fire Fighter I, NFPA 5.3.14)

1 Hold one wedge in each hand.

2 Insert the two wedges, one from each side, between the discharge orifice and the sprinkler deflector.

3 Bump the wedges securely into place to stop the water flow.

3. If the valve is locked in the open position with a chain and padlock, and the key is readily available, unlock and remove the chain. If no key is available, cut the lock or the chain with a pair of bolt cutters. Cut a link close to the padlock so that the chain can be reused. (STEP 1)
4. Turn the valve handle clockwise to close the valve. Keep turning until resistance is strong and little of the stem is visible. (STEP 2)
5. To open the OS&Y valve, turn the handle counter-clockwise until resistance is strong and the stem is visible again.
6. Lock the valve in the open position. (STEP 3)

To close and open a post indicator valve, follow the steps in SKILL DRILL 19-6:

1. Locate the PIV (it is usually found outside the building). The location of the valve should be indicated on a preincident plan.
2. Most PIVs will be locked in the open position by a padlock. Unlock the padlock if the key is readily available. If no key is available, cut the lock with a pair of bolt cutters. (STEP 1)
3. Remove the handle from its storage position on the PIV and place it on top of the valve, similar to the use of a hydrant wrench.

SKILL DRILL 19-5 Closing and Reopening a Main Control Valve (OS&Y)

(Fire Fighter I, NFPA 5.3.14)

1 Locate the OS&Y valve as indicated on the preincident plan. Identify the valve that controls sprinklers in the fire area. If the valve is locked in the open position with a chain and padlock, and the key is readily available, unlock and remove the chain. If no key is available, cut the lock or the chain with a pair of bolt cutters.

2 Turn the valve handle clockwise to close the valve. Keep turning until resistance is strong and little of the stem is visible.

3 To open the OS&Y valve, turn the handle counterclockwise until resistance is strong and the stem is visible again. Lock the valve in the open position.

4. Turn the valve stem in the direction indicated on top of the valve to close the valve. Keep turning until resistance is strong and the visual indicator changes from "Open" to "Shut." (STEP 2)
5. To reopen the PIV, turn the valve stem in the opposite direction until resistance is strong and the indicator changes back to "Open."
6. Lock the valve in the open position. (STEP 3)

Before a sprinkler system can be restored to normal operation, the sprinkler heads that have been activated must be replaced. Every sprinkler system should have spare heads stored somewhere, usually near the main control valves. An activated sprinkler head must be replaced with another head of the same design, size, and temperature rating.

The main sprinkler control valve or the appropriate zone valve must be closed, and the system must be drained before a sprinkler head can be changed. Special wrenches must be used when replacing sprinklers to prevent damage to the operating mechanism.

After the activated heads are replaced, the sprinkler system can be restored to service. Restoring a sprinkler system to service takes special training and should be performed only by qualified individuals.

SKILL DRILL 19-6 Closing and Reopening a Main Post Indicator Control Valve

(Fire Fighter I, NFPA 5.3.14)

1 Locate the PIV. Unlock the padlock with a key or cut the lock with a pair of bolt cutters.

2 Remove the handle from its storage position on the PIV and place it on top of the valve, similar to the use of a hydrant wrench. Turn the valve stem in the direction indicated on top of the valve to close the valve. Keep turning until resistance is strong and the visual indicator changes from "Open" to "Shut."

3 To reopen the PIV, turn the valve stem in the opposite direction until resistance is strong and the indicator changes back to "Open." Lock the valve in the open position.

Removing Water

Water that accumulates within a building or drips down from higher levels should be channeled to a drain or to the outside of the building to prevent or limit water damage to the structure. A salvage pump may be needed to help remove the water in some cases.

Some buildings have floor drains that will funnel water into the below-ground sewer system. These floor drains should be kept free from debris so that the water can drain freely. Fire fighters can use squeegees to direct the water into the drain.

In buildings or houses without floor drains, it may be possible to shut off the water supply to a floor-mounted toilet, remove the nuts that hold the toilet bowl to the floor, and remove the toilet from the floor flange. This maneuver creates a large drain capable of handling large quantities of water, as long as someone keeps the opening from becoming clogged with debris.

Water on a floor at ground level can often be channeled to flow outside of the structure through a doorway or other opening. Water on a floor above ground level can sometimes be drained to the outside by making an opening at floor level in an exterior wall of the building.

Water chutes or water catch-alls are often used to collect water leaking down from firefighting operations on higher floor levels and to help protect property on calls involving burst pipes or leaking roofs. A water chute catches dripping water and directs it toward a drain or to the outside through a window or doorway. A water catch-all is a temporary pond that holds dripping water in one location. The accumulated water must then be drained to the outside of the building. Water chutes and catch-alls can be constructed quickly using salvage covers.

A water chute can be constructed with a single salvage cover, or with a cover and two pike poles for support. To construct a water chute, follow the steps in SKILL DRILL 19-7:

1. Fully open a large salvage cover flat on the ground. (STEP 1)
2. Roll the cover tightly from one edge toward the middle. If using pike poles, lay one pole on the edge and roll the cover around the handle. (STEP 2)
3. Roll the opposite edge tightly toward the middle in the same manner. Stop when the rolls are 1 to 3 feet (30 to 91 centimeters) apart. (STEP 3)
4. Turn the cover upside down. Position the chute so that it collects the dripping water and channels it toward a drain or outside opening. Place the chute on the floor, with one end propped up by a chair or other object. (STEP 4)
5. Use a stepladder or other tall object to support chutes constructed with pike poles. (STEP 5)

SKILL DRILL 19-7 Constructing a Water Chute

(Fire Fighter I, NFPA 5.3.14)

1 Fully open a large salvage cover flat on the ground.

2 Roll the cover tightly from one edge toward the middle. If using pike poles, lay one pole on the edge and roll the cover around the handle.

3 Roll the opposite edge tightly toward the middle until the two rolls are 1 to 3 feet (30 to 91 centimeters) apart.

4 Turn the cover upside down. Position the chute so that it collects dripping water and channels it toward a drain or outside opening. Place the chute on the floor, with one end propped up by a chair or other object.

5 Use a stepladder or other tall object to support chutes constructed with pike poles.

To construct a water catch-all, follow the steps in SKILL DRILL 19-8:

1. Open a large salvage cover on the ground, and roll two edges inward from the opposite sides (approximately 3 feet (91 centimeters) on each side). (STEP 1)
2. Fold each corner over at a 90-degree angle, starting each fold approximately 3 feet (91 centimeters) in from the edge. (STEP 2)
3. Roll the remaining two edges inward approximately 2 feet (61 centimeters). (STEP 3)
4. Lift the rolled edge over the corner flaps, and tuck it in under the flaps to lock the corners in place. (STEP 4)

Water Vacuum

Special vacuum cleaners that suck up water also can be used during salvage operations. Two types of water vacuums are available: a small-capacity, backpack type and a larger, wheeled unit. The backpack vacuum cannot be used by someone who is wearing SCBA. Wet/dry shop vacuums may also be used as a low-cost alternative to water vacuums.

Fire Fighter Safety Tips

You cannot operate a gasoline-powered pump safely inside a structure without wearing SCBA because the operation of the pump generates deadly carbon monoxide gas.

SKILL DRILL 19-8 Constructing a Water Catch-All

(Fire Fighter I, NFPA 5.3.14)

1 Open a large salvage cover on the ground, and roll each edge of the salvage cover towards the opposite side.

2 Fold each corner over at a 90-degree angle, starting each fold approximately 3 feet (91 centimeters) in from the edge.

3 Roll the remaining two edges inward approximately 2 feet (61 centimeters).

4 Lift the rolled edge over the corner flaps, and tuck it in under the flaps to lock the corners in place.

VOICES OF EXPERIENCE

Some time ago, I responded to a structural fire in a single family dwelling. With three stories and an attached multi-car garage, this was a luxurious home. The fire started in a room on the second floor and extended to the third floor and attic. All of the occupants had escaped the structure, so operations concentrated on fire suppression and property conservation. Hoselines were advanced to the upper stories, the roof was opened, and crews were dispatched to protect and remove as much personal property as possible.

This home was furnished with many antiques and valuable art. The Incident Commander (IC) instructed the company officers to take care with overhaul and to salvage as much as possible. A lieutenant asked the homeowner what she most wanted protected or removed first. The homeowner directed the lieutenant to a specific drawer in a piece of furniture on the second floor. In the drawer the lieutenant would find a wicker basket.

The lieutenant directed his fire fighters to the location and a few minutes later they handed the wicker basket to the homeowner, who then became very emotional. After the fire and during the post-incident interview, the homeowner thanked the IC repeatedly, especially for the caring request of the lieutenant, which resulted in the safe return of her most precious possession. In the wicker basket were the pictures and mementoes of her child who had passed away some years earlier.

This was a real teaching moment for me and it made an impression that I will never forget. When we got back to the station, my lieutenant asked us, "If I walked into your home today, could I determine which possession you value the most?" Salvage and overhaul is about what is of value to the homeowner and protecting it as if it were priceless.

F. Patrick Marlatt
Maryland Fire and Rescue Institute
College Park, Maryland

Drainage Pumps

Drainage pumps remove water that has accumulated in basements or below ground level. Portable electric submersible pumps can be lowered directly into the water to pump it out of a building. Gasoline-powered portable pumps must be placed outside a building because they exhaust poisonous carbon monoxide gas. These pumps use a hard-suction hose, which is lowered into the basement through a window and drafts water out of the building.

Using Salvage Techniques to Limit Smoke and Heat Damage

One way to reduce property loss is to keep heat and smoke out of areas that are not involved in the fire. A closed door can effectively keep smoke and heat out of a room or area. Given this fact, fire fighters doing search and rescue operations should remember to close doors after searching a room.

Properly timed and effective ventilation practices also limit smoke and heat damage. Rapid ventilation will often reduce smoke damage in an area already filled with smoke. The visible components of smoke are mostly soot particles and other products of combustion, including corrosive chemicals that will settle on any horizontal surface when the smoke cools. Blowing the smoke out of the building is a better option than allowing contaminants to settle on the contents. All members of the fire suppression team—not just the salvage crews—should recognize opportunities to prevent smoke and heat damage. Ventilation is discussed in detail in the Ventilation chapter.

Salvage Covers

The most common method of protecting building contents is to cover them with salvage covers. Salvage covers are large square or rectangular sheets of heavy canvas or plastic material that are used to protect furniture and other items from water runoff, falling debris, soot, and particulate matter in smoke residue.

Salvage crews usually begin their work on the floor immediately below the fire, with the goal being to prevent water damage to furniture and other contents on lower floors. If the fire is in an attic, fire fighters may have enough time to spread covers over the furniture in the rooms directly below the fire before pulling the ceilings to attack the flames.

The most efficient way to protect a room's contents is to move all the furniture to the center of the room, away from the walls, where water could damage the backs of the furniture. This approach reduces the total area that must be covered, enabling one or two fire fighters to cover the pile quickly and move on to the next room FIGURE 19-7.

Remove any pictures from the walls and place them with the furniture. Put smaller pictures and valuable objects in drawers or wherever they will be protected from breakage. If you have enough time, roll up any rugs and place them on the pile.

Some departments use rolls of construction-grade polyethylene film for protecting structures' contents instead of salvage covers. This material comes in rolls as long as 120 feet (36.5 meters), so it can be unrolled over the room's contents and cut to the correct length with a box cutter. Polyethylene film is particularly useful for covering long surfaces, such as retail display counters. This material is disposable and can be left behind after the fire; in contrast, traditional salvage covers must be picked up, washed, dried, and properly folded for storage in fire apparatus compartments after each use. Occasionally, salvage covers may be left behind as protection for the building contents, but they should be picked up when they are no longer needed.

FIGURE 19-7 Move furniture to the center of the room so that a single salvage cover can protect all of the contents.

Special folding and rolling techniques are used to store salvage covers so that they can be deployed quickly by one or two fire fighters. Be certain to follow your fire department's SOPs/SOGs when folding or rolling a salvage cover.

To fold a salvage cover to prepare it for one fire fighter to deploy, follow the steps in SKILL DRILL 19-9:

1. Spread the salvage cover flat on the ground with a partner facing you. (STEP 1)
2. On the right side of the salvage cover, make a fold at the quarter point of the cover. Next, make a second fold that ends in the middle of the cover. (STEP 2)
3. On the left side of the salvage cover, make a fold at the quarter point of the cover. Next, make a second fold that ends in the middle of the cover. (STEP 3)
4. Fold the two halves together, and flatten the salvage cover to remove any trapped air. (STEP 4)
5. Make 1-foot (30-centimeter) folds from each end of the cover until you reach the center of the salvage cover. (STEP 5)
6. Fold the two halves together (STEP 6).

To fold a salvage cover to prepare it for two fire fighters to deploy, follow the steps in SKILL DRILL 19-10:

1. Spread the salvage cover flat on the ground with a partner facing you. Together, fold the cover in half. (STEP 1)
2. Together, grasp the unfolded edge and fold the cover in half again. Flatten the salvage cover to remove any trapped air. (STEP 2)

SKILL DRILL 19-9 Folding a Salvage Cover for use by One Fire Fighter

(Fire Fighter I, NFPA 5.3.14)

1. Spread the salvage cover flat on the ground with a partner facing you.

2. On the right side of the salvage cover, make a fold at the quarter point of the cover. Next, make a second fold that ends in the middle of the cover.

3. On the left side of the salvage cover, make a fold at the quarter point of the cover. Next, make a second fold that ends in the middle of the cover.

4. Fold the two halves together, and flatten the salvage cover to remove any trapped air.

5. Make 1-foot (30-centimeter) folds from each end of the cover until you reach the center of the salvage cover.

6. Fold the two halves together.

3. Move to the newly created narrow ends of the salvage cover and fold the salvage cover in half lengthwise. (STEP 3)
4. Fold the salvage cover in half lengthwise again. Make certain that the open end is on top. (STEP 4)
5. Fold the cover in half a third time. (STEP 5)

To fold and roll a salvage cover, follow the steps in SKILL DRILL 19-11:

1. Spread the salvage cover flat on the ground with a partner facing you. (STEP 1)
2. Together, fold the outside edge in to the middle of the cover, creating a fold at the quarter point. (STEP 2)
3. Fold the outside fold in to the middle of the cover, creating a second fold. (STEP 3)
4. Repeat Steps 2 and 3 from the opposite side of the cover so the folded edges meet at the middle of the cover, with the folds touching but not overlapping. (STEP 4)
5. Tightly roll up the folded salvage cover from the end. (STEP 5)

Although it takes two people to fold and roll a salvage cover for storage, this technique enables one person to unroll the cover easily and quickly. To perform the one-person salvage cover roll, follow the steps in SKILL DRILL 19-12:

1. Stand in front of the end of the object that you are going to cover. (STEP 1)
2. Start to unroll the cover over one end of the object. (STEP 2)
3. Continue unrolling the cover over the top of the object. Allow the remainder of the roll to settle at the other side of the object. (STEP 3)
4. Spread the cover, unfolding each side outward over the object to the first fold. (STEP 4)
5. Unfold the second fold on each side to drape the cover completely over the object. (STEP 5)
6. Tuck in all loose edges of the cover around the object. (STEP 6)

SKILL DRILL 19-10 Performing a Salvage Cover Fold for Two-Fire Fighter Deployment

(Fire Fighter I, NFPA 5.3.14)

1. Spread the salvage cover flat on the ground with a partner facing you. Together, fold the cover in half.

2. Together, grasp the unfolded edge and fold the cover in half again. Flatten the salvage cover to remove any trapped air.

3. Move to the newly created narrow ends of the salvage cover and fold the salvage cover in half lengthwise.

4. Fold the salvage cover in half lengthwise again. Make certain that the open end is on top.

5. Fold the cover in half a third time.

A single fire fighter can also use a shoulder toss to spread a salvage cover. To perform a shoulder toss, follow the steps in SKILL DRILL 19-13:

1. Place the folded salvage cover over one arm. (STEP 1)
2. Toss the cover over the salvaged object with a straight-arm movement. (STEP 2)
3. Unfold the cover until it completely drapes the object. (STEP 3)

Two fire fighters can use a balloon toss to cover a pile of building contents quickly. To perform a balloon toss, follow the steps shown in SKILL DRILL 19-14:

1. Place the cover on the ground beside the object. (STEP 1)
2. Unfold the cover so that it runs along the entire base of the object. Each fire fighter grabs one edge of the cover and brings it up to waist height. (STEP 2)
3. Together, lift the cover quickly so that it fills with air like a balloon. (STEP 3)
4. Move quickly to the other side of the item, using the air to help support the cover, and spread the entire cover over the object. (STEP 4)

Salvage Cover Maintenance

Salvage covers must be adequately maintained to preserve their shelf life. Salvage cover maintenance depends on the type of cover used. A canvas cover can usually be cleaned with a scrub brush and clean water. If the cover becomes particularly dirty, however, cleaning with a mild detergent may be necessary. Covers should be adequately rinsed when a detergent is used.

Canvas covers must be properly dried before being returned to service. Effectively drying a canvas cover will reduce mildewing. Vinyl type covers are easily maintained by rinsing and do not mildew as easy as canvas covers.

Once dried, salvage covers should be inspected for tears and holes. Any damage found can be patched by using duct tape or a sewn-on patch.

SKILL DRILL 19-11 Folding and Rolling a Salvage Cover

(Fire Fighter I, NFPA 5.3.14)

1. Spread the salvage cover flat on the ground with a partner facing you.

2. Together, fold the outside edge in to the middle of the cover, creating a fold at the quarter point.

3. Fold the outside fold in to the middle of the cover, creating a second fold.

4. Repeat Steps 2 and 3 from the opposite side of the cover so the folded edges meet at the middle of the cover, with the folds touching but not overlapping.

5. Tightly roll up the folded salvage cover from the end.

SKILL DRILL 19-12 Performing a One-Person Salvage Cover Roll

(Fire Fighter I, NFPA 5.3.14)

1 Stand in front of the end of the object that you are going to cover.

2 Start to unroll the cover over one end of the object.

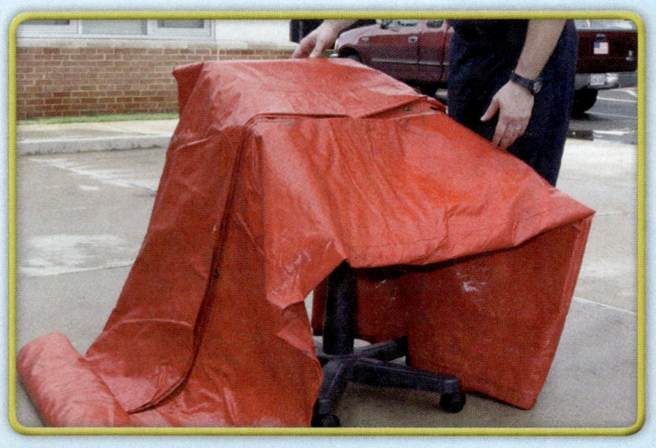

3 Continue unrolling the cover until you reach the top of the object. Allow the remainder of the cover to unroll and settle at the end of the object.

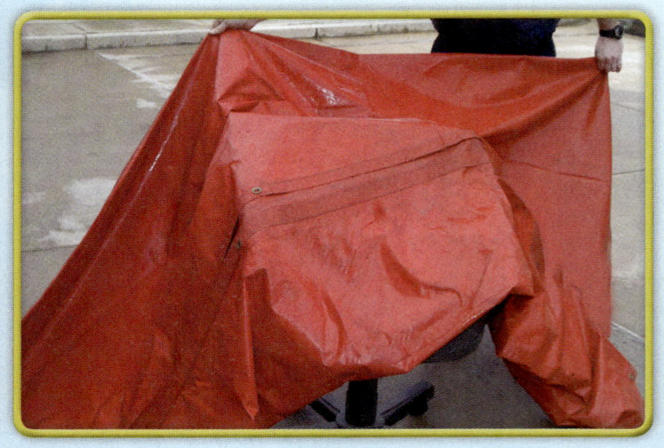

4 Spread the cover, unfolding each side outward over the object to the first fold.

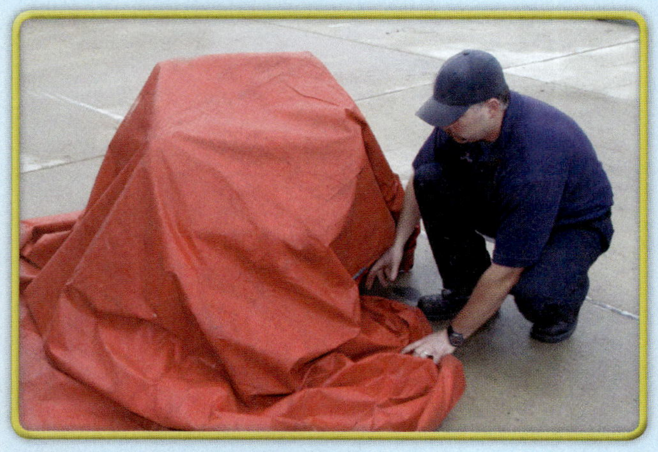

5 Unfold the second fold on each side and drape the cover completely over the object.

6 Tuck in all loose edges of the cover around the object.

SKILL DRILL 19-13 Performing a Shoulder Toss

(Fire Fighter I, NFPA 5.3.14)

1 Place the folded salvage cover over one arm.

2 Toss the cover over the salvaged object with a straight-arm movement.

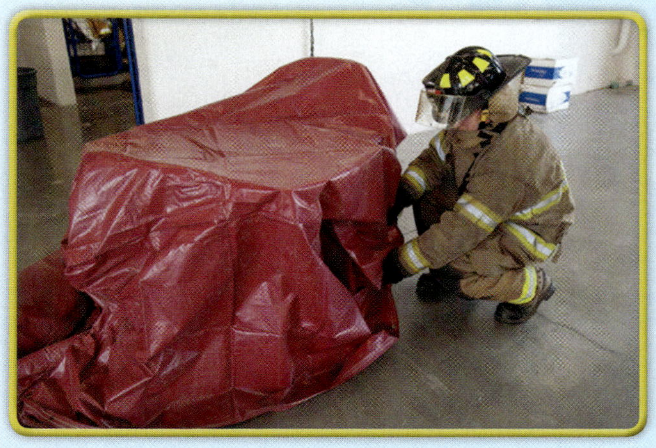

3 Unfold the cover until it completely drapes the object.

Floor Runners

A floor runner is a long section of protective material used to cover a section of flooring or carpet. Floor runners protect carpets or hardwood floors from water, debris, fire fighters' boots, and firefighting equipment. Fire fighters entering an area for salvage operations should unroll the floor runner ahead of themselves and stay on the floor runner while working in the area.

Other Salvage Operations

The best way to protect the contents of a building may be to move them to a safe location. The IC will make this determination. Any items removed from the building should be placed in a dry, secure area—preferably a single location. In some cases, salvaged items can be moved to a suitable location within the same building. If items are moved outside, they should be protected from further damage caused by firefighting operations or the weather. Valuable items should be placed in the care of a law enforcement officer if the property owner is not present.

Sometimes the building contents can provide clues about the cause or spread of the fire. In these situations, fire investigators should be consulted and supervise the removal process.

Salvage operations sometimes extend outside the building to include valuable property such as vehicles or machinery. Use a salvage cover to protect outside property or move vehicles if this action can be done without compromising the fire suppression efforts.

SKILL DRILL 19-14 Performing a Balloon Toss

(Fire Fighter I, NFPA 5.3.14)

1 Place the cover on the ground beside the object.

2 Unfold the cover so that it runs along the entire base of the object. Each fire fighter grabs one edge of the cover and brings it up to waist height.

3 Together, lift the cover quickly so that it fills with air like a balloon.

4 Move quickly to the other side of the item, using the air to help support the cover, and spread the entire cover over the object.

Overhaul Overview

Overhaul is the process of searching for and extinguishing any pockets of fire that remain after a fire has been brought under control. A new fire remains a possibility even if 99 percent of a fire is out and just 1 percent is smoldering. Indeed, a single pocket of embers can rekindle after fire fighters leave the scene and cause even more damage and destruction than the original fire. A fire cannot be considered fully extinguished until the overhaul process is complete.

The process of overhaul begins after the fire is brought under control. Overhaul can be a time-consuming, physically demanding process. The greatest challenge during this phase of fire operations is to identify and open any void spaces in a building where the fire might be burning undetected. If the fire extends into any void spaces, fire fighters must open the walls and ceilings to expose the burned area. Any materials that are still burning must be soaked with water or physically removed from the building. This process must continue until all of the burned material is located and unburned areas are exposed. Overhaul is also required for nonstructure fires such as those occurring in automobiles, bulk piles (tires or woodchips), vegetation, and even garbage.

Safety Considerations During Overhaul

Many injuries can occur during overhaul. As mentioned earlier, overhaul is strenuous work conducted in an area already damaged by fire. Fire fighters who were involved in the fire suppression efforts may be physically fatigued and so overlook hazards. Fire departments should summon fresh crews to conduct overhaul, giving the crews that fought the fire time to recover from their efforts. Provide adequate breaks for rehabilitation during the overhaul process.

Several hazards may be present in the overhaul area. Notably, the structural safety of the building is often compromised. Catastrophic building collapses have occurred during overhaul. Heavy objects could lead to roof or ceiling collapse, debris could litter the area, and there could be holes in the floor. Visibility is often limited, so fire fighters may have to depend on portable lighting. The presence of wet or icy surfaces makes falls more likely. Smoldering areas may burst into flames, and the air is probably not safe to breathe. In addition, during overhaul operations, dangerous equipment—including axes, pike poles, and power tools—is used in close quarters.

Fire fighters must be aware of these hazards and proceed with caution. When necessary, take extra time to evaluate the hazards and determine the safest way to proceed. There is no need to rush during overhaul, and there is no excuse for risking injury during this phase of operations.

A charged hose line must always be ready for use during overhaul operations because flare-ups may occur. For example, a smoldering fire in a void space could reignite suddenly when fire fighters open the space and allow fresh air to enter; such a fire could quickly become intense and dangerous. In addition, fine-grained combustible materials such as sawdust can smolder for a long time and then ignite explosively when they are disturbed and oxygenated.

Fire fighters must wear full PPE during overhaul. SCBA use is mandatory until the air is tested and found safe to breathe. Because working for extended periods in full PPE creates heat stress and dehydration, overhaul crews should work for short periods and take frequent breaks. Fresh crews or crews that have been properly rehabbed should replace fatigued crews.

A safety officer should always be present during overhaul operations to note any hazards and ensure that operations are conducted safely. Company officers should supervise operations, look for hazards, and make sure that all crew members work carefully. The work should proceed at a cautious pace, with an appropriate number of crew members: Too many people working in a small area creates chaos.

Always evaluate the structural condition of a building before beginning overhaul. Look for the following indicators of possible structural collapse:

- Lightweight and/or truss construction
- Cracked walls, out-of-alignment walls, or sagging floors
- Heavy mechanical equipment on the roof
- Overhanging cornices or heavy signs
- Accumulations of water

During the overhaul work, do not compromise the structural integrity of the building. When opening walls and ceilings, remove only the outer coverings and leave the structural members in place. If more invasive overhaul is required, be careful not to compromise the structure's load-bearing members. Avoid cutting lightweight wood trusses, load-bearing wall studs, and floor or ceiling joists, especially when using power tools.

If fire damages the structural integrity of a building, the IC may call for a "hydraulic overhaul" rather than a standard overhaul. In this situation, large-caliber hose streams are used to completely extinguish a fire from the exterior. This strategy is appropriate if the site poses excessive risks to fire fighters and damage to the property is so extensive that it has no salvage value. Heavy mechanical equipment may be used to demolish unsafe buildings and expose any remaining hot spots. A condemned or abandoned building that is going to be torn down is not a place to risk the life or health of a fire fighter.

If a complete overhaul cannot be conducted, the IC can establish a fire watch. The fire watch team remains at the fire scene and watches for signs of rekindling. This team can then request additional help if the fire reignites.

Fire Fighter Safety Tips

A charged hose line must always be available during overhaul.

Coordinating Overhaul with Fire Investigators

Overhaul crews must work with fire investigators to ensure that important evidence is not lost or destroyed as a result of their efforts to extinguish all remnants of the fire. Ideally, a fire investigator should examine the area before overhaul operations begin, identifying evidence and photographing the scene before it is disturbed.

If a fire investigator is not immediately available, the overhaul crews should make careful observations and report to the investigator later. When performing overhaul in or near the suspected area of fire origin, note burn patterns on the walls or ceilings that could indicate the exact site of origin or the path of fire travel. A piercing nozzle can be used to extinguish hidden fire in these void areas; its use limits the damage in and around the suspected area of origin. When moving appliances or other electrical items, note whether they were plugged in or turned on. Always look for evidence that the fire investigator can use to determine the cause of the fire.

If you observe anything suspicious—particularly indications of arson—you should delay the overhaul operations until an investigator can examine the scene. Ensure that the fire will not rekindle, but do not go any further until the investigator arrives. The Fire Cause Determination chapter provides additional information on preservation of evidence in case of fire.

Where to Overhaul

Determining when, where, and how much property needs to be overhauled requires good judgment. Overhaul must be thorough and extensive enough to ensure that the fire is completely out. At the same time, fire fighters should try not to destroy any more property than necessary.

Generally, it is better to make sure that the fire is definitely out than to be too careful about damaging property. If the fire rekindles, more property damage will occur, and the fire department could be held responsible for the additional losses.

The area that must be overhauled depends on the building's construction, its contents, and the size of the fire. All areas directly involved in the fire must be overhauled. If the fire was confined to a single room, all of the furniture in that room must be checked for smoldering fire. The overhaul process also must ensure that the fire did not extend into any void spaces in the walls, above the ceiling, or into the floor. If signs indicate that the fire spread into the structure itself or into the void spaces, all suspect areas must be opened to expose any hidden fire.

If the fire involved more than one room, overhaul must include all possible paths of fire extension. The paths available for a fire to spread within a building are directly related to the type of construction. As a fire fighter, you must learn to anticipate where and how a fire is likely to spread in different types of buildings to find any hidden pockets of fire.

Fire-resistive construction can help contain a fire and keep it from spreading within a building, although this outcome is not guaranteed. Look for any openings that would allow the fire to spread, including utility shafts, pipe chases, and fire doors or dampers that failed to close tightly.

Wood-frame and ordinary construction buildings may contain several areas where a hidden fire could be burning. In particular, these structures often have void spaces under the wall-covering materials. When a serious fire strikes a building of ordinary or wood-frame construction, fire fighters may need to open every wall, ceiling, and void space to check for fire extension. Neighboring buildings may also need to be overhauled if the fire could have spread into them.

In balloon-frame construction, a fire can extend directly from the basement to the attic, without obvious signs of fire on any other floor FIGURE 19-8. For this reason, these buildings require a thorough floor-by-floor overhaul. The problems could be compounded if the attic insulation consists of blown-in cellulose materials—these materials can smolder for a long time. Additional information on how building construction may affect fire spread is found in the Building Construction chapter.

If a building has been extensively remodeled, overhaul presents special challenges. Fire can hide in the space between a dropped ceiling and the original ceiling, or it may extend into a different section of the building through doors and windows covered by new construction. Some buildings have two roofs—one original and one added later—with a void space between them. A fire in this void space presents especially difficult overhaul problems.

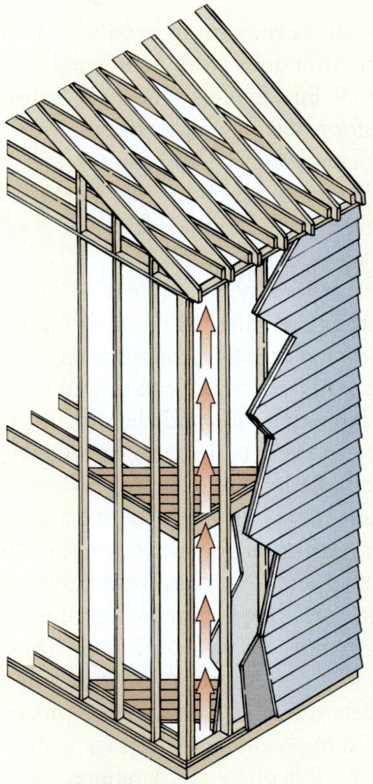

FIGURE 19-8 A fire in a balloon-frame construction building can spread directly from the basement to the attic.

The cause of the fire also can indicate the extent of necessary overhaul. A kitchen stove fire will probably extend into the kitchen exhaust duct and could ignite combustible materials in the immediate vicinity. Follow the path of the duct and open the areas around it to locate any residual fire. A lightning strike releases enormous energy through the wiring and piping systems, which can start multiple fires in different parts of the building. Overhaul after this incident must be quite extensive.

Using Your Senses

Efficient and effective overhaul requires the use of all your senses. Look, listen, and feel to detect signs of potential burning.

Look for these signs:

- Smoke seeping from cracks or from around doors and windows
- Fresh or new smoke
- Red, glowing embers in dark areas
- Burned areas
- Discolored material
- Peeling paint or cracked plaster
- "Hot spots" on a thermal imager

Listen for these sounds:

- Crackling sounds that indicate active fire
- Hissing sounds that indicate water has touched hot objects

Feel for this condition:

- Heat, using the back of your hand (but only if it is safe to remove your glove)

Experience is a valuable trait during overhaul situations. By using his or her senses in a systematic manner, an experienced fire fighter can often determine which parts of the structure need to be opened and which areas remain untouched by the fire.

Thermal Imaging

The thermal image device is a valuable high-tech tool that is often used during overhaul. The same type of thermal imager used for search and rescue (as discussed in the Search and Rescue chapter) can also be used to locate hidden hot spots or residual pockets of fire. It can quickly differentiate between unaffected areas and areas that need to be opened. Using a thermal imager can decrease the amount of time needed to overhaul a fire scene and reduce the amount of physical damage to the building.

The thermal imager distinguishes between objects or areas with different temperatures and displays hot areas and cold areas as different colors on the video screen. Because it is very sensitive, this device can "see" a hot spot even if the heat source is located behind a wall. It can also show the pattern of fire within a wall, even if there is only a few degrees' difference in the wall's surface temperature.

Interpreting the readings from a thermal image device requires practice and training FIGURE 19-9. This device displays relative temperature differences, so an object will appear to be warmer or cooler than its surroundings. If the room has been superheated by fire, all of the contents—including the walls, floor, and ceiling—will appear hot. In these circumstances, an extra-hot spot behind a wall could be difficult to distinguish. In such a case, it might be better to look at the wall from the side that was not exposed to the fire to identify any hot spots.

FIGURE 19-9 Thermal imaging devices can help locate hot spots.

Fire Fighter Safety Tips

Before overhaul operations begin, the safety officer should confirm that electric power and gas service in the overhaul area are shut off. Be careful: Even though the gas supply may be turned off, gas under pressure might still be present in the lines that needs to be bled off.

Do not rely solely on the information from a thermal imaging device to identify areas of persistent fire, because anything that acts as an insulator can hide hot spots. Hot spots behind heavy padding and carpeting or in insulated walls, for example, might not show up as hot spots on the thermal imager screen.

The thermal imager, then, is just a tool to complement the rest of the overhaul operation; it cannot completely rule out the possibility of concealed fire. The only way to ensure that no hidden fire exists is to open a ceiling or wall and do a direct inspection.

Overhaul Techniques

The objective of overhaul is to find and extinguish any fires that could still be burning after a fire is brought under control. Any fire uncovered during overhaul must be thoroughly extinguished. Overhaul operations should continue until the IC is satisfied that no smoldering fires are left.

During overhaul operations, a charged hose line should be available to douse any sudden flare-ups or exposed pockets of fire. If necessary, use a direct stream from the hose line, but avoid unnecessary water damage. Extinguish small pockets of fire or smoldering materials with the least possible amount of water by using a short burst from the hose line or simply drizzling water from the nozzle directly onto the fire.

Extinguish smoldering objects that can be safely picked up by dropping them into a bathtub or bucket filled with water. Remove materials prone to smoldering, such as mattresses and cushioned furniture, from the building and thoroughly soak them outside. Roll mattresses and secure them with rope or webbing to move them out of the building and decrease the possibility of rekindling.

Place debris that is moved outside far enough away from the building to prevent any additional damage if it reignites. Do not allow debris to block entrances or exits. In some cases, a window opening can be enlarged so that debris can be removed more easily. Heavy machinery, such as a front-end loader, may be used to remove large quantities of debris from commercial buildings.

Overhaul Tools

Tools used during overhaul are designed for cutting, prying, and pulling, thereby ensuring that fire fighters can access spaces that might contain hidden fires. Many of the tools used for overhaul are also used for ventilation and forcible entry. The following tools are frequently required in overhaul operations:

- Pike poles and ceiling hooks: for pulling ceilings and removing gypsum wallboard
- Crowbars and Halligan-type tools: for removing baseboards and window or door casings
- Axes: for chopping through wood, such as floor boards and roofing materials
- Power tools such as battery-powered saws: for opening up walls and ceilings
- Pitchforks and shovels: for removing debris
- Rubbish hooks and rakes: for pulling things apart
- Thermal imaging cameras: for identifying hot spots

Because overhaul situations usually do not require high pressures or large volumes of water, a 1½- or 1¾-inch (38- or 44-mm) hose line is usually sufficient to extinguish hot spots. Follow your department's SOPs when choosing a hose line during overhaul.

Buckets, tubs, wheelbarrows, and carryalls (rubbish-carriers) are used to remove debris from a building. A carryall is a 6-foot-square (1.8-meter-square) piece of heavy canvas material with rope handles in each corner FIGURE 19-10.

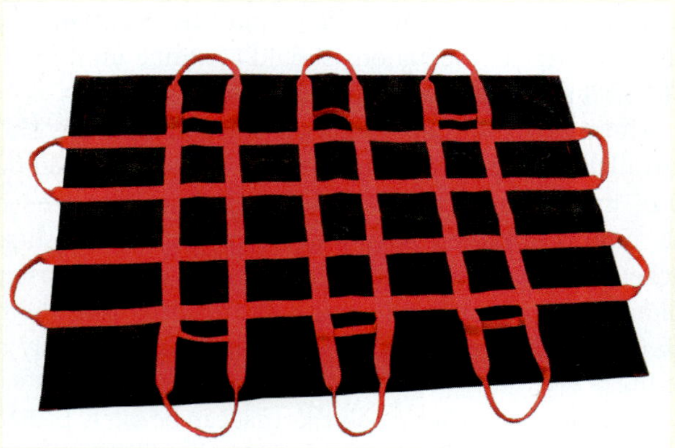

FIGURE 19-10 A carryall is used to remove debris during overhaul.

FIRE FIGHTER Tips

Overhaul is physically demanding work. Pace yourself and use proper technique to increase your efficiency. Ask experienced members of your team for tips on improving your technique.

Opening Walls and Ceilings

Pike poles are used to open ceilings and walls to expose hidden fire. To pull down a ceiling with a pike pole, follow the steps in SKILL DRILL 19-15:

1. Select the appropriate length of pike pole based on the height of the ceiling.
2. Determine which area of the ceiling will be opened. Open the most heavily damaged areas first, followed by the surrounding areas.
3. Position yourself to begin work with your back toward a door, so the debris you pull down will not block your access to the exit. (STEP 1)
4. Using a strong, upward-thrusting motion, penetrate the ceiling with the tip of the pike pole. Face the hook side of the tip away from you. (STEP 2)
5. Pull down and away from your body, so the ceiling material falls away from you. (STEP 3)
6. Continue pulling down sections of the ceiling until the desired area is open. Pull down any insulation, such as rolled fiberglass, found in the ceiling. (STEP 4)

Use the pike pole to break through and pull down large sections of ceilings made with gypsum board. Pulling down laths and breaking through plaster ceilings requires more force. Power saws may be required to cut through ceilings made with plywood or solid boards.

Pike poles, axes, power saws, and handsaws can all be used to open a hole in a wall. Use the same technique to open a wall with a pike pole as you do to open a ceiling. When using an axe, make vertical cuts with the blade of the axe, and then pull the wallboard away from the studs by hand or with the pick end of the axe. A power saw also can be used to make vertical cuts. Pull the wall section away with another tool or by hand.

To open an interior wall with a pick-head axe, follow the steps in SKILL DRILL 19-16:

1. Determine which area of the wall will be opened. Open those areas most heavily damaged by the fire first, followed by the surrounding areas. (STEP 1)
2. Use the axe blade to begin cutting near the top of the wall. Cut downward between wall studs. Be alert for electrical switches or receptacles, as they indicate the presence of electrical wires behind the wall. (STEP 2)
3. Make two vertical cuts, using the pick end of the axe to pull the wall material away from the studs and open the wall. Work from top to bottom. Remove items such as baseboards or window and door trim with a Halligan tool or axe. (STEP 3)
4. Continue opening additional sections of the wall until the desired area is open. Pull out any insulation, such as rolled fiberglass, found behind the wall. (STEP 4)

SKILL DRILL 19-15 Pulling a Ceiling Using a Pike Pole

(Fire Fighter I, NFPA 5.3.13)

1 Select the appropriate length of pike pole based on the height of the ceiling. Determine which area of the ceiling will be opened. Typically, the most heavily damaged areas are opened first, followed by the surrounding areas. Position yourself to begin work with your back toward a door, so the debris you pull down will not block your access to the exit.

2 Using a strong, upward-thrusting motion, penetrate the ceiling with the tip of the pike pole. Face the hook side of the tip away from you.

3 Pull down and away from your body, so the ceiling material falls away from you.

4 Continue pulling down sections of the ceiling until the desired area is opened. Pull down any insulation, such as rolled fiberglass, found in the ceiling.

SKILL DRILL 19-16 Opening an Interior Wall

(Fire Fighter I, NFPA 5.3.13)

1 Determine which area of the wall will be opened. Open those areas most heavily damaged by the fire first, followed by the surrounding areas.

2 Use the axe blade to begin cutting near the top of the wall. Cut downward between wall studs. Be alert for electrical switches or receptacles, as they indicate the presence of electrical wires behind the wall.

3 Make two vertical cuts, using the pick end of the axe to pull the wall material away from the studs and open the wall. Work from top to bottom. Remove items such as baseboards or window and door trim with a Halligan tool or axe.

4 Continue opening additional sections of the wall until the desired area is open. Pull out any insulation, such as rolled fiberglass, found behind the wall.

Wrap-Up

Chief Concepts

- Salvage and overhaul limit and reduce property losses from a fire. Salvage efforts protect property and belongings from damage, particularly from the effects of smoke and water. Overhaul ensures that a fire is completely extinguished by finding and exposing any smoldering or hidden pockets of fire in an area that has been burned.
- Because many emergency incidents occur at night, lighting is required to illuminate the scene and enable safe, efficient operations. Inside fire buildings, heavy smoke can completely block out natural light, and electrical service is often interrupted or disconnected for safety purposes. As a consequence, interior lighting is essential for fire fighters to conduct search and rescue, ventilation, fire suppression, salvage, and overhaul operations.
- Spotlights project a narrow concentrated beam of light. Floodlights project a more diffuse light over a wide area.
- All electrical equipment must be properly grounded. Electrical cords must be well insulated, without cuts or defects, and properly sized to handle the required amperage.
- Lights can be portable or mounted on apparatus. Large hand lights that project a powerful beam of light are preferred for search and rescue activities and interior fire suppression operations when fire fighters must penetrate smoke-filled areas quickly.
- Electricity for lighting equipment is supplied by a generator, an inverter, or a building's electrical system. An inverter is powered by a vehicle, whereas a generator is powered by gasoline or diesel fuel.
- Portable electrical equipment should be cleaned and properly maintained to ensure that it will work when needed. Test and run generators on a weekly or monthly basis to confirm that they will start, run smoothly, and produce power. Run gasoline-powered generators for 15 to 30 minutes to reduce any deposit build-up that could foul the spark plugs and make the generator hard to start.
- Salvage efforts usually are aimed at preventing or limiting secondary losses that result from smoke and water damage, fire suppression efforts, and other causes. Salvage operations include ejecting smoke, removing heat, controlling water runoff, removing water from the building, securing a building after a fire, covering broken windows and doors, and providing temporary patches for ventilation openings in the roof to protect the structure and contents.
- Safety is a primary concern during salvage operations. Structural collapse is always a possibility because of fire damage to the building's structural components. In addition, the water used to douse the fire can add extra weight to the damaged structural components.
- The best way to prevent water damage at a fire scene is to limit the amount of water used to fight the fire. One way to do so is to shut down the sprinkler system. Sprinklers should be shut down as soon as the IC declares that the fire is under control.
- Sprinklers can be shut down by using sprinkler wedges or sprinkler stops, or by shutting off the main control valve.
- Water may be removed from a structure with a salvage pump, water chutes, water catch-alls, water vacuums, or drainage pumps.
- Salvage techniques to limit smoke and heat damage include closed doors, ventilation, and salvage covers. Salvage covers can be used to protect floors and furniture from smoke and water damage.
- The best way to protect an object from water and smoke damage is to remove it from the structure and place it in a safe location.
- Overhaul is the process of searching for and extinguishing any pockets of fire that remain after a fire has been brought under control. A single pocket of embers can rekindle after fire fighters leave the scene and cause even more damage and destruction than the original fire.
- Many injuries can occur during overhaul. During this time, fire fighters may be physically fatigued and require rehabilitation. In addition, the structure may be compromised and visibility limited. Be aware of these hazards and proceed with caution.
- A charged hose line must always be ready for use during overhaul operations to suppress flare-ups and explosions.
- Always evaluate the structural condition of a building before beginning overhaul. Look for the following indicators of possible structural collapse:
  - Lightweight and/or truss construction
  - Cracked walls, out-of-alignment walls, or sagging floors
  - Heavy mechanical equipment on the roof
  - Overhanging cornices or heavy signs
  - Accumulations of water
- During overhaul operations, do not compromise the structural integrity of the building.
- The area that must be overhauled depends on the building's construction, its contents, and the size of the fire. All areas directly involved in the fire must be overhauled.

- Use your senses of sight, hearing, and touch to determine where overhaul is needed. You can also use a thermal imaging device to find hot spots.

Hot Terms

Carryall A piece of heavy canvas with handles, which can be used to tote debris, ash, embers, and burning materials out of a structure.

Floodlight A light that can illuminate a broad area.

Floor runner A piece of canvas or plastic material, usually 3 to 4 feet (91 to 122 centimeters) wide and available in various lengths, that is used to protect flooring from dropped debris and dirt from shoes and boots.

Generator An electromechanical device for the production of electricity. (NFPA 1901)

Inverter Equipment that is used to change the voltage level or waveform, or both, of electrical energy. Commonly, an inverter [also known as a power conditioning unit (PCU) or power conversion system (PCS)] is a device that changes DC input to AC output. Inverters may also function as battery chargers that use alternating current from another source and convert it into direct current for charging batteries. (NFPA 70)

Junction box A device that attaches to an electrical cord to provide additional outlets.

Outside stem and yoke (OS&Y) valve A sprinkler control valve with a valve stem that moves in and out as the valve is opened or closed.

Overhaul The process of final extinguishment after the main body of a fire has been knocked down. All traces of fire must be extinguished at this time. (NFPA 402)

Post indicator valve (PIV) A sprinkler control valve with an indicator that reads either open or shut depending on its position.

Salvage A firefighting procedure for protecting property from further loss following an aircraft accident or fire. (NFPA 402)

Salvage cover A large square or rectangular sheet made of heavy canvas or plastic material that is spread over furniture and other items to protect them from water runoff and falling debris.

Secondary loss Property damage that occurs due to smoke, water, or other measures taken to extinguish the fire.

Spotlight A light designed to project a narrow, concentrated beam of light.

Sprinkler stop A mechanical device inserted between the deflector and the orifice of a sprinkler head to stop the flow of water.

Sprinkler wedge A piece of wedge-shaped wood placed between the deflector and the orifice of a sprinkler head to stop the flow of water.

Water catch-all A salvage cover that has been folded to form a container to hold water until it can be removed.

Water chute A salvage cover that has been folded to direct water flow out of a building or away from sensitive items or areas.

Water vacuum A device similar to a wet/dry shop vacuum cleaner that can pick up liquids. It is used to remove water from buildings.

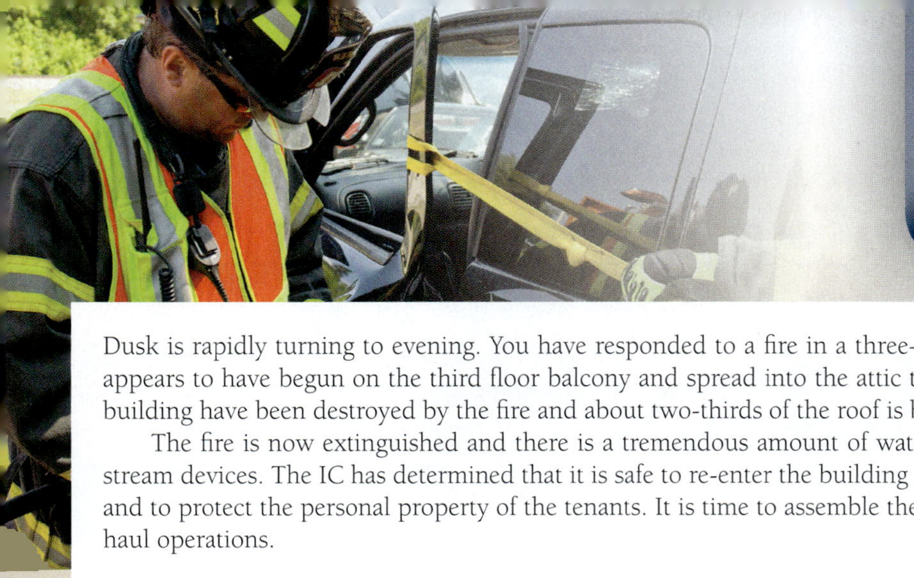

FIRE FIGHTER *in action*

Dusk is rapidly turning to evening. You have responded to a fire in a three-story, wood-frame apartment building. The fire appears to have begun on the third floor balcony and spread into the attic through the soffit. Four units on one end of the building have been destroyed by the fire and about two-thirds of the roof is burned away.

The fire is now extinguished and there is a tremendous amount of water on the upper floors due to the use of master stream devices. The IC has determined that it is safe to re-enter the building in order to knock down any remaining hotspots and to protect the personal property of the tenants. It is time to assemble the equipment that you will need to perform overhaul operations.

1. A _____ is a piece of canvas or plastic material, usually 3 to 4 feet (91 to 122 centimeters) wide and available in various lengths, that is used to protect flooring from dropped debris and/or dirt from shoes and boots.
   - **A.** floor runner
   - **B.** water chute
   - **C.** water catch-all
   - **D.** carryall
2. A _____ is a salvage cover that has been folded to direct water flow out of a building or away from sensitive items or areas.
   - **A.** floor runner
   - **B.** water chute
   - **C.** water catch-all
   - **D.** carryall
3. Which is the first choice for stopping water from flowing through a sprinkler system that has activated?
   - **A.** Close the OS&Y valve.
   - **B.** Close the PIV.
   - **C.** Use a sprinkler wedge to deactivate the sprinklers.
   - **D.** Close the sprinkler system control valve.
4. Which of the following should you perform during overhaul operations?
   - **A.** Clear off all floors so that when the fire investigator arrives, he can check for pour patterns.
   - **B.** Open all interior walls in the structure.
   - **C.** Remove all furniture from the room to prevent rekindling.
   - **D.** Allow the fire investigator to inspect the structure before beginning any overhaul operations.

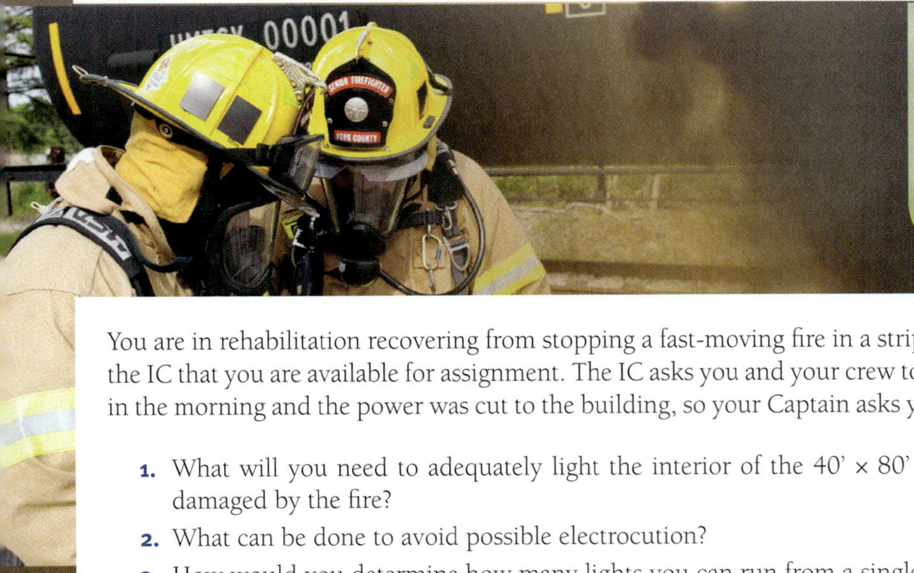

FIRE FIGHTER II *in action*

You are in rehabilitation recovering from stopping a fast-moving fire in a strip mall. After rehabilitation, your Captain notifies the IC that you are available for assignment. The IC asks you and your crew to begin the overhaul of the scene. It is one o'clock in the morning and the power was cut to the building, so your Captain asks you to set up lighting for the overhaul operations.

1. What will you need to adequately light the interior of the 40' × 80' (12 × 24 meters) unit of the building that was damaged by the fire?
2. What can be done to avoid possible electrocution?
3. How would you determine how many lights you can run from a single generator?

Fire Fighter Rehabilitation

Fire Fighter I

NFPA 1001 contains no Fire Fighter I Job Performance Requirements for this chapter.

Knowledge Objectives

After studying this chapter, you will be able to:

- Define rehabilitation. (p 644)
- Describe the factors and causes that require rehabilitation for fire fighters. (p 645)
- Explain how heat stress and personal protective equipment tax the fire fighter's body. (p 645)
- Describe the hazards of dehydration and explain how dehydration can be prevented. (p 646)
- List the signs of dehydration. (p 647)
- Explain why the body needs rehabilitation during an extended or specialized incident. (p 647)
- Describe the types of extended fire incidents during which fire fighters need rehabilitation. (p 648–651)
- List the steps in rehabilitation. (p 651–656)
- Describe the types of fluids that are ideal for fire fighters to drink during rehabilitation. (p 653)
- Describe the types of food that are ideal for fire fighters to eat during rehabilitation. (p 654–655)
- Explain what the individual fire fighter's personal responsibilities are in rehabilitation. (p 656)

Skills Objectives

There are no skill objectives for Fire Fighter I candidates. NFPA 1001 contains no Fire Fighter I Job Performance Requirements for this chapter.

CHAPTER 20

Fire Fighter II — FFII

Knowledge Objectives

There are no knowledge objectives for Fire Fighter II candidates. NFPA 1001 contains no Fire Fighter II Job Performance Requirements for this chapter.

Skills Objectives

There are no skill objectives for Fire Fighter II candidates. NFPA 1001 contains no Fire Fighter II Job Performance Requirements for this chapter.

Additional NFPA Standards

- NFPA 1500, *Standard on Fire Department Occupational Safety and Health Program*
- NFPA 1561, *Standard on Emergency Services Incident Management System*
- NFPA 1584, *Standard on the Rehabilitation Process for Members During Emergency Operations and Training Exercises*

You Are the Fire Fighter

It is shaping up to be a hot day, so you decide to get your workout in early. You have just completed three miles on the treadmill when you are dispatched to a house fire in an old section of town that dates back to the late 1890s. Upon arrival, you are assigned to fire attack and are told that the fire is in the basement. You and your partner battle through the heat as you make your way down the stairs. You have trouble locating the seat of the fire in the dirty brown smoke. When you finally reach it, you realize it is a large fire that is spreading quickly to the upper floors. Just as you are able to get a knockdown of the fire in the basement, you notice you are getting low on air. You notify command of your status, and you are asked to get your cylinder changed out quickly and go to the first floor to begin opening up walls. You change out your cylinder and head back in. You begin pulling walls and notice that you are breathing much harder than usual. You begin getting mild cramps in your arms as you tire from swinging the axe. You strive to keep up with your partner. You are relieved when you notice you are getting low on air, knowing you will be sent to rehabilitation.

1. What is the purpose of rehabilitation?
2. Should an incident commander consider sending personnel to rehabilitation sooner than after two bottles of air have been consumed?
3. Which actions could you have taken to reduce the heat stress on your body?

Introduction

A common saying within the fire service is "Take care of yourself first, take care of the rest of your team second, and take care of the people involved in the incident third." At first glance, this statement might seem self-centered. After all, fire fighters have a deep commitment to help others. If you were to put your personal comfort first, you probably would not want to get up in the middle of a cold winter night to answer a call. But this statement actually has a deeper meaning: You must take care of yourself physically and mentally so you can continue to help others. You must place a high priority on the health and well-being of your teammates for the good of the fire department as a whole. Only when you and your teammates are physically fit and mentally alert will you be able to perform the tasks necessary to save lives and protect property.

To rehabilitate means to restore someone or something to a condition of health or to a state of useful and constructive activity. Fighting fires is a job that requires excellent physical conditioning to combat the rigors of heat, cold, smoke, flames, physical exertion, and emotional stress that are routinely encountered by team members. Even a seasoned, well-conditioned fire fighter can quickly become fatigued when battling a tough fire **FIGURE 20-1**. New fire fighters are frequently amazed at the amount of energy expended during fire suppression activities. This exertion takes its toll on your body, leaving you hot, dehydrated, hungry, and tired. The goal of rehabilitation (usually called simply "rehab") is to take time to "recharge your body's batteries" so you can continue to be a productive member of the team. Rehabilitation is a critical factor in maintaining your health and well-being. It is so critical, in fact, that it is the focus of several NFPA standards, including NFPA 1500, *Standard on Fire Department Occupational Safety and Health Program*; NFPA 1561, *Standard on Emergency Services Incident Management System*; and NFPA 1584, *Standard on the Rehabilitation Process for Members During Emergency Operations and Training Exercises*.

FIGURE 20-1 Firefighting often involves extreme physical exertion and results in rapid fatigue.

Emergency incident rehabilitation is part of the overall emergency effort. This aspect of the emergency response ensures that fire fighters and other emergency workers who are exhausted, thirsty, hungry, ill, injured, or emotionally upset can take a break for rest, fluids, food, medical monitoring, and treatment of illnesses or injuries **FIGURE 20-2**. Without the opportunity to rest and recover, you may develop physical symptoms such as fatigue, headaches, gastrointestinal problems, or collapse. Rehabilitation gives you the opportunity to evaluate and address these issues.

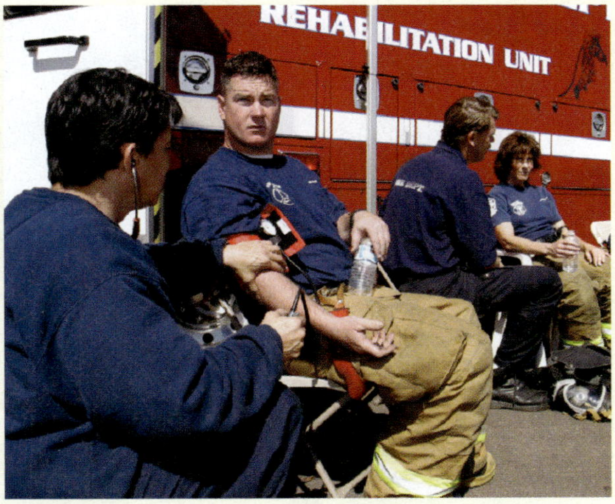

FIGURE 20-2 Rehabilitation provides fire fighters with an opportunity to take a break for rest, cooling or rewarming, rehydration, calorie replacement, medical monitoring, and ensuring accountability.

Rehabilitation enables fire fighters to perform more safely and effectively at an emergency scene. A tired or dehydrated fire fighter is not able to accomplish as much work, is more likely to be injured, and runs the risk of collapsing. The effort that is required to rescue a collapsed or injured fire fighter takes time and resources away from fire suppression activities. For all these reasons, rehabilitation is essential to fight fires safely and effectively.

Factors, Causes, and Need for Rehabilitation

Many conditions come together during a firefighting operation to produce a stressful environment. Consider the stresses involved in a typical middle-of-the-night call for a working fire. The loud, jarring sound of the alarm jolts your sleeping body awake. Without hesitation, you must immediately get up, get dressed, and put on your personal protective equipment (PPE). You do not have time to eat or get something to drink. You may have to drive an emergency vehicle, haul hoses, position a ladder, and climb to the roof to cut a ventilation hole. All of these tasks require a significant amount of energy and concentration, and you must be able to move into action quickly with no time to warm up your muscles as athletes do before an event.

FIRE FIGHTER Tips

The amount of rest needed to recover from physical exertion is directly related to the intensity of the work performed. Fire fighters who have expended a tremendous amount of energy will require a longer recovery period than those who have performed moderate work.

You may be called to a fire on the hottest day of the year, the coldest day of the year, and under all types of adverse circumstances. Because you know that lives and property are at risk, you may feel an added emotional stress, which in turn affects your body. Finally, fire fighters often work in unfamiliar, smoke-filled environments. All of these factors make firefighting very difficult and stressful.

Fire Fighter Safety Tips

A tired or dehydrated fire fighter is more likely to be injured and runs the risk of collapsing. Rehabilitation is essential to correct imbalances in the body that, if left untreated, could endanger you, your crew, and others at the scene.

FIRE FIGHTER Tips

The SAID principle—Specific Adaptation to Imposed Demands—identifies a need for training that mimics the type of work to be performed. In other words, the best way to prepare for physical work is to do activities that match the type, intensity, and duration of the work. The optimal training program should include both high- and low-intensity cardiovascular training and muscular work that involves strength, endurance, and power.

Personal Protective Equipment

Wearing PPE—the protective clothing and breathing apparatus used by fire fighters to reduce and prevent injuries—adds heat stress on the body FIGURE 20-3. PPE can weigh 40 pounds (18 kilogram) or more, and the extra weight increases the amount of energy needed simply to move around. PPE creates a protective envelope around a fire fighter that protects his or her body from the smoke, flames, heat, and steam of a fire. At the same time, however, it traps almost all body heat inside the protective envelope.

Normally, evaporating perspiration helps cool your body so that it does not overheat. PPE contains a vapor barrier that keeps hot liquids and steam from getting in. Depending on your vapor barrier and thermal liner, evaporation may take place on a limited basis. When a fire fighter is wearing PPE, perspiration soaks the inner clothing but evaporative cooling—such as might take place from bare skin—does not occur. Because evaporative cooling cannot occur, a fire fighter's core temperature quickly rises to dangerous levels.

Regardless of the temperature and relative humidity in the ambient air, the temperature and relative humidity inside your PPE will quickly rise to high levels. These high temperatures and high relative humidity produce heat stress. Heat stress reduces your ability to do work, and excessive amounts can lead to collapse if not treated. Because your body continues to perspire in an attempt to cool your body, you can quickly become dehydrated as well as overheated under conditions of heat stress.

FIGURE 20-3 The personal protective equipment that is worn to protect a fire fighter contributes to heat stress.

FIGURE 20-4 When perspiring, fire fighters can lose as much as 1 quart (1 liter) of fluid in 20 minutes.

Fire Fighter Safety Tips

The heat stress index combines temperature and relative humidity to measure the degree of heat stress. TABLE 20-1 shows the heat stress index.

Dehydration

Dehydration is a state in which fluid losses are greater than fluid intake into the body. If left untreated, this imbalance can lead to shock and even death. Fighting fires is a very strenuous activity, and the large amounts of muscular energy required in firefighting activities produce a significant amount of heat. During this exertion, the body loses a substantial amount of water through perspiration. In fact, fire fighters in action can lose as much as 2 quarts (2 liters) of fluid in less than 1 hour FIGURE 20-4.

Dehydration reduces strength, endurance, and mental judgment, as evidenced by a variety of signs and symptoms TABLE 20-2. Extreme dehydration can lead to confusion and total collapse. It is important to prevent or correct dehydration as quickly as possible. To prevent dehydration, drink 16 ounces (0.5 liter) of water two hours before physical exertion. It is simple to plan for and accomplish this fluid intake when you have a scheduled physical conditioning session or in-service drill. Fires are unscheduled events, however, so you must try to keep your body pre-hydrated when you are on duty. In addition, replenishing fluids during rehabilitation is essential to correct any fluid imbalance in your body.

Energy Consumption

Food provides the fuel your muscles need to work properly. In times of strenuous activity, the body burns carbohydrates and fats for energy. These energy sources then need to be replenished. Without a sufficient supply of the right food for energy, your body cannot continue to perform at peak levels for extended periods. The number of calories you need depends on the duration of the activity, the time since your last meal, and your general physical condition. For example, a 6'3" (2 meters) muscular athlete who has not eaten in four hours will need more calories to fuel his body than a 5'2" (1.5 meters) video gamer who has been munching on potato chips. The type of

TABLE 20-1 Heat Index

Relative Humidity (%)	Air Temperature (°F) 70	75	80	85	90	95	100	105	110	115	120
				Apparent Temperature (°F)							
0	64	69	73	78	83	87	91	95	99	103	107
10	65	70	75	80	85	90	95	100	105	111	110
20	66	72	77	82	87	93	99	105	112	120	130
30	67	73	78	84	90	96	104	113	123	135	148
40	68	74	79	86	93	101	110	123	137	151	
50	69	75	81	88	96	107	120	135	150		
60	70	76	82	90	100	114	132	149			
70	70	77	85	93	106	124	144				
80	71	78	86	97	113	136	157				
90	71	79	88	102	122	150	170				
100	72	80	91	108	133	166					

Apparent Temperature (°F)	Danger Category	Injury Threat
Below 80	None	Little or no danger under normal circumstances
80–90	Caution	Fatigue possible if exposure is prolonged and there is physical activity
91–100	Extreme Caution	Heat cramps and heat exhaustion possible if exposure is prolonged and there is physical activity
100–130	Danger	Heat cramps or exhaustion likely, heat stroke possible if exposure is prolonged and there is physical activity
Above 130	Extreme Danger	Heat stroke imminent

TABLE 20-2 Signs and Symptoms of Dehydration

Percentage of Body Weight (Lost)	Signs and Symptoms
1%	Increased thirst
2%	Loss of appetite Dry skin Dark urine Fatigue Dry mouth
3%	Increased heart rate
4–5%	Decreased work capacity by up to 30%
5%	Increased respiration Nausea Decreased sweating Decreased urine output Increased body temperature Markedly increased fatigue Muscle cramps Headache
10%	Muscle spasms Markedly elevated pulse rate Vomiting Dim vision Confusion Altered mental status

Source: IAFC, *A Guide for Best Practices: An Introduction to NFPA 1584 (2008 Standards).*

Fire Fighter Safety Tips

You should always sanitize your hands before you eat.

calories that you consume is also important—it is important to refuel the body with a balanced diet consisting of nutritious food.

Tolerance for Stress

Each individual has a different tolerance level for the stresses encountered when fighting fires. For example, younger individuals tend to have greater endurance and can tolerate higher levels of stress. Likewise, a person who is well rested and in good condition will have a greater level of endurance than someone who is tired and in poor condition. In addition, carrying extra weight and performing strenuous tasks will strain the cardiovascular system and greatly increase the risk of heart attack.

Conditioning plays a significant role in a fire fighter's level of endurance. A well-conditioned person with good cardiovascular capacity, good flexibility, and well-developed muscles will be better able to tolerate the stresses of fighting fires than a person who is out of shape **FIGURE 20-5**. Nevertheless, even the most impressive conditioning will not keep a fire fighter from becoming exhausted under physically stressful situations.

The Body's Need for Rehabilitation

Rehabilitation provides fire fighters with periods of rest and time to recover from the fatigue and stresses of fighting fires and participating in emergency operations. Studies have shown that proper rehabilitation is one way to prevent fire fighters from collapsing or suffering injuries during fire suppression activities and emergency operations. Taking short breaks, replacing fluids, ingesting healthy food, and cooling or rewarming are all measures that reduce the risks of injury and illness **FIGURE 20-6**. Rehabilitation even helps to improve the quality of decision making, because people who are tired tend to make poor decisions.

FIGURE 20-5 A well-conditioned fire fighter will have a greater tolerance for the stresses encountered when fighting fires.

FIGURE 20-6 Taking short breaks to rehabilitate lowers your risk of injury and illness.

Types of Incidents Affecting Fire Fighter Rehabilitation

The concept of rehabilitation needs to be addressed at all incidents, but it will not be necessary to implement all components of a rehabilitation center for every incident. For example, fire departments should always have fluids and high-quality foods available for rehabilitation. Fire fighters who put out a small fire in a single room might require only water for rehydration, whereas those who are involved in extinguishing a major wildland fire may need a full-fledged rehabilitation station.

FIRE FIGHTER Tips

Engaging in a variety of activities can help develop optimal stair-climbing endurance. At the health club, the fire fighter may participate in step aerobics classes or use stair-climbing devices (with or without gear) to develop his or her leg muscles and lung capacity. Outside of the health club, a fire fighter can use step boxes or stairwells to enhance his or her climbing capacity.

FIRE FIGHTER Tips

The introduction of structured rehabilitation procedures has contributed to a reduction in the number of injuries caused by heat stress and exhaustion. The old philosophy was to "fight hard until you drop," and it was not unusual to see exhausted fire fighters being carried out of burning buildings to waiting ambulances. Of course, this procedure disrupted effective firefighting operations because the fire fighters who could still function were busy rescuing their injured and exhausted teammates instead of attacking the fire.

Extended Fire Incidents

Structure Fires

Large emergency incidents require full-scale rehabilitation efforts. Major structure fires that involve extended time on the scene will be hard on crews FIGURE 20-7. Crews working on the interior will become dehydrated and fatigued quickly because of the intense heat and stressful conditions brought on by working in full PPE. They will require rehabilitation so they can continue working at a healthy and safe level. Rotating crews off the fire ground and bringing in fresh crews also promotes health and safety and helps get the job done in an efficient manner.

FIGURE 20-7 Major fires often require a strong effort that lasts for an extended period of time. Rehabilitation and crew rotation are important measures that limit the risk of exhaustion and injuries to fire fighters in such situations.

High-Rise Fires

High-rise fires place increased stress on fire fighters at the same time that they drain crew members' energy resources. It takes considerable energy to walk up many flights of stairs and then work in PPE and self-contained breathing apparatus (SCBA). Hose packs, hand tools, and extra oxygen cylinders must be carried up staircases that may be crowded with people exiting the building. Just getting everyone and everything up to the location of the fire may require a truly marathon effort, even before the attack on the fire begins. It is no wonder that fatigue and dehydration can soon occur in these circumstances.

During a high-rise fire, the incident commander (IC) may assign three companies to do the work that is normally done by one company. This strategy enables the companies to rotate between attacking the fire, replenishing their air supplies, and

FIRE FIGHTER Tips

Reducing the amount of clothing worn while at a safe distance from the fire can help to avoid heat stress before actually engaging in firefighting activities.

VOICES OF EXPERIENCE

Fire fighter rehabilitation is often overlooked by many experienced fire fighters, as it is a relatively new component of the fire service. At the Connecticut Fire Academy, we stress to our fire recruits the importance of rehabilitation, and provide a formal rehabilitation during their live fire training. Each recruit is evaluated after every live fire training session. In addition, the instructors follow a two-in, two-out rotation.

When conversing with a colleague, he told me about a fire that he had worked on with three recent graduates of our recruit fire fighter program. He stated that they came out of the fire after consuming their second cylinder of oxygen and said "Cap, it's time to go to rehab." The Captain looked at them and said, "Yeah, okay." He was very impressed that these new fire fighters knew and understood the need for rehabilitation, and I was impressed that the new fire fighters were able to take their seasoned captain into rehab with them.

The importance of rehabilitation is being spread by the new fire fighters graduating from fire academies. They understand that rehabilitation is not a means to pull them out of the fire, but instead a means to keep them in the fight. Through our training program, our recruits become very comfortable with the rehab process, and my hope is that they continue to educate seasoned fire fighters on the importance of rehabilitation.

Debra L. Burch
Connecticut Fire Academy
Windsor Locks, Connecticut

meeting their needs for rest, fluids, and nutrition. At high-rise fires, the rehabilitation center will often be located two or three floors below the level of the fire to reduce the time and energy expended walking up and down the stairs.

FIRE FIGHTER Tips

During active operations, fire fighters may be very reluctant to admit that they need a rest. Asking an obviously exhausted fire fighter if he or she needs to go to rehabilitation almost invariably brings the response, "No—I'm okay." All company members need to watch out for one another, and company officers must monitor their crews for indications of fatigue. It is better to go to rehabilitation a few minutes early than to wait too long and risk the consequences of injury or exhaustion. The IC should always plan ahead so that a fresh or rested crew is ready to rotate with a crew that needs rehabilitation.

Wildland Fires

Emergency incident rehabilitation was first used on a widespread basis during firefighting operations carried out in wildland areas such as grasslands and forests. Given the size and intensity of major wildland fires, crews need to work in shifts so that their bodies can recover from the heat, smoke, and stresses of the fire. Minimal rehabilitation efforts may be all that is needed for a small ground fire that is easily extinguished. By contrast, a large forest fire may require hundreds of fire fighters and take several weeks to control and extinguish; such a fire requires a significant and ongoing rehabilitation effort for all crews involved **FIGURE 20-8**.

Proper nutrition and rehydration are essential in keeping the crews healthy and ready to resume duty. Some studies have shown that wildland fire fighters may require as much as 5000 to 6000 calories (21 to 25 joules) per day.

FIGURE 20-8 Wildland fires often involve hundreds of fire fighters who work for days or weeks at the site.

Other Types of Incidents Requiring Rehabilitation

Other incidents may also require extensive rehabilitation efforts to maintain the health and safety of fire fighters. For example, hazardous materials incidents that require responders to wear a fully encapsulated suit (a protective suit that fully covers the responder, including the breathing apparatus) are especially stressful **FIGURE 20-9**. These kinds of incidents can expose responders to strenuous conditions for extended periods of time. They may also require that the staging area be located at significant distance from the hazardous materials, such that responders have to walk for long distances while wearing the heavy PPE. As a result, hazardous materials incidents require adequate rehabilitation for responders.

On other occasions, the fire department may be involved in long-duration search and rescue activities or other incidents that require the presence of public safety agencies for extended periods of time. These situations can be both mentally and physically taxing. The establishment of a rehabilitation center where fire fighters can recuperate during these stressful incidents is essential.

FIGURE 20-9 Responders wearing fully encapsulated suits must be carefully monitored for symptoms of heat stress.

The need for rehabilitation is not limited to emergency situations. Training exercises, athletic events, and even standby assignments may also require rehabilitation. Whenever fire fighters are required to be ready for action for an extended period of time, some provision should be made for providing replenishing fluids, nourishing foods, and rewarming or cooling for crew members.

Large-scale training activities, including live fire exercises, generate the same concerns as major fire incidents, and rehabilitation should likewise be incorporated in the planning

FIGURE 20-10 The same type of rehabilitation procedures should be implemented for training activities as are used for actual emergency incidents.

for these activities **FIGURE 20-10**. Training exercises may be conducted over a full day and involve a series of activities as well as time to set up each exercise. As part of this process, time should be set aside for the participants to go to rehabilitation between strenuous activities.

Weather conditions can significantly affect the need for rehabilitation. Fire fighters should always dress appropriately for the weather, and plans for rehabilitation procedures should take into account the anticipated environmental conditions. Whether it is hot or cold, returning the body's temperature back to normal is one of the primary goals of rehabilitation.

Emergencies that occur when the temperature is very hot will increase the need to rotate crews and allow extensive rehabilitation. Crews working on the interior fire attack may not notice much difference, because their environment during the attack will be hot regardless of the external temperature. However, hot weather will definitely affect crews working on the outside. These fire fighters will tend to become dehydrated and fatigued much faster than they would if the ambient temperature was in a more comfortable range.

Another factor that must be considered is the humidity of the air—humidity plays an important role in evaporative cooling. High humidity reduces evaporative cooling, making it more difficult for the body to regulate its internal temperature. The United States Fire Administration (USFA) recommends that rehabilitation operations be initiated whenever the heat stress index exceeds 90°F (32°C).

Fire Fighter Safety Tips

PPE is designed to protect fire fighters from hazards encountered in emergency operations. Fire fighters who are overheated should remove their PPE as soon as possible to permit evaporative cooling. PPE must be removed only in a safe place, outside the hazard area.

Cold weather also increases the need for rehabilitation and crew rotation. Indeed, cold temperatures can be just as dangerous as heat. Hypothermia, a condition in which the internal body temperature falls below 95°F (35°C), can lead to loss of coordination, muscle stiffness, coma, and death **FIGURE 20-11**. Even in cold weather, the weight of PPE and the physical exertion of fighting a fire will cause the body to sweat inside the protective clothing. The combination of damp clothing and cold temperatures can quickly lead to hypothermia. The USFA recommends that rehabilitation operations be initiated whenever the wind chill factor drops to 10°F (–12°C) or lower.

FIGURE 20-11 Winter weather presents a different set of problems for fire fighters than summer weather. Rehabilitation procedures must be conducted in a warm space, and fire fighters should be given hot liquids and food as well as dry clothing.

FIRE FIGHTER Tips

At most fire department incidents, only one rehabilitation center will be established. In contrast, at large-scale incidents, it may be necessary to set up multiple rehabilitation centers. In some communities, rehabilitation is initiated entirely by fire department personnel. In other communities, establishing the rehabilitation facility may be a combined effort between the Salvation Army, the American Red Cross, the fire department auxiliary unit, or a separate Emergency Medical Services (EMS) agency.

How Does Rehabilitation Work?

One way to understand how an emergency incident rehabilitation center works is to look at the functions it is designed to perform. A widely used model of rehabilitation consists of seven parts:

- Relief from climatic conditions
- Rest and recovery
- Active or passive cooling or warming
- Rehydration and calorie replacement
- Medical monitoring
- Member accountability
- Release and reassignment

Although these seven activities are not necessarily completed in the order listed, it is important that each step be addressed.

Relief from Climatic Conditions

Few days provide the perfect climatic conditions for rehabilitation. When the temperature is between approximately 68°F and 72°F (20°C and 22°C), the humidity is low, and plenty of shade is available, indoor rehabilitation may not be needed. Conversely, on days when the temperature is warmer or colder than 68°F to 72°F (20°C and 22°C), or when there is high humidity or wind, rehabilitation needs to take place in a climate-controlled location. In that case, a rehabilitation center may be set up in a nearby building such as a church, school, business, or shopping center, or rehabilitation services may be provided in a transit bus that is brought to the emergency scene.

When entering the rehabilitation center, it is important to take off your PPE to help your body warm up or cool down as needed FIGURE 20-12. If your inner clothing is wet, it may be necessary to change clothing. Many fire fighters carry an extra pair of socks and a second pair of gloves to handle such emergencies. Use a towel to dry off if needed.

Rest and Recovery

Rest should begin as soon as you arrive for rehabilitation. The rehabilitation center should be located away from the central activity of the emergency, so you can disengage from all other stressful activities and remove personal protective clothing. Rest continues as you go through the other parts of revitalization. Rehabilitation centers are equipped with chairs or cots so fire fighters can sit or lie down and relax in a comfortable position.

FIRE FIGHTER Tips

The rehabilitation center will provide some type of restroom facility. For example, the existing restrooms in the building may be used for the rehabilitation center, or portable restrooms may be brought to the emergency scene.

FIGURE 20-12 In the rehabilitation center, fire fighters should be able to remove personal protective clothing and get some rest before returning to action.

Active or Passive Cooling or Warming

The third step of rehabilitation is stabilizing the body's internal temperature. Interior firefighting requires wearing full PPE including SCBA. This encapsulating gear and the rigorous exercise required during firefighting generate huge amounts of body heat that cannot be dissipated by normal means. Performing strenuous external firefighting, especially in times of hot temperatures and high relative humidity, but even without the use of SCBA, also generates huge amounts of body heat. At other times, fire fighters may suffer from cold temperatures. Mounting a defensive attack on a large outdoor lumber yard in 5°F (–15°C) temperatures, for example, can lead to hypothermia. Rehabilitation centers must be ready to provide means to warm cold fire fighters and cool overheated fire fighters.

Active and Passive Cooling

For the overheated fire fighter, PPE should be removed as soon as possible to allow the body to cool. Two types of cooling are possible: active and passive.

Passive cooling is performed by removing PPE and other clothing and moving to a cooler environment. These measures allow the body to cool through effective sweating. Passive cooling works best when both the ambient temperature and the relative humidity are low, which is why stabilizing a fire fighter's body temperature through passive cooling becomes even more complicated during periods of hot weather or high humidity. During these periods, rehabilitation centers need to be climate controlled so that fire fighters can achieve a normal body temperature before resuming active duty. Some fire departments have air-conditioned vehicles available for rehabilitation. Others use buses or establish rehabilitation centers in an already-cooled area such as a shopping center.

If passive cooling is not adequate to produce a reduction in body temperature, active cooling may be required. Active cooling is performed by placing something cool on the fire fighter. This can be done by engaging a misting fan, by placing cool, wet towels around the head and neck, or by dipping both arms in cool water. Active cooling can produce a more rapid reduction in the body temperature.

Passive and Active Warming

Fighting fires in cold temperatures can cause problems as well. Fire fighters responding to incidents during cold weather are subject to both hypothermia and frostbite (damage to tissues resulting from prolonged exposure to cold) FIGURE 20-13. In these cases, the rehabilitation center needs to be heated so that fire fighters can warm up before returning to the cold environment. Passive warming occurs when the fire fighter gets out of the cold, and the body restores itself to an equilibrium temperature. Active heating may be needed when a fire fighter's body is unable to restore itself to the normal temperature balance. Active warming is achieved by medical professionals through the use of heat packs, warming blankets, or warmed IV fluids.

Fire fighters who are wet or severely chilled should be wrapped in warming blankets and moved into a well-heated area before they remove their PPE. As soon as possible, all

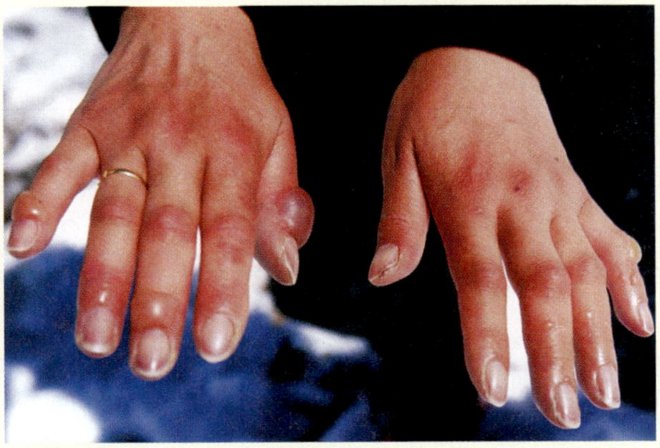

FIGURE 20-13 Prolonged exposures to freezing weather can result in severe frostbite.

wet clothing should be removed and replaced with warm, dry clothing.

Rehydration and Calorie Replacement

Rehydration

When the body becomes overheated, it sweats so that evaporative cooling can reduce body temperature. Perspiration is composed of water and other dissolved substances such as salt. During fire suppression activities, a fire fighter can perspire enough to lose 2 quarts (2 liters) of water in the time it takes to go through two cylinders of air. Because 1 pint (1 liter) of water weighs 1 pound (0.45 kilogram), losing 2 quarts (2 liters) of water is equivalent to a 2 percent loss in body weight for a 200-pound (91 kilogram) fire fighter. The loss of that much body fluid can result in impaired body temperature regulation. Further dehydration can result in reduced muscular endurance, reduced strength, and heat cramps. In addition, severe cases of dehydration can contribute to heat stroke and death. Because a fire fighter can lose so much water during fire suppression activities, it is important to replace fluids before severe dehydration takes place FIGURE 20-14.

When the body perspires, it loses electrolytes (certain salts and other chemicals that are dissolved in body fluids and cells) as well as water. If a fire fighter loses large amounts of water through perspiration, his or her electrolytes must be replenished in addition to the water. Sports-type drinks are often used in rehabilitation for this purpose; they contain water, balanced electrolytes, and some sugars.

Rehydrating the body is not as simple as just drinking lots of fluids quickly. Drinks such as colas, coffee, and tea should be avoided because they contain caffeine; caffeine acts as a diuretic that causes the body to excrete more water. Sugar-rich carbonated beverages are not tolerated or absorbed as well as straight water or sports drinks, because too much sugar is difficult to digest and causes swings in the body's energy level. In addition, drinks that are too cold or too hot may be difficult to consume and may prevent the fire fighter from ingesting enough liquids.

FIGURE 20-14 The rehabilitation center should have plenty of fluids available to rehydrate fire fighters. Plain water or diluted sports drinks are preferred.

An additional concern is the rate at which fluids are absorbed from the stomach. Drinking too much too quickly can cause bloating, a condition in which air fills the stomach. It causes a feeling of fullness and can lead to discomfort, nausea, and even vomiting.

Studies have demonstrated that the stomach can absorb approximately 1 quart (1 liter) of fluid per hour but, as previously noted, the body can lose up to 2 quarts (2 liters) of fluid per hour during intense activity. Thus the body may lose fluids much more rapidly than they can be replaced. Once the initial 2 quarts (2 liters) of water are lost, the body will require 1 to 2 hours to recover. One key to remaining hydrated while fighting fires is to make sure you are hydrated before you reach the fire ground. To do so, keep your body hydrated whenever you are on duty. This will reduce the need for playing catch-up with fluids during rehabilitation.

Try to drink enough fluids to keep your body properly hydrated at all times, particularly during hot weather.

FIRE FIGHTER Tips

The sensation of thirst is not a reliable indicator of the amount of water the body has lost. Thirst develops only after the body is already dehydrated. Do not wait until you feel thirsty to focus on your hydration. Instead, start to drink before you get thirsty, and remember to drink early and often. Fire fighters should strive to drink a quart of water per hour during periods of work.

Near Miss REPORT

Report Number: 10-0000880

Synopsis: Heat stress overcomes fire fighter.

Event Description: The morning of my shift the weather conditions were as follows: temperature 101 degrees, heat index 110, humidity 80%+, clear, sunny, and dry and no rain in the past 12–24 hrs.

I was assigned to the driver/operator position of my engine company, which is staffed only by a driver and officer. Approaching [time deleted], my company was dispatched for a working structure fire in a subdivision. Homes in this subdivision are approx 1800–2200 square foot (550 to 670 meter) wood frame construction built within the past 15 years.

Upon arrival to the scene we found a 2000-square-foot (610 meter) home with 25% involvement with exposures on the two sides. I hand-jacked 400 feet (122 meters) of 5" LDH (13 centimeter LDH) to the hydrant and then proceeded to charge two 1.75" (45 centimeter) cross lays. There was one engine (staffed with 2), one truck (staffed with 2), and one heavy rescue (staffed with 3). I assisted with pulling a 2.5" (76 centimeter) pre-connect portable master stream to the rear of the building where the fire was located. My pump was taken over by the next arriving apparatus and I was reassigned to interior operations. I performed interior operations for approximately 12 minutes and went to rehab with the rest of my crew. I was in rehab for about 10 minutes before I began to feel chest palpitations, weakness, and dizziness. According to medical crews on scene, I was also lethargic, disoriented, and hypotensive (BP 80 palp). I had hydrated prior to and on shift and while in rehab. At this point, I do not know the actual events that happened as this is all secondhand accounts from people on the scene. I was moved to the medic unit and an IV was established with normal saline. The cardiac monitor was applied, which showed sinus tachycardia. All my gear was removed and active cooling took place. After several minutes of cooling and IV fluids, my core temperature was 99. I was in the hospital for several hours and was administered three one liter bags of saline. I was sent home from the ER at approximately 1600 and placed on sick leave for the remainder of my shift.

Lessons Learned: The entire shift (two on the engine and two on the ambulance) were hydrating prior to and during the shift to prevent this very problem. Nobody else had this happen during this call, so does this mean it is simply the way one person's body reacted? I believe so. Hydration is important and must be constant during periods of high heat and humidity. However, hydration is not the only way to ensure our safety and health in this weather. We must monitor ourselves and not be afraid to say, "I need a break" earlier than a normal day. I was not pushing myself any harder than normal. But then again, who can say since I had this problem. Initial staffing levels would have allowed me to perform less work than I did. I delivered hose to the hydrant by hand (since the hydrant was not on our direction of travel), and I also pulled the master streamline because the fire fighters operating in the house were not in a position to perform this task. Proper staffing would have meant some of the tasks I did would be performed by somebody else.

This habit will decrease your risk of dehydration when an emergency incident occurs. Adequate hydration also ensures peak physical and mental performance.

Calorie Replacement

A fire fighter performs intense physical work and burns many calories during active firefighting, and it is necessary to replace those calories during rehabilitation. In much the same way that an engine runs on diesel fuel, the human body runs on glucose. Glucose (also known as blood sugar) is carried throughout the body by the bloodstream and is needed to burn fat efficiently and release energy.

For the body to work properly, its glucose levels need to be in balance. If blood sugar drops too low, the body becomes weak and shaky. If blood sugar is too high, the body becomes sluggish. Blood sugar levels can be balanced by eating a proper diet of carbohydrates, proteins, and fats.

Carbohydrates—a major source of fuel for the body—can be found in grains, vegetables, and fruits. The body converts carbohydrates into glucose, so "carbs" are an excellent energy source. A common dietary myth is that carbohydrates are fattening and should be avoided. In fact, carbohydrates should account for 55 to 65 percent of calories in a balanced diet. Carbohydrates have the same number of calories per gram as proteins and fewer calories than fat. Additionally, carbohydrates are the only fuel that the body can readily use during high-intensity physical activities such as fighting fires.

Proteins perform many vital functions within the body. Most protein comes from meats and dairy products, but smaller amounts are found in grains, nuts, legumes, and vegetables. The body uses proteins (amino acids) to grow and repair tissues; proteins are used as a primary fuel source only in extreme conditions, such as during starvation. Like other nutrients, excess proteins are converted and stored in the body as fat. A balanced diet will get 10 to 12 percent of its calories from proteins.

Fats are also essential for life. These nutrients are used as a source of energy, for insulating and protecting organs, and for breaking down certain vitamins. Some fats, such as those found in fish and nuts, are healthier and more beneficial to the body than others, such as the fats found in margarine. No more than 25 to 30 percent of the diet should come from fats, however. Excess fat consumption—particularly of saturated fats (which come mostly from animal products)—is linked to high cholesterol, high blood pressure, and cardiovascular disease.

Both candy and soft drinks contain sugar (unless they are specifically labeled as "sugar-free"), so the body can quickly absorb and convert these foods to fuel. Unfortunately, simple sugars also stimulate the production of insulin, which reduces blood glucose levels. As a consequence, eating a lot of sugar can actually result in lower energy levels.

For a fire fighter to sustain peak performance levels, it is necessary for him or her to "refuel" during rehabilitation. During short-duration incidents, low-sugar, high-protein sports bars can be used to keep the glucose balance steady. During extended-duration incidents, however, a fire fighter should eat a more complete meal FIGURE 20-15. The proper balance of carbohydrates, proteins, and fats will maintain energy levels throughout the emergency. To ensure peak performance, the meal should include complex carbohydrates such as whole-grain breads, whole-grain pasta, rice, and vegetables. It is also better to eat a series of smaller meals over time, rather than one or two large meals, because larger meals can increase glucose levels and slow down the body.

Healthy, balanced eating is the recommended lifestyle for all fire fighters. Proper nutrition reduces stress, improves health, and provides more energy. It is just as important to keep blood sugar levels balanced throughout the day as it is to remain hydrated. If you follow this regimen, you will be ready to react to an emergency at any time.

FIRE FIGHTER Tips

During rehabilitation, it is important to eat foods that contain carbohydrates. Foods such as energy bars or meal replacement bars are readily available and can be stored for extended periods of time, as opposed to perishable foods such as fruit. Fruit, however, provides carbohydrates and contains a lot of water, which helps maintain a good degree of hydration.

FIGURE 20-15 Plans for rehabilitation at extended-duration incidents should include complete balanced meals.

Medical Monitoring

Medical monitoring during rehabilitation is important to identify potentially dangerous health problems. The checking of vital signs is one way to identify potential problems. Many rehabilitation protocols specify the following types of medical monitoring:

- Checking the fire fighter's heart rate, respiratory rate, blood pressure, and temperature
- Determining how much oxygen is in the fire fighter's bloodstream by taking a pulse oximetry reading
- Performing a carbon monoxide assessment

If the fire fighter's vital signs or test results are abnormal upon entering the rehabilitation center, the fire fighter should be reexamined before being permitted to return to firefighting.

In addition, signs of illness or injury should be assessed in the rehabilitation center before the fire fighter resumes active duty. As a fire fighter, you should be alert for any of the following signs and symptoms:

- Chest pain
- Dizziness
- Shortness of breath
- Weakness
- Headache
- Cramps
- Mental status changes

- Behavioral changes
- Changes in speech
- Changes in gait

Report these signs to medical personnel at the rehabilitation center if you notice any of them in yourself or in one of your crew members.

FIRE FIGHTER Tips

All fire fighters entering rehabilitation should be quickly checked by medical personnel before returning to active firefighting.

Transportation to a Hospital

Another function of the rehabilitation center is to arrange for transportation of ill or injured fire fighters. An ambulance should be available to ensure that ill or injured fire fighters receive prompt transport.

FIRE FIGHTER Tips

Many fire departments have found that assigning a fire officer to be in charge of the rehabilitation center ensures that fire fighters stay in rehabilitation as long as needed and that those fire fighters who need to be transported to a hospital for further evaluation are transported rapidly.

Member Accountability

Crew integrity should be maintained during rehabilitation. Crew members should be released to rehabilitation together and should be released from rehabilitation together. This practice makes accountability on the fire ground much easier for the IC to control. It also ensures that all members of the crew report to rehabilitation to receive the rest, cooling or warming, hydration, food, and medical monitoring that they require. In the event that one member of a crew is not able to return to the active fire scene, the IC must determine how to handle the situation.

Release and Reassignment

Once they are rested, rehydrated, refueled, and rechecked to make certain that they are fit for duty, fire fighters can be released from rehabilitation and reassigned to active duty. Reassignments may be to the same job performed before the trip to rehabilitation or to a different task, depending on the decision of the IC.

Personal Responsibility in Rehabilitation

The goal of every firefighting team is to save lives first and property second. To achieve this goal, you need to take care of yourself first, take care of the rest of your crew second, and then take care of the people involved in the incident. Another way of looking at this issue is to remember that safety begins with you. The other members of your crew depend on you to bear your share of the load. To meet their expectations, you need to maintain your body in peak condition.

Part of your responsibility is to know your own limits. No one else can know what you ate, whether you are lightheaded or dehydrated, whether you are feeling ill, or whether you need a breather. You are the only person who knows these things. Therefore, you may be the only person who knows when you need to request rehabilitation. It may be difficult to say, "I need a break," while the rest of the crew is still hard at work. Even so, it is better to take a break when you need it than to push yourself too far and have to be rescued by other members of your crew.

Regular rehabilitation enables fire fighters to accomplish more work during a major incident, and it decreases the risk of stress-related injuries, illness, and death. Remember—safety begins and ends with you.

Fire Fighter Safety Tips

Stress management skills are crucial for high-quality performance, regardless of the type of stressor involved and the role that the person plays in emergency assistance. A fire fighter needs to implement basic stress management skills to deliver high-quality performance. Exercise, proper nutrition, and emotional support are all helpful in managing stress.

FIRE FIGHTER Tips

You must take care of yourself first if you are to be able to help others.

FIRE FIGHTER Tips

It is important to consider when and how often rehabilitation is necessary. NFPA 1584, *Standard on the Rehabilitation Process for Members During Emergency Operations and Training Exercises*, indicates fire fighters should undergo rehabilitation after exhausting a single 45-minute or 60-minute SCBA cylinder, after every second 30-minute SCBA cylinder, or after 40 minutes of intense work without SCBA. Another recommendation is that a team should have at least 20 minutes of rest during its period of rehabilitation. Additional rehabilitation time may be needed after completing very stressful activities.

Wrap-Up

Chief Concepts

- Rehabilitation is an important part of firefighting. It is designed to ensure that emergency personnel can rest, cool off or warm up, receive fluids and nourishment, and be evaluated for potential medical problems.
- Conditions such as stress, physical exertion, and weather extremes all take a toll on the body.
- Rehabilitation is essential to correct physical imbalances, such as overheating and dehydration, that could endanger members of the firefighting crew.
- Due to their protective qualities and weight, PPE and SCBA add heat stress on the body.
- Dehydration is a state in which fluid losses are greater than fluid intake into the body. If left untreated, it can lead to shock and even death. To avoid dehydration, frequently drink small amounts of nonsugary and noncaffeinated beverages throughout the shift.
- To work at peak capacity, the body needs to be fueled with the right amount of nutrient-rich foods. A physically active fire fighter needs more calories than an office worker.
- Rehabilitation provides fire fighters with periods of rest and time to recover from the physical fatigue and mental stresses of fighting fires. Taking short breaks, replacing fluids, ingesting healthy food, and cooling or rewarming are all measures that can reduce the risk of injury and illness. They also improve mental focus.
- Rehabilitation needs to be addressed at all incidents, but establishing a rehabilitation center may not be necessary for every incident. Types of incidents that require a dedicated rehabilitation center include structure fires, high-rise fires, and wildland fires. Rehabilitation is also needed for hazardous materials incidents, search and rescue operations, and training activities.
- Emergencies that occur during weather extremes require a rehabilitation center for cooling or rewarming.
- Rehabilitation includes seven components:
  - Relief from climatic conditions
  - Rest and recovery
  - Active or passive cooling or warming
  - Rehydration and calorie replacement
  - Medical monitoring
  - Member accountability
  - Release and reassignment
- Report the following signs to medical personnel at the rehabilitation center if you notice any of them in yourself or in one of your crew members:
  - Chest pain
  - Dizziness
  - Shortness of breath
  - Weakness
  - Headache
  - Cramps
  - Mental status changes
  - Behavioral changes
  - Changes in speech
  - Changes in gait
- Your responsibilities in rehabilitation are to know your limits, to listen to your body, and to use the rehabilitation facilities when needed.

Hot Terms

Dehydration A state in which fluid losses are greater than fluid intake into the body. If left untreated, dehydration may lead to shock and even death.

Electrolytes Certain salts and other chemicals that are dissolved in body fluids and cells. Proper levels of electrolytes need to be maintained for good health and strength.

Emergency incident rehabilitation A function on the emergency scene that cares for the well-being of the fire fighters. It includes relief from climatic conditions, rest, cooling or warming, rehydration, calorie replacement, medical monitoring, member accountability, and release.

Frostbite A localized condition that occurs when the layers of the skin and deeper tissue freeze. (NFPA 704)

Fully encapsulated suit A protective suit that completely covers the fire fighter, including the breathing apparatus, and does not let any vapor or fluids enter the suit. It is commonly used in hazardous materials emergencies.

Glucose The source of energy for the body. One of the basic sugars, it is the body's primary fuel, along with oxygen.

Hypothermia A condition in which the internal body temperature falls below 95°F (35°C), usually a result of prolonged exposure to cold or freezing temperatures.

Rehabilitate To restore someone or something to a condition of health or to a state of useful and constructive activity.

FIRE FIGHTER *in action*

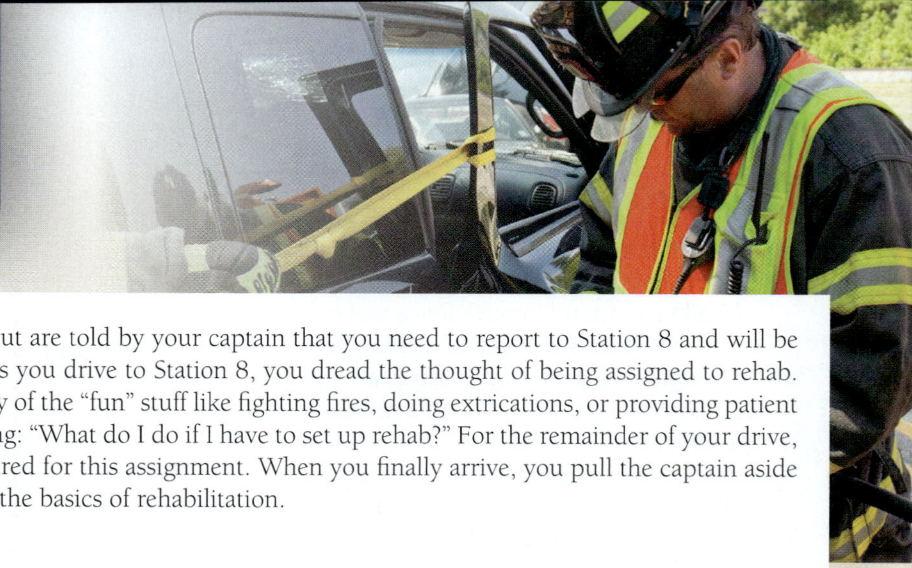

You arrive at your station for your tour of duty, but are told by your captain that you need to report to Station 8 and will be assigned to the rehabilitation unit for the shift. As you drive to Station 8, you dread the thought of being assigned to rehab. Your first thought is that you will not get to do any of the "fun" stuff like fighting fires, doing extrications, or providing patient care. Your second thought is even more concerning: "What do I do if I have to set up rehab?" For the remainder of your drive, you have a nagging feeling that you are not prepared for this assignment. When you finally arrive, you pull the captain aside and express your concerns. He begins to go over the basics of rehabilitation.

1. What is a state in which fluid losses are greater than fluid intake into the body?

A. Homeostasis
B. Dehydration
C. Hypothermia
D. Hyperthermia

2. According to the United States Fire Administration, when should rehabilitation operations be initiated?

A. Whenever the heat stress index exceeds 90°F (32°C)
B. Whenever the heat stress index exceeds 70°F (21°C)
C. Whenever the heat stress index is below 32°F (0°C)
D. Whenever the heat stress index is below 10°F (–12°C)

3. Where should the rehabilitation center be located?

A. Near the SCBA bottle changing station
B. Away from the central activity of the emergency
C. Close to the ancillary activities of the emergency
D. Close to the central activity of the emergency

4. Many rehabilitation medical monitoring protocols recommend all of the following except

A. Vital signs
B. Glucometry
C. Pulse oximetry
D. Carbon monoxide assessment

5. Which of the following is *not* one of the seven parts of rehabilitation?

A. Active or passive heating or cooling
B. Member accountability
C. Rest and recovery
D. PPE replacement

6. Removing PPE and other clothing and moving to a cooler environment is considered ______________.

A. active cooling
B. passive cooling
C. active warming
D. passive warming

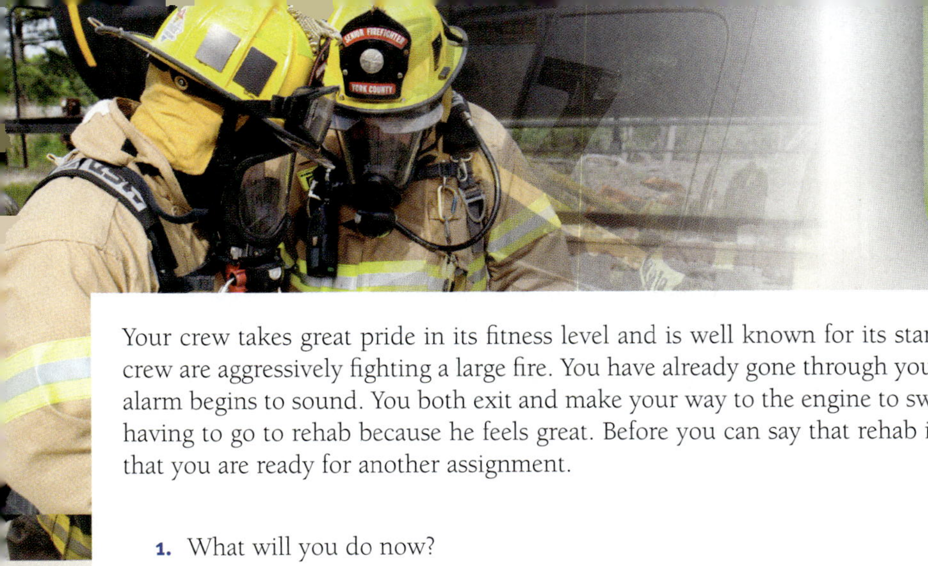

FIRE FIGHTER II *in action*

Your crew takes great pride in its fitness level and is well known for its stamina on the fire ground. One day, you and your crew are aggressively fighting a large fire. You have already gone through your first cylinder when your partner's low-pressure alarm begins to sound. You both exit and make your way to the engine to swap out the cylinder. He mentions how he dreads having to go to rehab because he feels great. Before you can say that rehab is for your own good, your partner radios the IC that you are ready for another assignment.

1. What will you do now?
2. What do you say when you get back to the station?
3. What do you say when your captain says that he noticed you did not go through rehabilitation?
4. If you had it to do over again, what would you do differently?

Wildland and Ground Fires

Fire Fighter I

Knowledge Objectives

After studying this chapter, you will be able to:

- Define the terms *wildland fires* and *ground fires*. (NFPA 5.3.19.A, p 662)
- Explain how the three elements of the wildland fire triangle affect each side of the fire triangle. (p 662–665)
- Describe light fuels, heavy fuels, subsurface fuels, surface fuels, and aerial fuels. (NFPA 5.3.19.A, p 663)
- Explain the relationship between a fuel's properties and the speed at which the fuel ignites and the ensuing fire spreads. (NFPA 5.3.19.B, p 663–664)
- Describe how weather factors and topography influence the growth of wildland fires. (NFPA 5.3.19.A, p 665)
- Label the parts of a wildland fire. (NFPA 5.3.19.A, p 666)
- Describe the methods and tools used to cool a fuel with water. (NFPA 5.3.19.A, p 666)
- Describe the methods and tools used to remove a fuel from wildland fires. (NFPA 5.3.19.A, p 666–667)
- Describe the methods and tools used to smother wildland fires. (NFPA 5.3.19.A, p 667)
- Describe how a direct attack is mounted on wildland fires. (NFPA 5.3.19.A, p 667, 669)
- Describe how an indirect attack is mounted on wildland fires. (NFPA 5.3.19.A, p 669)
- Itemize the characteristics of the fire apparatus used to suppress wildland fires. (NFPA 5.3.19.A, p 669–670)
- Describe the hazards associated with wildland and ground firefighting. (NFPA 5.3.19, p 671)
- Describe the personal protective equipment needed for wildland firefighting. (NFPA 5.3.19, p 671–672)
- Explain the problems created by the wildland–urban interface. (p 673–674)

Skills Objectives

After studying this chapter, you will be able to perform the following skills:

- Deploy a fire shelter. (NFPA 5.3.19.B, p 672)
- Suppress a ground cover fire. (NFPA 5.3.19.B, p 673)

CHAPTER

21

Fire Fighter II

FFII

Knowledge Objectives

There are no knowledge objectives for Fire Fighter II candidates. NFPA 1001 contains no Fire Fighter II Job Performance Requirements for this chapter.

Skills Objectives

There are no skill objectives for Fire Fighter II candidates. NFPA 1001 contains no Fire Fighter II Job Performance Requirements for this chapter.

Additional NFPA Standards

- NFPA 1051, *Standard for Wildland Fire Fighter Professional Qualifications*
- NFPA 1901, *Standard for Automotive Fire Apparatus*
- NFPA 1906, *Standard for Wildland Fire Apparatus*
- NFPA 1977, *Standard on Protective Clothing and Equipment for Wildland Fire Fighting*

You Are the Fire Fighter

You have been watching the national news with awe as you see homes with only their foundation remaining as another major wildland fire sweeps through the formerly pristine forest. Lately it seems that these types of fires have become a regular nightly news item. You think about your own community and whether it is possible that you would ever face a wildland fire since you are part of a large municipal fire department that is not surrounded by forests.

1. How is wildland firefighting similar to structural firefighting?
2. Are urban/suburban fire fighters likely to face wildland fires? Why or why not?
3. How are wildland vehicles and equipment similar to the equipment used in structural firefighting?

Introduction

Wildland fires are defined by the National Fire Protection Association (NFPA) as unplanned and uncontrolled fires burning in vegetative fuel (grass, leaves, crop fields, and trees) that sometimes includes structures. Wildland fires can consume grasslands, brush, and trees of all sizes. Some of these fires burn vegetation that is located many feet above the ground. Their incidence varies from season to season and from one year to the next. The threat and intensity of wildland fires are closely tied to the moisture content and the combustibility of the vegetative fuels, as well as to weather conditions and topography.

Wildland fires are sometimes referred to by different terminology, including brush fires, forest fires, grass fires, ground cover fires, ground fires, natural cover fires, and wildfires. While these names provide descriptions of specific types of wildland fires, all are properly classified as wildland fires.

Ground cover fires burn loose debris on the surface of the ground. This debris includes vegetation such as grass, as well as dead leaves, needles, and branches that have fallen from shrubs and trees. In most fire departments, ground cover fires are the most common type of wildland fire encountered.

Some fire departments respond to more wildland and ground fires than to structure fires. Although many calls for wildland and ground fires involve small incidents, some calls will escalate into major incidents. Most fire departments are responsible for at least some areas that can be classified as wildland. Wildland is land in an uncultivated natural state that is covered by timber, woodland, brush, or grass. Most large cities have some parklands that are covered with natural vegetation.

Most structural fire fighters will be called on to extinguish wildland and ground fires at some point in their careers. The information in this chapter is intended to teach you how to extinguish wildland and ground fires safely. It will not train you to be a wildland fire fighter, however. Fire fighters whose primary responsibility is suppressing wildland fires need to complete separate wildland firefighting training.

Large wildland fires are handled by agencies that specialize in this type of firefighting. Each state has an agency designated to coordinate wildland firefighting. Federal government agencies such as the U.S. Forest Service, Bureau of Land Management, National Wildfire Coordinating Group, Bureau of Indian Affairs, National Park Service, and U.S. Fish and Wildlife Service are also responsible for coordinating firefighting activities at large incidents and incidents that occur on federal lands. The incident command system (ICS) is also activated at these incidents. See the Incident Command System chapter for more information on working within the ICS during a wildland fire.

Wildland and Ground Fires and the Fire Triangle

To safely and effectively extinguish wildland and ground fires, fire fighters need to understand the various factors that cause fire ignition and affect the growth and spread of wildland fires. The fire triangle provides a good model for explaining the behavior of wildland and ground fires FIGURE 21-1.

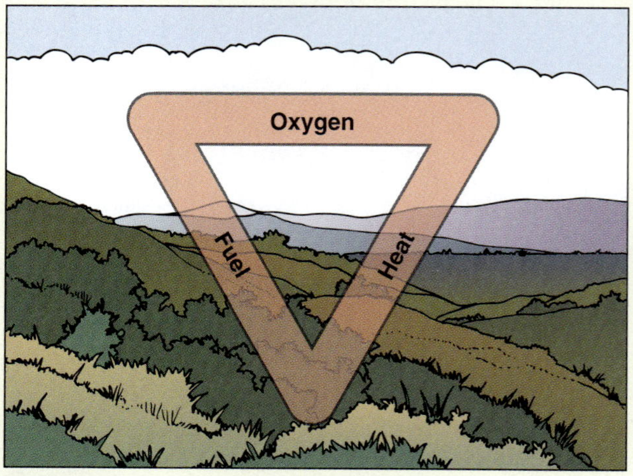

FIGURE 21-1 The wildland fire triangle.

Fire Fighter Safety Tips

All fires involving fine or heavy fuels should be treated with the same caution and safety. Many fire fighters have died because they underestimated fire behavior. Treat all wildland and ground fires with respect.

The fire triangle, as discussed in the Fire Behavior chapter, consists of three elements: fuel, oxygen, and heat. Wildland and ground fires require the same three elements as structural fires to enable them to start and continue, but the difference between these types of fires is the conditions under which the fuel, oxygen, and heat come together to produce an uncontrolled fire. Because weather conditions have a major impact on the behavior of wildland fires, we will also consider how each side of the fire triangle is influenced by certain weather conditions.

Fire Fighter Safety Tips

Heavy fuels such as trees and stumps require close attention. Spraying water on the outside of these materials will not extinguish the fire, because heavy fuels have a tendency to continue to burn on the inside. Heavy fuels need to be cut out to confirm that fire fighters have extinguished all of the fire and to prevent rekindling.

Fuel

The primary fuel for wildland and ground fires is the vegetation (grasslands, brush, and trees) in the area. The amount of fuel in an area may range from sparse grass to heavy underbrush and large trees. Some fuels ignite readily and burn rapidly when dry, whereas others are more difficult to ignite and burn more slowly. Vegetative fuels can be located under the ground (roots), on the surface (grass and fallen leaves), or above the ground (tree branches). These fuels are further classified as fine fuels or heavy fuels. TABLE 21-1 summarizes the types of fuels that are typically associated with wildland and ground fires.

Fine Fuels

Fine fuels and light fuels include dried vegetation such as twigs, leaves, needles, grass, moss, and light brush. Ground duff—the partly decomposed organic material on a forest floor—is another type of fine fuel. Fine fuels have a large surface area relative to their volume, which causes them to ignite easily, burn quickly, and produce a lot of heat. Such fuels are usually the main type of fuel present in ground cover fires. They facilitate the ignition of heavier fuels and cause fire to burn and spread with great intensity. In addition, they burn rapidly and unpredictably. As a consequence of these characteristics, fires spread more quickly in fine fuels than in heavy timber and brush.

Heavy Fuels

Heavy fuels are of a larger diameter than fine fuels. They include large brush, heavy timber, stumps, branches, and dead timber on the ground. Another type of heavy fuel is slash—the leftovers of logging and land-clearing operations. Slash includes both large and small pieces of logs, branches, bark, stumps, and other vegetative debris. Although fires involving heavy fuels do not spread as rapidly as fires involving fine fuels do, heavy fuels can burn with a high intensity.

TABLE 21-1 Fuel Classification

Fine fuels	1-hour fuels; are ¼ inch (0.6 centimeter) in diameter or less; consist of grass, leaves, tree moss, and loose surface litter. Fine fuels can absorb or lose moisture very rapidly, and their flammability can change very quickly from no flammability to very high flammability in just a short period of time.
Light fuels	10-hour fuels; range from ¼ inch to 1 inch (0.6 to 2.5 centimeters) in diameter; consist of small twigs and stems, both living and dead. Light fuels dry out very quickly and can become highly flammable.
Medium fuels	100-hour fuels; range from 1 inch to 3 inches (2.5 to 8 centimeters) in diameter; consist of sticks, branch wood, and medium duff. Medium fuels gain and lose moisture relatively slowly and cumulatively. They also add to wildfire intensity.
Coarse and heavily compacted fuels	1000-hour fuels; are 3 inches (8 centimeters) or more in diameter; include large logs, deep duff, and peaty material. The moisture content of coarse and heavily compacted fuels changes very slowly and cumulatively over long periods of time.

Subsurface, Surface, and Aerial Fuels

Subsurface fuels are those fuels located under the ground. Roots, moss, duff, and decomposed stumps are examples of subsurface fuels. Fires involving subsurface fuels are difficult to locate and challenging to extinguish.

Surface fuels are located close to the surface of the ground. They include grass, leaves, twigs, needles, small trees, and logging slash. Brush less than 6 feet (2 meter) above the ground is also classified as a surface fuel. Surface fuels, which are sometimes called ground fuels, are involved in ground cover fires.

Aerial fuels (also called canopy fuels) are located more than 6 feet (2 meters) above the ground. They usually consist of trees, including tree limbs, leaves and needles on limbs, and moss attached to the tree limbs.

Other Fuel Characteristics

Other characteristics of wildland fuels may determine how quickly the fuel ignites, how rapidly it burns, and how readily it spreads to other areas. These properties include the fuel's size and shape, its compactness, its continuity, its volume, and its moisture level.

The size and shape of a fuel influence how it burns. For example, fine fuels burn more quickly than heavy fuels and require less heat to reach their ignition temperature. This difference arises because fine fuels have a greater surface area relative to the total volume of the fuel.

Fuel compactness influences the rate at which a wildland fire burns. When fuels are tightly compressed, adequate oxygen and heat cannot get to large surfaces of the fuel in a short period of time. Therefore compact fuels often burn more slowly and over a long period of time. For example, a compact fuel such as a large tree trunk burns more slowly than a less compact fuel such as dry grass.

Fuel continuity refers to wildland fuels that have uninterrupted connections. For example, intertwined tree branches provide for fuel continuity. Fuels with continuity touch one another or are located close enough together that a fire can spread from one location to another—a characteristic that contributes to a rapidly spreading wildland fire. In situations where there is no fuel continuity, a fire may burn out at the spot where it started.

Fuel volume refers to the quantity of fuel available in a specific area. The amount of fuel in a given area influences the growth and intensity of the fire.

Fuel moisture refers to the amount of moisture contained in a fuel. This characteristic influences the speed of ignition, the rate of spread, and the intensity of the fire. Fuels with high moisture content will not ignite and burn as readily as fuels with low moisture content. Some types of vegetation naturally contain more moisture than others. The amount of moisture in a fuel is also related to the season of the year, the amount of rain that has fallen, and the relative humidity in the air. In a year with decreased rainfall, less moisture will be present in vegetative fuels compared with the same vegetation during a rainy year. The capacity of fuels to vary in fuel moisture content depends on the type and condition of the fuels.

Oxygen

The second side of the fire triangle is oxygen. Oxygen is needed to initiate and support the process of combustion. In a structure fire, a limited supply of oxygen results in a slow-burning fire. If additional oxygen becomes available to a fire, it will begin to burn more rapidly.

In wildland and ground fires, which are usually free burning, oxygen is rarely an important variable in the ignition or spread of the fire. Unlike in structure fires, oxygen is available in unlimited quantities in these open-air fires. Nevertheless, air movement around wildland and ground fires will influence the speed with which the fire moves. Wind blowing on a wildland and ground fire brings more oxygen to the fire, accelerates the process of combustion, and influences the direction in which the fire travels.

Heat

The third side of the fire triangle is heat. Sufficient heat must be applied to fuel in the presence of adequate oxygen to produce a fire. Three types of factors may ignite wildland and ground fires: natural causes, accidental causes, and intentional causes.

Lightning is the source of almost all naturally caused wildland and ground fires. Some storms produce dry lightning, in which lightning strikes occur without rain. A single lightning storm can produce many lightning strikes and start many fires.

Near Miss REPORT

Report Number: 11-0000349

Synopsis: Wildland crew saved by wind shift.

Event Description: We were dispatched to a brush/wildland fire in a neighboring district with 458 acres (185 hectares) of woods on fire. The fire was moving higher and faster than a person can move. We arrived in brush units with about 100 gallons (378 liters) of water at most. From the time we arrived, we were teamed up with other departments. We never once made contact with command concerning which radio channels we would be communicating on. We stopped at the top of the hill and saw several brush units flying down the hill to the right flank. We heard what sounded like a freight train coming over the hill. There were no trains nearby. No communications were made to tell us that the fire was coming right at us. We were just able to back up, turn around, and start to fly down the hill.

The brush truck engine died. The fire was moving faster than we could run. Our only saving grace was the wind changing direction and blowing the fire away from us. The wind blew hard enough to stop the fire and stand it straight up. The wind saved 19 lives that day. The fire was further than you could see, with nowhere to go.

Lessons Learned:

We need to establish good communications prior to entering a fire scene.

Equipment needs to be checked every day.

Communication is key. We need to make sure we have communication with all mutual aid companies.

A variety of accidental causes may also result in wildland and ground fires, such as discarded smoking materials and improperly extinguished campfires. In addition, downed electric wires may provide a heat source that can ignite vegetative fuels.

Finally, some wildfires and ground fires are the result of intentional acts of arson.

Other Factors That Affect Wildland Fires

Once a wildland and ground fire has started, the growth of the fire is influenced by weather factors and by the topography of the land involved in the fire. Weather conditions and topography have a greater effect on wildland and ground fires than they do on structure fires.

Weather

Weather conditions have a major impact on the course of a wildland fire. The two most critical aspects of weather that influence a wildland fire are moisture and wind.

> **Fire Fighter Safety Tips**
>
> The National Weather Service issues "Red Flag Warnings" and "Fire Weather Watches" for particularly dangerous weather conditions. Even small fires can rapidly spin out of control during these times, so extra vigilance must be used. *Do not* expect normal fire behavior when these warnings are issued!

Moisture

Moisture can be present in the form of either relative humidity or precipitation. Relative humidity is the ratio of the amount of water vapor present in the air compared to the maximum amount of water vapor that the air can hold at a given temperature. Warm air has a higher capacity for moisture content than cool air does. If the air contains the maximum amount of water vapor possible, the relative humidity is said to be 100 percent. On a rainy day, the relative humidity is usually 100 percent. In dry climates, the relative humidity may be 20 percent or less, which means the water vapor contained in the air is only 20 percent of the maximum amount that the air is capable of holding. If the relative humidity drops and stays at 10 percent, for example, eventually all fuels will equalize at that level—but light fuels will reach that level much more quickly than heavy fuels do.

Relative humidity is a major factor in the behavior of wildland and ground fires. When relative humidity is low, vegetative fuels dry out, making them more susceptible to ignition. Conversely, when relative humidity is high, the moisture from the air is absorbed by the vegetative fuels, making them less susceptible to ignition. Changes in relative humidity affect light fuels more dramatically than they affect heavy fuels.

Relative humidity typically varies with the time of day. As the temperature increases throughout the day, the relative humidity generally drops. Conversely, as the temperature decreases during the evening and night hours, the relative humidity rises. A rise in the relative humidity can help to slow the spread of a wildland fire and aid the work of fire crews who are trying to control the fire.

In some parts of the United States, the average relative humidity also varies with the season of the year. These changes contribute to variations in the moisture content of vegetative fuels—a factor that partly explains why many regions of the United States demonstrate seasonal patterns of wildland and ground fires. The seasons with the most wildland and ground fires are often the seasons with the lowest relative humidity.

The second type of moisture that influences the behavior of wildland and ground fires is precipitation. Moisture falling from the sky increases the relative humidity. In addition, much precipitation is absorbed by plants, which increases their moisture content and renders the plants less susceptible to combustion. In seasons with adequate precipitation, the incidence of wildland and ground fires is typically much lower than in years with below-average precipitation.

Wind

Air movement also has an effect on the spread and direction of wildland and ground fires. On a still day, a wildland or ground fire can burn and grow in size. A fire with similar conditions will spread more rapidly, however, if wind pushes the fire so that it comes in contact with additional fuels. Wind has the ability to move a fire at great speed. The effect of wind on a wildland and ground fire is similar to fanning a flame to help it burn more rapidly.

Topography

The movement of wildland and ground fires also reflects the topography of the land. Topography, which refers to the changes of elevation in the land as well as the positions of natural and human-made features, has a major effect on the behavior of wildland and ground fires. Heat from these fires rises just like the heat in a structure fire. When a wildland or ground fire burns on flat land, much of the heat from the fire will rise into the air, thereby preheating the fuels above the main body of fire. For example, when a fire is burning on a hillside, convective and radiant heat from the fire will preheat the fuels on the slope above, causing them to dry out and be more flammable. This factor may result in more rapid rates of spread over a larger area as the fire moves up the slope.

Other features of the topography influence the spread of a fire as well. Natural barriers, such as streams and lakes, may help contain fires. Human-built barriers, such as highways, also make it easier to contain a fire.

> **Fire Fighter Safety Tips**
>
> Certain topographical features, such as valleys, steep hillsides, and saddles, can cause sudden and rapid fire progression. Many fire fighter fatalities and near misses have occurred in these types of terrain. Lookouts, communications, escape routes, and safety zones (LCES) must be rigidly enforced when operating in these areas.

Extinguishing Wildland Fires

To safely attack and effectively extinguish wildland and ground fires, fire fighters must consider how the factors of weather, fuels, and topography come together to influence the behavior of the fire. They also need to know the terms that are commonly used to describe different parts of the fire so that they can understand directions from command and report information back to fire officers.

Anatomy of a Wildland Fire

The location where a wildland or ground fire begins is called the area of origin. As the fire grows and moves into new fuel, the most rapidly moving area—the traveling edge of the fire—is called the head of the fire FIGURE 21-2. As the fire gets bigger, the area close to the area of origin may be referred to as the heel of the fire or the rear of the fire.

As the fire grows, a change in wind or topography may cause it to move in such a way that a long, narrow extension of fire projects out from the fire. This type of expansion is called a finger. A finger can grow because of a change in topography, fuel, or wind direction and produce a secondary direction of travel for the fire. The unburned area between a finger and the main body of the fire is called a pocket. A pocket is a dangerous place for fire fighters because this area of unburned fuel is surrounded on three sides by fire. An area of land that is left untouched by the fire, but is surrounded by burned land, is called an island.

A spot fire is a new fire that starts outside the perimeter of the main fire. Spot fires usually begin when flaming vegetation, in the form of an ember or spark, is picked up by the convection currents generated by the fire and dropped some distance away from the original fire.

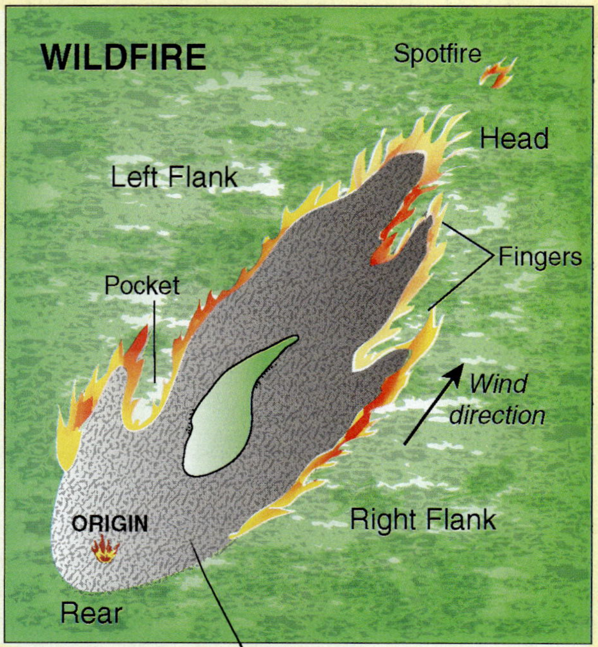

FIGURE 21-2 The anatomy of a wildland fire.

Fight from inside already burnt area

Green and black are terms often used by wildland fire fighters. "Green" describes areas containing unburned fuels; these areas are more likely to become involved in the fire than are black areas. "Black" describes areas that have already been burned. Because black areas contain few unburned fuels, they are often relatively safe for fire fighters.

Methods of Extinguishment

The methods used to extinguish wildland and ground fires focus on disrupting the conditions represented by the fire triangle. Wildland and ground fires can be controlled and extinguished by cooling the fuel, by removing fuel from the fire, or by smothering the fire.

Cooling the Fuel

Water is used to cool the fuels that drive wildland and ground fires. Although the principles involved in cooling wildland and ground fires are the same as those used for structural fires, the methods used to apply the water may be different.

For small fires with a light fuel load, backpack pump extinguishers may be effective. These devices can be transported to the fire and quickly put into operation before the fire has an opportunity to grow. Portable backpack pumps hold between 4 and 8 gallons of water. Fire fighters expel the water by using a hand pump.

Water from booster tanks carried on structural fire apparatus or on special wildland fire trucks is another means of cooling a wildland fire's fuel. The hoses used for wildland fires may be smaller than the hoses required for structural firefighting. Unfortunately, it is not always possible to get trucks into a position where they can attack a wildland fire. The water supply may also be limited to the amount of water carried by the apparatus, unless static water sources can be tapped by fire pumpers or portable pumps. Backpack pumps and small hose lines are used to extinguish most ground cover fires.

A third means of applying water to a wildland fire is from aircraft. Fixed-wing aircraft can take on a load of water from a lake and apply it to the fire. Rotary-wing aircraft (helicopters) can carry a container of water and drop it on the fire. The use of aircraft can be an effective means to control wildland fires, but it requires close communication between different agencies to coordinate this activity. Aircraft are most commonly used for fires involving heavy fuels.

Removing the Fuel

The second method of controlling and extinguishing wildland and ground fires is to remove the fuel from the fire. This can be done by fire fighters using hand tools or mechanized equipment.

In terms of hand tools, removal of fine fuels can be accomplished with a fire broom to sweep away light grass or weeds. Steel fire rakes can be used to create a fire line in light brush. A combination hoe and rake called a McLeod tool is also used to create a fire line. When working in heavier brush, fire fighters can use an adze to grub out brush to create a fire break. The Pulaski axe combines an adze and an axe for brush removal. A council rake is a long-handled rake constructed with hardened triangular-shaped steel teeth that is used for raking a fire line down to soil with no subsurface fuel (mineral soil), for digging, for rolling burning logs, and for cutting grass and small brush.

In some situations, saws are used to remove heavy brush and trees from the fire. The types of saws employed for this purpose range from handsaws to gasoline-powered chainsaws. Other powered equipment includes tractors, plows, and bulldozers.

Backfiring is another technique used to remove the fuel from a wildland fire. When properly set, a backfire can burn an area of vegetation in front of the fire, thereby creating a wide area devoid of vegetation. Backfires must be set at the right time and at the proper place to work correctly. When improperly set, a backfire may worsen a wildland fire or create a new one. These fires should be set only by fire fighters who have had special training in the control of wildland and ground fires.

FIRE FIGHTER II Tips **FFII**

Only specially trained personnel should operate chainsaws. When operating a chainsaw, you must wear the following protective gear: helmet, eye protection, hearing protection, gloves, and leather chaps to protect the legs. In addition, always have a lookout and a preplanned escape route when trees are being cut.

FIRE FIGHTER II Tips **FFII**

When operating around motorized apparatus or bulldozers, stay at least 50 feet away from the equipment. Wear a light on your helmet and a reflective vest at night. Avoid standing downhill from motorized apparatus. Finally, make sure that the equipment operator can see you.

Smothering the Fire (Removing Oxygen)

The third method that can be used to control and extinguish wildland fires is to remove oxygen from the fire. The smothering technique is most commonly used when fire fighters are overhauling the last remnants of a wildland and ground fire. In these circumstances, mineral soil (dirt with no organic material) and water are often thrown on smoldering vegetation until it is cool to the touch to prevent flare-ups. Subsurface fuels are also removed to prevent flare-ups. Smothering is less useful during the more active phases of a fire, although some vegetative fires in grasses can be smothered using a special fire-beater sheet constructed of heavy rubber.

Compressed-air foam systems (CAFS) combine foam concentrate, water, and compressed air to produce a foam that will stick to vegetation and structures that are in a wildland and ground fire's path. When the heat of the fire reaches the foam, the foam absorbs the fire's heat and breaks down, thereby cooling the fuel. CAFS can also extinguish a fire by coating the fuel and preventing oxygen from combining with the fuel. CAFS can be delivered from a backpack extinguisher or mounted on apparatus. Such systems help to extinguish fires with less water because the compressed air foam achieves better penetration of fuels and clings to the fuel, thereby increasing the effectiveness of the water used. The use of CAFS reduces the incidence of fires rekindling.

Fire Fighter Safety Tips

Before attacking a wildland fire, ensure that lookouts, communication systems, escape routes, and safety zones (LCES) are in place.

Types of Attacks

Wildland and ground fires can advance quickly and can change directions quickly. Given their unpredictability, they present a very hazardous environment for fire fighters. For these reasons, it is important for the incident commander (IC) to match the type of firefighting attack to the conditions present at the fire. Two general types of attacks are used to contain and extinguish wildland fires: direct attacks and indirect attacks.

An attack on a wildland fire needs to ensure the safety of fire fighters and citizens, and to be as effective as possible in minimizing the property losses from the fire. Fire attacks should start from an anchor point close to the place where the fire started. The anchor point provides an area of safety where the vegetation has already burned. Ideally, it should be located close to a highway or road. This strategy helps prevent a situation in which the fire might change direction, circle back on fire fighters, and trap them in a point of danger.

FIRE FIGHTER Tips

Two primary methods of attack are used for fighting wildland and ground fires: the pincer attack and the flanking attack.

- Pincer attack: Two crews of fire fighters mount a two-pronged attack, with one crew suppressing one side of the fire and the other crew suppressing the other side of the fire. The two crews work together to "pinch out" the fire.
- Flanking attack: Only one crew is required. The IC positions the crew on one side of the fire and the crew suppresses it.

Direct Attack

A direct attack on a wildland and ground fire is mounted by containing and extinguishing the fire at its burning edge. A direct attack can be made with a fire crew and hand tools, a structural engine company, a wildland engine company, a plow, or aircraft. Fire fighters might smother the fire with dirt, use hoses to apply water or compressed air foam to cool the fire, or remove the fuel. A direct attack is typically used on a small wildland fire before it has grown to a large size.

Any direct attack needs to be coordinated through the ICS. Before a direct attack is started, the IC will assess the resources available and determine the best way to attack the fire.

A direct attack is dangerous to fire fighters because they must work in smoke and heat close to the fire. Such an attack has the advantage of accomplishing quick containment of a fire, however. By extinguishing the fire while it is small, fire fighters reduce the risk posed by the fire. Most wildland and ground fires are kept small because the initial crews used an aggressive direct attack on the fire.

VOICES OF EXPERIENCE

In 2003, I was given a Strike Team Leader Trainee assignment on a large wildland fire in northern Montana. For three days I supervised two 20-person hand crews on a quiet flank of the fire. On the fourth day, I had two more crews assigned to me, as well as a tracked wildland engine. My Strike Team had turned into a large Task Force, but conditions were such that I was able to coordinate all the tactics and tasks needed to accomplish our goals.

As the next couple days went by, the fire changed direction and started heading in our direction. More complex water supply operations were needed, requiring coordination with tenders. Another crew was added and my area suddenly became a crucial spot for both providing a lookout and a radio communication relay site.

I soon became overwhelmed with trying to coordinate the activities of seven or eight different crews in an increasingly complex environment. I requested that the Division Supervisor try to find a dedicated lookout and radio relay person as well a water supply specialist. As the initial section of fire line was starting to come under control, I also released two of the four hand crews that had been assigned to me.

The day that my Strike Team returned to a normal size of two hand crews, the fire made a major push over the line, causing us to retreat to our safety zone. During the time the fire made its run, radio traffic became very chaotic. Fortunately, I was able to meet face-to-face with my hand crews. The radio traffic was so jammed that if I was still supervising an additional six crews, I would not have been able to contact them. By asking for help and narrowing the span of my supervision, I was able to maintain accountability of my crews charge and keep them safe.

The lesson for me was to not get complacent when all is going well. Anticipate and plan for potential changes. Be aware of your surroundings and capabilities and ask for help BEFORE a hostile event happens.

Derek Rosenquist
Estes Valley Fire Protection District
Estes Park, Colorado

Two major types of direct attacks are used on wildland and ground cover fires: the pincer attack and the flanking attack.

Pincer Attack

The pincer attack requires two teams of fire fighters. One team establishes an anchor point on the left side of the fire near the point of origin and mounts a direct attack along the left flank of the fire, working toward the head of the fire. A second team of fire fighters establishes an anchor point on the right side of the fire near the point of origin and mounts a direct attack along the right flank of the fire, also working toward the head of the fire. As the teams advance, the fire gets "pinched" between them, which reduces its growth. By starting at a safe anchor point, the risk to fire fighters is minimized. A successful pincer attack requires the availability of sufficient personnel to be able to mount a two-pronged attack **FIGURE 21-3**.

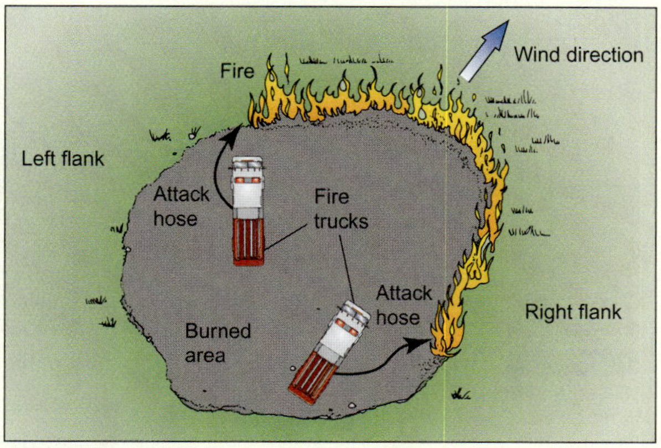

FIGURE 21-3 A pincer attack is made on both flanks of a wildland fire.

Flanking Attack

The flanking attack requires only one team of fire fighters. This team starts at an anchor point and attacks along one flank of the fire. Such an attack is used when sufficient resources are not available to mount two simultaneous attacks or when the risks posed by the fire are very high on one flank of the fire, but lower on the other flank of the fire. The decision regarding which flank of the fire to attack is based on the determination of which side of the fire poses the greatest risk **FIGURE 21-4**.

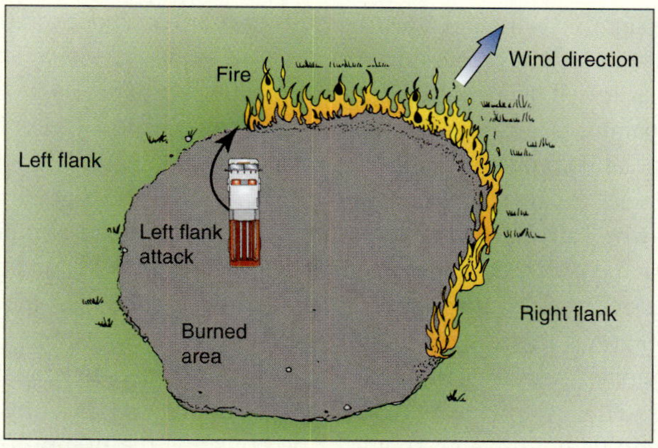

FIGURE 21-4 A flank attack is made on the left flank or right flank of a wildland fire.

Indirect Attack

An indirect attack is most often used for large wildland and ground fires that are too dangerous to approach through a direct attack. An indirect attack is mounted by building a fire line along natural fuel breaks, at favorable breaks in the topography, or at considerable distance from the fire and then burning out the intervening fuel. This strategy is similar to using a defensive attack on a structure fire. The fire line may be constructed fairly close to the fire, or it may be constructed several miles away. An indirect attack can be mounted using either hand tools or mechanized machinery.

Indirect attacks are employed when not enough equipment or personnel are available to mount a direct attack. They are also appropriate when the topography is so rough that a direct attack is dangerous or when the weather conditions pose a high risk for fire fighter safety.

Priorities of Attack

Before determining how to attack a wildland fire, the IC must assess and evaluate the priorities for preserving lives and property. The top priority, of course, is to ensure the safety of both fire fighters and civilians. All fire fighters should make sure that they know the escape route and escape plan before attacking the fire. The second priority is to preserve any structures threatened by the blaze.

Fire Apparatus Used for Wildland Fires

The fire apparatus used for wildland fire suppression include both structural fire apparatus and specialized wildland fire apparatus. Structural fire apparatus are commonly used because most small wildland fires are fought by departments whose primary mission is structural firefighting. Structural fire pumpers carry enough water to mount an attack on many small wildland fires. Booster lines provide enough water for attacking small wildland fires. For larger fires, preconnected 1¾-inch (44-mm) hose lines may be needed.

Using structural fire apparatus for wildland fires has several drawbacks. In particular, most structural fire apparatus are not designed to pump water at the same time that the truck is moving forward. The lack of a pump-and-roll capacity means that the apparatus must stop and be shifted into pump gear before a stream of water can be pumped. It also makes the apparatus less efficient when a continuous stream of water is needed while the truck moves along the flank of a fire. In addition, most structural fire apparatus is designed for on-the-road use. As a consequence, trucks that lack all-wheel drive and underside protection may become stuck or damaged when used in off-road situations. Structural fire apparatus that meet the requirements of NFPA 1901, *Standard for Automotive Fire Apparatus*, are not primarily designed for fighting wildland fires.

By contrast, wildland fire apparatus are especially designed for fighting wildland fires. Because the terrain and conditions vary from one geographic location to another, many variations in wildland fire apparatus exist. The design of any such apparatus needs to be suited to the local wildland conditions. Wildland apparatus range from small pickup trucks or Jeep-type vehicles to large trucks. Most of this apparatus is designed with all-wheel drive, underside protection, and high road clearance, which enables the vehicles to travel off the road. Wildland apparatus normally pump between 10 gallons per minute (gpm) and 550 gpm (38 liters per minute and 2100 liters per minute), are equipped with a water tank, and have pump-and-roll capabilities. This pump-and-roll capacity is achieved by powering the pump with a separate engine or by powering the pump from a power take-off shaft. Wildland fire apparatus should meet the requirements of NFPA 1906, *Standard for Wildland Fire Apparatus*.

Small apparatus for fighting wildland fires typically carry 200 to 300 gallons (758 to 1135 liters) of water. Most such small apparatus have a pump that is powered by a small gasoline or diesel engine, an arrangement that gives the vehicle a pump-and-roll capacity **FIGURE 21-5**. Small brush trucks, for example, can respond quickly and contain many fires before they grow into larger blazes **FIGURE 21-6**.

FIGURE 21-5 This apparatus has a pump-and-roll capacity.

FIGURE 21-6 A smaller wildland or brush truck.

Larger wildland fire apparatus are often built on a midsized or large truck chassis. These trucks carry more water than small apparatus, which enables them to operate for a longer period of time without being refilled. Many of these vehicles have crew cabs that enable an entire crew of fire fighters to travel to the fire scene in the truck.

A third type of wildland fire apparatus is a tanker or water tender **FIGURE 21-7**. These trucks are designed to transport water to the scene of a wildland fire. They are usually equipped with a pump that enables them to fill from a drafting site and to offload water when needed. Many tenders are equipped with a dump valve for rapidly unloading water into a portable tank. Some tenders are suited for off-road use, whereas others must remain on a road.

FIGURE 21-7 A water tender (tanker) is used to transport water to wildland fires.

Wildland fire apparatus also include plows and bulldozers **FIGURE 21-8**. These vehicles are used to create fire breaks.

As mentioned earlier, fixed-wing and rotary-wing aircraft are valuable resources for dropping water where the terrain makes it dangerous or impossible to fight a fire with ground-based fire fighters **FIGURE 21-9**. Aircraft can be used to attack the head of a fire, thereby slowing the growth and spread of the fire.

FIGURE 21-8 A wildland bulldozer is used to create fire breaks.

FIGURE 21-9 Aircraft can be used for making water drops on wildland fires.

Safety in Wildland Firefighting

Fighting wildland and ground fires is hazardous duty. The wildland and ground fire environment shares many of the hazards of structural firefighting but also includes other hazards associated with driving, falls, smoke, fire, electricity, and falling trees.

Hazards of Wildland Firefighting

Responses for wildland and ground fires involve all of the hazards encountered by personnel responding to structural fires. In addition, they may involve some unique hazards associated with driving fire apparatus on poorly maintained roads or trails, and driving apparatus off-road. Driving on unimproved roads and steep terrain greatly increases the chance of fire apparatus rollovers. Given these dangers, drivers must thoroughly understand the operating characteristics of their fire apparatus and operate the apparatus within the safe limits specific to its design. All fire fighters must keep their seat belts fastened whenever the apparatus is moving.

When working in rough terrain, wildland and ground fire fighters are at increased risk for falls. Rough ground often contains holes that are difficult to see in smoky conditions. Steep terrain also increases the likelihood of falls. Because wildland and ground firefighting involves working with sharp tools, it is important to prevent injuries caused by these tools.

Other hazards of fighting wildland and ground fires include burns, smoke inhalation, dehydration, and heat stroke. Because the personal protective equipment (PPE) worn by wildland and ground fire fighters provides less protection than the PPE worn by structural fire fighters, fire fighters must keep far enough from the heat of the fire to prevent burns. Also, because much wildland and ground firefighting is done without self-contained breathing apparatus (SCBA), fire fighters must avoid inhaling poisonous gases and suspended smoke particles. Use SCBA in any conditions where it is needed.

When engaged in wildland and ground firefighting, be alert for the hazards posed by falling trees. During a fire, the lower parts of trees may burn away and weaken the support for the rest of the tree. As a result, trees of all sizes can fall with little warning.

Also be alert for the presence of electrical hazards. Electrical transmission lines and other electrical wires may be present in the location of a wildland fire. Wires that drop on vegetation may ignite a wildland and ground fire and pose an electrical hazard to fire fighters. Many of these safety hazards can be difficult to see at night and in smoky conditions. Electricity from a downed electrical wire, however, may spark a ground fire. Be alert for electrical hazards even in the middle of an area covered with vegetation with no buildings around.

Because many wildland fires occur during hot weather and because combating them involves strenuous physical activity, all fire fighters must be alert to the possibility of dehydration and heat stroke. Make sure you are drinking enough water, and take breaks when needed to give your body time to recover.

Fire Fighter Safety Tips

An ash-covered pit with hot embers can be left behind when stumps and large logs burn out. Watch your step!

Personal Protective Equipment

It is important for wildland fire fighters to wear appropriate PPE FIGURE 21-10. Specifically, they should be equipped with a one-piece jumpsuit, or a coat, shirt, and trousers that meet the

FIGURE 21-10 PPE for a wildland fire fighter.

requirements of NFPA 1977, *Standard on Protective Clothing and Equipment for Wildland Fire Fighting*. These garments should be constructed of a fire-resistant material, such as Nomex. Such construction ensures that the clothing will stop burning as soon as the heat and flame are removed, which reduces the chance of fire fighters being burned. Wildland fire fighters should also wear an approved helmet with a protective shroud, eye protection, gloves, and protective footwear. All of this equipment should meet the provisions of NFPA 1977.

Respiratory protection for wildland and ground firefighting is usually limited to a filter mask that filters out small particles of smoke. Such masks provide a very limited degree of protection and should not be relied upon beyond the limits for which they are approved. SCBA should be used in conditions of heavy smoke and poisonous gases.

The PPE that is provided for wildland and ground firefighting varies from one location to another, depending on the types of wildland fires encountered and on the climate in a given location. Follow the local fire department's standards regarding PPE. The use of structural PPE for wildland fires can quickly result in rapid body overheating and dehydration. Given this risk, fighting wildland fires in structural PPE for long periods of time is not recommended.

Fire Fighter Safety Tips

Avoid the use of structural PPE for wildland firefighting. Wearing structural PPE for long periods of time will result in the retention of body heat and will greatly increase the chance of dehydration, heat exhaustion, and heat stroke.

Fire Shelters

One of the most important pieces of PPE for wildland and ground fire fighters is the fire shelter. This life-saving piece of equipment should be issued to all wildland and ground fire fighters. Fire shelters are made of a thin, reflective-foil layer that is attached to a layer of fiberglass. They are designed to reflect approximately 95 percent of a fire's radiant heat for a short period of time. This property allows a rapidly moving fire to pass over a fire fighter who has deployed a fire shelter.

Fires shelters are carefully folded and carried in a protective pouch on the fire fighter's belt. When fire fighters are in danger of being overrun by rapidly moving fire, they should try to get to a safe location. Only when they cannot escape should they use their fire shelters. To use a fire shelter, the fire fighter finds the largest available clearing, scrapes away flammable litter, opens the shelter, lies face down on the shelter, and covers himself or herself with the shelter. When properly used by well-trained personnel, this equipment can save lives. As with all equipment, however, it is important to receive proper training to use a fire shelter safely. Follow the steps in SKILL DRILL 21-1 FIGURE 21-11 (Fire Fighter I, NFPA 5.3.19):

1. Pick the largest available clearing. Wear gloves and a hardhat, and cover your face and neck if possible.
2. Scrape away flammable litter if you have time.
3. Pull the red ring to tear off the plastic bag covering the shelter.
4. Grasp the black handle in your left hand and the red handle in your right hand.
5. Shake the shelter until it is unfolded. If it is windy, lie on the ground to unfold the shelter.
6. Lie down in the shelter with your feet toward the oncoming fire. Slip your arms through the hold-down straps, push the sides out for more protection, and keep your mouth near the ground.

FIGURE 21-11 A wildland fire shelter.

Fire Fighter Safety Tips

When working with fire shelters, remember the following key points:

- Inspect the shelter regularly.
- Do not open the shelter's clear plastic bag during inspections.
- Be sure you are on the ground in the shelter when the fire arrives.
- After the fire has cooled, pick the safest area and wait for help.
- Watch for falling snags.

Fire Fighter Safety Tips

Safety zones are areas of refuge you can retreat to where you will not need to use a fire shelter. The size of the zone will be based on the behavior of the fire and your location relative to the fire. A basic guideline is to choose an area that provides a separation of four times the flame height. This guideline is based on a three-person engine crew on flat ground with no wind. The size of the safety zone would need to be adjusted to take into account wind, location on a slope, and the number of resources.

Suppressing a Ground Cover Fire

To suppress a ground cover fire, follow the steps in SKILL DRILL 21-2:

1. Don appropriate PPE. (STEP 1)
2. Identify safety and exposure risks.
3. Protect exposures if necessary.
4. Construct a fire line by removing fuel with hand tools. (STEP 2) OR
5. Extinguish the fire with a backpack pump extinguisher or a hand line. (STEP 3)
6. Overhaul the area completely to ensure complete extinguishment of the ground cover. (STEP 4)

The Challenge of the Wildland–Urban Interface

When people think of wildland fires, they tend to think of burning grass, brush, and forests. In fact, wildland fires are defined by the NFPA as "unplanned and uncontrolled fires burning in vegetation, including structures and other improvements." In other words, many wildland fires threaten not only vegetative fuels, but also buildings and other improvements that have been made on the land.

Today many people want a house that is surrounded by the beauty of nature—that is, they want a house that has modern conveniences, but is set in the middle of the woods.

SKILL DRILL 21-2 Suppressing a Ground Cover Fire
(Fire Fighter I, NFPA 5.3.19)

1 Don appropriate PPE. Identify safety and exposure risks. Protect exposures if necessary.

2 Construct a fire line by removing fuel with hand tools. **OR**

3 Extinguish the fire with a backpack pump extinguisher or a hand line.

4 Overhaul the area completely to ensure complete extinguishment of the ground cover.

The mixing of wildlands and developed areas in this way has created massive problems for fire departments in many parts of the United States. Wildland fires regularly ignite buildings and become structure fires. Structure fires, conversely, regularly ignite vegetative fuels and become wildland fires. In many cases, it is not possible to clearly separate one type of fire from the other. This phenomenon explains why most fire departments are involved in fighting some types of wildland fires.

The term wildland–urban interface is used to describe the mixing of wildland with developed areas. It is defined as an area where undeveloped land with vegetative fuels is mixed with human-made structures. The wildland–urban interface presents a major challenge for fire fighters. Because many people live in the wildland–urban interface, fires in this zone present a significant life-safety hazard. Also, because so many structures are built in this zone, there is a huge potential for property loss whenever a fire occurs. Finally, many parts of the wildland–urban interface do not have adequate municipal water systems.

The wildland–urban interface represents a unique challenge for fire fighters and emphasizes the importance of having structural fire fighters learn the basic principles of fighting wildland fires. While the material presented in this chapter will help fire fighters understand the basic principles of fighting wildland fires, it is not intended to provide all of the skills and knowledge needed by wildland fire fighters. These types of fires require special training on what to look for during a size-up and which tactical and strategic considerations to take into account. If your department is regularly involved in fighting wildland fires, you should complete a separate course on wildland firefighting based on NFPA 1051, *Standard for Wildland Fire Fighter Professional Qualifications*.

Much of the effort geared toward reducing the loss from wildland fires needs to be directed at prevention. Many communities work to involve homeowners by educating them about the risks their homes pose and the steps that they can take to reduce this risk. One important measure that has been implemented in many jurisdictions is to outlaw the use of flammable cedar shake roofs in houses in the wildland urban interface. Another important step is to encourage or require homeowners to maintain a defensible space around their structures. A defensible space is an area around a structure where combustible vegetation that can spread fire has been cleared, reduced, or replaced. It acts as a barrier between a structure and an advancing fire. Clearly, our best offense against the huge problem of fire in the wildland urban interface is often vigorous prevention.

Fire Fighter Safety Tips

Many areas that are prone to wildland fires have limited numbers of roads leading into and out of them. Be aware of the routes by which you can escape a rapidly moving fire or a fire that changes direction quickly. Do not let yourself get caught in a wildland fire with no escape route.

Wildland Fire Safety

The ten standard firefighting orders were developed by the U.S. Department of Agriculture Forest Service Chief after a task force investigated 16 tragic wildland fires. These orders were developed in an effort to prevent further tragedies while fighting these fires. The ten orders are as follows:

1. Keep informed on fire weather conditions and forecasts.
2. Know what your fire is doing at all times.
3. Base all actions on the current and expected behavior of the fire.
4. Identify escape routes and safety zones and make them known.
5. Post lookouts when there is possible danger.
6. Be alert. Keep calm. Think clearly. Act decisively.
7. Maintain prompt communications with your forces, your supervisor, and adjoining forces.
8. Give clear instructions and ensure they are understood.
9. Maintain control of your forces at all times.
10. Fight fire aggressively, having provided for safety first.

Wrap-Up

Chief Concepts

- Wildland fires are unplanned and uncontrolled fires burning in vegetative fuel that sometimes includes structures.
- Wildland fires are sometimes referred to as brush fires, forest fires, grass fires, ground cover fires, ground fires, natural cover fires, and wildfires.
- Wildland is land in an uncultivated natural state that is covered by timber, woodland, brush, or grass.
- Large wildland fires are handled by specialized agencies, such as the U.S. Forest Service.
- The wildland fire triangle consists of fuel, oxygen, and heat. These three elements affect the intensity of the fire.
- The primary fuel for wildland fires is vegetation, such as grasslands, brush, and trees. Vegetative fuels range from fine fuels (dried twigs, leaves, or grass) to heavy fuels (large brush, dead timber, or large stumps).
- The size and shape of a fuel influence how it burns. Fine fuels burn more quickly than heavy fuels and require less heat to reach their ignition temperature.
- In wildland fires, air movement influences the speed and direction which the fire moves. Wind blowing on a wildland fire brings more oxygen to the fire and speeds the process of combustion.
- Lightning is the source of almost all naturally caused wildland fires. A single storm can produce many lightning strikes and many fires. Discarded smoking materials and campfires are common human-made causes of wildland fires.
- Weather conditions and topography have a greater effect on wildland fires than they do on structure fires.
- The two most critical weather conditions that influence a wildland fire are moisture and wind. When the relative humidity is low, vegetative fuels dry out, making them more susceptible to ignition. Wind can then push the fire into additional fuels.
- Topography can influence the spread of the fire and the speed of the fire. A wildland fire will spread more quickly over dry, flat land than over a river valley.
- The location where the wildland fire begins is the area of origin.
- The most rapidly moving area of a wildland fire is the head of the fire.
- A long, narrow extension of fire that projects from the head of the fire is called a finger. A finger can grow and produce a secondary direction of travel for the fire.
- A pocket is an unburned area between a finger and the head of the fire. It is a dangerous place for fire fighters because it is surrounded by three sides of fire, leaving no options for escape.
- Green areas contain unburned fuels, whereas black areas have already been burned. Green areas are more likely to become involved in a wildland fire.
- Wildland fires can be controlled and extinguished by cooling the fuel, by removing fuel from the fire, or by smothering the fire.
- Backpack pump extinguishers may be effective against small wildland fires with light fuel loads. Water may also be applied by hoses on special wildland fire trucks or by aircraft.
- Hand tools such as rakes and power tools such as chainsaws may be used to remove fuel from the fire.
- CAFS foam may be used to smother wildland fires.
- The two general types of attacks used to contain and extinguish wildland fires are direct attacks and indirect attacks. A direct attack is mounted by containing and extinguishing the fire at its burning edge. An indirect attack is mounted by building a fire line along natural fuel breaks, at favorable breaks in the topography, or at considerable distance from the fire and then burning out the intervening fuel.
- All fire fighters should ensure that they know the escape route and escape plan before attacking a wildland fire.
- The hazards of wildland fires include fire apparatus rollovers, falls, burns, smoke inhalation, dehydration, and heat stroke.
- The fire shelter is one of the most important pieces of PPE for wildland fire fighters. Fire shelters are designed to reflect 95 percent of a fire's radiant heat for a short period of time.
- The wildland–urban interface is the mixing of wildland and developed areas. Because many people live in the wildland–urban interface, fires in this zone present a significant risk to both lives and structures.

Hot Terms

Adze A hand tool constructed of a thin, arched blade set at right angles to the handle. It is used to chop brush for clearing a fire line, or to mop up a wildland fire.

Aerial fuels Fuels located more than 6 feet (2 meters) off the ground, usually part of or attached to trees.

Area of origin The room or area where a fire began. (NFPA 921)

Backfiring A planned operation to remove fuel from a wildland fire by burning out large selected areas. This operation should be carried out only under the close supervision of experienced, authorized fire officers.

Backpack pump extinguisher A portable fire extinguisher consisting of a 4- to 8-gallon (15- to 30-liter) water tank that is worn on the user's back and features a hand-powered piston pump for discharging the water.

Black An area that has already been burned.

Compressed-air foam system (CAFS) A system that combines foam concentrate, water, and compressed air to produce foam that can stick to both vegetation and structures.

Council rake A long-handled rake constructed with hardened triangular-shaped steel teeth that is used for raking a fire line down to soil with no subsurface fuel, for digging, for rolling burning logs, and for cutting grass and small brush.

Defensible space An area as defined by the authority having jurisdiction [typically a width of 9.14 meters (30 feet) or more] between an improved property and a potential wildland fire where combustible materials and vegetation have been removed or modified to reduce the potential for fire on improved property spreading to wildland fuels or to provide a safe working area for fire fighters protecting life and improved property from wildland fire. (NFPA 1051)

Direct attack (wildland and ground fire) A method of fire attack in which fire fighters focus on containing and extinguishing the fire at its burning edge.

Fine fuels Fuels that ignite and burn easily, such as dried twigs, leaves, needles, grass, moss, and light brush.

Finger A narrow point of fire whose extension is created by a shift in wind or a change in topography.

Fire shelter An item of protective equipment configured as an aluminized tent utilized for protection, by means of reflecting radiant heat, in a fire entrapment situation. (NFPA 1500)

Flanking attack A direct method of suppressing a wildland or ground fire that involves placing a suppression crew on one flank of a fire.

Fuel compactness The extent to which fuels are tightly packed together.

Fuel continuity The relative closeness of wildland fuels, which affects a fire's ability to spread from one area of fuel to another.

Fuel moisture The amount of moisture present in a fuel, which affects how readily the fuel will ignite and burn.

Fuel volume The amount of fuel present in a given area.

Green An area of unburned fuels.

Ground cover fire A fire that burns loose debris on the surface of the ground.

Ground duff Partly decomposed organic material on a forest floor; a type of light fuel.

Head of the fire The main or running edge of a fire; the part of the fire that spreads with the greatest speed.

Heavy fuels Fuels of a large diameter, such as large brush, heavy timber, snags, stumps, branches, and dead timber on the ground. These fuels ignite and are consumed more slowly than light fuels.

Heel of the fire The side opposite the head of the fire, which is often close to the area of origin.

Indirect attack A method of attack in which the control line is located along natural fuel breaks, at favorable breaks in the topography, or at considerable distance from the fire, and the intervening fuel is burned out.

Island An unburned area surrounded by fire.

McLeod tool A hand tool used for constructing fire lines and overhauling wildland fires. One side of the head consists of a five-toothed to seven-toothed fire rake; the other side is a hoe.

Pincer attack A direct method of suppressing a wildland or ground fire, in which one suppression crew attacks the right flank of the fire and a second suppression crew attacks the left flank of the fire.

Pocket A deep indentation of unburned fuel along the fire's perimeter, often found between a finger and the head of the fire.

Power take-off shaft A supplemental mechanism that enables a fire engine to operate a pump while the engine is still moving.

Pulaski axe A hand tool that combines an adze and an axe for brush removal.

Rear of the fire The side opposite the head of the fire. Also called the heel of the fire.

Relative humidity The ratio between the amount of water vapor in the air (a gas) at the time of measurement and the amount of water vapor that could be in the air (gas) when condensation begins, at a given temperature. (NFPA 79)

Slash Debris resulting from natural events such as wind, fire, snow, or ice breakage, or from human activities such as building or road construction, logging, pruning, thinning, or brush cutting. (NFPA 1144)

Spot fire A new fire that starts outside areas of the main fire, usually caused by flying embers and sparks.

Subsurface fuels Partially decomposed matter that lies beneath the ground, such as roots, moss, duff, and decomposed stumps.

Surface fuels Fuels that are close to the surface of the ground, such as grass, leaves, twigs, needles, small trees, logging slash, and low brush. Also called ground fuels.

Topography The land surface configuration. (NFPA 1051)

Wildland Land in an uncultivated, more or less natural state and covered by timber, woodland, brush, and/or grass. (NFPA 901)

Wildland fire An unplanned fire burning in vegetative fuels. (NFPA 1051)

Wildland–urban interface An area where undeveloped land with vegetative fuels mixes with human-made structures.

FIRE FIGHTER *in action*

The alarm squawks out, "Grass Fire. Two miles east of Highway 125." You jump to your feet and scramble to the engine. Your captain directs you to the brush unit.

Highway 125 is a hilly road that winds through the countryside. The road is lined with pastures that are interspersed with groves of trees. You notice a large plume of billowing white smoke in the distance. This is your first real wildland fire and your mind begins to race. To calm yourself, you quickly cycle through what you learned in your basic training,

1. A grass fire is characterized by what type of fuel?
   - **A.** Fine
   - **B.** Light
   - **C.** Medium
   - **D.** Coarse and heavily compacted
2. Which of the following would be considered surface fuels?
   - **A.** Roots
   - **B.** Duff
   - **C.** Brush
   - **D.** Trees
3. Which of the following is a leg of the fire triangle?
   - **A.** Weather
   - **B.** Fuel
   - **C.** Humidity
   - **D.** Slash
4. A "pocket" is ___________.
   - **A.** a safe place for fire fighters
   - **B.** a dangerous place for fire fighters
   - **C.** at the heel of the fire
   - **D.** at the head of the fire
5. A McLeod tool is a ___________.
   - **A.** axe with an adz
   - **B.** tool that combines a hoe and an adz
   - **C.** tool that combines a hoe and a rake
   - **D.** long-handled rake with hardened, triangular-shaped steel teeth
6. A backfire is a(n) ___________.
   - **A.** pincer attack
   - **B.** flanking attack
   - **C.** direct attack
   - **D.** indirect attack
7. The basic guideline for a safety zone should include a minimum separation area of at least ___ times the height of the fire.
   - **A.** two
   - **B.** three
   - **C.** four
   - **D.** five

FIRE FIGHTER II *in action*

You are normally assigned to a wildland engine and have gained some valuable experience during your first fire season. As the season draws to a close, many of the fire fighters have returned to school and their off-season jobs. To fill the void, a senior fire fighter who normally is assigned to structural firefighting is assigned to your crew. Upon first getting a chance to visit with him, you realize he has little training and no experience in wildland firefighting.

1. How would you build trust with the newly assigned fire fighter?
2. How will you approach the subject and provide guidance?
3. How will you approach your supervisor about the issue of this fire fighter's lack of training?
4. What would be your first priority to teach him?

Fire Suppression

Fire Fighter I

Knowledge Objectives

After studying this chapter, you will be able to:

- Describe the objectives of an offensive attack. (NFPA 5.3.10, p 680–681)
- Describe the operations performed during an offensive attack. (NFPA 5.3.10, p 680–681)
- Describe the objectives of a defensive attack. (NFPA 5.3.10, p 680–681)
- Describe the operations performed during a defensive attack. (NFPA 5.3.10, p 680–681)
- Describe the characteristics of a fog stream. (NFPA 5.3.8, p 683)
- Describe the characteristics of a straight stream. (NFPA 5.3.8, p 683–684)
- Describe the characteristics of a solid stream. (NFPA 5.3.8, p 683–684)
- Describe the objectives of an interior fire attack. (NFPA 5.3.10, p 684)
- Describe the objectives of a direct attack. (NFPA 5.3.10, p 685)
- Describe the objectives of an indirect attack. (NFPA 5.3.10, p 685)
- Describe the objectives of a combination attack. (NFPA 5.3.10, p 686)
- Describe the techniques used to advance a large handline. (NFPA 5.3.1.A, p 686–689)
- Describe the characteristics of a master stream device. (NFPA 5.3.8.A, p 687–688)
- Describe the characteristics of a deck gun. (NFPA 5.3.8.A, p 687)
- Describe the characteristics of a portable monitor. (NFPA 5.3.8.A, p 688)
- Describe the characteristics of elevated master streams. (NFPA 5.3.8.A, p 690)
- Describe the tactics used to protect exposures. (NFPA 5.3.10.A, p 691)
- Describe the characteristics of concealed-space fires. (NFPA 5.3.10, p 691–692)
- Describe the characteristics of basement fires. (NFPA 5.3.10, p 692)
- Describe the tactics used to suppress basement fires. (NFPA 5.3.10, 5.3.10.A, 5.3.10.B, p 692)
- Describe the tactics used to suppress fires above ground level. (NFPA 5.3.10, 5.3.10.A, 5.3.10.B, p 693)
- Describe the tactics used to suppress fires in large buildings. (NFPA 5.3.10, p 694)
- Describe the tactics used to suppress fires in buildings under construction, renovation, or demolition. (NFPA 5.3.10, p 694)
- Describe the tactics used to suppress fires in lumberyards. (NFPA 5.3.8A, p 695)
- Describe the tactics used to suppress fires in stacked or piled materials. (NFPA 5.3.8, 5.3.8.A, p 695)
- Describe the tactics used to suppress fires in trash containers. (NFPA 5.3.8, p 695)
- Describe the tactics used to suppress fires in confined spaces. (NFPA 5.3.8.A, p 696)
- Describe the tactics used to suppress fires in vehicles. (NFPA 5.3.7, p 696–697)
- Describe the tactics used to suppress fires in alternative-fuel vehicles. (NFPA 5.3.7.A, p 697)
- Describe the tactics used to suppress fires in the passenger compartment of a vehicle. (NFPA 5.3.7, p 698)
- Describe the tactics used to suppress fires in the engine compartment of a vehicle. (NFPA 5.3.7.A, p 698)
- Describe the tactics used to suppress fires in the trunk of a vehicle. (NFPA5.3.7.A, p 698)
- Describe how to overhaul a vehicle fire. (NFPA 5.3.7.A, p 699)
- Describe the hazards presented by flammable-liquid fires. (NFPA 5.3.7, p 699)
- Describe the tactics used to suppress flammable-liquid fires. (NFPA 5.3.7, p 699)
- Discuss when gas service should be shut off. (NFPA 5.3.18, 5.3.18A, p 701)
- Describe when the electrical system should be shut off. (NFPA 5.3.18, 5.3.18A, p 701–702)
- Describe the hazards posed by electrical fires. (NFPA 5.3.18.A, p 701–702)
- Describe the tactics used to suppress an electrical fire. (p 701–702)

CHAPTER 22

Skills Objectives

After studying this chapter, you will be able to perform the following skills:

- Perform a direct attack. (NFPA 5.3.10.B, p 685)
- Perform an indirect attack. (NFPA 5.3.10.B, p 685)
- Perform a combination attack. (NFPA 5.3.10.B, p 686)
- Perform the one-fire fighter method for operating a large handline. (NFPA 5.3.8.B, p 686–688)
- Perform the two-fire fighter method for operating a large handline. (NFPA 5.3.8.B, p 687, 689)
- Operate a deck gun. (NFPA 5.3.8.B, p 690)
- Deploy and operate a portable monitor. (NFPA 5.3.8.B, p 688–689)
- Locate and suppress concealed-space fires. (NFPA 5.3.8.B, p 691)
- Extinguish an outside trash fire or other outside Class A fire. (NFPA 5.3.8.B, p 695)
- Extinguish a vehicle fire. (NFPA 5.3.7.B, p 697)
- Shut off gas utilities. (NFPA 5.3.18.B, p 701)
- Control the electric utility system. (NFPA 5.3.18.B, p 701)

Fire Fighter II — FFII

Knowledge Objectives

After studying this chapter, you will be able to:

- List the factors that the incident commander evaluates when determining whether to perform a defensive attack or an offensive attack. (NFPA 6.3.2, p 682)
- Describe the characteristics of a fog stream. (NFPA 6.3.2A, p 683)
- Describe the characteristics of a straight stream. (NFPA 6.3.2A, p 683–684)
- Describe the characteristics of a solid stream. (NFPA 6.3.2A, p 683–684)
- Describe the objectives of an interior fire attack. (NFPA 6.3.2B, p 683–684)
- Describe the objectives of a direct attack. (NFPA 6.3.2B, p 685)
- Describe the objectives of an indirect attack. (NFPA 6.3.2B, p 685)
- Describe the objectives of a combination attack. (NFPA 6.3.2B, p 686)
- Explain how ventilation is coordinated with fire suppression operations. (NFPA 6.3.2, p 691)
- Describe the characteristics of concealed-space fires. (NFPA 6.3.2, p 691)
- Describe the characteristics of basement fires. (NFPA 6.3.2, p 692)
- Describe the tactics used to suppress basement fires. (NFPA 6.3.2,B, p 692)
- Describe the tactics used to suppress fires above ground level. (NFPA 6.3.2.B, p 693)
- Describe the characteristics of flammable-gas cylinders. (NFPA 6.3.3A, p 699–700)
- Describe the hazards presented by flammable-gas fires. (NFPA 6.3.3A, p 699–700)
- Describe a boiling liquid/expanding vapor explosion (BLEVE). (NFPA 6.3.3A, p 700)
- Describe the tactics used to suppress flammable-gas fires. (NFPA 6.3.3, 6.3.3A, p 700–701)

Skills Objectives

After studying this chapter, you will be able to perform the following skills:

- Coordinate an interior attack. (NFPA 6.3.2.B, p 684)
- Suppress a flammable-gas cylinder fire. (NFPA 6.3.3.B, p 699–701)

You Are the Fire Fighter

This is it—your first night at the fire station. You lie in your bed wide awake, just hoping for a call. Just past midnight, the alarm sounds for a fire in a house. You slide down the pole and quickly slip on your turnout gear before climbing into the cab. Your heart is racing as the other crew members mount the apparatus like it is just another day at the office. As you get close to the scene, you can see the orange glow: The incident is a working fire.

1. What is the difference between an offensive attack and a defensive attack?
2. When would you use an indirect attack rather than a direct attack?
3. Which type of nozzle would you use for an indirect attack?

Introduction

The term "fire suppression" refers to all of the tactics and tasks that are performed on the fire scene to achieve the final goal of extinguishing the fire. Fire suppression can be accomplished through a variety of methods that will stop the combustion process. All of these methods involve removal of one of the three components of the fire triangle—that is, a fire can be extinguished by removing the oxygen, the fuel, or the heat from the combustion process.

This chapter presents the methods that fire fighters most frequently use to extinguish fires. The apparatus and equipment employed by most fire departments are designed to apply large volumes of water to a fire in an attempt to cool the fuel below its ignition temperature. Although fire departments typically deploy a variety of extinguishing agents for different situations, water is used more often than any other agent. Recent fire experiments conducted by NIST and UL have increased our understanding of fire behavior. They have also demonstrated safer and more efficient ways in which to attack certain fires. The results of these studies are incorporated throughout this chapter and the rest of this book. Many of these changes are incorporated into the sections on transitional fire attack, attic fires, and basement fires. These advances in fire fighting science may require revision of certain procedures in some fire departments. Water is a very effective extinguishing agent for many different types of fires, because tremendous quantities of heat energy are required to convert water into steam. When water is applied to a fire, all of the heat that is used to create steam is removed from the combustion process. If a sufficient quantity of water is applied, the fuel is cooled below its ignition temperature and the fire is extinguished.

Offensive, Defensive, and Transitional Operations

The first two types of fire suppression operations are offensive and defensive. When fire fighters advance hose lines toward a building to attack a fire, the strategy is offensive. By contrast, defensive operations are conducted from the exterior, or while moving away from the building or fire, by directing water streams toward the fire from a safe distance.

Offensive operations expose fire fighters to the heat and smoke of the fire inside the building as well as to several other risk factors, such as the possibility of being trapped by a structural collapse. The objective in an offensive operation is for fire fighters to get close enough to the fire to apply extinguishing agents, to the burning surfaces. This allows the extinguishing agent to be applied directly where it is needed to overpower the fire. When an offensive attack is successful, the fire can be controlled with the least amount of property damage.

An offensive strategy is used in situations where the fire is not too large or too dangerous to be extinguished by applying water using handlines. Small handlines (1½- or 1¾-inch [38- or 45-mm] lines) are used for most interior fire attack operations. Large handlines (2½-inch [65-mm] lines) are often used to conduct an interior attack on a large fire, such as a fire in a large commercial building or warehouse. Although they are much heavier and more difficult to maneuver inside a building than their smaller-diameter counterparts, the volume of fire may require the use of these larger-diameter hoses.

Large handlines and master streams are more often used in defensive operations. A defensive strategy is used in situations where the fire is too large to be controlled by an offensive attack and in situations where the level of risk to fire fighters conducting an interior attack would be unacceptable. The primary objective in a defensive operation is to prevent the fire from spreading. Thus water is directed into the building through doorways, windows, and openings in roofs, or onto exposures to keep the fire from spreading. During this type of fire suppression operation, fire fighters remain outside the building and operate from safe positions.

A third type of fire attack is the transitional attack. A transitional attack is an offensive fire attack initiated by an exterior, indirect handline operation into the fire compartment to initiate cooling while transitioning into an interior direct fire attack in coordination with ventilation operations. The purpose of this attack is to "soften the target" or cool the fire gases to reduce the risk of flashover before entering the fire compartment.

The decision on what type of fire attack to implement is made by the incident commander. This decision is made at the beginning of the fire suppression operation and is continuously evaluated throughout the incident. This decision must be made before operations begin and must be clearly communicated and understood by everyone who is involved in the operation. There is no room for confusion or errors: It could be a dangerous situation if one group of fire fighters initiated defensive operations while another group of fire fighters was conducting an offensive attack inside the building.

If the decision is made to switch from one strategy to another at any point during an operation, this change must be clearly communicated and understood by all fire fighters. A change from offensive to defensive operations could be warranted if an interior attack is unsuccessful or if the risk factors are determined to be too great to justify having fire fighters work inside the building. Sometimes the strategy switches from defensive to offensive after an exterior attack has reduced the volume of fire inside a building to the point at which fire fighters can enter and complete extinguishments with handlines. An offensive fire attack requires well-planned coordination among crews performing different tasks, such as ventilating, operating hose lines, and conducting aggressive search and rescue.

A defensive strategy should be implemented when the IC determines that it would be impossible to enter the burning building due to structural compromise or volume of fire and control the fire with handlines, as well as in situations where the risk of injury to or death of a fire fighter is excessive. A defensive operation involves the use of large-diameter hoses and master stream devices from the exterior of the structure in an attempt to confine the fire. Exposure protection should be a high priority during a defensive operation.

Transitional Fire Attack

Our primary means of suppressing fires is to apply water. Recent research conducted by NIST and UL provide valuable knowledge about more effective means of applying water to a fire. It has been widely accepted that attacking a fire from the outside may push the fire through other parts of the building. Traditionally, it has been taught that a residential fire should be attacked from the unburned side. This has led to the assumption that offensive attacks must be made exclusively from the inside, and unburned side, of the structure, effectively "pushing the fire" away from the uninvolved portion of the building and any occupants who might be behind the advancing hose team. This technique requires fire fighters to enter a burning building, often with low visibility and little idea of where the fire is located. NIST and UL research demonstrates that these assumptions are not always accurate.

As NIST and UL measured the movement of the fire, they determined that a properly-applied straight or solid stream does not "push" fire. Creating a higher pressure with improper stream applications may cause a high-pressure point at the exhaust, causing an internal flow path. They determined that the movement of the fire is dependent on the flow path of the heated smoke and gases. The flow path is the volume in a structure between an inlet and an outlet that allows the movement of heat and smoke from the higher pressure within the fire area towards the lower-pressure areas accessible via doorways, halls, stairs, and window openings. After verifying these results with multiple experiments, they realized that if water was introduced into the fire from outside, it produced cooling not just in the fire compartment, but also in other areas of the fire building. Temporary cooling was achieved even when the water entered the structure at a distance from the seat of the fire, as long as the stream was applied to a body of fire, not only the gases. In one experiment, a straight stream of water flowing 180 gallons of water per minute (gpm) (11 liters per second) was applied for 28 seconds and flowed 84 gallons (318 liters) of water into the house at some distance from the fire. The water reduced the temperature in the front room from 1200°F (649°C) to 300°F (149°C). These gains are only temporary unless additional water can be applied to the seat of the fire. (*Note*: These numbers are from fire behavior experiments conducted at Governor's Island by FDNY, NIST, and UL.)

These experiments showed that a transitional fire attack, an offensive exterior attack—introducing water from the outside to a body or volume of fire—reduced temperatures in other parts of the house at some distance from the fire, but it did not completely extinguish the fire. It slowed the growth of the fire by cooling huge quantities of very hot gaseous fuel and solid fuel below its ignition temperature **FIGURE 22-1**. This offensive exterior attack is sometimes referred to as a blitz attack, a transitional attack, softening the target, or resetting the fire.

The effect of this offensive exterior is significant but short-lived. To prevent the fire from regaining its former state, to complete extinguishment, and to search for and remove any occupants, it is necessary to quickly enter the building and to complete extinguishment of the remaining fire immediately

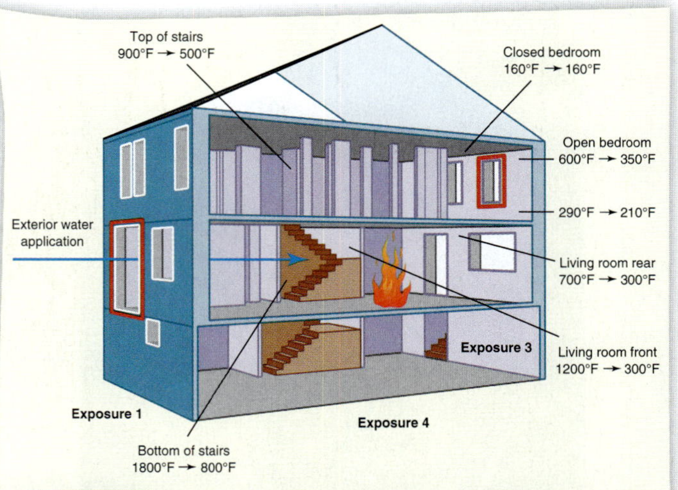

FIGURE 22-1 An offensive exterior attack some distance from the fire reduces temperatures throughout the house and partially suppresses the fire. The temperatures listed are from before and after the exterior application of water.

following the knock down from the exterior. This combined attack is safer for fire fighters because when entering the building, the internal temperature is cooler, visibility is increased, and the threat of a flashover is greatly reduced. The transitional attack can often effectively place water on the seat of the fire sooner, thus making conditions tenable for any trapped occupants FIGURE 22-2. With the fire knocked down, more aggressive ventilation can follow without the risk of inducing flashover. As visibility improves, fire fighters can move through the building more effectively, while efficiently and more safely conducting rescue operations, final suppression, and overhaul.

An initial exterior attack can be started using a preconnected hose line or a deck gun. Because this attack is conducted from the outside, it does not place fire fighters in a dangerous position in the fire compartment. Therefore, it can be started using water from the booster tank on the apparatus even before a permanent water supply has been established. The results of these experiments require us to reconsider our long-standing approaches to ventilation and the application of water into the fire building.

TABLE 22-1 shows the conclusions on exterior fire attacks that were drawn from the fire experiments conducted by NIST, UL, and FDNY.

FIRE FIGHTER II

Command Considerations

The IC must evaluate a whole range of factors to decide whether an offensive strategy (interior attack) or a defensive strategy (exterior attack) should be used at a particular fire. If the risk factors are too great, an exterior attack is the only acceptable option. If the decision is made to launch an interior attack, the IC must determine where and how to attack, after considering both safety issues and the potential effectiveness of the operation. The factors to be evaluated when considering whether to enter the structure to mount an attack include the following:

- What are the risks versus the potential benefits?
- Is it safe to send fire fighters into the building?
- What are the structural concerns?
- Is this building made of lightweight construction?
- Are there any lives at risk?
- Does the volume of the fire prohibit entry?
- Are enough fire fighters on the scene to mount an interior attack? (Remember the two-in/two-out rule.)
- Is an adequate water supply available?
- Can proper ventilation be carried out to support offensive operations?

TABLE 22-1 Exterior Fire Attack

• A properly-applied offensive exterior fire attack through a window or door, even when it is the only exterior vent, will not push fire. • Water application is most effective if a straight stream is aimed through the smoke into the ceiling of the fire compartment. This technique allows heated gases to continue to vent from the fire compartment while cooling the hot fuel inside. Fog patterns should not be used in this application. The fog pattern entrains large volumes of air and pushes air into the building. A fog stream can also block a ventilation opening, effectively changing the fire flow path. • Applying a straight or solid hose stream through a window or door into a room involved in a fire resulted in improved conditions throughout the structure. • Even in cases where the front and rear doors were open and windows had been vented, application of water through one of the vents improved conditions throughout the structure. • Applying water directly into the compartment as soon as possible resulted in the most effective means of suppressing the fire. • Transitional attack is an offensive exterior fire attack that occurs just prior to entry, search, and tactical ventilation. This technique is also known as a blitz attack, a transitional attack, or softening the target. • The transitional attack should begin from the outside, but it is necessary to finish it from the inside. • Coordinate the fire attack with any ventilation—do not ventilate before an attack stream is ready.

Courtesy of NIST.

FIGURE 22-2 Fire fighters using a transitional attack to reduce the burning and cool the interior before entering the burning building. A. Second-story window. B. Exterior attack.

Fire Fighter Safety Tips

The acronym S.L.I.C.E.R.S. has been developed by fire service instructors to help fire fighters remember the steps to be taken when conducting a transitional fire attack. The steps are as follows:

Sequential Actions

S Size Up
L Locate
I Identify and control the fire flow path
C Cool the space from the safest location
E Extinguish the fire

Actions of Opportunity

R Rescue
S Salvage

When using any acronym, keep in mind the objectives of your actions. It may be necessary to modify your actions to fit a specific situation.

After sizing up the situation, the IC must determine which type of attack is appropriate. As a new fire fighter, you are not responsible for determining the type of fire attack that will be used. Nevertheless, you should understand the various factors that go into making these decisions and recognize why the IC orders different types of fire attacks for different types of fires.

FIRE FIGHTER II

Operating Hose Lines

Some of the most basic skills that must be mastered by every fire fighter involve the use of hose lines to apply water onto a fire. Put simply, a fire fighter must be able to advance and operate a hose line effectively to extinguish a fire. The proper operation of a hose line is also essential to protect yourself, your crew, and any trapped victims from the fire. Fire attack operations are often conducted under extremely stressful conditions, including high heat conditions, limited or zero visibility, and unfamiliar surroundings. Care should be taken not to have opposing hose lines, such that two crews are working "against" each other, due to potential injury from the water streams and debris.

Fire fighters must learn how to operate both large and small handlines, as well as master stream appliances. Small handlines can be as large as 2 inches (50 mm) in diameter. The most frequently used size for interior fire attack is 1¾-inch (45 mm) handlines. Although one fire fighter can usually operate the nozzle on a small hose line, a second fire fighter will provide valuable assistance when a small hose line must be advanced and maneuvered.

Large handlines are defined as hoses that are at least 2½ inches (65 mm) in diameter. Because water can flow through these hoses at a rate of more than 250 gallons per minute (gpm) (1136 liters per minute [lpm]), large handlines are heavier and less maneuverable than smaller lines. At least two fire fighters are required to advance and control a large handline, although one fire fighter can control a large handline if it is firmly anchored. This task can be accomplished by utilizing a webbing strap or by looping the line and sitting down on it FIGURE 22-3.

Master streams are used when large quantities of water are needed to control a large fire. Such a stream can deliver water at a rate of at least 350 gpm (1591 lpm), and some master stream devices can flow more than 2000 gpm (9092 lpm). The most commonly used master stream devices deliver flows between 350 (1591) and 1500 (6819) gpm/lpm. Master stream devices are operated from a fixed position—either on the ground, on top of a piece of fire apparatus, or on an aerial ladder or elevated platform. They are typically used for defensive operations, although master streams can also be used to "blitz" a fire before beginning an offensive attack. This fire suppression method knocks the main body of fire down with a heavy stream; crews can then stretch handlines into the site and extinguish the remaining fire.

Fire Streams

Different types of fire streams are produced by using different types of nozzles. As described in the Fire Attack and Foam chapter, the nozzle defines the pattern and form of the water that is discharged onto the fire. A fire stream can be produced with a smooth-bore nozzle or an adjustable nozzle. Fire department policies and standard operating procedures (SOPs) usually dictate the types of nozzles that are used with different types of hose lines. The nozzle operator must know which type of nozzle should be used in the specific situation at hand. When an adjustable nozzle is used, the nozzle operator must know how to set the discharge pattern to produce different kinds of streams.

The first major distinction in nozzle discharge patterns is between a fog stream and a straight stream FIGURE 22-4. A fog stream divides water into droplets, which have a very large surface area and can absorb heat efficiently. When heat levels in a building need to be lowered quickly, a combination

FIGURE 22-3 One fire fighter can control a large handline if it is firmly anchored. This task can be accomplished by utilizing a webbing strap or by looping the line and sitting down on it.

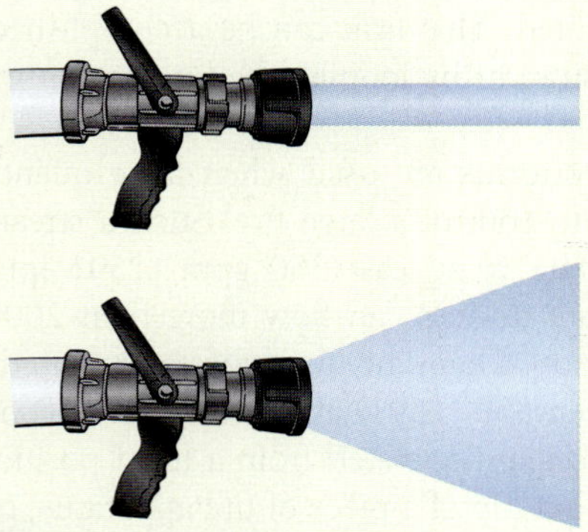

FIGURE 22-4 A straight stream and a fog stream are both produced using a combination nozzle.

of ventilation and a fog stream may be the fire suppression method of choice. A fog stream can also be used to protect fire fighters from the radiant heat of a large fire. Most adjustable nozzles can be adjusted from a straight stream, to a narrow fog pattern, to a wide fog pattern, depending on the reach that is required and how the stream will be used.

A straight stream has a greater reach than a fog stream, given similar water flow rates and nozzle pressures. A straight stream also keeps the water concentrated in a small area, so it can penetrate through a hot atmosphere to reach and cool the burning surfaces. To produce a straight stream, the fire fighter sets the adjustable nozzle to the narrowest pattern it can discharge. This type of stream is made up of a highly concentrated pattern of droplets that are all discharged in the same direction.

A solid stream is produced by a smooth-bore nozzle FIGURE 22-5. A solid stream has a greater reach and penetrating power than a straight stream, because it is discharged as a continuous column of water.

One consideration when selecting and operating nozzles is the amount of air that is moved along with the water. A fog stream naturally moves a large quantity of air along with the mass of water droplets. This air flows into the fire area along with the water. When this air movement is combined with steam production as the water droplets encounter a heated atmosphere, the thermal balance is likely to be disrupted quickly. In such a case, the hot fire gases and steam may be displaced back toward the nozzle operator. Straight and solid streams move little air in comparison with a fog stream, so fewer concerns with displacement and disruption of the thermal balance arise when these types of streams are used for fire suppression.

When applied correctly, the air movement created by a fog stream can be used for ventilation. Discharging a fog stream out through a window or doorway, for example, will draw smoke and heat out in the same manner as an exhaust fan. This operation must be performed carefully to prevent accidentally drawing hidden fire toward the nozzle operator. The use of a water stream to provide ventilation is called hydraulic ventilation and is discussed further in the Ventilation chapter.

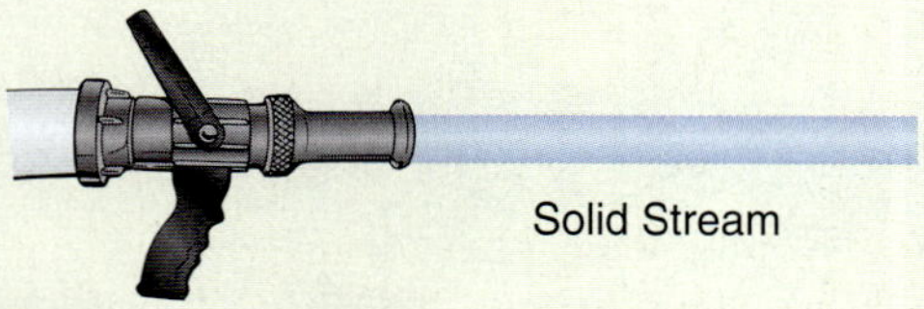

FIGURE 22-5 A solid stream is produced by a smooth-bore nozzle.

Interior Fire Attack

An interior fire attack is an offensive operation that requires fire fighters to enter a building and discharge an extinguishing agent (usually water) onto the fire. An interior structure fire is a fire that occurs inside a building or structure. Its fuel could be the contents of the building, or the structure itself might be burning. The larger the fire, the greater the challenge in suppressing it, and the more ominous the risks that are involved in interior fire suppression.

Interior fire attack can be conducted on many different scales. In many cases, such an attack is geared toward a fire that is burning in only one room; this kind of fire may be controlled quickly by one attack hose line. Larger fires require more water, which could be provided by two or more small handlines working together or by one or more larger handlines. Fires that involve multiple rooms, large spaces, or concealed spaces are more complicated and require more extensive coordination; nevertheless, the basic techniques for attacking these fires are similar to the techniques used when extinguishing smaller fires.

FIRE FIGHTER II

As a trained Fire Fighter II, you should be able to understand and coordinate an interior fire attack. To coordinate an interior attack, follow the steps in SKILL DRILL 22-1 (Fire Fighter II, NFPA 6.3.2):

1. Don full personal protective equipment (PPE), including self-contained breathing apparatus (SCBA). Enter the personnel accountability system, and proceed to work as a team.
2. Perform size-up and give an arrival report. Call for additional resources if needed.
3. Ensure that an adequate water supply and backup resources are available.
4. Select the appropriate attack technique.
5. Communicate the attack technique to the team.
6. Maintain constant team coordination.
7. Evaluate conditions on an ongoing basis.
8. Communicate and manage search, rescue, and ventilation requirements.
9. Report hazards.

FIRE FIGHTER II

10. Inform incident command of changing conditions.
11. Ensure complete extinguishment of fire during overhaul.
12. Exit the hazard area, account for all members of the team, and report to incident command.

Direct attack and indirect attack are two different methods of discharging water onto a fire. A combination attack is performed in two stages, beginning with an indirect attack and then continuing with a direct attack.

Fire Fighter Safety Tips

It is imperative to extinguish all fire as you proceed to the seat of the fire. Failure to do so may allow the fire to grow behind you, entrapping you by cutting off your escape route.

Direct Attack

An effective means of fire suppression in some situations is a direct attack. This kind of attack uses a straight or solid hose stream to deliver water directly onto the base of the fire. The water cools the burning surfaces until it is below its ignition temperature. Water should be applied until flame is no longer visible. To perform a direct attack, follow the steps in SKILL DRILL 22-2 (Fire Fighter I, NFPA 5.3.10):

1. Exit the fire apparatus wearing full PPE, including SCBA.
2. Select the proper hose line to fight the fire based on the fire's size, location, and type.
3. Advance the hose line from the apparatus to the entry point of the structure. Flake out excess hose in front of the door.
4. Don a face piece and activate the SCBA and personal alert safety system (PASS) device prior to entering the building.
5. Signal the pump operator/driver that you are ready for water.
6. Open the nozzle to purge air from the system and make sure water is flowing.
7. Make sure the ventilation team is being coordinated with the attack team.
8. Apply water to the base of the fire or the ceiling of the fire compartment.
9. Watch for changes in fire conditions.
10. Shut down the nozzle and listen.
11. Locate and extinguish hot spots.

Indirect Attack

Indirect application of water can be used in situations where the interior temperature is increasing and it appears that the room or space is ready to flash over. This application can be accomplished from a safer location, such as a doorway or hallway, to reduce the thermal exposure to operating fire fighters. With this fire suppression method, the fire fighter aims water at the ceiling to cool the superheated gases in the upper levels of the room or space. This action can prevent or delay flashover long enough for fire fighters to apply water directly to the seat of the fire or to make a safe exit. Follow your department's SOPs regarding the application of water.

The objective of an indirect attack is to apply water to the hot gases by sweeping the ceiling with a straight stream or a solid stream to cool the gases and droplets from the broken stream. This action may also hit and cool some of the solid fuel. This cooling lasts for only a short period of time unless the stream is sustained. The intention of this indirect attack is not to generate significant amounts of steam as much as it is to absorb the heat in the fire compartment. As soon as the water has reduced the temperature, the fire stream can transition into a direct attack.

Today's indirect attack is different from the indirect method that Lloyd Laymen advocated starting in the 1950s. In Lloyd Layman's indirect attack, the idea was to keep the fire compartment closed, introduce a low flow of water fog (25 to 30 gpm) into a window, and fill the compartment with steam. Today's indirect attack uses a straight stream or a solid stream, flowing between 120 to 180 gpm, to cool everything quickly, and does not generate significant amounts of steam.

To perform an indirect attack, follow the steps in SKILL DRILL 22-3 (Fire Fighter I, NFPA 5.3.10):

1. Exit the fire apparatus wearing full PPE, including SCBA.
2. Select the correct hose line to be used to attack the fire depending on the type of fire, its location, and its size.
3. Advance the hose line from the apparatus to the opening in the structure where the indirect attack will be made safely.
4. Don a face piece, and activate the SCBA and PASS device.
5. Notify the operator/driver that you are ready for water.
6. Open the nozzle and make sure that air is purged from the hose line and that water is flowing. A straight stream directed toward the celling has been shown to be effective in cooling the fire gases in a room. If using

FIRE FIGHTER Tips

When using an indirect or combination attack, watch for droplets of water raining down; they indicate lowered ceiling temperatures. If no droplets fall, the ceiling is still too hot.

a fog nozzle, ensure that it is set to the proper nozzle pattern for entry. Shut down the nozzle until you are in a position to apply water.

7. Advance with a charged hose line to the location where you will apply water.
8. Direct the water stream toward the upper levels of the room and ceiling into the heated area overhead, and move the stream back and forth. Flow water until the room begins to darken. Shut the nozzle off, and reassess the fire conditions.
9. Watch for changes and a reduction in the amount of fire. Once the fire is reduced, shut down the nozzle.
10. Confirm that ventilation has been completed.
11. Attack any remaining fire and hot spots until the fire is completely extinguished.

Combination Attack

A combination attack employs both indirect attack and direct attack methods in a sequential manner. This strategy should be used when a room's interior has been heated to the point that it is nearing a flashover condition. Fire fighters should first use an indirect attack method, from the safest location such as a hallway, adjoining room, or doorway, to cool the fire gases down to safer temperatures and prevent flashover from occurring. This operation is followed with a direct attack on the main body of fire. In a combination attack, the fire fighter who is operating the nozzle should be given plenty of space to maneuver.

A transitional fire attack is an offensive fire attack initiated by an exterior, indirect handline operation into the fire compartment to initiate cooling while transitioning into interior direct fire attack in coordination with ventilation operations. Only enough water as is needed to control the fire should be used, so as to avoid unnecessary water damage.

To perform a combination attack, follow the steps in **SKILL DRILL 22-4** (Fire Fighter I, NFPA 5.3.10):

1. Don full PPE and SCBA. Select the correct hose line to accomplish the suppression task at hand.
2. Stretch the hose line to the entry point of the structure, and signal the operator/driver that you are ready to receive water.
3. Open the nozzle to get the air out and make sure that water is flowing.
4. Enter the structure, and locate the room or area where the fire originated.
5. From the safest location, such as a hallway, adjoining room, or doorway, aim the nozzle at the upper-left corner of the fire and make either a "T," "O," or "Z" pattern with the nozzle. Start high and then work the pattern down to the fire level.
6. Use only enough water to darken down the fire without upsetting the thermal layering.
7. Once the fire has been reduced, find the remaining hot spots and complete fire extinguishment using a direct attack.

Large Handlines

Large handlines can be used either for offensive fire attacks or for defensive operations. In an offensive attack situation, a 2½-inch (65 mm) attack line can be advanced into a building to apply a heavy stream of water onto a large volume of fire. The same direct and indirect attack techniques that were described for small hose lines can also be used with large handlines. A 2½-inch (65 mm) handline can overwhelm a substantial interior fire if it can be discharged directly into the involved area. The extra reach of the stream can also prove valuable when making an interior attack in a large building.

It is more difficult for fire fighters to advance and maneuver a large handline inside a building, particularly in tight quarters or around corners. At least three team members are usually needed to advance and maneuver a 2½-inch (65 mm) handline inside a building. These fire fighters must contend with the nozzle reaction force as well as the combined weight of the hose and the water. In situations where the hose line must be advanced over a considerable distance into a building, additional fire fighters will be required to move the line. The extra effort required to deploy such a line, however, is balanced by the powerful fire suppression capabilities of a large handline.

Fire Fighter Safety Tips

Indicators of possible building collapse include:

- Lightweight construction or trusses
- Leaning walls
- Free-standing walls
- Smoke emitting from cracks
- Creaking sounds
- Sagging roofs or floors

Considerations for possible building collapse include:

- Exposure to serious fire
- Water loading from use of master streams

Large handlines are often used in defensive situations to direct a heavy stream of water onto a fire from an exterior position. In these cases, the nozzle is usually positioned so that it can be operated from a single location by one or two fire fighters. The stream can be used to attack a large exterior fire or to protect exposures. It can also be directed into a building through a doorway or window opening to knock down a large volume of fire inside. If the exterior attack is successful in reducing the volume of fire, the IC might decide to switch to an offensive (interior) attack to complete extinguishment of the fire.

One-Fire Fighter Method

One fire fighter can control a large attack hose by forming a large loop of hose about 2 feet (0.6 m) behind the nozzle. When the loop is placed over the top of the nozzle, the weight of the hose stabilizes the nozzle and reduces the nozzle reaction. To add more stability, lash the hose loop to the hose behind the nozzle where they cross. This technique reduces the energy needed to control the line if it is necessary to maintain the water stream for a long period of time. Although this method does not allow the hose to be moved while water is flowing, it is a good choice for protecting exposures when fire fighters are operating in a defensive attack mode.

To perform the one-fire fighter method for operating a large handline, follow the steps in SKILL DRILL 22-5:

1. Select the correct size of fire hose for the task to be performed.
2. While wearing full PPE and SCBA, advance the hose into the position from which you plan to attack the fire
3. Signal the pump operator that you are ready for water.
4. Open the nozzle to allow air to escape the hose and to ensure that water is flowing.
5. Close the nozzle and then make a loop with the hose, ensuring that the nozzle is under the hose line that is coming from the fire apparatus. (STEP 1)
6. Lash the hose sections together where they cross, or use your body weight to kneel or sit on the hose line at the point where the hose crosses itself. (STEP 2)
7. Allow enough hose to extend past the section where the line crosses itself for maneuverability. (STEP 3)
8. Open the nozzle and direct water onto the designated area. (STEP 4)

Two-Fire Fighter Method

When two fire fighters are available to operate a large handline, one should act as the nozzle operator, while the other serves as a backup. The nozzle operator grasps the nozzle with one hand and holds the hose behind the nozzle with the other hand. The hose should be cradled across the fire fighter's hip for added stability. The backup fire fighter should be positioned approximately 3 feet (0.9 m) behind the nozzle operator. This person grasps the hose with both hands and holds the hose against a leg or hip. The backup fire fighter can also use a hose strap to maintain a better hand grip on a large handline. When the line is operated from a fixed position, the second fire fighter can kneel on the hose with one knee to stabilize it against the ground.

To perform the two-fire fighter method for operating a large handline, follow the steps in SKILL DRILL 22-6:

1. Don all PPE and SCBA.
2. Select the correct hose line for the task at hand.
3. Stretch the hose line from the fire apparatus into position. (STEP 1)
4. Signal the pump operator that you are ready for water.
5. Open the nozzle a small amount to allow air to escape and to ensure that water is flowing.
6. Advance the hose line as needed. (STEP 2)
7. Before attacking the fire, the fire fighter on the nozzle should cradle the hose on his or her hip while grasping the nozzle with one hand and supporting the hose with the other hand.
8. The second fire fighter should stay approximately 3 feet behind the fire fighter who is on the nozzle. The second fire fighter should grasp the hose with two hands and, if necessary, use a knee to stabilize the hose against the ground.
9. Use a hose strap to maintain a better grip on the hose. (STEP 3)
10. Open the nozzle in a controlled fashion and direct water onto the fire or designated exposure. (STEP 4)

If it is necessary to advance a flowing 2½-inch (65 mm) handline over a short distance and only two fire fighters are available, be aware of the large reaction force exerted by the flowing water. It is much easier to shut down the nozzle momentarily and move it to the new position than to relocate a flowing line. If the line must be moved while water is flowing, both fire fighters must brace the hose against their bodies to keep it under control. Three fire fighters can stabilize and advance a large handline more comfortably and safely than can two fire fighters.

Master Stream Devices

Master stream devices are used to produce high-volume water streams for large fires. Several types of master stream devices exist, including portable monitors, deck guns, ladder pipes, and other elevated stream devices. Most master streams discharge between 350 and 1500 gallons (1591 to 6819 lpm) of water per minute, although much larger capacities are available for special applications. In addition, the stream that is discharged from a master stream device has a greater range than the stream from a handline, so it can be effective from a greater distance.

A master stream device can be either manually operated or directed by remote control. Many of these devices can be set up and then left to operate unattended. This capability may prove extremely valuable in a high-risk situation, because it eliminates the need to leave a fire fighter in an unsafe location or a hazardous environment to operate the device.

Master streams are used mainly during defensive operations. They should never be directed into a building while fire fighters are operating inside the structure, because these streams can push heat, smoke, or fire onto the fire fighters. The force and impact of the stream can also dislodge loose materials or cause a structural collapse.

Fire Fighter Safety Tips

A nozzle flowing 1000 gallons (4546 lpm) of water per minute is adding 4.2 tons (8345 pounds) (3785 kg) of weight to the building every minute. Be aware of the increasing risk of building collapse in such a scenario.

Deck Guns

A deck gun is permanently mounted on a vehicle and equipped with a piping system that delivers water to the device. These devices are sometimes called turret pipes or wagon pipes FIGURE 22-6. If the vehicle is equipped with a pump, the pump operator can usually open a valve to start the flow of water. Sometimes, however, a hose must be connected to a special inlet to deliver water to the deck gun. If

SKILL DRILL 22-5 Performing the One-Fire Fighter Method for Operating a Large Handline

(Fire Fighter I, NFPA 5.3.8)

1. Advance the hose into position. Signal that you are ready for water and open the nozzle to allow air to escape and to ensure that water is flowing. Close the nozzle and then make a loop with the hose.

2. Lash the hose sections together where they cross, or use your body weight to kneel or sit on the hose line at the point where the hose crosses itself.

3. Allow enough hose to extend past the section where the line crosses itself for maneuverability.

4. Open the nozzle and direct water onto the designated area.

your fire apparatus is equipped with a deck gun, you need to learn your role when placing it in operation. To operate a deck gun, follow the steps in SKILL DRILL 22-7:

1. Make sure that all firefighting personnel are out of a structure before using a deck gun.
2. Place the deck gun in the correct position. (STEP 1)
3. Aim the deck gun at the fire or at the target exposure.
4. Signal the pump operator that you are ready for water. (STEP 2)
5. Once water is flowing, adjust the angle, aim, or water flow as necessary. (STEP 3)

Portable Monitor

A portable monitor is a master stream device that can be positioned wherever a master stream is needed FIGURE 22-7. It is placed on the ground, and hose lines are then connected to the portable monitor to supply the water. Most of these devices come equipped with either one, two, or three inlets. Portable monitors can be built with 2½-inch (65 mm) inlets or with a large-diameter hose inlet. Smaller portable monitors are sometimes set up attached to a preconnected hose. This arrangement enables one fire fighter to step away from the engine with the portable monitor and attached hose, walk to an assigned area, and quickly place the portable monitor in operation. Some master stream devices can be used as

SKILL DRILL 22-6 Performing the Two-Fire Fighter Method for Operating a Large Handline

(Fire Fighter I, NFPA 5.3.8)

1 Stretch the hose line from the fire apparatus into position.

2 Signal that you are ready for water and open the nozzle to allow air to escape and to ensure water is flowing. Advance the hose line as needed.

3 Before attacking the fire, the fire fighter on the nozzle should cradle the hose on his or her hip while grasping the nozzle with one hand and supporting the hose with the other hand. The second fire fighter should stay 3 feet behind him or her and grasp the hose with two hands.

4 Open the nozzle in a controlled fashion and direct water onto the fire or designated exposure.

deck guns or taken off the fire apparatus and used as portable monitors.

To deploy a portable monitor, remove it from the apparatus and carry it to the location where it will be used. Advance an adequate number of hose lines from the engine to the monitor. The number of hose lines needed depends on the volume of water to be delivered and the size of the hose lines. Form a large loop in the end of each hose line in front of the monitor, and then attach the male coupling to the inlets of the monitor. The loops serve to counteract the force created by the flow of the water through the nozzle.

To set up and operate a portable monitor, follow the steps in SKILL DRILL 22-8 (Fire Fighter I, 5.3.8):

1. Remove the portable monitor from the fire apparatus and move it into the desired position.
2. Attach the necessary hose lines to the monitor as per SOPs or the manufacturer's instructions.
3. Loop the hose lines in front of the monitor to counteract the force created by water flowing out of the nozzle.
4. Signal the pump operator that you are ready for water.
5. Aim the water stream at the fire or onto the designated exposure, and adjust the stream as necessary.

If the portable monitor is not adequately secured, the nozzle reaction force can cause it to move from the position where it was originally placed. A moving portable monitor poses a danger to anyone in its path. Many of these master stream devices are equipped with a strap or chain that may be secured to a fixed object to prevent the monitor from moving. Pointed feet on the base also help to keep a portable monitor from moving. If the stream is operated at a low angle, the reaction force will tend to make the monitor unstable. For this reason, a safety lock is usually provided to prevent the monitor from being lowered beyond a safe angle of 35 degrees. When setting up any portable monitor, always follow the manufacturer's instructions and your department's SOPs to ensure its safe and effective operation.

Elevated Master Streams

Elevated master stream devices are mounted on aerial ladders, aerial platforms, or special hydraulically operated booms **FIGURE 22-8**. A ladder pipe is an elevated master stream device that is mounted at the tip of an aerial ladder or tower ladder. On many aerial ladders, the ladder pipe is attached to the top of the ladder only when it is actually needed, and a hose is run up the ladder to deliver water to the device. Most newer

FIGURE 22-6 A deck gun is permanently mounted on a vehicle and equipped with a piping system that delivers water to the device.

FIGURE 22-7 A portable monitor is placed on the ground and supplied with water from one or more hose lines.

SKILL DRILL 22-7 Operating a Deck Gun

(Fire Fighter I, NFPA 5.3.8)

1. Make sure that all firefighting personnel are out of a structure. Place the deck gun in the correct position. Aim the deck gun at the fire or at the target exposure. Signal that you are ready for water.

2. Once water is flowing, adjust the angle, aim, or water flow as necessary.

FIGURE 22-8 Elevated master stream devices can be mounted on aerial apparatus.

aerial ladders and tower ladders are equipped with a fixed piping system to deliver water to a permanently mounted master stream device at the top. This arrangement saves valuable setup time at a fire scene. If your apparatus is equipped with a ladder pipe, you need to learn how to assist in its setup.

Protecting Exposures

Protecting exposures refers to actions that are taken to prevent the spread of a fire to areas that are not already burning. Exposure protection is a consideration at every fire; it becomes even more important with a large fire. If the fire is relatively small and contained within a limited area, the best way to protect exposures is usually to extinguish the fire; when the fire is extinguished, the exposure problem ceases to exist. Modern fires are producing heat flux levels that may affect exposures earlier in the event. In cases where the fire is too large to be controlled by an initial attack, exposure protection becomes a priority. In some cases, the best outcome that fire fighters can hope to obtain is to stop the fire from spreading.

The IC must consider the size of the fire and the risk to exposures in relation to how much firefighting capability is available and how quickly those resources can be assembled. In some cases, the IC will direct the first-arriving companies to protect exposures while a second group of companies prepares to attack the fire. At other times, the IC must identify a point where the progress of the fire can be stopped and direct all firefighting efforts toward that objective.

Protecting exposures involves very different tactics from offensive fire attacks. At a large free-burning fire, the first priority is to protect exposed buildings and property from a combination of radiant heat, convective heat, and burning embers FIGURE 22-9. The best option in these circumstances is usually to direct the first hose streams at the exposures rather than at the fire itself. Wetting the exposures will keep the fuel from reaching its ignition temperature. Master stream devices such as deck guns, portable monitors, and elevated master streams are excellent tools for protecting exposures. They also ensure

FIGURE 22-9 Protecting an exposure from radiant heat.

that large volumes of water can be directed onto the exposures from a safe distance without putting fire fighters in the path of excess heat or in danger of building collapse. Fog streams can sometimes be used to absorb some of the heat coming from the main body of fire.

FIRE FIGHTER II

Ventilation

Ventilation must be coordinated with the suppression efforts to ensure that both events occur simultaneously and in a manner that supports the attack plan. Proper ventilation is designed to allow hot gases and smoke to be removed from the building, thereby improving visibility and tenability in the building for any trapped victims and fire fighters. Because many residential fires are ventilation-limited fires, it is important to be sure that charged hose lines are in place before ventilation is initiated. When preparing for ventilation, consider which fire attack is appropriate. Remember that improper ventilation may create a flashover or produce conditions that allow a fire to burn more aggressively and make it more difficult for fire fighters to enter the structure and attack the fire.

Coordination is essential to ensure that the hose lines will be ready to attack when the ventilation openings are made. These openings must be located so that the hot smoke and gases will be drawn away from the attack crews.

Fire Fighter Safety Tips

With modern lightweight construction and highly combustible contents, many fires today are likely to be underventilated when you arrive. Remember that even the act of opening the front door can introduce additional oxygen to the fire, contributing to a violent flashover in a very short period of time!

Specific Fire-Ground Operations

Concealed-Space Fires

Fires in ordinary and wood-frame construction can burn in combustible void spaces behind walls and under subfloors and ceilings. To prevent the fire from spreading, these fires must be found and suppressed. To locate and suppress fires behind walls and under subfloors, follow the steps in SKILL DRILL 22-9 (Fire Fighter I, 5.3.8):

1. Locate the area of the building where a hidden fire is believed to exist.
2. Look for signs of fire such as smoke coming from cracks or openings in walls, charred areas with no outward evidence of fire, and peeling or bubbled paint or wallpaper. Listen for cracks and pops or hissing steam.
3. Use a thermal imaging camera to look for areas of heat that may indicate a hidden fire.
4. Use the back of your hand to feel for heat coming from a wall or floor.
5. If a hidden fire is suspected, use a tool such as an axe or Halligan tool to remove the building material over the area.
6. If fire is found, expose the area as much as possible without causing unnecessary damage and extinguish the fire using conventional firefighting methods.

Basement Fires

Fires in basements or below grade level present several different challenges. First and foremost they may be hard to recognize as basement fires. Basement fires may spread to areas above. This means that when a fire is discovered by a neighbor or when fire fighters arrive on the scene, the fire may have spread from the basement to the first floor. Firefighters may assume that a basement fire is a fire in one of the main rooms of the house, since that is where the fire is most visible.

Many vertical voids within a building originate or terminate in the basement, providing ample opportunities for fire gases to spread throughout the building. In many parts of the country, especially in newer subdivisions with lightweight construction, built prior to the adoption of the 2012 International Residential Code, basement ceilings are most likely left unfinished when the house is constructed. Under these circumstances, fires originating in the basement will quickly involve the floor and support system, resulting in a failure of the floor over the fire and early collapse. These floor failures have the potential for causing injury or death to fire fighters entering the building with conditions of limited visibility. If fire fighters do not identify the basement fire prior to entering the building above the basement fire, they are at an increased risk for falling through the damaged floor and ending up in the burning basement below. Also remember that basement fires can spread to upper floors in houses with balloon-frame construction.

Basements are difficult and dangerous spaces to access because they have limited routes for entering and exiting. They may be difficult to ventilate because of the limited openings to the outside. Basements are often used for storage, so fire fighters may find it difficult to keep their sense of orientation in the narrow, disorganized cluttered spaces. Many basements contain a high volume of flammable materials, which results in a high fuel load and the production of large quantities of heat. All of these risks highlight the importance of completing a 360-degree size-up prior to entering a structure below grade.

One of the findings of recent research on basement fires is that water applied via the interior stairs has a limited effect on cooling the basement or extinguishing the fire. Researchers also discovered that the temperatures at the bottom of the basement stairs are not sufficiently cool to enable fire fighters to operate safely in that environment. Therefore the traditional tactic of pushing down the interior stairs to attack the seat of a fire should no longer be considered a safe and effective option for suppressing basement fires. If the only point of entry into a basement is an interior stairway, fire fighters must look for alternative openings to apply an exterior stream to the basement, reducing temperatures and the fire's impact on the floor system.

Research investigating the characteristics of basement fires has provided us with valuable information about how we can use a safer and more effective approach and attack basement fires. If the basement has exterior windows or an external door, these should be considered as points for ventilation and suppression. Once an external window or basement door is opened, it acts as an inlet to the fire or an exhaust from the fire. The growth of these fires can be slowed and the basement temperatures can be reduced through the use of an initial exterior fire attack—a transitional attack. A straight stream of water applied through an external window or door quickly darkens down a fire and greatly reduces temperatures throughout the building. Furthermore, fire experiments have shown that this transitional attack, when applied properly, does not push fire or hot gases up the interior stairs, nor does it create dangerous steam conditions. The darkening of the fire lasts for several minutes before the fire grows back to the size it was before the application of water.

To conduct a safe and effective attack on a basement fire, identify any external openings from the basement to the outside and advance a charged hose line to those openings. At the same time, make sure a crew is prepared to ventilate through one or more openings. As soon as the ventilation crew is ready to ventilate, apply a straight stream through the opening. Following this, fire fighters can advance into the basement to complete the knockdown, ventilation, final extinguishment and overhaul of the fire. TABLE 22-2 shows the conclusions drawn from the fire experiments conducted by NIST, UL, and the FDNY regarding basement fires.

TABLE 22-2	Basement Fires

- Fire flows from basement fires developed in locations other than the stairs, as the floor assembly often failed close to the location where the fire started.
- Flowing water at the top of the interior stairs had limited impact on basement fires.
- Offensive exterior attack through a basement window was effective in cooling the fire compartment.
- Offensive fire attack through an exterior door was effective in cooling the fire compartment.

Courtesy of NIST.

FIRE FIGHTER Tips

Some points to remember regarding basement fires are:

- Thermal imagers may help indicate there is a basement fire but cannot be used to assess structural integrity or fire severity from above.
- Attacking a basement fire from a stairway places fire fighters in a high-risk location due to being in the path of hot gases flowing up the stairs and working over the fire on a flooring system that has the potential to collapse due to fire exposure.
- Coordinating ventilation is extremely important. Ventilating the basement may create a flow path up the stairs and out through the front door of the structure, almost doubling the speed of the hot gases and increasing temperatures of the gases to levels that could cause injury or death to a fully protected fire fighter.

Fires Above Ground Level

Advancing charged hose lines up stairs and along narrow hallways requires much more physical effort than advancing a charged hose line on a level surface. It is important to protect stairways and other vertical openings between floors when fighting a fire in a multiple-level structure. Specifically, hose lines must be placed to keep the fire from extending vertically and to ensure that exit paths remain available.

Interior fire crews must always look for a secondary exit path in case their entry route becomes blocked by the fire or by a structural collapse. This secondary exit could be a second interior stairway, an outside fire escape, a ground ladder placed to a window, or an aerial device.

In high-rise buildings, the standpipe system is typically used to supply water for hose lines. Fire fighters must practice connecting hose lines to standpipe outlets and extending lines from stairways into remote floor areas. Additional hose lines, tools, air cylinders, and Emergency Medical Services equipment should be staged one or two floors below the fire.

Fire Fighter Safety Tips

Be alert to the risk of structural instability and collapse. Remember that modern construction does not always provide clear signs of weakness before collapse. Studies have shown floor deflection is not a reliable measuring factor and fire fighters cannot rely on sounding the floor for accurate stability indicators.

Attic Fires

To understand how to extinguish attic fires, it is helpful to know where these fires start, how they receive a supply of oxygen, and how the fire flow path contributes to the spread of these fires. Fires that involve attics can start in several different locations. Some start in the attic area. An example of this is a fire that starts from an electrical short in the attic. A second location where attic fires can start is in a location apart from the attic. An example of this is an older house built using balloon-type construction. In this case, a fire that starts in the basement can quickly spread from the basement to the attic through the openings between the studs, which extend from the basement to the attic. A third location involving attic spaces are fires that start outside and spread via combustible siding or insulating materials. These fires often start as a ground cover fire or a grilling adventure gone awry. They then impinge on the vinyl or other combustible siding and spread up the side of the house. When the fire impinges on the vinyl siding, it will cause the siding to melt and fall away from the exterior wall, and in many cases, ignite the combustible foam insulation. These insulation fires can transition into the attic space through the exposed soffit very rapidly.

When a fire enters the attic space, it reaches a place where there are large quantities of wood and possibly other combustible materials, open air spaces, and limited or absent fire stops. Convection currents within the attic and radiated heat contribute to the growth of the fire. The attic spaces in most houses are designed to have a constant exchange of air to prevent excess moisture from building up in the attic and to remove excess heat during hot weather. Therefore, there is usually a steady but limited supply of air. In modern construction, fresh air enters the attic space through vents mounted in the soffit. These soffit vents usually run the length of the house on both sides and supply cooler fresh air to the house under normal conditions. These same soffit vents serve as a limited source of oxygen during a fire, which involves the attic space. In other words, soffit vents serve as a limited inlet during an attic fire, though a new trend in the building industry is to create conditioned attic spaces with the use of spray polyurethane foam to seal the attic and soffit.

During normal conditions, convection currents exhaust warmer air from the attic through vents, which are usually mounted along the peak or ridge of the roof. In newer houses, this vent runs in a continuous pathway along the peak of the roof. When there is a fire in the attic, this ridge vent serves as the exhaust for the fire gases. Most modern construction frequently uses plastic soffit vents and plastic ridge vents. Older houses may be constructed with vents at each end of the house instead of using ridge vents. In older houses constructed with balloon construction, part of the inlet flow will be through the open spaces running from the basement to the attic, which is located between the wall studs. It is important to understand how flow paths operate in attic fires. It is also important to understand that there are many different types of building construction. Each type of construction will affect how fires grow and spread.

Attic fires contain large quantities of fuel; however, because of the limited oxygen supply available through the soffit vents and stud spaces, they usually become ventilation-limited fires. Further limiting of the exhaust fire flow path is the fact that the ridge vent is often made of plastic. Heat from the fire will often melt the ridge vent and seal this opening, thereby limiting the exhaust of hot fire fuel gases from the fire. This means that ventilation without coordinated suppression will usually result in rapid fire growth and spread of the fire. Treat attic fires as ventilation-limited fires and use a transitional fire attack. Because venting the attic allows more air to reach the fire, water needs to be applied to the fire to remove fuel gases and to cool the compartment before or as ventilation is performed. Remember, if air can flow into the attic, water can be applied through the same openings.

To carry out a transitional fire attack on an attic fire, it is necessary to apply a stream of water into the attic space. Since the oxygen is entering the fire compartment through the soffit vents, it is logical that water applied through these openings should be effective. Recent live-fire experiments have indicated that this is true. When performing a transitional fire attack, quickly remove the soffit vents from the underside of the soffit. In most modern-constructed houses, these are made of plastic, and in the case of a single-story house, can be easily removed with a pike pole. Once the vents are removed, use a straight-stream nozzle to spray a stream of water into the soffit-vent opening, directing water all the way along one side of the house.

This will wet the underside of the sheathing on that side, deflect off the ridge line, and run down the other side of the interior roof sheathing. This application of water serves several purposes. It removes large quantities of the fire fuel gases, cools the fire compartment, and wets the fuel in the attic. All these actions serve to reduce the growth of the fire and partially extinguish it. Following this initial transitional attack, it will be necessary to ventilate the attic space and extinguish any remaining fire. In the case of a two-story house, a transitional fire attack may be accomplished by using aerial apparatus to apply a stream of water through the soffit vent openings.

Older houses are sometimes built with openings at each gable end of the building. It would seem that an application of water through these openings on the end should be an effective way to perform a transitional fire attack. However, experiments have shown that applying water through the gable ends is not the best option since the stream bounces off the joists and may reach only one-third the length of the house.

Remember that these fire flow paths change if there is wind blowing to enhance or retard the airflow. Understanding flow paths is an important step in understanding how to extinguish attic fires.

Fires in Large Buildings

Large buildings have many different uses and different layouts. "Big box" stores contain one very large, open space surrounded by smaller rooms and storage areas. These stores are usually one story tall. Other large buildings are constructed for multiple offices and may be many stories tall. The fire protection for most large buildings is dependent on fire-resistive construction, fire detection, and automatic sprinkler systems. In large buildings equipped with automatic sprinkler systems, it is important to augment the water supply by connecting to the fire department sprinkler connection to increase the water volume and pressure in the sprinkler system.

The fire load in large buildings varies greatly depending on the building contents. Large stores and home improvement centers contain an extensive amount of flammable contents ranging from lumber to flammable fuels and large quantities of home furnishings and clothing. Recognize that the fire load may be high in some of these buildings. Consider the need for long hose lays if there is not an adequate number of standpipe connections. Remember that big buildings may produce big fires and require the use of large hose lines with unlimited water supplies.

Many large buildings have floor plans that can cause fire fighters to become lost or disoriented while working inside, particularly in low-visibility or zero-visibility conditions. In such a setting, guide lines may be necessary to keep fire fighters from becoming lost and running out of air. A well-organized preincident plan of the structure can be essential when fighting this type of fire. Knowing the occupancy and the other hazards beforehand will help in determining the best strategy and tactics.

FIRE FIGHTER II Tips **FFII**

In large buildings that are equipped with automatic sprinkler systems, it is important to augment the water supply by connecting to the fire department connection to increase the water volume and pressure in the sprinkler system.

Fires in Buildings During Construction, Renovation, or Demolition

Buildings that are under construction, renovation, or demolition are all at increased risk for destruction by fire. These buildings often have large quantities of combustible materials exposed, while lacking the fire-resistant features of a finished building. If the building lacks windows and doors, an almost unlimited supply of oxygen is available to fuel a fire. Fire detection, fire alarm, and automatic fire suppression systems are often not installed or are inoperable. In addition, construction workers using torches and other flame-producing devices pose a notable fire risk. Moreover, these buildings are often unoccupied and can be easy targets for arsonists.

Under these conditions, a fire in a large building can be very difficult to extinguish. If no life-safety hazards are involved, fire fighters may need to use a defensive strategy for this type of fire. In such a case, no fire fighters should enter the building, and a collapse zone should be established. A defensive exterior operation should be conducted using master streams, aerial streams, and large handlines to protect exposures. Inspections of construction and demolition sites can give fire fighters indications if construction is following code and maintaining protection systems to the floor below the last completed floor.

Fire Fighter Safety Tips

It is important to maintain safety precautions during overhaul. Large unsupported or fire-damaged walls may collapse hours after the main body of fire has been knocked down.

Fires in Lumberyards

Lumberyard fires are often prime candidates for a defensive firefighting strategy. A typical lumberyard contains large quantities of highly combustible materials that are stored in the open or in sheds where plenty of air is available to support combustion. Given this rich fuel supply, a lumberyard fire will usually produce tremendous quantities of radiant heat and release burning embers that will cause the fire to spread quickly from stack to stack. Lumberyards may be sited in locations where there is an inadequate water supply. Such fires may quickly extend to nearby buildings and other structures. Large buildings at lumberyards are often constructed using trusses and other lightweight building techniques, so you must be aware of the potential for rapid building collapse at these incidents.

Protecting exposures is often the primary objective at lumberyard fires, as there may be little life-safety risk. Exposure protection should be dealt with early in the operation by placing large handlines and master stream devices where they can be most effective. A collapse zone must be established around any stacks of burning material and buildings to keep fire fighters out of dangerous positions.

Fires in Stacked or Piled Materials

Fires occurring in stacked or piled materials can present a variety of hazards. The greatest danger to fire fighters is the possibility that a stack of heavy material, such as rolled paper or baled rags, will collapse without warning. Such an event might occur, for example, if a fire has damaged the stacked materials or if water has soaked into them. Absorbed water can increase the weight of many materials and weaken cardboard and paper products. In fact, the water discharged by automatic sprinklers alone can be sufficient to make some stacked materials unstable. A tall stack of material that falls on top of a team of fire fighters can cause injury or death.

Given these risks, fires in stacked materials should be approached cautiously. All fire fighters must remain outside potential collapse zones. Mechanical equipment should be used to move material that has been partially burned or water soaked.

Conventional methods of fire attack can often be used to gain control of the fire; however, water must penetrate into the stacked material to fully extinguish the residual combustion. Class A foam and wetting agents can be applied to extinguish smoldering fires in tightly packed combustible materials. Overhaul will require fire fighters to separate the materials to expose any remaining deep-seated fire. This operation can be a labor-intensive process unless mechanical equipment can be used to dig through the material.

Trash Container and Rubbish Fires

Trash container (dumpster) fires usually occur outside of any structure and appear to present fewer challenges than fires inside buildings. Even so, fire fighters must be vigilant in wearing full PPE and using SCBA when fighting trash container fires or trash pile fires, because there is no way of knowing what might be included in a collection of trash. Some trash containers may contain hazardous materials or materials that are highly flammable or explosive. Manufacturers are now selling polyethylene dumpsters, which can burn very rapidly. These dumpsters are not permitted within 15 feet of a structure, but the fire service needs to be aware of their presence.

If the fire is deep seated, fire fighters will have to overhaul the trash to make sure that the fire is completely extinguished. Manual overhaul involves pulling the contents of a trash container apart with pike poles and other hand tools so that water can reach the burning material. This process can be labor intensive and involves considerable risk to fire fighters. The fire fighters are exposed to any contaminants in the container as well as to the risks of injury from burns, smoke, or other causes. Considering the low value of the contents of a trash container, it is difficult to justify any risk to fire fighters' safety.

Class A foam is useful for extinguishing many trashcontainer fires, because it allows water to soak into thematerials and, therefore, can eliminate the need for manual overhaul. Some fire departments use the deck gun on the top of an engine to extinguish large trash container fires and then complete extinguishment by filling the container with water. This is done by pointing the deck gun at the dumpster and slowly opening the discharge gate to lob water into the container.

Trash containers are often placed behind large buildings and businesses. If the container is close to the structure, be sure to check for fire extension. Also look around the container for the presence of telephone, cable, and power lines that might have been damaged by the fire.

To extinguish an outside trash fire or other outside class A fire, follow the steps in SKILL DRILL 22-10 (Fire Fighter I, 5.3.8):

1. Don full PPE, including SCBA; enter the accountability system; and work as a team.
2. Perform size-up and give an arrival report. Call for additional resources if needed.
3. Ensure that apparatus is positioned uphill and upwind of the fire and that it protects the scene from traffic.
4. Develop and implement a fire suppression strategy.
5. Protect the crew from hazards.
6. Deploy an appropriate attack line (at least 1½ inches [38 mm] in diameter).
7. Direct the crew to attack the fire in a safe manner—uphill and upwind from the fire.
8. Break up compact materials with hand tools or hose streams.
9. Evaluate and modify the water application technique if necessary.
10. Maintain good body mechanics during the fire attack.
11. Notify command when the fire is under control.
12. Investigate the origin and cause of the fire. Preserve any evidence of arson. Return the equipment and crew to service.

Confined Spaces

Both fires and other types of emergencies can occur in confined spaces. Fires in underground vaults and utility rooms such as transformer vaults are too dangerous to enter. In such cases, fire fighters should summon the utility company and keep the area around manhole covers and other openings clear while awaiting the arrival of utility company personnel. To deal with emergencies in these areas, the Occupational Safety and Health Administration (OSHA) requires specially trained entry teams. Learn your fire department's operational procedures for handling such incidents.

It is also important that fire fighters know about confined spaces that exist in the industries within their individual response areas. Fire fighters should visit these areas with plant personnel so that preincident plans can be developed.

Unique hazards may arise in confined spaces, including oxygen deficiencies, toxic gases, and standing water. Although conditions might appear to be safe, the confined space might not have enough oxygen to sustain life. It is common for a victim to enter these spaces and pass out, only to be followed by another victim—an individual who comes in to assist the first person and also passes out. When entering confined spaces to attempt rescue, fire fighters should wear full SCBA gear and be attached to a life line. An additional life line should be lowered or brought with the fire fighter to tie the victim in a bowline on a bight (rescue knot), which allows surface personnel to raise the victim out of the hole.

Owing to the lack of ventilation in most confined spaces, fire fighters may notice an intense amount of heat once they have entered the space. Fire fighters will tire quickly in this environment, so they must recognize the signs of heat exhaustion and heat stroke.

Because confined spaces commonly have low oxygen levels and high levels of combustible gases, such as methane, fire fighters without breathing apparatus can quickly be overcome in confined spaces with these conditions. For this reason, fire fighters who must enter a confined space should carry a monitoring device. Air quality must be checked constantly, looking for the build-up of explosive gases as well as any decline in oxygen levels.

Fire fighters must adhere to a strict accountability system when they enter into a confined space. This procedure ensures that only those personnel with proper training and equipment enter the space. It is important for a safety officer to track the movement of personnel and the amount of time that they remain in the confined space.

Fire suppression in confined spaces must not begin until all utilities and industrial processes have been turned off. Potential suppression agents include hose streams, high- and low-expansion foams, carbon dioxide (CO_2) flooding systems, and built-in sprinkler systems.

Vehicle Fires

Vehicle fires are one of the most common types of fires handled by fire departments. In 2012, the NFPA reported that approximately 184,000 highway vehicle fires occurred in the United States. These fires may result from a variety of causes. For example, discarded smoking materials can cause fires in upholstery. Electrical short-circuits may cause fires in many different parts of a vehicle. Friction caused by dragging brakes or defective wheel bearings may cause fires. Collisions may lead to ruptured fuel lines, resulting in fires.

Fire fighters save few cars. In fact, when active flames are seen at a vehicle fire, the vehicle will likely be classified as "totaled" by the insurance company. Given this fact, it is important to understand the hazards involved in fighting vehicle fires and to mount a safe attack on the fire. Only when a viable victim is trapped in a burning vehicle does the scenario become a life-or-death situation.

Many hazards are associated with vehicle fires—traffic hazards, fuel, pressurized cylinders and containers that can explode, fire, and toxic smoke. Because vehicle fires usually occur on streets and highways, one of the biggest hazards faced by fire fighters is the danger posed by traffic. Drivers are easily distracted by the sight of a burning vehicle, which may lead to subsequent collisions. To counteract this risk, you should use your apparatus to block traffic. Do not hesitate to work with law enforcement to shut down or divert traffic flow if necessary to ensure safety for fire fighters. Fire apparatus operators often place their vehicles 100 feet behind the burning vehicle to stop traffic and to position the apparatus a safe distance from the burning vehicle. Follow the steps outlined by your department for guarding against traffic hazards.

Modern vehicles contain a variety of gas-filled, pressurized cylinders and containers containing explosive materials. Hydraulic pistons are used to support hatch backs, trunks, tailgates, and automobile hoods. Energy-absorbing bumper systems contain hydraulic pistons. In addition, many modern vehicles use a MacPherson strut suspension system to absorb road shocks. When these gas-filled components are quickly heated to high temperatures in a fire, they can release pressure explosively, sending metal parts hurtling away from the vehicle. In addition, the supplemental restraint systems (SRS) found in most vehicles consist of air bags and air curtains containing chemicals that can ignite explosively during a fire.

Modern automobiles are constructed from hundreds of pounds of plastics, which give off large quantities of toxic smoke and heat when they burn. They also contain a variety of petroleum products, including gasoline or diesel fuel, motor oil, brake fluid, and automatic transmission fluid. These products ignite easily, burn with high intensity, and produce large quantities of toxic gases. For this reason, it is important to always wear full PPE, including SCBA, when fighting a vehicle fire.

If the owner or driver of the burning vehicle is present, ask about any specific hazards that may be present in the vehicle, such as portable propane cylinders, propane torches, medical oxygen equipment, cans of spray paint, and other hazardous materials. If no driver or occupant is present, do not assume that the vehicle is safe; always be cautious as you approach a vehicle fire. Perform a risk–analysis assessment—that is, look at the big picture and weigh the options. Do not risk injuring fire fighters in a vehicle fire.

Attacking Vehicle Fires

As you prepare to approach a vehicle fire, make sure that you have created a safe area around the vehicle. The only people closer than 50 feet (15 m) to the vehicle should be fire fighters in full PPE and SCBA who are extinguishing the fire. Use a hose line at least 1½ inches (38 mm) in diameter; such a hose will provide sufficient cooling power to overwhelm the fire and provide protection from a sudden flare-up. The use of compressed air foam or Class B foam aids in fire suppression. Charge the hose line while you are at least 50 feet (15 m) from the fire. Bleed all the air from the hose line. Set the nozzle to initially deliver a straight stream or a fog pattern that is no wider than a 30-degree angle. Approach the vehicle from an uphill and upwind position, moving in from the side and at a 45-degree angle FIGURE 22-10. This path will help you avoid debris in case of an exploding bumper.

From a point approximately 30 feet (9 m) away from the vehicle, open the nozzle and sweep the bottom part of the vehicle using a horizontal motion. Extinguish all visible fire while advancing toward the vehicle. Sweeping along the undercarriage helps to cool the bumper pistons, shock absorbers, and hydraulic struts; cool the tires before they explode; and cool the fuel tank before it fails. By applying a sufficient quantity of water to the lower part of the vehicle, you reduce the chance of an explosive event. Observe the area under the car during the approach for any sign of leaking flammable liquids. If burning flammable liquids are present, widen the spray pattern on the nozzle. Foam can be used to extinguish the burning liquid and provide a vapor barrier to prevent reignition. Once this sweep is complete, fire fighters can begin to attack any fire in the passenger compartment, the engine compartment, and the cargo compartment.

To extinguish a vehicle fire, follow the steps in SKILL DRILL 22-11 (Fire Fighter I, 5.3.7):

1. Don full PPE, including SCBA; enter the accountability system; and work as a team.
2. Perform size-up and give an arrival report. Call for additional resources if needed.
3. Ensure that the fire apparatus is positioned uphill and upwind, and that it protects the scene from traffic.
4. Ensure that the crew is protected from any hazards.
5. Identify the type of fuel used in the burning vehicle and look for fuel leaks.
6. Advance a fire attack line of at least 1½-inch (38 mm) diameter using water or foam.
7. Attack from uphill and upwind of the fire, and at a 45-degree angle to the side of the vehicle.
8. Suppress the fire.
9. Overhaul all areas of the vehicle, passenger compartment, engine compartment, and trunk.
10. Notify command when the fire is under control.
11. Investigate the origin and cause of the fire. Preserve any evidence of arson. Return the equipment and crew to service.

Alternative-Fuel Vehicles

Always be alert for signs that a burning vehicle could be powered by an alternative fuel, such as compressed natural gas (CNG) or liquefied petroleum gas (LPG). Fully involved fires in vehicles powered by either type of fuel should be fought with an unmanned master stream to prevent injuries from exploding gas cylinders.

Automobiles powered by CNG contain storage cylinders that are very similar to SCBA cylinders. These cylinders are usually located in the trunk and contain CNG at high pressures. When a fire occurs, they must be cooled and protected just like any gas cylinder. CNG is a nontoxic, lighter-than-air gas that will rise and dissipate if it is released into the atmosphere.

Propane is stored in the same types of cylinders as are used for heating or cooking purposes. Propane is heavier than air, so its vapors will pool or collect in low areas.

Hybrid automobiles have small gasoline-powered engines and large battery banks. The batteries power electric motors that drive the wheels, much like a train locomotive. The engine is used for auxiliary power and to recharge the batteries.

Electric vehicles are propelled solely by electric motors, relying on a large bank of batteries. Usually these batteries must be recharged from an external source. There are some considerations to keep in mind when dealing with fires involving hybrid or electric vehicles:

- When responding to fires involving either of these types of vehicles, you need to don your PPE, including SCBA.
- Apply chock blocks to the wheels to prevent the vehicle from moving forward or in reverse.
- Determine whether the vehicle is a hybrid vehicle or an electric vehicle. If possible, turn the power off by activating the vehicle shut-off switch, removing the key from its switch or removing the proximity key from the vehicle. Some vehicles have a disconnect switch or fuses that can be removed to disconnect the motors from the power source.
- Locate the batteries. In a hybrid vehicle, the batteries are often located at the rear of the vehicle under the trunk. In an electric vehicle, they are often located under the passenger floor boards. Do not force access

FIGURE 22-10 When possible, approach vehicle fires in the hood or engine area from the upwind side.

to the batteries; this may cause you to come into contact with high-voltage cables or damage the batteries. Use access doors or points; look for existing openings.
- Avoid direct contact with the batteries. Watch out for increased voltage in the battery compartments. High-voltage, direct-current cables connect the batteries to the electric motors that power the wheels. Avoid cutting these cables.

It is also important to establish a water supply, since fires in these vehicles require more water to extinguish than fires in conventional vehicles. If the battery is not involved, treat these fires like fires in conventionally powered vehicles. If the battery is involved, realize it will take more water and a longer period of time to extinguish these fires. Apply water even after the flames are no longer visible; this is necessary to continue to cool the batteries.

Overhauling these vehicles takes longer than overhauling a conventional car fire. Batteries can reheat and reignite a long period of time after the flames were extinguished. Listen for popping sounds, which may indicate that the battery is heating up. There is a possibility of reignition several hours after extinguishment. In cases where you can see the battery, you can use a thermal imager to see if the battery is cooling or heating up. After a fire, electric vehicles and hybrids should be stored at least 50 feet from any exposures because of the possibility of a late rekindle.

Controlled fires in hybrid and electric vehicles have not shown any electrical hazards to fire fighters while they are applying water to fires in these vehicles.

More information about these vehicles is available from automobile manufacturing companies and from the NFPA.

Follow the manufacturer's instructions for disconnecting the power to alternative-powered vehicles. More information on alternative-powered vehicles is presented in the Vehicle Rescue and Extrication chapter.

Fire in the Passenger Area

Fires in the passenger area of a vehicle are more visible and accessible than fires in the engine compartment or the cargo area. Usually it is logical to extinguish the fire in the passenger compartment before moving on to extinguish any fire in the engine and cargo compartments. Often the windows are broken or a door is ajar, presenting an opening through which fire fighters can direct a stream of water. If the doors will not open, stand upwind from the window and use a striking tool to break out one or more windows. Be cautious, because a backdraft may occur.

As you get closer to the vehicle, change the nozzle to produce a wider pattern that will cool a wider area and give you some protection from the heat of the fire. Pay special attention to cooling areas such as the steering column and the dashboard on the passenger side. Cool areas that contain side-curtain air bags; cooling will greatly reduce the chance for accidental deployment of the SRS. Once the fire in the passenger compartment has been knocked down, attack any fire in the engine compartment or in the cargo compartment.

Fire in the Engine Compartment

The engine compartment of a vehicle is filled with a variety of devices that use petroleum products to power or lubricate them. They also contain components made of plastics and rubber. As a consequence, these devices produce a large amount of smoke when they burn. Other hazards present in the engine compartment include suspension struts and hydraulic lift cylinders for the hood. Vehicle batteries contain sulfuric acid, which can cause serious burns. Some vehicles have magnesium parts within the engine compartment; when magnesium is present, you must use a Class D extinguishing agent.

A challenging part of extinguishing a fire in the engine compartment is gaining access to the fire. An initial attack can be made through the wheel well. The plastic liner between the engine compartment and the wheel well is often consumed during such a fire, so it may be possible to spray water into the engine compartment through this opening. An alternative initial approach is to spray water through the grille after thoroughly cooling the area around the bumper. When using this tactic, avoid standing directly in front of the bumper until this area is thoroughly cooled. Although neither of these two methods is effective at totally extinguishing the fire, they will help to diminish the volume of fire while fire fighters are gaining access to the engine compartment.

The fastest way to gain access to the engine compartment is to pull the hood-release lever inside the passenger compartment to open the hood. Unfortunately, this method rarely works during a vehicle fire, so other methods usually need to be tried. A quick means of gaining access to the engine compartment is to insert a pry bar along the side of the hood between the edge of the hood and the fender. Pry the side of the hood away from the fender to produce an opening big enough to apply a stream of water into the engine compartment.

A second method of gaining access to the engine compartment is to open the hood of the vehicle. In most fires, the cable that normally opens the hood is damaged and will not release from the inside of the vehicle. To open the hood of a burning vehicle, break out the plastic grille and find the hood-release cable. With a gloved hand, pull on the cable to release the hood or use the forked end of a Halligan tool to twist the cable until the hood releases. Fire fighters can also use vise grip pliers to grasp the cable and pull it.

Once the hood is raised, fire fighters should have good access to any remaining fire and hot spots. Use plenty of water or foam to cool this area. Once the vehicle's hood has been opened and any fire in the engine compartment extinguished, disconnect the power to the vehicle by cutting the battery cables if the battery is not under the hood, check the wheel wells or trunk. First, remove a section of the negative cable at least 6 inches (152 mm) long by making two cuts with wire cutters. Then, remove a similar section of the positive cable. This excision will disrupt the flow of power to the vehicle and ensure that nothing is powered accidentally.

Fire in the Trunk

A fire in the trunk of a vehicle may present unknown hazards because it is not possible to know what the trunk contains. In addition, fires in this area are challenging to access. Initial access can sometimes be made by knocking out the tail light assembly on one side, which enables fire fighters to direct a hose stream into the trunk to cool down and partially extinguish fire in this area. A fire in the trunk area of an automobile can also be accessed by first using the pike of a Halligan tool to force the lock into the trunk, and then using a screwdriver or key tool (K-tool lock puller) to turn the lock cylinder in a clockwise direction. A charged hose line must be ready when the trunk lid is raised.

Fires in the rear of light trucks and vans must always be approached cautiously. In addition, vehicles using alternative fuels (discussed earlier in this chapter) may contain compressed natural gas or propane cylinders in the trunk. Vans are often used by couriers and could contain medical waste, laboratory specimens, and radioactive material.

Overhauling Vehicle Fires

Overhaul of vehicle fires is just as important as overhaul of structure fires. As soon as it is safe to approach the vehicle, chock the wheels to prevent the vehicle from moving. If a vehicle fire erupts quickly, the driver may not have time to set the brake and parking gear. Also, fire can damage cables and wires that control the operation of parking and braking mechanisms. After all visible fire has been knocked down, allow a few minutes for the steam and smoke to dissipate before starting overhaul. This delay will allow visibility to improve so that overhaul can be completed safely.

During overhaul of interior fires, remember that air bags can deploy without warning in a burning automobile. Never place any part of your body in the path of a front or side air bag.

As you overhaul the vehicle, be systematic and thorough. Do not miss areas that may contain lingering sparks. Direct the hose stream under the dashboard, and soak and remove smoldering upholstery. Apply water over and under all parts of the engine compartment. Confirm that no fluids are leaking from the vehicle. If contents in the car might potentially be salvaged, treat them with respect. Continue to use your SCBA as long as smoke or fumes are present.

Fire Fighter Safety Tips

Some vehicle components (engines or body) may be constructed from magnesium or other flammable or explosive metals that can react violently when water is applied during fire suppression. If possible, a Class D extinguishing agent should be used in these cases instead of water. All areas of the vehicle must be checked for fire extension and for potential victims. There have been numerous vehicle fires used to cover a crime.

Flammable-Liquid Fires

Flammable-liquid fires can be encountered in almost any type of occupancy. Most fires involving a vehicle (e.g., airplane, train, ship, car, truck) will involve a combustible or flammable liquid. Special tactics must be used when attempting to extinguish a flammable-liquid fire, and special extinguishing agents such as foam or dry chemicals may be needed.

Hazards

Fires involving flammable liquids such as gasoline require special extinguishing agents. Most flammable liquids can be extinguished using either foam or dry chemicals. Class B extinguishing agents are approved for use on Class B (flammable liquids) fires. Flammable-liquid fires can be classified as either two-dimensional or three-dimensional. A two-dimensional fire refers to a spill, pool, or open container of liquid that is burning only on the top surface. A three-dimensional fire refers to a situation in which the burning liquid is dripping, spraying, or flowing over the edges of a container.

A two-dimensional flammable-liquid fire can usually be controlled by applying the appropriate Class B foam to the burning surface. Several different formulations of Class B foams are suitable for a variety of liquids and situations. The foam will flow across the surface of the liquid and create a seal that stops the fuel from vaporizing; this separates the fuel from the oxygen and extinguishes the fire. Foam will also cool the liquid and reduce the possibility of reignition.

When dealing with flammable-liquid fires, fire fighters should look for hot surfaces or open flames that could cause the vapors to reignite after a fire has been extinguished. It is important to determine the identity of the liquid that is involved so as to select the appropriate extinguishing agent and to determine whether the vapors are lighter or heavier than air.

A three-dimensional flammable-liquid fire is much more difficult to extinguish with foam, because the foam cannot establish an effective seal between the fuel and the oxygen. Either dry-chemical or gaseous extinguishing agents are usually more effective than foam in controlling these kinds of fires. These agents can also be used to extinguish two-dimensional fires, although they do not provide a long-lasting seal between the fuel and the oxygen. In some cases, a fire can be extinguished with a dry chemical, and then the surface can be covered with foam to prevent reignition.

Fire fighters should avoid standing in pools of flammable liquids or contaminated runoff from them, because their PPE will absorb the flammable product and become contaminated. In cases of serious contamination, the PPE itself can become flammable.

Suppression

The skills used in suppressing small flammable-liquid fires are presented in the Portable Fire Extinguishers chapter. Larger flammable liquid fires may require the use of Class B foam. The equipment and methods used to apply Class B foam are described in the Fire Attack and Foam chapter.

FIRE FIGHTER II

Flammable-Gas Cylinders

Flammable-gas cylinders can be found in many places. Many types of flammable gases are stored in many different types and sizes of containers. A variety of flammable gases can be found in industrial occupancies.

Propane Gas

The popularity of propane gas for heating and cooking has meant that these cylinders have become commonplace in residential areas and many industrial and commercial locations. In addition, propane is used as an alternative fuel for vehicles and is often stored to power emergency electrical generators. Fire fighters should be familiar with the basic hazards and characteristics of propane as well as with procedures for fighting propane fires.

Propane (LPG) exists as a gas in its natural state at temperatures higher than –44°F (–42.2°C). When the gas is placed into a storage cylinder under pressure, it is changed into a liquid. Storing propane as a liquid is very efficient, because it has an expansion ratio of 270:1 (i.e., 1 cubic foot (0.02 cubic meters) of liquid propane is converted to 270 cubic feet (7.64 cubic meters) of gaseous propane when it is released into the atmosphere). Put simply, a large quantity of propane fuel can be stored in a small container.

Inside a propane container, there is a space filled with propane gas above the level of the liquid propane. As the contents of the cylinder are used, the liquid level becomes lower and the vapor space increases. The internal piping is arranged so as to draw product from the vapor space.

Propane gas containers come in a variety of sizes and shapes, with capacities ranging from a few ounces to thousands of gallons. The cylinder itself is usually made of steel or aluminum. A discharge valve keeps the gas inside the cylinder from escaping into the atmosphere and controls the flow of gas into the system where it is used. This valve should be easily visible and accessible. In the event of a fire, closing the valve should stop the flow of the product and extinguish the fire. The valve should be clearly marked to indicate the direction in which it should be turned or moved to reach the closed position.

A connection to a hose, tubing, or piping allows the propane gas to flow from the cylinder to its destination. In the case of portable tanks, this connection is often the most likely place for a leak to occur. If the gas is ignited, this area could become involved in fire.

A propane cylinder is always equipped with a relief valve to allow excess pressure to escape, thereby preventing an explosion if the tank becomes overheated. Propane cylinders must be stored in an upright position so that the relief valve remains within the vapor space. If the cylinder is placed on its side, the relief valve could fall below the liquid level. If a fire were then to heat the tank and cause an increase in pressure, the relief valve would release liquid propane, which would expand by the 270:1 ratio and create a huge cloud of potentially explosive propane gas.

Propane Hazards

Propane is highly flammable. Although it is nontoxic, this gas can displace oxygen and cause asphyxiation. By itself, propane is odorless, so leaks of pure propane cannot be detected by a human sense of smell. For this reason, mercaptan is added to propane to give it a distinctive odor.

Propane gas is heavier than air, so it will flow along the ground and accumulate in low areas. When using meters to check for LPG, be sure to check storm drains, basements, and other low-lying areas for concentrations of the gas.

When responding to a reported LPG leak, fire fighters and their apparatus should be staged uphill and upwind of the scene. Because an explosion can happen at any time, fire fighters should wear full PPE and SCBA at this type of incident. Life safety should be the highest priority; depending on the type and size of the leak, an evacuation might be necessary.

The greatest danger with propane and similar products is a BLEVE. If an LPG tank is exposed to heat from a fire, the temperature of the liquid inside the container will increase. The fire could then be fueled by propane escaping from the tank or from an external source. As the temperature of the product increases, the vapor pressure will also increase. The increasing pressure creates added stress on the container. If this pressure exceeds the strength of the cylinder, the cylinder can rupture catastrophically. An exploding LPG cylinder can produce the same explosive power as dynamite.

To protect the container from rupture, the relief valve will open to release some of the pressure. This relief valve is designed to exhaust the vapor until the pressure drops to a preset level. When the pressure returns to a safe level, the valve will close. If the heating continues, however, the liquid will begin to boil within the container. If the flame impinges directly on the tank, the container can weaken and fail somewhere above the liquid line. When this happens, the container will rupture and release its contents with explosive speed. The boiling liquid will expand, vaporize, and ignite in a giant fireball, accompanied by flying fragments of the ruptured container. Fire fighters have been killed in these explosions.

The best method to prevent a BLEVE is to direct heavy streams of water onto the tank from a safe distance. The water should be directed at the area where the tank is being heated. Cool the upper part of the tank to cool the gas vapors. The fire fighters operating these streams should work from shielded positions or use remote-controlled or unmanned monitors. Horizontal tanks are designed to fail at the ends if a catastrophic failure occurs, so fire fighters should operate only from the sides of the tank.

Flammable-Gas Fire Suppression

Fighting fires involving LPG or other flammable-gas cylinders requires careful analysis and logical procedures. If the gas itself is burning because of a pipe or regulator failure, the best way to extinguish the fire is to shut off the main discharge valve at the cylinder. If the fire is extinguished and the fuel continues to leak, there is a high probability that it will reignite explosively. Do not attempt to extinguish the flames unless the source of the fuel has been shut off or all of the fuel has been consumed. If the fire is heating the storage tank, use hose streams to cool the cylinder, being careful not to extinguish the fire.

Unless a remote shut-off valve is available, the flow of propane can be stopped only if it is safe to approach the cylinder. Fire fighters should inspect the integrity of the cylinder from a distance before they make any attempt to approach and shut off the valve. If the container is damaged or the valve is missing, the fuel should be allowed to burn off, while hose streams continue to cool the tank from a safe distance.

Approach a flammable-gas fire with two 1¾-inch (45 mm) hose lines working together. When approaching a horizontal LPG tank, always approach it from the sides. The nozzles should be set on a wide fog pattern, with the discharge streams interlocked to create a protective curtain. The team leader should be located between the two nozzle operators. On the command of the leader, the crew should move forward, remaining together and never turning their backs to the burning product. Upon reaching the valve, the fire fighter in the center can turn off the valve, stopping the flow of gas. Any remaining fire may then be extinguished by normal means. Continue the flow of water as a protective curtain and to reduce sources of ignition.

If the fire is extinguished prematurely, the valve should still be turned off as soon as the team reaches it. Always approach and retreat from these types of fires while facing the objective with water flowing, in case of reignition.

Unmanned master streams should be used to protect flammable-gas containers that are exposed to a severe fire. Direct the stream so that it is one-third of the way down the container. This technique will allow half of the water to roll up and over the container, while the remainder projects downward FIGURE 22-11. The objective is to cover as much of the exposed tank as possible.

If the LPG container is located next to a fully involved building or a fire that is too large to control, evacuate the area and do not fight the fire. If there is nothing to save, risk nothing.

Keep in mind that if the relief valve is open, the flammable-gas container is under stress. Exercise extreme caution in this scenario. As the gas pressure is relieved, it will sound like the whistle on a teakettle; if the sound is rising in frequency, an explosion could be imminent and evacuation should be ordered.

To suppress a flammable-gas cylinder fire, follow the steps in SKILL DRILL 22-12 (Fire Fighter II, 6.3.3):

1. Cool the tank from a distance until the relief valve resets.
2. Wearing full PPE, two teams of fire fighters using a minimum of two 1¾-inch (45 mm) hose lines advance using an interlocking 90-degree-wide fog pattern for protection. Do not approach the cylinder from the ends. The team leader should be located between the two nozzle persons. The leader coordinates the advance toward the cylinder.
3. When the cylinder is reached, the two nozzle teams isolate the shut-off valve from the fire with their fog streams while the leader closes the tank valve, eliminating the fuel source.
4. After the burning gas is extinguished, the fire fighters continue to apply water to the cylinder to cool the metal, with the goal of preventing tank failure and a subsequent BLEVE.
5. As cooling continues, fire fighters slowly back away from the cylinder, never turning their backs to it.

Shutting Off Gas Service

Many structures use either natural gas or propane gas for heat or cooking. In addition, these two energy sources have many industrial applications. If a gas line inside a structure becomes compromised during a fire, the escaping gas can add fuel to the fire. The means by which the gas is supplied to the structure must be located to stop the flow.

FIGURE 22-11 Using a master stream to protect a flammable-gas container that is exposed to fire.

Most residential gas supplies are delivered through a gas meter connected to an underground utility network or from a storage tank located outside the building. If the gas is supplied by an underground distribution system, the flow can be stopped by closing a quarter-turn valve on the gas meter. If the gas is supplied from an outside LPG storage cylinder, closing the cylinder valve will stop the flow.

After the gas service has been shut off, use a lockout tag to ensure that it is not turned back on. Only a professional can reestablish the flow of gas to a structure.

To shut off gas utilities, follow the steps in SKILL DRILL 22-13 (Fire Fighter I, NFPA 5.3.18):

1. Don PPE, including SCBA; enter the personnel accountability system.
2. Acknowledge the assignment.
3. Locate the exterior gas shut-off valve.
4. Shut off the gas valve.
5. Attach a shut-off tag and lock if required.
6. Notify command that the gas is shut off.

Fires Involving Electricity

The greatest danger with most fires involving electrical equipment is electrocution. For this reason, only Class C extinguishing agents should be used when energized equipment is involved in a fire. All electrical equipment should be considered as potentially energized until the power company or a qualified electrical professional confirms that the power is off. Once the electrical service has been disconnected, most fires in electrical equipment can be controlled using the same tactics and procedures as are used for a Class A fire.

When a fire occurs in a building, the electrical service should be turned off quickly to reduce the risks of injury or death to fire fighters, even if there is no involvement of electrical equipment in the fire. If possible, this shut-off should take place at the main circuit breaker box, with a lockout tag being used to prevent someone from accidentally turning the electrical current back on.

If it is not possible to turn the electricity off at the breaker box, fire fighters must notify the electrical utility company to send a representative to the fire scene to disconnect the service from a location outside the building. In many cases, the local utility company is automatically notified of any working fire.

To control the electric utility system, follow the steps in SKILL DRILL 22-14 (Fire Fighter I, 5.3.18):

1. Don PPE, including SCBA; enter the personnel accountability system.
2. Acknowledge the assignment
3. Locate the exterior shut-off.
4. Shut off the power.
5. Attach a lockout tag and lock.
6. Notify command that the electricity is shut off.

■ Electrical Fire Suppression

Fire suppression methods for fires involving electrical equipment vary according to the type of equipment and the power supply. In many cases, the best approach is to wait until the power is disconnected and then use the appropriate extinguishing agents to control the fire. If the power cannot be disconnected or the situation requires immediate action, only Class C extinguishing agents—such as halon agents, CO_2, or dry chemicals—should be used.

When delicate electronic equipment is involved in a fire, either halons or CO_2 should be used to limit the damage as much as possible. These agents cause less damage to computers and sensitive equipment than do water or dry chemical agents.

When power distribution lines or transformers are involved in a fire, special care must be taken to ensure the safety of both emergency personnel and the public. No attempt should be made to attack these fires until the power has been disconnected. In some cases, fire fighters may need to protect exposures or extinguish a fire that has spread to other combustible materials, if this can be accomplished without coming in contact with the electrically energized equipment. If a hose stream comes in contact with the energized equipment, the current can flow back through the water to the nozzle and electrocute fire fighters who are in contact with the hose line.

Many electrical transformers contain a cooling liquid that includes polychlorinated biphenyls, a cancer-causing material. Do not apply water to a burning transformer. Water can cause the transformer's cooling liquid to spill or splash, contaminating both fire fighters and the environment. If the transformer is located on a pole, it should be allowed to burn until electrical utility professionals arrive and disconnect the power. Dry-chemical extinguishers can then be used to control the fire. Fires in ground-mounted transformers can also be extinguished with dry chemical agents after the power has been disconnected. Fire fighters should stay out of the smoke and away from any liquids that are discharged from a transformer, and they must wear full PPE and SCBA to attack the fire.

Some very large transformers contain large quantities of cooling oil and require foam for fire extinguishment. This foam can be applied only after the power has been disconnected. Until the power is off, fire fighters can still protect exposures while taking care to avoid contamination.

Underground power lines and transformers are often located in vaults beneath the ground's surface. Explosive gases can build up within these vaults. If a spark then ignites the gases, the resulting explosion can lift a manhole cover from a vault and hurl it for a considerable distance. Products of combustion can also leak into buildings through the underground conduits. Fire fighters should never enter an underground electrical vault while the equipment is energized. Even after the power has been disconnected, these vaults should be considered to be confined spaces containing potentially toxic gases, explosive atmospheres, or oxygen-deficient atmospheres. Special precautions are required to enter this type of confined space.

Large commercial and residential structures often have high-voltage electrical service connections and interior rooms containing transformers and distribution equipment. These areas should be clearly marked with electrical hazard signs, and fire fighters should not enter them unless there is a rescue to be made. Until the power has been disconnected, fire suppression efforts should be limited to protecting exposures. Fire fighters must wear full PPE and SCBA owing to the inhalation hazards presented by the burning of the plastics and cooling liquids that are often used with equipment of this size.

Wrap-Up

Chief Concepts

- Fire suppression refers to all of the tactics and tasks performed on the fire scene to extinguish a fire.
- All fire suppression operations are either offensive or defensive:
    - Offensive—Tasks performed while moving toward the fire or inside the fire structure. Used when the fire is not too large or dangerous to be extinguished using interior handlines.
    - Defensive—Tasks performed outside. Used when the fire is too large to be controlled by an offensive attack or when the level of risk is too high.
- The IC evaluates conditions constantly to determine the type of attack that should be used. An interior attack may be switched to a defensive attack if necessary, and vice versa.
- A fire fighter must be able to advance and operate a hose line effectively to extinguish a fire.
- Care should be taken not to have opposing hose lines. Opposing hose lines may harm fire fighters with water streams or debris.
- The most frequently used hose line size for interior fire attack is 1¾-inch (45 mm) handlines. Although one fire fighter can usually operate the nozzle on a small hose line, a second fire fighter will provide valuable assistance when a small hose line must be advanced and maneuvered.
- Different types of fire streams are produced by different types of nozzles. A fog stream divides water into droplets, which can absorb heat efficiently. Fog streams can be used to protect fire fighters from the intense heat of the fire and to assist in ventilation. A straight stream can hit a fire from a longer distance away; it also keeps the water concentrated in a small area. A solid stream has an even greater reach and penetrating power than a straight stream and is charged as a continuous column of water.

- An interior fire attack is an offensive operation that requires fire fighters to enter a building and discharge an extinguishing agent onto the fire. Interior fire attack can be conducted on many different scales. In many cases, an interior attack is conducted on a fire that is burning in only one room and may be controlled quickly by one attack hose line. Larger fires require more water, which could be provided by two or more small handlines working together or by one or more larger handlines.
- The most effective means of fire suppression in most situations is a direct attack. This kind of attack uses a straight or solid hose stream to deliver water directly onto the base of the fire. The water cools the fuel until it is below its ignition temperature.
- The objective of an indirect attack is to quickly remove as much heat as possible from the fire atmosphere. Such an approach is particularly effective at preventing flashover from occurring. This method of fire suppression should be used when a fire has produced a layer of hot gases at the ceiling level.
- A combination attack employs both indirect attack and direct attack methods in a sequential manner. This strategy should be used when a room's interior has been heated to the point that it is nearing a flashover condition. Fire fighters first use an indirect attack method to cool the fire gases, then make a direct attack on the main body of fire.
- Large handlines can be used either for offensive fire attacks or for defensive operations.
- Master stream devices are used to produce high-volume water streams for large fires. Several types of master stream devices exist, including portable monitors, deck guns, ladder pipes, and other elevated stream devices. Most master streams discharge between 350 (1591 lpm) and 1500 (6819 lpm) gallons of water per minute.
- Fires in ordinary and wood-frame construction can burn in combustible void spaces behind walls and under subfloors and ceilings.
- Fires in basements or below grade level may cause the floor on the ground level to collapse.
- During an interior fire attack, protect stairways and other vertical openings between floors when fighting a fire in a multiple-level structure. Hose lines must be placed to keep the fire from extending vertically and to ensure that exit paths remain available to fire fighters and victims.
- Many large buildings have floor plans that can cause fire fighters to become lost or disoriented while working inside, particularly in low-visibility or zero-visibility conditions.
- Buildings that are under construction, renovation, or demolition are all at increased risk for destruction by fire. These buildings often have large quantities of combustible materials exposed, while lacking the fire-resistant features of a finished building.
- A lumberyard fire will usually produce tremendous quantities of radiant heat and release burning embers that can cause the fire to spread quickly from stack to stack.
- Fires in stacked materials should be approached cautiously. All fire fighters must remain outside potential collapse zones. Mechanical equipment should be used to move material that has been partially burned or water soaked.
- Fires in trash containers may be fought with foam or a deck gun.
- Some unique hazards exist with fires in confined spaces, including oxygen deficiencies, toxic gases, and standing water.
- Vehicle fires are one of the most common types of fires handled by fire departments. These fires may result from a variety of causes, including discarded smoking materials, electrical short-circuits, friction caused by dragging brakes or defective wheel bearings, and ruptured fuel lines due to a collision.
- Many hazards related to vehicle fires are possible—traffic hazards, fuel, pressurized cylinders and containers that can explode, fire, and toxic smoke.
- Overhaul of vehicle fires is just as important as overhaul of structure fires. As soon as it is safe to approach the vehicle, chock the wheels to prevent the vehicle from moving. After all visible fire has been knocked down, allow a few minutes for the steam and smoke to dissipate before starting overhaul. As you overhaul the vehicle, be systematic and thorough.
- Always be alert for signs that a burning vehicle could be powered by an alternative fuel, such as compressed natural gas (CNG) or liquefied petroleum gas (LPG). Fully involved fires in vehicles powered by either type of fuel should be fought with an unmanned master stream to prevent injuries from exploding gas cylinders.
- Flammable-liquid fires can be encountered in almost any type of occupancy. Most fires involving a vehicle (e.g., airplane, train, ship, car, truck) will involve a combustible or flammable liquid. Special tactics must be used when attempting to extinguish a flammable-liquid fire, and special extinguishing agents such as foam or dry chemicals may be needed.
- Flammable-gas cylinders can be found in many places. Many types of flammable gases are stored in many different types and sizes of containers. A variety of flammable gases can be found in industrial occupancies.
- The greatest danger with propane and similar flammable-gas products is a BLEVE. If an LPG tank is exposed to heat from a fire, the temperature of the liquid inside the container will increase. The fire itself could be fueled by propane escaping from the tank or from an external source. As the temperature of the product increases, the vapor pressure will increase in tandem. The increasing pressure creates added stress on the container. If this pressure exceeds the strength of the cylinder, the cylinder can rupture catastrophically.
- The greatest danger with most fires involving electrical equipment is electrocution. Only Class C extinguishing agents should be used when energized equipment is involved in a

fire. All electrical equipment should be considered as potentially energized until the power company or a qualified electrical professional confirms that the power is off.

- Once the electrical service has been disconnected, most fires in electrical equipment can be controlled using the same tactics and procedures as are used for Class A fires.

Hot Terms

Combination attack A type of attack employing both direct attack and indirect attack methods.

Deck gun An apparatus-mounted master stream device that is intended to flow large amounts of water directly onto a fire or exposed building.

Direct attack (structural fire) Firefighting operations involving the application of extinguishing agents directly onto the burning fuel. (NFPA 1145)

Elevated master stream device A nozzle mounted on the end of an aerial device that is capable of delivering large amounts of water onto a fire or exposed building from an elevated position.

Indirect application of water The use of a solid object such as a wall or ceiling to break apart a stream of water, creating more surface area on the water droplets and thereby causing the water to absorb more heat.

Indirect attack (structural fire) Firefighting operations involving the application of extinguishing agents to reduce the build-up of heat released from a fire without applying the agent directly onto the burning fuel. (NFPA 1145)

Ladder pipe A monitor that is fed by a hose and that holds and directs a nozzle while attached to the rungs of a vehicle-mounted aerial ladder. (NFPA 1965)

Master stream device A large-capacity nozzle that can be supplied by two or more hose lines or fixed piping. Such devices include deck guns, portable ground monitors, and elevated streams, and commonly flow between 350 (1591 lpm) and 1500 (6819 lpm) gallons per minute.

Portable monitor A monitor that can be lifted from a vehicle-mounted bracket and moved to an operating position on the ground by not more than two people. (NFPA 1965)

Solid stream A stream made by using a smooth-bore nozzle to produce a penetrating stream of water.

Soffit The material covering the gap between the edge of the roof and the exterior wall of the house. The soffits in most modern houses contain a system of vents to maintain an air flow to the attic area. Some attics may have sealed soffits to create a conditioned air space for energy conservation.

Straight stream A stream made by using an adjustable nozzle to provide a straight stream of water.

Transitional attack An offensive fire attack initiated by an exterior, indirect handline operation into the fire compartment to initiate cooling while transitioning into interior direct fire attack in coordination with ventilation operations.

FIRE FIGHTER *in action*

You are just returning to the station from a medical emergency when your engine is dispatched to the report of a car on fire in a garage. The battalion chief arrives first and establishes command. The IC gives a size-up indicating that the structure is a single-story, wood-frame house with dark, turbulent smoke coming from the garage area. He then assigns your crew to fire attack.

1. A(n) _____ uses a solid object such as a wall or ceiling to break apart a stream of water, creating more surface area on the water droplets and thereby causing the water to absorb more heat.
- **A.** interior attack
- **B.** indirect attack
- **C.** direct attack
- **D.** combination attack

2. Which of the following is not a master stream device?
- **A.** Deck guns
- **B.** Portable ground monitors
- **C.** Elevated streams
- **D.** Solid stream

3. A _____ is made by using a smooth-bore nozzle to produce a penetrating stream of water.

- **A.** solid stream
- **B.** fog stream
- **C.** straight stream
- **D.** master stream

4. What is the expansion ratio of propane?

- **A.** 5:1
- **B.** 17:1
- **C.** 70:1
- **D.** 270:1

5. What should be done once the gas has been shut off to a building?

- **A.** Attach a shut-off tag
- **B.** Report to staging
- **C.** Report to rehabilitation
- **D.** Notify the gas company

6. Large handlines will flow at least ____ gpm/lpm.

- **A.** 150 (681)
- **B.** 200 (909)
- **C.** 250 (1136)
- **D.** 300 (1363)

7. Which challenge do fire fighters frequently face when extinguishing an engine compartment fire?

- **A.** Confining the fire
- **B.** Vehicle movement
- **C.** Gaining access
- **D.** Locating the seat of the fire

FIRE FIGHTER II *in action*

It is a beautiful Saturday morning, and you are out washing the engine on the front apron of your fire station. A car pulls in and the driver walks over and begins to chat with you. He says he is a fire fighter from a neighboring jurisdiction, so he decided to stop and visit when he saw you out washing the engine. It is evident that he has been a fire fighter for only a short period of time, but he is very enthusiastic about learning and seems to be a very likeable individual. You know that his department has very limited financial resources, so you engage him in an informative discussion about fire attack.

1. Which factors would you tell him affect the incident commander's decision on whether to make an interior attack rather than an exterior attack?
2. How would you describe to him the characteristics of the various types of nozzles?
3. Which other activities would you tell him need to be coordinated with fire attack?

Preincident Planning

Fire Fighter I

Knowledge Objectives

There are no knowledge objectives for Fire Fighter I candidates. NFPA 1001 contains no Fire Fighter I Job Performance Requirements for this chapter.

Skills Objectives

There are no skill objectives for Fire Fighter I candidates. NFPA 1001 contains no Fire Fighter I Job Performance Requirements for this chapter.

Fire Fighter II

FFII

Knowledge Objectives

After studying this chapter, you will be able to:

- Describe why and for which types of properties a preincident plan is created. (p 708–710)
- List the typical target hazards that may be found in a community. (p 710)
- Describe how a preincident survey is performed. (NFPA 6.5, p 710–711)
- List the information that is gathered during a preincident survey. (NFPA 6.5.3, p 711)
- Describe the information included in any sketches or drawings created during the preincident survey. (NFPA 6.5.3, 6.5.3A, 6.5.3B, p 711–713)
- Describe the symbols commonly used in preincident plans. (NFPA 6.5.3A, p 712)
- Describe how preincident planning for safe and rapid response is performed. (p 711–713)
- Describe the information that needs to be gathered to assist the incident commander in making a rapid and correct size-up during an emergency incident. (p 713–718)

CHAPTER FFII

23

- Explain how to identify built-in fire detection and suppression systems during a preincident survey. (NFPA 6.5.3B, p 717–718)
- Describe the tactical information that is collected during a preincident survey. (NFPA 6.5.3, p 718–724)
- Describe how the sources of water supply for fire suppression operations are identified. (NFPA 6.5.3A, p 718–720)
- Explain why the locations of utilities are noted on the preincident plan. (p 720–721)
- Describe how preincident planning for an efficient search and rescue is performed. (p 721)
- Describe how preincident planning for rapid forcible entry is performed. (p 722)
- Describe how preincident planning for safe ladder placement is performed. (p 722)
- Describe how preincident planning for effective ventilation is performed. (p 722)
- List the occupancy considerations to take into account when conducting a preincident survey. (NFPA 6.5.3, p 723–724)
- List the types of locations that require special considerations in preplanning. (NFPA 6.5.3, p 724–725)

Skills Objectives

After studying this chapter, you will be able to perform the following skills:

- Conduct a preincident survey. (NFPA 6.5.3B, p 718)

Additional NFPA Standards

- NFPA 1, *Fire code, 2009 Edition*
- NFPA 170, *Standard for Fire Safety and Emergency Symbols*
- NFPA 220, *Standard on Types of Building Construction*
- NFPA 704, *Standard System for the Identification of the Hazards of Materials for Emergency Response*
- NFPA 1620, *Recommended Practice for Pre-incident Planning*

You Are the Fire Fighter

Your chief advises that the Insurance Services Office (ISO) will be evaluating your department during the fall, so the chief wants to ensure that all of the commercial buildings in your community have an up-to-date preincident plan. He has laid out a plan in which the town is divided into smaller sections, and each section is assigned to a different fire fighter. You are given a section with a convenience store, a group of storage units, and an old shoe factory, which is now vacant.

1. Why do fire departments conduct preincident planning?
2. What should you do to prepare?
3. Which process will you use when you arrive at a building?

Introduction

Preincident planning gives you the tools and knowledge that you need to become a much more effective fire fighter. Without a preincident plan, you will go into an emergency situation "blind." You will not be familiar with the structure, the location of hydrants, or the potential hazards. By contrast, with a preincident plan, you will know where the hydrants and exits are and which hazards to anticipate. A preincident plan puts all of this information at your fingertips, either on paper, in a computer file, or online and available to you at the fire scene.

Preincident planning helps your fire department to make better command decisions because important information is assembled before the emergency occurs. At the emergency scene, the incident commander (IC) can use the information gathered prior to the incident to direct the emergency operations much more effectively **FIGURE 23-1**. Because a preincident plan identifies potentially hazardous situations before an emergency occurs, fire fighters can be made aware of hidden dangers and prepare for them.

FIGURE 23-1 The preincident information is supplied to the IC at the emergency scene.

Preincident Plan

Preincident planning is the process of obtaining information about a building or a property and storing the information in a system so that it can be retrieved quickly for future reference. It is performed under the direction of a fire officer. The completed preincident plan should be made available to all units that might respond to an incident at that location.

The preincident plan is intended to help the IC make informed decisions when an emergency incident occurs at the location **FIGURE 23-2**. Some of this information would be useful to any fire company or unit responding to an incident at that location. In addition, a preincident plan can be used in training activities to help fire fighters become familiar with the properties within their jurisdiction.

The objective of a preincident plan is to make valuable information available quickly during an emergency incident that otherwise would not be readily evident or easily determined. The amount and the nature of the information provided for specific properties will depend on the size and complexity of the property, the types of risks present, and the particular hazards or challenges likely to be encountered.

The use of modern information technology has greatly enhanced the ability of fire departments to capture, store, organize, update, and retrieve preincident planning information. Accurate and current information such as drawings, maps, satellite and aerial imagery, photographs, descriptive text, lists of hazardous materials, and Material Safety Data Sheets can be made available instantly to the fire fighters who need it during an emergency incident. Geographic information systems (GIS) collect data on a geographic environment, such as population characteristics, types of occupancies, water, other utilities, and road maps; organize those data; and present the information digitally to the entire emergency response system, including the fire department.

A preincident plan makes information immediately available to fire fighters who need it during an emergency incident. It is particularly important to inform fire fighters of potential safety hazards. The most critical information, such as hydrant locations and life hazards, should be instantly available,

Tactical Priorities

Address:			
Occupancy Name:			
Preplan #:	Number Drawings:	Revised Date:	
District:	Subzone:	By:	
Rescue Considerations: Yes ( ) No ( )			
Occupancy Load Day:	Occupancy Load Night:		
Building Size:	Best Access:		
Knox Box:	Knox Switch:	Opticom:	
Roof Type:	Attic Space: Yes ( ) No ( )	Attic Height:	
Ventilation Horizontal:	Ventilation Vertical:		
Sprinklers: Yes ( )	No ( )	Full ( )	Partial ( )
Standpipes: Yes ( )	No ( )	Wet ( )	Dry ( )
Gas: Yes ( )	No ( )	Lpg ( )	
Hazardous Materials: Yes ( ) No ( )			
Firefighter Safety Considerations:			
Property Conservation And Special Considerations:			

FIGURE 23-2 An example of a preincident plan.

with additional data being accessible as needed. All of the information must be presented in an uniform and understandable format.

A preincident plan usually includes one or more diagrams that show details such as the building location and layout, access routes, entry points, exposures, and hydrant locations or alternative water supplies. In addition, the location and nature of any special hazards to the public or to fire fighters should be highlighted on the diagrams. Information about the actual building should include its height and overall dimensions, its type of construction, the nature of the occupancy, and the types of contents in different areas of the building. Additional information should include interior floor plans, stairway and elevator locations, utility shut-off locations, and information about built-in fire detection and suppression systems. For more information, see NFPA 1620, *Recommended Practice for Preincident Planning*.

Target Hazards

Most fire departments are not able to create a preincident plan for every individual property in their jurisdiction. Instead, they identify properties that are particularly large or that present unusual risks. These properties are identified as target hazards TABLE 23-1. Target hazard properties pose an increased risk to fire fighters.

A preincident plan should be prepared for every property that poses a high life-safety hazard to its occupants or presents safety risks for responding fire fighters. Preincident plans should also be prepared for properties that have the potential to create a large fire or conflagration (a large fire involving multiple structures).

Properties that have an increased life-safety hazard include the following structures:

- Hospitals
- Nursing facilities
- Assisted-living facilities
- Large apartment buildings
- Hotels and rooming houses
- Schools
- Public-assembly occupancies

Developing a Preincident Plan

The information that goes into a preincident plan is gathered during a preincident survey. This survey is usually completed by one of the crews that would respond to an emergency incident at the location. Its performance enables the crew to visit the property and become familiar with the location as they collect the preincident information.

Survey information can be collected and updated automatically in real time, through mobile devices. It can be uploaded to a central system or submitted manually in a hard-copy preincident plan format. Drawings and other graphics are usually prepared at the fire station, where the fire officer takes the time to organize the information properly and to generate reports and forms that can be used effectively during an emergency.

If the preincident plan is stored on paper, copies should be made and distributed to all of the companies and command officers who might respond to the location on an initial alarm. The plans should be kept in binders or in a filing system on each vehicle FIGURE 23-3. In some jurisdictions, the communications center can send a copy of the preincident plan to a portable fax machine in a command unit at the scene of an incident.

Fire departments using an integrated information technology system to organize data and create preincident plans can manage access to these plans through a network. This network can be accessed in multiple ways—directly, through computer terminals at fire stations, and through mobile computer devices FIGURE 23-4. When this approach is used, the preincident plan for a specific property can be automatically transmitted to the

TABLE 23-1 Typical Target Hazard Properties

- Bulk oil facilities and refineries
- High-rise buildings
- Hospitals
- Hotels and rooming houses
- Large apartment buildings
- Lumberyards
- Manufacturing plants
- Nursing homes and assisted-living facilities
- Public-assembly occupancies
- Schools
- Shopping centers
- Storage structures for hazardous materials
- Warehouses

Fire Fighter Safety Tips

Preincident planning is not the same thing as fire prevention. With fire prevention, the goal is to identify hazards and minimize or correct them so that fires do not occur or have limited consequences. With preincident planning, the person creating the plan assumes that a fire will occur and compiles information that responding fire fighters would need.

A preincident survey can often identify the need for a preincident plan or a fire prevention inspection. For this reason, many departments conduct preincident surveys simultaneously with fire safety inspections. If this method is used, different members of the department should be assigned to conduct the inspection and to complete the preincident survey. This separation of duties allows each fire fighter to concentrate on the objectives of one specific assignment.

FIGURE 23-3 Copies of preincident plans can be kept in three-ring binders.

responding fire stations and to responding vehicles. Updating preincident plans is also easier when they are stored on a computer network. As soon as the database is updated, everyone has access to the new information, which eliminates the need to make and distribute new copies to everyone. No matter which system your department uses, every effort should be made to keep preincident plans current, consistent, and readily available to personnel in the communications center, in responding vehicles, and to command staff.

FIGURE 23-4 Computers allow access to preincident plans from any location.

Conducting a Preincident Survey

A preincident survey should be conducted with the knowledge and cooperation of the property owner or occupant. The property owner or a representative should be contacted before the preincident survey is conducted. Making this initial contact enables the fire department to schedule an acceptable time, to explain the purpose and the importance of the preincident survey, and to clarify that the information is needed to prepare fire fighters in the event that an emergency occurs at the location.

The team members who conduct the survey should dress and conduct themselves in a manner appropriate to the department's mission. A representative of the property should accompany the survey team to answer questions and provide access to different areas. Every effort should be made to obtain accurate, useful information.

The preincident survey is conducted in a systematic fashion, following a uniform format. Begin with the outside of the building, gathering all of the necessary information about the building's geographic location, external features, and access points. Then survey the inside of the building to collect information about every interior area. A good, systematic approach starts at the roof and works down through the building, covering every level of the structure, including the basement. If the property is large and complicated, it may be necessary to make more than one visit to ensure that all the required information is obtained and recorded accurately.

The same set of basic information must be collected for each property that is surveyed TABLE 23-2. Additional information should be gathered for properties that are unusually large, are complicated, or might be the site of a particularly hazardous situation. Most fire departments use standard forms to record the survey information. After the team returns to the fire station, they can use the information to develop the preincident plan.

The fire fighters conducting the survey should prepare sketches or drawings to show the building layout and the location of important features such as exits. It takes practice and experience to learn how to sketch the required information while doing the survey, and then to convert the information to a final drawing. In some cases, the building owner can provide the survey team with a copy of a plot plan or a floor plan. Some departments use computer-assisted software to create and store these diagrams. These systems have the advantage of creating graphics more quickly and accurately than is possible with hand drawings. Many fire departments also use digital cameras to record information during the survey.

The completed drawing should use standard, easily understood map symbols FIGURE 23-5 and FIGURE 23-6.

Preincident Planning for Response and Access

Building layout and access information is particularly important during the response phase of an emergency incident. The preincident plan should provide information that would be valuable to units en route to an incident. For instance, it might identify the most efficient route for the apparatus to take to the fire building and note an alternate route if the time of the day and local traffic patterns might affect the primary route. An alternate route

TABLE 23-2 Information Gathered During the Preincident Survey

- Building location
- Apparatus access to the exterior of the building
- Access points to the interior of the building
- Hydrant locations and alternative water supplies
- Size of the building (height, number of stories, length, width)
- Exposures to the fire building and separation distances
- Type of building construction
- Building use
- Type of occupancy (assembly, institutional, residential, commercial, industrial)
- Floor plan
- Life hazards
- Building exit plan and exit locations
- Stairway locations (note whether the stairways are enclosed or unenclosed)
- Elevator locations and emergency controls
- Built-in fire protection systems (sprinklers, sprinkler control valves, standpipes, standpipe connections)
- Fire detection and alarm systems and location of the fire alarm annunciator panel (part of the fire alarm system that indicates the location of an alarm within the building)
- Utility shut-off locations
- Ventilation locations
- Presence of hazardous materials
- Presence of unusual contents or hazards
- Type of incident expected
- Sources of potential damage
- Special resources required
- General firefighting concerns

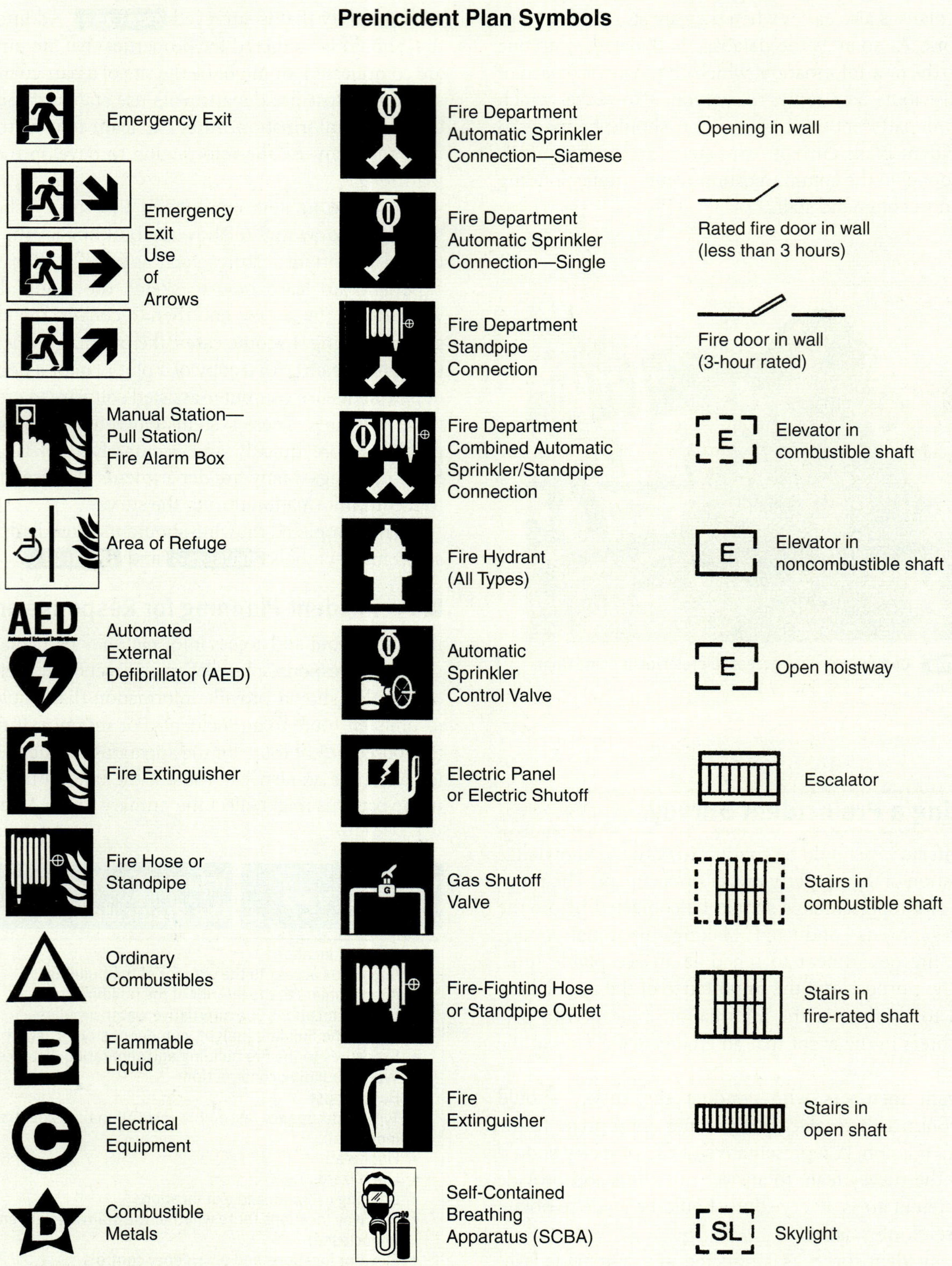

FIGURE 23-5 Preincident plan symbols.

should also be identified if the primary route requires crossing railroad tracks, drawbridges, or other potentially blocked routes.

When conducting a preincident survey, fire fighters should ensure that the building address is clearly visible to save time in locating the property during an emergency incident. If the building is part of a complex, the best route to each individual building, section, or apartment should be indicated on a map. The locations of hydrants and fire department connections (FDCs) for sprinkler and standpipe systems should always be identified.

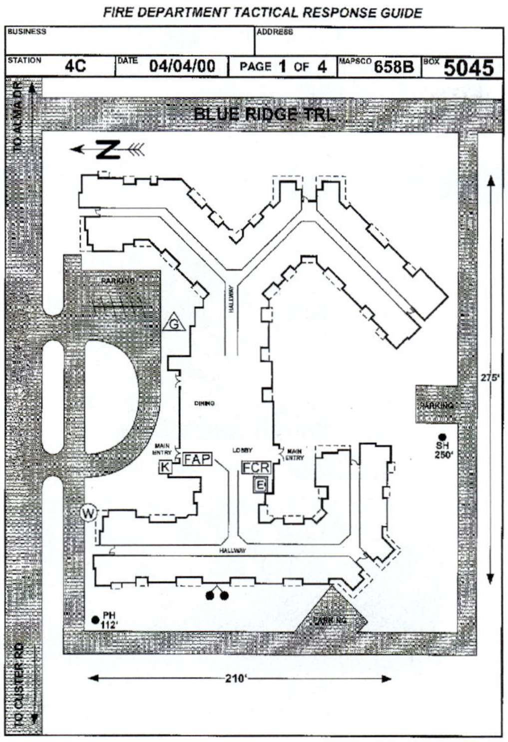

FIGURE 23-6 The completed drawing should use standard, easily understood map symbols.

Points of access into the building must also be noted, particularly if the building has multiple entrances or if access points are not easily seen from the street. In some cases, different entrances should be used, depending on the location of the emergency within the building. In other cases, the first-due unit should always respond to a designated entrance where the fire alarm annunciator panel is located or to meet a security guard who can act as a guide.

Diagrams should clearly indicate points at which gates, fences, or other barriers block access to parts of a building. The preincident survey should also note whether the topography makes one or more sides of a building inaccessible to fire apparatus or if an underground parking area will not support the weight of fire apparatus.

Access to the Exterior of the Building

The preincident plan should address all possible issues related to access to the exterior of the building. For example, you should ask the following questions during your survey:

- Do several roads lead to the building, or just a few?
- Where are the hydrants located?
- Where are the FDCs for automatic sprinkler and standpipe systems located?
- Do any security barriers limit access to the site?
- Are there fire lanes to provide access to specific areas?
- Are any barricades, gates, or other obstructions so narrow or low that they would prevent passage of fire apparatus?
- Are there bridges or underground structures that will not support the weight of apparatus?
- Do any gates require keys or a code to gain entry?
- Will it be necessary to cut fences to gain access to the site?
- Does the site include natural barriers such as streams, lakes, or rivers that might limit access?
- Does the topography limit access to any parts of the building?
- Might the landscaping or the presence of snow prevent access to certain parts of the building?

Access to the Interior of the Building

The preincident survey should also consider access to the interior of the building. The following questions are helpful in this regard:

- Is there a lockbox containing keys to the building? Do the keys work?
- Where is the lockbox located? Is it clearly visible?
- Are key codes needed to gain access to the building? Who has them? Are they in the lockbox?
- Does the building have security guards? Is a guard always on duty? Does the guard have access to all areas of the building?
- Is a key holder available to respond to the alarm within a reasonable amount of time? How can this person be reached?
- Where is the fire alarm annunciator panel located?
- Is the fire alarm annunciator panel properly programmed so you can quickly determine the exact location of an alarm?

Preincident Planning for Scene Size-Up

The preincident survey must also obtain essential information about the building that is important for size-up—that is, the ongoing observation and evaluation of factors that influence the objectives, strategies, and tactics for fire suppression. This information should include the building's construction, height, area, use, and occupancy, as well as the presence of hazardous materials or other risk factors. The locations of other buildings or structures that might be jeopardized by a fire in the building should be recorded as well.

Fire protection system information is also important for size-up. The preincident survey should identify areas that are protected by automatic sprinklers or other types of fire extinguishing systems, the locations of standpipes, and the locations of firewalls and other features designed to limit the spread of a fire. Likewise, the survey should note any areas where fire protection is lacking, such as an area without sprinklers in a building that otherwise has a sprinkler system.

Features inside a building that would allow a fire to spread but are not readily visible from the outside should be noted as well. A common attic space over several occupancies, unprotected openings between floors, or buildings connected by overhead passages or conveyor systems are examples of these features.

Construction

Building construction is an important factor to identify in the preincident survey. Although they might have similar appearances, two buildings might nevertheless be constructed differently and, therefore, behave differently during a fire. Five types of construction are defined in NFPA 220, *Standard on Types of Building Construction* TABLE 23-3. (Building construction is discussed in detail in this book in the Building

TABLE 23-3 Types of Building Construction

Type I: fire resistive	Buildings where the structural members are made of noncombustible materials that have a specified fire resistance FIGURE 23-7. Materials include concrete, protected steel beams, and masonry block walls, among others.	 FIGURE 23-7 A Type I building.
Type II: noncombustible	Buildings where the structural members are made of noncombustible materials but may not have fire resistance protection FIGURE 23-8. Includes unprotected steel beams.	 FIGURE 23-8 A Type II building.
Type III: ordinary	Buildings where the exterior walls are made of noncombustible or limited-combustible materials, but the interior floors and walls are made of combustible materials FIGURE 23-9.	 FIGURE 23-9 A Type III building.
Type IV: heavy timber	Buildings where the exterior walls are made of noncombustible or limited-combustible materials, but the interior walls and floors are made of combustible materials FIGURE 23-10. The dimensions of the interior materials are greater than those of ordinary construction (typically with minimum dimensions of 8 inches by 8 inches) (20 centimeter by 20 centimeter).	 FIGURE 23-10 A Type IV building.
Type V: wood frame	Buildings where the exterior wall, interior walls, floors, and roof are made of combustible wood materials FIGURE 23-11.	 FIGURE 23-11 A Type V building.

Construction chapter.) Preincident planning enables you to take the time that is needed to determine the exact type of building construction and identify any problem areas. Then, if an emergency occurs, the IC can check the preincident plan for information on construction, instead of hazarding a guess.

Lightweight Construction

It is important to identify whether the building contains lightweight construction. Lightweight construction uses assemblies of small components, such as trusses or fabricated beams, as structural support materials. This type of construction can be found in newer buildings as well as in older buildings with extensive remodeling. In many cases, the lightweight components are located in void spaces or concealed above ceilings and are not readily visible.

Fire Fighter Safety Tips

"Lightweight construction" is a relative term. A structure may be "lightweight" in comparison to other methods of construction, but its components are still heavy enough to cause serious injury or death if they collapse on a fire fighter.

FIGURE 23-12 A lightweight wood truss roof assembly.

Fire Fighter Safety Tips

Because lightweight construction uses truss supports or I-beams, a floor or roof in this type of structure may appear sturdier than it actually is. Two fire fighters in Houston, Texas, died while battling an early-morning blaze in a fast-food restaurant. The fire burned through the lightweight wood truss roof supports, and the roof-mounted air conditioner dropped into the building, trapping and killing the fire fighters. The fire was later determined to be arson.

FIGURE 23-13 A lightweight wood truss with gusset plates.

A wood truss constructed from 2-inch by 4-inch (5 centimeter by 10 centimeter), or 2-inch by 6-inch (5 centimeter by 15 centimeter), pieces of wood, for example, can be used to span a wide area and support a floor or roof. A truss uses the principle of a triangle to build a structure that can support a great deal of weight with much less supporting material than conventional construction methods FIGURE 23-12 and FIGURE 23-13.

Remodeled Buildings

Buildings that have been remodeled or renovated may present unique hazards. Remodeling can remove some of the original, built-in fire protection and create new hazards. Examples of remodeling-derived hazards include multiple ceilings with void spaces between them, new construction of concrete topping over a wooden floor, and new openings between floors or through walls that originally were fire resistant.

If you can conduct the preincident survey during construction or remodeling, you will be able to see the construction from the inside out, before everything is covered over. It is also important to realize that all buildings under construction—including those that are being remodeled or demolished—are especially vulnerable to fire. Unfinished

VOICES OF EXPERIENCE

While I have had my share of commercial and residential fires, extrications, and EMS incidents, the most satisfying part of my job is the ability to get into buildings and review their construction types and features before they are completed. Early into the job, an old timer told me, "Get into those buildings while they are being built, it will be your only chance to see it before it is covered." The only time that you will have free access to this wealth of structural information is during the construction process. Take the time to get into the building, move between the studs, around the block walls, make notes of special features, construction materials, and void spaces that are currently visible but will be secret after the drywall goes up. Observe the roof supports, truss or rafter, wood gussets, stamped gussets, or none. Take pictures for future reference and go back often throughout the construction process.

The information that you gather from a walk-through of a building under construction may save your life, the life of a co-worker, and/or a civilian life one day. Your visualization of the structural components during the construction process will prove invaluable in the years to come. You have an opportunity to view what may never be seen again unless the building is renovated.

After observing the construction process, it is incumbent upon you to share this information with the new generation of fire and emergency service personnel. You must pass this lesson on to all who follow you on the job and when you do, stress the importance of building construction. In the words of the late Francis Brannigan, "The building is your enemy, know your enemy." I equate Brannigan's words with the words passed on to me: by getting into the new construction before it is covered, I am able to see my enemy before it dons its armor.

R. Paul Long
York County Department of Fire and Life Safety
York County, Virginia

Fire Fighter Safety Tips

Buildings that are being remodeled or demolished are at higher risk for major fires. You should consider doing a special preincident survey for buildings undergoing substantial remodeling during the construction phase. This visit will enable you to gather more thorough information for a preincident plan of the completed building.

construction is open, so it lacks many of the fire-resistant features and fire detection/suppression systems that will be part of the finished structure.

FIRE FIGHTER II Tips FFII

A new component in lightweight building construction is the truss joist I-beam (or TGI beam). This framing material is made up of three wooden components. Channels may be drilled through TGI beams, thus providing another avenue for fire to spread.

Building Use

The second major consideration during size-up—and hence during a preincident survey—is the building's use and occupancy. Buildings are used for many purposes, such as residences, offices, stores, restaurants, warehouses, factories, schools, and churches. For each use, different types of problems, concerns, and hazards may be present.

For example, knowing the building use can help fire fighters determine the number of occupants and predict their ability to escape if a fire occurs. It is also a key factor in determining the probable contents of the building. These factors can have a major effect on the problems and hazards encountered by fire fighters responding to an emergency incident at the location.

A building is usually classified by its major use, which identifies the basic characteristics of the building TABLE 23-4. Within the major use classification, various occupancy subclassifications may provide a more specific description of the possible uses of sections of the building and their associated characteristics.

Many large building complexes contain multiple occupancy subcategories under one roof. For example, a shopping mall usually includes both retail stores and restaurants. The mall might also be part of an even larger complex that includes offices, residential condominiums, a hotel, a movie theater, an underground parking garage, and a subway station. Each area may have very different characteristics and, therefore, different risk factors.

TABLE 23-4 Classifications of Buildings

Major Use Classification	Occupancy Subcategories
Public assembly	Theaters, auditoriums, and churches Arenas and stadiums Convention centers and meeting halls Bars and restaurants
Institutional	Hospitals and nursing homes Schools
Commercial	Retail stores Industrial factories Warehouses Parking garages Offices

FIRE FIGHTER II Tips FFII

A building can have multiple uses during its lifetime or can incorporate more than one use within the same structure.

Occupancy Changes

Building use may change over time. An outdated factory may be transformed into a residential building, or an unused school may be converted into an office building. A warehouse that once stored concrete blocks may now be filled with flammable foam-plastic insulation or hazardous materials such as swimming pool chemicals. Given this propensity for change, preincident plans should be checked and updated on a regular schedule. A building's current occupancy information must always be determined during a preincident survey.

Exposures

An exposure is any other building or item that might be in danger if an incident occurs in another building or area. For example, an exposure could be an attached building, separated from the other structure by a common wall, or it could be a building across an alley or street. Exposures can include other buildings, vehicles, outside storage, or anything else that could be damaged by or involved in a fire. For example, heat from a burning warehouse could ignite adjacent buildings and spread the fire.

A preincident survey should identify any potential exposures for the property that is being evaluated. It should take into account the size, construction, and fire load (the amount of combustible material and the rate of heat release) of the property being evaluated, the distance to the exposure, and the ease of ignition of the exposure by radiation, convection, or conduction of heat.

Built-In Fire Protection Systems

The preincident survey should identify built-in fire detection and suppression systems on the property. These systems may include automatic sprinklers, standpipes, fire alarms, and fire detection systems, as well as systems designed to control or extinguish particular types of fires. Some buildings have automatic smoke control or exhaust systems. Most high-rise buildings have systems that control elevator function during an emergency. Each of these systems is covered in detail in the Fire Detection, Protection, and Suppression Systems chapter.

Automatic Sprinkler Systems

A properly designed and maintained automatic sprinkler system can help control or extinguish a fire before the arrival of the fire services unit. When properly designed and maintained, such systems are extremely effective and can play a major role in reducing the loss of life and property at an incident. They also create a safer situation for fire fighters.

The preincident survey should determine whether the building has a sprinkler system and which parts of the building that system covers. Note the locations of the valves that control water flow to different sections of the system. Normally, these valves are found in the open position.

In addition to the control valves, sprinkler systems should have a fire department connection (FDC) outside the building. The FDC is used to supplement the water supply to the sprinklers by pumping water from a hydrant or other water supply through fire hoses into this connection. The location of the FDC for the sprinkler system must be marked on the preincident plan.

Standpipe Systems

Standpipe systems are installed to deliver water to fire hose outlets on each floor of a building. Their use eliminates the need to extend hose lines from an engine at the street level up to the fire level. Standpipes may also be used in low-rise buildings, such as convention centers. In such structures, fire fighters can bring attack hose lines inside the building and connect them to a standpipe outlet close to the fire, while the engine delivers water to a FDC outside the building.

The location of the FDC, as well as the locations of the outlets on each floor, should be clearly marked on the preincident survey. The locations of nearby hydrants that will be used to supply water to the pumper that feeds the FDC must also be recorded. In a large building, multiple FDCs may be available to deliver water to different parts of the building. The area or floor levels served by each connection should be carefully noted on the preincident plan and labeled at the connection.

Fire Alarm and Fire Detection Systems

The primary role of a fire alarm system is to alert the occupants of a building so that they can evacuate or take appropriate action when an incident occurs. Some fire alarm systems are connected directly to the fire department; others are monitored by a service that calls the fire department when the system is activated.

In some cases, the fire alarm system must be manually activated. In other cases, it is activated automatically. A smoke or heat detection system can trigger the alarm, for example, or it may respond to a device that indicates when water is being discharged from the sprinkler system.

The annunciator panel, which is usually located close to a building entrance, indicates the location and type of device that activated the alarm system. In most cases, fire fighters who respond to the alarm must check the annunciator panel to determine the actual source of the alarm within the building or complex. The preincident survey should identify which type of system is installed, where the annunciator panel is located, and whether the system is remotely monitored.

Special Fire Extinguishing Systems

Several types of fixed fire extinguishing systems can be installed to protect areas where automatic sprinklers are not suitable. Most commercial kitchens and computer rooms are required to have installed fire suppression systems, for example. Special extinguishing systems are also found in many industrial buildings. The type of system and the area that is protected should be identified in the preincident survey.

Areas where flammable liquids are stored or used are often protected by sophisticated foam or dry-chemical fire suppression systems. The locations of these systems should be noted on the preincident plan. In addition, details about the method of operation, location of equipment, and foam supplies should be recorded during the preincident survey. More information about built-in systems is presented in the Fire Detection, Protection, and Suppression Systems chapter.

Conducting a Preincident Survey

To conduct a preincident survey, follow the steps in SKILL DRILL 23-1 (Fire Fighter II, NFPA 6.5.3):

1. Schedule the survey in advance.
2. Make contact with a responsible person.
3. Present a neat and professional image.
4. Identify yourself by name, title, and department.
5. Ensure that a representative from the facility accompanies you during the survey.
6. Take notes and pictures as needed, and start outside.
7. Note the building location.
8. Identify the building construction.
9. Identify the building use and occupancy.
10. Note any life hazards.
11. Note the access points to the interior of the building.
12. Note the utility shut-off locations.
13. Assess the apparatus access locations to the building.
14. Note hydrant locations and alternative water supplies.
15. Note ventilation concerns.
16. Record information about built-in fire detection and suppression systems.
17. Sketch floor plans.
18. Note the elevator and stairway locations.
19. Review exit plans and exit locations.
20. Identify any special hazards and hazardous materials.
21. Note the building exposures.
22. Anticipate the type of incident expected.
23. Identify any special resources needed.
24. Complete and file the preincident survey form.

Tactical Information

Many different types of tactical information can be obtained during a preincident survey and documented in a preincident plan. Such information would be of particular value to an IC in directing operations during an emergency incident or to fire fighters who are assigned to perform specific functions. When conducting a survey, all of the factors that could be significant when conducting emergency operations at that location should be considered.

Considerations for Water Supply

During the preincident survey, the amount of water needed to fight a fire in the building should be determined. The water supply source should be identified as well. The required flow rate, measured in gallons per minute or liters per minute, can be calculated based on the building's size, construction, contents, and exposures.

In most urban areas, the water supply will come from municipal hydrants. It is important to locate the hydrants closest to the building. In addition, for large buildings, it will be necessary to locate enough hydrants to supply the volume of water required to control a fire. The ability of the municipal water system to provide the required flow must be determined as well. In some cases, it may be necessary to use hydrants that are supplied by different water mains of sufficient size to achieve the needed water flow FIGURE 23-14.

In areas without municipal water systems, water may have to be obtained from a static water supply, such as a lake or stream, or be delivered by fire department tanker (or tenders).

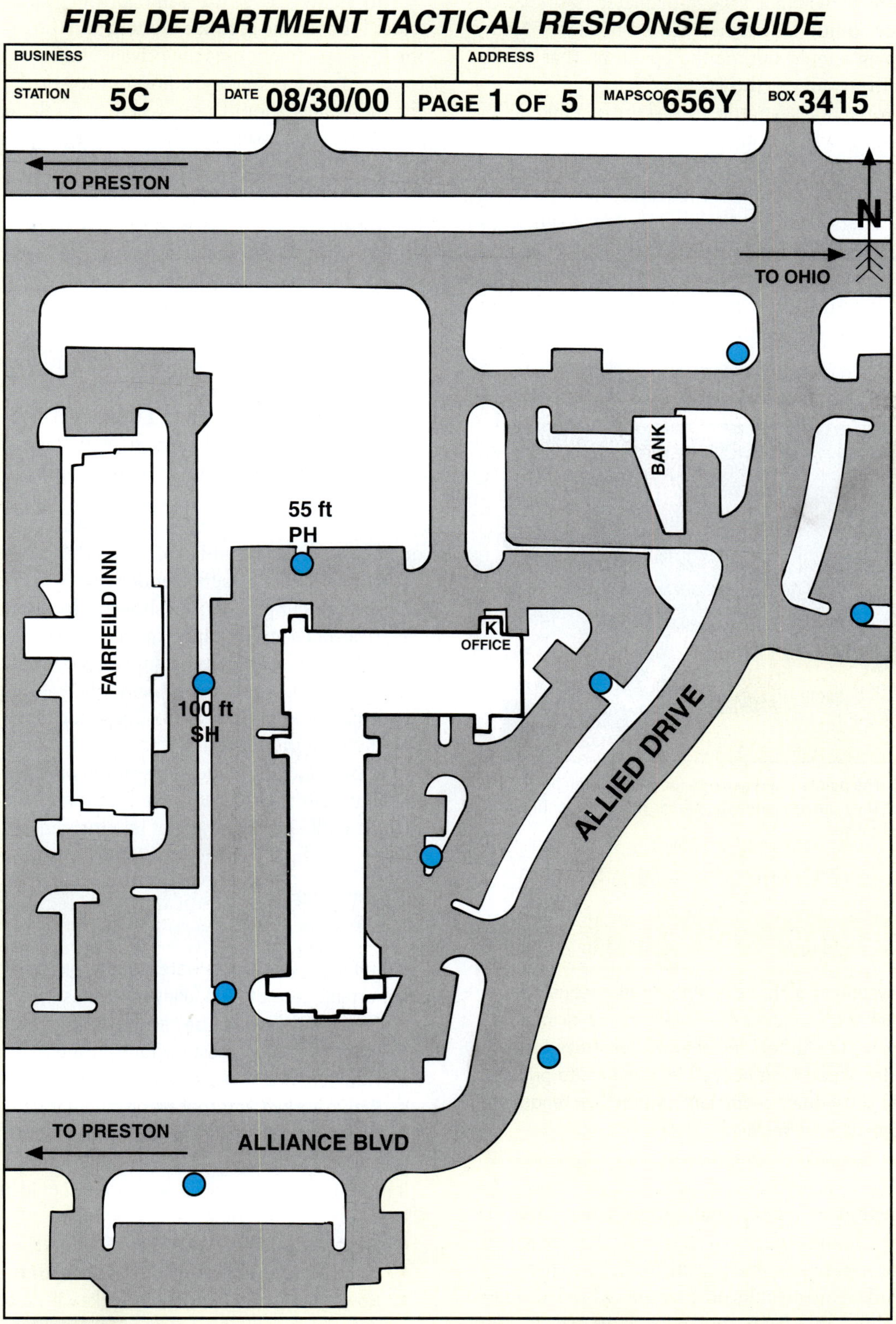

FIGURE 23-14 A diagram showing hydrant locations around a building. PH indicates the primary hydrant for connecting to the FDC. 55 indicates that the primary hydrant is located 55 feet (17 meter) from the FDC. SH indicates the secondary hydrant for connecting to the FDC. 100 indicates that the secondary hydrant is located 100 feet (30 meter) from the FDC.

When static water sources are used, the preincident plan must identify drafting sites (locations where an engine can draft water directly from the static source) or the locations of the nearest dry hydrants (an arrangement of pipes that is permanently connected to a static water supply). It is also important to measure the distance from the water source to the fire building to determine whether a large-diameter hose can be used and whether additional engines will be needed. Finally, the preincident plan should outline the operation that would be required to deliver water to the fire.

If a tanker (or tender) shuttle (a system of tankers [or tenders] transporting water from a water supply to the fire scene) will be used to deliver water, the preincident plan must include several additional details. For instance, sites for filling the tankers (or tenders) and for discharging their loads must be identified FIGURE 23-15. The preincident plan should also identify how many tankers (or tenders) will be needed based on the total distance they must travel, the quantity of water each vehicle can transport, and the total time it takes to empty and refill the vehicle.

FIGURE 23-15 The points where tankers (or tenders) can be filled and where they can discharge their loads must be identified in the preincident plan.

FIRE FIGHTER II Tips FFII

A private water system may not be able to simultaneously supply both sprinklers and fire hoses effectively. In one incident, a commercial property in Bluffton, Indiana, was destroyed when the responding fire department hooked its pumpers to private fire hydrants. The subsequent reduction in water flow rendered the building's sprinklers ineffective.

A large industrial or commercial complex may have its own water supply system that provides water for automatic sprinkler systems, standpipes, and private hydrants. These systems usually include storage tanks or reservoirs as well as fixed fire pumps to deliver the water under pressure. The details and arrangement of the private water supply system must be determined during the preincident survey.

The survey should indicate how much water is stored on the property and where the tanks are located. If a fixed fire pump is available, its location, capacity (in gallons or liters per minute), and power source (electricity or a diesel engine) should be noted. It is also important to confirm that these private water systems are being maintained in good operating condition. For example, the private water supply system on an abandoned property may be useless if it is no longer properly maintained.

In many cases, the same private water main provides water for both the sprinkler system and the private hydrants. If the fire department uses the private hydrants, the water supply to the sprinklers may be compromised. The preincident plan for these sites should note whether public, off-site hydrants should be used instead of the private hydrants.

Utilities

During an emergency incident, it may be necessary to turn off utilities such as electricity or natural gas as a safety measure. The preincident survey should note the locations of shut-offs for electricity, natural gas, propane gas, fuel oil, and other energy sources. These sites, as well as contact information for the appropriate utility company, should be included in the preincident plan. In some cases, special knowledge or equipment may be required to disconnect utilities. This information must be noted in the preincident plan, along with the procedure for contacting the appropriate individual or organization.

Electrical wires pose particular problems during firefighting operations. Electricity may be supplied by overhead wires or through underground cables. Overhead wires can be deadly if a ground ladder or aerial ladder comes in contact with them. The preincident plan should show the locations of high-voltage electrical lines and equipment that could prove dangerous to fire fighters. If electricity is supplied by underground cables, the shut-off may be located inside the basement of a building or in an underground vault. These locations should also be noted on the preincident plan.

If propane gas or fuel oil is stored on the property, the preincident plan should show the location and note the capacity of each tank. The presence of an emergency generator should

FIRE FIGHTER II Tips FFII

Identifying the available water supply for a particular building is important. Consider the following issues:

- How much water will be needed to fight a fire in the structure?
- Where are the nearest hydrants located?
- How much water is available and at what pressure?
- Where are hydrants that are on different water mains located?
- How many feet of hose would be needed to deliver the water to the fire?

If there are no hydrants, consideration should be given to these issues:

- What is the nearest water supply?
- Is the water readily available year-round? Is it seasonal?
- How many tanker (or tender) trucks will be needed to deliver the water?
- How long will it take to establish a tanker (or tender) shuttle?
- What is the total travel time from the scene to the fill site and then from the fill site back to the scene?

Near Miss REPORT

Report Number: 06-0000508

Synopsis: Fire fighter falls through skylight.

Event Description: My engine company (three-man crew) responded to an apartment fire. We had received multiple calls reporting the possibility of victims trapped.

We were the first unit on the scene and found heavy smoke and fire from the second story of an eight-unit apartment building. Upon exiting the apparatus, I was advised by sheriff's deputies that there were possibly two victims trapped. Another fire fighter and I pulled a 2-inch (5 centimeter) attack line, while the engine operator worked with the truck company to establish a supply line. The center of the building was fully involved and contained the entrances to all of the apartments.

My plan was to knock down the fire by the entrances to allow a rapid search for victims. I knew there was plenty of help on the way because I had called for a second alarm while responding to the scene. We knocked down the fire by the doors to the lower units and began moving up the stairs to the second floor. While advancing the attack line to the balcony/entrance area, I fell through the floor and landed on the enclosed concrete patio that was below me. This area was still fully involved with fire. The fire fighter who was with me attempted to call a "mayday" but could not get through owing to heavy radio traffic. I was able to exit the building on my own and sustained only a bruised shoulder and hip; my injuries could have been much worse.

I gathered my crew and exited out of the building. Command decided to switch to a defensive mode using master streams to knock down the majority of the fire. This incident eventually escalated to three alarms. There were no victims found.

After reviewing the incident and looking at some of the other buildings in the complex, it appeared that I actually fell through a 2-foot by 4-foot (0.6 by 1.2 meter) skylight/atrium located on the second floor. The railing had burned away and I could not see it owing to the smoke, fire, and steam. If we had not received reports of victims trapped, I would not have been as aggressive with our initial attack. In our jurisdiction (population of more than 45,000), we rarely find victims in the fire building. We have experienced only four fire fatalities in 30-plus years and have only a few "saves."

Lessons Learned: I fell through a hole that I did not know was there and could not see. This problem might have been prevented with better knowledge of preincident plans. I have been responding to this complex for 10 years but had never noticed the small atriums. My station has 11 large apartment complexes in our first-due area and more than 75 large apartment complexes in our first-alarm area. This makes it difficult to remember small details about each one of them.

FIRE FIGHTER II Tips — FFII

The locations of utility shut-offs are as varied as the layout of structures. Whereas some structures may have their utility shut-offs outside on the top of a utility pole, other structures may have their utility shut-offs inside in a basement closet. It is important to note these locations on the preincident plan to avoid lengthy delays in shutting down utilities during an emergency.

be noted as well, along with the fuel source and a list of equipment powered by the generator.

Preincident Planning for Search and Rescue

Fire fighters who are conducting search and rescue operations will need to know the locations of the occupants of a building as well as the locations of exits. The preincident survey should identify all entrances and exits to the building, including fire escapes and roof exits.

In addition to conducting a search for occupants who are unable to escape on their own, search and rescue teams may need to assist occupants who are trying to use the exits FIGURE 23-16. For example, fire fighters may need to assist occupants who are descending fire escapes or need to place ladders near windows or other locations for occupants to use.

During the preincident survey, team members should obtain the interior floor plan for the building. Information about the floor plan of a building can be life-saving knowledge when the structure is filled with smoke. It is also much easier to understand a floor plan when you can tour a building under nonemergency conditions. In large buildings, it may be necessary to plan for the use of ropes during search and rescue to prevent disorientation in conditions of limited visibility.

Preincident Planning for Forcible Entry

As previously noted, the preincident survey should consider both exterior and interior access problems. Locations where forcible entry may be required should be identified and marked

FIGURE 23-16 Fire fighters should plan to assist occupants who are trying to use the building exits.

FIGURE 23-17 Considerations for the use of ladders should include identifying the best locations to place ground ladders and use aerial apparatus.

on the site diagrams and building floor plan. Noting which tools would be needed to gain entry can save time during the actual emergency. The location of a lockbox and instructions on obtaining keys should be identified as well.

Preincident Planning for Ladder Placement

The preincident survey is an excellent time to identify the best locations for placing ground ladders or using aerial apparatus FIGURE 23-17. The length of ladder needed to reach a roof or entry point should be noted. When planning ladder placement, pay careful attention to any electrical wires and other obstructions that might not be visible at night or in a smoky atmosphere FIGURE 23-18. Plan to place ladders in locations that will not disturb other vital functions. For example, you should not place ladders in front of a door that will be used for entrance or exit from the fire building.

Preincident Planning for Ventilation

While performing a preincident survey, fire fighters should consider which information would be valuable to the members of a ventilation team during a fire. For example, what would be the best means to provide ventilation? How useful are the existing openings for ventilation? Are there windows and doors that would be suitable for horizontal ventilation? Where could fans be placed? Could the roof be opened to provide vertical ventilation? What is the best way to reach the roof? Is the roof construction safe? Are there ventilators or skylights that could be easily removed or bulkhead doors that could be opened easily? Will fire fighters need to use saws and axes to cut through the roof? Would multiple ceilings have to be punctured to allow smoke and heat to escape?

It is also important to determine whether the heating, ventilation, and air-conditioning (HVAC) system can be used to remove smoke without circulating it throughout the building. Many buildings with sealed windows have controls that enable the fire department to set the HVAC system to deliver outside air to some areas and exhaust smoke from other areas. The instructions for controlling the HVAC system should be included in the preincident plan.

FIGURE 23-18 The preincident survey should note overhead obstructions—in particular, electrical wires that might not be visible at night or in a smoky environment.

FIRE FIGHTER II Tips FFII

A church fire in Lake Worth, Texas, claimed the lives of three fire fighters. The fire spread into the roof, which was supported by lightweight wood trusses. Some fire fighters entered the building to initiate an interior attack, while other crews went to the roof to start ventilation operations. The fire burned through the trusses and the roof collapsed, trapping three fire fighters. An NFPA Fire Investigation Report is available on this fire.

Roof construction must also be evaluated to determine whether it would be safe for fire fighters to work on the roof when there is a fire below. If the roof is constructed with lightweight trusses, the risk of collapse is great. Many fire departments do not permit fire fighters on these roofs. The presence of an attic that might allow a fire to spread quickly under the roof should also be noted. In addition, common attics in townhouses and shopping malls can spread fire from one occupancy to another.

Occupancy Considerations

Each type of occupancy involves unique considerations that should be taken into account when preparing a preincident plan. Fire fighters should keep these factors in mind when conducting a preincident survey.

High-Rise Buildings

A high-rise building is generally defined as a structure that is more than 75 feet (23 meter) high, which is usually six or seven stories. High-rise buildings present special problems during an emergency because of the difficulty in gaining access and because of the large numbers of occupants. As a consequence, major fires in high-rise buildings often result in injuries, fatalities, and millions of dollars in property losses.

A preincident survey should identify both the building construction and any special features that have been installed. It should note all of the systems that are present in a particular building and identify how they are designed to function. This information, in turn, should be incorporated into the preincident plan.

The fire protection features of a high-rise building will vary depending on the age of the building and the specific building and fire code requirements in different jurisdictions. Older high-rise buildings are generally constructed of noncombustible materials and designed to confine a fire within a limited area. Newer buildings generally include automatic sprinklers, smoke detection and alarm systems, emergency generators, elevator control systems, smoke control systems, and building control stations where all these systems are monitored and controlled. The details of each system must be documented in the preincident plan. In many cases, expert consultants may be contacted to develop an emergency plan for such a building. These plans are intended to be used by the building tenants and management as well as the fire department.

Assembly Occupancies

Public-assembly venues—such as theaters, nightclubs, stadiums, churches, hotel meeting rooms, and arenas—present the possibility that large numbers of people could become involved in an emergency incident. These structures are often very large and complicated, and they may be equipped with complex systems intended to manage emergency situations. Gaining access to the location of the fire or emergency situation may prove difficult, however, when all of the occupants are trying to evacuate at the same time.

FIRE FIGHTER II Tips FFII

Modern high-rise buildings are often equipped with multiple fire protection systems—but these systems must be maintained properly to be effective. One Meridian Plaza (a modern high-rise office building in Philadelphia, Pennsylvania) had both a standpipe system and a partial sprinkler system on a few floors. When a fire broke out in this building in 1991, fire fighters attached their hose lines to the standpipe system. Because the pressure-reducing valves on the standpipe outlets had been set incorrectly, however, the fire fighters were unable to obtain sufficient pressure to control the fire. Three fire fighters died and the fire burned several floors of the building, which eventually had to be demolished. The fire was stopped only when it reached one of the floors that was equipped with a sprinkler system. An NFPA Fire Investigation Report is available on this fire.

Healthcare Facilities

Healthcare facilities—such as hospitals, nursing homes, assisted-living facilities, surgery centers, and ambulatory healthcare centers—require special preincident planning. Hospitals are often very large and include many different areas, ranging from operating rooms to walk-in clinics. The most challenging problem during an emergency incident at a healthcare facility is protecting nonambulatory patients.

A defend-in-place philosophy is used when designing fire protection for these facilities. This approach presumes that patients will not be able to escape from a fire without assistance and that there may not be enough staff present to move all of the patients. Therefore, the facility itself is designed to protect the patients from the fire. Most healthcare facilities utilize fire-resistant construction, have fire detection systems and sprinkler systems, and are compartmentalized.

If it becomes necessary to move patients from a dangerous area, the preferred approach is often horizontal evacuation (moving patients from a dangerous area to a safe area on the same floor). It is much easier to wheel a bed to another area on the same floor level than to carry patients down stairways.

Detention and Correctional Facilities

By their very nature, detention and correctional facilities are designed to ensure that the occupants cannot leave the building, even under normal conditions. The preincident plan must consider the problems involved in removing the inmates from a dangerous situation while protecting the fire fighters from angry or frightened inmates. Additionally, security concerns can make it difficult for fire fighters to gain rapid access to the building or for occupants and rescuers to exit the facility.

Residential Occupancies

Preincident plans are usually prepared for multifamily residential properties such as apartment complexes and condominiums. Most fire departments do not prepare individual preincident plans for one- and two-family residential buildings, unless there are unusual risk factors, such as a dwelling with handicapped occupants. In addition to the standard preincident plan information, the following information will be helpful when developing a preincident plan for a private residence:

- Detailed floor plan
- Location of sleeping areas
- Information about any handicapped occupants
- Escape routes and an outside rendezvous area
- Fire hydrant or water source location

Even if individual homes are not surveyed, many fire departments will survey different neighborhoods making special note of addresses, access routes, and hydrant locations. Knowing which styles and types of private residences are present in your jurisdiction will help you prepare for emergency incidents.

FIRE FIGHTER II Tips FFII

Several fires have occurred in home improvement stores. In one case, pool chemicals ignited, creating a special challenge for the fire department because of the hazardous materials involved. In another case, a fire started in a seat cushion display and spread rapidly through several aisles. The responding fire fighters were faced with a large fire, no visibility, and products dropping onto them from the upper shelves. Reports on these fires are available from NFPA Fire Investigations.

Some homeowners may request an individual fire safety survey of their property. These surveys are intended to identify fire hazards and to educate the homeowner about steps that can be taken to reduce the risk of fire.

Locations Requiring Special Considerations

Preincident planning should extend beyond planning for fires and other types of emergency situations that could occur in a building. Specifically, it should anticipate the types of incidents that could occur at locations such as airports, bridges, and tunnels, as well as incidents along highways or railroad lines, or at construction sites FIGURE 23-19.

Detailed preincident plans are prepared for airports, for example; these plans should note the various types of aircraft that use the airport. In addition, the airport facilities themselves often require extensive preincident planning because of their size, the number of people who may be present, and security issues. Similar planning should be done for bridges, tunnels, and any other locations where complicated situations could occur.

The following special locations would also require preincident planning:

- Gas or liquid fuel transmission pipelines
- Electrical transmission lines FIGURE 23-20
- Ships and waterways FIGURE 23-21
- Subways FIGURE 23-22
- Railroads FIGURE 23-23

Special Hazards

One of the most important reasons for developing a preincident plan is to identify any special hazards and to provide information that would be valuable during an emergency incident. This information includes chemicals or hazardous materials that are stored or used on the premises, structural conditions that could result in a building collapse, industrial processes that pose special hazards, high-voltage electrical equipment, and confined spaces. Preincident plans warn fire fighters of potentially dangerous situations and could include detailed instructions about what to do in those circumstances—information that could ultimately save the lives of fire fighters.

Fire fighters who are conducting a preincident survey should always look for special hazards and obtain as much information as possible at that time FIGURE 23-24. If the information is not readily available, it may be necessary to contact specialists for advice or conduct further research before completing the preincident plan.

FIGURE 23-19 Preincident planning should anticipate the types of incidents that are likely to occur at locations other than buildings, such as a tunnel.

FIGURE 23-20 Electrical transmission lines.

FIGURE 23-21 Ships and waterways.

Hazardous Materials

A preincident survey should obtain a complete description of all hazardous materials that are stored, used, or produced at the property. This assessment includes an inventory of the types and quantities of hazardous materials that are on the premises, as well as information about where they are located, how they are used, and how they are stored. The appropriate

FIGURE 23-22 Subway station.

FIGURE 23-23 Railroad.

FIGURE 23-24 During the preincident survey, look for special hazards and obtain as much information as possible.

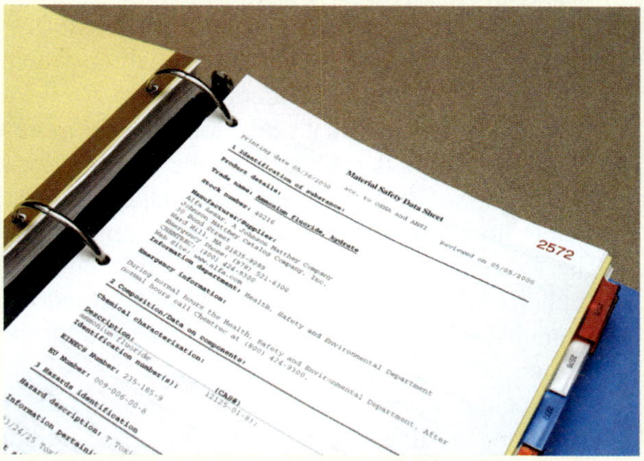

FIGURE 23-25 MSDS or SDS document.

actions and precautions for fire fighters in the event of a spill, leak, fire, or other emergency incident should also be listed.

Many jurisdictions require businesses that store or use hazardous materials to obtain a fire department permit. They may also require that hazardous materials specialists conduct special inspections. If the quantity of hazardous materials on hand exceeds a specified limit, federal and state regulations require a property owner to provide the local fire department with current inventories and Material Safety Data Sheet (MSDS) documents FIGURE 23-25. These documents are also called Safety Data Sheets (SDS). Fire fighters conducting preincident surveys should learn where the MSDS sheets are stored, ensure that the required information has been provided, and verify that it is up-to-date.

Fire fighters should expect to encounter hazardous materials at certain types of occupancies, such as chemical companies, garden centers, swimming pool supply stores, hardware stores, and laboratories. Of course, hazardous materials can also be found in unexpected locations, so fire fighters should always be alert for their presence.

Some communities put placards on the outside of any building that contains hazardous materials. These placards should use the marking system specified in NFPA 704, *Standard System for the Identification of the Hazards of Materials for Emergency Response*.

In case of fire, some hazardous materials require use of special suppression techniques or extinguishing agents. This information should be noted in the preincident plan. The plan should also contain contact information for individuals or organizations that can provide advice if an incident occurs. Hazardous materials are discussed in detail in the Hazardous Materials chapters.

FIRE FIGHTER II Tips FFII

Once the preincident plan is completed, train on it using a tabletop scenario. Invite neighboring fire departments to train with you and become familiar with the preincident plan.

FIRE FIGHTER II Tips FFII

All information in preincident plans, including MSDS, is confidential. The general public is not at liberty to see the information collected in your fire department's preincident plans. Keep all information that you learn about an occupancy during a preincident plan survey confidential.

Wrap-Up

Chief Concepts

- Preincident planning helps your department to make better command decisions because important information is assembled before the emergency occurs.
- Preincident planning is the process of obtaining information about a building or a property and storing the information in an accessible system. It is intended to help the IC make informed decisions when an emergency incident occurs at a specific location.
- The use of modern information technology has greatly enhanced the ability of fire departments to capture, store, organize, update, and retrieve preincident plans. Accurate and current information such as drawings, maps, satellite and aerial imagery, photographs, descriptive text, lists of hazardous materials, and MSDSs can be made available instantly to the fire fighters who need it during an emergency incident.
- A preincident plan usually includes one or more diagrams that show details such as the building location and layout, access routes, entry points, exposures, and hydrant locations and alternative water supplies.
- The location and nature of any special hazards to the public or to fire fighters should be highlighted on the diagrams. Information about the actual building should include its height and overall dimensions, its type of construction, the nature of the occupancy, and the types of contents in different areas of the building. Additional information should include interior floor plans, stairway and elevator locations, utility shut-off locations, and information about built-in fire detection and suppression systems.
- A preincident plan should be prepared for every property that poses a high life-safety hazard to its occupants or presents safety risks for responding fire fighters. Preincident plans should also be prepared for properties that have the potential to create a large fire or conflagration.
- The information that goes into a preincident plan is gathered during a preincident survey.
- A preincident survey should be conducted with the knowledge and cooperation of the property owner or occupant.
- The preincident survey is conducted in a systematic fashion, beginning with the outside of the building and moving inside. A good, systematic approach starts at the roof and works down through the building, covering every level of the structure, including the basement.
- The fire fighters conducting the preincident survey should prepare sketches or drawings to show the building layout and the location of important features such as exits.
- Building layout and access information is particularly important during the response phase of an emergency incident. The preincident plan should provide information that would be valuable to units en route to an incident. Points of access into the building must also be noted.
- The preincident survey must obtain essential information about the building that is important for size-up, including the building's construction, height, area, use, and occupancy; the presence of hazardous materials; locations of exposed structures; and fire protection system information.
- To conduct a preincident survey:
  - Schedule the survey in advance.
  - Make contact with a responsible person.
  - Present a neat and professional image.
  - Identify yourself by name, title, and department.
  - Ensure that a representative accompanies you during the survey.
  - Take notes and pictures as needed, and start outside.
  - Note the building location.
  - Identify the building construction.
  - Identify the building use and occupancy.
  - Note any life hazards.
  - Note the access points to the interior of the building.
  - Note the utility shut-off locations.
  - Assess the apparatus access to the building.
  - Note hydrant locations and alternative water supplies.
  - Note ventilation concerns.
  - Record information about built-in fire detection and suppression systems.
  - Sketch floor plans.
  - Note the elevator and stairway locations.
  - Review exit plans and exit locations.
  - Identify any special hazards and hazardous materials.
  - Note the building exposures.
  - Anticipate the type of incident expected.
  - Identify any special resources needed.
  - Complete and file the preincident survey form.
- During the preincident survey, collect tactical information about water supply considerations and locations of shut-offs for utilities.
- To facilitate a safe search and rescue operation at a site, the preincident survey should identify all entrances and exits to a building, including fire escapes and roof exits.
- To ensure a safe forcible entry operation, the preincident survey should identify any exterior and interior access issues and the locations where forcible entry may be required.
- To support safe ladder operations, the preincident survey should identify the best locations for placing ground ladders or aerial apparatus.

- To ensure safe ventilation operations, the preincident survey should identify the best locations for ventilation and determine whether the HVAC system can be used for ventilation.
- The following occupancies involve unique considerations:
  - High-rise buildings—Special issues include difficulty in gaining access and the large number of occupants to evacuate.
  - Public-assembly venues—These structures are often very large and contain large numbers of people to be evacuated.
  - Healthcare facilities—These structures are very large and contain nonambulatory occupants who need assistance in evacuating.
  - Detention and correctional facilities—Security concerns may make it difficult for fire fighters to gain rapid access to the building or for occupants to exit the facility.
  - Residential occupancies—Include apartment complexes and condominiums. The preincident plan should identify the locations of sleeping areas and the water supplies.
- Preincident planning should anticipate the types of incidents that could occur at locations such as airports, bridges, and tunnels, as well as incidents along highways or railroad lines, or at construction sites.

Hot Terms

Conflagration A large fire, often involving multiple structures.

Defend-in-place A strategy in which the victims are protected from the fire without relocating them.

Exposure Any person or property that could be endangered by fire, smoke, gases, runoff, or other hazardous conditions. (NFPA 402)

Fire alarm annunciator panel Part of the fire alarm system that indicates the source of an alarm within a building.

Fire load The weight of combustibles in a fire area (per ft^2 or m^2) or on a floor in buildings and structures, including either contents or building parts, or both. (NFPA 914)

Geographic information system (GIS) A system of computer software, hardware, data, and personnel to describe information tied to a spatial location. (NFPA 450)

Heating, ventilation, and air-conditioning (HVAC) system A system to manage the internal environment that is often found in large buildings.

Horizontal evacuation Moving occupants from a dangerous area to a safe area on the same floor level.

Lightweight construction Lightweight materials or advanced engineering, or both practices, which result in a weight saving without sacrifice of strength or efficiency. (NFPA 414)

Material Safety Data Sheet (MSDS) document A form, provided by manufacturers and compounders (blenders) of chemicals, containing information about chemical composition, physical and chemical properties, health and safety hazards, emergency response, and waste disposal of the material. (NFPA 472)

Nonambulatory A term describing individuals who cannot move themselves to an area of safety owing to their physical condition, medical treatment, or other factors.

Preincident plan A written document resulting from the gathering of general and detailed information to be used by public emergency response agencies and private industry for determining the response to reasonably anticipated emergency incidents at a specific facility.

Preincident survey The process used to gather information to develop a preincident plan.

Sprinkler system For fire protection purposes, an integrated system of underground and overhead piping designed in accordance with fire protection engineering standards. The installation includes at least one automatic water supply that supplies one or more systems. The portion of the sprinkler system above ground is a network of specially sized or hydraulically designed piping installed in a building, structure, or area, generally overhead, and to which sprinklers are attached in a systematic pattern. Each system has a control valve located in the system riser or its supply piping. Each sprinkler system includes a device for actuating an alarm when the system is in operation. The system is usually activated by heat from a fire and discharges water over the fire area. (NFPA 13, 2007)

Standpipe system An arrangement of piping, valves, hose connections, and allied equipment installed in a building or structure, with the hose connections located in such a manner that water can be discharged in streams or spray patterns through attached hose and nozzles, for the purpose of extinguishing a fire, thereby protecting a building or structure and its contents in addition to protecting the occupants. (NFPA 14, 2007)

Target hazard Any occupancy type or facility that presents a high potential for loss of life or serious impact to the community resulting from fire, explosion, or chemical release.

FIRE FIGHTER *in action*

It is a beautiful spring day, and your captain lays out the plan for the day: vehicle checks, trip to the store, ladder training, lunch, preincident planning, and then supper. Your captain goes on to say that he would like to knock out several buildings that are all in a row during preincident planning, so he will break the crew into two groups to speed up the process. You feel good when he looks at you and says that he wants you to act as the lead for the second group.

1. Which of the following would normally have a preincident plan because it is typically considered a target hazard?
  - **A.** Single-family residence
  - **B.** Lumberyard
  - **C.** Office building
  - **D.** Hardware store

2. Buildings where the exterior walls are made of noncombustible or limited-combustible materials, but the interior floors and walls are made of combustible materials, are termed ________.
  - **A.** fire resistive
  - **B.** noncombustible
  - **C.** ordinary
  - **D.** wood-frame

3. A ________ is a form, provided by manufacturers and compounders (blenders) of chemicals, containing information about chemical composition, physical and chemical properties, health and safety hazards, emergency response, and waste disposal of the material.
  - **A.** Material Safety Data Sheet (MSDS)
  - **B.** preincident survey
  - **C.** preincident plan
  - **D.** geographic information system (GIS)

4. In which type of occupancy would you most likely consider a shelter-in-place strategy?
  - **A.** Residential
  - **B.** Retail
  - **C.** Hospital
  - **D.** Factory

5. Over which height is a building normally considered to be a "high-rise"?
  - **A.** 32 feet (10 meter)
  - **B.** 50 feet (15 meter)
  - **C.** 75 feet (23 meter)
  - **D.** 100 feet (30 meter)

6. When preincident planning for ladder placement, which of the following is not normally a consideration?
  - **A.** Length of ladders required
  - **B.** Type of ladder construction
  - **C.** Locations of entrances and exits
  - **D.** Overhead obstructions

FIRE FIGHTER II *in action*

You are out conducting a preincident survey at a local strip shopping center. As you walk into one of the suites, the tenant becomes visibly upset about having the fire department in his business. He immediately begins complaining about the fire department trying to run businesses out of town through all of its rules. The tenant continues to raise his voice even more loudly as he goes on to complain about the fire department wasting fuel by driving the big apparatus for no good reason. He then goes on to state that he pays your "overpriced" salary and he is barely getting by.

1. How would you respond to this situation?
2. Which techniques can be used to defuse the situation?
3. Which comments would you choose to address?
4. How would you explain the difference between an inspection and a preincident plan?

Chapter 24: Fire and Emergency Medical Care and Chapter 25: Emergency Medical Care have been removed from this Edition.

Vehicle Rescue and Extrication

Fire Fighter I

Knowledge Objectives

After studying this chapter, you will be able to:

- Describe the types of motor vehicles. (NFPA 5.3.7.A, p 810–811)
- Describe the four different types of alternative fuels that power motor vehicles. (NFPA 5.3.7.A, p 811–814)
- Describe the extrication tools that are used for accessing locked compartments in a motor vehicle. (NFPA 5.3.7.A, p 821–823)
- Describe how to gain access to a victim of a motor vehicle collision. (NFPA 5.3.7.A, p 823)

Skills Objectives

There are no skill objectives for Fire Fighter I candidates. NFPA 1001 contains no Fire Fighter I Job Performance Requirements for this chapter.

Fire Fighter II

FFII

Knowledge Objectives

After studying this chapter, you will be able to:

- Describe a vehicle's anatomy. (NFPA 6.4.1A, p 811)
- List the hazards involved in responding to an emergency scene. (NFPA 6.4.1A, p 814–817)
- List the hazards to look for when arriving on the scene of a vehicle extrication situation. (NFPA 6.4.1A, p 816–817)
- Describe cribbing. (NFPA 6.4.1, p 817–819)
- Describe the extrication tools that are used for stabilizing, bending, cutting, and disassembling. (NFPA6.4.1,6.4.1A, p 821–827)
- Describe how to gain access to a victim of a motor vehicle collision. (NFPA6.4.1,6.4.1A, p 821–829)
- Describe how to disentangle a victim of a motor vehicle collision. (NFPA 6.4.1, p 827–834)
- Describe how to remove and transport victims of a motor vehicle collision. (NFPA 6.4.1, p 833–834)

CHAPTER 26

Skills Objectives

After studying this chapter, you will be able to perform the following skills:

- Disable the electrical system of a hybrid vehicle. (NFPA 6.4, 6.4.1B, p 813)
- Perform scene size-up at a motor vehicle crash. (NFPA 6.4, 6.4.1B, p 815–816)
- Mitigate the hazards at a motor vehicle crash. (NFPA 6.4, 6.4.1B, p 817–818)
- Stabilize a vehicle following a motor vehicle crash. (NFPA 6.4, 6.4.1B, p 819–820)
- Break tempered glass. (NFPA 6.4, 6.4.1B, p 824–825)
- Gain access to a vehicle following a motor vehicle crash. (NFPA 6.4, 6.4.1B, p 824, 826)
- Open a vehicle door. (NFPA 6.4, 6.4.1B, p 827–828)
- Gain access and provide medical care to a victim in a vehicle. (NFPA 6.4, 6.4.1B, p 827, 829)
- Displace the dashboard of a vehicle. (NFPA 6.4, 6.4.1B, p 831–832)
- Remove the roof of a vehicle. (NFPA 6.4, 6.4.1B, p 831, 833–834)

Additional NFPA Standards

- NFPA 1006, *Standard for Technical Rescuer Professional Qualifications*
- NFPA 1670, *Standard on Operations and Training for Technical Search and Rescue Incidents*

You Are the Fire Fighter

You have been assigned to the vehicle extrication company for the day. It begins in the usual manner with a detailed check of the vehicle extrication tools and equipment. As you look at all of the cribbing, the hydraulic spreaders, cutters, and rams, it dawns on you the gravity of your responsibility. You think of the past motor vehicle collisions that you have been to and how the vehicle extrication crew always seemed to know just the right tools and techniques to use to swiftly extricate the victims.

Then the station alerting system taps out the response for a person trapped. Bystanders are reporting that a semi-truck has crossed the median at mile marker 78 on Interstate 44, hitting a passenger van head-on. The driver and passenger of the van are trapped in the van.

1. What type of equipment will you most likely need in this situation?
2. What extrication techniques will be required?
3. What safety concerns do you have?

Introduction

Beginning fire fighters must understand the process of extrication and have some proficiency in extrication skills. Most fire departments respond to certain types of vehicle collisions. This chapter presents the knowledge and skills needed to assist rescuers in extricating victims from various types of motor vehicles involved in crashes.

While this chapter provides the knowledge and skills a beginning fire fighter needs, it does not cover all of the knowledge and skills needed by members of special extrication teams. Fire fighters who are members of special rescue teams or who will be expected to perform vehicle extrication should complete a course in rescue techniques that meets the requirements of NFPA 1006, *Standard for Technical Rescuer Professional Qualifications*, and NFPA 1670, *Standard on Operations and Training for Technical Search and Rescue Incidents*.

Types of Vehicles

Most of the vehicles on the road today are conventional vehicles, which use internal combustion engines for power. Internal combustion engines burn gasoline or diesel fuel to produce power. Fuel creates a hazard if it leaks after a crash. Other hazards associated with these vehicles include short circuits and battery acid leaks after a collision—not to mention the unknown of what people are carrying in their vehicles while they travel. The 12-volt electrical systems in these vehicles pose a minimal threat to rescuers, however.

Alternative-powered vehicles are powered by compressed natural gas (CNG), liquefied natural gas (LNG), or liquefied petroleum gas (LPG). These fuels are carried in cylinders that may be mounted in the trunk of smaller vehicles or on the roofs of buses. In particular, buses, delivery vans, and fleet sedans may be powered by CNG. CNG-powered vehicles should be identified by a "CNG" sticker mounted on the front and back of the vehicle **FIGURE 26-1**. After a crash, damage to the natural gas cylinders or fuel lines could result in the escape of flammable natural gas. In addition, a fire in a natural gas–powered vehicle poses the threat of a boiling liquid/expanding vapor explosion (BLEVE). Any sign of fire in a vehicle powered by CNG should be treated with extreme caution.

Electric vehicles are propelled by an electric motor that is powered by batteries. These vehicles contain a large number of batteries that provide the needed power. Hazards posed by these vehicles include the large amount of energy stored in the batteries, the potential for electrical shock or arcing and vehicle movement. Additional hazards include off-gassing and delayed-reaction electrical fire from damaged batteries. Be aware that these vehicles use much higher voltage systems in addition to the 12-volt system found in conventional vehicles.

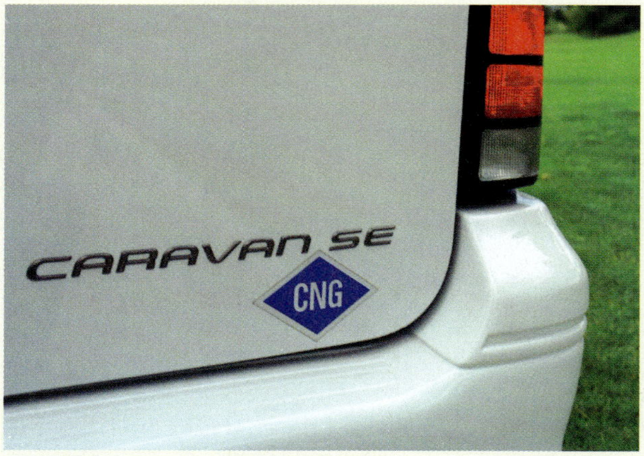

FIGURE 26-1 The CNG sticker identifies a vehicle that is powered by compressed natural gas.

Hybrid vehicles use both a battery-powered electric motor and a gasoline-powered engine. Hybrid vehicles contain all of the same hazards associated with conventional gasoline-powered vehicles, plus additional hazards linked to the electrical system. It is important for fire fighters to become familiar with the features of these vehicles, because their numbers are increasing. Vehicle manufacturers maintain Web sites that you can access to learn more about the operation and hazards of hybrid vehicles. More information on alternative-powered vehicles is found later in this chapter.

Fire Fighter Safety Tips

Some vehicles—for example, some mass-transit buses, forklifts, and school buses—run on liquefied petroleum gas, compressed natural gas, or liquefied natural gas. These vehicles, like gasoline- and diesel-powered vehicles, can produce flammable vapors. Given this potential threat, it is important to stretch a charged hose line at the scene of all motorized vehicle/equipment extrications.

FIRE FIGHTER II

Vehicle Anatomy

Parts of a Motor Vehicle

To reduce confusion and minimize the potential for mistakes at extrication scenes, it is important to use standardized terminology when referring to specific parts of vehicles. The front of a vehicle normally travels down the road first. The hood is located on the front of the vehicle. The trunk is usually located at the rear of the vehicle.

The left side of a vehicle is on your left as you sit in the vehicle. In the United States and Canada, the driver's seat is on the left side of the vehicle. The right side of a vehicle is where the passenger's seat is located. Always refer to left and right as they relate to the vehicle; do not refer to left and right from where you might be standing.

Vehicles contain "A," "B," and "C" posts, which are the vertical support members of a vehicle that hold up the roof and form the upright columns of the passenger compartment. The "A" posts are located closest to the front of the vehicle; they form the sides of the windshield. In four-door vehicles, the "B" posts are located between the front and rear doors of a vehicle. In some vehicles, these posts do not reach all of the way to the roof of the vehicle.

In four-door vehicles, the "C" posts are located behind the rear doors—specifically, behind the rear passenger windows. In two-door vehicles, the "C" post is the rear post.

The hood covers the engine compartment. The structure that divides the engine compartment from the passenger compartment is called the bulkhead or firewall. The passenger compartment includes the front and back seats. This part of a vehicle is also sometimes called the occupant cage or the occupant compartment (FIGURE 26-2). The trunk may be exposed to the back seats, as in a station wagon, or enclosed, as in a sedan.

FIRE FIGHTER II

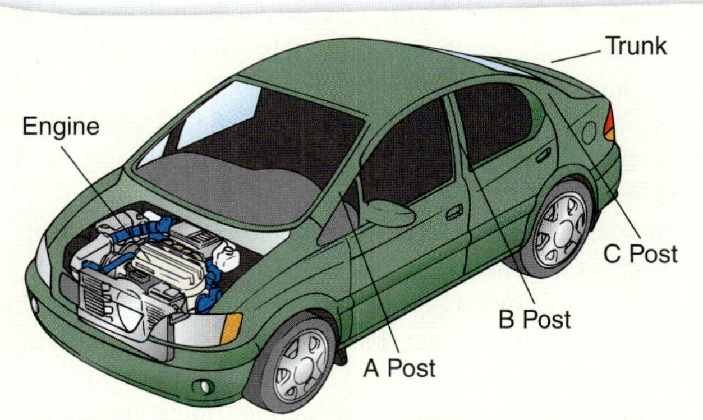

FIGURE 26-2 The anatomy of a vehicle.

Motor Vehicle Frames

Two types of vehicle frames are commonly found in today's motor vehicles: platform frame construction and unibody (unit body) construction.

Platform frame construction uses beams to fabricate the load-bearing frame of a vehicle. The engine, transmission, and body components are then attached to this basic platform frame. This type of frame construction is found primarily in trucks and larger SUVs, but is rarely present in smaller passenger cars. Platform frame construction provides a structurally sound place for stabilizing the vehicle and an anchor point for attaching cables or extrication tools.

Unibody construction, which is used for most modern passenger cars, combines the vehicle body and the frame into a single component. By folding multiple thicknesses of metal together, a column can be formed that is strong enough to serve as the frame for a lightweight vehicle. Unibody construction has the advantage of enabling auto manufacturers to produce lighter-weight vehicles. When extricating a person from a unibody vehicle, remember that these vehicles do not have the frame rails that are present with platform frame-constructed vehicles.

Alternative-Powered Vehicles

Several types of alternative-powered vehicles are currently either in production or in the planning stages. Alternative-powered vehicles may be powered by any of four different types of fuels: blended liquid fuels, compressed gases, electric and gasoline hybrid combination, and fuel cells. Following a crash or fire, these vehicles present hazards that are not encountered in incidents involving conventionally powered vehicles. It is important for rescuers to recognize the hazards these vehicles pose both to rescuers and to victims, and to be familiar with the additional steps needed to mitigate these hazards.

Blended Liquid Fuel–Powered Vehicles

The most common blended liquid fuel is a combination of gasoline and another flammable liquid. Two types of alcohol—methanol and ethanol—are commonly mixed with gasoline to produce a blended fuel. The mixture of 10 percent alcohol and 90 percent gasoline is sold in many cities in the United States as a measure to reduce pollution. Diesel fuel may also be blended with other combustibles to produce a blended diesel fuel.

Biofuels are derived from agriculturally produced products such as grains, grasses, or recycled animal oils. They are processed to produce a type of alcohol that is added to gasoline to produce a blended fuel. Biofuels can also be produced to power diesel vehicles.

Vehicles operating on these blended fuels are identified by special placards or designations. They are singled out for attention primarily because of their environmental benefits. Currently, biofuel-powered vehicles account for less than 3 percent of all vehicles in the United States.

Crashes involving vehicles that rely on blended fuels are handled in the same manner as crashes involving vehicles powered by conventional fuels. In the event of a fire involving blended fuel, it is necessary to use an alcohol-resistant foam as a fire-suppressing agent. Other types of foam will break down when exposed to alcohol.

Compressed Gas–Powered Vehicles

Three forms of compressed gas are used to power vehicles. Compressed natural gas (CNG) is commonly stored in steel, aluminum, or carbon fiber–wound tanks, under pressures ranging from 3000 to 3600 pounds per square inch (psi) (429 to 514 kilopascal [kPa]). Many cities power some of their municipal bus fleets with CNG. These vehicles can be identified by the "pregnant shape" of the roof top.

Liquefied natural gas (LNG) is stored at very low temperatures. Storage tanks for LNG are double walled and must be insulated to keep the liquefied gas cold. The use of LNG is not as common as the use of CNG.

Liquefied petroleum gas (LPG), also called propane, is the third type of compressed gas used to power vehicles. It is stored in a liquid state at a pressure of 50 psi (7 kPa). LPG-powered vehicles are often found in industrial settings—for example, LPG is used to power forklifts in factories.

When responding to crashes involving vehicles powered by compressed gases, rescuers need to able to identify and mitigate the special hazards posed by these vehicles. Look for signage or placards on the vehicle identifying it as being powered by a compressed gas. If a compressed gas is present, check for the presence of leaking cylinders or hoses. If you suspect that a leak is present, keep all sources of ignition away from the vehicle and take measures to stop the leak if possible. If LNG is leaking, avoid spraying a stream of water on the escaping liquid, as this action heats the very cold liquid; it may then convert to a gaseous state with explosive force. It is acceptable to spray a fog stream on an escaping cloud of vapor, however. Follow the procedures described in the Hazardous Materials chapters for mitigating hazardous materials. Ensure that a fire fighter in full protective equipment is standing by with a charged hose line that is least 1½ inches (38 millimeters) in diameter. In the event of a fire, keep the gas cylinders cool to avoid a BLEVE.

Hybrid and Electric Vehicles

A hybrid vehicle has two separate systems to power the drive train **FIGURE 26-3**. One part of the power system is a fuel-driven power train. This part of the drive train contains the same parts as a conventional vehicle. The second part of the power system is powered by an electrical motor. The electrical power system consists of an electrical motor, a high-voltage battery pack, a DC-DC converter (and possibly an AC-DC inverter), and electrical cables to connect these components. In some hybrid vehicles, the high-voltage batteries are charged by the onboard gasoline or diesel engine. In others, the batteries can also be charged from an external electrical power source. Usually hybrid vehicles are identified by placards such as the one shown in Figure 26-3.

The components of these vehicles continue to change, and the design and location of each component vary from one manufacturer to another. It is important to keep up with changes in vehicle architecture, as they affect the response to incidents involving the vehicles. Most manufacturers offer a wealth of information that can be viewed or downloaded from the manufacturer's Web site.

When responding to crashes involving hybrid vehicles, you will encounter the hazards normally associated with a gasoline- or diesel-powered vehicle—these hazards are still present in hybrid vehicles. The additional hazards posed by the electrical drive train are similar to those you would encounter around other high-voltage electrical hazards.

The high-voltage battery pack in a hybrid vehicle is usually located under or behind the rear seats or the trunk. High-voltage batteries are protected by a strong structural cage and located out of the way for most extrication evolutions. Hybrid vehicles have automatic shut-offs that are designed to disable the high-voltage electrical system in the event of a crash. According to the current manufacturers' recommendations, water remains the extinguishing agent of choice in case of a fire involving a hybrid vehicle.

FIGURE 26-3 A hybrid sticker indicates that this vehicle is powered by a gasoline engine and an electric motor, and it contains a high-voltage battery or batteries.

As part of the electrical power system, brightly colored high-voltage cables are routed from the high-voltage battery pack to the inverter and electric motor, which are located in the engine compartment. These cables are usually located beneath the passenger compartment toward the center of the vehicle, so they do not interfere with most extrication procedures. Because these cables carry high voltage, however, they pose a risk to rescuers if responders try to cut them without first disabling the electrical system. Look for placards and high voltage labels that will identify a potential shock hazard.

Although the procedures for disabling the electrical system of a hybrid vehicle vary from one manufacturer to another, generally they follow similar steps. To disable the electrical system of a hybrid vehicle, follow the steps in **SKILL DRILL 26-1**:

1. Identify the vehicle's propulsion system. Is it conventional, hybrid, or CNG?
2. Immobilize the vehicle by chocking the wheels. Set the parking brake. Place the vehicle in park. (**STEP 1**)
3. Disable both the high voltage and low voltage systems by shutting off the vehicle's ignition (power button or conventional key) and disconnecting or cutting the 12-volt battery cable (usually the negative). For vehicles equipped with a proximity key, remove the key at least 16 feet (5 meters) from the vehicle to prevent the possibility of unintentional restart. (**STEP 2**)

Alternate disabling methods involve removing fuses or the high voltage manual service disconnect. Be aware that some manufacturers require the use of electrical PPE for the removal of manual service disconnects while other manufacturers do not recommend the use of service disconnects at all for emergency responders. Check with the manufacturer's *Emergency Response Guide* or the NFPA's *Electric Vehicle Emergency Field Guide* for alternate disabling methods or before attempting to use the service disconnect.

Electric vehicles are powered by an electric motor and batteries, but do not contain a gasoline- or diesel-powered engine. Instead, these vehicles rely solely on batteries, which must be recharged by plugging them into an external power source. The procedures used to deal with electric vehicles involved in collisions are nearly the same as those used in dealing with a hybrid vehicle. Few such vehicles are currently in production and on the streets.

Review the manufacturers' fact sheets regularly to keep up-to-date with the steps needed to disable the electrical systems in hybrid vehicles. The NFPA also offers an *Electric Vehicle Emergency Field Guide* to assist fire fighters in responding safely to motor vehicle accidents involving electric vehicles.

Fuel Cell–Powered Vehicles

A fuel cell is a device that generates electricity through a chemical reaction between hydrogen gas and oxygen gas to

SKILL DRILL 26-1 Disable the Electrical System of a Hybrid Vehicle FFII

Fire Fighter II (NFPA 6.4.1)

1. Identify the vehicle's propulsion system. Is it conventional, hybrid, or CNG? Immobilize the vehicle by chocking the wheels. Set the parking brake. Place the vehicle in park.

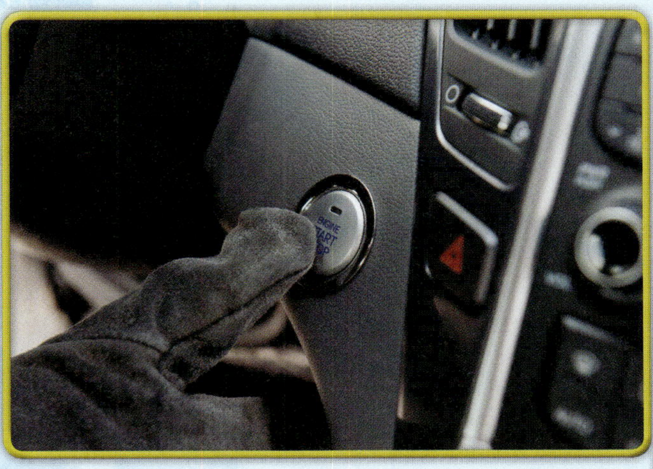

2. Disable both the high voltage and low voltage systems by shutting off the vehicle's ignition (power button or conventional key) and disconnecting or cutting the 12-volt battery cable (usually the negative). For vehicles equipped with a proximity key, remove the key at least 16 feet (5 meters) from the vehicle to prevent the possibility of unintentional restart.

produce water. The purpose of fuel cells is not to produce water, however, but rather to produce electricity that can then be used to drive an electric motor. Fuel cell–powered vehicles are essentially electric vehicles with a different device for generating the electricity. The major components needed for a fuel cell–powered vehicle are a hydrogen gas storage tank, a high-output battery, a group of fuel cells called a fuel cell stack, and an electric motor. The use of fuel cells is being advocated because they produce less pollution than internal combustion engines.

Several issues remain to be worked out before fuel cell–powered vehicles are likely to be widely produced. Because hydrogen gas contains one-third the amount of energy per volume that gasoline does, it is difficult to store enough hydrogen gas to propel the vehicle as far as gasoline vehicle can travel on a full tank. Several ways to store hydrogen gas are currently being explored.

Because fuel cell–powered vehicles are not currently in production, they do not present hazards to rescuers working in the field.

Fire Fighter Safety Tips

Hybrid vehicles and electric vehicles contain powerful batteries, and their electrical systems produce a higher voltage than would be encountered in a conventional-powered vehicle. Be alert for special electrical hazards when you respond to incidents involving these vehicles. Also be aware that batteries may be placed anywhere inside the vehicle—from under the hood to the trunk, to behind the backseat. See the section on alternative-powered vehicles.

Responding to the Scene

The first step in the vehicle extrication process is response. To perform extrication, you need to safely and efficiently arrive on the scene. Safe response includes picking the best route for the time of the day, driving in a safe manner, and keeping in mind the limitations of your vehicle and its warning devices. Always fasten your seat belt. Follow the same safety practices while responding to a motor vehicle crash that you would practice when responding to a structural fire call.

When responding to an emergency, evaluate the dispatch information that you have received, because it may help you to anticipate the necessary types of equipment and the procedures that you may need to perform. Use this time to mentally review the scenarios that you might encounter. Listen carefully for any directions that your officer may give during the response phase.

FIRE FIGHTER II

Arrival and Size-Up of the Scene

The next step in the extrication process is arrival and size-up. After arriving at the scene of a motor vehicle crash, it is important to assess the hazards present and to determine the scope of the incident and the need for any additional resources.

FIRE FIGHTER II

FIRE FIGHTER II Tips FFII

Before doing any type of work on a vehicle that has been involved in an accident, scan the entire area for hazards. Did the vehicle hit an electric box or rupture a gas line? Hazards are everywhere, and a quick scan might save everyone on the scene from further injury.

Traffic

The decision of where to position emergency vehicles should take into account the safety of emergency workers, the victims, and the motorists traveling along the road. Whenever possible, place emergency vehicles in a manner that will ensure safety and avoid disrupting traffic any more than necessary. However, do not hesitate to request that the road be closed if necessary. Sometimes the most important action you need to take at the scene of a vehicle crash is to slow, stop, or divert the flow of traffic before proceeding with additional actions. Remember—safety first!

Position large emergency vehicles so that they provide a barrier against motorists who fail to recognize or heed emergency warning lights. Many departments place apparatus at an angle to the crash and pointing away from oncoming traffic FIGURE 26-4. This position helps to push the apparatus to the side of a crash in the event that the emergency apparatus is struck from behind. Traffic cones or flares can be placed to direct motorists away from the crash FIGURE 26-5. Call for law enforcement to assist in traffic control whenever needed.

Fire fighters need to be readily visible at a crash scene. Personal protective equipment (PPE) should be bright to help ensure fire fighters' visibility during daylight hours; PPE that is used at night should be equipped with reflective material to increase visibility in the darkness. PPE must be worn at all motor vehicle crashes. Any time you are operating in a roadway or are close to traffic, you are required to wear a bright-colored reflective safety vest that meets the latest iteration of

FIGURE 26-4 Many fire departments place an apparatus at an angle to the crash.

FIGURE 26-5 Traffic cones or flares can be placed to direct motorists away from the crash.

ANSI Standard 207. This standard requires the vest to contain 450 square inches (11 meters) of high-visibility fabric and 201 square inches (5 meters) of retro-reflective material.

Before exiting fire apparatus at an emergency scene, be alert for any vehicles that might cause injury to fire fighters. Do not assume that motorists will always heed the warning lights. Let law enforcement personnel coordinate traffic control.

The incident commander (IC) will usually perform a size-up of the scene by conducting a 360-degree walk-around of the scene. During this size-up, the IC evaluates the hazards present and determines the number of crash victims. Using the information obtained from the scene size-up, the IC can create an action plan and call for additional resources, if needed.

Fire Hazards

Look for spilled fuel and other flammable substances. Motor vehicles carry a variety of fuels and lubricants that might pose fire hazards. Also look for the presence of fire, because a short in the electrical system or a damaged battery may cause a post-crash fire by releasing sparks and igniting spilled fuel. These fires may trap the occupants of the vehicle and require fire suppression.

Fire Fighter Safety Tips

The locations of vehicle batteries can vary widely. If you open the hood to cut the battery cables and do not find a battery, look in the not-so-obvious locations. Batteries may be behind the back seat, in the trunk, and inside rear wheel wells.

Electrical Hazards

Downed and low-hanging power lines represent an electrical hazard. Look closely to determine whether the crash has damaged any electrical power poles. Downed and low-hanging power lines may be either clearly visible or difficult to see, but they inevitably create a deadly hazard for rescuers, the victims of the crash, and bystanders.

As noted earlier, hybrid vehicles also contain high-voltage batteries and electrical cables that require special handling because they may pose an electrical hazard.

Fire Fighter Safety Tips

High-intensity discharge (HID) headlights are becoming more common on vehicles as standard equipment or as after-market modifications. HID headlights discharge several thousand volts of electricity. If the HID bulb is broken or removed, you may be exposed to an electrical hazard.

Other Hazards

Environmental conditions can lead to unique hazards at a crash scene. Crashes that occur in rain, sleet, or snow, for example, present an added hazard for rescuers and the victims of the crash.

Crashes that occur on hills are harder to handle than those that occur on level ground.

Assume all vehicles are carrying hazardous materials until proven otherwise. Assess the scene for hazardous material placards, unusual odors, or leaking liquids. More information about hazardous materials is found in the Hazardous Materials chapters.

Be especially alert for the presence of infectious bodily substances. Be prepared for the presence of blood and exercise universal precautions. Specifically, do not let blood or other bodily fluids come in contact with your skin, and wear gloves that will protect you from both contaminated fluids and sharp objects that are present at a crash site. If you or your clothes become contaminated, report the contamination, document it, and then clean and wash the affected clothes and equipment.

Threats of violence may be present at some crash scenes. In particular, intoxicated people or those who are upset with other motorists may pose a threat to you or to other people present at the scene. Be alert for weapons that are carried in vehicles.

Occasionally, animals become a hazard at crash scenes. Dogs and other family pets may be protective of their owners and threaten rescuers. Farm animals or horses that have been involved in a crash may need care. You may need to call in specialized resources to assist with this type of incident.

To perform the scene size-up at a motor vehicle crash, follow the steps in **SKILL DRILL 26-2**:

1. Position emergency vehicles to protect the crash scene and the rescuers. Take any additional actions needed to prevent further crashes. (**STEP 1**)
2. Perform a quick initial assessment from the first-arriving vehicle, establish command, and give a brief initial radio report. (**STEP 2**)
3. Perform a 360-degree walk-around looking for potential hazards. Look for overhead hazards and hazards under the vehicles, and determine the stabilization needed to prevent further movement of the vehicles involved in the incident. (**STEP 3**)

SKILL DRILL 26-2 Performing a Scene Size-up at a Motor Vehicle Crash FFII

Fire Fighter II (NFPA 6.4.1)

1 Position emergency vehicles to protect the crash scene and the rescuers. Take any additional actions needed to prevent further crashes.

2 Perform a quick initial assessment from the first-arriving vehicle, establish command, and give a brief initial radio report.

3 Perform a 360-degree walk-around looking for potential hazards. Look for overhead hazards and hazards under the vehicles, and determine the stabilization needed to prevent further movement of the vehicles involved in the incident.

4 Determine the number of patients, the severity of their injuries, and the amount of entrapment. Give an updated report and call for additional resources if needed.

5 Establish a secure working area and an equipment staging area.

6 Direct personnel to perform initial tasks.

4. Determine the number of patients, the severity of their injuries, and the amount of entrapment. Give an updated report and call for additional resources if needed. (**STEP 4**)
5. Establish a secure working area and an equipment staging area. (**STEP 5**)
6. Direct personnel to perform initial tasks. (**STEP 6**)

Stabilization of the Scene

The next step of the extrication process is scene stabilization. This step consists of reducing, removing, or mitigating the hazards at the scene. These hazards should have been identified during arrival and scene size-up. The order in which these hazards are addressed will depend on the specific conditions at a scene and the amount of risk that each hazard poses.

Traffic Hazards

The hazards posed by traffic need to be handled quickly before they lead to additional crashes. In some cases, the most critical issue is slowing or stopping traffic before further damage and injuries can occur. As mentioned previously, emergency vehicles can be used to block traffic from the crash scene. Traffic cones and flares can slow motorists and direct them in a safe pattern around the crash scene. Avoid positioning emergency vehicles in a manner that confuses oncoming motorists.

Traffic hazards are best handled by the appropriate law enforcement agency. This is their area of expertise, and giving

FIRE FIGHTER II

this duty to law enforcement personnel leaves fire department personnel free to handle other extrication tasks. All fire fighters need to work together with law enforcement officials to control traffic in a manner that is safe for emergency workers, victims of the crash, and motorists. If law enforcement officials are not on the scene when you arrive, verify that they are aware of the incident and that they have been dispatched.

Fire Hazards

Because there is a significant risk of spilled fuel in many motor vehicle crashes, the standard operating procedure (SOP) should be to advance a charged hose line close to the damaged vehicles to protect them from the threat of fire. This line, which should be at least a 1½ inches (38 mm) in diameter, should be staffed by a trained fire fighter in full turnout gear. It protects both rescuers and victims of the crash. Crashes that pose large fire hazards or actual fires may require additional fire suppression resources, which should be requested as soon as possible. Small fuel spills can be handled by using an absorbent material to remove the fuel from the area around the damaged vehicle. See the Fire Attack and Foam chapter for the steps required to advance an attack line. See the Fire Suppression chapter for more information on extinguishing vehicle fires.

To mitigate fire hazards at a motor vehicle crash, follow the steps in SKILL DRILL 26-3:

1. Don PPE, including self-contained breathing apparatus (SCBA). Attach a regulator if indicated. (STEP 1)
2. Assess the incident scene for hazards.
3. Communicate with other crews and command.
4. Advance a charged hose line to the proximity of the vehicle. Extinguish any fires. (STEP 2)
5. Open the vehicle's hood. (STEP 3)
6. Mitigate electrical sources of ignition by cutting the 12-volt battery cables. In hybrid vehicles, it is also necessary to disable the high-voltage electrical system. Follow the manufacturer's advice for disabling the high-voltage electrical system. (STEP 4)

Electrical Hazards

Disconnecting the vehicle's electrical system can mitigate electrical hazards posed by a damaged vehicle. Follow your department's SOP for disconnecting the electrical system. When deciding whether to pursue this path, you must weigh the need to disconnect the electrical system against the advantage of being able to operate electrically powered components of the vehicle. Additional information on handling the electrical hazards posed by hybrid and electric vehicles is presented later in this chapter.

Stabilizing electrical hazards posed by downed electrical wires is essential before fire fighters attempt to approach the crash site. At times it may be necessary to instruct victims of a crash to remain in their vehicles until the power can be turned off. Disconnecting the power to damaged electrical lines is a job for properly trained members of the power company. Do not attempt to approach downed electrical lines until the utility company personnel have turned off the electricity. Although it is frustrating to delay extrication and treatment because of downed power lines, this step is essential to avoid potentially life-threatening injuries to both rescuers and victims of the crash.

Other Hazards

Stabilize the environmental hazards present at a crash scene. At scenes where the temperature is hot, fire fighters may need to provide some shade for a victim during a prolonged extrication. In addition, rescue crews require rehabilitation more quickly in a hot environment. At scenes where the temperature is cold, provide covering to help the victim maintain body warmth.

Many crashes occur at night. At such crashes, it is important to provide adequate lighting so that rescuers can work quickly and safely. The Illuminating an Emergency Scene skill drill in the Salvage and Overhaul chapter describes the steps required to light an emergency scene.

Be careful when conditions are wet or icy, because it is important to prevent slips and falls. Sprinkling sand or a kitty litter type of material may help to give solid footing to rescuers.

Vehicle crashes are often characterized by a variety of sharp objects. For example, mangled pieces of sheet metal, plastic, and glass all have sharp edges that pose significant hazards for rescuers and victims of the crash. To reduce the chance of injury, rescuers should wear proper personal protective clothing. Another way to reduce the threat of injury from sharp edges is to cover them with some type of padding—for example, with short sections of old fire hose. Be especially careful when moving victims, making sure that they are protected from contact with any sharp objects.

In incidents involving animals, it may be necessary to deal with the animals before proceeding with other activities. In particular, dogs can be very possessive of their owners. If the dog's owner is injured, remove the dog from the crash site and place it with an uninjured family member or other responsible person. If a pet is injured, call the agency that is responsible for transporting animals to a veterinary clinic. As part of preincident planning, you should learn who in your community is responsible for caring for large animals at rescue scenes.

Cribbing

Unstable objects pose a threat to both rescuers and victims of a crash. Such objects need to be stabilized before fire fighters approach the scene. Most often, the objects that need to be stabilized are the damaged vehicles.

Vehicles that end up on their wheels need to be stabilized vertically with cribbing. Cribbing consists of short lengths of sturdy timber (4 inches by 4 inches [10 centimeters by 10 centimeters], 4 inches by 18 inches [10 centimeters by 46 centimeters], and 4 inches by 24 inches [10 centimeters by 61 centimeters], as well as other lengths and sizes) FIGURE 26-6. After cribbing has been placed against the wheels, a vehicle may still be able to move because of the give and motion that the suspension system causes when rescuers get into the vehicle and victims are extricated from the vehicle. This type of instability can cause further injuries to the victims of the crash.

FIRE FIGHTER II

SKILL DRILL 26-3 Mitigating the Hazards at a Motor Vehicle Crash FFII

Fire Fighter II (NFPA 6.4.1)

1 Don PPE, including self-contained breathing apparatus (SCBA). Attach a regulator if indicated.

2 Assess the incident scene for hazards. Communicate with other crews and command. Advance a charged hose line to the proximity of the vehicle. Extinguish any fires.

3 Open the vehicle's hood.

4 Mitigate electrical sources of ignition by cutting the 12-volt battery cables. Recognize and disable electrical systems in hybrid vehicles.

FIRE FIGHTER II

The suspension system of most vehicles can be stabilized with step chocks, which are shaped like stair steps and are placed under both sides of the vehicle. Place one step chock toward the front of the vehicle and a second step chock toward the rear of the vehicle FIGURE 26-7. Repeat this process on the other side of the vehicle. Once the step chocks are in place, the tires can be deflated by pulling out the valve stems to create a stable vehicle.

If step chocks are not available or are not the right size, you can build a box crib. To do so, place cribbing at right angles to the preceding layer of cribbing FIGURE 26-8.

After a crash, some vehicles might come to rest on their roofs or sides. These vehicles are very unstable, and placing the slightest amount of weight on them can cause them to move. Overturned vehicles can be stabilized using box cribs or step chocks on each end of the vehicle.

Besides cribbing, wedges are used to snug loose cribbing under the load or, when using rescue-lift air bags, to fill the void between the crib and the object as it is raised FIGURE 26-9. Only lift or raise the vehicle enough to place additional cribbing under it; remember the saying, "Lift an inch, crib an inch." Wedges should be the same width as the cribbing, with the tapered end no less than ¼ inch (6 millimeters) thick. When the ends are feathered to less than ¼ inch (6 millimeters), the end will commonly fracture under a load.

FIRE FIGHTER II

FIGURE 26-6 The application of cribbing in the front and back of the vehicle wheels stabilizes the vehicle.

FIGURE 26-7 Step chocks are used to stabilize a vehicle.

To stabilize a vehicle following a motor vehicle crash, follow the steps in SKILL DRILL 26-4:

1. Don PPE, including face protection.
2. Minimize the hazards to rescuers and victims.
3. Chock both sides of one tire to prevent the vehicle from rolling by placing one chock in front of a wheel and a second chock in back of the wheel. (STEP 1)
4. Consider deflating tires for added stability. Assess the need for step chocks for additional stability. (STEP 2)
5. Turn off the ignition and remove the key or fob. (STEP 3)
6. Place the gear shift in park, and apply the parking brake. (STEP 4)

Rescue-Lift Air Bags

Rescue-lift air bags are pneumatic-filled bladders made out of rubber or synthetic material FIGURE 26-10. They are used to lift an object or to spread one or more objects away from each other to assist in freeing a victim. Rescue-lift air bags are never used to shore or stabilize a vehicle by themselves; cribbing is always required in conjunction with rescue-lift air bags.

FIRE FIGHTER II

FIGURE 26-8 A box crib is constructed to stabilize a vehicle.

FIGURE 26-9 Wedges are used to snug loose cribbing when the void space is not big enough to accommodate a regular-sized crib.

FIGURE 26-10 A rescue-lift air bag.

Rescue-lift air bags are often used to lift a vehicle or object off a victim. Fire fighters should use extreme caution when

SKILL DRILL 26-4 Stabilizing a Vehicle Following a Motor Vehicle Crash FFII

Fire Fighter II (NFPA 6.4.1)

1. Don PPE, including face protection. Minimize hazards to rescuers and victims. Chock both sides of one tire to prevent the vehicle from rolling by placing one chock in front of a wheel and a second chock in back of the wheel.

2. Consider deflating tires for added stability. Assess the need for step chocks for additional stability.

3. Turn off the ignition and remove the key or fob.

4. Place the gear shift in park, and apply the parking brake.

applying this technique and should adhere to all safety precautions outlined in the manufacturer's instructions for the rescue-lift air bag. Cribbing must be used whenever fire fighters are lifting a load, because instability can occur from weight shifts or the rescue-lift air bags may fail under a load. Given these risks, it is a must—not an option—that cribbing be used in conjunction with rescue-lift air bags.

Several types of pneumatic rescue-lift air bags are in common use today: low pressure, medium pressure, and high pressure. None of these devices should be used without first properly blocking the wheels opposite of the lift air bag and cribbing the vehicle or object as it is lifted. These safety precautions will prove invaluable in case the rescue-lift air bag suffers a catastrophic failure.

As they age, rescue-lift air bags become more prone to failure. Given this fact, they should be tested regularly according to the manufacturer's recommendations. Rescue-lift air bags that are 10 years of age or older have generally reached the end of their useful life.

Low-Pressure Lift Air Bags

Low-pressure lift air bags are commonly used for recovery operations and are sometimes used for vehicle rescue operations. These devices come in many sizes and shapes; square air bags, however, offer greater stability. Because of their lightweight construction, low-pressure lift air bags can be less stable (at least until they are fully inflated) than high-pressure air bags, which have a lower height lift and stronger construction.

Medium-Pressure Lift Air Bags

Medium-pressure lift air bags are designed so that they contain either two or three cells. For general vehicle rescue, these devices are not appropriate. Medium-pressure lift air bags are more suitable for aircraft, medium or heavy truck, or bus rescue, and for recovery work.

High-Pressure Lift Air Bags

The rescue-lift air bags most frequently used by fire fighters are high-pressure lift air bags. These devices have a very sturdy construction and are generally made of vulcanized rubber mats that are reinforced by steel or other material woven into a fiber mat and then covered with rubber. When using this type of air bag, cribbing remains essential to provide a safe working environment.

Fire Fighter Safety Tips

- Never stack high-pressure lift air bags more than two units high.
- Do not use a rescue-lift air bag to pull a steering column.
- Do not use a rescue-lift air bag as the sole means to stabilize a vehicle; cribbing must be the primary stabilizer.
- Never operate the rescue-lift air bag system without having been properly trained and fully understanding how the system works.
- When stacking rescue-lift air bags, the largest size should be on the bottom and the smallest on top. The bottom rescue-lift air bag should be inflated first.
- Place a sheet of plywood on the ground under the air bag to protect it.
- Do not use boards or plywood between or above rescue-lift air bags.
- Clean rescue-lift air bags by following the manufacturer's recommendations.
- Test rescue-lift air bags regularly.
- Never store a rescue-lift air bag near gasoline.

Principles of Gaining Access and Disentangling the Victim

Before discussing the steps of gaining access and disentanglement, it is important to understand the principles involved with these steps of extrication. Although many different techniques and methods are used to gain access and disentangle victims from crashed vehicles, the objective of these techniques is accomplished by performing four functions:

- Stabilize or hold an object or vehicle. Vehicles are stabilized when cribbing is used to keep them from moving.
- Bend, distort, or displace. An example is bending a vehicle door back to get it out of the way.
- Cut or sever. An example is cutting a roof.
- Disassemble. An example is removing a vehicle door by unbolting the door hinges.

FIRE FIGHTER II Tips FFII

Extrications are dynamic events, especially on roadways, and fire fighters must be constantly aware of what is going on around them. Maintain situational awareness at all times.

Tools Used for Extrication

Many different tools can be used for extrication. It is important to understand the purpose of each tool, because a certain tool will be more appropriate in certain circumstances. Fire fighters need to know which tools are available in their department and how to operate each of these tools.

Cribbing, rescue-lift air bags, step clocks, high-lift jacks, and stabilization jacks are all types of stabilization devices. When using these tools, be certain that they are placed firmly under the vehicle and that the vehicle is stable before attempting to enter it.

Hand tools and powered tools are used to displace, bend, and distort metal parts of a vehicle to gain access to a victim and to disentangle a victim from the vehicle. Hand tools include pry axes, short pry bars, long pry bars, hacksaws, screw drivers, hammers, and Halligan tools; these tools can be used to bend sheet metal so that it can be pulled away from a victim. In addition, manual hydraulic rams and manual hydraulic spreaders can exert a large amount of force onto the vehicle. Powered hydraulic rams and powered hydraulic spreaders use a hydraulic pump powered by an electric motor or a gasoline engine. Because newer vehicles are designed to tear and rip (as a safety feature), sometimes hydraulic power is not the best option for extrication purposes.

Hand tools that can be used for cutting include axes, bolt cutters, cable cutters, hacksaws, and manual hydraulic cutters. Each of these cutting tools is designed to cut certain materials. If you try to use a tool on an inappropriate material, you will not only fail to cut the material, but also may damage the tool and injure the victim or yourself. In case of powered equipment failure, hand tools are an appropriate backup. Training sessions should include hand tools use and practice to show advantages and limitations.

A variety of power tools for cutting or severing are available—some are air powered, some are electrically powered, and others are hydraulically powered. Air chisels are powered by pressurized air from portable SCBA air cylinders or from a cascade system mounted on a vehicle FIGURE 26-11. Air chisels are efficient at cutting sheet metal and some types of plastic; they can also be used to cut metal posts.

A variety of powered saws are used for extrication. Some electric saws require power cords, whereas others operate on battery packs. Powered saws produce different types of motion.

VOICES OF EXPERIENCE

We treat many rescues as old hat. We have done rescues so many times that we want to get the victim out as quickly as possible. It is very important for all of us to just take a breath. Take a look at what is best for the victim. Do we need to rapidly extricate or begin treatment as we are attempting extrication?

A year back, we had a two car head-on collision with one fatality and four injuries. One victim was pinned by the steering wheel and dash. Some of the crew wanted to get the victim out by rapid extrication and pull her out on the backboard. The victim was in critical condition, but we decided that we could remove the roof quickly and start to care for the victim while removing the doors and rolling the dash.

As we continued the extrication, it became obvious that the victim's leg was injured badly and was entrapped by the seat and brake pedal. Removing the brake pedal freed the victim's leg enough that that we could slide her up the backboard. The victim had two obvious fractures to the left femur and her ankle was being held together by a small piece of skin. If we had not taken the time to look at the big picture, the victim's foot might have been lost. Remember to take your time and not cause other injuries that might not be fixed easily.

David Nix
Wilburton Fire Department
Wilburton, Oklahoma

FIGURE 26-11 An air chisel is useful for cutting sheet metal.

For example, reciprocating saws move back and forth very quickly, whereas rotary saws move in a circular motion. One of the most commonly used extrication tools is the powered hydraulic cutter, which can quickly cut through the roof posts of a vehicle. Whenever fire fighters use power and hydraulic tools, they should cover the victim to protect him or her.

Specialized equipment, such as the hand-operated come along, is also used during extrication. You can learn more about the proper use of this and other specialized equipment through special rescue courses or in-service training.

Sometimes the most effective way to remove a part of a vehicle is to disassemble it. This method of extrication has the advantage of being controlled; it may also provide an added degree of safety for the victim and rescuers. A disadvantage is that it is very slow and may not always be possible. The types of tools needed for disassembly are the same tools that are used for construction and repair of vehicles. A mechanic's toolkit that contains an assortment of wrenches and screwdrivers will enable a qualified rescuer to disassemble some parts of a vehicle efficiently and safely.

Gaining Access to the Victim

Open the Door

After stabilizing the vehicle, the simplest way to access a victim of a crash is to open a door. Try all of the doors first—even if they appear to be badly damaged. Remember the first rule of forcible entry: "Try before you pry." It is an embarrassing waste of time and energy to open a jammed door with heavy rescue equipment, only to discover that another door can be opened easily and without any special equipment.

Attempt to unlock and open the least damaged door first. Make sure the locking mechanism is released. Then try the outside and inside handles at the same time if possible.

Near Miss REPORT

Report Number: 10-0000452

Synopsis: Unsecured vehicle rolls down hill during extrication.

Event Description: We responded to a motor vehicle accident with injury. The victim in the driver's seat was restrained. A small amount of extrication was required. The fire crew was already on the scene, and stated they needed a medic in the vehicle. The vehicle was off the road on top of a hill, with a pond on both sides of the hill. I arrived on the scene and got in the backseat to hold the victim's c-spine and do patient care during extrication.

The vehicle had been chocked, but no one had checked whether it was in park or whether the emergency brake was on. As the extrication started, the vehicle started moving down the hill toward the pond. The patient and I were in the vehicle. It stopped prior to making it down the hill, and we finished the extrication and transported the patient to the hospital.

Lessons Learned

- Scene safety—Always be aware of surroundings.
- Teamwork—Work together and remind one another of possible safety issues.
- Command—Just because the scene has been established and units are on scene, it does not mean that all safety aspects have been taken care of.
- Training—Always chock vehicles and make sure they are in park with the emergency brake set.

FIRE FIGHTER II

Break Tempered Glass

If a victim's condition is serious enough to require immediate care and you cannot enter through a door, consider breaking a window. You cannot easily enter through the windshield, however; it is made of laminated windshield glass, which is more difficult to break. In contrast, the side and rear windows are made of tempered glass, which will break easily into small pieces when hit with a sharp, pointed object such as a spring-loaded center punch, a multipurpose tool, or the point of an axe FIGURE 26-12. Because these windows do not pose as great of a safety threat, they can be fire fighters' primary access route to the victim.

If you must break a window to unlock a door or gain access, cover the victim and try to break a window that is away from the victim. If the victim's condition warrants your immediate entry, however, do not hesitate to break the closest window. Small pieces of tempered glass do not usually pose a danger to victims trapped in cars. Advise Emergency Medical Services personnel if a victim is covered with broken glass so that they can notify the hospital emergency department.

After breaking the window, use your gloved hands to pull the remaining glass out of the window frame so that it does not fall onto any victims or injure any rescuers. If you use something other than a spring-loaded center punch to break the window, always aim for a low corner.

To break tempered glass, follow the steps in SKILL DRILL 26-5:

1. After donning complete PPE, including face and eye protection, select a tool. (STEP 1)
2. Ensure that the victim and other fire fighters are as properly protected as possible. (STEP 2)
3. Place the tool in the lower corner of the window and sharply apply pressure until the spring is activated. (STEP 3)
4. Remove loose glass around the window opening. (STEP 4)

Once you have broken the glass and removed the pieces from the frame, try to unlock the door again. Release the locking mechanism, and then use both the inside and outside door handles at the same time. This technique will often force a jammed locking mechanism to open the door, even when the door appears to be badly damaged.

Breaking the rear window will sometimes provide an opening large enough to enable a rescuer to gain access to the victim if there is no other rapid means for doing so. The simple techniques of opening a door and breaking the rear window will enable fire fighters to gain access to most vehicle crash victims, even those in an upside-down vehicle.

Follow the steps in SKILL DRILL 26-6 to gain access to a vehicle following a motor vehicle crash. The steps in this skill drill may have to be modified depending on the situation.

1. Don PPE including face protection.
2. Minimize hazards to rescuers and victims.
3. Ensure stability of the vehicle; chock wheels, turn off the ignition, remove smart keys away from the vehicle, place the gear shift in park, and apply the parking brake. Isolate the power by cutting the battery cables.
4. Determine the best access point, and then open a door, break needed glass, or distort metal. (STEP 1)
5. Enter the vehicle and approach the victim.
6. Communicate with the victim. (STEP 2)

FIRE FIGHTER II

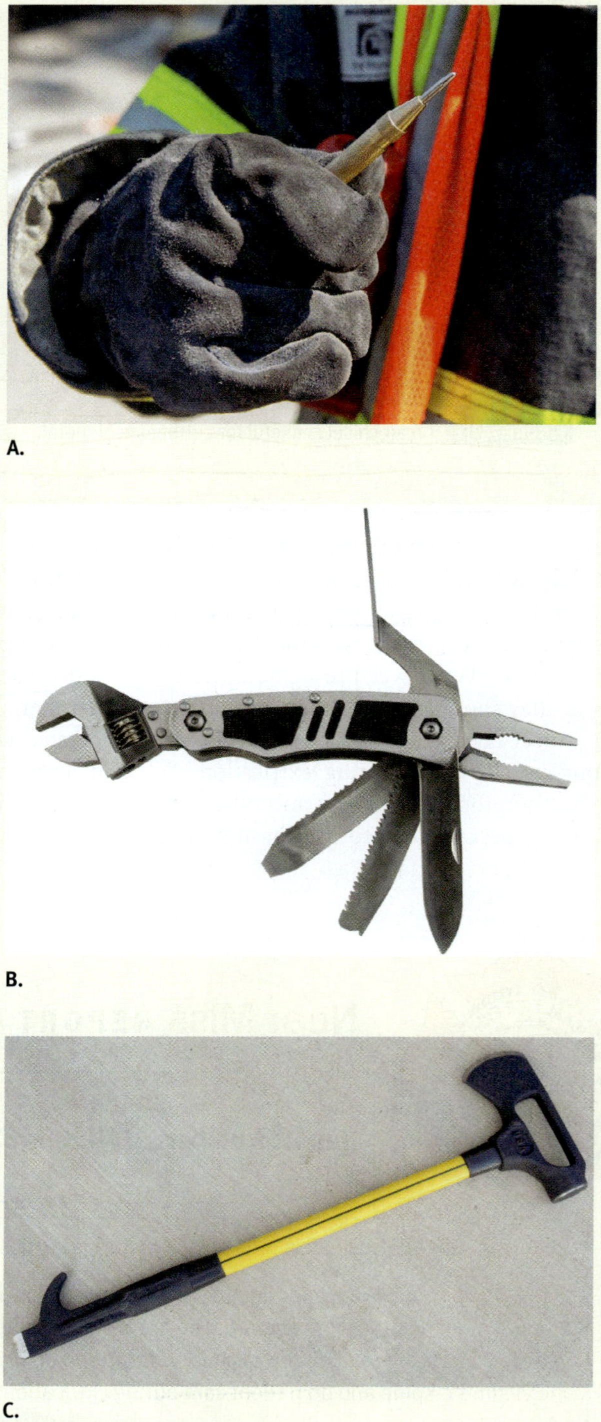

FIGURE 26-12 Three tools used to break tempered glass. A. Spring-loaded punch. B. Multipurpose tool. C. Axe.

FIRE FIGHTER II

SKILL DRILL 26-5 Breaking Tempered Glass FFII

Fire Fighter II (NFPA 6.4.1)

1. After donning complete PPE, including face and eye protection, select a tool.

2. Ensure that the victim and other fire fighters are as properly protected as possible.

3. Place the tool in the lower corner of the window and sharply apply pressure until the spring is activated.

4. Remove loose glass around the window opening.

Fire Fighter Safety Tips

Some newer vehicles may soon be equipped with side and rear windows made out of a new type of glass that cannot be broken with a sharp object. This glass promises to provide more protection during a crash and offers an increased level of protection against theft.

Force the Door

If fire fighters cannot gain access by the previously mentioned methods, they must use heavier extrication tools to gain access to the victim. The most commonly used technique for gaining access to a vehicle following a collision is door displacement FIGURE 26-13. The opening and displacement of vehicle doors may be difficult and somewhat unpredictable, however. As mentioned earlier, doors on newer vehicles will sometimes be hard to force, and several attempts may be necessary.

FIRE FIGHTER II

When it is necessary to force a door to gain access to the victim, choose a door that will not endanger the safety of the victim. For example, do not try to force a door open if the victim is leaning against it.

To force a door, fire fighters typically use a prying tool to bend the sheet metal at the edge of the door near the locking mechanism. Once this metal has been exposed, a hydraulic spreading tool can be inserted into the space between the door and the pillar of the vehicle and can pull the locking mechanism apart. Once a door has been removed, it should be placed away from the vehicle where it will not be a safety hazard to rescuers.

Powered hydraulic tools are the most efficient and widely used tools for opening jammed doors, which can be opened by releasing the door from the latch side or the hinge side of the

FIRE FIGHTER II

SKILL DRILL 26-6 Gaining Access to a Vehicle Following a Motor Vehicle Crash FFII

Fire Fighter II (NFPA 6.4.1)

1 Don PPE, including face protection. Minimize hazards to rescuers and victims. Ensure stability of the vehicle; chock wheels, turn off the ignition, place the gear shift in park, and apply the parking brake. Isolate the power by cutting the battery cables. Determine the best access point, and then open a door, break needed glass, or distort metal.

2 Enter the vehicle and communicate with the victim.

FIGURE 26-13 The most commonly used technique for gaining access to a vehicle following a collision is door displacement.

door. The decision regarding which method to use will depend on the structure of the vehicle and the type of damage the door has sustained.

The first step in opening a vehicle door is to use a pry tool to bend the sheet metal away from the edge of the door where the hydraulic tool is to be inserted. This action provides a place where the spreader of the hydraulic tool can be inserted.

Place the spreader in a position so that it is not in the pathway that the door will take when the hinges or latch break. Do not stand in a position that might put you in danger. To ensure that the door does not swing open violently, you can use a long pry bar to limit the movement of the door when it opens. Activate the hydraulic tool to push apart the outer sheet metal skin of the vehicle, exposing the hinges or the door latch. Once the outer sheet metal has been exposed, close the tips of the spreader and remove them.

FIRE FIGHTER II

Insert the closed tips onto the inner skin of the door and the doorjamb just above the latch or above the hinges. Activate the spreader to spread the tips until the latch or the hinge separates. When separating a door at the latch side, you can place 4-inch by 4-inch (10-centimeter by 10-centimeter) cribbing under the bottom of the door to hold it up and then start to separate the hinges of the door. Some hydraulic tools are capable of cutting door hinges. Check the manufacturer's recommendations for the proper operation of the hydraulic tool.

When separating a door from the hinge side, place the spreader on the top of the bottom hinge and then use the hydraulic spreader to separate the door from the hinge. Once the hinges have separated, place 4-inch by 4-inch (10-centimeter by 10-centimeter) cribbing underneath the door to hold it in place while you work on the latch side of the door.

Another way to expose the hinges of the door is to vertically pinch the door in the center of the door, from the bottom of the window down, using the hydraulic spreader. This will pull the sheet metal away from the edges of the door, exposing the latch and hinges.

FIRE FIGHTER II

FIRE FIGHTER II Tips FFII

As vehicle manufacturers strive to make new vehicles stronger and lighter, they have begun using increasing quantities of high-strength steel. These stronger steels, in turn, require more robust rescue tools. Tool manufacturers continue to increase the capacity of rescue tools to cut and distort the metals in newer vehicles. By studying individual vehicle manufacturers' specifications, you can determine the easiest places to cut or distort specific vehicle parts.

To open a door following a motor vehicle collision, follow the steps in SKILL DRILL 26-7:

1. Don PPE, including face protection.
2. Retrieve and set up the required tools. Check the equipment for readiness.
3. Assess the vehicle for hazards, including passive restraint devices. (STEP 1)
4. Communicate with the victim; minimize hazards to both the rescuers and the victim. (STEP 2)
5. Engage hand tools or a power unit to force the door, using good body mechanics. (STEP 3)
6. Ensure that the forced door has been stabilized. (STEP 4)

FIRE FIGHTER II Tips FFII

Having one leader coordinate the extrication and rescue effort will increase efficiency and help prevent the actions or one team from interfering with another.

Provide Initial Medical Care

As soon as you have secured access to the victim, begin to provide emergency medical care. A qualified emergency medical provider should perform this function. Being trapped in a damaged vehicle is a frightening experience, so one caregiver should remain with the victim and provide both emotional comfort and physical care. Victim care should be provided at the same time that extrication is ongoing. Although it might be necessary to delay one part of these processes for a short period, all personnel should work toward the goal of getting the victim stabilized and removed from the vehicle as quickly and safely as possible.

To gain access and provide medical care to a victim in a vehicle following a motor vehicle collision, follow the steps in SKILL DRILL 26-8:

1. Don PPE, including eye protection.
2. Communicate your actions with the victim.
3. Enter the vehicle through the best access point. Minimize the hazards present.
4. Perform an initial assessment of the victim, begin initial medical care, and communicate the victim's condition to the outside rescuer. (STEP 1)
5. Perform as complete an assessment as possible. (STEP 2)
6. Provide immediate care, and begin packaging the victim for extrication. (STEP 3)

FIRE FIGHTER II

Disentangling the Victim

The next step in the extrication sequence is disentangling the victim. The purpose of this step is to remove those parts of the vehicle that are trapping the victim. Note that the goal of this step is to remove the sheet metal and plastic from around the victim—not, as you may have read in the popular press, to "cut the victim out of the vehicle."

Before beginning disentanglement, study the situation. What is trapping the victim in the vehicle? Perform only those disentanglement procedures that are necessary to remove the victim safely from the vehicle. The order in which these procedures are carried out will be dictated by the conditions at the scene. Many times it will be necessary to perform one procedure before you can access the parts of the vehicle to perform another procedure.

As you work to disentangle the victim from the vehicle, be sure to protect the victim by covering him or her with a blanket or by using a backboard. Be sure that the victim understands what is being done. In addition, provide emotional support, because the sounds made by extrication procedures are frightening for victims.

To learn all of the methods of disentangling a victim, fire fighters need to take an approved extrication course. The five procedures presented here are the ones that are most commonly performed.

Displace the Seat

In frontal and rear-end crashes, the vehicle may become compressed FIGURE 26-14. That is, as the front of the vehicle is pushed back, the space between the steering wheel and the seat becomes smaller. In some cases, the driver may be trapped between the steering wheel and the back of the front seat.

FIGURE 26-14 In frontal crashes, the vehicle may become compressed.

FIRE FIGHTER II

SKILL DRILL 26-7 Opening a Vehicle Door FFII

Fire Fighter II (NFPA 6.4.1)

1 Don PPE, including face protection. Retrieve and set up the required tools. Check the equipment for readiness. Assess the vehicle for hazards, including passive restraint devices.

2 Communicate with the victim; minimize hazards to both the rescuers and the victim.

3 Engage hand tools or a power unit to force the door, using good body mechanics.

4 Ensure that forced door is stable.

Displacing the seat can relieve pressure on the driver and give rescuers more space for removal.

If it is necessary to displace a seat backward, start with the simplest steps. Many times you can gain some room by moving the seat backward on its track. This maneuver works well with short drivers who have their seat forward. To move a seat back, first make sure that the victim is supported. With manually operated seats, release the seat-adjusting lever and carefully slide the seat back as far as it will go. Attempt to use these simple methods first.

With seats that have adjustable backs and adjustable heights, use these features to move the seat back to give the victim more room or lower the seat to help disentangle the victim. Make sure that the victim is adequately supported before attempting these maneuvers.

FIRE FIGHTER II

If this technique is unsuccessful, perform a dash displacement or roll-up. As a last resort, use a manual hydraulic spreader or a powered hydraulic tool to move the seat back. Place one tip of the hydraulic tool on the bottom of the seat, but avoid pushing on the seat channel that is attached to the floor of the vehicle. Place the other tip of the spreader at the bottom of the "A" post doorjamb. Support the victim carefully. Engage the seat adjustment lever on manually operated seats, and open the spreader in a careful and controlled fashion.

In some cases, it may be helpful to remove the back of the seat. To accomplish this task, cut the upholstery away from the bottom of the seat back where it joins the main part of the seat. A reciprocating saw or a hydraulic cutter can then be used to cut the supports for the seat back. Be certain

FIRE FIGHTER II

SKILL DRILL 26-8 Gaining Access and Providing Medical Care to a Victim in a Vehicle FFII

Fire Fighter II (NFPA 6.4.1)

1 Don PPE, including eye protection. Communicate your actions with the victim. Enter the vehicle through the best access point. Minimize the hazards present. Perform an initial assessment of the victim, begin initial medical care, and communicate the victim's condition to the outside rescuer.

2 Perform as complete an assessment as possible.

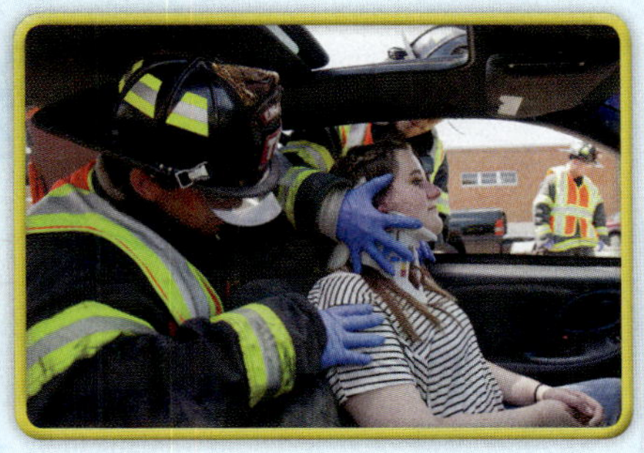

3 Provide immediate care, and begin packaging the victim for extrication.

that the victim is supported and protected throughout this procedure.

Displacing the seat or removing the seat back will give fire fighters more room to administer treatment and remove the victim from the vehicle.

Remove the Windshield

A second technique that is often part of the disentanglement process involves the removal of glass. Removing the rear window, the side glass, or the windshield may improve communication between rescue personnel inside the vehicle and personnel outside the vehicle. Sometimes all that rescuers need to do is to roll a window down to gain this advantage. Open windows provide a good route for passing medical care supplies to the inside caregiver. When the roof of a vehicle must be removed, it is necessary to first remove all of the glass from the vehicle.

The side windows and rear window on most vehicles are made of tempered glass, which can be removed by striking the glass in a lower corner with a sharp object, such as a spring-loaded center punch. Whenever this measure becomes necessary, fire fighters should protect the victim as much as possible by covering him or her to prevent the victim from being cut by the small pieces of glass.

Windshields, however, cannot be broken with a spring-loaded center punch. Windshields consist of plastic laminated glass, which has a "sandwich" type of construction: The two pieces of "bread" consist of thin sheets of specially constructed glass, and the "filling" of the sandwich is a thin layer of a flexible plastic. When laminated glass is struck by a sharp stone or by a spring-loaded center punch, a small mark appears, but the structure of the glass remains intact. Because of this special construction, it is necessary to remove the windshield in one large piece.

FIRE FIGHTER II

The windshields of most passenger vehicles are glued in place with plastic-type glue that is very strong. Nevertheless, there are several ways to remove a windshield after a collision, including cutting the windshield with a special glass saw or reciprocating saw, or carefully using an axe.

To remove a windshield, first make sure that the victim is protected from flying glass. One rescuer begins to cut the top of the windshield at the middle of the windshield. The rescuer continues cutting along the top of the windshield, and then cuts cut down the side of the windshield close to the "A" post FIGURE 26-15. Finally, the rescuer cuts along the bottom of the windshield. At this point, the first rescuer stabilizes the half of the windshield that has been cut free.

Once this operation is complete, a second rescuer starts at the top of the windshield where the initial cuts were made by the first rescuer and cuts the second half of the windshield, following the same sequence of cuts used by the first rescuer. When the second rescuer has completed cutting the windshield, the entire windshield is lifted out of its frame and placed under the vehicle or in another safe place, where it will not present a safety hazard.

Removing a windshield is an essential step prior to removing the roof of a vehicle. It also provides extra space when administering emergency medical care to an injured victim.

FIGURE 26-15 Removing the windshield by cutting close to the "A" post.

Fire Fighter Safety Tips

- On most recently manufactured vehicles, the steering wheel contains the driver's-side air bag, which is a life-saving safety feature for the driver of the vehicle. Front-passenger air bags may also be present.
- Air bags may present a hazard after deployment because some air bags deploy in two stages.
- If the air bag did not deploy during the crash, it presents a hazard both for the occupant of the vehicle and for rescue personnel. An undeployed air bag could deploy if wires are cut or if it becomes activated during the rescue operation.
- If the air bag did not deploy, disconnect the battery and allow the air bag capacitor to discharge. The time required to discharge the capacitor varies for different models of air bags.
- Do not place an object between the victim and an undeployed air bag.
- Do not attempt to cut the steering wheel if the air bag has not deployed.
- For your safety, never get in front of an undeployed air bag. You could suffer serious injury if it activates unexpectedly.
- Some vehicles contain side-mounted air bags or curtains that provide lateral protection for occupants. Check vehicles for the presence of these devices.

Remove the Steering Wheel

During normal driving, the steering wheel must be located close to the driver. During a crash, however, any compression of the front part of the vehicle may push the steering wheel back into the victim's abdomen or chest. Removing the steering wheel can be a major help in disentangling a victim from the vehicle in such a case.

Steering wheels can be cut with hand tools such as a hacksaw or bolt cutters. Hydraulic cutters and reciprocating saws will also cut steering wheel spokes and the steering wheel hoop. One method of removing a steering wheel is to cut the spokes as close to the hub of the steering wheel as possible. A second method is to cut the hoop or ring of the steering wheel. The steering wheel ring can be removed completely, or one section can be cut and removed. A third method is to cut out the column.

Note that cutting the steering wheel hoop or spokes leaves sharp edges that present a safety hazard. These sharp edges must be covered to prevent injury both to the victim and to rescuers.

Displace the Dashboard

During frontal crashes, the dashboard is often pushed down or backward. When a victim is trapped by the dash, it is necessary to remove this component FIGURE 26-16. The technique used to remove the dash is called dash displacement or dash roll-up. The objective with this procedure is to lift the dashboard up and move it forward. This evolution is also sometimes used to lift the steering column up and away from a victim.

Dash displacement requires a cutting tool such as a hacksaw, a reciprocating saw, an air chisel, or a hydraulic cutter. A cutting tool is needed to make a cut on the "A" post. A mechanical high-lift jack or a hydraulic ram is needed to push the dash forward, and cribbing is needed to maintain the opening made with these tools.

The first step of the dash displacement is to open both front doors. Tie them in the open position so that they will not move as the dash is being displaced. Alternatively, the doors can be removed, as discussed earlier in this chapter in the section on gaining access. Additionally, the roof could be removed.

Place a backboard or other protective device between the victim and the bottom part of the "A" post where the relief cut will be made. Do not place a backboard in this position if the driver's-side air bag has not deployed. Next, cut the bottom of the "A" post where it meets the sill or floor of the vehicle. It is critical to make the cut perpendicular to the "A" post—failure to do so may cut the fuel or power line located in the rocker panels.

Place the base of a high-lift mechanical jack or a hydraulically powered ram at the base of the "B" post where the sill and the "B" post meet. Place the tip of the jack or ram at the bend in

FIGURE 26-16 When a victim is trapped by the dash, it is necessary to displace this component.

FIRE FIGHTER II

the "A" post, which is located toward the top of the "A" post. In a controlled manner, extend the jack or ram to push the dash up and off of the victim. This evolution may be difficult if the crashed vehicle contains high-strength steel.

Once the dash has been removed from the victim, build a crib to hold the sill in the proper position and to prevent the dash from moving. Only then can the jack or ram be removed.

Performing dash displacement requires careful monitoring of the dashboard's movement to be certain that it is moving away from the victim and is not causing any additional harm to the victim. Throughout this evolution, make sure that the victim is protected and that someone keeps the victim informed about what is happening. Communication is key in this procedure.

To displace the dashboard of a vehicle following a motor vehicle collision, follow the steps in SKILL DRILL 26-9:

1. Don PPE, including face protection.
2. Retrieve and set up the required tools. Check the equipment for readiness.
3. Assess the vehicle for hazards, including passive restraint devices. (STEP 1)
4. Communicate with the victim and ensure that both the victim and rescuers are protected from hazards. (STEP 2)
5. Make relief cuts at the bottom of the "A" post. (STEP 3)
6. Insert the hydraulic spreader in the cut and lift the dashboard and firewall. Place cribbing under the dash as it is being lifted. (STEP 4)

FIRE FIGHTER II Tips FFII

Cribbing should be placed under the vehicle whenever downward pressure is expected. This will maximize the displacement of the dash.

Displace the Roof

Removing the roof of a vehicle has several advantages. First, it enables equipment to be easily passed to the emergency medical provider. In addition, it increases the amount of space available to perform medical care, and it increases the visibility and space for performing disentanglement. Likewise, visibility and fresh air supply are improved. The increased space also helps to reduce the feeling of panic caused by the confined space of the crashed vehicle. Finally, removing the roof provides a large exit route for the victim FIGURE 26-17.

One method of displacing the roof is to cut the "A" posts and fold the roof back toward the rear of the vehicle. This technique provides limited space. Given this fact, it is often preferable to remove the entire roof.

FIRE FIGHTER II Tips FFII

When removing the roof, remember to cut all of the seat belts so they will not interfere with the operation.

FIRE FIGHTER II

FIGURE 26-17 Removing the roof provides a large exit route for the victim.

Roof displacement can be accomplished with hand tools such as hacksaws, air chisels, and manual hydraulic cutters. Appropriate power tools include reciprocating saws and powered hydraulic cutters, especially the latter tools. Be aware that some "A" posts are made of high-strength steel and may be difficult to cut, especially with older cutting tools. In addition, most newer vehicles include additional reinforcement beams for the seat belts. Older hydraulic cutters may not be able to sever these beams, so cuts must be made above the seat belts.

The first step in displacing the roof is to ensure the safety of both rescuers and the victim inside of the vehicle. In particular, as rescuers cut the posts that support the roof, they must support the roof to keep it from falling on the victim.

To remove the roof, first remove the glass in the vehicle's windows and windshields to prevent it from falling on the victim. Cut the vehicle posts farthest away from the victim. Cut each post at a level that will ensure the least amount of post remains in place after roof removal. When cutting the wider rear posts, cut them at the narrowest point of the post. As each post is cut, a rescuer needs to support that post. Cut the post closest to the victim last, because it provides some protection for the victim.

Working together, several rescuers should remove the roof and place it away from the vehicle, where it does not pose a safety hazard. They should then cover the sharp ends of the cut posts with a protective device, thereby ensuring the safety of both rescuers and the victim.

To remove the roof of a vehicle following a motor vehicle collision, follow the steps in SKILL DRILL 26-10:

1. Don PPE, including face protection.
2. Retrieve and set up the required tools. Check the equipment for readiness.
3. Assess the vehicle for hazards, including passive restraint devices.
4. Communicate with the victim, and minimize hazards to both the rescuers and the victim.
5. Remove any remaining glass. (STEP 1)

SKILL DRILL 26-9 Displacing the Dashboard of a Vehicle FFII

Fire Fighter II (NFPA 6.4.1)

1 Don PPE, including face protection. Retrieve and set up the required tools. Check the equipment for readiness. Assess the vehicle for hazards, including passive restraint devices.

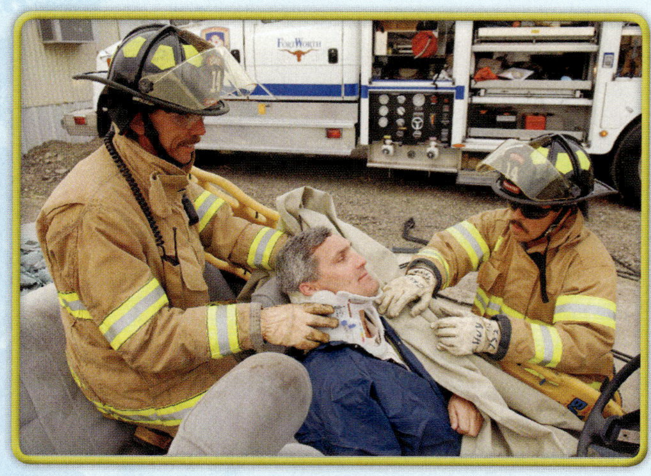

2 Communicate with the victim and ensure that both the victim and rescuers are protected from hazards.

3 Make relief cuts at the bottom of the "A" post.

4 Insert the hydraulic spreader in the cut and lift the dashboard and firewall. Place cribbing under the dash as it is being lifted.

6. Engage appropriate tools to remove the roof, while using good body mechanics. (STEP 2)
7. Control the vehicle roof and protect against sharp edges at all times. (STEP 3)
8. Remove the roof to a safe location.
9. Return the tools to the staging area upon completion of the tasks. (STEP 4)

Removing and Transporting the Victim

The final phase of the extrication process consists of removing the victim from the crashed vehicle. During this phase, the victim needs to be stabilized and packaged in preparation for removal. Follow the direction of the medical care provider who is in charge of victim care. The definitive treatment of trauma victims will occur at a hospital; therefore, the amount of stabilization that should be attempted while the victim is in the vehicle should be limited to only those steps that are needed to prevent further injury to the victim. See the Performing a Long Backboard Rescue skill drill in the Search and Rescue chapter for more information on using a long backboard to remove a victim from a vehicle.

Develop a plan for the victim's removal, and make sure that a clear exit pathway is available. The actual victim removal should be directed by a designated person—usually the rescuer controlling c-spine stabilization—and conducted with clear commands. Ensure that an adequate number of rescuers are able to assist with the victim removal, and confirm that everyone involved in this process understands the commands that will be given.

Try to make the removal process as seamless as possible. Ideally, the ambulance cot should be positioned close by so that the victim can be placed on the cot without any delay. In any event, the victim should be moved to an ambulance as soon as possible.

Transport the victim to an appropriate medical facility as soon as the initial stabilization has been completed. If your department uses a helicopter service to transport victims, you also need to be familiar with the procedures used to load victims into helicopters.

FIRE FIGHTER II Tips FFII

Always have a separate set of eyes, or a lookout, observing the extrication operation for safety.

Terminating an Incident

Terminating an incident includes securing the scene by removing the damaged vehicle and equipment from the scene and ensuring the scene is left in a safe condition. After you have secured the scene and packed your equipment, it is important to return to the station and fully inventory, clean, service, and maintain all the equipment (per the manufacturer's instructions) to prepare it for the next call. Some items will inherently need repair, but most will need simple maintenance before they are placed back on the fire apparatus and considered in service.

Securing the Scene

Just because the victim has been removed from the wreckage does not mean the scene can be abandoned. Law enforcement will secure the scene by conducting their on-scene investigation. A potential crime scene, such as a homicide, should be managed by law enforcement who will preserve and secure any evidence and close off the scene or roadway. Once a scene has been released by law enforcement, the removal of vehicles will be coordinated with a tow agency.

In some cases, a vehicle may need to be turned right side up. This may cause sparks or a fire. In addition, the high-voltage batteries in hybrid and electric vehicles may be a delayed reaction fire risk. Be aware of smoke, sparks, irritating fumes, or a gurgling sound from the battery. These are warning signs of a damaged high-voltage battery that is undergoing a thermal event and is at risk for fire hours or even days later. There is also a risk of toxic off-gassing. Being proactive by standing by with a charged hose line is always a good practice during one of these procedures.

Any fluid hazards that have spilled from the vehicle, such as gasoline, motor oil, transmission fluid, or radiator fluid that was either captured with an absorbent or not, will be removed by the towing agency. Towing agencies must be licensed to transport and dispose of any hazardous substances; fire rescue agencies are typically not licensed to do so. To assist the towing agency, place any vehicle parts, such as doors, roofs, and fenders, back in a heavily damaged vehicle. A word of advice is to always ask the tow agency representative if they need assistance before you start throwing items back into a vehicle. The tow agency may have procedures or a policy of their own for stowing and transporting loose objects. Being considerate of their preferences will help to maintain a good working relationship with the tow agency for future endeavors such as acquiring vehicles for use in an extrication class. It is also advisable to carry a contact list of various resources for any private and public organization or business that can offer a particular resource that can be utilized on the incident, such as public works/utilities, the department of transportation (DOT), or a heavy equipment company.

FIRE FIGHTER II Tips FFII

All electric and hybrid vehicles should be towed using a flatbed. Towing with the drive wheels on the ground could result in an electrical fire.

The exception to these procedures will be an incident involving a fatality. Law enforcement will acquire the scene immediately to conduct a fatality investigation. In the event that law enforcement investigators take possession of the vehicle, brief them on the hazards and warning signs of a damaged high-voltage battery.

SKILL DRILL 26-10 Removing the Roof of a Vehicle FFII

Fire Fighter II (NFPA 6.4.1)

1. Don PPE, including face protection. Retrieve and set up the required tools. Check the equipment for readiness. Assess the vehicle for hazards, including passive restraint devices. Communicate with the victim, and minimize hazards to both the rescuers and the victim. Remove any remaining glass.

2. Engage appropriate tools to remove the roof, while using good body mechanics.

3. Control the vehicle roof and protect against sharp edges at all times.

4. Remove the roof to a safe location. Return the tools to the staging area upon completion of the tasks.

FIRE FIGHTER II Tips FFII

To ensure your safety and the safety of every member of your crew:

- Follow the manufacturers' recommendations when using power tools.
- Inspect all power tools before and after use.

FIRE FIGHTER II Tips FFII

After the incident, conduct a brief meeting with the crew to discuss what went right during the incident and what could be done better.

Wrap-Up

Chief Concepts

- The design of motor vehicles is constantly evolving. As a fire fighter, you need to keep up-to-date on the design of motor vehicles.
- The most common types of vehicles on the road are conventional vehicles, which use internal combustion engines that burn gasoline for power. Alternative-powered vehicles may be powered by compressed natural gas (CNG), liquefied natural gas (LNG), or liquefied petroleum gas (LPG; propane). The hazards posed by these vehicles include fuel fires.
- Electric-powered vehicles are powered by an electric motor that is powered by batteries. These vehicles pose electrical hazards due to the large number of batteries needed to power the engine.
- Hybrid vehicles use both a battery-powered electric motor and a gasoline-powered engine. As a consequence, these vehicles pose both electrical hazards and fuel fire hazards.
- The main parts of a motor vehicle include the following components:
  - The "A" posts form the sides of the windshield.
  - The "B" posts are located between the front and rear doors of a vehicle.
  - The "C" posts are located behind the rear passenger windows.
  - The bulkhead divides the engine compartment from the passenger compartment.
  - The passenger compartment (also called the occupant cage) includes the front and back seats.
  - The two common types of vehicle frames are the platform frame and the unibody. The platform frame is used for SUVs and trucks; it uses beams to form the load-bearing frame of a vehicle. The unibody is used in most passenger vehicles; it combines the vehicle body and the frame into a single component.
- Extrication should follow a series of logical steps:
  - Safely respond to the scene—Pick the best route, drive in a safe manner, and wear your safety belt.
  - Arrival and scene size-up—Assess the hazards and determine whether additional resources are needed. Be aware of the traffic and any spilled fuel.
  - Stabilize the scene—Reduce, remove, or mitigate the hazards at the scene.
  - Stabilize the vehicle—Use cribbing and/or rescue-lift air bags to stabilize the vehicle before the victim can be removed.
  - Access the victim—The victim may be accessed (and assessed) through a door, window, or a removed roof.
  - Disentangle the victim—Using hand and power tools, remove the tangle of metal from around the trapped victim.
- The following tools are used for stabilizing a vehicle:
  - Cribbing
  - Step chocks
  - Wedges
  - Rescue-lift air bags
- Four techniques may be used for gaining access to and disentangling the victim of a motor vehicle collision:
  - Stabilize or hold an object or vehicle.
  - Bend, distort, or displace.
  - Cut or sever.
  - Disassemble.
- The following classes of tools are used for extrication:
  - Tools for stabilization: Cribbing, rescue-lift air bags, step chocks, high-left jacks, and stabilization jacks.
  - Tools to displace, bend, and distort metal: Pry axes, short pry bars, long pry bars, hacksaws, screw drivers, hammers, and Halligan tools.
  - Tools for cutting: Axes, bolt cutters, cable cutters, hacksaws, manual hydraulic cutters, air chisels, and power saws.
- To gain access to the victim, the following techniques are used:
  - Open the door.
  - Break a window.
  - Force open the door.
- To disentangle the victim from the vehicle, the following techniques are used:
  - Displace the seat.
  - Remove the windshield.
  - Remove the steering wheel.
  - Displace the dashboard.
  - Displace the roof.
- The final phase of the extrication process consists of removing and transporting the victim to the hospital. During this phase, the victim is stabilized, packaged, removed from the vehicle, and transported to a medical facility.

Hot Terms

Alternative-powered vehicle A vehicle that is powered by compressed natural gas or other fuel.

"A" posts Vertical support members that form the sides of the windshield of a motor vehicle.

"B" posts Vertical support members located between the front and rear doors of a motor vehicle.

Bulkhead (firewall) The wall that separates the engine compartment from the passenger compartment in a motor vehicle.

Wrap-Up, continued

Conventional vehicle A vehicle that uses an internal combustion engine and is fueled with either gasoline or diesel fuel.

"C" posts Vertical support members located behind the rear doors of a motor vehicle.

Cribbing Short lengths of timber/composite materials, usually 101.60 millimeters × 101.60 millimeters (4 inches × 4 inches) and 457.20 millimeters × 609.60 millimeters (18 inches × 24 inches) long that are used in various configurations to stabilize loads in place or while a load is moving. (NFPA 1006)

Electric vehicle An automotive-type vehicle for on-road use, such as passenger automobiles, buses, trucks, vans, neighborhood electric vehicles, and the like, primarily powered by an electric motor that draws current from a rechargeable storage battery, fuel cell, photovoltaic array, or other source of electric current. Electric motorcycles and similar-type vehicles and off-road, self-propelled electric vehicles, such as industrial trucks, hoists, lifts, transports, golf carts, airline ground support equipment, tractors, boats, and the like, are not included in this definition. (NFPA 70)

Fuel cell An electrochemical system that consumes fuel to produce an electric current. The main chemical reaction used in a fuel cell for producing electric power is not combustion, but sources of combustion may be used within the overall fuel cell system such as reformers/fuel processors. (NFPA 70)

Fuel cell stack Many fuel cells assembled into a group to produce larger quantities of electricity.

Hybrid vehicle A vehicle that uses a battery-powered electric motor and an internal combustion engine.

Laminated windshield glass A type of window glazing that incorporates a sheeting material that stops the glass from breaking into individual shards.

Platform frame A type of vehicle frame resembling a ladder, which is made up of two parallel rails joined by a series of cross members. This kind of construction is typically used for luxury vehicles, sport utility vehicles, and all types of trucks.

Post One of the vertical support members, or pillars, of a vehicle that holds up the roof and forms the upright columns of the passenger compartment.

Rescue-lift air bags Extrication tools consisting of air bags, filler hoses, air regulators, control valves, and a supply of compressed air. These tools are used to lift vehicles that have been involved in crashes.

Step chocks Specialized cribbing assemblies made out of wood or plastic blocks in a step configuration. They are typically used to stabilize vehicles.

Tempered glass A type of safety glass that is heat treated so that it will break into small, less dangerous pieces.

Unibody (unit body) The frame construction most commonly used in vehicles. The base unit is made of formed sheet metal; structural components are then added to the base to form the passenger compartment. Subframes are attached to each end. This type of construction does away with the rail beams used in platform-framed vehicles.

Wedges Material used to tighten or adjust cribbing and shoring systems. (NFPA 1006)

FIRE FIGHTER *in action*

It is midnight and you find yourself arriving at the scene of a serious motor vehicle collision where a pickup has t-boned the driver's door of a 2012 Honda Civic. The driver of the pickup, who appears highly intoxicated, is wandering around with minor cuts and scrapes. His passenger has her legs pinned between the seat and the dash. She has open fractures to her femurs.

The 19-year-old driver of the car is deceased, and her body is trapped between the door and the center console. The front-seat passenger is unconscious and her back is resting against the passenger door. The back-seat passenger is semi-conscious and is unable to get out of the back seat. Your captain looks at you and says, "I'll check out the pickup, you check the car."

FIRE FIGHTER *in action* (continued)

1. You decide to remove the roof and you begin with the vertical support member that runs along the windshield and the front of the passenger door. What is the term for this member?
   - **A.** "A" post
   - **B.** "B" post
   - **C.** "C" post
   - **D.** "D" post
2. Prior to making your first cut, you know you must first stabilize the vehicle. What will you use?
   - **A.** Rescue-lift air bag
   - **B.** Struts
   - **C.** Cribbing
   - **D.** Spreaders
3. What type of glass is generally used in the side and rear windows of automobiles?
   - **A.** Laminated
   - **B.** Tempered
   - **C.** Heat-resistant
   - **D.** Flat
4. Which of the following would precede the others in the extrication process?
   - **A.** Disentangle the patient
   - **B.** Stabilize the vehicle
   - **C.** Assess the patient
   - **D.** Transport the patient
5. Which of the following tools is used to displace, bend, or distort metal?
   - **A.** Hack saw
   - **B.** Axe
   - **C.** Step chock
   - **D.** Air chisel
6. After covering the patient, the next step in removing a windshield with an axe is to cut:
   - **A.** along the "A" post starting at the top.
   - **B.** in the center, starting at the top.
   - **C.** along the top.
   - **D.** along the bottom of the windshield.

FIRE FIGHTER II *in action*

You are the first to arrive at a motor vehicle collision in your privately owned vehicle (POV). It involves two vehicles, one of which has rolled down an embankment and is now resting on its side. You find two people in the overturned vehicle. One is pinned and has moderate injuries, while the other is in her seatbelt, but unharmed. In the other vehicle, one has a serious head injury, another has an amputated arm, while the last is yelling and screaming at the occupants of the overturned vehicle for causing the accident.

You recognize the need for more help than your small, volunteer department can provide. Your department has a single engine with a basic set of extrication tools. In addition, there is an ambulance responding. You quickly begin to process the information and then key up your radio.

1. What additional resources should you request?
2. How will you assign other fire fighters that arrive by POV?
3. How will you handle the irate occupant?
4. How will you balance your need to manage the incident with your need to provide care?

Assisting Special Rescue Teams

Fire Fighter I

Knowledge Objectives

There are no knowledge objectives for Fire Fighter I candidates. NFPA 1001 contains no Fire Fighter I Job Performance Requirements for this chapter.

Skills Objectives

There are no skill objectives for Fire Fighter I candidates. NFPA 1001 contains no Fire Fighter I Job Performance Requirements for this chapter.

Fire Fighter II FFII

Knowledge Objectives

After studying this chapter, you will be able to:

- Define the types of special rescues encountered by fire fighters. (NFPA 6.4.2.A, p 840–841)
- Describe the steps of a special rescue. (NFPA 6.4.2, 6.4.2.A, p 841–846)
- Describe the general procedures at a special rescue scene. (NFPA 6.4.2, 6.4.2.A, p 846–848)
- Describe how to safely approach and assist at a vehicle or machinery rescue incident. (NFPA 6.4.2, 6.4.2.A, p 850–851)
- Describe how to safely approach and assist at a confined-space rescue incident. (NFPA 6.4.2, 6.4.2.A, p 851–852)
- Describe how to safely approach and assist at a rope rescue incident. (NFPA 6.4.2, 6.4.2.A, p 852–853)
- Describe how to safely approach and assist at a trench and excavation rescue incident. (NFPA 6.4.2, 6.4.2.A, p 853–854)

CHAPTER 27

- Describe how to safely approach and assist at a structural collapse rescue incident. (NFPA 6.4.2, 6.4.2.A, p 854–855)
- Describe how to safely approach and assist at a water or ice rescue incident. (NFPA 6.4.2, 6.4.2.A, p 855–857)
- Describe how to safely approach and assist at a wilderness search and rescue incident. (NFPA 6.4.2, 6.4.2.A, p 857–858)
- Describe how to safely approach and assist at a hazardous materials rescue incident. (NFPA 6.4.2, 6.4.2.A, p 858–860)
- Describe how to safely respond to an elevator or escalator rescue. (NFPA 6.4.2, 6.4.2.A, p 859–860)

Skills Objectives

After studying this chapter, you will be able to perform the following skills:

- Establish a barrier. (NFPA 6.4, 6.4.2.B, p 847)
- Identify and retrieve rescue tools. (NFPA 6.4, 6.4.2.B, p 850)

Additional NFPA Standards

- NFPA 472, *Standard for Competence of Responders to Hazardous Materials/Weapons of Mass Destruction Incidents*
- NFPA 1006, *Standard for Technical Rescuer Professional Qualifications*
- NFPA 1500, *Standard on Fire Department Occupational Safety and Health Program*
- NFPA 1670, *Standard on Operations and Training for Technical Search and Rescue Incidents*
- NFPA 1951, *Standard on Protective Ensembles for Technical Rescue Incidents*

You Are the Fire Fighter

You hear a call come in that is only two blocks from your home. A construction worker has been buried in a trench collapse. You travel with your crew to the scene. On arrival, you find an idling backhoe at the end of the trench and two workers in the trench digging feverishly with hand shovels to find their co-worker.

1. What is the first priority?
2. What do you say to the workers?
3. What is your greatest concern?

Introduction

In recent years, the number of fires and the associated human and property losses from fires in the United States have generally decreased. This decrease is attributable to better fire prevention measures, better fire safety education, better building codes, and better fire safety design. At the same time, the mission of fire departments has expanded. As fire departments take on added roles, such as Emergency Medical Services (EMS), hazardous materials response, and technical rescue responses, they are generally seen as the agency that can handle any and all types of emergencies.

A technical rescue incident (TRI) is a complex rescue incident involving vehicles or machinery, water or ice, rope techniques, trench or excavation collapse, confined spaces, structural collapse, wilderness search and rescue, or hazardous materials that requires specially trained personnel and special equipment. This chapter shows you how to assist specially trained rescue personnel in carrying out the tasks for which they have been trained, but it will not make you an expert in the skills needed to handle these eight types of rescue situations. Specialized training is required to acquire true proficiency in TRIs. The more training you receive in these areas, the better you will understand how to conduct yourself at these scenes.

Training in technical rescue areas is conducted at three levels: awareness, operations, and technician.

- Awareness level: This training serves as an introduction to the topic, with an emphasis on recognizing hazards, securing the scene, and calling for appropriate assistance. There is no actual use of rescue skills at the awareness level.
- Operations level: This level is geared toward working in the "warm zone" of an incident (the area directly around the hazard area). Such training will allow you to directly assist those conducting the rescue operation and use certain skills and procedures to conduct the rescue.
- Technician level: This level allows you to be directly involved in the rescue operation itself. Such training includes use of specialized equipment, care of victims during the rescue, and management of the incident and of all personnel at the scene.

Most of the training you receive as a beginning fire fighter aims at preparing you at the awareness level so that you can identify hazards and secure the scene to prevent further injuries.

Types of Rescues Encountered by Fire Fighters

Most fire departments respond to a variety of special rescue situations FIGURE 27-1:

- Vehicle and machinery rescue
- Confined-space rescue
- Rope rescue
- Trench and excavation rescue
- Structural collapse rescue
- Water and ice rescue
- Wilderness rescue
- Hazardous materials incidents

FIGURE 27-1 Most fire departments respond to a variety of special rescue situations.

To become proficient in handling these situations, you must take a formal course to gain specialized knowledge and skills. NFPA 1670, *Standard on Operations and Training for Technical Search and Rescue Incidents*, identifies operational benchmarks and training needed to establish operational capacity for technical search and rescue events.

It is important for awareness-level responders to have an understanding of these special types of rescues. Often, the first emergency unit to arrive at a rescue incident is a fire department engine or truck company. As a consequence, the initial actions taken by that company may determine the safety of both the victims and the rescuers. These initial actions may also determine how efficiently the rescue is completed.

It is also important for you to understand and recognize the tools needed for rescue situations. You may not have to know how to use all of them, but you may be called upon to assist with a tool at any time.

Guidelines for Operations

When assisting rescue team members, keep in mind the five guidelines that you follow during other firefighting operations:

- Be safe.
- Follow orders.
- Work as a team.
- Think.
- Follow the Golden Rule of public service, which emphasizes the ethic of reciprocity: Treat others as you would like to be treated.

FIRE FIGHTER II Tips FFII

Emergency incidents are unpredictable. Don't just have one backup plan; have two or three.

Be Safe

Rescue situations may be full of hidden hazards, including oxygen-deficient atmospheres, weakened floors, electrical hazards, and strong water currents. Knowledge and training are required to recognize signs indicating that a hazardous rescue situation exists. Once the hazards are recognized, determine which actions are necessary to ensure your own safety as well as the safety of your team members, the victims of the incident, and bystanders. It requires experience and skill to determine that a rescue scene is not safe to enter, but that determination could ultimately save lives.

Follow Orders

When you begin your career as a fire fighter, you will have limited training and experience. By contrast, your officers and the rescue teams with whom you will work on special rescue incidents have received extensive specialized training. They have been chosen for their duties precisely because they have experience and skills in a particular area of rescue. It is critical to follow the orders of those who understand exactly what needs to be done to ensure safety and to mitigate the dangers involved in the rescue situation. Moreover, orders should be followed exactly as given. If you do not understand what is expected of you, ask for more information. Have the orders clarified so you will be able to complete your assigned task safely.

To do well as a fire fighter, you must have a strong appreciation for the basic philosophy of many firefighting organizations. Most fire departments are run like military organizations. Grasping the paramilitary structure will enable you to understand the command and control concept of fire departments.

- The fire officer's knowledge base and experience are greater than yours.
- Orders come from superiors. Legitimate orders are those given by a fire officer or other designated person.
- Follow rules and procedures. A fire fighter is required to follow rules, procedures, and guidelines regardless of his or her personal opinions.
- You must do your own job.
- Get the job done. In emergency situations, time is critical. Nevertheless, you must not act beyond your own skill and training level, and you must not violate any rules, procedures, standard operating guidelines, or orders of a superior in an effort to get the job done quickly.

Acting with these ideas clear in your head will enable you to deal with the orders of your commanding fire officers quickly and confidently.

Work as a Team

Rescue efforts often require many people to complete a wide variety of tasks. Some personnel have been trained in specific tasks, such as rope rescue or swift water rescue FIGURE 27-2. To do their jobs, however, they need the support and assistance of other fire fighters.

Firefighting requires members of the team to work together to complete the goal of fire suppression. Rescue is a different goal, but it requires the same kind of team effort. Your role in the rescue effort is an essential part of this team effort.

FIGURE 27-2 Some personnel are trained in specific rescue skills, such as rope rescue.

Think

As you are working on a rescue situation, you must constantly assess and reassess the scene. You might see something that your company officer does not see. If you think your assigned task may be unsafe, bring this matter to the attention of your superior officer. Do not try to reorganize the total rescue effort, as it is being directed by people who are highly trained and experienced—but do not ignore what is going on around you, either.

Observations that a fire fighter should bring to the attention of a superior fire officer include unsafe structures, changing weather conditions, suspicious packages, and broken equipment. In a hazardous materials rescue incident, the wind direction and speed are important to know. For example, an increase in wind speed can cause hazardous vapors to spread more quickly, which in turn can affect the response plan.

Follow the Golden Rule of Public Service

When you are involved in carrying out a rescue effort, it is easy to concentrate on the technical parts of the rescue and forget that there is a frightened person who needs your emotional support and encouragement. It is helpful to have a rescuer stay with the victim whenever possible. He or she can tell the victim which actions will be performed during the rescue process. This rescuer is often the one whom the victim remembers and the one who helps to fulfill the Golden Rule.

FIRE FIGHTER II Tips — FFII

In any TRI, it is important to know why rescuers fail. The reasons for rescue failures are most commonly referred to by the acronym FAILURE:

- F—Failure to understand the environment, or underestimating it
- A—Additional medical problems not considered
- I—Inadequate rescue skills
- L—Lack of teamwork or experience
- U—Underestimating the logistics of the incident
- R—Rescue versus recovery mode not considered
- E—Equipment not mastered

Steps of Special Rescue

Although special rescue situations may take many different forms, all rescuers take certain basic steps to perform special rescues in a safe, effective, and efficient manner:

1. Preparation
2. Response
3. Arrival and size-up
4. Stabilization
5. Access
6. Disentanglement
7. Removal
8. Transport
9. Security of the scene and preparation for the next call
10. Postincident analysis

Preparation

You can prepare for responses to emergency rescue incidents by training with the specialized rescue teams with whom you will be working. This kind of experience will enable you to learn about their techniques and equipment. It will help you to understand what you can do to assist them as well as what you should *not* do. It is also helpful to train with other fire departments in your area, as such training will better enable you to respond to a mutual aid call. In addition, you will be able to learn about the types of rescue equipment to which other departments have access as well as the training levels of their personnel.

Pay attention to terminology—knowing the terminology used in the field will make communicating with other rescuers easier and more effective. Also, learn to recognize the different types of rescue situations that you could encounter. Prior to a technical rescue call, your department must consider these issues:

- Does the department have the personnel and equipment to handle a technical rescue incident from start to finish?
- Does the department meet National Fire Protection Association (NFPA) and Occupational Safety and Health Administration (OSHA) standards for technical rescue calls?
- Which equipment and personnel will the department send on a technical rescue call?
- Do members of the department know the hazards located within the department's response area, and have they visited these areas with local representatives?

Response

A TRI should have its own dispatch protocol. If your agency has its own technical rescue team, it will usually respond with a rescue unit supported by other equipment and personnel, such as a medic unit, an engine company, a truck company, and a chief. The makeup of the response is based on the ability of the agency or department to provide resources for the response. Even if a technical rescue team is available, it may not be trained or equipped to handle all incidents that could be encountered. Your department should make provisions for additional resources and trained personnel to assist in handling incidents when the TRI is beyond the local technical rescue team's capabilities.

Often, it is necessary to notify power and utility companies during a TRI for possible assistance. Many technical rescues involve electricity, sewer pipes, or other factors that may create the need for additional heavy equipment, to which utility companies have ready access.

Arrival and Size-Up

The first company officer to arrive on the scene should assume command. A rapid and accurate size-up is needed to avoid placing would-be rescuers in danger and to identify additional resources that may be needed. Determine how many victims are involved, and note the extent of their injuries. This information will help to decide how many EMS units and other resources are needed.

When responding to a work site or industrial facility, the incident commander (IC) should make contact with the foreman or supervisor, also known as the responsible party. This individual can often provide valuable information about the work site and the emergency situation. An important part of any rescue is the identification of all hazards. It is also important to determine whether a TRI is a rescue event or a body recovery operation. After these determinations have been made, the IC can decide whether to call for additional resources and order other actions intended to stabilize the incident.

Do *not* rush into the incident scene until you have adequately assessed the situation, because you may inadvertently make the situation worse. For example, a fire fighter approaching a trench collapse may cause further collapse. A fire fighter entering a swiftly flowing river might be quickly knocked down and carried downstream. A fire fighter climbing down into a well to evaluate an unconscious victim may be overcome by an oxygen-deficient atmosphere. Stop and think about the dangers that may be present—do not make yourself part of the problem. Always be alert for invisible dangers such as electrical hazards and oxygen-deficient or poisonous atmospheres.

Fire Fighter Safety Tips

At many fire and emergency scenes, fire line tape is set up to keep bystanders back. Emergency responders operating at such a scene routinely ignore this barrier by simply ducking under or climbing over the tape. This complacency toward fire line tape as a warning device creates a potential hazard for responders at TRIs and other emergency scenes where the tape is placed to warn not just bystanders but responders as well. All fire line tape should be considered a hazard warning to fire fighters unless and until it is proved to be for bystanders only.

Stabilization

Once needed resources are on the way and the scene is safe to enter, begin to stabilize the incident. Establish an outer perimeter to keep the public and media out of the staging area, and establish a smaller perimeter directly around the rescue site. A rescue area is an area that surrounds the incident site (e.g., a collapsed structure, a collapsed trench, or a hazardous spill area) and whose size is proportional to the hazards that exist. Three controlled zones should be established:

- Hot zone: This area is for entry teams and rescue teams only. The hot zone immediately surrounds the dangers of the site (e.g., hazardous materials releases) and is established to protect personnel outside the zone.
- Warm zone: This area is for properly trained and equipped personnel only. The warm zone is where personnel and equipment decontamination and hot zone support take place.
- Cold zone: This area is for staging vehicles and equipment. The cold zone contains the command post. The public and the media should be kept clear of the cold zone at all times.

To figure out the scope of each of these zones, fire fighters should identify and evaluate the hazards that are discovered at the scene, observe the geographical area, note the routes of access and exit, observe weather and wind conditions, and consider evacuation problems and transport distances.

The most common method of establishing the control zones for an emergency incident site is to use police or fire line tape. Police or fire line tape is available in a variety of colors. The most commonly used colors are red, for the hot zone; orange, for the warm zone; and yellow, for the cold zone. For other methods of securing the perimeters of the controlled zones, see the Hazardous Materials: Personal Protective Equipment, Scene Safety, and Scene Control chapter.

Once the fire department has established the three controlled zones, responders should ensure that the zones of the emergency scene are enforced. Because of lack of firefighting personnel, scene control activities are sometimes assigned to law enforcement personnel.

Lockout and tagout systems should be used at this time to ensure an electrically safe environment **FIGURE 27-3**. Lockout and tagout systems are intended to ensure that the electricity has been shut off and that electrically powered equipment is not unintentionally turned on.

In an emergency situation, all power must be turned off and locked out. As part of this effort, switches or valves must be tagged with labels to protect personnel and emergency response workers from accidental machine start-up. Lockout and tagout procedures are used to warn personnel and ensure that the electrical power is disconnected. Authorized and

FIGURE 27-3 Lockout and tagout systems are methods of ensuring that the electricity has been shut down.

trained personnel can disconnect the source of power, lock it out, and tag it. Locks and tags are used for everyone's protection against electrical dangers. For your safety and that of others, never remove or ignore a lock or tag. Always be alert for electrical hazards, because they are not necessarily easy to recognize. Vehicle crashes, building collapses, entrapments, and natural disasters, for example, can all create electrical hazards to rescuers and to patients. These scenes require a careful overview of the scene and continual monitoring of the scene.

During stabilization, atmospheric monitoring should be started to identify any immediately dangerous to life and health (IDLH) environments for rescuers and victims. The next steps involve looking at the type of incident and planning how to rescue victims safely. For example, in a trench rescue, these measures might include setting up ventilation fans for air flow, setting up lights for visibility, or protecting a trench from further collapse.

Access

Once the scene is stabilized, rescuers need to consider how to gain access to the victim. How is the victim trapped? In a trench situation, a dirt pile may be to blame. In a rope rescue, scaffolding may have collapsed. In a confined space, a hazardous atmosphere may have caused the victim to collapse. To reach a victim who is buried or trapped beneath debris, it is sometimes necessary to dig a tunnel as a means of rescue and escape. Identify the actual reason for the rescue, and work toward freeing the victim safely.

Communicate with the victims at all times during the rescue to make sure they are not injured further by the rescue operation. Even if they are not injured, all victims need to be reassured that the team is working as quickly as possible to free them.

Emergency medical care should be initiated as soon as safe access is made to the victim. Technical rescue paramedics are vital resources at TRIs: Not only can these responders start intravenous lines and treat medical conditions, but they also know how to deal with the equipment and procedures that are going on around them. It is vitally important that EMS personnel be effectively integrated into ongoing operations during rescue incidents. Their main functions are to treat victims and to stand by in case a rescue team member needs medical assistance. As soon as a rescue area or scene is secured and stabilized, EMS personnel must be allowed access to the victims for medical assessment and stabilization. Some fire departments have trained paramedics to the technical rescue level so that they can enter hazardous areas and provide direct assistance to the victim. Throughout the course of the rescue operation (which can sometimes span many hours), EMS personnel must continually monitor and ensure the stability of the victims and, therefore, must be allowed access to the victims.

The strategy used to gain access to victims depends on the type of incident. For example, in an incident involving a motor vehicle, the location and position of the vehicle, the damage to the vehicle, and the position of the victim are important considerations. The means of gaining access to the victim must also take into account the victim's injuries and their severity. The chosen means of access may have to be changed during the course of the rescue as the nature or severity of the victim's injuries becomes apparent.

Disentanglement

Once precautions have been taken and the reason for entrapment has been identified, the victim needs to be freed as quickly and as safely as possible. A team member should remain with the victim to direct the rescuers who are performing the disentanglement. In a trench incident, this step would include digging either with a shovel or by hand to free the victim.

In a vehicle accident, the most important point to remember is that the vehicle is to be removed from around the victim, rather than trying to remove the victim through the wreckage. Many different parts of the vehicle may trap the occupants, including the steering wheel, seats, pedals, and dashboard. In this scenario, disentanglement entails the cutting of a vehicle (and or machinery) away from trapped or injured victims, which is accomplished by using extrication and rescue tools with a variety of extrication methods.

FIRE FIGHTER II Tips **FFII**

Technical rescue incidents may involve longer operational periods than structure fires. Pay attention to your physical state and that of your team members. If you are tired, go to the rehabilitation area.

Removal

Once the victim has been disentangled, efforts will be redirected toward removing the victim **FIGURE 27-4**. In some instances, this may simply amount to having someone assist the victim up a ladder. In other cases, it may require removal with spinal immobilization due to possible injuries. Stokes baskets, backboards, stretchers, and other immobilization devices are all commonly used to remove an injured victim from a trench, a confined space, or an elevated point.

FIGURE 27-4 Victim removal.

Preparing the victim for removal involves maintaining continued control of all life-threatening problems, dressing wounds, and immobilizing suspected fractures and spinal injuries. The application of standard splints in confined areas may be difficult or sometimes even impossible, but stabilization of the arms to the victim's trunk and of the legs to each other will often be adequate until the victim is positioned on a long spine board, which may serve as a splint for the whole body. A short spine immobilization device is sometimes used to stabilize a sitting victim.

Sometimes a victim has to be removed quickly (rapid extrication) because his or her general condition is deteriorating, and time will not permit meticulous splinting and dressing procedures. Quick removal may also be necessary if hazards—such as spilled gas or hazardous materials—pose a danger to the victim or the rescue personnel. The only time the victim should be moved prior to completion of initial care, assessment, stabilization, and treatment is when the victim's life or the emergency responder's life is in immediate danger.

Packaging is the process of preparing the victim for movement as a unit. It is often accomplished by means of a long spine board or similar device. Such boards are essential for moving victims with potential or actual spine injuries.

Rough-terrain rescues may require passing a stretcher up and over rough terrain, fording streams, or rock climbing. These kinds of rescues often require ingenuity to suspend or pad a stretcher so that the victim is provided with a reasonably comfortable ride. Padding such as inflated inner tubes, 4- to 5-inch (10- to 13-centimeter) foam padding, or loosely rolled blankets may be used in such circumstances. Try to achieve smooth movement of the stretcher to prevent excess swaying and bouncing of the patient.

The overriding objective for each rescue, transfer, and removal is to complete the process as safely and efficiently as possible. It is important that the rescuer utilize good body mechanics, victim-packaging, removal, and transportation skills.

Transport

Once the victim has been removed from the hazard area, transport to an appropriate medical facility is accomplished by EMS personnel. The type of transport will vary depending on the severity of the victim's injuries and the distance to the medical facility. For example, if a victim is critically injured or if the rescue is taking place some distance from the hospital, air transportation may be more appropriate than use of a ground ambulance.

In rough-terrain rescues, four-wheel drive, high-clearance vehicles may be required to transport victims on stretchers to an awaiting ambulance. Snowmobiles with attached sleds can be used to transport victims down a snowy mountain. Helicopters are increasingly used for quick evacuation from remote areas, but weather conditions may sometimes limit their use.

Transporting victims who have been injured in a hazardous materials incident may pose many problems. Prior to transport, it is extremely important that adequate decontamination occur. Placing a poorly decontaminated victim inside an ambulance or helicopter and then closing the doors puts both the victim and the rescue personnel in danger. Any toxic fumes given off by the victim or by the victim's clothing can contaminate the inside of the transport vehicle, causing injury to the EMS personnel. In such an incident, any bags containing contaminated clothing or other personal effects must be properly sealed if they are going with the victim. Receiving hospitals should be notified of the impending arrival of victims who have been involved in a hazardous materials exposure incident. This will alert the hospital's triage teams so that they can take the appropriate precautions once the victims begin to arrive.

Throughout any extended-duration technical rescue event, it is important to address the rehabilitation needs of rescue personnel. During many such situations, it will be necessary to rotate personnel. This course of action reduces the chance for injury or sickness to the rescuers and also provides a safer rescue for the victim. Rehabilitation is especially important when temperatures are very high or very low, and when rescuers are subjected to wet conditions. It is better to plan for rehab and not need it than to not have rehab available when it is needed. For more information on rehabilitation, see the Fire Fighter Rehabilitation chapter.

Postincident Duties

Security of the Scene and Preparation for the Next Call

Once the rescue is complete, the scene must be stabilized by the rescue crew to ensure that no one else becomes injured. In a trench incident, this includes filling the trench with dirt and roping off the surrounding area.

In a hazardous materials incident, the clean-up of equipment and personnel takes place after the hazardous materials incident has been completely controlled and all victims have been treated and transported. Trained disposal crews should be available to clean the site. All equipment, protective gear, and clothing, as well as the rescue personnel themselves, must be decontaminated.

In a vehicle rescue incident, the vehicle is removed; usually it is transported by a tow truck to a storage lot. The street or roadway is cleaned of all debris from the vehicle. This includes the clean-up of any spilled fluids, including gasoline, from the vehicle that was involved in the incident.

In an industrial setting, the supervisor of the facility is responsible for securing the scene. Nevertheless, the technical rescue team must follow up with the supervisor or the safety coordinator to ensure that further problems are prevented.

Once you have secured the scene and packed up your equipment, it is important to return to the station and fully inventory, clean, service, and maintain all of the equipment to prepare it for the next call. Although some items may need repair, most will require only simple maintenance before being placed back on the apparatus and considered in service.

Back at the station, as you prepare for the next call, complete all paperwork and document the rescue incident. Record keeping serves several important purposes. Adequate reporting and the keeping of accurate records ensure the continuity of quality care, guarantee proper transfer of responsibility, and fulfill the administrative needs of the fire department for local, state, and federal reporting requirements. In addition, the reports can be used to evaluate response times, equipment usage, and other areas of administrative interest.

Postincident Analysis

As with any type of call, the best way to prepare for the next rescue call is to review the last one and identify the strengths and weaknesses in your response. What did you do well? What could have been done better? Would any other equipment have made the rescue safer or easier? If a death or serious injury occurred during the call, a critical incident stress management session may occur to assist fire fighters. Reviewing a TRI with everyone involved will allow everyone to learn from the call and make the next call even more successful.

General Rescue Scene Procedures

At any scene you respond to—whether it is a fire, EMS, or technical rescue call—the safety of you, your company, and the public is paramount. At a TRI, many issues need to be considered during the response. While the temptation may be to immediately approach the victim or the accident area, it is critically important to slow down and properly evaluate the situation.

In confined-space rescue incidents, for example, potential hazards may include deep or isolated spaces, multiple complicating hazards (such as water or chemicals), failure of essential equipment or service, and environmental conditions (such as snow or rain). In all rescue incidents, fire fighters should consider the potential general hazards and risks posed by utilities, hazardous materials, confined spaces, and environmental conditions, as well as IDLH hazards.

Approaching the Scene

As you approach the scene of a TRI, you will not always know which kind of scene you are entering. Is it a construction scene? Do you see piles of dirt that would indicate a trench? Has a structural collapse occurred in a building? Which actions are the civilians taking? Are they attempting to rescue trapped people, possibly placing themselves at great danger?

Beginning with the initial dispatch of the rescue call, fire fighters should be compiling facts and factors about the call. Size-up begins with the information gained first from the person reporting the incident and then from the bystanders at the scene upon arrival.

The information received when an emergency call is received is important to the success of the rescue operation. It should include the following details:

- The location of the incident
- The nature of the incident (kinds, cause, what is involved, etc.)
- The conditions and positions of victims
- The conditions and positions of vehicles, building, structure, terrain, etc.
- The number of people trapped or injured, and the types of their injuries
- Any specific or special hazard information
- The name of the person calling and a number where that person can be reached

Many times the information received on the emergency call will be incomplete. You will have to evaluate the incident with the information available. Once on the scene, you can better identify life-threatening hazards and take the appropriate corrective measures to mitigate those dangers. If additional resources are needed, they should be ordered by the IC.

A size-up should include the initial and continuous evaluation of the following issues:

- Scope and magnitude of the incident
- Risk–benefit analysis
- Number of known and potential victims
- Hazards
- Access to the scene
- Environmental factors
- Available and necessary resources
- Establishment of control perimeters

Careful size-up and coordination of rescue efforts are a must to avoid further injuries to the victims and to provide for the safety of the fire fighters.

Dealing with Utility Hazards

Are there any downed electrical wires near the scene FIGURE 27-5? Is the equipment or machinery present electrically charged so that it might present a hazard to the victim or the rescuers? The IC should ensure that the proper procedures have been taken to shut off the utilities in the area where the rescuers will be working.

Utility hazards require the assistance of trained personnel. For electrical hazards, such as downed lines, park at least one utility pole span away. Watch for falling utility poles—a damaged pole may bring other poles down with it. Do not touch any wires, power lines, or other electrical sources until they have been deactivated by a power company representative. It is not just the wires that are hazardous; any metal that they touch is also energized. Metal fences that become energized,

FIGURE 27-5 Downed electrical wires present a hazard at many types of rescue scenes.

for example, are energized for their entire unbroken length. Be careful around running or standing water, because it is an excellent conductor of electricity.

Both natural gas and liquefied petroleum gas are nontoxic but are classified as asphyxiants because they displace breathing air. In addition, both gases are explosive. If a call involves leaking gas, call the gas company immediately. If a victim has been overcome by leaking gas, wear positive-pressure self-contained breathing apparatus (SCBA) when making the rescue. Remove the victim from the hazardous atmosphere before beginning treatment.

Providing Scene Security

Has the area been secured to prevent people from entering the area? Often co-workers, family members, and sometimes other rescuers will enter an unsafe scene and become additional victims. The IC should coordinate with law enforcement to help secure the scene and control access. A strict personnel accountability system should be implemented by the fire department to control access to the rescue scene.

One task you will often be called upon to perform is to establish barriers to provide scene security at a rescue incident. When emergency incidents occur on streets or highways, vehicle warning lights, traffic cones, and fuses are often used to ward off oncoming traffic. When crowd control is an issue, plastic barrier tape marked with the words "Caution" or "Fire Line" is often used to keep out nonessential personnel. The steps for establishing a barrier are listed in SKILL DRILL 27-1:

1. Respond safely to the emergency scene.
2. Place the emergency vehicle in a safe position that protects the scene. (STEP 1)
3. Perform a size-up to assess for hazards.

SKILL DRILL 27-1 Establishing a Barrier FFII

Fire Fighter II (NFPA 6.4.2)

1 Place the emergency vehicle in a safe position that protects the scene.

2 Perform a size-up to assess for hazards. Secure the scene.

3 Use appropriate devices to establish a barrier, following the IC's orders.

* Note: This is an artistic rendering. The fire apparatus should be positioned away from the building collapse field.

4. Secure the scene. (STEP 2)
5. Call for needed assistance.
6. Don the appropriate personal protective equipment (PPE).
7. Use appropriate devices to establish a barrier, following the IC's orders. (STEP 3)

Using Protective Equipment

Firefighting gear is designed to protect the body from the high temperatures of fire. It does, however, restrict movement. Technical rescues generally require the ability to move around freely, so most firefighting gear does not work well in a TRI. For this reason, most specialized teams also carry items such as harnesses; smaller, lighter helmets; and jumpsuits that are easier to move in than turnout gear FIGURE 27-6.

FIGURE 27-6 Technical rescue technicians may need to be able to move in their gear, depending on the type of incident.

For personnel operating in a water rescue hazard zone, the minimum PPE includes a Coast Guard–approved personal flotation device (PFD), thermal protection, a helmet appropriate for water rescue, a cutting device, a whistle that works in water, and contamination protection (if necessary).

Other equipment items that are easily carried by fire fighters include binoculars, chalk or spray paint for marking searched areas, a compass for wilderness rescues, first-aid kits, a whistle, a hand-held global positioning system (GPS), and cyalume-type light sticks. The IC and the technical rescue team will help determine which PPE you will need to wear while assisting in the rescue operation. Also see NFPA 1951, *Standard on Protective Ensembles for Technical Rescue Incidents*.

Using the Incident Command System

The first-arriving officer immediately assumes command and starts using the incident command system (ICS). The ICS's activation is critically important because many TRIs will eventually become very complex and require a large number of assisting units. Without the ICS in place, it will be difficult—if not impossible—to ensure the safety of the rescuers. For more information on ICS, see the Incident Command System chapter.

Ensuring Accountability

Accountability should be practiced at all emergencies, no matter how small. A personnel accountability system is important to ensure rescuers' safety. It tracks the personnel on the scene, including their identities, assignments, and locations. This system ensures that only rescuers who have been given specific assignments are operating within the area where the rescue is taking place. By using a personnel accountability system and working within the ICS, an IC can track the resources at the scene, make appropriate assignments, and ensure that every person at the scene operates safely. For more information on the personnel accountability system, see the Fire Fighter Survival chapter.

Making Victim Contact

At any rescue scene, every effort should be made to communicate with the victim if at all possible. Technical rescue situations often last for hours, with the victim being left alone for long periods of time. Sometimes making contact with the victim is just not possible, but do try to communicate via a radio, by a cell phone, or by yelling. Reassure the victim that everything possible is being done to ensure his or her safety.

For the fire fighter, it is important to stay in constant communication with the victim. If possible, assign someone to talk to the victim while other fire fighters carry out the rescue. Realize that the victim could be sick or injured and is probably frightened. If the fire fighter is calm, his or her reassuring demeanor will calm the victim. To help keep a victim calm, do the following:

- Make and keep eye contact with the victim.
- Tell the truth. Lying destroys trust and confidence. You may not always tell the victim everything, but if the victim asks a specific question, answer it truthfully.
- Communicate at a level that the victim can understand.
- Be aware of your own body language.
- Always speak slowly, clearly, and distinctly.
- Use the victim's name.
- If a victim is hard of hearing, speak clearly and directly to the person, so that the person can read your lips.
- Allow time for the victim to answer or respond to your questions.
- Try to make the victim comfortable and relaxed whenever possible.

Many victims at TRIs require medical care, but this care should be given only if it can be done safely. Do not become a victim yourself during a rescue attempt. Many fire fighters have been killed while entering a TRI in an attempt to help a dead victim.

Assisting Rescue Crews

Every TRI is different. If you have the role of assisting a technical rescue team, training with the team is probably the most important thing you can do to prepare for your participation in a TRI. By training with the team you will respond with, you will get a feel for how the team members operate and they will get an idea of what they can trust you with. The more knowledge you have, the more you will be able to do.

VOICES OF EXPERIENCE

Several years ago, I was involved in a technical rescue in one of the mountainous regions of our response area. A hiker had fallen an unknown distance while scrambling up a hillside. On our arrival, it became clear the rescue would be technically complex and involved. The victim appeared to have an unstable pelvic fracture and possible spinal injuries. Just getting to the victim involved a significant hike; climbing rocky ledges, negotiating tangled brush, and scrambling up loose hillsides.

While the victim was being packaged, we planned the rescue. We decided to carry the victim approximately 300 yards to where a high-angle rope system could be set up. The victim would be lowered down a sandstone slope, and then transitioned to a sloping highline to clear a cliff and thick underbrush. The bottom of the highline would bring the victim right to the ambulance.

Three teams of rescuers, hundreds of feet of rope, and many pounds of equipment were required to accomplish the rescue. It took nearly two hours before the victim was loaded into the ambulance. In the end, the victim only remembered one detail from the rescue. She remembered and appreciated greatly the rescuer who took the time to massage her feet when they became cold. I continue to use this story as an example of the importance of the "little things" during a rescue.

Shane Baird
Durango Fire Rescue
Durango, Colorado

At any TRI, follow the orders of the company officer who receives direction from the IC. Many of the tasks will involve moving equipment and objects from one place to another. Other tasks will involve protecting the team and victims. Do not take these tasks lightly; they might not seem important but they are essential to the effort. Without scene security, traffic control, protective hose lines, and people to maintain equipment, the rescue cannot and will not happen.

One task you will often be called upon to perform at any type of special rescue incident is to retrieve needed rescue tools. To do so, you must become familiar with the tools that are used by the specialized rescue teams in your department. The steps for identifying and retrieving rescue tools are listed in SKILL DRILL 27-2 (Fire Fighter II, NFPA 6.4.2):

1. Respond safely to the emergency scene.
2. Place the emergency vehicle in a safe position that protects the scene.
3. Perform a size-up to assess for hazards.
4. Secure the scene.
5. Call for needed assistance.
6. Don the appropriate PPE.
7. Receive the request from the special rescue team to retrieve rescue tools.
8. Identify and retrieve the rescue tools FIGURE 27-7.

FIGURE 27-7 Identify and retrieve the rescue tools.

As you read the information that follows about assisting at specific types of rescue scenes, bear in mind the three factors about safely approaching the scene, which apply to all TRIs:

- Approach the scene cautiously.
- Position the apparatus properly.
- Assist the specialized team members as needed.

Vehicles and Machinery

A wide variety of rescue situations involving vehicles and machinery are possible. By far, the most common type that fire fighters encounter is motor vehicle crashes. The actions commonly taken at the scene of motor vehicle crashes are covered in the Vehicle Rescue and Extrication chapter. This section focuses on other types of rescues involving vehicles and machinery, including motorcycles, trucks, buses, trains, mass-transit vehicles, aircraft, watercraft, farm machinery, agriculture implements, construction machinery, and industrial machinery. Rescues involving mass transit often involve many victims, whereas most other types of rescues involve single victims.

Safe Approach

Vehicle and machinery rescues present a wide variety of problems to rescuers and victims. Hazards present at these scenes may, for example, include flammable liquids, electrical hazards, and unstable machines or vehicles.

Electricity is an invisible hazard. Consider any machine you encounter to be electrically charged until proven otherwise. Because of this risk, it is important to use lockout and tagout systems with any potential electrical source before approaching the scene. The operator of the machine or a maintenance person who is familiar with the machine can be a valuable resource in this regard.

With vehicle-related calls, traffic control is a crucial issue. Every year, fire fighters and EMS workers are struck and killed at accident scenes by passing drivers who became distracted by the commotion. Placement of the apparatus is key to the protection of you and those around you. In addition, whenever you are in a traffic-related situation, you must wear an approved vest to increase your visibility and prevent injuries.

Stabilization of the rescue scene includes making sure that the machine cannot move. Gaining access to some victims will be easy. In other situations, access to the victim can occur only after extensive stabilization of the scene.

Unusual and special rescue situations include extrication and disentanglement operations at incidents involving vehicles on their tops or sides, trucks and large commercial vehicles, and vehicles on top of other vehicles. To ensure responder safety, be wary of fuel spilled at these scenes.

How You Can Assist

At a vehicle or machinery rescue call, you may be called upon to assist in the extrication and treatment of the victim. Adequate protection of the victim (e.g., with c-spine immobilization or holding of blankets) will allow the technical rescue team to operate hydraulic tools and cut the vehicle or machine apart without further hurting the victim. Many hydraulic tools are heavy, so you may need to help support the tools while technicians operate these pieces of equipment.

One skill with which you can assist is the removal of a patient from a vehicle using a long backboard. This skill is covered in the Search and Rescue chapter. You may also be called upon to do the following:

- Assist with controlling site security and the perimeter of the rescue incident
- Obtain information from witnesses
- Retrieve more rescue tools and equipment from the fire apparatus or rescue trucks
- Console a family member of a victim

- Keep bystanders out of the way
- Assist in moving items that are in the way of the rescue team
- Service equipment
- Set up power tools
- Stand by with a hose line
- Extinguish a fire

Tools Used

Many tools—including some inexpensive items—are required for a successful vehicle or machinery rescue situation. Simple tools such as a spring-loaded punch, for example, can safely break out a vehicle window to access a victim. Tools and items that will be required for a vehicle or machinery rescue include the following:

- Personal protective equipment
- Hydraulic tools (spreaders, cutters, rams)
- Halligan tool
- Cutting torch
- Air chisels
- Cribbing
- Saber saw
- Windshield cutter
- Spring-loaded punch
- Chains
- Air bags (high and low pressure)
- SCBA air cylinders
- Basic hand tools
- Come along
- Portable generator
- Seat belt cutters
- Hand lights and other scene lighting
- Hose lines (protection)
- Blanket

Confined Spaces

A **confined space** is an enclosed area that is not designed for people to occupy. Confined spaces have limited openings for entrance and exit and may have limited ventilation to provide air circulation and exchange. They can occur in natural, farm, commercial, and industrial settings. For example, grain silos, industrial pits, tanks, and below-ground structures are all confined spaces. Likewise, cisterns, well casings, sewers, and septic tanks are confined spaces that are found in residential settings.

Confined spaces present special hazards, as they may be oxygen deficient, contain poisonous gases, or have combustible atmospheres. Entering a confined space without testing the atmosphere for safety and without the proper breathing apparatus can result in death. Confined spaces may also contain hazardous chemicals as well as harmful dusts and molds. Other serious safety or health hazards present at these sites may include slips, trips, falls, noise, dust, poor lighting, and sensitizing agents. A confined-space call is sometimes dispatched as a heart attack or medical illness call, because the caller assumes that the person who entered a confined space and became unresponsive because of a heart attack or medical illness rather than because of a hazardous atmosphere.

All confined spaces should be considered to contain, or have the potential to contain, a hazardous atmosphere until proven otherwise.

Fire Fighter Safety Tips

The two greatest hazards in a confined space are a lack of oxygen and the presence of poisonous gases or asphyxiants. If rescuers fail to identify a confined space and to ensure that it contains a safe atmosphere, injury and death can occur to rescuers as well as to the original victim. In several cases, well-meaning rescuers have gone into a confined space, become incapacitated, and died because they did not properly evaluate the situation.

Safe Approach

As you approach a rescue scene, look for a bystander who might have witnessed the emergency. Information gathered prior to the technical rescue team's arrival will save valuable time during the actual rescue. Do not assume that a person in a pit has simply suffered a heart attack; instead, assume that an IDLH atmosphere is present at any confined-space call. An IDLH atmosphere can immediately incapacitate anyone who enters the area without breathing protection. Toxic or explosive gases may be present, or the confined space may not have enough oxygen to support life.

When a rescue involves a confined space, remember that it will take some time for qualified rescuers to arrive on the scene and prepare for a safe entry into the confined space. Given this delay, the victim of the original incident may have died before your arrival. Do not put your life in danger on behalf of a body recovery. The majority of deaths in confined spaces are would-be rescuers.

How You Can Assist

Your main role in confined-space rescues is to secure the scene, preventing other people from entering the confined space until additional rescue resources arrive. If the incident has occurred at an industrial site, ascertain whether a confined-space entry permit was prepared. If so, locate the entry supervisor and attendant listed on the permit and have them stand by at the space. As additional highly trained personnel arrive, your company may provide help by giving the rescuers a situation report.

The first-responding company must share whatever information it discovered at the rescue scene with other arriving crews. Anything that might be important to the response should be noted by the first-arriving unit. Observed conditions should be compared to reported conditions, and a determination should be made as to the relative change over the time period. Whether an incident appears to be stable or has changed greatly since the first report will affect the operation strategy for the rescue. A size-up should be completed quickly and immediately upon arrival, and this information should be relayed to the special rescue team members upon their arrival at the scene. Other items of importance that should be included in a situation report are a description of any rescue attempts that have been made, the exposures, any hazards present, the extinguishment of fires, the facts and probabilities of the scene, the situation and resources of the fire company, the identities of any hazardous materials present, and an evaluation of the progress made so far.

Many engine and ladder companies carry atmospheric or gas detection devices; if you have this capability, obtain readings to determine whether the situation presents a hazard. If you do encounter an oxygen-deficient atmosphere, set up a ventilation fan to help remove toxic gases and improve air flow to the victim. Sometimes you can help a victim without entering the confined space by passing down SCBA, oxygen, first-aid supplies, or even a ladder that the victim can use to climb out.

Confined-space rescues can be complex and take a long time to complete. You may be asked to assist by bringing rescue equipment to the scene, maintaining a charged hose line, or providing crowd control. By understanding the hazards associated with confined spaces, you will be better prepared to assist a specialized team that is dealing with an emergency involving a confined space.

Tools Used

One of the key tools used in any confined-space rescue is a supplied-air respirator system FIGURE 27-8. These devices are similar to SCBA, except that instead of carrying the air supply in a cylinder on your back, you are connected by a hose line to an air supply located outside the confined space. This kind of system provides the rescuer with a continuous supply of air that is not limited by the capacity of a back-mounted air cylinder.

A number of other tools and pieces of equipment can also be used during a confined-space rescue operation:

- Air lines
- Air carts
- Extra SCBA air cylinders
- Tripod and winch for raising and lowering rescuers and victims
- Pry bar or manhole cover remover
- Rescue rope
- Personnel harnesses
- Gas monitors
- Ventilation fans
- Explosion-proof lights
- Radios
- Hard-wired communications equipment
- Protective gear
- Lockout, tagout, and blankout kits
- Accountability boards
- Victim removal devices such as short backboard devices, Stokes baskets, backboards, and other commercially made devices
- Medical equipment
- Carabiners and locking "D" rings
- Descenders
- Safety goggles
- Hand light (battery-operated)
- Hearing protection headsets

Rope Rescue

Rope rescue skills are used in many types of TRIs. For example, an injured victim might be lifted from inside a tank using a raising system, or rescuers might need to lower themselves down to the area where victims are trapped in a structural collapse. Rope rescue skills are the most versatile and widely used technical rescue skills.

Rope rescue incidents are divided into low-angle and high-angle operations. Low-angle operations are situations where the slope of the ground over which the fire fighters are working is less than 45 degrees, fire fighters are dependent on the ground for their primary support, and the rope system becomes the secondary means of support FIGURE 27-9. An example of a low-angle system is a rope stretched from the top of an embankment and used for support by rescuers who are carrying a victim up an incline. In this scenario, the rope is the secondary means of support for the rescuers; the primary support is the rescuers' contact with the ground.

Low-angle operations are used when the scene requires ropes to be used only as assistance to pull or haul up a victim or rescuer. This step usually becomes necessary when adequate footing is not present in areas such as a dirt or rock embankment. In such an incident, a rope will be tied to the rescuer's harness and the rescuer will climb the embankment on his or

FIGURE 27-8 Supplied-air respirator system.

FIGURE 27-9 Low-angle operations.

her own, using the rope only to make sure he or she does not fall. Low-angle operations also include lifelines deployed during ice or water rescues.

Ropes can also be used to assist in carrying a Stokes basket. This use of ropes aids the rescuers and frees them from having to carry all of the weight over rough terrain. Rescuers at the top of the embankment can help to pull up or lower the basket using a rope system.

High-angle operations are situations where the slope of the ground is greater than 45 degrees, and rescuers or victims depend on life safety rope for support, rather than on a fixed surface such as the ground. High-angle rescue techniques are used to raise or lower a person when other means of raising or lowering are not readily available. Sometimes, a rope rescue is performed to remove a person from a position of peril; at other times, it is needed to remove an ill or injured victim.

In the Ropes and Knots chapter, you learned how to tie some basic knots and to use ropes to lift and lower selected tools. Rope rescue training courses will build on the skills you have already learned. There is much more to learn before you are ready to perform high-angle rescues, however, because you need to be able to perform complex tasks quickly and safely in these incidents. Successful rope rescue operations also require a cohesive team effort.

Safe Approach

If you respond to an incident that may require a rope rescue operation, consider your safety and the safety of those around you. Rope rescues are among the most time-consuming calls that you will encounter. Extensive setup is necessary, and a significant body of equipment must be assembled prior to initiating any rescue. Protect your safety by remaining away from the area under the victim and away from any loose materials that may fall or slide. Likewise, when you are above the victim, stay back from the edge unless you are securely tied off to keep from falling. Use extreme caution so you do not cause any loose debris to fall on the victim. Work to control the scene so that the bystanders and friends of the victim move to an area where they will not be injured.

You can do a lot to stabilize the scene and prevent further injuries by remaining calm and putting your skills to work. A rope can, for example, be used to secure a rescuer when making a rescue attempt, haul a stretcher or litter up an embankment, or lower a rescuer into a trench or cave.

How You Can Assist

Rope rescues are labor-intensive operations. By remembering the introductory rope skills you have learned, you may be able to assist with many phases of the rescue operation. For instance, a technical rescue team member might ask you to tie knots and get anchors ready. Do not be offended if the team members go back and check your work—it is protocol to check a system two or three times to ensure its safety.

Avoid stepping on ropes, because any damage or friction to the outer layers of a rope can decrease its tensile strength and possibly cause a catastrophic failure. Keep everyone clear of areas where falling objects could cause injuries. Keep in mind the importance of following the ICS. Work to complete your assigned tasks.

Tools Used

Rope rescues rely almost completely on the equipment used. When you are rappelling down to rescue a victim who is 300 feet (90 meters) off the ground, you need to have equipment that has been properly designed and maintained to keep you alive. Some of the tools and equipment that you will need to recognize and use are listed here:

- PPE (helmet, rescue gloves)
- Personnel harnesses
- Stokes basket
- Harness for Stokes basket
- Rescue ropes
- Carabiners
- Webbing and Prussic cord
- Miscellaneous hardware (racks, pulleys)

Trench and Excavation Collapse

Trench and excavation rescues become necessary when earth has been removed for placement of a utility line or for other construction and the sides of the excavation collapse, trapping a worker FIGURE 27-10. Entrapments can occur when children play around a pile of sand or earth that collapses. Many entrapments occur because the required safety precautions were not taken.

Whenever a collapse has occurred, you need to understand that the collapsed product is unstable and prone to further collapse. Earth and sand are heavy, and a person who is partly entrapped in these materials cannot simply be pulled out. Instead, the victim must be carefully dug out. This step can be taken only after shoring has stabilized the sides of the excavation.

FIGURE 27-10 Trench rescue.

Vibration or additional weight on top of displaced earth will increase the probability of a secondary collapse—that is, a collapse that occurs after the initial collapse. A secondary collapse can be caused by equipment vibration, personnel standing at the edge of the trench, or water eroding away the soil. Safe removal of trapped persons requires a special rescue team that is trained and equipped to erect shoring to protect the rescuers and the entrapped person from secondary collapse.

Safe Approach

Safety is of paramount importance when approaching a trench or excavation collapse. Walking close to the edge of a collapsed area can trigger a secondary collapse, so stay away from the edge of the collapse and keep all workers and bystanders at a distance from the site. Vibration from equipment and machinery can also cause secondary collapses, so shut off all heavy equipment. Likewise, vibrations caused by nearby traffic can cause collapse, so it may be necessary to stop or divert traffic near the scene.

Soil that has been removed from the excavation and placed in a pile is called the spoil pile. This material is unstable and may collapse if placed too close to the excavation. Avoid disturbing the spoil pile.

Make verbal contact with the trapped person if possible, but do not place yourself in danger while doing so. If you plan to approach the trench at all, do so from the narrow end, where the soil will be more stable. However, it is best not to approach the trench unless absolutely necessary. Stay out of a trench unless you have been properly trained in trench rescue techniques.

Provide reassurance by letting the trapped person know that a trained rescue team is on the way. By moving people away from the edges of the excavation, shutting down machinery, and establishing contact with the victim, you start the rescue process.

You can also size up the scene by looking for evidence that would indicate where the trapped victims may be located. Hand tools are an indicator of where victims may have been working, as are hardhats. By questioning the bystanders, you can also determine where the victims were last seen.

How You Can Assist

As the rescue team starts to work, your company will be assigned to carry out certain tasks, ranging from unloading lumber for shoring to assisting with cutting timbers at a safe distance away from the entrapment. This type of rescue can take a long time. If it is hot or cold, a rehabilitation sector may need to be set up. Early implementation of the ICS will help make this type of rescue go smoothly. If someone in your company has a specialty such as carpentry, he or she should make this fact known, because this skill is valuable at a trench rescue. Cutting and measuring timber and shores for a trench operation is an important aspect of the rescue.

Like confined spaces, trenches may have an IDLH atmosphere inside owing to the presence of poisonous gases such as methane or other sewer gases. Another hazard may be gasoline fumes from gas-powered equipment or gas cans in the trench. Buried power lines may also be present. Setting up ventilation fans can often make a difference in victim and rescuer survivability in these scenarios.

It may also be necessary to pump water out of a trench. If the collapse occurred because of a ruptured pipe or during a rainstorm, for example, rising water in the trench may endanger the trapped victim and cause additional soil to collapse into the trench. Removing this water can, therefore, stabilize the situation.

When extricating victims from the trench, it may be necessary to lift them out using a raising system, just as in a high-angle rope rescue operation. This effort will require that the rescuers have rescue ropes, harnesses, pulleys, carabiners, and all of the other associated equipment available. Personnel must be trained in how to use and apply these systems in this type of rescue scenario.

Tools Used

Trench and excavation rescue uses equipment, rescue techniques, and skills similar to those used with both confined-space rescue and structural collapse rescue. Tools and equipment used in trench and excavation rescue include the following:

- PPE: helmet, gloves, personal protective clothing, harness, flashlight, work boots, knee pads, elbow pads, eye protection, SCBA, and supplied-air breathing apparatus (SABA)
- Hydraulic, pneumatic, and wood shores
- Lumber and plywood for shoring
- Cribbing
- Power cutting tools and saws
- Carpentry hand tools
- Shovels
- Buckets for moving soil
- Rescue rope, harnesses, webbing, and associated hardware
- Utility rope
- Ventilation fans
- Water pumps
- Lighting
- Ladders
- Shovels
- Extrication equipment such as Stokes baskets or backboards
- Harness sets for Stokes baskets
- Medical equipment

Structural Collapse

Structural collapse is the sudden and unplanned fall of part or all of a building FIGURE 27-11. Such collapses may occur because of fires, removal of supports during construction or renovation, vehicle crashes, explosions, rain, wind, snowstorms, earthquakes, and tornados.

The Building Construction chapter emphasizes the importance of all building components working together to provide a stable building. Consider the type of building construction when determining the potential for collapse. When any part of a building becomes compromised, the dynamics of the building will change. As a consequence, fire fighters should always be alert for signs of a possible building collapse. A partial

FIGURE 27-11 Structural collapse is the sudden and unplanned fall of part or all of a building.

building collapse may also be hazardous to rescuers because of the potential for secondary collapse.

Safe Approach

Many different factors can lead to building collapses, so you must approach the scene carefully. The causes of some collapses—for example, a vehicle crashing through the wall of a house—will be readily evident. Other causes, such as a natural gas explosion, may require extensive investigation before they are definitively identified.

Fire Fighter Safety Tips

Fires often lead to structural collapses. Be alert for the possibility of a building collapse anytime a fire has damaged a building.

As you approach any building collapse, consider the need to shut off utilities. Entering a structure with escaping natural gas or propane is extremely hazardous. Electricity from damaged wiring can also present a deadly hazard.

A prime safety consideration is the stability of the building. Even a well-trained engineer cannot always determine the stability of a building simply by looking at its exterior. Therefore, you must operate as though the building might experience a secondary collapse at any time. The IC must make the decision regarding whether the building is safe to enter. This is a difficult decision, especially given that bystanders cannot always tell you whether people were present in the building before the collapse.

FIRE FIGHTER II Tips FFII

New fire fighters must learn when to offer helpful suggestions and when to remain quiet. Sometimes you might see something that the company officer does not see, or you might have an idea that the company officer has not considered. In such a case, the need to provide helpful information must be balanced against the risk of overwhelming a company or chief officer with "helpful" suggestions. No hard-and-fast rule covers all situations, of course, but stay mindful of the fact that remaining quiet is sometimes the most helpful thing you can do.

How You Can Assist

Rescue operations at a structure collapse vary depending on the size of the building and the amount of damage to the building. In cases of large building collapses, the rescue operation will be sizable. Urban search and rescue teams and structural collapse teams have received special training in dealing with these types of situations. Specifically, they are trained in shoring and specialized techniques for gaining access and extricating victims that enable them to systematically search the affected building. These types of rescue situations take a lot of time and require a lot of personnel.

Personnel who do not have this kind of special training will usually be assigned to support operations. For example, you might be assigned to be part of a bucket brigade that works to remove debris from the building. A substantial amount of manual digging and searching may be necessary, all of which is physically taxing. You may be able to assist with this activity if the work is not in an unstable location.

For a rescue effort of this type to work, there must be teamwork and a well-organized and well-implemented ICS. Without a solid command structure, most large-scale rescue efforts are doomed to fail.

Tools Used

Fire fighters should know the tools and equipment designed for structural collapse emergency rescue incidents:

- PPE: helmet, gloves, work boots, harness, elbow pads, knee pads, eye protection, SCBA, SABA, and dust masks
- Shoring equipment
- Lumber for shoring and cribbing
- Power tools
- Hand tools
- Lighting
- Rescue ropes, harnesses, webbing, and associated hardware
- Utility ropes
- Buckets
- Shovels

Water and Ice Rescue

Almost all fire departments have the potential for being called to perform a water rescue FIGURE 27-12. Water is present in small streams and large rivers, and it fills lakes, oceans, reservoirs, irrigation ditches, canals, and swimming pools. A static source such as a lake may have no current, whereas a whitewater stream or flooded river may have a swift current. During a flood, a dry wash in the desert can temporarily become a huge raging, swiftly flowing river.

Fire Fighter Safety Tips

Wearing turnout gear while conducting water rescue is dangerous and can be deadly. Do not wear turnout gear within 10 feet of water, especially around swift-moving water.

Near Miss REPORT

Report Number: 08-0000557

Synopsis: Embankment rescue could have ended tragically.

Event Description: We were called on a technical rescue response for a report of two dogs stuck on a cliff. The response consisted of an engine, a rescue company, a truck company, a chief, and a safety officer. The reported location was a rugged area of suburban/wildland at the mouth of a steep canyon. The first-due engine responded from the station approximately 1 mile (1.6 kilometer) away. The rest of the units came from substantially farther away and would have a delayed arrival.

Upon arrival, the first-due engine company met with the reporting person and were escorted up to the area where the dogs were. They were on small ledge approach, 75 feet (23 meter) down from a small promontory that served as their access point. The terrain below the access point was a series of steep loose rock and grass slopes ending in small 5–10 drop-offs to a ledge. The dogs were on the third ledge down at the end of a 10-foot (3 meter) drop-off. Below that ledge was an approximately 50-foot (15 meter) drop to a dirt access road.

The two fire fighters had grabbed a utility rope bag and a rope rescue pack before heading up. Upon reaching the access point, the first fire fighter began to scramble down to the dogs. As the dogs were initially agitated and growling, the second fire fighter felt that he should proceed down to assist his partner, who had now reached the dogs.

About halfway down the slope, the second firefighter realized the severity of the terrain and asked the bystander to throw down a rope. He threw down a utility rope while holding on to one end. The closest fire fighter tied a loop around his waist and tossed the end down to his partner, who did the same. The fire fighters asked the bystander to "belay" them.

The engine officer observed this part of the operation from a vantage point below the access point. He gave a size-up that said they had reached the dogs and it was a simple low-angle rescue.

When the rescue company arrived on scene, the rescue officer was assigned to rescue operations and proceeded up to the access point. He updated the size-up to include the fact that the rescue was a high-angle rescue and that the first priority was to remove the two fire fighters to safety. A lowering system with safety was established, and a technical rescue team member was lowered to the stuck fire fighter. He was put in a harness and safety and was hauled up. The technical rescue team member proceeded down to the first fire fighter and the dogs. He was placed in a harness and hauled up. The dogs were then assisted up to the access point and walked down to animal control officers, who checked them over and figured out their ownership. The technical rescue team member came up safely and the operation was broken down.

Lessons Learned: The main lesson learned was that you should not let emotion and adrenalin override judgment. The dogs had been on the cliff for at least several days and were in no immediate danger. There was plenty of time to assess the situation and plan for a proper, safe operation.

The first-in officer needed to be more of a presence and control his crew to prevent them from acting hastily.

- Awareness—The first fire fighters began to scramble down and were already in a dangerous situation before they realized that fact and asked a bystander for assistance.
- Training—Frequent training with the rescue equipment carried on the first-due engine would have allowed the team members to set up a safe lowering system in a timely manner and still achieve their objectives.

In North America, the most common swiftwater rescue scenario involves a car that has tried to drive through a pool of water created by a storm. When the vehicle stalls because of the depth of the water, it leaves the vehicle occupants stranded in a rising stream with a swift current. If the water is high enough, the vehicle can be swept away. These incidents are especially dangerous for rescue personnel, because it is difficult to determine the depth of the water around the vehicle.

Fire departments should be prepared for all types of water rescue incidents that may occur in their communities. Fixed bodies of water in your community are easy to identify. Identifying the areas in your community where flooding is likely to occur will require preplanning. Other common water rescue scenarios include cars rolling into lakes or streams, cars going off bridges, people falling into bodies of water, and swimmers getting into trouble.

FIGURE 27-12 Water rescue.

Safe Approach

When responding to water rescue incidents, your safety and the safety of other teammates is your primary concern. Your turnout gear is designed for structural firefighting, and it provides amazing protection for that purpose. Unfortunately, it is not designed for water rescue activities. Fire fighters who fall into the water while wearing turnout gear quickly find that their movements are severely limited by the bulky gear. For this reason, when working at a water rescue scene, you should use PPE designed for water rescue, not structural firefighting. Anytime you are within 10 feet of the water, you should be wearing a Coast Guard–approved PFD. Shoes that provide solid traction are preferred to fire boots in this situation.

If you are part of the first-arriving engine or truck company, and the endangered people are in a vehicle or holding on to a tree or other solid object, try to communicate with them. Let them know that additional help is on the way.

Do not exceed your level of training. If you cannot swim, operating around or near water is not recommended. A person who is trained as a lifeguard for still water is not prepared to enter flowing water with a strong current, such as is found in a river, stream, or ocean. Trained swift water rescuers may consider using a throw rope, a pike pole, or a ladder to reach a person. Many people who are in trouble in the water are close to shore. Do not use a boat unless you are trained in its use. Ensure that bystanders do not try to rescue the victim and place themselves in a situation where they need to be rescued as well.

Ice rescues are common in colder climates, and many departments in colder regions have developed specialized equipment for these scenarios. Throwing a rope or flotation device may be helpful initially. In addition, ladders can be extended to the victim or used to distribute the weight of the rescuer on ice-covered water. A fire hose that has been capped at each end and then inflated with an SCBA air cylinder can be used to create to create a flotation buoy. In addition, many other devices can be used to keep a victim above the ice and pull him or her to safety. Rescuers in this situation need special ice rescue suits to prevent hypothermia and provide flotation. If your department is involved in ice rescues, you need to receive training in these specialized procedures.

FIRE FIGHTER II Tips **FFII**

All fire fighters should be trained in the use of water rescue throw bags. This equipment contains a rope that is thrown to a struggling victim in water or on ice. The rescuer secures one end of the rope before throwing the water rescue throw bag. Once the victim has grasped the water rescue throw bag, the rescuer can pull the victim to shore.

How You Can Assist

Water and ice rescues are dangerous for rescuers. PFDs should be worn by everyone operating at the scene. Good scene control is needed to prevent additional people from becoming victims.

At water and ice rescues, you may be called upon to assist by keeping victims in sight, retrieving equipment, assisting with rescue ropes, and changing rescuers' air supplies. You can also assist special rescue teams by relaying communications.

Be alert for changing weather conditions, such as an increase in wind speed or a change in temperature that could affect the rescue operations.

Tools Used

Fire fighters should become familiar with the tools used in water rescue operations:

- PFD
- Self-contained underwater breathing apparatus (SCUBA)
- Helmet
- Waterproof whistle
- Throw bag with rope
- Boat
- Rescue rope, webbing, and associated hardware
- Lighting and flashlights
- Gloves
- Goggles or other eye protection
- Suitable footwear for the environmental and rescue conditions

Wilderness Search and Rescue

Wilderness search and rescue (SAR) is conducted by a limited number of fire departments. It is included in NFPA 1670, *Standard on Operations and Training for Technical Search and Rescue Incidents*, as part of technical rescue training. SAR missions consist of two parts: search and rescue. Search is defined as looking for a lost or overdue person. Rescue, in the SAR context, is defined as removing a victim from a hostile environment.

Several types of situations may result in the initiation of SAR missions. For example, small children may wander off and be unable to find their way back to a known place. Older adults who are suffering from disorders such as Alzheimer's disease may fail to remember where they are going and become lost. People who are hiking, hunting, or participating in other wilderness activities may become lost because they do not have the proper training or equipment for that activity, because the weather changes unexpectedly, or because they become sick or injured.

Safe Approach

As a beginning fire fighter, you may respond to a possible SAR mission as part of your fire company. In cases of a lost person, some small fire departments will request all available personnel to respond to assist with a search. It is important to respond to such a call as part of your department and to work together as a team. The IC for such a mission should be a person who is well trained in directing SAR missions. In some communities, SAR is the responsibility of law enforcement personnel. In others, it becomes the responsibility of volunteer search and rescue groups.

The term "wilderness" may be used to describe a variety of environments, such as forests, mountains, deserts, parks, and animal refuges. Depending on the terrain and environmental factors, the wilderness can be as little as a few minutes into the backcountry or a few feet off the roadway. An incident with a short access time could require an extended evacuation and, therefore, qualify as a wilderness incident. Examples of terrains found in wilderness areas include cliffs, steep slopes, rivers, streams, valleys, mountainsides, and beaches. Terrain hazards may include cliffs, caves, wells, mines, avalanches, and rock slides.

When you participate in SAR missions, prepare for the weather conditions by bringing suitable clothing. Make sure that you do not exceed your physical limitations, and do not get in situations that are beyond your ability to handle yourself in the wilderness. Call for a special wilderness rescue team, depending on the needs of the situation and your local protocols.

How You Can Assist

The fire service teaches its members to work in a buddy system; wilderness rescues are another situation in which you never go out alone. By working in teams of at least two and having a radio, teams can be methodically deployed and assigned to a search. The more knowledge you have about the search area, the more vital your role is during the search. Knowing your response area and the places where a child or lost person might hide is always beneficial to the search. A well-coordinated team can be an effective SAR force; conversely, an unorganized group of people will produce increasing chaos. Do not enter the search area before the search team arrives. If the team will be using dogs, your scent will distract them.

Tools Used

Each specialized rescue team will bring its own equipment to the scene. Many of the same tools, equipment, and personal protective clothing that you would use for other specialized rescue situations will be used in SAR emergencies:

- Personal clothing appropriate for the environment
- Water
- Food
- Lighting and flashlights
- Communications equipment, including radios
- Medical equipment
- Maps, compass, and GPS
- Shelter
- Rescue ropes, harnesses, webbing, and associated hardware
- Extrication equipment such as a Stokes basket and harness
- Flare gun and flares, whistles, or other signaling devices

Hazardous Materials Incidents

Hazardous materials are defined as any materials or substances that pose a significant risk to the health and safety of persons or to the environment if they are not properly handled during manufacture, processing, packaging, transportation, storage, use, or disposal FIGURE 27-13.

FIGURE 27-13 Hazardous materials incident.

Although hazardous materials incidents often involve petroleum products, many other chemicals in our society have toxic effects when they are not handled properly. We tend to think of hazardous materials incidents as occurring during transportation or at a large industrial setting, but it is important to consider that many retail businesses contain significant quantities of hazardous materials. Home and garden stores, farm cooperatives, hardware stores, and pool supply stores, for example, are places where hazardous materials can be found in significant quantities.

Most fire departments are trained and equipped to recognize these incidents, contain the hazards, and evacuate people from the area if necessary. See the Hazardous Materials chapters for complete coverage of hazardous materials incidents.

Safe Approach

Hazardous materials incidents are not always dispatched as hazardous materials incidents. Given this fact, you must be able to recognize the signs indicating that hazardous materials may be present as you approach the scene. Preincident plans may note that a fixed facility stores chemicals, or the use of a specific type of transport container may tip you off to the presence of hazardous materials. You might also see an escaping chemical or smell a suspicious odor. Warning placards are required for most hazardous materials for storage or in transit.

Once you have recognized the presence of a hazardous material, you must protect yourself by staying out of the area exposed to the material. If you have happened upon a hazardous materials incident, it is important to have the hazardous materials team dispatched as soon as possible. In addition to standard action and precautions, call for a special hazardous materials rescue team immediately upon your arrival at such an incident, implement site control and scene management, and assist specialized personnel after their arrival according to your training level.

How You Can Assist

To assist in operations at a hazardous materials incident, you need formal training in techniques to deal with this kind of threat. Training at the awareness level will provide you with the knowledge and skills you need to be able to recognize the presence of a hazardous material, protect yourself, call for appropriate assistance, and evacuate or secure the affected area. As part of your training, you will learn how to assist other responders to the hazardous materials incident.

The four major objectives of training at the operational level are to analyze the magnitude of the hazardous materials incident, plan an initial response, implement the planned response, and evaluate the progress of the actions taken to mitigate the incident. Being trained and equipped to perform emergency decontamination of victims may help to minimize the effects of chemical exposures while hazardous materials technicians are en route.

Tools Used

The following tools are used in a hazardous materials incident:

- PPE appropriate to the level of the hazard
- Two-way radios
- Lighting
- Gas monitors
- Sensing and monitoring tools to identify the materials involved
- Extensive research material
- Decontamination equipment
- A wide variety of hand tools (e.g., hammers, screwdrivers, axes, wrenches, pliers)
- Devices for sealing breached containers
- Leak-control devices
- Binoculars
- Fire line tape
- Control agents

Elevator and Escalator Rescue

Elevator and escalator rescues are technical responses that require specialized knowledge. Rescue crews should review the American Society of Mechanical Engineers' *A17.4 Guide for Emergency Personnel* and OSHA's lockout and tagout procedures.

All elevator and escalator responses begin with knowledge of the machinery found in your department's jurisdiction—information that can be gathered only through a preincident analysis of these structures. Specialized training classes are also an essential component of safe and successful operations.

Elevator Rescue

As metropolitan areas continue to grow, building heights are also increasing. Because of our aging population, more buildings are being equipped with elevators. In turn, elevator emergencies are increasing in frequency FIGURE 27-14.

Fire fighters should never attempt to move or relocate an elevator under any circumstances. Only professional elevator technicians who are thoroughly trained and authorized to do so should consider this step. When responding to elevator incidents, consider these recommendations:

1. Always cut power to a malfunctioning elevator, and secure the power supply through the appropriate lockout and tagout procedures. Remember that elevator technicians and building maintenance personnel will have been summoned to the same elevator incident; they may try to turn on the elevator equipment. To eliminate any possibility of this happening, use the lockout and tagout procedures at every incident. Turn off the power supply in the elevator machine room, and identify the location of

FIGURE 27-14 Elevator emergencies are a common emergency call for many fire departments.

(Continued)

Elevator and Escalator Rescue (*Continued*)

the stalled elevator within the hoist-way or elevator shaft. Are there victims in the stalled elevator? If so, communicate with them, and reassure them that the situation is under control, they are not in danger, and the situation will soon be corrected.

2. Implement incident risk management and risk assessment. Can the victims in the disabled elevator be removed without increasing the risk of injury or death to fire fighters as well as to the victims? Elevator cabs are passenger compartments that are securely locked to ensure that the victims are contained within the cab for their own safety. Fire fighters must carefully evaluate the situation and the inherent risks involved in a rescue and extrication operation, and understand that the victims are in a much safer environment inside the elevator than outside it. The best option may be to reassure the victims and wait for the elevator technician to arrive and correct the situation.
3. Are there enough trained personnel and the proper equipment on scene to perform the rescue and extrication operation safely? Elevator hall door release keys, safety harnesses, rope bags, ladders, forcible entry tools, portable lighting, and an assortment of hand tools will be necessary in such a case. Always have additional trained personnel on hand, and anticipate obstacles and problems. You need to know the procedure your department uses to call for assistance when confronted with an elevator rescue.
4. Once the incident has been resolved, leave the power supply off. Always leave the building with the elevator power supply turned off.

Escalator Rescue

Escalator emergencies can be challenging and complex. Fire fighters should never attempt to move an escalator under any circumstances—only professional escalator technicians who are thoroughly trained and authorized to do so should consider this step. This is particularly true if a victim (usually a child) is caught or entangled in the machinery.

Good incident risk management and risk assessment skills are crucial in this type of incident, because an error in judgment or a poorly executed reaction on the part of rescue personnel can easily turn an abrasion, laceration, or fractured bone into an amputated limb. Victims who are caught in machinery must be reassured, calmed, and stabilized medically by the rescuers.

Consider these recommendations:

1. The escalator machinery must be stopped. Stop switches are located at the top and bottom of most escalators. One member of your company must maintain constant pressure on the stop switch until the power supply can be shut down. If the escalator machinery is stopped upon your arrival, which is generally the case, secure the power source. Escalator power supplies are usually located under the top landing or step plate of the escalator. Always turn off the power supply to the escalator and secure the de-energized (power off) power source with the lockout and tagout procedures.
2. Are personnel competently trained for this type of incident? Is the proper equipment on the scene to perform the rescue and extrication operation safely? These are reasonable questions, as you might not always have the experienced, adequately trained personnel or the equipment needed to handle the situation. Again, exercise good judgment through risk management and risk assessment; simply reacting to the situation could have a very negative outcome. Unnecessary or excessive damage to the escalator equipment is also undesirable.
3. The best course may be to take no action until a trained professional arrives on the scene to apply special expertise regarding this type of incident. The machinery may have to be dismantled to reduce the possibility of causing more damage to the entangled victim (usually the victim's hands or feet) or the equipment involved.
4. Leave the power supply to the escalator turned off after the escalator incident has been resolved. Assume nothing. Remember—safety is the number one priority.

Wrap-Up

Chief Concepts

- A technical rescue incident (TRI) is a complex rescue incident involving vehicles or machinery, water or ice, rope techniques, trench or excavation collapse, confined spaces, structural collapse, wilderness search and rescue, or hazardous materials that requires specially trained personnel and special equipment.
- Training in technical rescue areas is conducted at three levels: awareness, operations, and technician. Awareness training serves as an introduction to the topic with an emphasis on recognizing hazards, securing the scene, and calling for appropriate assistance. There is no actual use of rescue skills at the awareness level. Most of the training that beginning fire fighters receive is at the awareness level.
- The following types of TRIs are often encountered by fire fighters:
  - Vehicle and machinery rescue
  - Confined-space rescue
  - Rope rescue
  - Trench and excavation rescue
  - Structural collapse rescue
  - Water and ice rescue
  - Wilderness rescue
  - Hazardous materials incidents
- To become proficient in handling these types of situations, you must take a formal course to gain specialized knowledge and skills.
- When assisting rescue team members, keep five guidelines in mind:
  - Be safe.
  - Follow orders.
  - Work as a team.
  - Think.
  - Follow the Golden Rule of public service: treat others as you would like to be treated.
- The basic steps of special rescue operations are outlined here:
  - Preparation—Train with the specialized rescue teams in your area and become familiar with the terminology used in the field.
  - Response—A technical rescue team will usually respond with a rescue unit supported by other units such as a medic unit, an engine company, a truck company, and a chief.
  - Arrival and size-up—The first company officer to arrive on the scene should assume command and rapidly assess the situation. Do not rush into the incident scene until the situation has been assessed and the hazards identified.
  - Stabilization—Once the needed resources are on the way and the scene is safe to enter, begin to stabilize the scene by establishing a perimeter around the rescue site.
  - Access—Identify the actual reason for the rescue, and work toward freeing the victim safely. Communicate with the victims at all times and initiate emergency medical care as soon as safe access is made to the victim.
  - Disentanglement—Once precautions have been taken and the reason for entrapment has been identified, the victim needs to be freed as safely as possible. Disentanglement removes what is confining the victim from around the victim.
  - Removal—This step could be as simple as assisting a victim up a ladder or as complex as packaging the victim in a Stokes basket and lifting him or her out of a trench.
  - Transport—Once the victim has been removed from the hazard area, transport to an appropriate medical facility is accomplished by EMS personnel.
  - Security of the scene and preparation for the next call—Once the victim has been transported, the scene must be stabilized by the rescue crew to ensure that no one else becomes injured.
  - Postincident analysis—Identify what worked well and which procedures could work better.
- Beginning with the initial dispatch of the rescue call, begin compiling facts and identifying factors pertinent to the call. The information received when an emergency call is received is important to the success of the rescue operation. The information should include the following details:
  - The location of the incident
  - The nature of the incident (kinds and number of vehicles)
  - The conditions and positions of victims
  - The conditions and positions of vehicles
  - The number of people trapped or injured, and the types of their injuries
  - Any specific or special hazard information
  - The name of the person calling and a number where that person can be reached
- A size-up should include the initial and continuous evaluation of the following issues:
  - Scope and magnitude of the incident
  - Risk–benefit analysis

- Number of known and potential victims
- Hazards
- Access to the scene
- Environmental factors
- Available and necessary resources
- Establishment of control perimeters

- After sizing up the scene, take the following precautions to avoid further injuries to the victims and to provide for the safety of other fire fighters:
  - Secure utility hazards.
  - Provide scene security.
  - Use the proper protective equipment for the emergency.
  - Activate the incident command system.
  - Activate the personnel accountability system.
  - Make contact with the victim and keep the victim calm.
- At any type of TRI, follow the orders of the company officer who receives direction from the IC. Many of the tasks assigned to fire fighters will involve moving equipment and objects from one place to another; others will involve protecting the team and victims.
- Vehicle and machinery rescues occur in many settings. These situations require responders to stabilize the machinery and ensure that the electricity is off.
- With a confined space, rescuers must ensure that a pure and adequate air supply is available before entering confined spaces.
- Low-angle and high-angle rope rescues require safe equipment, adequate training, and much caution.
- Trench and excavation rescues are hazardous and require responders to minimize the chance for secondary collapses.
- A damaged building is prone to structural collapse. Whenever a building has been damaged, assume that it may collapse. Consider fire scenes to be potential building collapses.
- Water rescue training is needed in almost all communities. Training, proper PFDs, and appropriate clothing are important to ensure rescuer safety.
- Wilderness rescue may be called for even when initial access to a lost or stranded individual occurs quickly.
- Hazardous materials incidents are not always dispatched as hazardous materials incidents, so be cautious when approaching any incident. Once the presence of a hazardous material is identified, protect yourself by staying out of the area exposed to the material.
- Never attempt to move or relocate an elevator under any circumstances. Always cut the power to a malfunctioning elevator and secure the power supply. Implement incident risk management and risk assessment. Always have additional trained personnel on hand and anticipate obstacles and problems. Once the incident has been resolved, leave the power supply off.

Hot Terms

Cold zone (technical rescue incident) The control zone of an incident that contains the command post and such other support functions as are deemed necessary to control the incident. (NFPA 1500)

Confined space An area large enough and so configured that a member can bodily enter and perform assigned work, but which has limited or restricted means for entry and exit and is not designed for continuous human occupancy. (NFPA 1500)

Entrapment A condition in which a victim is trapped by debris, soil, or other material and is unable to extricate himself or herself.

Hazardous material A substance that, when released, is capable of creating harm to people, the environment, and property. (NFPA 472)

High-angle operation A rope rescue operation where the angle of the slope is greater than 45 degrees. In this scenario, rescuers depend on life safety rope rather than a fixed support surface such as the ground.

Hot zone (technical rescue incident) The area immediately surrounding a hazardous materials spill/incident site that is directly dangerous to life and health. All personnel working in the hot zone must wear complete, appropriate protective clothing and equipment. Entry requires approval by the incident commander or a designated sector officer. Complete backup, rescue, and decontamination teams must be in place at the perimeter before operations begin.

Lockout and tagout systems Methods of ensuring that electricity and other utilities have been shut down and switches are "locked" so that they cannot be switched on, so as to prevent flow of power or gases into the area where rescue is being conducted.

Low-angle operation A rope rescue operation on a mildly sloping surface (less than 45 degrees) or flat land. In this scenario, fire fighters depend on the ground for their primary support, and the rope system is a secondary means of support.

Packaging The process of securing a victim in a transfer device, with regard to existing and potential injuries or illness, so as to prevent further harm during movement. (NFPA 1006)

Personal flotation device (PFD) A displacement device worn to keep the wearer afloat in water. (NFPA 1925)

Personnel accountability system A system that readily identifies both the locations and the functions of all members operating at an incident scene. (NFPA 1500)

Placards Signage required to be placed on all four sides of highway transport vehicles, railroad tank cars, and other forms of hazardous materials transportation that identifies the hazardous contents of the vehicle, using a standardized system: 10¾ inch by 10¾ inch diamond-shaped indicators.

Secondary collapse A subsequent collapse in a building or excavation. (NFPA 1006)

Shoring A structure such as a metal hydraulic, pneumatic/mechanical, or timber system that supports the sides of an excavation and is designed to prevent cave-ins. (NFPA 1670)

Spoil pile A pile of excavated soil next to the excavation or trench. (NFPA 1006)

Supplied-air respirator An atmosphere-supplying respirator for which the source of breathing air is not designed to be carried by the user. (NFPA 1404)

Technical rescue incident (TRI) A complex rescue incident requiring specially trained personnel and special equipment to complete the mission. (NFPA 1670)

Technical rescue team A group of rescuers specially trained in the various disciplines of technical rescue.

Warm zone (technical rescue incident) The area located between the hot zone and the cold zone at an incident. The decontamination corridor is located in the warm zone.

Wilderness search and rescue (SAR) The process of locating and removing a victim from the wilderness.

FIRE FIGHTER *in action*

You find yourself having to travel from your normal station assignment to a downtown station where the technical rescue team is located. You report to your captain who assigns you to the engine company; however, he advises you that the engine crew will be training on technical rescue with his rescue company. You are thrilled by the prospect of getting to learn more about these highly-specialized duties. As soon as the morning briefing is over, the captain asks you to follow him to the bays. He wants to get a feel for your current level of knowledge.

1. What level of responder takes offensive actions at technical rescue incidents?
   A. Technician
   B. First responder
   C. Operations
   D. Awareness
2. In what zone is the contamination corridor usually located?
   A. Cold
   B. Tepid
   C. Warm
   D. Hot
3. A rope rescue operation of greater than 45° is considered a __________ rescue.
   A. Low-angle
   B. High-angle
   C. Obtuse-angle
   D. Acute-angle
4. What must responders ensure is available before entry into a confined space incident?
   A. Ropes and rope hardware
   B. Confined space permit
   C. Operations level responders
   D. Pure and adequate air supply
5. Which of the following is true about elevator incidents?
   A. Do not move an elevator to rescue victims.
   B. Only move the elevator in an upward direction to rescue victims.
   C. Only move the elevator in a downward direction to rescue victims.
   D. Elevators can be moved in either direction to rescue victims.
6. When at the scene of a swift water incident should a PFD be worn?
   A. Whenever a rescuer is in a boat.
   B. Whenever a rescuer is in the water.
   C. Whenever a rescuer is within 10' of the water.
   D. Whenever a rescuer is on scene.
7. What task will you likely be called upon to perform at a technical rescue incident?
   A. Set up safety zones
   B. Lock out, tag out of utilities
   C. Carry tools and equipment
   D. Assist with the extrication of the victim

FIRE FIGHTER II *in action*

Since you first began your career, you have dreamed of getting on the rescue company so you can perform the technical rescue duties. While not exactly the same, you are thrilled when you are assigned to the engine company that is co-located with the heavy rescue company. You have heard that the rescue company crew is highly seasoned and is very competent at their jobs. However, they are also known to be a bit arrogant. At times, their behavior is downright demeaning to others.

During your first shift at the station, the rescue company was going over their equipment with the rookie fire fighter. When asked about the K-tool, his mind drew a blank and he froze. They immediately began riding him about being incompetent. You noticed that everything the rookie did, from cleaning the toilets to washing the Battalion vehicle, was not good enough for them and they all made fun of him. You could tell that it really bothered him.

1. How would you handle the situation?
2. With whom would you discuss the issue?
3. What would you do if your Captain was present when they were teasing him?
4. What responsibility do you have in addressing this?

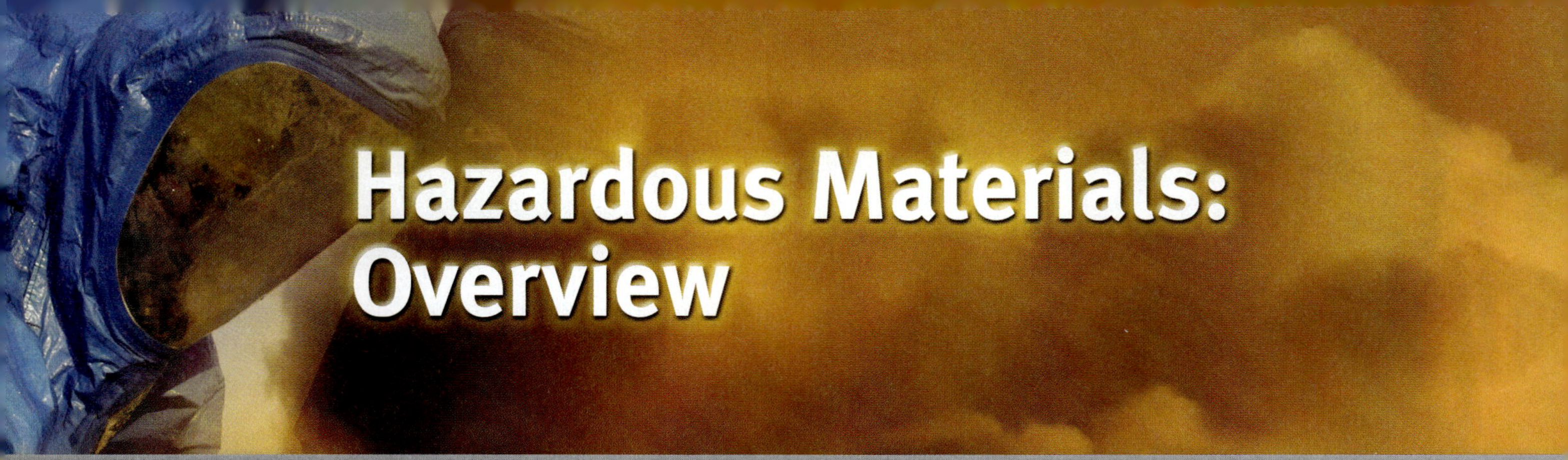

Hazardous Materials: Overview

Fire Fighter I

Knowledge Objectives

After studying this chapter, you will be able to:

- Define a hazardous material. (NFPA 472, 5.1, 5.1.1, 5.1.1.1, 6.6, 6.6.1, 6.6.1.1, 6.6.1.1.1, p 868–869)
- Define hazardous waste. (NFPA 5.1.1.1, p 869)
- List the common locations that may contain hazardous materials. (NFPA 5.1.1.2, p 869)
- Distinguish between a regulation and a standard. (p 869–870)
- Describe which regulations and standards govern how fire departments respond to hazardous materials incidents. (p 869–870)
- Describe the roles and responsibilities of awareness-level personnel. (NFPA 5.1.1.1, 5.1.1.2, 5.1.1.3, 6.6.1.1.2, 6.6.1.1.3, 6.6.1.1.4, p 870–871)
- Describe the roles and responsibilities of operations-level personnel. (NFPA 5.1.1.1, 5.1.1.2, 5.1.1.3, 6.6.1.1.2, 6.6.1.1.3, 6.6.1.1.4, 6.6.1.2, 6.6.1.2.1, p 871)
- Describe the roles and responsibilities of technician-level personnel. (NFPA 6.6.1.1.3, p 871)
- Describe the roles and responsibilities of specialist-level personnel. (p 871–873)
- Describe the roles and responsibilities of a hazardous materials officer. (p 873)
- Describe the roles and responsibilities of hazardous materials safety officers. (p 873)
- Describe the roles and responsibilities of hazardous material technicians with specialties. (p 873)
- List the laws that govern hazardous material response activities. (p 873)
- Explain the differences between hazardous materials incidents and other emergencies. (p 873–874)
- Explain the need for a planned response to a hazardous materials incident. (p 874)

Skills Objectives

There are no skill objectives for Fire Fighter I candidates. NFPA 1001 contains no Fire Fighter I Job Performance Requirements for this chapter.

CHAPTER 28

Fire Fighter II

FFII

Knowledge Objectives

There are no knowledge objectives for Fire Fighter II candidates. NFPA 1001 contains no Fire Fighter II Job Performance Requirements for this chapter.

Skills Objectives

There are no skill objectives for Fire Fighter II candidates. NFPA 1001 contains no Fire Fighter II Job Performance Requirements for this chapter.

Additional NFPA Standards

- NFPA 1072, *Hazardous Materials/Weapons of Mass Destruction Emergency Response Personnel Professional Qualifications*

You Are the Fire Fighter

You and your crew are dispatched for the report of an unusual odor in a multistory residential structure. Initial dispatch information is sketchy but indicates the residents have smelled the odor for several days, and it is causing irritation to their eyes and making them sick to their stomachs. As you arrive on scene, you and your crew observe several of the residents outside of the building; they have skin irritation, are vomiting, and complain of trouble breathing. You and your crew are asked to conduct an initial size-up and report back to the incident commander.

1. Which initial dispatch information would lead you to believe that this incident might involve some type of hazardous material?
2. Which other indicators or direct observations will help identify the presence of hazardous materials?
3. Which actions should be taken to immediately to protect other people in the area, assist in identification of the problem, and make appropriate notifications?

Introduction

The modern fire service must respond to both existing and emerging threats to lives and property. In addition to fighting fires, today's fire fighters may be called to incidents involving chemical spills, gas leaks, emergencies at industrial plants, or railroad or truck crashes with resultant material leaks or spills. These incidents frequently involve hazardous materials that threaten lives, property, and the environment.

This chapter is designed to give you the basic information required to recognize, respond to, and take initial action at a hazardous materials incident. The information is required for Fire Fighter I and II certification and is designed to build a foundation for safe and efficient response practices.

The following material describes the competency levels for first responders at the awareness and operations levels as found in NFPA 472, *Standard for Competence of Responders to Hazardous Materials/Weapons of Mass Destruction Incidents*, depending on your fire department's policies and procedures as they relate to hazardous materials response, and you likely will receive additional training in this area once you complete your basic training.

What Is a Hazardous Material?

A hazardous material, as defined broadly by the U.S. Department of Transportation (DOT), is a material that poses an unreasonable risk to the health and safety of operating emergency personnel, the public, and/or the environment if it is not properly controlled during handling, storage, manufacture, processing, packaging, use and disposal, or transportation. One working definition of hazardous materials is "any substance that stores potentially harmful energy when not contained in its intended container." For our purposes, both definitions are applied throughout this chapter. All fire fighters must be able to recognize the presence of hazardous materials, analyze the available information, and take appropriate action.

FIRE FIGHTER Tips

A concerted effort is being made to ensure that all fire fighters are trained to the operations core competencies.

The ability to recognize a potential hazardous materials incident is critical to ensuring your own safety FIGURE 28-1. To avoid entering a dangerous area, you must be alert to the possibility that hazardous materials are present at the scene. Next, you must attempt to identify and isolate the released material, and then prevent and address any injuries or exposures encountered. After all immediate life-safety hazards are addressed, you must decide which level of action is required and request any additional resources (i.e., appropriate equipment and expertise) necessary to mitigate the incident.

FIGURE 28-1 The ability to recognize a potential hazardous materials incident is critical to ensuring your safety.

FIRE FIGHTER Tips

In almost any incident involving hazardous materials, structural firefighting gear does *not* provide an adequate level of chemical protection.

Under both the DOT definition and the working definition, a hazardous material can be almost anything. Milk, for example, is not regarded as a dangerous chemical substance, but 5000 gallons (18,927 liters) of milk leaking into a creek does pose a risk to the environment. A large chlorine cloud fits these definitions, as does a sustained leak from a 55-gallon (208-liter) drum of sulfuric acid. A fire fighter must be able to identify the substances or hazards involved and place them in the proper context. In other words, ask yourself this question: What is the chemical or hazard that is identified and what are the conditions under which it was released?

Hazardous materials can be found anywhere. In sites ranging from hospitals to petrochemical plants, pure chemicals and chemical mixtures are used to create millions of consumer products. More than 80,000 chemicals are registered for use in commerce in the United States, and it is estimated that 2,000 new ones are introduced annually FIGURE 28-2. The bulk of the new chemical substances are industrial chemicals, household cleaners, and lawn care products.

In addition, manufacturing processes sometimes generate hazardous waste. Hazardous waste is a potentially harmful byproduct or residue that remains after a manufacturing process. This type of waste can be just as dangerous as pure chemicals. Hazardous waste can consist of mixtures of several chemicals, resulting in a hybrid substance. It may be difficult to determine how such a substance will react when it is released or comes into contact with other chemicals.

The good news is that compared to the volumes of hazardous materials used and the amounts of hazardous waste generated in the United States, very few hazardous material incidents occur that require intervention. When accidents and incidents do happen, the fire department is usually the agency that handles the initial emergency response phase.

FIGURE 28-2 Hazardous materials can be found anywhere. **A.** Residential. **B.** Industrial. **C.** Business. **D.** Transportation.

Fire Fighter Safety Tips

An awareness and an understanding of the surroundings is your first defense against danger. Be aware of the hazards in your response area. Most often, this is referred to as "situational awareness."

Levels of Training: Regulations and Standards

To understand where your training fits in, you must first understand some of the regulatory processes that govern hazardous materials response, beginning with the difference

Fire Fighter Safety Tips

Each state has the right to adopt and supersede any safety and health regulations put forth by the U.S. Occupational Safety and Health Administration (OSHA) as long as the state's regulations meet or exceed the federal standards. States that have exercised this right are called state-plan states. California, for example, is a state-plan state; its regulatory body is called CAL-OSHA. Approximately half of the states in the United States have state-plans. However, hazardous materials may be treated differently because many of the OSHA regulations contained in 29 CFR 1910.120 are repeated under regulations promulgated by the U.S. Environmental Protection Agency (EPA). In most cases, any federal regulations will be seen as the minimum standards for dealing with such safety issues. Be sure to check which regulations are in force in your state.

between a regulation and a standard. Regulations are issued and enforced by governmental bodies such as the U.S. Occupational Safety and Health Administration (OSHA) and the U.S. Environmental Protection Agency (EPA). In contrast, standards are issued by nongovernmental entities and are generally consensus based. Essentially, consensus organizations such as the National Fire Protection Association (NFPA) issue standards that the public and other interested parties can comment on before committee members agree to adopt them.

NFPA standards governing hazardous materials response come from the Technical Committee on Hazardous Materials Response Personnel. This group includes members from private industry, the fire service, and state and federal agencies as well as other stakeholders from outside of the fire service that may be called on to respond to emergencies that involve hazardous materials.

FIRE FIGHTER Tips

Many times a regulatory agency such as OSHA will refer to or actually cite a standard within the agency's regulation, which then gives the standard equal authority with the regulation.

NFPA has two current hazardous materials–specific standards that relate directly to how fire departments respond to hazardous materials incidents: NFPA 472, *Standard for Competence of Responders to Hazardous Materials/Weapons of Mass Destruction Incidents*, and NFPA 473, *Standard for Competencies for EMS Personnel Responding to Hazardous Materials/Weapons of Mass Destruction Incidents*. The NFPA has recently withdrawn NFPA 471, *Recommended Practice for Responding to Hazardous Materials Incidents*.

The NFPA 472 levels of competencies are awareness, operations (core with mission-specific competencies), technician, technician with specialty, and incident commander. In 2008, the operations level was divided into the basic core competencies that responders should have when responding to a hazardous materials incident and mission-specific competencies. These core competencies include understanding basic chemical and physical properties, concepts of chemical-protective clothing, and decontamination. The mission-specific competencies allow the authority having jurisdiction (AHJ) to decide which additional individual competencies responders should have based on their expected duties. For example, if a fire department has air monitoring equipment and expects to use it, then the responders must receive additional training and demonstrate competency in its use. Air monitoring then becomes a mission-specific skill added to the list of core competencies in which the fire fighter is expected to show proficiency.

A concerted effort is being made to ensure that all fire fighters are trained to the operations core competencies by having the committee that wrote NFPA 1001, *Standard for Fire Fighter Professional Qualifications*, work closely with the committee that developed NFPA 472. The NFPA Technical Committees on Hazardous Materials Response Personnel formed a task group including members of the Fire Fighter Professional Qualifications which meets on a regular basis and has drafted NFPA 1072, *Hazardous Materials/Weapons of Mass Destruction Emergency Response Personnel Professional Qualifications*. The task group is working to set forth Job Performance Requirements (JPRs) that will be used to demonstrate those skills needed to meet the question of qualifications.

In the United States, OSHA regulations are also important. The federal regulations detailing the hazardous materials response competencies are collectively known as HAZWOPER (Hazardous Waste Operations and Emergency Response). This set of regulations, issued in the late 1980s, standardized training for hazardous materials response and hazardous waste site operations. The regulations are found in 29 CFR 1910.120. Fire departments are primarily concerned with Subsection (q), Emergency Response. The training levels found in HAZWOPER, much like the NFPA competencies levels, are identified as awareness, operations, technician, specialist, and incident commander.

Fire fighters in the United States generally must be trained at the operations level. Fire departments must follow both the OSHA and the NFPA requirements for hazardous materials response. NFPA 472 clearly states, "The operations level responder shall receive additional training to meet applicable governmental occupational health and safety regulations."

The following descriptions provide a general overview (as found in NFPA 472 and the HAZWOPER regulations) of the different levels of hazardous materials training and competencies.

Awareness Level

According to the NFPA, awareness-level personnel are individuals who, in the course of their normal duties, could encounter an emergency involving hazardous materials or weapons of mass destruction (WMDs) and who are expected to recognize the presence of the hazardous materials or WMDs, protect themselves, call for trained personnel, and secure the scene.

According to OSHA, awareness-level personnel are first responders who are likely to witness or discover a hazardous substance release and who have been trained to initiate an emergency response sequence by notifying the proper authorities of the release. They do not take any further action beyond notifying the authorities of the release. First responders at the awareness level should have sufficient training or sufficient experience to objectively demonstrate competency.

In other words, any person who comes upon an incident and has been trained to recognize, identify, and notify is considered to be operating at the awareness level. This definition applies to personnel during their regular course of work. For example, the public works employee who is picking up road debris and comes upon a drum leaking a liquid with lots of dead foliage and animals around it should be able to protect the area and make a notification.

Operations Level

According to the NFPA, the operations-level responder responds to hazardous materials or WMD incidents for the purpose of protecting nearby persons, the environment, or property from the effects of the release. This responder should be trained to meet all competencies at the awareness level and the core competencies of the operational level. The operations-level responder receives additional training to meet applicable governmental occupational health and safety regulations.

According to OSHA, first responders at the operations level are individuals who respond to releases or potential releases of hazardous substances as part of the initial response to the site for the purpose of protecting nearby persons, property, or the environment from the effects of the release. They are trained to respond in a defensive fashion without actually trying to stop the release. Their function is to contain the release from a safe distance, keep it from spreading, and prevent exposures. First responders at the operational level must have received at least 8 hours of training or have sufficient experience to objectively demonstrate competency in the following areas in addition to those listed for the awareness level: basic hazard and risk assessment techniques; selection and use of personal protective equipment; basic hazardous materials terms; and control, containment, and/or confinement operations.

In other words, operations-level responders respond to hazardous materials and WMD incidents for the purpose of protecting nearby persons, the environment, or property from the effects of the release. Fire fighters in modern society are usually trained to the operations level because they should be able to recognize potential hazardous materials incidents, isolate and deny entry to the affected area by other responders and the public, evacuate persons in danger, and take defensive actions such as remotely shutting off valves and protecting drains without having contact with the product. Operations-level responders act in a defensive fashion FIGURE 28-3.

Many hazardous materials incidents require some form of product control—hence the need for product control training for the operations-level responder. These responders may be called upon to attempt to confine, contain, or otherwise control a released substance. The ultimate goal is to control the release and reduce or eliminate the threat to people, property, or the environment. Operations-level responders should work under the guidance of a hazardous material technician, an allied professional, or standard operating procedures (SOPs) and should be familiar with the policies and procedures of the local jurisdiction.

FIGURE 28-3 Operations-level responders.

Additional Levels

- Technician level: At this level, fire fighters are trained to enter heavily contaminated areas using the highest levels of personal protection. Hazardous materials technicians take offensive actions using a variety of specialized tools, equipment, and training in an effort to stop the release FIGURE 28-4.
- Specialist level: This level receives more specialized training than a hazardous materials technician. Practically speaking, however, the two levels are not very different. The majority of the specialized

FIGURE 28-4 Hazardous materials technicians.

VOICES OF EXPERIENCE

Fire fighters are often intimidated by the demands of an effective hazardous materials response. Remember that hazardous materials response is much like fire suppression: it requires you to draw upon all of the knowledge, skills, and abilities that you have learned and mastered. Then you must implement this body of knowledge and skills based on the hazards and conditions that you have identified and analyzed at the hazardous materials incident.

Our hazardous materials response team is a regional asset for our mutual aid group. We often respond with neighboring departments and with members of local industry to provide support for incidents ranging from refinery explosions to unidentified odor investigations. Some incidents are extremely challenging and present a great potential for a significant negative impact on the community if they are not managed correctly. I attribute our ability to successfully handle each of these incidents to an effective and ongoing hazardous materials response training program.

One thing I have learned through experience is, "If you train like you'll fight, you'll fight like you train." Training must be realistic and challenging. Training to mastery should be the goal. First concentrate on the basics, and then focus on the most likely scenarios you will encounter based on the target hazards found in your community. Then train to your maximum capability and strive for excellence.

Robert Havens
Lamar Institute of Technology/Port Arthur Fire Department
Beaumont, Texas

training relates to a specific product, such as chlorine, or to a specific mode of transportation, such as rail emergencies.

- Hazardous materials branch director/group supervisor: This level of training is intended for those persons assuming command of a hazardous materials incident beyond the operations level. Individuals trained as hazardous materials officers receive operations-level training as well as additional training specific to commanding a hazardous materials incident. The hazardous materials officer is trained to act as a branch director or group supervisor for the hazardous materials component of the incident.
- Incident commander (IC): The IC's primary responsibilities during an incident include the development of strategies and tactics and acquisition of the resources necessary for the response.
- Hazardous materials safety officer: This person is assigned to the Hazardous Materials Branch and is responsible for ensuring the safety of hazardous materials personnel and recognizing that appropriate hazardous materials/WMD practices are followed.
- Hazardous materials technician with specialties: This individual is a hazardous materials technician with training in specialized areas such railroad tank cars, highway cargo vehicles, intermodal containers, and others as listed in NFPA 472. Three levels of specialist employees are distinguished: A, B, and C. These designations are primarily used in the industrial world and describe the employee's capabilities with regard to his or her response during incidents on or off site. Such personnel may have specialties with regard to specialized chemicals, containers, and their uses.

In addition to the initial training requirements listed previously, OSHA regulations require annual refresher training. Consult your AHJ for more specific requirements on refresher training.

Fire Fighter Safety Tips

You can't do it all! The members of your department have different skill levels and capabilities. All personnel must use their collective talents and training at a hazardous materials incident.

Other Hazardous Materials Laws, Regulations, and Regulatory Agencies

Several government agencies are concerned with hazardous materials. The DOT, for example, enforces and publicizes laws and regulations that govern the transportation of goods by highways, rail, air, pipelines and, in some cases, marine transport.

The EPA regulates and governs issues relating to hazardous materials in the environment. The EPA's version of HAZWOPER can be found in *Title 40, Protection of the Environment, Part 311: Worker Safety*. As mentioned, OSHA is very concerned about hazardous materials response. In addition to the HAZWOPER regulations, OSHA issues guidance on respiratory protection, personal protective equipment, and a multitude of other topics regarding worker safety. Depending on your state, some of these regulations may apply to your fire department. Check with your department to determine how these regulations affect your ability to respond to hazardous materials incidents.

The Superfund Amendments and Reauthorization Act (SARA) was one of the first laws to affect how fire departments respond in a hazardous material emergency. Finalized in 1986, SARA was the original driver for the OSHA HAZWOPER regulation. Indirectly, it indicated that workers handling hazardous wastes should have a minimum amount of training. Additionally, this law laid the foundation that ultimately allowed local fire departments and the public to obtain information on how and where hazardous materials were stored in their community. The Emergency Planning and Community Right to Know Act requires a business that handles chemicals to report storage type, quantity, and storage methods to the fire department and the local emergency planning committee. Such documentation is known as tier reports.

Local emergency planning committees (LEPCs) gather and disseminate to the public information about hazardous materials. These committees comprise members of industry, transportation, media, fire and police agencies, and the public at large. Essentially, LEPCs ensure that local resources are adequate to respond to a chemical event in the community. Fire departments should be familiar with their LEPCs and know how their departments work with this committee. They are often a good resource for technical information, specialized training, and expertise in hazardous materials response.

LEPCs collect material safety data sheets (MSDSs). Under the Globally Harmonized System of Classification and Labeling of Chemicals (GHS), the MSDS term is changing to the term Safety Data Sheet (SDS) as the impacts of global harmonization become more integrated into the regulatory and environmental health and safety communities. Additional information on global harmonization and its effects on chemical classification and labeling can be found on the OSHA web-site. An MSDS is a detailed profile of a single chemical or mixture of chemicals provided by the manufacturer and/or supplier of a chemical. It details a specific chemical's properties and all pertinent information about it. These chemical-specific data sheets are an important source of information for fire fighters when dealing with hazardous materials incidents. Later chapters will discuss MSDSs in detail FIGURE 28-5.

Each state has a State Emergency Response Commission (SERC). The SERC serves as the liaison between local and state levels of authority. It involves agencies such as the fire service, police services, and elected officials in the collection and dissemination of information relating to hazardous materials. SERCs and LEPCs often work together closely to help a community prepare to deal with hazardous materials incidents.

Difference Between Hazardous Materials Incidents and Other Types of Emergencies

A critical distinction must be made between a hazardous materials event and other emergencies, such as a structure fire. Fire fighters cannot approach a hazardous materials incident with

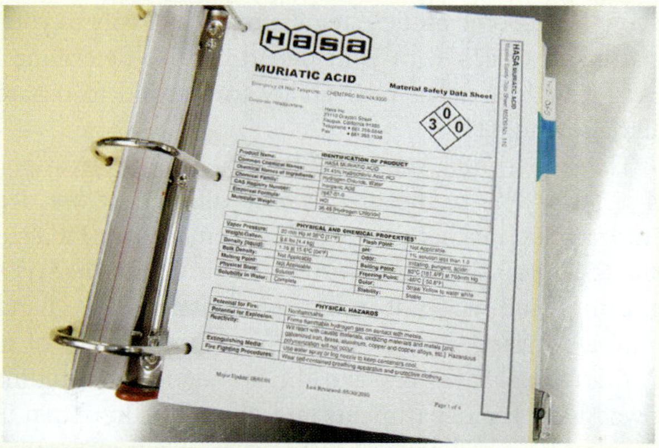

FIGURE 28-5 An MSDS or SDS details the properties of a specific chemical.

the same mindset used for a structure fire. Structural firefighting employs a fairly basic strategy, determined during the size-up phase, with the ultimate goal of extinguishing the fire. By comparison, hazardous materials incidents are more complex. Consequently, the strategies used in addressing such incidents must be thought out and planned in advance.

For example, most of the time, a hazardous materials emergency moves more slowly than a structure fire. If a rescue is required, or if the situation is imminently dangerous and requires rapid intervention, events may move quickly, but a hazardous materials emergency requires time, forethought, and planning to ensure that the incident is handled safely and effectively.

The actions taken at hazardous materials incidents are dictated largely by the chemicals involved and the available resources. In many cases, it takes considerable effort to identify them and assemble the appropriately trained personnel and equipment. It becomes much more difficult and time-consuming to handle an incident involving an unknown substance. Response objectives, the choice of personal protective equipment, and the type of decontamination needed are all complex decisions that depend on the chemical properties of the hazardous substance.

In the case study at the beginning of this chapter, there are some clues to the nature of the material involved but not enough to indicate all the hazards involved. Ask yourself these important questions:

- Is the material flammable, toxic or corrosive?
- What is the chemical name of the material involved?
- How will it respond when exposed to air, water, or other materials?

Fires are considerably more straightforward. Every fire will generate heat, smoke, flames, and toxic gases. Personal protective equipment needs do not change radically from fire to fire. The tactics and strategy change based on the type of building and degree of fire involvement, rather than because of some difference in the actual chemical nature of the fire.

When approaching hazardous materials events, make a conscious effort to change your perspective. Slow down, think about the problem and available resources, and take well-considered actions to solve the problem.

Fire Fighter Safety Tips

The goal of the fire fighter is to favorably change the outcome of the hazardous materials incident. This is accomplished by sound planning and by establishing safe and reasonable response objectives based on the level of training. Bottom line: Don't do it if you are not trained or properly equipped to do it, no matter what the situation—but even more so when a hazardous material is involved!

Planning a Response

It is a mistake to assume that the response to a hazardous materials incident begins when the alarm sounds. The response really begins with training, learning about the regulations and agencies involved, and finding out about potential hazards in your area. Response agencies also should conduct incident-planning activities at target hazards and other potential problem areas throughout the jurisdiction. Preplanning activities enable agencies to develop logical and appropriate response procedures for anticipated incidents. Planning should focus on the real threats that exist in your community or in adjacent communities where you could be assisting other departments at an incident FIGURE 28-6.

Once the threats have been identified, agencies and fire departments must determine how they will build their response. Some agencies establish standard operating procedures or guidelines that guide their response to particular hazardous material incidents. Those parameters outline incident severity based on the nature of the chemical, the amount released, or the type of occupancy involved in the incident.

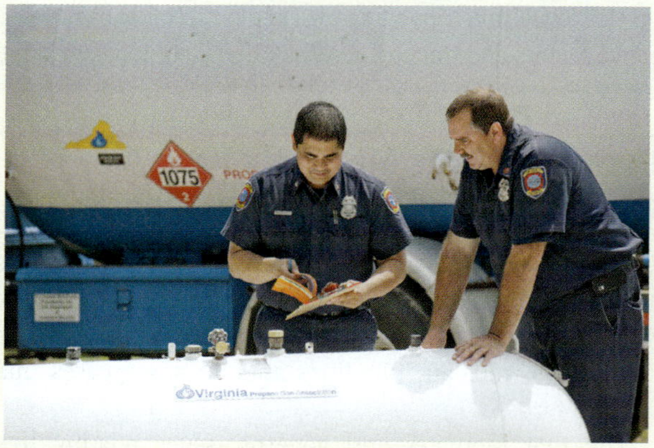

FIGURE 28-6 Conduct preincident planning activities at target hazards throughout the jurisdiction.

Wrap-Up

Chief Concepts

- A hazardous material is a material that poses an unreasonable risk to the health and safety of operating emergency personnel, the public, and/or the environment if it is not properly controlled during handling, storage, manufacture, processing, packaging, use and disposal, or transportation.
- All fire fighters must be able to recognize the presence of hazardous materials, analyze information, and take appropriate action.
- The ability to recognize a potential hazardous material incident is critical to ensuring your own safety.
- To avoid entering a dangerous area, you must be alert to the possibility that hazardous materials are present at the scene.
- At a hazardous materials incident, you must attempt to identify and isolate the released material, and then prevent and address any injuries or exposures encountered. After any life-safety hazards are addressed, you must decide which level of action and protection is required to alleviate or resolve the incident.
- Hazardous materials can be found anywhere. At sites ranging from hospitals to petrochemical plants, pure chemicals and chemical mixtures are used to create millions of consumer products. In addition, manufacturing processes sometimes generate hazardous waste.
- Regulations are issued and enforced by governmental bodies such as OSHA and the EPA. HAZWOPER is a set of OSHA regulations that cover hazardous materials response training and hazardous waste site operations.
- Standards are issued by nongovernmental entities, such as the NFPA, and are generally consensus based. The NFPA has two current standards that pertain to hazardous materials incidents: NFPA 472 and NFPA 473.
- Training or competencies levels of proficiency for responders to a hazardous materials incident are found in HAZWOPER and outlined by the NFPA: awareness, operations, operations mission–specialist technician, and incident commander.
- Any person who comes upon a hazardous materials incident and has been trained to recognize it, identify it, and notify the appropriate agencies is considered to be operating at the awareness level.
- Operations-level responders respond to hazardous materials and WMD incidents for the purpose of protecting nearby persons, the environment, or property from the effects of the release.
- Technician-level personnel are trained to enter heavily contaminated areas using the highest levels of personal protection.
- Specialist-level personnel have a similar level of training as technician-level personnel, but have additional training related to specific chemicals, container types, or transportation modes.
- Hazardous materials officers are trained to assume command of a hazardous materials incident beyond the operations level.
- Hazardous materials safety officers are responsible for ensuring the safety of hazardous materials personnel and the use of appropriate hazardous materials and WMD practices.
- LEPCs gather and disseminate to the public information about hazardous materials. These committees include members from industry, transportation, media, fire and police agencies, and the public at large. LEPCs ensure that local resources are adequate to respond to a chemical-related event in the community. LEPCs collect MSDSs, which serve as an important informational resource for fire fighters.
- Hazardous materials incidents cannot be approached in the same manner as a structural fire. These incidents are more complex and structural firefighting gear does not provide adequate protection from hazardous materials.
- The actions taken at hazardous materials incidents are dictated largely by the chemicals involved and the resources available.
- A hazardous materials response begins with preincident planning.

Hot Terms

Awareness level (29 CFR 1910.12: First Responder at the Awareness Level) Personnel who, in the course of their normal duties, could encounter an emergency involving hazardous materials and weapons of mass destruction (WMDs) and who are expected to recognize the presence of the hazardous materials and WMDs, protect themselves, call for trained personnel, and secure the scene. (NFPA 472)

Department of Transportation (DOT) The federal agency that publicizes and enforces rules and regulations that relate to the transportation of many hazardous materials.

Emergency Planning and Community Right to Know Act Federal legislation that requires a business that handles chemicals (depending on the quantity stored) to report storage type, quantity, and storage methods to the fire department and the local emergency planning committee.

Environmental Protection Agency (EPA) Established in 1970, the federal agency that ensures safe manufacturing, use, transportation, and disposal of hazardous substances.

Hazardous material A substance that, when released, is capable of creating harm to people, the environment, and property. (NFPA 472)

Hazardous materials branch director/group supervisor Commanders of hazardous materials incidents beyond the operations level.

Hazardous materials officer (NIMS: Hazardous Materials Branch Director/Group Supervisor) The person who is responsible for directing and coordinating all operations involving hazardous materials and weapons of mass destruction as assigned by the incident commander. (NFPA 472)

Hazardous materials safety officer The person who is responsible for ensuring the safety of hazardous materials personnel and ensuring that appropriate hazardous materials/weapons of mass destruction practices are followed.

Hazardous materials technician with specialties A hazardous materials technician with training in areas such as specialized chemicals, containers, and their uses.

Hazardous waste Waste that is potentially damaging to the environment or human health due to its toxicity, ignitability, corrosivity, or chemical reactivity or another cause. (NFPA 820)

HAZWOPER Hazardous Waste Operations and Emergency Response; the OSHA regulation that governs hazardous materials waste sites and response training. Specifics can be found in 29 CFR 1910.120. Subsection (q) is specific to emergency response.

Local emergency planning committee (LEPC) A group comprising members of industry, transportation, the public at large, media, and fire and police agencies; it gathers and disseminates information on hazardous materials stored in the community and ensures that there are adequate local resources to respond to a chemical event in the community.

Material safety data sheet (MSDS) Transitioning to Safety Data Sheets (SDS). A form, provided by manufacturers and compounders (blenders) of chemicals, containing information about chemical composition, physical and chemical properties, health and safety hazards, emergency response, and waste disposal of the material. (NFPA 472)

National Fire Protection Association (NFPA) A private organization that develops and maintains nationally recognized minimum consensus standards on many areas of fire safety and specific standards on hazardous materials.

Occupational Safety and Health Administration (OSHA) The federal agency that regulates worker safety and, in some cases, responder safety. OSHA is a part of the Department of Labor.

Operations level Persons who respond to hazardous materials/weapons of mass destruction incidents for the purpose of implementing or supporting actions to protect nearby persons, the environment, or property from the effects of the release. (NFPA 472)

Specialist level Persons who respond to hazardous materials/weapons of mass destruction incidents who have received more specialized training than hazardous materials technicians. Most of the training that specialist employees receive is either product or transportation mode specific.

State Emergency Response Commission (SERC) The liaison between local and state levels that collects and disseminates information relating to hazardous materials. The SERC involves agencies such as the fire service, police services, and elected officials.

Superfund Amendments and Reauthorization Act (SARA) One of the first federal laws to affect how fire departments respond in a hazardous material emergency.

Target hazard Any occupancy type or facility that presents a high potential for loss of life or serious impact to the community resulting from fire, explosion, or chemical release.

Technician level Persons who respond to hazardous materials/weapons of mass destruction incidents and are allowed to enter heavily contaminated areas using the highest levels of protection. This is the most aggressive level of training as identified in NFPA 472.

FIRE FIGHTER *in action*

It is 1300 hours, and your engine company is dispatched for an incident that involves several leaking drums at a local self-storage facility near the elementary school. En route to the scene, you start thinking about the potential scene: Are children outside the school playing and running around? Is there pedestrian traffic on the street? What does this facility look like?

As your apparatus approaches the dispatched address, you see a white cloud appearing from the doorway of one of the storage lockers. Your officer immediately stops the apparatus and notifies dispatch with a report of what he has observed. He starts requesting additional resources.

FIRE FIGHTER *in action* (continued)

1. Based on the type of incident, which level of hazardous materials training will be needed to mitigate it?
   A. Awareness
   B. Operations
   C. Technician
   D. Specialist employee
2. Which actions should the first arriving engine company consider as first priority?
   A. Life safety of responders
   B. Evacuation
   C. Property conservation
   D. Environmental considerations
3. Which actions, if any, should awareness-level personnel take in the incident described?
   A. None
   B. Identification of product
   C. Notification of emergency/trained personnel
   D. Make entry to assist in evacuation
4. Which actions, if any, should operations-level responders take in the incident described?
   A. Evacuation
   B. Entry to identify product
   C. Entry to stop leaking drum
   D. Act in defensive fashion
5. Because the incident has occurred in your community, which local committee would have an impact on your preplanning and knowledge of this facility?
   A. State Emergency Response Commission
   B. Local emergency planning committee
   C. OSHA HAZWOPER regulations
   D. NFPA Standards

FIRE FIGHTER II *in action*

Your engine company has been asked to give a presentation on a recent incident involving a suspected hazardous material. Your captain has assigned you the task of reporting on the activities and responsibilities of the hazardous materials technician during the incident. As you interview the rest of the company, which includes both Fire Fighter I and II members, you receive some suggestions on how to improve the safety of the operation. One particular fire fighter suggests better monitoring of personnel entering the hot zone.

1. Which safety considerations should be the primary concern of each responder on the scene of a hazardous materials incident?
2. As a Fire Fighter II, what are your responsibilities to ensure the safety of other responders?
3. Who should be responsible for the accountability of personnel in the hot zone?

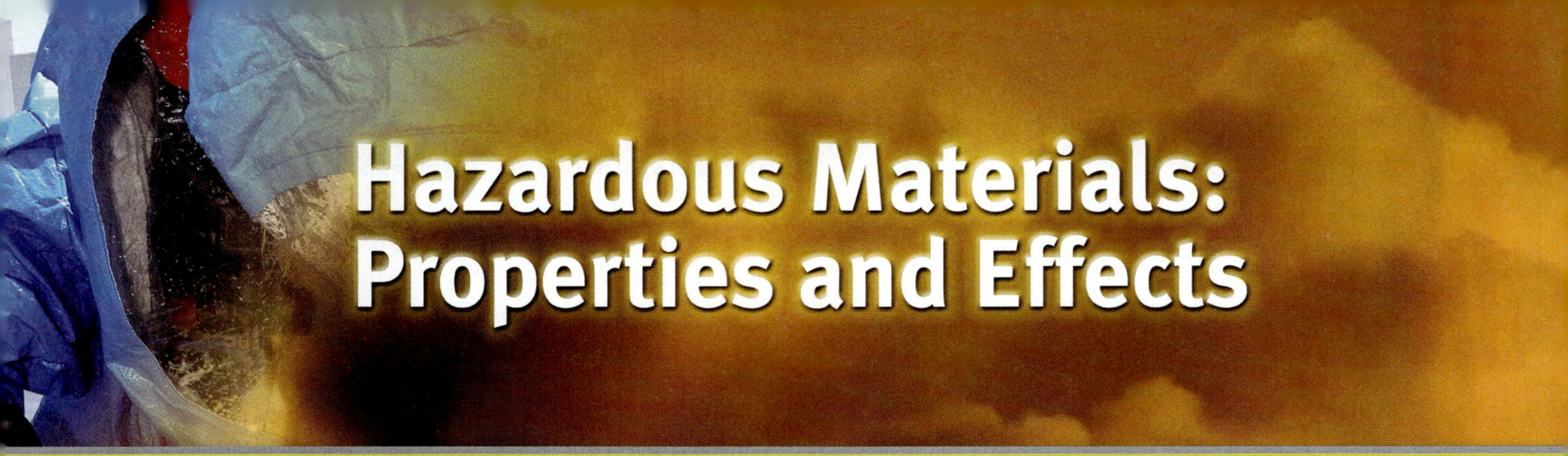

Hazardous Materials: Properties and Effects

Fire Fighter I

Knowledge Objectives

After studying this chapter, you will be able to:

- Describe how to identify a substance's state of matter. (NFPA472,5.2.3,6.6.1.1.2, p 880)
- Describe the process of chemical change. (NFPA472,5.2.3,6.6.1.1.2, p 881)
- Describe the process of physical change. (NFPA472,5.2.3,6.6.1.1.2, p 881)
- Define boiling point and explain how this principle affects hazardous materials. (NFPA472,5.2.3,6.6.1.1.2, p 881)
- Define flash point and explain how this principle affects hazardous materials. (NFPA472,5.2.3,6.6.1.1.2, p 881–882)
- Define fire point and explain how this principle affects hazardous materials. (NFPA472,5.2.3,6.6.1.1.2, p 882)
- Define ignition temperature and explain how this principle affects hazardous materials. (NFPA472,5.2.3,6.6.1.1.2, p 882)
- Define flammable range and explain how this principle affects hazardous materials. (NFPA472,5.2.3,6.6.1.1.2, p 882)
- Define vapor density and explain how this principle affects hazardous materials. (NFPA472,5.2.3,6.6.1.1.2, p 883)
- Define vapor pressure and explain how this principle affects hazardous materials. (NFPA472,5.2.3,6.6.1.1.2, p 883–884)
- Define specific gravity and explain how this principle affects hazardous materials. (NFPA472,5.2.3,6.6.1.1.2, p 884)
- Define water miscibility and explain how this principle affects hazardous materials. (NFPA472,5.2.3,6.6.1.1.2, p 884–885)
- Define corrosivity and explain how this principle affects hazardous materials. (NFPA472,5.2.3,6.6.1.1.2, p 885)
- Define pH and explain how this principle affects hazardous materials. (NFPA472,5.2.3,6.6.1.1.2, p 885)
- Describe how to determine a substance's pH in the field.
- Describe the physical hazards posed by the toxic products of combustion. (NFPA472,5.2.3,6.6.1.1.2, p 885)
- Describe the differences between non-ionizing and ionizing radiation. (NFPA472,5.2.3,6.6.1.1.2, p 886–887)
- Describe how radiation is detected in the field. (p. 886)
- Define alpha particles and describe how to avoid exposure. (NFPA472,5.2.2,5.2.3,5.2.4,6.6.1.1.2, p 887)
- Define beta particles and describe their potential effects on the human body. (NFPA472,5.2.3,6.6.1.1.2, p 887)
- Define gamma radiation. (NFPA472,5.2.3,6.6.1.1.2, p 887)
- Describe the differences between contamination and secondary contamination. (NFPA472,5.2.3,6.6.1.1.2, p 888)
- List the types of weapons of mass destruction. (NFPA472,5.2.3,6.6.1.1.2, p 888)
- Describe how nerve agents damage the human body. (NFPA472,5.2.3,6.6.1.1.2, p 888–889)
- List the signs and symptoms of nerve agent exposure. (NFPA472,5.2.3,6.6.1.1.2, p 888–889)
- Describe how blister agents damage the human body. (NFPA472,5.2.3,6.6.1.1.2, p 889)
- List the signs and symptoms of blister agent exposure. (NFPA472,5.2.3,6.6.1.1.2, p 889)
- Explain the route of exposure for cyanide. (NFPA472,5.2.3,6.6.1.1.2, p 889)
- List the signs and symptoms of cyanide exposure. (NFPA472,5.2.3,6.6.1.1.2, p 889)
- Describe how choking agents damage the human body. (NFPA472,5.2.3,6.6.1.1.2, p 889)
- Describe how irritants damage the human body. (NFPA472,5.2.3,6.6.1.1.2, p 889)
- List convulsant chemicals and describe how they damage the human body. (NFPA472,5.2.3,6.6.1.1.2, p 889–890)
- List the four ways chemicals can enter the human body. (NFPA472,5.2.3,6.6.1.1.2, p 890–894)
- Describe the precautions fire fighters take to avoid chemical exposure through inhalation. (NFPA472,6.6.1.1.2, p 891)
- Describe the precautions fire fighters take to avoid chemical exposure through absorption. (NFPA472,6.6.1.1.2, p 892)
- Explain the differences between chronic and acute health effects. (NFPA472,5.2.3,6.6.1.1.2, p 894)
- List the four major sections of the *Emergency Response Guidebook*. (NFPA472,5.2.3,6.6.1.1.2, p 894)

CHAPTER

29

Skills Objectives

After studying this chapter, you will be able to perform the following skill:

- Demonstrate the ability to properly utilize the current edition of the *DOT Emergency Response Guidebook* (p 894).

Fire Fighter II

FFII

Knowledge Objectives

There are no knowledge objectives for Fire Fighter II candidates. NFPA 1001 contains no Fire Fighter II Job Performance Requirements for this chapter.

Skills Objectives

There are no skill objectives for Fire Fighter II candidates. NFPA 1001 contains no Fire Fighter II Job Performance Requirements for this chapter.

You Are the Fire Fighter

Your fire department and hazardous materials team are responding to an incident involving a leaking drum at a local building supply store. Upon arrival, you obtain basic information from the manager of the store including location and type of container that is leaking. The manager states that they have stored several different chemicals in the area that include both solids and liquids and one of the drums is leaking after being hit by a forklift.

1. Which of the chemical properties of the product would be helpful in determining the type of hazards the liquid presents?
2. How could the product enter the body and how can you, the responder, prevent it?
3. Which information from the driver would be helpful in obtaining further information from the *Emergency Response Guidebook* for the involved chemicals?

Introduction

To understand hazardous materials incidents, fire fighters must understand the chemical and physical properties of the materials and substances involved. Chemical and physical properties are the characteristics of a substance that are measurable, such as vapor density, flammability, corrosivity, and water reactivity. This requirement does not mean that you need to be a chemist, however. In most cases, being an astute observer and correctly interpreting the visual clues presented provide enough information. The basis for interpreting those visual clues lies in having a solid grasp of some basic hazardous materials concepts.

The first step in understanding the hazard posed by any substance involves identifying its state of matter. The state of matter identifies the hazard as a solid, liquid, or gas **FIGURE 29-1**. Why is this information important? If you know the state of matter and physical properties of the chemical, you can predict what the substance will do if it escapes from its container. Remember that the "working definition" for a hazardous material is potential energy that could be harmful when released from its container—hence, the importance of being able to "predict" the behavior of a material when it escapes its container is obvious.

For example, it is vital to know whether a released gas is heavier or lighter than air. In a trench rescue, a gas that is heavier than air will sink to the lowest point, posing a significant problem to the victim. It is also critical to determine whether the released gas is flammable, is potentially toxic, or could displace oxygen.

Based on the various potential hazards associated with a material, you need to see the big picture before committing to a course of action. The bottom line is this: Think before you act. This simple statement could save your life or the lives of your team members.

This chapter identifies concepts and key terms to help you understand the hazards a chemical poses and the information contained in material safety data sheets (MSDSs) and other chemical reference sources.

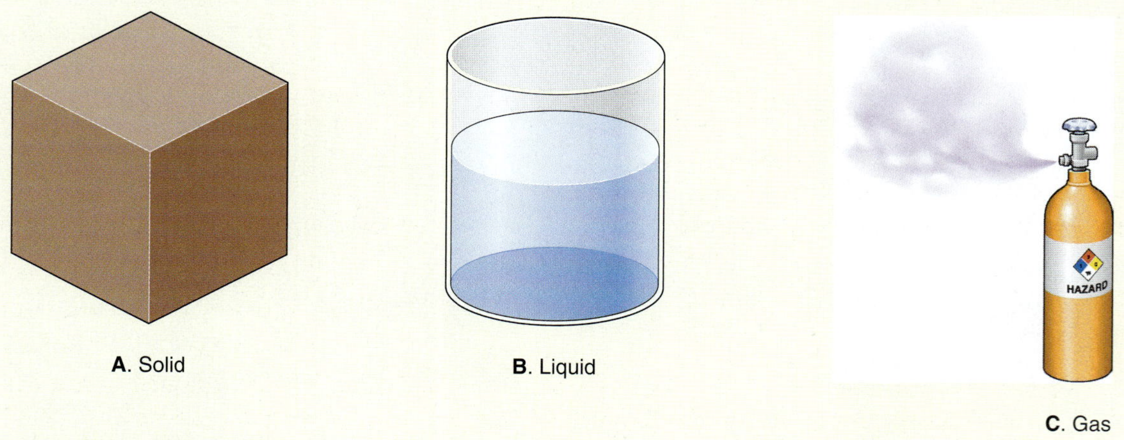

FIGURE 29-1 The state of matter identifies the hazard as a solid, liquid, or gas.

Characteristics of Hazardous Materials

Physical and Chemical Changes

Chemicals exist in three physical states: solid, liquid, or gas. The state that they are in at any given moment is a matter of the environment that they are in or a matter of how they are stored in a given container. Chemicals can undergo a physical change when subjected to outside influences such as heat, cold, and pressure. When water is frozen, for example, it undergoes a physical change from a liquid to a solid. The actual chemical makeup (H_2O) is still the same, but the state in which it exists is different.

Propane gas has the potential to undergo physical change based on how it is stored inside a vessel. Liquefied propane has a large expansion ratio. The expansion ratio is a description of the volume increase that occurs when a liquid material changes to a gas. For example, propane has an expansion ratio of 270:1. This means that for every 1 volume of liquid propane, there would be 270 times that amount of propane vapor. In other words, 1 gallon of liquid propane has the ability to vaporize to 270 gallons of propane vapor when released.

If the propane cylinder is exposed to heat, the liquid propane inside changes into gaseous propane, which increases the pressure inside the vessel. If this uncontrolled expansion takes place faster than the relief valve can vent the excess pressure, a catastrophic container failure could occur. Such a catastrophic failure in a pressurized cylinder of a liquid is referred to as a BLEVE (boiling liquid expanding vapor explosion). The pressure in the container exceeds the container's ability to hold the pressure and the container fails in a catastrophic fashion, but the chemical makeup of propane has not changed: The substance has just physically changed state from liquid to gas.

The term chemical change describes the ability of a chemical to undergo a change in its chemical makeup, usually with a release of some form of energy. The reactive nature of a chemical may be influenced in many ways, such as by mixing it with another chemical or by applying heat to it.

Chemical change is different from physical change. Physical change is essentially a change in state, whereas chemical change results in an alteration of the chemical nature of the material. Steel rusting and wood burning are examples of chemical changes. When the chemical reaction is complete, the substance itself is not the same as it was.

To understand this distinction, think of a Mr. Potato Head toy. Taking off its arms and legs and changing them around would be a physical change: Mr. Potato Head would still be the same toy—it would just look a bit different. A chemical change would be a transformation in which Mr. Potato Head changed from plastic to wood. Understanding how a given chemical reacts or can affect its environment when it is in the three different states of matter can be a matter of life and death for a fire fighter.

The concepts of physical and chemical changes can be applied to nearly any incident that involves chemicals. Assume that the owner of a landscape company left some rags soaked with linseed oil wadded up in a corner of a shed. The rags ignited spontaneously and started a fire. The heat from that fire ultimately caused a catastrophic failure of a 20-pound propane cylinder. The fire also caused the valve on a Type A (100-pound) chlorine cylinder to fail (frangible disk functions as designed), allowing chlorine gas to be released into the atmosphere.

Looking closely at each of these events, it is apparent how the various physical and chemical changes altered the course of the fire incident. Linseed oil is an organic material that generates heat as it decomposes (a chemical change). That heat, in the presence of oxygen in the surrounding air, ignited the rags, which in turn ignited other combustibles (a common occurrence). Propane is gas liquefied by pressure and temperature and stored in steel cylinders. The surrounding heat from the fire caused the propane to expand (physical change) until it overwhelmed the ability of the container and the relief valve to handle the buildup of pressure. The cylinder ultimately exploded, demolished the shed, and damaged the chlorine cylinder. That cylinder then leaked chlorine gas, which mixed with the moisture in the air and formed an acidic compound called hypochlorous acid (chemical change). The breeze carried the acid mist toward an apartment complex. Because chlorine vapor is heavier than air, most of the residents who complained about eye, nose, and throat irritation would likely be located on the bottom floor of the apartment complex.

Boiling Point

Boiling point is the temperature at which a liquid will continually give off vapors in sustained amounts and, if held at that temperature long enough, will eventually turn completely into a gas. The boiling point of water is 212°F. At this temperature, the water molecules have enough kinetic energy (energy in motion) to break free from the liquid and overcome the force of the surrounding atmospheric pressure. Standard atmospheric pressure at sea level is 14.7 pounds per square inch (psi), or 101.4 kilopascals (kPa). In other words, at sea level, 14.7 psi (101.4 kPa) of pressure are exerted on every surface of every object.

An example of boiling point can be observed in the preparation of popcorn. Popcorn pops because of water inside the kernel of corn. Water has an expansion ratio of 1,700:1 and a boiling point of 212°F (100°C). A kernel of corn has no relief valve, so a rapid buildup of pressure causes it to rupture. When popcorn kernels are heated, the small amount of water inside exceeds its boiling point, expands to 1,700 times its original water volume, and the kernel pops. Unpopped kernels are those with insufficient water inside the kernel to allow for this change.

Flammable liquids with low boiling points are dangerous because of their potential to produce large volumes of flammable vapor when exposed to relatively low temperatures.

Flash Point, Fire Point, Ignition (Autoignition) Temperature, and Flammable Range

When looking at the fire potential of a chemical, there are three important aspects to consider: flash point, ignition temperature, and flammable range. When it comes to combustion, materials must be in a gaseous or vapor state to burn. Neither solids nor liquids actually burn; rather, they must give off a gas or a vapor that must be ignited to sustain combustion.

Think of a log in a fireplace. If you look closely, the fire is not directly on the log, but rather slightly above the surface. The reason for this is that the wood must be heated to the point where it produces enough "wood gas." The log, then, does not burn; instead, the gas produced by heating the log is what starts the process of combustion. Liquids behave in the same way as a log in a fire. That is, one way or another, there must be vapor production before there can be fire.

Flash point is an expression of the temperature at which a liquid fuel gives off sufficient vapor that, when an ignition source is present, will result in a flash fire. The flash fire involves only the vapor phase of the liquid, so it goes out once the vapor fuel is consumed.

The flash point of gasoline is –43°F (–41°C). In other words, when the temperature of gasoline is above –43°F (–41°C), the gasoline gives off sufficient flammable vapors to support combustion. If those vapors reach an ignition source, the vapors will ignite in a ball of fire FIGURE 29-2. Diesel fuel has a much higher flash point—approximately 120°F (49°C). In either case, once the temperature of the liquid surpasses its flash point, the fuel will give off sufficient flammable vapors to support combustion.

Flash point is merely one aspect of how flammable and combustible liquids behave in a fire situation. The fire point is the temperature at which sustained combustion occurs. It is usually only slightly above the flash point for most materials.

FIGURE 29-2 Gasoline at –43°F gives off sufficient vapor to result in a flash point when an ignition source is present.

The ignition (autoignition) temperature is another important landmark in the burning process. When a liquid fuel is heated beyond its ignition temperature, the substance ignites without an external ignition source. Think of a pan full of cooking oil being heated on the stove. For illustration purposes, assume that the ignition temperature of the oil is 300°F (149°C). What would happen if the oil were heated past that point? Once the temperature exceeds the ignition (autoignition) temperature, the oil ignites; there is no need for an external ignition source. This is a common cause of stove fires.

Flammable range is another important concept. In broad terms, a substance's flammable range describes a mixture of fuel and air, defined by upper and lower flammable limits, and reflects an amount of vapor mixed with a volume of air. To illustrate this concept, think about gasoline. The flammable range for gasoline vapors is 1.4 percent to 7.6 percent. These two percentages, called the upper flammable limit (UFL = 7.6 percent) and the lower flammable limit (LFL = 1.4 percent), respectively, define the boundaries of the fuel/air mixture necessary for gasoline to burn properly. These terms are also known as the upper explosive limit and lower explosive limit (UEL/LEL). The upper flammable limit (UFL) defines the amount of flammable vapor needed to keep a fire burning; the lower flammable limit (LFL) defines the amount of vapor needed to cause ignition. If a given gasoline/air mixture falls between the upper and lower flammable limits, and that mixture is exposed to an ignition source, there will likely be a flash fire. This fire will appear to be explosive in nature, even though it is not technically an explosion owing to the speed of the flame front/pressure wave generated. Of course, this is a moot point should the fire fighter be near the event as it occurs—the result, though not an explosion by strict definition, will nevertheless be catastrophic for the fire fighter.

The concept underlying the functioning of an older-style automobile carburetor is based on flammable range. The carburetor is the place where gasoline and air are mixed. With the right mixture of gasoline vapors and air, the car runs smoothly. When there is too much fuel and not enough air, the mixture is considered to be "too rich." In the case of too much air and not enough fuel, the mix is called "too lean." In either case, optimal combustion is not achieved and the motor does not run correctly.

The flammable range in the hazardous materials environment works in much the same way. If gasoline is released from a rolled-over tanker, gasoline vapors and air will mix. If the mixture of vapors and air falls between the upper and lower flammable limits, and those vapors encounter an ignition source, a flash fire will occur. As with carburetion, if there is too much or too little fuel, the mixture will not adequately support combustion.

To accurately assess the flammable range of a released vapor or gas, specialized combustible gas indicators must be used. These indicators are generally used by hazardous materials technicians and help determine the presence of a dangerous flammable atmosphere.

Generally speaking, the wider the flammable range, the more dangerous the material TABLE 29-1. This relationship exists

TABLE 29-1	Flammable Ranges of Common Gases
Hydrogen	4.0–75%
Natural gas	5.0–15%
Propane	2.5–9.0%

because the wider the flammable range, the more opportunity there is for an explosive mixture to find an ignition source. The one major concern in the case of an unconfined vapor cloud is that as the cloud expands outward from the point of release, the likelihood of the edges of such a vapor cloud having a flammable mixture at one or more points becomes much greater.

When operating in any environment where a flammable or combustible vapor cloud is suspected to be present, always err on the side of safety and assume that at some point in the suspected vapor cloud the mixture is of such a concentration that the danger of an ignition exists. Always assume that any ignition source could ignite the vapor cloud and that, once ignited, that vapor cloud will continue to burn, possibly with explosive forces. For this reason, it is essential to always operate in a very defensive mode in such situations.

Vapor Density

In addition to addressing the flammability of a particular vapor or gas, fire fighters must determine whether the vapor or gas is heavier or lighter than air. In essence, you must know the vapor density of the substance in question. Vapor density is the weight of an airborne concentration (vapor or gas) as compared to an equal volume of dry air FIGURE 29-3.

Basically, vapor density compares the "weight" of a given vapor with the "weight" of air. What will the gas or vapor do when it is released into the air? Will it collect in low spots in the topography or somewhere inside a building, or will it float upward into the ventilation system? These are important response considerations for hazardous materials incidents, and the answers to these questions can influence the severity of the incident and directly affect fire fighters' ability to operate safely in the environment.

To find a chemical's vapor density, consult a reference source such as an MSDS. The vapor density is expressed in numerical fashion: 1.2, 0.59, 4, and so on. The vapor density of propane, for example, is 2.4. Air has an index value of 1. Gases such as propane or chlorine are heavier than air and, therefore, have a vapor density value greater than 1. By contrast, substances such as acetylene, natural gas, and hydrogen are lighter than air and, therefore, have a vapor density value less than 1.

In the absence of reliable reference sources in the field, you can use the HA HA MICEN mnemonic to remember a number of lighter-than-air gases:

H: Hydrogen
A: Acetylene
H: Helium
A: Ammonia
M: Methane
I: Illuminating gas (neon and hydrogen cyanide)
C: Carbon monoxide
E: Ethylene
N: Nitrogen

Although this list is not all inclusive, most materials not on this list are heavier than air (i.e., vapor density > 1). If you encounter a leaking propane cylinder, for example, just think of the HA HA MICEN mnemonic. Propane is not on that list, so by default, you can assume that propane is heavier than air.

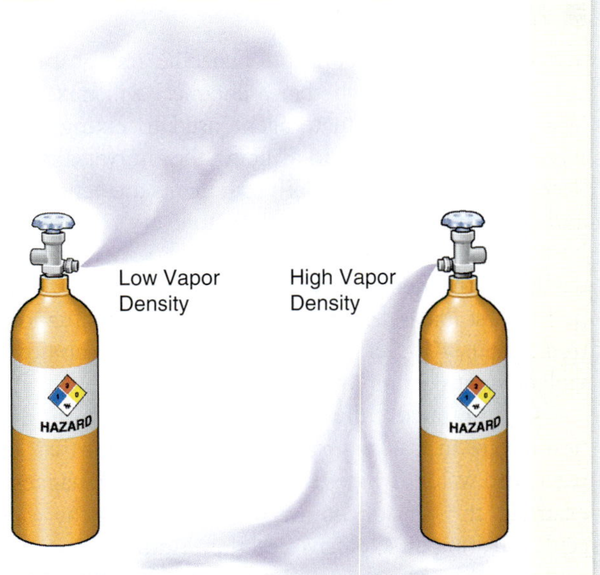

FIGURE 29-3 Vapor density.

Vapor Pressure

For fire fighters' purposes, the definition of vapor pressure can be assumed to pertain to liquids FIGURE 29-4. When liquids are held in a closed container, such as a 55-gallon (208-liter) drum, pressure will develop inside that container. All liquids—even water—develop a certain amount of pressure in the airspace between the top of the liquid and the container.

The key point to understanding vapor pressure is this: The vapors released from the surface of the liquid must be contained if they are to exert any pressure. Essentially, the liquid inside the container will vaporize until the molecules given off by the liquid reach equilibrium with the liquid itself. It is a balancing act between the liquid and the vapors in which some liquid molecules turn into vapor, even as other molecules leave the vapor phase and return to the liquid phase.

Carbonated beverages illustrate this concept very clearly. Inside the basic carbonated soft drink is the formula for the soda and a certain amount of carbon dioxide (CO_2) molecules. The CO_2 makes the bubbles in the soda. When the soda sits in an unopened plastic bottle on the shelf, the balancing act of CO_2 is happening: CO_2 from the liquid is turning into gaseous

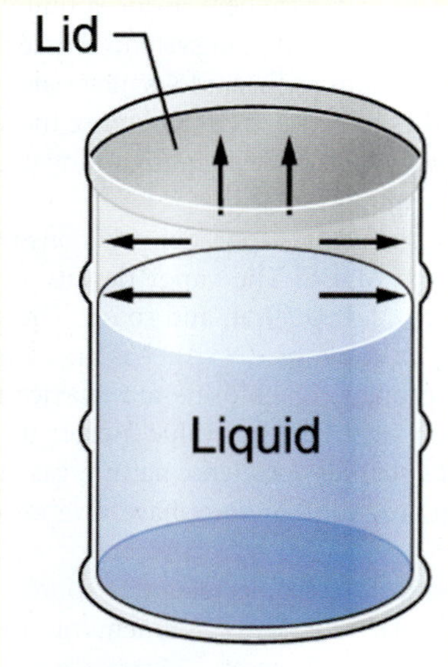

FIGURE 29-4 Vapor pressure.

CO_2 above the liquid, and vice versa. The pressure inside the bottle can be indirectly observed by feeling the rigidity of the unopened plastic bottle. When the CO_2 is not in a closed vessel (that is, the bottle is opened), however, it escapes into the atmosphere. With the pressure released, you can easily dent the sides of the bottle with your fingers. Ultimately, the opened bottle of soda goes "flat" because all of the CO_2 has escaped. If you put the cap back on the bottle tightly before all of the CO_2 has escaped, after a short time the bottle will feel rigid again. If you shake the closed bottle, you will release more of the CO_2 and the bottle (container) could fail as a result of the buildup of vapor pressure.

Armed with that knowledge, the technical definition of vapor pressure makes ready sense. The vapor pressure of a liquid is the pressure exerted by its vapor until the liquid and the vapor are in equilibrium. Again, this process occurs inside a closed container, and temperature has a direct influence on the vapor pressure. For example, if heat impinges on a drum of acetone, the pressure above the liquid will increase and perhaps cause the drum to fail. If the temperature is reduced dramatically, the vapor pressure will drop. This is true for any liquid held inside a closed container.

What happens if the container is opened or spilled onto the ground to form a puddle? The liquid still has a vapor pressure, but it is no longer confined to a container. In such a case, liquids with high vapor pressures evaporate much more quickly than do liquids with low vapor pressures. Vapor pressure directly correlates to the speed at which a material evaporates once it is released from its container.

For example, motor oil has a low vapor pressure. When it is released, it will stay on the ground for a long time. By contrast, chemicals such as isopropyl alcohol and diethyl ether have high vapor pressures. When either of these materials is released and collects on the ground, it evaporates rapidly. If ambient air temperature or pavement temperatures are elevated, the evaporation rate will increase. Wind speed, shade, humidity, and the surface area of the spill also influence how fast the chemical evaporates.

In reference sources, vapor pressure might be expressed in several different units—psi, atmospheres (atm), torr, or millimeters of mercury (mm Hg)—and is usually shown at a temperature of 68°F (20°C). For a frame of reference, use the following comparison:

14.7 psi = 1 atm = 760 torr = 760 mm Hg

The vapor pressure of water at room temperature is approximately 15 torr. Standard 40 weight motor oil, by comparison, has a vapor pressure of approximately less than 0.1 mm Hg at 68°F (20°C); it is practically vaporless at room temperature. Isopropyl alcohol has a very high vapor pressure of 30 mm Hg at room temperature. The temperature of the materials and the ambient temperature both have a direct correlation to vapor pressure.

Specific Gravity

Specific gravity is to liquids what vapor density is to gases and vapors—that is, it is a comparison value. In this case, the comparison is between the weight of a liquid and the weight of water, where water is assigned an index value of 1 as its specific gravity. Any material with a specific gravity less than 1 floats on water, whereas any material with a specific gravity greater than 1 sinks and remains below the surface of water. Consult a comprehensive reference source to find the specific gravity of the chemical in question.

Most flammable liquids float on water FIGURE 29-5. Gasoline, diesel fuel, motor oil, and benzene are examples of liquids that float on water. Carbon disulfide, however, has a specific gravity of approximately 2.6. Consequently, if water is applied to a puddle of carbon disulfide, the water will cover it completely.

Water Miscibility

When discussing the concept of specific gravity, it is also necessary to determine whether a chemical will mix with water. Water is the predominate agent used to extinguish fires; it has often been referred to as being the "universal solvent and decontamination agent." Both statements are general in nature and there are certainly exceptions to every rule; nevertheless, both are also relatively correct. Always remember that when dealing with chemical emergencies, water might not always be the best and safest choice to mitigate the situation. It is a tremendously aggressive solvent and has the ability to react violently with certain other chemicals.

Concentrated sulfuric acid, metallic sodium, and magnesium are just a few examples of substances that adversely react with water. If water is applied to burning magnesium, for example, the heat of the fire will break apart the water molecule, creating an explosive reaction. Adding water to sulfuric acid would be like throwing water on a pan full of hot oil, in that popping and spattering might occur.

Conversely, water can be a friendly ally when fire fighters are attempting to mitigate a chemical emergency. Depending on

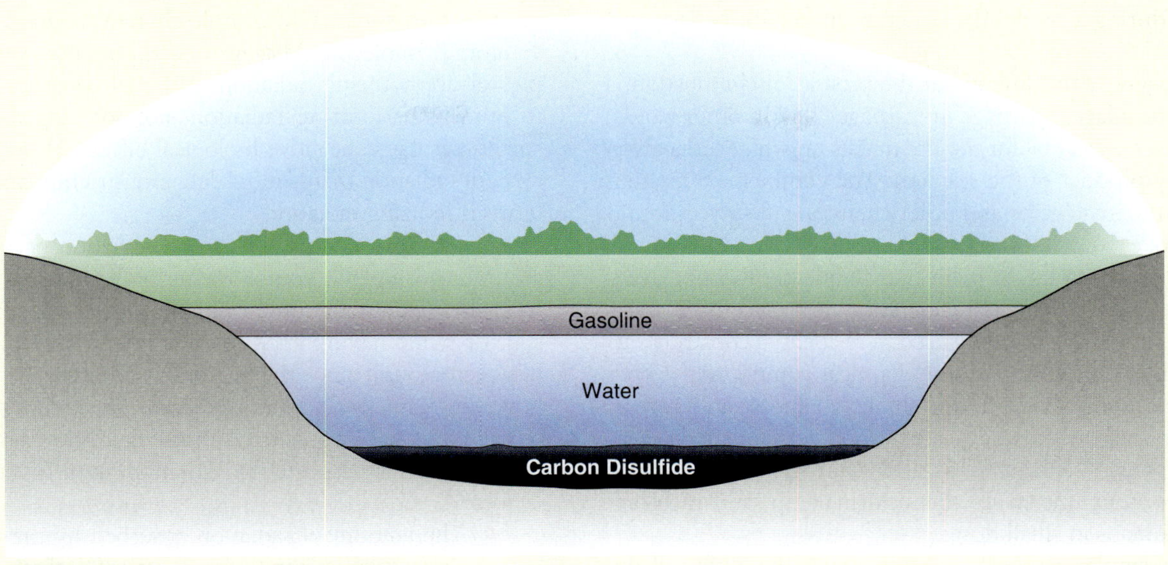

FIGURE 29-5 Gasoline floats on water, whereas carbon disulfide does not.

the chemical involved, fog streams operated from handlines can knock down vapor clouds. Additionally, fires involving heavier-than-water flammable liquids can be extinguished by gently applying water to the surface of the liquid. The term *water solubility* describes the ability of a substance to dissolve in water.

It is important to consider the application of water carefully in all cases involving spilled or released chemicals. Do not approach any problem with the "it is just water" mindset. The results could be something worse than the initial problem.

Corrosivity (pH)

Corrosivity is the ability of a material to cause damage (on contact) to skin, eyes, or other parts on the body. The technical definition, as written by the U.S. Department of Transportation (DOT), includes the following language: "destruction or irreversible damage to living tissue at the site of contact." Corrosive materials are often also damaging to clothing, rescue equipment, and other physical objects in the environment.

Corrosives constitute a complex group of chemicals and should not be taken lightly. Tens of thousands of corrosive chemicals are used in general industry, semiconductor manufacturing, and biotechnology, all of which can be further categorized into two classes: acid and base.

A number of technical definitions for acid and base have been put forth, but the most common way to define them is in terms of pH. Think of pH as the "potential of hydrogen." Essentially, pH is an expression of the concentration of hydrogen ions in a given substance. For fire fighters' purposes, pH serves as a measurement of corrosive strength, which loosely translates to the substance's degree of hazard. In simple terms, fire fighters can use pH to judge how aggressive a particular corrosive might be. To anticipate the seriousness of a particular hazard, they must understand the pH scale.

Acids such as sulfuric acid, hydrochloric acid, phosphoric acid, nitric acid, and acetic acid (vinegar) have pH values less than 7. Chemicals that are considered to be bases, such as sodium hydroxide, potassium hydroxide, sodium carbonate, and ammonium hydroxide, have pH values greater than 7. The midpoint of the pH scale (pH = 7) is where a chemical is considered to be neutral—that is, neither acidic nor basic. At this value, a chemical does not harm human tissue.

Generally, substances with pH values of 2.5 and lower or 12.5 and higher are considered to be "strong." Strong corrosives (both acids and bases) react more aggressively with metallic substances such as steel and iron; they cause more damage to unprotected skin; they react more adversely when contacting other chemicals; and they might react violently with water.

To determine pH in the field, fire fighters can use specialized pH test paper. They can also obtain this information from the MSDS or call for specialized hazardous materials teams to determine the pH level of a chemical. Hazardous materials technicians have more specialized ways of determining pH and can be a useful resource when handling corrosive incidents.

At the operations level, it is critical to understand that incidents involving corrosive substances might not be as straightforward as other types of chemical emergencies. The materials themselves are more complicated and, therefore, the tactics employed to deal with them could, in fact, be outside the fire fighter's scope of practice.

Toxic Products of Combustion

Toxic products of combustion are the hazardous chemical compounds that are released when a material decomposes under heat. The process of combustion is a chemical reaction and, like other reactions, is certain to generate a given amount of by-products. The smoke produced by a structure fire is obvious, but its composition and the toxic gases that are liberated during a residential structure fire might be less well known.

An easy way to think about the process of combustion is to apply the adage, "Garbage in, garbage out." In other words, whatever serves as fuel for the fire makes up whatever is given off as a by-product of the reaction: The combustion reaction produces the toxic gases and other chemical substances found in the smoke.

Burning wood might seem like simple combustion reaction, but consider that the incineration of a piece of Douglas fir (a type of wood commonly used in residential construction) gives off more than 70 harmful chemical compounds. Other substances found in most fire smoke include soot, carbon monoxide, carbon dioxide, water vapor, formaldehyde, cyanide compounds, and many oxides of nitrogen. Each of these substances is unique in its chemical makeup, and most are toxic to humans in small doses.

For example, carbon monoxide affects the ability of the human body to transport oxygen. When the body inhales too much of this gas, the red blood cells cannot get oxygen to the cells and the person dies from tissue asphyxiation. Cyanide compounds also affect oxygen uptake in the body and are a common cause of civilian deaths in structure fires. Formaldehyde is found in many plastics and resins and is one of the many components in smoke that causes eye and lung irritation. The oxides of nitrogen—including nitric oxide, nitrous oxide, and nitrogen dioxide—are deep lung irritants that can cause a serious medical condition called pulmonary edema, or fluid buildup in the lungs.

Fire fighters should always consider the origin of any fire they encounter. The conclusions reached might require evacuating affected citizens, dictating unique tactics (e.g., no water), or helping them better understand and treat injuries to civilians or fire-operations personnel.

Radiation

Most fire fighters have not been trained on the finer points of handling incidents involving radioactive materials and, therefore, have many misconceptions about radiation. Oddly enough, in many ways radiation incidents are some of the easiest hazardous materials events to deal with—assuming that fire fighters have the right training and the right equipment to respond to them.

Radiation is energy transmitted through space in the form of electromagnetic waves or energetic particles. Electromagnetic radiation (light or radio waves) has no mass or charge, whereas energetic particles do have mass.

Radiation can be described as either ionizing or non-ionizing. Ionizing radiation consists of energetic particles or waves that have the potential to ionize an atom or molecule through atomic interactions. Ionization is a function of the energy of the individual particles or waves, rather than a function of the number of particles or waves present. If enough of these ionizations occur, they can prove destructive to biological organisms by damaging the DNA in individual cells. In contrast, non-ionizing radiation refers to any type of electromagnetic radiation that does not carry enough energy to ionize atoms or molecules—that is, to completely remove an electron from an atom or molecule. Microwaves and radio waves are examples of non-ionizing radiation. Although it is not as harmful to humans as ionizing radiation, non-ionizing radiation can sometimes cause negative biological effects. Nevertheless, the form of radiation that is most dangerous to humans in general terms is ionizing radiation.

No one can completely escape radiation because it is all around. During the course of his or her life, every person receives a certain amount of radiation from the sun, from the soil, or while having an X-ray. Radiation is everywhere and has been around since the beginning of time. The focus here, however, is not background radiation, but rather the occupational exposures encountered in the field. For the most part, the health hazards posed by radiation are a function of two factors:

- The amount of radiation absorbed by the body has a direct relationship to the degree of damage done.
- The amount of exposure time ultimately determines the extent of the injury.

The periodic table lists all of the known elements that make up every known chemical compound. Those elements, in turn, are made up of atoms. In the nucleus of those atoms are protons, which have a positive (+) electrical charge, and neutrons, which do not have any electrical charge (neutral). Orbiting the nucleus are electrons, which have a negative (–) electrical charge.

All atoms of any given element have the same number of protons in their nucleus. Some elements, however, have varying numbers of neutrons in their nucleus. Each variation in the number of neutrons results in a different isotope of the element. For most purposes, different isotopes of the same element can be considered to behave identically in chemical reactions. A radioactive isotope has an unstable configuration of protons and neutrons in the nucleus of the atom. Carbon-14, for example, is a radioactive isotope of a carbon atom. Other examples include sulfur-35 and phosphorus-32, which are radioactive isotopes of sulfur and phosphorus atoms, respectively. Radioactivity is the natural and spontaneous process by which unstable atoms (isotopes) of an element decay to a different state and emit or radiate excess energy in the form of particles or waves.

Small radiation detectors are available that can be worn on turnout gear. These detectors sound an alarm when dangerous levels of radioactivity are encountered and alert fire fighters to leave the scene and call for more specialized assistance. The detection of radiation is usually a responsibility of hazardous materials technicians or specialist employees.

If the presence of radiation is suspected, steps must be taken to identify the type of radiation being emitted. In simple terms, radioactive isotopes give off energy from the nucleus of an unstable atom in an attempt to reach a stable state. Typically, a combination of alpha, beta, and gamma radiation is present. Each of these forms of energy given off by a radioactive isotope varies in intensity and consequently determines how fire fighters attempt to reduce the exposure potential FIGURE 29-6.

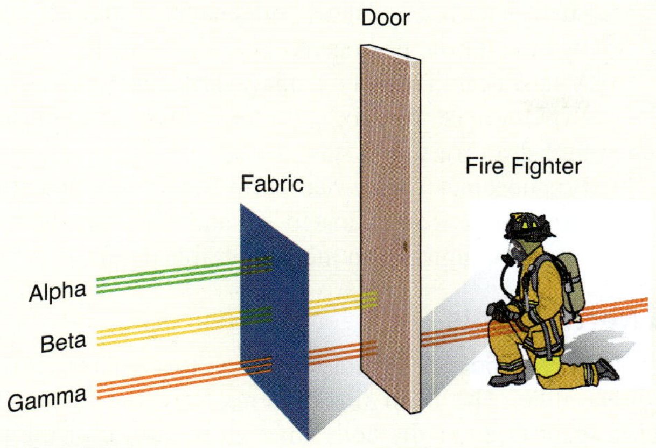

FIGURE 29-6 Alpha, beta, and gamma radiation.

Alpha Particles

As mentioned previously, radiation stems from an imbalance in the numbers of protons and neutrons contained in the nucleus of an atom. Alpha radiation is a reflection of that instability. This form of radiation energy occurs when an electrically charged particle is given off by the nucleus of an unstable atom. Alpha particles have weight and mass and, therefore, cannot travel very far (typically less than a few centimeters) from the nucleus of the atom. For the purpose of comparison, alpha particles are like dust particles.

Typical alpha emitters include americium (found in smoke detectors), polonium (identified in cigarette smoke), radium, radon, thorium, and uranium. Fire fighters can protect themselves from alpha emitters by staying several feet away from the radioactive material and by protecting their respiratory tract with a HEPA (high-efficiency particulate air) filter on a simple respirator or by using self-contained breathing apparatus (SCBA). HEPA filters catch particles as small as 0.3 micron—a size much smaller than a typical dust particle. Although fire fighters can protect themselves from exposure to alpha particles by wearing HEPA filters and proper personal protective equipment (PPE), this type of radiation still poses a serious health risk if it enters into the body, especially if it is inhaled.

Beta Particles

Beta particles are more energetic than alpha particles and, therefore, pose a greater health hazard. Essentially, beta particles are like electrons, except that a beta particle is ejected from the nucleus of an unstable atom. Depending on the strength of the source, beta particles can travel several feet in the open air. Beta particles themselves are not radioactive; rather, the radiation energy is generated by the speed at which the particles are emitted from the nucleus. As a consequence, beta radiation can break chemical bonds at the molecular level and cause damage to living tissue. This breaking of chemical bonds creates an ion, such that beta particles are considered ionizing radiation. Ionizing radiation has the capability to cause changes in human cells. This damage can ultimately create a mutation of the cell and become the root cause of cancer. Examples of ionizing radiation are X-rays and gamma rays (discussed later in this chapter).

Beta particles can redden (erythema) and burn skin; they can also be inhaled. Beta particles that are inhaled can directly damage the cells of the human body. Most solid objects can stop beta particles, and SCBA should provide adequate respiratory protection against them. Common beta emitters include tritium (luminous dials on gauges), iodine (medical treatment), and cesium.

Gamma Rays and Neutrons

Gamma radiation is the most energetic type of radiation that fire fighters might encounter. Gamma rays differ from alpha and beta particles in that they are not particles ejected from the nucleus, but rather pure electromagnetic energy. Gamma radiation has no mass, has no electrical charge, and travels at the speed of light. Gamma rays can pass through thick solid objects (including the human body) very easily and generally follow the emission of a beta particle. If the nucleus still has too much energy after ejecting beta particles, a photon (a packet of pure energy) is released.

Like beta particles, gamma radiation is ionizing radiation and can be deadly. Structural firefighting gear with SCBA does not protect fire fighters from gamma rays; if fire fighters are near the source, they will be exposed to gamma radiation. Typical sources of gamma radiation are cesium (cancer treatment and soil density testing at construction sites) and cobalt (medical instrument sterilization).

Neutrons are also penetrating particles found in the nucleus of the atom. They can be removed from the nucleus through nuclear fusion or fission. Although neutrons themselves are not radioactive, exposure to neutrons can create radiation, such as gamma radiation. Most fire fighters will never encounter neutrons in any form.

Hazard Exposure and Contamination

Hazard and Exposure

According to NFPA 472, *Standard for Competence of Responders to Hazardous Materials/Weapons of Mass Destruction Incidents*, a hazard is defined as a material capable of posing an unreasonable risk to health, safety, or the environment—that is, a material capable of causing harm. The same source defines exposure as the process by which people, animals, the environment, and equipment are subjected to or come into contact with a hazardous material.

For fire fighters' purposes, the two definitions are intertwined. When fire fighters respond to hazardous materials incidents, it is vital they determine the risk so that fire personnel can reduce their potential for exposure. This is done by taking the time to understand the problem at hand, including the physical properties of the chemical and the release scenario, and determining whether fire fighters have the right training, equipment, and protective gear to positively influence the outcome of the incident. All of the information presented to this point has been directed at understanding the nature of the chemicals involved.

Contamination

When a chemical has been released and physically contacts people, the environment, animals, tools, or other items, either intentionally or unintentionally, the residue is called contamination. In some cases, the process of removing those contaminants is complex and requires a significant effort to complete. In other cases, the contamination is easily removed and does not pose much of a threat to the overall success in handling the incident. For now, simply understand that when a dangerous chemical escapes its container, fire fighters might need to organize a system to remove it safely and efficiently or reduce its physical hazard.

Secondary Contamination

Secondary contamination occurs when a person or an object transfers the contamination or the source of contamination to another person or object by direct contact. In such cases, fire fighters might become contaminated and subsequently handle tools and equipment, grasp door handles, or touch other responders, thereby spreading the contamination. The possibility of spreading contamination brings up a point that all fire fighters should understand: The cleaner fire fighters stay during the response, the less decontamination they will have to do later. Many fire fighters have the misperception that PPE is worn to enable them to contact the product. In fact, quite the opposite is true: PPE, including specialized chemical protective gear, is worn to protect fire fighters in the event that they cannot avoid product contact. The less contamination encountered or spread, the easier the decontamination will be.

Types of Hazardous Materials: Weapons of Mass Destruction

In recent years, weapons of mass destruction (WMD) have become a highly relevant topic for fire fighters. WMD represent a real threat in the United States, and each fire fighter should have a basic knowledge of those potential threats. To that end, this section briefly addresses the threats posed by nerve, blister, blood, and choking agents, and other potential hazards associated with various irritant materials.

WMD events can be complex, and the harm inflicted can be the result of several factors—nerve agents dispersed by explosions, radioactive contamination spread by the detonation of conventional explosives (so-called dirty bombs), and so on. Generally speaking, however, the damages caused by a terrorist attack or any other hazardous materials incident can be broadly classified into seven categories, best remembered by the mnemonic TRACEMP:

- Thermal: Heat created from intentional explosions or fires, or cold generated by cryogenic liquids.
- Radiologic: Radioactive contamination from dirty bombs; alpha, beta, and gamma radiation.
- Asphyxiation: Oxygen deprivation caused by materials such as nitrogen; tissue asphyxiation from blood agents.
- Chemical: Injury and death caused by the intentional release of toxic industrial chemicals, nerve agents, vesicants, poisons, or other chemicals.
- Etiologic: Illness and death resulting from biohazards such as anthrax, plague, and smallpox; hazards posed by bloodborne pathogens.
- Mechanical: Property damage and injury caused by explosion, falling debris, shrapnel, firearms, explosives, and slips, trips, and falls.
- Psychogenic: The mental harm from being potentially exposed to, contaminated by, and even just being in close proximity to an incident of this nature.

Nerve Agents

Nerve agents pose a significant threat to civilians and fire fighters alike. The main threat stems from the ability of a nerve agent to enter the body through the lungs or the skin and then systemically affect the function of the human body. Nerve agents, such as many common pesticides, disrupt the central nervous system, possibly causing death or serious impairment.

The human central nervous system is laid out much like the electrical wiring in a house. The brain serves as the main panel. The wires—that is, the nerves—go to muscles, organs, glands, and other critical areas of the body. When a nerve agent, such as sarin or VX, gets into the body, it affects the ability of the body to transact nerve impulses at the junction points between nerves and muscles, nerves and vital organs, and nerves and glands. Subsequently, those critical areas do not function properly and associated body functions are affected.

Signs and symptoms of nerve agent exposure include pinpoint pupils, tearing of the eyes, twitching muscles, loss of bowel and bladder control, and slow or rapid heartbeat. Mild exposures can result in mild symptoms and include some or all of these symptoms. Significant exposures can result in rapid death, serious impairment, or a more pronounced manifestation of symptoms. The bottom line is that an exposure to a nerve agent is not an automatic death sentence. The dose absorbed directly correlates to the degree of damage done and the symptoms presented by the victim.

Recognition of the signs and symptoms of a nerve agent exposure is vital in determining the presence of such a threat. Nerve agents in general are reported to have fruity odors. In most cases, fire fighters do not have the specialized detection devices necessary to identify the presence of a nerve agent; the symptoms presented by the victims should tip them off, however. To that end, the mnemonic SLUDGEM briefly summarizes some of the more common signs and symptoms of nerve agent exposure:

S: Salivation
L: Lachrymation (tearing)
U: Urination
D: Defecation
G: Gastric disturbance
E: Emesis (vomiting)
M: Miosis (constriction of the pupils)

Many times, nerve agents are incorrectly referred to as nerve gas. Nerve agents are liquids, not gases. Each evaporates at a different rate. Sarin, for example, is a water-like liquid that evaporates more easily than VX, which is a thick, oily liquid. VX is the most toxic of the weapons-grade nerve agents.

Other nerve agents include soman and tabun. The organophosphate and carbamate classes of pesticides are similar to the weapons-grade agents but are less toxic in general terms.

Blister Agents

Blister agents (also known as vesicants) are classified as such because of their ability to cause blistering of the skin. Common blister agents include sulfur mustard and Lewisite, which interact in unique ways with the human body.

Sulfur mustard was first used as a weapon in World War I. Chemical casualties suffered horrific injuries to the skin and lungs; in some cases, those exposures were fatal. Although sulfur mustard is commonly referred to as mustard gas, it is typically found in a liquid state. This chemical is colorless and odorless when pure. When mixed with other substances or otherwise impure, however, it can look brownish in color and smell like garlic or onions.

Exposures to sulfur mustard might not be immediately apparent. A victim might not experience pain or any other sign indicating current contact with this agent. Indeed, redness and blistering might not appear for 2 to 24 hours after exposure. Once the blistering does occur, skin decontamination will not reduce its effects FIGURE 29-7.

If sulfur mustard exposure is suspected, however, immediate skin decontamination is indicated. Early skin decontamination can reduce the effects of exposure.

Sulfur mustard produces injuries similar to second-degree and third-degree thermal burns. In cases involving mild skin exposures, victims typically make a complete recovery. Inhalation exposures are much more serious, with the damage to lung tissues sometimes resulting in permanent respiratory function impairment. Additionally, sulfur mustard is very persistent in the environment when released. It can remain intact, sticking to the ground or other surfaces, for several days.

Lewisite has many of the same characteristics of sulfur mustard. Lewisite, however, contains arsenic; thus, when it is absorbed by the skin, this blister agent can produce some symptoms specific to arsenic poisoning, such as vomiting and low blood pressure. Unlike sulfur mustard, Lewisite exposures to the skin produce immediate pain. If a Lewisite exposure is suspected, immediate and aggressive skin decontamination is indicated. Like sulfur mustard, Lewisite does not occur naturally in the environment and is considered to be a chemical warfare agent.

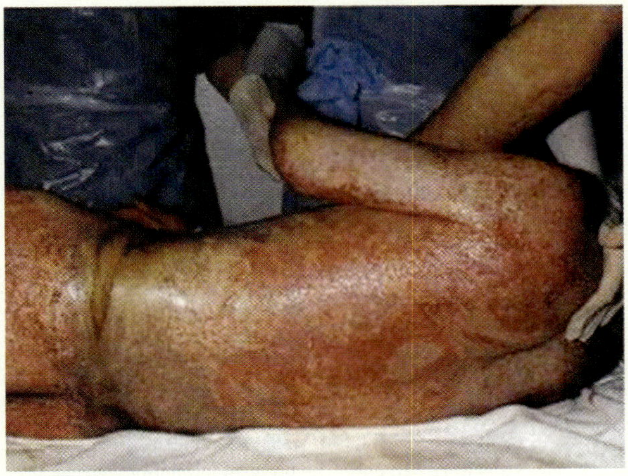

FIGURE 29-7 Blister agent exposure.

Cyanide

Cyanide compounds, including hydrogen cyanide and cyanogen chloride, prevent the body from using oxygen. Victims exposed to high levels of cyanide, for example, could be placed on supplemental oxygen but still not survive the exposure. The cause of death in this case is not a lack of oxygen, but rather the inability of the body to get oxygen to the cells.

The main route of exposure to cyanide compounds is through the lungs, but many of these agents can also be absorbed through the skin. Hydrogen cyanide has an odor of bitter almonds. Sixty percent of the general population cannot detect the presence of hydrogen cyanide by smell; thus, because the odor threshold and the amount of cyanide that would be lethal are very close, many times odor alone cannot be used to detect dangerous amounts of hydrogen cyanide. By contrast, cyanogen chloride, because of the chlorine atom within the chemical compound, has a much more irritating and pungent odor.

Some of the typical signs and symptoms of cyanide exposure include vomiting, dizziness, watery eyes, and deep and rapid breathing. High concentrations lead to convulsions, inability to breathe, loss of consciousness, and death.

Choking Agents

Choking agents are predominantly designed to incapacitate rather than to kill; nevertheless, death and serious injuries from these agents are certainly possible. The nature of choking agents (extremely irritating odor) alerts potential victims of their presence and allows for escape from the environment when possible. Some commonly recognized choking agents include chlorine, phosgene, and chloropicrin.

Each of these substances is associated with a specific odor. Chlorine smells like a swimming pool, whereas phosgene and chloropicrin have been reported to smell like freshly mown grass or hay. In either case, the odors are noticeable and, in many cases, strong enough and irritating enough to alert victims to the presence of an unusual situation.

In the event of a significant exposure, the main threat from these materials is pulmonary edema (also known as dry drowning because the victim's lungs fill with fluid). When inhaled, choking agents damage sensitive lung tissue and cause fluid to be released by the injury. That fluid results in chemically induced pneumonia, which is fatal (over time) in many cases. Additionally, these agents act as skin irritants and can cause mild to moderate skin irritation and significant burning when victims are exposed to them in high concentrations.

Irritants (Riot Control Agents)

Irritants (also known as riot control agents) include substances, such as mace, that can be dispersed to briefly incapacitate a person or groups of people. Chiefly, irritants cause pain and a burning sensation in exposed skin, eyes, and mucous membranes. The onset of symptoms occurs within seconds after contact and lasts several minutes to several hours, usually with no lasting effects. Irritants can be dispersed from canisters, hand-held sprayers, and grenades.

From a terrorism perspective, irritants can be employed to incapacitate rescuers or to drive a group of people into another area where a more dangerous substance can be released. Of all the groups of WMD, irritants pose the least amount of danger in terms of toxicity. Exposed patients can be decontaminated with clean water, and the residual effects of the exposure are rarely significant. Fire fighters are also likely to encounter these agents when they respond to calls in which law enforcement personnel have used these agents for riot control or suspect apprehension. It is important for fire fighters to know which specific agents are used by local law enforcement agencies and any specific decontamination techniques recommended for use with those agents.

Convulsants

Chemicals classified as convulsants are capable of causing convulsions or seizures when absorbed by the body. Chemicals that fall into this classification include nerve agents such as sarin, soman, tabun, VX, and the organophosphate and carbamate classes of pesticides. These substances interfere with the central nervous system and disrupt normal transmission of neuromuscular impulses throughout the body. Pesticides capable of functioning as convulsants include parathion, aldicarb, diazinon, and fonofos.

When dealing with these materials, it is important to identify their presence and to avoid breathing vapors or allowing liquid to contact skin. In some cases, even a small exposure can be fatal.

Harmful Substances' Routes of Entry into the Human Body

For a chemical or other harmful substance to injure a person, it must first get into his or her body. Throughout the course of his or her fire service career, a fire fighter will inevitably be bombarded by harmful substances such as diesel exhaust, bloodborne pathogens, smoke, and accidentally and intentionally released chemicals. Fortunately, most of these exposures are not immediately deadly. Unfortunately, repetitive exposures to these materials can have a negative health effect after a 20-year career. To protect yourself now and give yourself the best shot at a healthy retirement and long life, it is important to understand some basic concepts about toxicology and the four ways that chemicals enter the body FIGURE 29-8:

- Inhalation: Through the lungs
- Absorption: By permeating the skin
- Ingestion: By the gastrointestinal tract
- Injection: Through cuts or other breaches in the skin

Toxicology is the study of the adverse effects of chemical or physical agents on living organisms. The following sections discuss the potential routes of entry and methods fire fighters can employ to protect themselves against these agents.

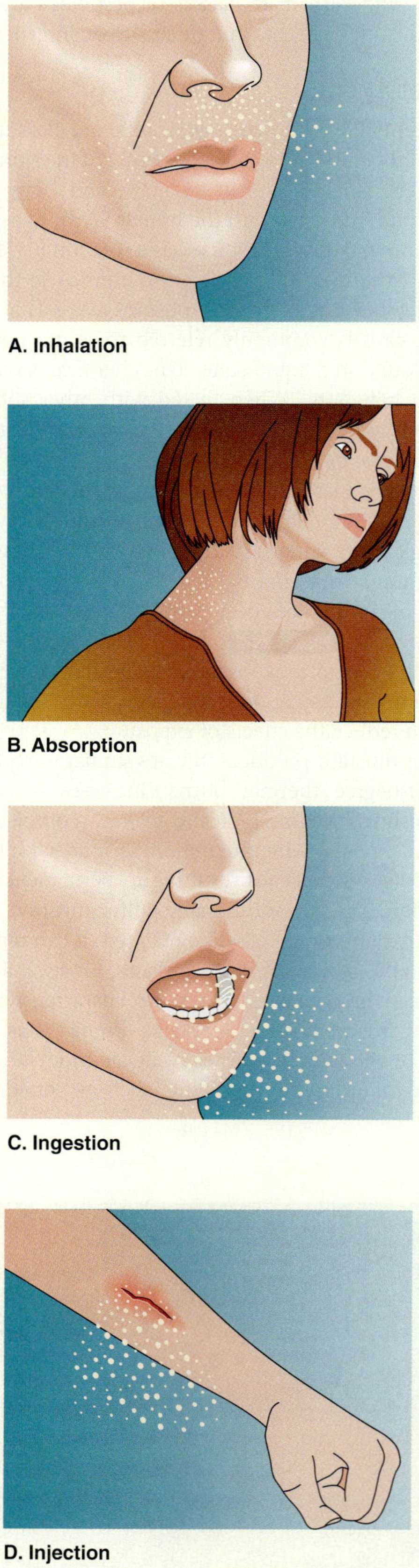

FIGURE 29-8 The four ways a chemical substance can enter the body. **A.** Inhalation. **B.** Absorption. **C.** Ingestion. **D.** Injection.

Inhalation

Inhalation exposures occur when harmful substances are brought into the body through the respiratory system. The lungs are a direct point of access to the bloodstream, so they can quickly transfer an airborne substance into the bloodstream.

The entire respiratory system is vulnerable to attack from most hazardous materials and WMD, corrosive materials such as chlorine and ammonia, solvent vapors such as gasoline and acetone, superheated air, and any other material finding its way into the air. In addition to gases and vapors, small particles of dust and smoke can lodge in sensitive lung tissue, causing substantial irritation.

Given these facts, it is imperative that fire fighters always consider using respiratory protection when they are operating in the presence of airborne contamination. Fortunately, fire fighters have ready access to an excellent form of respiratory protection—namely, the positive-pressure, open-circuit SCBA. This equipment is by far the single most important piece of PPE that fire fighters have at their disposal—so use it! Sometimes fire fighters might need to use other forms of respiratory protection depending on the specific respiratory hazard they are facing while performing a given task. For example, full-face and half-face air-purifying respirators (APRs) or battery-powered air-purifying respirators offer specific degrees of protection if the chemical hazard present is known and the appropriate filter canister is used FIGURE 29-9.

Respirators do not provide oxygen or breathing air, however. Thus, if the oxygen content of the work area is low, respirators are not a viable option. According to the Occupational Safety and Health Administration (OSHA), any work environment containing less than 19.5 percent oxygen is considered to be oxygen deficient and requires the use of an air-supplying respirator. Another drawback to APRs is that filters can be costly and are often needed in large quantities, depending on the nature of the response.

FIGURE 29-9 Air-purifying respirators offer some degree of protection against airborne chemical hazards.

Respirators are lighter than SCBA, are more comfortable to wear, and usually allow for longer work periods because they do not depend on a limited source of breathing air. SCBA certainly offers a higher level of respiratory protection, but in the right circumstances APRs represent a viable alternative to the former devices.

When considering protection against airborne contamination, it is important to understand the origin of the contamination, its concentration, and the impact that the contamination might potentially have on oxygen levels in the area. In short, determining the appropriate type of respiratory protection must be accomplished by looking at the overall situation, including the nature of the contaminant. In some cases, the anticipated particle size of the contamination determines the level of respiratory protection employed TABLE 29-2.

TABLE 29-2 Particle Sizes of Common Types of Respiratory Hazards

Fume	< 1 micron
Smoke	< 1 micron–1 micron
Dust	1 micron+
Fog	< 40 microns
Mist	> 40 microns

FIRE FIGHTER II Tip **FFII**

Air monitoring cannot be done haphazardly. It defines the zones of operations, so pay close attention to your air-monitoring equipment.

Anthrax spores offer an excellent example to illustrate this point. Weaponized anthrax spores typically vary in size from 0.5 micron to 1 micron. Based on that range, a typical full-face APR with a nuisance dust filter would not offer sufficient protection against this hazard. Anyone operating in an area contaminated with anthrax should wear SCBA or, at minimum, an air-purifying respirator with HEPA filtration (protection from 0.3-micron or larger particles).

Now consider this scenario: A container of cryogenic nitrogen is leaking inside a poorly ventilated storage room. The main threat is the cold vapor produced and the possibility of oxygen deficiency. In this case, SCBA is the appropriate type of respiratory protection based on the anticipated hazard. In many cases, hazardous materials technicians might need to respond; they will use air-monitoring devices to characterize the work area prior to entry. The choices made regarding respiratory protection then will be based on definitive and known information.

Particle size also determines where the inspired contamination ultimately ends up. Bigger particles that make up visible mists are captured in the nose and upper airway, for example, whereas smaller particles work their way deeper into the lung TABLE 29-3. Gaseous substances such as carbon monoxide and

TABLE 29-3 Location of Respiratory Trapping by Particle Size

>7 microns	Nose
5–7 microns	Larynx
3–5 microns	Trachea and bronchi
2–3 microns	Bronchi
1–2.5 microns	Respiratory bronchioles
0.5–1 micron	Alveoli

hydrogen cyanide can diffuse into the bloodstream and affect entire body systems.

Respiratory protection is one of the most important components of any fire fighter's PPE. In all cases where airborne contamination is encountered, take the time to understand the nature of the threat, evaluate the respiratory protection available, and decide whether it provides adequate protection. When protecting the lungs, "good enough" is not an option.

Absorption

The skin—the largest organ in the body—is susceptible to damage by a number of substances. In addition to serving as the body's protective shield against heat, light, and infection, the skin helps to regulate body temperature, stores water and fat, and serves as a sensory center for painful and pleasant stimulation. Without this important organ, humans would not be able to survive.

When discussing chemical exposures, however, absorption is not limited only to the skin. Absorption is the process by which hazardous materials travel through body tissues until they reach the bloodstream. The eyes, nose, mouth (any mucous membrane), and, to a certain degree, the intestinal tract are all part of this equation. The eyes, for example, absorb a high amount of liquid and vapor that come into contact with the sensitive tissues. This absorption is particularly problematic because the eyes connect directly to the optic nerve, which allows the chemical to follow a direct route to the brain and the central nervous system.

Although the skin functions as a shield for the body, that shield can be pierced by a number of chemicals. Aggressive solvents such as methylene chloride (found in paint stripper), for example, can be readily absorbed through the skin. A secondary hazard associated with this chemical occurs when the body attempts to metabolize the methylene chloride after it is absorbed. A by-product of that metabolization reaction is carbon monoxide, a cellular asphyxiant. Asphyxiants are substances that prevent the body from using oxygen, thus causing suffocation. In this scenario, the initial chemical is broken down to another substance that poses a greater health hazard than the original chemical. Methylene chloride is also suspected to be a human carcinogen.

Many other hydrocarbon-based substances can be absorbed through the skin and ultimately enter the bloodstream. A hydrocarbon is a substance consisting of carbon and hydrogen atoms, among other components. Examples include gasoline, diesel fuel, benzene, and other solvents.

Fire Fighter Safety Tips

When responding to incidents that involve large or small hydrocarbon spills, fire fighters should not walk through the product with leather boots. The leather will absorb the product and expose and contaminate the fire fighter.

Absorption hazards are not limited to hydrocarbons, however. Hydrofluoric acid, for example, poses a significant threat to life when it is absorbed through the skin. This unique corrosive has the ability to bind with certain substances in the body (predominately calcium). Secondary health effects occurring after exposure can include muscular pain and potentially lethal cardiac arrhythmias.

In the field, fire fighters must constantly evaluate the possibility of chemical contact with their skin and eyes. In many cases, structural firefighting turnout gear provides little or no protection against liquid chemicals. Consult the *Emergency Response Guidebook* (ERG) or an MSDS for response guidance when deciding whether turnout gear is appropriate given the hazards at hand.

In the event that turnout gear does not offer adequate protection, fire fighters might have to increase their level of protection to include specialized chemical protective clothing. Keep in mind that the operations level of training is primarily intended to allow fire fighters to take defensive actions while avoiding contact with the released product. If they have to wear something other than turnout gear, they might be exceeding their level of training. The use of SCBA and proper protective firefighting gear can reduce many of the standard routes of exposure, but not always! Refer to your fire department's standard operating procedures and guidelines for guidance in such situations.

Ingestion

In addition to absorption through the skin, chemicals can be brought into the body through the gastrointestinal tract, by the process of ingestion. The water, nutrients, and vitamins the body needs are predominately absorbed in this manner. This is an uncommon route of significant chemical exposure for most people—except for fire fighters. For example, at a structure fire, fire fighters generally have an opportunity to rotate out of the building for rest and refreshment. If they do not take the time to wash up prior to eating or drinking, they have a high probability of spreading contamination from the hands to the food and subsequently to the intestinal tract. If you do not think about every situation where you might become exposed, you put yourself in harm's way.

Injection

Chemicals brought into the body through open cuts and abrasions qualify as injection exposures. Protecting yourself from this route of exposure starts with realizing that you are going to work in a compromised state. Any cuts or open wounds should be addressed before reporting for duty. If they

VOICES OF EXPERIENCE

Our fire department was called one day to a fire in a storage shed/barn/garage building in a rural area. Upon arrival, the structure was fully involved. We set up suppression operations and soon had the bulk of the fire under control. We then started to mop up the scene. Because of the heat still in the building, we were applying streams from the outside, and when water was shot under a workbench, we got a small explosion and a fireball rolled out from underneath the bench.

When we asked the property owner what might be causing this, he had no idea. After a few minutes discussion with him we learned that under the bench was where he stored his calcium carbide for some lamps that he had owned. Knowing that calcium carbide is a solid that when combined with water creates acetylene gas, we then knew what had caused the explosion and fire. It was obvious that we had already knocked the lid off at least one of the containers, so our strategy then became to cool the rest of the structure while avoiding the area under the workbench.

By carefully placing our water in an effort to avoid the calcium carbide we were able to cool the rest of the structure without further incident. Once everything else cooled down we could get close enough to use a pike pole to reach in and pick up the buckets containing the calcium carbide and safely remove them from the structure.

Knowing and understanding the chemical and physical properties of calcium carbide allowed us to determine how to handle the incident while maintaining safe operations for our responders.

Allen Michel
Nebraska State Fire Marshal—Training Division
Lincoln, Nebraska

are significant, you might be excluded from operating in contaminated environments. Open wounds are a direct portal to the bloodstream and subsequently to muscles, organs, and other body systems. If a chemical substance contacts this open portal, the health effects could be immediate and pronounced. Remember—intact skin is a good protective shield. Do not go into battle if your shield is not up to the task.

Chronic versus Acute Health Effects

A chronic health hazard (chronic health effect) is an adverse health effect to a substance that occurs gradually over time. Chronic health effects can occur either after long-term exposures or after multiple short exposures over a shorter period. The exposures can occur by all routes of entry into the body and can result in health issues such as cancer, permanent loss of lung function, or repetitive skin rashes. For example, inhaling asbestos fibers for years without respiratory protection can result in a form of lung cancer called asbestosis.

Chemicals that pose a hazard to health after only relatively short exposure periods are said to cause acute health effects. Such an exposure period is considered to be subchronic, with effects occurring either immediately after a single exposure or up to several days or weeks after an exposure. Essentially, acute exposures are "right now" exposures that produce some observable condition such as eye irritation, coughing, dizziness, or skin burns.

Skin irritation and burning after a dermal exposure to sulfuric acid are classified as acute health effects. Other chemicals, such as formaldehyde, are also capable of causing acute health effects. Among other things, formaldehyde is a sensitizer that is capable of causing an immune system response when it is inhaled or absorbed through the skin. This reaction is similar to other allergic reactions in that an initial exposure produces mild health effects but subsequent exposures cause much more severe reactions such as breathing difficulties and skin irritation. OSHA defines a sensitizer as a chemical that causes a substantial proportion of exposed people or animals to develop an allergic reaction in normal tissue after repeated exposure to the chemical.

Using the *Emergency Response Guidebook*

Probably one of the most frequently consulted hazardous materials references used by emergency response agencies is the ERG. The ERG is jointly developed by the U.S. DOT, Transport Canada, the Secretariat of Transport and Communications of Mexico, and CIQUIME of Argentina. It is designed to provide basic information to keep both responders and the general public safe during the initial stages of a hazardous materials incident.

The ERG is divided into four major sections, plus some additional reference materials:

- Yellow section: Lists hazardous materials in numerical order by the chemical identification number. This identification number is followed by a guide number and the chemical name.
- Blue section: Lists hazardous chemicals in alphabetical order. The chemical name is followed by the guide number and the four-digit chemical identification number.
- Orange section: The most important section of the ERG. Its 62 "guides" provide basic response information for specific groups of chemicals—namely, potential hazards (health and fire/explosion), public safety (general information regarding chemical hazards, protective clothing requirements, and evacuation needs), and emergency response (fire situations, spill or leak situations, and first-aid requirements).
- Green section: Provides information on the evacuation/isolation distance for specific chemicals. This section lists the chemical identification number, the chemical name, and isolation and protection distances for both small and large spill situations.
- White pages: List reference resources, including who to call for assistance, placarding information, rail and highway vehicle identification charts, and a glossary of key terms.

Emergency responders also use numerous other chemical reference manuals. For example, *Firefighters' Handbook of Hazardous Materials, Seventh Edition*, is similar to the ERG. This handbook is divided into three major sections. The first section lists chemicals in alphabetical order and provides a listing of the basic properties for each chemical in a spreadsheet format. The second section provides emergency guides and is very similar to the guides section of the ERG. The third section lists chemicals by the DOT identification number.

To use the DOT's ERG, follow the steps in SKILL DRILL 29-1:

1. Identify the chemical name and/or the chemical identification number for the suspect material.
2. Look up the material name in the appropriate section. Use the yellow section to obtain information by the chemical identification number or the blue section to obtain information by the alphabetical chemical name.
3. Determine the correct emergency action guide to use for the chemical identified.
4. Identify the isolation distance and the protective actions required for the chemical identified.
5. Identify the potential hazards of the chemical identified.
6. Identify the emergency response actions of the chemical identified.

Wrap-Up

Chief Concepts

- The first step in understanding the hazard posed by a substance involves identifying its state of matter: solid, liquid, or gas.
- A physical change occurs when a substance is subjected to outside influences such as heat, cold, or pressure. For example, when water is exposed to cold, it becomes ice. A chemical change is a change in the substance's chemical makeup with a release of some form of energy. For example, if oxygen is removed from water (H_2O), the chemical makeup of the substance changes (H).
- The boiling point is the temperature at which a liquid eventually turns to gas. Flammable liquids with low boiling points are dangerous because of their potential to produce large volumes of flammable vapor when exposed to relatively low temperatures.
- The three aspects to consider when evaluating the fire potential of a chemical are the following:
  - Flash point: An expression of the temperature at which a liquid fuel gives off sufficient vapor that, when an ignition source is present, results in a flash fire.
  - Ignition temperature: When a liquid fuel is heated beyond its ignition temperature, the substance ignites without an external ignition source.
  - Flammable range: A mixture of fuel and air, defined by upper and lower flammable limits, that reflects an amount of vapor mixed with a volume of air. The wider the flammable range, the more dangerous the material.
- When operating in any environment where a flammable or combustible vapor cloud is suspected to be present, always err on the side of safety and assume that the danger of ignition exists.
- In addition to addressing the flammability of a particular vapor or gas, fire fighters must determine the vapor density (or weight) of the substance. To find a chemical's vapor density, consult a reference source such as the MSDS.
- In the absence of reliable reference sources in the field, fire fighters can use the HA HA MICEN mnemonic to remember a number of lighter-than-air gases:

  H: Hydrogen
  A: Acetylene
  H: Helium
  A: Ammonia
  M: Methane
  I: Illuminating gas (neon and hydrogen cyanide)
  C: Carbon monoxide
  E: Ethylene
  N: Nitrogen
- The key point to understanding vapor pressure is this: The vapors released from the surface of the liquid must be contained if they are to exert any pressure. Essentially, the liquid inside a container will vaporize until the molecules given off by the liquid reach equilibrium with the liquid itself. It is a balancing act between the liquid and the vapors in which some liquid molecules turn into vapor, even as other molecules leave the vapor phase and return to the liquid phase.
- Always remember that when dealing with chemical emergencies water might not always be the best and safest choice to mitigate the situation. It is a tremendously aggressive solvent and has the ability to react violently with certain other chemicals.
- Corrosivity is the ability of a material to cause damage to the body on contact. Corrosives constitute a complex group of chemicals and fall into two classes: acids and bases. Acids have pH values less than 7. Bases have pH values greater than 7. Generally, substances with pH values of 2.5 and lower and 12.5 and higher are considered to be strong.
- Radiation is the energy transmitted through space in the form of electromagnetic waves or energetic particles. It can be described as ionizing or non-ionizing. Ionizing radiation has the potential to ionize an atom or molecule through atomic interactions. Non-ionizing radiation does not carry enough energy to remove an electron from an atom or molecule. Radiation is everywhere and has been around since the beginning of time. For the most part, the health hazards posed by radiation are a function of two factors:
  - The amount of radiation absorbed by the body has a direct relationship to the degree of damage done.
  - The amount of exposure time ultimately determines the extent of the injury.
- Fire fighters can protect themselves from alpha particle exposure by wearing HEPA filters. Fire fighters can protect themselves from beta particle exposure by wearing SCBA. Structural firefighting gear and SCBA do not protect fire fighters from gamma rays.
- The residue from a chemical is contamination. Secondary contamination occurs when a person or object transfers the contamination to another person or object by direct contact. To prevent this, decontamination is necessary.
- The seven categories of WMD are as follows:
  - Thermal: Heat created from intentional explosions or fires, or cold generated by cryogenic liquids.
  - Radiological: Radioactive contamination from dirty bombs; alpha, beta, and gamma radiation.
  - Asphyxiation: Oxygen deprivation caused by materials such as nitrogen; tissue asphyxiation from blood agents.

- Chemical: Injury and death caused by the intentional release of toxic industrial chemicals, nerve agents, vesicants, poisons, or other chemicals.
- Etiological: Illness and death resulting from biohazards such as anthrax, plague, and smallpox; hazards posed by bloodborne pathogens.
- Mechanical: Property damage and injury caused by explosion, falling debris, shrapnel, firearms, explosives, and slips, trips, and falls.
- Psychogenic: The mental harm from being potentially exposed to, contaminated by, and even just being in close proximity to an incident of this nature.

- Nerve agents attack the body's central nervous system. The signs and symptoms of nerve agent exposure can be remembered through the mnemonic SLUDGEM:
  S: Salivation
  L: Lachrymation (tearing)
  U: Urination
  D: Defecation
  G: Gastric disturbance
  E: Emesis (vomiting)
  M: Miosis (constriction of the pupil)
- Blister agents include sulfur mustard and Lewisite. Immediate skin decontamination is required.
- Cyanide compounds prevent the body from using oxygen. The typical signs and symptoms include vomiting, dizziness, watery eyes, and deep and rapid breathing.
- Choking agents are mostly designed to incapacitate rather than to kill. Irritants can be dispersed to briefly incapacitate a person or groups of people.
- Convulsants are capable of causing seizures. They include nerve agents such as sarin, soman, and VX. It is critical to identify their presence and avoid exposure.
- The four ways that chemicals enter the body are the following:
  - Inhalation: Through the lungs
  - Absorption: By permeating the skin
  - Ingestion: By the gastrointestinal tract
  - Injection: Through cuts or other breaches in the skin
- A chronic health hazard is an adverse health effect that occurs gradually over time. Acute health effects occur after only relatively short exposure periods.
- One of the most frequently consulted hazardous materials references used by emergency response agencies is the *Emergency Response Guidebook* (ERG). It is designed to provide basic information to keep both responders and the general public safe during the initial stages of a hazardous materials incident. The ERG is divided into four major sections, plus some additional reference materials:
  - Yellow section: Lists hazardous materials in numerical order by the chemical identification number.
  - Blue section: Lists hazardous chemicals in alphabetical order.
  - Orange section: Provides basic response information for specific groups of chemicals.
  - Green section: Provides information on the evacuation/isolation distance for specific chemicals.
  - White pages: List reference resources, including who to call for assistance, placarding information, rail and highway vehicle identification charts, and a glossary of key terms.

Hot Terms

Absorption The process by which hazardous materials travel through body tissues until they reach the bloodstream.

Acid A material with a pH value less than 7.

Acute health effect A health problem caused by relatively short exposure periods to a harmful substance that produces observable conditions such as eye irritation, coughing, dizziness, and skin burns.

Alpha particle A positively charged particle emitted by certain radioactive materials, identical to the nucleus of a helium atom. (NFPA 801, *Standard for Fire Protection for Facilities Handling Radioactive Materials*)

Asphyxiant A material that causes the victim to suffocate.

Base A material with a pH value greater than 7.

Beta particle An elementary particle, emitted from a nucleus during radioactive decay, with a single electrical charge and a mass equal to 1/1837 that of a proton. (NFPA 801)

Blister agent A chemical that causes the skin to blister.

Blood agent A chemical that, when absorbed by the body, interferes with the transfer of oxygen from the blood to the cells.

Boiling point The temperature at which the vapor pressure of a liquid equals the surrounding atmospheric pressure. (NFPA 30, *Flammable and Combustible Liquids Code*)

Carcinogen A cancer-causing substance that is identified in one of several published lists, including, but not limited to, *NIOSH Pocket Guide to Chemical Hazards, Hazardous Chemicals Desk Reference*, and the ACGIH *2007 TLVs and BEIs*. (NFPA 1851, *Standard on Selection, Care, and Maintenance of Protective Ensembles for Structural Fire Fighting and Proximity Fire Fighting*)

Chemical change The ability of a chemical to undergo an alteration in its chemical makeup, usually accompanied by a release of some form of energy.

Chlorine A yellowish gas that is about 2.5 times heavier than air and slightly water-soluble. Chlorine has many industrial uses but also damages the lungs when inhaled; it is a choking agent.

Choking agent A chemical designed to inhibit breathing and typically intended to incapacitate rather than kill.

Chronic health hazard A health problem occurring after a long-term exposure to a substance.

Contamination The process of transferring a hazardous material from its source to people, animals, the environment, or equipment, all of which can act as carriers for the material.

Convulsant A chemical capable of causing convulsions or seizures when absorbed by the body.

Corrosivity The ability of a material to cause damage (on contact) to skin, eyes, or other parts on the body.

Expansion ratio The ratio of the volume of foam in its aerated state to the original volume of nonaerated foam solution. (NFPA 1901, *Standard for Automotive Fire Apparatus*)

Exposure The process by which people, animals, the environment, and equipment are subjected to or come into contact with a hazardous material.

Fire point The lowest temperature at which a liquid will ignite and achieve sustained burning when exposed to a test flame in accordance with ASTM D 92, *Standard Test Method for Flash and Fire Points by Cleveland Open Cup*. (NFPA 704, *Standard System for the Identification of the Hazards of Materials for Emergency Response*)

Flammable range The range of concentrations between the lower and upper flammable limits. (NFPA 68, *Standard on Explosion Protection by Deflagration Venting*)

Flammable vapor A concentration of constituents in air that exceeds 10 percent of its lower flammable limit (LFL). (NFPA 115, *Standard for Laser Fire Protection*)

Flash point The minimum temperature at which a liquid emits vapor in sufficient concentration to form an ignitable mixture with air near the surface of the liquid, with the container as specified by appropriate test procedures and apparatus. (NFPA 122, *Standard for Fire Prevention and Control in Metal/Nonmetal Mining and Metal Mineral Processing Facilities*)

Gamma radiation A type of radiation that can travel significant distances, penetrating most materials and passing through the body. Gamma radiation is the most destructive type of radiation to the human body.

Hazard A fuel complex defined by kind, arrangement, volume, condition, and location that determines the ease of ignition and/or of resistance to fire control. (NFPA 1144, *Standard for Reducing Structure Ignition Hazards from Wildland Fire*)

HEPA (high-efficiency particulate air) filter A filter capable of catching particles down to 0.3-micron size—much smaller than a typical dust or alpha radiation particle.

Ignition (autoignition) temperature The minimum temperature at which a fuel, when heated, will ignite in air and continue to burn. Also called autoignition temperature.

Ingestion Exposure to a hazardous material by swallowing it.

Inhalation Exposure to a hazardous material by breathing it into the lungs.

Injection Exposure to a hazardous material by it entering cuts or other breaches in the skin.

Ionizing radiation Radiation of sufficient energy to alter the atomic structure of materials or cells with which it interacts, including electromagnetic radiation such as X-rays, gamma rays, and microwaves and particulate radiation such as alpha and beta particles. (NFPA 1991, *Standard on Vapor-Protective Ensembles for Hazardous Materials Emergencies*)

Irritant A substance such as mace that can be dispersed to incapacitate a person or groups of people briefly.

Lewisite A blister-forming agent that is an oily, colorless to dark brown liquid with an odor of geraniums.

Lower flammable limit (LFL) The minimum concentration of combustible vapor or combustible gas in a mixture of the vapor or gas and gaseous oxidant above which propagation of flame will occur on contact with an ignition source. (NFPA 115)

Nerve agent A toxic substance that attacks the central nervous system in humans.

Neutron A penetrating particle found in the nucleus of the atom that is removed through nuclear fusion or fission. Although neutrons are not radioactive, exposure to neutrons can create radiation.

pH A measure of the acidity or basic nature of a material; more technically, an expression of the concentration of hydrogen ions in the substance.

Physical change A transformation in which a material changes its state of matter—for instance, from a liquid to a solid.

Pulmonary edema Fluid buildup in the lungs.

Radiation The emission and propagation of energy through matter or space by means of electromagnetic disturbances that display both wave-like and particle-like behavior. (NFPA 801)

Radioactive isotope An atom that has unequal numbers of protons and neutrons in the nucleus and that emits radioactivity.

Radioactivity The spontaneous decay or disintegration of an unstable atomic nucleus accompanied by the emission of radiation. (NFPA 801)

Sarin A nerve agent that when dispersed sends droplets in the air that when inhaled harm intended victims.

Secondary contamination The process by which a contaminant is carried out of the hot zone and contaminates people, animals, the environment, or equipment.

Sensitizer A chemical that causes a large portion of people or animals to develop an allergic reaction after repeated exposure to the substance.

Specific gravity As applied to gas, the ratio of the weight of a given volume to that of the same volume of air, both measured under the same conditions. (NFPA 54, *National Fuel Gas Code*)

State of matter The physical state of a material—solid, liquid, or gas.

Sulfur mustard A clear, yellow, or amber oily liquid with a faint, sweet odor of mustard or garlic that can be dispersed in an aerosol form. It causes blistering of exposed skin.

Toxic A property of any chemical that has the capacity to produce injury to workers, which is dependent on concentration, rate, and method and site of absorption. (NFPA 306, *Standard for the Control of Gas Hazards on Vessels*)

Toxicology The study of the adverse effects of chemical or physical agents on living organisms.

Toxic product of combustion A hazardous chemical compound that is released when a material decomposes under heat.

TRACEMP An acronym to help remember the effects and potential exposures to a hazardous materials incident: thermal, radiation, asphyxiant, chemical, etiologic, mechanical, and psychogenic.

Upper flammable limit (UFL) The highest concentration of a combustible substance in a gaseous oxidizer that will propagate a flame. (NFPA 68, *Standard on Explosion Protection by Deflagration Venting*)

Vapor The gas phase of a substance, particularly of those that are normally liquids or solids at ordinary temperatures. (NFPA 92, *Standard for Smoke Control Systems*)

Vapor density The weight of an airborne concentration (vapor or gas) as compared to an equal volume of dry air.

Vapor pressure The pressure, measured in pounds per square inch (or kilopascals), absolute (psia), exerted by a liquid, as determined by ASTM D 323, *Standard Method of Test for Vapor Pressure of Petroleum Products (Reid Method)*. (NFPA 385, *Standard for Tank Vehicles for Flammable and Combustible Liquids*)

VX A nerve agent.

Weapons of mass destruction (WMD) (1) Any destructive device, such as any explosive, incendiary, or poison gas bomb, grenade, rocket having a propellant charge of more than 4 ounces (113 grams), missile having an explosive or incendiary charge of more than ¼ ounce (7 grams), mine, or device similar to the above; (2) any weapon involving toxic or poisonous chemicals; (3) any weapon involving a disease organism; or (4) any weapon that is designed to release radiation or radioactivity at a level dangerous to human life. (NFPA 472, *Standard for Competence of Responders to Hazardous Materials/Weapons of Mass Destruction Incidents*)

FIRE FIGHTER *in action*

The station alarm goes off for a multi-vehicle incident on the major highway that runs through the center of the response area. Additional information comes in stating that there are five semitrailers involved in an accident at the off ramp and four of them have jack knifed. Two of the semitrailers were double trailers and all five trailers were placarded with various different hazard classes. Upon arrival on the scene, the captain hands you the binoculars and asks you to identify the hazardous materials, obtain information, and start establishing protective action distances to protect response personnel and citizens.

1. You start obtaining the four-digit identification number off of several of the trailers. Which section of the *Emergency Response Guidebook* (ERG) could you consult to obtain the chemical name and appropriate guide to action for the specific chemical family?
   - **A.** White section
   - **B.** Blue section
   - **C.** Yellow section
   - **D.** Green section
2. Once you have the basic chemical name and identification number, which section of the ERG would be most beneficial in assisting to determine the initial isolation and protection distances necessary for the chemical?
   - **A.** Orange section
   - **B.** Blue section
   - **C.** Yellow section
   - **D.** Green section

FIRE FIGHTER *in action* (continued)

3. While obtaining the placard information on the last trailer, you find that the four-digit number is highlighted in the section you are looking. Based on this notation, what section of the ERG should you go to immediately?
   A. White section
   B. Blue section
   C. Yellow section
   D. Green section
4. Once you have obtained basic chemical information, the driver hands you paperwork on one of the loaded chemicals involved and you find that the product is a gas with a vapor density of 2.4. Based on this information, what would you expect a leak of this material to do?
   A. Rise quickly and dissipate in normal air
   B. Rise slowly and hang in the air
   C. Sink to all low-lying areas
   D. Sink and dissipate
5. Another trailer has placards that denote that it is carrying several low-level radioactive materials. The driver hands you the Bill of Lading shipping paper and information that explains each material. It shows that one material gives off "Alpha" energy. How far will this product travel if released from its container?
   A. Miles
   B. Yards
   C. Feet
   D. Inches
6. When using the ERG, where would you find information with regard to water-reactive materials?
   A. White section
   B. Blue section
   C. Yellow section
   D. Green section
7. When using the ERG, where would you find basic information with regard to protective clothing?
   A. White section
   B. Blue section
   C. Yellow section
   D. Green section

FIRE FIGHTER II *in action*

It is the first week of December and you arrive at the station for your first day of duty. All too soon, the alarm bells dispatch your engine company to a derailment of a freight train in your first-due response area. It is 95°F (35°C) outside, with a humidity of 25%. There is a slight wind from the northeast at 3 miles per hour (5 kilometers per hour) and the sun has just come up over the horizon. As one of the first-due companies arriving on scene, your engine is immediately assigned to start evacuation in the neighborhood closest to the incident scene. As you and your personnel start your assignment, you notice several police officers sitting in their vehicles in a low-lying area, downhill and downwind from your position. When you walk over, you notice that they are speaking with slurred speech and are complaining of nausea as well as burning in their eyes. You notice what appears to be a cloud moving toward your site from the derailment scene.

1. As the senior fire fighter of your company, which decisions should you consider immediately to protect the personnel in your company?
2. Which decision should you make that will impact the safety of other responders and the residents?
3. Based on the current conditions, what would you consider to be the current incident priorities?

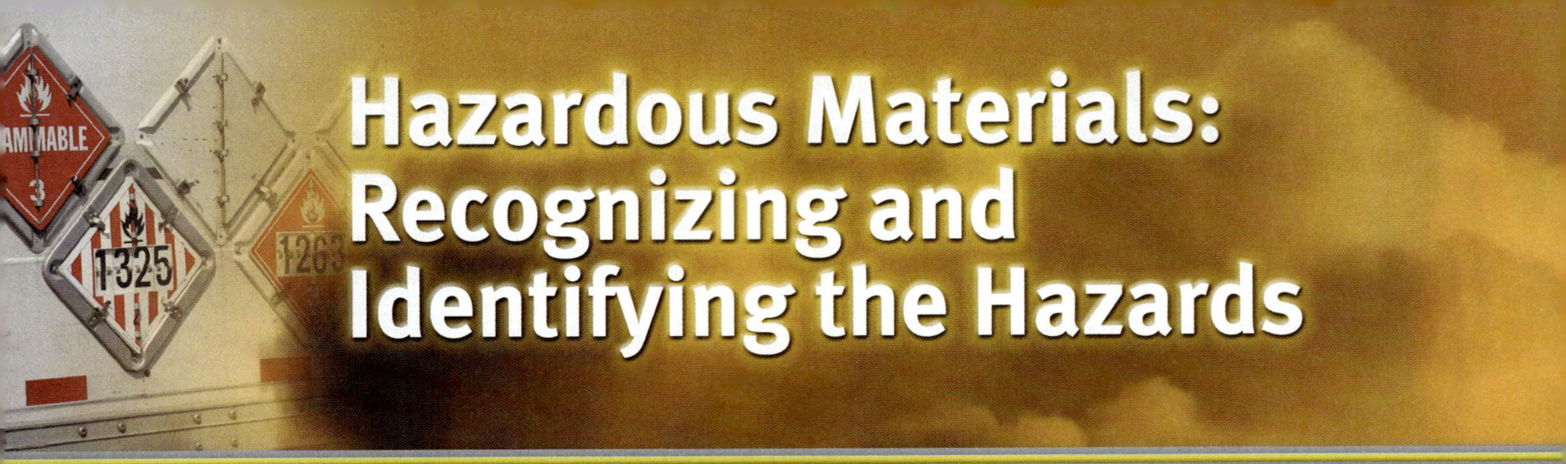

Hazardous Materials: Recognizing and Identifying the Hazards

Fire Fighter I

Knowledge Objectives

After studying this chapter, you will be able to:

- Describe occupancies that may contain hazardous materials. (NFPA 472, 5.1, 5.1.2, 5.1.2.2, 5.2, 5.2.1, 6.6.1.1.2, p 902)
- Describe how your senses can be used to detect the presence of hazardous materials safely. (NFPA 472, 5.1.2.2, 6.6.1.1.2, p 903)
- Describe the general characteristics of a bulk storage container. (NFPA 472, 5.2.1.1, 5.2.1.2.2, 6.6.1.1.2, p 903)
- Describe the common types of bulk storage containers. (NFPA 472, 5.2.1.1, 5.2.1.2.2, 6.6.1.1.2, p 903)
- List the hazardous materials commonly stored in each type of bulk storage container. (NFPA 472, 5.2.1.3, 6.6.1.1.2, p 904)
- Describe the general characteristics of nonbulk storage vessels. (NFPA 472 5.2.1.1, 5.2.1.1.5, 5.2.1.2, 6.6.1.1.2, p 904)
- Describe the general characteristics of drums. (NFPA 472, 5.2.1.1, 5.2.1.1.5, 6.6.1.1.2, p 904)
- Describe the general appearance of intermediate bulk containers and ton containers. (NFPA 472, 5.2.1.1, 5.2.1.1.6, 6.6.1.1.2, p 903–904)
- Describe the general characteristics of bags. (NFPA 472, 5.2.1.1, 5.2.1.1.5, 6.6.1.1.2, p 905)
- List the information found on a pesticide label. (NFPA 472, 5.2.1.1, 5.2.1.3.2, 6.6.1.1.2, p 905)
- Describe the general characteristics of carboys. (NFPA 472, 5.2.1.1, 5.2.1.1.5, 6.6.1.1.2, p 905–906)
- Describe the general characteristics of cylinders. (NFPA 472, 5.2.1.1, 5.2.1.1.5, 6.6.1.1.2, p 906)
- Describe the common types of cylinders. (NFPA 472, 5.2.1.1, 5.2.1.1.5, 6.6.1.1.2, p 906)
- Describe the types of chemical tankers that transport hazardous materials and their common characteristics. (NFPA 472, 5.2.1.1, 5.2.1.1.3, 5.2.1.1.4, 6.6.1.1.2, p 906–908)
- Describe the types of railroad cars that transport hazardous materials and their common characteristics. (NFPA 472, 5.2.1.1, 5.2.1.1.1, 5.2.1.1.2, 5.2.1.2.1, 6.6.1.1.2, p 907–910)
- Describe how to identify the product, owner, and emergency telephone number on a pipeline marker. (NFPA 472, 5.2.1.1, 5.2.1.3.1, 6.6.1.1.2, p 910)
- Describe how to identify a placard and label. (NFPA 472, 5.1.2.2, 6.6.1.1.2, p 910)
- List the nine Department of Transportation chemical families. (NFPA 472, 5.2.2, 6.6.1.1.2, p 911–912)
- Explain how to use the *North American Emergency Response Guidebook*. (NFPA 472, 5.2.2, 6.6.1.1.2, p 912)
- Describe the NFPA 704 hazard identification system. (NFPA 472, 5.2.2, 6.6.1.1.2, p 912–913)
- Describe markings used by the military to indicate hazardous materials and weapons of mass destruction markings. (NFPA 472, 5.2.2, 6.6.1.1.2, p 913–914)
- List the information about a hazardous material found on material safety data sheets. (NFPA 472, 5.2.1.5, 5.2.2, 6.6.1.1.2, p 914)
- Explain when shipping papers are utilized. (NFPA 472, 5.2.1.5, 5.2.2, 6.6.1.1.2, p 914)
- Describe the role of CHEMTREC during a hazardous materials incident. (NFPA 472, 5.1.2.2, 5.2.2, 6.6.1.1.2, p 915)
- List the information CHEMTREC needs to assist with a hazardous materials incident. (NFPA 472, 5.1.2.2, 5.2.2, 6.6.1.1.2, p 915)
- Describe the role of the National Response Center during a hazardous materials incident. (NFPA 472, 5.1.2.2, 5.2.2, 6.6.1.1.2, p 915)
- Describe the common containers used to hold radioactive materials. (NFPA 472 5.2.1.1, 5.2.1.1.7, 5.2.1.3.3, 6.6.1.1.2, p 915–916)
- Describe the signs of a potential terrorist incident. (NFPA 472, 5.2.1.6, 6.6.1.1.2, p 916)

Skills Objectives

There are no skill objectives for Fire Fighter I candidates. NFPA 1001 contains no Fire Fighter I Job Performance Requirements for this chapter.

CHAPTER 30

Fire Fighter II — FFII

Knowledge Objectives

There are no knowledge objectives for Fire Fighter II candidates. NFPA 1001 contains no Fire Fighter II Job Performance Requirements for this chapter.

Skills Objectives

There are no skill objectives for Fire Fighter II candidates. NFPA 1001 contains no Fire Fighter II Job Performance Requirements for this chapter.

Other NFPA Standards

- NFPA 70, *National Electric Code*
- NFPA 434, *Code for the Storage of Pesticides*
- NFPA 472, *Standard for Competence of Responders to Hazardous Materials/Weapons of Mass Destruction Incidents*
- NFPA 498, *Standard for Safe Havens and Interchange Lots for Vehicles Transporting Explosives*
- NFPA 704, *Standard System for the Identification of the Hazards of Materials for Emergency Response*

You Are the Fire Fighter

The engine and truck from your station have been dispatched for several visible leaks from railcars and highway cargo vehicles at an industrial facility in your first-due territory. En route to the site, your officer receives an update that a strong chemical odor has been detected in the air near the intersection of the facility. On one corner of the intersection is the entrance to the rail yard, which includes several tracks and nonpressurized tank cars. On the other corner is a trucking company, which has several different highway cargo tank trucks. On the other side of the street is the industrial facility that stores large amounts of chemicals. Upon your company's arrival on the scene, your officer requests that you recognize and identify the area for potential indicators of the source of the chemical odor.

1. Which markings would you look for on the outside of the facility to indicate whether any potential hazardous materials are present at the site?
2. Which Department of Transportation placard information would you look for on the transportation vehicles, railcars, and highway cargo tank trucks that would indicate the products they might carry and hazards they might present?
3. Which other markings would you look for to indicate other potential hazards on site?

Introduction

Scene size-up is important in any emergency but is especially so in hazardous materials incidents. Situational awareness, or the ability to "read" the scene, is a critical skill. Fire fighters must be able to identify sensory clues to be useful and safe.

In hazardous materials incidents, it is not always possible to know the whole story before action is required. Even so, you may be able to identify the incident as involving hazardous materials based on information from the dispatcher, knowledge of the response area, and visual, auditory, or olfactory (odorous) clues. A leaking cargo trailer, colored placards on abandoned drums, and certain odors are all indications that hazardous materials may be present. Departmental standard operating procedures (SOPs) in concert with standard operating guidelines (SOGs) and your level of training should guide any initial actions. Following departmental SOPs/SOGs will help you take a consistent approach to the size-up of hazardous materials incidents as well as the material identification process. SOPs/SOGs serve as a road map to lead properly trained personnel to an endpoint.

This chapter focuses on how to identify and obtain necessary information on hazardous materials incidents. Recognizing the potential for hazardous materials based on the containers involved or the available shipping information will help you take proper action at the scene.

Recognizing a Hazardous Materials Incident

A hazardous material is any material that poses an unreasonable risk of damage or injury to persons, property, or the environment if it is not properly controlled during handling, storage, manufacture, processing, packaging, use and disposal, or transportation. Recognizing a hazardous materials event and identifying the materials involved often require some detective work. Take the time to look at the whole scene, to identify the critical visual indicators, and to fit them into what is known about the problem.

If the dispatch information indicates that an incident involves hazardous materials, it may be a good idea to use binoculars and view the scene from a safe distance. Be sure to question anyone involved in the incident, such as the truck driver or the plant owner. Take time to scan the scene and interpret visual clues such as dead animals near the release, discolored pavement, dead grass, visible vapors or puddles, or labels that might help identify the presence of a hazardous material.

Occupancy and Location

Hazardous materials incidents are not confined to places like chemical facilities or nuclear power plants—they can occur almost anywhere. Hazardous materials are stored in warehouses, hospitals, laboratories, industrial occupancies, residential garages, bowling alleys, home improvement centers, garden supply stores, restaurants, and many other facilities and businesses in your response area. So many different chemicals are used in daily life that you could encounter several types of hazardous materials as you respond to emergencies. For example, a fire in a pesticide storage facility could release dangerous chemicals; a person exposed to a solvent while cleaning automotive parts could require emergency medical assistance; a rollover on the interstate could involve leaking diesel fuel. Nearly any emergency or even nonemergency response might potentially include a hazardous materials component.

The location and type of occupancy are two indicators of the presence of hazardous materials. The previous chapter outlined the basic reporting requirements for occupancies that use hazardous materials. By reporting the kinds and quantities of hazardous materials present to the local fire department, facilities that pose a significant threat to the community become part of a comprehensive community plan. These reports also alert fire departments to the need for a preincident plan for the site so that fire fighters are aware of the location and types of hazardous materials involved.

Senses as a Hazardous Materials Detector

Another way to detect the presence of hazardous materials is to use your senses, although this must be done carefully to avoid exposure. Many hazardous materials can be recognized by smell, taste, or touch. Getting close enough to exercise these senses, however, may expose you to the hazard. The farther you are from the incident when you notice a problem, the safer you will be.

Clues that are seen or heard provide warning information from a distance, enabling you to take precautionary steps. Vapor clouds at the scene, for example, are a signal to move away to safety; the sound of an alarm from a toxic gas cabinet is a warning to retreat. This chapter provides you with information on visual clues such as containers, placards, labels, and shipping information that signal possible hazardous materials incidents.

Fire Fighter Safety Tips

Package delivery services such as UPS, DHL, and FedEx carry packages that could contain hazardous materials, although small amounts might not require placards. However, the package may require labels. If a package is damaged, it could become a hazardous material. If a delivery transport vehicle is involved in an accident, multiple packages could open, allowing their contents to mix and produce a hazardous material.

Containers

A container is any vessel or receptacle that holds material, including storage vessels, pipelines, and packaging. Often, container type, size, and material provide important clues about the nature of the substance inside. Fire fighters should not rely solely on the type of container, however, when making a determination about hazardous materials.

Red phosphorus from a drug lab, for example, might be found in an unmarked plastic jug. Acetone or other solvents might be stored in 55-gallon (208 liters) steel drums with two capped openings on the top. Sulfuric acid, at 97 percent concentration, is typically found in a polyethylene drum, but that drum could be black, red, white, or blue. The same sulfuric acid might also be found in a 1-gallon (3.8 liters) amber glass container. Hydrofluoric acid, in contrast, is incompatible with silica (glass) and would be stored in a plastic container.

Container Type

Hazardous materials can be found in many different types of containers, ranging from 1-gallon (3.8 liters) glass containers to 5000-gallon (18,927 liters) steel storage tanks. Steel or polyethylene plastic drums, bags, high-pressure gas cylinders, railroad tank cars, plastic buckets, above-ground and underground storage tanks, truck tankers, and pipelines are all types of containers used with hazardous materials.

Some very recognizable chemical containers, such as 55-gallon (208 liters) drums and compressed gas cylinders, can be found in almost every commercial building. Materials in cardboard drums are usually in solid form. Stainless steel containers are used to hold particularly dangerous chemicals, and cold liquids are kept in thermos-like Dewar containers designed to maintain the appropriate temperature.

Containers may be large or small. Sometimes, smaller containers may be stored and shipped in larger containers. One way to distinguish containers is to divide them into two separate categories: bulk storage containers and nonbulk storage vessels.

FIRE FIGHTER Tips

When you consider locations for possible hazardous materials incidents, do not limit your thinking to chemical plants. Many hazards can be found in residential structures in the kitchen, laundry room, and both attached and stand-alone garages. In exploring your response district, you may be surprised by how many bulk diesel, oxygen, nitrogen, and high-volume corrosive storage tanks you find.

Container Volume

Bulk storage containers, or large-volume containers, are defined by their internal capacity based on the following measures:

- Liquids: more than 119 gallons (450 liters)
- Solids: more than 882 pounds (400 kilograms)
- Gases: more than 882 pounds (400 kilograms)

Bulk storage containers include fixed tanks, large transportation tankers, totes, and intermodal tanks. In general, these kinds of containers are found in occupancies that rely on and need to store large quantities of a particular chemical. Most manufacturing facilities have at least one bulk storage container. Often, these bulk storage containers are surrounded by a supplementary containment system to help control an accidental release. Secondary containment is an engineered method to control spilled or released product if the main containment vessel fails. A 5000-gallon (18,927 liters) vertical storage tank, for example, might be surrounded by a series of short walls that form a catch basin around the tank. This basin typically can hold the entire volume of the tank and accommodate water from hose lines or sprinkler systems in the event of fire. Many storage vessels, including 55-gallon (208 liters) drums, may have secondary containment systems. If bulk storage containers in your response area have secondary containment systems, large leaks in them will be easier to deal with.

Large-volume horizontal tanks are also common. When stored above ground, these tanks are referred to as above-ground storage tanks (ASTs); if they are placed underground, they are known as underground storage tanks (USTs). These tanks can hold amounts ranging from a few hundred gallons to several thousand gallons (several hundred liters to tens of thousands of liters) of product and are usually made of aluminum, steel, or plastic. USTs and ASTs can be pressurized or nonpressurized. Nonpressurized horizontal tanks are usually constructed from steel or aluminum. Because it is difficult to relieve internal pressure in these tanks, they are dangerous when exposed to fire. Typically, they hold flammable or combustible materials such as gasoline, oil, or diesel fuel.

Pressurized horizontal tanks have rounded ends and large vents or pressure-relief stacks. The most common aboveground pressurized tanks are liquid propane and liquid ammonia tanks, which can hold volumes ranging from a few hundred gallons to several thousand gallons (several hundred liters to tens of thousands of liters) of product. These tanks also contain a small vapor space; 10 to 15 percent of total capacity can be vapor. Refer to the discussion of liquefied gases in the Hazardous Materials and Effects chapter for more information on these materials.

Another common bulk storage vessel is the tote. Totes are portable plastic tanks surrounded by a stainless steel web that adds both structural stability and protection. They can hold a few hundred gallons (several hundreds of liters) of product and may contain any type of chemical, including flammable liquids, corrosives, food-grade liquids, and oxidizers.

Shipping and storing totes can be hazardous. These containers are often stacked atop one another and are moved with a forklift. A mishap with the loading or moving process can compromise the container. Because totes have no secondary containment mechanism, any leak will create a large puddle. Additionally, the steel webbing around the tote makes leaks difficult to patch.

Intermodal tanks are both shipping and storage vehicles TABLE 30-1. They hold between 5000 and 6000 gallons (18,927 to 22,712 liters) of product and can be either pressurized or nonpressurized. In most cases, an intermodal tank is shipped to a facility, where it is stored and used, and then it is returned to the shipper for refilling. Intermodal tanks can be shipped by any mode of transportation: air, sea, or land. These horizontal round tanks are surrounded by, or are part of, a boxlike steel framework for shipping. Three basic types of intermodal (IM or IMO) tanks are routinely encountered:

- IM-101 containers have a 6000-gallon (22,712 liters) capacity, with internal working pressures between 25 pounds per square inch (psi) (172 kilopascals [kPa]) and 100 psi (689 kPa). These containers typically carry mild corrosives, food-grade products, or flammable liquids.
- IM-102 containers have a 6000-gallon (22,712 liters) capacity, with internal working pressures between 14 psi (97 kPa) and 30 psi (207 kPa). They primarily carry flammable liquids and corrosives.
- Spec 51 or IMO Type 5 (internationally) containers are high-pressure vessels with internal pressures of several hundred psi (kilopascals) that carry liquefied gases such as propane and butane.

TABLE 30-1 Common Bulk Storage Vessels, Locations, and Contents

Tank Shape	Common Locations	Hazardous Materials Commonly Stored
Underground tanks	Residential, commercial	Fuel oil and combustible liquids
Covered floating roof tanks	Bulk terminals and storage	Highly volatile flammable liquids (especially ethanol-blended fuels)
Cone roof tanks	Bulk terminal and storage	Combustible liquids
Open floating roof tanks	Bulk terminal and storage	Flammable and combustible liquids
Dome roof tanks	Bulk terminal and storage	Combustible liquids
High-pressure horizontal tanks	Industrial storage and terminal	Flammable gases, chlorine, ammonia
High-pressure spherical tanks	Industrial storage and terminal	Liquid propane gas, liquid nitrogen gas
Cryogenic liquid storage tanks	Industrial and hospital storage	Oxygen, liquid nitrogen gas

Nonbulk Storage Vessels

Essentially, nonbulk storage vessels are all other types of containers. Nonbulk storage vessels can hold volumes ranging from a few ounces to many gallons (fewer than 100 milliliters to hundreds of liters); they include drums, bags, carboys, compressed gas cylinders, cryogenic containers, and more. Nonbulk storage vessels hold commonly used commercial and industrial chemicals such as solvents, industrial cleaners, and compounds. This section describes the most common nonbulk storage vessels.

Drums

Drums are easily recognizable, barrel-like containers. They are used to store a wide variety of substances, including food-grade materials, corrosives, flammable liquids, and grease. Drums may be constructed of low-carbon steel, polyethylene, cardboard, stainless steel, nickel, or other hybrid materials FIGURE 30-1. Generally, the nature of the material dictates the construction of the storage drum. Steel utility drums, for example, hold flammable liquids, cleaning fluids, oil, and other noncorrosive chemicals. Polyethylene drums are used for corrosives such as acids, bases, oxidizers, and other materials that cannot be stored in steel containers. Cardboard drums hold solid materials such as soap flakes, sodium hydroxide pellets, and food-grade materials. Stainless steel or other heavy-duty drums generally hold materials too aggressive for either plain steel or polyethylene.

Closed-head drums have a permanently attached lid with one or more small openings called bungs. Typically, these openings are threaded holes sealed by caps that can be removed only by using a special wrench called a bung wrench. Closed-head drums usually have one 2-inch (5 centimeters) bung and one ¾-inch (2 centimeters) bung. The larger hole is used to pump product from the drum, while the smaller bung functions as a vent.

An open-head drum has a removable lid fastened to the drum with a ring. The ring is tightened with a clasp or a threaded nut-and-bolt assembly. The lid of an open-head drum may or may not have bung-type openings.

FIRE FIGHTER Tips

A ton (907 kilograms) container is a metal container that may resemble a large cylinder. It may carry gases such as chlorine. An intermediate bulk container is cube-shaped and may be made of plastic or steel. It is used to carry bulk items.

A. B. C.

FIGURE 30-1 Drums may be constructed of **A.** cardboard, **B.** polyethylene, or **C.** stainless steel.

Bags

Bags are commonly used to store solids and powders such as cement powder, sand, pesticides, soda ash, and slaked lime. Storage bags may be constructed of plastic, paper, or plastic-lined paper. Bags come in different sizes and weights, depending on their contents.

Pesticide bags must be labeled with specific information **FIGURE 30-2**. Fire fighters can learn a great deal from the label:

- Name of the product
- Statement of ingredients
- Total amount of product in the container
- Manufacturer's name and address
- U.S. Environmental Protection Agency (EPA) registration number, which provides proof that the product was registered with the EPA
- The EPA establishment number, which shows where the product was manufactured
- Signal words to indicate the relative toxicity of the material:
    - Danger: Poison: Highly toxic by all routes of entry.
    - Danger: Severe eye damage or skin irritation.
    - Warning: Moderately toxic.
    - Caution: Minor toxicity and minor eye damage or skin irritation.
- Practical first-aid treatment description
- Directions for use
- Agricultural use requirements
- Precautionary statements such as mixing directions or potential environmental hazards
- Storage and disposal information
- Classification statement on who may use the product

In addition, every pesticide label must include the following statement: "Keep out of reach of children."

FIGURE 30-2 A pesticide container must be labeled with specific information.

Carboys

Some corrosives and other types of chemicals are transported and stored in vessels called carboys **FIGURE 30-3**. A carboy is a glass, plastic, or steel container that holds 5 to 15 gallons (19 to 57 liters) of product. Glass carboys often have a protective wood or fiberglass box to help prevent breakage.

FIGURE 30-3 A carboy is used to transport and store corrosive chemicals.

Nitric, sulfuric, and other strong acids are transported and stored in thick glass carboys protected by a wooden or Styrofoam crate to shield the glass container from damage during normal shipping.

Cylinders

Several types of cylinders are used to hold liquids and gases FIGURE 30-4. Uninsulated compressed gas cylinders are used to store gases such as nitrogen, argon, helium, and oxygen. They have a range of sizes and pressure readings. A typical oxygen cylinder used for medical purposes, for example, has a gas pressure reading of approximately 2000 pounds per square inch (psi) (13,790 kPa). Very large compressed gas cylinders found at a fixed facility may have pressure readings of 5000 psi (34,474 kPa) or greater.

The high pressures exerted by these cylinders are potentially dangerous to both fire fighters and others. If the cylinder is punctured or the valve assembly fails, the rapid release of compressed gas will turn the cylinder into a deadly missile. Additionally, if the cylinder is heated rapidly, it could explode with tremendous force, spewing product and metal fragments over long distances. Compressed gas cylinders do include pressure-relief valves, but those valves might not be sufficient to relieve the pressure created during a fast-growing fire. The result would be a catastrophic explosion.

A propane cylinder is another type of compressed gas cylinder. Propane cylinders have lower pressures (200 to 300 psi [1,379 to 2,068 kPa]) and contain a liquefied gas. Liquefied gases such as propane are subject to the phenomenon called a boiling liquid/expanding vapor explosion (BLEVE). For more information on BLEVEs, refer to the Fire Suppression chapter.

The low-pressure Dewar is another common cylinder type. Dewars are thermos-like vessels designed to hold cryogenic liquids (cryogens)—that is, gaseous substances that have been chilled until they liquefy FIGURE 30-5. Typical cryogens are oxygen, helium, hydrogen, argon, and nitrogen. Under normal conditions, each of these substances is a gas. A complex process turns them into liquids that can be stored in Dewar containers. Nitrogen, for example, becomes a liquid at –320°F (–160°C) and must be kept that cold to remain a liquid.

Cryogens pose a substantial threat if the Dewar fails to maintain the low temperature. Cryogens have large expansion ratios—even larger than the expansion ratio of propane. Cryogenic helium, for example, has an expansion ratio of approximately 750 to 1. Thus, if one volume of liquid helium is warmed to room temperature and vaporized in a totally enclosed container, it can generate a pressure of more than 14,500 psi (99,974 kPa). To offset this risk, cryogenic containers usually have two pressure-relief devices: a pressure-relief valve and a frangible (easily broken) disk.

Transporting Hazardous Materials

Although rail, air, and sea transport is used to deliver chemicals and other hazardous materials to their destinations, the most common method of transport is over land, by highway transportation vehicles. Even when another method of transport is used, motor vehicles often carry the shipments from the station, airport,

A. B.

FIGURE 30-4 Cylinders. **A.** 100-pound (45 kilograms) cylinder. **B.** 150-pound (68 kilograms) cylinder.

FIGURE 30-5 A small cryogenic Dewar.

or dock to the factory or plant. For this reason, fire fighters must become familiar with the types of chemical transport vehicles, or tankers, they might encounter during a traffic emergency.

One of the most common chemical tankers is the MC-306/DOT406 flammable liquid tanker FIGURE 30-6A. It typically carries gasoline or other flammable and combustible materials. The oval-shaped tank is pulled by a diesel tractor and can carry between 6000 and 10,000 gallons (22,712 and 37,854 liters). The MC-306/DOT406 is nonpressurized, usually made of aluminum, and offloaded through valves at the bottom of the tank. It is a common highway sight and a reliable way to transport chemicals.

A similar vehicle is the MC-307/DOT407 chemical hauler, which has a round or horseshoe-shaped tank and typically carries 6000 to 7000 gallons (22,712 to 26,498 liters) FIGURE 30-6B. The MC-307/DOT407 is used to transport flammable liquids, mild corrosives, and poisons. Tanks that transport corrosives may have a rubber lining.

The MC-312/DOT412 corrosives tanker is used for concentrated sulfuric and nitric acids and other corrosive substances FIGURE 30-6C. This tanker has a smaller diameter than either the MC-306/DOT406 or the MC-307/DOT407 and is often (not always) characterized by several reinforcing rings around the tank. These rings provide structural stability during transportation and in the event of a rollover. The inside of an MC-312/DOT412 tanker operates at a pressure of approximately 75 psi (517 kPa) and holds approximately 6000 gallons (22,712 liters).

The MC-331 pressure cargo tanker carries materials such as ammonia, propane, and butane FIGURE 30-6D. The tank has rounded ends, typical of a pressurized vessel, and is commonly constructed of steel with a single tank compartment. The MC-331 operates at a pressure of approximately 300 psi (2068 kPa) and could be a significant explosion hazard if it accidentally rolls over or is threatened by fire. Fire fighters must use great care when dealing with this type of highway emergency.

The MC-338 cryogenic tanker operates much like the Dewar containers described earlier and carries many of the same substances FIGURE 30-6E. This low-pressure tanker relies on tank insulation to maintain the low temperatures required by the cryogens it carries. A boxlike structure containing the tank control valves is typically attached to the rear of the tanker. Special training is required to operate valves on this and any other tanker. An untrained individual who attempts to operate the valves may disrupt the normal operation of the tank, thereby compromising its ability to keep the liquefied gas cold and creating a potential explosion hazard.

Tube trailers carry compressed gases such as hydrogen, oxygen, helium, and methane FIGURE 30-6F. Essentially, they are high-volume transportation vehicles comprising several individual cylinders banded together and affixed to a trailer. The individual cylinders on the tube trailer work much like the smaller compressed gas cylinders discussed earlier. These large-volume cylinders operate at a pressure of 3000 to 5000 psi (20,684 to 34,474 kPa); one trailer may carry several different gases in individual tubes. Typically, a valve control box can be found toward the rear of the trailer, and each individual cylinder has its own relief valve. These trailers frequently can be seen at construction sites or at facilities that use great quantities of these materials.

Dry bulk cargo tanks also are commonly seen on the road and carry dry bulk goods such as powders, pellets, fertilizers, and grain FIGURE 30-6G. These tanks are not pressurized, but they may use pressure to offload product. Dry bulk cargo tanks are generally V-shaped and have rounded sides that funnel the contents to the bottom-mounted valves. Intermodal tanks can be laid out in a wide variety of configurations and are found in all modes of transportation FIGURE 30-6H.

Railroad Transportation

Railroads move almost 2 million carloads of freight per year, yet are involved in relatively few hazardous materials incidents. Railway tank cars may carry volumes up to 30,000 gallons (113,562 liters) and have the potential to create large leaks or vapor clouds if their contents leak. Hazardous materials incidents involving railroad transportation, although relatively rare, are extremely dangerous.

Fortunately, there are only three basic railcar configurations that fire fighters should recognize: nonpressurized, pressurized, and special use. Each type of railcar has a distinct profile that can be recognized from a long distance. Additionally, railcars are usually labeled on both sides with the volume and maximum working pressure inside the tank. Dedicated haulers often have the chemical name clearly visible.

DOT 111 nonpressurized (general service) railcars typically carry general industrial chemicals, consumer products such as corn syrup, flammable and combustible liquids, and mild corrosives. Nonpressurized railcars are easily identified by

FIGURE 30-6 **A.** An MC-306 flammable liquid tanker. **B.** An MC-307 chemical hauler. **C.** An MC-312 corrosives tanker. **D.** An MC-331 pressure-cargo tanker. **E.** An MC-338 cryogenic tanker. **F.** A tube trailer. **G.** A dry bulk cargo tank. **H.** Intermodal tanks.

looking at the top of the car: Nonpressurized railcars will have visible valves and piping without a dome cover **FIGURE 30-7**.

DOT 105, DOT 112 pressurized railcars feature an enclosed dome on the top of the railcar. These cars transport materials such as propane, ammonia, ethylene oxide, and chlorine. Pressurized cars have internal working pressures ranging from 100 to 500 psi (689 to 3447 kPa) and are equipped with relief valves, similar to those on bulk storage tankers. Unfortunately, the high volumes carried in these cars can generate long-duration, high-pressure leaks that may be impossible to stop **FIGURE 30-8**.

Special-use railcars include boxcars, flat cars, and corrosive tank cars, and high-pressure compressed gas tube cars. DOT 113 cryogenic rail tank cars carry products that have varying

VOICES OF EXPERIENCE

My experience proves that a hazardous materials incident may be encountered in a residential garage or basement. While operating on an engine company for the Cincinnati Fire Department, I was dispatched to a report of a fire in a basement or a garage. The dispatch sounded vague and uncertain with very little concrete information. Upon arrival, there appeared to be a basement fire in a private residence. A young man stood in front of the building, holding an elderly gentleman covered in soot. The young man stated that he thought that the older gentleman was working with ammunition in his basement when a fire broke out.

Realizing that we could not enter the basement safely, we advanced a hose line and continued to investigate the extent of the fire. There was a side entrance to the house that was near the basement stairway. As the fire officer, I entered the side door with full gear on and attempted to gather further information.

I could hear a hissing sound and radioed command that there may be a gas leak. In hindsight, the hissing sound was probably a black powder canister heating up. Suddenly an explosion pushed me out of the side door.

I received second degree burns to my head, ears, and hands. The explosion blew my helmet off and pulled my hood back, burning my head and ears. My hands were burned even though I was wearing my fire gloves.

As quickly as the orange glow appeared, it was gone. The company managed to extinguish the remainder of the fire in the basement before most of the other responding companies had arrived.

Howard Harper
Cincinnati State College
Cincinnati, Ohio

temperatures that are less than 1300°F (700°C) FIGURE 30-9. Products such as oxygen, helium, hydrogen, argon, and others are shipped via this method to larger industrial facilities that use these materials in everyday industrial processes. In each case, the hazard will be unique to the particular railcar and its contents. Do not assume that only the chemical tank cars pose a threat; until you know what is in a particular car, assume the incident is a hazardous situation.

The train's engineer or conductor will have information about all the cars on that train and should be consulted during a railway emergency. Details about the shipping papers unique to railway transportation are presented later in this chapter.

FIGURE 30-7 DOT 111 nonpressurized (general service) railcar.

FIGURE 30-8 DOT 105, DOT 112 pressurized railcar.

FIGURE 30-9 DOT 113 cryogenic tank car.

Pipelines

Of the various methods used to transport hazardous materials, the high-volume pipeline is rarely involved in emergencies. However, every fire fighter should be aware of these structures and know where they run in their communities. Although the number of incidents involving pipelines is low, the incidents that have occurred have resulted in large losses of both property and lives and have affected many communities. In many areas, large-diameter pipelines transport natural gas, gasoline, diesel fuel, and other products from delivery terminals to distribution facilities. Pipelines are often buried underground, but they may be above ground in remote areas. The pipeline right-of-way is an area, patch, or roadway that extends a certain number of feet (meters) on either side of the pipe itself. This area is maintained by the company that owns the pipeline. The company is also responsible for placing warning signs at regular intervals along the length of the pipeline.

Pipeline warning signs include a warning symbol, the pipeline owner's name, and an emergency contact phone number FIGURE 30-10. Pipeline emergencies are complex events that require specially trained responders. If you suspect an emergency involving a pipeline, contact the owner of the line immediately. The company pipeline will dispatch a crew to assist with the incident.

FIGURE 30-10 Information about the pipe's contents and owner is often found attached to vent pipes.

Information about the pipe's contents and owner is also often found at the vent pipes. These inverted J-shaped tubes provide pressure relief or natural venting during maintenance and repairs. Vent pipes are clearly marked and are approximately 3 feet (91 centimeters) above the ground.

Facility and Transportation Markings and Colorings

Labels, placards, and other markings on buildings, packages, boxes, and containers often enable fire fighters to identify a spilled or released chemical. When used correctly, marking systems indicate the presence of a hazardous material and provide clues about the substance. This section provides an introduction to various marking systems being used. It does not cover all the intricacies and requirements for every marking system; it will, however, acquaint you with the most common systems.

Department of Transportation System

The Department of Transportation (DOT) marking system is characterized by a system of labels and placards. The DOT's *Emergency Response Guidebook* (ERG) is also part of this system and offers a certain amount of guidance for fire fighters operating at a hazardous materials incident.

Labels and Placards

Placards are diamond-shaped indicators (10¾ inches [3.3 meters] on each side) that must be placed on all four sides of highway transport vehicles, railroad tank cars, and other forms of transportation carrying hazardous materials **FIGURE 30-11**. Labels are smaller versions (4-inch [10 centimeters] diamond-shaped indicators) of placards that are used on the four sides of individual boxes and smaller packages being transported.

Placards and labels are intended to give fire fighters a general idea of the hazard inside a particular container. A placard may identify the broad hazard class (e.g., flammable, poison, corrosive) that a tanker contains, while the label on a box inside a delivery truck relates only to the potential hazard inside that package **FIGURE 30-12**.

FIGURE 30-11 A DOT placard.

The ERG can be used to identify types of threats during the initial phase of a hazardous materials incident. The book organizes chemicals into nine basic hazard classes, or families, the members of which exhibit similar properties:

- DOT Class 1: Explosives
- DOT Class 2: Gases
- DOT Class 3: Flammable combustible liquids
- DOT Class 4: Flammable solids
- DOT Class 5: Oxidizers

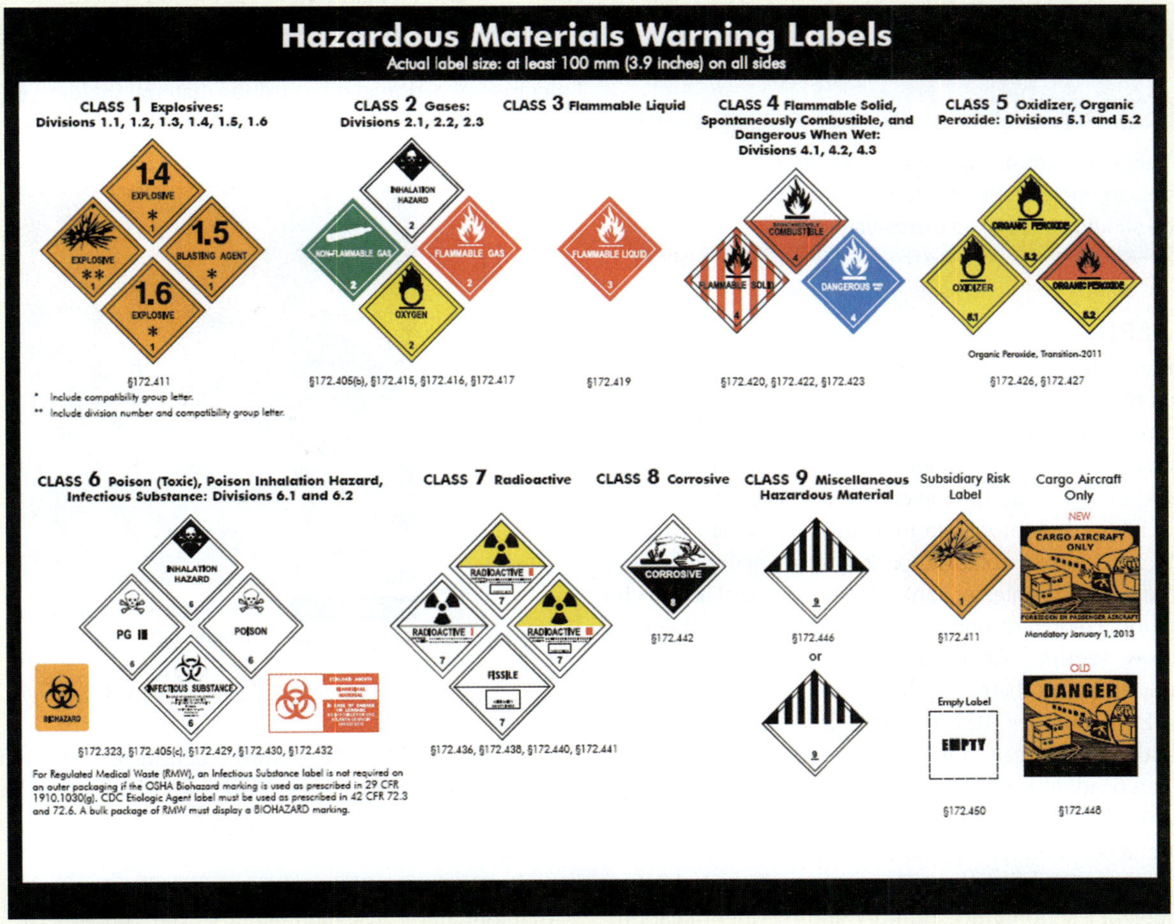

FIGURE 30-12 A DOT hazardous materials warning label chart.

- DOT Class 6: Poisons (including blood agents and choking agents)
- DOT Class 7: Radioactive materials
- DOT Class 8: Corrosives
- DOT Class 9: Other regulated materials

There also is a "Dangerous" placard, which indicates that more than one hazard class is contained in the same load. The DOT system is a broad-spectrum look at chemical hazards and a valuable resource to fire fighters.

Other Considerations

The DOT system does not require that all chemical shipments be marked with placards or labels. In most cases, the package or tank must contain a certain amount of hazardous material before a placard is required. For example, the "1000-pound (453 kilograms) rule" applies to blasting agents, flammable and nonflammable gases, flammable or combustible liquids, flammable solids, air reactive solids, oxidizers and organic peroxides, poison solids, corrosives, and miscellaneous (Class 9) materials. Placards are required for these materials only when the shipment weighs more than 1000 pounds (453 kilograms).

In contrast, some chemicals are so hazardous that shipping any amount of them requires the use of labels or placards. These include Class 1.1, 1.2, or 1.3 explosives, poison gases, water-reactive solids, and high-level radioactive substances. A four-digit United Nations (UN) number may be required on some placards. This number identifies the specific material being shipped; a list of UN numbers is included in the ERG.

Fire Fighter Safety Tips

The fire fighter should always attempt to identify hazardous materials using more than one source. The ERG is just one source and should not be used exclusively.

Using the ERG

The ERG is a preliminary action guide for the initial response to a hazardous materials incident. When the ERG refers to a small spill, it means a leak from one small package, or a small leak in a large container (up to a 55-gallon [208 liters] drum), a small cylinder leak, or any small leak, even one in a large package. A large spill is a large leak or spill from a larger container or package, a spill from a number of small packages, or anything from a 1-ton (907 kilograms) container, tank truck, or railcar. The ERG is divided into four colored sections: yellow, blue, orange, and green.

- **Yellow section**: Chemicals in this section are listed numerically by their four-digit UN identification number. Entry number 1017, for example, identifies chlorine. Use the yellow section when the UN number is known or can be identified. The entries include the name of the chemical and the emergency action guide number.
- **Blue section**: Chemicals in the blue section are listed alphabetically by name. The entry will include the emergency action guide number and the identification number. The same information, organized differently, appears in both the blue and yellow sections.
- **Orange section**: This section contains the emergency action guides. Guide numbers are organized by general hazard class and indicate which basic emergency actions should be taken, based on the hazard class.
- **Green section**: This section is organized numerically by UN identification number and provides the initial isolation distances for specific materials. Chemicals included in this section are highlighted in the blue or yellow sections. Any materials listed in the green section are always extremely hazardous. This section also directs the reader to consult the tables listing toxic inhalation hazard (TIH) materials—gases and volatile liquids that are extremely toxic to humans and pose a hazard to health during transportation. In addition, the green section includes a table of water-reactive materials that produce toxic gases and the associated protective actions to be taken.

NFPA 704 Marking System

The DOT hazardous materials marking system is used when materials are being transported from one location to another. The National Fire Protection Association (NFPA) NFPA 704 hazard identification system is designed for fixed-facility use. The use of the NFPA 704 marking system is purely voluntary in most jurisdictions, but local fire and building codes may require its use in some instances. Check with your code enforcement agencies for your local situation.

NFPA 704 diamonds are found on the outside of buildings, on doorways to chemical storage areas, and on fixed storage tanks. Fire fighters can use the NFPA diamonds to determine a course of action at a hazardous material incident FIGURE 30-13.

FIGURE 30-13 Example of a placard using the NFPA 704 hazard identification system.

The NFPA 704 hazard identification system uses a diamond-shaped symbol of any size, which is itself broken into four smaller diamonds, each representing a particular property or characteristic. The blue diamond at the nine o'clock position indicates the health hazard posed by the material. The top red diamond indicates flammability. The yellow diamond at the three o'clock position indicates reactivity. The bottom white diamond is used for special symbols and handling instructions.

The blue, red, and yellow diamonds will each contain a numerical rating of 0 to 4, with 0 being the least hazardous and 4 being the most hazardous for that type of hazard. The white quadrant will not have a number but may contain special symbols. Among the symbols used are a burning O (oxidizing capability), a three-bladed fan (radioactivity), and a W with a slash through it (water reactive). **TABLE 30-2** provides a description of the numerical ratings in each category.

For more complete information on the NFPA 704 system, consult NFPA 704, *Standard System for the Identification of the Hazards of Materials for Emergency Response*.

FIRE FIGHTER Tips

NFPA 704 placards may not have accurate information, due to business owners placing them on their buildings and guessing which information should be included on the placard.

Hazardous Materials Information System Marking System

Since 1983, the Hazardous Materials Information System (HMIS) hazard communication program has helped employers comply with the Hazard Communication Standard developed by the U.S. Occupational Safety and Health Administration (OSHA). The HMIS is similar to the NFPA 704 marking system and uses a numerical hazard rating with similarly colored horizontal columns.

The HMIS is more than just a label; it is a method used by employers to give their personnel necessary information to work safely around chemicals. This system includes training materials that are used to educate workers about chemical hazards in their workplace. The HMIS is not required by law but is a voluntary system that employers choose to use to comply with OSHA's Hazard Communication Standard. In addition to describing the chemical hazards posed by a particular substance, the HMIS provides guidance about the personal protective equipment that employees need to use to protect themselves from workplace hazards. Letters and icons, which are explained in the text, are used to specify the different levels and combinations of protective equipment.

Fire fighters must understand the fundamental difference between NFPA 704 and the HMIS. NFPA 704 is intended for responders; the HMIS is intended for the employees of a facility. Although the HMIS is not a response information tool, it

FIRE FIGHTER Tips

More than 4 billion tons (3.6 billion tonnes) of hazardous materials are shipped annually by land, sea, air, and rail. While it is not possible to know each and every hazardous material by name or hazard class, you can learn to identify many of the more common hazardous materials by understanding the available identification systems and using other methods to identify the presence of a hazardous material. Properly identifying the hazardous material is the first step in a safe and successful response.

FIRE FIGHTER Tips

Beware of skin exposure to cryogens. Exposed skin can suffer significant burn injuries when it comes in contact with these hazardous materials.

TABLE 30-2 Hazard Levels in the NFPA Hazard Identification System

Flammability Hazards	
4	Materials that will rapidly or completely vaporize at atmospheric pressure and normal ambient temperature, or that are readily dispersed in air and that will burn readily. Liquids with a flashpoint below 73°F (22°C) and a boiling point below 100°F (38°C).
3	Liquids and solids that can be ignited under almost all ambient temperature conditions. Liquids with a flashpoint below 73°F (22°C) and a boiling point above 100°F (38°C) or liquids with a flashpoint above 73°F (22°C) but not exceeding 100°F (38°C) and a boiling point below 100°F (38°C).
2	Materials that must be moderately heated or exposed to relatively high ambient temperatures before ignition can occur. Liquids with flashpoint above 100°F (38°C) but not exceeding 200°F (93°C).
1	Materials that must be preheated before ignition can occur. Liquids that have a flashpoint above 200°F (93°C).
0	Materials that will not burn.
Health Hazards	
4	Materials that on very short exposure could cause death or major residual injury.
3	Materials that on short exposure could cause serious temporary or residual injury.
2	Materials that on intense or continued, but not chronic, exposure could cause incapacitation or possible residual injury.
1	Materials that on exposure would cause irritation but only minor residual injury.
0	Materials that on exposure under fire conditions would offer no hazard beyond that of ordinary combustible material.
Reactivity Hazards	
4	Materials that in themselves are readily capable of detonation or of explosive decomposition or reaction at normal temperatures and pressures.
3	Materials that in themselves are capable of detonation or explosive decomposition or reactions but require a strong initiating source, or that must be heated under confinement before initiation, or that react explosively with water.
2	Materials that readily undergo violent chemical change at elevated temperatures and pressures, or that react violently with water, or that may form explosive mixtures with water.
1	Materials that in themselves are normally stable, but can become unstable at elevated temperatures and pressures.
0	Materials that in themselves are normally stable, even under fire exposure conditions, and are not reactive with water.
Special Hazards	
OX	Oxidizer
~~W~~	Reacts with water

can be used as a clue to the presence and nature of the hazardous materials present in a facility.

Military Hazardous Materials/WMD Markings

The U.S. military has developed its own marking system for hazardous materials. The military system has been developed primarily to identify detonation, fire, and special hazards.

In general, hazardous materials within the military marking system are divided into four categories based on the relative detonation and fire hazards. Division 1 materials are considered mass detonation hazards and are identified by a number 1 printed inside an orange octagon. Division 2 materials have explosion-with-fragment hazards and are identified by a number 2 printed inside an orange X. Division 3 materials are mass fire hazards and are identified by a number 3 printed inside an inverted orange triangle. Division 4 materials are moderate fire hazards and are identified by a 4 printed inside an orange diamond.

Chemical hazards in the military system are depicted by colors. Toxic agents (such as sarin and mustard gas) are identified by the color red. Harassing agents (such as tear gas and smoke producers) are identified by yellow. White phosphorus is identified by white. Specific personal protective gear requirements are indicated by pictograms.

Other Reference Sources

Labels and placards may be helpful in identifying hazardous materials, but other sources of information are also available. Among these reference sources are material safety data sheets (MSDSs) or Safety Data Sheets, shipping papers, and staffed national resources such as CHEMTREC and the National Response Center. First-responding fire fighters should place a high priority on identifying the released material and finding a reliable source of information about the chemical and physical properties of the released substance. The most appropriate source will depend on the situation; use your best judgment in making your selection.

MSDS

A commonly used source of information about a particular chemical is the MSDS specific to that substance. Essentially, an MSDS provides basic information about the chemical makeup of a substance, the potential hazards it presents, appropriate first aid in the event of an exposure, and other pertinent data for safe handling of the material. Generally, an MSDS will include the following data:

- Physical and chemical characteristics
- Physical hazards of the material
- Health hazards of the material
- Signs and symptoms of exposure
- Routes of entry
- Permissible exposure limits
- Responsible party contact
- Precautions for safe handling (including hygiene practices, protective measures, and procedures for cleaning up spills or leaks)
- Applicable control measures, including personal protective equipment
- Emergency and first aid procedures
- Appropriate waste disposal

When responding to a hazardous materials incident at a fixed facility such as a factory or plant, fire fighters should ask the site manager for an MSDS for the spilled material. All facilities that use or store chemicals are required, by law, to have an MSDS on file for each chemical used or stored in the facility. Although the MSDS is not a definitive response tool, it is a piece of the puzzle. Fire fighters should investigate as many sources as possible (preferably at least three) to obtain emergency information about a released substance.

Shipping papers are required whenever materials are transported from one place to another. They include the names and addresses of the shipper and the receiver, identify the material being shipped, and specify the quantity and weight of each part of the shipment. Shipping papers for road and highway transportation are called bills of lading or freight bills and are located in the cab of the vehicle. Drivers transporting chemicals are required by law to have a set of shipping papers on their person or within easy reach inside the cab at all times.

A bill of lading may include additional information about a hazardous substance, such as its packaging group designation. The packaging group designation is also used by shippers to identify special handling requirements or hazards. Some DOT hazard classes require shippers to assign packaging groups based on the material's flash point and toxicity. A packaging group designation might, for example, signal that the material poses a greater hazard than similar materials in a hazard class. Three packaging group designations are used:

- Packaging Group I: high danger
- Packaging Group II: medium danger
- Packaging Group III: minor danger

Shipping papers for railroad transportation are called waybills; the list identifying every car on the train is called a consist. The conductor, engineer, or a designated member of the train crew will have a copy of the consist when the train is traveling in mainline service. They may have the waybill as well, but it is more often used in the rail yards. On a marine vessel, shipping papers are called the dangerous cargo manifest. The dangerous cargo manifest is generally kept in a tube-like container in the wheelhouse in the custody of the captain or master. For air transport, the air bill is the shipping paper; it is kept in the cockpit and is the pilot's responsibility. The responsible person in each situation should maintain the information about hazardous cargo and provide it in an emergency situation.

Where to Find an MSDS

MSDSs are stored in several different formats depending on the number of chemicals stored within a facility. Facilities with small chemical inventories typically store the MSDS forms in a three-ring binder. In some cases, fire departments require the binder to be stored in a lockbox system located near an entrance. In facilities with large chemical inventories, the MSDS forms may be stored in file cabinets or in multiple binders in a library system format. In some cases, large MSDS libraries may be stored electronically on a computer system with a paper backup.

CHEMTREC

Located in Arlington, Virginia, established by the Chemical Manufacturers Association, and now operated by the American Chemical Council, the Chemical Transportation Emergency Center (CHEMTREC) is a clearinghouse of emergency response information. Essentially, this emergency call center is an information resource for fire fighters responding to chemical incidents. The toll-free number for CHEMTREC is 1-800-424-9300. Many fire fighters write this number inside their helmets so that they always have it in case of need. When calling CHEMTREC, be sure to have the following basic information ready:

- Name of the caller and callback telephone number
- Location of the actual incident or problem
- Name of the chemical involved in the incident (if known)
- Shipper or manufacturer of the chemical (if known)
- Container type
- Railcar or vehicle markings or numbers
- Shipping carrier's name
- Recipient of material
- Local conditions and exact description of the situation

When speaking with CHEMTREC, spell out all chemical names; if using a third party, such as a dispatcher, it is vital that you confirm all spellings so that there is no chance of a misunderstanding of any spelling or numbers. One number or letter could be vital—when in doubt, be sure to obtain clarification.

CHEMTREC is a free service that connects fire fighters with chemical manufacturers, chemists, and other product specialists who can help during a chemical incident. The Canadian equivalent of CHEMTREC is CANUTEC; the Mexican counterpart is SETIQ. Phone numbers for all of these resources can be found in the ERG. Fire fighters who are at or near the U.S. border may need to be able to access these international counterparts to CHEMTREC and can do so both directly and through CHEMTREC itself.

National Response Center

Whenever a significant hazardous incident occurs, the National Response Center (NRC) must be notified. The NRC is operated by the U.S. Coast Guard and serves as a central notification point, rather than a guidance center. Once the NRC is notified, it will alert the appropriate state and federal agencies. The NRC must be notified if any spilled material could enter a navigable waterway, through any method ranging from a small spill into a local stream to a release at a sewerage treatment plant or a leak into an underground water table. The toll-free number for the NRC is 1-800-424-8802. The NRC specifies complex reporting requirements for different chemicals based on the reportable quantity (RQ) for that chemical. The shipper and/or the owner of the chemical has the ultimate legal responsibility to make this call, but by doing so response agencies will have their reporting bases covered.

Radiation

Fire fighters must be able to recognize the situations in which radioactive materials could be encountered. Industries that routinely use radioactive materials include food testing labs, hospitals, medical research centers, biotechology facilities, construction sites, and medical laboratories. Usually, some visual indicators (signs or placards) will indicate the presence of radioactive substances in a structure, but this is not always the case.

If you suspect a radiation incident at a fixed facility, you should initially ask for the facility's radiation safety officer. This person is responsible for the use, handling, and storage procedures for all the radioactive material at the site. He or she likely will be a tremendous resource to fire fighters and will know exactly which materials are being used at the facility.

If the incident occurs somewhere other than a fixed site, the presence of radiation might never be apparent. Radioactive materials cannot be detected by sight, smell, taste, or the other senses. Therefore, if you have any suspicion that the incident involves radiation, it will be necessary to call a hazardous materials team or some other resource with radiation detection capabilities.

Significant incidents involving radiation are few and far between. This is not to say accidents cannot or will not happen, but regulations for using, storing, and transporting significant radioactive sources are so comprehensive that the entire process has become quite safe. Most of the incidents you may encounter will involve low-level radioactive sources and can be handled safely. These low-level sources are typically found in radioactive packaging, industrial radioactive packaging, and Type A packaging **FIGURE 30-14**. Type A packaging method is unique to radioactive substances and contains materials such as radiopharmaceuticals and other low-level emitters.

Type A packaging is characterized by having an inner containment vessel of glass, plastic, or metal and packaging materials made of polyethylene, rubber, or vermiculite. This type of packaging is designed to protect the contents from damage during normal shipping and handling. The key is

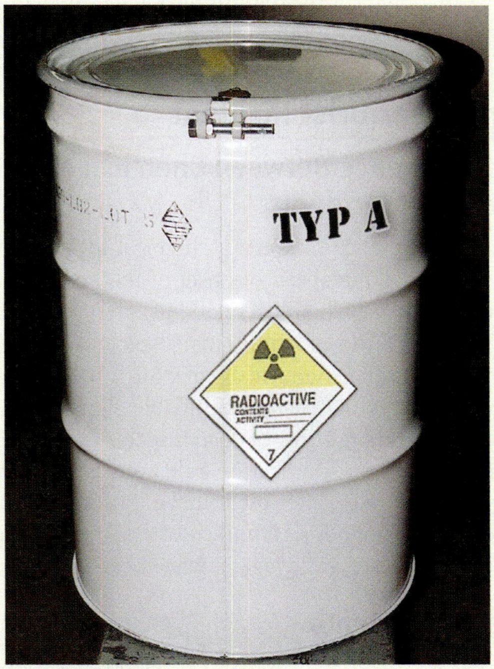

FIGURE 30-14 Type A package.

to be able to suspect, recognize, and understand when and where you may encounter radioactive sources involving Type A packaging.

More dangerous radioactive sources might be found in Type B packaging FIGURE 30-15. This type of containment vessel contains materials such as spent radioactive waste and other high-level emitters. Type B packages are designed to prevent greater exposure from the contents. The amount of protection is based on the potential severity of the hazard. Type B packages include small drums and heavily shielded casks weighing more than 100 metric tons.

FIGURE 30-15 Type B package.

Type C packages are authorized for international shipments of radioactive material by air. They are designed to withstand severe accident conditions associated with air transport. The Type C package performance requirements are significantly more stringent than those for Type B packages. Because Type C packages are not authorized for domestic use, it is unlikely that you would encounter them in the United States.

Potential Terrorist Incident

The Difference Between Chemical and Biological Incidents

The biggest difference between a chemical incident and a biological incident is the speed of onset of the health effects from the involved agents. Chemical incidents are typically characterized by a rapid onset of symptoms (usually minutes to hours after exposure to the chemical agent). Such incidents also usually have some type of easily identifiable signature such as an odor, liquid or solid residue, or dead insects or foliage. Biological incidents, by comparison, feature a delayed onset of symptoms (usually days to weeks after the initial exposure). Because most biological agents are odorless and colorless, there are usually no outward indicators that the agents have been spread.

Chemical Agents

Indicators of possible criminal or terrorist activity involving chemical agents may vary depending on the complexity of the operation. The perpetrators may have left behind protective equipment such as rubber gloves, chemical suits, and respirators. Chemical containers of various shapes and sizes may be present, especially glass containers. The chemicals that are involved in the incident may provide unexplained odors that are out of character for the surroundings. Residual chemicals (in liquid, powder, or gas form) might also be observed in the area. Chemistry books or other reference materials may be present as well, along with equipment that is used to manufacture chemical weapons (such as scales, thermometers, or torches). Finally, personnel working around the materials may exhibit symptoms of chemical exposure—for example, irritation to the eyes, nose, and throat; difficulty breathing; tightness in the chest; nausea and vomiting; dizziness; headache; blurred visions; blisters or rashes; disorientation; or even convulsions.

Biological Agents

Indicators of incidents that potentially involve biological agents are somewhat similar to those that involve chemical agents. For example, production or containment equipment may be present, such as Petri dishes, vented hoods, Bunsen burners, pipettes, microscopes, and incubators. Microbiology or biology textbooks or reference manuals may be found. Containers used to transport biological agents may be found at the site, including metal cylindrical cans or red plastic boxes or bags with biohazard labels. Personal protective equipment (e.g., respirators, chemical or biological suits, and latex gloves) may be observed as well. Excessive amounts of antibiotics may be present as a means to protect personnel working with the agents. Other potential indicators may include abandoned spray devices and unscheduled or unusual sprays being disseminated (especially if outdoors at night). Finally, personnel working in the area of the lab may exhibit symptoms consistent with contact with the biological weapons with which they are working.

Radiological Agents

Because radiological agents are difficult to access, you are less likely to see incidents involving them. Indicators of radiological agents typically will include production or containment equipment, such as lead or stainless steel containers (with nuclear or radiological labels), and equipment that can be used to detonate the radioactive source, including containers such as pipes, caps, fuses, gunpowder, timers, wire, and detonators. Personal protective equipment present may include radiological-protective suits and respirators. Radiation monitoring equipment such as Geiger counters may be present. Personnel working around the radiological agents may exhibit exposure symptoms such as burns or difficulty breathing.

Illicit Laboratories

There are many indicators of possible criminal or terrorist activity involving illicit labs. One set of indicators may be terrorist paraphernalia such as terrorist training manuals, ideological propaganda, and documents indicating affiliation with known terrorist groups. Certain locations are also commonly associated with clandestine labs—for example, basements with unusual or multiple vents, buildings with heavy security, buildings with windows obscured, and buildings with odd or unusual odors.

Personnel working in illegal lab settings will exhibit a certain degree of suspicion. They may be nervous and have a high level of anxiety. They may be very protective of the lab area and not want to allow anyone to access the area for any reason. They may rush people to leave the lab area as soon as possible.

Other equipment that may be present in lab areas includes surveillance materials (such as videotapes, photographs, maps, blueprints, or time logs of the target hazard locations), nonweapon supplies (such as identification badges, uniforms, and decals that might be used to allow the terrorist to access target hazards), and weapon supplies (such as timers, switches, fuses, containers, wires, projectiles, and gun powder or fuel). Finally, security weapons such as guns, knives, and booby trap systems may be present.

Drug labs are by far the most common type of clandestine labs encountered. Most are very primitive. Many materials used to manufacture the drugs will be common, everyday items (jars, bottles, glass cookware, coolers, and tubing) modified to produce the illicit drugs. Specific chemicals and materials present will include large quantities of cold tablets (ephedrine or pseudoephedrine), hydrochloric or sulfuric acid, paint thinner, drain cleaners, iodine crystals, table salt, aluminum foil, and camera batteries. The strong smell of urine or unusual chemical smells like ether, ammonia, or acetone are also commonly encountered in drug labs. Clandestine drug labs should be considered to represent significant hazardous materials scenes. The inexperienced chemists who run these labs take many shortcuts and disregard typical safety protocols in an effort to increase production.

Possible Criminal or Terrorist Activity Involving Explosives

Indicators of possible criminal or terrorist activity involving explosives will typically include materials that fit into four major categories—protective equipment, production and containment materials, explosive materials, and support materials. Protective equipment may include rubber gloves, goggles and face shields, and fire extinguishers. Production and containment equipment may include funnels, spoons, threaded pipe, caps, fuses, timers, wires, detonators, and concealment containers such as briefcases, backpacks, or other common packages. Explosive materials may include gunpowder, gasoline, fertilizer, and similar materials. Support materials may include explosive reference manuals, Internet reference materials, and military information.

Indicators of Secondary Devices

Indicators of potential secondary devices may include trip devices such as timers, wires, or switches. In addition, they may include common concealment containers, such as briefcases, backpacks, boxes, or other common packages, and uncommon concealment containers, such as pressure vessels (propane tanks) or industrial chemical containers (chlorine storage containers). Personnel may be watching the site at which the primary devices are placed, waiting to manually activate the secondary devices when first responders arrive at the scene.

Wrap-Up

Chief Concepts

- To ensure your safety and the safety of your crew, you must be able to recognize a potential hazardous material as soon as possible.
- Hazardous materials incidents can occur almost anywhere. Hazardous materials are stored in warehouses, hospitals, laboratories, industrial occupancies, residential garages, bowling alleys, home improvement centers, garden supply stores, restaurants, and many other businesses.
- Many hazardous materials can be recognized by smell, taste, or touch. Getting close enough to exercise these senses, however, may expose you to the hazard. The farther you are from the incident, the safer you will be.
- A container is any vessel or receptacle that holds material, including storage vessels, pipelines, and packaging. Often the container type, size, and material provide important clues about the nature of the substance inside.
- Bulk storage containers include the following types:
    - Fixed tanks—Include vertical storage tanks, aboveground storage tanks, and underground storage tanks. Often these containers are surrounded by a supplementary containment system to help control an accidental release.
    - Large transportation tankers—Used to transport hazardous materials.
    - Totes—Portable plastic tanks surrounded by a stainless steel web that adds both structural stability and protection. Both shipping and storing totes can be hazardous operations.
    - Intermodal tanks—Both shipping and storage vehicles. They hold between 5000 and 6000 gallons

(18,927 and 22,712 liters) of product and can be either pressurized or nonpressurized.

- Bulk storage containers, or large-volume containers, are defined by their internal capacity based on the following measures:
  - Liquids: more than 119 gallons (450 liters)
  - Solids: more than 882 pounds (400 kilograms)
  - Gases: more than 882 pounds (400 kilograms)
- Nonbulk storage vessels include the following types:
  - Drums—Barrel-like containers used to store a wide variety of substances.
  - Bags—Commonly used to store solids and powders. Pesticide bags must be labeled with specific information.
  - Carboys—A glass, plastic, or steel container that holds 5 to 15 gallons (19 to 57 liters) of product, such as some corrosives and other types of chemicals.
  - Compressed gas cylinders—Used to hold liquids and gases.
  - Cryogenic containers—Include Dewar containers that hold gaseous substances that have been chilled until they liquefy.
- Chemical transport vehicles or tankers include the following types:
  - MC-306/DOT406 flammable liquid tanker—Typically carries gasoline or other flammable and combustible materials.
  - MC-307/DOT407 chemical hauler—Round or horseshoe-shaped tank that typically carries 6000 to 7000 gallons (22,712 to 26,498 liters) of flammable liquids, mild corrosives, and poisons.
  - MC-312/DOT412 corrosives tanker—Used to transport concentrated sulfuric and nitric acids and other corrosive substances.
  - MC-331 pressure cargo tanker—Carries materials such as ammonia, propane, and butane.
  - MC-338 cryogenic tanker—Low-pressure tanker that relies on tank insulation to maintain low temperatures to transport cryogenic gases.
  - Tube trailers—Carry compressed gases such as hydrogen, oxygen, helium, and methane.
  - Dry bulk cargo tanks—Commonly seen on the road and carry dry bulk goods such as powders, pellets, fertilizers, and grain.
- Railway tank cars can carry volumes up to 30,000 gallons (113,562 liters) and have the potential to create large leaks or vapor clouds. There are three basic railcar configurations:
  - DOT 111 nonpressurized railcars—Carry general industrial chemicals, consumer products, flammable and combustible liquids, and mild corrosives.
  - DOT 105, DOT 112 pressurized railcars—Transport materials such as propane, ammonia, ethylene oxide, and chlorine.
  - Special-use railcars—Include boxcars, flat cars, corrosive tank cars, and high-pressure compressed gas tube cars.
- Pipelines often transport hazardous materials. They may be located either underground or above ground.
- The DOT marking system is characterized by a system of labels and placards. The DOT's ERG is also part of this system and offers a certain amount of guidance for fire fighters operating at a hazardous materials incident.
- Labels, placards, and other markings on buildings, packages, boxes, and containers often enable fire fighters to identify a spilled or released chemical.
- Placards are diamond-shaped indicators (10¾ inches [27.3 centimeters] on each side) that must be placed on all four sides of highway transport vehicles, railroad tank cars, and other forms of transportation carrying hazardous materials.
- Labels are smaller versions of placards (4-inch [10 centimeters] diamond-shaped indicators) and are used on the four sides of individual boxes and smaller packages being transported.
- Nine DOT chemical families are recognized in the ERG:
  - DOT Class 1: Explosives
  - DOT Class 2: Gases
  - DOT Class 3: Flammable combustible liquids
  - DOT Class 4: Flammable solids
  - DOT Class 5: Oxidizers
  - DOT Class 6: Poisons (including blood agents and choking agents)
  - DOT Class 7: Radioactive materials
  - DOT Class 8: Corrosives
  - DOT Class 9: Other regulated materials
- The ERG is a preliminary action guide for hazardous materials incidents. It is divided into four colored sections:
  - Yellow section—Used when the UN number is known or can be identified.
  - Blue section—Lists chemicals alphabetically by name.
  - Orange section—Contains the emergency action guides.
  - Green section—Organized numerically by UN identification number and provides the initial isolation distances for specific materials.
- The NFPA hazard identification system is designed for fixed-facility use. The NFPA 704 hazard identification system uses a diamond-shaped symbol of any size, which is itself broken into four smaller diamonds, each representing a particular property or characteristic. The blue diamond at the nine o'clock position indicates the health hazard posed by the material; the top red diamond indicates flammability; the yellow diamond at the three o'clock position indicates reactivity; and the bottom white diamond is used for special symbols and handling instructions.
- Additional reference sources for hazardous materials include material safety data sheets (MSDS), shipping papers, and staffed national resources such as CHEMTREC and the National Response Center. The MSDS provides basic information about the chemical makeup of a substance, the potential hazards it presents,

appropriate first aid in the event of an exposure, and other pertinent data for safe handling of the material. MSDSs may be kept in a three-ring binder or a file cabinet, or a computer system.

- Industries that routinely use radioactive materials include food testing labs, hospitals, medical research centers, biotechnology facilities, construction sites, and medical laboratories. Radioactive materials are not detected by sight, smell, taste, or other senses. If you have any suspicion that the incident involves radiation, it will be necessary to call a hazardous materials team or some other resource with radiation detection capabilities.

Hot Terms

Air bill Shipping papers on an airplane.

Bill of lading Shipping papers for roads and highways.

Bulk storage containers Large-volume containers that have an internal volume greater than 119 gallons (450 liters) for liquids and greater than 882 pounds (400 kilograms) for solids and a capacity of greater than 882 pounds (400 kilograms) for gases.

Bungs One or more small openings in closed-head drums.

Carboys Glass, plastic, or steel containers, ranging in volume from 5 to 15 gallons (19 to 57 liters).

Chemical Transportation Emergency Center (CHEMTREC) A national call center for basic chemical information, established by the Chemical Manufacturers Association and now operated by the American Chemical Council.

Consist A list of every car on a train.

Container Any vessel or receptacle that holds material, including storage vessels, pipelines, and packaging.

Cryogenic liquids (cryogens) Gaseous substances that have been chilled to the point at which they liquefy; a liquid having a boiling point lower than −150°F (−101°C) at 14.7 psia (an absolute pressure of 101 kPa).

Cylinder A portable compressed gas container.

Dangerous cargo manifest Shipping papers on a marine vessel, generally located in a tube-like container.

Department of Transportation (DOT) marking system A unique system of labels and placards that, in combination with the *Emergency Response Guidebook*, offers guidance for first responders operating at a hazardous materials incident.

Dewar containers Containers designed to preserve the temperature of the cold liquid held inside.

Drums Barrel-like containers built to DOT Specification 5P (1A1).

Dry bulk cargo tanks Tanks designed to carry dry bulk goods such as powders, pellets, fertilizers, or grain; they are generally V-shaped with rounded sides that funnel toward the bottom.

Emergency Response Guidebook (ERG) A reference book, written in plain language, to guide emergency responders in their initial actions at the incident scene. (NFPA 472)

Freight bills Shipping papers for roads and highways.

Hazardous material A substance that, when released, is capable of creating harm to people, the environment, and property. (NFPA 472)

Hazardous Materials Information System (HMIS) A color-coded marking system by which employers give their personnel the necessary information to work safely around chemicals.

Intermodal tanks Bulk containers that can be shipped by all modes of transportation—air, sea, or land.

Labels Smaller versions (4-inch [10 centimeters] diamond-shaped markings) of placards, placed on four sides of individual boxes and smaller packages.

MC-306/DOT406 flammable liquid tanker Commonly known as a gasoline tanker; a tanker that typically carries gasoline or other flammable and combustible materials.

MC-307/DOT407 chemical hauler A tanker with a rounded or horseshoe-shaped tank.

MC-312/DOT412 corrosives tanker A tanker that often carries aggressive acids such as concentrated sulfuric and nitric acid, which features reinforcing rings along the side of the tank.

MC-331 pressure cargo tanker A tank commonly constructed of steel with rounded ends and a single open compartment inside; there are no baffles or other separations inside the tank.

MC-338 cryogenics tanker A low-pressure tanker designed to maintain the low temperature required by the cryogens it carries.

National Response Center (NRC) An agency maintained and staffed by the U.S. Coast Guard; it should always be notified if any spilled material could possibly enter a navigable waterway.

NFPA 704 hazard identification system A hazardous materials marking system designed for fixed-facility use.

Nonbulk storage vessels Containers other than bulk storage containers.

Pipeline A length of pipe including pumps, valves, flanges, control devices, strainers, and/or similar equipment for conveying fluids. (NFPA 70)

Pipeline right-of-way An area, patch, or roadway that extends a certain number of feet on either side of the pipe itself and that may contain warning and informational signs about hazardous materials carried in the pipeline.

Placards Signage required to be placed on all four sides of highway transport vehicles, railroad tank cars, and other forms of hazardous materials transportation that identifies the hazardous contents of the vehicle, using a standardization system with 10¾-inch (27.3 centimeters) diamond-shaped indicators.

Secondary containment Any device or structure that prevents environmental contamination when the primary container or its appurtenances fail. Secondary containment shall be designed and constructed to intercept and contain pesticide spills and leaks and to prevent runoff or leaching of pesticides into the environment. Examples of secondary containment include dikes, curbing, and double-walled tanks. (NFPA 434)

Shipping papers A shipping order, bill of lading, manifest, or other shipping document serving a similar purpose and containing the information required by regulations of the U.S. Department of Transportation. (NFPA 498)

Signal words Information on a pesticide label that indicates the relative toxicity of the material.

Special-use railcars Boxcars, flat cars, cryogenic and corrosive tank cars, and high-pressure compressed gas tube-type cars.

Totes Portable tanks, which usually hold a few hundred gallons (several hundred liters) of product and are characterized by a unique style of construction.

Toxic inhalation hazard (TIH) materials Gases or volatile liquids that are extremely toxic to humans.

Tube trailers High-volume transportation devices made up of several individual compressed gas cylinders banded together and affixed to a trailer.

Vent pipes Inverted J-shaped tubes that allow for pressure relief or natural venting of the pipeline for maintenance and repairs.

Waybill Shipping papers for trains.

FIRE FIGHTER in action

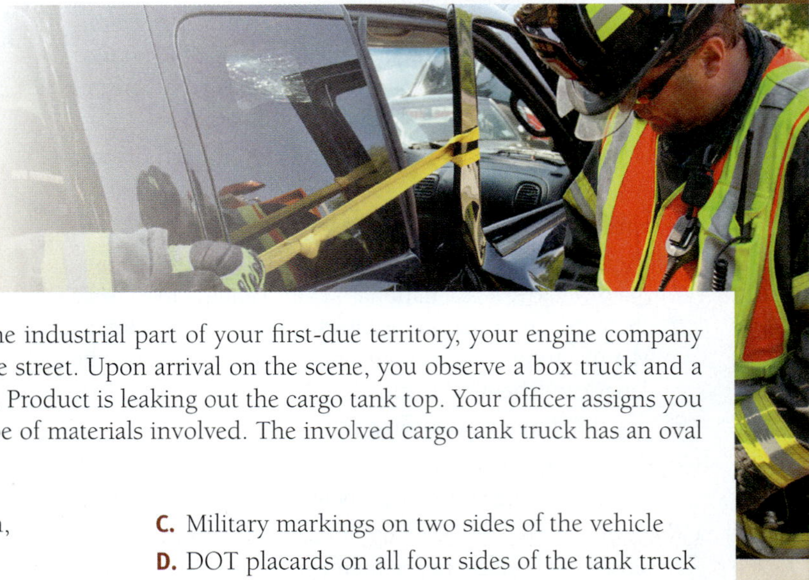

While you are conducting an commercial inspection in the industrial part of your first-due territory, your engine company is dispatched for a multiple-vehicle accident just down the street. Upon arrival on the scene, you observe a box truck and a highway cargo tank overturned in the middle of the street. Product is leaking out the cargo tank top. Your officer assigns you the task of sizing up the situation and determining the type of materials involved. The involved cargo tank truck has an oval shape and is red.

1. Which type of tank truck, based on the description, can you see?
   - **A.** MC-307/DOT407
   - **B.** MC-312/DOT412
   - **C.** MC-331
   - **D.** MC-306/DOT406
2. Based on the type of tanker involved, which product is most likely being released?
   - **A.** Gasoline
   - **B.** Sulfuric acid
   - **C.** Ethanol-blended fuel
   - **D.** Both a and c
3. Which type of label or placard would you most likely find on this tank truck?
   - **A.** NFPA 704 labels located on all four sides of the tank
   - **B.** HMIS labels located on all four sides of the vehicle
   - **C.** Military markings on two sides of the vehicle
   - **D.** DOT placards on all four sides of the tank truck
4. Which shipping papers would be present on the tank truck and where would they be located?
   - **A.** A bill of lading would be located in the cab within reach of the driver.
   - **B.** A waybill would be located in a tube on the side of the tank.
   - **C.** A bill of lading would be located in a tube on the side of the tank.
   - **D.** A consist would be carried by the driver.

FIRE FIGHTER II *in action*

It is the start of the Fourth of July weekend. You and your company are sitting in the station discussing your plans for the weekend when the tones go off. There is a Hazardous Materials Incident in progress on the major interstate that cuts through your county.

You are assigned to the engine company along with two other fire fighters and a fire officer. As the engine turns onto the entrance ramp to the interstate, you notice that traffic is already backed up. As you approach the scene, you see three semi-tractor trailers involved in a multiple vehicle accident. Two of the drivers are walking around, but the third is trapped in his truck. You see that the last truck has several placards and is leaking a large amount of liquid from its trailer.

1. What key decision-making tools would you use to identify the hazards to personnel on the scene?
2. To whom would you report your observations?
3. How much detail should you provide when reporting your observations?
4. Which resources would you be able to use to assist your fire officer in identifying the product(s), shipper, and emergency contacts for the truck that is leaking product?

Hazardous Materials: Implementing a Response

Fire Fighter I

Knowledge Objectives

After studying this chapter, you will be able to:

- List and describe the hazardous materials incident levels. (NFPA472,5.4.3,6.6.1.1.2, p 924)
- List the information to provide when reporting a hazardous materials incident. (NFPA472,5.4.3,5.5.2,6.6.1.1.2, p 924–925)
- Describe how to plan an initial response. (NFPA472,5.1.2.2,5.2.1.4,5.3,5.3.2, p 925–926)
- List and describe defensive objectives. (NFPA472,5.1.2.2, p 926)
- List and describe defensive actions. (NFPA472,5.1.2.2, p 926)
- Describe how to estimate the size and scope of a hazardous materials/weapons of mass destruction incident. (NFPA472,5.2.4, p 928)
- Describe the resources available for determining the concentrations of a released hazardous material. (NFPA472,5.2.4, p 928)
- Describe a secondary attack and its impact on responders. (NFPA472,5.3.1, p 928–930)
- List the signs that indicate the possible presence of a secondary device. (NFPA472,5.3.1, p 928–930)
- Explain how the Incident Command System (ICS) is adapted for hazardous materials incidents. (NFPA472,5.1.2.2, p 930)
- Identify the considerations for determining the location of the incident command post (ICP). (NFPA472,5.4.3, p 930)

Skills Objectives

There are no skill objectives for Fire Fighter I candidates. NFPA 1001 contains no Fire Fighter I Job Performance Requirements for this chapter.

CHAPTER 31

Fire Fighter II | FFII

Knowledge Objectives

There are no knowledge objectives for Fire Fighter II candidates. NFPA 1001 contains no Fire Fighter II Job Performance Requirements for this chapter.

Skills Objectives

There are no skill objectives for Fire Fighter II candidates. NFPA 1001 contains no Fire Fighter II Job Performance Requirements for this chapter.

You Are the Fire Fighter

While you are at the volunteer station performing hose testing, a car drives up on the ramp and reports that a truck parked up the street is possibly leaking a chemical gas. You notify the officer on duty and call dispatch to notify it of the response. As the first apparatus gets closer to the scene, you notice an odor of chlorine in the area. As you approach the parking lot where the truck is stopped, you see a cloud of gas emitting from several cylinders in the bed of the truck.

1. Once you have arrived on scene, what should be your first actions?
2. Which immediate communications should be made?
3. Which command elements should be activated immediately?

Introduction

The first step after determining that a hazardous material is present is to identify those resources that can assist you in the process of reporting the hazardous materials incident. Fire fighters must follow their local standard operating procedures and guidelines.

Who to Contact

Fire fighters should make a series of contacts after determining that they are responding to a hazardous materials incident. Some of these notifications are related to operational concerns; others are made for legal reasons TABLE 31-1.

First and foremost, when a hazardous materials incident is detected, there should be an initial call for additional resources FIGURE 31-1. The resources sought should include support personnel, trained hazardous materials technicians, and technical specialists who will help identify the hazardous material and control the incident. Additional calls should then request decontamination personnel and equipment. Other notifications could include the police, highway department, Chemical Transportation Emergency Center (CHEMTREC), the National Response Center, local and state environmental agencies, and the local emergency planning commission. A predetermined list of contact names, agencies, and numbers should be established and maintained at the dispatch center.

No offensive action should be implemented until the identity of the hazardous material involved is confirmed. It makes no sense to try to deal with a hazardous material until it has been properly identified. This identification should be confirmed in a minimum of three references. For example, the hazardous material could first be identified by a placard, which may be used to look up information in the *Emergency*

FIGURE 31-1 When a hazardous materials incident is detected, there should be an initial call for additional resources.

TABLE 31-1 Hazardous Materials Incident Levels

Level I	Level II	Level III
Lowest level of threat. Small amount of hazardous material involved. Can usually be handled by the local fire department. Fire fighters must wear turnout gear and SCBA. Example: a small gasoline spill from a motor vehicle accident	A hazardous materials team is needed at this level. Fire fighters simply support the hazardous materials team. Additional PPE required will be specialized and carried only by the hazardous materials team. Civilian evacuations may be required. Decontamination may need to be performed. Example: a gasoline tanker overturns in a tunnel and begins spilling gasoline onto the highway.	The highest level of threat. Large-scale evacuations may be needed. Federal agencies will be called. Example: a ship in a highly populated harbor catches fire and begins to release chlorine vapors from its cargo area.

Response Guidebook (ERG). Then the shipping papers or Material Safety Data Sheets (MSDSs) could be checked to ensure that they list the same hazardous material as the placard. Finally, the National Institute for Occupational Safety and Health references and CHEMTREC could be consulted for a final confirmation of the identity of the hazardous material.

A variety of sources of information should be compared for consistency. If the information presented in the reference books is not consistent, the information that reflects a more conservative course of action should be used. Protection and safety of fire fighters is always the first priority in any hazardous materials incident and cannot be compromised for any reason. Therefore, if two references indicate that the use of water is acceptable but one reference cautions against it, the use of water should not be considered until further information is obtained. On-scene research should continue throughout the course of the incident.

After the material is identified, an operations-level responder should perform only actions that do not involve contact with the material. The responder must use full personal protective equipment (PPE) or chemical protective equipment (CPE) during any activity and must complete decontamination procedures prior to leaving any area where the hazardous material is present.

FIRE FIGHTER Tips

If the hazardous material or the hazards of the product cannot be clearly identified, the most conservative defense strategy and tactics must be employed.

What to Report

When additional agencies are contacted regarding a potential hazardous materials incident, it is important that the information given be as clear, concise, and accurate as possible. An error in spelling, an incorrect measurement, a mispronunciation of a chemical name, or incorrect identification of a hazardous substance may prove disastrous. The change or omission of just one letter in a chemical name, for example, could lead to incorrect identification of the material at hand. If a chemical is incorrectly identified, the proper response and safety procedures will not be carried out. This type of error can be deadly to both responders and the public. The fire fighter should keep information as simple as possible, spell names that are complex or potentially confusing, and confirm that the receiver of the message has heard the correct information by repeating back what was heard.

If these agencies are to plan, prepare, and begin assisting fire fighters, they must have as much information as can be obtained. This will help ensure that every possible situation can be taken into account and planned for, so as to prevent any further injury, property loss, or environmental damage. If possible, provide all of the following information:

- The exact address and specific location of the leak or spill
- Identification of indicators and markers of hazardous materials
- All color or class information obtained from placards
- Four-digit United Nations/North American Hazardous Materials Code numbers for the hazardous materials
- Hazardous material identification obtained from shipping papers or MSDSs and the potential quantities of hazardous materials involved
- Description of the container, including its size, capacity, type, and shape
- The amount of chemical that could leak and the amount that has already leaked
- Exposures of people and the presence of special populations (children or elderly)
- The environment in the immediate area
- Current weather conditions, including wind direction and speed
- A contact or callback telephone number and two-way radio frequency or channel

Plan an Initial Response

When planning an initial hazardous materials incident response, the first priority is to consider the safety of the responding personnel. Responders are there to isolate, contain, and remedy the problem—not to become part of it. Proper incident planning will keep responders safe and provide a means to control the incident effectively, thereby preventing further harm to persons or property.

Planning the response begins with the initial call for help. Information obtained during this call is used to determine the safest, most effective, and fastest route to the hazardous materials scene. Choose a route that approaches the scene from an upwind and upgrade direction, so that natural wind currents blow the hazardous material vapors away from arriving responders. A route that places the responders uphill as well as upwind of the site is also desirable, so that any liquid or vapor flows away from responders.

Responders need to know the type of material involved. Is the material a solid, a liquid, or a gas? Is it contained in a drum, a barrel, or a pressurized tank? Response to a spill of a solid hazardous material will differ from response to a liquid-release or vapor-release incident. A solid is local and can be easily contained, whereas a released gas can be widespread and constantly moving, depending on the gas characteristics and weather conditions **FIGURE 31-2**.

The characteristics of the affected area near the location of the spill or leak are also important factors in planning the response to an incident. If an area is heavily populated, evacuation procedures and a decontamination process must be established very early in the course of the incident. If the area is sparsely populated and rural, isolating the area from anyone trying to enter the location may be the top priority. A high-traffic area such as a major highway would necessitate immediate rerouting of traffic.

When responding to an incident, the more information that can be obtained the better. If information is unknown or is unconfirmed, assume the worst-case scenario. When planning

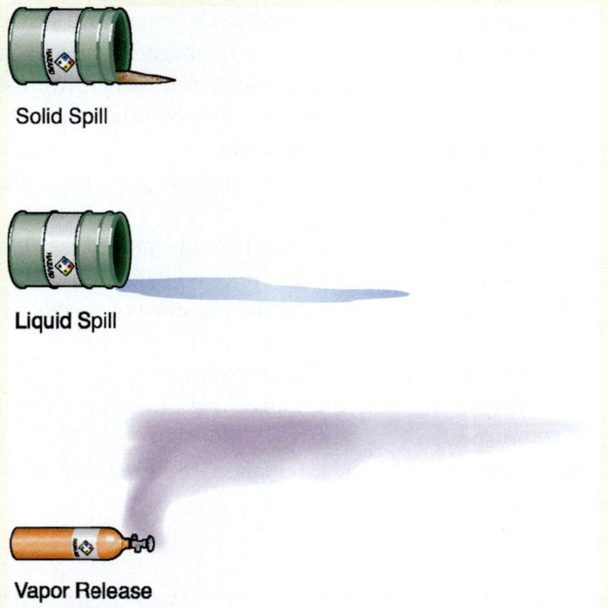

FIGURE 31-2 Response to a spill of a solid hazardous material will differ from response to a liquid-release or vapor-release incident.

for hazardous materials incidents, always provide the largest margin of safety possible.

A number of solutions to a given problem may exist. The challenge in dealing with hazardous materials incidents is arriving at a solution that can be employed quickly and safely, while minimizing the potential negative effects to people, property, and the environment. Handling a hazardous materials incident is a bit like a chess game—one move sets up and then positively or negatively influences the next move or series of moves.

For example, in the case of an anhydrous ammonia release, the incident commander (IC) might decide to use a water fog to knock down the vapor cloud. If the cloud is not immediately threatening lives or property and the water runoff is heading toward a drain that goes to a wastewater treatment facility, this might not be a sound plan. Conversely, if the ammonia cloud is immediately threatening to envelope an elementary school, the water runoff may be an acceptable risk. It is vital to use a thought process that incorporates the risk–reward trade-off when choosing to employ any product control option. An initial response objective should be realistic and well planned, taking into account all potential positive and negative effects.

Sometimes no action is the safest course of action to choose. That is, on some occasions, the incident is so extreme that it may be prudent to create a safe perimeter and let the problem stabilize on its own. In firefighting, structures with inaccessible fires may be encountered. In these cases, an IC may choose to pull back, go defensive, and protect exposures. A similar mentality should be considered when dealing with extreme or unusual hazardous materials incidents. A fire in a gasoline tanker may have to burn itself out if not enough foam is available to fight the fire. When incompatible chemicals are mixed, it might be wise to let the reaction run its course before intervening. Highly volatile flammable liquids may be left to evaporate rather than being the subject of an offensive attack.

It is important that you understand the concepts behind the various product control options available to you. It is equally important that you take the time to do enough hands-on training to become proficient with these skills. You should never learn a new skill under the pressure of a real incident.

It is also vitally important to identify the physical and chemical properties of the released substance. Such knowledge facilitates the selection of the proper PPE. It also provides operations-level responders with enough information to make a wise decision when selecting the appropriate control option. Lastly, as with any other facet of hazardous materials response, decontamination should be performed when and where appropriate.

Response Objectives

At the operational level of training, all response objectives should be primarily defensive in purpose. That is, personnel do not actually come in contact with the hazardous material. Some effective defense actions can be taken safely at a distance. Defensive objectives are objectives that do not involve stopping the leak or release of a hazardous material:

- Isolate the area affected by the leak or spill, and evacuate victims who could become exposed to the hazardous material if the leak or spill were to progress.
- Control where the spill or release is spreading.
- Contain the spill to a specific area.

Defensive Actions

Defensive actions that can be taken include diking and damming, absorbing or adsorbing the hazardous material, stopping the flow remotely from a valve or shut-off, diluting or diverting the material, and suppressing or dispersing vapor. These defensive actions are covered in detail in the Hazardous Materials: Response Priorities and Actions chapter.

Proper Personal Protective Equipment

The determination of which PPE is needed is based on the hazardous material involved, specific hazards, and the physical state of the material, along with a consideration of the tasks to be performed by the operations-level responder. At a minimum, fire fighters should wear full protective gear with no skin exposed and use self-contained breathing apparatus (SCBA). Although this gear provides some respiratory protection, note that standard structural firefighting PPE offers only limited hazardous material protection depending on the type of hazardous material involved. The Hazardous Materials: Personal Protective Equipment, Scene Safety, and Scene Control chapter discusses proper PPE for a hazardous materials incident in detail.

VOICES OF EXPERIENCE

One spring morning at approximately 0900 hours, we responded, along with multiple other agencies, to a neighboring department for an overturned truck tanker that failed to negotiate a turn in the road. At the time of our response, it was unknown what type of chemical we were dealing with.

After I checked in with the incident commander, he assigned me as incident safety officer. Once we were in place, all operational individuals were briefed on the suspected chemical and its properties, along with the response plan. Our first responsibility was to verify the identity of the chemical involved. Next we established the hot, warm, and cold zones. A decontamination corridor was also established, with constant air monitoring.

An entry team, with backup teams in place, identified the damage to the truck tanker and the approximate amount of the chemical spilled. They also conducted diking and absorption activities in containing the spilled chemical.

Through the post-briefing of the entry team, it was determined that the safest and most efficient means to mitigate the situation would be to transfer the chemical to another truck tanker. The incident commander, along with the operations section chief, developed an incident action plan, and the incident was mitigated exactly as planned. Although this incident lasted approximately 15 hours and involved many agencies including private cleanup contractors, we were able to ensure life safety by establishing control zones, delegating tasks through use of the incident command system, and taking protective actions such as conducting a safety briefing and utilizing backup teams. Remember to utilize all of your resources when dealing with a potential hazardous materials release.

Jeffry J. Harran
Lake Havasu City Fire Department
Lake Havasu City, Arizona

Emergency Decontamination Procedures

Even though there should be no intentional contact with the hazardous material involved, there must be a procedure or a plan in place to decontaminate any responder who accidentally becomes contaminated FIGURE 31-3. Victims who are removed from a contaminated zone must be decontaminated by personnel who have donned the appropriate protective gear and who have been trained in the proper methods of victim decontamination.

FIGURE 31-3 There must be a plan for decontamination at every hazardous materials incident.

The decontamination methods employed for a specific incident are dictated by the hazardous material, the physical state of the material, and the hazards involved. Decontamination can be as simple as removing clothing and flushing material away with water or as complex as using drug therapy. It is covered in detail in the Hazardous Materials: Decontamination Techniques chapter.

Gauging the Potential Harm or Severity of the Incident

When attempting to evaluate a hazardous materials event's potential to cause injury to people or destruction of property and the environment, responders need to consider factors such as the size of the container, the nature of the hazardous material involved, the amount released, and the area exposed. Based on the toxicity and the concentration of the hazardous material, they may be able to gauge how the incident might progress. For example, the difference between a 20-pound (9 kilograms) propane cylinder and an 18,000-gallon (68,137 liters) propane tank will determine how extensive and lengthy an incident will become.

Resources for Determining the Size of the Incident

To determine the size of the incident, two resources should be used. The first is the ERG, which is published by the U.S. Department of Transportation. This reference book identifies and outlines predetermined protective action zones and basic action plans, based on spill size estimates, for thousands of chemicals. The second resource is a computerized or hard-copy preincident plan, which should include both reports submitted by the fire department and topographical mapping information.

Monitoring devices such as wind direction and weather forecasting equipment are critical resources for the IC in formulating response plans. Computer modeling programs such as Aerial Location of Hazardous Atmospheres Model can plot the predicted movements of vapor clouds and plumes. The availability of monitoring and portable detection devices will allow the IC to determine the hot, warm, and cold zones and the evacuation distances required.

Reporting the Size and Scope of the Incident

Reporting the estimated physical size of the area affected by a hazardous materials incident is accomplished by using information available at the scene. If a vehicle is transporting a known amount of material, an estimate of the size of the release can be made by subtracting the amount remaining in the container. For example, suppose a gasoline tanker containing 9000 gallons (34,000 liters) overturns, and 4500 gallons (17,000 liters) remain in the tanker. In this case, an estimated 4500 gallons (17,000 liters) of gasoline has spilled. The actual spill area can then be estimated in square feet. Depending on its size, the affected area may be expressed in units as small as square feet or as large as square miles.

When it is initially unsafe to approach a vehicle and shipping papers are not immediately available, other methods must be used to determine the original amount of hazardous material and estimate the remaining contained amount. Remember that the safety of responders is paramount to maintaining an effective response to any hazardous materials incident.

Determining the Concentration of a Released Hazardous Material

The MSDS or Safety Data Sheet (SDS) usually states the concentration of the hazardous material and can be used to estimate the concentration of a chemical release FIGURE 31-4. Litmus paper (pH strips) can be used to determine the hazardous material's pH. To determine the concentration of a gaseous hazardous material, monitors are used to analyze the atmosphere from a safe distance.

Once the hazardous material's concentration is known, the IC can evaluate the incident response plan. A high concentration of an acid, for example, would call for a higher level of personal protection. The revised plan might also require the evacuation of civilians owing to the concentration of the hazardous material in vaporized form.

Secondary Attacks

Implementing a secondary attack on responders is a tactic that many terrorist organizations use to make their attacks even more dramatic. It is essential that responders be able to assess the potential for secondary attacks and devices. This may be much easier said than done, however, because the indicators of secondary attacks and devices can be difficult to locate.

syngenta

MATERIAL SAFETY DATA SHEET

Syngenta Crop Protection, Inc.
Post Office Box 18300
Greensboro, NC 27419

In Case of Emergency, Call
1-800-888-8372

1. PRODUCT IDENTIFICATION

Product Name: **ENVOKE** Product No.: A9842A

EPA Signal Word: Caution

Active Ingredient(%): Trifloxysulfuron-Sodium (75.0%) CAS No.: 199119-58-9

Chemical Name: N-[(4,6-Dimethoxy-2-pyrimidinyl)amino]carbonyl-3-(2,2,2-trifluoro-ethoxy)-pyridin-2-sulfonamide sodium salt

Chemical Class: Sulfonylurea Herbicide

EPA Registration Number(s): 100-1132 **Section(s) Revised: 14**

2. HAZARDS IDENTIFICATION

Health and Environmental

Presents a low hazard during normal handling.

Hazardous Decomposition Products

May decompose at high temperatures forming toxic gases.

Physical Properties

Appearance: Light beige to brown granules

Odor: Not determined

Unusual Fire, Explosion and Reactivity Hazards

During a fire, irritating and possibly toxic gases may be generated by thermal decomposition or combustion.

3. COMPOSITION/INFORMATION ON INGREDIENTS

Material	OSHA PEL	ACGIH TLV	Other	NTP/IARC/OSHA Carcinogen
Diatomaceous Earth	80 mg/m^3/%SiO2 (20 mppcf) TWA	Not Established	6 mg/m^3 TWA **	IARC 3
Crystalline Silica, Quartz	10 mg/m^3/(%SiO2+2) (respirable dust)	0.025 mg/m^3 (respirable silica)	0.05 mg/m^3 (respirable dust) **	IARC 1; ACGIH A2
Surfactant	Not Established	Not Established	15 mg/m^3 TWA (total dust) *	No
Sodium Sulfite	Not Established	Not Established	Not Established	IARC Group 3
Trifloxysulfuron-Sodium (75.0%)	Not Established	Not Established	10 mg/m^3 TWA ***	No

* recommended by manufacturer
** recommended by NIOSH
*** Syngenta Occupational Exposure Limit (OEL)

Ingredients not precisely identified are proprietary or non-hazardous. Values are not product specifications.
Syngenta Hazard Category: B

Product Name: **ENVOKE** Page: 1

FIGURE 31-4 The MSDS or Safety Data Sheet (SDS) contains critical information.

Terrorists who want to injure responding personnel with a secondary device or attack will typically make the initial attack very dramatic to draw responders into very close proximity to the scene. Indeed, the primary attack may purposely injure members of the public to draw responders into the scene. As they begin to treat victims, the secondary attack then takes place. Signs of secondary devices may include trip wires, timers, and ordinary everyday containers (e.g., luggage, briefcases, boxes) found in very close proximity to the initial incident site. The smell of chemicals or the sighting of chemical dispersion devices or other containers that may hold chemical, biological, or even radioactive agents could be an indicator that a secondary attack is possible.

Incident Command System

As described in the Incident Command System chapter, an ICS can be expanded to handle an incident of any size and complexity. Hazardous materials incidents can be complex, and local, state, and federal responders and agencies may all become involved in many cases of long duration. The basic ICS consists of five functions: command, operations, logistics, planning/intelligence, and finance/administration.

In an incident involving hazardous materials, the application of an ICS takes on two unique characteristics. First, U.S. federal law requires that an ICS be used in the response to all hazardous materials incidents, including the use of a written incident action plan. Second, a special technical group may be developed under the Operations Section, known as the Hazardous Materials Branch. The Hazardous Materials Branch consists of some or all of the following positions as needed for the safe control of the incident:

- A second safety officer reporting directly to the hazardous materials officer as well as to the incident safety officer. This hazardous materials safety officer (the National Incident Management System [NIMS] calls this position Assistant Safety Officer—Hazardous Materials) is responsible for the hazardous materials team's safety only.
- A hot-zone entry team.
- A decontamination team.
- A backup entry team.
- A hazardous materials information research team.

The Incident Command Post

The main hub of the ICS is the incident command post (ICP), which serves as the collection point for all information and resources. The ICP must be located upwind and upgrade from the spill or leak in the cold zone to keep it from becoming contaminated or unsafe. If the ICP, including those personnel housed there, were to become contaminated, the personnel inside the ICP would no longer be able to control the operation. The officers and members in that location would become victims, the ICP would become a part of the hot zone, and all operations would have to be reestablished elsewhere using a completely different pool of personnel, equipment, and supplies. In this scenario, the overall efficiency of command would be negatively affected, and control of the scene would be temporarily lost during the transfer of command.

Based on the potential threat or severity of the hazard, the ICP could be as close as one block from the incident site or as far away as several miles from the hot zone. When choosing a site for the ICP, the maximum margin of safety must be used.

Wrap-Up

Chief Concepts

- After identifying that a hazardous material is present, you must identify the resources needed to properly respond to the incident.
- There are three levels of hazardous materials incidents:
  - Level I—Small amount of hazardous material involved and usually handled by the local fire department.
  - Level II—A hazardous materials team is required with fire fighters providing assistance.
  - Level III—The largest and most serious scale. Federal agencies will be called and large-scale evacuations may be needed.
- The resources sought for a hazardous materials incident should include support personnel, trained hazardous materials technicians, and technical specialists who will help identify the hazardous material and control the incident. Additional calls should then request

decontamination personnel and equipment. Other notifications could include the CHEMTREC, the National Response Center, local and state environmental agencies, and the local emergency planning commission.

- A predetermined list of contact names, agencies, and numbers should be established and maintained at the dispatch center.
- No offensive action should be taken until the hazardous material has been properly identified. After the material is identified, an operations-level responder should perform only actions that do not involve contact with the material. The responder must use full PPE during any activity and must complete decontamination procedures prior to leaving any area where the hazardous material is present.
- When reporting the hazardous material incident, if possible, provide all of the following information:
  - The exact address and specific location of the leak or spill
  - Identification of indicators and markers of hazardous materials
  - All color or class information obtained from placards
  - Four-digit United Nations/North American Hazardous Materials Code numbers for the hazardous materials
  - Hazardous material identification obtained from shipping papers or MSDS and the potential quantities of hazardous materials involved
  - Description of the container, including its size, capacity, type, and shape
  - The amount of chemical that could leak and the amount that has already leaked
  - Exposures of people and the presence of special populations (children or elderly)
  - The environment in the immediate area
  - Current weather conditions, including wind direction and speed
  - A contact or callback telephone number and two-way radio frequency or channel
- The highest priority of an initial response plan is to consider the safety of the responding personnel. Responders are there to isolate, contain, and remedy the problem—not to become a part of it.
- Sometimes no action is the safest course of action.
- At the operational level, all response objectives should be primarily defensive (personnel do not come in contact with the hazardous material).
- Defensive objectives do not involve stopping the leak or release of a hazardous material and include the following:
  - Isolating the area affected by the leak or spill, and evacuating victims who could become exposed to the hazardous material if the leak or spill were to progress
  - Controlling where the spill or release is spreading
  - Containing the spill to a specific area
  - Diking and damming
  - Absorbing or adsorbing a hazardous material
  - Stopping the flow remotely
  - Diluting or diverting a material
  - Suppressing or dispersing a vapor
- Terrorists who want to injure responding personnel with a secondary device or attack will typically make the initial attack very dramatic to draw responders into very close proximity to the scene. As responders begin to treat victims, the secondary attack then takes place.
  - Signs of secondary devices may include trip wires, timers, and ordinary everyday containers (e.g., luggage, briefcases, boxes) found in very close proximity to the initial incident site.
  - The smell of chemicals or the sighting of chemical dispersion devices or other containers that may hold chemical, biological, or even radioactive agents could be an indicator that a secondary attack is possible.
- The ICS can be expanded to handle a hazardous materials incident. A special technical group may be developed under the Operations Section, known as the Hazardous Materials Branch.

Hot Terms

Backup entry team A dedicated team of fully qualified and equipped responders who are ready to enter the hot zone at a moment's notice to rescue any member of the hot zone entry team.

Decontamination team The team responsible for reducing and preventing the spread of contaminants from persons and equipment used at a hazardous materials incident. Members of this team establish the decontamination corridor and conduct all phases of decontamination.

Defensive objectives Actions that do not involve the actual stopping of the leak or release of a hazardous material, or contact of responders with the material; these include preventing further injury and controlling or containing the spread of the hazardous material.

Hazardous Materials Branch The function within an overall incident management system that deals with the mitigation and control of the hazardous materials/weapons of mass destruction portion of an incident. (NFPA 472)

Hazardous materials information research team A dedicated team of responders who serve as an information gathering and referral point for the incident commander as well as the hazardous materials officer.

Hazardous materials officer (NIMS: Hazardous Materials Branch Director/Group Supervisor.) The person who is responsible for directing and coordinating all operations involving hazardous materials/weapons of mass

Wrap-Up, continued

destruction as assigned by the incident commander. (NFPA 472)

Hazardous materials safety officer (NIMS: Assistant Safety Officer—Hazardous Material.) The person who works within the Incident Management System (specifically, the Hazardous Materials Branch/Group) to ensure that recognized hazardous materials/weapons of mass destruction (WMD) safe practices are followed at hazardous materials/WMD incidents. (NFPA 472)

Hot-zone entry team The team of fire fighters assigned to the entry into the designated hot zone.

FIRE FIGHTER *in action*

Your engine company is dispatched to a new-home construction site for the report of a pipeline struck by a backhoe. Additional information comes in while you are en route that states the damaged pipeline is a transmission line that supplies several gasoline tank farms in the area. Upon arrival at the scene, you see a stream of product coming from the ground near the backhoe. You conduct a quick survey and find the operator has escaped the backhoe and is safe, but the product is flowing to the street and into both the storm drains and a creek. After further investigation, you find a pipeline marker.

1. Based on the initial incident information, which immediate contacts would you consider making?
   - **A.** The emergency contact number identified on the pipeline marker
   - **B.** CHEMTREC
   - **C.** State Environmental Protection Agency/Department of Natural Resources
   - **D.** All of the above
2. Which of the following pieces of information is essential information that needs to be immediately determined?
   - **A.** The approximate amount of product that has been released
   - **B.** The exact location of the leak
   - **C.** The location of ignition sources
   - **D.** All of the above
3. How would you determine the boundaries of the hot, warm, and cold zones for this specific incident?
   - **A.** Use a detection device to determine the lower explosive limit of the area
   - **B.** Use on-scene personnel's sense of smell to determine where the gasoline odor can be determined
   - **C.** Use the protective action zones recommended by a bystander
   - **D.** Base the zone distance on the area where gasoline can actually be seen
4. What is the primary purpose of the hazardous materials safety officer?
   - **A.** Ensuring the safety of the general public
   - **B.** Ensuring the safety of the hazardous materials team
   - **C.** Ensuring the safety of all on-scene responders
   - **D.** All of the above
5. Although a view of the incident is extremely helpful, the primary goal in locating the incident command post should be to find an area uphill and upwind of the incident that provides a safe location for incident command operations.
   - **A.** True
   - **B.** False

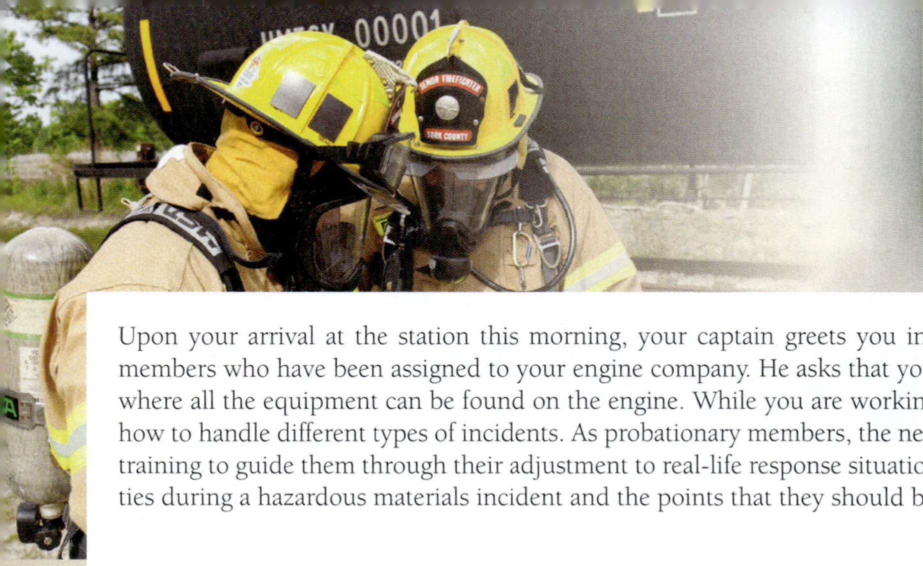

FIRE FIGHTER II *in action*

Upon your arrival at the station this morning, your captain greets you in the engine room with three new probationary members who have been assigned to your engine company. He asks that you familiarize them with company operations and where all the equipment can be found on the engine. While you are working with the new crew, a question comes up about how to handle different types of incidents. As probationary members, the new fire fighters are relying on your experience and training to guide them through their adjustment to real-life response situations. Your discussion focuses on their responsibilities during a hazardous materials incident and the points that they should be observing during the incident.

1. When arriving first on the scene of a hazardous materials incident, which initial response objectives would be established by your engine company?
2. Which on-site information is available to assist in the process of planning the initial response?
3. Which part of the Incident Command System (ICS) will play a role in the initial decision-making process, and where do these probationary fire fighters fit into the process?
4. Based on your knowledge of the ICS, what should you tell your probationary members about the best reporting practices that should be used?

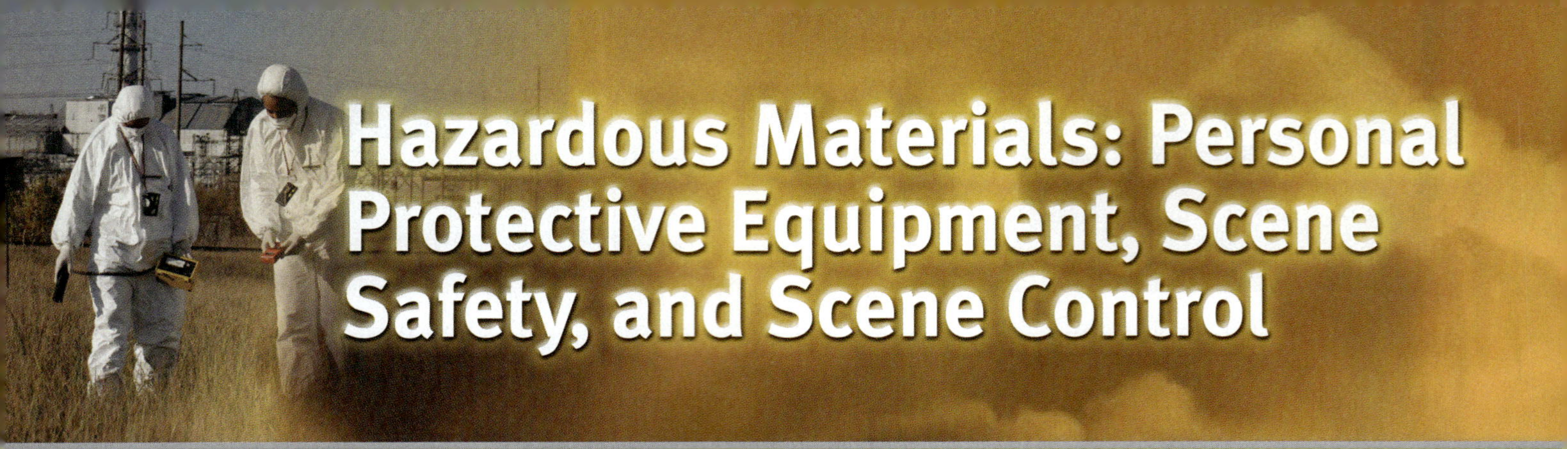

Hazardous Materials: Personal Protective Equipment, Scene Safety, and Scene Control

Fire Fighter I

Knowledge Objectives

After studying this chapter, you will be able to:

- Explain how a hazardous material's threshold limit value determines the level of protection required for responders. (NFPA 472, 5.3.3, 6.6.1.1.2, 6.6.3.2, p 936)
- List the three categories of threshold limit values. (NFPA 472, 5.3.3, 6.6.1.1.2, 6.6.3.2, p 936)
- List and describe the regulatory measures set by the Occupational Safety and Health Administration (OSHA). (NFPA 472, 5.3.3, 6.6.1.1.2, 6.6.3.2, p 936)
- List the subcategories of Immediately Dangerous to Life and Health atmospheres. (NFPA 472, 5.3.3, 6.6.1.1.2, 6.6.3.2, p 936)
- List and define the three basic types of atmospheres at a hazardous materials incident. (NFPA 472, 5.3.3, 6.6.1.1.2, 6.6.3.2, p 937)
- List the categories of personal protective equipment. (NFPA 472, 5.3.3, 6.6.1.1.2, 6.6.3.2, p 937)
- Describe the purpose and components of street clothing and work uniforms. (NFPA 472, 5.3.3, 6.6.1.1.2, 6.6.3.2, p 937)
- Describe the purpose and components of structural firefighting protective clothing. (NFPA 472, 5.3.3, 6.6.1.1.2, 6.6.3.2, p 938)
- Describe the purpose and components of high-temperature protective equipment. (NFPA 472, 5.3.3, 6.6.1.1.2, 6.6.3.2, p 938)
- Describe the purpose and components of chemical-protective clothing and equipment. (NFPA 472, 5.3.3, 6.6.1.1.2, 6.6.3.2, p 938–939)
- Describe the purpose and components of liquid splash-protective clothing. (NFPA 472, 5.3.3, 6.6.1.1.2, 6.6.3.2, p 939)
- Describe the purpose and components of vapor-protective clothing. (NFPA 472, 5.3.3, 6.6.1.1.2, 6.6.3.2, p 939)
- Discuss respiratory protection in a hazardous materials incident. (NFPA 472, 5.3.3, 6.6.1.1.2, 6.6.3.2, p 939–940)
- Describe the levels of hazardous materials personal protective equipment. (NFPA 472, 5.3.3, 6.6.1.1.2, 6.6.3.2, p 941–949)
- List the ratings of chemical-protective clothing. (NFPA 472, 5.3.3, 6.6.1.1.2, 6.6.3.2, p 941–949)
- Describe the purpose and components of Level A protection. (NFPA 472, 5.3.3, 6.6.1.1.2, 6.6.3.2, p 941)
- Describe the purpose and components of Level B protection. (NFPA 472, 5.3.3, 6.6.1.1.2, 6.6.3.2, p 941–942)
- Describe the purpose and components of Level C protection. (NFPA 472, 5.3.3, 6.6.1.1.2, 6.6.3.2, p 945–947)
- Describe the purpose and components of Level D protection. (NFPA 472, 5.3.3, 6.6.1.1.2, 6.6.3.2, p 947–949)
- Identify the potential skin-contact hazards encountered at hazardous materials incidents. (NFPA 472, 5.3.3, 6.6.1.1.2, 6.6.3.2, p 950)
- Describe the safety precautions to be observed, including those for heat and cold stress, when approaching and working at hazardous materials incidents. (NFPA 472, 5.3.3, 6.6.1.1.2, 6.6.3.2, p 950–953)
- Describe the signs and symptoms of heat cramps. (NFPA 472, 5.4.4, 6.6.1.1.2, p 950)
- Describe the signs and symptoms of heat exhaustion. (NFPA 472, 5.4.4, 6.6.1.1.2, p 950)
- Describe how to prevent cold injuries. (NFPA 472, 5.4.4, 6.6.1.1.2, p 952–953)
- Describe the physical capabilities required and limitations of personnel working in PPE. (NFPA 472, 5.4.4, 6.6.1.1.2, p 953)
- List the three control zones. (NFPA 472, 5.4.1, 6.6.1.1.2, p 954)
- Define hot zone and describe the tasks performed in the zone. (NFPA 472, 5.4.1, 6.6.1.1.2, p 954)
- Define warm zone and describe the tasks performed in the zone. (NFPA 472, 5.4.1, 6.6.1.1.2, p 954)
- Define cold zone and describe the tasks performed in the zone. (NFPA 472, 5.4.1, 6.6.1.1.2, p 955)
- Describe the importance of the buddy system and backup personnel. (NFPA 472, 5.4.4, 6.6.1.1.2, p 955)

Skills Objectives

After studying this chapter, you will be able to perform the following skills:

- Don a Level B encapsulated chemical-protective clothing ensemble. (NFPA 472, 5.3.3, 6.6.1.1.2, 6.6.3.2, p 942–944)
- Don a Level B nonencapsulated chemical-protective clothing ensemble. (NFPA 472, 5.3.3, 6.6.1.1.2, 6.6.3.2, p 942, 944–945)
- Doff a Level B encapsulated chemical-protective clothing ensemble. (NFPA 472, 5.3.3, 6.6.1.1.2, 6.6.3.2, p 945–946)
- Doff a Level B nonencapsulated chemical-protective clothing ensemble. (NFPA 472, 5.3.3, 6.6.1.1.2, 6.6.3.2, p 945, 947)
- Don a Level C chemical-protective clothing ensemble. (NFPA 472, 5.3.3, 6.6.1.1.2, 6.6.3.2, p 946–947, 949)
- Doff a Level C chemical-protective clothing ensemble. (NFPA 472, 5.3.3, 6.6.1.1.2, 6.6.3.2, p 947)
- Don a Level D chemical-protective clothing ensemble. (NFPA 472, 5.3.3, 6.6.1.1.2, 6.6.3.2, p 948–949)
- Doff a Level D chemical-protective clothing ensemble. (NFPA 472, 5.3.3, 6.6.1.1.2, 6.6.3.2, p 949)

CHAPTER 32

Fire Fighter II FFII

Knowledge Objectives

There are no knowledge objectives for Fire Fighter II candidates. NFPA 1001 contains no Fire Fighter II Job Performance Requirements for this chapter.

Skills Objectives

There are no skill objectives for Fire Fighter II candidates. NFPA 1001 contains no Fire Fighter II Job Performance Requirements for this chapter.

Additional NFPA Standards

- NFPA 1951, *Standard on Protective Ensembles for Technical Rescue Incidents*
- NFPA 1981, *Standard on Open-Circuit Self-Contained Breathing Apparatus (SCBA) for Emergency Services*
- NFPA 1991, *Standard on Vapor-Protective Ensembles for Hazardous Materials Emergencies*
- NFPA 1992, *Standard on Liquid Splash-Protective Ensembles and Clothing for Hazardous Materials Emergencies*
- NFPA 1994, *Standard on Protective Ensembles for First Responders to CBRN Terrorism Incidents*
- NFPA 1999, *Standard on Protective Clothing for Emergency Medical Operations*
- NIOSH Standard for Chemical, Biological, Radiological, and Nuclear (CBRN) *Open-Circuit Self-Contained Breathing Apparatus*
- NIOSH Standard for Chemical, Biological, Radiological, and Nuclear (CBRN) *Full Facepiece Air Purifying Respirator (APR)*
- NIOSH Standard for Chemical, Biological, Radiological, and Nuclear (CBRN) *Air-Purifying Escape Respirator and CBRN Self-Contained Escape Respirator*

You Are the Fire Fighter

It is the middle of July and the temperatures have soared to 95°F (35°C) and a unit train of denatured ethanol, which is a flammable liquid used as a motor fuel additive, is traveling through the center of your city when one of the tank cars is struck at a crossing by a car and has damage that has now caused a leak. Your engine company is dispatched and upon arrival on the scene you are directed by your officer to grab shovels from the apparatus compartment to build a dam around the storm drain to prevent the product from entering the storm sewer system. Prior to defensive operations, some key questions concerning personal protective equipment (PPE) need to be answered.

1. Which properties of the spilled chemical and the hazards that might be encountered will be important in helping you determine the correct type of PPE?
2. Which type of PPE should be used, based on the hazards of this material?
3. What concerns, based on the hazards of the material and the weather, might be raised by the prospect of personnel wearing the appropriate PPE and working in and around the spill area?

Introduction

The proper use of PPE and proper scene control is important at all emergencies. At a hazardous materials incident, however, PPE, scene control, site management, and personnel accountability are critical issues that have a direct bearing on the life safety of all responders charged with handling such incidents. The safe handling of the incident is often determined in the first 5 to 15 minutes, based on the actions of the fire fighters who are the initial responders. Mistakes made by those responders can lead to disaster.

Levels of Damage Caused by Chemicals to Humans

The damage that a hazardous material inflicts on a human depends on the material's threshold limit value (TLV). The TLV, which is the point at which the material begins to affect a person, can be defined in several ways. The various definitions indicate the levels of protection that are required.

The threshold limit value/time-weighted average (TLV/TWA) is the maximum concentration of material to which a worker could be exposed for 8 hours a day, 40 hours a week, with no ill effects. The lower the TLV/TWA, the more toxic the substance.

The threshold limit value/short-term exposure limit (TLV/STEL) is the maximum concentration of material to which a person can be exposed for 15-minute intervals, up to four times a day, without experiencing irritation or chronic or irreversible tissue damage. There should be a minimum one-hour rest period between any exposures to this concentration of the material. The lower the TLV/STEL, the more toxic the substance.

The threshold limit value/ceiling (TLV/C) is the maximum concentration of material to which a worker should not be exposed, even for an instant. Again, the lower the TLV/C, the more toxic the substance.

The threshold limit value/skin indicates the concentration at which direct or airborne contact with a material could result in possible and significant exposure from absorption through the skin, mucous membranes, and eyes. This designation is intended to suggest that appropriate measures be taken to minimize skin absorption so that the TLV/TWA is not exceeded.

The Occupational Safety and Health Administration (OSHA) has set the following by the regulatory measures that are comparable to the TLV/TWA:

- The permissible exposure limit (PEL) measures the maximum, time-weighted concentration of material to which 95 percent of healthy adults can be exposed without suffering any adverse effects over a 40-hour workweek. PELs are set by OSHA for certain chemicals and are enforceable as federal law as set exposure limits for those chemicals.
- Immediately Dangerous to Life and Health (IDLH) means that an atmospheric concentration of a toxic, corrosive, or asphyxiant substance poses an immediate threat to life or could cause irreversible or delayed adverse health effects. Three general IDLH atmospheres are distinguished: toxic, flammable, and oxygen-deficient.

The last term is the recommended exposure level (REL). The National Institute for Occupational Safety and Health (NIOSH) sets RELs for other chemicals; these limits do not have the force of law, however.

Individuals who are exposed to atmospheric concentrations below the IDLH value could (in theory) escape from the atmosphere without experiencing irreversible damage to their health, even if their respiratory protection fails. Individuals who might be exposed to atmospheric concentrations at or above the IDLH value must use positive-pressure self-contained breathing apparatus (SCBA) or equivalent respiratory protection to guard against harm from the substance.

At most hazardous materials emergencies, fire fighters are not able to measure concentrations of specific chemicals.

Measuring concentrations of chemicals requires specific monitoring instruments and trained response personnel to interpret the results.

Exposure guidelines are intended to minimize the possibility that either emergency responders or the public will be exposed to hazardous materials or atmospheres that will lead to harm. Once these exposure guidelines are understood, they can be applied at the scene of a hazardous materials emergency. For example, exposure guidelines can be used to define the three basic atmospheres at a hazardous materials emergency:

- Safe atmosphere: No harmful hazardous materials effects exist, so personnel can handle routine emergencies without donning specialized PPE.
- Unsafe atmosphere: A hazardous material that is no longer contained has created an unsafe condition or atmosphere. A person who is exposed to the material for long enough will probably experience some form of acute or chronic injury.
- Dangerous atmosphere: Serious, irreversible injury or death can occur in the environment.

All exposure guidelines share a common goal: to ensure the safety and health of people exposed to a hazardous material. As long as the exposure remains below these values, it is considered safe to the average healthy adult, based on current information. As additional research and studies on the toxic effects of various chemicals are completed, these values might be adjusted.

Personal Protective Equipment

PPE is the clothing and protection provided to shield or insulate a person from chemical, physical, and thermal hazards. It is essential that a fire fighter's PPE meet both National Fire Protection Association (NFPA) standards and OSHA regulations and that it be properly maintained and used.

Selecting the appropriate PPE for a hazardous materials incident is an important, even potentially life-saving decision. Without the proper protective clothing and respiratory protection, fire fighters might be injured. Adequate PPE should protect the fire fighter's respiratory system, as well as the skin, eyes, face, hands, feet, head, body, and hearing. Each member of the responding team must know the shielding capabilities and limitations of his or her PPE. When product-specific chemical suits are unavailable, fire fighters must know how much and what kind of protection their regular firefighting clothing provides. Remember—protective clothing is your last line of defense, not your first.

Which type of PPE is appropriate for the incident depends on the activities that a fire fighter is expected to perform—specifically, the tasks that the fire fighter is expected to complete. Standard structural firefighting gear might be suitable for support operations at many hazardous materials incidents, but personnel who are involved in decontamination activities can require chemical-protective clothing because of the possibility of cross-contamination issues. The incident commander (IC) has the ultimate authority and responsibility to approve the level of PPE required for a given activity.

Fire fighters should operate only at the incident level that matches their knowledge, training, and equipment. If conditions indicate a need for a higher response level, additional personnel with the appropriate training and PPE should be summoned.

Fire Fighter Safety Tips

TRACEM is a common acronym used to sum up a collection of potential hazards that an emergency responder might face.

- Thermal
- Radiologic
- Asphyxiating
- Chemical
- Etiologic/biologic
- Mechanical

Hazardous Materials–Specific Personal Protective Equipment

Different levels of PPE are required at a hazardous materials incident. This section reviews the protective qualities of various outfits, from those offering the least protection to those providing the greatest protection.

Street Clothing and Work Uniforms

Street clothing and work uniforms offer the least amount of protection in a hazardous materials emergency. Normal clothing can prevent a noncaustic powder from coming into direct contact with the skin, but it offers no more protection. A one-piece flame-resistant or lightly flame-resistant coverall can enhance protection slightly. Such PPE is often used in industrial applications such as oil refineries as a general work uniform for this very reason FIGURE 32-1.

FIGURE 32-1 Nomex jumpsuit.

Structural Firefighting Protective Clothing

The next level of protection is provided by structural firefighting protective equipment, also known as turnouts or bunker gear. Such an outfit includes a helmet, a bunker coat, bunker pants, boots, gloves, a hood, and a personal alert safety system (PASS) device. Standard firefighting turnout gear offers no chemical protection, although it does have some abrasion resistance and prevents direct skin contact. However, it easily absorbs product, breaks down when exposed to chemicals, and does not provide adequate protection from the harmful gases, vapors, liquids, or dusts that are encountered during hazardous materials incidents FIGURE 32-2.

FIGURE 32-2 Standard turnout gear.

FIGURE 32-3 High temperature–protective equipment protects the wearer from high temperatures during a short exposure.

High Temperature–Protective Equipment

High temperature–protective equipment protects the wearer during short-term exposures to high temperatures FIGURE 32-3. It allows the properly trained fire fighter to work in extreme fire conditions (i.e., it provides protection against high temperatures) and is not designed to protect the fire fighter from hazardous materials.

Depending on the radioactive isotope, this same gear and SCBA can be a good choice for alpha emitters and perhaps beta emitters. Remember that this equipment is not intended to be radiation protection—it is designed to shield the wearer against radioactive particles, perhaps from a radiological dispersal device.

Chemical-Protective Clothing and Equipment

Chemical-protective clothing is designed to prevent chemicals from coming in contact with the body and can offer varying degrees of resistance. It comprises a material or combination of materials used in an item of clothing to isolate parts of the wearer's body from contact with a hazardous chemical. Other performance requirements that must be considered when selecting chemical-protective materials include chemical resistance, flexibility, abrasion, temperature resistance, shelf life, and sizing criteria.

Chemical resistance is the ability to resist chemical attack. Time, temperature, stress, and reagent are all factors that affect the chemical resistance of materials. Chemical-resistant materials are specifically designed to inhibit or resist the passage of chemicals into and through the material by the processes of penetration, permeation, or degradation:

- Penetration is the flow or movement of a hazardous chemical through closures (e.g., zippers), seams, porous materials, pinholes, or other imperfections in the material. Although liquids are most likely to penetrate material, solids (e.g., asbestos) can also penetrate protective clothing materials.
- Permeation is the process by which a hazardous chemical moves through a given material on the molecular level. It differs from penetration in that permeation occurs through the material itself rather than through openings in the material.
- Degradation is the physical destruction or decomposition of a clothing material owing to chemical exposure, general use, or ambient conditions (e.g., storage in sunlight). It might be evidenced by visible signs such as charring, shrinking, swelling, color changes, or dissolving. Materials can also be tested for weight changes, loss of fabric tensile strength, and other properties to measure degradation.

Chemical-protective clothing can be constructed as a single- or multipiece garment. A single-piece garment completely encloses the wearer and is known as an encapsulated suit. Encapsulated protective clothing offers full body protection from hostile environments and requires the use of supplied-air respiratory protection devices, such as SCBA. Some materials are so caustic or corrosive that they could destroy SCBA after only a short exposure, however. A multipiece garment works with the wearer's respiratory protection, an attached or detachable hood, gloves, and boots to protect against a specific hazard.

A variety of materials is used in both encapsulated and nonencapsulated protective clothing. The most common include butyl rubber, Tyvek, Saranex, polyvinyl chloride, Viton, as well as many other materials that are protected by the manufacturer, either singly or in multiple layers of several materials. Special chemical-protective clothing is adequate for some chemicals yet useless for other chemicals; no single material provides satisfactory protection from all chemicals. Protective clothing materials must be compatible with the chemical substances involved in the incident, and use must be consistent with the manufacturer's instructions. The manufacturer's guidelines and recommendations should be consulted for material compatibility information.

Personnel trained to the operations level should not be operating in encapsulated suits. Operational-level personnel can wear some nonencapsulated chemical-protective clothing while providing support for those entering the hazardous materials area.

Liquid splash–protective clothing consists of several pieces of clothing and equipment that protects the skin and eyes from chemical splashes FIGURE 32-4. It does not provide total body protection from gases or vapors. For this reason, liquid splash–protective clothing should not be used for incidents involving liquids that emit vapors known to affect or be absorbed through the skin.

FIGURE 32-4 Liquid splash–protective clothing must be worn when there is the danger of chemical splashes.

Vapor-protective clothing must be used when hazardous vapors are present FIGURE 32-5. This type of clothing traps the fire fighter's body heat, so fire fighters and support personnel must be aware of the possibility of heat-related emergencies. SCBA or air-line hose units must also be used for respiratory protection.

FIGURE 32-5 Vapor-protective clothing retains body heat and increases the possibility of heat-related emergencies.

Fire fighters can wear liquid splash–protective clothing over or under structural firefighting clothing in some situations. This multiple-PPE approach provides limited chemical splash and thermal protection. Fire fighters trained to the operational level can wear liquid splash–protective clothing when they are assigned to enter the initial site, protect decontamination personnel, or construct isolation barriers such as dikes, diversions, retention areas, or dams.

Respiratory Protection

Positive-Pressure Self-Contained Breathing Apparatus

Respiratory protection, which is usually provided by a SCBA, is an important—and separate—level of protection at a hazardous materials incident. Using positive-pressure SCBA prevents both inhalation and ingestion (two primary routes of exposure) and should be mandatory for fire service personnel. Although it is an excellent barrier in many hazardous environments, positive-pressure SCBA is not designed for, and has never been tested or approved for use in, atmospheres that might contain unknown hazards. Fire fighters must be aware of the limitations of SCBA in the hazardous atmosphere.

New standards are now in effect for the testing of SCBA in environments that are contaminated with certain concentrations of chemical, biological, radiological, and nuclear (CBRN) materials. Check with your fire department to see whether it has adopted any of these specialized SCBA units for day-to-day or special unit use.

Other Respiratory Protection

Supplied-air respirators (SARs) have an external air source such as a compressor or storage cylinder. A hose connects the user to the air source and provides air to the face piece. SARs are useful during extended operations such as decontamination, clean-up, and remedial work. These devices might be less bulky and weigh less than SCBA, but the length of the air hose can limit the fire fighter's movement FIGURE 32-6.

Air-purifying respirators (APRs) and powered air-purifying respirators (PAPRs) are filtering devices that remove particulates and contaminants from the air. The difference between the two devices is that the PAPR has a built-in fan that assists in drawing air into a hose to the face piece. This action provides the wearer with a positive flow of air while it is operating. They should be worn only in atmospheres where the type and quantity of the contaminants are known and where sufficient oxygen for breathing is available. APRs/PAPRs might be appropriate for operations involving volatile solids and for remedial clean-up and recovery operations where the type and concentration of contaminants are verifiable FIGURE 32-7.

FIGURE 32-6 A supplied-air respirator is less bulky than an SCBA but is limited by the length and structural integrity of the air hose.

APRs/PAPRs use filtration devices to remove particulate matter, gases, or vapors from the atmosphere. PAPRs include a fan motor and pack where the filters mount and a battery pack for operation as well as a hose from the fan to a full face piece. APRs range from full-face-piece, dual-cartridge masks with eye protection to half-mask, face piece–mounted cartridges with no eye protection. APRs/PAPRs do not have a separate source of air but rather filter and purify ambient air before it is inhaled. The models used in environments containing hazardous gases or vapors are commonly equipped with an absorbent material that soaks up or reacts with the gas. Consequently, cartridge selection is based on the expected contaminants. Particle-removing respirators use a mechanical filter to separate the contaminants from the air. Both types of devices require that the ambient atmosphere contain a minimum of 19.5 percent oxygen.

APRs are easy to wear and offer no hose line restriction, but they do have some drawbacks. Because filtering cartridges are specific to expected contaminants, they will be ineffective if the contaminant changes suddenly, endangering the lives of fire fighters. The air must be continually monitored for both the known substance and the ambient oxygen level throughout the incident. For these reasons, APRs should not be employed at hazardous materials incidents until qualified personnel have tested the ambient atmosphere and determined that such devices can be used safely.

A. B.

FIGURE 32-7 Air-purifying respirators and powered air-purifying respirators can be used only by trained personnel. **A.** Air-purifying respirator. **B.** Powered air-purifying respirator.

Chemical-Protective Clothing Ratings

Chemical-protective clothing is rated for its effectiveness in several different ways. Testing measures whether chemicals will permeate the suit and, if so, how quickly and to what degree. This information can be obtained from the manufacturer of the clothing. The U.S. Environmental Protection Agency defines levels of protection using an alphabetic system. Also consult NFPA 1992, *Standard on Liquid Splash–Protective Ensembles and Clothing for Hazardous Materials Emergencies*, and NFPA 1994, *Standard on Protective Ensembles for First Responders to CBRN Terrorism Incidents*.

Level A

Level A protection consists of a heavy, encapsulating suit that envelops both the wearer and his or her SCBA totally. It should be used when the hazardous material identified requires the highest level of protection for skin, eyes, and respiration. Level A protection is effective against vapors, gases, mists, and even dusts FIGURE 32-8. To warrant this rating, chemical-protective clothing must meet the requirements for vapor-gas protection outlined in NFPA 1991, *Standard on Vapor-Protective Ensembles for Hazardous Materials Emergencies*. Level A protection also requires open-circuit, positive-pressure SCBA or an SAR for respiratory protection.

FIGURE 32-8 Level A protection envelops the wearer in a totally encapsulating suit.

Recommended PPE for Level A protection includes the following components:

- SCBA or SAR
- Fully encapsulating chemical-resistant suit
- Inner chemical-resistant gloves
- Chemical-resistant safety boots/shoes
- Two-way radio
- Hard hat

Optional PPE for Level A protection includes the following components:

- Coveralls
- Long cotton underwear
- Disposable gloves and boot covers

Fire Fighter Safety Tips

Mechanical hazards are best addressed by understanding the NFPA performance requirements for chemical-protective ensembles. Typical performance requirements include tests for durability, barrier integrity after flex and abrasion challenges, cold temperature flex, and flammability. Essentially, each part of the suit must pass a particular set of challenges prior to receiving certification based on the NFPA testing standards. Users of any garment that meets the performance requirements set forth by the NFPA can rest assured that the garment will withstand reasonable insults from most mechanical-type hazards encountered on the scene. This does not mean the suit is "bullet proof" and cannot fail. It is up to the user to be aware of the hazards and avoid situations that might cause the garment to fail.

Level B

Level B protection consists of chemical-protective clothing, boots, gloves, and SCBA FIGURE 32-9. This type of PPE should be used when the type and atmospheric concentration of identified substances require a high level of respiratory protection but less skin protection. The type of gloves and boots worn depends on the identified chemical; wrists and ankles must be properly sealed to prevent splashed liquids from contacting skin.

Recommended PPE for Level B protection includes the following components:

- SCBA or SAR
- Chemical-resistant clothing
- Inner and outer chemical-resistant gloves
- Chemical-resistant safety boots/shoes
- Hard hat
- Two-way radio

Optional PPE for Level B protection includes the following components:

- Coveralls
- Long cotton underwear
- Disposable gloves and boot covers

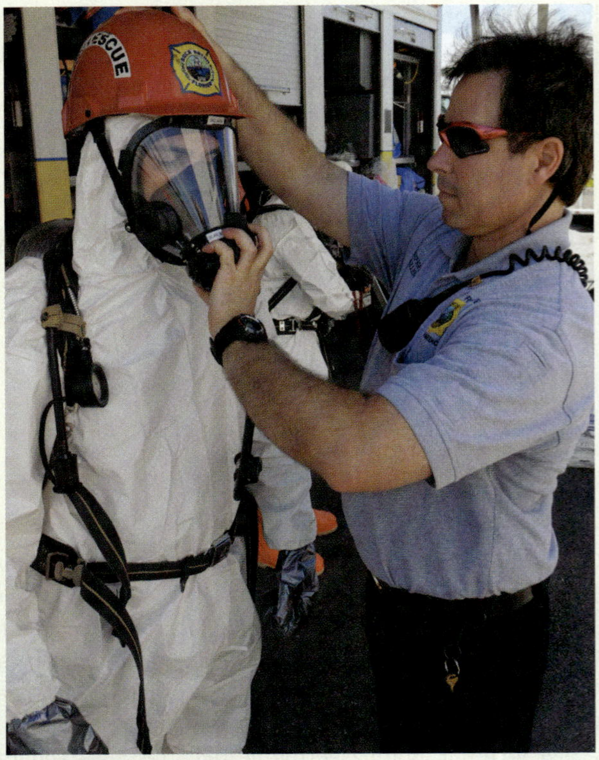

FIGURE 32-9 Level B protection provides a high level of respiratory protection but less skin protection.

Fire Fighter Safety Tips

Chemicals and/or biological agents can challenge a responder with a mixed bag of hazards resulting from the unique properties of each and every substance. The following list of NFPA and NIOSH documents offers an overview of the current testing and certification standards affecting the current PPE on the market.

- NFPA 1951, *Standard on Protective Ensembles for Technical Rescue Incidents*
- NFPA 1981, *Standard on Open-Circuit Self-Contained Breathing Apparatus (SCBA) for Emergency Services*
- NFPA 1991, *Standard on Vapor-Protective Ensembles for Hazardous Materials Emergencies*
- NFPA 1994, *Standard on Protective Ensembles for First Responders to CBRN Terrorism Incidents*
- NFPA 1999, *Standard on Protective Clothing for Emergency Medical Operations*
- NIOSH Chemical, Biological, Radiological, and Nuclear (CBRN) *Standard for Open-Circuit Self-Contained Breathing Apparatus*
- NIOSH Standard for Chemical, Biological, Radiological, and Nuclear (CBRN) *Full Facepiece Air Purifying Respirator (APR)*
- NIOSH Standard for Chemical, Biological, Radiological, and Nuclear (CBRN) *Air-Purifying Escape Respirator and CBRN Self-Contained Escape Respirator*

To don a Level B encapsulated chemical-protective clothing ensemble, follow the steps in in SKILL DRILL 32-1:

Fire Fighter Safety Tips

CBRN-certified SCBA refers to SCBA and APR units that are safe to use during a chemical, biological, radiological, or nuclear incident. CBRN-certified SCBA and APRs have gone through rigorous testing to ensure their integrity during such incidents. Be sure to understand the specific capabilities and limitations of any such devices your fire department issues to you for both day-to-day and special uses.

1. Conduct a pre-entry briefing, medical monitoring, and equipment inspection.
2. While seated, pull on the suit to waist level; pull on the chemical boots over the top of the chemical suit. Pull the suit boot covers over the tops of the boots. (STEP 1)
3. Stand up and don SCBA and the SCBA face piece, but do not connect the regulator to the face piece. (STEP 2)
4. Place the helmet on the head. (STEP 3)
5. Don the inner gloves. (STEP 4)
6. With assistance, complete donning the suit by placing both arms in the suit, pulling the expanded back piece over the SCBA, and placing the chemical suit over the head.
7. Instruct the assistant to connect the regulator to the SCBA face piece and ensure that air flow is working. Note when you begin to use the SCBA. (STEP 5)
8. Instruct the assistant to close the chemical suit by closing the zipper and sealing the splash flap. (STEP 6)
9. Review hand signals and indicate that you are okay. (STEP 7)

To doff a Level B encapsulated chemical-protective clothing ensemble, follow the steps in SKILL DRILL 32-2:

1. After completing decontamination, proceed to the clean area for suit doffing.
2. Pull the hands and arms out of the suit gloves and sleeves, and cross the arms in front inside the suit. (STEP 1)
3. Instruct the assistant to open the chemical splash flap and suit zipper. (STEP 2)
4. Instruct the assistant to begin at the head and roll the suit down and away until the suit is below waist level. (STEP 3)
5. Sit and instruct the assistant to complete rolling down the suit and remove the outer boots and suit. Rotate on the bench to the direction that will allow you to place feet on dry, clean area. (STEP 4)
6. Stand and doff the SCBA using the quick release method. The face piece should be kept in place while the SCBA frame is placed on the ground. (STEP 5)
7. Take a deep breath, doff the SCBA mask, and walk away from the clean area.
8. Go to the rehabilitation area for medical monitoring, rehydration, and personal decontamination shower. (STEP 6)

SKILL DRILL 32-1 Donning a Level B Encapsulated Chemical-Protective Clothing Ensemble

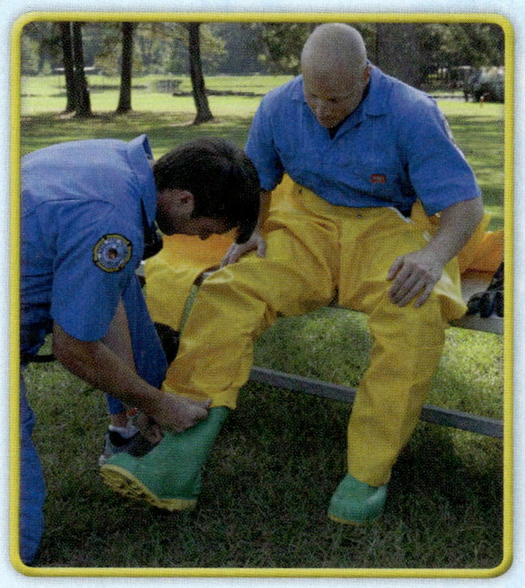

1 While seated, pull on the suit to waist level; put on the chemical boots over the chemical suit. Pull the suit boot covers over the tops of the boots.

2 Stand up and don SCBA and the SCBA face piece, but do not connect the regulator to the face piece.

3 Place the helmet on the head, if required.

4 Don the inner gloves.

5 With assistance, complete donning the suit. Instruct the assistant to connect the regulator to the SCBA face piece and ensure air flow.

(Continued)

SKILL DRILL 32-1 Donning a Level B Encapsulated Chemical-Protective Clothing Ensemble (*Continued*)

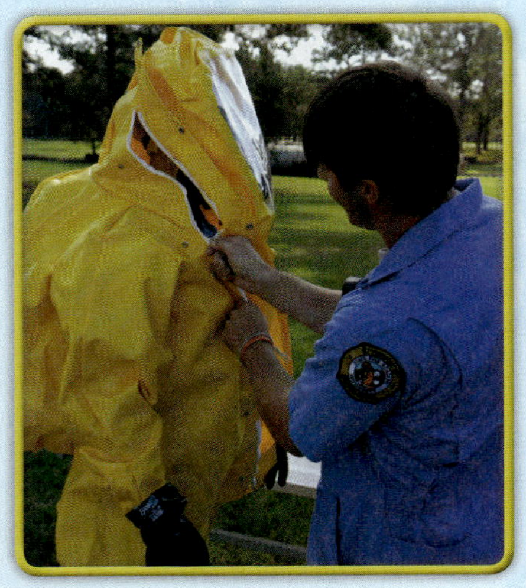

6 Instruct the assistant to close the chemical suit by closing the zipper and sealing the splash flap.

7 Review hand signals and indicate that you are ready to operate.

SKILL DRILL 32-2 Doffing a Level B Encapsulated Chemical-Protective Clothing Ensemble

1 After completing decontamination, proceed to the clean area. Remove the hands and arms from the suit gloves and sleeves, and cross the arms in front inside the suit.

2 Instruct the assistant to open the chemical splash flap and open suit zipper.

3 Instruct the assistant to begin at the head and roll the suit down and away from you until the suit is below waist level.

(*Continued*)

SKILL DRILL 32-2 Doffing a Level B Encapsulated Chemical-Protective Clothing Ensemble (*Continued*)

4 Sit and instruct the assistant to complete rolling down the suit and remove the outer boots and suit. Rotate on the bench to the direction that will allow you to place feet on dry, clean area.

5 Stand and doff the SCBA using the quick-release method. Keep the face piece in place while the SCBA frame is placed on the ground.

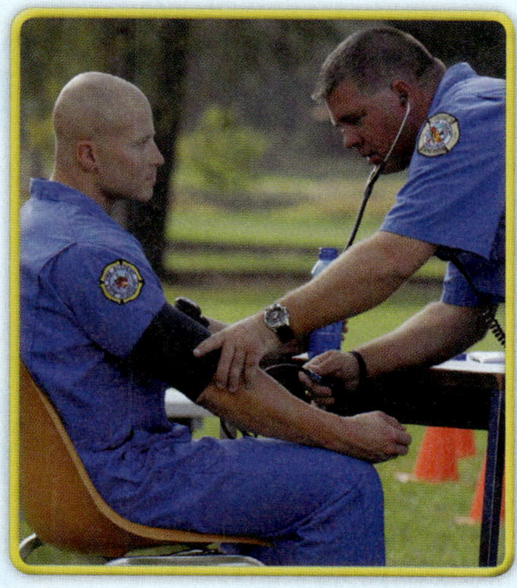

6 Take a deep breath, doff the SCBA mask, and walk away from the clean area. Go to rehabilitation area for medical monitoring, rehydration, and personal decontamination shower.

To don a Level B nonencapsulated chemical-protective clothing ensemble, follow the steps in SKILL DRILL 32-3:

1. Conduct a pre-entry briefing, medical monitoring, and equipment inspection.
2. Sit down, pull on the suit to waist level; pull on the chemical boots over the top of the chemical suit. Pull the suit boot covers over the tops of the boots or tape if needed. (STEP 1)
3. Don the inner gloves. (STEP 2)
4. With assistance, complete donning the suit by placing both arms in the suit and pulling the suit over the shoulders.
5. Instruct the assistant to close the chemical suit by closing the zipper and sealing the splash flap. (STEP 3)
6. Stand up and don SCBA and the SCBA face piece, but do not connect the regulator to the face piece.
7. With assistance, pull the hood over the head and SCBA face piece. (STEP 4)
8. Place the helmet on the head.
9. Pull the gloves over and/or under the sleeves, depending on the situation.
10. Instruct the assistant to connect the regulator to the SCBA face piece and ensure that air flow is working.
11. Review hand signals and indicate that you are okay. (STEP 5)

To doff a Level B nonencapsulated chemical-protective clothing ensemble, follow the steps in SKILL DRILL 32-4:

1. After completing decontamination, proceed to the clean area for suit doffing.
2. Stand and doff the SCBA. Keep the face piece in place while the SCBA frame is placed on the ground. (STEP 1)
3. Instruct the assistant to open the chemical splash flap and suit zipper. (STEP 2)
4. Remove your hands and arms from the suit gloves (except the inner gloves) and sleeves, and cross your arms in front inside the suit.
5. Instruct the assistant to begin at the head and roll the suit down and away until the suit is below waist level. (STEP 3)
6. Sit down and instruct the assistant to complete rolling down the suit and remove the outer boots and suit. Rotate on the bench to the direction that will allow you to place feet on dry, clean area. The assistant helps remove the inner gloves. (STEP 4)
7. Remove the SCBA face piece.
8. Go to the rehabilitation area for medical monitoring, rehydration, and personal decontamination shower. (STEP 5)

Level C

Level C protection consists of standard work clothing plus chemical-protective clothing, chemical-resistant gloves, and a form of respiratory protection. Level C protection is appropriate when the type of airborne substance is known, its

SKILL DRILL 32-3 Donning a Level B Nonencapsulated Chemical-Protective Clothing Ensemble

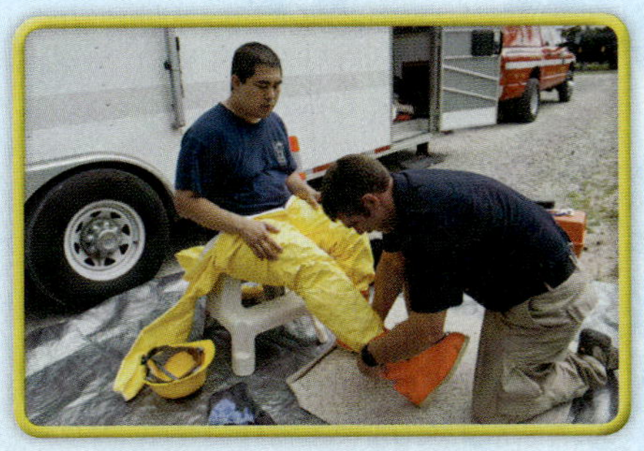

1. While seated, pull on the suit to waist level; pull on the chemical boots over the top of the chemical suit. Pull the suit boot covers over the tops of the boots or tape if needed.

2. Don the inner gloves.

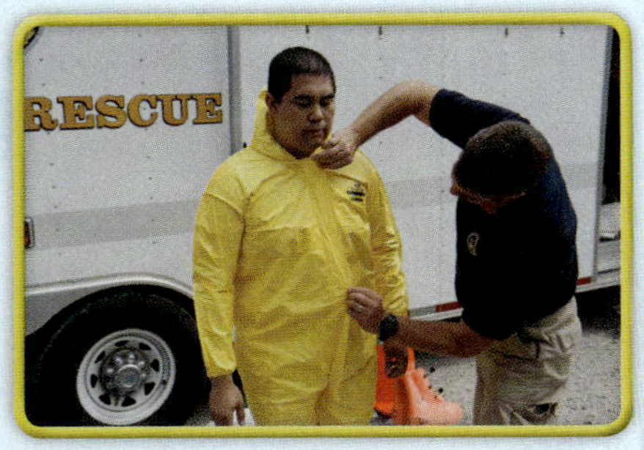

3. With assistance, complete donning the suit by placing both arms in suit and pulling suit over shoulders. Instruct the assistant to close chemical suit by closing zipper and sealing splash flap.

4. Stand up and don SCBA and SCBA face piece but do not connect regulator to the face piece. With assistance, pull hood over head and SCBA face piece.

5. Place helmet on head. Instruct the assistant to connect the regulator to the SCBA face piece and ensure you have air flow.

concentration is measured, the criteria for using APRs are met, and skin and eye exposure is unlikely FIGURE 32-10.

Level C respiratory protection can be provided with a filter or cartridge mask or an APR. The APR could be a half-face or a full-face mask that includes eye protection. Head protection and goggles are also required.

Recommended PPE for Level C protection includes the following components:

- Full-face APR
- Chemical-resistant clothing
- Inner and outer chemical-resistant gloves
- Chemical-resistant safety boots/shoes
- Two-way radio
- Hard hat

Optional PPE for Level C protection includes the following components:

- Coveralls
- Long cotton underwear
- Disposable gloves and boot covers

To don a Level C chemical-protective clothing ensemble, follow the steps in SKILL DRILL 32-5:

1. Conduct a pre-entry briefing, medical monitoring, and equipment inspection.
2. While seated, pull on the suit to waist level; pull on the chemical boots over the top of the chemical suit. Pull the suit boot covers over the tops of the boots. (STEP 1)
3. With assistance, complete donning the suit by placing both arms in the suit and pulling the suit over the shoulders.
4. The assistant closes the chemical suit by closing the zipper and sealing the splash flap. (STEP 2)
5. Don the inner gloves. (STEP 3)
6. Stand up and don the APR/powered air-purifying respirator (PAPR) and APR/PAPR face piece.
7. With assistance, pull the hood over the head and the APR/PAPR face piece.

SKILL DRILL 32-4 Doffing a Level B Nonencapsulated Chemical-Protective Clothing Ensemble

1 After completing decontamination, proceed to clean area for suit doffing. Stand and doff the SCBA. Face piece should be kept in place while the SCBA frame is placed on the ground.

2 Instruct the assistant to open the chemical splash flap and suit zipper.

3 Remove your hands and arms from suit gloves (except the inner gloves) and sleeves and cross your arms in front inside the suit. Instruct the assistant to begin at the head and roll the suit down and away until the suit is below waist level.

4 Sit down and instruct the assistant to complete rolling down the suit and remove the outer boots and suit. Rotate on the bench to the direction that will allow you to place feet on dry, clean area.

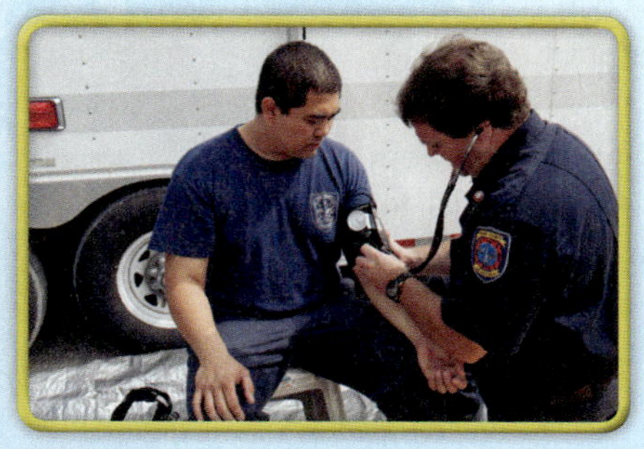

5 Remove the SCBA face piece. Go to the rehabilitation area for medical monitoring, rehydration, and personal decontamination shower.

8. Place the helmet on the head.
9. Review hand signals and indicate that you are okay. (STEP 4)

To doff a Level C chemical-protective clothing ensemble (CPE), follow the steps in SKILL DRILL 32-6 FIGURE 32-11:

1. After completing decontamination, proceed to the clean area for suit doffing.
2. Instruct the assistant to open the chemical splash flap and suit zipper.
3. Remove the hands and arms from the suit gloves (except the inner gloves) and sleeves, and cross the arms in front inside the suit.
4. Instruct the assistant to begin at the head and roll the suit down and away until the suit is below waist level.
5. Sit down. Instruct the assistant to complete rolling down the suit and take the outer boots and suit away.
6. Instruct the assistant to help remove the inner gloves.
7. Rotate on the bench to the direction that allows the feet to be placed on a dry, clean area.
8. Remove the APR/PAPR.
9. Go to the rehabilitation area for medical monitoring, rehydration, and personal decontamination shower.

Level D

Level D protection is the lowest level of protection, which includes coveralls, work shoes, hard hats, gloves, and standard work clothing FIGURE 32-12. It should be used when

FIGURE 32-10 Level C protection includes chemical-protective clothing and gloves as well as respiratory protection.

FIGURE 32-11 An assistant helps in doffing Level C CPE.

the atmosphere contains no known hazard, and when work functions preclude splashes, immersion, or the potential for unexpected inhalation of or contact with hazardous levels of chemicals. Level D PPE should be used for nuisance contamination (such as dust) only; it should not be worn at any site where respiratory or skin hazards exist.

Fire Fighter Safety Tips

It should be noted that the fire fighter structural clothing is *not* considered to be Level D because it is not rated as chemical-protective clothing.

Recommended PPE for Level D protection includes the following components:

- Coveralls
- Safety boots/shoes
- Safety glasses or chemical-splash goggles
- Hard hat
- Gloves

Optional PPE for Level D protection includes the following components:

- Face shield
- Escape mask

To don a Level D chemical-protective clothing ensemble, follow the steps in **SKILL DRILL 32-7** **FIGURE 32-13**:

1. Conduct a pre-entry briefing, medical monitoring, and equipment inspection.
2. Don the Level D suit.
3. Don the boots.

FIGURE 32-12 Level D protection is primarily a work uniform that includes coveralls and provides minimal protection.

SKILL DRILL 32-5 Donning a Level C Chemical-Protective Clothing Ensemble

1. Conduct a pre-entry briefing, medical monitoring, and equipment inspection. While seated, pull on the suit to waist level; pull on the chemical boots over the top of the chemical suit. Pull the suit boot covers over the tops of the boots.

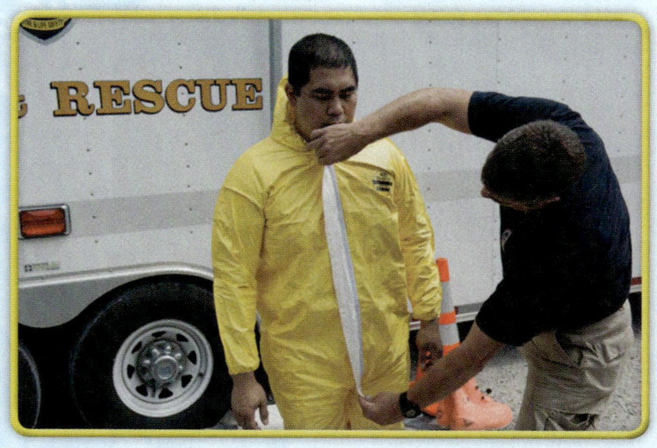

2. With assistance, complete donning the suit by placing both arms in the suit and pulling the suit over the shoulders. The assistant closes the chemical suit by closing the zipper and sealing the splash flap.

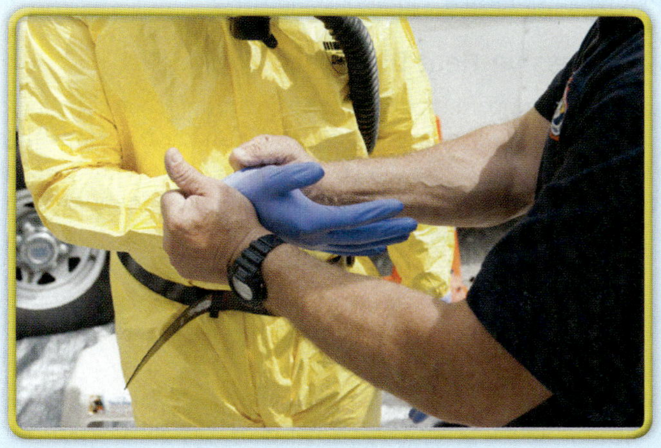

3. Don the inner gloves.

4. Stand up and don the APR/PAPR and APR/PAPR face piece. With assistance, pull the hood over the head and the APR/PAPR face piece. Place the helmet on the head.

4. Don safety glasses or chemical goggles.
5. Don a hard hat.
6. Don gloves, a face shield, and any other required equipment.

To doff a Level D chemical-protective clothing ensemble, follow the steps in SKILL DRILL 32-8:

1. After completing decontamination, proceed to the clean area for suit doffing.
2. Remove the gloves and hard hat.
3. Begin at the head and roll the suit down and away until the suit is below waist level.
4. Sit down. The assistant completes rolling down the suit and takes the boots and suit away.
5. Instruct the assistant to help remove the inner gloves.
6. Rotate on the bench to the direction that allows the feet to be placed on a dry, clean area.
7. Go to the rehabilitation area for medical monitoring and rehydration.

FIGURE 32-13 Level D CPE.

Skin Contact Hazards

The principal dangers of hazardous materials are toxicity, flammability, and reactivity. However, many hazardous materials have more than one dangerous characteristic. These materials can be extremely dangerous and produce harmful effects on the unprotected or inadequately protected human body. The best action to take when dealing with an unknown chemical or with a chemical that has more than one dangerous characteristic is to assume the worst and to provide the largest safety margin possible.

The skin can absorb harmful toxins without any sensation to the skin itself. Given this fact, fire fighters should not rely on pain or irritation as a warning sign of absorption. Some poisons are so concentrated that just a few drops placed on the skin can result in death.

Skin absorption is enhanced by abrasions, cuts, heat, and moisture. This relationship can create critical problems for fire fighters who are working at incidents that involve any form of chemical or biological agents. Fire fighters with large open cuts, rashes, or abrasions should be prohibited from working in areas where they might be exposed to hazardous materials. Smaller cuts or abrasions should be covered with nonporous dressings.

The rate of absorption can vary depending on the body part that is exposed. For example, chemicals can be absorbed through the skin on the scalp much faster than they are absorbed through the skin on the forearm.

The high absorbency rate of the eyes makes them one of the fastest means of exposure. This type of exposure can occur when a chemical is splashed directly into the eye, when a chemical is carried from a fire by toxic smoke particles, or when gases or vapors are absorbed through the eyes. Absorption of a chemical through the eyes also can be seen as an early warning signal for either PPE or SCBA failure. This type of exposure is also a major problem because the eyes are directly connected to the optic nerve, which serves as a direct route to the brain.

Chemicals such as corrosives immediately damage skin or body tissues upon contact. Acids, for example, have a strong affinity for moisture and can create significant skin and respiratory tract burns. In addition, acid injuries create a clot-like barrier that blocks deep skin penetration of the chemical. In contrast, alkaline materials dissolve the fats and lipids that make up skin tissue and change solid tissue into a soapy liquid. This process is similar to the way caustic chemical drain cleaners dissolve grease and other materials in sinks and drains. As a result, alkaline burns are often much deeper and more destructive than acid burns are.

Safety Precautions

Fire fighters need to take several types of safety precautions during a hazardous materials incident. In addition to taking the standard safety precautions of using PPE and respirator protection, they must guard against the dangers posed by temperature-related conditions and stress.

Excessive-Heat Disorders

Fire fighters operating in protective clothing should be aware of the signs and symptoms of heat rash, heat cramps, heat exhaustion, heat stroke, and dehydration. If the body is unable to disperse heat because it is covered by a closed, impermeable garment, serious short- and long-term medical situations could result.

Heat rash can occur for a variety of reasons. In fire fighters it occurs as a result of PPE rubbing the skin when it is hot. Heat rash can also occur if there are folds or creases in the skin where sweat can collect and create clear, raised bumps on the skin. These bumps are usually not itchy. However, repeated or prolonged outbreaks can lead to infection.

Heat cramps are painful, muscle spasms that can occur during physical exertion in high temperatures. The cause of heat cramps is electrolyte imbalance caused by excess sweating. Consuming a sports drink can rebalance the body.

Heat exhaustion is a mild form of shock that occurs when the circulatory system begins to fail because the body cannot dissipate excessive heat and becomes overheated. The body's core temperature rises, an event that is followed by weakness and profuse sweating. The fire fighter might become weak or dizzy and have episodes of blurred vision. Other signs and symptoms of heat exhaustion include rapid, shallow breathing; weak pulse; cold, clammy skin; and, sometimes, loss of consciousness. Although heat exhaustion is not an immediately life-threatening condition, the affected individual should be removed at once from the source of heat, rehydrated with electrolyte solutions, and kept cool. If not properly treated, heat exhaustion can progress to heat stroke.

VOICES OF EXPERIENCE

A page was received which indicated a hazardous materials incident at the local paper mill. Dispatch was vague with the details, but the local fire department commander had decided to call us in.

Upon arrival, the team leader assigned the team members our tasks. I was in charge of choosing the correct PPE for the entry team. The plant personnel informed us that "black liquor" was leaking from a fitting in a storage tank that was above 200°F (93°C). I was informed that the material was a highly caustic liquid and a by-product in the digestion of wood pulp. The plant captures this liquid, concentrates it, and then recycles it through the use of a recovery boiler.

I determined right away that we would need to perform a Level A entry because any skin exposure would not be acceptable. The "black liquor" would cause skin damage almost immediately. I wasn't able to find any specific information in the suit selection guides that were provided by the manufacturers, so I decided to call the toll-free number listed in the guide. The technician who answered the call was very knowledgeable and asked if I had one of the Level A suits with flash protection that the company sold. I informed him that we had several in our inventory. Upon his recommendation, we chose those suits to provide the entry team not only with chemical protection, but also protection from the temperature of the liquid should a rupture or catastrophic failure occur while the team members were stopping the leak.

Everything went well. The entry team was able to stop the leak and get out without any problems. I was glad that I had called the manufacturer because I had been completely overlooking the thermal hazard and was focused only on the chemical properties of the liquid. If something had happened during the mitigation of the incident, I believe that the entry team would have been better prepared because of the flash and chemical protection the recommended suits offered.

It is easy to be focused on one item and completely overlook the obvious. Calling the manufacturers of your PPE might not be something that you have to do very often, but they are a great resource and shouldn't be overlooked.

Jon Mink
Muskegon County Hazmat Team
Muskegon, Michigan

Heat stroke is a severe and potentially fatal condition resulting from the failure of the temperature-regulating capacity of the body. It is caused by exposure to the sun or high temperatures. Reduction or cessation of sweating is an early symptom. The body temperature can rise to 105°F (40.5°C) or higher, with the victim experiencing a rapid pulse, hot skin, headache, confusion, unconsciousness, and (possibly) convulsions. Heat stroke is a true medical emergency that requires immediate transport to a medical facility, preferably with advanced life support (ALS) in place.

The issue of dehydration should also be addressed. Fire fighters should be encouraged to drink 8 to 16 ounces (237 to 473 milliliters) of water before donning any protective clothing. During rehabilitation, they should be required to rehydrate with 16 ounces (473 milliliters) of water for each SCBA tank used or consumed.

Cooling Technologies

In an effort to combat heat stress while wearing PPE, many response agencies employ some form of cooling technology under the garment. These technologies include but are not limited to air, ice, and water-cooled vests, along with phase change cooling technology. Many studies have been conducted on each form of cooling technology. Each is designed to accomplish the same goal: to reduce the impact of heat stress on the human body.

The human body is constantly seeking to reach an acceptable equilibrium with the external environment. To that end, heat generated by the body's core is carried through the circulatory system to the skin, where it is wicked off as cooler ambient air is passed over the skin. This process, called convective cooling, is how the body cools itself under normal conditions.

As the outside temperature rises, the difference between normal body heat and the surrounding environment narrows, and the normal process of convective cooling is ineffective. To compensate, the body switches to a process called evaporative cooling. Perspiration is generated during this process, and the ensuing evaporation of sweat produces a tremendous cooling effect. This method of cooling is effective only when the surrounding air is dry. When the moisture content in the surrounding environment rises (humidity), perspiration cannot evaporate because the air is equally moist. Unfortunately, the human body has no other alternative when these natural mechanisms fail to work efficiently. The same suit that seals you up against the hazardous materials also seals in the heat, defeating the body's natural cooling mechanisms.

Forced air-cooling systems operate by forcing prechilled air through a system of hoses worn close to the body. As the cooler air passes by the skin, heat is drawn away (by convection) from the body and released into the atmosphere. Forced-air systems are designed to function as the first level of cooling the body would naturally employ. Typically, these systems are lightweight and provide long-term cooling benefits, but mobility is limited because the hose is attached to an external fixed compressor.

Fluid-chilled systems operate by pumping ice-chilled liquids from a reservoir through a series of tubes held within a vest-like garment and back to the reservoir. Mobility might be limited with some varieties of this system because the pump might be located away from the garment. Some units have a battery-operated unit worn on the hip, but the additional weight increases the workload, which generates more heat, thereby defeating the purpose of the cooling vest.

Passive systems, ice- or gel-packed vests, are commonly used because of the low cost, unlimited portability, and unlimited recharging by refreezing the packs. These garments are vest-like in their design and intended to be worn around the torso. The principle is that the ice-chilled vest absorbs the heat generated by the body. This technology is bulkier and heavier than the aforementioned systems and can cause discomfort to the wearer because of the nature of an ice-cold vest near the skin. Additionally, the cold temperature near the skin can actually fool the body into thinking it is cold instead of hot, thereby causing it to retain even more heat.

Phase change cooling technology operates in a similar fashion to the passive ice- or gel-packed vests. The main difference is that the temperature of the material in the packs is chilled to approximately 60°F, and the fabric of the vest is designed to wick perspiration away from the body. The packs typically recharge more quickly than ice- or gel-packed vests.

Fire Fighter Safety Tips

Remember to take rehabilitation breaks throughout the hazardous materials incident. Wearing any type of PPE requires a great deal of physical energy and mental concentration.

Cold-Temperature Exposures

Fire fighters at hazardous materials incidents can be exposed to two types of cold temperatures: those caused by the materials and those caused by weather. Hazardous materials such as liquefied gases and cryogenic liquids expose fire fighters to the same low-temperature hazards as those created by cold-weather environments. Exposure to severe cold for even a short period can cause severe injury to body surfaces, particularly to the ears, nose, hands, and feet.

Two factors in particular influence the development of cold-related injuries: temperature and wind speed. Because still air is a poor heat conductor, fire fighters working in low temperatures with little wind can endure these conditions for longer periods, assuming that their clothing remains dry. When low temperatures are combined with faster winds, however, wind chill occurs. As an example, if the temperature with no wind is 20°F (–7°C), it will feel like –5°F (–20°C) when the wind speed is 15 miles/hour (54 kilometers per hour).

Regardless of the temperature, fire fighters perspire heavily when they are wearing impermeable chemical-protective clothing. When this wet inner clothing is removed during the decontamination process, particularly in a cold environment, the body can cool rapidly. Wet clothing extracts heat from the body as much as 240 times faster than dry clothing does. This phenomenon can also lead to hypothermia, a true medical emergency in which the body temperature falls below 95°F (35°C).

Fire fighters must be aware of the dangers of frostbite, hypothermia, and impaired ability to work when temperatures are low, when the wind chill factor is low, or when they are working in wet clothing. All personnel should wear appropriate, layered clothing and should be able to warm themselves in heated shelters or vehicles such as buses.

The layer of clothing next to the skin—especially the socks—should always be kept dry. Trench foot can result when wet socks are worn at long-term emergencies in cool (not cold) environments **FIGURE 32-14**. Fire fighters should carefully schedule their work and rest periods and monitor their physical working conditions. Warm shelters should be available for donning and doffing protective clothing.

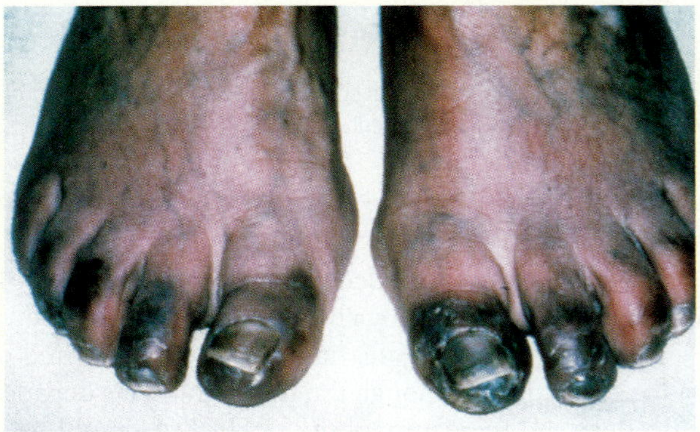

FIGURE 32-14 Trench foot can result when wet socks are worn at long-term emergencies in cool environments.

Physical Capability Requirements

Hazardous materials response operations put a great deal of both physiologic and psychological stress on responders. During the incident, these personnel can be exposed to both chemical and physical hazards. They might face life-threatening emergencies, such as fire and explosions, or they might develop heat stress while wearing protective clothing or working under extreme temperatures.

For these reasons, every emergency response organization should have a health and safety management program. The components of a health and safety management system for hazardous materials responders are outlined in OSHA 29 CFR 1910.120, *Hazardous Waste Site Operations and Emergency Response*. Among the components of such a program are medical surveillance, PPE, and site safety practices and procedures.

A medical surveillance program is the cornerstone of an effective health and safety management system for responders. The two primary objectives of a medical surveillance program are to determine whether an individual can perform his or her assigned duties, including the use of personal protective clothing and equipment and to detect any changes in body system functions caused by physical and/or chemical exposures.

Medical monitoring is the ongoing, systematic evaluation of response personnel who might experience adverse effects because of exposure to heat, cold, stress, or hazardous materials. Regular medical monitoring can quickly identify problems so that they can be treated in a timely fashion, thereby preventing severe adverse effects and maintaining the optimal health and safety of on-scene personnel.

Fire fighters should be examined on a regular basis as determined by the fire department's medical director. For hazardous materials personnel, this might be on an annual basis. During this examination, the physician performs a respiratory function test to measure lung capacity and lung function and to evaluate an individual's fitness for wearing SCBA and other respiratory protection devices. The specific requirements for any fire fighter who is assigned to any duty where any form of respiratory protection will be used can be found in 29 CFR 1910.134, OSHA's respiratory protection standard.

Whenever possible and immediately before fire fighters don PPE to enter a hazardous materials environment, they should have their vital signs taken as part of the pre-entrance medical monitoring. This examination should also include an inspection of the individual's skin for rashes and open sores or wounds and an evaluation of the individual's mental status. Fire fighters at a hazardous materials incident should be alert and oriented to time and place, with clear speech, a normal gait, and the ability to respond appropriately. A recent medical history should be obtained, including current weight, medications taken within the past 72 hours, alcohol consumption within the past 24 hours, any new medical treatment or diagnosis made within the past two weeks, and any symptoms of fever, nausea, vomiting, diarrhea, or cough within the past 72 hours.

This pre-entrance medical monitoring should be performed on all individuals who will wear liquid splash– or vapor-protective clothing and execute hazardous materials operations. It should be completed as soon as possible and within the hour before the fire fighter enters the hazardous environment.

Hazardous materials PPE puts significant stress on the body, particularly on the cardiovascular system. If a member of the hazardous materials team exhibits changes in gait, speech, or behavior, he or she should leave the environment for immediate decontamination, doffing of protective clothing, and assessment. Anyone who complains of chest pain, dizziness, shortness of breath, weakness, nausea, or headache should also undergo immediate decontamination, doffing of protective clothing, and assessment by Emergency Medical Services (EMS) personnel on the scene, preferably at the ALS level.

After the hazardous materials team members undergo decontamination, they should once again have their vital signs checked and monitored. Components of postevent medical monitoring should include history, vital signs, weight, skin evaluation, and mental status. Normal baseline values should be attained within 15 to 20 minutes after a person leaves the environment. Anyone who does not return to normal baseline entry levels should be treated and transported to a definitive care facility for follow-up monitoring.

During this postevent medical monitoring, vital signs should be checked every 5 to 10 minutes. The monitoring team should obtain information from medical control on latent reactions or symptoms and communicate this information to response personnel. If any of these symptoms appear, medical control should be contacted for direction and the affected individual should be prepared for possible transport to a medical facility. Some hazardous materials teams maintain their own hazardous materials–specific medical control.

Response Safety Procedures

One of the first things that first responders at the awareness level should do at a hazardous materials incident scene is to isolate the area and prevent anyone from entering the scene. These basic protective actions should be taken immediately. For the first several minutes, not even the first responders themselves should enter the area. During this time, first responders can gather necessary scene information and identify the materials involved. Any actions taken should follow the predetermined local response plan or guideline. For example, local and state officials, the local emergency planning committee, and other federal response agencies should be notified as required.

Any possible ignition sources should be eliminated, particularly if the incident involves the release or probable release of flammable materials. Whenever possible, electrical devices used in the immediate area of the hazard should be certified as safe by recognized organizations. Note that some flammable vapors are heavier than air and, therefore, can travel along the ground to an ignition source.

The first step in gaining control of a hazardous materials incident is to isolate the problem and keep people away from it **FIGURE 32-15**. The IC cannot begin extended operations until the hazard area is identified and the perimeter is secured. Fire fighters can help to isolate the area by stretching banner tape across roads to divert traffic away from a spill and by coordinating their efforts with those of police, EMS, and other agencies.

FIGURE 32-15 Isolate the area by using banner tape.

Control Zones

Isolating the incident helps to limit the number of civilian and public service personnel who might be exposed to the hazardous materials. This can be done through a series of control zones around the incident site. Control zones are designated areas at a hazardous materials incident based on safety and the degree of hazard. These zones should not be defined too narrowly **FIGURE 32-16**. Indeed, as the IC gets more information about the specifics of the chemical or material involved, the control zones can be reevaluated. If the incident is not as severe as first reported, the IC can readjust the control zones. It is always easier and safer to reduce the size of the control zones than to discover that they need to be larger.

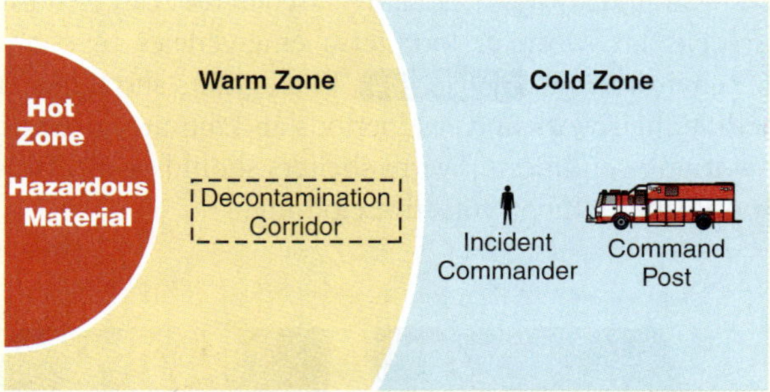

FIGURE 32-16 Control zones spread outward from the center of a hazardous materials incident.

Control zones are labeled as hot, warm, or cold. They might also be color-coded, such as red, yellow, and blue.

The hot zone is the area immediately around and adjacent to the incident. Although the size of this zone depends on the incident and the nature of the product, it should be large enough that adverse effects from the hazardous materials will not affect people outside the zone. An incident that involves a gas, for example, requires a larger hot zone than one involving a liquid leak. The hot zone should cover the incident site as well as the personnel and equipment necessary to control it. All personnel and equipment in this zone must be decontaminated when they leave the area, and personnel who work within it should be closely monitored. Access to the hot zone must be limited to only those persons necessary to control the incident. Individuals must log in at the access control point, recording their entry and exit times. Only individuals trained to the hazardous materials technician level should be permitted in the hot zone.

The warm zone (also referred to as the contamination reduction zone) is the area where personnel and equipment are staged before they enter and after they leave the hot zone. It contains control points for access corridors as well as the decontamination corridor (also referred to as the contamination reduction corridor). Designating a warm zone helps to reduce the spread of contamination. Only the minimal amount of personnel and equipment necessary to perform a task or support those personnel who are operating in the hot zone should be permitted in the warm zone.

Personnel trained to the operations level can work with moderate supervision in the warm zone. Generally, personnel working in the warm zone can use PPE that is one level below the PPE used in the hot zone. If personnel in the hot zone are using Level A encapsulated chemical-protective suits, for example, personnel in the warm zone might dress in Level B protection. Ultimately, only the IC can determine which level of protection is needed for each responder.

Beyond the warm zone is the cold zone. In this "safe area," personnel do not need to wear any special protective clothing for safe operation. Personnel staging, the incident command post, EMS providers, and an area for medical treatment after decontamination are all located in the cold zone.

FIRE FIGHTER Tips

In addition to the control zones listed here, law enforcement organizations use the terms *inner* and *outer perimeters* for their control zones. Law enforcement organizations consider the hot, warm, and cold zones as part of the inner perimeter, and from the cold zone to where the law enforcement agency has set up the exclusion area is the cold zone.

Fire Fighter Safety Tips

The first step in establishing control zones is to immediately isolate the incident scene and deny entry to all personnel.

Isolation Techniques

Whenever possible, approach a hazardous materials incidents cautiously from upwind and uphill FIGURE 32-17. Resist the urge to rush in; do not attempt to help others until the situation has been fully assessed. Without entering the immediate hazard area (the hot zone), isolate the scene so that the safety of both people and the environment are ensured. Keep people away from the scene and outside the safety perimeter. Only those directly involved in the emergency response operations should be permitted in the area, and even they must wait until control has been established and the situation assessed.

There are several ways to isolate the hazard area and create the control zones. The first step that a first responder should take is to identify the area verbally over the radio. Providing this information helps ensure that those who arrive later will remain in a safe area and will not accidentally enter the hot zone. Cordon off the area using standard traffic cones, police, or fire-line barrier tape to control access.

If the incident takes place inside a structure, the best place to control access is at the normal points of entry—that is, the doors. Once the doors are secured so that no unauthorized personnel can enter the facility, appropriately trained emergency response crews can begin to isolate areas above and below the hazard. The same concept applies to outdoor incidents. Secure the entry points and the area around the hazard. Begin by controlling intersections, on/off ramps, service roads, or other access routes to the scene.

Police officers should assist fire fighters in this operation by diverting traffic at a safe distance outside the hazard area. They should block off streets, close intersections, or use their vehicles to redirect traffic. During a long-term incident, the highway department can set up traffic barriers. Crowd-control devices used by the police department can also be used to identify the hazard zones and restrict access. No matter which devices are used to restrict access, they should not limit or inhibit rapid withdrawal from the area by personnel working inside the hot zone.

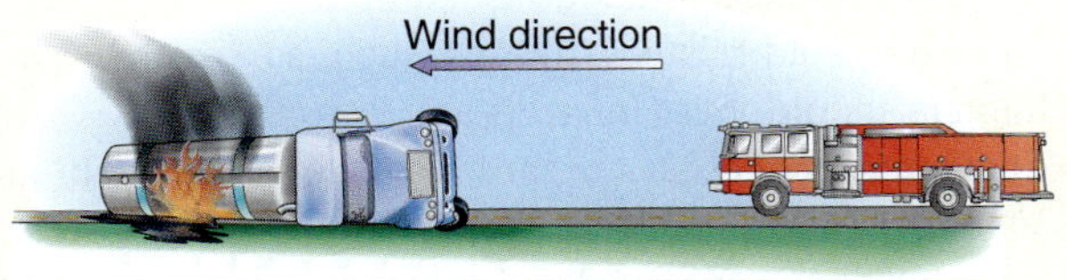

FIGURE 32-17 Approach a hazardous materials incident cautiously.

The Buddy System and Backup Personnel

The PASS device and the use of personal accountability and buddy systems are essential for fire fighter safety. On a hazardous materials incident, backup personnel are also used to ensure the safety of emergency crews. They stand by at the scene, ready to remove personnel who are working in the hot zone if an emergency occurs. Backup personnel must be in place and prepared whenever personnel are operating in the hot zone.

All personnel must be fully briefed before they approach the hazard area or enter the hot zone. In addition, the decontamination team must be in place before anyone enters the hot zone. No one should enter the hot zone alone; instead, at least two team members should suit up and enter the hot zone together. At least two additional team members should suit up and stand by to assist those who entered first. The IC should ensure that backup personnel are prepared for immediate entry into the hot zone.

Team members should always remain within sight, sound, or touch of each other. Unfortunately, electronic communication devices sometimes fail, or the individual needing rescue might not be physically able to summon assistance. Radios should be used for communications at hazardous materials incident sites. They cannot, however, replace visual or voice contact or be the only way to contact an individual working in the hot zone.

Fire Fighter Safety Tips

It is easy to focus all of your attention on the hazardous materials component of the scene. In reality, many other hazards are associated with emergency incident scenes. Watch out for downed power lines, jagged edges of metal, and tripping hazards. Do not get tunnel vision; consider all types of hazards.

FIRE FIGHTER Tips

Be certain to document and properly report the PPE worn during the hazardous materials incident according to your standard operating procedures.

Wrap-Up

Chief Concepts

- The proper use of PPE and scene control are critical to ensuring the safety of fire fighters, victims, and bystanders.
- The levels of TLVs are the following:
  - Threshold limit value/short-term exposure limit (TLV/STEL): The maximum concentration of material to which a person can be exposed for 15-minute intervals, up to four times a day, without experiencing irritation or chronic or irreversible tissue damage.
  - Threshold limit value/ceiling (TLV/C): The maximum concentration of material to which a worker should not be exposed, even for an instant.
  - Threshold limit value/skin: The concentration at which direct or airborne contact with a material could result in possible and significant exposure from absorption through the skin, mucous membranes, and eyes.
- OSHA's regulatory measures that are comparable to the TLV/TWA are as follows:
  - The permissible exposure limit (PEL): Measures the maximum, time-weighted concentration of material to which 95 percent of healthy adults can be exposed without suffering any adverse effects over a 40-hour workweek.
  - Immediately Dangerous to Life and Health (IDLH): An atmospheric concentration of a toxic, corrosive, or asphyxiant substance poses an immediate threat to life or could cause irreversible or delayed adverse health effects.
- Exposure guidelines are intended to minimize the possibility that either responders or the public will be exposed to hazardous materials or atmospheres that will lead to harm. Exposure guidelines can be used to define the three basic atmospheres at a hazardous materials emergency:
  - Safe atmosphere: No harmful hazardous materials effects exist, so personnel can handle routine emergencies without donning specialized PPE.
  - Unsafe atmosphere: A hazardous material that is no longer contained has created an unsafe condition or atmosphere. A person who is exposed to the material for long enough will probably experience some form of acute or chronic injury.
  - Dangerous atmosphere: Serious, irreversible injury or death can occur in the environment.
- Selecting the appropriate PPE for a hazardous materials incident is a potentially life-saving decision. Adequate PPE should protect the fire fighter's respiratory system, as well as the skin, eyes, face, hands, feet, head, body, and hearing.
- Which type of PPE is required for the incident depends on the activities that the fire fighter is expected to perform.
- The categories of hazardous materials–specific PPE include the following:
  - Street clothing and work uniforms: Offer the least amount of protection
  - Structural firefighting protective clothing: Offers no chemical protection but protects against abrasion and direct skin contact
  - High temperature–protective equipment: Protects the wearer during short-term exposures to high temperatures
  - Chemical-protective clothing and equipment: Includes liquid splash–protective clothing and vapor-protective clothing
- Respiratory protection includes the following:
  - SCBA
  - Supplied-air respirators (SARs): Have an external air source such as a compressor or storage cylinder
  - Air-purifying respirators (APRs): Filtering devices
  - Powered air-purifying respirators (PAPRs): Filtering devices
- The chemical-protective clothing ratings are as follows:
  - Level A protection: Provides the highest level of protection for skin, eyes, and respiration
  - Level B protection: Used when the type and atmospheric concentration of identified substances require a high level of respiratory protection but less skin protection
  - Level C protection: Used when the type of airborne substance is known, its concentration is measured, the criteria for using APRs are met, and skin and eye exposure is unlikely
  - Level D protection: Used when the atmosphere contains no known hazard, and when work functions preclude splashes, immersion, or the potential for unexpected inhalation of or contact with hazardous levels of chemicals
- Temperature-related conditions and stress pose dangers. If the body cannot disperse heat because it is covered by a closed, impermeable garment, serious short- and long-term medical situations could result. These medical conditions include the following:
  - Heat rash: Can lead to skin infections.
  - Heat cramps: Painful muscle spasms caused by electrolyte imbalance.
  - Heat exhaustion: Mild form of shock that occurs when the body overheats.
  - Heat stroke: Potentially fatal condition. Reduction or cessation of sweating is the first symptom. Immediate medical attention is required.
- To avoid cold-related injuries such as frostbite, fire fighters should dress in layers and try to keep dry.
- The control zones at a hazardous materials incident are as follows:
  - Hot zone: Area immediately around the incident. Only properly trained and equipped personnel should enter this zone.

- Warm zone: Contains the decontamination corridor.
- Cold zone: No special PPE is required. This zone contains the incident command post, personnel staging area, and rehabilitation area.

- Team members should always remain within sight, sound, or touch of each other.

Hot Terms

Air-purifying respirator (APR) A respirator with an air-purifying filter, cartridge, or canister that removes specific air contaminants by passing ambient air through the air-purifying element. (NFPA 1404, *Standard for Fire Service Respiratory Protection Training*)

Backup personnel Individuals who remove or rescue those working in the hot zone in the event of an emergency.

Chemical-resistant material A material used to make chemical-protective clothing, which can maintain its integrity and protection qualities when it comes into contact with a hazardous material. These materials also resist penetration, permeation, and degradation.

Cold zone The control zone of hazardous materials/weapons of mass destruction incidents that contains the incident command post and such other support functions as are deemed necessary to control the incident. (NFPA 472, *Standard for Competence of Responders to Hazardous Materials/Weapons of Mass Destruction Incidents*)

Control zone An area at hazardous materials/weapons of mass destruction incidents within an established perimeter that is designated based upon safety and the degree of hazard. (NFPA 472)

Degradation (1) A chemical action involving the molecular breakdown of a protective clothing material or equipment caused by contact with a chemical. (2) The molecular breakdown of the spilled or released material to render it less hazardous during control operations. (NFPA 472)

Heat cramps Painful muscle spasms usually associated with vigorous activity in a hot environment.

Heat exhaustion A mild form of shock caused when the circulatory system begins to fail as a result of the body's inadequate effort to give off excessive heat.

Heat rash Clear, raised bumps on the skin caused by friction and sweat.

Heat stroke A severe and sometimes fatal condition resulting from the failure of the temperature-regulating capacity of the body. It is caused by prolonged exposure to the sun or high temperatures. Reduction or cessation of sweating is an early symptom; body temperature of 105°F (40°C) or higher, rapid pulse, hot and dry skin, headache, confusion, unconsciousness, and convulsions can occur. Heat stroke is a true medical emergency requiring immediate transport to a medical facility.

High temperature–protective equipment Protective clothing designed to shield the wearer during short-term exposures to high temperatures.

Hot zone The control zone immediately surrounding hazardous materials/weapons of mass destruction incidents, which extends far enough to prevent adverse effects of hazards to personnel outside the zone. (NFPA 472)

Level A protection Personal protective equipment that provides protection against vapors, gases, mists, and even dusts. The highest level of protection, it requires a totally encapsulating suit that includes self-contained breathing apparatus.

Level B protection Personal protective equipment that is used when the type and atmospheric concentration of substances have been identified. It generally requires a high level of respiratory protection but less skin protection: chemical-protective coveralls and clothing, chemical protection for shoes, gloves, and self-contained breathing apparatus outside of a nonencapsulating chemical-protective suit.

Level C protection Personal protective equipment that is used when the type of airborne substance is known, the concentration is measured, the criteria for using air-purifying respirators are met, and skin and eye exposure is unlikely. It consists of standard work clothing with the addition of chemical-protective clothing, chemically resistant gloves, and a form of respirator protection.

Level D protection Personal protective equipment that is used when the atmosphere contains no known hazard, and work functions preclude splashes, immersion, or the potential for unexpected inhalation of or contact with hazardous levels of chemicals. It is primarily a work uniform that includes coveralls and affords minimal protection.

Liquid splash–protective clothing The garment portion of a chemical-protective clothing ensemble that is designed and configured to protect the wearer against chemical liquid splashes but not against chemical vapors or gases. (NFPA 472)

Penetration The movement of a material through a suit's closures, such as zippers, buttonholes, seams, flaps, or other design features of chemical-protective clothing, and through punctures, cuts, and tears. (NFPA 472)

Permeation A chemical action involving the movement of chemicals, on a molecular level, through intact material. (NFPA 472)

Permissible exposure limit (PEL) The maximum permitted eight-hour, time-weighted average concentration of an airborne contaminant. (NFPA 5000, *Building Construction and Safety Code*)

Powered air-purifying respirator (PAPR) Air-purifying respirator with a hood or helmet, breathing tube, canister, cartridge, filter, and a blower that passes ambient air through the purifying element. The blower can be stationary or portable. It can also be designed and equipped with a full face piece, in which case it is referred to as a full-face-piece powered air-purifying respirator (FFPAPR). (NFPA 1404)

Recommended exposure level (REL) The established standard limit of exposure to a hazardous material. It is based on the maximum time-weighted concentration at which 95 percent of exposed, healthy adults suffer no adverse effects over a 40-hour workweek. RELs are set by the National Institute for Occupational Safety and Health but do not have the force of federal law.

Supplied-air respirator (SAR) An atmosphere-supplying respirator for which the source of breathing air is not designed to be carried by the user. (NFPA 1404)

Threshold limit value/ceiling (TLV/C) The maximum concentration of hazardous material to which a worker should not be exposed, even for an instant.

Threshold limit value/short-term exposure limit (TLV/STEL) The maximum concentration of hazardous material to which a worker can sustain a 15-minute exposure not more than four times daily, with a 60-minute rest period required between each exposure. The lower the value, the more toxic the substance.

Threshold limit value/skin A designation indicating that direct or airborne contact with a material could result in significant exposure from absorption through the skin, mucous membranes, and eyes.

Threshold limit value/time-weighted average (TLV/TWA) The airborne concentration of a material to which a worker can be exposed for 8 hours a day, 40 hours a week, and not suffer any ill effects.

Vapor-protective clothing The garment portion of a chemical-protective clothing ensemble that is designed and configured to protect the wearer against chemical vapors or gases. (NFPA 472)

Warm zone The control zone at hazardous materials/weapons of mass destruction incidents where personnel and equipment decontamination and hot zone support take place. (NFPA 472)

FIRE FIGHTER *in action*

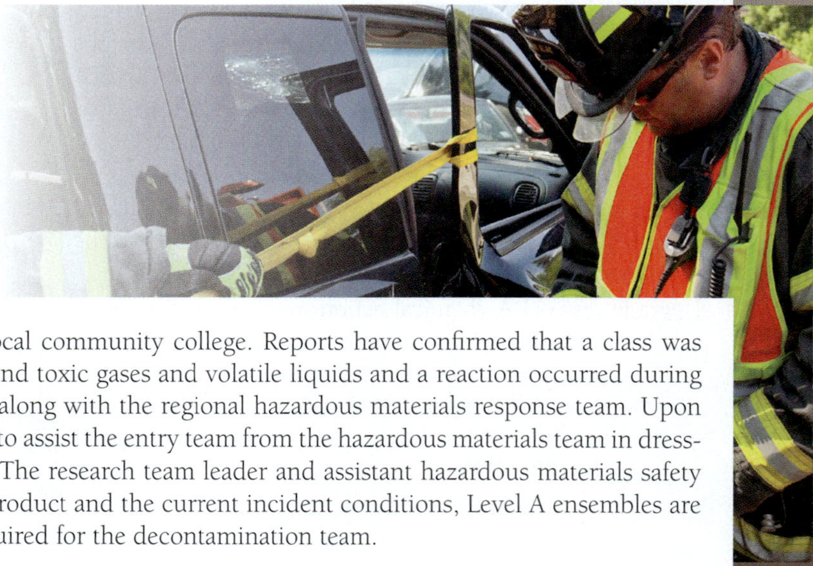

A chemical release has been reported in the lab at the local community college. Reports have confirmed that a class was conducting experiments using several different corrosive and toxic gases and volatile liquids and a reaction occurred during one of those experiments. Your fire department responds along with the regional hazardous materials response team. Upon arrival at the scene, you and your crew have been assigned to assist the entry team from the hazardous materials team in dressing out for entry into the hot zone to control the release. The research team leader and assistant hazardous materials safety officer have determined that, based on the hazard of the product and the current incident conditions, Level A ensembles are required for the entry team and Level B ensembles are required for the decontamination team.

1. Based on the hazard and chemical information provided, Level A provides the best protection for the entry team.
   A. True
   B. False
2. Level A protection requires the wearer to use some type of respiratory protection. Which type of protection is required for use in this level of protection?
   A. SCBA
   B. Supplied air respirator
   C. Air-purifying respirator
   D. Powered air-purifying respirator
3. After 15 minutes in the hot zone, one of the entry team members calls on the radio stating he is sweating profusely and feeling dizzy. Which type of heat stress might he be suffering from?
   A. Heat stroke
   B. Heat cramps
   C. Heat rash
   D. Heat exhaustion

4. During a hazardous materials incident, the first-due engine arrives and establishes protective zones. In which zone is the command post established?
   - **A.** Hot zone
   - **B.** Cold zone
   - **C.** Warm zone
   - **D.** Outer perimeter
5. When an entry team enters the hot zone, how many personnel with be in the backup position for life safety actions in the event the entry team has a problem?
   - **A.** 2
   - **B.** 4
   - **C.** 3
   - **D.** 5
6. What types of failure might chemical-protective clothing incur?
   - **A.** Permeation
   - **B.** Degradation
   - **C.** Penetration
   - **D.** All of the above

FIRE FIGHTER II *in action*

It is an unusually hot day for this time of year. Over the radio, you hear a dispatch for a fire in a large industrial complex on the other side of the county. Approximately 45 minutes into the incident, your company is dispatched to assist the Hazardous Materials Branch. The fire has raced through the complex, destroying several larger warehouses that store several types of hazardous chemicals. Many of the containers have been breached and have released product throughout the site.

Upon arrival at the staging site, you are sent directly to the hazardous materials officer, who requests that your personnel assist in both the entry group and the site control group. You are assigned to the entry group and asked to assist in the dressing of the hazardous materials technicians, while your long-time partner is assigned to the site control group. You both take a Fire Fighter I with you to help with your assigned tasks.

1. Who should be notified of your assignments?
2. What should be the primary concern for you and your assistant while assisting the hazardous materials personnel in getting dressed for entry?
3. While working in the site control group, what equipment should your partner and his partner use and which safety parameters should they consider?

Hazardous Materials: Response Priorities and Actions

Fire Fighter I

Knowledge Objectives

After studying this chapter, you will be able to:

- Define exposures in regard to a hazardous materials incident. (NFPA 472, 5.2.4, 6.6.1.1.2, p 962)
- List two protective actions that may be taken in a hazardous materials incident. (NFPA 472, 5.4, 5.4.1, 6.6.1.1.2, p 962–963)
- Describe the role of fire fighters during an evacuation. (NFPA 472, 5.4.1, 6.6.1.1.2, p 962–963)
- Describe how and when a fire department conducts shelter-in-place operations with a local population. (NFPA 472, 5.4.1, 6.6.1.1.2, p 963)
- Describe the monitoring methods used to detect the presence of a hazardous material. (NFPA 472, 6.6.2, p 963–964)
- Explain the special considerations to follow when performing search and rescue operations in a hazardous materials incident. (p 964)
- List the methods of protecting exposures from hazardous materials. (p 965–966)
- Describe confinement and containment operations. (NFPA 472, 6.6.3, 6.6.3.1, 6.6.4, 6.6.4.1, 6.6.5, p 965)
- Describe the methods and tools utilized to extinguish flammable liquid fires. (NFPA 472, 6.6.1.2.2, p 965–966)
- Describe the actions taken during a pressurized-gas cylinder leak. (NFPA 472, 6.6.3, 6.6.3.1, 6.6.4, 6.6.4.1, 6.6.5, p 966)
- List the common hazardous materials control activities. (NFPA 472, 6.6.3, 6.6.3.1, 6.6.4, 6.6.4.1, 6.6.5, p 966–972)
- Describe how process of absorption can mitigate a hazardous materials incident. (NFPA 472, 6.6.3, 6.6.3.1, 6.6.4, 6.6.4.1, 6.6.5, p 966–968)
- Describe how the process of adsorption can mitigate a hazardous materials incident. (NFPA 472, 6.6.3, 6.6.3.1, 6.6.4, 6.6.4.1, 6.6.5, p 966–968)
- Describe how the process of diking can mitigate a hazardous materials incident. (NFPA 472, 6.6.3, 6.6.3.1, 6.6.4, 6.6.4.1, 6.6.5, p 968)
- Describe how the process of damming can mitigate a hazardous materials incident. (NFPA 472, 6.6.3, 6.6.3.1, 6.6.4, 6.6.4.1, 6.6.5, p 968–969)
- List the three types of dams that may be constructed during a hazardous materials incident. (NFPA 472, 6.6.3, 6.6.3.1, 6.6.4, 6.6.4.1, 6.6.5, p 969)
- Describe how the process of diversion can mitigate a hazardous materials incident. (NFPA 472, 6.6.3, 6.6.3.1, 6.6.4, 6.6.4.1, 6.6.5, p 969)
- Describe how the process of retention can mitigate a hazardous materials incident. (NFPA 472, 6.6.3, 6.6.3.1, 6.6.4, 6.6.4.1, 6.6.5, p 969–970)
- Describe how the process of vapor dispersion can mitigate a hazardous materials incident. (NFPA 472, 6.6.3, 6.6.3.1, 6.6.4, 6.6.4.1, 6.6.5, p 970–971)
- Describe how the process of vapor suppression can mitigate a hazardous materials incident. (NFPA 472, 6.6.3, 6.6.3.1, 6.6.4, 6.6.4.1, 6.6.5, p 971)
- List the types of containers and tanks with remote shut-off valves. (NFPA 472, 6.6.3, 6.6.3.1, 6.6.4, 6.6.4.1, 6.6.5, p 971–972)
- Explain why the incident commander might withdraw personnel from the hazardous materials incident. (NFPA 472, 5.5, 5.5.1, 5.6, p 972)
- Describe the recovery phase of a hazardous materials incident.
- Explain the factors that enter into the decision to terminate a hazardous materials incident. (NFPA 472, 6.6.6, p 972)
- Describe the precautions to take if the hazardous materials incident involves potential criminal or terrorist activity. (NFPA 472, 5.4.2, p 972)

CHAPTER 33

Skills Objectives

After studying this chapter, you will be able to perform the following skills:

- Use a multi-gas meter to provide atmospheric monitoring. (p 964)
- Utilize absorption/adsorption to manage a hazardous materials incident. (NFPA 472, 6.6.3, 6.6.3.1, 6.6.4, 6.6.4.1, 6.6.5, p 968)
- Construct a dike. (NFPA 472, 6.6.3, 6.6.3.1, 6.6.4, 6.6.4.1, 6.6.5, p 968)
- Construct an overflow dam. (NFPA 472, 6.6.3, 6.6.3.1, 6.6.4, 6.6.4.1, 6.6.5, p 969)
- Construct an underflow dam. (NFPA 472, 6.6.3, 6.6.3.1, 6.6.4, 6.6.4.1, 6.6.5, p 969)
- Construct a diversion. (NFPA 472, 6.6.3, 6.6.3.1, 6.6.4, 6.6.4.1, 6.6.5, p 969)
- Construct a retention system. (NFPA 472, 6.6.3, 6.6.3.1, 6.6.4, 6.6.4.1, 6.6.5, p 970)
- Perform vapor dispersion to manage a hazardous materials incident. (NFPA 472, 6.6.3, 6.6.3.1, 6.6.4, 6.6.4.1, 6.6.5, p 971)
- Perform vapor suppression to manage a hazardous materials incident. (NFPA 472, 6.6.3, 6.6.3.1, 6.6.4, 6.6.4.1, 6.6.5, p 971)

Fire Fighter II FFII

Knowledge Objectives

There are no knowledge objectives for Fire Fighter II candidates. NFPA 1001 contains no Fire Fighter II Job Performance Requirements for this chapter.

Skills Objectives

There are no skill objectives for Fire Fighter II candidates. NFPA 1001 contains no Fire Fighter II Job Performance Requirements for this chapter.

You Are the Fire Fighter

A 55-gallon (208-liter) steel drum has fallen off a loading dock at an abandoned warehouse in your first-due response area. The warehouse borders a small residential area with garden apartments. Your engine company is dispatched as part of the initial response and is the first unit to arrive on the scene. A police officer meets you at the front gate and points out the location of the drum on the ground. From the gate, you can see the drum, and it is obviously leaking from the closure on the top. The police officer has already checked the area and there are no occupants in the area, except for the playground in the residential complex, where children are playing. You and your crew must manage the incident until the department's hazardous materials team arrives.

1. Which immediate actions can you and your crew safely take based on the level of training that you currently have?
2. Which additional information should you obtain while waiting for the hazardous materials team to arrive?
3. Which resources do you currently have available to use and which resources will you request to respond to the incident?

Introduction

During the initial size-up at a hazardous materials incident, the first decision concerns personnel safety. The incident commander (IC) needs to decide whether any response would be effective to control and minimize the hazardous materials incident, or if such action would unnecessarily risk harming fire fighters and other responders for little or no benefit. Basically, the question asked is this: Will doing nothing make the situation worse than if fire fighters respond with some sort of action?

If an unknown hazardous material is covering a large area and already negatively affecting large numbers of residents, the appropriate answer to this question is to withdraw responding personnel for their safety. This decision goes against the basic nature of fire fighters, who are willing to take risks to help others. It will be difficult to accept the idea that no response action can be taken because the risk is too great or because there are no living victims to save. This situation can be compounded if family members or friends are located in affected areas. Nevertheless, if fire fighters or other responders expose themselves to unnecessary risk, injury, exposure, or contamination, such action merely complicates the problem.

It has long been held that fire fighters "must" do something. In the case of an unknown or uncontrolled hazardous materials incident, doing "nothing" may well be the best course of action and is often the correct response.

Exposures

In case of a hazardous materials incident, exposures can consist of people, property, structures, or the environment that are subject to influence, damage, or injury as a result of contact with the hazardous material. The amount of exposures that remain is determined both by the location of the incident and by the amount of progress that has been made in protecting those exposures via isolation and other indirect responses. Incidents in urban areas will likely have more exposures and, therefore, will likely need more resources to protect those exposures from the hazardous materials.

Protective Actions

The IC must consider many factors before deciding on protective actions for responders and the general public, such as evacuation or sheltering-in-place. The IC must first evaluate the risk assessment of the hazardous material, the container holding the hazardous material, and the environment surrounding the incident, as well as the resources available to handle the incident.

Evacuating the exposed area is one of the first response priorities at a hazardous materials incident **FIGURE 33-1** Evacuation involves some significant risks even when it is properly planned. Once the evacuation order is given, fire fighters are charged with assisting in the physical evacuation of residents to a safe area. This duty might include traveling to homes and informing residents that they must leave their homes for a shelter.

Before an evacuation order is given, a safe area and suitable shelter should be established. In many cases, schools are used for shelters.

Depending on the time of day and the season, it may take a considerable amount of time to evacuate even a small area. The evacuation area should be located close enough to the exposure for the evacuation to be practical, yet far enough away from the incident to be safe. Temporary evacuation areas may be needed to shelter residents until longer-term evacuation sites or structures indicated in the community's emergency plan can be accessed and opened.

Transportation to temporary evacuation areas must be arranged for all populations. The needs of handicapped and special-needs persons should be considered as well. The initial areas that need to be evacuated can be determined from the *Emergency Response Guidebook*. This guidebook provides the minimum basic guidelines for the initial response, but it does not give complete detailed information for all conditions that responders might potentially encounter, such as weather

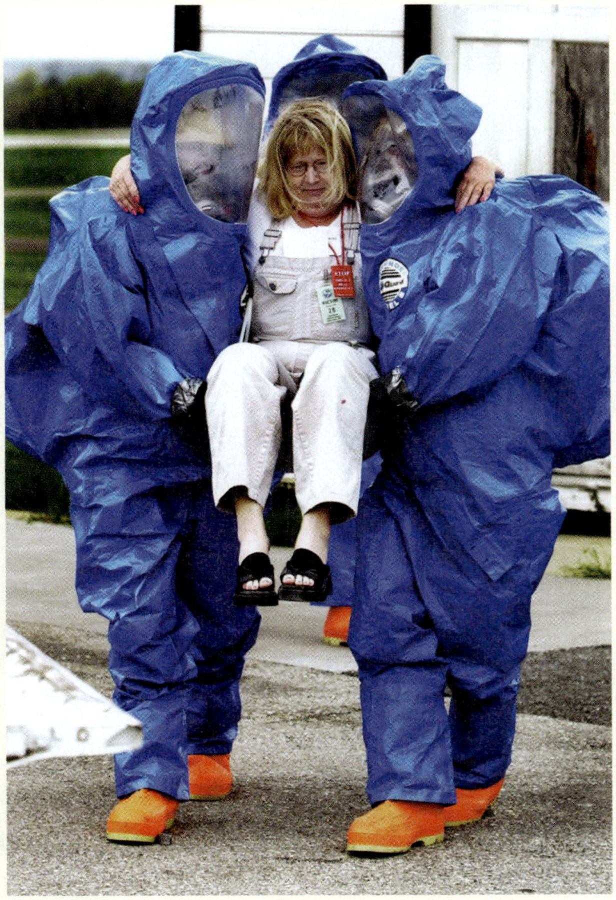

FIGURE 33-1 Evacuating the exposed area is one of the first response priorities at the hazardous materials incident.

extremes, road conditions, or actual conditions encountered at the scene.

The best way for the IC to determine the appropriate area of evacuation and decide how far to extend actual evacuation distances is to have hazardous materials technicians with detection devices monitor both the concentrations of hazardous material and the direction and speed that the material is moving because of wind currents and terrain. Access to local weather conditions is invaluable in this decision-making process.

Another concern when considering evacuation is the potential for exposure to the hazardous material during the evacuation attempt. Personnel may have to go into or through the hazardous material to remove residents. Also, when the evacuated residents are removed from protected areas, they may pass into or through the hazardous material. These individuals must be adequately protected from the hazardous material during their egress or they will become victims.

If the area to be evacuated is more than a few blocks or if it involves any special hazards such as jails, hospitals, or nursing homes, the IC will need to activate local emergency management plans and involve agencies other than the fire department for the safe and orderly evacuation of these types of facilities.

Shelter-in-place is a method of safeguarding people located near or in a hazardous area by keeping them in a safe atmosphere, usually inside structures. Local emergency planning should address local facilities that have large populations to evacuate, such as schools, churches, hospitals, nursing homes, and apartment complexes. When residents are sheltered-in-place, they remain indoors with windows and doors closed. All ventilation systems are turned off to prevent outside air from being drawn into the structure.

The toxicity of the hazardous material is a major factor in the decision whether to evacuate or to shelter-in-place. The more toxic the substance, the less likely a shelter-in-place strategy is to be appropriate. The expected duration of the incident is another factor in determining whether shelter-in-place is a viable option. The longer the incident, the more time the material has to enter or permeate into protected areas.

Flammability is another consideration in the decision whether to evacuate or to shelter-in-place. If the risk of flammability is high, then sheltering-in-place may no longer be the safest option.

In areas that are home to many chemical facilities, it is not uncommon for fire departments and emergency management agencies to provide preincident information on shelter-in-place actions. The dissemination of this type of information may be a function of the fire department's Fire Prevention Bureau.

The ability to evacuate an area and to staff an adequate emergency shelter will be determined by the local resources available. There is no mathematical formula to calculate how many residents can be evacuated at any given time. In severe weather, evacuation can be challenging at best. Even in perfect circumstances, evacuation of residents will be a process that is measured in hours, not minutes. Local emergency planning for evacuations is part of a safe and successful emergency evacuation.

Fire Fighter Safety Tips

When evacuating resident populations, you should be properly protected in case the wind shifts or the hazardous material engulfs you suddenly. Don the proper personal protective equipment (PPE) before you begin assisting with evacuation.

Detecting the Hazard

When responding to hazardous material incidents, responders will be required to take appropriate steps to monitor the atmosphere for potential hazards. In general, atmospheric monitoring should be conducted in a prioritized manner. Monitoring priorities should include radiation, corrosive vapors, oxygen levels, combustibility (flammability), and final toxicity.

Radiation monitoring is conducted using a direct-reading radiation monitor such as a Geiger counter. Corrosive vapor monitoring can be conducted using either pH paper or a pH meter. For example, a stick with a dry piece of pH paper may be used to test liquids and a wet piece of pH paper may be used to test vapors. Detector/colorimetric tubes are another method that can be used to monitor for a wide variety of chemicals.

FIRE FIGHTER Tips

Combustible gas indicators are not direct-reading instruments unless they are monitoring for the gas or vapor for which the combustible gas indicator is calibrated.

Oxygen levels and combustible levels are typically monitored with the use of a two-, three-, four-, or five-gas monitor. These direct-reading monitors provide a direct indication of the oxygen level (as a percentage), combustible gas levels [typically expressed as a percentage of the lower flammability level (LFL)], and the levels of specific gases such as carbon monoxide, hydrogen sulfide, and sulfur dioxide. Other specialized chemical monitors can be used to detect individual chemicals. Photo-ionization and flame-ionization detectors can be used to detect an even wider range of organic and inorganic gases. Finally, specialized instruments are available to detect chemical agents.

To use a multi-gas meter to provide atmospheric monitoring, follow the steps in SKILL DRILL 33-1 FIGURE 33-2:

1. Prior to use of the multi-gas meter, make sure you understand the manufacturer's recommendations and local standard operating procedures for multi-gas analyzer use. The instruments should be calibrated and tested prior to use to verify that they are working properly.
2. Turn on the analyzer, allow the instrument to warm up, and zero it in a clean atmospheric environment.
3. Make sure you understand which hazards and conditions to avoid.
4. Approach the hazardous material and monitor the atmosphere.
5. Note the readings and interpret the meaning of the following conditions:
   a. Oxygen increase or decrease
   b. Lower explosive limit (LEL)/LFL increase
   c. Carbon monoxide increase
   d. Hydrogen sulfide increase
6. When mission is completed, allow meter to run in fresh air. Follow the appropriate procedures to turn the meter off and return it to service.

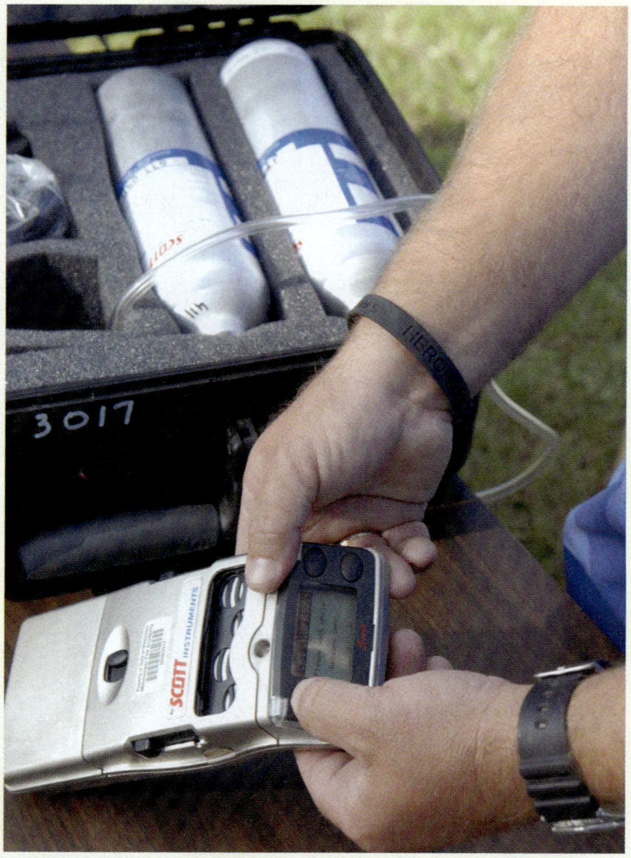

FIGURE 33-2 A multi-gas meter can provide atmospheric monitoring.

FIRE FIGHTER II Tip FFII

When using detection equipment, you must understand the capabilities and the limitations of each instrument or detection device. Remember that the detection equipment provides you with a piece of information and you must understand how to interpret that information properly. Verification of the presence of a hazardous material should not be based solely on the response of a single detection device, but rather must be based on a complete and thorough assessment of the scene.

Search and Rescue

The protection of life is the first priority in any emergency response situation, and a hazardous materials incident is no exception. The search for and rescue of people involved in a hazardous materials event is the primary concern. However, this process becomes more complicated in a hazardous materials incident than in a fire situation.

In a structure fire, search and rescue operations are done quickly to remove victims from the smoke-filled environment. By contrast, in a hazardous materials incident, all emergency response personnel (e.g., fire, law enforcement, emergency medical services) must first recognize and identify which hazardous materials may be present. After this task is accomplished, the IC must slowly and cautiously evaluate the conditions before sending in personnel to attempt a search and rescue.

After identifying the hazards present, the IC must determine whether hazardous materials technicians can safely enter the hot zone to perform search and rescue. If the hazardous material is highly toxic, victims may not survive the exposure to the material; in this case, no search would be necessary if a hazardous material exposure is not survivable. If it is determined that a search can be safely performed, however, rescue teams wearing proper PPE may enter the hot zone to retrieve victims. These victims would then be removed to the warm zone, where they can be decontaminated and given emergency medical attention in preparation for transport to a medical facility.

FIRE FIGHTER Tips

When a fire fighter is injured in a hazardous materials incident, he or she is not the only fire fighter who must leave the scene. One or more fire fighters are needed to treat the injured fire fighter and/or transport that individual to medical care unless Emergency Medical Services personnel are on-scene to perform those duties.

Exposure Protection

Exposures can be protected in different ways. For example, the exposure from the threat can be eliminated by evacuating residents. In addition, a barrier might be constructed to prevent the hazardous materials from reaching the exposure. Alternatively, the hazardous material could be neutralized chemically, such as by treating an acid with a basic material to weaken the strength of the acid. Only a responder at the hazardous materials technician level should perform this kind of neutralization technique. The IC is ultimately responsible for determining which offensive and or defensive actions are to be taken and who will take those actions on the scene of a hazardous materials incident.

Confinement and Containment

Confinement is the process of attempting to keep the hazardous material on the site or within the immediate area of the release. It can usually be accomplished by damming or diking a material, or by confining vapors to a specific area.

Containment refers to actions that stop the hazardous material from leaking or escaping from its container. Examples of containment include patching or plugging a breached container, or righting an overturned container. Only a responder at the hazardous materials technician level should patch or plug a container, given that direct product contamination is likely to occur while the responder is undertaking such an operation. A responder operating at the hazardous materials operations level can perform several different actions where no direct contact with the product would be likely to occur.

When considering a containment response, a number of factors must be evaluated, including the maximum quantity of material that can be released, the level of risk posed by the hazard, and the likely duration of the incident if no intervention is attempted. Ultimately, the IC—and only the IC—is responsible for directing any offensive or defensive operations and for managing who is doing what at the scene.

Flammable Liquids Vapor Control and Fire Extinguishment

Fires associated with a hazardous materials incident should be addressed with great caution, and the principles of extinguishment should be dealt with in a systematic manner. First and foremost is the need to recognize and identify the hazardous material or materials involved. Some hazardous materials react violently if water is applied—important knowledge for fire fighters given that water is the most commonly used firefighting agent. After identification of the material is made and confirmed, its physical and chemical properties—such as its specific gravity, flash point, and flammable ranges—should be researched.

Most flammable and combustible liquid fires can be extinguished by the use of foam. A complete discussion of foam appears in the Fire Attack and Foam chapter.

Aqueous film-forming foam (AFFF) can be used at 1 percent, 3 percent, or 6 percent concentrations. This type of foam is designed to form a blanket over spilled flammable liquids to suppress vapors, or on actively burning pools of flammable liquids. Foam blankets are designed to prevent a fire from reigniting once extinguished. Most AFFF concentrates are biodegradable. AFFF is usually applied through in-line foam eductors, foam sprinkler systems, and portable or fixed proportioning systems.

Alcohol-resistant aqueous film-forming foam is specially formulated to form a polymeric membrane or barrier between the fuel surface and the foam blanket. This foam is intended to be used at proportioning ratios of 1 percent and 3 percent for hydrocarbon fuels and 3 percent or 6 percent for polar fuels such as ethanol-blended fuels.

FIRE FIGHTER Tips

Today the number one product being transported by rail tank cars and highway cargo tank trucks is ethanol-blended fuel. The widespread transport of this hazardous material has raised safety concerns.

Fluoroprotein foams contain protein products mixed with a synthetic fluorinated surfactant. Fluoroprotein foams can be used on fires or spills involving gasoline, oil, or similar products. Fluoroprotein foams rapidly spread over the fuel, causing fire knockdown and vapor suppression. These foams, like many of the synthetic foam concentrates, are resistant to polar solvents such as ketones and ethers.

Protein foam concentrates are made from hydrolyzed proteins, along with stabilizers and preservatives. These types of foams are very stable with good expansion properties. They are quite durable and resistant to reignition when used on Class B fires or spills involving nonpolar substances such as gasoline, toluene, oil, and kerosene.

High-expansion foams are used when large volumes of foam are required for spills or fires in warehouses, tank farms, and hazardous waste facilities. With these products, it is not uncommon to have the ability to produce more than 1000 gallons of finished foam for every gallon of foam concentrate used. This is accomplished by pumping large volumes of air through a small screen coated with a foam solution. Because of the large amount of air in the foam, high-expansion foam is referred to as "dry" foam. By excluding oxygen from the fire environment, high-expansion foam smothers the fire, yet leaves very little water residue.

Remember that as foam is applied, more volume will be added to the spill.

Foam should be gently applied or bounced off another adjacent object so that it flows down across the liquid and does not directly upset the burning surface. Foam can also be applied in a rain-down method by directing the stream into the air over the material and letting the foam fall gently, much as rain would fall to the ground.

This technique is less effective when the fire is creating an intense thermal column. When the heat from the thermal column is high, the water content of the foam will turn to steam before it contacts the liquid surface.

Foam can also be applied through the roll-on method, by bouncing the stream directly into the front of the spill area and allowing it to gently push forward into the pool rather than directly "splashing" it in.

Pressurized-Gas-Cylinder Vapor Control and Fire Extinguishment

Incidents involving pressurized-gas cylinders may consist of fires or releases of the cylinders' contents. Some materials stored in pressurized cylinders may be flammable, others may be flammable and toxic, and still other gases may pose asphyxiant hazards to personnel. The precise nature of the threat depends on the contents of the cylinder, the area in which the cylinder is stored in at the time of the emergency, and the design and integrity of the cylinder itself.

Response to incidents involving cylinders may be very routine, extremely hazardous, or anywhere in between the two extremes. Owing to the unique nature of the hazards presented by cylinders involved in hazardous materials incidents, responders need to err on the side of caution at all times. Evacuation to a safe distance may well be the best response.

By definition, an emergency involving a cylinder or compressed gas comprises the actual or potential release of a hazardous material that cannot be stopped by closing the product's cylinder or container valve. Leaks involving compressed gases require immediate evacuation of the contaminated area. Isolate the cylinder in a well-ventilated and secure area, but move it only if the transport can be done in a safe manner. It is important to recognize all hazards associated with such a material so that proper action can be taken without risk to anyone:

- Evacuation of all nonessential personnel
- Rescue of injured people by crews equipped with the necessary PPE
- Corrective action to minimize the leak or at least minimize exposure to people and equipment
- Assurance that all necessary resources are available for the final resolution of the situation
- Firefighting action
- Decontamination
- Written documentation and critique

These procedures describe the type of actions to take when the leak is of a minimal size and corrective measures can be taken without risk to personnel and/or direct product exposure. Leaks of a large nature, however, require more sophisticated response efforts. Emergency action plans must be based on the nature of the product and the exact situation.

In some ways, fires are less of a problem with pressurized-gas-cylinder incidents because the escaping gas is consumed by the fire. No fire should be extinguished unless fire fighters are certain that they will be able to control the escaping gas once the fire is out. Otherwise, allowing the fire to burn itself out is likely the best course of action.

Some fire departments may believe that the previously described procedures are too offensive in nature and prefer to take a defensive approach to all incidents that involve compressed gases. Always consult your fire department's standard operating policies and procedures and follow them as well as the advice of the local hazardous materials team.

Hazardous Materials Control Activities

As fire fighters approach a hazardous materials incident, they should look for natural control points—that is, areas in the terrain or places in a structure where materials such as liquids and powders or solids can be contained. Hazardous materials technicians should use monitoring equipment to determine where a material has traveled. Some natural barriers that might be used to contain the material include doors to a room, doors to a building, and curb areas of roadways; these natural barriers make for an effective way to establish easily identifiable control zones. In addition, operational personnel can use a number of defensive control measures to manage the incident effectively.

Absorption/Adsorption

Absorption is the process of applying a material (an absorbent) that will soak up and hold the hazardous material in much the same way that a sponge gathers up water. This technique makes collection and disposal of a hazardous material in liquid form more manageable. For example, a dry, granular, clay-based material or dry sand may be added to a spill to help contain the spilled product. To apply the absorbent material, fire fighters should distribute it from a distance using shovels. Any tools used for dispersal and pickup of the absorbent material (e.g., shovels) need to be nonsparking to prevent ignition of the spilled material.

The technique of absorption is often difficult for operational personnel because it generally requires them to be in close proximity to the spill. It also involves the application of material to a spilled product, which adds volume to the spill. The absorbent material may react with certain hazardous substances as well, so it is important to determine whether an absorbent substance is compatible with the hazardous material before it is applied to the spill. The addition of the absorbent material does not change the hazard of the spilled material, and appropriate PPE must still be worn.

Adsorption is a process that occurs when a liquid hazardous material accumulates on the surface of a solid, forming a molecular or atomic film. It is different from absorption, in which a substance passes into a material to form a solid.

VOICES OF EXPERIENCE

As emergency responders, always be mindful that responses we may think of as routine may turn into a hazardous materials incident. For example, when I was serving as an emergency management specialist, a "routine" garbage truck on fire became a hazardous materials incident. The single engine company that responded first encountered a fire in the rear area of the garbage compartment of the truck. While overhauling the fire area, the engine company found a closed metal ammunition-type can. The personnel hooked the can to a pike pole to remove it from the compartment. During removal from the compartment, the lid opened and the product inside started to react and ignite. Water was applied to the fire and the material reacted. After the application of massive amounts of water, the fire was extinguished. At that time, a hazardous materials response was requested and I made my way to the scene.

One of the key elements of a hazardous materials response was undertaken: the identification of the products involved. The product involved was determined to be phosphorous. This raised an eyebrow because phosphorous is sometimes used in the illicit manufacture of methamphetamine. Various law enforcement agencies then became involved at the local, state, and federal levels. Clean-up operations of the dried runoff in the roadway were conducted, as it is possible for phosphorous residue to ignite from friction caused by vehicle tires.

For me, this experience underscores the need for situational awareness at any incident and the need to expect the unexpected. Remember that products that initially seem relatively innocuous may turn out to be harmful.

Glenn C. Clapp
Fire Training Commander
High Point, North Carolina Fire Department

Disposal is generally not the responsibility of a fire department, but rather is carried out by the responsible party. If a responsible party cannot be quickly located, however, fire departments need to plan for the proper disposal of any contaminated materials collected from these types of incidents. Notify the appropriate local or state health or environmental regulatory agency to coordinate the removal and disposal of the spilled material.

To use absorption/adsorption to manage a hazardous materials incident, follow the steps in **SKILL DRILL 33-2**:

1. Collect the basic materials:
   a. Absorbents
   b. Adsorbents
   c. Absorbent pads
   d. Absorbent booms
2. Decide which material is best suited for use with the spilled product.
3. Assess the location of the spill and stay clear of any spilled product. (**STEP 1**)
4. Apply the appropriate material to control and contain the spilled material. (**STEP 2**)
5. Maintain materials and take appropriate steps for their disposal. (**STEP 3**)

Diking, Damming, Diversion, and Retention

Controlling liquid spills can be difficult because liquids tend to take the shape of their container regardless of where that container is located. Several techniques can be effective in managing these types of incidents.

Diking

Diking is the placement of impervious materials to form a barrier that will keep a hazardous material in liquid form from entering an area, or that will hold the material in a specific area. To construct a dike, follow the steps in **SKILL DRILL 33-3** **FIGURE 33-3**:

1. Based on the spilled material and the topography conditions, determine the best location for the dike.
2. Dig a depression in the channel 6 to 8 inches deep if needed.
3. Use plastic to line the bottom of the depression, and apply sufficient plastic to cover the dike wall.
4. Build a dam with sandbags or other available materials.
5. Install plastic to the top of the dike and secure it with additional sandbags.
6. Complete the dike installation and ensure the size will contain all of the spilled product.

Damming

Damming is used when a liquid is flowing in a natural channel or depression and its progress can be stopped by blocking the channel. Three kinds of dams are typically used in such circumstances: a complete dam, an overflow dam, or an underflow dam.

A complete dam is placed across a small stream or ditch to completely stop the flow of materials through the channel. This type of dam is used only in areas where the stream or ditch is basically dry and the amount of material that needs to be controlled is relatively small.

For streams or ditches with continuous flow of water, an overflow or underflow dam will need to be constructed. An overflow dam is used to contain materials that are not water soluble and have a specific gravity greater than 1. It is constructed by building the dam base up to a level that hold back the flow and then installing polyvinyl chloride pipe or hard

SKILL DRILL 33-2 Using Absorption/Adsorption to Manage a Hazardous Materials Incident

1. Decide which material is best suited with the spilled product. Assess location of spill and stay clear of spilled product.
2. Apply appropriate material to control and contain the spilled material.
3. Maintain materials and take appropriate steps for their disposal.

FIGURE 33-3 Dikes keep a hazardous material from entering an area.

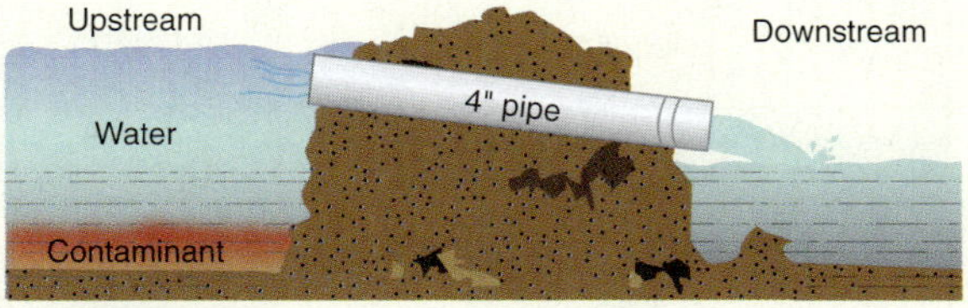

FIGURE 33-4 An overflow dam.

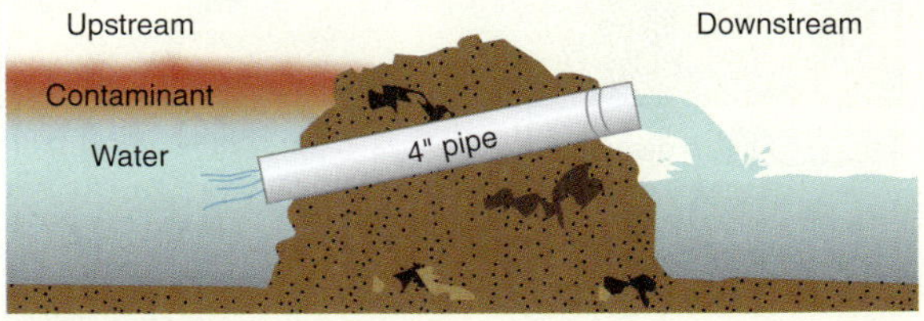

FIGURE 33-5 An underflow dam.

suction hose at an angle that allows the water to flow "over" the dam, trapping the heavier material at a low level at the base of the dam. An underflow dam is used to contain materials that are not water soluble and have a specific gravity less than 1. It is basically constructed in the opposite manner of the overflow dam. The piping on the underflow dam is installed near the bottom of the dam so that waters flow "under" the dam, allowing the materials floating on the water to accumulate at the top of the dam area. It is critical to have sufficient pipes or hoses to allow sufficient water to flow past the dam without overflowing over the top of the dam.

To construct an overflow dam, follow the steps in SKILL DRILL 33-4 FIGURE 33-4:

1. Based on the spilled material having a specific gravity greater than 1, determine the need for an overflow dam.
2. Dig a depression in the channel 6 to 8 inches (15 to 20 centimeters) deep.
3. Use plastic to line the bottom of the depression. (Step 1)
4. Build a dam with sandbags or other available materials. (Step 2)
5. Install two to three lengths of 3- to 4-inch (7- to 10-centimeter) plastic pipe horizontally on top of the dam. Add more sandbags on top of the dam if available.
6. Complete the dam installation and ensure its size will allow the proper flow. (Step 3)

To construct an underflow dam, follow the steps in SKILL DRILL 33-5 FIGURE 33-5:

1. Based on the spilled material having a specific gravity less than 1, determine the need for an underflow dam.
2. Dig a depression in the channel 6 to 8 inches (15 to 20 centimeters) deep.
3. Use plastic to line the bottom of the depression.
4. Build a dam with sandbags or other available materials.
5. Install two to three lengths of 3- to 4-inch (7- to 10-centimeter) plastic pipe at a 20- to 30-degree angle at a level suitable to ensure that clean water passes through the pipes.
6. Add more sandbags on top of the dam if available.
7. Complete the dam installation and ensure its size will allow the proper flow.

Diversion

A diversion technique is used to redirect the flow of a liquid away from an area. Liquid hazardous materials spills can be controlled by using existing barriers such as curbs and channeling liquids away from storm drains that would otherwise allow them to leave the site. Diversion can also be used to protect environmentally sensitive areas.

To use diversion to manage a hazardous materials incident, follow the steps in SKILL DRILL 33-6 FIGURE 33-6:

1. Based on the spilled material and the topography conditions, determine the best location for the diversion.
2. Use sandbags or other materials to divert the product flow to an area of fewer hazards.
3. Stay clear of the product flow.
4. Monitor the diversion channel to ensure the integrity of the system.

Retention

Retention is the process of creating an area to hold hazardous materials. For instance, fire fighters might dig a depression and allow material to pool in the depression. The hazardous material would then be held there until a clean-up contractor could recover it. A hazardous material that is leaking in a room of an industrial facility might be safely approachable after its proper recognition and identification by operational personnel wearing the proper level of PPE, who could then close a door and contain the spill within one room. As another example, the

FIGURE 33-6 A diversion.

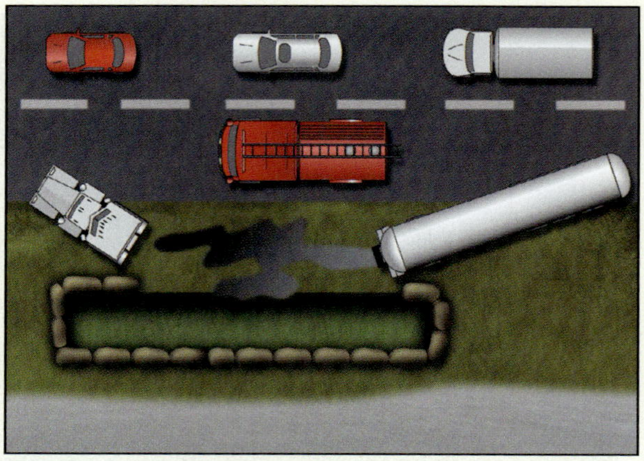

FIGURE 33-7 A retention system.

public works department or an independent contractor might use a backhoe to create a retention area for runoff at a safe downhill distance from the release.

To use retention, follow the steps in SKILL DRILL 33-7 FIGURE 33-7:

1. Based on the spilled material and the topography conditions, determine the best location for the retention system.
2. Dig a depression in the channel to serve as a retention area.
3. Use plastic to line the bottom of the depression.
4. Use sandbags or other materials to hold the plastic in place.
5. Stay clear of the product flow.
6. Monitor the retention system to ensure its integrity.

Dilution

Another protective action that can be taken from a safe distance is dilution. Dilution entails the addition of water or another liquid to weaken the strength or concentration of a hazardous material. This tactic can be used only when the identity and properties of the hazardous material are known with certainty. One concern with dilution relates to the fact that water applied to dilute a hazardous material increases the total volume; if the volume increases too much, it may overwhelm containment measures. Dilution is considered a risky operation because of these potential pitfalls and the possibility that it might exacerbate the contamination. As a consequence, it should be used with extreme caution and only on the advice of those persons who are knowledgeable regarding the nature of the chemicals involved in the incident.

To use dilution to manage a hazardous materials incident, follow the steps in SKILL DRILL 33-8 FIGURE 33-8:

1. Based on the spilled material's characteristics and the size of the spill, determine the viability of a dilution operation.

FIGURE 33-8 Using dilution.

2. Obtain guidance from a hazardous materials technician.
3. Ensure that the water used will not overflow and affect other product-control activities.
4. Add small amounts of water from a distance to dilute the product.
5. Contact the hazardous materials technician if any additional issues arise.

Vapor Dispersion and Suppression

When a hazardous material produces a vapor that collects in an area or increases in concentration, vapor dispersion or suppression is required.

Vapor dispersion is the process of lowering the concentration of vapors by spreading them out. Vapors can be dispersed with hose streams set on fog patterns, large displacement fans, or heating and cooling systems found in fixed hazardous materials handling systems.

Before dispersing vapors, the consequences of this action should be considered. If the vapors are highly flammable, an attempt at dispersing the gases might ignite them. If the vapors are corrosive, the runoff will be corrosive and must be confined. The dispersed vapors might also contaminate areas outside the hot zone. Vapor dispersion with fans or a fog stream should be attempted only after the hazardous material is safely identified and only if the weather conditions will promote the diversion of vapors away from any populated areas. Conditions must be constantly monitored during the entire operation.

Vapor suppression is the technique of controlling fumes or vapors that are given off by certain materials, such as flammable liquid fuels. For example, gasoline gives off vapors, which are dangerous because they are flammable. To control these vapors, a blanket of firefighting or vapor-suppressing foam is layered on the surface of the liquid. The foam seals the surface of the liquid, prevents vapors from escaping, and reduces the danger of vapor ignition.

Reducing the temperature of some hazardous materials will also suppress vapor formation. However, there is no really effective way to accomplish this goal except in cases of the smallest spills.

To use vapor dispersion to manage a hazardous materials incident, follow the steps in SKILL DRILL 33-9 FIGURE 33-9:

1. Based on the characteristics of the released material, the size of the release, and the topography, determine the viability of a dispersion operation. Stay upwind and uphill.
2. Use a four-gas monitor to identify a safe work area.
3. Ensure that ignition sources in the area have been removed if the material is flammable.
4. Use water from a distance to disperse vapors.
5. Monitor the environment until the vapors have been adequately dispersed.

FIGURE 33-9 Using vapor dispersion to manage a hazardous materials incident.

To use vapor suppression to manage a hazardous materials incident, follow the steps in SKILL DRILL 33-10 FIGURE 33-10:

1. Based on the characteristics of the released material, the size of the leak, and the topography, determine the viability of a vapor suppression operation.
2. Use a four-gas monitor to identify a safe work area.
3. Ensure that ignition sources in the area have been removed.
4. Use a foam stream from a distance to suppress vapors.
5. Monitor the environment until the vapors have been adequately dispersed.

FIGURE 33-10 Using vapor suppression to manage a hazardous materials incident.

Fire Fighter Safety Tips

Remember to take rehabilitation breaks throughout the hazardous materials incident. Participating in defensive operations takes a great deal of physical energy and mental concentration. Maintain hydration levels and watch your fellow responders for signs of fatigue.

Remote Shut-off

A protective action that should always be considered, especially with transportation emergencies or incidents at fixed facilities, is the identification and isolation of remote shut-off valves. Many chemical processes, or piped systems that carry chemicals, have a way to remotely shut down a system or isolate a valve. This may prove to be a much safer way to mitigate the problem. Fixed ammonia systems provide good examples of the effectiveness of using remote shut-off valves. In the event of a large-scale ammonia release, most current systems have a main dump valve that will immediately redirect anhydrous ammonia into a water deluge system. This deluge system is designed to knock down the vapors and collect the resulting liquid in a designated holding system.

Many types of cargo tanks have emergency remote shut-off valves. MC-306/DOT406 cargo tanks, for example, are certified to carry flammable and combustible liquids and Class B poisons. These single-shell aluminum tanks can hold 9000 gallons or more of product at atmospheric pressures as high as 4 pounds per square inch (psi). The tanks have an oval/elliptical cross-section and overturn protection that serves to protect the top-mounted fittings and valves in the event of a roll-over. In the event of fire, a fusible vent is designed to melt at a predetermined temperature, which in turn causes the closure of the internal product discharge valve. A baffle system within the cargo tank compartment reduces the surge of the liquid contents. With these containers, remote shut-off valves are often found near the front (adjacent to the driver's door) or located at the rear of the cargo tank.

MC-307/DOT407 cargo tanks are certified to carry chemicals that are transported at low pressure (less than 25 psi), such as flammable and combustible liquids and other mild corrosives. These vehicles can carry as much as 7000 gallons of product and are outfitted with similar safety features as are found on the MC-306/DOT406 cargo tanks.

MC-331 cargo tanks carry substances such as anhydrous ammonia, propane, butane, and liquefied petroleum gas. The tanks are not insulated and feature rounded ends and a circular cross-section. MC-331 cargo tanks have a carrying capacity of between 2500 and 11,500 gallons. These containers have emergency shut-off valves either at both ends of the tank (tractor-trailer type) or at one location with possible radio control (bobtail or straight-truck types); internal shut-off valves; internal volume and pressure gauges; and two top-mounted vents.

Decision to Withdraw

After all factors have been considered, the IC may decide that the hazardous materials incident cannot be handled without incurring unnecessary risks to all personnel. Perhaps the resources are not available to mitigate the incident or there are no exposures to be saved. In such a case, the IC may decide to withdraw to a safe distance, set a defensive perimeter, and wait for additional resources to arrive. He or she may also simply allow the hazardous materials incident to run its course. This situation arises with certain chemicals, such as pesticides; they are consumed in a hot fire and, therefore, may be allowed to burn. The decision to extinguish a fire involving pesticides could result in runoff contamination of surrounding areas, worsening the impact of the incident.

FIRE FIGHTER Tips

When deciding to withdraw from a hazardous materials incident, the IC will consult with technical advisors, local and state health agencies, and environmental agencies.

Recovery

The recovery phase of a hazardous materials incident occurs when the imminent danger to people, property, and the environment has passed or is controlled, and clean-up begins. During the recovery phase, both state and federal agencies may become involved in cleaning up the site, determining the responsible party, and implementing cost-recovery methods. In addition, a transition is made between on-scene public safety responders and private-sector commercial clean-up companies. The recovery phase also includes completion of the records necessary for documenting the incident.

It must be stressed that the hazardous materials incident is not over at this point. As long as public safety personnel and responders remain on the scene, they should maintain their vigilance for safety to ensure that no new hazards are created and no injuries are incurred during this final phase. Always remember that fire fighters' primary responsibility is to ensure the safety of themselves and other responders.

When to Terminate the Incident

The decision to terminate the hazardous materials incident is made by the IC. The recovery phase in large-scale incidents can go on for days, weeks, or even months. The recovery phase and clean-up will likely require amounts of resources and equipment that go far beyond the capabilities of local responders. The ultimate goal of incident termination is to return the property or site of the incident to a preincident condition, and then return the facility or mode of transportation to the responsible party. This handoff often takes place long after fire fighters are released from the hazardous materials scene.

Crime or Terrorist Incident

When an incident involves potential criminal or terrorist activity, responders must take extra precautions and implement appropriate actions to ensure their own safety as well as the safety of the general public. One of the critical actions that must be taken is notifying the appropriate law enforcement agencies of the situation. These notifications may include not only local law enforcement but also appropriate federal agencies depending on the circumstances. Other actions that go hand in hand with law enforcement notification are the preservation of the potential crime scene and good documentation of the initial scene activities by responders. Proper crime scene preservation and strong documentation may prove invaluable to law enforcement during subsequent legal proceedings related to the incident.

In addition, specific activities should be implemented to ensure safety. For example, proper measures should be implemented to limit the potential for cross-contamination. These steps include isolating the scene from the general public and unneeded responders by establishing control zones, isolating people or animals that were potentially contaminated during the initial incident, and ensuring that potentially contaminated materials are not removed from the scene. Responders should also exercise extreme caution during incidents that involve potential criminal or terrorist activity and be aware of the potential risk of secondary devices that may be intended to injure arriving personnel.

As noted earlier, during hazardous materials incidents, especially those involving criminal or terrorist activity, responders should take appropriate measures to preserve incident evidence. Upon arriving on the scene, responders should note key information regarding the scene itself. This step will include documenting information such as who was present, where specific pieces of evidence were located, when time-specific elements of the incident took place, and any other information pertinent to the incident.

The use of photography to document the scene is encouraged if it can be done safely and will not create additional hazards. Document the date and time when the photographs were taken as well as the name of the person who took the pictures. Do not move potential evidence if at all possible. If evidence must be moved to mitigate the incident, document and photograph it if possible.

Limit the number of responders working within the area. All responders who work in the immediate incident area should be asked to document their specific actions in the area as well as their findings regarding evidence in the area. Work with law enforcement agencies to coordinate the interviewing of witnesses, victims, responders, and other individuals who may have important information about the incident.

Wrap-Up

Chief Concepts

- Exposures can consist of people, property, structures, or the environment. The number of exposures is determined by the location and the amount of progress made in protecting the exposures.
- The two main protective actions utilized to protect human exposures are evacuation and sheltering-in-place.
- With evacuation, the role of the fire fighter is to assist in the physical evacuation of residents to a safe area or shelter such as a school. Potential shelters should be determined during preincident planning.
- Shelter-in-place involves keeping residents near a hazardous area safe by keeping them in a healthy atmosphere, usually inside structures and with the ventilation system turned off.
- The toxicity of the hazardous material is a major factor in deciding whether to evacuate or shelter-in-place. The more toxic the substance, the more likely evacuation will be needed.
- Responders are required to take appropriate steps to monitor the atmosphere for potential hazards. Monitoring priorities should include radiation, corrosive vapors, oxygen levels, combustibility, and final toxicity. Radiation monitoring is conducted using a Geiger counter. Oxygen levels and combustible levels are typically monitored with the use of a two-, three-, four-, or five-gas monitor. Other specialized chemical monitors can be used to detect individual chemicals.
- The search and rescue process is more complicated in a hazardous materials incident than in a structure fire. Before beginning search and rescue operations, all emergency response personnel must first recognize and identify when hazardous materials may be present. After this is accomplished, the IC must evaluate the conditions. If hazardous materials are present, the IC must determine whether hazardous materials technicians can safely enter the hot zone to perform search and rescue operations.
- The IC determines which offensive or defensive actions are to be taken and who will take those actions on the scene of a hazardous materials incident.
- Confinement is the process of attempting to keep the hazardous materials on the site or within the immediate area of release. Containment refers to actions that stop the hazardous material from leaking or escaping.
- Fires associated with a hazardous materials incident should be addressed with great caution because some hazardous materials react violently if water is applied. Most flammable and combustible liquid fires can be extinguished through the application of foam.
- Incidents involving pressurized-gas cylinders may involve fires or releases of the cylinders' contents. Some materials stored in pressurized cylinders may be flammable, others may be flammable and toxic, and still other gases may pose asphyxiant hazards to personnel. If a pressurized-gas cylinder is leaking, responders should take these actions:
  - Evacuation of all nonessential personnel
  - Rescue of injured people by crews equipped with the necessary PPE
  - Corrective action to minimize the leak or at least minimize exposure to people and equipment
  - Assurance that all necessary resources are available for the final resolution of the situation
  - Firefighting action
  - Decontamination
  - Written documentation and critique

- Hazardous materials control activities include the following:
    - Absorption/adsorption—Techniques for holding a hazardous material to make collection and disposal easier.
    - Diking—Emplacement of an impervious material to form a barrier that will keep a hazardous material from entering an area.
    - Damming—Includes complete dams, overflow dams, and underflow dams. A complete dam completely stops the flow of materials through a channel. An overflow dam uses piping to trap the material at the base of the dam. An underflow dam allows water to flow through piping under the dam and collects the hazardous material at the top of the dam.
    - Diversion—Redirection of the flow of a liquid from an area.
    - Retention—Creation of an area to hold hazardous materials.
    - Dilution—Addition of water or another liquid to weaken the strength or concentration of a hazardous material.
    - Vapor dispersion—Lowering the concentration of vapors by spreading them out.
    - Vapor suppression—Controlling fumes or vapors that are given off by certain materials.
    - Identify and isolate a remote shut-off valve—Many systems have a way to remotely shut down a system.
- After all of the factors have been considered, the IC may decide that the hazardous materials incident cannot be handled without unnecessary risks and that all personnel should withdraw.
- The recovery phase of a hazardous materials incident occurs when the imminent danger to people, property, and the environment has passed or is controlled, and clean-up begins. During this phase, state and federal agencies may become involved in cleaning up the site, determining the responsible party, and implementing cost-recovery methods.
- The recovery phase in large-scale incidents can last for days, weeks, or even months and may require resources and equipment beyond those available to local responders.
- If a hazardous materials incident involves potential criminal or terrorist activity, responders must notify law enforcement and appropriate federal agencies. Specific activities should be implemented to ensure safety, such as isolating the scene.

Hot Terms

Absorption The process of applying a material (an absorbent) that will soak up and hold a hazardous material in a sponge-like manner, for collection and disposal.

Adsorption A process that occurs when a liquid hazardous material accumulates on the surface of a solid.

Confinement Actions taken to keep a material, once released, in a defined or local area. (NFPA 472)

Containment Actions taken to keep a material in its container (e.g., stop a release of the material or reduce the amount being released). (NFPA 472)

Damming A process used when liquid is flowing in a natural channel or depression and its progress can be stopped by blocking the channel.

Diking The placement of materials to form a barrier that will keep a hazardous material in liquid form from entering an area, or that will hold the material in an area.

Dilution The process of adding some substance—usually water—to a product to weaken its concentration.

Diversion Redirecting spilled or leaking material to an area where it will have less impact.

Recovery phase The stage of a hazardous materials incident after the imminent danger has passed, when clean-up and the return to normalcy have begun.

Remote shut-off valve A device designed to retain a hazardous material in its container.

Retention The process of purposefully collecting hazardous materials in a defined area. It may include creating that area—by digging, for instance.

Shelter-in-place A method of safeguarding people in a hazardous area by keeping them in a safe atmosphere, usually inside structures.

Vapor dispersion The process of lowering the concentration of vapors by spreading them out.

Vapor suppression The process of controlling vapors given off by hazardous materials, thereby preventing vapor ignition, by covering the product with foam or other material or by reducing the temperature of the material.

FIRE FIGHTER *in action*

During your annual department dinner, the alarm sounds, sending your engine company along with the rest of the department to the report of a large yellow cloud coming from the wastewater treatment plant. As your engine approaches the scene, you can see the cloud and smell what appears to be chlorine. In addition, a stream of liquid is coming down the driveway and heading toward the storm drain that empties into the nearby creek. You see the cloud staying low to the ground and moving toward the mall approximately one-fourth mile downwind. Upon seeing the cloud movement, your officer immediately requests additional resources as well as the hazardous materials team to respond.

1. Based on the initial size-up, which type of population-protection strategies could you impose for the mall?
   - **A.** Evacuation
   - **B.** Shelter-in-place
   - **C.** Both a and b
   - **D.** No population protection strategies would be required
2. Based on the liquid moving downhill from the incident, the best method for product control would be damming and retention of the product.
   - **A.** True
   - **B.** False
3. If you use portable monitor streams to provide vapor dispersion because of the gas cloud, you may produce an additional hazard. Which spill control methods could be used to control the runoff?
   - **A.** Diking
   - **B.** Damming
   - **C.** Retention
   - **D.** All of the above
4. Based on the first-due response company's protective clothing, it may be prudent to withdraw until additional hazardous materials resources arrive.
   - **A.** True
   - **B.** False
5. Based on the product involved, which monitoring equipment would be best suited to be used to check the levels of the hazard?
   - **A.** Geiger counter
   - **B.** Multi-gas monitor (LEL/O_2/CO/H_2S)
   - **C.** pH paper
   - **D.** Chemical agent detector
6. Vapor suppression is a relatively harmless defensive operation and should readily be deployed at every hazardous materials scene.
   - **A.** True
   - **B.** False

FIRE FIGHTER II *in action*

Your company has just finished a run to a local chemical facility to respond to an injured employee. While en route back to quarters, you notice a rash on your forearm. You start scratching and remember that you touched the employee with your bare skin accidently while lifting him onto the stretcher.

1. What actions are appropriate for you to take now?
2. Considering the facility and the injuries to the employee, was there anything that should have been done differently to avoid this situation?
3. What policies or procedures should you be aware of when responding to chemical facilities within your response territory?

Hazardous Materials: Decontamination Techniques

Fire Fighter I

Knowledge Objectives

After studying this chapter, you will be able to:

- Describe the purpose of decontamination during a hazardous materials incident. (NFPA 472, 5.3.4, p 978)
- Explain how cross-contamination can occur. (NFPA 472, 5.3.4, p 978)
- List the types of decontamination. (NFPA 472, 5.3.4, p 978–980)
- Describe the process of emergency decontamination. (NFPA 472, 5.3.4, p 978–979)
- Describe the process of gross decontamination. (NFPA 472, 5.3.4, p 979)
- Describe the process of technical decontamination. (NFPA 472, 5.3.4, p 980)
- Describe the process of mass decontamination. (NFPA 472, 5.3.4, p 980)
- List the methods of decontamination. (NFPA 472, 5.3.4, p 980–983)
- Describe the process of absorption. (NFPA 472, 5.3.4, p 982)
- Describe the process of adsorption. (NFPA 472, 5.3.4, p 982)
- Describe the process of dilution. (NFPA 472, 5.3.4, p 982)
- Describe the process of disinfection. (NFPA 472, 5.3.4, p 983)
- Describe the proper procedures to follow to safely dispose of items that cannot be decontaminated. (NFPA 472, 5.3.4, p 983)
- Describe the process of solidification. (NFPA 472, 5.3.4, p 983)
- Describe the process of emulsification. (NFPA 472, 5.3.4, p 983)
- Describe the process of vapor dispersion. (NFPA 472, 5.3.4, p 983)
- Describe the process of soil removal. (NFPA 472, 5.3.4, p 983)
- Describe the process of vacuuming. (NFPA 472, 5.3.4, p 983)
- List the general steps of decontamination. (NFPA 472, 5.3.4, p 984)

Skills Objectives

After studying this chapter, you will be able to perform the following skills:

- Perform emergency decontamination. (NFPA 472, 5.3.4, p 979)
- Perform mass decontamination. (NFPA 472, 5.3.4, p 980, 982)
- Perform responder decontamination. (NFPA 472, 5.3.4, 6.6.4.2, p 984–985)

CHAPTER

34

Fire Fighter II

FFII

Knowledge Objectives

There are no knowledge objectives for Fire Fighter II candidates. NFPA 1001 contains no Fire Fighter II Job Performance Requirements for this chapter.

Skills Objectives

There are no skill objectives for Fire Fighter II candidates. NFPA 1001 contains no Fire Fighter II Job Performance Requirements for this chapter.

You Are the Fire Fighter

The local community college is just completing a clean-up of the science building in your locality when one of the professors knocks over a box containing both labeled and unlabeled bottles. Several of the other teachers and helpers are exposed and contaminated by the released chemicals. The station bells ring, and you are dispatched to the incident along with the hazardous materials team. When your engine company arrives, you see the victims standing near the doorway showing signs and symptoms of some type of chemical exposure and contamination.

1. Which types of decontamination systems might be required for this incident?
2. Which type of personnel resources would be required to staff the decontamination operations as well as to fulfill the fire protection needs?
3. Which type of personal protective equipment and resources would be needed?

Introduction

Decontamination is the physical or chemical process of reducing and preventing the spread of hazardous materials by persons and equipment. Decontamination makes personnel, equipment, and supplies safe by removing or reducing the amount of hazardous materials or weapons of mass destruction (WMDs). Proper decontamination is essential at every hazardous materials/WMD incident to ensure the safety of both personnel and property. In some ways, proper decontamination is the most vital function performed by the fire department at the scene of such an incident.

Contamination

Contamination (secondary contamination) is the process of transferring a hazardous material from its source to people, animals, the environment, or equipment, any of which may act as a carrier of the contaminant. Although contamination can occur though a variety of processes, at hazardous materials incidents it often results from the release of a hazardous material from its intended container. Once a person, animal, or surface is contaminated, that person, animal, or surface can cause other persons, animals, and surfaces to become contaminated. This sort of spreading contamination is known as cross-contamination or secondary contamination.

Cross-contamination

Cross-contamination (secondary contamination) is the process by which a contaminant is carried out of the hot zone and contaminates people, animals, the environment, or equipment. Cross-contamination may occur in several ways:

- A contaminated victim comes into physical contact with an emergency responder.
- A bystander or other emergency responder comes into contact with a contaminated object from the hot zone.
- A decontaminated responder comes into contact with another contaminated responder or object.

Because cross-contamination can occur in so many ways, control zones should be established, clearly marked, and enforced at all hazardous materials incidents.

Types of Decontamination

As noted earlier, decontamination as a process makes personnel, equipment, and supplies safe by removing or eliminating the hazardous materials. Fire fighters are often responsible for establishing a decontamination corridor at a hazardous materials incident for the initial emergency response crews, for the victims of the incident, and in support of the hazardous materials response team. The decontamination corridor is a controlled area, usually located within the warm zone of the incident, where decontamination procedures take place. Everyone who leaves the hot zone is required to pass through the decontamination corridor as he or she exits through the warm zone and eventually into the cold zone.

The major categories of decontamination are emergency decontamination, gross decontamination, technical decontamination, and mass decontamination. Each category provides for a different level of decontamination.

Emergency Decontamination

Emergency decontamination is used in potentially life-threatening situations to rapidly remove the bulk of the contamination from an individual, regardless of the presence or absence of a technical decontamination corridor. A more formal and detailed decontamination process usually follows later. Emergency decontamination usually involves dousing of the victim with adequate quantities of water usually via a hand-held hose line from an engine company using a fog pattern, removal of contaminated clothing, and a second round of dousing. This process is known as "flush, strip, and flush." If a decontamination corridor has not yet been established, fire fighters can isolate the exposed victims in a contained area and establish an emergency decontamination area.

Do not allow the water runoff from emergency decontamination to flow into drains, streams, or ponds; instead, try to divert it into an area where it can be treated or disposed of later. Nevertheless, do not delay decontamination: Human life always comes first. If any extension of the contaminated area occurs as a result of this emergency decontamination, be sure to notify command as soon as possible to ensure an appropriate response to the newly contaminated area.

Whenever possible, identify the hazardous material before beginning decontamination. Some hazardous materials are water reactive and will require a special decontamination process. When in doubt, proceed with water-based decontamination, albeit with caution.

The best method of decontamination for fire fighters and first responders is to avoid contact with the hazardous material. If you do not become contaminated, there is no need to decontaminate!

To perform emergency decontamination, follow the steps in SKILL DRILL 34-1:

1. Ensure that you have the appropriate personal protective equipment (PPE) to protect against the chemical threat.
2. Stay clear of the product, and do not make physical contact with it.
3. Make an effort to contain runoff by directing victims out of the hazard zone and into a suitable location for decontamination.
4. Flush the victim to remove the product from the victim's clothing. (STEP 1)
5. Instruct the victim in removing contaminated clothing. (STEP 2)
6. Flush the contaminated victim.
7. Assist or obtain medical treatment for the victim, and arrange for the victim's transport. (STEP 3)

Gross Decontamination

Gross decontamination, like emergency decontamination, aims to significantly reduce the amount of surface contaminant by delivery of a continuous shower of water and removal of outer clothing or PPE. It differs from emergency decontamination, however, in that gross decontamination is controlled through the decontamination corridor. Hazardous materials teams use a large shower system for gross decontamination as the first part of a multistep decontamination process or line.

The shower system works well if water is appropriate for removing the hazardous material. With a shower-type arrangement, gross decontamination begins when a responder, still fully clothed in PPE and using the appropriate level of respiratory protection, steps into a portable shower where high-pressure, low-volume water flow (or spray) is used to rinse off the contaminants. This step will remove most solid or liquid contaminants from the PPE. Runoff water should be diverted to a safe area for treatment or disposal; it should not be allowed to flow into drains, streams, and ponds.

Do not confuse emergency decontamination with gross decontamination—the two are completely separate concepts. Emergency decontamination is performed for life-safety reasons. Gross decontamination is intended to carry out a brief and rapid contamination reduction—a sort of pre-wash before technical decontamination.

FIRE FIGHTER Tips

Because runoff water from gross decontamination is likely to contain contaminants, make some effort to contain it. Be aware of any obvious storm drains, streams, or ponds in the area, and divert contaminated runoff away from them if possible.

SKILL DRILL 34-1 Performing Emergency Decontamination

1 Ensure that you have the appropriate PPE. Stay clear of the product, and do not make physical contact with it. Make an effort to contain runoff by directing the victim out of the hazard zone and into a suitable location. Flush the victim to remove the product from the victim's clothing.

2 Instruct the victim in removing contaminated clothing.

3 Flush the contaminated victim. Assist or obtain medical treatment for the victim, and arrange for the victim's transport.

Technical Decontamination

Technical decontamination is performed after gross decontamination and is a more thorough cleaning process. This type of decontamination may involve several stations or steps. During technical decontamination, multiple personnel, known as the decontamination team, typically use brushes to scrub and wash the person or object to remove contaminants.

To clean a responder wearing chemical-protective clothing, for example, the decontamination team might use a water spray, long-handled scrub brushes, and a special cleaning solution. The type of cleaning solution will depend on the hazardous material involved. In some situations, special chemicals may be needed for decontamination.

In some cases, law enforcement or search canines may require decontamination after a mission. Search dogs used by Federal Urban Search and Rescue (USAR) teams have particular ways to perform canine decontamination. Protocols will be available from the USAR team or the canine handlers. Law enforcement will have similar protocols for their equipment. Before entering the hot zone, the decontamination team needs to know or be supervised on these procedures.

Additionally, it may be necessary to perform decontamination on a criminal subject, firearms, or other specialized pieces of equipment. Unless life safety is a priority, it is wise to consult the owner of the piece of equipment, or an otherwise knowledgeable person, before doing anything that might have an adverse effect. Unusual items might require unusual methods of decontamination.

It may also be necessary to decontaminate a piece of evidence prior to turning it back over to law enforcement officials. In these cases, maintaining the chain of custody, even through the decontamination process, is important. The item should arrive at decontamination at least double-bagged. Protocols will need to be in place prior to the incident. The outer bag can then be washed, rendering it safe for transportation, laboratory analysis, or other further handling. Care should be taken not to compromise the integrity of any of the containers holding evidence. Additionally, the method used for decontamination should not alter the evidence in any way.

Mass Decontamination

Mass decontamination is often used in incidents involving unknown agents or in the case of a contamination of large groups of people. It takes place in the field and is a way of quickly performing gross decontamination on a large number of victims who have escaped from a hazardous materials incident. Because water is a generally effective solvent, washing off as much of the contaminant as possible with a massive water spray is the best and quickest way to decontaminate a large group of people.

The process itself relies on the "strip, flush, and cover" principle. The victims must first be "herded" into an area where a corridor has been set up and, if possible, has been excluded from public view. Next, the victims should be instructed to remove the outermost layer of clothing necessary to remove the contaminant from themselves. They should then enter the shower area. If fire apparatus is used for this purpose, it should be set up to distribute large quantities of water at a low pressure so as not to injure victims or embed the contaminant in the skin. Once victims exit the shower area, they should be given a cover for protection. The cover could be a blanket, gown, Tyvek suit, or even a black trash bag.

In mass decontamination, the primary concern is to remove a hazardous material from a large number of victims. This process can take place on any street, parking lot, or area where fire apparatus can be deployed with a continuous, uncontaminated water supply. Environmental concerns are secondary; there is rarely enough time to build structures to contain runoff. Although notifying command about the situation remains a priority, mass decontamination should not be delayed to allay environmental concerns.

Victims who go through mass decontamination will need further decontamination and medical monitoring after they emerge from the shower system. When more responders arrive, they can establish a decontamination corridor to clean and thoroughly decontaminate the victims before transporting them to a medical facility.

Mass decontamination can be set up in a variety of ways, and your fire department may have a specific procedure for it. Generally, this operation requires a minimum of two fire apparatus placed side by side. Place fog nozzles on the side discharge outlets, and use a spray pattern to douse the victims. An aerial ladder device can provide a complete overhead spray. The nozzle pressure should be reduced substantially. Be aware that the water temperature may cause hypothermia. Check with your department regarding its specific policy and procedures for mass decontamination.

To set up and use mass decontamination, follow the steps in SKILL DRILL 34-2:

1. Ensure that you have the appropriate PPE to protect against the chemical threat.
2. Stay clear of the product, and do not make physical contact with it.
3. Make an effort to contain runoff by directing victims out of the hazard zone and into a suitable location for decontamination. (STEP 1)
4. Set up the appropriate type of mass-decontamination system based on the type of apparatus available.
5. Instruct victims to walk through the decontamination area and remove their contaminated clothing. (STEP 2)
6. Flush the contaminated victims.
7. Direct the decontaminated victims to the triage area. (STEP 3)

Methods of Decontamination

The previous descriptions of decontamination focused on using water to dilute or wash away the hazardous material. Some hazardous materials have chemical properties that require different methods of decontamination, or circumstances may dictate that fire fighters use an alternative to water as a decontamination measure. Although most of the methods described here can be carried out by the operations-level responder, you should consult the local hazardous materials response team) before attempting to implement any of them.

VOICES OF EXPERIENCE

Today there are many concerns raised with regard to the decontamination of mass-casualty victims. We must prepare ourselves for the task of effectively decontaminating these persons by developing, training, and practicing methods with which we can safely remove the hazardous products from the victims so that they may be triaged, treated, and transported to medical facilities.

The first-arriving emergency response units may be faced with an overwhelming number of victims, and the officer in charge will be faced with many decisions. The decontamination of multiple civilian casualties requires the implementation of effective methods with due consideration of citizen safety, modesty, and sensibility. The idea that first-arriving units will simply open up on the crowd with a deck gun is wrong and must be dispelled.

The mass decontamination of civilians is a critical public safety capability. Whether these people are the victims of a terrorist act or an industrial accident, the effective and efficient decontamination of exposed persons has been demonstrated to result in fewer and less severe casualties. Mass decontamination, however, is only one part of a rational decontamination plan. To protect exposed persons and provide for the survivability of the emergency health care system, we cannot rely exclusively on field mass decontamination of victims at a single place. The concept of "decontamination in depth" provides for multiple opportunities to decontaminate victims—implemented in the field; in hospital emergency rooms, clinics, and trauma centers; and at other casualty collection points. "Decontamination in depth," with field mass-casualty decontamination as a critical centerpiece, is the only system imaginable for dealing effectively with a mass contamination incident.

Glen Rudner
Virginia Department of Emergency Management, Retired
Spotsylvania, Virginia

SKILL DRILL 34-2 Performing Mass Decontamination

1 Ensure that you have the appropriate PPE to protect from the chemical threat. Stay clear of the product, and do not make physical contact with it. Make an effort to contain runoff by directing victims out of the hazard zone and into a suitable location.

2 Set up the appropriate type of mass-decontamination system based on the type of apparatus available. Instruct victims to walk through decontamination area and remove contaminated clothing.

3 Flush the contaminated victims. Direct the decontaminated victims to the triage area.

Alternative decontamination procedures include the following techniques:

- Absorption
- Adsorption
- Dilution
- Disinfection
- Disposal
- Solidification
- Emulsification
- Vapor dispersion
- Removal
- Vacuuming

Absorption

In absorption, a spongy material is mixed with a liquid hazardous material, and the contaminated mixture is then collected and disposed of. This technique is primarily used for decontaminating equipment and property; it has limited application for decontaminating personnel. Absorption may be used, for example, in a shuffle pit for the boots of responders, so that they are cleaned prior to entering the rest of the decontamination line.

Absorption minimizes the surface areas of liquid spills but is effective only on flat surfaces. Although absorbent materials such as soil and sawdust are inexpensive and readily available, they become hazardous themselves after coming in contact with the product and must be disposed of properly after the incident. One problem with this process is that it creates a hazardous waste that must be disposed of by someone (usually the "generator"), but the fire department may inadvertently become the generator of waste. Both the federal government and most states have laws and regulations that dictate how to dispose of used absorbent materials. Consult a hazardous materials authority before disposing of any materials suspected of being contaminated.

Adsorption

The opposite of absorption is adsorption. In adsorption, the contaminant adheres to the surface of an added material, including sand and activated carbon, rather than combining with it, as in absorption. Similar issues related to disposing of the contaminated materials arise whether you use an absorbent or an adsorbent.

Dilution

Dilution uses plain water or a soap-and-water mixture to lower the concentration of a hazardous material while flushing it off a contaminated person or object. Dilution is both fast and economical—water is a readily available solvent that usually generates no toxic fumes. Gross decontamination, technical decontamination, and mass decontamination processes all use dilution.

Fire fighters tend to use dilution as the first decontamination method because water is readily available from the tank on an engine or from a hydrant. Before using water, consider whether the contaminant will react with water, be soluble in water, or spread to a larger area. The more water used, the more hazardous waste generated that may spread the contamination and later must be disposed of safely. Dilution also can be effective in removing some solids such as dusts and fibers; however, care must be used to prevent the spread of these contaminants to larger areas.

Disinfection

Disinfection is a process used to destroy recognized disease-carrying (pathogenic) microorganisms. Commercial disinfectants are packaged with a detailed brochure that describes the limitations and capabilities of each product. Fire fighters who have medical research laboratories, hospitals, clinics, mortuaries, medical waste disposal facilities, blood banks, or universities in their response areas should be familiar with the specific types of biological hazards present and the best disinfectants to use for each hazard.

Take special precautions when responding to an emergency at a facility where bloodborne pathogens may be involved in an accident or spill. Also, be aware that the overuse of disinfectants contributes to the development of drug-resistant organisms, so always follow the manufacturer's instructions when such products are used.

Disposal

Disposal is a two-step removal process for items that cannot be properly decontaminated. First, the contaminated article is removed and isolated in a designated area. Then, the item is packaged in a suitable container and transported to an approved waste facility, where it is either incinerated or buried in a hazardous waste landfill. Contaminated items such as disposable coveralls should be collected, bagged, and tagged. Contaminated tools and equipment should be placed in bags, barrels, or buckets.

Solidification

Solidification is the chemical process of treating a hazardous liquid so as to turn it into a solid material. This transformation makes the material easier to handle but does not change its inherent chemical properties.

Emulsification

Emulsification is the process of neutralizing or reducing the harmful effects of a hazardous material by changing its chemical properties. This process can be used with some petroleum-based products. Although emulsification may change the chemical nature of the hazardous material, the resultant product still needs to be disposed of properly. Federal, state, and local regulations may apply to the use of emulsifier products.

The principal advantage of emulsification is that it renders the hazardous material less harmful, thereby decreasing risks to emergency responders. It can also help limit clean-up costs. Unfortunately, the chemicals required for emulsification may be harmful to fire fighters. In addition, it may take some time to determine which chemicals should be used and whether they are available.

Vapor Dispersion

Vapor dispersion is the process of separating and diminishing harmful vapors. A water spray is commonly used to disperse or move vapors away from certain areas or materials. Vapor dispersion should be used with extreme caution because the rapid introduction of a large volume of air can create dynamic results.

Removal

Removal as a mode of decontamination applies specifically to contaminated soil that can be taken away from the scene and disposed of properly. Removal may be necessary when highly toxic materials that cannot be rendered harmless are involved, when on-site treatment presents unacceptable risks to fire fighters, or when on-site treatment costs exceed disposal costs.

For example, it may be more cost-effective and less risky to pump the hazardous material into a drum and transport it to a hazardous waste dumpsite than to neutralize a spill and decontaminate emergency equipment. Removing the contaminants reduces clean-up time and limits the exposure risk to fire fighters. It is not always the most economical decision, however, and often requires the purchase of hazardous waste and transportation permits.

Vacuuming

Vacuuming is the removal of dusts, particles, and some liquids by sucking them up into a container. A filtering system prevents the contaminated material from recirculating and reentering the atmosphere. A special high-efficiency particulate air (HEPA) vacuum cleaner is used to remove hazardous dusts, powders, or fibers that are 0.3 micron or larger. HEPA filters allow air to pass through but capture the particulate matter in the air. The filters must be replaced regularly if they are to maintain their effectiveness.

The Decontamination Process

The decontamination process takes place in the decontamination corridor which is set up in the warm zone between the hot and cold zones. Within this corridor, fire fighters pass through several stations to complete the decontamination process.

A clearly marked, easily seen, readily accessible entry point to the decontamination corridor should be established. If the decontamination operation occurs at night, the area should be well lit. All personnel leaving the hot zone must pass through the decontamination corridor.

The decontamination team must wear self-contained breathing apparatus (SCBA) and a level of PPE equal to those being decontaminated. All decontamination team members must undergo decontamination before they leave the area.

Steps in Decontamination

Personnel leaving the hot zone should place any tools that they used in a tool drop area located near the decontamination corridor. This tool drop area can consist of a container, a recovery drum, or a special tarp. The tools can then be either used by another team entering the hot zone or collected for decontamination.

Next, the fire fighter, still wearing full PPE and SCBA, steps into the decontamination corridor for gross decontamination. The first step usually involves a portable shower, where medium-pressure, low-volume water flow rinses off and dilutes the contaminants on the PPE.

The decontamination team is responsible not only for scrubbing and swabbing the PPE worn by personnel, but also for containing runoff of the water used for washing the equipment. This step generally involves one to three wash-and-rinse stations. At each station, one decontamination team member handles the scrubbing and another does the rinsing. The decontamination team member who is scrubbing should pay special attention to the gloves, the kneecaps, behind the knees, under the arms, under the boot bottoms, and any areas where significant folds in the personal protective gear are noted, because these are the places with the highest contamination levels.

After outerwear is thoroughly scrubbed and rinsed, it can be removed along with chemical-protective clothing. SCBA, however, should remain in place. The decontamination team member then unzips the PPE and peels back the suit, exposing the SCBA. The team should fold the PPE back, so that the contaminated side contacts only itself. If the procedure is done properly, the contaminated side of the suit will not touch either the interior of the suit or the person wearing it. If the fire fighter is wearing outer gloves, they can be removed at this point; the inner gloves will be removed later. The fire fighter then moves down the decontamination corridor to remove the remainder of the PPE, SCBA, and other support equipment.

When removing protective clothing, turn as much of the PPE inside-out as possible. This tactic keeps any remaining trace contaminants inside the suit or glove. The fire fighter should be seated and allow the decontamination team members to remove PPE. Removing outside clothing properly is especially important when a fire fighter is wearing structural PPE, because these clothing items are porous and can absorb the hazardous material.

Remove the helmet, boots, and other support equipment along with the SCBA. The SCBA mask is usually the last item removed because it serves as an additional safeguard against trace contaminants. Place equipment on the contaminated side of the decontamination corridor area. Deposit the SCBA in a plastic bag or place it on a tarp for bagging later. Highly contaminated SCBA should be removed and isolated until it can undergo complete decontamination. Remove the inner gloves, turn them inside out, and sort them into individual containers for clean-up or disposal. Heavy plastic bags can be used as containers for these items because they usually provide sufficient temporary protection from most materials. The bags should be sealed with tape and transported elsewhere for clean-up or disposal. Place the bags in a properly marked recovery drum when disposing of them.

After removing personal clothing, the fire fighter should wash his or her entire body. An overhead shower produces much better results than a hose line. Apply ample soap to all areas of the body, but especially to the head and the groin. Liquid surgical soaps in plastic squeeze bottles give the best results; local hospitals may be able to provide guidance about obtaining prepackaged cleaning kits. Use small brushes and sponges for scrubbing the body surface. Afterward, bag and mark all cleaning items for disposal. The water runoff from this step should be controlled and contained, just as in previous decontamination steps.

After going through the decontamination corridor, the fire fighter should dry off using towels or sheets. Each towel can be used only once; it is then placed in bags on the contaminated side for disposal. Plastic garbage cans lined with plastic bags are good storage containers for this purpose.

Decontaminated personnel can now don clean clothes. Disposable cotton coveralls, hospital gowns, hospital booties, slippers, and flip-flops are inexpensive and easy-to-use options for this step. They can be prepackaged according to size and stored for easy access. Provisions should be made to bag clothing in the event that uncontaminated clothing can be salvaged. During mass-decontamination operations, agencies have also used large opaque plastic bags with holes cut out for the head and arms as a means to cover personnel after their contaminated clothes have been removed. The use of the bags provides a very cost-effective solution to provide coverage to personnel of many different sizes.

Fire Fighter Safety Tips

Contact lenses may trap contaminants under the lens. Anyone wearing contact lenses should remove them and flush the eyes thoroughly. If any discomfort or irritation remains, the eye should be evaluated.

To perform technical decontamination at a hazardous materials incident, follow the steps in SKILL DRILL 34-3:

1. Drop any tools and equipment into a tool drum or onto a designated tarp. (STEP 1)
2. Perform gross decontamination. (STEP 2)
3. Wash and rinse the entry team member. (STEP 3)
4. Remove outer hazardous materials–protective clothing and isolate PPE. (STEP 4)
5. Remove SCBA and the face piece. (STEP 5)
6. Remove personal clothing. Bag and tag all personal clothing. (STEP 6)
7. Shower and wash the body. Dry off the body and put on clean clothing.
8. The entry team member should proceed to medical monitoring.
9. At the station, the entry members should fill out any record-keeping information about the incident to include exposure reporting. (STEP 7)

Medical Follow-Up

After personnel are thoroughly decontaminated, they should proceed to an Emergency Medical Services (EMS) station for a medical evaluation. EMS personnel will take vital signs for each person leaving decontamination and compare those data to baseline data (taken prior to the person's entry into the hot or warm zone). Any open wounds or breaks in skin surfaces should be reported immediately to medical control. Unless advised otherwise by the medical command physician, all open wounds should be cleaned on the scene.

SKILL DRILL 34-3 Performing Technical Decontamination

1 Drop any tools and equipment into a tool drum or onto a designated tarp.

2 Perform gross decontamination.

3 Wash and rinse the entry team member.

4 Remove outer hazardous materials–protective clothing and isolate PPE.

5 Remove SCBA and the face piece.

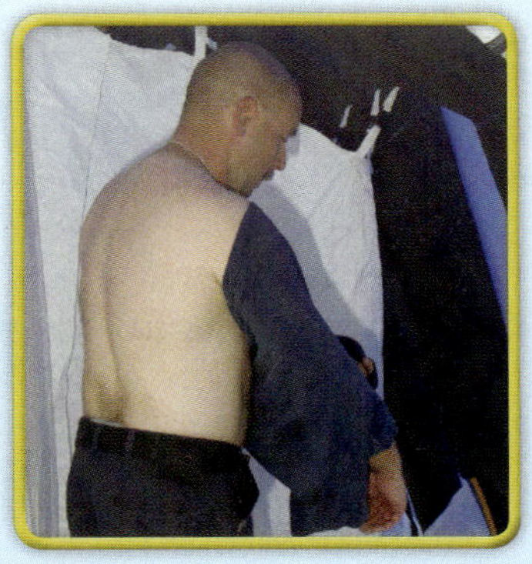

6 Remove personal clothing. Bag and tag all personal clothing.

7 Shower and wash the body. Dry off the body and put on clean clothing. The entry team member should proceed to medical monitoring. At the station, the entry members should fill out any record-keeping information about the incident to include exposure reporting.

Wrap-Up

Chief Concepts

- Decontamination makes personnel, equipment, and supplies safe by removing or reducing hazardous materials or WMDs. Proper decontamination is essential at every hazardous material/WMD incident to ensure the safety of both personnel and property.
- Cross-contamination is the process by which a contaminant is carried out of the hot zone and contaminates people, animals, the environment, or equipment. Cross-contamination may occur in several ways:
  - A contaminated victim comes into physical contact with an emergency responder.
  - A bystander or other emergency responder comes into contact with a contaminated object from the hot zone.
  - A decontaminated responder comes into contact with another contaminated responder or object.
- Fire fighters are often responsible for establishing a decontamination corridor.
- There are four major categories of decontamination:
  - Emergency decontamination—Aims to reduce the amount of surface contaminant. Involves dousing the victim with water via a hose line outside of a decontamination corridor. Formal decontamination is performed later.
  - Gross decontamination—Uses a large shower system in a decontamination corridor. This type of decontamination is used for a brief and rapid contamination reduction.
  - Technical decontamination—Performed after gross decontamination and may involve several steps.
  - Mass decontamination—Used to decontaminate large groups of people in the field. Victims will need to undergo further and more thorough decontamination.
- Alternative decontamination procedures include the following options:
  - Absorption—Use of a spongy material to soak up the hazardous material. This technique is used to decontaminate equipment and property.
  - Adsorption—Binding of the contaminant to the surface of an added material, which is disposed of.
  - Dilution—Use of soap and water to flush the hazardous material off a person or object.
  - Disinfection—Process used to destroy disease-carrying organisms.
  - Disposal—Two-step removal process for items that cannot be decontaminated properly.
  - Solidification—Chemical process of treating a hazardous liquid to become a solid.
  - Emulsification—Process of neutralizing or reducing the harmful effects of a hazardous material by changing its chemical properties.
  - Vapor dispersion—Process of separating and diminishing harmful vapors, often with a water spray.
  - Removal—Process of removing contaminated soil from a hazardous materials site.
  - Vacuuming—Removal of dusts, particles, and liquids by sucking them up into a container.

Hot Terms

Absorption Decontamination technique in which a spongy material is mixed with a liquid hazardous material. The contaminated mixture is collected and disposed of.

Adsorption The process of adding a material such as sand or activated carbon to a contaminant, which then adheres to the surface of the material. The contaminated material is then collected and disposed of.

Decontamination The physical and/or chemical process of reducing and preventing the spread of contaminants from people, animals, the environment, or equipment involved at hazardous materials/weapons of mass destruction incidents. (NFPA 472)

Decontamination corridor The area usually located within the warm zone where decontamination is performed. (NFPA 472)

Decontamination team The team responsible for reducing and preventing the spread of contaminants from persons and equipment used at a hazardous materials incident. Members of this team set up the decontamination corridor and conduct all phases of decontamination.

Disinfection The process used to inactivate virtually all recognized pathogenic microorganisms but not necessarily all microbial forms, such as bacterial endospores. (NFPA 1581)

Disposal A two-step removal process for contaminated items that cannot be properly decontaminated. Items are bagged and placed in appropriate containers for transport to a hazardous waste facility.

Emergency decontamination The physical process of immediately reducing contamination of individuals in potentially life-threatening situations with or without the formal establishment of a decontamination corridor. (NFPA 472)

Emulsification The process of changing the chemical properties of a hazardous material to reduce its harmful effects.

Gross decontamination The phase of the decontamination process during which the amount of surface contaminants is significantly reduced. (NFPA 472)

Mass decontamination The physical process of reducing or removing surface contaminants from large numbers of victims in potentially life-threatening situations in the fastest time possible. (NFPA 472)

Removal A mode of decontamination that applies specifically to contaminated soil, which is taken away from the scene.

Solidification The process of chemically treating a hazardous liquid so as to turn it into a solid material, thereby making the material easier to handle.

Technical decontamination The planned and systematic process of reducing contamination to a level that is as low as reasonably achievable.

Vacuuming The process of cleaning up dusts, particles, and some liquids using a vacuum with high-efficiency particulate air filtration to prevent recontamination of the environment.

FIRE FIGHTER *in action*

Your department is requested, as part of a mutual aid agreement, to send additional apparatus to assist at a large nursery/garden center fire just before Christmas. The nursery/garden center is at the peak of its growing season for holiday plants, which require the facility to construct large greenhouses made of plastic sheeting. The ensuing fire has enveloped the entire area in flames and has impinged on the chemical and fuel storage areas of the complex. Your engine company has been assigned to set up a decontamination area for all personnel operating at the scene. Hazardous materials personnel direct you to set up both a technical decontamination and an emergency decontamination. While setting up the decontamination area, a report comes in stating that a fire fighter has fallen near the chemical building and landed in several bags of open fertilizer.

1. Based on the reports, the fire fighter should go through emergency decontamination because the process would allow the victim to receive treatment as soon as possible.

- **A.** True
- **B.** False

2. As fire fighters exit the hot zone, what is the first thing they should do as they begin to enter the decontamination corridor?

- **A.** Report to emergency decontamination
- **B.** Drop their tools in the tool drop area prior to entering the decontamination area
- **C.** Report to the mass decontamination wash site with their tools
- **D.** Drop their tools, gloves, and boots in the drop area before entering the decontamination area

3. Because the likelihood of contamination is very small, decontamination team members need not pass through the decontamination process after they carry out the decontamination process for victims and other responders.

- **A.** True
- **B.** False

4. When conducting decontamination operations, decontamination team members should pay close attention to which of the following areas when scrubbing?

- **A.** Gloves
- **B.** Knee caps
- **C.** Boot bottoms
- **D.** All of the above

5. What is the purpose of the gross decontamination in a decontamination system?

- **A.** To remove small amounts of contaminants using small brushes
- **B.** To decontaminate multiple personnel at the same time
- **C.** To conduct final cleaning of the contaminated person after formal decontamination
- **D.** To remove large amounts of contaminants in a controlled manner

6. What is the next step in the decontamination process after all protective clothing and SCBA have been removed?

- **A.** Remove personal clothing and shower
- **B.** Report to staging for reassignment
- **C.** Assist the decontamination team in bagging the contaminated clothing
- **D.** Have a medical evaluation completed

FIRE FIGHTER II *in action*

You arrive on the scene of a motor vehicle accident where a tanker has overturned. The tank is round and it has reinforcing ribs visible on the outside. There is a placard on the side that says "Corrosive." There is a large amount of liquid that is pooling around the vehicle. Bystanders are screaming that the driver is trapped in the cab.

1. How would you manage the bystanders?
2. How would you provide emergency decontamination?
3. What are your greatest risks with this incident?

Terrorism Awareness

Fire Fighter I

Knowledge Objectives

After studying this chapter, you will be able to:

- Describe the threat posed by terrorism. (p 992)
- Identify potential terrorist targets in your jurisdiction. (p 993–994)
- Describe how to respond to a terrorist incident. (p 1005–1008)
- Describe the dangers posed by explosive devices. (p 996–998)
- Explain the difference between chemical and biological agents. (p 998–1004)
- Describe the dangers posed by radiological incidents. (p 1004–1005)
- List the types of terrorism. (p 994)
- Describe the need for decontamination of exposed victims and response personnel. (p 1007–1008)

Skills Objectives

There are no skill objectives for Fire Fighter I candidates. NFPA 1001 contains no Fire Fighter I Job Performance Requirements for this chapter.

CHAPTER

35

Fire Fighter II FFII

Knowledge Objectives

There are no knowledge objectives for Fire Fighter II candidates. NFPA 1001 contains no Fire Fighter II Job Performance Requirements for this chapter.

Skills Objectives

There are no skill objectives for Fire Fighter II candidates. NFPA 1001 contains no Fire Fighter II Job Performance Requirements for this chapter.

Additional NFPA Standards

- NFPA 472, *Standard for Competence of Responders to Hazardous Materials/Weapons of Mass Destruction Incidents*

You Are the Fire Fighter

It is 10:12 am and you are dispatched to a water flow alarm in the basement of a 22-story office building. The basement consists of four floors of parking as well as the HVAC system. Once enroute, the dispatcher advises that there are reports of smoke coming from the basement and that the entire building shook just before the alarm sounded. Currently the building is being evacuated.

1. What has most likely occurred and how would you approach this incident?
2. How would your thinking change if you found out that the FBI has an office in the building?
3. How would you approach this incident if you received news reports that a bomb had gone off in the basement?

What Is Terrorism?

Terrorism can be described as the unlawful use of violence or threats of violence to intimidate or coerce a government, the civilian population, or any segment thereof, to further political or social objectives. This broad definition encompasses a wide range of acts committed by different groups for different purposes.

The goal of terrorism is to produce feelings of terror in a population or a group. Terrorism is not limited to certain tools or weapons, however—almost anything can potentially be used as a tool by terrorists. For example, a gasoline tanker could be intentionally wrecked and burned to destroy property or lives; such an event would be classified as a terrorist incident. Yet, if an identical gasoline tanker wrecked and caught fire because of driver error, we would call that event an accident. The steps the fire department needs to take to mitigate each of these incidents might be the same. Many times first responders will not know whether an incident is caused by accidental circumstances or by intentional actions until a thorough investigation has been completed.

The Federal Bureau of Investigation (FBI) classifies terrorism as either domestic or international. Domestic terrorism refers to acts that are committed within the United States by individuals or groups that operate entirely within the United States and are not influenced by any foreign interests. International terrorism includes any acts that transcend international boundaries.

Within the United States, only 24 terrorist incidents occurred between 1994 and 1999; three other crimes that occurred during this period were also suspected of being terrorist in nature. Most of these incidents were classified as domestic terrorism. On September 11, 2001, the largest terrorist event in the history of the United States—the successful attacks on the World Trade Center and the Pentagon and the crash of a hijacked plane in Pennsylvania that aborted another attack—resulted in almost 3000 deaths FIGURE 35-1. Although terrorism can occur at any time, thankfully it is not a common event in the United States. In 2010, the U.S. State Department reported no deaths in the United States from terrorist events.

FIGURE 35-1 The September 11 attacks on the World Trade Center and the Pentagon, along with the crash of a hijacked plane in Pennsylvania, was the largest terrorist attack in the history of the United States.

Fire Service Response to Terrorist Incidents

The role of the fire service in handling terrorist events consists of the same functions that fire fighters perform on a day-to-day basis. The fire service role includes supplying emergency medical services, hazardous materials mitigation, technical rescue, and fire suppression. All of these functions may be needed when a terrorist incident occurs. Terrorism also presents new challenges for the fire service, because it requires fire fighters both to perform the tasks they have been doing successfully and to learn

FIRE FIGHTER Tips

Emergency responders at a terrorist event may later be called to testify in court. Make mental notes if you must disturb or move any evidence to rescue someone in immediate danger. Document this information as soon as possible.

FIRE FIGHTER Tips

The Code of Federal Regulations defines terrorism as "the unlawful use of force and violence against persons or property to intimidate or coerce a government, the civilian population, or any segment thereof in furtherance of political or social objectives" (28 C.F.R. Section 0.85).

to handle some new tasks. In this sense, responding to terrorist incidents represents an extension of the threats fire fighters encounter on a routine basis.

The terrorist threat requires fire fighters to work closely with local, state, and federal law enforcement agencies; emergency management agencies; allied health agencies; and the military. It is critical that all of these agencies work together in a coordinated and cooperative manner. Emergency responders and law enforcement agencies must be prepared to face a wide range of potential situations.

One threat posed by terrorists is the use of weapons of mass destruction (WMD)—devices that are designed to cause maximum damage to property or people. These weapons could be made from chemical, biological, or radiological agents, or they could involve conventional weapons and explosives. They pose a grave danger to life, such that mass casualties could result from a WMD attack in an urban area. An incident involving a WMD could quickly overwhelm emergency response agencies and the local healthcare system.

To prepare for the threat of terrorist attacks, fire fighters need to understand the technologies used to manage WMD incidents. In addition, first responders need to know their role in handling incidents involving releases of chemical, biological, and radiological agents.

Potential Targets and Tactics

Terrorists are motivated by a cause and choose targets they believe will help them achieve their goals and objectives. Terrorist incidents aim to instill fear and panic among the general population and to disrupt daily ways of life. Given this goal, terrorists tend to choose symbolic targets, such as a place of worship, an embassy, a monument, or a prominent government building. Sometimes the objective is sabotage—that is, to destroy or disable a facility that is significant to the terrorist cause. The ultimate goal could be to cause economic turmoil by interfering with transportation, trade, or commerce.

Terrorists choose a method of attack they think will make the desired statement or achieve the maximum results. They may vary their methods or change them over time. Explosive devices have remained popular weapons for terrorists, however, and have been used in thousands of terrorist attacks. In recent years, the number of suicide bombings has increased dramatically.

The type and location of terrorist incidents also change over time. Political instability in a country or region may result in an increase in terrorist incidents. For example, in 2011, the lack of a stable government in Somalia resulted in 25 incidents of ships being hijacked off the coast of that country. The purpose of these terrorist incidents appeared to be financial gain rather than political change.

Terrorism can occur in any community and in connection with many different issues. There are many causes with supporters who range from peaceful, nonviolent organizations to fanatical fringe groups; the latter groups may be the wellspring of terrorism. For example, a rural ski lodge tucked away in the mountains might be attacked by an environmental group that is upset with its plans for expansion. A small retailer in an upscale suburban community could become the target of an animal rights group that objects to the sale of fur coats. An antiabortion group might plant a bomb at a local community health clinic.

It is often possible to anticipate likely targets and potential attacks. Law enforcement agencies routinely gather intelligence about terrorist groups, threats, and potential targets. The fire service is increasing its capacity to engage in intelligence sharing so that preincident plans can be developed for possible targets and scenarios. Even if no specific threats have been made, certain types of occupancies are known to be potential targets; preincident planning for those locations should include the possibility of both a terrorist attack and an accidental fire.

News accounts of terrorist incidents abroad can help keep fire fighters up-to-date with trends related to terrorist tactics. They can provide useful information about situations that could occur in your jurisdiction in the future. A variety of Internet resources can also help you to keep abreast of current threats. Periodic e-mails sent by list servers, which are available through federal response agencies, can provide useful information. By becoming familiar with potential targets and current tactics, emergency responders can plan appropriate strategies and tactics for potential attacks.

Potential targets include both natural landmarks and human-made structures. These sites can be classified into three broad categories: infrastructure targets, symbolic targets, and civilian targets.

Terrorism can also be classified according to the part of society that is targeted—for example, agroterrorism, ecoterrorism, and cyberterrorism. Incidents of terrorism involving these three tactics will probably not have as direct an impact on fire departments as incidents involving natural landmarks and human-made structures, however. Incidents involving cyberterrorism, for instance, will have a significant indirect impact on the fire department and could significantly disrupt its operations, but would probably not require a fire department response.

FIRE FIGHTER Tips

Potential Terrorist Targets

Infrastructure Targets
Bridges and tunnels
Emergency facilities
Hospitals
Oil refineries
Pipelines
Power plants
Railroads
Telecommunications systems
Water reservoirs and treatment plants

Symbolic Targets
Embassies
Government buildings
Military bases
National monuments
Places of worship

Civilian Targets
Arenas and stadiums
Airports and railroad stations
Mass-transit systems
Schools and universities
Shopping malls
Theme parks

Ecoterrorism Targets
Controversial development projects
Environmentally sensitive areas
Research facilities

Agroterrorism Targets
Crops
Feed storage
Grain elevators
Livestock and poultry

Cyberterrorism Targets
Banking and finance computer systems
Business computer systems
Court computer systems
Government computer systems
Law enforcement computer systems
Military computer systems

Infrastructure Targets

Terrorists might select bridges, tunnels, or subways as targets in an attempt or disrupt transportation and inflict a large number of casualties **FIGURE 35-2**. They might also plan to attack the public water supply or try to disable the electrical power distribution system, telephones, or the Internet. Disruption of a community's 911 system or public safety radio network would directly affect emergency response agencies.

Symbolic Targets

National monuments such as the Lincoln Memorial, Washington Monument, or Statue of Liberty could be targeted by groups who want to attack symbols of national pride **FIGURE 35-3**. Foreign embassies and institutions might be attacked by groups promoting revolution within those countries or protesting their international policies. Religious institutions are potential targets of hate groups. By attacking these symbols, terrorist groups seek to make people aware of their demands and to create a sense of fear in the public.

Civilian Targets

Civilian targets include places where large numbers of people gather, such as shopping malls, schools, or stadiums. In incidents involving these sites, the goal of terrorists is to indiscriminately kill or injure large numbers of people, thereby creating fear in the society as a whole **FIGURE 35-4**. Attacking civilian targets tends to produce a high degree of fear in the general population.

Ecoterrorism

Ecoterrorism refers to illegal acts committed by groups supporting environmental or related causes. Examples include spiking trees to sabotage logging operations, vandalizing a university research laboratory that is conducting experiments on animals, and firebombing a store that sells fur coats. Unless these incidents involve arson or explosions, fire departments are rarely directly involved in responding to ecoterrorism.

FIRE MARKS

The Earth Liberation Front claimed credit for arson fires that destroyed several mountaintop buildings at a ski resort in Vail, Colorado, in 1998. The group was trying to prevent the development of additional ski trails on the mountain.

Agroterrorism

Agroterrorism includes the use of chemical or biological agents to attack the agricultural industry or the food supply **FIGURE 35-5**. The deliberate introduction of an animal disease such as foot-and-mouth disease to the livestock population could result in major losses to the food industry and produce fear among members of the general population.

Cyberterrorism

Groups of terrorists could engage in cyberterrorism by electronically attacking government or private computer systems. In the past, several attempts have been made to disrupt the Internet or to attack government computer systems and other critical networks. This type of terrorism would disrupt many day-to-day activities in our society, and disrupt the notification, dispatch, response, and operation of emergency services.

Agents and Devices

Terrorists can turn the most ordinary objects into powerful weapons. We tend to think of gasoline tankers and commercial airliners as valuable parts of our society, but in the hands of determined terrorists they could become WMD. Acts of terrorism reflect the instigator's intentions more than the use of specific devices. For example, demolition companies use

A. Subway.

B. Airport.

C. Bridge.

D. Hospital.

FIGURE 35-2 Subways, airports, bridges, and hospitals are all vulnerable to attack by terrorists who seek to interrupt a country's infrastructure.

FIGURE 35-3 Terrorists might attempt to destroy visible national icons.

FIGURE 35-4 By attacking a civilian target such as a crowded stadium, terrorists might make citizens feel vulnerable in their everyday lives.

FIRE MARKS

Major Terrorist Incidents in the United States

September 1984: Dalles, Oregon

To influence local elections, a religious sect spread *Salmonella* on salad bars in four restaurants, resulting in 750 cases of *Salmonella* poisoning.

February 1993: New York City

A large explosive device was detonated in a van parked in the underground garage of the World Trade Center. Six workers were killed and more than 1000 people were injured.

April 1995: Oklahoma City, Oklahoma

The Alfred E. Murrah Federal Building was demolished by a truck bomb that also killed 167 people. (An NFPA Fire Investigations report is available on this incident.)

1978–1995: United States

Over a period of 17 years, the Unabomber (Theodore Kaczynski) mailed at least 16 packages containing explosives to university professors, corporate executives, and other targeted individuals. These attacks killed 3 individuals and injured 23 others.

July 1996: Atlanta, Georgia

A pipe bomb exploded in Centennial Olympic Park, killing one person and injuring 111 others.

January 1997: Atlanta, Georgia

Following the bombing of an abortion clinic in suburban Atlanta, a secondary device exploded, wounding several emergency responders. A month later, another secondary device was found and disarmed at the scene of a bombing at a gay nightclub in Atlanta.

January 1998: Birmingham, Alabama

A bomb killed a police officer who was providing security at an abortion clinic.

October 1998: Vail, Colorado

Arson destroyed eight buildings at a ski resort. An extremist environmental group opposed to expansion of the resort claimed responsibility.

September 11, 2001: New York City/Washington, D.C./Pennsylvania

Terrorists hijacked four commercial jets. Two were flown into the World Trade Center, one struck the Pentagon, and the fourth crashed into a field in rural Pennsylvania. More than 3000 people died in the various incidents.

Fall 2001: United States

Five people died after letters containing anthrax virus were sent to various locations in the eastern United States.

explosives to bring down unneeded structures quickly and safely; terrorists could use the same agents to bring down an occupied building, thereby killing many people.

Terrorists use several different kinds of weapons. Although bombings are the most frequent international terrorist acts, fire fighters also must be aware of the threat posed by other potential weapons. Shooting into a crowd at a shopping mall or train station with an automatic weapon could cause devastating carnage. The release of a microscopic biological agent into a subway system could cause numerous people to become ill and die, thereby creating a public panic response that might overwhelm the resources of the fire service and Emergency Medical Services (EMS) system. A computer virus that attacks the banking industry could lead to tremendous economic losses. Given the breadth of these threats, planning should consider the full range of possibilities.

FIGURE 35-5 Agroterrorism might affect the food supply.

Fire Fighter Safety Tips

A terrorist event may be designed to target emergency responders as well as civilians. Explosives do not discriminate—anyone nearby can be severely injured or killed when they detonate.

Explosives and Incendiary Devices

Explosives and incendiary devices are a common tool of terrorists. Incendiary devices account for 20 to 25 percent of all bombing incidents in the United States. These weapons are of special concern to fire fighters because they are capable of causing large fires in highly flammable settings.

Groups or individuals have used explosives for many different purposes: to further a cause, to intimidate a co-worker or former spouse, to take revenge, or simply to experiment with a recipe found in a book or on the Internet.

Each year, thousands of pounds of explosives are stolen from construction sites, mines, and military facilities. How much of this material makes its way to criminals and terrorists is not known **FIGURE 35-6**. Terrorists can also use commonly available materials, such as ammonium nitrate fertilizer and fuel oil (ANFO), to create their own explosives.

An improvised explosive device (IED) is any explosive device that is fabricated from readily available materials. An IED could be contained in almost any type of package—from a letter bomb to a truckload of explosives. The Unabomber, for example, constructed at least 16 bombs that were delivered in small packages through the U.S. Postal Service. By contrast, the bombings of the World Trade Center in 1993 and the Alfred E. Murrah Federal Building in 1995 both involved delivery vehicles loaded with ANFO and detonated via a simple timer **FIGURE 35-7**.

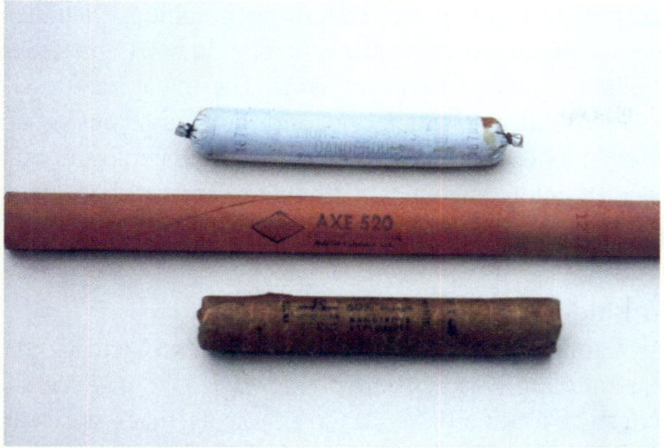

FIGURE 35-6 Every year, thousands of pounds of explosives are stolen from their rightful owners.

FIGURE 35-7 The Alfred E. Murrah Federal Building in Oklahoma City was destroyed by a truck bomb.

FIRE FIGHTER Tips

A radio-free area, spreading out for at least 1000 feet, should be established at any incident for which a suspected explosion area is identified. The establishment of such an area should be noted in an operating procedure. Within the radio-free area, all personnel should be ordered to cease any and all transmissions through any radio-wave-transmitting device, including mobile radios, mobile data terminals, and cell phones.

Pipe Bombs

The most common IED is the pipe bomb. A pipe bomb is simply a length of pipe filled with an explosive substance and rigged with some type of detonator **FIGURE 35-8**. Most pipe bombs are simple devices, made with black powder or smokeless powder and ignited by a hobby fuse. More sophisticated pipe bombs may use a variety of chemicals and incorporate electronic timers, mercury switches, vibration switches, photocells, or remote control detonators as triggers.

FIGURE 35-8 Pipe bombs come in many shapes and sizes.

Pipe bombs are sometimes packed with nails or other objects that act as shrapnel, reflecting the bomber's goal of inflicting as much injury as possible on anyone in the vicinity. A chemical or biological agent or radiological material could also be added to a pipe bomb to create a much more complicated and dangerous incident.

Secondary Devices

Emergency responders must realize that terrorists sometimes place secondary devices in the area where an initial event has occurred. These devices, which are intended to explode some time after the initial device detonates, are designed to kill or injure emergency responders, law enforcement personnel, spectators, and news reporters. Terrorists have used this tactic to attack the best-trained and most-experienced investigators and emergency responders, or simply to increase the levels of fear and chaos following an attack.

The use of secondary devices is a common tactic in incidents abroad and has occurred at a few incidents in North America. For example, in 1998, a bomb exploded outside a Georgia abortion clinic. Approximately an hour later, a second explosion injured seven people, including two emergency responders. A similar secondary device was discovered nearly a month later at the scene of a bombing in a nearby community, although responders were able to disable this device before it detonated.

Although the possibility of a secondary device exists at every terrorist incident, do not forget that the hazards of a secondary explosion or collapse exist at any incident involving an explosion. As a consequence, fire fighters and other first responders must be alert to all hazards at the scene of *any* explosion. The hazards at an accidental explosion can be just as deadly as secondary devices at an incident caused by terrorists.

Potentially Explosive Devices

All fire fighters should know their department's standard operating procedures for handling incidents involving explosives. Fire department units responding to an incident that involves a potentially explosive device—that is, a device that has not yet exploded—should move all civilians away from the area and establish a perimeter at a safe distance. At no time should fire fighters handle a potentially explosive device unless they have received special training in doing so. Instead, trained explosive ordnance disposal (EOD) personnel should assess the device and render it inoperative.

While waiting for the properly trained EOD personnel to arrive, the fire department should establish a command post and a staging area FIGURE 35-9. Because secondary devices may be present, the staging area should be located at least 3000 feet from the incident site. In this type of incident, a unified command must be established, with multiagency coordination including all agencies involved in the incident.

FIGURE 35-9 If the fire department is first to arrive at the scene of an explosion, it should establish a unified command post.

If the bomb disposal team decides to disarm the device, an emergency action plan must be developed in case of accidental detonation. The incident commander (IC) should work in concert with law enforcement and EOD personnel to determine a safe perimeter where fire fighters and emergency medical personnel will be staged.

In some cases, a forward staging area will be established. A rapid intervention team then stands by in this area, ready to provide immediate assistance to the bomb disposal team if something goes wrong. Other fire fighters and emergency medical personnel would remain in a remote staging area in this scenario.

Fire Fighter Safety Tips

Radio waves can trigger electric blasting caps, initiating an explosion. For this reason, radio transmitters should not be used near a suspicious device.

Actions Following an Explosion

In an explosion, as in any type of incident, your first priority should be to ensure the safety of the scene. During the initial stages of an incident, you will not know whether the event was caused by an intentional act or by accidental circumstances. In any incident involving an explosion, follow departmental procedures to ensure the safety of rescuers, victims, and bystanders. Consider the possibility that a secondary device and additional hazards might be present in the vicinity. Quickly survey the area for any suspicious bags, packages, or other items.

It is also possible that chemical, biological, or radiological agents might be involved in a terrorist bombing. Qualified personnel with monitoring instruments should be assigned to check the area for potential contaminants. The initial size-up should also include an assessment of hazards and dangerous situations. The stability of any building involved in the explosion must be evaluated before anyone is permitted to enter it, because entering an unstable area without proper training and equipment may complicate rescue and recovery efforts.

Working with Other Agencies

Joint training with the local, state, and federal agencies charged with handling incidents involving explosive devices should occur on a routine basis. Among these agencies are local and state police; the FBI; the Bureau of Alcohol, Tobacco, Firearms and Explosives (ATF); and military EOD units. In some jurisdictions, the fire department is responsible for handling explosive devices. In many other communities, this function is handled by specially trained law enforcement personnel.

Chemical Agents

Chemical agents have the potential to kill or injure great numbers of people. Many chemical agents are readily available because they are produced in the United States and used in a wide variety of industrial and commercial processes. For instance, ammonia is used as a fertilizer for crops; chlorine is used for many purposes including purification of drinking water and swimming pools. These same chemicals, if released in a terrorist event, could cause injury or death to many people. Mitigating an accidental release of chlorine gas, however, requires the same safety precautions as handling an intentional release of chlorine gas by terrorists.

Chemical agents are not new; they have been used as weapons for at least 100 years. For example, phosgene, chlorine, and mustard agents were used as weapons in World War I, resulting in thousands of battlefield deaths and permanent injuries. Chemical weapons were also used during the Iran–Iraq War (1980 to 1988). During the same period, the Iraqi government used these weapons against the minority Kurdish people in that country. Although international agreements prohibit the use of chemical and biological agents on the battlefield, concern remains that chemical weapons could be used by terrorists.

In 1995, the religious cult Aum Shinrikyo released a nerve agent, sarin, into the Tokyo subway system. Although the attack resulted in 12 deaths and more than 1000 injuries, many experts believe that it was ineffective because there were relatively few casualties despite the potential exposure of thousands of subway riders to the nerve agent. Even so, this attack achieved one of the major goals of terrorism: It instilled fear in a large population.

The basic instructions for making chemical weapons can be readily found through the Internet, books, and other publications, and the chemicals needed can be obtained fairly easily. Many of these chemicals are routinely used in legitimate industrial processes and, therefore, are widely available. For example, chlorine gas is used in swimming pools, at water treatment facilities, and in industry FIGURE 35-10. Despite its

VOICES OF EXPERIENCE

As I arrived on the scene of the New York City World Trade Center (WTC) attacks in September 2001, I saw that the city was quiet—eerily quiet. I walked to the command post to report for duty, and the scene was surreal. I looked north, south, east, and west, and it was like a scene from a bad Hollywood movie. I looked around, found my bearings, and walked through the ankle-deep dust and debris. It covered the streets like snow or ash.

I was sent to the WTC attacks as a member of the then-U.S. Department of Justice, Office of State and Local Domestic Preparedness Support (OSLDPS), to provide support to the New York City Fire Department in any way possible. I reported to the Chief of Logistics, Chief Charlie Blaich, and worked side-by-side with him and fire fighter Lee Morris until the day before Christmas 2001.

What did I see there, and what did I learn? I had to use every skill that I have ever learned as a fire fighter and EMS officer. I provided safety support as I had done while a safety officer at the Federal Emergency Management Agency (FEMA), where I had been a Disaster Safety Officer for the previous eight years.

One thing is certain: When it comes to terrorism, we don't know everything, and we must be constantly alert. During an event involving terrorism, nothing is as it seems—nothing. Revert to your training and practice, practice, practice your skills.

Remember that during such an event, other factors may be in play that do not necessarily equate to why you were brought there in the first place. There will be many unknowns, and your skills—your professionalism, your analytical qualities, and your demeanor—may well be the only thing that makes any sense during the early stages of the incident.

What you read someplace doesn't matter during such an event; the only thing that matters is what you can do to help the situation by being an engaged member of the team. There is no place for lone wolves and no place for grandstanding—it's not the TV show *24*. It's real people needing assistance in making informed decisions, and your task may be to help them make the right ones, as part of a team. It's up to you to be well trained, well informed, and well intentioned, and to be a part of the solution. Be safe, be proactive, and stay alive.

Michael J. Fagel
University of Chicago—Master of Threat Risk Program
Chicago, Illinois

FIGURE 35-10 Although chlorine is regularly used in swimming pools and water treatment facilities, it is also classified as a pulmonary agent that can be used in a terrorist attack.

legitimate uses, chlorine is classified as a pulmonary irritant (a choking agent) because its inhalation causes severe pulmonary damage.

Like chlorine, cyanide has many legitimate uses, but it kills quickly once it enters the body. Cyanide compounds such as hydrogen cyanide are used in the production of paper and synthetic textiles as well as in photography and printing. These chemicals are stored and transported in containers of various sizes, ranging from small containers and pressurized cylinders to railroad tank cars.

Chemical weapons can be disseminated in several ways. Simply releasing chlorine gas from a storage tank in an unguarded rail yard, for example, might cause many injuries and deaths. To ensure broader distribution of the chemical agent, it might be added to an explosive device. Crop-dusting aircraft, truck-mounted spraying units, or hand-operated pump tanks could also be used to disperse an agent over a wide area FIGURE 35-11.

Dissemination of a toxic gas or suspended particles depends on wind direction, wind speed, air temperature, and humidity. Because these factors can change quickly, it is difficult to predict the exact direction that might be taken by a chemical release cloud. Hazardous materials teams use computer models to predict the pathway of a toxic cloud. They also have the training and equipment to safely handle these situations.

FIGURE 35-11 Crop-dusting equipment could be used to distribute chemical agents.

Protection from Chemical Agents

Air flow is the most commonly used dispersal method for chemical agents. Agents released outdoors will follow wind currents; agents released inside will be dispersed through a building's air circulation system. Some chemical agents have distinctive odors. Hence, if any unusual odor is noticed at an emergency scene, fire fighters must use full personal protective equipment (PPE) including self-contained breathing apparatus (SCBA) FIGURE 35-12. The presence of some chemicals requires the use of fully encapsulated suits. In addition, the area should be monitored with the appropriate detection devices.

FIGURE 35-12 If an unusual odor is reported at the scene, a fire fighter must don full PPE including SCBA.

It is never sufficient to rely strictly on odor to determine the presence of a chemical agent. Instead, definitive detection and identification of a chemical agent require the activation of a well-trained hazardous materials team. As a fire fighter, your job is to prevent further injuries to rescuers and citizens—not to neutralize the chemical agent.

Nerve Agents

Nerve agents are toxic substances that attack the central nervous system. These weapons were first developed in Germany before World War II. Today, several countries maintain stockpiles of these agents. Nerve agents are similar to some pesticides but are much more toxic—in some cases, 100 to 1000 times more toxic than their pesticide counterparts. Exposure to these agents can result in injury or death within minutes.

In their normal states, most nerve agents are liquids FIGURE 35-13. As liquids, these agents are unlikely to contaminate large numbers of people because direct contact with the agent is required for the chemical to inflict its health-damaging effects. To be an effective weapon, the liquid must be dispersed in aerosol form or broken down into fine droplets that can then be inhaled or absorbed by the skin.

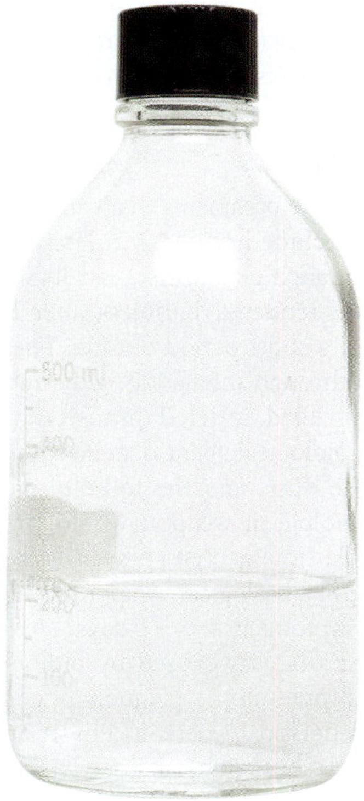

FIGURE 35-13 In their normal states, nerve agents are liquids. They must be dispersed in aerosol form if they are to be inhaled or absorbed by the skin.

Pouring a liquid nerve agent onto the floor of a crowded building would probably not immediately contaminate large numbers of people, but it would produce fear and panic. In this scenario, the effectiveness of the agent would depend on how long it remained in the liquid state and how widely it became dispersed throughout the building. Sarin, the most volatile nerve agent, evaporates at the same rate as water and is not considered persistent. The most stable nerve agent, V-agent, is considered persistent because it takes several days or weeks to evaporate. Common nerve agents, their methods of contamination, and specific characteristics are listed in TABLE 35-1.

When a person is exposed to a nerve agent, symptoms will become evident within minutes. These symptoms include pinpoint pupils, runny nose, drooling, difficulty breathing, tearing, twitching, diarrhea, convulsions or seizures, and loss of consciousness. The same symptoms are seen in victims who have been exposed to pesticides. Several mnemonics may be used to help you remember these symptoms, including the mnemonic used most often by the fire service: SLUDGEM TABLE 35-2.

TABLE 35-1 Common Nerve Agents

Nerve Agent	Method of Contamination	Characteristics
Tabun (GA)	Skin contact Inhalation	Disables the chemical connections between nerves and target organs
Soman (GD)	Skin contact Inhalation	Odor of camphor
Sarin (GB)	Inhalation	Evaporates quickly
V-agent (VX)	Skin contact	Oily liquid that can persist for weeks

TABLE 35-2 Symptoms of Nerve Agent Exposure

S—salivation (drooling)
L—lacrimation (tearing)
U—urination
D—defecation
G—gastric upset (upset stomach, vomiting)
E—emesis (vomiting)
M—miosis (constriction of the pupils)

The U.S. military has developed a nerve agent antidote kit, which contains two medications, atropine and pralidoxime chloride. These antidotes can be quickly injected into a person who has been contaminated by a nerve agent. The first antidote kit to be widely distributed was the Mark 1 Nerve Agent Antidote Kit, which was issued to many fire departments' hazardous materials teams. The DuoDote Kit has since been introduced to replace expired Mark 1 kits. One cautionary note about use of the DuoDote Kit: The DuoDote auto-injector should be administered by EMS personnel who have received adequate training in the recognition and treatment of nerve agent or insecticide intoxication. The DuoDote auto-injector is intended as an initial treatment of the symptoms of organophosphorous nerve agent or insecticide poisoning. Definitive medical care should be sought immediately.

Blister Agents

Two chemical agents are generally used as blistering agents, so called because contact with these chemicals causes the skin to blister:

- Sulfur mustard is a clear, yellow, or amber, oily liquid with a faint sweet odor of mustard or garlic. It vaporizes slowly at temperate climates and may be dispersed in an aerosol form.
- Lewisite is an oily, colorless-to-dark-brown liquid with an odor of geraniums.

Blister agents produce painful burns and blisters with even minimal exposure FIGURE 35-14. The major difference between these two blister agents is that lewisite causes pain

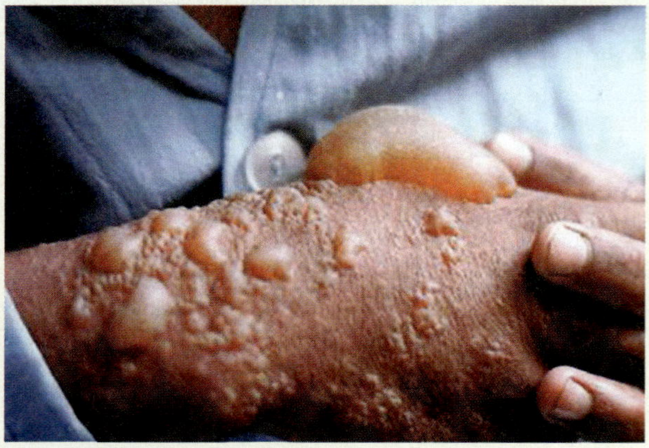

FIGURE 35-14 Typical effects of a blistering agent.

immediately upon contact with the skin, whereas the signs and symptoms from exposure to mustard gas may not appear for several hours. The victim will complain of burning at the site of the exposure, the skin will redden, and blisters will appear. The eyes may also itch, burn, and turn red. Inhalation of mustard gas produces significant respiratory damage.

If you suspect exposure to a chemical agent, treat victims as if they are contaminated. Use full PPE and prevent other people from being exposed to the chemical agent.

Pulmonary Agents

Pulmonary agents (also referred to as choking agents) cause severe damage to the lungs and lead to asphyxia. Two chemicals that might be used as pulmonary agents in terrorist attacks are phosgene and chlorine. Both of these agents were used extensively as weapons in World War I, and both have several industrial uses.

Phosgene and chlorine are heavier than air, so they tend to settle in low areas. Thus subways, basements, and sewers are prime areas for these agents to accumulate. Fire fighters wearing appropriate PPE should quickly evacuate people from such areas when the possibility of phosgene or chlorine exposure exists.

Exposure to high concentrations of phosgene or chlorine will immediately irritate the eyes, nose, and upper airway. Within hours, the exposed individual will begin to develop pulmonary edema (fluid in the lungs). Individuals who are exposed to lower concentrations may not exhibit any initial symptoms but can still experience respiratory damage.

Although neither pulmonary agent is absorbed through the skin, both phosgene and chlorine can cause skin burns on contact. Decontamination consists of removing exposed individuals from the area and flushing the skin with water.

Metabolic Agents

Metabolic agents, such as cyanide, interfere with the utilization of oxygen by the cells of the body. As mentioned earlier, cyanide compounds are highly toxic poisons that can cause death within minutes of exposure. The two most common cyanide compounds are hydrogen cyanide and cyanogen chloride, both of which have legitimate uses in industry.

Cyanide can be inhaled or ingested. Historically, the gaseous form was used as a means of carrying out the death penalty (hence the term "gas chamber"). Liquid cyanide mixed with fruit punch was used in the mass suicide/murder of 913 members of a religious cult in Guyana in 1978.

The symptoms associated with cyanide exposure appear quickly. A person who is exposed to the gas will begin gasping for air. In addition, if enough agent is inhaled, the skin may begin to appear red. Seizures are also possible.

In the face of cyanide exposure, a complete and rapid evacuation of the area is the most appropriate course of action.

Biological Agents

Biological agents are organisms that cause diseases and attack the body. They include bacteria, viruses, and toxins. Some of these organisms, such as anthrax, can live in the ground for years; others are rendered harmless after being exposed to sunlight for only a short period of time. The highest potential for infection is through inhalation, although some biological agents can be absorbed, injected through the skin, or ingested. The effects of a biological agent depend on the specific organism or toxin, the dose, and the route of entry. Most experts believe that a biological weapon would probably be spread by a device similar to a garden sprayer or crop-dusting plane. After the release of a biological weapon, residents would begin experiencing symptoms in 1 to 17 days.

Some of the diseases caused by biological agents, such as smallpox and pneumonic plague, are contagious and can be passed from person to person. Physicians are concerned about the use of contagious diseases as weapons, because the resulting epidemic could overwhelm the healthcare system. Experts have different opinions about how difficult it would be to infect large numbers of people with one of these naturally occurring organisms. Because of their incubation period, it would take between 2 and 17 days after exposure to these organisms before people would show signs of being infected.

Anthrax

Anthrax is an infectious disease caused by the bacterium *Bacillus anthracis*. These bacteria are typically found around farms and infect livestock. When anthrax is to be used as a weapon, the bacteria are cultured to develop anthrax spores. The spores, in powdered form, can then be dispersed in a variety of ways FIGURE 35-15. Approximately 8000 to 10,000 spores are typically required to cause an anthrax infection. Spores infecting the skin cause cutaneous anthrax, ingested spores cause gastrointestinal anthrax, and inhaled spores cause inhalational anthrax. Experts are not sure whether it would be possible to infect large numbers of people with anthrax.

In 2001, four letters containing anthrax were mailed to locations in New York City; Boca Raton, Florida; and Washington, D.C. Five people died after being exposed to the

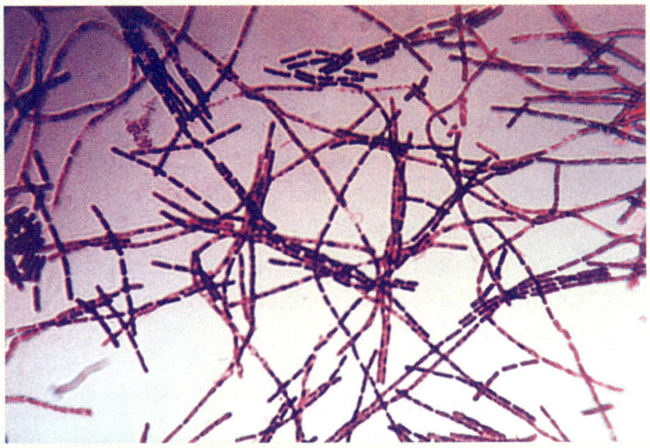

FIGURE 35-15 Anthrax spores can be dispersed in a variety of ways.

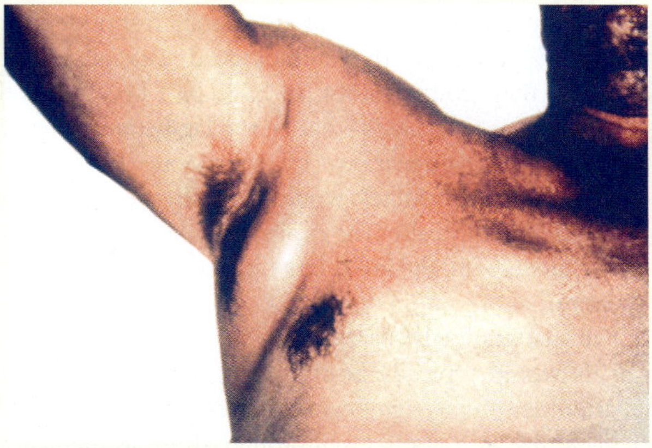

FIGURE 35-16 A bubo—one of the symptoms of plague—consists of a swollen, painful lymph node.

contents of these letters, including two postal workers who came in contact with the anthrax as the letters passed through postal sorting centers. Additional exposures resulted in more than 20 injuries that have had lasting effects. Several major government buildings had to be shut down for months to be decontaminated.

These incidents followed shortly after the terrorist attacks of September 11, 2001, and they caused tremendous public concern. In their wake, emergency personnel had to respond to many incidents involving suspicious packages, and many citizens believed that they might have been exposed to anthrax. Today, simple tests are available that hazardous materials teams can perform in the field to determine whether the threat of anthrax is real.

Anthrax has an incubation period of 2 to 6 days. The disease can be successfully treated with a variety of antibiotics if it is diagnosed early enough.

Plague

Plague is caused by *Yersinia pestis*, a bacterium that is commonly found on rodents. These bacteria are most often transmitted to humans by fleas that feed on infected animals and then bite humans.

Three forms of plague are distinguished: bubonic, septicemic, and pneumonic. Individuals who are bitten generally develop bubonic plague, which attacks the lymph nodes FIGURE 35-16. Pneumonic plague can be contracted by inhaling the bacterium.

Y. pestis can survive for weeks in water, moist soil, or grains. These bacteria have been cultured for distribution as a weapon in aerosol form. Inhalation of the aerosol form would put the target population at risk for pneumonic plague.

The incubation period for the plague ranges from 2 to 6 days. This disease can be treated with antibiotics.

Smallpox

Smallpox is a highly infectious and often fatal disease caused by *Variola*, a virus; it kills approximately 30 percent of those persons infected with this pathogen. Although smallpox was once routinely encountered throughout the world, by 1980 it had been successfully eradicated as a public health threat through the use of an extremely effective vaccine. Two countries maintained cultures of the disease for research purposes, however, and international terrorist groups may have acquired the virus.

The smallpox virus could potentially be dispersed over a wide area in an aerosol form. Infection of only a small number of people might lead to a rapid spread of the disease throughout a targeted population given smallpox's highly contagious nature: The disease is easily spread by direct contact, droplet, and airborne transmission. Patients are considered highly infectious and should be quarantined until the last scab has fallen off FIGURE 35-17. The incubation period for smallpox is between 4 and 17 days (average = 12 days).

There is no treatment for smallpox except for supportive care. Although an effective vaccine against this disease exists, it has not been widely used in recent decades because smallpox was deemed to be eliminated in 1980. Today millions of people have never been vaccinated against this infection, and millions more have reduced immunity because decades have passed since their last immunization.

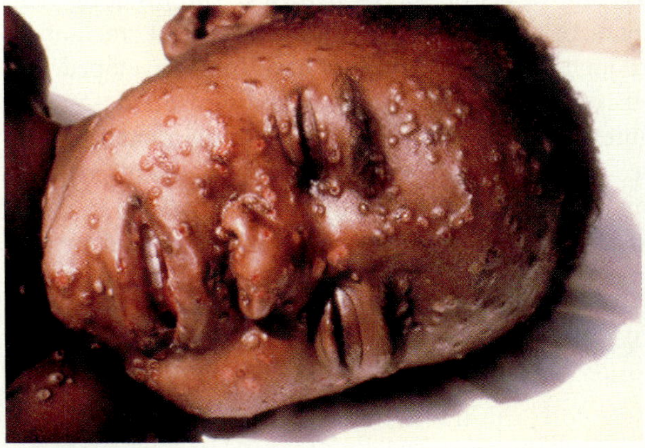

FIGURE 35-17 Smallpox is a highly contagious disease with a mortality rate of approximately 30 percent.

Dealing with Biological Agents

It is unlikely that fire fighters or emergency responders would immediately recognize that a biological weapon had been released in their communities, in large part because of these agents' incubation period—that is, the lag between the actual infection and the appearance of symptoms. The symptoms of a biological agent attack would typically become manifest over a period of days after the exposure. If left untreated, many of these diseases might kill a large proportion of the infected population.

The Centers for Disease Control and Prevention and area hospitals would typically be the first to recognize the situation, after identifying significant numbers of people arriving at their facilities with similar symptoms. Multiple medical calls about patients exhibiting similar symptoms could provide a clue about the use of a biological weapon, especially if their location is considered to be a potential target.

Fire fighters or emergency responders treating people who may have been exposed to a biological agent should follow the recommendations of their departments regarding universal precautions (protective measures for use in dealing with objects, blood, body fluids, or other items associated with potential exposure risks of communicable diseases). These measures include wearing gloves, masks, eye protection, and surgical gowns when treating patients.

Emergency responders who exhibit flu-like symptoms after a potential terrorist incident should seek medical care immediately. It is important that fire fighters report any possible exposure to department officials. Postincident actions may include medical screening, testing, or vaccinations.

Radiological Agents

A third type of agent that might potentially be used by terrorists is radiation. The most probable scenarios involve the use of various approaches to disperse radiological agents and contaminate an area with radioactive materials. This threat is very different from that posed by a nuclear detonation.

Radiation

Radioactive materials release energy in the form of electromagnetic waves or energy particles. The resulting radiation cannot be detected by the normal senses of smell and taste, although several instruments have been developed that can detect the presence of radiation and measure dose rates. Fire fighters should become familiar with the radiation detectors used by their departments. Do not enter any place that might contain radiation unless you have been trained in measuring radiation, have a working radiation detector, and are properly protected.

Three types of radiation exist: alpha particles, beta particles, and gamma rays.

- Alpha particles quickly lose their energy and, therefore, can travel only 1 to 2 inches from their source. Clothing or a sheet of paper can stop this type of energy. If ingested or inhaled, alpha particles can damage a number of internal organs.
- Beta particles are more powerful, capable of traveling 10 to 15 feet. Heavier materials such as metal, plastic, and glass can stop this type of energy. Beta radiation can harm both the skin and the eyes, and its ingestion or inhalation will damage internal organs.
- Gamma radiation can travel significant distances, penetrate most materials, and pass through the body. This type of radiation is the most destructive to the human body. The only materials with sufficient mass to stop gamma radiation are concrete, earth, and dense metals such as lead.

Effects of Radiation

Symptoms of low-level radioactive exposure might include nausea and vomiting. Exposure to high levels of radiation can cause vomiting and digestive system damage within a short time. Other symptoms of high-level radiation exposure include bone marrow destruction, nerve system damage, and radioactive skin burns. An extreme exposure may cause death rapidly; more typically, it takes a considerable amount of time before the signs and symptoms of radiation poisoning become obvious. As the effects progress over the years, a prolonged spiral into death caused by leukemia or carcinoma is likely.

A contaminated person will have radioactive materials on the exposed skin and clothing. In severe cases, this contamination could penetrate to the internal organs of the body. For this reason, decontamination procedures are required to remove the radioactive materials. Decontamination must be complete and thorough because the situation could escalate rapidly if this procedure is inadequate. Specifically, the contamination could quickly spread to ambulances, hospitals, and other locations where contaminated individuals or items are transported.

Radioactive materials both expose rescuers to radiation injury and continue to affect the exposed person. Anyone who comes in contact with a contaminated person must be protected by appropriate PPE and shielding, depending on the particular contaminant. Rescuers and medical personnel must all be decontaminated after interacting with a contaminated person because the only way to know whether radiation is present on their bodies or clothing is to use radiation detectors.

A person can be exposed to radiation without coming into direct contact with a radioactive material through inhalation, skin absorption, or ingestion. A person can also be injured by exposure to radiation without being contaminated by the radioactive material itself. Someone who has been exposed to radiation, but has not been contaminated by radioactive materials, requires no special handling.

There are three ways to limit exposure to radioactivity:

- Keep the time of the exposure as short as possible.
- Stay as far away from the source of the radiation as possible.
- Use shielding to limit the amount of radiation absorbed by the body.

In addition, do not enter any area that might contain radiation until trained personnel have assessed the radiation level using approved detection devices.

If radioactive contamination is suspected, everyone who enters the area should be equipped with a personal dosimeter to measure the amount of radioactive exposure FIGURE 35-18. Rescue attempts should be made only by properly trained and equipped personnel, each of whom must carry an approved radiation monitor to measure the amount of radiation present.

The Dirty Bomb

In recent years, the radiation dispersal device or "dirty bomb" has emerged as a source of concern in terms of terrorism. Packing radioactive material around a conventional explosive device could contaminate a wide area, with the size of the affected area ultimately depending on the amount of radioactive material and the power of the explosive device.

To limit this threat, radioactive materials, even in small amounts, are kept secure and protected. Such materials are widely used in industry and health care, particularly in conjunction with X-ray machines. A terrorist could potentially construct a dirty bomb with just a small quantity of stolen radioactive material.

Security experts are also concerned that a large explosive device, such as a truck bomb, might potentially be detonated near a nuclear power plant. Such an explosion could damage the reactor containment vessel and release radioactive material into the atmosphere, allowing it to spread over a large area.

FIGURE 35-18 A personal dosimeter measures the amount of radioactive exposure received by an individual.

Operations

Responding to a terrorist incident puts fire fighters and other emergency personnel at risk. Although responders must ensure their own safety at every incident, a terrorist incident may carry an extra dimension of risk. Because the terrorist's objective is to cause as much harm as possible, emergency responders are just as likely to be targets as ordinary civilians.

In most cases, the first emergency units will not be dispatched for a known WMD or terrorist incident. Rather, the initial dispatch might be for an explosion, for a possible hazardous materials incident, for a single person with difficulty breathing, or for multiple victims with similar symptoms. Emergency responders will usually not know that a terrorist incident has occurred until personnel on the scene begin to piece together information gained from their own observations and from witnesses.

If appropriate precautions are not taken, initial responders may find themselves in the middle of a dangerous situation before they realize what has happened. Initial responders should be aware of any factors that suggest the possibility of a terrorist incident and immediately implement appropriate procedures. The possibility of a terrorist incident should be considered when responding to any location that has been identified as a potential terrorist target. It could be difficult to determine the true nature of the situation until a scene size-up is conducted.

Fire Fighter Safety Tips

Casualty Management After Detonation of a Radiological Weapon

- Do not enter an area that is suspected to be contaminated until radiation levels have been measured and found to be safe.
- Enter the area only to save lives; radiation levels will be very high.
- Wear PPE including SCBA.
- All personnel and equipment must be decontaminated after leaving the contaminated area. Because radioactive dust collects on clothing, all clothing should be removed and discarded after leaving the area. Failure to do so will result in continued radiation exposure to yourself and to others.
- Wash your entire body thoroughly with lukewarm water as soon as possible after leaving the scene, even if you have been through a decontamination process.
- Do not eat, drink, or smoke while exposed to potential radioactive dust or smoke.
- Use radiation monitoring devices to map the locations of areas with high radioactivity. These monitoring devices should be wrapped in plastic bags to prevent their contamination.
- Vehicles should be washed before they leave the scene. The only exception is emergency units transporting critical victims.
- Life-threatening injuries are more serious than radioactive contamination. Treat such injuries before decontaminating the individual.
- Alert local hospitals that they may receive radioactive-contaminated victims.

Initial Actions

Fire fighters should approach a known or potential terrorist incident just as they would a hazardous materials incident. The possibility that chemical, biological, or radiological agents are involved cannot be ruled out until a hazardous materials team checks the area with detection and monitoring instruments. Apparatus and personnel should, if possible, approach the scene from a position that is uphill and upwind. Emergency responders should don PPE, including SCBA. Additional arriving units should be staged an appropriate distance away from the incident.

The first units to arrive should establish an outer perimeter to control access to and from the scene. They should deny access to all but emergency responders, and they should prevent potentially contaminated individuals from leaving the area before they have been decontaminated. The perimeter must completely surround the affected area, with the goal of keeping people who were not initially involved from becoming additional victims.

The incident command post should be established in a safe location, which could be as far as 3000 feet from the actual incident scene. It must be set up outside the area of possible contamination and beyond the distance where a secondary device might be planted. The initial priority should be to determine the nature of the situation, the types of hazards that could be encountered, and the magnitude of the problems faced by the responders.

An initial reconnaissance ("recon") team should be sent out to quickly examine the involved area and to determine how many people are involved. Proper use of PPE, including SCBA, is essential for the recon team, and the initial survey must be conducted very cautiously, albeit as rapidly as possible. The possibility that chemical, biological, or radiological agents are involved cannot be ruled out until qualified personnel with appropriate instruments and detection devices have surveyed the area. Responders should take special care not to touch any liquids or solids or to walk through pools of liquid. Any emergency responders who become contaminated must not leave the area until they have been decontaminated.

A process of elimination may be required to determine the nature of the situation. Occupants and witnesses should be asked if they observed any unusual packages or detected any strange odors, mists, or sprays. The presence of a large number of casualties who have no outward signs of trauma might indicate a chemical agent exposure. In such a case, victims' symptoms might include difficulty breathing, skin irritations, or seizures.

A visible vapor cloud would be a strong indicator of a chemical release, which may have occurred as the result of either an accident or a terrorist attack. The presence of dead or dying animals, insects, or plant life might also signal a chemical agent release.

When approaching the scene of an explosion, responders should consider the possibility of a terrorist bombing incident and remain vigilant for secondary explosive devices. If they note any suspicious packages, responders must notify the IC immediately. Fire fighters who have not been specially trained should never approach a suspicious object. Instead, EOD personnel should examine any suspicious articles and disable them.

FIRE FIGHTER Tips

In 1993, the first units to respond to a call at the World Trade Center thought a transformer explosion had occurred somewhere beneath the huge complex. As responders explored the smoke-filled interior, they realized that a large explosion had occurred in the underground parking area. As they pieced together the information, they determined that terrorists had detonated a truck bomb, causing tremendous destruction and filling the twin towers with smoke.

At the same location in 2001, many of the first responders saw a plane crash into one of the towers and believed that they were responding to a terrible accident. They realized the true nature of the situation only when a second plane crashed into the other tower.

Interagency Coordination

If a terrorist incident is suspected, the IC should consult immediately with local law enforcement officials. If a mass-casualty situation is evident, the IC should notify area hospitals and activate the local medical response system. Local technical rescue teams should be requested to evaluate structural damage and initiate rescue operations.

State emergency management officials should be notified as soon as possible. This step will help ensure a quick response by both state and federal resources to a major incident. Large-scale search and rescue incidents could require the response of Urban Search and Rescue task forces as activated through the Federal Emergency Management Agency (FEMA). Medical response teams, such as Disaster Medical Assistance Teams, may be needed for incidents involving large numbers of people.

An emergency operations center (EOC) can help to coordinate the actions of all involved agencies in a large-scale incident, particularly if terrorism is involved. The EOC is usually set up in a predetermined remote location and is staffed by experienced command and staff personnel FIGURE 35-19. The IC, who remains at the scene of the incident, should provide detailed situation reports to the EOC and request additional resources as needed.

Fire fighters must remember that a terrorist incident is also a crime scene. To avoid destroying important evidence that could lead to a conviction of those parties responsible for the attack, responders should not disturb the scene any more than is necessary. Where possible, law enforcement personnel should be consulted prior to overhaul and before any material is removed from the scene. Fire fighters should also realize that one or more terrorists could be among the injured victims.

FIGURE 35-19 An EOC is set up in a predetermined location for large-scale incidents.

Be alert for threatening behavior, and make note of anyone who seems determined to leave the scene.

TEAMWORK Tips

A hazardous materials team should respond to any potential or suspected terrorist incident. These personnel will test for chemical, biological, or radiological contaminants using appropriate test instruments.

Decontamination

Those persons who have been directly contaminated to chemical, biological, or radiological agents must be decontaminated to remove or neutralize any chemicals or substances on their bodies or clothing. Those who have been exposed to these materials should be evaluated first before decontamination takes place to avoid unnecessary exposure and waste of resources. If in doubt, then decontaminate. Decontamination should occur as soon as possible to prevent further absorption of a contaminant and to reduce the possibility of spreading the contamination. Equipment must also be decontaminated before it leaves the scene. The Hazardous Materials: Decontamination Techniques chapter explains these decontamination procedures in detail.

The perimeter must fully surround the area of known or suspected contamination. Every effort must be made to avoid contaminating any additional areas, particularly hospitals and medical facilities. Qualified personnel must monitor the perimeter with instruments or detection devices to ensure that contaminants are not spread.

Standard decontamination procedures usually involve a series of stations. At each station, clothing and protective equipment are removed and the individual is cleaned. Although some contaminants require only soap and water for their removal, chemical and biological agents may require special neutralizing solutions.

A terrorist incident that results in a large number of casualties may require a somewhat different decontamination process. Special procedures for mass decontamination have been devised for incidents involving hundreds or thousands of people. These procedures use master stream devices from engine companies and aerial apparatus to create high-volume, low-pressure showers. This approach allows large numbers of people to be decontaminated rapidly **FIGURE 35-20**.

FIGURE 35-20 Mass decontamination procedures may be required to handle a large group of contaminated victims.

FIRE FIGHTER Tips

Additional Resources for Terrorist Incidents

Resources	Action
Air evacuation	Air transportation of injured
Air supply unit	Refill SCBA cylinders
Bomb squad	Render explosive devices safe
Emergency medical services	Emergency medical care
Hazardous materials unit	Neutralize hazardous substances, including chemical agents
Health department	Technical; advice on chemical and biological agents
Local law enforcement	Law enforcement, perimeter control, and evidence recovery
Structural collapse teams	Search and stabilize collapsed structures

FIRE FIGHTER Tips

Emergency responders at a terrorist event may later be called to testify in court. Make mental notes if you must disturb or move any evidence to rescue someone in immediate danger. Document this information as soon as possible.

Mass Casualties

A terrorist or WMD incident may result in a large number of casualties. As part of the overall planning process, mass-casualty plans are essential to manage this type of situation, which would quickly overwhelm the normal capabilities of most emergency response systems. Mass-casualty plans typically involve using the resources of multiple agencies in an effort to handle large numbers of victims efficiently.

The basic principles of mass-casualty operations are described in the Emergency Medical Care chapter. A terrorist incident, however, might potentially involve several additional complications and considerations, including the possibility of contamination by chemical, biological, or radiological agents. The mass-casualty plan must be expanded to address these problems.

If contamination is suspected, the plan must ensure that it does not spread beyond a defined perimeter. In some cases, victims may need to be decontaminated as they are moved to the triage and treatment areas FIGURE 35-21. This step will keep these areas free of contaminants. In other cases, the triage and treatment areas may be considered contaminated zones. In this scenario, victims would be decontaminated before they are transported from the scene.

During the early stages of an incident, it may be difficult to determine which agent was used and, therefore, which treatment and decontamination procedures are appropriate. In such a case, it may be necessary to quarantine exposed individuals within an area that is assumed to be contaminated until laboratory test results are evaluated. The strategic plan for this situation must assume that any personnel who provide assistance or treatment to victims during this period will also become contaminated, as will any equipment or vehicles used in the contaminated area.

FIGURE 35-21 Victims should be decontaminated before they are placed in triage and treatment areas.

Additional Resources

A terrorist incident would likely result in a massive response by local, state, and federal government agencies. The FBI is the lead agency for crisis management, and FEMA is the lead agency for consequence management during a terrorist incident. Other federal agencies such as the ATF and the Department of Homeland Security may also be involved. Several federal agencies have specific responsibilities regarding terrorist incidents.

The National Terrorism Advisory System (NTAS) is designed to effectively communicate information about terrorist threats by providing timely, detailed information to the public, government agencies, first responders, airports and other transportation hubs, and the private sector. It recognizes that first responders need to be aware of heightened risk or terrorist attacks in the United States to know what to do to in the event of an emergency incident.

National Terrorism Advisory System Alerts are issued when credible information is received that indicates a terrorist threat has been discovered that poses a significant danger to the United States. This system replaced the color-coded system that was originally put in place in 2002. The NTAS issues two types of alerts: the Imminent Threat Alert and the Elevated Threat Alert. An Elevated Threat Alert warns of a credible terrorist threat against the United States. An Imminent Threat Alert warns of a credible, specific, and impending terrorist threat against the United States. Both types of alerts provide a concise summary of the potential threat. In some cases, alerts are sent directly to law enforcement or specific areas of the private sector; in other situations, alerts are issued more broadly to the American people through public media. Each threat alert is issued for a specific time and then expires. It may be extended if new information becomes available or if the threat evolves.

Chief Concepts

- Terrorism can be described as the unlawful use of violence or threats of violence to intimidate or coerce a government, the civilian population, or any segment thereof, to further political or social objectives.
- This broad definition can include a wide range of acts committed by different groups for different purposes.
- The goal of terrorism is to produce feelings of terror in a population or a group.
- Terrorism can occur in any community. Be aware of potential targets in your area.
- Terrorists can turn ordinary objects into weapons.
- Responding to a terrorist incident puts fire fighters and other emergency personnel at risk.
- A call for a terrorist incident may first appear to be a routine incident.
- When responding to a potential terrorist-related incident, fire fighters should establish a staging area at a safe distance from the scene and follow the direction of the IC.

Hot Terms

Agroterrorism The intentional act of using chemical or biological agents against the agricultural industry or food supply.

Alpha particles A type of radiation that quickly loses energy and can travel only 1 to 2 inches from its source. Clothing or a sheet of paper can stop this type of energy. Alpha particles are not dangerous to plants, animals, or people unless the alpha-emitting substance has entered the body.

Ammonium nitrate fertilizer and fuel oil (ANFO) An explosive made of commonly available materials.

Anthrax An infectious disease spread by the bacterium *Bacillus anthracis*; it is typically found around farms, infecting livestock.

Beta particles A type of radiation that is capable of traveling 10 to 15 feet. Heavier materials, such as metal and glass, can stop this type of energy.

Biological agents Disease-causing bacteria, viruses, and other agents that attack the human body.

Blistering agents Chemicals that cause the skin to blister.

Chlorine A yellowish gas that is approximately 2.5 times heavier than air and slightly water soluble. It has many industrial uses but also damages the lungs when inhaled (it is a choking agent).

Choking agent A chemical designed to inhibit breathing. It is typically intended to incapacitate rather than kill.

Cyanide A highly toxic chemical agent that attacks the circulatory system.

Cyberterrorism The intentional act of electronically attacking government or private computer systems.

Decontamination The physical or chemical process of removing any form of contaminant from a person, an animal, an object, or the environment.

DuoDote Kit A prefilled auto-injector that provides two medications (atropine and pralidoxime chloride) for use in treating exposure to nerve agents and insecticide poisoning.

Ecoterrorism Terrorism directed against causes that radical environmentalists think would damage the earth or its creatures.

Forward staging area A strategically placed area, close to the incident site, where personnel and equipment can be held in readiness for rapid response to an emergency event.

Gamma radiation A type of radiation that can travel significant distances, penetrating most materials and passing through the body. Gamma rays are the most destructive type of radiation to the human body.

Improvised explosive device (IED) An explosive or incendiary device that is fabricated in an improvised manner.

Incubation period The time period between the initial infection by an organism and the development of symptoms by a victim.

Lewisite A blister-forming agent that is an oily, colorless-to-dark brown liquid with an odor of geraniums.

Mass decontamination The process of flushing a large number of victims with water for rapid decontamination in the field when the contaminating agent is unknown.

National Terrorism Advisory System Alerts A national system to communicate information about terrorist threats by providing timely information to the public, government agencies, and first responders. It consists of Imminent Threat Alerts and Elevated Threat Alerts.

Nerve agents Toxic substances that attack the central nervous system in humans.

Personal dosimeter A device that measures the amount of radioactive exposure to an individual.

Phosgene A chemical agent that causes severe pulmonary damage; it is a by-product of incomplete combustion.

Pipe bomb A device created by filling a section of pipe with an explosive material.

Plague An infectious disease caused by the bacterium *Yersinia pestis*, which is commonly found on rodents.

Radiation dispersal device Any device that causes the purposeful dissemination of radioactive material without a nuclear detonation; a dirty bomb.

Wrap-Up, continued

Radiological agents Materials that emit radioactivity.

Sarin A nerve agent that is primarily a vapor hazard.

Secondary device An explosive device designed to injure emergency responders who have responded to an initial event.

Smallpox A highly infectious disease caused by the virus *Variola*.

Soman A nerve gas that is both a contact hazard and a vapor hazard; it has the odor of camphor.

Sulfur mustard A clear, yellow, or amber oily liquid with a faint sweet odor of mustard or garlic that may be dispersed in an aerosol form. It causes blistering of exposed skin.

Tabun A nerve gas that is both a contact hazard and a vapor hazard; it operates by disabling the chemical connection between the nerves and their target organs.

Triage The process of sorting victims based on the severity of their injuries and medical needs to establish treatment and transportation priorities.

Universal precautions Procedures for infection control that treat blood and certain body fluids as capable of transmitting bloodborne diseases.

V-agent A nerve agent, principally a contact hazard; an oily liquid that can persist for several weeks.

Weapons of mass destruction (WMD) Weapons whose use is intended to cause mass casualties, damage, and chaos.

FIRE FIGHTER *in action*

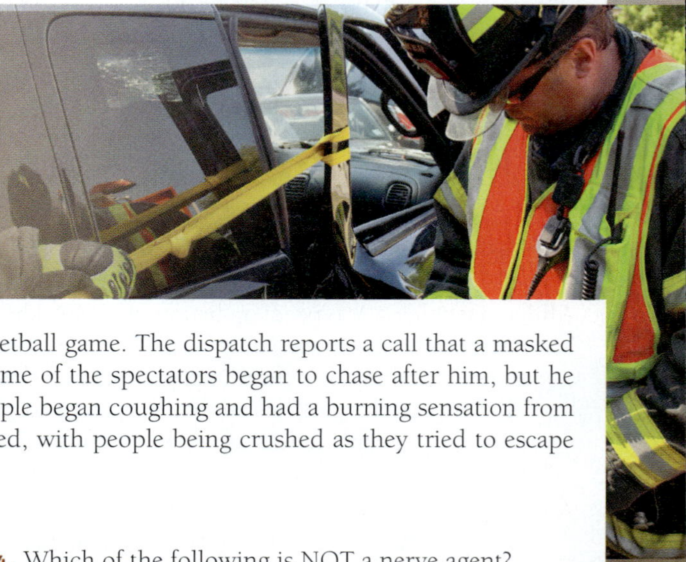

You are dispatched along with the police to the local high school basketball game. The dispatch reports a call that a masked man ran out onto the court spraying some type of fog into the air. Some of the spectators began to chase after him, but he pulled out a handgun, injuring two of them, and escaped the area. People began coughing and had a burning sensation from the spray that had a sweet garlic smell. The exits were quickly jammed, with people being crushed as they tried to escape the gym.

1. The agent most likely was _____.
   - **A.** sulfur mustard
   - **B.** a radiological agent
   - **C.** Lewisite
   - **D.** Anthrax
2. Which of the following is an infectious disease that can be used in a WMD attack?
   - **A.** Anthrax
   - **B.** Soman
   - **C.** V-agent
   - **D.** Sarin
3. A type of radiation that quickly loses energy and can travel only 1 to 2 inches from its source is __________.
   - **A.** alpha
   - **B.** beta
   - **C.** gamma
   - **D.** neutron
4. Which of the following is NOT a nerve agent?
   - **A.** Tabun
   - **B.** Phosgene
   - **C.** Soman
   - **D.** Sarin
5. The intentional act of using chemical or biological agents against the agricultural industry or food supply is termed _____.
   - **A.** Ecoterrorism
   - **B.** Cyberterrorism
   - **C.** Nutriterrorism
   - **D.** Agroterrorism
6. Because of the risk of a secondary device in incidents involving explosive devices, the staging area should be at least _____ feet away.
   - **A.** 300
   - **B.** 1000
   - **C.** 1500
   - **D.** 3000

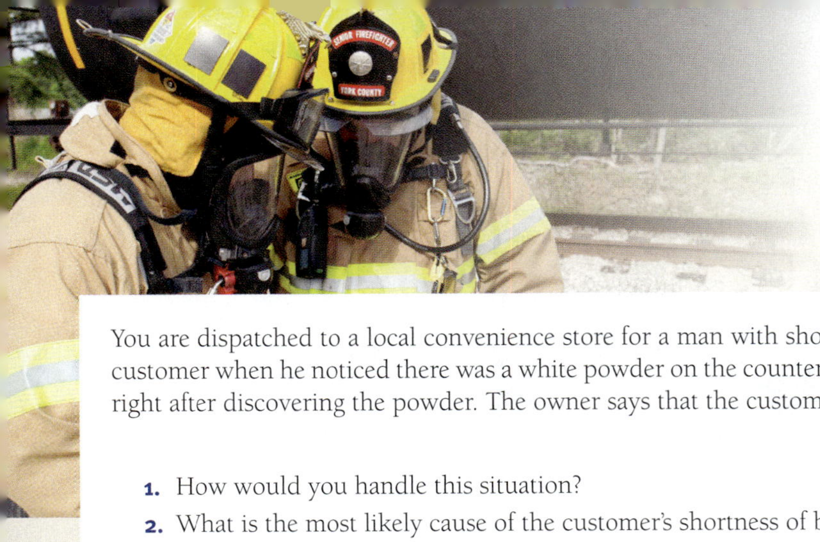

FIRE FIGHTER II *in action*

You are dispatched to a local convenience store for a man with shortness of breath. The owner says that he was waiting on a customer when he noticed there was a white powder on the counter. The owner says that the customer became short of breath right after discovering the powder. The owner says that the customer also became nauseated and broke out in a cold sweat.

1. How would you handle this situation?
2. What is the most likely cause of the customer's shortness of breath?
3. How will you calm and reassure the owner?
4. What concerns do you have about this incident?

Fire Prevention and Public Education

Fire Fighter I

Knowledge Objectives

There are no knowledge objectives for Fire Fighter I candidates. NFPA 1001 contains no Fire Fighter I Job Performance Requirements for this chapter.

Skills Objectives

There are no skill objectives for Fire Fighter I candidates. NFPA 1001 contains no Fire Fighter I Job Performance Requirements for this chapter.

Fire Fighter II — FFII

Knowledge Objectives

After studying this chapter, you will be able to:

- Describe the activities that prevent fires and limit their consequences if fire occurs. (NFPA 6.5.2, p 1014)
- Identify elements of public fire safety education programs covering stop, drop, and roll; exit drills in the home; the selection and use of portable fire extinguishers; and the importance of smoke alarms and residential sprinkler systems in preventing fire deaths. (NFPA 6.5.1.A, p 1015–1022)
- Explain the importance of residential sprinkler systems in preventing residential fire deaths. (NFPA 6.5.1.B, p 1020–1021)
- Stress the importance of having portable fire extinguishers. (NFPA 6.5.1.B, p 1021)
- Recognize hazards during a fire safety survey of an occupied structure. (NFPA 6.5, 6.5.1, p 1022–1026)
- Describe the steps in conducting a fire station tour. (NFPA 6.5.2.A, p 1025–1028)

FFII

CHAPTER 36

Additional NFPA Standards

- NFPA 1, *Fire Code*
- NFPA 101, *Life Safety Code*
- NFPA 1452, *Guide for Training Fire Service Personnel to Conduct Dwelling Fire Safety Surveys*

Skills Objectives

After studying this chapter, you will be able to perform the following skills:

- Complete an occupied structure fire safety survey. (NFPA 6.5.1.B, p 1025–1026)
- Perform a public fire safety education presentation on stop, drop, and roll. (NFPA 6.5.2, p 1017)
- Execute a public fire safety education presentation on exit drills in the home. (NFPA 6.5.2, p 1018–1019)
- Install and maintain a smoke alarm. (NFPA 6.5.2, p 1019, 1021)
- Give a public education tour of a fire station. (NFPA 6.5.2.B, p 1027–1028)

You Are the Fire Fighter

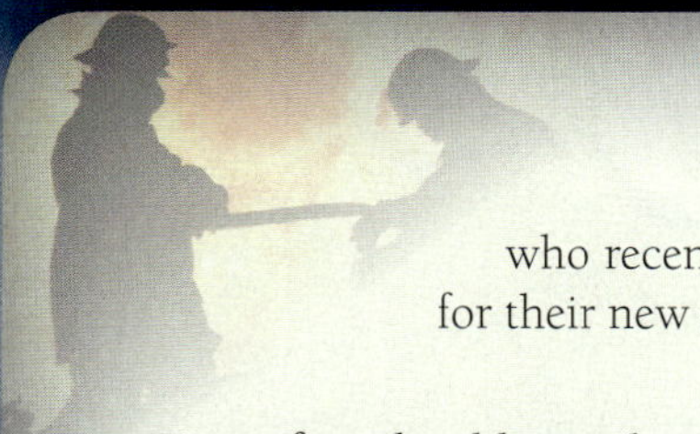

You are attending a family reunion and the conversation turns to your new job as a fire fighter. You beam with pride as you tell your family about your training and your new assignment. The discussion turns to your role in fire prevention. Your cousin and his wife, who recently had a new baby boy begin to question you on what they can do to ensure their home is safe for their new addition. You think back to what you learned in your Fire Fighter I and II courses.

1. How often should you change the batteries in your smoke alarm?
2. Does your family have an evacuation plan in case of a fire?
3. What actions would you recommend to your cousin?

Introduction

Fire prevention is one of the most important activities performed by fire fighters. Our most effective means of avoiding the loss of lives and property from fires is to prevent fires rather than extinguish them. Most fires are caused by unsafe or careless acts, equipment failure, arson, or acts of nature. Countless fires can be prevented by teaching safe behaviors and eliminating unsafe conditions. This kind of education ultimately saves lives and reduces the losses caused by fires.

FIRE FIGHTER II Tips **FFII**

According to the U.S. Fire Administration, 1,331,500 fires were reported in the United States in 2010. Of that number, 384,000 fires were in residential occupancies, and 98,000 fires were in nonresidential buildings. Clearly, the majority of structural fires occur in residential occupancies. In 2010, U.S. fire fighters also responded to 215,000 vehicle fires and 634,000 outdoor fires.

Fire prevention duties for fire fighters include teaching children how to stop, drop and roll; teaching families to conduct exit drills in the home (EDITH); delivering public safety education on smoke alarms; educating the public about the role of residential fire sprinklers; stressing the role of portable fire extinguishers; performing fire safety inspections in residential occupancies; and conducting fire station tours. This chapter covers the information necessary to accomplish these tasks.

FIRE FIGHTER II Tips **FFII**

Fires that are prevented do not cause any deaths, injuries, or property loss. Most fires are caused by actions or situations that could have been avoided. A key element in reducing the number of fire fighter line-of-duty deaths and civilian deaths is proactive fire prevention.

What Is Fire Prevention?

Fire prevention includes a range of activities that are intended to prevent the outbreak of fires and to limit the consequences if a fire does occur. These activities include enacting and enforcing fire codes, conducting property inspections, and presenting public fire safety education programs. Investigating fire causes is also included in the overall scope of fire prevention activities. All of these activities share a common objective—limiting life loss, injuries, and property damage. These activities increase fire fighter safety as well. Your highest priority as a fire fighter should always be to prevent fires.

Every fire fighter must work to prevent fires and to educate the public about fire risks and hazards, even if a separate bureau or division within the department has primary responsibility for fire prevention. Some departments share the responsibility for conducting inspections and enforcing safety codes with the jurisdiction's building department or other agency. Some departments use on-duty suppression personnel to perform basic fire inspections within their response area. Although Fire Fighter I and Fire Fighter II personnel may have limited responsibilities for formal fire prevention activities, it is imperative that they understand the objectives of fire prevention, the delineation of responsibilities within their department, and the importance of their role in fire prevention.

Enactment of Fire Codes

Preventing fires and protecting lives and property in the event of a fire are important community concerns and responsibilities. Fire codes—regulations that have been legally adopted by a governmental body with the authority to pass laws and enforce safety regulations—are enacted to ensure a minimum level of fire safety in the home and workplace environments. These codes are enforced through a legal process similar to the way in which driving-speed limits and other traffic regulations are enforced. A fire code can be adopted by a local government or by a state or province. In some cases, one basic set of regulations applies throughout a state, but local jurisdictions are free to adopt more stringent regulations.

The Fire Codes are created by the National Fire Protection Association (NFPA) and the International Code Council (ICC). The Fire Codes are intended to address a wide range of issues relating to fire and safety. Many jurisdictions have adopted or incorporated these documents into their local codes and enforce them just as they would any other government regulation.

Most communities adopt and enforce a full set of codes that cover a range of areas and that are designed to establish basic health and safety standards. Generally, these codes include a building code, a fire code, an electrical code, a plumbing code, a mechanical code, and other local rules and regulations. These codes and regulations must be coordinated to avoid conflicting requirements and to ensure a safe community. For example, the electrical code includes requirements that are intended to prevent electrical fires as well as requirements geared toward reducing the risk of accidental electrocution.

The fire code generally includes regulations designed to prevent fires from occurring, eliminate fire hazards, protect lives, and limit fire losses. Fire department personnel are usually involved in the enforcement of this code. Every fire fighter needs to understand the effects of the relevant codes and their involvement in establishing public safety requirements.

Fire codes are also closely related to building codes. Building codes establish safety requirements that apply to the construction of new buildings and almost always include fire protection requirements. When a new building is constructed, the building code specifies the basic requirements for the type of construction, allowable heights and areas, distances between firewalls, and the building materials. In addition, this code usually includes requirements for safe exiting and determines when fire alarm systems, automatic sprinklers, and standpipes are required. The Building Construction chapter covers issues that are closely related to building codes.

Generally, fire codes apply to all buildings, new or old, and to many different situations related to fire risks and hazardous conditions. They require unobstructed, unlocked exits; properly maintained sprinkler systems; and fire alarm systems in operating condition. They may limit the number of people who can be present in places of public assembly (that is, occupancy limits). Fire codes also include regulations for dispensing flammable liquids, for storing them in approved storage containers, and for prohibiting their use or storage in some buildings. Regulations that prohibit parking in a fire lane or obstructing a fire hydrant are usually found in the fire code as well. All of these requirements are intended to promote fire safety and protect the community.

Inspection and Code Enforcement

After a jurisdiction adopts a fire code, it must be able to enforce it. Citizens have a legal obligation to comply with all rules and regulations set by such an authority. Regular inspections ensure compliance, note any violations, and require their correction. If violations are not corrected, a series of increasing penalties can be imposed, beginning with a verbal warning or a written notice. An individual who fails to correct the violation within a specified time can be fined. In some cases, a jail sentence can be imposed for very serious violations.

Fire codes usually specify the types of occupancies that can be inspected and the frequency of the inspections. For example, some fire codes require all schools to be inspected twice each year. In other cases, the schedule of inspections is discretionary. The agency responsible for performing inspections and enforcing codes is usually named in the fire code; it may vary in different communities. Sometimes the responsibility belongs to specific individuals or positions within the fire department, such as fire inspectors or fire marshals. Other fire departments train all fire fighters to conduct inspections and take code enforcement actions. In some communities, building inspectors or special code enforcement personnel are given this responsibility. Generally, newly assigned fire fighters are not directly involved in code enforcement activities, but they should nevertheless understand how code enforcement is handled in the community.

In many cases, the fire code does not apply to the interior of a private dwelling, and the fire department cannot require an occupant to submit to an inspection once a certificate of occupancy is issued. The occupant can request a voluntary inspection to identify hazardous conditions, however, and many fire departments offer this service. Such a voluntary inspection is called a home fire safety survey, and the legal requirements of the fire code are not applied when it is conducted. Voluntary inspections of private dwellings are one way that fire departments can educate citizens about the risks of fire in their homes. Such inspections are an important community service offered by the fire department, and the diplomacy and professionalism of fire fighters conducting home safety surveys cannot be underestimated in furthering the mission of the fire department.

Every fire fighter should know how to conduct a home fire safety survey in a private dwelling—including knowing what to look for during the survey, how to contact the occupant, and how to communicate the survey findings. This information is presented in detail later in this chapter.

Public Fire and Life Safety Education

The goals of public fire and life safety education are to help people understand how to prevent fires from occurring and to teach them how to react in an appropriate manner if a fire occurs. Making people aware of common fire risks and hazards and providing information about reducing or eliminating those dangers can prevent many fires. Public education teaches techniques to reduce the risks of death or injury in the event of a fire. The programs are also designed to prevent other types of accidents and injury.

Public fire safety education programs include those focused on the following topics:

- Stop, drop, and roll FIGURE 36-1
- Exit drills in the home (EDITH)

FIGURE 36-1 The Stop, Drop, and Roll program teaches young children what to do if their clothing catches fire.

- Installation and maintenance of smoke alarms
- Advantages of residential sprinkler systems
- Selection and use of portable fire extinguishers
- Learn Not to Burn®
- Change Your Clock—Change Your Battery
- Fire safety for special populations
- Bike safety
- Fall prevention
- Wildland fire prevention programs

Most fire and life safety programs are presented to groups such as school classes, scout troops, church groups, senior citizens groups, civic organizations, hospital staff, and business employees. Presentations are often made at community events and celebrations as well as during Fire Prevention Week activities. Five public fire safety education programs—stop, drop, and roll; EDITH; installation and maintenance of smoke alarms; importance of residential fire sprinklers; and the selection and use of portable fire extinguishers—are described here.

Another popular fire safety education activity is a fire station tour. Both children and adults enjoy the opportunity to tour the local fire station, and a fire station tour is an excellent opportunity to promote fire prevention. Every fire fighter must understand how to conduct an effective fire station tour. The skills and knowledge for conducting a fire station tour are presented later in this chapter.

Stop, Drop, and Roll

The Stop, Drop, and Roll public fire safety education program can be taught to children as young as preschoolers, but it is also a valuable educational program for adults. This program is designed to instruct people what to do if their clothing catches fire. When clothing catches fire, the normal reaction is to run for help. Unfortunately, running tends to fan the flames and increases the chance for severe burns. Teaching the mnemonic "stop, drop, and roll" and demonstrating each step helps to reduce burn injuries.

When teaching the Stop, Drop, and Roll program to children, ask them to discuss situations that might result in their clothing catching fire. These circumstances might include being too close to a kitchen stove or fireplace, or playing with matches or lighters. The fire fighter should talk about ways to reduce these hazards, such as keeping matches and lighters out of the reach of children. He or she should then stress each step of the sequence:

- **Stop** means just that: Stay in place; do not run. Stress the fact that running will fan the flames and spread the fire.
- **Drop** means getting down on the ground or on the floor. The person should cover his or her face with the hands to help protect the airway and the eyes.
- **Roll** involves encouraging children to tuck in their elbows and keep their legs together and roll like a bug. They should then roll over and over, and then roll back in the other direction. This helps to smother the flames.

When teaching adults this exercise, stress prevention techniques such as wearing short-sleeved clothing when cooking. Adults can be told that a stream of water will extinguish fire on their sleeves. A good prevention tactic is to talk about safe use of gasoline when filling lawn equipment; the safe use of charcoal lighter and other flammable liquids should be stressed as well.

Adults should be instructed that they can give directions to another person if their own clothing catches fire. Remind students that a blanket can be used to smother a fire and that a garden hose both extinguishes the fire and cools the burned area. By teaching ways to prevent clothing from catching fire and steps to smother the fire, you can help to reduce serious burns.

To teach the Stop, Drop, and Roll program, follow the steps in SKILL DRILL 36-1:

1. Explain the purpose of stop, drop, and roll.
2. Instruct students to **stop**: Stay in place. (STEP 1)
3. Instruct students to **drop** to the floor or ground and cover their faces with their hands. (STEP 2)
4. Instruct students to tuck their elbows, keep their legs together, and **roll** over and then back to extinguish the flames. (STEP 3)
5. Demonstrate the stop, drop, and roll procedure.
6. Have students practice the procedure. (STEP 4)

FIRE FIGHTER II Tips — FFII

There are many good reasons to have a form of a written plan for fire safety presentations. A written plan offers the following benefits:

- Guides the presentation
- Outlines the main points the presentation will cover
- Ensures consistency when multiple presenters are used

SKILL DRILL 36-1 Teaching Stop, Drop, and Roll — FFII

(Fire Fighter II, NFPA 6.5.2)

1 Instruct students to STOP: Stay in place.

2 Instruct students to DROP to the floor or ground and cover their faces with their hands.

3 Instruct students to tuck their elbows, keep their legs together, and ROLL over and then back to extinguish the flames.

4 Have students practice the procedure.

FIRE FIGHTER II Tips FFII

Whenever you are dealing with the public, pay attention to the language that you use. Make sure that it will be understood by your audience. Avoid using firefighting terms and jargon. Speak at a slower pace for young children or the elderly.

Exit Drills in the Home

A second public fire safety education program, called Exit Drills in the Home (EDITH), teaches residents how to safely get out of their homes in the event of a fire or other emergency. Many fire deaths could be prevented if everyone learned a few simple rules about what to do in the event of a fire in the home. As part of this exercise, residents need to understand the importance of having properly installed and properly working smoke alarms. Working smoke alarms alert residents to a fire and give them more time to escape from a burning building. A public fire safety program about smoke alarms is described later in this chapter. More information about the operation of smoke alarms is also found in the Fire Detection, Protection, and Suppression Systems chapter.

As part of your EDITH presentation, stress the importance of keeping bedroom doors closed during sleeping hours, as this step prevents smoke and heat from reaching bedrooms. Emphasize the need to have two escape routes from each bedroom. If smoke, heat, or flames are present in the primary escape route, use the secondary escape route. The secondary escape route may be through a window or through a different door. Stress the importance of alerting other occupants to the presence of a fire.

When alerted to the presence of a fire, all occupants should roll out of bed, stay low, and crawl toward the designated exit. By staying low, individuals decrease the amount of smoke and gases they inhale.

Before opening any door, touch the closed door to see if it is hot. Teach students to use the back of a bare hand to sense the temperature of the door. If the door is hot, do not open it; instead, use a window or another door for escape.

Once outside the house, gather all family members at a preestablished meeting spot. Do not go back into the house. Also, make sure that the fire department has been called. If any doubt arises about whether the fire department has been called, use a neighbor's phone or a cell phone to call the fire department again.

Once the steps for emergency drills have been taught, families should practice EDITH regularly. Some drills should be held in the daytime, and some should take place at night.

To teach EDITH, follow the steps in SKILL DRILL 36-2:

1. **Plan**. Describe the importance of making a quick exit when a fire occurs.
2. Stress the importance of having a sufficient number of working smoke alarms on each level of the house.
3. Explain why you should sleep with bedroom doors closed.
4. Draw a diagram of your house.
5. Plan two escape routes from each room. (STEP 1)
6. **Practice**. When you hear a smoke alarm, roll out of bed and crawl to the exit. (STEP 2)
7. Check exit doors for heat. Do not open a door if it is hot. (STEP 3)
8. Follow the escape route to the preestablished meeting place.
9. Ensure that the fire department has been called. (STEP 4)
10. Do *not* reenter a burning building.

Smoke Alarms

A third type of public fire safety presentation covers the proper installation and maintenance of smoke alarms. Smoke alarms can save lives—but only if they are working properly. The goal of this presentation is to help people properly install and maintain approved smoke alarms in their homes. More information about this topic is presented in the Fire Detection, Protection, and Suppression Systems chapter.

During your presentation on this topic, stress the importance of having a working smoke alarm on every floor of a house. The area outside each sleeping area should contain a smoke alarm. Remind home residents that smoke alarms should not be installed close to kitchens, fireplaces, or garages, or near windows, exterior doors, or heating or air-conditioning ducts, as placing smoke alarms in these locations may result in false alarms. Smoke alarms should be mounted on the ceiling at least 4 inches (10 centimeters) from a wall. If ceiling mounting is impractical, mount the detector 4 to 12 inches (10 to 30.5 centimeters) below the ceiling as high as possible on the wall.

Stress the importance of keeping smoke alarms in working order. Alarms should be tested once a month by pressing the test button. Identify the type of battery in the smoke alarm. Smoke alarms powered by alkaline batteries should have their batteries changed every six months. Many communities participate in the Change Your Clock—Change Your Battery campaign, which encourages regular battery replacement. Many new smoke alarms contain a 10 year battery. Once the smoke alarm signals a low battery alarm, replace the smoke alarm, not the battery. Remind homeowners to keep circuit breakers supplying current to smoke alarms on to ensure that these alarms remain operational. Vacuum the detector chamber regularly to prevent the buildup of dust or insects, which can cause false alarms. Some alarms are manufactured to detect the presence of smoke as well as carbon monoxide; it is important to explain the function of each alarm when these dual-purpose alarms are present.

Teaching these simple facts about the installation and maintenance of smoke alarms will help families keep their homes safe in the event of a fire. To present a public fire safety

SKILL DRILL 36-2 Teaching EDITH FFII

(Fire Fighter II, NFPA 6.5.2)

1. Plan. Draw a diagram of your house. Plan two escape routes from each room.

2. Practice. When you hear a smoke alarm or smell smoke, roll out of bed and crawl to the exit.

3. Check exit doors for heat. Do not open a door if it is hot.

4. Follow the escape route to the preestablished meeting place. Ensure that the fire department has been called.

FIRE FIGHTER II Tips FFII

Many manufacturers of smoke alarms recommend replacing smoke detectors every 10 years. Given that lithium batteries need to be changed only every 10 years, these smoke detectors are considered disposable.

education program on installation and maintenance of smoke alarms, follow the steps listed in SKILL DRILL 36-3:

1. Explain the importance of properly installing and maintaining smoke alarms.
2. Stress the importance of installing a smoke alarm on each floor and outside each sleeping area.
3. Remind students to avoid installing smoke alarms in kitchens and garages, or near fireplaces, windows, and exit doors.
4. Mount smoke alarms on the ceiling or as high as possible on walls. (STEP 1)
5. Test smoke alarms once a month using the test button. (STEP 2)
6. Change alkaline batteries in smoke detectors every six months, if not 10 year batteries. (STEP 3)
7. Clean alarms regularly to prevent false alarms. (STEP 4)

Residential Fire Sprinkler Systems

A fourth type of public fire safety presentation covers the importance of residential fire sprinkler systems in preventing

VOICES OF EXPERIENCE

Although I had been a volunteer in my community for a few years, I never understood fully the impact that fire prevention programs can have on a community. Once I was hired full-time, my first week on duty was during Fire Prevention Week. For our fire department, Fire Prevention Weeks means three weeks of presenting fire safety programs to students from kindergarten to third grade in all ten of the town's elementary schools.

On my first day, a veteran fire fighter said, "Okay, kid, watch me do one of these this morning. You're going to get really good at these." It was an example of "see one, do one, and then teach one." I watched the veteran fire fighter open up to the kids at the school, a totally different personality from who he was at the fire station, and I saw how the kids reacted.

True to his word, starting that afternoon and for the next two weeks everyone looked to me to perform the fire safety presentations. I learned quickly on the best ways to present fire safety to young children. Early on I saw that elementary school children were easily frightened by fully dressed fire fighters approaching them unexpectedly. I began to incorporate donning PPE in front of the children while explaining how each part of the PPE helps to protect fire fighters during a fire.

Over time many fire fighters from our department have incorporated their ideas into our fire safety talks. Today we try to relate our equipment to things the kids already use. For example SCBA is like a backpack, turnout gear is like a snow suit, and helmets are like bike helmets.

Alan Zygmunt
Southington Connecticut Fire Department
Southington, Connecticut

FIRE FIGHTER II Tips FFII

Always follow the manufacturer's instructions for mounting and maintenance of smoke alarms.

residential fire deaths. Statistics show that the combination of working smoke alarms and residential fire sprinklers can reduce the risk of fire death by 82 percent. With these statistics in mind, there is a national movement in the fire service to have fire sprinklers installed in all new residential construction. For more information, see the Fire Detection, Protection, and Suppression Systems chapter.

Selection and Use of Portable Fire Extinguishers

Many groups visiting your fire station will have questions about which types of fire extinguishers they should have in their home, kitchen, garage, car, or boat. You should be familiar with the information you learned in the Portable Fire Extinguishers chapter, so you can answer any questions that arise.

Fire Cause Determination

Fire cause determination is the process of establishing the cause of a fire through careful investigation and analysis of available evidence. It is considered part of fire prevention because finding the causes of fires can help prevent similar

SKILL DRILL 36-3 Installing and Maintaining Smoke Alarms FFII

(Fire Fighter II, NFPA 6.5.1)

1 Ensure that smoke alarms are mounted on the ceiling or as high as possible on walls.

2 Test smoke alarms once a month using the test button.

3 Change alkaline batteries in the smoke detector every six months if not a 10 year battery.

4 Clean alarms regularly to prevent false alarms.

fires from occurring in the future. Fire cause determination is equally important for both accidental and intentional fires. The fire fighter's role in fire cause determination is covered in the Fire Cause Determination chapter.

FIRE FIGHTER II Tips FFII

According to the NFPA, the top five causes of fire in homes are cooking equipment, heating equipment, intentional acts, electrical distribution, and open flames.

Conducting a Fire Safety Survey in a Private Dwelling

Many fire departments offer to conduct fire safety surveys in private dwellings. A home fire safety survey helps identify fire and life-safety hazards and provides the occupants with recommendations on making their home safer.

Residential fire safety surveys cannot be conducted without the occupant's permission. Usually, the survey comes in response to a request by the occupant for an appointment. Fire departments may use public service announcements on radio and television to increase public awareness of the need for a fire-safe home and the availability of a fire safety survey. In many communities, the fire department offers to check home smoke alarms and to conduct a home fire safety survey at the same time. Some fire departments will install smoke alarms free of charge in a dwelling as a community service.

A home fire safety survey is a joint effort by the fire department and the occupant to prevent fires and reduce fire fatalities. If the occupant accompanies you during the survey, point out hazards, explain the reasons for making different recommendations, and answer questions as they arise. Many fire departments provide educational materials that address safety issues such as installation and maintenance of smoke alarms, the role in residential fire sprinklers in preventing fire deaths, and selection and use of portable fire extinguishers. These materials can be left with the occupant.

Getting Started

During a home inspection, you should present a neat, professional image FIGURE 36-2. Identify yourself by name, title, and organization. Inform the occupant of the purpose of your visit and the survey format. Always remember that you are a guest in a private home, and respect the privacy of the people who live there. You are there to help teach this person how to protect his or her family and property from fire.

During the survey, you will need to consider several types of hazards and issues. Concentrate on the hazard categories that most often cause residential fires—for example, cooking equipment, heating equipment, electrical wiring and appliances, smoking materials, candles and other open flames, and children playing with fire. Some of the hazards you encounter will depend on the local climate. For example, wood-burning stoves and fireplaces are more common in colder climates. Other hazards may be related to the age of the occupants. The fire hazards in a residence with young children, occupants with a physical or mental disability, and only senior citizens would be quite different, for example.

Always look for fire protection equipment such as smoke alarms, fire extinguishers, and residential fire sprinkler systems. Make sure that any fire alarm or suppression equipment in the home is properly installed, maintained, and fully operational FIGURE 36-3. Reference your department's policies before making recommendations on fire protection equipment. Most jurisdictions do not allow recommending a specific manufacturer's product.

FIGURE 36-2 Present a neat, professional image during a home inspection.

FIGURE 36-3 Ensure that smoke alarms are properly installed, maintained, and fully operational during a home inspection.

Some departments address additional safety issues during a home survey, such as swimming pool safety, slip-and-fall hazards, and the value of carbon monoxide alarms. Mentioning these additional safety items to the occupant can be very helpful.

Conduct the survey in a systematic fashion for both the inside and the outside of the home. Some fire fighters begin by inspecting the outside of the house, whereas others prefer to start with the interior. Many fire inspectors will start at the top of a building and systematically work their way down to ensure that all areas are covered. Inside the house, systematically inspect each room for fire hazards. Each section will have its own hazards.

Outside Hazards

On the exterior of a residential occupancy, you need to check several items. Make sure that the house number, or address, is clearly visible from the street so emergency responders can identify the correct house quickly **FIGURE 36-4**. Look for accumulated trash that could present a fire hazard or obstruct an exit. Note any flammable materials that are located close to the house or cars that are parked so close that a vehicle fire could spread to the structure. Point out shrubs and vegetation that need to be trimmed or removed.

FIGURE 36-4 The home's address should be clearly visible from the street.

For example, in an urban–wildland interface zone, no dead vegetation should be found within 30 feet (9 meter) of a house.

Locate any chimneys and try to determine the condition of the mortar. Ask the occupant whether the chimney has been cleaned within the past year, and recommend a cleaning if it has not.

Fire Fighter Safety Tips

Security bars and double-cylinder door locks are great deterrents to burglars but can prove deadly for occupants when trying to escape from a house fire. These measures can also waste precious time if fire fighters must enter to rescue the occupants or control a fire. Occupants should always be able to release security bars and open locked doors from the inside.

Inside Hazards

During the interior inspection, explain why different situations are considered potential fire risks and hazards. A person who understands why an overloaded electrical circuit is a fire hazard will be more likely to avoid overloading circuits in the future. Your goal should be to educate the occupant, not simply to point out fire hazards.

As you walk through the house, help the occupant identify alternate escape routes that can be used in case of fire. Look for a primary exit and an alternate exit from each room, and discuss these points of egress with the occupant. Be sure to mention the importance of EDITH involving all family members.

Smoke Alarms

Studies show that the presence of working smoke alarms on each level of a house reduces the risk of death from fire by 50 percent. Take the time to verify and test all smoke alarms. Some fire departments keep donated smoke alarms available and install one in any dwelling that lacks this important protection. Give the residents a copy of the NFPA fact sheet on smoke alarms and request that they read it. More information about the proper maintenance of smoke alarms is presented in the Fire Detection, Protection, and Suppression Systems chapter.

The following important smoke alarm tips should be shared with the home's occupants:

- Smoke alarms should be installed on every floor and in or near every sleeping room of the home.
- Smoke alarms should be mounted on the ceiling, at least 4 inches (10 centimeters) from a wall, or high on the wall, 4 to 12 inches (10 to 30 centimeters) below the ceiling.
- Smoke alarms should not be located near windows, exterior doors, or duct vents that could interfere with their operation.

- Only qualified electricians should install or replace AC-powered smoke alarms.
- Smoke alarms should be tested at least once a month by using the test button on the device FIGURE 36-5.
- Smoke alarms should be dusted and vacuumed regularly.
- Alkaline batteries in smoke alarms should be replaced twice each year, or as soon as the intermittent "chirp" warning that the battery is low sounds. Remind homeowners to "change the battery when you change your clock." If the smoke alarm is equipped with lithium batteries, the smoke alarm batteries need to be replaced only once every 10 years.

FIGURE 36-5 Smoke alarms should be tested at least monthly using the test button.

Bedrooms

The most common causes of fires in bedrooms are defective wiring, improper use of heating devices, improper use of candles, children playing with matches, and smoking in bed. Although smoking in bed causes only 5.4 percent of all fires in the United States, these fires are responsible for 23 percent of civilian fire deaths. Fire fighters conducting home safety surveys should stress the fire hazard that is associated with smoking in bed. In houses with children, it is equally important to discuss the risk of children playing with lighters and matches.

Kitchens

Kitchen fires are responsible for 22 percent of residential fires and usually result from cooking accidents. In particular, these fires are often caused by leaving cooking food on the stove unattended and by faulty electric appliances. In the kitchen, look for signs of frayed wires on appliances, flammable towels or potholders stored too close to the stove, and flammable cooking oils stored near the stove. Nothing that can be ignited—such as a plastic food container or paper packaging—should ever be placed on a cooking surface, even if the burners are turned off and the surface is cold. Many fires have resulted from someone inadvertently turning on the wrong burner and igniting something that was left on the stovetop.

Every kitchen should be equipped with an approved ABC-rated fire extinguisher FIGURE 36-6. Kitchens that are equipped with a deep-fat fryer should also have a Class K fire extinguisher. Check whether this extinguisher is visible and properly charged. Additional recommendations for kitchen safety include the following measures:

- Do not leave anything cooking unattended on a stove.
- Keep all flammable materials, cleaning supplies, cooking oils, and aerosols away from the stove.
- Do not place anything that could ignite on a cooking surface, even when it is turned off and cold.
- Do not place towel racks near the stove.
- Store smoking materials out of the reach of children.
- Do not overload electrical outlets or extension cords.
- Keep electric cords properly maintained; replace any frayed cords.

FIGURE 36-6 Encourage homeowners to purchase an approved ABC-rated fire extinguisher.

Living Rooms

The "living room" category includes formal living rooms, family rooms, and entertainment rooms. The primary causes of fires in these areas are smoking and electrical equipment. Look for indications of careless smoking or improperly installed home entertainment equipment, such as multiple devices plugged into a light-duty extension cord. Also look for overloaded electrical outlets and extension cords in home offices with computers, printers, scanners, fax machines, and other devices. Stress the importance of using power strips that contain circuit breakers to prevent overloading of extension cords.

If the room contains a fireplace, a wood stove, or a portable heater, ensure that no flammable or easily ignited materials are stored nearby. A fireplace should have a screen to keep sparks and hot embers from escaping. If solid fuels are used, the chimney or flue pipe should be professionally cleaned at least once each year.

Garages, Basements, and Storage Areas

Explain the importance of good housekeeping and the need to clear junk out of garages, basements, and storage areas. These areas are often filled with large quantities of unneeded materials that can easily ignite, such as old newspapers and trash. Furnaces and hot-water heaters that operate with an open flame are often located in a basement or garage as well; they can become ignition sources if flammable materials are stored too close to them.

Storage of gasoline and other flammable substances is a major concern, because an open flame or pilot light can easily ignite leaking vapors. Gasoline and other flammable liquids should be stored only in approved containers and in outside storage areas or outbuildings. Propane tanks, such as those used in gas grills, also should be stored outside or in outbuildings. Small quantities of flammable and combustible liquids (such as paint, thinners, varnishes, and cleaning fluids) should be stored in closed metal containers away from heat sources. Oily or greasy rags should also be stored in closed metal containers. Recommend that homeowners maintain fully charged fire extinguishers (2A:10B:C classification) for basements and garages.

Closing Review

During the home fire safety survey, listen carefully to any questions from the occupants, because they will enable you to address any special concerns or correct inaccurate information. At the end of the survey, complete the inspection form and give a copy to the occupants. Carefully review the findings and describe the steps that need to be taken to eliminate any identified hazards. Talk to the occupants and take the time to answer their questions fully. Emphasize the importance of smoke alarms, home exit plans, and fire drills.

After you have completed a home fire safety survey or any other type of inspection, you must file your report according to the standard operating procedures of your department. If you identified hazards that require further action or follow-up, discuss them with your company officer or a designated representative from the code enforcement office.

Conducting a home fire safety survey is a complex and important task. Initially, you may accompany your company officer, a more experienced fire fighter, or an inspector to observe how it is done. You will work with a more experienced partner until you are ready to conduct home surveys on your own. Remember—a home fire safety survey is important in helping occupants prevent fires to reduce loss of lives and property.

To conduct a home fire safety survey, follow the steps in **SKILL DRILL 36-4**:

1. Explain that the purpose of the home fire safety survey is to make a home safer for the occupants.
2. Check for a plainly visible house number.
3. Look outside the house for accumulated trash, overgrown shrubs, and blocked exits. (**STEP 1**)
4. Check inside the house for properly working smoke alarms, fire extinguishers, and fire sprinkler systems.
5. Help building occupants identify primary and alternate escape routes.
6. Look for improper wiring, use of candles, and evidence of smoking in bedrooms.
7. Explain how improper cooking procedures can start kitchen fires. Stress the safe storage of cooking oils and flammable objects away from the stove. (**STEP 2**)
8. Check living rooms for improper wiring and hazards from smoking.
9. Explain the safe use of fireplaces, heating stoves, and portable heaters.
10. Stress the importance of good housekeeping practices and safe storage of flammable liquids.
11. Review the results of the fire safety survey with the building occupants. (**STEP 3**)
12. Leave fire and life-safety brochures with the building occupants. (**STEP 4**)

Conducting Fire Station Tours

Fire station tours present a unique opportunity to help people learn how the fire department operates and how they can help prevent fires. Most fire departments have a set format for conducting fire station tours. Your station commander is in charge of all activities and will teach you how to conduct tours in your station.

Some tours start in the dayroom, or in a meeting room, with a general overview of the fire station, which is then followed by a tour of the fire apparatus. A tour may also start with an inspection of the apparatus and end in a meeting room with a question-and-answer session and information about the department and fire prevention techniques. Showing a fire prevention video or a video about the operation of your department is a good way to start or finish a tour and stimulate questions from your audience.

In preparing for a fire station tour, remember that in conducting a tour, you are representing your department to the visitors. Your professional appearance and presentation project a positive image of the department to the community. One of the first things you should cover in your welcoming remarks is to tell the visitors what to expect and what they should do if the station receives an alarm during the tour. Visitors should not wander around the station while fire fighters respond to an emergency call. This is particularly important if the tour guide also might have to leave in response to a call.

SKILL DRILL 36-4 Home Safety Survey FFII

(Fire Fighter II, NFPA 6.5.1)

1 Look outside the house for accumulated trash, overgrown shrubs, and blocked exits.

2 Check inside the house for properly working smoke alarms, fire extinguishers, and fire sprinkler systems. Explain how improper cooking procedures can start kitchen fires. Explain the safe use of fireplaces, heating stoves, and portable heaters.

3 Review the results of the fire survey with the building occupants.

4 Leave fire and life-safety brochures with the building occupants.

The tour format will vary depending on the age and interests of the group, which could consist of small children, teenagers, adults, or senior citizens. The tour plan, including your presentation and fire prevention messages, should be designed to meet the needs of each group. The weather may also be an important factor if you are planning any outdoor demonstrations.

Young children like to see action, so squirting some water or raising an aerial ladder can be a real crowd-pleaser FIGURE 36-7. Fire prevention messages for young people include use of the stop, drop, and roll procedure; learning how to call 911; the importance of EDITH; selection and use of portable fire extinguishers; the importance of residential fire sprinklers; and the dangers of playing with lighters and matches.

Teenagers are ready for lessons that they can apply in everyday life. Many teenagers work as babysitters, so prevention messages can be geared toward cooking safety and evacuating children in the event of a fire. Teenagers are also a high-risk group for motor vehicle collisions, so messages about seat belts, proper cell phone use, safe driving, the dangers of drinking and driving, and actions to take if they see or hear an emergency vehicle while driving are appropriate. Some teenagers might be interested in becoming fire fighters or joining

FIGURE 36-7 Young children always like to see fire fighters.

FIRE FIGHTER II Tips FFII

Fire stations can be potentially dangerous for the general public. Before conducting a tour, ask yourself:

- Is the area safe?
- Am I showing the fire service in a positive light?
- Will there be enough fire fighters to supervise the tour group?
- Are the planned activities age-appropriate?

a fire fighter explorer group. This age group can be a good source of recruits for both career and volunteer departments.

Adults are probably more interested in home fire safety, whereas senior citizens are often more interested in the Emergency Medical Services available. Remind smokers not to smoke in bed or throw burning cigarettes into wastepaper baskets. Many seniors use oxygen concentrators or supplemental oxygen, so stress the fire hazards associated with the use of medical oxygen. Some seniors may forget to turn off burners or may leave cooking food unattended. They also may be worried about their ability to move quickly if a fire starts in their home. Tailor your prevention messages to the specific concerns of the age group you are addressing.

FIRE FIGHTER II Tips FFII

Adults often ask about funding sources for the fire department. If you give station tours, you should be aware of how your fire department is funded. Be prepared to answer questions on how taxpayer money is used to provide the myriad of services the fire department provides.

Try to leave every tour group with both a message and materials. First, reinforce a simple prevention or safety message that expands their awareness of fire and safety and helps them to deal with potential emergency situations. Second, give them printed fire prevention materials FIGURE 36-8. These materials could range from coloring books for young children to a home fire prevention checklist for adults.

FIGURE 36-8 Always leave visitors with printed fire prevention materials.

The steps for conducting a fire station tour are listed in SKILL DRILL 36-5:

1. Introduce yourself to the tour participants and explain the mission of the fire department.
2. Explain the layout and purpose of the fire station.
3. Consider the use of an educational video presentation. (STEP 1)
4. Tour the living areas of the station. (STEP 2)
5. Explain the purpose of each piece of fire apparatus. (STEP 3)
6. Conclude the tour in the meeting room.
7. Encourage questions from participants.
8. Hand out appropriate fire safety materials. (STEP 4)

SKILL DRILL 36-5 Conducting a Fire Station Tour FFII

Fire Fighter II (NFPA 6.5.2)

1 Introduce yourself to the tour participants and explain the mission of the fire department. Explain the layout and purpose of the fire station. Consider the use of an educational video presentation.

2 Tour the living areas of the station.

3 Explain the purpose of each piece of fire apparatus.

4 Conclude the tour in the meeting room. Encourage questions from participants. Hand out appropriate fire safety materials.

Near Miss REPORT

Report Number: 06-0000531

Synopsis: Smoke from peanut oil causes severe reaction.

Event Description: We were holding our annual fire prevention night activities which included the smoke house trailer. The kids and adults were having a good time. We were teaching the children about safety in the home and bedroom. The trailer then filled with smoke and everything seemed to be fine. After our first few tours a woman screamed for help and stated that her child was having trouble breathing. We did not realize at first that it was an allergic reaction until we were en route to the hospital. We asked the mother if the child had any allergies. She stated yes, he was severely allergic to peanuts. We did not put two and two together until after we were returning from the hospital. We kept thinking of how his allergic reaction could have occurred at the station. We tried to think where the child could have gotten peanuts. Then one of us thought about the smoke trailer. The smoke is made from peanut oil, which started his reaction. This is one of those things you just don't think of until something goes wrong.

Lessons Learned: We learned that we need to have signs and announcements made regarding the smoke in the trailer. We now ask prior to all people entering if they or their children are allergic to peanuts as the smoke is made from peanut oil.

Wrap-Up

Chief Concepts

- Fire prevention is as important as fire suppression. Countless fires could be prevented by teaching safe behaviors and eliminating unsafe conditions. This kind of education ultimately saves lives and reduces the losses caused by fires.
- Fire prevention includes a range of activities that are intended to prevent the outbreak of fires and to limit the consequences if a fire does occur. These activities include enacting and enforcing fire codes, conducting property inspections, and presenting public fire safety education programs.
- Your highest priority as a fire fighter should always be to prevent fires. Although Fire Fighter I and Fire Fighter II personnel may have limited responsibilities for formal fire prevention activities, it is imperative that you understand the objectives of fire prevention, the delineation of fire prevention responsibilities within your department, and the importance of your role in fire prevention.
- Fire codes are enacted to ensure a minimum level of fire safety in the home and workplace environments. After a jurisdiction adopts a fire code, it must be able to enforce it. Regular inspections ensure compliance, note any violations, and require their correction.
- Public education teaches techniques to reduce the risks of death or injury in the event of fire.
- Public fire and life safety education programs include those focused on the following topics:
  - Stop, drop, and roll
  - Exit drills in the home (EDITH)
  - Installation and maintenance of smoke alarms
  - Advantages of residential sprinkler systems
  - Selection and use of portable fire extinguishers

- Learn Not to Burn®
- Change Your Clock—Change Your Battery
- Fire safety for special populations
- Bike safety
- Fall prevention
- Wildland fire prevention programs

- The Stop, Drop, and Roll program is designed to instruct people what to do if their clothing catches fire.
  - **Stop** means just that: Stay in place; do not run. Stress the fact that running will fan the flames and spread the fire.
  - **Drop** means getting down on the ground or on the floor. Cover the face with the hands to help protect the airway and the eyes.
  - **Roll** involves tucking in the elbows and keeping the legs together while rolling like a bug. Individuals should then roll over and over, and then roll back in the other direction. This helps to smother the flames.
- EDITH teaches residents how to safely get out of their homes in the event of a fire or other emergency.
- Another type of public fire safety presentation covers the proper installation and maintenance of smoke alarms. In discussing this topic, stress the importance of having a working smoke alarm on every floor of a house. Also stress the importance of keeping smoke alarms in good working order.
- Many fire departments offer to conduct fire safety surveys in private dwellings. A home fire safety survey helps identify fire and life-safety hazards and provides occupants with recommendations on making their home safer.
- During a home fire safety survey, consider several types of hazards and issues. Concentrate on the hazard categories that most often cause residential fires, such as cooking equipment, heating equipment, electrical wiring, smoking materials, candles and other open flames, and children playing with fire.
- Most fire departments have a set format for conducting fire station tours. For example, a tour might start with an inspection of the apparatus and end in a meeting room with a question-and-answer session and information about the department and fire prevention techniques.
- Try to leave every tour group with both a fire prevention message and materials.

Hot Terms

Exit Drills in the Home (EDITH) A public safety program designed to teach occupants how to safely exit a house in the event of a fire or other emergency. It stresses the importance of having a preestablished meeting place for all family members.

Fire code A set of legally adopted rules and regulations designed to prevent fires and protect lives and property in the event of a fire.

Fire prevention Activities conducted to prevent fires and protect lives and property in the event of a fire. Fire prevention activities include the enactment and enforcement of fire codes, the inspection of properties, the presentation of fire safety education programs, and the investigation of the causes of fires.

FIRE FIGHTER *in action*

You are a volunteer fire fighter in a small rural community. One night after training, the chief asks if you would be willing to serve as the department's public educator. You know there is a desperate need and you enjoy talking about the fire department, but you just aren't sure where you would begin. You think back to the programs and concepts you learned in Fire Fighter I & II training.

1. Regulations designed to prevent fires from occurring, eliminate fire hazards, protect lives, and limit fire losses are called __________ codes.
   - **A.** building
   - **B.** electrical
   - **C.** fire
   - **D.** model

2. Where do fire codes not generally apply?
   - **A.** Interior of residential dwellings
   - **B.** Abandoned property
   - **C.** New construction
   - **D.** Commercial property

3. Which program focuses on instructing children what to do when their clothes catch fire?
   - **A.** Learn Not to Burn®.
   - **B.** Stop, drop, and roll.
   - **C.** Get low and go.
   - **D.** EDITH

4. Which of the following is accurate about fire safety in the home?
   - **A.** Pre-identify an exit.
   - **B.** Keep bedroom doors closed.
   - **C.** Keep bedroom doors open.
   - **D.** Keep a heavy object next to the window.

5. Where should smoke alarms be installed?
   - **A.** Garage
   - **B.** Kitchen
   - **C.** Near an exterior door
   - **D.** All of the above

6. By what percent does a working smoke alarm and sprinkler system reduce the risk of fire death?
   - **A.** 23%
   - **B.** 50%
   - **C.** 82%
   - **D.** 94%

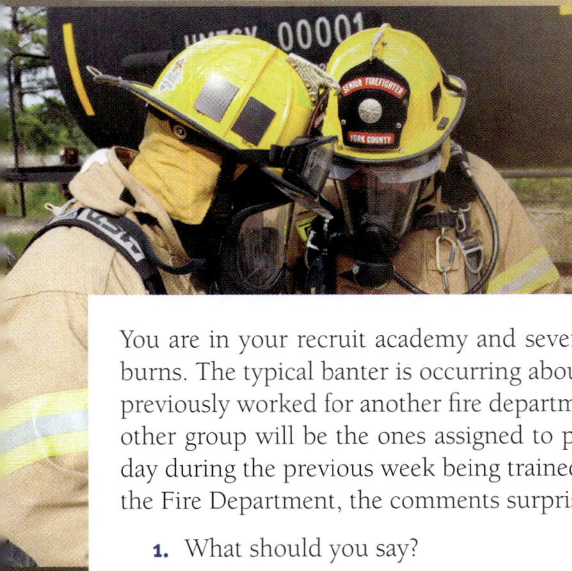

FIRE FIGHTER II *in action*

You are in your recruit academy and several of the other recruits are talking as they clean the room after a day of live fire burns. The typical banter is occurring about how much better one group did over another. Eventually, a fire fighter who had previously worked for another fire department makes the comment that his group will be the ones fighting the fires while the other group will be the ones assigned to public education. Others quickly jump in with similar comments. Having spent a day during the previous week being trained on the importance of public education, including a personal visit by the Chief of the Fire Department, the comments surprise you.

1. What should you say?
2. Should you mention this discussion to the Chief?
3. How would you go about changing the attitudes of the group?
4. How do you maintain your attitude when others around you do not share your views?

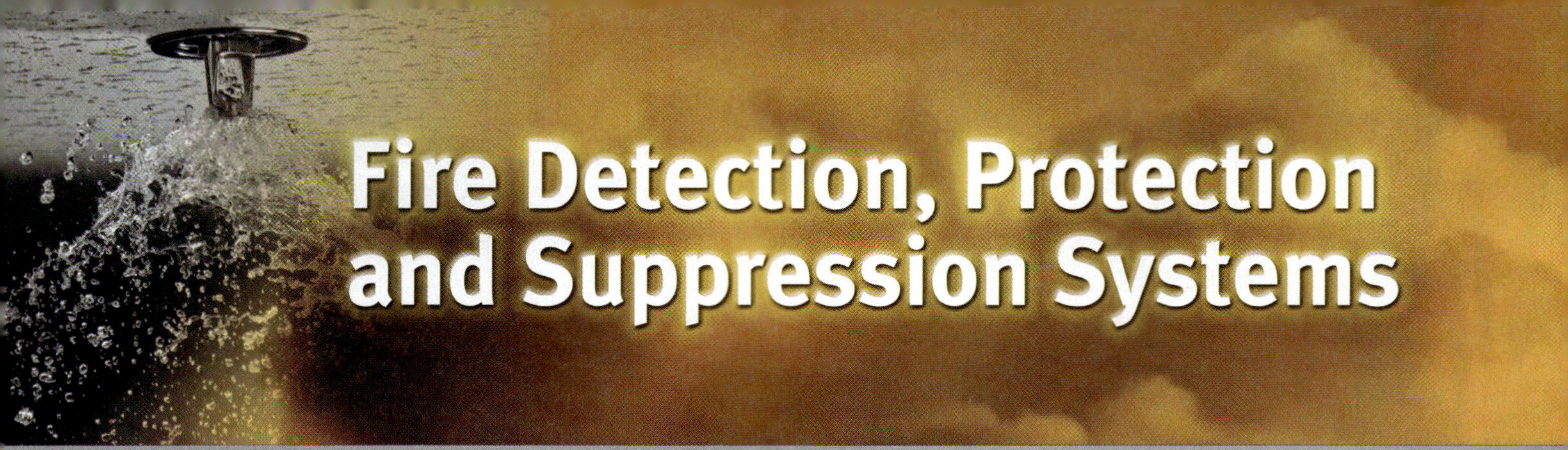

Fire Detection, Protection and Suppression Systems

Fire Fighter I

Knowledge Objectives

After studying this chapter, you will be able to:

- Explain why all fire fighters should have a basic understanding of fire protection systems. (NFPA5.3.14.A, p 1034)
- Describe when and how water is shut off to a building's sprinkler system and how to stop water at a single sprinkler head. (NFPA5.3.14.A, p 1058)

Skills Objectives

There are no skill objectives for Fire Fighter I candidates. NFPA 1001 contains no Fire Fighter I Job Performance Requirements for this chapter.

Fire Fighter II

FFII

Knowledge Objectives

After studying this chapter, you will be able to:

- Describe the basic components and functions of a fire alarm system. (NFPA6.5.3.A, p 1034–1038)
- Describe the basic types of fire alarm initiation devices and indicate where each type is most suitable. (NFPA6.5.3.A, p 1038–1044)
- Describe the fire department's role in resetting fire alarms. (NFPA6.5.3.A, p 1038–1044)
- Describe the basic types of alarm notification appliances. (p 1044)
- Describe the basic types of fire alarm annunciation systems. (p 1045)
- Explain the different ways that fire alarms may be transmitted to the fire department. (NFPA6.5.3.A, p 1045–1047)
- Identify the four types of sprinkler heads. (NFPA6.5.3.A, p 1047–1050)
- Identify the different styles of indicating valves. (NFPA6.5.3.A, p 1051–1054)

CHAPTER 37

- Describe the operation and application of the following types of automatic sprinkler systems:
  - Wet-pipe system
  - Dry-pipe system
  - Preaction system
  - Deluge system (NFPA 6.5.3.A, p 1056–1058)
- Describe the differences between commercial and residential sprinkler systems. (NFPA 6.5.3.A, p 1058)
- Identify the three types of standpipes and point out the differences among them. (NFPA 6.5.3.A, p 1059–1062)
- Describe two problems that fire fighters could encounter when using a standpipe in a high-rise building. (NFPA 6.5.3.A, p 1060–1062)
- Identify the hazards that specialized extinguishing systems can pose to responding fire fighters. (NFPA 6.5.3.A, p 1062–1064)

Skills Objectives

There are no skill objectives for Fire Fighter II candidates. NFPA 1001 contains no Fire Fighter II Job Performance Requirements for this chapter.

You Are the Fire Fighter

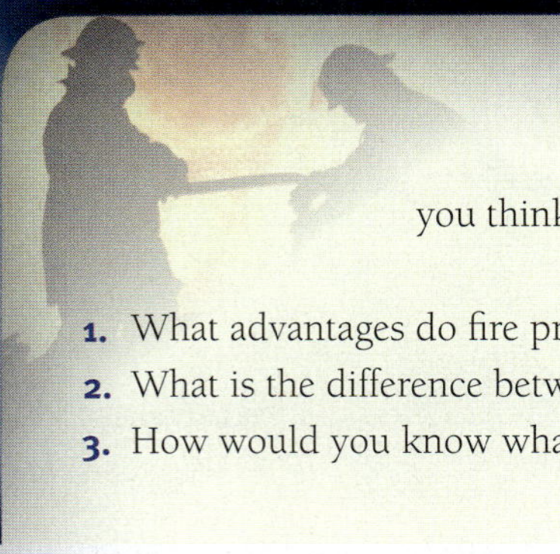

You are riding in the back seat as you arrive at your first fire in a high-rise structure. As you step out of the cab, you see fire coming out of the 18th floor window. Over the radio, you hear the IC direct your crew to connect to the standpipe. Your heart rate quickens as you think about the task ahead and worry about getting things right.

1. What advantages do fire protection systems offer when fighting high-rise fires?
2. What is the difference between a standpipe system and a sprinkler system?
3. How would you know what systems the fire department connection serves?

Introduction

Today, fire prevention and building codes require that almost all new structures have some sort of fire protection system installed. This requirement makes it more important than ever for the line fire fighter to have a working knowledge of these systems. Fire protection systems include fire alarm, automatic fire detection, and fire suppression systems. Understanding how these systems operate is both important for the fire fighter's safety and necessary to provide effective customer service to the public.

From a safety standpoint, fire fighters need to understand the operations and limitations of fire detection and suppression systems. A building with a fire protection system will have very different working conditions during a fire than an unprotected building. Fire fighters need to know how to fight a fire in a building with a working fire suppression system and how to shut down the system after the fire is extinguished.

From a customer service standpoint, fire fighters who understand how fire protection systems work can help dispel misconceptions about these systems and advise building owners and occupants appropriately after an alarm is sounded. Most people have no idea of how the fire protection or detection systems in their building work, and there are often more false alarms in buildings with fire protection systems than actual fires. Fire fighters can help the owners and occupants determine what activated the system, how they can prevent future false alarms, and what needs to be done to restore a system to service.

The term "false alarm" is not always completely accurate. In such events, the sensors within the fire alarm system have usually recorded an error somewhere within the system. In the modern construction of multifamily apartments, smoke detectors may be placed too close to the kitchen, which results in numerous "false alarms." In such a case, the alarm was not raised in error; smoke or heat has truly activated the system, though no crisis is in process. The major problem with these nuisance alarms is that people become used to hearing them and no longer respond to them with any urgency. Such was the case in January 2000, when three students died in their dormitory rooms at Seton Hall University. The students ignored the alarm, which signaled a real fire emergency, because they had become attuned to hearing the alarm on a regular basis for nonemergencies.

Fire alarms should be treated as true emergencies until someone from the fire department arrives on the scene and dictates otherwise. This chapter discusses the basic design and operation of fire alarm, fire detection, and fire suppression systems, each of which could be the subject of an entire book. It is not necessary to understand all of the design criteria for a fire protection system, but simply to appreciate how the system is intended to function. Fire protection systems have fairly standardized design requirements across North America; most areas follow the applicable National Fire Protection Association (NFPA) standards. Local fire and building codes may require different types of systems for different buildings.

Fire Alarm and Detection Systems

Practically all new construction includes some sort of fire alarm and detection system. In most cases, both fire detection and fire alarm components are integrated into a single system. A fire detection system recognizes when a fire is occurring and activates the fire alarm system, which alerts the building occupants and, in some cases, the fire department. Some fire detection systems also automatically activate fire suppression systems to control or extinguish the fire.

Fire alarm and detection systems range from simple, single-station smoke alarms for private homes, to complex fire detection and control systems for high-rise buildings. Many fire alarm and detection systems in large buildings also control other systems to help protect occupants and control the spread of fire and smoke. Although these systems can be complex, they generally include the same basic components as the simpler versions.

Fire Alarm System Components

A fire alarm system has three basic components: an alarm initiation device, an alarm notification device, and a control panel. The alarm initiation device is either an automatic or manually operated device that, when activated, causes the system to indicate an alarm. The alarm notification device is generally an audible device, often accompanied by a visual device that alerts the building occupants when the system is activated. The control panel links the initiation device to the notification device and performs other essential functions.

A single-station smoke alarm in a residential dwelling has the same basic components as a fire alarm system in a multistory office building. That is, a single-station smoke alarm includes a smoke detector, an audible notification device, and a control unit.

Fire Alarm System Control Panels

Most fire alarm systems in buildings other than single residential units have several alarm initiation devices in different areas and use both audible and visible devices to notify the occupants of an alarm. The fire alarm control panel serves as the "brain" of the system, linking the initiation devices to the notification devices.

The control panel manages and monitors the proper operation of the system. It can indicate the source of an alarm, so that responding fire personnel will know what activated the alarm and where the initial activation occurred. The control panel also manages the primary power supply and provides a backup power supply for the system. It may perform additional functions, such as notifying the fire department or central station monitoring company when the alarm system is activated, and may interface with other systems and facilities.

Control panels vary greatly depending on the age of the system and the manufacturer. For example, an older system might simply indicate that an alarm has been activated, whereas a newer system might indicate that an alarm occurred in a specific zone within the building **FIGURE 37-1**. The most modern panels actually specify the exact location of the activated initiation device.

Fire alarm control panels are used to silence the alarm and reset the system. These panels should always be locked. Alarms should not be silenced or reset until the activation source has been found and checked by fire fighters to ensure that the situation is under control. In some systems, the individual initiating devices must be reset before the entire system can be reset; others reset themselves after the problem is resolved.

Many buildings have an additional display panel in a separate location, usually near the front door of the building. This panel, which is called a remote annunciator, enables fire fighters to ascertain the type and location of the activated alarm device as they enter the building **FIGURE 37-2**.

The fire alarm control panel should also monitor the condition of the entire alarm system to detect any faults. Faults within the system are indicated by a trouble alarm, which shows that a component of the system is not operating properly and requires attention. Trouble alarms do not activate the building's fire alarm, but should make an audible sound and illuminate a special light at the alarm control panel. They may also transmit a notification to a remote service location.

A fire alarm system is usually powered by a 110-volt line, even though the system's appliances may use a lower voltage.

FIGURE 37-1 Most modern fire alarm control panels indicate the zone where an alarm was initiated.

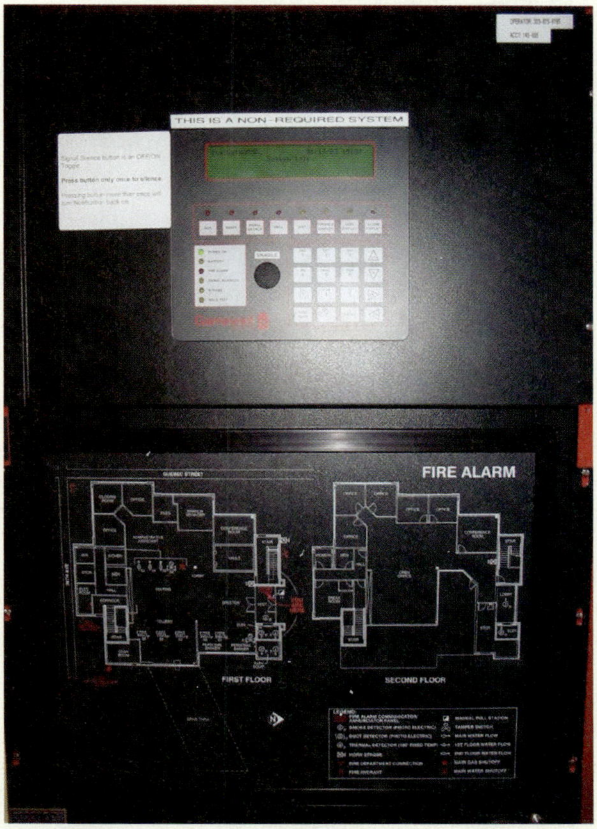

FIGURE 37-2 A remote annunciator allows fire fighters to quickly determine the type and location of the activated alarm device.

In addition to the normal power supply, fire codes call for a backup power supply for all alarm systems FIGURE 37-3. In some systems, a battery in the fire alarm control panel becomes activated automatically when the external power is interrupted. Fire codes will specify how long the system must be able to function on the battery backup. In large buildings, the backup power supply could consist of an emergency generator or a secondary power source. If either the main power supply or the backup power source fails, the trouble alarm should sound.

Depending on the building's size and floor plan, an activated alarm may sound throughout the building or only in particular areas. In high-rise buildings, fire alarm systems are often programmed to alert only the occupants on the same floor as the activated alarm as well as those on the floors immediately above and below the affected floor. Alarms on the remaining floors can be manually activated from the system control panel. Some systems include a public address feature, which enables a fire department officer to provide specific instructions or information for occupants in different areas.

The control panel in a large building may be programmed to perform several additional functions. For example, it may automatically shut down or change the operation of air-handling systems, recall elevators to the ground floor, and unlock stairwell doors so that a person in an exit stairway can reenter an occupied floor.

FIGURE 37-3 Fire prevention codes require that alarm systems have a backup power supply, which becomes activated automatically when the normal electrical power is interrupted.

Residential Fire Alarm Systems

Current building and fire codes usually require the installation of a fire detection and alarm system in all residential dwelling units. The most common type of residential fire alarm system is a single-station smoke alarm FIGURE 37-4. This life-safety device includes a smoke detection device, an automatic control unit, and an audible alarm within a single unit. It alerts occupants quickly when a fire occurs. Millions of single-station smoke alarms have been installed in private dwellings and apartments.

Smoke alarms can be either battery-powered or hard-wired to a 110-volt electrical system or both. Most building codes require hard-wired, AC-powered smoke alarms in all newly constructed dwellings; battery-powered units are popular for existing residencies. The major concern with a battery-powered smoke alarm is ensuring that the battery is replaced on a regular basis. Newer battery-powered smoke alarms are available with lithium batteries that will last for 10 years.

The most up-to-date codes require new homes to have a smoke alarm in every bedroom and on every floor level. They also require a battery backup in the event of a power failure. In newer installations, all smoke alarms in the dwelling are interconnected; as a consequence, all of them will sound if one is activated. For example, if the basement smoke alarm is activated, alarms will sound on all levels to alert the occupants.

Many home smoke alarms are part of security systems. Like commercial systems, these home fire alarm/security systems have an alarm control panel and require a passcode to set or reset the system. They may also be monitored by a central station.

Ionization versus Photoelectric Smoke Detectors

Two types of fire detection devices may be used in a smoke alarm to detect combustion:

- Ionization detectors are triggered by the invisible products of combustion.
- Photoelectric detectors are triggered by the visible products of combustion.

FIGURE 37-4 A single-station smoke alarm.

Ionization smoke detectors work on the principle that burning materials release many different products of combustion, including electrically charged microscopic particles. An ionization detector senses the presence of these invisible charged particles (ions).

An ionization smoke detector contains a very small amount of radioactive material inside its inner chamber. This radioactive material releases charged particles into the chamber, and a small electric current flows between two plates FIGURE 37-5. When smoke particles enter the chamber, they neutralize the charged particles and interrupt the current flow. The detector then senses this interruption and activates the alarm.

Photoelectric smoke detectors use a light beam and a photocell to detect larger visible particles of smoke FIGURE 37-6. They operate by reflecting the light beam either away from or onto the photocell, depending on the design of the device. When visible particles of smoke pass through the light beam, they reflect the beam onto or prevent it from striking the photocell, thereby activating the alarm.

Ionization smoke detectors are more common and less expensive than photoelectric smoke detectors. They react more quickly than photoelectric smoke detectors to fast-burning fires, such as a fire in a wastepaper basket, which may produce little visible smoke FIGURE 37-7. On the downside, fumes and steam from common activities such as cooking and showering may trigger unwanted alarms.

By comparison, photoelectric smoke detectors are more responsive to slow-burning or smoldering fires, such as a fire caused by a cigarette caught in a couch, which usually produces a large quantity of visible smoke FIGURE 37-8. They are less prone to false alarms from regular cooking fumes and shower steam than are ionization smoke detectors. Most photoelectric smoke alarms require more current to operate, however, so they are typically connected to a 110-volt power source.

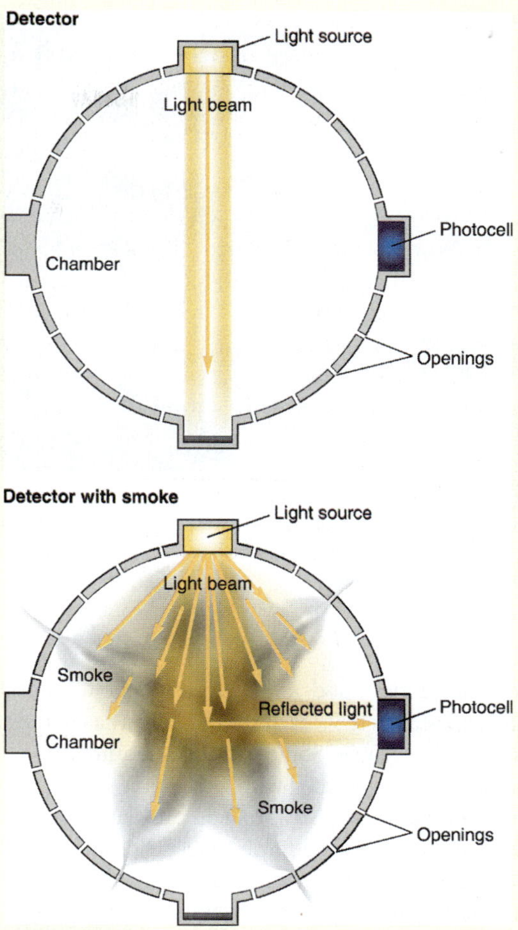

FIGURE 37-6 A photoelectric smoke alarm.

FIGURE 37-5 An ionization smoke alarm.

FIGURE 37-7 Ionization smoke detectors react more quickly than photoelectric smoke detectors to a fast-burning fire, such as a fire in a wastepaper basket, which may produce little visible smoke.

FIGURE 37-8 Photoelectric smoke alarms react more quickly than ionization smoke detectors to a slow-burning or smoldering fire, such as a cigarette in furniture, which usually produces a large quantity of visible smoke.

FIGURE 37-9 Several types of manual fire alarm boxes (also known as manual pull-stations) are available.

Both ionization and photoelectric smoke detectors are acceptable life-safety devices. Indeed, in most cases, they are considered interchangeable. Some fire codes require that photoelectric smoke detectors be used within a certain distance of cooking appliances or bathrooms, so as to help prevent unwanted alarms.

Combination ionization/photoelectric smoke alarms are also available. These alarms quickly react to both fast-burning and smoldering fires. They are not suitable for use near kitchens or bathrooms, because they are prone to the same unwanted alarms as regular ionization smoke detectors.

Most ionization and photoelectric smoke alarms look very similar to each other. The only sure way to identify the type of alarm is to read the label, which is often found on the back of the case. An ionization alarm must have a label or engraving stating that it contains radioactive material.

Alarm Initiation Devices

Alarm initiation devices begin the fire alarm process either manually or automatically. Manual initiation devices require human activation; automatic devices function without human intervention. Manual fire alarm boxes are the most common type of alarm initiation devices that require human activation. Automatic initiation devices include various types of heat and smoke detectors and other devices that automatically recognize the evidence of a fire.

Manual Initiation Devices

Manual initiation devices are designed so that building occupants can activate the fire alarm system on their own if they discover a fire in the building. Many older alarm systems could be activated only manually. The primary manual initiation device is the manual fire alarm box, or manual pull-station **FIGURE 37-9**. Such a station has a switch that either opens or closes an electrical circuit to activate the alarm.

Pull-stations come in a variety of sizes and designs, depending on the manufacturer. They can be either single-action or double-action devices. Single-action pull-stations require a person to pull down a lever, toggle, or handle to activate the alarm; the alarm sounds as soon as the pull-station is activated. Double-action pull-stations require a person to perform two steps before the alarm will activate. They are designed to reduce the number of false alarms caused by accidental or intentional pulling of the alarm. The person must move a flap, lift a cover, or break a piece of glass to reach the alarm activation device. Designs that use glass are no longer in favor, both because the glass must be replaced each time the alarm is activated and because the broken glass poses a risk of injury.

Once activated, a manual pull-station should stay in the "activated" position until it is reset. This status enables responding fire fighters to determine which pull-station

Troubleshooting Smoke Alarms

Most problems with smoke alarms are caused by a lack of power, dirt in the sensing chamber, or defective alarms.

Power Problems

Smoke alarms require power from a battery or from a hard-wired 110-volt power source. Some smoke alarms that are hard-wired to a 110-volt power source are equipped with a battery that serves as a backup power source.

The biggest problem with battery-powered alarms is that they may contain a dead or missing battery. Some people do not change batteries when recommended, resulting in inoperable smoke alarms. People may also remove smoke alarm batteries to use the battery for another purpose or to prevent nuisance alarm activations. The solution in this situation is to install a new battery in the alarm. New lithium batteries will last for 10 years in a smoke alarm.

Hard-wired smoke detectors become inoperable if someone turns the power off at the circuit breaker. The solution in this situation is to turn the circuit breaker back on.

Most smoke alarms containing batteries will signal a low-battery condition by emitting a chirp every few seconds. This chirp indicates that the battery needs to be replaced to keep the smoke alarm functional. Many hard-wired smoke alarms that are equipped with battery backups will chirp to indicate that the backup battery is low, even when the 110-volt power source is operational. This feature ensures that the smoke alarm always has an operational backup system.

Dirt Problems

A second problem with smoke alarms is the increased sensitivity that results when dust or an insect becomes lodged in the photoelectric or ionization chamber. Such an obstruction causes the alarm to become activated when a small quantity of water vapor or smoke enters the chamber, leading to unnecessary alarms. The solution in this situation is to remove the cover of the chamber and gently vacuum out the chamber. Alternatively, you can use a puff of air from a can of compressed air to remove the dust or insect. Follow the manufacturer's instructions for this process.

Alarm Problems

The third common problem you may encounter with smoke alarms is worn-out detectors. It is recommended that smoke alarms be replaced every 10 years because the sensitivity of the alarm can change. Some worn-out alarms may become overly sensitive and emit false alarms. Others may develop decreased sensitivity and fail to emit an alarm in the event of a fire. The date of manufacture should be stamped on each smoke alarm. If a detector is more than 10 years old, it should be replaced with a new alarm.

Understanding the basic functions and troubleshooting of simple smoke alarms enables you to respond to citizens who call your fire department when they encounter problems with their smoke alarms. Most importantly, it enables you to keep smoke alarms operational. Remember—only operational smoke alarms can save lives.

initiated the alarm. Resetting the pull-station requires a special key, screwdriver, or Allen wrench. In many systems, the pull-station must be reset before the building alarm can be reset at the alarm control panel.

A variation on the double-action pull-station, designed to prevent malicious false alarms, is covered with a piece of clear plastic **FIGURE 37-10**. These covers are often used in areas where malicious false alarms occur frequently, such as high schools and college dormitories. The plastic cover must be opened before the pull-station can be activated. Lifting the cover triggers a loud tamper alarm at that specific location, but does not activate the fire alarm system. Snapping the cover back into place resets the tamper alarm. The intent is that a person planning to initiate a false alarm will drop the cover and run when the tamper alarm sounds. The pull-station tamper alarm is not connected to the fire alarm system in any way.

Even automatic fire alarm systems can be manually activated by pressing a button or flipping a switch on the alarm system control panel. In large buildings (and particularly high-rise buildings), the control panel has separate manual activation switches for different areas or on particular floors.

Automatic Initiation Devices

Automatic initiation devices are designed to function without human intervention and will activate the alarm system when they detect evidence of a fire. These systems can be programmed to transmit the alarm to the fire department, even if the building is unoccupied, and to perform other functions when a detector is activated.

Automatic initiation devices can use any of several types of detectors. Some detectors are activated by smoke or by the invisible products of combustion; others react to heat, the

FIGURE 37-10 A variation on the double-action pull-station, designed to prevent malicious false alarms, has a clear plastic cover and a separate tamper alarm.

light produced by an open flame, or specific gases. Automatic initiation devices sense fire in the same manner as a single-station alarm does in a residence—that is, they use either an ionization or a photoelectric chamber to detect products of combustion.

Smoke Detectors

A smoke detector is designed to sense the presence of smoke. Among fire fighters, the term "smoke detector" generally refers to a sensing device that is part of a fire alarm system. This type of device is commonly found in school, hospital, business, and commercial occupancies that are equipped with fire alarm systems FIGURE 37-11.

Smoke detectors come in a variety of designs and styles geared toward different applications. The most commonly used smoke detectors are ionization and photoelectric models, which operate in the same way that residential smoke alarms do. However, the smoke detectors used in commercial fire alarm systems are usually more sophisticated and more expensive than residential smoke alarms.

Each detection device is rated to protect a certain floor area, so in large areas the detectors are often placed in a grid pattern. Newer smoke detectors also have a visual indicator, such as a steady or flashing light, that indicates when the device has been activated.

FIGURE 37-11 Commercial ionization smoke detector.

A beam detector is a type of photoelectric smoke detector used to protect large open areas such as churches, auditoriums, airport terminals, and indoor sports arenas. In these facilities, it would be difficult or costly to install large numbers of individual smoke detectors, but a single beam detector could be used for the entire area FIGURE 37-12.

A beam detector has two components: a sending unit, which projects a narrow beam of light across the open area, and a receiving unit, which measures the intensity of the light when the beam strikes the receiver. When smoke interrupts the light beam, the receiver detects a drop in the light intensity and activates the fire alarm system. Most photoelectric beam detectors are set to respond to a certain obscuration rate, or percentage of light blocked. If the light is completely blocked, such as when a solid object is moved across the beam, the trouble alarm will sound, but the fire alarm will not be activated.

Smoke detectors are usually powered by a low-voltage circuit and send a signal to the fire alarm control panel when they

FIGURE 37-12 Beam detectors are used in large open spaces.

are activated. Both ionization and photoelectric smoke detectors are self-restoring. After the smoke condition clears, the alarm system can be reset at the control panel.

Heat Detectors

Heat detectors are also commonly used as automatic alarm initiation devices. These devices can provide property protection but cannot provide reliable life-safety protection because they do not react quickly enough to incipient fires. They are generally used in situations where smoke alarms cannot be used, such as dusty environments and areas that experience extreme cold or heat. Heat detectors are often installed in unheated areas, such as attics and storage rooms, as well as in boiler rooms and manufacturing areas.

Heat detectors are generally very reliable and less prone to false alarms than are smoke alarms. You may come across heat detectors that were installed 30 or more years ago and are still in service. Unfortunately, older units lack a visual trigger that indicates which device was activated, so tracking down the cause of an alarm may be very difficult. Newer models have an indicator light that shows which device was activated.

Several types of heat detectors are available, each of which is designed for specific situations and applications. Single-station heat alarms are sometimes installed in unoccupied areas of buildings that do not have fire alarm systems, such as attics or storage rooms. Spot detectors are individual units that can be spaced throughout an occupancy, so that each detector covers a specific floor area. Spot detectors are usually installed in light commercial and residential settings; the units may be in individual rooms or spaced at intervals along the ceiling in larger areas.

Heat detectors can be designed to operate at a fixed temperature or to react to a rapid increase in temperature. Either fixed-temperature or rate-of-rise devices can be configured as spot or line detectors.

Fixed-Temperature Heat Detectors Fixed-temperature heat detectors, as the name implies, are designed to operate at a preset temperature FIGURE 37-13. A typical temperature for a light-hazard occupancy, such as an office building, would be 135°F (57°C). Fixed-temperature detectors typically include a metal alloy that will melt at the preset temperature. This melting alloy releases a lever-and-spring mechanism, which either opens or closes a switch. Most fixed-temperature heat detectors must be replaced after they have been activated, even if the activation was accidental.

Rate-of-Rise Heat Detectors Rate-of-rise heat detectors become activated when the temperature of the surrounding air increases by more than a set amount in a given period of time. A typical rating might be "greater than 12°F (–11°C) in 1 minute." If the temperature increase occurs more slowly than this rate, the rate-of-rise heat detector will not activate. By contrast, a temperature increase at a pace greater than this rate will activate the detector and set off the fire alarm. Rate-of-rise heat detectors should not be located in areas that normally experience rapid changes in temperature, such as near garage doors in heated parking areas.

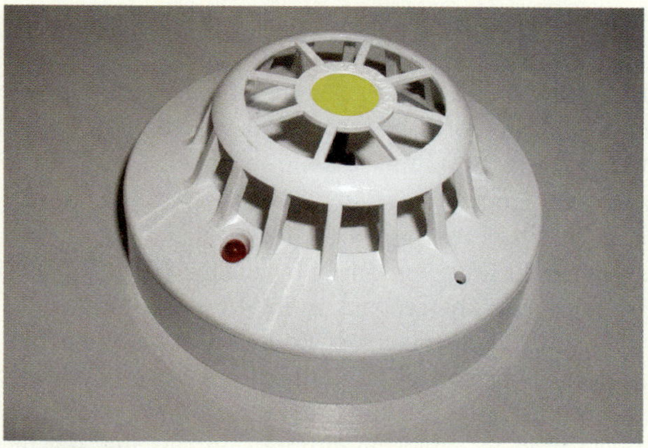

FIGURE 37-13 A fixed-temperature heat detector initiates an alarm at a preset temperature.

Some rate-of-rise heat detectors have a bimetallic strip made of two metals that respond differently to heat: A rapid increase in temperature causes the strip to bend unevenly, which opens or closes a switch. Another type of rate-of-rise heat detector uses an air chamber and diaphragm mechanism: As air in the chamber heats up, the pressure increases. Gradual increases in pressure are released through a small hole, but a rapid increase in pressure will press upon the diaphragm and activate the alarm. Most rate-of-rise heat detectors are self-restoring, so they do not need to be replaced after an activation unless they were directly exposed to a fire.

Rate-of-rise heat detectors generally respond more rapidly to most fires than do fixed-temperature heat detectors. However, a slow-burning fire, such as a smoldering couch, might not activate a rate-of-rise heat detector until the fire is well established.

Combination rate-of-rise and fixed-temperature heat detectors are available. These devices balance the faster response of the rate-of-rise detector with the reliability of the fixed-temperature heat detector.

Line Heat Detectors

Line detectors use wire or tubing strung along the ceiling of large open areas to detect an increase in heat. An increase in temperature anywhere along the line will activate the detector. Line detectors are often found in churches, warehouses, and industrial or manufacturing applications.

Line detectors use wires or a sealed tube to sense heat. One wire-type model has two wires inside, separated by an insulating material. When heat melts the insulation, the wires short out and activate the alarm. The damaged section of insulation must be replaced with a new piece after activation of the detector.

Another wire-type model measures changes in the electrical resistance of a single wire as it heats up. This device is self-restoring and does not need to be replaced after activation unless it is directly exposed to a fire.

The tube-type line heat detector contains a sealed metal tube filled with air or a nonflammable gas. When the tube is heated, the internal pressure increases and activates the alarm. Like the single-wire line heat detector, this device is self-restoring and does not need to be replaced after activation, unless it is directly exposed to a fire.

Flame Detectors

Flame detectors are specialized devices that detect the electromagnetic light waves produced by a flame FIGURE 37-14. These devices can quickly recognize even a very small fire.

Typically, flame detectors are found in places such as aircraft hangars or specialized industrial settings in which early detection and rapid reaction to a fire are critical. Such detectors are also used in explosion suppression systems, where they detect and suppress an explosion as it is occurring.

Flame detectors are both complicated and expensive. A key disadvantage of these models is that other infrared or ultraviolet sources, such as the sun or a welding operation, can set off an unwanted alarm. Flame detectors that combine infrared and ultraviolet sensors are sometimes used to lessen the chances of a false alarm.

Gas Detectors

Gas detectors are calibrated to detect the presence of a specific gas that is created by combustion or that is used in the facility. Depending on the system, a gas detector may be programmed to activate either the building's fire alarm system or a separate alarm. Some of these specialized instruments need regular calibration if they are to operate properly. Gas detectors for specific gases are often found in commercial or industrial applications. The most common type of gas detector you will encounter in residential settings is a carbon monoxide detector.

Air Sampling Detectors

Air sampling detectors continuously capture air samples and measure the concentrations of specific gases or products of combustion. These devices draw in air samples through a sampling unit and analyze them using an ionization or photoelectric smoke detector FIGURE 37-15. Duct detectors (air sampling detectors) are often installed in the return air ducts of large buildings. They will sound an alarm and shut down the air-handling system if they detect smoke.

FIGURE 37-14 Flame detectors are specialized devices that detect the electromagnetic light waves produced by a flame.

FIGURE 37-15 Air sampling detector.

More complex systems are sometimes installed in special hazard areas to draw air samples from rooms, enclosed spaces, or equipment cabinets. These samples pass through gas analyzers that can identify smoke particles, products of combustion, and concentrations of other gases associated with a dangerous condition. Air sampling detectors are most often used in areas that hold valuable contents or sensitive equipment and, therefore, where it is important to detect problems early.

Residential Carbon Monoxide Detectors

Residential carbon monoxide detectors are designed to sound an audible or visual alarm when the concentration of carbon monoxide becomes high enough to pose a health risk to the occupants of the building. Such an alarm is triggered by a very high concentration of carbon monoxide that accumulates in a short period of time or by a lower concentration of carbon monoxide that occurs over a longer period of time. Many carbon monoxide alarms look like smoke alarms to most people. Some of these units are combined with smoke alarms to provide warning from a fire or from elevated levels of carbon monoxide FIGURE 37-16. These combination alarms have one audible or visible alarm for fire and a different audible or visible alarm for elevated levels of carbon monoxide.

Carbon monoxide alarm activations need to be investigated by personnel who can use proper gas detection devices to determine the level of the gas and the cause of the alarm activation. Anytime you respond to a carbon monoxide alarm, you need to check all building occupants to determine whether they are suffering from the following signs or symptoms of carbon monoxide poisoning:

- Headache
- Nausea
- Disorientation
- Unconsciousness
- Multiple people with flulike symptoms in the same location

FIGURE 37-16 One type of combination carbon monoxide and smoke alarm.

Be especially alert for multiple people who are all suffering flulike symptoms, as they may be suffering from carbon monoxide poisoning.

Alarm Initiation by Fire Suppression Systems

Other fire protection systems in a building may be used to activate the fire alarm system. Automatic sprinkler systems are usually connected to the fire alarm system and will activate the alarm if a water flow occurs FIGURE 37-17. Such a system not only alerts the building occupants and the fire department to a possible fire but also ensures that someone is made aware that water is flowing, in case of an accidental discharge. Any other fire-extinguishing systems in a building, such as those found in kitchens or those containing halogenated agents, should also be tied into the building's fire alarm.

FIGURE 37-17 Automatic sprinkler systems use an electric flow switch to activate the building's fire alarm system.

False, Unwanted, and Nuisance Alarms

As the number of fire detection and alarm systems increases, so does the number of false, unwanted, or nuisance alarms. Knowing how to handle these alarms is just as important as knowing how to deal with an actual fire. Fire department personnel who are informed about fire alarm and detection systems can advise building owners and operators when these systems experience problems.

The term "false alarm" is generally used by the public to describe all fire alarm activations that are not associated with a true emergency. In reality, three distinct types of false alarms are possible: malicious false alarms, unwanted alarms, and nuisance alarms. It is important to distinguish among the three types in determining the root cause of the fire alarm activation, because the problem must be recognized before it can be rectified.

Regardless of the underlying cause, all three types of false alarms have the same results: They waste fire department resources and may delay legitimate responses. In addition, frequent false alarms at the same site can desensitize building occupants to the alarm system so that they might not respond appropriately to a real emergency.

Malicious False Alarms

Malicious false alarms occur when individuals deliberately activate a fire alarm when there is no fire, causing a disturbance. Manual fire alarm boxes are popular targets for pranksters. A malicious false alarm is an illegal act.

Unwanted Alarms

An unwanted alarm occurs when an alarm system is activated by a condition that is not really an emergency. For example, a smoke alarm placed too close to a kitchen may be triggered by normal cooking activities. An unwanted alarm could also occur if a person who is smoking unknowingly stands under a smoke detector. The smoke detector does its job—recognizing smoke and activating the alarm system—but there is no real emergency.

Nuisance Alarms

Nuisance alarms are caused by improper functioning of an alarm system or one of its components. Alarm systems must be properly maintained on a continual basis. A mechanical failure or a lack of maintenance that causes an alarm system to become activated when no emergency is actually occurring means that the system might not be activated when there is a real fire.

Preventing Unwanted and Nuisance Alarms

Systems that experience unwanted and nuisance alarms should be examined to determine the cause of the alarms and correct the problem. Many unwanted alarms could be avoided by relocating a detector or using a different type of detector in a particular location. Proper design, installation, and system maintenance are essential to prevent nuisance alarms.

If an alarm system is activated whenever it rains, water is probably leaking into the wiring. Detectors may also respond to a buildup of dust, dirt, or other debris by becoming more sensitive and needlessly going off. That risk explains why smoke alarms and smoke detectors must be cleaned periodically.

Several methods can be used to reduce unwanted and nuisance alarms caused by smoke detection systems:

- In a cross-zoned system, the activation of a single smoke detector will not sound the fire alarm, although it will usually set off a trouble alarm. In this kind of system, a second smoke detector must be activated before the actual fire alarm will sound.
- In a verification system, a delay of 30 to 60 seconds separates activation and notification. During this time, the system may show a trouble or "pre-alarm" condition at the system control panel. After the preset interval, the system rechecks the detector. If the condition has cleared, the system returns to normal. If the detector is still sensing smoke, however, the fire alarm is activated.

Both cross-zoned and verification systems are designed to prevent brief exposures to common occurrences such as cigarette smoke from activating the alarm system. For more information, see the NFPA's *Fire Service Guide to Reducing Unwanted Fire Alarms*.

FIGURE 37-18 A speaker fire alarm notification appliance.

Alarm Notification Appliances

Audible and visual alarm notification devices such as bells, horns, and electronic speakers produce an audible signal when the fire alarm is activated. Newer systems also incorporate visual alerting devices. These audible and visual alarms alert occupants of a building to a fire.

Older systems used a variety of sounds as notification devices, but this inconsistency often led to confusion over whether the sound was actually an alarm. More-recent fire codes have adopted a standardized audio pattern, called the temporal-3 pattern, that must be produced by any audio device used as a fire alarm. The temporal-3 pattern consists of three tones followed by a short pause. Even single-station smoke alarms designed for residential occupancies now use this sound pattern. As a consequence, people can recognize a fire alarm immediately.

Some public buildings also play a recorded evacuation announcement in conjunction with the temporal-3 pattern. This recorded message is played through the fire alarm speakers and provides instructions on safe evacuation of the site FIGURE 37-18. In facilities such as airport terminals, this announcement is recorded in multiple languages. Such a system may also include a public address feature that fire department or building security personnel can use to provide specific instructions, information about the situation, or notice when the alarm condition is terminated.

Many new fire alarm systems incorporate visual notification devices such as high-intensity strobe lights or other types of flashing lights as well as audio devices FIGURE 37-19. Visual devices alert hearing-impaired occupants to a fire alarm and are very useful in noisy environments where an audible alarm might not be heard.

FIGURE 37-19 This alarm notification appliance has both a loud horn and a high-intensity strobe light.

Other Fire Alarm Functions

In addition to alerting occupants and summoning the fire department, fire alarm systems may control other building functions, such as air-handling systems, fire doors, and elevators. To control smoke movement through the building, the system may either shut down or start up air-handling systems. Fire doors that are normally held open by electromagnets may be released to compartmentalize the building and confine the fire to a specific area. Doors allowing reentry from exit stairways into occupied areas may be unlocked. Elevators may be summoned to a predetermined floor, usually the main lobby, so they can be used by fire crews.

Responding fire personnel must understand which building functions are being controlled by the fire alarm, for both safety and fire suppression reasons. This information should be gathered during preincident planning surveys and should be available in printed form or on a graphic display at the control panel location.

Fire Alarm Annunciation Systems

Some fire alarm systems give little information at the alarm control panel; others specify exactly which initiation device activated the fire alarm. The systems can be further subdivided based on whether they are zoned or coded systems.

In a zoned system, the alarm control panel indicates where in the building the alarm was activated. Today, almost all alarm systems are zoned to some extent; only the most rudimentary alarm systems give no information at the alarm control panel about where the alarm was initiated. In a coded system, the zone is identified not only at the alarm control panel, but also through the audio notification device. Using these two variables, systems can be broken down into four categories: noncoded alarm, zoned noncoded alarm, zoned coded alarm, and master-coded alarm TABLE 37-1.

Noncoded Alarm System

In a noncoded alarm system, the control panel does not provide any information indicating where in the building the fire alarm was activated. The alarm typically sounds a bell or horn. Fire department personnel must then search the entire building to find which initiation device was activated. This type of system is generally found only in older, small buildings.

Zoned Noncoded Alarm System

The zoned noncoded alarm system is the most common type of system encountered by fire fighters, particularly in newer buildings. With this model, the building is divided into multiple zones, often by floor or by wing. The alarm control panel indicates in which zone the activated device is located and, sometimes, which type of device was activated. Responding personnel can go directly to that part of the building to search for the problem and check the activated device.

Many zoned noncoded alarm systems maintain an individual indicator light for each zone. When a device in that area is activated, the indicator light goes on. Computerized alarm systems also may use "addressable devices." In these systems, each individual initiation device—whether it is a smoke detector, heat detector, or pull-station—has its own unique identifier. When the device is activated, the identifier is indicated on a display or printout at the control panel. Responding personnel can then determine exactly which device or devices have been activated.

Zoned Coded Alarm

In addition to having all the features of a zoned alarm system, a zoned coded alarm system indicates which zone has been activated over the announcement system. Hospitals often use this type of system, because it is not possible to evacuate all staff and patients for every fire alarm. The audible notification devices give a numbered code that indicates which zone was activated. A code list tells building personnel which zone is in an alarm condition and which areas must be evacuated.

More modern systems in these occupancies use speakers as their alarm notification devices. As a consequence, a voice message indicating the nature and location of the alarm accompanies the audible alarm signal.

Master-Coded Alarm

In a master-coded alarm system, the audible notification devices for fire alarms also are used for other purposes. For example, a school may use the same bell to announce a change in classes, to signal a fire alarm, to summon the janitor, or to make other notifications. Most of these systems have been replaced by modern speaker systems that use the temporal-3 pattern fire alarm signal and have public address capabilities. This type of system is rarely installed in new buildings.

Fire Department Notification

The fire department should always be notified when a fire alarm system is activated. In some cases, a person must make a telephone call to the fire department. In other cases, the fire alarm system may be connected directly to the fire department or to a remote location where someone on duty calls the fire department. Fire alarm systems can be classified in five categories, based on how the fire department is notified of an alarm TABLE 37-2.

Local Alarm Systems

A local alarm system does not notify the fire department. Instead, the alarm sounds only in the building to notify the occupants. Buildings with this type of system should have notices posted requesting occupants to call the fire department and report the alarm after they exit FIGURE 37-20.

Remote Station Systems

A remote station system sends a signal directly to the fire department or to another monitoring location via a telephone line or a radio signal. This type of direct notification can be installed only in jurisdictions where the fire department is equipped to handle direct alarms. If the signal goes to a monitoring location, that site must be continually staffed by someone who will call the fire department.

TABLE 37-1 Categories of Alarm Annunciation Systems

Category	Description
Noncoded alarm	No information is given on which device was activated or where it is located.
Zoned noncoded alarm	The alarm system control panel indicates the zone in the building that was the source of the alarm. It may also indicate the specific device that was activated.
Zoned coded alarm	The system indicates over the audible warning device which zone has been activated. This type of system is often used in hospitals, where it is not feasible to evacuate the entire facility.
Master-coded alarm	The system is zoned and coded. The audible warning devices are also used for other emergency-related functions.

TABLE 37-2 Fire Department Notification Systems

Type of System	Description
Local alarm	The fire alarm system sounds an alarm only in the building where it was activated. No signal is sent out of the building. Someone must call the fire department to respond.
Remote station	The fire alarm system sounds an alarm in the building and transmits a signal to a remote location. The signal may go directly to the fire department or to another location where someone is responsible for calling the fire department.
Auxiliary system	The fire alarm system sounds an alarm in the building and transmits a signal to the fire department via a public alarm box system.
Proprietary system	The fire alarm system sounds an alarm in the building and transmits a signal to a monitoring location owned and operated by the facility's owner. Depending on the nature of the alarm and arrangements with the local fire department, facility personnel may respond and investigate, or the alarm may be immediately retransmitted to the fire department. These facilities are monitored 24 hours a day.
Central station	The fire alarm system sounds an alarm in the building and transmits a signal to an off-premises alarm monitoring facility. The off-premises monitoring facility is then responsible for notifying the fire department to respond.

FIGURE 37-20 Buildings with a local alarm system should post notices requesting occupants to call the fire department and report the alarm after they exit.

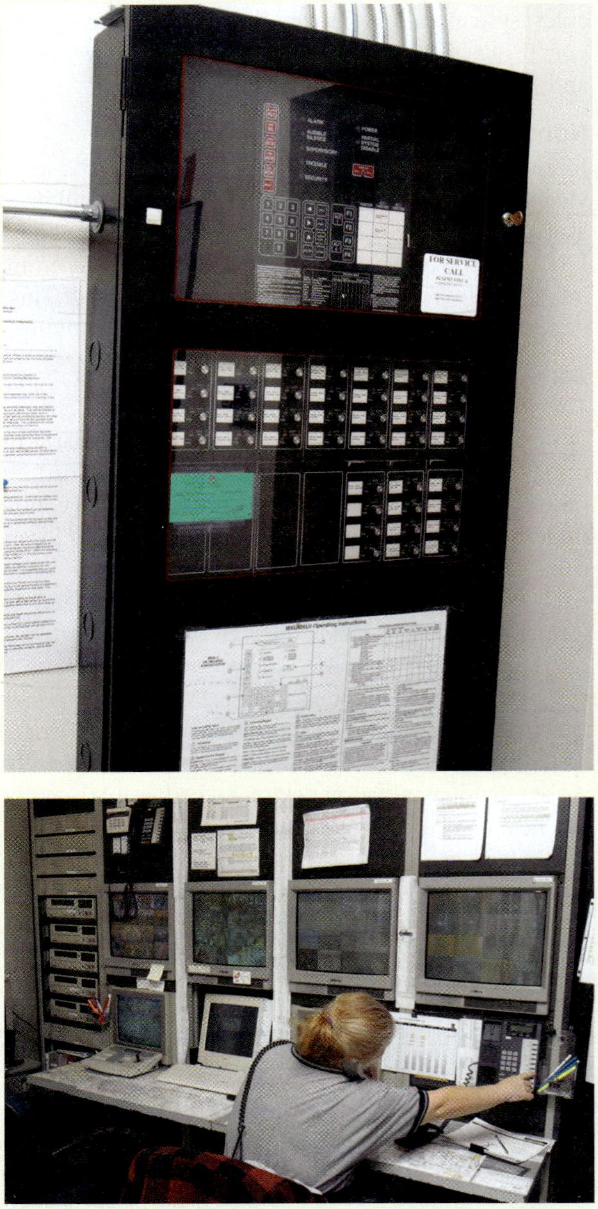

FIGURE 37-21 In a proprietary system, fire alarms from several buildings are connected to a single monitoring site owned and operated by the buildings' owner.

Auxiliary Systems

Auxiliary systems can be used in jurisdictions with a public fire alarm box system. With this approach, the building's fire alarm system is tied into a master alarm box located outside the building. When the alarm activates, it trips the master box, which transmits the alarm directly to the fire department communications center. Most cities have phased out public fire alarm boxes because they are expensive to maintain. They have been largely replaced by more modern communication devices.

Proprietary Systems

In a proprietary system, the building's fire alarms are connected directly to a monitoring site that is owned and operated by the building's owner. Proprietary systems are often installed at facilities where multiple buildings belong to the same owner, such as universities or industrial complexes. Each building is connected to a monitoring site on the premises (usually the security center), which is staffed at all times **FIGURE 37-21**. When an alarm sounds, the staff at the monitoring site report the alarm to the fire department, usually by telephone or a direct line.

Central Stations

A central station is a third-party, off-site monitoring facility that monitors multiple alarm systems. Individual building owners contract with and pay the central station to monitor their facilities FIGURE 37-22. When an activated alarm at a covered building transmits a signal to the central station by telephone or radio, personnel at the central station notify the appropriate fire department of the fire alarm. The central station facility may be located in the same city as the facility or in a different part of the country.

FIGURE 37-22 A central station monitors alarm systems at many locations.

Usually, building alarms are connected to the central station through leased or standard telephone lines. The use of either cellular telephone frequencies or radio frequencies, however, is becoming more common. Cellular or radio connections may be used to back up regular telephone lines in case they fail; in remote areas without telephone lines, they may be the primary transmission method.

Fire Suppression Systems

Fire suppression systems include automatic sprinkler systems, standpipe systems, and specialized extinguishing systems such as dry-chemical systems. Most newly constructed commercial buildings incorporate at least one of these systems, and increasing numbers of single-family dwellings are being built with residential sprinkler systems as well.

Understanding how these systems work is important because they can affect fire behavior. In addition, fire fighters should know how to interface with the fire suppression system and how to shut down a system to prevent unnecessary damage.

Automatic Fire Sprinkler Systems

The most common type of fire suppression system is the automatic sprinkler system. Automatic sprinklers are reliable and effective, with a history of more than 100 years of successfully controlling fires. When properly installed and maintained, these systems can help control fires and protect lives and property.

Unfortunately, few members of the general public have an accurate understanding of how automatic sprinklers work. In movies and on television, when one sprinkler is activated, the entire system begins to discharge water. This inaccurate portrayal of how automatic sprinkler systems operate has made people hesitant to install automatic sprinklers, fearing that they will cause unnecessary water damage.

The reality is quite different. In most automatic sprinkler systems, the sprinkler heads open one at a time as they are heated to their operating temperature. Usually, only one or two sprinkler heads open before the fire is controlled.

The basic operating principles of an automatic sprinkler system are simple. A system of water pipes is installed throughout a building to deliver water to every area. Depending on the design and occupancy of the building, these pipes may be placed above or below the ceiling. Automatic sprinkler heads are located along this system of pipes, such that each sprinkler head covers a particular floor area. A fire in that area will activate the sprinkler head, which then discharges water on the fire. This kind of system is analogous to having a fire fighter in every room with a charged hose line, just waiting for a fire.

One of the major advantages of a sprinkler system is that it can function as both a fire detection system and a fire suppression system. An activated sprinkler head not only discharges water on the fire but also triggers a water-motor gong, a flow alarm that signifies water is flowing in the system. In addition, the system prompts electric flow or pressure switches to activate the building's fire alarm system, notifying the fire department and occupants. This kind of system is so effective that by the time fire fighters arrive, the sprinklers have often extinguished the fire.

Automatic Sprinkler System Components

The overall design of automatic sprinkler systems can be complex, especially in large buildings. This complexity is somewhat deceiving, however, because even the largest systems include just four major components: the automatic sprinkler heads, piping, control valves, and a water supply, which may or may not include a fire pump FIGURE 37-23. Most sprinkler systems are directly connected to a public water supply system and do not need electricity to function.

Automatic Sprinkler Heads

Automatic sprinkler heads, which are commonly referred to as sprinkler heads, are the working ends of a sprinkler system. In most systems, the heads serve two functions: They activate the sprinkler system and they apply water to the fire. Sprinkler heads are composed of a body, which includes the orifice (opening); a release mechanism, which holds a cap in place over the orifice; and a deflector, which directs the water in a spray pattern FIGURE 37-24. Standard sprinkler heads have a ½-inch (12.7-mm) orifice, but several other sizes are available for special applications.

Although sprinkler heads come in several styles, all of them are categorized according to the type of release mechanism used and the intended mounting position—upright, pendant, or horizontal. They are also rated according to their release temperature. The release mechanisms hold the cap in

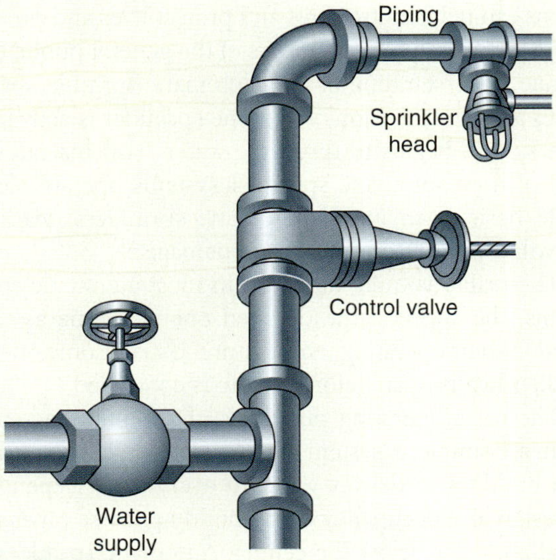

FIGURE 37-23 The basic components of an automatic sprinkler system include sprinkler heads, piping, control valves, and a water supply.

FIGURE 37-24 Automatic sprinkler heads have a body with an opening, a release mechanism, and a water deflector.

place until the release temperature is reached. At that point, the mechanism is released, and the water pushes the cap out of the way as it discharges onto the fire **FIGURE 37-25**.

Fusible-link sprinkler heads use a metal alloy, such as solder, that melts at a specific temperature **FIGURE 37-26**. This alloy links two pieces of metal that keep the cap in place. When the designated operating temperature is reached, the solder melts and the link breaks, releasing the cap. Fusible-link sprinkler heads come in a wide range of styles and temperature ratings.

Frangible-bulb sprinkler heads use a glass bulb filled with glycerin or alcohol to hold the cap in place **FIGURE 37-27**. This bulb also contains a small air bubble. As the bulb is heated, the liquid absorbs the air bubble and expands until it breaks the glass, releasing the cap. The volume and composition of the liquid and the size of the air bubble determine the temperature at which the head activates as well as the speed with which it responds.

Chemical-pellet sprinkler heads use a plunger mechanism and a small chemical pellet to hold the cap in place. The chemical pellet liquefies when the temperature reaches a

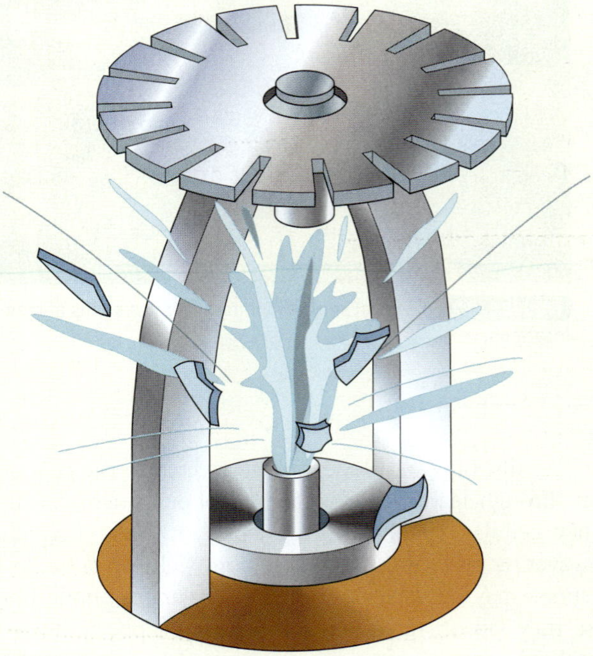

FIGURE 37-25 Automatic sprinkler heads are designed to activate at a wide range of temperatures. They may react slowly or quickly, depending on their settings.

FIGURE 37-26 In fusible-link sprinkler heads, two pieces of metal are linked together by an alloy such as solder.

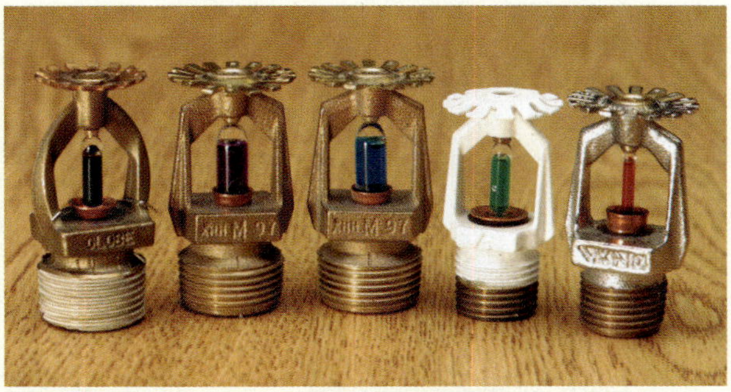

FIGURE 37-27 Frangible-bulb sprinkler heads activate when the liquid in the bulb expands and breaks the glass.

FIRE FIGHTER Tips

Here are some popular misconceptions about sprinklers:

- The entire building will be drowned when the sprinklers go off.
- The manual alarm will set off every sprinkler in the building.
- Water does more damage than fire.
- Activation of smoke detectors sets off all of the sprinklers in the building.

These beliefs are all wrong!

preset point. When the pellet melts, the liquid compresses the plunger, releasing the cap and allowing water to flow.

Special Sprinkler Heads

Sprinkler heads can also be designed for special applications, such as covering large areas or discharging the water in extra-large droplets or as a fine mist. Some sprinkler heads have protective coatings to help prevent corrosion. Builders and installers should consider these characteristics when designing the system and selecting appropriate heads. It is important to ensure that the proper heads are installed and that any replacement heads inserted in subsequent years are of the same type.

Sprinkler heads that are intended for residential occupancies are manufactured with a release mechanism that provides for a fast response.

By contrast, early-suppression fast-response sprinkler heads have improved heat collectors that speed up the response and ensure rapid release of water **FIGURE 37-28**. They are used in large warehouses and distribution facilities where early fire suppression is important. These heads often have large orifices to discharge large volumes of water onto a fire.

Deluge Heads

Deluge heads are easily identifiable by the fact that they have no cap or release mechanism. The orifice is always open **FIGURE 37-29**. Deluge heads are used only in deluge sprinkler systems, which are covered later in this chapter.

Temperature Ratings

Sprinkler heads are rated according to their release temperature. A typical rating for sprinkler heads in a light-hazard

FIGURE 37-28 An early-suppression fast-response sprinkler head.

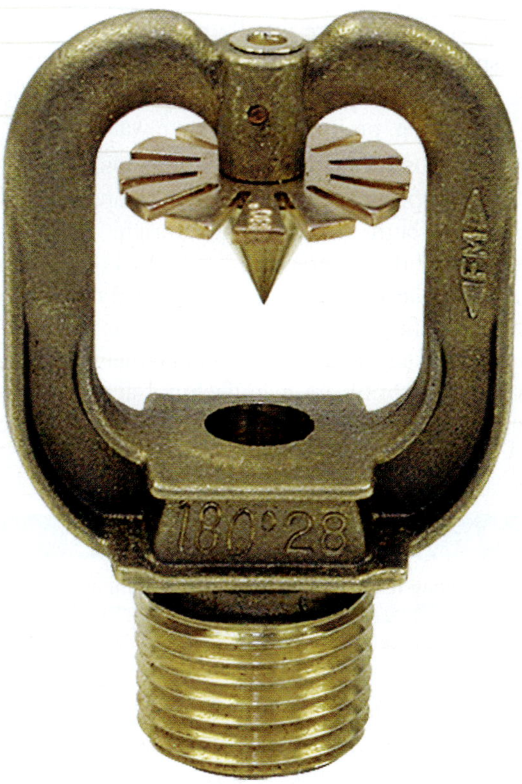

FIGURE 37-29 A deluge sprinkler head has no release mechanism.

occupancy, such as an office building, would be 165°F (74°C). Sprinkler heads that are used in areas with warmer ambient air temperatures would have higher ratings. The rating should be stamped on the body of the sprinkler head. Frangible-bulb sprinkler heads use a color-coding system to identify the temperature rating TABLE 37-3. Some fusible-link and chemical-pellet sprinklers also use this system.

The temperature rating on a sprinkler head must match the anticipated ambient air temperatures. If the rating is too low for the ambient air temperature, accidental alarms may occur. Conversely, if the rating is too high, the system will be slow to react to a fire, and the fire may be able to establish itself and grow before the system becomes activated.

Spare heads that match those used in the system should always be available on-site. Usually the spare heads are kept in a clearly marked box near the main control valve FIGURE 37-30. Having spare heads handy enables sprinkler systems to be returned to full service quickly, whether they were set off by a fire or by accident.

FIGURE 37-30 Spare sprinkler heads should be kept in a special box near the main sprinkler system valve.

Mounting Position

Sprinkler heads with different mounting positions are not interchangeable, because each mounting position has deflectors specifically designed to produce an effective water stream down or out toward the fire. Each automatic sprinkler head is designed to be mounted in one of three positions FIGURE 37-31:

- Pendant sprinkler heads are designed to be mounted on the underside of the sprinkler piping, hanging down toward the room. These heads are commonly marked SSP, which stands for "standard spray pendant."
- Upright sprinkler heads are designed to be mounted on top of the supply piping, as their name suggests. Upright heads are usually marked SSU, for "standard spray upright."
- Sidewall sprinkler heads are designed for horizontal mounting, projecting out from a wall.

Old-Style versus New-Style Sprinkler Heads

In automatic sprinklers manufactured before the mid-1950s, deflectors in both pendant and upright sprinkler heads directed part of the water stream up toward the ceiling. At the time, it was believed that this action helped cool the area and extinguish the fire. Sprinkler heads with this design are called old-style sprinklers, and many of them remain in service today.

Automatic sprinklers manufactured after the mid-1950s deflect the entire water stream down toward the fire. These types of heads are referred to as new-style heads or standard spray heads. New-style heads can replace old-style heads, but the reverse is not true. Due to different coverage patterns, old-style heads should not be used to replace any new-style heads.

Sprinkler Piping

Sprinkler piping, the network of pipes that delivers water to the sprinkler heads, includes the main water supply lines, risers, feeder lines, and branch lines. Although sprinkler pipes are usually made of steel, other metals can be used as well FIGURE 37-32. For example, plastic pipe is sometimes used in residential sprinkler systems.

Sprinkler system designers use hydraulic calculations to determine the size of pipe and the layout of the "grid." Older systems were designed using piping schedules. Most new systems are designed using computer software. Near the main control valve, pipes have a large diameter; as the pipes approach the sprinkler heads, the diameter generally decreases.

Valves

A sprinkler system includes several different valves, such as the main water supply control valve, the alarm valve, and other, smaller valves used for testing and service. Many large systems have zone valves, which enable the water supply to different areas to be shut down without turning off the entire system. All of these valves play a critical role in the design and function of the system.

TABLE 37-3 Temperature Rating Determined by Color of Sprinkler Head

Maximum Ceiling Temperature (°F)	Temperature Rating (°F)	Color Code	Glass Bulb Colors
100	135 to 170	Uncolored or black	Orange or red
150	175 to 225	White	Yellow or green
225	250 to 300	Blue	Blue
300	325 to 375	Red	Purple
375	400 to 475	Green	Black
475	500 to 575	Orange	Black
625	650	Orange	Black

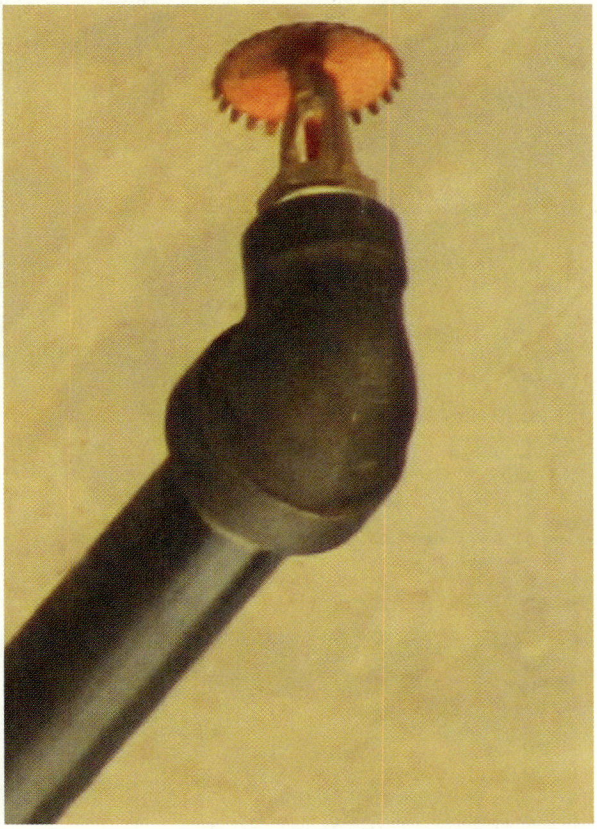

A.

B.

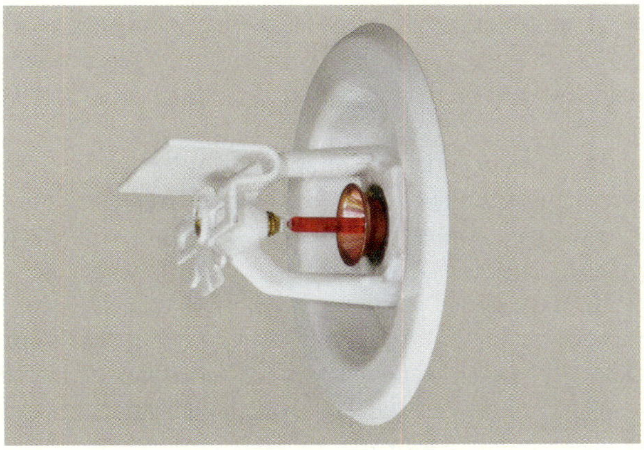

C.

FIGURE 37-31 Sprinkler head mounting positions. **A.** Upright. **B.** Pendant. **C.** Sidewall.

FIGURE 37-32 Most new sprinkler systems use steel pipes.

Water Supply Control Valves

Every sprinkler system must have at least one main control valve that allows water to enter the system. This water supply control valve must be of the "indicating" type, meaning that the position of the valve itself indicates whether it is open or closed. Two examples are the outside stem and yoke (OS&Y) valve and the post indicator valve (PIV).

An OS&Y valve has a stem that moves in and out as the valve is opened or closed **FIGURE 37-33**. If the stem is out, the

FIGURE 37-33 An outside stem and yoke (OS&Y) valve is often used to control the flow of water into a sprinkler system.

valve is open; if the stem is in, the valve is closed. OS&Y valves are often found in a mechanical room in the building, where water to supply the sprinkler system enters the building. In warmer climates, they may be found outside.

The PIV has an indicator that reads either open or shut depending on its position **FIGURE 37-34**. A PIV is usually located in an open area outside the building and controls an underground valve. Opening or closing a PIV requires a wrench, which is usually attached to the side of the valve.

A wall post indicator valve (WPIV) is similar to a PIV but is designed to be mounted on the outside wall of a building **FIGURE 37-35**.

The main control valve, whether it is an OS&Y valve or a PIV, should be locked in the open position. This practice ensures that the water supply to the sprinkler system is never shut off unless the proper people are notified that the system is out of service. It is critically important that the sprinkler system always remain charged with water and ready to operate if needed.

As an alternative to locking the valves open, the valves may be equipped with tamper switches **FIGURE 37-36**. These devices monitor the position of the valve. If someone opens or closes the valve, the tamper switch sends a signal to the fire alarm control panel, indicating a change in the valve position. If the change has not been authorized, the cause of the signal can be investigated and the problem can be corrected.

FIGURE 37-34 A post indicator valve (PIV) is used to open or close an underground valve.

FIGURE 37-35 A wall post indicator valve (WPIV) controls the flow of water from an underground pipe into a sprinkler system.

FIGURE 37-36 A tamper switch activates an alarm if someone attempts to close a valve that should remain open.

Main Sprinkler System Valves

The type of main sprinkler system valve used depends on the type of sprinkler system installed. Options include an alarm valve, a dry-pipe valve, or a deluge valve. These valves are usually installed on the main riser, above the water supply control valve.

The primary functions of an alarm valve are to signal an alarm when a sprinkler head is activated and to prevent nuisance alarms caused by pressure variations and surges in the water supply to the system. This valve has a clapper mechanism that remains in the closed position until a sprinkler head opens. The closed clapper prevents water from flowing out of the system and back into the public water mains when water pressure drops.

When a sprinkler head is activated, the clapper opens fully and allows water to flow freely through the system. The open clapper also allows water to flow to the water-motor gong, sounding an alarm. Electrical flow and pressure switches then activate connections to external alarm systems.

Near Miss REPORT

Report Number: 05-0000472

Synopsis: Responded to an automatic fire alarm. Fire found on second floor. Officer trapped.

Event Description: Our company responded to an automatic fire alarm. The occupancy was a renovated grade/middle school building that now housed the local Park District offices and the area headquarters for a State Representative. The building was of brick and block construction with large windows. The still-alarm assignment consisted of two engines, one tower ladder with the shift commander aboard, and one ambulance.

The first-arriving engine's lieutenant and [department-specific name deleted] number 3 fire fighter entered the building, leaving the engineer with the rig. The ambulance crew finished donning SCBA, gathered tools, and staged. Both the lieutenant and fire fighter had portable radios, and the number 3 fire fighter had irons. The crew located the annunciator and found it indicating a smoke alarm on the second floor. The lieutenant radioed command and advised him that it smelled like paper burning, "like a trash can."

Command acknowledged this information and requested to be advised of developments/needs. The crew proceeded to the second floor and, upon gaining the landing, reported "light haze of smoke" at the ceiling. Again, command acknowledged the message. The lieutenant and fire fighter began searching numerous offices and cubicles in the State Representative's office area.

The second-due engine arrived and requested orders. Command ordered the lieutenant and number 3 fire fighter to enter the building from the "C" side and meet up with initial crew to assist in the search. Hearing this order via the radio, the first-due lieutenant sent his number 3 fire fighter down to the rear entrance to open the door for the second crew, as the Knox-box was on the front of the building and had already been accessed to enter the building. Not long after the number 3 fire fighter left, the lieutenant, alone on the second floor, radioed command stating that conditions had deteriorated greatly, with thick black, hot smoke. Command acknowledged the message. This was then followed by another transmission from the lieutenant: "It's getting hot up here and I can't find the door."

Command then exited the tower ladder and, dressed only in his bunker pants and coat, entered the building to attempt to locate the lieutenant. He did not remain inside long. When he did emerge, he finally radioed our dispatch center to upgrade to a full-still and then a general alarm. In the meantime, the second-due engine crew, along with the first-due number 3 fire fighter, were attempting to retrace the fire fighter's steps back to the lieutenant's last location. The doors in this government office, however, automatically locked upon closing. About this time the lieutenant made a third radio transmission repeating that it was "hot up here." Command acknowledged the message and reported that a crew was on its way up to get him.

The second-due crew never reported to command the trouble they were encountering with the locked doors. The lost lieutenant eventually made his way out of the office on his own after locating a receptionist's desk and remembering that the door was just in front of it. He emerged into the hallway as the second-due engine crew and his number 3 fire fighter were coming down the hall. The lieutenant suffered no injuries as a result of this incident. The fire eventually went to a second [alarm; department-specific name deleted].

Lessons Learned:

Lesson 1: An automatic fire alarm can still be a fire. Treat every one like it is.

Lesson 2: There are very few things on the fire ground that can safely be done alone.

Lesson 3: Be aware of building construction and features (i.e., doors that lock behind you).

Lesson 4: Locate second and even third exits when you enter a space. If there is only one way in and out, ensure you know where you are in relation to that exit at all times.

Lesson 5: Call the mayday early and give as much information as you can regarding your whereabouts.

Lesson 6: Command is just that—command. The incident commander is not a line-participant.

Lesson 7: Communicate your needs/difficulties/assessments to command so that other plans can be made if needed.

Lesson 8: Complacency kills. This incident was the exception.

Without a properly functioning alarm valve, sprinkler system flow alarms would occur frequently. The normal pressure changes and surges in a public water system will not open the clapper, however, which prevents water from flowing to the water-motor gong or tripping the electrical water flow or pressure switches unnecessarily.

In dry-pipe and deluge systems, the main valve functions both as an alarm valve and as a dam, holding back the water until the sprinkler system is activated. When the system is activated, the valve opens fully so that water can enter the sprinkler piping. Both dry-pipe and deluge systems are described later in this chapter.

Additional Valves

Sprinkler systems are equipped with a variety of other control valves. Several smaller valves are usually located near the main control valve, and still others are located elsewhere in the building. These smaller valves include drain valves, test valves, and connections to alarm devices. All of these valves should be properly labeled.

In larger facilities, the sprinkler system may be divided into zones, with a specific valve controlling the flow of water to each zone. This design makes maintenance easy and can prove extremely valuable when a fire occurs. After the fire is extinguished, water flow to the affected area can be shut off so that the activated heads can be replaced. Fire protection in the rest of the building, however, is unaffected by this shutdown.

Water Supplies

The water used in an automatic sprinkler system may come from a municipal water system, from on-site storage tanks, or from static water sources such as storage ponds or rivers. Whatever its source, the water supply must be able to satisfy the demands of the sprinkler system as well as meet the needs of the fire department in the event of a fire.

The preferred water source for a sprinkler system is a municipal water supply, if one is available. If the municipal supply cannot meet the water pressure and volume requirements of the sprinkler system, alternative supplies must be established.

Fire pumps are used when the water comes from a static source. They may also be deployed to boost the pressure in some sprinkler systems, particularly for tall buildings **FIGURE 37-37**. Because most municipal water supply systems do not provide enough pressure to control a fire on the upper floors of a high-rise building, fire pumps will turn on automatically when the sprinkler system activates or when the pressure drops to a preset level. In high-rise buildings, a series of fire pumps on upper floors may be needed to provide adequate pressure.

A large industrial complex could have more than one water source, such as a municipal system and a backup storage tank **FIGURE 37-38**. Multiple fire pumps can provide water to the sprinkler and standpipe systems in different areas through underground pipes. Private hydrants may also be connected to the same underground system.

Each sprinkler system should also have a fire department connection (FDC). This connection allows the department's engine to pump water into the sprinkler system **FIGURE 37-39**. The FDC may be used as either a supplement or the main source of water to the sprinkler system if the regular supply is interrupted or a fire pump fails.

The FDC usually has two or more 2½-inch (64-mm) female couplings or one large-diameter hose coupling mounted on an outside wall or placed near the building. It ties directly into the sprinkler system after the main control valve or alarm

FIGURE 37-37 In tall buildings, a fire pump may be needed to maintain appropriate pressure in the sprinkler system.

FIGURE 37-38 An elevated storage tank ensures that sufficient water and adequate pressure will be available to fight a fire.

valve. Each fire department should establish a standard operating procedure for first-arriving companies that specifies how to connect to the FDC and when to charge the system.

In large facilities, a single FDC may be used to deliver water to all fire protection systems in the complex. The water from this connection flows into the private underground water mains instead of into each system. Water pumped into this type of FDC should come from a source that does not service the complex, such as a public hydrant on a different grid.

Water Flow Alarms

All sprinkler systems should be equipped with a method for sounding an alarm whenever water begins flowing in the pipes. This type of warning is important both in an actual fire and in an accidental activation. Without these alarms, the occupants or the fire department might not be aware of the sprinkler activation. If a building is unoccupied, the sprinkler system could continue to discharge water long after a fire is extinguished, leading to extensive water damage.

Most systems incorporate a mechanical flow alarm called a water-motor gong **FIGURE 37-40**. When the sprinkler system is activated and the main alarm valve opens, some water is fed through a pipe to a water-powered gong located on the outside of the building. This gong alerts people outside the building that there is water flowing. This type of alarm will function even if there is no electricity.

Accidental soundings of water-motor gongs are rare. If a water-motor gong is sounding, water is probably flowing from the sprinkler system somewhere in the building. Fire companies that arrive and hear the distinctive sound of a water-motor gong know that there is a fire or that something else is causing the sprinkler system to flow water.

Sprinkler systems also are connected to the building's fire alarm system by either an electric flow switch or a pressure switch. This connection triggers the alarm to alert the building's occupants of its activation; a monitored system will notify the fire department as well. Unlike water-motor gongs, flow and pressure switches can be accidentally triggered by water pressure surges in the system. To reduce the risk of accidental activations, these devices usually have a time delay before they will sound an alarm.

FIGURE 37-39 A fire department connection delivers additional water and boosts the pressure in a sprinkler system.

FIGURE 37-40 A water-motor gong sounds when water is flowing in a sprinkler system.

Types of Automatic Sprinkler Systems

Automatic sprinkler systems are classified into four categories, depending on which type of sprinkler head is used and how the system is designed to activate:

- Wet sprinkler systems
- Dry sprinkler systems
- Preaction sprinkler systems
- Deluge sprinkler systems

Although many buildings may use the same type of system to protect the entire facility, it is not uncommon to see two or three systems combined in one building. Some facilities use a wet sprinkler system to protect most of the structure but implement a dry sprinkler or preaction system in a specific area. In many cases, a dry sprinkler or preaction system will branch off from the wet sprinkler system.

Wet Sprinkler Systems

A wet sprinkler system is the most common and least expensive type of automatic sprinkler system. As its name implies, the piping in a wet system is always filled with water. When a sprinkler head activates, water is immediately discharged onto the fire. The major drawback to a wet sprinkler system is that it cannot be used in areas within the building where temperatures drop below freezing. Water will also begin to flow in such a system if a sprinkler head is accidentally opened or a leak occurs in the piping.

If only a small, unheated area needs to be protected, two options are available. A dry-pendant sprinkler head can be used in very small areas, such as walk-in freezers FIGURE 37-41. The bottom part of a dry-pendant head, which resembles a standard sprinkler head, is mounted inside the freezer. The head has an elongated neck, usually 6 to 18 inches (15 to 65 centimeters) long, that extends up and connects to the wet sprinkler piping in the heated area above the freezer. The vertical neck section is filled with air and capped at each end. The top cap prevents water from entering the lower section, where it would freeze; the bottom cap acts just like the cap on a standard sprinkler head. When the head is activated and the lower cap drops out, a device inside the neck releases the upper cap, so water can flow down. The entire dry-pendant head assembly must be replaced after it has been activated.

Larger unheated areas, such as a loading dock, can be protected with an antifreeze loop. This small section of the wet sprinkler system is filled with glycol or glycerin instead of water. A check valve separates the antifreeze loop from the rest of the sprinkler system. When a sprinkler head in the unheated area is activated, the antifreeze sprays out first, followed by water. These areas can also be protected by dry sprinkler systems.

Dry Sprinkler Systems

A dry sprinkler system operates much like a wet sprinkler system, except that the pipes are filled with pressurized air instead of water. A dry-pipe valve keeps water from entering the pipes until the air pressure is released. Dry systems are used in facilities that may experience below-freezing temperatures, such as unheated warehouses, garages, or attics.

The air pressure in this kind of sprinkler system is set high enough to hold a clapper inside the dry-pipe valve in the closed position FIGURE 37-42. When a sprinkler head opens, the air escapes. As the air pressure drops, the water pressure on

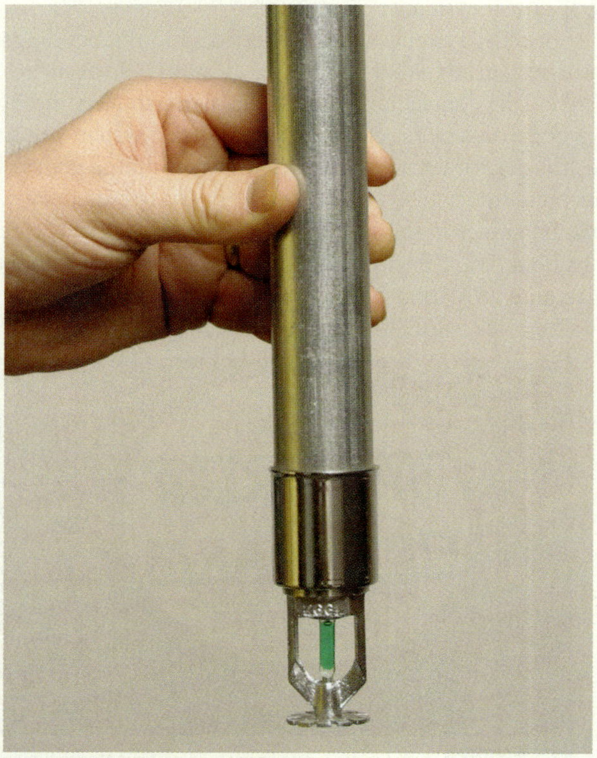

FIGURE 37-41 A dry-pendant sprinkler head can be used to protect a freezer box.

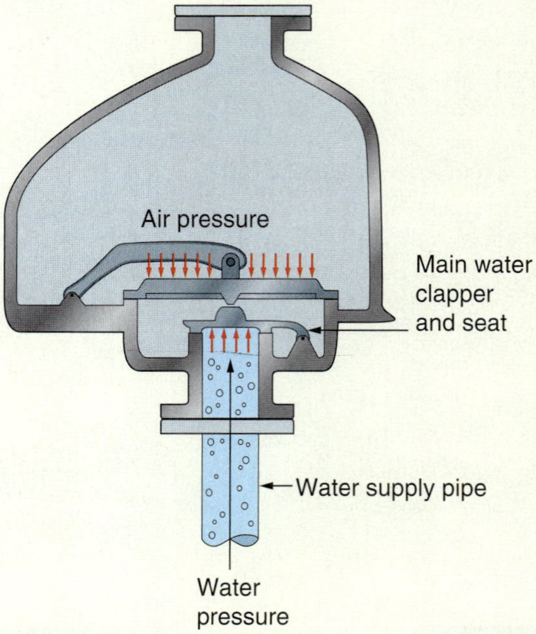

FIGURE 37-42 Water pressure on one side of the dry-pipe valve is held down by air pressure on the other side of the valve.

the other side of the clapper forces it open and water begins to flow into the pipes. When the water reaches the open sprinkler head, it is discharged onto the fire.

Dry sprinkler systems do not eliminate the risk of water damage from accidental activation. If a sprinkler head breaks, the air pressure will drop and water will flow, just as in a wet sprinkler system.

The action of the clapper assembly located inside most dry-pipe valves relies on a pressure differential. The system (or air) side of the clapper has a larger surface area than the supply (or wet) side. As a consequence, a lower air pressure can hold back a higher water pressure. A small compressor is used to maintain the air pressure in the system.

Dry sprinkler systems should have an air pressure alarm to alert building management personnel if the air pressure drops. The activation of this alarm could mean one of two things: The compressor is not working or there is an air leak in the system. If the air pressure in the system is too low, the clapper will open and the system will fill with water. At that point, the system would essentially function as a wet sprinkler system, which could cause it to freeze in low temperatures. In this scenario, the system would have to be drained and reset to prevent the pipes from freezing.

Dry sprinkler systems must be drained after every activation so the dry-pipe valve can be reset. The clapper also must be reset, and the air pressure must be restored before the water is turned back on.

Accelerators and Exhausters

One problem encountered in dry sprinkler systems is the delay between the activation of a sprinkler head and the actual flow of water out of the head. The pressurized air that fills the system must escape through the open head before the water can flow. For personal safety and property protection reasons, any delay longer than 90 seconds is unacceptable. Large systems, however, can take several minutes to empty out the air and refill the pipes with water. To compensate for this problem, two additional devices are used: accelerators and exhausters.

An accelerator is installed at the dry-pipe valve. The rapid drop in air pressure caused by an open sprinkler head triggers the accelerator, which allows air pressure to flow to the supply side of the clapper valve. This quickly eliminates the pressure differential, opening the dry-pipe valve and allowing the water pressure to force the remaining air out of the piping.

An exhauster is installed on the system side of the dry-pipe valve, often at a remote location in the building. Like an accelerator, the exhauster monitors the air pressure in the piping. If it detects a drop in pressure, it opens a large-diameter portal, allowing the air in the pipes to escape. The exhauster closes when it detects water, diverting the flow to the open sprinkler heads. Large systems may have multiple exhausters located in different sections of the piping.

Preaction Sprinkler Systems

A preaction sprinkler system is similar to a dry sprinkler system with one key difference: In a preaction sprinkler system, a secondary device—such as a smoke detector or a manual-pull alarm—must be activated before water is released into the sprinkler piping. When the system is filled with water, it functions as a wet sprinkler system.

A preaction system uses a deluge valve instead of a dry-pipe valve. The deluge valve will not open until it receives a signal that the secondary device has been activated. Because a detection system usually becomes activated more quickly than a sprinkler system does, water in a preaction system will generally reach the sprinklers before a head is activated.

The primary advantage of a preaction sprinkler system is its ability to prevent accidental water discharges. If a sprinkler head is accidentally broken or the pipe is damaged, the deluge valve will prevent water from entering the system. This feature makes preaction sprinkler systems well suited for locations where water damage is a major concern, such as computer rooms and electrical rooms.

Deluge Sprinkler Systems

A deluge sprinkler system is a type of dry sprinkler system in which water flows from all of the sprinkler heads as soon as the system is activated FIGURE 37-43. A deluge system does not have closed heads that open individually at the activation temperature; instead, all of the heads in a deluge system are always open.

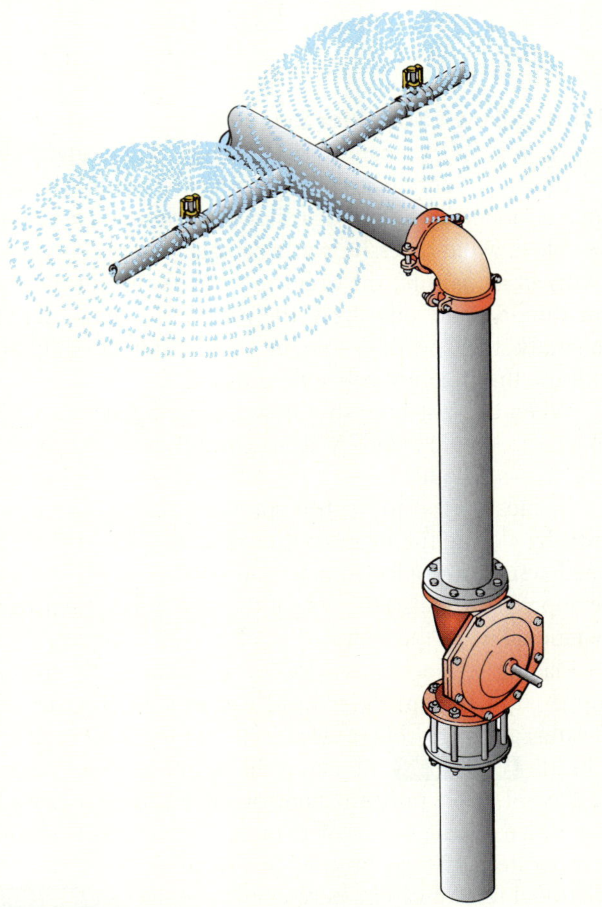

FIGURE 37-43 Water flows from all of the heads in a deluge system as soon as the system is activated.

FIRE FIGHTER Tips

Some dry-pipe valves will not open if there is water above the clapper valve, a mechanical device that allows the water to flow in only one direction. Unless the main valve is designed to operate as either a wet- or dry-pipe valve, a dry system that has been filled with water should be immediately drained and reset for dry-pipe operation.

Deluge systems can be activated in three ways:

1. A detection system can release the deluge valve when a detector is activated.
2. The deluge system can be connected to a separate pilot system of air-filled pipes with closed sprinkler heads. When a head on the pilot line is activated, the air pressure drops, opening the deluge valve.
3. Most deluge valves can be released manually.

Deluge systems are used in special applications such as aircraft hangars or industrial processes, where rapid fire suppression is critical. In some cases, foam concentrate is added to the water, so that the system will discharge a foam blanket over the hazard. Deluge systems are also used for special hazard applications, such as liquid propane gas loading stations. In these situations, a heavy deluge of water is needed to protect exposures from a large fire that ignites very rapidly.

Shutting Down Sprinkler Systems

Responding fire companies should know where and how to shut down any automatic sprinkler system if needed. If the system is accidentally activated, it should be shut down as soon as possible to avoid excessive water damage.

In an actual fire, the order to shut off the sprinkler system should come only from the incident commander (IC). Generally, the sprinkler system should not be shut down until the fire is completely extinguished.

When the system is shut down, a fire fighter with a portable radio should stand by at the control valve, in case it has to be reopened quickly.

In most cases, the entire sprinkler system can be shut down by closing the main control valve (OS&Y or PIV). In a zoned system, the affected zone can be shut down by closing the appropriate valve; the rest of the system can then remain operational while the activated heads are being replaced.

Placing a wooden wedge or a commercial sprinkler stopper into the sprinkler head can quickly stop the flow of water, although this measure will not work with all types of heads FIGURE 37-44. Blocking the sprinkler head shuts off the flow of water until the control valve can be located and shut down. Some types of commercial sprinkler stoppers incorporate a fusible link so the sprinkler head becomes functional until it can be permanently replaced FIGURE 37-45. After the main valve is closed, opening the drain valve allows the system to drain so that the activated head can be replaced.

FIGURE 37-44 A sprinkler stopper can be used to stop the flow of water from a sprinkler head that has been activated.

FIGURE 37-45 This shut-off device shuts off the sprinkler and provides a temporary fusible link to keep the sprinkler head in service.

Residential Sprinkler Systems

Residential sprinkler systems are becoming more accepted, with many recently constructed homes including them. Some communities require residential sprinkler systems in every newly constructed home.

The design of and theory underlying residential sprinkler systems are similar to those of commercial systems, but with some significant variations. The primary objective of a residential sprinkler system is life safety, so the sprinkler heads are designed to respond quickly FIGURE 37-46. A fast response helps protect the occupants and control incipient-stage fires.

Residential systems typically use smaller piping and sprinkler heads with smaller orifices and less water discharge. To control costs, plastic pipe may be used instead of metal pipe. These systems typically protect high-risk areas such as rooms

FIGURE 37-46 Special sprinkler heads are used in residential systems. They open more quickly than commercial heads and discharge less water.

and corridors; small areas such as closets and bathrooms may not be covered.

Residential systems are usually wet sprinkler systems. They are typically connected to the domestic water supply of a public water main. Residential sprinkler systems do not have a fire department connection or have to be tied into a separate fire alarm system, unless required locally. But they usually do require the system to include a water flow alarm.

The responding fire companies in the jurisdiction may not be familiar with every home that has a sprinkler system, but fire fighters should know both how these systems work and how to shut down a system that has been activated. Usually the main shut-off valve for the sprinklers can be found near the location where water enters the house. If the house has a basement, the shut-off valve may be located near other utility controls. If a separate shut-off valve for the sprinkler system cannot be found, the main water shut-off for the house can be used to deactivate the sprinkler system. Installing working smoke alarms and a residential sprinkler system reduces the chance of death from a home fire by 82 percent.

Fire Fighter Safety Tips

Do not shut down the water to a sprinkler system until ordered to do so by the IC. The fire must be completely out and hose lines must be available should the fire reignite. A fire fighter should be stationed at the valve with a portable radio, ready to reopen the valve if necessary.

Standpipe Systems

A standpipe system consists of a network of inlets, pipes, and outlets for fire hoses that are built into a structure to provide water for firefighting purposes. It includes one or more inlets supplied by a municipal water supply or fire department hoses, piping to carry the water closer to the fire, and one or more outlets equipped with valves to which fire hoses can be connected. Standpipe systems are required in many high-rise buildings, and they are found in many other structures as well. Standpipes are also installed to carry water to large bridges and to supply water to limited-access highways that are not equipped with fire hydrants FIGURE 37-47.

FIGURE 37-47 Standpipe outlets allow fire hoses to be connected inside a building.

Standpipes are found in buildings both with and without sprinkler systems. In many newer buildings, sprinklers and standpipes are combined into a single system. In older buildings, the sprinkler and standpipe systems are typically separate entities. Three categories of standpipes—Class I, Class II, and Class III—are distinguished based on their intended use.

Class I Standpipe

A Class I standpipe is designed for use by fire department personnel only. Each outlet has a 2½-inch (64-mm) male coupling and a valve to open the water supply after the attack line is connected FIGURE 37-48. Often the connection is located inside a cabinet, which may or may not be locked. Responding fire personnel carry the hose into the building with them, usually in some sort of roll, bag, or backpack. A Class I standpipe system must be able to supply an adequate volume of water with sufficient pressure to operate fire department attack lines.

Class II Standpipes

A Class II standpipe is designed for use by the building occupants. The outlets are generally equipped with a length of 1½-inch (38-mm) single-jacket hose preconnected to the system FIGURE 37-49. These systems are intended to enable occupants to attack a fire before the fire department arrives, but their safety and effectiveness are questionable. After all, most building occupants are not trained to attack fires safely. If a fire cannot be controlled with a regular fire extinguisher, it is usually safer for the occupants to simply evacuate the building and call the fire department. Class II standpipes may be useful at facilities such as refineries and military bases, where workers are trained as an in-house fire brigade.

Responding fire personnel should not use Class II standpipes for fighting a fire, because the hoses and nozzles may be of inferior quality and lack maintenance and testing. The water flow may not be adequate to control a fire. Class II standpipe outlets are frequently connected to the domestic water piping system in the building rather than an outside main or a separate system. Instead of using equipment (hose and nozzle) that may not be reliable or adequate, fire fighters should always use department-issued equipment.

FIGURE 37-48 A Class I standpipe provides water for fire department hose lines.

FIGURE 37-49 A Class II standpipe is intended to be used by building occupants to attack incipient-stage fires.

Class III Standpipes

A Class III standpipe has the features of both Class I and Class II standpipes in a single system. This kind of system has 2½-inch (64-mm) outlets for fire department use as well as smaller outlets with attached hoses for occupant use. The occupant hoses may have been removed—either intentionally or by vandalism—in many facilities, so the system basically becomes a Class I system. Fire fighters should use only the 2½-inch (64-mm) outlets, even if they employ an adapter to connect a smaller hose. The 1½-inch (38-mm) outlets may have pressure-reducing devices to limit the flow and pressure for use by untrained civilians.

Water Flow in Standpipe Systems

Standpipes are designed to deliver a minimum amount of water at a particular pressure to each floor. The design requirements depend on the code requirements in effect when the building was constructed.

Flow-restriction devices FIGURE 37-50 or pressure-reducing valves FIGURE 37-51 are often installed at the outlets to limit the pressure and flow. A vertical column of water, such as the water in a standpipe riser, exerts a backpressure (also called head pressure). In a tall building, this backpressure can amount to hundreds of pounds per square inch (psi) at lower floor levels. If a hose line is connected to an outlet without a flow restrictor or a pressure-reducing valve, the water pressure could rupture the hose, and the excessive nozzle pressure could make the line difficult or dangerous to handle.

If flow restrictors or pressure-reducing valves are not properly installed and maintained, these devices can cause problems for fire fighters. An improperly adjusted pressure-reducing valve, for example, could severely restrict the flow to a hose line. Likewise, a flow-restriction device could limit the flow of water to fight a fire.

The flow and pressure capabilities of a standpipe system should be determined during preincident planning. Many standpipe systems deliver water at a pressure of only 65 psi (448 kilopascals) at the top of the building. The combination fog/straight-stream nozzles used by many fire departments are designed to operate at 100 psi (689 kilopascals). For this reason, many fire departments use low-pressure combination nozzles for fighting fires in high-rise buildings or require the use of only smooth-bore nozzles when operating from a standpipe system.

VOICES OF EXPERIENCE

One of the most valuable skills I have learned is being calm while under extreme duress. A fire fighter who is traumatized and panics has a detrimental effect on his or her performance, endangers the lives of team members, and can endanger everyone on the incident. Fire fighters must be able to function as a team, which is partially why the camaraderie is so strong in the fire service. Literally, you must be willing to trust your life with your fellow fire fighters. You must train extensively to develop a sense of calm so that you can function effectively when everyone and everything around you appears to be in a chaotic state. Fire fighters routinely operate in environments that endanger life and property. The old adage that, "fire fighters run in, when everyone else is running away," still holds true.

The best fire fighters are always seeking knowledge to enhance their ability to perform. This knowledge should be obtained from a variety of sources:

- College-based instruction and formal education
- Local, Regional, State, National Fire schools and academies
- Institutional knowledge such as thorough knowledge of the department and local mutual aid resources available
- Experiential learning by hands-on practical application of techniques
- Community knowledge: pertinent information about community demographics, water mains, water sources, utilities, building design and construction, potential community hazards, traffic routes, evacuation routes, and location of fire protection systems

The importance of each fire fighter's contribution to the overall mission of public safety and the department's role must be extolled. Fire fighters are called when people do not know what to do; it may involve a seemingly insignificant operation such as water in the basement, or a devastating fire with multiple fatalities. The only way a fire fighter can effectively respond to this task is through being calm and constantly learning.

Charles Garrity
Berkshire Community College
Pittsfield, Massachusetts

FIGURE 37-50 A flow-restriction device on a standpipe outlet can cause problems for fire fighters.

FIGURE 37-51 A pressure-reducing valve on a standpipe outlet may be necessary on lower floors to avoid problems caused by backpressure.

Preincident planning for high-rise buildings should include an evaluation of the building's standpipe system and a determination of the anticipated flows and pressures. This information should be used to make decisions about the appropriate nozzles and tactics for those buildings.

Engine companies that respond to buildings equipped with standpipes should carry a kit that includes the appropriate hose and nozzle, a spanner wrench, and any required adapters. This kit should also include tools to adjust the settings of pressure-reducing valves or to remove restrictors that are obstructing flows.

Water Supplies

Both standpipe systems and sprinkler systems are supplied with water in essentially the same way. That is, many wet standpipe systems in modern buildings are connected to a public water supply and equipped with an electric or diesel fire pump to provide additional pressure. Some also have a water storage tank that serves as a backup supply. In these systems, the FDC on the outside of the building can be used to increase the flow, boost the pressure, or obtain water from an alternative source.

Dry standpipe systems are found in many older buildings. If freezing weather is a problem, such as in open parking structures, bridges, and tunnels, dry standpipe systems are still acceptable. Most of these systems do not have a permanent connection to a water supply, so the FDC must be used to pump water into the system. If a fire occurs in a building with dry standpipes, connecting the hose lines to the FDC and charging the system with water are high priorities.

Some dry standpipe systems are connected to a water supply through a dry-pipe or deluge valve, similar to a sprinkler system. In such systems, opening an outlet valve or tripping a switch next to the outlet releases water into the standpipes.

High-rise buildings often incorporate complex systems of risers, storage tanks, and fire pumps to deliver the needed flows to upper floors. The details of these systems should be obtained during preincident planning surveys. Department procedures should dictate how responding units will supply the standpipes with water as well as how crews should use the standpipes inside the building.

Specialized Extinguishing Systems

Automatic sprinkler systems are used to protect whole buildings, or at least major sections of buildings. Nevertheless, in certain situations, more specialized extinguishing systems are needed. These kinds of systems are often used in areas where water would not be an acceptable extinguishing agent FIGURE 37-52. For example, water is not the agent of choice for areas containing sensitive electronic equipment or contents such as computers, valuable books, or documents. Water is also incompatible with materials such as flammable liquids or water-reactive chemicals. Given these facts, areas where these materials are stored or used may have a separate extinguishing system.

Dry-Chemical and Wet-Chemical Extinguishing Systems

Dry-chemical and wet-chemical extinguishing systems are the most common specialized agent systems. In commercial kitchens, they are used to protect the cooking areas and exhaust systems. In addition, some gas stations have dry-chemical systems that protect the dispensing areas. These systems are also installed inside buildings to protect areas where flammable liquids are stored or used. Both dry-chemical and wet-chemical extinguishing systems are similar in basic design and arrangement.

Dry-chemical extinguishing systems use the same types of finely powdered agents as dry-chemical fire extinguishers FIGURE 37-53. The agent is kept in self-pressurized tanks or in tanks with an external cartridge of carbon dioxide or nitrogen that provides pressure when the system is activated.

Wet-chemical extinguishing systems are used in most new commercial kitchens FIGURE 37-54. They use a liquid extinguishing agent, which is much more effective on vegetable oils than are the dry chemicals used in older kitchen systems. Wet-chemical systems are also easier to clean up after a discharge, so the kitchen can resume operations more quickly after the system has discharged.

FIGURE 37-52 Specialized extinguishing systems are used in areas where water would not be effective or desirable.

FIGURE 37-53 Dry-chemical extinguishing systems are installed at many self-service gasoline filling stations.

Wet-chemical extinguishing agents are not compatible with normal all-purpose dry-chemical extinguishing agents, however. Only wet agents or B:C-rated dry-chemical extinguishers should be used where these systems are installed.

FIGURE 37-54 Wet-chemical extinguishing systems are used in most new commercial kitchens.

With both dry-chemical and wet-chemical extinguishing agent systems, fusible-link or other automatic initiation devices are placed above the target hazard to activate the system FIGURE 37-55. A manual discharge button is also provided so that workers can activate the system if they discover a fire FIGURE 37-56. Open nozzles are located over the target areas to discharge the agent directly onto a fire. When the system is activated, the extinguishing agent flows out of all the nozzles.

Many kitchen systems discharge the extinguishing agent into the ductwork above the exhaust hood as well as onto the cooking surface. This approach helps prevent a fire from igniting any grease build-up inside the ductwork and spreading throughout the system. Although the ductwork should be cleaned regularly, it is not unusual for a kitchen fire to extend into the exhaust system.

Most dry- and wet-chemical extinguishing systems are tied into the building's fire alarm system. Kitchen extinguishing systems should also shut down the gas or electricity to the cooking appliances and exhaust fans.

Clean-Agent Extinguishing Systems

Clean-agent extinguishing systems are often installed in areas where computers or sensitive electronic equipment are used or where valuable documents are stored. The agents used in these systems are nonconductive and leave no residue. Notably, halogenated agents or carbon dioxide are used for this purpose because they will extinguish a fire without causing significant damage to the contents of a room.

Clean-agent systems operate by discharging a gaseous agent into the atmosphere at a concentration that will extinguish a fire. Smoke detectors or heat detectors installed in these areas activate the system, although a manual discharge button is also provided with most installations. Discharge is usually delayed 30 to 60 seconds after the detector is activated to allow workers to evacuate the area.

FIGURE 37-55 Fusible links can be used to activate a specialized extinguishing system.

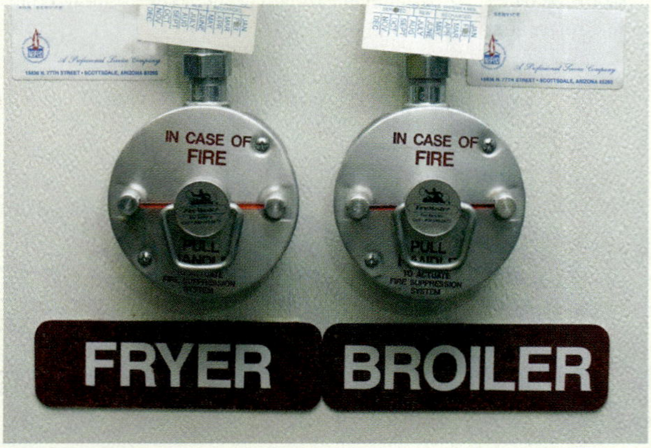

FIGURE 37-56 Most special extinguishing systems can also be manually activated.

During this delay (the pre-alarm period), an abort switch can be used to stop the discharge. In some systems, the abort button must be pressed until the detection system is reset; releasing the abort button too soon causes the system to discharge. If there is a fire, the clean-agent system should be completely discharged before fire fighters arrive. Whether the fire was successfully extinguished or not, fire fighters entering the area should use self-contained breathing apparatus (SCBA) until the area has been properly ventilated. Although the gaseous agents used in clean-agent extinguishing systems are not considered immediately dangerous to life and health, it is better to avoid any unnecessary exposure to them. Toxic products or by-products of combustion could be present in the atmosphere as well, or the oxygen level could be reduced.

Clean-agent extinguishing systems should be tied to the building's fire alarm system and indicated as a zone on the control panel. This notification scheme alerts fire fighters that they are responding to a situation where a clean agent has discharged. If the system has a preprogrammed delay, the pre-alarm should activate the building's fire alarm system.

Until the 1990s, Halon 1301 was the agent of choice for protecting areas such as computer rooms, telecommunications rooms, and other sensitive areas. Halon 1301 is a nontoxic, odorless, colorless gas that leaves behind no residue. It is very effective at extinguishing fires because it interrupts the chemical reaction of combustion. Unfortunately, Halon 1301 is also damaging to the environment—a fact that led to its discontinuation as a firefighting agent. Alternative agents have since been developed for use in new systems; they are also replacing Halon 1301 in many existing systems. A stockpile of Halon 1301 is still available to recharge existing systems when no suitable substitute agent exists.

Carbon Dioxide Extinguishing Systems

Carbon dioxide extinguishing systems are similar in design to clean-agent systems. The primary difference is that carbon dioxide extinguishes a fire by displacing the oxygen in the room and smothering the fire. Large quantities of carbon dioxide are required for this purpose, because the area must be totally flooded to extinguish a fire FIGURE 37-57.

FIGURE 37-57 Carbon dioxide extinguishes a fire by displacing the oxygen in the room and smothering the fire.

Carbon dioxide systems may be designed to protect either a single room or a series of rooms. They usually have the same series of pre-alarms and abort buttons as are found in Halon 1301 systems. Because the carbon dioxide discharge creates an oxygen-deficient atmosphere in the room, the activation of such a system is immediately dangerous to life. Any occupant who is still in the room when the agent is discharged is likely to be rendered unconscious and asphyxiated. Fire fighters responding to a carbon dioxide extinguishing system discharge must use SCBA protection until the area is fully vented.

Carbon dioxide extinguishing systems should be connected to the building's fire alarm system. Responding fire fighters should see that a carbon dioxide system discharge has been activated. Using this knowledge, they can deal with the situation safely.

Wrap-Up

Chief Concepts

- Fire protection systems include fire alarms, automatic fire detection, and fire suppression systems.
- Fire alarm and detection systems range from simple, single-station smoke alarms for private homes, to complex fire detection and control systems for high-rise buildings. Many fire alarm and detection systems in large buildings also control other systems to help protect occupants and control the spread of fire and smoke.
- A fire alarm system has three basic components: an alarm initiation device, an alarm notification device, and a control panel. The alarm initiation device is either an automatic or manually operated device that, when activated, causes the system to indicate an alarm. The alarm notification device is generally an audible device, but is often accompanied by a visual device, which alerts the building occupants. The control panel links the initiation device to the notification device and performs other essential functions.
- The most common type of residential fire alarm system is a single-station smoke alarm. This life-safety device includes a smoke detection device, an automatic control unit, and an audible alarm within a single unit. It alerts occupants quickly when a fire occurs.
- The most up-to-date codes require new homes to have a smoke alarm in every bedroom and on every floor level. They also require a battery backup for smoke alarms in the event of a power failure.
- Two types of fire detection devices may be used in a smoke alarm to detect combustion:
  - Ionization detectors use radioactive material within the device to detect invisible products of combustion.
  - Photoelectric detectors use a light beam to detect the presence of visible particles of smoke.
- Alarm initiation devices begin the fire alarm process either manually or automatically. Manual alarm initiation devices include single-action pull-stations and double-action pull-stations. Automatic initiation devices sense fire using an ionization detector or a photoelectric chamber to detect products of combustion and trigger an audible alarm.
- Various types of automatic initiation devices are available:
  - Smoke detectors—Designed to sense the presence of smoke.
  - Beam detectors—Photoelectric smoke detectors used to protect large open areas.
  - Heat detectors—Designed to sense the presence of too much heat.
  - Fixed-temperature heat detectors—Designed to activate at a preset temperature.
  - Rate-of-rise heat detectors—Designed to activate if the temperature of the surrounding air increases by more than a set amount in a given period of time.
  - Line detectors—Use wire or tubing strung along the ceiling of large open areas to detect an increase in heat.
  - Flame detectors—Specialized devices that detect the electromagnetic light waves produced by a flame.
  - Gas detectors—Designed to detect the presence of a specific gas that is created by combustion or that is used in the facility.
  - Air sampling detectors—Designed to continuously capture air samples and measure the concentrations of specific gases or products of combustion.
- Residential carbon monoxide detectors are designed to sound an audible or visual alarm when the concentration of carbon monoxide is high enough to pose a health risk to the occupants of the building.
- Knowing how to handle false, unwanted, and nuisance alarms is just as important as knowing how to deal with an actual fire.
- Three types of false alarms are possible:
  - Malicious false alarms—Occur when individuals deliberately activate a fire alarm when there is no fire, causing a disturbance.
  - Unwanted alarms—Occur when an alarm system is activated by a condition that is not really an emergency, such as smoke from normal cooking activities.
  - Nuisance alarms—Occur when an alarm system malfunctions.
- Several methods can be used to reduce unwanted and nuisance alarms caused by smoke detection systems:
  - Cross-zoned system—The activation of a single smoke detector will not sound the fire alarm, although it will usually set off a trouble alarm. In this kind of system, a second smoke detector must be activated before the actual fire alarm will sound.
  - Verification system—A delay of 30 to 60 seconds separates activation and notification. During this time, the system may show a trouble or "pre-alarm" condition at the system control panel. After the preset interval, the system rechecks the detector. If the condition has cleared, the system returns to normal. If the detector is still sensing smoke, the fire alarm is activated.
- Some fire alarm systems give little information at the alarm control panel, while others specify exactly which initiation device activated the fire alarm. Alarm annunciation systems are classified into four categories:
  - Noncoded alarm—No information is given on the control panel.
  - Zoned noncoded alarm—The control panel indicates the zone in the building that was the source of the alarm.

- Zoned coded alarm—The system indicates over the audible warning device which zone has been activated.
- Master-coded alarm—The audible notification device is also used for other purposes, such as a public address system.

■ There are five categories of fire department notification systems:
- Local alarm—Sounds an alarm only in the building where it was activated.
- Remote station—Sounds an alarm in the building and transmits a signal to a remote location.
- Auxiliary system—Sounds an alarm in the building and transmits a signal to the fire department via a public alarm box system.
- Proprietary system—Sounds an alarm in the building and transmits a signal to a monitoring location owned and operated by the facility's owner.
- Central station—Sounds an alarm in the building and transmits a signal to an off-premises alarm monitoring facility.

■ Fire suppression systems include sprinkler systems, standpipe systems, and specialized extinguishing systems.

■ The most common type of fire suppression system is the automatic sprinkler system.

■ The basic operating principles of an automatic sprinkler system are simple. A system of water pipes is installed throughout a building to deliver water to every area where a fire might occur. Depending on the design and occupancy of the building, these pipes may be placed above or below the ceiling. Automatic sprinkler heads are located along this system of pipes, such that each sprinkler head covers a particular floor area. A fire in that area will activate the sprinkler head, which then discharges water on the fire.

■ Every automatic sprinkler system includes four major components:
- Automatic sprinkler heads—Commonly called sprinkler heads and available in several styles.
- Piping—Carries the water or extinguishing agent to the sprinkler heads.
- Control valves—Control the flow of water to the piping.
- Water supply—May come from a municipal water system, from on-site storage tanks, or from static water sources.

■ Automatic sprinkler systems are divided into four categories:
- Wet sprinkler systems—The most common type of automatic sprinkler system, in which the pipes are always filled with water.
- Dry sprinkler systems—The pipes are filled with pressurized air instead of water until the system is activated; used when the pipes may be exposed to freezing temperatures.
- Preaction sprinkler systems—Similar to a dry sprinkler system, in which a secondary device must be activated before water is released into the sprinkler piping.
- Deluge sprinkler systems—A type of dry sprinkler system in which water flows from all of the sprinkler heads as soon as the system is activated.

■ A standpipe system consists of a network of inlets, pipes, and outlets for fire hoses that are built into a structure to provide water for firefighting purposes. Three types of standpipes are distinguished based on their intended use:
- Class I—Designed for use by fire department personnel only.
- Class II—Designed for use by the building occupants.
- Class III—Has the features of both Class I and Class II standpipes in a single system.

■ Specialized extinguishing systems may be installed to protect areas where water may not be used, such as computer rooms.

Hot Terms

Accelerator A device that speeds up the removal of the air from a dry-pipe or preaction sprinkler system. An accelerator reduces the time required for water to start flowing from sprinkler heads.

Air sampling detector A system that captures a sample of air from a room or enclosed space and passes it through a smoke detection or gas analysis device.

Alarm initiation device An automatic or manually operated device in a fire alarm system that, when activated, causes the system to indicate an alarm condition.

Alarm notification device An audible and/or visual device in a fire alarm system that makes occupants or other persons aware of an alarm condition.

Alarm valve A valve that signals an alarm when a sprinkler head is activated and prevents nuisance alarms caused by pressure variations.

Automatic sprinkler heads The working ends of a sprinkler system, which serve to activate the system and to apply water to the fire.

Automatic sprinkler system A system of pipes filled with water under pressure that discharges water immediately when a sprinkler head opens.

Auxiliary system A fire alarm system that sounds an alarm in the building and transmits a signal to the fire department via a public alarm box system.

Beam detector A smoke detection device that projects a narrow beam of light across a large open area from a sending unit to a receiving unit. When the beam is interrupted by smoke, the receiver detects a reduction in light transmission and activates the fire alarm.

Bimetallic strip A device with components made from two distinct metals that respond differently to heat. When heated, the metals will bend or change shape.

Carbon dioxide extinguishing system A fire suppression system designed to protect either a single room or series of rooms by flooding the area with carbon dioxide.

Central station An off-premises facility that monitors alarm systems and is responsible for notifying the fire department of an alarm. These facilities may be geographically located some distance from the protected building(s).

Chemical-pellet sprinkler head A sprinkler head activated by a chemical pellet that liquefies at a preset temperature.

Clapper mechanism A mechanical device installed within a piping system that allows water to flow in only one direction.

Class I standpipe A standpipe system designed for use by fire department personnel only. Each outlet should have a valve to control the flow of water and a 2½-inch (64-mm) male coupling for fire hose.

Class II standpipe A standpipe system designed for use by occupants of a building only. Each outlet is generally equipped with a length of 1½-inch (38-mm) single-jacket hose and a nozzle, which are preconnected to the system.

Class III standpipe A combination system that has features of both Class I and Class II standpipes.

Coded system A fire alarm system design that divides a building or facility into zones and has audible notification devices that can be used to identify the area where an alarm originated.

Cross-zoned system A fire alarm system that requires activation of two separate detection devices before initiating an alarm condition. If a single detection device is activated, the alarm control panel will usually show a problem or trouble condition.

Deluge head A sprinkler head that has no release mechanism; the orifice is always open.

Deluge sprinkler system A sprinkler system in which all sprinkler heads are open. When an initiation device, such as a smoke detector or heat detector, is activated, the deluge valve opens and water discharges from all of the open sprinkler heads simultaneously.

Deluge valve A valve assembly designed to release water into a sprinkler system when an external initiation device is activated.

Double-action pull-station A manual fire alarm activation device that requires two steps to activate the alarm. The user must push in a flap, lift a cover, or break a piece of glass before activating the alarm.

Dry-chemical extinguishing system An automatic fire extinguishing system that discharges a dry chemical agent.

Dry-pipe valve The valve assembly on a dry sprinkler system that prevents water from entering the system until the air pressure is released.

Dry sprinkler system A sprinkler system in which the pipes are normally filled with compressed air. When a sprinkler head is activated, it releases the air from the system, which opens a valve so the pipes can fill with water.

Early-suppression fast-response sprinkler head A sprinkler head designed to react quickly and suppress a fire in its early stages.

Exhauster A device that accelerates the removal of the air from a dry-pipe or preaction sprinkler system.

False alarm The activation of a fire alarm system when no fire or emergency condition exists.

Fire alarm control panel The component in a fire alarm system that controls the functions of the entire system.

Fire department connection (FDC) A fire hose connection through which the fire department can pump water into a sprinkler system or standpipe system.

Fixed-temperature heat detector A sensing device that responds when its operating element is heated to a predetermined temperature.

Flame detector A sensing device that detects the radiant energy emitted by a flame.

Flow switch An electrical switch that is activated by water moving through a pipe in a sprinkler system.

Frangible-bulb sprinkler head A sprinkler head with a liquid-filled bulb. The sprinkler head becomes activated when the liquid is heated and the glass bulb breaks.

Fusible-link sprinkler head A sprinkler head with an activation mechanism that incorporates two pieces of metal held together by low-melting-point solder. When the solder melts, it releases the link and water begins to flow.

Gas detector A device that detects and/or measures the concentration of dangerous gases.

Halon 1301 A liquefied gas extinguishing agent that puts out a fire by chemically interrupting the combustion reaction between fuel and oxygen. Halon agents leave no residue.

Heat detector A fire alarm device that detects abnormally high temperature, an abnormally high rate of rise in temperature, or both.

Ionization smoke detector A device containing a small amount of radioactive material that ionizes the air between two charged electrodes to sense the presence of smoke particles.

Line detector Wire or tubing that can be strung along the ceiling of large open areas to detect an increase in heat.

Local alarm system A fire alarm system that sounds an alarm only in the building where it was activated; that is, no signal is sent out of the building.

Malicious false alarm A fire alarm signal sent when there is no fire, usually initiated by individuals who wish to cause a disturbance.

Manual pull-station A device with a switch that either opens or closes a circuit, activating the fire alarm.

Master-coded alarm An alarm system in which audible notification devices can be used for multiple purposes, not just for the fire alarm.

Noncoded alarm An alarm system that provides no information at the alarm control panel indicating where the activated alarm is located.

Nuisance alarm A fire alarm signal caused by malfunction or improper operation of a fire alarm system or component.

Obscuration rate A measure of the percentage of light transmission that is blocked between a sender and a receiver unit.

Outside stem and yoke (OS&Y) valve A sprinkler control valve with a valve stem that moves in and out as the valve is opened or closed.

Pendant sprinkler head A sprinkler head designed to be mounted on the underside of sprinkler piping so that the water stream is directed in a downward direction.

Photoelectric smoke detector A device to detect visible products of combustion using a light source and a photosensitive sensor.

Post indicator valve (PIV) A sprinkler control valve with an indicator that reads either open or shut depending on its position.

Preaction sprinkler system A dry sprinkler system that uses a deluge valve instead of a dry-pipe valve and requires activation of a secondary device before the pipes will fill with water.

Proprietary system A fire alarm system that transmits a signal to a monitoring location owned and operated by the facility's owner.

Rate-of-rise heat detector A fire detection device that responds when the temperature increases at a rate that exceeds a predetermined value.

Remote annunciator A secondary fire alarm control panel in a different location than the main alarm panel; it is usually located near the front door of a building.

Remote station system A fire alarm system that sounds an alarm in the building and transmits a signal to the fire department or an off-premises monitoring location.

Residential sprinkler system A sprinkler system designed to protect dwelling units.

Sidewall sprinkler head A sprinkler that is mounted on a wall and discharges water horizontally into a room.

Single-action pull-station A manual fire alarm activation device in which the user takes a single step—such as moving a lever, toggle, or handle—to activate the alarm.

Single-station smoke alarm A single device that both detects visible and invisible products of combustion (smoke) and sounds an alarm.

Smoke detector A device that detects visible and invisible products of combustion (smoke) and sends a signal to a fire alarm control panel.

Spot detector A single heat-detector device. These devices are often spaced throughout an area.

Sprinkler piping The network of piping in a sprinkler system that delivers water to the sprinkler heads.

Standpipe system An arrangement of piping, valves, and hose connections installed in a structure to deliver water for fire hoses.

Tamper alarm An alarm that sounds when the clear plastic cover of a pull-station is lifted; it is intended to prevent malicious false alarms.

Tamper switch A switch on a sprinkler valve that transmits a signal to the fire alarm control panel if the normal position of the valve is changed.

Temporal-3 pattern A standard fire alarm audible signal for alerting occupants of a building.

Unwanted alarm A fire alarm signal caused by a device reacting properly to a condition that is not a true fire emergency.

Upright sprinkler head A sprinkler head designed to be installed on top of the supply piping; it is usually marked SSU ("standard spray upright").

Verification system A fire alarm system that does not immediately initiate an alarm condition when a smoke detector activates. The system will wait a preset interval (generally 30 to 60 seconds) before checking the detector again. If the condition is clear, the system returns to normal status. If the detector still senses smoke, the system activates the fire alarm.

Wall post indicator valve A sprinkler control valve that is mounted on the outside wall of a building. The position of the indicator tells whether the valve is open or shut.

Water-motor gong An audible alarm notification device that is powered by water moving through the sprinkler system.

Wet-chemical extinguishing system An extinguishing system that discharges a proprietary liquid extinguishing agent. It is often installed over stoves and deep-fat fryers in commercial kitchens.

Wet sprinkler system A sprinkler system in which the pipes are normally filled with water.

Zoned coded alarm A fire alarm system that indicates which zone was activated both on the alarm control panel and through a coded audio signal.

Zoned noncoded alarm A fire alarm system that indicates the activated zone on the alarm control panel.

Zoned system A fire alarm system design that divides a building or facility into zones so that the area where an alarm originated can be identified.

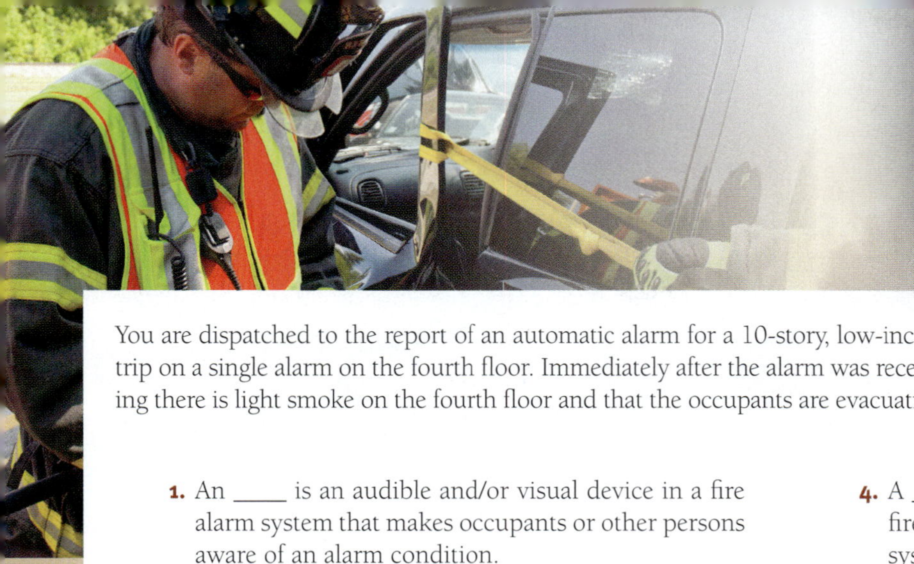

FIRE FIGHTER *in action*

You are dispatched to the report of an automatic alarm for a 10-story, low-income housing facility. The alarm company reports a trip on a single alarm on the fourth floor. Immediately after the alarm was received, calls began coming into the 911 center advising there is light smoke on the fourth floor and that the occupants are evacuating. You arrive on scene and prepare to go to work.

1. An ____ is an audible and/or visual device in a fire alarm system that makes occupants or other persons aware of an alarm condition.

- **A.** alarm initiating device
- **B.** accelerator
- **C.** auxiliary system
- **D.** alarm notification device

2. A type ___ standpipe system is designed for use by occupants of a building only.

- **A.** I
- **B.** II
- **C.** III
- **D.** IV

3. A(n) _____ fire alarm system transmits a signal to a monitoring location owned and operated by the facility's owner.

- **A.** proprietary system
- **B.** auxiliary
- **C.** central station
- **D.** local

4. A ____ is a fire hose connection through which the fire department can pump water into a sprinkler system or standpipe system.

- **A.** fire department connection
- **B.** post indicator valve
- **C.** dry-pipe valve
- **D.** outside stem and yoke

5. A(n) ____ is a sprinkler head designed to be mounted on the underside of sprinkler piping so that the water stream is directed down.

- **A.** frangible-bulb sprinkler head
- **B.** fusible-link sprinkler head
- **C.** pendant sprinkler head
- **D.** upright sprinkler head

6. A(n) _____ alarm occurs when an alarm system is activated by a condition that is not really an emergency.

- **A.** nuisance
- **B.** malicious false
- **C.** undesired
- **D.** unwanted

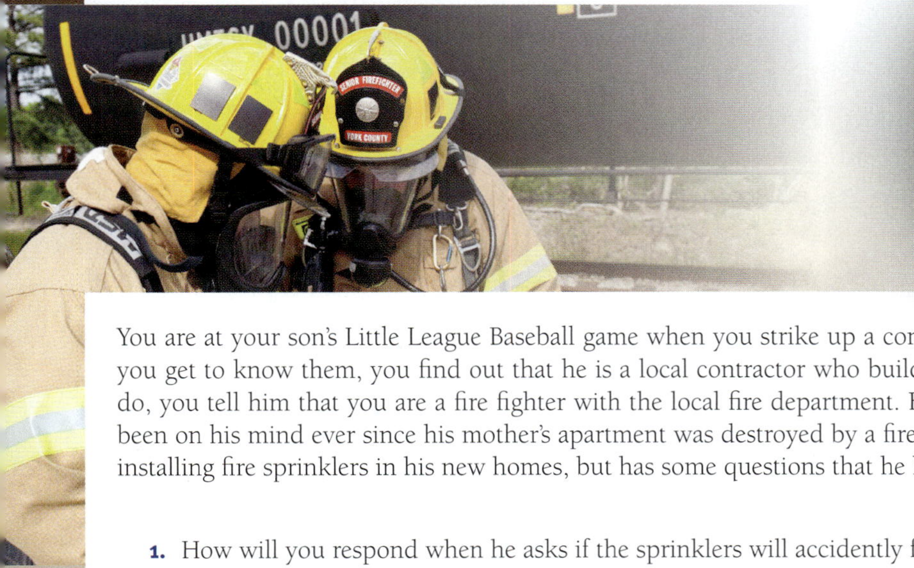

FIRE FIGHTER II *in action*

You are at your son's Little League Baseball game when you strike up a conversation with the couple sitting next to you. As you get to know them, you find out that he is a local contractor who builds houses in your district. When asked what you do, you tell him that you are a fire fighter with the local fire department. He mentions that trying to make homes safer has been on his mind ever since his mother's apartment was destroyed by a fire caused by a neighbor's cigar. He is committed to installing fire sprinklers in his new homes, but has some questions that he hopes you can answer.

1. How will you respond when he asks if the sprinklers will accidently flood the house?
2. How will you respond if he asks how he can justify the cost to his customers?
3. How will you respond if he asks if you have sprinklers in your home?
4. How will you encourage this contractor to continue to promote fire sprinklers in residential structures?

Fire Cause Determination

Fire Fighter I

Knowledge Objectives

After studying this chapter, you will be able to perform the following skills:

- Differentiate accidental fires from incendiary fires. (NFPA 5.3.8, p 1074)
- Describe the point of origin. (NFPA 5.3.8, 5.3.8.A, 5.3.8.B, p 1073–1074)
- Define the chain of custody. (NFPA 5.3.13, 5.3.13.A, 5.3.13.B, p 1077–1078)
- Describe techniques for preserving fire-cause evidence. (NFPA 5.3.14.B, p 1077–1078)
- Describe the observations that fire fighters should make during fire-ground operations. (NFPA 5.3.13.B, p 1080–1083)

Skills Objectives

There are no skills objectives for Fire Fighter I candidates. NFPA 1001 contains no Fire Fighter I Job Performance Requirements for this chapter.

Fire Fighter II FFII

Knowledge Objectives

After studying this chapter, you will be able to perform the following skills:

- Describe the role and relationship of the Fire Fighter II to criminal investigators and insurance investigators. (NFPA 6.3.4A, p 1073)
- Describe how the origin and cause of a fire are determined. (NFPA 6.3.4.A, p 1074)
- Describe how to assist the fire investigators in the process of digging out the fire scene. (NFPA 6.3.4.B, p 1075–1076)
- Describe the types of evidence that may be uncovered at a fire scene. (NFPA 6.3.4.A, p 1076)
- Describe techniques for preserving fire-cause evidence. (NFPA 6.3.4.B, p 1076–1077)
- Describe the steps needed to secure a property. (NFPA 6.3.4, 6.3.4A, 6.3.4B, p 1083–1084)
- Explain the importance of protecting a fire scene to aid in cause determination. (NFPA 6.3.4, 6.3.4A, 6.3.4B, p 1083–1084)
- Describe the common motives of an arsonist. (p 1084–1086)

CHAPTER 38

Additional NFPA Standards

- NFPA 921, *Guide for Fire and Explosion Investigations*

Skills Objectives

There are no skill objectives for Fire Fighter II candidates. NFPA 1001 contains no Fire Fighter II Job Performance Requirements for this chapter.

You Are the Fire Fighter

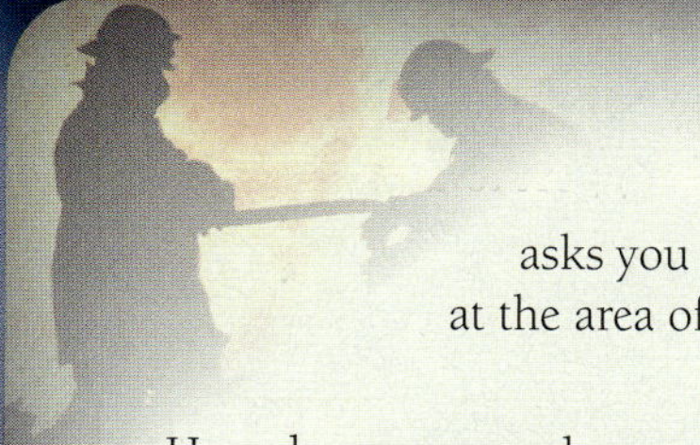

It is 2:00 a.m. on Saturday morning and you have just put out a fire in a church. After going through rehab, the Chief of your small department asks you to assist him as he begins an investigation. As he begins taking photos of the exterior of the building and making notes, he asks you to set up a fire line and once you are done with that, to get some tools to "dig it out" and meet him at the area of origin.

1. How do you properly secure a scene?
2. What tools would you bring to dig it out?
3. How would you know where the area of origin is located?

Introduction

Fire fighters usually reach a fire scene before a trained fire investigator arrives. As a consequence, fire fighters are able to observe important signs and patterns that the fire investigators can use in determining how and where the fire started. By identifying and preserving possible evidence, as well as recalling and reporting objective findings, fire fighters provide essential assistance to fire investigators. In some cases, the observations and actions of fire fighters could be significant in apprehending and convicting arsonists.

Determining the causes of fires allows fire departments to take steps to prevent future fires. For example, a fire department might develop an education program to reduce accidental fires, such as those caused when food is left unattended on a hot stove. A series of fires could point to a product defect, such as a design error in a chimney flue or the improper installation of chimney flues. Identifying fires that were intentionally set could lead to the arrest of the person responsible for the crimes.

Fire fighters must understand the basic principles of fire investigation and participate in taking care of this important departmental responsibility. Fire cause determination is often difficult, because important evidence may be consumed by the fire or destroyed during fire suppression operations or during salvage and overhaul. Fire investigators must rely on fire fighters to observe, capture, and retain information, as well as to preserve evidence until it can be examined. Evidence that is lost can never be replaced.

FIRE FIGHTER Tips

The preliminary investigation report will include names, dates, and witnesses. Make a mental note of what you witnessed—that is, who you saw, who opened or closed doors, and so on. Although the investigation report will be written by a trained fire investigator, your information is vital to its creation. A report is only as good as the information in it.

FIRE FIGHTER II Tips FFII

Fire departments can use fire cause data to support ordinances or create policies designed to eliminate major fires from occurring. For example, such data can lead to development of ordinances related to removing combustible landscape materials such as pine needles, removing construction materials from the outside of multiple-family buildings, or requiring the use of residential sprinklers.

Who Conducts Fire Investigations?

In most jurisdictions, the chief of the fire department has a legal responsibility to determine the causes of fires. In large fire departments, the fire chief usually delegates this responsibility to a special unit staffed with trained fire investigators. In smaller jurisdictions, one or two fire department members may be trained to conduct these investigations. Sometimes a special unit conducts fire investigations for several area fire departments, under an established policy that determines when those unit members should be called to conduct an investigation. In very small communities, the fire chief may personally examine the scene to determine the cause.

Many fire departments automatically dispatch a fire investigator to all working structure fires as well as to any other fire where the incident commander (IC) suspects something unusual. Other jurisdictions have policies that require the use of a fire investigator only when the damage exceeds a predetermined level or when the incident involves injuries or fatalities. In other departments, the IC may be expected to conduct a preliminary investigation and decide whether a fire investigator is needed. If the cause of the fire is evident and accidental, the IC would be responsible for gathering the information and filing the necessary reports. If the cause cannot be determined or appears to be intentional, the IC will summon a fire investigator.

When a fire investigator is not on the scene while the fire is being extinguished or overhauled, fire fighters must be particularly careful to preserve evidence. In such a situation, they serve as the eyes and ears of the fire investigator and must pass on their observations after extinguishing the fire. In some cases, fire fighters may need to be interviewed or to prepare a written report documenting their observations. These tasks are part of the fire fighter's role and responsibility in the fire investigation process.

Fire-cause investigation should not be confused with a criminal investigation of arson, which is the malicious burning of property with a criminal intent. The objective of a fire-cause investigation is to identify the cause of the fire; the fire investigator will never have a preconceived idea about what started the fire. The evidence and the facts will indicate whether the fire was started deliberately or resulted from an accidental cause. Only if the fire investigator determines that arson is the cause will a criminal investigation begin to find the person responsible.

FIRE FIGHTER II

Law Enforcement Authority

An arson investigation must determine not only the origin and cause of the fire, but also the person who was responsible for starting the fire and the sequence of events that led up to it. Determination of origin and cause is a small, yet important step in the overall investigation. To prove that a crime took place, fire investigators must establish a link from the fire to a particular origin and cause, eliminating any other cause as a possibility.

Whether the fire investigators have police powers and can conduct a criminal investigation depends on state and local laws. In some jurisdictions, fire investigators have police powers during fire investigations. In other areas, police officers are trained as fire investigators. Sometimes fire investigators determine the origin and cause of the fire but then turn the investigation over to a law enforcement agency if the fire is determined to be intentionally set. In other jurisdictions, fire department and police department personnel work together throughout the investigation. Successful fire investigations often depend on good working relationships between different organizations. You need to understand arrangements for fire investigation in your community.

Investigation Assistance

Most fire departments and government agencies have limited time, personnel, and resources available for fire investigations. To provide for more extensive investigations, a state fire marshal or similar authority may establish an investigations unit that concentrates on major incidents and supports local investigators on larger fires. Because these investigators cannot always reach the fire scene quickly, the local fire department must be prepared to conduct a thorough preliminary investigation and to protect the scene and preserve evidence.

Federal resources are also available for major investigations. The U.S. Bureau of Alcohol, Tobacco, Firearms and Explosives (ATF) has individual agents who can assist local jurisdictions with fire investigations. The ATF also maintains response teams spread across the country that respond to large-scale incidents in the United States. Each of these teams includes approximately 15 agents and uses equipment that ranges from simple tools for digging out a fire scene to laser-surveying devices for fully documenting the scene.

Insurance companies often investigate fires to determine the validity of a claim for damages or to identify factors that might help prevent future fires. The cost of an investigation is more than offset by the savings the company would realize by identifying a fraudulent claim. Some insurance companies employ their own fire investigators, whereas others retain the services of independent fire investigators. These outside fire investigators often have valuable experience and can provide critical technical support to determine the cause of a fire.

FIRE FIGHTER II

Causes of Fires

Every fire has a cause, which the fire investigator tries to uncover. An origin and cause investigation determines where, why, and how the fire originated. Some fires have simple causes that are easily identified and understood; others result from a complex set of circumstances that must be examined carefully to determine what actually happened. In some cases, the cause of a fire will never be determined with absolute certainty.

Every fire has a starting point where ignition occurs and fuel begins to burn. At this point of origin, an ignition source comes into contact with a fuel supply, such as a lighted match touching a piece of paper. To find the point of origin, fire investigators first search for the general area of origin and then attempt to locate the exact point of origin. The cause of the fire is the particular set of circumstances that brought the ignition source into contact with the fuel.

Fires result when a competent ignition source and a fuel come together for long enough to ignite. A competent ignition source must have enough heat energy to ignite the fuel and must remain in contact with the fuel until the fuel reaches its ignition temperature—a period that could range from a fraction of a second to hours, days, or even weeks.

A fire can be caused either by an act or by an omission. Igniting a piece of paper with a match is an act; leaving a pot of grease unattended on a hot stove is an omission. The cause of a fire can also be classified as either incendiary or accidental. Arson is an incendiary cause. According to NFPA 921, *Guide for Fire and Explosion Investigations*, arson is a crime of maliciously and intentionally, or recklessly, starting a fire or causing an explosion. By contrast, accidental fires do not involve a criminal or malicious intent, even if they are caused by human error or carelessness. Falling asleep with a lit cigarette, for example, would be considered an accidental cause.

Fire investigators always consider a fire to have an undetermined cause until the specific cause is established. The evidence from both incendiary and accidental fires can be very similar, so untrained individuals should not attempt to categorize fires as one or the other of these types. Instead, fire fighters should focus on helping to identify and preserve possible evidence for the fire investigator to examine.

FIRE FIGHTER Tips

NFPA 921, *Guide for Fire and Explosion Investigations*, contains information and guidance for fire investigators.

Fire Cause Statistics

According to the National Fire Protection Association (NFPA), 370,000 home structure fires were reported in the United States in 2011 (the latest year for which statistics are available). These fires caused 2520 fatalities and more than $6.9 billion in direct property damage.

Most structure fires occur in residential occupancies. The leading causes of fires in the home in 2010 are shown in TABLE 38-1.

TABLE 38-1	The Leading Causes of Residential Fires in the United States in 2011
Cooking	49.8%
Intentional	15%
Heating appliances	14%
Electrical malfunctions	14%
Smoking materials	5%

Source: U.S. Fire Administration.

Accidental Fire Causes

Accidental fires have hundreds of possible causes and involve multiple factors and circumstances. The most important reason for investigating and determining the causes of accidental fires is to prevent future fires. To reduce the number of fires, efforts must concentrate on the most frequent causes and those involving the greatest risks of death, injury, and property damage.

Most fires, fire deaths, and injuries occur in residential occupancies. The most commonly reported accidental causes of fire in these occupancies involve cooking, heating equipment, electrical equipment, and smoking. These statistics provide a foundation for fire prevention and public education efforts. Additional analysis can provide more specific information. For instance, fires caused by electrical equipment can be divided into four groups: those caused by worn-out or defective equipment, those caused by improper use of approved equipment, those caused by defective installations, and those caused by other accidents. A proper, thorough investigation of an electrical fire would identify and classify the specific cause within one of these groups.

For example, worn-out or defective electrical equipment that could cause fires would include deteriorating 50-year-old wiring circuits or a computer circuit board with an internal defect. Other worn-out equipment that might still be in use and cause fires includes electrical motors, switches, appliances, and extension cords. Properly used and maintained equipment that has been tested and listed by a recognized laboratory rarely causes a fire, but it should be replaced when it wears out.

Placing a portable heater too close to combustible materials or using a toaster oven to heat a container of flammable glue are examples of the improper use of electrical equipment. Defective installations are those not acceptable under electrical codes or printed instructions, such as using a light-duty extension cord to connect a heavy-duty appliance to a wall outlet. It is important to stress to the residents of a house the importance of reading and following the manufacturer's instructions when using any electrical device.

Some electrical fires result from accidental misuse or oversight, such as unintentionally leaving a cooking appliance turned on. Natural events can also cause accidental electrical fires, such as when a tree falls on a wire. A fire caused by an electrical overload or a short circuit can start wherever electricity is present, such as in the electrical panel, fuses, fuse boxes, circuit breakers, wiring, and appliances.

Sometimes it may be difficult or impossible to identify the specific source of ignition because the fire destroyed every trace of evidence. Fire investigators will classify these fires as having an undetermined cause, rather than making a "best guess." For example, low-temperature ignition can occur when wood is subjected to low heat, such as that generated by steam pipes or incandescent light bulbs, over a long period of time. In this scenario, the wood can gradually deteriorate and eventually ignite through a process called pyrolysis. If the resulting fire destroys the building, its point of origin would be difficult to identify.

Incendiary Fire Causes

People may set fires for several reasons and in many different ways. The same type of origin and cause investigation that is employed for an accidental fire is needed to identify where and how an incendiary fire started. With incendiary fires, it is particularly important to rule out possible accidental causes to prove beyond doubt that the fire was deliberately set.

A fire caused by arson requires a second phase of investigation to identify the person responsible. All of the evidence relating to the cause of the fire must be handled in a way that ensures it will be admissible as evidence in a criminal trial. A trained, qualified fire investigator should always be called to determine the cause of any fire that may have been deliberately set. Arson and the factors that could indicate an incendiary fire cause are discussed in more detail later in this chapter.

FIRE FIGHTER II

Determining the Origin and Cause of a Fire

According to NFPA 921, a scientific method and a systematic analysis are needed to determine the origin and cause of a fire. As part of his or her responsibilities, the investigator must determine where the fire started and how it was ignited. He or she must look at the situation objectively to be sure that the evidence is convincing and fully explains the situation. If more than one explanation for the observations exists, each possibility must be considered. The cause cannot be determined with absolute certainty until all alternative explanations have been ruled out.

Identifying the Point of Origin

One of the first steps in a fire investigation is identifying the point of origin. At this location, the fire investigator can look for clues indicating the specific cause of the fire.

The investigation process usually begins with an examination of the building's exterior. The fire investigator will look for indications that the fire originated outside the building before looking inside, and he or she will size up the building to identify important information. The overall size, construction, layout, and occupancy of the building will be noted, as well as the extent of damage that is visible from the exterior. The fire investigator will look for any openings that might have created drafts that influenced the fire spread and will examine the condition of outside utilities, such as the electrical power connection and gas meter.

The search for indications of the fire's point of origin continues inside the building, beginning with the area of lightest damage and moving systematically to the area of heaviest damage. The area of heaviest damage is probably the area that was burning for the longest time; areas with less damage were probably not as heavily involved in the fire or were involved for shorter periods.

The depth of char can be used to help determine how intensely and how long a fire burned in a particular location FIGURE 38-1. The depth of char is most closely related to the intensity of the fire at a particular location. In other words, an area that was exposed to a lower intensity of fire or heat impingement will display shallow charring. Conversely, areas that were exposed to higher intensities of heat and fire will display deeper charring and greater destruction of material. The depth of the char also reflects the amount of time the fuel remains in contact with the flame. Some research shows that an analysis of the depth of char is more reliable for evaluating fire spread than for establishing specific burn times or the intensity of heat from burning materials. Although charring may be deepest at the point of origin, the presence of ignitable liquids and combustible materials, as well as the effects of ventilation, can also influence charring.

Burn patterns and smoke residue can be helpful in identifying the point of origin, but—like depth of char—are not conclusive evidence. Burn patterns and damage will often spread outward from the room or area where the damage is most severe. Because heat rises, the flow of heated gases from a fire will often be up and out from the point of origin. This upward, outward flow can usually be recognized, even when all of the building's contents were involved in the fire. Often the point of origin is found directly below the most damaged area on the ceiling, where the heat of the fire was most intensely concentrated.

FIGURE 38-1 Depth of char.

A charred V, hourglass, or U-shaped burn pattern on a wall indicates that fire spread up and out from something at the base. The best place to start looking for a specific fire cause in a room with a burn pattern on the wall, a pile of charred debris at the base, and minimal damage to the other room contents is in the pile of debris FIGURE 38-2.

FIGURE 38-2 Often the point of a burn pattern is near or at the point of origin.

An experienced fire investigator knows that many factors can influence burn patterns, including ventilation, fire suppression efforts, and the burning materials themselves. A burn pattern on a wall could indicate that an easily ignited and intensely burning fuel source was present at that location. It might also indicate that something fell from a higher level and burned on the floor. Such patterns may be caused by direct flame contact, heat, smoke, soot, or a combination of these elements. The patterns produced by the fire may be apparent on any surface that displays damage indicating a movement from the point of origin outward. Examples include wall and ceiling surfaces, appliances and furnishings, doors and windows, and even fire victims.

An inverted cone may be present when short-duration fires do not spread horizontally from the point of origin. An inverted cone on a wall may also indicate that an ignitable liquid was spread along the base of the wall as someone set the fire intentionally.

Once the fire investigator identifies the exact or approximate location of the point of origin, the search for indications of a specific cause can begin. The fire investigator must determine what happened at that location to cause the fire. This effort involves identifying both the source of ignition and the fuels that were involved. The fire investigator must ultimately determine how the source of ignition and fuel came together, whether accidentally or intentionally.

Digging Out

FIRE FIGHTER II

After the fire investigator identifies the area of origin, fire fighters could be asked to assist in digging out the fire scene. "Digging out" is a term used to describe the process of carefully looking for evidence within the debris. Sometimes the entire

FIRE FIGHTER II

fire scene must be closely examined to determine the cause of the fire and gather evidence.

The fire investigator will take extensive photographs of the fire scene as it first appears. He or she will then begin to remove and inspect the debris, layer by layer, working from the top of the pile down to the bottom. The type of evidence uncovered and its location within the layers of debris can provide important indications of how the fire originated and progressed.

Removing and inspecting the layers of debris enables the fire investigator to determine the sequence in which items burned, whether an item burned from the top down or from the bottom up, and how long it burned. Did the fire start at a low point and burn up, or did burning items fall down from above and ignite combustible materials below? Did the fire spread along the ceiling or along the floor? Is a residue of rags containing an ignitable liquid found under the furniture? Why are papers that should have been in a metal filing cabinet stacked on the floor and partially burned?

Systematically digging out through the debris often can uncover the exact point of origin and point to the cause of both accidental and deliberate fires. If circumstances or eyewitness accounts indicate that a deliberate fire is a possibility, the fire investigator may request that fire fighters help examine the entire area. Common search methods include the "grid" search (also known as the "double strip" search) and the "strip" search. The investigator will explain generally what to look for, how to search, and what to do with any potential evidence.

When you find possible evidence, stop and inform the fire investigator so that he or she can examine it in place. It is the fire investigator's job to document, photograph, and remove any potential evidence, whether or not it supports the suspected cause.

To determine the origin and cause of a fire, the fire investigator must evaluate all potential causes of the fire. The process of eliminating alternative causes and documenting the reasons for rejecting them is as important as properly documenting the ultimate cause of the fire. For example, if the fire started in the kitchen, potential causes could include these:

- The stove
- Any of the electrical appliances
- Light fixtures
- Smoking materials
- Cleaning supplies
- Other factors, such as candles

Collecting and Processing Evidence

Evidence refers to all of the information gathered and used by a fire investigator in determining the cause of a fire. Evidence can be used in a legal process to establish a fact or prove a point. To be admissible in court, it must be gathered and processed under strict procedures.

Types of Evidence

Physical evidence consists of items that can be observed, photographed, measured, collected, examined in a laboratory, and presented in court to prove or demonstrate a point FIGURE 38-3. Fire investigators can gather physical evidence at the fire scene,

FIRE FIGHTER II

FIGURE 38-3 Physical evidence at a fire scene.

such as a burn pattern on a wall or an empty gasoline can, to explain how the fire started or to document how it burned.

Trace (transfer) evidence consists of a minute quantity of physical evidence that is conveyed from one place to another. For example, a suspect's clothing may contain the residue of the same ignitable liquid found at the scene of a fire.

Demonstrative evidence is anything that can be used to validate a theory or to show how something could have occurred. To demonstrate how a fire could spread, for example, a fire investigator might use a computer model of the burned building.

Evidence used in court can be classified as either direct or circumstantial. Direct evidence includes facts that can be observed or reported first-hand. Testimony from an eyewitness who saw a person actually ignite a fire and a videotape from a security camera showing the person starting the fire are examples of direct evidence. Whereas direct evidence is rare in arson cases, eyewitnesses often can describe the circumstances that led to an accidental fire.

Circumstantial evidence is information that can be used to prove a theory, based on facts that were observed directly. For example, an investigation might show that the gasoline from a container found at the scene was used to start a fire. Two different witnesses might testify that the suspect purchased a container of gasoline before the fire and walked away from the fire scene without the container a few minutes before the fire department arrived. Such circumstantial evidence clearly places the suspect at the fire site with an ignition source at the time the fire started. Fire investigators must often work with circumstantial evidence to connect an arson suspect to a fire.

Preservation of Evidence

Fire fighters have a responsibility to preserve evidence that could indicate the cause or point of origin of a fire. Fire fighters who discover something that could be evidence while digging out a fire scene or performing other activities should leave it in place, make sure that no one interferes with the item or the surrounding area, and notify a fire officer or fire investigator immediately.

Evidence is most often found during the salvage and overhaul phases of a fire. Salvage and overhaul should always be performed carefully and can often be delayed until a fire investigator has examined the scene. Do not move debris any more than is absolutely necessary, and never discard debris until the fire investigator gives his or her approval to do so. It is the fire investigator's job to decide whether the evidence is relevant, not the fire fighter's.

Fire fighters at the scene are not in the position to decide whether the evidence they find will be admissible in court and, therefore, is worthy of preservation; that is also the fire investigator's decision. As a fire fighter, your responsibility is to make sure that potential evidence is not destroyed or lost. Too much evidence is better than too little, so no piece of potential evidence should be considered insignificant.

If evidence could be damaged or destroyed during fire suppression activities, cover it with a salvage cover or some other type of protection, such as a garbage can. Use barrier tape to keep others from accidentally walking through evidence. These methods may prove to be ineffective, however, if no indication of their purpose is given. For this reason, if evidence is at risk of being damaged or altered by fire suppression activities, it should be attended to by personnel or recovered. Before moving an object to protect it from damage, be sure that witnesses are present, that a location sketch is drawn, and that a photograph is taken.

Evidence should not be contaminated (i.e., altered from its original state) in any way. Fire investigators use special containers to store evidence and prevent contamination from any other products.

FIRE FIGHTER Tips

Gasoline-powered tools and equipment, including saws and generators, are often used during firefighting operations. Gasoline from such equipment could contaminate the area and lead to an erroneous assumption that accelerants (materials used to initiate or increase the spread of fire) were used in the fire. Limit the use of gasoline-powered equipment inside the structure, particularly in the area of the fire's origin. At a minimum, refuel such equipment completely outside the investigation area and the structure itself to prevent contamination of the site by spilled fuel.

FIRE FIGHTER Tips

To avoid contaminating evidence, fire investigators always wash their tools between taking samples. This step ensures that material from one piece of evidence will not be unintentionally transferred to another piece. Fire investigators also change their gloves each time they take a sample of evidence and place evidence only in absolutely clean containers. Before entering the fire scene, fire investigators often wash their boots to keep from transporting any contamination into the fire scene.

FIRE FIGHTER Tips

Ensure that your personal protective equipment (PPE) does not retain flammable liquids. Wash your PPE anytime it has come in contact with a flammable liquid or smoke.

Chain of Custody

To be admissible in a court of law, physical evidence must be handled according to certain prescribed standards. Because the cause of the fire may not be known when evidence is first collected, all evidence should be handled according to the same procedure. In an incendiary or arson fire, evidence relating to the cause of the fire will probably have to be presented in court. The same is true for accidental fires, because lawsuits might be filed to claim or recover damages.

Chain of custody (also known as *chain of evidence* or *chain of possession*) is a legal term that describes the process of maintaining continuous possession and control of the evidence from the time it is discovered until it is presented in court. Every step in the capture, movement, storage, and examination of the evidence must be properly documented. For example, if a gasoline can is found in the debris of a fire suspected to involve arson, documentation must record the person who found the can, along with where and when the can was found. Photographs should be taken to show the site where the gasoline can was found and its condition. In court, the fire investigator must be able to show that the gas can presented is the same can that was found at the fire site.

The person who takes initial possession of the evidence must keep it under his or her personal control until the item is turned over to another official. Each successive transfer of possession must be properly recorded. If evidence is stored, documentation must indicate where and when it was placed in storage, whether the storage location was secure, and when the evidence was removed from that location. Often, evidence is maintained in a secured evidence locker to ensure that only authorized personnel have access to it.

Everyone who had possession of the evidence must be able to attest that it has not been contaminated, damaged, or changed in any way. If evidence is examined in a laboratory, the laboratory tests must be documented. The documentation for chain of custody must establish that the evidence was never out of the control of the responsible agency and that no one could have tampered with it.

Fire fighters are frequently the first link in the chain of custody. The fire fighter's responsibility in protecting the integrity of this chain is relatively simple: Report everything to a supervisor and disturb nothing needlessly. The individual who finds the evidence should remain with it until the material is turned over to a company officer or to the fire investigator FIGURE 38-4. Remember—you could be called as a witness to state that you are the fire fighter who discovered this particular piece of evidence.

FIGURE 38-4 Evidence should remain where you find it until you can turn it over to your fire officer or the fire investigator.

The fire investigator's standard operating procedures (SOPs) for collecting and processing evidence generally include the following steps:

- Take photographs of each piece of evidence as it is found and collected. If possible, photograph the item exactly as it was found, before it is moved or disturbed.
- On the fire scene, sketch, mark, and label the location of the evidence. Sketch the scene as near to scale as possible.
- Place evidence in appropriate containers to ensure its safety and prevent contamination. Unused paint cans with lids that automatically seal when closed are good containers for transporting evidence. Glass mason jars sealed with a sturdy sealing tape are appropriate for transporting smaller quantities of materials. Plastic containers and plastic bags should not be used to hold evidence containing petroleum products, as these chemicals may lead to deterioration of the plastic. Paper bags can be used for collecting dry clothing or metal articles, matches, or papers. Soak up small quantities of liquids with either a sterile cellulose sponge or sterile cotton batting. Protect partially burned paper and ash by placing them between layers of glass.
- Tag all evidence at the fire scene. The container for evidence being transported to the laboratory should be labeled with the date, time, location, discoverer's name, and witnesses' names.
- Record the time when the evidence was found, the location where it was found, and the name of the person who found it. Keep a record of each person who handled the evidence.
- Keep a constant watch on the evidence until it can be stored in a secure location. Evidence that must be moved temporarily should be put in a secure place that is accessible only to authorized personnel.
- Preserve the chain of custody in handling all the evidence. A broken chain of custody may result in a court ruling that the evidence is inadmissible.

Only one person should be responsible for collecting and taking custody of all evidence at a fire scene, no matter who discovers it. If someone other than the assigned evidence collector must seize the evidence, that person must photograph, mark, and contain the evidence properly and turn it over to the evidence collector as soon as possible. The evidence collector must also document all evidence that is collected, including the date, time, and location of discovery; the name of the finder; the known or suspected nature of the evidence; and the evidence number from the evidence tag or label. A log of all photographs taken should be recorded at the scene as well.

Identifying Witnesses

Although fire fighters may not interview witnesses, you can identify potential witnesses to the fire investigator. People who were on the scene when fire fighters arrived could have invaluable information about the fire. If a fire fighter learns something that might be related to the cause of the fire, he or she should pass this information on to a supervisor or to the fire investigator.

Interviews with witnesses should be conducted by the fire investigator or by a police officer. If the fire investigator is not on the scene or does not have the opportunity to interview the witness, the fire fighter should obtain the witness's name, address, and telephone number and give it to the fire investigator. A witness who leaves the scene without providing this information could be difficult or impossible to locate later.

Fire fighters have a primary responsibility to save lives and property. Until the fire is under control, they must concentrate on fighting the fire, not investigating its cause. Even so, fire fighters should pay attention to the situation and make mental notes about any observations. They must tell the fire investigator about any odd or unusual happenings. Information, suspicions, or theories about the fire should be shared only with the fire investigator and only in private.

Do not make any statements of accusation, personal opinion, or probable cause to anyone other than the fire investigator. Comments that are overheard by the property owner, the occupant, a news reporter, or a bystander can impede the efforts of the fire investigator to obtain complete and accurate information. A witness who is trying to be helpful might report an overheard comment as a personal observation. In this way, inaccurate information can generate a rumor, which then becomes a theory, which then turns into a reported "fact" as it passes from person to person.

VOICES OF EXPERIENCE

A few years ago, while I was assigned to a fire apparatus for the Forsyth County Fire Department in North Carolina, we were dispatched to a working house fire within a small community of our county jurisdiction. Once on the fire scene, we located a female who identified herself as the occupant of the burning dwelling. She stated that her son had attempted to injure her by setting fire to her home and then tried to set her on fire as well. Even though the female occupant stated that the son had left the scene, the alert incident commander felt the need to have law enforcement respond as we extinguished the fire.

Once we completed our task of extinguishing the fire and searching the structure, I exited the front door of the structure. While exiting, I saw a young man running across the yard towards the structure. I tried to get his attention, but he ignored me and continued toward the basement door. As the young man approached the structure, he picked up a gasoline container stored near the basement door. After grabbing the gasoline container, the young man entered the basement. While fire fighters were working above, the young man emptied the contents of the container onto the basement floor carpeting.

The young man then pulled a lighter from his pocket and was in the process of igniting the materials when several of us stopped him. After we removed the young man from the dwelling, law enforcement personnel arrived to secure the young man.

There were several lessons learned that day. Always consider the fire scene as a potential crime scene until proven otherwise. Your fire department should have policies in place to contact law enforcement authorities upon the confirmation of any incident. Both will aid in the successful outcome of the incident and increase the level of safety for all of those involved.

In most cases law enforcement personnel may only direct traffic or secure the perimeter of a fire scene. In those rare cases, like the one we experienced that day, you may need immediate law enforcement assistance to prevent fire fighter injuries. After all, anything can happen at any given time on the fire scene.

Gary Styers
Fire Marshal
Mooresville Fire-Rescue

Never make jesting remarks or jokes at the scene FIGURE 38-5. Careless, unauthorized, or premature remarks could embarrass the fire department. Statements to news reporters about the fire's cause should be made only by an official spokesperson after the fire investigator and ranking fire officer have agreed on their accuracy and validity. Until then, "The fire is under investigation" is a sufficient reply to any questions concerning the cause of the fire.

FIGURE 38-5 Jesting remarks and jokes should never be made at the scene, because your comments could be overheard by the media and other bystanders.

Observations During Fire-Ground Operations

Although a fire fighter's primary concern is saving lives and property, you will inevitably make observations and gather information as you perform your duties. What you observe could prove significant in the subsequent investigation of the incident. Remember, the fire investigator is the only person who can determine if something is suspicious and needs to be investigated further.

Dispatch and Response

During dispatch and response, form a mental image of the scene you expect to encounter. Note the time of day, the weather conditions, and any route obstructions. As the incident progresses, pay attention to things that do not match your expectations—they could help the fire investigator determine the origin and cause of the fire.

Time of Day

Time of day and type of occupancy can indicate the number and type of people at an incident. Offices, stores, and other business places are filled with people during the day but are mostly empty at night. Conversely, restaurants and nightclubs are likely to be crowded after dark. Residential occupancies are more likely to be occupied at night and on weekends and holidays.

As victims evacuate the building, some may stand out, either because they are fully dressed when everyone else is in pajamas, or because their behavior and demeanor are quite different from those of the other victims.

Weather Conditions

Note whether the day is hot, cold, cloudy, or clear, and whether conditions in the burning structure match the weather. On a cold day, windows should be closed; on a hot day, the furnace should be off.

Lightning, heavy snow, ice, flooding, fog, or other hazardous conditions can help cover an arsonist's activities because they delay the fire department's arrival and make a fire fighter's job more difficult. Because wind direction and velocity help determine the natural path of fire spread, being aware of these conditions will help to determine if the fire behaved in an unnatural way.

Route Obstructions

Unusual traffic patterns or barriers blocking the route to the scene may be early indications of an incendiary fire. Be sure to note these and any other obstructions, such as barricades, felled trees, downed cables, or trash containers that cause delays.

Arrival and Size-Up

Size-up operations can provide valuable information for fire investigators. Pay attention to the fire conditions, the building characteristics, positions of doors and windows, and any vehicles and people at or leaving the scene.

Description of the Fire

The fire investigator will compare the dispatcher's description with the actual fire conditions. If the fire has intensified dramatically in a short time, an accelerant could have been used. During fire suppression operations, note whether flames are visible or whether only smoke is apparent. Also observe the quantity, color, and source of the smoke. Does the fire appear to be burning in one place or in multiple locations?

People on the Scene

The appearance and behavior of people and vehicles at the scene of a fire can provide valuable clues. Note the attitude and dress of the owner and/or the occupants of the building, as well as any other individuals at the scene of a fire. Normally, the owner or occupants of a building will be distressed and wearing clothing appropriate to the time of day. For example, at 4 a.m., you would expect to see residents in pajamas.

Anyone who seems out of place at a fire or someone who has been observed at several fires in various locations should be reported to the fire investigator. Some arsonists are emotionally disturbed individuals who receive personal satisfaction in watching a "working" fire. Sometimes fire investigators will photograph the crowd at a fire scene, particularly if a series of similar fires has occurred recently, and look for the same faces at different incidents.

Most people at a fire scene are serious and intent on watching the drama unfold FIGURE 38-6. Someone who is talking loudly, laughing, or making light of the situation should be considered suspicious. Fire investigators may also want to know about someone who eagerly volunteered multiple theories or too much information.

FIGURE 38-6 Most persons at the fire scene are intent on watching the fire fighters at work.

Unusual Items or Conditions

Always note any unusual items or conditions about the property, such as whether windows and doors are open or closed. Arsonists commonly draw the shades or cover windows and doors with blankets to delay the discovery of a fire. Gasoline cans, forcible-entry tools, and a damaged hydrant or sprinkler connection might all suggest an intentionally set fire.

Entry

As you prepare to enter the burning structure, note if you observe any evidence of prior entry, such as shoeprints leading into or out of the structure or tracks from vehicle tires. Note whether the windows and doors are intact, whether they are locked or unlocked, and whether any unusual barriers limit access to the structure. Also note any signs of forced entry by others.

Forced entry could leave impressions of tools on the windows and doors. Likewise, cut or torn edges of wood, metal, or glass may indicate forced entry. Fire investigators might be able to determine whether the glass had been broken by heat or by mechanical means. Note if you observe any signs of a burglary, too; the fire may have been set to destroy evidence of another crime.

FIRE FIGHTER Tips

Photographs—whether taken by film or digital cameras—can document conditions before the arrival of fire investigators, but specific procedures must be followed for these photographs to be admissible in court as evidence. Follow your department's SOPs regarding picture taking.

Search and Rescue

As you enter the building to perform search and rescue or interior fire suppression activities, consider the location and extent of the fire. Be especially alert for signs of multiple fires in a building. Arsonists often set more than one fire in an attempt to ensure the destruction of a building. The presence of multiple fires increases the risk of injury and death both to the building's occupants and to responding fire fighters.

The location and condition of the building's contents may provide clues for the fire investigator as well. Unusual building contents or conditions—such as barriers in doors and windows, the absence of inventory from a warehouse, or insurance papers or deeds left out on a desk—should be noted to the fire investigator.

The team responsible for shutting off electric power to the building should identify whether the circuit breakers were on or off when they arrived. The location of any people found in the building should be noted as well. When cutting off electrical panels, try to cut off main breakers and do not disturb breakers that are "tripped."

Ventilation

The ventilation crew should note whether the windows and doors of the building were open or closed, locked or unlocked. They should also note the color and quantity of the smoke as well as the presence of any unusual odors.

Color of Smoke

The color of smoke sometimes indicates which kind of material is burning. Unusual smoke might indicate that more fuel was added to the mix.

Unusual Odors

Self-contained breathing apparatus protects fire fighters from hazardous fumes and toxic odors. Nevertheless, sometimes an odor is so strong that it can be detected en route or lingers after the fire has been extinguished. Cooking fires and overheated light ballasts produce distinctive, identifiable odors that are highly familiar to experienced fire fighters. Other common odors familiar to fire fighters include gasoline, kerosene, paint thinner, lacquers, turpentine, linseed oil, furniture polish, or natural gas.

Odors often linger in the soil under a building without a basement, particularly if an accelerant has been used and if the ground is wet. Concrete, brick, and plaster will all retain vapors after a fire has been extinguished. These observations should be passed along to the fire investigator.

Effects of Ventilation

Fire department ventilation operations can dramatically influence the behavior of a fire and alter burn patterns. Certain ventilation situations can cause a fire to become more involved. Relay information about these operations to the fire investigator so that he or she can correctly interpret burn patterns.

Suppression

Fire behavior, the presence of incendiary devices (materials used to start a fire), obstacles encountered during fire suppression operations, and charring and burn patterns are among the factors that might help the fire investigator determine the origin and cause of the blaze.

Behavior of Fire

During the fire attack, observe the behavior of the fire and note how it reacts when an extinguishing agent is applied. Look for unusual flame colors, sounds, or reactions. For example, most ignitable liquids will float, continue to burn, and spread the fire when water is applied to them. Rekindled fire in the same area or a flare-up when water is applied could indicate the presence of an ignitable liquid. Fires set in multiple locations and accelerants present an increased risk to you and your fellow fire fighters during suppression activities.

Incendiary Devices, Trailers, and Accelerants

While fighting the fire, be aware of streamers or trailers (combustible materials positioned to spread the fire). Note if there are any combustible materials such as wood, paper, or rags in unusual locations. Note containers of ignitable liquids that are normally not found in the type of occupancy. Very intense heat or rapid fire spread might indicate the use of an accelerant to increase the fire spread.

Incendiary devices can include unusual items in unlikely places, such as a packet of matches tied to a bundle of combustible fibers or attached to a mechanical device. Often, incendiary devices fail to ignite or burn out without igniting other materials.

Condition of Fire Alarm or Suppression Systems

If the building is equipped with a fire alarm or fire suppression system, fire fighters should note whether the system operated properly or was disabled. Note if the OS&Y valve was open or closed in an occupied structure. If the fire suppression system failed to work, the culprit could be poor maintenance or deliberate tampering. Notify the fire investigator of any findings. Removal of the batteries from a smoke detector is a common way that an arsonist attempts to delay discovery of a fire.

Obstacles

An arsonist may place obstacles to hinder the efforts of fire fighters. Note whether any furniture was obviously moved to block entry. Arsonists also may prop open fire doors, pull down plaster to expose the wood structure, or punch holes in walls and ceilings to increase the rate of fire spread. In such circumstances, be alert for unexpected openings in floors.

Fire Fighter Safety Tips

If you encounter an incendiary device that has not ignited, notify your fire officer or fire investigator in the area immediately! Do not make any attempt to move or disable the device.

Contents

Fire fighters who are involved in interior fire suppression, like the members of the search and rescue team, should make note of anything unusual about the contents of the building. The absence of personal items in a residence may indicate that they were removed before the incident and that the fire was intentionally set. Empty boxes in a warehouse may belong there, or they may indicate that valuable contents were removed prior to the fire. If only one item or a particular stock of products burns, and there is no reasonable explanation for the fire, arson could be suspected. Sometimes the contents will be removed and replaced by less expensive or nonfunctioning items. Neighbors may report that they observed items being removed or replaced prior to the fire. Check whether appliances are plugged in at the time of the fire.

In some cases of arson, clothing or other valuables may be placed in areas of a structure considered to be protected from the fire. In addition, the arsonist may place clothing in vinyl or plastic bags to reduce its absorption of smoke. Valuable documents or items may be placed under floors, in crawl spaces, or in detached garages. The fire investigator will consider these factors with other evidence before reaching any conclusion.

Charring and Burn Patterns

Charring in unusual places—such as open floor space away from any likely accidental ignition source—could indicate that the fire was deliberately set. Char on the underside of doors or on the underside of a low horizontal surface, such as a tabletop, could indicate that a pool of an ignitable liquid was placed below the item.

FIRE FIGHTER Tips

As a result of the *Michigan v. Tyler* court decision, fire investigators are allowed to remain at a fire for a reasonable time without a search warrant to determine the cause of the fire. Care should be taken to secure only evidence that is in plain view. Searching for evidence in drawers and closets where there is no reasonable likelihood of evidence would violate the spirit of this decision.

Normally, once fire investigators leave the premises, they should obtain a search warrant prior to reentering the site. Failure to do so may mean that any evidence gathered in later searches is inadmissible in court.

Overhaul

During overhaul, the smoke and steam should begin to dissipate, enabling both fire fighters and fire investigators to get a better look at the surroundings. Fire fighters should continue to look for the signs, patterns, and evidence previously discussed while conducting overhaul in a way that allows evidence to be identified and preserved. The overhaul process, if not done carefully, can quickly destroy valuable evidence.

If possible, the fire investigator will take a good look at an area before overhaul begins. He or she can often quickly identify potential evidence and can help direct or guide the overhaul operation so that this evidence is properly preserved. Evidence located during overhaul should be left where it is found, untouched and undisturbed, until the fire investigator examines it. Evidence that must be removed from the scene should be properly identified, documented, photographed, packaged, and placed in a secure location.

Fire suppression personnel and fire investigators must work as a team to ensure that the fire is completely extinguished and properly overhauled while continually searching for and preserving signs, patterns, and evidence. Taking photographs during this phase of the fire operation is a good idea.

In short, be careful not to destroy evidence during overhaul. Avoid throwing materials into a pile. Use low-velocity hose streams to avoid breaking up and scattering debris needlessly. Thermal imaging devices can be used to find hot spots without tearing apart the interior structure.

Watch for evidence that was shielded from the fire and is lying beneath burned debris. For example, a wall clock may have fallen during the fire and been covered with debris. If the clock is near the area of origin and stopped approximately when the fire broke out, it could be an important piece of evidence.

A fire investigator will always try to determine whether the building contents were changed or removed prior to a fire. If all of a business's computers are missing, for example, the fire might have been set to conceal a theft. In other cases, missing portraits or other sentimental items might have been removed by the building's owner before setting the fire.

The cause of wildland and vehicle fires should be investigated as well, although the evidence collected at these fires may be somewhat different from the evidence gathered at a building fire. An accidental wildland fire might have started from an untended campfire, for example, whereas intentional wildland fires can be started from a book of matches or other incendiary device.

Automobiles, recreational vehicles, and boats are common targets for arson. Some people believe that starting a fire in a vehicle or boat will destroy evidence of an incendiary fire. Fire investigators, however, will gather much evidence about the area of origin and seek to establish whether the fire was accidental or intentional. Take care at the scene of a vehicle fire to preserve the scene as much as possible.

A vehicle fire may also start from an accidental electrical short or from a combination of gasoline and matches placed in the passenger compartment. Signs of spilled fuel on the ground around a vehicle may be evidence of an intentional fire. Do not move the car without documenting this spilled fuel; otherwise, the evidence will be lost.

FIRE FIGHTER Tips

Some of the best friends of fire investigators have four legs. Specially trained dogs are invaluable in searching for accelerants.

FIRE FIGHTER II Tips FFII

While looking for hidden fires, leave as much of the room of origin intact by breeching walls on the opposite side of the burned areas from the room of origin. This approach will leave valuable burn patterns intact and aid the fire investigator in his or her scene examination.

Injuries and Fatalities

Any fire that results in an injury or fatality must be thoroughly investigated and the fire scene documented. A search and rescue operation should never be compromised, but it is important to document the location and position of any victims, especially in relation to the fire and the exits.

Clothing removed from any victim should be preserved as evidence. It may contain traces of ignitable liquids, and burn patterns on the clothing can indicate the fire flow. If the clothing is removed in the ambulance or at the hospital, personnel should be assigned to collect it and keep this evidence as intact as possible.

Document anything lying under a victim's body after it is removed, because this is often a protected area and may reveal important evidence. Victims who do not escape the fire should not be disturbed, and no attempts should be made to try to identify them. Fire investigators will work with other investigators, including the coroner or medical examiner, to examine the victims and the surrounding area for items of evidentiary value.

FIRE FIGHTER II

Securing and Transferring the Property

Fire departments have the legal authority to maintain control of a building after a fire to determine the cause of the fire. Maintaining site integrity is critical to the fire investigation. To meet this requirement, the building and premises must be properly secured and guarded until the fire investigator has finished gathering evidence and documenting the fire scene. This could be accomplished by posting fire fighters or police personnel. Otherwise, any efforts to determine the cause of a malicious or incendiary fire, no matter how efficient or complete, may be wasted.

Fire fighters must be aware of any state or local laws pertaining to right of access by the owners or occupants while the property is under the control of the fire department. The fire department should keep a log of the name of anyone who enters the property, the times of his or her entry and departure, and a description of any items taken from the scene. After receiving permission from the fire investigator, a fire officer or a fire fighter should accompany anyone who enters the premises for any reason until the scene is released.

If a fire investigator is not immediately available, the premises should be guarded and maintained under the control of the fire department until the investigation takes place and all evidence is collected. In the interim, take the following steps:

- Suspend salvage and overhaul, and secure the scene. Keep nonessential personnel out of the area. Deny entry to all unauthorized and unnecessary persons.
- Photograph the fire scene extensively. Start with the area with the least amount of damage, and work your way toward the area of possible origin. Take several pictures of the point of origin from various angles. Photograph any incendiary devices on the premises exactly where they were found.
- If weather, traffic, or other factors could destroy evidence, take steps to preserve these items in the best way possible. Protect tire tracks or footprints by placing boxes over them to prevent dust accumulation. Use barricades to block off the area to further traffic. Rope off areas surrounding plants, trailers, and devices, and post a guard at the site.

To secure the property, cordon off the area with fire- or police-line tape. A member of the fire department or law enforcement agency should remain at the scene to ensure

FIRE FIGHTER II

that no unauthorized persons cross the line. As previously mentioned, the fire department has the authority to deny access to a building for as long as necessary to conduct a thorough investigation and ensure the safety of the public.

Before leaving the scene, make sure that the building is properly secured and that no hazards to public safety exist. Shut off all utilities and seal any openings in the roof to prevent additional water damage. Board up and secure windows and doors to prevent unauthorized entry.

Fire departments can secure and protect the premises in several ways. Lock and guard gates if necessary. Rope off dangerous areas and mark them with signs. Most fire departments have contracts with local companies that provide 24-hour board-up services, and they sometimes hire private security guards to secure a property after a fire.

Eventually, when fire department operations are over, the property will be returned to the owner. This handover should not occur until the investigation is complete and all evidence is collected. The fire department's authority ends when the property is formally released to the property owner. Afterward, the fire investigator might need a search warrant or written consent to search the site, and unguarded evidence could be contaminated by subsequent visitors to the property.

Incendiary Fires

The term "incendiary fires" refers to all fires that are deliberately started for malicious or criminal intent. In many jurisdictions, arson has a narrower, specific legal definition. This chapter uses the term "arson" to mean the malicious burning of property with criminal intent.

Fire fighters must be aware of factors that could indicate an intentionally set fire, report any observations to a supervisor or fire investigator, and help protect evidence.

Indications of Arson

Arson fires have several distinct, recognizable patterns or indications. For example, a deliberate fire might have multiple points of origin or include multiple simultaneous fires. An arsonist may ignite fires in many different areas so as to create a large fire as quickly as possible and involve the entire building. Arsonists may also use trailers made from combustible materials, such as paper, rags, clothing, curtains, kerosene-soaked rope, or other fuels, to enhance the fire's spread. Such trailers often leave distinctive char and burn patterns that fire investigators can use to trace the spread of a fire.

An incendiary device is a device or mechanism, such as a candle, timer, or electrical heater, that is used to start a fire or explosion. Many incendiary devices leave behind valuable evidence—for example, metal parts, electrical components, or mechanical devices—at the point of origin. Often an arsonist will use more than one incendiary device; sometimes, a failed incendiary device is found at the fire scene. If trailers were used, investigators may be able to trace their burn patterns back to an incendiary device.

Evidence consisting of an ignitable liquid often indicates the presence of an incendiary fire but does not necessarily establish arson as the cause: An accident could also explain the presence of an ignitable liquid spill. To prove arson, the fire investigator must determine that an ignitable liquid was used and that no explanation other than arson justifies its presence.

Extensive burn damage on a floor's surface could indicate that an ignitable liquid was poured and ignited. For example, a wood floor might demonstrate a pool-shaped burn pattern, or floor tiles could be discolored or blistered with an irregular burn pattern. On concrete and masonry floors, an ignitable liquid pattern is usually irregular and consists of various shades of gray to black. Liquids flow to the lowest level possible, so pooled liquids could be visible in corners and along the base of walls, where low levels of charring might also be apparent.

Sometimes the first indications of arson are entirely circumstantial. For example, a particular fire might fit an existing pattern, such as a series of fires in the same area, at about the same time on the same days of the week. Alternatively, a series of fires might occur in the same type of business or in properties with the same owner. A fire fighter who notices a pattern or a set of similar circumstances should immediately report that observation to a supervisor or a fire investigator.

Cause Determination

Fire investigations involve much more than simply determining what caused the fire, where it started, and whether it was accidental or incendiary (intentionally set). Many fire investigations also examine many other aspects of a fire incident. If the fire results in injuries or fatalities, the investigation often will study every factor that could have contributed to or prevented those losses. In a fire that causes millions of dollars in damage, a whole range of factors could contribute to the large loss. Fire investigators typically examine the construction and contents of the building to determine whether fire and building codes were followed, if the built-in fire protection systems functioned properly, or if lessons from the fire could be applied by amending codes or inspection procedures. A fire investigation can teach many lessons, even if the specific cause and origin of the fire are never determined.

Arsonists

An arsonist is a person who deliberately sets a fire with criminal intent. Arsonists fall into several categories, with various explanations for their behavior. The fire service has identified two groups who are responsible for a large number of fires: pyromaniacs and juvenile fire-setters. Many other arsonists start fires for a wide range of motives.

Pyromaniacs

A pyromaniac is a pathological fire-setter. Most are adult males, and they are often loners. These individuals are usually introverted, are polite but timid, and have difficulty relating to other people. The fires set by pyromaniacs have the following characteristics:

- Fires are set in easily accessible locations, such as immediately inside entrances, on basement stairs, in trash bins, or on porches.

FIRE FIGHTER II

- Fires are set in structures such as occupied residences of all types, barns, and vacant buildings.
- Accelerants are rarely used. The pyromaniac is impulsive, so materials readily at hand are used.
- Each pyromaniac follows a unique pattern—for example, setting fires at the same time of day or night, using the same method, and in similar locations.

Juvenile Fire-Setters

Juvenile fire-setters are usually divided into three groups according to age: 8 years old and younger, 9 to 12 years old (preadolescent), and 13 to 17 years old (adolescent). Preadolescent and adolescent fire-setters often exhibit the same personality traits as adult pyromaniacs: They are introverted, they have difficulty with interpersonal relationships, and they are extremely polite when questioned.

Children younger than 8 years old are seldom criminally motivated when they set fires; they usually are just curious and experimenting. Children of this age do not really understand the danger of fire. They typically set fires in or near their homes: in attics, basements, bathrooms, garages, closets, or nearby fields or vacant lots. They start fires with matches or by sticking combustible material into equipment such as electric heaters that provide an ignition source. The remains of matches, matchboxes, or matchbooks are often found at the point of origin when these young fire-setters are responsible for a fire incident.

Preadolescent fire-starters do not venture far from home, yet members of this group set most of the fires that involve schools and churches. Preadolescents have motivations other than idle curiosity. The motivations of boys range from spite to revenge and disruptive behavior. They typically set fires in closets, classrooms, and wastebaskets, and they may try burning hallway bulletin boards. Girls usually set less aggressive fires and are motivated by a need for attention or as a response to a particular stress, such as a test they do not want to take.

Preadolescent fire-setters rarely use elaborate trailers or incendiary devices. Instead, they use common, readily available accelerants such as gasoline, kerosene, or lighter fluid to fuel the fire. Preadolescents often take advantage of whatever materials are at hand and on site, including trash container contents, loose papers, and rags. The fire set by a preadolescent will show a lack of planning, preparation, and sophistication, particularly if it was a group effort.

In some cases, preadolescents set fires in an attempt to cover vandalism and theft. The items taken from the scene are typically personal, rather than especially valuable. For example, a child may take athletic equipment, which is later displayed to classmates, or may reclaim an item that had been confiscated by a teacher.

The fires set by adolescents are similar to those set by adults. Adolescents have better access to transportation and can travel farther, so they have access to a wider variety of buildings. They also have many of the same motivations of adult fire-setters, such as revenge or attempts to hide larceny, and they often use accelerants. Two-thirds of all fires set in vacant buildings are initiated by adolescents, but no class of property is exempt from harm. Often a large amount of vandalism at the scene serves as a clue that the fire was set by an adolescent.

FIRE FIGHTER II Tips FFII

Arson Facts

- Intentionally set home-structure fires are most likely to be set between 3:00 p.m. and midnight.
- Arson is the leading cause of property damage in the United States, resulting in $1.1 billion in property damage annually.
- In the 2005–2009 period, 440 civilian deaths annually resulted from fires started by arson.

Source: NFPA.

FIRE FIGHTER II Tips FFII

Serial Arson

NFPA 921 identifies three types of serial (repetitive) arson:

- Serial arson involves an offender who sets three or more fires, with a cooling-off period between fires.
- Spree arson involves an arsonist who sets three or more fires at separate locations, with no emotional cooling-off period between fires.
- Mass arson involves an offender who sets three or more fires at the same site or location during a limited period of time.

Source: NFPA 921, *Guide for Fire and Explosion Investigations, 2011 Edition*, Section 22.4.9.3.1.2.

Arsonist Motives

Arsonists set all types of fires and have a range of motives. Male arsonists are usually motivated by profit, revenge, or vanity, and they often use accelerants. In the past, female fire-setters were seldom motivated by profit or used accelerants, but this pattern seems to be changing.

Six common motives of arsonists are listed in NFPA 921:

1. Vandalism
2. Excitement
3. Revenge
4. Crime concealment
5. Profit
6. Extremism

Arsonists who are motivated by excitement are relatively easy to apprehend because they usually have some identifiable connection with their targets, such as a failing student who sets fire to a school or a disgruntled employee who sets fire to a business. One category of excitement fires includes fires started by a would-be hero, who can be either a discoverer or an assister. The discoverer sets the fire, "discovers" it, and initiates the alarm. The assister sets the fire, sometimes discovers it

Near Miss REPORT

Report Number: 09-0000909

Event Description: Units responded to a report of a natural gas odor. Upon arrival, the deputy chief located the caller who stated that the vacant house next door has a strong odor of gas coming from it. After a complete 360 degree walk-around, the deputy chief confirmed the dispatch.

The house had a "FOR SALE" sign in the front yard and was reported to be vacant. The gas line going into the structure was shut off by the fire department. The gas utility company was notified and responded. Once the gas company arrived, the fire department gained access to the structure but did not enter. After venting the structure for approximately twenty minutes, the fire department entered. Once in the house, they entered the utility room to find a gas line for a gas clothes dryer was left in the "on" position. The dryer had been removed. Right next to the gas dryer was a wash sink with water slowly running from the faucet. The water was hot water. In the same room and in close proximity to the gas line was a hot water heater. This created the perfect recipe to blow up a house. Upon further investigation the house just went into foreclosure.

We consider this a close call because, if we had just entered this house, we potentially could have lost an entire company. The decision to not enter this vacant building was important and communication was a key factor.

Lessons Learned:

- Always conduct a 360 degree walk-around.
- Notice "FOR SALE" signs, identify vacant properties, and talk to neighbors.
- Step back and proceed cautiously when responding to situations like this.
- Good decision making and communications skills are always key factors.
- Maintain situational awareness when planning what to do.

Remember to conduct a complete size-up and use caution when responding to reported gas odors. Good communication and command structure protected the responders from a potential disaster. When approaching a structure with a suspected gas leak inside, use a meter and check explosive levels before entering. To help mitigate this situation, we utilized a gas powered fan to ventilate. However, for safety purposes, we started the fan at a remote location and then walked it to the position just outside the front door. Remember, if a scene seems suspicious, do not tamper with evidence. In this case, we notified the police department and the Fire Investigation Unit for further investigation.

FIRE FIGHTER II

and reports it, and is always on hand to help fire fighters pull hose, raise ladders, and rescue people.

Another category of excitement arsonists is closely related to the would-be heroes: fire fighters and would-be fire fighters (persons who have tried to become fire fighters and failed) who intentionally start fires. Some just want the opportunity to operate a particular piece of apparatus or equipment; others want to get some practice. Some of these individuals have even started fires in nearby jurisdictions in an attempt to harm another fire department's reputation.

Revenge arsonists are often very careful and make detailed plans for setting the fire and escaping detection afterward. Such persons motivated by revenge seldom consider the extent of damage that fire can inflict; instead, their intent is to harm a particular person or group so as to get revenge for a perceived injustice. This type of fire is dangerous for fire fighters and occupants because the revenge arsonist will often set fires near doorways, thereby blocking an exit point. Revenge is often a factor for arsonists in the other motive categories as well.

FIRE FIGHTER II

Arsonists also may set fires to conceal a crime. In such a case, the fire is intended to destroy evidence of another crime, such as burglary, vandalism, fraud, or murder. Fires have also been set to divert attention while another crime is being committed or during attempted escapes from jails, hospitals, and other institutions.

Arson for profit occurs because the arsonist hopes to benefit from the fire either directly or indirectly. Fires that directly benefit the arsonist often involve businesses. For example, a business owner might set a fire or arrange to have one set, with the goal being to collect an insurance settlement that can be used to cover other financial needs. An increasing problem connected with this kind of "arson for profit" is fraud. For example, a person may buy and insure a property, only to burn it for the sole purpose of defrauding an insurance company.

FIRE FIGHTER II

During times of economic hardship, the incidence of arson may increase. People who owe more on a house than the house is worth may set fire to try to get insurance money. When large numbers of houses are unoccupied in a neighborhood, the incidence of arson often increases. Given this relationship, keeping vacant houses secure is often a priority for fire departments. Firefighters must be aware of the added risks that occur in vacant houses, especially when houses have been vandalized.

The other type of profit-motivated arsonist sets a fire to benefit from another person's loss. Included in this motive category are the following individuals:

- Insurance agents who want to sell more insurance to the victim's neighbors
- Contractors who want to secure a contract for rebuilding or wrecking the burned building
- Competitors who want to drive the victim out of business
- Owners of adjoining property who would like to buy the property and expand their own holdings
- Owners of nearby property who want to prevent an occupancy change, such as the conversion of a building to a drug rehabilitation center

Some individuals with extreme beliefs set fires in an attempt to aid their cause. These fires are prompted by political, religious, and racial differences. Churches, abortion clinics, and government buildings have been targets of arsonists who belong to the extremism category.

Wrap-Up

Chief Concepts

- Fire fighters usually reach a fire scene before a trained fire investigator and are able to observe important signs and patterns that the fire investigators can use in determining how and where the fire started.
- Determining the causes of fires allows fire departments to take steps to prevent future fires.
- The size of the fire department and the size of the fire determine who performs the fire investigation and when a fire is investigated.
- When a fire investigator is not on the scene while the fire is being extinguished or overhauled, fire fighters must be particularly careful to preserve evidence.
- Basic fire investigation includes locating the point of origin, determining the fuel used, and identifying the ignition source for the fire.
- Fire investigation should be performed by one of the following: trained fire department investigators, the fire marshal's office, insurance company investigators, or a law enforcement agency.
- An origin and cause investigation determines where, why, and how the fire originated.
- At the point of origin, an ignition source comes into contact with a fuel supply. As part of their examination of a fire scene, fire investigators first search for the general area of origin and then attempt to locate the exact point of origin.
- The cause of a fire can be classified as either incendiary or accidental. Arson is an incendiary cause, whereas accidental fires do not involve a criminal or malicious intent.
- Accidental fires have hundreds of possible causes and involve multiple factors and circumstances. The most important reason for investigating and determining the causes of accidental fires is to prevent future fires. To reduce the number of fires, efforts must concentrate on the most frequent causes and those involving the greatest risks of death, injury, and property damage.
- Most fires, fire deaths, and injuries occur in residential occupancies. The most commonly reported accidental causes of fire in these occupancies involve cooking, heating equipment, electrical equipment, lightning strikes, and smoking.
- Sometimes it may difficult or impossible to identify the specific source of ignition because the fire destroyed every trace of evidence. Fire investigators will classify these fires as having an undetermined cause.
- According to NFPA 921, a scientific method and a systematic analysis are needed to determine the origin and cause of a fire. The fire investigator must determine where the fire started and how it was ignited.
- At the location of the fire, the fire investigator will look for clues indicating the specific cause of the fire. This fire investigation will begin on the exterior of the structure and move to the interior.
- Depth of char, burn patterns, and smoke residue can be helpful in identifying the point of origin but are not conclusive evidence.

- After the fire investigator identifies the area of origin, fire fighters may be asked to assist in digging out the fire scene to help look for evidence within the debris.
- Evidence refers to all of the information gathered and used by fire investigators in determining the cause of a fire:
  - Physical evidence—Items that can be observed, photographed, measured, collected, examined in a laboratory, and presented in court to prove or demonstrate a point
  - Trace or transfer evidence—A minute quantity of physical evidence that is conveyed from one place to another
  - Demonstrative evidence—Anything that can be used to validate a theory or to show how something could have occurred
  - Direct evidence—Facts that can be observed or reported first-hand
  - Circumstantial evidence—Information that can be used to prove a theory, based on facts that were observed directly
- Fire fighters have a responsibility to preserve evidence that could indicate the cause or origin of a fire. If fire fighters discover something that could be evidence, they should leave it in place and notify a supervisor or fire investigator immediately. Evidence is most often found during the salvage and overhaul phases.
- Physical evidence must be handled in a manner that protects the integrity of the chain of custody. Every step in the capture, movement, storage, and examination of the evidence must be properly documented.
- The fire investigator's standard operating procedures (for collecting and processing evidence generally include the following steps:
  - Take photographs of each piece of evidence as it is found and collected.
  - On the fire scene, sketch, mark, and label the location of the evidence.
  - Place evidence in appropriate containers to ensure its safety and prevent contamination. Tag all evidence at the fire scene.
  - Keep a constant watch on the evidence until it can be stored in a secure location.
  - Preserve the chain of custody in handling all the evidence.
- Although fire fighters may not interview witnesses, they can identify potential witnesses to the fire investigators.
- The fire fighter's role in identifying and preserving evidence continues throughout the fire suppression sequence, and takes into account the following factors:
  - The time of day and the weather conditions
  - People leaving the scene
  - The extent of the fire and the number of locations in which fire is found
  - The security of the building
  - Any signs of property break-in
  - Vehicles or people in the area
  - Indications of unusual fire situations
  - Unusual color of smoke
  - The positions of the windows and roof
  - The reaction of the fire during initial attack
  - Any signs of abnormal behavior of fire
  - The condition of the building contents
  - The need to coordinate overhaul and evidence preservation activities
- The building and premises must be properly secured and guarded until the fire investigator has finished gathering evidence and documenting the fire scene—for example, by posting fire fighters or police personnel at the site. Fire fighters must be aware of any state or local laws pertaining to right of access by the owners or occupants while the property is under the control of the fire department.
- If a fire investigator is not immediately available, the premises should be guarded and maintained under the control of the fire department until the investigation takes place and all evidence is collected. In the interim, take the following steps:
  - Suspend salvage and overhaul, and secure the scene.
  - Keep nonessential personnel out of the area. Deny entry to all unauthorized and unnecessary persons.
  - Photograph the fire scene extensively.
  - If weather, traffic, or other factors could destroy the evidence, take steps to preserve it in the best way possible.
- Incendiary fires refer to all fires that are deliberately started for malicious or criminal intent. You must be aware of factors that could indicate an intentionally set fire, report any relevant observations to a supervisor or fire investigator, and help protect evidence in such a scenario.
- Arson fires have several distinct, recognizable patterns or indications. For example, a deliberately set fire might have multiple points of origin or include multiple simultaneous fires.
- Fire investigations involve much more than simply determining what caused the fire, where it started, and whether it was accidental or incendiary (intentionally set). Many fire investigations examine other aspects of a fire incident. For example, if the fire results in injuries or fatalities, the investigation often will study every factor that could have contributed to or prevented those losses.
- Fire investigators typically examine the construction and contents of the building to determine whether fire and building codes were followed, if the built-in fire protection systems functioned properly, or if lessons from the fire could be applied by amending codes or inspection procedures.
- An arsonist is someone who deliberately sets a fire with criminal intent. Arsonists fall into several categories, with various explanations for their behavior. The fire service

has identified two groups who are responsible for a large number of fires: pyromaniacs and juvenile fire-setters.

- The six common motives of arsonists are listed in NFPA 921:
  - Vandalism
  - Excitement
  - Revenge
  - Crime concealment
  - Profit
  - Extremism

Hot Terms

Accidental Fire-cause classification that includes fires with a proven cause that does not involve a deliberate human act.

Arson The crime of maliciously and intentionally, or recklessly, starting a fire or causing an explosion. (NFPA 921)

Arsonist A person who deliberately sets a fire to destroy property with criminal intent.

Chain of custody A legal term used to describe the paperwork or documentation describing the movement, storage, and custody of evidence, such as the record of possession of a gas can from a fire scene.

Circumstantial evidence The means by which alleged facts are proven by deduction or inference from other facts that were observed directly.

Competent ignition source An ignition source that can ignite a fuel under the existing conditions at the time of the fire. It must have sufficient heat and be in close enough proximity to the fuel for a sufficient amount of time to ignite the fuel.

Contaminated A term used to describe evidence that may have been altered from its original state.

Demonstrative evidence Materials that are used to demonstrate a theory or explain an event.

Depth of char The thickness of the layer of a material that has been consumed by a fire. The depth of char on wood can be used to help determine the intensity of a fire at a specific location.

Direct evidence Evidence that is reported first-hand, such as statements from an eyewitness who saw or heard something.

Ignitable liquid Classification for liquid fuels including both flammable and combustible liquids.

Incendiary A fire that is intentionally ignited under circumstances in which the person knows that the fire should not be ignited.

Incendiary device A device or mechanism used to start a fire or explosion.

Mass arson Arson in which an offender sets three or more fires at the same site or location during a limited period of time.

Physical evidence Items that can be observed, photographed, measured, collected, examined in a laboratory, and presented in court to prove or demonstrate a point.

Point of origin The exact location where a heat source and a fuel come in contact with each other and a fire begins. (NFPA 921)

Pyromaniac A pathological fire-setter.

Serial arson A series of fires set by the same offender, with a cooling-off period between fires.

Spree arson A series of fires started by an arsonist who sets three or more fires at separate locations, with no emotional cooling-off period between fires.

Trace (transfer) evidence A minute quantity of physical evidence that is conveyed from one place to another.

Trailers Combustible materials (such as rolled rags, blankets, and newspapers or ignitable liquid) that are used to spread fire from one point or area to other points or areas, often used in conjunction with an incendiary device.

Undetermined Fire-cause classification that includes fires for which the cause has not or cannot be proven.

FIRE FIGHTER *in action*

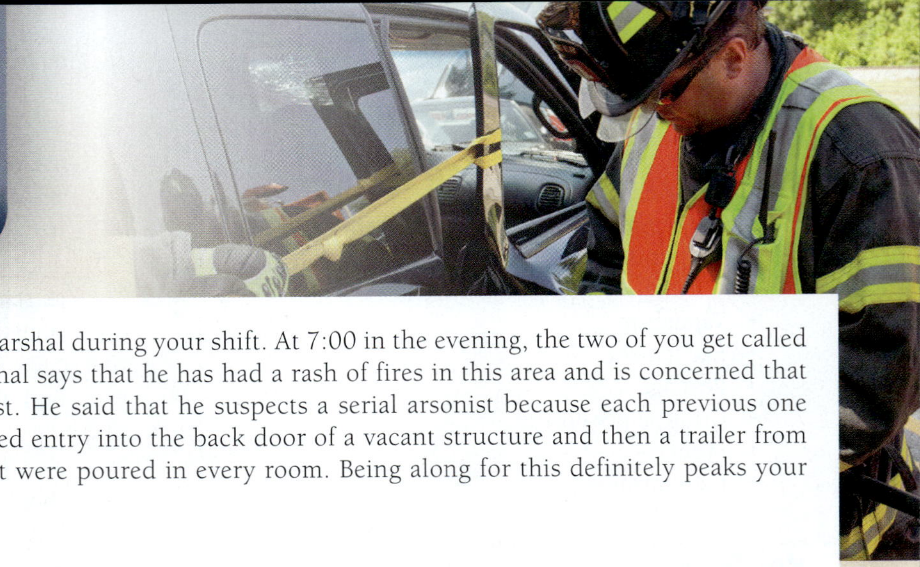

You are excited to get to job shadow the fire marshal during your shift. At 7:00 in the evening, the two of you get called out for a fire in a vacant house. The fire marshal says that he has had a rash of fires in this area and is concerned that this will be another one caused by an arsonist. He said that he suspects a serial arsonist because each previous one has been set in exactly the same manner: forced entry into the back door of a vacant structure and then a trailer from the back door leading to ignitable liquids that were poured in every room. Being along for this definitely peaks your curiosity.

1. What is a "trailer"?
   - **A.** A device or mechanism used to start a fire or explosion.
   - **B.** Evidence of a minute quantity of physical evidence that is conveyed from one place to another.
   - **C.** Combustible materials that are used to spread fire from one point or area to other points or areas.
   - **D.** The exact location where a heat source and a fuel come in contact with each other and a fire begins.
2. Which of the following is a fire-cause classification that includes fires that are deliberately set?
   - **A.** Arson
   - **B.** Natural
   - **C.** Incendiary
   - **D.** Accidental
3. Arson in which an offender sets three or more fires at the same site or location during a limited period of time is considered a ______.
   - **A.** mass arson
   - **B.** spree arson
   - **C.** serial arson
   - **D.** pyromaniac
4. Which type of evidence is where alleged facts are proven by deduction or inference from other facts that were observed directly?
   - **A.** Demonstrative
   - **B.** Circumstantial
   - **C.** Direct
   - **D.** Indirect
5. Which pattern might indicate a short-duration fire that does not spread horizontally from the point of origin?
   - **A.** U-pattern
   - **B.** Y-pattern
   - **C.** V-pattern
   - **D.** Inverted V-pattern
6. When securing a scene, a fire department should ______.
   - **A.** allow anyone to enter the premises for any reason.
   - **B.** not allow anyone to enter the premises for any reason.
   - **C.** keep a log of the name of non-firefighting personnel who enters the property, the times of his or her entry and departure.
   - **D.** keep a log of the name of anyone who enters the property, the times of his or her entry and departure.
7. After the property is officially turned over to the owner, the fire department ______ to search the house.
   - **A.** will not have to obtain a search warrant
   - **B.** may have to obtain a search warrant
   - **C.** must have a search warrant
   - **D.** may not return under any condition

FIRE FIGHTER II *in action*

Having just returned to the station from an overnight fire in a vacant structure, several of you sit around the table discussing the incident. The crews made a great stop; particularly considering the time of night and that there was heavy fire involvement upon your arrival. When the discussion naturally turns to how the fire started, you notice that one fire fighter doesn't say much, but sits there with a smirk on his face as it is discussed. At one point, he mumbles that this is only the beginning, but no one else seems to notice the comment. You become suspicious that he may have something to do with this and other recent fires.

1. How would you handle the situation?
2. To whom would you report your suspicions?
3. Is there a moral obligation to report this? Why or why not?
4. How would you handle it if you reported it and you were wrong?

An Extract from: NFPA 1001, *Standard for Fire Fighter Professional Qualifications*, 2013 Edition

Chapter 5 Fire Fighter I

5.1 General. For qualification at Level I, the fire fighter candidate shall meet the general knowledge requirements in 5.1.1; the general skill requirements in 5.1.2; the JPRs defined in Sections 5.2 through 5.5 of this standard; and the requirements defined in Chapter 5, Core Competencies for Operations Level Responders, and Section 6.6, Mission-Specific Competencies: Product Control, of NFPA 472, *Standard for Competence of Responders to Hazardous Materials/Weapons of Mass Destruction Incidents.*

5.1.1 General Knowledge Requirements. The organization of the fire department; the role of the Fire Fighter I in the organization; the mission of fire service; the fire department's standard operating procedures (SOPs) and rules and regulations as they apply to the Fire Fighter I; the value of fire and life safety initiatives in support of the fire department mission and to reduce fire fighter line-of-duty injuries and fatalities; the role of other agencies as they relate to the fire department; aspects of the fire department's member assistance program; the importance of physical fitness and a healthy lifestyle to the performance of the duties of a fire fighter; the critical aspects of NFPA 1500, *Standard on Fire Department Occupational Safety and Health Program*.

5.1.2 General Skill Requirements. The ability to don personal protective clothing, doff personal protective clothing and prepare for reuse, hoist tools and equipment using ropes and the correct knot, and locate information in departmental documents and standard or code materials.

5.2 Fire Department Communications. This duty shall involve initiating responses, receiving telephone calls, and using fire department communications equipment to correctly relay verbal or written information, according to the JPRs in 5.2.1 through 5.2.4.

5.2.1 Initiate the response to a reported emergency, given the report of an emergency, fire department SOPs, and communications equipment, so that all necessary information is obtained, communications equipment is operated correctly, and the information is relayed promptly and accurately to the dispatch center.

(A) Requisite Knowledge. Procedures for reporting an emergency; departmental SOPs for taking and receiving alarms, radio codes, or procedures; and information needs of dispatch center.

(B) Requisite Skills. The ability to operate fire department communications equipment, relay information, and record information.

5.2.2 Receive a telephone call, given a fire department phone, so that procedures for answering the phone are used and the caller's information is relayed.

(A) Requisite Knowledge. Fire department procedures for answering nonemergency telephone calls.

(B) Requisite Skills. The ability to operate fire station telephone and intercom equipment.

5.2.3 Transmit and receive messages via the fire department radio, given a fire department radio and operating procedures, so that the information is accurate, complete, clear, and relayed within the time established by the AHJ.

(A) Requisite Knowledge. Departmental radio procedures and etiquette for routine traffic, emergency traffic, and emergency evacuation signals.

(B) Requisite Skills. The ability to operate radio equipment and discriminate between routine and emergency traffic.

5.2.4 Activate an emergency call for assistance, given vision obscured conditions, PPE, and department SOPs, so that the fire fighter can be located and rescued.

(A) Requisite Knowledge. Personnel accountability systems, emergency communication procedures, and emergency evacuation methods.

(B) Requisite Skills. The ability to initiate an emergency call for assistance in accordance with the AHJ's procedures, the ability to use other methods of emergency calls for assistance.

5.3 Fireground Operations. This duty shall involve performing activities necessary to ensure life safety, fire control, and property conservation, according to the JPRs in 5.3.1 through 5.3.19.

5.3.1 Use self-contained breathing apparatus (SCBA) during emergency operations, given SCBA and other personal protective equipment, so that the SCBA is correctly donned, the SCBA is correctly worn, controlled breathing techniques are used, emergency procedures are enacted if the SCBA fails, all low-air warnings are recognized, respiratory protection is not intentionally compromised, and hazardous areas are exited prior to air depletion.

(A) Requisite Knowledge. Conditions that require respiratory protection, uses and limitations of SCBA, components of SCBA, donning procedures, breathing techniques, indications for and emergency procedures used with SCBA, and physical requirements of the SCBA wearer.

(B) Requisite Skills. The ability to control breathing, replace SCBA air cylinders, use SCBA to exit through restricted passages, initiate and complete emergency procedures in the event of SCBA failure or air depletion, and complete donning procedures.

5.3.2* Respond on apparatus to an emergency scene, given personal protective clothing and other necessary personal protective equipment, so that the apparatus is correctly mounted and dismounted, seat belts are used while the vehicle is in motion, and other personal protective equipment is correctly used.

(A) Requisite Knowledge. Mounting and dismounting procedures for riding fire apparatus, hazards and ways to avoid hazards associated with riding apparatus, prohibited practices, and types of department personal protective equipment and the means for usage.

(B) Requisite Skills. The ability to use each piece of provided safety equipment.

5.3.3 Establish and operate in work areas at emergency scenes, given protective equipment, traffic and scene control devices, structure fire and roadway emergency scenes, traffic hazards and downed electrical wires, an assignment, and SOPs, so that procedures are followed, protective equipment is worn, protected work areas are established as directed using traffic and scene control devices, and the fire fighter performs assigned tasks only in established, protected work areas.

(A) Requisite Knowledge. Potential hazards involved in operating on emergency scenes including vehicle traffic, utilities, and environmental conditions; proper procedures for dismounting apparatus in traffic; procedures for safe operation at emergency scenes; and the protective equipment available for members' safety on emergency scenes and work zone designations.

(B) Requisite Skills. The ability to use personal protective clothing, deploy traffic and scene control devices, dismount apparatus, and operate in the protected work areas as directed.

5.3.4 Force entry into a structure, given personal protective equipment, tools, and an assignment, so that the tools are used as designed, the barrier is removed, and the opening is in a safe condition and ready for entry.

(A) Requisite Knowledge. Basic construction of typical doors, windows, and walls within the department's community or service area; operation of doors, windows, and locks; and the dangers associated with forcing entry through doors, windows, and walls.

(B) Requisite Skills. The ability to transport and operate hand and power tools and to force entry through doors, windows, and walls using assorted methods and tools.

5.3.5 Exit a hazardous area as a team, given vision-obscured conditions, so that a safe haven is found before exhausting the air supply, others are not endangered, and the team integrity is maintained.

(A) Requisite Knowledge. Personnel accountability systems, communication procedures, emergency evacuation methods, what constitutes a safe haven, elements that create or indicate a hazard, and emergency procedures for loss of air supply.

(B) Requisite Skills. The ability to operate as a team member in vision-obscured conditions, locate and follow a guideline, conserve air supply, and evaluate areas for hazards and identify a safe haven.

5.3.6 Set up ground ladders, given single and extension ladders, an assignment, and team members if needed, so that hazards are assessed, the ladder is stable, the angle is correct for climbing, extension ladders are extended to the necessary height with the fly locked, the top is placed against a reliable structural component, and the assignment is accomplished.

(A) Requisite Knowledge. Parts of a ladder, hazards associated with setting up ladders, what constitutes a stable foundation for ladder placement, different angles for various tasks, safety limits to the degree of angulation, and what constitutes a reliable structural component for top placement.

(B) Requisite Skills. The ability to carry ladders, raise ladders, extend ladders and lock flies, determine that a wall and roof will support the ladder, judge extension ladder height requirements, and place the ladder to avoid obvious hazards.

5.3.7 Attack a passenger vehicle fire operating as a member of a team, given personal protective equipment, attack line, and hand tools, so that hazards are avoided, leaking flammable liquids are identified and controlled, protection from flash fires is maintained, all vehicle compartments are overhauled, and the fire is extinguished.

(A) Requisite Knowledge. Principles of fire streams as they relate to fighting automobile fires; precautions to be followed when advancing hose lines toward an automobile; observable results that a fire stream has been properly applied; identifying alternative fuels and the hazards associated with them; dangerous conditions created during an automobile fire; common types of accidents or injuries related to fighting automobile fires and how to avoid them; how to access locked passenger, trunk, and engine compartments; and methods for overhauling an automobile.

(B) Requisite Skills. The ability to identify automobile fuel type; assess and control fuel leaks; open, close, and adjust the flow and pattern on nozzles; apply water for maximum effectiveness while maintaining flash fire protection; advance 1½ in. (38 mm) or larger diameter attack lines; and expose hidden fires by opening all automobile compartments.

5.3.8 Extinguish fires in exterior Class A materials, given fires in stacked or piled and small unattached structures or storage containers that can be fought from the exterior, attack lines, hand tools and master stream devices, and an assignment, so that exposures are protected, the spread of fire is stopped, collapse hazards are avoided, water application is effective, the fire is extinguished, and signs of the origin area(s) and arson are preserved.

(A) Requisite Knowledge. Types of attack lines and water streams appropriate for attacking stacked, piled materials and outdoor fires; dangers—such as collapse—associated with stacked and piled materials; various extinguishing agents and their effect on different material configurations; tools and methods to use in breaking up various types of materials; the difficulties related to complete extinguishment of stacked and piled materials; water application methods for exposure protection and fire extinguishment; dangers such as exposure to toxic or hazardous materials associated with storage building and container fires; obvious signs of origin and cause; and techniques for the preservation of fire cause evidence.

(B) Requisite Skills. The ability to recognize inherent hazards related to the material's configuration, operate handlines or master streams, break up material using hand tools and water streams, evaluate for complete extinguishment, operate hose lines and other water application devices, evaluate and modify water application for maximum penetration, search for and expose hidden fires, assess patterns for origin determination, and evaluate for complete extinguishment.

5.3.9 Conduct a search and rescue in a structure operating as a member of a team, given an assignment, obscured vision conditions, personal protective equipment, a flashlight, forcible entry tools, hose lines, and ladders when necessary, so that ladders are correctly placed when used, all assigned areas are searched, all victims are located and removed, team integrity is maintained, and team members' safety—including respiratory protection—is not compromised.

(A) Requisite Knowledge. Use of forcible entry tools during rescue operations, ladder operations for rescue, psychological effects of operating in obscured conditions and ways to manage them, methods to determine if an area is tenable, primary and secondary search techniques, team members' roles and goals, methods to use and indicators of finding victims, victim removal methods (including various carries), and considerations related to respiratory protection.

(B) Requisite Skills. The ability to use SCBA to exit through restricted passages, set up and use different types of ladders for various types of rescue operations, rescue a fire fighter with functioning respiratory protection, rescue a fire fighter whose respiratory protection is not functioning, rescue a person who has no respiratory protection, and assess areas to determine tenability.

5.3.10 Attack an interior structure fire operating as a member of a team, given an attack line, ladders when needed, personal protective equipment, tools, and an assignment, so that team integrity is maintained, the attack line is deployed for advancement, ladders are correctly placed when used, access is gained into the fire area, effective water application practices are used, the fire is approached correctly, attack techniques facilitate suppression given the level of the fire, hidden fires are located and controlled, the correct body posture is maintained, hazards are recognized and managed, and the fire is brought under control.

(A) Requisite Knowledge. Principles of fire streams; types, design, operation, nozzle pressure effects, and flow capabilities of nozzles; precautions to be followed when advancing hose lines to a fire; observable results that a fire stream has been properly applied; dangerous building conditions created by fire; principles of exposure protection; potential longterm consequences of exposure to products of combustion; physical states of matter in which fuels are found; common types of accidents or injuries and their causes; and the application of each size and type of attack line, the role of the backup team in fire attack situations, attack and control techniques for grade level and above and below grade levels, and exposing hidden fires.

(B) Requisite Skills. The ability to prevent water hammers when shutting down nozzles; open, close, and adjust nozzle flow and patterns; apply water using direct, indirect, and combination attacks; advance charged and uncharged 1½ in. (38 mm) diameter or larger hose lines up ladders and up and down interior and exterior stairways; extend hose lines; replace burst hose sections; operate charged hose lines of 1½ in. (38 mm) diameter or larger while secured to a ground ladder; couple and uncouple various handline connections; carry hose; attack fires at grade level and above and below grade levels; and locate and suppress interior wall and subfloor fires.

5.3.11 Perform horizontal ventilation on a structure operating as part of a team, given an assignment, personal protective equipment, ventilation tools, equipment, and ladders, so that the ventilation openings are free of obstructions, tools are used as designed, ladders are correctly placed, ventilation devices are correctly placed, and the structure is cleared of smoke.

(A) Requisite Knowledge. The principles, advantages, limitations, and effects of horizontal, mechanical, and hydraulic ventilation; safety considerations when venting a structure; fire behavior in a structure; the products of combustion found in a structure fire; the signs, causes, effects, and prevention of backdrafts; and the relationship of oxygen concentration to life safety and fire growth.

(B) Requisite Skills. The ability to transport and operate ventilation tools and equipment and ladders, and to use safe procedures for breaking window and door glass and removing obstructions.

5.3.12 Perform vertical ventilation on a structure as part of a team, given an assignment, personal protective equipment, ground and roof ladders, and tools, so that ladders are positioned for ventilation, a specified opening is created, all ventilation barriers are removed, structural integrity is not compromised, products of combustion are released from the structure, and the team retreats from the area when ventilation is accomplished.

(A) Requisite Knowledge. The methods of heat transfer; the principles of thermal layering within a structure on fire; the techniques and safety precautions for venting flat roofs, pitched roofs, and basements; basic indicators of potential collapse or roof failure; the effects of construction type and elapsed time under fire conditions on structural integrity; and the advantages and disadvantages of vertical and trench/strip ventilation.

(B) Requisite Skills. The ability to transport and operate ventilation tools and equipment; hoist ventilation tools to a roof; cut roofing and flooring materials to vent flat roofs, pitched roofs, and basements; sound a roof for integrity; clear an opening with hand tools; select, carry, deploy, and secure ground ladders for ventilation activities; deploy roof ladders on pitched roofs while secured to a ground ladder; and carry ventilation-related tools and equipment while ascending and descending ladders.

5.3.13 Overhaul a fire scene, given personal protective equipment, attack line, hand tools, a flashlight, and an assignment, so that structural integrity is not compromised, all hidden fires are discovered, fire cause evidence is preserved, and the fire is extinguished.

(A) Requisite Knowledge. Types of fire attack lines and water application devices most effective for overhaul, water application methods for extinguishment that limit water damage, types of tools and methods used to expose hidden fire, dangers associated with overhaul, obvious signs of area of origin or signs of arson, and reasons for protection of fire scene.

(B) Requisite Skills. The ability to deploy and operate an attack line; remove flooring, ceiling, and wall components to expose void spaces without compromising structural integrity; apply water for maximum effectiveness; expose and extinguish hidden fires in walls, ceilings, and subfloor spaces; recognize and preserve obvious signs of area of origin and arson; and evaluate for complete extinguishment.

5.3.14 Conserve property as a member of a team, given salvage tools and equipment and an assignment, so that the building and its contents are protected from further damage.

(A) Requisite Knowledge. The purpose of property conservation and its value to the public, methods used to protect property, types of and uses for salvage covers, operations at properties protected with automatic sprinklers, how to stop the flow of water from an automatic sprinkler head, identification of the main control valve on an automatic sprinkler system, forcible entry issues related to salvage, and procedures for protecting possible areas of origin and potential evidence.

(B) Requisite Skills. The ability to cluster furniture; deploy covering materials; roll and fold salvage covers for reuse; construct water chutes and catch-alls; remove water; cover building openings, including doors, windows, floor openings, and roof openings; separate, remove, and relocate charred material to a safe location while protecting the area of origin for cause determination; stop the flow of water from a sprinkler with sprinkler wedges or stoppers; and operate a main control valve on an automatic sprinkler system.

5.3.15 Connect a fire department pumper to a water supply as a member of a team, given supply or intake hose, hose tools, and a fire hydrant or static water source, so that connections are tight and water flow is unobstructed.

(A) Requisite Knowledge. Loading and off-loading procedures for mobile water supply apparatus; fire hydrant operation; and suitable static water supply sources, procedures, and protocol for connecting to various water sources.

(B) Requisite Skills. The ability to hand lay a supply hose, connect and place hard suction hose for drafting operations, deploy portable water tanks as well as the equipment necessary to transfer water between and draft from them, make hydrant-to-pumper hose connections for forward and reverse lays, connect supply hose to a hydrant, and fully open and close the hydrant.

5.3.16 Extinguish incipient Class A, Class B, and Class C fires, given a selection of portable fire extinguishers, so that the correct extinguisher is chosen, the fire is completely extinguished, and correct extinguisher-handling techniques are followed.

(A) **Requisite Knowledge.** The classifications of fire; the types of, rating systems for, and risks associated with each class of fire; and the operating methods of and limitations of portable extinguishers.

(B) **Requisite Skills.** The ability to operate portable fire extinguishers, approach fire with portable fire extinguishers, select an appropriate extinguisher based on the size and type of fire, and safely carry portable fire extinguishers.

5.3.17 Illuminate the emergency scene, given fire service electrical equipment and an assignment, so that designated areas are illuminated and all equipment is operated within the manufacturer's listed safety precautions.

(A) **Requisite Knowledge.** Safety principles and practices, power supply capacity and limitations, and light deployment methods.

(B) **Requisite Skills.** The ability to operate department power supply and lighting equipment, deploy cords and connectors, reset ground-fault interrupter (GFI) devices, and locate lights for best effect.

5.3.18 Turn off building utilities, given tools and an assignment, so that the assignment is safely completed.

(A) **Requisite Knowledge.** Properties, principles, and safety concerns for electricity, gas, and water systems; utility disconnect methods and associated dangers; and use of required safety equipment.

(B) **Requisite Skills.** The ability to identify utility control devices, operate control valves or switches, and assess for related hazards.

5.3.19 Combat a ground cover fire operating as a member of a team, given protective clothing, SCBA (if needed), hose lines, extinguishers or hand tools, and an assignment, so that threats to property are reported, threats to personal safety are recognized, retreat is quickly accomplished when warranted, and the assignment is completed.

(A) **Requisite Knowledge.** Types of ground cover fires, parts of ground cover fires, methods to contain or suppress, and safety principles and practices.

(B) **Requisite Skills.** The ability to determine exposure threats based on fire spread potential, protect exposures, construct a fire line or extinguish with hand tools, maintain integrity of established fire lines, and suppress ground cover fires using water.

5.3.20 Tie a knot appropriate for hoisting tool, given personnel protective equipment, tools, ropes, and an assignment, so that the knots used are appropriate for hoisting tools securely and as directed.

(A) **Requisite Knowledge.** Knot types and usage; the difference between life safety and utility rope; reasons for placing rope out of service; the types of knots to use for given tools, ropes, or situations; hoisting methods for tools and equipment; and using rope to support response activities.

(B) **Requisite Skills.** The ability to hoist tools using specific knots based on the type of tool.

5.4 Rescue Operations. This duty shall involve no requirements for Fire Fighter I.

5.5 Preparedness and Maintenance. This duty shall involve performing activities that reduce the loss of life and property due to fire through response readiness, according to the JPRs in 5.5.1 and 5.5.2.

5.5.1 Clean and check ladders, ventilation equipment, SCBA, ropes, salvage equipment, and hand tools, given cleaning tools, cleaning supplies, and an assignment, so that equipment is clean and maintained according to manufacturer's or departmental guidelines, maintenance is recorded, and equipment is placed in a ready state or reported otherwise.

(A) **Requisite Knowledge.** Types of cleaning methods for various tools and equipment, correct use of cleaning solvents, and manufacturer's or departmental guidelines for cleaning equipment and tools.

(B) **Requisite Skills.** The ability to select correct tools for various parts and pieces of equipment, follow guidelines, and complete recording and reporting procedures.

5.5.2 Clean, inspect, and return fire hose to service, given washing equipment, water, detergent, tools, and replacement gaskets, so that damage is noted and corrected, the hose is clean, and the equipment is placed in a ready state for service.

(A) **Requisite Knowledge.** Departmental procedures for noting a defective hose and removing it from service, cleaning methods, and hose rolls and loads.

(B) **Requisite Skills.** The ability to clean different types of hose; operate hose washing and drying equipment; mark defective hose; and replace coupling gaskets, roll hose, and reload hose.

Chapter 6 Fire Fighter II

6.1 General. For qualification at Level II, the Fire Fighter I shall meet the general knowledge requirements in 6.1.1, the general skill requirements in 6.1.2, the JPRs defined in Sections 6.2 through 6.5 of this standard, and the requirements defined in Chapter 5.

6.1.1 General Knowledge Requirements. Responsibilities of the Fire Fighter II in assuming and transferring command within an incident management system, performing assigned duties in conformance with applicable NFPA and other safety regulations and AHJ procedures, and the role of a Fire Fighter II within the organization.

6.1.2 General Skill Requirements. The ability to determine the need for command, organize and coordinate an incident management system until command is transferred, and function within an assigned role in an incident management system.

6.2 Fire Department Communications. This duty shall involve performing activities related to initiating and reporting responses, according to the JPRs in 6.2.1 and 6.2.2.

6.2.1 Complete a basic incident report, given the report forms, guidelines, and information, so that all pertinent information is recorded, the information is accurate, and the report is complete.

(A) **Requisite Knowledge.** Content requirements for basic incident reports, the purpose and usefulness of accurate reports, consequences of inaccurate reports, how to obtain necessary information, and required coding procedures.

(B) **Requisite Skills.** The ability to determine necessary codes, proof reports, and operate fire department computers or other equipment necessary to complete reports.

6.2.2 Communicate the need for team assistance, given fire department communications equipment, SOPs, and a team, so that the supervisor is consistently informed of team needs, departmental SOPs are followed, and the assignment is accomplished safely.

(A) **Requisite Knowledge.** SOPs for alarm assignments and fire department radio communication procedures.

(B) Requisite Skills. The ability to operate fire department communications equipment.

6.3 Fireground Operations. This duty shall involve performing activities necessary to ensure life safety, fire control, and property conservation, according to the JPRs in 6.3.1 through 6.3.4.

6.3.1 Extinguish an ignitible liquid fire, operating as a member of a team, given an assignment, an attack line, personal protective equipment, a foam proportioning device, a nozzle, foam concentrates, and a water supply, so that the correct type of foam concentrate is selected for the given fuel and conditions, a properly proportioned foam stream is applied to the surface of the fuel to create and maintain a foam blanket, fire is extinguished, reignition is prevented, team protection is maintained with a foam stream, and the hazard is faced until retreat to safe haven is reached.

(A) Requisite Knowledge. Methods by which foam prevents or controls a hazard; principles by which foam is generated; causes for poor foam generation and corrective measures; difference between hydrocarbon and polar solvent fuels and the concentrates that work on each; the characteristics, uses, and limitations of fire-fighting foams; the advantages and disadvantages of using fog nozzles versus foam nozzles for foam application; foam stream application techniques; hazards associated with foam usage; and methods to reduce or avoid hazards.

(B) Requisite Skills. The ability to prepare a foam concentrate supply for use, assemble foam stream components, master various foam application techniques, and approach and retreat from spills as part of a coordinated team.

6.3.2 Coordinate an interior attack line for a team's accomplishment of an assignment in a structure fire, given attack lines, personnel, personal protective equipment, and tools, so that crew integrity is established; attack techniques are selected for the given level of the fire (e.g., attic, grade level, upper levels, or basement); attack techniques are communicated to the attack teams; constant team coordination is maintained; fire growth and development is continuously evaluated; search, rescue, and ventilation requirements are communicated or managed; hazards are reported to the attack teams; and incident command is apprised of changing conditions.

(A) Requisite Knowledge. Selection of the nozzle and hose for fire attack, given different fire situations; selection of adapters and appliances to be used for specific fireground situations; dangerous building conditions created by fire and fire suppression activities; indicators of building collapse; the effects of fire and fire suppression activities on wood, masonry (brick, block, stone), cast iron, steel, reinforced concrete, gypsum wallboard, glass, and plaster on lath; search and rescue and ventilation procedures; indicators of structural instability; suppression approaches and practices for various types of structural fires; and the association between specific tools and special forcible entry needs.

(B) Requisite Skills. The ability to assemble a team, choose attack techniques for various levels of a fire (e.g., attic, grade level, upper levels, or basement), evaluate and forecast a fire's growth and development, select tools for forcible entry, incorporate search and rescue procedures and ventilation procedures in the completion of the attack team efforts, and determine developing hazardous building or fire conditions.

6.3.3 Control a flammable gas cylinder fire, operating as a member of a team, given an assignment, a cylinder outside of a structure, an attack line, personal protective equipment, and tools, so that crew integrity is maintained, contents are identified, safe havens are identified prior to advancing, open valves are closed, flames are not extinguished unless the leaking gas is eliminated, the cylinder is cooled, cylinder integrity is evaluated, hazardous conditions are recognized and acted upon, and the cylinder is faced during approach and retreat.

(A) Requisite Knowledge. Characteristics of pressurized flammable gases, elements of a gas cylinder, effects of heat and pressure on closed cylinders, boiling liquid expanding vapor explosion (BLEVE) signs and effects, methods for identifying contents, how to identify safe havens before approaching flammable gas cylinder fires, water stream usage and demands for pressurized cylinder fires, what to do if the fire is prematurely extinguished, valve types and their operation, alternative actions related to various hazards, and when to retreat.

(B) Requisite Skills. The ability to execute effective advances and retreats, apply various techniques for water application, assess cylinder integrity and changing cylinder conditions, operate control valves, and choose effective procedures when conditions change.

6.3.4 Protect evidence of fire cause and origin, given a flashlight and overhaul tools, so that the evidence is noted and protected from further disturbance until investigators can arrive on the scene.

(A) Requisite Knowledge. Methods to assess origin and cause; types of evidence; means to protect various types of evidence; the role and relationship of Fire Fighter IIs, criminal investigators, and insurance investigators in fire investigations; and the effects and problems associated with removing property or evidence from the scene.

(B) Requisite Skills. The ability to locate the fire's origin area, recognize possible causes, and protect the evidence.

6.4 Rescue Operations. This duty shall involve performing activities related to accessing and disentangling victims from motor vehicle accidents and helping special rescue teams, according to the JPRs in 6.4.1 and 6.4.2.

6.4.1 Extricate a victim entrapped in a motor vehicle as part of a team, given stabilization and extrication tools, so that the vehicle is stabilized, the victim is disentangled without further injury, and hazards are managed.

(A) Requisite Knowledge. The fire department's role at a vehicle accident, points of strength and weakness in auto body construction, dangers associated with vehicle components and systems, the uses and limitations of hand and power extrication equipment, and safety procedures when using various types of extrication equipment.

(B) Requisite Skills. The ability to operate hand and power tools used for forcible entry and rescue as designed; use cribbing and shoring material; and choose and apply appropriate techniques for moving or removing vehicle roofs, doors, windshields, windows, steering wheels or columns, and the dashboard.

6.4.2 Assist rescue operation teams, given standard operating procedures, necessary rescue equipment, and an assignment, so that procedures are followed, rescue items are recognized and retrieved in the time as prescribed by the AHJ, and the assignment is completed.

(A) Requisite Knowledge. The fire fighter's role at a technical rescue operation, the hazards associated with technical rescue operations, types and uses for rescue tools, and rescue practices and goals.

(B) Requisite Skills. The ability to identify and retrieve various types of rescue tools, establish public barriers, and assist rescue teams as a member of the team when assigned.

6.5 Fire and Life Safety Initiatives, Preparedness, and Maintenance. This duty shall involve performing activities related to reducing the loss of life and property due to fire through hazard identification, inspection, and response readiness, according to the JPRs in 6.5.1 through 6.5.5.

6.5.1 Perform a fire safety survey in an occupied structure, given survey forms and procedures, so that fire and life safety hazards are identified, recommendations for their correction are made to the occupant, and unresolved issues are referred to the proper authority.

(A) Requisite Knowledge. Organizational policy and procedures, common causes of fire and their prevention, the importance of a fire safety survey and public fire education programs to fire department public relations and the community, and referral procedures.

(B) Requisite Skills. The ability to complete forms, recognize hazards, match findings to preapproved recommendations, and effectively communicate findings to occupants or referrals.

6.5.2 Present fire safety information to station visitors or small groups, given prepared materials, so that all information is presented, the information is accurate, and questions are answered or referred.

(A) Requisite Knowledge. Parts of informational materials and how to use them, basic presentation skills, and departmental standard operating procedures for giving fire station tours.

(B) Requisite Skills. The ability to document presentations and to use prepared materials.

6.5.3 Prepare a preincident survey, given forms, necessary tools, and an assignment, so that all required occupancy information is recorded, items of concern are noted, and accurate sketches or diagrams are prepared.

(A) Requisite Knowledge. The sources of water supply for fire protection; the fundamentals of fire suppression and detection systems; common symbols used in diagramming construction features, utilities, hazards, and fire protection systems; departmental requirements for a preincident survey and form completion; and the importance of accurate diagrams.

(B) Requisite Skills. The ability to identify the components of fire suppression and detection systems; sketch the site, buildings, and special features; detect hazards and special considerations to include in the preincident sketch; and complete all related departmental forms.

6.5.4 Maintain power plants, power tools, and lighting equipment, given tools and manufacturers' instructions, so that equipment is clean and maintained according to manufacturer and departmental guidelines, maintenance is recorded, and equipment is placed in a ready state or reported otherwise.

(A) Requisite Knowledge. Types of cleaning methods, correct use of cleaning solvents, manufacturer and departmental guidelines for maintaining equipment and its documentation, and problem-reporting practices.

(B) Requisite Skills. The ability to select correct tools; follow guidelines; complete recording and reporting procedures; and operate power plants, power tools, and lighting equipment.

6.5.5 Perform an annual service test on fire hose, given a pump, a marking device, pressure gauges, a timer, record sheets, and related equipment, so that procedures are followed, the condition of the hose is evaluated, any damaged hose is removed from service, and the results are recorded.

(A) Requisite Knowledge. Procedures for safely conducting hose service testing, indicators that dictate any hose be removed from service, and recording procedures for hose test results.

(B) Requisite Skills. The ability to operate hose testing equipment and nozzles and to record results.

An Extract from NFPA 472, *Standard for Competence of Responders to Hazardous Materials/Weapons of Mass Destruction Incidents*, 2013 Edition

Chapter 4 Competencies for Awareness Level Personnel

4.1 General.

4.1.1 Introduction.

4.1.1.1 Awareness level personnel shall be persons who, in the course of their normal duties, could encounter an emergency involving hazardous materials/weapons of mass destruction (WMD) and who are expected to recognize the presence of the hazardous materials/WMD, protect themselves, call for trained personnel, and secure the area.

4.1.1.2 Awareness level personnel shall be trained to meet all competencies of this chapter.

4.1.1.3 Awareness level personnel shall receive additional training to meet applicable governmental occupational health and safety regulations.

4.1.2 Goal.

4.1.2.1 The goal of the competencies in this chapter shall be to provide personnel already on the scene of a hazardous materials/WMD incident with the knowledge and skills to perform the tasks in 4.1.2.2 safely and effectively.

4.1.2.2 When already on the scene of a hazardous materials/WMD incident, the awareness level personnel shall be able to perform the following tasks:

(1) Analyze the incident to determine both the hazardous materials/WMD present and the basic hazard and response information for each hazardous materials/WMD agent by completing the following tasks:
   (a) Detect the presence of hazardous materials/WMD.
   (b) Survey a hazardous materials/WMD incident from a safe location to identify the name, UN/NA identification number, type of placard, or other distinctive marking applied for the hazardous materials/WMD involved.
   (c) Collect hazard information from the current edition of the DOT *Emergency Response Guidebook*.
(2) Implement actions consistent with the authority having jurisdiction (AHJ), and the current edition of the DOT *Emergency Response Guidebook* by completing the following tasks:
   (a) Initiate protective actions
   (b) Initiate the notification process

4.2 Competencies —Analyzing the Incident.

4.2.1 Detecting the Presence of Hazardous Materials/WMD. Given examples of various situations, awareness level personnel shall identify those situations where hazardous materials/WMD are present by completing the following requirements:

(1) Identify the definitions of both hazardous material (or dangerous goods, in Canada) and WMD

(2) Identify the UN/DOT hazard classes and divisions of hazardous materials/WMD and identify common examples of materials in each hazard class or division

(3) Identify the primary hazards associated with each UN/DOT hazard class and division

(4) Identify the difference between hazardous materials/WMD incidents and other emergencies

(5) Identify typical occupancies and locations in the community where hazardous materials/WMD are manufactured, transported, stored, used, or disposed of

(6) Identify typical container shapes that can indicate the presence of hazardous materials/WMD

(7) Identify facility and transportation markings and colors that indicate hazardous materials/WMD, including the following:

(a) Transportation markings, including UN/NA identification number marks, marine pollutant mark, elevated temperature (HOT) mark, commodity marking, and inhalation hazard mark

(b) NFPA 704, *Standard System for the Identification of the Hazards of Materials for Emergency Response*, markings

(c) Military hazardous materials/WMD markings

(d) Special hazard communication markings for each hazard class

(e) Pipeline markings

(f) Container markings

(8) Given an NFPA704 marking, describe the significance of the colors, numbers, and special symbols

(9) Identify U.S. and Canadian placards and labels that indicate hazardous materials/WMD

(10) Identify the following basic information on material safety data sheets (MSDS) and shipping papers for hazardous materials:

(a) Identify where to find MSDS

(b) Identify major sections of an MSDS

(c) Identify the entries on shipping papers that indicate the presence of hazardous materials

(d) Match the name of the shipping papers found in transportation (air, highway, rail, and water) with the mode of transportation

(e) Identify the person responsible for having the shipping papers in each mode of transportation

(f) Identify where the shipping papers are found in each mode of transportation

(g) Identify where the papers can be found in an emergency in each mode of transportation

(11) Identify examples of clues (other than occupancy/location, container shape, markings/color, placards/labels, MSDS, and shipping papers) to include sight, sound, and odor of which indicate hazardous materials/WMD

(12) Describe the limitations of using the senses in determining the presence or absence of hazardous materials/WMD

(13) Identify at least four types of locations that could be targets for criminal or terrorist activity using hazardous materials/WMD

(14) Describe the difference between a chemical and a biological incident

(15) Identify at least four indicators of possible criminal or terrorist activity involving chemical agents

(16) Identify at least four indicators of possible criminal or terrorist activity involving biological agents

(17) Identify at least four indicators of possible criminal or terrorist activity involving radiological agents

(18) Identify at least four indicators of possible criminal or terrorist activity involving illicit laboratories (clandestine laboratories, weapons lab, ricin lab)

(19) Identify at least four indicators of possible criminal or terrorist activity involving explosives

(20) Identify at least four indicators of secondary devices

4.2.2 Surveying Hazardous Materials/WMD Incidents. Given examples of hazardous materials/WMD incidents, awareness level personnel shall, from a safe location, identify the hazardous material(s)/WMD involved in each situation by name, UN/NA identification number, or type placard applied by completing the following requirements:

(1) Identify difficulties encountered in determining the specific names of hazardous materials/WMD at facilities and in transportation

(2) Identify sources for obtaining the names of, UN/NA identification numbers for, or types of placard associated with hazardous materials/WMD in transportation

(3) Identify sources for obtaining the names of hazardous materials/WMD at a facility

4.2.3 Collecting Hazard Information. Given the identity of various hazardous materials/WMD (name, UN/NA identification number, or type placard), awareness level personnel shall identify the fire, explosion, and health hazard information for each material by using the current edition of the DOT *Emergency Response Guidebook* by completing the following requirements:

(1) Identify the three methods for determining the guidebook page for a hazardous material/WMD

(2) Identify the two general types of hazards found on each guidebook page

4.3 Competencies — Planning the Response. (Reserved)

4.4 Competencies — Implementing the Planned Response.

4.4.1 Initiating Protective Actions. Given examples of hazardous materials/WMD incidents, the emergency response plan, the standard operating procedures, and the current edition of the DOT *Emergency Response Guidebook*, awareness level personnel shall be able to identify the actions to be taken to protect themselves and others and to control access to the scene by completing the following requirements:

(1) Identify the location of both the emergency response plan and/or standard operating procedures

(2) Identify the role of the awareness level personnel during hazardous materials/WMD incidents

(3) Identify the following basic precautions to be taken to protect themselves and others in hazardous materials/WMD incidents:

(a) Identify the precautions necessary when providing emergency medical care to victims of hazardous materials/WMD incidents

(b) Identify typical ignition sources found at the scene of hazardous materials/WMD incidents

(c) Identify the ways hazardous materials/WMD are harmful to people, the environment, and property

(d) Identify the general routes of entry for human exposure to hazardous materials/WMD

(4) Given examples of hazardous materials/WMD and the identity of each hazardous material/WMD (name, UN/NA identification number, or type placard), identify the following response information:

(a) Emergency action (fire, spill, or leak and first aid)
(b) Personal protective equipment necessary
(c) Initial isolation and protective action distances

(5) Given the name of a hazardous material, identify the recommended personal protective equipment from the following list:
(a) Street clothing and work uniforms
(b) Structural fire-fighting protective clothing
(c) Positive pressure self-contained breathing apparatus
(d) Chemical-protective clothing and equipment

(6) Identify the definitions for each of the following protective actions:
(a) Isolation of the hazard area and denial of entry
(b) Evacuation
(c) Shelter-in-place

(7) Identify the size and shape of recommended initial isolation and protective action zones

(8) Describe the difference between small and large spills as found in the Table of Initial Isolation and Protective Action Distances in the DOT *Emergency Response Guidebook*

(9) Identify the circumstances under which the following distances are used at hazardous materials/WMD incidents:
(a) Table of Initial Isolation and Protective Action Distances
(b) Isolation distances in the numbered guides

(10) Describe the difference between the isolation distances on the orange-bordered guidebook pages and the protective action distances on the green-bordered ERG (*Emergency Response Guidebook*) pages

(11) Identify the techniques used to isolate the hazard area and deny entry to unauthorized persons at hazardous materials/WMD incidents

(12) Identify at least four specific actions necessary when an incident is suspected to involve criminal or terrorist activity

4.4.2 **Initiating the Notification Process.** Given scenarios involving hazardous materials/WMD incidents, awareness level personnel shall identify the initial notifications to be made and how to make them, consistent with the AHJ.

4.5 **Competencies — Evaluating Progress. (Reserved)**

4.6 **Competencies — Terminating the Incident. (Reserved)**

Chapter 5 Core Competencies for Operations Level Responders

5.1 **General.**

5.1.1 **Introduction.**

5.1.1.1 The operations level responder shall be that person who responds to hazardous materials/weapons of mass destruction (WMD) incidents for the purpose of protecting nearby persons, the environment, or property from the effects of the release.

5.1.1.2 The operations level responder shall be trained to meet all competencies at the awareness level (*see Chapter 4*) and the competencies of this chapter.

5.1.1.3 The operations level responder shall receive additional training to meet applicable governmental occupational health and safety regulations.

5.1.2 **Goal.**

5.1.2.1 The goal of the competencies in this chapter shall be to provide operations level responders with the knowledge and skills to perform the core competencies in 5.1.2.2 safely.

5.1.2.2 When responding to hazardous materials/WMD incidents, operations level responders shall be able to perform the following tasks:

(1) Analyze a hazardous materials/WMD incident to determine the scope of the problem and potential outcomes by completing the following tasks:
(a) Survey a hazardous materials/WMD incident to identify the containers and materials involved, determine whether hazardous materials/WMD have been released, and evaluate the surrounding conditions
(b) Collect hazard and response information from MSDS; CHEMTREC/CANUTEC/SETIQ; local, state, and federal authorities; and shipper/manufacturer contacts
(c) Predict the likely behavior of a hazardous materials/WMD and its container
(d) Estimate the potential harm at a hazardous materials/WMD incident

(2) Plan an initial response to a hazardous materials/WMD incident within the capabilities and competencies of available personnel and personal protective equipment by completing the following tasks:
(a) Describe the response objectives for the hazardous materials/WMD incident
(b) Describe the response options available for each objective
(c) Determine whether the personal protective equipment provided is appropriate for implementing each option
(d) Describe emergency decontamination procedures
(e) Develop a plan of action, including safety considerations

(3) Implement the planned response for a hazardous materials/WMD incident to favorably change the outcomes consistent with the emergency response plan and/or standard operating procedures by completing the following tasks:
(a) Establish and enforce scene control procedures, including control zones, emergency decontamination, and communications
(b) Where criminal or terrorist acts are suspected, establish means of evidence preservation
(c) Initiate an incident command system (ICS) for hazardous materials/WMD incidents
(d) Perform tasks assigned as identified in the incident action plan
(e) Demonstrate emergency decontamination

(4) Evaluate the progress of the actions taken at a hazardous materials/WMD incident to ensure that the response objectives are being met safely, effectively, and efficiently by completing the following tasks:
(a) Evaluate the status of the actions taken in accomplishing the response objectives
(b) Communicate the status of the planned response

5.2 **Core Competencies—Analyzing the Incident.**

5.2.1 **Surveying Hazardous Materials/WMD Incidents.** Given scenarios involving hazardous materials/WMD incidents, the operations level responder shall collect information about the incident to identify the containers, the materials involved, the surrounding conditions, and whether hazardous materials/WMD have been released by completing the requirements of 5.2.1.1 through 5.2.1.6.

5.2.1.1 Given three examples each of liquid, gas, and solid hazardous material or WMD, including various hazard classes, operations level personnel shall identify the general shapes of containers in which the hazardous materials/WMD are typically found.

5.2.1.1.1 Given examples of the following tank cars, the operations level responder shall identify each tank car by type, as follows:

(1) Cryogenic liquid tank cars

(2) Nonpressure tank cars (general service or low pressure cars)

(3) Pressure tank cars

5.2.1.1.2 Given examples of the following intermodal tanks, the operations level responder shall identify each intermodal tank by type, as follows:

(1) Nonpressure intermodal tanks

(2) Pressure intermodal tanks

(3) Specialized intermodal tanks, including the following:

(a) Cryogenic intermodal tanks

(b) Tube modules

5.2.1.1.3 Given examples of the following cargo tanks, the operations level responder shall identify each cargo tank by type, as follows:

(1) Compressed gas tube trailers

(2) Corrosive liquid tanks

(3) Cryogenic liquid tanks

(4) Dry bulk cargo tanks

(5) High pressure tanks

(6) Low pressure chemical tanks

(7) Nonpressure liquid tanks

5.2.1.1.4 Given examples of the following storage tanks, the operations level responder shall identify each tank by type, as follows:

(1) Cryogenic liquid tank

(2) Nonpressure tank

(3) Pressure tank

5.2.1.1.5 Given examples of the following nonbulk packaging, the operations level responder shall identify each package by type, as follows:

(1) Bags

(2) Carboys

(3) Cylinders

(4) Drums

(5) Dewar flask (cryogenic liquids)

5.2.1.1.6 Given examples of the following packaging, the operations level responder shall identify the characteristics of each container or package by type as follows:

(1) Intermediate bulk container (IBC)

(2) Ton container

5.2.1.1.7 Given examples of the following radioactive material packages, the operations level responder shall identify the characteristics of each container or package by type, as follows:

(1) Excepted

(2) Industrial

(3) Type A

(4) Type B

(5) Type C

5.2.1.2 Given examples of containers, the operations level responder shall identify the markings that differentiate one container from another.

5.2.1.2.1 Given examples of the following marked transport vehicles and their corresponding shipping papers, the operations level responder shall identify the following vehicle or tank identification marking:

(1) Highway transport vehicles, including cargo tanks

(2) Intermodal equipment, including tank containers

(3) Rail transport vehicles, including tank cars

5.2.1.2.2 Given examples of facility containers, the operations level responder shall identify the markings indicating container size, product contained, and/or site identification numbers.

5.2.1.3 Given examples of hazardous materials incidents, the operations level responder shall identify the name(s) of the hazardous material(s) in 5.2.1.3.1 through 5.2.1.3.3.

5.2.1.3.1 The operations level responder shall identify the following information on a pipeline marker:

(1) Emergency telephone number

(2) Owner

(3) Product

5.2.1.3.2 Given a pesticide label, the operations level responder shall identify each of the following pieces of information, then match the piece of information to its significance in surveying hazardous materials incidents:

(1) Active ingredient

(2) Hazard statement

(3) Name of pesticide

(4) Pest control product (PCP) number (in Canada)

(5) Precautionary statement

(6) Signal word

5.2.1.3.3 Given a label for a radioactive material, the operations level responder shall identify the type or category of label, contents, activity, transport index, and criticality safety index as applicable.

5.2.1.4 The operations level responder shall identify and list the surrounding conditions that should be noted when a hazardous materials/WMD incident is surveyed.

5.2.1.5 The operations level responder shall describe ways to verify information obtained from the survey of a hazardous materials/WMD incident.

5.2.1.6 The operations level responder shall identify at least three additional hazards that could be associated with an incident involving terrorist or criminal activities.

5.2.2 Collecting Hazard and Response Information. Given scenarios involving known hazardous materials/WMD, the operations level responder shall collect hazard and response information using MSDS, CHEMTREC/CANUTEC/SETIQ, governmental authorities, and shippers and manufacturers by completing the following requirements:

(1) Match the definitions associated with the UN/DOT hazard classes and divisions of hazardous materials/WMD, including refrigerated liquefied gases and cryogenic liquids, with the class or division

(2) Identify two ways to obtain an MSDS in an emergency

(3) Using an MSDS for a specified material, identify the following hazard and response information:

(a) Physical and chemical characteristics

(b) Physical hazards of the material

(c) Health hazards of the material

(d) Signs and symptoms of exposure

(e) Routes of entry

(f) Permissible exposure limits

(g) Responsible party contact

(h) Precautions for safe handling (including hygiene practices, protective measures, and procedures for cleanup of spills and leaks)

(i) Applicable control measures, including personal protective equipment

(j) Emergency and first-aid procedures

(4) Identify the following:

(a) Type of assistance provided by CHEMTREC/CANUTEC/SETIQ and governmental authorities

(b) Procedure for contacting CHEMTREC/CANUTEC/SETIQ and governmental authorities

(c) Information to be furnished to CHEMTREC/CANUTEC/SETIQ and governmental authorities

(5) Identify two methods of contacting the manufacturer or shipper to obtain hazard and response information

(6) Identify the type of assistance provided by governmental authorities with respect to criminal or terrorist activities involving the release or potential release of hazardous materials/WMD

(7) Identify the procedure for contacting local, state, and federal authorities as specified in the emergency response plan and/or standard operating procedures

(8) Describe the properties and characteristics of the following:

(a) Alpha radiation

(b) Beta radiation

(c) Gamma radiation

(d) Neutron radiation

5.2.3 **Predicting the Likely Behavior of a Material and Its Container.** Given scenarios involving hazardous materials/WMD incidents, each with a single hazardous material/WMD, the operations level responder shall describe the likely behavior of the material or agent and its container by completing the following requirements:

(1) Use the hazard and response information obtained from the current edition of the DOT Emergency Response Guidebook, MSDS, CHEMTREC/CANUTEC/SETIQ, governmental authorities, and shipper and manufacturer contacts, as follows:

(a) Match the following chemical and physical properties with their significance and impact on the behavior of the container and its contents:

i. Boiling point

ii. Chemical reactivity

iii. Corrosivity (pH)

iv. Flammable (explosive) range [lower explosive limit (LEL) and upper explosive limit (UEL)]

v. Flash point

vi. Ignition (autoignition) temperature

vii. Particle size

viii. Persistence

ix. Physical state (solid, liquid, gas)

x. Radiation (ionizing and non-ionizing)

xi. Specific gravity

xii. Toxic products of combustion

xiii. Vapor density

xiv. Vapor pressure

xv. Water solubility

(b) Identify the differences between the following terms:

i. Contamination and secondary contamination

ii. Exposure and contamination

iii. Exposure and hazard

iv. Infectious and contagious

v. Acute effects and chronic effects

vi. Acute exposures and chronic exposures

(2) Identify three types of stress that can cause a container system to release its contents

(3) Identify five ways in which containers can breach

(4) Identify four ways in which containers can release their contents

(5) Identify at least four dispersion patterns that can be created upon release of a hazardous material

(6) Identify the time frames for estimating the duration that hazardous materials/WMD will present an exposure risk

(7) Identify the health and physical hazards that could cause harm

(8) Identify the health hazards associated with the following terms:

(a) Alpha, beta, gamma, and neutron radiation

(b) Asphyxiant

(c) Carcinogen

(d) Convulsant

(e) Corrosive

(f) Highly toxic

(g) Irritant

(h) Sensitizer, allergen

(i) Target organ effects

(j) Toxic

(9) Given the following, identify the corresponding UN/DOT hazard class and division:

(a) Blood agents

(b) Biological agents and biological toxins

(c) Choking agents

(d) Irritants (riot control agents)

(e) Nerve agents

(f) Radiological materials

(g) Vesicants (blister agents)

5.2.4 **Estimating Potential Harm.** Given scenarios involving hazardous materials/WMD incidents, the operations level responder shall describe the potential harm within the endangered area at each incident by completing the following requirements:

(1) Identify a resource for determining the size of an endangered area of a hazardous materials/WMD incident

(2) Given the dimensions of the endangered area and the surrounding conditions at a hazardous materials/WMD incident, describe the number and type of exposures within that endangered area

(3) Identify resources available for determining the concentrations of a released hazardous materials/WMD within an endangered area

(4) Given the concentrations of the released material, describe the factors for determining the extent of physical, health, and safety hazards within the endangered area of a hazardous materials/WMD incident

(5) Describe the impact that time, distance, and shielding have on exposure to radioactive materials specific to the expected dose rate

5.3 Core Competencies—Planning the Response.

5.3.1 Describing Response Objectives. Given at least two scenarios involving hazardous materials/WMD incidents, the operations level responder shall describe the response objectives for each example by completing the following requirements:

(1) Given an analysis of a hazardous materials/WMD incident and the exposures, describe the number of exposures that could be saved with the resources provided by the AHJ

(2) Given an analysis of a hazardous materials/WMD incident, describe the steps for determining response objectives

(3) Describe how to assess the risk to a responder for each hazard class in rescuing injured persons at a hazardous materials/WMD incident

(4) Describe the potential for secondary attacks and devices at criminal or terrorist events

5.3.2 Identifying Action Options. Given examples of hazardous materials/WMD incidents (facility and transportation), the operations level responder shall identify the options for each response objective and shall meet the following requirements:

(1) Identify the options to accomplish a given response objective

(2) Describe the prioritization of emergency medical care and removal of victims from the hazard area relative to exposure and contamination concerns

5.3.3 Determining Suitability of Personal Protective Equipment. Given examples of hazardous materials/WMD incidents, including the names of the hazardous materials/WMD involved and the anticipated type of exposure, the operations level responder shall determine whether available personal protective equipment is applicable to performing assigned tasks by completing the following requirements:

(1) Identify the respiratory protection required for a given response option and the following:

(a) Describe the advantages, limitations, uses, and operational components of the following types of respiratory protection at hazardous materials/WMD incidents:

i. Positive pressure self-contained breathing apparatus (SCBA)

ii. Positive pressure air-line respirator with required escape unit

iii. Closed-circuit SCBA

iv. Powered air-purifying respirator (PAPR)

v. Air-purifying respirator (APR)

vi. Particulate respirator

(b) Identify the required physical capabilities and limitations of personnel working in respiratory protection

(2) Identify the personal protective clothing required for a given option and the following:

(a) Identify skin contact hazards encountered at hazardous materials/WMD incidents

(b) Identify the purpose, advantages, and limitations of the following types of protective clothing at hazardous materials/WMD incidents:

i. Chemical-protective clothing such as liquid splash–protective clothing and vapor-protective clothing

ii. High temperature–protective clothing such as proximity suit and entry suits

iii. Structural fire-fighting protective clothing

5.3.4 Identifying Decontamination Issues. Given scenarios involving hazardous materials/WMD incidents, the operations level responder shall identify when decontamination is needed by completing the following requirements:

(1) Identify ways that people, personal protective equipment, apparatus, tools, and equipment become contaminated

(2) Describe how the potential for secondary contamination determines the need for decontamination

(3) Explain the importance and limitations of decontamination procedures at hazardous materials incidents

(4) Identify the purpose of emergency decontamination procedures at hazardous materials incidents

(5) Identify the methods, advantages, and limitations of emergency decontamination procedures

5.4 Core Competencies—Implementing the Planned Response.

5.4.1 Establishing Scene Control. Given two scenarios involving hazardous materials/WMD incidents, the operations level responder shall explain how to establish and maintain scene control, including control zones and emergency decontamination, and communications between responders and to the public by completing the following requirements:

(1) Identify the procedures for establishing scene control through control zones

(2) Identify the criteria for determining the locations of the control zones at hazardous materials/WMD incidents

(3) Identify the basic techniques for the following protective actions at hazardous materials/WMD incidents:

(a) Evacuation

(b) Shelter-in-place

(4) Demonstrate the ability to perform emergency decontamination

(5) Identify the items to be considered in a safety briefing prior to allowing personnel to work at the following:

(a) Hazardous material incidents

(b) Hazardous materials/WMD incidents involving criminal activities

(6) Identify the procedures for ensuring coordinated communication between responders and to the public

5.4.2 Preserving Evidence. Given two scenarios involving hazardous materials/WMD incidents, the operations level responder shall describe the process to preserve evidence as listed in the emergency response plan and/or standard operating procedures.

5.4.3 Initiating the Incident Command System. Given scenarios involving hazardous materials/WMD incidents, the operations level responder shall implement the incident command system as required by the AHJ by completing the following requirements:

(1) Identify the role of the operations level responder during hazardous materials/WMD incidents as specified in the emergency response plan and/or standard operating procedures

(2) Identify the levels of hazardous materials/WMD incidents as defined in the emergency response plan

(3) Identify the purpose, need, benefits, and elements of the incident command system for hazardous materials/WMD incidents

(4) Identify the duties and responsibilities of the following functions within the incident management system:

(a) Incident safety officer

(b) Hazardous materials branch or group

(5) Identify the considerations for determining the location of the incident command post for a hazardous materials/WMD incident

(6) Identify the procedures for requesting additional resources at a hazardous materials/WMD incident

(7) Describe the role and response objectives of other agencies that respond to hazardous materials/WMD incidents

5.4.4 **Using Personal Protective Equipment.** Given the personal protective equipment provided by the AHJ, the operations level responder shall describe considerations for the use of personal protective equipment provided by the AHJ by completing the following requirements:

(1) Identify the importance of the buddy system

(2) Identify the importance of the backup personnel

(3) Identify the safety precautions to be observed when approaching and working at hazardous materials/WMD incidents

(4) Identify the signs and symptoms of heat and cold stress and procedures for their control

(5) Identify the capabilities and limitations of personnel working in the personal protective equipment provided by the AHJ

(6) Identify the procedures for cleaning, disinfecting, and inspecting personal protective equipment provided by the AHJ

(7) Describe the maintenance, testing, inspection, and storage procedures for personal protective equipment provided by the AHJ according to the manufacturer's specifications and recommendations

5.5 **Core Competencies—Evaluating Progress.**

5.5.1 **Evaluating the Status of Planned Response.** Given two scenarios involving hazardous materials/WMD incidents, including the incident action plan, the operations level responder shall determine the effectiveness of the actions taken in accomplishing the response objectives and shall meet the following requirements:

(1) Identify the considerations for evaluating whether actions taken were effective in accomplishing the objectives

(2) Describe the circumstances under which it would be prudent to withdraw from a hazardous materials/WMD incident

5.5.2 **Communicating the Status of Planned Response.** Given two scenarios involving hazardous materials/WMD incidents, including the incident action plan, the operations level responder shall report the status of the planned response through the normal chain of command by completing the following requirements:

(1) Identify the procedures for reporting the status of the planned response through the normal chain of command

(2) Identify the methods for immediate notification of the incident commander and other response personnel about critical emergency conditions at the incident.

5.6 **Competencies—Terminating the Incident. (Reserved)**

Chapter 6 Competencies for Operations Level Responders Assigned Mission-Specific Responsibilities

6.2 **Mission-Specific Competencies: Personal Protective Equipment.**

6.2.1 **General.**

6.2.1.1 **Introduction.**

6.2.1.1.1 The operations level responder assigned to use personal protective equipment shall be that person, competent at the operations level, who is assigned to use personal protective equipment at hazardous materials/WMD incidents.

6.2.1.1.2 The operations level responder assigned to use personal protective equipment at hazardous materials/WMD incidents shall be trained to meet all competencies at the awareness level (*see Chapter 4*), all core competencies at the operations level (*see Chapter 5*), and all competencies in this section.

6.2.1.1.3 The operations level responder assigned to use personal protective equipment at hazardous materials/WMD incidents shall operate under the guidance of a hazardous materials technician, an allied professional, or standard operating procedures.

6.2.1.1.4 The operations level responder assigned to use personal protective equipment shall receive the additional training necessary to meet specific needs of the jurisdiction.

6.2.1.2 Goal. The goal of the competencies in this section shall be to provide the operations level responder assigned to use personal protective equipment with the knowledge and skills to perform the following tasks safely and effectively:

(1) Plan a response within the capabilities of personal protective equipment provided by the AHJ in order to perform mission-specific tasks assigned

(2) Implement the planned response consistent with the standard operating procedures and site safety and control plan by donning, working in, and doffing personal protective equipment provided by the AHJ

(3) Terminate the incident by completing the reports and documentation pertaining to personal protective equipment

6.2.2 **Competencies —Analyzing the Incident. (Reserved)**

6.2.3 **Competencies — Planning the Response.**

6.2.3.1 **Selecting Personal Protective Equipment.** Given scenarios involving hazardous materials/WMD incidents with known and unknown hazardous materials/WMD and the personal protective equipment provided by the AHJ, the operations level responder assigned to use personal protective equipment shall select the personal protective equipment required to support mission-specific tasks at hazardous materials/WMD incidents based on local procedures by completing the following requirements:

(1) Describe the types of personal protective equipment that are available for response based on NFPA standards and how these items relate to EPA levels of protection

(2) Describe personal protective equipment options for the following hazards:

(a) Thermal

(b) Radiological

(c) Asphyxiating

(d) Chemical

(e) Etiological/biological

(f) Mechanical

(3) Select personal protective equipment for mission-specific tasks at hazardous materials/WMD incidents based on local procedures

(a) Describe the following terms and explain their impact and significance on the selection of chemical protective clothing:

i. Degradation

ii. Penetration

iii. Permeation

(b) Identify at least three indications of material degradation of chemical-protective clothing

(c) Identify the different designs of vapor-protective and splash-protective clothing and describe the advantages and disadvantages of each type

(d) Identify the relative advantages and disadvantages of the following heat exchange units used for the cooling of personnel operating in personal protective equipment:

i. Air cooled

ii. Ice cooled

iii. Water cooled

iv. Phase change cooling technology

(e) Identify the physiological and psychological stresses that can affect users of personal protective equipment

(f) Describe local procedures for going through the technical decontamination process

6.2.4 Competencies — Implementing the Planned Response.

6.2.4.1 Using Protective Clothing and Respiratory Protection. Given the personal protective equipment provided by the AHJ, the operations level responder assigned to use personal protective equipment shall demonstrate the ability to don, work in, and doff the equipment provided to support mission specific tasks by completing the following requirements:

(1) Describe at least three safety procedures for personnel wearing protective clothing

(2) Describe at least three emergency procedures for personnel wearing protective clothing

(3) Demonstrate the ability to don, work in, and doff personal protective equipment provided by the AHJ

(4) Demonstrate local procedures for responders undergoing the technical decontamination process

(5) Describe the maintenance, testing, inspection, storage, and documentation procedures for personal protective equipment provided by the AHJ according to the manufacturer's specifications and recommendations

6.2.5 Competencies — Terminating the Incident.

6.2.5.1 Reporting and Documenting the Incident. Given a scenario involving a hazardous materials/WMD incident, the operations level responder assigned to use personal protective equipment shall document use of the personal protective equipment by completing the documentation requirements of the emergency response plan or standard operating procedures regarding personal protective equipment.

6.6 Mission-Specific Competencies: Product Control.

6.6.1 General.

6.6.1.1 Introduction.

6.6.1.1.1 The operations level responder assigned to perform product control shall be that person, competent at the operations level, who is assigned to implement product control measures at hazardous materials/WMD incidents.

6.6.1.1.2 The operations level responder assigned to perform product control at hazardous materials/WMD incidents shall be trained to meet all competencies at the awareness level (*see Chapter 4*), all core competencies at the operations level (*see Chapter 5*), all mission-specific competencies for personal protective equipment (*see Section 6.2*), and all competencies in this section.

6.6.1.1.3 The operations level responder assigned to perform product control at hazardous materials/WMD incidents shall operate under the guidance of a hazardous materials technician, an allied professional, or standard operating procedures.

6.6.1.1.4 The operations level responder assigned to perform product control at hazardous materials/WMD incidents shall receive the additional training necessary to meet specific needs of the jurisdiction.

6.6.1.2 Goal.

6.6.1.2.1 The goal of the competencies in this section shall be to provide the operations level responder assigned to product control at hazardous materials/WMD incidents with the knowledge and skills to perform the tasks in 6.6.1.2.2 safely and effectively.

6.6.1.2.2 When responding to hazardous materials/WMD incidents, the operations level responder assigned to perform product control shall be able to perform the following tasks:

(1) Plan an initial response within the capabilities and competencies of available personnel, personal protective equipment, and control equipment and in accordance with the emergency response plan or standard operating procedures by completing the following tasks:

(A) Describe the control options available to the operations level responder

(B) Describe the control options available for flammable liquid and flammable gas incidents

(2) Implement the planned response to a hazardous materials/WMD incident

6.6.2 Competencies—Analyzing the Incident. (Reserved)

6.6.3 Competencies—Planning the Response.

6.6.3.1 Identifying Control Options. Given examples of hazardous materials/WMD incidents, the operations level responder assigned to perform product control shall identify the options for each response objective by completing the following requirements as prescribed by the AHJ:

(1) Identify the options to accomplish a given response objective

(2) Identify the purpose for and the procedures, equipment, and safety precautions associated with each of the following control techniques:

(A) Absorption

(B) Adsorption

(C) Damming

(D) Diking

(E) Dilution

(F) Diversion

(G) Remote valve shutoff

(H) Retention

(I) Vapor dispersion

(J) Vapor suppression

6.6.3.2 Selecting Personal Protective Equipment. Given the personal protective equipment provided by the AHJ, the operations level responder assigned to perform product control shall select the personal protective equipment required to support product control at hazardous materials/WMD incidents based on local procedures (*see Section 6.2*).

6.6.4 Competencies—Implementing the Planned Response.

6.6.4.1 Performing Control Options. Given an incident action plan for a hazardous materials/WMD incident, within the capabilities and equipment provided by the AHJ, the operations level responder assigned to perform product control shall demonstrate control functions set out in the plan by completing the following requirements as prescribed by the AHJ:

(1) Using the type of special purpose or hazard suppressing foams or agents and foam equipment furnished by the AHJ, demonstrate the application of the foam(s) or agent(s) on a spill or fire involving hazardous materials/WMD

(2) Identify the characteristics and applicability of the following Class B foams if supplied by the AHJ:

- **(A)** Aqueous film-forming foam (AFFF)
- **(B)** Alcohol-resistant concentrates
- **(C)** Fluoroprotein
- **(D)** High-expansion foam

(3) Given the required tools and equipment, demonstrate how to perform the following control activities:

- **(A)** Absorption
- **(B)** Adsorption
- **(C)** Damming
- **(D)** Diking
- **(E)** Dilution
- **(F)** Diversion
- **(G)** Retention
- **(H)** Remote valve shutoff
- **(I)** Vapor dispersion
- **(J)** Vapor suppression

(4) Identify the location and describe the use of emergency remote shutoff devices on MC/DOT-306/406, MC/DOT- 307/407, and MC-331 cargo tanks containing flammable liquids or gases

(5) Describe the use of emergency remote shutoff devices at fixed facilities

6.6.4.2 The operations level responder assigned to perform product control shall describe local procedures for going through the technical decontamination process.

6.6.5 Competencies—Evaluating Progress. (Reserved)

6.6.6 Competencies—Terminating the Incident. (Reserved)

NFPA 1001 and 472 Correlation Guide

Fire Fighter I

NFPA 1001, *STANDARD FOR FIRE FIGHTER PROFESSIONAL QUALIFICATIONS*, 2013 EDITION	CORRESPONDING CHAPTER(S)	CORRESPONDING PAGE(S)
5.1	1	5–6
5.1.1	1, 2	6–7, 8–10, 11, 28–31
5.1.2	1, 3, 10	8, 44–49, 50–53, 73, 272–274, 277–293
5.2	4	92–93
5.2.1	4	93–96, 98–99
5.2.1(A)	4	96
5.2.1(B)	4	102, 111
5.2.2	4	96–97, 111
5.2.2(A)	4	111
5.2.2(B)	4	111
5.2.3	4	103–104, 105–108
5.2.3(A)	4	107–108
5.2.3(B)	4	107
5.2.4	18	591–592
5.2.4(A)	18	591–592
5.2.4(B)	18	591–592
5.3	3	49, 51, 54
5.3.1	3	62, 73
5.3.1(A)	3, 22	54–56, 58–59, 59–64, 73, 686–689
5.3.1(B)	3, 16, 18	62, 64–65, 66–72, 75–76, 78–80, 504, 596–597
5.3.2	11	301–302
5.3.2(A)	11	302–303
5.3.2(B)	11	302–303
5.3.3	11	303–306
5.3.3(A)	11	303–304
5.3.3(B)	11	303
5.3.4	9, 12	248–249, 320–321
5.3.4(A)	12	325–344
5.3.4(B)	12	321, 327–332, 334, 337–348
5.3.5	18	588–589, 596–600
5.3.5(A)	18	587, 589–591, 593, 596–600
5.3.5(B)	18	594
5.3.6	13	366–368
5.3.6(A)	13	357–362, 369–370

NFPA 1001, *STANDARD FOR FIRE FIGHTER PROFESSIONAL QUALIFICATIONS*, 2013 EDITION	CORRESPONDING CHAPTER(S)	CORRESPONDING PAGE(S)
5.3.6(B)	13	371–396
5.3.7	22	696–697
5.3.7(A)	22, 26	697–699, 810–814, 821–823
5.3.7(B)	22	697
5.3.8	9, 22, 38	249–251, 682, 695, 1073–1074
5.3.8(A)	22, 38	687–690, 695–696, 1073–1074
5.3.8(B)	22, 38	683–691, 692, 695, 1073–1074
5.3.9	14	404–407, 413–414
5.3.9(A)	14	405–418, 427–428
5.3.9(B)	14, 18	414–415, 418–435, 596–597, 601–604
5.3.10	22	680–684, 690–694
5.3.10(A)	6, 17, 22	143, 152–154, 552–553, 559–562, 566–572, 690, 692–693
5.3.10(B)	16, 17, 22	508, 510–511, 537–541, 554, 557–559, 561–564, 566–572, 684, 692–693
5.3.11	9, 15	250–251, 451–458
5.3.11(A)	6, 15	144–145, 149–156, 442–451, 455–460
5.3.11(B)	15	451–459
5.3.12	15	459–463, 467–468
5.3.12(A)	6, 7, 15	146–147, 152–153, , 174–180, 446, 463–467, 470–478
5.3.12(B)	13, 15	391-393, 395, 453–481
5.3.13	9, 19, 38	251, 632–636, 1077–1078
5.3.13(A)	17, 19, 38	552–553, 633–636, 1077–1078
5.3.13(B)	38	1077–1078, 1080–1083
5.3.14	19	632, 636
5.3.14(A)	12, 19, 37	349, 616–623, 1034, 1058
5.3.14(B)	19	618–638
5.3.15	16	505–506, 508–512
5.3.15(A)	16	489–491, 495–498, 504–505
5.3.15(B)	16, 17	491, 497, 499, 507, 523–537, 567–568
5.3.16	8	216–218, 225–227
5.3.16(A)	8	199–216
5.3.16(B)	8	216, 218–226
5.3.17	19	612–615
5.3.17(A)	19	612, 614–615
5.3.17(B)	19	614–615
5.3.18	11, 22	305–306, 701
5.3.18(A)	11, 22	305–306, 701–702
5.3.18(B)	11, 22	305–306, 701
5.3.19	21	671–672
5.3.19(A)	21	662–663, 665–670
5.3.19(B)	21	663–664, 672–673
5.3.20	10	272–279, 286
5.3.20(A)	10	260–265, 268–272
5.3.20(B)	10	272–274, 277–293
5.4	General	General
5.5	General	General
5.5.1	3, 9, 10, 13, 19	73, 73–78, 81–82, 252–253, 268–271, 362–366, 628

NFPA 1001, *STANDARD FOR FIRE FIGHTER PROFESSIONAL QUALIFICATIONS*, 2013 EDITION	CORRESPONDING CHAPTER(S)	CORRESPONDING PAGE(S)
5.5.1(A)	9, 10	252–253, 268
5.5.1(B)	9, 10	236–237, 252–253, 268
5.5.2	16	513–519, 522–532, 544
5.5.2(A)	16, 17	512–513, 543, 567, 568
5.5.2(B)	16, 17	513–514, 554–558, 560

Fire Fighter II

NFPA 1001, *STANDARD FOR FIRE FIGHTER PROFESSIONAL QUALIFICATIONS*, 2013 EDITION	CORRESPONDING CHAPTER(S)	CORRESPONDING PAGE(S)
6.1	1	6
6.1.1	1, 5	6, 119–134
6.1.2	5	130, 131, 134
6.2	4	109–110
6.2.1	4	109–110
6.2.1(A)	4	109–110
6.2.1(B)	4	110
6.2.2	4	109
6.2.2(A)	4	109
6.2.2(B)	4	107
6.3	17	573
6.3.1	17	573
6.3.1(A)	17	573–578
6.3.1(B)	17	576–578
6.3.2	15, 22	450, 682, 691
6.3.2(A)	15, 22	464, 683–684
6.3.2(B)	15, 22	450, 683–686, 692–693
6.3.3	22	700–701
6.3.3(A)	22	699–701
6.3.3(B)	22	699–701
6.3.4	38	1083–1084
6.3.4(A)	38	1073–1074, 1076, 1083–1084
6.3.4(B)	38	1075–1076, 1083–1084
6.4	26, 27	813, 815–834, 847, 850
6.4.1	26	817–819, 821–834
6.4.1(A)	26	811, 814–817, 821–829
6.4.1(B)	26	813–834
6.4.2	10, 27	265–268, 841–846, 850–860
6.4.2(A)	9, 27	250, 840–846, 850–860
6.4.2(B)	9, 27	250, 847, 850
6.5	23, 36	710–711, 1022–1026
6.5.1	36	1022–1026
6.5.1(A)	36	1015–1022
6.5.1(B)	36	1020–1021, 1025–1026
6.5.2	36	1014, 1017–1019, 1021
6.5.2(A)	36	1025–1028

NFPA 1001, *STANDARD FOR FIRE FIGHTER PROFESSIONAL QUALIFICATIONS*, 2013 EDITION	CORRESPONDING CHAPTER(S)	CORRESPONDING PAGE(S)
6.5.2(B)	36	1027–1028
6.5.3	23	711, 718–725
6.5.3(A)	23, 37	711–713, 718–720, 1034–1064
6.5.3(B)	23	711–713, 717–718
6.5.4	9, 19	252–253, 615
6.5.4(A)	9, 19	252–253, 615
6.5.4(B)	9, 19	525, 615
6.5.5	16	513–514
6.5.5(A)	16	513–514
6.5.5(B)	16	513–514

NFPA 472: Core Competencies for Awareness Level Responders

NFPA 472, *STANDARD FOR COMPETENCE OF RESPONDERS TO HAZARDOUS MATERIALS/ WEAPONS OF MASS DESTRUCTION INCIDENTS*, 2013 EDITION	CORRESPONDING CHAPTER(S)	CORRESPONDING PAGE(S)
4.1	General	General
4.1.1	General	General
4.1.1.1	28	869–871
4.1.1.2	28	869–871
4.1.1.3	28	869–871
4.1.2	General	General
4.1.2.1	28, 29, 30, 31	868–874, 880–894, 902–917, 924–930
4.1.2.2	28, 29, 30, 31	868–874, 880–894, 902–917, 924–930
4.2	General	General
4.2.1	28, 29, 30, 31	868–869, 873–874, 888–894, 902–916, 924, 928–930
4.2.2	30	910–914
4.2.3	30, 31	912, 914–915, 924
4.3	Reserved	Reserved
4.4	General	General
4.4.1	28, 31, 32	870–871, 873–874, 924–930, 936–955, 962–963
4.4.2	31	924–926
4.5	General	General
4.6	General	General

NFPA 472: Core Competencies for Operations Level Responders

NFPA 472, *STANDARD FOR COMPETENCE OF RESPONDERS TO HAZARDOUS MATERIALS/WEAPONS OF MASS DESTRUCTION INCIDENTS*, 2013 EDITION	CORRESPONDING CHAPTER(S)	CORRESPONDING PAGE(S)
5.1	28, 30	868–869
5.1.1	28	868–869
5.1.1.1	28	868–869, 871
5.1.1.2	28	869–871
5.1.1.3	28	870–871
5.1.2	30	902

NFPA 472, *STANDARD FOR COMPETENCE OF RESPONDERS TO HAZARDOUS MATERIALS/WEAPONS OF MASS DESTRUCTION INCIDENTS*, 2013 EDITION	CORRESPONDING CHAPTER(S)	CORRESPONDING PAGE(S)
5.1.2.1	General	General
5.1.2.2	30, 31	902–903, 910, 915, 925–930
5.2	30	902
5.2.1	30	902
5.2.1.1	30	903–910, 915–916
5.2.1.1.1	30	907–910
5.2.1.1.2	30	907–910
5.2.1.1.3	30	906–908
5.2.1.1.4	30	906–908
5.2.1.1.5	30	904–906
5.2.1.1.6	30	903–904
5.2.1.1.7	30	915–916
5.2.1.2	30	904
5.2.1.2.1	30	907–910
5.2.1.2.2	30	903
5.2.1.3	30	904
5.2.1.3.1	30	910
5.2.1.3.2	30	905
5.2.1.3.3	30	915–916
5.2.1.4	31	925–926
5.2.1.5	30	914
5.2.1.6	30	916
5.2.2	29, 30	887, 911–915
5.2.3	29	880–894
5.2.4	29, 31, 33	887, 928, 962
5.3	31	925–926
5.3.1	31	928–930
5.3.2	31	925–926
5.3.3	32	936–953
5.3.4	34	978–985
5.4	33	962–963
5.4.1	32, 33	954, 962–963
5.4.2	33	972
5.4.3	31	924–925, 930
5.4.4	32	950–953, 955
5.5	33	972
5.5.1	33	972
5.5.2	31	924–925
5.6	33	972

NFPA 472: Core Competencies for Operations Level Responders Assigned Mission-Specific Responsibilites

NFPA 472, *STANDARD FOR COMPETENCE OF RESPONDERS TO HAZARDOUS MATERIALS/WEAPONS OF MASS DESTRUCTION INCIDENTS,* 2013 EDITION	CORRESPONDING CHAPTER(S)	CORRESPONDING PAGE(S)
6.2	General	General
6.2.1	General	General
6.2.1.1	General	General
6.2.1.1.1	32	936–955
6.2.1.1.2	28, 29, 30, 31, 32, 33	870–871, 880–894, 902–916, 924–925, 936–955, 962–963
6.2.1.1.3	32	936–955
6.2.1.1.4	32	936–955
6.2.1.2	32	936–955
6.2.2	General	General
6.2.3	General	General
6.2.3.1	32	936–955
6.2.4	General	General
6.2.4.1	32	936–955
6.2.5	General	General
6.2.5.1	31	924–930
6.6	28	868–869
6.6.1	28	868–869
6.6.1.1	28	868–869
6.6.1.1.1	28	868–869
6.6.1.1.2	28, 29, 30, 31, 32, 33	870–871, 880–894, 902–916, 924–925, 936–955, 962–963
6.6.1.1.3	28	870–871
6.6.1.1.4	28	870–871
6.6.1.2	28	871
6.6.1.2.1	28	871
6.6.1.2.2	33	965–966
6.6.2	33	963–964
6.6.3	33	965–972
6.6.3.1	33	965–972
6.6.3.2	32	936–953
6.6.4	33	965–972
6.6.4.1	33	965–972
6.6.4.2	34	984–985
6.6.5	33	965–972
6.6.6	33	972

Pro Board Assessment Methodology Matrices for NFPA 1001

NFPA 1001 - Fire Fighter I - 2013 Edition

INSTRUCTIONS: In the column titled 'Cognitive/Written Test' place the number of questions from the Test Bank that are used to evaluate the applicable JPR, RK, RS, or objective. In the column titled 'Manipulative/Skills Station' identify the skill sheets that are used to evaluate the applicable JPR, RS, or objective. When the Portfolio or Projects method is used to evaluate a particular JPR, RK, RS, or objective, identify the applicable section in the appropriate column and provide the procedures to be used as outlined in the NBFSPQ Operational Procedures, COA-5. Evaluation methods that are not cognitive, manipulative, portfolio, or project based should be identified in the 'Other' column.

OBJECTIVE / JPR, RK, RS		COGNITIVE	MANIPULATIVE			
SECTION	ABBREVIATED TEXT	WRITTEN TEST	SKILLS STATION	PORTFOLIO	PROJECTS	OTHER
5.1	Requirements of NFPA 472, Operations Level; Chapter 5, Sec 6.6 (to include Sec 6.2)					Chapter 1 (p 5–6), Chapter 3 (p 44–54), Chapter 10 (p 260–293), Chapters 28–34 (p 868–930)
5.2.1	Initiate the response to a reported emergency					Chapter 4 (p 93–96, 98–99)
5.2.1(A)	RK: reporting an emergency					Chapter 4 (p 96)
5.2.1(B)	RS: operate communications equipment					Chapter 4 (p 102, 111)
5.2.2	Receive a business or personal telephone call					Chapter 4 (p 111)
5.2.2(A)	RK: Fire department procedures for answering nonemergency telephone calls					Chapter 4 (p 111)
5.2.2(B)	RS: fire station telephone and intercom equipment					Chapter 4, Skill Drill 4-5 (p 111)
5.2.3	Transmit and receive messages via the fire department radio					Chapter 4 (p 103–104, 105–108)
5.2.3(A)	RK: Departmental radio procedures					Chapter 4 (p 107–108)
5.2.3(B)	RS: radio equipment and discriminate between routine and emergency traffic					Chapter 4, Skill Drill 4-3 (p 107)
5.2.4	Activate an emergency call for assistance					Chapter 18 (p 591–592)
5.2.4(A)	RK: Personal Accountability, emergency comms					Chapter 18 (p 591–592)
5.2.4(B)	RS: ability to initiate an emergency call for assistance					Chapter 18, Skill Drill 18-1 (p 591–592)
5.3.1	Use SCBA during emergency operations					Chapter 3 (p 62, 73)
5.3.1(A)	RK: respiratory protection					Chapter 3 (p 54–73)
5.3.1(B)	RS: use SCBA					Chapter 3 (p 70)
5.3.2	Respond on apparatus to an emergency scene					Chapter 11 (p 301–302)
5.3.2(A)	RK: riding fire apparatus					Chapter 11 (p 302–303)
5.3.2(B)	RS: use each piece of provided safety equipment					Chapter 11 (p 302–303)

OBJECTIVE / JPR, RK, RS		COGNITIVE	MANIPULATIVE			
SECTION	ABBREVIATED TEXT	WRITTEN TEST	SKILLS STATION	PORTFOLIO	PROJECTS	OTHER
5.3.3	Operate in established work areas at emergency					Chapter 11 (p 303–306)
5.3.3(A)	RK: Potential hazards at emergency scene					Chapter 11 (p 303–304)
5.3.3(B)	RS: Operate in protected area					Chapter 11 (p 303)
5.3.4	Force entry into a structure					Chapter 12 (p 320–321)
5.3.4(A)	RK: Basic construction					Chapter 12 (p 325–344)
5.3.4(B)	RS: transport and operate hand and power tools and to force entry					Chapter 12, Skill Drill 12-1, 12-2, 12-3, 12-4, 12-5, 12-6, 12-7, 12-8, 12-9, 12-10, 12-11, 12-12, 12-13 (p 327–332, 334, 337–348)
5.3.5	Exit a hazardous area as a team					Chapter 18 (p 588–589, 596–600)
5.3.5(A)	RK: Personnel accountability systems					Chapter 18 (p 589–590)
5.3.5(B)	RS: operate as a team					Chapter 18 (p 594)
5.3.6	Set up ground ladders					Chapter 13 (p 366–368)
5.3.6(A)	RK: parts and hazards of ladders					Chapter 13 (p 357–362)
5.3.6(B)	RS: carry raise and extend ladders					Chapter 13, Skill Drill 13-2, 13-3, 13-4, 13-5, 13-6, 13-7, 13-8, 13-9, 13-10, 13-11, 13-12, 13-13, 13-14, 13-15, 13-16, 13-17, 13-18, 13-19, 13-20 (p 371–395)
5.3.7	Attack a passenger vehicle fire operating as a member of a team					Chapter 22 (p 696–697)
5.3.7(A)	RK: fire streams					Chapter 22 (p 697–699)
5.3.7(B)	RS: automobile fire					Chapter 22 (p 697)
5.3.8	Extinguish fires in exterior Class A materials					Chapter 9, 22, 38 (p 249–251, 683, 684, 1073–1074)
5.3.8(A)	RK: attack lines and water streams					Chapter 22 (687–690, 695–696)
5.3.8(B)	RS: handlines or master streams					Chapter 22 (p 686–691, 695)
5.3.9	Conduct a search and rescue in a structure operating as a member of a team					Chapter 14 (p 404–407, 413–414)
5.3.9(A)	RK: forcible entry tools					Chapter 9 (p 248–249)
5.3.9(B)	RS: SCBA, set up ladders, rescue, and assess areas to determine tenability					Chapter 14, Skill Drill 14-3, 14-4, 14-5, 14-6, 14-7, 14-8, 14-9, 14-10, 14-11, 14-12, 14-13, 14-14, 14-15, 14-16, 14-17, 14-18, 14-19 (p 414–435)
5.3.10	Attack an interior structure fire operating as a member of a team					Chapter 22 (p 684–686, 692–694)
5.3.10(A)	RK: fire streams					Chapter 17, 22 (p 552–553, 559–562, 566–572, 691–694)
5.3.10(B)	RS: water hammers; flow and patterns; apply water; and suppress interior wall and subfloor fires					Chapter 17, 22, Skill Drill 22-2, 22-3, 22-4, 22-9 (p 559, 561–564, 566–572, 685–686, 692)
5.3.11	Perform horizontal ventilation on a structure operating as part of a team					Chapter 15 (p 452–460)
5.3.11(A)	RK: ventilation					Chapter 15 (p 449–460)

OBJECTIVE / JPR, RK, RS		COGNITIVE	MANIPULATIVE			
SECTION	ABBREVIATED TEXT	WRITTEN TEST	SKILLS STATION	PORTFOLIO	PROJECTS	OTHER
5.3.11(B)	RS: ventilation tools and equipment					Chapter 15 (p 453–461)
5.3.12	Perform vertical ventilation on a structure operating as part of a team					Chapter 15 (p 461–464, 467–468)
5.3.12(A)	RK: heat transfer					Chapter 6, 15 (p 146–147, 152, 153, 464–468)
5.3.12(B)	RS: ventilation tools and equipment					Chapter 13, 15 (p 391–393, 395, 453–461)
5.3.13	Overhaul a fire scene					Chapter 19 (p 632–636)
5.3.13(A)	RK: attack lines and water application devices					Chapter 17 (p 552–553)
5.3.13(B)	RS: attack line; expose void spaces; apply water; signs of area of origin and arson; and complete extinguishment					Chapter 17, 19, 38 (p 552–553, 636–638, 683–684, 1077–1078, 1080–1083)
5.3.14	Conserve property as a member of a team					Chapter 19 (p 632, 636)
5.3.14(A)	RK: property conservation					Chapter 19, 37 (p 616–623, 1034, 1058)
5.3.14(B)	RS: covering materials					Chapter 19, Skill Drill 19-11, 19-12, 19-13, 19-14 (p 628–632)
5.3.15	Connect a fire department pumper to a water supply as a member of a team					Chapter 16 (p 505–506, 508–512)
5.3.15(A)	RK: mobile water supply apparatus; fire hydrant operation; and suitable static water supply					Chapter 16 (p 489–491, 495–498, 504–505)
5.3.15(B)	RS: lay a supply hose					Chapter 16 (p 523–537)
5.3.16	Extinguish incipient Class A, Class B, and Class C fires					Chapter 8 (p 216–218, 225–227)
5.3.16(A)	RK: classifications of fire					Chapter 8 (p 199–216)
5.3.16(B)	RS: operate portable fire extinguishers					Chapter 8, Skill Drill 8-1, 8-2, 8-3, 8-4, 8-5, 8-6, 8-7, 8-8, 8-9 (p 216, 218–226)
5.3.17	Illuminate the emergency scene					Chapter 19 (p 612–615)
5.3.17(A)	RK: Safety principles					Chapter 19 (p 612, 614–615)
5.3.17(B)	RS: department power supply					Chapter 19, Skill Drill 19-1 (p 614–615)
5.3.18	Turn off building utilities					Chapter 11, 22 (p 305–306, 701)
5.3.18(A)	RK: electricity, gas, and water systems;					Chapter 11, 22 (p 305–306, 701–702)
5.3.18(B)	RS: utility control devices					Chapter 11, 22 Skill Drill 22-13, 22-14 (p 305–306, 701)
5.3.19	Combat a ground cover fire operating as a member of a team					Chapter 21 (p 671–672)
5.3.19(A)	RK: ground cover fires					Chapter 21 (p 662–663, 665–670)
5.3.19(B)	RS: exposure threats					Chapter 21 (p 672–673)
5.3.20	tie a knot appropriate for hoisting tool					Chapter 10 (p 272–279, 286)
5.3.20(A)	RK: knot types and usage, diff between life safety and utility rope					Chapter 10 (p 260–265, 268–272)

OBJECTIVE / JPR, RK, RS		COGNITIVE	MANIPULATIVE			
SECTION	ABBREVIATED TEXT	WRITTEN TEST	SKILLS STATION	PORTFOLIO	PROJECTS	OTHER
5.3.20(B)	RS: ability to hoist tools using specific knots					Chapter 10, Skill Drill 10-5, 10-6, 10-7, 10-8, 10-9, 10-10, 10-11, 10-12, 10-13, 10-14, 10-15, 10-16, 10-17, 10-18, 10-19, 10-20, 10-21 (p 272–274, 276–293)
5.5.1	Clean and check ladders, ventilation equipment, self-contained breathing apparatus (SCBA), ropes, salvage equipment, and hand tools					Chapter 3, 9, 10, 13, 19 (p 73, 73–78, 81–82, 252–253, 268–271, 362–366, 628)
5.5.1(A)	RK: cleaning methods					Chapter 9, 10 (p 252–253, 268)
5.5.1(B)	RS: correct tools					Chapter 9, 10 (p 236–237, 252–253, 268)
5.5.2	Clean, inspect, and return fire hose to service					Chapter 16 (p 513–519, 522–532, 544)
5.5.2(A)	RK: noting a defective hose					Chapter 16 (p 512–513)
5.5.2(B)	RS: clean hose;					Chapter 16, Skill Drill 16-12 (p 513)

NFPA 1001 - Fire Fighter II - 2013 Edition

INSTRUCTIONS: In the column titled 'Cognitive/Written Test' place the number of questions from the Test Bank that are used to evaluate the applicable JPR, RK, RS, or objective. In the column titled 'Manipulative/Skills Station' identify the skill sheets that are used to evaluate the applicable JPR, RS, or objective. When the Portfolio or Projects method is used to evaluate a particular JPR, RK, RS, or objective, identify the applicable section in the appropriate column and provide the procedures to be used as outlined in the NBFSPQ Operational Procedures, COA-5. Evaluation methods that are not cognitive, manipulative, portfolio, or project based should be identified in the 'Other' column.

OBJECTIVE / JPR, RK, RS		COGNITIVE	MANIPULATIVE			
SECTION	ABBREVIATED TEXT	WRITTEN TEST	SKILLS STATION	PORTFOLIO	PROJECTS	OTHER
6.1.1	Fire Fighter I					Chapter 1, (p 6)
6.2.1	Complete a basic incident report					Chapter 5 (p 131)
6.2.1(A)	RK: basic incident reports					Chapter 5 (p 131)
6.2.1(B)	RS: complete reports					Chapter 5, Skill Drill 5-2 (p 131)
6.2.2	Communicate the need for team assistance					Chapter 4 (p 109)
6.2.2(A)	RK: Standard operating procedures					Chapter 4 (p 109)
6.2.2(B)	RS: operate fire department communications equipment					Chapter 4 (p 107)
6.3.1	Extinguish an ignitible liquid fire					Chapter 17 (p 573)
6.3.1(A)	RK: foam					Chapter 17 (p 573–578)
6.3.1(B)	RS: prepare a foam concentrate					Chapter 17, Skill Drill 17-17, 17-18, 17-19 (p 576–578)
6.3.2	Coordinate an interior attack line team's accomplishment of an assignment in a structure fire					Chapter 15, 22 (p 450, 682, 691, 692)
6.3.2(A)	RK: proper nozzle and hose					Chapter 15, 22 (p 464, 683–684)
6.3.2(B)	RS: assemble a team					Chapter 22, Skill Drill 22-1 (p 684)
6.3.3	Control a flammable gas cylinder fire operating as a member of a team					Chapter 22 (p 700–701)
6.3.3(A)	RK: Characteristics of pressurized flammable gases					Chapter 22 (p 699–700)

OBJECTIVE / JPR, RK, RS		COGNITIVE	MANIPULATIVE			
SECTION	ABBREVIATED TEXT	WRITTEN TEST	SKILLS STATION	PORTFOLIO	PROJECTS	OTHER
6.3.3(B)	RS: advances and retreats					Chapter 22, Skill Drill 22-12 (p 700–701)
6.3.4	Protect evidence of fire cause and origin					Chapter 38 (p 1083–1084)
6.3.4(A)	RK: assess origin and cause					Chapter 38 (p 1073–1074, 1076, 1083–1084)
6.3.4(B)	RS: locate the fire's origin area					Chapter 38 (p 1075–1076, 1083–1084)
6.4.1	Extricate a victim entrapped in a motor vehicle as part of a team					Chapter 26 (p 817–819, 821–834)
6.4.1(A)	RK: fire department's role at a vehicle accident					Chapter 26 (p 811, 814–817, 821–829)
6.4.1(B)	RS: operate hand and power tools					Chapter 26 Skill Drill 26-3, 26-4, 26-5, 26-6, 26-7, 26-8, 26-9, 26-10 (p 818–834)
6.4.2	Assist rescue operation teams					Chapter 10, 27 (p 265–268, 841–846, 850–860)
6.4.2(A)	RK: fire fighter's role at a technicial rescue operation					Chapter 9, 27 (p 250, 840–846, 850–860)
6.4.2(B)	RS: identify and retrieve various types of rescue tools					Chapter 9, 27 (p 250, 847, 850)
6.5.1	Perform a fire safety survey in a occupied structure					Chapter 36 (p 1022–1026)
6.5.1(A)	RK: Organizational policy and procedures					Chapter 36 (p 1015–1022)
6.5.1(B)	RS: complete forms					Chapter 36 (p 1020–1021, 1025–1026)
6.5.2	Present fire safety information to station visitors or small groups					Chapter 36 (p 1014, 1017–1019, 1021)
6.5.2(A)	RK: informational materials					Chapter 36 (p 1025–1028)
6.5.2(B)	RS: document presentations and using prepared materials					Chapter 36 (p 1027–1028)
6.5.3	Prepare a preincident survey					Chapter 23 (p 711, 718–725)
6.5.3(A)	RK: sources of water supply					Chapter 23, 37 (p 711–713, 718–720, 1034–1064)
6.5.3(B)	RS: components of fire suppression and detection systems					Chapter 37 (p 1034–1064)
6.5.4	Maintain power plants, power tools, and lighting equipment					Chapter 9, 19 (p 252–253, 615)
6.5.4(A)	RK: Types of cleaning methods					Chapter 9, 19 (p 252–253, 615)
6.5.4(B)	RS: correct tools					Chapter 19 (p 615)
6.5.5	Perform an annual service test on fire hose					Chapter 16 (p 513–514)
6.5.5(A)	RK: hose service testing					Chapter 16 (p 513–514)
6.5.5(B)	RS: hose testing equipment and nozzles and to record results					Chapter 16, Skill Drill 16-13, 16-14 (p 513–514)

Pro Board Assessment Methodology Matrices for NFPA 472

NFPA 472 - HazMat Awareness - 2013 Edition

INSTRUCTIONS: In the column titled 'Cognitive/Written Test' place the number of questions from the Test Bank that are used to evaluate the applicable JPR, RK, RS, or objective. In the column titled 'Manipulative/Skills Station' identify the skill sheets that are used to evaluate the applicable JPR, RS, or objective. When the Portfolio or Projects method is used to evaluate a particular JPR, RK, RS, or objective, identify the applicable section in the appropriate column and provide the procedures to be used as outlined in the NBFSPQ Operational Procedures, COA-5. Evaluation methods that are not cognitive, manipulative, portfolio, or project based should be identified in the 'Other' column.

OBJECTIVE / JPR, RK, RS		COGNITIVE	MANIPULATIVE			
SECTION	ABBREVIATED TEXT	WRITTEN TEST	SKILLS STATION	PORTFOLIO	PROJECTS	OTHER
4.2.1	Detecting the Presence of Hazardous Materials					Chapter 30 (p 902–903)
4.2.1(1)	definition of hazardous materials and WMD					Chapter 28 (p 868–869)
4.2.1(2)	hazard classes					Chapter 31 (p 924)
4.2.1(3)	primary hazards					Chapter 31 (p 924)
4.2.1(4)	difference between hazardous materials incidents and other emergencies					Chapter 28 (p 873–874)
4.2.1(5)	typical occupancies and locations					Chapter 30 (p 902)
4.2.1(6)	typical container shapes					Chapter 30 (p 903–916)
4.2.1(7)	facility and transportation markings and colors					Chapter 30 (p 903–916)
4.2.1(8)	NFPA 704					Chapter 30 (p 912–913)
4.2.1(9)	placards and labels					Chapter 30 (p 910)
4.2.1(10)	information on material safety data sheets					Chapter 30 (p 914)
4.2.1(11)	examples of clues					Chapter 30 (p 903)
4.2.1(12)	limitations of using the senses					Chapter 30 (p 903)
4.2.1(13)	four types of locations					Chapter 30 (p 902)
4.2.1(14)	difference between a chemical and a biological incident					Chapter 29 (p 888–894)
4.2.1(15)	indicators of possible criminal or terrorist activity involving chemical agents					Chapter 30 (p 916)
4.2.1(16)	indicators of possible criminal or terrorist activity involving biological agents					Chapter 30 (p 916)
4.2.1(17)	indicators of possible criminal or terrorist activity involving radiological agents					Chapter 30 (p 916)

OBJECTIVE / JPR, RK, RS		COGNITIVE	MANIPULATIVE			
SECTION	ABBREVIATED TEXT	WRITTEN TEST	SKILLS STATION	PORTFOLIO	PROJECTS	OTHER
4.2.1(18)	indicators of possible criminal or terrorist activity involving illicit laboratories					Chapter 30 (p 916)
4.2.1(19)	indicators of possible criminal or terrorist activity involving explosives					Chapter 30 (p 916)
4.2.1(20)	indicators of secondary device					Chapter 30, 31 (p 916, 928–930)
4.2.2	Surveying the Hazardous Materials Incident from a safe location					Chapter 30 (p 910)
4.2.2(1)	determining the specific names					Chapter 30 (p 910–914)
4.2.2(2)	obtaining the names of, UN/NA identification numbers					Chapter 30 (p 910–914)
4.2.2(3)	obtaining the names of hazardous materials					Chapter 30 (p 910–914)
4.2.3	Collecting Hazard Information					Chapter 30 (p 914–915)
4.2.3(1)	determining the appropriate guide page					Chapter 30 (p 912)
4.2.3(2)	general types of hazards					Chapter 31 (p 924)
4.4.1	Initiating Protective Actions					Chapter 31 (p 924–926)
4.4.1(1)	location of both the local emergency response plan and the organization's					Chapter 31 (p 924–926)
4.4.1(2)	role of the first responder					Chapter 28 (p 870–871)
4.4.1(3)	basic precautions					Chapter 28 (p 873–874)
4.4.1(4)	identification of response information					Chapter 31 (p 924–925)
4.4.1(5)	identify the recommended personal protective equipment					Chapter 32 (p 936–953)
4.4.1(6)	identify the definitions					Chapter 32 (p 936–953)
4.4.1(7)	identify the shapes of recommended initial isolation and protective action zones					Chapter 32 (p 954–955)
4.4.1(8)	describe the difference between small and large spills					Chapter 32 (p 954–955)
4.4.1(9)	identify circumstances under which distances are used at a hazmat incident					Chapter 32 (p 954–955)
4.4.1(10)	describe the difference between the isolation distances and protective action distances					Chapter 31 (p 924–930)
4.4.1(11)	identify the techniques used to isolate the hazard area and deny entry					Chapter 32 (p 955)
4.4.1(12)	identify at least four specific actions necessary when an incident is suspected					Chapter 33 (p 962–963)
4.4.2	Initiating the Notification Process					Chapter 31 (p 924–926)

NFPA 472 - HazMat Operational - Core Competencies - 2013 Edition

INSTRUCTIONS: In the column titled 'Cognitive/Written Test' place the number of questions from the Test Bank that are used to evaluate the applicable JPR, RK, RS, or objective. In the column titled 'Manipulative/Skills Station' identify the skill sheets that are used to evaluate the applicable JPR, RS, or objective. When the Portfolio or Projects method is used to evaluate a particular JPR, RK, RS, or objective, identify the applicable section in the appropriate column and provide the procedures to be used as outlined in the NBFSPQ Operational Procedures, COA-5. Evaluation methods that are not cognitive, manipulative, portfolio, or project based should be identified in the 'Other' column.

OBJECTIVE / JPR, RK, RS		COGNITIVE	MANIPULATIVE			
SECTION	ABBREVIATED TEXT	WRITTEN TEST	SKILLS STATION	PORTFOLIO	PROJECTS	OTHER
5.1.1.2	Awareness level					Chapter 28 (p 869–871)
5.2.1	survey the incident					Chapter 30 (p 902)
5.2.1.1	identify shapes of containers					Chapter 30 (p 903–910, 915–916)
5.2.1.1.1	identify each type of tank car					Chapter 30 (p 907–910)
5.2.1.1.2	identify each type of intermodal tank					Chapter 30 (p 907–910)
5.2.1.1.3	identify each type of cargo tank					Chapter 30 (p 906–908)
5.2.1.1.4	identify each type of storage tank					Chapter 30 (p 906–908)
5.2.1.1.5	identify each package by type					Chapter 30 (p 904–906)
5.2.1.1.6	identify each container/ package by type					Chapter 30 (p 903–904)
5.2.1.2	identify the markings that differentiate one container from another					Chapter 30 (p 904)
5.2.1.2.1	identify the vehicle or tank identification markings					Chapter 30 (p 907–910)
5.2.1.2.2	identify the markings indicating container size, product contained, and/or site identification numbers					Chapter 30 (p 903)
5.2.1.3	identify the name(s) of hazmats					Chapter 30 (p 904)
5.2.1.3.1	identify pipeline marker information					Chapter 30 (p 910)
5.2.1.3.2	identify pesticide label information					Chapter 30 (p 905)
5.2.1.3.3	identify radioactive material label information					Chapter 30 (p 915–916)
5.2.1.4	identify conditions surrounding a hazmat incident					Chapter 31 (p 925–926)
5.2.1.5	ways to verify information					Chapter 30 (p 914)
5.2.1.6	additional hazards associated with criminal or terrorist activities					Chapter 30 (p 916)
5.2.2	collect hazard and response information					Chapter 29, 30 (p 887, 911–915)
5.2.2(1)	definitions associated with the DOT hazard classes					Chapter 30 (p 911–912)
5.2.2(2)	ways to obtain a material safety data sheet					Chapter 30 (p 914)
5.2.2(3)	identify hazard and response information					Chapter 30 (p 914)
5.2.2(4)	Identify type, procedures, and information regarding CHEMTREC					Chapter 30 (p 915)
5.2.2(5)	methods of contacting the manufacturer					Chapter 30 (p 914)
5.2.2(6)	identify type of assistance available with respect to criminal or terrorist activities					Chapter 30 (p 915–916)

OBJECTIVE / JPR, RK, RS		COGNITIVE	MANIPULATIVE			
SECTION	ABBREVIATED TEXT	WRITTEN TEST	SKILLS STATION	PORTFOLIO	PROJECTS	OTHER
5.2.2(7)	identify procedure for contacting local, state, and federal authorities					Chapter 30 (p 915–916)
5.2.2(8)	describe properties or radioactive materials					Chapter 29 (p 886–887)
5.2.3	Predicting the Behavior of a Material and Its Container					Chapter 29 (p 880–894)
5.2.3(1)	interpret the hazard and response information					Chapter 29 (p 894)
5.2.3(1)(a)	significance of chemical and physical properties on behavior of the container and/or its contents					Chapter 29 (p 880–885)
5.2.3(1)(b)	differences between terms					Chapter 29 (p 880–894)
5..2.3(2)	types of stress					Chapter 29 (p 880–885)
5.2.3(3)	ways in which containers can breach					Chapter 29 (p 880–885)
5.2.3(4)	ways in which containers can release their contents					Chapter 29 (p 880–885)
5.2.3(5)	dispersion patterns					Chapter 29 (p 880–885)
5.2.3(6)	time frames for predicting the length of time that exposures					Chapter 29 (p 894)
5.2.3(7)	health and physical hazards					Chapter 29 (p 885)
5.2.3(8)	health hazards					Chapter 29 (p 885)
5.2.3(9)	warfare agents					Chapter 29 (p 885)
5.2.4	Estimating the Potential Harm					Chapter 33 (p 962–964)
5.2.4(1)	resource for determining the size of an endangered area					Chapter 33 (p 962–964)
5.2.4(2)	number and type of exposures					Chapter 33 (p 962–964)
5.2.4(3)	resources available for determining the concentrations					Chapter 33 (p 962–964)
5.2.4(4)	factors for determining the extent of physical, health, and safety hazards					Chapter 33 (p 962–964)
5.2.4(5)	impact of time, distance and shielding					Chapter 33 (p 962–964)
5.3.1	Describing Response Objectives for Hazardous Materials Incidents					Chapter 31 (p 928–930)
5.3.1(1)	steps for determining the number of exposures					Chapter 31 (p 928–930)
5.3.1(2)	steps for determining defensive response objectives					Chapter 31 (p 926)
5.3.1(3)	assess risk to responder					Chapter 31 (p 928–930)
5.3.1(4)	assess potential for secondary attack					Chapter 31 (p 928–930)
5.3.2	Identifying Action Options					Chapter 31 (p 925–926)
5.3.2(1)	options to accomplish objectives					Chapter 31 (p 925–926)
5.3.2(2)	prioritization of emergency medical care					Chapter 31 (p 925–926)
5.3.3	Determining Suitablity of Personal Protective Equipment					Chapter 32 (p 936–953)
5.3.3(1)	appropriate respiratory protection					Chapter 32 (p 937–953)

OBJECTIVE / JPR, RK, RS		COGNITIVE	MANIPULATIVE			
SECTION	ABBREVIATED TEXT	WRITTEN TEST	SKILLS STATION	PORTFOLIO	PROJECTS	OTHER
5.3.3(2)	appropriate personal protective clothing					Chapter 32 (p 937–953)
5.3.4	Identifying Decontamination Issues					Chapter 34 (p 978–985)
5.3.4(1)	ways that personnel, personal protective equipment, apparatus, and tools and equipment become contaminated					Chapter 34 (p 978)
5.3.4(2)	potential for secondary contamination					Chapter 34 (p 978)
5.3.4(3)	importance and limitations of decontamination procedures					Chapter 34 (p 978–985)
5.3.4(4)	purpose of emergency decontamination procedures					Chapter 34 (p 978–979)
5.3.4(5)	factors to be considered in emergency decontamination					Chapter 34 (p 978–979)
5.4.1	Establishing and Enforcing Scene Control Procedures					Chapter 32, 33 (p 954, 962–963)
5.4.1(1)	procedures for establishing scene control					Chapter 32 (p 954)
5.4.1(2)	criteria for determining the locations					Chapter 32 (p 954)
5.4.1(3)	basic techniques for the following protective actions					Chapter 33 (p 962–963)
5.4.1(4)	demonstrate emergency decontamination					Chapter 34 (p 978–979)
5.4.1(5)	identify safety items					Chapter 32, 33 (p 954, 962–963)
5.4.1(6)	procedures for ensuring coordinated communication					Chapter 32, 33 (p 954, 962–963)
5.4.2	Preserving Evidence					Chapter 33 (p 972)
5.4.3	Initiating ICS					Chapter 31 (p 924–925, 955)
5.4.3(1)	role of the first responder					Chapter 28 (p 870–874)
5.4.3(2)	levels of hazardous materials incidents					Chapter 31 (p 924)
5.4.3(3)	purpose, need, benefits, and elements of an incident management system					Chapter 31 (p 930)
5.4.3(4)	identify duties and responsibilities					Chapter 28 (p 870–874)
5.4.3(5)	considerations for determining location of command post					Chapter 31 (p 930)
5.4.3(6)	procedures for requesting additional resources					Chapter 31 (p 924–925, 955)
5.4.3(7)	role and objectives of other agencies					Chapter 31 (p 924–925, 955)
5.4.4	Using Personal Protective Equipment					Chapter 32 (p 950–953, 955)
5.4.4(1)	importance of the buddy system					Chapter 32 (p 955)
5.4.4(2)	importance of the backup personnel					Chapter 32 (p 955)
5.4.4(3)	identify safety precautions					Chapter 32 (p 950–953, 955)
5.4.4(4)	symptoms of heat and cold stress					Chapter 32 (p 950–953)

OBJECTIVE / JPR, RK, RS		COGNITIVE	MANIPULATIVE			
SECTION	ABBREVIATED TEXT	WRITTEN TEST	SKILLS STATION	PORTFOLIO	PROJECTS	OTHER
5.4.4(5)	physical capabilities required for, and the limitations of, personnel working in the personal protective equipment					Chapter 32 (p 953)
5.4.4(6)	procedures for cleaning, disinfecting, and inspecting personal protective equipment					Chapter 34 (p 984)
5.4.4(7)	maintenance, testing, inspections, storage of PPE					Chapter 32 (p 950–953, 955)
5.5.1	Evaluating the Status of Planned Response					Chapter 33 (p 972)
5.5.1(1)	considerations for evaluating whether actions were effective					Chapter 33 (p 972)
5.5.1(2)	circumstances under which it would be prudent to withdraw					Chapter 33 (p 972)
5.5.2	Communicating the Status of the Planned Response					Chapter 31 (p 924–925)
5.5.2(1)	methods for communicating the status of the planned response					Chapter 31 (p 924–925)
5.5.2(2)	methods for immediate notification					Chapter 31 (p 924–925)

NFPA 472 - HazMat Operational w/Personal Protective Equipment Mission Specific Competencies - 2013 Edition

INSTRUCTIONS: In the column titled 'Cognitive/Written Test' place the number of questions from the Test Bank that are used to evaluate the applicable JPR, RK, RS, or objective. In the column titled 'Manipulative/Skills Station' identify the skill sheets that are used to evaluate the applicable JPR, RS, or objective. When the Portfolio or Projects method is used to evaluate a particular JPR, RK, RS, or objective, identify the applicable section in the appropriate column and provide the procedures to be used as outlined in the NBFSPQ Operational Procedures, COA-5. Evaluation methods that are not cognitive, manipulative, portfolio, or project based should be identified in the 'Other' column.

OBJECTIVE / JPR, RK, RS		COGNITIVE	MANIPULATIVE			
SECTION	ABBREVIATED TEXT	WRITTEN TEST	SKILLS STATION	PORTFOLIO	PROJECTS	OTHER
6.2.1.1.2	awareness level (Chapter 4), all core competencies at the operations level (Chapter 5), and all competencies in this section.					Chapter 28, 29, 30, 31, 32, 33 (p 870–871, 880–894, 902–916, 924–925, 936–955, 962–963)
6.2.3.1	select PPE					Chapter 32 (p 936–955)
6.2.3.1(1)	Describe the types of protective clothing and equipment					Chapter 32 (p 936–955)
6.2.3.1(2)	Describe personal protective equipment options for hazards					Chapter 32 (p 936–955)
6.2.3.1(3)	Select personal protective equipment for mission-specific tasks at hazardous materials					Chapter 32 (p 936–955)
6.2.3.1(3)(a)	Describe terms and explain their impact and significance					Chapter 32 (p 936–955)
6.2.3.1(3)(b)	identify three indications of material degradation					Chapter 32 (p 936–955)
6.2.3.1(3)(c)	identify the different designs of vapor-protective and splash-protective clothing					Chapter 32 (p 939–940)
6.2.3.1(3)(d)	identify the relative advantages and disadvantages of heat exchange units					Chapter 32 (p 950–953)
6.2.3.1(3)(e)	identify physiological and psychological stresses					Chapter 32 (p 953)

OBJECTIVE / JPR, RK, RS		COGNITIVE	MANIPULATIVE			
SECTION	ABBREVIATED TEXT	WRITTEN TEST	SKILLS STATION	PORTFOLIO	PROJECTS	OTHER
6.2.3.1(3)(f)	describe local procedures for technical decontamination					Reserved
6.2.4.1	Using Protective Clothing and Respiratory Protection					Chapter 32 (p 936–955)
6.2.4.1(1)	Describe at least three safety procedures					Chapter 32 (p 936–955)
6.2.4.1(2)	Describe at least three emergency procedures					Chapter 32 (p 936–955)
6.2.4.1(3)	Demonstrate the ability to don, work in, and doff personal protective equipment					Chapter 32 (p 938–949)
6.2.4.1(4)	Demonstrate local procedures					Reserved
6.2.4.1(5)	Describe the maintenance, testing, inspection, storage, and documentation procedures					Chapter 32 (p 936–955)
6.2.5.1	Reporting and Documenting the Incident					Chapter 31 (p 924–930)

NFPA 472 - HazMat Operational w/Product Control Mission Specific Competencies - 2013 Edition

INSTRUCTIONS: In the column titled 'Cognitive/Written Test' place the number of questions from the Test Bank that are used to evaluate the applicable JPR, RK, RS, or objective. In the column titled 'Manipulative/Skills Station' identify the skill sheets that are used to evaluate the applicable JPR, RS, or objective. When the Portfolio or Projects method is used to evaluate a particular JPR, RK, RS, or objective, identify the applicable section in the appropriate column and provide the procedures to be used as outlined in the NBFSPQ Operational Procedures, COA-5. Evaluation methods that are not cognitive, manipulative, portfolio, or project based should be identified in the 'Other' column.

OBJECTIVE / JPR, RK, RS		COGNITIVE	MANIPULATIVE			
SECTION	ABBREVIATED TEXT	WRITTEN TEST	SKILLS STATION	PORTFOLIO	PROJECTS	OTHER
6.6.1.1.2	awareness level (Chapter 4), all core competencies at the operations level (Chapter 5), all mission-specific competencies for personal protective equipment (Section 6.2), and all competencies in this section					Chapter 28, 29, 30, 31, 32, 33 (p 870–871, 880–894, 902–916, 924–925, 936–955, 962–963)
6.6.3.1	identify control options					Chapter 33 (p 965–972)
6.6.3.1(1)	Identify the options to accomplish a given response objective					Chapter 33 (p 965–972)
6.6.3.1(2)	Identify the purpose for and the procedures					Chapter 33 (p 965–972)
6.6.3.2	select personal protective equipment					Chapter 32 (p 936–953)
6.6.4.1	perform control options					Chapter 33 (p 965–972)
6.6.4.1(1)	Using special purpose or hazard suppressing foams or agents					Chapter 33 (p 965–972)
6.6.4.1(2)	Identify the characteristics and applicability of Class B foams					Chapter 17 (p 575)
6.6.4.1(3)	demonstrate how to perform control activities					Chapter 33 (p 965–972)
6.6.4.1(4)	Identify location and describe use of emergency remote shutoff devices					Chapter 33 (p 971–972)
6.6.4.1(5)	Describe the use of emergency remote shutoff devices					Chapter 33 (p 971–972)
6.6.4.2	describe local procedures for going through the technical decontamination process					Reserved

Glossary

"A" posts Vertical support members that form the sides of the windshield of a motor vehicle.

"B" posts Vertical support members located between the front and rear doors of a motor vehicle.

"C" posts Vertical support members located behind the rear doors of a motor vehicle.

911 dispatcher/telecommunicator From the communications center, the dispatcher takes the calls from the public, sends appropriate units to the scene, assists callers with treatment instructions until the EMS unit arrives, and assists the incident commander with needed resources.

A tool A cutting tool with a pry bar built into the cutting part of the tool.

Absorption (decontamination) Technique in which a spongy material is mixed with a liquid hazardous material. The contaminated mixture is collected and disposed of.

Absorption (contamination) The process by which hazardous materials travel through body tissues until they reach the bloodstream.

Absorption (hazardous materials response) The process of applying a material (an absorbent) that will soak up and hold a hazardous material in a sponge-like manner, for collection and disposal.

Accelerator A device that speeds up the removal of the air from a dry-pipe or preaction sprinkler system. An accelerator reduces the time required for water to start flowing from sprinkler heads.

Accidental Fire-cause classification that includes fires with a proven cause that does not involve a deliberate human act.

Accordion hose load A method of loading hose on a vehicle that results in a hose appearance that resembles accordion sections. It is achieved by standing the hose on its edge, then placing the next fold on its edge, and so on.

Acid A material with a pH value less than 7.

Activity logging system A device that keeps a detailed record of every incident and activity that occurs.

Acute health effect A health problem caused by relatively short exposure periods to a harmful substance that produces observable conditions such as eye irritation, coughing, dizziness, and skin burns.

Adaptor A device that joins hose couplings of the same type, such as male to male or female to female.

Adjustable-gallonage fog nozzle A nozzle that allows the operator to select a desired flow from several settings.

Adsorption The process of adding a material such as sand or activated carbon to a contaminant, which then adheres to the surface of the material. The contaminated material is then collected and disposed of.

Advanced Emergency Medical Technician (AEMT) An EMS provider who has training in specific aspects of Advanced Life Support care, such as intravenous therapy, interpretation of heart rhythms, and defibrillation.

Advanced Life Support (ALS) Functional provision of advanced airway management including intubation, advanced cardiac monitoring, manual defibrillation, establishment and maintenance of intravenous access, and drug therapy. (NFPA 1584)

Adz The prying part of the Halligan tool.

Adze A hand tool constructed of a thin, arched blade set at right angles to the handle. It is used to chop brush for clearing a fire line, or to mop up a wildland fire.

Aerating The process of inducing air into the foam solution, which expands and finishes the foam.

Aerial fuels Fuels located more than 6 feet (2 meters) off the ground, usually part of or attached to trees.

Aerial ladder A self-supporting, turntable-mounted, power-operated ladder of two or more sections permanently attached to a self-propelled automotive fire apparatus and designed to provide a continuous egress route from an elevated position to the ground. (NFPA 1901)

Aerosols Small solid particles or liquid droplets that are suspended in air.

Agroterrorism The intentional act of using chemical or biological agents against the agricultural industry or food supply.

Air bill Shipping papers on an airplane.

Air cylinder The component of the SCBA that stores the compressed air supply.

Air line The hose through which air flows, either within an SCBA or from an outside source to a supplied air respirator.

Air management The use of a limited air supply in such a way as to ensure that it will last long enough to enter a hazardous area, accomplish needed tasks, and return safely.

Air sampling detector A system that captures a sample of air from a room or enclosed space and passes it through a smoke detection or gas analysis device.

Air-purifying respirator (APR) A respirator with an air-purifying filter, cartridge, or canister that removes specific air contaminants by passing ambient air through the air-purifying element. (NFPA 1404, Standard for Fire Service Respiratory Protection Training)

Aircraft/crash rescue fire fighter (ARFF) An individual with specialized training in aircraft fires, extrication of victims from aircraft, and extinguishing agents. These fire fighters wear special types of turnout gear and respond in fire apparatus that protect them from high-temperature fires.

Alarm initiation device An automatic or manually operated device in a fire alarm system that, when activated, causes the system to indicate an alarm condition.

Alarm notification device An audible and/or visual device in a fire alarm system that makes occupants or other persons aware of an alarm condition.

Alarm valve A valve that signals an alarm when a sprinkler head is activated and prevents nuisance alarms caused by pressure variations.

Alcohol-resistant foam Used for fighting fires involving water-soluble materials or fuels that are destructive to other types of foams. Some alcohol-resistant foams are capable of forming a vapor-suppressing aqueous film on the surface of hydrocarbon fuels. (NFPA 412)

Alpha particle A positively charged particle emitted by certain radioactive materials, identical to the nucleus of a helium atom. (NFPA 801, Standard for Fire Protection for Facilities Handling Radioactive Materials)

Alternative-powered vehicle A vehicle that is powered by compressed natural gas or other fuel.

Ammonium nitrate fertilizer and fuel oil (ANFO) An explosive made of commonly available materials.

Ammonium phosphate An extinguishing agent used in dry-chemical fire extinguishers that can be used on Class A, B, and C fires.

Annealed The process of forming standard glass.

Anthrax An infectious disease spread by the bacterium *Bacillus anthracis*; it is typically found around farms, infecting livestock.

Aqueous film-forming foam (AFFF) A concentrated aqueous solution of one or more hydrocarbon and/or fluorochemical surfactants that forms a foam capable of producing a vapor-suppressing, aqueous film on the surface of hydrocarbon fuels. (NFPA 403)

Arched roof A rounded roof usually associated with a bowstring truss.

Area of origin The room or area where a fire began. (NFPA 921)

Arson The crime of maliciously and intentionally, or recklessly, starting a fire or causing an explosion. (NFPA 921)

Arsonist A person who deliberately sets a fire to destroy property with criminal intent.

Asphyxiant A material that causes the victim to suffocate.

Assistant or division chief A midlevel chief who often has a functional area of responsibility, such as training, and who answers directly to the fire chief.

Atom The smallest particle of an element, which can exist alone or in combination.

Attack engine An engine from which attack lines have been pulled.

Attack hose (attack line) Hose designed to be used by trained fire fighters and fire brigade members to combat fires beyond the incipient stage. (NFPA 1961)

Automatic location identification (ALI) An enhanced 911 service feature that displays where the call originated or where the phone service is billed.

Automatic number identification (ANI) An enhanced 911 service feature that shows the calling party's phone number on the display screen at the telecommunicator's terminal.

Automatic sprinkler heads The working ends of a sprinkler system, which serve to activate the system and to apply water to the fire.

Automatic sprinkler system A system of pipes filled with water under pressure that discharges water immediately when a sprinkler head opens.

Automatic-adjusting fog nozzle A nozzle that can deliver a wide range of water stream flows. It operates by means of an internal spring-loaded piston.

Auxiliary system A fire alarm system that sounds an alarm in the building and transmits a signal to the fire department via a public alarm box system.

Awareness level (29 CFR 1910.12: First Responder at the Awareness Level) Personnel who, in the course of their normal duties, could encounter an emergency involving hazardous materials and weapons of mass destruction (WMDs) and who are expected to recognize the presence of the hazardous materials and WMDs, protect themselves, call for trained personnel, and secure the scene. (NFPA 472)

Awning windows Windows that have one large or two medium-size panels, which are operated by a hand crank from the corner of the window.

Backdraft A phenomenon that occurs when a fire takes place in a confined area, such as a sealed aircraft fuselage, and burns undetected until most of the oxygen within is consumed. The heat continues to produce flammable gases, mostly in the form of carbon monoxide. These gases are heated above their ignition temperature and when a supply of oxygen is introduced, as when normal entry points are opened, the gases could ignite with explosive force. (NFPA 402)

Backfiring A planned operation to remove fuel from a wildland fire by burning out large selected areas. This operation should be carried out only under the close supervision of experienced, authorized fire officers.

Backpack pump extinguisher A portable fire extinguisher consisting of a 4- to 8-gallon (15- to 30-liter) water tank that is worn on the user's back and features a hand-powered piston pump for discharging the water.

Backpack The harness of the SCBA, which supports the components worn by a fire fighter.

Backup entry team A dedicated team of fully qualified and equipped responders who are ready to enter the hot zone at a moment's notice to rescue any member of the hot zone entry team.

Backup personnel Individuals who remove or rescue those working in the hot zone in the event of an emergency.

Ball valves Valves used on nozzles, gated wyes, and engine discharge gates. They consist of a ball with a hole in the middle of the ball.

Balloon-frame construction An older type of wood frame construction in which the wall studs extend vertically from the basement of a structure to the roof without any fire stops.

Bam-bam tool A tool with a case-hardened screw, which is secured in the keyway of a lock and used to remove the keyway from the lock.

Bangor ladder A ladder equipped with tormentor poles or staypoles that stabilize the ladder during raising and lowering operations.

Banked Covering a fire to ensure low burning.

Bankshot (or bank-down) method A method that applies the stream onto a nearby object, such as a wall, instead of directly aiming at the fire.

Base (bed) section The lowest or widest section of an extension ladder. (NFPA 1931)

Base A material with a pH value greater than 7.

Base station A stationary radio transceiver with an integral AC power supply. (NFPA 1221)

Batch mixing Pouring foam concentrate directly into the fire apparatus water tank, thereby mixing a large amount of foam at one time.

Battalion chief Usually the first level of fire chief; also called a district chief. These chiefs are often in charge of running calls and supervising multiple stations or districts within a city. A battalion chief is usually the officer in charge of a single-alarm working fire.

Battering ram A tool made of hardened steel with handles on the sides used to force doors and to breach walls. Larger versions may be used by as many as four people; smaller versions are made for one or two people.

Beam detector A smoke detection device that projects a narrow beam of light across a large open area from a sending unit to a receiving unit. When the beam is interrupted by smoke, the receiver detects a reduction in light transmission and activates the fire alarm.

Beam The main structural side of a ground ladder. (NFPA 1931)

Bend A knot used to join two ropes together.

Beta particle An elementary particle, emitted from a nucleus during radioactive decay, with a single electrical charge and a mass equal to 1/1837 that of a proton. (NFPA 801)

Bight A U shape created by bending a rope with the two sides parallel.

Bill of lading Shipping papers for roads and highways.

Bimetallic strip A device with components made from two distinct metals that respond differently to heat. When heated, the metals will bend or change shape.

Biological agents Disease-causing bacteria, viruses, and other agents that attack the human body.

Bite A small opening made to enable better tool access in forcible entry.

Black An area that has already been burned.

Black fire A hot, high-volume, high-velocity, turbulent, ultra-dense black smoke that indicates an impending flashover or auto ignition.

Blister agent A chemical that causes the skin to blister.

Block creel construction Rope constructed without knots or splices in the yarns, ply yarns, strands or braids, or rope. (NFPA 1983)

Blood agent A chemical that, when absorbed by the body, interferes with the transfer of oxygen from the blood to the cells.

Boiling liquid/expanding vapor explosion (BLEVE) An explosion that occurs when a tank containing a volatile liquid at the bottom of the tank and a flammable gas at the top of the tank is heated to the point where the tank ruptures.

Boiling point The temperature at which the vapor pressure of a liquid equals the surrounding atmospheric pressure. (NFPA 30, Flammable and Combustible Liquids Code)

Bolt cutter A cutting tool used to cut through thick metal objects, such as bolts, locks, and wire fences.

Booster hose (booster line) A noncollapsible hose used under positive pressure having an elastomeric or thermoplastic tube, a braided or spiraled reinforcement, and an outer protective cover. (NFPA 1962)

Bowstring truss A truss that is curved on the top and straight on the bottom.

Box A burning structure.

Box-end wrench A hand tool used to tighten or loosen bolts. The end is enclosed, as opposed to an open-end wrench. Each wrench is a specific size, and most have ratchets for easier use.

Braided rope Rope constructed by intertwining strands in the same way that hair is braided.

Branch director A person in a supervisory level position in either the operations or logistics function to provide a span of control. (NFPA 1561)

Branch The organizational level having functional, geographical, or jurisdictional responsibility for major aspects of incident operations. (NFPA 1026)

Breakaway type nozzle A nozzle with a tip that can be separated from the shut-off valve.

Bresnan distributor nozzle A nozzle that can be placed in confined spaces. The nozzle spins, spreading water over a large area.

Bulk storage containers Large-volume containers that have an internal volume greater than 119 gallons (450 liters) for liquids and greater than 882 pounds (400 kilograms) for solids and a capacity of greater than 882 pounds (400 kilograms) for gases.

Bulkhead (firewall) The wall that separates the engine compartment from the passenger compartment in a motor vehicle.

Bungs One or more small openings in closed-head drums.

Bunker coat The protective coat worn by a fire fighter for interior structural firefighting; also called a turnout coat.

Bunker pants The protective trousers worn by a fire fighter for interior structural firefighting; also called turnout pants.

Butt plate An alternative to a simple butt spur; a swiveling plate with both a spur and a cleat or pad that is attached to the butt of the ladder.

Butt spurs That component of ground ladder support that remains in contact with the lower support surface to reduce slippage. (NFPA 1931)

Butt The end of the beam that is placed on the ground, or other lower support surface, when ground ladders are in the raised position. (NFPA 1931)

Butterfly valves Valves that are found on the large pump intake valve where the suction hose connects to the inlet of the fire pump.

Call box A system of telephones connected by phone lines, radio equipment, or cellular technology to a communications center or fire department.

Captain The second rank of promotion in the fire service, between the lieutenant and the battalion chief. Captains are responsible for managing a fire company and for coordinating the activities of that company among the other shifts.

Carabiner An auxiliary equipment system item; load-bearing connector with a self-closing gate used to join other components of life safety rope. (NFPA 1983)

Carbon dioxide (CO2) fire extinguisher A fire extinguisher that uses carbon dioxide gas as the extinguishing agent. It is rated for use on Class B and C fires.

Carbon dioxide extinguishing system A fire suppression system designed to protect either a single room or series of rooms by flooding the area with carbon dioxide.

Carbon monoxide (CO) A toxic gas produced through incomplete combustion.

Carboys Glass, plastic, or steel containers, ranging in volume from 5 to 15 gallons (19 to 57 liters).

Carcinogen A cancer-causing substance that is identified in one of several published lists, including, but not limited to, NIOSH Pocket Guide to Chemical Hazards, Hazardous Chemicals Desk Reference, and the ACGIH 2007 TLVs and BEIs. (NFPA 1851)

Carpenter's handsaw A saw designed for cutting wood.

Carryall A piece of heavy canvas with handles, which can be used to tote debris, ash, embers, and burning materials out of a structure.

Cartridge/cylinder fire extinguisher A fire extinguisher that has the expellant gas in a separate container from the extinguishing agent storage container. The storage container is pressurized by a mechanical action that releases the expellant gas.

Cascade system A method of piping air tanks together to allow air to be supplied to the SCBA fill station using a progressive selection of tanks, each with a higher pressure level. (NFPA 1901)

Case-hardened steel Steel created in a process that uses carbon and nitrogen to harden the outer core of a steel component, while the inner core remains soft. Case-hardened steel can be cut only with specialized tools.

Casement windows Windows in a steel or wood frame that open away from the building via a crank mechanism.

Ceiling hook A tool with a long wooden or fiberglass pole that has a metal point with a spur at right angles at one end. It can be used to probe ceilings and pull down plaster lath material.

Cellar nozzle A nozzle used to fight fires in cellars and other inaccessible places. The device works by spreading water in a wide pattern as the nozzle is lowered through a hole into the cellar.

Central station An off-premises facility that monitors alarm systems and is responsible for notifying the fire department of an alarm. These facilities may be geographically located some distance from the protected building(s).

Chain of command A rank structure, spanning the fire fighter through the fire chief, for managing a fire department and fire-ground operations.

Chain of custody A legal term used to describe the paperwork or documentation describing the movement, storage, and custody of evidence, such as the record of possession of a gas can from a fire scene.

Chainsaw A power saw that uses the rotating movement of a chain equipped with sharpened cutting edges. It is typically used to cut through wood.

Chase Open space within walls for wires and pipes.

Chemical change The ability of a chemical to undergo an alteration in its chemical makeup, usually accompanied by a release of some form of energy.

Chemical energy Energy that is created or released by the combination or decomposition of chemical com-pounds.

Chemical Transportation Emergency Center (CHEMTREC) A national call center for basic chemical information, established by the Chemical Manufacturers Association and now operated by the American Chemical Council.

Chemical-pellet sprinkler head A sprinkler head activated by a chemical pellet that liquefies at a preset temperature.

Chemical-resistant material A material used to make chemical-protective clothing, which can maintain its integrity and protection qualities when it comes into contact with a hazardous material. These materials also resist penetration, permeation, and degradation.

Chief of the department The top position in the fire department. The fire chief has ultimate responsibility for the fire department and usually answers directly to the mayor or other designated public official.

Chief's trumpet An obsolete amplification device that enabled a chief officer to give orders to fire fighters during an emergency. It was a precursor to a bullhorn and portable radios.

Child Anyone between 1 year of age and the onset of puberty (12 to 14 years of age).

Chisel A metal tool with one sharpened end that is used to break apart material in conjunction with a hammer, mallet, or sledgehammer.

Chlorine A yellowish gas that is about 2.5 times heavier than air and slightly water-soluble. Chlorine has many industrial uses but also damages the lungs when inhaled; it is a choking agent.

Choking agent A chemical designed to inhibit breathing. It is typically intended to incapacitate rather than kill.

Chronic health hazard A health problem occurring after a long-term exposure to a substance.

Circumstantial evidence The means by which alleged facts are proven by deduction or inference from other facts that were observed directly.

Clapper mechanism A mechanical device installed within a piping system that allows water to flow in only one direction.

Class A fire A fire in ordinary combustible materials, such as wood, cloth, paper, rubber, and many plastics. (NFPA 10)

Class A foam Foam intended for use on Class A fires. (NFPA 1150)

Class B fire A fire in flammable liquids, combustible liquids, petroleum greases, tars, oils, oil-based paints, solvents, lacquers, alcohols, and flammable gases. (NFPA 10)

Class B foam Foam intended for use on Class B fires. (NFPA 1901)

Class C fire A fire that involves energized electrical equipment. (NFPA 10)

Class D fire A fire in combustible metals, such as magnesium, titanium, zirconium, sodium, lithium, and potassium. (NFPA 10)

Class I standpipe A standpipe system designed for use by fire department personnel only. Each outlet should have a valve to control the flow of water and a 2½-inch (64-mm) male coupling for fire hose.

Class II standpipe A standpipe system designed for use by occupants of a building only. Each outlet is generally equipped with a length of 1½-inch (38-mm) singlejacket hose and a nozzle, which are preconnected to the system.

Class III standpipe A combination system that has features of both Class I and Class II standpipes.

Class K fire A fire in a cooking appliance that involves combustible cooking media (vegetable or animal oils and fats). (NFPA 10)

Claw bar A tool with a pointed claw-hook on one end and a forked- or flat-chisel pry on the other end. It is often used for forcible entry.

Claw The forked end of a tool.

Clean agent Electrically nonconducting, volatile, or gaseous fire extinguishant that does not leave a residue upon evaporation. (NFPA 2001)

Clemens hook A multipurpose tool that can be used for several forcible entry and ventilation applications because of its unique head design.

Closed-circuit breathing apparatus SCBA designed to recycle the user's exhaled air. This system removes carbon dioxide and generates fresh oxygen.

Closet hook A type of pike pole intended for use in tight spaces, commonly 2 to 4 feet (0.6 to 1 meter) in length.

Cockloft The concealed space between the top-floor ceiling and the roof of a building.

Coded system A fire alarm system design that divides a building or facility into zones and has audible notification devices that can be used to identify the area where an alarm originated.

Cold zone (technical rescue incident) The control zone of an incident that contains the command post and such other support functions as are deemed necessary to control the incident. (NFPA 1500)

Cold zone The control zone of hazardous materials/weapons of mass destruction incidents that contains the incident command post and such other support functions as are deemed necessary to control the incident. (NFPA 472, Standard for Competence of Responders to Hazardous Materials/Weapons of Mass Destruction Incidents)

Combination attack A type of attack employing both direct attack and indirect attack methods.

Combination ladder A ground ladder that is capable of being used both as a stepladder and as a single or extension ladder. (NFPA 1931)

Combustibility The property describing whether a material will burn and how quickly it will burn.

Combustion A chemical process of oxidation that occurs at a rate fast enough to produce heat and usually light in the form of either a glow or a flame. (NFPA 101)

Come along A hand-operated tool used for dragging or lifting heavy objects that uses pulleys and cables or chains to multiply a pulling or lifting force.

Command Staff The public information officer, safety officer, and liaison officer, all of whom report directly to the incident commander and are responsible for functions in the incident management system that are not a part of the function of the line organization. (NFPA 1561)

Command The first component of the ICS. It is the only position in the ICS that must always be staffed.

Communications center A building or portion of a building that is specifically configured for the primary purpose of providing emergency communications services or public safety answering point services to one or more public safety agencies under the authority or authorities having jurisdiction. (NFPA 1221)

Company officer The officer, or any other position of comparable responsibility in the department, in charge of a fire department company or station. (NFPA 1143)

Compartment A space completely enclosed by walls and a ceiling. The compartment enclosure is permitted to have openings in walls to an adjoining space if the openings have a minimum lintel depth of 8 inches (203 mm) from the ceiling and the openings do not exceed 8 feet (2.44 m) in width. A single opening of 36 inches (914 mm) or less in width without a lintel is permitted when there are no other openings to adjoining spaces. (NFPA 13)

Competent ignition source An ignition source that can ignite a fuel under the existing conditions at the time of the fire. It must have sufficient heat and be in close enough proximity to the fuel for a sufficient amount of time to ignite the fuel.

Compressed air foam (CAF) Class A foam produced by injecting compressed air into a stream of water that has been mixed with 0.1 percent to 1.0 percent foam.

Compressed air foam system (CAFS) A foam system that combines air under pressure with foam solution to create foam. (NFPA 1901)

Compressor A device used for increasing the pressure and density of a gas. (NFPA 853)

Computer-aided dispatch (CAD) A combination of hardware and software that provides data entry, makes resource recommendations, notifies and tracks those resources before, during, and after fire service alarms, preserving records of those alarms and status changes for later analysis. (NFPA 1221)

Conduction Heat transfer to another body or within a body by direct contact. (NFPA 921)

Confined space An area large enough and so configured that a member can bodily enter and perform assigned work, but which has limited or restricted means for entry and exit and is not designed for continuous human occupancy. (NFPA 1500)

Confinement Actions taken to keep a material, once released, in a defined or local area. (NFPA 472)

Conflagration A large fire, often involving multiple structures.

Consensus document A code document developed through agreement between people representing different organizations and interests. NFPA codes and standards are consensus documents.

Consist A list of every car on a train.

Container Any vessel or receptacle that holds material, including storage vessels, pipelines, and packaging.

Containment Actions taken to keep a material in its container (e.g., stop a release of the material or reduce the amount being released). (NFPA 472)

Contaminated A term used to describe evidence that may have been altered from its original state.

Contamination The process of transferring a hazardous material from its source to people, animals, the environment, or equipment, all of which can act as carriers for the material.

Contemporary construction Buildings constructed since about 1970 that incorporate lightweight construction techniques and engineered wood components. These buildings exhibit less resistance to fire than older buildings.

Control zone An area at hazardous materials/weapons of mass destruction incidents within an established perimeter that is designated based upon safety and the degree of hazard. (NFPA 472)

Convection Heat transfer by circulation within a medium such as a gas or a liquid. (NFPA 921)

Conventional vehicle A vehicle that uses an internal combustion engine and is fueled with either gasoline or diesel fuel.

Convulsant A chemical capable of causing convulsions or seizures when absorbed by the body.

Coping saw A saw designed to cut curves in wood.

Corrosivity The ability of a material to cause damage (on contact) to skin, eyes, or other parts on the body.

Council rake A long-handled rake constructed with hardened triangular-shaped steel teeth that is used for raking a fire line down to soil with no subsurface fuel, for digging, for rolling burning logs, and for cutting grass and small brush.

Coupling One set or pair of connection devices attached to a fire hose that allow the hose to be interconnected to additional lengths of hose or adapters and other fire-fighting appliances. (NFPA 1963)

Crew A team of two or more fire fighters. (NFPA 1500)

Cribbing Short lengths of timber/composite materials, usually 101.60 millimeters × 101.60 millimeters (4 inches × 4 inches) and 457.20 millimeters × 609.60 millimeters (18 inches × 24 inches) long that are used in various configurations to stabilize loads in place or while a load is moving. (NFPA 1006)

Critical incident stress debriefing (CISD) A postincident meeting designed to assist rescue personnel in dealing with psychological trauma as the result of an emergency. (NFPA 1006)

Critical incident stress management (CISM) A program designed to reduce both acute and chronic effects of stress related to job functions. (NFPA 450)

Cross-zoned system A fire alarm system that requires activation of two separate detection devices before initiating an alarm condition. If a single detection device is activated, the alarm control panel will usually show a problem or trouble condition.

Crowbar A straight bar made of steel or iron with a forked-like chisel on the working end that is suitable for performing forcible entry.

Cryogenic liquids (cryogens) Gaseous substances that have been chilled to the point at which they liquefy; a liquid having a boiling point lower than −150°F (−101°C) at 14.7 psia (an absolute pressure of 101 kPa).

Curtain wall Nonbearing walls that separate the inside and outside of the building but are not part of the support structure for the building.

Curved roof A roof with a curved shape.

Cutting tools Tools that are designed to cut into metal or wood.

Cutting torch A torch that produces a high-temperature flame capable of heating metal to its melting point, thereby cutting through an object. Because of the high temperatures (5700°F [3148°C]) that these torches produce, the operator must be specially trained before using this tool.

Cyanide A highly toxic chemical agent that attacks the circulatory system.

Cyberterrorism The intentional act of electronically attacking government or private computer systems.

Cylinder A portable compressed gas container.

Cylinder (fire extinguisher) The body of the fire extinguisher where the extinguishing agent is stored.

Cylindrical locks The most common fixed locks in use today. The locks and handles are placed into predrilled holes in the doors. One side of the door will usually have a key-in-the-knob lock; the other will have a keyway, a button, or some other type of locking/unlocking mechanism.

Damming A process used when liquid is flowing in a natural channel or depression and its progress can be stopped by blocking the channel.

Dangerous cargo manifest Shipping papers on a marine vessel, generally located in a tube-like container.

Dead load The weight of the aerial device structure and all materials, components, mechanisms, or equipment permanently fastened thereto. (NFPA 1901)

Deadbolt Surface- or interior-mounted lock on or in a door with a bolt that provides additional security.

Decay stage The stage of fire development in which the fire has consumed either the available fuel or oxygen and is starting to die down.

Deck gun An apparatus-mounted master stream device that is intended to flow large amounts of water directly onto a fire or exposed building.

Decontamination corridor The area usually located within the warm zone where decontamination is performed. (NFPA 472)

Decontamination team The team responsible for reducing and preventing the spread of contaminants from persons and equipment used at a hazardous materials incident. Members of this team establish the decontamination corridor and conduct all phases of decontamination.

Decontamination The physical and/or chemical process of reducing and preventing the spread of contaminants from people, animals, the environment, or equipment involved at hazardous materials/weapons of mass destruction incidents. (NFPA 472)

Defend-in-place A strategy in which the victims are protected from the fire without relocating them.

Defensible space An area as defined by the authority having jurisdiction [typically a width of 9.14 meters (30 feet) or more] between an improved property and a potential wildland fire where combustible materials and vegetation have been removed or modified to reduce the potential for fire on improved property spreading to wildland fuels or to provide a safe working area for fire fighters protecting life and improved property from wildland fire. (NFPA 1051)

Defensive attack Exterior fire suppression operations directed at protecting exposures.

Defensive objectives Actions that do not involve the actual stopping of the leak or release of a hazardous material, or contact of responders with the material; these include preventing further injury and controlling or containing the spread of the hazardous material.

Degradation (1) A chemical action involving the molecular breakdown of a protective clothing material or equipment caused by contact with a chemical. (2) The molecular breakdown of the spilled or released material to render it less hazardous during control operations. (NFPA 472)

Dehydration A state in which fluid losses are greater than fluid intake into the body. If left untreated, dehydration may lead to shock and even death.

Deluge head A sprinkler head that has no release mechanism; the orifice is always open.

Deluge sprinkler system A sprinkler system in which all sprinkler heads are open. When an initiation device, such as a smoke detector or heat detector, is activated, the deluge valve opens and water discharges from all of the open sprinkler heads simultaneously.

Deluge valve A valve assembly designed to release water into a sprinkler system when an external initiation device is activated.

Demonstrative evidence Materials that are used to demonstrate a theory or explain an event.

Department of Transportation (DOT) marking system A unique system of labels and placards that, in combination with the *Emergency Response Guidebook*, offers guidance for first responders operating at a hazardous materials incident.

Department of Transportation (DOT) The federal agency that publicizes and enforces rules and regulations that relate to the transportation of many hazardous materials.

Depressions Indentations felt on a kernmantle rope that indicate damage to the interior (kern) of the rope.

Depth of char The thickness of the layer of a material that has been consumed by a fire. The depth of char on wood can be used to help determine the intensity of a fire at a specific location.

Designated incident facilities Assigned locations where specific functions are always performed.

Dewar containers Containers designed to preserve the temperature of the cold liquid held inside.

Diking The placement of materials to form a barrier that will keep a hazardous material in liquid form from entering an area, or that will hold the material in an area.

Dilution The process of adding some substance—usually water—to a product to weaken its concentration.

Direct attack (structural fire) Firefighting operations involving the application of extinguishing agents directly onto the burning fuel. (NFPA 1145)

Direct attack (wildland and ground fire) A method of fire attack in which fire fighters focus on containing and extinguishing the fire at its burning edge.

Direct evidence Evidence that is reported first-hand, such as statements from an eyewitness who saw or heard something.

Direct line A telephone that connects two predetermined points.

Discipline The guidelines that a department sets for fire fighters to work within.

Disinfection The process used to inactivate virtually all recognized pathogenic microorganisms but not necessarily all microbial forms, such as bacterial endospores. (NFPA 1581)

Dispatch To send out emergency response resources promptly to an address or incident location for a specific purpose. (NFPA 450)

Disposal A two-step removal process for contaminated items that cannot be properly decontaminated. Items are bagged and placed in appropriate containers for transport to a hazardous waste facility.

Distributors Relatively small-diameter underground pipes that deliver water to local users within a neighborhood.

Diversion Redirecting spilled or leaking material to an area where it will have less impact.

Division A supervisory level established to divide an incident into geographic areas of operations. (NFPA 1561)

Division of labor Breaking down an incident or task into a series of smaller, more manageable tasks and assigning personnel to complete those tasks.

Division supervisor A person in a supervisory-level position who is responsible for a specific geographic area of operations at an incident. (NFPA 1561)

Doff To take off an item of clothing or equipment.

Don To put on an item of clothing or equipment.

Door An entryway; the primary choice for forcing entry into a vehicle or structure.

Double-action pull-station A manual fire alarm activation device that requires two steps to activate the alarm. The user must push in a flap, lift a cover, or break a piece of glass before activating the alarm.

Double-female adaptor A hose adaptor that is used to join two male hose couplings.

Double-hung windows Windows that have two movable sashes that can go up and down.

Double-jacket hose A hose constructed with two layers of woven fibers.

Double-male adaptor A hose adaptor that is used to join two female hose couplings.

Double-pane glass A window design that traps air or inert gas between two pieces of glass to help insulate a house.

Drag rescue device (DRD) A component integrated within the protective coat element to aid in the rescue of an incapacitated fire fighter. (NFPA 1851)

Drums Barrel-like containers built to DOT Specification 5P (1A1).

Dry bulk cargo tanks Tanks designed to carry dry bulk goods such as powders, pellets, fertilizers, or grain; they are generally V-shaped with rounded sides that funnel toward the bottom.

Dry hydrant An arrangement of pipe permanently connected to a water source other than a piped, pressurized water supply system that provides a ready means of water supply for firefighting purposes and that utilizes the drafting (suction) capability of a fire department pump. (NFPA 1142)

Dry sprinkler system A sprinkler system in which the pipes are normally filled with compressed air. When a sprinkler head is activated, it releases the air from the system, which opens a valve so the pipes can fill with water.

Dry-barrel hydrant A type of hydrant used in areas subject to freezing weather. The valve that allows water to flow into the hydrant is located underground, and the barrel of the hydrant is normally dry.

Dry-chemical extinguishing system An automatic fire extinguishing system that discharges a dry chemical agent.

Dry-chemical fire extinguisher An extinguisher that uses a mixture of finely divided solid particles to extinguish fires. The agent is usually sodium bicarbonate, potassium bicarbonate, or ammonium phosphate based, with additives being included to provide resistance to packing and moisture absorption and to promote proper flow characteristics. These extinguishers are rated for use on Class B and C fires, although some are also rated for Class A fires.

Dry-pipe valve The valve assembly on a dry sprinkler system that prevents water from entering the system until the air pressure is released.

Dry-powder extinguishing agent An extinguishing agent used in putting out Class D fires. Common examples include sodium chloride and graphite-based powders.

Dry-powder fire extinguisher A fire extinguisher that uses an extinguishing agent in powder or granular form. It is designed to extinguish Class D combustible-metal fires by crusting, smothering, or heat-transferring means.

Drywall hook A specialized version of a pike pole that can remove drywall more effectively because of its hook design.

Dual-path pressure reducer A feature that automatically provides a backup method for air to be supplied to the regulator of an SCBA if the primary passage malfunctions.

Duck-billed lock breaker A tool with a point that can be inserted into the shackles of a padlock. As the point is driven farther into the lock, it gets larger and forces the shackles apart until they break.

Dump valve A large opening from the water tank of a mobile water supply apparatus for unloading purposes. (NFPA 1901)

DuoDote Kit A prefilled auto-injector that provides two medications (atropine and pralidoxime chloride) for use in treating exposure to nerve agents and insecticide poisoning.

Duplex channel A radio system that uses two frequencies per channel; one frequency transmits and the other receives messages. Such a system uses a repeater site to transmit messages over a greater distance than is possible with a simplex system.

Dutchman A short fold placed in a hose when loading it into the bed; the fold prevents the coupling from turning in the hose bed.

Dynamic rope A rope generally made from synthetic materials that is designed to be elastic and stretch when loaded. Dynamic rope is often used by mountain climbers.

Early-suppression fast-response sprinkler head A sprinkler head designed to react quickly and suppress a fire in its early stages.

Ecoterrorism Terrorism directed against causes that radical environmentalists think would damage the earth or its creatures.

Egress To go or come out; to exit from an area or a building.

Ejectors Also called smoke ejectors, electric fans used in negative-pressure ventilation.

Electric vehicle An automotive-type vehicle for on-road use, such as passenger automobiles, buses, trucks, vans, neighborhood electric vehicles, and the like, primarily powered by an electric motor that draws current from a rechargeable storage battery, fuel cell, photovoltaic array, or other source of electric current. Electric motorcycles and similar-type vehicles and off-road, self-propelled electric vehicles, such as industrial trucks, hoists, lifts, transports, golf carts, airline ground support equipment, tractors, boats, and the like, are not included in this definition. (NFPA 70)

Electrical energy Heat that is produced by electricity.

Electrolytes Certain salts and other chemicals that are dissolved in body fluids and cells. Proper levels of electrolytes need to be maintained for good health and strength.

Elevated master stream device A nozzle mounted on the end of an aerial device that is capable of delivering large amounts of water onto a fire or exposed building from an elevated position.

Elevated water storage tower An above-ground water storage tank that is designed to maintain pressure on a water distribution system.

Elevation pressure The amount of pressure created by gravity. Also known as head pressure.

Emergency decontamination The physical process of immediately reducing contamination of individuals in potentially life-threatening situations with or without the formal establishment of a decontamination corridor. (NFPA 472)

Emergency incident rehabilitation A function on the emergency scene that cares for the well-being of the fire fighters. It includes relief from climatic conditions, rest, cooling or warming, rehydration, calorie replacement, medical monitoring, member accountability, and release.

Emergency Medical Services (EMS) company A company that may be made up of medical units and first-response vehicles. Members of this company respond to and assist in the transport of medical and trauma victims to medical facilities. They often have medications, defibrillators, and paramedics who can stabilize a critical patient.

Emergency Medical Services (EMS) personnel Personnel who are responsible for administering prehospital care to people who are sick and injured. Prehospital calls make up the majority of responses in most fire departments, and EMS personnel are cross-trained as fire fighters.

Emergency Medical Technician (EMT) An EMS provider who has training in Basic Life Support care, including automated external defibrillation, simple airway techniques, and controlling external bleeding.

Emergency Planning and Community Right to Know Act Federal legislation that requires a business that handles chemicals (depending on the quantity stored) to report storage type, quantity, and storage methods to the fire department and the local emergency planning committee.

Emergency Response Guidebook (ERG) A reference book, written in plain language, to guide emergency responders in their initial actions at the incident scene. (NFPA 472)

Emergency traffic An urgent message, such as a call for help or evacuation, transmitted over a radio that takes precedence over all normal radio traffic.

Employee assistance programs (EAPs) Fire service programs that provide confidential help to fire fighters with personal issues.

Emulsification The process of changing the chemical properties of a hazardous material to reduce its harmful effects.

End-of-service-time-indicator (EOSTI) A warning device on a SCBA that alerts the user that the end of the breathing air is approaching.

Endothermic Reactions that absorb heat or require heat to be added.

Engine company A group of fire fighters who work as a unit and are equipped with one or more pumping engines that have rated capacities of 2840 L/min (750 gpm) or more. (NFPA 1410)

Entrapment A condition in which a victim is trapped by debris, soil, or other material and is unable to extricate himself or herself.

Environmental Protection Agency (EPA) Established in 1970, the federal agency that ensures safe manufacturing, use, transportation, and disposal of hazardous substances.

Escape rope A system component; a single-purpose, one-time use, emergency self-escape (self-rescue) rope; not classified as a life safety rope. (NFPA 1983)

Evacuation signal A distinctive signal intended to be recognized by the occupants as requiring evacuation of the building. (NFPA 72)

Exhauster A device that accelerates the removal of the air from a dry-pipe or preaction sprinkler system.

Exit Drills in the Home (EDITH) A public safety program designed to teach occupants how to safely exit a house in the event of a fire or other emergency. It stresses the importance of having a preestablished meeting place for all family members.

Exothermic Reactions that result in the release of energy in the form of heat.

Expansion ratio The ratio of the volume of foam in its aerated state to the original volume of nonaerated foam solution. (NFPA 1901, Standard for Automotive Fire Apparatus)

Exposure Any person or property that could be endangered by fire, smoke, gases, runoff, or other hazardous conditions. (NFPA 402)

Extension Fire that moves into areas not originally involved in the incident, including walls, ceilings, and attic spaces; also, the movement of fire into uninvolved areas of a structure.

Extension ladder A non-self-supporting ground ladder that consists of two or more sections traveling in guides, brackets, or the equivalent arranged so as to allow length adjustment. (NFPA 1931)

Exterior wall A wall—often made of wood, brick, metal, or masonry—that makes up the outer perimeter of a building. Exterior walls are often load bearing.

Extinguishing agent A material used to stop the combustion process. Extinguishing agents may include liquids, gases, dry chemical compounds, and dry powder compounds.

Extra (high) hazard locations Occupancies where the total amounts of Class A combustibles and Class B flammables are greater than expected in occupancies classed as ordinary (moderate) hazards.

Face piece A component of SCBA that fits over the face.

False alarm The activation of a fire alarm system when no fire or emergency condition exists.

Federal Communications Commission (FCC) The federal regulatory authority that oversees radio communications in the United States.

Film-forming fluoroprotein (FFFP) A protein-based foam concentrate incorporating fluorinated surfactants that forms a foam capable of producing a vapor-suppressing aqueous film on the surface of hydrocarbon fuels. This foam can show an acceptable level of compatibility with dry chemicals and might be suitable for use with those agents. (NFPA 412)

Finance/Administration Section Section responsible for all costs and financial actions of the incident or planned event, including the time unit, procurement unit, compensation/claims unit, and the cost unit. (NFPA 1026)

Fine fuels Fuels that ignite and burn easily, such as dried twigs, leaves, needles, grass, moss, and light brush.

Finger A narrow point of fire whose extension is created by a shift in wind or a change in topography.

Fire A rapid, persistent chemical reaction that releases both heat and light.

Fire alarm annunciator panel Part of the fire alarm system that indicates the source of an alarm within a building.

Fire alarm control panel The component in a fire alarm system that controls the functions of the entire system.

Fire and life safety education specialist A member of the fire department who deals with the public on education, fire safety, and juvenile fire safety programs.

Fire apparatus driver/operator A fire department member who is authorized by the authority having jurisdiction to drive, operate, or both drive and operate fire department vehicles. (NFPA 1451)

Fire apparatus maintenance personnel The people who repair and service the fire and EMS vehicles, ensuring that they are always ready to respond to emergencies.

Fire code A set of legally adopted rules and regulations designed to prevent fires and protect lives and property in the event of a fire.

Fire department connection (FDC) A fire hose connection through which the fire department can pump water into a sprinkler system or standpipe system.

Fire department EMS system A system in which the fire department both provides medical first response and transports the patient to the hospital emergency room.

Fire enclosure A fire-rated assembly used to enclose a vertical opening such as a stairwell, elevator shaft, and chase for building utilities.

Fire Fighter I A person, at the first level of progression as defined in Chapter 5, who has demonstrated the knowledge and skills to function as an integral member of a firefighting team under direct supervision in hazardous conditions. (NFPA 1001)

Fire Fighter II A person, at the second level of progression as defined in Chapter 6, who has demonstrated the skills and depth of knowledge to function under general supervision. (NFPA 1001)

Fire helmet Protective head covering worn by fire fighters to protect the head from falling objects, blunt trauma, and heat.

Fire hook A tool used to pull down burning structures.

Fire hydraulics The physical science of how water flows through a pipe or hose.

Fire load The weight of combustibles in a fire area (measured in ft2 or m2) or on a floor in buildings and structures, including either contents or building parts, or both. (NFPA 914)

Fire mark Historically, an identifying symbol on a building informing fire fighters that the building was insured by a company that would pay them for extinguishing the fire.

Fire marshal/fire inspector/fire investigator A member of the fire department who inspects businesses and enforces laws that deal with public safety and fire codes. A fire investigator may also respond to fire scenes to help incident commanders investigate the cause of a fire. Investigators may have full police powers of arrest and deal directly with investigations and arrests.

Fire partition An interior wall extending from the floor to the underside of the floor above.

Fire point The lowest temperature at which a liquid will ignite and achieve sustained burning when exposed to a test flame in accordance with ASTM D 92, *Standard Test Method for Flash and Fire Points by Cleveland Open Cup*. (NFPA 704, Standard System for the Identification of the Hazards of Materials for Emergency Response)

Fire police Members of the fire department who protect fire fighters by controlling traffic and securing the scene from public access. Many fire police are sworn peace officers as well as fire fighters.

Fire prevention Activities conducted to prevent fires and protect lives and property in the event of a fire. Fire prevention activities include the enactment and enforcement of fire codes, the inspection of properties, the presentation of fire safety education programs, and the investigation of the causes of fires.

Fire protection engineer A member of the fire department who is responsible for reviewing plans and working with building owners to ensure that the design of and systems for fire detection and suppression will meet applicable codes and function as needed.

Fire shelter An item of protective equipment configured as an aluminized tent utilized for protection, by means of reflecting radiant heat, in a fire entrapment situation. (NFPA 1500)

Fire tetrahedron A geometric shape used to depict the four components required for a fire to occur: fuel, oxygen, heat, and chemical chain reactions.

Fire triangle A geometric shape used to depict the three components of which a fire is composed: fuel, oxygen, and heat.

Fire wall A fire division assembly with a fire resistance rating of 3 test hours or longer, built to permit complete burnout and collapse of the structure on one side without extension of fire through the fire wall or collapse of the fire wall. (NFPA 901)

Fire wardens Individuals who were charged with enforcing fire regulations in colonial America.

Fire window A window assembly rated in accordance with NFPA 257 and installed in accordance with NFPA 80. (NFPA 5000)

Fire-ground command (FGC) An incident management system developed in the 1970s for day-to-day fire department incidents (generally handled with fewer than 25 units or companies).

Fire-resistive construction Buildings that have structural components of noncombustible materials with a specified fire resistance. Materials can include concrete, steel beams, and masonry block walls. Fire-resistive construction is also known as Type I building construction, as defined in NFPA 220, *Standard on Types of Building Construction*.

Fireplug A valve installed to control water accessed from wooden pipes.

FIRESCOPE (Fire Resources of California Organized for Potential Emergencies) An organization of agencies established in the early 1970s to develop a standardized system for managing fire resources at large-scale incidents such as wildland fires.

Fixed-gallonage fog nozzle A nozzle that delivers a set number of gallons per minute (liters per minute) as per the nozzle's design, no matter what pressure is applied to the nozzle.

Fixed-temperature heat detector A sensing device that responds when its operating element is heated to a predetermined temperature.

Flame detector A sensing device that detects the radiant energy emitted by a flame.

Flame point (fire point) The lowest temperature at which a substance releases enough vapors to ignite and sustain combustion.

Flammability limits (explosive limits) The upper and lower concentration limits (at a specified temperature and pressure) of a flammable gas or vapor in air that can be ignited, expressed as a percentage of the fuel by volume.

Flammable range The range of concentrations between the lower and upper flammable limits. (NFPA 68, Standard on Explosion Protection by Deflagration Venting)

Flammable vapor A concentration of constituents in air that exceeds 10 percent of its lower flammable limit (LFL). (NFPA 115, Standard for Laser Fire Protection)

Flanking attack A direct method of suppressing a wildland or ground fire that involves placing a suppression crew on one flank of a fire.

Flash point The lowest temperature at which a liquid produces a flammable vapor.

Flash point The minimum temperature at which a liquid emits vapor in sufficient concentration to form an ignitable mixture with air near the surface of the liquid, with the container as specified by appropriate test procedures and apparatus. (NFPA 122, Standard for Fire Prevention and Control in Metal/Nonmetal Mining and Metal Mineral Processing Facilities)

Flashover A transition stage in the development of a compartment fire in which surfaces exposed to thermal radiation reach ignition temperature more or less simultaneously and fire spreads rapidly throughout the space, resulting in full room involvement or total involvement of the compartment or enclosed space. (NFPA 921)

Flat bar A specialized type of prying tool made of flat steel with prying ends suitable for performing forcible entry.

Flat hose load A method of putting a hose on a vehicle in which the hose is laid flat and stacked on top of the previous section.

Flat roof A horizontal roof; often found on commercial or industrial occupancies.

Flat-head axe A tool that has a head with an axe on one side and a flat head on the opposite side.

Floodlight A light that can illuminate a broad area.

Floor runner A piece of canvas or plastic material, usually 3 to 4 feet (91 to 122 centimeters) wide and available in various lengths, that is used to protect flooring from dropped debris and dirt from shoes and boots.

Flow path The volume in a structure between an inlet and an outlet that allows the movement of heat and smoke from the higher pressure within the fire area toward the lower-pressure areas accessible via doorways, halls, stairs, and window openings.

Flow pressure The amount of pressure created by moving water.

Flow switch An electrical switch that is activated by water moving through a pipe in a sprinkler system.

Fluoroprotein foam A protein-based foam concentrate to which fluorochemical surfactants have been added. This has the effect of giving the foam a measurable degree of compatibility with dry chemical extinguishing agents and an increase in tolerance to contamination by fuel. (NFPA 402)

Fly section Any section of an aerial telescoping device beyond the base section. (NFPA 1901)

Foam concentrate A concentrated liquid foaming agent as received from the manufacturer. (NFPA 11)

Foam eductor A device placed in the hose line that draws foam concentrate from a container and introduces it into the fire stream.

Foam injector A device installed on a fire pump that meters out foam by pumping or injecting it into the fire stream.

Foam proportioner A device or method to add foam concentrate to water to make foam solution. (NFPA 1901)

Foam solution A homogeneous mixture of water and foam concentrate in the proper proportions. (NFPA 1901)

Fog-stream nozzle A nozzle that is placed at the end of a fire hose and separates water into fine droplets to aid in heat absorption.

Folding ladder A single-section ladder with rungs that can be folded or moved to allow the beams to be brought into a position touching or nearly touching each other. (NFPA 1931)

Forcible entry Techniques used by fire personnel to gain entry into buildings, vehicles, aircraft, or other areas of confinement when normal means of entry are locked or blocked. (NFPA 402)

Forward lay A method of laying a supply line where the line starts at the water source and ends at the attack engine.

Forward staging area A strategically placed area, close to the incident site, where personnel and equipment can be held in readiness for rapid response to an emergency event.

Four-way hydrant valve A specialized type of valve that can be placed on a hydrant and that allows another engine to increase the supply pressure without interrupting flow.

Frangible-bulb sprinkler head A sprinkler head with a liquid-filled bulb. The sprinkler head becomes activated when the liquid is heated and the glass bulb breaks.

Freelancing The dangerous practice of acting independently of command instructions.

Freight bills Shipping papers for roads and highways.

Fresno ladder A narrow, two-section extension ladder that has no halyard. Because of its limited length, it can be extended manually.

Friction loss The reduction in pressure resulting from the water being in contact with the side of the hose. This contact requires force to overcome the drag that the wall of the hose creates.

Frostbite A localized condition that occurs when the layers of the skin and deeper tissue freeze. (NFPA 704)

Fuel A material that will maintain combustion under specified environmental conditions. (NFPA 53)

Fuel cell An electrochemical system that consumes fuel to produce an electric current. The main chemical reaction used in a fuel cell for producing electric power is not combustion, but sources of combustion may be used within the overall fuel cell system such as reformers/fuel processors. (NFPA 70)

Fuel cell stack Many fuel cells assembled into a group to produce larger quantities of electricity.

Fuel compactness The extent to which fuels are tightly packed together.

Fuel continuity The relative closeness of wildland fuels, which affects a fire's ability to spread from one area of fuel to another.

Fuel moisture The amount of moisture present in a fuel, which affects how readily the fuel will ignite and burn.

Fuel volume The amount of fuel present in a given area.

Fuel-limited A fire that has sufficient oxygen for fire growth, but has a limited amount of fuel available for burning.

Fully developed stage The stage where the fire is consuming the maximum amount of fuel possible. It achieves the maximum heat release rate possible for the fuel supply and oxygen present.

Fully encapsulated suit A protective suit that completely covers the fire fighter, including the breathing apparatus, and does not let any vapor or fluids enter the suit. It is commonly used in hazardous materials emergencies.

Fusible-link sprinkler head A sprinkler head with an activation mechanism that incorporates two pieces of metal held together by low-melting-point solder. When the solder melts, it releases the link and water begins to flow.

Gamma radiation A type of radiation that can travel significant distances, penetrating most materials and passing through the body. Gamma radiation is the most destructive type of radiation to the human body.

Gas A material that has a vapor pressure greater than 300 kPa absolute (43.5 psia) at 50°C (122°F) or is completely gaseous at 20°C (68°F) at a standard pressure of 101.3 kPa absolute (14.7 psia). (NFPA 30A)

Gas detector A device that detects and/or measures the concentration of dangerous gases.

Gate valves Valves found on hydrants and sprinkler systems.

Gated wye A valved device that splits a single hose into two separate hoses, allowing each hose to be turned on and off independently.

General use life safety rope A life safety rope that is no larger than 5/8" (16 mm) and no smaller than 7/16" (11 mm), with a minimum breaking strength of 8992 lbf (40 kN).

Generator An electromechanical device for the production of electricity. (NFPA 1901)

Geographic information system (GIS) A system of computer software, hardware, data, and personnel to describe information tied to a spatial location. (NFPA 450)

Glass blocks Thick pieces of glass that are similar to bricks or tiles.

Glazed Glass or transparent or translucent plastic sheet used in windows, doors, skylights, or curtain walls. [ASCE/SEI 7:6.2] (NFPA 5000)

Glucose The source of energy for the body. One of the basic sugars, it is the body's primary fuel, along with oxygen.

Governance The process by which an organization exer-cises authority and performs the functions assigned to it.

Grade A measurement of the angle used in road design and expressed as a percentage of elevation change over distance. (NFPA 1901)

Gravity-feed system A water distribution system that depends on gravity to provide the required pressure. The system storage is usually located at a higher elevation than the end users.

Green An area of unburned fuels.

Gripping pliers A hand tool with a pincer-like working end that can be used to bend wire or hold smaller objects.

Gross decontamination The phase of the decontamination process during which the amount of surface contaminants is significantly reduced. (NFPA 472)

Ground cover fire A fire that burns loose debris on the surface of the ground.

Ground duff Partly decomposed organic material on a forest floor; a type of light fuel.

Group A supervisory level established to divide an incident into functional areas of operation. (NFPA 1561)

Group supervisor A person in a supervisory-level position who is responsible for a functional area of operation. (NFPA 1561)

Growth stage The stage of fire development in which the fire is spreading beyond the point of origin and beginning to involve other fuels in the immediate area.

Guideline A rope used for orientation when fire fighters are inside a structure where there is low or no visibility. The line is attached to a fixed object outside the hazardous area.

Guides Strips of metal or wood that serve to guide a fly section during extension. Channels or slots in the bed or fly section may also serve as guides.

Gusset plate A connecting plate used in trusses, typically made of wood or lightweight metal.

Gypsum A naturally occurring material consisting of calcium sulfate and water molecules.

Gypsum board The generic name for a family of sheet products consisting of a noncombustible core primarily of gypsum with paper surfacing. (NFPA 5000)

Hacksaw A cutting tool designed for use on metal. Different blades can be used for cutting different types of metal.

Halligan tool A prying tool that incorporates a pick and a fork, specifically designed for use in the fire service.

Halogenated extinguishing agent A liquefied gas extinguishing agent that puts out fires by chemically interrupting the combustion reaction between the fuel and oxygen.

Halogenated-agent fire extinguisher An extinguisher that uses a halogenated extinguishing agent.

Halon 1211 A halogenated agent whose chemical name is bromochlorodifluoromethane, CBrClF2; it is a multipurpose, Class ABC-rated agent effective against flammable liquid fires. (NFPA 408)

Halon 1301 A liquefied gas extinguishing agent that puts out a fire by chemically interrupting the combustion reaction between fuel and oxygen. Halon agents leave no residue.

Halyard Rope used on extension ladders for the purpose of raising a fly section(s). (NFPA 1931)

Hammer A striking tool.

Hand light A small, portable light carried by fire fighters to improve visibility at emergency scenes; it is often powered by rechargeable batteries.

Handle The grip used for holding and carrying a portable fire extinguisher.

Handline nozzle A nozzle with a rated discharge of less than 1591 L/min (350 gpm). (NFPA 1964)

Handsaw A manually powered saw designed to cut different types of materials. Examples include hacksaws, carpenter's handsaws, keyhole saws, and coping saws.

Hard suction hose A hose used for drafting water from static supplies (lakes, rivers, wells, and so forth). It can also be used for supplying pumps on fire apparatus from hydrants if designed for that purpose. The hose contains a semirigid or rigid reinforcement. (NFPA 1963)

Hardware The parts of a door or window that enable it to be locked or opened.

Harness A piece of equipment worn by a rescuer that can be attached to a life safety rope.

Hazard A fuel complex defined by kind, arrangement, volume, condition, and location that determines the ease of ignition and/or of resistance to fire control. (NFPA 1144, Standard for Reducing Structure Ignition Hazards from Wildland Fire)

Hazardous material A substance that, when released, is capable of creating harm to people, the environment, and property. (NFPA 472)

Hazardous materials branch director/group supervisor Commanders of hazardous materials incidents beyond the operations level.

Hazardous Materials Branch The function within an overall incident management system that deals with the mitigation and control of the hazardous materials/weapons of mass destruction portion of an incident. (NFPA 472)

Hazardous materials company A fire company that responds to and controls scenes where hazardous materials have spilled or leaked. Responders wear special suits and are trained to deal with most chemicals.

Hazardous materials information research team A dedicated team of responders who serve as an information gathering and referral point for the incident commander as well as the hazardous materials officer.

Hazardous Materials Information System (HMIS) A color-coded marking system by which employers give their personnel the necessary information to work safely around chemicals.

Hazardous materials officer (NIMS: Hazardous Materials Branch Director/Group Supervisor.) The person who is responsible for directing and coordinating all operations involving hazardous materials/weapons of mass destruction as assigned by the incident commander. (NFPA 472)

Hazardous materials safety officer (NIMS: Assistant Safety Officer—Hazardous Material.) The person who works within the Incident Management System (specifically, the Hazardous Materials Branch/Group) to ensure that recognized hazardous materials/weapons of mass destruction (WMD) safe practices are followed at hazardous materials/WMD incidents. (NFPA 472)

Hazardous materials technician A person who responds to hazardous materials/weapons of mass destruction incidents using a risk-based response process by which they analyze the problem at hand, select applicable decontamination procedures, and control a release while using specialized protective clothing and control equipment. (NFPA 472)

Hazardous materials technician with specialties A hazardous materials technician with training in areas such as specialized chemicals, containers, and their uses.

Hazardous waste Waste that is potentially damaging to the environment or human health due to its toxicity, ignitability, corrosivity, or chemical reactivity or another cause. (NFPA 820)

HAZWOPER Hazardous Waste Operations and Emergency Response; the OSHA regulation that governs hazardous materials waste sites and response training. Specifics can be found in 29 CFR 1910.120. Subsection (q) is specific to emergency response.

Head of the fire The main or running edge of a fire; the part of the fire that spreads with the greatest speed.

Heads-up display A visual display of information and system conditions status that is visible to the wearer of the SCBA.

Heat cramps Painful muscle spasms usually associated with vigorous activity in a hot environment.

Heat detector A fire alarm device that detects abnormally high temperature, an abnormally high rate of rise in temperature, or both.

Heat exhaustion A mild form of shock caused when the circulatory system begins to fail as a result of the body's inadequate effort to give off excessive heat.

Heat flux The measure of the rate of heat transfer from one surface to another. It is measured in kilowatts per meter squared (kW/m2) or Btu/ft2.

Heat rash Clear, raised bumps on the skin caused by friction and sweat.

Heat sensor label A label that changes color at a preset temperature to indicate a specific heat exposure. (NFPA 1931)

Heat stroke A severe and sometimes fatal condition resulting from the failure of the temperature-regulating capacity of the body. It is caused by prolonged exposure to the sun or high temperatures. Reduction or cessation of sweating is an early symptom; body temperature of 105°F (40°C) or higher, rapid pulse, hot and dry skin, headache, confusion, unconsciousness, and convulsions can occur. Heat stroke is a true medical emergency requiring immediate transport to a medical facility.

Heating, ventilation, and air-conditioning (HVAC) system A system to manage the internal environment that is often found in large buildings.

Heavy fuels Fuels of a large diameter, such as large brush, heavy timber, snags, stumps, branches, and dead timber on the ground. These fuels ignite and are consumed more slowly than light fuels.

Heel of the fire The side opposite the head of the fire, which is often close to the area of origin.

HEPA (high-efficiency particulate air) filter A filter capable of catching particles down to 0.3-micron size—much smaller than a typical dust or alpha radiation particle.

Higbee indicators Indicators on the male and female threaded couplings that indicate where the threads start. These indicators should be aligned before fire fighters start to thread the couplings together.

High temperature–protective equipment Protective clothing designed to shield the wearer during short-term exposures to high temperatures.

High-angle operation A rope rescue operation where the angle of the slope is greater than 45 degrees. In this scenario, rescuers depend on life safety rope rather than a fixed support surface such as the ground.

Hitch A knot that attaches to or wraps around an object so that when the object is removed, the knot will fall apart. (NFPA 1670, Standard on Operations and Training for Technical Search and Rescue Incidents)

Hockey puck lock A type of lock with hidden shackles that cannot be forced open through conventional methods.

Hollow-core door A door made of panels that are honeycombed inside, creating an inexpensive and lightweight design.

Horizontal evacuation Moving occupants from a dangerous area to a safe area on the same floor level.

Horizontal ventilation The process of making openings on the same level as the fire so that smoke, heat, and gases can escape horizontally from a building through openings such as doors and windows.

Horizontal-sliding windows Windows that slide open horizontally.

Horn The tapered discharge nozzle of a carbon dioxide–type fire extinguisher.

Horseshoe hose load A method of loading hose in which the hose is laid into the bed along the three walls of the bed, so that it resembles a horseshoe.

Hose appliance A piece of hardware (excluding nozzles) generally intended for connection to fire hose to control or convey water. (NFPA 1962)

Hose clamp A device used to compress a fire hose so as to stop water flow.

Hose jacket A device used to stop a leak in a fire hose or to join hoses that have damaged couplings.

Hose liner (hose inner jacket) The inside portion of a hose that is in contact with the flowing water.

Hose roller A device that is placed on the edge of a roof and is used to protect hose as it is hoisted up and over the roof edge.

Hot zone (technical rescue incident) The area immediately surrounding a hazardous materials spill/incident site that is directly dangerous to life and health. All personnel working in the hot zone must wear complete, appropriate protective clothing and equipment. Entry requires approval by the incident commander or a designated sector officer. Complete backup, rescue, and decontamination teams must be in place at the perimeter before operations begin.

Hot zone The control zone immediately surrounding hazardous materials/weapons of mass destruction incidents, which extends far enough to prevent adverse effects of hazards to personnel outside the zone. (NFPA 472)

Hot-zone entry team The team of fire fighters assigned to the entry into the designated hot zone.

Hux bar A multipurpose tool that can be used for several forcible entry and ventilation applications because of its unique design. It may also be used as a hydrant wrench.

Hybrid vehicle A vehicle that uses a battery-powered electric motor and an internal combustion engine.

Hydrant wrench A hand tool that is used to operate the valves on a hydrant; it may also be used as a spanner wrench. Some models are plain wrenches, whereas others have a ratchet feature.

Hydraulic shears A lightweight, hand-operated tool that can produce up to 10,000 pounds (4,500 kilogram) of cutting force.

Hydraulic spreader A lightweight, hand-operated tool that can produce up to 10,000 pounds (4,500 kilogram) of prying and spreading force.

Hydraulic ventilation Ventilation that relies on the movement of air caused by a fog stream that is placed 2 to 4 feet (0.6 to 1.2 meters) in front of the open window.

Hydrogen cyanide A toxic gas produced by the combustion of materials containing cyanide.

Hydrostatic testing Pressure testing of the extinguisher to verify its strength against unwanted rupture. (NFPA 10)

Hypothermia A condition in which the internal body temperature falls below 95°F (35°C), usually a result of prolonged exposure to cold or freezing temperatures.

Hypoxia A state of inadequate oxygenation of the blood and tissue sufficient to cause impairment of function. (NFPA 99B)

I-beam A ladder beam constructed of one continuous piece of I-shaped metal or fiberglass to which the rungs are attached.

ICS general staff The chiefs of each of the four major sections of ICS: Operations, Planning, Logistics, and Finance/Administration.

Ignitable liquid Classification for liquid fuels including both flammable and combustible liquids.

Ignition (autoignition) temperature The minimum temperature at which a fuel, when heated, will ignite in air and continue to burn. Also called autoignition temperature.

Ignition point The minimum temperature at which a substance will burn.

Ignition temperature Minimum temperature a substance should attain to ignite under specific test conditions. (NFPA 921)

Ignition The action of setting something on fire.

Immediately dangerous to life and health (IDLH) Any condition that would pose an immediate or delayed threat to life, cause irreversible adverse health effects, or interfere with an individual's ability to escape unaided from a hazardous environment. (NFPA 1670, 2004)

Improvised explosive device (IED) An explosive or incendiary device that is fabricated in an improvised manner.

Incendiary A fire that is intentionally ignited under circumstances in which the person knows that the fire should not be ignited.

Incendiary device A device or mechanism used to start a fire or explosion.

Incident action plan (IAP) The objectives reflecting the overall incident strategy, tactics, risk management, and member safety that are developed by the incident commander. Incident action plans are updated throughout the incident. (NFPA 1500)

Incident command post (ICP) The field location at which the primary tactical-level, on-scene incident command functions are performed. (NFPA 1026)

Incident command system (ICS) The combination of facilities, equipment, personnel, procedures, and communications operating within a common organizational structure that has responsibility for the management of assigned resources to effectively accomplish stated objectives pertaining to an incident or training exercise. (NFPA 1670)

Incident commander (IC) The person who is responsible for all decisions relating to the management of the incident and is in charge of the incident site. (NFPA 1500)

Incipient stage The initial or beginning stage of a fire, in which it can be controlled or extinguished by portable extinguishers or small amounts of dry extinguishing agents, without the need for protective clothing or breathing apparatus. (NFPA 484)

Incomplete combustion A burning process in which the fuel is not completely consumed, usually due to a limited supply of oxygen.

Incubation period The time period between the initial infection by an organism and the development of symptoms by a victim.

Indirect application of water The use of a solid object such as a wall or ceiling to break apart a stream of water, creating more surface area on the water droplets and thereby causing the water to absorb more heat.

Indirect attack (structural fire) Firefighting operations involving the application of extinguishing agents to reduce the build-up of heat released from a fire without applying the agent directly onto the burning fuel. (NFPA 1145)

Information management Fire fighters or civilians who take care of the computer and networking systems that a fire department needs to operate.

Ingestion Exposure to a hazardous material by swallowing it.

Inhalation Exposure to a hazardous material by breathing it into the lungs.

Injection Exposure to a hazardous material by it entering cuts or other breaches in the skin.

Integrated communications The ability of all appropriate personnel at the emergency scene to communicate with their supervisor and their subordinates.

Interior attack The assignment of a team of fire fighters to enter a structure and attempt fire suppression.

Interior finish The exposed surfaces of walls, ceilings, and floors within buildings. (NFPA 101)

Interior wall A wall inside a building that divides a large space into smaller areas; also known as a partition.

Intermodal tanks Bulk containers that can be shipped by all modes of transportation—air, sea, or land.

Inverter Equipment that is used to change the voltage level or waveform, or both, of electrical energy. Commonly, an inverter [also known as a power conditioning unit (PCU) or power conversion system (PCS)] is a device that changes DC input to AC output. Inverters may also function as battery chargers that use alternating current from another source and convert it into direct current for charging batteries. (NFPA 70)

Ionization smoke detector A device containing a small amount of radioactive material that ionizes the air between two charged electrodes to sense the presence of smoke particles.

Ionizing radiation Radiation of sufficient energy to alter the atomic structure of materials or cells with which it interacts, including electromagnetic radiation such as X-rays, gamma rays, and microwaves and particulate radiation such as alpha and beta particles. (NFPA 1991, Standard on Vapor-Protective Ensembles for Hazardous Materials Emergencies)

Irons A combination of tools, usually consisting of a Halligan tool and a flat-head axe, that are commonly used for forcible entry.

Irritant A substance such as mace that can be dispersed to incapacitate a person or groups of people briefly.

Island An unburned area surrounded by fire.

J tool A tool that is designed to fit between double doors equipped with panic bars.

Jalousie windows Windows made of small slats of tempered glass, which overlap each other when the window is closed. Often found in trailers and mobile homes, jalousie windows are held together by a metal frame and operated by a small hand wheel or crank found in the corner of the window.

Jamb The part of a doorway that secures the door to the studs in a building.

Junction box A device that attaches to an electrical cord to provide additional outlets.

K tool A tool that is used to remove lock cylinders from structural doors so the locking mechanism can be unlocked.

Kelly tool A steel bar with two main features: a large pick and a large chisel or fork.

Kerf cut A cut that is only the width and depth of the saw blade. It is used to inspect cockloft spaces from the roof.

Kernmantle rope Rope made of two parts—the kern (interior component) and the mantle (the outside sheath).

Kevlar® A strong synthetic material used in the construction of protective clothing and equipment.

Keyhole saw A saw designed to cut circles in wood for keyholes.

Knot A fastening made by tying together lengths of rope or webbing in a prescribed way. (NFPA 1670)

Labels Smaller versions (4-inch [10 centimeters] diamond-shaped markings) of placards, placed on four sides of individual boxes and smaller packages.

Ladder belt A compliant equipment item that is intended for use as a positioning device for a person on a ladder. (NFPA 1983)

Ladder gin An A-shaped structure formed with two ladder sections. It can be used as a makeshift lift when raising a trapped person. One form of the device is called an A-frame hoist.

Ladder halyards Rope used on extension ladders to raise a fly section.

Ladder pipe A monitor that is fed by a hose and that holds and directs a nozzle while attached to the rungs of a vehicle-mounted aerial ladder. (NFPA 1965)

Laminar smoke flow Smooth or streamlined movement of smoke, which indicates that the pressure in the building is not excessively high.

Laminated glass Safety glass. The lamination process places a thin layer of plastic between two layers of glass, so that the glass does not shatter and fall apart when broken.

Laminated windshield glass A type of window glazing that incorporates a sheeting material that stops the glass from breaking into individual shards.

Laminated wood Pieces of wood that are glued together.

Large-diameter hose (LDH) A hose 3.5 inches (89 mm) or larger that is designed to move large volumes of water to supply master stream appliances, portable hydrants, manifolds, standpipe and sprinkler systems, and fire department pumpers from hydrants and in relay. (NFPA 1410)

Latch A spring-loaded latch bolt or a gravity-operated steel bar that, after release by physical action, returns to its operating position and automatically engages the strike plate when it is returned to the closed position. (NFPA 80)

Lath Thin strips of wood used to make the supporting structure for roof tiles.

Legacy construction An older type of construction that used sawn lumber and was built before about 1970.

Level A protection Personal protective equipment that provides protection against vapors, gases, mists, and even dusts. The highest level of protection, it requires a totally encapsulating suit that includes self-contained breathing apparatus.

Level B protection Personal protective equipment that is used when the type and atmospheric concentration of substances have been identified. It generally requires a high level of respiratory protection but less skin protection: chemical-protective coveralls and clothing, chemical protection for shoes, gloves, and self-contained breathing apparatus outside of a nonencapsulating chemical-protective suit.

Level C protection Personal protective equipment that is used when the type of airborne substance is known, the concentration is measured, the criteria for using air-purifying respirators are met, and skin and eye exposure is unlikely. It consists of standard work clothing with the addition of chemical-protective clothing, chemically resistant gloves, and a form of respirator protection.

Level D protection Personal protective equipment that is used when the atmosphere contains no known hazard, and work functions preclude splashes, immersion, or the potential for unexpected inhalation of or contact with hazardous levels of chemicals. It is primarily a work uniform that includes coveralls and affords minimal protection.

Lewisite A blister-forming agent that is an oily, colorless to dark brown liquid with an odor of geraniums.

Liaison officer A member of the Command Staff who serves as the point of contact for assisting or coordinating agencies. (NFPA 1026)

Lieutenant A company officer who is usually responsible for a single fire company on a single shift; the first in line among company officers.

Life line A rope secured to a fire fighter that enables the fire fighter to retrace his or her steps out of a structure.

Life safety rope Rope dedicated solely for the purpose of supporting people during rescue, firefighting, other emergency operations, or during training evolutions. (NFPA 1983)

Light (low) hazard locations Occupancies where the total amount of combustible materials is less than expected in an ordinary hazard location.

Light-emitting diode (LED) An electronic semiconductor that emits a single-color light when activated.

Lightweight construction Lightweight materials or advanced engineering, or both practices, which result in a weight saving without sacrifice of strength or efficiency. (NFPA 414)

Line detector Wire or tubing that can be strung along the ceiling of large open areas to detect an increase in heat.

Liquid Any material that (1) has a fluidity greater than that of 300 penetration asphalt when tested in accordance with ASTM D5, *Standard Test Method for Penetration of Bituminous Materials*, or (2) is a viscous substance for which a specific melting point cannot be determined but that is determined to be a liquid in accordance with ASTM D4359, *Standard Test Method for Determining Whether a Material Is a Liquid or a Solid*. (NFPA 30)

Liquid splash–protective clothing The garment portion of a chemical-protective clothing ensemble that is designed and configured to protect the wearer against chemical liquid splashes but not against chemical vapors or gases. (NFPA 472)

Live load The load produced by the use and occupancy of the building or other structure, which does not include construction or environmental loads such as wind load, snow load, rain load, earthquake load, flood load, or dead load. Live loads on a roof are those produced (1) during maintenance by workers, equipment, and materials; and (2) during the life of the structure by movable objects such as planters and by people. [ASCE/SEI 7:4.1] (NFPA 5000)

Load-bearing wall A wall that is designed to provide structural support for a building.

Loaded-stream fire extinguisher A water-based fire extinguisher that uses an alkali metal salt as a freezing-point depressant.

Local alarm system A fire alarm system that sounds an alarm only in the building where it was activated; that is, no signal is sent out of the building.

Local emergency planning committee (LEPC) A group comprising members of industry, transportation, the public at large, media, and fire and police agencies; it gathers and disseminates information on hazardous materials stored in the community and ensures that there are adequate local resources to respond to a chemical event in the community.

Lock body The part of a padlock that holds the main locking mechanisms and secures the shackles.

Locking mechanism A device that locks an extinguisher's trigger to prevent its accidental discharge.

Lockout and tagout systems Methods of ensuring that electricity and other utilities have been shut down and switches are "locked" so that they cannot be switched on, so as to prevent flow of power or gases into the area where rescue is being conducted.

Logistics Section Chief The general staff position responsible for directing the logistics function. It is generally assigned on complex, resource-intensive, or long—duration incidents.

Logistics Section Section responsible for providing facilities, services, and materials for the incident or planned event, including the communications unit, medical unit, and food unit within the service branch and the supply unit, facilities unit, and ground support unit within the support branch. (NFPA 1026)

Loop A piece of rope formed into a circle.

Louver cut A cut that is made using power saws and axes to cut along and between roof supports so that the sections created can be tilted into the opening.

Low-angle operation A rope rescue operation on a mildly sloping surface (less than 45 degrees) or flat land. In this scenario, fire fighters depend on the ground for their primary support, and the rope system is a secondary means of support.

Low-volume nozzle A nozzle that flows 40 gallons per minute (182 lpm) or less.

Lower flammable limit (LFL) That concentration of a combustible material in air below which ignition will not occur; also known as the lower explosive limit (LEL). Mixtures below this limit are said to be "too lean." (NFPA 329)

Malicious false alarm A fire alarm signal sent when there is no fire, usually initiated by individuals who wish to cause a disturbance.

Mallet A short-handled hammer.

Manual pull-station A device with a switch that either opens or closes a circuit, activating the fire alarm.

Manufactured (mobile) home A factory-assembled structure or a structure transportable in one or more sections that is built on a permanent chassis and designed to be used as a dwelling without a permanent foundation when connected to the required utilities, including the plumbing, heating, air-conditioning, and electric systems contained therein.

Masonry Built-up unit of construction or combination of materials such as clay, shale, concrete, glass, gypsum, tile, or stone set in mortar. (NFPA 5000)

Mass arson Arson in which an offender sets three or more fires at the same site or location during a limited period of time.

Mass decontamination The physical process of reducing or removing surface contaminants from large numbers of victims in potentially life-threatening situations in the fastest time possible. (NFPA 472)

Mass-casualty incident An emergency situation involving more than one victim, which can place such great demand on equipment or personnel that the system is stretched to its limit or beyond.

Master stream device A large-capacity nozzle that can be supplied by two or more hose lines or fixed piping. Such devices include deck guns, portable ground monitors, and elevated streams, and commonly flow between 350 (1591 lpm) and 1500 (6819 lpm) gallons per minute.

Master stream nozzle A nozzle with a rated discharge of 1591 lpm (350 gpm) or greater. (NFPA 1964)

Master-coded alarm An alarm system in which audible notification devices can be used for multiple purposes, not just for the fire alarm.

Material safety data sheet (MSDS) Transitioning to Safety Data Sheets (SDS). A form, provided by manufacturers and compounders (blenders) of chemicals, containing information about chemical composition, physical and chemical properties, health and safety hazards, emergency response, and waste disposal of the material. (NFPA 472)

Matter A substance made up of atoms and molecules.

Maul A specialized striking tool, weighing 6 pounds (3 kilogram) or more, with an axe on one end and a sledgehammer on the other end.

Mayday A code indicating that a fire fighter is lost, missing, or trapped, and requires immediate assistance.

MC-306/DOT406 flammable liquid tanker Commonly known as a gasoline tanker; a tanker that typically carries gasoline or other flammable and combustible materials.

MC-307/DOT407 chemical hauler A tanker with a rounded or horseshoe-shaped tank.

MC-312/DOT412 corrosives tanker A tanker that often carries aggressive acids such as concentrated sulfuric and nitric acid, which features reinforcing rings along the side of the tank.

MC-331 pressure cargo tanker A tank commonly constructed of steel with rounded ends and a single open compartment inside; there are no baffles or other separations inside the tank.

MC-338 cryogenics tanker A low-pressure tanker designed to maintain the low temperature required by the cryogens it carries.

McLeod tool A hand tool used for constructing fire lines and overhauling wildland fires. One side of the head consists of a five-toothed to seven-toothed fire rake; the other side is a hoe.

Mechanical energy A form of potential energy that can generate heat through friction.

Mechanical saw A saw that is usually powered by an electric motor or a gasoline engine. The three primary types of mechanical saws are chainsaws, rotary saws, and reciprocating saws.

Mechanical ventilation A process of removing heat, smoke, and gases from a fire area by using exhaust fans, blowers, air-conditioning systems, or smoke ejectors. (NFPA 402, Guide for Aircraft Rescue and Fire-Fighting Operations)

Medium-diameter hose (MDH) Hose with a diameter of 2½ or 3 inches (65 mm or 76 mm).

Mildew A fungus that can grow on hose if the hose is stored wet. Mildew can damage the jacket of a hose.

Mobile data terminal Technology that allows fire fighters to receive information while in the fire apparatus or at the station.

Mobile radio A two-way radio that is permanently mounted in a fire apparatus.

Mobile water supply apparatus A vehicle designed primarily for transporting (pickup, transporting, and delivering) water to fire emergency scenes to be applied by other vehicles or pumping equipment. (NFPA 1901)

Mortise locks Door locks with both a latch and a bolt built into the same mechanism; the two locking mechanisms operate independently of each other. Mortise locks are often found in hotel rooms.

Multipurpose dry-chemical fire extinguisher A fire extinguisher rated to fight Class A, B, and C fires.

Multipurpose hook A long pole with a wooden or fiberglass handle and a metal hook on one end used for pulling.

Municipal water system A water distribution system that is designed to deliver potable water to end users for domestic, commercial, industrial, and fire protection purposes.

National Fire Incident Reporting System (NFIRS) A system used by fire departments to report and maintain computerized records of fires and other fire department incidents in a uniform manner.

National Fire Protection Association (NFPA) A private organization that develops and maintains nationally recognized minimum consensus standards on many areas of fire safety and specific standards on hazardous materials.

National Incident Management System (NIMS) A system mandated by Homeland Security Presidential Directive 5 (HSPD-5) that provides a systematic, proactive approach guiding government agencies at all levels, the private sector, and nongovernmental organizations to work seamlessly to prepare for, prevent, respond to, recover from, and mitigate the effects of incidents, regardless of cause, size, location, or complexity, so as to reduce the loss of life or property and harm to the environment. (NFPA 1026)

National Institute for Occupational Safety and Health (NIOSH) The U.S. federal agency responsible for research and development on occupational safety and health issues.

National Response Center (NRC) An agency maintained and staffed by the U.S. Coast Guard; it should always be notified if any spilled material could possibly enter a navigable waterway.

National Terrorism Advisory System Alerts A national system to communicate information about terrorist threats by providing timely information to the public, government agencies, and first responders. It consists of Imminent Threat Alerts and Elevated Threat Alerts.

Natural ventilation The flow of air or gases created by the difference in the pressures or gas densities between the outside and inside of a vent, room, or space. (NFPA 853, Standard for the Installation of Stationary Fuel Cell Power Systems)

Negative-pressure ventilation Ventilation that relies on electric fans to pull or draw the air from a structure or area.

Nerve agent A toxic substance that attacks the central nervous system in humans.

Neutral plane The interface at a vent, such as a doorway or a window opening, between the hot gas flowing out of a fire compartment and the cool air flowing into the compartment where the pressure difference between interior and exterior is zero.

Neutron A penetrating particle found in the nucleus of the atom that is removed through nuclear fusion or fission. Although neutrons are not radioactive, exposure to neutrons can create radiation.

NFPA 704 hazard identification system A hazardous materials marking system designed for fixed-facility use.

Nomex® A fire-resistant synthetic material used in the construction of personal protective equipment for firefighting.

Nonambulatory A term describing individuals who cannot move themselves to an area of safety owing to their physical condition, medical treatment, or other factors.

Nonbearing wall Any wall that is not a bearing wall. [ASCE/SEI 7:11.2] (NFPA 5000)

Nonbulk storage vessels Containers other than bulk storage containers.

Noncoded alarm An alarm system that provides no information at the alarm control panel indicating where the activated alarm is located.

Normal operating pressure The observed static pressure in a water distribution system during a period of normal demand.

Nose cups An insert inside the face piece of an SCBA that fits over the user's mouth and nose.

Nozzle A constricting appliance attached to the end of a fire hose or monitor to increase the water velocity and form a stream. (NFPA 1965)

Nozzle shut-off A device that enables the fire fighter at the nozzle to start or stop the flow of water.

Nuisance alarm A fire alarm signal caused by malfunction or improper operation of a fire alarm system or component.

Obscuration rate A measure of the percentage of light transmission that is blocked between a sender and a receiver unit.

Occupancy The purpose for which a building or other structure, or part thereof, is used or intended to be used. (NFPA 5000)

Occupational Safety and Health Administration (OSHA) The federal agency that regulates worker safety and, in some cases, responder safety. OSHA is a part of the Department of Labor.

Offensive attack An advance into the fire building by fire fighters with hose lines or other extinguishing agents that are intended to overpower the fire.

Open-circuit breathing apparatus SCBA in which the exhaled air is released into the atmosphere and is not reused.

Open-end wrench A hand tool that is used to tighten or loosen bolts. The end is open, as opposed to a box-end wrench. Each wrench is a specific size.

Operations level Persons who respond to hazardous materials/weapons of mass destruction incidents for the purpose of implementing or supporting actions to protect nearby persons, the environment, or property from the effects of the release. (NFPA 472)

Operations Section Chief The general staff position responsible for managing all operations activities. It is usually assigned when complex incidents involve more than 20 single resources or when Command cannot be involved in the details of tactical operations.

Operations Section Section responsible for all tactical operations at the incident or planned event, including up to 5 branches, 25 divisions/groups, and 125 single resources, task forces, or strike teams. (NFPA 1026)

Operator lever The handle of a door that turns the latch to open it.

Ordinary (moderate) hazard locations Occupancies that contain more Class A and Class B materials than are found in light hazard locations.

Ordinary construction Buildings whose exterior walls are made of noncombustible or limited-combustible materials, but whose interior floors and walls are made of combustible materials. Ordinary construction is also known as Type III building construction, as defined in NFPA 220, *Standard on Types of Building Construction*.

Outside stem and yoke (OS&Y) valve A sprinkler control valve with a valve stem that moves in and out as the valve is opened or closed.

Overhaul The process of final extinguishment after the main body of a fire has been knocked down. All traces of fire must be extinguished at this time. (NFPA 402, 2002)

Oxidation Reaction with oxygen either in the form of the element or in the form of one of its compounds. (NFPA 53)

Oxygen deficiency Any atmosphere where the oxygen level is less than 19.5 percent. Low oxygen levels can have serious effects on people, including adverse reactions such as poor judgment and lack of muscle control.

Packaging The process of securing a victim in a transfer device, with regard to existing and potential injuries or illness, so as to prevent further harm during movement. (NFPA 1006)

Padlocks The most common types of locks on the market today, built to provide regular-duty or heavy-duty service. Several types of locking mechanisms are available, including keyways, combination wheels, and combination dials.

Parallel chord truss A truss in which the top and bottom chords are parallel.

Paramedic An EMS provider who has extensive training in Advanced Life Support care, including intravenous therapy, pharmacology, endotracheal intubation, and other advanced assessment and treatment skills.

Parapet walls Walls on a flat roof that extend above the roofline.

Partition A nonstructural interior wall that spans horizontally or vertically from support to support. The supports may be the basic building frame, subsidiary structural members, or other portions of the partition system. [ASCE/SEI 7:11.2] (NFPA 5000)

Party wall A wall constructed on the line between two properties.

PASS Acronym for the steps involved in operating a portable fire extinguisher: Pull pin, Aim nozzle, Squeeze trigger, Sweep across burning fuel.

Pawls Devices attached to a fly section(s) to engage ladder rungs near the beams of the section below for the purpose of anchoring the fly section(s). (NFPA 1931)

PBI® A fire-retardant synthetic material used in the construction of personal protective equipment.

Peak cut A ventilation opening that runs along the top of a peaked roof.

Pendant sprinkler head A sprinkler head designed to be mounted on the underside of sprinkler piping so that the water stream is directed in a downward direction.

Penetration The movement of a material through a suit's closures, such as zippers, buttonholes, seams, flaps, or other design features of chemical-protective clothing, and through punctures, cuts, and tears. (NFPA 472)

Permeation A chemical action involving the movement of chemicals, on a molecular level, through intact material. (NFPA 472)

Permissible exposure limit (PEL) The maximum permitted eight-hour, time-weighted average concentration of an airborne contaminant. (NFPA 5000, Building Construction and Safety Code)

Personal alert safety system (PASS) A device that continually monitors for lack of movement of the wearer and, if no movement is detected, automatically activates an alarm signal indicating the wearer is in need of assistance. The device can also be manually activated to trigger the alarm signal. (NFPA 1982)

Personal dosimeter A device that measures the amount of radioactive exposure to an individual.

Personal flotation device (PFD) A displacement device worn to keep the wearer afloat in water. (NFPA 1925)

Personal protective equipment (PPE) The basic protective equipment for wildland fire suppression includes a helmet, protective footwear, gloves, and flame-resistant clothing as defined in NFPA 1977, *Standard on Protective Clothing and Equipment for Wildland Fire Fighting*. (NFPA 1051)

Personnel accountability report (PAR) Periodic reports verifying the status of responders assigned to an incident or planned event. (NFPA 1026)

Personnel accountability system A system that readily identifies both the locations and the functions of all members operating at an incident scene. (NFPA 1500)

Personnel accountability tag (PAT) An identification card used to track the location of a fire fighter on an emergency incident.

pH A measure of the acidity or basic nature of a material; more technically, an expression of the concentration of hydrogen ions in the substance.

Phosgene A chemical agent that causes severe pulmonary damage; it is a by-product of incomplete combustion.

Photoelectric smoke detector A device to detect visible products of combustion using a light source and a photosensitive sensor.

Physical change A transformation in which a material changes its state of matter—for instance, from a liquid to a solid.

Physical evidence Items that can be observed, photographed, measured, collected, examined in a laboratory, and presented in court to prove or demonstrate a point.

Pick The pointed end of a pick axe, which can be used to make a hole or bite in a door, floor, or wall.

Pick-head axe A tool that has a head with an axe on one side and a pointed end ("pick") on the opposite side.

Piercing nozzle A nozzle that can be driven through sheet metal or other material to deliver a water stream to that area.

Pike pole A pole with a sharp point ("pike") on one end coupled with a hook. It is used to make openings in ceilings and walls. Pike poles are manufactured in different lengths for use in rooms of different heights.

Pincer attack A direct method of suppressing a wildland or ground fire, in which one suppression crew attacks the right flank of the fire and a second suppression crew attacks the left flank of the fire.

Pipe bomb A device created by filling a section of pipe with an explosive material.

Pipe wrench A wrench having one fixed grip and one movable grip that can be adjusted to fit securely around pipes and other tubular objects.

Pipeline A length of pipe including pumps, valves, flanges, control devices, strainers, and/or similar equipment for conveying fluids. (NFPA 70)

Pipeline right-of-way An area, patch, or roadway that extends a certain number of feet on either side of the pipe itself and that may contain warning and informational signs about hazardous materials carried in the pipeline.

Pitched chord truss A type of truss typically used to support a sloping roof.

Pitched roof A roof with sloping or inclined surfaces.

Pitot gauge A type of gauge that is used to measure the velocity pressure of water that is being discharged from an opening. It is used to determine the flow of water from a hydrant or nozzle.

Placards Signage required to be placed on all four sides of highway transport vehicles, railroad tank cars, and other forms of hazardous materials transportation that identifies the hazardous contents of the vehicle, using a standardization system with 10¾-inch (27.3 centimeters) diamond-shaped indicators.

Plague An infectious disease caused by the bacterium *Yersinia pestis*, which is commonly found on rodents.

Planning Section Chief The general staff position responsible for planning functions. It is assigned when Command needs assistance in managing information.

Planning Section Section responsible for the collection, evaluation, dissemination, and use of information related to the incident situation, resource status, and incident forecast. (NFPA 1026)

Plaster hook A long pole with a pointed head and two retractable cutting blades on the side.

Plate glass A type of glass that has additional strength so it can be formed in larger sheets but will still shatter upon impact.

Platform frame A type of vehicle frame resembling a ladder, which is made up of two parallel rails joined by a series of cross members. This kind of construction is typically used for luxury vehicles, sport utility vehicles, and all types of trucks.

Platform-frame construction Construction technique for building the frame of the structure one floor at a time. Each floor has a top and bottom plate that acts as a firestop.

Plume The column of hot gases, flames, and smoke rising above a fire; also called convection column, thermal updraft, or thermal column. (NFPA 921)

Pocket A deep indentation of unburned fuel along the fire's perimeter, often found between a finger and the head of the fire.

Point of origin The exact location where a heat source and a fuel come in contact with each other and a fire begins. (NFPA 921)

Polar solvent A water-soluble flammable liquid such as alcohol, acetone, ester, and ketone.

Policies Formal statements that provide guidelines for present and future actions. Policies often require personnel to make judgments.

Portable ladder A ladder carried on fire apparatus but designed to be removed from the apparatus and deployed by fire fighters where needed.

Portable monitor A monitor that can be lifted from a vehicle-mounted bracket and moved to an operating position on the ground by not more than two people. (NFPA 1965)

Portable radio A battery-operated, hand-held transceiver. (NFPA 1221)

Portable tanks Folding or collapsible tanks that are used at the fire scene to hold water for drafting.

Positive-pressure ventilation Ventilation that relies on fans to push or force clean air into a structure.

Post indicator valve (PIV) A sprinkler control valve with an indicator that reads either open or shut depending on its position.

Post One of the vertical support members, or pillars, of a vehicle that holds up the roof and forms the upright columns of the passenger compartment.

Pounds per square inch (psi) The standard unit for measuring pressure.

Power take-off shaft A supplemental mechanism that enables a fire engine to operate a pump while the engine is still moving.

Powered air-purifying respirator (PAPR) Air-purifying respirator with a hood or helmet, breathing tube, canister, cartridge, filter, and a blower that passes ambient air through the purifying element. The blower can be stationary or portable. It can also be designed and equipped with a full face piece, in which case it is referred to as a full-face-piece powered air-purifying respirator (FFPAPR). (NFPA 1404)

Preaction sprinkler system A dry sprinkler system that uses a deluge valve instead of a dry-pipe valve and requires activation of a secondary device before the pipes will fill with water.

Preincident plan A written document resulting from the gathering of general and detailed information to be used by public emergency response agencies and private industry for determining the response to reasonably anticipated emergency incidents at a specific facility.

Preincident survey The process used to gather information to develop a preincident plan.

Premixed foam Solution produced by introducing a measured amount of foam concentrate into a given amount of water in a storage tank. (NFPA 11)

Pressure gauge A device that measures and displays pressure readings. In an SCBA, the pressure gauges indicate the quantity of breathing air that is available at any time.

Pressure indicator A gauge on a pressurized portable fire extinguisher that indicates the internal pressure of the expellant.

Primary cut The main ventilation opening made in a roof to allow smoke, heat, and gases to escape.

Primary feeders The largest-diameter pipes in a water distribution system, which carry the greatest amounts of water.

Primary search A rapid search to locate living victims.

Private water system A privately owned water system that operates separately from the municipal water system.

Projected windows Windows that project inward or outward on an upper hinge; also called factory windows. They are usually found in older warehouses or commercial buildings.

Proprietary system A fire alarm system that transmits a signal to a monitoring location owned and operated by the facility's owner.

Protection plates Reinforcing material placed on a ladder at chafing and contact points to prevent damage from friction and contact with other surfaces.

Protective hood A part of a fire fighter's personal protective equipment that is designed to be worn over the head and under the helmet; it provides thermal protection for the neck and ears.

Protein foam A protein-based foam concentrate that is stabilized with metal salts to make a fire-resistant foam blanket. (NFPA 402)

Pry axe A specially designed hand axe that serves multiple purposes. Similar to a Halligan bar, it can be used to pry, cut, and force doors, windows, and many other types of objects. Also called a multipurpose axe.

Pry bar A specialized prying tool made of a hardened steel rod with a tapered end that can be inserted into a small area.

Public information officer (PIO) A member of the Command Staff who is responsible for interacting with the public and media or with other agencies with incident-related information requirements. (NFPA 1026)

Public safety answering point (PSAP) A facility in which 911 calls are answered. (NFPA 1221)

Pulaski axe A hand tool that combines an adze and an axe for brush removal.

Pulley A device with a free-turning, grooved metal wheel (sheave) used to reduce rope friction. Side plates are available for a carabiner to be attached. (NFPA 1670)

Pulmonary edema Fluid buildup in the lungs.

Pulse The wave of pressure created by the heart as it contracts and forces blood out into the major arteries.

Pump tank fire extinguisher A nonpressurized, manually operated water extinguisher that is rated for use on Class A fires. Discharge pressure is provided by a hand-operated, double-acting piston pump.

Pyrolysis The destructive distillation of organic compounds in an oxygen-free environment that converts the organic matter into gases, liquids, and char. (NFPA 820)

Pyromaniac A pathological fire-setter.

Rabbet A type of door frame in which the stop for the door is cut into the frame.

Rabbet tool A hydraulic spreading tool designed to pry open doors that swing inward.

Radiation dispersal device Any device that causes the purposeful dissemination of radioactive material without a nuclear detonation; a dirty bomb.

Radiation The emission and propagation of energy through matter or space by means of electromagnetic disturbances that display both wave-like and particle-like behavior. (NFPA 801)

Radio repeater system A radio system that automatically retransmits a radio signal on a different frequency.

Radioactive isotope An atom that has unequal numbers of protons and neutrons in the nucleus and that emits radioactivity.

Radioactivity The spontaneous decay or disintegration of an unstable atomic nucleus accompanied by the emission of radiation. (NFPA 801)

Radiological agents Materials that emit radioactivity.

Rafters Joists that are mounted in an inclined position to support a roof.

Rail The top or bottom piece of a trussed beam assembly used in the construction of a trussed ladder. Also, the top and bottom surfaces of an I-beam ladder. Each beam has two rails.

Rapid intervention company/crew (RIC) A minimum of two fully equipped personnel on site, in a ready state, who are assigned to immediate rescue of injured or trapped fire fighters. Also called a rapid intervention team (RIT).

Rapid intervention crew/company universal air connection (RIC UAC) A system that allows emergency replenishment of breathing air to the SCBA of disabled or entrapped fire or emergency services personnel. (NFPA 1981)

Rapid oxidation A chemical process that occurs when a fuel is combined with oxygen, resulting in the formation of ash or other waste products and the release of energy as heat and light.

Rate-of-rise heat detector A fire detection device that responds when the temperature increases at a rate that exceeds a predetermined value.

Rear of the fire The side opposite the head of the fire. Also called the heel of the fire.

Reciprocating saw A saw that is powered by an electric motor or a battery motor, and whose blade moves back and forth.

Recommended exposure level (REL) The established standard limit of exposure to a hazardous material. It is based on the maximum time-weighted concentration at which 95 percent of exposed, healthy adults suffer no adverse effects over a 40-hour workweek. RELs are set by the National Institute for Occupational Safety and Health but do not have the force of federal law.

Reconnaissance report The inspection and exploration of a specific area for the purpose of gathering information for the incident commander.

Recovery phase The stage of a hazardous materials incident after the imminent danger has passed, when clean-up and the return to normalcy have begun.

Reducer A fitting used to connect a small hose line or pipe to a larger hose line or pipe. (NFPA 1142)

Regulations Rules, usually issued by a government or other legally authorized agency, that dictate how something must be done. Regulations are often developed to implement a law.

Rehabilitate To restore someone or something to a condition of health or to a state of useful and constructive activity.

Rehabilitation An intervention designed to mitigate against the physical, physiological, and emotional stress of fire fighting so as to sustain a member's energy, improve performance, and decrease the likelihood of on-scene injury or death. (NFPA 1521)

Rekindle A return to flaming combustion after apparent but incomplete extinguishment. (NFPA 921)

Relative humidity The ratio between the amount of water vapor in the air (a gas) at the time of measurement and the amount of water vapor that could be in the air (gas) when condensation begins, at a given temperature. (NFPA 79)

Remote annunciator A secondary fire alarm control panel in a different location than the main alarm panel; it is usually located near the front door of a building.

Remote shut-off valve A device designed to retain a hazardous material in its container.

Remote station system A fire alarm system that sounds an alarm in the building and transmits a signal to the fire department or an off-premises monitoring location.

Remote-controlled hydrant valve A valve that is attached to a fire hydrant to allow the operator to turn the hydrant on without flowing water into the hose line.

Removal A mode of decontamination that applies specifically to contaminated soil, which is taken away from the scene.

Rescue company A group of fire fighters who work as a unit and are equipped with one or more rescue vehicles. (NFPA 1410)

Rescue Those activities directed at locating endangered persons at an emergency incident, removing those persons from danger, treating the injured, and providing for transport to an appropriate healthcare facility. (NFPA 402)

Rescue-lift air bags Extrication tools consisting of air bags, filler hoses, air regulators, control valves, and a supply of compressed air. These tools are used to lift vehicles that have been involved in crashes.

Reservoir A water storage facility.

Residential sprinkler system A sprinkler system designed to protect dwelling units.

Residual pressure The pressure that exists in the distribution system, measured at the residual hydrant at the time the flow readings are taken at the flow hydrants. (NFPA 24)

Resource management Under NIMS, includes mutual-aid agreements; the use of special federal, state, local, and tribal teams; and resource mobilization protocols. (NFPA 1026)

Respirator A device that provides respiratory protection for the wearer. (NFPA 1994)

Response The deployment of an emergency service resource to an incident. (NFPA 901)

Retention The process of purposefully collecting hazardous materials in a defined area. It may include creating that area—by digging, for instance.

Reverse lay A method of laying a supply line where the supply line starts at the attack engine and ends at the water source.

Rim locks Surface- or interior-mounted locks located on or in a door with a bolt that provide additional security.

Risk–benefit analysis An assessment of the risk to rescuers versus the benefits that can be derived from their intended actions. (NFPA 1006)

Rocker lugs (rocker pins) Fittings on threaded couplings that aid in coupling the hoses.

Rollover The condition where unburned fuel (pyrolysate) from the originating fire has accumulated in the ceiling layer to a sufficient concentration (i.e., at or above the lower flammable limit) that it ignites and burns. This can occur without ignition of, or prior to, the ignition of other fuels separate from the origin. (NFPA 921)

Roof covering The membrane, which can also be the roof assembly, that resists fire and provides weather protection to the building against water infiltration, wind, and impact. (NFPA 5000, Building Construction and Safety Code)

Roof decking The rigid component of a roof covering.

Roof hooks The spring-loaded, retractable, curved metal pieces that allow the tip of a roof ladder to be secured to the peak of a pitched roof. The hooks fold outward from each beam at the top of a roof ladder.

Roof ladder A single ladder equipped with hooks at the top end of the ladder. (NFPA 1931)

Roofman's hook A long pole with a solid metal hook used for pulling.

Rope bag A bag used to protect and store rope so that the rope can be easily and rapidly deployed without kinking.

Rope record A record for each piece of rope that includes a history of when the rope was placed in service, when it was inspected, when and how it was used, and which types of loads were placed on it.

Rotary saw A saw that is powered by an electric motor or a gasoline engine, and that uses a large rotating blade to cut through material. The blades can be changed depending on the material being cut.

Round turn A piece of rope looped to form a complete circle with the two ends parallel.

Rubber-covered hose (rubber-jacket hose) Hose whose outside covering is made of rubber, which is said to be more resistant to damage.

Run cards Cards used to determine a predetermined response to an emergency.

Rung The ladder crosspieces, on which a person steps while ascending or descending. (NFPA 1931)

Running end The part of a rope used for lifting or hoisting.

Safe location A location remote or separated from the effects of a fire so that such effects no longer pose a threat. (NFPA 101)

Safety knot A knot used to secure the leftover working end of the rope.

Safety officer A member of the Command Staff who is responsible for monitoring and assessing safety hazards and unsafe situations, and for developing measures for ensuring personnel safety. (NFPA 1561)

Salvage A firefighting procedure for protecting property from further loss following an aircraft accident or fire. (NFPA 402)

Salvage cover A large square or rectangular sheet made of heavy canvas or plastic material that is spread over furniture and other items to protect them from water runoff and falling debris.

San Francisco hook A multipurpose tool that can be used for several forcible entry and ventilation applications because of its unique design, which includes a built-in gas shut-off and directional slot.

Saponification The process of converting the fatty acids in cooking oils or fats to soap or foam; the action caused by a Class K fire extinguisher.

Sarin A nerve agent that when dispersed sends droplets in the air that when inhaled harm intended victims.

SCBA harness The part of SCBA that allows fire fighters to wear it as a "backpack."

SCBA regulator The part of the SCBA that reduces the high pressure in the cylinder to a usable lower pressure and controls the flow of air to the user.

Screwdriver A tool used for turning screws.

SCUBA dive rescue technician A responder who is trained to handle water rescues and emergencies, including recovery and search procedures, in both water and under-ice situations. (SCUBA stands for "self-contained underwater breathing apparatus.")

Search and rescue The process of searching a building for a victim and extricating the victim from the building.

Search rope A guide rope used by fire fighters that allows them to maintain contact with a fixed point.

Search The process of locating victims who are in danger.

Seat belt cutter A specialized cutting device that cuts through seat belts.

Seat of the fire The main area of the fire.

Secondary collapse A subsequent collapse in a building or excavation. (NFPA 1006)

Secondary containment Any device or structure that prevents environmental contamination when the primary container or its appurtenances fail. Secondary containment shall be designed and constructed to intercept and contain pesticide spills and leaks and to prevent runoff or leaching of pesticides into the environment. Examples of secondary containment include dikes, curbing, and double-walled tanks. (NFPA 434)

Secondary contamination The process by which a contaminant is carried out of the hot zone and contaminates people, animals, the environment, or equipment.

Secondary cut An additional ventilation opening made for the purpose of creating a larger opening or limiting fire spread.

Secondary device An explosive device designed to injure emergency responders who have responded to an initial event.

Secondary feeders Smaller-diameter pipes that connect the primary feeders to the distributors.

Secondary loss Property damage that occurs due to smoke, water, or other measures taken to extinguish the fire.

Secondary search A methodical search designed to locate deceased victims after the fire has been suppressed, searching areas not covered during the primary search.

Self-contained breathing apparatus (SCBA) An atmosphere-supplying respirator that supplies a respirable air atmosphere to the user from a breathing air source that is independent of the ambient environment and designed to be carried by the user. (NFPA 1981)

Self-contained underwater breathing apparatus (SCUBA) A respirator with an independent air supply that is used by underwater divers.

Self-expelling fire extinguisher A fire extinguisher in which the agents have sufficient vapor pressure at normal operating temperatures to expel themselves. (NFPA 10)

Self-rescue Escaping or exiting a hazardous area under one's own power. (NFPA 1006)

Sensitizer A chemical that causes a large portion of people or animals to develop an allergic reaction after repeated exposure to the substance.

Serial arson A series of fires set by the same offender, with a cooling-off period between fires.

Shackles The U-shaped part of a padlock that runs through a hasp and then is secured back into the lock body.

Shelter-in-place A method of safeguarding people in a hazardous area by keeping them in a safe atmosphere, usually inside structures.

Shipping papers A shipping order, bill of lading, manifest, or other shipping document serving a similar purpose and containing the information required by regulations of the U.S. Department of Transportation. (NFPA 498)

Shock load An instantaneous load that places a rope under extreme tension, such as when a falling load is suddenly stopped as the rope becomes taut.

Shoring A structure such as a metal hydraulic, pneumatic/mechanical, or timber system that supports the sides of an excavation and is designed to prevent cave-ins. (NFPA 1670)

Shut-off valve Any valve that can be used to shut down water flow to a water user or system.

Siamese connection A device that allows two hoses to be connected together and flow into a single hose.

Sidewall sprinkler head A sprinkler that is mounted on a wall and discharges water horizontally into a room.

Signal words Information on a pesticide label that indicates the relative toxicity of the material.

Simplex channel A radio system that uses one frequency to transmit and receive all messages.

Single command A command structure in which a single individual is responsible for all of the strategic objectives of the incident. It is typically used when an incident is within a single jurisdiction and is managed by a single discipline.

Single resource An individual, a piece of equipment and its personnel, or a crew or team of individuals with an identified supervisor that can be used on an incident or planned event. (NFPA 1026)

Single-action pull-station A manual fire alarm activation device in which the user takes a single step—such as moving a lever, toggle, or handle—to activate the alarm.

Single-station smoke alarm A single device that both detects visible and invisible products of combustion (smoke) and sounds an alarm.

Size-up The observation and evaluation of existing factors that are used to develop objectives, strategy, and tactics for fire suppression. (NFPA 1051)

Slash Debris resulting from natural events such as wind, fire, snow, or ice breakage, or from human activities such as building or road construction, logging, pruning, thinning, or brush cutting. (NFPA 1144)

Sledgehammer A long, heavy hammer that requires the use of both hands.

Small-diameter hose (SDH) Hose with a diameter ranging from 1 to 2 inches (25 mm to 50 mm).

Smallpox A highly infectious disease caused by the virus *Variola*.

Smoke color The attribute of smoke that reflects the stage of burning of a fire and the material that is burning in the fire.

Smoke density The thickness of smoke. Because it has a high mass per unit volume, smoke is difficult to see through.

Smoke detector A device that detects visible and invisible products of combustion (smoke) and sends a signal to a fire alarm control panel.

Smoke explosion A violent release of confined energy that occurs when a mixture of flammable gases and oxygen are present, usually in a void or other area separate from the fire compartment, and come in contact with a source of ignition. In a smoke explosion, there is no change to the ventilation profile, such as an open door or window; rather it occurs from the travel of smoke within the structure to an ignition source.

Smoke inversion The condition in which smoke hangs low to the ground because of the presence of cold air.

Smoke particles Airborne solid materials consisting of ash and unburned or partially burned fuel released by a fire.

Smoke The airborne solid and liquid particulates and gases evolved when a material undergoes pyrolysis or combustion, together with the quantity of air that is entrained or otherwise mixed into the mass. (NFPA 318)

Smoke velocity The speed of smoke leaving a burning building.

Smoke volume The quantity of smoke, which indicates how much fuel is being heated.

Smooth-bore nozzle A nozzle that produces a straight stream that is a solid column of water.

Smooth-bore tip A nozzle device that is a smooth tube, which is used to deliver a solid column of water.

Socket wrench A wrench that fits over a nut or bolt and uses the ratchet action of an attached handle to tighten or loosen the nut or bolt.

Soffit The material covering the gap between the edge of the roof and the exterior wall of the house.

Soft suction hose A large-diameter hose that is designed to be connected to the large port on a hydrant (steamer connection) and into the engine.

Solid beam A ladder beam constructed of a solid rectangular piece of material (typically wood), to which the ladder rungs are attached.

Solid One of the three stages of matter; a material that has three dimensions and is firm in substance.

Solid stream A stream made by using a smooth-bore nozzle to produce a penetrating stream of water.

Solid-core door A door design that consists of wood filler pieces inside the door. This construction creates a stronger door that may be fire rated.

Solidification The process of chemically treating a hazardous liquid so as to turn it into a solid material, thereby making the material easier to handle.

Soman A nerve gas that is both a contact hazard and a vapor hazard; it has the odor of camphor.

Sounding The process of striking a roof with a tool to determine where the roof supports are located.

Spalling Chipping or pitting of concrete or masonry surfaces. (NFPA 921)

Span of control The maximum number of personnel or activities that can be effectively controlled by one individual (usually three to seven). (NFPA 1006)

Spanner wrench A type of tool used to couple or uncouple hoses by turning the rocker lugs on the connections.

Special-use railcars Boxcars, flat cars, cryogenic and corrosive tank cars, and high-pressure compressed gas tube-type cars.

Specialist level Persons who respond to hazardous materials/weapons of mass destruction incidents who have received more specialized training than hazardous materials technicians. Most of the training that specialist employees receive is either product or transportation mode specific.

Specific gravity As applied to gas, the ratio of the weight of a given volume to that of the same volume of air, both measured under the same conditions. (NFPA 54, *National Fuel Gas Code*)

Split hose bed A hose bed arranged to enable the engine to lay out either a single supply line or two supply lines simultaneously.

Split hose lay A scenario in which the attack engine forward lays a supply line from an intersection to the fire, and the supply engine reverse lays a supply line from the hose left by the attack engine to the water source.

Spoil pile A pile of excavated soil next to the excavation or trench. (NFPA 1006)

Spot detector A single heat-detector device. These devices are often spaced throughout an area.

Spot fire A new fire that starts outside areas of the main fire, usually caused by flying embers and sparks.

Spotlight A light designed to project a narrow, concentrated beam of light.

Spree arson A series of fires started by an arsonist who sets three or more fires at separate locations, with no emotional cooling-off period between fires.

Spring-loaded center punch A spring-loaded punch used to break automobile glass.

Sprinkler piping The network of piping in a sprinkler system that delivers water to the sprinkler heads.

Sprinkler stop A mechanical device inserted between the deflector and the orifice of a sprinkler head to stop the flow of water.

Sprinkler system For fire protection purposes, an integrated system of underground and overhead piping designed in accordance with fire protection engineering standards. The installation includes at least one automatic water supply that supplies one or more systems. The portion of the sprinkler system above ground is a network of specially sized or hydraulically designed piping installed in a building, structure, or area, generally overhead, and to which sprinklers are attached in a systematic pattern. Each system has a control valve located in the system riser or its supply piping. Each sprinkler system includes a device for actuating an alarm when the system is in operation. The system is usually activated by heat from a fire and discharges water over the fire area. (NFPA 13, 2007)

Sprinkler wedge A piece of wedge-shaped wood placed between the deflector and the orifice of a sprinkler head to stop the flow of water.

Stack effect The vertical air flow within buildings caused by the temperature-created density differences between the building interior and exterior or between two interior spaces. (NFPA 92B, Standard for Smoke Management Systems in Malls, Atria, and Large Spaces)

Staging area A prearranged, strategically placed area, where support response personnel, vehicles, and other equipment can be held in an organized state of readiness for use during an emergency. (NFPA 424)

Standard operating procedure (SOP) A written organizational directive that establishes or prescribes specific operational or administrative methods to be followed routinely for the performance of designated operations or actions. (NFPA 1521)

Standing part The part of a rope between the working end and the running end.

Standpipe system An arrangement of piping, valves, and hose connections installed in a structure to deliver water for fire hoses.

Standpipe system An arrangement of piping, valves, hose connections, and allied equipment installed in a building or structure, with the hose connections located in such a manner that water can be discharged in streams or spray patterns through attached hose and nozzles, for the purpose of extinguishing a fire, thereby protecting a building or structure and its contents in addition to protecting the occupants. (NFPA 14, 2007)

State Emergency Response Commission (SERC) The liaison between local and state levels that collects and disseminates information relating to hazardous materials. The SERC involves agencies such as the fire service, police services, and elected officials.

State of matter The physical state of a material—solid, liquid, or gas.

Static pressure The pressure that exists at a given point under normal distribution system conditions measured at the residual hydrant with no hydrants flowing. (NFPA 24)

Static rope A rope generally made out of synthetic material that stretches very little under load.

Static water source A water source such as a pond, river, stream, or other body of water that is not under pressure.

Staypoles Poles attached to each beam of the base section of extension ladders, which assist in raising the ladder and help provide stability of the raised ladder. (NFPA 1931)

Steamer port The large-diameter port on a hydrant.

Step chocks Specialized cribbing assemblies made out of wood or plastic blocks in a step configuration. They are typically used to stabilize vehicles.

Stop A piece of material that prevents the fly sections of a ladder from becoming overextended, leading to collapse of the ladder.

Stored-pressure fire extinguisher A fire extinguisher in which both the extinguishing material and the expellant gas are kept in a single container, and that includes a pressure indicator or gauge. (NFPA 10)

Stored-pressure water-type fire extinguisher A fire extinguisher in which water or a water-based extinguishing agent is stored under pressure.

Storz-type (nonthreaded) hose coupling A hose coupling that has the property of being both the male and the female coupling. It is connected by engaging the lugs and turning the coupling a one-third turn.

Straight stream A stream made by using an adjustable nozzle to provide a straight stream of water.

Strike team leader The person in charge of a strike team. This individual is responsible to the next higher level in the incident organization and serves as the point of contact for the strike team within the organization.

Strike team Specified combinations of the same kind and type of resources, with common communications and a leader. (NFPA 1026)

Striking tools Tools designed to strike other tools or objects such as walls, doors, or floors.

Strip cut Another term for a trench cut.

Subsurface fuels Partially decomposed matter that lies beneath the ground, such as roots, moss, duff, and decomposed stumps.

Suggested operating guideline (SOG) Another term for standard operating procedure.

Sulfur mustard A clear, yellow, or amber oily liquid with a faint, sweet odor of mustard or garlic that can be dispersed in an aerosol form. It causes blistering of exposed skin.

Superfund Amendments and Reauthorization Act (SARA) One of the first federal laws to affect how fire departments respond in a hazardous material emergency.

Supplied-air respirator (SAR) An atmosphere-supplying respirator for which the source of breathing air is not designed to be carried by the user. (NFPA 1404)

Supply hose (supply line) Hose designed for the purpose of moving water between a pressurized water source and a pump that is supplying attack lines. (NFPA 1961)

Surface fuels Fuels that are close to the surface of the ground, such as grass, leaves, twigs, needles, small trees, logging slash, and low brush. Also called ground fuels.

Survivability profiling The process of using available data and observations to determine if there is a reasonable chance that victims trapped inside might still be alive. This process is used to decide whether to send fire fighters into a fire or other hazardous environment.

Sweep (or roll-on) method A method of applying foam that involves sweeping the stream just in front of the target.

Tabun A nerve gas that is both a contact hazard and a vapor hazard; it operates by disabling the chemical connection between the nerves and their target organs.

Talk-around channel A simplex channel used for on-site communications.

Tamper alarm An alarm that sounds when the clear plastic cover of a pull-station is lifted; it is intended to prevent malicious false alarms.

Tamper seal A retaining device that breaks when the locking mechanism is released.

Tamper switch A switch on a sprinkler valve that transmits a signal to the fire alarm control panel if the normal position of the valve is changed.

Tanker shuttle A method of transporting water from a source to a fire scene using a number of mobile water supply apparatus.

Target hazard Any occupancy type or facility that presents a high potential for loss of life or serious impact to the community resulting from fire, explosion, or chemical release.

Task force Any combination of single resources assembled for a particular tactical need, with common communications and a leader. (NFPA 1051)

Task force leader The person in charge of a task force. This individual is responsible to the next higher level in the incident organization and serves as the point of contact for the task force within the organization.

TDD/TTY/text phone system User devices that allow speech- and/or hearing-impaired citizens to communicate over a telephone system. TDD stands for telecommunications device for the deaf; TTY stands for -teletype; text phones visually display text. The displayed text is the equivalent of a verbal conversation between two hearing persons.

Technical decontamination The planned and systematic process of reducing contamination to a level that is as low as reasonably achievable.

Technical rescue incident (TRI) A complex rescue incident requiring specially trained personnel and special equipment to complete the mission. (NFPA 1670)

Technical rescue team A group of rescuers specially trained in the various disciplines of technical rescue.

Technical rescue technician A fire fighter who is trained in special rescue techniques for incidents involving structural collapse, trench rescue, swiftwater rescue, confined-space rescue, and other unusual rescue situations.

Technical use life safety rope A life safety rope with a diameter that is 3/8" (9.5 mm) or greater, but is less than 1/2" (12.5mm), with a minimum breaking strength of 4496 lbf (20 kN). Used by highly trained rescue teams that deploy to very technical environments such as mountainous and/or wilderness terrain.

Technician level Persons who respond to hazardous -materials/weapons of mass destruction incidents and are allowed to enter heavily contaminated areas using the highest levels of protection. This is the most aggressive level of training as identified in NFPA 472.

Telecommunicator A trained individual who is responsible for answering requests for emergency and nonemergency assistance from citizens. This individual assesses the need for a response and alerts responders to the incident.

Telephone interrogation The phase in a 911 call during which the telecommunicator asks questions to obtain vital information such as the location of the emergency.

Temperature The degree of sensible heat of a body as measured by a thermometer or similar instrument.

Tempered glass A type of safety glass that is heat treated so that it will break into small, less dangerous pieces.

Temporal-3 pattern A standard fire alarm audible signal for alerting occupants of a building.

Ten-codes A system of predetermined coded messages, such as "What is your 10-20?", used by responders over the radio.

Thermal column A cylindrical area above a fire in which heated air and gases rise and travel upward.

Thermal conductivity A property that describes how quickly a material will conduct heat.

Thermal imager (TI) An electronic device that detects differences in temperature based on infrared energy and then generates images based on those data. This device is used in smoky environments to locate victims.

Thermal layering The stratification (heat layers) that occurs in a room as a result of a fire.

Thermal radiation The means by which heat is transferred to other objects.

Thermoplastic material Plastic material capable of being repeatedly softened by heating and hardened by cooling and, that in the softened state, can be repeatedly shaped by molding or forming. (NFPA 5000)

Thermoset material Plastic material that, after having been cured by heat or other means, is substantially infusible and cannot be softened and formed. (NFPA 5000)

Threaded hose coupling A type of coupling that requires a male fitting and a female fitting to be screwed together.

Threshold limit value/ceiling (TLV/C) The maximum concentration of hazardous material to which a worker should not be exposed, even for an instant.

Threshold limit value/short-term exposure limit (TLV/STEL) The maximum concentration of hazardous material to which a worker can sustain a 15--minute exposure not more than four times daily, with a 60--minute rest period required between each exposure. The lower the value, the more toxic the substance.

Threshold limit value/skin A designation indicating that direct or airborne contact with a material could result in significant exposure from absorption through the skin, mucous membranes, and eyes.

Threshold limit value/time-weighted average (TLV/TWA) The airborne concentration of a material to which a worker can be exposed for 8 hours a day, 40 hours a week, and not suffer any ill effects.

Tie rod A metal rod that runs from one beam of the ladder to the other to keep the beams from separating. Tie rods are typically found in wood ladders.

Time marks Status updates provided to the communications center every 10 to 20 minutes. Such an update should include the type of operation, the progress of the incident, the anticipated actions, and the need for additional resources.

Tip The very top of the ladder.

Topography The land surface configuration. (NFPA 1051)

Totes Portable tanks, which usually hold a few hundred gallons (several hundred liters) of product and are characterized by a unique style of construction.

Toxic A property of any chemical that has the capacity to produce injury to workers, which is dependent on concentration, rate, and method and site of absorption. (NFPA 306, Standard for the Control of Gas Hazards on Vessels)

Toxic inhalation hazard (TIH) materials Gases or volatile liquids that are extremely toxic to humans.

Toxic product of combustion A hazardous chemical compound that is released when a material decomposes under heat.

Toxicology The study of the adverse effects of chemical or physical agents on living organisms.

Trace (transfer) evidence A minute quantity of physical evidence that is conveyed from one place to another.

TRACEMP An acronym to help remember the effects and potential exposures to a hazardous materials incident: thermal, radiation, asphyxiant, chemical, etiologic, mechanical, and psychogenic.

Trailers Combustible materials (such as rolled rags, blankets, and newspapers or ignitable liquid) that are used to spread fire from one point or area to other points or areas, often used in conjunction with an incendiary device.

Training officer The person designated by the fire chief as having authority for overall management and control of the organization's training program. (NFPA 1401)

Transfer of command The formal procedure for transferring the duties of an incident commander at an -incident scene. (NFPA 1026)

Transitional attack An offensive fire attack initiated by an exterior, indirect handline operation into the fire compartment to initiate cooling while transitioning into interior direct fire attack in coordination with ventilation operations.

Trench cut A roof cut that is made from one load-bearing wall to another load-bearing wall and that is intended to prevent horizontal fire spread in a building.

Triage The process of sorting victims based on the severity of their injuries and medical needs to establish treatment and transportation priorities.

Triangular cut A triangle-shaped ventilation cut in the roof decking that is made using a saw or an axe.

Trigger The button or lever used to discharge the agent from a portable fire extinguisher.

Triple-layer load A hose loading method in which the hose is folded back onto itself to reduce the overall length to one-third before loading the hose in the bed. This load method reduces deployment distances.

Truck company A group of fire fighters who work as a unit and are equipped with one or more pieces of aerial fire apparatus. (NFPA 1410)

Trunking system A radio system that uses a shared bank of frequencies to make the most efficient use of radio resources.

Truss A collection of lightweight structural components joined in a triangular configuration that can be used to support either floors or roofs.

Truss block A piece of wood or metal that ties the two rails of a trussed beam ladder together and serves as the attachment point for the rungs.

Trussed beam A ladder beam constructed of top and bottom rails joined by truss blocks that tie the rails together and support the rungs.

Tube trailers High-volume transportation devices made up of several individual compressed gas cylinders banded together and affixed to a trailer.

Turbulent smoke flow Agitated, boiling, angry-movement smoke, which indicates great heat in the burning building. It is a precursor to flashover.

Turnout coat Protective coat that is part of a protective clothing ensemble for structural firefighting; also called a bunker coat.

Turnout pants Protective trousers that are part of a protective clothing ensemble for structural firefighting; also called bunker pants.

Twisted rope Rope constructed of fibers twisted into strands, which are then twisted together.

Two-in/two-out rule A safety procedure that requires a minimum of two personnel to enter a hazardous area and a minimum of two backup personnel to remain outside the hazardous area during the initial stages of an incident.

Two-way radio A portable communication device used by fire fighters. Every firefighting team should carry at least one radio to communicate distress, progress, changes in fire conditions, and other pertinent information.

Type I construction (fire resistive) Buildings with structural members made of noncombustible materials that have a specified fire resistance.

Type II construction (noncombustible) Buildings with structural members made of noncombustible materials without fire resistance.

Type III construction (ordinary) Buildings with the exterior walls made of noncombustible or limited-combustible materials, but with interior floors and walls made of combustible materials.

Type IV construction (heavy timber) Buildings constructed with noncombustible or limited-combustible exterior walls, and interior walls and floors made of large-dimension combustible materials.

Type V construction (wood frame) Buildings with exterior walls, interior walls, floors, and roof structures made of wood.

Underventilated fire A fire that has an adequate supply of fuel available but is not receiving an adequate supply of oxygen.

Underwriters Laboratories, Inc. (UL) The U.S. organization that tests and certifies that fire extinguishers (among many other products) meet established standards. The Canadian equivalent is Underwriters Laboratories of Canada.

Undetermined Fire-cause classification that includes fires for which the cause has not or cannot be proven.

Unibody (unit body) The frame construction most commonly used in vehicles. The base unit is made of formed sheet metal; structural components are then added to the base to form the passenger compartment. Subframes are attached to each end. This type of construction does away with the rail beams used in platform-framed vehicles.

Unified command An application of the incident command system that allows all agencies with jurisdictional responsibility for an incident or planned event, either geographical or functional, to manage an incident or planned event by establishing a -common set of -incident objectives and strategies. (NFPA 1561)

Unity of command The concept according to which each person within an organization reports to one, and only one, designated person. (NFPA 1026)

Universal precautions Procedures for infection control that treat blood and certain body fluids as capable of transmitting bloodborne diseases.

Unlocking mechanism A keyway, combination wheel, or combination dial.

Unwanted alarm A fire alarm signal caused by a device reacting properly to a condition that is not a true fire emergency.

Upper flammable limit (UFL) The highest concentration of a combustible substance in a gaseous oxidizer that will propagate a flame. (NFPA 68)

Upright sprinkler head A sprinkler head designed to be installed on top of the supply piping; it is usually marked SSU ("standard spray upright").

Utility rope Rope used for securing objects, for hoisting equipment, or for securing a scene to prevent bystanders from being injured. Utility rope must never be used in life safety operations.

V-agent A nerve agent, principally a contact hazard; an oily liquid that can persist for several weeks.

Vacuuming The process of cleaning up dusts, particles, and some liquids using a vacuum with high-efficiency particulate air filtration to prevent recontamination of the environment.

Vapor density The weight of an airborne concentration (vapor or gas) as compared to an equal volume of dry air.

Vapor dispersion The process of lowering the concentration of vapors by spreading them out.

Vapor pressure The pressure, measured in pounds per square inch (or kilopascals), absolute (psia), exerted by a liquid, as determined by ASTM D 323, *Standard Method of Test for Vapor Pressure of Petroleum Products (Reid Method)*. (NFPA 385, Standard for Tank Vehicles for Flammable and Combustible Liquids)

Vapor suppression The process of controlling vapors given off by hazardous materials, thereby preventing vapor ignition, by covering the product with foam or other material or by reducing the temperature of the material.

Vapor The gas phase of a substance, particularly of those that are normally liquids or solids at ordinary temperatures. (NFPA 92, Standard for Smoke Control Systems)

Vapor-protective clothing The garment portion of a chemical-protective clothing ensemble that is designed and configured to protect the wearer against chemical vapors or gases. (NFPA 472)

Vent pipes Inverted J-shaped tubes that allow for pressure relief or natural venting of the pipeline for maintenance and repairs.

Ventilation The fire-fighting operation of the systematic and coordinated removal of smoke and heat from a structure by opening windows and doors or making holes in the roof. (NFPA 923)

Ventilation-limited A fire in an enclosed building that is restricted because there is insufficient oxygen available for the fire to burn as rapidly as it would with an unlimited supply of oxygen.

Verification system A fire alarm system that does not immediately initiate an alarm condition when a smoke detector activates. The system will wait a preset interval (generally 30 to 60 seconds) before checking the detector again. If the condition is clear, the system returns to normal status. If the detector still senses smoke, the system activates the fire alarm.

Vertical ventilation The process of making openings so that smoke, heat, and gases can escape vertically from a structure.

Voice recording system Recording devices or computer equipment connected to telephone lines and radio equipment in a communications center to record telephone calls and radio traffic.

Volatility The ability of a substance to produce combustible vapors.

VX A nerve agent.

Wall post indicator valve A sprinkler control valve that is mounted on the outside wall of a building. The position of the indicator tells whether the valve is open or shut.

Warm zone (technical rescue incident) The area located between the hot zone and the cold zone at an incident. The decontamination corridor is located in the warm zone.

Warm zone The control zone at hazardous materials/weapons of mass destruction incidents where personnel and equipment decontamination and hot zone support take place. (NFPA 472)

Water catch-all A salvage cover that has been folded to form a container to hold water until it can be removed.

Water chute A salvage cover that has been folded to direct water flow out of a building or away from sensitive items or areas.

Water curtain nozzle A nozzle used to deliver a flat screen of water that forms a protective sheet of water to protect exposures from fire.

Water hammer The surge of pressure that occurs when a high-velocity flow of water is abruptly shut off. The pressure exerted by the flowing water against the closed system can be seven or more times that of the static pressure. (NFPA 1962)

Water knot A knot used to join the ends of webbing together.

Water main A generic term for any underground water pipe.

Water supply A source of water for firefighting activities. (NFPA 1144)

Water thief A device that has a 2½-inch (65 mm) inlet and a 2½-inch (65 mm) outlet in addition to two 1½-inch (38 mm) outlets. It is used to supply many hoses from one source.

Water vacuum A device similar to a wet/dry shop vacuum cleaner that can pick up liquids. It is used to remove water from buildings.

Water-motor gong An audible alarm notification device that is powered by water moving through the sprinkler system.

Waybill Shipping papers for trains.

Weapons of mass destruction (WMD) (1) Any destructive device, such as any explosive, incendiary, or poison gas bomb, grenade, rocket having a propellant charge of more than 4 ounces (113 grams), missile having an explosive or incendiary charge of more than ¼ ounce (7 grams), mine, or device similar to the above; (2) any weapon involving toxic or poisonous chemicals; (3) any weapon involving a disease organism; or (4) any weapon that is designed to release radiation or radioactivity at a level dangerous to human life. (NFPA 472, Standard for Competence of Responders to Hazardous Materials/Weapons of Mass Destruction Incidents)

Wedges Material used to tighten or adjust cribbing and shoring systems. (NFPA 1006)

Wet sprinkler system A sprinkler system in which the pipes are normally filled with water.

Wet-barrel hydrant A hydrant used in areas that are not susceptible to freezing. The barrel of the hydrant is normally filled with water.

Wet-chemical extinguishing agent An extinguishing agent for Class K fires. It commonly consists of solutions of water and potassium acetate, potassium carbonate, potassium citrate, or any combination thereof.

Wet-chemical extinguishing system An extinguishing system that discharges a proprietary liquid extinguishing agent. It is often installed over stoves and deep-fat fryers in commercial kitchens.

Wet-chemical fire extinguisher A fire extinguisher for use on Class K fires that contains a wet-chemical extinguishing agent.

Wetting-agent water-type fire extinguisher An extinguisher that expels water combined with a chemical or chemicals to reduce its surface tension; it is used for Class A fires.

Wheeled fire extinguisher A portable fire extinguisher equipped with a carriage and wheels, which is intended to be transported to the fire by one person. (NFPA 10)

Wilderness search and rescue (SAR) The process of -locating and removing a victim from the wilderness.

Wildland fire An unplanned fire burning in vegetative fuels. (NFPA 1051)

Wildland Land in an uncultivated, more or less natural state and covered by timber, woodland, brush, and/or grass. (NFPA 901)

Wildland–urban interface An area where undeveloped land with vegetative fuels mixes with human-made structures.

Wildland/brush company A fire company that is dispatched to woods and brush fires where larger engines cannot gain access. Wildland/brush companies have four-wheel drive vehicles and special firefighting equipment.

Wired glass A glazing material with embedded wire mesh. (NFPA 80)

Wood truss An assembly of small pieces of wood or wood and metal.

Wood-frame construction Buildings whose exterior walls, interior walls, floors, and roofs are made of combustible wood materials. Wood-frame construction is also known as Type V building construction, as defined in NFPA 220, *Standard on Types of Building Construction*.

Wooden beam Load-bearing member assembled from individual wood components.

Working end The part of the rope used for forming a knot.

Wye A device used to split a single hose into two separate lines.

Zoned coded alarm A fire alarm system that indicates which zone was activated both on the alarm control panel and through a coded audio signal.

Zoned noncoded alarm A fire alarm system that indicates the activated zone on the alarm control panel.

Zoned system A fire alarm system design that divides a building or facility into zones so that the area where an alarm originated can be identified.

Index

E

F

M

N

O

T

U

W

Y

Z

Credits

Voices of Experience Courtesy of David Jackson, Saginaw Township Fire Department; **You are the Fire Fighter** © Duncan Noakes/Dreamstime.com.

Chapter 1

Opener © Bettmann/Corbis; **1-3** Courtesy of Captain David Jackson, Saginaw Township Fire Department; Courtesy of Captain David Jackson, Saginaw Township Fire Department; **1-4** © Patrick Kane/AP Photos; **1-6** © Brad Sauter/ShutterStock, Inc.; **1-15** © PURPLE MARBLES YORK 1/Alamy Images.

Chapter 2

Opener © Patricia Marks/ShutterStock, Inc.; **2-2** © Joseph Kaczmarek/AP Photos.

Chapter 4

4-6 © Chase Swift/Corbis.

Chapter 5

Opener © Aled Llywelyn/Alamy Images; **5-1** © Peter Weber/ShutterStock, Inc.; **5-2** © Mark C. Ide; **5-3** © Rick McClure/AP Photos; **5-8** Courtesy of Captain David Jackson, Saginaw Township Fire Department; **5-17** © Keith D. Cullom.

Chapter 6

Opener © Darin Echelberger/ShutterStock, Inc; **6-16** Courtesy of NIST; **6-19** Courtesy of NIST; **6-22** Courtesy of Courtesy of UL.

Chapter 7

Opener © PD Loyd/ShutterStock, Inc.; **7-1A** © MitarArt/ShutterStock, Inc.; **7-1B** © Ken Durden/ShutterStock, Inc.; **7-2** Courtesy of Captain David Jackson, Saginaw Township Fire Department; **7-4** © AbleStock; **7-5** © John Foxx/Alamy Images; **7-6** © Michael Doolittle/Alamy Images; **7-8** © Brandon Bourdages/ShutterStock, Inc.; **7-11** © Dorn1530/ShutterStock, Inc.; **7-12** © Lourens Smak/Alamy Images; **7-13A** © AbleStock; **7-13B** Courtesy of Pacific Northwest National Laboratory; **7-13C** Courtesy of Doyal Dum/US Army Corps of Engineers; **7-13D** © AbleStock; **7-14A** © Paul Springett/Up the Resolution/Alamy Images; **7-14B** © Ron Chapple/Thinkstock/Alamy Images; **7-14C** © TH Photo/Alamy Images; **7-15** Courtesy of Captain David Jackson, Saginaw Township Fire Department; **7-19** Courtesy of Captain David Jackson, Saginaw Township Fire Department; **7-20** © M Stock/Alamy Images; **7-24** Courtesy of Securalldoors.com; **7-26** © Steven Townsend/Code 3 Images.

Chapter 8

8-1A Courtesy of Amerex Corporation; **8-1B** Courtesy of Amerex Corporation; **8-6** Courtesy of Tyco Fire and Building Products; **8-7** © Andrew Lambert Photography/Photo Researchers, Inc.; **8-8** © Scott Leman/ShutterStock, Inc.; **8-18** © craftvision/iStockphoto.com; **8-19** Courtesy of Amerex Corporation; **Skill Drill 9-1** Courtesy of Tyco (ANSUL); **Skill Drill 9-2** Courtesy of Tyco (ANSUL); **Skill Drill 9-3** Courtesy of Tyco (ANSUL); **Skill Drill 9-4** Courtesy of Tyco (ANSUL); **Table 8-2** Reproduced with permission from NFPA's *Fire Protection Handbook*®, Vol. 2, 20th ed., Copyright © 2008, National Protection Association.

Chapter 9

9-11 Courtesy of Donald M. Colarusso, AllHandsFire.com; **9-12** © 2003, Berta A. Daniels; **9-29** © 2003, Berta A. Daniels; **9-32** Courtesy of ISG/INFRASYS.

Chapter 10

10-16 Courtesy of Captain David Jackson, Saginaw Township Fire Department; **10-17** © Mike Meadows/AP Photos; © Mike Meadows/AP Photos; **10-18** Courtesy of Captain David Jackson, Saginaw Township Fire Department; **10-21** © AbleStock.

Chapter 11

Opener © Keith Muratori/ShutterStock, Inc.; **11-5** © Carlos Arroyo/ShutterStock, Inc.; **11-7** © 2003, Berta A. Daniels; **11-8** © Gregory Price, Lewiston Son Journal/AP Photos; **11-10** © Glen E. Ellman.

Chapter 12

12-4 Courtesy of Paratech Inc.; **12-8** © AbleStock; **12-13** Courtesy of Horton Automatics; **12-14A** © Dave White/iStockphoto.com; **12-14B** © Photodisc/Creatas; **12-23** Courtesy of IR Safety and Security—Americas; **12-25** Courtesy of Master Lock Company LLC; **12-26** Courtesy of IR Safety and Security—Americas; **12-27** Courtesy of IR Safety and Security—Americas; **12-28** Courtesy of DOE/NREL, Credit—Ed Hancock.

Chapter 13

13-5 Courtesy of Dennis Wetherhold, Jr.; **13-6** © SteveStone/iStockphoto.com; **13-7** Courtesy of Duo-Safety Ladder Corporation; **13-9** Courtesy of Duo-Safety Ladder Corporation; **13-12** Courtesy of Duo-Safety Ladder Corporation; **13-17** Courtesy of Underwriters Laboratories; **13-24** Courtesy of Dennis Wetherhold, Jr; **Table 13-2** Reproduced with permission from NFPA's 1901–1909, *Automotive Fire Apparatus*, Copyright © 2008, National Fire Protection Association. This reprinted material is not the complete and official position of the NFPA on the referenced subject, which is represented only by the standard in its entirety.

Chapter 14

Opener © Glen E. Ellman; **14-1** © Irene Teesalu/ShutterStock, Inc.; **14-2B** Courtesy of Captain David Jackson, Saginaw Township Fire Department; **14-10** Courtesy of District Chief Chris E. Mickal/New Orleans Fire Department, Photo Unit.

Chapter 15

Opener © James Quigg/AP Photos; **15-5** © Steven Townsend/Code 3 Images; **15-6** Courtesy of District Chief Chris E. Mickal/New Orleans Fire Department, Photo Unit; **15-7** Courtesy of District Chief Chris E. Mickal/New Orleans Fire Department, Photo Unit; **15-9** Courtesy of Super Vacuum Mfg. Co., Inc.; **15-14** Courtesy of Captain David Jackson, Saginaw Township Fire Department; **15-15** © craig robinson/iStockphoto.com; **15-17** Courtesy of Captain David Jackson, Saginaw Township Fire Department; **15-18** © NicVW/Alamy Images; **15-20** Courtesy of Four Seasons Solar Products LLC; **15-26** © olaf schlueter/ShutterStock, Inc.

Chapter 16

16-2 © Peter Kim/ShutterStock, Inc.; **16-3** © Will Powers/AP Photos; **16-5** © Maran Garai/ShutterStock, Inc.; **16-6** © IAN MARLOW/Alamy Images; **16-7** Courtesy of Captain David Jackson, Saginaw Township Fire Department; **16-9** © Paul Glendell/Alamy Images; **16-13** Courtesy of American AVK Company; **16-14A** Courtesy of American AVK Company; **16-16** Courtesy of Captain David Jackson, Saginaw Township Fire Department; **16-31** Courtesy of Akron Brass Company; **16-32** Courtesy of Akron Brass Company; **16-33** Courtesy of Akron Brass Company; **16-35** © 2003, Berta A. Daniels; **16-36** © 2003, Berta A. Daniels; **16-37** © 2003, Berta A. Daniels; **16-39A** Courtesy of Akron Brass Company; **16-39B** Courtesy of Akron Brass Company; **16-39C** Courtesy of Akron Brass Company; **16-43** Courtesy of Scott Dornan, ConocoPhillips Alaska; **Table 16-1** Reprinted with permission from NFPA 24-2013, *Installation of Private Fire Service Mains and Their Appurtenances*,

Copyright © 2012, National Fire Protection Association. This reprinted material is not the complete and official position of the NFPA on the referenced subject, which is represented only by the standard in its entirety.

Chapter 17

Opener Courtesy of the University of Nevada, Reno Fire Science Academy; **17-5** © Goodman Photography/ShutterStock, Inc.; **17-6** Courtesy of Akron Brass Company; **17-7** Courtesy of POK of North America, Inc.

Chapter 18

Opener © MarcoCoda/iStockphoto.com; **18-1** © Condor 36/ShutterStock, Inc.; **18-5** © Mark C. Ide.

Chapter 19

19-1 © Lusoimages/ShutterStock, Inc.; **19-2** © Glen E. Ellman; **19-3** © 2003, Berta A. Daniels; **19-6** Courtesy of Tyco Fire and Building Products; **19-10** Courtesy of Cascade Fire Equipment Company.

Chapter 20

20-7 © Jerry S. Mendoza/AP Photos; **20-8** Courtesy of Michael Rieger/FEMA News Photo; **20-11** © Jack Dagley Photography/ShutterStock, Inc.; **20-12** © Tom Carter/PhotoEdit, Inc.; **20-13** Courtesy of Neil Malcom Winkelmann; **20-15** © Photodisc/Creatas; Table 1 Courtesy of the International Association of Fire Chiefs.

Chapter 21

Opener © Gene Blevins/Landov; **21-5** © Steven Townsend/Code 3 Images; **21-6** © Michael Routh/Alamy Images; **21-7** © Steven Townsend/Code 3 Images; **21-8** Courtesy of Bob McMillan/FEMA; **21-9** Courtesy of Andrea Booher/FEMA; **21-10** © Michael Routh/Alamy Images; **21-11** Courtesy of Anchor Industries, Inc.

Chapter 22

Opener Courtesy of the University of Nevada, Reno Fire Science Academy.; **22-1** Courtesy of NIST; **22-2A** Courtesy of Courtesy of UL; **22-2B** Courtesy of Courtesy of UL; **22-3** © Jack Dagley Photography/ShutterStock, Inc.; **22-9** © SteveStone/iStockphoto.com; **22-13** Courtesy of Barry Hyvarinen, Massachusetts Fire Academy.

Chapter 23

Opener Courtesy of L. Charles Smeby, Jr./University of Florida; **23-2** Courtesy of Tempe Fire Department; **23-7** © John Foxx/Alamy Images; **23-8** © Rick's Photography/ShutterStock, Inc.; **23-9** © Brandon Bourdages/ShutterStock, Inc.; **23-10** © Helen Filatova/ShutterStock, Inc.; **23-12** Courtesy of Underwriters Laboratories; **23-13** © David Huntley/ShutterStock, Inc.; **23-15** © Jim Smalley; © Jim Smalley; **23-16** © Keith D. Cullom; **23-17** © Colin Archer/AP Photos; **23-18** © Photos.com; **23-19** © Corbis/PictureQuest; **23-21** © Photodisc; **23-22** © Arthur S. Aubry/Photodisc/Getty Images; **23-23** © JJJ/ShutterStock, Inc; **23-5** Reproduced with permission from NFPA 170-2012, *Standard for Fire Safety and Emergency Symbols*, Copyright © 2011, National Fire Protection Association. This reprinted material is not the complete and official position of the NFPA on the referenced subject, which is represented only by the standard in its entirety.

Chapter 24

Opener © BrandyTaylor/iStockphoto.com; **24-3** © Murray Wilson/Fotolia.com; **24-4** © Mark C. Ide; **24-5** LIFEPAK® 1000 Defibrillator courtesy of Physio-Control. Used with permission of Physio-Control, Inc.; **24-8** © Tony Freeman/PhotoEdit, Inc.; **24-11** © Andrei Malov/Dreamstime.com.

Chapter 25

Opener © Anthony Behar/Sipa Press/firesipatb.015/1107120615/AP Photos; **25-2** Courtesy of Kimberly Smith/CDC; **25-40** LIFEPAK® 1000 Defibrillator courtesy of Physio-Control. Used with permission of Physio-Control, Inc.; **25-49** Courtesy of Rhonda Beck; **25-54A** © E. M. Singletary, M.D. Used with permission.; **25-55A** Courtesy of RhondaBeck; **25-63** © Amy Walters/ShutterStock, Inc.; **25-64** © E. M. Singletary, M.D. Used with permission.; **25-70A** © Chuck Stewart, MD.; **25-70B** © Chuck Stewart, MD.; **Skill Drill 9-1** Courtesy of Jennifer and Marc Lemaire; **Skill Drill 9-2** Courtesy of Jennifer and Marc Lemaire; **Skill Drill 9-3** Courtesy of Jennifer and Marc Lemaire; **Skill Drill 9-4** Courtesy of Jennifer and Marc Lemaire; **Skill Drill 14-1** Courtesy of Jennifer and Marc Lemaire; **Skill Drill 14-2** Courtesy of Jennifer and Marc Lemaire; **Skill Drill 14-3** Courtesy of Jennifer and Marc Lemaire; **Skill Drill 14-4** Courtesy of Jennifer and Marc Lemaire; **Skill Drill 14-5** Courtesy of Jennifer and Marc Lemaire; **Skill Drill 14-6** Courtesy of Jennifer and Marc Lemaire.

Chapter 26

Opener © Monkey Business Images/ShutterStock, Inc.; **26-1** Courtesy of DOE/NREL, Credit—Warren Gretz; **26-5** Courtesy of David Jackson, Saginaw Township Fire Department; **26-12B** © sbarabu/ShutterStock, Inc.; **26-14** © EuroStyle Graphics/Alamy Images.

Chapter 27

Opener © Samuel Hoffman, The Journal Gazette/AP Photos; **27-5** Courtesy of Lara Shane/FEMA Photo News; **27-9** © Keith D. Cullom; **27-10** Courtesy of Captain David Jackson, Saginaw Township Fire Department; **27-11** Courtesy of Dave Gatley/FEMA; **27-12** © Nate Rawner; **27-13** Courtesy of Photographer's Mate 2nd Class Daniel R. Mennuto/U.S. Navy; **27-14** © Dennis Wetherhold, Jr.

Chapter 28

Opener © Lon C. Diehl/PhotoEdit, Inc.; **28-1** © FIORENTINI MASSIMO/ShutterStock, Inc.; **28-2B** © Richard Goldberg/ShutterStock, Inc.; **28-2C** © Harris Shiffman/ShutterStock, Inc.; **28-2D** © michael rubin/ShutterStock, Inc.; **28-4** Courtesy of Staff Sgt. Brandie Session/U.S. Air Force.

Chapter 29

Opener Courtesy of Photographer's Mate 1st Class James Schauer/U.S. Navy; **29-7** Courtesy of Dr. Saeed Keshavarz/RCCI (Research Center of Chemical Injuries)/IRAN; **29-9** Courtesy of Survivair.

Chapter 30

Opener © mikeledray/ShutterStock, Inc.; **30-1A** © Joe Potato Photo/iStockphoto.com; **30-1B** © Natalia7/ShutterStock, Inc.; **30-1C** © sherwood/ShutterStock, Inc.; **30-3** Courtesy of EMD Chemicals, Inc.; **30-4A** Courtesy of Glen Rudner; **30-4B** Courtesy of Glen Rudner; **30-5** Courtesy of Cryofab, Inc.; **30-6A** Courtesy of Polar Tank Trailer L.L.C.; **30-6B** Courtesy of Polar Tank Trailer L.L.C.; **30-6D** Courtesy of Rob Schnepp; **30-6E** Courtesy of Jack B. Kelley, Inc.; **30-6F** Courtesy of Jack B. Kelley, Inc.; **30-6G** Courtesy of Polar Tank Trailer L.L.C.; **30-6H** Courtesy of Eurotainer; **30-9** © Mark Gibson/Index Stock/Alamy Images; **30-10** © Photodisc; **30-12** Courtesy of DOT; **30-14** Courtesy of Sandia National Laboratories; **30-15** Courtesy of Sandia National Laboratories; **Table 30-2** Reprinted with permission from NFPA 704-2012, *System for the Identification of the Hazards of Materials for Emergency Response*, Copyright © 2011, National Fire Protection Association. This reprinted material is not the complete and official position of the NFPA on the referenced subject, which is represented only by the standard in its entirety.

Chapter 31

Opener © Huntstock, Inc./Alamy Images; **31-3** © Bill Karrow, Dunkirk Observer/AP Photos; **31-4** Courtesy of Syngenta Crop Protection, Inc.

Chapter 32

Opener © Steve Morgan/Alamy Images; **32-1** © KellyBoreson/ShutterStock, Inc.; **32-3** © Photodisc; **32-5** Courtesy of Lakeland Industries, Inc.; **32-6** © Jim Ruymen/Reuters/Landov; **32-7A** Courtesy of Survivair; **32-14** © NMSB/Custom Medical Stock Photo; **32-15** © Jim Ruymen/Reuters/Landov.

Chapter 33

Opener © Dennis Wetherhold, Jr.; **33-1** © Ric Dugan, Herald-Mail/AP Photos.

Chapter 35

Opener © ED BETZ/AP Photos; **35-1** © Todd Hollis/AP Photos; **35-2A** © Arthur S. Aubry/Photodisc/Getty Images; **35-2B** © Steve Allen/Brand X Pictures/Alamy Images; **35-2C** © AbleStock; **35-2D** © Michael Dwyer/Alamy Images; **35-3** © James P. Blair/Photodisc/Getty Images; **35-4** © Photodisc/Getty Images; **35-5** © Larry Rana/USDA; **35-6** From When Violence Erupts: A Survival Guide for Emergency Responders, courtesy of Dennis R. Krebs; **35-7** Courtesy of FEMA News Photo; **35-8** Courtesy of Captain David Jackson, Saginaw Township Fire Department; **35-9** Courtesy of Captain David Jackson, Saginaw Township Fire Department; **35-10** © TFoxFoto/ShutterStock, Inc.; **35-11** © Tim McCabe/USDA; **35-13** © AbleStock; **35-14** Courtesy of Dr. Saeed Keshavarz/RCCI (Research Center of Chemical Injuries)—Iran; **35-15** Courtesy of CDC; **35-16** Courtesy of CDC; **35-17** Courtesy of CDC; **35-18** Courtesy of Fluke Biomedical; **35-20** © Jim Mone/AP Photos; **35-21** © Scott Fisher/MCT/Landov.

Chapter 36

Opener © Jones & Bartlett Learning. Photographed by Glen E. Ellman; **36-3** Courtesy of Nampa Fire Department—Nampa, Idaho; **36-4** © David Papazian/Corbis; **36-5** © Monkey Business Images/ShutterStock, Inc.; **36-6** Photography courtesy of the Pacific Northwest National Laboratory (PNNL), proudly operated by Battelle since 1965 for the U.S. Department of Energy.; **36-7** Courtesy of Captain David Jackson, Saginaw Township Fire Department.

Chapter 37

Opener © Ocean/Corbis; **37-1** Courtesy of Honeywell/Gamewell; **37-2** Courtesy of Aaron Fire & Safety and Honeywell/Gamewell; **37-3** Courtesy of Honeywell/Gamewell; **37-4** Courtesy of Kidde Residential and Commercial Division; **37-5** © AbleStock; **37-7** © Brendan Byrne/age footstock; **38-8** Courtesy of Kevin Reed; Courtesy of Kevin Reed; **37-9A** © rob casey/Alamy Images; **37-9B** Courtesy of Honeywell/Fire-Lite Alarms; **37-10** Courtesy of STI-USA; **37-12** Courtesy of Fire Fighting Enterprises Ltd.; **37-13** Courtesy of Firetronics Pte Ltd.; **37-16** © kvisel/iStockphoto.com; **37-18** Courtesy of Honeywell/Fire-Lite Alarms; **37-22** Courtesy of www.acimonitoring.com, Doug Beaulieu.; Courtesy of www.acimonitoring.com, Doug Beaulieu.; **37-29** Courtesy of Tyco Fire & Building Products; **37-31C** Courtesy of Tyco Fire & Building Products; **37-32** © ImageState/Alamy Images; **37-45** Courtesy of Scott Dornan, ConocoPhillips Alaska; **37-46** © AbleStock; **37-48** © Stephen Coburn/ShutterStock, Inc.; **37-49** Courtesy of Ralph G. Johnson (nyail.com/fsd); **37-50** Courtesy of Dixon Valve and Coupling; **37-51** Courtesy of Dixon Valve and Coupling; **37-57** Courtesy of Chemetron Fire Systems; **Table 37-4** Reprinted with permission from NFPA 13-2013, *Installation of Sprinkler Systems*, Copyright © 2012, National Fire Protection Association. This reprinted material is not the complete and official position of the NFPA on the referenced subject, which is represented only by the standard in its entirety.

Chapter 38

Opener © James Quine/Alamy Images; **38-1** Courtesy of Charles B. Hughes/Unified Investigations & Sciences, Inc.; **38-2** Courtesy of Charles B. Hughes/Unified Investigations & Sciences, Inc.; **38-3** Courtesy of Charles B. Hughes/Unified Investigations & Sciences, Inc.; **38-5** © Mark C. Ide; **38-6** © Louis Brems, The Herald-Dispatch/AP Photos; **Table 38-1** Reprinted with permission from *"Home Structure Fires"* by Marty Aherns, Copyright © 2012, National Fire Protection Association.

Appendix A

Reproduced with permission from NFPA 1001-2013, *Standard for Fire Fighter Professional Qualifications*, Copyright © 2012, National Fire Protection Association. This reprinted material is not the complete and official position of the NFPA on the referenced subject, which is represented only by the standard in its entirety.

Reproduced with permission from NFPA 472-2013, *Standard for Competence of Responders to Hazardous Materials/Weapons of Mass Destruction Incidents*, Copyright © 2012, National Fire Protection Association. This reprinted material is not the complete and official position of the NFPA on the referenced subject, which is represented only by the standard in its entirety.

Unless otherwise indicated, all photographs and illustrations are under copyright of Jones & Bartlett Learning, courtesy of Maryland Institute for Emergency Medical Services Systems, or photographed by Glen E. Ellman. Photographs and illustrations have also been provided by the National Fire Protection Association, the American Academy of Orthopaedic Surgeons, or the author(s).

Use of released U.S. Navy imagery does not constitute product or organizational endorsement of any kind by the U.S. Navy.